KT-216-384

GEOCHIMICA ET COSMOCHIMICA ACTA

SUPPLEMENT 3

# PROCEEDINGS OF THE THIRD LUNAR SCIENCE CONFERENCE

Houston, Texas, January 10–13, 1972

GEOCHIMICA ET COSMOCHIMICA ACTA

Journal of The Geochemical Society and the Meteoritical Society

SUPPLEMENT 3

# PROCEEDINGS OF THE THIRD LUNAR SCIENCE CONFERENCE

## Houston, Texas, January 10–13, 1972

Sponsored by
The Lunar Science Institute

VOLUME 2

CHEMICAL AND ISOTOPE ANALYSES

ORGANIC CHEMISTRY

Edited by
DIETER HEYMANN
Rice University, Houston, Texas

THE MIT PRESS

Cambridge, Massachusetts, and London, England

It was set in Monotype Times Roman
by Wolf Composition Company, Inc.
Printed and bound by The Colonial Press Inc.
in the United States of America.



Library of Congress Cataloging in Publication Data

Lunar Science Conference, 3d, Houston, Tex., 1972. Proceedings.

(Geochimica et cosmochimica acta. Supplement 3)
"Sponsored by the Lunar Science Institute."
Includes bibliographies.
CONTENTS: v. 1. Mineralogy and petrology, edited by Elbert A. King, Jr.—v. 2. Chemical and isotope analyses, organic chemistry, edited by Dieter Heymann.—v. 3. Physical properties, edited by David R. Criswell.
1. Lunar geology—Congresses. 2. Moon—Congresses. I. King, Elbert A., ed. II. Heymann, Dieter, ed. III. Criswell, David R., ed. IV. Lunar Science Institute. V. Series.
QB592.L85 1972 559.9′1 72–5496
ISBN 0–262–12060–7 (vol. 1)
0–262–12062–3 (vol. 2)
0–262–12063–1 (vol. 3)
0–262–12064–X (3 vol. set)



Supplement 3
GEOCHIMICA ET COSMOCHIMICA ACTA
Journal of The Geochemical Society and The Meteoritical Society

## BOARD OF ASSOCIATE EDITORS

# Preface

THE STUDIES REPORTED in this volume of the *Proceedings of the Third Lunar Science Conference*, which was held 10–13 January 1972 at the Manned Spacecraft Center near Houston, are unique in at least one respect, namely that the overwhelming majority of the experimental work resulted in the permanent destruction of lunar material. Because of this aspect, I firmly believe that this volume will always remain a prime source of data for future generations of lunar and solar system scientists. They will modify and perhaps reject some of the theories and hypotheses that we have expressed in this book; they will probably perform similar experiments on some of the same samples with more advanced techniques. However, I believe that nearly all of the data and many of the theories contained in this volume will survive the test of time surprisingly well.

In Volume 2 the papers are arranged by specialties. Alphabetical ordering by first author usually applies thereafter. Three cross-reference sections have been added to enhance the usefulness of the *Proceedings*. Two "Lunar Sample Cross References" are presented (Volumes 2 and 3) that list for a given sample number every article in all three *Proceedings* sets in which that sample is mentioned. A complete and annotated "Sample Inventory" for Apollos 11, 12, 14, and 15 samples is given in Volume 1. Each volume contains an author index and a subject index, prepared in part on the basis of key words suggested by the authors. J. L. Warner of the Manned Spacecraft Center has provided substantial assistance in the preparation of all of these sections.

Thanks are due the collective authors who have chosen the *Proceedings* for the publication of their papers. Without the great expenditure of time and effort by the Associate Editors and Technical Reviewers, the *Proceedings* would not have appeared in print. Mmes. Ann Geisendorff, Olene Edwards, Mildred Armstrong, and Jewel White deserve thanks for their fine support. I thank Dr. S. Lakatos for his assistance with the final proofreading.

On behalf of The Lunar Science Institute, sponsor of these *Proceedings*, I thank NASA for its continued interest in the scientific aspects of the Apollo program. I also thank Pergamon Press, Ltd. for allowing these *Proceedings* to appear as Supplement 3 to *Geochimica et Cosmochimica Acta* with The MIT Press as the publisher.

Dieter Heymann

*Rice University*
*Houston, Texas*
*May 1972*

# Contents

GEOCHIMICA ET COSMOCHIMICA ACTA

SUPPLEMENT 3

# PROCEEDINGS
# OF THE
# THIRD LUNAR SCIENCE CONFERENCE

Houston, Texas, January 10–13, 1972

Proceedings of the Third Lunar Science Conference
(Supplement 3, *Geochimica et Cosmochimica Acta*)
Vol. 2, pp. 1133–1147
The M.I.T. Press, 1972

# Distribution of elements between different phases of Apollo 14 rocks and soils

A. O. BRUNFELT, K. S. HEIER, B. NILSSEN, and B. SUNDVOLL
Mineralogical-Geological Museum, Oslo, Norway

and

E. STEINNES
Institutt for Atomenergi, Kjeller, Norway

**Abstract**—Neutron activation analysis data are given for samples 14163 (soil), 14276 (breccia), 14303 (breccia) and 14310 (crystalline rock). Two polished thin sections of breccia 14303 were studied and electron-microprobe data are given on the constituent minerals. Orthopyroxene separate from rock 14310 was also analyzed by electron microprobe. Data are given on plagioclase, orthopyroxene, three glass fractions, and two rock fractions from sample 14163, and on plagioclase and orthopyroxene fractions from rock 14310. The different glass and rock fractions of sample 14163 have similar compositions. Published modal analyses of 14310 show that plagioclase plus pyroxene constitute about 95% of the rock. Most of the trace elements are concentrated in the remaining 5%. Iron is also concentrated in this fraction being probably related to the presence of metallic iron in the sample.

## INTRODUCTION

### *Samples received*

THE FOLLOWING SAMPLES were received: 14163,154 (<1 mm fines, 4.960 g); 14276,8 (breccia, 0.236 g); 14303,14 (breccia, 32.31 g) with two polished sections; and 14310,123 (rock, 4.04 g).

Sample 14276 was part of a consortium sample with Professor G. J. Wasserburg as consortium leader.

Data on sample 15601,75 (<1 mm fines, 1.01 g) were reported by us at the Third Lunar Science Conference (Brunfelt *et al.*, 1972). The analytical data are presented in the Appendix, but will not be discussed further in this paper.

### *Previously published papers on the received material*

In addition to our abstract in the proceedings of the Third Lunar Science Conference (Brunfelt *et al.*, 1972), a determination of 36 elements in sample 14163,154 was reported by Brunfelt *et al.* (1971). A discussion of REE distributions in apatite and whitlockite from lunar rock 14310,123 and from a terrestrial occurrence (Ödegaarden, Norway) was published by Griffin *et al.* (1972) and will not be commented on further in this paper.

### *Sample preparation and analytical methods*

Only bulk analyses have been carried out on sample 14276 while both the bulk breccia and a separated matrix fraction have been analyzed in the case of 14303. Two polished sections (14303,48 and 14303,49) were studied microscopically and analyzed by electron microprobe.

Mineral separations were carried out on samples 14163 and 14310 after a bulk sample had been split off. In the case of 14163 a fine fraction (<0.12 mm grain size) was washed with ethanol and removed by sieving before the mineral separation. This fine fraction was analyzed separately. Plagioclase, orthopyroxene, light and dark rock fragments, glass spheres, irregular dark glass fragments, and twisted glass fragments were concentrated by handpicking from 14163 and analyzed.

Sample 14310 was split into a light and a heavy fraction by the use of acetylene tetrabromide. Plagioclase and orthopyroxene fractions of high purities were obtained from these fractions by magnetic separation and handpicking.

All the separated fractions and bulk samples were analyzed by neutron activation following the procedure of Brunfelt and Steinnes (1971). Some of the 14163 fractions were too small to permit determination of elements other than those that could be assayed without using radiochemical separation steps.

Orthopyroxene from sample 14310 was also analyzed by electron microprobe, using an ARL-EMX probe at the Central Institute for Industrial Research, Oslo. Various natural minerals were used as standards for these analyses (see Griffin *et al.* (1972) for REE determinations in whitlockite and apatite).

## Presentation and Discussion of Data

### *Regolith 14163,154 (<1 mm fines)*

The analytical data obtained by neutron activation analysis on the regolith and separated fractions are given in Table 1. There is no significant difference between the bulk sample and the less than 0.12 mm size fraction (columns 1 and 2). Our data compare well with emission spectrographic and spark source mass spectrometric determinations by Taylor *et al.* (1971) for 19 elements. Significant differences exist only for Ga and W (4.5 and 0.66 ppm, respectively, according to Taylor *et al.*). Our data for W agree well with that of Wänke *et al.* (1972), who reported 1.95 ppm W. Baedecker *et al.* (1972), Helmke and Haskin (1972), and Wänke *et al.* (1972) found 8.7, 7.5, and 8.3 ppm Ga, respectively, in this sample. Baedecker *et al.* (1972) obtained values for Zn and Cd in good agreement with those reported by us. Of the number of elements determined both by Helmke and Haskin (1972) and ourselves, we find significant disagreement only for Ce (157 versus 203 ppm). Taylor *et al.* (1971) and Wakita *et al.* (1972) both obtained 200 ppm Ce. Hubbard and Gast (1972) obtained 176 ppm Ce. The other data given by Hubbard and Gast (1972) agrees well with our own except for Ba, which they reported as 926 ppm versus 748 ppm in our analysis. Taylor *et al.* (1971) and Wakita *et al.* (1972) found 710 and 730 ppm Ba, respectively. Other determinations of trace elements in 14163 are by Jackson *et al.* (1972); Keith *et al.* (1972); Laul *et al.* (1972); Morgan *et al.* (1972). A study of the mean values from these analyses indicates that our determination of Th (11.3 ppm) is somewhat low and that the content is more likely to be 13 and 14 ppm.

The bulk composition of sample 14163 is compared with the average composition of the bulk regolith from the Apollo 11 and 12 landing sites in Fig. 1. The higher

Table 1. Compositions of lunar regolith 14163,154 by neutron activation analysis.

| Element | 14163 Regolith | 14163 Fine fraction | 14163/1 Plagio-clase | 14163/2 Glass (spheres) | 14163/3 Glass (dark) | 14163/4–5 Glass (twisted) | 14163/9 Ortho-pyroxene | 14163/12 Light rock fragments | 14163/13 Dark rock fragments |
|---|---|---|---|---|---|---|---|---|---|
| Na (%) | 0.54 | 0.55 | 0.88 | 0.39 | 0.46 | 0.45 | 0.11 | 0.60 | 0.56 |
| Mg (%) | 5.5 | 6.7 | 1.7 | 5.5 | 6.7 | 5.4 | 18.8 | 7.2 | 7.4 |
| Al (%) | 9.1 | 9.6 | 16.5 | 8.9 | 9.0 | 10.0 | 2.6 | 10.2 | 8.9 |
| Cl (ppm) | 47 | | | | | | | | |
| K (%) | 0.41 | 0.33 | | | | | | | |
| Ca (%) | 7.2 | 7.3 | 11.5 | 8.3 | 7.6 | 7.6 | 1.7 | 8.6 | 7.2 |
| Sc (ppm) | 21 | 20.5 | 2.5 | 23.1 | 22.8 | 20.3 | 25.7 | 16.5 | 19.7 |
| Ti (%) | 0.90 | 1.11 | 0.13 | 0.82 | 1.05 | 1.08 | 0.35 | 0.71 | 0.84 |
| V (ppm) | 48 | 45 | < 10 | 29 | 34 | 39 | 50 | 18 | 38 |
| Cr (%) | 0.137 | 0.143 | 0.019 | 0.194 | 0.154 | 0.156 | 0.256 | 0.116 | 0.147 |
| Mn (%) | 0.103 | 0.102 | 0.132 | 0.114 | 0.102 | 0.093 | 0.132 | 0.085 | 0.098 |
| Fe (%) | 8.1 | 8.16 | 0.95 | 9.16 | 8.65 | 7.75 | 12.0 | 6.62 | 7.92 |
| Co (ppm) | 34 | 34.7 | 4.35 | 77.2 | 34.6 | 25.5 | 36.6 | 19.8 | 24.6 |
| Ni (ppm) | | | 34 | 990 | 330 | 190 | 140 | 220 | 230 |
| Cu (ppm) | 10.4 | 13.4 | | | | | | | |
| Zn (ppm) | 33 | 40 | | | | | | | |
| Ga (ppm) | 7.7 | 8.2 | | | | | | | |
| As (ppm) | 0.02 | 0.10 | | | | | | | |
| Se (ppm) | 0.29 | | | | | | | | |
| Rb (ppm) | 16 | 13 | 8.5 | 8.4 | 10 | 16 | 28 | 18 | 15 |
| Sr (ppm) | 185 | 170 | | | | | | | |
| Pd (ppm) | 0.11 | ~0.1 | | | | | | | |
| Sb (ppm) | 0.003 | 0.01 | 1.3 | 0.11 | | 0.01 | | | |
| Cs (ppm) | 0.68 | 0.56 | | | | | | | |
| Ba (ppm) | 748 | 740 | 447 | 515 | 689 | 583 | 163 | 647 | 753 |
| La (ppm) | 67 | 61 | 14 | 49 | 68 | 54 | 23 | 64 | 81 |
| Ce (ppm) | 203 | | 40 | 162 | 214 | 183 | | 189 | 214 |
| Pr (ppm) | | 26 | | | | | | | |
| Sm (ppm) | 30 | 27.3 | 4.4 | 20.1 | 32.1 | 27.5 | 8.0 | 30.0 | 37.3 |
| Eu (ppm) | 2.5 | 2.8 | 4.5 | 2.3 | 2.4 | 2.4 | 1.7 | 2.8 | 2.8 |
| Tb (ppm) | 6.3 | 5.8 | 0.9 | 4.2 | 6.4 | 5.3 | 1.2 | 5.8 | 7.1 |
| Dy (ppm) | | 33.5 | 5.8 | 24.5 | 35.3 | 30.2 | 8.3 | 33.4 | 41.5 |
| Ho (ppm) | | 8.2 | | | | | | | |
| Er (ppm) | | 17.3 | | | | | | | |
| Yb (ppm) | 21 | 15 | 3.7 | 16 | 19 | 18 | 8.6 | 19 | 25 |
| Lu (ppm) | 3.1 | | | | | | | | |
| Hf (ppm) | 19 | 20.6 | 13.4 | 19.0 | 25.4 | 20.7 | | 23.3 | |
| Ta (ppm) | 2.7 | 2.9 | 0.4 | 2.3 | 3.4 | 2.6 | 0.6 | 2.7 | 3.6 |
| W (ppm) | 1.8 | 1.40 | | | | | | | |
| Au (ppb) | | 2.3 | | | | | | | |
| Th (ppm) | 11.3 | 10.6 | 2.2 | 8.1 | 11.1 | 10.5 | 2.7 | 10.9 | 13.6 |
| U (ppm) | 3.4 | 3.4 | 0.69 | 2.8 | 3.7 | 3.2 | 1.2 | 4.0 | 4.4 |

concentration of most trace elements in the Apollo 14 soil is readily apparent from this figure.

The different types of glass and rock fragments separated from the bulk sample all have rather similar compositions (Fig. 2). The glass spheres tend to have the lowest concentrations of most elements, notable exceptions are Fe, Co, and Ni.

Plagioclase and orthopyroxene were the major crystalline mineral constituents in the sample. They differ significantly in composition from the other separated fragments. The chondrite normalized REE distribution patterns illustrate this point rather well, Fig. 3.

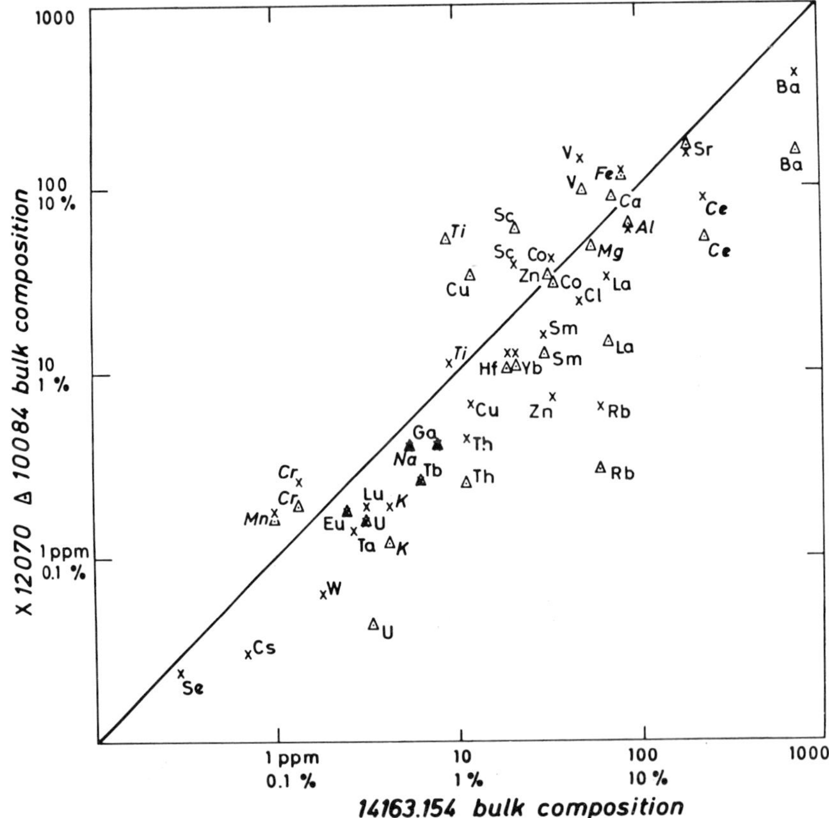


Fig. 1. Bulk composition of regolith 14163 compared with regoliths 10084 and 12070. Na, Mg, Al, K, Ca, Ti, Cr, Mn, Fe in percent; rest in ppm. Solid line indicates 1 : 1 distribution ratio.

## *Breccias 14276 and 14303*

The neutron activation data on the breccias are listed in Table 2. The elemental compositions of the 14303 ground mass and bulk sample are very similar. The groundmass is slightly enriched in Fe, Co, and Ni. The compositions of the breccias are compared with the regolith in Fig. 4. The general similarity of regolith and breccias is apparent from this figure.

Two polished thin sections of breccia 14303 (14303,48 and 14303,49) have been studied. Electron-microprobe data on constituent minerals are shown in Table 3. The large variation in FeO/MgO ratios between individual grains indicates the complexity of mineral studies of this material.

We are not aware of any other chemical data available on samples 14276 and 14303. No information appears to be published in *Lunar Science—III* (editor C. Watkins) Lunar Science Institute Contr. 88.

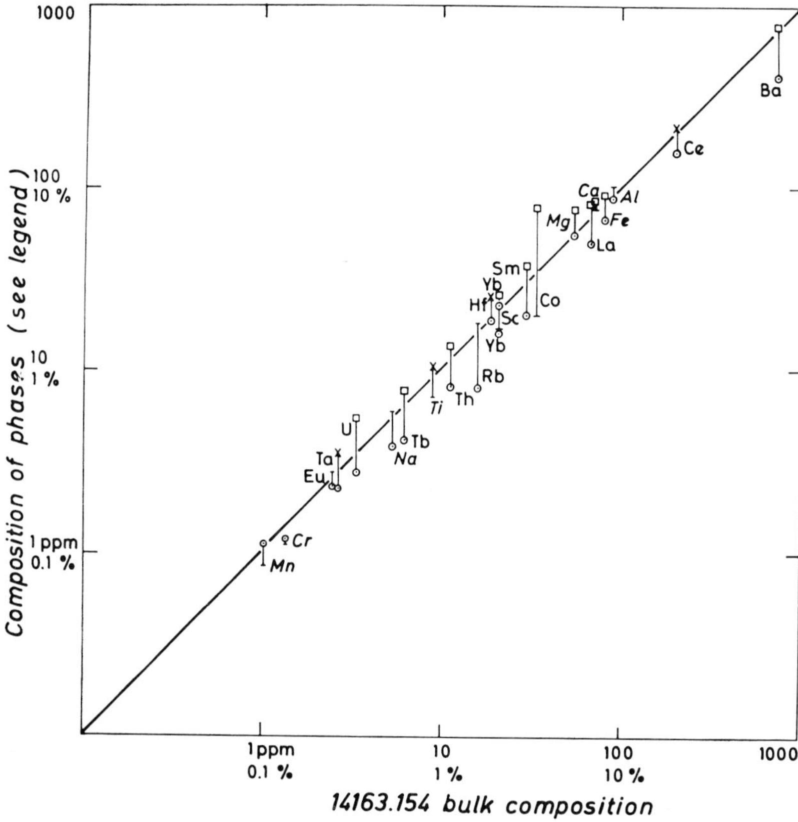


Fig. 2. Distribution of elements between different fragments of 14163,154 and bulk sample. Phases showing maximum and minimum values are indicated, remaining phases plot in the indicated range (□ dark rock fragments, - light rock fragments, x dark irregular glass fragments, v twisted glass fragments, ⊙ glass spheres).

*Rock 14310,123*

This is the one of the two homogeneous crystalline rocks collected on the Apollo 14 mission. Chemical data on bulk rock and separated plagioclase and orthopyroxene fractions are shown in Table 4. Electron-microprobe analyses of polished mounts of orthopyroxene separates are given in Table 5. No thin sections were made available to us for electron-microprobe analyses or microscopic studies. The sizes of the orthopyroxene grains range between 0.1 and 0.2 mm and the mineral appears to be only slightly heterogeneous. The orthopyroxene identification was verified by optical and x-ray diffraction techniques. Clinopyroxene appears to be present in only subordinate amounts in the chip received by us. This is at variance with the findings of Gancarz *et al.* (1971), who reported no orthopyroxene in the sample. Hollister *et al.* (1972) reported 23.8% augite, 11.4% hypersthene, and 1.6% pigeonite, while Melson *et al.* (1972) reported 18% low-calcium and 36% high-calcium pyroxene in 14310.

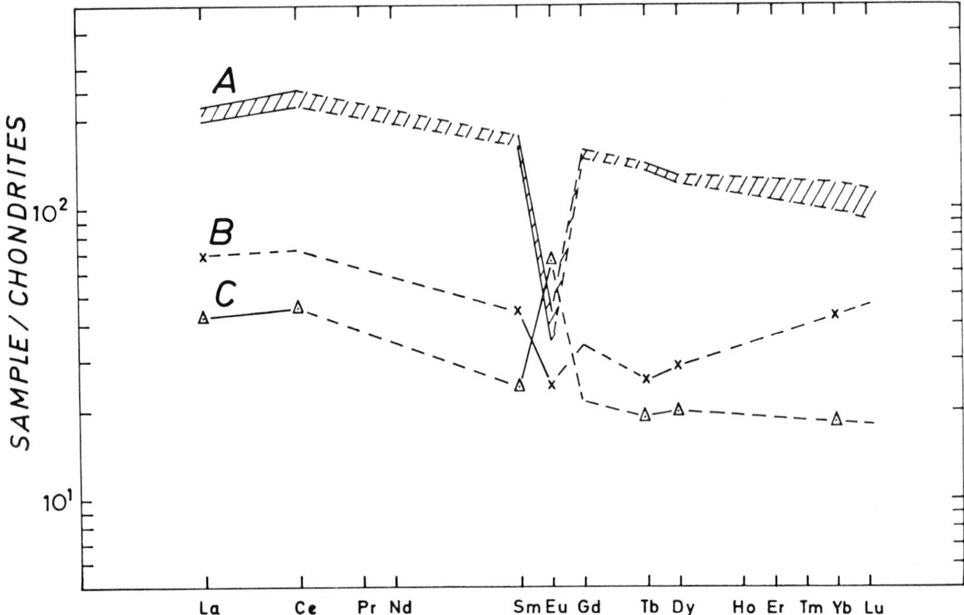


Fig. 3. Chondrite normalized REE-distribution patterns in 14163,154 pyroxene (B), plagioclase (C), bulk fines and other separated fractions (A, shaded area).

Neither of these are in agreement with our impression. Battey *et al.* (1972), on the basis of Mössbauer studies, claimed that clinopyroxene, with a composition close to $En_{46}Fs_{28}Wo_{26}$, is the major iron mineral in rock 14310, and Jagodzinski and Korokawa (1972) wrote that "orthopyroxenes are very rare in samples 14310,106." The petrology of rock 14310 and its pyroxenes is also discussed in some detail by Bence and Papike (1972); Ford *et al.* (1972); Kushiro (1972); Ridley *et al.* (1972); Ringwood *et al.* (1972); Takeda and Ridley (1972); and Walter (1972). The presence of orthopyroxene, augitic clinopyroxene, and pigeonite is reported by all these authors but without any statements being made about their relative amounts.

These very different findings regarding the nature and relative amounts of pyroxenes in rock 14310 could indicate heterogeneity in the rock. However, studies of the other published whole rock analyses by Baedecker (1972), Helmke and Haskin (1972), Hubbard and Gast (1972), Keith *et al.* (1972), Laul *et al.* (1972), and Morgan *et al.* (1972) give very little support for the idea of a heterogeneous sample. We report significantly higher concentrations of some of the heavy REE than, for instance, Helmke and Haskin, and Hubbard and Gast. Morgan *et al.* report much higher values for Au (4.31 versus 0.3 ppb) and In (130 versus 30 ppb) than we do, but again analytical error or sample contamination are the most likely explanations.

The concensus of opinion among petrologists and geochemists, who have studied this rock, is that it represents the crystallization product of a surface impact melt, i.e., Ringwood *et al.* (1972); Walker *et al.* (1972). Its similar composition to the regolith 14163 is illustrated in Fig. 5. Modal analyses by LSPET (1971) and Gancarz *et al.*

Table 2. Elemental composition of samples 14303,14 and 14276,8 (4) by neutron activation analysis.

| Element | 14303 Bulk rock | 14303 Groundmass | 14276 Split B |
|---|---|---|---|
| Na (%) | 0.62 | 0.60 | 0.52 |
| Mg (%) | 6.5 | 6.6 | 5.2 |
| Al (%) | 8.77 | 8.53 | 11.7 |
| Cl (ppm) | 18 | | 23 |
| K (%) | 1.33 | | 0.36 |
| Ca (%) | 7.1 | 6.3 | 9.5 |
| Sc (ppm) | 23.2 | 23.9 | 18.2 |
| Ti (%) | 1.09 | 1.01 | 0.66 |
| V (ppm) | 45 | 47 | 25 |
| Cr (%) | 0.137 | 0.142 | 0.1450 |
| Mn (%) | 0.1085 | 0.1075 | 0.08 |
| Fe (%) | 8.07 | 8.54 | 6.04 |
| Co (ppm) | 30.5 | 34.2 | 13 |
| Ni (ppm) | 260 | 320 | |
| Cu (ppm) | 75 | | 38 |
| Zn (ppm) | 0.8–3.7* | | 1.7 |
| Ga (ppm) | 5.3 | | 3.8 |
| As (ppm) | 0.07 | | 0.16 |
| Rb (ppm) | 20 | 27 | 14 |
| Sr (ppm) | 160 | | 170 |
| Cd (ppm) | | | <0.02 |
| In (ppb) | | | 56 |
| Sb (ppm) | <0.03 | | 0.024 |
| Cs (ppm) | 0.86 | 1.1 | 0.57 |
| Ba (ppm) | 890 | 830 | 540 |
| La (ppm) | 72 | 71 | 43 |
| Ce (ppm) | 210 | 200 | |
| Sm (ppm) | 34.6 | 33.3 | 20.0 |
| Eu (ppm) | 2.5 | 2.3 | 1.9 |
| Tb (ppm) | 7.0 | 7.0 | 4.8 |
| Dy (ppm) | 50.8 | 50.9 | 23 |
| Ho (ppm) | 9.5 | | 7.3 |
| Er (ppm) | 30 | | |
| Yb (ppm) | 28 | 29 | 11.7 |
| Lu (ppm) | 4.4 | 4.5 | |
| Hf (ppm) | 25.6 | 25.4 | 16 |
| Ta (ppm) | 3.2 | 3.4 | 2.1 |
| W (ppm) | 0.85 | | 2.5 |
| Au (ppb) | | | 0.3 |
| Th (ppm) | 12.6 | 12.9 | 8.0 |
| U (ppm) | 3.6 | 3.4 | 2.5 |

* Probably inhomogeneities.

(1971) indicate 61% plagioclase and 31% pyroxene, versus 50% plagioclase and 40% pyroxene, respectively, in this rock. In Fig. 6 we have plotted the sum of the element concentrations in plagioclase and pyroxene (calculated as $\frac{2}{3}$ plagioclase and $\frac{1}{3}$ pyroxene) versus bulk rock (K in pyroxene and Ti in plagioclase taken as zero; Sr and Ca in pyroxene as 10 ppm and 1.7%, respectively). If we assume plagioclase and pyroxene to make up 95% of the rock, the average composition of the remaining 5% is (in percent): Mg (30.2), Fe (49.2), Ti (11.6), K (0.48), Mn (0.6), Cr (0.20); in ppm: Sc (132), Co (166), Rb (216), Ba (6640), La (854), Sm (360), Tb (159), Dy (134), Yb (172), Hf (266), Ta (41), Th (148), U (48.6). The high Fe content in the remaining

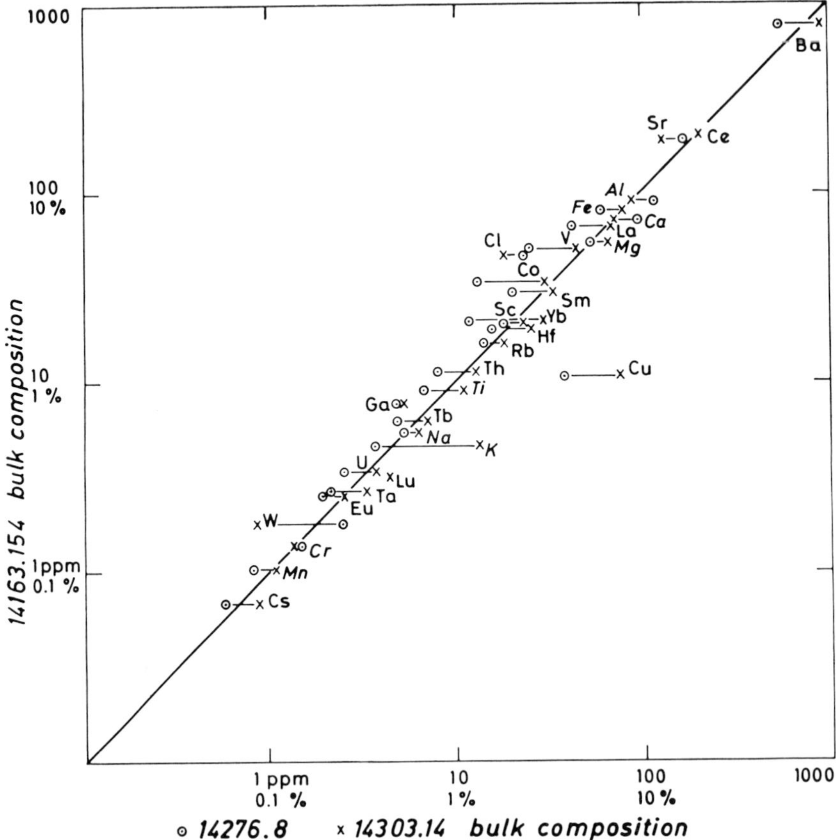


Fig. 4. Comparison of breccias 14276 and 14303 with bulk regolith 14163.

phases could be related to the presence of metallic iron (i.e., El Goresy *et al.*, 1972), while the high Mg content may partly be related to analytical error. Compston *et al.* (1972), Hubbard and Gast (1972), and Kushiro (1972) all report between 7.6 and 7.9% MgO while we have found 8.8% MgO. It is interesting that the bulk of the potassium is not to be found in the plagioclase concentrate which represents the lightest mineral fraction.

It must be emphasized that what we have termed "remaining phases" could include outer rims of the plagioclase and pyroxene grains which may have been selectively removed during the grinding and mineral separation. Some of the trace elements could be strongly concentrated in these rims.

The chondrite normalized REE patterns of the plagioclase and orthopyroxene concentrate are compared with that of the bulk rock in Fig. 7. The importance of minor phases as hosts for a number of trace elements is strikingly demonstrated by the lunar rocks.

Table 3. Electron-microprobe data on minerals in breccia 14303 (nos. 1 to 12 from 14303,48; nos. 13 to 20 from 14303,49).

| | Olivines | | | | | | |
|---|---|---|---|---|---|---|---|
| | 1 | 8 | 14 | 16 | 17 | 18 | 20 |
| $SiO_2$ | 37.78 | 38.72 | 39.48 | 40.65 | 37.84 | 40.34 | 37.59 |
| $TiO_2$ | 0.0 | 0.0 | 0.0 | 0.0 | 0.0 | 0.0 | 0.0 |
| $Cr_2O_3$ | 0.13 | 0.12 | 0.15 | 0.16 | 0.13 | 0.07 | 0.11 |
| $Al_2O_3$ | 0.38 | 0.54 | 0.21 | 0.39 | 0.41 | 0.39 | 0.46 |
| FeO | 30.21 | 23.44 | 18.36 | 13.55 | 28.03 | 15.17 | 29.30 |
| MnO | 0.32 | 0.23 | 0.18 | 0.13 | 0.30 | 0.14 | 0.28 |
| MgO | 28.97 | 37.00 | 41.67 | 45.80 | 32.96 | 44.01 | 32.38 |
| CaO | 0.51 | 0.36 | 0.0 | 0.06 | 0.39 | 0.15 | 0.25 |
| $Na_2O$ | 0.0 | 0.0 | 0.0 | 0.0 | 0.0 | 0.0 | 0.0 |
| | 98.30 | 100.41 | 100.05 | 100.74 | 100.06 | 100.27 | 100.37 |

| | Pigeonites | | | | | Spinel 14303,49 |
|---|---|---|---|---|---|---|
| | 3 | 4 | 12 | 13 | 19 | |
| $SiO_2$ | 52.72 | 53.15 | 54.04 | 48.58 | 50.29 | 0.0 |
| $TiO_2$ | 0.43 | 0.38 | 0.31 | 0.39 | 0.31 | 0.0 |
| $Ca_2O_3$ | 0.80 | 0.82 | 0.84 | 0.80 | 0.30 | 5.61 |
| $Al_2O_3$ | 2.00 | 1.91 | 1.57 | 1.17 | 0.82 | 62.91 |
| FeO | 16.56 | 15.55 | 16.78 | 17.85 | 30.72 | 13.41 |
| MnO | 0.34 | 0.33 | 0.34 | 0.27 | 0.55 | 0.0 |
| MgO | 20.57 | 21.60 | 23.34 | 24.74 | 10.75 | 18.17 |
| CaO | 5.11 | 5.15 | 3.39 | 5.15 | 5.98 | 0.0 |
| $Na_2O$ | 0.0 | 0.0 | 0.0 | 0.0 | 0.04 | — |
| | 98.53 | 98.89 | 100.61 | 98.95 | 99.76 | 100.10 |

| | Orthopyroxenes | | | | Clinopyroxenes | | |
|---|---|---|---|---|---|---|---|
| | 7 | 9 | 10 | 11 | 2 | 6 | 15 |
| $SiO_2$ | 54.30 | 55.81 | 56.77 | 54.50 | 52.31 | 51.00 | 52.73 |
| $TiO_2$ | 0.33 | 0.35 | 0.17 | 0.65 | 0.58 | 1.50 | 0.91 |
| $Cr_2O_3$ | 0.34 | 0.56 | 0.29 | 0.33 | 0.01 | 0.65 | 0.18 |
| $Al_2O_3$ | 1.18 | 0.98 | 0.90 | 1.16 | 1.56 | 1.79 | 1.42 |
| FeO | 17.91 | 12.02 | 11.30 | 18.63 | 9.77 | 8.47 | 15.64 |
| MnO | 0.30 | 0.21 | 0.22 | 0.30 | 0.22 | 0.20 | 0.30 |
| MgO | 23.58 | 29.31 | 30.38 | 22.16 | 14.50 | 16.10 | 12.65 |
| CaO | 1.79 | 1.74 | 0.68 | 2.25 | 21.12 | 19.80 | 16.34 |
| $Na_2O$ | 0.0 | 0.0 | 0.0 | 0.0 | 0.08 | 0.05 | 0.11 |
| | 99.73 | 100.98 | 100.71 | 99.98 | 100.15 | 99.56 | 100.28 |

Analyses by Dr. W. L. Griffin.

## *Comparison of plagioclase and orthopyroxene from regolith 14163 and rock 14310*

A rigorous comparison between these mineral separates is hampered by the larger amounts of impurities in the minerals separated from sample 14163. The 14310 orthopyroxene has the highest Al, Ca, Na contents and the highest Fe/Mg ratio. The compositions of the plagioclase are virtually identical, 14163 plagioclase being slightly more sodic. The chondrite normalized REE patterns of the minerals are compared in Fig. 8. The 14163 orthopyroxene is significantly enriched in the light REE compared with 14310 orthopyroxene while the contents of heavy REE are approximately the same in both minerals. The plagioclase have similar REE distribution patterns, the 14163 plagioclase having slightly higher absolute concentrations and a markedly higher Eu content.

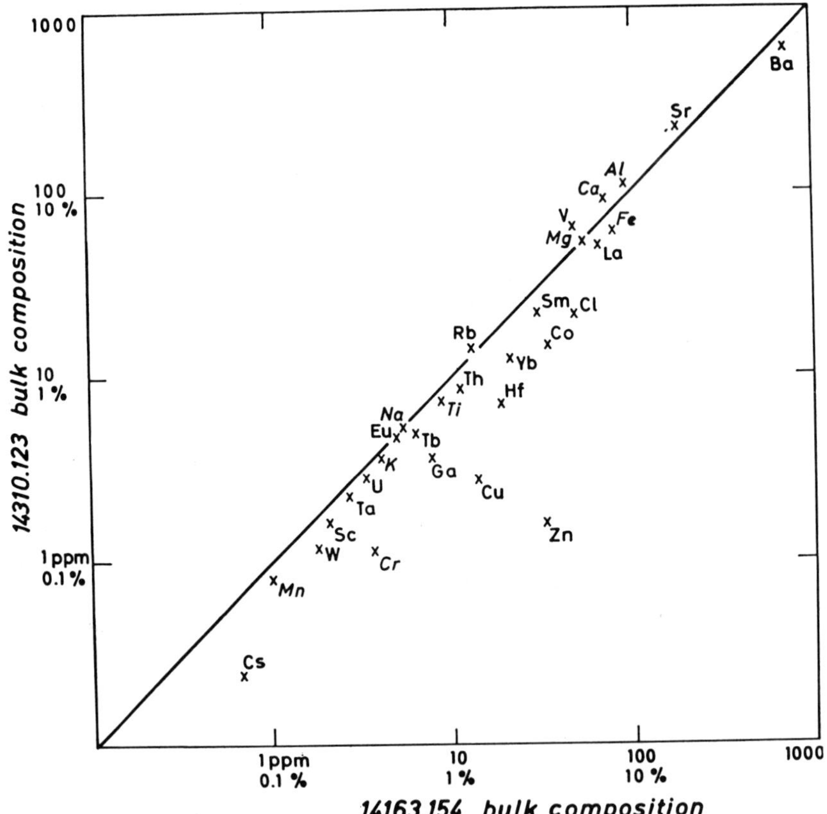


Fig. 5. Comparison of 14163,154 and 14310,123 bulk compositions.

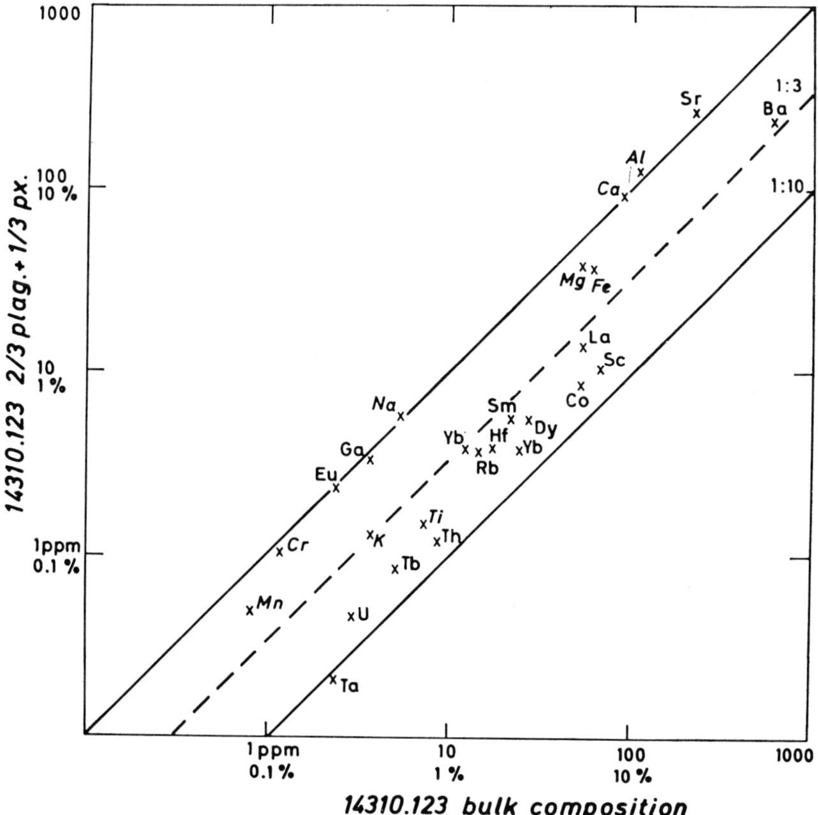


Fig. 6. Distribution of elements between plagioclase plus pyroxene and bulk rock 14310,123.

Table 4. Composition of rock 14310,123 by neutron activation analysis.

| Element | 14310 Bulk rock | 14310/1 Plagioclase | 14310/2 Pyroxene (orthorombic) |
|---|---|---|---|
| Na (%) | 0.54 | 0.78 | 0.20 |
| Mg (%) | 5.3 | 0.5 | 11.3 |
| Al (%) | 11.0 | 16.5 | 3.9 |
| Cl (ppm) | 22 | | |
| K (%) | 0.37 | 0.20 | |
| Ca (%) | 9.0 | 11.8 | 3.9 |
| Sc (ppm) | 16.7 | 1.0 | 28.2 |
| Ti (%) | 0.73 | 0.05 | 0.44 |
| V (ppm) | 56 | < 50 | 150 |
| Cr (%) | 0.116 | 0.014 | 0.288 |
| Mn (%) | 0.0842 | 0.0067 | 0.1382 |
| Fe (%) | 6.17 | 0.92 | 9.30 |
| Co (ppm) | 15.1 | 1.1 | 17.8 |
| Ni (ppm) | | 5 | 20 |
| Cu (ppm) | 2.8 | 15 | |
| Zn (ppm) | 1.6 | 12 | |
| Ga (ppm) | 3.7 | 4.6 | |
| As (ppm) | 0.03 | 0.6 | |
| Rb (ppm) | 15 | 4.1 | 2.8 |
| Sr (ppm) | 220 | 390 | |
| Cd (ppm) | < 0.02 | | |
| In (ppb) | 30 | | |
| Sb (ppm) | 0.004 | < 0.01 | 0.09 |
| Cs (ppm) | 0.4 | | |
| Ba (ppm) | 595 | 350 | 90 |
| La (ppm) | 53 | 12 | 7.0 |
| Ce (ppm) | | 32 | 17 |
| Pr (ppm) | 17 | | |
| Sm (ppm) | 22.7 | 4.7 | 4.8 |
| Eu (ppm) | 2.4 | 3.0 | 1.1 |
| Tb (ppm) | 5.1 | 0.73 | 1.1 |
| Dy (ppm) | 27.3 | 4.3 | 8.1 |
| Ho (ppm) | 6.5 | | |
| Er (ppm) | 16 | | |
| Yb (ppm) | 12.5 | 2.3 | 7.2 |
| Hf (ppm) | 17.2 | 2.2 | 4.3 |
| Ta (ppm) | 2.3 | 0.22 | 0.23 |
| W (ppm) | 1.20 | 0.30 | |
| Au (ppb) | 0.3 | | |
| Th (ppm) | 8.6 | 1.5 | 1.0 |
| U (ppm) | 2.9 | 0.50 | 0.40 |

Table 5. Electron-microprobe analysis of orthopyroxene in rock 14310,123 (average composition).

| | Wt. % |
|---|---|
| $SiO_2$ | 52.28 |
| $TiO_2$ | 0.58 |
| $Cr_2O_3$ | 0.30 |
| $Al_2O_3$ | 1.88 |
| FeO | 15.35 |
| MnO | 0.30 |
| MgO | 26.96 |
| CaO | 2.33 |
| $Na_2O$ | 0.10 |
| | 100.08 |

Analyses by Dr. W. L. Griffin.

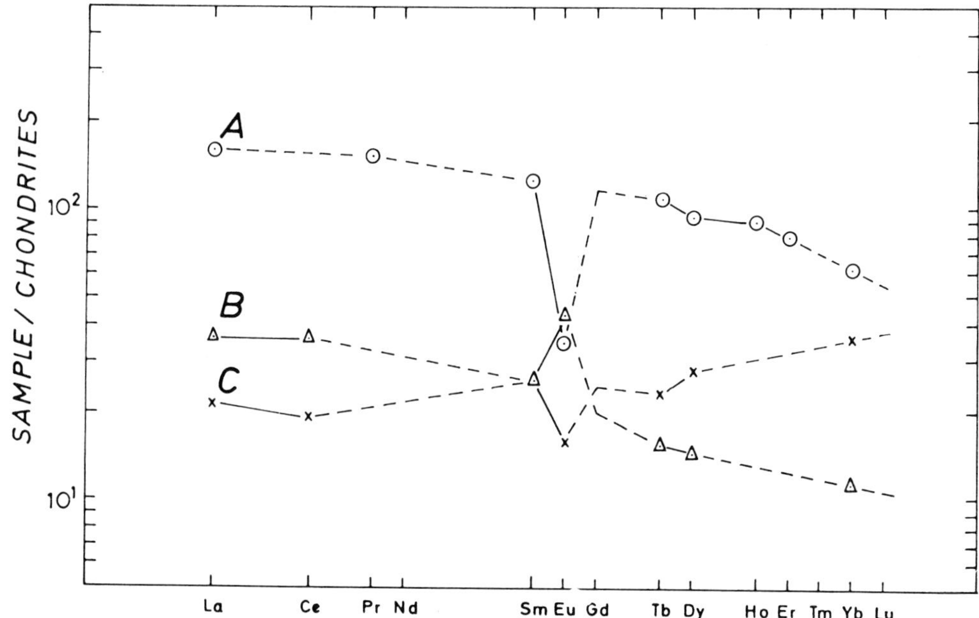


Fig. 7. REE distribution in 14310,123 bulk rock (A), plagioclase (B), and pyroxene (C).

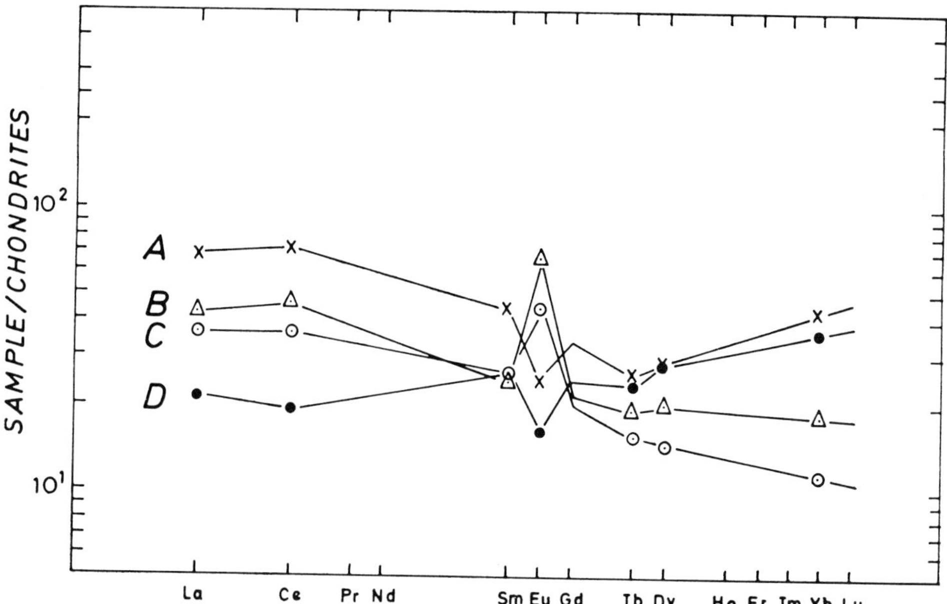


Fig. 8. Comparison of REE distribution patterns of 14310,123 and 14163,154 plagioclase and orthopyroxene. (A) Pyroxene 14163. (B) Plagioclase 14163. (C) Plagioclase 14310. (D) Pyroxene 14310.

*Acknowledgments*—We thank Dr. W. L. Griffin for undertaking the electron-microprobe analyses and Mrs. B. Jensen for correcting the English. Financial support by the Royal Norwegian Council for Scientific and Industrial Research (Research contract B 1206 3070) is gratefully acknowledged.

## References

Baedecker P. A., Chou C.-L., Kimberlin J., and Wasson J. T. (1972) Trace element studies of lunar rocks and soils (abstract). In *Lunar Science—III* (editor C. Watkins), pp. 35–37, Lunar Science Institute Contr. No. 88.

Battey M. H., Gibb T. C., Greatrex R., and Greenwood N. N. (1972) Mössbauer studies of Apollo 14 lunar samples (abstract). In *Lunar Science—III* (editor C. Watkins), pp. 44–46, Lunar Science Institute Contr. No. 88.

Bence A. E. and Papike J. (1972) Crystallization histories of pyroxenes from lunar basalts (abstract). In *Lunar Science—III* (editor C. Watkins), pp. 59–61, Lunar Science Institute Contr. No. 88.

Brunfelt A. O., Heier K. S., Nilssen B., Steinnes E., and Sundvoll B. (1972) Distribution of elements between different phases of Apollo 14 rocks and soils (abstract). In *Lunar Science—III* (editor C. Watkins), pp. 99–101, Lunar Science Institute Contr. No. 88.

Brunfelt A. O., Heier K. S., Steinnes E., and Sundvoll B. (1971) Determination of 36 elements in Apollo 14 bulk fines 14163 by activation analysis. *Earth Planet. Sci. Lett.* **11,** 351–353.

Brunfelt A. O. and Steinnes E. (1971) A neutron activation scheme developed for the determination of 42 elements in lunar material. *Talanta* **18,** 1197–1208.

Compston W., Vernon M. J., Berry H., Rudowski R., Gray C. M., and Ware N. (1972) Age and petrogenesis of Apollo 14 basalts (abstract). In *Lunar Science—III* (editor C. Watkins), pp. 151–153, Lunar Science Institute Contr. No. 88.

El Goresy A. and Ramdohr P. (1972) Fra Mauro crystalline rocks: Petrology, geochemistry, and subsolidus reduction of the opaque minerals (abstract). In *Lunar Science—III* (editor C. Watkins), pp. 224–226, Lunar Science Institute Contr. No. 88.

Ford C. E., Humphries D. J., Wilson G., Dixon D., Biggar G. M., and O'Hara M. J. (1972) Experimental petrology of high alumina basalt, 14310, and related compositions (abstract). In *Lunar Science—III* (editor C. Watkins), pp. 274–276, Lunar Science Institute Contr. No. 88.

Gancarz A. J., Albee A. L., and Chodos A. A. (1971) Petrologic and mineralogic investigation of some crystalline rocks returned by the Apollo 14 mission. *Earth Planet. Sci. Lett.* **12,** 1–18.

Griffin W. L., Åmli R., and Heier K. S. (1972) Whitlockite and apatite from lunar rock 14310 and Ödegärden, Norway. *Earth Planet. Sci. Lett.* **15,** 53–88.

Helmke P. A. and Haskin L. A. (1972) Rare earths and other trace elements in Apollo 14 lunar samples (abstract). In *Lunar Science—III* (editor C. Watkins), pp. 366–368, Lunar Science Institute Contr. No. 88. See all their papers in this volume.

Hollister L., Trzcienski W. Jr., Dymek R., Kulick C., Weigand P., and Hargraves R. (1972) Igneous fragment 14310,21 and the origin of the Mare basalts (abstract). In *Lunar Science—III* (editor C. Watkins), pp. 386–388, Lunar Science Institute Contr. No. 88.

Hubbard N. J. and Gast P. (1972) Chemical composition of Apollo 14 materials and evidence for alkali volatilization (abstract). In *Lunar Science—III* (editor C. Watkins), pp. 407–409, Lunar Science Institute Contr. No. 88.

Jackson P. F. S., Coetzee J. H. J., Strasheim A., Strelow F. W. E., Gricius A. J., Wybenga F., and Kokot M. L. (1972) The analysis of lunar material returned by Apollo 14 (abstract). In *Lunar Science—III* (editor C. Watkins), pp. 424–426, Lunar Science Institute Contr. No. 88.

Jagodzinski H. and Korekawa M. (1972) X-ray-studies of plagioclases and pyroxenes (abstract). In *Lunar Science—III* (editor C. Watkins), pp. 427–429, Lunar Science Institute Contr. No. 88.

Keith J. E., Clark R. S., and Richardson K. A. (1972) Gamma ray measurements of Apollo 12, 14, and 15 lunar samples (abstract). In *Lunar Science—III* (editor C. Watkins), pp. 446–448, Lunar Science Institute Contr. No. 88.

Kushiro I. (1972) Petrology of lunar high-alumina basalt (abstract). In *Lunar Science—III* (editor C. Watkins), pp. 466–468, Lunar Science Institute Contr. No. 88.

Laul J. C., Boynton W. V., and Schmitt R. A. (1972) Bulk, REE, and other elemental abundances in four Apollo 14 clastic rocks and three core samples, two Luna 16 breccias and four Apollo 15 soils (abstract). In *Lunar Science—III* (editor C. Watkins), pp. 480–482, Lunar Science Institute Contr. No. 88.

Melson W. G., Mason B., and Nelen J. (1972) Apollo 14 basaltic rocks (abstract). In *Lunar Science—III* (editor C. Watkins), pp. 535–536, Lunar Science Institute Contr. No. 88.

Morgan J. W., Laul J. C., Krähenbühl U., Ganapathy R., and Anders E. (1972) Major impacts on the moon: Chemical characterization of projectiles (abstract). In *Lunar Science—III* (editor C. Watkins), pp. 552–554, Lunar Science Institute Contr. No. 88.

Ridley W. I., Williams R. J., Brett R., and Takeda H. (1972) Petrology of lunar basalt 14310 (abstract). In *Lunar Science—III* (editor C. Watkins), pp. 648–650, Lunar Science Institute Contr. No. 88.

Ringwood A. E., Green D. H., and Ware N. G. (1972) Experimental petrology and petrogenesis of Apollo 14 basalts (abstract). In *Lunar Science—III* (editor C. Watkins), pp. 654–656, Lunar Science Institute Contr. No. 88.

Takeda H. and Ridley W. I. (1972) Crystallography and mineralogy of pyroxenes from Fra Mauro soil and rock 14310 (abstract). In *Lunar Science—III* (editor C. Watkins), pp. 738–740, Lunar Science Institute Contr. No. 88.

Taylor S. R., Muir P., and Kaye M. (1971) Trace element chemistry of Apollo 14 lunar soils from Fra Mauro. *Geochim. Cosmochim. Acta* **35,** 975–981.

Wakita H., Showalter D. L., and Schmitt R. A. (1972) Bulk, REE, and other abundances in Apollo 14 soils (3), clastic (1), and igneous (1) rocks (abstract). In *Lunar Science—III* (editor C. Watkins), pp. 767–769, Lunar Science Institute Contr. No. 88.

Walker D., Longhi J., and Hays F. J. (1972) Experimental petrology and origin of Fra Mauro rocks and soil (abstract). In *Lunar Science—III* (editor C. Watkins), pp. 770–772, Lunar Science Institute Contr. No. 88.

Walter L. S., French B. M., and Doan A. S. Jr. (1972) Petrographic analysis of lunar samples 14171 and 14305 (breccias) and 14310 (melt rock) (abstract). In *Lunar Science—III* (editor C. Watkins), pp. 773–775, Lunar Science Institute Contr. No. 88.

Wänke H., Baddenhausen H., Balacescu A., Teschke F., Spettel B., Dreibus G., Quijano M., Kruse H., Wlotzka F., and Begemann F. (1972) Multielement analyses of lunar samples (abstract). In *Lunar Science—III* (editor C. Watkins), pp. 779–781, Lunar Science Institute Contr. No. 88.

## Appendix

*Elemental composition of sample 15601,75 by neutron activation analysis.*

| Element | 15601 Bulk regolith | Element | 15601 Bulk regolith |
|---|---|---|---|
| Na (%) | 0.24 | | |
| Mg (%) | 6.9 | Sb (ppm) | < 0.03 |
| Al (%) | 5.83 | Cs (ppm) | 0.30 |
| Cl (ppm) | 7.6 | Ba (ppm) | 135 |
| K (%) | 0.18 | La (ppm) | 11.3 |
| Ca (%) | 6.8 | Ce (ppm) | 29 |
| Sc (ppm) | 35.1 | Pr (ppm) | 3.0 |
| Ti (%) | 1.18 | Sm (ppm) | 6.32 |
| V (ppm) | 200 | Eu (ppm) | 1.0 |
| Cr (%) | 0.351 | Tb (ppm) | 1.3 |
| Mn (%) | 0.1962 | Dy (ppm) | 9.7 |
| Fe (%) | 14.6 | Ho (ppm) | 1.8 |
| Co (ppm) | 48.9 | Er (ppm) | 5.7 |
| Ni (ppm) | 170 | Yb (ppm) | 5.2 |
| Cu (ppm) | 6.4 | Lu (ppm) | 0.9 |
| Zn (ppm) | 9.8 | Hf (ppm) | 4.86 |
| Ga (ppm) | 3.1 | Ta (ppm) | 0.60 |
| As (ppm) | < 0.005 | W (ppm) | 0.066 |
| Rb (ppm) | 5.3 | Th (ppm) | 1.52 |
| Sr (ppm) | 120 | U (ppm) | 0.46 |

Proceedings of the Third Lunar Science Conference
(Supplement 3, *Geochimica et Cosmochimica Acta*)
Vol. 2, pp. 1149–1160
The M.I.T. Press, 1972

# Oxygen and bulk element composition studies of Apollo 14 and other lunar rocks and soils

WILLIAM D. EHMANN

Department of Chemistry, University of Kentucky, Lexington, Kentucky 40506
and Visiting Professor, Department of Oceanography, Florida State University,
Tallahassee, Florida 32306

and

DAVID E. GILLUM and JOHN W. MORGAN*

Department of Chemistry, University of Kentucky,
Lexington, Kentucky 40506

**Abstract**—Abundances of O, Si, Al, Mg, and Fe have been determined in 13 rock samples and 3 soils from the Apollo 14 mission. Most of the crystalline rocks and breccias fall closely along a Si-rich extension of the Si–O regression line reported for Apollo 11 crystalline rocks: O% = 0.98 Si% + 20.8 (Ehmann and Morgan, 1970). This slope is significantly different than that for terrestrial igneous rocks, and the lunar rocks are depleted by 1–2% O with respect to terrestrial rocks of comparable Si contents. The lunar fines exhibit an excess O deficiency in addition to that determined by the mineralogy. Various models for this depletion are discussed in the light of recent data. The bulk chemistry of rock 14053 is similar to that found for a clast in breccia 14321 and also resembles that for Si-rich basalt 12064. Rock 14310 has the highest Al abundance and Al/Si ratio of any bulk lunar material we have analyzed and may represent remelted soil from near the mountain fronts. Additional data are presented for abundances in Apollo 11 and Lunar 16 sieved fines and Apollo 12 rocks and fines.

## INTRODUCTION

INSTRUMENTAL ACTIVATION ANALYSIS employing 14 MeV and thermal neutrons has been used to determine a number of major, minor, and trace elements in Apollo 14, Luna 16, and other lunar rocks and fines. The method used is essentially non-destructive, except for minor amounts of long-lived radioactivities induced in the samples and radiation damage. The samples used in these studies have been returned to the Lunar Sample Curator and could be used for a variety of other studies.

We have previously reported major element abundance studies on Apollo 11 rocks and fines (Ehmann and Morgan, 1970), Apollo 12 rocks and fines (Ehmann and Morgan, 1971), rock 12013 (Morgan and Ehmann, 1970a), and Luna 16 fines (Gillum *et al.*, 1972). This paper presents new data derived from the analyses of Apollo 14 rocks and fines and additional analyses of samples collected on previous missions. The data are discussed with reference to our previous findings and the reports of other investigators, as presented at the Third Lunar Science Conference.

* Present address: Enrico Fermi Institute for Nuclear Studies, University of Chicago, Chicago, Illinois 60637.

It should be noted that our samples (with a few exceptions, as noted later) have been furnished to us and processed in our laboratory under a dry nitrogen atmosphere. The oxygen data of our group are the only direct oxygen determinations reported on lunar materials that have not been exposed to the terrestrial atmosphere and should closely represent the true oxygen abundances in these materials on the lunar surface. Grossman *et al.* (1972) and other scientists have suggested that the surfaces of lunar samples are highly reactive with respect to $O_2$ and $H_2O$ vapor. As will be discussed later, our experience suggests that lunar fines that have been exposed to the terrestrial atmosphere are enriched by 1–2% O over samples which have been maintained under dry $N_2$.

## Analytical Method

Detailed procedures for the determination of major elements by 14 MeV neutron activation have been reported previously by Morgan and Ehmann (1970b; 1971) and will not be repeated here. The instruments, sample preparation procedures, and standards used specifically for the analyses of lunar samples are described by Ehmann and Morgan (1971).

The techniques used in the thermal neutron activation analyses for minor and trace elements have also been reviewed previously by Ehmann (1970) and in other publications from our laboratory. The thermal neutron activation data presented in this paper were obtained by irradiation of the samples in sealed polyethylene vials for 30 minutes at a flux of $10^{13}$ n cm$^{-2}$ sec$^{-1}$ at the University of Missouri Research Reactor Facility. Primary standards were used for Sc and Co determinations, while three splits of U.S.G.S. standard rock BCR–1 were used as secondary standards in the determination of Na, Cr, Eu, Hf, and Ta. The composition of BCR–1 was taken from the average values reported by Flanagan (1969). The samples were counted with an ORTEC Ge(Li) detector with an efficiency of 3.9% and a resolution of 2.7 keV (fwhm) at $^{60}$Co.

Sample sizes ranged from 17 mg for the Luna 16 fines to as much as 2 g for some of the Apollo 14 samples. The majority of the samples analyzed were 0.5 to 1.0 g. All samples were maintained under a dry nitrogen atmosphere throughout our packaging and analytical procedures. However, the two Luna 16 samples and Apollo 14 fines 14163 and 14259 were reported by the Lunar Sample Curator to have been exposed to the terrestrial atmosphere prior to receipt in our laboratories. In the case of the Apollo 14 samples, the exposure was apparently due to a failure of the primary vacuum box seal.

The Luna 16 and Apollo 14 fines had been sieved prior to receipt in our laboratories. The Apollo 11 fines (10084,50) were sieved in our laboratories with nylon screens under a dry nitrogen atmosphere. The mass distribution of the various size fractions with respect to the total fines was:

| Size Fraction | % of Total Fines |
|---|---|
| >0.25 mm | 23.93 |
| 0.125–0.250 mm | 67.17 |
| 0.084–0.125 mm | 8.64 |
| <0.084 mm | 0.26 (not analyzed) |

The largest size fraction consisted mainly of glass beads and slaggy glass fragments. Analyses for Zr, Hf, and a number of other trace elements in these sieved fines are currently in progress and will be reported later.

## Results

The results of over 400 replicate analyses for O, Si, Al, Mg, and Fe in Apollo 14 rocks and fines are presented in Table 1. Normally, 5 to 10 replicate analyses were done for O and Si, and 3 to 4 replicate analyses for Al, Mg, and Fe in each sample.

Table 1. Major elements in Apollo 14 rocks and fines.*

| Sample | %O | %Si | %Al | %Mg | %Fe | %Ca Other Work | %Ti (Ref)† |
|---|---|---|---|---|---|---|---|
| Cryst. Rocks | | | | | | | |
| 14053,42 | 42.1 | 22.2 | 6.8 | 5.0 | 13.0 | 8.0(1) | 1.8(1) |
| 14310,113 | 43.1 | 21.9 | 10.4 | 3.6 | 6.5 | — | — |
| | — | [22.8] | [10.8] | [3.8] | [6.8] | 8.9(1) | 0.7(1) |
| Breccias | | | | | | | |
| 14047,29 | 41.3 | 23.0 | 9.1 | 5.9 | 7.9 | 8.2(4) | 1.1(4) |
| 14303,15–3 | 43.5 | 23.1 | 8.6 | 7.2 | 8.6 | — | — |
| 14303,15–2,4,5 | 43.6 | 23.1 | 8.8 | 6.1 | 7.9 | — | — |
| 14303,15–6 | 42.9 | 23.0 | 8.8 | 5.5 | 7.7 | — | — |
| Avg. 14303 | 43.3 | 23.1 | 8.7 | 6.3 | 8.1 | — | — |
| 14305,76 | 41.1 | 20.4 | 7.5 | 6.5 | 8.0 | — | — |
| | — | [22.2] | [8.2] | [7.1] | [8.7] | 6.0(2) | 1.0(2) |
| 14311,66 | 42.9 | 22.8 | 9.2 | 6.6 | 9.1 | 7.3(3) | 1.1(3) |
| 14321,64 | 43.6 | 21.8 | 7.4 | 5.7 | 9.7 | — | — |
| 14321,171A | 41.5 | 21.5 | 8.8 | 6.4 | 7.8 | — | — |
| 14321,171B | 42.1 | 22.5 | 8.9 | 5.0 | 8.5 | — | — |
| 14321,225A | 43.1 | 22.7 | 8.0 | 6.1 | 9.5 | — | — |
| 14321,225B | 42.8 | 21.5 | 6.7 | 4.9 | 12.0 | — | — |
| Avg. 14321 | 42.6 | 22.0 | 8.0 | 5.6 | 9.5 | 7.1(3) | 1.2(3) |
| Avg. breccias | 42.2 | 22.3 | 8.5 | 6.2 | 8.5 | 7.2 | 1.1 |
| Fines | | | | | | | |
| 14003,13 | 43.2 | 23.2 | 9.2 | 4.6 | 8.4 | 7.8(3) | 1.2(3) |
| 14163,87 | 44.5 | 23.4 | 9.1 | 4.9 | 7.9 | 6.6(2) | 0.9(2) |
| 14259,65 | 44.8 | 23.1 | 9.3 | — | 8.4 | 7.7(2) | 0.9(2) |
| Avg. fines | 44.2 | 23.2 | 9.2 | 4.8 | 8.2 | 7.4 | 1.0 |

* Standard deviations of the means for the replicate analyses are typically ±0.3% O, 0.2% Si, 0.1% Al, 0.4% Mg, and 0.3% Fe.

† Summations of major element abundances determined in this work together with data from other work for Ca and Ti, where available, equaled 98 ± 1% for all samples except saw chip and sawdust samples 14310,113 and 14305,76. Dilution decrements of 4% for 14310 and 8% for 14305 are estimated from these summations, and corrected data are given in brackets. References for the Ca and Ti data are: (1) Compston *et al.* (1972), (2) Wänke *et al.* (1972), (3) Scoon (1972), (4) Wakita *et al.* (1972).

*Note:* Other sample descriptions—14053, 14311, and 14321,64 were furnished to us as homogeneous fines; 14047 was a dark gray chip; 14303,15–3 was a largely white matrix chip; 14303,15–2,4,5 was a mixture of small gray-white chips; 14303,15–6 appeared to be a medium gray lithic fragment (14303 was separated and furnished to us by Dr. V. R. Murthy); 14321,171A and B were medium grained heterogeneous chips; 14321,225A was a coarser grained heterogeneous chip; 14321,225B appeared to be largely a lithic clast; 14003, 14163, and 14259 were less than 1 mm sieve fractions of these fines. The 14163 and 14259 fines had been exposed to the terrestrial atmosphere, and we estimate the O abundances reported here may reflect a pick up of 1–2% O, based on comparisons with pristine fines sample 14003.

*Note added in proof:* Analyses for Ti and Mn in a number of these samples were completed after submittal of this manuscript. The following data are based on 2 to 5 replicate analyses using standard rocks BCR-1 and W-1 for comparators. The relative standard deviations of the means for the replicate determinations were 1 to 2%. The determinations are based on $^{252}$Cf instrumental neutron activation analysis.

| Sample | %Ti | %Mn |
|---|---|---|
| 14047,29 | 1.15 | 0.098 |
| 14053,42 | 1.93 | 0.190 |
| 14310,113 | 0.64** | 0.087** |
| 14321,171A | 0.83 | 0.089 |
| 14321,171B | 0.83 | 0.099 |
| 14321,225A | 1.24 | 0.119 |
| 14321,225B | 1.40 | 0.157 |

Analysts: M. Janghorbani and J. Storm.

** Corrected for 4% dilution.

The accuracy of the determinations was checked by analyzing various standard rocks along with the lunar samples. The results obtained for U.S.G.S. standard rock BCR–1 over the course of these analyses are given in Table 2 together with the compilation values of Flanagan (1969). The descriptions given are based on our observations of the samples received and on information provided by the Lunar Sample Curator. The samples 14310,113 (saw cut chips) and 14305,76 (sawdust powder) clearly exhibited evidence of contamination. Fibers resembling cellulose from filter papers were found in both samples but were particularly abundant in sample 14305,76. Summations of major element abundances determined in this work together with abundances of Ca and Ti from other work presented at the Third Lunar Science Conference equaled 98 ± 1% for all our samples except the two contaminated samples just noted. The reported abundances of Na, K, Cr, and Mn account almost exactly for the residual 2 ± 1% in the samples. Dilution decrements of approximately 4% for 14310,113 and 8% for 14305,76 are estimated from the summed major element abundance data. Showalter *et al.* (1972) have shown a dilution decrement of approximately 27% in sawdust derived from cutting rock 12013. Apparently the contamination is largely C and Cu derived from the diamond saw wire and from the fibers described previously. Abundances of Si, Al, Mg, and Fe corrected for dilution are

Table 2. Some major element abundances in sieved Apollo 11 and Luna 16 fines.*

| Sample | Size (mm) | %O | %Si | %Al | %Mg | %Fe |
|---|---|---|---|---|---|---|
| 10084,50 | >0.25 | 41.0 | 20.4 | 7.0 | 4.6 | 12.3 |
| 10084,50 | 0.125–0.250 | 40.3 | 19.9 | 7.2 | 5.0 | 11.8 |
| 10084,50 | 0.084–0.125 | 43.4 | 20.3 | 8.3 | 4.9 | 11.1 |
| Luna–16–A–5 | <0.125 | 40.2 | 20.5 | 9.2 | — | — |
| Luna–16–G–5 | <0.125 | 43.0 | 22.2 | 10.3 | — | — |
| Rock BCR–1 | This work | 44.8 | 25.6 | 7.2 | 2.1 | 9.5 |
| | Flanagan (1969) | 44.6 | 25.4 | 7.2 | 2.0 | 9.5 |

* Standard deviations of the means for the replicate analyses are typically ±0.2% O, 0.2% Si, 0.3% Al, 0.3% Mg, and 0.3% Fe. Luna 16 fines had been exposed to the atmosphere prior to receipt. Apollo 11 fines and BCR–1 were maintained in a dry $N_2$ atmosphere.

Table 3. New determinations of some minor and trace elements in Apollo 12 rocks and fines.*

| Sample | Type | %Na | %Cr | ppm Sc | ppm Co | ppm Eu | ppm Hf | ppm Ta |
|---|---|---|---|---|---|---|---|---|
| Rocks | | | | | | | | |
| 12004,32 | A | — | 0.42 | 43.3 | 51.7 | 0.86 | 2.2 | — |
| 12063,60 | A | 0.21 | 0.34 | 60.8 | 43.4 | 1.31 | 4.6 | — |
| 12051,46 | AB | — | 0.24 | 58.0 | 35.1 | 1.21 | 3.1 | — |
| 12021,81 | B | 0.20 | 0.25 | 53.9 | 29.5 | 1.08 | 2.2 | — |
| 12022,53 | B | 0.19 | 0.33 | 53.8 | 42.5 | 0.99 | 3.4 | 0.7 |
| 12035,04 | B | — | 0.39 | 33.7 | 53.2 | 0.66 | <1 | — |
| Fines | | | | | | | | |
| 12032,34 | | 0.45 | 0.24 | 36.5 | 35.6 | 2.20 | 13.1 | 1.8 |
| 12044,18 | | 0.37 | 0.28 | 43.0 | 42.1 | 1.67 | 10.5 | 1.4 |
| 12057,79 | | — | 0.31 | 44.7 | 42.5 | 1.56 | 14.0 | 1.5 |
| 12070,70 | | — | 0.28 | 42.7 | — | — | 11.2 | — |

* Co and Sc analyses are based on primary standards, and these data are assigned an accuracy of ±5%. The remaining elements were determined relative to three splits of U.S.G.S. standard rock BCR–1 using the values of Flanagan (1969), and these data are assigned on accuracy of ±10%. Major element abundances in these samples have been published previously in Ehmann and Morgan (1971).

given in brackets in Table 1. A corrected value is not given for O, since the contamination with cellulose fibers would already have made the value of the O determinations questionable. A more accurate estimate of the dilution decrement could probably be obtained by direct determination of C and Cu in these samples.

Some major element abundances in sieved Apollo 11 fines 10084,50 and two samples of Luna 16 fines are presented in Table 2. The sample Luna–16–A–5 (NASA #) is from level A (~7 cm) and sample Luna–16–G–5 (NASA #) is from level D (~30 cm depth). These Luna 16 data together with other elemental abundances determined at Oregon State University are discussed in detail elsewhere (Gillum *et al.*, 1972).

A variety of trace element abundances in Apollo 12 rocks and fines, which were determined following our previous report on Apollo 12 major element abundances (Ehmann and Morgan, 1971), are given in Table 3. These elements were determined by instrumental thermal neutron activation and are the only determinations that have been reported for some of these elements in several of the samples. Where data comparisons are possible, the agreement with published data is quite satisfactory. The major discrepancy is in the case of Hf, where our analyses agree well with those of Wakita *et al.* (1971) and Smales *et al.* (1971) but are significantly lower than the data presented by Wänke *et al.* (1971) and Brunfelt *et al.* (1971). Our Hf data suggest that additional confidence should be ascribed to the lower Hf values.

## Discussion

### *Major element relationships in Apollo 14 materials*

In our previous studies (Ehmann and Morgan, 1970; 1971) we have shown that the lunar crystalline rocks are depleted by approximately 1–2% O with respect to terrestrial igneous rocks of comparable Si contents. The lunar rocks define a Si–O regression line with a slope (0.96–0.98) that is distinctly different than that found for terrestrial igneous rocks (0.415). This difference must be due in part to highly reducing conditions at the time of crystallization of the lunar basalts. The Si–O relationships for the Apollo 14 crystalline rocks and breccias are consistent with these previous observations. The majority of the Apollo 14 rocks analyzed in this work fall closely along a Si-rich extension of the Si–O regression line for the Apollo 11 crystalline rocks: $O\% = 0.98\ Si\% + 20.8$ (Fig. 1). Actually, only 14053 and 14310 are crystalline rocks, and the remainder of the rocks analyzed are breccias. Rock 14310,113 was furnished to us in the form of saw chips and use of the dilution-corrected Si abundance would move it even closer to the Apollo 11 regression line. LSPET (1971) has noted that rock 14053 was collected from the side of a large boulder at Station $C_2$ and may be a clast weathered from this boulder. Indeed, the bulk element abundances for this rock almost exactly equal the abundances we found in sample 14321, 225B, which appeared to be largely a fragment of a clast from this heterogeneous breccia.

Rock 14053 has been described as being more primitive than the mare basalts and may be representative of an early-stage liquid suitable as parent material for the

Apollo 12 basalts (Biggar *et al.*, 1972). In Fig. 1, rock 14053 falls between the Apollo 11 and 12 regression lines, and its bulk element abundances are very similar to those we reported previously for the Si-rich Apollo 12 basalt, 12064,34 (Ehmann and Morgan, 1971). Several investigators have suggested that 14053 is highly reduced and may be a good candidate for the occurrence of the species $Ti^{3+}$ (Bence *et al.*, 1972; Haggerty, 1972; Pavícevíc *et al.*, 1972). Our data confirm that 14053 is somewhat O-deficient with respect to other Apollo 14 rocks (14310 and 14321) with comparable Si contents.

Numerous investigators (for example, Helmke and Haskin, 1972; Schnetzler *et al.*, 1972; Morgan *et al.*, 1972; Ridley *et al.*, 1972) have suggested that rock 14310 may have been formed by fusion of material similar to Apollo 14 soils or breccias. In comparison with any of the bulk lunar materials we have analyzed to date, it is true that the bulk element abundances in 14310,113 most closely resemble those

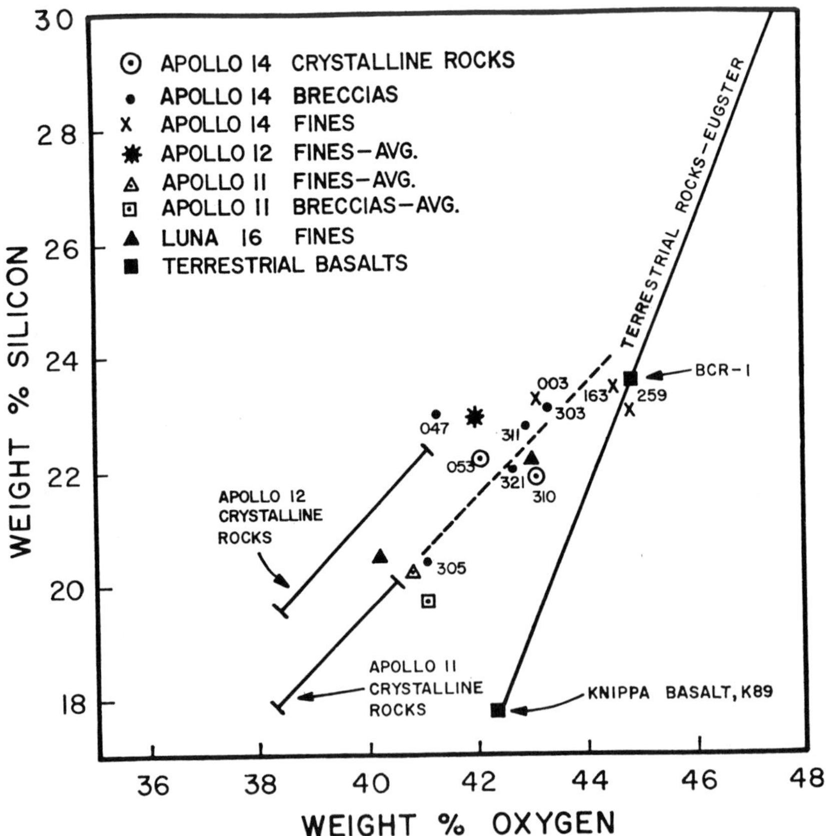


Fig. 1. Si–O relationship in Apollo 14 rocks and fines. The line for terrestrial igneous rocks is taken from Eugster (1969). The points for terrestrial basalts BCR–1 and K 89 were taken from data obtained in our laboratory. The data for Apollo 11 and 12 materials were taken from our previous work (Ehmann and Morgan, 1970; 1971).

found for the Apollo 14 soils. However, the Al abundance in this rock is the highest we have found in any bulk lunar sample we have analyzed (LSPET, 1972; Al data for anorthositic Apollo 15 rocks 15415 and 15418 are still higher). Figure 2 indicates that the Al/Si weight ratio for this rock (0.48) greatly exceeds the values for the three Apollo 14 fines we analyzed (0.39 $\pm$ 0.01) and is approached only by our values for the two Luna 16 fines (0.45–0.46). The x-ray fluorescence data obtained during the Apollo 15 mission (Adler *et al.*, 1972) suggest Al/Si concentration ratios for the lunar highlands are in the range 0.42 to 0.78, while the mare values range from 0.29 to 0.38. These mare values are in reasonable agreement with the values obtained in our analyses of mare material, as shown in Fig. 2. It should be noted that our Luna 16 samples were less than 0.125 mm fines, and the true bulk Al values in these layers may be lower than we report for these sieve fractions. In Table 2 we present data for sieved Apollo 11 fines 10084,50, which clearly show an increase in the Al/Si weight

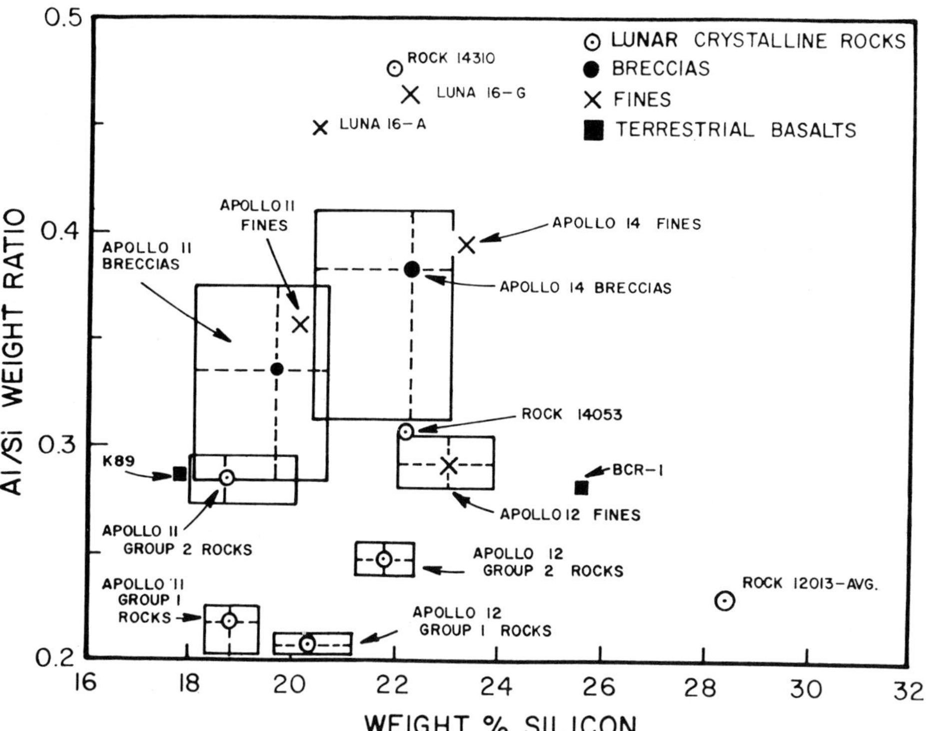


Fig. 2. The Al/Si weight ratio in lunar materials, as a function of weight percent Si, based on samples analyzed in this laboratory. The symbols represent the mean values, and the boxes represent the total range of analytical data for each type material. The range of values for the Apollo 11 and Apollo 14 fines is approximately the size of the symbol used for the mean value. Points for terrestrial basalts BCR-1 and K 89 are based on data obtained in this laboratory. For comparison the Al/Si ratios for the Type I carbonaceous chondrites, ordinary chondrites, and Ca-rich achondrites are 0.08, 0.06, and 0.2, respectively.

ratio from 0.34 to 0.41 for the coarsest to the finest sieve fractions. Similar abundance trends in sieve fractions have been noted for Apollo 11 fines by Goles *et al.* (1970), and for Apollo 12 fines by Wänke *et al.* (1972). The high Al/Si ratio for rock 14310,113 suggests to us that it is unlikely that this rock could have been formed by simple fusion of the Apollo 14 soils we have analyzed, unless it was derived from an aggregate of the finest particles in a region where some size sorting process had taken place. It is more likely that 14310 is derived from a soil formed closer to the Al-rich units, which must occur in the mountain fronts. This is supported by the observations of LSPET (1972) that front soil samples collected during the Apollo 15 mission are consistently higher in Al and lower in Fe than the mare-type soil collected near Hadley Rille. If we exclude rock 14310, our data for all lunar sites show an Al enrichment, a Fe depletion, and little change in Si in the fines as compared to the associated crystalline rocks. These major element variations are consistent with those predicted by mixing material similar to anorthosite rock 15415 (Si = 20.6%, Al = 18.8%, Fe = 0.18%; LSPET, 1972) with fragmentary material derived from the mare basalts.

LSPET (1972) has also pointed out a strong inverse correlation of Al with Fe in the Apollo 15 fines and breccias. In fact, their data for the Apollo 15 crystalline rocks (with the exception of anorthositic rocks 15415 and 15418) also fall close to their Al–Fe regression line for the fines and breccias. Using our analytical data on a large group of lunar samples including Apollo 11 and 12 Type B basalts, Apollo 14 rocks 14053 and 14310, all lunar fines and breccias, and the LSPET Apollo 15 data (excluding the anorthosites), we find the data fit surprisingly well an Al–Fe regression line given by the equation $Al\% = -0.56\ Fe\% + 13.7$. Only the Type A Apollo 11 and 12 basalts (Al-deficient) and the anorthosites (Al-rich) deviate appreciably from this correlation. We hope to investigate this interesting relationship in more detail when we have completed our own analyses of the Apollo 15 samples.

In our previous papers (Ehmann and Morgan, 1970; 1971) we pointed out an apparent excess O deficiency of approximately 1.6% O in Apollo 11 and 12 fines and breccias, as compared to the amount of O required for simple stoichiometry based on total silicate analyses. This O deficiency with respect to simple stoichiometry was smaller, or nonexistent, in most of the crystalline rocks. Similar results expressed in terms of "excess reducing capacity" have been reported by Rose *et al.* (1970) and Cuttitta *et al.* (1971). They correctly point out that this effect cannot be completely accounted for by the metallic Fe or the S content of the samples and suggest the presence of $Ti^{3+}$. Based on our previous work, we suggested that at least a part of this effect might be due to reduction of the surface fines and breccias by interaction with solar-wind H. Indeed, most of the lunar fines and breccias have high abundances of solar-wind rare gases.

It is interesting to note that soil 12033, which was recovered from approximately 15 cm below the lunar surface, is reported by Cuttitta *et al.* (1971) to have an "excess reducing capacity" of only 0.2–0.3%, which is far less than values they obtained for the surface lunar soils they analyzed. Epstein and Taylor (1971) state that 12033 soil has the lowest solar-wind $H_2$ concentration among the soils and breccias that they analyzed. Figure 1 shows that breccia 14047 lies to the O-deficient side of the regression line defined by most of the other Apollo 14 rocks, including the large breccia rock 14321. LSPET (1972) reports that rock 14047 (and several other rocks not analyzed here) is enriched in solar-wind noble gases by approximately two orders of magnitude over a group of four other rocks, which includes rock 14321. Epstein and Taylor (1972) detected almost no solar-wind H in the interior of 14321, which suggests that this rock could not be classed as a polymict breccia of lunar soil and clastic fragments. Silver (1972) and Wakita *et al.* (1972) suggest 14047 is merely compacted soil, which would be consistent with the apparent excess O depletion we find. Unfortunately, our Apollo 14 analyses were not able to shed much additional light on this problem, since a number of our samples were unsuitable for such investigations (sawdust, saw chips, fines exposed to the terrestrial atmosphere prior to our receipt), and abundance data on several additional major and minor elements required for the stoichiometry calculations were not available to us at this writing. Using our available data and some preliminary data selected from reports at the Third Lunar Science Conference, we calculated an apparent O deficiency with respect to stoichiometry of 1.3% for pristine fines sample 14003,13 and $\sim 1\%$ O for crystalline rock 14053. This latter rock has been reported by Haggerty (1972) to be intensely reduced and to contain veinlets of Fe and FeS.

Epstein and Taylor (1971; 1972), Bibring *et al.* (1972), and others have detected ultrathin coatings on particles of lunar fines, which could be due to solar-wind effects and/or redeposition from a transient lunar atmosphere resulting from vaporization of silicate material during meteorite, micrometeorite, or particle bombardment of the lunar surface. Epstein and Taylor have determined that this surface layer material is more depleted in O than in Si and may have O/Si ratios as low as 0.8 to 1.3. We feel the relation of the O depletion to the solar-wind gas contents of the lunar fines and breccias, as reviewed here, supports our original suggestion of solar-wind H reduction of these materials, although the existence of $Ti^{3+}$ and metallic Fe in the soil precursors and thin film redeposition from a transient lunar atmosphere may also play important roles in the observed effect.

### *Minor and trace element relationships in some Apollo 12 materials*

In our previous work we reported major element abundances in a group of Apollo 12 rocks and fines (Ehmann and Morgan, 1971) and here report the presence of a group of minor and trace element abundances obtained later in these same samples. We previously reported that rocks 12004, 12022, and 12063 exhibit lower Al abundances and higher Fe and Mg abundances than Type B rocks 12021 and 12051. Rock 12035 is anomalous, having low Al but somewhat higher Fe and Mg abundances than the low-Al Type A rocks. This is a reflection of its unusual mineralogy, which would classify it terrestrially as a troctolite. It is of interest to examine briefly some of the trace element relationships in the light of these classifications based on major element abundances.

Chromium and Co are higher in the low-Al group (and in the troctolite) than in the high-Al group. If the anomalous troctolite is excluded, the distribution of both elements is strongly correlated with Mg. This is not too surprising, since Cr enrichment (as chromite) is often associated with ultrabasic rocks, and presumably the increase in Mg in the lunar rocks indicates a trend of increasing olivine. There is a strong correlation between Cr and Co, and the troctolite fits well into this trend. It appears that these two elements may be associated in the same mineral host and that this was not enriched to the same extent as olivine during the formation of the troctolite.

In the low-Al rocks the trace elements Hf, Eu, and Sc are correlated with mafic index [Fe/(Fe + Mg)], though the trend is largely defined by the troctolite. There are only two points for the high-Al rocks, so a trend cannot be defined. However, they do not coincide with the low-Al rock correlation. Among the trace elements themselves, Eu correlates with Sc, and both rock types fall closely on the same correlation line. In the case of Hf in the low-Al rocks, there is a strong correlation with Sc, and the point for the troctolite fits in well. The two points for the high-Al group do not fall on the trend but lie parallel to it at somewhat lower Hf values.

The elements Na, Eu, and Ta are enriched in the fines as compared to the typical Apollo 12 igneous rocks by a factor of approximately 2, while Hf is enriched by a factor of 4. The elements Cr, Sc, and Co exhibit similar abundance ranges in the fines and igneous rocks, while Fe is slightly depleted in the fines. As with the rocks, Cr and Co are strongly correlated in the fines and plot quite closely to a correlation line derived from the rock data. In distinction to the rocks, both elements in the fines are positively correlated with Sc. More surprisingly, the fines data suggest that Eu is *negatively* correlated with Sc (although the trend is determined largely by the fines sample 12032), which is quite the opposite to the positive trend found in the crystalline rocks.

*Acknowledgments*—Financial support for this work was provided by NASA grant NGR 18-001-058 and the University of Kentucky Research Foundation. We also wish to acknowledge the technical assistance in computer data processing of Mr. John Jones and the University of Kentucky Computer Center.

## References

Adler I., Trombka J., Gerard J., Lowman P., Yin L., Blodgett H., Gorenstein P., and Bjorkholm P. (1972) Preliminary results from the S–161 x-ray fluorescence experiment (abstract). In *Lunar Science—III* (editor C. Watkins), pp. 4–6, Lunar Science Institute Contr. No. 88.

Bence A. E. and Papike J. J. (1972) Crystallization histories of pyroxenes from lunar basalts (abstract). In *Lunar Science—III* (editor C. Watkins), pp. 59–61, Lunar Science Institute Contr. No. 88.

Bibring J. P., Maurette M., Meunier R., Durieu L., Jouret C., and Eugster O. (1972) Solar wind implantation effects in the lunar regolith (abstract). In *Lunar Science—III* (editor C. Watkins), pp. 71–73, Lunar Science Institute Contr. No. 88.

Biggar G. M., Ford C. E., Humphries D. J., Wilson G., and O'Hara M. J. (1972) Melting relations of more primitive mare-type basalts 14053 and M (Reid 1970); and of breccia 14321 and soil 14162 (average lunar crust?) (abstract). In *Lunar Science—III* (editor C. Watkins), pp. 74–76, Lunar Science Institute Contr. No. 88.

Brunfelt A. O., Heier K. S., and Steinnes E. (1971) Determination of 40 elements in Apollo 12 materials by neutron activation analysis. *Proc. Second Lunar Sci. Conf., Geochim. Cosmochim. Acta* Suppl. 2, Vol. 2, pp. 1281–1290. MIT Press.

Compston W., Vernon M. J., Berry H., Rudowski R., Gray C. M., Ware N., Chappel B. W., and Kaye M. (1972) Age and petrogenesis of Apollo 14 basalts (abstract). In *Lunar Science—III* (editor C. Watkins), pp. 151–153, Lunar Science Institute Contr. No. 88.

Cuttitta F., Rose H. J. Jr., Annell C. S., Carron M. K., Christian R. P., Dwornik E. J., Greenland L. P., Heiz A. W., and Ligon D. T. Jr. (1971) Elemental composition of some Apollo 12 lunar rocks and soils. *Proc. Second Lunar Sci. Conf., Geochim. Cosmochim. Acta* Suppl. 2, Vol. 2, pp. 1217–1229. MIT Press.

Ehmann W. D. (1970) Non-destructive techniques in activation analysis. *Fortschr. chem. Forsch.* **14/1,** 49–91.

Ehmann W. D. and Morgan J. W. (1970) Oxygen, silicon, and aluminum in Apollo 11 rocks and fines by 14 MeV neutron activation. *Proc. Apollo 11 Lunar Sci. Conf., Geochim. Cosmochim. Acta* Suppl. 1, Vol. 2, pp. 1071–1079. Pergamon.

Ehmann W. D. and Morgan J. W. (1971) Major element abundances in Apollo 12 rocks and fines by 14 MeV neutron activation. *Proc. Second Lunar Sci. Conf., Geochim. Cosmochim. Acta* Suppl. 2, Vol. 2, pp. 1237–1245. MIT Press.

Epstein S. and Taylor H. P. Jr. (1971) $O^{18}/O^{16}$, $Si^{30}/Si^{28}$, and $C^{13}/C^{12}$ ratios in lunar samples. *Proc. Second Lunar Sci. Conf., Geochim. Cosmochim. Acta* Suppl. 2, Vol. 2, pp. 1421–1441. MIT Press.

Epstein S. and Taylor H. P. Jr. (1972) $O^{18}/O^{16}$, $Si^{30}/Si^{28}$, $C^{13}/C^{12}$, and D/H studies of Apollo 14 and 15 samples (abstract). In *Lunar Science—III* (editor C. Watkins), pp. 236–238, Lunar Science Institute Contr. No. 88.

Eugster H. P. (1969) Oxygen, abundance in common igneous rocks. In *Handbook of Geochemistry* (editor K. H. Wedepohl), Vol. II, part I, Chap. 8, p. E–1. Springer-Verlag.

Flanagan F. J. (1969) U.S. Geological Survey standards—II. First compilation of data for the new U.S.G.S. rocks. *Geochim. Cosmochim. Acta* **33,** 81–120.

Gillum D. E., Ehmann W. D., Wakita H., and Schmitt R. A. (1972) Bulk and rare earth abundances in the Luna 16 soil levels A and D. *Earth Planet. Sci. Lett.* **13,** 444–449.

Goles G. G., Randle K., Osawa M., Lindstrom D. J., Jérome D. Y., Steinborn T. L., Beyer R. L., Martin M. R., and McKay S. M. (1970) Interpretations and speculations on elemental abundances in lunar samples. *Proc. Apollo 11 Lunar Sci. Conf., Geochim. Cosmochim. Acta* Suppl. 1, Vol. 2, pp. 1177–1194. Pergamon.

Grossman J. J., Mukherjee N. R., and Ryan J. A. (1972) Microphysical, microchemical, and adhesive properties of lunar material, III. Gas interaction with lunar material (abstract). In *Lunar Science—III* (editor C. Watkins), pp. 344–346, Lunar Science Institute Contr. No. 88.

Haggerty S. E. (1972) Subsolidus reduction and compositional variations of lunar spinels (abstract). In *Lunar Science—III* (editor C. Watkins), pp. 347–349, Lunar Science Institute Contr. No. 88.

Helmke P. A. and Haskin L. A. (1972) Rare earths and other trace elements in Apollo 14 lunar samples (abstract). In *Lunar Science—III* (editor C. Watkins), pp. 366–368, Lunar Science Institute Contr. No. 88.

LSPET (Lunar Sample Preliminary Examination Team) (1971) Preliminary examination of lunar samples from Apollo 14. *Science* **173,** 681–693.

LSPET (Lunar Sâmple Preliminary Examination Team) (1972) The Apollo 15 lunar samples: A preliminary description. *Science* **175,** 363–375.

Morgan J. W. and Ehmann W. D. (1970a) Lunar rock 12013; O, Si, Al, and Fe abundances. *Earth Planet. Sci. Lett.* **9,** 164–168.

Morgan J. W. and Ehmann W. D. (1970b) Precise determination of oxygen and silicon in chondritic meteorites by 14 MeV neutron activation using a single transfer system. *Anal Chim. Acta* **49,** 287–299.

Morgan J. W. and Ehmann W. D. (1971) 14 MeV neutron activation analysis of rocks and meteorites. *Activation Analysis in Geochemistry and Cosmochemistry* (editors A. O. Brunfelt and E. Steinnes), pp. 81–97. Universitetsforlaget, Oslo.

Morgan J. W., Krähenbuhl U., Ganapathy R., and Anders E. (1972) Volatile and siderophile elements in Apollo 14 and 15 rocks (abstract). In *Lunar Science—III* (editor C. Watkins), pp. 555–557, Lunar Science Institute Contr. No. 88.

Pavícevíc M., Ramdohr P., and El Goresy A. (1972) Microprobe investigations of the oxidation state of Fe and Ti in ilmenite in Apollo 11, Apollo 12, and Apollo 14·crystalline rocks (abstract). In *Lunar Science—III* (editor C. Watkins), pp. 596–598, Lunar Science Institute Contr. No. 88.

Ridley W. I., Williams R. J., Brett R., Takeda H., and Brown R. W. (1972) Petrology of lunar basalt 14310 (abstract). In *Lunar Science—III* (editor C. Watkins), pp. 648–650, Lunar Science Institute Contr. No. 88.

Rose H. J., Cuttitta F., Dwornik E. J., Carron M. K., Christian R. P., Lindsay J. R., Ligon D. T., and Larson R. R. (1970) Semimicro x-ray fluorescence analysis of lunar samples. *Proc. Apollo 11 Lunar Sci. Conf., Geochim. Cosmochim. Acta* Suppl. 1, Vol. 2, pp. 1493–1497. Pergamon.

Schnetzler C. C., Philpotts J. A., Nava D. F., Thomas H. H., Bottino M. L., and Barker J. L. Jr. (1972) Chemical compositions of Apollo 14, Apollo 15, and Luna 16 material (abstract). In *Lunar Science—III* (editor C. Watkins), p. 682, Lunar Science Institute Contr. No. 88.

Scoon J. H. (1972) Chemical analyses of lunar samples 14003, 14311, and 14321 (abstract). In *Lunar Science—III* (editor C. Watkins), pp. 690–691, Lunar Science Institute Contr. No. 88.

Showalter D. L., Wakita H., Smith R. H., Schmitt R. A., Gillum D. E., and Ehmann W. D. (1972) Chemical composition of sawdust from lunar rock 12013 and comparison of a Java tektite with the rock. *Science* **175**, 170–172.

Silver L. T. (1972) U–Th–Pb abundances and isotopic characteristics in some Apollo 14 rocks and soils and an Apollo 15 soil (abstract). In *Lunar Science—III* (editor C. Watkins), pp. 704–706, Lunar Science Institute Contr. No. 88.

Smales A. A., Mapper D., Webb M. S. W., Webster R. K., Wilson J. D., and Hislop J. S. (1971) Elemental composition of lunar surface material (part 2). *Proc. Second Lunar Sci. Conf., Geochim. Cosmochim. Acta* Suppl. 2, Vol. 2, pp. 1253–1258. MIT Press.

Wakita H., Rey P., and Schmitt R. A. (1971) Abundances of the 14 rare-earth elements and 12 other trace elements in Apollo 12 samples: Five igneous and one breccia rocks and four soils. *Proc. Second Lunar Sci. Conf., Geochim. Cosmochim. Acta* Suppl. 2, Vol. 2, pp. 1319–1329. MIT Press.

Wakita H., Showalter D. L., and Schmitt R. A. (1972) Bulk, REE and other abundances in Apollo 14 soils (3), clastic (1), and igneous (1) rocks (abstract). In *Lunar Science—III* (editor C. Watkins), pp. 767–769, Lunar Science Institute Contr. No. 88.

Wänke H., Wlotzka F., Baddenhausen H., Balcescu A., Spettel B., Teschke F., Jagoutz E., Kruse H., Quijano-Rico M., and Reider R. (1971) Apollo 12 samples: Chemical composition and its relation to sample locations and exposure ages, the two component origin of the various lunar soil samples and studies on lunar metallic particles. *Proc. Second Lunar Sci. Conf., Geochim. Cosmochim. Acta* Suppl. 2, Vol. 2, pp. 1187–1208. MIT Press.

Wänke H., Baddenhausen H., Balacescu A., Teschke F., Spettel B., Dreibus G., Quijano-Rico M., Kruse H., Wlotzka F., and Begemann F. (1972) Multielement analyses of lunar materials (abstract). In *Lunar Science—III* (editor C. Watkins), pp. 779–781, Lunar Science Institute Contr. No. 88.

Proceedings of the Third Lunar Science Conference
(Supplement 3, *Geochimica et Cosmochimica Acta*)
Vol. 2, pp. 1161–1179
The M.I.T. Press, 1972

# Nonmare basalts: Part II

N. J. HUBBARD and P. W. GAST
NASA Manned Spacecraft Center, Houston, Texas 77058

J. M. RHODES, B. M. BANSAL, and H. WIESMANN
Lockheed Electronics Company, Houston, Texas 77058

and

S. E. CHURCH
National Research Council, Houston, Texas 77058

**Abstract**—Three clearly distinguishable rock types are found on the moon. They are (1) iron-rich mare basalts, (2) high-Al basaltic rocks with relatively high abundances of trace elements (KREEP basalts), and (3) a group of plagioclase-rich basaltic to anorthositic rocks with low abundances of trace elements (low-K nonmare basalts, etc.). The latter two types are extensively metamorphosed and considered to be ancient material from the lunar highlands. Among the highland materials two "boundary" chemical compositions are recognized: (1) the trace element-rich KREEP basalts and (2) the trace element-poor low-K nonmare basalts. Both types give rise to anorthosites, although low-K anorthosites appear to be far more common. It is possible to derive both types of nonmare basalts from a common source by varying the extent of melting. A combined consideration of partial melting and thermal models indicates that the aluminous and refractory element-rich source for nonmare basalts must extend to $>100$ km depth.

## INTRODUCTION

THE RESULTS of extensive chemical studies of rocks from the lunar surface along with the chemical characterization of the lunar surface from lunar orbit (Adler *et al.*, 1972; Metzger *et al.*, 1972) suggest that three clearly distinguishable rock types are commonly found in the lunar crust, or upper 50 km of the moon. First, there are iron-rich mare basalts that dominate the collections returned from the Apollo 11, 12, and 15 sites. Abundant macroscopic specimens of this type have been studied. Thus, there is little ambiguity in the classification and description of these rocks. Second, there are aluminum-rich basaltic rocks with relatively high abundances of large ion lithophile elements (KREEP basalts). Very few unmetamorphosed fragments of these rocks have been described. Thus, their characterization is more ambiguous and dependent on their chemical composition, which we will show to be rather well defined. Third, there are plagioclase-rich anorthositic rocks that range from almost pure plagioclase to basaltic rocks with about two-thirds plagioclase. These plagioclase-rich rocks have very low trace element concentrations (Hubbard *et al.*, 1971b) and may be derived from a trace element-poor, plagioclase-rich basalt that is presently more of a concept than a known rock type. The characterization of these rocks is also primarily based on the study of extensively metamorphosed regolith fragments and on chemical compositions.

The widespread abundance of the latter two types has been recognized from the observation that large areas of the lunar highlands consist of Al-rich material (Adler *et al.*, 1972) and that the region around Mare Imbrium and the regolith in the northwestern Oceanus Procellarum region must have an unusually high Th content (Metzger *et al.*, 1972). The metamorphic nature of the latter two rock types is consistent with their apparent old age and the extensive cratering of the lunar surface in premare time. It should also be noted that this threefold classification of lunar rocks is undoubtedly oversimplified. A number of rocks with intermediate characterstics have already been found, suggesting that more detailed exploration of the lunar highlands will produce a greater variety of rock types than that found in the mare regions.

The objective of this paper is to characterize further the chemical characteristics of the large ion lithophyle (LIL) element-rich or KREEP basalts that were first identified as an important basalt type at the Apollo 12 site (Hubbard *et al.*, 1971a). It was anticipated from these early data that the Apollo 14 or Fra Mauro samples would have similar characteristics. The data presented in this paper clearly establish this similarity. In addition, some rocks with intermediate characteristics are described, and the origin of plagioclase-rich, trace element-poor nonmare materials will be considered from the viewpoint of LIL element abundance patterns. Finally, the effects of high temperatures during the deposition of the Fra Mauro formation are considered in terms of their influence on several trace elements.

## MAJOR ELEMENT DATA

The broad features of the major element chemistry of lunar rocks are readily discerned from a plot of $Al_2O_3$ versus FeO (Fig. 1). These two major components account for the bulk of the variance in the major element chemistry of lunar samples and, as a consequence, provide the best discriminatory criteria. Figure 1 and Table 1 show that mare basalts are distinguished from nonmare basalts (KREEP and low-K) by higher FeO ($>14\%$) and lower $Al_2O_3$ ($<12.0\%$) concentrations. Furthermore, the covariation of $Al_2O_3$ and FeO in mare basalts is distinctly different from the covariation of these elements in nonmare basalts and their anorthositic derivatives. Although the slopes of the regression lines for the mare and nonmare basalts differ markedly, both trends intersect the $Al_2O_3$ axis at about 35%, indicating the effect of plagioclase control. This control is much more pronounced, and probably more real, for the nonmare basalts and their associates than it is for the mare basalts. There are some mare-like basalts (14053, 14072, L–16, 12038, and vitrophyre) with FeO concentrations greater than 14% that extend the mare basalt $Al_2O_3$–FeO trend to $Al_2O_3$ concentrations near those of KREEP basalts (Fig. 1 and Table 1). On nearly all major element plots these mare-like basalts occupy a position intermediate in chemistry between mare and KREEP basalts. It is possible that these mare-like basalts may constitute a third basalt type or indicate a continuum between KREEP and mare basalts. It should be noted that mare basalts, mare-like basalts, and KREEP basalts have overlapping CaO and MgO concentrations, with a tendency for the mare basalts to extend to lower CaO and higher MgO concentrations than KREEP basalts

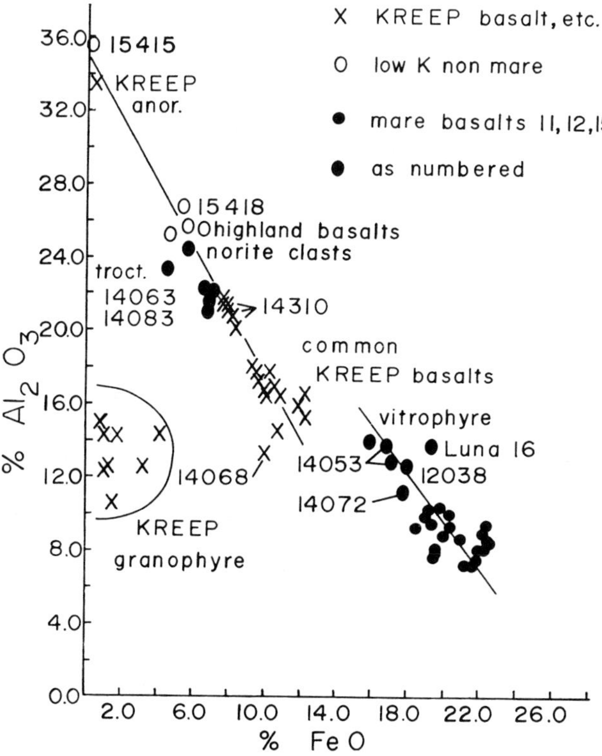


Fig. 1. FeO versus $Al_2O_3$ variation diagram for lunar basalts, anorthositic materials, and KREEP granophyre. Data are from sources referenced in text or Table 1.

Table 1. Chemical compositions of mare basalts, mare-like basalts, the "boundary types" of nonmare basalts (KREEP and low-K), and two types of anorthosites.

| | Average Mare Basalts | | | | Mare-like Basalts | | | | |
|---|---|---|---|---|---|---|---|---|---|
| | Apollo 11[a] Low Rb, K | Apollo 11[a] High Rb, K | Apollo 12[b] Mare Basalts/ Sigma | Apollo 15[c] Mare Basalts/ Sigma | 14053 | 14072 | 12038 | Vitrophyre from 14321 | Luna 16 Basalt |
| No. of Samples: | N = 6 | N = 4 | N = 16 | N = 7 | N = 2 | N = 1 | N = 2 | N = 7 | N = 1 |
| $SiO_2$ | 40.6 | 40.3 | 44.9 ± 1.3 | 46.1 ± 2.0 | 46.3 | 45.2 | 46.8 | 46.8 | 43.8 |
| $TiO_2$ | 10.2 | 11.5 | 3.36 ± 0.92 | 2.13 ± 0.46 | 2.79 | 2.57 | 3.24 | 2.33 | 4.9 |
| $Al_2O_3$ | 10.2 | 8.72 | 8.95 ± 1.17 | 8.95 ± 0.40 | 13.7 | 11.1 | 12.7 | 13.8 | 13.7 |
| FeO | 19.0 | 19.6 | 20.8 ± 0.8 | 21.2 ± 1.6 | 17.0 | 17.8 | 17.7 | 15.9 | 19.4 |
| MgO | 6.92 | 7.64 | 11.2 ± 3.5 | 9.51 ± 1.25 | 8.54 | 12.2 | 6.76 | 6.30 | 7.05 |
| CaO | 11.6 | 11.5 | 9.70 ± 1.34 | 10.2 ± 0.58 | 11.2 | 9.84 | 11.5 | 11.6 | 10.4 |
| $Na_2O$ | 0.41 | 0.52 | 0.27 ± 0.07 | 0.26 ± 0.03 | 0.44 | 0.32 | 0.65 | 0.80 | 0.38 |
| $K_2O$ | 0.08 | 0.30 | 0.063 ± 0.011 | 0.034 ± 0.01 | 0.11 | 0.08 | 0.07 | 0.25 | 0.15 |
| $P_2O_5$ | 0.09 | 0.18 | 0.11 ± 0.06 | 0.066 ± 0.02 | 0.11 | 0.08 | 0.16 | 0.25 | — |
| $Cr_2O_3$ | — | — | 0.51 ± 0.15 | 0.70 | 0.37 | 0.51 | — | — | 0.28 |
| Method: | XRF & WC | XRF & WC | XRF & WC | XRF | XRF[d] | XRF[e] | XRF[f] | Microprobe Grieve *et al.* (1972) | XRF Vinogradov (1971) |

Table 1.—*Continued.*

| Nonmare Basalts and Anorthosites | | | | | | | | | | |
|---|---|---|---|---|---|---|---|---|---|---|
| | Apollo 14[g] Common KREEP Breccia/Sigma | Apollo 14 Small KREEP fragments | Apollo 12 Broad Beam Microprobe Analyses of KREEP fragments | Apollo 14 Soil Survey Glass B&C | 14310 Felds-pathic KREEP Basalt | Average "Highland Basalt" Glass | 15418 | Calculated Low-K Nonmare Basalt | 15415 Low-K Anorthosite | 12033,97,7 KREEP Anorthosite |
| No. of Samples: | N = 14 | N = 8 | N = 29 | N = 0 | N = 1 | N = 0 | N = 1 | n.a. | N = 1 | N = 1 |
| $SiO_2$ | 47.5 ± 0.62 | — | 47.6 | 48.0 | 47.2 | 45.5 | 45.0 | — | 44.1 | 46.5 |
| $TiO_2$ | 1.82 ± 0.26 | 1.75 | 1.7 | 2.1 | 1.24 | 0.5 | 0.27 | — | 0.02 | 0.02 |
| $Al_2O_3$ | 16.4 ± 1.3 | — | 17.2 | 17.1 | 20.1 | 25.3 | 26.7 | — | 35.5 | 33.5 |
| FeO | 10.8 ± 1.1 | — | 9.4 | 10.5 | 8.38 | 5.8 | 5.37 | — | 0.23 | 0.7 |
| MgO | 10.6 ± 1.3 | 10.9 | 9.1 | 8.7 | 7.87 | 8.0 | 5.38 | 6.0–9.0 | 0.09 | 0.4 |
| CaO | 10.5 ± 0.7 | 9.83 | 10.2 | 10.7 | 12.3 | 14.3 | 16.1 | 13.0–15.0 | 19.7 | 16.5 |
| $Na_2O$ | 0.82 ± 0.11 | 0.78 | 1.0 | 0.7 | 0.63 | 0.21 | 0.21 | 0.2–0.4 | 0.34 | 1.18 |
| $K_2O$ | 0.57 ± 0.11 | 0.60 | 0.7 | 0.5 | 0.49 | 0.06 | 0.05 | 0.03–0.07 | 0.015 (I.D.) | 0.24 (I.D.) |
| $P_2O_5$ | 0.56 ± 0.11 | — | — | — | 0.34 | — | 0.93 | — | 0.01 | — |
| $Cr_2O_3$ | 0.21 ± 0.02 | — | 0.17 | — | 0.18 | — | 0.11 | — | — | — |
| Method: | XRF & WC | ID & AA; This report table 3 | Microprobe Keil *et al.* (1971) Meyer *et al.* (1971) Smith *et al.* (1970) | Microprobe Apollo Soil Survey (1971) | XRF Analyst: J. M. Rhodes this report table 2 | Microprobe analyses of glasses Reid *et al.* (1972) | XRF Analyst: J. M. Rhodes LSPET (1972) | | XRF Analyst: J. M. Rhodes LSPET (1972) | Microprobe Analysis Analyst: Roy Brown; this report |

[a] Analysts: Chappell, Peck, Maxwell, Engel, Scoon, and Wiik, *Science* Moon Issue, Vol. 167, No. 3918 (1970).
[b] Analysts: Rose, Chappell, Scoon, Engel, Ahrens, and Maxwell (*Proc. Second Lunar Sci. Conf.*, Vol. 1 and 2 (1971).
[c] Analyst: J. M. Rhodes, LSPET (1972).
[d] Analysts: Rhodes and Chappell; Compston *et al.* (1972); and this report, Table 2.
[e] Analysts: Chappell; Compston *et al.* (1972).
[f] Analysts: Rose and Chappell; Compston *et al.* (1971); Cuttatta *et al.* (1971)
[g] Analysts: Rose, Rhodes, and Scoon. *Revised Abstracts of the Third Lunar Science Conference* (1972).

(not shown). Consequently, they are indistinguishable on the basis of these two elements. The group of rocks designated low-K nonmare basalts, etc., (Fig. 1) have lower MgO ($\sim$6.0%) and higher CaO ($\sim$15%) than the other basalt types. The chemical composition of low-K nonmare basalts is currently known only from the results of two highly interpretive studies (Reid *et al.*, 1972; and Hubbard *et al.*, 1971b). Sample 15418 may be considered a large crystalline fragment of this material and thus may have a role similar to that of 15415 in the understanding of lunar anorthosites.

The KREEP basalt chemical composition is as well defined as that of Fe-rich mare basalts for any single Apollo site (Table 1), if one ignores 14310, 14068, and similar samples. Four separate types of analytical data yield nearly identical average chemical compositions (Table 1). The XRF and wet chemical analyses of Apollo 14 common KREEP breccias (Fig. 1) show that these samples have a very well-defined chemical composition (Table 1). The one sigma variation of these 14 analyses overlaps the average chemical composition of Apollo 12 KREEP fragments, except for FeO, MgO, $Cr_2O_3$, and $Na_2O$. The differences in FeO, MgO, $Cr_2O_3$, and $Na_2O$ are insignificant relative to the dispersion in the 29 analyses averaged. Eight coarse fine fragments, breccia clasts, and groundmass (Table 3) have average $TiO_2$, MgO, CaO, $Na_2O$, and $K_2O$ concentrations that are completely within the standard deviation of the 14 larger samples. Glass fragments analyzed by the Apollo Soil Survey (1971) provide a highly interpretive fourth approach to the chemical composition of KREEP basalts. Two preferred glass compositions recognized at that time (B and C) are close to the common KREEP average (Table 1). Even though

Table 2. X-ray fluorescence data for major, minor, and trace elements in Apollo 14 samples.

| | 006,3 Breccia | 053,50 Basalt | 068,3 Breccia | 163,65 < 1 mm Soil | 301,101 Sawdust, Breccia | 307,58 Chips, Breccia | 310,101 Basalt |
|---|---|---|---|---|---|---|---|
| $SiO_2$ | 47.0 | 46.4 | 47.2 | 47.2 | 47.6 | 46.9 | 47.2 |
| $TiO_2$ | 1.77 | 2.64 | 1.39 | 1.79 | 1.77 | 1.84 | 1.24 |
| $Al_2O_3$ | 16.4 | 13.6 | 13.3 | 17.2 | 15.9 | 16.5 | 20.1 |
| FeO | 10.9 | 16.8 | 10.0 | 10.4 | 11.9 | 12.3 | 8.38 |
| MnO | 0.14 | 0.26 | 0.13 | 0.14 | 0.14 | 0.14 | 0.11 |
| MgO | 10.7 | 8.48 | 17.6 | 9.37 | 10.4 | 9.76 | 7.87 |
| CaO | 10.5 | 11.2 | 8.28 | 11.0 | 10.1 | 10.8 | 12.3 |
| $Na_2O$ | 0.79 | — | 0.75 | 0.66 | 0.74 | 0.70 | 0.63 |
| $K_2O$ | 0.35 | 0.10 | 0.59 | 0.58 | 0.69 | 0.55 | 0.49 |
| $P_2O_5$ | 0.75 | 0.09 | 0.55 | 0.46 | 0.58 | 0.51 | 0.34 |
| S | 0.11 | 0.14 | 0.07 | 0.08 | 0.09 | 0.10 | 0.02 |
| $Cr_2O_3$ | — | — | — | 0.22 | 0.20 | 0.23 | 0.18 |
| Total | 99.4 | 99.7 | 99.9 | 99.1 | 100.1 | 100.3 | 98.9 |
| Sr | 192 | — | — | 186 | 175 | 163 | 193 |
| Rb | 6.5 | — | — | 15 | 18 | 14 | 13 |
| Y | 276 | — | — | 213 | 238 | 188 | 174 |
| Th | 18 | — | — | 13 | 15 | 12 | 11 |
| Zr | 1376 | — | — | 978 | 1215 | 842 | 842 |
| Nb | 81 | — | — | 65 | 73 | 53 | 52 |
| Ni | 263 | — | — | 322 | 203 | 251 | 64 |

trace element data on such glass fragments are not available, the major element comparisons clearly indicate that KREEP basalt and Fra Mauro basalt are equivalent.

The variation trend for $Al_2O_3$ between KREEP basalts and KREEP anorthosites suggests simple plagioclase enrichment. The major element data for 14063, 14083 (Rose, personal communication), norite clasts (Grieve *et al.*, 1972), troctolite clast (Compston *et al.*, 1972), and 14310 (also 14073 and 14276), all having the minor mineralogy of KREEP materials, allow insight into differentiation of nonmare basalts. MgO and $K_2O$ vary along the $Al_2O_3$–FeO trend line as follows: common KREEP basalts, 10.7% MgO; 14310, 8.5% MgO; 14063/14083, 11.0% MgO; troctolite, 15% MgO; norite clasts, 6.2% MgO; KREEP anorthosite, 0.1% MgO. Potassium concentrations drop from ~0.6% $K_2O$ for KREEP basalts to 0.4% $K_2O$ for 14310 (also 14073, 14276), to 0.3% in 14063/14083, to 0.06% in the troctolite, and then increase to 0.6% $K_2O$ in the norite clasts. Obviously, simple plagioclase enrichment has not generated the entire basalt-anorthosite association—although 14310, 14073, 14276, and the norite clasts may be so derived from common KREEP basalts. Samples 14063, 14083, and the troctolite clast show that at least one other differentiation mechanism exists or that these samples were not derived from KREEP basalts, even though they possess similar chemical features. Basaltic breccias 14063 and 14083 also contain a pink Mg–Al spinel (Warner, 1972), and thus their high $Al_2O_3$ and MgO concentrations are consistent with a plagioclase, olivine, spinel assemblage of igneous origin (Walker *et al.*, 1972). Because spinel is present, it can be argued that 14063 and 14083 represent partial melts produced at higher temperatures than KREEP basalts. This higher temperature also implies more extensive melting, thus accounting for the lower $K_2O$ values. If 14063 and 14083 represent igneous liquids produced by larger degrees of partial melting than common KREEP basalts, then these samples will have

Table 3. Isotope dilution data for Apollo 14 samples.

| Sample No. | 006,3 | 053,50 | 068,3 | Individual 2–4 mm Coarse Fines 161,35,2 | 161,35,3 | 161,35,4 | 161,35,5 | 161,35,6 | 163,65 | 163,65,1 | 163,65,2 | 301,48,1 | 307,26,1 | 307,26,2 | 310,130 | 15023,2,5 |
|---|---|---|---|---|---|---|---|---|---|---|---|---|---|---|---|---|
| | 263 mg Brec. | 13.1 mg Basalt | 106 mg Brec. | 55.0 mg Brec. | 27.8 mg Brec. | 27.6 mg Brec. | 30.1 mg Brec. | 39.8 mg Brec. | 300 mg Soil < 100 mm | 14.7 mg 33 Similar Soil Frag. | 5.5 mg 13 Similar Soil Frag. | 64.8 mg Brec. | 6.8 mg White Clast | 68.4 mg Matrix | 177 mg Basalt | 5.5 mg Apollo 15 KREEP Basalt |
| La (ppm) | 84.7 | 13.0 | — | 55.6 | — | — | — | — | 68.2 | — | — | 71.8 | — | — | 56.4 | — |
| Ce (ppm) | 214 | 34.5 | 157 | 252 | 205 | 266 | 165 | 212 | 176 | 227 | 188 | 201 | 230 | 164 | 144 | 193 |
| Nd (ppm) | 131 | 21.9 | 93.4 | 149 | 122 | 158 | 106 | 132 | 103 | 129 | 106 | 121 | 138 | 99.2 | 87.0 | 114 |
| Sm (ppm) | 36.0 | 6.56 | 28.1 | 42.8 | 34.4 | 44.3 | 29.7 | 38.7 | 29.0 | 36.5 | 30.0 | 34.7 | 38.8 | 28.0 | 24.0 | 32.0 |
| Eu (ppm) | 2.73 | 1.21 | 2.01 | 2.76 | 2.74 | 3.04 | 2.49 | 2.76 | 2.54 | 2.92 | 2.42 | 2.69 | 2.74 | 2.25 | 2.15 | 2.60 |
| Gd (ppm) | 42.5 | 8.59 | 29.1 | 49.1 | 43.0 | — | 34.9 | 43.6 | — | 42.7 | 35.9 | 40.3 | — | 34.0 | 28.1 | — |
| Dy (ppm) | 47.1 | 10.5 | 35.1 | 55.8 | 45.6 | 56.8 | 40.3 | 49.3 | 38.3 | 46.9 | 39.7 | 46.0 | 52.0 | 37.2 | 32.7 | 44.0 |
| Er (ppm) | 28.8 | 6.51 | — | — | 31.2 | — | 24.6 | — | 23.8 | 28.5 | 24.5 | 28.0 | 30.1 | 22.9 | 19.7 | 28.3 |
| Yb (ppm) | 25.8 | 6.00 | 20.0 | — | 26.1 | — | 23.4 | 27.4 | 23.6 | 30.6 | 24.6 | 25.5 | 28.0 | 20.6 | 18.4 | 21.5 |
| Na (ppm) | 0.68 | 0.46 | 0.56 | 0.60 | 0.58 | — | 0.54 | 0.56 | 0.51 | 0.67 | 0.95 | — | 0.67 | 0.56 | 0.53 | 0.61 |
| K (ppm) | 2700 | 912 | 4508 | 4733 | 2372 | 5107 | 4700 | 5699 | 4840 | 6010 | 5300 | 6874 | 5300 | 4940 | 4250 | 4110 |
| Rb (ppm) | 6.07 | 2.19 | 14.5 | 12.9 | 3.38 | 15.2 | 14.7 | 16.9 | 15.3 | 17.9 | 18.1 | 21.7 | 16.0 | 15.3 | 12.8 | 13.2 |
| Mg (%) | 6.38 | 5.09 | 10.6 | 7.45 | 6.84 | 5.88 | 7.44 | 6.88 | 5.44 | 6.20 | 4.96 | 5.33 | 6.63 | 5.71 | 4.37 | 5.16 |
| Ca (%) | 7.40 | 7.92 | 5.62 | 6.51 | 7.02 | 7.24 | 6.21 | 6.72 | 7.83 | 7.17 | 7.35 | 7.37 | 7.68 | 7.47 | 8.93 | 6.74 |
| Sr (ppm) | 180 | 98 | 139 | 171 | 180 | 197 | 170 | 182 | 186 | — | — | 185 | 192 | — | 188 | — |
| Ba (ppm) | 781 | 146 | 780 | 1022 | 775 | 811 | 817 | 916 | 926 | 1076 | 1634 | 959 | 890 | 735 | 617 | 683 |
| U (ppm) | 4.07 | 0.60 | 3.47 | 5.03 | 4.08 | 4.71 | 3.92 | 4.61 | — | — | — | 4.32 | 4.90 | 3.28 | — | 3.09 |
| Ti (%) | — | 1.63 | 0.82 | 1.10 | 0.94 | 1.18 | 0.97 | 0.96 | — | — | — | 1.04 | 1.05 | 1.19 | — | 1.28 |
| K/Rb | 445 | 416 | 311 | 373 | 702 | 336 | 319 | 337 | 3.16 | 336 | 293 | 317 | 331 | 323 | 332 | 311 |
| K/U | 663 | 1520 | 1300 | 914 | 582 | 1085 | 1200 | 1236 | 1370 | — | — | 1591 | 1082 | 1500 | — | 1330 |

lower K, REE, P, etc., than KREEP basalts and thus represent an intermediate nonmare basalt type. Note that troctolite is a logical crystal cumulate from such high-$Al_2O_3$, high-MgO liquids.

Only one of the chemically recognized basalt types (mare basalts, including mare-like basalts) is clearly igneous and represents extrusive vulcanism. The other chemical types are commonly brecciated and metamorphosed, although igneous textures are found for some KREEP basalts (14310, 15023,2,5, 14276). Walker *et al.* (1972) have shown common KREEP basalts to be reasonable partial melts, and they indicate that 14310 can be either a partial melt at greater depths than common KREEP basalts or, alternatively, a plagioclase-enriched common KREEP basalt. We consider the previously noted features and the distinctive trace and minor element concentrations of KREEP materials as firm evidence that KREEP basalts represent an independent composition and not a derivative composition produced by fractional crystallization of a parental mare basalt composition.

Derivation of KREEP basalt from a mare basalt composition by fractional crystallization can be further restricted by comparing the analyzed $Cr_2O_3$ and $K_2O$ concentration in KREEP basalts to those calculated for the derivation of high-K liquids from mare basalts using the theoretical approach of Gast (1968). Under the most favorable conditions, in which $K_2O$ liquid/$K_2O$ crystals = 20, the high concentration of $K_2O$ in KREEP basalts (average 0.57%) can only be obtained from a parental mare basalt, with $K_2O$ about 0.03% to 0.08%, by a tenfold increase in potassium, requiring about 90% crystallization of the magma. At this stage the $Cr_2O_3$ remaining in the liquid would be reduced to well below 0.05% from about 0.51% in the parental mare basalt, a value completely at variance with the observed concentration of 0.21% $Cr_2O_3$ in KREEP basalts. This conclusion holds whether one uses the conservative revised distribution coefficient $Cr_2O_3$ pyrox/$Cr_2O_3$ liquid of 3.4 given by Ringwood and Essene or higher values based on terrestrial models and conditions (Gast, 1968; Ringwood and Essene, 1970; Smith *et al.*, 1970) or whether one takes an initial $Cr_2O_3$ content in the supposed parental mare basalt composition of 0.51%, as in Apollo 12 basalts, or as high as 0.7%, as in Apollo 15 basalts (LSPET, 1972). A graphic illustration of the rapid depletion of $Cr_2O_3$ in mare basalts with crystallization is provided by the sharp decrease of this element in pyroxenes and olivines with decreasing Mg/Mg + Fe ratios (Ringwood and Essene, 1970; Keil *et al.*, 1971). Furthermore, one would expect that early removal of pyroxenes and olivines with high Mg/Fe + Mg ratios, together with spinel, would also substantially reduce the MgO content of the derived liquid, rather than maintaining a value similar to the parental magma.

The Apollo 12 sample return provided one rock (12013) that is high in $SiO_2$, K, Ba, U, Th, etc., (LSPET, 1970) and has been found to be a KREEP basaltic breccia (Hubbard *et al.*, 1971a) intruded by a granite-like material (Drake *et al.*, 1970). Similar granitic material (Fig. 1) has been recognized in samples from all missions (Reid *et al.*, 1972) and is common (~1–3%) in KREEP breccias (Meyer *et al.*, 1971, and personal communication). Mare basalts contain a nearly identical material that represents the high $SiO_2$ liquid generated when mare basalts crystallize and produce a residual liquid that becomes immiscible and splits into high-$SiO_2$ and high-$FeO_2$

liquid phases (Roedder and Weiblen, 1971). The high-$SiO_2$ liquids in mare basalts are distinct from the granitic material in 12013 and other KREEP materials because they have lower Ba concentrations (Roedder and Weiblen, 1971; Drake *et al.*, 1970) reflecting the lower Ba concentrations of mare basalts. The KREEP basalt composition is a particularly suitable parent for this granitic material because of its high $K_2O$ and relatively high $SiO_2$ concentrations. We argue that the granitic material in KREEP breccias is directly associated with the KREEP chemical composition because it is implausible that the small volumes of immiscible liquid in mare basalts would be efficiently extracted and then added to the Apollo 14 breccias. Further, the granitic material in 12013 and other KREEP breccias could not have been derived from most mare basalts because KREEP breccia materials are older ($>3.9 \times 10^9$ years, Wasserburg and Papanastassiou, 1971; Huneke *et al.*, 1972) than currently dated mare basalts (Wasserburg and Papanastassiou, 1971), except for mare-like basalts 14053 and 14072. Nyquist *et al.* (1972) argue that the granitic material in 12013 and other, more normal, KREEP breccias had a common genesis at $4.4 \times 10^9$ years of age. In view of these facts, we propose that the KREEP associated granitic material be known as KREEP granophyre and postulate that much of it was originally produced during slow cooling of thick or ponded lava flows or during slow cooling of dykes or sills (Meyer, 1972).

## LARGE ION LITHOPHILE ELEMENT ABUNDANCES

Perhaps the most distinctive characteristic of the KREEP basalts is the abundance pattern of large ion lithophile elements, that is, the group of elements strongly concentrated into a liquid during crystallization of mafic liquids or during early stages of partially melting a mafic assemblage. A number of Apollo 12 soil fragments relatively rich in Al and poor in Fe, in some cases containing orthopyroxene as the dominant mafic mineral, were analyzed for K, Rb, Sr, Ba, and the REE (Hubbard and Gast, 1971). These analyses clearly identified a distinct LIL element-rich basaltic rock. We have studied similar fragments and clasts from Apollo 14 soil sample 14163 and accompanying coarse fine sample 14161. For the Apollo 14 samples, we have also determined U, Ca, Mg, and Ti by stable isotope dilution methods. High-precision Sr isotope determinations were also made on many of these samples (Nyquist *et al.*, 1972). Analytical results for seven soil fragments, two breccia clasts, three total breccia samples, one soil sample, and two igneous rocks from the Apollo 14 site are shown in Table 3. Data for a single igneous soil fragment from the Apollo 15 site are also given there. The chondrite normalized abundance patterns for REE, Ba, and U are remarkably similar for these samples and overlap almost exactly the range of concentrations found for KREEP fragments from the Apollo 12 site and dark portions of rock 12013 (Hubbard and Gast, 1971).

The average patterns for typical Apollo 12, Apollo 14, and the single Apollo 15 KREEP fragment are illustrated in Fig. 2. The most significant variation seen in these patterns is the variation in Ba relative to U and Ce. Three Apollo 14 samples, 14068, 14310, and a clast from 14321, have REE abundances generally similar to the average KREEP fragments and deserve special mention. We have already noted that sample 14310 is enriched in Al and slightly depleted in Mg. In addition, we note

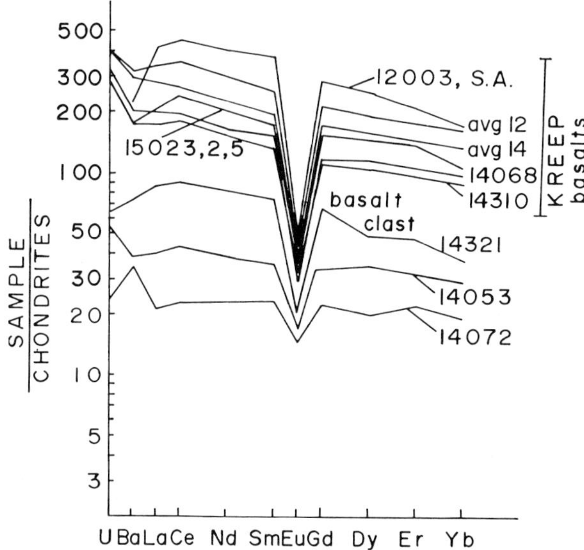


Fig. 2. REE, Ba, and U diagram for KREEP basalts from Apollo 12, 14, and 15 and for three other Apollo 14 samples (14053, 14072, and a basalt clast from 14321). Data for 14072 are an average of data from Helmke and Haskin (1972) and Taylor *et al.* (1972). Data for the basalt clast from 14321 are from Taylor *et al.* (1972) and have been corrected for interlab differences in Ce and La.

here that it is somewhat depleted in LIL elements and has a distinctly higher Sr/Eu ratio than the other KREEP fragments. All of these characteristics can be explained by the addition of plagioclase phenocrysts or xenocrysts as suggested by Ridley *et al.* (1972). Sample 14068 is also depleted in LIL elements. It is, however, enriched in MgO, and its chemical characteristics can be explained by addition of olivine to a normal KREEP basalt. A basalt fragment from sample 14321 also has lower concentrations of LIL elements (Fig. 2), but due to a lack of major element data little more can be learned.

Three other Apollo 14 samples, 14053, 14072, and 14063, are markedly different from KREEP basalts in either or both major element and LIL element abundances. Samples 14072 (data from Helmke *et al.*, 1972; and Taylor *et al.*, 1972) and 14053 are obviously distinguished from typical KREEP basalts by their high FeO and low $Al_2O_3$ and Ree values. In fact, 14072 is very similar to Fe-rich mare basalts in terms of both major and LIL element abundances. If this sample is derived from the Fra Mauro formation, then basalts with characteristics very similar to those of typical mare basalts must have existed on the lunar surface prior to 3.9 b.y. ago. Sample 14063, on the other hand, has a major element composition (Rose, personal communication) somewhat similar to that of KREEP basalts like 14310, while its LIL element pattern —in particular, the abundance of Eu—is quite different from that of the KREEP basalts (Fig. 3).

The K/U ratios of individual fragments of KREEP basalt analyzed in this study are summarized in Fig. 4 along with K/U ratios measured on mare basalts. This summary shows that KREEP basalts have a relatively narrow range of K/U ratios

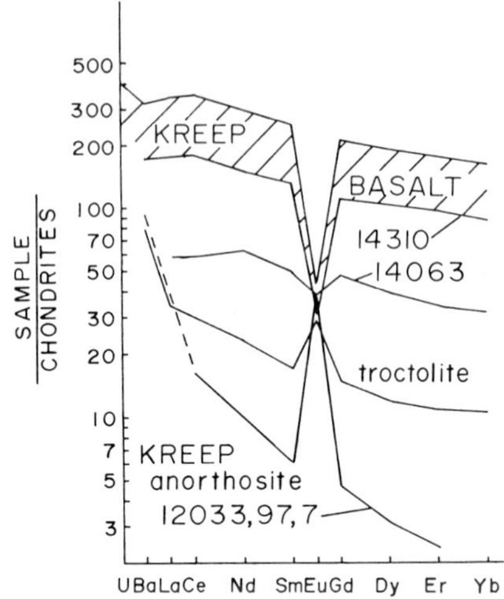


Fig. 3. A comparison of U, Ba, and REE data for KREEP basalts, 14063, the troctolite from 14321, and KREEP anorthosite. Data are from sources referenced in text.

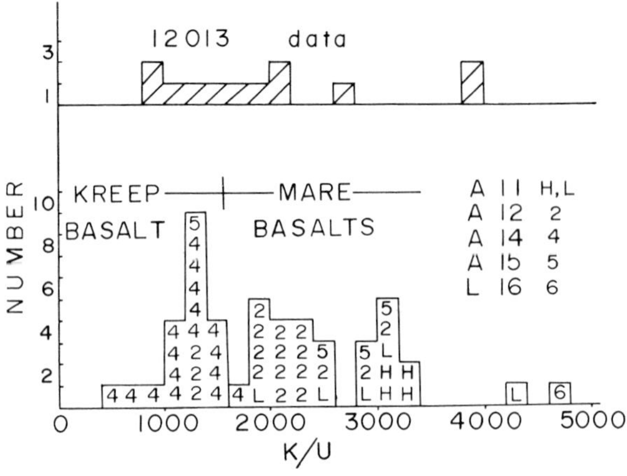


Fig. 4. Histogram of K/U ratios in KREEP basalts and mare basalts. Note that the KREEP data are from three sites, Apollo 12, 14, and 15. Data are from Gast *et al.* (1970), Gopalan *et al.* (1970), Wakita *et al.* (1970), Tatsumoto (1970), Silver (1970), O'Kelley *et al.* (1970), Hubbard and Gast (1971), Compston *et al.* (1971), O'Kelley *et al.* (1971), Cliff *et al.* (1971), Tatsumoto *et al.* (1971; 1972), Hubbard *et al.* (1970), Wakita and Schmitt (1970), Eldridge *et al.* (1972), O'Kelley *et al.* (1972), Albee *et al.* (1972), and this report.

that is clearly resolved from mare basalts. Moreover, rock 14053, which is intermediate in other characteristics, also has an intermediate K/U ratio (Table 3). Three KREEP basalts have K/U ratios below 1000 and are also readily distinguished in terms of other element ratios, for example, K/Ba and K/Rb. With the exception of these three samples, the K/U ratio of KREEP basalts is 1300 $\pm$ 300. In addition, our data give an average U content of 4.32 ppm, which leads to an average Th content of

16.4 ppm, assuming a Th/U ratio of 3.8. The K/U ratios of Apollo 12 mare basalts are between those of KREEP basalts and high-K Apollo 11 and Apollo 15 basalts. The low-K basalts of Apollo 12 have a wide range of K/U ratios. In addition, the K/U ratios for individual fragments from rock 12013 also cover an extremely wide range, although the dark portions of this rock have ratios similar to those of the KREEP basalts. The range in K/U ratios observed for this specimen may represent the mixing of two end members, namely, the dark, KREEP-like, low K/U material and white material with a high K/U ratio.

Hubbard *et al.* (1972) have determined the LIL element abundances for several anorthositic or plagioclase-rich fragments. They suggest that at least two types of lunar anorthosites can be distinguished from these data. The first and most common type is apparently derived from an LIL element-poor parent liquid. Other fragments clearly are derived from liquids with KREEP-like LIL element abundance patterns. Taylor *et al.* (1972) have analyzed a single troctolite fragment that, by similar arguments, can clearly be associated with KREEP basalts. This similarity is illustrated in Fig. 3.

The high $Al_2O_3$ and low FeO of 14063 and its Eu value suggest that it must somehow be related to nonmare materials. Interpretations of the data for 14063 may be tested by noting that 14063 has only slightly more $Al_2O_3$ than 14310 (Rose, personal communication) but about one-third of the REE, except for Eu. Clearly the depletion in REE is not due to dilution with plagioclase. The amount of olivine dilution that can be inferred from the MgO difference between 14063 and 14310 is grossly insufficient to account for the depletion in REE, and thus we rule out addition of low REE minerals as the only cause of the low REE in 14063. The low REE in 14063 may be due to removal of high REE minerals. If so, this removal must be accomplished by REE-rich minor minerals, because the major minerals are all REE poor. Early crystallization of REE-rich minor minerals is difficult to substantiate because of sampling inadequacy and the extensive metamorphism and brecciation that has taken place. If we assume that 14063 represents a magma produced from the same source material as KREEP basalts, then simple partial melting models (batch melting) predict the observed overall pattern and concentration at $\sim 10\%$ melting but a deeper negative Eu anomaly. The smaller observed Eu anomaly can be explained by Rayleigh melting models that maintain a constant $Eu^{+3/+2}$ ratio in the source as liquid is removed or if the source of 14063 has a higher $Eu^{+3/+2}$ ratio. Thus, two possible alternatives remain: (1) 14063 was derived from KREEP basalts by an igneous differentiation process that removed trivalent REE by precipitation of unknown REE-rich minor minerals and preserved the original MgO concentrations while enriching the liquid in plagioclase, or (2) 14063 was directly produced by 10–15% of partial melting from the source of KREEP basalts but with elevated $Eu^{+3/+2}$ ratios or by a Rayleigh process involving oxidation of Eu in the source.

## Metamorphism of KREEP Basalts

All KREEP basalts except 14310, 14073, 14276, and a few small fragments have been brecciated and thermally metamorphosed (Warner, 1972; Meyer *et al.*, 1971). The effect of this metamorphic event on the chemical composition of these basalts can be investigated because of the abundance of relevant data and the recent establishment of a preliminary body of experimental data directly applicable to the problem (Gibson and Hubbard, 1972). An understanding of this metamorphic event should allow a more accurate appraisal of the original chemical features of KREEP basalts.

During thermal metamorphism the groundmass of a basalt is most susceptible to chemical alteration. When a basalt crystallizes a wide variety of chemical elements are rejected by the major minerals and quenched to unstable glassy or cryptocrystalline refuse. Thermal metamorphism provides a second chance for this refuse to form a stable mineral assemblage. One of the characteristic features of brecciated and metamorphosed KREEP basalts is the absence of this chemical refuse and the presence of abundant exotic minor minerals (Meyer *et al.*, 1971; Marvin *et al.*, 1971). Certain of these minor minerals contain the trace elements that are characteristic of KREEP. The REE are contained in certain phosphates, Ba, K, and Rb are in K-feldspar or a granophyric material similar to the white portion of 12013, and U is concentrated in several minor minerals. Because the characteristic trace and minor elements of KREEP have been segregated into several different minor minerals, one must establish the extent to which elements have been transported into or out of the sample volumes used for analyses. This problem can be approached by plotting the elements of interest, particularly those of differing volatility, against one another. If one plots the involatile elements Ce versus U (not shown, Table 2), a good correlation is found, indicating no differential movement and yielding a well-defined Ce/U ratio of 47 $\pm$ 6. A plot of Ce versus Ba (not shown, Table 2; also Hubbard and Gast, 1971, Table 2) shows an almost total lack of correlation in metamorphosed KREEP basalts. Plotting Ba versus K/Rb (Fig. 5) for KREEP basalts and various fragments from 12013 shows that both have the same K/Rb ratio over a wide range in Ba (also K and Kb) con-

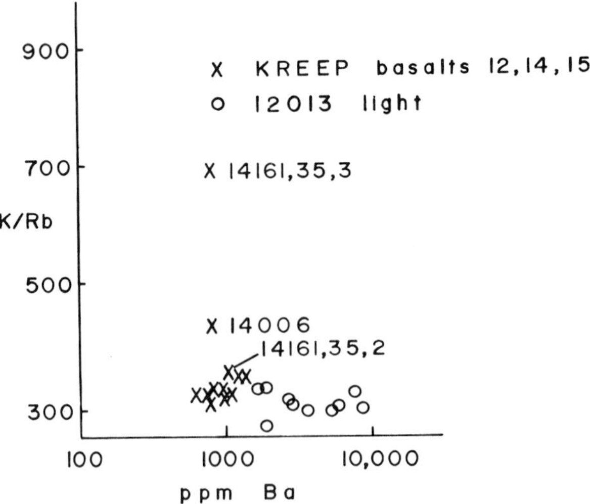


Fig. 5. K/Rb versus Ba for KREEP basalts and 12013 light material. The samples with elevated K/Rb ratios are identified as samples that have lost vaporized K and Rb.

centrations. There are, however, six samples that have elevated K/Rb ratios but similar Ba concentrations. Two of these six samples have greatly reduced K and Rb concentrations and the highest K/Rb ratios yet found for KREEP basalts. Experimental Studies (Gibson and Hubbard, 1972) have shown that the observed changes in K, Rb, and K/Rb can be reproduced by thermal volatilization under vaccum at temperatures between 950 and 1050°C and in a system open to loss of K and Rb vapor. Under these conditions K and Rb are found to be volatile, but Ba, Ce, and U are not. Textural evidence (Warner, 1972; Meyer *et al.*, 1971) indicates that the granophyre material is molten at these temperatures, and thus a mechanism exists for generating the independent variation of Ba and Ce (also U). Furthermore, both 12013 and KREEP basalts have a poorly defined K/Ba ratio ($6.0 \pm 1.0$) inconsistent with the migration of a granophyre melt with a well-defined K/Ba ratio. It now seems plausible that migration of granophyric material could occur as liquid + vapor in a system closed to loss of Rb and K vapor and thus produce variations in K and Rb, relative to Ba, without producing changes in K/Rb. The 12013 K/U data show a wide range in K/U ratios (Fig. 4) where the ratios increase with K concentration, suggesting that this granophyric material contains very little U relative to K.

We thus arrive at the following characteristic premetamorphism trace element ratios for KREEP basalts: $K/Rb = 330 \pm 20$, $Ce/U = 47 \pm 6$, $K/Ba = 6.0 \pm 1.0$, and $K/U = 1300 \pm 300$. The REE abundance patterns (Fig. 2) are equally unaffected by the metamorphic event and thus are characteristic of KREEP basalts.

## Chemical Composition of Nonmare Source Material

It has been recognized that the source of nonmare basalts is chemically different from the source material for mare basalts (Hubbard and Gast, 1971; Ringwood *et al.*, 1972). The nonmare source is enriched about $3\times$ in Al, Ca, and REE relative to the mare source and contains these elements at $7–10\times$ the chondritic concentrations (Hubbard and Gast, 1971). The nonmare source material also has much lower Fe/Mg than the mare basalt source (Ringwood *et al.*, 1972).

The relative variation of volatile and refractory elements can be inferred from the K/U ratios in lunar basalts (Fig. 4) because of the volatility differences between K and U. Lunar basalts show a wide range in K/U ratios over a wide range of basalt types (Fig. 4), suggesting that measured K/U ratios must be corrected for the effect of partial melting before they can be used as indicators of original K/U ratios. This is easily done using the partial melting model of Hubbard and Gast (1971) and by noting that uranium and cerium are strongly covariant and assuming that U and Ce have identical distributions during partial melting and that all K goes into the liquid. These calculations show that the K/U ratio is $\sim 1500$ in the source of KREEP and $\sim 2300$ in the source of Apollo 12 mare basalts—almost exactly the mean measured ratios (Fig. 4). From this we conclude that the wide range in K/U ratios seen in Fig. 4 reflects considerable alkali/refractory variation in the source regions of lunar basalts. Inspection of K and U data for lunar basalts shows no obvious correlation of K and U concentrations and K/U ratios. Comparison of the K and U concentrations for KREEP and Apollo 12 mare basalt sources indicates why this may be so.

The back-calculated K concentrations are similar (~130.ppm for the KREEP source and ~90 ppm for the Apollo 12 mare basalt source), while the U concentrations are ~3X higher in the KREEP source than in the mare basalt source (~0.09 versus ~0.03 ppm U). These results suggest that the approximately threefold variation in refractory elements is accompanied by only a slight increase (~50%) in K and perhaps other volatiles. The low K concentration of the KREEP source is consistent with that expected if low-K nonmare basalts are the product of 30–40% melting of the same nonmare source as KREEP, that is, if low-K nonmare basalts have 300 to 600 ppm K (Hubbard *et al.*, 1971b; Reid *et al.*, 1972). We thus conclude that the primary zonation of the outer part of the moon is essentially restricted to refractory elements and that this enrichment in refractory elements was accompanied by small enrichments in volatiles.

## DEFINITION OF KREEP AND LOW-K NONMARE BASALTS AND COMMENTS ON THEIR GENESIS

KREEP was originally defined as a high $Al_2O_3$, low FeO basalt with distinctively high K, REE, and P concentration relative to mare basalts (Hubbard *et al.*, 1971a; Meyer *et al.*, 1971; Hubbard and Gast, 1971). The high concentrations of K, REE, and P are accompanied by high concentrations of U, Zr, Nb, and Ba, high Rb/Sr, and $Sr^{87/86}$. This material is a major portion of some Apollo 12 soils and a significant fraction of the Apollo 11 soil (Hubbard *et al.*, 1971; Meyer *et al.*, 1971). The moderate abundance of KREEP in Apollo 11 and 12 soils and its high concentrations of trace elements satisfied the needs expressed by terms like "magic component," "cryptic component," "transferred component," etc. The chemical compositions found or predicted by Hubbard *et al.* (1971a), Meyer *et al.* (1971), and Hubbard and Gast (1971) are substantiated by the Apollo 14 samples. Most analyzed KREEP basalts (now meta-igneous rocks) show a rather tight cluster and are denoted "common" KREEP basalts (Table 1 and Fig. 1). KREEP basalts like 14310 are recognized as a plagioclase-rich subtype (feldspathic KREEP basalts) that is enriched in $Al_2O_3$ and CaO and depleted in FeO, MgO, and REE, etc., relative to "common" KREEP basalts. An olivine enriched subtype (14068) is also recognized and has ~30% lower REE, etc.

KREEP basalts are distinct from the inferred low-K nonmare basalts, because KREEP basalts are highly enriched in those elements that are strongly partitioned into basaltic liquids during partial melting. Thus, KREEP and low-K nonmare basalts can be viewed as the compositional extremes of nonmare basalts produced by partial melting of a common source to varying degrees. The trace element concentrations of low-K nonmare basalts are predicted to be those calculated by Hubbard *et al.* (1971) or within a factor of 2 higher. It may be assumed that low-K nonmare basalts have the average major element composition of "highland basalt" (Reid *et al.*, 1972), because this composition is consistent with the trace and major element data of Hubbard *et al.* (1971b) (also Table 1) and is close to the analyzed composition of 15418 (LSPET, 1972).

It has been postulated that KREEP and low-K nonmare basalts have a common or similar source material (Hubbard *et al.*, 1971b) and that the fundamental differences between the two basalt types are directly related to large differences in the extent of partial melting that occurred when they were produced. The concept of a common parent can now be put on a firmer basis. The first step is to show that both compositions can be derived from a common source or that both compositions can be back-calculated to a common source via partial melting models. The calculation approach was used earlier and obtained a common K concentration. If the KREEP source is assumed to have 7–10× chondritic REE, Ba, U, Sr, Ca, Al, etc., (Hubbard and Gast, 1971) concentrations, then the 30–40% melting used to get the common K concentration can also be used to get the REE concentration independently calculated by Hubbard *et al.* (1971b), or within a factor of 2 higher, and with chondritic relative abundances. Under conditions of 30–40% melting, the Eu and Sr concentrations in liquid and coexisting plagioclase residue are nearly equal and buffered by the large amounts of both solid and liquid, thus producing the constant Eu and Sr concentrations observed by Hubbard *et al.* (1971b). Because plagioclase would be a major residual mineral, low-K anorthosite cumulates should be common.

Thermal calculations (Gast and McConnell, 1972) suggest that the extensive melting predicted for production of low-K nonmare basalts can only be the result of external heating and thus could affect the entire lunar surface. If extensive and early melting of the lunar surface is correct, we suggest that the $Al_2O_3$-rich nonmare crust found by Adler *et al.* (1972) may have been generated by the gravitational separation of low-K nonmare basalt magma, which also carried as xenocrysts the plagioclase fraction of the residue. This suggestion is akin to the early model of Wood *et al.* (1970) but is fundamentally different in that the ultramafic residue has no genetic relationship to mare basalts.

Intermediate chemical compositions similar to those represented by 14063 and 14083 and the intermediate degrees of partial melting implied by these compositions suggest thermal conditions intermediate between those for low-K nonmare basalts and KREEP basalts. The thermal model of Gast and McConnell (1972) suggests that this is improbable. However, the postulated molten surface layer must interface with solid material below, and thus a zone with decreasing fractions of molten material is to be expected. We postulate that chemical compositions like 14063 and 14083 were derived from this zone. It may be further postulated that this zone of decreasing liquid may be enriched in Mg-rich minerals, which contribute to the high MgO concentrations of 14063 and 14083.

The thermal model of Gast and McConnell (1972) suggests that KREEP basalts must be generated at depths greater than 100 km in order for heat from radioactive decay to raise temperatures above the melting point. If this suggestion is coupled with the partial melting model used earlier and the conclusion that all nonmare materials can be derived from a single source material, then one can further conclude that the nonmare source material must extend to depths gretear than 100 km.

## Conclusions

1. Nonmare basalts are chemically distinct from mare basalts, primarily in FeO and $Al_2O_3$ concentrations and $Al_2O_3$/FeO ratios.

2. The chemical compositions of nonmare basalts are essentially bounded by the

well-defined KREEP composition and the inferred low-K nonmare composition. The spectrum of nonmare basalt compositions can be primarily related to the degree of partial melting. To a first approximation, all nonmare basalts can be derived from a single source material enriched 7–10× in refractory element (Ca, Al, REE, U, Ba, Sr, etc.) concentrations relative to chondrites.

3. The chemical composition of KREEP basalts from Apollo 12, 14, and 15 shows very little chemical variation and substantiates the major and trace element concentrations found and predicted for KREEP basalts (Hubbard *et al.*, 1971a; Hubbard and Gast, 1971). The KREEP type of nonmare basalt is distinct from all other lunar basalts because of its high REE, etc., concentrations.

4. The source of nonmare basalts is approximately threefold richer in refractory elements than the mare basalt source. The enrichment in refractories is accompanied by a small (~50%) enrichment in K and perhaps other volatiles.

*Acknowledgments*—Jeff Warner, Ian Ridley, and Chuck Meyer provided valuable discussions. We greatly appreciate Beverly Atkinson's rapid and diligent typing of interim drafts and final copy. We also thank Roy Brown for the microprobe analysis of KREEP anorthosite.

## REFERENCES

Adler I., Trombka J., Gerard J., Lowman P., Yin L., and Blodgett H. (1972) Preliminary results from the S–161 x-ray fluorescence experiment (abstract). In *Lunar Science—III* (editor C. Watkins), pp. 4–7, Lunar Science Institute Contr. No. 88.

Albee A. L., Chodos A. A., Gancarz A. J., Haines E. L., Papanastassiou D. A., Ray L., Tera F., Wasserburg G. J., and Wen T. (1972) Mineralogy, petrology and chemistry of a Luna 16 basaltic fragment, sample B–1. *Earth Planet. Sci. Lett.* **13,** 353–367.

Apollo Soil Survey (1971) Apollo 14: Nature and origin of rock types in soil from the Fra Mauro formation. *Earth Planet. Sci. Lett.* **12,** 49–50.

Cliff R. A., Lee-Hu C., and Wetherill G. W. (1971) Rb–Sr and U–Th–Pb measurements on Apollo 12 material. *Proc. Second Lunar Sci. Conf., Geochim. Cosmochim. Acta* Suppl. 2, Vol. 2, pp. 1493–1502. MIT Press.

Compston W., Berry H., Vernon M. J., Chappell B. W., and Kaye M. J. (1971) Rubidium–Strontium chronology and chemistry of lunar materials from the Ocean of Storms. *Proc. Second Lunar Sci. Conf., Geochim. Cosmochim. Acta* Suppl. 2, Vol. 2, pp. 1471–1485. MIT Press.

Compston W., Vernon M. J., Berry H., Rudowski R., Gray C. M., Ware N., Chappell B. W., and Kaye M. J. (1972) Age and petrogenesis of Apollo 14 basalts (abstract). In *Lunar Science—III* (editor C. Watkins), pp. 151–153, Lunar Science Institute Contr. No. 88.

Cuttitta F., Rose H. J. Jr., Annell C. S., Carron M. K., Christian R. P., Dwornik E. J., Greenland L. P., Helz A. W., and Ligon D. T. Jr. (1971) Elemental composition of some Apollo 12 lunar rocks and soils. *Proc. Second Lunar Sci. Conf., Geochim. Cosmochim. Acta* Suppl. 2, Vol. 2, pp. 1217–1229. MIT Press.

Drake M. J., McCallum I. S., McKay G. A., and Weill D. F. (1970) Mineralogy and petrology of Apollo 12 sample 12013: A progress report. *Earth Planet. Sci. Lett.* **9,** 103–123.

Eldridge J. S., O'Kelley G. D., and Northcutt K. J. (1972) Abundances of primordial and cosmogenic radionuclides in Apollo 14 rocks and fines (abstract). In *Lunar Science—III* (editor C. Watkins), pp. 221–223, Lunar Science Contr. No. 88.

Gast P. W. (1968) Trace element fractionation and the origin of tholeiitic and alkaline magma types. *Geochim Cosmochim. Acta* **32,** 1057–1086.

Gast P. W. and Hubbard N. J. (1970) Abundance of alkali metals, alkaline and rare earths and strontium-87/strontium-86 ratios in lunar samples. *Science* **167,** 485–486.

Gast P. W., Hubbard N. J., and Wiesmann H. (1970) Chemical composition and petrogenesis of basalts from Tranquility Base. *Proc. Apollo 11 Lunar Sci. Conf., Geochim. Cosmochim. Acta* Suppl. 1, Vol. 2, pp. 1143–1163. Pergamon.

Gast P. W. and McConnell R. K. Jr. (1972) Evidence for initial chemical layering of the moon (abstract). In *Lunar Science—III* (editor C. Watkins), pp. 289–290, Lunar Science Institute Contr. No. 88.

Gibson E. K. and Hubbard N. J. (1972) Thermal volatilization studies on lunar samples. Submitted to *Proc. Third Lunar Sci. Conf., Geochim. Cosmochim. Acta* Suppl. 3.

Gopalan K., Kaushal S., Lee-Hu C., and Wetherill G. W. (1970) Rb–Sr and U–Th–Pb ages of lunar materials. *Proc. Apollo 11 Lunar Sci. Conf., Geochim. Cosmochim. Acta* Suppl. 1, Vol. 2, pp. 1195–1205. Pergamon.

Grieve G., McKay G., Smith H., and Weill D. (1972) Mineralogy and petrology of polymict breccia 14321 (abstract). In *Lunar Science—III* (editor C. Watkins), pp. 338–340, Lunar Science Institute Contr. No. 88.

Helmke P. A. and Haskin L. A. (1972) Rare earths and other trace elements in Apollo 14 lunar samples (abstract). In *Lunar Science—III* (editor C. Watkins), pp. 366–368, Lunar Science Institute Contr. No. 88.

Hubbard N. J., Gast P. W., and Wiesmann H. (1970) Rare earth, alkaline and alkali metal and $Sr^{87/86}$ data for subsamples of lunar sample 12013. *Earth Planet. Sci. Lett.* **9,** 181–184.

Hubbard N. J. and Gast P. W. (1971) Chemical composition and origin of nonmare lunar basalts. *Proc. Second Lunar Sci. Conf., Geochim. Cosmochim. Acta* Suppl. 2, Vol. 2, pp. 999–1020. MIT Press.

Hubbard N. J., Meyer C. Jr., Gast P. W., and Wiesmann H. (1971a) The composition and derivation of Apollo 12 soils. *Earth Planet. Sci. Lett.* **10,** 341–350.

Hubbard N. J., Gast P. W., Meyer C., Nyquist L. E., Shih C., and Wiesmann H. (1971b) Chemical composition of lunar anorthosites and their parent liquids. *Earth Planet. Sci. Lett.* **13,** 71–75.

Huneke J. C., Podosek F. A., Turner G., and Wasserburg G. J. (1972) $^{40}Ar$–$^{39}Ar$ systematics in lunar rocks and separated minerals of lunar rocks from Apollo 14 and 15 (abstract). In *Lunar Science—III* (editor C. Watkins), pp. 413–414, Lunar Science Institute Contr. No. 88.

Keil K., Prinz M., and Bunch T. E. (1971) Mineralogy, petrology, and chemistry of some Apollo 12 samples. *Proc. Second Lunar Sci. Conf., Geochim. Cosmochim. Acta* Suppl. 2, Vol. 1, pp. 319–341. MIT Press.

LSPET (Lunar Sample Preliminary Examination Team) (1970) Preliminary examination of lunar samples from Apollo 12. *Science* **167,** 1325–1339.

LSPET (Lunar Sample Preliminary Examination Team) (1972) The Apollo 15 lunar samples: A preliminary description. *Science* **175,** 363–375.

Marvin U. B., Wood J. A., Taylor G. J., Reid J. B., Powell B. N., Dickey J. S. Jr., and Bower J. F. (1971) Relative proportions and probable sources of rock fragments in the Apollo 12 soil samples. *Proc. Second Lunar Sci. Conf., Geochim. Cosmochim. Acta* Suppl. 2, Vol. 1, pp. 679–701. MIT Press.

Metzger A. E., Trombka J. I., Peterson R. C., Reedy R. C., and Arnold J. A. (1972) A first look at the lunar orbital gamma ray data (abstract). In *Lunar Science—III* (editor C. Watkins), pp. 540–541, Lunar Science Institute Contr. No. 88.

Meyer C. Jr., Brett R., Hubbard N. J., Morrison D. A., McKay D. S., Aitken F. K., Takeda H., and Schonfeld E. (1971) Mineralogy, chemistry, and origin of the KREEP component in soil samples from the Ocean of Storms. *Proc. Second Lunar Sci. Conf., Geochim. Cosmochim. Acta* Suppl. 2, Vol. 1, pp. 393–411. MIT Press.

Nyquist L. E., Hubbard N. J., Gast P. W., Church S. E., Bansal B., and Wiesmann H. (1972) Rb–Sr systematics for chemically defined Apollo 14 breccias. Submitted to *Proc. Third Lunar Sci. Conf., Geochim. Cosmochim. Acta* Suppl. 3.

O'Kelley G. D., Eldridge J. S., Schonfeld E., and Bell P. R. (1970) Primordial radionuclide abundances, solar proton, and cosmic ray effects and ages of Apollo 11 lunar samples by nondestructive gamma-ray spectrometry. *Proc. Apollo 11 Lunar Sci. Conf., Geochim. Cosmochim. Acta* Suppl. 1, Vol. 2, pp. 1407–1423. Pergamon.

O'Kelley G. D., Eldridge J. S., Schonfeld E., and Bell P. R. (1971) Abundances of the primordial radionuclides K, Th, and U in Apollo 12 lunar samples by nondestructive gamma-ray spectrometry: Implications for origin of lunar soils. *Proc. Second Lunar Sci. Conf., Geochim. Cosmochim. Acta* Suppl. 2, Vol. 2, pp. 1159–1168. MIT Press.

O'Kelley G. D., Eldridge J. S., Schonfeld E., and Northcutt K. J. (1972) Concentrations of primordial radioelements and cosmogenic radionuclides in Apollo 15 samples by nondestructive gamma-ray spectrometry (abstract). In *Lunar Science—III* (editor C. Watkins), pp. 587–589, Lunar Science Institute Contr. No. 88.

Reid A. M., Ridley W. I., Warner J. L., Harmon R. S., Brett R., Jakes P., and Brown R. W. (1972) Chemistry of highland and mare basalts as inferred from glass in the lunar soils (abstract). In *Lunar Science—III* (editor C. Watkins), pp. 640–642, Lunar Science Institute Contr. No. 88.

Ridley W. I., Williams R. J., Brett R., Takeda H., and Brown R. (1972) Petrology of lunar basalt 14310 (abstract). In *Lunar Science—III* (editor C. Watkins), pp. 648–650, Lunar Science Institute Contr. No. 88.

Ringwood A. E. and Essene E. (1970) Petrogenesis of Apollo 11 basalts, internal constitution and origin of the moon. *Proc. Apollo 11 Lunar Sci. Conf., Geochim. Cosmochim. Acta* Suppl. 1, Vol. 1, pp. 769–799. Pergamon.

Ringwood A. E., Green D. H., and Ware N. G. (1972) Experimental petrology and petrogenesis of Apollo 14 basalts (abstract). In *Lunar Science—III* (editor C. Watkins), pp. 654–656, Lunar Science Institute Contr. No. 88.

Roedder E. and Weiblen P. W. (1971) Petrology of silicate melt inclusions, Apollo 11 and Apollo 12 and terrestrial equivalents. *Proc. Second Lunar Sci. Conf., Geochim. Cosmochim. Acta* Suppl. 2, Vol. 1, pp. 507–529. MIT Press.

Silver L. T. (1970) Uranium–thorium–lead isotopes in some Tranquility Base samples and their implications for lunar history. *Proc. Apollo 11 Lunar Sci. Conf., Geochim. Cosmochim. Acta* Suppl. 1, Vol. 2, pp. 1533–1574. Pergamon.

Smith J. V., Anderson A. T., Newton R. C., Olsen E. J., Wyllie P. J., Crewe A. V., Isaacson M. S., and Johnson D. (1970) Petrologic history of the moon inferred from petrography, mineralogy, and petrogenesis of Apollo 11 rocks. *Proc. Apollo 11 Lunar Sci. Conf., Geochim. Cosmochim. Acta* Suppl. 1, Vol. 1, pp. 897–925. Pergamon.

Tatsumoto M. (1970) Age of the moon: An isotopic study of U–Th–Pb systematics of Apollo 11 lunar samples—II. *Proc. Apollo 11 Lunar Sci. Conf., Geochim. Cosmochim. Acta* Suppl. 1, Vol. 2, pp. 1595–1612. Pergamon.

Tatsumoto M., Knight R. J., and Doe B. R. (1971) U–Th–Pb systematics of Apollo 12 lunar samples. *Proc. Second Lunar Sci. Conf., Geochim. Cosmochim. Acta* Suppl. 2, Vol. 2, pp. 1521–1546. MIT Press.

Tatsumoto M., Hedge C. E., Doe B. R., and Unruh D. (1972) U–Th–Pb and Rb–Sr measurements on some Apollo 14 lunar samples (abstract). In *Lunar Science—III* (editor C. Watkins), pp. 741–743, Lunar Science Institute Contr. No. 88.

Taylor S. R., Muir P., Nance W., Rudowski R., and Kaye M. (1972) Composition of the lunar uplands, I. Chemistry of Apollo 14 samples from Fra Mauro (abstract). In *Lunar Science—III* (editor C. Watkins), pp. 744–746, Lunar Science Institute Contr. No. 88.

Vinogradov A. P. (1971) Preliminary data on lunar ground brought to Earth by automic probe "Luna-16". *Proc. Second Lunar Sci. Conf., Geochim. Cosmochim. Acta* Suppl. 2, Vol. 1, pp. 1–16. MIT Press.

Wakita H. and Schmitt R. A. (1970) Elemental abundances in seven fragments from lunar rock 12013. *Earth Planet. Sci. Lett.* **9**, 169–176.

Wakita H., Schmitt R. A., and Rey P. (1970) Elemental abundances of major, minor, and trace elements in Apollo 11 lunar rocks, soil, and core samples. *Proc. Apollo 11 Lunar Sci. Conf., Geochim. Cosmochim. Acta* Suppl. 1, Vol. 2, pp. 1685–1717. Pergamon.

Walker D., Longhi J., and Hays J. F. (1972) Experimental petrology and origin of Fra Mauro rocks and soils (abstract). In *Lunar Science—III* (editor C. Watkins), pp. 770–772, Lunar Science Institute Contr. No. 88.

Warner J. L. (1972) Apollo 14 breccias: Metamorphic origin and classification (abstract). In *Lunar Science—III* (editor C. Watkins), pp. 782–784, Lunar Science Institute Contr. No. 88.

Wasserburg G. J. and Papanastassiou D. A. (1971) Age of an Apollo 15 mare basalt; lunar crust and mantle evolution. *Earth Planet. Sci. Lett.* **13,** 97–104.

Wood J. A., Dickey J. S. Jr., Marvin U. B., and Powell B. N. (1970) Lunar anorthosites and a geophysical model of the moon. *Proc. Apollo 11 Lunar Sci. Conf., Geochim. Cosmochim. Acta* Suppl. 1, Vol. 1, pp. 965–988. Pergamon.

Proceedings of the Third Lunar Science Conference
(Supplement 3, *Geochimica et Cosmochimica Acta*)
Vol. 2, pp. 1181–1200
The M.I.T. Press, 1972

# Bulk, rare earth, and other trace elements in Apollo 14 and 15 and Luna 16 samples

J. C. LAUL, H. WAKITA,* D. L. SHOWALTER,†
W. V. BOYNTON, and R. A. SCHMITT

Department of Chemistry and the Radiation Center,
Oregon State University, Corvallis, Oregon 97331

**Abstract**—The chemical abundances of 24 and 34 bulk, minor and trace elements have been measured by instrumental (INAA) and radiochemical (RNAA) neutron activation analysis, respectively, in a variety of lunar specimens. Apollo 14 soils are characterized by significant enrichments of $Al_2O_3$, $Na_2O$, and $K_2O$ and depletions of $TiO_2$, FeO, MnO, and $Cr_2O_3$ relative to Apollo 11 and to most of Apollo 12 soils. The uniform abundances in 14230 core tube soils and three other Apollo 14 soils indicate that the regolith is uniform to at least 22 cm depth and within ~200 m from the lunar module. The chondritic normalized REE distribution patterns of Apollo 14 soils resemble those of KREEP-norite and Apollo 12 soils. A Sm/Eu ratio of 11.6 observed in the average Apollo 14 soil is compared to 7.4, 9.7, and 3.9 in Apollo 11, 12, and Luna 16 soils, respectively. This indicates a larger KREEP-noritic component in Apollo 14 soils. Elemental abundances in the igneous rock 14073 are quite similar to igneous rock 14310 and to clastic rocks such as 14318. Apollo 14 soils may be produced by pulverizing ~80% rocks like igneous 14073 or breccia 14318 and ~20% KREEP. Clastic rock 14047 is indistinguishable in composition from Apollo 14 soils and, therefore, 14047 is merely a compacted soil. Elemental abundances in four clastic rocks, 14063 and 14083 (STA Cl), 14318 (STA H) and 14066 (STA F) reveal a large variety of KREEP and variable quantities (25–100%) of KREEP components in individual clasts as well as in the overall rock composition. Two Luna 16 breccias are similar in composition to Luna 16 soils. Four Apollo 15 soils (LM, STA 4, 9, and 9a) have variable compositions. The REE abundances reveal ~6–14% KREEP components in these soils compared to ~25–80% KREEP in Apollo 12 and 14 soils. Although the Apollo 15 site is in the ESE edge of Mare Imbrium, the presumed source of ejected KREEP-norite matter, the mare lavas at Palus Putredinis (3.2 AE) have flowed to tens of meters in depths over the older (3.9 AE) Mare Imbrium event. Interelement correlations between MnO–FeO, Sc–FeO, V–$Cr_2O_3$, and $K_2O$–Hf negate the hypothesis that howardite achondrites may be primitive lunar matter, argue against the fission hypothesis for the origin of the moon, and precludes any selective large scale volatilization of alkalies during lunar magmatic events.

## EXPERIMENTAL

The detailed analytical procedure used here has been described by Schmitt *et al.* (1970), Rey *et al.* (1970), and Wakita *et al.* (1970). In our sequential INAA procedure for 24 elements, sample weights ranged from 1 to 500 mg. To increase the sensitivity of elemental detection, samples were grouped into four batches depending on their weights, i.e., <20 mg, 20–80 mg, 80–200 mg, and 200–500 mg weight groups. A duplicate set of U.S.G.S. standards BCR–1 and one sample GSP–1 were also analyzed with each batch. Aliquants of two soils, 14003 and 14163, were also subjected to RNAA.

* Present address: Dept. of Chemistry, University of Tokyo, Japan.
† Present address: Dept. of Chemistry, Wisconsin State University, Stevens Pt., Wisconsin.

## RESULTS AND DISCUSSION

RESULTS ARE SHOWN in Tables 1 and 2. In general, the elemental abundance data of this work agree well with those obtained by Brunfelt *et al.* (1971), Schnetzler and Nava (1971), Taylor *et al.* (1971), Helmke and Haskin (1972), Hubbard and Gast (1972), Schnetzler *et al.* (1972), and Rose *et al.* (1972). The overall reproducibility of our procedure was tested from replicate analyses of BCR–1. Only the average value of eight replicates of BCR–1 is listed in Table 2.

Table 1. Elemental abundances in three Apollo 14 soils, one clastic and one igneous rock.[a]

| | Soils | | | | | | Clastic rock | | Igneous rock |
|---|---|---|---|---|---|---|---|---|---|
| | 14003,31 | | 14163,55 | | 14240,10 (SESC) | | 14047,30 | | 14073,3 0.0308g powdered aliquant[d] |
| Element | 0.528g[b] | 0.478g[c] | 0.537g[b] | 0.392g[c] | 0.227g[b] | 0.187g[b] | 0.402g[b] | 0.553g[b] | |
| $TiO_2$ (%) | 1.8 | | 1.9 | | 1.8 | 2.0 | 1.8 | 1.9 | 1.2 |
| $Al_2O_3$ (%) | 18.3 | | 18.4 | | 19.4 | 18.5 | 18.5 | 18.8 | 20.0 |
| FeO (%) | 11.0 | | 10.4 | | 10.9 | | 10.9 | | 8.9 |
| CaO (%) | 11 | | 11 | | 12 | 12 | 12 | 11 | 13 |
| $Na_2O$ (%) | 0.689 | | 0.711 | | 0.717 | 0.745 | 0.670 | 0.662 | 0.76 |
| $K_2O$ (%) | 0.57 | 0.60 | 0.52 | 0.54 | 0.57 | 0.60 | 0.48 | 0.47 | 0.48 |
| MnO (%) | 0.128 | | 0.124 | | 0.125 | 0.130 | 0.125 | 0.123 | 0.120 |
| $Cr_2O_3$ (%) | 0.214 | | 0.197 | | 0.209 | | 0.204 | | 0.196 |
| Rb (ppm) | | 13 | | 13 | | | | | |
| Cs (ppm) | | 0.5 | | 0.5 | | | | | |
| Sc (ppm) | 23 | | 21 | | 22 | | 22 | | 21 |
| V (ppm) | 53 | | 57 | | 55 | 40 | 40 | 50 | 45 |
| Co (ppm) | 39 | | 38 | | 36 | | 38 | | 18 |
| Cd (ppb) | | (139) | | (189) | | | | | |
| In (ppb) | | (104) | | (85) | | | | | |
| Zr (ppm) | 780 | | 900 | | 790 | | 880 | | 810 |
| Hf (ppm) | 20 | | 20 | | 20 | | 20 | | 17 |
| Th (ppm) | 14 | | 13 | | 14 | | 14 | | 8 |
| Ba (ppm) | 810 | | 730 | | 790 | | 730 | | 660 |
| La (ppm) | 68 | 66 | 69 | 68 | 70 | | 69 | | 60 |
| Ce (ppm) | 200 | 193 | 200 | 200 | 214 | | 204 | | 196 |
| Pr (ppm) | | 20.5 | | 24.4 | | | | | |
| Nd (ppm) | | 103 | | 103 | | | | | |
| Sm (ppm) | 30 | 31.2 | 29 | 32.2 | 31 | | 29 | | 26.4 |
| Eu (ppm) | 2.60 | 2.56 | 2.70 | 2.78 | 2.65 | | 2.70 | | 2.2 |
| Gd (ppm) | | 36 | | 37 | | | | | |
| Tb (ppm) | | 6.1 | | 6.4 | | | | | |
| Dy (ppm) | | 41 | | 41 | | | | | |
| Ho (ppm) | | 9.7 | | 10.2 | | | | | |
| Er (ppm) | | 23.9 | | 24.5 | | | | | |
| Tm (ppm) | | 4.0 | | 4.1 | | | | | |
| Yb (ppm) | 23 | 22 | 22 | 24 | 24 | | 22 | | 21 |
| Lu (ppm) | 3.2 | 3.3 | 3.1 | 3.6 | 3.2 | | 3.0 | | 2.9 |
| Y (ppm) | | 192 | | 204 | | | | | |
| ΣREE + Y (ppm) | | 754 | | 784 | | | | | |
| Sm/Eu | 11.5 | 12.2 | 10.7 | 11.6 | 11.7 | | 10.7 | | 12.0 |

[a] One standard deviation due to counting statistics and other errors for single determinations are approximately ±2–3% for Al, Na, Mn, and Cr; ±5% for Ti, Fe, Mg, Ca, Sc, Co, Cd, In, 14 rare earth elements and Y; ±10% for K, Rb, Cs, and Hf; ±15% for V, Zr, Th, and Ba, and ±10% for Ce and Eu in 14073 analysis. Cd and In abundances are probably high due to LRL contamination.

[b] Abundances were obtained via instrumental neutron activation analysis.

[c] Abundances were obtained via radiochemical neutron activation analysis.

[d] This sample was prepared by D. S. Burnett as a member of the G. J. Wasserburg consortium for analysis of the 14073 rock.

Table 2. Elemental abundances in one igneous and four Apollo 14 clastic rocks, three Apollo 14 core tube soils, two Luna 16 breccias and four Apollo 15 soils.[a]

| | Element | | | | | | | | | | | | | |
|---|---|---|---|---|---|---|---|---|---|---|---|---|---|---|
| | $TiO_2$ (%) | $Al_2O_3$ (%) | FeO (%) | CaO (%) | $Na_2O$ (%) | $K_2O$ (%) | MnO (%) | $Cr_2O_3$ (%) | Sc (ppm) | V (ppm) | Co (ppm) | Zr (ppm) | Hf (ppm) | Th (ppm) |
| 14073,3A (STA G) Igneous 269 mg | 1.3 | 20.8 | 8.2 | 12.8 | 0.737 | 0.49 | 0.108 | 0.150 | 18.7 | 29 | 18 | 660 | 21 | 11 |
| 14063,37 Clastic (STA C1) | | | | | | | | | | | | | | |
| A10 Finest matrix 81 mg | 1.5 | 22.0 | 7.0 | 13.1 | 0.835 | 0.17 | 0.081 | 0.180 | 13.6 | 29 | 20 | — | — | — |
| All Coarse matrix int. 59 mg | 1.4 | 21.5 | 6.8 | 13.2 | 0.782 | — | 0.081 | 0.171 | 12.4 | 27 | 18 | — | — | — |
| A12 Black clasts 61 mg | 1.3 | 20.4 | 7.7 | 11.6 | 0.755 | 0.11 | 0.085 | 0.233 | 13.5 | 31 | 27 | — | — | — |
| BS Scrap. frag. int. 80 mg | 1.8 | 22.8 | 6.5 | 13.3 | 0.782 | 0.12 | 0.079 | 0.160 | 14.7 | 17 ± 6 | 17 | — | — | — |
| B11–3 Coarse matrix int. 72 mg | 1.6 | 23.5 | 6.4 | 14.6 | 0.795 | 0.13 | 0.076 | 0.147 | 14.1 | 18 ± 6 | 17 | — | — | — |
| 14083 Clastic (STA C1) | | | | | | | | | | | | | | |
| 2,1 Black and white frag. 73 mg | 1.6 | 16.4 | 10.4 | 10.4 | 0.984 | 0.36 | 0.116 | 0.180 | 21.5 | 45 | 34 | 830 | 31 | 19 |
| 2, White A, 1/16 Pure white matter 31 mg | 0.72 | 21.8 | 6.8 | 14.1 | 0.657 | 0.21 | 0.076 | 0.100 | 10.1 | 25 | 17 | 450 | 13.0 | 6.1 |
| 2,4 Int. Dark 1/8 Pure dark matter 28 mg | 1.7 | 17.0 | 9.1 | 12.4 | 0.973 | 0.44 | 0.110 | 0.700 | 19.2 | 48 | 26 | 1100 | 31 | 19 |
| 14318 Clastic (STA H) | | | | | | | | | | | | | | |
| 40 Sawdust[b] 56 mg | 1.6 | 16.3 | 9.9 | 10.0 | 0.728 | 0.58 | 0.105 | 0.193 | 17.9 | 46 | 31 | 800 | 20 | 12 |
| Sawdust[c] 31 mg | 1.2 | 16.1 | 10.7 | 11.3 | 0.728 | 0.53 | 0.109 | 0.180 | 17.1 | 30 | 28 | 600 | 21 | 13 |
| 26-A clast 14.2 mg | 1.8 | 17.8 | 9.8 | 10.8 | 0.821 | 0.63 | 0.113 | 0.143 | 18.6 | 46 | 26 | 1400 | 42 | 24 |
| 26-B clast 5.2 mg | 1.7 | 18.9 | 11.5 | 10.3 | 0.976 | 0.61 | 0.107 | 0.148 | 17.5 | 45 | 86 | 950 | 32 | 18 |
| 26-C clast 12.8 mg | 1.0 | 15.9 | 7.7 | 9.4 | 0.852 | 3.3 | 0.099 | 0.121 | 16.6 | 24 | 28 | 600 | 23 | 24 |
| 27-B clast 7.6 mg | 1.1 | 19.3 | 7.5 | 10.3 | 0.837 | 2.1 | 0.101 | 0.097 | 15.3 | 25 | 12 | 1300 | 27 | 17 |
| 14066 Clastic (STA F) | | | | | | | | | | | | | | |
| 31,3 Sawdust 269 mg | 1.8 | 15.3 | 9.5 | 9.5 | 0.764 | 0.77 | 0.112 | 0.165 | 17.7 | 33 | 28 | 950 | 28 | 15 |
| 21,2.02 Many clasts like 2.03 226 mg | 1.0 | 23.1 | 6.2 | 12.8 | 1.07 | 0.33 | 0.073 | 0.128 | 9.8 | 21 | 20 | 640 | 19 | 10 |
| 21,1.01 Matrix small clasts 33 mg | 1.6 | 16.3 | 8.4 | 11.0 | 0.864 | 1.3 | 0.111 | 0.150 | 17.6 | 59 | 23 | 970 | 33 | 16 |
| 21,2.01 Breccia clast 27 mg | 2.0 | 16.6 | 11.7 | 8.9 | 0.935 | 0.30 | 0.115 | 0.240 | 16.1 | 55 | 38 | 550 | 23 | 15 |
| 21,2.04,2 White igneous clast 10.6 mg | 1.0 | 17.9 | 8.4 | 9.1 | 0.759 | 1.4 | 0.107 | 0.124 | 15.3 | 68 | 17 | 800 | 33 | 13 |
| 14230 Core Tube (STA G) | | | | | | | | | | | | | | |
| 110 13—14.3 cm 56 mg | 1.7 | 18.1 | 9.3 | 11.2 | 0.714 | 0.60 | 0.122 | 0.180 | 18.8 | 50 ± 7 | 32 | 550 | 21 | 13 |
| 116 18—19.8 cm 74 mg | 1.6 | 18.0 | 10.1 | 12.5 | 0.741 | 0.48 | 0.125 | 0.200 | 20 | 64 ± 9 | 33 | 600 | 21 | 12 |
| 127 21.5–22.6 cm 74 mg | 1.7 | 18.7 | 9.7 | 11.6 | 0.714 | 0.55 | 0.129 | 0.190 | 20 | 34 ± 7 | 34 | 700 | 22 | 14 |
| Average in 3 Ap. 14 soils, 3,163,240 | 1.9 | 18.6 | 10.8 | 11.3 | 0.74 | 0.56 | 0.127 | 0.207 | 22 | 52 | 38 | — | 20 | — |
| Luna 16 breccias | | | | | | | | | | | | | | |
| A-36 1.07 mg | 3.7 | 16.8 | 16.5 | 14.0 | 0.99 | 0.21 | 0.215 | 0.314 | 49 | 106 | 33 | — | 6.5 | — |
| G-41 2.32 mg | 3.8 | 16.8 | 16.4 | 9.0 | 0.76 | 0.20 | 0.213 | 0.314 | 49 | 70 | 32 | — | 8.8 | — |
| 15021,24 soil, < 1 mm LM cont. 261 mg | 1.8 | 14.1 | 15.0 | 10.8 | 0.434 | 0.22 | 0.190 | 0.400 | 28 | 114 | 40 | 350 | 9.9 | 4.9 |
| 15471,49 soil, < 1 mm STA 4 Dune c. 500 mg[d] | 1.6 | 13.3 | 15.9 | 10.9 | 0.364 | 0.12 | 0.204 | 0.430 | 31 | 147 | 44 | 160 | 6.3 | 2.4 |
| 15501,40 soil, < 1 mm STA 9 Scraplet c. 497 mg[d] | 1.6 | 12.7 | 16.5 | 10.0 | 0.377 | 0.17 | 0.207 | 0.430 | 31 | 139 | 43 | 370 | 7.1 | 3.4 |
| 15531,52 soil, < 1 mm STA 9a Rille 498 mg[d] | 2.1 | 10.0 | 19.5 | 10.5 | 0.301 | 0.09 | 0.248 | 0.490 | 36 | 182 | 50 | 130 | 4.4 | 1.8 |
| BCR-1 Basalt | 2.23 | 13.7 | 12.3 | 7.0 | 3.32 | 1.71 | 0.174 | 0.0022 | 32 | 430 | 36 | — | 4.7 | — |
| F.R.[e] Basalt | 1.5 | 14.8 | 10.3 | 11.3 | 2.15 | 0.65 | 0.171 | 0.290 | 39 | 280 | 45 | 170 | 2.9 | 0.5 |
| KREEP or Noritic matter[f] Average (range) | 1.6 (1.3–2.9) | 19 (15–21) | 9 (9–12) | 10.5 (10–11) | 1.0 (0.8–1.0) | 0.9 (0.5–1.1) | 0.18 (0.16–0.22) | 0.19 (0.17–.021) | 25 | 50 | 35 | 800 | 36 | 19 |

[a] Abundances were determined by INAA. Estimated errors due to counting statistics are: $TiO_2$, $\sim \pm 7\%$; $Al_2O_3$, $Na_2O$, MnO, $Cr_2O_3$, Sc, La and Sm, $\sim \pm 1$–$3\%$; FeO, $K_2O$, Co, Yb, Lu and Hf, $\sim \pm 5\%$; CaO Eu, Ta and Th, $\sim \pm 5$–$10\%$; V, Zr, Ba, Ce, Tb, and U $\sim \pm 10$–$30\%$. Standards and duplicate BCR-1 samples were activated with all samples. Whenever BCR-1 values differed from published values, sample abundances for a given activation analysis were recalculated using BCR-1 values in bottom line of this table. Average values in Apollo 14 soils were taken from Wakita *et al.* (1971). Abundances of Zr and Th were determined relative to 540 and 110 ppm, respectively, in GSP-1.

[b] This sawdust sample, obtained by the LRL in sawing 14318, was scooped from the container before the sawdust was mixed; therefore, this sample is not representative of the rock sawdust.

[c] This sample was obtained from well-mixed sawdust and, therefore, should be representative of the whole rock composition after correcting for sawing contamination.

Table 2.—*Continued.*

| | Element | | | | | | | | | | | | | |
|---|---|---|---|---|---|---|---|---|---|---|---|---|---|---|
| | U (ppm) | Ba (ppm) | La (ppm) | Ce (ppm) | Sm (ppm) | Eu (ppm) | Tb (ppm) | Yb (ppm) | Lu (ppm) | Ta (ppm) | Sm/Eu | K/U | K/Ba | Th/U |
| 14073,3A (STA G) Igneous 269 mg | 2.6 | 600 | 61 | 142 | 27 | 2.1 | 4.8 | 22 | 2.9 | 2.4 | 11.3 | 1560 | 6.8 | 4.2 |
| 14063,37 Clastic (STA C1) | | | | | | | | | | | | | | |
| A10 Finest matrix 81 mg | 1.1 | 360 | 26.5 | — | 10.6 | 2.4 | — | 10 | 1.4 | — | 4.4 | 1280 | 3.9 | — |
| A11 Coarse matrix int. 59 mg. | 1.8 | 550 | 34.2 | — | 15.6 | 2.4 | — | 12 | 1.7 | — | 6.5 | — | — | — |
| A12 Black clasts 61 mg | 0.5 | 280 | 18.6 | — | 8.7 | 2.4 | — | 7.0 | 1.0 | — | 3.6 | 1830 | 3.3 | — |
| BS Scrap. frag. int. 80 mg | 0.9 | 310 | 22.6 | — | 10.6 | 2.4 | — | 8.6 | 1.2 | — | 4.4 | 1110 | 3.2 | — |
| B11–3 Coarse matrix int. 72 mg | 0.8 | 260 | 21.8 | — | 10.1 | 2.5 | — | 8.0 | 1.1 | — | 4.0 | 1350 | 4.2 | — |
| 14083 Clastic (STA C1) | | | | | | | | | | | | | | |
| 2,1 Black and white frag. 73 mg | 5.0 | 900 | 109 | 253 | 44 | 3.2 | 9.4 | 36 | 5.0 | 3.7 | 13.8 | 600 | 3.3 | 3.8 |
| 2, White A, 1/16 Pure white matter 31 mg | 1.6 | 450 | 41 | 94 | 18.0 | 2.1 | 3.6 | 13.4 | 2.0 | 1.3 | 8.6 | 1090 | 3.9 | 3.8 |
| 2,4 Int. Dark 1/8 Pure dark matter 28 mg | 4.5 | 1100 | 92 | 249 | 42 | 3.2 | 8.6 | 32 | 4.4 | 3.5 | 13.1 | 810 | 3.3 | 4.2 |
| 14318 Clastic (STA H) | | | | | | | | | | | | | | |
| 40 Sawdust[b] 56 mg | 4.1 | 700 | 66 | 151 | 26 | 2.0 | 5.8 | 23 | 3.1 | 2.3 | 13.0 | 1170 | 6.9 | 2.9 |
| Sawdust[c] 31 mg | 2.1 | 600 | 65 | 170 | 26 | 2.0 | 5.9 | 22 | 3.1 | 2.4 | 13.0 | 2100 | 7.3 | 6.2 |
| 26-A clast 14.2 mg | 6.1 | 1100 | 129 | 290 | 51 | 3.0 | 11 | 37 | 5.6 | 5.1 | 17.0 | 860 | 4.8 | 3.9 |
| 26-B clast 5.2 mg | 6.0 | 1000 | 110 | 255 | 53 | 2.7 | 9.6 | 30 | 4.2 | 4.6 | 19.6 | 840 | 5.1 | 3.0 |
| 26-C clast 12.8 mg | 6.3 | 2600 | 58 | 115 | 22 | 2.7 | 5.7 | 24 | 4.3 | 5.0 | 8.1 | 4350 | 10.5 | 3.8 |
| 27-B clast 7.6 mg | 5.4 | 1700 | 95 | 215 | 40 | 2.2 | 8.1 | 30 | 4.5 | 3.7 | 18.1 | 3230 | 10.3 | 3.1 |
| 14066 Clastic (STA F) | | | | | | | | | | | | | | |
| 31,3 Sawdust 269 mg | 4.0 | 920 | 79 | 178 | 34 | 2.6 | 6.3 | 22 | 3.8 | 3.2 | 13.1 | 1600 | 6.9 | 3.8 |
| 21,2.02 Many clasts like 2.03 226 mg | 2.1 | 770 | 58 | 130 | 24 | 4.5 | 4.2 | 14.2 | 2.4 | 2.0 | 5.3 | 1300 | 3.6 | 4.8 |
| 21,1.01 Matrix small clasts 33 mg | 4.1 | 1000 | 78 | 200 | 40 | 2.5 | 7.8 | 28 | 4.0 | 3.4 | 16.0 | 2630 | 10.8 | 3.9 |
| 21,2.01 Breccia clast 27 mg | 2.7 | 800 | 61 | 180 | 32 | 3.0 | 6.3 | 21 | 2.7 | 3.4 | 10.7 | 920 | 3.1 | 5.6 |
| 21,2.04,2 White igneous clast 10.6 mg | 3.3 | 1350 | 57 | 140 | 28 | 2.3 | 5.4 | 17 | 3.2 | 3.2 | 12.2 | 3520 | 8.6 | 3.9 |
| 14230 Core tube (STA G) | | | | | | | | | | | | | | |
| 110 13–14.3 cm 56 mg | 3.5 | 850 | 64 | 162 | 29 | 2.7 | 6.0 | 22 | 3.2 | 2.8 | 10.7 | 1420 | 5.9 | 3.7 |
| 116 18–19.8 cm 74 mg | 3.0 | 800 | 66 | 160 | 32 | 2.5 | 5.8 | 23 | 3.2 | 2.5 | 12.8 | 1330 | 5.0 | 4.0 |
| 127 21.5–22.6 cm 74 mg | 3.4 | 800 | 68 | 170 | 30 | 2.6 | 6.4 | 24 | 3.3 | 2.8 | 11.5 | 1340 | 5.7 | 4.1 |
| Average in 3 Ap. 14 soils, 3,163,240 | — | 780 | 68 | 203 | 31 | 2.7 | 6.2 | 23 | 3.3 | — | 11.5 | — | 6.0 | — |
| Luna 16 Breccias | | | | | | | | | | | | | | |
| A-36 1.07 mg | — | — | 14 | 30 | 8.4 | 2.9 | 1.6 | 6.2 | 0.91 | 1.5 | 2.9 | — | — | — |
| G-41 2.32 mg | — | — | 13 | 23 | 7.6 | 3.6 | 1.3 | 6.0 | 0.90 | 1.3 | 2.1 | — | — | — |
| 15021,24 Soil, < 1 mm LM cont. 261 mg | 1.5 | 320 | 26 | 73 | 12.9 | 1.4 | 2.3 | 9.5 | 1.3 | 1.2 | 9.2 | 1220 | 5.7 | 3.3 |
| 15471,49 Soil, < 1 mm STA 4 Dune c. 500 mg[d] | 0.7 | 200 | 15.0 | 42 | 7.4 | 0.64 | 1.5 | 5.7 | 0.81 | 0.75 | 11.5 | 1420 | 5.0 | 3.4 |
| 15501,40 Soil, < 1 mm STA 9 Scraplet c. 497 mg[d] | 1.0 | 120 | 20 | 53 | 9.7 | 0.77 | 1.8 | 6.2 | 1.1 | 0.83 | 12.6 | 1410 | 11.8 | 3.4 |
| 15531,52 Soil, < 1 mm STA 9a Rille 498 mg[d] | — | 110 | 10.9 | 35 | 5.7 | 1.0 | 1.0 | 4.8 | 0.66 | 0.60 | 5.7 | — | 6.8 | — |
| BCR-1 Basalt | 1.4 | 700 | 26 | 53 | 7.0 | 2.0 | 0.96 | 3.6 | 0.55 | 0.74 | 3.5 | 10100 | 20 | — |
| F.R.[e] Basalt | — | 90 | 6.7 | 15 | 3.2 | 1.2 | 0.62 | 2.1 | 0.43 | 0.43 | 2.7 | 18000 | 60 | — |
| KREEP or Noritic matter[f] | 6 | 1100 | 108 | 297 | 49 | 3.2 | 10 | 39 | 5.7 | 3.9 | 15.3 | 1250 | 6.8 | 3.2 |
| Average (range) | | (600–1600) | (93–134) | (243–374) | (38–62) | (2.6–3.7) | | (34–47) | (4.7–7.0) | | | | | |

[d] Sample was split into two aliquants that were analyzed independently. Average of two analyses are given.

[e] F.R. basalt is an Oregon coastal mountain basalt, picked up near the east boundary of the range on the Fred B. Ramsey property (123° 19′ × 44° 39′).

[f] Overall average and range values for $TiO_2$, $Al_2O_3$, FeO, CaO, $Na_2O$, and $K_2O$ are taken from Hubbard and Gast (1971). Values for anorthositic basalt are not included in range. All other values from MnO to Lu are taken from Hubbard and Gast (1971) for five KREEP fragments from 12003 and 12033 and the dark material from 12013, 10–5 and also from Schnetzler *et al.* (1970) for dark 12013,10,15, and 8 matrix and from Wakita and Schmitt (1970) for 12013,06. Brown *et al.* (1971) determined MnO and $Cr_2O_3$ in noritic glass. Assuming 12034 breccia rock is ≈80% KREEP (Wänke *et al.*, 1971), values for Zr, Tb, and Ta in KREEP or noritic matter were calculated from abundances of these three elements in 12034 (Goles *et al.*, 1971 and Wakita *et al.*, 1971).

*Apollo 14 soils*

Apollo 14 soils are characterized by significant enrichment of $Al_2O_3$, $Na_2O$, and $K_2O$ and depletions of $TiO_2$, FeO, MnO, and $Cr_2O_3$, relative to Apollo 11 and to most of Apollo 12 soils. Chemical compositions, including trace elements, of these three Apollo 14 soils are essentially the same (Table 1). This suggests that the regolith seems to be uniform for distances of $\sim$100 meters both east and west of North Crater. Abundances of Hf, Th, Ba, and the REE greatly exceed those observed in Apollo 11 soil (Wakita *et al.*, 1970) and Luna 16 soil (Gillum *et al.*, 1972) by a factor of $\sim$5, and exceed those in some Apollo 12 soils by a factor of $\sim$2 (Wakita *et al.*, 1971). However, abundances of bulk, minor and trace elements in Apollo 14 soils are nearly equal to those measured in Apollo 12 soil 12033 and breccia rocks 12010 and 12034 (Goles *et al.*, 1971 and Wakita *et al.*, 1971). Total REE + Y abundances in Apollo 14 soils with an average value of 770 ppm are higher by factors of $\sim$2.5 and $\sim$4, compared to Apollo 11 and Luna 16 soils and are slightly higher than REE + Y abundances of Apollo 12 soil, 12033. These observations are consistent with the suggestion of Schnetzler and Nava (1971) that breccia rocks, such as 12010 and 12034, may have been ejected from cratering events on the Fra Mauro formation onto the Apollo 12 mare site followed by comminution to Apollo 12 soils such as 12033.

The chondritic normalized REE distribution patterns of Apollo 14 soils (Fig. 1) resemble those of Apollo 12 soils and these distribution patterns are quite different from those of Apollo 11 and Luna 16 soils. Apollo 14 soils have the greatest negative Eu anomalies. Apparent positive Ce anomalies ($\sim$15%) are observed in all Apollo 14 soils, one clastic and on igneous rocks that we analyzed. The Ce abundance in soil 14163 agrees with those reported by Brunfelt *et al.* (1971) and Taylor *et al.* (1971) and these Ce values are $\simeq$20% higher than those obtained by Schnetzler and Nava (1971) and Hubbard and Gast (1971). Masuda *et al.* (1972) also found the same degree of the positive Ce anomalies using isotope dilution technique. However, since we did not observe positive Ce anomalies in 14230 core tube soils, the alleged Ce anomaly may not be real.

The Rb/Cs ratio is fairly constant for Apollo 11, 12, 14, and Luna 16 soils. The Rb/Cs ratio of 26 observed in the average Apollo 14 soils is compared with 27 (Tera *et al.*, 1970), 15–29 (Wakita *et al.*, 1971) and 23 (Morgan *et al.*, 1972) found in Apollo 11, 12 and Luna 16 soils, respectively. A Sm/Eu ratio of 11.6 observed in the average Apollo 14 soils is compared to those of 7.4, 9.7, and 3.9 in Apollo 11, 12 and Luna 16 soils, respectively. This indicates a larger KREEP or noritic component in Apollo 14 soils. For KREEP-norite, the average Sm/Eu is 15.3 (Table 2).

*Core samples 14230, STA G*

Abundances of 24 elements are essentially identical for the three layers and for soil 14240 (STA G), scooped at $\sim$2 cm depth near the core tube and other soils 14003 and 14163 (Tables 1 and 2). In Apollo 12 double core Laul *et al.* (1971) reported a peculiar layer 12028,66 (13.2 cm) which was spectacularly enriched in Bi, Cd, and relatively less abundant in other trace elements with respect to neighboring layers. This led them to conclude that little mixing had occurred and that the

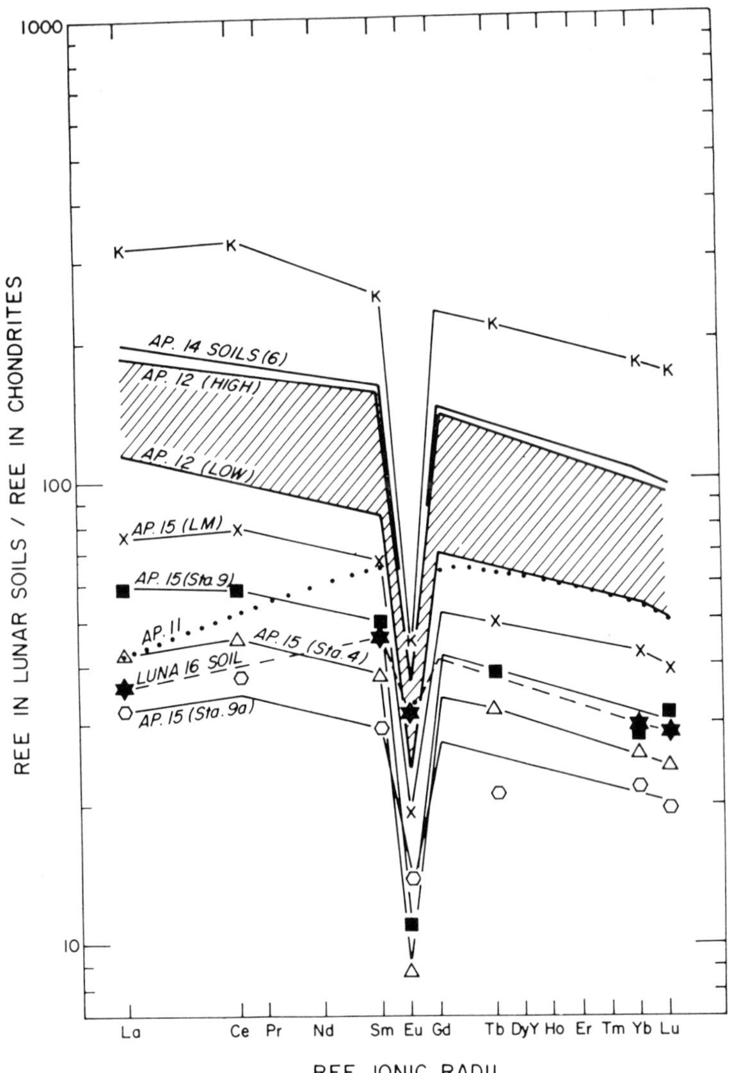


Fig. 1. Abundances of REE (La, Ce, Sm, Eu, Tb, Yb, and Lu) in soils normalized to chondritic abundances (La 0.34, Ce 0.91, Sm 0.195, Eu 0.073, Tb 0.047, Yb 0.22, Lu 0.034, Ba 3.6). Values of Apollo 11 and 12 and Luna 16 soils are taken from Wakita *et al.* (1970, 1971) and Gillum *et al.* (1972). Apollo 15 soils are 15021 (LM), 15471 (STA 4), 15501 (STA 9), and 15531 (STA 9a). Average values for KREEP (K) are listed in Table 2.

turn-over rate was of the order of magnitude slower than previously estimated. Based on our trace data, we see no evidence for vertical or horizontal nonmixing and we predict similar behavior for other trace elements. If such a chemical homogenity is typical of the Fra Mauro site, the regolith has been well-mixed with a turnover rate of few $10^6$ Y/cm.

Derivation of these soils can be explained by a simple two-component mixing model of KREEP and crystalline rock (Schnetzler *et al.*, 1970 and Hubbard and Gast, 1971). About 20% KREEP-noritic like component and 80% comminuted basaltic rocks like 14073 are required to account for the mass-balance of the soil composition. In addition, about 2% type C1 carbonaceous chondrites are needed to explain the enrichment of trace elements (Laul *et al.*, 1971); anorthosite, considered to be an highland material, and microbreccia are additional components (Duncan *et al.*, 1972).

*Igneous rock 14073, STA G*

Elemental abundances of one igneous rock 14073 are similar to those observed in Apollo 14 soils, with some exceptions such as ~10–15% higher $Al_2O_3$, CaO, and $Na_2O$ and 20–30% lower $TiO_2$ and FeO and ~50% lower Co abundances in 14073 relative to the soils. The REE abundances in 14073 are about 90% of those in Apollo 14 soils; the same degree of the negative Eu anomaly was observed. These REE abundances are significantly higher than those of Apollo 11, 12, and Luna 16 igneous rocks. The chondritic normalized REE distribution pattern of 14073 is quite different from those of Apollo 11 and 12 and Luna 16 igneous rocks; the 14073 pattern is very similar to those observed in KREEP or noritic fragments. The chemical compositions and ages of the alkali-rich rocks 14073 and 14310 are the same (LSPET, 1971; Helmke and Haskin, 1972; Wasserburg *et al.*, 1972). This suggests a common lava flow. On the other hand, Morgan *et al.* (1972) showed that the trace elemental abundances in 14310 are similar to soil and breccia and concluded that the so-called igneous 14310 may actually be a melted breccia rock. Our data for clastic rock 14318 (Table 2) whose overall chemical composition can be represented by the sawdust (Showalter *et al.*, 1972) matches closely to 14310. This infers some genetic relationship between them.

*Rock 14047*

This rock has been reported to be a fine-grained gas rich, clastic rock and has a glass coating (LSPET, 1971). Our analyses for bulk and trace elemental composition and the Sm/Eu ratio of this work are identical to those of Apollo 14 soils (Table 1), except for slightly lower $Na_2O$ and $K_2O$ abundances in the clastic rock. Such a close chemical similarity suggests that the rock 14047 is merely a compacted soil sample. Similarities in noble gas contents (LSPET, 1971) between 14047 and 14259 soil support this conclusion. Our conclusion is further strengthened by the close correspondence for siderophilic and volatile trace data in 14047 and soils (Morgan *et al.*, 1972).

*Clastic rocks*

Clastic rocks are more prevalent than crystalline rocks on the Apollo 14 area (LSPET, 1971). Petrographic and stratigraphic studies of these clastic rocks are complex. Presumably, KREEP-norite is associated with these clasts. The average

KREEP-norite composition and its range are listed in the bottom line of Table 2. We have also tabulated ratios of Sm/Eu, K/U, and K/Ba as KREEP-norite indicators. Sm/Eu is an indicator for Eu depletion and Sm, for the total REE abundance. The ratio of Th/U in the last column (Table 2) merely reflects a typical ratio of two highly refractory elements in high-temperature lunar minerals.

*Clastic rock 14063, STA C1 (Tatsumoto Consortium)*

This rock is a thermally metamorphosed microbreccia F3 (Jackson and Wilshire, 1972). Its parent rock is 14065, an alkali-rich rock collected from the continuous Cone Crater ejecta blanket. Five samples of 59–81 mg each were analyzed by us before they were analyzed via RNAA by Morgan *et al.* (1972). We could not report Zr, Hf, Th, Tb, and Ta because of insufficient time for decay of short-lived radionuclides.

These samples have in general high plagioclase content. Overall, our data suggests that the composition of all five $Al_2O_3$ rich (~20%) clasts generally are rather similar in composition, especially BS and B11–3 which are quite similar. However, detailed inspection shows some differences in their REE content (Fig. 2). Coarse matrix A11 has the highest REE content and Sm/Eu ratio of 6.5, and therefore, the largest fraction of KREEP. The upper limit of KREEP in this clast is ~30%. The distribution patterns of other REE are identical except that they have lower KREEP components.

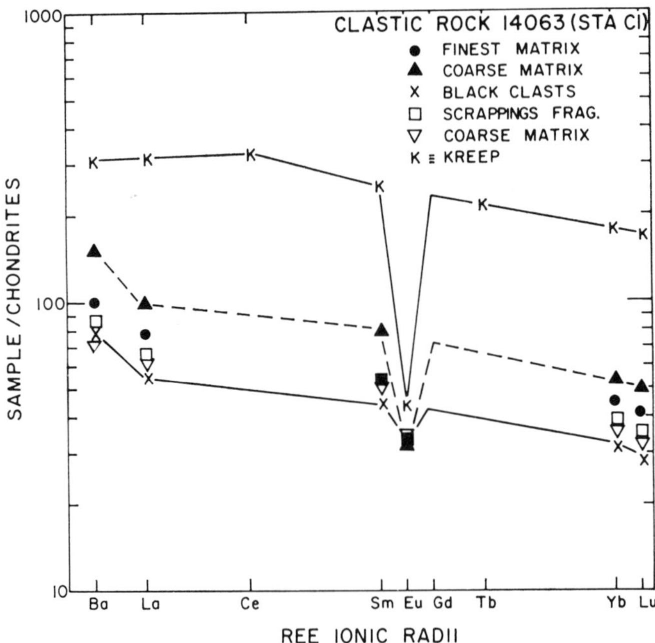


Fig. 2. Chondritic normalized abundances of 7 REE and Ba in clastic rock 14063 (STA C1). Clasts are A10 finest matrix, A11 coarse matrix, A12 black clasts, BS scrappings and B11-3 coarse matrix.

The compositions of the clasts from 14083, also picked up from STA C1, vary markedly (see below for discussion). The average value of K/U is 1390 and K/Ba, 3.6 for 14063. Potassium correlates linearily with U but not with Ba.

*Clastic rock 14083, STA C1 (Wasserburg Consortium)*

This rock is also a thermally metamorphosed microbreccia like 14063 (Jackson and Wilshire, 1972) and has been dated via Rb–Sr at 3.95 AE (Wasserburg *et al.*, 1972). This rock was a piece of 14082 rock, which was chipped from the top surface of the "White" rock, ~20 m near Cone Crater rim. Abundances in a black and white fragment and in a pure dark interior sample are quite similar to the composition of KREEP or noritic matter for all elements, except for the high $Cr_2O_3$ abundance in the pure dark matter that we attribute to a mineral grain of chromite. The higher CaO and $Al_2O_3$ contents and lower FeO, $Na_2O$, MnO, V, and $TiO_2$ suggest higher anorthitic plagioclase and lower pyroxene and ilmenite contents in the pure white matter. REE distribution patterns (Fig. 3) are similar for the three clasts. The black white fragment 2,1 and pure dark matter 2,4 match the average composition of KREEP. The close correspondence between these two clasts suggests that the pure dark matter dominates the black-white fragment. The pure white matter is relatively depleted in KREEP; i.e., <40% maximum KREEP. The Sm/Eu for the black-white

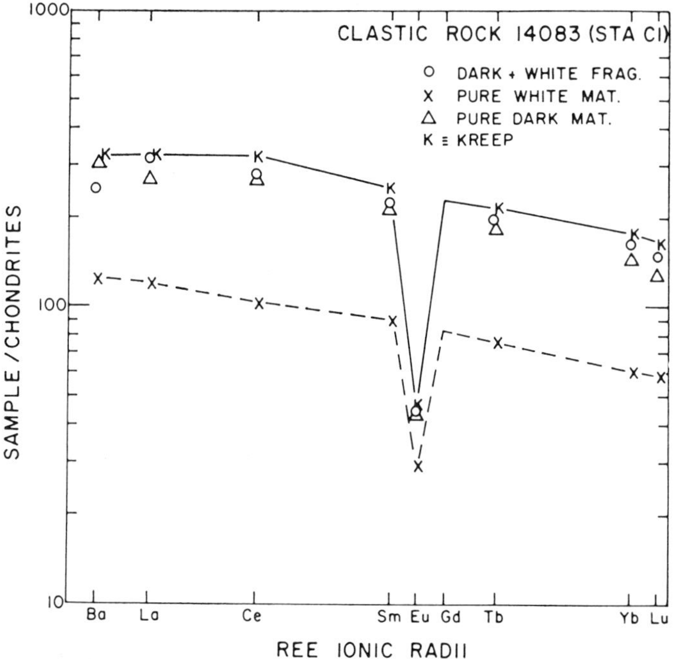


Fig. 3. Chondritic normalized values of 7 REE and Ba in clastic rock 14083 (STA C1). Clasts are 2,1 black and white; 2, white-A, $\frac{1}{16}$ pure white matter, and 2,4 interior dark $\frac{1}{8}$ pure dark matter.

fragment, pure dark and pure white matter are 14, 13, and 9, respectively. The mean K/U is 830 and K/Ba is 3.5. Potassium correlates linearly with Ba but not with U whereas 14063 shows the reverse correlations.

Since the clastic rocks 14063 and 14083 are Cone Crater ejecta, the chemical dissimilarity between these two rocks suggest a very complex and heterogeneous stratigraphy for the Fra Mauro formation. The same conclusion was reached by LSPET (1971), who analyzed two rocks, 14321 and 14065, both of which were on the Cone Crater ejecta blanket and which also yielded quite dissimilar chemical compositions.

*Clastic rock 14318, STA H (Tatsumoto Consortium)*

This rock, although picked up ~100 m NW of the LM, was also in the Cone Crater ejecta blanket (Swann *et al.*, 1971). We analyzed four clasts (5–15 mg) and two sawdust samples. No specific petrographic information for these four analyzed clasts is available at this writing. In general the overall rock composition for 24 elements of 14318, as exemplified by the well-mixed 31 mg sawdust sample (Showalter *et al.*, 1972), match the composition of igneous rock 14073 with the exception of slightly lower $Al_2O_3$ and higher FeO content for the sawdust. This suggests a genetic relationship between igneous and microbreccia rocks at the Apollo 14 site. Also the similarity in composition between 14318 and the average Apollo 14 soil suggests that either the soil is pulverized from breccia rocks such as 14318 or that 14318 was compacted from the soil in a meteoritic impact event. The latter suggestion may be valid since 14318 rock is a shocked compressed microbreccia (Jackson and Wilshire, 1972). If ejected from Cone Crater, the chemical composition of 14318 underscores the complexity of the Fra Mauro formation as noted in the 14083 discussion. The REE distribution is shown in Fig. 4. Clasts 26A and 26B are largely KREEP-noritic matter, while 27B contains ~80–90% KREEP and a large content of orthoclase. Although clast 26C has ~3.5 times more orthoclase and ~2.5 times more Ba than average KREEP-norite, 26C appears to contain ≤50% KREEP. Normal KREEP abundances of U were found in these two clasts. The Sm/Eu ratios of 17,20, and 18 for 26A, 26B, and 27B seem to be the highest values obtained so far for any lunar sample. If we compare the average composition of 14318 and the Apollo 12 samples, 12034 breccia, 12033 and 12032 soils, which contain about 80%, 70%, and 60% KREEP (Wänke *et al.*, 1971), respectively, we find no close correspondence. The mean K/U is 2300 for clasts; K/Ba is 7.3. These high average values are attributed to high K abundances of the clasts 26C and 27B. Potassium and Ba correlate fairly well, while K and U show no correlation at all.

*Clastic rock 14066, STA F (Reynolds Consortium)*

This rock was picked up from smooth terrain, ~100 m SSE of Weird Crater. The rock is a thermally metamorphosed microbreccia (Jackson and Wilshire 1972) and contains cosmogenic and extremely low solar wind content (Kaiser, 1972). Its Rb–Sr age is 4.0 AE (Cliff *et al.*, 1972). We analyzed one sawdust and four different clasts.

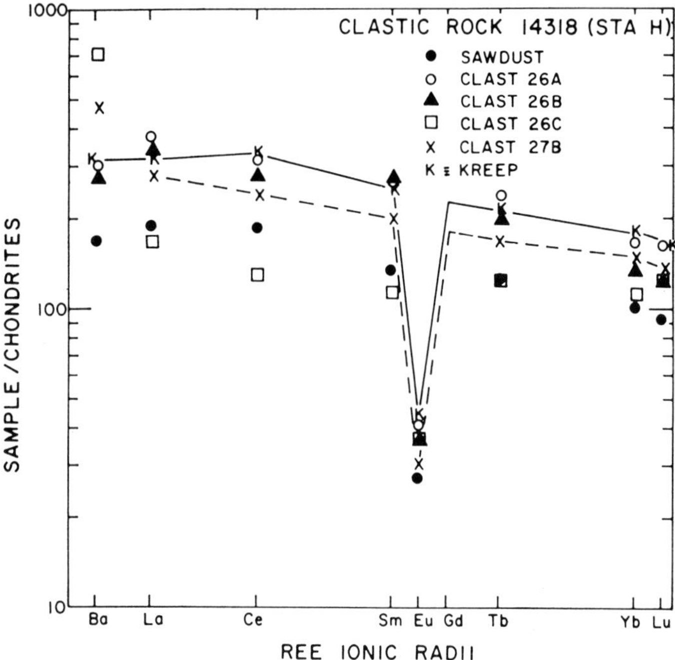


Fig. 4. Chondritic normalized abundances of 7 REE and Ba in clastic rock 14318 (STA H). Data for sawdust (31 mg) is plotted.

The compositions of the sawdust and four clasts are all different. The overall rock (sawdust) composition is slightly lower in $Al_2O_3$ and higher in $K_2O$ and REE compared to the average in Apollo 14 soil. On the other hand, the whole rock (sawdust) composition of 14318 is higher in plagioclase, lower in orthoclase, the REE, Zr, Hf, Th, U, and Ta abundances, relative to rock 14066. Among the clasts, "matrix small clasts" and the "white igneous clast" have high $K_2O$ abundance of 1.3 and 1.4%, respectively. An enrichment of anorthitic plagioclase is noted in the "many clasts" relative to the other fragments.

REE patterns for the five specimens are shown in Fig. 5. The matrix has $\leq 75\%$ KREEP component; the remaining fragments cluster at $\leq 50\%$ KREEP. The KREEP component in clastic rock 14066 is higher than that observed in clastic rock 14318. The $K_2O$ abundances of 1.3 and 1.4% are not correlated with the REE, Ba, and U abundances. Six fragments in rock 12013 also had high $K_2O$ abundances ranging from 0.7 to 3.6% and low REE abundances that were $\leq 50\%$ KREEP composition (Wakita and Schmitt, 1970). We may conclude that high orthoclase contents ($> 1.1\%$ $K_2O$) are not always associated with REE abundances and distribution patterns that are indicative of average KREEP-norite. This underscores the complexity of lunar crustal matter as would be expected for chemical differentiation of a large planetary body. The mean K/U is 1990 and K/Ba is 7. Potassium shows no correlation with U and Ba.

The REE patterns and the tabulated Sm/Eu ratios in different clasts clearly demonstrate that KREEP is not a homogeneous component, i.e., large variety of

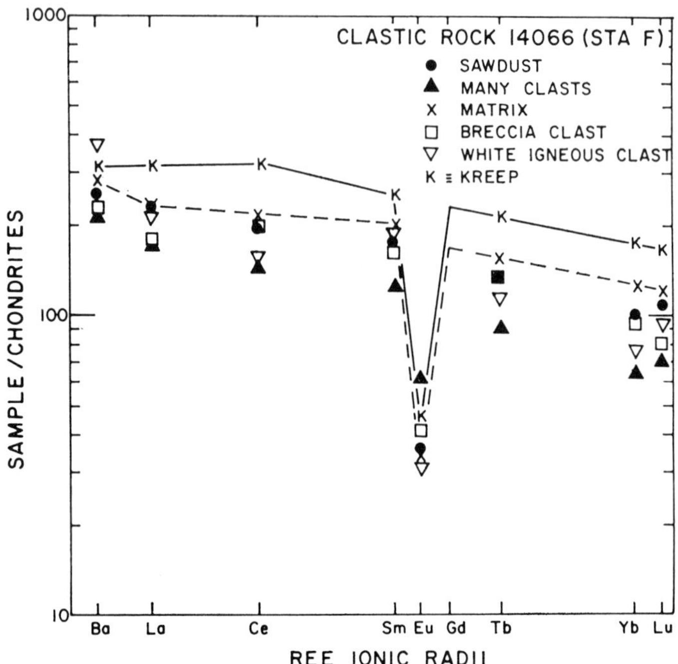


Fig. 5. Chondritic normalized abundances of 7 REE and Ba in clastic rock 14066 (STA F). Clasts are 21,2.02 many clasts, 21,1.01 matrix, 21,2.01 breccia, and 21.2.04,2 white igneous rock. Rock sawdust is for overall rock composition.

KREEP exists (Hubbard and Gast, 1971; and Meyer *et al.*, 1971). Our work supports this conclusion and further suggests that KREEP is associated to variable degrees, i.e., from 25–100%, in clasts from the Apollo 14 clastic rocks. Our analyses show high enrichments of the refractory and large ionic elements Zr, Hf, Ta, Th, and U elements in these clasts. This suggests that the crust of the moon is rich in refractory elements. KREEP is considered to be derived from an exotic source by impact (Hubbard and Gast, 1971; Meyer *et al.*, 1971; Marvin *et al.*, 1972). Ganapathy *et al.* (1972) suggested that an impact event on the Mare Imbrium (3.9 AE) was caused by a cyprus-size body (190 km) with low velocity (5 km/s) and that the associated ejecta thrown onto the Apollo 14 area is now the Fra Mauro formation. If valid, the average composition of its counterpart (Mare Imbrium) should possess similar KREEP components; i.e., the Imbrium basin should be rich in refractories.

*Luna 16 breccias*

The chemical compositions of these small 1–2 mg breccia samples (Table 2) are similar to the Luna 16 soil composition with the exceptions of the alkalies; i.e., the feldspar contents are higher in these breccias by factors of two or greater. This is attributed to sample heterogenity. The basic conclusion reached here is the same as was reported for the soil, i.e., the overall Luna 16 composition matches more closely to Apollo 11 soil than to any other lunar soil (Gillum *et al.*, 1972; Laul *et al.*, 1972a).

*Apollo 15 soils: 15021,21 (LM); 15471,49 (STA 4); 15501,40 (STA 9); and 15531,52 (STA 9a)*

These soils are $<1$ mm fines and represent a wide sampling area. Our major elemental results (Table 2) agree well with x-ray fluorescence data of LSPET (1972). Their 15601 soil matches closely to our 15531, both from the same Rille site. Our REE data for 15531 are in agreement with isotopic dilution data of Schnetzler *et al.* (1972). In general, elemental abundances of all four soils are different. Soils 15531 (STA 9a) and 15501 (STA 9) are only 0.3 km apart, yet the bulk compositions are widely different. Comparing Apollo 11 and 14 soils, we note that none of the soils match in their major elemental content. Among Apollo 12 soils, 12032 comes close to 15021 in FeO, $Al_2O_3$, CaO, $Cr_2O_3$, and MnO contents; however, the alkalies, REE, and refractories are depleted in 15021, indicating much less KREEP matter. Likewise, soil 12070 comes closer to 15501 (STA 9) in bulk composition, but 12070 soil is more abundant for many other elements.

The REE abundance patterns (Fig. 1) for the four soils vary considerably, e.g., soil 15021 (LM) has the highest REE content while 15531 (STA 9a) has the lowest REE abundances of all lunar soils measured to date. These REE abundances are lower than those measured in Apollo 11, 12, and 14 soils. The REE abundances in 15531 soil are also less than the REE abundances in Luna 16 soil. The chondritic normalized REE patterns of the Apollo 15 soils parallel the observed distribution in 12070 soil. The Sm/Eu ratios though low, are quite variable 6–13 and do not correlate with the KREEP content as expected. The ratio $Th/U = 3.4$ is constant for the soils.

It is apparent from Fig. 1 that the KREEP component in Apollo 15 soils is much less than observed in Apollo 12 and 14 soils. Schnetzler *et al.* (1972) studied separated minerals and whole rock 15555 (STA 9a) and estimated 6% KREEP, 88% contribution of 15555 basalt, and 6% anorthositic components for the composition of 15531 soil. Our upper limit, neglecting the basaltic rock contribution, for KREEP is 10%, which is close to their value of 6%. If we use 6% as the KREEP component value for 15531 and assume that the average REE abundances in all Apollo 15 basalts are similar to that observed in 15555, then we calculate that the KREEP components for the soils 15021, 15471, and 15501 are about 14%, 8%, and 11%, respectively, all of which are very low compared to KREEP in Apollo 12 and 14 soils. Even though the Apollo 15 site is on the eastern edge of Mare Imbrium, the presumed source of ejected KREEP matter, the low KREEP content in Apollo 15 soils indicates that basaltic components dominate the soil constituency. This suggests that the Palus Putredinis mare basalts (3.2 AE), perhaps derived by melting processes at depths $>200$ km, have covered the older (3.9 AE) Mare Imbrium basin with many lava flows of considerable thicknesses.

*Inter-element correlations*

We have attempted to establish a few inter-element correlations with the wealth of data from five lunar missions. In our comparison, we have included primitive carbonaceous chondrite type 1 (C1), meteoritic "basalts" (Ca-rich achondrites: eucrites and howardites) and terrestrial basalts (continental and oceanic ridge).

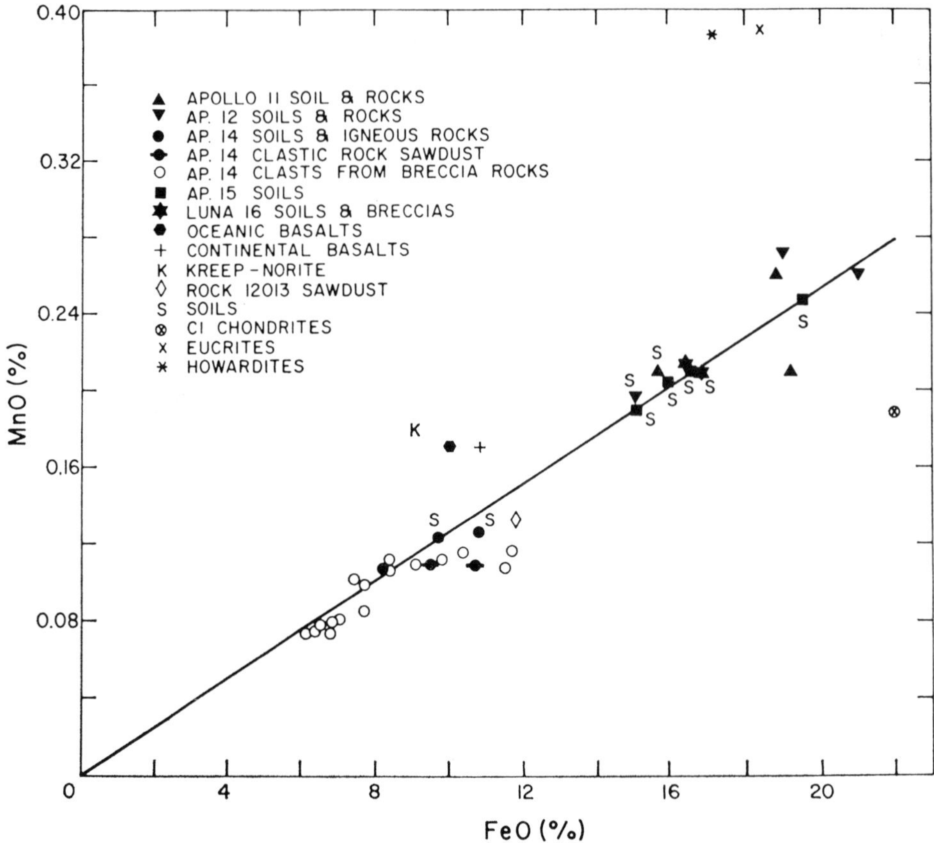


Fig. 6. Abundances of FeO and MnO in lunar, meteoritic and terrestrial matter. Sources for these data are Wakita *et al.* (1970), Wakita and Schmitt (1971), this work, Gillum *et al.* (1972), Wakita *et al.* (1972), and Corliss (1970). For KREEP, see references in Table 2. The correlation coefficient is 0.95 (>99% confidence level) for all lunar values.

The unique breccia rock 12013 and dominant component KREEP-norite are included as well. Symbols in the Figs. 6–9 are unified.

Figure 6 shows a correlation between MnO versus FeO. Such a strong correlation suggests that FeO and MnO are associated together in pyroxenes. Extrapolation of FeO and MnO through zero implies that Fe and Mn did not fractionate significantly during lunar differentiation processes. Marvin *et al.* (1972) noted howardite-like composition for lunar green glasses and suggested howardites may have been some of the raw material for planetary accretion of the moon. The howardite data for Fe and Mn are narrow in range (Schmitt *et al.*, 1972). It is evident from Fig. 6 that howardites and eucrites both fall far away from the lunar correlation line. Morgan *et al.* (1972) also reported that the abundances of 15 trace elements in lunar green glass do not approximate the corresponding abundances observed in six howardites by Laul *et al.* (1972b). Both evidences support our conclusion that no genetic relationship exists between howardites and primitive lunar composition.

Figure 7 indicates that Sc is approximately correlated with FeO and resides mainly in pyroxenes (Goles *et al.*, 1970). However, we note considerable Sc enrichments in Apollo 11 low and high alkali rocks and to a lesser degree in Apollo 12 (Mg-poor) rocks. Mass balance calculations indicate that ilmenite in Apollo 11 rocks (Goles *et al.*, 1970) is not responsible for the high Sc content in Apollo 11 rocks. Apparently either the magmatic processes or the primitive matter responsible for generation of Apollo 11 basalts differed considerably from the corresponding factors that produced other lunar basalts.

In Fig. 8, we note that V and $Cr_2O_3$ correlate strongly over a factor of six in abundances. It seems that V and Cr are tied together in spinels. Extrapolation to zero V content yields ~800 ppm $Cr_2O_3$ in the whole rock; this quantity of $Cr_2O_3$ may be attributed to the Cr content in pyroxenes. Howardites and C1 chondrites lie below the correlation line and eucrites tend to approach the line. The average value for oceanic ridge basalts is spectacularly high. Spinels are high-temperature minerals and should reflect early condensates from a hot cooling nebular gas. Such

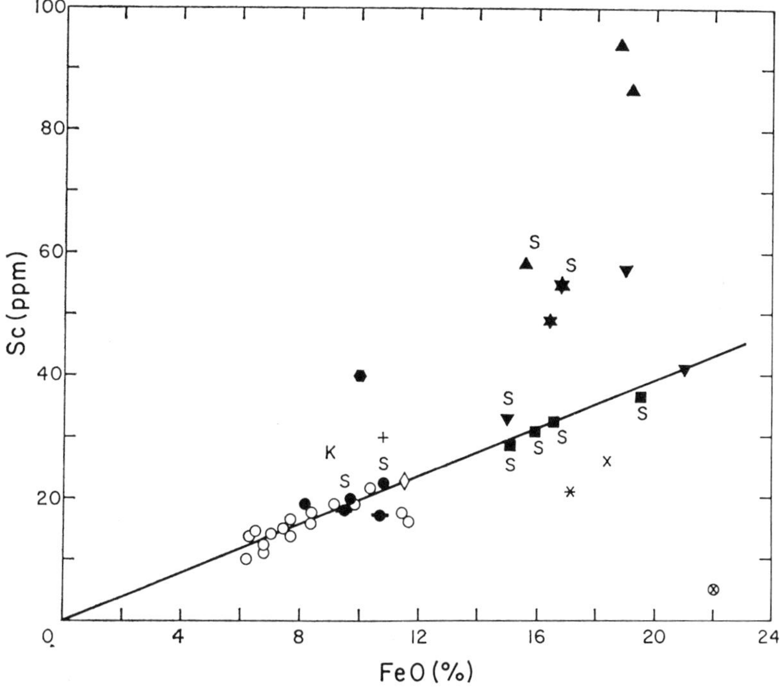


Fig. 7. Sc–Fe correlation in lunar, meteoritic and terrestrial matter. Symbols are identical to those found in Fig. 6. The correlation coefficient is 0.81 (> 99% confidence level) for all lunar values and 0.93 for lunar values excluding Apollo 11 soils and rocks and Apollo 12 low-Mg rocks.

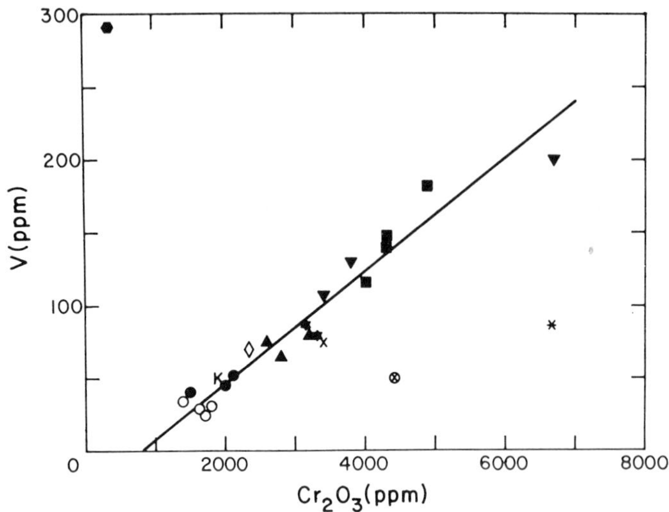


Fig. 8. Correlation of V–$Cr_2O_3$. Same symbols as found in Fig. 6. The correlation coefficient is 0.97 (>99% confidence level) for all lunar values.

a hiatus between lunar and terrestrial correlations for V and $Cr_2O_3$ argues against the fission hypothesis for the origin of the moon (Wise, 1963; O'Keefe, 1969).

Siderophile and volatile trace patterns by Ganapathy *et al.* (1970), Laul *et al.* (1971), and Morgan *et al.* (1972) have been used to characterize the nature of impacting meteoritic bodies on the lunar surface, but this work has not escaped criticism of relative volatilization. The authors argued against selective volatilization on the constancy of Rb/Cs, Cs/U and Tl/U ratios in all lunar rocks and soils. Hubbard and Gast (1972) from laboratory experiments presented evidence for alkali loss from basaltic rocks that were differentially volatilized to temperatures of 850–1050°C at $10^{-3}$ atm. However, Erlank *et al.* (1972) reported a relative constant ratio for K/Zr (4.5) for lunar matter. This observation in lunar magmatic processes argues against any large-scale volatilization of the alkalies from lunar lavas or by exotic impact processes. To test it further, we have plotted $K_2O$ (volatile) versus Hf, a refractory element-like Zr, in Fig. 9. A good correlation exists. Only the clasts from Apollo 14 clastic rocks fall off the line. These clastic deviations may be ascribed to a large variety of KREEP. Therefore high or low K (in KREEP) values in clasts may account for the observed dispersion. The sawdusts, which represent the overall rock composition of clastic rocks, fall on the line. Our data precludes selective large-scale volatilization of alkalies during lunar magmatic events.

In Table 3, we have tabulated correlation coefficients for 11 pairs of elements from data obtained from sample analyses of five different lunar sites. The first eight pairs $Na_2O$–$K_2O$ to Ba–La (Table 3) except Sc–FeO show strong linear relationships within the 95% confidence level in both basalts and basalts plus soils and clastic rocks. This suggests that these genetically related pairs do not fractionate appreciably during magmatic events. The last three pairs, CaO–$Na_2O$, CaO–$Al_2O_3$ and FeO–$TiO_2$

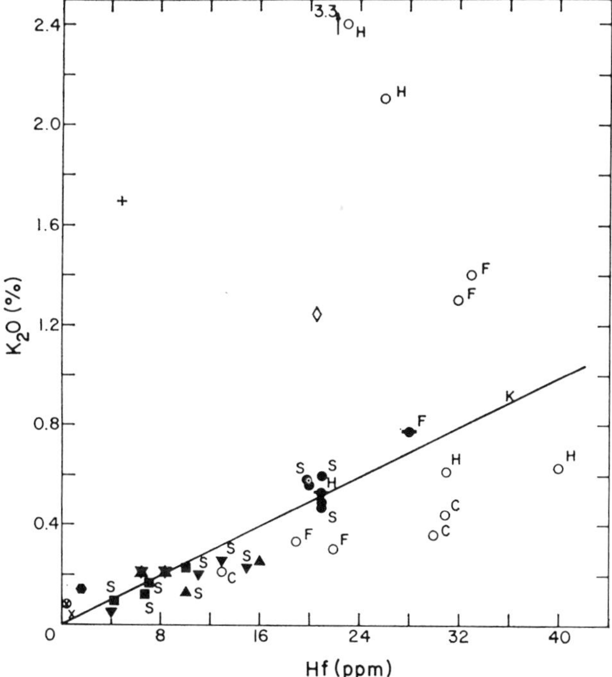


Fig. 9. Correlation of $K_2O$–Hf. Same symbols as found in Fig. 6. The correlation coefficient is 0.48 and 0.96 (both >99% confidence level) for all lunar values and for lunar values excluding the individual clasts, respectively.

Table 3. Correlation coefficients for 11 elemental pairs in lunar samples.[a]

| | $Na_2O$ $K_2O$ | MnO FeO | V $Cr_2O_3$ | Sc FeO | $K_2O$ Hf | Zr Hf | Nb Ta | Ba La | CaO $Na_2O$ | CaO $Al_2O_3$ | FeO $TiO_2$ |
|---|---|---|---|---|---|---|---|---|---|---|---|
| C.C. for igneous | 0.93 | 0.76 | 0.98 | 0.60 | 0.88 | 0.95 | 1.00 | 0.98 | 0.74 | 0.66 | 0.30 |
| basalts[b] | (9) | (8) | (6) | (6) | (6) | (5) | (4) | (7) | (8) | (8) | (7) |
| C.C. for igneous plus | 0.94 | 0.95 | 0.97 | 0.81 | 0.96 | 0.97 | 0.98 | 0.98 | 0.52 | 0.67 | 0.58 |
| clastic plus soils[c] | (15) | (35) | (21) | (37) | (20) | (11) | (7) | (14) | (11) | (12) | (11) |

[a] Best average data of our work and that available from the literature for Apollo 11, 12, 14, and 15, and Luna 16 samples.

[b] Numbers in parenthesis represent the number of points used for C.C. computations. Normally, one best average value each for Ap. 11 high and Ap. 11 low alkali rocks, Ap. 12 Mg-rich and Mg-poor rocks, Ap. 14 igneous, Ap. 15 igneous and Luna 16 igneous basalts.

[c] Numbers in parenthesis represents the number of points used for computations. In addition to numbers in a, the best average values each for Apollo 11, 12, 14, and 15, and Luna 16 soils and the average value for Ap. 14 clastic rocks are included. In MnO-FeO, V-$Cr_2O_3$, Sc-FeO, and $K_2O$-Hf correlations, individual clasts and rock sawdust abundances were also included.

are poorly correlated. This may be attributed to the fact that Ca does not always reside in plagioclase; because of the variable ilmenite content in lunar basalts, FeO and $TiO_2$ are not well correlated. Average abundance ratios of Zr and Hf are $28 \pm 6$ and $31 \pm 4$ for Apollo 11 igneous rocks and soils; likewise, $36 \pm 6$ and $40 \pm 4$ were obtained for Apollo 12 igneous rocks and soils (Wakita *et al.*, 1971). In general, the Zr/Hr ratio for Apollo 12 samples were 20–30% higher than for Apollo 11 samples. This was attributed to different chemical fractionizations at the two sites. The Zr/Hf ratios at the Apollo 14 site are 33 ± 3 for the soils, 31 for igneous rock 14073, and 31 ± 2 for the clastic rocks. These ratios indicate no evidence for significant fractionization for this geochemically coherent pair in Apollo 11, 12, and 14 igneous basalts.

*Acknowledgments*—This study was supported by NASA grants 38-002-039 and 38-002-020. We acknowledge the assistance of the Oregon State University TRIGA reactor group for neutron activations and T. Cooper for assistance in some phases of this study, and Mrs. Susan R. Shurtliff for typing the manuscript.

## REFERENCES

Brown G. M., Emeleus C. H., Holland J. G., Peckett A., and Phillips R. (1971) Picritic basalts, ferrobasalts, feldspathic norites, and rhyolites in a strongly fractionated lunar crust. *Proc. Second Lunar Sci. Conf., Geochim. Cosmochim. Acta* Suppl. 2, Vol. 1, pp. 583–600. MIT Press.

Brunfelt A. O., Heier K. S., Steinnes E., and Sundvoll B. (1971) Determination of 36 elements in Apollo 14 bulk fines 14163 by activation analysis. *Earth Planet. Sci. Lett.* **11,** 351–353.

Cliff R. A., Lee-Hu C., and Wetherill G. W. (1972) K, Rb, and Sr measurements in Apollo 14 and 15 material (abstract). In *Lunar Science—III* (editor C. Watkins), pp. 146–147, Lunar Science Institute Contr. No. 88.

Corliss J. (1970) Mid-ocean ridge basalts. Ph.D. thesis, University of California, San Diego.

Duncan A. R., Lindstrom M. M., Lindstrom D. J., McKay S. M., Stoeser J. W., Goles G. G., and Fruchter J. S. (1972) Comments on the Genesis of Breccia 14321 (abstract). In *Lunar Science—III* (editor C. Watkins), pp. 192–194, Lunar Science Institute Contr. No. 88.

Erlank A. J., Willis J. P., Ahrens L. H., Gurney J. J., and McCarthy T. S. (1972) Inter-element-relationships between the moon and stony meteorites with particular reference to some refractory elements (abstract). In *Lunar Science—III* (editor C. Watkins), pp. 239–241, Lunar Science Institute Contr. No. 88.

Ganapathy R., Keays R. R., Laul J. C., and Anders E. (1970) Trace elements in Apollo 11 lunar rocks: Implications for meteorite influx and origin of moon. *Proc. Apollo 11 Lunar Sci. Conf., Geochim. Cosmochim. Acta* Suppl. 1, Vol. 2, pp. 1117–1142. Pergamon.

Ganapathy R., Laul J. C., Morgan J. W., and Anders E. (1972) Moon: Possible nature of the body that produced the Imbrium Basin, from the composition of Apollo 14 samples. *Science* **175,** 55–59.

Gillum D. E., Ehmann W. D., Wakita H., and Schmitt R. A. (1972) Bulk and rare earth abundances in the Luna 16 soil levels A and D. *Earth Planet. Sci. Lett.* **13,** 444–449.

Goles G. G., Duncan A. R., Lindstrom D. J., Martin M. R., Beyer R. L., Osawa M., Randle K., Meek L. T., Steinborn T. L. ,and McKay S. M. (1971) Analyses of Apollo 12 specimens: Compositional variations, differentiation processes, and lunar soil mixing models. *Proc. Second Lunar Sci. Conf., Geochim. Cosmochim. Acta* Suppl. 2, Vol. 2, pp. 1063–1081. MIT Press.

Goles G. G., Randle K., Osawa M., Lindstrom D. J., Jerome D. Y., Steinborn T. L., Beyer R. L., Martin M. R., and McKay S. M. (1970) Interpretations and speculations on elemental abundances in lunar samples. *Proc. Apollo 11 Lunar Sci. Conf., Geochim. Cosmochim. Acta* Suppl. 1, Vol. 2, pp. 1177–1194. Pergamon.

Helmke P. A. and Haskin L. A. (1972) Rare earths and other trace elements in Apollo 14 lunar samples (abstract). In *Lunar Science—III* (editor C. Watkins), pp. 366–368, Lunar Science Institute Contr. No. 88.

Hubbard N. and Gast P. W. (1971) Chemical composition and origin of nonmare lunar basalts. *Proc. Second Lunar Sci. Conf., Geochim. Cosmochim. Acta* Suppl. 2, Vol. 2, pp. 999–1020. MIT Press.

Hubbard N. J. and Gast P. W. (1972) Chemical composition of Apollo 14 materials and evidence for alkali volatilization (abstract). In *Lunar Science—III* (editor C. Watkins), pp. 407–409, Lunar Science Institute Contr. No. 88.

Jackson E. D. and Wilshire H. G. (1972) Classification of the samples returned from the Apollo 14 landing site (abstract). In *Lunar Science—III* (editor C. Watkins), pp. 418–420, Lunar Science Institute Contr. No. 88.

Kaiser W. A. (1972) Rare gas measurements in three Apollo 14 samples (abstract). In *Lunar Science—III* (editor C. Watkins), pp. 442–443, Lunar Science Institute Contr. No. 88.

Laul J. C., Ganapathy R., Morgan J. W., and Anders E. (1972a) Meteoritic and nonmeteoritic trace elements in Luna 16 samples. *Earth Planet. Sci. Lett.* **13,** 450–454.

Laul J. C., Keays R. R., Ganapathy R., Anders E., and Morgan J. W. (1972b) Chemical fractionations in meteorites—V. Volatile and siderophile elements in achondrites and ocean ridge basalts. *Geochim. Cosmochim. Acta* **36,** 329–345.

Laul J. C., Morgan J. W., Ganapathy R., and Anders E. (1971) Meteoritic material in lunar samples: Characterization from trace elements. *Proc. Second Lunar Sci. Conf., Geochim. Cosmochim. Acta* Suppl. 2, Vol. 2, pp. 1139–1159. Pergamon.

LSPET (Lunar Sample Preliminary Examination Team) (1971) Preliminary examination of lunar samples from Apollo 14. *Science* **173,** 681–693.

LSPET (Lunar Sample Preliminary Examination Team) (1972) The Apollo 15 lunar samples: A preliminary description. *Science* **175,** 363–374.

Marvin U. B., Reid J. B. Jr., Taylor G. J., and Wood J. A. (1972) A survey of lithic and vitreous types in the Apollo 14 soil samples (abstracts). In *Lunar Science—III* (editor C. Watkins), pp. 507–509, Lunar Science Institute Contr. No. 88.

Masuda A., Nakamura N., Kurasawa H., and Tanaka T. (1972) Chondrite-normalized rare earth patterns of the Apollo 14 samples (abstract). In *Lunar Science—III* (editor C. Watkins), pp. 517–519, Lunar Science Institute Contr. No. 88.

Meyer C. Jr., Brett R., Hubbard N. J., Morrison D. A., McKay D. S., Aitken F. K., Takeda H., and Schonfeld E. (1971) Mineralogy, chemistry, and origin of the KREEP component in soil samples from the Ocean of Storms. *Proc. Second Lunar Sci. Conf., Geochim. Cosmochim. Acta* Suppl. 2, Vol. 2, pp. 393–411. MIT Press.

Morgan J. W., Laul J. C., Krahenbühl U., Ganapathy R. and Anders E. (1972) Major impacts on the moon: Chemical characterization of projectiles (abstract). In *Lunar Science—III* (editor C. Watkins), pp. 552–554, Lunar Science Institute Contr. No. 88.

O'Keefe J. A. (1969) Origin of the moon. *J. Geophys. Res.* **74,** 2758–2767.

Rey P., Wakita H., and Schmitt R. A. (1970) Radiochemical activation analysis of In, Cd, and the 14 rare earth elements and Y in rocks. *Anal. Chim. Acta* **57,** 163–178.

Rose H. S., Cuttitta F. Jr., Annell C. S., Carron M. K., Christian R. D., Dwornik E. J., and Ligon D. T. Jr. (1972) Compositional data for 15 Fra Mauro lunar samples (abstract). In *Lunar Science—III* (editor C. Watkins), pp. 660–662, Lunar Science Institute Contr. No. 88.

Schmitt R. A., Goles G. G., Smith R. H., and Osborn T. W. III (1972) Elemental abundances in stony meteorites. *Meteoritics*, in press.

Schmitt R. A., Linn T. A. Jr., and Wakita H. (1970) The determination of 14 common elements in rocks via sequential instrumental activation analysis. *Radiochim. Acta* **13,** 200–212.

Schnetzler C. C. and Nava D. F. (1971) Chemical composition of Apollo 14 soils 14163 and 14259. *Earth Planet. Sci. Lett.* **11,** 345–350.

Schnetzler C. C., Philpotts J. A., and Bottino M. L. (1970) Li, K, Rb, Sr, Ba, and rare earth concentrations, and Rb–Sr age of lunar rock 12013. *Earth and Planet. Sci. Lett.* **9,** 185–192.

Schnetzler C. C., Philpotts J. A., Nava D. F., Schuhmann S., and Thomas H. H. (1972) Geochemistry of Apollo 15 basalt 15555 and soil 15531. *Science* **175,** 426–428.

Showalter D. L., Wakita H., Smith R. H., Schmitt R. A., Gillum D. E., and Ehmann W. D. (1972) Chemical composition of sawdust from lunar rock 12013 and comparison of a Java tektite with the rock. *Science* **175,** 170–172.

Swann G. G., Trask N. J., Hait M. H., and Sutton R. L. (1971) Geologic setting of the Apollo 14 samples. *Science* **173,** 716–719.

Taylor S. R., Muir P., and Kaye M. (1971) Trace element chemistry of Apollo 14 lunar soil from Fra Mauro. *Geochim. Cosmochim. Acta* **35,** 975–981.

Tera F., Eugster O., Burnett D. S., and Wasserburg G. J. (1970) Comparative study of Li, Na, K,

Rb, Cs, Ca, Sr, and Ba abundances in achondrites and in Apollo 11 lunar samples. *Proc. Apollo 11 Lunar Sci. Conf., Geochim. Cosmochim. Acta* Suppl. 1, Vol. 2, pp. 1637–1657. Pergamon.

Wakita H., Rey P., and Schmitt R. A. (1971) Abundances of the 14 rare earth elements and 12 other trace elements in Apollo 12 samples: Five igneous and one breccia rocks and four soils. *Proc. Second Lunar Sci. Conf., Geochim. Cosmochim. Acta* Suppl. 2, Vol. 2, pp. 1319–1329. MIT Press.

Wakita H. and Schmitt R. A. (1970) Elemental abundances in seven fragments from lunar rock 12013. *Earth and Planet. Sci. Lett.* **9**, 169–176.

Wakita H. and Schmitt R. A. (1971) Bulk elemental composition of Apollo 12 samples: Five igneous and one breccia rocks and four soils. *Proc. Second Lunar Sci. Conf., Geochim. Cosmochim. Acta* Suppl. 2, Vol. 2, pp. 1231–1236. MIT Press.

Wakita H., Schmitt R. A., and Rey P. (1970) Elemental abundances of major, minor, and trace elements in Apollo 11 lunar rocks, soil, and core samples. *Proc. Apollo 11 Lunar Sci. Conf., Geochim. Cosmochim. Acta* Suppl. 1, Vol. 2, pp. 1685–1717. Pergamon.

Wakita H., Showalter D. L., and Schmitt R. A. (1972) Bulk, REE, and other abundances in Apollo 14 soils (3), clastic (1), and igneous (1) rocks (abstract). In *Lunar Science—III* (editor C. Watkins), pp. 767–769, Lunar Science Institute Contr. No. 88.

Wänke H., Wlotzka F., Baddenhausen H., Balacescu A., Spettel B., Teschke F., Jagoutz E., Kruse H., Quijano-Rico M., and Rieder R. (1971) Apollo 12 samples: Chemical composition and its relation to sample locations and exposure ages, the two component origins of the various soil samples and studies on lunar metallic particles. *Proc. Second Lunar Sci. Conf., Geochim. Cosmochim. Acta* Suppl. 2, Vol. 2, pp. 1187–1208, MIT Press.

Wasserburg G. J., Turner G., Tera F., Podosek F. A., Papanastassiou D. A., and Huneke J. C. (1972) Comparison of Rb–Sr, K–Ar, and U–Th–Pb ages, lunar chronology and evolution (abstract). In *Lunar Science—III* (editor C. Watkins), pp. 788–790, Lunar Science Institute Contr. No. 88.

Wise D. U. (1963) An origin of the moon by rotational fission during formation of the earth's core. *J. Geophys. Res.* **68**, 1547–1554.

Proceedings of the Third Lunar Science Conference
(Supplement 3, *Geochimica et Cosmochimica Acta*)
Vol. 2, pp. 1201–1214
The M.I.T. Press, 1972

# Compositional characteristics of some Apollo 14 clastic materials

M. M. LINDSTROM, A. R. DUNCAN, J. S. FRUCHTER, S. M. MCKAY,
J. W. STOESER, G. G. GOLES, and D. J. LINDSTROM

Center for Volcanology, University of Oregon, Eugene, Oregon 97403

**Abstract**—Eighty-two subsamples of Apollo 14 materials have been analyzed by instrumental neutron activation analysis techniques for as many as 25 elements. In many cases, it was necessary to develop new procedures to allow analyses of small specimens. Compositional relationships among Apollo 14 materials indicate that there are small but systematic differences between regolith from the valley terrain and that from Cone Crater ejecta. Fragments from 1–2 mm size fractions of regolith samples may be divided into compositional classes, and the "soil breccias" among them are very similar to valley soils. Multicomponent linear mixing models have been used as interpretive tools in dealing with data on regolith fractions and subsamples from breccia 14321. These mixing models show systematic compositional variations with inferred age for Apollo 14 clastic materials.

## INTRODUCTION

SPECIMENS RETURNED from the Apollo 14 site afford us an opportunity to study the relationships of KREEP-like materials (Meyer *et al.*, 1971) to other compositional varieties of lunar materials. In many respects, these relationships are well exhibited in Apollo 14 clastic rocks and soils. Most of the results of our detailed studies of breccia 14321 will be reported elsewhere (Duncan *et al.*, 1972b); here we are concerned with comparing the results of some of those studies with compositions of other clastic materials returned by the Apollo 14 mission.

## ANALYTICAL DATA

Abundances of up to 25 major and trace elements were determined in samples of Apollo 14 clastic materials by instrumental neutron activation analysis (INAA). Analyses were performed on 4 samples of surficial soils, 3 trench samples, and 3 core samples, together with 5 fragments from breccia 14270 and 50 samples of clasts and matrix from breccia 14321. In addition, we analysed 17 fragments hand-picked by Dr. J. A. Wood from the 1–2 mm fractions of soils and were able to determine abundances of 16 elements in those fragments. Compositional data for soil, trench, and core samples appear in Table 1, those for soil fragments in Table 2, and those for breccia samples in Table 3.

Our INAA procedures were similar to those of Gordon *et al.* (1968) with the addition of procedures to determine Ti, Al, V, and Mn by means of short-lived radioisotopes (Schmitt *et al.*, 1970). Modifications of these basic procedures allowed us to analyse either large (>150 mg) or small (2–150 mg) samples. Small samples were subjected to a greater neutron flux and counted for longer times so that estimates of precision, based mainly on Poisson counting statistics, were not markedly increased

Table 1. Apollo 14 regolith samples.

| Sample | Soils | | | | | Trench | | | Core | | | Representative |
|---|---|---|---|---|---|---|---|---|---|---|---|---|
| | 14163,89 | 14259,34 | 14049,31A | 14049,31B | 14141,33 | 14148,26 | 14156,26 | 14149,42 | 14230,112 | 14230,119 | 14230,129 | % Error |
| Wt. (mg) | 162.8 | 95.7 | 512.5 | 473.9 | 191.2 | 156.0 | 235.6 | 197.5 | 132.6 | 148.9 | 103.4 | |
| Al (%) | — | — | — | — | 8.75 | 9.13 | 8.97 | 9.16 | — | — | — | 2 |
| Na | 5630 | 5020 | 5780 | 6070 | 5870 | 5050 | 5060 | 5260 | 5020 | 5310 | 5410 | 2 |
| K | 4300 | — | — | — | 5300 | 4150 | 4000 | 4200 | 3950 | 4200 | 3800 | 5–10 |
| Cs | 0.78 | 0.75 | — | — | 0.62 | 0.40 | 0.54 | 0.66 | — | — | — | 10–20 |
| Ba | 950 | 740 | 900 | 830 | 900 | 750 | 780 | 740 | 750 | 780 | 700 | 8 |
| La | 67.3 | 57.8 | 69.8 | 70.2 | 71.4 | 64.3 | 65.1 | 65.1 | 62.3 | 66.1 | 65.1 | 2 |
| Ce | 194 | 178 | 216 | 185 | 200 | 176 | 175 | 177 | 167 | 170 | 176 | 3 |
| Nd | 100 | — | 98 | 95 | 104 | 98 | — | 100 | — | — | 100 | 10 |
| Sm | 29.6 | 26.5 | 31.1 | 31.8 | 34.7 | 31.5 | 31.4 | 31.6 | 27.8 | 29.9 | 29.4 | 2 |
| Eu | 2.75 | 2.63 | 2.97 | 3.04 | 2.82 | 2.68 | 2.66 | 2.76 | 2.91 | 3.26 | 3.44 | 3 |
| Tb | 7.1 | 5.9 | 7.0 | 7.0 | 7.4 | 6.6 | 6.6 | 6.7 | 5.0 | 6.7 | 6.9 | 5 |
| Yb | 22.0 | 21.4 | 21.0 | 23.8 | 23.8 | 21.7 | 21.5 | 21.7 | 22.5 | 22.8 | 21.5 | 3 |
| Lu | 3.21 | 3.05 | 3.26 | 3.26 | 3.35 | 3.18 | 3.05 | 3.08 | 3.15 | 3.28 | 3.28 | 2 |
| Th | 15.2 | 14.0 | — | — | 15.3 | 13.8 | 13.8 | 13.6 | — | — | — | 5 |
| U | — | 3.0 | 3.6 | 3.4 | — | — | — | — | — | — | 2.1 | 10 |
| Hf | 25.3 | 22.5 | 24.3 | 24.0 | 25.0 | 25.7 | 23.2 | 23.0 | 22.4 | 23.4 | — | 3 |
| Zr | 720 | 590 | 750 | 860 | 760 | 690 | 700 | 660 | 990 | 900 | — | 8 |
| Ta | 4.3 | 3.9 | 5.0 | 4.8 | 5.7 | 4.7 | 4.8 | 4.8 | 3.9 | 4.5 | 5.0 | 5 |
| Fe (%) | 8.3 | 8.3 | 8.1 | 8.3 | 7.9 | 8.0 | 8.1 | 7.9 | 8.1 | 8.3 | 8.0 | 2 |
| Ti (%) | — | — | — | — | 0.98 | 1.01 | 0.99 | 0.93 | — | — | — | 4 |
| Sc | 21.4 | 21.9 | 21.7 | 21.9 | 21.5 | 21.0 | 20.9 | 20.5 | 20.9 | 22.0 | 21.5 | 2 |
| V | — | — | — | — | — | 44 | 36 | — | — | — | — | 8 |
| Cr | 1280 | 1290 | 1260 | 1280 | 1350 | 1310 | 1350 | 1300 | 1290 | 1350 | 1260 | 2 |
| Mn | — | — | — | — | 960 | 985 | 965 | 910 | — | — | — | 3 |
| Co | 36.0 | 37.5 | 35.6 | 36.4 | 31.0 | 34.4 | 36.2 | 40.0 | 42.4 | 36.2 | 36.4 | 2 |

Table 2. Fragments separated from 1–2 mm Apollo 14 soils.

| | Glasses | | | | | |
|---|---|---|---|---|---|---|
| Sample | 14262,2, 251,3 | 14262,2, 251,4 | 14258,36, 251,13 | 14258,36, 250,16 | 14258,36, 251,10 | 14258,36, 251,17 |
| Wt. (mg) | 2.60 | 4.06 | 2.01 | 2.02 | 1.91 | 2.96 |
| Na | 5150 | 5280 | 5220 | 4580 | < 500 | 4310 |
| K | 2460 | — | 2520 | 1940 | — | — |
| Ba | 740 | 970 | 1000 | 780 | — | 880 |
| La | 71 | 73 | 77 | 65 | 4.0 | 77 |
| Ce | 187 | 171 | 192 | 163 | 7 | 183 |
| Sm | 34.4 | 34.9 | 37.1 | 30.2 | 1.92 | 36.5 |
| Eu | 2.6 | 2.5 | 4.1 | 2.6 | — | 2.4 |
| Tb | 7.8 | 7.1 | 7.1 | 6.9 | — | 7.0 |
| Yb | 21.2 | 23.0 | 19.3 | 21.0 | 1.4 | 21.9 |
| Lu | 3.41 | 3.47 | 3.27 | 2.90 | — | 3.32 |
| Th | 12.6 | 13.8 | 13.1 | 12.9 | — | 13.1 |
| Hf | 20.9 | 24.4 | 22.3 | 25.2 | — | 25.4 |
| Fe(%) | 8.49 | 8.39 | 8.31 | 8.56 | 4.64 | 10.45 |
| Sc | 23.2 | 22.9 | 22.6 | 22.1 | 11.2 | 23.1 |
| Cr | 1390 | 1395 | 1380 | 1410 | 930 | 1680 |
| Co | 34.2 | 44.7 | 42.4 | 47.7 | 23.2 | 99.5 |

| | Soil Breccias | | | | |
|---|---|---|---|---|---|
| Sample | 14262,2, 250,1 | 14262,2, 250,2 | 14262,2, 249,5 | 14258,36, 249,14 | 14258,36, 250,15 |
| Wt. (mg) | 4.26 | 4.39 | 2.92 | 2.47 | 5.47 |
| Na | 4830 | 5140 | 6720 | 4830 | 5500 |
| K | 2430 | — | 4430 | — | 4330 |
| Ba | — | 1130 | 1220 | 590 | 670 |
| La | 70 | 73 | 78 | 68 | 70 |
| Ce | 178 | 187 | 206 | 173 | 169 |
| Sm | 33.3 | 34.8 | 40.3 | 31.5 | 34.5 |
| Eu | 2.5 | 2.9 | 3.1 | 2.2 | 2.6 |
| Tb | 6.3 | 6.8 | 7.7 | 5.9 | 6.6 |
| Yb | 21.1 | 21.6 | 23.1 | 20.3 | 20.5 |
| Lu | 3.38 | 3.19 | 3.68 | 3.19 | 3.27 |
| Th | 14.0 | 12.6 | 14.8 | 13.2 | 13.4 |
| Hf | 22.9 | 27.0 | 25.1 | 21.5 | 25.7 |
| Fe(%) | 8.31 | 8.41 | 7.96 | 8.03 | 8.93 |
| Sc | 22.9 | 23.0 | 22.1 | 21.6 | 24.1 |
| Cr | 1400 | 1425 | 1478 | 1360 | 1560 |
| Co | 37.5 | 38.4 | 40.3 | 38.6 | 39.7 |

| | Annealed Breccias | | | | | | |
|---|---|---|---|---|---|---|---|
| Sample | 14262,2, 249,6 | 14262,2, 250,7 | 14258,36, 249,8 | 14258,36, 249,9 | 14258,36, 249,11 | 14258,36, 250,12 | Representative % Error |
| Wt. (mg) | 5.58 | 3.49 | 4.66 | 4.58 | 5.97 | 9.46 | |
| Na | 6290 | 5900 | 6230 | 7750 | 5900 | 6980 | 2 |
| K | — | 2300 | 5850 | 5720 | 5260 | — | 10 |
| Ba | 1020 | 1220 | 650 | 710 | 1320 | 830 | 10 |
| La | 89 | 89 | 98 | 68 | 79 | 102 | 3 |
| Ce | 234 | 244 | 247 | 162 | 198 | 243 | 3 |
| Sm | 42.0 | 44.1 | 45.5 | 32.3 | 37.0 | 49.0 | 2 |
| Eu | 2.9 | 3.4 | 2.8 | 3.3 | 3.7 | 3.0 | 10 |
| Tb | 8.7 | 9.2 | 8.9 | 6.6 | 7.3 | 9.5 | 5 |
| Yb | 24.6 | 24.7 | 24.1 | 20.6 | 26.2 | 27.0 | 5 |
| Lu | 3.79 | 4.17 | 3.71 | 3.11 | 3.94 | 4.10 | 3 |
| Th | 15.8 | 18.4 | 12.2 | 14.5 | 22.2 | 17.6 | 8 |
| Hf | 24.9 | 30.0 | 34.0 | 20.5 | 20.9 | 29.9 | 3 |
| Fe(%) | 9.24 | 9.61 | 9.43 | 6.88 | 6.89 | 7.80 | 2 |
| Sc | 24.8 | 27.9 | 26.8 | 16.9 | 19.0 | 20.4 | 2 |
| Cr | 1400 | 1580 | 1300 | 1027 | 980 | 1145 | 2 |
| Co | 29.8 | 58.3 | 21.4 | 33.8 | 24.7 | 27.2 | 3 |

Table 3. Subsamples of Apollo 14 breccias.

| | 14270 Fragments | | | | | 14321,184 Materials | | | | | | | | |
|---|---|---|---|---|---|---|---|---|---|---|---|---|---|---|
| | | | | | | Basalt | Microbreccia-2 | | Microbreccia-3 | | | Matrix | | |
| Sample | 14270, 1,1 | 14270, 1,2 | 14270, 1,3 | 14270, 1,4 | 14270, 1,5 | Average | 14321, 184,15 | 14321, 184,42 | Average | 14321, 184,14A | 14321, 184,19A | 14321, 184,9A | 14321, 184,13 | Representative % Error |
| Wt. (mg) | 540.5 | 497.5 | 305.8 | 168.4 | 295.8 | | 78.7 | 12.3 | | 60.0 | 65.3 | 77.9 | 27.0 | |
| Al (%) | — | — | — | — | — | 6.66 | 8.07 | 8.66 | 8.29 | 8.89 | 8.02 | 7.02 | 7.80 | 2 |
| Na | 6540 | 6120 | 5440 | 5890 | 5960 | 4063 | 6560 | 7090 | 6215 | 6010 | 6050 | 4430 | 5180 | 2 |
| K | 5000 | 3530 | — | 3300 | 4600 | 1303 | 7970 | 5870 | 5102 | 4300 | 3800 | 1450 | 2560 | 5–10 |
| Cs | — | — | — | — | — | — | 1.29 | 2.6 | — | 0.54 | 0.42 | — | — | 10–20 |
| Ba | 970 | 800 | 750 | 770 | 820 | 250 | 1140 | 1110 | 1062 | 1070 | 730 | — | 600 | 8 |
| La | 101.6 | 65.5 | 64.8 | 62.5 | 69.2 | 20.7 | 97.1 | 112 | 92.2 | 88.6 | 77.7 | 27.3 | 51.0 | 2 |
| Ce | 265 | 169 | 181 | 175 | 199 | 72 | 260 | 334 | 255 | 260 | 211 | 82 | 138 | 3 |
| Nd | 158 | 101 | 93 | 115 | 114 | — | 150 | 162 | — | 125 | 147 | — | 55 | 10 |
| Sm | 48.1 | 31.1 | 30.3 | 29.3 | 32.2 | 11.6 | 46.9 | 53.9 | 41.8 | 42.2 | 37.6 | 14.7 | 23.7 | 2 |
| Eu | 3.50 | 2.71 | 3.04 | 3.27 | 2.80 | 1.44 | 3.34 | 3.06 | 3.22 | 3.42 | 2.70 | 1.55 | 1.96 | 3 |
| Tb | 9.9 | 6.2 | 5.3 | 6.6 | 6.6 | — | 9.4 | 9.5 | — | 9.6 | 7.6 | 3.1 | 4.6 | 5 |
| Yb | 32.0 | 22.0 | 22.9 | 22.9 | 23.8 | 6.90 | 32.6 | 32.5 | 30.7 | 30.5 | 25.5 | 9.3 | 15.8 | 3 |
| Lu | 4.60 | 3.12 | 3.38 | 3.30 | 3.50 | 1.28 | 4.35 | 4.60 | 4.35 | 4.30 | 3.50 | 1.6 | 2.3 | 2 |
| Th | — | — | — | — | — | — | 21.9 | 19.4 | — | 19.0 | 15.4 | 3.6 | 9.2 | 5 |
| U | 4.40 | 2.56 | — | — | — | — | — | — | — | — | — | — | — | 10 |
| Hf | 35.7 | 22.0 | 24.0 | 21.0 | 22.4 | 8.3 | 32.1 | 38.7 | 33.4 | 29.5 | 24.1 | 9.8 | 18.1 | 3 |
| Zr | 1200 | 760 | 850 | 680 | 960 | — | 1070 | 755 | — | 720 | 820 | — | — | 8 |
| Ta | 5.9 | 4.5 | 4.6 | 4.2 | 4.8 | 1.2 | 7.3 | 5.5 | 4.8 | 6.0 | 6.0 | — | 3.4 | 5 |
| Fe (%) | 8.1 | 8.9 | 9.3 | 9.0 | 9.0 | 13.0 | 8.2 | 8.4 | 8.7 | 8.3 | 9.6 | 12.0 | 10.0 | 2 |
| Ti (%) | — | — | — | — | — | 1.34 | 1.04 | 1.20 | 1.08 | 0.95 | 1.24 | 1.36 | 1.25 | 4 |
| Sc | 19.9 | 26.5 | 25.8 | 26.2 | 26.3 | 55.9 | 21.4 | 21.7 | 20.9 | 20.3 | 29.6 | 52.8 | 38.9 | 2 |
| V | — | — | — | — | — | — | 38 | — | — | 39 | 56 | 104 | 86 | 8 |
| Cr | 1090 | 1570 | 1750 | 1690 | 1570 | 2905 | 1180 | 1025 | 1285 | 1280 | 1620 | 2920 | 2160 | 2 |
| Mn | — | — | — | — | — | 1802 | 1040 | 1070 | 1077 | 870 | 1220 | 1660 | 1480 | 3 |
| Co | 32.4 | 37.3 | 38.6 | 36.4 | 36.8 | 32.4 | 31.4 | 19.8 | 37.5 | 39.0 | 37.9 | 33.2 | 33.4 | 2 |

over those of larger samples, except where sample sizes were less than 10–15 mg. During collection of gamma spectra, small samples were placed inside gelatin capsules lining our standard counting vials, so that all samples were in comparable geometrical relationship to the detector.

Several aliquots of U.S.G.S. standard rocks W–1 and BCR–1 (varying in weight from 10 to 500 mg) were analysed and showed no significant variation outside $2\sigma$ counting statistics; thus we conclude that the disagreements between duplicate samples of Apollo 14 materials are an artifact of sampling and not a result of analytical error. Estimates of precision are given in percentages in the last column of each table. The accuracy of our determinations can be checked by comparison of standard rock abundances given in Goles *et al.* (1971).

## Compositional Relationships among Apollo 14 Materials

The Apollo 14 landing site as described by Swann *et al.* (1971) consists of a smooth valley terrain and a ridge terrain, onto which the Cone Crater impact has splashed an ejecta blanket of underlying bedrock.

Compared to other Apollo 14 soil samples, soil 14141,33 from the edge of Cone Crater contains the highest concentrations of Na, K, rare earth elements (REE), and Hf, constituents that are generally characteristic of KREEP material as defined by Meyer *et al.* (1971). The valley soils, typified by sample 14259,34, differ compositionally from 14141,33 only with regard to the abundances of these KREEP-related elements. Other Apollo 14 soils can be classified as being similar either to 14141,33 or to 14259,34. Samples 14049,31 and 14163,89, collected from probable Cone Crater ray material, are very similar to 14141,33; trench and core samples collected from the smooth terrain strongly resemble 14259,34.

Among the 25 elements determined in the six soils of the trench and core (both taken at Station G), the only significant variation in composition is shown by Co. The deepest trench and the shallowest core samples have higher Co contents than do the other trench and core samples. Although it is impossible to determine the actual depth of sampling from the "disturbed" core 14230, our data suggest that a layer (or layers) some 15 to 30 cm deep in the regolith is slightly enriched in Co relative to layers above and below.

Most of the rocks at the Apollo 14 site are breccias. Breccia 14321, collected at the rim of Cone Crater, contains many clasts of varied lithology incorporated in a fragmental light-colored matrix. The large basaltic clasts (up to 5 cm in greatest dimension) are relatively rare in our sample 14321,184. Most of the analysed subsamples that make up our "average basalt" are probably from the same large clast. The numerous microbreccia clasts in 14321,184 vary markedly in lithology and composition. The average composition of the most abundant type of microbreccia (designated microbreccia-3 by Grieve *et al.*, 1972a, 1972b, and Duncan *et al.*, 1972a, 1972b) is given in Table 3 along with several analyses of individual microbreccia clasts. The less abundant and older microbreccia-2 clasts show considerable variability in composition. Data for two of these microbreccia-2 clasts (14321,184,15 and 14321,184,42) indicate that they are very similar in composition to KREEP material (Meyer *et al.*, 1971). The light-colored matrix samples, 14321,184,9A and 14321,184,13, are compositionally somewhat varied but are intermediate between basalts and microbreccias.

Breccia 14270 is strongly annealed (Warner, 1972a) without distinguishable clasts; consequently our subsamples of 14270,1 are random fragments broken from the rock. Four of the fragments (14270,1,2,3,4,5) are quite similar in composition and do not match in detail the composition of any other Apollo 14 clastic material we have studied (although they most resemble a microbreccia clast, 14321,184,19A). Fragment 14270,1,1 is compositionally distinct from the other four and matches the composition of average microbreccia-3 clasts from 14321,184.

Another useful approach to outlining the range of compositional variations among Apollo 14 clastic materials is to study lithic fragments from regolith samples. As mentioned above, we have analysed 17 fragments from 1–2 mm size fractions of soil specimens, supplied to us by Dr. J. A. Wood. The soil fragments (Table 2) have been grouped on petrographic grounds into glasses, soil breccias, and annealed breccias (Wood, personal communication). The soil breccias and four dark-colored glasses (14262,2,251,3, 14262,2,251,4, 14258,36,251,13, and 14258,36,250,16) form a coherent compositional group and are almost identical in composition to the valley soils from which they were taken. Compositions of the soil breccias and their related glasses show somewhat more variability than do those of the soils but cluster around the soil compositions. The suggestion, based on major element analyses and petrography, of Wood *et al.* (1971) that soil breccias are partially-fused soils seems to be confirmed by our data.

The remaining two glass fragments (14258,36,251,10, 14258,36,251,17) show little compositional similarity to any of our other analysed clastic materials. 14258,36,251,10 is a colorless glass rich in Ca, Al, and Mg (Wood, personal communication), with a norm of about 67% anorthite and 30% clinopyroxene, suggesting that it is a fragment of an anorthositic gabbro that has been vitrified. 14258,36,251,17 is a yellow-colored glass with a mafic composition, but it appears compositionally distinct from other Apollo 14 materials.

The annealed breccias have highly variable compositions, distinct from those of the soil breccias, and thus they are best compared to microbreccia clasts. Fragments 14262,2,249,6, 14262,2,250,7, and 14258,36,249,8 are similar to the series 14270,1,2 through 5 and to 14321,184,19A, containing more Fe, Sc, Cr, Na, and REE than the valley soils. Fragments 14258,36,249,9 and 14258,36,249,11 are unlike any other clastic materials we have analysed. They have very low Fe contents and moderately high Na, K, and REE contents. One sample of an annealed breccia that we have analysed (14258,36,250,12) resembles typical microbreccia-3 clasts from 14321,184. Clearly among both the glass fragments and the annealed breccia fragments in the soils, there are some materials with compositions that are not represented among the clasts of even such complex breccias as 14321 and 14270.

## COMPARISON WITH OTHER LUNAR SITES

Clastic materials from the Apollo 14 site are compositionally distinct from most of the materials obtained from other lunar sites. Soil samples from Fra Mauro are generally richer in Al, Na, K, Ba, Hf, and rare earths, and poorer in Fe, Sc, and Cr than are mare soils, Apollo 11 breccias, and many Apollo 12 samples (Warner, 1972b). The Apollo 14 clastic materials also show a similar compositional relationship to the soils from the Apollo 15 site for which data are available (LSPET, 1972; Wänke *et al.*, 1972; and Brunfelt *et al.*, personal communication).

There are, however, definite similarities between the Apollo 14 clastic materials and those Apollo 12 samples containing considerable KREEP component. 14321,184 microbreccia-3 (see Table 3) is quite similar in composition to the dark portion of 12013 (Wakita and Schmitt, 1970a; Hubbard *et al.*, 1970), although it is slightly richer in Al, K, and Cr and poorer in Na than is the dark portion of 12013. Breccia samples 14270,1,2 to 14270,1,5 (Table 3) are poorer in Na, K, Ba, Hf, and REE than microbreccia-3, and they closely match the compositions of Apollo 12 breccia 12010 and soil sample 12033 (Warner, 1972b). In general these materials all seem related to one another in terms of the amount of a KREEP component incorporated in each. This relationship will be discussed next in the section on mixing models.

## Mixing Models for Apollo 14 Clastic Materials

Multicomponent mixing models for some of the lunar soils have been presented by a number of authors (for example, Goles *et al.*, 1971; Meyer *et al.*, 1971). Studies of individual particles in some lunar soils (for example, Wood *et al.*, 1971; Apollo Soil Survey, 1971) have provided considerable information on their individual constituents, as well as indications concerning how the proportions of these constituents may affect the overall compositional characteristics of any particular soil sample. There are, however, statistical and sampling problems involved in making estimates of the proportions of different rock types in lunar soils from counts of identifiable rock fragments or of analysed glasses in a restricted size range. For this reason the practice of interpreting compositional data for bulk samples of lunar soils by the use of multicomponent mixing models has become increasingly common. Such mixing models may be used to describe the compositional parameters of lunar soil samples and to contrast the compositional characteristics of soils from different lunar sites (for example, Schonfeld, 1972). In this paper we extend the use of these models to a consideration of Apollo 14 clastic materials in a more general sense, and we present, together with models for soils, models for two different types of clastic materials from Apollo 14 breccia 14321.

In order to evaluate the linear mixing models discussed below we have used a least-squares analysis technique similar to that described by Goles *et al.* (1971) but with a modified weighting procedure. Previously our data were weighted by dividing each mixing component, element by element, by the *elemental abundances* in the component whose least-squares approximation was sought. We now weight our data by dividing each mixing component, element by element, by the *analytical error* at the one standard deviation level for the elemental abundances in the component whose approximation is sought (a technique similar to that used by Schonfeld, 1972). This normalization technique gives the greatest weight in calculating the model to those elements that have been most precisely determined. We have found that in general these two weighting procedures produce results that are very similar to one another. It would clearly be preferable to weight by the *variances* of the elements in the component whose approximation is sought (if analyses of multiple samples were available) in order to make allowance for both sampling errors and analytical precision, but the required information is generally not available for the models described here. We have been able to calculate mixing models only for those samples in which we determined Ti and Al; without these elements the computed component ratios are very poorly controlled.

## *Soils*

Mixing models have been calculated for four of the Apollo 14 soils (14141, 14148, 14149, 14156). The mixing proportions are given in Table 4, and the mixes are illustrated in histogram form in Fig. 1. All the mixes have low residuals for the 15 elements used, with very few elements showing differences between the analysed and computed soil abundances greater than three standard deviations in terms of estimated analytical precision. The components used in these mixing models are anorthositic gabbro, average Type 1 carbonaceous chondrite, a KREEP component and a local basalt (local in the sense that samples of this material were found at the Apollo 14 landing site). Additional details of the mixes are available on request.

For the local basalt component we have tried three alternative compositions, 14053, 14072, and the average of what seem to be several subsamples of a large (~20 gm) basalt clast from 14321,184. We have found that mixing models of the soil using 14072 as the basaltic component have markedly lower residuals than do comparable models using the other two basalt compositions and thus conclude that 14072-type basalt is probably the most abundant compositional variety of basaltic material that has been incorporated into the Apollo 14 soils. The composition of 14072 (Taylor *et al.*, 1972; Compston *et al.*, 1972) is similar to that of Apollo 12 group

Table 4. Mixing models for Apollo 14 clastic materials.

| | Estimated Proportions of Components (%) | | | | |
|---|---|---|---|---|---|
| Soils | 14072 | Anorthosite | CC-1 | 14321,184,15 | |
| 14141 | 10.9 | 13.2 | 0.8 | 75.1 | |
| 14148 | 12.6 | 17.0 | 1.8 | 68.6 | |
| 14149 | 11.0 | 18.1 | 3.1 | 67.8 | |
| 14156 | 13.2 | 16.7 | 2.3 | 67.9 | |
| 14321,184, Microbreccia-3 | Basalt | Anorthosite | CC-1 | 14321,184,15 | 14321,184,42 |
| 14321,184,19A | 29.3 | 3.3 | 3.4 | — | 64.0 |
| 14321,184,14A | 2.8 | 8.5 | 2.9 | 57.9 | 27.9 |
| 14321,184,AVMB | 6.7 | 4.1 | 3.3 | 25.9 | 60.0 |
| 14321,184 Matrix | Basalt | Anorthosite | 14321,184,AVMB | | |
| 14321,184,9A | 90.0 | 1.0 | 9.0 | | |
| 14321,184,13 | 53.6 | 5.0 | 41.4 | | |

Elements used to determine models: Fe, Sc, Ti, Cr, Mn, Co, Al, Na, K, La, Ce, Sm, Eu, Yb, and Lu.
Identification of components: The data for basalt 14072 are from Compston *et al.* (1972) and Taylor *et al.* (1972). Anorthosite data are from Wood *et al.* (1970) and Wakita and Schmitt (1970b). Average Type 1 carbonaceous chondrite (CC-1) data are from Schmitt *et al.* (1964), Schmitt *et al.* (1971), Keil (1969), and Goles (1971). Data for 14321,184,15 and 14321,184,42, KREEP microbreccias, are in Table 3. 14321,184, AVMB is an average of analyses of 14 microbreccia-3 clasts from 14321; the data are in Table 3. Basalt is an average of 14 analyses of typical basalt clasts from 14321; these data are in Table 3.

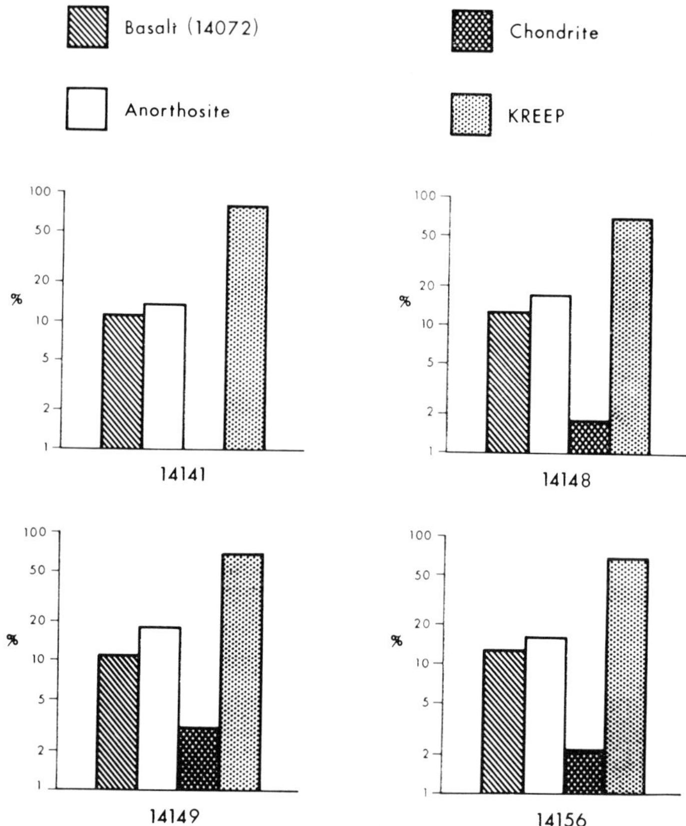


Fig. 1. Histograms to illustrate the proportions of components in some Apollo 14 soils. The height of the bar represents the percentage of each component (note log scale) in the computed mixing model. Those components present below the 1% level are not plotted. The details of the mixing models and information on the data sources for the different components are given in Table 4.

1 igneous rocks (Goles *et al.*, 1971) but is slightly poorer in Fe; it is also very similar to the composition of a small basalt clast from 14321,184 (Duncan *et al.*, 1972b). Thus it is possible that the predominant basalt type in the Apollo 14 soils has been derived both from the Fra Mauro formation and from adjacent maria.

The KREEP component used in these mixing models is a local one, with compositional data taken from sample 14321,184,15, an early-stage microbreccia clast in 14321 (see Grieve *et al.*, 1972a, and Duncan *et al.*, 1972a). The composition of this material is shown in Table 3. One should note that it has rather higher K and REE abundances than do most of the KREEP materials described by Meyer *et al.* (1971). We have tried two other KREEP compositions in our soil mixing models, 14321,184,42 (Table 3 and Duncan *et al.*, 1972b) and our estimate of "average KREEP" at the Apollo 12 site (Goles *et al.*, 1971, Table 2). In all of the soil mixing models reported here we find that the KREEP material 14321,184,15 gives lower residuals than do the other two KREEP materials mentioned earlier.

In a previous section we discussed the qualitative observation that soils from the mapped ejecta blanket of Cone Crater appear to have higher abundances of those elements that are characteristically enriched in KREEP (K, REE, Hf, etc.), when compared to soils from locations away from the Cone Crater ejecta blanket. By use of the mixing models (Table 4) it is possible to quantify this and other differences.

The Cone Crater soil (14141) is enriched in a KREEP component relative to the soils collected away from the Cone Crater ejecta blanket (14148, 14149, 14156) and contains lesser amounts of basaltic, anorthositic, and chondritic components. Cone Crater has been dated by exposure ages on the ejected material as being approximately 27 m.y. old (Lugmair and Marti, 1972), and it has been suggested that material in the Cone ejecta blanket may have come from depths of up to 80 m below the surface of the Fra Mauro formation. Based on these data we consider that there are two reasonable hypotheses to explain the observed compositional differences between the Cone Crater soil and the other three soils. Either the Fra Mauro formation is inhomogeneous in a lateral or vertical sense, or the compositional difference is due to compositional evolution with time of soils at the Fra Mauro site, presumably because of the secular addition of one or more extra-Fra Mauro components. Owing to the considerable distance between the Apollo 14 site and the putative source of the Fra Mauro formation in the Imbrium basin (about 1200 km), its probable mode of transport as an ejecta plume or turbulent jet (Anderson *et al.*, 1972), and the size of the sample provided by Cone Crater, it seems unlikely that the Fra Mauro formation would show compositional differences that are this large over such a small vertical or lateral distance.

Consequently, we believe that the differences between the Cone Crater soil and the other soils are best interpreted to show the approximate amounts and types of material added to the surface of the Fra Mauro formation at the Apollo 14 site since its deposition. We thus conclude that there has been an addition of 1–2% of non-KREEP basaltic material, 3–5% of anorthosite or anorthositic gabbro, and possibly 1–2% of a chondritic component. The amount of chondritic material added is very speculative because our models provide rather low-precision estimates of the proportion of this component and because we have not included in our mixing models an ultramafic component that may be present (Schonfeld, 1972). For the elements that we have determined, such an ultramafic component would behave very similarly to a chondritic component in our computed mixing models.

### *Breccia materials*

Mixing models for individual Apollo 14 breccias on a whole-rock basis are unlikely to be meaningful owing to the extreme heterogeneity shown by most of these rocks on the scale at which they were sampled. However, it is possible to derive reasonable mixing models for some of the fine-grained microbreccia clasts and for the fine-grained matrix material present in some of these rocks. The compositional relationships between microbreccia clasts in 14321 and 14270 have already been discussed.

In this paper we discuss only a few of the mixing models for 14321 materials in order to elucidate relationships between soils and some breccia materials at the Apollo 14 site. Specific mixing models for a larger number of 14321 subsamples are given in Duncan *et al.* (1972b).

The components used in these mixing models for breccia materials (Fig. 2, Tables 3 and 4) are: (1) average composition of large basalt clasts (which may themselves

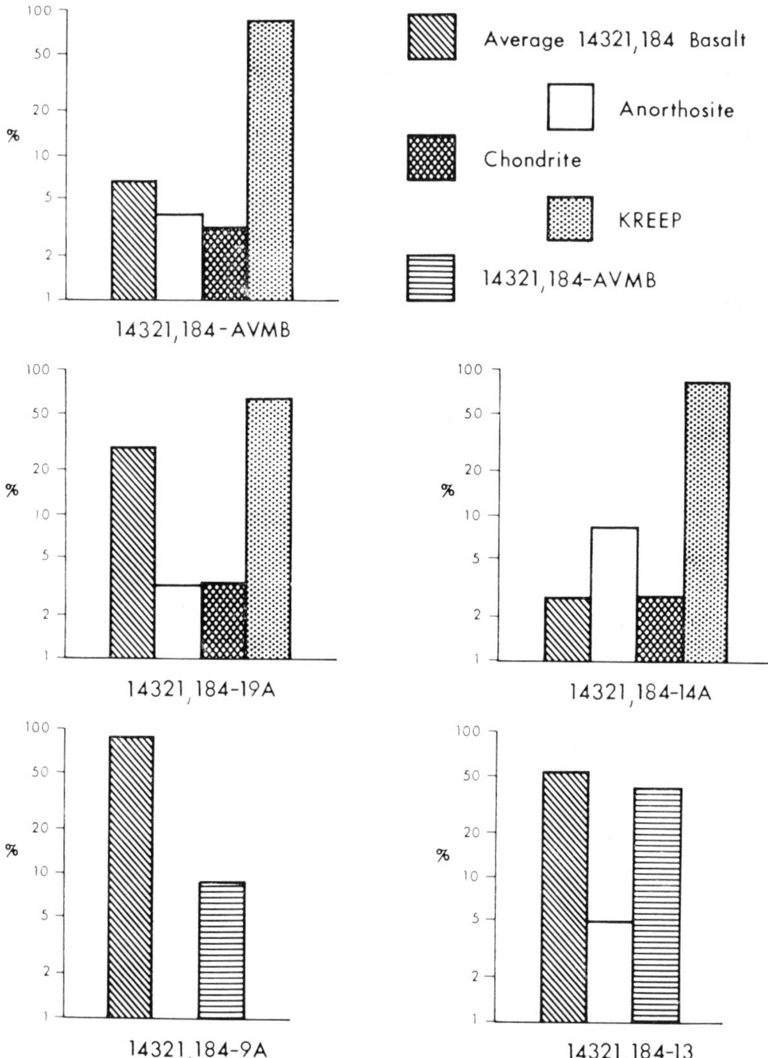


Fig. 2. Histograms to illustrate the proportions of components in some clastic materials sampled from breccia 14321,184. The height of the bar represents the percentage of each component (note log scale) in the computed mixing model. Those components present below the 1% level are not plotted. The details of the mixing models and information on the data sources for the different components are given in Table 4.

be fragments of a single 20 gm clast) from 14321,184; (2) anorthositic gabbro; (3) average Type 1 carbonaceous chondrite; and (4) KREEP materials of two types, as represented by early-stage microbreccia clasts 14321,184,15 and 14321,184,42 (see Grieve *et al.*, 1972b, and Duncan *et al.*, 1972b for a discussion of the relative age relationships of clasts within breccia 14321).

Residuals are low in the three mixing models for microbreccia-3 materials from 14321 but not as low as those in the soil mixing models discussed earlier. This contrast is probably due to an incomplete set of components in these mixes. We are aware of both potash microgranite and early-stage dunitic microbreccia components in 14321 (Grieve *et al.*, 1972b) for which trace element data are too poorly known to allow their use as explicit components in the mixing models. In addition, when mixing models for clastic rock samples of 50–200 mg size are being devised, it may be necessary to consider the possibility of using discrete mineral compositions as components. Microbreccia-3 materials from 14321 are rather varied in composition; hence their mixing models also show a rather marked variability. The two specific microbreccia clasts for which mixing models are given (14321,184,14A and 14321, 184,19A) are representative of the extremes of the compositional range shown by the main group of 14321 microbreccias. A model is also given for the average composition of this main group of microbreccia-3 materials.

It is notable that the mixing model for the average microbreccia-3 (Table 4, Fig. 2) indicates lower contents of non-KREEP basalt and anorthositic gabbro and a higher KREEP content than was present in either group of Apollo 14 soils. Since the microbreccias clearly predate the soils, it appears that the physical mixing system in which these clastic materials formed started with a very high initial KREEP content, which has undergone dilution by successive additions of anorthosite and basalt, with the ratio of anorthosite to basalt apparently increasing with time. It is certainly possible that this change in the anorthosite/basalt ratio is an artifact of our sampling or of the assumptions inherent in our model. However, it should be possible to extend this approach as data on more breccia samples become available and to determine if any general conclusions can be drawn from such a study. We consider that secular change in such a ratio indicates either a change in the ratio of these materials exposed on the lunar surface or a change in the location of the physical mixing system, or a combination of these hypotheses.

We have included mixing models for two samples of the matrix material produced in the final assembly stage of breccia 14321 to illustrate a different type of mixing system (Table 4, Fig. 2). The residuals for these mixes are extremely small, and we conclude that the matrix material in 14321 was produced by mutual abrasion of the three major clast types. The absence of any additional chondritic component, other than that already present in microbreccia-3, indicates that the matrix material was not pre-existing lunar regolith similar to that sampled by the Apollo missions.

## CONCLUSIONS

We have developed both our analytical and interpretive tools beyond their previous limits. Our capabilities for analyses of small specimens are now limited more by sampling errors than by the INAA techniques themselves. We have shown that linear mixing models are useful tools in relating the history of breccias and soils, provided that breccias can be adequately characterized. We consider that there has been a progressive addition of anorthosite and basalt components to successive generations of clastic materials at the Apollo 14 site and an increasing dominance of anorthosite with time. This conclusion should be tested by examination of additional samples of the Apollo 14 clastic materials.

*Acknowledgments*—We gratefully acknowledge the assistance of Dr. J. A. Wood of the Smithsonian Astrophysical Observatory, who provided the selected samples required for our study of soil particles, together with microprobe data for these samples. Ernest Schonfeld of NASA provided valuable advice that led to the improvement of our computer technique for calculating mixing models. Lauren Dunn aided our data reduction with the development of additional computer programs, and Jean Duncan and Shirley Sales performed the typing chores. This work was supported in large part by NASA grant NGL 38-003-024.

## References

Anderson A. T., Braziunas T. F., Jacoby J., and Smith J. V. (1972) Breccia populations and thermal history: Nature of pre-Imbrian crust and impacting body (abstract). In *Lunar Science—III* (editor C. Watkins), pp. 24–26, Lunar Science Institute Contr. No. 88.

Apollo Soil Survey (1971) Apollo 14: Nature and origin of rock types in soil from the Fra Mauro formation. *Earth Planet. Sci. Lett.* **12,** 49–54.

Compston W., Vernon M. J., Berry H., Rudowski R., Gray C. M., and Ware N. (1972) Age and petrogenesis of Apollo 14 basalts (abstract). In *Lunar Science—III* (editor C. Watkins), pp. 151–153, Lunar Science Institute Contr. No. 88.

Duncan A. R., Lindstrom M. M., Lindstrom D. J., McKay S. M., Stoeser J. W., Goles G. G., and Fruchter J. S. (1972a) Comments on the genesis of breccia 14321 (abstract). In *Lunar Science—III* (editor C. Watkins), pp. 192–194, Lunar Science Institute Contr. No. 88.

Duncan A. R., McKay S. M., Fruchter J. S., Lindstrom M. M., Lindstrom D. J., Stoeser J. W., and Goles G. G. (1972b) Subsamples of breccia 14321,184: Their compositions and inter-relationships (in preparation).

Goles G. G., Duncan A. R., Lindstrom D. J., Martin M. R., Beyer R. L., Osawa M., Randle K., Meek L. T., Steinborn T. L., and McKay S. M. (1971) Analyses of Apollo 12 specimens: Compositional variations, differentiation processes, and lunar soil mixing models. *Proc. Second Lunar Sci. Conf., Geochim. Cosmochim. Acta* Suppl. 2, Vol. 2, pp. 1063–1081. MIT Press.

Gordon G. E., Randle K., Goles G. G., Corliss J. B., Beeson M. H., and Oxley S. S. (1968) Instrumental activation analysis of standard rocks with high-resolution $\gamma$-ray detectors. *Geochim. Cosmochim. Acta* **32,** 369–396.

Grieve R., McKay G., Smith H., Weill D., and McCallum S. (1972a) Mineralogy and petrology of polymict breccia 14321 (abstract). In *Lunar Science—III* (editor C. Watkins), pp. 338–340, Lunar Science Institute Contr. No. 88.

Grieve R., McKay G., Smith H., Weill D., and McCallum S. (1972b) Mineralogy and petrology of lunar polymict breccia 14321 (in preparation).

Hubbard N. J., Gast P. W., and Wiesmann H. (1970) Rare earth, alkaline and alkali metal, and $^{87/86}$Sr data for subsamples of lunar sample 12013. *Earth Planet. Sci. Lett.* **9,** 181–184.

Keil K. (1969) Meteorite composition. In *Handbook of Geochemistry* (V. 1) (editor K. H. Wedepohl), pp. 78–115. Springer-Verlag.

LSPET (Lunar Sample Preliminary Examination Team) (1972) The Apollo 15 lunar samples: A preliminary description. *Science* **175,** 363–375.

Lugmair G. W. and Marti K. (1972) Neutron and spallation effects in Fra Mauro regolith (abstract). In *Lunar Science—III* (editor C. Watkins), pp. 495–497, Lunar Science Institute Contr. No. 88.

Meyer C., Brett R., Hubbard N. J., Morrison D. A., McKay D. S., Aitken F. K., Takeda H., and Schonfeld E. (1971) Mineralogy, chemistry, and origin of the KREEP component in soil samples from the Ocean of Storms. *Proc. Second Lunar Sci. Conf., Geochim. Cosmochim. Acta* Suppl. 2, Vol. 1, pp. 393–412. MIT Press.

Schmitt R. A., Smith R. H., and Olehy D. A. (1964) Rare-earth, yttrium, and scandium abundances in meteoritic and terrestrial matter, II. *Geochim. Cosmochim. Acta* **28,** 67–86.

Schmitt R. A., Linn T. A., and Wakita H. (1970) The determination of 14 common elements in rocks via sequential instrumental activation analysis. *Radiochim. Acta* **13**, 200–212.

Schmitt R. A., Goles G. G., Smith R. H., and Osborn T. W. III (1972) Elemental abundances in stone meteorites (to be published).

Schonfeld E. (1972) Component abundance and ages in soils and breccia (abstract). In *Lunar Science—III* (editor C. Watkins), pp. 683–685, Lunar Science Institute Contr. No. 88.

Swann G. A., Trask N. J., Hait M. H., and Sutton R. L. (1971) Geologic setting of the Apollo 14 samples. *Science* **173,** 716–719.

Taylor S. R., Kaye M., Muir M., Nance W., Rudowski R., and Ware N. (1972) Composition of the lunar uplands, I. Chemistry of Apollo 14 samples from Fra Mauro (this volume).

Wakita H. and Schmitt R. A. (1970a) Elemental abundances in seven fragments from lunar rock 12013. *Earth Planet. Sci. Lett.* **9,** 169–176.

Wakita H. and Schmitt R. A. (1970b) Lunar anorthosites: Rare-earth and other elemental abundances. *Science* **170,** 969–974.

Wänke H., Baddenhausen A., Balacescu A., Teschke F., Spettel B., Dreibus G., Quijano M., Kruse H., Wlotzka F., and Begemann F. (1972) Multielement analyses of lunar samples (abstract). In *Lunar Science—III* (editor C. Watkins), pp. 779–781, Lunar Science Institute Contr. No. 88.

Warner J. L. (1972a) Apollo 14 breccias: Metamorphic origin and classification (abstract). In *Lunar Science—III* (editor C. Watkins), pp. 782–784, Lunar Science Institute Contr. No. 88.

Warner J. L. (1972b) Continuing compilation of Apollo chemical, age, and modal information. M.S.C. Curator's Office (unpublished).

Wood J. A., Marvin U. B., Powell B. N., and Dickey J. S. Jr. (1970) Mineralogy and petrology of the Apollo 11 lunar sample. Smithsonian Astrophysical Observatory, Spec. Rep. No. 307.

Wood J. A., Marvin U. B., Reid J. B., Taylor G. J., Bower J. F., Powell B. N., and Dickey J. S. Jr. (1971) Mineralogy and petrology of the Apollo 12 lunar sample. Smithsonian Astrophysical Observatory, Spec. Rep. No. 333.

Proceedings of the Third Lunar Science Conference
(Supplement 3, *Geochimica et Cosmochimica Acta*)
Vol. 2, pp. 1215–1229
The M.I.T. Press, 1972

# Compositional data for twenty-one Fra Mauro lunar materials*

H. J. ROSE, JR., FRANK CUTTITTA, C. S. ANNELL, M. K. CARRON,
R. P. CHRISTIAN, E. J. DWORNIK, L. P. GREENLAND,
and D. T. LIGON, JR.

U.S. Geological Survey, Washington, D.C. 20242

**Abstract**—Major, minor, and trace element analyses are presented for two igneous rocks, six breccias, four microbreccias, two breccia clasts, and six soils, as well as a sample of sawdust from rock 14066. Evaluation of the data suggests that the samples from the Fra Mauro highlands have the same non-terrestrial characteristics shown previously by the samples returned from the mare regions by Apollo 11 and 12—a high refractory element content, a lower volatile element content, and an excess reducing capacity above that due to FeO.

The Fra Mauro soils have higher concentrations of $Al_2O_3$, $Na_2O$, and $K_2O$ and lower amounts of FeO and $TiO_2$ than do the mare soils. They also show a bimodal distribution of Ni ($\approx 300$ and $>400$ ppm), B ($<10$ and $\approx 18$ ppm), and Nb ($\approx 40$ and $\approx 70$ ppm).

The highland breccias are richer in $SiO_2$, $Al_2O_3$, MgO, $Na_2O$, and $K_2O$ than those returned from the mare lowlands. FeO, $TiO_2$, and MnO are lower in concentration at Fra Mauro, and the highland breccias are more complex mineralogically than those collected previously.

The Apollo 14 basalts are richer in $SiO_2$, $Al_2O_3$, CaO, $Na_2O$, $K_2O$, $P_2O_5$, Pb, Rb, Sr, Li, Be, Nb, La, Y, and Zr when compared to the mare basalts.

## INTRODUCTION

THE MAJOR, MINOR, and trace element composition of two basaltic rocks, six breccias, four microbreccias, two breccia clasts, and six soils collected by the Apollo 14 mission at the Fra Mauro formation (Apollo 14 Preliminary Science Report, 1971) were determined by combined semimicro chemical, x-ray fluorescence, and optical emission methods. Details of the analytical procedures and accuracy of the methods have been described previously (Annell and Helz, 1970a, 1970b; Cuttitta *et al.*, 1971; Rose *et al.*, 1970a, 1970b, 1970c, 1971) and will not be reported here. Some of the samples were provided through consortium groups, each of which was given the responsibility of planning mineralogic and petrographic study, disaggregating the rock, and distributing portions of the sample to each of its members for more detailed study or analysis.

The Apollo 14 samples were collected in an area approximately 50 km south of the southern rim crest of the Imbrium Basin and the formation is believed to be partly the ejecta blanket resulting from the catastrophic excavation of that Basin (Apollo 14 Preliminary Science Report, 1971). Compared to the Apollo 11 (Tranquility Base) and Apollo 12 (Ocean of Storms) rocks, the Fra Mauro rocks have similar mineral assemblages but different proportions of major components, such as

* Publication authorized by the Director, U.S. Geological Survey.

plagioclase, pyroxene, olivine, and ilmenite, and of minor mineralogic constituents, such as cristobalite, apatite, spinel, and troilite. The soils from the highland area differ physically from those returned from the lunar maria both in their generally lighter color and in their greater range of particle size.

Preliminary reports on the Apollo 14 soils (Annell *et al.*, 1972) and rocks (Cuttitta *et al.*, 1972) have been made. In comparing the data for the Apollo 14 samples with

Table 1. Composition of some Apollo 14 basalts (oxides in weight percentage, elements* in ppm).

| Constituent | 14310,114 | 14276,8 |
|---|---|---|
| $SiO_2$ | 47.81 | 47.60 |
| $Al_2O_3$ | 21.54 | 21.34 |
| $Fe_2O_3$ | 0.00 | 0.00 |
| FeO | 7.62 | 7.94 |
| MgO | 7.48 | 7.10 |
| CaO | 12.92 | 13.18 |
| $Na_2O$ | 0.68 | 0.72 |
| $K_2O$ | 0.48 | 0.48 |
| $TiO_2$ | 1.11 | 1.20 |
| $P_2O_5$ | 0.43 | 0.40 |
| MnO | 0.10 | 0.12 |
| $Cr_2O_3$ | 0.25 | 0.26 |
| Total | 100.42 | 100.34 |
| ΔRC† | +0.53 | +0.11 |
| Pb | 13.0 | 9.0 |
| Zn | <4 | <4 |
| Ga | 3.2 | 4.2 |
| Cu | 9.0 | 42 |
| Rb | 15 | 13 |
| Li | 23 | 21 |
| Co | 17.0 | 9.0 |
| Ni | 120 | 113 |
| Ba | 780 | 700 |
| Sr | 175 | 165 |
| V | 38 | 37 |
| Be | 4.2 | 4.0 |
| Nb | 29 | 33 |
| Sc | 25 | 20 |
| La | 59 | 59 |
| Y | 185 | 200 |
| Yb | 16 | 16 |
| Zr | 610 | 620 |

* The elements in the table are listed in order of decreasing volatility.

† ΔRC = Total reducing capacity less the reducing capacity attributable to the FeO content of the sample, in % FeO.

*Note:* The following elements were looked for but not detected. If present, they would be in concentrations below the amount (in ppm) indicated in parentheses: Ag (0.2), As (4), Au (0.2), B (10), Bi (1), Cd (8), Ce (100), Cs (1), Ge (1), Hf (20), Hg (8), In (1), Mo (2), Nd (100), P (2,000), Pt (3), Re (30), Sb (100), Sn (10), Ta (100), Te (300), Th (100), Tl (1), U (500), and W (200).

those for the samples collected by the Apollo 11 and 12 missions, we have attempted whenever possible to use the data gathered in this laboratory for reasons of consistency.

In general, we are in agreement with the sample classifications given in *Apollo 14: Preliminary Science Report* (1971) or with the descriptions provided by the consortium groups. Optical, microscopic, and photographic examination of all samples were made prior to analysis. All the soils were found to contain varying amounts of breccia and crystalline rock fragments; ilmenite, pyroxene, and plagioclase grains; glass spheres or nearly spherical glass particles ranging from colorless to green to black opaque. No metallic fragments were found.

## Composition of Basalts

### *Elemental variations*

The analyses for the two basalts are given in Table 1. The two samples (14276 and 14310) are similar in chemical composition. The high CaO and $Al_2O_3$ values support the petrographic observation that plagioclase ($\sim$An 90) is the major mineral component. Only slight variations are evident in some of the trace elements. Although we believe the variation in Pb, Mn, and Co to be real, we feel strongly that the five-fold increase in Cu in sample 14276,8 compared to other lunar basalts is due mainly to contamination incurred during preparation of that sample elsewhere.

Basalt 14310 was also analyzed by others (LSPET, 1971; Compston *et al.*, 1972; Kushiro, 1972; Engel *et al.*, 1972). Although there is some variation among the chips analyzed, the average composition derived from the available data is: $SiO_2$, 47.36%; $Al_2O_3$, 20.87%; FeO, 8.11%; MgO, 7.59%; CaO, 12.58%; $Na_2O$, 0.69%; $K_2O$, 0.49%; $P_2O_5$, 0.34%; $TiO_2$, 1.20%; MnO, 0.12%; and $Cr_2O_3$, 0.21%. The analysis of Engel *et al.* (1972) could not be included because it was averaged with two other samples. The $SiO_2$ and $P_2O_5$ values reported by Kushiro were somewhat lower than those of the other analysts, perhaps attesting to the inhomogeneity of the rock, but the data were included in the calculation of the average values above.

### *Comparison with mare basalts*

Comparison of the elemental data of the highland basalts (Apollo 14) with the mare samples returned previously (Apollo 11 and Apollo 12) is given in Table 2.

The following observations are made:

(1) The elements Si, Al, Ca, Na, P, and Ni show an increasing trend (Apollo 11 $<$ 12 $<$ 14).

(2) The elements Fe, Ti, Cr, and Sc show a decreasing trend (Apollo 11 $>$ 12 $>$ 14).

(3) The highland basalts are enriched in K, Pb, Rb, Sr, Li, Ba, Be, Nb, La, Y, and Zr.

(4) The lowland basalts have higher concentrations of Co, Ga, Mn, and V.

Table 2. Comparison of average composition of Apollo 11, 12, and 14 basalts (oxides in weight percentage, elements in ppm).

| Constituent | Apollo 11 | Apollo 12 | Apollo 14 |
|---|---|---|---|
| $SiO_2$ | 40.10 | 47.10 | 47.70 |
| $Al_2O_3$ | 8.60 | 12.80 | 21.44 |
| $Fe_2O_3$ | 0.00 | 0.00 | 0.00 |
| FeO | 18.90 | 17.40 | 7.78 |
| MgO | 7.74 | 6.80 | 7.29 |
| CaO | 10.70 | 11.40 | 13.05 |
| $Na_2O$ | 0.46 | 0.64 | 0.70 |
| $K_2O$ | 0.30 | 0.07 | 0.48 |
| $TiO_2$ | 12.20 | 3.17 | 1.16 |
| $P_2O_5$ | <0.2 | 0.17 | 0.42 |
| MnO | 0.25 | 0.24 | 0.11 |
| $Cr_2O_3$ | 0.37 | 0.31 | 0.25 |
| Total | 99.82 | 100.10 | 100.38 |
| ΔRC* | +2.1 | +0.37 | +0.32 |
| Pb | <2 | <2 | 11 |
| Ag | <0.2 | <0.2 | 0.2 |
| Zn | <4 | <4 | <4 |
| Cu | 8.8 | 10.5 | 9.0 |
| Ga | 4.8 | 4.9 | 3.7 |
| Rb | 5.1 | 1.4 | 14.0 |
| Li | 17.0 | 5.9 | 22.0 |
| Co | 30.8 | 64.0 | 13.0 |
| Ni | 6.6 | 70.0 | 116 |
| Ba | 440 | 63.0 | 740 |
| Sr | 135 | 64.0 | 170 |
| V | 72.8 | 160 | 37.5 |
| Be | 3.1 | 1.4 | 4.1 |
| Nb | 24.0 | 13.0 | 31.0 |
| Sc | 96.5 | 40.0 | 22.5 |
| La | 26.2 | <20 | 59 |
| Y | 162 | 39 | 192 |
| Yb | 20.0 | 5.2 | 16.0 |
| Zr | 594 | 110 | 615 |

* ΔRC = Total reducing capacity less the reducing capacity attributable to the FeO content of the sample, in % FeO.

### *Significant geochemical ratios*

A comparison of some elemental ratios calculated for Apollo 11, 12, and 14 basalts is presented in Table 3. The data reveal that some of the elemental ratios distinguish the highland basalts of Apollo 14 from those returned from the lunar maria:

(1) Si/Al, Fe/Ni, and Cr/Ni decrease (Apollo 11 > 12 > 14). The Fe/Ni ratio shows marked decreases in the sequence of missions (21,000, 2520, and 504 for Apollo 11, 12, and 14, respectively). This effect is due to both decreasing Fe and increasing Ni contents.

(2) Ni/Co, Al/Ti, and Nb/Ti increase markedly (Apollo 11 < 12 < 14).

(3) The highland basalts are characterized by high Mg/Fe, Ba/Sr, Ba/V, Rb/Sr, and Cr/V.

(4) K/Rb is higher in the mare basalts.

Table 3. Comparison of some elemental ratios of Apollo 11, 12, and 14 samples.

| | Basalt | | | Breccia | | | Soil | | |
|---|---|---|---|---|---|---|---|---|---|
| | Apollo 11 | 12 | 14 | Apollo 11 | 12 | 14 | Apollo 11 | 12 | 14 |
| Si/Al | 4.12 | 3.25 | 1.96 | 2.82 | 2.81 | 2.52 | 2.67 | 3.04 | 2.40 |
| Ca/Si | 0.41 | 0.37 | 0.42 | 0.43 | 0.37 | 0.34 | 0.43 | 0.35 | 0.31 |
| Mg/Fe | 0.63 | 0.39 | 0.92 | 0.42 | 0.65 | 0.80 | 0.50 | 0.41 | 0.89 |
| Fe/Ni | 21,000 | 2520 | 504 | 612 | 640 | 254 | 661 | 617 | 218 |
| Cr/Ni | 357 | 39.0 | 15.0 | 16.3 | 20.3 | 4.47 | 9.4 | 14.8 | 4.7 |
| Ni/Co | 0.23 | 0.98 | 8.92 | 6.31 | 4.59 | 9.78 | 7.70 | 3.36 | 10.6 |
| Al/Ti | 0.62 | 3.56 | 16.3 | 1.36 | 5.95 | 8.80 | 1.64 | 4.46 | 8.95 |
| $10^3 \times$ Nb/Ti | 0.33 | 0.68 | 4.40 | 0.45 | — | 5.06 | 0.40 | 2.38 | 5.32 |
| K/Rb | 588 | 500 | 342 | 533 | 388 | 400 | 518 | 390 | 423 |
| K/Ba | 5.7 | 9.2 | 5.3 | 5.2 | 6.6 | 5.05 | 5.3 | 4.5 | 4.4 |
| Ba/V | 6.04 | 0.48 | 19.7 | 4.17 | 5.75 | 19.0 | 4.2 | 5.3 | 18.6 |
| Ba/Sr | 3.25 | 0.95 | 4.35 | 1.91 | 2.82 | 5.69 | 1.6 | 4.3 | 5.5 |
| Rb/Sr | 0.35 | 0.008 | 0.82 | 0.023 | 0.058 | 0.088 | 0.015 | 0.062 | 0.069 |
| Cr/V | 34.3 | 13.5 | 46.4 | 53.3 | 37.9 | 28.6 | 34.8 | 26.9 | 30.9 |

Table 4. Composition of some Apollo 14 breccias (oxides in weight percentage, elements in ppm).

| Constituent | 14047, 27* | 14066,31(4) Sawdust (1) | 14066, 21,203 | 14301, 62 | 14303, 34 | 14318 Misc. Frag. | 14318, 27A |
|---|---|---|---|---|---|---|---|
| $SiO_2$ | 47.45 | 46.31 | 47.59 | 48.26 | 47.49 | 47.94 | 47.97 |
| $Al_2O_3$ | 17.75 | 14.80 | 14.61 | 16.52 | 16.05 | 17.95 | 17.80 |
| $Fe_2O_3$ | 0.00 | 0.00 | 0.00 | 0.00 | 0.00 | 0.00 | 0.00 |
| FeO | 10.36 | 9.59 | 10.82 | 10.29 | 10.96 | 9.43 | 9.62 |
| MgO | 9.35 | 10.78 | 13.67 | 9.98 | 10.99 | 9.63 | 9.79 |
| CaO | 11.19 | 9.32 | 9.07 | 10.29 | 10.03 | 11.13 | 11.16 |
| $Na_2O$ | 0.75 | 1.10 | 1.01 | 0.84 | 0.87 | 0.81 | 0.79 |
| $K_2O$ | 0.49 | 0.87 | 0.45 | 0.75 | 0.46 | 0.62 | 0.60 |
| $H_2O^-$ | 0.00 | 0.06 | 0.00 | 0.00 | 0.00 | 0.00 | 0.00 |
| $TiO_2$ | 1.48 | 1.49 | 1.68 | 2.06 | 1.98 | 1.44 | 1.48 |
| $P_2O_5$ | 0.39 | 0.58 | 0.55 | 0.64 | 0.56 | 0.55 | 0.56 |
| MnO | 0.13 | 0.12 | 0.13 | 0.14 | 0.15 | 0.13 | 0.13 |
| $Cr_2O_3$ | 0.22 | 0.22 | 0.26 | 0.21 | 0.21 | 0.18 | 0.19 |
| Cu | — | 2.30 | — | — | — | — | — |
| L.O.I. | — | 2.94 | — | — | — | — | — |
| Total | 99.76 | 100.48 | 99.84 | 99.98 | 99.75 | 99.81 | 100.09 |
| ΔRC† | — | +8.29 | +1.38 | — | — | +1.32 | +1.38 |
| Pb | — | 6.4 | 12 | 15 | 8.5 | 18 | 18 |
| Zn | — | — | — | 39 | — | 15 | 15 |
| Cu | — | >1% | 22 | 43 | 20 | 150 | 170 |
| Ga | — | 6.3 | 5.2 | 7.7 | 3.8 | 4.4 | 4.5 |
| Rb | — | 26 | 12 | 17 | 10 | 16 | 14 |
| Li | — | 29 | 28 | 28 | 26 | 24 | 24 |
| Co | — | 32 | 38 | 27 | 28 | 30 | 38 |
| Ni | — | 315 | 285 | 255 | 245 | 330 | 420 |
| Ba | — | 1400 | 1000 | 1280 | 980 | 760 | 640 |
| Sr | — | 150 | 140 | 195 | 175 | 160 | 140 |
| V | — | 47 | 53 | 49 | 46 | 50 | 47 |
| Be | — | 12 | 5.4 | 10 | 8.0 | 8.0 | 7.6 |
| B | — | 27 | 15 | 24 | 24 | 17 | 16 |
| Nb | — | 51 | 44 | 64 | 53 | 48 | 52 |
| Sc | — | 22 | 20 | 26 | 26 | 24 | 22 |
| La | — | 97 | 87 | 92 | 88 | 75 | 66 |
| Y | — | 280 | 240 | 335 | 320 | 260 | 260 |
| Yb | — | 27 | 22 | 23 | 23 | 22 | 23 |
| Zr | — | 700 | 830 | 940 | 940 | 720 | 820 |

* Insufficient samples for trace element analysis.

† ΔRC = Total reducing capacity less the reducing capacity attributable to the FeO content of the sample, in % FeO.

## Composition of Breccias

### *Elemental variations*

The lunar breccias are composed of rock fragments and mineral grains of the types mentioned previously, but their complexity is magnified by the presence of glass and microbreccia fragments that resulted from earlier impact events. Results of the analyses of five breccia samples are given in Table 4. Trace elements were not determined for sample 14047,27 because of insufficient amount of sample, about 100 mg. Sample 14318 was composed of miscellaneous fragments, whereas sample 14318,27A was a single chip from the parent rock. Two samples of 14066 were received, one a chip 14066 (21,203) and the other the LRL sawdust 14066 (31,4).

The data (Table 4) show that the breccias have a somewhat wider range of composition than do the soils. Sample 14066 (21,203) is considerably lower in $Al_2O_3$ and CaO but, conversely, contains the highest amount of MgO of all the samples analyzed. The trace element composition is also indicative of the complex nature of the breccias; many of the elements have ranges that vary by twofold (Pb, Zn, Ga, Rb, Ba, and Be). The higher Cu in the two fragments of 14318 is considered due to contamination during cutting of that rock prior to receipt.

### *Comparison with mare breccias*

A comparison of the average elemental composition of the highland breccias (Apollo 14) with those from the lunar maria (Apollo 11 and 12) is given in Table 5. As we were assigned no breccias from the Apollo 12 mission, column 2 presents an average of data gathered from the literature (Morrison *et al.*, 1971; Schnetzler *et al.*, 1971; Wakita *et al.*, 1971; and Wanke *et al.*, 1971). Examination of the data shows that:

(1) Si, Al, Mg, Na, Ba, K, Li, Rb, La, Yb, and Zr show an increasing trend (Apollo 11 < 12 < 14).
(2) Fe, Ca, Ti, Mn, and Sc are progressively lower with each set of returned breccias (Apollo 11 > 12 > 14).
(3) Ni and Y are characteristically higher in the highland breccia.
(4) V has a low concentration in the highland breccias as compared to those from the maria.

### *Significant geochemical ratios*

The average compositions for the breccias (Table 4) were used to calculate a number of element ratios, and these are reported in Table 3. The following relationships are noted:

(1) Mg/Fe, Al/Ti, Ba/V, Ba/Sr, and Rb/Sr show an increasing trend (Apollo 11 < 12 < 14).
(2) Ca/Si and Cr/V decrease (Apollo 11 > 12 > 14).

Table 5. Comparison of average composition of Apollo 11, 12, and 14 breccias (oxides in weight percentage, elements in ppm).

| Constituent | Apollo 11 | Apollo 12 | Apollo 14 |
|---|---|---|---|
| $SiO_2$ | 41.80 | 46.52 | 47.78 |
| $Al_2O_3$ | 13.10 | 14.64 | 16.76 |
| $Fe_2O_3$ | 0.00 | 0.00 | 0.00 |
| FeO | 15.90 | 13.85 | 10.24 |
| MgO | 7.70 | 9.06 | 10.57 |
| CaO | 11.80 | 11.15 | 10.48 |
| $Na_2O$ | 0.46 | 0.61 | 0.83 |
| $K_2O$ | 0.16 | 0.40 | 0.56 |
| $TiO_2$ | 8.49 | 2.17 | 1.68 |
| $P_2O_5$ | $<0.2$ | — | 0.54 |
| MnO | 0.22 | 0.19 | 0.13 |
| $Cr_2O_3$ | 0.32 | 0.33 | 0.21 |
| Total | 99.95 | 98.92 | 99.78 |
| ΔRC* | +3.75 | — | +1.36 |
| Pb | $<2$ | $<2$ | 13 |
| Zn | 24 | 6.9 | 23 |
| Cu | 15 | 5.7 | 28 |
| Ga | 5.0 | 4.6 | 5.0 |
| Rb | 3.0 | 10.3 | 14 |
| Li | 12 | 22 | 26 |
| Co | 32 | 36 | 32 |
| Ni | 202 | 165 | 313 |
| Ba | 250 | 500 | 910 |
| Sr | 131 | 177 | 160 |
| V | 60 | 87 | 49 |
| Sc | 68 | 35 | 23 |
| Nb | 23 | — | 51 |
| La | 19 | 54 | 81 |
| Y | 102 | 107 | 278 |
| Yb | 14 | 19 | 22 |
| Zr | 401 | 583 | 840 |

* ΔRC = Total reducing capacity less the reducing capacity attributable to the FeO content of the sample, in % FeO.

(3) The lower Fe/Ni and Cr/Ni in the highland breccias distinguish them from those in the Mare basins.

(4) Ni/Co is higher in the Apollo 14 breccias.

*Comments on sawdust sample 14066 (31,4)*

Breccia sample 14066 (31,4) is a sample of sawdust collected during original cutting of the rock. It was felt that if the contamination introduced during processing was insignificant, the sample could reflect the bulk composition of the rock and would be useful for some scientific studies. Optical and microscopic examination revealed the presence of fine copper chips and wires, diamond chips, and a large number of matted organic fibers. Based on these observations, the difficulty encountered in homogenizing the sample, the comparison of the chemical analyses of the sawdust with chip sample 14066 (21,203), and the general disagreement among the analysts given a split of the sawdust, we strongly recommend that the sample not be used for any further lunar studies.

### *Microbreccias and clasts*

Breccia sample 14083 was received as two small separates of white and dark clasts taken from the host rock. Some of the dark clasts were coated with a fine layer of white material identical to the white clasts. Further effort to remove the white layer was discontinued because of the small quantity of sample ($\sim$200 mg). Breccia 14063, a friable, feldspathic microbreccia, is the white material observed in photographs of the large grey breccia boulders near Cone Crater. Because they are considered to be separates, microbreccias 14063 (three chip samples) and 14315 have been included with the 14083 clasts in Table 6.

Table 6. Composition of some microbreccias and clast separates (oxides in weight percentage, elements in ppm).

| Element Oxide | 14063,46 | 14063, 37A11–2 | 14063, 37B11–2 | 14083,2d Dark Clasts | 14083,2w White Clasts | 14315,4 |
|---|---|---|---|---|---|---|
| $SiO_2$ | 44.69 | 45.02 | 45.22 | 46.19 | 44.20 | 47.76 |
| $Al_2O_3$ | 22.31 | 21.53 | 21.02 | 17.16 | 22.06 | 21.31 |
| $Fe_2O_3$ | 0.00 | 0.00 | 0.00 | 0.00 | 0.00 | 0.00 |
| FeO | 6.71 | 7.00 | 6.94 | 9.66 | 7.16 | 7.82 |
| MgO | 10.80 | 10.79 | 10.40 | 11.55 | 11.48 | 8.28 |
| CaO | 12.70 | 12.40 | 12.76 | 10.76 | 12.66 | 12.77 |
| $Na_2O$ | 0.76 | 0.93 | 0.93 | 1.01 | 0.65 | 0.76 |
| $K_2O$ | 0.15 | 0.20 | 0.16 | 0.40 | 0.23 | 0.35 |
| $TiO_2$ | 1.48 | 1.58 | 1.87 | 2.21 | 0.80 | 0.80 |
| $P_2O_5$ | 0.22 | 0.29 | 0.23 | 0.63 | 0.45 | 0.23 |
| MnO | 0.08 | 0.09 | 0.09 | 0.11 | 0.06 | 0.11 |
| $Cr_2O_3$ | 0.21 | 0.19 | 0.17 | 0.21 | 0.14 | 0.23 |
| Total | 100.11 | 100.02 | 99.79 | 99.89 | 99.89 | 100.42 |
| ΔRC* | — | 0.0 | 0.0 | 0.0 | 0.0 | — |
| Pb | 2.9 | 4.2 | 3.0 | 11.0 | 6.0 | 15.0 |
| Zn | <4 | 4.0 | <4 | 4.0 | 4.0 | 34 |
| Cu | 2.3 | 3.3 | 5.0 | 11.0 | 6.0 | 10 |
| Ga | 4.8 | 6.0 | 5.5 | 4.4 | 4.6 | 5.2 |
| Li | 24 | 23 | 21 | 26 | 18 | 16 |
| Rb | 5.0 | 6.0 | 4.0 | 8.8 | 5.2 | 10 |
| Co | 16 | 17 | 16 | 25 | 17 | 30 |
| Ni | 110 | 110 | 110 | 180 | 73 | 355 |
| Ba | 250 | 380 | 315 | 940 | 500 | 410 |
| Sr | 205 | 220 | 200 | 140 | 160 | 165 |
| V | 33 | 38 | 36 | 44 | 31 | 50 |
| Be | 2.2 | 3.7 | 3.7 | 6.7 | 3.3 | 3.0 |
| B | 20 | 16 | 15 | — | — | 18 |
| Nb | 16 | 20 | 16 | 56 | 26 | 30 |
| Sc | 13 | 15 | 18 | 23 | 12 | 18 |
| La | 27 | 30 | 26 | 97 | 61 | 41 |
| Y | 94 | 130 | 110 | 300 | 190 | 155 |
| Yb | 6.8 | 9.5 | 8.2 | 31 | 16 | 11 |
| Zr | 260 | 300 | 340 | 800 | 380 | 400 |

*ΔRC = Total reducing capacity less the reducing capacity attributable to the FeO content of the sample, in % FeO.

The three samples of 14063 (46, 37A11-2, and 37B11-2) are nearly identical in composition. They are high in CaO and $Al_2O_3$ and are very similar compositionally to 14083,2w, the white clasts.

The white and dark clasts of 14083 are different. The dark clasts are enriched in Si, Fe, Na, K, Ti, P, Mn, Cr, Pb, Cu, Li, Rb, Co, Ni, Ba, V, Be, Nb, Sc, La, Y, Yb, and Zr. The white clasts are enriched in only $Al_2O_3$ and CaO, which is attributed to a larger feldspathic component.

Another interesting feature is shown by these materials. In contrast to all the other lunar samples analyzed, the microbreccias and clasts have no measurable excess reducing capacity. This unusual feature may be attributable either to a multiplicity of meteoritic events leading to their formation under unusual environmental circumstances or to a magmatic sequence different from those that generally characterize mare breccias.

Table 7. Element composition of some Apollo 14 fines (oxides in weight percentage, elements in ppm).

| Constituent | 14003,30 | 14049,37 | 14163,54 | 14240,9 | 14259,12 | 14421,23 |
|---|---|---|---|---|---|---|
| $SiO_2$ | 48.08 | 47.81 | 47.97 | 47.77 | 48.16 | 47.80 |
| $Al_2O_3$ | 17.59 | 17.44 | 17.57 | 17.99 | 17.60 | 17.40 |
| $Fe_2O_3$ | 0.00 | 0.00 | 0.00 | 0.00 | 0.00 | 0.00 |
| FeO | 10.45 | 10.44 | 10.41 | 10.02 | 10.41 | 10.48 |
| MgO | 9.27 | 9.08 | 9.18 | 9.47 | 9.26 | 9.36 |
| CaO | 11.12 | 11.13 | 11.15 | 11.25 | 11.25 | 11.26 |
| $Na_2O$ | 0.65 | 0.75 | 0.68 | 0.70 | 0.61 | 0.68 |
| $K_2O$ | 0.54 | 0.56 | 0.58 | 0.54 | 0.51 | 0.49 |
| $TiO_2$ | 1.77 | 1.79 | 1.77 | 1.67 | 1.73 | 1.74 |
| $P_2O_5$ | 0.58 | 0.56 | 0.52 | 0.55 | 0.53 | 0.44 |
| MnO | 0.14 | 0.14 | 0.14 | 0.14 | 0.14 | 0.14 |
| $Cr_2O_3$ | 0.26 | 0.22 | 0.26 | 0.23 | 0.26 | 0.27 |
| Total | 100.45 | 99.92 | 100.23 | 100.33 | 100.46 | 100.06 |
| ΔRC* | +2.70 | +3.40 | +2.37 | +2.50 | +2.94 | +3.20 |
| Pb | 10 | 11 | 11 | 12 | 8 | 8 |
| Zn | 28 | 23 | 28 | 23 | 24 | 23 |
| Cu | 16 | 18 | 17 | 16 | 19 | 21 |
| Ga | 5.0 | 4.4 | 5.5 | 7.5 | 4.4 | 6.2 |
| Rb | 13 | 14 | 13 | 13 | 12 | 13 |
| Li | 24 | 25 | 24 | 23 | 22 | 19 |
| Co | 38 | 28 | 36 | 33 | 38 | 38 |
| Ni | 430 | 295 | 400 | 320 | 440 | 335 |
| Ba | 1000 | 990 | 1100 | 1170 | 1100 | 920 |
| Sr | 135 | 150 | 140 | 390 | 150 | 170 |
| V | 58 | 42 | 57 | 52 | 62 | 62 |
| Be | 6.0 | 5.1 | 7.0 | 7.2 | 6.0 | 8.0 |
| B | 18.0 | nd† | 17.0 | nd | 17.0 | nd |
| Nb | 70 | 44 | 70 | 42 | 67 | 39 |
| Sc | 27 | 22 | 25 | 28 | 28 | 30 |
| La | 75 | 70 | 79 | 67 | 77 | 75 |
| Y | 300 | 240 | 290 | 250 | 285 | 290 |
| Yb | 27 | 19 | 28 | 23 | 30 | 21 |
| Zr | 790 | 900 | 820 | 930 | 800 | 640 |

* ΔRC = Total reducing capacity less the reducing capacity attributable to the FeO content of the sample, in % FeO.

† nd = Not detected.

## Composition of Soils

### *Elemental variation*

Analyses of six lunar soils are given in Table 7. The most striking feature of the major element composition is the uniformity of the data. $SiO_2$, for example, has a range between 47.77 and 48.16%, averaging 47.93% (standard deviation of the method is 0.22%). MnO is 0.14% in all the samples. The apparent homogeneous nature of the lunar fines permits the calculation of an average composition for the Fra Mauro soil from the data reported at the Third Lunar Conference (January 10–13, 1972, NASA, Houston, Texas). These calculations have been made for 23 different samples and splits (Jackson *et al.*, 1972; Scoon, 1972; Compston *et al.*, 1972; Klein *et al.*, 1972; Wakita *et al.*, 1972; Wanke *et al.*, 1972; and Rose *et al.*, 1972). The data give the following average values: $SiO_2$, 47.89%; $Al_2O_3$, 17.56%; FeO, 10.43%; MgO, 9.26%; CaO, 11.07%; $Na_2O$, 0.70%; $K_2O$, 0.56%; $P_2O_5$, 0.49%; $TiO_2$, 1.78%; MnO, 0.14%; and $Cr_2O_3$, 0.21%. Although Compston *et al.* and Rose *et al.*, combined, provide data for more than half the soil samples reported, the average value for each was used as a single entry into calculation of the average values.

Although little variation is shown for the major elements, the trace elements show discernible differences (Table 7). Ni, B, Nb, and Yb have uniformly higher values in samples 14003, 14163, and 14259 than those shown by the other three soils. The data suggest a bimodal distribution for Ni ($\approx$300 and >400 ppm), B (<10 and $\approx$18 ppm), Nb ($\approx$40 and $\approx$70 ppm) and Yb (<23 and >27 ppm).

### *Comparison with mare soils*

A comparison of the average composition of Apollo 14 highland soils with those of the Apollo 11 and 12 mare soils (Table 8) indicates a higher feldspar component in the Fra Mauro samples. The data show:

(1) Si, Na, K, P, Ga, Li, Ba, Be, Nb, La, Y, and Zr increase (Apollo 11 < 12 < 14).
(2) Fe, Ti, and Sc decrease (Apollo 11 > 12 > 14).
(3) Al, Pb, Zn, Cu, Ni, and Zr are all enriched in the highland soils.
(4) Mn and Cr are enriched in the mare soils. The trend of increase in concentration of the largest or highly charged cations is evident when going from a mare basaltic region to a highland area where a greater feldspathic component exists in the soils (Taylor *et al.*, 1971).

### *Significant geochemical ratios*

Some element ratios calculated from the average data for Apollo 11, 12 and 14 soils have been given in Table 3. They show that:

(1) Ca/Si and Fe/Ni decrease (Apollo 11 > 12 > 14).
(2) Al/Ti, Nb/Ti, Ba/V, Ba/Sr, and Rb/Sr increase (Apollo 11 < 12 < 14).
(3) Mg/Fe and Ni/Co are highest in the highland soils.
(4) Si/Al and Cr/Ni are higher in the mare soils.

Table 8. Comparison of average composition of Apollo 11, 12, and 14 fines (oxides in weight percentage, elements in ppm).

| Constituent | Apollo 11 | Apollo 12 | Apollo 14 |
|---|---|---|---|
| $SiO_2$ | 42.04 | 46.40 | 47.93 |
| $Al_2O_3$ | 13.92 | 13.50 | 17.60 |
| $Fe_2O_3$ | 0.00 | 0.00 | 0.00 |
| FeO | 15.74 | 15.50 | 10.37 |
| MgO | 7.90 | 9.73 | 9.24 |
| CaO | 12.01 | 10.50 | 11.19 |
| $Na_2O$ | 0.44 | 0.59 | 0.68 |
| $K_2O$ | 0.14 | 0.32 | 0.55 |
| $TiO_2$ | 7.48 | 2.66 | 1.74 |
| $P_2O_5$ | 0.12 | 0.40 | 0.53 |
| MnO | 0.21 | 0.21 | 0.14 |
| $Cr_2O_3$ | 0.30 | 0.40 | 0.25 |
| Total | 100.30 | 100.21 | 100.22 |
| *ΔRC | +4.1 | +1.3 | +2.8 |
| Pb | <2 | <2 | 10 |
| Zn | 19.0 | 6.7 | 24.9 |
| Cu | 10.0 | 10.6 | 17.9 |
| Ga | 3.8 | 4.9 | 5.5 |
| Rb | 2.7 | 8.2 | 13.0 |
| Li | 11.0 | 18.0 | 23.0 |
| Co | 24.0 | 58.0 | 35.0 |
| Ni | 185 | 195 | 370 |
| Ba | 210 | 563 | 1030 |
| Sr | 130 | 131 | 189 |
| V | 50.0 | 107 | 55.6 |
| Be | 1.6 | 5.2 | 6.6 |
| Nb | 18.0 | 38.0 | 55.3 |
| Sc | 56.0 | 40.0 | 26.7 |
| La | 16.0 | 54.0 | 74.0 |
| Y | 18.0 | 164 | 276 |
| Zr | 273 | 548 | 813 |

* ΔRC = Total reducing capacity less the reducing capacity attributable to the FeO content of the sample, in % FeO.

## Comparison of the Apollo 14 Basalts with the Breccias and Soils

Within the Fra Mauro formation, characteristic features distinguish the igneous rocks from the breccia and soils. The data given in Tables 4, 5, and 8 indicate that Apollo 14 basalts are enriched in Ca and Al when compared to the Apollo 14 breccias and soils. The breccias and soils are higher in Fe, Mg, K, Ti, P, Mn, Zn, Cu, Ga, Co, Ni, Ba, V, Nb, La, Y, and Zr. Compared to the highland igneous rocks, the breccias and soils appear to be enriched in ferromagnesian silicates.

The notable distinctions between the element ratio data for the basalts and those of the breccias and soils at Fra Mauro (Table 3) are: Rb/Sr is tenfold higher in the basalts, and Fe/Ni, Al/Ti, and Cr/Ni are at least twice as high in the basalts. Si/Al, Ni/Co, Nb/Ti, K/Rb, and Ba/Sr are somewhat lower in the basalts, whereas Ca/Si and Cr/V are lower in the breccias and soils.

The higher concentrations of Fe and Ni in the soils and breccias compared to basalts are probable indicators of a meteoritic contribution.

The soils are more homogeneous than the breccias at Fra Mauro, suggesting mixing or turnover of the soil particles by a succession of meteoritic impacts.

## Discussion

Baedecker *et al.* (1972), Dence *et al.* (1972), Kushiro (1972), Morgan *et al.* (1972), and Schnetzler *et al.* (1972) have suggested that basalt 14310 was derived from a simple melting of the soil. However, comparison of our analyses of 14310 and closely similar 14276 (Table 1) with the soil analyses (Table 7) does not support this conclusion. These basalts are distinctly richer in Ca and Al than the associated soil. The soils are richer than 14310 in those elements that would be attributable to meteoritic and KREEP components. It is also unlikely that 14310 was derived by a mixing of soil and plagioclase. Ga and Sr, which would be expected to be enriched by the addition of plagioclase, are actually depleted in the basalt relative to the soil.

Several authors have noted the difficulty of deriving 14310 from the magmas of primitive basalt 14053 (Biggar *et al.*, 1972; Kushiro, 1972; and Melson *et al.*, 1972). Basalt 14310 probably represents a more primitive magma modified by assimilation of lunar crust. The igneous nature of 14310 is indicated by the presence of xenocrysts (Hollister *et al.*, 1972; Ridley *et al.*, 1972). A near-surface intrusion of an ordinary basalt magma could result in extensive assimilation and a product having the composition suggested by Anderson *et al.* (1972), Gast (1972), and Wasserburg *et al.* (1972). If assimilation processes have been widespread (Wasserburg *et al.*, 1972), we should expect to find other basalts transitional between 14053 and 14310.

Comparison of the average breccia (Table 5) with the average soil (Table 8) at Fra Mauro shows them to be nearly identical in composition. Assuming that some were formed as early as the Imbrium event, the breccias then present a closed system, and the equal abundances of the meteoritic (Ni, Co) and KREEP elements in the breccias and soils imply that little change in the soil composition has occurred in the last $3.9 \times 10^9$ yr. This observation provides an indirect confirmation of the short-half-life, exponentially decreasing particle influx rate suggested by Shoemaker (1972) and Cliff *et al.* (1972). The greater Ni content of the soils at Fra Mauro relative to the soils in the mare suggests that the Imbrium impacting body was an iron meteorite rather than the moonlet suggested by Anderson *et al.* (1972).

## Conclusions

Preliminary evaluation of these and previously published chemical, petrographic, and photographic data suggests:

(1) The Fra Mauro highland basalts have the same nonterrestrial composition found previously in the lowland materials at the Sea of Tranquility and Ocean of Storms. Except for the relatively high Pb content of Apollo 14 materials, all the lunar samples are distinctly enriched in refractory elements and depleted in volatile elements when compared with average terrestrial basalts.

(2) The greater mineralogical complexity of the Fra Mauro highland rocks and soils implies a different mode of formation or a history different from rocks returned from lunar mare regions. The samples returned from the three missions are not members of a simple differentiation sequence as shown by the covariation of antipathetic elements. The substantial increase in Al, Pb, and Rb in Apollo 14 basalts coupled with small changes in Mg, K, and Cr is also incompatible with a simple differentiation process.

(3) The high reducing capacity of all the lunar materials strongly suggests that the elemental-oxygen stoichiometry generally accepted for terrestrial rocks may be invalid for the lunar materials and (or) a major element other than iron capable of existing in more than one valence state.

(4) The higher concentrations of Fe and Ni in the soils and breccias as compared to the basalts are probable indicators of a meteoritic contribution.

(5) Differences in composition of the Fra Mauro soils compared to mare soils include the marked enrichment of Be (Apollo 14 : 11, Be 4 : 1). The greater concentration of K, Na, Al, and P is consistent with the lighter colored soil and the higher concentrations of large or highly charged cations. In comparison to the basalts at Fra Mauro, the higher concentrations of Fe and Mg in the soils indicate a high proportion of ferromagnesian minerals.

(6) The Apollo 11, 12, and 14 basalts could not have been derived from magmas of the same composition. It is necessary to assume that the different magma source regions differ in chemical composition. The composition of the Apollo 14 basalts of the 14310 type may reflect assimilation processes.

(7) The Si/Al ratio of the mare basalts is nearly twice as high as that of the highland basalts. To a lesser degree, the mare breccias and soils also have higher Si/Al ratios than the highland materials. These data are in agreement with Apollo 15 orbiter x-ray fluorescence experiments (Adler *et al.*, 1972), which show that on a global scale the mare regions are higher in Si/Al than the highland regions.

*Acknowledgments*—This research was undertaken on behalf of the National Aeronautics and Space Administration under requisition No. 0-365-036, Order No. T-2360-A. We wish to thank our colleagues David Gottfried and Thomas Wright for critical review of the manuscript.

## References

Adler I., Trombka J., Gerard J., Lowman P., Schmadabeck R., Blogett H., Eller E., Yin L., and Lamothe R. (1972) Apollo 15 geochemical x-ray fluorescence experiments: Preliminary report. *Science* **175,** 436–440.

Anderson A. T. Jr., Braziunas T. F., Jacoby J., and Smith J. V. (1972) Breccia populations and thermal history: Nature of pre-imbrian crust and impacting body (abstract). In *Lunar Science—III* (editor C. Watkins), pp. 24–27, Lunar Science Institute Contr. No. 88.

Annell C. S. and Helz A. W. (1970a) Emission spectrographic determination of trace elements in lunar samples. *Science* **167,** 521–523.

Annell C. S. and Helz A. W. (1970b) Emission spectrographic determination of trace elements in lunar samples from Apollo 11. *Proc. Apollo 11 Lunar Sci. Conf., Geochim. Cosmochim. Acta* Suppl. 1, Vol. 2, pp. 991–994. Pergamon.

Annell C. S., Carron M. K., Christian R. P., Cuttitta F., Dwornik E. J., Ligon D. T. Jr., and Rose

H. J. Jr. (1972) Preliminary studies of six Apollo 14 lunar soils. U.S. Geological Survey Prof. Paper 800-C (in press).

Apollo 14 Preliminary Science Report (1971) NASA SP-272, National Aeronautics and Space Administration, Washington, D.C.

Baedecker P. A., Chou C. L., Kimberlin J., and Wasson, J. T. (1972) Trace element studies of lunar soils (abstract). In *Lunar Science—III* (editor C. Watkins), pp. 35–38, Lunar Science Institute Contr. No. 88.

Biggar G. M., Ford C. E., Humphries D. J., Wilson G., and O'Hara M. J. (1972) Melting relations of more primitive mare-type basalts 14053 and M (Reid, 1970); and of breccia 14321 and soil 14162 (average lunar crust?) (abstract). In *Lunar Science—III* (editor C. Watkins), pp. 74–77, Lunar Science Institute Contr. No. 88.

Cliff R. A., Lee-Hu C., and Wetherill G. W. (1972) K, Rb, and Sr measurements in Apollo 14 and 15 material (abstract). In *Lunar Science—III* (editor C. Watkins), pp. 146–149, Lunar Science Institute Contr. No. 88.

Compston W., Vernon M. J., Berry H., Rudowski R., Gray C. M., Ware N., Chappell B. W., and Kaye M. (1972) Age of petrogenesis of Apollo 14 basalts (abstract). In *Lunar Science—III* (editor C. Watkins), pp. 151–154, Lunar Science Institute Contr. No. 88.

Cuttitta F., Rose H. J. Jr., Annell C. S., Carron M. K., Christian R. P., Dwornik E. J., Greenland L. P., Helz A. W., and Ligon D. T. Jr. (1971) Elemental composition of some Apollo 12 lunar rocks and soils. *Proc. Second Lunar Sci. Conf., Geochim. Cosmochim. Acta* Suppl. 2, Vol. 2, pp. 1217–1229. MIT Press.

Cuttitta F., Rose H. J. Jr., Annell C. S., Carron M. K., Christian R. P., Dwornik E. J., and Ligon D. T. Jr. (1972) Preliminary studies of some Apollo 14 lunar rocks. U.S. Geological Prof. Paper 800-C (in press).

Dence M. R., Plant A. G., and Traill R. J. (1972) Impact-generated shock and thermal metamorphism in Fra Mauro samples (abstract). In *Lunar Science—III* (editor C. Watkins), pp. 174–177, Lunar Science Institute Contr. No. 88.

Engel A. E. J., Engel C. G., and Sutton A. L. (1972) Earth-moon genesis (abstract). In *Lunar Science—III* (editor C. Watkins), pp. 230–233, Lunar Science Institute Contr. No. 88.

Gast P. W. and McConnell R. K. Jr. (1972) Evidence for initial layering of the moon (abstract). In *Lunar Science—III* (editor C. Watkins), pp. 289–292, Lunar Science Institute Contr. No. 88.

Hollister L., Tryzcienski W. Jr., Dymek R., Kulick C., Wiegand P., and Hargraves R. (1972) Igneous fragment 14310,21 and the origin of the mare basalts (abstract). In *Lunar Science—III* (editor C. Watkins), pp. 386–389, Lunar Science Institute Contr. No. 88.

Jackson P. F. S., Coetzee J. H. J., Strasheim A., Strelow F. W. E., Gricius A. J., Wybenga F., and Kokot M. L. (1972) The analysis of lunar material returned by Apollo 14 (abstract). In *Lunar Science—III* (editor C. Watkins), pp. 424–427, Lunar Science Institute Contr. No. 88.

Klein C. Jr., Drake J. C., Frondel C., and Ito J. (1972) Mineralogy and petrology of several Apollo 14 rock types and chemistry of the soil (abstract). In *Lunar Science—III* (editor C. Watkins), pp. 455–458, Lunar Science Institute Contr. No. 88.

Kushiro I. (1972) Petrology of lunar high-alumina basalt (abstract). In *Lunar Science—III* (editor C. Watkins), pp. 477–480, Lunar Science Institute Contr. No. 88.

LSPET (Lunar Science Preliminary Examination Team) (1971) Preliminaiy examination of lunar samples from Apollo 14. *Science* **173,** 681–693.

Melson W. G., Mason B., Nelen J., and Jacobson S. (1972) Apollo 14 basaltic rocks (abstract). In *Lunar Science—III* (editor C. Watkins), pp. 535–538, Lunar Science Institute Contr. No. 88.

Morgan J. W., Krähenbühl U., Ganapathy R., and Anders E. (1972) Volatile and siderophile elements in Apollo 14 and 15 rocks (abstract). In *Lunar Science—III* (editor C. Watkins), pp. 555–558, Lunar Science Institute Contr. No. 88.

Morrison G. H., Gerard J. T., Potter N. M., Gangadharam E. V., Rothenberg A. M., and Burdo R. A. (1971) Elemental abundances of lunar soils and rocks from Apollo 12. *Proc. Second. Lunar Sci. Conf., Geochim. Cosmochim. Acta* Suppl. 2, Vol. 2, pp. 1169–1185. MIT Press.

Ridley W. I., Williams R. J., Brett R., and Takeda H. (1972) Petrology of lunar basalt 14310 (abstract). In *Lunar Science—III* (editor C. Watkins), pp. 648–651, Lunar Science Institute Contr. No. 88.

Rose H. J. Jr., Cuttitta F., Dwornik E. J., Carron M. K., Christian R. P., Lindsay J. R., Ligon D. T. Jr., and Larson R. R. (1970a) Semimicro chemical and x-ray fluorescence analysis of lunar samples. *Science* **167**, 520–521.

Rose H. J. Jr., Cuttitta F., Dwornik E. J., Carron M. K., Christian R. P., Lindsay J. R., Ligon D. T. Jr., and Larson R. R. (1970b) Semimicro x-ray fluorescence analysis of lunar samples. *Proc. Apollo 11 Lunar Sci. Conf., Geochim. Cosmochim. Acta* Suppl. 1, Vol. 2, pp. 1493–1497. Pergamon.

Rose H. J. Jr., Cuttitta F., Dwornik E. J., Carron M. K., Christian R. P., Lindsay J. R., Ligon D. T. Jr., and Larson R. R. (1970c) Errata and addenda. *Geochim. Cosmochim. Acta* **34**, p. 1367.

Rose H. J. Jr., Cuttitta F., Annell C. S., Carron M. K., Christian R. P., Dwornik E. J., Helz A. W., and Ligon D. T. Jr. (1971) Semimicro analysis of Apollo 12 lunar samples. U.S. Geological Survey Prof. Paper 750-C, pp. 182–184.

Rose H. J. Jr., Cuttitta F., Annell C. S., Carron M. K., Christian R. P., Dwornik E. J., and Ligon D. T. Jr. (1972) Compositional data for fifteen Fra Mauro Lunar samples (abstract). In *Lunar Science—III* (editor C. Watkins), pp. 660–663, Lunar Science Institute Contr. No. 88.

Schnetzler C. C. and Philpotts J. A. (1971) Alkali, alkaline earth, and rare earth element concentrations of some Apollo 12 soils, rocks and separated phases. *Proc. Second Lunar Sci. Conf., Geochim. Cosmochim. Acta* Suppl. 2, Vol. 2, pp. 1101–1122. MIT Press.

Schnetzler C. C., Philpotts J. A., Nava D. F., Thomas H. H., Bottino M. L., and Barker J. K. Jr. (1972) Chemical compositions of Apollo 14, Apollo 15, and Luna 16 material (abstract). In *Lunar Science—III* (editor C. Watkins), pp. 682–685, Lunar Science Institute Contr. No. 88.

Scoon J. H. (1972) Chemical analyses of lunar samples 14003, 14311, and 14321 (abstract). In *Lunar Science—III* (editor C. Watkins), pp. 690–693, Lunar Science Institute Contr. No. 88.

Shoemaker E. M. (1972) Cratering history and early evolution of the moon (abstract). In *Lunar Science—III* (editor C. Watkins), pp. 696–699, Lunar Science Institute Contr. No. 88.

Taylor S. R., Rudowski R., Muir Patricia, and Graham A. (1971) Trace element chemistry of lunar samples from the Ocean of Storms. *Proc. Second Lunar Sci. Conf., Geochim. Cosmochim. Acta* Suppl. 2, Vol. 2, pp. 1083–1099. MIT Press.

Wakita H. and Schmitt R. A. (1971) Bulk elemental composition of Apollo 12 samples: Five igneous and one breccia rocks and four soils. *Proc. Second Lunar Sci. Conf., Geochim. Cosmochim. Acta* Suppl. 2, Vol. 2, pp. 1231–1236. MIT Press.

Wakita H., Showalter D. L., and Schmitt R. A. (1972) Bulk REE and other abundances in Apollo 14 soils (3), clastic (1) and igneous (1) rocks (abstract). In *Lunar Science—III* (editor C. Watkins), pp. 767–770, Lunar Science Institute Contr. No. 88.

Wanke H., Wlotzka F., Baddenhausen H., Balacescu A., Spettel B., Teschke F., Jagoutz E., Kruse H., Quijano-Rico M., and Rieder R. (1971) Apollo 12 samples: Chemical composition and its relation to sample location and exposure ages, the two component origin of the various soil samples and studies on lunar metallic particles. *Proc. Second Lunar Sci. Conf., Geochim. Cosmochim. Acta* Suppl. 2, Vol. 2, pp. 1187–1208. MIT Press.

Wanke H., Baddenhausen H., Balacescu A., Teschke E., Spettel B., Quijano-Rico M., Kruse H., Wlotzka F., and Begemann R. (1972) Multielement analyses of lunar samples (abstract). In *Lunar Science—III* (editor C. Watkins), pp. 779–782, Lunar Science Institute Contr. No. 88.

Wasserburg G. J., Turner G., Tera F., Podosek A., Papanastassiou D. A., and Huneke J. C. (1972) Comparison of Rb–Sr, K–Ar, and U–Th–Pb ages: lunar chronology and evolution (abstract). In *Lunar Science—III* (editor C. Watkins), pp. 788–791, Lunar Science Institute Contr. No. 88.

Proceedings of the Third Lunar Science Conference
(Supplement 3, *Geochimica et Cosmochimica Acta*)
Vol. 2, pp. 1231–1249
The M.I.T. Press, 1972

# Composition of the lunar uplands: Chemistry of Apollo 14 samples from Fra Mauro

S. R. TAYLOR, MAUREEN KAYE, PATRICIA MUIR,
W. NANCE, R. RUDOWSKI, and N. WARE

Department of Geophysics and Geochemistry, Australian National University
Canberra, Australia

**Abstract**—Data are presented for 45 elements in two lunar soils (14003, 14163); two crystalline rocks (14072, 14310); a breccia (14047); basalt and troctolite clasts from 14321; matrix, light and dark fragments from 14306; and matrix from 14063. The most striking feature is the high concentration of alkali elements, REE, Th, U, Zr, Hf, and so forth. The samples from the smooth terrain unit of the Fra Mauro formation form a uniform compositional group. This, coupled with the diversity in composition of the fragments and clasts, suggests a thorough mixing process, identified with pre-Imbrium cratering, and the Imbrium event. Crystalline rock 14310 is interpreted as formed by impact melting. Basalt 14072 resembles Apollo 12 basalts and basalt 14321 has analogies with Apollo 11 high-K basalts, indicating that mare-type igneous activity preceded the Imbrium collision. The high concentrations of elements in the Fra Mauro samples is matched only by rock 12013 and the KREEP components of the Apollo 12 soil. This composition does not resemble tektites. The Fra Mauro chemistry differs from primitive solar nebula abundances, being strongly enriched in involatile elements, and depleted in volatiles. It is high in Ni due to addition of a meteoritic component. The enrichment of the *nonvolatile* elements over chondritic abundances is shown to depend in detail on crystal chemical factors of ionic radius and valency. This is interpreted as indicating extensive melting of the outer regions of the moon, with cooling and solidification occurring before the final stages of upland cratering and the Imbrium event.

## INTRODUCTION

UNUSUAL INTEREST attaches to the Apollo 14 landing on February 5, 1971 on the Fra Mauro Formation at 3° 40′ south latitude, 17° 27′ West longitude, since this was the first lunar mission to sample lunar upland material directly. The geochemical investigation reported here was undertaken to provide information on the nature of the pre-Imbrium surface, whether this was fractionated relative to the whole moon, or to primitive solar nebula abundances, and the effect of the Imbrium collision and of probable pre-Imbrium cratering. A preliminary discussion of the geochemistry has been given by Taylor *et al.* (1971).

## FRA MAURO FORMATION AND THE IMBRIUM COLLISION

The Fra Mauro Formation (Wilhelms, 1970) is interpreted as throwout produced during the collision that formed the Imbrium basin. This giant impact structure has at least three concentric rings with diameters of 670 km, 950 km, and 1340 km and presumably resembled the Mare Orientale structure before extensive flooding and filling by mare basalts. The Apollo 14 landing site is 550 km south of the Carpathian-Apennine rim. Impact crater theory (e.g., as developed for Copernicus by Shoemaker,

1962), based on ballistic trajectories, and ignoring the complicating effects of base surge, indicates that the Fra Mauro material would be derived from surface layers midway between the center of the Imbrium basin and its southern rim, or from deeper layers closer to the center. The heavily cratered uplands (e.g., around Clavius) provide a probable model for the condition of the pre-Imbrium surface. If we assume that brecciation extended to depths of 0.2–0.3 crater diameter, as in terrestrial examples, any simple pre-Imbrium stratigraphy would have been destroyed before the Imbrium impact. The age of the Imbrium event appears to be not older than 4 billion years (Compston, personal communication). This allows 500 million years to develop the pre-Imbrium surface, which allows adequate time for cooling and solidification of a molten crust, as postulated by Baldwin (1970). Accordingly, there is adequate opportunity for crystal-liquid fractionation to occur.

## SAMPLE LOCATION

At the Apollo 14 site, the Fra Mauro Formation is divided into three units (Fig. 1).

(1) Smooth terrain, labeled (Is), exposed in the western part of the traverse area, on which the Lunar Module (LM) landed. The samples from this unit (two soils: 14003 and 14163; two breccias: 14047 and 14306; and one of the two large (>50 g) crystalline rocks, 14310) cover the range of sample types collected. Station locations are shown on Fig. 1.

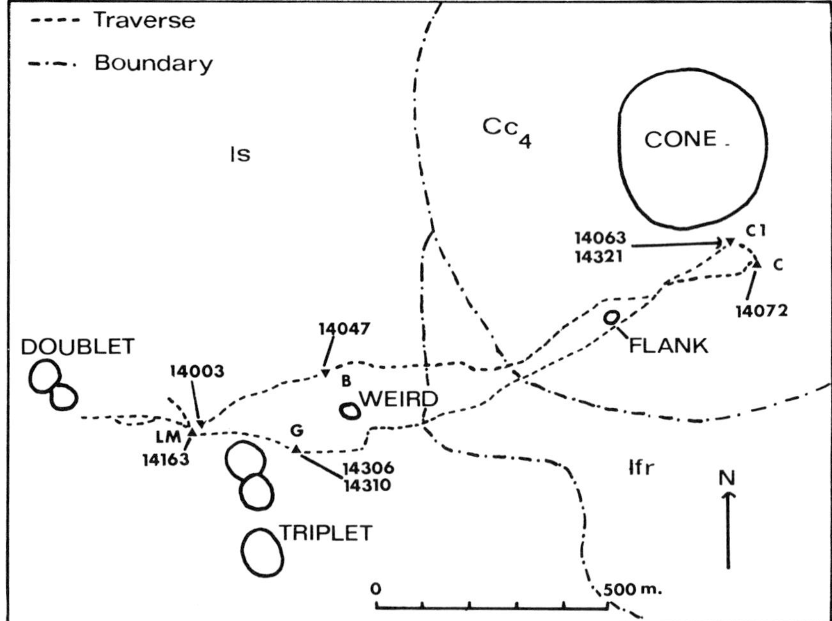


Fig. 1. Sketch map of Apollo 14 traverse area [3½°S. Lat. 17½°W. Long.] Location of stations (B) and samples (14163) is shown. Cc4 = Cone Crater materials. Is = smooth terrain material of Fra Mauro formation. Ifr = ridge material of Fra Mauro formation.

(2) Ridge material, labeled (Ifr), typical of the prominent elongate north-south ridges of the Fra Mauro formation.

(3) Material excavated by Cone Crater (340 m in diameter), labeled (Cc4). Sample 14063 and consortium samples 14072 (basalt) 14321 (breccia) were from this unit. Station locations are shown on Fig. 1.

## Analytical Methods and Results

Up to 45 elements were determined in the samples by spark source mass spectrography (SSMS) emission spectrography (ES) x-ray fluorescence (XRF) and electron microprobe (EP) techniques. The method used for each element is identified in Tables 1a and 1b. The analytical methods have been described by Taylor (1971) and Taylor *et al.* (1971). A new technique, by which amounts down to 5 mg can be

Table 1a. Analytical data for Apollo 14 samples from smooth terrain Unit (Is) of Fra Mauro Formation. Data for large cations and rare earth elements. All data in parts per million (ppm wt.) except where indicated in wt.% (K, Ca, Na) (—: No data available).

| Method | | 14163,136 | 14003,38 | 14047,32 | 14310,125 | 14306,45 Matrix | 14306,45 Dark | 14306,45 White |
|---|---|---|---|---|---|---|---|---|
| S | Cs | 0.7 | — | 1.0 | 0.7 | 1.3 | 0.75 | 1.2 |
| S | Rb | 13 | 14 | 16 | 12 | 33 | 14 | 32 |
| XP | K% | 0.43 | 0.43 | 0.40 | 0.41 | 0.66 | 0.55 | — |
| S | Ba | 710 | 760 | 730 | 610 | 1650 | 1350 | 1300 |
| S | Eu | 2.25 | 2.38 | 2.60 | 2.28 | 2.27 | 2.90 | 2.12 |
| S | Pb | 9 | — | 7 | — | 4 | 7.6 | 2 |
| E | Sr | 180 | 180 | 180 | 185 | 200 | 195 | 230 |
| XP | Ca% | 7.4 | 7.4 | 8.21 | 8.78 | 8.7 | 6.4 | 8.1 |
| XP | Na% | 0.42 | 0.42 | 0.50 | 0.47 | 0.47 | 0.77 | 0.24 |
| | K/Rb | 331 | 307 | 250 | 342 | 200 | 393 | — |
| | K/Cs | 6140 | — | 4000 | — | 5080 | 7330 | — |
| | Rb/Cs | 18 | — | 16 | — | 25 | 18 | 26 |
| | Ba/Rb | 54 | 54 | 45 | 50 | 50 | 96 | 40 |
| | Rb/Sr | 0.072 | 0.078 | 0.089 | 0.065 | 0.165 | 0.072 | 0.140 |
| | Sr/Eu | 80 | 76 | 70 | 79 | 88 | 67 | 108 |
| S | La | 64 | 81 | 80 | 72 | 100 | 110 | 55 |
| S | Ce | 200 | 228 | 235 | 207 | 299 | 300 | 148 |
| S | Pr | 26 | 29 | 26 | 23 | 32 | 41 | 19 |
| S | Nd | 102 | 121 | 102 | 91 | 119 | 151 | 71 |
| S | Sm | 29 | 36 | 28 | 23 | 26 | 42 | 20 |
| S | Eu | 2.25 | 2.38 | 2.60 | 2.28 | 2.27 | 2.90 | 2.12 |
| S | Gd | 33 | 33 | 31 | 29 | 33 | 49 | 24 |
| S | Tb | 5.0 | 4.9 | 4.7 | 4.2 | 5.1 | 8.0 | 4.0 |
| S | Dy | 32 | 33 | 33 | 29 | 41 | 46 | 29 |
| S | Ho | 8.0 | 8.3 | 8.0 | 6.8 | 9.2 | 12.3 | 6.3 |
| S | Er | 23 | 23 | 19 | 19 | 25 | 34 | 18 |
| S | Tm | 3.5 | 3.7 | 3.3 | 3.0 | 3.6 | 5.3 | 2.8 |
| S | Yb | 18.5 | 19.3 | 17 | 15 | 19 | 27 | 14 |
| S | Lu | — | — | — | — | — | — | — |
| | Σ REE | 546 | 623 | 590 | 524 | 714 | 829 | 413 |
| E | Y | 190 | 193 | 191 | 160 | 210 | 297 | 148 |
| | Σ REE+Y | 736 | 816 | 781 | 684 | 924 | 1126 | 561 |
| | La/Yb | 3.46 | 4.20 | 4.71 | 4.8 | 5.26 | 4.07 | 3.93 |
| | Gd/Eu | 14.7 | 13.9 | 11.9 | 12.7 | 14.5 | 16.9 | 11.3 |
| | Eu/Eu* | 0.25 | 0.25 | 0.25 | 0.30 | 0.25 | 0.22 | 0.33 |

Method: S = Spark source mass spectrography; E = Emission spectrography; X = X-ray fluorescence; P = Electron microprobe.

Table 1b. Data for large, high-valency cations and ferromagnesian elements for smooth terrain unit (Is) of Fra Mauro Formation. Data in parts per million (ppm wt.) except where listed as % (Ti, Fe, Mg, Al, Si). (—: no data). Major element analyses for 14163, 14003, 14310 from LSPET (1971).

| Method | | 14163,136 | 14003,38 | 14047,32 | 14310,125 | 14306,45 Matrix | 14306,45 Dark | 14306,45 White |
|---|---|---|---|---|---|---|---|---|
| S | Th | 12 | 12 | 12 | 12 | 12.7 | 18.5 | 9.1 |
| S | U | 3.2 | 3.1 | 3.2 | 3.0 — | 3.2 | 4.2 | 2.3 |
| E | Zr | 850 | 800 | 780 | 890 | 836 | 1370 | 1150 |
| S | Hf | 19.5 | 23 | 17 | 16 | 14.8 | 22 | 17 |
| S | Sn | — | — | 0.4 | — | — | — | — |
| S | Nb | 46 | 45 | 45 | 36 | 82 | 76 | 43 |
| XP | Ti% | 1.1 | 1.1 | 1.05 | 0.74 | 0.70 | 0.95 | — |
| S | W | 0.7 | 0.9 | 0.5 | 0.6 | — | 1.0 | 0.5 |
| | Th/U | 3.75 | 3.9 | 3.75 | 4.0 | 3.97 | 4.40 | 3.96 |
| | K/U | 1340 | 1390 | 1250 | 1370 | 2060 | 1300 | — |
| | Zr/Hf | 44 | 35 | 46 | 55 | 56 | 62 | 67 |
| E | Cr | 1400 | 1400 | 1220 | 1080 | 900 | 950 | 300 |
| E | V | 49 | 48 | 43 | 36 | 23 | 28 | — |
| E | Sc | 21 | 24 | 23 | 20 | 14 | 21 | — |
| E | Ni | 340 | 370 | 370 | Var.* | 200 | 430 | — |
| E | Co | 34 | 34 | 34 | 17 | 19 | 39 | — |
| E | Cu | 8 | 10 | [150] | 5 | [175] | 5 | — |
| XP | Fe% | 8.0 | 8.0 | 8.18 | 6.51 | 5.3 | 8.3 | 6.1 |
| E | Mn | 1400 | 1200 | 1100 | 930 | 700 | 1000 | 1300 |
| | Zn | — | — | — | — | — | — | — |
| X | Mg% | 5.9 | 5.9 | 5.36 | 4.75 | 3.5 | 6.1 | 5.9 |
| E | Li | 32 | 26 | — | 22 | — | — | — |
| E | Ga | 4.5 | 4.6 | 4.1 | 3.2 | 3.2 | 4.0 | — |
| XP | Al% | 9.3 | 9.3 | 9.6 | 10.6 | 11.4 | 8.0 | 10.9 |
| XP | Si% | 22.5 | 22.5 | 22.0 | 22.0 | 23.5 | 23.6 | 23.1 |
| X | P | — | — | 2200 | 1500 | — | — | — |
| | V/Ni | 0.14 | 0.13 | 0.12 | — | 0.12 | 0.065 | — |
| | Cr/V | 28 | 29 | 28 | 30 | 39 | 33 | — |
| | Ni/Co | 10.0 | 10.9 | 10.9 | — | 10.5 | 11.0 | — |
| | Fe/Ni | 235 | 216 | 221 | — | 265 | 193 | — |
| | Al/Ga ($\times 10^4$) | 2.1 | 2.0 | 2.4 | 3.3 | 3.6 | 2.0 | — |

Method: S = Spark source mass spectrography; E = Emission spectrography; X = X-ray fluorescence; P = Electron microprobe.

* Nickel content of 14310 is variable. (See Table 5).

analyzed by SSMS using tipped electrodes, will be described separately. The samples were treated as follows:

*Soils* (14003, 14163): Bulk analyses, and analysis of separated glass spherules. The latter data will be reported elsewhere.

*Crystalline rocks* (14310): Bulk sample, plagioclase, and pyroxene separates were analyzed for sample 14310. The minerals were separated by handpicking of density fractions, separated by centrifuging with bromoform. The plagioclase fraction was judged 98% pure by examination by binocular microscope. The pyroxene (pigeonite) is intimately intergrown with ilmenite and other phases and is not more than 90% pure. (14072): Two separate basalt whole rock samples were analyzed from this consortium sample.

*Breccias* (14047): The sample weight supplied (0.53 g) of this fine grained clastic rock was used to carry out a bulk analysis for major and trace elements. (14306): This polymict breccia was separated into three components (1) very fine grained matrix (<20 microns) separated by decantation following ultrasonic washing in acetone (2) melanocratic rock fragments 99% pure (3) leucocratic rock fragments 85% pure. (14063): The fine grained matrix (<20 micron) was separated by ultrasonic washing with acetone and decantation. A plagioclase separate, 99% pure was separated. (14321): A basalt clast and a troctolite (50% $An_{95}$ plagioclase, 50% olivine $Fo_{86}$) clast was analyzed from this consortium sample.

The analytical data, as elemental abundances, are presented in Tables 1a, 1b, for the western area (Is) samples and in Tables 2a, 2b, for the Cone Crater (Cc4) samples. The major element data, expressed as oxides, are given in Table 3. The mineral data are given in Table 4.

Table 2a. Analytical data for large cations and rare earth elements for Apollo 14 samples from Cone Crater unit (Cc4) of Fra Mauro Formation. (Format as for Table 1a).

| | 14063,48 Matrix | 14072 (1) | 14072 (2) | 14321,88 Basalt 2 | 14321,88 Troc. |
|---|---|---|---|---|---|
| Cs | — | — | — | 0.38 | 0.08 |
| Rb | 3.5 | 1.5 | 1.3 | 5.7 | 0.90 |
| K% | 0.09 | 0.066 | 0.066 | — | 0.05 |
| Ba | 460 | 135 | 120 | 280 | 300 |
| Eu | 3.6 | 1.02 | 0.97 | 1.50 | 2.0 |
| Pb | 1.7 | 0.8 | 1.3 | 3.5 | 0.4 |
| Sr | 235 | 110 | 106 | 120 | 150 |
| Ca% | 9.3 | 7.03 | 7.03 | — | 8.77 |
| Na% | 0.52 | 0.24 | 0.24 | — | 0.21 |
| K/Rb | 260 | 440 | 510 | — | — |
| K/Cs | — | — | — | — | — |
| Rb/Cs | — | — | — | 15 | 11 |
| Ba/Rb | 131 | 90 | 92 | 49 | 333 |
| Rb/Sr | 0.015 | 0.014 | 0.012 | 0.048 | 0.006 |
| Sr/Eu | 65 | 108 | 109 | 80 | 75 |
| La | 47 | 8.7 | 8.7 | 28 | 14 |
| Ce | 133 | 26 | 27 | 84 | 34 |
| Pr | 16.5 | 3.4 | 3.2 | 12 | 3.6 |
| Nd | 70 | 13 | 13 | 46 | 13 |
| Sm | 19 | 4.3 | 4.4 | 14 | 3.2 |
| Eu | 3.6 | 1.02 | 0.97 | 1.50 | 2.0 |
| Gd | 24 | 5.3 | 6.4 | 17 | 3.8 |
| Tb | 3.1 | 0.88 | 0.93 | 2.5 | 0.50 |
| Dy | 20 | 6.3 | 5.9 | 15 | 3.6 |
| Ho | 4.6 | 1.9 | 1.6 | 3.7 | 0.73 |
| Er | 12.5 | 4.4 | 4.7 | 9.8 | 2.3 |
| Tm | 1.8 | 0.79 | 0.76 | 1.5 | 0.43 |
| Yb | 8.2 | 4.0 | 4.0 | 7.7 | 2.2 |
| Lu | — | — | — | — | — |
| Σ REE | 363 | 80 | 82 | 243 | 83 |
| Y | 90 | 40 | 36 | 74 | 22 |
| Σ REE + Y | 453 | 120 | 118 | 317 | 105 |
| La/Yb | 5.73 | 2.18 | 2.18 | 3.64 | 6.36 |
| Gd/Eu | 6.67 | 5.19 | 6.60 | 11.3 | 1.90 |
| Eu/Eu* | 0.60 | 0.35 | 0.35 | 0.30 | 2.1 |

Table 2b. Analytical data for large high-valency cations and ferromagnesian elements for Apollo 14 samples from Cone Crater unit (Cc4) of Fra Mauro Formation. (Format as for Table 1b).

| | 14063,48 Matrix | 14072 (1) | 14072 (2) | 14321 Basalt 2 | 14321 Troc. |
|---|---|---|---|---|---|
| Th | 3.2 | 0.78 | 1.04 | 2.9 | 0.56 |
| U | 0.82 | 0.22 | 0.29 | 0.71 | 0.16 |
| Zr | 325 | 160 | 172 | 440 | 110 |
| Hf | 6.8 | 3.0 | 3.2 | 7.5 | 2.8 |
| Sn | — | — | 0.3 | 0.2 | — |
| Nb | 20 | 9.9 | 13 | 22 | 3.2 |
| Ti% | 0.76 | 1.54 | 1.54 | — | 0.11 |
| W | 0.4 | 0.2 | 0.1 | 0.2 | 0.1 |
| Th/U | 3.90 | 3.55 | 3.59 | 4.08 | 3.50 |
| K/U | 1100 | 3000 | 2300 | — | 3125 |
| Zr/Hf | 48 | 53 | 54 | 59 | 53 |
| Cr | 1100 | 2500 | 2500 | — | — |
| V | 23 | — | — | — | — |
| Sc | 12 | — | — | — | — |
| Ni | 99 | — | — | — | — |
| Co | 17 | — | — | — | — |
| Cu | 10 | — | — | — | — |
| Fe% | 4.5 | 13.9 | 13.9 | — | 3.54 |
| Mn | 650 | 2100 | 2100 | — | 4.60 |
| Zn | — | — | — | — | — |
| Mg% | 5.8 | 5.18 | 5.18 | — | 9.54 |
| Li | — | — | — | — | — |
| Ga | 3 | — | — | — | — |
| Al% | 12.2 | 5.86 | 5.86 | — | 12.3 |
| Si% | 21.3 | 21.1 | 21.1 | — | 20.3 |
| P | — | 570 | 570 | — | 130 |
| V/Ni | 0.23 | — | — | — | — |
| Cr/V | 48 | — | — | — | — |
| Ni/Co | 5.8 | — | — | — | — |
| Fe/Ni | 455 | — | — | — | — |
| Al/Ga ($\times 10^4$) | 4.1 | — | — | — | — |

Table 3. Major element data expressed as oxides, (wt.%) for Apollo 14 samples.

| | Smooth Terrain Unit (Is) | | | | Cone Crater (Cc4) | | |
|---|---|---|---|---|---|---|---|
| | 14047 | 14306 Matrix | 14306 Dark | 14306 White | 14063 Matrix | 14072 | 14321 Troctolite |
| $SiO_2$ | 47.16 | 50.3 | 50.4 | 49.4 | 45.5 | 45.15 | 43.5 |
| $TiO_2$ | 1.75 | 1.16 | 1.58 | — | 1.27 | 2.57 | 0.19 |
| $Al_2O_3$ | 18.22 | 21.6 | 15.2 | 20.7 | 23.0 | 11.07 | 23.3 |
| FeO | 10.52 | 6.83 | 10.66 | 7.87 | 5.82 | 17.82 | 4.56 |
| MnO | 0.14 | 0.13 | 0.18 | 0.17 | 0.10 | 0.27 | 0.06 |
| MgO | 8.89 | 5.76 | 10.17 | 9.79 | 9.67 | 12.16 | 15.82 |
| CaO | 11.49 | 12.12 | 9.00 | 11.28 | 13.0 | 9.84 | 12.27 |
| $Na_2O$ | 0.68 | 0.63 | 1.04 | 0.32 | 0.70 | 0.32 | 0.28 |
| $K_2O$ | 0.48 | 0.79 | 0.66 | — | 0.11 | 0.08 | 0.06 |
| $P_2O_5$ | 0.50 | — | — | — | — | 0.08 | 0.03 |
| S | 0.08 | — | — | — | — | 0.51 | — |
| $Cr_2O_3$ | 0.15 | 0.13 | 0.15 | 0.05 | 0.16 | 0.12 | — |
| Σ | 100.06 | 99.46 | 99.05 | 99.6 | 99.3 | 99.99 | 100.06 |
| Analyst | BWC | NW | NW | NW | NW | BWC | NW |
| Method | XRF | EP | EP | EP | EP | XRF | EP |

—: No data; XRF = X-ray fluorescence; EP = electron microprobe.

Table 4. Mineral data, in parts per million, for Apollo 14 samples. Analyses by spark source mass spectrograph.

| | Plagioclase 14063 | Plagioclase 14310 | Pyroxene 14310 |
|---|---|---|---|
| Rb | 1.4 | 2.7 | — |
| Ba | 500 | 500 | — |
| Pb | — | 0.7 | — |
| Sr | 500 | 320 | 49 |
| La | 10 | 14 | 24 |
| Ce | 29 | 40 | 70 |
| Pr | 2.8 | 4.3 | 7.4 |
| Nd | 11 | 16 | 31 |
| Sm | 2.9 | 4.4 | 9.8 |
| Eu | 4.9 | 2.7 | 0.68 |
| Gd | 2.7 | 5.7 | 12 |
| Tb | 0.44 | 0.80 | 2.1 |
| Dy | 2.7 | 5.5 | 17 |
| Ho | 0.44 | 1.1 | 4.3 |
| Er | 1.4 | 3.1 | 12 |
| Tm | 0.26 | 0.50 | 2.0 |
| Yb | 1.4 | 2.6 | 11.2 |
| Σ REE | 70 | 101 | 204 |
| Y | 13 | 38 | 130 |
| Σ REE + Y | 83 | 139 | 334 |
| Th | — | 1.6 | 2.7 |
| U | — | 0.45 | 0.74 |
| Zr | 39 | 196 | 310 |
| Hf | — | 2.1 | 6.6 |
| Th/U | — | 3.6 | 3.6 |
| Zr/Hf | — | 93 | 47 |

## Major Elements

### *Smooth terrain unit (Is)*

The most striking geochemical feature is the high abundance of most elements relative to the concentrations, reported for the Apollo 11 and 12 mare basalts (see summary by Levinson and Taylor, 1971). The initial conclusion from the data in Tables 1a and 1b is that all the bulk samples analyzed (soils, breccia, crystalline rock) show the same distinctive features and are generally similar in bulk composition. If these samples are typical, the smooth terrain unit is well mixed, for the separated fragments of 14306 show diverse compositions. These Apollo 14 samples define a different type of lunar nonmare basic rock, here termed Fra Mauro basalt following the usage of Reid (1971). In terms of major element composition, it is characterized by high Al ($\sim 18\%$ $Al_2O_3$) low Fe ($\sim 10\%$ FeO) and high K ($\sim 0.5\%$ $K_2O$) in comparison with Apollo 11 or 12 mare basalts. The bulk analysis of 14047 given in Table 1a, Table 1b, and Table 3 appears to be typical. Such material was identified in minor amount in Apollo 11 soil and was the significant exotic component in Apollo 12 soil called KREEP, norite, nonmare basalt, and so forth. The widespread distribution of this material indicates that it was a significant component of the pre-Imbrium crust.

*Cone Crater (Cc4)*

No bulk samples were analyzed from this unit and the clasts show considerable diversity in composition. The fine grained matrix from 14063 is plagioclase rich, and resembles the composition of the feldspathic glasses of probable highland derivation (Type E basalts) except for higher Na, K, and Ti. The basalt sample 14072 closely resembles the Apollo 12 mare basalts, indicating that material equivalent to mare basalts was being formed prior to the Imbrium collision. The basalt clast from 14321 resembles the Apollo 11 high-K basalts, while the troctolite sample is unique.

## TRACE ELEMENTS

Most of the discussion in this section refers to the samples from the smooth terrain unit (Is) which is considered to be typical of the Fra Mauro formation. Occasional reference is made to the Cone Crater clasts.

*Large cations*

A major distinction between the Fra Mauro basalts and the mare basalts is the high content of potassium and related elements. The K is typically 4000–4300 ppm, a factor of 8 higher than in the Apollo 12 basalts and nearly twice the 2400 ppm average of the Apollo 11 high-K suite. The K/Rb ratios are lower, averaging 300 compared to 430 in Apollo 11 high-K, 890 in Apollo 11 low-K and 470 average in Apollo 12 basalts. The Cs is relatively more abundant with Rb/Cs ratios of 20, contrasting with 35 in Apollo 11, 25 in Apollo 12, but the same as in rock 12013 (Anders *et al.*, 1971). Barium concentrations are likewise high, typically 700–800 ppm, compared with 100 in Apollo 11 low-K, 280 ppm in Apollo 11 high-K and about 60 ppm in Apollo 12 rocks, but comparable with the abundances in rock 12013. All these elements are concentrated by processes of fractional crystallization, or in the early stages of partial melting, and the Apollo 14 material is highly evolved geochemically compared to primitive abundances (see later). A striking feature is the constancy of the Sr/Eu ratio at about 80, extending the relationship noted by Willis *et al.* (1971), further confirming the parallel behavior of these two elements. Note that Eu enriches slightly relative to Sr in comparison with chondrites and achondrites, where the ratio is about 110. The parallel or near-parallel behavior of Sr and Eu is evidence that they are contained in the same phases and that Eu is mainly divalent. Since Sr is involatile, this relationship suggests that the depletion of Eu in the surface regions of the moon is not due to pre-accretion loss as a volatile compound, and that the whole moon is not depleted in Eu.

*Rare earth elements (REE)*

The smooth terrain unit soils, breccias and rocks show extreme enrichment in REE. Figure 2 shows the REE abundances, divided element by element by the REE abundances in chondrites (Haskin *et al.*, 1966). The soils, breccias, and rock 14310

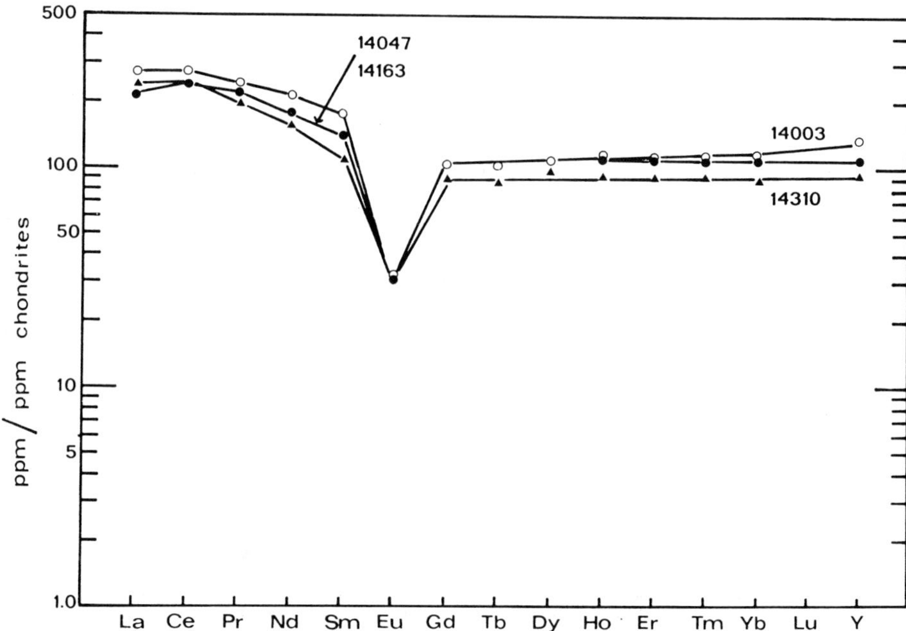


Fig. 2. Rare earth elements in Apollo 14 soils (14003, 14163), breccia (14047), and crystalline rock (14310), divided element by element, by the abundances in chondritic meteorites, from Haskin *et al.* (1966).

all show the same relative patterns. The small REE (Gd − Yb + Y) show monotonic enrichment by factors of about 100 over chondrites. Eu shows a marked depletion (30X chondrites) and the large REE (Sm-La) show a smooth progressive enrichment from Sm to Ce (240X chondrites) with a slight fall to La. These patterns are similar to those observed for rock 12013 and are close to the absolute abundances for the light (most abundant) part of 12013 and to the light grey fines 12033. These Apollo 14 samples are enriched absolutely (and relatively for the large REE La–Sm) compared to the Apollo 11 and 12 basalts. The parallel enrichment of large REE in Apollo 12 soil 12070 is due to the addition of a similar component. A comparison for the Apollo 11, 12, and 14 soils is given in Fig. 1 of Taylor *et al.* (1971). The three phases separated from 14306 all show parallel patterns, with the matrix pattern close to the Apollo 14 soil (Fig. 3). Figure 4 shows the plagioclase and clinopyroxene patterns relative to the total rock, showing the extreme enrichment in Eu in the plagioclase, and the flat pattern for the other REE. The basalt from Cone Crater 14072 (Fig. 5) although generally resembling Apollo 12 basalts has a low Eu anomaly, with some enrichment of the large REE. In contrast, the basalt clast from 14321 has a pattern parallel to 14310 but with lower abundances. It differs also from the Apollo 11 high-K basalts in being enriched in the large REE (La–Sm), and this feature appears typical of all the Apollo 14 samples, including the troctolite clast from 14321.

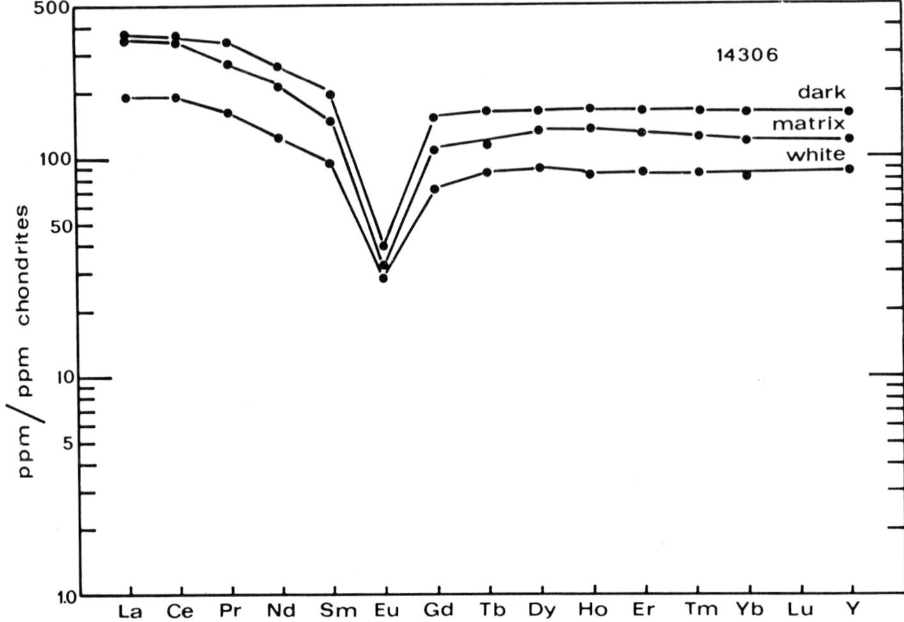


Fig. 3. Rare earth abundances in matrix, light and dark fragments of breccia 14306, ratioed to chondrites.

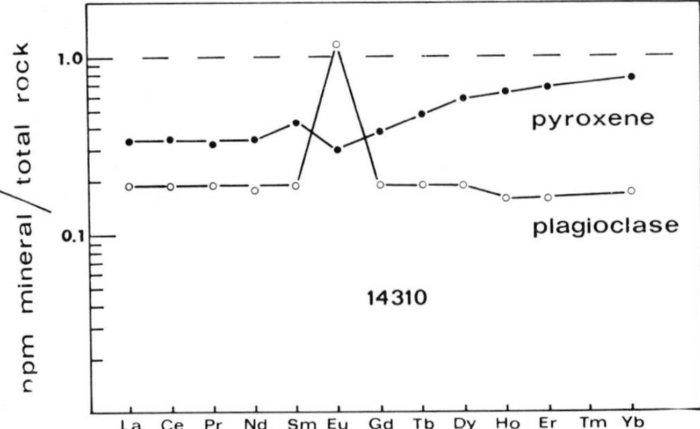


Fig. 4. Rare earth abundances in plagioclase and pyroxene from 14310, ratioed to whole rock abundances.

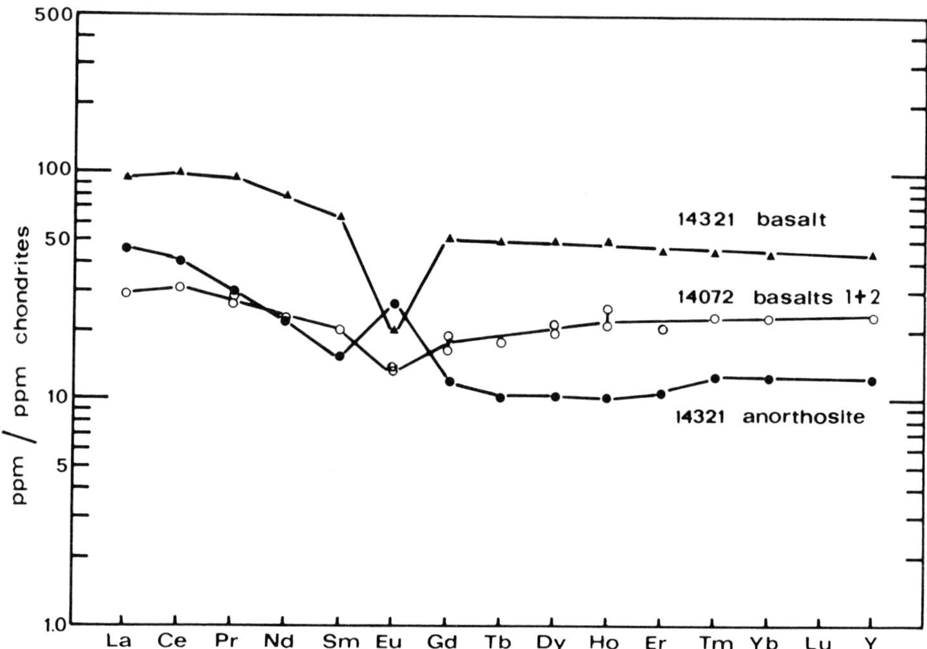


Fig. 5. Rare earth abundances in mare type basalt 14072 from Cone Crater, and in basalt and troctolite (= anorthosite) clasts from breccia 14321.

*Large high-valency cations*

In the Apollo 14 samples Th and U are enriched, but the Th/U ratio is not different from other lunar samples, averaging about 3.7. Zr/Hf ratios, averaging 50, are higher than those in Apollo 11 (average 27) or 12 (average 43) rocks. The significance of these variations will be discussed elsewhere, but it is suggested that the cause of the variation is that Zr is partly trivalent, is present in different phases from Hf in the lunar interior, so that the Zr/Hf ratio reflects the degree of partial melting. The low content of volatiles (e.g., F) the absence of water, and the great fluidity of the lunar magmas will inhibit complex formation. Thus conventional explanations for terrestrial variations in the ratio are probably not applicable to the lunar environment.

The K/U ratios are low, similar to the Apollo 12 values of ~ 1400 but much lower than the Apollo 11 values of 3000 except for 14072 basalt.

*Ferromagnesian elements*

The Fra Mauro samples contain the same characteristic lunar features as do the Apollo 11 and 12 samples. These include high Cr, high Cr/V and Cr/Ni ratios. A notable feature of the Apollo 14 samples is the higher content of nickel in all Apollo 14 samples discussed here. The nickel content is typically in the range 300–400 ppm, with Fe/Ni ratios of about 240. The distribution of nickel in the crystalline rock 14310 is exceedingly sporadic. The data are shown in Table 5 for eight replicate analyses for

Table 5. Distribution of Ni, Cr, and Sc in 8 replicate analyses of crystalline rock 14310. (All data in ppm.) (Method Es.)

| | Cr | Sc | Ni | Cr/Ni | Cr/Sc |
|---|---|---|---|---|---|
| 1 | 1110 | 20.2 | 100 | 11 | 55 |
| 2 | 1000 | 19.6 | 163 | 6.1 | 51 |
| 3 | 1110 | 19.6 | 210 | 5.3 | 57 |
| 4 | 1130 | 20.1 | 107 | 10.6 | 56 |
| 5 | 1070 | 19.6 | 430 | 2.5 | 55 |
| 6 | 1040 | 19.1 | 115 | 9.0 | 54 |
| 7 | 1120 | 20.0 | 164 | 6.8 | 56 |
| 8 | 1130 | 20.5 | 164 | 6.9 | 55 |

Cr, Sc, and Ni. Cr, Sc, and the Cr/Sc ratio are uniform, but Ni varies by a factor of 4.3, as does the Cr/Ni ratio. Probe analyses indicate that metallic Fe is distributed sporadically in 14310. It is suggested that the Ni in 14310 is distributed in a metal phase which has failed to homogenize during mixing. This is contributory evidence for 14310 being an impact melt. The overall high abundance of Ni in Apollo 14 samples is attributed to derivation from meteoritic impact, consistent with the extensive multiple brecciation of the samples. If all the Ni is derived from chondritic meteorites, there is an overall meteoritic component of up to 3% (assuming 1.1% Ni in Type 1 carbonaceous chondrites). This is about double the meteoritic component in Apollo 11 and 12 soils.

The Cr values of 1000–1400 ppm, although about half the values in Apollo 11 and 12 rocks, are high enough to preclude much removal of Cr-containing phases by fractional crystallization.

### *Comparison of Apollo 14 and 12 soils*

Figure 6 shows a comparison between Apollo 14 soil 14163 and Apollo 12 soil 12070, showing the uniform enrichment by a factor of two for a wide variety of elements: Li, Na, K, Rb, Cs, Ba, REE, Th, U, Zr, and Hf.

This is consistent with the derivation of the Apollo 12 soil component containing these elements by dilution of Apollo 14 type compositions. Note that nickel is enriched in Apollo 14, due to a greater meteoritic component, but that the other ferromagnesian elements Cr, Sc, and V are depleted by a factor of about 2 in Apollo 14. Cobalt, which is partly added by meteoritic addition, is less depleted. Note that it would not be possible to produce this rather uniform difference for so many elements by simple removal or addition of individual major mineral phases (olivine, pyroxene, plagioclase), so that the Apollo 12 and 14 samples cannot be linked by simple fractional crystallization models.

## Comparison with Chondritic Abundances

What is the relationship of the Fra Mauro basaltic material to primitive solar nebula abundances? Figure 7 compares the abundances in Apollo 14 fines, with those in Type 1 carbonaceous chondrites, considered to be the closest approximation to

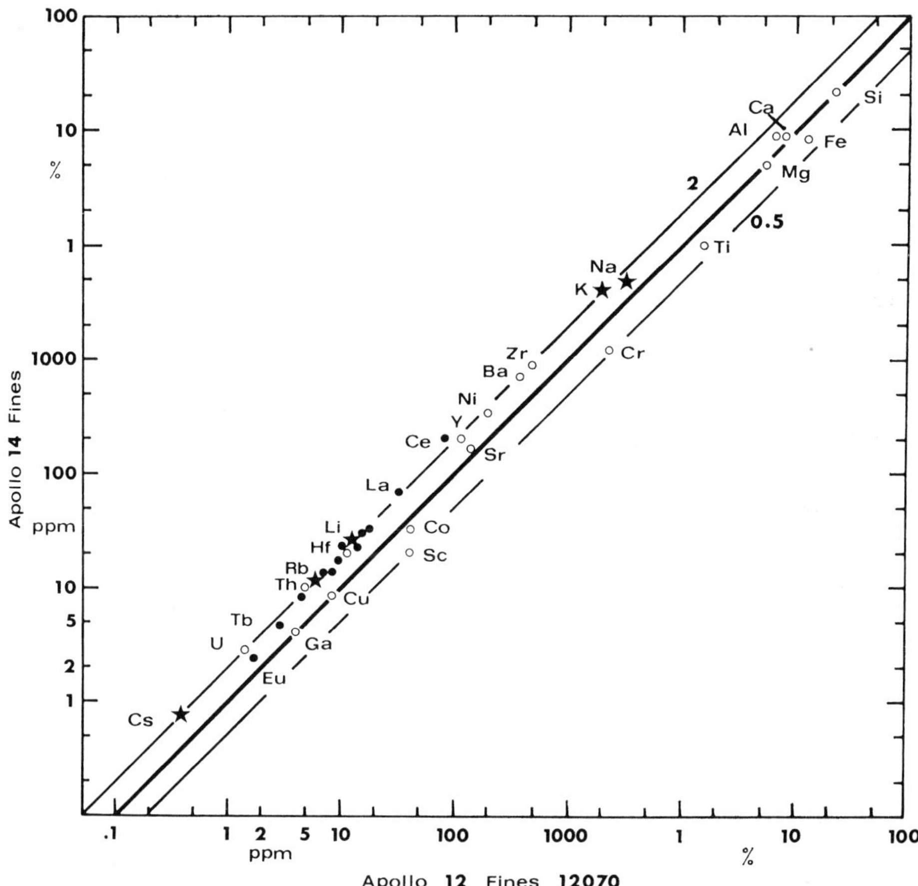


Fig. 6. Comparison of the abundances in Apollo 12 and 14 soils, showing uniform enrichment by a factor of two for many elements in Apollo 14. Filled circles indicate rare earth elements.

primitive solar nebula composition. The similarity between the solar and these meteoritic abundances (excluding H, He, and the noble gases) constitutes the rationale for this choice (Anders, 1971). Since the Imbrium event excavated material which probably resembled the lunar uplands, and since the Fra Mauro material predates the mare basalts as sampled by Apollo 11 and 12, it is of considerable interest to look for evidence of more primitive compositions. None are found. As is clear from Fig. 7, the Apollo 14 samples display, generally in more extreme form, the same compositional characteristics as the mare basalts. There is the same striking enrichment or depletion depending on volatility. The data are given in Table 6, showing the enrichment or depletion factors relative to the Type 1 carbonaceous chondrites. The dependence in detail, on volatility is shown by the relative order of depletion of the

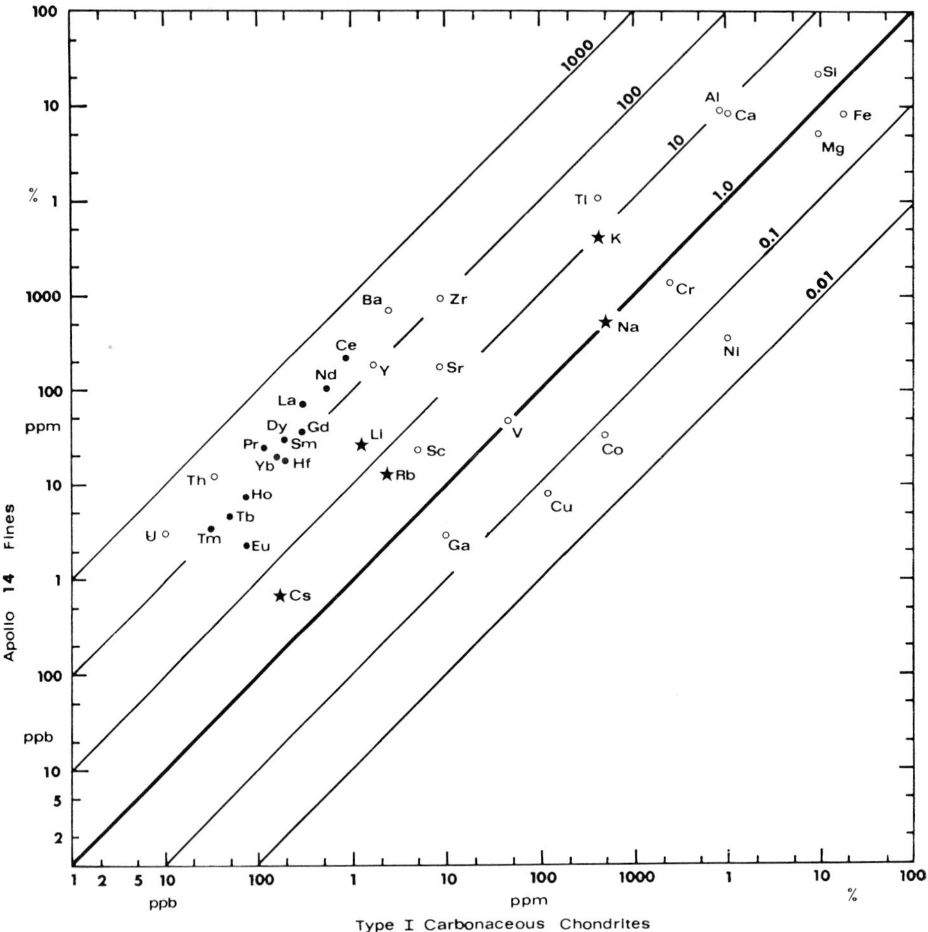


Fig. 7. Comparison of element abundances in Apollo 14 with abundances in Type 1 carbonaceous chondrites (Data from Mason, 1971). Points lying to the left of the diagonal line are enriched in Apollo 14 and those lying to the right are depleted, relative to the meteoritic abundances. The other lines indicate enrichment factors of 10,100, and 1000 and depletion factors of 0.1 and 0.01.

alkali elements Cs > Rb > K > Li. This effect has not been masked by subsequent fractionation processes that probably included partial melting. Note, however that Na does not fit the sequence. This is probably due to the fact that Na will enter major mineral phases in the lunar interior, while the other alkalies will be in minor and interstitial phases. Thus they will be released as a group during partial melting, while Na will be selectively retained. The extreme enrichments and depletions observed in Apollo 14 element abundances indicate that (a) the Fra Mauro formation material is not primitive, and (b) it cannot be representative of the whole moon (e.g., Th, and U are too high).

Table 6. Average composition at Apollo 14 site, Fra Mauro Formation. (1. Abundances in parts per million (wt.) or wt.%. 2. Values in column 1 divided by abundances in Type 1 carbonaceous chondrites giving relative enrichment or depletion. Meteoritic data from Mason (1971).)

| | 1 | 2 | | 1 | 2 |
|---|---|---|---|---|---|
| Cs | 0.7 | 3.8 | La | 74 | 246 |
| Rb | 13 | 5.4 | Ce | 217 | 258 |
| K | 0.40% | 10 | Pr | 26 | 217 |
| K/Rb | 310 | — | Nd | 104 | 179 |
| Ba | 700 | 280 | Sm | 29 | 138 |
| Sr | 180 | 21 | Eu | 2.4 | 32 |
| Rb/Sr | 0.072 | — | Gd | 32 | 100 |
| Ca | 8.2% | 7.7 | Tb | 4.7 | 96 |
| Na | 0.50% | 1.0 | Dy | 32 | 103 |
| Li | 27 | 21 | Ho | 7.8 | 107 |
| Th | 12 | 270 | Er | 21 | 100 |
| U | 3.2 | 260 | Tm | 3.4 | 103 |
| Th/U | 3.75 | — | Yb | 17.5 | 103 |
| K/U | 1250 | — | Σ REE | 571 | — |
| Zr | 900 | 100 | Y | 190 | 106 |
| Hf | 19 | 100 | Σ REE + Y | 761 | -- |
| Zr/Hf | 47 | — | La/Yb | 4.2 | — |
| Nb | 44 | — | Gd/Eu | 13.4 | — |
| W | 0.7 | — | Eu/Eu* | 0.27 | — |
| Ti | 1.05% | 24 | | % | — |
| Cr | 1400 | 0.58 | $SiO_2$ | 47.2 | — |
| V | 44 | 1.0 | $TiO_2$ | 1.8 | — |
| Sc | 22 | 4.3 | $Al_2O_3$ | 18.2 | — |
| Ni | 340 | — | FeO | 10.5 | — |
| Co | 34 | 0.07 | MgO | 8.9 | — |
| Fe | 8.2% | 0.44 | CaO | 11.5 | — |
| Mg | 5.4% | 0.56 | $Na_2O$ | 0.50 | — |
| Cu | 8 | 0.06 | $K_2O$ | 0.48 | — |
| Ga | 4 | 0.4 | $P_2O_5$ | 0.50 | — |
| Al | 9.6% | 11 | S | 0.08 | — |
| Si | 22% | 2.1 | $Cr_2O_3$ | 0.15 | — |

*Behavior of involatile elements*

The involatile elements are enriched as a group relative to the primordial abundances. In addition to this order of magnitude enrichment, a second trend appears when the data are considered from the basis of factors significant in crystal chemistry. In Table 6 the data are divided, element by element, by the abundances from Mason (1971) in Type 1 carbonaceous chondrites. The data for the nonvolatile elements only, are shown on Fig. 8, plotted according to ionic radius and valency. Contour intervals are drawn for successive enrichment factors over the primitive abundances. Those elements close in radius and valency to $Fe^{2+}$ and $Mg^{2+}$, major constituents of the lunar interior, are depleted. The other involatile elements are progressively enriched in the Apollo 14 Fra Mauro material, as their ionic radius and valency differ from those of $Fe^{2+}$ and $Mg^{2+}$. The order of enrichment, for the divalent elements is Ba > Sr > Ca; for the trivalent elements, large REE > small REE + Y > Sc; for the quadrivalent elements Th = U > Zr = Hf. There are no exceptions to this rule for the elements considered. Note that these effects appear between elements which do not differ appreciably in volatility.

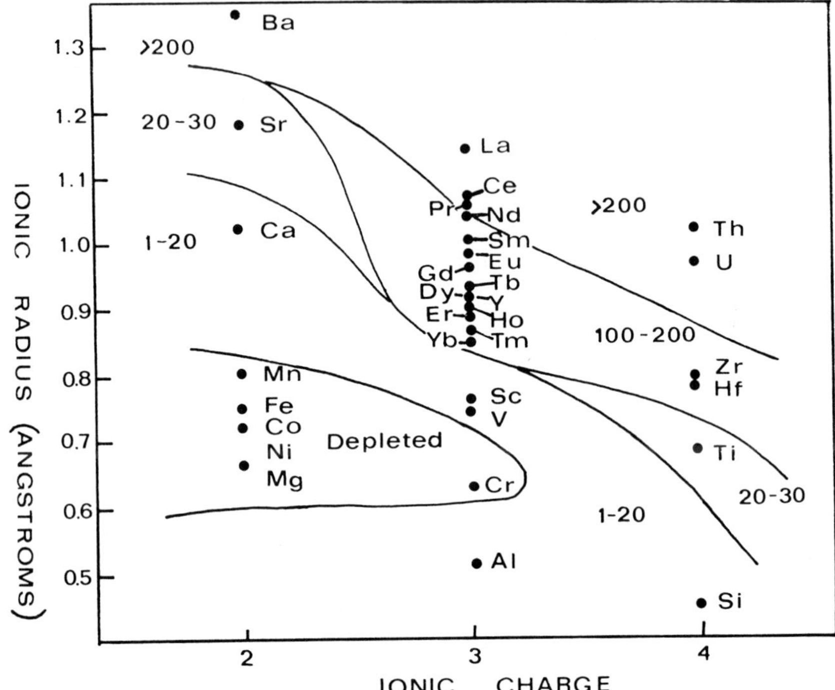


Fig. 8. Relative enrichment or depletion of involatile elements in Apollo 14 samples relative to abundances in Type 1 carbonaceous chondrites, based on ionic radius and valency. Contour lines separate areas of differing degrees of enrichment. Note the progressive enrichment as radius and valency differ from those of Mg and Fe. Data from Table 6.

The assumption is made, for the *nonvolatile* elements considered here, that there has been no, or little *relative* fractionation between the whole moon and the primitive abundances (e.g., they both have the same relative REE abundances). If this assumption of no relative fractionation of involatile elements between the moon and the primordial abundances is valid, then the data in Table 6 and Fig. 8 suggest that there has been a strong upward concentration in the moon of involatile elements on the basis of properties which are important in crystal-liquid fractionation. Thus in addition to the loss of volatile elements at or before accretion, there has been an internal primitive lunar fractionation, resulting in strong surficial concentration of elements based on crystal chemical properties. This is strong evidence for extensive melting of the outer layers of the moon to a depth of a few hundred kilometers at least.

## Comparison of Fra Mauro Material with Tektites

The suggestion by O'Keefe (1971) that tektite glass was present in sample 12013 was refuted by King *et al.* (1971). Taylor (1971) and Levinson and Taylor (1971) have discussed the tektite problem in the light of the Apollo 11 and 12 data, concluding, from the differences in chemistry, age, oxygen and lead isotopes, that tektites do not come from the moon. The different chemical composition of the Fra Mauro samples makes it worthwhile to repeat the comparison. Figure 9 shows a plot of the Fra Mauro abundances, divided element by element, by average australite abundances from Taylor (1968). The same relative pattern emerges if data from other tektite groups or microtektites (Frey *et al.*, 1970) are used, but it is appropriate to use the australite data since they have been considered a prime candidate for a lunar origin. It is clear from Fig. 9 that there are many discrepancies in composition. Thus for example, Cr is enriched by a factor of 20 in the lunar samples, while Rb is depleted by an order of magnitude. There are wide discrepancies for nearly all the elements considered, both major and minor and no sign of any first order similarities in composition, which would form the necessary basis for the lunar hypothesis. The REE data have been discussed by Frey *et al.* (1970) and Taylor *et al.* (1971). The characteristic Eu depletion in the lunar sample is not present in the tektites. It should be noted here that the small amounts of "potash granite" glass noted in the Apollo 14 fines by Reid (1971) does not resemble the composition of any tektite group being notably low in Fe and Mg, and high in K, a comparison also remarked of the terrestrial granites. It appears from all scientifically measurable parameters that the lunar samples differ decisively from tektites.

## Conclusions

(1) The Fra Mauro formation, although containing clasts of many differing rocks, is uniform in overall composition, and has thus been well mixed, consistent with heavy pre-Imbrium cratering, in addition to mixing during the Imbrium event.

(2) The element abundances for the alkalies, Ba, REE, Th, U, Zr and Hf are greatly enriched in the Fra Mauro formation, compared to the mare basalts. They resemble the concentrations in rock 12013, and the KREEP components of the Apollo 12 soil. The same distinctive features of lunar chemistry (high Cr, Eu depletion, low content of volatile elements) are present.

(3) The Apollo 11 and 12 mare basalts are not derived from material of Fra Mauro composition, or vice versa. The Fra Mauro basalt is a distinctive chemical type.

(4) The crystalline rock 14310 is probably an impact melt.

(5) Mare-type basalts are present as clasts, and hence some were forming prior to the Imbrium collision.

(6) The object responsible for the Imbrium collision does not appear to differ from the other meteoritic component in the lunar soil with regard to nickel content.

(7) Material of tektite composition is not present at the Fra Mauro site.

(8) The Fra Mauro material shows major differences from primitive solar nebula abundances related to relative volatility.

(9) There is evidence for extensive pre-Imbrium melting and differentiation. Impact melting appears unlikely to produce extensive fractionation. The involatile elements have been fractionated according to size and valency. This is consistent with

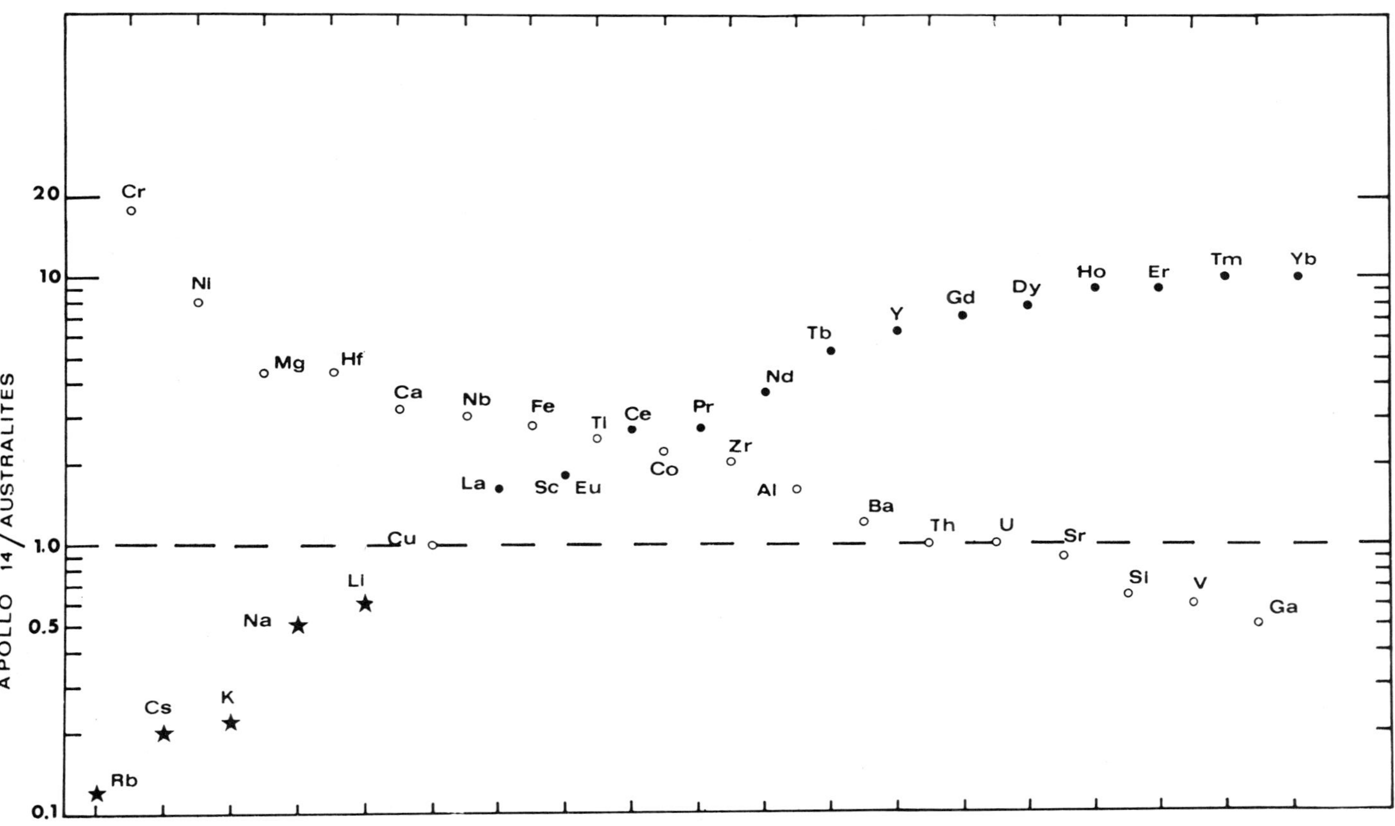


Fig. 9. Comparison of Apollo 14 composition with that of the australite group of tektites. Data from Taylor (1968). Note the wide divergence in composition.

crystal-liquid fractionation probably during partial melting. The high-Cr contents argue against extensive crystal fractionation during solidification. The overall evidence is consistent with melting of the outer layers of the moon, but these must have cooled and solidified before the end of the major cratering and the Imbrium collision.

## References


Anders E. (1971) How well do we know "cosmic" abundances. *Geochim. Cosmochmi. Acta* **35,** 516–522.

Anders E. (1971) Volatile and siderophile elements in lunar rocks: comparison with terrestrial and meteoritic basalts. *Proc. Second Lunar Sci. Conf., Geochim. Cosmochim. Acta* Suppl. 2, Vol. 2, pp. 1021–1036. MIT Press.

Baldwin R. B. (1970) Summary of arguments for a hot moon. *Science* **170,** 1297–1300.

Frey F. A., Spooner C. M., and Baedecker P. A. (1970) Microtektites and tektites: a chemical comparison. *Science* **170,** 845–847.

Haskin L. A., Frey F. A., Schmitt R. A., and Smith R. H. (1966) Meteoritic, solar, and terrestrial rare earth distribution. *Phys. Chem. Earth* **VII,** 169–321.

King E. A., Martin R., and Nance W. B. (1970) Tektite glass not in Apollo 12 sample. *Science* **170,** 199–200.

Levinson A. A. and Taylor S. R. (1971) *Moon Rocks and Minerals.* Pergamon.

LSPET (Lunar Sample Preliminary Examination Team) (1971) Preliminary examination of lunar samples from Apollo 14. *Science* **173,** 681–698.

Mason B. (1971) *Handbook of Elemental Abundances in Meteorites.* Gordon and Breach.

O'Keefe J. A. (1970) Tektite glass in Apollo 12 sample. *Science* **168,** 1209–1210.

Reid A. M. (1971) Apollo 14: Nature and origin of rock types in soil from the Fra Mauro Formation. *Earth Planet. Sci. Lett.* (in press).

Reid A. M., Ridley W. I., Harmon R. S., Warner J., Brett R., Jakes P., and Brown R. W. (1971) Feldspathic basalts in lunar soils and the nature of the lunar highlands. *Earth Planet. Sci. Lett.* (in press).

Shoemaker E. M. (1962) Interpretation of lunar craters. *Physics and Astronomy of the Moon*, pp. 283–359. Academic.

Taylor S. R. (1968) Geochemistry of Australian impact glasses and tektites (australites). *Origin and Distribution of the Elements* (editor L. H. Ahrens), pp. 533–541. Pergamon.

Taylor S. R. (1971) Geochemical analysis by spark source mass spectrography II photoplate data processing. *Geochim. Cosmochim. Acta* **35,** 1187–1196.

Taylor S. R., Muir P., and Kaye M. (1971a) Trace element chemistry of Apollo 14 lunar soil from Fra Mauro. *Geochim. Cosmochim. Acta* **35,** 975–981.

Taylor S. R., Rudowski R., Muir P., Graham A., and Kaye M. (1971b) Trace element chemistry of lunar samples from the Ocean of Storms. *Proc. Second Lunar Sci. Conf., Geochim. Cosmochim. Acta* Suppl. 2, Vol. 2, pp. 1083–1099. MIT Press.

Wilhelms D. E. (1970) Summary of lunar stratigraphy—telescopic observations. U.S. Geol. Surv. Prof. Paper 599-F, pp. 1–47.

Willis J. P., Ahrens L. H., Danchin R. V., Erlank A. J., Gurney J. J., Hofmeyr P. K., McCarthy T. S., and Orren M. J. (1971) Some interelement relationships between lunar rocks and fines, and stony meteorites. *Proc. Second Lunar Sci. Conf., Geochim. Cosmochim. Acta* Suppl. 2, Vol. 2, pp. 1123–1138. MIT Press.

Proceedings of the Third Lunar Science Conference
(Supplement 3, *Geochimica et Cosmochimica Acta*)
Vol. 2, pp. 1251–1268
The M.I.T. Press, 1972

# Multielement analyses of lunar samples and some implications of the results

H. Wänke, H. Baddenhausen, A. Balacescu, F. Teschke,
B. Spettel, G. Dreibus, H. Palme, M. Quijano-Rico,
H. Kruse, F. Wlotzka, and F. Begemann

Max-Planck-Institut für Chemie (Otto-Hahn-Institut),
Mainz, Germany

**Abstract**—The concentrations of 48 major, minor, and trace elements were determined in 10 lunar samples from Apollo 14 and 15 and in the achondrites, Juvinas, and Kapoeta. Among these elements F was studied more extensively to clarify some discrepancies in the literature about this element.

A three-component mixing model for the lunar soils and breccias was set up. In this way we proved large amounts of KREEP and of anorthosite to be present in all Apollo 15 soils and breccias. The anorthosite content is highest for samples from the mountain front and decreases with increasing distance from the front.

It is shown that besides U, Th, and Hf also Ta and W are related with KREEP. KREEP seems to have been added to all mare basalts; noticeable especially for the elements K and La.

La, U, Th, and many other trace elements are found to be enriched in the Apollo 14 samples up to a factor 400 as compared to chondritic values. The nearly identical composition of Apollo 12 KREEP and the Apollo 14 soils and breccias and their distribution as Imbrium ejecta is taken as evidence that the layer of KREEP rocks (norite layer see Wood *et al.*, 1971) must be at least a factor of 10 thicker than the upper limit of 1.5 km that one would get for a layer enriched 400 times in certain elements. We therefore strongly support the conclusion of Gast and McConnell (1972) about an initial chemical layering of the moon. Specifically we suggest the upper 200 km to have had initially an about eucritic or two times eucritic abundance of the elements in question including K. With this assumption the thermal history of the moon could also be brought in accord with the observations. Furthermore, it is shown that with such a composition even small planetesimals would melt within a few hundred million years after accretion.

A number of elements that are of meteoritic origin were studied also and attempts were made to distinguish between Imbrian and post-Imbrian components.

## Introduction and Experimental Results

The major and many of the minor and trace elements have been determined in seven Apollo 14 and three Apollo 15 samples. For comparison studies on trace element concentrations two basaltic achondrites, the eucrite Juvinas and the howardite Kapoeta, were analyzed in addition. We also measured the distribution of the rare earth elements in the silicate phase of mesosiderites; we will report on these measurements in detail elsewhere. For nearly all measurements neutron activation methods were used. About half of the data of all 48 elements analyzed were obtained via intrumental activation analyses using fast and thermal neutrons. For the other half of the elements radio-chemical separations were carried out after additional neutron activation. Most of the measurements were carried out on one and the same sample.

Only Li and Cl were measured on separate small aliquants. The techniques applied were nearly identical to those described previously (Rieder and Wänke, 1969; Wänke

Table 1. Major and trace elements in lunar samples and in basaltic achondrites.

| | Fines | | Clastic rocks: micro breccias | | | Clastic rocks: igneous fragm. | | Fines | | | Achondrites | | |
|---|---|---|---|---|---|---|---|---|---|---|---|---|---|
| *Percent* | 14163,126 | 14259,24 | 14066,31 | 14305,81 | 14321 184,25 | 14321, 184,1E | 14321,223 | 15021,80 | 15471,29 | 15601,45 | Eucrite Juvinas | Howardite Kapoeta | Accuracy (%) |
| O | 43.7 | 43.8 | 44.4 | 44.2 | 44.0 | 42.2 | 42.3 | 43.8 | 42.7 | 41.8 | 42.8 | 43.3 | 1 |
| Mg | 5.6 | 5.6 | 6.8 | 6.2 | 6.8 | 5.4 | 5.3 | 6.3 | 7.0 | 6.8 | 4.0 | 9.5 | 4 |
| Al | 9.6 | 9.2 | 8.4 | 8.6 | 8.7 | 6.5 | 6.4 | 7.5 | 7.1 | 5.7 | 7.1 | 4.4 | 2 |
| Si | 22.6 | 22.2 | 23.0 | 22.6 | 22.3 | 22.3 | 22.2 | 22.0 | 22.6 | 21.8 | 23.0 | 23.5 | 1 |
| Ca | 7.3 | 7.7 | 7.2 | 7.1 | 6.7 | 7.4 | 8.1 | 6.4 | 7.1 | 6.7 | 7.7 | 3.72 | 15 |
| Ti | 0.87 | 0.85 | 0.60 | 0.91 | 0.78 | 1.07 | 1.08 | 1.05 | 0.69 | 0.94 | 0.38 | 0.18 | 15 |
| Fe | 8.1 | 8.0 | 7.8 | 8.1 | 8.3 | 12.8 | 12.3 | 11.6 | 12.8 | 15.0 | 14.5 | 13.6 | 3 |
| *ppm* | | | | | | | | | | | | | |
| Li | 27 | 14 | | | | | | | | | 5.1 | | 15 |
| F | 145 | 106 | — | 119 | — | — | 131 | 59 | — | 45 | 19 | 13 | 8 |
| Na | 5000 | 4710 | 6230 | 5670 | 5880 | 4060 | 3720 | 2890 | 2470 | 2260 | 2800 | 2050 | 5 |
| Cl | 280 | 40 | | | | | | | | | 18 | | 20 |
| K | 4430 | 4020 | 7870 | 5300 | 4630 | 1430 | 1050 | 1650 | 1000 | 870 | 222 | 180 | 5 |
| Sc | 22.8 | 23.0 | 20 | 24 | 20 | 61 | 55 | 26.6 | 31 | 36.3 | 28.5 | 20.7 | 5 |
| Cr | 1290 | 1310 | 1190 | 1330 | 1070 | 3070 | 2800 | 2500 | 2980 | 3540 | 2090 | 4750 | 5 |
| Mn | 1010 | 1040 | 920 | 1040 | 970 | 1820 | 1720 | 1420 | 1560 | 1880 | 3990 | 3830 | 5 |
| Co | 43 | 36 | 30 | 31 | 39 | 30 | 28 | 42 | 45 | 51 | 5.8 | 28 | 5 |
| Ni | 400 | 380 | — | 200 | — | — | 39 | — | — | 90 | | 410 | 15 |
| Cu | 15.6 | 12.3 | — | 10.9 | — | — | 8.2 | — | — | 8.2 | 1.65 | 4.83 | 10 |
| Zn | — | 22 | — | 2.1 | — | — | 5.1 | — | — | 1.33 | 1.1 | 4.2 | 10 |
| Ga | 8.3 | 7.6 | — | 5.0 | — | — | 4.0 | — | — | 3.4 | 2.16 | 1.04 | 10 |
| Ge | — | 0.59 | — | 0.44 | — | — | 0.47 | — | — | 0.20 | 0.06 | 0.31 | 10 |
| As | 0.087 | 0.076 | — | 0.077 | — | — | 0.0087 | — | — | 0.0153 | 0.18 | 0.092 | 15 |
| Rb | 23 | 19 | — | 25 | — | — | 6.7 | — | — | | — | | 15 |
| Sr | 180 | — | — | 190 | — | — | 100 | — | — | 104 | | — | 10 |
| Pd | 0.028 | 0.020 | — | 0.015 | — | — | 0.001 | — | — | 0.0062 | ≤0.001 | 0.016 | 20 |
| In | 1.01 | 0.034 | — | 0.0047 | — | — | 0.0065 | — | — | — | ~0.0015 | ~0.002 | 10 |
| Cs | 0.74 | 0.67 | — | 1.36 | — | — | 0.32 | — | — | 0.27 | | | 20 |
| Ba | 775 | — | — | 830 | — | — | 100 | — | — | 120 | — | | 15 |
| La | 68 | 66 | 75 | 109 | 91 | 21 | 22 | 25 | 14.6 | 12.9 | 2.58 | 1.39 | 5 |
| Ce | 180 | 170 | 200 | 200 | 230 | 65 | 60 | 65 | 49 | 35 | | — | 15 |
| Pr | 22 | 21 | — | 26 | — | — | 7.4 | — | — | 4.6 | 0.94 | 0.83 | 10 |
| Nd | 130 | 120 | — | 140 | — | — | | | — | — | 5.07 | | 20 |
| Sm | 28 | 26 | — | 23 | — | — | 8.6 | 8.8 | — | 4.6 | 1.48 | 0.54 | 10 |
| Eu | 2.45 | 2.29 | 2.76 | 2.60 | 3.03 | 1.40 | 1.17 | 1.34 | 1.12 | 1.03 | 0.61 | 0.32 | 5 |
| Gd | 35 | 34 | — | 38 | — | — | 14.4 | — | — | 9.4 | 2.3 | 1.25 | 10 |
| Tb | 6.6 | 6.3 | 7.8 | 7.4 | 8.9 | 2.5 | 2.5 | | 1.78 | 1.60 | 0.60 | 0.29 | 10 |
| Dy | 40 | 38 | 39 | 43 | 48 | 13 | 13 | 15 | 8.8 | 8.6 | 3.2 | 1.26 | 10 |
| Ho | 6.6 | 6.0 | — | 6.5 | — | — | 2.2 | 2.3 | — | 1.4 | 0.42 | 0.23 | 10 |
| Er | 28 | 26 | — | 32 | — | — | 9.3 | — | — | 6.1 | 2.3 | 0.79 | 10 |
| Yb | 23.5 | 21.5 | 25.1 | 24.2 | 28 | 7.5 | 6.8 | 8.3 | 5.05 | 4.71 | 1.72 | 0.89 | 5 |
| Lu | 2.7 | 2.7 | 3.6 | 3.5 | 3.9 | 1.20 | 0.94 | 1.20 | 0.72 | 0.77 | 0.28 | 0.14 | 10 |
| Hf | 23 | 21 | 30 | 26 | 31 | 8.0 | 7.1 | 8.8 | 5.5 | 4.4 | 1.3 | 0.6 | 10 |
| Ta | 3.2 | 2.8 | 3.6 | 3.2 | 4.0 | 1.2 | 1.0 | 1.20 | 0.68 | 0.46 | 0.12 | 0.10 | 15 |
| W | 1.95 | 1.70 | — | 1.94 | — | — | 0.55 | — | — | 0.28 | 0.041 | 0.036 | 10 |
| Ir | 0.019 | 0.016 | — | 0.010 | — | — | 0.0011 | — | — | 0.0041 | 0.0270 | 0.020 | 20 |
| Au | 0.0061 | 0.0055 | — | 0.0067 | — | — | 0.0006 | — | — | 0.0016 | 0.0079 | 0.0068 | 15 |
| Th | 15.9 | 14.3 | — | 17.4 | — | — | 2.6 | — | — | 1.8 | — | | 10 |
| U | 4.07 | 3.79 | — | 5.15 | — | — | 0.54 | — | — | 0.57 | 0.089 | 0.051 | 10 |
| Σ (%) | 99.0 | 98.4 | 99.5* | 99.3 | 98.9 | 98.7 | 98.7 | 99.5 | 100.8 | 99.6 | 100.4 | 99.4 | |

*The values for sample 14066 were corrected for 6% contamination, see text. No entry means not determined. Samples 14066,31,14321,184,25, 14321,184,1E, 15021,80, and 15471,29 were analyzed by INAA only.

*et al.*, 1970). We only made a slight change in the chemical processing in order to obtain in addition data on Zn. Another new element which we included in our work is F. Only very few data of this element on lunar samples exist and the discrepancies between different authors are very large. For our fluorine determinations on 50 mg aliquots we used pyrohydrolysis for the separation and a specific ion electrode for the determinations.

For the samples 14066,31, 14321,184,1E, and 14321,184,25, only instrumental neutron activation was applied prior to destructive analysis by Baedecker *et al.* (1972).

The results are given in Tables 1 and 2. As we have also determined oxygen, we could calculate the sums directly. We did not determine P, S, and Zr, which elements amount in the KREEP-rich samples up to about 0.5%. In these cases we can theoretically only reach sums of about 99.5%.

For the sawdust sample 14066,31 the total sum found was only 93%, excluding about 3% of copper. The remaining 3% must come from elements not in our list, most probably organic contaminations. We also saw in the instrumental neutron analysis a substantial $\gamma$-peak from gold. Previously, we had noted that all our rock samples had been highly contaminated with gold. We suspect that copper and gold comes from the sawing at L.R.L., Houston. In Table 1 we have for sample 14066 corrected all elements to a sum of 99.5%.

Only a few comparisons with the data of other authors could be made; the agree-

Table 2. Fluorine content in lunar samples. A $H_2O$ leach on sample 14163 yielded 9 ppm, the residue 135 ppm given a total of 144 ppm fluorine.

| Sample | Fluorine (in ppm) Indiv. run | This work mean | Literature | Sample | Fluorine (in ppm) Indiv. run | This work mean | Literature |
|---|---|---|---|---|---|---|---|
| *Igneous rocks* | | | | *Fines* | | | |
| 10017 | 75 | 78 | | 10084 | 87 | 86 | 96[b] |
| | 80 | | | | 84 | | 254–900[d] |
| 10044 | 82 | 83 | | 12001 | 58 | 61 | |
| | 84 | | | | 64 | | |
| 10057 | 92 | 88 | 90[c] | 12070 | 60 | 60 | 60[a] |
| | 84 | | | | 60 | | 63[b] |
| 12053 | 35 | 33 | 51[a] | | | | 241[d] |
| | 31 | | | 14163 | 150 | 145 | |
| | 33 | | | | 140 | | |
| 12063 | 51 | 53 | 84[b] | 14259 | 102 | 106 | |
| | 55 | | | | 109 | | |
| 14321* | 130 | 131 | | 15021 | 58 | 59 | |
| | 132 | | | | 59 | | |
| *Breccias* | | | | | | | |
| 10018 | 101 | 101 | | 15601 | 44 | 45 | |
| 14305 | 114 | 119 | | | 46 | | |
| | 124 | | | | | | |

* Igneous fragment of clast (14321,223).
[a] Morrison *et al.* (1971).
[b] Bouchet *et al.* (1971).
[c] Smales *et al.* (1971).
[d] Reed and Jovanovic (1971).

ment is as previously excellent for most of the elements especially with the results given by Brunfelt *et al.* (1971). From the three samples of 14321 that we analyzed, two were igneous fragments with nearly identical chemical composition highly differing from the third one, a microbreccia, whose composition was found to be similar to that of the soil samples.

In Table 2 all the fluorine data on lunar samples obtained by us are listed. It can be seen that our data agree reasonably with that of Morrison *et al.* (1971), Bouchet *et al.* (1971), and Smales *et al.* (1971), but the F-values given by Reed and Jovanovic (1971) appear to be far too high.

## MIXING MODEL FOR SOILS AND BRECCIAS OF APOLLO 12 AND APOLLO 15

Mixing models for the lunar soils have been presented by various authors (for example, Goles *et al.* (1971), Hubbard *et al.* (1971a), Lindsay (1971), Schonfeld (1972), and Wänke *et al.* (1971)). In principle two types of methods seem to be feasible:

Method I: Calculation of the various mixtures from known endmembers (single fit).

Method II: Calculation of the unknown endmembers from the composition of the various mixtures (multiple fit).

Most of the mixing ratios published are derived in a way described here as Method I. We prefer Method II, because the mixtures (i.e., the soils and breccias) are rather well represented by millions of grains as compared to the endmembers being single rocks. The three components denoted by A, B, and K (for example anorthosite, basalt, and KREEP) are represented by the corners of a triangle. Normal to this plane we plot the concentration values for each element in each corner. In this way, every element fixes an individual element plane (Fig. 1a). To find the mixing ratio of a particular soil, we compute *that* point within the triangle, whose concentration values (in perpendicular direction) make the best fit to the individual element planes. Now the mixing ratios are given by the three distances from the computed point D to the

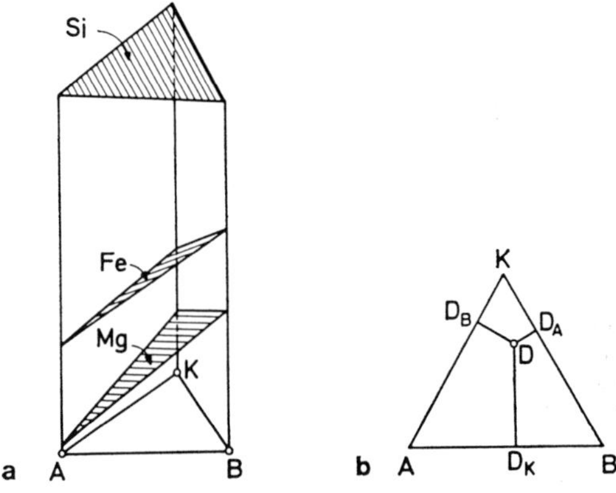


Fig. 1. The three dimension mixing model.

opposite sides of the triangle (Fig. 1b). So far, the calculation is identical to Method I (single fit).

Method II: We now take the computed mixing points of the various samples of Method I and their concentration values as a first approximation, which should be improved by a further least squares method of fitting. This seems to be necessary, because the standard deviations, obtained by the above method, are rather high, showing especially that the element concentrations of selected basalt and KREEP samples do not represent the ideal composition.

Subsequently, we compute a new series of element planes, fixing only the concentration values of anorthosite and fitting the "best" planes to those concentration points in space, which belong to the same element. In this computation the values of basalt and KREEP have been omitted.

To these newly computed planes we fit new concentration perpendiculars. Those of basalt and KREEP, together with anorthosite, fix a new triangle in the base plane.

This method of alternating computation of element planes and concentration perpendiculars is repeated until the desired accuracy is gained. The deviations between measured and calculated values obtained by this method are much smaller than those of the first method.

The results of our computations for Apollo 12 and 15 samples are compiled in Table 3 and in Fig. 2. For the Apollo 12 soils the use of the third component anorthosite compared to the two component model presented in our Apollo 12 paper results in a slight reduction of the mean standard deviations. The calculated anorthosite contents range only between 2 and 7%.

The following elements were used for the Apollo 15 soils, mainly from LSPET (1972): Si, Ti, Al, Fe, Mg, Ca, Na, K, P, Mn, Cr, Y, and Zr. Unfortunately, we do not have data of the REE for most of the samples. Also our Method II is in this case

Table 3. Computed compositions of Apollo 12 and 15 soils and breccias. For the Apollo 12 samples the "multiple fit" data are considered to be more accurate as the "single fit" data. For comparison the results of the two component model from our Apollo 12 paper are also given. The Apollo 12 samples are listed according to increasing KREEP content, the Apollo 15 samples according to increasing anorthosite content. $V(z)$ stands for mean standard deviation of the computed values for all elements from the experimental values.

| | Three components | | | | | | | | Two components | | |
|---|---|---|---|---|---|---|---|---|---|---|---|
| | Method II (multiple fit) | | | | Method I (single fit) | | | | Multiple fit | | |
| | $V(z)$ (%) | KREEP (%) | Basalt (%) | Anorthosite (%) | $V(z)$ (%) | KREEP (%) | Basalt (%) | Anorthosite (%) | $V(z)$ (%) | KREEP (%) | Basalt (%) |
| 12037 | 10 | 20 | 78 | 2 | 22 | 16 | 77 | 7 | 14 | 25 | 75 |
| 12070 | 7 | 31 | 63 | 6 | 16 | 25 | 64 | 11 | 9 | 40 | 60 |
| 12044 | 8 | 31 | 65 | 4 | 11 | 27 | 65 | 8 | 8 | 40 | 60 |
| 12042 | 15 | 32 | 61 | 7 | 23 | 26 | 62 | 12 | 16 | 42 | 58 |
| 12032 | 9 | 48 | 49 | 3 | 16 | 44 | 49 | 7 | 9 | 62 | 38 |
| 12073 | 10 | 50 | 43 | 7 | 19 | 42 | 46 | 12 | 11 | 65 | 35 |
| 12033 | 7 | 58 | 38 | 4 | 19 | 48 | 42 | 10 | 7 | 74 | 26 |
| 12034 | 5 | 63 | 33 | 4 | 10 | 52 | 38 | 10 | 6 | 80 | 20 |
| | | | | | | | | | *Station* | | |
| 15601 | 6 | 20 | 79 | 1 | 14 | 9 | 86 | 5 | Rille | | |
| 15558 | 6 | 43 | 54 | 3 | 9 | 19 | 68 | 13 | Rille | | |
| 15501 | 4 | 35 | 60 | 5 | 9 | 15 | 71 | 14 | Scarplet Crater | | |
| 15021 | 6 | 45 | 48 | 7 | 10 | 20 | 62 | 18 | LM | | |
| 15471 | 11 | 25 | 62 | 13 | 14 | 11 | 71 | 18 | Dune Crater | | |
| 15301 | 11 | 33 | 52 | 15 | 15 | 15 | 62 | 23 | Spur Crater | | |
| 15265 | 3 | 60 | 25 | 15 | 15 | 28 | 41 | 31 | Front | | |
| 15271 | 3 | 49 | 35 | 16 | 15 | 22 | 48 | 30 | Front | | |
| 15101 | 8 | 37 | 40 | 23 | 15 | 17 | 49 | 34 | St. George | | |

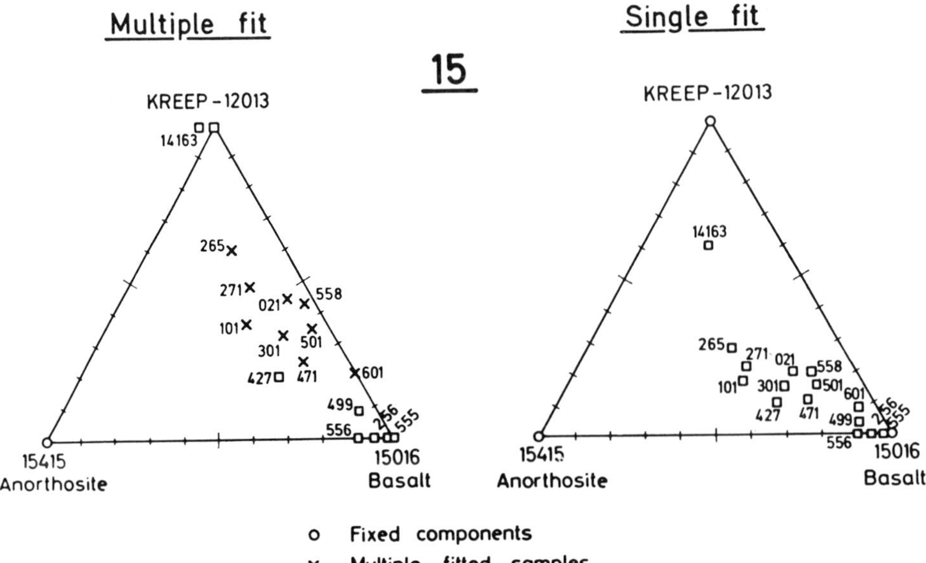


Fig. 2. Mixing diagrams for Apollo 15 soils and breccias.

no real advantage compared with Method I, as most of the data are from single measurements with limited accuracy. The computed anorthosite concentrations vary here between 1 and 23%. The highest anorthosite concentrations were found for the samples from locations close to the Apennine front decreasing to almost zero with increasing distance from this front. The results can be understood by assuming the Apennine to be rich in anorthosite. Besides anorthosite large amounts of KREEP are present in the Apollo 15 soils, but no clear cut distribution pattern seems visible. Qualitatively our compositions agree with the finding of Schonfeld (1972), but his anorthosite values are higher compared with our values derived via Method II. They agree almost exactly with those from Method I.

From Table 4 the advantage of Method II is clearly observable for the Apollo 12 samples where all the data used are mean values of many determinations of a large number of authors. For KREEP we had used the same elemental concentrations as in our Apollo 12 paper (data from Hubbard *et al.*, 1971b). According to the present knowledge our assumed composition of KREEP*, i.e., the mean of KREEP, rock

Table 4. Comparison of the measured composition of Apollo 12 KREEP* with the computed composition and the Apollo 14 samples 14301 and 14163. KREEP* is the mean of KREEP and rock 12013 total and 12013 dark. (Data from Hubbard *et al.* (1971)). The Zr value for 14163 is the mean of the data given by Jackson *et al.* (1972), Wakita *et al.* (1972) and Gast *et al.* (1972); all others from this paper.

| | Si | Fe | Mg (%) | Al | Ca | Na | K | Mn | Sc | La (ppm) | Zr | Ba | W | U |
|---|---|---|---|---|---|---|---|---|---|---|---|---|---|---|
| KREEP* (measured) | 25.2 | 8.4 | 4.6 | 7.7 | 6.4 | 7670 | 7650 | 1200 | 27 | 91 | — | 1460 | — | 7.9 |
| KREEP (computed) | 21.6 | 8.1 | 5.0 | 9.0 | 8.0 | 6380 | 5580 | 1245 | 26 | 99 | 933 | 1070 | 2.3 | 4.5 |
| 14305 | 22.6 | 8.1 | 6.2 | 8.6 | 7.1 | 5670 | 5300 | 1040 | 24 | 109 | — | — | 1.9 | 5.2 |
| 14163 | 22.6 | 8.1 | 5.6 | 9.6 | 7.3 | 5000 | 4430 | 1010 | 23 | 68 | 880 | 1100 | 2.0 | 4.1 |

12013, and dark portion of rock 12013, was not quite correct as the light portion of rock 12013 represents a different type of rock with no obvious relation to KREEP. The computed concentrations for Apollo 12 KREEP end member are therefore considerably different and they all lie much closer to the concentrations found in 14163 and 14305, proving once more that the soil and breccias from the Fra Mauro site are nearly identical with the Apollo 12 KREEP.

## The KREEP Elements

Besides K, the REE and P a number of elements show a clear correlation with KREEP. Obviously KREEP itself consists of a number of mineral components. From the work of Keil *et al.* (1971), Brown *et al.* (1971), Schnetzler and Philpotts *et al.* (1971), Taylor *et al.* (1971), and others, it is evident that in lunar basalts La and the other lighter REE have to be attributed mainly to the phosphate minerals apatite and whitlockite, while for the heavier REE the contribution from pyroxene becomes important (for chemical evidence see Figs. 3 and 4); Eu and K reside predominantly in feldspars. (See Fig. 5.) In the KREEP rocks probably only the phosphate and feldspar minerals contribute significantly to the K, REE, and P.

With our data we find that fluorine clearly correlates with La (Fig. 6). Assuming all F and Cl found in sample 14259 to be derived from apatite, the corresponding amount of P would be 623 ppm. As Jackson *et al.* (1972) found 1839 ppm P in 14259, one would

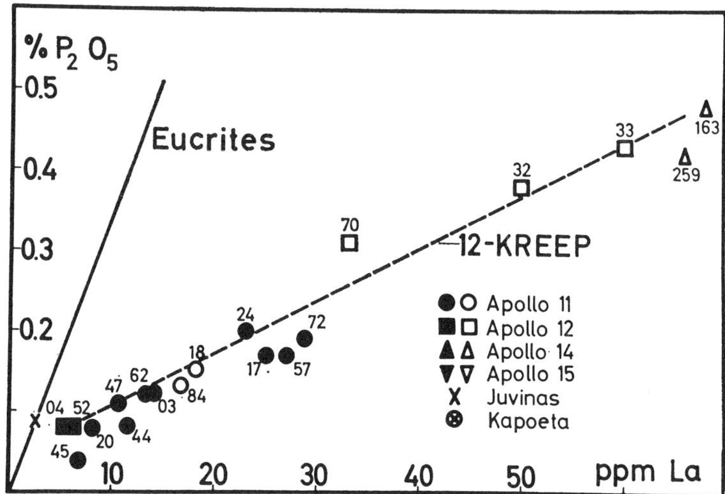


Fig. 3. $P_2O_5$ versus La. A linear correlation is observed for nearly all types of lunar samples. Data for $P_2O_5$ from Compston *et al.* (1970), Compston *et al.* (1971), Engel *et al.* (1971), Hubbard *et al.* (1972), Jackson *et al.* (1972), LSPET (1972), Scoon (1971), and Willis *et al.* (1971). Data for La of Apollo 11 samples are mean values of all authors (see compilation Warner, 1971); of Apollo 12, 14, and 15 from Wänke *et al.* (1971) and this paper. Full symbols refer to igneous rocks, open symbols to soils and breccias. Apollo 11 high-K rocks: 17, 22, 24, 48, 49, 57, 69, 71, and 72. Apollo 11 low-K rocks: 03, 20, 44, 45, 47, 50, 58, and 62.

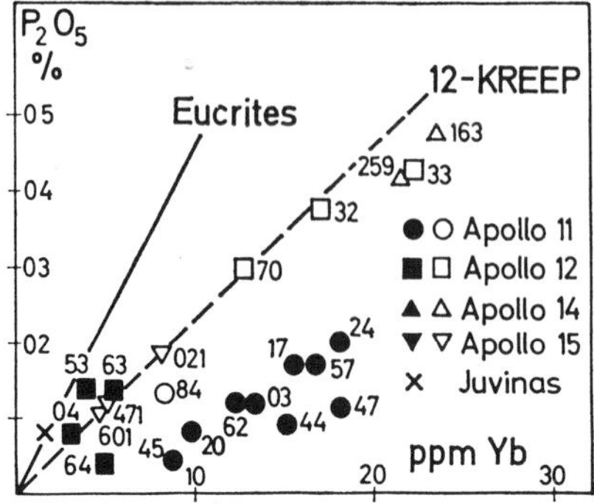


Fig. 4. $P_2O_5$ versus Yb. For the Apollo 12 rocks and all soils and breccias rich in KREEP a linear correlation is observed. The relative enrichment of Yb in the Apollo 11 rocks indicates that here the phosphate minerals are not the only important contributors.

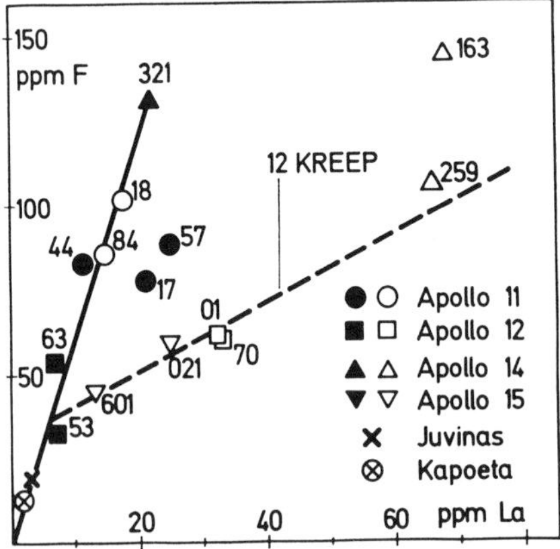


Fig. 5. F versus La. A good correlation is found for igneous rocks, while KREEP has a different F/La ratio. Data for F from this paper, those for La see legend of Fig. 3. Also see Fig. 3 for symbols.

estimate an apatite to whitlockite ratio of about 1:2. The distribution of the various lunar samples in the F–La plot is similar to that in the Eu–La plot (Fig. 6). (See below in Genesis of Lunar Rocks.)

As noticed by various authors among the elements determined by us U, Th, and Hf correlate with the KREEP elements in all lunar samples. U correlates best with La, while U–Yb ratios vary between different lunar rock classes. On the contrary the Hf–Yb ratios (Fig. 7) are more uniform than the Hf–La ratios.

We can add tantalum and tungsten to the long list of elements enriched in KREEP (Figs. 8 and 9). Here again the ratios W/La and Ta/Yb are more uniform than the ratios W/Yb and Ta/La. We have previously found a large tungsten excess in the lunar metal (Wänke *et al.*, 1971) and had attributed this excess to some tungsten rich rock class on the moon, from which tungsten could have entered the metal phase

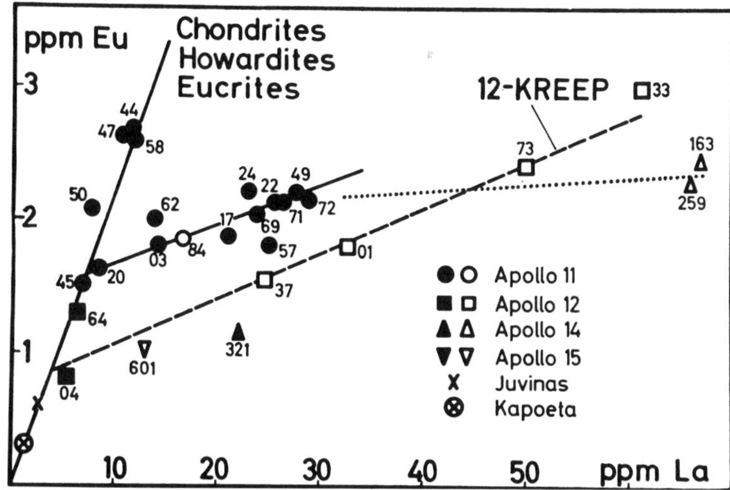


Fig. 6. Eu versus La. As in the F–La plot, a good correlation is found for the Apollo 12 and the low-K Apollo 11 rocks, having the unchanged chondritic Eu/La ratio. For KREEP a completely different ratio is observed. The high K-rocks of Apollo 11 were probably formed from low-K rocks by adding of KREEP. For data and symbols, see legend of Fig. 3.

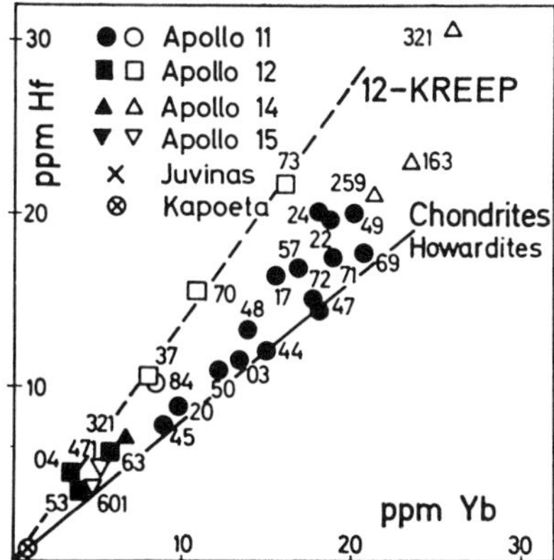


Fig. 7. Hf versus Yb. Hf and Yb correlate perfectly for all lunar samples. The Hf/Yb ratio is the same as in chondrites. For data and symbols, see legend of Fig. 3.

under reducing conditions. Obviously KREEP type material can act as the required source for W.

All Apollo 14 samples showed large negative europium anomalies (Fig. 10). The overall REE enrichment is highest for the microbreccias 14321,184,25 and 14305,81. The REE abundance patterns of all soils and breccias are very similar except for sample 14305,81, which has a marked lanthanum excess.

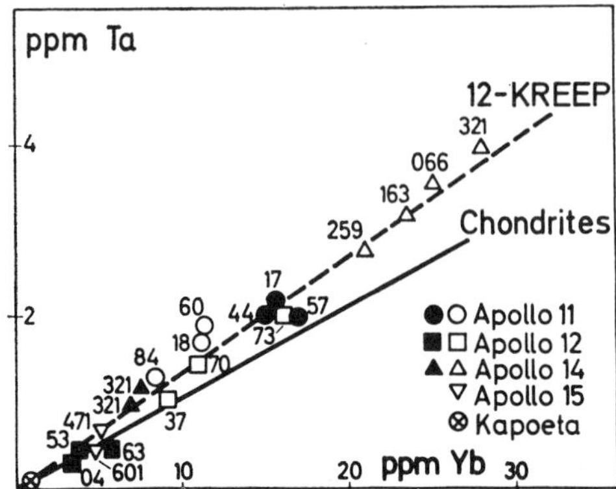


Fig. 8. Ta versus Yb. Ta is definitely correlated with the KREEP elements. As the data on Ta in the literature scatter widely we have only used measurements from this laboratory Wänke *et al.* (1970), Wänke *et al.* (1971), and this paper.

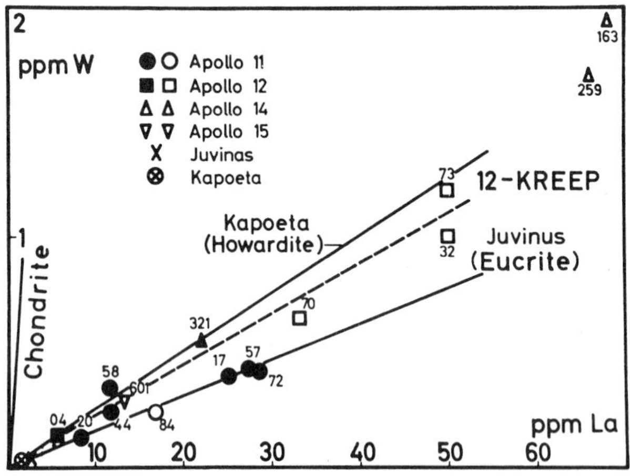


Fig. 9. W versus La. W also seems to belong to the KREEP elements. Only the data points for the high-K rocks of Apollo 11 deviate somewhat from the others. For data and symbols, see legend of Fig. 3.

## METEORITIC COMPONENTS

For the elements determined by us in soils and breccias the meteoritic contribution is dominant for the following ones: Ni, Ge, As, Pd, Au, and Ir. W is also moderately correlated with these elements in the *Apollo 12 and 14 bulk soils* as well as in their metal particles, but *not* in Apollo 11 soils and breccias and *not* in Apollo 12 and 14 breccias. We therefore think W to be of lunar origin mixed and partly equilibrated with the meteoritic component during the Imbrian event. Only very few determinations on Pd and As have been reported so far; our data on these elements are strongly correlated with gold (Figs. 11 and 12).

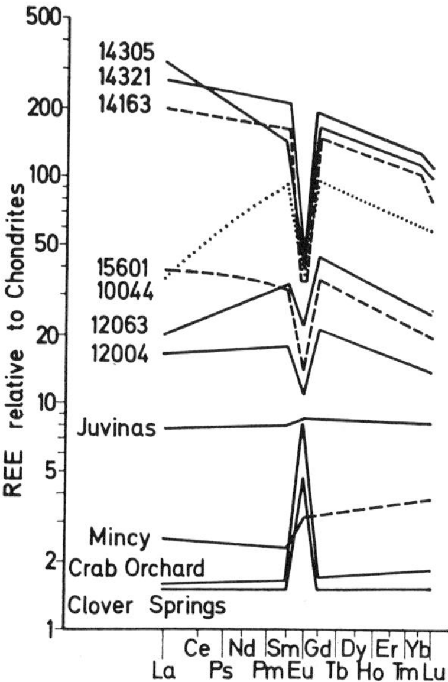


Fig. 10. REE abundances normalized to chondritic values. Data for chondrites are taken from Wakita and Zellmer (1970). All other data from Wänke *et al.* (1970), Wänke *et al.* (1971), and this work. Note the excess of La in sample 14305.

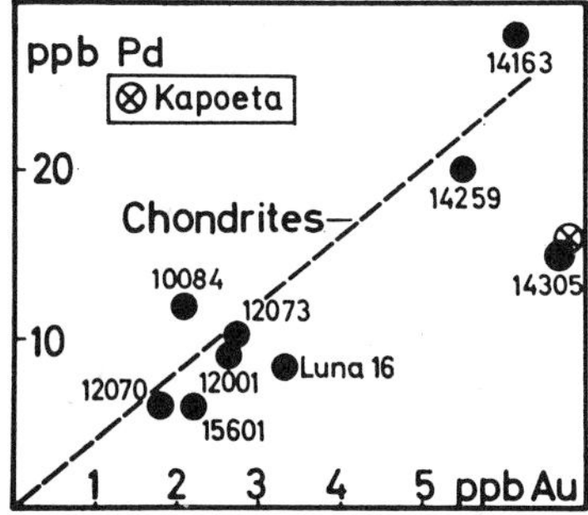


Fig. 11. Pd versus Au. The meteoritic character of palladium is clearly visible. Note that the point for the microbreccia 14305 falls far off the meteorite curve. Data on Luna 16 from Vinogradov (1971), all others from this laboratory.

The data points in the Pd/Au and As/Au diagrams for the microbreccia 14305 fall far below the chondritic line. As also the Ir/Au and the Ni/Au ratio is below the chondritic value Au is obviously enriched in this sample. It has been emphasized by Ganapathy *et al.* (1972) that the object which excavated the Imbrium basin was different in composition compared with most meteoritic classes, except iron meteorites of class IVa.

The deviation of the element ratios in question from the chondritic values for samples related to the Imbrian event (i.e., the soils and breccias of Apollo 14 and

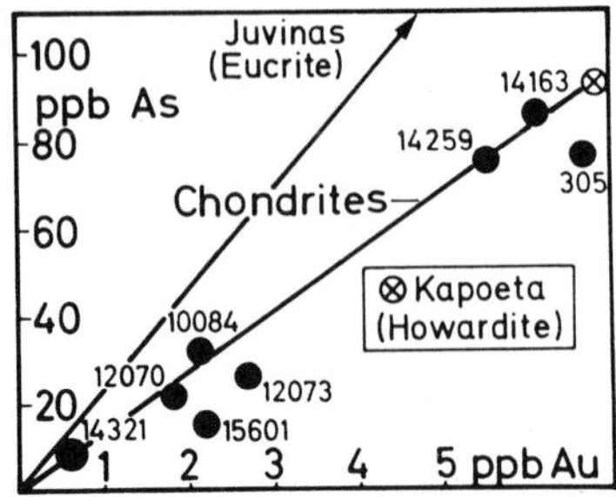


Fig. 12. As versus Au. All As found is definitely of meteoritic origin. For data, see legend of Fig. 11.

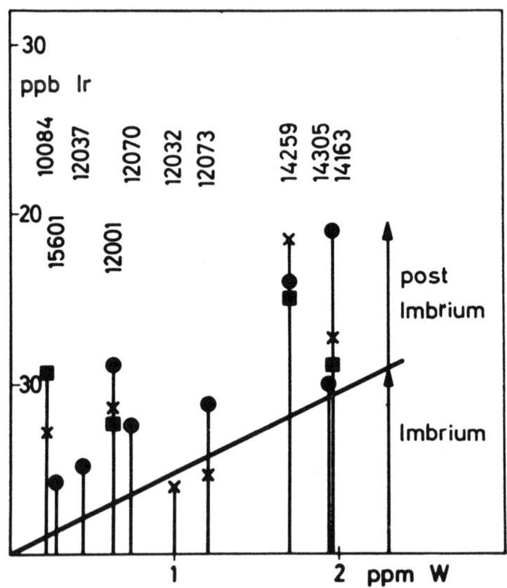


Fig. 13. Ir versus W. Full circles Ir data from this laboratory; crosses Ir data from Morgan *et al.* (1972) and Laul *et al.* (1971); full squares Ir data from Baedecker *et al.* (1972). All W values from this laboratory. The strong full line connects the breccias from Apollo 12 and 14. As W seems to be supplied to all sites by the Imbrian ejecta and behaves on the lunar surface like the siderophile elements (i.e., it is predominantly concentrated in metal particles), it can serve as an indicator for Imbrian and post-Imbrian meteoritic components. According to this model only the portion of the Ir found at the various soils and breccias above the dividing line should be post-Imbrian.

those rich in KREEP from Apollo 12) is in most cases larger for the microbreccias 14305 and 14321 (Morgan *et al.*, 1972) than for the soils. Obviously these breccias were formed at the time of the Imbrian event itself or shortly thereafter and therefore preserved the element distribution of the Imbrium object, having not been contaminated with post-Imbrian meteoritic components. Similarly, the material excavated by the Copernicus event was kept until 850 m.y. ago (Silver, 1971; Eberhardt *et al.*, 1972) at a depth which was not reached by the post-Imbrian meteoritic component.

In Fig. 13 we have plotted Ir against W. As mentioned above, the W content represents the percentage of Imbrian ejecta material in each sample. The line connecting the points of microbreccia 14305 and the nearly pure Copernicus ejecta 12032 and 12073 separates the Imbrian component (or less likely the pre-Imbrian component) from the post-Imbrian normal meteoritic component. As can be seen from Fig. 13 only about half of the Ir of the Apollo 14 soils is post-Imbrian. On the other hand the soils from Apollo 11 and those from Apollo 12 which are free of, or low in, ejecta material from Copernicus contain mainly the post-Imbrian meteoritic material.

## Genesis of Lunar Rocks

The plots Eu versus La and La versus K in Figs. 6 and 14 give some information about the genesis of lunar rocks. There is a very sharp distinction between both the low-K rocks of Apollo 11 and the Apollo 12 rocks, on one hand, to the high-K rocks of Apollo 11 on the other hand. Keeping in mind the proportionality of $P_2O_5$ to La (Fig. 3), it seems evident that the elements strongly enriched in KREEP were contributed to the mare basalts by inclusion of material rich in KREEP by the upwelling magmas. Such a conclusion was already reached by Urey *et al.* (1971) and for Rb by Wasserburg *et al.* (1972). Among the REE this is especially the case for La. The proportionality of La to K is even better (Fig. 14); it also holds for most of the Apollo 12 rocks and the low-K rocks of Apollo 11. Hence, K and La were probably supplied from the same source material (KREEP) to the lunar basalts.

Studying Figs. 3 and 7 in detail, one also comes to the conclusion that the high-K Apollo 11 rocks were derived from the low-K rocks by addition of KREEP. In this

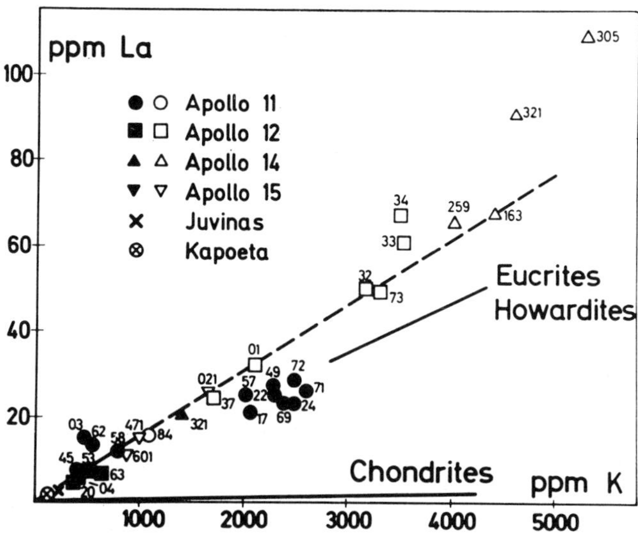


Fig. 14. La versus K. A nearly perfect correlation is observed for all lunar samples. Only the low-K rocks 10003 and 10062 have a considerable La excess. For data and symbols, see legend of Fig. 3.

connection one should remember that according to Turner (1971) the high-K rocks are more than 0.2 b.y. younger than the low-K rocks. The average content of strongly siderophile elements is about a factor of 3 higher in the high-K rocks. One might speculate therefore that the high-K rocks were formed from the low-K rocks by taking up material of an earlier regolith enriched in KREEP and meteoritic elements.

## HEAT PRODUCTION AND THE COMPOSITION OF THE LUNAR INTERIOR

Relative to chondrites a number of elements, for example, La, U, and Th, are enriched in some near-surface layers of the moon by factors up to 400. If the average composition of the moon were *chondritic* a simple material balance consideration shows that any such layer can have a maximum thickness of only 1.5 km. On the other hand, the homogeneity of the Imbrian ejecta (see Tables 1 and 4), especially the identical composition of the Fra Mauro formation and the ray material of Copernicus (KREEP in Apollo 12 soil, Morgan *et al.*, 1971) which also represents Imbrian ejecta underlying the Copernicus area, clearly indicates that this enriched layer must be 10 times thicker at least. Thus, we conclude that a considerable enrichment of the elements in question must have occurred prior to the accumulation of the moon. It is improbable that the total enrichment occurred in the material of the outer layer before accumulation, because no such material is known to exist in the solar system at present. A two-step process, however, with an initial enrichment of about a factor of 10 and subsequent enrichment on the moon is quite plausible, because in this case the initial material would have been of about eucritic composition, especially as far as the elemental ratios are concerned. It is important to note that a *homogeneous* eucritic moon is impossible because it would have been completely molten 4 b.y. ago—and stayed so ever since—but that the heat flow of $0.79 \times 10^{-6}$ cal/cm$^2$ sec (Langseth *et al.*, 1972) is exactly that obtained for the concentrations of radioelements as measured in the eucrite Juvinas (Table 1).

Hence, we strongly support the conclusion reached by Gast and McConnel (1972) of an initial chemical layering of the moon. We propose the upper 200–300 km to be enriched in U, Th, K, La, and other trace elements by a factor of 2–3 relative to Juvinas that, incidentally, would approximately correspond to the concentrations found in another eucrite, namely Stannern. With such enrichments of U and Th (but not of K)—about 20–30-fold relative to chondrites—the upper 200 km of the moon would be molten within less than 500 m.y. after accretion even assuming an initially cool moon. In this calculation the high insulation against heat loss from the surface of a planet covered with a layer of dust is essential (Urey *et al.*, 1971).

This suggestion has some interesting implications. If the upper 200 km or so of the moon had originally eucritic or slightly higher U and Th contents, then other planetesimals with similar concentrations should have existed—and perhaps still exist—as well. Such a differentiation prior to accumulation—however it may have occurred—would circumvent the difficulties one encounters with finding an appropriate heat source necessary to produce basaltic achondrites by differentiation of chondritic material. The subsequent melting of the eucrites that has quite clearly happened to Ibitira, e.g., as shown by its macroscopic vesicular structure very similar to that of lunar basalts (Fig. 15) could then have occurred on this eucrite parent body. An object

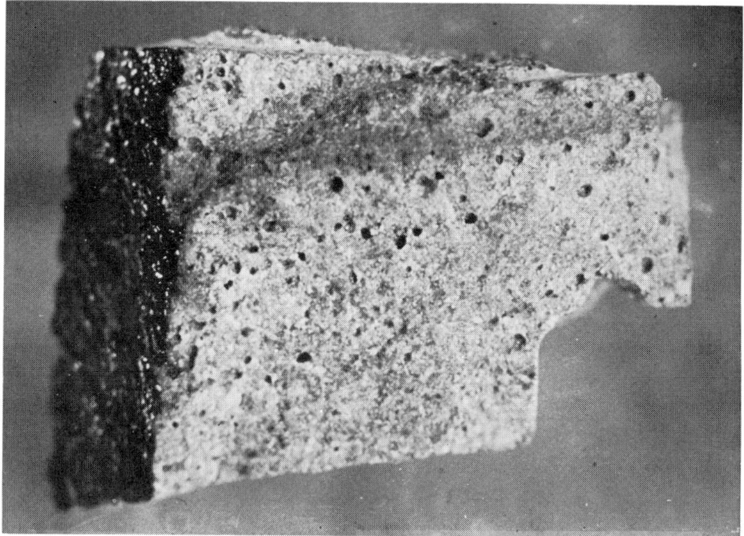

Fig. 15. The eucrite Ibitira. The specimen clearly shows a vesicular structure like the lunar basalts.

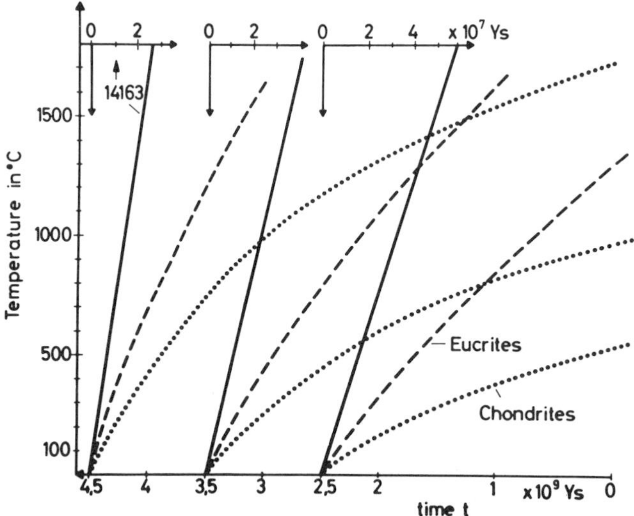


Fig. 16. Temperature increase in perfectly insulated material (chondrites, eucrites, and 14163 fines) due to internal radioactivity integrated over the time *t*, starting from 4.5, 3.5, and 2.5 billion years. (Chondrites: 11 ppb U, 40 ppb Th, and 845 ppm K; eucrites (Juvinas): 89 ppb U, 270 ppb Th and 220 ppm K; 14163: 4.07 ppm U, 15.9 ppm Th, and 4430 ppm K; a specific heat of 0.3 cal/g°C has been assumed).

with a diameter of even less than 100 km consisting of a loose agglomerate of such material would melt by itself, partly because the material accumulated under low gravity will have a heat conductivity even lower than that measured for lunar soil on earth (Cremers and Birkebak, 1971). For such a body the insulation against heat loss would be almost perfect as long as the surface layers are not yet molten resulting in very fast melting indeed (Fig. 16) and once they are molten due to its small size the cooling would be rather rapid.

That such a melting process for Ca-rich achondrites may have happened later than 4.6 b.y. ago is indicated by the work of Podosek (1972) who reports for Lafayette and Petersburg gas-retention formation times of 350 and 146 m.y., respectively, after that of the chondrite St. Severin.

*Acknowledgments*—We are grateful to NASA for making available the lunar material for this investigation. The samples were irradiated in TRIGA-research reactor of the Institut für Anorganische Chemie und Kernchemie der Universität Mainz. We thank Prof. J. T. Wasson and his group for their cooperation which made it possible for us to analyze their samples via INAA prior to their destructive determinations. We wish to thank the staff of the TRIGA-research reactor and the members of the staff of our Institute who are not listed as authors, in particular Mr. P. Deibele, Mr. H. Engler, and Miss H. Prager. The financial support by the Bundesministerium für Bildung und Wissenschaft is gratefully acknowledged.

## References

Baedecker P. A., Chou C.-L., Kimberlin J., and Wasson J. T. (1972) Trace element studies of lunar rocks and soils (abstract). In *Lunar Science—III* (editor C. Watkins), pp. 35–37, Lunar Science Institute Contr. No. 88.

Bouchet M., Kaplan G., Voudon A., and Bertoletti M. J. (1971) Spark mass spectrometric analysis of major and minor elements in six lunar samples. *Proc. Second Lunar Sci. Conf., Geochim. Cosmochim. Acta* Suppl. 2, Vol. 2, pp. 1247–1252. MIT Press.

Brown G. M., Emeleus C. H., Holland J. G., Peckett A., and Phillips R. (1971) Picrite basalts, ferrobasalts, feldspatic norites, and rhyolites in a strongly fractionated lunar crust. *Proc. Second Lunar Sci. Conf., Geochim. Cosmochim. Acta* Suppl. 2, Vol. 1, pp. 583–600. MIT Press.

Brunfelt A. O., Heier K. S., Steinnes E., and Sundvoll B. (1971) Determination of 36 elements in Apollo 14 bulk fines 14163 by activation analysis. *Earth Planet. Sci. Lett.* **11,** 351–353.

Compston W., Chappell B. W., Arriens P. A., and Vernon M. J. (1970) The chemistry and age of Apollo 11 lunar material. *Proc. Apollo 11 Lunar Sci. Conf., Geochim. Cosmochim. Acta* Suppl. 1, Vol. 2, pp. 1007–1027. Pergamon.

Compston W., Berry H., Vernon M. J., Chappell B. W., and Kaye M. J. (1971) Rubidium-strontium chronology and chemistry of lunar material from the Ocean of Storms. *Proc. Second Lunar Sci. Conf., Geochim. Cosmochim. Acta* Suppl. 2, Vol. 2, pp. 1471–1485. MIT Press.

Cremers C. J. and Birkebak R. C. (1971) Thermal conductivity of fines from Apollo 12. *Proc. Second Lunar Sci. Conf., Geochim. Cosmochim. Acta* Suppl. 2, Vol. 3, pp. 2311–2315. MIT Press.

Eberhardt P., Eugster O., Geiss J., Schwarzmüller J., Stettler A., and Weber L. (1972) When was the Apollo 12 KREEP ejected? (abstract). In *Lunar Science—III* (editor C. Watkins), pp. 206–208, Lunar Science Institute Contr. No. 88.

Engel A. E. J., Engel C. G., Sutton A. L., and Myers A. T. (1971) Composition of five Apollo 11 and Apollo 12 rocks and one Apollo 11 soil and some petrogenic considerations. *Proc. Second Lunar Sci. Conf., Geochim. Cosmochim. Acta* Suppl. 2, Vol. 1, pp. 439–448. MIT Press.

Ganapathy R., Laul J. C., Morgan J. W., and Anders E. (1972) Moon: Possible nature of the body that produced the Imbrian basin, from the composition of Apollo 14 studies. *Science* **175,** 55–59.

Gast P. W. and McConnell R. K. Jr. (1972) Evidence for initial chemical layering of the moon (abstract). In *Lunar Science—III* (editor C. Watkins), pp. 289–290, Lunar Science Institute Contr. No. 88.

Goles G. G., Duncan A. R., Lindstrom D. J., Martin M. R., Beyer R. L., Osawa M., Randle K., Meek L. T., Steinborn T. L., and McKay S. M. (1971) Analyses of Apollo 12 specimens: Compositional variations, differentiation processes, and lunar soil mixing models. *Proc. Second Lunar Sci. Conf., Geochim. Cosmochim. Acta* Suppl. 2, Vol. 2, pp. 1063–1081. MIT Press.

Hubbard N. J., Meyer C. Jr., Gast P. W., and Wiesmann H. (1971a) The composition and derivation of Apollo 12 soils. *Earth Planet. Sci. Lett.* **10,** 341–350.

Hubbard N. J., Gast P. W., and Meyer C. (1971b) The origin of the lunar soil based on REE, K, Rb, Ba, Sr, P, and $Sr^{87/88}$ data. Second Lunar Science Conference (unpublished proceedings).

Hubbard N. J., Gast P. W., Rhodes M., and Wiesmann H. (1972) Chemical composition of Apollo 14 materials and evidence for alkali volatilization (abstract). In *Lunar Science—III* (editor C. Watkins), pp. 407–409, Lunar Science Institute Contr. No. 88.

Jackson P. F. S., Coetzee J. H. J., Strasheim A., Strelow F. W. E., Gricius A. J., Wybenga F., and Kokot M. L. (1972) The analysis of lunar material returned by Apollo 14 (abstract). In *Lunar Science—III* (editor C. Watkins), pp. 424–426, Lunar Science Institute Contr. No. 88.

Keil K., Prinz M., and Bunch T. E. (1971) Mineralogy, petrology, and chemistry of some Apollo 12 samples. *Proc. Second Lunar Sci. Conf., Geochim. Cosmochim. Acta* Suppl. 2, Vol. 1, pp. 319–341. MIT Press.

Langseth M. G. Jr., Clark S. P. Jr., Chute J. Jr., and Keihm S. (1972) The Apollo 15 lunar heat flow measurement (abstract). In *Lunar Science—III* (editor C. Watkins), pp. 475–477, Lunar Science Institute Contr. No. 88.

Laul J. C., Morgan J. W., Ganapathy R., and Anders E. (1971) Meteoritic material in lunar samples: Characterization from trace elements. *Proc. Second Lunar Sci. Conf., Geochim. Cosmochim. Acta* Suppl. 2, Vol. 2, pp. 1139–1158. MIT Press.

Lindsay J. F. (1971) Mixing models and the recognition of end-member groups in Apollo 11 and 12 soils. *Earth Planet. Sci. Lett.* **12,** 67–72.

LSPET (Lunar Sample Preliminary Examination Team) (1972) The Apollo 15 lunar samples: A preliminary description. *Science* **175,** 363–375.

Morgan J. W., Ganapathy R., Laul J. C., and Anders E. (1971) Lunar crater Copernicus: Possible nature of impact. Preprint.

Morgan J. W., Laul J. C., Krähenbühl U., Ganapathy R., and Anders E. (1972) Major impacts on the moon: Chemical characterization of projectiles (abstract). In *Lunar Science—III* (editor C. Watkins), pp. 552–554, Lunar Science Institute Contr. No. 88.

Morrison G. H., Gerard J. T., Potter N. M., Gangadharam E. V., Rothenberg A. M., and Burdo R. A. (1971) Elemental abundances of lunar soil and rocks from Apollo 12. *Proc. Second Lunar Sci. Conf., Geochim. Cosmochim. Acta* Suppl. 2, Vol. 2, pp. 1169–1185. MIT Press.

Podosek, F. A. (1972) Gas retention chronology of Petersburg and other meteorites. *Geochim. Cosmochim. Acta* (in press).

Reed G. W. and Jovanovic S. (1971) The halogens and other trace elements in Apollo 12 samples and the implications of halides, platinum metals, and mercury on surfaces. *Proc. Second Lunar Sci. Conf., Geochim. Cosmochim. Acta* Suppl. 2, Vol. 2, pp. 1261–1276. MIT Press.

Rieder R. and Wänke H. (1969) Study of trace element abundance in meteorites by neutron activation. In *Meteorite Research* (editor P. Millman), pp. 75–86, D. Reidel.

Schnetzler C. C. and Philpotts J. A. (1971) Alkali, alkaline earth, and rare-earth element concentrations in some Apollo 12 soils, rocks, and separated phases. *Proc. Second Lunar Sci. Conf., Geochim. Cosmochim. Acta* Suppl. 2, Vol. 2, pp. 1101–1122. MIT Press.

Schonfeld E. (1972) Component abundance and ages in soils and breccia (abstract). In *Lunar Science—III* (editor C. Watkins), pp. 683–685, Lunar Science Institute Contr. No. 88.

Scoon J. H. (1971) Chemical analyses of lunar samples 12040 and 12064. *Proc. Second Lunar Sci. Conf., Geochim. Cosmochim. Acta* Suppl. 2, Vol. 2, pp. 1259–1260. MIT Press.

Silver L. T. (1971) U–Th–Pb isotopes systems in Apollo 11 and 12 regolithic materials and a possible age for the Copernicus impact event. Paper presented at Amer. Geophys. Union 52nd Ann. Meeting, Washington D.C.

Smales A. A., Mapper D., Webb M. S. W., Webster R. K., Wilson J. D., and Hislop J. S. (1971) Elemental composition of lunar surface material (part 2). *Proc. Second Lunar Sci. Conf., Geochim. Cosmochim. Acta* Suppl. 2, Vol. 2, pp. 1253–1258. MIT Press.

Taylor S. R., Rudowski R., Muir P., and Graham A. (1971) Trace element chemistry of lunar samples from the Ocean of Storms. *Proc. Second Lunar Sci. Conf., Geochim. Cosmochim. Acta* Suppl. 2, Vol. 2, pp. 1083–1099. MIT Press.

Turner G. (1971) $^{40}$Ar–$^{39}$Ar ages from the lunar maria. *Earth Planet. Sci. Lett.* **11,** 169–191.

Urey H. C., Marti K., Hawkins J. W., and Liu M. K. (1971) Model history of the lunar surface. *Proc. Second Lunar Sci. Conf., Geochim. Cosmochim. Acta* Suppl. 2, Vol. 2, pp. 987–998. MIT Press.

Vinogradov A. P. (1971) Preliminary data on lunar ground brought to earth by automatic probe Luna 16. *Proc. Second Lunar Sci. Conf., Geochim. Cosmochim. Acta* Suppl. 2, Vol. 1, pp. 1–16. MIT Press.

Wakita H. and Zellmer D. (1970) unpublished work. Data used quoted in Wakita *et al.* (1971).

Wakita H., Rey P., and Schmitt R. A. (1971) Abundances of the 14 rare-earth elements and 12 other trace elements in Apollo 12 samples: Five igneous and one breccia rocks and four soils. *Proc. Second Lunar Sci. Conf., Geochim. Cosmochim. Acta* Suppl. 2, Vol. 2, pp. 1319–1329. MIT Press.

Wänke H., Rieder R., Baddenhausen H., Spettel B., Teschke F., Quijano-Rico M., and Balacescu A. (1970) Major and trace elements in lunar material. *Proc. Apollo 11 Lunar Sci. Conf., Geochim. Cosmochim. Acta* Suppl. 1, Vol. 2, pp. 1719–1727. Pergamon.

Wänke H., Wlotzka F., Baddenhausen H., Balacescu A., Spettel B., Teschke F., Jagoutz E., Kruse H., Quijano-Rico M., and Rieder R. (1971) Apollo 12 samples: Chemical composition and its relation to sample locations and exposure ages and studies on lunar metallic particles. *Proc. Second Lunar Sci. Conf., Geochim. Cosmochim. Acta* Suppl. 2, Vol. 2, pp. 1187–1208. MIT Press.

Warner J. L. (1971) A summary of Apollo 11 chemical, age, and modal data. Extracted from *Proc. Apollo 11 Lunar Sci. Conf., Geochim. Cosmochim. Acta* Suppl. 1. Pergamon.

Wasserburg G. J., Turner G., Tera F., Podosek F. A., Papanastassiou D. A., and Huneke J. C. (1972) Comparison of Rb–Sr, K–Ar, and U–Th–Pb-ages; lunar chronology and evolution (abstract). In *Lunar Science—III* (editor C. Watkins), pp. 788–790, Lunar Science Institute Contr. No. 88.

Willis J. P., Ahrens L. H., Danchin R. V., Erlank A. J., Gurney J. J., Hofmeyr P. K., McCarthy T. S., and Orren M. J. (1971) Some interelement relationships between lunar rocks and fines, and stony meteorites. *Proc. Second Lunar Sci. Conf., Geochim. Cosmochim. Acta* Suppl. 2, Vol. 2, pp. 1123–1138. MIT Press.

Wood J. A., Marvin U. B., Reid J. B. Jr., Taylor G. J., Bower J. F., Powell B. N., and Dickey J. S. Jr. (1971) Mineralogy and petrology of the Apollo 12 lunar sample. Research in Science. Smithsonian Institution, Astrophysical Observatory, Cambridge, Mass., U.S.A., Special Report No. 333.

Proceedings of the Third Lunar Science Conference
(Supplement 3, *Geochimica et Cosmochimica Acta*)
Vol. 2, pp. 1269–1273
The M.I.T. Press, 1972

# Major, minor, and trace element data for some Apollo 11, 12, 14, and 15 samples

J. P. WILLIS, A. J. ERLANK, J. J. GURNEY, R. H. THEIL, and L. H. AHRENS

Department of Geochemistry, University of Cape Town,
Rondebosch, Cape, South Africa

**Abstract**—One Apollo 15 fines (15101), two Apollo 14 fines (14259, 14163), two rocks (14310, 14053), and one breccia sawdust (14305), one Apollo 12 rock (12038) and an Apollo 11 rock (10017) and fines (10084) have been analyzed for eighteen elements by x-ray fluorescence spectrometry. Revised data are given for Nb, Zr, Y, and Sr in the Apollo 12 rocks (12002, 12053, 12063) and fines (12032 and 12070). Summations of the elemental concentrations derived from our data together with available oxygen analyses are briefly discussed.

WE HAVE DETERMINED by x-ray fluorescence analysis (XRF) several major, minor, and trace elements in a variety of lunar materials. For the purposes of the present paper, we wish only to report analytical data obtained since our last report (Willis *et al.*, 1971). Detailed discussions, mainly involving inter-element relationships in lunar materials, stony meteorites, and terrestrial basaltic rocks, will be presented elsewhere.

Major element data for a variety of lunar rocks, fines, and a breccia are listed in Table 1. The method of sample preparation and analysis has been described previously (Willis *et al.*, 1971); this applies also to the trace element data presented in Table 2. However, recalibration of our technique has resulted in slightly higher values for Sr, slightly lower values for Zr, and appreciably lower values for Nb and Y than those previously obtained. We have improved the quality of our data for these elements by increased counting times and more careful corrections for line and background interferences. These comments are of importance in view of the very close coherence existing between Zr and Nb in lunar materials, and also in the eucritic and howarditic stony meteorites (Erlank *et al.*, 1972). For comparison purposes, average values obtained during our various lunar runs for four U.S.G.S. standard rocks are given in Table 2. For completeness, we present in Table 3 revised data for Nb, Y, Sr, and Zr in samples we have previously analyzed; these data replace that reported by Willis *et al.* (1971). It may be noted that the Nb data in Tables 2 and 3 are some 10% lower than those reported by Erlank *et al.* (1972).

Small quantities (0.1 to 0.3%) of $H_2O^-$ (at 120°C) were found in all samples, except for breccia 14305,121 (0.93%). This sample was supplied as sawdust by C. C. Schnetzler (Consortium Chief). The presence of $H_2O^-$ was considered to be due to atmospheric contamination, and all analyses are presented on an $H_2O^-$-free basis.

The data presented in Tables 1 and 2 agree very well with the XRF data presented by LSPET (1972), where the same elements and samples have been analyzed (10084, 14163, 14310, and 15101), and with data presented by several other authors in the

Table 1. Major element data (%).

| Sample | s.d.* | Fines 15101,67 | Rock 14310,117 | Breccia† 14305,121 | Fines 14259,59 | Fines 14163,56 | Rock 14053,43 | Rock 12038,77 | Rock 10017,70 | Fines 10084,173 |
|---|---|---|---|---|---|---|---|---|---|---|
| $SiO_2$ | 0.23 | 45.97 | 47.16 | 47.92 | 46.94 | 47.25 | 46.08 | 46.61 | 40.31 | 41.97 |
| $TiO_2$ | 0.01 | 1.26 | 1.25 | 1.71 | 1.75 | 1.79 | 2.91 | 3.25 | 11.66 | 7.47 |
| $Al_2O_3$ | 0.07 | 17.58 | 20.35 | 15.47 | 17.31 | 17.34 | 12.54 | 12.45 | 8.09 | 13.76 |
| FeO‡ | 0.08 | 11.54 | 8.31 | 11.34 | 10.60 | 10.32 | 16.97 | 17.75 | 19.49 | 15.89 |
| MnO | 0.002 | 0.157 | 0.113 | 0.144 | 0.139 | 0.137 | 0.255 | 0.251 | 0.242 | 0.210 |
| MgO | 0.08 | 10.32 | 7.83 | 11.14 | 9.55 | 9.36 | 8.97 | 6.83 | 7.82 | 7.99 |
| CaO | 0.05 | 11.71 | 12.43 | 9.96 | 11.06 | 10.97 | 11.07 | 11.48 | 10.71 | 12.12 |
| $Na_2O$ | 0.02 | 0.37 | 0.72 | 0.73 | 0.60 | 0.66 | 0.44 | 0.67 | 0.49 | 0.47 |
| $K_2O$ | 0.002 | 0.165 | 0.485 | 0.681 | 0.484 | 0.563 | 0.097 | 0.067 | 0.291 | 0.130 |
| $P_2O_5$ | 0.006 | 0.158 | 0.385 | 0.575 | 0.460 | 0.486 | 0.114 | 0.12 | 0.16 | 0.09 |
| $Cr_2O_3$ | 0.01 | 0.33 | 0.17 | 0.22 | 0.20 | 0.20 | 0.42 | 0.32 | 0.39 | 0.32 |
| S | 0.001 | 0.085 | 0.066 | 0.094 | 0.101 | 0.101 | 0.132 | 0.07 | 0.21 | 0.12 |
| Subtotal | | 99.645 | 99.269 | 99.984 | 99.194 | 99.177 | 99.998 | 99.868 | 99.863 | 100.540 |
| O ≡ S | | 0.043 | 0.033 | 0.047 | 0.051 | 0.051 | 0.066 | 0.035 | 0.105 | 0.060 |
| Total | | 99.602 | 99.236 | 99.937 | 99.143 | 99.126 | 99.932 | 99.833 | 99.758 | 100.480 |

*Average experimental standard deviation ($1\sigma$).
† Sawdust.
‡ Total Fe as FeO.

Table 2. Trace element data (ppm).

| | Fines 15101, 67 | Rock 14310, 117 | Breccia* 14305, 121 | Fines 14259, 59 | Fines 14163, 56 | Rock 14053, 43 | Rock 12038, 77 | Rock 10017, 70 | Fines 10084, 173 | G-1 | W-1 | BCR-1 | GSP-1 |
|---|---|---|---|---|---|---|---|---|---|---|---|---|---|
| Ba | 233 | 666 | 913 | 808 | 855 | 163 | 107 | 324 | 177 | 1038 | 160 | — | — |
| Nb | 17.8 | 54.5 | 70.7 | 61.3 | 63.4 | 15.7 | 9.3 | 27.4 | 19.4 | 21.7 | 6.8 | 11.3 | 25.4 |
| Zr | 313 | 852 | 1158 | 961 | 1022 | 215 | 182 | 499 | 309 | 219 | 95.1 | 192 | 543 |
| Y | 67.1 | 174 | 238 | 200 | 209 | 54.7 | 50.5 | 172 | 104 | 11.9 | 20.3 | 33.4 | 25.4 |
| Sr | 134 | 177 | 162 | 173 | 177 | 98.6 | 185 | 163 | 160 | 252 | 192 | 334 | 234 |
| Rb | 5.1 | 12.1 | 17.9 | 13.8 | 16.0 | 2.2 | n.d. | 5.4 | 2.4 | 210 | 21.3 | 47.3 | 252 |

Average experimental standard deviation ($1\sigma$): Ba = 2 ppm, Nb = 1.0 ppm, Zr = 2.0 ppm, Y = 1.3 ppm, Sr = 1.3 ppm, Rb = 0.5 ppm.
* Sawdust; n.d. = not detected, < 0.5 ppm.

Table 3. Revised trace element data (ppm).

| | Rock 12002,113 | Rock 12053,24 | Rock 12063,52 | Fines 12032,38 | Fines 12070,88 |
|---|---|---|---|---|---|
| Nb | 7.1 | 8.2 | 6.4 | 43.0 | 33.2 |
| Zr | 102 | 133 | 128 | 705 | 523 |
| Y | 33.5 | 44.2 | 54.8 | 166 | 126 |
| Sr | 89.0 | 114 | 149 | 160 | 140 |

Revised Abstracts of the Third Lunar Science Conference. Within our quoted precisions we have not noted any systematic errors, except for Sr where our data tends to be low by about 5% when compared with LSPET (1972) data and those obtained by isotope dilution methods. We are unable to account for this apparent systematic discrepancy, as the U.S.G.S. standard rock data in Table 2 are in good agreement with isotope dilution data, some of which are summarized in Cherry *et al.* (1970).

Oxygen abundances, by neutron activation analysis, are available for most lunar samples analyzed by us, and we have summed these abundances and our elemental abundances for the major and trace elements (including sulfur) listed in Tables 1, 2, 3, and Willis *et al.* (1971). Details and references are presented in Table 4. It is apparent that the summations derived by using the oxygen data of Wänke *et al.* (1970, 1971, 1972) more consistently approach the utopian 100% than do those using the oxygen analyses of Ehmann and Morgan (1970, 1971) and Ehmann and Gillum (1972).

Table 4. Summated elemental and oxygen abundances (%).

| Sample | | Σ Elements | Oxygen | Total | Reference for oxygen |
|---|---|---|---|---|---|
| 10084 | F | 58.28 | 40.8 | 99.08 | Ehmann and Morgan (1970) |
| | | | 41.5 | 99.78 | Wänke *et al.* (1970) |
| | | | | 100.58* | This paper |
| 12032 | F | 57.26 | 41.9 | 99.16 | Ehmann and Morgan (1971) |
| | | | | 100.37* | Willis *et al.* (1971) |
| 12070 | F | 57.62 | 41.4 | 99.02 | Ehmann and Morgan (1971) |
| | | | 42.6 | 100.22 | Wänke *et al.* (1971) |
| | | | | 100.37 | Willis *et al.* (1971) |
| 14163 | F | 55.59 | 43.7 | 99.29 | Wänke *et al.* (1972) |
| | | | 44.5 | 100.09 | Ehmann and Gillum (1972) |
| | | | | 99.42 | This paper |
| 14259 | F | 55.66 | 43.8 | 99.46 | Wänke *et al.* (1972) |
| | | | 44.8 | 100.46 | Ehmann and Gillum (1972) |
| | | | | 99.42 | This paper |
| 10017 | R | 59.70 | 40.7 | 100.40 | Wänke *et al.* (1970) |
| | | | | 99.91 | This paper |
| 12002 | R | 59.02 | 41.7 | 100.72 | Wänke *et al.* (1971) |
| | | | | 100.53 | Willis *et al.* (1971) |
| 12053 | R | 58.10 | 42.1 | 100.20 | Wänke *et al.* (1971) |
| | | | | 99.95 | Willis *et al.* (1971) |
| 12063 | R | 59.00 | 38.7 | 97.70 | Ehmann and Morgan (1971) |
| | | | 41.4 | 100.40 | Wänke *et al.* (1971) |
| | | | | 100.38 | Willis *et al.* (1971) |
| 14053 | R | 58.00 | 42.1 | 100.10 | Ehmann and Gillum (1972) |
| | | | | 100.00 | This paper |
| 14310 | R | 57.71 | 43.1 | 100.81 | Ehmann and Gillum (1972) |
| | | | | 99.48 | This paper |
| 14305 | B | 56.23 | 44.2 | 100.43 | Wänke *et al.* (1972) |
| | | | 41.1 | 97.33 | Ehmann and Gillum (1972) |
| | | | | 100.26 | This paper |

Σ Elements = sum of elemental concentrations listed in Tables 1, 2, and 3 and Willis *et al.* (1971).
F = Fines; R = Rock; B = Breccia.
* Totals of XRF analyses (majors + trace elements as oxides) from Tables 1, 2 and 3 (this paper) and Willis *et al.* (1971).

However, Ehmann and Morgan (1970) have noted that their lunar sample oxygen analyses are generally lower than those of Wänke *et al.* (1970), and point out that they have taken special care to handle their lunar materials under high-purity dry nitrogen at all times, to exclude the possibility of water contamination. Accordingly, they suggest that their oxygen data closely reflect the true abundances of oxygen in the samples under lunar conditions. The samples analyzed by Wänke *et al.* and those analyzed by us have been exposed to, and fine-ground under, normal laboratory atmospheric conditions and hence it might be expected that the summations derived from our data and those of Wänke *et al.* would be satisfactory, and that these are in agreement with the XRF summations derived by stoichiometry given in Table 4.

Ehmann and Morgan (1970, 1971) have observed Si–O correlations in Apollo 11 and 12 rocks which are different from those found in terrestrial rocks. An apparent oxygen deficiency of approximately 1.6% O was noted in the Apollo 11 fines, as compared to the amount of oxygen required for conventional stoichiometry based on total silicate analyses; a similar oxygen deficiency was noted in Apollo 12 fines and rocks (an oxygen deficiency in Apollo 11 rocks appears to be minimal or nonexistent). Rose *et al.* (1970) and Cuttitta *et al.* (1971) have found that the total reducing capacity of Apollo 11 and 12 samples is higher than the values determined for total Fe as FeO, with the highest "excess reducing capacity" being present in the lunar fines. They interpret these data, together with high summations they have obtained for lunar materials, to be suggestive of the presence of some Ti as Ti (III). Hence these observations, and also those of Maxwell *et al.* (1970), are in qualitative agreement with the postulated oxygen deficiency in lunar materials.

In the light of the above comments, it would seem that further discussion is unnecessary; nevertheless it is instructive to consider the summations given in Table 4. Breccia 14305 should perhaps not be considered as the same sample has not been analyzed by all three groups concerned. With regard to the fines, samples 10084, 12032 and 12070 would appear to support the postulated oxygen deficiency if the data of Ehmann and Morgan (1970, 1971) are used: this is not so for samples 14163 and 14259. However, Ehmann and Gillum (1972) have reported that these fines may have been subjected to terrestrial atmospheric contamination. It is noteworthy that the oxygen abundances reported by them for these two samples are distinctly higher than those reported by Wänke *et al.* (1972). The situation with regard to the rock samples listed in Table 4 is confusing. If the oxygen data of Ehmann and Morgan (1971) are used, sample 12063 would appear to have an oxygen deficiency greater than that indicated by these authors for Apollo 12 rocks. In considering the low total (97.7%) derived by using their data, and the total (100.4%) obtained using the results of Wänke *et al.* (1972),* we note the following details with respect to our own analyses. First, our major element data have been corrected for loss at 120°C ($H_2O^- = 0.15\%$ for 12063). Second, our samples have been heated to 1050°C prior to fusion for major element analysis, and we can compare the actual with the expected gain on ignition if all FeO is oxidized to $Fe_2O_3$ (the actual analysis of Fe by XRF is unaffected by the oxidation states involved). The shortfall for 12063 is 0.3% (i.e., expected gain minus actual gain). This could represent incomplete oxidation during preheating, adsorbed $H_2O$ not released at 120°C, oxidation of some of the FeO during grinding, or some or all of these. Our data for this sample could therefore be increased by as much as 0.3%. This shows that adsorbed water by itself cannot explain the discrepancy in the summations for 12063. In contrast, the summations for rocks 14053 and 14310 do not show any oxygen deficiency, as suggested for Apollo 14 rocks by Ehmann and Gillum (1972). In fact, our total for 14310 in Table 4, using stoichiometry, is lower than that derived by summation with the oxygen abundance.

We are unable to account for these discrepancies but have presented this discussion to show how incomplete is our knowledge regarding lunar stoichiometry and oxidation states and the response of lunar samples to atmospheric handling.

Ehmann and Morgan (1972) have emphasized that the calculated oxygen deple-

* Wänke *et al.* (1972) obtained a high total of 102.05% for this sample, using their own major element abundances, but note that this is due to inaccurate Ca measurements; this observation is supported by the Ca abundance we have obtained for 12063.

tions involve subtraction of large numbers to obtain small differences. We concur, and stress further that current chemical and instrumental techniques are still not sufficiently accurate and precise to resolve these problems.

*Acknowledgments*—We wish to acknowledge the assistance of Mrs. D. C. Corrigall and Mr. H. Fortuin in obtaining the data presented here. We are grateful to both the C.S.I.R., Pretoria, and the U.C.T. Staff Research Fund for their encouragement and generous financial support.

## References

Cherry R. D., Hobbs J. B. M., Erlank A. J., and Willis J. P. (1970) Thorium, uranium, potassium, lead, strontium, and rubidium in silicate rocks by gamma spectroscopy and/or x-ray fluorescence. *Canadian Spectr.* **15**, 1–8.

Cuttitta F., Rose H. J. Jr., Annell C. S., Carron M. K., Christian R. P., Dwornik E. J., Greenland L. P., Helz A. W., and Ligon D. T. Jr. (1971) Elemental composition of some Apollo 12 lunar rocks and soils. *Proc. Second Lunar Sci. Conf., Geochim. Cosmochim. Acta* Suppl. 2, Vol. 2, pp. 1217–1229. MIT Press.

Ehmann W. D. and Morgan J. W. (1970) Oxygen, silicon, and aluminium in Apollo 11 rocks and fines by 14 MeV neutron activation. *Proc. Apollo 11 Lunar Sci. Conf., Geochim. Cosmochim. Acta* Suppl. 1, Vol. 2, pp. 1071–1079. Pergamon.

Ehmann W. D. and Morgan J. W. (1971) Major element abundances in Apollo 12 rocks and fines by 14 MeV neutron activation. *Proc. Second Lunar Sci. Conf., Geochim. Cosmochim. Acta* Suppl. 2, Vol. 2, pp. 1237–1245. MIT Press.

Ehmann W. D. and Gillum D. E. (1972) Oxygen and other major elements in Apollo 14 rocks and some lunar soils (abstract). In *Lunar Science—III* (editor C. Watkins), pp. 209–211, Lunar Science Institute Contr. No. 88.

Erlank A. J., Willis J. P., Ahrens L. H., Gurney J. J., and McCarthy T. S. (1972) Inter-element relationships between the moon and stony meteorites with particular reference to some refractory elements (abstract). In *Lunar Science—III* (editor C. Watkins), pp. 239–241, Lunar Science Institute Contr. No. 88.

LSPET (Lunar Sample Preliminary Examination Team) (1972) The Apollo 15 lunar samples: A preliminary description. *Science* **175**, 363–375.

Maxwell J. A., Peck L. C., and Wiik H. B. (1970) Chemical composition of Apollo 11 lunar samples 10017, 10020, 10072, and 10084. *Proc. Apollo 11 Lunar Sci. Conf., Geochim. Cosmochim. Acta* Suppl. 1, Vol. 2, pp. 1369–1374. Pergamon.

Rose H. J. Jr., Cuttitta F., Dwornik E. J., Carron M. K., Christian R. P., Lindsay J. R., Ligon D. T. Jr., and Larson R. R. (1970) Semimicro x-ray fluorescence analysis of lunar samples. *Proc. Apollo 11 Lunar Sci. Conf., Geochim. Cosmochim. Acta* Suppl. 1, Vol. 2, pp. 1493–1497. Pergamon.

Wänke H., Rieder R., Baddenhausen H., Spettel B., Teschke F., Quijano-Rico M., and Balacescu A. (1970) Major and trace elements in lunar material. *Proc. Apollo 11 Lunar Sci. Conf., Geochim. Cosmochim. Acta* Suppl. 1, Vol. 2, pp. 1719–1727. Pergamon.

Wänke H., Wlotzka F., Baddenhausen H., Balacescu A., Spettel B., Teschke F., Jagoutz E., Kruse H., Quijano-Rico M., and Rieder R. (1971) Apollo 12 samples: Chemical composition and its relation to sample locations and exposure ages, the two component origin of the various soil samples and studies on lunar metallic particles. *Proc. Second Lunar Sci. Conf., Geochim. Cosmochim. Acta* Suppl. 2, Vol. 2, pp. 1187–1208. MIT Press.

Wänke H., Baddenhausen H., Balacescu A., Teschke F., Spettel B., Dreibus G., Quijano-Rico M., Kruse H., Wlotzka F., and Begemann F. (1972) Multielement analyses of lunar samples (abstract). In *Lunar Science—III* (editor C. Watkins), pp. 779–781, Lunar Science Institute Contr. No. 88.

Willis J. P., Ahrens L. H., Danchin R. V., Erlank A. J., Gurney J. J., Hofmeyr P. K., McCarthy T. S., and Orren M. J. (1971) Some inter-element relationships between lunar rocks and fines, and stony meteorites. *Proc. Second Lunar Sci. Conf., Geochim. Cosmochim. Acta* Suppl. 2, Vol. 2, pp. 1123–1138. MIT Press.

Proceedings of the Third Lunar Science Conference
(Supplement 3, *Geochimica et Cosmochimica Acta*)
Vol. 2, pp. 1275–1292
The M.I.T. Press, 1972

# Rare earths and other trace elements in Apollo 14 samples

PHILIP A. HELMKE, LARRY A. HASKIN, RANDY L. KOROTEV, and KAREN E. ZIEGE

Department of Chemistry, University of Wisconsin,
Madison, Wisconsin 53706

**Abstract**—REE and other trace elements have been determined in igneous rocks 14053, 14072, and 14310, in breccias 14063 and 14313, and in fines 14163. All materials analyzed have typical depletions of Eu except for feldspar fragments from the breccias and igneous fragments from 14063. Igneous rocks 14072 and 14053 have REE concentrations very similar to Apollo 12 basalts; 14310 has the highest REE concentrations yet observed for a large fragment of lunar basalt. The effects of crystallization of a basaltic liquid as a closed system on the concentrations of Sm and Eu in feldspar are considered. Small anorthositic fragments may have originated by simple crystallization from very highly differentiated basalt (KREEP) or by closed-system crystallization in a less differentiated starting material. Application of independent models of igneous differentiation to Sm and Eu in massive anorthosite 15415 and to Sm and Eu in lunar basalts suggests a common starting material with a ratio of concentrations of Sm and Eu about the same as that in chondrites and with concentrations of those elements about 15 times enriched over chondrites.

## INTRODUCTION

RARE EARTHS AND SEVERAL OTHER trace elements have been determined in three Apollo 14 igneous rocks (14053, 14072, and 14310), one sample of fines (14163), and portions of two breccias (14063 and 14313) by neutron activation analyses. The procedures used for the analyses are modifications of those used for our earlier work (Haskin *et al.*, 1970, 1971) and reported by Denechaud *et al.* (1970) and Allen *et al.* (1970).

The igneous rocks were received as 1 g chips; they were powdered and sieved to finer than 100 mesh. Approximately 0.5 g each of the igneous rocks and soil was used for analyses. A portion of breccia 14313 was powdered and sieved; the less than 200-mesh fraction was taken to represent the composition of the whole rock, and a magnetic separate of feldspar was obtained from the 100–200 mesh size fraction. From a 68 mg portion of disaggregated material from breccia 14063, we hand-picked two pure (as determined by inspection under a binocular microscope) plagioclase crystals and six fragments of igneous rock (mafic minerals plus feldspar). The remaining material was then sieved. A portion of the $<200$ mesh fraction was analyzed as "fines," and small crystals of feldspar were handpicked from the $>200$ mesh fraction to provide a feldspar "separate."

The results of the analyses of the various materials are given in Tables 1 and 2.

Rocks 14072 and 14053 are within a factor of 2 of each other in trace element abundances. Their REE abundances are compared with those of chondrites in Fig. 1. On the graph of Sm/Eu versus Sm (Fig. 2) the points for those basalts fall among the

Table 1. Trace element concentrations (in ppm) for three Apollo igneous rocks, one fines, and one breccia.

| | 14053* | 14072 | 14310 | 14163 | 14313 |
|---|---|---|---|---|---|
| La | 12.8 ± 0.2 | 6.76 | 57.0 | 70.4 | 65.4 |
| Ce | 36.4 ± 0.8 | 17.9 | 135 | 157 | 171 |
| Nd | 22 ± 1 | 13 | 93 | 101 | 114 |
| Sm | 6.50 ± 0.06 | 3.93 | 25.6 | 30.8 | 30.8 |
| Eu | 1.23 ± 0.03 | 0.88 | 2.08 | 2.57 | 2.65 |
| Gd | 8.5 ± 0.2 | 4.2 | — | 36 | 40 |
| Tb | 1.62 ± 0.08 | 0.98 | 5.3 | 6.4 | 6.3 |
| Dy | 11.1 ± 0.4 | 6.0 | 36.2 | 44.8 | 38 |
| Ho | 2.1 ± 0.2 | 1.50 | 6.7 | 8.6 | 8.0 |
| Er | — | 3.5 | 20 | 25 | 22 |
| Yb | 6.1 ± 0.3 | 4.05 | 18.6 | 24.6 | 21.8 |
| Lu | 0.89 ± 0.01 | 0.61 | 2.76 | 3.16 | 3.17 |
| Co | 25 ± 1 | 32 | 16.1 | 27 | 31.5 |
| Cr | 2860 ± 80 | 3880 | 1440 | 1570 | 2450 |
| Hf | 9.8 ± 0.4 | 6.9 | 18 | 22 | 21 |
| Ga | 4.8 ± 0.2 | 3.8 | 4.3 | 7.5 | 11.0 |
| Mn | 1360 ± 20 | 1840 | 705 | — | 1020 |
| Ni | 14 ± 3 | 31 | 150 | 331 | 273 |
| Sc | 55 ± 2 | 47.1 | 18.7 | 20.5 | 24.6 |
| Zn | 3.4 ± 0.3 | 8 | <7 | 34 | 56 |

* The uncertainties given for 14053, if converted to percentages, correspond approximately to the uncertainties, which are standard deviations based on counting statistics, for the other data in this table. A systematic uncertainty of about ±2% should be included for the entire group of REE from a given sample.

Table 2. Trace element concentrations (in ppm) and sample weights in milligrams for portions of two Apollo 14 breccias.

| | 14063* "Fines" | 14063 Igneous fragments | 14063 Feldspar crystals | 14063 Feldspar separate | 14313 Feldspar separate |
|---|---|---|---|---|---|
| La | 19.4 ± 2 | 15.9 | 2.6 | 4.1 | 15.1 |
| Ce | 47 ± 1 | 49 | 5.0 | — | 34.3 |
| Nd | 36 ± 8 | 17 | — | — | 26 |
| Sm | 9.17 ± 0.05 | 6.84 | 1.06 | 1.16 | 4.82 |
| Eu | 2.55 ± 0.02 | 2.89 | 3.18 | 3.49 | 4.98 |
| Gd | 11.6 ± 1.4 | 8.3 | 2.7 | — | 6.1 |
| Tb | 1.96 ± 0.12 | 1.49 | 0.22 | — | 0.97 |
| Dy | 12.0 ± 0.2 | 8.3 | 1.52 | 0.93 | 6.12 |
| Ho | — | — | — | — | 1.4 |
| Er | 7 ± 3 | — | — | — | 6 |
| Yb | 6.8 ± 0.2 | 4.7 | 0.86 | 0.83 | 4.6 |
| Lu | 0.99 ± 0.01 | 0.71 | 0.113 | 0.16 | 0.60 |
| Co | 21 ± 3 | 15 | <2 | <10 | 1.4 |
| Hf | 11 ± 2 | 12 | <0.4 | 38 | 24 |
| Sc | 14.2 ± 0.4 | 12.6 | 4.0 | 13.9 | 2.48 |
| (mg) | 7.32 | 3.55 | 2.88 | 2.14 | 7.27 |

* The uncertainties given for 14063 "fines," if converted to percentages, correspond approximately to the uncertainties, which are standard deviations based on counting statistics, for the other data in this table. A systematic uncertainty of about ±2% should be included for the entire group of REE from a given sample.

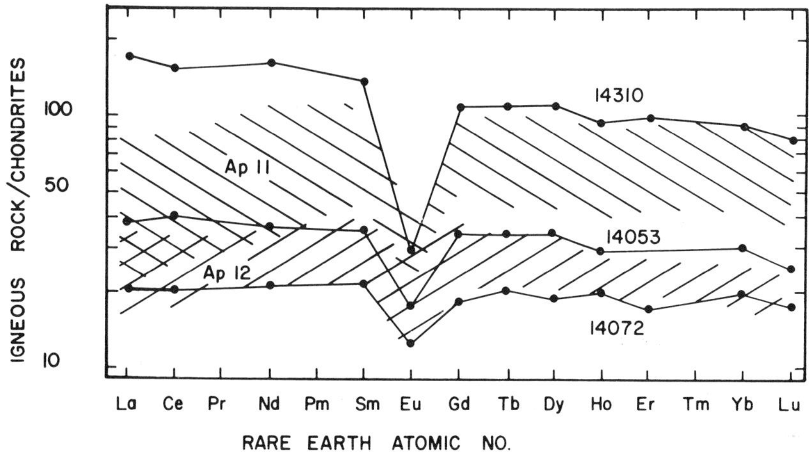


Fig. 1. Comparison diagram for 14053, 14072, and 14310. Ranges for Apollos 11 and 12 basalts are included.

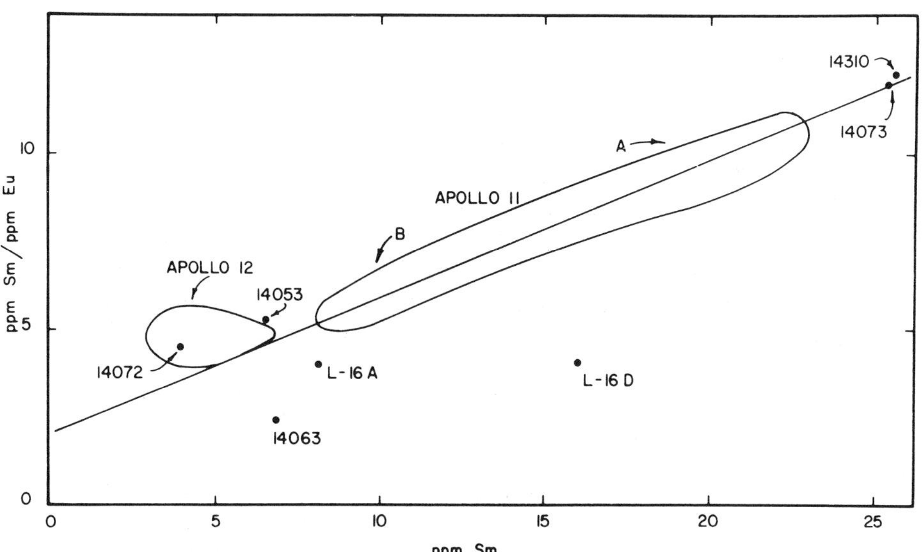


Fig. 2. Ratios of concentrations of Sm to concentrations of Eu are plotted against concentrations of Sm as required by equations describing partial melting. The data fit a straight line equally well when logarithmic coordinates are used, as required by equations describing fractional crystallization (Haskin *et al.*, 1970).

points for Apollo 12 basalts (Haskin *et al.*, 1971). Rock 14072 has a smaller Eu depletion and 0.5 to 0.8 times the total REE concentrations found in 14053. Rock 14072 contains more Co, Mn, and Ni, but less Hf and Ga than 14053, the same trends as observed among Apollo 12 basalts. In analogy with the Apollo 12 basalts, accumulation of olivine in a liquid with the composition of 14053 would be expected to produce a rock like 14072. Cumulative olivine is found in 14072 (Walker *et al.*, 1972).

The concentrations of REE in 14310 are the highest yet observed for a large fragment of lunar basalt. The relative REE abundances (Fig. 1) are essentially those found in the fines sample 14163. Concentrations for most of the other trace elements in 14310 resemble those in 14163 much more than those in 14053 or 14072. Apparently, basalt 14073 is very similar to 14310 in composition (Wakita *et al.*, 1972).

Do rocks such as 14310 and 14073 represent fused fines or are they the parent materials of fines such as 14163? The high concentration of Ni in 14310 is suggestive of addition of that element to a soil by meteorite infall (e.g. Ganapathy *et al.*, 1970). However, in comparison with soil 14163, the concentration in 14310 of REE, Cr, Hf, and Sc are 0.8 to 0.95 times as high, whereas those for the siderophilic elements Co, Ga, and Ni are in the range 0.45 to 0.6 times as high, and that for the relatively volatile element Zn is very low, $<0.2$ times as high (Morgan *et al.*, 1972). These values tend to argue against a large chondritic component in 14310. Since rock 14310 is very old ($3.87 \pm 0.04$ AE, Papanastassiou and Wasserburg, 1971), it presumably has not had the opportunity to receive the same contribution from chondritic meteorites as that received by 14163. Furthermore, it is not clear that the bombarding meteoroids that long ago might have produced the fine material from which 14310 might have been formed were of chondritic composition. A chondritic component was not found for old Apollo 14 breccias (Morgan *et al.*, 1972). Thus, the chemical evidence bearing on the mode of origin of rock 14310 is ambiguous.

It would appear that, in order to produce soil as highly differentiated from primodial matter as is 14163, rocks of the required composition must be made first. The abundances of many trace elements in 14310 and 14073 (Wakita *et al.*, 1972) appear to be in line with the trends that characterize the igneous differentiation of lunar matter that produced the basalts from the other Apollo missions. For example, they extend the line found for Apollo 11 and 12 basalts to higher values of Sm/Eu and Sm (Fig. 2).

The concentrations of the trace elements measured in the "whole rock" sample of breccia 14313 are very similar to those of 14163, confirming the general nature of that rock as agglomerated soil (Floran *et al.*, 1972). In the feldspar concentrate from 14313, the high concentration of Eu is typical of that for feldspars. The concentrations of Co and Sc are very low, indicating that mafic mineral fragments were efficiently excluded by the magnetic separation. The rather high value of Hf suggests that 0.1 to 0.2% of the feldspar separate was zircon. Because of this, the concentrations of the REE heavier than Dy are believed to be anomously high for the feldspar separate (Fig. 3) but not for the feldspar crystals.

The concentrations of REE in the white breccia 14063 are significantly lower than those in the Apollo 14 fines and in 14313. The Eu depletion in the "fines" fraction is very small, and the igneous chips are slightly enriched in Eu, relative to the value

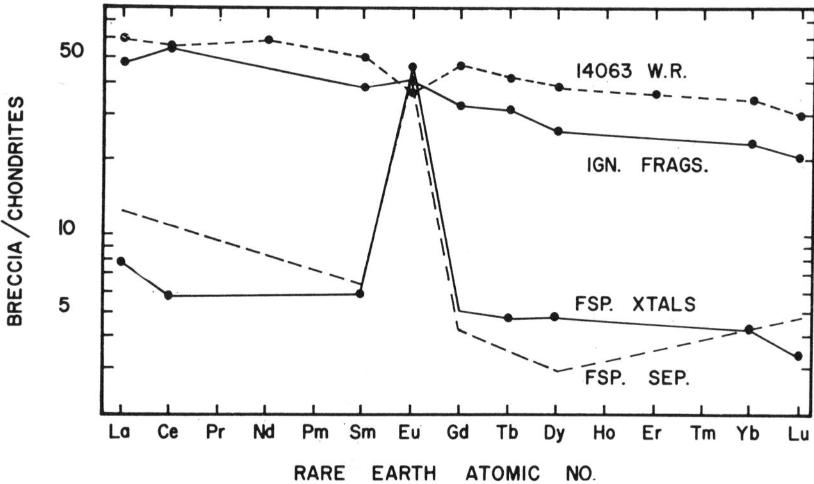


Fig. 3. Comparison diagram for fragments from breccia 14063.

expected by interpolation on a comparison diagram with chondrites (Fig. 3). The feldspar crystals and the feldspar "separate" have similar REE abundances and Eu enrichments. The feldspar separate for 14063, like that for 14313, is rich in Hf and, presumably, in zircon.

The points for the igneous fragments from 14063, like those for the igneous fragments from Luna 16 samples, do not fall on the diagram of ppm Sm/ppm Eu versus ppm Sm (Fig. 2) (Helmke and Haskin, 1972). These igneous fragments may indicate that their parent liquids were in some manner unrelated to the starting materials for the other lunar basalts. The finer materials associated with both 14063 and the Luna 16 fragments are similar in relative REE abundances to their respective igneous fragments. This suggests that the relative REE abundances in the igneous fragments are not merely artifacts of some accidental, statistically improbable sampling of very few grains of different minerals with highly selective affinities for the different REE. The positions of the points for these fragments on a graph of Sm/Eu to Sm could be seriously in error if there were fluctuations in REE concentrations but not *relative* abundances among small fragments from the same rock. The REE concentrations for the two sets of Luna 16 fragments differ by a factor of 2 although the relative abundances are identical.

## Closed System Crystallization

It would be useful to interpret the Sm and Eu concentrations and ratios found in a single crystal of a mineral, or in a concentrate of a particular mineral, in terms of the Sm and Eu concentrations and ratios of concentrations in the parent liquid from which those minerals came. It is first necessary to consider the effects of complete crystallization of a liquid as a closed system on the average concentrations of the elements in the crystal. This can be done in terms of the logarithmic model of trace element partition, beginning with the equations used previously (Haskin *et al.*, 1970) and considering each of the minerals to correspond to a derived phase. Many changes in composition of the initial liquid occur as crystallization proceeds. Compositions and proportions of the minerals crystallized also change as solidification of the melt progresses. Earlier formed phases may cease to crystallize. Residual liquid may finally yield to exotic accessory minerals (mesostasis). Despite these many real complications, it is adequate for illustrating the problem to assume a simple model in which the proportions of the various minerals crystallizing and the distribution coefficients for the elements of interest in those minerals are fixed throughout the entire crystallization of the liquid.

In the derived phase (minerals) at any stage of crystallization, where $C_{p,\mathrm{Sm}}$ represents the average concentration of Sm (taken as an example) in the entire solidified portion (mixture of minerals), $C_{i,\mathrm{Sm}}$ is the average concentration of Sm in mineral $i$, and $f_i$ is the fraction by weight of the solidified portion that is represented by mineral $i$,

$$C_{P,\mathrm{Sm}} = \sum_i C_{i,\mathrm{Sm}} f_i. \tag{1}$$

For each increment of solid formed, the concentration of Sm in one mineral $C_{i,\mathrm{Sm}}$ is related to that in another mineral $C_{j,\mathrm{Sm}}$ through their respective distribution coefficients $D_i$ and $D_j$. ($D$ is defined as the ratio of the equilibrium concentration of Sm in a solid phase to its concentration in its parent liquid.) The relative concentrations of a trace element in those minerals will stay the same throughout crystallization, i.e.,

$$C_{j,\mathrm{Sm}}/C_{i,\mathrm{Sm}} = D_{j,\mathrm{Sm}}/D_{i,\mathrm{Sm}}. \tag{2}$$

Rewriting equation (1) as follows, we obtain

$$C_{P,\mathrm{Sm}} = C_{j,\mathrm{Sm}} f_j + \sum_{i \neq j} C_{i,\mathrm{Sm}} f_i; \tag{3}$$

and substituting for $C_{i,\mathrm{Sm}}$ from (2) we have

$$C_{P,\mathrm{Sm}} = C_{j,\mathrm{Sm}} \left[ f_j + \left( \sum_{i \neq j} \frac{D_i f_i}{D_j} \right) \right]. \tag{4}$$

But since $C_{P,\mathrm{Sm}}$ must also equal $C_{A,\mathrm{Sm}}[1 - (1 - x)^{D_{\mathrm{Sm}}}]/x$ (Haskin *et al.*, 1970; equation (2)) in which $C_{A,\mathrm{Sm}}$ is the average concentration of Sm in the entire crystallizing system, $x$ is the fraction crystallized and $D_{\mathrm{Sm}}$ is the average distribution coefficient for the solid portion; i.e.,

$$D_{\mathrm{Sm}} = \sum_i D_{i,\mathrm{Sm}} f_i; \tag{5}$$

then

$$C_{j,\mathrm{Sm}} = \frac{C_{A,\mathrm{Sm}}[1 - (1 - x)^{D_{\mathrm{Sm}}}]/x}{f_j + (\sum_{i \neq j} D_{i,\mathrm{Sm}} f_i)/D_{j,\mathrm{Sm}}} \tag{6}$$

If the distribution coefficients for the minerals favor retention of the trace element by the liquid (i.e., $D_{\mathrm{Sm}} < 1$), the concentration of the trace element in the liquid will increase. As the crystals of each mineral grow, their average concentrations will continue to increase, too, because each new increment of solid added has a higher concentration of that element than did the previous increment. If the process of solidification is allowed to reach completion with no changes in mineralogy for the solid phase, all of the Sm must eventually be taken up by those minerals. Thus, for $x = 1$,

$$C_{j,\mathrm{Sm}} = \frac{C_{A,\mathrm{Sm}} D_{j,\mathrm{Sm}}}{\sum_i D_{i,\mathrm{Sm}} f_i}. \tag{7}$$

Note that the initial concentration of $C_{j,\mathrm{Sm},0}$ for any mineral is given by the following equation:

$$C_{j,\mathrm{Sm},0} = D_{j,\mathrm{Sm}} C_{A,\mathrm{Sm}}. \tag{8}$$

Thus, the increase in concentration, i.e., the ratio of the final concentration to the initial concentration, is

$$C_{j,\mathrm{Sm},1}/C_{j,\mathrm{Sm},0} = 1/\sum_i D_{i,\mathrm{Sm}} f_i. \tag{9}$$

Finally, the maximum change in the ratio of two trace elements to each other, say Sm to Eu, with crystallization, is

$$(C_{j,\mathrm{Sm},1}/C_{j,\mathrm{Sm},0})/(C_{j,\mathrm{Eu},1}/C_{j,\mathrm{Eu},0}) = \sum_i D_{i,\mathrm{Eu}} f_i / \sum_i D_{i,\mathrm{Sm}} f_i. \tag{10}$$

Note also that the extent of change in the ratio of Sm to Eu is the same for every mineral.

In a real rock, some minerals may have phenocrystic cores of essentially uniform composition. The proportions of the minerals change as solidification progresses. A major mineral may not start crystallizing before a significant fraction of the parent liquid has solidified. The final dregs of liquid crystallize as mesostasis. The compositions of the minerals and their parent liquid change as solidification progresses and so, therefore, may the $D$ values. Portions of mineral crystals may be lost during separation from the rock. In view of these differences between real rocks and the model rocks described above, can we draw any conclusions about the parent liquid from which a mineral crystal came? Before answering this question, let us examine some predictions of the model.

Several choices of minerals and values for distribution coefficients are given in Table 3. Values for distribution coefficients are based on those found for natural and

Table 3. Values of parameters for and results of calculation of extents of changes in concentrations of Sm, Eu, and in the ratios of concentrations of Sm and Eu as a result of closed-system crystallization of a silicate liquid.

| Case | $f_{\mathrm{pl}}$ | $f_{\mathrm{opx}}$ | $f_{\mathrm{cpx}}$ | $D_{\mathrm{pl,Sm}}$ | $D_{\mathrm{opx,Sm}}$ | $D_{\mathrm{cpx,Sm}}$ | $D_{\mathrm{pl,Eu}}$ | $D_{\mathrm{opx,Eu}}$ | $D_{\mathrm{cpx,Eu}}$ | $C_{\mathrm{Sm}}$ Final / $C_{\mathrm{Sm}}$ Initial | $C_{\mathrm{Eu}}$ Final / $C_{\mathrm{Eu}}$ Initial | Sm/Eu Final / Sm/Eu Initial |
|---|---|---|---|---|---|---|---|---|---|---|---|---|
| 1 | 0.33 | 0.33 | 0.33 | 0.05 | 0.05 | 0.3 | 0.8 | 0.06 | 0.4 | 7.5 | 2.5 | 3.1 |
| 2 | 0.5 | 0.5 | — | 0.05 | 0.05 | — | 0.8 | 0.06 | — | 20 | 2.3 | 8.6 |
| 3 | 0.5 | — | 0.5 | 0.05 | — | 0.3 | 0.8 | — | 0.4 | 5.7 | 1.7 | 3.4 |
| 4 | 0.5 | 0.5 | — | 0.2 | 0.05 | — | 0.8 | 0.06 | — | 6.7 | 2.3 | 2.9 |
| 5 | 0.33 | 0.33 | 0.33 | 0.2 | 0.05 | 0.3 | 0.8 | 0.06 | 0.4 | 5.8 | 2.5 | 2.4 |
| 6 | 0.33 | 0.33 | 0.33 | 0.02 | 0.05 | 0.3 | 1.2 | 0.06 | 0.4 | 8.1 | 1.8 | 4.5 |
| 7 | 0.5 | 0.5 | — | 0.02 | 0.05 | — | 1.2 | 0.06 | — | 29 | 1.6 | 18 |
| 8 | 0.7 | 0.2 | 0.1 | 0.02 | 0.05 | 0.3 | 1.2 | 0.06 | 0.4 | 19 | 1.1 | 17 |
| 9 | 0.3 | 0.6 | 0.1 | 0.02 | 0.05 | 0.3 | 1.2 | 0.06 | 0.4 | 5.8 | 2.3 | 2.5 |

experimental mafic systems (e.g., Schnetzler and Philpotts, 1970; Green *et al.*, 1971; Paster, Schauwecker, and Haskin, unpublished). Some discussion of these choices was made previously (Haskin *et al.*, 1970). Only values of distribution coefficients for plagioclase were varied. We regard values of

$$D_{\mathrm{pl,Sm}} = 0.2 \text{ and } D_{\mathrm{pl,Eu}} = 1.2$$

as being too high; the values we have found for $D_{\mathrm{pl,Sm}}$ in layered intrusions are near 0.02 and those of $D_{\mathrm{pl,Eu}}$ for calcic plagioclase, when corrected for effects of oxygen fugacity (Haskin *et al.*, 1970) are about 0.8; we regard these as the most probable values.

We regard the absence of calcic clinopyroxene (or some other fairly abundant mineral with values for REE distribution coefficients of the order of 0.3–0.5) from the suite of crystallizing minerals as an extreme condition for crystallization. We have used orthopyroxene as the mineral with low values for REE distribution coefficients. Values for olivine or opaques would yield roughly similar results.

The changes in average concentrations for the minerals in case 1 from Table 3 as a function of the fraction of the original liquid crystallized are shown in Fig. 4. It is

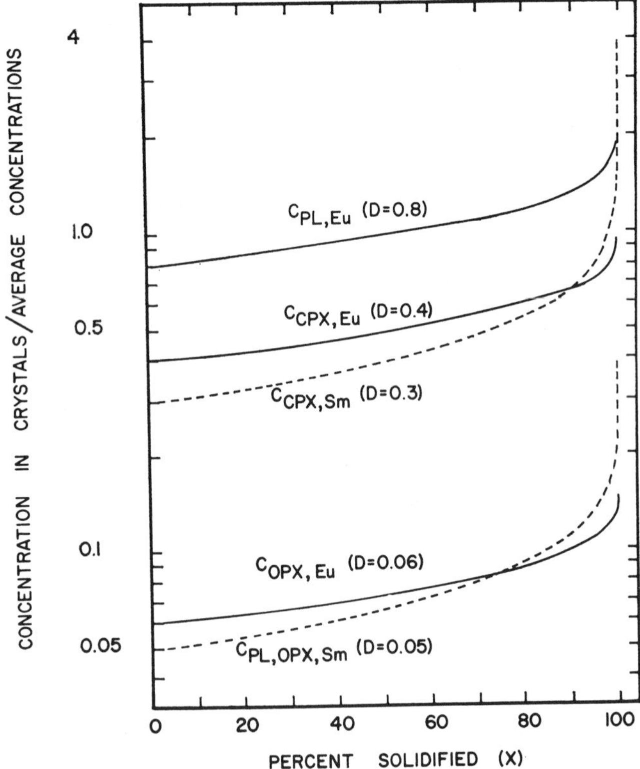


Fig. 4. Changes in concentrations of Sm and Eu in plagioclase, orthopyroxene, and clinopyroxene as described by the model for closed system crystallization of a silicate liquid. Parameters used correspond to case 1, Table 3.

seen that the most drastic changes in average concentrations occur within the last few percent of solidification. Table 3 gives the ratio of the final to the initial concentrations for Sm and for Eu, and it gives the ratio of the final ratio of the concentration of Sm to that of Eu to the initial ratio of concentrations for these elements.

As seen in Table 3, the largest changes in the average ratio of concentration of Sm to Eu occur for those systems in which crystallization of plagioclase is not accompanied by crystallization of another mineral with fairly high values (>0.1) of *D* for Eu and Sm. In such systems, depending on the values used for *D*, changes in the ratio of Sm to Eu of nearly 30 times were calculated. With clinopyroxene present, this change was much less, amounting to only 17 times. From these results, we conclude that the ratio of Sm to Eu, provided that clinopyroxene or some equivalent mineral is present, would be unlikely to change during crystallization by more than 15 times, and probably not more than five times.

With no clinopyroxene present, changes in concentrations of Sm of up to 30 times were estimated. With even as little as 10% clinopyroxene in the system, the maximum change in concentration of Sm is less than 20 times the initial concentration. We conclude that changes exceeding a factor of 20 are very unlikely and that changes of a factor of 8 are probable.

For Eu, no change in concentration was generated that was greater than a factor of 2.5, regardless of the amount of clinopyroxene or the choice of values of *D*. This results because the value of *D* for plagioclase is nearly one.

Two further cases are considered. It is mainly the plagioclase that separates Eu from Sm. Suppose we use the *D* and *f* values for case 1 but let the plagioclase crystallize entirely before the ortho- and clinopyroxenes begin to form. The changes in concentrations of Sm and Eu in the plagioclase can then be obtained through the use of equation (2) of our earlier paper (Haskin *et al.*, 1970). The change in concentration in the plagioclase for Sm is a factor of 1.2 and for Eu is a factor of 1.04. The change in the ratio is a factor of 1.15. Thus, early crystallization of plagioclase has a relatively small effect on the concentrations of Sm and Eu in the plagioclase and in the residual liquid.

At the other extreme, let the ortho and clinopyroxenes crystallize prior to any feldspar. Then the feldspar, because it must accept all the Sm and Eu in the remaining liquid, will end up with the same ratio of Sm to Eu as the liquid following crystallization of the pyroxenes. (The crystal will be strongly zoned with respect to its ratio of Sm to Eu.) Thus, we see that the maximum separation of Sm from Eu through mineral paragenesis occurs when feldspar crystallizes simultaneously with other minerals throughout solidification of the liquid, and especially during the final stages of crystallization.

Some effects of crystallization of real liquids not included in the model must be evaluated before the results of calculations based on the model can be applied to lunar rocks. Suppose that phenocrysts are present. If we presume that the concentrations of Sm and Eu in phenocrystic cores (of those mineral grains that have such cores) have solidified from the parent liquid at an early stage of solidification, then the effect of preferential incorporation of such cores into the analyzed mineral separate will be to lower the average concentrations of Sm and Eu and to decrease the average ratio of Sm to Eu in comparison with "typical" crystals of that mineral. The best estimate of both the initial concentrations and *D* values if *D* values are to be measured will be obtained from the cores of phenocrysts. In the case of a mineral separate, grains of all ages represented in the rock will be obtained. To the extent that grains of all ages are equally well represented, this will not affect the measurement. If the grains that began crystallizing after a significant fraction of the initial liquid had solidified are preferentially included in the mineral separate, or if crystallization of the mineral started late, the measured concentrations of Sm and Eu and the ratio of Sm to Eu will be too high compared with the true average expected from the model. If the latest portions crystallized are lost from the mineral crystals during separation, or if crystallization of the mineral stopped early, then the measured concentrations of Sm and Eu and the ratio of Sm to Eu will be lower than the average for the model. Even while the final dregs are producing a new suite of minerals (mesostasis) in a real rock, some further growth of older, major minerals probably continues. Note from Fig. 4 that the last percent of the entire mass of a mineral to crystallize contains such a high concentration of Sm or Eu that the average concentrations for the entire crystals are changed several fold. Mesostasis seldom comprises as much as 1% of the mass of a rock. Nor can this effect necessarily be avoided by grinding away the outermost layers of minerals to be analyzed, since growth is probably uneven, and high concentrations may occur only at interfaces with the last liquids, in tension cracks in cooling minerals, etc.

These effects are large enough to account for much of the spread in *D* values derived for REE from natural systems, for the apparent deficiencies of Eu in the *D* values for minerals other than feldspars, and for *D* values for Eu in feldspars that appear to be too low relative to the values for the REE (e.g., Schnetzler and Philpotts, 1970; Philpotts, 1970). Note, for example, the lower *D* values and absence of an anomalous value for Eu in the core of an augite phenocryst as compared with those based on analyses of the rim (Schnetzler and Philpotts, 1968). They are also sufficient to explain the distribution and concentrations of the REE and several other trace elements in separates of ilmenite and other minerals from lunar rocks (e.g., Philpotts and Schnetzler, 1970; Goles *et al.*, 1970). They probably account for the difficulties often encountered in trying to make a mass balance for trace elements for a rock from modes and mineral data. The presence of mesostasis with minerals in which the REE are essential components decreases somewhat the high Sm and Eu concentrations and the ratio of Sm to Eu predicted by the model. An increase in the relative proportion of feldspar as crystallization progressed would increase the observed ratio of Sm to Eu above the predicted value. The effects of changes in composition of the minerals and the liquids on values of distribution coefficients are hard to evaluate. There is some evidence that even fairly large changes in composition of the liquid phase may not severely affect values of *D* (Cullers, 1971). For Eu, it appears that values of *D* increase with increasing Na content in feldspars (Schnetzler and Philpotts, 1970).

Despite these effects, we can estimate limits from the model on the concentrations of Sm and Eu in the liquid from which an analyzed mineral grain (or separate) crystallized. The upper limit, obtained by merely dividing the concentration for the element by the value of the distribution coefficient for that element, assumes that the entire measured mineral was not part of a closed system. Within the framework of the model, this corresponds to presuming that the mineral grains as analyzed were produced at the very onset of crystallization. A virtually certain lower limit can be estimated by considering the extremes of conditions of crystallization and the values for the distribution coefficients, because all effects of closed-system crystallization increase the concentrations of elements that have values of $\sum D_i f_i < 1$ in crystals as solidification continues. The "virtually certain" upper limits used here were obtained from the more extreme values in Table 3 and are those listed above. "Probable" lower limits can be established if the extreme values of $D$ and conditions of crystallization are ignored. The values used here are those discussed above. In setting these lower limits, the mineral grains or separates analyzed are regarded as whole end products of crystallization of a liquid as a closed system.

We have used the upper and lower limits suggested by the model to estimate the range of concentrations for the parent liquids from which several analyzed lunar feldspar crystals were produced. The results of the calculations, along with independent evidence for the concentrations of Sm and Eu in the parent material, are given in Table 4. The first two samples considered are feldspar concentrated from two Apollo 11 fine-grained rocks analyzed by Philpotts and Schnetzler (1970). The concentrations in the whole rocks should represent those of the parent liquids from which these feldspars crystallized under conditions of a closed system. Note that the probable lower limits agree very well, within the limitations of the model, with the actual concentrations in both rocks. (Differences of a factor of 2 between predicted lower limits and true values would not exceed the uncertainties of this model.)

On observing that the model makes reasonable predictions for known cases, we consider the significance of the predictions for feldspars from rocks 14063 and 14313. The probable lower limits for feldspar from 14063 are the same for the mineral separate and the two grains. They are about the same as the observed concentrations in the igneous fragments separated from that breccia, and not too different from the values obtained by analysis of "fines" sieved from the whole rock. This suggests that these feldspars may be mineral grains crystallized as parts of igneous rocks similar to those that provided fragments and matrix of the breccia. The probable lower limits for feldspar grains from 14313 are very close to the concentrations and their ratio found in the "whole rock" sample, and an interpretation similar to that for the feldspars from 14063 seems plausible.

If, on the other hand, we regard the feldspars from these breccias as fragments of massive anorthosite, the upper limit to the concentrations and ratios of Sm and Eu would apply. This is best illustrated in the case of the anorthosite 15415. The large grain size and near absence of any mineral other than feldspar (e.g., Hargraves and Hollister, 1972; James, 1972) suggest that this rock is a portion of massive feldspar separated efficiently from a parent liquid, and containing very little trapped liquid from closed system crystallization. Thus, the composition of the parent liquid can best be predicted by dividing the Sm and Eu concentrations by their respective values of $D_{\text{pl}}$. This results in concentrations of Sm and Eu that are 14 times greater than the average for chondrites, whose REE concentrations can be considered as a first

Table 4. Estimates of upper and lower limits for concentrations of Sm and Eu (in ppm) and comparison with parent or possibly related materials.

| | | Probable upper limit | Probable lower limit | Virtually certain lower limit | Observed for "parent" rock | "Parent" |
|---|---|---|---|---|---|---|
| 10062 | Sm[a] | 100 | 13 | 5.0 | 14.7 | 10062 rock |
| | Eu | 3.4 | 1.4 | 0.9 | 2.07 | |
| | Sm/Eu | 18 | 5.9 | 2.5 | 7.1 | |
| 10044 | Sm[a] | 220 | 27 | 5.8 | 23.4 | 10044 rock |
| | Eu | 6.0 | 2.4 | 1.6 | 2.21 | |
| | Sm/Eu | 36 | 7.3 | 2.4 | 10.6 | |
| 14063 | Sm | 53 | 6.7 | 2.7 | 6.8 | 14063 ign. frag. |
| | Eu | 4.0 | 1.6 | 1.1 | 2.9 | |
| | Sm/Eu | 13 | 2.7 | 0.9 | 2.4 | |
| 14313 | Sm | 240 | 30 | 12 | 31 | 14313 whole rock |
| | Eu | 6.2 | 2.5 | 1.7 | 2.65 | |
| | Sm/Eu | 39 | 7.8 | 2.6 | 11.6 | |
| 15415 | Sm[b] | 2.4 | 0.20 | 0.15 | — | |
| | Eu | 1.0 | 0.41 | 0.26 | — | |
| | Sm/Eu | 2.4 | 0.48 | 0.16 | — | |
| 10085 | Sm[c] | 35 | 3.4 | 1.72 | 13.7 | 10084 fines |
| | Eu | 1.11 | 0.45 | 0.30 | 1.75 | |
| | Sm/Eu | 31 | 6.0 | 2.1 | 7.8 | |
| 10085 | Sm[b] | 17.5 | 2.2 | 0.85 | 13.7 | 10084 fines |
| | Eu | 0.81 | 0.33 | 0.22 | 1.75 | |
| | Sm/Eu | 21.6 | 4.4 | 1.5 | 7.8 | |
| 10085 | Sm[b] | 77.5 | 9.7 | 4.4 | 13.7 | 10084 fines |
| | Eu | 1.05 | 0.43 | 0.28 | 1.75 | |
| | Sm/Eu | 74 | 15 | 4.9 | 7.8 | |
| 12033 | Sm[b] | 57 | 7.1 | 2.8 | 21.5 | 12033 fines |
| | Eu | 3.3 | 1.3 | 0.88 | 2.31 | |
| | Sm/Eu | 17.3 | 3.5 | 1.2 | 9.3 | |
| 12033 | Sm | | | | 38–62 | 12033 KREEP |
| | Eu | | | | 2.6–3.7 | |
| | Sm/Eu | | | | 12–1.8 | |

[a] Data from Philpotts and Schnetzler (1970).
[b] Data from Hubbard *et al.* (1972).
[c] Data from Wakita and Schmitt (1970).

approximation for the starting material from which a silicate planetary body might form. The predicted ratio of concentrations of Sm to Eu of 2.4 is nearly equal to that of the chondrites (2.6). Similar estimates were made by Hubbard *et al.* (1972). This estimate of concentration supports our suggestion based on REE concentrations of Apollo 11 basalts (Haskin *et al.*, 1970) that the materials from which lunar surface rocks derived were already considerably enriched in REE (and other refractory trace elements) over the REE concentrations in chondrites before significant separation among members of the REE group began.

The high concentrations of Sm (compared with the chondritic values) in the parent liquids for the feldspars from breccias 14063 and 14313 as predicted by the probable upper limits are so uncomfortably high that it is easier to regard the feldspars as minerals from closed system crystallization. Nevertheless, chilled liquids (KREEP) of sufficiently high REE concentrations have been observed (Hubbard and Gast,

1971) that an origin as massive anorthosite for the feldspars from 14063 and 14313 cannot be ruled out. Data for fragments of anorthosite reported by Hubbard *et al.* (1972) and by Wakita and Schmitt (1970) are included in Table 4. The accompanying values for possible parent materials are analytical results for fine-fines samples 10084 and 12033 (Haskin *et al.*, 1970, 1971) and 12033 KREEP (Hubbard and Gast, 1971). As pointed out by Hubbard *et al.* (1972), the treatment of the anorthosite fragments as pieces of massive anorthosite does indicate a parent liquid with Sm and Eu concentrations in the range found for the KREEP fragments. We find no clear evidence to favor an origin as massive anorthosite over an origin as crystals from closed system crystallization for the anorthosite fragments from the Apollos 11 and 12 coarse fines or from the Apollo 14 breccias. It is difficult to imagine that large quantities of massive anorthosite on the lunar surface could have been derived from KREEP because of the extreme extent of differentiation of more primative material to produce KREEP in the first place.

## Eu Deficiencies in Lunar Basalts

Now that a large fragment of massive anorthosite (15415) has been analyzed for REE (Hubbard *et al.*, 1972), it is clear that there exists feldspar on the moon that contains significant quantities of Eu (0.8 ppm) but very little Sm (0.048 ppm). Suppose that, as a crude first estimate to the starting material from which lunar basalts came, we calculate the mass of anorthosite with Sm and Eu concentrations like those of 15415 that would have to be added to those basalts to produce a final product with a ratio of Sm concentration to Eu concentration of 2.6, the same as found in chondritic meteorites. The results of this calculation are given in Table 5. The averages for the Apollo 11 rocks are from Haskin *et al.* (1970), those for Apollo 12 rocks are from Haskin *et al.* (1971) and include only those Apollo 12 rocks that are low in olivine phenocrysts, those for 14053 and 14310 are from this paper, and those for KREEP are the upper and lower limits of the data of Hubbard and Gast (1971).

The interesting result of this calculation is found in the next to the last column of Table 5, in which the concentration of Sm in the mixture has been divided by the mass of the mixture. The value for ppm Sm in all cases falls between 2.4 and 2.6 ppm. This suggests strongly a starting material of common composition for the average Apollo

Table 5. Average concentrations of Sm and Eu (in ppm) for lunar basalts, values of Eu′ corresponding to ppm Sm/ppm Eu = 2.6, quantities of 15415 anorthosite containing the "missing" Eu per gram of basalt, concentrations of $Sm_0$ in the resulting mixture of 15415 and basalt, and fractions of the mixtures that are 15415.

| | Sm | Eu | Eu′ | Δ Eu | $\frac{\Delta Eu}{0.806}$ | $Sm_0$ | $1 - f_b = x$ |
|---|---|---|---|---|---|---|---|
| 11A | 20 | 2.1 | 7.6 | 5.5 | 6.8 | 2.6 | 0.88 |
| 11B | 10 | 1.4 | 3.8 | 2.4 | 3.0 | 2.5 | 0.75 |
| 12 | 5.8 | 1.23 | 2.2 | 0.97 | 1.2 | 2.6 | 0.57 |
| 14310 | 25 | 2.1 | 9.5 | 7.4 | 9.2 | 2.5 | 0.90 |
| 14053 | 6.5 | 1.2 | 2.5 | 1.3 | 1.6 | 2.5 | 0.62 |
| 12033* | 38–62 | 2.6–3.7 | 14.5–23.6 | 12–20 | 15–25 | 2.4 | 0.93–0.96 |

* Data from Hubbard and Gast (1971).

11, 12, 14 basalts, and KREEP (which according to its major element composition is also basaltic). Taking 2.5 ppm as the Sm concentration for the starting material, the data fit roughly the linear equation.

$$C_{\mathrm{Sm},b} = 17.3\, C_{\mathrm{Eu},b} - 14.0, \tag{11}$$

in which $C_{\mathrm{Sm},b}$ and $C_{\mathrm{Eu},b}$ are the Sm and Eu concentrations of the basalts.

Equation (11) can be derived if it is assumed that a common starting material is always divided into two parts, the parts varying in relative sizes. One part contains no Sm and a constant amount of Eu (0.8 ppm), the other part contains all the Sm and the rest of the Eu. Using subscripts 0 to represent this hypothetical starting material, $a$ to represent the material with constant concentration of Eu, $b$ to represent the remainder (basalt), and $f_b$ to represent the fraction of the total mass of the system that the basalt portion makes up, we get from considerations of mass balance

$$C_{\mathrm{Sm},0} = C_{\mathrm{Sm},b} f_b + C_{\mathrm{Sm},a}(1 - f_b) \tag{12}$$

and

$$C_{\mathrm{Eu},0} = C_{\mathrm{Eu},b} f_b + C_{\mathrm{Eu},a}(1 - f_b). \tag{13}$$

Combining, and letting $C_{\mathrm{Sm},a} \approx 0$, we obtain

$$C_{\mathrm{Sm},b} = \frac{C_{\mathrm{Sm},0}}{C_{\mathrm{Eu},a} - C_{\mathrm{Eu},0}} C_{\mathrm{Eu},b} + \frac{C_{\mathrm{Sm},0}}{C_{\mathrm{Eu},a} - C_{\mathrm{Eu},0}} C_{\mathrm{Eu},a}. \tag{14}$$

Adding the constraints that $C_{\mathrm{Eu},a} = 0.8$ and $C_{\mathrm{Sm},0}/C_{\mathrm{Eu},0} = 2.6$ (chondritic ratio), we obtain equation (11). The parameters of slope and intercept for any straight line are, of course, unique. Thus, to the extent that the data for lunar basalts fit this line it can be implied that those basalts and 15415 are complementary parts of starting material with the concentrations of Sm and Eu that are determined by the slope and intercept of the line. Based on this assumption, the fraction of the starting material that has crystallized to produce each group of basalts $f_b$ can be calculated from equation (13). The results are given in the last column of Table 5.

The line corresponding to equation (11) is shown in Fig. 5, along with the points corresponding to the data given in the first two columns of Table 5. Clearly, the fit is not perfect. The samples with the highest concentrations really require removal of more Eu than is implied by the equation. More surprising is that the model implied by equation (11) fits the data as well as it does. From considerations of partial melting and fractional crystallization, separation of one phase with a constant concentration of Eu is not reasonable; the concentration in that phase should increase as the concentration in the residual phase increases (Haskin *et al.*, 1970).

The fraction of the starting material that had to crystallize to form each of the groups of basalts (last column, Table 5) corresponds to the fraction of the system solidified ($x$) in fractional crystallization (equation (1), Haskin *et al.*, 1970). The values for $x(= 1 - f_b)$ in Table 5 can be compared with the values predicted by Fig. 2 of Haskin *et al.* (1970) for the corresponding increase in Sm concentration over the original, using $D_{\mathrm{Sm}} = 0.02$ (which corresponds closely to the curve for $D = 0.01$ in

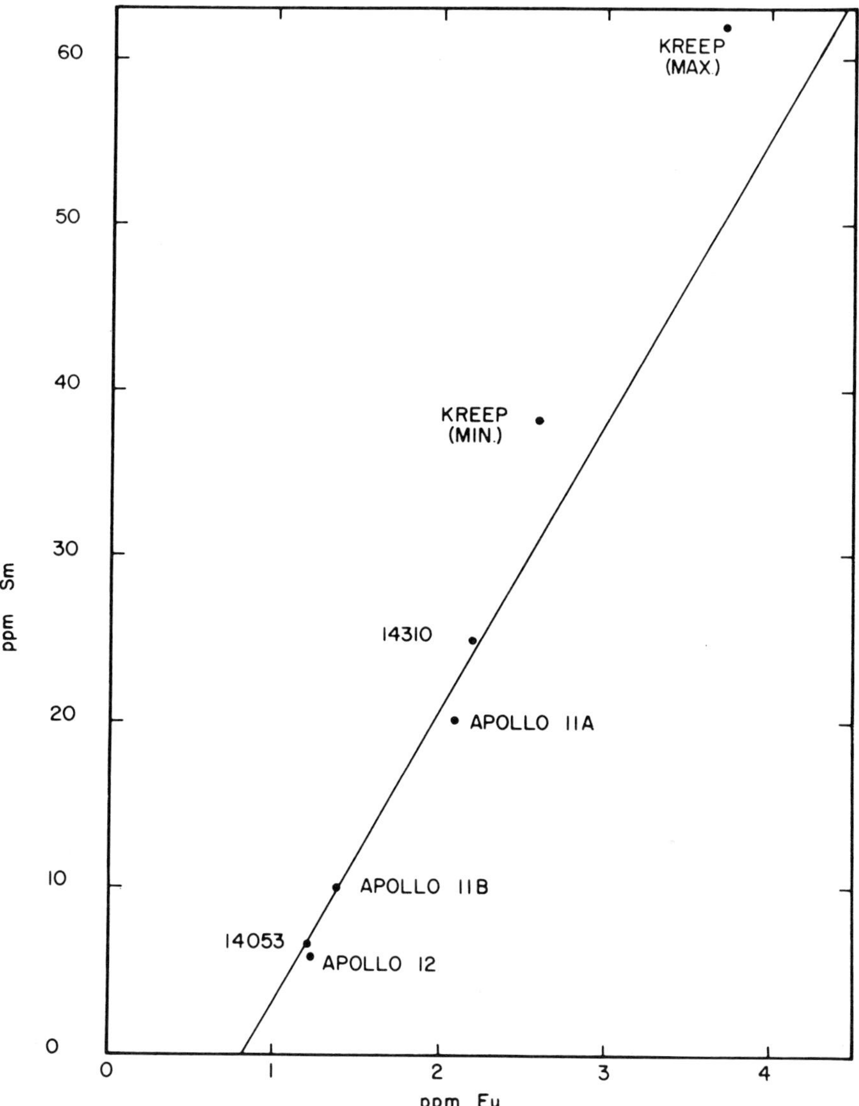


Fig. 5. Graph of ppm Sm versus ppm Eu and line corresponding to equation (11).

Fig. 2 of Haskin *et al.*, 1970). The agreement is excellent. Apparently, the linear equation (11) gives a good approximation to the results of the logarithmic equations for Sm over the range of values of $x$ considered. Extrapolating to pure starting material on the logarithmic curve (Fig. 2, Haskin *et al.*, 1970), instead of by the use of equation (11), would not lower the concentrations of Sm and Eu in the hypothetical starting material by as much as a factor of 2.

## Some Conclusions and Difficulties

Several independent lines of evidence lead to similar conclusions about the parent liquids for lunar basalts and massive anorthosites.

The parent material for 15415 feldspar was estimated to contain about 2.4 ppm Sm and 1 ppm Eu. This result was based upon theoretical considerations of fractional crystallization and on values of distribution coefficients for Sm and Eu of 0.02 and 0.8. Those $D$ values were derived from studies of terrestrial layered intrusions. The approximately chondritic ratio of the concentration of Sm to that of Eu in the estimated starting material is a conclusion, not an assumption, in this model. No considerations of lunar basalts enter into these estimates.

The parent material for Apollo 11, 12, and 14 lunar basalts and KREEP can be estimated from the linear equation (14). This equation uses data for Sm and Eu from the lunar basalts, plus two more conditions, namely, that these basalts could all be derived from a common starting material with a chondritic ratio of the concentration of Sm to that of Eu, and that the basalts were residues following removal of varying amounts of a single material with a constant concentration of Eu and essentially no Sm. Although the idea for this model was arrived at through data for 15415, it is completely independent of the concentrations of Sm and Eu found in 15415. It nevertheless predicts concentrations of Sm and Eu for the starting material that are essentially the same as those predicted by the previous model which was based on entirely independent assumptions and data. This latter model makes no use of $D$ values.

Finally, when the extent of crystallization based on the above model for the basalts is compared with results for a fractional crystallization model in which $D$ values of 0.02 for Sm and 0.8 for Eu are used, the results are equivalent. This means that, within the limits of uncertainty, those values for $D$ are essentially correct for the derivation of the basalts from the above common starting material. They are the same values of $D$ chosen on the basis of terrestrial studies (plus considerations of oxygen fugacities) and used in the estimate based on the data for the anorthosite 15415. An extrapolation of the logarithmic model for fractional crystallization that would extend the results from equation (11) to $x = 0$ would not decrease the Sm and Eu concentrations of the starting material as derived from the two above models by as much as a factor of 2.

Thus, we conclude that there may have been a common material near the lunar surface that produced both massive anorthosite and the lunar basalts. This material had approximately 2.5 ppm Sm, approximately 1 ppm Eu, and an approximately chondritic ratio of concentration of Sm to Eu. It was thus about 14 times richer in REE than chondrites. This suggests a very efficient extraction of REE (and many other trace elements) with no appreciable fractionation of Sm from Eu from a minimum amount of primordial matter of about 15 times the mass of the proposed parent material of the basalts and anorthosite. The residue would presumably contain little or no feldspar, and would be more mafic than the lunar basalts. It may be a residue from partial melting, somewhat less probably (Haskin *et al.*, 1970) from fractional crystallization, or possibly both. It may contain sufficient amounts, in the form of trapped liquid, of the parent material for the basalts that the total amount of residue may significantly exceed 15 times the amount of starting material from which the basalts and anorthosite formed. This quantity of residue is based on the notion that the concentrations of Sm and Eu in the matter from which planetary bodies formed are similar to those in chondrites. Thus, the bulk of the residue from the starting material may not be in the moon but elsewhere in the solar system.

This suggestion of a common parent material for the anorthosite and the lunar basalts is based mostly on considerations of equilibrium partial melting and fractional crystallization, and is based only on considerations of Sm and Eu and, particularly, incorporation into feldspar as the only efficient mechanism for the separation of Eu from the other REE. It is offered here not as a definite or final answer to the way in which the basalts are related to each other, to the anorthosite, and to primordial matter. Rather, it is given because it demonstrates some very self-consistent relationships among Sm and Eu in several lunar (and terrestrial) materials. Taken in its present form, the relationship proposed for the basalts and anorthosite has some severe flaws.

The major element concentrations of the basalts (e.g., Hubbard and Gast, 1971; Compston *et al.*, 1972) do not reflect increasing loss of feldspar from an original liquid. Except for 14310 (Walker *et al.*, 1972) and possibly KREEP the basalts as liquids apparently were not in equilibrium with plagioclase (e.g. Ringwood and Essene, 1970; Green *et al.*, 1971). For this reason, some investigators have proposed models of nonequilibrium partial melting to explain the Eu depletions in lunar basalts (Green *et al.*, 1971; Goles *et al.*, 1971).

*Acknowledgments*—We thank the crew of the University of Wisconsin Nuclear Reactor for irradiating the samples with neutrons. We thank the Apollo 14 astronauts and the personnel of NASA whose efforts provided these samples for analysis. We thank D. Schauwecker for helpful discussions about models for fractional crystallization. This work was supported by the National Aeronautics and Space Administration under Grant NGL-50-002-148.

## References

Allen R. O., Haskin L. A., Anderson M. R., and Muller O. (1970) Neutron activation analysis for 39 elements in small or precious geologic samples. *J. Radioanal. Chem.* **6,** 115–137.

Compston W., Vernon M. J., Berry H., Rudowski R., Gray C. M., and Ware N. (1972) Age and petrogenesis of Apollo 14 basalts. (abstract). In *Lunar Science—III* (editor C. Watkins), pp. 151–153, Lunar Science Institute Contr. No. 88.

Cullers R. (1971) The partitioning of the rare-earth elements among rock-forming silicate phases and water. Ph.D. thesis, University of Wisconsin, Madison, Wisconsin.

Denechaud E. B., Helmke P. A., and Haskin L. A. (1970) Analysis for the rare-earth elements by neutron activation and Ge(Li) spectrometry. *J. Radioanal. Chem.* **6,** 97–113.

Floran R. J., Cameron K., Bence A. E., and Papike J. J. (1972) The 14313 consortium: A mineralogic and petrologic report (abstract). In *Lunar Science—III* (editor C. Watkins), pp. 268–270, Lunar Science Institute Contr. No. 88.

Ganapathy R., Keays R. R., Laul J. C., and Anders E. (1970) Trace elements in Apollo 11 lunar rocks: Implications for meteorite influx and origin of moon. *Proc. Apollo 11 Lunar Sci. Conf., Geochim. Cosmochim. Acta* Suppl. 1, Vol. 2, pp. 1117–1142. Pergamon.

Goles G. G., Randle K., Osawa M., Lindstrom D. J., Jerome D. Y., Steinborn T. L., Beyer R. L., Martin M. R., and McKay S. M. (1970) Interpretations and speculations on elemental abundances in lunar samples. *Proc. Apollo 11 Lunar Sci. Conf., Geochim. Cosmochim. Acta* Suppl. 1, Vol. 2, pp. 1177–1194. Pergamon.

Goles G. G., Duncan A. R., Lindstrom D. J., Martin M. R., Beyer R. L., Osawa M., Randle K., Meek L. T., Steinborn T. L., and McKay S. M. (1971) Analyses of Apollo 12 specimens: Compositional variations, differentiation processes, and lunar soil mixing models. *Proc. Second Lunar Sci. Conf., Geochim. Cosmochim. Acta* Suppl. 2, Vol. 2, pp. 1063–1081. MIT Press.

Green D. H., Ringwood A. E., Ware N. G., Hibberson W. O., Major A., and Kiss E. (1971) Experimental petrology and petrogenesis of Apollo 21 basalts. *Proc. Second Lunar Sci. Conf., Geochim. Cosmochim. Acta* Suppl. 2, Vol. 1. pp. 601–615. MIT Press.

Hargraves R. B. and Hollister L. S. (1972) Mineralogic and petrologic study of lunar anorthosite slide 15415, 18. *Science* **175**, 430–432.

Haskin L. A., Allen R. O., Helmke P. A., Paster T. P., Anderson M. L., Korotev R. L., and Zweifel K. A. (1970) Rare earths and other trace elements in Apollo 11 lunar samples. *Proc. Apollo 11 Lunar Sci. Conf., Geochim. Cosmochim. Acta* Suppl. 1, Vol. 2, pp. 1213–1231. Pergamon.

Haskin L. A., Helmke P. A., Allen R. O., Anderson M. R., Korotev R. L., and Zweifel K. A. (1971) Rare-earth elements in Apollo 12 lunar materials. *Proc. Second Lunar Sci. Conf., Geochim. Cosmochim. Acta* Suppl. 2, Vol. 2, pp. 1307–1317. MIT Press.

Helmke P. A. and Haskin L. A. (1972) Rare earths and other trace elements in Luna 16 soil. *Earth Planet. Sci. Lett.* **13,** 441–443.

Hubbard N. J. and Gast P. W. (1971) Chemical composition and origin of nonmare lunar basalts *Proc. Second Lunar Sci. Conf., Geochim. Cosmochim. Acta* Suppl. 2, Vol. 2, pp. 999–1020. MIT Press.

Hubbard N. J., Gast P. W., Meyer C., Nyquist L. E., and Shih C. (1972) Chemical composition of lunar anorthosites and their parent liquids. *Earth Planet. Sci. Lett.* (in press).

James O. (1972) Lunar anorthosite 15415: Texture, mineralogy, and metamorphic history. *Science* **175,** 432–436.

Morgan J. W., Laul J. C., Krahenbuhl U., Ganapathy R., and Anders E. (1972) Major impacts on the moon: Chemical characterization of projectiles (abstract). In *Lunar Science—III* (editor C. Watkins), pp. 552–554, Lunar Science Institute Contr. No. 88.

Papanastassiou D. A. and Wasserburg G. J. (1971) Rb–Sr ages of igneous rocks from the Apollo 14 mission and the age of the Fra Mauro formation. *Earth Planet. Sci. Lett.* **12,** 36–48.

Philpotts J. A. (1970) Redox estimation from a calculation of $Eu^{2+}$ and $Eu^{3+}$ concentration in natural phases. *Earth Planet Sci. Lett.* **9,** 257–268.

Philpotts J. A. and Schnetzler C. C. (1970) Apollo lunar samples: K, Rb, Sr, Ba, and rare-earth concentration in some rocks and separated phases. *Proc. Apollo 11 Lunar Sci. Conf., Geochim. Cosmochim. Acta* Suppl. 1, Vol. 2, pp. 1471–1486. Pergamon.

Ringwood A. E. and Essene E. (1970) Petrogenesis of lunar basalts and the internal constitution and origin of the moon. *Science* **167,** 607–610.

Schnetzler C. C. and Philpotts J. A. (1968) Partition coefficients of rare-earth elements and barium between igneous matrix materials and rock-forming minerals phenocrysts. *Origin and Distribution of the Elements* (editor L. H. Ahrens), pp. 929–938. Pergamon.

Schnetzler C. C. and Philpotts J. A. (1970) Partition coefficients of rare-earth elements between igneous matrix material and rock-forming mineral phenocrysts, II. *Geochim. Cosmochim. Acta* **34,** 331–340.

Wakita H. and Schmitt R. A. (1970) Lunar anorthosites: Rare-earth and other elemental abundances. *Science* **170,** 969–974.

Wakita G., Showalter D. L., and Schmitt R. A. (1972) Bulk, REE, and other abundances in Apollo 14 soils (3), clastic (1), and igneous (1) rocks (abstract). In *Lunar Science—III* (editor C. Watkins), pp. 767–769, Lunar Science Institute Contr. No. 88.

Walker D., Longhi J., and Hays J. F. (1972) Experimental petrology and origin of Fra Mauro rocks and soil (abstract). In *Lunar Science—III* (editor C. Watkins), pp. 770–772, Lunar Science Institute Contr. No. 88.

Proceedings of the Third Lunar Science Conference
(Supplement 3, *Geochimica et Cosmochimica Acta*)
Vol. 2, pp. 1293–1305
The M.I.T. Press, 1972

# Apollo 14: Some geochemical aspects

JOHN A. PHILPOTTS, C. C. SCHNETZLER, DAVID F. NAVA,
MICHAEL L. BOTTINO,* PAUL D. FULLAGAR,* HERMAN H. THOMAS,
SHUFORD SCHUHMANN, and CHARLES W. KOUNS

Planetology Branch, Goddard Space Flight Center,
Greenbelt, Maryland 20771

**Abstract**—Chemical analyses have been obtained for five samples of Apollo 14 regolith fines, three 14230 core samples, 14049 soil clod, 14305 and 14319 breccias, 14310 basalt, and some separated phases. The chemical uniformity of these Apollo 14 samples indicates thorough mixing and/or uniform source rocks. Basalt 14310 can be matched well in composition by a four to one mixture of soil and plagioclase. $Eu^{2+}/Eu^{3+}$ ratios calculated for 14310 pigeonite and plagioclase are similar to those for Apollo 12 and 15 mare-type basalt phases; this indicates similar redox conditions. Our Apollo 14 samples are chemically similar to Apollo 12 and 15 KREEP as distinct from Apollo 11, 12, and 15, and Luna 16 mare-type basalts. A relationship, relating the two types of basalt and in which mare-basalts would represent fused cumulates, is suggested.

## INTRODUCTION

APOLLO 14 WAS THE FIRST MISSION to visit an uplands (i.e., nonmare) region of the moon. The Fra Mauro formation, which underlies the landing site, is thought to represent throwout produced during an impact that formed the Imbrium Basin (LSPET, 1971). Thus the Apollo 14 returned lunar samples were expected to represent material sampled at some depth within the moon and to be older than most mare-basalts, at least those at the surface. Subsequent dating of these samples (Turner *et al.*, 1971; Papanastassiou and Wasserburg, 1971; Compston *et al.*, 1971) bore out this expected age relationship. Preliminary chemical studies of the Apollo 14 samples, almost all of which are fragmental, revealed affinities to the dark component of 12013 breccia and the exotic component in the Apollo 12 soils rather than to the Apollo 11 and 12 mare-basalts (LSPET, 1971). The purpose of the present paper is to report chemical data we have obtained to date on Apollo 14 samples and to discuss briefly some of the implications. Data obtained in this laboratory for two Apollo 14 soils have been discussed previously (Schnetzler and Nava, 1971).

## SAMPLES

*Soils:*

14003,32 $<1$ mm (90% of total mass), contingency fines.
14148,29 $<1$ mm (93% of total mass), surface of trench fines, Station G.
14149,45 $<1$ mm (72% of total mass), bottom of trench fines, Station G.
14156,29 $<1$ mm (91% of total mass), middle of trench fines, Station G.
14421,24 $<10$ mm, reserve from unsieved bulk fines.

---

* Presently at University of North Carolina, Chapel Hill, North Carolina.

*Core:* 14230,105 fines 12 cm below top of drive-tube liner, Station G.
14230,114 fines 18 cm below top of liner, Station G.
14230,123 fines 22 cm below top of liner, Station G.

*Soil clod:* 14049,32 extremely friable breccia (coherent soil), 80 m east of Station B. This sample was separated into 4 sieve fractions: > 100, < 100, < 200, and < 325 mesh.

*Breccias:* 14305,121 fine sawdust, 40 m west of the comprehensive sample site.
14319,8 Station H. Sample received as chip and subsequently separated on the basis of color and mineralogy. Fraction C represents dark breccia clasts and fraction E, light colored breccia.

*Crystalline:* 14310,131 basalt, Station G. Sample received as chips. 430 mg were taken for a whole-rock sample. Mineral separates of the remainder yielded 370 mg of pyroxene and 200 mg of plagioclase.

Phase separations were carried out utilizing heavy-liquid, electromagnetic, and hand-picking techniques. Any sieving of samples was through nylon mesh. Any grinding of samples was on boron carbide.

## Analytical Procedures and Results

Determination of most major (>1%) and minor elements (0.01–1%) in cosmochemical materials by atomic absorption spectrophotometry (AAS) using samples as small as 0.05 g has been demonstrated (Nava, 1970; Maxwell *et al.*, 1970; Nava *et al.*, 1971; Schnetzler and Nava, 1971) to yield analytical results having precisions and accuracies equal to, or better than, those obtained with classical chemical methods, owing to the main atomic absorption features of high analytical sensitivity and relative freedom from chemical interferences. Due to the limited quantities available, a combined semi-micro analytical scheme of atomic absorption and colorimetric spectrophotometry was devised to provide accurate bulk chemical analysis for all of the major (Si, Ti, Al, Mg, Ca, Fe) and most of the minor (Na, K, Mn, P, Cr) constituents of these Apollo 14 samples.

Prior to undertaking the elemental analyses, each Apollo 14 sample was optically examined to check for obvious contaminants from grinding, sawing, and/or packaging (plastic or aluminium vial particles). All samples appeared to be acceptable and aliquants of approximately 50 mg (duplicates when available, as listed in Table 1) were weighed on a precision microbalance and prepared without further grinding (either as received from MSC or as yielded by the mineral and grain size fraction separations in our laboratory) by the teflon decomposition, HF–$H_3BO_3$ matrix procedure described by Bernas (1968). All samples were quantitatively diluted to a volume of 100 ml, which included 0.5 ml aqua regia, 3 ml HF, and 2.8 g $H_3BO_3$. The AAS analysis method (Bernas, 1968) was modified, however, in the following ways in order to maximize accuracy: Close bracketing sets of synthetically prepared (from Matthey Specpure and Spex Industries spectrographic grade metals and reagent compounds and J. T. Baker acids) composite analytical standards, containing the equivalent fluoborate acid matrix concentration, were used for calibration of the double-beam atomic absorption and colorimetric spectrophotometers. All of the AAS determinations were made on the original sample solution volume, without further dilution, by means of appropriate selection of burner heads and rotations. Samples were analyzed in small-numbered sets, each with their own set of composite analytical standards, in order to minimize instrumental variables such as drift. The solution matrix of fluoborate, in addition to providing a common interference suppressant matrix for AAS determination of Si, Al, and Ca (acetylene-nitrous oxide flame) and of Fe (total), Mg, Na, K, Mn, and Cr (acetylene-air flame), served as a complexing agent to eliminate iron interference in colorimetric determinations of Ti by $H_2O_2$. Another aliquot of the AAS solution was taken for colorimetric P via molybdivanadophosphoric acid complex. Due to limited sample availability and anticipated low concentrations in these lunar materials, $H_2O^-$ and $H_2O^+$ were not determined.

Sample analysis sets consisted of: (a) 14003, 14148, 14149, 14156, 14421, and 14310 whole rock: (b) 14305, 14049 sized fractions, 14230 cores, and 14319 light and dark fractions; (c) 14310 feldspar; and (d) 14310 pyroxene. Reagent blanks were monitored during all analyses. Accuracy of this combined semi-micro analytical scheme was monitored by simultaneous analyses of U.S. Geological Survey reference silicates W-1, BCR-1, and AGV-1, and can be assessed from the results listed in Table 1, along with recent recommended literature values for comparison. Relative total precision (decomposition, volumetric, plus instrumental) of this method for the Apollo 14 duplicates and U.S.G.S. reference replicates was $\pm 1\%$, or better, for most elements.

Sample 14305, a breccia sawdust, which by optical observation appeared to be clean, was analyzed for copper by AAS (and found to be $<0.01\%$) because some sawdust samples were known to include segments from a copper-wire saw (Rose, personal communication, 1972).

The trace-element results given in Table 2 were obtained by mass-spectrometric stable-isotope dilution. Chemical procedures utilized for the soil samples were those used previously in our Apollo 11 and 12 work (Philpotts and Schnetzler, 1970; Schnetzler and Philpotts, 1971). Rock 14310 and its separates underwent the slightly modified chemistry utilized in the analysis of Luna 16 soils and 14305 that used for the Luna 16 rock chips (Philpotts *et al.*, 1972). Blanks prepared along with the Apollo 14 samples have not yet been analyzed. However, blank values obtained using each of the chemical techniques indicate generally negligible corrections; the largest corrections (using the previously determined blank values) were for Zr ($\sim 2\%$), Lu ($\sim 1.4\%$), and K ($\sim 1\%$) values for the plagioclase separated from 14310. All the values reported in Table 2 are for single determinations. Precision and accuracy of previous analyses (Philpotts *et al.*, 1972) of standard rock monitors indicate that all the Apollo 14 data are probably good to within less than $\pm 3\%$ relative.

## Homogeneity of Apollo 14 Samples

It is apparent from the data in Tables 1 and 2 and those for 14163 and 14259 (Schnetzler and Nava, 1971), that the Apollo 14 soil samples analyzed in this laboratory are quite homogeneous in chemical composition. For these surficial fines, trench and core samples, and the soil clod, the largest ranges in element concentrations, expressed in terms of the percent deviations from the mean of the extreme values, were: Na 12; P 11; K 9; Ti, Mn, Cr 8; Zr 7; Al, Li 6; Mg, Rb 5; and Fe 4. Ranges

Table 1. Major element composition (in wt.%) of Apollo 14 samples and standard rock monitors (analyst: D. F. Nava).

| | Soils | | | | | Core | | | Soil clod 14049,32 | | |
|---|---|---|---|---|---|---|---|---|---|---|---|
| Constituent | 14421,24 | 14003,32 | 14148,29 | 14156,29 | 14149,45 | 14230,105 | 14230,114 | 14230,123 | < 325 mesh | < 200 mesh | < 100 mesh |
| $SiO_2$ | 48.4 | 48.2 | 48.5 | 48.1 | 48.0 | 48.2 | 48.4 | 48.0 | 48.4 | 48.4 | 48.3 |
| $TiO_2$ | 1.70 | 1.75 | 1.71 | 1.71 | 1.61 | 1.66 | 1.73 | 1.62 | 1.77 | 1.75 | 1.75 |
| $Al_2O_3$ | 17.52 | 17.69 | 17.38 | 17.15 | 17.45 | 17.27 | 17.03 | 17.60 | 16.74 | 15.65 | 16.20 |
| MgO | 9.41 | 9.52 | 9.66 | 9.55 | 9.54 | 9.43 | 9.54 | 9.45 | 9.32 | 10.23 | 10.00 |
| CaO | 10.49 | 10.32 | 10.40 | 10.16 | 10.54 | 10.78 | 10.94 | 10.80 | 10.96 | 10.64 | 10.63 |
| $Na_2O$ | 0.68 | 0.69 | 0.71 | 0.72 | 0.75 | 0.70 | 0.70 | 0.70 | 0.73 | 0.71 | 0.74 |
| $K_2O$ | 0.52 | 0.55 | 0.53 | 0.53 | 0.58 | 0.56 | 0.56 | 0.56 | 0.58 | 0.55 | 0.60 |
| FeO | 10.54 | 10.53 | 10.55 | 10.55 | 9.95 | 10.17 | 10.21 | 10.09 | 10.37 | 10.86 | 10.83 |
| MnO | 0.14 | 0.13 | 0.13 | 0.14 | 0.13 | 0.12 | 0.13 | 0.13 | 0.14 | 0.14 | 0.14 |
| $P_2O_5$ | 0.48 | 0.49 | 0.50 | 0.49 | 0.48 | 0.48 | 0.48 | 0.49 | 0.50 | 0.40 | 0.44 |
| $Cr_2O_3$ | 0.19 | 0.18 | 0.19 | 0.19 | 0.19 | 0.19 | 0.19 | 0.19 | 0.20 | 0.21 | 0.20 |
| $H_2O^-$ | — | — | — | — | — | — | — | — | — | — | — |
| Sum of constituents determined | 100.07 | 100.05 | 100.26 | 99.29 | 99.22 | 99.56 | 99.91 | 99.63 | 99.71 | 99.54 | 99.83 |
| Sample weights, mg | 49.8 | 50.1 | 49.8 | 50.0 | 49.9 | 50.1 | 49.9 | 50.0 | 50.0 | 50.1 | 49.9 |
| | 50.1 | 49.9 | | | | | | | 49.8 | | |

Table 1—*continued*.

Table 1. Major element composition (in wt.%) of Apollo 14 samples and standard rock monitors (analyst: D. F. Nava).

| | Breccias | | | Basalt 14310,131 | | | Standard rock monitors | | | | | | | |
|---|---|---|---|---|---|---|---|---|---|---|---|---|---|---|
| | | 14319,8 | | | | | BCR-1 | | | W-1 | | | AGV-1 | |
| Constituent | 14305,121 | -L, dark | -E, light | Whole rock | Feldspar | Pyroxene | Split 22 * | Position 22 † | Position 22 ‡ | * | † | ‡ | Split 70 † | Position 12 ‡ |
| $SiO_2$ | 48.2 | 47.7 | 49.3 | 48.3 | 46.6 | 50.6 | 54.6 | 54.6 | 54.58 | 52.7 | 52.7 | 52.64 | 59.4 | 58.99 |
| $TiO_2$ | 1.71 | 1.63 | 1.82 | 1.25 | 0.10 | 1.28 | 2.19 | 2.20 | 2.23 | 1.06 | 1.11 | 1.07 | 1.06 | 1.08 |
| $Al_2O_3$ | 15.16 | 16.40 | 15.76 | 20.74 | 32.84 | 8.84 | 13.65 | 13.61 | 13.66 | 15.04 | 15.01 | 14.85 | 17.32 | 17.01 |
| MgO | 11.12 | 10.71 | 10.12 | 8.00 | 0.63 | 17.57 | 3.45 | 3.44 | 3.28 | 6.81 | 6.73 | 6.62 | 1.69 | 1.49 |
| CaO | 10.13 | 10.33 | 9.67 | 11.61 | 17.57 | 7.11 | 7.11 | 7.11 | 6.95 | 10.86 | 10.81 | 10.96 | 4.87 | 4.98 |
| $Na_2O$ | 0.87 | 0.86 | 0.84 | 0.76 | 0.93 | 0.34 | 3.33 | 3.25 | 3.31 | 2.09 | 2.16 | 2.15 | 4.35 | 4.33 |
| $K_2O$ | 0.73 | 0.69 | 1.05 | 0.52 | 0.31 | 0.27 | 1.72 | 1.70 | 1.68 | 0.71 | 0.72 | 0.64 | 2.85 | 2.89 |
| FeO | 10.88 | 10.03 | 9.76 | 7.78 | 0.84 | 13.68 | 11.84 | 11.85 | 12.22 | 9.94 | 9.89 | 9.98 | 5.98 | 6.11 |
| MnO | 0.13 | 0.12 | 0.12 | 0.11 | < 0.01 | 0.20 | 0.18 | 0.18 | 0.18 | 0.17 | 0.17 | 0.17 | 0.09 | 0.10 |
| $P_2O_5$ | 0.64 | 0.83 | 0.60 | 0.38 | 0.06 | 0.25 | 0.35 | 0.35 | 0.36 | 0.14 | 0.13 | 0.14 | 0.47 | 0.49 |
| $Cr_2O_3$ | 0.19 | 0.16 | 0.16 | 0.17 | 0.01 | 0.35 | < 0.003 | < 0.003 | ~ 0.002 | 0.018 | 0.015 | 0.016 | < 0.003 | ~ 0.002 |
| $H_2O^-$ | — | — | — | — | — | — | 0.86 | 0.86 | 0.83 | 0.17 | 0.17 | 0.13 | 0.99 | 1.03 |
| Sum of constituents determined | 99.76 | 99.46 | 99.20 | 99.62 | 99.90 | 100.49 | 99.28 | 99.15 | 99.18 | 99.71 | 99.61 | 99.33 | 99.07 | 98.49 |
| Sample weights, mg | 50.0 | 50.0 | 49.8 | 50.1 | 50.0 | 50.0 | 50.2 | 50.0 | | 49.8 | | | | |
| | 50.1 | 50.1 | 50.1 | 50.2 | | | 50.2 | 50.2 | | 50.1 | 50.0 | | 49.7 | |
| | | | | | | | 49.9 | | | 50.0 | 50.1 | | 49.7 | |

* Analyzed with 14230, 14049, 14305, 14319, and 14310 separates.
† Analyzed with 14421, 14003, 14148, 14156,14149, and 14310 whole-rock.
‡ Literature values: BCR-1 and AGV-1, Table 4 of Flanagan (1969); W-1, Table 1, column 3b of Fleischer (1969). Total iron expressed as FeO.

Table 2. Trace element concentrations (in ppm by wt.) in Apollo 14 samples.

| | Soils | | | | | Breccia 14305 | Basalt 14310 | | | Normalizing values |
|---|---|---|---|---|---|---|---|---|---|---|
| | 14421 | 14003 | 14148 | 14156 | 14149 | Sawdust | Whole-rock | Feldspar | Pyroxene | |
| Li | 31.7 | 32.1 | 30.3 | 30.5 | 34.0 | 38.4 | 27.5 | 24.0 | 28.9 | 1.8 |
| K | 3980 | 4260 | 4190 | 4190 | 4730 | 5940 | 3920 | 2230 | 2040 | 1000 |
| Rb | 13.9 | 14.6 | 14.8 | 14.7 | 14.5 | 19.0 | 12.7 | 4.39 | 6.78 | 3 |
| Sr | 180.5 | 176.5 | 177.4 | 179.2 | 177.5 | 165.9 | 180.9 | 281.6 | 72.6 | 11 |
| Ba | 806 | 823 | 824 | 813 | 827 | 924 | 649 | 373 | 343 | 3.66 |
| Ce | 170 | 170 | 171 | 169 | 169 | 193 | 143 | 35.0 | 89.6 | 0.787 |
| Nd | 105 | 107 | 107 | 107 | 108 | 121 | 87.9 | 20.0 | 59.3 | 0.580 |
| Sm | 29.7 | 30.2 | 30.6 | 29.9 | 30.1 | 34.4 | 24.6 | 5.06 | 17.5 | 0.185 |
| Eu | 2.68 | 2.63 | 2.55 | 2.57 | 2.56 | 2.56 | 2.09 | 2.87 | 0.971 | 0.071 |
| Gd | — | 35.5 | 34.4 | 34.9 | 34.7 | 40.1 | — | — | 22.4 | 0.256 |
| Dy | 38.8 | 38.9 | 38.9 | 38.9 | 39.3 | 46.7 | 31.7 | 5.79 | 26.8 | 0.303 |
| Er | 23.0 | 22.9 | 23.2 | 23.1 | 23.4 | 28.3 | 19.3 | 3.29 | 16.7 | 0.182 |
| Yb | 21.7 | 21.6 | 21.6 | 21.8 | 22.0 | 23.3 | 18.1 | 2.97 | 17.0 | 0.188 |
| Lu | 3.27 | — | 3.31 | 3.27 | 3.34 | 3.82 | 2.66 | 0.440 | — | 0.034 |
| Zr | 858 | 877 | 992 | 913 | 907 | 1060 | 893 | 155 | 709 | 7 |
| Hf | 20.9 | 20.7 | 20.6 | — | 20.8 | 28.6 | 21.0 | — | 16.3 | 0.2 |
| K/Rb | 286 | 291 | 283 | 285 | 326 | 313 | 308 | 508 | 300 | 333 |
| K/Ba | 4.94 | 5.18 | 5.08 | 5.15 | 5.71 | 6.42 | 6.04 | 5.98 | 5.94 | 273 |
| Rb/Sr | 0.0770 | 0.0828 | 0.0839 | 0.0820 | 0.0817 | 0.1142 | 0.0704 | 0.0156 | 0.0934 | 0.273 |
| Ce/Yb | 7.85 | 7.87 | 7.92 | 7.75 | 7.68 | 8.28 | 7.90 | 11.8 | 5.27 | 4.19 |
| Sm/Eu | 11.1 | 11.5 | 12.0 | 11.6 | 13.4 | 13.4 | 11.8 | 1.76 | 18.0 | 2.61 |
| Sr/Eu | 67.4 | 67.1 | 69.6 | 69.7 | 69.3 | 65.2 | 86.6 | 98.1 | 74.8 | 155 |
| Zr/Hf | 41.1 | 42.4 | 48.2 | — | 43.6 | 37.0 | 42.5 | — | 43.5 | 35.0 |
| $Eu^{2+}/Eu^{3+}$ | — | — | — | — | — | — | 5 | 41 | 2.9 | — |
| Sample weight | 0.100g | 0.10076g | 0.11739g | 0.10318g | 0.09049g | 0.09353g | 0.11283g | 0.11709g | 0.15835g | |

for other elements were smaller and close to the expected precisions of the analytical techniques. The apparent ranges for Mn and Cr may be analytical, in part, because of the low concentrations. The variations of Na and K are thought to be largely real. Na and K tend to vary sympathetically. The fact that Zr also appears to be positively correlated with Na and K (as does Li) suggests that selective volatility is not the cause of these variations. Ti shows some positive correlation with Na and K but there appear to be other factors (ilmenite?) influencing its concentration. The P variations appear to be relatively random and may reflect somewhat heterogeneous distribution of phosphates. Al is low in the soil clod samples 14049, two of which have relatively high Mg and Fe; this might reflect a somewhat lower plagioclase/mafics ratio for these samples. The sympathetic variations in alkalis and Zr most probably reflect somewhat heterogeneous distribution of a phase rich in these elements, possibly resembling the mesostasis material of lunar basalts. Our data do not indicate any definite general variations in soil composition correlated either with grain size or with collection depth. Na and K (and correlated elements) are lowest in 14259 (Schnetzler and Nava, 1971), and highest in 14149 and 14049 (100–200 mesh fraction). Sample 14259 is the finest grained and 14149 the coarsest grained of the soil samples (LSPET, 1971) and this suggests a possible correlation of alkalis with grain size. However, 14421 a coarser grained sample from the same site as 14259 is also low in K. The data indicate a possible geographic variation in alkalis, inasmuch as 14259 was collected farthest away from Cone Crater, and except for 14049, 14149 was the closest. However, with the limited number of samples, and lacking material from the ridge Fra Mauro formation and Cone Crater throwout, firm conclusions are not possible at this time.

Although there is some heterogeneity as discussed above, the most interesting conclusion resulting from intercomparison of Apollo 14 soil samples is that they are, in fact, remarkably homogeneous. This uniformity is brought out in Fig. 1 for the

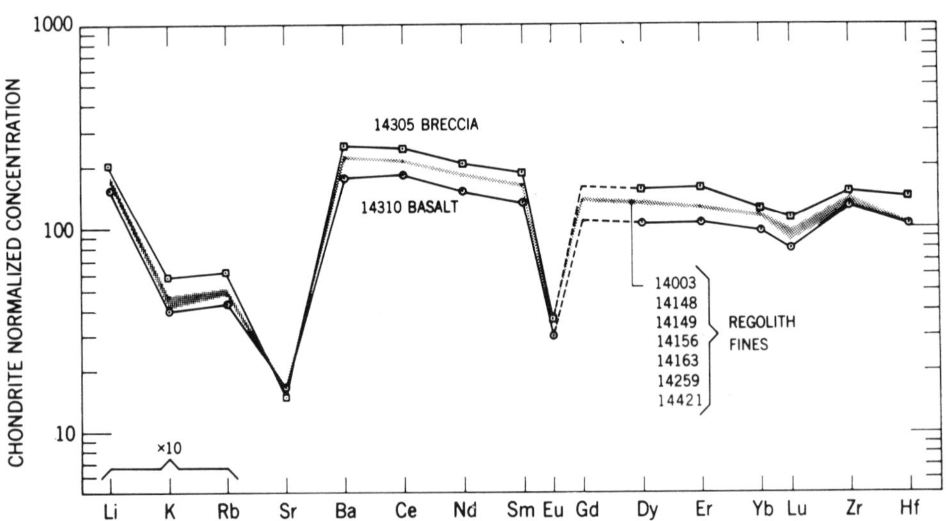


Fig. 1. Chondrite normalized trace element concentrations for 14305 breccia sawdust, 14310 whole-rock basalt, and seven Apollo 14 soil samples.

trace elements; concentrations have been normalized to the chondritic values given in the last column of Table 2. The situation is in marked contrast to that holding for the Apollo 12 soils (Schnetzler and Philpotts, 1971). The homogeneity of Apollo 14 soils indicates thorough mixing of the regolith at this site and/or relatively homogeneous source rocks with chemical composition very similar to that of the soils.

Breccia 14305 sawdust is similar in many respects in chemical composition to the soils (Table 1). However, $Al_2O_3$ is distinctly lower and MgO distinctly higher in the breccia. Also most of the trace elements are enriched in the breccia although their relative concentrations are not much affected (Table 2; Fig. 1). These differences relative to the soil could be accounted for in terms of a deficiency of feldspar and the presence of an accessory phase(s) rich in trace elements. For breccia 14319 we have at present only the major element analyses of light and dark portions reported in Table 1. In terms of $Al_2O_3$ and MgO, 14319 is intermediate between 14305 breccia and the soils. The high $P_2O_5$, $Na_2O$, and $K_2O$ of the 14319 materials indicate that other trace elements might also be relatively enriched. The light and dark portions of 14319 show some chemical differences, particularly for $SiO_2$ and $TiO_2$; it appears likely, however, that the bulk concentrations of these oxides in 14319 might be close to those for 14305 breccia. The variations in composition within and between these breccias tempt the speculation that the average chemical composition of the Fra Mauro breccias might be very similar to the soil composition (except for solar wind components, etc.).

## 14310 Basalt

Basalt 14310 is the only Apollo 14 basalt that we have studied. It is apparent from Tables 1 and 2 and Fig. 1 that 14310 is similar in chemical composition to the Apollo 14 soils. However, the composition of 14310 is not within our soil range. Whereas the major element data for the breccias indicate an excess of mafic phases relative to the soil, the 14310 data indicate an excess of plagioclase. Caution is warranted, of course, in comparing 4 b.y.-old rocks (Turner *et al.*, 1971; Papanastassiou and Wasserburg, 1971; Compston *et al.*, 1971) with soil samples that were evolving up to the time of their collection. Nevertheless, the fact that the 14310 and breccia compositions bracket those of the soils suggests that any addition of foreign components to the soils may not have greatly affected the concentrations of the elements we have determined. The 14310 composition is matched remarkably well by a mixture of soil and feldspar: the data appear to preclude addition or subtraction of significant amounts of other components. These points are well illustrated in Fig. 2 in which the compositions of the soil average, 14310 separated plagioclase (Tables 1 and 2), and a 4:1 soil-feldspar mixture are plotted, all data being normalized to the composition of 14310 whole-rock. As can be seen, the concentrations in the mixture are within $\pm 10\%$ of those in 14310 for almost all of the 25 elements. The major discrepancy is for Eu that is 25% higher in the mixture than in 14310. The agreement among the soil Eu values indicates that this discrepancy is not due to error in the average soil value. It is possible, however, that the soils contain, as a foreign component, plagioclase (anorthosite?) with a very large positive Eu anomaly. If this is the explanation of the discrepancy then the amount of feldspar needed in a mixture with uncontaminated

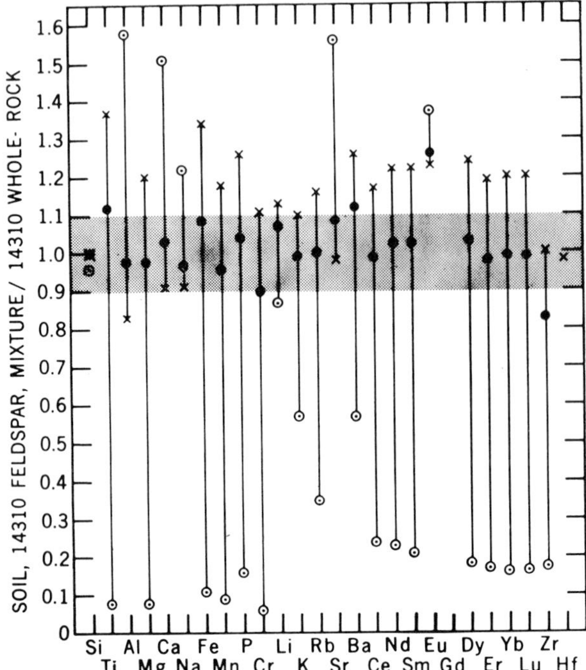


Fig. 2. 14310 whole-rock normalized elemental concentrations for average Apollo 14 regolith fines (x), 14310 plagioclase (⊙), and a 4 : 1 soil : feldspar mixture (●).

soil to match the 14310 composition would be greater than the indicated 20%. Analytical error is possible in the determination of the 14310 whole-rock and plagioclase Eu concentrations but we see no reason at the present time to doubt these values. Helmke and Haskin (1972) report a 14310 whole-rock Eu concentration of 2.08 ppm and Hubbard *et al.* (1972) one of 2.15 ppm, in good agreement with our value of 2.09 ppm. The Eu discrepancy would be reduced if feldspar with less Eu than 14310 plagioclase were used in the mixture. A lower feldspar Eu concentration could be explained by one or both of two possibilities. Feldspar from a more rapidly cooled chemical equivalent of 14310 would be expected to have a smaller Eu anomaly because of kinetics, inherent contamination, etc. However, if this were the explanation Sr would be expected to show a discrepancy similar to that for Eu, and it does not. The second possibility entails having in the mixture a feldspar with lower Eu/Sr than 14310 plagioclase. Eu/Sr in lunar basalts and constituent feldspars tends to increase with increasing overall trace element concentration which presumably reflects extent of differentiation. The lowest Eu/Sr yet reported for a lunar plagioclase is for 15415 anorthosite (Hubbard *et al.*, 1971); use of this material in the 20% plagioclase-soil mixture would result in a Eu excess of only 7%.

The second largest discrepancy in modeling 14310 in terms of a soil-feldspar mixture is for Zr. This discrepancy is probably not due to analytical error; the soil data are internally consistent, and 14310 whole-rock has a reasonable Zr/Hf ratio. It could well be due to the presence in the 14310 sample analyzed of an excess of an accessory phase rich in Zr. A deficiency in 14310 of ilmenite, etc., could similarly explain the Ti discrepancy.

Although there are some discrepancies, the overall compositional match of 14310 whole-rock and the 4:1 soil-plagioclase mixture is good. It seems entirely possible that the soils come close to representing the average composition of Apollo 14 rocks, at least for the elements we have studied, and that 14310 deviates from this typical composition in containing an excess of plagioclase. If 14310 represents a fused soil or breccia as has been suggested on the basis of its "meteoritic" component (Morgan *et al.*, 1972a, b) then the feldspar excess indicates local heterogeneity in surficial fragmental material at that time. If 14310 represents an igneous rock then both the excess feldspar and the "meteoritic" component may have been added by assimilation and/or by mechanical incorporation of xenolithic material in accord with the suggestion of Ridley *et al.* (1972).

Analytical data for feldspar ($An_{84}$) and pyroxene (pigeonite) separated from 14310 are given in Tables 1 and 2. The phases were separated utilizing heavy liquid, electromagnetic, and hand-picking techniques; microscopic examination indicated that the feldspar separate was about 99% pure whereas the pyroxene separate contained about 10% contaminants. Except for Li, Sr, and Eu, which are largely accounted for, the whole-rock, plagioclase, and pyroxene data indicate that about half of the whole-rock trace-element content is concentrated in accessory phases; this is in accord with previous observations (Philpotts and Schnetzler, 1970; Schnetzler and Philpotts, 1971; Schnetzler *et al.*, 1972).

The $Eu^{2+}/Eu^{3+}$ values reported for 14310 feldspar and pyroxene in Table 2 were calculated according to the method of Philpotts (1970). $Eu^{2+}/Eu^{3+}$ ratios calculated using the whole-rock data are within 20% of these mineral-mineral values. The plagioclase ratio is very similar to those calculated for plagioclases from two Apollo 11 basalts (Philpotts, 1970) and lower than those for 12035, 12040 (Schnetzler and Philpotts, 1971), and 15555 (Schnetzler *et al.*, 1972) plagioclases. The pyroxene $Eu^{2+}/Eu^{3+}$ ratio is higher than any previously determined for this mineral in lunar samples. The difference between the 14310 pyroxene and feldspar ratios is smaller than any previously determined for lunar samples with the exception of that for 12021 that shows a similar spread. The spread in pyroxene-feldspar ratios tends to be smaller in finer-grained rocks and may be due to kinetics, inherent contamination of phases, or other factors. Taking this effect into account and assuming (a) equilibrium in all cases, (b) the validity of the calculation, and (c) that $Eu^{2+}/Eu^{3+}$ reflects redox conditions, then it appears that 14310 crystallized under conditions slightly more reducing than those holding for the Apollo 11 basalts and perhaps similar to those holding for 12035, 12040, and 15555. These relationships appear to be in accord with petrographic observations (Ridley *et al.*, 1972).

## Petrogenesis

The Apollo 14 materials appear to be identical in composition to the exotic fragments in the Apollo 12 regolith that were termed KREEP (Hubbard and Gast, 1971). KREEP-type material also appears to be abundant in the Apollo 15 soils judging from their chemistry (LSPET, 1972; Schnetzler *et al.*, 1972; our unpublished data). In contrast, the compositions of the soils from Apollo 11 (Philpotts and Schnetzler, 1970) and Luna 16 (Philpotts *et al.*, 1972) indicate very little KREEP component. These bulk chemistry conclusions are in accord with the studies by Reid *et al.* (1972) of regolith glasses. The abundance of KREEP at the Apollo 12, 14, and 15 sites and its paucity at the Apollo 11 and Luna 16 sites, suggest that this material may be related to the Imbrium Basin or its throwout. The Apollo 15 $\gamma$-ray experiment is of interest in this respect in that it indicated high Th, U, and K concentrations (as in KREEP) in Mare Imbrium and Oceanus Procellarum and a smooth decrease in concentration away from these regions (Metzger *et al.*, 1972). It appears, therefore, that Apollo 14 type material may have a restricted geographical distribution.

Apollo 14 material shares with the mare basalts certain lunar characteristics such as low volatile content. However, relative to the mare basalts, Apollo 14 material has higher concentrations of large-ion lithopile trace elements and Al, higher Mg/Fe, and lower Fe. Some of the difficulties encountered in attempting to relate these two types of material have been touched upon by Hubbard and Gast (1971). It appears most unlikely that they are directly related to each other in terms of various degrees of partial fusion or fractional crystallization of the same source material. If they were so related it is difficult to explain the equivalent concentrations of Mg considering the distinctly different Al, Fe, and trace-element concentrations, for example. It appears that different source material is required. There appears to be general agreement among experimental petrologists that Apollo 14 material probably resulted from partial fusion of an olivine-plagioclase-pyroxene assemblage (Biggar *et al.*, 1972a; Ford *et al.*, 1972; Kushiro, 1972; Ringwood *et al.*, 1972; Walker *et al.*, 1972). The mare-basalts are of less certain origin (Biggar *et al.*, 1972b).

It has been suggested that the different source materials of the Fra Mauro and the mare basalts reflect primary accretion layering of the moon, with the more refractory source of the (older) Fra Mauro basalts being at shallower depths than the source of the mare basalts (Hubbard and Gast, 1971; Gast and McConnell, 1972; Ringwood *et al.*, 1972). One possible problem with this hypothesis is that it requires material being added during the later stages of accretion to be more refractory than the earlier materials, whereas if any heterogeneity in accretion is expected at all, it might generally be expected in the reverse direction. Further, ratios of volatile to involatile elements for both the high alumina and the mare basalts do not appear to support this hypothesis. K and Ba may be the best pair to consider in that they show good geochemical coherence in igneous processes (hence K/Ba for a basalt is expected to be quite similar to K/Ba of its source rock) but have quite different volatilities. K/Ba ratios of 8.4 and 4.5 for trace element enriched and depleted Apollo 11 mare-type basalts, respectively (Philpotts and Schnetzler, 1970), bracket the ratios for all other bulk lunar samples that we have analyzed. The Apollo 14 samples have K/Ba ratios from 4.9 to 6.4 (Table 2); 12013 samples have ratios from 4.6 to 6.6 (Schnetzler *et al.*, 1970). Ratios for Apollo 11 enriched-type, normal Apollo 12 (Schnetzler and Philpotts, 1971), and Apollo 15555 (Schnetzler *et al.*, 1972) mare-type basalts are higher than those for the KREEP-type materials. But ratios for Apollo 11 depleted-type, Apollo 12038, and Luna 16 (Philpotts *et al.*, 1972) mare-type basalts are at or below the lower values of the KREEP-type material. Thus no good case can be made for K/Ba being distinctly different in the source materials of the two basalt types. On the contrary, the evidence tends to indicate that K/Ba may have been identical, initially, in both source materials. This appears to be difficult to reconcile with a simple model of volatility dependent, heterogeneous accretion. Any simple volatility related process sufficient to effect sizable increases in the concentrations of refractory elements such as Al might be expected to effect very extensive depletion of volatile elements such as K. It is possible, of course, that all the accreting material was depleted in volatile elements to the extent that they were at essentially equilibrium values prohibiting further loss. However, before giving exhaustive consideration to such possibilities, homogeneous accretions models merit appraisal.

O'Hara and colleagues (Biggar *et al.*, 1972a; Ford *et al.*, 1972) have suggested that Apollo 14 types aluminous liquids might be produced by partial melting in systems rich in alkalis and/or water, and that these volatiles have subsequently been lost. The mare-type basalts might then be produced by partial melting of the volatile and feldspar depleted residue. However, as far as alkali loss is concerned there is first the problem of the mechanism whereby significant loss could be effected and second the fact that the K/Ba (etc.) data and the Rb–Sr data show no evidence that any significant alkali loss has taken place. Similarly, the available data on aluminous lunar rocks show no evidence (such as hydrous phases, deuteric alteration, etc.) of significant concentrations of water in their evolution. At the present time these ideas, therefore, appear to be interesting but unsupported speculations.

An apparently more attractive homogeneous accretion model is one in which Apollo 14 and other aluminous lunar materials represent primary aluminous liquid and plagioclase-rich cumulates in various proportions formed in an igneous event taking place during the first few hundred million years of lunar evolution, whereas mare-type basalts result from later fusion of mafic-rich cumulates formed during the initial event(s). This model seems to fit adequately the chemical and geochronologic constraints. Detailed thermal calculations for this model have not yet been made but it appears that modifications of Hays' (1972) third model would suffice.

*Acknowledgments*—We express our most profound gratitude to our respective families, present and future, for their patience during the course of this work. We also sincerely thank Miss Marianne Sandilands and Mr. M. Leatherwood for hurried preparation of the manuscript and illustrations, respectively.

## References

Bernas B. (1968) A new method for decomposition and comprehensive analysis of silicates by atomic absorption spectrometry. *Anal. Chem.* **40,** 1682–1686.

Biggar G. M., Ford C. E., Humphries D. J., Wilson G., and O'Hara M. J. (1972a) Melting relations of more primitive mare-type basalts 14053 and M (Reid 1970); and of breccia 14321 and soil 14162 (average lunar crust?) (abstract). In *Lunar Science—III* (editor C. Watkins), pp. 74–76, Lunar Science Institute Contr. No. 88.

Biggar G. M., O'Hara M. J., Humphries D. J., and Pecket A. (1972b) Experimental crystallization of Apollo 11 and 12 lavas—a re-examination (abstract). In *Lunar Science—III* (editor C. Watkins), pp. 77–79, Lunar Science Institute Contr. No. 88.

Compston W., Vernon M. J., Berry H., and Radowski R. (1971) The age of the Fra Mauro formation: A radiometric older limit. *Earth Plan. Sci. Lett.* **12,** 55–58.

Flanagan F. J. (1969) U.S. Gelogical Survey Standards—II, First compilation of data for the new U.S.G.S. rocks. *Geochim. Cosmochim. Acta* **33,** 81–120.

Fleischer M. (1969) U.S. Geological Survey Standards—I. Additional data on rocks G-1 and W-1, 1965–1967. *Geochim. Cosmochim. Acta* **33,** 65–79.

Ford C. E., Humphries D. J., Wilson G., Dixon D., Biggar G. M., and O'Hara M. J. (1972) Experimental petrology of high alumina basalt, 14310, and related compositions (abstract). In *Lunar Science—III* (editor C. Watkins), pp. 274–276, Lunar Science Institute Contr. No. 88.

Gast P. W. and McConnell R. K. Jr. (1972) Evidence for initial chemical layering of the moon (abstract). In *Lunar Science—III* (editor C. Watkins), pp. 289–290, Lunar Science Institute Contr. No. 88.

Hays J. F. (1972) Radioactive heat sources in the lunar interior. *Phys. Earth Planet. Interiors* **5,** 77–84.

Helmke P. A. and Haskin L. A. (1972) Rare earths and other trace elements in Apollo 14 lunar samples (abstract). In *Lunar Science—III* (editor C. Watkins), pp. 366–368, Lunar Science Institute Contr. No. 88.

Hubbard N. J. and Gast P. W. (1971) Chemical composition and origin of nonmare lunar basalts. *Proc. Second Lunar Sci. Conf., Geochim. Cosmochim. Acta* Suppl. 2, Vol. 2, pp. 999–1020. MIT Press.

Hubbard N. J., Gast P. W., Meyer C., Nyquist L. E., Shih C., and Wiesmann H. (1971) Chemical composition of lunar anorthosites and their parent liquids. *Earth Planet. Sci. Lett.* **13,** 71–75.

Hubbard N. J., Gast P. W., Rhodes M., and Wiesmann H. (1972) Chemical composition of Apollo 14 materials and evidence for alkali volatilization (abstract). In *Lunar Science—III* (editor C. Watkins), pp. 407–409, Lunar Science Institute Contr. No. 88.

Kushiro I. (1972) Petrology of lunar high-alumina basalt (abstract). In *Lunar Science—III* (editor C. Watkins), pp. 466–468, Lunar Science Institute Contr. No. 88.

LSPET (Lunar Sample Preliminary Examination Team) (1971) Preliminary examination of lunar samples from Apollo 14. *Science* **173,** 681–693.

LSPET (Lunar Sample Preliminary Examination Team) (1972) The Apollo 15 lunar samples: A preliminary description. *Science* **175,** 363–375.

Maxwell J. A., Peck L. C., and Wiik H. B. (1970) Chemical composition of Apollo 11 lunar samples 10017, 10020, 10072, and 10084. *Proc. Apollo 11 Lunar Sci. Conf., Geochim. Cosmochim. Acta* Suppl. 1, Vol. 2, pp. 1369–1374. Pergamon.

Metzger A. E., Trombka J. I., Peterson L. E., Reedy R. C., and Arnold J. R. (1972) A first look at the lunar orbital gamma ray data (abstract). In *Lunar Science—III* (editor C. Watkins), pp. 540–541, Lunar Science Institute Contr. No. 88.

Morgan J. W., Krähenbühl U., Ganapathy R., and Anders E. (1972a) Volatile and siderophile elements in Apollo 14 and 15 rocks (abstract). In *Lunar Science—III* (editor C. Watkins), pp. 555–557, Lunar Science Institute Contr. No. 88.

Morgan J. W., Laul J. C., Krähenbühl U., Ganapathy R., and Anders E. (1972b) Major impacts on the moon: Chemical characterization of projectiles (abstract). In *Lunar Science—III* (editor C. Watkins), pp. 552–554, Lunar Science Institute Contr. No. 88.

Nava D. F. (1970) Atomic absorption and classical chemical analyses of the Lost City meteorite (abstract). *E.O.S. Transactions of the American Geophysical Union*, Vol. 51, p. 580.

Nava D. F., Walter L. S., and Doan A. S. Jr. (1971) Chemistry and mineralogy of the Lost City meteorite. *J. Geophys. Res.* **76,** 4067–4071.

Papanastassiou D. A. and Wasserburg G. J. (1971) Rb–Sr ages of igneous rocks from the Apollo 14 mission and the age of the Fra Mauro formation. *Earth Planet. Sci. Lett.* **12,** 36–48.

Philpotts J. A. (1970) Redox estimation from a calculation of $Eu^{2+}$ and $Eu^{3+}$ concentrations in natural phases. *Earth Planet. Sci. Lett.* **9,** 257–268.

Philpotts J. A. and Schnetzler C. C. (1970) Apollo 11 lunar samples: K, Rb, Sr, Ba, and rare-earth concentrations in some rocks and separated phases. *Proc. Apollo 11 Lunar Sci. Conf., Geochim. Cosmochim. Acta* Suppl. 1, Vol. 2, pp. 1471–1486. Pergamon.

Philpotts J. A., Schnetzler C. C., Bottino M. L., Schuhmann S., and Thomas H. H. (1972) Luna 16: Some Li, K, Rb, Sr, Ba, rare-earth, Zr, and Hf concentrations. *Earth Planet. Sci. Lett.* **13,** 429–435.

Reid A. M., Ridley W. I., Warner J., Harmon R. S., and Brett R. (1972) Chemistry of highland and mare basalts as inferred from glasses in the lunar soils (abstract). In *Lunar Science—III* (editor C. Watkins), pp. 640–642, Lunar Science Institute Contr. No. 88.

Ridley W. I., Williams R. J., Brett R., and Takeda H. (1972) Petrology of lunar basalt 14310 (abstract). In *Lunar Science—III* (editor C. Watkins), pp. 648–650, Lunar Science Institute Contr. No. 88.

Ringwood A. E., Green D. H., and Ware N. G. (1972) Experimental petrology and petrogenesis of Apollo 14 basalts (abstract). In *Lunar Science—III* (editor C. Watkins), pp. 654–656, Lunar Science Institute Contr. No. 88.

Rose H. J. (1972) Private communication.

Schnetzler C. C., Philpotts J. A., and Bottino M. L. (1970) Li, K, Rb, Sr, Ba, and rare-earth concentrations, and Rb–Sr age of lunar rock 12013. *Earth Planet. Sci. Lett.* **9,** 185–192.

Schnetzler C. C. and Nava D. F. (1971) Chemical composition of Apollo 14 soils 14163 and 14259. *Earth Planet. Sci. Lett.* **11,** 345–350.

Schnetzler C. C. and Philpotts J. A. (1971) Alkali, alkaline earth, and rare-earth element concentrations in some Apollo 12 soils, rocks, and separated phases. *Proc. Second Lunar Sci. Conf., Geochim. Cosmochim. Acta* Suppl. 2, Vol. 2, pp. 1101–1122. MIT Press.

Schnetzler C. C., Philpotts J. A., Nava D. F., Schuhmann S., and Thomas H. H. (1972) Geochemistry of Apollo 15 basalt 15555 and soil 15531. *Science* **175,** 426–428.

Turner G., Huneke J. C., Podosek F. A., and Wasserburg G. J. (1971) $^{40}$Ar–$^{39}$Ar ages and cosmic ray exposure ages of Apollo 14 samples. *Earth Planet. Sci. Lett.* **12,** 19–35.

Walker D., Longhi J., and Hays J. F. (1972) Experimental petrology and origin of Fra Mauro rocks and soil (abstract). In *Lunar Science—III* (editor C. Watkins), pp. 770–772, Lunar Science Institute Contr. No. 88.

Proceedings of the Third Lunar Science Conference
(Supplement 3, *Geochimica et Cosmochimica Acta*)
Vol. 2, pp. 1307–1313
The M.I.T. Press, 1972

# Precise determination of rare-earth elements in the Apollo 14 and 15 samples

AKIMASA MASUDA and NOBORU NAKAMURA
Department of Chemistry, Science University of Tokyo,
Kagurazaka, Shinjuku-ku, Tokyo, Japan

and

HAJIME KURASAWA and TSUYOSHI TANAKA
Geological Survey of Japan, Hisamoto, Kawasaki, Japan

**Abstract**—Rare earth (REE) abundances not only in lunar samples but also in chondrites were determined precisely. It was found that there is a positive cerium anomaly when normalized against the REE abundances in a representative L chondrite (Leedey) and that the extent of the anomaly is variable from site to site.

In contrast with lanthanum in terrestrial rocks, the behavior of La in lunar materials is regular in association with Nd and Sm. This is in accord with the lunar magmas being dry. The chondrite-normalized REE patterns of lunar samples studied have a small gap between Sm and Gd, apart from an Eu negative anomaly.

It is inferred that Fra Mauro materials and Apennine front soils are the most representative highland materials and are originally the quenched liquids which were left after nearly total melting and subsequent extensive differential solidification.

## INTRODUCTION

AS REVIEWED BY LSPET (1971), the Fra Mauro region investigated by the Apollo 14 mission is a kind of an island of premare material. Namely, the Fra Mauro formation might be regarded as composed of materials of the ancient highland crust of the moon. The Hadley-Apennine site on which the Apollo 15 mission landed is geologically complex; some materials may be of nonmare origin and some others of mare origin.

Our results on rare earth elements (REE) in 14163, 14310, 14321, 15101, and 15401 are reported here together with the REE abundances in two L chondrites, Leedey and Peace River. Sample 14163 is the fines; 14310 is an only crystalline rock returned by Apollo 14 mission; and 14321 is breccia (fragmental rock). Both of the Apollo 15 samples reported here are fines.

## EXPERIMENTAL

Ten rare-earth elements, i.e., La, Ce, Nd, Sm, Eu, Gd, Dy, Er, Yb, and Lu were determined by stable isotope dilution, according to the experimental procedure described by one of us (Masuda, 1968a), with some modifications. When a rhenium single-filament system was employed, silica gel was used (Akishin *et al.*, 1957; Cameron *et al.*, 1969) as preliminary mounting onto the filament or addition to the separated REE sample. The silica gel was prepared by a procedure worked out by Gensho (1971), i.e., hydrolysis of $SiF_4$ by ammonia water. The use of silica gel has an intensifying effect on some REE ions, especially, on $Eu^+$ when using a single filament. Also the triple filament ion source was employed. The use of this type of ion source was desirable for determination of Ce.

Thus some rare-earth elements were determined with both the single-filament system and the triple-filament one.

Prior to these determinations, the standard solutions to calibrate the spike solutions were prepared with much caution, because the REE oxides are subject to absorption or adsorption of water and carbon dioxide (Gast *et al.*, 1970) in the air. Our examination showed that the adsorbed component is mainly water and is removable by ignition and that the oxide weight does not change several hours after the ignition. Therefore, the oxide reagents were heated strongly until the weights became constant. The concentrations of the standard solutions were also determined titrimetrically. The differences between the concentrations determined titrimetrically and those determined gravimetrically were 0.6% or less, with the exception of Eu (the difference of 1.3%). Insofar as the stated purities (99.9 to 99.99%) of reagent oxides are reliable, it is believed that the spike solutions have been calibrated with the accuracies better than 1.0%.

## CHONDRITIC ABUNDANCES

In order that the fine structure of the REE pattern of the Masuda-Coryell plotting (Masuda, 1962; Coryell *et al.*, 1963) can be discussed with certainty, REE abundances not only in lunar materials but also in chondrites need to be precise. For this purpose, REE in nine ordinary chondrites have been determined (Masuda *et al.*, 1972) using the same spike solutions as employed for investigation of lunar materials. Our results on ordinary chondrites show that the cerium abundance in ordinary chondrites is subject to more or less variation relative to La, Nd, and Sm, but appears to be "regular" in L chondrites rather than in H chondrites. Moreover, the results show that the mutually normalized REE patterns of ordinary chondrites have individualities in fine structures. This leaves some doubt as to whether the composite material prepared by mixing of randomly chosen chondrites can tell the most basic and pristine data on rare-earth relative abundances for chondrites. A survey of our data has led us to a judgment that Leedey is a safe choice to represent chondrites free from individuality. This is in line with Dr. C. B. Moore's personal suggestion that Leedey be taken as the reference chondritic material.

## RESULTS AND DISCUSSION

The results of abundance determination are presented in Table 1. The chondrite-normalized REE patterns for studied samples are shown in Fig. 1. It can be seen that the chondrite-normalized REE patterns for three Fra Mauro materials are similar

Table 1. REE abundances (ppm) in five lunar samples (Apollo 14 and 15) and in two L chondrites.

| | 14163 Fines | 14310 Igneous | 14321 Breccia | 15101 Fines | 15401 Fines | Leedey | Peace River |
|---|---|---|---|---|---|---|---|
| La | 66.6 | 54.9 | 70.6 | 21.00 | 19.40 | 0.378 | 0.412 |
| Ce | 177.8 | 151.1 | 193.4 | 55.1 | 51.0 | 0.976 | 1.052 |
| Nd | 106.5 | 88.8 | 114.2 | 34.37 | 31.91 | 0.716 | 0.744 |
| Sm | 30.21 | 25.06 | 31.75 | 9.91 | 9.20 | 0.230 | 0.228 |
| Eu | 2.655 | 2.330 | 2.647 | 1.189 | 1.003 | 0.0866 | 0.0862 |
| Gd | 34.78 | 29.04 | 37.01 | 11.74 | 10.67 | 0.311 | 0.309 |
| Dy | 40.6 | 33.62 | 41.7 | 12.87 | 12.50 | 0.390 | 0.375 |
| Er | 24.23 | 20.28 | 25.63 | 7.86 | 7.33 | 0.255 | 0.253 |
| Yb | 21.93 | 18.65 | 22.78 | 6.88 | 6.41 | 0.249 | 0.245 |
| Lu | 3.17 | 2.60 | 3.32 | 1.021 | 0.952 | 0.0387 | 0.0382 |

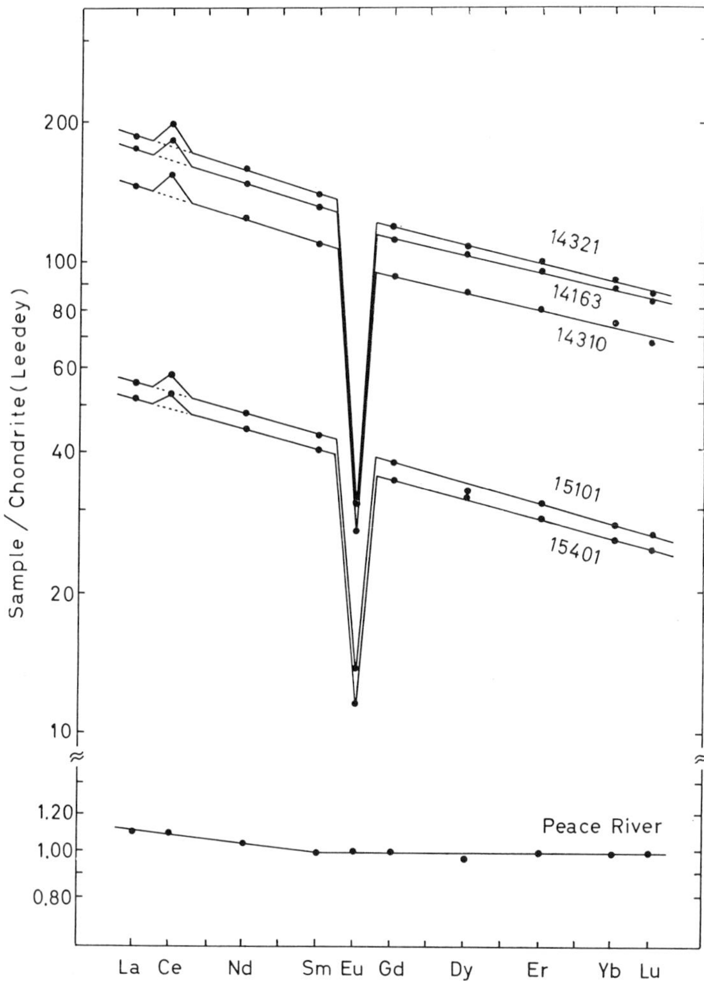


Fig. 1. Chondrite-normalized REE patterns for Apollo 14 and 15 samples (Table 1) and for the Peace River chondrite; the Leedey chondrite is used as a reference for normalization.

to each other and are logarithmically linear as a whole with a few fine irregularities as discussed below. For 14163, quick reports on REE along with other elements have been given by Schnetzler and Nava (1971), Brunfelt *et al.* (1971), and Taylor *et al.* (1971). Although Taylor *et al.* (1971) mentioned that the small heavy REE are enriched "monotonically" by a factor of 105 $\pm$ 10 over the chondrites, Fig. 1 shows that the apparent enrichment is not monotonic but shows an inclination. The patterns presented by Philpotts, Schnetzler, and their co-workers show, usually, irregularities for Lu; this is considered to be due to too high a value of the chondritic Lu abundance employed by them for normalization.

All of the above three reports pointed out that 14163 closely resembles in chemical and/or rare earth composition the exotic fragments, such as KREEP (Hubbard *et al.*, 1971) or norite (Wood *et al.*, 1971), found in Apollo 11 and 12 soils, or the dark portion of rock 12013 (Schnetzler *et al.*, 1970; Hubbard *et al.*, 1970; Wakita and Schmitt, 1970). Attention should be paid to the fact that the pattern of 14310 is quite similar to that of 14163 and rock 14310 is a fine-grained basaltic rock (LSPET, 1971).

Apart from the Eu anomaly and the fine structures involving Ce anomaly, the logarithmic rectilinearity of chondrite-normalized REE patterns for Fra Mauro samples is remarkable, a fact suggesting that the Fra Mauro formation is composed of typical primary-liquid-type materials (Masuda and Matsui, 1966). In this sense, our inference is parallel with the suggestion by Schnetzler and Nava (1971) that "the moon apparently underwent extreme differentiation very early in its history and/or accumulated inhomogeneously." If we are right in believing that the Apollo 14 samples are primary-liquid-type materials, this differentiation is inferred to have been produced by the total melting of the original lunar materials and the subsequent solidification. If so, there must be primary-solid-type materials somewhere within the moon. It would be worth pointing out that rock 10045 of Mare Tranquillitatis appears to be solid-type material which can be associated with sample 14163 in REE pattern in a conjugate manner (Masuda, 1972).

As mentioned above, the Apollo 15 samples reported in this paper are fines. One of them, 15101, was collected at Station 2 which is located near the Apennine front approximately 600 m north of the northeastern part of the St. George crater rim crest, and another soil 15401 is from Station 6a which is the highest location visited on the Apennine front (Swann *et al.*, 1971). As seen in Fig. 1, the Apennine front soils also have the characteristics of primary-liquid-type materials in that their chondrite-normalized REE patterns are logarithmically rectilinear. The REE patterns of Apennine front soils are similar in shape to those of Apollo 14 samples, but the absolute abundances are lower than these by a factor of about 3. This suggests either that the rare earth absolute abundances in original materials for Apollo 15 soils were lower than those for Apollo 12 and 14 materials by a factor of 3, or that the bulk partition coefficients pertaining to the differentiation which generated the Apollo 15 soils are greater than those pertaining to Apollo 12 and 14 materials by a certain factor, say, 1.6. Anyway, this fact means that the Apollo 15 soils, probably, the Apennine front crustal materials belong to a differentiation series somewhat different from those of Apollo 12 and 14 materials; this difference, however, is not so essential.

One of the remarkable features seen in Fig. 1 is a cerium positive anomaly. Needless to say, the reality of a rather small degree of the anomaly in chondrite-normalized pattern rests on chondritic abundance employed for normalization. (Note that, if erroneously too low chondritic abundance is employed for normalization, apparent positive anomaly will arise in the resultant pattern.) As emphasized above, the Leedey chondrite has been employed as a reference, based on very cautious analyses of ordinary chondrites for REE. The cerium anomalies are not the result of experimental error. In order to evince this, the data on Peace River are also presented in Fig. 1 and Table 1. (Strangely, Dy for Peace River (Fig. 1) deviates from the smooth curve; this deviation has been confirmed by the repetition of the experiment.) When the REE abundances in Leedey are applied to terrestrial basic rocks, the positive Ce anomaly like this is seldom observed in the resultant normalized patterns. This fact would be worth bearing in mind when discussing the genetic relationship between the earth and the moon.

In Table 2 is presented the extent of the positive cerium anomaly for five samples studied by us. On the average, the cerium anomalies for Apollo 14 samples and Apollo 15 samples under consideration are 12.0 and 7.1%, respectively. It is evident that the cerium anomalies are variable from site to site (or from sample to sample). This anomaly reminds us of two factors, namely, possible tetravalency of the cerium ion and the exceptionally low melting point of cerium sesquioxide. If the tetravalency is associated with the observed cerium anomaly, it would mean that the lunar original materials passed through a highly oxidizing stage. If the tetravalent state of cerium is denied for any solar materials at any stages preceding the formation of celestial nongaseous bodies, the exceptionally low melting point of cerium oxide (Samsonov, 1969) or some other irregularities in physical properties such as volatility might have to be invoked. It is also an intriguing problem whether the cerium anomaly in question is due to prelunar process or postlunar process. Although we prefer the former possibility, we cannot help admitting at present that our knowledge about the physico-chemical properties of cerium oxide and/or cerium is too poor to give conclusive words.

A scrutiny of the chondrite-normalized REE patterns in Fig. 1 reveals that there is a small gap between Sm and Gd. This discontinuity is small, but its significance could be great, because it is understood to provide us with a clue to the genesis of the lunar materials.

It is also noticeable that the behavior of La appears to be always regular in association with Nd and Sm. In terrestrial basaltic rocks, La is not so "regular" as in lunar materials. This lanthanum irregularity in terrestrial igneous materials has been ascribed to the high solubility product of lanthanum hydroxide (Masuda, 1968b), namely, the effect of water. The regularity of La in lunar materials is considered to be in accord with the lunar magmas being dry.

The inclination of the chondrite-normalized REE pattern for the span from Gd through Lu is identical for Apollo 14 and Apollo 15 samples. The Apollo 12 samples have substantially the same inclinations for the same span. Also the Apollo 11 samples

Table 2. Cerium positive anomalies for five lunar samples.

| Sample | Anomaly (%)* |
|---|---|
| 14321 | +12.2 |
| 14163 | +10.9 |
| 14310 | +13.0 |
| 15101 | +7.1 |
| 15401 | +7.2 |

* Relative to "regular" values estimated from rectilinear lines fitting La, Nd, and Sm.

show the similar inclinations for the corresponding REE span. This fact indicates that the bulk partition coefficients are almost the same for REE from Gd through Lu.

LSPET (1971) mentions, although preliminarily, that some premare lunar materials have been produced by extensive and efficient chemical differentiation process and they may represent a sample of the premare lunar surface. Schnetzler and Nava (1971) made a similar suggestion. However, Schnetzler and Nava and LSPET appear to be prudent in concluding finally that the Fra-Mauro-like material is representative of the lunar highlands. If the logarithmic rectilinearity of chondrite-normalized REE patterns for Fra Mauro materials and Apennine front soils is interpreted straightforward, it means that those samples are the most representative highland materials and are the quenched liquids which were left after nearly total melting and subsequent extensive differential solidification.

### *Note Added in Proof*

Our subsequent considerations have led us to inferences that Apollo 12 rock such as 12035 is relatively close to original lunar magma (Masuda and Tanaka, 1972), and the original lunar magma may have been produced by a partial melting of chondritic material (Nakamura and Masuda, unpublished).

*Acknowledgments*—We thank Prof. M. Honda, University of Tokyo, for his generous donation of ordinary chondrites and his encouragement of our study. Personal communication from Dr. C. B. Moore, Arizona State University, is acknowledged gratefully. This work was supported in part by Grant-in-Aid for Scientific Research from the Ministry of Education, Japan, and in part by Grant of the Mitsubishi Foundation.

## References

Akishin P. A., Nikitin O. T., and Panchekov G. M. (1957) A new effective ionic emitter for the isotopic analysis of lead. *Geochemistry* (USSR), 500–505.

Brunfelt A. O., Heier K. S., Steinnes E., and Sundvoll B. (1971) Determination of 36 elements in Apollo 14 bulk fines 14163 by activation analysis. *Earth Planet. Sci. Lett.* **11,** 351–353.

Cameron A. E., Smith D. H., and Walker R. L. (1969) Mass spectrometry of nanogram-size samples of lead. *Anal. Chem.* **41,** 525–526.

Coryell C. D., Chase J. W., and Winchester J. W. (1963) A procedure for geochemical interpretation of terrestrial rare-earth patterns. *J. Geophys. Res.* **68,** 559–566.

Gast P. W., Hubbard N. J., and Wiesmann H. (1970) Chemical composition and petrogenesis of basalts from Tranquility Base. *Proc. Apollo 11 Lunar Sci. Conf., Geochim. Cosmochim. Acta* Suppl. 1, Vol. 2, pp. 1143–1163. Pergamon.

Gensho R. (1971) (personal communication).

Hubbard N. J., Gast P. W., and Wiesmann H. (1970) Rare earth, alkaline and alkali metal and $^{87}Sr/^{86}Sr$ data for subsamples of lunar sample 12013. *Earth Planet. Sci. Lett.* **9,** 181–184.

Hubbard N. J., Meyer C. Jr., Gast P. W., and Wiesmann H. (1971) The composition and derivation of Apollo 12 soils. *Earth Planet. Sci. Lett.* **10,** 341–350.

LSPET (1971) Preliminary examination of lunar samples from Apollo 14. *Science* **173,** 681–693.

Masuda A. (1962) Regularities in variation of relative abundances of lanthanide elements and an attempt to analyze separation-index patterns of some minerals. *J. Earth. Sci., Nagoya Univ.* **10,** 173–187.

Masuda A. (1968a) Lanthanides in the Norton County achondrite. *Geochem. J.* **2,** 111–135.

Masuda A. (1968b) Nature of the experimental Mohole basalt—Redetermination of lanthanides. *J. Geophys. Res.* **73,** 5425–5428.

Masuda A. (1972) Lunar solid-type and liquid-type materials and rare earth abundances. *Nature* **235,** 132.

Masuda A. and Matsui Y. (1966) The difference in lanthanide abundance pattern between the crust and the chondrite and its possible meaning to the genesis of crust and mantle. *Geochim. Cosmochim. Acta* **30,** 239–250.

Masuda A., Nakamura N., and Tanaka T. (1972) Fine structures of mutually normalized rare-earth patterns of chondrites (in preparation).

Masuda A. and Tanaka T. (1972) Possible europium-normal rare-earth abundances estimated from the lunar samples, and the terrestrial analogue. *Contr. Mineral. Petrol.* (in press).

Samsonov G. V. (1969) Physico-chemical properties of oxides (hand-book, Japanese translation). *Metallurgy*, Publishing Office, Moscow.

Schnetzler C. C., Philpotts J. A., and Bottino M. L. (1970) Li, K, Rb, Sr, Ba, and rare-earth concentrations, and Rb–Sr age of lunar rock 12013. *Earth Planet. Sci. Lett.* **9,** 185–192.

Schnetzler C. C. and Nava D. F. (1971) Chemical composition of Apollo 14 soils 14163 and 14259. *Earth Planet. Sci. Lett.* **11,** 345–350.

Swann G. A., Hait M. H., Schaber G. G., Freeman V. L., Ulrich G. E., Wolfe E. W., Reed V. S., and Sutton R. L. (1971) Preliminary description of Apollo 15 sample environments. Interagency Report: 36 (prepared by the Geological Survey for NASA).

Taylor S. R., Muir P., and Kaye M. (1971) Trace element chemistry of Apollo 14 lunar soil from Fra Mauro. *Geochim. Cosmochim. Acta* **35,** 975–981.

Wakita H. and Schmitt R. A. (1970) Elemental abundances in seven fragments from lunar rock 12013. *Earth Planet. Sci. Lett.* **9,** 169–176.

Wood J. A., Marvin U. B., Reid J. B. Jr., Taylor G. J., Bower J. F., Powell B. N., and Dickey J. S. Jr. (1971) Mineralogy and petrology of the Apollo 12 lunar samples. Smithsonian Astrophysical Observatory, Special Report 333.

Proceedings of the Third Lunar Science Conference
(Supplement 3, *Geochimica et Cosmochimica Acta*)
Vol. 2, pp. 1315–1326
The M.I.T. Press, 1972

# Provenance of Apollo 12 KREEP

JOHN T. WASSON and PHILIP A. BAEDECKER

Department of Chemistry and Institute of Geophysics and Planetary Physics,
University of California, Los Angeles, California 90024

**Abstract**—Four possible origins for Apollo 12 KREEP can be distinguished: (1) ejecta from Copernicus; (2) ejecta from Reinhold; (3) ejecta from the local highlands; and (4) material excavated from Fra Mauro deposits buried below the landing site. The first three sources do not appear able to account for the substantial deposits of KREEP found near the crests of Head and Bench Craters. This material appears to be buried Fra Mauro material brought to a near surface location (as demanded by the high neutron dosage) as ejecta from Middle Crescent Crater. KREEP in the common soil (e.g., 12070) at the Apollo 12 site appears to be a mixture of roughly equal proportions of Fra Mauro materials from each of the four proposed sources. Insofar as the U–Pb metamorphism ages of Apollo 12 soils date real events, the most likely candidates are (1) the formation of Middle Crescent Crater for the 12032 and 12033 soils, and (2) the formation of Copernicus and Reinhold for the 12070 soils.

## INTRODUCTION

SOILS AND BRECCIAS at the Apollo 12 landing site contain varying amounts of KREEP, a basaltic material highly enriched in certain of the incompatible elements. The acronym was proposed by Hubbard *et al.* (1971) and is a mnemonic emphasizing the high K, P, and rare earth content of this material. KREEP occurs mainly as glass and microbreccia fragments. Although distinct compositional variations are observed, surveys of soil glasses show that the KREEP population is well resolved from other major soil components, such as anorthositic gabbro and mare-type basalts (Reid *et al.*, 1972). KREEP is a major soil component at the Apollo 14 site. There are minor differences in the mean composition of KREEP between the two landing sites, but plots of individual analyses show a large degree of overlap in compositional fields.

Figure 1 shows a portion of Orbiter photo IV-125H3 which includes the Apollo 12 site.* The most striking local features are the Copernican ray that traverses the landing area, and the large amounts of highlands material which surround the site at distances of 20–100 km.

KREEP at the Apollo 14 site appears to originate in the main from the local highlands, i.e., from the local Fra Mauro formation, which presumably is ejecta from the Imbrium Basin. The local rocks at the Apollo 12 site consist of mare-type basalt, however, and the KREEP is therefore an exotic component (Baedecker *et al.*, 1971a; Hubbard *et al.*, 1971; Schnetzler *et al.*, 1970). Origins which have been proposed for Apollo 12 KREEP are (1) ejecta from Copernicus, which is located 370 km to the north; (2) ejecta from Reinhold, which is located 163 km north of the landing site;

* The Apollo 12 site in ACIC coordinates is 23.34°W, 2.99°S (L. Jaffe, private communication). The coordinates given by LSPET (1970) and Shoemaker *et al.* (1970a) are in error.

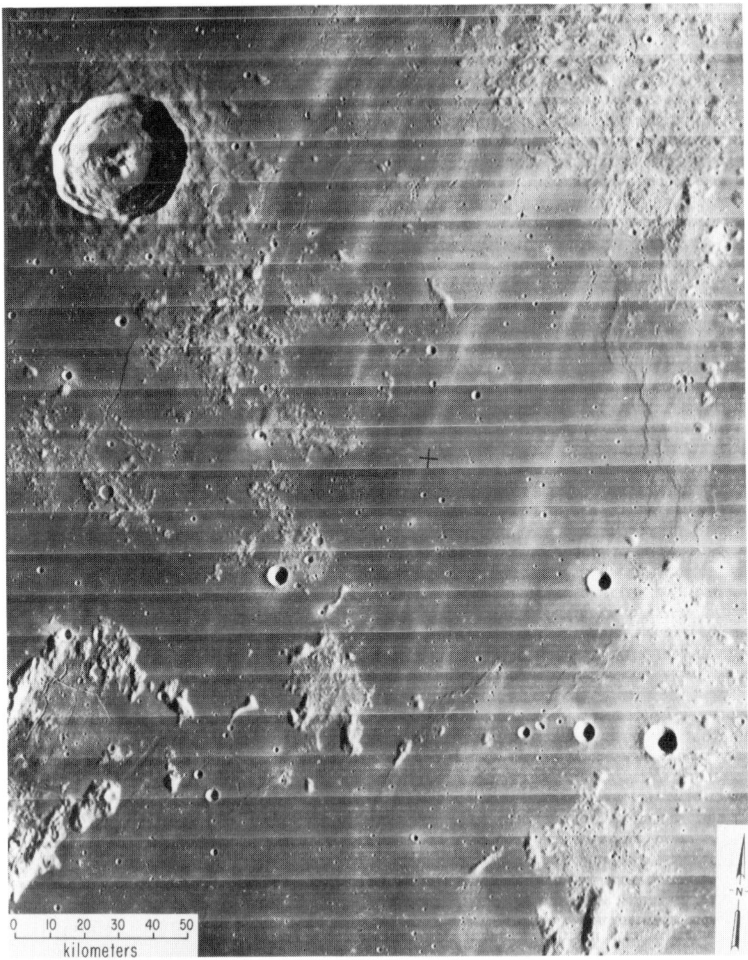


Fig. 1. A portion of Lunar Orbiter photo IV-125H3, showing the environs of the Apollo 12 landing site (marked with a cross). The most striking features are the Copernican Ray, which crosses the site, and the extensive outcrops of the Fra Mauro formation. The large impact crater, Lansberg, in the upper left, predates the filling of the local mare.

(2a) ejecta from Lansberg, which is located about 120 km to the northwest; (3) Fra Mauro material ejected from smaller cratering events from nearer outcrops, the nearest of which is about 20 km to the west; (4) Fra Mauro material underlying the mare basaltic material at the landing site; and (4a) premare, presumably Fra Mauro regolith underlying the mare-basalt. Source 2a is not considered to be viable, because Lansberg is stratigraphically older than the mare surface at the landing site (Schmitt *et al.*, 1967). Origin 4a is essentially an extension of origin 4, and will not be discussed other than to show that it cannot account for the KREEP in soil 12033.

The idea that KREEP could be Copernican in origin is one that captures the imagination, and the consensus to date seems to favor this viewpoint. However, we find that there are several reasonably strong arguments against such an origin. The purpose of this paper is to review available evidence and summarize the boundary conditions that must be met by any source of the Apollo 12 KREEP.

## Amount of KREEP in the Apollo 12 Regolith

The bulk composition of the Apollo 12 soils and breccias can be fitted satisfactorily by mixing models consisting of basalts, KREEP, and anorthositic materials (Lindsay, 1971; Wänke *et al.*, 1972). Goles *et al.* (1971) have also added a component consisting of local breccia. Since these breccias are themselves mixtures, we find no special merit to their inclusion. Addition of an extralunar component (Laul *et al.*, 1971; Baedecker *et al.*, 1971b) is only necessary to account for a few siderophilic elements. Studies of mixing models show that two components dominate: a member similar to Mg-rich Apollo 12 basalts, and a component with a composition to that of KREEP, as defined by Hubbard *et al.* (1971). Wänke *et al.* (1972) show that all soils and breccias contain KREEP components lying between 30%, as observed in the mature Apollo 12 soils 12070 and 12044 and 62%, as observed in the breccia 12034. H. Wänke (private communication) states that the soil 12037, for which they calculate a KREEP component of only 20%, should be disregarded, since it was returned to earth in the same sample bag as a more massive, high-Mg basaltic rock. The KREEP-rich soil 12033 contains 58% KREEP, according to Wänke *et al.* (1972).

Other research groups have estimated KREEP components which are comparable to those of Wänke *et al.* (1972). For example, Hubbard *et al.* (1971) estimate the KREEP components of 12070 and 12033 to be 25% and 65%, respectively. It appears that a mean KREEP concentration of 30% in soil 12070 is an accurate estimate. There are several reasons to believe that 12070 is representative of the mean soil at the Apollo 12 site. The great bulk of soil samples taken at random are similar to 12070, whereas the soils richer in KREEP were normally sampled because Conrad and Bean were struck by their higher albedo. That the 12070-type soils contain consistently larger extralunar components is also an indication that they are more representative of mature soils at the site (Baedecker *et al.*, 1971b).

The regolith depth at the Apollo 12 site is not well characterized. The most careful estimate would appear to be that of Oberdeck and Quaide (1968), who assigned a depth of 4.6 m to the regolith at Orbiter site III-P9c, which included the Apollo 12 site. These authors give the same estimate for Orbiter site II-P6b, which is near the Apollo 11 site. Shoemaker *et al.* (1970b) estimate the regolith depth at Tranquility Base to range from 3 to 6 m. Some authorities (e.g., LSPET, 1970) estimate that the regolith should be only about half as thick at the Apollo 12 as at the Apollo 11 site, but this is not in keeping with the fact that the cratering ages at the two sites are comparable. For example, Soderblom and Lebofsky (1972) estimate that the crater density at the Apollo 11 site is 1.5 times that at the Apollo 12 site, which, according to equation (1) of Baedecker *et al.* (1971b) indicates that the regolith depth at the Apollo 11 site should be greater by a factor of about 1.3. A reasonable compromise on the Apollo 12 regolith depth would appear to be about 3.5 m. If we assume that the mean density of the regolith is 2.15 g $cm^{-3}$, as measured in the lowest section of the 2.4-m Apollo 15 core, we calculate a regolith depth of 750 g $cm^{-2}$.

Although the striking inhomogeneities in the Apollo 12 regolith make it clear that it is not well mixed on a short time scale, there is evidence that it is well mixed on a time scale of the order of 2 g.y. The $He^3$ contents of the light components in the soil can be used to set upper limits on the time that these materials have been near the surface of the regolith. These are typically about 100 m.y. for particles separated from 12033, and are 260 and 520 m.y. for particles from the 15 and 23 cm layers of the 12028 core, respectively (Funkhouser, 1971). These upper limits on the time of emplacement of the observed inhomogeneities indicate that events capable of producing substantial stirring occur on a time scale of the order of a few hundred m.y. or less. The most powerful evidence for evaluating the mixing of the regolith appear to be the data of Burnett *et al.* (1971) and Marti and Lugmair (1971) regarding neutron-capture effects in Gd. According to Burnett *et al.*, these data indicate that the 12028 core has been stratified for only about 50 m.y., and that, on the basis of these and other Gd data, the maria appear to be "uniformly well-mixed seas." The KREEP content of a well-mixed Apollo 12 regolith is calculated to be about 230 g $cm^{-2}$.

## SOURCE STRENGTHS

Of the four postulated sources for Apollo 12 KREEP, the one involving the greatest lateral transport of material is the Copernicus event. Presumably, the other three sources are potentially stronger. Let us therefore examine the expected yields of Copernican material. Copernicus is located about 370 km from the Apollo 12 site. Shoemaker *et al.* (1970b) give a curve which shows that, for a single cratering event, 0.25% of the ejecta falls beyond a radial distance of 428 km, whereas 0.35% falls beyond a radial distance of 331 km. Thus, 0.1% of the ejecta should fall within an annular ring with an area of about $2.3 \times 10^{11}$ $m^2$. For Copernicus, this ring would include the Apollo 12 site. The volume of material ejected from Copernicus can be calculated from data given by Baldwin (1963). He gives the apparent diameter of Copernicus as 91.2 km and notes that the true excavated diameter is generally about 0.83 times the apparent (rim crest) diameter of a crater. The true excavated depth of a crater as large as this is not known, but a conservatively large depth would be 0.4 radius, as observed in artificial cratering events. Baldwin (1963) calculates the excavated volume from the formula: volume $= 0.5\pi r^2 d$, which with the above assumption regarding depth becomes: volume $= 0.2\pi r^3$. Combining the above, one calculates that 0.14 m of Copernican ejecta was deposited at the Apollo 12 site. According to Wood (1972), the density of KREEP is about 3.0 g $cm^{-3}$. Thus the Copernican ejecta should amount to about 42 g $cm^{-2}$ at this distance. This is more than a factor of 5 lower than the estimate of 230 g $cm^{-2}$ KREEP at the Apollo 12 site.

The presence of a Copernican ray may indicate a larger than average ejecta deposit at the Apollo 12 site. However, Oberbeck's (1971) investigations of Copernican rays indicate that these are bright because they are rich in bright secondary craters from the Copernican event. It is not known whether the smaller ejecta fragments that account for most of the mass will show the same distribution as the large ejecta masses which are capable of producing the bright haloed craters. An additional complexity at the Apollo 12 site results from the assertion by Quaide *et al.* (1971) that the brightness of the ray here results from tertiary material ejected from a field of secondary craters some 45 km to the north. We will arbitrarily estimate that the maximum amount of Copernican deposita is 84 g $cm^{-2}$, twice the nominal value estimated above.

This estimate is in agreement with the study of Oberbeck *et al.* (1972). Their computer simulation of regolith growth shows that the thickness expected at the Apollo 12 site based on the observed crater density is the same as that estimated from crater profiles, to within an accuracy of roughly 0.5 m. It conflicts with observations of E. M. Shoemaker (private communication) that the Copernican ejecta accounts for 2 m of the local regolith.

Shoemaker (private communication) states that Reinhold postdates the Apollo 12 site, although the ages are similar. The latter is considered to be the uppermost member of the Imbrian period, whereas Reinhold is assigned to the younger Eratosthenian period. We note in passing that Soderblom and Lebofsky (1972) assign Eratosthenes an older age than the Apollo 12 site. They do not present data on Reinhold. We have calculated the amount of Reinhold deposita expected at the Apollo 12 site in the same fashion as shown for Copernicus. The average depth of deposit at 163 km from Reinhold is estimated to be 69 g $cm^{-2}$, comparable to the amount estimated for Copernican deposita. The ray pattern of Reinhold is no longer visible, therefore no means are available for estimating whether the deposition at the Apollo 12 site was greater or less than the nominal value.

Although it is difficult to provide anything resembling a quantitative calculation, it seems clear that, because of its nearness, the highland material which Fig. 1 shows within 20 to 100 km in all directions from the Apollo 12 site should provide a source of exotic material which exceeds that estimated for Copernicus or Reinhold. Whether the buried Fra Mauro is capable of producing a high KREEP yield depends on the burial depth of such a hypothesized layer below the local surface. This problem will be discussed in a later section.

## Metamorphism Ages of KREEP Fragments

The three chief techniques for dating the formation or metamorphism of lunar samples are those based on the decay of $Rb^{87}$, $K^{40}$, and the isotopes of U. Samples of Apollo 12 KREEP consist, unfortunately, of very small fragments, making the application of these techniques very difficult. The one sample that was large enough to allow accurate study was the unusual breccia 12013, the dark portion of which is scarcely distinguishable from KREEP. It yielded a Rb–Sr age of 4.0 g.y. (Albee *et al.*, 1970) and an $Ar^{39}$–$Ar^{40}$ age of 4.03 g.y. (Turner, 1971). The latter study is of special significance, because it revealed that (a) outgassing of Ar was complete at the time of formation, and (b) no gas loss has occurred since formation.

Funkhouser (1971) reported K–Ar ages for a number of KREEP fragments separated from Apollo 12 soil samples, particularly 12033 and 12070. Of 16 particles, 14 give apparent ages lying between 1.0 and 1.9 g.y.; the other two were lower. Funkhouser assumes that the particles were outgassed at the time of deposition and that the maximum apparent age corresponds to the minimum age of a single depositionary event. Eberhardt *et al.* (1972) report $Ar^{39}$–$Ar^{40}$ ages of 0.82 and 0.87 g.y. for two density fractions separated from 12033. Since each of these fractions consisted of about seven particles (P. Eberhardt, private communication), there is no unique interpretation that can be applied to these ages. They may measure an event, or they may reflect the combination of material of higher age with some which has experienced recent outgassing. That the apparent age climbs in their high-temperature gas fractions tends to support the latter viewpoint, because this effect could be due to residual gas remaining in some recently outgassed fragments. The Eberhardt *et al.* (1972) data would seem to provide lower limits on the emplacement of KREEP by a single event.

The most striking evidence for metamorphism is provided by U–Pb dating techniques (Cliff *et al.*, 1971; Tatsumoto *et al.*, 1971). In contrast to the Apollo 11 soils, which yielded data plotting on the concordia curve at about 4.7 g.y., data on the Apollo 12 soils plot on or near a two-stage mixing line extending between 4.7 g.y. and a point at about 1.2–2.0 g.y. on the concordia curve. This observation seems to indicate that a portion of the sample lost Pb at about 1.2–2.0 g.y. ago. The 12070 data lie nearer to the 4.7 g.y. intercept than do the data from the KREEP-rich soils 12032 and 12033. If one assumes that a two-stage model is the correct interpretation of the data, the data plotted by Tatsumoto *et al.* (1971) can be interpreted to show that 30% of the 12070 soil and 50% of the 12033 soil lost their Pb 2.0 g.y. ago. Similarly, the data of Cliff *et al.* (1971) are consistent with 36% of 12070 and 39% of 12032 being Pb-free about 1.5 g.y. ago. Since most of the U even in 12070 is associated with the KREEP, these are estimates of the fraction of KREEP which has lost Pb during high temperature events.

Studies of the Apollo 14 soils (Silver, 1972; Tatsumoto *et al.*, 1972) again show some tendency of the points to lie along a chord indicating that a portion of the material lost Pb about 2.2–3.0 g.y. ago. The data plot much nearer to the high-age concordia intercept, however, indicating that the fraction of material experiencing Pb loss is smaller than that observed at the Apollo 12 site.

## AGE OF COPERNICUS

The age of Copernicus can be estimated from crater counts (Gault, 1970; Hartmann, 1970) or by observing the effects of crater erosion (Soderblom and Lebofsky, 1972). The latter study seems to provide the better precision, mainly because it involves integration over smaller areas. The Soderblom and Lebofsky data indicate that the integrated flux of crater producing projectiles at Copernicus is 0.47 times that at the Apollo 12 site. They interpret their data to indicate that the flux of these projectiles has been decreasing with a half-life of 0.8 g.y. for the past 3.6 g.y. We (Baedecker *et al.*, 1972) find that our trace-element data indicate that the flux has been falling off more slowly, with a half-life of about 2.0 g.y. Taking the age of the Apollo 12 site to be 3.26 g.y., fluxes with these half-lives yield ages of Copernicus of 2.5 and 2.0 g.y., respectively. We believe that the latter is the more accurate estimate.

As indicated earlier, the age of Reinhold appears to be roughly the same as that of the Apollo 12 site.

## Geography and Cosmic-Ray Ages

According to the data of Wänke *et al.* (1972), the most KREEP-rich samples other than 12013 are the soils 12033 and 12032 and the breccias 12034 and 12073. Sutton and Schaber (1971) assign the first three definite locations at the Apollo 12 site, whereas only a tentative assignment is made for 12073. The locations are shown on Fig. 2. It seems particularly significant that the two soils were taken from the crests of the two most recent craters which are large enough to have penetrated the regolith. Cosmic-ray ages for the soils are relatively short. The $He^3$ data of Funkhouser (1971) indicate cosmic-ray ages for KREEP fragments ranging from 32 to 200 m.y. Eberhardt *et al.* (1972) report $Ne^{21}$ ages for individual KREEP fragments which range from 100 to 290 m.y. Crozaz *et al.* (1971) argue that 12033 was deposited $\leq$40 m.y. ago, based on the lowest track densities observed in their sample (found in 8% of their sample). On the average, fragments from 12070 are older than those from 12033. The cosmic-ray age of breccia 12013 is about 55 m.y. (Albee *et al.*, 1970); its particle-track surface-residence age is about 15 m.y. (Burnett *et al.*, 1970).

Wänke *et al.* (1971) point out that the ages of lunar craters can often be determined on the basis of data from rocks picked up in their vicinity. They note that a peak in the crystalline-rock cosmic-ray ages at about 90 m.y. probably represents the age of

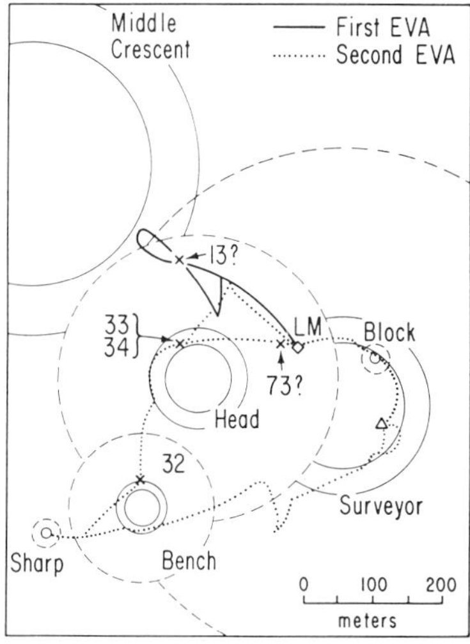


Fig. 2. Schematic map showing the detailed crater stratigraphy at the Apollo 12 site. The two inner solid concentric circles show the inner rim and the ridge crest of each crater. The outer, broken circle shows the probable limit of the continuous ejecta blanket, and is used to indicate superposition. The locations of KREEP-rich soils and breccias are shown by *X* and the last two digits of the sample number. [After Sutton and Schaber (1971).]

Head Crater or the stratigraphically younger Bench Crater. They suggest that a peak at about 200 m.y. represents the Surveyor Crater event. Funkhouser (1971) suggests that the 40 m.y. age reported for the 12033 soil by Crozaz *et al.* (1971) represents the age of Head Crater. Since there are no rocks of this age, this interpretation seems questionable. This is more likely to be the date of the smaller Bench Crater event, which surely added material to the deposit on the rim of Head Crater. We suggest that the best dates that can be assigned at present are 40, 90, and 200 m.y. for the events which produced Bench, Head, and Surveyor craters. It seems more likely that these err on the low side than on the high. E. M. Shoemaker (private communication) finds stratigraphic evidence indicating that both Surveyor and the older Middle Crescent Crater predate Copernicus.

The effects of cosmic-ray-secondary neutrons in generating changes in isotopic effects provide especially important constraints regarding the origin of Apollo 12 KREEP. The data of Burnett *et al.* (1971) show that soil 12033 has received a neutron dosage about 0.9 times that observed in soils 12070 and 12042 and in the core 12028, which are quite high, and similar to those in Apollo 11 soils. Baedecker *et al.* (1971a) interpret the siderophilic-element data in 12033 to indicate that it consists of a mixture of about 60% mature local soil with about 40% KREEP-like material which was not previously part of the regolith. According to this interpretation, the fresh KREEP has received a neutron dosage about 75% as great as the mean value in the local soil, or about $1.6 \times 10^{16}$ neutrons $cm^{-2}$. Unpublished calculations by Lingenfelter, Canfield, and Hample indicate that the maximum depth that material could be stored on the moon and still receive this total flux within 4 g.y. is about 550–700 g $cm^{-2}$, i.e., of the order of the mean depth estimated for the local regolith. Reducing the storage age to 2 g.y. increases the storage depth by about 110 g $cm^{-2}$, or by about 40 cm of crystalline KREEP.

## REGOLITH HISTORY OF KREEP IN APOLLO 12 SOILS

Mention was made in the previous paragraph of the interpretation of data on Ni, Ge, and Ir by Baedecker *et al.* (1971a) to indicate that about 60% of 12033 is old regolith material, whereas 40% has spent a negligible portion of its history as regolith. If the latter material were ever part of a preexisting regolith, it must have been covered within a period short relative to 3.3 g.y. Quaide *et al.* (1971) do not attempt to separate 12033 into old and young components but point out that the total sample shows minimal evidence of reworking. For example, the proportion of shocked grains is small, while the mean grain size is large relative to a normal soil sample. Although it seems very unlikely that 12033 KREEP was excavated from a preexisting regolith, Quaide *et al.* believe that the material has been moved two or three times since it was formed as crystalline rock.

Oberbeck *et al.* (1972) give results of a Monte Carlo calculation that shows the fraction of regolith material produced by craters within given radius intervals. It is of interest to use this calculation to estimate the burial depth of a hypothetical subsurface KREEP stratum. The KREEP component of the mature soils at the Apollo 12 site amounts to about 30%. The data of Oberbeck *et al.* (1972) show that 40% of the regolith is produced by craters having radii greater than 140 m. The depth of a crater with a 140 m radius is about 0.4 r, or 56 m. If we make the gross assumption that 75% of the material ejected from craters greater in radius than 140 m comes from below 56 m, then 30% of the total regolith comes from below this depth. If the bulk of the KREEP at the Apollo 12 site has been excavated from Fra Mauro material lying below a thin layer of mare basalt, this is a rough estimate of the mean depth of this basaltic material.

## Discussion of the Proposed Origins

The possible sources of KREEP mentioned in the Introduction were (1) ejecta from Copernicus; (2) ejecta from Reinhold; (3) ejecta from cratering events occurring in the Fra Mauro outcrops within about 100 km of the Apollo 12 site; (4) material excavated from Fra Mauro deposits buried below the basaltic material at the Apollo 12 site.

It is clear that the first two sources cannot account for certain properties of the KREEP-rich soils in a simple fashion. The noble-gas and particle-track ages of 12033 KREEP clearly indicate that this material has not been at its last location near the surface for more than 40–200 m.y. Since both cratering evidence and neutron-dosage measurements on Apollo 12 and Apollo 14 soils indicate extensive mixing during periods comparable to the age of Copernicus or of the Apollo 12 regolith, it is very unlikely that this KREEP-rich material could have been stored in the upper 1–3 m of the regolith and still have retained its compositional and textural uniqueness. Nor does it seem reasonable to transport Copernican or other Fra Mauro material from any of the surrounding highlands regions within the past 40–200 m.y. and still have preserved the compositional characteristics of 12033. The only reasonable way to understand these properties seems to involve storage below the surface of the Apollo 12 site in deposits which were not involved in regolith-mixing. The neutron-dosage data indicate that such a deposit could not be much lower than the bottom of the regolith, i.e., of the order of 700 g $cm^{-2}$.

It is possible in principle that a local accumulation of Copernican material could have been deposited and then quickly covered with enough regolith to prevent mixing prior to its reexcavation by the Head Crater or Bench Crater event. However, the impact velocity of Copernican ejecta must have been greater than or equal to 0.77 km $sec^{-1}$. Since the (loosely consolidated) Apollo 12 regolith had achieved 0.6–0.7 of its present depth prior to the Copernican event, and since the amount of Copernican material must have been small compared to the regolith volume, it is hard to conceive that the Copernican material could have formed a local massive deposit retaining its compositional uniqueness for 2 g.y. Further, if this deposition occurred after the formation of Middle Crescent Crater, we are forced to assume that a local concentration of the exotic material occurred on a topographic high (the Middle Crescent ejecta blanket) rather than the more likely occurrence on a topographic low. Much the same arguments apply in only slightly reduced strength to an origin from the crater Reinhold. A particularly difficult problem is understanding how the Reinholdian material managed to stay within 3–4 m of the surface despite being buried under the Middle

Crescent ejecta blanket. We consider these processes to be highly unlikely and effectively to exclude these origins for the 12033 KREEP.

A source consisting of buried Fra Mauro material involves no problems in lateral transport. It must, however, be located near enough to the surface to account for the high neutron dosage observed in 12033. A model for bringing this material to the surface is given below.

The low cosmic-ray age of 12013, its comparably low neutron dosage, and its completely flat temperature release pattern for $Ar^{39}$–$Ar^{40}$ rule out the possibility that it has been transported to the Apollo 12 site together with Copernican or Reinholdian ejecta, or that from the surrounding highlands. It must be of local provenance.

The frequent outcrops of Fra Mauro material in the neighborhood of the Apollo 12 site make it clear that such material is also buried below the Apollo 12 site. The only question is how deep. Since the relief of the local outcrops is not pronounced, it would seem reasonable that the basaltic cover is only some tens of meters thick.

There are two ways in which such material might occur within about 3 m of the surface, as demanded by the neutron dosage evidence. The Fra Mauro outcrops may be irregular in thickness, and a topographic high may occur immediately below Head and Bench Craters. A second, more plausible idea is suggested by Fig. 6 in the paper by Warner (1971). He notes that Head Crater occurs in the ejecta blanket of Middle Crescent Crater. Bench Crater must also penetrate into this material. Ejecta blankets are known to be stratified inversely relative to the local bedrock. Middle Crescent Crater has a true excavated radius of about 360 m, and we therefore estimate the excavated depth to be about 144 m. If the lowest stratum penetrated was Fra Mauro material, then the upper layer in the ejecta blanket is of KREEP composition. Subsequent formation of Surveyor Crater and others in the vicinity would have provided both surficial mixing of and coverage by regolith. Head and Bench Craters are estimated to have excavated depths of about 18 and 10 m. Both of these could have easily penetrated the regolith, and probably the ejecta blanket of Surveyor Crater as well. The two miraculous aspects of this model are (1) that the 12033 material was ejected from only about 3 m depth rather than from a greater depth in one of these craters; and (2) the U–Pb evidence that a sizable fraction of this material has lost Pb. This Pb loss may have occurred during the ejection of material from Middle Crescent Crater, but the situation is complicated by the apparent loss of Pb from the 12070 soil as well.

The KREEP in 12070 probably comes from all the postulated sources—Copernicus, Reinhold, local Fra Mauro outcrops, and buried Fra Mauro. It seems reasonable that the amounts of KREEP from these sources are roughly comparable. If only about 10% of the regolith KREEP comes from buried Fra Mauro materials, the calculations of Oberbeck *et al.* (1972) can be used to estimate the mean burial depth to be roughly of the order of 80–100 m.

The U–Pb metamorphism "ages" apparently indicate severe Pb loss by the KREEP about 1.2–2.0 g.y. ago. The 12033 data can be understood if Pb was lost from the Middle Crescent ejecta, with the discordant chord representing a mixing line between this material and the normal regolith. The position of 12070 on the concordia plot is much more difficult to explain. If we presume that Pb volatilized during the Middle

Crescent event traveled only a short distance ($\leq 1$ km) before falling out, it ought to be in 12070, and this soil should lie near concordia. If such Pb travels much greater distances on the average, the presence of KREEP-rich highlands surrounding the Apollo 12 site should lead to an excess rather than a deficiency of parentless Pb. Because of their high ejection velocities, Copernican and Reinholdian ejecta are prime candidates for 12070 materials that have lost Pb. If this explanation is correct, the similarity in "metamorphism ages" for 12033 and 12070 KREEP is fortuitous.

*Acknowledgments*—We are indebted to R. Greeley, L. Jaffe, V. R. Oberbeck, E. M. Shoemaker, H. Wänke, and G. W. Wetherill for stimulating discussions. This research was supported in part by NASA Grant NAS-05-007-291.

## References

Albee A. L., Burnett D. S., Chodos A. A., Haines E. L., Huneke J. C., Papanastassiou D. A., Podosek F. A., Russ G. P. III, and Wasserburg G. J. (1970) Mineralogic and isotopic investigations on lunar rock 12013. *Earth Planet. Sci. Lett.* **9,** 137–163.

Baedecker P. A., Cuttitta F., Rose H. J., Schaudy R., and Wasson J. T. (1971a) On the origin of lunar soil 12033. *Earth Planet. Sci. Lett.* **10,** 361–364.

Baedecker P. A., Schaudy R., Elzie J. L., Kimberlin J., and Wasson J. T., (1971b) Trace element studies of rocks and soils from Oceanus Procellarum and Mare Tranquilitatis. *Proc. Second Lunar Sci. Conf., Geochim. Cosmochim. Acta* Suppl. 2, Vol. 2, pp. 1037–1061. MIT Press.

Baedecker P. A., Chou C.-L., and Wasson J. T. (1972) The extralunar component in lunar soils and breccias. *Proc. Third Lunar Sci. Conf., Geochim. Cosmochim. Acta* Suppl. 3, Vol. 2, MIT Press.

Baldwin R. B. (1963) *The Measure of the Moon.* University of Chicago Press.

Burnett D. S., Monnin M., Seitz M., Walker R., Woolum D., and Yuhas D. (1970) Charged particle track studies in lunar rock 12013. *Earth Planet. Sci. Lett.* **9,** 127–137.

Burnett D. S., Huneke J. C., Podosek F. A., Russ G. P., and Wasserburg G. J. (1971) The irradiation history of lunar samples. *Proc. Second Lunar Sci. Conf., Geochim. Cosmochim. Acta* Suppl. 2, Vol. 2, pp. 1671–1679. MIT Press.

Cliff R. A., Lee-Hu C., and Wetherill G. W. (1971) Rb–Sr and U, Th–Pb measurements on Apollo 12 material. *Proc. Second Lunar Sci. Conf., Geochim. Cosmochim. Acta* Suppl. 2, Vol. 2, pp. 1493–1502. MIT Press.

Crozaz G., Walker R., and Woolum D. (1971) Nuclear track studies of dynamic surface processes on the moon and the constancy of solar activity. *Proc. Second Lunar Sci. Conf., Geochim. Cosmochim. Acta* Suppl. 2, Vol. 3, pp. 2543–2558. MIT Press.

Eberhardt P., Eugster P., Geiss J., Grögler N., Schwarzmüller J., Stettler A., and Weber L. (1972) When was the Apollo 12 KREEP ejected? (abstract). In *Lunar Science—III* (editor C. Watkins), pp. 206–208, Lunar Science Institute Contr. No. 88.

Funkhouser J. (1971) Noble gas analysis of KREEP fragments in lunar soil 12033 and 12070. *Earth Planet. Sci. Lett.* **12,** 263–272.

Gault D. E. (1970) Saturation and equilibrium conditions for impact cratering on the lunar surface: Criteria and implications. *Radio Sci.* **5,** 273–291.

Goles G. G., Duncan A. R., Lindstrom D. J., Martin M. R., Beyer R. L., Osawa M., Randle K., Meek L. T., Steinborn T. L., and McKay S. M. (1971) Analyses of Apollo 12 specimens: Compositional variations, differentiation processes, and lunar soil mixing models. *Proc. Second Lunar Sci. Conf., Geochim. Cosmochim. Acta* Suppl. 2, Vol. 2, pp. 1063–1081. MIT Press.

Hartmann W. K. (1970) Lunar cratering chronology. *Icarus* **13,** 299–301.

Hubbard N. J., Meyer C., Gast P. W., and Wiesmann H. (1971) The composition and derivation of Apollo 12 soils. *Earth Planet. Sci. Lett.* **10,** 341–350.

Laul J. C., Morgan J. W., Ganapathy R., and Anders E. (1971) Meteoritic material in lunar samples: Characterization from trace elements. *Proc. Second Lunar Sci. Conf., Geochim. Cosmochim. Acta* Suppl. 2, Vol. 2, pp. 1139–1158. MIT Press.

Lindsay J. F. (1971) Mixing models and the recognition of end-member groups in Apollo 11 and 12 soils. *Earth Planet. Sci. Lett.* **12,** 67–72.

LSPET (Lunar Sample Preliminary Examination Team) (1970) Preliminary examination of lunar samples from Apollo 12. *Science* **167,** 1325–1339.

Marti K. and Lugmair G. W. (1971) $Kr^{81}$–Kr and K–$Ar^{40}$ ages, cosmic-ray spallation products, and neutron effects in lunar samples from Oceanus Procellarum. *Proc. Second Lunar Sci. Conf., Geochim. Cosmochim. Acta* Suppl. 2, Vol. 2, pp. 1591–1605. MIT Press.

Oberbeck V. R. (1971) A mechanism for the production of lunar crater rays. *The Moon* **2,** 263–278.

Oberbeck V. R. and Quaide W. L. (1968) Genetic implications of lunar regolith thickness variations. *Icarus* **9,** 446–465.

Oberbeck V. R., Quaide W. L., Mahon M., and Paulson J. (1972) Monte Carlo calculations of lunar regolith thickness distributions. Preprint.

Quaide W., Oberbeck V., Bunch T., and Polkowski G. (1971) Investigations of the natural history of the regolith at the Apollo 12 site. *Proc. Second Lunar Sci. Conf., Geochim. Cosmochim. Acta* Suppl. 2, Vol. 1, pp. 701–718. MIT Press.

Reid A. M., Ridley W. I., Warner J., Harmon R. S., Brett R., Jakes P., and Brown R. W. (1972) Chemistry of highland and mare basalts as inferred from glasses in the lunar soils (abstract). In *Lunar Science—III* (editor C. Watkins), pp. 640–642, Lunar Science Institute Contr. No. 88.

Schmitt H. H., Trask N. J., and Shoemaker E. M. (1967) Geologic map of the Copernicus Quadrangle of the moon. U.S. Geol. Surv. Map 1-515.

Schnetzler C. C., Philpotts J. A., and Bottino M. L. (1970) Li, K, Rb, Sr, Ba, and rare-earth concentrations, and Rb–Sr age of lunar rock 12013. *Earth Planet. Sci. Lett.* **9,** 185–192.

Shoemaker E. M., Batson R. M., Bean A. L., Conrad C., Dahlem D. H., Goddard E. N., Hait M. H., Larson K. B., Schaber G. G., Schleicher D. L., Sutton R. L., Swann G. A., and Waters A. A. (1970a) Preliminary geologic investigation of the Apollo 12 landing site. *Apollo 12 Preliminary Science Report*, NASA Document NASA SP-235, pp. 113–156.

Shoemaker E. M., Hait M. H., Swann G. A., Schleicher D. L., Schaber G. G., Sutton R. L., Dahlem D. H., Goddard E. N., and Waters A. C. (1970b) Origin of the lunar regolith at Tranquility Base. *Proc. Apollo 11 Lunar Sci. Conf., Geochim. Cosmochim. Acta* Suppl. 1, Vol. 3, pp. 2399–2412. Pergamon.

Silver L. T. (1972) U–Th–Pb abundances and isotopic characteristics in some Apollo 14 rocks and soils and an Apollo 15 soil (abstract). In *Lunar Science—III* (editor C. Watkins), pp. 704–706, Lunar Science Institute Contr. No. 88.

Soderblom L. A. and Lebofsky L. A. (1972) Technique for rapid determination of relative ages of lunar areas from orbital photography. *J. Geophys. Res.* **77,** 279–296.

Sutton R. L. and Schaber G. G. (1971) Lunar locations and orientations of rock samples from Apollo missions 11 and 12. *Proc. Second Lunar Sci. Conf., Geochim. Cosmochim. Acta* Suppl. 2, Vol. 1, pp. 17–26. MIT Press.

Tatsumoto M., Knight R. J., and Doe B. R. (1971) U–Th–Pb systematics of Apollo 12 lunar samples. *Proc. Second Lunar Sci. Conf., Geochim. Cosmochim. Acta* Suppl. 2, Vol. 2, pp. 1521–1546. MIT Press.

Tatsumoto M., Hedge C. E., Doe B. R., and Unruh D. (1972) U–Th–Pb and Rb–Sr measurements on some Apollo 14 lunar samples (abstract). In *Lunar Science—III* (editor C. Watkins), pp. 741–743, Lunar Science Institute Contr. No. 88.

Turner G. (1971) $^{40}Ar$–$^{39}Ar$ ages from the lunar maria. *Earth Planet. Sci. Lett.* **11,** 169–191.

Wänke H., Baddenhausen H., Balacescu A., Teschke F., Spettel B., Dreibus G., Quijano M., Kruse H., Wlotzka F., and Begemann F. (1972) Multielement analyses of lunar samples (abstract). In *Lunar Science—III* (editor C. Watkins), pp. 779–781, Lunar Science Institute Contr. No. 88.

Warner J. (1971) Lunar crystalline rocks: Petrology and geology. *Proc. Second Lunar Sci. Conf., Geochim. Cosmochim. Acta* Suppl. 2, Vol. 1, pp. 469–480. MIT Press.

Wood J. A. (1972) The nature of the lunar crust and composition of undifferentiated lunar material. *The Moon*, in press.

Proceedings of the Third Lunar Science Conference
(Supplement 3, *Geochimica et Cosmochimica Acta*)
Vol. 2, pp. 1327–1333
The M.I.T. Press, 1972

# Beryllium and chromium abundances in Fra Mauro and Hadley-Apennine lunar samples

K. J. EISENTRAUT, M. S. BLACK, F. D. HILEMAN, and R. E. SIEVERS

Aerospace Research Laboratories, ARL/LJ, Air Force Systems Command
Wright-Patterson Air Force Base, Ohio 45433

and

W. D. ROSS

Monsanto Research Corporation, Dayton, Ohio 45407

**Abstract**—Beryllium and chromium contents of lunar rocks and soils have been determined in fourteen samples returned from the Apollo 14 and Apollo 15 landing sites using gas chromatography. Lunar rock and soil samples from the Fra Mauro region have much higher beryllium contents than those of lunar samples from the other Apollo mission sites. Lunar crystalline rocks from the Apollo 15 site contain less beryllium than those returned from any of the previous Apollo missions. Lunar breccia 14321 is inhomogeneous with respect to both beryllium and chromium contents. The analytical determination of the inhomogeneity of this sample has shown that the beryllium and chromium concentrations vary inversely in each of the 65 mg fragments from the same breccia chip. In general, the beryllium content of the lunar crystalline rocks is lower than that of the soils or breccias. The high beryllium content of the Fra Mauro samples supports the possibility that highland material was translocated and contaminated the soil in the maria regions, Mare Tranquillitatis and Oceanus Procellarum, thus explaining the higher beryllium content of soils relative to that of the crystalline rocks at the Apollo 11 and 12 sites. A mass spectrometric study to determine whether organosiloxanes reported by others are present in representative Apollo 11, 12, and 14 samples has been completed; organosiloxanes were not detected in the samples analyzed by us.

## INTRODUCTION

APOLLO 14 AND 15 lunar samples have been analyzed for beryllium and chromium using electron capture gas chromatography. The beryllium content of various U.S. Geological Survey standard rock samples and of a St. Severin meteorite sample has also been determined. The analytical method consists of converting the metal to a volatile chelate which is extracted into benzene, followed by separation and detection by electron capture gas chromatography as described previously (Sievers *et al.*, 1971; and Eisentraut *et al.*, 1971).

In addition to the beryllium and chromium analyses reported herein a method was devised using special dissolution procedures and mass spectrometry to determine whether organosiloxanes could be detected in representative Apollo 11, 12, and 14 samples. The method employed was quite different from that of Gehrke *et al.* (1970) who reported the presence of organosiloxanes in Apollo 11 fines.

## ANALYTICAL METHODS

### *Beryllium and chromium analyses*

The analytical method for the analysis of beryllium and chromium consists of solubilization of the lunar samples using a sodium carbonate fusion followed by dissolution in dilute HCl. A rough pH adjustment is made with dilute NaOH and an acetic acid-sodium acetate buffer solution is added to pH 5.0 to 5.5. A $Na_2EDTA$ solution is then added to the samples for beryllium analysis in order to mask other metals present in the sample. ($Na_2EDTA$ is not added to the sample solutions for the chromium analysis). The beryllium or chromium aqueous solutions are then shaken with a solution of trifluoroacetylacetone, H(tfa), in benzene in order to cause reaction of the metals of interest with the ligand. The respective beryllium or chromium volatile chelate thus formed is extracted into the benzene phase and one or two microliters of this solution is injected into the gas chromatograph equipped with an electron capture detector and analyzed. Three separate *ca.* 65-mg portions of each lunar sample are fused independently and each of these portions is divided into five distinct aliquots with each of the five aliquots being injected into the chromatograph 5 or 6 times. This provides between 75 and 90 separate measurements for each lunar sample and accounts for the high precision of the analyses as evidenced by the standard deviations obtained. The details of the technique have been reported earlier (Eisentraut *et al.*, 1971; Sievers *et al.*, 1971).

### *Organosiloxane analysis*

The method for organosiloxane analysis consists of partial dissolution of representative lunar samples in freshly distilled neat trifluoroacetylacetone at room temperature. The organic extract is then vacuum pumped at room temperature and inserted via the direct probe into the DuPont, CEC Model 21-491 mass spectrometer. Numerous mass spectra are taken at probe temperatures ranging from 30° to 200°C. The mass range between 12 and 600 is examined for organosiloxane peaks that were reported by Gehrke *et al.* (1970).

The following lunar samples were examined using this method:

(1) Three different samples containing approximately 25 mg each of Apollo 11 fines, 10084,145 were separately shaken with 50 microliters of H(tfa) in a Kimax melting point capillary tube. The tubes were then sealed and allowed to react for 24, 120, and 168 hours, respectively, at room temperature.

(2) A quantity (587 mg) of Apollo 12 rock, 12002,141 was submerged in 3 ml of H(tfa) and the reaction was allowed to proceed for two weeks at room temperature in a closed Kimax borosilicate vessel. At the end of the reaction period the color was deep red-orange, characteristic of the iron(III) chelate of trifluoroacetylacetone which is formed together with other metal chelates just by allowing the ligand to stand in contact with lunar material. The volume of the solution was reduced by evaporation at room temperature and the last remaining drops of solution were transferred to a melting point capillary tube. Most of the remaining highly volatile liquid was removed in vacuo and the residue in the capillary tube was then subjected to mass spectrometric analysis.

(3) A quantity (505 mg) of Apollo 14 fines, 14163,149 was allowed to react with 2 ml of H(tfa) for one week at room temperature in a Kimax vessel sealed with a Teflon-lined cap. At the conclusion of the reaction period the volume of solution was reduced by evaporation at room temperature and the last remaining drops of solution were transferred to a melting point capillary tube which was treated as before.

Precautions were taken in sample reaction and analysis to preclude any contamination. Blanks were run for all samples analyzed. Appropriate background and blank corrections were made in the data interpretation.

## DISCUSSION OF RESULTS

### *Beryllium analysis*

The beryllium concentrations in the U.S. Geological Survey standard rocks as determined by the electron capture gas chromatographic method are shown in

Table 1. Beryllium concentration of U.S. Geological Survey standard rocks.

| Sample | Beryllium Conc. (ppm) |
|---|---|
| Basalt, BCR–1 | 1.53 ± 0.18 |
| Granite, G–2 | 2.07 ± 0.13 |
| Peridotite, PCC–1 | 0.0019 ± 0.0001 |
| Dunite, DTS–1 | 0.0041 ± 0.0002 |
| Andesite, AGV–1 | 2.24 ± 0.08 |
| Granodiorite, GSP–1 | 1.27 ± 0.08 |
| Diabase, W–1 | 0.63 ± 0.05 |

Table 2. Beryllium concentration of some meteorites.

| Meteorite | Beryllium Conc. (ppm) |
|---|---|
| St. Severin | 0.062 ± 0.002 |
| Allende | 0.03 ± 0.004 to |
| | 0.12 ± 0.01 |
| Murchison | 0.031 ± 0.002 |

Table 1 and beryllium concentrations in some chondritic meteorites are shown in Table 2. The values for W-1, BCR-1, Allende meteorite, and Murchison meteorite have been reported previously (Sievers *et al.*, 1971).

The beryllium content of the meteorites shown in Table 2 and the values reported for chondritic, achondritic, and iron meteorites by Sill and Willis (1962) indicate that meteorites contain much less beryllium than the lunar samples returned to date.

The beryllium concentrations in the Apollo 14 and 15 samples are shown in Table 3 and Table 4, respectively. The phenomenon that the beryllium content of the lunar fines is much larger than that of the crystalline rocks observed previously (Sievers *et al.*, 1971) for the Mare Tranquillitatis and Oceanus Procellarum samples is also observed in the case of the Apollo 15 samples.

Table 3. Beryllium analysis of Apollo 14 lunar samples.

| NASA Sample No. | Sample Type | Beryllium Conc. (ppm) |
|---|---|---|
| 14163,105 | Apollo 14 fines | 6.64 ± 0.21 |
| 14311,33 | Apollo 14 breccia | 11.72 ± 0.57 |
| 14301,64 | Apollo 14 breccia | 7.09 ± 0.33 |
| 14047,46 | Apollo 14 breccia | 5.97 ± 0.29 |
| 14310,135 | Apollo 14 crystalline rock | 5.78 ± 0.24 |
| 14321,192* | Apollo 14 breccia, | 6.99 ± 0.14 |
| | Interior fragment | 4.23 ± 0.08 |
| | | 1.77 ± 0.13 |
| | | 4.84 ± 0.14 |
| | | 5.28 ± 0.36 |
| 14321,188* | Apollo 14 breccia, | 3.31 ± 0.13 |
| | Exterior fragment | 6.09 ± 0.10 |

* Inhomogeneous rock.

Table 4. Beryllium analysis of Apollo 15 lunar samples.

| NASA Sample No. | Type of Sample | Beryllium Conc. (ppm) |
|---|---|---|
| 15555,143 | Apollo 15 crystalline rock | $0.28 \pm 0.02$ |
| 15058,86 | Apollo 15 crystalline rock | $0.45 \pm 0.03$ |
| 15601,66 | Apollo 15 fines | $1.31 \pm 0.04$ |
| 15471,30 | Apollo 15 fines | $1.59 \pm 0.04$ |
| 15421,26 | Apollo 15 fines | $1.02 \pm 0.06$ |
| 15301,78 | Apollo 15 fines | $1.85 \pm 0.06$ |
| 15021,96 | Apollo 15 fines | $2.66 \pm 0.11$ |

The samples returned from the lunar highland, Fra Mauro, have the highest beryllium content of any lunar samples yet observed from the Apollo missions. This finding, together with the work of many other investigators dealing with other elements, suggests that highland type material could well have contaminated the fines at the other Apollo sites. Certain crystalline rock samples returned from the Apollo 15 site are specimens having the lowest beryllium content observed for any of the previous Apollo mission samples.

Lunar breccia 14321 shows significant inhomogeneity in beryllium content in both the interior and exterior fragments analyzed. This sample is unique in that it is the first lunar sample that we have found to be clearly inhomogeneous (65 mg fragments of 1 g samples) with respect to beryllium concentration. As mentioned in the chromium section, we have found this breccia to be inhomogeneous with respect to chromium content as well, and the beryllium and chromium contents vary inversely in the fragments analyzed.

Crystalline rock 14310 is the first example of a lunar crystalline rock that we have analyzed from any of the Apollo missions that contains more than 1 ppm Be. This rock has been described by LSPET (1971) as being of special interest because its chemistry and modal mineralogy differ from those of previously described lunar basaltic rocks and that it is very similar to the fragmental rocks in composition. These facts raise interesting questions about its origin. The beryllium content of 14310 is much larger than one might have expected from our other analyses of crystalline rocks from the Apollo 11, 12, and 15 sites. Whether this fact is due to a manifestation of the above mentioned peculiarities of 14310 compared with other basaltic crystalline rocks or rather Apollo 14 highland material just contains more beryllium in relation to the other Apollo sites is presently unknown.

The results of our analyses for beryllium in the lunar samples have shown that not only do the lunar fines contain more beryllium than the crystalline rocks, but also, as might be expected, the breccias also contain considerably more beryllium than do the crystalline rocks. It has been shown previously (Sievers *et al.*, 1971) that neither meteoritic contamination nor the solar wind can account for the higher beryllium content of soils found in the maria regions, Mare Tranquillitatis and Oceanus Procellarum, and it was speculated that material uncommon to the Apollo 11 and 12 sites could have contaminated the soil with additional beryllium. Since samples containing much more beryllium have been returned from Fra Mauro than from any other site, it seems plausible that highland material could have been translocated to the Apollo 11 and 12 maria regions and could have preferentially contaminated the fines with beryllium.

*Chromium analysis*

The Apollo 14 and 15 missions visited two distinctly different regions of the moon, both of which have the common ancestry of being related to the Imbrium impact. The Fra Mauro region is representative of the blanket material ejected from the impact while the Hadley-Apennine region is representative of a mare region lying on the edge of the Imbrium impact basin. The chromium concentration in Apollo 14 and 15 samples as shown in Tables 5 and 6, respectively, reflects the distinctive differences in these two areas. Chromium concentrations in the Apollo 14 highland samples were approximately one-half to one-third the values found for samples taken from the mare-like Apollo 15 site and at the mare regions visited by Apollo 11 and 12. The only exception to this generalization was the previously mentioned inhomogeneous breccia (14321) collected at the Apollo 14 site. This rock showed widely fluctuating concentration values for both chromium and beryllium thus suggesting a complex formation process.

A possible explanation for the observed data might involve a premare differentiation process similar to that hypothesized for iron containing minerals (Murthy

Table 5. Chromium analysis for Apollo 14 samples.

| Sample No. | Type of Sample | Chromium Conc. (%) |
|---|---|---|
| 14163,105 | Apollo 14 fines | 0.148 ± 0.006 |
| 14311,33 | Apollo 14 breccia | 0.102 ± 0.008 |
| 14301,64 | Apollo 14 breccia | 0.136 ± 0.006 |
| 14047,46 | Apollo 14 breccia | 0.138 ± 0.002 |
| 14310,135 | Apollo 14 crystalline rock | 0.115 ± 0.002 |
| 14321,192* | Apollo 14 breccia | 0.177 ± 0.005 |
| | Interior fragment | 0.196 ± 0.005 |
| | | 0.410 ± 0.006 |
| | | 0.265 ± 0.004 |
| | | 0.250 ± 0.002 |
| 14321,188* | Apollo 14 breccia | 0.289 ± 0.008 |
| | Exterior fragment | 0.200 ± 0.002 |

* Inhomogeneous rock.

Table 6. Chromium analysis for Apollo 15 samples.

| Sample No. | Type of Sample | Chromium Conc. (%) |
|---|---|---|
| 15601,66 | Apollo 15 fines | 0.338 ± 0.011 |
| 15471,30 | Apollo 15 fines | 0.309 ± 0.009 |
| 15421,26 | Apollo 15 fines | 0.301 ± 0.008 |
| 15301,78 | Apollo 15 fines | 0.247 ± 0.007 |
| 15021,96 | Apollo 15 fines | 0.277 ± 0.004 |
| 15555,143 | Apollo 15 crystalline rock | 0.354 ± 0.019 |
| | | 0.383 ± 0.006 |
| | | 0.439 ± 0.006 |
| 15058,86 | Apollo 15 crystalline rock | 0.587 ± 0.009 |
| | | 0.380 ± 0.015 |
| | | 0.479 ± 0.003 |

*et al.*, 1972). This could include fractional crystallization which depletes the premare lunar crust of minerals rich in chromium. When the Imbrium impact occurred this crust may have been thrown out forming the highlands which are consequently low in chromium content. The sublayers of molten material which may have then formed the present mare were correspondingly enriched in chromium thus resulting in the observed higher chromium concentration for the mare samples over that of the highland samples.

In comparing the beryllium and chromium contents in the samples, a trend was observed in which the concentrations of these two elements tended to vary inversely. It should be emphasized that the same 65 mg samples were used for both the chromium and the beryllium analyses. This trend was observed for a wide range of different rocks and soil samples (Fig. 1) as well as within small fragments from a single particular inhomogeneous breccia 14321 (Table 7).

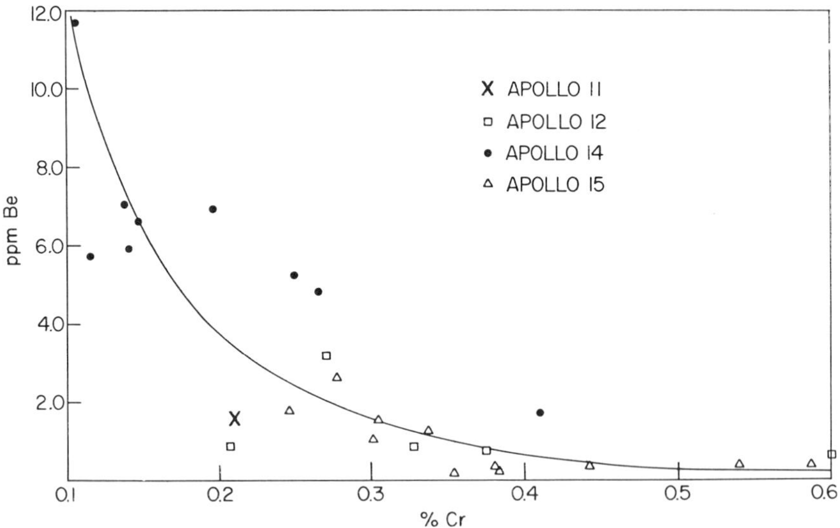


Fig. 1. Variation of beryllium and chromium in the lunar samples.

Table 7. Comparison of beryllium and chromium concentrations in Apollo 14 samples.

| Sample No. and Type | Be Concentration (ppm) | Cr Concentration (%) |
|---|---|---|
| Apollo 14–14321,192 inhomogeneous breccia | | |
| Sample 1 | 6.99 | 0.196 |
| Sample 2 | 5.28 | 0.250 |
| Sample 3 | 4.84 | 0.265 |
| Sample 4 | 1.77 | 0.410 |

### *Organosiloxane analysis*

Preliminary studies in our laboratory demonstrated that the lunar samples were partially dissolved by the organic compound, trifluoroacetylacetone. This compound acts both as a chelating agent and solvent. This finding is significant because it affords an entirely new approach to leaching materials from a lunar sample matrix. It was decided to use this technique in an attempt, entirely independent from dissolution by strong mineral acids, to determine if the presence of organosiloxanes containing the dimethylpolysiloxane structure $[R(OSi[CH_3]_2)n]^+$ as reportedly found by Gehrke *et al.* (1970) in the hydrochloric acid hydrolysates of lunar fines 10086 using gas chromatography-mass spectrometry could be confirmed. In our search for the organosiloxanes we decided to employ mass spectrometry alone and not to make a preliminary gas chromatographic separation in order to eliminate the possibility of introducing contamination into the samples since siloxanes are commonly encountered in gas chromatographic work.

Samples of Apollo 11 fines, 10084, Apollo 12 crystalline rock, 12002, and Apollo 14 fines, 14163 were examined. Sample weights taken for analysis ranged from 25 mg for the Apollo 11 samples to 587 mg for the Apollo 12 sample. Direct probe mass spectrometry, at probe temperatures from 30°C to 200°C, was used to observe the material dissolved from the lunar samples.

No peaks corresponding to the organosiloxanes identified by Gehrke *et al.* (1970) could be detected above the blanks or background. We concluded that the samples analyzed by us do not contain detectable amounts that can be leached out by partial dissolution. In the addendum of the article by Gehrke *et al.*, the authors recognized that their siloxane peaks may have been an artifact of the particular analytical scheme used.

*Acknowledgments*—We would like to thank Drs. Michael L. Taylor and B. Mason Hughes for their assistance with the mass spectrometric measurements. The authors would also like to thank Drs. Brian H. Mason and Roy S. Clarke, Jr. for the meteorite samples and Dr. Francis J. Flanagan for the U.S.G.S. standard rock samples.

## References

Eisentraut K. J., Griest D. J., and Sievers R. E. (1971). Ultratrace analysis for beryllium in terrestrial, meteoritic, and Apollo 11 and 12 lunar samples using electron capture gas chromatography. *Anal. Chem.* **43,** 2003–2007.

Gehrke C. W., Zumwalt R. W., Aue W. A., Stalling D. L., Duffield A., Kvenvolden K. A., and Ponnamperuma C. (1970) Carbon compounds in lunar fines from Mare Tranquillitatis, III. Organosiloxanes in hydrochloric acid hydrolysates. *Proc. Apollo 11 Lunar Sci. Conf., Geochim. Cosmochim. Acta* Suppl. 1, Vol. 2, pp. 1845–1856. Pergamon.

LSPET (1971) (Lunar Sample Preliminary Examination Team) Preliminary examination of lunar samples from Apollo 14. *Science* **173,** 681–693.

LSPET (1972) (Lunar Sample Preliminary Examination Team) The Apollo 15 lunar samples: A preliminary description. *Science* **175,** 363–375.

Murthy V. R., Evensen N. M., Jahn B., and Coscio M. R. (1972) Rb–Sr ages, trace elements, and speculation on lunar differentiation (abstract). In *Lunar Science—III* (editor C. Watkins), pp. 499–500, Lunar Science Institute Contr. No. 88.

Sill C. W. and Willis C. P. (1962) The beryllium content of some meteorites. *Geochim. Cosmochim. Acta* **26,** 1209–1214.

Sievers R. E., Eisentraut K. J., Griest D. J., Richardson M. F., Wolf W. R., Ross W. D., Frew N. M., and Isenhour T. L. (1971) Variations in beryllium and chromium contents in lunar fines compared with crystalline rocks. *Proc. Second Lunar Sci. Conf., Geochim. Cosmochim. Acta* Vol. 2, pp. 1451–1459. MIT Press.

Proceedings of the Third Lunar Science Conference
(Supplement 3, *Geochimica et Cosmochimica Acta*)
Vol. 2, pp. 1335–1336
The M.I.T. Press, 1972

# Chemical analyses of lunar samples 14003, 14311, and 14321

J. H. Scoon

Department of Mineralogy and Petrology,
University of Cambridge, Cambridge, England

**Abstract**—Three samples, 14003,33, 14311,67, and 14321,65 of the Apollo 14 collection were analyzed by classical chemical methods for major elements. The results of these analyses and C.I.P.W. norms are given. A small but significant amount of water is reported on 14003,33 and 14321,65. Sample 14311,67 appears to contain a comparatively large amount of combined water. Iron is present entirely in the ferrous state. $P_2O_5$ is more abundant than in the Apollo 12 samples.

The composition of lunar samples 14003,33, 14311,67, and 14321,65, together with their C.I.P.W. norms are given in Table 1. Classical methods of chemical analysis were used, as described by Scoon (Agrell *et al.*, 1970), except that PbO was used as a flux in the determination of total water by the Penfield method in place of $PbCrO_4$. An attempt was made to estimate the total reducing capacity of these samples. The Pratt method for the determination of ferrous iron was repeated, with the addition of a known excess of a ferric salt to the crucible before heating with HF and 1 : 1 $H_2SO_4$. These results were calculated as FeO. The total iron in the sample was also

Table 1.

| Chemical Analyses (wt.%) of Lunar Rocks | 14003,33 | 14311,67 | 14321,65 | C.I.P.W. Norms | 14003,33 | 14311,67 | 14321,65 |
|---|---|---|---|---|---|---|---|
| $SiO_2$ | 47.99 | 47.24 | 47.78 | Q | — | — | — |
| $Al_2O_3$ | 17.54 | 18.05 | 15.20 | Or | 3.07 | 4.77 | 3.66 |
| $Fe_2O_3$ | nil | nil | nil | Ab | 6.34 | 7.16 | 6.60 |
| FeO | 10.68 | 11.13 | 12.25 | An | 42.90 | 42.86 | 36.14 |
| MnO | 0.13 | 0.14 | 0.17 | Di | 7.26 | 1.95 | 8.59 |
| MgO | 9.24 | 9.59 | 10.73 | En | 2.08 | 0.55 | 2.44 |
| CaO | 10.99 | 10.16 | 9.94 | Fs | 1.47 | 0.40 | 1.76 |
| $Na_2O$ | 0.75 | 0.85 | 0.78 | Wo | 3.70 | 1.00 | 4.38 |
| $K_2O$ | 0.52 | 0.81 | 0.62 | Hy | 35.62 | 32.22 | 35.18 |
| $H_2O^+$ | 0.04 | 0.29 | 0.05 | En | 20.88 | 18.62 | 20.44 |
| $H_2O^-$ | 0.01 | 0.03 | nil | Fs | 14.74 | 13.60 | 14.74 |
| $TiO_2$ | 1.94 | 1.81 | 2.06 | Ol | 0.02 | 5.83 | 4.81 |
| $P_2O_5$ | 0.41 | 0.76 | 0.41 | Fo | 0.01 | 3.23 | 2.68 |
| $Cr_2O_3$ | 0.19 | 0.15 | 0.26 | Fa | 0.01 | 2.60 | 2.13 |
| S | 0.08 | 0.06 | 0.07 | Il | 3.68 | 3.42 | 3.91 |
| F | nil | nil | nil | Ap | 0.97 | 1.79 | 0.97 |
| | 100.51 | 101.07 | 100.32 | Pr | 0.15 | — | 0.13 |
| Less S ≡ O | 0.02 | 0.02 | 0.02 | | | | |
| | 100.49 | 101.05 | 100.30 | | | | |
| Total iron as $Fe_2O_3$ | 11.76 | 11.91 | 13.59 | | | | |
| ΔRC | 1.15 | 1.27 | 0.34 | | | | |

calculated as FeO and the difference between these two figures is shown as ΔRC in Table 1.

Lunar sample 14311,67, which was a sample obtained from the dust of the sawing operation, is unusual in containing an appreciable amount of combined water. This sample also contained numerous small shiny white metallic fragments. Some fibrous material was also noted on microscopic examination. It was assumed that the metallic fragments arose from the aluminum foil used in the cutting operation. An attempt was made to get some idea of the extent of this contamination by extracting a portion of the sample with warm 5% NaOH solution. It was found that approximately 1.3% of Al was removed in this way. The unacceptably high summation for this sample is probably due to the metallic fragments being reported as oxide. Similar contamination was noted on 14321,65 (homogenized fines) but to a much smaller extent.

## Reference

Agrell S. O. and Scoon J. H. *et al.* (1970) Observations on the chemistry, mineralogy, and petrology of some Apollo 11 lunar samples. *Proc. Apollo 11 Lunar Sci. Conf., Geochim. Cosmochim. Acta* Suppl. 1, Vol. 1, pp. 93–128. Pergamon.

Proceedings of the Third Lunar Science Conference
(Supplement 3, *Geochimica et Cosmochimica Acta*)
Vol. 2, pp. 1337–1342
The M.I.T. Press, 1972

# Analysis of lunar samples 14163, 14259, and 14321 with isotopic data for $^7$Li/$^6$Li

A. Strasheim, P. F. S. Jackson, J. H. J. Coetzee, F. W. E. Strelow,*
F. T. Wybenga, A. J. Gricius, M. L. Kokot, and R. H. Scott

National Physical Research Laboratory (*National Chemical Research Laboratory),
C.S.I.R., Pretoria, South Africa

**Abstract**—The lunar samples 14163,151, 14259,77, and 14321,222 were analyzed comprehensively by several analytical methods, including wet chemical techniques and various instrumental methods. The values obtained by these methods are presented, including the spark-source mass spectrographic values obtained for the forsteritic olivine which was heterogeneously distributed in sample 14321. The $^7$Li/$^6$Li ratio have also been determined using a laser-source mass spectrometer.

## Introduction

The 3.0 g bulk soil sample 14163,151, the 1.0 g sample 14163,77 (a sample of the comprehensive soil sample that was skimmed from the upper one centimeter of the surface), and the 5.6 g rock sample 14321,222 received, were analyzed for major and minor elements using wet chemical and various instrumental methods. The sample 14321 was found to contain heterogeneously distributed forsteritic olivine. Three milligrams of separated inclusion, which was isolated under a microscope, was analysed by spark-source mass spectrography after identification by x-ray diffraction. This forsteritic olivine proved to be almost free from impurities.

## Analytical Methods

The methods used included the following:

### *X-ray spectrography*

Various approaches were used to determine the concentrations of major and minor elements, using a computer controlled Philips 1220c x-ray spectrometer with a 100 kV generator. These were:

(i) *Pressed powder method* (x-ray P). The sample (0.8 g) was firmly pressed onto a makrofol window.

(ii) *Pellet method* (x-ray PT). The sample mixture (0.6 g sample: 0.25 g urea) was pressed into a pellet using urea as a backing.

(iii) *Double melt technique* (x-ray D). Discs were prepared of 0.6 g samples fused in lithium tetraborate and sodium tetraborate, respectively, (1 : 4).

(iv) *Fusion technique* (x-ray F). Discs were prepared of 0.3 g samples fused with lithium tetraborate (1 : 9).

### *Direct electron excitation*

Fluorine and, in some cases, sodium were determined using a Jeol JP X − 3 Primary x-ray Analyzer. The excitation was at 30 kV with specimen currents of approximately 90 $\mu$A, using a KAP analyzing crystal.

A 0.3 g sample was ground together with 0.1 g graphite for 2 hours and pressed into a disc at a pressure of 3.45 kbar with a bakelite backing. The disc's diameter was 25 mm.

*Atomic absorption spectroscopy*

Analyses were carried out using a Techtron AA5 instrument. About twenty absorbance readings were taken at 5 sec intervals, for each sample and standard, and the values were averaged. The methods of sample dissolution used, were:

(i) *Acid dissolution.* The sample was transferred to a 200 cm³ Pt/Au crucible, slightly wetted with water, and 10 cm³ of acid mixture (consisting of HF and $HClO_4$ in a 2 : 1 ratio) was added. After the sample was fumed to dryness at about 70°C, a second 10 cm³ portion of acid mixture was added and the fuming process was repeated. The residue was then dissolved in 10–20 cm³ 2N HCl and made up to volume to give a 1% w/v solution.

(ii) *Fusion.* The glass bead prepared for the x-ray spectrographic analysis was remelted at 1100° for 15 minutes in a graphite crucible. The molten sample was then added directly to an aqueous solution containing $HNO_3$(2N), 2000 ppm La and 1000 ppm Cs (Butler and Kokot, 1969).

Natural rock standards were prepared in the same way as the samples. When necessary, samples and solutions were diluted with the La/Cs solution.

*Wet chemical analysis*

A wet chemical determination of Al, total Fe, Ti, Ca, Mg, Na, K, Mn, Zr, and V was carried out using a complete ion exchange separation system. Details of the separation and the methods of determination employed have been described in the literature (Strelow *et al.*, 1969).

*Mass spectrography*

(i) *Spark-source mass spectrography.* The mass spectrographic analysis of the rock samples has been carried out using a Varian-Mat SM 1 BF mass spectrograph. This makes use of the low voltage discharge source as a source of positive ions and the double-focussing geometry of Mattauch Hertzog. The optics of the system are so arranged that a broad (110 $\mu$) rectilinear line profile is used and the translation of the photoblackening data into concentration is carried out according to the method described by Franzen and Schuy (1967).

The electrode systems used to obtain these analyses were mixtures of rock and silver powder on the one hand and rock and graphite on the other. Mixtures containing 90% conductive matrix were mixed carefully by hand with 10% rock powder. The conductive matrices chosen were ultrapure chemicals, with very low blank values. As some are subjected to particle segregation due to the action of a vibrating end-mill, hand mixing was preferred. One of the minor isotopes of iron was taken as primary internal standard as this could be measured accurately in nearby laboratories. In Apollo 14 material with iron present at a concentration of about 1500–2000 ppm, the $^{57}Fe$ isotope gave lines of suitable intensities. All elemental analyses were determined relative to this internal standard. Two photoplates from each electrode system were exposed together with the W1 standard.

(ii) *Lithium isotope ratios.* The laser source mass spectrograph as designed by Scott *et al.* (1971) was used to evaluate the lithium isotope ratios in the lunar samples. The extreme sensitivity of the technique to the alkali metals makes this technique eminently suitable for this purpose. Photographic recording was used and the transparency curves for the two isotopes constructed by a computer. The difference in intensity of the two curves is directly attributable to the difference in the isotopic ratio within the sample. Minor differences were found in the bulk samples of the three samples received by us but one small crystal of forsterite was found to have a very low value. These differences at this moment remain inexplicable.

## PROCEDURE

The analytical methods were chosen and developed to obtain values by as many techniques as possible. A general sample flow diagram is given in Fig. 1. The methods

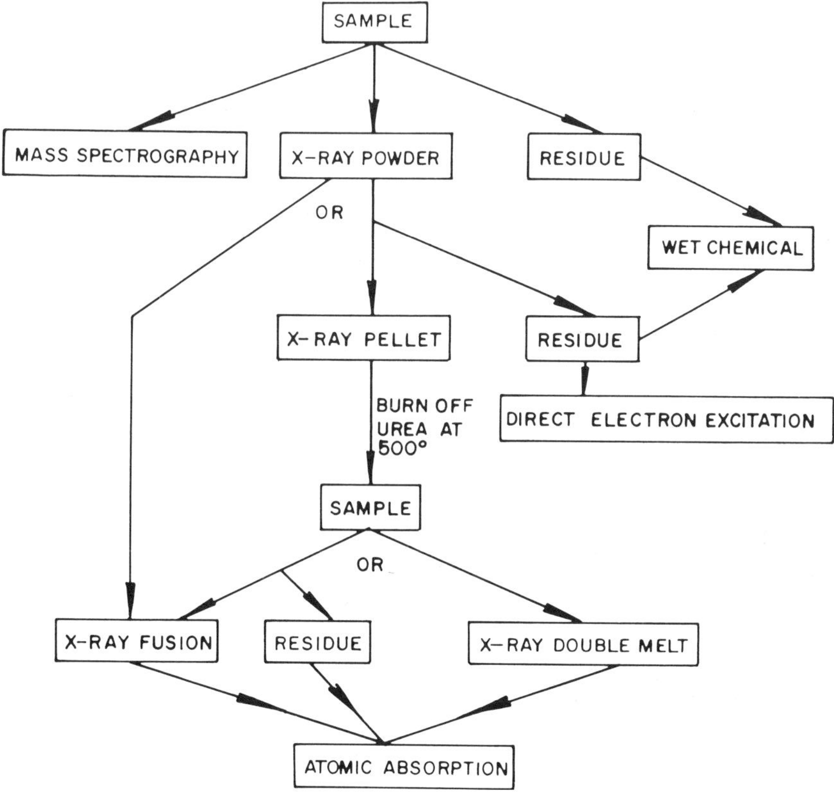


Fig. 1. General "sample flow."

that could be used for the analysis of each sample were determined by the quantity of sample available. The techniques used for each sample are stated in Table 1.

## Results and Discussion

Table 1 presents the average values obtained using the different techniques and the arithmetic means of these values where applicable. Table 2 demonstrates the precision of some of the methods at different concentration levels.

For the major elements, the precision obtained with the wet chemical method is normally better than 0.5% with the x-ray spectrographic methods better than 1%, and with the atomic absorption spectroscopy better than 2%.

The spark-source mass spectrometer was mainly used to determine elements in the concentration ranges 1%–1 ppm. Two systems were used, namely a rock-silver powder and a rock-graphite powder. The rock-silver powder system has been shown to yield precisions approaching ±2%. The sensitivity for most elements is adequate but for some elements, notably those of which the oxides are stable at high temperature (1000°C +), the sensitivity was extremely poor. (For barium a factor of between 100 and 1000 had to be applied to the results to obtain the correct value.) In the

Table 1. Chemical analyses of lunar samples.

| | 14259,77 | | | | | | 14163,151 | | | | | 14321,222 | | | | | | | | 14321,22 |
|---|---|---|---|---|---|---|---|---|---|---|---|---|---|---|---|---|---|---|---|---|
| Major (%) | Chem | X-ray P | X-ray F | MS | AA | Average | Chem | X-ray P | MS | AA | Average | Chem | X-ray P | X-ray PT | X-ray F | X-ray D | MS | AA | Average | Forsterite MS |
| $SiO_2$ | — | 47.3 | 49.0 | — | — | 48.15 | — | 47.3 | — | — | 47.3 | 47.21 | 47.4 | 47.4 | 49.5 | 49.1 | — | — | 48.1 | Major |
| $Al_2O_3$ | 17.21 | 16.70 | 17.15 | — | 16.90 | 16.99 | 17.20 | 17.00 | — | — | 17.10 | 14.01 | 14.85 | 14.50 | 14.8 | 15.1 | — | 14.45 | 14.62 | — |
| FeO | 10.54 | 10.56 | 10.57 | — | 10.66 | 10.58 | 10.27 | 10.17 | — | 10.01 | 10.15 | 13.10 | 12.9 | 12.9 | 12.8 | — | — | 12.91 | 12.9 | 7.5 |
| MgO | 9.07 | — | 9.56 | — | — | 9.32 | 8.99 | — | — | 9.01 | 9.00 | 10.90 | 10.9 | 10.9 | — | — | 10 | 11.21 | 11.0 | Major |
| CaO | 10.70 | 10.95 | 10.69 | 10.5 | — | 10.71 | 10.66 | 10.85 | 10.5 | 10.45 | 10.65 | 9.99 | 10.1 | 10.0 | 10.33 | 9.63 | 10 | 9.80 | 10.0 | 0.35 |
| $K_2O$ | 0.51 | 0.43 | 0.45 | 0.41 | 0.55 | 0.47 | 0.54 | 0.51 | 0.43 | 0.64 | 0.53 | 0.443 | 0.41 | 0.41 | 0.42 | 0.35 | 0.47 | 0.43 | 0.42 | 0.044 |
| $Na_2O$ | 0.64 | — | — | — | 0.65 | 0.65 | 0.64 | — | — | 0.66 | 0.65 | 0.706 | — | — | — | — | — | 0.68 | 0.69 | — |
| $TiO_2$ | 1.79 | 1.84 | 1.76 | 1.90 | — | 1.82 | 1.82 | 1.82 | 1.87 | — | 1.84 | 2.00 | 2.04 | 2.01 | 1.99 | 2.15 | 2.0 | 2.04 | 2.03 | 0.085 |
| MnO | 0.133 | 0.125 | — | 0.150 | — | 0.136 | 0.130 | 0.125 | 0.125 | — | 0.127 | 0.175 | 0.16 | 0.16 | — | — | 0.15 | 0.182 | 0.17 | 0.080 |
| $Cr_2O_3$ | — | 0.17 | — | 0.165 | 0.16 | 0.165 | — | 0.17 | 0.145 | 0.21 | 0.18 | 0.23 | 0.24 | 0.21 | — | — | 0.245 | 0.24 | 0.24 | 0.080 |
| $P_2O_5$ | — | — | 0.49 | 0.42 | — | 0.46 | — | — | 0.50 | — | 0.50 | 0.33 | 0.37 | 0.36 | 0.45 | — | 0.38 | — | 0.38 | 0.010 |
| Total (%) | | | | | | 99.18 | | | | | 98.1 | | | | | | | | 100.55 | |
| Trace (ppm) | | | | | | | | | | | | | | | | | | | | |
| Ni | 376 | 345 | — | 270 | — | 330 | 316 | 280 | 240 | — | 279 | 144 | 125 | 130 | — | — | 130 | — | 132 | 3 |
| Ba | — | 995 | 907 | 1020 | — | 974 | — | 1050 | 1080 | — | 1065 | — | 630 | 625 | — | — | 630 | — | 628 | 5 |
| Nb | — | 57 | — | 50 | — | 54 | — | 57 | 67 | — | 62 | — | 60 | 60 | — | — | 56 | — | 59 | < 2 |
| Zr | 599 | 785 | — | 775 | — | 720 | 648 | 805 | 845 | — | 766 | — | 715 | 700 | — | — | 710 | — | 708 | < 5 |
| Y | — | 200 | — | 220 | — | 210 | — | 210 | 260 | — | 235 | — | 190 | 170 | — | — | 200 | — | 187 | < 3 |
| Sr | — | 225 | — | 270 | — | 248 | — | 230 | 240 | — | 235 | — | 200 | 160 | — | — | 270 | 108 | 185 | 24 |
| Rb | — | 10 | — | 17 | — | 14 | — | 10 | 15 | 12 | 12 | — | 11 | 11 | — | — | 14 | — | 12 | < 1 |
| Co | — | 36 | — | 30 | — | 33 | — | 34 | 31 | — | 33 | — | 36 | 36 | — | — | 33 | 50 | 39 | 25 |
| V | 33 | — | — | 34 | — | 34 | 36 | — | 47 | — | 42 | — | 82 | 82 | — | — | 68 | — | 77 | 20 |
| Cu | — | — | — | 17 | 15 | 16 | — | — | 15 | 12 | 14 | — | 13 | 12 | — | — | 7.4 | 7 | 9.9 | < 1 |
| B | — | — | — | 20 | — | — | — | — | 20 | — | — | — | — | — | — | — | 26 | — | — | 1 |
| Sc | — | — | — | 15 | — | — | — | — | 19 | — | — | — | — | — | — | — | — | — | — | — |
| Ga | — | — | — | 1 | — | — | — | — | 1 | — | — | — | — | — | — | — | 2 | — | — | < 1 |
| As | — | — | — | 2 | — | — | — | — | 1.6 | — | — | — | — | — | — | — | 2.5 | — | — | < 1 |
| Ta | — | — | — | < 1 | — | — | — | — | < 1 | — | — | — | — | — | — | — | — | — | — | — |
| W | — | — | — | < 5 | — | — | — | — | < 5 | — | — | — | — | — | — | — | — | — | — | — |
| Pb | — | — | — | < 10 | — | — | — | — | < 10 | < 30 | — | — | — | — | — | — | < 5(3) | — | — | < 5 |
| Zn | — | 20 | — | 80 | 60 | 53 | — | 25 | 60 | 45 | 43 | — | < 20 | < 20 | — | — | 30 | 7 | — | 12 |
| Li | — | — | — | — | — | — | — | — | — | 19 | — | — | — | — | — | — | — | — | — | — |
| Ratio $^7Li/^6Li$ | — | — | — | 11.4 | — | — | — | — | 12.4 | 11.8 | — | — | — | — | — | — | 12.2 | — | — | — |

Chem: wet chemical; X-ray P: X-ray on powder sample; X-ray F: X-ray on fused sample; X-ray PT: X-ray on pellet; X-ray D: X-ray with double melt technique; MS: mass spectrographic; AA: atomic absorption.

Direct electron excitation: Sample 14321,22: $Na_2O$: 0.73%, F: 0.15%.

Table 2. Precisions of some of the methods employed.

| | Wet chemical | Atomic absorption | Mass spectrography | X-ray powder | X-ray pellet |
|---|---|---|---|---|---|
| % $Al_2O_3$ | 13.98 | 14.24 | — | 14.85 | 14.50 |
| | 14.03 | 14.53 | — | 14.85 | 14.60 |
| | 14.03 | 13.84 | — | — | — |
| % $TiO_2$ | 2.00 | 2.0 | 2.0 | 2.05 | 2.01 |
| | 1.99 | 1.7 | 2.0 | 2.04 | 2.00 |
| | 2.00 | 2.2 | — | — | — |
| % MnO | 0.177 | 0.182 | 0.15 | 0.16 | 0.16 |
| | 0.174 | 0.182 | 0.155 | 0.16 | 0.16 |
| | 0.175 | 0.182 | — | — | — |

rock-graphite system this phenomenon was not observed. Previous work by Scott, Jackson and Strasheim, using a laser to excite the plasma, also showed this effect. Their explanation of the reducing rate of the graphite in promoting the chemical reduction of these very stable oxides in the plasma or at the sample surface, would seem to be fitting in this case and the presence, even in small amounts (3%), of carbon seems essential to the promotion of the existence of these ion-types from such a matrix. (This may not apply to other types of spark-sources.) Thus even in the rock-silver system 5% by weight graphite was added to the electrode mixture.

The two systems were used concurrently with little cross contamination and then only in the case of silver appearing in the silver-free system. In the silver system, complex polyatomic species of almost identical mass to these elements in the first transition series do not occur whereas in the graphite system the "blank" equivalent concentration due to these ions may be a few hundred parts per million. The primary standard used in the setting up of the calibration functions and the determination of relative sensitivity factors was the U.S.G.S. standard W1. Estimates of precision and accuracy were made using portions of the rock standard as standard or unknown. In most cases the precision obtained was considerably better than 5% with an average ±2%. The best estimate for the average accuracy was ±7%.

From the results given in Tables 1 and 2 the following conclusions seem valid:

(1) With about 5 grams of material a comprehensive comparison between a number of instrumental and chemical techniques is possible.

(2) The comparative results listed for the major elements are very satisfactory. The mass spectrometer results recorded for Ca at 10% is significant. It seems that if a low relative concentration isotope is available analysis at high concentrations is possible using this technique.

(3) The trace element values recorded for using the x-ray powder technique and the spark-source mass spectrometer are very concurrent. This gives confidence to the mass spectrometric (MS) values recorded for such elements as Sc, Ga, Ta, W, etc. The low values recorded chemically for zirconium in relation to the XRF and MS values warrant further investigation. The same applies to zinc where low values were recorded for the x-ray powder technique and high values for the MS and AA techniques. The very low values listed for strontium using the AA technique for sample

14321,222 seem at fault. The MS values always seem to be higher than that found using the x-ray technique.

(4) Copper is an element that can be determined very satisfactorily using the AA technique. The corresponding values recorded for the MS and AA techniques has indicated that the MS values for this element in rocks can be accepted with confidence.

(5) The MS technique for the isotopic ratio for $^{7}Li/^{6}Li$ seems to compare favorably with that recorded using the AA technique. The small amount of material needed for the MS technique should make this technique invaluable to study this relation in other known and meteorite material of which only limited quantities are available.

(6) As the x-ray powder technique only requires limited sample preparation and as the results recorded were very favorable in comparison to the other techniques tested, it can be used with confidence for general rock analysis for the major and minor elements.

In a following publication the results listed above will be compared to those determined for ten of the oldest rocks on earth.

## REFERENCES

Butler L. R. P. and Kokot M. L. (1969) The application of atomic absorption to the analysis of upper mantle and other rocks. "Upper Mantle Project," Geological Society of South Africa, Special Publications No. 2.

Franzen J. and Schuy K. D. (1967) Advances in precision of mass-spectroscopic spark-source analysis of conducting materials. *Z. Anal. Chem.* **225,** 295–323.

Strasheim A., Scott R. H., and Jackson P. F. S. (1971) Laser-source mass spectrography. Lecture given at the 10th S.A.S. Meeting, St. Louis.

Strelow F. W. E., Liebenberg C. J., and Toerien F. von S. (1969) Accurate silicate analysis based on separation by ion exchange chromatography. *Anal. Chim. Acta* **47,** 251–260.

Proceedings of the Third Lunar Science Conference
(Supplement 3, *Geochimica et Cosmochimica Acta*)
Vol. 2, pp. 1343–1359
The M.I.T. Press, 1972

# The extralunar component in lunar soils and breccias

PHILIP A. BAEDECKER, CHEN-LIN CHOU, and JOHN T. WASSON
Department of Chemistry and Institute of Geophysics and Planetary Physics,
University of California, Los Angeles, California 90024

**Abstract**—Concentrations of Ni, Zn, Ga, Ge, Cd, In, and Ir in soils and breccias returned by Apollo 14 and Apollo 15 are reported. The integrated flux of extralunar material at each lunar landing site is estimated on the basis of the siderophilic elements Ni, Ge, Ir, and Au. It falls with time between 3.95 and 3.26 g.y. before the present. The flux versus time relationship indicated by the trace element data is consistent with estimates of the same relationship based on crater statistics and morphologies, as would be expected if the same populations of interplanetary objects were responsible for both properties. The data are consistent with the presence of two populations of material bombarding the moon early in its history, one with a short half-life that was dominant until about 3.8 g.y. ago, and another with a half-life of about 2.0 g.y. that dominated during the more recent era.

## INTRODUCTION

AGE DATA OBTAINED from analysis of lunar material returned by the Apollo and Luna missions show that most igneous activity on the moon ceased at the end of the first 1.5 g.y. of its history. For the past three g.y. the most important dynamic process on the moon has been the bombardment of the lunar surface by extralunar materials. Aside from the extensive cratering of the lunar surface, the accumulation of extralunar material is also recorded in the composition of the lunar soils and in some breccias (Ganapathy *et al.*, 1970; Wasson and Baedecker, 1970; Baedecker *et al.*, 1971a; Laul *et al.*, 1971). A number of elements that appear in low concentrations in lunar and terrestrial crustal rocks either because of their geochemical affinity for metallic iron or their relatively high volatility are enriched in lunar soil relative to the igneous rocks. Since most meteorites have much higher concentrations of the siderophile elements than lunar crustal rocks, the enrichment of these elements in the soil is thought to reflect the amount of extralunar material. Although, as discussed below, the evidence is equivocal, the nearly equal enrichment of volatile elements relative to the C1 chondrites suggests that the extralunar component in the soil is predominantly of highly primitive character.

Öpik (1969) has estimated the size distribution of interplanetary material and suggested that the major flux of material onto the lunar surface is in the smallest size range. A study of the extralunar component in lunar fines should therefore provide information regarding both the composition of interplanetary dust and the integrated flux of these materials onto the lunar surface since the cessation of igneous activity at a particular site. The flux of larger crater-forming projectiles is obtained by crater population (Gault, 1970; Hartmann, 1970) and crater morphology (Soderblom and Lebofsky, 1972) studies. In this paper we will discuss the chemical composition, concentration, and time history of the extralunar material.

## Experimental

### *Samples and sample preparation*

Approximately 2 g each of the comprehensive and bulk soil samples 14259 and 14163 were provided for analysis along with 1 g of contingency soil sample 14003. In addition, a 2 g sample of the largest igneous rock recovered at the Apollo 14 site, 14310, was also provided. We have analyzed three separated clasts as well as matrix material from the breccia 14321, as members of the Goles consortium. The clastic materials picked from this rock were described by Duncan *et al.* (1972). A microbreccia clast and sawdust from the cutting of rock 14066 completed our suite of Apollo 14 rocks. These samples were allocated to us as members of the Reynolds-Alexander consortium. In addition to the data obtained on these samples recovered on the Apollo 14 mission, we report preliminary data on four Apollo 15 samples: 15091, 15211, 15231, and 15531.

Sample preparation, packaging, flux monitor preparation, and irradiation procedures were identical to those described in reports of our previous work (Wasson and Baedecker, 1970; Baedecker *et al.*, 1971a). A split of one exterior sample of 14310 was treated using the microsandblasting procedure described by Baedecker *et al.* (1971a) in an attempt to remove possible surficial contamination.

### *Analytical and radiometric procedures*

In the previous studies of the returned lunar material we determined the six elements Zn, Ga, Ge, Cd, In, and Ir. The analytical and radiometric procedures for these six elements have been described in the references cited above. We have subsequently added a seventh element, Ni, to our analytical scheme. The details of our analytical procedure for Ni have previously been described by Müller *et al.* (1971), although the scheme has been modified slightly for our current work. Whereas separate aliquots of the solution resulting from sample dissolution were processed for Ni and Ir in the work described by Müller *et al.*, in the scheme employed in this study, Ni as well as the other six elements were separated from a single solution. Following sample dissolution, Ga, Ge, In, and Ir were precipitated as the hydroxides by addition of $NH_4OH$, while Zn and Cd were only partially carried by the precipitate; Ni was then precipitated from the supernatant by the addition of dimethylglyoxime, and the supernatant retained for Zn and Cd chemistry. The Ni-dimethylglyoxime precipitate was dissolved and the resulting solution processed in essentially the same manner as described by Müller *et al.*

### *Precision and accuracy*

In order to assess the precision and accuracy of our procedures we have analyzed several replicates of seven standard rock powders distributed by the United States Geological Survey, as well as a powdered sample of the Allende meteorite provided by the United States National Museum for purposes of interlaboratory comparison. The data are summarized in Table 1, where means and 95% confidence limits on the mean are reported. For the elements Ga, Ge, and In where five or more analyses have been carried out on each rock, obvious outliers were discarded before calculating a mean. Our Ir data for rocks containing less than 0.1 ppb Ir generally show poor precision. We attribute this to post irradiation contamination, which is a serious problem in our laboratory. For these rocks an average of the two or three lowest values are presented. Although the relative standard deviations for the various rocks tended to show a fairly wide variation, the following would appear to be reasonable estimates of the precisions for each of our elements: Zn, $\pm 3.4\%$; Ga, $\pm 4\%$; Ge, $\pm 6\%$; Cd, $\pm 16\%$; In, $\pm 10\%$; Ir, $\pm 30\%$. We have analyzed too few replicates to adequately estimate the precision of our Ni procedure.

The accuracy of our results can best be assessed by comparing them with data obtained by other workers on the same rocks. Such a comparison shows no significant systematic differences between our data and data published by other analysts.

Table 1. Means of replicate analyses of standard rock and meteorite samples.

| Sample | Ni (ppm) | | Zn (ppm) | | Ga (ppm) | | Ge (ppm) | | Cd (ppb) | | In (ppb) | | Ir (ppb) | |
|---|---|---|---|---|---|---|---|---|---|---|---|---|---|---|
| | Anal. | Mean | Anal. | Mean | Anal. | Mean | Anal. | Mean | Anal. | Mean | Anal. | Mean | Anal. | Mean |
| W-1 | | | 3 | 91 ± 5 | 6 | 18.5 ± 0.6 | 5 | 1.87 ± 0.18 | 4 | 146 ± 34 | 6 | 64 ± 4 | 6 | 0.34 ± 0.09 |
| G-2 | | | 5 | 91 ± 5 | 4 | 25.1 ± 0.5 | 5 | 1.23 ± 0.14 | 4 | 26 ± 5 | 6 | 28 ± 4 | 2 | ≤0.16 |
| GSP-1 | 1 | 8.4 | 4 | 131 ± 59 | 6 | 25.1 ± 1.7 | 4 | 1.38 ± 0.12 | 4 | 61 ± 19 | 6 | 52 ± 7 | 3 | 0.05 ± 0.04 |
| AGV-1 | 1 | 18 | 3 | 102 ± 5 | 6 | 22.3 ± 1.2 | 5 | 1.32 ± 0.27 | 4 | 75 ± 7 | 5 | 45 ± 4 | 2 | ≤0.08 |
| PCC-1 | | | 3 | 47 ± 25 | 6 | 0.68 ± 0.03 | 6 | 1.02 ± 0.14 | 3 | 23 ± 16 | 5 | 3.5 ± 0.6 | 5 | 5.1 ± 2.2 |
| DTS-1 | 2 | 2,500 ± 80 | 4 | 53 ± 4 | 6 | 0.46 ± 0.02 | 5 | 0.83 ± 0.04 | 3 | 9.6 ± 1.1 | 5 | 2.4 ± 0.4 | 4 | 0.58 ± 0.21 |
| BCR-1 | 1 | 11 | 4 | 139 ± 12 | 5 | 23.7 ± 0.4 | 7 | 1.57 ± 0.05 | 5 | 148 ± 59 | 6 | 95 ± 3 | 2 | ≤0.24 |
| Allende (USNM) | 1 | 14,000 | 3 | 110 ± 35 | 3 | 6.0 ± 0.6 | 3 | 17.9 ± 2.0 | 2 | 730 ± 370 | 3 | 28 ± 3 | 3 | 800 ± 41 |

Table 2. Replicate and mean concentrations of seven trace elements in rocks and soils from Fra Mauro and Hadley Apennines.

| Sample | Description | Ni (ppm) | | Zn (ppm) | | Ga (ppm) | | Ge (ppb) | | Cd (ppb) | | In (ppb) | | Ir (ppb) | |
|---|---|---|---|---|---|---|---|---|---|---|---|---|---|---|---|
| | | Repl. | Mean | Repl. | Mean | Repl. | Mean | Repl. | Mean | Repl. | Mean | Repl. | Mean | Repl. | Mean |
| 14163,115 | Bulk fines | 360, 340 | | 34, 39 | | 9.0, 9.0 | | 670, 710 | | 200, 200 | | 250*, 120 | | 9.4, 13 | |
| | | | 350 | 37 | 36 | 8.1 | 8.7 | 780 | 720 | 190 | 196 | 120 | 120 | 9.6 | 11 |
| 14259,89 | Comp. fines | 410, 390 | | 27, 29 | | 7.2, 6.8 | | 770, 820 | | 100, 86 | | 46, 30 | | (6.2), 15 | |
| | | | 400 | 28 | 28 | 6.5 | 6.8 | 1010 | 860 | | 93 | 43 | 39 | 14 | 15 |
| 14003,4 | Cont. fines | 380 | 380 | 31, 28 | 29 | 7.7, 7.4 | 7.6 | 810, 770 | 790 | 170, $\leq$220* | 170 | 44, 38 | 41 | 12, 11 | 12 |
| 14310,120 | Basalt | 250, 203 | | 1.4, 3.3* | | 4.2, 4.1 | | 210*, 94 | | 11, 6.7 | | 57, 21† | | 7.6, 9.4 | |
| | | 188 | 210 | 1.5 | 1.5 | 4.4 | 4.2 | 86 | 90 | 7.5 | 8.4 | 18† | 20 | 6.5 | 7.8 |
| 14321,184,1E | Basalt clast | 26, 46 | 36 | 2.9, 4.5 | 3.7 | 4.0, 3.9 | 4.0 | 1100, 700 | 880 | 12, 3.8 | 7.9 | 3.3, 4.1 | 3.7 | 0.33, 0.55 | 0.4 |
| 14321,184,25A | Breccia clast | 350, 410 | 380 | 4.2, 3.9 | 4.1 | 6.3, 6.6 | 6.5 | 720, 870 | 800 | 20, 12 | 16 | 1.4, 1.4 | 1.4 | 4.9, 25* | 4.9 |
| 14321,184,31 | Breccia clast | 290, 314 | 300 | 5.4, 3.4 | 4.4 | 6.5, 5.6 | 6.0 | 540, 660 | 600 | 37, 15 | 26 | 2.6, 3.0 | 2.8 | 10.5, 7.6 | 9.0 |
| 14321,184,32 | Matrix | 200 | 200 | 35 | 35 | 5.2 | 5.2 | 160 | 160 | 84 | 84 | 3.4 | 3.4 | 5.2 | 5.2 |
| 14066,21,1.01 | Breccia clast | 200 | 200 | 2.3 | 2.3 | 6.1 | 6.1 | 580 | 580 | 54 | 54 | 2.7 | 2.7 | 52* | |
| 14066,32,2 | Sawdust | 510, 350 | 430 | 51*, 14 | 14 | 5.9, 5.5 | 5.7 | 600, 530 | 560 | 175, 2200* | 175 | 17, 88* | 17 | 6.2, 6.1 | 6.1 |
| 15231,51 | Fines-St. George | 270 | | 15 | | 4.0 | | 460 | | 40 | | 11 | | 11 | |
| 15091,36 | Fines-St. George | 250 | | 15 | | 4.3 | | 380 | | 44 | | 8.0 | | 8.5 | |
| 15211,8 | Fines-St. George | 240 | | 16 | | 4.4 | | 420 | | 46 | | 3.2 | | 8.2 | |
| 15531,46 | Fines-rille | 140 | | 7.7 | | 4.0 | | 120 | | 20 | | 410 | | 3.6 | |

* Value shows evidence of contamination. Not included in mean. † Surface of rock sample removed with SiC abrasive powder prior to analysis.

## Results

The results of our analyses are presented in Table 2. All Cd results have been corrected for the production of $Cd^{115}$ by neutron induced fission of U, since this produces a significant interference in the analysis of lunar rocks which have high U and low Cd concentrations. This correction becomes largest in the case of the igneous rock samples, but even in these cases does not exceed 35% of the reported value. Uranium data was taken from the following sources: Eldridge *et al.* (1972), LSPET (1972), Laul *et al.* (1972), Tatsumoto *et al.* (1972), and Wänke *et al.* (1972).

As in our previous work on the Apollo 11 and 12 samples, contamination of the rock and soil samples due to the continued use of In seals on the sample return containers frustrated our attempts to obtain meaningful data for this cosmochemically important element. Our microsandblasting procedure appeared to be only moderately successful in reducing the In level observed for sample 14310. The lowest In results were obtained for the hand-picked clasts from 14321 and 14066: These data were comparable to the lowest concentrations we observed for igneous rocks recovered from the Apollo 11 and 12 sites (Wasson and Baedecker, 1970; Baedecker *et al.*, 1971a).

## Diagnostic Elements

The character of the extralunar component in soil and breccia samples can best be understood in terms of the observed enrichment of certain diagnostic trace elements in these materials relative to lunar crystalline rocks. We have examined the published trace element data on the Apollo 12 material in order to assess the usefulness of various elements as tracers of the extralunar component. The Apollo 12 data were chosen for two reasons: (1) They are somewhat more extensive and more accurate than the other large body of data, that on Apollo 11; and (2) The concentration of the extralunar component is apparently the lowest observed on the first five successful lunar missions, thus providing the best test for the usefulness of individual elements. Ratios of the average concentration of an element in the mature soil (12070 or equivalent) to its average concentration in both high-Mg and low-Mg rocks were calculated for various elements which were considered potentially useful. Most were either relatively stable in the metallic state, or of relatively high volatility.

The soil : basalt ratios for these elements are plotted in Fig. 1 in order of increasing atomic number. The bars are based on the spread of the data for the crystalline rocks and are drawn so as to include 95% of the basalt values. To establish the diagnostic usefulness of a particular element, we first set the criterion that the soil : basalt ratio must be greater than 3. Some elements that met this criterion were rejected either because the results appeared to be biased due to sample contamination (e.g., In and Ag), because the analytical data were sparse or of poor quality (e.g., Pd, As, and $Pb^{204}$), or because there was evidence that the soil enrichment was due to exotic components indigenous to the moon (e.g., P and the halogens). All available data for the remaining elements were tabulated and critically examined. A few results that differed greatly from the more typical values were discarded. In general, the data which were discarded were results by a given analyst which were systematically higher

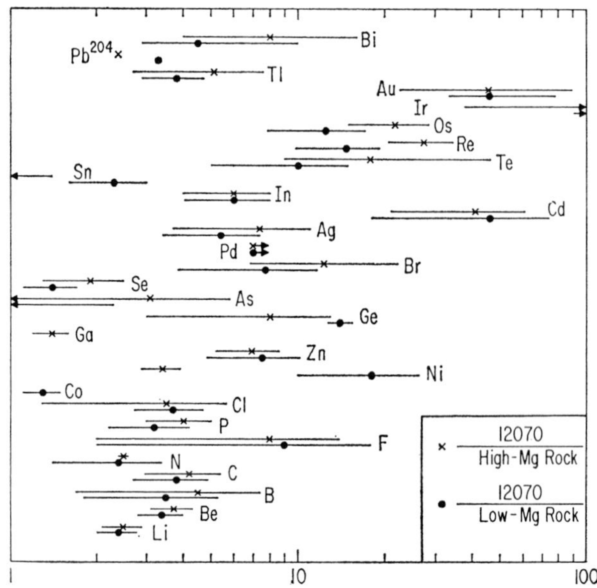


Fig. 1. The ratios of trace element concentration in lunar fines 12070 relative to low-Mg and high-Mg crystalline rocks collected at the Apollo 12 site for 29 elements that are potentially useful for characterizing the extralunar component. The error bars are drawn to include 95% of the data for the crystalline rocks. Data by the following authors was taken from the *Proceedings of the Second Lunar Science Conference:* Anders *et al.*, Baedecker *et al.*, Bouchet *et al.*, Brunfelt *et al.*, Cliff *et al.*, Compston *et al.*, Cuttitta *et al.*, Engel *et al.*, Epstein and Taylor, Friedman *et al.*, Gibson and Johnson, Goles *et al.*, Herr *et al.*, Kaplan and Petrowski, Kharkar and Turekian, Laul *et al.*, Lovering and Hughes, Moore *et al.*, Morrison *et al.*, Reed and Jovanovic, Schnetzler and Philpotts, Scoon, Sievers *et al.*, Smales *et al.*, Taylor *et al.*, Travesi *et al.*, Volbecky *et al.*, Wakita *et al.*, Wänke *et al.*, Willis *et al.* Data from the following sources was also employed: Ganapathy *et al.* (1970) *Science* **170,** 533; Eugster (1971) *Earth Planet. Sci. Lett.* **12,** 273.

than results obtained by other analysts, and which were thought to reflect either sample contamination or a poor analytical method.

Net concentrations of the selected elements (i.e., bulk concentration minus crystalline rock concentration) were calculated for mature soils from the first four Apollo landing sites and for Luna 16 soil. These, divided by concentrations observed in water-free C1 chondrites, are plotted in Fig. 2. Data is also shown for rock 14310.

Few crystalline rocks were recovered from the Fra Mauro site. Rock 14310 is the only large one, but it has a sizable extralunar component of its own and may be a remelted soil. Therefore, we have arbitrarily used mean concentration data on Apollo 12 mare basalts as an estimate of the amounts of indigenous lunar material in the Apollo 14 regolith. Similarly, data for Apollo 12 crystalline rocks were employed where none was available for Apollo 15 mare basalts, and the indigenous components in the Luna 16 soil were estimated from Apollo 11 data. It is possible that the alkali-rich material which appears to be the main contributor to the regolith at the Apollo 14

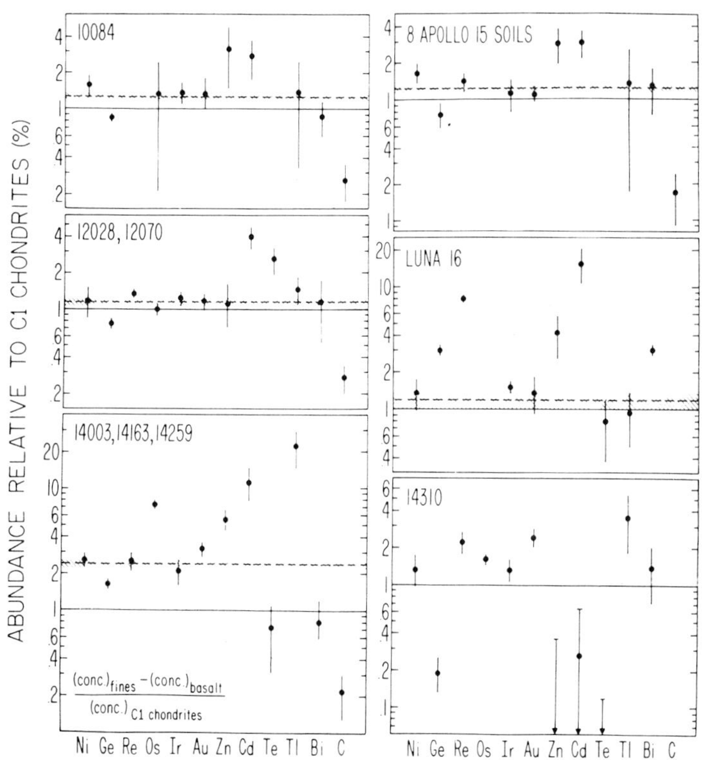


Fig. 2. Abundance pattern of 12 trace elements in the extralunar component obtained by subtracting average abundances for crystalline rocks from those of the soils and normalizing to C1 chondrites. The error bars represent estimates of $\pm$ one standard deviation on the means for both soils and crystalline rocks. The dashed lines represent the concentration of extralunar component as determined from an analysis of the data as presented in Fig. 4, with bands drawn to indicate an uncertainty of $\pm$ one standard deviation. Data by the following authors was taken from the *Proceedings of the Apollo 11 Lunar Science Conference:* Annell and Helz, Compston *et al.*, Epstein and Taylor, Friedman *et al.*, Ganapathy *et al.*, Haskin *et al.*, Kaplan *et al.*, Lovering and Butterfield, Maxwell *et al.*, Moore *et al.*, Morrison *et al.*, Smales *et al.*, Taylor *et al.*, Turekian and Kharkar, Wänke *et al.*, Wasson and Baedecker; *Proceedings of the Second Lunar Science Conference:* Anders *et al.*, Baedecker *et al.*, Bouchet *et al.*, Brunfelt *et al.*, Compston *et al.*, Cuttitta *et al.*, Epstein and Taylor, Friedman *et al.*, Herr *et al.*, Kaplan and Petrowski, Laul *et al.*, Lovering *et al.*, Moore *et al.*, Morrison *et al.*, Reed and Jovanovic, Smales *et al.*, Taylor *et al.*, Vinogradov, Wänke *et al.*, Willis *et al.*,; revised abstracts of the third lunar science conference: Baedecker *et al.*, Barnes *et al.*, Hamilton *et al.*, Helmke and Haskin, Herpers *et al.*, Hubbard and Gast, Jackson *et al.*, Moore *et al.*, Morgan *et al.*, Reed *et al.*, Rose *et al.*, Sakai *et al.*, Wakita *et al.*, Wänke *et al.* Data from the following sources was also employed: Brunfelt *et al.* (1971) *Earth Planet. Sci. Lett.* **11,** 351; Jerome *et al.* (1972) *Earth Planet. Sci. Lett.* **13,** 436; Laul *et al.* (1972) *Earth Planet. Sci. Lett.* **13,** 450. The shaded bands are calculated from the renormalized Ni, Ge, Ir, and Au values, and are the same as those shown in Fig. 4.

site has a higher content of the diagnostic elements than the mare basalts. In this case we will have underestimated the indigenous contribution and this possibility will be considered in the subsequent analysis of the data.

The soils we have plotted are those that appear to be representative of each of the sampling sites. They were generally chosen on a rather random basis by the astronauts, they show extensive reworking (see, e.g., Quaide *et al.*, 1971) and they have high concentrations of extralunar diagnostic elements. Other soils (e.g., 12033) that show lower concentrations of the extralunar component generally show less reworking and were chosen because the astronauts felt they were unusual (Wasson and Baedecker 1972). The high relief near the Apollo 15 site makes it likely that most soils differ from the composition they would have had were the site planar.

Both the fractionation pattern and the concentration of the extralunar component is similar in mature soils at the five landing sites. All show extralunar-component concentrations near 1% except the Apollo 14 site, where the concentration is about twice as high. The abundance pattern of the siderophilic elements Ni, Ge, Ir, and Au is remarkably uniform from site to site. The other elements vary in an unsystematic fashion which limits their usefulness as diagnostic elements.

Data are plotted in Fig. 2 for eight different Apollo 15 soil samples. An additional soil sample recovered 20 m from the rim of Hadley Rille (15331) contains an extralunar component that is a factor of two lower than is present in these other fines recovered from other stations (see Table 2). The explanation for the lower concentration in 15531 appears to reflect the fact that the gradual loss of soil into the rille results in a lower effective bombardment age for this regolith. Swann and Schaber (1972) have pointed out that the regolith is particularly thin near the rille rim. The major-element data indicate that the lower extralunar component in the rille soil does *not* result from dilution by *exotic* lunar materials.

The siderophilic-element data for the rock 14310 (Fig. 2) is best interpreted in terms of an extralunar component. However, Ge and other more volatile elements are low, resulting in a strongly fractionated abundance pattern. Although some (e.g., LSPET, 1971) have suggested that 14310 is a primary igneous rock typical of bedrock at the Fra Mauro site, our data are not consistent with this interpretation. For example, the extralunar Ge/Ni ratio is nearly constant independent of landing site. Subtraction of 14310 concentrations from Apollo 14 soil data would result in a Ge/Ni ratio at this site that is twice as high as that observed at any other site. Further, the bedrock material should be KREEP of Fra Mauro origin. Data from the KREEP-rich soil 12033 (Baedecker *et al.*, 1971b) indicates that the siderophilic-element content of KREEP is very low compared to that in 14310.

Our data for Ga (Table 2) also indicate that material similar to 14310 is not a major component in the soil at Fra Mauro and is distinctly different from the KREEP component, although the major element compositions are similar. Gallium is strongly correlated with Al in most lunar materials, with Apollo 14 soil samples falling on the main trend line in a plot of Ga versus Al, and having the highest Ga concentrations observed for lunar materials. However, 14310 has a low Ga content and a low Ga/Al ratio. With the exceptions of Bi and Tl, the elements that are present in low concentrations in 14310 are those with the highest volatilities. The best interpretation of rock 14310 appears to be that it is a lithified soil that has been hot enough to vaporize off a portion of these elements, although not sufficiently hot to have affected the K/Rb ratio (Hubbard *et al.*, 1972).

Although the enrichment of siderophile elements in the soil appears to be best explained by the presence of an extralunar component, the situation with regard to the volatile elements is more complex. The C1 chondrite normalized abundances of Zn and Cd (with the exception of Zn in the Apollo 12 soils) are consistently greater than those of the siderophile elements. In the soils recovered at the Apollo 15 site, the ratios of Zn and Cd to siderophile element abundances are constant, while the absolute abundances of these elements vary by a factor of two. The latter observation suggests that the Zn and Cd enrichment is due to an extralunar component having a higher abundance of these elements than C1 chondrites. However, this suggestion is not supported by data obtained on soil samples from other sites. First, the Zn/Cd ratio for Apollo 12 soils is distinctly lower than those from other sampling locales. Second, although there is a rough correlation between the levels of Zn and Cd and the siderophile elements among lunar soils, the ratio of Zn and Cd to siderophiles varies by as much as a factor of 2. Zinc and Cd are strongly correlated in lunar material, although the Zn/Cd ratio varies by over a factor of 5. The enrichment of Zn and Cd in soils and breccias might also be ascribed to the presence of one or more Zn- and/or Cd-rich exotic components, which have not been represented in the igneous rocks and clasts which have been analyzed to date (Baedecker *et al.*, 1971a; Laul *et al.*, 1971). Unfortunately, there is also some possibility that an unknown contamination source is contributing.

The abundance data on the other three volatile elements plotted in Fig. 2 are no less confusing. Tellurium is enriched relative to siderophiles in the Apollo 12 soil but shows a relative depletion in soils from Fra Mauro and Mare Fecunditatis. Bismuth and Tl appear to be present at levels consistent with an enrichment due to C1-like material in soil recovered by Apollo 11 and 12 and Luna 16. However, Tl was found to be spectacularly enriched in soil at the Apollo 14 site, while Bi showed a smaller enrichment than at other sites in spite of a larger apparent extralunar component. Anders *et al.* (1971) find a correlation between Tl and the alkali elements in lunar basalts, and the high Tl content of the Apollo 14 soil may be due to the high content of alkali-rich material.

Because of the enigmatic data on the distribution of volatile elements in lunar fines, their use in characterizing the extralunar component should be accepted with caution. In particular, attempts to identify individual cratering projectiles as belonging to a particular meteorite type based on volatile element data should be considered temerarious.

We attribute the minor variations observed in the relative abundances of the elements Ni, Ge, Ir, and Au in mature soils to analytical error. The near constancy in the relative abundances of these elements between sites would tend to rule out the possibility that the level of their enrichment in the soil is appreciably affected by variations in the compositions of the rock types which make up the bulk of the soil. The large apparent enrichments of Re in Luna 16 soil and Os in the Apollo 14 soils may also be due to analytical problems, because analyses for these two elements are subject to a number of pitfalls, including loss of volatile $OsO_4$ and the inherent difficulties involved in assaying the indicator radionuclides by $\beta$ counting (Morgan, 1965).

As shown by Fig. 2, the abundance of C in the lunar soil is nearly an order of magnitude lower than would be expected by a 1% addition of C1-like material to the lunar regolith. Either the extralunar component is significantly poorer in C than are C1 chondrites, or C has been depleted from the soil by direct volatilization or by escape of $CH_4$ following reaction with solar-wind H (Kaplan and Smith, 1970). Most researchers assign most of the observed soil C to origins from the solar wind (e.g., Abell *et al.*, 1971) or to lunar crystalline rocks (e.g., Kaplan and Petrowski, 1971). The data in Fig. 2 show that the C abundance varies over only a small range of about 0.19 to 0.29%. This mainly indicates that C in the lunar soil is highly mobile, whatever its source.

## Apollo 14 Breccias

In Table 2 we present data for our suite of elements for the breccias 14321 and 14066. The sawdust from 14066 and the matrix material from 14321 are similar in their trace element abundance patterns to the mature local soil, except for an apparent depletion of Ge in the latter. The data for the material separated from 14321, normalized to C1 chondrite abundances, has been plotted in Fig. 3. The concentrations of

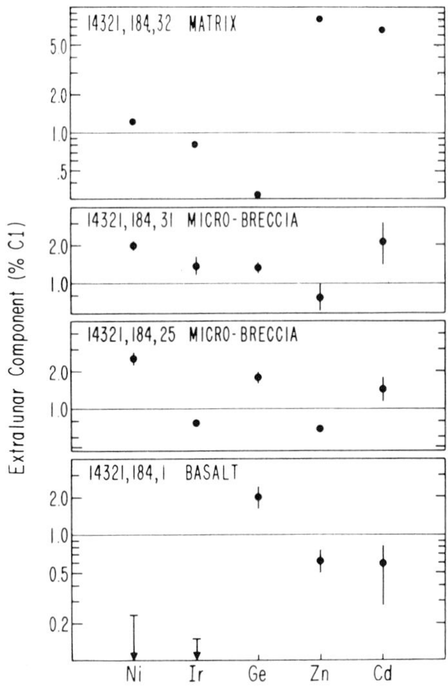


Fig. 3. Abundance pattern of five trace elements in the extralunar component in separated clasts and matrix material from the breccia 14321. The error bars represent the ranges in analyses of replicate samples.

Ni and Ir in basalt clast 1E are similar to those observed for mare basalts recovered from Oceanus Procellarum. Germanium appears to be anomalously high, and the absence of similar high-Ge basalts among the material obtained thus far leads us to suspect sample contamination. Zinc and Cd appear to be moderately enriched relative to Apollo 12 basalts in this clast. The microbreccia clasts from 14321 have an apparent extralunar component less than or equal to that in the local soil based on data for the siderophile elements. The most striking feature of the data for 14321 is the much lower levels of Zn and Cd observed in the microbreccia clasts relative to the local soil.

## Concentration of Extralunar Component at Five Lunar Sites

For reasons outlined above, the siderophile elements are the most useful indicators of the level of extralunar material in lunar fines and breccias. In particular the elements Ni, Ge, Ir, and Au appear to be those determined with the greatest precision and accuracy.

An equation relating the integrated flux of extralunar material to the concentrations of the extralunar component at two sites on the lunar surface was developed by Baedecker *et al.* (1971a). They assumed that the amount of extralunar material in the regolith at any site was directly proportional to the total integrated flux and inversely proportional to regolith depth. Utilizing data presented by Oberbeck and Quaide (1968), who estimated the depth of the regolith at four different sites based on crater profiles, and the integrated flux at each site based on crater frequency distributions, Baedecker *et al.* obtained an expression relating regolith depth to total integrated flux. This approach yielded the following expression:

$$\log \frac{\chi_2}{\chi_1} = 2.41 \log \frac{[X]_2}{[X]_1} \tag{1}$$

where $[X]$ is the concentration of the extralunar component at a given site, and $\chi$ is the total integrated flux. An extralunar-component concentration ratio of 1.10 for the first two Apollo sites yielded a ratio of 1.25 in integrated flux flow.

As observed above, the relative abundance patterns of the siderophile elements Ni, Ge, Ir, and Au are nearly the same among the different lunar soil samples thus far sampled. However, the pattern deviates significantly from a C1 abundance pattern to some degree, particularly in the low abundance of Ge in the lunar fines relative to the other three elements. It is desirable to choose a normalization standard that more accurately reflects the true pattern in the extralunar component. We have done this in the following fashion: Ir is an element that is currently being accurately determined and that, with a C1 chondrite normalization, yields extralunar-component concentrations which are intermediate between the low values based on Ge and the slightly higher values estimated from Ni abundances. We have arbitrarily defined new concentrations of these elements in the extralunar component such that the mean estimates of the concentration of the extralunar component in the regolith at the Apollo 11, 12, 14, and 15 sites are the same for each element. The assigned concentrations are Ni, 1.49%; Ge, 29.5 ppm; and Au, 204 ppb. We have, of course, retained the water-free C1 chondrite concentration for Ir, 634 ppb.

The renormalized extralunar-component concentrations based on Ni, Ge, Ir, and Au concentrations in mature soils at the five lunar landing sites are plotted in Fig. 4. Mean values were then calculated by weighting the concentration calculated for each element by its estimated variance, and are shown as dashed lines in Figs. 2 and 4. The stippled bands show 70% confidence limits on the means (based on Student's *t* distributions) for each landing site.

## RELATIVE INFLUX OF CRATER-PRODUCING AND SIDEROPHILE-RICH MATTER

By applying equation (1) we have used data on the siderophiles Ni, Ge, Ir, and Au to calculate integrated fluxes at the Apollo 11, 14, 15, and Luna 16 sites relative to that at the Apollo 12 site.

In Fig. 5a we compare our estimates with those of Soderblom and Lebovsky (1972), which are based on crater morphology (except that for Apollo 14, which they take from crater-count data of Swann *et al.*, 1971), and with those of Hartmann (1970), which are based on crater populations. The ages of the different landing sites are means of the crystalline rock ages at each site. In effect, we are assuming that the rocks are a random selection of the material which has been comminuted to make the soils at each site. For clarity, the flux estimates of Soderblom and Lebofsky are plotted 0.01 g.y. high and those of Hartmann 0.01 g.y. low.

The vertical bars attached to the estimates of Soderblom and Lebofsky (1972) reflect the error limits they assign to their values. The vertical bars attached to our estimates are the 70% confidence limits shown by the bands in Fig. 4 increased for all sites except Apollo 12 to allow for an estimated uncertainty of 5% in the constant in equation (1).

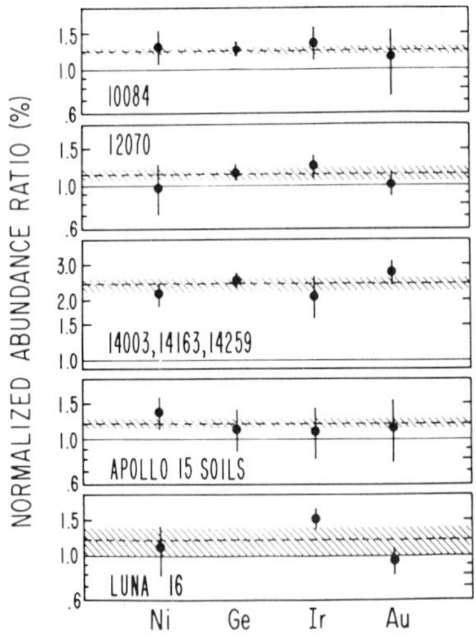


Fig. 4. Cl chondrite normalized enrichments of four siderophile elements in lunar soils renormalized to Ir in order to correct for uncertainties in the abundance pattern of the extralunar component. The concentration of extralunar material is represented by the dashed line, which is the mean of the individual values of the normalized abundance ratio for each of the four elements, weighted by the reciprocal of the variance. The shaded band represents 70% confidence limits about the mean.

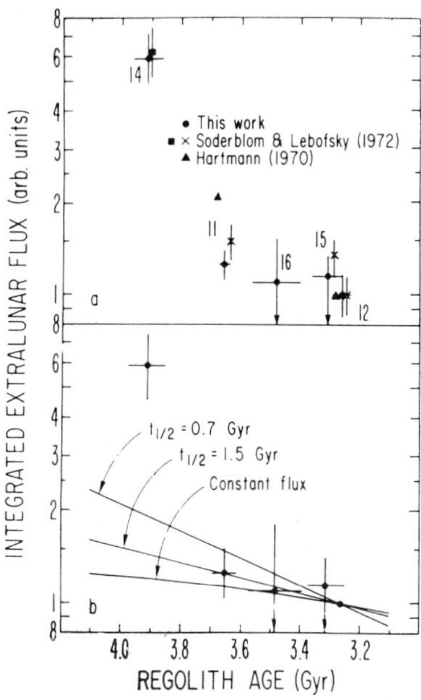


Fig. 5. The variation of the total integrated flux of the extralunar component at five lunar landing sites as a function of the regolith age at each site: (a) comparison of estimates based on siderophilic-element data with those based on crater observations; (b) siderophilic element-based estimates with exponential curves fitted to the Apollo 11 and Apollo 12 points. See text for details.

The Apollo 15 site is characterized by its high relief. This complicates its use for a study such as this. As noted earlier, the regolith near Hadley Rille has a low apparent bombardment age as a result of the loss of material into the rille. For this reason we have extended the error bar on the Apollo 15 flux estimate down toward the apparent value of 0.35 estimated on the basis of our Rille soil 15531. On the other hand, talus from the Apennines probably makes a substantial contribution to the soil near the Apennine Front. The Front is probably of Fra Mauro age (ca. 4.0 g.y.) and must have participated in the higher flux of extralunar matter postulated below to explain the higher extralunar-component concentrations observed at the Apollo 14 site.

The ages of the Apollo 11, 12, and 14 sites are based on the Rb–Sr data of Papanastassiou and Wasserburg (1970, 1971a, 1971b). The errors assigned these values are 70% confidence limits based on the scatter in the reported isochron ages. The age of the regolith at the Luna 16 site is based on the $Ar^{39}$–$Ar^{40}$ data obtained for fragment B-1 by Huneke *et al.* (1972). We have calculated a mean age of 3.48 g.y. from their data on individual thermal release fractions by weighting each fraction by the reciprocal of its variance. We have arbitrarily assigned an error of 0.08 g.y. to this value to reflect the large uncertainty attached to dating a regolith by analyzing a single 62 mg particle. The age (3.31 $\pm$ 0.05 g.y.) assigned the Apollo 15 site is the $Ar^{39}$–$Ar^{40}$ and Rb–Sr age measured in rock 15555 by Wasserburg *et al.* (1972). We have assumed that their error assignment applies to the uncontaminated, mature local regolith since similar ages have been reported by other groups (e.g., Schaeffer *et al.*, 1972; York *et al.*, 1972).

The similar trends observed in the three sets of data plotted in Fig. 5a indicate that the flux of large, crater- and regolith-producing objects onto the surface of the moon paralleled that of the total mass flux, which is probably dominated by material in the small size range. This, in turn, implies that both types of material either have the same source, or that the sources are different but have similar orbital characteristics.

## Extralunar Matter: Variation With Time

In Fig. 5b we plot only our estimates of the integrated flux. Here we have combined the age and flux errors previously assigned to the Apollo 12 point into the errors associated with the values for the other lunar sites. The errors attached to the Luna 16 and Apollo 15 data are large, but those attached to Apollo 11 and Apollo 14 are small enough to suggest that it is worthwhile attempting to assess the time history of the extralunar flux during the period from 4.0 to 3.3 g.y.

It seems reasonable to assume that the influx of extralunar matter to the moon decreases exponentially with time. The data in Fig. 5 are clearly inconsistent with the exponential decay of a single population of matter, however. As has been pointed out by Hartmann (1970), Shoemaker and Lebofsky (1972), Shoemaker (1972), and others, a minimum of two populations are necessary to explain the observed trends. We shall assume that only two populations are involved, one with a short half-life that accounts for the rapid decrease in the flux between 4.0 and 3.7 g.y., and one with a much longer half-life that dominated during the past 3.7 g.y.

The three values for the integrated fluxes at the Apollos 11, 12, and 14 sites can be used to construct two linear equations relating these ratios to the half-lives of the two populations of extralunar matter, and to their intensity ratio at some arbitrarily defined time. With two equations and three unknowns, no unique solution is possible. It is nonetheless informative to assume values of the half-life of one component and calculate that of the other. In doing this we have neglected possible errors in the integrated flux values. In Table 3 are shown four possible combinations of the half-lives of the long-lived and short-lived components, and of the relative contributions of the two populations of extralunar matter during the interval between the initiation of regolith growth at the Apollo 11 and 12 sites. Note that the assumption of a longer half-life for one population results in a longer estimated half-life for the other.

Hartmann (1970) has interpreted his cratering data to indicate that the flux of

Table 3. Comparison of half-live estimates for the long-lived and short-lived populations of extralunar matter.

| Half-life long-lived (m.y.) | Half-life short-lived (m.y.) | Ratio short-lived/ long-lived fluxes 3.65–3.26 g.y. ago |
|---|---|---|
| $\infty$ | 50 | 1.2 |
| 2000 | 40 | 0.24 |
| 1700 | 35 | 0.11 |
| 1500 | $\leq 30$ | $\leq 0.04$ |

extralunar matter has been increasing during the last few g.y. Although our data could be fitted with such a model, we do not consider it very plausible and have only considered models involving a constant flux, or one undergoing exponential decay. As a result, the maximum half-life we estimate for the short-lived population is 50 m.y. The minimum half-life of the short-lived flux that affects the estimate of the long-lived population is 30 m.y.; below this value the contribution of the short-lived population to the integrated extralunar flux at the Apollo 11 site is negligible. Shoemaker (1972) argued that the half-life of the short-lived component cannot have been less than 45 m.y., else the total mass accreted to the earth since 4.65 g.y. would have exceeded the mass of the earth. We are not convinced of the quantitativeness of this argument, since the gravitational cross-section of the earth was much smaller at the beginning of this period, and the accretion rate correspondingly smaller. It would appear that the half-life might be as short as 10 m.y., as estimated for meteorites in earth-crossing orbits (Wetherill, 1969).

For purposes of illustration we have drawn curves on Fig. 5b illustrating half-lives of the long-lived component of 0.7, 1.5, and $\infty$ m.y. All of these provide a reasonable fit to the data on the four mare sites samples to date. Our error limits on the Apollo 11 value allow the assignment of a rather stringent lower limit of 0.7 m.y. on the half-life of the long-lived flux (assuming a negligible contribution from the short-lived population).

As best estimates of the half-lives of the short- and long-lived populations we would choose 40 and 2000 m.y., respectively. Limits on these estimates are given by the first and last lines of Table 3.

## Relative Compositions of Extralunar Matter Populations

The long-lived population contributes no more than 25% of the extralunar matter observed at the Fra Mauro site. It is therefore of great interest to reexamine the Apollo 14 data to see whether a systematic difference in the composition of the extralunar component exists between this site and the sites of the other four lunar missions. A reexamination of the siderophilic element data in Fig. 4 suggests that Ge and Au might be enriched relative to Ir and Ni if the comparison is limited to the Apollo 11 and 12 soils, for which the most accurate data are available. The differences are at most about 30%, however, and are probably within experimental error ranges.

A second glance at Fig. 2 shows that the Apollo 14 extralunar component has its own volatile element fingerprint. However, the variations are mainly in Te, Bi, and Tl, which are elements subject to rather large experimental errors. Further, the high Bi is associated with a high Cd value, and the high Tl with a high alkali content. Laul *et al.* (1971) find evidence that there are indigenous lunar materials which are rich in Cd and Bi, and Anders *et al.* (1971) suggest that lunar Tl follows alkalis. We conclude that, at this time, there is no basis for assigning anomalies in the volatile element spectrum to differences in composition between the long-lived and the short-lived extralunar materials.

Ganapathy *et al.* (1972) ascribe the high concentration of extralunar materials at the Apollo 14 site to the planetesimal which created the Imbrium Basin. A third alternative is that the Preimbrian regolith accounts for these high concentrations. The absence of an appreciable extralunar component in relatively unaltered Fra Mauro material, such as the KREEP portion of the 12033 fines (Baedecker *et al.*, 1971b) speaks strongly against these alternative explanations for the high concentration of the extralunar component at the Apollo 14 site.

*Acknowledgments*—We are indebted to LSAPT and the MSC curatorial staff for the lunar samples, and to R. Bild, R. Glimp, J. F. Kaufman, J. Kimberlin, S. Krasner, L. Pickering, H. Quimby, K. Robinson, and L. Sundberg. Neutron irradiations at the UCLA, Ames Laboratory and Northrop reactors were capably handled by J. Brower, A. F. Voight, D. Rusling, and their associates. This research was supported in part by NASA Grant NAS-05-007-291.

## REFERENCES

Abell P. I., Cadogan P. H., Eglinton G., Maxwell J. R., and Pillinger C. T. (1971) Survey of lunar carbon compounds, I. The presence of indigenous gases and hydrolysable carbon compounds in Apollo 11 and Apollo 12 samples. *Proc. Second Lunar Sci. Conf., Geochim. Cosmochim. Acta* Suppl. 2, Vol. 2, pp. 1843–1863. MIT Press.

Anders E., Ganapathy R., Keays R. R., Laul J. C., and Morgan J. W. (1971) Volatile and siderophile elements in lunar rocks. *Proc. Second Lunar Sci. Conf., Geochim. Cosmochim. Acta* Suppl. 2, Vol. 2, pp. 1024–1036. MIT Press.

Baedecker P. A., Schaudy R., Elzie J. L., Kimberlin J., and Wasson J. T. (1971a) Trace element studies of rocks and soils from Oceanus Procellarum and Mare Tranquilitatis. *Proc. Second Lunar Sci. Conf., Geochim. Cosmochim. Acta* Suppl. 2, Vol. 2, pp. 1037–1061. MIT Press.

Baedecker P. A., Cuttitta F., Rose H. J., Schaudy R., and Wasson J. T. (1971b) On the origin of lunar soil 12033. *Earth Planet. Sci. Lett.* **10,** 361–364.

Duncan A. R., Lindstrom M. M., Lindstrom D. J., McKay S. M., Stoeser J. W., Goles G. G., and Fruchter J. S. (1972) Comments on the genesis of breccia 14321 (abstract). In *Lunar Science—III* (editor C. Watkins), pp. 192–194, Lunar Science Institute Contr. No. 88.

Eldridge J. S., O'Kelley G. D., and Northcutt K. J. (1972) Abundances of primordial and cosmogenic radionuclides in Apollo 14 rocks and fines (abstract). In *Lunar Science—III* (editor C. Watkins), pp. 221–223, Lunar Science Institute Contr. No. 88.

Ganapathy R., Keays R. R., Laul J. C., and Anders E. (1970) Trace elements in Apollo 11 lunar rocks: Implications for meteorite influx and origin of moon. *Proc. Apollo 11 Lunar Sci. Conf., Geochim. Cosmochim. Acta* Suppl. 1, Vol. 2, pp. 1117–1142. Pergamon.

Gault D. E. (1970) Saturation and equilibrium conditions for impact cratering on the lunar surface: Criteria and implications. *Radio Sci.* **5,** 273–291.

Hartmann W. K. (1970) Lunar cratering chronology. *Icarus* **13,** 299–301.

Hubbard N. J., Gast P. W., Rhodes M., and Wiesmann H. (1972) Chemical composition of Apollo 14 materials as evidence for alkali volatilization (abstract). In *Lunar Science—III* (editor C. Watkins), pp. 407–409, Lunar Science Institute Contr. No. 88.

Huneke J. C., Podosek F. A., and Wasserburg G. J. (1972) Gas retention and cosmic-ray exposure ages of a basalt fragment from Mare Fecunditatis. *Earth Planet. Sci. Lett.* **13,** 375–383.

Kaplan I. R. and Petrowski C. (1971) Carbon and sulfur isotope studies on Apollo 12 lunar samples. *Proc. Second Lunar Sci. Conf., Geochim. Cosmochim. Acta* Suppl. 2, Vol. 2, pp. 1397–1406. MIT Press.

Kaplan I. R. and Smith J. W. (1970) Concentration and isotopic composition of carbon and sulfur in Apollo 11 lunar samples. *Science* **167,** 541–543.

Laul J. C., Boynton W. V., and Schmitt R. A. (1972) Bulk, REE, and other elemental abundances in four Apollo 14 clastic rocks and three core samples, two Luna 16 breccias and four Apollo 15 soils (abstract). In *Lunar Science—III* (editor C. Watkins), pp. 480–482, Lunar Science Institute Contr. No. 88.

Laul J. C., Morgan J. W., Ganapathy R., and Anders E. (1971) Meteoritic material in lunar samples: Characterization from trace elements. *Proc. Second Lunar Sci. Conf., Geochim. Cosmochim. Acta* Suppl. 2, Vol. 2, pp. 1139–1158. MIT Press.

LSPET (Lunar Sample Preliminary Examination Team) (1972) The Apollo 15 lunar samples: A preliminary description. *Science* **175,** 363–375.

Morgan J. W. (1965) The simultaneous determination of Re and Os in rocks by neutron activation analysis. *Anal. Chim. Acta* **32,** 8–16.

Müller O., Baedecker P. A., and Wasson J. T. (1971) Relationship between siderophilic-element content and oxidation state of ordinary chondrites. *Geochim. Cosmochim. Acta* **35,** 1121–1137.

Oberbeck V. R. and Quaide W. L. (1968) Genetic implications of lunar regolith thickness variations. *Icarus* **9,** 446–465.

Öpik E. J. (1969) The moon's surface. *Ann. Rev. Astron. Astrophys.* **7,** 473–526.

Papanastassiou D. A. and Wasserburg G. J. (1971a) Lunar chronology and evolution from Rb–Sr studies of Apollo 11 and 12 samples. *Earth Planet. Sci. Lett.* **11,** 37–62.

Papanastassiou D. A. and Wasserburg G. J. (1971b) Rb–Sr ages of igneous rocks from the Apollo 14 mission and the age of the Fra Mauro formation. *Earth Planet. Sci. Lett.* **12,** 36–48.

Quaide W., Oberbeck V., Bunch T., and Polkowski G. (1971) Investigations of the natural history of the regolith at the Apollo 12 site. *Proc. Second Lunar Sci. Conf., Geochim. Cosmochim. Acta* Suppl. 2, Vol. 1, pp. 701–718. MIT Press.

Schaeffer O. A., Husain L., Sutter J., and Funkhouser J. (1972) The ages of lunar material from Fra Mauro and the Hadley Rille-Apennine Front area (abstract). In *Lunar Science—III* (editor C. Watkins), pp. 675–677, Lunar Science Institute Contr. No. 88.

Shoemaker E. M. (1972) Cratering history and early evolution of the moon (abstract). In *Lunar Science—III* (editor C. Watkins), pp. 696–698, Lunar Science Institute Contr. No. 88.

Soderblom L. A. and Lebofsky L. A. (1972) Technique for rapid determination of relative ages of lunar areas from orbital photography. *J. Geophys. Res.* **77,** 279–296.

Swann G. A., Bailey N. G., Batson R. M., Eggleton R. E., Hair M. H., Holt H. E., Larson R. B., McEwen M. C., Mitchell E. D., Schaber G. G., Schafer J. P., Shepard A. B., Sutton R. L., Trask N. J., Ulrich G. E., Wilshire H. G., and Wolfe E. W. (1971) Preliminary geologic investigations of the Apollo 14 landing site. *Apollo 14 Preliminary Science Report*, NASA Document NASA SP-272, 39–86.

Swann G. A. and Schaber G. G. (1972) Geology of the Apollo 15 landing site (abstract). In *Lunar Science—III* (editor C. Watkins), pp. 735–737, Lunar Science Institute Contr. No. 88.

Tatsumoto M., Hedge C. E., Doe B. R., and Unruh D. (1972) U–Th–Pb and Rb–Sr measurements on some Apollo 14 lunar samples (abstract). In *Lunar Science—III* (editor C. Watkins), pp. 741–743, Lunar Science Institute Contr. No. 88.

Wänke H., Baddenhausen H., Balacescu A., Teschke F., Spettel B., Dreibus G., Quijano M., Kruse H., Wlotzka F., and Begemann F. (1972) Multielement analyses of lunar samples (abstract). In *Lunar Science—III* (editor C. Watkins), pp. 779–781, Lunar Science Institute Contr. No. 88.

Wasserburg G. J., Turner G., Podosek F. A., Papanastassiou D. A., and Huneke J. C. (1972) Comparison of Rb–Sr, K–Ar, and U–Th–Pb ages: Lunar chronology and evolution (abstract). In *Lunar Science—III* (editor C. Watkins), pp. 788–790, Lunar Science Institute Contr. No. 88.

Wasson J. T. and Baedecker P. A. (1970) Ga, Ge, In, Ir, and Au in lunar, terrestrial, and meteoritic basalts. *Proc. Apollo 11 Lunar Sci. Conf., Geochim. Cosmochim. Acta* Suppl. 1, Vol. 2, pp. 1741–1750. Pergamon.

Wasson J. T. and Baedecker P. A. (1972) Provenance of Apollo 12 KREEP. *Proc. Third Lunar Sci. Conf., Geochim. Cosmochim. Acta* Suppl. 3, Vol. 2. MIT Press.

Wetherill G. W. (1969) Relationships between orbits and sources of chondritic meteorites. In *Meteorite Research*, (editor P. M. Millman), pp. 573–589, Reidel.

York D., Kenyon W. J., and Doyle R. J. (1972) $^{40}$Ar–$^{39}$Ar ages of Apollo 14 and 15 samples (abstract). In *Lunar Science—III* (editor C. Watkins), pp. 822–824, Lunar Science Institute Contr. No. 88.

Proceedings of the Third Lunar Science Conference
(Supplement 3, *Geochimica et Cosmochimica Acta*)
Vol. 2, pp. 1361–1376
The M.I.T. Press, 1972

# Trace elements in Apollo 15 samples: Implications for meteorite influx and volatile depletion on the moon

JOHN W. MORGAN, URS KRÄHENBÜHL, R. GANAPATHY, and EDWARD ANDERS

Enrico Fermi Institute and Department of Chemistry,
University of Chicago, Chicago, Illinois 60637

**Abstract**—Five Apollo 15 soils and four rocks were analyzed by neutron activation analysis for Ag, Au, Bi, Br, Cd, Co, Cs, Ge, In, Ir, Ni, Sb, Se, Rb, Re, Te, Tl, and Zn. Elbow Crater soil 15081, collected 65 m from the rim, contains a meteoritic component equivalent to 1.72% Cl material, similar to that at other lunar sites. Other soils are lower, owing to dilution by fresh bedrock or talus from the Apennine Front. A component rich in Rb, Cs, Cd, and Zn (KREEP?) seems to be present in all soils, but though the abundance of these elements is highest in soil 15431 from the front, it does not decrease regularly with distance from the front. Hence the front may not be the only or even the principal source of KREEP.

Anorthosite 15415 is outstandingly low in these elements. Apollo 11 anorthosites, on the other hand, contain a sizable meteoritic component, similar to that in Apollo 14 dark norites. This component may represent debris of the Imbrian or Serenitatis projectiles or mixed planetesimal debris from the pre-Imbrium regolith.

Apennine Front basalt 15256 is high in Tl, In, Br, Sb, and especially Cd, but otherwise it resembles Apollo 15 mare basalts, which in turn are similar to Apollo 11, 12 mare basalts. Apollo 14 norites and basalts are enriched in most elements that we measured.

Apollo 14 rocks may have a slightly higher Tl/Cs ratio [$(1.6 \pm 1.2) \times 10^{-2}$ versus $(1.0 \pm 0.9) \times 10^{-2}$] than Apollo 11, 12, 15 basalts, anorthosites, "granites" (12013), or Luna 16 soil. The mean Tl/Cs ratio of all Apollo rocks, $(1.2 \pm 1.0) \times 10^{-2}$, still is more than an order of magnitude below the terrestrial and eucritic ratios. The mean Cs/U ratio is $0.23 \pm 0.11$, 1.5% the cosmic ratio. In terms of the two-component model of planetary accretion, this suggests that the moon contains only ~1.5% low-temperature material, with a nominal accretion temperature of ~450°K. Comparison with the earth and 6 meteorite classes shows that the moon's depletion is unique. Possibly this reflects accretion in the earth's neighborhood.

## INTRODUCTION

FIVE APOLLO 15 SOILS and four rocks were analyzed by neutron activation analysis for 18 volatile and siderophile elements. We shall first interpret these results in terms of local geologic problems at the Apollo 15 site and then review their bearing on two broader issues: meteorite influx and volatile element depletion on the moon. Our earlier work on this subject was summarized in the Apollo 11 and 12 Conference Proceedings (Ganapathy *et al.*, 1970; Anders *et al.*, 1971; Laul *et al.*, 1971). Data on Apollo 14 are given in the companion paper by Morgan *et al.* (1972a), and on Luna 16, by Laul *et al.* (1972a).

## RESULTS

The data are shown in Table 1. Our chemical procedure has not been published but can be obtained from the sources listed in Anders *et al.* (1971). The precision and

Table 1. Trace elements in Apollo 15 samples (ppb; Ni, Co, Zn, Rb ppm).

| Sample No. and Location | Ir | Re | Au | Ni | Co | Sb | Ge | Se | Te | Ag* | Br | In* | Bi | Zn | Cd | Tl | Rb | Cs |
|---|---|---|---|---|---|---|---|---|---|---|---|---|---|---|---|---|---|---|
| *Basalts* | | | | | | | | | | | | | | | | | | |
| 15256,6 Apennine Front | 0.022 | 0.0049 | 0.019 | — | 46.2 | 0.43 | 3.8 | 119 | 2.0 | 0.78 | 51 | 6.8 | 0.41 | 0.92 | 104 | 1.45 | 0.67 | 32 |
| 15555,25 Rille | 0.006 | 0.0013 | 0.139 | — | — | 0.067 | 8.5 | 156 | 3.4 | 1.0 | 6.0 | 0.55 | 0.089 | 0.78 | 2.1 | 0.20 | 0.65 | 30 |
| 15597,6 Rille | 0.0072 | 0.0081 | 0.045 | — | 39.6 | 1.49 | 6.5 | 117 | 1.9 | 0.90 | 24 | 0.59 | 0.21 | 1.2 | 1.7 | 0.32 | 0.72 | 40 |
| *Anorthosite* | | | | | | | | | | | | | | | | | | |
| 15415,12 Spur Crater | ≤0.010 | 0.00084 | 0.117 | — | — | 0.067 | 1.2 | 0.23 | 2.1 | 1.73 | 2.3 | 0.178 | 0.097 | 0.26 | 0.57 | 0.09 | 0.11 | 23 |
| *< 1 mm Soils* | | | | | | | | | | | | | | | | | | |
| 15471,32 Dune Crater | 5.9 | 0.47 | 1.79 | 150 | 44.8 | 1.24 | 310 | 184 | 8.7 | *61.8* | 90 | *330* | 0.87 | 11 | 37 | 1.46 | 3.04 | 123 |
| 15501,32 Rille | 6.3 | 0.64 | 2.09 | 150 | 43.6 | 1.51 | 317 | 195 | 6.2 | 30.7 | 131 | *7.4* | 2.15 | 14 | 36 | 1.86 | 4.18 | 175 |
| 15071,31 Elbow Crater, 25 m | 5.6 | 0.71 | 1.99 | 170 | 46.1 | 2.66 | 269 | 162 | 7 | 6.5 | 118 | 3.4 | 2.70 | 10 | 22 | 1.53 | 3.2 | 125 |
| 15431,33 Apennine Front | 4.5 | 0.58 | 1.63 | 150 | 35.4 | 1.11 | 271 | 181 | 11 | 12.1 | 160 | *5.6* | 2.38 | 25 | 49 | 4.23 | 5.2 | 245 |
| 15081,26 Elbow Crater, 65 m | 7.1 | 0.76 | 2.32 | 200 | 46 | 2.67 | 233 | 217 | 12 | 12.8 | 115 | *42* | 2.92 | 14 | 28 | 1.80 | 3.9 | 153 |
| *BCR–1* | 0.004 | 0.81 | 0.47 | — | 35.7 | 556 | 1310 | 81 | 4 | 25 | 110 | 88 | 45.6 | 143 | 142 | 305 | 46 | 920 |

* Italicized values presumably reflect contamination from In–Ag vacuum gaskets in ALSRC or LRL.

accuracy of our results can be judged from comparison of the BCR–1 analyses with earlier values for this standard and replicate analyses of Apollo 11, 12 rocks and soils (Laul *et al.*, 1971, and references cited therein). A new element, Ge, has been added.

## Soils

### *Apollo 15*

Data for five Apollo 15 soils are shown in Fig. 1. Elements on the left (Ir to Br) are largely meteoritic, those on the right, largely nonmeteoritic.

*Abundance levels; dilution by bedrock.* Of the two Elbow Crater soils, that collected farther from the rim (15081, solid triangles) is consistently higher in nearly all elements than the other (15071, open triangles). Apparently the latter has been diluted by ~20% freshly excavated bedrock, low in meteoritic as well as nonmeteoritic elements. A similar dilution effect was observed in Apollo 12: Soils from crater rims were consistently lower in meteoritic elements than soils from intercrater regions (Laul *et al.*, 1971). Apparently surface soils in undisturbed, intercrater areas acquire micrometeorites up to a steady-state level of ~2% Cl equivalent. Lower levels imply dilution by fresh rubble or soil from deeper layers of shorter surface exposure.

It seems that of the five soils studied in this work, only 15081 comes close to being an intercrater soil in this sense. All others appear to have been diluted by freshly excavated material, judging from the fact that they fall below 15081 in Fig. 1. This is to be expected from their location. Dune Crater soil 15471 was collected less than 30 m from the rim of the crater. Rille soil 15501 came from the ejecta blanket of a 15 m crater, 10 m from the crater rim. Though this crater did not penetrate to bedrock,

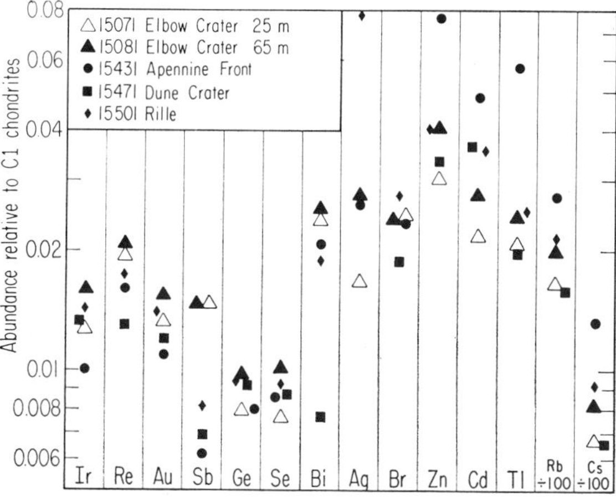


Fig. 1. Of the two Elbow Crater soils, the one collected farther from the rim (15081) is consistently higher in nearly all elements than the other (15071). Presumably the latter was diluted by freshly excavated bedrock. The remaining soils, all collected near crater rims, show a similar dilution effect. Apennine Front soil 15431, though low in meteoritic elements (Ir to Br), is high in Zn, Cd, Tl, Rb, Cs. Probably this implies dilution by an alkali-rich rock type from the front.

it apparently excavated compact soil breccias from some depth (Swann *et al.*, 1971). Possibly these had short surface exposure ages and, hence, low micrometeorite content.

Apennine Front soil 15431 was collected *inside* Spur Crater. Hence it is not surprising that it is lowest in meteoritic elements. However, it happens to be highest in nonmeteoritic elements, Zn to Cs. Neither mare basalts nor Apennine Front basalt are high enough in these elements to account for their high abundance in 15431 and, for that matter, in all other Apollo 15 soils. Apparently other rock types, rich in alkalis, Zn, and Cd (KREEP?) have contributed to these soils. Inasmuch as the contribution seems to be greatest for the soil nearest the front, it seems likely that these rocks are derived from the front.

Schonfeld (1972) and Wänke *et al.* (1972) have interpreted the soils as a three-component mixture of mare basalts, anorthositic gabbro (or anorthosite), and KREEP. Though the first two components correlate smoothly with distance from the front, the third one does not. Thus the distribution of KREEP may not bear a simple relation to that of anorthosite. The front may not be the only or even the principal source of KREEP.

*Ray material.* Two of the soils (15431 and 15471) lie on the Aristillus-Autolycus ray crossing the site; three others lie off it. Yet no systematic difference was seen that could be attributed to ray material. Presumably ray material is not greatly enriched in any of the elements that we measured.

### *Meteoritic component in lunar soils*

Our estimate for Apollo 15 will be based on the only intercrater soil in the present suite, 15081. The net meteoritic component can be found by correcting the gross abundances in Fig. 1 for the indigenous component. The latter is at least a three-component mixture of mare basalt, anorthositic gabbro, and KREEP. However, for the present purpose, the first two components can probably be treated as one, in view of the very low trace-element content of 15415 anorthosite (Table 1). Using the composition of meteorite-free KREEP from Morgan *et al.* (1972b), we estimated the KREEP content of 15081 as 23% via the Cs content. The net meteoritic component was then calculated by subtracting a nonmeteoritic component of 77% Apollo 15 basalts and 23% KREEP (Table 2). Similar estimates for three other landing sites are also shown. Only those values where the indigenous contribution was less than one-third of the total are considered meaningful.* The indigenous component for Luna 16 is not known and was therefore assumed to be equal to that for Apollo 11. Both soils appear to be rich in anorthosite and poor in KREEP (Reid *et al.*, 1972). Apollo 14 soils are not shown at all, because they contain an atypical meteoritic component of ancient origin (Morgan *et al.*, 1972a).

Of the nine elements listed in Table 2, the first four are siderophile in cosmochemical fractionations (Anders, 1972), while the next five are volatile, with volatility

* We have omitted Zn, which is largely meteoritic at some landing sites but not at others, as well as Se and Te, which seem to be partially lost from soils, like their congener S.

Table 2. Meteoritic components in lunar soils (percent CI material or equivalent).*

| | | Siderophiles | | | | Volatiles | | | | | Mean |
|---|---|---|---|---|---|---|---|---|---|---|---|
| | | Ir | Re | Au | Ni | Sb | Ge | Ag | Br | Bi | Siderophiles |
| Gross | Apollo 11 | 1.70 | 1.95 | 1.83 | 2.03 | 1.22 | 1.14 | *(4.31)* | 1.77 | 1.42 | — |
| | Apollo 12 | 1.71 | 1.62 | 1.82 | 2.00 | — | 1.06 | *(2.64)* | 2.60 | 1.79 | — |
| | Apollo 15 | 1.61 | 2.09 | 1.55 | 1.94 | 1.48 | 0.69 | *(6.27)* | 2.39 | 2.56 | — |
| | Luna 16 | 2.19 | *9.89* | 1.87 | 1.80 | 2.11 | 4.08 | (29.4) | 2.81 | (4.25) | — |
| Net | Apollo 11 | 1.69 | 1.91 | 1.79 | 1.88 | (0.79) | 0.97 | *(3.72)* | 1.23 | 1.11 | 1.82 |
| | Apollo 12 | 1.70 | 1.59 | 1.77 | 1.80 | — | 1.00 | *(1.89)* | (1.46) | 1.43 | 1.72 |
| | Apollo 15 | 1.60 | 2.07 | 1.48 | — | 1.05 | 0.67 | *(5.81)* | 2.08 | 2.43 | 1.72 |
| | Luna 16 | 2.18 | *9.85* | 1.83 | 1.66 | 1.68 | (3.90) | (28.8) | 2.27 | (3.94) | 1.89 |

* Italicized values: contamination suspected; parenthesized values: indigenous contribution estimated (or suspected) to be larger than one-third of gross abundance.

*Notes on sources (see following list of letter designations)*:

Apollo 11 soils: mean of 6 determinations on 10084 from *e*, *g*, *h*, except Re and Sb (single determination from *h*), Ni (*c*, *f*, *n*), and Ge (*m*). Indigenous correction: mean of all rock analyses from *a*, *e*, except Re (*i*), Ni (*f*), Ge (*m*).

Apollo 12 soils: mean of all analyses on 12070 and 12028 (*g*), except Re (*i*), Ni [mean of all analyses on 12070 and 12057 from Annell, Bouchet, Compston, Rose, Smales, Wänke, Willis from Proceedings of the Second Lunar Science Conference and LSPET (1971)], and Ge (*b*). Indigenous correction: linear combination of 38% meteorite-free KREEP (*j*) and 62% crystalline rocks [mainly *a*, except Re (*i*), Ni (*d*), and Ge (*b*)].

Apollo 15 soils: this work. Indigenous correction: 23% meteorite-free KREEP (*j*) and 77% mare basalts (this work).

Luna 16 soils: all values from *h*, except Ge (average of *h* and *k*) and Ag (*k*). Indigenous correction: same as Apollo11.

*a*. Anders *et al.* (1971).
*b*. Baedecker *et al.* (1971).
*c*. Compston *et al.* (1970).
*d*. Compston *et al.* (1971).
*e*. Ganapathy *et al.* (1970).
*f*. Gast and Hubbard (1970).
*g*. Laul *et al.* (1971).
*h*. Laul *et al.* (1972a).
*i*. Lovering and Hughes (1971).
*j*. Morgan *et al.* (1972b).
*k*. Vinogradov (1971).
*m*. Wasson and Baedecker (1970).
*n*. Wiik and Ojanperä (1970).

increasing roughly in the order listed. All are "largely meteoritic" in lunar soils, but from the vantage point of Apollo 15, it seems that siderophiles are more reliable than volatiles as indicators of meteoritic material. The intrinsic abundance of siderophiles in lunar rocks is consistently low at all landing sites, and thus the indigenous correction is small (a few percent) and reliable. The abundance of volatiles is more variable and often larger, apparently due to contributions by unknown rock types rich in volatiles such as Ag, Bi, Cd, Ge, Zn (Ganapathy *et al.*, 1970; Laul *et al.*, 1971, 1972a).

*Siderophiles.* Soils from all four landing sites show very similar gross and net abundances of the four siderophile elements. The high Re value in Luna 16 is probably due to contamination (Laul *et al.*, 1972a). The Ir/Au and Ir/Re ratios show slight variations of marginal significance, but the mean meteoritic components are virtually the same at all four sites: 1.72 to 1.89% Cl chondrite equivalent.

*Volatiles.* The volatiles present a much less tidy picture. Most Ag values are unreasonably high, owing to contamination from In–Ag vacuum gaskets and, possibly, contributions from a silver-rich rock type similar to Unit VI in core 12028 (Laul *et al.*, 1971). The values for Sb, Ge, Br, and Bi do not look too unreasonable, but since these elements are overabundant in at least one soil (Luna 16), the indigenous corrections may have been generally underestimated. Though most lunar rocks are quite low in these elements, a few fall near or above the level of soils. With limited sampling, their importance cannot be accurately determined.

In view of these uncertainties, we have not computed the mean abundance of meteoritic volatiles, as it would be a less reliable indicator of total meteorite content than the mean abundance of siderophiles (Table 2, last column).

*Comparison with other authors.* Wasson and Baedecker (1970) and Baedecker *et al.* (1971) have published estimates of the lunar meteoritic component that are considerably lower than ours, 1.14 ± 0.25% for Apollo 11 and 1.04 ± 0.24% for Apollo 12. Part of the discrepancy is due to a difference in units. Whereas we express our values as "% Cl chondrite material (or equivalent)," Baedecker *et al.* give theirs as "% *water-free* Cl chondrite." Since Cl chondrites have a water content of 20%, recalculation to a water-free basis lowers all values by a factor of 1.25. Baedecker *et al.* justify their preference on the grounds that it reflects more closely the actual chemical state of the meteoritic material in the soil. But if that is the objective, several additional corrections must be applied. Cl chondrites also contain 6% S and 5–6% organic matter, most of which is not retained in lunar soil according to isotopic studies of S and C. And as Wänke *et al.* (1972) have shown, some reduction of $Fe^{3+}$, $Fe^{2+}$ to metal apparently takes place on impact. Most meteoritic noble metals are found in metal grains in lunar soil, and since these grains also contain large amounts of tungsten (of evidently lunar origin), they must have been made on impact.

Thus it seems difficult to characterize the "actual chemical state" of meteoritic material in lunar soil. Consequently, we prefer to use the actual composition of Cl chondrites as the norm. It is the closest known analog to the composition of the material before infall, which is more important for many purposes than the composition after infall.

A second reason for the discrepancy is that Baedecker *et al.* compute the meteoritic component essentially from only two elements, Ir and Ge. Germanium consistently gives a value lower by a factor of 1.6, a trend that is also evident in Table 2. Since Ge is volatile, this might imply a composition less primitive than Cl, with volatiles depleted relative to siderophiles. But other volatile elements (Br, Bi) do not show this trend, and it therefore seems conceivable that the Ge value for Cl's is too high. Similar errors may be expected for some of the other elements whose abundances in Cl chondrites are based on only a few analyses. Combined analytical and sampling errors may approach 50% in such cases. Ultimately, this problem will have to be resolved by careful redetermination of Cl abundances; in the meantime, the best way to minimize errors from this source is to use as many elements as possible (preferably siderophile) for determination of the meteoritic component.

Baedecker *et al.* have, in fact, attempted to broaden the base of their estimate by considering seven to eight additional elements determined by us (Pd, Au, Ag, Zn, Cd, Br, Bi, Tl), three of which were also measured by them (Au, Zn, Cd). However, as we have noted in our papers, Zn, Cd, and Tl often are largely nonmeteoritic. Bi and Pd are mainly meteoritic, but Baedecker *et al.* applied too large an indigenous correction based on values that we ourselves had disowned. They also failed to take into account the great bulk of our Apollo 12 data (Anders *et al.*, 1971).

Baedecker *et al.* attribute the slight difference between the Apollo 11 and Apollo 12 meteoritic components (1.14 ± 0.25% and 1.04 ± 0.24% water-free Cl material) to the greater age of the former site and construct a model in which the abundance of meteoritic material is related to regolith thickness. However, as the data on Apollo 12 and 15 show, the abundance of meteoritic elements in the soil depends very strongly on proximity to a crater, owing to dilution by freshly excavated bedrock, or juvenile soil. Though we agree that older soils should be enriched in meteoritic material, we believe this to be a second-order effect, smaller than predicted by the model of Baedecker *et al.* Such variations as are evident in Table 2 probably represent mainly dilution, rather than age effects. Hadley Rille soil 15501 and Luna 16 soil both come from locations where the regolith is thin, yet both are rather high in meteoritic material.

## Local Craters

Of the five soils investigated, only 15471 (Dune Crater) shows any evidence of a meteoritic contribution from the local craters. It is low in Bi and Sb, which suggests the presence of fractionated meteoritic material along with the ubiquitous Cl component. Correcting the Bi abundance for an estimated indigenous contribution of 0.27 ppb and attributing the remainder to the Cl component, we obtain a fractionated meteoritic component equivalent to 0.37% H-chondrite or 0.10% Group I iron meteorite. Such amounts of meteoritic material would be expected at impact velocities of 16 and 35 km/sec, respectively. The lower value falls near the observed mean for meteorites (Millman, 1969), while the higher value falls nearly outside the observed range. On these grounds, a chondritic projectile seems more probable than an iron projectile.

### *Meteoritic component in Apollo 11 anorthosite*

In an earlier paper (Laul *et al.*, 1971), we reported results on two anorthosite fractions from Apollo 11 soil. Both were enriched in "meteoritic" elements relative to mare basalts, and we therefore suggested that they contained a meteoritic component. Dickey (1970) had previously noted that some of the metal particles in Apollo 11 anorthosites might be meteoritic. However, neither the amount nor the abundance pattern of this meteoritic component could be assessed with confidence, because the indigenous correction was not known.

It is interesting to re-examine these data, now that results for the 15415 anorthosite are available (Fig. 2). Insofar as 15415 is representative of the Apollo 11 anorthosites, the indigenous correction would seem to be small, generally less than 10%. (Actually, the two anorthosites are not completely alike. While 15415 is a true anorthosite consisting almost entirely of plagioclase, the Apollo 11 anorthosites appear to be anorthositic gabbros with 20 to 40% mafics, judging from their iron content.)

Both Apollo 11 samples contain nearly equal amounts of siderophiles (Ir, Ni, Au), but the coarser-grained sample is notably poorer in volatiles (Zn, Bi, Se, Ag). Presumably the difference reflects lesser contamination by Cl material in the coarse-grained fraction, as observed in Apollo 11 and 14 samples (Ganapathy *et al.*, 1970; Morgan *et al.*, 1972a). Thus 10085,107,1 probably gives the best approximation to the meteoritic component in Apollo 11 anorthosite.*

* In this sample at least, Zn and Se appear to be good indicators of meteoritic material. Zinc falls below siderophiles even without an indigenous correction, so the meteoritic component must fall lower still. Selenium is lower in the coarse-grained than in the fine-grained sample, which suggests that volatilization losses were not important in these samples.

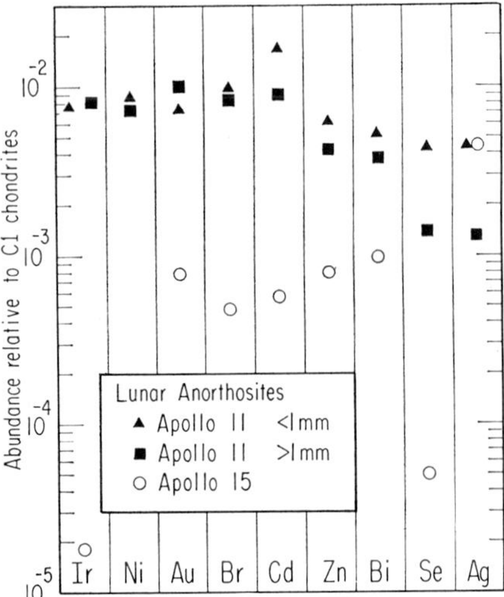


Fig. 2. Compared to 15415, Apollo 11 anorthosites are enriched in meteoritic elements. This meteoritic component resembles that in Apollo 14 dark norites (Morgan *et al.*, 1972a) and may represent debris of the Imbrium or Serenitatis projectiles or mixed planetesimal debris from the pre-Imbrian regolith.

We can try to identify this component by comparing it to known meteorite classes. For such a comparison, we subtract an indigenous correction based on 15415 and normalize the data first to Cl abundances and then to Au. The purpose of this second normalization is to permit comparison of materials of different metal contents (Ganapathy *et al.*, 1972). The results are as follows:

Meteoritic component in 10085,107,1 anorthosite (in percentages).

| | Ir | Ni | Au | Br | Cd | Zn | Bi | Se | Ag |
|---|---|---|---|---|---|---|---|---|---|
| Norm. to Cl | 0.81 | 0.74 | 0.92 | 0.83 | 0.83 | 0.34 | 0.28 | 0.14 | $<0.13$ |
| Norm. to Cl and Au | 0.88 | 0.81 | 1.00 | 0.90 | 0.90 | 0.38 | 0.30 | 0.15 | $<0.14$ |

No indigenous correction was applied for Ag, because the Ag abundance in 15415 was higher than that in 10085 and would have led to a negative Ag abundance.

These data may now be compared with meteorite compositions, similarly normalized (Morgan *et al.*, 1972a). Since we did not correct for residual Cl contamination and may have underestimated the indigenous contribution, all these values must be regarded as upper limits. Accordingly, we shall look for those instances where the anorthosite values fall below the meteoritic ones.

We can immediately eliminate carbonaceous chondrites, L, LL, and E4 chondrites, because they are too high in Ag and Se. Perhaps Se should be disregarded because it seems to be partially volatilized from lunar soil (Laul *et al.*, 1971). In that case, H-chondrites, E5,6 chondrites, and irons remain as possible candidates, though none have quite the right Ir/Au ratio.

Interestingly, the anorthosite pattern resembles one of the meteoritic components at the Apollo 14 site, found in dark norites, ropy glass, and a light norite (Morgan *et al.*, 1972a). Both have similar abundances of Ir, Ag, Se, and Bi. Three sources have been considered for the meteoritic material in dark norite: (1) Imbrian projectile debris, ejected directly from the basin to Fra Mauro; (2) Serenitatis projectile debris, deposited on the pre-Imbrian surface and then ejected by the Imbrian impact; (3) mixed planetesimal debris from the pre-Imbrian regolith.

All three origins are equally acceptable for the Apollo 11 anorthosite. The highlands northwest to south of the Apollo 11 site contain pre-Imbrian regolith, mantled by Imbrian and Serenitatis ejecta. A recent impact (Theophilus?) may have ejected this material to the Apollo 11 site. Of course, this recent impact must have added some meteoritic material of its own, possibly all of that observed if the impact velocity was low. Clearly, more anorthosite samples from other landing sites will have to be studied to see whether their meteoritic components show any systematic regional variations.

## Rocks

A summary of our data is shown in Figs. 3 and 4. Ranges for Apollo 11 and 12 rocks are shown by vertical bars.

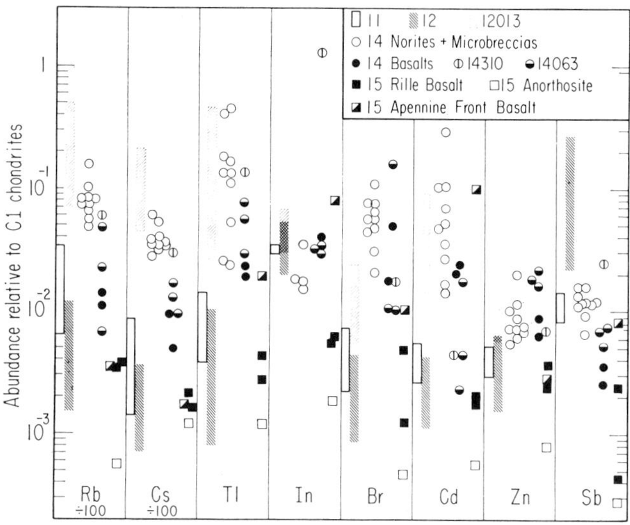


Fig. 3. Apollo 15 anorthosite has some of the lowest trace element contents seen in lunar rocks to date. Apollo 15 rille basalts are similar to Apollo 11, 12 basalts, but Apennine Front basalt is distinctly higher in some elements (Tl, In, Br, Cd, and Sb). Apollo 14 rocks show a varied pattern, norites and impact melt 14310 being highest, and basalts, lowest. The norites contain a substantial meteoritic component, probably related to the Imbrian impact (Ganapathy *et al.*, 1972; Morgan *et al.*, 1972a).

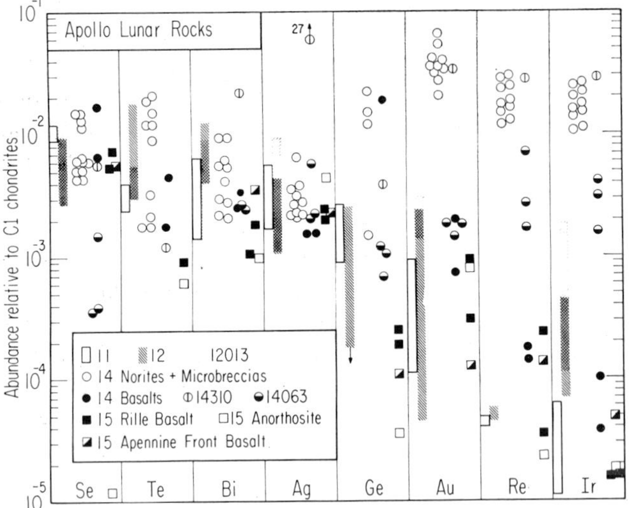


Fig. 4. Apollo 15 anorthosite has some of the lowest trace element contents seen in lunar rocks to date. Apollo 15 rille basalts are similar to Apollo 11, 12 basalts, but Apennine Front basalt is distinctly higher in some elements (Tl, In, Br, Cd, and Sb). Apollo 14 rocks show a varied pattern, norites and impact melt 14310 being highest, and basalts, lowest. The norites contain a substantial meteoritic component, probably related to the Imbrian impact (Ganapathy *et al.*, 1972; Morgan *et al.*, 1972a).

*Apollo 15*

Rille basalts (filled squares) are similar to Apollo 11, 12 basalts, except for their lower content of In, Sb, Ge, and Ir. Some of these differences (for example, Sb, Ge) may represent systematic interlaboratory errors, because the Apollo 11, 12 data for these elements were determined by other authors, and the later ones by us.

The Apennine Front basalt (half-filled squares) is similar to the mare basalts, except for its higher content of Tl, In, Br, Cd, and Sb. But although it is shock-melted, it is not enriched in meteoritic elements.

The anorthosite (open squares) is exceedingly low in most elements measured by us, as predicted by Laul *et al.* (1971). Often it falls well below the range for other lunar samples.

*Apollo 14*

Norites (soil separates and microbreccia fractions from 14321, open circles) fall very high, often in the range of rock 12013. They are outstandingly enriched in meteoritic elements (Ge, Au, Re, Ir), as discussed by Morgan *et al.* (1972a).

Basalts (14053 and a clast from 14321, filled circles) are lower though occasionally still above the range for mare basalts. Note, for example, Tl, Br, Cd, Zn.

Breccia 14063 (half-filled circles) falls in an intermediate position with respect to alkalis and meteoritic elements, which might be taken to mean that it is a mixture of basalt and a small amount of norite ($\sim 10\%$). However, it is very low in a few elements, notably, Cd, Se, and Ge. Hence a simple mixing model is not adequate. Either selective volatilization or a different source material must be assumed.

Rock 14310 (circle with vertical line) is classified as an igneous rock but contains meteoritic elements in the same amounts and proportions as noritic breccias. As argued by Morgan *et al.* (1972a), it is probably an impact melt, solidified too quickly to permit gravitational segregation of metal. We have no explanation for its exceedingly high silver content.

## Accretion Temperature of the Moon

One of the objectives of our work is to estimate the moon's overall depletion in volatile elements. As explained in our earlier papers, this depletion seems to have taken place in the solar nebula (Larimer and Anders, 1967; Ganapathy *et al.*, 1970; Anders, 1971; Anders *et al.*, 1971). When dust accretes from a cosmic gas, elements condense until their partial pressure in the gas equals their vapor pressure at that temperature (Urey, 1952, 1954; Larimer, 1967). The remainder stay behind in the nebula and are subsequently lost with $H_2$, He, and other uncondensed material. Thus volatile elements (Pb, Bi, Tl, In) can serve as cosmothermometers, their abundance in a planetary body being a function of accretion temperature (Larimer and Anders, 1967; Keays *et al.*, 1971; Laul *et al.*, 1972b, 1972c).

In a differentiated body such as the moon, the effects of planetary differentiation are superimposed on the early, nebular fractionation. We therefore compare three "large-ion, lithophile" elements, Tl, Cs, and U, that differ greatly in volatility but are almost quantitatively concentrated in the crust during planetary differentiation. Thallium condenses around 450°K, Cs somewhat higher, and U still higher, around 1300°K. Since these elements do not fractionate much from each other in magmatic processes, their ratios in lunar surface rocks should be close to their ratios in the whole moon.

The Tl/Cs ratio in lunar rocks (Fig. 5) is nearly constant over a 300-fold variation in Cs content, as noted in our earlier papers (Ganapathy *et al.*, 1970; Anders *et al.*, 1971). To be sure, some differences do exist. Apollo 14 rocks fall above rock 12013, and the Apennine Front basalt is also high. But the overall range of variation is quite small for samples from five landing sites, including rocks as diverse as basalts, norites, anorthosites, and granites. The lunar points fall consistently below the terrestrial points, which in turn fall below the cosmic ratio. The mean value of the Tl/Cs mass ratio in lunar rocks is $(1.2 \pm 1.0) \times 10^{-2}$.

The Cs/U ratio is shown in Fig. 6. The Cs values were taken from our work and that of Brunfelt *et al.* (1971); the U values were taken from papers in the first three Lunar Science Conference Proceedings.

Again, the ratio remains nearly constant over a 100-fold range in Cs content. Only the 15415 anorthosite (not plotted) falls out of line, having a high Cs/U ratio. However, since it contains both elements in very low abundance, it probably makes only a minor contribution to the moon's total inventory of these elements. The mean Cs/U ratio of the rocks plotted in Fig. 6 is $0.23 \pm 0.11$.

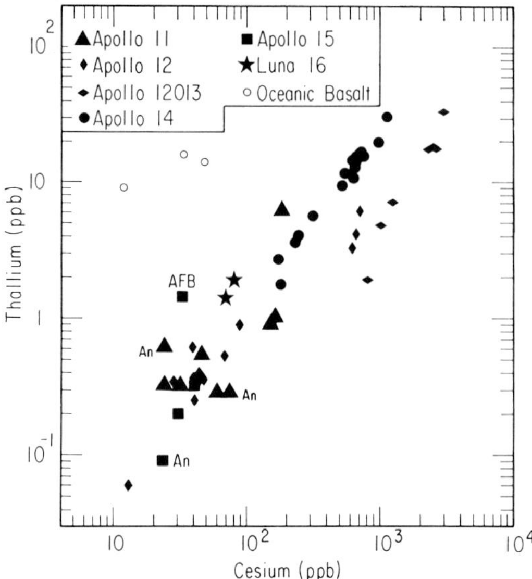


Fig. 5. Thallium and cesium from five lunar landing sites show a fairly good linear correlation. The mean Tl/Cs ratio, $(1.2 \pm 1.0) \times 10^{-2}$, is markedly lower than the cosmic, terrestrial, or meteoritic ratios. An = anorthosite; AFB = Apennine Front basalt.

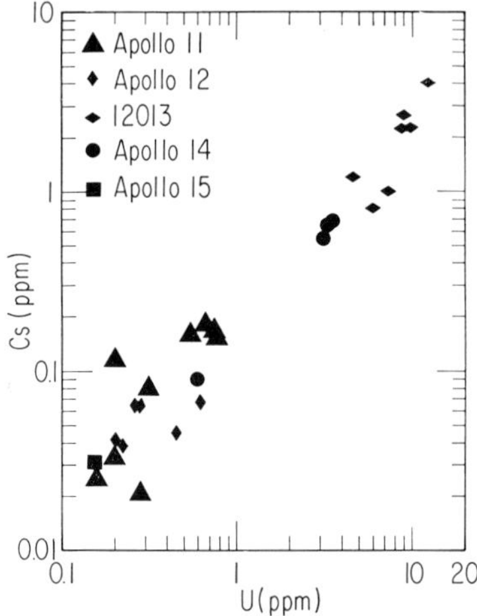


Fig. 6. Cesium and uranium in lunar rocks show a good correlation. The mean Cs/U ratio, $0.23 \pm 0.11$, is markedly lower than cosmic, terrestrial, or chondritic ratios.

We now plot these ratios (normalized to the cosmic ratio) against each other for the moon, earth, and six classes of meteorites studied at Chicago (Fig. 7; Cs and Tl data from Keays *et al.*, 1971; Laul *et al.*, 1972b, 1972c). This plot has a straightforward physical meaning under the two-component model of meteorite and planet formation (Larimer and Anders, 1967, 1970; Anders, 1971). According to this model, each planet consists of a high-temperature fraction that lost *all* its volatiles (both Tl and Cs) and a low-temperature fraction that condensed most of them, with the partial exception of the most volatile elements such as Tl. Thus the Cs/U ratio gives the amount of low-temperature fraction; the Tl/Cs ratio gives the accretion temperature. The low-temperature fraction presumably is a fine-grained ($\sim 10^{-5}$ cm) condensate that equilibrated with the gas to relatively low temperatures. The high-temperature fraction (exemplified by meteoritic chondrules) may represent remelted low-temperature material (Whipple, 1966; Cameron, 1966) or a coarse-grained initial condensate (Blander and Abdel-Gawad, 1969), which ceased to equilibrate with the gas at a higher temperature.

Apparently most objects in Fig. 7 contained more than 10% low-temperature material. The moon and eucrites, on the other hand, contain much less, on the order of 1%. This may reflect a depletion in Cs but perhaps also an enrichment in U (Larimer and Anders, 1970; Anders, 1971). Many authors now believe that the moon is enriched in refractory elements (Ca, Ti, Al, U, Th, etc.) relative to Si (Gast, 1971, and references cited therein).

The moon is also unique in having a very low Tl/Cs ratio. Nominally, this corresponds to a slightly higher accretion temperature for the low-temperature fraction, for example, 450°K versus 430°K for the earth. But this difference could equally well

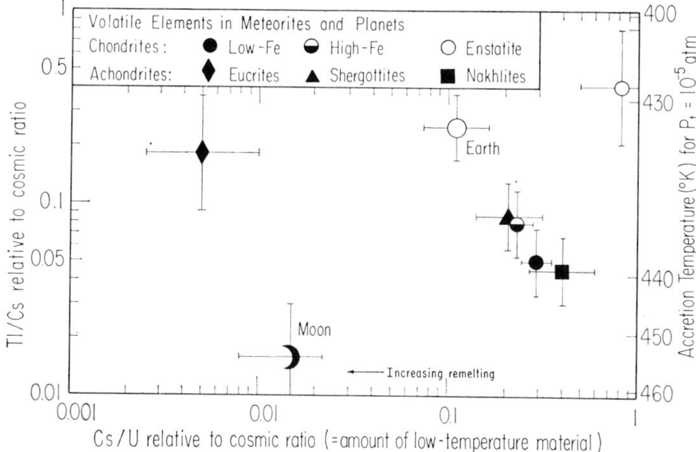


Fig. 7. In terms of the two-component model of planet formation, the Cs/U ratio indicates the fraction of low-temperature material, and the Tl/Cs ratio, the accretion temperature of this material. The moon falls into an unoccupied portion of this diagram, being strongly depleted in both Cs and Tl. This may imply a lower accretion rate than that of the earth in the final stages of accretion, after volatiles had condensed on the nebular dust.

represent a lower accretion rate in the final stages of accretion, when the volatiles had largely condensed on the remaining dust (Ganapathy *et al.*, 1970; Anders *et al.*, 1971; Wasson, 1971). This is particularly likely because $O^{18}$ thermometry shows no difference between the earth and moon (Onuma *et al.*, 1972). The $O^{18}$ content of accreting dust varies much less steeply with temperature than the Tl content, and thus the mean $O^{18}$ content of a planet will not be much influenced by the composition of the last material to accrete, while the Tl content will.

It seems likely, for several reasons, that the bulk of the volatiles were acquired toward the end of accretion (Anders, 1968; Turekian and Clark, 1969). Temperatures probably fell during accretion, and so the material last to accrete would be richest in volatiles. Moreover, material from colder regions (above the median plane or farther from the center of the nebula) may have reached the inner parts of the disk in later stages. Thus the moon-earth difference may imply nothing more than a lower accretion efficiency in the final stages. One possible reason for such low accretion efficiency might be proximity to the earth. Singer and Bandermann (1970) have shown that bodies large enough to move independently of the gas ($>10$ m at $P_{\text{total}} = 10^{-4}$ atm), would strike the moon preferentially, having been concentrated near the earth by the gravitational effect of the latter. But for accretion of volatiles, only particles small enough to equilibrate with the gas are relevant ($\sim 10^{-5}$ cm). They tend to move with the gas. Whipple (1972) has considered this problem in detail and finds that small particles are accreted less efficiently by small bodies. This selectivity can reach factors of $10^{-2}$ to $10^{-3}$ at favorable combinations of radius, particle size, and velocity. The fact that the moon ended up with only $\frac{1}{81}$ the mass of the earth suggests that the moon in fact had a lower overall accretion efficiency.

It must be emphasized that Fig. 7 is not tied to any particular model. The Tl/Cs and Cs/U ratios are remarkably constant and well defined for each planetary body or *differentiated* meteorite class. This supports the notion that these elements are quantitatively concentrated in the crust during planetary differentiation and that these ratios therefore are a fundamental and direct clue to the bulk composition of the planets. Any model with a claim to serious attention should therefore strive to explain these ratios.

*Acknowledgments*—We are indebted to Rudy Banovich for preparation of the drawings on exceedingly short notice. This work was supported in part by NASA Grant NGL 14-001-167. Some of the counting equipment had been provided by the AEC under Contract AT(11-1)-382.

## REFERENCES

Anders E. (1968) Chemical processes in the early solar system, as inferred from meteorites. *Acc. Chem. Res.* **1**, 289–298.

Anders E. (1971) Meteorites and the early solar system. *Ann. Rev. Astron. Astrophys.* **9**, 1–34.

Anders E. (1972) Conditions in the early solar system, as inferred from meteorites. In *Nobel Symposium 21 "From Plasma to Planet"* (in press). Almqvist and Wiksell.

Anders E., Ganapathy R., Keays R. R., Laul J. C., and Morgan J. W. (1971) Volatile and siderophile elements in lunar rocks: Comparison with terrestrial and meteoritic basalts. *Proc. Second Lunar Sci. Conf., Geochim. Cosmochim. Acta* Suppl. 2, Vol. 2, pp. 1021–1036. MIT Press.

Baedecker P. A., Schaudy R., Elzie J. L., Kimberlin J., and Wasson J. T. (1971) Trace element studies of rocks and soils from Oceanus Procellarum and Mare Tranquillitatis. *Proc. Second Lunar Sci. Conf., Geochim. Cosmochim. Acta* Suppl. 2, Vol. 2, pp. 1037–1061. MIT Press.

Blander M. and Abdel-Gawad M. (1969) The origin of meteorites and the constrained equilibrium condensation theory. *Geochim. Cosmochim. Acta* **33,** 701–716.

Brunfelt A. O., Heier K. S., and Steinnes E. (1971) Determination of 40 elements in Apollo 12 materials by neutron activation analysis. *Proc. Second Lunar Sci. Conf., Geochim. Cosmochim. Acta* Suppl. 2, Vol. 2, pp. 1281–1290. MIT Press.

Cameron A. G. W. (1966) The accumulation of chondritic material. *Earth Planet. Sci. Lett.* **1,** 93–96.

Compston W., Chappell B. W., Arriens P. A , and Vernon M. J. (1970) The chemistry and age of Apollo 11 lunar material. *Proc. Apollo 11 Lunar Sci. Conf., Geochim. Cosmochim. Acta* Suppl. 1, Vol. 2, pp. 1007–1027. Pergamon.

Compston W., Berry H., Vernon M. J., Chappell B. W., and Kaye M. J. (1971) Rubidium–strontium chronology and chemistry of lunar material from the Ocean of Storms. *Proc. Second Lunar Sci. Conf., Geochim. Cosmochim. Acta* Suppl. 2, Vol. 2, pp. 1471–1485. MIT Press.

Dickey J. S. Jr. (1970) Nickel–iron in lunar anorthosites. *Earth Planet. Sci. Lett.* **8,** 387–392.

Ganapathy R., Keays R. R., Laul J. C., and Anders E. (1970) Trace elements in Apollo 11 lunar rocks: Implications for meteorite influx and origin of moon. *Proc. Apollo 11 Lunar Sci. Conf., Geochim. Cosmochim. Acta* Suppl. 2, Vol. 2, pp. 1117–1142. Pergamon.

Ganapathy R., Laul J. C., Morgan J. W., and Anders E. (1972) Moon: Possible nature of the body that produced the Imbrian basin, from the composition of Apollo 14 samples. *Science* **175,** 55–59.

Gast P. W. (1971) The chemical composition of the earth, the moon, and chondritic meteorites. In *The Nature of the Solid Earth* (editors E. C. Robertson, J. F. Hays, and L. Knopoff) (in press). McGraw-Hill.

Gast P. W. and Hubbard N. J. (1970) Abundance of alkali metals, alkaline and rare earths, and strontium-87/Strontium-86 ratios in lunar samples. *Science* **167,** 485–487.

Keays R. R., Ganapathy R., and Anders E. (1971) Chemical fractionations in meteorites—IV. Abundances of fourteen trace elements in L-chondrites; implications for cosmothermometry. *Geochim. Cosmochim. Acta* **35,** 337–363.

Larimer J. W. (1967) Chemical fractionations in meteorites—I. Condensation of the elements. *Geochim. Cosmochim. Acta* **31,** 1215–1238.

Larimer J. W. and Anders E. (1967) Chemical fractionations in meteorites—II. Abundance patterns and their interpretation. *Geochim. Cosmochim. Acta* **31,** 1239–1270.

Larimer J. W. and Anders E. (1970) Chemical fractionations in meteorites—III. Major element fractionations in chondrites. *Geochim. Cosmochim. Acta* **34,** 367–388.

Laul J. C., Morgan J. W., Ganapathy R., and Anders E. (1971) Meteoritic material in lunar samples: Characterization from trace elements. *Proc. Second Lunar Sci. Conf., Geochim. Cosmochim. Acta* Suppl. 2, Vol. 2, pp. 1139–1158. MIT Press.

Laul J. C., Ganapathy R., Morgan J. W., and Anders E. (1972a) Meteoritic and non-meteoritic trace elements in Luna 16 samples. *Earth Planet. Sci. Lett.* **13,** 450–454.

Laul J. C., Keays R. R., Ganapathy R., Anders E., and Morgan J. W. (1972b) Chemical fractionations in meteorites—V. Volatile and siderophile elements in achondrites and ocean ridge basalts. *Geochim. Cosmochim. Acta* **36,** 329–345.

Laul J. C., Ganapathy R., Anders E., and Morgan J. W. (1972c) Chemical fractionations in meteorites—VI. Accretion temperatures of H-, LL-, and E-chondrites, from abundances of volatile trace elements. Submitted to *Geochim. Cosmochim. Acta.*

Lovering J. F. and Hughes T. C. (1971) Rhenium and osmium abundance determinations and meteoritic contamination levels in Apollo 11 and Apollo 12 lunar samples. *Proc. Second Lunar Sci. Conf., Geochim. Cosmochim. Acta* Suppl. 2, Vol. 2, pp. 1331–1335. MIT Press.

Millman P. M. (1969) Astronomical information on meteorite orbits. In *Meteorite Research* (editor P. M. Millman), pp. 541–551. Reidel.

Morgan J. W., Laul J. C., Krähenbühl U., Ganapathy R., and Anders E. (1972a) Major impacts on the moon: Characterization from trace elements in Apollo 12 and 14 samples. *Proc. Third Annual Lunar Sci. Conf., Geochim. Cosmochim. Acta* Suppl. 3, Vol. 2. MIT Press.

Morgan J. W., Ganapathy R., Laul J. C., and Anders E. (1972b) Lunar crater Copernicus: Nature of impacting body. Submitted to *Geochim. Cosmochim. Acta.*

Onuma N., Clayton R. N., and Mayeda T. K. (1972) Oxygen isotope cosmothermometer. *Geochim. Cosmochim. Acta* **36,** 169–188.

Reid J. B. Jr., Taylor G. J., Marvin U. B., and Wood J. A. (1972) Luna 16: Relative proportions and petrologic significance of particles in the soil from Mare Fecunditatis. *Earth Planet. Sci. Lett.* **13,** 286–298.

Schonfeld E. (1972) Component abundance and ages in soils and breccia (abstract). In *Lunar Science—III* (editor C. Watkins), p. 683, Lunar Science Institute Contr. No. 88.

Singer S. F. and Bandermann L. W. (1970) Where was the moon formed? *Science* **170,** 438–439.

Swann G. A., Hait M. H., Schaber G. G., Freeman V. L., Ulrich G. E., Wolfe E. W., Reed V. S., and Sutton R. L. (1971) Preliminary description of Apollo 15 sample environments. U.S.G.S. Interagency Report No. 36.

Turekian K. K. and Clark S. P. Jr. (1969) Inhomogeneous accumulation of the earth from the primitive solar nebula. *Earth Planet. Sci. Lett.* **6,** 346–348.

Urey H. C. (1952) Chemical fractionation in the meteorites and the abundance of the elements. *Geochim. Cosmochim. Acta* **2,** 269–282.

Urey H. C. (1954) On the dissipation of gas and volatilized elements from protoplanets. *Astrophys. J.* Suppl. **1** [6], 147–173.

Vinogradov A. P. (1971) Preliminary data on lunar ground brought to earth by automatic probe "Luna-16." *Proc. Second Lunar Sci. Conf., Geochim. Cosmochim. Acta* Suppl. 2, Vol. 1, pp. 1–16. MIT Press.

Wänke H., Baddenhausen H., Balacescu A., Teschke F., Spettel B., Dreibus G., Quijano M., Kruse H., Wlotzka F., and Begemann F. (1972) Multielement analyses of lunar samples. In *Lunar Science—III* (editor C. Watkins), pp. 779–781, Lunar Science Institute Contr. No. 88.

Wasson J. T. (1971) Volatile elements on the Earth and the Moon. *Earth Planet. Sci. Lett.* **11,** 219–225.

Wasson J. T. and Baedecker P. A. (1970) Ga, Ge, In, Ir, and Au in lunar, terrestrial, and meteoritic basalts. *Proc. Apollo 11 Lunar Sci. Conf., Geochim. Cosmochim. Acta* Suppl. 1, Vol. 2, pp. 1741–1750, Pergamon.

Whipple F. L. (1972) On certain aerodynamic processes for asteroids and comets. In *Proceedings of Nobel Symposium 21 "From Plasma to Planet"* (in press). Almqvist and Wiksell.

Wiik H. B. and Ojanperä P. (1970) Chemical analyses of lunar samples 10017, 10072 and 10084. *Science* **167,** 531–532.

Proceedings of the Third Lunar Science Conference
(Supplement 3, *Geochimica et Cosmochimica Acta*)
Vol. 2, pp. 1377–1395
The M.I.T. Press, 1972

# Major impacts on the moon: Characterization from trace elements in Apollo 12 and 14 samples

JOHN W. MORGAN, J. C. LAUL,* URS KRÄHENBÜHL, R. GANAPATHY, and EDWARD ANDERS

Enrico Fermi Institute and Department of Chemistry, University of Chicago, Chicago, Illinois 60637

**Abstract**—Seventeen trace elements (Ag, Au, Bi, Br, Cd, Co, Cs, Ge, In, Ir, Rb, Re, Sb, Se, Te, Tl, and Zn) have been determined by neutron activation analysis in 33 lunar samples from Apollo 14, 5 from Apollo 12, and 2 from Luna 16. Apollo 14 soils and breccias contain at least two, and possibly three, ancient meteoritic components of unusual composition, probably derived from the Imbrian and Serenitatis impacts, and mixed planetesimal debris from the pre-Imbrian regolith. These components have a lower ratio of volatiles (Ag, Ge, Sb, Se) to siderophiles (Ir, Re, Au, Ni) than any known class of chondrites. They also have Ir/Au, Ge/Au ratios outside the range for most iron meteorites, except Groups IVA and possibly IIIA. One of these components, of very low Ir/Au, Re/Au ratio, occurs in light norites, 14321 microbreccias, and KREEP separates from 12033 soil. Another is found in dark norites, glasses, and several other Apollo 14 samples, as well as rock 12013 and Apollo 11 anorthosite. From these compositional clues it appears that the Imbrian body and the pre-Imbrian planetesimals, like the earth, were relatively rich in iron ($\geq 20\%$ Fe), but depleted in volatiles ($< 10\%$ of cosmic abundance). Such a composition is consistent with the Imbrian body originating as an earth-crossing planetesimal. Other arguments (impact velocity, age of Imbrian basin), support this view. An asteroidal origin may be less likely, but still possible.

## INTRODUCTION

THE INFLUX OF METEORITIC MATERIAL can be estimated chemically using those siderophile and volatile elements which are abundant in meteorites, but are depleted in lunar rocks (Ganapathy *et al.*, 1970a). Additionally, the composition of this material can be characterized from the relative abundances of these elements. Thus fractionated meteorites (for example, ordinary chondrites or iron meteorites) may be readily distinguished from more primitive types (e.g. carbonaceous chondrites) by a large depletion of volatile relative to siderophile elements (Laul *et al.*, 1971).

Soils and breccias from Apollo 11 and soils from intercrater areas of the Apollo 12 site contain a meteoritic component predominantly derived from micrometeorites of primitive (C1-chondrite-like) composition, equivalent to $\sim 1.9\%$ C1 material (Ganapathy *et al.*, 1970a, b; Laul *et al.*, 1971). The micrometeorite material seems to be associated chiefly with the finer soil fractions (Ganapathy *et al.*, 1970a). By contrast, three Apollo 12 soils collected from crater rims, and two breccias contain smaller amounts of meteoritic material of a fractionated type, corresponding to 0.9% L-chondrite, 0.6% H-chondrite, or 0.2% Group I iron (Laul *et al.*, 1971). This appears to be associated with the KREEP component in these soils (Morgan *et al.*, 1972a).

* Present address: Radiation Center, Oregon State University, Corvallis, Oregon 97331.

Two coarse-grained anorthosite fractions (+100 mesh and 1–10 mm) from Apollo 11 soil also have their own meteoritic component, again of fractionated composition (Laul *et al.*, 1971). It seems that coarse soil fractions and at least some breccias carry the record of discrete impacts, unobscured by the ubiquitous C1 component. Accordingly we have given preference to such samples, in the hope of characterizing the projectiles involved in major impacts.

In the present work, we have studied 33 Apollo 14 samples: 7 bulk soils (< 1 mm), 12 hand-picked separates from the 1–2 mm fraction of these soils, 11 samples from breccias 14047, 14063, and 14321, and 2 samples of the two putative igneous rocks from the Apollo 14 mission, 14053 and 14310. Eight of these measurements had been previously reported by Ganapathy *et al.* (1972). For the record, we have included data on bulk soil 12037 (< 1 mm) and four KREEP separates from soil 12033 (Morgan *et al.*, 1972a) and also on two Luna 16 soils (Laul *et al.*, 1972a).

## EXPERIMENTAL

### *Samples*

*Bulk soils.* Apollo 14 (< 1 mm) and Luna 16 (< 0.125 mm) soils were analyzed as received. Rocks were coarsely crushed in an agate mortar before analysis.

*KREEP.* Two KREEP fragments (one lithic, one glassy), 10 mg total weight, from the 2–4 mm fraction of 12033,105 were kindly provided by N. J. Hubbard of NASA Manned Spacecraft Center and were combined for analysis. A +30 mesh fraction of < 1 mm soil 12033,20 was cleaned ultrasonically in double-distilled reagent grade acetone, to remove adhering powder. From this, samples of 21 mg glassy KREEP (12033,20,4) and of 52 mg of cindery KREEP (12033,20,7) were selected. The 30–100 mesh fraction was similarly cleaned and a magnetic fraction (35 mg) was extracted with a small hand magnet.

*Apollo 14 soil separates.* Seven soils (1–2 mm) were ultrasonically cleaned in double-distilled acetone. Petrologically distinct fractions were hand-picked by Ursula B. Marvin and John A. Wood of the Smithsonian Astrophysical Observatory, Cambridge.

*Breccia 14321,184.* Selected clasts and a sample of "matrix" (light-colored material from the clast-groundmass interface) were provided by A. R. Duncan and G. G. Goles of the University of Oregon, Eugene. The samples were analyzed by the same group for major, minor, and trace elements by non-destructive neutron activation.

*Breccia 14063,37.* Samples of clasts and groundmass from this rock were supplied by B. R. Doe and M. Tatsumoto of the U.S. Geological Survey, Denver. Nondestructive neutron activation analyses on the samples were carried out by the group led by R. A. Schmitt of Oregon State University, Corvallis.

### *Analytical procedure*

Our original procedure is unpublished but may be obtained from sources given by Anders *et al.* (1971). The separation scheme has been streamlined somewhat, and extended to include Se, Te, and more recently, Re, Ge, and Sb.

## RESULTS

Analytical data for 17 elements in 41 lunar samples are given in Table 1.

The accuracy and precision of our analytical method has been discussed previously (Anders *et al.*, 1971; Laul *et al.*, 1971), though not for the three new elements Re, Sb, and Ge. Since we routinely analyze a sample of U.S.G.S. standard basalt BCR-1 with each irradiation, we can compare the results for these three new elements with previous work by other groups. For five analyses we find a mean of

Table 1. Abundances in Apollo 14, Apollo 12, and Luna 16 samples (ppb; Zn, Rb, Co, ppm).*

| Sample no. | Type | Ir | Re | Au | Sb | Ge | Se | Te | Ag | Br | In | Bi | Zn | Cd | Tl | Rb | Cs | Co |
|---|---|---|---|---|---|---|---|---|---|---|---|---|---|---|---|---|---|---|
| *Soils, < 1 mm* | | | | | | | | | | | | | | | | | | |
| 14003,17 | Contingency | 11.0 | 0.97 | 4.4 | 2.4 | | 310 | 25 | 11.5 | 330 | | 1.3 | 25 | 94 | 19 | 13.5 | 575 | |
| 14163,57 | Bulk | 13.6 | 0.93 | 5.4 | | | 335 | 70 | 16.6 | | *72* | 1.7 | 31 | 140 | 30 | 15.8 | 645 | |
| 14163,57 | Bulk | 11.7 | 1.07 | 5.3 | 5.7 | | 350 | 30 | 18.4 | 490 | | 1.6 | 31 | 139 | 35 | 16.1 | 730 | |
| 14259,20 | Comprehensive | 18.6 | 1.3 | 6.6 | | | 350 | 50 | 26.5 | | *34* | 1.8 | 22 | 83 | 18 | 15.4 | 620 | |
| 14141,32 | Cone Crater | 12.6 | 1.26 | 11.0 | 3.1 | | 270 | ≤20 | 30 | 480 | | 1.7 | 31 | 461 | 22 | 18.3 | 790 | |
| 14148,25 | Surface trench | 13.7 | 1.34 | 6.9 | 2.8 | | 340 | 25 | 12.6 | 360 | | 1.8 | 22 | 111 | 19 | 16.8 | 695 | |
| 14156,25 | Middle trench | 12.7 | 1.11 | 5.3 | 3.6 | | 280 | 25 | 11.7 | 290 | | 1.3 | 20 | 77 | 19 | 13.5 | 570 | |
| 14149,41 | Bottom trench | 11.1 | 1.05 | 7.5 | 2.4 | | 290 | 15 | 11.8 | 430 | | 2.2 | 19 | 199 | 26 | 13.9 | 605 | |
| *Fractions 1–2 mm* | | | | | | | | | | | | | | | | | | |
| 14002,3,33 | Magnetic, contingency | 13.3 | 1.03 | 8.5 | 2.0 | | 357 | 55 | 10.6 | 470 | *80* | 1.10 | 22 | 143 | 39 | 16.5 | 670 | |
| 14258,36, | Magnetic, comprehensive | 13.5 | 1.10 | 5.3 | 1.8 | | 285 | 20 | 8.5 | 330 | *24* | 0.95 | 19 | 77 | 16 | 13.6 | 580 | |
| 14146,2,3 | Magnetic, surface of trench | 17.7 | 1.47 | 7.6 | 2.2 | | 351 | 30 | 12.6 | 450 | *40* | 1.3 | 28 | 114 | 21 | 16.0 | 670 | |
| 14154,2,3 | Magnetic, middle of trench | 11.9 | 0.99 | 10.0 | 2.4 | | 273 | 110 | 10 | 310 | *26* | 1.5 | 23 | 78 | 20 | 13.2 | 580 | |
| 14154,2,7 | Dark norite, middle of trench | 8.9 | 0.92 | 4.8 | 2.2 | | 130 | ≤10 | 1.4 | 270 | *12* | 0.63 | 2.9 | 48 | 33 | 16.5 | 735 | |
| 14151,12,7 | Dark norite, bottom of trench | 6.6 | 0.63 | 3.1 | 2.9 | | 130 | 30 | 0.94 | 220 | *20* | 0.63 | 2.4 | 36 | 30 | 16.4 | 710 | |
| 14146,2,9 | Light norite, surface of trench | 4.9 | 0.43 | 5.4 | 1.7 | | 91 | ≤70 | 1.2 | 150 | *10.6* | ≤0.7 | 1.8 | 14 | 9.9 | 15.2 | 620 | |
| 14142,1,12 | Light norite, Cone Crater | 4.8 | 0.47 | 4.1 | 1.2 | | 110 | 100 | 0.85 | 100 | | 0.48 | 2.4 | 27 | 10 | 20.0 | 980 | |
| 14151,12,12 | Light norite, bottom of trench | 13.7 | 1.09 | 4.9 | 0.99 | 152 | 97 | ≤3 | 0.65 | 287 | *12.7* | 0.36 | 1.6 | 11 | 8.8 | 12.8 | 511 | 28 |
| 14258,36,5 | Ropy glass, comprehensive | 13.1 | 0.91 | 5.1 | 0.87 | | 121 | 30 | 2.5 | 70 | 5.9 | 0.49 | 5.6 | 20 | 5.2 | 4.1 | 185 | |
| Yellow glass | From: 14002, 14258, 14154, 14151 | 3.6 | 0.32 | 1.2 | 0.63 | | 88 | ≤70 | 0.74 | 190 | 3.6 | 0.72 | 2.8 | 16 | 3.0 | 25.1 | 935 | |
| Green glass | From: 14002, 14258, 14146, 14154, 14262 | 7.6 | 0.53 | 2.0 | 0.53 | | 62 | ≤3 | 0.78 | 100 | 7.8 | 0.26 | 1.5 | 16 | 3.5 | 7.1 | 250 | |
| *Rocks* | | | | | | | | | | | | | | | | | | |
| 14047,38 | Glass coating | 11.7 | 1.12 | 5.2 | 2.0 | | 315 | 35 | 10.3 | 360 | *27* | 1.1 | 23 | 78 | 16 | 16.2 | 670 | |
| 14047,25 | Fragment of rock | 11.2 | 1.06 | 5.4 | 2.1 | | 320 | 85 | 11.0 | 370 | *50* | 1.1 | 20 | 102 | 17 | 14.6 | 650 | |
| 14053,26 | Basalt | 0.017 | 0.0066 | 0.11 | 0.64 | | 141 | 15 | 0.60 | 50 | *15* | 0.29 | 2.1 | 20 | 1.4 | 2.1 | 90 | |
| 14063,37,A10 | Fine matrix | 0.71 | 0.26 | 0.22 | 1.4 | 23 | 8 | ≤3 | 0.80 | 49 | 2.8 | 0.17 | 5.9 | 4.5 | 4.0 | | 238 | 19 |
| 14063,37,A11 | Coarse matrix | 1.37 | 0.064 | 0.28 | 1.3 | 36 | 8 | ≤3 | 0.87 | 52 | 3.3 | 0.28 | 5.3 | 18 | 5.6 | | 309 | 18 |
| 14063,37,A12 | Dark clasts | 1.82 | 0.099 | 0.28 | 0.46 | 42 | 31 | ≤3 | 2.3 | 840 | 3.1 | 0.30 | 7.2 | 2.3 | 1.8 | | 178 | 26 |
| 14310,119 | "Basalt" | 10.5 | 1.02 | 4.31 | 4.5 | 130 | 120 | 4 | 10780 | 235 | 130 | 2.5 | 2.3 | 2.6 | 10 | 11.8 | 540 | |
| 14321,184,1B | Basaltic clast | 0.044 | 0.0051 | 0.30 | 0.78 | 640 | 338 | 6 | 0.60 | 85 | 1.84 | 0.39 | 2.9 | 24 | 1.7 | 2.7 | 170 | |
| 14321,184,9A | Matrix | 0.71 | 0.056 | 0.70 | 15.3 | 240 | 162 | | 1.10 | 85 | 2.74 | 0.55 | 2.8 | 7.3 | 1.7 | 3.6 | 230 | |
| 14321,184,14A | Microbreccia | 7.8 | 0.70 | 6.06 | 2.2 | (50) | 128 | 11 | 0.88 | 300 | 3.40 | 0.34 | 6.6 | 298 | 8.1 | 12.9 | 650 | |
| 14321,184,15 | Dark clast (microbreccia) | 6.9 | 0.64 | 8.08 | 2.1 | (500) | 139 | 8 | 1.49 | 520 | 1.69 | 0.32 | 3.8 | 17 | 3.9 | 30.8 | 1150 | |
| 14321,184,16A | Microbreccia | 9.7 | 0.86 | 9.96 | 2.9 | 760 | 110 | | 2.63 | 270 | 1.77 | 0.25 | 2.5 | 106 | 1.8 | 10.9 | 640 | |
| 14321,184,19A | Microbreccia | 6.1 | 0.55 | 6.41 | 2.4 | (400) | 92 | 6 | 0.83 | 215 | 1.45 | 0.24 | 3.3 | 52 | 1.9 | 9.3 | 510 | |
| *Apollo 12* | | | | | | | | | | | | | | | | | | |
| 12033,105 | 2 KREEP fragments (2–4 mm) | 1.6 | 0.097 | 0.99 | 0.43 | | 56 | | 3.3 | 70 | 1.25 | *4.6* | 1.1 | 8.0 | 6.0 | 19.4 | 710 | |
| 12033,20,4 | Glassy KREEP (+30 mesh) | 1.5 | 0.15 | 1.5 | 0.79 | | 84 | | 2.2 | 160 | 1.83 | 0.70 | 1.6 | 3.8 | 4.1 | 17.0 | 655 | |
| 12033,20,7 | Cindery KREEP (+30 mesh) | 2.2 | 0.19 | 2.1 | 1.5 | | 99 | | 2.0 | 140 | 1.21 | 0.4 | 2.3 | 7.7 | 3.3 | 17.0 | 625 | |
| 12033,20,1 | Magnetic (30–100 mesh) | 6.7 | 0.57 | 5.4 | 3.1 | | 172 | | 2.5 | 130 | 1.31 | 0.4 | 3.2 | 13 | 3.3 | 14.9 | 535 | |
| 12037,25 | < 1 mm soil | 4.2 | 0.25 | 2.2 | 0.77 | | 166 | | 25 | 90 | *222* | 0.70 | 6.1 | 56 | 2.6 | 4.9 | 210 | |
| *Luna 16* | | | | | | | | | | | | | | | | | | |
| L-16-A-5 | < 0.125 mm soil | 9.6 | 3.7 | 2.7 | 3.3 | | 340 | 18 | *1340* | 120 | 3.0 | 4.6 | 24 | 200 | 1.4 | 1.5 | 68 | |
| L-16-G-5 | < 0.125 mm soil | 9.8 | 3.4 | 2.9 | 4.2 | 1300 | 370 | 20 | *540* | 150 | 1.9 | 5.1 | 29 | 10900 | 1.9 | 1.9 | 82 | |
| *Apollo 11* | | | | | | | | | | | | | | | | | | |
| 10084,49 | < 1 mm soil | 8.7 | 0.71 | 2.4 | 2.2 | | 270 | 17 | 9.0 | 85 | — | 1.5 | 19 | 37 | 1.7 | 2.7 | 106 | |
| BCR-1 | Columbia River basalt | < 0.009 | 0.78 | 0.75 | 530 | 1370 | 92 | 4 | 26 | 60 | 88 | 46 | 116 | 129 | 300 | 46 | 925 | |

* Italicized values presumably reflect contamination from Ir–Ag vacuum gaskets in ALSRC or LRL. Parenthesized Ge values are less accurate; see text.

0.78 ± 0.11 ppb Re, which is in good agreement with earlier determinations averaging 0.84 ppb (Morgan and Lovering, 1967; Lovering and Hughes, 1971). Previous values for Sb of 500 ppb (Tanner and Ehmann, 1967) and 580 ppb (Brunfelt and Steinnes, 1969) compare well with our value of 530 ± 45 ppb (means of 3 determinations). We find 1370 ± 160 ppb Ge (3 determinations) in BCR-1, in reasonable accord with other values of 1700 ppb (Tandon and Wasson, 1968) and 1550 ppb (Baedecker *et al.*, 1971). Analytical difficulties were experienced with some of our earlier Ge determinations, and in several instances radiochemical yields were very low; considerably less than 1%. These less reliable results are shown in parentheses in Table 1. In later runs, yields tended to fall between 5 and 35%.

## DISCUSSION

### *Meteoritic material at Fra Mauro*

Figure 1 shows the abundances of four siderophile elements in bulk soils (< 1 mm), for Apollo 14 and four other lunar landing sites. Crater-rim soils are marked by vertical ticks. Data for igneous rocks (open symbols) are also shown. Abundances have been normalized to C1 chondrites to provide a convenient frame of reference.

As at previous sites, soils are greatly enriched in these elements relative to rocks,

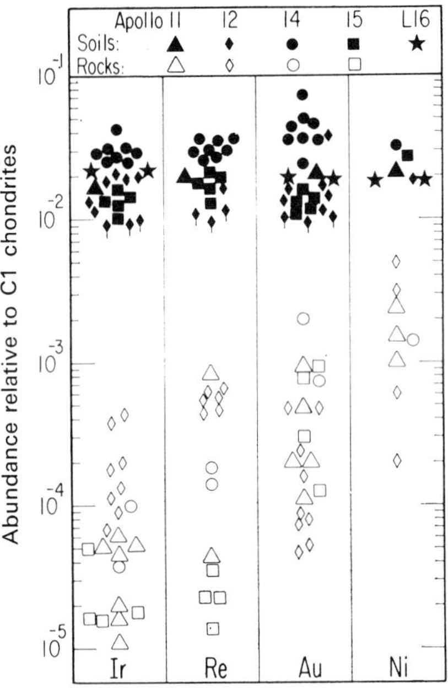


Fig. 1. The abundances (normalized to C1 chondrites) of 4 siderophile elements are higher in Apollo 14 soils than in those from other lunar landing sites. Crater rim soils from Oceanus Procellarum, marked with vertical ticks, have lower than average abundances. Abnormally high points for Re in Luna 16 soils have been omitted. All soils (solid symbols) are substantially enriched in these elements over the abundances in crystalline igneous rocks (open symbols). The Apollo 14 igneous rocks plotted are 14053 and basaltic clast 14321,184,1B. Rock 14310 is apparently an impact melt, contaminated with meteoritic material, and has not been included.

apparently due to addition of meteoritic material. At Apollo 14, however, the amounts are considerably larger: 3–5% C1 equivalent compared to 1.5–2% at the other sites. We cannot immediately conclude that the siderophiles in Apollo 14 soils are largely meteoritic, rather than indigenous, because the only igneous rocks from that site are (alkali-poor) highland basalts, whereas the soils consist largely of (alkali-rich) norites or KREEP. The norite fractions from 1–2 mm soils and the 14321 microbreccia-clasts are almost as rich in siderophiles as the bulk soils (Fig. 2), and it is possible that these elements are indigenous. Direct proof is not available, because no pristine norite rock has yet been found on the moon.* Nonetheless, three lines of evidence suggest that norite has a low intrinsic siderophile element content, which is not much higher than that of other lunar rocks in Fig. 1. First, the siderophile content of our Apollo 12 KREEP separates is quite variable while abundances of alkali metals are almost constant. A regression analysis indicates that the indigenous siderophile content is low, less than $\sim 10^{-3}$ C1 equivalent (Morgan *et al.*, 1972a). Second, most samples of rock 12013, from both the granitic and KREEP portions, show low siderophile contents, equivalent to $1\text{–}3 \times 10^{-3}$ C1 material (Laul *et al.*, 1970). Third, it seems unlikely that a differentiation process extreme enough to enrich U, lanthanides, and

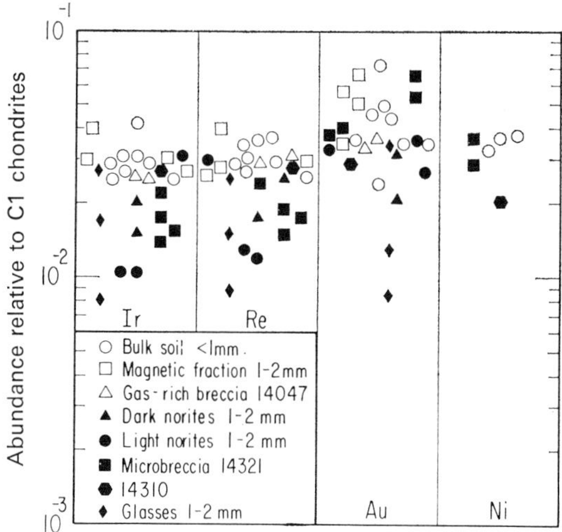


Fig. 2. The abundances of siderophile elements in many Apollo 14 soil separates are similar to those of the bulk soils themselves. The data have been normalized to C1 chondrites to facilitate inter-element comparisons. Ni values taken from Watkins (1972).

* Rock 14310 was first thought to be such a pristine norite, but petrologic data suggest that it is an impact melt (Dence *et al.*, 1972; Ringwood *et al.*, 1972; Hollister *et al.*, 1972). Trace element data are consistent with this interpretation, because 14310 is as rich in most meteoritic elements as the soils (Morgan *et al.*, 1972b; Baedecker *et al.*, 1972; Helmke and Haskin, 1972). Breccia 14065 was reported to have low Ni but high K, Rb (LSPET, 1971), suggesting that it might be a meteorite-free norite. Our analysis of 14063 (parent rock of 14065) shows it to be low in both siderophiles and alkalis, however, (Table 1) and this rock seems to represent a new, spinel-rich type (Christophe-Michel-Lévy *et al.*, 1972), with a small, ill-defined meteoritic component.

other lithophiles more than 100-fold over cosmic levels should have retained a sizeable complement of unfractionated siderophiles. Various Ca-rich achondrites and terrestrial rocks are typically depleted in siderophiles to $10^{-3}$–$10^{-4}$ of C1 abundance (Laul *et al.*, 1972b).

It seems highly likely, therefore, that the siderophiles in Apollo 14 soils are more than 95% meteoritic. Since the indigenous correction is apparently small (<5%) and hard to estimate precisely, we shall neglect it altogether for siderophile elements. For volatile elements, the indigenous correction is also uncertain, but probably larger. Accordingly we shall use the gross abundances of volatiles as upper limits of the meteoritic component. Like all upper limits, they are interesting only if unexpectedly low.

## *The micrometeorite component*

It appears that two distinct meteoritic components are present at the Apollo 14 site. One is related to surface exposure and probably consists of micrometeorites. The other, unrelated to surface exposure, seems to consist of projectile material from one or more major impacts. This difference is brought out most clearly in Fig. 3. In order to permit comparison of samples with different absolute contents of meteoritic material, we have taken the abundances normalized to C1 chondrites and normalized these a second time to Au, a siderophile element of negligible indigenous abundance. The first two elements in Fig. 3 are siderophile; the next five are volatile, arranged roughly

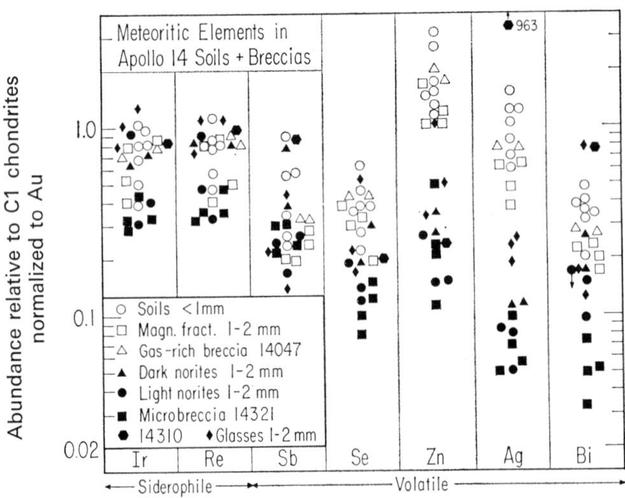


Fig. 3. Separated 1–2 mm soil fragments (solid symbols) show little evidence of micrometeorite contamination of primitive (C1-chondrite-like) composition. In contrast, samples of significant surface exposure (open symbols) are relatively enriched in volatile elements. In order to compare meteoritic elements in samples containing differing absolute amounts, abundances have been doubly normalized, to C1 chondrites and then internally to Au. It appears that the soil fragments contain a relatively uncontaminated ancient component.

in order of increasing volatility. Samples of significant surface exposure, as indicated by solar-wind gases, are shown by open symbols; all others, by filled symbols.

The samples clearly separate into two groups. Those of high surface exposure (fine soils, soil breccias, and magnetic fractions consisting of glass-bonded agglutinates) are markedly higher in volatiles, especially Zn and Ag. Presumably this reflects the presence of micrometeorites of primitive (C1 chondrite-like) composition, acquired with solar-wind gases during surface exposure. For our present purposes, micrometeorite material represents an undesirable contaminant, which obscures the record of major impacts. We shall therefore focus our attention on samples of low surface exposure (filled symbols in Figs. 2 and 3). These samples apparently contain a meteoritic component of their own, manifested by their high content of siderophiles (Fig. 2).

*Ancient meteoritic component(s)*

We can at once rule out the (trivial) possibility that the high siderophile meteoritic component observed at the Apollo 14 site is derived from the local craters. Four of our samples (14141, 14142, 14321, and 14063) were collected on the rim of Cone Crater; all others were collected at least 5.3 crater radii away, outside the ejecta blanket of any of the local craters (LSPET, 1971). The bulk soil and the trench samples were collected near Triplet North Crater, about 2.7 and 1.5 crater radii away; the comprehensive soil was taken about 3.8 crater radii from Doublet South Crater. Our work on Apollo 12 and 15 shows that at these distances the local component is no longer prominent, being swamped by micrometeorites and ray material from distant, large craters. Moreover, nuclear-track and noble-gas measurements show that Cone Crater formed about 24 m.y. ago (Burnett *et al.*, 1972). Only two of our soil samples, Cone Crater and trench-bottom, consist largely of material of such young exposure age. Others have higher ages (400–500 m.y.), suggesting that they come mainly from earlier impacts. It seems unlikely that each of these impacts at the Apollo 14 site should have deposited much larger amounts of meteoritic material than analogous impacts at the Apollo 11, 12, 15, and Luna 16 sites. It is more likely that this meteoritic material is common to the Fra Mauro formation at the Apollo 14 site, and possibly elsewhere. In this case, it must date back either to the impact that made the Imbrium Basin, or to an earlier generation of impacts recorded in the pre-Imbrian regolith. Let us then attempt to characterize this ancient meteoritic material.

*Chemical characterization of ancient meteoritic component(s)*

*Number of components.* Ganapathy *et al.* (1972) had noticed some compositional variation among the seven Apollo 14 samples analyzed by them but put special emphasis on a light norite of extreme composition, on the assumption that it represented the ancient component in purest form. Now that data on 25 additional samples are available, it appears that at least two types of ancient meteoritic material are present which differ mainly in their Ir/Au and Re/Au ratio. Among the samples of low surface exposure (filled symbols in Fig. 3), light norites and 14321 microbreccias have low ratios, both Ir/Au and Re/Au being around 0.3–0.4 when normalized to C1 chondrites, whereas dark norites, glasses, impact melt 14310, and a peculiar light norite

(14151) have higher ratios, 0.6–1.1, when similarly normalized. We shall call these groups LN and DN, for "light norite" and "dark norite". It is possible that the LN group is composite. Two samples from Cone Crater (14141 $<1$ mm soil and 14142 light norite) and 14321,184,14A microbreccia have slightly higher ratios than the other LN samples and form a tight little grouping at Ir/Au $\approx$ Re/Au $\approx$ 0.4. We shall use the symbols LN1 and LN2 for the lower and upper of these subgroups.

The meteoritic component in matrix 14321,184,9A matches the LN1 subgroup in siderophile element ratios, but it is much less abundant than in the microbreccias. The matrix is largely composed of fragments of basaltic and microbreccia clasts with the former predominating (Duncan *et al.*, 1972; Grieve *et al.*, 1972). The siderophile elements in our sample of 14321 matrix could be accounted for by 9% of microbreccia 14321,184,15 diluted with basalt similar to clast 14321,184,1B, although such a mixture would provide too little Sb and too much Se and Cd. Another sample of matrix (14321,184,32) analyzed by Baedecker *et al.* (1972) appears to be much richer in siderophile elements, and may be predominantly composed of microbreccia.

Interestingly, the division into the LN and DN groups is recognizable even among samples of high surface exposure (open symbols), although their ratios must have been shifted upward by addition of C1-like material with Ir/Au, Re/Au = 1.

It is remotely possible that these groupings are spurious, being caused by a high indigenous Au content of light norites and 14321 microbreccias. In order to give the observed low and constant ratios, however, the "indigenous" Au would have to correlate quite precisely with the "meteoritic" Ir, Re. In addition the same two groups have been seen in Apollo 12 material. Four KREEP separates from 12033 soil have Ir/Au, Re/Au that resemble those of the LN group (Morgan *et al.*, 1972a), whereas a siderophile-rich, noritic sample from breccia 12013 (Laul *et al.*, 1970) has Ir/Au that falls into the DN group.

Both LN and DN components show a *fractionated* abundance pattern, being depleted in volatiles relative to siderophiles (Fig. 3). We must bear in mind, of course, that the volatile abundances are only upper limits, because they contain an unknown and occasionally large indigenous contribution. It is therefore the *low* values which are important. High values of Ag and Bi in 14310 may reflect the presence of the peculiar, Ag-, Bi-rich material first seen in core 12028 and subsequently elsewhere (Ganapathy *et al.*, 1970b; Laul *et al.*, 1971, 1972a; Morgan *et al.*, 1972c). The general tendency for glasses to lie above crystalline samples may indicate admixture of soil or projectile material in the glass-forming events.

The breccia 14063 does not fit easily into this discussion. It is of a rare type at the Apollo 14 site, F3, which comprises only about 5% of the rocks weighing > 1 gram examined by Jackson and Wilshire (1972). Unlike most other Apollo 14 samples, 14063 is, on the whole, rather alkali-poor (Laul *et al.*, 1972). (An atypical matrix sample, 14063,37,A12, is quite alkali-rich, and was analyzed by us for this very reason). The breccia is very poor in meteoritic elements, containing only $1 - 2 \times 10^{-3}$ Cl equivalent estimated from Au, whereas other Apollo 14 breccias appear to contain at least 10 times this amount (LSPET, 1971; Baedecker *et al.*, 1972; Wänke *et al.*, 1972). Mössbauer studies indicate that 14063 is apparently devoid of metallic Fe, unlike other Apollo 14 breccias investigated by this method (Schwerer *et al.*, 1972). The meteoritic component shows rather variable amounts of Ir and particularly Re. In general, however, the Ir/Au and Re/Au ratios (normalized to Cl chondrites) tend to be higher than in the DN group. The high Re content, relative to Au, in fine matrix 14063,37,A10 may possibly be due to contamination (Re mass spectrometer filaments?). Preliminary results indicate that alkali-poor material with a low meteoritic component also may be present at the Apollo 15 site, and it seems appropriate to defer a full discussion of 14063 until more data are available for similar material.

*Comparison with meteorites.* It is interesting to see whether these ancient meteoritic components resemble any known meteorite classes. The high abundance of siderophiles points to iron-rich types: chondrites, stony irons, and irons. Figure 4 compares the lunar samples with the major chondrite classes. (LL-chondrites have been omitted; they fall slightly above the L-chondrites.) Let us again emphasize that the lunar values are upper limits, especially for the volatile elements. Thus a value *above* the meteoritic level is merely a possible match. A value *below* the meteoritic level, on the other hand, is a definite mismatch.

Clearly, the ancient lunar components do not match any of the known chondrite classes. They are too low in Sb, Se, and Ag; the LN component also in Ir and Re. These deficiencies can hardly be due to volatilization during impact or metamorphism. Chondrites have been metamorphosed at temperatures similar to those experienced by the Apollo 14 breccias, 600–1000°C (Onuma *et al.*, 1972; Warner, 1972; Anderson *et al.*, 1972), but for much longer times: $\sim 10^7$ yr versus $\sim 10^2$ yr. Chondrites show no losses of Sb, Se, and Ag attributable to metamorphism (Keays *et al.*, 1971; Laul *et al.*, 1972c), however, and the heating of the breccias by impact can scarcely have been more intense (certainly not enough to melt the samples), or as sustained.

Let us now extend our comparison to iron meteorites. The elements most useful for such a comparison are Ir, Au, and Ge; all are siderophile, but of increasing volatility. Wasson and coworkers have developed a classification scheme of iron meteorites based on Ir and Ge (see Baedecker, 1971, for a review). We have remolded their scheme by normalization to Cl chondrite abundances and Au, so that lunar samples, chondrites, and iron meteorites may be compared on the same diagram.

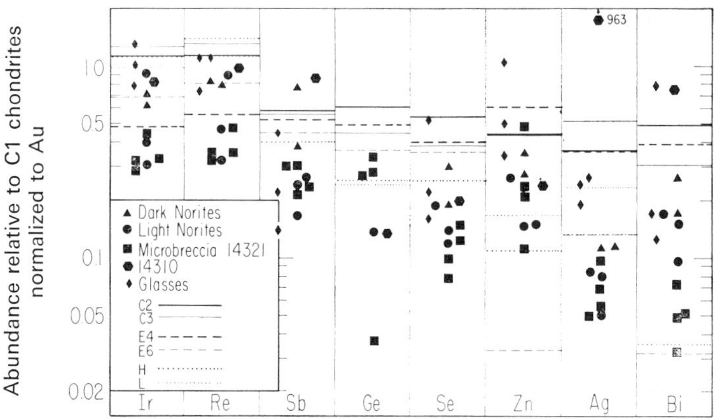


Fig. 4. The meteoritic component in Apollo 14 samples does not resemble any major chondrite class. (LL-chondrites have been omitted; they fall somewhat above L-chondrites in volatile elements.) Abundances are doubly normalized to C1 chondrites and to Au. The lunar values for volatile elements are upper limits only; those which are lower than corresponding meteoritic values are the most significant. In particular, Sb, Se, and Ag are lower than *any* chondrite class.

(Gold is especially suitable for normalization. It does not readily separate from Fe in cosmochemical fractionations, as shown by the fact that Au/Fe ratios are constant within a factor of 2 in most meteorite classes. Thus Au provides a reliable link between the abundances of trace and major elements in meteorites.)

The results are shown in Fig. 5. The six lunar samples appear to fall into two, or possibly three, groupings. Three 14321 microbreccias lie very closely together, just above the IIIA irons. (Two other microbreccias from this rock, analyzed by Baedecker *et al.*, 1972, have similar Ge and Ir contents.) The Ir is probably largely meteoritic, while the Ge may contain a substantial indigenous contribution, and accordingly we look for meteorite groups directly *below* the lunar points. For this group of microbreccias, IIIA or IVA irons are *possible* matches. All others are excluded.

A second grouping, consisting of 14310 and 14151 light norite, lies just below the IIIA irons. By the upper limit criterion, this group is comparable only to IVA irons, or perhaps to some unsampled meteorite group of intermediate composition.

The final sample, which appears to have a very low Ge content, is 14321,184,14A microbreccia. Its Ge value was parenthesized in Table 1, because the chemical yield was exceptionally low. If this analysis is correct, it may indicate the presence of a third grouping. Again, of the known iron meteorite groups IVA is the only possible match.

It is unfortunate that more Ge data are not available on our Apollo 14 separates,

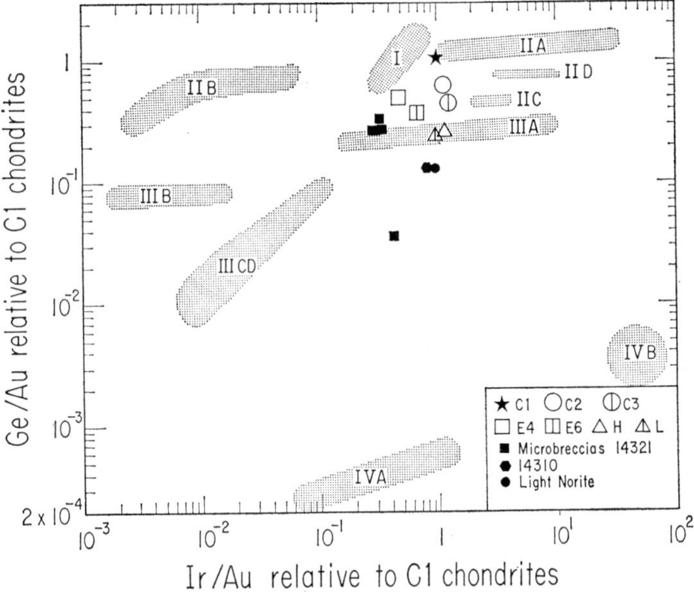


Fig. 5. Iron meteorites form distinct groupings on a Ge versus Ir plot. By normalizing each element to Au, we are able to compare patterns of iron meteorites, chondrites, and lunar samples. Iridium and gold are largely meteoritic; however, some Ge may be indigenous. Accordingly, the iron meteorite groups which lie below the lunar sample points must be considered. There appear to be 2 or possible 3 groups of samples, perhaps related to IIIA and IVA irons.

to verify the apparent Ge groups. It is interesting nevertheless that the Ge contents correlate with our classification based on Ir/Au and Re/Au ratios. The three microbreccias of high Ge content all fall in the LN1 group, at Ir/Au $\approx$ 0.3. The microbreccia of very low Ge content falls in the LN2 subgroup (Ir/Au $\approx$ 0.4). Finally, rock 14310 and norite 14151,12,12, of intermediate Ge content, fall in the DN group (Ir/Au $\approx$ 1.0).

*Spatial distribution of ancient meteoritic components*

*Apollo 12.* It appears that the ancient meteoritic components are widely distributed on the lunar surface. As mentioned above, the four KREEP fractions from 12033 soil (Table 1) have the same, peculiarly low Ir/Au, Re/Au ratios as the LN component from Apollo 14 (Morgan *et al.*, 1972a), while the most noritic and siderophile-rich portion of breccia 12013 has a higher ratio, in the range of the DN component (Laul *et al.*, 1970). These samples are thought to be related to a light-colored Copernican ray which lies across the Apollo 12 site (LSPET, 1970; Adams and McCord, 1971; Dollfus *et al.*, 1971). All show evidence of a thermal disturbance about 0.85 AE ago, which apparently represents the date of the Copernican impact (Tatsumoto, 1970; Silver, 1971; Eberhardt *et al.*, 1972).

Since Copernicus was excavated in the Imbrian ejecta blanket, the unusually low Ir/Au, Re/Au ratios may be typical of large areas of the Fra Mauro formation. The similarity in siderophile element patterns between Apollo 12 KREEP on the one hand, and the two LN groups on the other, extends in a general way to the volatile elements (Fig. 6; see also Morgan *et al.*, 1972a). In all three cases we see Sb depleted relative to Au to about the same extent as Ir and Re ($\sim$0.3–0.5 Au) and the other volatiles (Se, Zn, Ag, Bi) much more strongly depleted ($\leq$0.2 Au). It is curious that no conspicuous chemical difference shows up that could be attributed to the Copernican projectile.

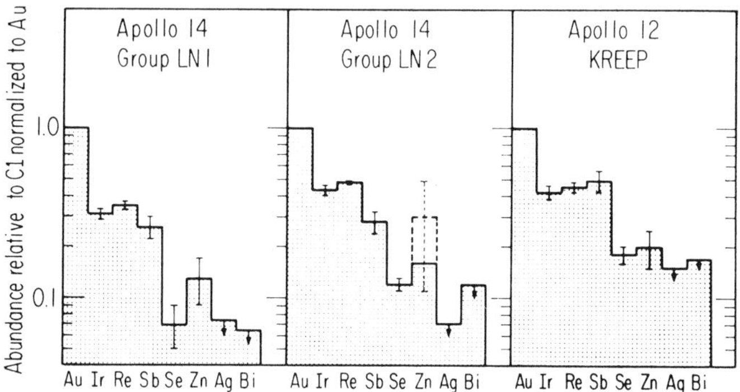


Fig. 6. Meteoritic patterns in Apollo 14 LN1 and LN2 groups are very similar to those of Apollo 12 KREEP samples. In all 3 cases, when normalized to C1 chondrites, the refractory siderophile elements Re and Ir are depleted (0.3–0.5 Au), and Sb is depleted by about the same factor. The volatile elements Se, Zn, Ag, and Bi are more strongly depleted ($\leq$0.2 Au). In the LN2 group, the unusually high Zn value for 14321,184,14A has been omitted from the mean. The average, if this value is included, is shown as a dotted line.

Only a small enhancement of volatiles in Apollo 12 KREEP is evident in Fig. 6 and this may simply represent slight contamination with soil or indigenous volatiles magnified disproportionately by the low absolute Au abundance. In this case, one must assume that the Copernican projectile was chemically inconspicuous (achondrite?), or underrepresented in the ejecta at the Apollo 12 site. On the other hand, the slight enhancement of volatiles may be due to the Copernican projectile. If the Apollo 12 KREEP samples are assumed to be representative of the entire ejecta from the impact, then a high-velocity projectile (~30–40 km/sec) of primitive composition (~C1) is indicated (Morgan *et al.*, 1972a), which may have been a cometary nucleus.

*Apollo 11.* Material resembling the DN component has also been found at the Apollo 11 site. Two anorthosite fractions hand-picked from 10084 or 10085 soil contain Ir, Au, and Ni at levels corresponding to ~0.9% C1 equivalent (Laul *et al.*, 1971), and, judging from the composition of 15415 anorthosite, the indigenous correction should be small (Morgan *et al.*, 1972c). When normalized to Au, the abundance pattern of Apollo 11 anorthosites resembles that of the DN component, in that both have Ir/Au ratios in the chondritic range, but are markedly lower in Ag and Se.

*Origin of ancient meteoritic components*

Three major sources must be considered for the ancient meteoritic components:

(1) Imbrian projectile debris, broadcast directly from the basin to Fra Mauro, Copernicus, and the highlands near Tranquility Base.

(2) Serenitatis projectile debris, deposited on the pre-Imbrian surface and then ejected by the Imbrian impact, as in (1).

(3) Mixed planetesimal debris from the pre-Imbrian regolith.

We have not yet found any decisive clues that permit assignment of each component to a particular source. Only some very tentative assignments can be made at present. Let us first compare the frequencies of these components at each of the three sites.

| Component | Apollo 11 | Apollo 12/ Copernicus | Apollo 14 |
|---|---|---|---|
| LN | | 4 | 6 (2) |
| DN | 2 | 1 | 11 (5) |
| Host rock | Anorthosite | Norite | Norite |

Parenthesized numbers refer to bulk soils; all others to separated fractions.

The most widespread component is DN, found at three sites and in two different host rocks. On this basis one might suspect it to be mixed planetesimal debris from the pre-Imbrian regolith. But this is only a working hypothesis, not a firm conclusion. The highlands south and west of Mare Tranquilitatis, the probable source of Apollo 11 anorthosite, are still within range of Imbrian and Serenitatis ejecta.

On the basis of woefully inadequate statistics, the LN component seems to become less common with increasing distance from the Imbrium Basin. It is present at the Apollo 12 site chiefly in material that seems to represent Imbrian ejecta originally deposited at Copernicus. If the usual rule of thumb holds, the material at Copernicus comes from greater depth in the Imbrium Basin than the material at the Apollo 14 or 11 sites, and thus is less apt to be contaminated by Serenitatis ejecta or pre-Imbrian regolith material. Again, this is a most tentative working hypothesis, subject to instant burial if evidence to the contrary is found.

We can test the elasticity of our working hypothesis by making the reverse assumption: that LN is planetesimal debris, and that DN is Imbrian projectile debris. The DN component in Apollo 11 anorthosite must then be regarded as Imbrian projectile debris, which implies anorthositic composition for some of the pre-Imbrian surface. This is not unreasonable, but it is surprising that LN, regarded as mixed planetesimal debris under the revised hypothesis, should occur mainly at Copernicus, the site closest to the basin. Thus the assignment of LN to Imbrium and DN to the regolith seems more plausible by a slight margin than the reverse assumption.

In the ensuing discussion we will tentatively adopt the working hypothesis that the LN component represents material from the Imbrian body. This assignment may well prove to be incorrect; nevertheless the discussion itself may still be of value. The alternative sources for the bulk of the ancient meteoritic component in Apollo 14 soils (the Serenitatis object and other pre-Imbrian planetesimals) were probably themselves members of an early population of basin-forming planetesimals of which the Imbrian body was a late survivor (Cliff *et al.*, 1972; Dence *et al.*, 1972; Wilshire and Jackson, 1972). Thus, our conclusions concerning the origin of the Imbrian body may still be generally applicable to the source of the Apollo 14 meteoritic material, even though our present working hypothesis should be in error.

### *Some properties of the Imbrian body*

For most of the following discussion it does not matter whether the planetesimal is identified with the LN or DN components. Both are unusually abundant in the soil (judging from the high siderophile content; Fig. 1) and both are strongly depleted in volatiles (Figs. 3 and 4). Let us assume, for the sake of definiteness, that it had LN composition, and try to infer some of its properties from the information available.

*Impact velocity.* The high siderophile content of Apollo 14 samples may be a significant clue. From cratering theory (Öpik, 1961) the impact velocity $w$ is related to the ratio of eroded mass, $M$, to projectile mass $\mu$, by

$$M/\mu = kw/(\rho/s)^{1/2}$$

where $s$ is the crushing strength of the target rock (assumed to be $9 \times 10^8$ dynes/cm$^2$), $\rho$ is its density, and $k$ is a dimensionless number varying slowly with velocity, which can be graphically evaluated from data given by Öpik.

If the *mean* value of $M/\mu$ is known for a given impact, the projectile velocity $w$ can be calculated; however, projectile and target probably do not mix uniformly, and extensive sampling would be required to obtain the true $M/\mu$. When only a few samples are available, the results will be fairly reliable if these samples are reasonably representative of the mean. This is usually not a bad assumption when a few members of a large population are drawn at random.

Before we can apply the above equation, we need to deduce the ratio $M/\mu$ from the trace element data. A convenient link between trace and major elements is the Au/Fe ratio. Under equilibrium conditions, these two elements condense together from the solar nebula (Larimer, 1967; Anders, 1971), and do not separate from each other as long as most of the iron remains in metallic form. In metal-rich meteorites (irons, chondrites) the Au/Fe ratio remains within a factor of 2 of the cosmic ratio.

Accordingly, we shall assume that the Au/Fe (mass) ratio in the planetesimal was identical to the cosmic ratio. Using subscripts $E$ for ejecta, $I$ for the Imbrian body, and $C$ for cosmic, we can write:

$$\mu/(M + \mu) \approx \mu/M \approx (Au_E/Fe_I)(Fe_C/Au_C).$$

We can put $(Au/Fe)_C = 7.9 \times 10^{-7}$ and $Au_E = 6.7$ ppb (mean of LN samples). We do not know $Fe_I$, but can consider four possibilities running the gamut of likely compositions: "lunar," "terrestrial," cosmic, and IVA iron. The results are given in Table 2, along with masses and diameters estimated from Öpik's (1969) formulas.

The results obviously are sensitive to $\mu/M$. Had we used the Au contents of Apollo 12 KREEP (average, 2.5 ppb), the velocities would have been higher by a factor of ~3, and the masses, lower by the same factor. The Apollo 12 values may be less representative, however, since the Imbrian ejecta blanket in the Copernicus region is only 1.5 km deep (Short and Foreman, 1970), and the Imbrian overlay ejected to Apollo 12 may have been diluted by freshly excavated, meteorite-free bedrock. If we choose to ignore this dilution effect, we find the average of all Apollo 14 and 12 values for $Au_E$ is 4.6 ppb. By coincidence this is the identical value used by Ganapathy *et al.* (1972), on the basis of 7 analyses of Apollo 14 alone, and leads to velocities about 1.5 times higher than those of the LN group (Table 2).

With either value of $Au_E$, a "lunar" composition is clearly inadmissible, as it gives a velocity below lunar escape velocity, 2.38 km/sec. A pure iron projectile seems unlikely also, because an iron object this size would require differentiation of a chondritic body of 3 to 10 times its mass, and then efficient removal of the silicate mantle. This seems improbable, since even after 4.6 AE of fragmentation, there are few, if indeed any, asteroids whose spectral reflectivity is compatible with an iron surface (Chapman, 1972). Even so, the iron model does set a rather firm upper limit, and we can put the impact velocity between 2.4 and 10 km/sec with some degree of confidence. This supports Urey's view that the Imbrian body impacted at low velocity, between 2.4 and 6 km/sec (Urey, 1952, 1961; Urey and MacDonald, 1971).

*Size and internal structure.* If the planetesimal was indeed as small as ~100 km, then it probably was not internally differentiated. Fricker *et al.* (1970) have compared the observed cooling rates of iron meteorites with thermal models of asteroids, and conclude that the IVA irons, with a wide range of cooling rates, apparently came from a 200-km body whose metal phase did not coalesce into a single core, but remained scattered throughout the body. Differentiation would be even less complete in a 100-km body, and it seems likely that the siderophile-rich pattern seen in noritic samples at the Apollo 12 and 14 sites is typical of the whole body.

*Composition.* Three compositional traits of the planetesimal can be specified. The first and third of these also apply to the body or bodies responsible for the DN component.

(1) It apparently contained nearly its cosmic complement of siderophiles. Thus it

Table 2. Properties of Imbrian body, for four possible compositions.

| Composition | Weight fraction Fe | Density ($g/cm^3$) | Velocity* (km/sec) A | Velocity* (km/sec) B | Diameter (km) A | Diameter (km) B | Mass* ($10^{21}$ g) A | Mass* ($10^{21}$ g) B |
|---|---|---|---|---|---|---|---|---|
| Lunar | 0.13 | 3.41 | 1.2 | 1.8 | — | — | — | — |
| Terrestrial | 0.36 | 4.0 | 3.5 | 4.9 | 99 | 83 | 2.0 | 1.2 |
| Cosmic | 0.28 | 3.76 | 2.7 | 3.9 | 113 | 95 | 2.8 | 1.7 |
| IVA iron | 0.92 | 7.91 | 7.6 | 10.0 | 54 | 45 | 0.63 | 0.94 |

* Based on apparent amount of meteoritic material, as estimated from Au content and cosmic Au/Fe ratio. A: $Au_E = 6.7$ ppb; average of LN group. B: $Au_E = 4.6$ ppb; average of Apollo 14 and Apollo 12 KREEP samples.

must have been an independently formed body, not a cast-off of a larger body that had undergone a "planetary" segregation of metal and silicate. Such segregation usually depletes siderophiles by factors of $10^{-3}$ to $10^{-4}$ (Anders *et al.*, 1971; Laul *et al.*, 1972b). Hence it cannot represent material spun off the earth after core formation (Wise, 1963; O'Keefe, 1969, 1970), a condensate or volatilization residue from a hot earth (Ringwood, 1966, 1970), or a fragment of a differentiated proto-moon disrupted during capture (Urey and MacDonald, 1971). If its Fe content deviated at all from cosmic, this must be attributed to a "nebular" metal-silicate fractionation, by analogy with meteorites and planets (Urey, 1954; Wood, 1962; Larimer and Anders, 1967, 1970; Anders, 1971). This process, presumably based on the ferromagnetism of metal grains, has fractionated metal from silicate by factors of up to 3 in the inner solar system.

(2) Refractory metals (Ir, Re) were depleted by 60–70%. This raises the possibility that refractory oxides (CaO, $Al_2O_3$, etc.) were likewise depleted, because these elements tend to correlate in cosmochemical fractionation processes. However, no direct evidence on this point is available from the Apollo 14 samples, because lunar surface rocks themselves are enriched in refractory oxides.

(3) Volatile elements were depleted to $<0.1$ their cosmic abundance, as in the eucrites and the earth and moon as a whole. In terms of the two-component model of planet formation (Larimer and Anders, 1967, 1970), the planetesimal thus contained less than 10% of low-temperature, volatile-rich material; the remainder consisted of high-temperature, volatile-poor material.

*Possible nature of Imbrian body*

The Imbrian body struck the moon less than 3.9 b.y. ago, 700 m.y. or more after formation of the solar system (Sutter *et al.*, 1971; Turner *et al.*, 1971; Papanastassiou and Wasserburg, 1971). This cannot have been a statistical fluke, because at least one body (Orientale) fell still later. The Imbrian projectile must have been stored in an orbit of low selenocentric velocity, and yet with a long enough collision lifetime to permit survival for 700 m.y. There are four origins which may be considered as possibilities.

(*1*) *A planetesimal in earth-crossing orbit.* It could have been one of the last survivors of a population of planetesimals. The mean lifetime for such objects is $\sim$40 m.y., according to Monte Carlo calculations by Arnold (1965) and Wetherill (1968). This would give a survival fraction of only $10^{-7}$ after 700 m.y. However, the actual lifetime may be longer, because secular variations in eccentricity, inclination, and node will put the object out of range of the earth for part of the time. Unpublished calculations by Mellick and Anders that make some allowance for these effects give a mean life of 100 m.y., corresponding to a survival fraction of $6 \times 10^{-4}$. This may be adequate, since the Imbrian body had a mass of only $2 \times 10^{-5}$ M .

The velocity of such "tardy" planetesimals is 6 to 21 km/sec according to Mellick and Anders, and in approximately the same range according to Arnold (1965) and Wetherill (1968). The lower end of the range is marginally consistent with the higher values in Table 2, and if we are dealing with only one object, this origin must be rated as acceptable.

(*2*) *An earth-grazing planetesimal.* It may have been an interplanetary object whose orbit barely grazed that of the earth. Öpik (1951) has shown that a large fraction of such objects, in orbits of low eccentricity, could survive indefinitely between the inner planets, only those trespassing into the space swept by the planets being eliminated by collisions, with lifetimes of $>10^8$ yr. However, if the Imbrian body was a member of such a population, evenly distributed between earth and Mars, other members of similar size should have survived to this day. No such objects have ever been found in asteroid searches, although they would be among the easiest to observe.

(*3*) *It may have been an earth satellite, swept up by the moon during tidal recession or capture.* This possibility, favored by Urey and MacDonald (1971) and Öpik (1969), readily accounts for the low velocity and long collision lifetime. Probably no decisive test of this hypothesis will be possible until the ages of several more lunar basins have been determined. It generally stands and falls with the capture theory.

(*4*) *It may have been an asteroid initially in a Mars-crossing orbit, later perturbed into an earth-crossing orbit.* The elimination of asteroids by this process roughly follows an exponential law, with mean lives ranging from a few hundred million to a few billion years, depending on the initial orbital elements (Arnold, 1965; P. J. Mellick and E. Anders, unpublished work). Velocities are of the right order, about 3 to 24 km/sec. The shorter of the above lifetimes agrees with the date of the Imbrian collision, though it is still too long to explain the steep decline in the flux of crater- and basin-forming objects between 3 and 4 AE ago (Hartmann and Wood, 1971). Perhaps some extinct variety of Mars-crossing asteroid was responsible. Williams (1971) has noted that the distribution of asteroid perihelia shows an abrupt drop in the region swept by Mars, suggesting that a large fraction of initially Mars-crossing asteroids have been eliminated in the course of time. If this extinct population had small perihelia and low inclinations, it might have been eliminated with about the right mean life, 100–200 m.y.

At present, alternative (1) seems most attractive, followed by (4). It is certainly suggestive that all three ancient meteoritic components show the sort of volatile depletion that has been proposed for the earth and moon as a whole (Gast, 1971; Anders *et al.*, 1971). It seems plausible that earth-like or moon-like planetesimals should originate in the earth's neighborhood. But we cannot rule out the possibility that similar planetesimals originated in the neighborhood of Mars and comprised the main source of crater-forming objects in terrestrial space 4 AE ago.

*Acknowledgment*—We are indebted to Rudy Banovich for preparation of the drawings on exceedingly short notice. Mr. Gary Snyder and the reactor staff at Plum Brook Station were most helpful in arranging irradiations. This work was supported in part by NASA Grant NGL 14-001-167. Some of the counting equipment had been provided by the AEC under Contract AT(11-1)-382.

## REFERENCES

Adams J. B. and McCord T. B. (1971) Optical properties of mineral separates, glass, and anorthositic fragments from Apollo mare samples. *Proc. Second Lunar Sci. Conf., Geochim. Cosmochim. Acta* Suppl. 2, Vol. 3, pp. 2183–2195. MIT Press.

Anders E. (1971) Meteorites and the early solar system. *Ann. Rev. Astron. Astrophys.* **9,** 1–34.

Anders E., Ganapathy R., Keays R. R., Laul J. C., and Morgan J. W. (1971) Volatile and siderophile elements in lunar rocks: comparison with terrestrial and meteoritic basalts. *Proc. Second Lunar Sci. Conf., Geochim. Cosmochim. Acta* Suppl. 2, Vol. 2, pp. 1021–1036. MIT Press.

Anderson A. T. Jr., Braziunas T. F., Jacoby J., and Smith J. V. (1972) Breccia populations and thermal history: Nature of pre-Imbrian crust and impacting body (abstract). In *Lunar Science—III* (editor C. Watkins), pp. 24–26, Lunar Science Institute Contr. No. 88.

Arnold J. R. (1965) The origin of meteorites as small bodies, II. The model. *Astrophys J.* **141,** 1536–1547.

Baedecker P. A. (1971) Iridium (77). In *Handbook of Elemental Abundances in Meteorites* (editor B. Mason), pp. 463–472. Gordon and Breach.

Baedecker P. A., Schaudy R., Elzie J. L., Kimberlin J., and Wasson J. T. (1971) Trace element studies of rocks and soils from Oceanus Procellarum and Mare Tranquilitatis. *Proc. Second Lunar Sci. Conf., Geochim. Cosmochim. Acta* Suppl. 2, Vol. 2, pp. 1037–1061. MIT Press.

Baedecker P. A., Chou C.-L., Kimberlin J., and Wasson J. T. (1972) Trace element studies of lunar rocks and soils (abstract). In *Lunar Science—III* (editor C. Watkins), pp. 35–37, Lunar Science Institute Contr. No. 88.

Burnett D. S., Huneke J. C., Podosek F. A., Russ G. P. III, Turner G., and Wasserburg G. J. (1972) The irradiation history of lunar samples (abstract). In *Lunar Science—III* (editor C. Watkins), pp. 105–107, Lunar Science Institute Contr. No. 88.

Chapman C. R. (1972) Surface properties of asteroids. Ph.D. Thesis, Massachusetts Institute of Technology.

Christophe-Michel-Lévy M., Lévy C., Caye R., and Pierrot R. (1972) The magnesian spinel-bearing rocks from the Fra Mauro formation. *Proc. Third Lunar Sci. Conf., Geochim. Cosmochim. Acta* Suppl. 3, Vol. 2. MIT Press.

Cliff R. A., Lee-Hu C., and Wetherill G. W. (1972) K, Rb, and Sr measurements in Apollo 14 and 15 material (abstract). In *Lunar Science—III* (editor C. Watkins), pp. 146–147, Lunar Science Institute Contr. No. 88.

Dence M. R., Plant A. G., and Traill R. J. (1972) Impact-generated shock and thermal metamorphism in Fra Mauro lunar samples (abstract). In *Lunar Science—III* (editor C. Watkins), pp. 174–176, Lunar Science Institute Contr. No. 88.

Dollfus A., Geake J. E., and Titulaer C. (1971) Polarimetric properties of the lunar surface and its interpretation. Part 3: Apollo 11 and Apollo 12 lunar samples. *Proc. Second Lunar Sci. Conf., Geochim. Cosmochim. Acta* Suppl. 2, Vol. 3, pp. 2285–2300. MIT Press.

Duncan A. R., Lindstrom M. M., Lindstrom D. J., McKay S. M., Stoeser J. W., Goles G. G., and Fruchter J. S. (1972) Comments on the genesis of breccia 14321 (abstract). In *Lunar Science—III* (editor C. Watkins), pp. 192–194, Lunar Science Institute Contr. No. 88.

Eberhardt P., Eugster O., Geiss J., Grögler N., Schwarzmüller J., Stettler A., and Weber L. (1972) When was the Apollo 12 KREEP ejected? (abstract). In *Lunar Science—III* (editor C. Watkins), pp. 206–208, Lunar Science Institute Contr. No. 88.

Fricker P. E., Goldstein J. I., and Summers A. L. (1970) Cooling rates and thermal histories of iron and stony-iron meteorites. *Geochim. Cosmochim. Acta* **34,** 475–491.

Ganapathy R., Keays R. R., Laul J. C., and Anders E. (1970a) Trace elements in Apollo 11 lunar rocks: implications for meteorite influx and origin of moon. *Proc. Apollo 11 Lunar Sci. Conf., Geochim. Cosmochim. Acta* Suppl. 1, Vol. 2, pp. 1117–1142. Pergamon.

Ganapathy R., Keays R. R., and Anders E. (1970b) Apollo 12 lunar samples: Trace element analysis of a core and the uniformity of the regolith. *Science* **170,** 533–535.

Ganapathy R., Laul J. C., Morgan J. W., and Anders E. (1972) Moon: Possible nature of the body that produced the Imbrian basin, from the composition of Apollo 14 samples. *Science* **175,** 55–59.

Gast P. W. (1971) The chemical composition of the earth, the moon, and chondritic meteorites. In *The Nature of the Solid Earth* (editors E. C. Robertson, J. F. Hays, and L. Knopoff). McGraw-Hill.

Grieve R., McKay G., Smith H., and Weill D. (1972) Mineralogy and petrology of polymict breccia 14321 (abstract). In *Lunar Science—III* (editor C. Watkins), pp. 338–340, Lunar Science Institute Contr. No. 88.

Hartmann W. K. and Wood C. A. (1971) Moon: Origin and evolution of multi-ring basins. *The Moon* **3,** 3–78.

Helmke P. A. and Haskin L. A. (1972) Rare earths and other trace elements in Apollo 14 lunar samples (abstract). In *Lunar Science—III* (editor C. Watkins), pp. 366–368, Lunar Science Institute Contr. No. 88.

Hollister L., Trzcienski W. Jr., Dymek R., Hulick C., Wiegand P., and Hargraves R. (1972) Igneous fragment 14310,21 and the origin of the mare basalts (abstract). In *Lunar Science—III* (editor C. Watkins), pp. 386–388, Lunar Science Institute Contr. No. 88.

Jackson E. D. and Wilshire H. G. (1972) Classification of the samples returned from the Apollo 14 landing site (abstract). In *Lunar Science—III* (editor C. Watkins), pp. 418–420, Lunar Science Institute Contr. No. 88.

Keays R. R., Ganapathy R., and Anders E. (1971) Chemical fractionations in meteorites—IV. Abundances of fourteen trace elements in L-chondrites; implications for cosmothermometry. *Geochim. Cosmochim. Acta* **35,** 337–363.

Larimer J. W. (1967) Chemical fractionations in meteorites—I. Condensation of the elements. *Geochim. Cosmochim. Acta* **31,** 1215–1238.

Larimer J. W. and Anders E. (1967) Chemical fractionations in meteorites—II. Abundance patterns and their interpretation. *Geochim. Cosmochim. Acta* **31,** 1239–1270.

Larimer J. W. and Anders E. (1970) Chemical fractionations in meteorites—III. Major element fractionations in chondrites. *Geochim. Cosmochim. Acta* **34,** 367–388.

Laul J. C., Keays R. R., Ganapathy R., and Anders E. (1970) Abundance of 14 trace elements in lunar rock 12013,10. *Earth Planet. Sci. Lett.* **9,** 211–215.

Laul J. C., Morgan J. W., Ganapathy R., and Anders E. (1971) Meteoritic material in lunar samples: Characterization from trace elements. *Proc. Second Lunar Sci. Conf., Geochim. Cosmochim. Acta* Suppl. 2, Vol. 2, pp. 1139–1158. MIT Press.

Laul J. C., Ganapathy R., Morgan J. W., and Anders E. (1972a) Meteoritic and non-meteoritic trace elements in Luna 16 samples. *Earth Planet. Sci. Lett.* **13,** 450–454.

Laul J. C., Keays R. R., Ganapathy R., Anders E., and Morgan J. W. (1972b) Chemical fractionations in meteorites—V. Volatile and siderophile elements in achondrites and ocean ridge basalts. *Geochim. Cosmochim. Acta* **36,** 339–345.

Laul J. C., Ganapathy R., Anders E., and Morgan J. W. (1972c) Chemical fractionations in meteorites—VI. Accretion temperatures of H-, LL-, and E-chondrites, from abundances of volatile trace elements. *Geochim. Cosmochim. Acta* (submitted).

Laul J. C., Boynton W. V., and Schmitt R. A. (1972d) Bulk, REE, and other elemental abundances in four Apollo 14 clastic rocks and three core samples, two Luna 16 breccias and four Apollo 15 soils (abstract). In *Lunar Science—III* (editor C. Watkins), pp. 480–482, Lunar Science Institute Contr. No. 88.

Lovering J. F. and Hughes T. C. (1971) Rhenium and osmium abundance determinations and meteoritic contamination levels in Apollo 11 and Apollo 12 lunar samples. *Proc. Second Lunar Sci. Conf., Geochim. Cosmochim. Acta* Suppl. 2, Vol. 2, pp. 1331–1335. MIT Press.

LSPET (Lunar Sample Preliminary Examination Team) (1970) Preliminary examination of lunar samples from Apollo 12. *Science* **167,** 1325–1339.

LSPET (Lunar Sample Preliminary Examination Team) (1971) Preliminary examination of lunar samples from Apollo 14. *Science* **173,** 681–693.

Morgan J. W. and Lovering J. F. (1967) Rhenium and osmium abundances in some igneous and metamorphic rocks. *Earth Planet. Sci. Lett.* **3,** 219–224.

Morgan J. W., Ganapathy R., Laul J. C., and Anders E. (1972a) Lunar crater Copernicus: Nature of impacting body. *Geochim. Cosmochim. Acta* (submitted).

Morgan J. W., Laul J. C., Krähenbühl U., Ganapathy R., and Anders E. (1972b) Major impacts on the moon: Chemical characterization of projectiles (abstract). In *Lunar Science—III* (editor C. Watkins), pp. 552–554, Lunar Science Institute Contr. No. 88.

Morgan J. W., Krähenbühl U., Ganapathy R., and Anders E. (1972c) Trace elements in Apollo 15 samples: Implications for meteorite influx and volatile depletion on the moon. *Proc. Third Lunar Sci. Conf., Geochim. Cosmochim. Acta* Suppl. 3, Vol. 2. MIT Press.

O'Keefe J. A. (1969) Origin of the moon. *J. Geophys. Res.* **74,** 2758–2767.

O'Keefe J. A. (1970) The origin of the moon. *J. Geophys. Res.* **75,** 6565–6574.

Onuma N., Clayton R. N., and Mayeda T. K. (1972) Oxygen isotope temperatures of "equilibrated" ordinary chondrites. *Geochim. Cosmochim. Acta* **36,** 157–168.

Öpik E. J. (1951) Collision probabilities with the planets and distribution of interplanetary matter. *Proc. Roy. Irish Acad.* **54,** Sec. A, 165–199.

Öpik E. J. (1961) Notes on the theory of impact craters. *Proc. Geophys. Lab.-Lawrence Radiation Lab. Cratering Symposium*, Washington, D.C., March 28–29, **2,** Paper S, 1–28. UCRL Report No. 6438.

Öpik E. J. (1969) The moon's surface. *Ann. Rev. Astron. Astrophys.* **7,** 473–526.

Papanastassiou D. A. and Wasserburg G. J. (1971) Rb–Sr ages of igneous rocks from the Apollo 14 mission and the age of the Fra Mauro formation. *Earth Planet. Sci. Lett.* **12,** 36–48.

Ringwood A. E. (1966) Chemical evolution of the terrestrial planets. *Geochim. Cosmochim. Acta* **30,** 41–104.

Ringwood A. E. (1970) Petrogenesis of Apollo 11 basalts and implications for lunar origin. *J. Geophys. Res.* **75,** 6453–6479.

Ringwood A. E., Green D. H., and Ware N. G. (1972) Experimental petrology and petrogenesis of Apollo 14 basalts (abstract). In *Lunar Science—III* (editor C. Watkins), pp. 654–656, Lunar Science Institute Contr. No. 88.

Schwerer F. C., Huffman G. P., Fisher R. M., and Nagata T. (1972) D.C. electrical conductivity of lunar surface rocks with complementary Mössbauer studies (abstract). In *Lunar Science—III* (editor C. Watkins), pp. 686–687, Lunar Science Institute Contr. No. 88.

Short N. M. and Forman M. L. (1970) Thickness of impact crater ejecta on the lunar surface. Goddard Space Flight Center Preprint No. X-652-70-336.

Silver L. T. (1971) U–Th–Pb isotope systems in Apollo 11 and 12 regolithic materials and a possible age for the Copernicus impact event. Paper presented at American Geophysical Union 52nd Annual Meeting, April 12–16, Washington, D.C.

Sutter J. E., Husain L., and Schaeffer O. A. (1971) $^{40}Ar/^{39}Ar$ ages from Fra Mauro. *Earth Planet. Sci. Lett.* **11,** 249–253.

Tandon S. N. and Wasson J. T. (1968) Gallium, germanium, indium, and iridium variations in a suite of L-group chondrites. *Geochim. Cosmochim. Acta* **32,** 1087–1109.

Tanner J. T. and Ehmann W. D. (1967) The abundance of antimony in meteorites, tektites and rocks by neutron activation analysis. *Geochim. Cosmochim. Acta* **31,** 2007–2026.

Tatsumoto M. (1970) U–Th–Pb age of Apollo 12 rock 12013. *Earth Planet. Sci. Lett.* **9,** 193–200.

Turner G., Huneke J. C., Podosek F. A., and Wasserburg G. J. (1971) $^{40}Ar$–$^{39}Ar$ ages and cosmic ray exposure ages of Apollo 14 samples. *Earth Planet. Sci. Lett.* **12,** 19–35.

Urey H. C. (1952) *The Planets.* Yale University Press.

Urey H. C. (1954) On the dissipation of gas and volatilized elements from protoplanets. *Astrophys. J.* Suppl. **1,** [6], 147–173.

Urey H. C. (1961) The origin and nature of the moon. In *Smithsonian Report for 1960*, pp. 251–265, Smithsonian Institution, Washington, D.C.

Urey H. C. and MacDonald G. J. F. (1971) Origin and history of the moon. In *Physics and Astronomy of the Moon*, 2nd edition (editor Z. Kopal), pp. 213–289. Academic Press.

Wänke H., Baddenhausen H., Balacescu A., Teschke F., Spettel B., Dreibus G., Quijano M., Kruse H., Wlotzka F., and Begemann F. (1972) Multielement analyses of lunar samples (abstract). In *Lunar Science—III* (editor C. Watkins), pp. 779–781, Lunar Science Institute Contr. No. 88.

Warner J. L. (1962) Apollo 14 breccias: Metamorphic origin and classification (abstract). In *Lunar Science—III* (editor C. Watkins), pp. 782–784, Lunar Science Institute Contr. No. 88.

Watkins C. (editor) (1972) *Lunar Science—III.* Lunar Science Institute Contr. No. 88.

Wetherill G. W. (1968) Dynamical studies of asteroidal and cometary orbits and their relation to the origin of meteorites. In *Origin and Distribution of the Elements* (editor L. H. Ahrens), pp. 423–443. Pergamon.

Williams J. G. (1971) Proper elements, families, and belt boundaries. In *Physical Studies of Minor Planets* (editor T. Gehrels), in press.

Wilshire H. G. and Jackson E. D. (1972) Petrology of the Fra Mauro formation at the Apollo 14 landing site (abstract). In *Lunar Science—III* (editor C. Watkins), pp. 803–805, Lunar Science Institute Contr. No. 88.

Wise D. U. (1963) An origin of the moon by rotational fission during formation of the earth's core. *J. Geophys. Res.* **68,** 1547–1554.

Wood J. A. (1962) Chondrules and the origin of the terrestrial planets. *Nature* **194,** 127–130.

Proceedings of the Third Lunar Science Conference
(Supplement 3, *Geochimica et Cosmochimica Acta*)
Vol. 2, pp. 1397–1420
The M.I.T. Press, 1972

# The abundances of components of the lunar soils by a least-squares mixing model and the formation age of KREEP

ERNEST SCHONFELD and CHARLES MEYER, Jr.
NASA Manned Spacecraft Center,
Houston, Texas 77058

**Abstract**—The chemical composition of lunar soils and breccias have been closely matched by a least-squares mixing model that includes mare basalts, KREEP basalts, anorthosites, anorthositic gabbros, ultramafics, granites, and meteorites as component materials. Each of these end member components have been directly observed in the coarse fraction of the soil samples and have previously been considered as fundamental rock types on the basis of their distinctive chemical compositions and mineral associations. The chemical compositions for the soils, breccias, and each of the components used in this mixing model include data for about 30 elements (Si, Ti, Al, Ca, Fe, Mg, P, Cr, Mn, Na, K, Rb, Ba, U, Th, La, Ce, Sm, Eu, Sr, Yb, Y, Sc, V, Zr, Nb, Co, Ni, Li, Au, Ir).

The results of the chemical mixing model show that iron-rich mare basalt ($< 15\%$ $Al_2O_3$) is the dominant material (35–80%) at the Apollo 11, 12, 15, Luna 16, and Surveyor 5 and 6 sites. KREEP basalt (15–22% $Al_2O_3$) is found to be present at all sites, but is most abundant at the Apollo 12, 14, 15, and possibly Surveyor 7 sites. Anorthositic material ($> 22\%$ $Al_2O_3$) is found to be an important component ($\sim 20\%$) of the Apollo 11, Luna 16, some Apollo 15 and Surveyor 7 soils. Meteoritic, granitic, and ultramafic components are only present in small amounts ($< 5\%$). Basalt 14310 can be considered to be a mixture of $\sim 8\%$ mare basalt, $\sim 65\%$ KREEP basalt, $\sim 25\%$ anorthosite, and $\sim 3\%$ meteorite; and should, therefore, be considered a melted soil. Basalt 14053 can be considered to be a mare basalt contaminated at depth with $\sim 7\%$ KREEP basalt. The component percentages of KREEP basalt and granite are found to be correlated, suggesting a common source for these two rock types. Breccias and core tube samples have variable component abundances indicating that the lunar regolith is not well mixed at any site.

The KREEP basalt component is the most important contributor of K, U, Th, Rb, REE, and radiogenic Sr in the soils (about 30% in the Apollo 11 soil, about 90% in the Apollo 12 soils, about 95–99% in the Apollo 14 soils, and about 80–90% in the Apollo 15 soils). In the Apollo 11 and 12 soils, there is no evidence of Rb or $^{87}Sr^*$ transfer; whereas the excess radiogenic lead in the Apollo 11 and 14 soils and the depletion of such lead in the Apollo 12 soils suggest volatile transfer of lead.

A two-stage Rb–Sr evolution model gives a formation age for the KREEP basalt of $4.41 \pm 0.06$ billion years.

## INTRODUCTION

LUNAR SOILS ARE the end products of extensive bombardment of the moon's surface by cosmic projectiles and accreting planetesimals. Such impacts have excavated lunar material from as deep as about 20 km and dispersed it as ejecta and rays for many crater diameters. For example, the rays of Tycho extend as much as one-quarter of the circumference of the moon. Consequently, the lunar soil samples represent *mixtures* of local surface material with large amounts of other rock types from the region surrounding the sample collection site, and a few percent of each soil may even come from great distances, or great depths, in the moon, or from the projectiles themselves. In the work presented here, we have used a least-squares mixing model technique to determine the abundance and distribution of "fundamental" rock types from the bulk chemical composition of the lunar soil and breccia samples. In order to perform such a mixing model calculation on soil samples, it is important to have properly recognized the "end member" rock types to be used as components and to have chosen reasonable elemental compositions for each rock type. It is also important to treat the soils at each site as slightly different problems because a proper average for the local component is required if correct amounts of the exotic components are to result. Mixing model calculations, such as the one presented here, are possible because a large number of high quality analyses are now available for many lunar samples. Future analyses should be aimed at filling in the matrix of data and further defining the rock types.

One of the important results of the lunar sample program is that a relatively small number of "fundamental" rock types have been recognized in the returned samples. These rock types have been identified by direct observation of lunar rocks and coarse fragments in lunar soils. For example, the mare basalts from all sites are found to be quite similar to each other in chemical composition. They are characterized by high iron ($\sim 20\%$ FeO), titanium, and magnesium, and low aluminum ($<15\%$ $Al_2O_3$). This material has been found in abundance at all of the mare sites and is certain to be the local rock type at mare sites. Another basaltic material, termed KREEP basalt, is abundant in Apollo 12, 14, and 15 samples. It is lower in iron ($\sim 10\%$ FeO) and higher in aluminum (15–22% $Al_2O_3$) than mare basalt and is highly enriched in the geochemically incompatible trace elements (Rb, K, U, Th, Zr, REE, etc.) (Meyer *et al.*, 1971; Hubbard *et al.*, 1971a). A third type of lunar material is very high in aluminum ($>22\%$ $Al_2O_3$) and calcium and has been called anorthositic material (Wood, 1970).

These three major lunar rock types also appear to represent three major, rock-forming, episodes in lunar history. The orbiting x-ray fluorescence experiment (Adler *et al.*, 1972) and gamma-ray experiment (Metzger *et al.*, 1972) show that both of the basaltic rock types fill the lunar mare regions and that a large portion of the highland areas and the backside of the moon are covered with high Al, low K, U, Th materials. So it appears that an anorthositic crust of the moon is stratigraphically older than the KREEP and mare basalts. It is also apparent that most KREEP materials predate the Imbrium event (about 4 b.y.) and that most mare basalts are younger (3.1–4 b.y.) than most KREEP basalts.

The results of the chemical mixing model along with the average isotopic properties of each component allow a calculation of radiogenic material balances for Sr and Pb isotopes. In this way, consideration of the soils as *mixtures* of rock types provides the basis for a better understanding of the "ages" of soils and breccias. Papanastassiou and Wasserburg (1971) and others, have, in the past, claimed that a "magic" component with isotopic properties equivalent to 4.6 b.y. was responsible for the old ages of lunar soils. In this paper, we wish to show that the isotopic properties and abundances of the KREEP component are such that a "magic" component is no longer required.

## THE MIXING MODEL TECHNIQUE

The method used is a linear mixing model of chemical elements according to the weighted least-squares method of Gauss (Kendall and Stuart, 1961). It is assumed that the chemical composition of a soil is the result of adding the contribution of each of the rock types present. Each element is represented by an equation as follows:

$$E_c = E_1 f_1 + E_2 f_2 + \cdots + E_n f_n \tag{1}$$

where $E_c$ is the calculated concentration of an element in a soil (such as Fe, Al, etc.), $E_1$ is the concentration of the respective element in the rock Type 1, $f_1$ is the fraction of rock Type 1 in the soil, and $n$ is the number of rock types used in the model. The fractions $f_1, f_2, \cdots f_n$ are obtained by the weighted least-squares method by minimizing the following expression:

$$\sum_{i=1}^{n} w_i (E_o - E_c)^2 \tag{2}$$

where $w_i$ is the statistical weight for each chemical element (inversely proportional to the square of the estimated standard deviation of $E_o$), $E_o$ is the observed concentration for each element in a soil, and $E_c$ is given by equation (1). The weighted least-squares method of Gauss also provides an estimate of the standard error for the fractions for each of the components.

In most cases, the chemical elements used in this analysis are Na, K, Rb, Sr, Ba, La, Ce, Sm, Eu, Yb, Y, Nb, Si, Al, Ca, Fe, Ti, Cr, Mn, Zr, U, Th, V, Co, Au, Ir, Ni, Sc, and P. In the case of Apollo 15 soils, a shorter list of elements was used. Several criteria were used in the selection of this set of chemical elements. Volatile elements such as Pb, Bi, and Cd were excluded. Elements that have very similar chemical properties, such as the rare earths, were represented by a selected few in order to avoid giving too much weight to a particular group in comparison with the rest of the elements. The availability of analytical data for the majority of the samples was also taken into account in the selection.

Since not all the elements have been determined with the same degree of accuracy, the use of statistical weights is very important. The least-squares method has been already applied to soils without using weights (Bryan *et al.*, 1969; Lindsay, 1971) or by using a first approximation in weighting each element (Goles *et al.*, 1971). In this work, the statistical weights have the role of putting each element on an equal basis according to the estimated degree of accuracy of determination. Of course, the elements that are the most fractionated between rock types are still the most important in this calculation. The natural variability or standard error for each element in each component is not well known, and, in our model, it is assumed to be zero. When large variability was found for a component, we subdivided it into two components (i.e., high and low K Apollo 11 basalts and high and low Mg Apollo 12 basalts). In the application of this mixing model to soils, the effect of the natural variability in the elemental composition in the components is minimized because of the very nature of the averaging process in generating soils. The fact that the residuals (as will be shown later) of the soils 10084 and 12070 are small also indicates that our assumption is reasonable.

A very useful part of the mixing model is the residual analysis. For each element the residual quantity $(E_o - E_c)/\sigma$ is computed where $\sigma$ is the estimated standard deviation of $E_o$. If the model fits the data, this expression should be around $\pm 1$ for

Table 1. Chemical composition and ages of the component rock types.

| | Mare basalts | | | | | | | | Anorthosite clan | | | | | |
|---|---|---|---|---|---|---|---|---|---|---|---|---|---|---|
| | Apollo 11 | | Apollo 12 | | Apollo[f] 14 | Apollo 15 | Luna[b] 16 | KREEP[e] | Anortho-site | Gabbroic anortho-site | Anortho-sitic gabbro | Granite[c] | Meteorite[d] (cc-1) | Ultra-mafic |
| | low K | high K | low Mg | high Mg | | | | | | | | | | |
| Si (%) | 18.9 | 19.37 | 21.80 | 20.70 | 21.60 | 21.5 | 20.5 | 22.50 | 20.7 | 20.9 | 21.0 | 30.0 | 10.8 | 20.0 |
| Ti | 6.43 | 7.00 | 2.20 | 1.58 | 1.76 | 1.28 | 2.94 | 1.14 | 0.10 | 0.18 | 0.24 | 0.60 | 0.05 | 1.16 |
| Al | 5.48 | 4.30 | 5.50 | 6.73 | 6.80 | 4.74 | 7.23 | 9.26 | 17.73 | 14.8 | 13.50 | 6.50 | 0.94 | 2.80 |
| Ca | 8.49 | 7.76 | 8.03 | 6.20 | 7.95 | 7.3 | 7.43 | 7.87 | 12.7 | 11.6 | 10.4 | 4.8 | 1.08 | 6.60 |
| Fe | 14.71 | 14.79 | 15.50 | 16.35 | 13.30 | 17.6 | 15.03 | 8.33 | 1.3 | 3.1 | 4.7 | 7.7 | 17.1 | 16.6 |
| Mg | 4.30 | 4.48 | 4.40 | 7.80 | 5.10 | 5.7 | 4.25 | 4.82 | 1.33 | 3.80 | 4.64 | 2.0 | 9.39 | 10.5 |
| $P_2O_5$ | 0.10 | 0.17 | 0.1 | 0.1 | 0.13 | 0.07 | 0.11 | 0.90 | | 0.02 | 0.1 | | 0.32 | |
| Cr (ppm) | 1915 | 2394 | 2630 | 4210 | 2530 | 4600 | 1920 | 1300 | 400 | 200 | 700 | 1300 | 2430 | 3240 |
| Mn | 2140 | 1843 | 2090 | 2120 | 2020 | 2250 | 1550 | 1300 | — | — | 700 | 1150 | 1880 | 2080 |
| Na | 2957 | 3580 | 1560 | 2080 | 3260 | 1930 | 2920 | 6520.0 | 5200 | 2600 | 2200 | 9000 | 5110 | 1400 |
| K | 600 | 2500 | 510.0 | 450.0 | 912 | 380 | 1200.0 | 5200.0 | 150 | 400 | 450 | 30000 | 560 | 250 |
| Rb | 0.812 | 5.65 | 1.1 | 0.9 | 2.1 | 0.6 | 1.90 | 16.3 | 0.32 | 1.27 | (1.35) | 100 | 2.38 | |
| Ba | 93 | 293 | 76.0 | 65.0 | 146 | 33 | 222 | 1030.0 | 30 | 61 | (70) | 5000 | 3.0 | |
| U | 0.25 | 0.86 | 0.27 | 0.23 | 0.59 | 0.13 | | 4.50 | 0.058 | — | — | 15.0 | 0.02 | |
| Th | 0.97 | 3.30 | 1.10 | 0.81 | 2.10 | 0.50 | | 17.2 | | | | 50.0 | 0.075 | |
| La | 12.2 | 28.1 | 5.55 | 7.35 | 13.0 | — | 7.7 | 93.0 | 1.58 | (7) | (8) | 60 | 0.19 | |
| Ce | 40.3 | 78.1 | 20.8 | 16.7 | 34.5 | 8.1 | 34.0 | 250 | 2.0 | 9.6 | (10) | 100.0 | 0.60 | |
| Sm | 13.7 | 22.3 | 5.7 | 4.3 | 6.6 | 2.1 | 8.5 | 42.0 | 0.5 | 1.55 | (1.65) | 15.0 | 0.20 | |
| Eu | 2.02 | 2.24 | 1.28 | 0.91 | 1.21 | 0.7 | 2.5 | 3.25 | 0.75 | 0.75 | (0.75) | 2.2 | 0.074 | |
| Sr | 170.3 | 166.8 | 115.0 | 90.0 | 101.0 | 98.0 | 370 | 200 | 160 | 160 | (160) | 150.0 | 8.2 | |
| Yb | 12.7 | 19.05 | 3.77 | 4.75 | 10.0 | 1.5 | 5.6 | 30.0 | 0.40 | 1.71 | (1.8) | 30 | 0.13 | |
| Y | 103 | 160 | 50.0 | 38.0 | 90 | 23 | | 300 | 5.0 | | | 280.0 | 1.55 | |
| Sc | 81 | 79.0 | 52.0 | 42.0 | 90 | — | | 28.0 | 5.5 | | | 25.0 | 5.1 | |
| V | 81 | 74.0 | 130.0 | 210.0 | 135 | — | | 50.0 | | | | 85.0 | 73.0 | |
| Zr | 310 | 410 | 140 | 100 | 310 | 88.0 | 296.0 | 1300 | | 150 | 150 | 3400 | 10 | 390 |
| Nb | 20 | 25 | 7.0 | 6.0 | 19 | 4.3 | — | 85.0 | | | | 200 | — | |
| Co | 16.1 | 28 | 40.0 | 60.0 | 48 | | | 27.0 | 6.4 | | | — | 520.0 | |
| Ni | 8.0 | 27.0 | 30.0 | 80.0 | 14 | 42.0 | | ~20 | (0) | (0) | (0) | 100 | 10000.0 | ~400 |
| Li | 11.4 | 17.7 | 8.5 | 7.4 | 11 | 6.4 | 10.0 | 44.0 | | | | 100 | 1.50 | |
| Au (ppb) | 0.04 | 0.04 | 0.04 | 0.04 | 0.017 | 0.05 | | ~0.1 | (0) | (0) | (0) | 0.7 | 150 | (0) |
| Ir | 0.07 | 0.07 | 0.10 | 0.10 | 0.11 | 0.01 | | ~0.1 | (0) | (0) | (0) | 0.5 | 442 | (0) |
| Age (b.y.) | 3.65 ±0.05 | 3.65 ±0.05 | 3.25 ±0.08 | 3.25 ±0.08 | 3.95 ±0.05 | 3.30 ±0.06 | 3.42 ±0.18 | 4.42 ±0.06 | | | | (4.42) | 4.6 ±0.1 | |
| $T_{BABI}$[a] | 4.7 ±0.2 | 3.87 ±0.1 | 4.5 ±0.1 | 4.5 ±0.1 | 4.6 ±0.1 | 4.4 ±0.3 | 3.8 ±0.3 | 4.41 ±0.06 | ~4.5 ±0.6 | ~4.5 ±0.6 | ~4.5 ±0.6 | 4.40 ±0.1 | 4.6 ±0.1 | |
| $T_{ADOR}$ | 5.1 ±0.2 | 3.96 ±0.1 | 5.0 ±0.1 | 5.0 ±0.1 | 4.8 ±0.1 | 4.8 ±0.3 | 4.5 ±0.4 | 4.45 ±0.06 | ~5.0 ±0.6 | ~5.0 ±0.6 | ~5.0 ±0.6 | 4.40 ±0.1 | 4.6 ±0.1 | |

[a] Model ages (b.y.); BABI = 0.69898; ADOR = 0.69884. [b] Vinogradov (1971). [c] Rock 12013. [d] Goles *et al.* (1971).
[e] Meyer *et al.* (1971), Hubbard *et al.* (1971a), Hubbard and Gast (1971), Keil *et al.* (1971), and the Apollo Soil Survey (1971). [f] Rock 14053.

each element (Kendall and Stuart, 1961). If this is not the case, it is possible that an unpredicted rock type is present and/or that the compositions chosen for the rock types used in the least-squares analysis are not correct. Residual analysis also provides a crude estimate of the composition of any unpredicted rock types.

*Choice of components*

A minimum number of "fundamental" lunar rock types have been recognized in the many studies of small rock and glass fragments in the soil samples (Meyer, *et al.*, 1971; Wood, 1970; Marvin *et al.*, 1971; Keil *et al.*, 1970; Keil and Prinz, 1971; Reid *et al.*, 1972; and many others). We have chosen the following rock types as components: mare basalts (slightly different at each site), KREEP basalt, anorthosite, gabbroic anorthosite, anorthositic gabbro, granite, ultramafic component, and Type I carbonaceous chondrite. This choice of components was found to give small residuals when the mixing model was applied to many soil compositions, giving independent evidence that we have picked realistic components. Table 1 gives the compositions we have used for each component. These compositions are averages of selected analytical data taken from many sources (*Proc. Apollo 11 Lunar Sci. Conf.*, 1970; *Proc. Second Lunar Sci. Conf.*, 1971; *Proc. Third Lunar Sci. Conf.*, 1972). We have favored data obtained by isotopic dilution and x-ray fluorescence.

Figure 1 is an oversimplified example of the natural separation that has been observed in lunar sample studies. The elements Fe, Al, and K show some of the important differences that characterize the end member rock types. The mare basalts from all sites have a high Fe/Al ratio ($\sim 2$). The consequence of their mafic composition is that mare basalts generally have mineral assemblages rich in opaque minerals, olivine, and pyroxene and poor in plagioclase. The pyroxenes in mare basalts are characteristically high in calcium. At all sites except Apollo 14, the mare basalts have been well analyzed because many large specimens have been returned. Although there is considerable range in the abundances of some elements (Ti, Sr, K, and Mg), mare basalts are all characterized by a high value for FeO ($\sim 20\%$). In a later section we will discuss the possibility of contamination of some of these mare basalts, but most of them are certain to be primary or near primary liquids derived from deep in the moon (Ringwood and Essene, 1971; Hubbard and Gast, 1971).

Another end member rock type that has been identified in the fragments of lunar soil has been termed KREEP basalt because of its high concentration in potassium, rare earth elements, and phosphorous (Meyer and Hubbard, 1970; Hubbard *et al.*, 1971a; Meyer *et al.*, 1971). The Fe/Al ratio for KREEP is $\sim 1$ which gives the crystalline version of this rock type a mafic to plagioclase ratio of 1 : 1. The relative absence of opaque minerals and the abundance of low calcium pyroxenes also help to distinguish fragments of this rock type from mare basalts. The annealed brecciated variety of this rock type (Meyer *et al.*, 1971; Anderson and Smith, 1971) has abundant orthopyroxene and has been called "norite" by some authors (Marvin *et al.*, 1971). The KREEP glasses at the Apollo 14 site were called Fra Mauro basalt by the Apollo Soil Survey (1971). The KREEP fragments that were found in Apollo 11, 12, and 14 were all breccias and glasses. (As we will show later, 14310 is a mixture of anorthosite,

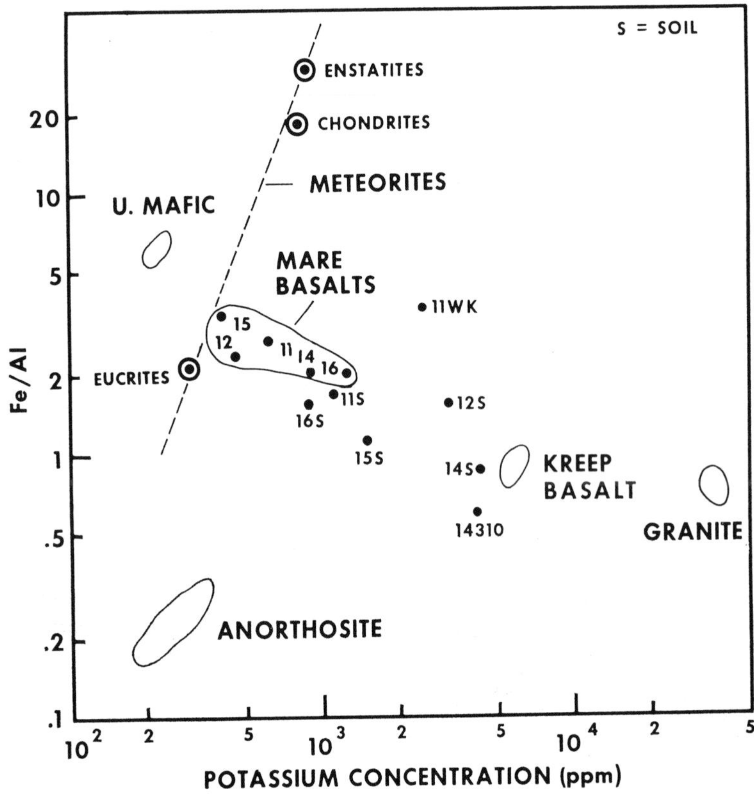


Fig. 1. Fe, Al, K concentrations of component rock types used in chemical mixing model for soils. The Fe/Al ratio is numerically similar to the mafic mineral to plagioclase ratio. The log K helps in resolving the different components. The soil samples (S) are intermediate to the major components. The high K Apollo 11 basalt is marked 11 HK. Lunar glasses are most densely populated in the circled regions.

local mare basalt, meteorite, and KREEP and is, hence, *not* a KREEP basalt.) However, in Apollo 15, there are fresh subophitic fragments of KREEP basalt in the coarse fine fraction of the soil (15023) (Meyer, 1972). Thin sections of rake sample 15382 (3.2 gm) have similar textures and mineralogy indicating that it may be a KREEP basalt. KREEP basalt is greatly enriched in many elements (K, Rb, Ba, U, Th, REE, Zr, P, etc.) and is an important lever in our model. We have obtained the composition of KREEP from the data of Meyer *et al.* (1971), Hubbard *et al.* (1971a), Hubbard and Gast (1971), Keil *et al.* (1971), and the Apollo Soil Survey (1971). The Apollo 12 and 14 glass and breccia fragments could be mixtures of KREEP with other rock types. However, those fragments which have the highest REE values cannot be diluted by very much of any other rock types, and we have used the values for the other elements also measured on these same fragments.

The return of sample 15415 has conclusively shown that lunar anorthosites exist. Wood (1970) and others had previously shown that many anorthositic fragments were present in Apollo 11 samples. The plagioclase-rich mineral assemblages of these fragments have characteristically exhibited brecciated and annealed textures indicating that they have had a complicated history. The anorthositic class of lunar material is subdivided into anorthosite (>90% plagioclase), gabbroic anorthosite (77–90% plagioclase), and anorthositic gabbro (77–60% plagioclase). For each of these components, we have used data from Wood (1970), Hubbard *et al.* (1971b), Keil *et al.* (1970), Keil and Prinz (1971), and Wakita and Schmitt (1971). A careful study of natural groupings in glass analyses by Reid *et al.* (1972) has led them to conclude that anorthositic gabbro (called highland basalt by them) is probably the most abundant type of material in the lunar highlands. However, it is not yet certain whether their uniform composition for anorthositic gabbro (highland basalt) at each site represents a fundamental rock type or perhaps instead a well-mixed regolith.

In KREEP-rich soils from Apollo 12 and 14, there are also a small number of anorthosite fragments containing potassium-rich plagioclase (Meyer, 1972; Hubbard *et al.*, 1971b). These anorthosites are apparently derived from KREEP magma; and in this mixing model, we have not treated this potassium-rich anorthosite as a component.

A small number of granitic glasses and lithic fragments have been found in soil samples and because of the unusual chemistry of this rock type, it is important to include it as a component in the mixing model. Granitic material is best characterized by high $SiO_2$ (~70%), high $K_2O$ (~6%), and BaO (~2%), and an extremely high Fe/Mg ratio (Meyer, 1972; Apollo Soil Survey, 1971). For trace element data on this component, we have used data from the light portion of 12013 (*Earth Planet. Sci. Lett.* 9, No. 2). The mineral assemblage of this component is K, Ba-feldspar, plagioclase, quartz, fayalitic olivine, and iron-rich pyroxene. Quite a few patches of this material can be found as melted clasts in Fra Mauro breccias.

Anders *et al.* (1971) and others have demonstrated the existence of a meteorite component in lunar soils even though fragments of this component have rarely been found. Wänke *et al.* (1971) and Goldstein *et al.* (1970) and others have reported numerous metallic iron grains with high Ni, Au, Ir contents. However, we follow Goles *et al.* (1971) in assuming that the composition of Type I carbonaceous meteorites is appropriate for the meteorite component.

An ultramafic component might be expected to be present in lunar soils as a companion to the anorthosite because whatever processes formed anorthosite should have also formed some residual of ultramafic composition and the deep impacts on the moon should have brought some of this material to the surface. However, examples of ultramafic components in lunar samples have proved to be elusive. Reid *et al.* (1972) give an average composition for a high $TiO_2$, low $Al_2O_3$ glass type called Fecunditatis B. Chao *et al.* (1970) and Keil and Prinz (1971) give data for some lunar ultramafic materials. The abundant green (mafic) glasses in Apollo 15 soil may represent this component. All of these materials are characterized by high FeO and MgO and low $Al_2O_3$ and $SiO_2$ contents.

The Ni, Au, and Ir contents for each component present special problems because they are excellent indicators of the amount of meteoritic component in lunar samples as long as no significant amount of these elements exists in any "fundamental" lunar rock type. Figure 2 is a plot of Ni versus Au for several KREEP-rich and anorthosite rich samples. The Ni/Au ratio of this plot is the same as that of carbonaceous chondrites. For this reason Morgan *et al.* (1972) consider all these samples to be contaminated with small amounts of meteorite. We accept this argument and assume that the Ni, Au, and Ir concentration of the original KREEP basalt and anorthosite was very small. Laul *et al.* (1971) report some high values of Ni and Au for some anorthositic gabbro samples, supporting our interpretation that anorthositic gabbro is actually melted highland regolith material. Chao *et al.* (1970) report relatively low values for NiO in their ultramafic glasses such that small amounts of ultramafic component in lunar soils cannot, by itself, explain the high Ni concentration of lunar soils.

*Results of the mixing model*

The results of the mixing model for the percentage abundance of each component in the soil and breccia samples are given in Table 2. The KREEP component is found to be more abundant in the Apollo 11 soil (~5%) than in breccias that are variable (1–3%). The abundance of combined anorthositic components (about 19%) is the same for both the soils and breccias. This combined anorthositic component is composed of about 70% anorthositic gabbro and smaller amounts of true anorthosite and gabbroic anorthosite.

There are more low-potassium basalts than high-potassium basalts in the Apollo 11 soil, suggesting that the low-potassium variety is the local surface type. The ratio

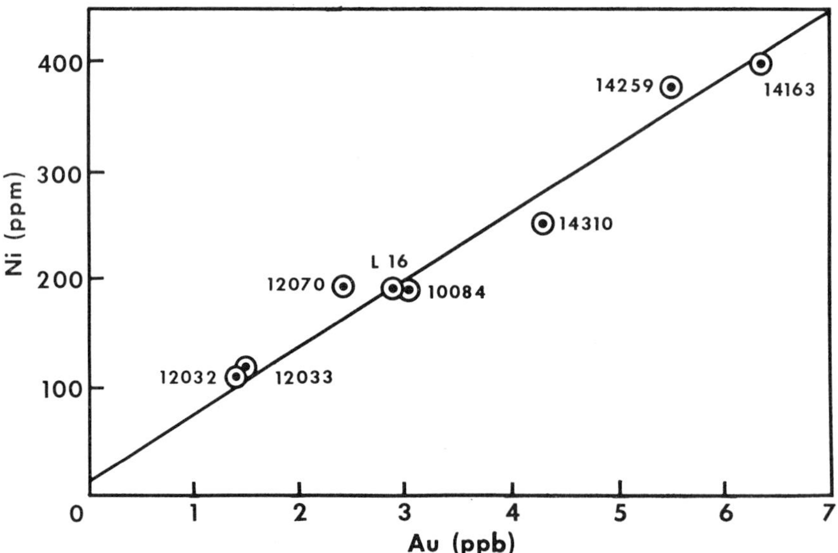


Fig. 2. Ni versus Au concentration for selected lunar samples. (Data from Anders *et al.*, 1971; Baedecker *et al.*, 1972; Laul *et al.*, 1971; Morgan *et al.*, 1972; and Wänke *et al.*, 1971). The Ni/Au ratio of this plot is the same as the value for carbonaceous chondrites. 14310 has a meteorite component.

of low-K to high-K Apollo 11 basalts in the breccia is quite variable (ranging from 1.8 to 0.38) and in all cases is lower than the soil (2.9). Eberhardt *et al.* (1971) have shown that the high-K basalts were located in a shielded place from cosmic rays about 60–240 m.y. before they were brought up (recently) to the surface of the regolith. The low-K have been in a less shielded area for ~130 m.y. This difference in location of these two types of local basalt may be responsible for the variable ratio in the breccias.

The Apollo 12 soil and breccia samples are mixtures of the Apollo 12 mare basalts with large amounts of the KREEP component (30–68%) and with only minor meteoritic (<2%), anorthositic (<10%), ultramafic (<5%), and granitic components (1–2%) (Meyer *et al.*, 1971). The preliminary data indicate that there is important stratification in the core tubes. It is highly desirable that more analyses for 30 elements be made along the length of lunar core tubes.

The Apollo 14 soils are made up predominantly by the KREEP component (60–97%), with minor contribution of mare basalt (0–15%), anorthositic gabbro (0–30%), meteoritic component (~3%), granite (~2.5%), and ultramafic component (<4%). The breccias at the Apollo 14 site are very heterogeneous, but there are not yet enough published results on the chemical composition of these samples.

The mixing model applied to the preliminary data on soils and breccia from the Apollo 15 mission (LSPET, 1972) using the elements Si, Ti, Al, Fe, Mn, Ca, Na, K, P, Cr, Sr, Zr, Nb, Rb, Th, Ni, and Y gives the results shown in Fig. 3. The samples obtained closest to the Apennine Front contain an increased amount of "anorthositic gabbro." The abundance of mare basalts follows an opposite trend. Most of the soils at this site are rich in KREEP (see also O'Kelley *et al.*, 1972). Samples that represent deeper locations such as sample 15601 (from the edge of Hadley Rille) and 15471 (from the rim of Dune Crater) have the smallest amount of KREEP, suggesting that the mare area was covered with a blanket rich in KREEP. The high concentration of anorthosite close to the Apennine Front is consistent with an anorthositic model of the lunar highlands (Wood, 1970). In all Apollo 15 samples, the granitic component is equal to or less than about 1%, and the meteoritic component is about 2%. The highest

Table 2. Component abundance in Apollo 11 and Luna 16 soils and breccia[b]

| Sample | | Apollo 11 mare basalts Low K (%) | High K (%) | "Anorthosites"[a] (%) | KREEP (%) | Meteorites (cc-1) (%) | Granites (%) | Basalts Low K/High K |
|---|---|---|---|---|---|---|---|---|
| 10084 | Soil | 58 ± 10 | 20 ± 7 | 19.0 ± 2.4 | 4.7 ± 1.2 | 1.8 ± 0.22 | 0.2 ± 0.5 | 2.9 |
| 10073 | Breccia | 53 ± 11 | 29 ± 9 | 21 ± 3 | 1 ± 1.7 | 1.9 ± 0.22 | 0.5 ± 0.5 | 1.8 |
| 10019 | Breccia | 48 ± 17 | 32 ± 10 | 19 ± 4 | 3.5 ± 2 | 1.5 ± 0.4 | 0.4 ± 0.8 | 1.5 |
| 10061 | Breccia | 43 ± 11 | 37 ± 11 | 17 ± 3 | 3.1 ± 1.6 | 2.0 ± 0.25 | 0.4 ± 0.4 | 1.16 |
| 10018 | Breccia | 40 ± 10 | 41 ± 10 | 17 ± 3 | 3.0 ± 1.4 | 2.3 ± 0.23 | 0.2 ± 0.4 | 1.0 |
| 10059 | Breccia | 29 ± 6 | 50 ± 8 | 16 ± 3 | 1.6 ± 1.6 | 2.1 ± 0.2 | 0 ± 0.5 | 0.58 |
| 10060 | Breccia | 23 ± 7 | 61 ± 5 | 17 ± 3 | 1 ± 2 | 1.2 ± 0.3 | 0 ± 0.7 | 0.38 |
| 10021 | Breccia | 20 ± 8 | 47 ± 7 | 18 ± 3 | 1.7 ± 1.6 | 1.7 ± 0.3 | 0 ± 0.5 | 0.42 |
| Luna 16 | Soil | 71 ± 5 (L-16 basalt) | | 22 ± 7 | 2 ± 1 | 1.9 ± 0.3 | 0.5 ± 1 | |

[a] Mainly (~70%) anorthositic gabbro type.
[b] All samples indicated the possibility of ~1–5% ultramafic component.

value for an ultramafic component is about 20% in samples 15923 (green clod) and 15301 (close to Spur Crater). This is the first evidence for any significant ultramafic component in lunar samples.

The results of the mixing model for the Luna 16 soil, with the Luna 16 basalt as the representative mare component, indicate about 20% anorthositic gabbro and 2% KREEP as components which is very similar to the results for Apollo 11.

An important result of this calculation is that the meteoritic, granitic, and ultramafic components can only be present in very small amounts in the Apollo 11, 12, 14, and Luna 16 soils and breccia.

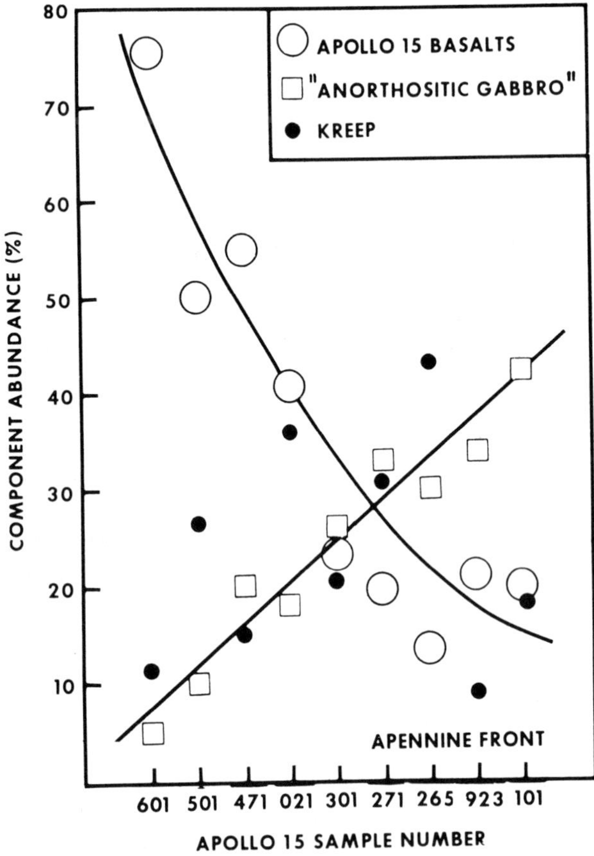


Fig. 3. The preliminary results for the component abundance of Apollo 15 soils and breccias (LSPET data 1971). The "anorthositic gabbro" component is greatest in samples that were collected closest to the Apennine Front and least in samples collected at distance from the Front. The mare basalt component shows an inverse trend than the "anorthositic gabbro" component. A large, variable KREEP component is predicted. Not shown are the granitic, meteoritic, and ultramafic components. Their abundances are low with the exception of the ultramafic component in samples 15923 and 15301 (about 20%).

The results for granite in Apollo 11, 12, and 14 soils are shown in Fig. 4. Although the granitic component is small in these sites, there is an interesting correlation between the abundances of granite and the KREEP components. The ratio of KREEP component to granite component is about 35 ± 10.

In this paper we will make only a few comments in comparison of this mixing model and the results with other workers (Goles *et al.*, 1971; Lindsay, 1971; Wänke *et al.*, 1971). To compare and evaluate our results, we listed in Table 3 the composition used for three soils, the residuals, and the percent differences between the observed and calculated component abundances. The residuals are excellent, and the percent deviation is in good agreement with the estimated experimental error, showing that we have chosen reasonable components for our analysis of these soils. The important factors that enter in the least-squares mixing model are the choice of components, the elemental composition of the components, the list of elements, and the statistical weights. For example, Goles *et al.* (1971), Lindsay (1971), and Wänke *et al.* (1971) did not consider granitic and ultramafic components (Goles *et al.* used the high-Mg Apollo 12 mare basalt as a source of Mg for the Apollo 11 soil analysis). If granites are excluded in the analysis, the residuals will increase for elements such as Rb, K, and Ba. For the soil 12032, Goles *et al.* had a −25.6% error in the Rb concentration that can be compared with our 1.8% error for Rb in the same sample. Differences in the selection of elemental composition data (like x-ray fluorescence and isotopic dilution data versus neutron activation data) will also give significant deviations in the residuals. Probably the −40.3% error in the Na analysis in sample 12032 by Goles *et al.* (compared to our 3.3% error) can be attributed to the low-Na concentration assigned by them to KREEP. We also found that by not using statistical

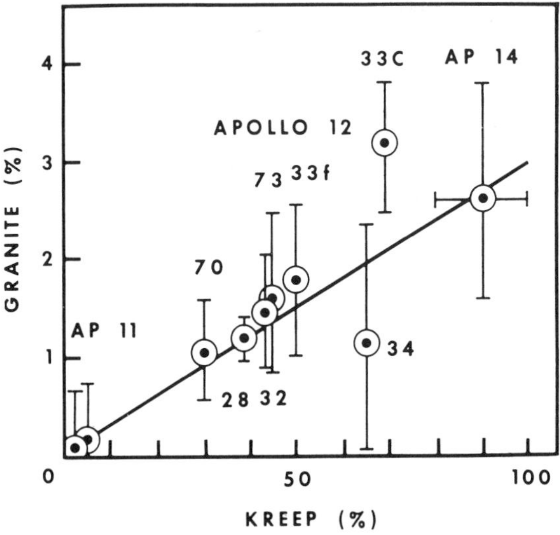


Fig. 4. The results of the chemical mixing model for the KREEP and granite components. The ratio of KREEP to granite is 35 ± 10. This correlation suggests a geographic and/or petrologic relationship between these two rock types.

Table 3. Concentrations and residuals for lunar soils 10084, 12070, and 12032.

| | | Soil 10084 | | | Soil 12070 | | | Soil 12032 | | |
|---|---|---|---|---|---|---|---|---|---|---|
| Elements | Units | Concentration | Res Obs-cal/$\sigma$ | Obs-cal/Obs × 100 | Concentration | Res Obs-cal/$\sigma$ | Obs-cal/Obs × 100 | Concentration | Res Obs-cal/$\sigma$ | Obs-cal/Obs × 100 |
| Si | % | 19.7 | 0.07 | 0.2 | 21.6 | 0.4 | 0.9 | 21.75 | 0.1 | 0.2 |
| Ti | % | 4.69 | −0.2 | −0.9 | 1.72 | 1.0 | 5.8 | 1.53 | 0.6 | 3.9 |
| Al | % | 7.25 | 0.1 | 0.3 | 6.72 | −0.8 | −2.4 | 7.62 | −1.1 | −3.2 |
| Ca | % | 8.58 | −0.5 | −1.1 | 7.50 | 0.5 | 1.3 | 7.69 | 0.6 | 1.7 |
| Fe | % | 12.23 | −0.4 | −0.6 | 12.80 | −1.7 | −2.7 | 11.74 | −0.5 | −0.9 |
| Mg | % | 4.76 | 0.3 | 1.3 | 5.8 | 1.4 | 4.8 | 5.81 | 1.1 | 3.8 |
| $P_2O_5$ | % | 0.13 | −0.3 | −11 | 0.32 | −0.4 | −6.2 | 0.29 | −0.8 | −13 |
| Cr | ppm | 1870 | 1.0 | 5.4 | 2850 | 1.3 | 6.8 | 2350 | 0.1 | 0.6 |
| Mn | ppm | 1630 | −1.1 | −7 | 1710 | −0.8 | −4.7 | 1630 | −0.2 | −1.6 |
| Na | ppm | 3190 | 0.5 | 1.6 | 3210 | −0.4 | −1.7 | 4267 | 1.0 | 3.3 |
| K | ppm | 1100 | −1.3 | −4.7 | 2030 | −0.8 | −3.9 | 3030 | 0.9 | 3.0 |
| Rb | ppm | 2.77 | 0.3 | 1.1 | 6.41 | 0.2 | 0.6 | 9.21 | 0.8 | 1.8 |
| Ba | ppm | 172.6 | −0.2 | −0.6 | 390 | 0.1 | 0.3 | 529 | −1.8 | −0.34 |
| U | ppm | 0.54 | −0.1 | −0.6 | 1.65 | 0.3 | 1.8 | 2.35 | 0.1 | 3.0 |
| Th | ppm | 2.09 | 0.35 | 1.7 | 6.30 | 0.5 | 2.3 | 8.72 | 0.3 | 1.4 |
| La | ppm | 16.6 | −0.3 | −2.0 | 30 | −1.1 | −6.5 | 47 | 0.6 | 2.3 |
| Ce | ppm | 46.6 | −0.4 | −2.9 | 90 | −0.2 | −1.4 | 117 | −0.9 | −4.6 |
| Sm | ppm | 13.7 | 0.4 | 2.8 | 16.4 | 0.2 | 2.2 | 20.7 | −0.4 | −3.4 |
| Eu | ppm | 1.82 | 0.3 | 1.6 | 1.73 | 0.1 | 1.0 | 2.12 | 0.5 | 4.2 |
| Sr | ppm | 166.4 | 0.6 | 1.1 | 142.5 | 1.0 | 2.1 | 157 | 0.7 | 0.6 |
| Yb | ppm | 11.05 | −1.1 | −4.9 | 12.5 | 0.2 | 1.4 | 15.2 | −0.7 | −4.1 |
| Y | ppm | 99 | 0.1 | 0.5 | 130 | 1.0 | 6.1 | 175 | 2.1 | 8.0 |
| Sc | ppm | 65 | 1.3 | 6.4 | 42 | 1.2 | 7.1 | 36 | −0.1 | −1.0 |
| V | ppm | 70 | −0.25 | −2.9 | 110 | −0.6 | −5.4 | 110 | 0.1 | 0.9 |
| Zr | ppm | 320 | −0.9 | −5.6 | 510 | 0.1 | 0.8 | 680 | −0.1 | −0.6 |
| Nb | ppm | 20 | −0.1 | −2 | 30 | −0.3 | −4.0 | 45 | 0.4 | 3 |
| Co | ppm | 29 | 0.7 | 6 | 40 | 0.1 | 0.8 | 39 | −0.5 | −1.9 |
| Ni | ppm | 185 | 0.1 | 0.5 | 200 | 0.1 | 0.5 | 117 | 0.1 | 0.9 |
| Li | ppm | 12.5 | 0.4 | 6 | 18 | −0.4 | −5.2 | 23.4 | −0.4 | −5.1 |
| Au | ppb | 2.7 | 0.2 | 2.9 | 24 | −0.5 | −10 | 1.4 | 0.4 | 14 |
| Ir | ppb | 7.4 | −0.4 | −3.2 | 8.5 | 1.5 | 9 | 4.0 | 1.0 | 12 |

weights higher apparent values of the anorthositic component and of the residuals will result; this may explain why Lindsay obtained higher abundances of anorthosites in the Apollo 12 samples, when compared to this work. We have included in our results an estimate of the standard error of the abundances of the components in each soil.

## MODEL "AGES" OF SOILS

A least-squares fit to all Rb–Sr data for lunar soils gives a slope of 4.5 $\pm$ 0.2 b.y. with an intercept close to 0.6990. Also model ages for soils actually range from 4.3 to 4.8 b.y. The model ages for lunar rocks are given in Table 1, and are generally about 4.6 b.y. except for the high potassium Apollo 11 rocks which are about 3.8 b.y. On the other hand, the accurately measured crystallization ages of these lunar rocks as determined by Rb–Sr and/or $Ar^{39}/Ar^{40}$ methods range from about 3.2 to 4 b.y. (Compston *et al.*, 1970, 1971a, 1971b, 1972; Cliff *et al.*, 1971; Lunatic Asylum, 1970; Murthy *et al.*, 1971; Papanastassiou and Wasserburg, 1971a, 1971b; Turner, 1971).

In order to understand the model ages of the soils and breccia, it is necessary to know the average isotropic properties of each component. The isotropic properties of the KREEP component are significant because we will show later in this section that KREEP is radiogenically the most significant component in the soils. A summary of the Rb–Sr isotopic properties of KREEP fragments (Table 4) shows that KREEP material from the Apollo 11, 12, and 14 has the same average isotopic properties of Rb and Sr. The average model age of KREEP is 4.41 $\pm$ 0.06 b.y. and this value is the same for each site. Table 4 also shows that the model age of the granite component ($\sim$4.4 b.y.) is similar to the average model age of the KREEP component.

The average isotopic properties of lead for KREEP are more difficult to obtain

Table 4. Average Rb–Sr isotopic properties and model age of KREEP.

| | $^{87}Sr/^{86}Sr$ | $^{87}Rb/^{86}Sr$ | Model age[d] (b.y.) |
|---|---|---|---|
| 5 KREEP fragments, Apollo 12 (Hubbard and Gast, 1971) | 0.7141 | 0.241 | 4.37 |
| 11 KREEP fragments, Apollo 14 (Nyquist *et al.*, 1972) | 0.7140 | 0.235 | 4.47 |
| Luny Rock 1, Apollo 11 (Papanastassiou *et al.*, 1970) | 0.7140 | 0.236 | 4.45 |
| 5 rocks[a] and 4 soils,[b] Apollo 14 (Papanastassiou and Wasserburg, 1971b) | 0.7133 | 0.227 | 4.41 |
| 30 KREEP material[c] | 0.7141 | 0.239 | 4.41 |
| 7 Granitic fragments from rock 12013 (Lunatic Asylum, 1970) | 0.7920 | 1.485 | 4.40 |

Note: In some cases samples like 14310 and Apollo 14 soils are not pure KREEP but mixtures very rich in that component, and therefore will give somewhat lower Rb/Sr and $^{87}Sr/^{86}Sr$ ratios.

[a] Samples 14001,7,3, 140073, 14310, 14276, 14321,144 dark matrix.

[b] Samples 14141, 14149, 14163, 14259.

[c] All previous samples from this table plus samples 12070 frag. A, 14001,7,1 frag., 12013,10 comp. B, and 12013,10,9 dark (data from Papanastassiou and Wasserburg, 1971a, 1971b; and Lunatic Asylum, 1970).

[d] Referred to BABI = 0.69898.

directly, because of the high volatility of lead and because not many KREEP fragments have been directly measured. One way to obtain these properties for lead is from the "formation age" as determined from Rb–Sr data (discussed in a later section) and the U and Th concentrations as measured for KREEP fragments. Another way is to study the U, Th, Pb systematics of the KREEP-rich soils of Apollo 12 and 14. A summary of the U, Th, Pb data from Cliff *et al.* (1971), Silver (1971a, b), Tatsumoto *et al.* (1971), and Wasserburg *et al.* (1972) is shown in Fig. 5. In order to obtain the U, Pb systematics for KREEP, these data were corrected by subtracting the small contribution of U–Pb from mare basalts using the results of the chemical mixing model. It is seen that the Apollo 12 fragments and breccia can be explained by either a continuous diffusion (Tilton, 1960) or by an episodic event. The gas retention ages of samples 12033 and 12034 of about $1 \pm 0.2$ b.y. (Pepin *et al.*, 1972; Schonfeld, 1971) and the $^{39/40}$A ages of $0.85 \pm 0.05$ b.y. (Eberhardt *et al.*, 1972) support the episodic interpretation. The other intercept of the chord defined by these samples with Concordia is about 4.4 b.y., which will be shown in a later section to be consistent with the initial age of formation of KREEP materials as determined from Rb–Sr data. Two rocks from Apollo 14 (14310 and 14073 data from Wasserburg *et al.*, 1972, and Silver, 1972a) when coupled with their known crystallization ages of 3.90 b.y. (Papanastassiou and Wasserburg, 1971b; Compston *et al.*, 1972) also give an apparent initial age of 4.4 b.y. in very good agreement with the Apollo 12 data and the results of the Rb–Sr systematics. Apparently 14310 and 14073 did not loose all their old lead when they recrystallized at 3.90 b.y. Thus it is reasonable to assign radiogenic lead concentrations equivalent to a formation age of $\sim$4.4 b.y. to KREEP.

Using the results shown in Table 2 of the mixing model on samples 10084, 10059, 12070, 14310, and 14053, the material balance of the elements Rb, Sr, and the radiogenic $^{87}Sr$ was calculated. The results are shown in Table 5. The agreement between

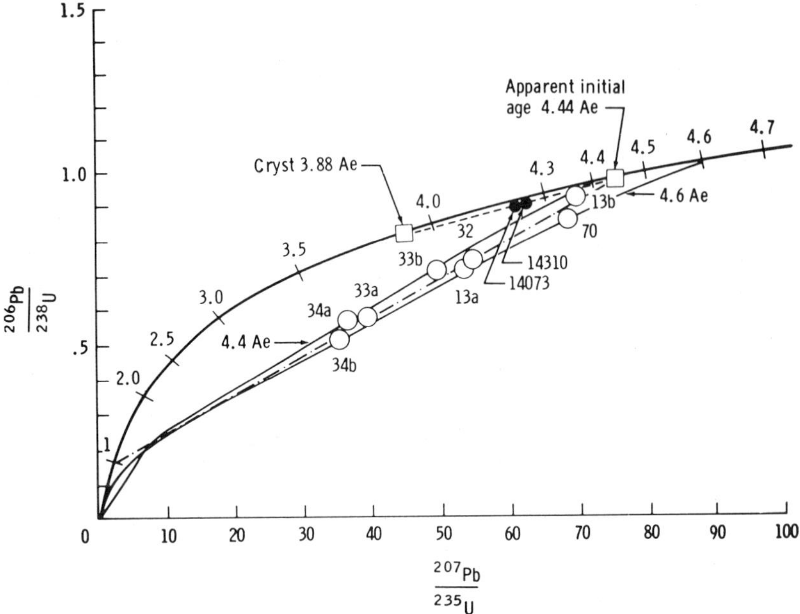


Fig. 5. U–Pb systematics for the KREEP component of Apollo 12 soils and breccias. A small correction for the amount of each isotope in the non-KREEP components has been subtracted from the bulk soil or breccia value. (Data from Cliff *et al.*, 1971; Silver, 1971a, b; and Tatsumoto *et al.*, 1971). This figure shows two curves (4.4 and 4.6 AE = $10^9$ yr) for continuous diffusion of lead (Tilton, 1960). A lead loss episode at 0.9 billion years and an apparent initial age of 4.4 billion years are indicated for the KREEP component. 14310 and 14073 lie on a chord between their crystallization age 3.9 billion years and the apparent initial age of KREEP at 4.4 billion years similar to the Apollo 12 KREEP material. (Data from Silver, 1972a; and Wasserburg *et al.*, 1972).

Table 5. Radiogenic $^{87}Sr^*$ material balance in soils, breccia, and rocks.

| | $^{87}Sr^*$ ($10^{-11}$ mole/g) | | | | |
|---|---|---|---|---|---|
| | Soil 10084 | Breccia 10059 | Soil 12070 | Rock 14310 | Rock 14053 |
| KREEP | 17 | 6 | 103.5 | 221 | 22 |
| Apollo 11 high K basalt | 20.3 | 53.2 | | | |
| Apollo 11 low K basalt | 9.4 | 5.4 | | | |
| Apollo 12 basalt | | | 13 | | |
| Apollo 14 basalt | | | | 2 | 23 |
| Anorthosites | 5.5 | 5.2 | 1 | 7 | 0 |
| Granites | 2.7 | 1 | 12 | 27 | 0 |
| Meteorite | 1 | 1 | 1 | 1 | 0 |
| Total | 55.9 | 71.8 | 130.5 | 258 | 45 |
| Measured[a] | 56.1 | | 134.5 | 265 | |
| Measured[b] | 57.3 | 73.2 | 132.5 | 250 | 44.7 |

[a] Compston *et al.*, 1970, 1971a, 1972.
[b] Papanastassiou and Wasserburg, 1971a, 1971b.

the calculated values by the mixing model and the observed radiogenic $^{87}Sr$ values (Papanastassiou *et al.*, 1970; Papanastassiou and Wasserburg, 1971a, b; Compston *et al.*, 1970, 1971a, 1972) is excellent. Similar material balances for these samples were about 1% for Rb and about 2% for Sr. The errors for the radiogenic $^{87}Sr$, Rb, and Sr are all in the same direction (observed larger than calculated). Since these are the elements that determine the model ages of the soils by the Rb–Sr method, the excellent material balance indicates that the model ages can be explained by a simple linear addition of these elements from each rock type. This result also shows that in the Apollo 11 and 12 sites, there is little or no bulk volatilization loss of Rb. In the Apollo 11 soil, KREEP contributes 30% of the total radiogenic $^{87}Sr^*$ and 87% in the Apollo 12 soil 12070. In general, the KREEP component contributes about 90% of the radiogenic $^{87}Sr^*$ in the Apollo 12 soils, about 95% in Apollo 14 soils, and about 80% in the Apollo 15 soils. Thus the "age" of the soils rich in KREEP are dominated by the model age of KREEP of 4.41 $\pm$ 0.06 b.y. and do not represent the formation age of the moon. The lower model age of the Apollo 11 breccia 10059 of 4.27 b.y. and of the Luna 16 soil of 4.2 b.y. can be explained by the fact that they are dominated by the local mare basalts with low model ages of about 4 b.y.

Using the abundances found for each of the components by the mixing model and the U, Th, and radiogenic lead properties of each of the components, it was found that in the Apollo 11 soil there is about 34% excess of $^{206}Pb$, about 73% excess $^{207}Pb$, and about 30% excess $^{208}Pb$. In the Apollo 12 site (see Fig. 5) the episode at 0.9 b.y. ago released large amounts of radiogenic lead. Using the graphical method of Wetherill (1956) for samples 12032, 12033, and 12034, the Pb losses by volatilization are, respectively, 29, 36, and 60% at the time of the episode. Laboratory experiments on lunar samples by Silver (1971a, b) showed that the radiogenic lead can easily be volatilized at temperatures as low as 600°C. This volatilized lead on the moon produces apparent $^{207}Pb/^{206}Pb$ ages for the Apollo 12 samples of 4.7 b.y., and probably has been deposited at other places on the moon. The Apollo 14 soils show parentless lead (Silver, 1972b; Tatsumoto *et al.*, 1972) of an apparent age of about 4.8. b.y.

## Formation Age of KREEP

As we have shown in Table 4, the average model age of KREEP-rich fragments is 4.41 b.y. for all sites. However, the real age of formation of the KREEP component may be considerably younger. In the following discussion, we will calculate what we will call "the formation age" of the KREEP component by using a two-stage evolution model (Schonfeld, 1971, 1972). This model has the advantage of being the simplest model that can be constructed that is consistent with lunar data. The first stage simply assumes that the moon was formed as a closed system at 4.65 b.y. ago and had an initial Sr isotope composition (MABI*) close to the value found in meteorites.

$$\left(\frac{^{87}\mathrm{Sr}}{^{86}\mathrm{Sr}}\right)_b - (\mathrm{MABI}) = \left(\frac{^{87}\mathrm{Rb}}{^{86}\mathrm{Sr}}\right)_b \left(e^{\lambda(4.65-t)} - 1\right). \tag{3}$$

* MABI (Moon Average Best Initial strontium isotopic ratio.) Values for MABI may be BABI = 0.69898 or ADOR = 0.69884 (Papanastassiou *et al.*, 1970).

The second stage began when KREEP was formed from the source material with a new Rb/Sr ratio.

$$\left(\frac{^{87}\mathrm{Sr}}{^{86}\mathrm{Sr}}\right)_p - \left(\frac{^{87}\mathrm{Sr}}{^{86}\mathrm{Sr}}\right)_b = \left(\frac{^{87}\mathrm{Rb}}{^{86}\mathrm{Sr}}\right)_p \left(e^{\lambda t} - 1\right). \tag{4}$$

In both equations, $b$ refers to the time just before KREEP was formed from its source, and $p$ refers to the present time; $t$ is the "age of formation" of KREEP and $\lambda$ is the decay constant of $^{87}$Rb ($0.0139 \times 10^{-9}\ \mathrm{yr}^{-1}$).

For convenience, we also define a fractionation factor $F$ which represents a change in the Rb/Sr ratio when KREEP formed from its source:

$$F = \frac{\left(\frac{^{87}\mathrm{Rb}}{^{86}\mathrm{Sr}}\right)_a}{\left(\frac{^{87}\mathrm{Rb}}{^{86}\mathrm{Sr}}\right)_b} = \frac{\left(\frac{^{87}\mathrm{Rb}}{^{86}\mathrm{Sr}}\right)_p e^{\lambda t}}{\left(\frac{^{87}\mathrm{Rb}}{^{86}\mathrm{Sr}}\right)_b} \tag{5}$$

where the subscript $a$ refers to the time right after the formation of KREEP.

The equation for the model age ($T$) of KREEP is

$$\left(\frac{^{87}\mathrm{Sr}}{^{86}\mathrm{Sr}}\right)_p - (\mathrm{MABI}) = \left(\frac{^{87}\mathrm{Rb}}{^{86}\mathrm{Sr}}\right)_p \left(e^{\lambda T} - 1\right). \tag{6}$$

Addition of equations (3) and (4) and substitution for equations (5) and (6) gives the following expression

$$e^{\lambda T} = e^{\lambda t} + \frac{(e^{\lambda 4.65} - e^{\lambda t})}{F}. \tag{7}$$

Figure 6 is a plot of this expression for $t$ versus $F$ for $T_{\mathrm{BABI}} = 4.41$ b.y. and $T_{\mathrm{ADOR}} = 4.45$ b.y. It is seen from equation (7) (since $t$ will range only between 0–4.65 that $t$ approaches $T_{\mathrm{BABI}}$ or $T_{\mathrm{ADOR}}$ as $F$ becomes large. In the case of KREEP, all the incompatible elements—K, U, Th, Zr, REE, as well as Rb—are highly enriched, and the Rb/Sr ratio is large. Thus, *if* KREEP was formed from a primitive source material such as eucrites (Rb/Sr $\sim$ 0.0045) or a source material similar to that of mare basalts (Rb/Sr $\approx$ 0.003–0.01, Papanastassiou and Wasserburg, 1971a) then the fractionation factor $F$ is in the range of 10–20, which gives a formation age for KREEP very close to its average model age of 4.41 b.y. (Fig. 6). This age is in excellent agreement with the intercept of the chord with Concordia on the modified Pb–Pb diagram for Apollo 12 soils (Fig. 5). This age is also supported by a similar two-stage model calculation using the isochron for KREEP breccias of Nyquist *et al.* (1972) with the initial value of 0.7004 at 3.95 b.y. If the assumptions in this model hold then KREEP is the oldest moon rock that has so far been recognized and the widespread occurrence of this component suggests, that a large portion of the moon underwent melting in its very early history.

Using the same assumptions as in the case of KREEP the two-stage evolution model shows a formation age for granite (Rb–Sr isotopic data at present and at 4 b.y. from Lunatic Asylum, 1970) very close to its model age of about 4.4 b.y. In this case

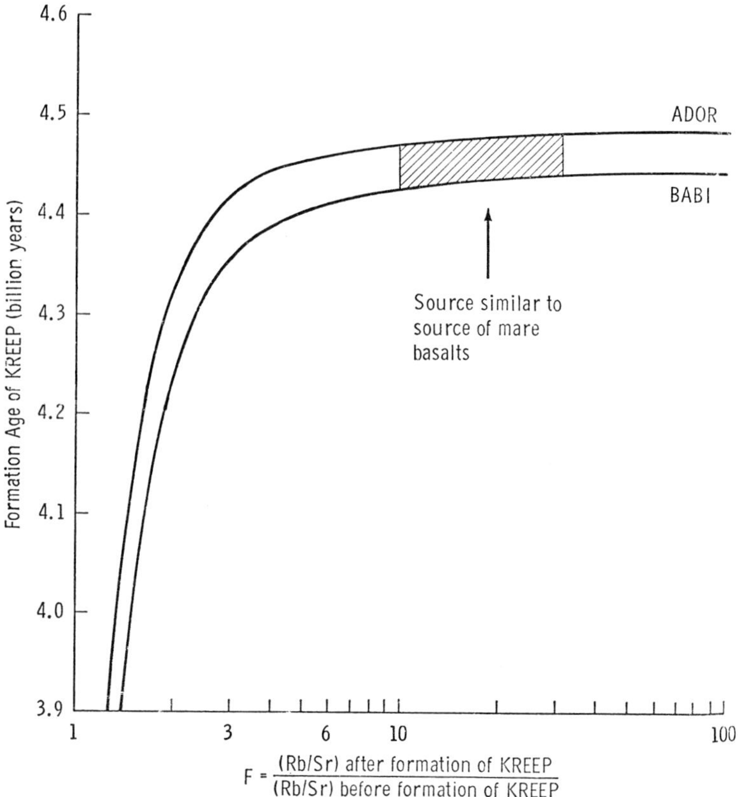


Fig. 6. Graphical solution of equation (7) for the "formation" age ($t$) of KREEP as a function of the fractionation factor ($F$) according to a two-stage model for the formation of KREEP. If the Rb/Sr ratio for the source of KREEP is low (i.e., similar to the source of mare basalt), then the formation age of KREEP is the same as the average model age of KREEP. This age is also slightly dependent on the initial Sr isotope ratio for the moon (BABI = 0.69898, ADOR = 0.69884).

the amount of data is less than in the case of KREEP. On the other hand, application of this two-stage model to mare basalts (which have very old model ages) is difficult to do because mare basalts have much smaller Rb/Sr ratios than KREEP and $F$ is not easy to estimate.

## Discussion

Simple mixing of lunar rock types is one of the most universal of lunar processes because of the extensive cratering of the lunar regolith. However, this mixing is not complete. The regolith at the Apollo 12 site is not at all well mixed because of the close presence of a ray from Copernicus (Meyer *et al.*, 1971). Therefore, a single soil analysis from any one site might or might not be completely representative of the composition of that part of the moon (i.e., Luna 16, Luna 20, Apollo 11). The natural groupings in the composition of small ($\sim 100\ \mu$) glass particles in the lunar soils also show that mixing is not complete, especially in the small size range (Reid *et al.*, 1972; Apollo Soil Survey, 1971). The philosophy that the average of the compositions of these natural groupings is equal to the composition of a lunar rock is questionable, because there are always some glasses in each group that are diluted by mixing with other rock types. This would have the effect of displacing the average from the center toward composition of the other major rock types. For this reason, the composition of end member components is hard to obtain from glass analysis alone. Such a technique might be expected to work best in being able to define the composition of the most local rock type, but the composition of the natural grouping of glasses in Apollo 11 soil fails to distinguish the high-K, low-K subdivision observed in the composition of the large rocks. Conversely, the observed groupings of KREEP glasses into several subdivided rock types at Apollo 14 (Apollo Soil Survey, 1971) might be controlled by an equal number of small, glass-forming craters into local regoliths with slightly variable compositions. The highland basalt ("anorthositic gabbro") composition of Reid *et al.* (1972) and Jakeš *et al.* (1972) is claimed by them to be the composition of a major rock type. However, Adams and McCord (1962) observe infrared reflection spectra from highland regoliths that can be explained by addition of high-Ca pyroxene derived from the mare regions. Thus, the composition of highland basalt ("anorthositic gabbro") is likely to be a mixture of highland and mare rock types.

The relative abundance of small glasses may be some sort of indication of the relative abundance of rock types in the vicinity of a landing site, but no direct comparison is expected with the abundance of rock types as derived by our mixing model because the average of all the glass analyses does not equal the measured composition of the same bulk soil (Jakeš *et al.*, 1972). Perhaps Fe-rich glasses are more often devitrified and not included in such a study, or perhaps the Apollo Soil Survey and mixing model techniques are "looking" at different diameter areas surrounding the sample collection site.

The mixing process is not only responsible for the chemical composition of soils and breccias, but has also played a part in producing some igneous rocks. When 14310 is analyzed by the mixing model, it is found to contain $\sim 8\%$ mare basalt, $\sim 65\%$ KREEP, $\sim 25\%$ anorthosite, and $\sim 3\%$ meteorite and should, therefore, be considered a melted soil. The oldest mare basalt, 14053, does not contain a meteorite component (see Table 1) but could be considered as a mare basalt that has been contaminated at depth by $\sim 7\%$ KREEP. However, the high-K basalts from Apollo 11 cannot be mixtures of low-K basalts, KREEP, and/or granite. The high-K Apollo 11 basalts may have been contaminated by a process that did not simply involve mixing of rock types, or they may represent natural variations in the partial melting process.

Our mixing model shows that lunar granite is not as important a component to lunar soils as has been claimed by the Lunatic Asylum (1970). One wonders whether large bodies of granite really exist on the moon. So far, there has been no evidence for them from lunar orbit. The high Fe/Mg ratio, high $SiO_2$, and high K, U, Th, Ba contents for granite are consistent with late stage crystallization of a basaltic magma. If so, then granitic liquid is much more likely to have been derived from KREEP than mare basalt. One of the tentative results of our mixing model is that the amount of granitic component is correlated with the amount of KREEP in a soil. In addition, both of these rock types are found in breccia 12013, and they both have the same "formation age" of about 4.4 b.y. (Table 4). Thus, these two rock types are certainly related to each other. However the texture and mineral assemblage of most KREEP fragments is that of a high grade metamorphic rock which suggests a second mechanism for the formation of granite from KREEP. It is possible that the granitic fragments formed by separation of a liquid rich in $SiO_2$, $K_2O$, and BaO that recrystallized as veins in a thick ejecta blanket of annealed KREEP breccia (Meyer, 1972).

## The Distribution of Lunar Rock Types

Recognition of the important rock types and their representative chemical composition helps in the interpretation of the remote and orbiting chemistry experiments such as the alpha scattering experiment (Turkevich, 1971), the x-ray fluorescence experiment (Adler *et al.*, 1972), the gamma-ray experiment (Metzger *et al.*, 1972), and the infrared reflection spectra (Adams and McCord, 1972). These experiments have the advantage of covering large areas of the lunar surface but interpretation of remote analyses requires ground truth data from the sampling sites. Table 6 is a summary of the results of the mixing model for five widely separated sampling sites. Although some KREEP is apparently present at all sites, it is apparent that the Apollo 12, 14, and 15 regoliths have much more KREEP than do Apollo 11 or Luna 16 regoliths. This region of high KREEP content was extended by the Apollo 15 gamma-ray experiment to include much of southern Imbrium and northern Procellarum. Likewise, the gamma-ray experiment verified the low KREEP content of the lunar highlands and Maria Fecunditatis and Tranquilitatis.

A mixing model analysis using lunar rock types as identified at landing sites indicates that the Surveyor 5 and 6 and Lunakod sites are all crudely 70–80% mare basalt with about 20% anorthosite, as was the case for Apollo 11 and Luna 16. The Surveyor 7 analysis at Tycho indicates ~60% anorthosite, but the presence of up to 30% KREEP cannot be excluded at present because the alpha scattering experiment is not sensitive to elements like K, Th, and U. These elements, however, can be measured by gamma-ray spectroscopy from lunar orbit. The Surveyor 7 analysis does indicate the presence of fluorine, which is present in relatively high concentration in KREEP basalt, but not in anorthosite (Marvin *et al.*, 1971).

The KREEP-rich, Fra Mauro materials seem to support the suggestion by Meyer

Table 6. Distribution and abundance of components on the moon.

| Location | Percent component | | |
|---|---|---|---|
| | KREEP | Local mare basalt | "Anorthosite" |
| Apollo 11 | 1–6 | 78 | 19 |
| Luna 16 | ~2 | 71 | 22 |
| Apollo 12 | 30–60 | 40–60 | ~5 |
| Apollo 14 | 60–100 | 0–10 | ~5–15 |
| Apollo 15 | 10–35 | 20–80 | 10–45 |
| Granites ~ ≤1% | Meteorites ~2% | | U. Mafic ~0–5% |

*et al.* (1971) that KREEP is Imbrium ejecta. However, a surprising result of the Apollo 15 gamma-ray experiment was that it did not find high radioactivity in the Imbrium ejecta covering the Haemus Mountains south of Serenitatis. For this reason we conclude that KREEP and other Fra Mauro materials were originally surface materials on the southern and western sides of the Imbrium impact that were radially distributed further from Imbrium by a base surge mechanism. In other words, KREEP basalt is likely to be a pre-Imbrian *mare* material.

Both Apollo 11 and Luna 16 were near lunar highlands that have been observed from lunar orbit to be composed of very high aluminum material. The results that the soils from these missions contain a large abundance of anorthosite component indicates that the high-Al material that has been mapped from lunar orbit by the x-ray experiment is largely anorthosite and gabbroic anorthosite as predicted by Wood (1970). The lack of ultramafic samples and the small abundance that can be accounted for by the chemical composition of the soil is surprising due to the abundance of anorthosites. One explanation for this lack of ultramafic samples may be that the anorthositic layer is very thick such that none of the large impacts have sampled beneath it. The high Al/Si ratios determined by the orbiting x-ray experiments in the Crisium ejecta blanket provide further evidence for the extreme thickness of this aluminum rich crust of the moon.

## CONCLUSIONS

The conclusions in order of importance are:

(1) The mixing model shows that the most important components on the moon are the mare basalts, KREEP, and the anorthosites. Minor components are granitic, meteoritic, and ultramafic components.

(2) KREEP is the most important contributor of $^{87}Sr^*$, Rb, K, U, Th, Zr, and REE in most of the soils and breccia. It is the so-called "magic" component in the soils and the "ages" of soils rich in KREEP are dominated by the model age of this component and do not represent the formation age of the moon.

(3) The formation age of KREEP is 4.41 $\pm$ 0.06 b.y. as determined by a two-stage evolution model. Thus it is the oldest lunar material found so far. Its rather wide spread distribution suggests that a considerably large part of the moon melted in its early history. The Imbrian event at about 4 b.y. and probably the Copernicus event at about 0.9 b.y. excavated and dispersed this component over the lunar surface.

(4) There is no evidence of Rb or $^{87}Sr^*$ bulk transfer, whereas the excess radiogenic lead in the Apollo 11 and 14 soils and the depletion in the Apollo 12 soils suggest volatile transfer of lead.

(5) Sample 14310 can be considered as a mixture of anorthosite, mare basalt, KREEP, and meteorite components. Sample 14053 can be considered to be a mare basalt contaminated at depth by about 7% KREEP.

(6) Breccias and core tube samples have variable component abundances indicating that the lunar regolith is not well mixed at any site.

(7) There is a correlation between the amounts of KREEP component and granitic component in the Apollo 12 and 14 soils and breccia. This correlation plus

the similarity of the average model ages of these rocks of about 4.4 b.y. suggest a common origin of KREEP and granite.

(8) The lack of ultramafics and the presence of large amount of anorthosites suggests that the anorthositic crust is quite thick.

*Acknowledgments*—We thank W. Compston, P. W. Gast, G. G. Goles, and L. E. Nyquist for their critical review, and B. A. Atkinson and D. L. Sanders for typing many drafts.

## REFERENCES

Adams J. B. and McCord T. B. (1972) Optical evidence for regional cross-contamination of highland and mare soils (abstract). In *Lunar Science—III* (editor C. Watkins), pp. 1–3, Lunar Science Institute Contr. No. 88.

Adler I., Trombka J., Gerard J., Lowman P., Schmadebeck R., Blodget H., Eller E., Yin L., Lamothe R., Gorenstein P., and Bjorkholm P. (1972) Apollo 15 geochemical x-ray fluorescence experiment: Preliminary report. *Science* **175,** 436–440.

Anders E., Ganapathy R., Keays R. R., Laul J. C., and Morgan J. W. (1971) Volatile and siderophile elements in lunar rocks: Comparison with terrestrial and meteoritic basalts. *Proc. Second Lunar Sci. Conf., Geochim. Cosmochim. Acta* Suppl. 2, Vol. 2, pp. 1021–1036. MIT Press.

Anderson A. T. Jr. and Smith J. V. (1971) Nature, occurrence, and exotic origin of "gray mottled" (Luny Rock) basalts in Apollo 12 soils and breccias. *Proc. Second Lunar Sci. Conf., Geochim. Cosmochim. Acta* Suppl. 2, Vol. 1, pp. 431–438. MIT Press.

Apollo Soil Survey (1971) Apollo 14: Nature and origin of rock types in soil from the Fra Mauro formation. *Earth Planet. Sci. Lett.* **12,** 49–54.

Baedecker P. A., Chou C. L., Kimberlin J., and Wasson J. T. (1972) Trace element studies of lunar rocks and soils (abstract). In *Lunar Science—III* (editor C. Watkins), pp. 35–37, Lunar Science Institute Contr. No. 88.

Bryan W. B., Finger L. W., and Chayes F. (1969) Estimating proportions in petrographic mixing equations by least-squares approximations. *Science* **163,** 926–927.

Chao E. C. T., Boreman J. A., Minkin J. A., and James O. B. (1970) Lunar glasses of impact origin: Physical and chemical characteristics and geological implications. *J. Geophys. Res.* **75,** 7445–7479.

Cliff, R. A., Lee-Hu C., and Wetherill G. W. (1971) Rb–Sr and U, Th–Pb measurements on Apollo 12 material. *Proc. Second Lunar Sci. Conf., Geochim. Cosmochim. Acta* Suppl. 2, Vol. 2, pp. 1493–1502. MIT Press.

Compston W., Chappell B. W., Arriens P. A., and Vernon M. J. (1970) The chemistry and age of the Apollo 11 lunar material. *Proc. Apollo 11 Lunar Sci. Conf., Geochim. Cosmochim. Acta* Suppl. 1, Vol. 2, pp. 1007–1027. Pergamon.

Compston W., Berry H., Vernon M. J., Chappell B. W., and Kaye M. J. (1971a) Rubidium-strontium chronology and chemistry of lunar material from the Ocean of Storms. *Proc. Second Lunar Sci. Conf., Geochim. Cosmochim. Acta* Suppl. 2, Vol. 2, pp. 1471–1485. MIT Press.

Compston W., Vernon M. J., Berry H., and Rudowski R. (1971b) The age of the Fra Mauro formation: A radiometric older limit. *Earth Planet. Sci. Lett.* **12,** 55–58.

Compston W., Vernon M. J., Berry H., Rudowski R., Gray C. M., and Ware N. (1972) Apollo 14 mineral age and the thermal history of the Fra Mauro formation. *Proc. Third Lunar Sci. Conf., Geochim. Cosmochim. Acta* Suppl. 3, Vol. 2. MIT Press.

Eberhardt P., Geiss J., Graf H., Grogler N., Krahenbuhl U., Schwaller H., Schwarzmuller J., and Stettler A. (1971) Correlation between rock type and irradiation history of Apollo 11 igneous rocks. *Earth Planet. Sci. Lett.* **10,** 67–72.

Eberhardt P., Eugster O., Geiss J., Grogler N., Schwarzmuller J., Stettler A., and Weber L. (1972) When was the Apollo 12 KREEP ejected? (abstract). In *Lunar Science—III* (editor C. Watkins), pp. 206–208, Lunar Science Institute Contr. No. 88.

Gast P. W., Hubbard N. J., and Wiesmann H. (1970) Chemical composition and petrogenesis of basalts from Tranquility Base. *Proc. Apollo 11 Lunar Sci. Conf., Geochim. Cosmochim. Acta* Suppl. 1, Vol. 2, pp. 1143–1163. Pergamon.

Goldstein J. I., Henderson E. P., and Yakowitz H. (1970) Investigation of lunar metal particles. *Proc. Apollo 11 Lunar Sci. Conf., Geochim. Cosmochim. Acta* Suppl. 1, Vol. 1, pp. 499–512. Pergamon.

Goles G. G., Duncan A. R., Lindstrom D. J., Martin R. M., Beyer R. L., Osawa M., Randle K., Meek L. T., Steinborn T. L., and McKay S. M. (1971) Analyses of Apollo 12 specimens: Compositional variations, differentiation processes, and lunar soil mixing models. *Proc. Second Lunar Sci. Conf., Geochim. Cosmochim. Acta* Suppl. 2, Vol. 2, pp. 1063–1081. MIT Press.

Hubbard N. J. and Gast P. W. (1971) Chemical composition and origin of nonmare lunar basalts. *Proc. Second Lunar Sci. Conf., Geochim. Cosmochim. Acta* Suppl. 2, Vol. 2, pp. 999–1020. MIT Press.

Hubbard N. J., Meyer C. Jr., Gast P. W., and Wiesmann H. (1971a) The composition and derivation of Apollo 12 soils. *Earth Planet. Sci. Lett.* **10,** 341–350.

Hubbard N. J., Gast P. W., Meyer C. Jr., Nyquist L. E., Shih C., and Wiesmann H. (1971b) Chemical composition of lunar anorthosites and their parent liquids. *Earth Planet. Sci. Lett.* **13,** 71–75.

Jakes P., Warner J., Ridley W. I., Reid A. M., Harmon R. S., and Brett R. (1972) Petrology of a portion of the Mare Fecunditatis regolith. *Earth Planet. Sci. Lett.* **13,** 257–271.

Keil K., Bunch T. E., and Prinz M. (1970) Mineralogy and composition of Apollo 11 lunar samples. *Proc. Apollo 11 Lunar Sci. Conf., Geochim. Cosmochim. Acta* Suppl. 1, Vol. 1, pp. 561–598. Pergamon.

Keil K. and Prinz M. (1971) Mineralogy, petrology, and chemistry of some Apollo 12 samples. *Proc. Second Lunar Sci. Conf., Geochim. Cosmochim. Acta* Suppl. 2, Vol. 1, pp. 319–344. MIT Press.

Kendall M. G. and Stuart A. (1961) *The Advanced Theory of Statistics.* Hafner, New York.

Laul J. C., Morgan J. W., Ganapathy R., and Anders E. (1971) Meteoritic material in lunar samples. *Proc. Second Lunar Sci. Conf., Geochim. Cosmochim. Acta* Suppl. 2, Vol. 2, pp. 1139–1158. MIT Press.

Lindsay J. F. (1971) Mixing models and the recognition of end-member groups in Apollo 11 and 12 soils. *Earth Planet. Sci. Lett.* **12,** 67–72.

LSPET (Lunar Sample Preliminary Examination Team) (1972) The Apollo 15 lunar samples: A preliminary description. *Science* **175,** 363–375.

Lunatic Asylum (1970) Mineralogic and isotopic investigations on lunar rock 12013. *Earth Planet. Sci. Lett.* **9,** 137–163.

Marvin U. B., Wood J. A., Taylor G. J., Reid J. B. Jr., Powell B. N., Dickey J. S. Jr., and Bower J. F. (1971) Relative proportions and probable sources of rock fragments in the Apollo 12 soil samples. *Proc. Second Lunar Sci. Conf., Geochim. Cosmochim. Acta* Suppl. 2, Vol .1, pp. 679–699. MIT Press.

Metzger A. E., Trombka J. I., Peterson L. E., Reedy R. C., and Arnold J. R. (1972) A first look at the orbital gamma ray data (abstract). In *Lunar Science—III* (editor C. Watkins), pp. 540–541, Lunar Science Institute Contr. No. 88.

Meyer C. Jr. (1972) Mineral assemblages and the origin of non-mare lunar rock types (abstract). In *Lunar Science—III* (editor C. Watkins), pp. 542–544, Lunar Science Institute Contr. No. 88.

Meyer C. Jr. and Hubbard N. J. (1970) High potassium high phosphorous glass as an important rock type in the Apollo 12 soil sample (abstract). *Meteoritics* **5,** 210–211.

Meyer C. Jr., Brett R., Hubbard N. J., Morrison D. A., McKay D. S., Aitken F. K., Takeda H., and Schonfeld E. (1971) Mineralogy, chemistry, and origin of the KREEP component in soil samples from the Ocean of Storms. *Proc. Second Lunar Sci. Conf., Geochim. Cosmochim. Acta* Suppl. 2, Vol. 1, pp. 393–411. MIT Press.

Morgan J. W., Laul J. C., Krahenbuhl U., Ganapathy R., and Anders E. (1972) Major impacts on the moon: Chemical characterization of projectiles (abstract). In *Lunar Science—III* (editor C. Watkins), pp. 552–557, Lunar Science Institute Contr. No. 88.

Murthy V. R., Evensen N. M., John B., and Coscio M. R. Jr. (1971) Rb–Sr ages and elemental abundances of K, Rb, Sr, and Ba in samples from the Ocean of Storms. *Geochim. Cosmochim. Acta* **35,** 1139–1154.

Nyquist L. E., Hubbard N. J., Gast P. W., and Church S. E. (1972) Rb–Sr relationships for some chemically defined lunar materials (abstract). In *Lunar Science—III* (editor C. Watkins), pp. 584–586, Lunar Science Institute Contr. No. 88.

O'Kelley G. D., Eldridge J. S., Schonfeld E., and Northcutt K. J. (1972) Primordial radioelements and cosmogenic radionuclides in lunar samples from Apollo 15. *Science* **175,** 440–443.

Papanastassiou D. A., Wasserburg G. J., and Burnett D. S. (1970) Rb–Sr ages of lunar rocks from the Sea of Tranquility. *Earth Planet. Sci. Lett.* **8,** 1–18.

Papanastassiou D. A. and Wasserburg G. J. (1971a) Lunar chronology and evolution from Rb–Sr studies of Apollo 11 and 12 samples. *Earth Planet. Sci. Lett.* **11,** 37–62.

Papanastassiou D. A. and Wasserburg G. J. (1971b) Rb–Sr ages of igneous rocks from the Apollo 14 samples. *Earth Planet. Sci. Lett.* **12,** 36–48.

Pepin R. O., Bradley J. G., Dragon J. C., and Nyquist L. E. (1972) K–Ar dating of lunar soils: Apollo 12, Apollo 14, and Luna 16 (abstract). In *Lunar Science—III* (editor C. Watkins), pp. 602–604, Lunar Science Institute Contr. No. 88.

Reid A. M., Ridley W. I., Warner J., Harmon R. S., Brett R., Jakes P., and Brown R. W. (1972) Chemistry of highland and mare basalts as inferred from glasses in lunar soils (abstract). In *Lunar Science—III* (editor C. Watkins), pp. 640–642, Lunar Science Institute Contr. No. 88.

Ringwood A. E. and Essene E. (1971) Petrogenesis of Apollo 11 basalts, internal constitution, and origin of the moon. *Proc. Apollo 11 Lunar Sci. Conf., Geochim. Cosmochim. Acta* Suppl. 1, Vol. 1, pp. 769–799. Pergamon.

Schonfeld E. (1971) The formation age of the lunar material KREEP. NASA Report MSC-4975, Rev. 1, Oct. 1971.

Schonfeld E. (1972) Component abundance and ages in soils and breccia (abstract). In *Lunar Science—III* (editor C. Watkins), pp. 683–685, Lunar Science Institute Contr. No. 88.

Silver L. T. (1971a) U–Th–Pb isotope relations in Apollo 11 and 12 lunar samples. Second Lunar Science Conference (unpublished proceedings).

Silver L. T. (1971b) U–Th–Pb isotope systems in Apollo 11 and 12 regolithic materials and a possible age for the Copernicus impact event (abstract). *Trans. Am. Geophys. Union* **52,** 534.

Silver L. T. (1972a) U–Th–Pb abundances and isotopic characteristics in some Apollo 14 rocks and soils and Apollo 15 soil (abstract). In *Lunar Science—III* (editor C. Watkins), pp. 704–706, Lunar Science Institute Contr. No. 88.

Silver L. T. (1972b) Lead volatilization and volatile transfer on the moon (abstract). In *Lunar Science —III* (editor C. Watkins), pp. 701–703, Lunar Science Institute Contr. No. 88.

Tatsumoto M., Knight R. J., and Doe B. R. (1971) U–Th–Pb systematics of Apollo 12 lunar samples. *Proc. Second Lunar Sci. Conf., Geochim. Cosmochim. Acta* Suppl. 2, Vol. 2, pp. 1521–1546. MIT Press.

Tatsumoto M., Hedge C. E., Doe B. R., and Unruh D. (1972) U–Th–Pb and Rb–Sr measurements on some Apollo 14 lunar samples (abstract). In *Lunar Science—III* (editor C. Watkins), pp. 741–743, Lunar Science Institute Contr. No. 88.

Tilton G. R. (1960) Volume diffusion as a mechanism for discordant lead ages. *J. Geophys. Res.* **65,** 2933–2945.

Turkevich A. (1971) Comparison of the analytical results from the Surveyor, Apollo and Luna missions. *Proc. Second Lunar Sci. Conf., Geochim. Cosmochim. Acta* Suppl. 2, Vol. 2, pp. 1209–1215. MIT Press.

Turner G. (1971) $^{40}Ar$–$^{39}Ar$ ages from the lunar maria. *Earth Planet. Sci. Lett.* **11,** 169–191.

Vinogradov A. P. (1971) Preliminary data on lunar ground brought to Earth by automatic probe "Luna 16." *Proc. Second Lunar Sci. Conf., Geochim. Cosmochim. Acta* Suppl. 2, Vol. 1, pp. 1–16. MIT Press.

Wakita H. and Schmitt R. A. (1970) Lunar anorthosites: Rare-earth and other elemental abundances. *Science* **170,** 969–974.

Wanke H., Wlotzka F., Baddenhausen H., Balacescu A., Spettel B., Teschke F., Jagoutz E., Kruse H., Quijano-Rico M., and Rieder R. (1971) Apollo 12 samples: Chemical composition and its relation to sample locations and exposure ages, the two component origin of the various soil samples and studies on lunar metallic particles. *Proc. Second Lunar Sci. Conf., Geochim. Cosmochim. Acta* Suppl. 2, Vol. 2, pp. 1187–1208. MIT Press.

Wasserburg G. J., Turner F., Tera F., Podosek F. A., Papanastassiou D. A., and Huneke J. C. (1972) Comparison of Rb–Sr, K–Ar, and U–Th–Pb ages; lunar chronology and evolution (abstract). In *Lunar Science—III* (editor C. Watkins), pp. 788–790, Lunar Science Institute Contr. No. 88.

Wetherill G. W. (1956) Discordant uranium-lead ages, I. *Trans. Am. Geophys. Union* **37,** 320–326.

Wood J. A. (1970) Petrology of the lunar soil and geophysical implications. *J. Geophys. Res.* **75,** 6497–6513.

Proceedings of the Third Lunar Science Conference
(Supplement 3, *Geochimica et Cosmochimica Acta*)
Vol. 2, pp. 1421–1427
The M.I.T. Press, 1972

# ESCA-Investigation of lunar regolith from the Seas of Fertility and Tranquility

A. P. Vinogradov, V. I. Nefedov, V. S. Urusov, and N. M. Zhavoronkov

V. I. Vernadsky Institute of Geochemistry and Analytical Chemistry
and N. S. Kuzhakov Institute of General and Inorganic Chemistry,
USSR Academy of Sciences, Moscow

**Abstract**—The x-ray photoelectron spectra Fe2p, Si2p, Ti2p, Al2p, Mg2p, O1s have been investigated in the samples of lunar regolith from the Seas of Fertility and Tranquility. The same spectra were obtained for approximately 30 minerals, oceanic gabbro, meteorite-eucrite, and several iron meteorites. The spectra were measured by means of a VIEE-15 spectrometer. The analysis of results shows that all investigated elements have a usual oxidation degree and in their nearest environment there are oxygen atoms. The prevailing coordination number of Al is 4.

In the Fe2p spectrum of regolith, together with lines of oxidized iron (from various mineral phases), a peak belonging to metallic iron and equaling 10–15% of the line intensity of oxidized iron was detected. Careful analysis and comparison of the Fe2p spectra in regolith, various iron meteorites and stainless steels leads to the following conclusion: Metallic iron in lunar regolith is present in a finely dispersed state and is extremely stable with respect to oxygen of the earth atmosphere.

## Introduction

The x-ray photoelectron spectroscopy can be used for the determination of the oxidation degree of elements in chemical compounds and for the study of different structural aspects (Siegbahn *et al.*, 1967). We have applied this new method for investigation of lunar regolith.

The principle of the ESCA-investigation is quite simple. The inner shell electron of the element under investigation is expelled by an x-ray quantum of a known energy $h\nu$. The kinetic energy $E_{kin}$ of this electron is measured in the spectrometer. The binding energy $E_b$ of the inner shell electron can then be calculated as the difference between the x-ray energy and the kinetic energy of the electron from the equation

$$E_b = h\nu - E_{kin} - \varphi$$

Where $\varphi$ is the work function of the spectrometer material.

The value of the binding energy $E_b$ depends on the chemical environment of the element under investigation. For example, this value for 2p-electron of iron is different for metallic iron and for compounds with iron in oxidation states II or III.

## Results and Discussion

The spectra were recorded by a Varian IEE-15 spectrometer. Alk$\alpha$-line was used for excitation of spectra (voltage: 8 kV; current: 80mA; vacuum: $10^{-6}$ Torr). With help of retarding field the value $E_{kin}$ was reduced to 100 eV. These conditions lead to the spectrum intensity maximum. The experimental procedure was quite similar to that described previously (Nefedov *et al.*, 1972).

The samples were prepared by rubbing the powder into a golden net placed on a metallic cylinder covered with an organic film not containing oxygen. The samples were also prepared by putting the powder on a sticky scotch-tape. Meteorites were investigated in the form of fillings. The spectra of the Sichote-Alyn meteorite and of some steels were recorded from whole samples in cylinder form. All spectra were recorded from several independent samples. For example, three regolith samples from the Sea of Fertility Luna-16-A3 ($<0.083$ mm) and two samples from the Sea of Tranquility 10005,35,2 were prepared (Table 1).

We also have investigated the electron binding energies of O, Mg, Al, Si, Fe, and Ti in about 30 minerals (Nefedov *et al.*, 1972), in gabbro from the mid-Indian Mountain ridge of the Indian Ocean, in several meteorites (including eucrite, kamacite, and taenite) and various steels. In the table there are the major element compositions in some of our samples (Vinogradov, 1971).

ESCA-spectra of regolith from different sites are quite similar, although as it was already established previously that the regolith from the Sea of Tranquility contains more Ti than the regolith from the Sea of Fertility. O1s-, Si2p-, Al2p- and Mg2p-spectra in lunar regolith are similar to those of the eucrite Chervony Koot and oceanic gabbro. It can be concluded that Mg, Al, and Si have the usual oxidation states in regolith and the nearest neighbors of these elements are oxygen atoms (Nefedov *et al.*, 1972).

A comparison of Al2p- and Si2p-lines in regolith and in well-known minerals allowed a qualitative evaluation of the mineral composition of lunar regolith. In accordance with many other methods (Vinogradov, 1971), it was shown that the coordination number of aluminum atom in most regolith minerals is four because the shift of Al2p-line in regolith (relative to the line in aluminum) is equal to 2.3 eV (Nefedov *et al.*, 1972). The Si2p-line shift in regolith is about 3.4 eV relative to the line in silicon. The shift of these lines in silicates varies from 4.0 eV in quartz $SiO_2$ to 3.0 eV in nepheline $NaAlSiO_4$. It can be concluded that the extreme basic and acid mineral phases are not present in essential amounts in lunar regolith. These results

Table 1. The compositions of lunar regoliths, eucrite and oceanic gabbro (wt. %).

| Component | The Regolith from the Sea of Fertility Luna–16–A3 | The Regolith from the Sea of Tranquility 10005,35,2 | Eucrite Chervony Koot (Chirvinsky, Sokolova, 1946) | Oceanic Gabbro |
|---|---|---|---|---|
| $SiO_2$ | 42 | 43 | 49 | 27 |
| $Al_2O_3$ | 15 | 13 | 13 | 12 |
| $Fe_2O_3$ | — | — | — | 14 |
| FeO | 17 | 16 | 19 | 17 |
| MgO | 12 | 12 | 11 | 5 |
| CaO | 8.7 | 8 | 7 | 10 |
| $TiO_2$ | 3.4 | 7 | 0.7 | 8 |
| $Na_2O$ | 0.4 | 0.5 | 0.5 | 0.6 |
| $K_2O$ | 0.1 | 0.1 | 0.1 | 0.9 |
| S | 0.2 | — | — | — |
| Ni | 0.02 | 0.007 | — | — |

are in good agreement with the data of other methods (Vinogradov, 1971); the regolith consists mainly of plagioclase, pyroxene, and olivine. The shift of Si2p-line in these minerals and regolith is practically the same.

Most interesting results can give a comparison of the Fe2p-lines in lunar and terrestrial samples. Three independent Fe2p-spectra of two samples from the Sea of Tranquility and seven independent Fe2p-spectra of three samples from the Sea of Fertility were recorded (Figs. 1 and 2). In all regolith spectra there are lines of metallic iron. These lines are absent in spectra of eucrite and oceanic gabbro. The lines of metallic iron are of course also absent in oxygen-containing minerals of usual origin.

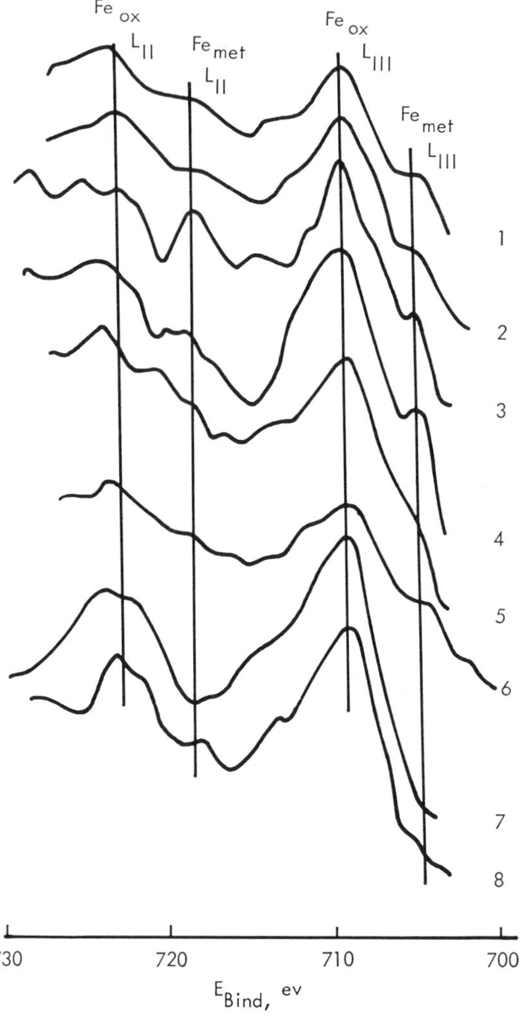


Fig. 1. Fe2p-lines. Curves: 1, 2 = the regolith from the Sea of Tranquility; 3–6 = the regolith from the Sea of Fertility; 7 = oceanic gabbro; 8 = eucrite Chervony Koot.

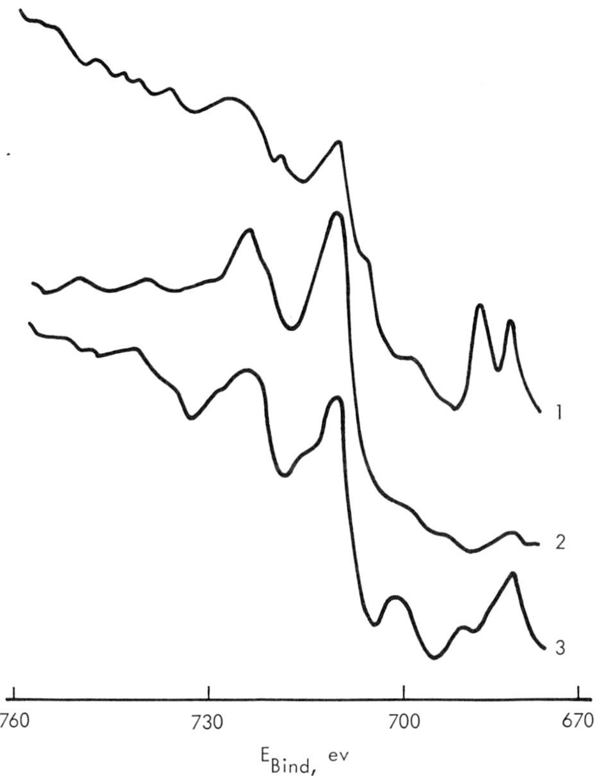


Fig. 2. Fe2p-lines (the larger spectral range). Curves: 1 = the regolith from the Sea of Fertility; 2 = oceanic gabbro; 3 = eucrite.

In pyrite $FeS_2$ and in pyrrhotite $Fe_{1-x}S$ there is a line in the same region (in $Fe_{1-x}S$ probably owing to the presence of $FeS_2$ in the surface layer, because in stoichiometric synthetic troilite FeS as well as in the sample of troilite from Sichote-Alyn iron meteorite this line is absent) (Fig. 3). We believe that this line in regolith spectra is mainly due to the presence of metallic iron and not to that of pyrite. There are several arguments for such a conclusion. The regolith contains little sulphur (only about two-tenths of a percent). The minerals pyrite and pyrrhotite have also not been found by other physical or chemical methods. It could be supposed that pyrite is distributed as a thin layer, therefore it was not possible to determine it. But in this case we must have a very intensive sulphur line. The latter was not proved. Thus, we must conclude that the recorded unusual maximum belongs mainly to metallic iron.

The presence of metallic iron in lunar regolith is in accordance with data of other methods. However ESCA-spectra allow to find out some unusual properties of this iron. For example, Mössbauer and electron spin resonance spectra of regolith have shown that the concentration of metallic iron constitutes only about 2–5% of all iron (Vinogradov, 1971). But from our Fe2p-spectra we must conclude that the line intensity of metallic iron is not less than 10% of the line intensity of the oxidized iron.

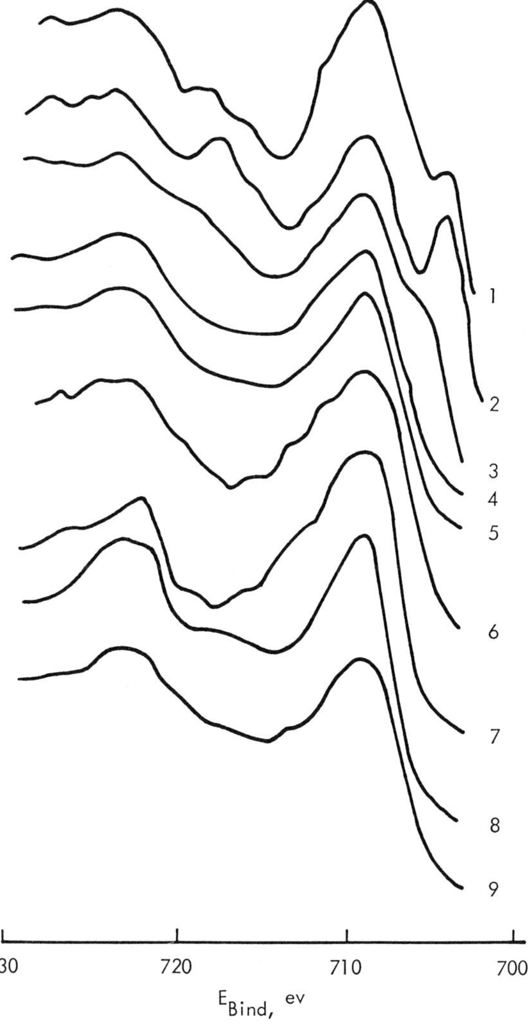


Fig. 3. Fe2p-lines. Curves: 1 = the regolith from the Sea of Fertility; 2 = pyrite $FeS_2$; 3 = pyrrhotite $Fe_{1-x}S$; 4 = troilite FeS; 5 = ilmenite $FeTiO_3$; 6 = olivine $(Mg, Fe)_2 SiO_4$; 7 = biotite $K(Mg, Fe)_3 (AlSi_3O_{10}) (OH)_2$; 8 = hematite $Fe_2O_3$; 9 = magnetite $Fe_3O_4$.

It is necessary to take into account that x-ray-photoelectron spectra can be obtained only from the thin surface layer of a sample (about 100 Å), therefore the relative surface of the metallic iron constitutes at least 10% of the oxidized iron surface although its volume and weight concentration is noticeably less. Besides that it is necessary to take into consideration that some part of the metallic iron surface is oxidized. From this we conclude that the metallic iron is present in regolith in a more dispersed state in comparison with other iron-containing minerals.

The second unusual property of lunar iron is its very high stability in the earth air atmosphere. Indeed the Fe2p-line of stainless steel shows that the surface layer is entirely oxidized (Fig. 4). Only after a careful cleaning of this sample surface the metallic iron line appears in the spectrum (Fig. 4). The time of air action is several minutes. After 24 hours of a sample being in the air, the line intensity of metallic iron sharply decreases and after three weeks practically disappears (Fig. 4).

The Fe2p-spectra of several iron meteorites were also investigated. In fresh kamacite fillings of the Gress meteorite, kamacite of the meteorite Sichote-Alyn and taenite of the meteorite Santa-Katarina, we observed the lines of metallic iron (Fig.

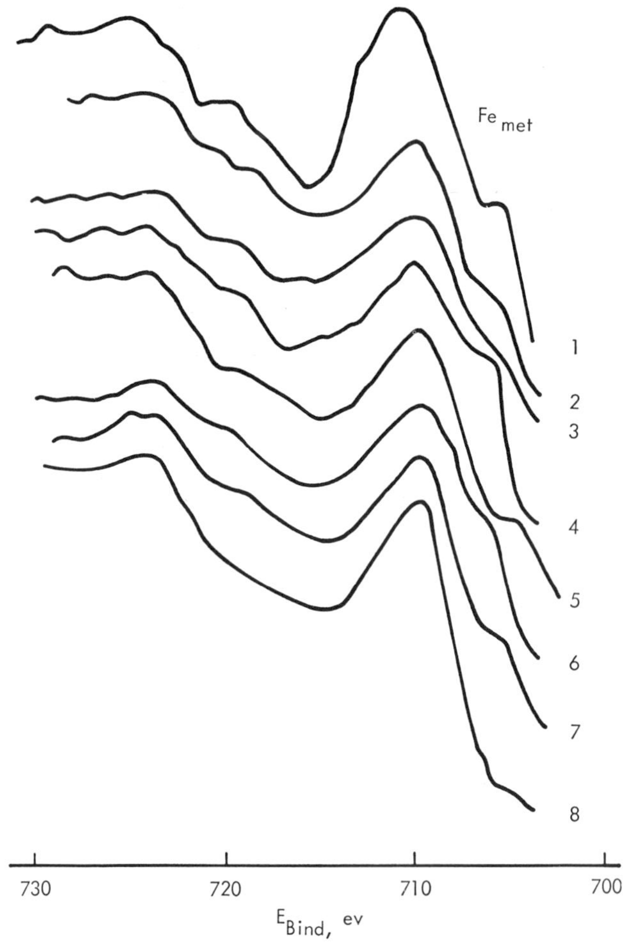


Fig. 4. Fe2p-lines. Curves: 1 = the regolith from the Sea of Fertility; 2 = kamacite, Santa-Katarina; 3 = taenite, Gressk; 4 = kamacite, Sichote-Alyn; 5 = kamacite, Sichote-Alyn, after 2 days of being in the air; 6 = stainless steel, after cleaning of surface; 7 = stainless steel, after 24 hours of being in the air; 8 = stainless steel, after 3 weeks of being in the air.

4). These line intensities decreases sharply after a stay in the air during two days, however after the following ten days or three weeks in the air they did not sufficiently change. These facts show that meteoritic iron is relatively resistant to air oxygen.

The investigation regolith samples had been in the atmosphere approximately during half a year. Nevertheless the ratio of intensity maxima of metallic and oxidized iron in lunar regolith is approximately equivalent to this ratio in samples of iron meteorites or steels, though metallic iron in regolith is a very small part of the oxidized iron. Thus, these observations are in accordance with each other only in the case of an unusual resistance of lunar metallic iron to oxidation by oxygen of the earth atmosphere.

Indeed let us suppose that lunar metallic iron is noticeably oxidized. Then the observed ratio of the intensity maxima of oxidized and metallic iron would be possible only at a very high dispersity degree of the metallic iron in regolith. In this case the surface of metallic iron would be much larger than the surface of the iron-containing minerals. However then we had to expect a sharp increase of relative Fe2p-line intensity from oxidized forms of Fe, for example, relative to the Si2p-line in comparison with eucrite or oceanic gabbro. But this is not the case in our experiments.

Thus, the experimental data lead to the conclusions that metallic iron in lunar regolith is in a finely dispersed state and the oxide film on its surface is very thin or even partly absent. The possible reasons for such unusual surface properties are the following: (1) the high purity of the metallic phase (or unusual set of dopants); (2) the condensation of the metallic phase under conditions of a very high vacuum; (3) the irradiation (and cleaning) of the metallic iron surface by ionic flow (solar wind) during a long time.

Now all these factors are under investigation in our laboratories.

## References

Chirvinsky P. N. and Sokolova A. I. (1946) Petrographical and chemical characteristics of the Chervony Koot eucrite (fall on June 23, 1939). *Meteoritica* **III,** 37.

Nefedov V. I., Urusov V. S., and Kakhana M. M. (1972) ESCA-investigation of chemical bond in minerals of Na, Mg, Al, and Si. *Geochemistry* (Russian), N 1, 11.

Siegbahn K., Nordling C., Fahlman A., Nordberg R., Hamrin K., Hedman J., Johansson G., Bergmark T., Karlsson S. E., Lindgren I., and Lindberg B. (1967) ESCA atomic, molecular, and solid state structure studied by means of electron spectroscopy. *Nova Acta Reg. Soc. Scient. Upsaliensis*, ser. IV, v. 20, Uppsala.

Vinogradov A. P. (1971) Preliminary data on lunar ground brought to earth by the automatic probe "Luna-16." *Geochemistry* (Russian), N 3, 261; also *Proc. Second Lunar Sci. Conf., Geochim. Cosmochim. Acta* Suppl. 2, Vol. 1, pp. 1–16. Pergamon.

Proceedings of the Third Lunar Science Conference
(Supplement 3, *Geochimica et Cosmochimica Acta*)
Vol. 2, pp. 1429–1454
The M.I.T. Press, 1972

# $O^{18}/O^{16}$, $Si^{30}/Si^{28}$, $C^{13}/C^{12}$, and D/H studies of Apollo 14 and 15 samples

S. EPSTEIN and H. P. TAYLOR, Jr.
Division of Geological and Planetary Sciences,
California Institute of Technology,
Pasadena, California 91109

**Abstract**—New $\delta O^{18}$ and $\delta Si^{30}$ data on Apollo 14 and 15 samples further confirm the remarkable isotopic similarity of lunar rocks from different sites. As in the case of Apollo 11 and 12 samples, the lunar fines are enriched in $O^{18}$ and $Si^{30}$ relative to the crystalline rocks from which they are derived. Certain Apollo 14 breccias (F-2) are also enriched in $O^{18}$, and thus may in large part be derived from lunar soil. The Apollo 14 bottom trench samples are depleted in $O^{18}$ relative to surface soils. The anorthosite 15415 has a $\delta O^{18}$ very similar to that obtained for plagioclases in the lunar basalts. Utilizing successive partial fluorinations, large $O^{18}$ and $Si^{30}$ enrichments (up to $+55$ and $+30‰$, respectively) were observed on the surfaces of grains of all Apollo 14 and 15 fines, but no such enrichment exists for 14321, a rock breccia (class F-4). These large $O^{18}$- and $Si^{30}$-enrichments are directly related to the amount of solar wind hydrogen present in the sample. These isotopic effects must be a result of exposure of the lunar fines to bombardment by meteorites, micrometeorites, and/or nuclear particles, with preferential loss of $Si^{28}$ and $O^{16}$ by fractional vaporization, or preferential gain of $Si^{30}$ and $O^{18}$ by fractional condensation. These effects are accompanied by a reduction in the overall oxygen/silicon ratio, indicating that the $O_2/Si$ ratio of material carried away during vaporization is $>2$.

Further confirmation has been obtained that the deuterium concentration in hydrogen gas extracted from the lunar fines is less than about 3 ppm. Even this small amount of deuterium could be due to spallation because an abnormally deuterium-rich fraction of gas was extracted during the melting of the lunar samples. The bulk of the hydrogen in the lunar fines is clearly of solar wind origin; the latter apparently has a $\delta D$ value close to $-1000$ (0 ppm deuterium). The small amounts of $H_2O$ ($\sim 10$ $\mu$mole/g) extracted from one lunar soil and one rock breccia have $\delta O^{18}$ values of $-5.9$ and $-18.2$, well within the $\delta O^{18}$ range of terrestrial atmospheric $H_2O$. Taken together with other considerations, this strongly indicates that most of this water is a terrestrial contaminant taken to the moon by the Apollo spacecraft or added to the samples after their return to earth. The carbon concentrations and the $\delta C^{13}$ values in lunar samples are both highly variable, about 10 to 260 ppm and about $-30$ to $+20$, respectively. Nonetheless, there is a progressive enrichment in $\delta C^{13}$ with increasing carbon content, and all the data can be explained by mixing solar wind carbon having a variable $\delta C^{13}$ of $+10$ to $+25$ with lunar basalts containing about 25 ppm carbon and having a $\delta C^{13} = -30$ to $-20$. However, because of possible addition of meteoritic carbon, and because of possible $C^{13}$ enrichment due to particle bombardment, we cannot at this time definitely determine whether the bulk of the lunar carbon is of solar wind or meteoritic origin.

## OXYGEN AND SILICON ISOTOPE DATA

THE NEW $\delta O^{18}$ and $\delta Si^{30}$ data obtained on various Apollo 14 and 15 lunar samples are given in Table 1 and Fig. 1. These data supplement those already published on Apollo 11 and 12 rocks (Taylor and Epstein, 1970a; Epstein and Taylor, 1970; O'Neil and Adami, 1970; Onuma *et al.*, 1970; Epstein and Taylor, 1971; Clayton *et al.*, 1971) and on the unusual, high-$SiO_2$ lunar rock 12013 (Taylor and Epstein, 1970b).

Table 1. Oxygen and silicon isotope analyses of Apollo 14 and 15 samples.

| Sample | $\delta O^{18}$ (‰)‡ | $\delta Si^{30}$ (‰) |
|---|---|---|
| *Apollo 14 samples* | | |
| *Fines* | | |
| 14298 | 6.62 (1) | |
| 14422 | 6.53 ± 0.09 (2) | |
| 14148 trench top | 6.56 ± 0.05 (3) | |
| 14156 trench middle | 6.45 ± 0.13 (2) | |
| 14149 trench bottom | 6.22 ± 0.11 (3) | |
| 14240 trench bottom SESC | 6.18 (1) | |
| *Breccias* | | |
| 14321, type F4, whole rock§ | 5.82 (1) | |
| 14307, type F2, black matrix | 6.58 ± 0.07 (2) | |
| 14307, type F2, D-2 brown clast | 5.84 ± 0.12 (2) | |
| 14068 matrix | 6.43 (1) | |
| 14063, type F3, whole rock | 5.87 (1) | |
| *Apollo 15 samples* | | |
| *Fines* | | |
| 15021 | 6.30 ± 0.09 (2) | |
| 15251 | 6.12 ± 0.03 (2) | |
| 15301 | 6.25 (1) | |
| *Basalt (microgabbro)* | | |
| 15058 clinopyroxene | 5.73 ± 0.04 (2) | −0.31 ± 0.05 (2) |
| 15058 plagioclase | 6.13 ± 0.05 (2) | −0.14 ± 0.10 (2) |
| *Anorthosite* | | |
| 15415 | 6.05 ± 0.04 (4) | −0.18 (1) |
| *Breccias* | | |
| 15455 white* | 5.83 (1) | |
| 15455 dark† | 5.90 (1) | −0.26 (1) |
| 15205 gray matrix | 5.92 (1) | |
| 15205 black glass | 6.07 (1) | |

* Brecciated norite, containing plagioclase and pale-green pyroxene.
† Vesicular basaltic breccia that veins the white material.
‡ Analytical error shown is average deviation from the mean. Numbers in parentheses indicate number of separate determinations.
§ F2, F4, D-2 notation is after Jackson and Wilshire (1972).

The most striking feature of all these data is that the Apollo 14 and 15 samples are essentially isotopically indistinguishable from the Apollo 11 and 12 rocks. The total range of $O^{18}/O^{16}$ in *all* whole-rock lunar samples is from $\delta O^{18} = 5.53$ to 6.62, and for $\delta Si^{30}$, from −0.49 to +0.31. In comparing $O^{18}$ results from different laboratories, cognizance must be taken of the difference in SMOW standards between the Chicago and Pasadena laboratories; our results are 0.2–0.3‰ higher than those of Onuma *et al.* (1970) and Clayton *et al.* (1971).

Note that we are reporting no silicon isotope results on Apollo 14 samples. We have obtained $\delta Si^{30}$ analyses for all the samples listed in Table 1, and many of these analyses fall in the above-quoted range of $\delta Si^{30}$. However, some of the samples exhibit apparent $\delta Si^{30}$ values as low as −1.5‰. Because of the relatively high abundance of the high-phosphorus KREEP component in all the Apollo 14 rocks and soils, we feel that many of our silicon isotope data on these rocks may be in error by as much as 1‰. As was pointed out previously for rock 12013 (Epstein and Taylor, 1971) the presence of appreciable amounts of phosphorus interferes with the $SiF_4$ analysis on our mass spectrometer; this effect leads to abnormally negative $\delta Si^{30}$ values. We will

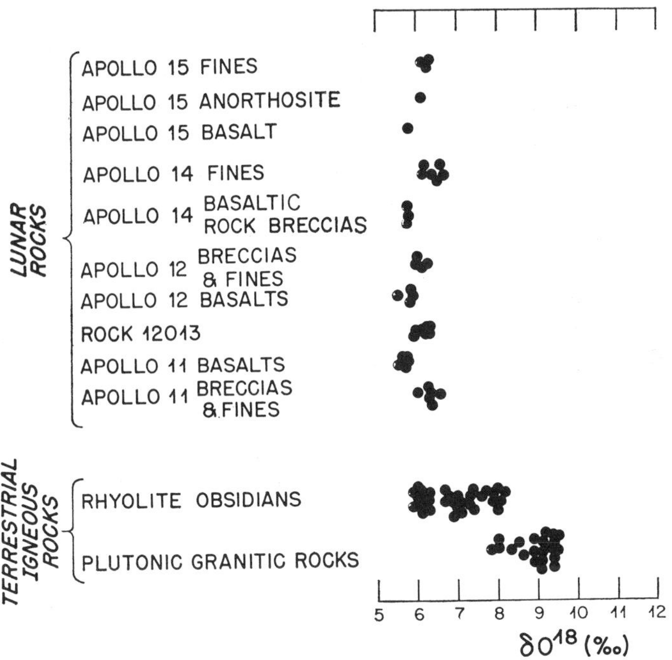


Fig. 1. Plot of $\delta O^{18}$ values of all lunar whole-rock samples analyzed by us (Taylor and Epstein, 1970a; 1970b; Epstein and Taylor, 1971; and Table 1 in the present paper). The $\delta O^{18}$ values of the lunar samples are essentially indistinguishable from terrestrial basalts, suggesting a genetic affinity between the earth and moon. For comparison, the $\delta O^{18}$ values of a number of "normal" $SiO_2$-rich terrestrial igneous rocks (Taylor, 1968) are also shown on the diagram. Very large $\delta O^{18}$ differences exist in silicate minerals on earth ($\delta O^{18}$ varies from $-10$ to $+40‰$). Much of this $O^{18}$ variation on earth is due to low-temperature sedimentary and metamorphic reactions involving $H_2O$; obviously such types of reactions have not been important on the moon.

have to treat separate aliquots of the Apollo 14 samples with HCl, thereby eliminating the phosphate, and then perform a chemical separation of the silicon for isotopic analysis. Unfortunately, such a procedure introduces $O^{18}/O^{16}$ contamination, so we cannot obtain valid $\delta O^{18}$ analyses on samples treated in this fashion. Therefore, for KREEP-rich samples, we cannot simultaneously obtain both valid $\delta O^{18}$ and valid $\delta Si^{30}$ data on the same aliquot of sample material. Preliminary experiments indicate that these difficulties can be resolved, but it obviously means that more lengthy experimental procedures are required to obtain useful silicon isotope data.

Lunar soils and soil breccias have similar $\delta O^{18}$ and $\delta Si^{30}$ for both the Apollo 11 and 12 sites, but they are enriched in $\delta O^{18}$ and $\delta Si^{30}$ relative to the local basalts from which they are presumably largely derived: the same relationship holds for the Apollo 14 and 15 samples (see Fig. 1). These $\delta O^{18}$ and $\delta Si^{30}$ enrichments in the whole-rock samples are much more readily apparent in the finest size fractions and/or the grain surfaces of the lunar fines, as is discussed more fully below. Among the samples of Apollo 14 fines are some of the most $O^{18}$-rich samples yet analyzed from the moon. Samples 14298, 14422, 14148 are all richer in $O^{18}$ than any other lunar samples except the soil breccia 10061. The $O^{18}$ enrichment of Apollo 14 soils relative to other lunar soils, particularly those from Apollo 12 and 15, may be simply a result of higher $\delta O^{18}$ values in the parent material from which the Apollo 14 fines were derived (perhaps more plagioclase-rich?). It is interesting to compare the average $\delta O^{18}$ values of the lunar soils and soil breccias from the four Apollo landing sites: Apollo 11: +6.31 (six samples); Apollo 12: +6.08 (four); Apollo 14: +6.43 (six); Apollo 15: +6.22 (three). Other than differences in mineralogical composition in the parent rocks from which these lunar fines were derived, there is no obvious reason for the small differences in $O^{18}$ among the various sites. All surface soil samples at each site are enriched in solar wind $H_2$ and carbon, and all the soils show grain surface $\delta O^{18}$ and $\delta Si^{30}$ enrichments. (This statement must be qualified because we have not yet carried out partial fluorination experiments on Apollo 12 surface soil samples.)

Note in Table 1 that the Apollo 14 trench fines show a systematic downward depletion of $O^{18}$. Two separate trench bottom samples (14149 and 14240) are both lower in $O^{18}$ than all other Apollo 14 surface fines, further substantiating the $O^{18}$ depth effect.

The anorthosite, 15415, has a $\delta O^{18}$ value at the low end of the spectrum of $\delta O^{18}$ values previously obtained on plagioclase from Apollo 11 and 12 basalts and microgabbros. It is also similar in $\delta O^{18}$ to plagioclase from the 15058 microgabbro. All the lunar plagioclase separates analyzed by us have $\delta O^{18}$ values in the narrow range +6.05 to +6.33. The plagioclase-clinopyroxene fractionation in 15058 is 0.4, similar to the values obtained in other lunar basalts and appropriate for isotopic equilibrium at magmatic temperatures.

The Apollo 14 and 15 rock breccias exhibit some interesting $O^{18}/O^{16}$ variations. The two Apollo 15 breccias and the Type F-3 and F-4 Apollo 14 breccias are almost identical in $O^{18}$ to the lunar basalts and microgabbros. They therefore contain little or none of the "exotic" $O^{18}$-rich component present in the lunar soils and soil breccias. As will become apparent in the discussion that follows, this implies that these rock breccias probably were not formed from materials that resided in a fragmented condition on the lunar surface for appreciable lengths of time. This conclusion is valid irrespective of whether or not there has been closed-system (internal) chemical re-equilibration in the breccias during metamorphism or cooling, because even though the large grain-surface $O^{18}$ effects might disappear, the whole-rock sample should still show as much as 0.5 to 0.8‰ $O^{18}$ enrichment relative to the lunar basalts.

Another class of Apollo 14 breccia (F-2 of Jackson and Wilshire, 1972) definitely does show $O^{18}$ enrichment. The black matrix, which constitutes the major portion of sample 14307, has a $\delta O^{18}$ very similar to that of the Apollo 14 surface fines. This suggests that 14307 may be composed in large part of consolidated lunar soil. The high-$O^{18}$ character of these F-2 breccias conceivably also could be due to the greater abundance of glass fragments in such rocks. It will be interesting to further test these concepts by seeing whether $Si^{30}$ is also enriched in these Apollo 14 F-2 breccias. Note also that rock 14307 as a whole probably cannot have undergone appreciable metamorphism at high temperature, because it contains clasts that have much lower $\delta O^{18}$

Table 2. $\delta O^{18}$ and $\delta Si^{30}$ Data obtained by successive partial fluorinations of Apollo 14 lunar samples.

| Sample | Cum. $\mu$mole $O_2$ / mg original sample | $\delta O^{18}$ | $\mu$mole $O_2$ | $\delta Si^{30}$ | $\mu$mole $SiF_4$ | Ratio $O_2/SiF_4$ |
|---|---|---|---|---|---|---|
| 14298, 14 fines (0.772 g) | | | | | | |
| a. 30 min/80°C | 0.036 | +57.7 | 28 | +24.8 | 28 | 1.00 |
| b. 60 min/121°C | 0.049 | +34.9 | 9.5 | +7.65 | ~7.5 | 1.27 |
| c. 30 min/143°C | 0.070 | +30.9 | 16.5 } | +4.20 | 39 | 1.39 |
| d. 30 min/197°C | 0.118 | +23.8 | 37 } | | | |
| e. 30 min/209°C | 0.177 | +15.66 | 46 | +1.16 | 31 | 1.48 |
| f. 30 min/247°C | 0.303 | +11.93 | 97 | +1.31 | 65 | 1.49 |
| Total | | | 234 | | 170.5 | 1.37 |
| 14422, 12 fines (1.265 g) | | | | | | |
| a. 20 min/87°C | 0.020 | +56.6 | 25 | +23.2 | 25 | 1.00 |
| b. 60 min/121°C | 0.033 | +30.0 | 16.5 } | +6.51 | 30 | 1.48 |
| c. 45 min/146°C | 0.055 | +26.7 | 28 } | | | |
| d. 25 min/190°C | 0.113 | +20.6 | 73 | +3.07 | 55 | 1.33 |
| e. 25 min/217°C | 0.177 | +15.40 | 82 | +2.87 | 55 | 1.49 |
| f. 30 min/247°C | 0.336 | +11.29 | 201 | +0.47 | 154 | 1.31 |
| g. 30 min/279°C | 0.559 | +8.24 | 281 | | 167 | 1.68 |
| h. 30 min/299°C | 0.774 | +8.13 | 277 | | 175 | 1.58 |
| i. 30 min/345°C | 1.097 | +7.38 | 404 | | 258 | 1.57 |
| Total | | | 1387.5 | | 919 | 1.51 |
| 14148, 7 trench fines, top (0.6639 g) | | | | | | |
| a. 30 min/77°C | 0.035 | +51.4 | 23 | +22.3 | 23 | 1.00 |
| b. 30 min/120°C | 0.049 | +27.6 | 9.5 } | +5.97 | 15 | 1.37 |
| c. 30 min/145°C | 0.066 | +28.9 | 11 } | | | |
| d. 30 min/200°C | 0.132 | +19.8 | 44 | +2.62 | 29 | 1.52 |
| e. 30 min/247°C | 0.341 | +11.31 | 139 | | 93 | 1.49 |
| f. 50 min/266°C | 0.645 | +8.40 | 202 | | 138 | 1.46 |
| g. 30 min/296°C | 0.935 | +7.01 | 192 | | 127 | 1.51 |
| h. 30 min/303°C | 1.186 | +6.63 | 167 | | 111 | 1.50 |
| Total | | | 787.5 | | 536 | 1.47 |
| 14156, 4 trench fines, middle (0.6273 g) | | | | | | |
| a. 30 min/77°C | 0.021 | +56.3 | 13 | +26.4 | 12 | 1.17 |
| b. 30 min/121°C } | 0.040 | +33.3 | 12 | +8.97 | 12 | 1.00 |
| c. 30 min/147°C } | | | | | | |
| d. 30 min/198°C | 0.096 | +26.3 | 35 | +4.50 | 24 | 1.46 |
| Total | | | 60 | | 48 | 1.25 |
| 14149,23 trench fines, bottom (0.7696 g) | | | | | | |
| a. 30 min/77°C | 0.017 | +42.4 | 13 | +20.6 | 12 | 1.08 |
| b. 30 min/121°C | 0.038 | +25.9 | 16 | +8.91 | 10 | 1.60 |
| c. 30 min/147°C | 0.064 | +24.4 | 20 | +6.36 | 9 | 2.22 |
| d. 30 min/200°C | 0.146 | +15.47 | 63 | | ~45 | ~1.4 |
| e. 30 min/247°C | 0.326 | +10.31 | 139 | | 96 | 1.45 |
| f. 50 min/266°C | 0.585 | +7.79 | 199 | | 138 | 1.44 |
| g. 30 min/297°C | 0.823 | +6.95 | 183 | | 130 | 1.41 |
| h. 30 min/303°C | 1.046 | +6.73 | 172 | | 114 | 1.51 |
| Total | | | 805 | | 554 | 1.45 |
| 14321,59 rock breccia (0.963 g) | | | | | | |
| a. 30 min/80°C | 0.021 | +7.81 | 20 | +1.48 | 11 | 1.82 |
| b. 30 min/123°C | 0.032 | +9.12 | 11 } | −0.13 | 17 | 1.21 |
| c. 30 min/156°C | 0.042 | +8.09 | 9.5 } | | | |
| d. 30 min/202°C | 0.066 | +7.78 | 23 | −1.32 | 17 | 1.35 |
| Total | | | 63.5 | | 45 | 1.41 |

values; a brown clast in this rock (D-2 of Jackson and Wilshire?) is isotopically identical to the lunar basalts and 0.7‰ lower in $O^{18}$ than the matrix.

## Large $O^{18}$ and $Si^{30}$ Enrichments in the Lunar Fines

We have continued our investigations of the $O^{18}$-rich and $Si^{30}$-rich component(s) in the lunar soils and breccias. Previously we had shown by a series of partial fluorination experiments that the grain surfaces and/or the finest size fractions of the Apollo 11 and 12 lunar fines and soil breccias were as much as 50 and 25‰ enriched in $O^{18}$ and $Si^{30}$, respectively (Epstein and Taylor, 1971). Utilizing the same analytical procedures, we have obtained the data shown in Tables 2 and 3; the data are illustrated graphically in Figs. 2, 3, 4, 5, 6, and 7. The isotopic enrichments in all these figures are plotted versus the fraction of sample reacted (denoted on a log scale to emphasize that the extreme isotopic enrichments are observed only in the initial two or three extractions). A smoothed curve has been drawn through the data points obtained on each sample. Note that a value of 0.1 $\mu$mole $O_2$ per mg means that only about 0.8% of the oxygen has been removed from the sample, and that even in the most extreme cases the isotopic effects disappear after about 8% of the oxygen has been removed.

Comparison of Figs. 2–7 in the present paper with Fig. 3 in Epstein and Taylor (1971) shows the similarity of the various curves obtained on soils 10084, 14298,

Table 3. $\delta O^{18}$ and $\delta Si^{30}$ Data obtained by successive partial fluorinations of Apollo 15 lunar samples.

| Sample | Cum. $\mu$ mole $O_2$ / mg original sample | $\delta O^{18}$ | $\mu$mole $O_2$ | $\delta Si^{30}$ | $\mu$mole $SiF_4$ | Ratio $O_2/SiF_4$ |
|---|---|---|---|---|---|---|
| 15021,25 fines (0.8811 g) | | | | | | |
| a. 30 min/80°C | 0.040 | +49.7 | 35 | +24.5 | 29 | 1.07 |
| b. 60 min/118°C } | 0.042 | +24.6 | 20 } | +7.74 | 24 | 1.17 |
| c. 45 min/144°C } | | | 20 } | | | |
| d. 25 min/190°C | 0.085 | +26.0 | | | | |
| e. 25 min/217°C | 0.125 | +18.64 | 35 | +4.15 | 23 | 1.35 |
| f. 30 min/247°C | 0.187 | +15.16 | 55 | +2.17 | 40 | 1.38 |
| g. 30 min/294°C | 0.439 | +8.33 | 222 | −0.02 | 166 | 1.34 |
| h. 30 min/303°C | 0.691 | +6.32 | 222 | −0.74 | 168 | 1.32 |
| i. 30 min/311°C | 0.928 | +5.28 | 209 | −0.92 | 129 | 1.62 |
| j. 30 min/342°C | 1.239 | +5.07 | 274 | −0.37 | 175 | 1.57 |
| k. 30 min/371°C | 1.546 | +6.25 | 270 | −0.20 | 141 | 1.91 |
| Total | | | 1362 | | 895 | 1.52 |
| 15251,24 fines (0.8847 g) | | | | | | |
| a. 30 min/67°C | 0.020 | +51.8 | 18 | +25.4 | 12 | 1.50 |
| b. 60 min/119°C | 0.037 | +37.0 | 15 | +12.9 | 8 | 1.88 |
| c. 45 min/144°C | 0.052 | +35.0 | 13 | +8.87 | 6 | 2.17 |
| d. 25 min/190°C | 0.094 | +26.7 | 37 | +6.24 | 31 | 1.19 |
| e. 25 min/216°C | 0.144 | +18.2 | 44 | +2.66 | 36 | 1.22 |
| f. 25 min/259°C | 0.270 | +12.40 | 112 | +1.04 | 80 | 1.40 |
| g. 30 min/291°C | 0.527 | +8.01 | 227 | −0.06 | 154 | 1.50 |
| h. 30 min/303°C | 0.729 | +6.86 | 179 | −0.24 | 139 | 1.32 |
| i. 35 min/319°C | 0.931 | +6.13 | 179 | −0.41 | 143 | 1.28 |
| j. 30 min/343°C | 1.241 | +5.06 | 274 | −0.33 | 152 | 1.83 |
| k. 30 min/369°C | 1.552 | +5.66 | 275 | −0.10 | 141 | 1.98 |
| Total | | | 1373 | | 902 | 1.52 |

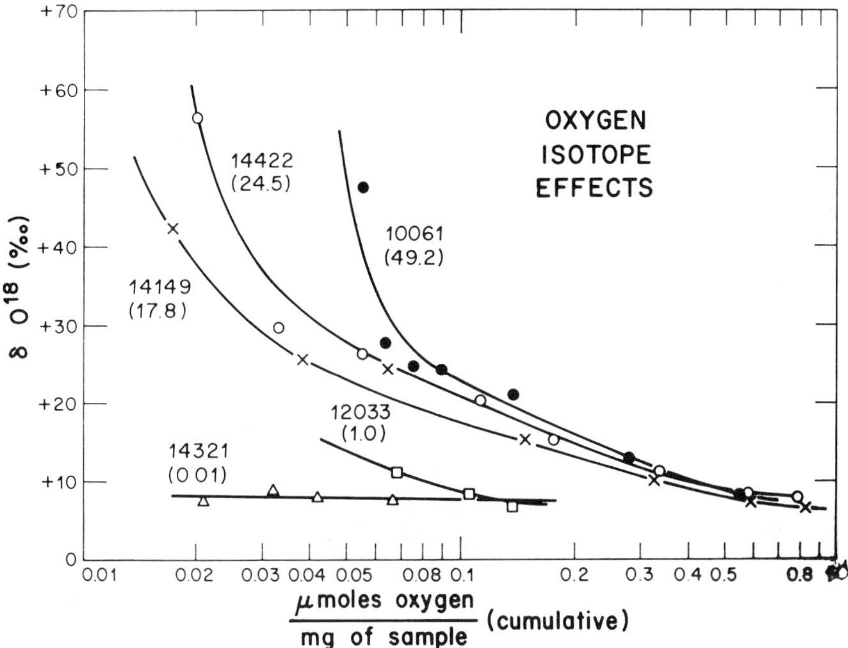


Fig. 2. Comparison of partial fluorination $O^{18}$ data obtained on samples 14422, 14149, and 14321 in the present work (Table 2) with similar data on 10061 and 12033 obtained by Epstein and Taylor (1971). The $\delta O^{18}$ values measured on each successive fraction are plotted versus the cumulative number of micromoles of oxygen produced per milligram of original sample. The numbers in parentheses represent measured hydrogen concentrations in the same samples (in micromoles of $H_2$ per gram); the value of 17.8 was not determined directly on bottom-trench sample 14149, but is an analysis of another bottom-trench sample from the same location, namely, 14240 (see Table 4). Note the correlation between the $H_2$ concentrations and the positions of the various curves.

15021, 14422, 15251, and the top (14148) and middle (14156) of the Apollo 14 trench fines, both for $\delta O^{18}$ and $\delta Si^{30}$. None of the above samples shows quite such extreme isotopic enrichments as does 10061, an Apollo 11 soil breccia (see Fig. 2 and 3), and all show distinctly *greater* isotopic enrichments than either 14149 or 12033, both of which are bottom-trench samples. All of these samples of lunar fines, however, display greater isotopic enrichments than does an interior fragment of an Apollo 14 rock breccia from the rim of Cone Crater, 14321 (see Figs. 2 and 3). The latter shows only very small $O^{18}$ or $Si^{30}$ enrichments, and is thus somewhat analogous to the fresh Kilauea basalt from Hawaii previously analyzed by us (Epstein and Taylor, 1971); by inference, the various lunar basalts also ought to show negligible isotopic enrichments, but none has as yet been studied in this fashion. The very sm all$O^{18}$ and $Si^{30}$ enrichment shown in the initial fractions extracted from 14321 conceivably could indicate the presence of a minor lunar soil component (very ancient?) in this breccia. In fact, the $\delta Si^{30}$ results on 14321 shown in Table 2 probably should be raised to take into

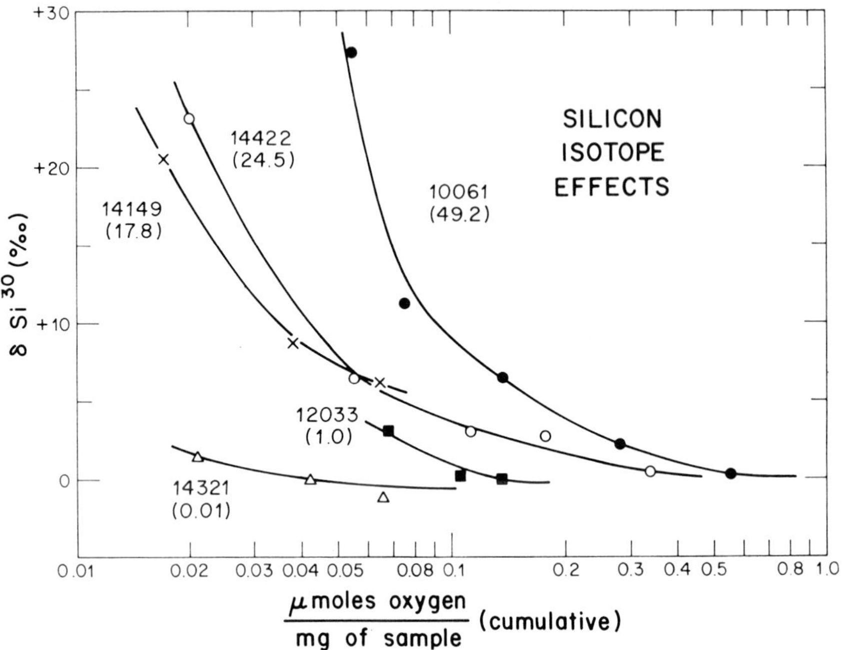


Fig. 3. Comparison of partial fluorination $Si^{30}$ data obtained on the same lunar samples shown in Fig. 2. All terminology is the same as given in Fig. 2. Note the correlation between $H_2$ concentrations and the positions of the various curves.

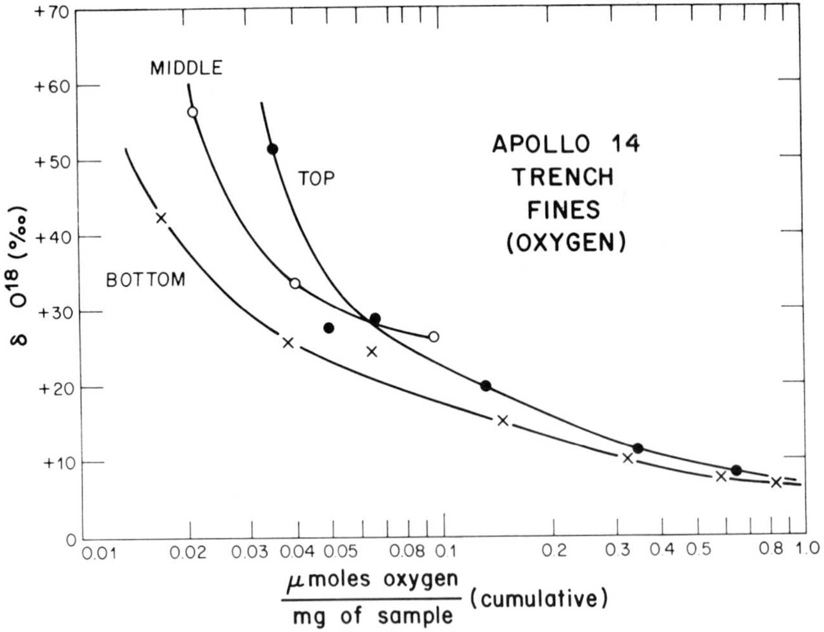


Fig. 4. Comparison of partial fluorination $O^{18}$ data on the top (14148), middle (14156), and bottom (14149) of the Apollo 14 trench fines. Note the systematic changes in position of the curves from the bottom of the trench to the surface.

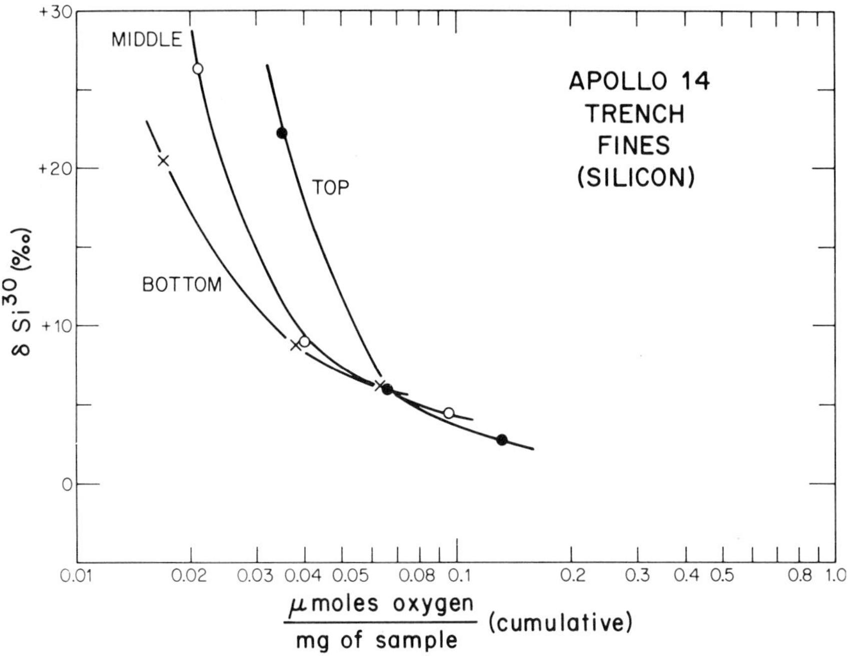


Fig. 5. Comparison of partial fluorination $Si^{30}$ data on the top (14148), middle (14156), and bottom (14149) of the Apollo 14 trench fines. Note the systematic changes in position of the curves from the bottom of the trench to the surface.

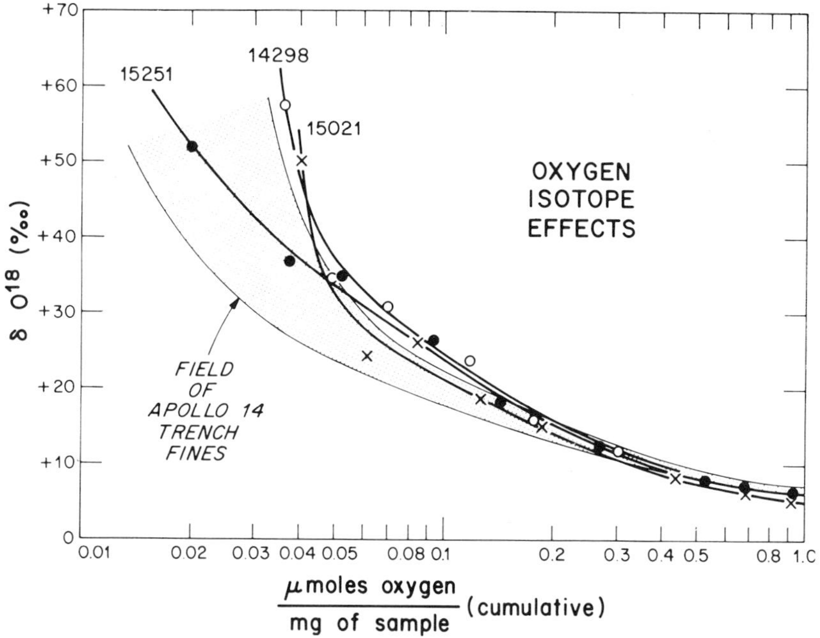


Fig. 6. Comparison of partial fluorination $O^{18}$ data on lunar soils 15251, 14298, and 15021 with the $O^{18}$ data for Apollo 14 trench samples shown in Fig. 4.

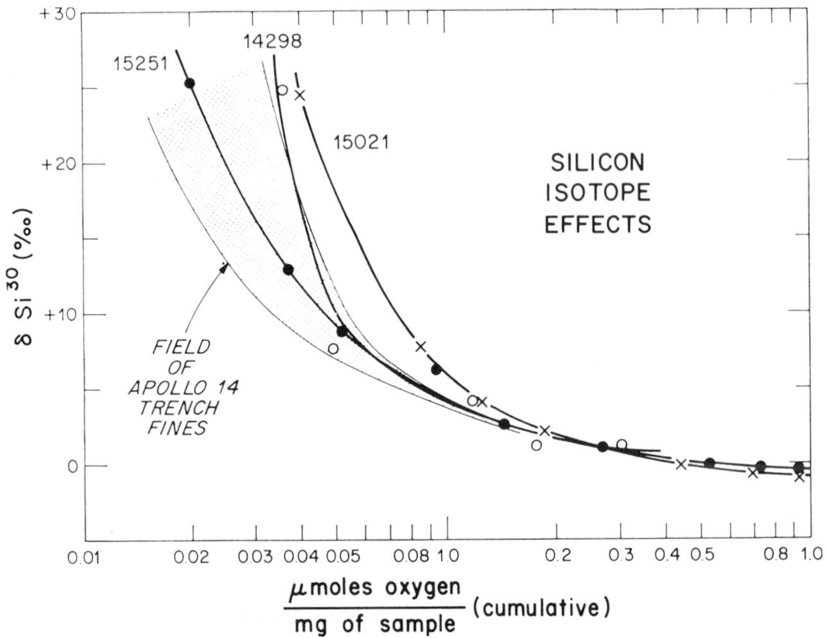


Fig. 7. Comparison of partial fluorination $Si^{30}$ data obtained on lunar soils 15251, 14298, and 15021 with the $Si^{30}$ data for Apollo 14 trench samples shown in Fig. 5.

account the presence of phosphorus impurities during the mass spectrometric measurements.

It is apparent from an examination of the data in the present paper and in Epstein and Taylor (1971) that the $O^{18}$ and $Si^{30}$ enrichments correlate very well with the amounts of solar wind hydrogen and rare gases in the lunar soils. This is illustrated in Figs. 2 and 3 where the concentrations of hydrogen in $\mu$mole/g are given along side each isotopic enrichment curve. The soil breccia, 10061, has the highest solar wind hydrogen content (49.2 $\mu$/g) and 14321 contains negligible solar wind $H_2$ ($<0.01$ $\mu$/g). Our sample of 14321 is not only an interior portion of a large rock fragment and thus largely shielded from solar wind or particle bombardment, but particle track data indicate that it is part of the deep ejecta from Cone Crater and has been residing on the lunar surface for only about 25 m.y. (Crozaz *et al.*, 1972; Burnett *et al.*, 1972).

It is remarkable that both the $O^{18}$ and $Si^{30}$ enrichments from the top, middle, and bottom samples of the Apollo 14 trench fines vary so systematically (see Figs. 4 and 5). The isotopic effects are markedly smaller in the bottom trench sample, 14149. We do not yet have solar wind $H_2$ data on these particular samples, but another bottom trench sample, 14240, contains appreciably less solar wind $H_2$ than the Apollo 14 surface soil sample. Assuming that the three trench samples all have similar grain-size distributions, these data may indicate that "gardening" and homogenization of the lunar soil take place relatively slowly. Probably, the turnover time to a depth equal to the bottom of trench is at least 25 m.y. or more. Arguments can be presented that indicate that the grain size distributions in all the trench samples are very similar.

Note in Table 2 that the *amounts* of $O_2$ and $SiF_4$ produced during the successive partial fluorinations of the three samples of trench fines are almost identical. Care was taken to make the times and temperatures of each successive extraction for 14148 (top) and 14149 (bottom) as similar as possible, so that the uniformity in the amounts of $O_2$ and $SiF_4$ obtained in each and every cut from (a) through (h) argues against there being any significant differences in grain-size or other physical properties between the two samples. This suggests that the consistently greater isotopic enrichments on the grain surfaces in 14148 (top) must reflect a greater exposure of the grains to the event(s) which produce the $\delta O^{18}$ and $\delta Si^{30}$ enrichments of the grain surfaces, or the top contains a greater fraction of "exposed" grains. The latter is more probable if $\delta O^{18}$ effects are subject to saturation. Both this and the correlation with solar wind $H_2$ are in agreement with data of Crozaz *et al.* (1972) who have shown from track studies that the Apollo 14 bottom trench sample has suffered far less irradiation than has the top or middle. We thus now feel certain that the $\delta O^{18}$ and $\delta Si^{30}$ enrichments we have observed are correlated with the degree of exposure of the sample at the lunar surface.

The data in Tables 2 and 3 also confirm and amplify the conclusions of Epstein and Taylor (1971) that the grain-surface coatings of the lunar fines are all depleted in oxygen relative to silicon. The $O_2/SiF_4$ ratios in the initial fluorination cut (a) are typically about 1, whereas the whole-rock samples typically exhibit values of about 1.7 to 1.8. Note in Tables 2 and 3 that for a given sample, if one averages all the fluorination cuts which show significant $O^{18}$ and/or $Si^{30}$ enrichments, the overall $O_2/SiF_4$ ratios are generally about 1.3 to 1.5. This means that the outer 5 to 8% and/or the finest 5 to 8% of the lunar dust grains are not only enriched in $Si^{30}$ and $O^{18}$, but are about 20% depleted in total oxygen relative to total silicon.

Other workers have studied the "coatings" on the lunar dust grains, showing them to be rather opaque and amorphous. Leaching and volatilization experiments have shown that these "coatings" and/or the finest grain-size fractions are also probably the locus of the old, parentless, lead component that has been added to lunar soils (Silver, 1972). There are definitely some new materials added to the grain surfaces and/or the finest size fractions. This added material in part represents solar wind and/or solar flare implantation (i.e., the $H_2$, rare gases, and at least some of the carbon). In part this material is probably a deposit of "volatilized" material from other parts of the moon (i.e., some of the Pb and possibly some Hg, Cd, Bi, etc.). In addition, some meteoritic debris may have been added. The questions immediately arise as to what role each of these possible additions of material play in explaining the unusual O and Si isotopic and concentration data. We have previously proposed that the latter represents fractional condensation onto the grain surfaces and/or fractional vaporization of material from the surface. Either process can in principle explain the gross aspects of the observed isotopic enrichments if we assume that some of the vaporized $Si^{28}$ and $O^{16}$ preferentially escaped completely from the moon's gravitational field. Such vaporization with attendant formation of Si, SiO, O, and $O_2$ species in the vapor is probably happening continuously on the lunar surface during micrometeorite, solar wind, solar flare, and cosmic ray bombardment.

In Fig. 8, the $\delta Si^{30}$ results for the partially fluorinated samples are plotted against

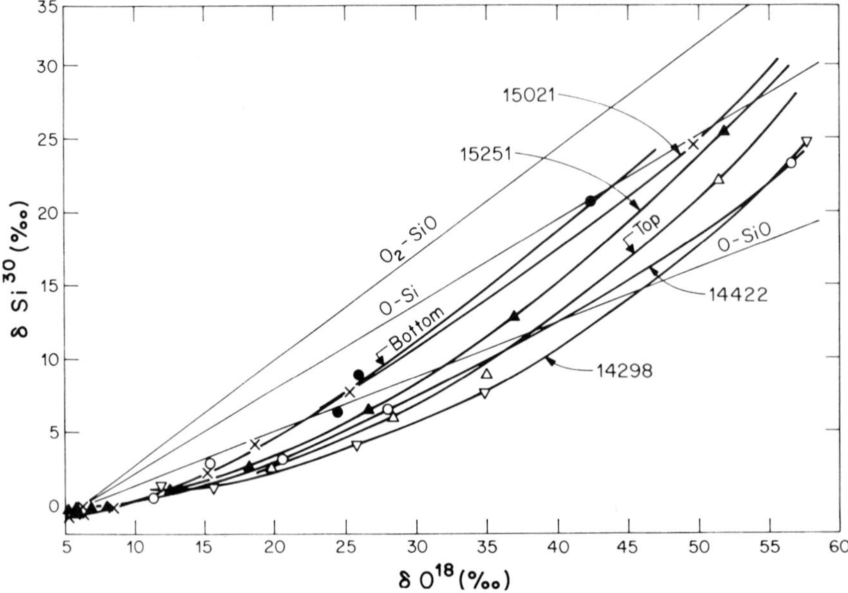


Fig. 8. Plot of $\delta Si^{30}$ versus $\delta O^{18}$ for all the partial fluorination data obtained on lunar samples in the present study (Tables 2 and 3). Curves have been drawn through the data points obtained on each sample; note that all the curves are concave upward and have similar shapes. The curves labeled Top and Bottom are 14148 and 14149, respectively, from the Apollo 14 trench fines. The diagonal lines labeled $O_2$–SiO, O–Si, and O–SiO represent theoretical 1 : 1 fractional vaporization lines for these various pairs of chemical species (see Fig. 4 in Epstein and Taylor, 1971, for a more complete discussion).

$\delta O^{18}$. See Fig. 3 in Epstein and Taylor (1971) and its accompanying figure caption for a discussion of the straight lines labeled $O_2$–SiO, O–Si, and O–SiO; these are theoretical fractional vaporization trends assuming that for each pair equal proportions of the oxygen- and silicon-species are vaporized. Obviously, if only $O_2$ or O was removed from the grain surfaces there would be no $Si^{30}$ enrichment and the trend lines would be horizontal. We know from the $O_2/SiF_4$ ratios given in Tables 2 and 3 that the $O_2$/Si ratio of the material vaporized or removed from the condensate must have been greater than 1.8, and probably was greater than 2. Vaporization of equal portions of SiO and $O_2$ alone would not produce this result; at least 1.5 times more $O_2$ than SiO would have to be removed. Qualitatively, such excess loss of $O_2$ relative to SiO fits the data shown on Fig. 8 because the observed data points all plot well beneath the theoretical 1:1 $O_2$–SiO line. However, the mechanism(s) must be more complicated than this because none of the curves obtained from actual analysis of the lunar samples is a simple straight line. Assuming that the concave-upward shape of the curves shown in Fig. 8 is not due to isotopic fractionation during the analytical procedure (very unlikely because neither 14321 or the aforementioned Kilauea basalt show such effects), the curves indicate that if fractional condensation has occurred, we are not dealing with a simple two-component mixture of condensate and rock, for which a linear trend would result. Also, because of the curvature, if fractional vaporization is the explanation, the proportions and types of species vaporized cannot have been uniform. It is possible that the curves shown in Fig. 8 result from a combination of fractional vaporization and fractional condensation. Certainly, if vaporization occurs during particle bombardment, some re-condensation of this vapor must occur, both nearby and at some distance from the impact site.

Even though a completely definitive interpretation of the curves shown in Fig. 8 cannot as yet be given it is worth pointing out that their concave-upward shape and position can be explained by a simple fractional vaporization model in which during the early stages of particle bombardment, oxygen is principally lost relative to silicon (perhaps with an $O_2/Si$ ratio in the vapor greater than 4 or 5); this could cause the isotopic enrichments to lie along a curve with a very shallow slope, perhaps less steep than the O–SiO curve shown in Fig. 8. Later, after appreciable depletion of the surface in oxygen relative to silicon, the relative amounts of Si species vaporized would have to increase simply because other things being equal the $O_2/Si$ ratio in the vapor should be roughly proportional to the $O_2/Si$ ratio in the "target." Thus, the outermost portions of the bombarded grains should have lower $O_2/Si$ ratios (as they in fact do, see Tables 2 and 3), and the $\delta Si^{30}$ fractionation effects should be relatively greater. The pattern of isotopic enrichments on a diagram such as Fig. 8 would then follow a curve with a progressively steepening slope; this is exactly what is observed.

Alternately, it is possible to explain the curves in Fig. 8 by a combination of fractional vaporization and condensation. If, as seems reasonable, the condensate contains a significantly higher Si/O ratio than the vapor from which it condenses, then the right-hand portions of the curves in Fig. 8 might largely represent condensate, whereas the left-hand portions with a shallower slope would represent fractionally vaporized surfaces. Thus, the concave upward appearance of the curves might simply be due to the joining of separate mixing lines in a three-component system (i.e., isotopically normal material, condensate, and fractionally vaporized surface).

Obviously, further isotopic studies and more detailed investigations of the grain-surfaces in the lunar fines will be necessary before we can decide between the alternative explanations outlined above. It is also necessary to obtain more data before we can decide whether meteorite, micrometeorite, solar wind, solar flare, or cosmic ray bombardment (or some combination of these processes) is principally responsible for producing the observed $O^{18}$ and $Si^{30}$ effects.

Note that either or both of the above models conceivably could produce $O_2/Si$ depletions in the lunar fines. Utilizing the data of Tables 2 and 3, one can estimate that the $O_2/Si$ ratio in the bulk lunar fines ought to be lowered by about 1 to 1.5% relative to the ratio in the lunar basalts from which the fines are largely derived. In other words, the overall $O_2/Si$ ratio might be lowered from a value such as 1.75 to 1.73. Our volumetric measuring techniques were not designed to allow us to make such fine distinctions, and judging from the quoted analytical precision, neither do the data of Ehmann and Morgan (1970, 1971). Ehmann and Morgan (1970) do suggest, however, that the Apollo 11 breccias and fines are about 1.6% depleted in oxygen relative to the crystalline rocks and to the amount required for simple stoichiometry. It is possible that they are observing the same effects in the bulk samples that we are observing in our partial fluorination experiments. However, it is difficult to utilize arguments based on stoichiometry, because we do not know how much Ca, Al, Mg, etc., have been vaporized or condensed on the grain surfaces of the lunar fines. Also, for Apollo 12 samples, Ehmann and Morgan (1971) did not observe any distinctions between the soils and the basalts. In the future it should prove useful to carry out chemical leaching experiments in conjunction with our isotopic "stripping" experiments, in order to more fully characterize the chemical and isotopic nature of the "coatings" on the grains of lunar fines.

## ISOTOPIC COMPOSITION AND CONCENTRATION OF HYDROGEN

Hydrogen and water were extracted from three lunar soil samples and a rock breccia returned by the Apollo 14 and 15 missions. The procedures used were those described previously (Epstein and Taylor, 1971); namely, the samples were heated in vacuum and the evolved water and hydrogen gas were measured volumetrically and their $\delta D$ values determined. The results are given in Table 4 in permil relative to SMOW (standard mean ocean water).

The results for sample 15301 are similar to those obtained previously for Apollo 11 and 12 soils. About four times more hydrogen than water is given off by the soil upon heating. Essentially all the $H_2O$ is given off at lower temperatures (fractions A and B, see p. 1434, Epstein and Taylor, 1971). The hydrogen gas, however, is principally given off at higher temperatures (fractions B and C). Somewhat different results were obtained for sample 14422, in that this sample contains a little more $H_2O$. The higher $H_2O$ content is probably responsible for the larger $\delta D$ value of $H_2$ in 14422 ($-765$ versus $-833$ for 15301), because we know from previous studies (Epstein and Taylor, 1970, 1971) that isotopic cross contamination of $H_2$ and $H_2O$ takes place during the extraction procedure.

Sample 14240 is from the bottom of a trench. Similar to our previously analyzed

Table 4. The isotopic composition and concentration of $H_2$ and $H_2O$ from lunar samples.

| | | Hydrogen | | Water | |
|---|---|---|---|---|---|
| Sample | Fraction | Conc. $\mu$mole/g | $\delta D$ ($^0/_{00}$) | Conc. $\mu$mole/g | $\delta D$ ($^0/_{00}$) |
| 14240 | A | — | — | | |
| fines | B | 13 | $-823$ | 17.4 | $-227$ |
| (bottom of trench) | C | 3 | $-619$ | | |
| | D | 1.8 | $-255$* | | |
| | Total | 17.8 | $-731$ | 17.4 | $-227$ |
| 14422 | A | — | — | 7.7 | $-313$ |
| fines | B | 10 | $-881$ | | |
| | C | 12.1 | $-770$ | 2.7 | $-419$ |
| | D | 2.4 | $-250$* | | |
| | Total | 24.5 | $-765$ | 10.4 | $-341$ |
| 15301 | A | — | — | 7.8 | $-230$ |
| fines | B | 21.3 | $-902$ | | |
| | C | 5.8 | $-600$ | | |
| | D | | | | |
| | Total | 26.1 | $-833$ | 7.8 | $-230$ |

* These $\delta D$ values of the high-temperature fraction are remarkably heavy and probably represent a contribution of spallation deuterium (see text).

trench sample 12033, its hydrogen gas content is lower than in surface samples from all the other Apollo sites. These depth effects on the concentration of solar wind hydrogen can only occur if the soil mixing or "gardening" rate is sufficiently slow that the deep samples were less affected by solar wind bombardment than the near-surface samples.

Sample 14240 was inadvertently exposed to atmospheric moisture in the lunar module and on the earth by the failure of its vacuum seal (Burlingame *et al.*, 1971). As the results in Table 4 show, this sample contains a larger concentration of water than any other lunar sample analyzed, suggesting that much of this water is a terrestrial contaminant. Small amounts of $H_2O$ are apparently readily adsorbed on the grain surfaces of the lunar fines. This adsorption may occur by reaction of water on chemically active centers, activated by solar wind or radiation bombardment.

In a previous study (Epstein and Taylor, 1971), we were able to estimate the $\delta D$ value of the lunar hydrogen gas to be about $-980$ by utilizing a technique in which we eliminated the deuterium from the contaminating $H_2O$ by isotopic exchange with deuterium-free water. A $\delta D$ of $-980$ is equivalent to about 3 ppm deuterium.

The relatively large variation in the hydrogen to water ratio in the lunar samples we have studied to date permits us to make an additional independent estimate of the $\delta D$ value for uncontaminated solar wind hydrogen. Let us make the reasonable assumption that the $\delta D$ value of water contaminating the various lunar soils and breccias is very approximately the same for all samples. During extraction, contamination of the hydrogen gas takes place through addition of deuterium from this coexisting water. Therefore, a plot of the relationship between the $\delta D$ value for the hydrogen gas versus the mole fraction of water in the $H_2O$–$H_2$ mixture ought to allow us to extrapolate to the conditions that would prevail if no contaminating water was present. Figure 9 shows this plot, and a calculated least squares line extrapolates to about 1 ppm D or a $\delta D$ value of $-995$ for the hypothetical $H_2O$-free sample. Considering the errors inherent in such a plot, this value is compatible with our previous $\delta D$ estimate of $-980$ for lunar hydrogen gas. The two points in Fig. 9 which represent an unusually high mole fraction of water are the samples 14321 and 12033. These samples are not markedly enriched in $H_2O$; the high mole fractions of $H_2O$ are simply due to the low concentrations of $H_2$, only 0.01 $\mu$mole/g in 14321, the interior fragment of breccia from Cone Crater, and 1 $\mu$mole/g of $H_2$ in 12033, an Apollo 12 trench sample.

One of the unusual results obtained from samples 14240 and 14422 is the abnormally high $\delta D$ value for the hydrogen extracted in fraction D (i.e., at the melting temperature and above). Such hydrogen, evolved only after melting of the sample, must in part come from the interiors of the soil grains. The $\delta D$ values for fraction D are unusually positive compared with most of the hydrogen gas extracted from lunar samples (the $\delta D$ values are $-250$ and $-255$). It is impossible to explain such high $\delta D$ values by cross-contamination of deuterium from the coexisting $H_2O$, because at least in the case of 14422, none of the water fractions extracted contains a sufficiently high concentration of deuterium. Therefore, it follows that these minor amounts of hydrogen extracted during the melting of lunar samples 14240 and 14422 are themselves markedly enriched in deuterium relative to the major solar wind hydrogen component.

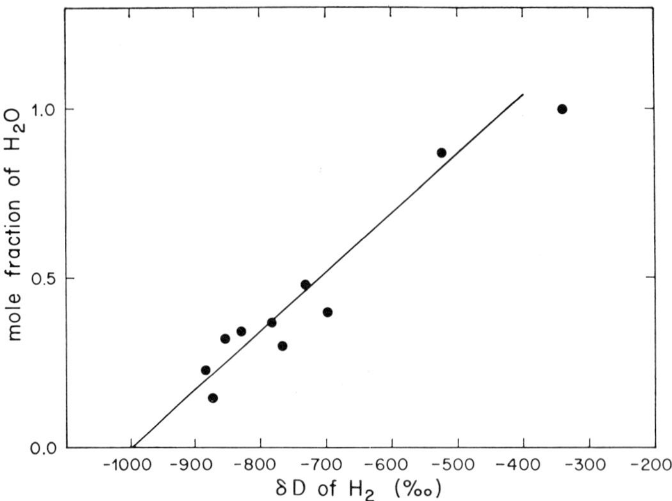


Fig. 9. The relationship between the δD value of the hydrogen gas and the mole fraction of $H_2O$ in the total $H_2O$–$H_2$ mixture extracted from all samples of lunar fines and soil breccias that we have analyzed to date (Epstein and Taylor, 1970; 1971; and Table 4 of the present study). A least-squares line that in the absence of $H_2O$ extrapolates to a δD value of $-995$ has been drawn through the data points; except for small amounts of spallation deuterium this should represent the approximate δD value of uncontaminated solar wind $H_2$. The point in the upper right hand corner of the diagram was excluded from the least-squares calculation because it represents sample 14321, which contains hydrogen that is mainly of nonsolar wind origin (see text).

In fact, this high-temperature hydrogen almost certainly is exceedingly deuterium-rich, because there is probably at least some solar wind $H_2$ evolved in fraction D as well. This result is further evidence that in some of the soil samples cosmic ray spallation deuterium is present. This was previously suggested by data of Friedman, *et al.* (1971) and by Epstein and Taylor (1971), on the basis of a calculation utilizing data on spallation xenon, but this is the first *direct* evidence of the presence of clearly measurable amounts of spallation deuterium in lunar *soil* samples. Thus, the actual solar wind $H_2$ component in the soils must have a δD value less than $-980$, perhaps very close to $-1000$ (0 ppm deuterium).

Sample 14321 is a large fragment of rock breccia from the rim of Cone Crater. Very little hydrogen ($<0.01$ μmole/g) or carbon was found in this sample, which is understandable since this is an interior piece that originally resided at least 1 cm inward from the surface. An approximate δD value of $-350$ was obtained on this small amount of $H_2$, indicating that in this case little or no spallation deuterium is present ($<1 \times 10^{-6}$ μmole/g of D). The lack of spallation deuterium is compatible with the results of other workers (Burnett *et al.*, 1972) who have shown that rock 14321 possesses a short exposure age, ~24 m.y., and was probably originally derived from about 80-m depth in Cone Crater (Wilshire and Jackson, 1972). The 0.01 μmole of hydrogen per g of sample may be attributed to ~10 to 20 MeV protons from solar flares because ordinary solar wind protons would not penetrate so deeply into the rock. It would be interesting to ascertain if hydrogen concentration measurements could be used to monitor the distribution of solar flare protons throughout the interior of a lunar rock, such as rock 14321. In turn, the proton distribution in the rock could give information regarding the energy distribution of the impacting protons.

Our procedure of extracting $H_2$ and $H_2O$ from lunar samples is accompanied by exchange of hydrogen between the two compounds. This cross-contamination appears to be most pronounced at high extraction temperatures. Inasmuch as it was the objective of our experiment to determine the concentration and $\delta D$ values of both the $H_2$ and $H_2O$ as they exist in the lunar samples, we felt that if extraction of the two gases could be done at room temperature, then the cross-contamination effects might be eliminated. We therefore tried to extract hydrogen from sample 14422 by using the method that Yaniv and Heymann (1971) used to extract helium gas from lunar samples. Their method involves the pounding of a metal rod upon the grains of soil, at room temperature, in a vacuum, causing a release of helium gas. We successfully duplicated their results in that we obtained appreciable amounts of helium gas, but no measurable hydrogen gas was given off by this method. Thus, from our point of view, this experiment was a failure. The result of this experiment, however, suggests that the helium and hydrogen may occupy positions of differing depth within an individual soil grain. Possibly the hydrogen may be buried deeper into the grain. On the other hand, it may be that for chemical reasons the hydrogen is just inherently more difficult to release at room temperature; the activation energy for release of hydrogen is apparently higher than that for the release of helium gas.

## The $O^{18}/O^{16}$ Ratio of "Lunar Water"

Up to the present time no $\delta O^{18}$ determinations have been made on the small amounts of $H_2O$ that occur in the samples of lunar fines; only $\delta D$ values have been determined for this $H_2O$ (Epstein and Taylor, 1970, 1971; Friedman *et al.*, 1970). Our interpretations regarding the origin of this water have heretofore been based solely on the concentrations and D/H ratios of the water. These data suggested that the water extracted from the lunar samples is primarily terrestrial in origin. Briefly, these arguments were as follows: (1) Almost all the $H_2O$ escapes much more readily during heating than any of the solar wind gases, including the very loosely bound helium; (2) Making allowances for cross contamination with the coexisting solar wind $H_2$, the estimated $\delta D$ values of this $H_2O$ are very similar to those of temperate zone atmospheric water vapor; (3) Most of the water in the lunar samples can, in a matter of a few hours, be isotopically exchanged with $H_2O$ vapor at about 300°C; (4) The grain surfaces in the lunar fines have all undergone a great deal of radiation, probably of sufficient magnitude to produce chemically active surfaces that would be expected to readily adsorb $H_2O$; (5) The concentrations of $H_2O$ in the lunar samples are very small, about 8 to 17 $\mu$mole/g, and the highest concentration was observed in sample 14240, which was inadvertently exposed to a moist atmosphere on earth.

It was of interest to ascertain if the $O^{18}/O^{16}$ analyses could add some pertinent information regarding the origin of this "lunar water." Samples 15301 and 14321 were heated in the usual way and the water collected in a liquid nitrogen cooled trap (this corresponds to the fractions A and B shown in Table 4). The extracted waters were then reacted with fluorine gas and the resultant oxygen converted to $CO_2$ by passing it over hot graphite (O'Neil and Epstein, 1966). The $\delta O^{18}$ of the $H_2O$ in 15301 is $-5.9$ and in 14321 is $-18.2$. Inasmuch as only very small amounts of $H_2O$ are present, the $\delta O^{18}$ values are probably accurate to only $\pm 2$ to 3‰. These data are plotted in Fig. 10, together with the $\delta D$ value of $H_2O$ from another aliquot of sample 15301 and for 14321 a "typical" $\delta D$ value for "lunar water" obtained previously on Apollo 11 and 12 samples. On the graph we also include an estimated $\delta D$ and $\delta O^{18}$ range of values for Pasadena water vapor. This only represents a rough average because at times this range may extend to both more positive and more negative values. The graph also shows the $\delta D$ and $\delta O^{18}$ values for terrestrial ocean water and for terrestrial primary magmatic waters (Sheppard *et al.*, 1969). The values for magmatic waters are estimated from the known $O^{18}/O^{16}$ and D/H fractionation factors between igneous minerals and water, and from measured $\delta D$ and $\delta O^{18}$ values for igneous hydrous minerals (Suzuoki and Epstein, 1972; O'Neil and Taylor, 1969). The arrows emitting from the data-points indicate the direction in which the $\delta O^{18}$ and $\delta D$ values for the "lunar water" should be displaced in order to correct for isotopic exchange effects that occur during the extraction procedures. The original, unexchanged water would almost certainly have had a lower $\delta O^{18}$ value than the measured value because of exchange with the $O^{18}$-rich lunar soil at about 300–500°C.

The sample that is shown as being further displaced from the true value (longest horizontal arrow) is an ordinary lunar soil and is thus one in which the surfaces are probably highly enriched in $O^{18}$ (15301): this $H_2O$ was thus probably more enriched in $O^{18}$ by exchange. The other sample (14321) does not show surface $O^{18}$ enrichment (see Fig. 2), and thus should have contributed less $O^{18}$ to the $H_2O$ during the extraction; it is shown with a shorter horizontal arrow. We also know that during the extraction, the evolved $H_2O$ exchanges with the coexisting solar wind hydrogen, and thus becomes depleted in deuterium. The measured $\delta D$ values shown in Fig. 10 must therefore be corrected for this effect (shown by vertical arrows), but we are not as yet able to calculate the exact amount of cross contamination. Note that both the $\delta O^{18}$ and $\delta D$ corrections will shift the measured data-points closer to the values typical of terrestrial atmospheric water vapor.

What $\delta O^{18}$ and $\delta D$ values might we expect for true lunar water? It would be most difficult to assign a $\delta D$ value because of the undoubtedly complicated history of water during the formation and evolution of the moon. Isotopic fractionations probably occurred during vaporization and outgassing of $H_2O$ from the moon, but it is difficult to predict either their magnitude or direction. Also, isotopic equilibration with a variety of possible hydrogen-containing chemical compounds could have resulted in a wide range of D/H ratios. Added to these difficulties is the problem of knowing what role solar wind and solar flare activity (perhaps greatly enhanced over the present) played in adding hydrogen to the moon during its early history. The $\delta D$ value for possible lunar water therefore cannot be predicted. However, it would certainly be coincidental and highly fortuitous for true lunar water to have a $\delta D$ value that is so similar to meteoric waters in the earth's temperature zone.

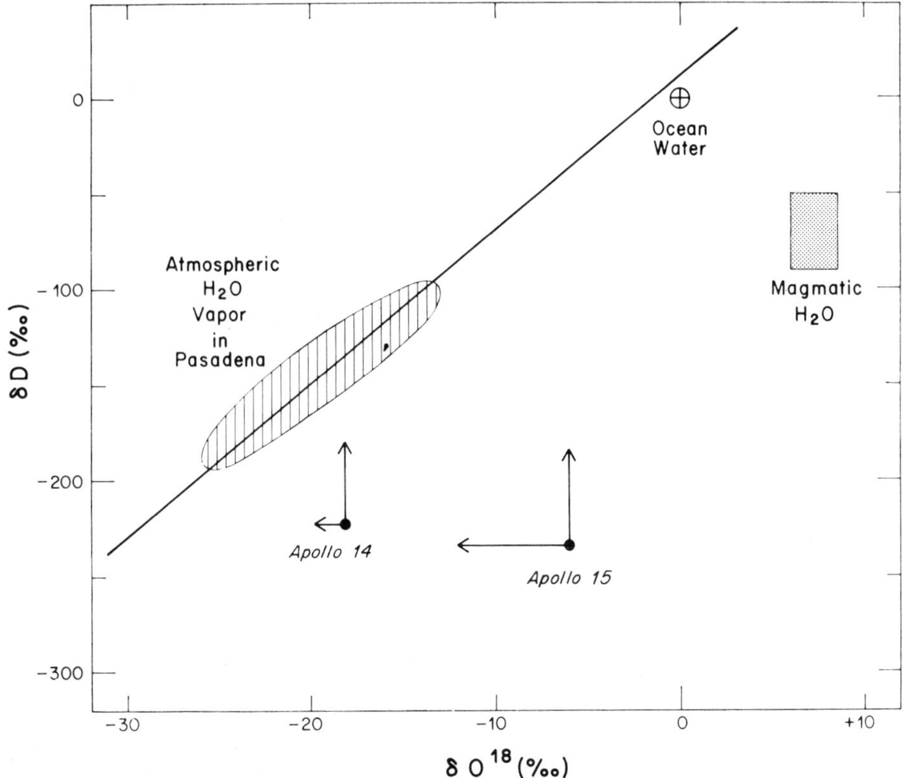


Fig. 10. A plot of $\delta D$ versus $\delta O^{18}$ showing the isotopic analyses of $H_2O$ extracted from samples 14321 and 15301. The plotted $\delta D$ value for sample 14321 was not actually measured on that sample, it was estimated from the average $\delta D$ value of $H_2O$ in all lunar soils we have studied to date ($\delta D \approx -225$). For comparison, we also show the position of terrestrial ocean water, "primary magmatic $H_2O$," and Pasadena $H_2O$ vapor; the diagonal line shown is the locus of terrestrial meteoric waters. The arrows emanating from the lunar data-points indicate the direction in which the isotopic data should be shifted to take into account contamination during sample extraction (see text).

The $\delta O^{18}$ value for lunar water is, however, more predictable. Irrespective of what the $\delta O^{18}$ values for lunar water may have been during the formation of the moon, the $H_2O$ probably equilibrated with the hot lunar interior or surface volcanic rocks and should have acquired a $\delta O^{18}$ value of about +6 to +8. If some reequilibration with the rocks occurred at lower temperatures, the $\delta O^{18}$ of the $H_2O$ could perhaps become a few permil lighter than this. It would be very difficult to visualize an exchange mechanism in which the $\delta O^{18}$ could become as low as −10 or −20; this could only occur at temperatures lower than about 100°C, where rates of exchange are known to be very slow. In view of the arguments outlined above that the $H_2O$ resides on or close to the surface of the lunar soil grains, we might also expect that by analogy with the observed $\delta O^{18}$, $\delta Si^{30}$, and $\delta C^{13}$ (see below) enrichments on these grain-surfaces, any such $H_2O$ would also be abnormally enriched in $\delta O^{18}$ (and $\delta D$?). In fact, just the reverse is true, the "lunar water" is unusually low in $O^{18}$. Also if appreciable $H_2O$ has escaped from the moon, then one would expect that isotopically light $H_2O$ would have preferentially been lost; the remaining water would thus be even heavier than any original magmatic lunar water.

Summing up, it seems unlikely that true lunar water would acquire the $\delta O^{18}$ and $\delta D$ values we experimentally obtained from our lunar samples. Combined with all the other arguments outlined above, we are still forced to consider the "lunar waters" we extracted to be primarily water of terrestrial origin, which was probably brought to the moon by the Apollo spacecraft and the astronauts. Some water may also have been added to the samples after their return to earth. In the light of our isotopic data, we conclude that before accepting the possibility that measurable amounts of lunar water exist, it is important to obtain much more conclusive evidence than that presented by Freeman *et al.* (1972), who detected an $H_2O$ "event" on the moon on 7 March 1971, by means of a cold cathode gauge.

## LUNAR CARBON

The $\delta C^{13}$ and carbon concentrations were previously measured on Apollo 11 and 12 lunar samples by ourselves (Epstein and Taylor, 1970, 1971) and by other investigators (Friedman *et al.*, 1970; Kaplan *et al.*, 1970; Kaplan and Petrowski, 1971). We have now extended these studies to include a few Apollo 14 and 15 samples. Our analytical procedure is to collect both the $CO_2$ and CO given off during heating of the lunar samples from about 300°C to well above their melting point. The CO is passed over hot CuO at approximately 750–800°C and converted to $CO_2$. The total $CO_2$ is then separated from $SO_2$ and purified by passing the gases over $MnO_2$. The resulting gas sample is then analyzed mass spectrometrically, with the $\delta C^{13}$ results given relative to PDB (see Table 5).

Experiments were made to determine the factors which might affect the $C^{13}/C^{12}$ ratio of carbon extracted from lunar samples. A sample of lunar breccia 10061 was heated at about 1200°C in a circulating atmosphere ($<10$ mm Hg) of purified oxygen gas. The resulting gases, such as $CO_2$, $SO_2$, and $H_2O$, were condensed in a liquid $N_2$ cold trap. After exposure to oxygen, the hot lunar sample was heated above its melting temperature without circulating oxygen to further extract carbon and possible trace amounts of hydrogen. This two-stage extraction procedure resulted in more than a three-fold increase in the extracted carbon and more than doubling of the hydrogen as compared to the amounts extracted by the direct pyrolysis method that we have used on all our previous analyses of lunar samples. The $\delta D$ of the resulting hydrogen was

Table 5. Isotopic and concentration data for carbon in Apollo 14 and 15 samples.

| Sample | Conc. (ppm) | $\delta C^{13}$ (‰) |
|---|---|---|
| 15301 fines | 60 | +5.1 |
| 14240 fines (SESC sample) | 43 | −1.5 |
| 14422 fines | 89 | +0.4 |
| 14321 rock breccia | 24 | −18.6 |

much greater and the $C^{13}/C^{12}$ decreased markedly, indicating terrestrial contamination. There is an approximate 1 to 1 ratio between the increase in carbon and $H_2O$, indicating that there must have been oxidation of a hydrocarbon, such as the stopcock grease present in other parts of the vacuum extraction system. Similar blank experiments were made using a melted lunar sample from which *all* the hydrogen and carbon had been previously extracted. There were 47 $\mu$mole of $CO_2$ and 44 $\mu$mole of $H_2O$ produced, again indicating that some oxidation of stopcock grease can take place, provided the sample and its container is maintained at a high temperature with circulation of oxygen over the sample. In summary, in trying to analyze the very small amounts of C and H in the lunar samples, serious contamination problems can arise during procedures where oxygen is used to combust the samples. The portions of the extraction system that come into contact with the circulating oxygen gas must be completely free of hydrocarbons such as stopcock grease. This type of contamination will always produce a decrease in $\delta C^{13}$ and an increase of $\delta D$ in the carbon and hydrogen extracted from lunar breccias and fines, reinforcing our conclusions that the simple vacuum pyrolysis method that we have been using for lunar sample analysis is a valid one. We have therefore continued to use the pyrolysis method on Apollo 14 and 15 soil samples.

There are several general statements that can be made from the data obtained so far on lunar samples. A much greater amount of carbon is present in lunar soils and breccias than in lunar igneous rocks. Carbon must have been added to the lunar fines from the solar wind, from carbonaceous meteorites or comets, or from some unknown source elsewhere on the moon. Abundance measurements alone cannot distinguish between these various possibilities. A knowledge of the isotopic composition of the carbon is critical to any evaluation of the possible origin of the lunar carbon. The $\delta C^{13}$ of the carbon in the lunar fines and soil breccias is unlike any reduced carbon on earth, the lunar carbon being anywhere from 15 to 40‰ enriched in $C^{13}$ compared to the $\delta C^{13}$ of the reduced terrestrial carbon. Indeed, there is no common terrestrial carbon from any source, be it reduced or oxidized, which is as $C^{13}$-rich as the carbon in the lunar fines, particularly the Apollo 11 soils and breccias. The carbon of other possible extraterrestrial sources such as meteorites also has a much lower $\delta C^{13}$ than that in the lunar fines.

In Fig. 11, we plot the carbon isotope and concentration data obtained on lunar samples by ourselves (Epstein and Taylor, 1970, 1971) and by Kaplan *et al.* (1970) and Kaplan and Petrowski (1971). Note that there is a 50‰ variation in $\delta C^{13}$, from $-30$ to $+20$. An approximate curve (solid line) has been visually drawn through the data points. Although the relationship as shown on the graph is approximate, there is an unmistakable increase in $\delta C^{13}$ with increasing carbon concentration. The scatter of the data on the right-hand side of the diagram probably indicates a heterogeneity in the soils and breccias both in the distribution of carbon and in its isotopic composition.

The data points on the left-hand side of Fig. 11 ($\delta C^{13} = -18.6$ to $-29.8$) all represent determinations on lunar basalts, except for one analysis of an interior fragment of rock breccia sample 14321 (Table 5), which as described previously, is known to have suffered very little exposure to solar wind or other particle bombardment. There is a pronounced gap in $\delta C^{13}$ values to the right of the 14321 point; no samples have as yet been analyzed with $\delta C^{13}$ values in the range $-18.6$ to $-3.6$. It is

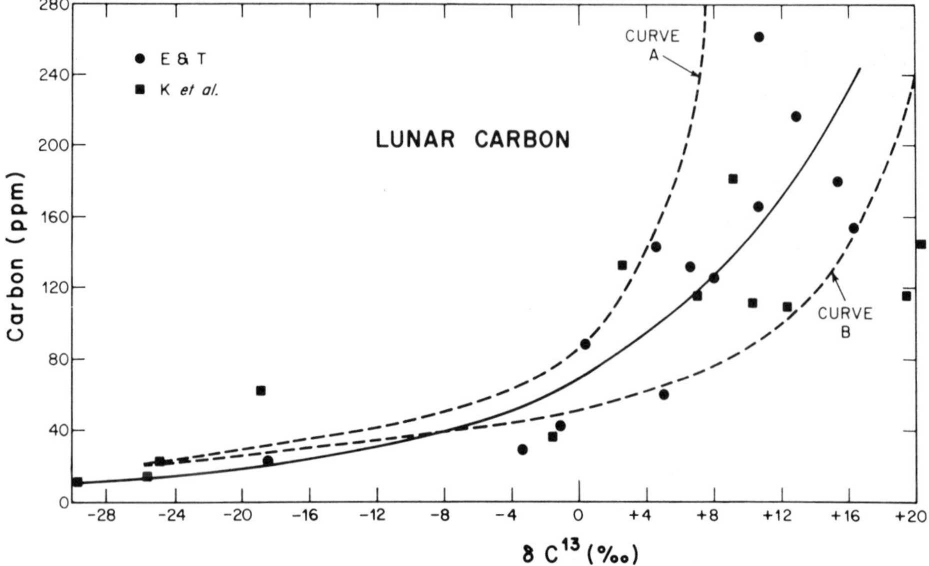


Fig. 11. The relationship between concentration and $\delta C^{13}$ values for carbon in all lunar samples analyzed in the present work and by Kaplan *et al.* (1970); Kaplan and Petrowski (1971); and Epstein and Taylor (1970; 1971). The solid curve was visually drawn through the data points. Curve A represents a hypothetical mixing curve obtained by successively adding increments of carbon with $\delta C^{13} = +10.8$ to the average lunar basalt, which contains about 25 ppm carbon having a $\delta C^{13}$ of $-26.0$. Curve B is a similar mixing curve, but one in which the added carbon is assumed to have $\delta C^{13} = +25$.

interesting to note that this happens to be precisely the $\delta C^{13}$ range of reduced carbon from carbonaceous chondrites (Boato, 1954). All the samples with $\delta C^{13} > -3.6$ are either soil breccias or fines. Within this $C^{13}$-rich grouping, the lowest $\delta C^{13}$ values and carbon concentrations are found in samples 12033 and 14240, both of which are trench samples that have abnormally low solar wind $H_2$ contents and relatively small $\delta Si^{30}$ and $\delta O^{18}$ enrichments (see Figs. 2 and 3). The remaining samples in the $C^{13}$-rich grouping show a good deal of scatter, but the samples with highest carbon contents and highest $\delta C^{13}$ are all Apollo 11 soil breccias or fines with very high solar wind hydrogen contents. To further graphically illustrate these relationships, in Fig. 12 we plot carbon concentration versus hydrogen concentration for all samples analyzed in our laboratories. Note that an approximate linear relationship exists, and that the Apollo 14 and 15 fines, even though having $H_2$ contents similar to the Apollo 11 and 12 fines, contain significantly less carbon.

The above discussion strongly points toward a solar wind origin for much of the carbon in the lunar fines. The data in Fig. 11 indicate that this solar wind carbon must be very enriched in $C^{13}$, but with variable $\delta C^{13}$ values. The simplest assumption one can make to explain the data in Fig. 11 is a two-component mixing model involving this solar wind carbon and the carbon that is intrinsic to the lunar basalts. On

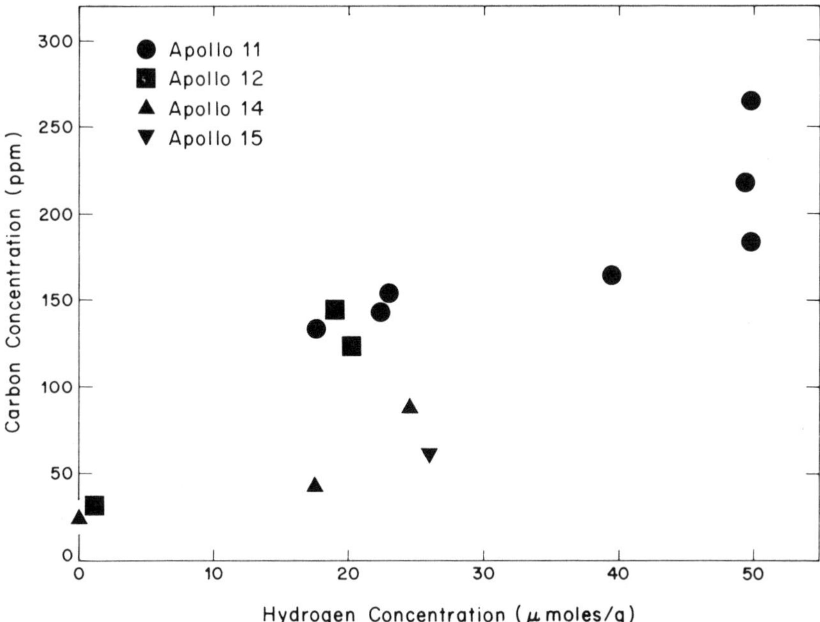


Fig. 12. The relationship between the hydrogen gas concentration (in micromoles per gram) and carbon concentration (ppm) in lunar fines and breccias analyzed in the present work and by Epstein and Taylor (1970; 1971).

the basis of determinations by Kaplan *et al.* (1970) and Kaplan and Petrowski (1971) we obtain average values of $-26‰$ and 25 ppm for the basaltic $\delta C^{13}$ and carbon concentration, respectively. Let us now consider the situation wherein we successively add increments of high $C^{13}$ solar wind carbon to a material that contains 25 ppm carbon of $\delta C^{13} = -26$. Two such mixing curves, Curve A and Curve B, are shown on Fig. 11. Curve A was constructed using a $\delta C^{13}$ of $+10.8$ for the solar wind carbon. This $\delta C^{13}$ value is that of the 10061 soil breccia, which contains the highest carbon and solar wind $H_2$ concentrations of any sample we have studied. Curve B was constructed using a $\delta C^{13}$ of $+25‰$. The particular curves chosen encompass nearly all the measured data points. This illustrates that if the solar wind carbon varies in $\delta C^{13}$ from about $+10$ to $+25$, it is possible to explain all our data in terms of a simple two-component mixing model.

Is there a way to explain the data in Fig. 11 without appealing to an inherent $\delta C^{13}$ variation of 15‰ in the solar wind carbon in the lunar fines? There is, after all, no *a priori* reason to expect such variation. In the light of the pronounced $\delta O^{18}$ and $\delta Si^{30}$ enrichments on the grain-surfaces of the lunar fines, it is reasonable that $C^{13}$ enrichment of the deposited solar wind carbon might take place during particle bombardment. If such isotopic fractionation occurs, it must be accompanied by an overall depletion in the carbon content as various carbon species are isotopically fractionated and lost from the lunar fines. This concept is illustrated schematically in Fig. 13. Curve A in Fig. 13 is the same as shown in Fig. 11; we use it to represent

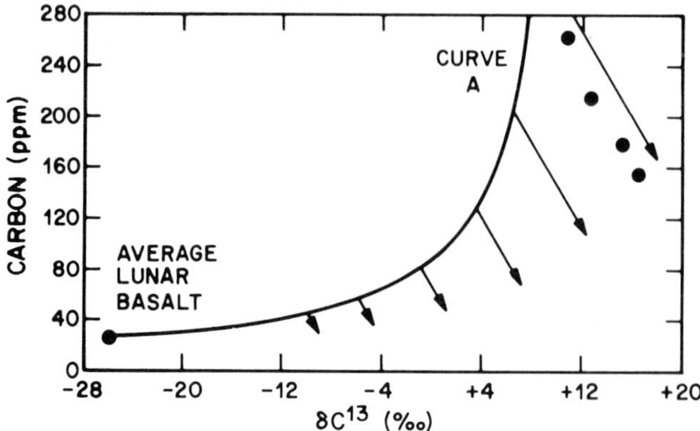


Fig. 13. A schematic diagram similar to that shown in Fig. 11. Curve A is the Curve A of Fig. 11. The diagonal arrows indicate the hypothetical increase in $\delta C^{13}$ that might accompany depletion in the concentration of lunar carbon as a result of particle bombardment. The unlabeled points in the upper right corner represent data obtained on 3 different aliquots of 10061 breccia and a single sample of 10084 soil; of the data points shown, the latter sample has the lowest carbon content and is the most $C^{13}$-rich.

mixing between the lunar basalts and an arbitrary solar wind carbon with $\delta C^{13} = +10.8$. If no isotopic fractionations occur, all samples of lunar fines should plot along such a curve. However, if particle-bombardment induced fractionation takes place the data-points should be displaced from the curve in the rough directions indicated by the arrows. The lengths of the arrows are meant to represent the degree of particle bombardment; the shortest arrows are drawn in the vicinity of the data points obtained on buried Apollo 12 and 14 trench samples. The slopes of the arrows are arbitrary, but have been drawn to approximately conform to Apollo 11 data obtained from three separate samples of 10061 soil breccia and one sample of 10084 fines (these data-points are shown on the figure). This variation in $\delta C^{13}$ and carbon concentration in the Apollo 11 breccia is well outside analytical error, and is difficult to explain on the basis of any reasonable mixing model involving known sources of carbon.

In the above models, we have arbitrarily utilized $\delta C^{13} > +10$ for the solar wind carbon and have ignored the possible contribution of a third component of carbon, namely, from meteorites. It is possible that the solar wind carbon is actually lighter than this, or that we are in fact dealing entirely with meteoritic carbon having an initial $\delta C^{13} \approx -5$ to $-15$, and that none of the samples of lunar fines yet analyzed is free of $C^{13}$-enrichment due to bombardment isotopic fractionation effects. Ganapathy *et al.* (1970) suggested that as much as 2% of the lunar fines may be meteoritic debris. It is necessary to point out that if the solar wind carbon all has a $\delta C^{13} \approx +25$ to $+30$, then all the data points on Fig. 11 conceivably could be explained by meteoritic contamination. Carbonaceous chondrites typically contain about 2–3 wt.% carbon with $\delta C^{13} = -5$ to $-15$ (Boato, 1954). Addition of such meteoritic carbon to material that initially fell along Curve B (Fig. 11) would cause a shift of the sample points upward and to the left along another set of mixing lines.

It is interesting to note in this connection that the trace element data of Ganapathy *et al.* (1970) suggest that the high solar wind $H_2$ breccia 10061 contains more meteoritic material than does 10084; it also contains more carbon, compatible with a greater content of carbonaceous chondrite debris. Thus, the trend in the sequence of points in the upper right-hand corner of Fig. 13 is explicable in terms of greater amounts of meteoritic material in the more carbon-rich samples. Although the present data do not warrant further speculation, it is useful to point out that if the model is correct, then for those samples with $\delta C^{13} > -5$, the ones lying close to Curve A should generally contain more meteoritic debris than the samples that plot in the vicinity of Curve B. It is most important to obtain much more trace element and carbon isotope data on the same lunar samples, to see if this idea has any validity.

Summarizing, we have proposed three separate models to explain the distribution and scatter of the lunar carbon isotope analyses obtained to date: (1) two-component mixing of fragmented lunar igneous rocks with solar wind carbon of variable $\delta C^{13}$ in the range +10 to +30; (2) two-component mixing of the lunar rocks with solar wind carbon or meteoritic carbon of $\delta C^{13} \approx -10$ to $+10$, with superimposition of $C^{13}$ enrichment and carbon loss due to particle bombardment; (3) three-component mixing of lunar rocks, meteoritic debris, and solar wind carbon of $\delta C^{13} \approx +25$ to $+30$. We cannot as yet distinguish among these various possibilities, and all three models may have some validity.

*Acknowledgments*—We wish to thank L. T. Silver and D. S. Burnett for fruitful discussions, and Paul Yanagisawa for carrying out much of the laboratory work. This research was supported by the National Aeronautics and Space Administration, NASA Grant No. NGL-05-002-190. This paper is Contribution No. 2164, Publications of the Division of Geological and Planetary Sciences, California Institute of Technology, Pasadena, California 91109.

## References

Boato G. (1954) The isotopic composition of hydrogen and carbon in the carbonaceous chondrites. *Geochim. Cosmochim. Acta* **6,** 209–220.

Burlingame A. L. *et al.* (1971) Simulation-3 sand transfer; transfer of Apollo 14 SESC lunar material. Report, University of California, Berkeley, Space Science Laboratory, p. 11.

Burnett D. S., Huneke J. C., Podosek F. A., Russ G. P. III, Turner G., and Wasserburg G. J. (1972) The irradiation history of lunar samples (abstract). In *Lunar Science—III* (editor C. Watkins), pp. 105–107, Lunar Science Institute Contr. No. 88.

Clayton R. N., Onuma N., and Mayeda T. (1971) Oxygen isotope fractionation in Apollo 12 rocks and soils. *Proc. Second Lunar Sci. Conf., Geochim. Cosmochim. Acta* Suppl. 2, Vol. 2, pp. 1417–1420. MIT Press.

Crozaz G., Drozd R., Hohenberg C. M., Hoyt H. P. Jr., Ragan D., Walker R. M., and Yuhas D. (1972) Solar flare and galactic cosmic ray studies of Apollo 14 samples (abstract). In *Lunar Science—III* (editor C. Watkins), pp. 167–169, Lunar Science Institute Contr. No. 88.

Ehmann W. D. and Morgan J. W. (1970) Oxygen, silicon, and aluminum in Apollo 11 rocks and fines by 14 MeV neutron activation. *Proc. Apollo 11 Lunar Sci. Conf., Geochim. Cosmochim. Acta* Suppl. 1, Vol. 2, pp. 1071–1079. Pergamon.

Ehmann W. D. and Morgan J. W. (1971) Major element abundances in Apollo 12 rocks and fines by 14 MeV neutron activation. *Proc. Second Lunar Sci. Conf., Geochim. Cosmochim. Acta* Suppl. 2, Vol. 2, pp. 1237–1245. MIT Press.

Epstein S. and Taylor H. P. Jr. (1970) The concentration and isotopic composition of hydrogen, carbon, and silicon in Apollo 11 lunar rocks and minerals. *Proc. Apollo 11 Lunar Sci. Conf., Geochim. Cosmochim. Acta* Suppl. 1, Vol. 2, pp. 1085–1096. Pergamon.

Epstein S. and Taylor H. P. Jr. (1971) $O^{18}/O^{16}$, $Si^{30}/Si^{28}$, D/H, and $C^{13}/C^{12}$ ratios in lunar samples. *Proc. Second Lunar Sci. Conf., Geochim. Cosmochim. Acta* Suppl. 2, Vol. 2, pp. 1421–1441. MIT Press.

Freeman S. W. Jr., Hills H. K., and Vondrak R. R. (1972) Water vapor, whence comest thou? (abstract). In *Lunar Science—III* (editor C. Watkins), pp. 283–285, Lunar Science Institute Contr. No. 88.

Friedman I., Gleason J. D., and Hardcastle K. G. (1970) Water, hydrogen, deuterium, carbon, and $C^{13}$ content of selected lunar material. *Proc. Apollo 11 Lunar Sci. Conf., Geochim. Cosmochim. Acta* Suppl. 1, Vol. 2, pp. 1103–1109. Pergamon.

Friedman I., O'Neil J. R., Gleason J. D., and Hardcastle K. (1971) The carbon and hydrogen content and isotopic composition of some Apollo 12 materials. *Proc. Second Lunar Sci. Conf., Geochim. Cosmochim. Acta* Suppl. 2, Vol. 2, pp. 1407–1415. MIT Press.

Ganapathy R., Keays R. R., Laul J. C., and Anders E. (1970) Trace elements in Apollo 11 lunar rocks: Implications for meteorite influx and origin of moon. *Proc. Apollo 11 Lunar Sci. Conf., Geochim. Cosmochim. Acta* Suppl. 1, Vol. 2, pp. 1117–1142. Pergamon.

Heymann D. and Yaniv A. (1971) Breccia 10065: Release of inert gases by vacuum crushing at room temperature. *Proc. Second Lunar Sci. Conf., Geochim. Cosmochim. Acta* Suppl. 2, Vol. 2, pp. 1681–1692. MIT Press.

Jackson E. D. and Wilshire H. G. (1972) Classification of the samples returned from the Apollo 14 landing site (abstract). In *Lunar Science—III* (editor C. Watkins), pp. 418–420, Lunar Science Institute Contr. No. 88.

Kaplan I. R., Smith J. W., and Ruth E. (1970) Carbon and sulfur concentrations and isotopic composition in Apollo 11 lunar samples. *Proc. Apollo 11 Lunar Sci. Conf., Geochim. Cosmochim. Acta* Suppl. 1, Vol. 2, pp. 1317–1329. Pergamon.

Kaplan I. R. and Petrowski C. (1971) Carbon and sulfur isotope studies on Apollo 12 lunar samples. *Proc. Second Lunar Sci. Conf., Geochim. Cosmochim. Acta* Suppl. 2, Vol. 2, pp. 1397–1406. MIT Press.

O'Neil J. R. and Epstein S. (1966) A method for oxygen isotope analysis of milligram quantities of water and some of its applications. *J. Geophys. Res.* **71,** 4955–4961.

O'Neil J. R. and Taylor H. P. Jr. (1969) Oxygen isotope equilibrium between muscovite and water. *J. Geophys. Res.* **74,** 6012–6022.

O'Neil J. R. and Adami L. H. (1970) Oxygen isotope analyses of selected Apollo 11 materials. *Proc. Apollo 11 Lunar Sci. Conf., Geochim. Cosmochim. Acta* Suppl. 1, Vol. 2, pp. 1425–1427. Pergamon.

Onuma N., Clayton R. N., and Mayeda T. (1970) Apollo 11 rocks: Oxygen isotope fractionation between minerals and an estimate of the temperature of formation. *Proc. Apollo 11 Lunar Sci. Conf., Geochim. Cosmochim. Acta* Suppl. 1, Vol. 2, pp. 1429–1434. Pergamon.

Sheppard S. M. F., Nielsen R. L., and Taylor H. P. Jr. (1969) Oxygen and hydrogen isotope ratios of clay minerals from porphyry copper deposits. *Econ. Geol* **64,** No. 7, 755–777.

Silver L. T. (1972) Lead volatization and volatile transfer processes on the moon (abstract). In *Lunar Science—III* (editor C. Watkins) pp. 701–703, Lunar Science Institute Contr. No. 88.

Suzuoki T. and Epstein S. (1972) Hydrogen isotope fractionation between OH-bearing silicate minerals and water. In preparation.

Taylor H. P. Jr. (1968) The oxygen isotope geochemistry of igneous rocks. *Contrib. Mineral. Petrol.* **19,** 1–71.

Taylor H. P. Jr. and Epstein S. (1970a) $O^{18}/O^{16}$ ratios of Apollo 11 lunar rocks and minerals. *Proc. Apollo 11 Lunar Sci. Conf., Geochim. Cosmochim. Acta* Suppl. 1, Vol. 2, pp. 1613–1626. Pergamon.

Taylor H. P. Jr. and Epstein S. (1970b) Oxygen and silicon isotope ratios of lunar rock 12013. *Earth Planet. Sci. Lett.* **9,** 208–210.

Wilshire H. G. and Jackson E. D. (1972) Petrology of the Fra Mauro formation at the Apollo 14 landing site (abstract). In *Lunar Science—III* (editor C. Watkins), pp. 803–805, Lunar Science Institute Contr. No. 88.

Proceedings of the Third Lunar Science Conference
(Supplement 3, *Geochimica et Cosmochimica Acta*)
Vol. 2, pp. 1455–1463
The M.I.T. Press, 1972

# Oxygen isotopic compositions and oxygen concentrations of Apollo 14 and Apollo 15 rocks and soils

ROBERT N. CLAYTON,* JULIE M. HURD, and TOSHIKO K. MAYEDA

The Enrico Fermi Institute, University of Chicago
Chicago, Illinois 60637

**Abstract**—Some Apollo 14 igneous rocks, and basaltic clasts from fragmental rocks, have intermineral oxygen isotopic fractionations larger than typical igneous values, resulting from some post-crystallization thermal event. Apollo 15 basalt 15016 has normal igneous fractionation factors. The $\delta O^{18}$ of microbreccia clasts from Apollo 14 fragmental rocks is correlated with the textural classification of the rock, showing $O^{18}$ enrichment for class F2, and none for class F4. Accurate determination of departures of oxygen content from normal stoichiometric values may aid in understanding the nature and origin of the $O^{18}$ and $Si^{30}$ rich component in lunar soils.

## INTRODUCTION

OXYGEN ISOTOPE ANALYSES of separated mineral phases of Apollo 11 and 12 crystalline rocks have previously shown that these phases were in isotopic equilibrium at the time of crystallization and that their isotopic compositions have not changed significantly since then, so that mineral-pair fractionations provide a measure of the crystallization temperature. It was of interest to see if the isotopic distributions in Apollo 14 crystalline rocks and breccias provided evidence for a more complex thermal history than that of the mare basalts.

Both the Apollo 14 and Apollo 15 collections provided interesting glassy rocks and glass spheres and fragments in the soils and breccias. It might be expected that the processes of their formation could have produced isotopic variations with respect to the lunar crystalline rocks, which have been found to be very uniform in oxygen isotopic composition.

Lunar soils and soil-related microbreccias have been found to be enriched in $O^{18}$ relative to crystalline rocks. The effect is mainly associated with the finest grains and/or grain surfaces. Since several types of breccia were returned from the Fra Mauro site, we may expect to find variable amounts of this $O^{18}$-enriched component, depending on the extent of incorporation of earlier surficial material.

Another problem, possibly related to the $O^{18}$-enriched component of soils and microbreccias, is the apparent oxygen deficiency in some soils and rocks (Ehmann and Morgan, 1970; 1971). Sufficiently accurate determinations of oxygen contents of various fractions may aid in determining the nature of the isotopic enrichment process.

* Also Departments of Chemistry and Geophysical Sciences.

## ANALYTICAL PROCEDURES

The procedures for mineral separation, oxygen extraction, and mass spectrometric analysis were the same as those used for Apollo 11 and Apollo 12 samples (Onuma *et al.*, 1970). Oxygen isotopic compositions are reported in the $\delta$-terminology, as permil ($^0/_{00}$) deviations from the SMOW standard. Standard error in the mean of the duplicate measurements on each sample is estimated to be $\pm 0.07$ permil.

Several of the samples analyzed were from consortia as follows: 14073 (G. J. Wasserburg), 14276 (G. J. Wasserburg), 14306 (E. Anders), 14313 (L. A. Haskin), 14321 (G. G. Goles).

## RESULTS AND DISCUSSION

### *Oxygen isotopes*

Oxygen isotopic analyses of separated phases from crystalline rocks, from crystalline clasts of fragmental rocks, and glassy igneous rocks are presented in Table 1. Mean $\delta$-values for minerals from several Apollo 12 basalts are shown for comparison. For a given mineral phase, there are no differences from rock to rock greater than $0.4^0/_{00}$, which is only a few times greater than the analytical uncertainty. However, the mineral-pair fractionations for some of the Apollo 14 crystalline rocks are con-

Table 1. Oxygen isotopic compositions of minerals from Apollo 14 and Apollo 15 rocks.

| Sample | Gl* | Pc | Pg | Cpx | Opx | Ol | Il |
|---|---|---|---|---|---|---|---|
| Apollo 14 Rocks | | | | | | | |
| Basalts | | | | | | | |
| 14053,25 | — | 5.97† | 5.55 | 5.43 | — | — | 3.79 |
| 14072,7 | — | 5.98 | 5.39 | 5.53 | — | — | 3.67 |
| Basaltic clasts | | | | | | | |
| 14321,184–1A | — | 5.93 | 5.56 | — | — | 4.89 | — |
| 14321,184–3C | — | 5.83 | 5.59 | — | — | 4.96 | — |
| Orthopyroxene basalts | | | | | | | |
| 14073 | — | 5.94 | 5.52 | — | — | — | — |
| 14276,8 | — | 5.86 | 5.51 | — | 5.66 | — | — |
| 14310,129 | — | 5.94 | — | — | 5.52 | — | — |
| Anorthosite clast | | | | | | | |
| 14321,184–7B | — | 5.97 | — | — | — | — | — |
| Apollo 15 Rocks | | | | | | | |
| Basalts | | | | | | | |
| 15016,43 | — | 5.81 | 5.56 | 5.42 | — | 5.00 | 4.04 |
| 15597,3 | 5.60 | — | 5.41 | — | — | — | — |
| Anorthosite | | | | | | | |
| 15415,13 | — | 5.78 | — | — | — | — | — |
| Glassy rock | | | | | | | |
| 15426,42, whole rock | 5.41 | — | — | — | — | — | — |
| 15426,42 spheres | 5.27 | — | — | — | — | — | — |
| Apollo 12 Rocks | | | | | | | |
| Means of 7 basalts | — | 5.98 | 5.81 | 5.56 | — | 5.15 | 4.07 |

* Abbreviations: Gl = glass; Pc = plagioclase; Pg = pigeonite; Cpx = calcic clinopyroxene; Opx = orthopyroxene; Ol = Olivine; Il = ilmenite.

† Oxygen isotopic compositions in $\delta$-notation, expressed as permil (parts per thousand) deviation of $O^{18}/O^{16}$ from SMOW standard.

sistently larger than corresponding fractionations for Apollo 12 and 15 basalts. The petrographically similar rocks 14053 and 14072 have indistinguishable isotopic patterns, and yield a plagioclase-ilmenite isotopic temperature of 960°C, about 100° lower than the value for Apollo 12 crystalline rocks. The basaltic clasts 14321,184–1A and 14321,184–3C also have isotopic patterns similar to 14053, and have a plagioclase-olivine fractionation of 0.95, which may be significantly larger than the value of 0.78 ± 0.04 determined for three Apollo 12 basalts (Clayton *et al.*, 1971) and 0.63 for 15016.

If the Apollo 14 rocks were excavated from some considerable depth by the Imbrium impact event, the possibility must be considered that they might have experienced a thermal metamorphism before impact, or a substantial heating in the impact itself. The oxygen isotopic data indicate that there has been no re-equilibration among minerals with respect to oxygen at temperatures below 960°C. Thus there is no record of a pre-Imbrium metamorphism. A rapid impact heating to ≥1000°C, followed by rapid quenching, as suggested by textural evidence in the fragmental rocks, would be compatible with the oxygen isotopic evidence. The K-Ar and Rb-Sr data (Papanastassiou *et al.*, 1971; Turner *et al.*, 1971) imply that basaltic fragments from rock 14321 have had a similar thermal history to crystalline rock 14053. The oxygen isotope results support this conclusion.

The petrographic and chemical studies of Grieve *et al.* (1972) and Duncan *et al.* (1972) on rock 14321 lead to the conclusions that basaltic clasts, such as those analyzed in this study, were incorporated into the fragmental rock relatively late in the series of breccia-forming events and imply that these clasts were not subjected to intense reheating. It is possible that the isotopic effects observed result from a minor heating event causing a small nonequilibrium increase of the isotopic fractionations from their primary igneous values.

The orthopyroxene-bearing rocks 14073, 14276, and 14310 all exhibit typical "igneous" values of oxygen isotopic fractionations, although the absence of measurable amounts of ilmenite prevents assignment of an isotopic temperature. Sample 14321,184,7B was a cluster of millimeter-sized plagioclase crystals within the fragmental rock 14321. This feldspar has the same isotopic composition as that in the crystalline rocks.

All of the Apollo 15 igneous rocks analyzed to date are slightly depleted in $O^{18}$ relative to Apollo 12 basalts, which in turn were found to be $0.2^0/_{00}$ lower in $O^{18}$ than Apollo 11 rocks (Clayton *et al.*, 1971). Nevertheless, the entire range for a given mineral or whole-rock $\delta$-value is only about $0.4^0/_{00}$, which is much smaller than the corresponding range in terrestrial basaltic rocks. Isotopic fractionations among minerals in the porphyritic basalt 15016 are typical of igneous values, the plagioclase-ilmenite fractionation corresponding to an isotopic temperature of 1090°C.

The anorthosite 15415 has the isotopic composition to be expected if the rock formed in igneous processes. "Anorthositic" fragments of indistinguishable $O^{18}/O^{16}$ ratio were found in Apollo 11 soil (Onuma *et al.*, 1970).

The glass-rich rock 15426, collected from the rim of Spur Crater, has a lower $\delta O^{18}$ than any lunar rock previously analyzed. Green glass spheres separated from this rock are almost $1^0/_{00}$ lower in $O^{18}$ than glass spheres in Apollo 11 and 14 soils.

Further interpretation of these isotopic compositions must await the results of chemical and petrographic studies of the glasses.

Oxygen isotopic compositions of soils (<1 mm fines) from the Apollo 14 and 15 sites, and of microbreccia fragments of Apollo 14 fragmental rocks are presented in Table 2. Mean $\delta O^{18}$ values of Apollo 11 and 12 and Luna 16 soils are given for comparison. At all lunar sites studied, the soils are enriched in $O^{18}$ relative to local igneous rocks, typically by 0.2–0.4$^0/_{00}$ (Clayton *et al.*, 1971; Epstein and Taylor, 1971; O'Neil and Adami, 1970; Onuma *et al.*, 1970; Taylor and Epstein, 1970). Detailed studies on the nature of this enrichment have been made by Epstein and Taylor (1971, 1972). The soils tend to fall into two groups: Apollo 11, Apollo 14, and Luna 16, with $\delta O^{18} \approx 6.2$, and Apollo 12 and Apollo 15, with $\delta O^{18} \approx 5.8$. These differences probably result both from the lower $\delta O^{18}$ of the Apollo 12 and 15 crystalline rocks, and from differences in the extent of $O^{18}$-enrichment occurring either in the soil-forming process or in later bombardment of the regolith.

The Apollo 11 microbreccias showed an $O^{18}$ enrichment relative to crystalline rocks equal to that of the Apollo 11 soils (O'Neil and Adami, 1970; Onuma *et al.*, 1970; Taylor and Epstein, 1970), in accord with the hypothesis that the microbreccias were derived by induration of soils. In this regard, it is interesting that one of the Apollo 14 breccias (14313) also shows this enrichment, whereas others (14306, 14321) do not. Epstein and Taylor (1972) have also shown that an interior fragment of breccia 14321 contains negligible amounts of solar wind hydrogen, and no measurable $O^{18}$ enrichment in fluorine stripping experiments. Thus there seems to be a correspondence between $O^{18}$ enrichment and petrologic class of the Apollo 14 fragmental rocks (Jackson and Wilshire, 1972): Class F2 (and perhaps F1) are similar to Apollo 11 microbreccias in containing large amounts of solar wind gases (Kaiser, 1972) and showing $O^{18}$-enrichments; Class F4 (and perhaps F3) containing little solar wind gas and showing no $O^{18}$ enrichment. If this relationship is verified, it will help in under-

Table 2. Oxygen isotopic compositions of Apollo 14 and Apollo 15 soils and breccias.

| | Sample | $\delta O^{18}$ ($^0/_{00}$ rel. SMOW) |
|---|---|---|
| Apollo 14 | | |
| Soils | 14003,11 | 6.23 |
| | 14163,62 | 6.14 |
| | 14163,62 glass spheres | 6.07 |
| | 14259,60 | 6.20 |
| Microbreccias* | 14306,35–10 | 5.72 |
| | 14306,35–11 | 5.64 |
| | 14313,34–E | 6.14 |
| | 14313,34–E (<2.77)† | 6.34 |
| | 14321,184–4 | 5.77 |
| | 14321,184–7B | 5.71 |
| Apollo 15 | | |
| Soils | 15270,1 | 5.96 |
| | 15421,20 | 5.80 |
| | 15600,1 | 5.68 |
| Other Soils | | |
| Apollo 11 mean value | — | 6.18 |
| Apollo 12 mean value | — | 5.81 |
| Luna 16 | — | 6.20 |

* Fragments or clasts from fragmental rocks.
† Plagioclase rich separate, specific gravity <2.77.

standing the nature and timing of the processes leading to enrichment of $O^{18}$ in lunar fines.

*Oxygen concentrations*

Direct measurement of the oxygen content of lunar rocks and soils may be useful in determining departures from normal stoichiometry due to the presence of ions in unusual oxidation states (such as $Ti^{3+}$) or to the presence of components in fine materials that have suffered radiation damage or that have received an external surface deposit. Ehmann and Morgan (1970, 1971) have discussed these possibilities and report an oxygen deficiency of 1.6 wt.%, relative to the stoichiometric amount calculated from major element analysis for Apollo 11 soils and for Apollo 12 soils and crystalline rocks.

Another experimental observation which may be related to a deficiency of oxygen in lunar soils is that of Epstein and Taylor (1971, 1972) on the composition of gases produced in the initial stages of the reaction of fluorine with these materials. They observed that the ratio of $O_2/SiF_4$ liberated in the first few percent of reaction was near 1, whereas the ratio for the whole sample is about 1.8. Furthermore, the initial oxygen and silicon released were strongly enriched in their heavier stable isotopes, $O^{18}$ and $Si^{30}$, respectively. The extent of isotopic enrichment is correlated with the amount of solar wind gases implanted in the soil. Epstein and Taylor have suggested that these effects result from the removal of oxygen and silicon from the surfaces of soil grains, leaving a residue depleted in the light isotopes, and depleted in oxygen relative to silicon. They also discussed the possibility of deposition of a layer of heavy-isotope-enriched material on the surfaces of the lunar fines, as an alternative explanation for their observations. It is of interest to investigate whether accurate determination of departures from normal stoichiometry with respect to oxygen can be of use in choosing between these alternatives.

Epstein and Taylor (1971) presented detailed results of their partial fluorination experiments on the Apollo 11 breccia 10061, which illustrates all the isotopic effects typical of solar-wind irradiated soils and microbreccias. With the exception of one data point (for the first reaction), all the data for $\delta O^{18}$, $\delta Si^{30}$, and $O_2/SiF_4$ ratio are compatible with a two-component mixing model, of which one component is "ordinary" Apollo 11 basalt, and the other component, characterized by very high $\delta O^{18}$ and $\delta Si^{30}$ values, and a low $O_2/Si$ ratio is interpreted either as an exotic addition to the soil, or as a residue after fractional vaporization. The data do not uniquely determine the composition of this second component, but yield the product: $R(\delta O^{18}-5.9)$, where $R$ is the molar ratio $O_2/Si$, and $\delta O^{18}$ is its oxygen isotopic composition, and 5.9 is the $\delta O^{18}$ value of the basaltic component.

Let us consider first the possibility of an addition of material to the surfaces. The observed isotopic compositions and $O_2/SiF_4$ ratios of the earliest reacting portions require a value of $R < 1$ and $\delta O^{18} > 48‰$. A reasonable minimum value for $R$ is 0.5, corresponding to the composition of SiO, which might have been deposited from the vapor phase following some high-temperature event. A component with $R = 0.5$, $\delta O^{18} = 51.5‰$, $\delta Si^{30} = 23‰$ fits the mixing model satisfactorily. It can then be shown that the observations of Epstein and Taylor imply an overall oxygen deficiency with respect to stoichiometric composition for rock 10061 of 0.3 wt.% of the rock.

In the case in which the isotopically heavy component is assumed to be the residue of a fractional vaporization process, the net oxygen deficiency depends on the proportions of silicon and oxygen lost by vaporization. In order to produce a residue with $O_2/Si < 1.8$, this ratio in the vapor must have been $> 1.8$. As an extreme case, if no silicon is lost, the data for 10061 imply an oxygen deficiency, relative to the stoichiometric composition, of 0.8 wt.% of the rock. However, this extreme case is inadmissable since it cannot account for the silicon isotope enrichment. If silicon is also lost by vaporization, a larger fraction of oxygen must also be lost, increasing the net oxygen deficiency with respect to normal stoichiometry. For example, if the isotopically enriched residue were material from which one-half of the original silicon had been removed, the fraction evaporated must have had an $O_2/Si$ ratio of about 3 in order to account for the low $O_2/Si$ ratio in the residue. The net oxygen deficiency for this case amounts to 1.3 wt.% for the bulk rock of 10061. Thus the expected oxygen deficiencies, relative to normal stoichiometry, show the lowest value (0.8 wt.%) for no silicon loss, and increase for increasing loss of silicon.

The two categories of model for accounting for the $O^{18}$ and $Si^{30}$ enriched material differ in that the fractional vaporization model results in a substantially greater oxygen deficiency than the model of accretion of an exotic component.

The experimental data are equivocal as to the presence or magnitude of oxygen deficiencies relative to stoichiometric compositions. Our data on oxygen contents of various lunar soils and breccias are listed in Table 3 and are compared with oxygen contents calculated from published analyses of major elements in Fig. 1. The points

Table 3. Oxygen contents of lunar soils and breccias.

| Sample No. | U.C.[a] | E.[b] | W.[c] | Stoich.[d] |
|---|---|---|---|---|
| 10060 (br) | 41.3 | 40.3 | 41.4 | 41.7 |
| 10084 | 42.6 | 40.8 | 41.5 | 42.1 |
| 12033 | 42.7 | — | — | 43.2 |
| 12037 | 41.4 | — | 42.4 | 42.9 |
| 12057 | 42.1 | 42.8 | — | 42.5 |
| 12070 | 42.9 | 41.4 | 42.6 | 42.6 |
| 14003 | 44.5 | 43.2 | — | 44.2 |
| 14163 | 44.2 | 44.5 | 43.7 | 44.2 |
| 14259 | 43.8 | 44.8 | 43.8 | 44.2 |
| 14313 (br) | 44.3 | — | — | — |
| 15270 | 43.2 | — | — | 43.7 |
| 15421 | 42.6 | — | — | — |
| 15600 | 42.2 | — | 41.8 | 41.9 |
| L-16-A | 41.9 | 40.2 | — | 41.9 |

[a] University of Chicago analyses by volumetric measurement of $O_2$ liberated by reaction with $BrF_5$.

[b] Ehmann and Morgan (1970; 1971), Ehmann and Gillum (1972), neutron-activation analysis.

[c] Wänke *et al.* (1970, 1971, 1972) neutron-activation analysis.

[d] Mean values calculated from major-element analyses of: Compston *et al.* (1970; 1971; 1972); Cuttitta *et al.* (1971); Frondell *et al.* (1971); Hubbard *et al.* (1972); Klein *et al.* (1972); Maxwell and Wiik (1971); Maxwell *et al.* (1970); Morrison *et al.* (1970); Rose *et al.* (1970); Wakita and Schmitt (1971).

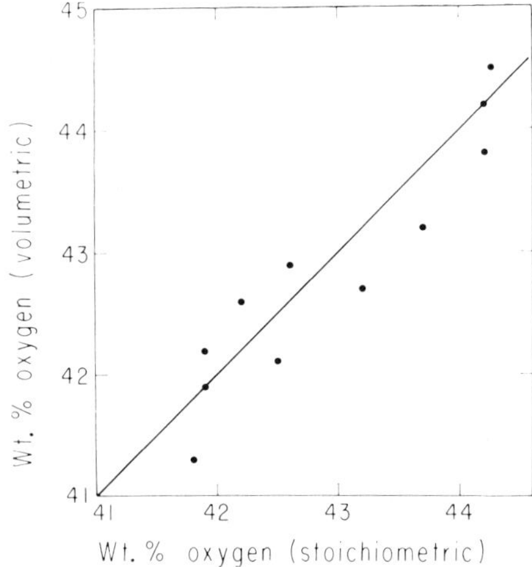

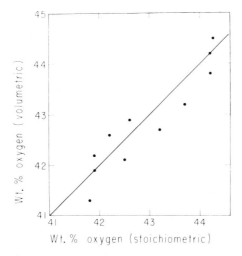

Fig. 1. Comparison of oxygen contents of soils determined by volumetric measurement of $O_2$ liberated by reaction of samples with $BrF_5$ with the stoichiometric oxygen content calculated from major element analyses, using conventional assignment of oxidation states.

scatter about the 45° line, with no systematic difference evident for the two types of analysis. Only in the case of sample 12037 is there a serious discrepancy between the two types of analysis. This may be a result of sample heterogeneity, since the elemental analysis (Frondell *et al.*, 1971) was done on a small (120 mg) sample of a fraction with grain-size $<37\ \mu$. The elemental analysis also has a rather high summation of 100.6%. Excluding this one point, the relative standard deviation of the volumetric results from the line is 0.8%. The maximum apparent oxygen deficiency is 0.5 wt.% (for samples 12033 and 15270). This observation is in contrast with that of Ehmann and Morgan (1970, 1971), and Ehmann and Gillum (1972), who found an average oxygen deficiency of 1.6% relative to the stoichiometrically calculated amount. Comparisons between the volumetric oxygen results in Table 3 and the activation analysis results of Ehmann and co-workers on the same samples reveals rather poor agreement, with differences between the two laboratories ranging from +1.8 to −1.0 wt.%. A similar comparison between our data and the activation analysis results of Wänke *et al.* (1970, 1971, 1972) shows rather good agreement. Of all the analysts whose results are discussed here, only Ehmann and co-workers took care to avoid contact of the samples with air. Hence it may be true, as they claim, that the freshly returned lunar samples are oxygen-deficient by one or two percent, and that they react with atmospheric water or oxygen on exposure to air. If this is the case, then the relationship between the bulk oxygen deficiency and the surface-correlated isotope effects observed by Epstein and Taylor becomes obscure, since the latter workers observed $O^{18}$ and $Si^{30}$ enrichments whether or not their samples had been exposed to laboratory air (S. Epstein and H. P. Taylor, Jr., personal communication). Furthermore, it is surprising that Ehmann and co-workers found an oxygen deficiency of similar magnitude in the Apollo 12 crystalline rocks. Such a result is unexpected if the oxygen deficiency is correlated with solar wind bombardment as Epstein and Taylor have observed for the surface effects. It is apparent that there are still experimental difficulties which must be resolved before the nature and source of non-stoichiometric oxygen and silicon components can be established.

*Acknowledgment*—This work was supported in part by NASA grant NGL 14-001-169.

## REFERENCES

Clayton R. N., Onuma N., and Mayeda T. K. (1971) Oxygen isotope fractionation in Apollo 12 rocks and soils. *Proc. Second Lunar Sci. Conf., Geochim. Cosmochim. Acta* Suppl. 2, Vol. 2, pp. 1417–1420. MIT Press.

Compston W., Chappell B. W., Arriens P. A., and Vernon M. J. (1970) The chemistry and age of Apollo 11 lunar material. *Proc. Apollo 11 Lunar Sci. Conf., Geochim. Cosmochim. Acta* Suppl. 1, Vol. 2, pp. 1007–1027. Pergamon.

Compston W., Berry H., Vernon M. J., Chappell B. W., and Kaye M. J. (1971) Rubidium-strontium chronology and chemistry of lunar material from the Ocean of Storms. *Proc. Second Lunar Sci. Conf., Geochim. Cosmochim. Acta* Suppl. 2, Vol. 2, pp. 1471–1485. MIT Press.

Compston W., Vernon M. J., Berry H., Rudowski R., Gray C. M., and Ware N. (1972) Age and petrogenesis of Apollo 14 basalts (abstract). In *Lunar Science—III* (editor C. Watkins), pp. 151–153, Lunar Science Institute Contr. No. 88.

Cuttitta F., Rose H. J. Jr., Annell C. S., Carron M. K., Christian R. P., Dwornik E. J., Greenland L. P., Helz A. W., and Ligon D. T. Jr. (1971) Elemental composition of some Apollo 12 lunar rocks and soils. *Proc. Second Lunar Sci. Conf., Geochim. Cosmochim. Acta* Suppl. 2, Vol. 2, pp. 1217–1229. MIT Press.

Duncan A. R., Lindstrom M. M., Lindstrom D. J., McKay S. M., Stoeser J. W., Goles G. G., and Fruchter J. S. (1972) Comments on the genesis of breccia 14321 (abstract). In *Lunar Science—III* (editor C. Watkins) pp. 192–194, Lunar Science Institute Contr. No. 88.

Ehmann W. D. and Gillum D. E. (1972) Oxygen and other major elements in Apollo 14 rocks and some lunar soils (abstract). In *Lunar Science—III* (editor C. Watkins), pp. 209–211, Lunar Science Institute Contr. No. 88.

Ehmann W. D. and Morgan J. W. (1970) Oxygen, silicon, and aluminum in Apollo 11 rocks and fines by 14 Mev neutron activation. *Proc. Apollo 11 Lunar Sci. Conf., Geochim. Cosmochim. Acta* Suppl. 1, Vol. 2, pp. 1071–1079. Pergamon.

Ehmann W. D. and Morgan J. W. (1971) Major element abundances in Apollo 12 rocks by 14 Mev neutron activation. *Proc. Second Lunar Sci. Conf., Geochim. Cosmochim. Acta* Suppl. 2, Vol. 2, pp. 1237–1245. MIT Press.

Epstein S. and Taylor H. P. Jr. (1971) $O^{18}/O^{16}$, $Si^{30}/Si^{28}$, D/H, and $C^{13}/C^{12}$ ratios in lunar samples. *Proc. Second Lunar Sci. Conf., Geochim. Cosmochim. Acta* Suppl. 2, Vol. 2, pp. 1421–1441. MIT Press.

Epstein S. and Taylor H. P. Jr. (1972) $O^{18}/O^{16}$, $Si^{30}/Si^{28}$, $C^{13}/C^{12}$, and D/H studies of Apollo 14 and 15 samples (abstract). In *Lunar Science—III* (editor C. Watkins), pp. 236–238, Lunar Science Institute Contr. No. 88.

Frondell C., Klein C. Jr., and Ito J. (1971) Mineralogical and chemical data on Apollo 12 lunar fines. *Proc. Second Lunar Sci. Conf., Geochim. Cosmochim. Acta* Suppl. 2, Vol. 1, pp. 719–726. MIT Press.

Grieve R., McKay G., Smith H., Weill D., and McCallum S. (1972) Mineralogy and petrology of polymict breccia 14321 (abstract). In *Lunar Science—III* (editor C. Watkins), pp. 338–340, Lunar Science Institute Contr. No. 88.

Hubbard N. J., Gast P. W., Rhodes M., and Wiesmann H. (1972) Chemical composition of Apollo 14 materials and evidence for alkali volatilization (abstract). In *Lunar Science—III* (editor C. Watkins), pp. 407–409, Lunar Science Institute Contr. No. 88.

Jackson E. D. and Wilshire H. G. (1972) Classification of the samples returned from the Apollo 14 landing site (abstract). In *Lunar Science—III* (editor C. Watkins), pp. 418–420, Lunar Science Institute Contr. No. 88.

Kaiser W. A. (1972) Rare gas measurements in three Apollo 14 samples (abstract). In *Lunar Science—III* (editor C. Watkins), pp. 442–443, Lunar Science Institute Contr. No. 88.

Klein C. Jr., Drake J. C., Frondell C., and Ito J. (1972) Mineralogy and petrology of several Apollo 14 rock types and chemistry of the soil (abstract). In *Lunar Science—III* (editor C. Watkins), pp. 455–457, Lunar Science Institute Contr. No. 88.

Maxwell J. A. and Wiik H. B. (1971) Chemical composition of Apollo 12 lunar samples 12004, 12033, 12051, 12052, and 12065. Second Lunar Science Conference (unpublished proceedings).

Maxwell J. A., Peck L. C., and Wiik H. B. (1970) Chemical composition of Apollo 11 lunar samples 10017, 10020, 10072, and 10084. *Proc. Apollo 11 Lunar Sci. Conf., Geochim. Cosmochim. Acta* Suppl. 1, Vol. 2, pp. 1369–1374. Pergamon.

Morrison G. H., Gerard J. T., Kashuba A. T., Gangadharam E. V., Rothenberg A. M., Potter N. M., and Miller G. B. (1970) Elemental abundances of lunar soil and rocks. *Proc. Apollo 11 Lunar Sci. Conf., Geochim. Cosmochim. Acta* Suppl. 1, Vol. 2, pp. 1383–1392. Pergamon.

Onuma N., Clayton R. N., and Mayeda T. K. (1970) Apollo 11 rocks: Oxygen isotope fractionation between minerals and an estimate of the temperature of formation. *Proc. Apollo 11 Lunar Sci. Conf., Geochim. Cosmochim. Acta* Suppl. 1, Vol. 2, pp. 1429–1434. Pergamon.

O'Neil J. R. and Adami L. H. (1970) Oxygen isotope analyses of selected Apollo 11 materials. *Proc. Apollo 11 Lunar Sci. Conf., Geochim. Cosmochim. Acta* Suppl. 1, Vol. 2, pp. 1425–1427. Pergamon.

Papanastassiou D. A. and Wasserburg G. J. (1971) Rb–Sr ages of igneous rocks from the Apollo 14 mission and the age of the Fra Mauro formation. *Earth Planet. Sci. Lett.* **12,** 36–48.

Rose H. J. Jr., Cuttitta F., Dwornik E. J., Carron M. K., Christian R. P., Lindsay J. R., Ligon D. T., and Larson R. R. (1970) Semimicro x-ray fluorescence analysis of lunar samples. *Proc. Apollo 11 Lunar Sci. Conf., Geochim. Cosmochim. Acta* Suppl. 1, Vol. 2, pp. 1493–1497. Pergamon.

Taylor H. P. Jr. and Epstein S. (1970) $O^{18}/O^{16}$ ratios of Apollo 11 lunar rocks and minerals. *Proc. Apollo 11 Lunar Sci. Conf., Geochim. Cosmochim. Acta* Suppl. 1, Vol. 2, pp. 1613–1626. Pergamon.

Turner G., Huneke J. C., Podosek F. A., and Wasserburg G. J. (1971) $^{40}Ar$-$^{39}Ar$ ages and cosmic ray exposure ages of Apollo 14 samples. *Earth Planet. Sci. Lett.* **12,** 19–35.

Wakita H. and Schmitt R. A. (1971) Bulk elemental composition of Apollo 12 samples: Five igneous and one breccia rocks and four soils. *Proc. Second Lunar Sci. Conf., Geochim. Cosmochim. Acta* Suppl. 2, Vol. 2, pp. 1231–1236. MIT Press.

Wänke H., Rieder R., Baddenhausen H., Spettel B., Teschke F., Quijano-Rico M., and Balacescu A. (1970) Major and trace elements in lunar material. *Proc. Apolle 11 Lunar Sci. Conf., Geochim. Cosmochim. Acta* Suppl. 1, Vol. 2, pp. 1719–1727. Pergamon.

Wänke H., Wlotzka F., Baddenhausen H., Balacescu A., Spettel B., Teschke F., Jagoutz E., Kruse H., Quijano-Rico M., and Rieder R. (1971) Apollo 12 samples: Chemical composition and its relation to sample locations and exposure ages, the two component origin of the various soil samples and studies on lunar metallic particles. *Proc. Second Lunar Sci. Conf., Geochim. Cosmochim. Acta* Suppl. 2, Vol. 2, pp. 1187–1208. MIT Press.

Wänke H., Baddenhausen H., Balacescu A., Teschke F., Spettel B., Dreibus G., Quijano M., Kruse H., Wlotzka F., and Begemann F. (1972) Multielement analyses of lunar samples (abstract). In *Lunar Science—III* (editor C. Watkins), pp. 779–781, Lunar Science Institute Contr. No. 88.

Proceedings of the Third Lunar Science Conference
(Supplement 3, *Geochimica et Cosmochimica Acta*)
Vol. 2, pp. 1465–1472
The M.I.T. Press, 1972

# Isotopic abundance ratios and concentrations of selected elements in Apollo 14 samples

I. L. Barnes, B. S. Carpenter, E. L. Garner, J. W. Gramlich, E. C. Kuehner, L. A. Machlan, E. J. Maienthal, J. R. Moody, L. J. Moore, T. J. Murphy, P. J. Paulsen, K. M. Sappenfield, and W. R. Shields

National Bureau of Standards, Institute for Materials Research, Analytical Chemistry Division, Washington, D.C. 20234

**Abstract**—Absolute or relative isotopic abundance ratios have been determined for U, Pb, Rb, Sr, Ca, and Cu on a representative fraction of the bulk fines 14163,159 and for a breccia 14321,221. No significant variations from terrestrial values were noted for the nonradiogenic isotopes. Concentrations were determined for the above elements as well as for Th, B, Ag, Cd, Fe, Ti, and Ni.

$^{207}Pb/^{206}Pb$ ages of about 4860 m.y. (million years) and 4420 m.y. were calculated for 14163 and 14321, respectively. The Pb–U and Pb–Th ages are very slightly discordant, the soils exhibit a reversed discordancy and the breccia a normal discordancy. Extreme inhomogeneity of Rb and Sr in the fines sample was found.

## Introduction

We are reporting the results of analyses on one Apollo 14 soil sample and one breccia sample. The fines were the less than 1 mm fraction from the bulk fines sample 14163,159 and the breccia sample was a sawn piece from the large breccia 14321,221 from Station C1 at Cone Crater. The materials have been described in detail by LSPET (1971).

The objective of this study was to perform a series of isotopic analyses and concentration determinations for a number of elements *all on the same sample* so that, for example, the lead, uranium, thorium, rubidium, and strontium data would all be directly related to the identical material. An effort was made to ensure that this sample would be representative as nearly as practically possible to the entire "whole rock" or fines sample. Most importantly, however, the elements to be analyzed were chosen on the basis of the availability of well characterized standard reference materials, many of which have had absolute abundance ratios determined and have been accompanied by a comprehensive mineral survey to establish the consistency of these ratios for terrestrial material. All the elements determined by thermal ionization mass spectrometry except thorium were analyzed along with, and in the identical manner as these standard reference materials. Standard reference materials are not available for thorium.

## Analytical Procedures

### *Chemical preparation*

After careful mixing to ensure a representative sample, a 1 gram sample and two 0.5 gram samples were removed from the fines material 14163. The 1 gram sample was used to determine isotopic compositions of the various elements while the half-gram samples were used for duplicate isotopic dilution concentration determinations.

The samples were put into solution in covered Teflon beakers using a perchloric-hydrofluoric acid mixture as described previously (Shields, 1970a). Before dissolution an isotopic tracer containing a suitable amount of the enriched isotope for each element to be determined was added to the isotope dilution samples. The procedure for preparing, storing and adding spikes has been discussed in detail in various NBS Technical Notes (Shields, 1966, 1967, 1968, 1970).

The dissolution procedure for the breccia, 14321, was somewhat different. A 9.6 gram piece of this material was examined. After microscopic examination no gross contamination from the sawing was apparent and the sample was then supported on two Teflon rods and broken in half with a blow from a steel pestle which had been wrapped in a Teflon sheet. A 4.3 gram piece was placed in solution in a sealed Teflon-lined bomb of the type described by Krogh (1971). The bomb containing a perchloric-hydrofluoric acid mixture was placed in an oven at 150°C for 16 hours. At the end of that period, the contents were transferred to a Teflon beaker and the insoluble fluorides were converted to soluble perchlorates as described previously (Shields, 1970a). After adding sufficient hydrochloric acid so that the resulting solution was 1*N* in HCl, the solution was transferred to a Teflon bottle and diluted to 150 ml. A clear solution with no visible insoluble matter resulted. A weighed aliquot representing 1 gram of material was taken for compositional analysis and two aliquots representing 0.5 gram of material were withdrawn and spiked with weighed portions of the enriched isotope solutions. From this point on all samples were treated in the same way. Three blanks were prepared for each aliquot or sample for each element to be determined and were treated exactly like the samples through all subsequent steps of the procedure. The separation and purification procedures used for each element have been previously published (Shields, 1970a).

### *Mass spectrometric procedures*

Five different single sector magnetic deflection mass spectrometers were used during the course of the analyses reported here. Two of these are 6-inch radius–60° instruments, two are 12-inch–90° spectrometers and one is a 12-inch–68° instrument. Except for the size and radius all instruments are identical, having identical sources, collectors, and measuring circuits. The details of each of these components have been published (Shields, 1966, 1967, 1968, 1970a, 1970b). In addition, a two-stage computer controlled spectrometer equipped with a high-speed electronic pulse counting circuit was used. This instrument was particularly designed to have a very high abundance sensitivity and was used to check the analyses for those isotopes present in very small amount such as the $^{234}$U content.

The analytical mass spectrometric procedures for each element have also been published (see references of Shields mentioned above). In general, the techniques used for the certification of various standard reference materials (SRM) were used, the appropriate SRM was analyzed with the samples and the quoted error limits are applicable to the lunar materials. An exception to this is the lead analyses where the samples were too small for the very precise triple filament procedure of Catanzaro (1967). The silica gel procedure reported by Cameron *et al.* (1969) and modified by us (Shields, 1970b) was used in this case. Long-time laboratory experience gives error limits (95% L.E. for a single analysis) of $\pm 0.1\%$, $\pm 0.07\%$, and $\pm 0.1\%$ for the 208/206, 207/206, and 204/206 ratios for common lead and these error terms are applicable to the lunar samples except for the 204/206 ratio measurement which expands to plus or minus the effect of the 204 blank. The average filament fractionation is 0.04% per mass unit. NBS standard reference materials 981, common lead, and 983, radiogenic lead, were analyzed along with the lunar materials.

### *Reagents*

The water, nitric acid, perchloric acid, hydrochloric acid, and sulfuric acid used were prepared by distillation in all quartz subboiling stills. Concentrated HF (27*N*) was prepared in a still of the same design made entirely of Teflon. The stills are contained within an enclosure which provides a continuous class 100 air environment. An analysis of each reagent was performed for some 17 elements including those of interest here so that an indication could be obtained of the total reagent blank to be expected for each element. The details of this procedure including the maximum impurity levels obtained by a single distillation of all elements analyzed will be published elsewhere; however, an indication of results that were obtained, the maximum lead level in 27*N* HF was found to be 0.05 ng/gm and in the distilled water was 0.008 ng/gm.

*Laboratory environment*

A novel vertical flow class 100 laboratory was designed and constructed which contained provisions for isolated dissolution, evaporation and other preparation of many samples at one time. Details of the features of this new laboratory will be published in the near future. All chemical operations were carried out in this facility. All mounting of filaments etc., were performed in portable class 100 clean air hoods.

In addition to the elements determined by thermal ionization mass spectrometry, several were determined by other techniques that were particularly applicable to the elements in question. The other analytical methods used were isotope dilution spark source mass spectrometry, differential cathode ray polarography and nuclear track counting. Each of these has been previously described (Paulsen *et al.*, 1969; Maienthal, 1972; and Carpenter, 1972). The reagents used and the laboratory environment for these other methods were the same as described above.

## Analytical Results

To date we have obtained absolute or relative isotopic composition and/or concentrations for lead, uranium, thorium, rubidium, strontium, copper, calcium, boron, silver, cadmium, iron, titanium, and nickel. The results of these analyses along with the results for the appropriate standard reference material are shown in Tables 1 through 5.

## Discussion

*Sample inhomogeneity*

The extreme range of possible sample inhomogeneity is shown by the data for lead, uranium, and thorium for the duplicate sampling of the fines sample 14163.

Table 1. Absolute isotope ratios and concentration of lead, uranium and thorium.

| | | Lead composition | | | Lead concentration | |
|---|---|---|---|---|---|---|
| Sample | 208/206 | 207/206 | 204/206 | blank | | blank |
| SRM 981 | 2.1681 ±0.0008 | 0.91464 ±0.00033 | 0.059042 ±0.000037 | | | |
| 14163,159 | 0.9491 | 0.7378 | 0.001945 | 5.63 ng | (a) 9.893 ppm<br>(b) 9.622 ppm | 14 ng |
| 14321,221 | 0.9884 | 0.5477 | 0.001755 | 7.72 ng | 3.312 ppm | |
| | | **Uranium composition** | | | **Uranium concentration** | |
| Sample | 235/234 | 235/236 | 235/238 | | | blank |
| Belgian Congo pitchblende | 131.87 ± 1.20 | | 137.88 ± 0.14 | | | |
| 14163,159 | 133.89 | $>10^4$ | 137.82 | | (a) 3.596 ppm<br>(b) 3.704 ppm | 0.8 ng |
| 14321,221 | 133.96 | $>10^4$ | 137.80 | | 1.550 ppm | 0.7 ng |
| | | **Thorium composition** | | | **Thorium concentration** | |
| Sample | 230/232 | | | | | blank |
| 14163,159 | $<10^{-4}$ | | | | (a) 13.22 ppm<br>(b) 13.22 ppm | 0.1 ng |
| 14321,221 | $<10^{-4}$ | | | | 5.775 ppm | 0.1 ng |

Table 2. Isotope ratios and concentration of rubidium and strontium.

| | Rubidium composition (absolute) | | | Rubidium concentration | |
|---|---|---|---|---|---|
| Sample | | 85/87 | | | blank |
| SRM 984 | | 2.5926 ± 0.0017 | | | |
| 14163,159 | | 2.5921 | | 15.41 ppm | 3.1 ng |
| 14321,221 | | 2.5916 | | 5.964 ppm | 3.1 ng |
| | Strontium composition[a] | | | Strontium concentration | |
| Sample | 88/86[b] | 87/86 | 84/86 | | blank |
| SRM 987 | 8.3752 | 0.71014 | 0.05655 | | |
| 14163,159 | 8.3752 | 0.71484 | 0.05651 | 185.50 ppm | 117 ng |
| 14321,221 | 8.3752 | 0.70799 | 0.05655 | 120.29 ppm | 117 ng |

[a] Normalized to 88/86 = 8.3752.
[b] To within 0.05% lunar samples have the same 88/86 ratio as SRM 987 before normalization.

Table 3. Absolute isotope ratio and concentration of copper.

| | Composition | Concentration | |
|---|---|---|---|
| Sample | 63/65 | | blank |
| SRM 976 | 2.2440 ± 0.0021 | | |
| 14163,159 | 2.2417 | 9.981 ppm | 34 ng |
| 14321,221 | 2.2458 | 77.91 ppm | 140 ng |

Table 4. Relative isotope ratios for calcium.

| Sample | 40/44 | 42/44 | 43/44 | 46/44 | 48/44 |
|---|---|---|---|---|---|
| SRM 915 | 46.480 | 0.3104 | 0.06478 | 0.0017 | 0.08976 |
| | ±0.087 | ±0.0011 | ±0.00090 | ±0.0005 | ±0.00055 |
| 14163,159 | 46.448 | 0.3100 | 0.06447 | 0.0016 | 0.08958 |
| 14321,221 | 46.426 | 0.3100 | 0.06451 | 0.0016 | 0.08961 |

Experimentally observed values, not absolute values.

Table 5. Concentrations of various elements.

| Element | 14163,159 | 14321,221 |
|---|---|---|
| Iron[a] | 7.74% | 11.50% |
| Titanium[a] | 0.98% | 1.31% |
| Nickel[a] | 333 ppm | 145 ppm |
| Boron[b] | 2.19 ppm | — |
| Cadmium[c] | ≤0.3 ppm | 0.50 ppm |
| Silver[c] | 0.018 ppm | <0.007 ppm |

[a] By differential cathode ray polarography.
[b] By nuclear track counting.
[c] By isotope dilution spark source mass spectrometry.

Even though reasonable precautions were taken to obtain homogeneous samples the differences range from 0% for thorium to 2.8% for lead to ~4% for uranium, about 10 to 40 times the predicted experimental error.

These limited data indicate that the thorium seems rather uniformly distributed while both the lead and the uranium seem concentrated in one or more phase and/or size distribution. This was evident in the small sample analyzed for uranium and boron by the nuclear track method and has been reported by numerous workers (Fleischer *et al.*, 1970; Burnett *et al.*, 1971). However, ages calculated for the samples separately agree with each other to ~0.05% indicating that the uranium and its associated lead do not separate.

Sample inhomogeneity is even more dramatically illustrated by a comparison of the values for rubidium and strontium reported by various workers for 14163. These data are shown in Table 6. The concentrations of rubidium range over ±3.5% from the average, those for strontium vary ±1.2%, and the calculated ages by 1.1% from the average of 4595 m.y. which represents a total range of about 100 m.y.

A significant part of this variation was expected since prior to our work on these lunar samples we had initiated an interlaboratory analysis program utilizing a set of standard reference materials (SRM 610 to 619) which are silicate glasses doped with 61 elements at the nominal 500, 50, 1, and 0.02 ppm levels. With the cooperation of a number of workers data have been obtained on many elements among which are lead, uranium, thorium, rubidium, and strontium. For the lead, uranium, thorium analyses, there was excellent agreement among the cooperating laboratories on all the samples at all levels. These data suggest that adequate chemical and mass spectrometric procedures and standards are available and in general use for these elements and barring actual sample inhomogeneities results should be reliable.

A completely different situation, however, was observed for rubidium and strontium in these glasses. It was readily apparent that, as a minimum, a 0.7% systematic bias exists for strontium analyses and a 1% systematic bias exists for rubidium despite satisfactory precision.

In attempting to define the source of this bias, four different commercially available samples of high purity strontium carbonate were examined. It was found that based on acidimetry all contained about 99.3% strontium carbonate with the remainder consisting of $NH_4^+$, $HCO_3^-$, and $NO_3^-$ ions resulting from the generally used method of manufacture. Heating to 800°C in an atmosphere of $CO_2$ resulted in a stoichiometric compound containing 99.98 ± 0.02% $SrCO_3$. The systematic bias of 0.7% could and probably does result from the general use of the nonstoichiometric material as a standard for portions of the analyses. No explanation for the apparent 1% bias in the rubidium analyses is offered.

Regardless of the cause it would appear that the lunar analysis program would benefit from the use of a common standard for rubidium and strontium analyses. We suggest that SRM 984, rubidium chloride and SRM 987, strontium carbonate would be eminently satisfactory for this purpose.

Table 6. Comparison of rubidium and strontium data on 14163.

| Author | Rubidium (ppm) | Strontium (ppm) | $^{87}Sr/^{86}Sr$ | Age (m.y.) |
|---|---|---|---|---|
| Tatsumoto *et al.* (1972) | 14.61 | 182.1 | 0.71448 | 4647 |
| Papanastassiou and Wasserburg (1972) | 14.68 | 182.4 | 0.71414 | 4535 |
| Nyquist *et al.* (1972) | 15.1 | 185 | 0.71454 | 4540 |
| This work A | 15.57 | 186.45 | 0.71484 | 4571 |
| This work B | 15.24 | 184.04 | 0.71484 | 4608 |

$^{88}Sr/^{86}Sr$ = 8.3752. $(^{87}Sr/^{86}Sr)_I$ = 0.69898 (BABI). $\lambda = 1.39 \times 10^{-11}\ yr^{-1}$.

*Age calculations*

The data presented above, of course, contain all the requisite information for the calculation of ages. Corrections are necessary for the small amount of blank lead and for the "primordial" lead present. These corrections were made using experimentally determined values for the blank lead and the troilite values of Oversby (1970) for the initial lead correction and are shown in Table 7. The ages calculated for the U–Pb and Th–Pb systems as well as the rubidium-strontium ages are shown in Table 8. These are "whole-rock" ages only. No attempt was made to obtain data from which to construct an isochron age. The initial $^{87}Sr/^{86}Sr$ values were those reported by Papanastassiou and Wasserburg (1971). Compston *et al.* (1971) obtained a nearly identical value for $(^{87}Sr/^{86}Sr)_I$ for sample 14321 and the ages agree nominally with those obtained by these two groups. The ages from the U–Pb and Th–Pb systems show, in general, the same effects as found by Tatsumoto *et al.* (1971) and others, that the ages are discordant and that the fines ages are older than the rocks or breccias. However, the ages are only very slightly discordant as shown in Fig. 1. These very limited data do not permit any speculations as to the "crystallization" ages of these samples; however, several observations may be made.

The positive discordant ages of the fines sample do indicate that "extra" lead is present in this material. Whether this is the result of volatile transfer as has been suggested by Silver (1970, 1972) and Tatsumoto (1970) cannot be determined from this limited information; however, the data would imply that this lead must be

Table 7. Isotope ratios for lead.

| Ratio | Observed | After blank[a] correction | After troilite[b] correction |
|---|---|---|---|
| | Sample 14163,159 (av.) | | |
| 208/206 | 0.94909 | 0.94857 | 0.90922 |
| 207/206 | 0.73775 | 0.73742 | 0.73090 |
| 204/206 | 0.001945 | 0.001930 | — |
| | Sample 14321,221 | | |
| 208/206 | 0.98837 | 0.98671 | 0.95736 |
| 207/206 | 0.54773 | 0.54719 | 0.53960 |
| 204/206 | 0.001755 | 0.001467 | — |

[a] Blank lead: 208 52.34%; 207 22.08%; 206 24.14%; 204 1.42%.
[b] Troilite lead: 208 58.47%; 207 20.63%; 206 18.87%; 204 2.02%.

Table 8. Apparent ages (m.y.).

| Sample | $^{207}Pb/^{206}Pb$ | $^{206}Pb/^{238}U$ | $^{207}Pb/^{235}U$ | $^{208}Pb/^{232}Th$ | $^{87}Rb/^{87}Sr$ |
|---|---|---|---|---|---|
| 14163,159 | 4861 | 4923 | 4881 | 4910 | 4600[a] |
| 14321,221 | 4416 | 4375 | 4404 | 4379 | 4180[b] |

[a] Initial $^{87}Sr/^{86}Sr = 0.69898$ assumed.
[b] Initial $^{87}Sr/^{86}Sr = 0.69942$ assumed.
$\lambda_{87} = 1.39 \times 10^{-11}$ yr$^{-1}$.
$\lambda_{238} = 1.54 \times 10^{-10}$ yr$^{-1}$.
$\lambda_{235} = 9.72 \times 10^{-10}$ yr$^{-1}$.
$\lambda_{232} = 4.88 \times 10^{-11}$ yr$^{-1}$.
$^{238}U/^{235}U = 137.7$.

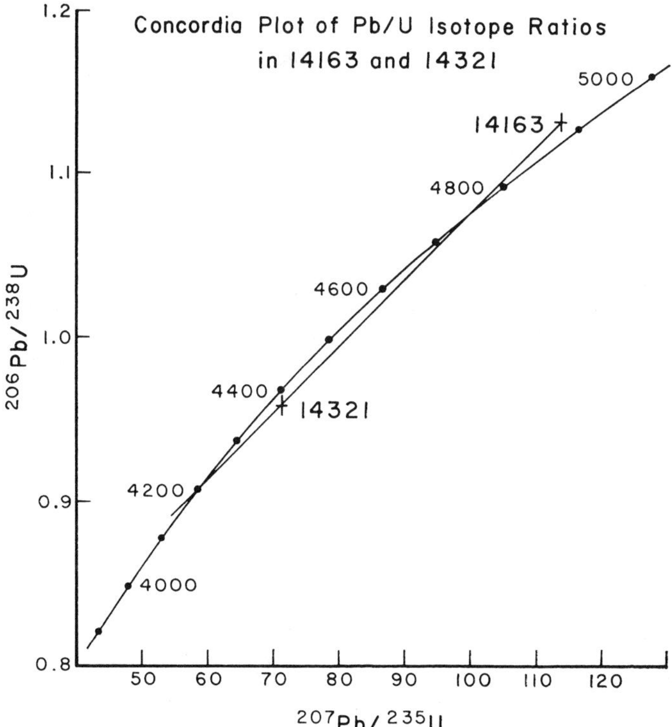


Fig. 1. A Pb/U concordia diagram with the positions of samples 14321 and 14163 shown as determined by the data reported here. The length of lines forming the cross for the position of 14163 shows the approximate error limits.

extremely well distributed if present for the two separate samples to show identical ages with lead-uranium concentration differing by 3–4%.

As shown in Fig. 1 a line joining the two data points intersects concordia at about 4730 m.y. An extension of the line to a lower intercept intersects at about 4200 m.y. and, within limits, is consistent with the age of 3800 to 4000 m.y. found by Tatsumoto *et al.* (1972) and others.

*Acknowledgments*—We thank W. A. Bowman, III for his assistance in the design, construction, and maintenance of much of the instrumentation used in this work. We are indebted to T. W. Stern, U.S. Geological Survey, who read the manuscript and whose comments and suggestions were of great help.

## References

Burnett D., Monnin M., Seitz M., Walker R., and Yuhas D. (1971) Lunar astrology—U–Th distributions and fission-track dating of lunar samples. *Proc. Second Lunar Sci. Conf., Geochim. Cosmochim. Acta* Suppl. 2, Vol. 3, pp. 1503–1519. MIT Press.

Cameron A. E., Smith D. H., and Walker R. L. (1969) Mass spectrometry of nanogram-size samples of lead. *Anal. Chem.* **41,** 525–526.

Carpenter B. S. (1972) The determination of trace concentrations of boron and uranium in glass by the nuclear track technique. *Anal. Chem.* in press, March 1972.

Catanzaro E. J. (1967) Triple-filament method for solid-sample lead isotope analysis. *J. Geophys. Res.* **72,** 1325–1327.

Compston W., Vernon M. J., Berry H., and Rudowski R. (1971) The age of the Fra Mauro formation: A radiometric older limit. *Earth Planet. Sci. Lett.* **12,** 55–58.

Fleischer R. L., Haines E. L., Hart H. R. Jr., Woods R. T., and Comstock G. M. (1970) The particle track record of the Sea of Tranquility. *Proc. Apollo 11 Lunar Sci. Conf., Geochim. Cosmochim. Acta* Suppl. 1, Vol. 3, pp. 2103–2120. Pergamon.

Garner E. L., Machlan L. A., and Shields W. R. (1971) Uranium isotopic reference materials. *NBS Special Publication 260-17*, 65–74. U.S. Government Printing Office, Washington, D.C.

Krogh T. E. (1971) A simplified technique for the dissolution of zircons and the isolation of uranium and lead. In *Carnegie Institution Year Book* **69,** 341–344. Carnegie Institution, Washington, D.C.

LSPET (Lunar Sample Preliminary Examination Team) (1971) Preliminary examination of lunar samples from Apollo 14. *Science* **173,** 681–693.

Maienthal E. J. (1972) The determination of iron, titanium, and nickel in Apollo 14 samples by cathode ray polarography. In preparation.

Nyquist L. E., Hubbard N. J., Gast P. W., Wiesmann H., and Church S. E. (1972) Rb–Sr relationships for some chemically defined lunar materials (abstract). In *Lunar Science—III* (editor C. Watkins), pp. 584–586, Lunar Science Institute Contr. No. 88.

Oversby V. M. (1970) The isotopic composition of lead in iron meteorites. *Geochim. Cosmochim. Acta* **34,** 65–75.

Papanastassiou D. A. and Wasserburg G. J. (1971) Rb–Sr ages of igneous rocks from the Apollo 14 mission and the age of the Fra Mauro formation. *Earth Planet. Sci. Lett.* **12,** 36–48.

Paulsen P. J., Alvarez R., and Kelleher D. E. (1969) Determination of trace elements in zinc by isotope dilution spark source mass spectrometry. *Spectrochimica Acta* **24,** Sec. B, 535–544.

Shields W. R. (1966) Analytical Mass Spectrometry Section, *NBS Technical Note 277*, U.S. Government Printing Office, Washington, D.C.

Shields W. R. (1967) Analytical Mass Spectrometry Section, *NBS Technical Note 426*, U.S. Government Printing Office, Washington, D.C.

Shields W. R. (1968) Analytical Mass Spectrometry Section, *NBS Technical Note 456*, U.S. Government Printing Office, Washington, D.C.

Shields W. R. (1970a) Analytical Mass Spectrometry Section, *NBS Technical Note 546*, U.S. Government Printing Office, Washington, D.C.

Shields W. R. (1970b) Analytical Mass Spectrometry Section, *NBS Technical Note 506*, U.S. Government Printing Office, Washington, D.C.

Silver L. T. (1970) Uranium–thorium–lead isotopes in some Tranquility Base samples and their implications for lunar history. *Proc. Apollo 11 Lunar Sci. Conf., Geochim. Cosmochim. Acta* Suppl. 1, Vol. 2, pp. 1533–1574. Pergamon.

Silver L. T. (1972) Lead volatilization and volatile transfer processes on the moon (abstract). In *Lunar Science—III* (editor C. Watkins), pp. 701–703, Lunar Science Institute Contr. No. 88.

Tatsumoto M. (1970) Age of the moon: An isotopic study of U–Th–Pb systematics of Apollo 11 lunar samples—II. *Proc. Apollo 11 Lunar Sci. Conf., Geochim. Cosmochim. Acta* Suppl. 1, Vol. 2, pp. 1595–1612. Pergamon.

Tatsumoto M., Knight R. J., and Doe B. R. (1971) U–Th–Pb systematics of Apollo 12 lunar samples. *Proc. Second Lunar Sci. Conf., Geochim. Cosmochim. Acta* Suppl. 2, Vol. 3, pp. 1521–1546. MIT Press.

Tatsumoto M., Hedge C. E., Doe B. R., and Unruh D. (1972) U–Th–Pb and Rb–Sr measurements on some Apollo 14 lunar samples (abstract). In *Lunar Science—III* (editor C. Watkins), pp. 741–743, Lunar Science Institute Contr. No. 88.

Proceedings of the Third Lunar Science Conference
(Supplement 3, *Geochimica et Cosmochimica Acta*)
Vol.2, pp. 1473–1477
The M.I.T. Press, 1972

# Deuterium content of lunar material

L. MERLIVAT, G. NIEF, and E. ROTH
Centre d'Etudes Nucléaires de Saclay,
D.R.A.—Boîte Postale n° 2-91-Gif-sur-Yvette, France

**Abstract**—The hydrogen and deuterium content of lunar samples 10019, 14003, and 14259, have been measured. Water extracted from the samples 10019 and 14259 has also been analyzed. Equivalent amounts of $H_2$ and $H_2O$ equal to about 20 $\mu$mole/g have been found. The measured concentration of deuterium is 20 ppm in hydrogen and 100 ppm in water. The conditions of isotopic exchange between hydrogen and water vapor have been investigated and are reported. Consideration of the influence of these exchange processes leads to an assignment of an upper limit of 13 ppm to the D content of hydrogen. In addition, constraints are placed on the origin of the water. The fact that no hydrogen is released from the fines or breccias at temperatures above 650°C in this work is explained by experiments showing that oxidation starts at 575°C.

## INTRODUCTION

DEUTERIUM ABUNDANCE IN LUNAR MATERIAL has been measured by several laboratories (Epstein and Taylor, 1970; Friedman *et al.*, 1970; Hintenberger *et al.*, 1970; Epstein and Taylor, 1972) since moon samples have been returned to earth. We report here measurements made on Apollo 11 and 14 samples.

## EXPERIMENTAL

About 250 mg aliquots of a sample are transferred from the shipment container to a gold foil wrapping in a glove box under dry nitrogen. They are inserted in a silica tube connected to a ground joint and a stopcock. This assembly is sealed to a vacuum line where the silica tube can be heated to 1100°C by a dc furnace.

Prior to gas extraction the quartz tube is degassed, the moon sample being kept at room temperature. This is made possible by turning the tube around the ground joint.

After the sample is dropped in the furnace, it is pumped out while heating until a temperature of 275°C. We start then to collect the extractable gases. The condensable gases and vapors are trapped by liquid nitrogen. Temperature in the furnace and pressure, as measured by a Pirani gauge, are recorded.

The temperature is increased step by step in such a way that a sufficient amount of hydrogen is released for isotopic analysis. Each fraction extracted is brought in contact with vanadium.

The vanadium acts as a trap for hydrogen. At 250°C the vanadium reacts with $H_2$ to form vanadium hydride. At high temperature (800°C), this compound is decomposed and hydrogen is liberated. Preliminary studies of the adsorption and desorption isotherms of hydrogen by the vanadium have shown that it is necessary to use 150 mg of metal in order to transfer 20 to 50 mm$^3$ STP of hydrogen.

A liquid nitrogen trap just ahead of the vanadium tube has been found necessary to avoid reduction of the residual water of the glass line by the hot vanadium. Blank experiments with hydrogen containing only 1 part per million (ppm) of deuterium have been run. They have shown that samples of 20 $mm^3$ STP of hydrogen can be enriched in deuterium in our apparatus by 3 ppm under our conditions of extraction. We have corrected by this amount the isotopic content measured in the lunar samples.

During the extraction, the gases are first in contact with vanadium at 250°C, then its temperature is brought down to ambient. Residual gases (mainly rare gases) are pumped out. By heating again the vanadium at 900°C, the hydrogen is liberated and transferred to ampoules filled with charcoal at liquid nitrogen temperature. At −180°C the charcoal retains completely the hydrogen.

The gas is desorbed at room temperature on the line of the mass spectrometer. The hydrogen sample is attached to the double inlet system and compared to the standard at the same pressure. Precision ($2\sigma$) of ratio measurements is $\pm 0.1\ 10^{-6}$ for 20 $mm^3$ samples, $\pm 0.5\ 10^{-6}$ for 2 $mm^3$, and $\pm 2.10^{-6}$ for 0.5 $mm^3$ samples.

## RESULTS

We have made five extractions from three different moon samples. The results are given in Tables 1 and 2. In two cases we have also measured the amount and the isotopic content of the water collected in the trap during heating of the lunar samples from 200°C to 800°C. This water comes mainly from the moon sample, but it is not excluded that some contamination of the glass line has been added too. Measured values can be found in Table 2.

Duplicate extractions have been made for the fines samples 14259a, b and 14003a, b. Between two different extractions from fractions of the same sample, we observe noticeable variations in the quantity of hydrogen collected. These differences can be related to small changes in extraction procedure, mainly to variations in the lapse of time during which the extracted gases are kept in contact with the hot moon sample before they are transferred to the vanadium and adsorbed. This observation is supported by experiments showing that hydrogen is oxidized to water by lunar material. At room temperature we have introduced over a moon sample, previously degassed at 800°C, a volume of hydrogen gas of about 100 $mm^3$ STP. When we raise the temperature, we observe a decrease of the pressure in the apparatus and water is formed. This oxidation starts at 575°C and the combustion is complete after 90 min, the temperature of the moon sample being raised from ambient to 800°C.

This oxidation also explains why hydrogen is not collected at temperatures higher than 640°C. In this range, the oxidation reaction is faster than the degassing of the hydrogen of the sample. We must conclude that our measurements represent lower limits of the hydrogen content of the lunar samples we have analyzed. It is interesting to note that the isotopic results are not markedly affected by varying conditions in the extraction procedure.

## DISCUSSION AND INTERPRETATION OF THE RESULTS

Keeping in mind that lunar hydrogen is mostly of solar wind origin that theoretically should be deuterium free (D/H about $10^{-17}$), we have tried to interpret the value of $2.10^{-5}$ found for deuterium concentration in the lunar gas.

Table 1.

| Temperature variation (°C) | 10019 $m = 0.245$ g | | 14259 a $m = 0.277$ g | | 14259 b $m = 0.249$ g | | 14003 a* $m = 0.235$ g | | 14003 b $m = 0.247$ g | | 14003 c† $m = 0.256$ g | |
|---|---|---|---|---|---|---|---|---|---|---|---|---|
| | $F$(%) | (D/H) $10^6$ | $F$(%) | (D/H) $10^6$ | $F$(%) | (D/H) $10^6$ | $F$(%) | (D/H) $10^6$ | $F$(%) | (D/H) $10^6$ | $F$(%) | (D/H) $10^6$ |
| 275–370 | 9.2 | 33.1 | | | 7.6 | 26.5 | 15.4 | 24.3 | | | 33.5 | 28.0 |
| | | | 28.3 | 16.2 | | | | | 7.1 | 39.6 | | |
| | | | | | | | | | 29.0 | 19.5 | | |
| 370–485 | 34.5 | 23.0 | 21.9 | 13.5 | 22.8 | 17.1 | 18.0 | 17.9 | 8.6 | 22.7 | 24.2 | 23.8 |
| | 13.3 | 22.4 | | | 20.7 | 15.7 | 30.4 | 17.0 | 10.8 | 17.6 | | |
| 485–540 | 24.0 | 22.9 | 26.7 | 19.2 | 22.7 | 17.7 | 23.5 | 19.1 | 20.1 | 18.1 | 26.2 | 24.5 |
| 540–640 | 19.0 | 35.0 | 23.1 | 22.6 | 26.2 | 27.7 | 12.7 | 26.5 | 24.4 | 22.0 | 16.1 | 25.8 |

* First extraction performed. The temperature variation is different of the indications of the 1st column.
† Sample purposedly contaminated with water vapor of deuterium content equal to 700 ppm.
*F* is the fraction of gas extracted in the given interval of temperature in percent.

Table 2.

| | $H_2$ ($\mu$ mole/g) | (D/H) (ppm) | $H_2O$ ($\mu$ mole/g) | (D/H) (ppm) |
|---|---|---|---|---|
| 10019 | | | | |
| Breccia | 17.6 | 26.1 | 24.8 | 101.4 |
| 14259 | | | | |
| Fines a | 17.1 | 17.9 | | |
| Fines b | 19.9 | 20.4 | 14.4 | 96.8 |
| 14003 | | | | |
| Fines a* | 13.4 | 20.0 | | |
| Fines b | 19.9 | 21.3 | | |
| Fines c† | 16.2 | 25.7 | | |

* First extraction performed. The temperature variation is different of the indications of the 1st column.

† Sample purposedly contaminated with water vapor of deuterium content equal to 700 ppm.

As previously communicated by Epstein and Taylor (1972), we have investigated the exchange of deuterium between hydrogen and water in the presence of lunar material. Three experiments have been performed at 200°C, 400°C, and 525°C. Volumes of about 100 mm$^3$ STP of water vapor and hydrogen are introduced in the extraction line with a moon sample previously degassed at 800°C. After one hour of contact at 200°C, no exchange is detectable. After the same period of time, 29 and 75% of the total exchange has taken place at 400° and 525°C. A rough estimation of the activation energy of about 12 kcal/mole/0°C can be calculated from these experiments. We must stress that these exchange experiments involve free hydrogen and water in the gas phase, in the presence of lunar material, and not gases inbedded in this material. The experiments show that at temperatures of about 400°C, at which a large part of the water contained in the moon sample is evolved (Epstein and Taylor, 1970; Friedman *et al.*, 1970; Epstein and Taylor, 1972) some isotopic exchange between water and hydrogen has occurred. The deuterium content we have measured in lunar hydrogen is necessarily affected by this isotopic exchange.

Second, the order of magnitude of the activation energy shows that if no free deuterium "lunar water" is present, the hydrogen and the water should be in isotopic equilibrium on the surface of the moon, where temperatures of 150°C are reached, after a period of only a few years. However, hydrogen and water evolved in experiments reported in Table 2 are further from equilibrium than we would expect from the previous measurements. It is to be noticed that for sample 10019, containing more water, the D/H content of both water and hydrogen is higher than for sample 14259. With more samples available differing in the ratio $H_2/H_2O$ it would be possible to evaluate deuterium content for both species under the assumptions that they undergo some exchange and that their initial values are the same in every case. From such considerations, we estimate that the measured value of 13 ppm is an upper limit for the deuterium content of hydrogen nonaffected by exchange and zero as lower limit is not excluded. Within limits of incertitude, these values are in agreement with the 5 ppm figure quoted by Epstein and Taylor (1972).

We have performed experiments to investigate whether the D/H content of water in a sample may have been altered during collection by exchange with moisture of nonlunar origin. For this purpose, sample 140003c prior to extraction was exposed for 24 hours in a closed vessel to air saturated with 700 ppm deuterium water vapor. Results are given in Tables 1 and 2. They show an increase of 5–6 ppm in the isotopic content of the subsequently extracted hydrogen, roughly constant in the whole range of temperature. This small variation can be explained either by a contribution of the water adsorbed on the sample envelope or by the contamination of the line. Because of the limited enrichment of hydrogen, most of the additional water must have been pumped out below 200°C and does not exchange further.

Another indication of the fact that water extracted from the lunar samples does not originate essentially from contamination undergone after the material has been returned to earth is derived from a comparison between results of sample 10019 and 14259. The first sample was exposed to air for more than a year and the second was carefully kept under dry nitrogen since its return to earth. The D content of both extracted hydrogen gases and waters are comparable. This would not be the case if exchange between water of the sample and surrounding vapor (approximately 135 ppm of D in midlatitude regions) had, even slowly, taken place.

Thus, if this water were of terrestrial origin, it must have been fixed by the lunar material at its very first exposure to moisture, because it can neither be easily removed nor does it exchange at room temperature by exposure to water vapor. We tend to believe that this strongly bound water is mainly contamination of primarily terrestrial origin because of its high deuterium content compared to that which would be in equilibrium with the hydrogen. However, it has not been conclusively proved that an equilibrium state would always be reached under lunar conditions.

In conclusion, samples returned from the moon hold equivalent amounts of water and hydrogen around 20 $\mu$mole/g, extractable in the same range of temperature. Because of the isotopic exchange that takes place between hydrogen and this strongly bound water and because it is very difficult to displace, the original deuterium content of the solar wind hydrogen cannot be directly measured. The lowest value we have measured is 13.5 ppm. Estimates within the range 0–13 ppm for deuterium content of solar wind hydrogen are compatible with our measurements. The isotopic ratio of the water is lower than most terrestrial samples, but too high to be in isotopic equilibrium with hydrogen at a temperature of about 100°C on the moon's surface. A mixing of terrestrial origin moisture and oxidized solar wind would account for the isotopic content of the water, but further experiments are desirable.

*Acknowledgments*—The authors wish to thank Mrs. M. Lelu for her valuable help in preparing the samples and her assistance in the analyses. We acknowledge P. Lohez for making improvements in the performance of the mass spectrometer. We are indebted to J. Robert, G. Defaye, H. Nguyen Ghy for their participation in preliminary experimental part of the work. We thank Prof. O. Schaffer for the loan of the sample.

## References

Epstein S. and Taylor H. P. Jr. (1970) The concentration and isotopic composition of hydrogen, carbon, and silicon in Apollo 11 lunar rocks and minerals. *Proc. Apollo 11 Lunar Sci. Conf., Geochim. Cosmochim. Acta* Suppl. 1, Vol. 2, pp. 1085–1096. Pergamon.

Epstein S. and Taylor H. P. (1972) Private communication, paper in press.

Friedman I., Gleason J. D., and Hardcastle K. (1970) Water, hydrogen, deuterium, carbon, and $C^{13}$ content of selected lunar material. *Proc. Apollo 11 Lunar Sci. Conf., Geochim. Cosmochim. Acta* Suppl. 1, Vol. 2, pp. 1103–1109. Pergamon.

Hintenberger H., Weber H. W., Voshage H., Wänke H., Begemann F., and Wlotzka F. (1970) Concentrations and isotopic abundances of the rare gases hydrogen and nitrogen in lunar matter. *Proc. Apollo 11 Lunar Sci. Conf., Geochim. Cosmochim. Acta* Suppl. 1, Vol. 2, pp. 1269–1282. Pergamon.

Proceedings of the Third Lunar Science Conference
(Supplement 3, *Geochimica et Cosmochimica Acta*)
Vol. 2, pp. 1479–1485
The M.I.T. Press, 1972

# Sulphur concentrations and isotope ratios in lunar samples

C. E. Rees and H. G. Thode

Department of Chemistry, McMaster University,
Hamilton, Ontario, Canada

**Abstract**—Sulphur concentrations for the Apollo 12 and 14 samples analyzed lie in the range 549 ppm to 1149 ppm. Fines, and to a lesser extent breccias, are enriched in heavy sulphur isotopes relative to the rocks. The abundance patterns for $S^{33}$, $S^{34}$, and $S^{36}$ relative to $S^{32}$ for Apollo 11, 12, and 14 samples are consistent with derivation from a primitive sulphur source via mass dependent isotope fractionation. There is inhomogeneity of both sulphur concentration and sulphur isotope abundances within individual samples and there are indications that in an Apollo 14 fines sample the individual troilite grains may be isotopically inhomogeneous.

## Introduction

Determinations have been made of sulphur concentrations and $\delta S^{34}$*, $\delta S^{33}$, and $\delta S^{36}$ values for Apollo 12 and 14 samples (12002, 12018, 12021, 12022, 12053, 12070, 14321, 14163) and of $\delta S^{33}$ and $\delta S^{36}$ values for Apollo 11 samples (10086, 10002, 10049, 10057). In addition, an attempt has been made to extract isotopically distinct sulphur components from an Apollo 14 fines sample.

## Analytical Procedures

Sulphur is released, as hydrogen sulphide, from fines and crushed rock samples by treatment with hydrochloric acid. Successive conversions are made from hydrogen sulphide to cadmium sulphide to silver sulphide to sulphur hexafluoride, the latter compound being the isotopic analysis gas. Sulphur concentrations are determined gravimetrically at the silver sulphide step. Details of these procedures and of the mass spectrometry have been given by Thode and Rees (1971). Reported sulphur concentrations have an error of $\pm 3\%$, while the standard deviations for individual determinations of $\delta S^{34}$, $\delta S^{33}$, and $\delta S^{36}$ are estimated as $\pm 0.07‰$, $\pm 0.07‰$, and $\pm 0.7‰$, respectively.

## Results and Discussion

### *Sulphur concentrations and $\delta S^{34}$ values*

The data are shown in Table 1. For convenience of comparison the present data have been combined with those of Thode and Rees (1971). The pattern of sulphur

* $\delta S^M$, ‰ $= \{[(S^M/S^{32})_{\text{sample}}/(S^M/S^{32})_{\text{standard}}] - 1\} \times 1000$; where M = 33, 34, 36. Standard is sulphur from the troilite of the Canyon Diablo meteorite.

Table 1. Sulphur concentrations and isotope ratio data for Apollo 11, 12, and 14 samples.

| Sample | Sulphur content (ppm) | $\delta S^{34}$ (‰) | $\Delta S^{33}$ (‰) | $\Delta S^{36}$ (‰) |
|---|---|---|---|---|
| 12002,136* | 632 | 0.46 | −0.06 | −0.4 |
| 12002,136† | — | 0.47 | — | — |
| 12018,33* | 549 | 0.68 | +0.05 | −0.7 |
| 12021,60* | 892 | 0.38 | +0.04 | −1.5 |
| 12021,60 | 768 | 0.56 | 0.00 | −0.4 |
| 12021,60 | 683 | 0.65 | −0.05 | −0.4 |
| 12022,50* | 914 | 0.37 | +0.01 | −0.1 |
| 12053,71* | 700 | 0.68 | 0.11 | 0.0 |
| 12053,71 | 745 | 0.47 | −0.09 | +0.7 |
| 12070,57* | 698 | 8.70 | −0.17 | −1.6 |
| 14321,220 | 750 | 1.44 | +0.09 | +0.1 |
| 14321,220 | 1149 | 0.90 | −0.01 | +0.7 |
| 14163,132 | 682 | 9.47 | −0.12 | 0.0 |
| 14163,132 | 788 | 9.76 | −0.08 | −0.7 |
| 14163,132‡ | 701 | 9.68 | — | — |
| 10086 | — | — | −0.13 | −1.1 |
| 10086 | — | — | −0.18 | −1.2 |
| 10002 | — | — | −0.13 | −0.9 |
| 10049 | — | — | −0.05 | +2.3 |
| 10057 | — | — | −0.10 | +1.2 |
| Mean value (‰) | | | −0.06 | −0.2 |
| Standard deviation of mean (‰) | | | ±0.02 | ±0.2 |

$\delta S^{34}$ values are relative to Canyon Diablo troilite.
$\Delta S^{33} = \delta S^{33} - 0.515\ \delta S^{34}$.
$\Delta S^{36} = \delta S^{36} - 1.90\ \delta S^{34}$.
* Previously reported in Thode and Rees (1971).
† Repeat determination on $Ag_2S$.
‡ Mean value from fines fractionation experiment, see text.
The Apollo 11 determinations were made on $Ag_2S$ separates provided by I. R. Kaplan.

concentrations and $\delta S^{34}$ values is in general accord with other determinations (Kaplan and Smith, 1970; Ponnamperuma *et al.*, 1970; Kaplan *et al.*, 1970; Kvenvolden *et al.*, 1970; Kaplan and Petrowski, 1971; Thode and Rees, 1971; Sakai *et al.*, 1972). The $\delta S^{34}$ values reported here for Apollo 12 rocks are tightly grouped in the range +0.37‰ to +0.68‰, while the fines sample 12070 has a $\delta S^{34}$ value of +8.7‰. The Apollo 14 fines sample 14163 is still more highly enriched in $S^{34}$, having a range of $\delta S^{34}$ values of from +9.47‰ to +9.76‰. This latter value is the highest so far encountered for a bulk sample. Two separate samples of the breccia 14321 have $\delta S^{34}$ values of +0.9‰ and +1.44‰.

Sulphur concentration values for the rocks and the breccia are lower by about a factor of three than those reported for Apollo 11 samples and lie in the same range as those for fines from the Apollo 11, 12, and 14 sites. There is inhomogeneity of both sulphur concentrations and $\delta S^{34}$ values within individual samples. For determinations on different chips of the rocks 12021 and 12053 and on the breccia 14321 there appears to be a negative correlation between sulphur concentration and $\delta S^{34}$ value. For the fines sample 14163 this trend is reversed and high $\delta S^{34}$ values are associated with high sulphur concentrations.

*$\delta S^{33}$ and $\delta S^{36}$ values*

Measurements of $\delta S^{33}$ and $\delta S^{36}$ were made in order to investigate possible mass independent sulphur isotope abundance variations. Hulston and Thode (1965) have shown that for mass dependent isotope fractionation processes, such as evaporation, diffusion, chemical equilibration, and chemical reaction, the relations

$$\delta S^{33} = 0.515\ \delta S^{34}$$

and

$$\delta S^{36} = 1.9\ \delta S^{34}$$

will hold to within 1% for $\delta S^{34}$ values of $\lesssim 15‰$. The quantities $\Delta S^{33}$ and $\Delta S^{36}$, shown in columns 4 and 5 of Table 1 are defined as

$$\Delta S^{33} = \delta S^{33} - 0.515\ \delta S^{34}$$

and

$$\Delta S^{36} = \delta S^{36} - 1.9\ \delta S^{34},$$

respectively. They represent residues after allowance has been made for mass dependent isotope fractionation processes. Thode and Rees (1971) indicated that deviations from zero of $\Delta S^{33}$ and $\Delta S^{36}$ for lunar sulphur, due to cosmic ray spallation processes, will be below present detection limits.

The data in Table 1 show that all the $\Delta S^{33}$ and $\Delta S^{36}$ values are indeed small. The slight deviation from zero of the mean value of $\Delta S^{33}$ is almost certainly not significant and probably represents a systematic error in the conversion of mass spectrometric ion current data to $\delta S^{33}$ values. For the lunar material examined thus far sulphur isotope abundance variations can be attributed to mass dependent isotope fractionation processes.

*"Fractionation" experiment on the fines sample 14163*

An attempt has been made to separate isotopically distinct sulphur components from a 1 g bulk fines (<1 mm) sample (14163). The sample was treated with hydrochloric acid in the normal way except that the released hydrogen sulphide was collected in three successive fractions. The first two fractions were collected with the acid at room temperature while the third fraction was generated by boiling the acid. A similar experiment was performed on a crushed sample of troilite from the Canyon Diablo meteorite in order to monitor possible isotope fractionation effects in the acid dissolution of troilite. The results of the experiments are summarized in Table 2 and Figs. 1 and 2.

Contrary to expectations an isotope effect was observed in the generation of hydrogen sulphide from the meteoritic troilite. The $\delta S^{34}$ values of the three sulphur fractions are consistent with a first-order chemical reaction in which the ratio of rate constants for the reaction of $S^{32}$ and $S^{34}$ is $1.0008 \pm 0.0001$. It is not understood why this is so. It was anticipated that the reaction of hydrochloric acid with the troilite grains would be a surface effect and that as successive layers of troilite reacted they would expose fresh layers with the same isotopic composition as that of the bulk material. Whatever

Table 2. Fractionation experiments on lunar and meteoritic troilite.

| | 14163,132 | | Canyon Diablo | |
|---|---|---|---|---|
| | Total sulphur (%) | $\delta S^{34}$ (‰)* | Total sulphur (%) | $\delta S^{34}$ (‰)* |
| Fraction 1 (room temperature) | 30.1 | −2.41 | 16.4 | −0.81 |
| Fraction 2 (room temperature) | 36.2 | +0.10 | 53.0 | −0.23 |
| Fraction 3 (~95°C) | 33.7 | +2.04 | 30.6 | +0.84 |

* $\delta S^{34}$ values are relative to the bulk material.

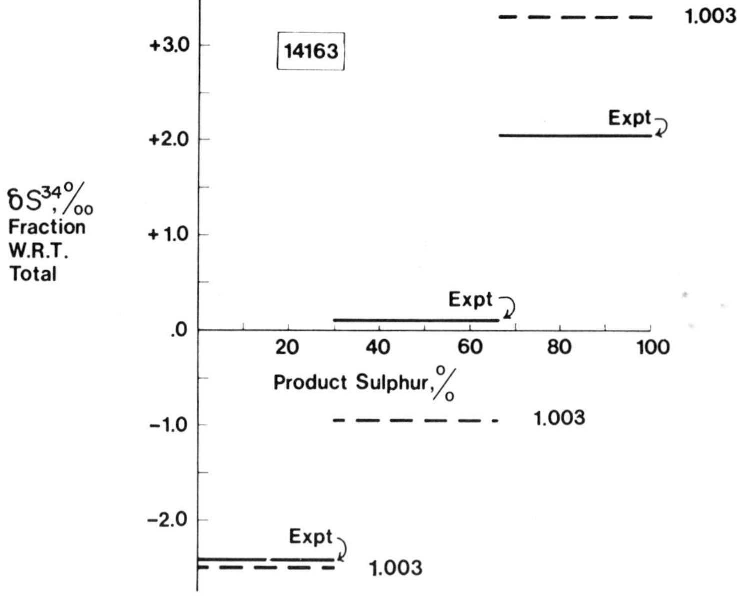


Fig. 1. The isotopic composition of successive fractions of $H_2S$ generated from troilite in fines sample 14163. The solid lines represent the experimental determinations, and the dashed lines are the isotopic compositions for an isotopically competitive reaction with ratio of rate constants 1.003.

the causes of the observed isotope effect, it is clear that they must be taken into account when considering the results of the parallel experiment on the lunar sample.

The $\delta S^{34}$ pattern obtained for successive fractions of hydrogen sulphide generated from the lunar troilite is somewhat different from that obtained in the Canyon Diablo experiment. The $\delta S^{34}$ difference between the first and third fractions is about a factor of 3 greater and the behavior of $\delta S^{34}$ with fraction of reaction is no longer consistent with a first-order reaction. The trend of the experimental results is in good agreement with that of a very similar experiment reported by Sakai *et al.* (1972). It appears likely that the results can be explained by a combination of two mechanisms. The first of these is the isotope fractionation process observed in the Canyon Diablo experiment,

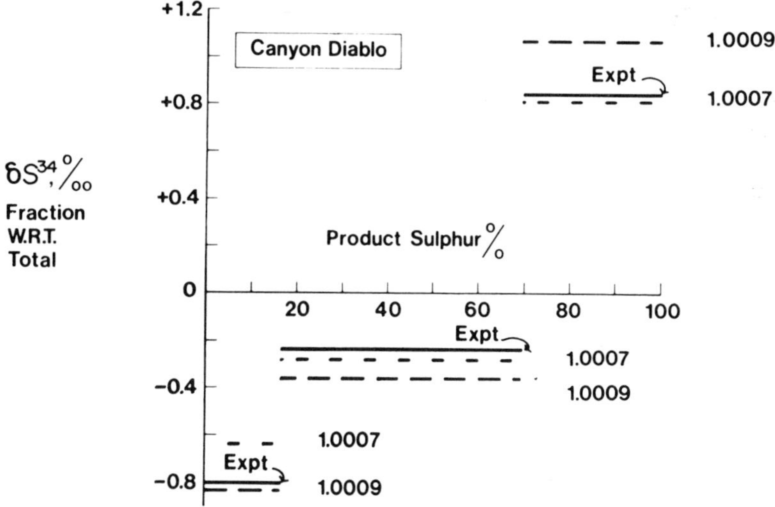


Fig. 2. The isotopic composition of successive fractions of $H_2S$ generated from Canyon Diablo troilite. The solid lines represent the experimental determinations, and the dashed lines are the isotopic compositions for isotopically competitive reactions with ratios of rate constants 1.0007 and 1.0009.

while the second is the sampling of isotopically distinct troilite in the fines material. This latter mechanism is further discussed below.

### *Origin of the $S^{34}$ enrichment of lunar fines*

Shortly after the Apollo 11 sulphur isotope results became available the solar wind was discussed as a possible agent for the $S^{34}$ enrichment of lunar fines material (Kaplan and Smith, 1970; Kaplan *et al.*, 1970; Berger, 1970). It was proposed that solar wind protons could react with troilitic sulphur to form hydrogen sulphide and that some or all of the hydrogen sulphide could then be lost to space. Isotope fractionation in the formation or loss of hydrogen sulphide could then enrich the residual sulphur in $S^{34}$. Such a mechanism seemed particularly attractive because of the factor of three depletion of sulphur in fines relative to rocks at the Apollo 11 site.

Sulphur concentration data for rocks, breccias, and fines from subsequent Apollo sites indicate that the Apollo 11 rocks have unusually high sulphur contents. While there is considerable variation of sulphur concentration for other returned samples there appears to be no clear cut distinction between rocks, breccias, and fines. It should be carefully noted however that this does not necessarily invalidate the solar wind hypothesis. If it is supposed that the sulphur in lunar fines is derived from a source sulphur that has $\delta S^{33}$, $\delta S^{34}$, and $\delta S^{36}$ all lying close to zero then the data presented above for $\Delta^{33}$ and $\Delta^{36}$ indicate either (a) the removal of sulphur by a "well behaved" mass dependent isotope fractionation mechanism, or (b) the addition of sulphur which was itself derived via a mass dependent process from a source having $\delta S^{33}$, $\delta S^{34}$, and $\delta S^{36}$ close to zero.

It is instructive to contrast the sulphur isotope patterns for fines with the isotope patterns for oxygen and silicon found by Epstein and Taylor (1971, 1972). Bulk fines are slightly enriched in the heavy isotopes of oxygen and silicon and grossly enriched in the heavy isotopes of sulphur. For oxygen and silicon the slight bulk enrichment is a consequence of a gross enrichment at grain surfaces, and the degree of surface enrichment is correlated with trapped solar wind gas concentrations. Taking the simplest possible interpretation of our sulphur isotope fractionation experiment described above and of the similar experiment described by Sakai *et al.* (1972), the gross heavy sulphur isotope enrichment persists throughout the bulk of the fines and in fact seems somewhat less for the surface than for the interior of troilite.

Although the isotope pattern for sulphur is quite different from those for oxygen and silicon, it is possible that the same process has been in operation in all three cases. Numerous workers (see for example, Carter (1971, 1972) and references therein) have noted that associations of metallic iron and troilite in fines commonly occur with troilite partially or completely surrounding the iron. This morphology strongly indicates that troilite in the fines has been at some time melted or even vaporized. This suggests the following mechanism for the establishment of the fines $\delta S^{34}$ pattern: During the vaporization of troilite there is significant loss of sulphur, as elemental sulphur, with an appreciable isotope effect. This leads to the overall enrichment in $S^{34}$ of the bulk fines material. Following the loss of sulphur, and during rapid cooling, troilite grows around metallic iron centers and does so with $S^{34}$ favored over $S^{32}$. This leads to isotopically inhomogeneous troilite with the interior enriched in $S^{34}$ relative to the exterior. This mechanism is clearly conjectural. More experiments will be carried out on lunar soil samples to test this and other possible mechanisms.

*Acknowledgments*—We would like to thank the National Research Council of Canada and the Geological Survey of Canada for financial support of this research. We have benefited greatly from discussions with Jan Monster and H. P. Schwarcz.

Silver sulphide separates of Apollo 11 samples were kindly supplied to us by I. R. Kaplan of U.C.L.A.

## REFERENCES

Berger R. (1970) Reaction of carbon and sulphur isotopes in Apollo 11 samples with solar hydrogen atoms. *Nature* **226,** 738–739.

Carter J. L. (1971) Chemistry and surface morphology of fragments from Apollo 12 soil. *Proc. Second Lunar Sci. Conf., Geochim. Cosmochim. Acta* Suppl. 2, Vol. 1, pp. 873–892. MIT Press.

Carter J. L. (1972) Morphology and chemical composition of metallic mounds produced by $H_2$ and C reduction of material of simulated lunar composition (abstract). In *Lunar Science—III* (editor C. Watkins), pp. 125–127, Lunar Science Institute Contr. No. 88.

Epstein S. and Taylor H. P. Jr. (1971) $O^{18}/O^{16}$, $Si^{30}/Si^{28}$, D/H, and $C^{13}/C^{12}$ ratios in lunar samples. *Proc. Second Lunar Sci. Conf., Geochim. Cosmochim. Acta* Suppl. 2, Vol. 2, pp. 1421–1441. MIT Press.

Epstein S. and Taylor H. P. Jr. (1972) $O^{18}/O^{16}$, $Si^{30}/Si^{28}$, $C^{13}/C^{12}$, and D/H studies of Apollo 14 and 15 samples (abstract). In *Lunar Science—III* (editor C. Watkins), pp. 236–238, Lunar Science Institute Contr. No. 88.

Hulston J. R. and Thode H. G. (1965) Variations in the $S^{33}$, $S^{34}$, and $S^{36}$ contents of meteorites and their relation to chemical and nuclear effects. *J. Geophys. Res.* **70,** 3475–3484.

Kaplan I. R. and Smith J. W. (1970) Concentration and isotopic composition of carbon and sulfur in Apollo 11 lunar samples. *Science* **167,** 541–543.

Kaplan I. R., Smith J. W., and Ruth E. (1970) Carbon and sulfur concentration and isotopic composition of Apollo 11 lunar samples. *Proc. Apollo 11 Lunar Sci. Conf., Geochim. Cosmochim. Acta* Suppl. 1, Vol. 2, pp. 1317–1330. Pergamon.

Kaplan I. R. and Petrowski C. (1971) Carbon and sulfur isotope studies on Apollo 12 lunar samples. *Proc. Second Lunar Sci. Conf., Geochim. Cosmochim. Acta* Suppl. 2, Vol. 2, pp. 1397–1406. MIT Press.

Kvenvolden K. A., Chang S., Smith J. W., Flores J., Pering K., Saxinger C., Woeller F., Keil K., Breger I., and Ponnamperuma C. (1970) Carbon compounds in lunar fines from Mare Tranquillitatis —*I*. Search for molecules of biological significance. *Proc. Apollo 11 Lunar Sci. Conf., Geochim. Cosmochim. Acta* Suppl. 1, Vol. 2, pp. 1813–1828. Pergamon.

Ponnamperuma C., Kvenvolden K., Chang S., Johnson R., Pollock G., Philpott C., Kaplan I., Schopf J. W., Gehrke C., Hodgson G., Breger I. A., Halpern B., Duffield A., Krauskopf K., Barghoorn E., Holland H., and Keil K. (1970) Search for organic compounds in the lunar dust from the Sea of Tranquility. *Science* **167,** 760–762.

Sakai H., Petrowski C., Goldhaber M. B., and Kaplan I. R. (1972) Distribution of carbon, sulfur, and nitrogen in Apollo 14 and 15 material (abstract). In *Lunar Science—III* (editor C. Watkins), pp. 673–675, Lunar Science Institute Contr. No. 88.

Thode H. G. and Rees C. E. (1971) Measurement of sulphur concentrations and the isotope ratios $^{33}S/^{32}S$, $^{34}S/^{32}S$, and $^{36}S/^{32}S$ in Apollo 12 samples. *Earth Planet. Sci. Lett.* **12,** 434–438.

Proceedings of the Third Lunar Science Conference
(Supplement 3, *Geochimica et Cosmochimica Acta*)
Vol. 2, pp. 1487–1501
The M.I.T. Press, 1972

# Apollo 14 mineral ages and the thermal history of the Fra Mauro formation

W. COMPSTON, M. J. VERNON, H. BERRY, R. RUDOWSKI, C. M. GRAY, and N. WARE
Department of Geophysics and Geochemistry,
Australian National University

and

B. W. CHAPPELL and M. KAYE
Department of Geology,
Australian National University

**Abstract**—Rb–Sr mineral isochrons for two basalt clasts from the partially annealed breccia 14321 (4.06 b.y.) are detectably older than 14310 (3.93 b.y.). Less precise ages for a troctolite clast from 14321 and the igneous chip 14072 can agree with either of these. A directionally consistent thermoremanent magnetism in 14321 reported by Pearce *et al.* shows that the temperature of this breccia was at least 800°C after final assembly. If the annealing of the breccia occurred within the Fra Mauro formation after deposition, temperatures may have remained above 600°C for as long as $\sim 10^2$ yr owing to thermal blanketing. During such a thermal event, the estimated diffusion ranges of Rb and Sr in glass (mesostasis), plagioclase, and possibly pyroxene would exceed the basalt grainsize and probably produce local equilibration of Sr isotopes between minerals. The age of the 14321 clasts would therefore represent the age of the Imbrium impact on this model, and 14310 would be a post-Fra Mauro rock rather than a Fra Mauro breccia clast. On the other hand, if the Fra Mauro formation was deposited at low temperatures, the ages of the clasts and their magnetization would represent pre-Imbrian events, and only a limiting age of younger than 3.9 b.y. can be set for the Imbrium impact.

## INTRODUCTION

IN THIS PAPER we document and give multiple interpretations of the isotopic ages of Apollo 14 igneous rocks cited previously at the Third Lunar Science Conference (Compston *et al.*, 1972). Our major-element analyses for the latter were given at the same conference, and major- and trace-element analyses on these and of the remainder of our allocation, various Apollo 14 fines, will be reported and discussed later.

### *Analytical methods*

The complete fragment received for each of the crystalline rocks, 14310 and 14072, was crushed and representative aliquots of 0.9 g and 0.3 g were taken, for trace-element and major-element analysis by x-ray fluorescence, respectively, using standard methods (Compston *et al.*, 1970). Part of the remainder was used for mineral separations with the aid of heavy liquids, and hand picking gave final concentrates of plagioclase free from ilmenite (low Rb/Sr) and ilmenite and pyroxene free from plagioclase (high Rb/Sr owing to included alkali-rich mesostasis). Representative Rb–Sr total rock samples were also prepared. All handling and mineral separations were done in a clean-air work station. Three igneous clasts were extracted from faces of the breccia 14321,88 and mineral concentrates made, as described for one of them (basalt 4A) by Compston *et al.* (1971a). Representative total-rock aliquots of two of these, basalt 6A and troctolite 8A, and minerals from 8A were analyzed for trace-elements (Taylor *et al.*, 1972a), and the troctolite for major-elements (Compston *et al.*, 1972). Electron-microprobe analysis of minerals from all three clasts will be reported by Ware and Green (in preparation).

Some details of the mineral concentrates and the Rb, Sr, and $^{87}Sr/^{86}Sr$ analytical results are given for all samples in Table 1. The chemical procedures and mass-spectrometer used were the same as those described by Compston *et al.* (1971a).

## CALIBRATION OF MASS SPECTROMETERS

Our previous measurements of $^{87}Sr/^{86}Sr$ in lunar samples (Compston *et al.*, 1971b), using a Nuclide 12-60-SU machine with Faraday-cup collector and Cary

Table 1. Isotope dilution analyses of Apollo 14 igneous rocks and minerals. Concentrates prepared solely by hand-picking are denoted (*H*) and purity is given in percent.

| | Weight (mg) | Rb (ppm) | Sr (ppm) | $^{87}Rb/^{86}Sr$* | $^{87}Sr/^{86}Sr$* |
|---|---|---|---|---|---|
| *14310,118, basalt* | | | | | |
| Plagioclase 1, (*H*), ~100% | 10.6 | 2.47 | 287.2 | 0.0248 ± 2 | 0.70190 ± 4 |
| Plagioclase 2, 98%, ~2% mesostasis | 19.4 | 7.00 | 309.1 | 0.1312 ± 6 | 0.70782 ± 6 |
| Plagioclase-rich mesostasis A | 13.4 | 22.35 | 304.3 | 0.2122 ± 11 | 0.71258 ± 5 |
| Plagioclase-rich mesostasis B | 14.9 | 16.75 | 232.2 | 0.2084 ± 10 | 0.71231 ± 5 |
| Total-rock A | 22.3 | 12.96 | 192.3 | 0.1948 ± 10 | 0.71144 ± 3 |
| Total-rock B | 12.2 | 12.83 | 190.6 | 0.1946 ± 10 | 0.71157 ± 3 |
| Pyroxene 3, (*H*), ~100% | 9.8 | 1.38 | 46.13 | 0.0866 ± 3 | 0.70540 ± 9 |
| Pyroxene 2, with ilmenite | 9.5 | 5.46 | 50.8 | 0.3105 ± 15 | 0.71848 ± 4 |
| Pyroxene 1, 98%, 2% mesostasis | 41.3 | 4.54 | 31.1 | 0.4212 ± 21 | 0.72484 ± 7 |
| *14072,2, basalt* | | | | | |
| Plagioclase 1, (*H*), ~100% | 13.0 | 0.66 | 210.7 | 0.0090 ± 1 | 0.69980 ± 8 |
| Plagioclase 2, 98%, ~2% mesostasis | 15.8 | 1.69 | 227.0 | 0.0215 ± 2 | 0.70049 ± 3 |
| Pyroxene, pigeonite/augite, 99% | 51.6 | 0.95 | 10.2 | 0.0270 ± 1 | 0.70069 ± 5 |
| Total-rock | 20.5 | 1.21 | 81.6 | 0.0428 ± 4 | 0.70183 ± 3 |
| Ilmenite 1, (*H*), ~100% | 8.1 | 1.80 | 19.5 | 0.265 ± 3 | 0.71480 ± 2 |
| Ilmenite 2, ~90%, 10% pyroxene | 4.8 | 1.62 | 20.8 | 0.223 ± 3 | 0.71205 ± 5 |
| *14321,88, basalt 6A* | | | | | |
| Plagioclase 1, (*H*), ~100% | 1.8 | 2.08 | 279.6 | 0.0215 ± 7 | 0.70062 ± 6 |
| 2.6 < ρ < 2.9, plagioclase + plagioclase mesostasis | 3.1 | 5.40 | 197.0 | 0.0791 ± 9 | 0.70392 ± 5 |
| ρ < 2.6, plagioclase + trace K-feldspar | 1.4 | 4.83 | 162.5 | 0.0857 ± 18 | 0.70427 ± 6 |
| Plagioclase-rich mesostasis | 9.2 | 5.79 | 129.2 | 0.1294 ± 1 | 0.70689 ± 5 |
| ρ > 3.3, pyroxene, ~95% | 15.4 | 2.19 | 29.35 | 0.2154 ± 17 | 0.71200 ± 5 |
| Ilmenite, 35%, + pyroxene + olivine | 1.7 | 7.16 | 71.4 | 0.290 ± 4 | 0.7170 ± 1 |
| *14321,88, troctolite 8A* | | | | | |
| Plagioclase, ~94% + olivine | 31.4 | 1.39 | 220.4 | 0.0183 ± 1 | 0.70047 ± 7 |
| Total-rock 2 | 29.6 | 1.22 | 162.8 | 0.0217 ± 2 | 0.7006 ± 1 |
| Total-rock 1 | 22.4 | 0.85 | 161.3 | 0.0152 ± 2 | 0.70042 ± 4 |
| K-feldspar + plagioclase A | 11.4 | 3.34 | 205.5 | 0.0466 ± 2 | 0.70209 ± 2 |
| K-feldspar + plagioclase B | 12.2 | 3.06 | 191.2 | 0.0463 ± 2 | 0.70222 ± 6 |
| Olivine, ~98% | 46.7 | 0.16 | 3.93 | 0.114 ± 2 | 0.7056 ± 1 |

* Uncertainties refer to the last digits and represent the standard error for precision. In addition, all values for $^{87}Rb/^{86}Sr$ should be increased by 1.8% according to recalibration of spikes by De Laeter *et al.* (1972).

model 31 electrometer (serial number 1588), have been systematically ~0.0002 higher (0.03%) than Papanastassiou and Wasserburg (1971a) and ~0.0001 higher than Murthy *et al.* (1971). The only reasonable source of such biases appears to be nonlinearity in the measurement system: for example, if our $^{88}Sr/^{86}Sr$ were erroneously low by 0.06% owing to the system characteristic, then the operation of normalizing $^{88}Sr/^{86}Sr$ to a standard value will introduce a fixed error of 0.03% in $^{87}Sr/^{86}Sr$ because the $^{88}Sr/^{86}Sr$ error will be treated as isotopic fractionation. We have previously monitored nonlinearity by injecting highly accurate voltage ratios (from 0.02 to 1) into the feedback loop of the electrometer. The corresponding output ratios as registered via a digital system have consistently agreed to better than 0.01%. However this procedure does not completely simulate the measurement of input ion-currents: It is necessary instead to feed voltage ramps in accurately known ratios to the electrometer input via a stable capacitor. A satisfactory ramp-generator has now been constructed and is being applied to the calibration of a number of Cary electrometers. Figures 1 and 2 show the characteristics obtained for two Cary 31 electrometers, over the range in ion currents used in measuring $^{88}Sr/^{86}Sr$ and $^{87}Sr/^{86}Sr$. For electrometer #1588, a nonlinearity of 0.06% for a ratio of ~10:1 is strikingly revealed, in contrast to #1565 that has no detectable nonlinearity, at least for the 3 V output range. The

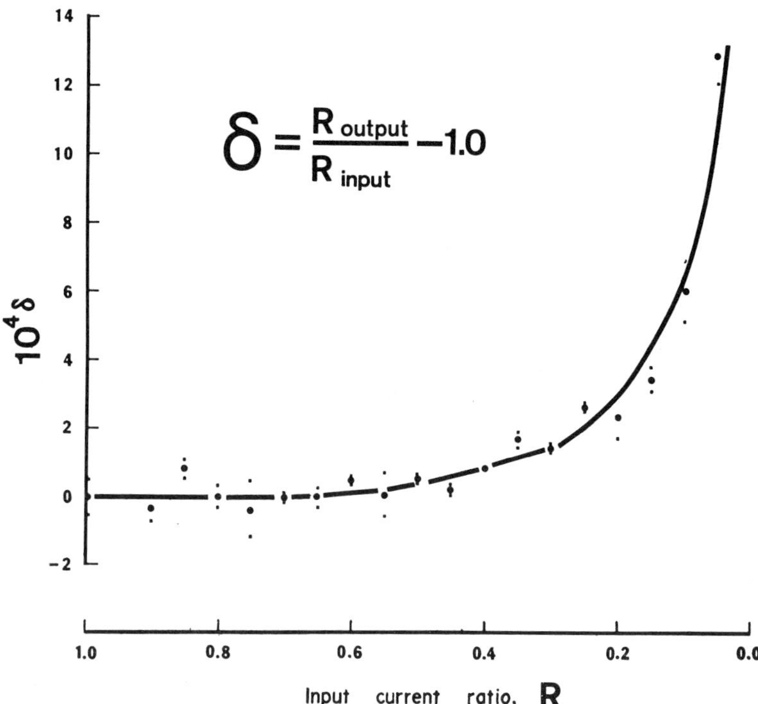


Fig. 1. Nonlinearity detected in Cary model 31 electrometer #1588 using a ramp generator. The input resistor was 2 × $10^{10}$ Ω (Victoreen Hi-meg) and electrometer output voltage 1.6 V on the 3V range for the reference input current. Input ratios are the fractions shown of the reference input current, of accuracy about 10 ppm.

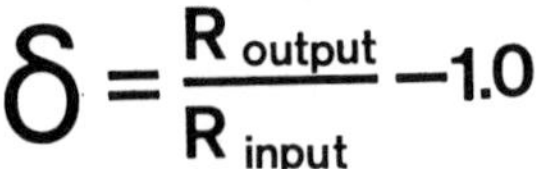

$$\delta = \frac{R_{output}}{R_{input}} - 1.0$$

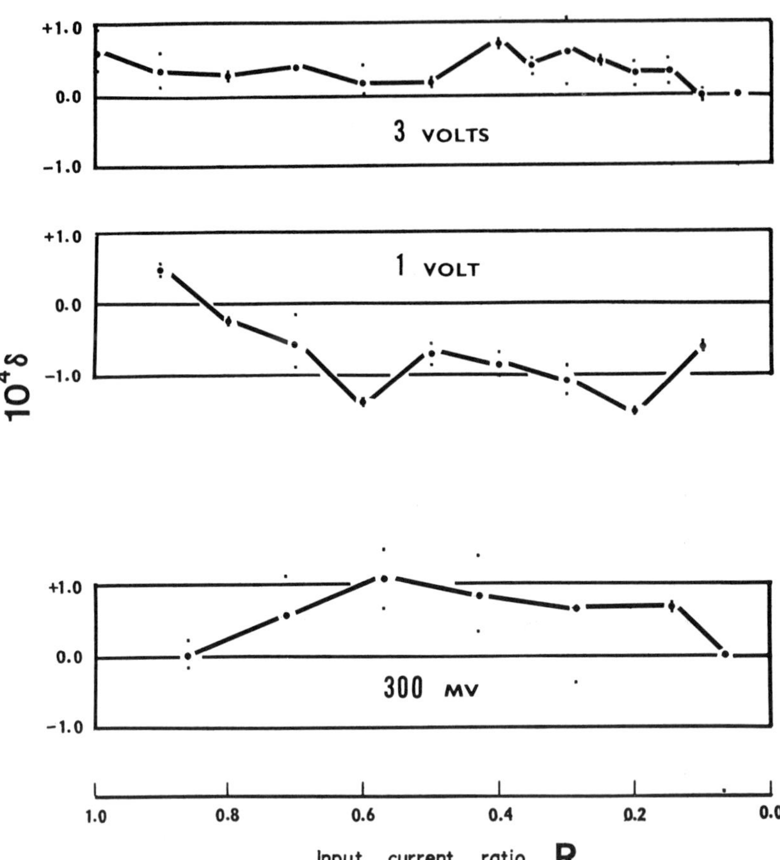


Fig. 2. Limits to non-linearity of Cary model 31 electrometer #1565 employed for Apollo 14 analyses, for three different output ranges. The input resistor was Victoreen $2 \times 10^{10}$ Ω.

interlaboratory bias in $^{87}Sr/^{86}Sr$ is thus located as originating within our electrometer (which nevertheless is working to within manufacturers' specifications) rather than within Papanastassiou and Wasserburg's. The cause of the nonlinearity will be examined later: It is not simple polarization of the input resistor, as it is present no matter which particular resistor is used.

We do not know yet whether the apparently systematic variations in Fig. 2 within the 1 V and 300 mV ranges of electrometer #1565 are real. If they are, then the $^{87}Sr/^{86}Sr$ measurements reported here and by us previously for 14321,88,4A (Compston *et al.*, 1971a), will be high by approximately 0.01%.

A small number of measurements for the C.I.T. seawater standard made so far on #1565 give 0.70918 ± 6 (without correction for the above possible bias of about 0.01%). More adequate data for the same standard on #1588 give 0.70931 ± 4, which becomes 0.70910 after correction for the known nonlinearity of this electrometer on the 3 V range. As we have not yet checked the 300 mV and 1 V ranges of #1588 (on which $^{87}Sr/^{86}Sr$ was measured) with the ramp generator, a further correction for nonlinearity may be needed. However, it is clear that our results for $^{87}Sr/^{86}Sr$ in the Apollo 11 and 12 samples, which were obtained solely with electrometer #1588, should be reduced by 0.00021 for direct comparison with those of Papanastassiou and Wasserburg (1971a).

## Thermal History of the Fra Mauro Formation

The crystallinity of the Apollo 14 fragmental rocks that constitute the Fra Mauro formation has been described as "compatible with a single very large impact event in which annealing took place within a thick, hot ejecta blanket" (LSPET, 1971). This view is disputed by Chao *et al.* (1972), who believe that the Fra Mauro formation was deposited at low temperatures. Quantitative information on the thermal history of such an ejecta blanket is vital to the interpretation of the mineral ages of the basaltic clasts in breccia 14321, and probably also of 14072 and 14310 which may be fragments of bigger clasts. If the clasts had a large enough temperature-time integral during and after final deposition, their original pre-Fra Mauro mineral ages could have been reset. On the other hand, if the Imbrium debris had cooled sufficiently by the time it eroded and deposited the igneous clasts, the original ages of the latter may be completely unchanged.

Petrographic evidence for recrystallization of lithic fragments and glass in Apollo 14 soils and breccias has been given by Anderson *et al.* (1972), Chao *et al.* (1972), Grieve *et al.* (1972), Kurat *et al.* (1972), Lindsay (1972), Quaide (1972), Taylor *et al.* (1972b), Warner (1972), and Wilshire and Jackson (1972). Warner (1972) describes a range in metamorphic grade within the breccias. From the distribution of ejecta around Cone Crater, Wilshire and Jackson (1972) and Quaide (1972) suggest that this range is associated with the depth of particular samples within the Fra Mauro formation, due at least in part to differences in their thermal blanketing during post-depositional cooling of the (hot) debris flow. Taylor *et al.* (1972b) refer to noritic fragments and to matrices of clasts in the Apollo 14 soils that have been remelted: They believe that this occurred after the deposition of the Fra Mauro formation. Anderson *et al.* (1972) infer that temperatures of some breccia components were as high as 900°C during at least the early part of the Imbrium impact, and suggest the presence of vapor during compaction (of the Fra Mauro formation). The growth of crystals in vuggy recrystallized breccias (McKay *et al.*, 1972) is compelling evidence for high postdepositional temperatures and vapor transport. The origin of the Fra Mauro formation as a single, very hot debris-flow has been detailed by Williams (1972), who integrates petrographic with diverse physical and chemical properties of the breccias.

Grieve *et al.* (1972) in a detailed analysis of textural ordering in breccia 14321, see different degrees of metamorphism in its different components. In their view, norites and other lithic fragments were subjected to an early period of severe thermal metamorphism; rebrecciation followed, accompanied by a second less-severe thermal metamorphism; final assembly of the sample occurred with the incorporation of basaltic and other clasts in an unrecrystallized light matrix. Multiple periods of heating and brecciation are also seen by Chao *et al.* (1972), who in addition use the sample distribution to infer the co-existence of unannealed with highly annealed breccias, and to conclude that the Fra Mauro formation was deposited at low temperatures with little or no subsequent annealing. They see the annealed breccias as pre-Fra Mauro clasts and adduce support for their interpretation from the high porosity of the Fra Mauro formation found by Kovachs *et al.* (1971). Dence *et al.* (1972), Hörz, and Chao (personal communication) believe that a low temperature of deposition for the Fra Mauro formation must be expected by analogy with annealing evidence found in terrestrial ejecta blankets.

Independent and quantitative evidence for elevated postassembly temperatures for breccia 14321 is available from studies of the remanent magnetization of lunar samples. Pearce *et al.* (1972) have discovered that three adjacent fragments of breccia 14321 have the same stable and directionally consistent magnetization component, which they regard as a thermoremanent magnetization carried by metallic iron. Hargraves and Dorety (1972) find the same TRM orientation for a more distant fourth fragment of 14321 (Gose, personal communication). This implies that the temperature of this particular breccia, which contains both annealed and unannealed components, was above the Curie point, ~780°C, *during or after its final assembly*. Either the TRM was acquired before deposition in the Fra Mauro formation and the entire rock 14321 is a single clast, or the final assembly of 14321 occurred, and its TRM was acquired, within the deposited Fra Mauro formation, and the temperature of the latter was at least 780°C.

The thermal inertia of the Fra Mauro formation limits its rate of cooling. If we assume that the formation is 100 m thick at the Apollo 14 site (Offield, 1970), and for simplicity, that its temperature was uniform throughout at the time of emplacement, we may use the calculations of Jaeger (1961) for the cooling of an extrusive sheet to estimate the time required for its central temperature to drop by 10%. This time is approximately 45 yr, using the Horai *et al.* (1970) measurement of thermal diffusivity for lunar breccias. For a 90% temperature drop, the time will be about 200 yr. Even though the assumption of a uniform emplacement temperature may be only a poor approximation, and the Fra Mauro formation may be a factor of 2 less in thickness (Kovach *et al.*, 1971), the order of magnitude of the above cooling times will be correct, and applicable to breccia 14321, believed to be excavated from a depth of ~30 m (dark clast-dominated group, Wilshire and Jackson, 1972).

## DIFFUSION OF Rb AND Sr WITHIN LUNAR BASALT

During contact metamorphism, biotite rapidly loses its Rb–Sr age above 300°C and other minerals, including K-feldspar, lose their ages at temperatures less than 600°C (Hanson and Gast, 1967). If the temperature within the Fra Mauro formation was 780°C or more immediately after its final deposition at the Apollo 14 site, then at least its central portion must have remained *above* 600°C for a period of the order of $10^2$ yr. Whether the minerals of the basaltic clasts lost their pre-Fra Mauro ages during this time depends critically on the rate at which Rb and Sr diffuse within and between the minerals at the raised temperature.

Table 2 lists some measurements for the diffusion coefficients $D$ of alkalis in glass and in albite. The composition of the mesotasis glass (Gancarz *et al.*, 1971) is not matched by any of the experimentally determined glasses. However the range in $D$ observed is quite small for large changes in glass composition, and the values for Na, K, and Rb are similar. The $D$ for Sr in alkali glasses will be smaller but probably comparable; $D$ for Ar also is within the range observed for alkalis.

The time $t$ required for effectively complete diffusion ($\geq 98\%$) into or out of a sphere of radius $L$ can be calculated using the relationship $\sqrt{DT}/L = 0.75$ (Kingery, 1960).

Taking $10^{-6}$ as the mean $D$ for alkalis in glass above 600°C, and 0.01 cm for $L$ as the order of magnitude of the mesostasis grain size, we find that this time is only 56 sec. Obviously the diffusion of alkalis in the mesostasis is extremely rapid, and even if $D$ for Sr in glass is taken as several orders of magnitude lower, Sr diffusion also will be rapid over the time scale available ($10^2$ yr). A flux of Rb and Sr must be expected across the glass boundaries owing to intergranular diffusion, which is usually faster than volume diffusion. Thus the K-rich mesostasis can hardly be viewed as chemically closed during the Fra Mauro thermal event.

To change the mesostasis age, there must be a decrease (or increase) in its ratio radiogenic $Sr^{87}/Sr^{87}$. By analogy with terrestrial metamorphism, a decrease can be expected, owing to Sr isotopic exchange between radiogenic $Sr^{87}$ and common Sr diffusing from adjacent Sr-rich minerals, which may result in Sr isotopic equilibration within small volumes of the rock. Such equilibration will be limited for the lunar basalts by the diffusion coefficients of Sr in plagioclase and pyroxene, which are not known. However an upper limit for plagioclase can be taken from the diffusion of Na in albite and a lower limit for the pyroxene from the diffusion of Ca in CaO and $Ca_3Si_2O_7$ (Table 2).

Calculating as before the time required for complete diffusion, only a short time is required for plagioclase ($\leq 6$ days) if the mean $D^{Sr}$ above 600°C is taken as $10^{-12}$ $cm^2\ sec^{-1}$, equal to the value for Na in albite at 600°C. For 45 yr above 700°C, as required for breccia 14321 from the Curie point datum and the cooling rate, diffusion would be complete for $D^{Sr}$ as small as $4 \times 10^{-16}$ $cm^2\ sec^{-1}$, which is unreasonably

Table 2. Some typical measured values for diffusion coefficients of alkalis in glass.

| Phase* | Species | log D 800°C | 700°C | 600°C | Reference |
|---|---|---|---|---|---|
| Glass, 19% $Na_2O$, 6% $Rb_2O$ | Rb | −6.2 | −6.8 | −8.1 | McVay and Day (1970) |
| Glass, 19% $Na_2O$, 6% $Rb_2O$ | Na | −5.4 | −5.8 | −6.8 | McVay and Day (1970) |
| Glass, 20% $K_2O$ | K | −7.0 | −7.3 | −7.9 | Doremus (1962) |
| Glass, 26% $K_2O$, 5% CaO, 5% $Al_2O_3$ | Ar | −6.6 | −7.1 | −8.4 | Reynolds (1957) |
| Albite | Na | −10.5 | −11.0 | −12.2 | Sippel (1963) |
| CaO, $Ca_3Si_2O_7$ | Ca | −17.0 | | | Kingery (1960) |

* Balance is $SiO_2$ for glasses.

low. Thus it seems certain that the Sr of the plagioclase would equilibrate with that in adjacent mesostasis during the time available. We can be much less sure about equilibration of the pyroxene. However as only a small fraction of the total Sr in the Apollo 14 basalts is located in the pyroxene, this is less important.

To summarize, local Sr isotope equilibration must be expected between minerals in igneous clasts deeply buried within a thick, hot ejecta blanket. It is likely also that small clasts would be open systems, especially for alkali diffusion. For the basalts within breccia 14321, internal Rb–Sr isochrons should be completely reset to the age of the Fra Mauro formation *provided the final assembly of this breccia took place at high temperatures deep within the formation.* On the other hand, if its final assembly occurred elsewhere, very little can be said of the cooling time for 14321 and the interpretation of the ages of its clasts becomes ambiguous.

## MINERAL AGES

### *14310*

The isotopic data for this rock and its various mineral separates (Table 1) are plotted in Fig. 3. The points cover a wide range in $^{87}Rb/^{86}Sr$ and all fit the regression

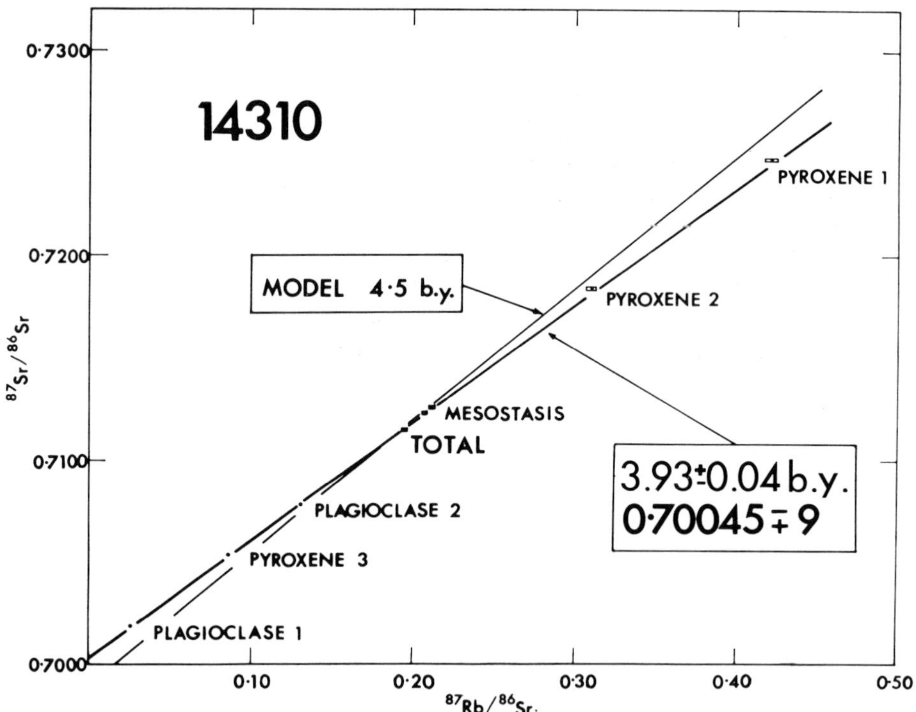


Fig. 3. Isochron diagram for igneous rock 14310. The fit of points in this and other isochron diagrams is within experimental precision, which is shown as 95% confidence limits. The "Model 4.5 b.y." isochron shown for comparison is drawn through initial $^{87}Sr/^{86}Sr$ of 0.6990. The $^{87}Rb$ decay constant is taken as 0.0139 b.y.$^{-1}$. Errors in isochron parameters are 95% limits.

line to within experimental error, so that the slope of the line is precisely measured ($\pm 1\%$, at 95% confidence limits). The mineral age is 3.93 $\pm$ 0.04 b.y., in agreement with the results of Papanastassiou and Wasserburg (1971b), Murthy *et al.* (1972), and Tatsumoto *et al.* (1972) for the same rock. (We have reduced our previous values for lunar ages by 1.8% following recalibration of spikes using the National Bureau of Standards stoichiometric Rb and Sr salts, as reported by De Laeter *et al.*, 1972.)

The variation in $^{87}Rb/^{86}Sr$ for the pyroxene concentrates in Fig. 3 is plainly due to variation in their content of the alkali-rich mesostasis, which has $^{87}Rb/^{86}Sr$ of at least 2.4 (Papanastassiou and Wasserburg, 1971b). However the mesostasis does not wholly control the slope of the isochron. The total rock-plagioclase tieline is also well-defined at 3.88 b.y., with limits as small as $\pm$0.08 b.y., so that if the mesostasis has lost any radiogenic $^{87}Sr$ by postcrystallization diffusion, the plagioclase has absorbed it. The good fit of all samples to the regression line favors an isochron interpretation either as the original age of crystallization or as a later thermal event involving complete isotopic equilibration. However this is not a compelling argument. Partial isotopic equilibration might have occurred, making the plagioclase and mesostasis each internally heterogeneous in $^{87}Sr/^{86}Sr$, but a good "isochron" alignment could still be produced by adequate mixing and representative sampling of the plagioclase and mesostasis during the operations of mineral separation. The high value for "initial" $^{87}Sr/^{86}Sr$ of 14310 compared with mare basalts is likewise suggestive of metamorphic equilibration of isotopes but not compelling, as mare basalts themselves have a range in initial $^{87}Sr/^{86}Sr$.

*14321,88, basalt clast 6A*

This was a ~0.5 cm diameter, rounded clast within the light grey matrix that Grieve *et al.* (1972) describe as the youngest textural element of breccia 14321. The clast weighed ~150 mg after cleaning, and was then crushed and divided into two equal and representative portions, one for mineral separation and Rb–Sr dating (Table 1) and the other for determination of trace-elements by spark-source mass spectrometry (Taylor, 1972a).

Figure 4 shows that the data for six different mineral separates give well-defined values for isochron slope, 4.05 b.y., and intercept. They are not distinguishable from our (revised) result for the basalt clast 4A (4.08 $\pm$ 0.1 b.y., Compston *et al.*, 1971a), which is from the opposite face of the breccia slab within the same light grey matrix. However both basalts are significantly older than 14310 and lower in initial $^{87}Sr/^{86}Sr$, the difference between the pooled basalt age and 14310 being 0.13 $\pm$ 0.08 b.y. for 95% confidence limits. Papanastassiou and Wasserburg (1971b) detect a similar difference, 0.08 $\pm$ 0.06 b.y., between the mineral ages of 14310 and several igneous chips of similar type, and a basalt clast (191 X1) extracted from 14321. Turner *et al.* (1971) likewise detect a difference in the $^{40}Ar$–$^{39}Ar$ ages of these groups.

Interpretation of the isochron as a geological process is again ambiguous: The good alignment suggests either the original crystallization or complete postcrystallization resetting, although the data also allow an interpretation as partial resetting if the making and sampling of mineral concentrates were representative.

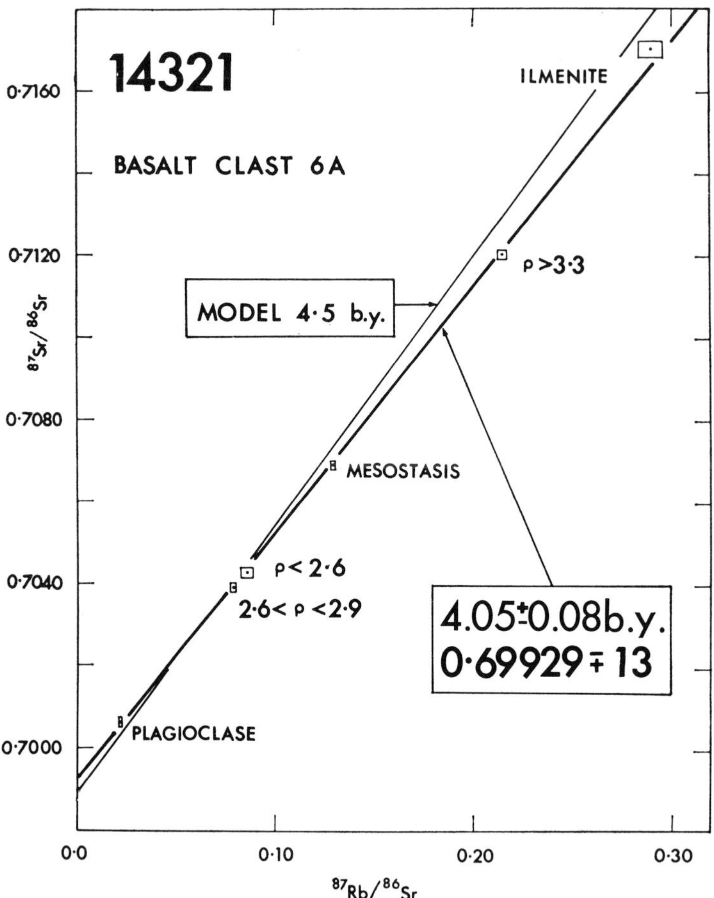


Fig. 4. Isochron diagram for basalt 6A, a clast from breccia 14321,88.

*14072*

The apparent age and initial $^{87}Sr/^{86}Sr$ of this basalt (Fig. 5) are not detectably different to those given by our analyses of basalt clasts 4A and 6A, but the initial $^{87}Sr/^{86}Sr$ for clast 191 X1 and rock 14053 by Papanastassiou and Wasserburg (1971b) is slightly higher. Our age is in good agreement with the $^{39}Ar$–$^{40}Ar$ result of York *et al.* (1972) for the same rock.

*14321,88 troctolite clast 8A*

Like the basalts 4A and 6A, this clast also occurs within the (youngest) light grey matrix of the breccia. Its apparent age (Fig. 6) is less precisely determined because a satisfactory separation of the traces of K-feldspar within this sample from Sr-rich plagioclase could not be achieved, with the result that the dispersion in Rb/Sr between the mineral concentrates was comparatively small. Furthermore, the Rb and Sr

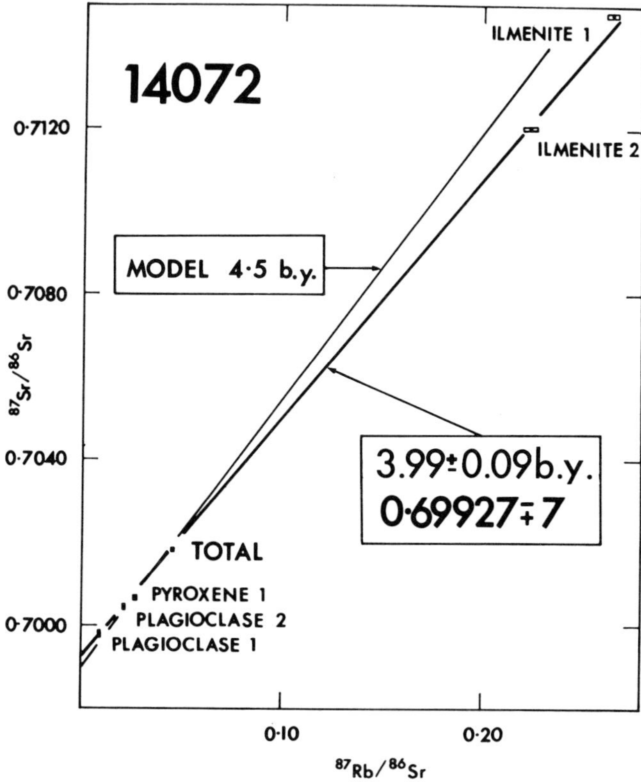


Fig. 5. Isochron diagram for basalt chip 14072.

contents of the highest $^{87}Rb/^{86}Sr$ sample, the olivine, were so low that the normal variation in the Rb blank and short-term Sr beam noise limited the precision of the analysis. Nevertheless, the mineral age is in good agreement with the overall Fra Mauro formation results, and the initial $^{87}Sr/^{86}Sr$ of the troctolite is clearly placed with the "mare" type basalts rather than rock 14310 type. A mineral age of $3.9 \pm 0.3$ b.y. is given by the K-feldspar and total-rock samples alone. This clast could belong either to the 14310 age group or to the 14053, 4A, 6A, and 191 X1 clast group.

## Discussion

A good quality fit of Rb–Sr mineral points to a regression line is necessary to interpret the line as an isochron. But no event thus identified can be equated to a particular geological process without additional information. Previously we used the well-preserved igneous mineralogy of clast 4A from 14321 and lack of shock effects as evidence that the dated event was the pre-Imbrian crystallization of an original lava flow. Papanastassiou and Wasserburg (1971b) cited the good agreement of their Rb–Sr ages with the $^{40}Ar$–$^{39}Ar$ ages of Turner *et al.* (1971) as additional evidence to support the same conclusion. However, neither preservation of petrography nor age

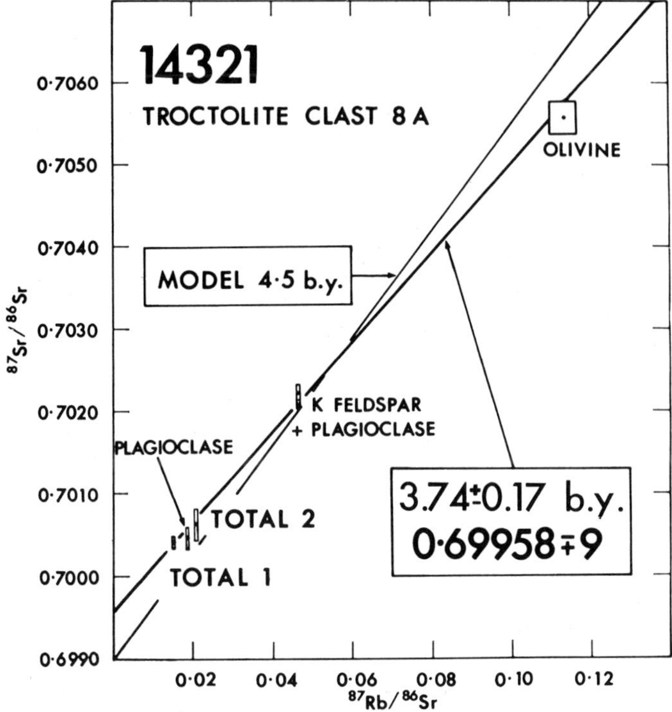


Fig. 6. Isochron diagram for troctolite clast 8A, from breccia 14321,88.

concordance are conclusive arguments. Instead we must now recognize the possibility of high and prolonged temperatures within the (deposited) Fra Mauro formation and the consequent diffusion of trace-elements within the basalt clasts which this would imply. For such a model, the good quality isochron fits are best interpreted as evidence for total re-equilibration of Sr isotopes between the minerals of each clast during the thermal event, and the concordance with K–Ar ages as due to complete Ar degassing at the same time. All clasts would be expected to give identical mineral ages; for the group of known basaltic clasts (4A, 6A, 191 X1) and probable clasts (14053) dated so far, this appears to be so. All have ages close to 4 b.y. In addition, Cliff *et al.* (1972) find the same value for the Rb–Sr mineral age of igneous fragments within a clast of older breccia included in the highly annealed breccia 14066, which is in close agreement with the result of these authors for another basalt clast from 14321. Thus, for the model of high-temperature deposition for the Fra Mauro formation, the age of deposition, i.e., the Imbrian event, is directly registered as 4 b.y. Schaeffer *et al.* (1972) have given this interpretation for their $^{40}Ar$–$^{39}Ar$ results.

It also follows that the slightly younger group of basalts that also show perfectly fitted mineral isochrons (14310, 14073, and 14001,7,3) cannot be regarded on this model as clasts belonging to the Fra Mauro formation. Their present mineral ages must have formed slightly later, and probably elsewhere. There is no geological evidence to preclude this: Rock 14310 for example, is a large isolated boulder completely free of adhering breccia-type matrix. Schaeffer *et al.* (1972) report still younger "nonmare" type basaltic fragments in Apollo 14 soils, between 3.5 and 3.6 b.y., that they also interpret as post-Fra Mauro formation.

In contrast to the above, if the Fra Mauro formation was deposited at low temperature, then the age of the Imbrian event is much less well determined. If the 14310 type rocks are assumed to be clasts, then the Imbrian event can be taken as 3.9 b.y. or younger. The significance of the 4 b.y. ages in the mare-type basalt clasts is now unknown: They could be unreset crystallization ages if the pre-Imbrian magnetization and annealing of the breccias occurred in a very short heating and cooling pulse, or they could be partially or completely reset mineral ages associated with a sufficiently long period of pre-Imbrian metamorphism.

*Acknowledgments*—We thank E. C. T. Chao, W. Gose, F. Hörz, O. James, and J. F. Lindsay for valuable discussion and D. H. Green for petrographic advice. M. W. Gamlen designed and constructed the ramp generator, and D. J. Millar assisted with the mass spectrometry.

This paper was written while the senior author was a Visiting Scientist at the Lunar Science Institute, which is operated by the Universities Space Research Association under Contract No. NSR 09-051-001 with the National Aeronautics and Space Administration.

## References

Anderson A. T., Braziunas T. F., Jacoby J., and Smith J. V. (1972) Breccia populations and thermal history (abstract). In *Lunar Science—III* (editor C. Watkins), pp. 24–26, Lunar Science Institute Contr. No. 88.

Chao E. C. T., Minkin J. A., and Best J. B. (1972) Apollo 14 breccias: General characteristics and classification. *Proc. Third Lunar Sci. Conf., Geochim. Cosmochim. Acta* Suppl. 3, Vol. 1. MIT Press.

Cliff R. A., Lee-Hu C., and Wetherill G. W. (1972) K, Rb, and Sr measurements in Apollo 14 and 15 material (abstract). In *Lunar Science—III* (editor C. Watkins), pp. 146–147, Lunar Science Institute Contr. No. 88.

Compston W., Chappell B. W., Arriens P. A., and Vernon M. J. (1970) The chemistry and age of Apollo 11 lunar material. *Proc. Apollo 11 Lunar Sci. Conf., Geochim. Cosmochim. Acta* Suppl. 1, Vol. 2, pp. 1007–1027. Pergamon.

Compston W., Vernon M. J., Berry H., and Rudowski R. (1971) The age of the Fra Mauro formation: A radiometric older limit. *Earth Planet. Sci. Lett.* **12,** 55–58.

Compston W., Berry H., Vernon M. J., Chappell B. W., and Kaye M. J. (1971b) Rb–Sr chronology and chemistry of lunar material from the Ocean of Storms. *Proc. Second Lunar Sci. Conf., Geochim. Cosmochim. Acta* Suppl. 2, Vol. 2, pp. 1471–1485. MIT Press.

Compston W., Vernon M. J., Berry H., Rudowski R., Gray C. M., Ware N., Chappell B. W., and Kaye M. (1972) Age and petrogenesis of Apollo 14 basalts (abstract). In *Lunar Science—III* (editor C. Watkins), pp. 151–153, Lunar Science Institute Contr. No. 88.

De Laeter J. R., Vernon M. J., and Compston W. (1972) Revision of lunar Rb–Sr ages. Submitted to *Geochim. Cosmochim. Acta.*

Dence M. R., Plant A. G., and Traill R. J. (1972) Impact-generated shock and thermal metamorphism in Fra Mauro lunar samples (abstract). In *Lunar Science—III* (editor C. Watkins), pp. 174–176, Lunar Science Institute Contr. No. 88.

Doremus R. H. (1962) Diffusion in non-crystalline silicates. In *Modern Aspects of the Vitreous State* (editor J. D. Mackenzie), Vol. 2, pp. 1–71, Butterworth & Co.

Gancarz A. J., Albee A. L., and Chodos A. A. (1971) Petrologic and mineralogic investigation of some crystalline rocks returned by the Apollo 14 mission. *Earth Planet. Sci. Lett.* **12,** 1–18.

Grieve R., McKay G., Smith H., Weill D., and McCallum S. (1972) Mineralogy and petrology of polymict breccia 14321 (abstract). In *Lunar Science—III* (editor C. Watkins), pp. 338–340, Lunar Science Institute Contr. No. 88.

Hanson G. N. and Gast P. W. (1967) Kinetic studies in contact metamorphic zones. *Geochim. Cosmochim. Acta* **31,** 1119–1153.

Hargraves R. B. and Dorety N. (1972) Magnetic property measurements on several Apollo 14 rock samples (abstract). In *Lunar Science—III* (editor C. Watkins), pp. 357–359, Lunar Science Institute Contr. No. 88.

Horai K., Simmons G., Kanamori H., and Wones D. (1970) Thermal diffusivity and conductivity of lunar material. *Science* **167,** 730–731.

Jaeger J. C. (1961) The cooling of irregularly shaped igneous bodies. *Amer. J. Sci.* **259,** 721–734.

Kingery W. D. (1960) *Introduction to Ceramics.* Wiley.

Kovach R. L., Watkins J. S., and Landers T. (1971) Active seismic experiment. In *Apollo 14 Preliminary Science Report*, NASA SP-272.

Kurat G., Keil K., Prinz M., and Bunch T. E. (1972) A "chondrite" of lunar origin: Textures, lithic fragments, glasses and chondrules (abstract). In *Lunar Science—III* (editor C. Watkins), pp. 463–465, Lunar Science Institute Contr. No. 88.

Lindsay J. F. (1972) Sedimentology of clastic rocks from the Fra Mauro region of the moon. *J. Sediment Petrol.*, in press.

LSPET (Lunar Sample Preliminary Examination Team) (1971) Preliminary examination of lunar samples from Apollo 14. *Science* **173,** 681–693.

McKay D. S., Clanton U. S., Heiken G. H., Morrison D. A., Taylor R. M., and Ladle G. (1972) Vapor phase crystallization in Apollo 14 breccias and size analysis of Apollo 14 soils (abstract). In *Lunar Science—III* (editor C. Watkins), pp. 529–531, Lunar Science Institute Contr. No. 88.

McVay G. L. and Day D. E. (1970) Diffusion and internal friction in Na–Rb silicate glasses. *J. Amer. Ceram. Soc.* **53,** 508–513.

Murthy V. Rama, Evensen N. M., Jahn B.-M., and Coscio M. R. (1971) Rb–Sr ages and elemental abundances of K, Rb, Sr, and Ba in samples from the Ocean of Storms. *Geochim. Cosmochim. Acta* **35,** 1193–1153.

Murthy V. R., Evensen N. M., Jahn B.-M., and Coscio M. R. (1972) Rb–Sr ages, trace-elements, and speculations on lunar differentiation (abstract). In *Lunar Science—III* (editor C. Watkins), pp. 571–572, Lunar Science Institute Contr. No. 88.

Offield T. W. (1970) Geologic map of the Fra Mauro site—Apollo 13, Scale 1 : 5000. USGS.

Papanastassiou D. A. and Wasserburg G. J. (1971a) Lunar chronology and evolution from Rb–Sr studies of Apollo 11 and 12 samples. *Earth Planet. Sci. Lett.* **11,** 37–62.

Papanastassiou D. A. and Wasserburg G. J. (1971b) Rb–Sr ages of igneous rocks from the Apollo 14 mission and the age of the Fra Mauro formation. *Earth Planet. Sci. Lett.* **12,** 36–48.

Pearce G. W., Strangway D. W., and Gose W. A. (1972) Remanent magnetization of lunar samples (abstract). In *Lunar Science—III* (editor C. Watkins), pp. 599–601, Lunar Science Institute Contr. No. 88.

Quaide W. (1972) Mineralogy and origin of Fra Mauro fines and breccias (abstract). In *Lunar Science—III* (editor C. Watkins), pp. 627–629, Lunar Science Institute Contr. No. 88.

Reynolds M. B. (1957) Diffusion of argon in a potassium-lime-silica glass. *J. Amer. Ceram. Soc.* **40,** 395–398.

Schaeffer O. A., Husain L., Sutter J., Funkhouser J., Kirsten T., and Kaneoka I. (1972) The ages of lunar material from Fra Mauro and the Hadley Rille-Apennine Front area (abstract). In *Lunar Science—III* (editor C. Watkins), pp. 675–677, Lunar Science Institute Contr. No. 88.

Sippel R. F. (1963) Sodium self-diffusion in natural minerals. *Geochim. Cosmochim. Acta* **27,** 107–120.

Taylor G. J., Marvin U. B., Reid J. B., and Wood J. A. (1972b) Noritic fragments in the Apollo 14 and 12 soils and the origin of Oceanus Procellarum. *Proc. Third Lunar Sci. Conf., Geochim. Cosmochim. Acta* Suppl. 3, Vol. 1. MIT Press.

Taylor S. R., Kaye M., Muir P., Nance W., Rudowski R., and Ware N. (1972a) Composition of the lunar uplands—I. Chemistry of Apollo 14 samples from Fra Mauro. *Proc. Third Lunar Sci. Conf., Geochim. Cosmochim. Acta* Suppl. 3, Vol. 2. MIT Press.

Turner G., Huneke J. C., Podosek F. A., and Wasserburg G. J. (1971) $^{40}$Ar–$^{39}$Ar ages and cosmic ray exposure ages of Apollo 14 samples. *Earth Planet. Sci. Lett.* **12,** 19–35.

Warner J. L. (1972) Apollo 14 breccias: Metamorphic origin and classification (abstract). In *Lunar Science—III* (editor C. Watkins), pp. 782–784, Lunar Science Institute Contr. No. 88.
Williams R. J. (1972) The petrogenesis of lunar breccias. Submitted to *Earth Planet. Sci. Lett.*
Wilshire H. G. and Jackson E. D. (1972) Petrology of the Fra Mauro formation at the Apollo 14 landing site (abstract). In *Lunar Science—III* (editor C. Watkins), pp. 803–805, Lunar Science Institute Contr. No. 88.
York D., Kenyon W. J., and Doyle R. J. (1972) $^{40}Ar$–$^{39}Ar$ ages of Apollo 14 and 15 samples (abstract). In *Lunar Science—III* (editor C. Watkins), pp. 822–824, Lunar Science Institute Contr. No. 88.

Proceedings of the Third Lunar Science Conference
(Supplement 3, *Geochimica et Cosmochimica Acta*)
Vol. 2, pp. 1503–1514
The M.I.T. Press, 1972

# Apollo 14 and 15 samples: Rb-Sr ages, trace elements, and lunar evolution

V. RAMA MURTHY, N. M. EVENSEN, BOR-MING JAHN, and M. R. COSCIO, JR.

Department of Geology and Geophysics, University of Minnesota,
Minneapolis, Minnesota 55455

**Abstract**—The Apollo 14 basalt sample 14310 has been dated by the Rb–Sr internal isochron method at 3.93 ± 0.06 AE (1 Aeon = AE = $10^9$ years), making it the oldest lunar basalt measured to date. The unusually high initial $^{87}Sr/^{86}Sr$ ratio (0.70033 ± 19) and K, Rb, Ba abundances in this rock are probably due to contamination of primitive basaltic material by a trace-element rich layer near the lunar surface. Further interpretation depends upon the question of whether the formation of 14310 basalt pre- or post-dates the Mare Imbrium excavation.

Chemical and isotopic characteristics of the Apollo 14 soils permit their derivation by comminution of local rocks, in contrast to the situation for Apollo 12 and 15 soils. To date, all lunar soils show variable fractions of an "exotic component" that must be an early differentiated material on the moon, rich in trace elements and radiogenic Sr. From the size fraction analyses of some Apollo 14 and 15 soil samples and the inferences made on the abundances of the exotic component in Apollo 11, 12, and Luna 16 soils, we suggest that this ubiquitous component is the Fra Mauro material itself or the trace-element rich layer from which it is derived, excavated by the Imbrian impact event from considerable depths in the moon, and distributed widely both at the time of the Imbrian event and at later times by impacts such as the Copernican event.

The Apollo 15 basalt 15555 has been dated by the Rb–Sr internal isochron method at 3.30 ± 0.08 AE, in complete agreement with a K–Ar date on the plagioclase separate. Both in age and trace-element contents this basalt is very similar to mare basalts from the Ocean of Storms. Thus, it appears that extensive volcanism and mare flooding occurred on the lunar surface at ~3.3 AE ago, suggesting that mare basalt generation is due to the internal thermal regime of the moon.

Total samples and size fractions of soils from the Apollo 14 site (14149, 14259) and the Apollo 15 site (15531) uniformly show model ages close to 4.6 AE despite almost an order of magnitude variation in their Rb–Sr ratios. Apollo 11 and 12 soils show similar characteristics. The Rb–Sr systematics of the soils point to a primordial differentiation of the moon at 4.6 AE ago. This is inferred directly from the nature of the exotic component and indirectly from the mare basalt material, whose Rb–Sr behavior may have been controlled by partial melting of early formed Rb–rich phases in the interior of the moon.

## INTRODUCTION

IN THIS PAPER we report the results of our work to date on Rb–Sr isotopic relations and K, Rb, Sr, Ba abundance measurements in Apollo 14 and 15 rocks and soils. These studies include age determination by the internal isochron technique on one basaltic rock sample from each of the two missions (14310 and 15555). In addition, total sample measurements have been made on samples of fines material from both missions (14003, 14149, 14163, 14259, 15091, and 15531). Finally we have analyzed trace element abundances and Rb–Sr systematics in the various grain size fractions of three of the above fines (14149, 14259, and 15531). This last type of study, which we plan to continue on other lunar fines, is being undertaken in conjunction with rare gas measurements on the same materials by Professor R. O. Pepin of the School of Physics, University of Minnesota. We expect these studies to shed light on the apparently complex nature and history of the lunar soil.

The Apollo 14 landing site at Fra Mauro is believed to be dominated by material ejected from the Mare Imbrium basin during its excavation. The rock samples returned from this mission were predominantly very complex multigeneration breccias; only two large basaltic rocks were collected (LSPET, 1971). The landing site of Apollo 15 was between the Appenine front and the Hadley Rille at the eastern edge of the Imbrium basin. Sample collection during this mission was done on the area of the landing site, the Appenine front and the Hadley Rille area. This has produced the largest and most varied sample collection thus far returned from the moon. Study of materials from these two missions provides interesting comparisons with the mare materials returned from the previous Apollo 11 and 12 missions.

## EXPERIMENTAL PROCEDURES

The experimental techniques used in these studies have previously been described in Murthy *et al.* (1971a). Here we present a brief summary and a description of some new techniques related to size fraction separation of the fines.

The rock samples were crushed to $-200 + 325$ mesh size, and from this size fraction mineral separations were carried out. In the case of sample 14310 density fractions obtained by heavy liquid separation were used. For sample 15555 a combination of density separations, magnetic separation and hand picking was used. A plagioclase fraction was hand picked from a fraction with density $\rho < 2.89$ g/cm$^3$ of $-100$, $+200$ mesh size from 15555,12. Aliquots of this hand picked plagioclase were used both for the Rb–Sr and for the K–Ar dating reported earlier (Murthy *et al.*, 1972). Magnetic separation techniques for the fraction with $\rho < 3.33$ g/cm$^3$ yielded moderately pure ($\sim 90\%$) olivine and pyroxene fractions. Ilmenite constitutes about 1% of rock 15555 (LSPET, 1972), and sufficient sample size of pure ilmenite could not be obtained from our small sample allocation. However, a highly magnetic fraction containing olivine, pyroxene and ilmenite was used to obtain the highest $^{87}$Rb/$^{87}$Sr we observed in this rock. Total rock analyses were made on two small chips from each of the rock samples rather than on splits of homogenized samples.

Particle size separations of samples 14149, 14259, and 15531 were made in high purity reagent grade acetone. For separations above 25 $\mu$ size, nylon screens supported in electropolished stainless steel holders were employed. For sizes below this and down to 2 $\mu$ size, separations were carried out in an acetone medium and by the use of special Buckbee-Mears electroformed nickel screens supported in stainless steel holders. Particles of less than 2 $\mu$ size were separated by sedimentation techniques in acetone and, after centrifugation to remove the acetone, were dried in an oven at 110°C for 24 hours. All tools and containers used in this work were repeatedly cleaned ultrasonically in acetone and double quartz-distilled water and dried with filtered high purity nitrogen. All operations were carried out in the clean room facilities of our laboratories.

The samples thus prepared were dissolved, aliquots taken, and spiked as appropriate. Elemental separations were then made by ion exchange. All procedures are generally designed to minimize contamination and are carried out in an ultra clean laboratory. The consistent blank levels remain at K = 0.02 $\mu$g, Rb = 0.1 ng, Sr = 1.0 ng, and Ba = 0.07 $\mu$g. Blank corrections are negligible in all but the smallest sample sizes used.

All mass spectrometry was done on a 30 cm radius single focusing machine with associated magnetic field stepping and automatic data collection system using the techniques described earlier (Murthy *et al.*, 1971a). With these techniques we have previously found that measurements of $^{87}$Sr/$^{86}$Sr could be made to a precision of $\sim 0.01\%$. Our chief index of the long term precision and reproducibility of our data has been the continuing measurements of the seawater provided by Professor Wasserburg of the California Institute of Technology. The results of several measurements of this standard, made during the period of the work reported here, are in complete agreement with those we have previously reported. The mean of these measurements on the sea water standard to date is $^{87}Sr/^{86}Sr = 0.70916 \pm 0.00005(2\sigma)$.

## Analytical Results

We were allocated one basaltic rock sample from Apollo 14, sample 14310,133. Five density separates were prepared from this rock and analyzed (Table 1). No

Table 1. Rb–Sr isotopic data on Apollo 14 and 15 crystalline rocks.

| Sample | Wt. (mg) | Rb (μg/g) | Sr (μg/g) | $^{87}Rb/^{86}Sr$[a] (× $10^2$) | $^{87}Sr/^{86}Sr$[b] |
|---|---|---|---|---|---|
| *14310,133* | | | | | |
| Total rock | 28.30 | 10.99 | 158.8 | 20.039 | 0.71159 ± 7 |
| Density fractions | | | | | |
| ($\rho < 2.78$) | 36.75 | 14.19 | 274.6 | 14.963 | — |
| ($\rho = 2.78$–$2.80$) | 52.65 | 13.71 | 272.7 | 14.558 | — |
| "A" ($\rho = 2.80$–$2.89$) | 32.90 | 14.03 | 238.6 | 17.026 | 0.70984 ± 6 |
| "B" ($\rho = 2.89$–$2.96$) | 35.30 | 10.36 | 118.2 | 25.379 | 0.71454 ± 5 |
| "C" ($\rho = 2.96$–$3.33$) | 36.50 | 12.07 | 122.6 | 28.507 | 0.71627 ± 8 |
| *15555,12* | | | | | |
| Total rock | 23.87 | 0.538 | 74.11 | 2.102 | 0.70005 ± 5 |
| Plagioclase (*H*) | 29.78 | 0.198 | 319.8 | 0.180 | 0.69917 ± 7 |
| Pyroxene (*H*) | 36.10 | 0.167 | 13.50 | 3.575 | 0.70076 ± 13 |
| Olivine (*H*) | 46.50 | 0.166 | 11.29 | 4.265 | 0.70110 ± 5 |
| "Ilmenite" (*D*, *M*) | 41.10 | 0.982 | 29.23 | 9.730 | 0.70362 ± 9 |
| *15555,15* | | | | | |
| Total rock | 39.25 | 0.700 | 85.32 | 2.375 | 0.70009 ± 6 |

[a] Errors of ±2% are assigned.
[b] Statistical errors correspond to the last figures given and are equal to $\pm 2\sigma_M$ where

$$\sigma_M = \left[\frac{\sum_{i=1}^{n} (\bar{X}_i - m)^2}{n(n-1)}\right],$$

$n$ is the number of sets of 10 scans each and $(X_i - m)$ is the difference between the mean value of the $i$th set and mean of all sets. See Murthy *et al.* (1971a) for a discussion of data analysis and errors. $D$ = Density fraction $\rho > 3.33$; $H$ = Hand picked; $M$ = Magnetic fraction.

significant amount of material with $\rho > 3.33$ was obtained, in contrast with all other lunar basalts we have analyzed. Also unusual was the fact that the light density fractions ($\rho < 2.89$) failed to yield a nonradiogenic plagioclase separate. This suggests that a significant proportion of the total Rb in the sample may lie in the low density glassy interstitial material and also reflects the highly radiogenic nature of this rock as compared with the basalts from other regions of the moon.

However, the total spread in $^{87}Sr/^{86}Sr$ in the separates is enough to define a relatively precise age of 3.93 ± 0.06 AE (Fig. 1), in good agreement with the ages obtained for this rock by Papanastassiou and Wasserburg (1971b) and by Compston *et al.* (1972). Because of the extrapolation required, an imprecise intercept was obtained, corresponding to an initial $^{87}Sr/^{86}Sr$ of 0.70033 ± 0.00019. Despite the large error, this is also in good agreement with the results of Papanastassiou and Wasserburg (1971b).

We received samples 15555,12 and 15555,15 as part of the Apollo 15 rapid sample allocation program and have previously reported our studies on this rock (Murthy

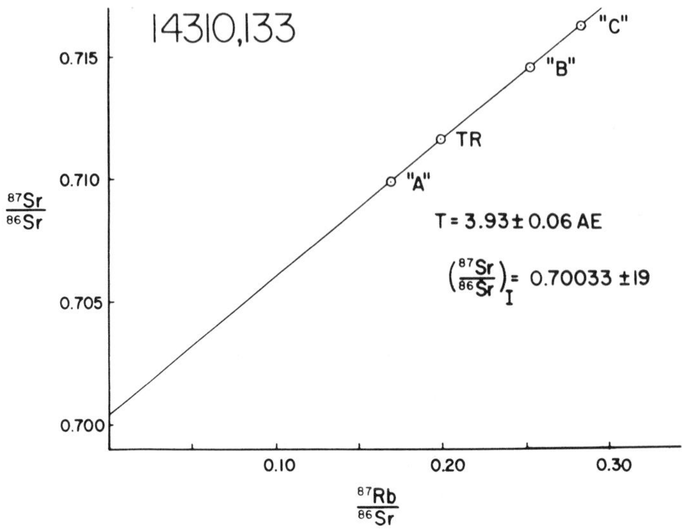


Fig. 1. $^{87}$Rb–$^{87}$Sr internal isochron for rock 14310,133. TR = total rock; density fractions A, B, C are described in Table 1. Errors for the $^{87}$Rb/$^{86}$Sr are $\pm 2\%_0$. The errors for $^{87}$Sr/$^{86}$Sr are noted in Table 1. Best fit line obtained by a York (1966) type of weighted regression analysis. All points shown deviated less than 1.5 parts $10^4$ from the best fit line. All errors are $2\sigma$.

*et al.*, 1972), which will be briefly summarized here. We analyzed four mineral separates obtained by combined density and magnetic separations and hand picking, as described above. Total rock measurements were made on each of the two chips received (Table 1). The data yield an internal isochron age of 3.30 ± 0.08 AE, and an initial $^{87}$Sr/$^{86}$Sr of 0.69906 ± 0.00004. The initial ratio we observed in 15555 is lower by about 3 parts in 7000 than that obtained by Chappell *et al.* (1972) and by Wasserburg and Papanastassiou (1971). This difference is significantly beyond experimental errors and also cannot be attributed to interlaboratory bias, as can be seen from the comparative measurements on sea water standard (Murthy *et al.*, 1971a), and agreement for low initial ratios on other lunar rocks. Contrary to LSPET (1972) descriptions, the sample of 15555 that we received was highly friable and disintegrated during shipment. Our present interpretation of this discrepancy is that the initial Sr in this rock may be variable. We have previously observed similar behavior in rock 12004, where two separate chips measured in our laboratory yielded similar ages but initial Sr values differing by about 1 part in 7000 (Murthy *et al.*, 1971a). Our experience with lunar rocks is not sufficient to unequivocally rule out small variations of this magnitude. A K–Ar age of 3.31 ± 0.07 AE was obtained for this sample, with the Ar analysis performed on the plagioclase separate by Prof. R. O. Pepin of the University of Minnesota School of Physics. The excellent agreement of these two ages demonstrates the very high Ar retentiveness of the plagioclase in lunar basalts, as was first shown by Eberhardt *et al.* (1971). In age, initial Sr and trace element abundances, sample 15555 is virtually indistinguishable from the Apollo 12 basalts measured in our laboratories (Murthy *et al.*, 1971a).

Data on the size fractions separated from the Apollo 14 and 15 soils are shown in Table 2. The two Apollo 14 soils show considerable scatter in K, Rb, and Ba abundances in the various size fractions, but no trend is apparent with decreasing grain size. In sample 14259,14 the maximum K, Rb, and Ba values are found in the $>147\ \mu$ size fraction. Pepin *et al.* (1972) found that this same size fraction lay far off the $^{40}Ar$–$^{36}Ar$ correlation line for the $<147\ \mu$ fractions in this soil, and falls precisely on a mixing line defined by Apollo 12 soils. This mixing line has an intercept at a K–Ar age of 0.95 AE, and may represent a contribution to the soil from Copernican ejecta. It is interesting that this fraction should prove distinguishable both in rare gas systematics and trace element abundances.

The data for Apollo 15 soil 15531,42 are not yet complete; so far analyses are available for the total sample and the three size fractions $>37\ \mu$. Even in this limited sampling, however, there appears to be a definite trend toward higher K, Rb, Ba abundances and $^{87}Sr/^{86}Sr$ ratios with decreasing grain size. From material balance considerations it appears that the $<37\ \mu$ fraction as a whole should strongly continue this trend. Possible interpretations of this Apollo 14 and 15 soil data will be discussed below.

Table 2. Elemental abundances and Rb–Sr isotopic data on some Apollo 14 and 15 fines and size fractions.

| Sample | Wt. (mg) | K ($\mu$g/g) | Rb ($\mu$g/g) | Sr ($\mu$g/g) | Ba ($\mu$g/g) | $^{87}Rb/Sr^{86}$[a] ($\times 10^2$) | $^{87}Sr/^{86}Sr$[b] |
|---|---|---|---|---|---|---|---|
| *14003,41* | | | | | | | |
| Total sample | 36.70 | 3723 | 14.61 | 173.8 | 767 | 24.34 | 0.71561 ± 8 |
| *14149,44* | | | | | | | |
| Total sample | 10.24 | 4786 | 17.87 | 183.5 | 817 | 28.20 | 0.71681 ± 13 |
| $>147\mu$ | 9.11 | 4481 | 13.29 | 181.6 | 699 | 21.20 | 0.71181 ± 7 |
| 74–147$\mu$ | 10.08 | 4438 | 14.11 | 171.3 | 670 | 23.85 | 0.71451 ± 9 |
| 37–74$\mu$ | 9.27 | 4251 | 15.34 | 169.1 | 688 | 26.27 | 0.71461 ± 8 |
| 16–37$\mu$ | 10.40 | 4497 | 14.19 | 169.1 | 694 | 24.30 | 0.71438 ± 9 |
| 4–16$\mu$ | 9.78 | 5431 | 15.08 | 184.9 | 788 | 23.62 | 0.71363 ± 11 |
| *14163,139* | | | | | | | |
| Total sample | 30.95 | 3750 | 14.32 | 169.6 | 770 | 24.45 | 0.71594 ± 9 |
| *14259,14* | | | | | | | |
| Total sample | 12.75 | 4268 | 13.88 | 180.7 | 753 | 22.24 | 0.71418 ± 7 |
| $>147\mu$ | 12.62 | 5659 | 16.88 | 180.0 | 961 | 26.72 | 0.71609 ± 6 |
| 74–147$\mu$ | 11.26 | 4335 | 14.31 | 181.1 | 726 | 22.89 | 0.71355 ± 6 |
| 37–74$\mu$ | 12.35 | 4188 | 13.47 | 178.2 | 703 | 21.89 | 0.71390 ± 7 |
| $<37\mu$ | 13.66 | 4684 | 15.08 | 191.8 | 791 | 22.77 | 0.71390 ± 11 |
| *15091,37* | | | | | | | |
| Total sample | 21.41 | 1632 | 5.05 | 139.7 | 208 | 10.47 | 0.70535 ± 15 |
| *15531,42* | | | | | | | |
| Total sample | 24.19 | 907.8 | 2.618 | 104.32 | 117.99 | 7.267 | 0.70390 ± 8 |
| $>147\mu$ | 26.16 | 583.8 | 1.448 | 97.21 | 66.09 | 4.313 | 0.70173 ± 10 |
| 74–147$\mu$ | 24.15 | 635.8 | 1.767 | 84.79 | 73.41 | 6.034 | 0.70293 ± 10 |
| 37–74$\mu$ | 24.12 | 737.1 | 2.052 | 87.74 | 98.71 | 6.772 | 0.70331 ± 14 |

[a] Errors of ±2% are assigned.
[b] Statistical errors correspond to the last figures given and are $\pm 2\sigma$ as described in Table 1.

Finally it is worth commenting upon the remarkable uniformity of Sr abundances in the lunar materials. This uniformity, within a factor of 2–3, is seen in materials with widely varying K, Rb, Ba abundances, ranging from the primitive mare basalts of Apollo 11 and 12 to highly evolved material such as rock 12013, KREEP, and the Apollo 14 materials. We can also see the remarkable constancy in the different soil fractions shown here. It appears highly probable that however complex the mixture of materials making up the soil may be, it consists predominantly of materials having Sr abundances similar to those so far sampled in lunar materials.

## DISCUSSION

The Apollo 14 basaltic rock 14310 has been dated at 3.93 $\pm$ 0.06 AE and is the oldest basaltic rock returned from the moon to date. The primary question in the interpretation of the age of 14310 is whether its formation pre- or postdated the Imbrium basin excavation and is only answerable if a unique interpretation can be made of the origin of this basalt in view of its high abundances of various trace elements and the radiogenic initial $^{87}Sr/^{86}Sr$ (Table 1 and Fig. 1). Such a unique interpretation is not possible at this time and several possibilities remain.

The very primitive petrologic nature of 14310 (e.g., Ford *et al.*, 1972; El Goresy *et al.*, 1972) combined with the high abundances of various trace elements strongly suggests contamination at the time of melting, perhaps during upward migration of the magma. Furthermore, the high initial $^{87}Sr/^{86}Sr$ ratio of this rock (0.70033) implies that the trace element rich and radiogenic strontium containing material which produced the contamination was already present in the moon well before 3.9 AE ago. This is compatible with the early differentiation models of the moon discussed by several workers, in which many lithophilic and radioactive elements are enriched in the outer layers of the moon during a melting event involving the outer 200–400 km of the moon (e.g., Smith *et al.*, 1970; Papanastassiou and Wasserburg, 1971a; Gast *et al.*, 1972; Murthy *et al.*, 1971b) occurring very soon after its formation. Whether this layer, rich in trace elements and containing highly evolved Sr, is a primitive crust on the lunar surface or a layer at some depth which is sampled only in deep excavations by large meteoroid impacts is not known at this time. These trace-element and isotopic characteristics are shown in both the regolith and basalts from the Fra Mauro area. If the Fra Mauro ejecta are considered to have been produced by Imbrium basin excavation, this suggests that the enriched layer must occur at some depth in the moon rather than as a crustal material in the lunar highlands.

If 14310 was formed *prior* to the Imbrium collision, the contamination could have occurred during migration of a primitive magma formed at depth upward through this trace element enriched layer. This hypothesis, however, would require that the Imbrium excavation was later than 3.9 AE ago, and that impacting bodies of the size required were still available at this time (Papanastassiou and Wasserburg, 1971b). Alternatively, 14310 could have formed *subsequent* to Imbrium excavation, either by melting at depth and subsequent contamination during injection into the Fra Mauro ejecta, or by *in situ* melting of the ejecta blanket itself (Dence *et al.*, 1972). In this latter case the "contamination" would be purely mechanical, consisting of the mixing of primitive and differentiated material derived from varying depths within the moon during the excavation, and prior to the actual formation of the basalt.

The Rb–Sr internal isochron age for Apollo 15 basalt 15555 is 3.30 ± 0.08 AE (Fig. 2) and is in essential agreement with the K–Ar age obtained by us on a plagioclase separate (Murthy *et al.*, 1972) as well as numerous other ages reported in the literature. Both the trace element content and the Rb–Sr isotopic systematics of this sample highly resemble the basaltic material from the Ocean of Storms we have studied earlier (Murthy *et al.*, 1971a). These data on basaltic rocks from widely separated mare regions show the contemporaneity of basalt flooding over extensive areas on the moon and indicate that mare basalt production is probably due to the internal thermal regime of the moon, as well as illustrating an underlying homogeneity in lunar igneous processes.

The fines sampled at the Apollo 14 site are invariably trace element rich and high in $^{87}Sr/^{86}Sr$ (Table 2) compared to soils from the other mare sites sampled. A comparison of the soil trace element chemistry with the breccias and basaltic material at Fra Mauro is compatible with production of soil by comminution of local rocks. However, it was first suggested in the case of Apollo 11 soils (Papanastassiou *et al.*, 1970) and later strikingly borne out in the studies of Apollo 12 soils (Papanastassiou and Wasserburg, 1971a; Murthy *et al.*, 1971a) that the soils are not simply derived from the local rocks but contain an additional component characterized as old and enriched in radiogenic Sr and several lithophilic trace elements. We have previously reported leaching studies on Apollo 11 and 12 soils (Murthy, 1970; Murthy *et al.*, 1971a) that showed possible radiogenic components with $^{87}Sr/^{86}Sr$ as high as 0.744

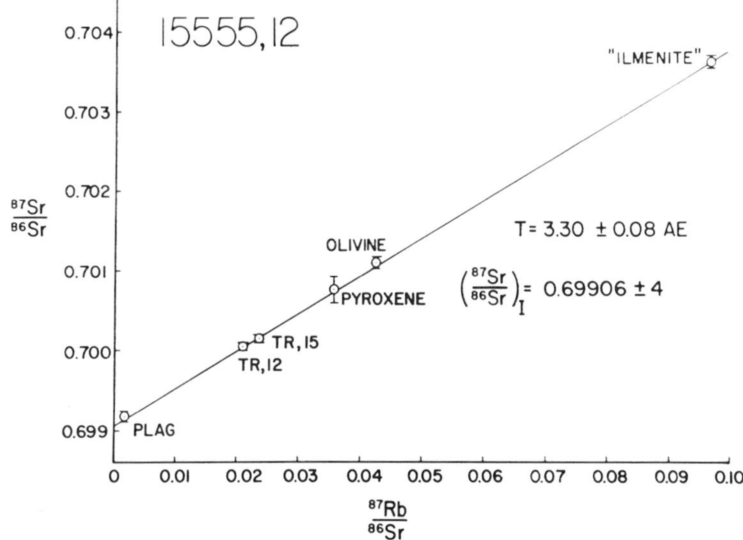


Fig. 2. $^{87}Rb$–$^{87}Sr$ internal isochron for rock 15555,12. TR,12 and TR,15 are two total rock chips. Olivine, pyroxene, and plagioclase are hand picked. "Ilmenite" is a highly magnetic fraction from a density separate $\rho > 3.3$ g/cm$^3$. Other comments as in Fig. 1.

and model ages ~4.5 AE. It is clear from all these considerations that whatever the provenance of this old and highly differentiated material ("exotic component") may be, it is easily available for incorporation into the lunar soils.

A detailed study of the various size fractions of two fines from the Fra Mauro site shows that the high trace element content is common to all size fractions (Table 2). Elemental abundances of K and Rb are much more variable (20–50%), but the Sr and Ba abundances vary by only ~10% in the various size fractions. Assuming that the trace element abundances are largely dependent upon the fraction of the exotic component present in the soils, we would expect a grain size dependency of trace element abundances if the exotic component had been transported over long distances. This, however, is not the case in those Fra Mauro soil samples and size fractions that we have analyzed. A reasonable interpretation, therefore, is that a wholesale transport of material occurred in the Imbrium collision event and that from this material the soil was locally produced by lunar surficial degradation processes.

There is some evidence for a component due to the Copernican ejecta, similar to that observed at the Apollo 11 site, in the Fra Mauro fines sample 14259, >147 $\mu$ size fraction (Pepin *et al.*, 1972). Since the trace element characteristics of this fraction largely resemble those of other fines samples and their size fractions at Fra Mauro and since the Copernican event penetrated the Fra Mauro formation (Schmitt *et al.*, 1967), it seems reasonable that the exotic component may be the Fra Mauro formation itself, but transported into the Apollo 14 site at the time of the Copernican event, about a billion years ago.

Because of the characteristics of the Fra Mauro material and the probability that this material underlies both the Procellarum site and Copernicus crater, it is tempting to suggest that the ultimate source of the exotic component observed in the lunar soils in general is the Fra Mauro material, excavated from considerable depths (~50 km) in the moon at the time of Imbrium basin excavation and scattered by smaller impacts into the mare regions following their flooding. The progressively smaller amounts of the exotic material inferred in lunar soils from Oceanus Procellarum, Mare Tranquilitatis, and Mare Fecunditatis are consistent with this point of view.

In contrast to the soil samples from the Fra Mauro site, the two soil samples from the Apollo 15 site are deficient in the lithophilic trace elements and resemble in general the soil samples from the mare sites of Tranquilitatis and Procellarum. However, sample 15091 from the St. George crater area is distinctly richer in these trace elements than the basaltic rock 15555 and the soil 15531 from the Hadley Rille area. It is clear that both these soils also contain a fraction of the exotic component. Size fraction analyses of sample 15531 and material balance considerations show that this component is preferentially in the finer grain size fractions. We interpret this as resulting from transport of this component from a distance by meteorite impacts such as the Copernican event, as we have discussed above. The K–Ar studies now in progress on these size fractions should further clarify the situation. Also, sieving studies on soil samples from other Apollo landing sites are being planned to investigate the consistency of this behavior.

Figure 3 shows all of the soil analyses performed to date in our laboratory, plotted on a Sr evolution diagram along with several reference isochrons passing through

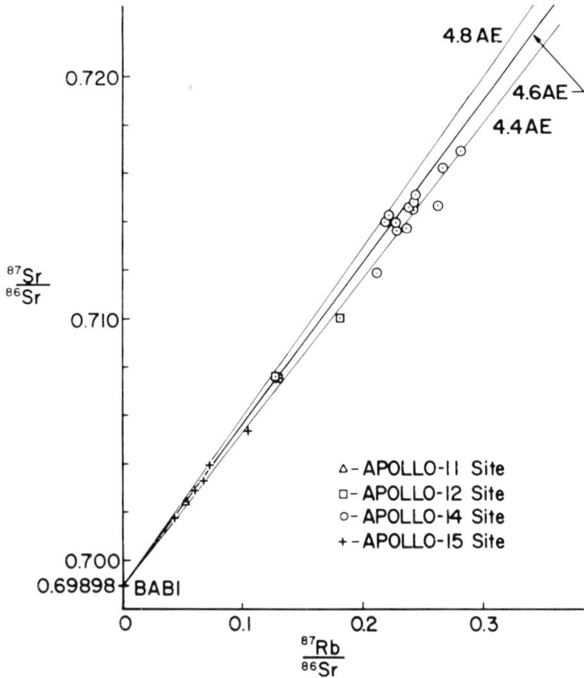


Fig. 3. Rb–Sr evolution diagram for lunar soils analyzed in our laboratories from Apollo missions 11, 12, 14, and 15. Total samples and size fraction data as shown in Table 2. Solid lines show reference $T_{BABI}$ isochrons of 4.6 ± 0.2 AE.

BABI. This plot demonstrates the remarkable tendency of the lunar soils to lie near the 4.6 AE reference isochron, as can be seen even in the various size fractions of Apollo 14 and 15 soils. It is highly unlikely that the lunar soils have been closed systems since ~4.6 AE ago; rather they appear to have been derived in part from rocks which have themselves presumably undergone differentiation at the time of their formation much less than 4.6 AE ago, and in part from an exotic component of unknown origin, but almost certainly not directly related to the rocks. Thus, the consistent 4.6 AE model ages of the soil imply that the major components present must themselves have ~4.6 AE model ages, as has been shown for many of the lunar rocks (e.g., Compston *et al.*, 1971). This in turn implies that either the individual components have themselves been closed systems since ~4.6 AE or they have been derived from other materials in a way which gives them the *appearance* of being closed systems. The first possibility is highly unlikely for the rocks but quite possible in the case of the exotic component. The second possibility is an attractive one for the rocks, and might operate through a mechanism such as the partial melting model proposed by Graham and Ringwood (1971), in which the Rb–Sr systematics of the partial melts are controlled by the accessory phases formed at the time of primary differentiation of the source region, in this case presumably at the "magic" age of 4.6 AE.

As lunar studies have progressed with the samples returned by the Apollo missions, it has become clear that at least the outer layers of the moon underwent a major differentiation very near the time of the moon's formation 4.6 AE ago. It is not clear, however, what mechanism may account for such an early intense heating event. Heating by extinct radioactivity seems to be ruled out as no traces of their decay products have been found. In addition, unless accretion was inhomogeneous, extinct radioactivity would have heated the deep interior of the moon even more strongly than the surface regions. In such an event, the departure of the moon's figure from equilibrium (Kaula, 1969) and its very low seismicity (Latham *et al.*, 1971) cannot be easily explained.

The most attractive explanation for early lunar differentiation seems to be provided by late stage accretion energy. Both accretion rate and gravitational potential energy of the accreting particles are dependent upon the mass that the accreting planet has attained. These effects should be accentuated if the moon accreted in earth orbit (Kaula, 1971) and, provided that accretion is sufficiently rapid, might well lead to large scale melting of the outer regions of the moon.

Whatever caused the initial melting on the moon, a number of models have been proposed to explore the possible consequences of such an early melting history (e.g., Smith *et al.*, 1970; Papanastassiou and Wasserburg, 1971a; Gast and McConnell, 1972) in terms of the petrologic, trace-element, and isotopic characteristics of lunar materials. Most of these models require an upward concentration of radioactivities of $^{40}$K, U, and Th in the first extensive heating event. One of the salient features of any thermal history model of the moon is the chronology represented by the mare basalts and their generation. As earlier noted, basalt production appears to be due to internal heat generation in the moon. The range of mare basalt ages of ~3.1–3.7 AE requires that a *second* melting episode on the moon reach a peak about a billion years after the formation of the moon and be terminated at about 3 AE ago, with a time span of approximately 0.6 AE. In the thermal regime of a moon where U and Th play a more dominant role than in chondrites or the earth, as can be inferred by the low K/U ratios of lunar materials, the thermal history required by the mare basalt ages can only be met by simultaneously satisfying several assumptions (initial temperature distribution, extent of differentiation, amount of radioactivity and its distribution).

In a recent model of differentiation of the moon (Murthy *et al.*, 1971b) we have suggested that the early surficial melting of the moon would result in the sinking of an Fe–FeS eutectic melt into the lunar interior, until prevented in the cold interior at a depth of ~250 km. In this case thermodynamic calculations suggest (Hall and Murthy, 1971) that appreciable amounts of $^{40}$K may be carried down in the Fe–FeS liquids. Thus, in the initial melting of the outer layers of the moon, a decoupling between the radioactivities of U, Th, and $^{40}$K can result. The heat from U and Th would be produced near the surface and would be more effectively lost by conduction, whereas the heat from buried $^{40}$K in the trapped Fe–FeS layer could cause melting and production of basaltic liquids on the time scale represented by the mare basalt ages. We therefore suggest that mare basalt generation is related to the presence of $^{40}$K in lenses of Fe–FeS, which we term "fescons," underneath the mare areas. A consequence of this model which we suggested prior to the Apollo 15 mission is that mare areas should be characterized by high heat flow. We note with satisfaction the high heat flow measurements in the Hadley Rille area at the edge of Mare Imbrium (Langseth, 1972) and await future heat flow measurements.

*Acknowledgments*—This research was performed under support of NASA grant NGR 24-005-223. We are indebted to Professor R. O. Pepin for many valuable discussions on lunar science. To our wives, who patiently and sympathetically assumed the positions of "lab-widows," we are especially grateful.

## References

Chappell B. W., Compston W., Green D. H., and Ware N. G. (1972) Chemistry, geochronology, and petrogenesis of lunar sample 15555. *Science* **175,** 415–416.

Compston W., Berry H., Vernon M. J., Chappell B. W., and Kaye M. J. (1971) Rubidium-strontium chronology and chemistry of lunar material from the Ocean of Storms. *Proc. Second Lunar Sci. Conf., Geochim. Cosmochim. Acta* Suppl. 2, Vol. 2, pp. 1471–1485. MIT Press.

Compston W., Vernon M. J., Berry H., Rudowski R., Gray C. M., Ware N., Chappell B. W., and Kay M. (1972) Age and petrogenesis of Apollo 14 basalts (abstract). In *Lunar Science—III* (editor C. Watkins), pp. 151–154, Lunar Science Institute Contr. 88.

Dence M. R., Plant A. G., and Traill R. J. (1972) Impact-generated shock and thermal metamosphism in Fra Mauro samples (abstract). In *Lunar Science—III* (editor C. Watkins), pp. 174–177, Lunar Science Institute Contr. No. 88.

Eberhardt P., Geiss J., Grögler N., Krähenbühl U., Mörgeli M., and Stettler A. (1971) Potassium-argon age of Apollo 11 rock 10003. *Earth Planet. Sci. Lett.* **11,** 245–247.

El Goresy A., Ramdohr P., and Taylor L. A. (1972) Fra Mauro crystalline rocks: Petrology, geochemistry, and subsolidus reduction of the opaque minerals (abstract). In *Lunar Science—III* (editor C. Watkins), pp. 224–227, Lunar Science Institute Contr. No. 88.

Ford C. E., Humphries D. J., Wilson G., Dixon D., Biggar G. M., and O'Hara M. J. (1972) Experimental petrology of high alumina basalt, 14310, and related compositions (abstract). In *Lunar Science—III* (editor C. Watkins), pp. 274–277, Lunar Science Institute Contr. No. 88.

Gast P. W. and McConnell R. K. Jr. (1972) Evidence for initial chemical layering in the moon (abstract). In *Lunar Science—III* (editor C. Watkins), pp. 289–292, Lunar Science Institute Contr. No. 88.

Graham A. L. and Ringwood A. E. (1971) Lunar basalt genesis: the origin of the europium anomaly. *Earth Planet. Sci. Lett.* **13,** 105–115.

Hall H. T. and Murthy V. R. (1971) The early chemical history of the earth: Some critical elemental fractionations. *Earth Planet. Sci. Lett.* **11,** 239–244.

Kaula W. M. (1969) Interpretation of lunar mass concentrations. *Phys. Earth Planet. Interiors* **2,** 123–137.

Kaula W. M. (1971) Dynamical aspects of lunar origin. (*Abstr.*) *Trans. Amer. Geophys. Union* **52,** 266.

Langseth M. G. (1972) The Apollo 15 lunar heat flow measurement (abstract). In *Lunar Science—III* (editor C. Watkins) pp. 475–478, Lunar Science Institute, Contr. No. 88.

Latham G., Ewing M., Dorman J., Lammlein D., Press F., Töksoz N., Sutton G., Duennebier F., and Nakamura Y. (1971) Moonquakes. *Science* **174,** 687–692.

LSPET (Lunar Sample Preliminary Examination Team) (1971) Preliminary examination of lunar samples from Apollo 14. *Science* **173,** 681–693.

LSPET (Lunar Sample Preliminary Examination Team) (1972) The Apollo 15 lunar samples: A preliminary description. *Science* **175,** 363–375.

Murthy V. R. (1970) Model ages and leaching studies in the Rb–Sr system in Apollo 11 and 12 lunar fines materials. (*Abstr.*) *Trans. Amer. Geophys. Union* **51,** 584.

Murthy V. R., Evensen N. M., Jahn Bor-ming, and Coscio M. R. Jr. (1971a) Rb–Sr ages and elemental abundances of K, Rb, Sr, and Ba in samples from the Ocean of Storms. *Geochim. Cosmochim. Acta* **35,** 1139–1153.

Murthy V. R., Evensen N. M., and Hall H. T. (1971b) A model of early lunar differentiation. *Nature* **234,** 267.

Murthy V. R., Evensen N. M., Jahn Bor-ming, Coscio M. R. Jr., Dragon J. C., and Pepin R. O. (1972) Rubidium-strontium and potassium-argon age of lunar sample 15555. *Science* **175,** 419–421.

Papanastassiou D. A. and Wasserburg G. J. (1970) Rb–Sr ages from the Ocean of Storms. *Earth Planet. Sci. Lett.* **8,** 269–278.

Papanastassiou D. A., Wasserburg G. J., and Burnett D. S. (1970) Rb–Sr ages of lunar rocks from the Sea of Tranquility. *Earth Planet. Sci. Lett.* **8,** 1–19.

Papanastassiou D. A. and Wasserburg G. J. (1971a) Lunar chronology and evolution from Rb–Sr studies of Apollo 11 and 12 samples. *Earth Planet. Sci. Lett.* **11,** 37–62.

Papanastassiou D. A. and Wasserburg G. J. (1971b) Rb–Sr ages of igneous rocks from the Apollo 14 mission and the age of the Fra Mauro formation. *Earth Planet. Sci. Lett.* **12,** 36–48.

Papanastassiou D. A. and Wasserburg G. J. (1972) Rb–Sr age of a Luna 16 basalt and the model age of luna soils. *Earth Planet. Sci. Lett.* **13,** 368–374.

Pepin R. O., Bradley J. G., Dragon J. C., and Nyquist L. E. (1972) K–Ar dating of lunar soils: Apollo 12, Apollo 14, and Luna 16 (abstract). In *Lunar Science—III* (editor C. Watkins), pp. 602–605, Lunar Science Institute Contr. No. 88.

Schmitt H. H., Trask N. J., and Shoemaker E. M. (1967) Geologic map of the Copernicus quadrangle of the moon. Map I-515, U.S. Geological Survey.

Smith J. V., Anderson A. T., Newton R. C., Olsen E. J., and Wyllie P. J. (1970) A petrologic model for the moon based on petrogenesis, experimental petrology, and physical properties. *J. Geol.* **78,** 381–405.

Wasserburg G. J. and Papanastassiou D. A. (1971) Age of an Apollo 15 mare basalt; lunar crust and mantle evolution. *Earth Planet. Sci. Lett.* **13,** 97–104.

York D. (1966) Least square fitting of a straight line. *Can. J. Phys.* **44,** 1079–1086.

Proceedings of the Third Lunar Science Conference
(Supplement 3, *Geochimica et Cosmochimica Acta*)
Vol. 2, pp. 1515–1530
The M.I.T. Press, 1972

# Rb-Sr systematics for chemically defined Apollo 14 breccias

L. E. NYQUIST, N. J. HUBBARD, and P. W. GAST
NASA Manned Spacecraft Center, Houston, Texas 77058

and

S. E. CHURCH
National Research Council, Houston, Texas 77058

and

B. M. BANSAL and H. WIESMANN
Lockheed Electronics Company, Houston, Texas 77058

**Abstract**—Chemical analyses and whole rock Rb–Sr systematics of selected Apollo 14 breccias and breccia fragments from soil 14161 show that at least breccias of high metamorphic grade underwent thermal metamorphism 3.94 ± 0.12 AE ago (1 Aeon = AE = $10^9$ years). As the Imbrium impact is a known source of energy we identify the 3.9 AE age with this event. It is suggested that the Rb–Sr systematics of these breccias are dominated by granophyre patches rich in K, Rb, and Ba, prompting a comparison of chronologies for KREEP breccias and rock 12013. A two-stage model comparison of KREEP breccias and 12013 "granites" using initial $^{87}Sr/^{86}Sr$ ratios yields total model ages of 4.40 AE and 4.47 AE, respectively. This age difference vanishes if equal metamorphic ages of 3.95 AE are assumed. In this case coincident two-stage model ages of 4.42 AE result. A whole rock age of 4.0 ± 0.2 AE and model age of 4.43 AE is obtained from 24 KREEP materials analyzed to date. The possibility that the chronologies of KREEP and "granites" are related merits serious consideration.

## INTRODUCTION

THE RB AND SR SYSTEMATICS were determined for eight lithic fragments from the coarse fines of Apollo 11 (10085), Apollo 12 (12033), and Apollo 14 (14161); for bulk soils from Luna 16, Apollo 14, and Apollo 15 and for "whole rocks" from Apollo 14 and the Apollo 15 anorthosite 15415. These samples were analyzed as part of our program of chemical and isotopic analysis of lunar materials. In this survey approach, it is our goal to identify important lunar rock types by their major and trace element chemistry and utilize Rb and Sr systematics to obtain a chronology for events which establish or modify their chemical composition. Thus, to obtain the most comprehensive sampling we have emphasized analysis of small lithic fragments (3–50 mg) picked from samples of coarse fines from the various mission sites. Chemical data on these samples are reported in the accompanying paper (Hubbard *et al.*, 1972). This report applies the "whole rock" Rb–Sr technique to these chemically defined Apollo 14 breccias. Whereas the Rb–Sr systematics are inherently less precisely defined than for the internal mineral isochron technique, important complementary information is obtained as a result of examining a larger number of individual rock fragments. We have established that the Apollo 14 breccias were open to Rb migration 3.94 ± 0.12 AE ago and identify this event with thermal metamorphism and brecciation produced by the Imbrium impact.

## ANALYTICAL PROCEDURE AND RESULTS

Rb and Sr analytical results are shown in Table 1. Basic chemical procedures and ion exchange chromatography were the same as described by Gast *et al.* (1970). The Sr fraction was eluted with

Table 1 Rb and Sr analytical results.

| Sample | Wt. (mg) | Rb[a] (ppm) | Sr[b] (ppm) | $^{87}Rb/^{86}Sr$ (× 100) | $^{87}Sr/^{86}Sr$[c] | $T_{BABI}$ |
|---|---|---|---|---|---|---|
| I. KREEP materials | | | | | | |
| 14161,35,2 | 55.0 | 13.1 ± 0.4 | 173 ± 5 | 21.9 ± 0.9 | 0.71322 ± 7 | 4.51 ± 0.18 |
| 14161,35,3 | 27.8 | 3.43 ± 0.11 | 182 ± 5 | 5.46 ± 0.25 | 0.70383 ± 7 | 5.98 ± 0.28 |
| 14161,35,4 | 27.6 | 15.4 ± 0.5 | 198 ± 5 | 22.5 ± 0.9 | 0.71402 ± 6 | 4.63 ± 0.18 |
| 14161,35,5 | 30.1 | 14.9 ± 0.2 | 172 ± 5 | 24.2 ± 0.8 | 0.71547 ± 8 | 4.54 ± 0.14 |
| 14161,35,6 | 39.8 | 17.2 ± 0.2 | 184 ± 5 | 27.0 ± 0.9 | 0.71654 ± 6 | 4.51 ± 0.15 |
| 14006,3 | 263.4 | 6.16 ± 0.08 | 181 ± 2 | 9.89 ± 0.16 | 0.70616 ± 7 | 4.98 ± 0.10 |
| 14310,130 | 176.7 | 13.0 ± 0.2 | 190 ± 2 | 19.7 ± 0.3 | 0.71177 ± 32 | 4.49 ± 0.13 |
| 14310[d] | 13.2 | 12.1 | 192 | 18.2 | 0.71041 ± 6 | 4.37 |
| 14307,26,2 | 68.4 | 15.5 ± 0.2 | 166 ± 5 | 27.0 ± 0.9 | 0.71573 ± 15 | 4.30 ± 0.14 |
| 14068,3 | 106.4 | 14.7 ± 0.2 | 140 ± 2 | 30.4 ± 0.4 | 0.71780 ± 43 | 4.30 ± 0.11 |
| 14301,48,1 | 64.8 | 22.0 ± 0.3 | 187 ± 2 | 33.9 ± 1.1 | 0.72042 ± 24 | 4.39 ± 0.15 |
| 14307,26,1 | 6.8 | 16.2 ± 0.5 | 194 ± 3 | 24.1 ± 0.9 | 0.71532 ± 20 | 4.68 ± 0.18 |
| 14163,65,2 | 5.5 | 18.5 ± 0.5 | 190 ± 3 | 28.3 ± 1.0 | 0.71736 ± 56 | 4.50 ± 0.20 |
| II. Bulk soils | | | | | | |
| 14163,65,II | 218.7 | 15.1 ± 0.2 | 187 ± 2 | 23.4 ± 0.3 | 0.71454 ± 22 | 4.59 ± 0.09 |
| 14163[d] | 30.0 | 14.68 | 182.4 | 23.3 | 0.71414 ± 8 | 4.54 ± 0.04 |
| 14163[e] | — | 14.61 | 182.1 | 23.2 | 0.71448 | 4.65 |
| 14163[f] | — | 15.4 | 185.5 | 24.0 | 0.71485 | 4.60 |
| Luna 16 | | | | | | |
| L16-A3 | 15.47 | 1.82 ± 0.08 | 280 ± 3 | 1.88 ± 0.08 | 0.70020 ± 14 | 4.09 ± 0.58 |
| L16-G3 | 15.70 | 1.90 ± 0.08 | 298 ± 3 | 1.85 ± 0.08 | 0.70028 ± 12 | 4.45 ± 0.54 |
| L16-A37 | 21.49 | 1.87 ± 0.04 | 282 ± 3 | 1.92 ± 0.05 | 0.70029 ± 15 | 4.35 ± 0.40 |
| L16-G37 | 24.00 | 1.94 ± 0.04 | 306 ± 3 | 1.83 ± 0.05 | 0.70022 ± 5 | 4.28 ± 0.33 |
| A-2[d] | 3.2 | 1.79 | 271 | 1.91 | 0.70009 ± 1 | 4.06 ± 0.16 |
| G-2[d] | 3.4 | 1.88 | 273 | 2.00 | 0.70016 ± 8 | 4.12 ± 0.30 |
| 15101,90 | 53.7 | 4.52 ± 0.04 | 146 ± 1 | 9.0 ± 0.1 | 0.70514 ± 31 | 4.67 ± 0.24 |
| III. Whole rock | | | | | | |
| 14053,50 B-1 | 13.1 | 2.22 ± 0.07 | 99 ± 3 | 6.50 ± 0.29 | 0.70351 ± 35 | 4.72 ± 0.42 |
| 14053[d] | 40.0 | 2.04 | 103 | 5.72 | 0.70276 ± 7 | 4.60 |
| IV. Anorthosites | | | | | | |
| 10085,11,47 | 3.05 | 0.27 ± 0.27 | 151 ± 6 | 0.51 ± 0.52 | 0.69897 ± 29 | — |
| 10085,11,146 | 5.17 | 1.29 ± 0.17 | 161 ± 7 | 2.32 ± 0.32 | 0.70110 ± 110 | — |
| 12033,97,7 | 38.17 | 2.91 ± 0.10 | 345 ± 8 | 2.44 ± 0.09 | 0.69968 ± 11 | — |
| 14161,35,1 | 11.07 | 0.32 ± 0.08 | 164 ± 2 | 0.69 ± 0.17 | 0.69960 ± 23 | — |
| 15415,1 | 53.25 | 0.17 ± 0.02 | 179 ± 2 | 0.28 ± 0.03 | 0.69938 + 9 − 18 | — |
| 15415-B[d] | 3.7 | 0.16 | 173 | 0.27 | 0.69914 ± 5 | — |

[a] Spike $^{87}Rb > 99\%$; concentration calibrated with SRM-727; uncertainties include spiking, blanks, spike composition, and concentration.

[b] Spike $^{84}Sr > 99\%$; concentration calibrated with SRM-987.

[c] Normalized to $^{88}Sr/^{86}Sr = 8.3752$ assuming $(^{87}Sr/^{86}Sr)_N = {}^{87}Sr/^{86}Sr\ (8.3752\ {}^{86}Sr/^{88}Sr)^{1/2}$, N = normalized, other ratios are measured values corrected for spike.

[d] Papanastassiou and Wasserburg (1971a; 1971b; 1972).

[e] Tatsumoto *et al.* (1972).

[f] Barnes *et al.* (1972).

dilute $HNO_3$ through a Biorad ZP-1 column to improve separation from Rb (M. Coscio, personal communication, 1971). Rb detected at the beginning of the Sr analyses was normally of natural composition and is attributed to a small handling blank. Samples were separately spiked for Rb and Sr and analyzed using an NBS single-focusing, 6-in. 60° sector mass spectrometer. Triple Re filaments were used for the Rb analyses. Analyses of SRM-727 prior to and after the sample sequence yielded $^{85}Rb/^{87}Rb = 2.608 \pm 0.002$ in exact agreement with the direct measurements of Catanzaro *et al.* (1969) indicating an absolute instrument bias of 0.6%. Error contributions from variations in instrumental bias are certainly negligible compared to those from other sources. The largest error contribution for the Rb concentrations derives from an uncertainty of $\pm 0.03$ ml in pipetting the Rb spike. Blanks averaged $1 \pm 1$ ng.

The Sr spike concentration was initially calibrated with a shelf standard which had been titrated against an EDTA solution previously calibrated against a primary standard $ZnCl_2$ solution (Gast *et al.*, 1970). The concentration of the shelf standard determined with the recently available SRM-987 differed from the titration value by 1.1%. The entries in Table 1 have been corrected for this bias and thus differ from those in the original report (Nyquist *et al.*, 1972). The Sr blanks averaged $30 \pm 20$ ng. Uncertainties for Rb and Sr concentrations in Table 1 include contributions from pipetting and blanks as well as from spike composition and concentration.

The Sr analyses were performed in a manner similar to that described by Papanastassiou and Wasserburg (1971b). Normally, samples were loaded onto an oxidized Ta filament and left for 10–12 hours in the instrument at a filament temperature of 1100°C to reduce $^{85}Rb$ below the detection limits of the instrument ($^{85}Rb/^{87}Sr \lesssim 0.02\%$ at the end of an analysis). Occasionally the Sr fraction contained significant impurities and a $^{85}Rb$ signal was detected throughout the analysis. In this case the $^{87}Sr/^{86}Sr$ was corrected assuming $^{85}Rb/^{87}Rb$ equal to that measured as the filament was being initially brought to the 1100°C temperature.

The mass spectrum was scanned by voltage stepping (NBS Tech. Note 546). Although voltage scanning introduced relatively large mass discriminations ($\sim 0.5\%$/mass unit) this effect was adequately compensated for by normalizing to $^{88}Sr/^{86}Sr = 8.3752$. Apparent voltage drift during a typical set of ten scans was $< 20\%$ of the flat top peak width. Tailing of mass-88 towards mass-87 was found to be pressure dependent; analyses were performed at analyzer section pressures $< 10^{-8}$ Torr for which the baseline between masses 88 and 87 differed by $< 0.02\%$ of the mass-87 peak from that between masses 87 and 86.

Ion beam intensities were registered by a vibrating reed electrometer, digitized, and recorded on punched paper tape for subsequent computer analysis. The $10^{11}$ ohm input resistor was used and data recorded on a single electrometer range at $^{88}Sr$ beam intensities of $3 \times 10^{-12}$ amp to $3 \times 10^{-11}$ amp. Beam integration times of five seconds were used with 5 second delay between $^{86}Sr$ and $^{87}Sr$ and 25 second delay between $^{88}Sr$ and $^{86}Sr$. With these delay times transient voltage contributions to succeeding ion beams were negligible. Secondary electron contributions to the measured currents were not detected.

Nominal operating procedure called for recording at least ten sets of ten $^{87}Sr/^{86}Sr$ and $^{88}Sr/^{86}Sr$ ratios at beam intensities of $\sim 10^{-11}$ amp. When these conditions were met, precision as defined by twice the standard deviation of the mean

$$\sigma_m = \left(\frac{\Sigma(r_i - \bar{r})^2}{n(n - 1)}\right)^{1/2}$$

where $r$ = average ratio for $i$th set, and $\bar{r}$ = mean of $n$ sets was about 0.01–0.02%. Unstable ion beams were encountered for some samples producing significantly greater uncertainties. Analyses of the C.I.T. seawater standard (Table 2) show our results to be biased with respect to those of Papanastassiou and Wasserburg (1971b), and Compston *et al.* (1971a) by about $(+0.00012 \pm 0.00007)$ and $(-0.00011 \pm 0.00007)$, respectively, and agree within the analytical uncertainty with those of Murthy *et al.* (1971). Our results on the E. and A. standard as well as on lunar samples analyzed in the other laboratories are consistent with these biases.

Table 2. Strontium isotopic standard analyses

| Standard | Date | $^{87}Sr/^{86}Sr$ | $2\sigma m$ ($\times 10^5$) |
|---|---|---|---|
| Bowen Seawater[b] | 6–09–71 | 0.70927[a] | ±12 |
| (C.I.T.) | 6–10–71 | 0.70916 | ± 7 |
| | 6–11–71 | 0.70922 | ± 8 |
| | 9–15–71 | 0.70914 | ±10 |
| | 12–07–71 | 0.70920 | ±14 |
| Mean | | 0.70920 | ± 5 |
| C.I.T.[d] | | 0.70908 | ± 5 |
| A.N.U.[e] | | 0.70931 | ± 3 |
| U.M.[f] | | 0.70917 | ± 2 |
| Eimer and Amend | 7–07–71 | 0.70801 | ±14 |
| Standard[c] | 7–13–71 | 0.70808 | ±10 |
| U.S.G.S.[g] | | 0.70798 | ± 6 |
| A.N.U.[e] | | 0.70818 | ± 2 |
| U.M.[f] | | 0.70810 | ± 2 |

[a] Corrected for spike, measured $^{84}Sr/^{86}Sr = 0.667$; spike composition: $^{84}Sr = 99.14\%$, $^{86}Sr = 0.0015\%$, $^{87}Sr = 0.0007\%$, $^{88}Sr = 0.0064\%$.
[b] Courtesy of Dr. G. J. Wasserburg.
[c] Courtesy of Dr. C. Hedge.
[d] Papanastassiou and Wasserburg (1971b).
[e] Compston *et al.* (1971).
[f] Murthy *et al.* (1971)
[g] Hildreth and Henderson (1971).

## Discussion

The Rb–Sr systematics for the Apollo 14 KREEP breccias are shown in Fig. 1. One important conclusion which can be drawn is that these whole rock samples do not represent closed system evolution from $(^{87}Sr/^{86}Sr)_0 = 0.6991$ (BABI adjusted for our bias). This argument is not significantly affected by the ambiguity in initial $^{87}Sr/^{86}Sr$ ratios arising from possibly sampling different systems. A least squares fit to all the data, using a York (5) program (York, 1966) yields as intercept $(^{87}Sr/^{86}Sr)_I = 0.7005 \pm 0.0003$. This result is largely determined by four samples which lie distinctly off the 4.6 AE reference isochron through BABI. Thus assuming the moon was formed about 4.6 AE ago these samples show clear evidence of fractionation of Rb from Sr on a scale of at least 1–5 mm. In particular, samples 14161,35,3 and 14006,3 have model ages $T_{BABI} = 6.0 \pm 0.3$ AE and $5.0 \pm 0.1$ AE, respectively. As direct measurement of the $^{87}Sr/^{86}Sr$ ratio of the Apollo 15 anorthosite 15415 show $(^{87}Sr/^{86}Sr)_0$ for the moon to be indistinguishable from BABI (Papanastassiou and Wasserburg, 1971c) these high model ages must be the result of a decrease in the Rb/Sr ratio in these samples after their initial crystallization. As shown in Table 1, the Sr concentrations in these samples are the same as in the other KREEP materials while the Rb concentrations are a factor 3–5 lower. This pattern of volatile element depletion is discussed by Gibson and Hubbard (1972). Similarly, the Rb/Sr ratio must have increased for samples 14068,3 and 14301,48,1.

Figure 2 shows concentrations of the volatile elements K and Rb in representative lunar materials normalized to the geochemically coherent but refractory Ba. With the exception of three Apollo 14 samples, 14161,35,3, 14006,3, and 14161,35,2 the KREEP materials trend along the line K/Rb = 325. Gibson and Hubbard (1972) discuss the

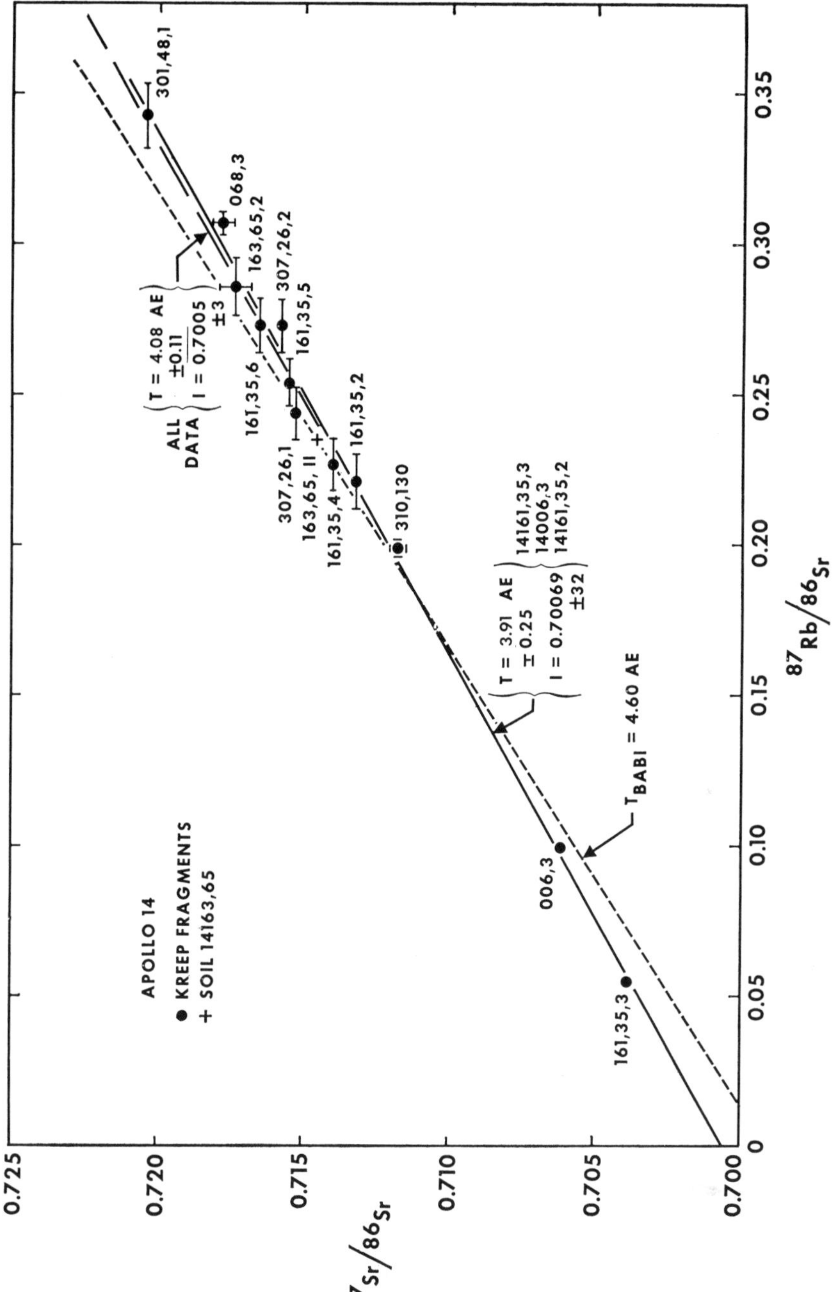


Fig. 1 Rb–Sr diagram for KREEP breccias. Isochrons shown are the best fit to the data as determined by the York program for (a) all data and (b) three samples evidencing volatile loss of Rb. A 4.6 AE reference isochron is also shown.

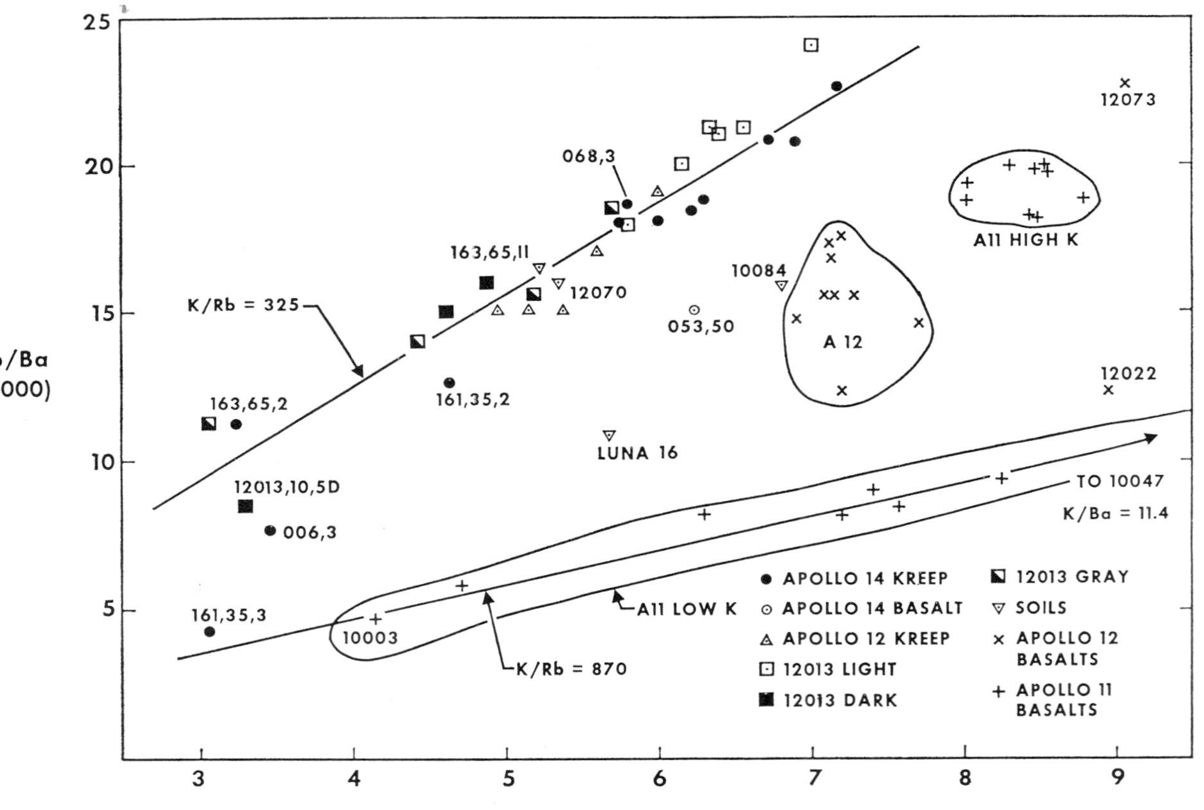


Fig. 2. Rb/Ba versus K/Ba for representative lunar materials. KREEP data from Hubbard *et al.* (1972), 12013 data from Schnetzler *et al.* (1970) and Hubbard *et al.* (1970). The K concentrations measured by Schnetzler *et al.* (1970) are systematically ~10% lower than those measured in this laboratory. The other data are from several authors reporting at the First and Second Lunar Science Conferences.

evidence for volatilization loss of K and Rb from the latter samples during thermal metamorphism; we point out another aspect of the K, Rb, and Ba systematics seen especially in Fig. 2. Straight lines on a diagram of this type are mixing lines for two components containing the three elements in different *proportions* as distinct from *concentrations* (Reynolds, 1963; cf. Funk *et al.*, 1967). This is illustrated in Fig. 2 by the striking coincidence of 12013 data with the KREEP data although the concentrations of K, Rb, and Ba in the light phase of 12013 are a factor of 10 higher than in KREEP. This similarity between KREEP and 12013 in Fig. 2 may suggest a genetic relationship between these two types of material; discussion of this point is deferred.

Here we note that with the exception of the three low Rb, high K/Rb samples, the pattern of Fig. 2 suggests domination by two phases. Bulk soil sample 14163,65,II plots to the left of most of the samples with normal K and Rb concentrations suggesting that our sampling was biased in favor of the high K/Ba end member. Sample 14163,65,2 consisting of several large fragments and having an unusually high Ba concentration plots near the extreme left of the diagram and is apparently enriched in the second end member. Thus the two phases must have K/Ba $\lesssim$ 3.3, and K/Ba $\gtrsim$ 7, respectively.

The similarity between the KREEP and 12013 patterns in Fig. 2 suggests the nature of the required two phases. The light phase of 12013 is permeated by an alkali feldspar and silica-rich granitic component that is also common in the dark phase as spherical to ovoid fillings (Drake *et al.*, 1970). Similar granophyre patches are present in our KREEP breccias (C. Meyer, personal communication). The chemical similarity between 12013 "granites" and the high silica glass mesostasis phase (Roedder and Weiblen, 1970) also called "quintessence" (Papanastassiou and Wasserburg, 1971a), was noted by Drake *et al.* (1970). Gancarz *et al.* (1971) have identified "silica-poor" glasses with high K concentrations and K/Ba $\sim$2 and "silica-rich" glasses with factor of $\sim$3 lower K concentrations and K/Ba $\sim$20 in Apollo 14 crystalline rocks. Thus we suggest that the K, Rb, and Ba systematics for Apollo 14 breccias can be explained by heterogeneity of two types of granophyres on a scale of at least several millimeters. This heterogeneity, in turn, can be explained by flowing of the quintessence melt or by vapor deposition (Gibson and Hubbard, 1972). In this interpretation, the three low Rb, high K/Rb samples are depleted in both granophyre types.

The above considerations and those of Gibson and Hubbard (1972) show that the Rb–Sr systematics of Fig. 1 have been affected by a metamorphic event that caused redistribution of Rb. Complete resetting of the Rb–Sr clock requires homogenization of the Sr isotopic composition which is most likely to have occurred in breccias of high metamorphic grade. Thin sections of our samples were classified by J. Warner according to degree of metamorphism. Figure 2 shows the Rb–Sr data for four dark breccias of grade 6 or above in Warner's classification (Warner, 1972). Included are the three breccias showing Rb depletion and breccia 14068. A least squares fit to the data yields an apparent age of 3.94 $\pm$ 0.12 AE and $(^{87}Sr/^{86}Sr)_I = 0.70067 \pm 0.00019$.

An ambiguity in age remains because $(^{87}Sr/^{86}Sr)_I$ may have differed among the various breccias. Although the data fall along a straight line within the analytical uncertainty, fulfilling the usual criteria for a whole rock isochron, this is not sufficient

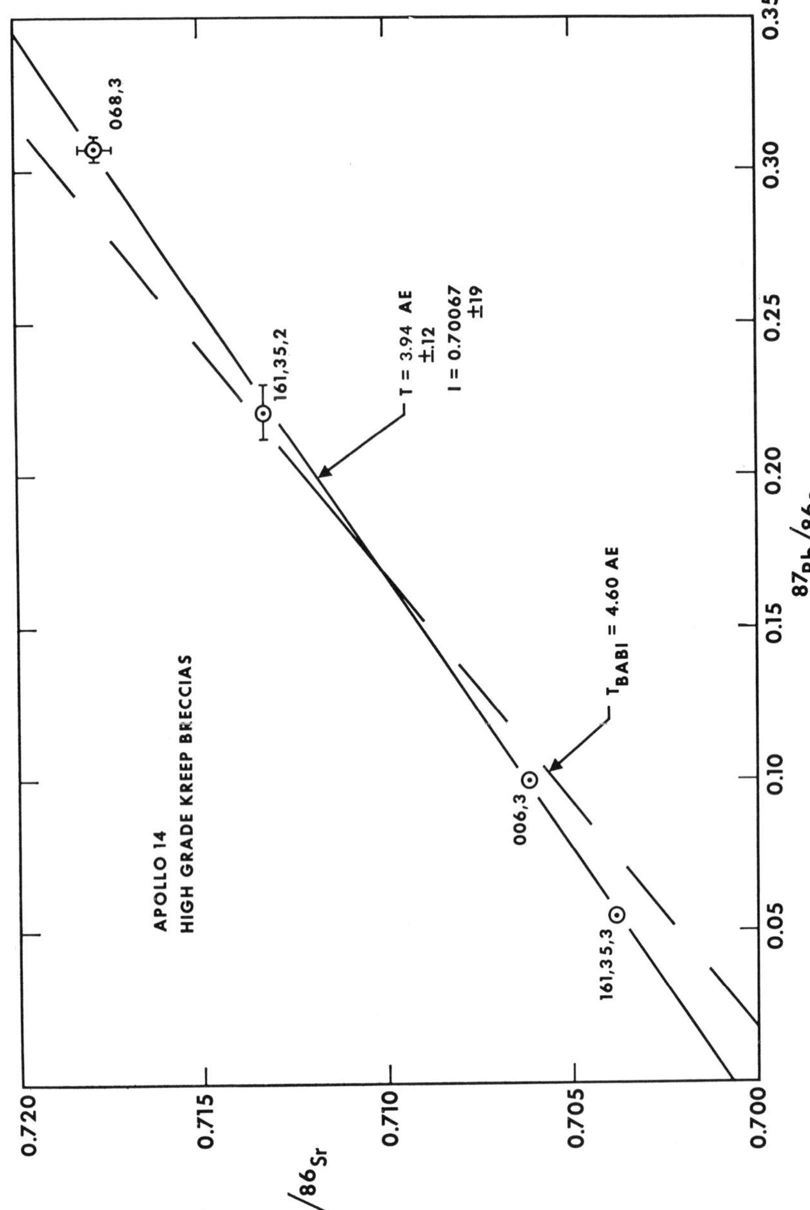


Fig. 3. Rb–Sr evolution diagram for four high grade KREEP breccias. A 4.6 AE reference isochron is also shown.

for a real age. As discussed by the Lunatic Asylum (1970), an erroneous "age" can be generated by evolution of a system which was originally a mixture of two systems of differing $(^{87}Sr/^{86}Sr)_I$. However, it can be strongly argued that this effect probably has not produced an error in excess of the stated uncertainty. Although the Apollo 14 crystalline rocks fall into two populations of $(^{87}Sr/^{86}Sr)_I$ with values of about 0.6995 and 0.7004 (Papanastassiou and Wasserburg, 1971b), the chemical composition of these breccias definitely show they are to be associated with the latter group. Further, $(^{87}Sr/^{86}Sr)_I$ obtained from our whole rock isochron (0.7006 $\pm$ 0.0003 corrected for bias) is characteristic of this group. If our isochron had resulted from a mixed system it is likely that the end member with the higher Rb/Sr ratio would also have had the higher $^{87}Sr/^{86}Sr$ ratio in which case the "isochron" generated would have an intercept below that of either of the two mixing end members. Moreover, although the Rb–Sr data of the four high grade breccias is most internally consistent, the age is not particularly affected by the choice of data as long as samples 14161,35,3 and 14006 are included in the least squares analysis. We obtain 3.91 AE, 3.94 AE, 4.08 AE, and 4.02 AE, respectively, by considering the data in the following groups: (a) three samples with chemical evidence of Rb depletion (Fig. 1); (b) four high-grade breccias (Fig. 3); (c) all Apollo 14 KREEP data (Fig. 1); and (d) all Apollo 12 and 14 data as shown in Fig. 4. The latter two data sets show more scatter about the correlation line but contain sufficient data that differences in $(^{87}Sr/^{86}Sr)_I$ are expected to be averaged out. Finally, the total Ar gas retention age of 14006 is 3.8 AE (Bogard and Nyquist, 1972) which, although a lower limit because of possible gas loss, is probably not greatly in error as shown by K–Ar systematics of Apollo 14 samples (Huneke *et al.*, 1972). Breccias 14006 and 14161,35,3 are, of course, of most interest in the present context as they show the clearest evidence of K and Rb migration. Thus we identify the age of 3.94 $\pm$ 0.12 AE with a metamorphic event most likely associated with the Imbrium event and an accompanying brecciation of these rocks.

There is increasing evidence for a close association between KREEP basalts and lunar "granites" as exemplified by the light portion of 12013. As shown earlier (Hubbard *et al.*, 1971) KREEP and the dark portion of 12013 are chemically indistinguishable. From consideration of Li, K, Rb, Ba, and REE abundances in the dark and light phases of 12013 Schnetzler *et al.* (1970) suggested that the two phases may be related. Figure 2 shows the striking similarity between the relative proportions of K, Rb, and Ba in Apollo 14 KREEP materials and rock 12013; in this representation, the similarity between KREEP and 12013 is extended to the granitic phase. As discussed above, this is probably a consequence of granophyre dominating the abundances of these elements in the KREEP breccias. The mixing model of Schonfeld (1972) shows a correlation between the KREEP and granitic components in Apollo 11, 12, and 14 soils with KREEP/"granite" $\simeq$ 35 $\pm$ 10. Meyer (1972) has pointed out that "granites" could be formed from KREEP magma or by thermal metamorphism of KREEP parent material; the second process has also been suggested by Hubbard *et al.* (1972).

In view of these chemical considerations suggesting a relationship between KREEP and "granites" a comparison of chronologies seems appropriate; particularly as the metamorphic ages (3.9–4.0 AE) of these two rock types suggest that the major

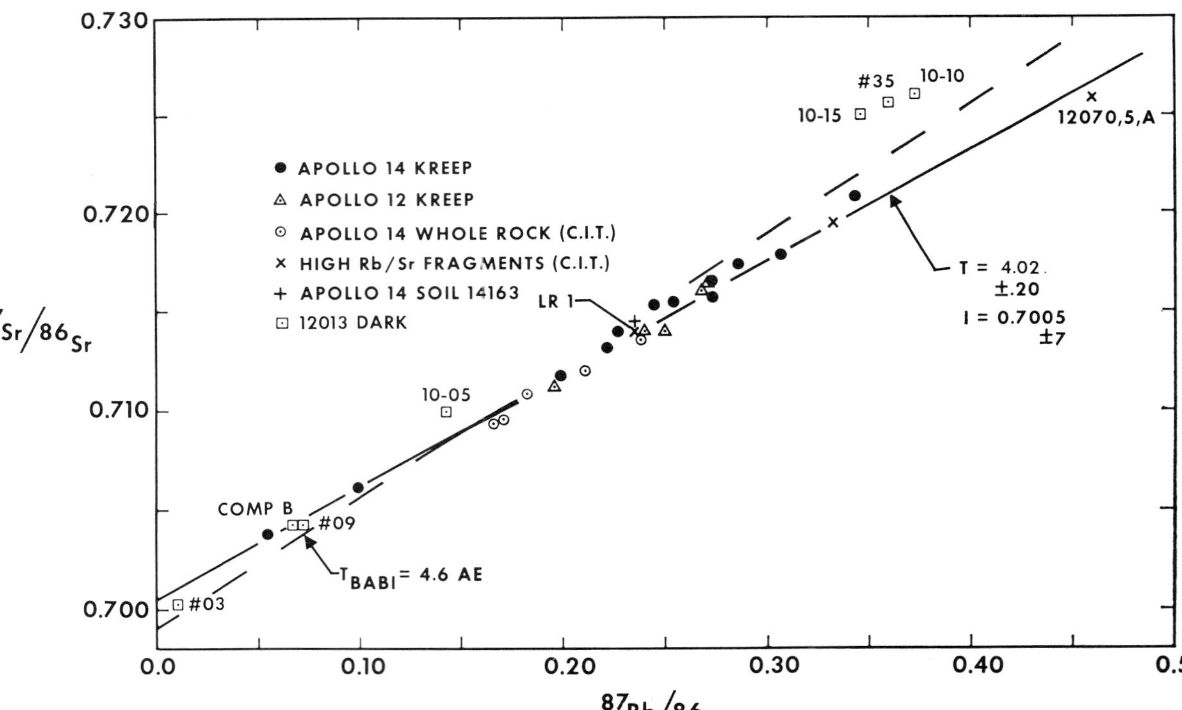


Fig. 4. Rb–Sr diagram for KREEP materials including lithic fragments from Apollo 11 and 12, Apollo 12 glasses, and Apollo 14 breccias and basalts. Data from the dark phase of 12013 are shown for comparison but were not included in the least squares regression. Data are from this investigation and Hubbard *et al.* (1970) and Papanastassiou and Wasserburg (1970a; 1970b; 1971a; 1971b). The average $^{87}Sr/^{86}Sr$ and $^{87}Rb/^{86}Sr$ yields a model age of 4.43 AE.

chemical composition was fixed even earlier. The $^{40}Ar$–$^{39}Ar$ ages of Turner (1970; 1971) provide the most precise comparison of the ages of dark and light lithologies interior to rock 12013. Turner's work yields an identical result (uncertainty $\pm 0.03$ AE) for light and dark phase; the K and Ca abundances obtained by him indicate his dark phase sample to have been essentially pure KREEP with little or no admixture of the K-rich granitic component. The internal Rb–Sr isochrons obtained by the Lunatic Asylum (1970) pertain primarily to the granitic portion; that of Bottino *et al.* (1971) to a mixture of phases; thus the Rb–Sr systematics of 12013 dark are largely undefined.

Data from whole rock and plagioclase samples from 12013 dark are plotted in Fig. 4 (Lunatic Asylum, 1970; Bottino, 1971; Hubbard *et al.*, 1971). These data do not give meaningful age information as evidenced by a whole rock age of $\simeq 5.0$ AE. Samples #10-10, #10-15, and #10-05 have essentially the KREEP trace element composition (Schnetzler *et al.*, 1970; Hubbard *et al.*, 1970). Comparatively high-Rb concentrations and $^{87}Rb/^{86}Sr$ ratios for #10-10, #10-15, and #35 suggest the high whole rock "age" results from migration of radiogenic $^{87}Sr$ during the formation of this rock. This would also explain the generally low model ages of samples from the granitic fraction ($T_{BABI} \leq 4.4$ AE in 7 of 8 cases) and is consistent with the local nature of the Sr isotope homogenization in this rock (Lunatic Asylum, 1970). The low Rb concentrations in #09 and #10-05 relative to other KREEP materials are consistent with our observations on 14161,35,3 and 14006 and suggest thermal metamorphism and Rb volatilization $\simeq 4.0$ AE ago. While it may be accidental that #09 falls near our KREEP isochron, this sample could represent material essentially uninfluenced by radiogenic $^{87}Sr$ from the light phase. Indeed this sample is described as "closest in character to the black end member" (Lunatic Asylum, 1970). Likewise Sr in plagioclase #03 may not have been completely equilibrated with that in surrounding area 9 dark material.

In Fig. 5 we attempt to use the Rb–Sr systematics of KREEP and 12013 granite to obtain information concerning the pre-3.9 AE history of these materials. We have plotted $^{87}Sr/^{86}Sr$ and $^{87}Rb/^{86}Sr$ for "granite," KREEP breccias, and Apollo 14 basalts as they would have been 3.88 AE ago. That is, we have corrected the present-day "whole rock" $^{87}Rb/^{86}Sr$ ratios for $^{87}Rb$ decay in the last 3.88 AE and plotted them against initial $^{87}Sr/^{86}Sr$ ratio corrected for $^{87}Sr$ evolution in the time interval between the internal isochron age of the rock and 3.88 AE ago. For the KREEP breccias we have chosen to plot the $(^{87}Sr/^{86}Sr)_I$ obtained from the four high grade breccias versus the average $^{87}Rb/^{86}Sr$ ratio for all analyzed KREEP fragments (Fig. 4). We justify this choice on the grounds that of the breccias analyzed by us, $^{87}Sr/^{86}Sr$ is most likely to be homogenized in those of high grade, whereas the average $^{87}Rb/^{86}Sr$ for KREEP fragments should most nearly approximate the closed system value. Further, the $^{87}Rb/^{86}Sr$ thus obtained is identical with that measured by us for soil 14163, and identical within the error limits shown in Fig. 5, to the average for four soils analyzed by Papanastassiou and Wasserburg (1971b). The correction for $^{87}Rb$ decay is about 5.5% for all samples. $(^{87}Sr/^{86}Sr)_I$ was adjusted by $+0.0002 \pm 0.0004$ for the KREEP breccias corresponding to the difference between $3.94 \pm 0.12$ and 3.88 AE. The magnitude of the corresponding correction for the 12013 data is shown in Fig. 5.

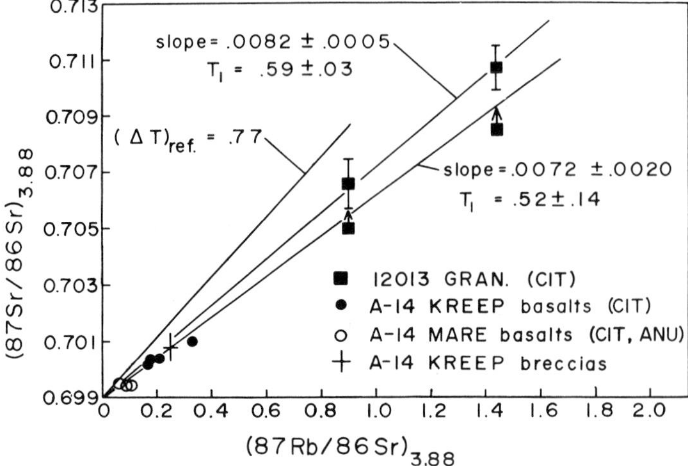


Fig. 5. $^{87}Rb/^{86}Sr$–I diagram for Apollo 14 crystalline rocks (Papanastassiou and Wasserburg, 1970; Papanastassiou and Wasserburg, 1971b; Compston *et al.*, 1971), KREEP breccias (this investigation) and granite samples 29 and 40 of 12013 (Lunatic Asylum, 1970). Data are normalized to values 3.88 AE ago. Arrows show the direction of the correction for 12013. $(^{87}Sr/^{86}Sr)_I$ for KREEP breccias has been corrected for a bias of +0.00012.

Isochrons on Rb–Sr diagrams similar to Fig. 5 yield the duration $T_1$ of first stage evolution of closed systems of total age $T_0 = T_1 + T_2$. The duration $T_2$ of the second stage is known from internal isochrons used to determine the $(^{87}Sr/^{86}Sr)_I$ values. Colinear systems on this diagram extrapolating to $(^{87}Sr/^{86}Sr)_I >$ BABI would imply cogenetic systems formed at $T_0$ incorporating Sr with an evolved isotopic composition and thus an earlier stage prior to $T_0$. The data in Fig. 5 do not allow the resolution of a third stage; thus we assume $T_0 = 4.65$ AE and $(^{87}Sr/^{86}Sr)_0 =$ BABI yielding a reference isochron $\Delta T = 4.65\text{ AE} - T_2$. Consequently, assuming $T_2 = 3.88$ AE means that closed system evolution starting from BABI at 4.65 AE ago would generate the $\Delta T = 0.77$ AE reference line in Fig. 5. Isochron slopes on this diagram can be given a time significance only if the data represent closed systems. In most respects this representation contains the same information as the traditional "whole rock" isochron but focuses attention on the first stage evolution. Another advantage is that failure of a particular "whole rock" point to satisfy the closed system criteria is reflected in the $^{87}Rb/^{86}Sr$ axis only.

Because it is not known whether the experimental data represent two stage closed systems the conclusions obtained from Fig. 5 are not absolute. Thus if it is assumed that the duration of the first stage is given by the reference isochron, $\Delta T = 0.77$ AE, it follows that the experimental data do not represent closed systems. For the KREEP breccias, one would conclude that the Rb/Sr ratio was increased by about a factor 1.5 during the 3.9 AE event. Similarly enrichments of Rb/Sr of a factor 1.3 at 4.0 AE would be deduced for 12013 "granites."

On the other hand, if it can be argued that the experimental data approximate closed system evolution, i.e., that no fractionation in Rb/Sr occurred during the 3.9–4.0 AE event(s), the model isochrons through BABI and the data for the various rock types yields an estimate of $T_1$, the duration of first stage evolution. The argument for no fractionation can best be made for 12013 and the KREEP breccias where we interpret the event at $T_2$ (4.0 AE or 3.94 AE, respectively) as metamorphic in nature. Thus averaging a sufficiently large number of "whole rock" data should minimize the effect of Rb redistribution. For the KREEP breccias one finds $T_1 = 0.52 \pm 0.14$. The "granite" data are more subject to violation of the closed system criteria but are colinear with BABI well within analytical uncertainty, suggesting a good approximation to closed system evolution. Thus for 12013 "granites" we estimate $T_1 = 0.59 \pm 0.03$ AE yielding total two stage model ages of 4.40 AE and 4.47 AE for KREEP breccias and "granites," respectively.

There is a suggestion in Fig. 5 that $T_1$ may be the same for KREEP and 12013 "granites." The data for these rock types are colinear within the error limits. The linearity is improved if no difference in $T_2$ is assumed for these materials. Whereas the resolution in age between the Apollo 14 mare-like basalts (3.95 AE) and crystalline KREEP rocks (3.88 AE) is well documented (Papanastassiou and Wasserburg, 1971b; Turner *et al.*, 1971), the resolution between the metamorphic ages of KREEP breccias and 12013 is less clear. The internal Rb–Sr ages of 12013 granitic fragments #29 and #40 is $4.00 \pm 0.05$ AE but is subject to a "mixing" ambiguity (Lunatic Asylum, 1970). The $^{40}Ar$–$^{39}Ar$ age of 12013 was originally reported as $3.87 \pm 0.07$ AE but was subsequently corrected to $4.03 \pm 0.07$ (Turner, 1970; 1971). Our whole rock isochron for all samples analyzed is $4.08 \pm 0.11$ AE (Fig. 4); Cliff *et al.* (1972) report an age of 4.01 AE for a radiogenic clast from Apollo 14 high grade breccia 14066. Huneke *et al.* (1972) report a disturbed $^{40}Ar$–$^{39}Ar$ age spectrum for fragmental rocks (breccias) but total $^{40}Ar$ ages of 3.92–3.93 AE. Crozaz *et al.* (1972) have confirmed the LSPET (1971) suggestion of $^{244}Pu$ fission Xe in breccia 14301 that, if produced by in situ decay, indicates this rock to be older than 12013 for which no $^{244}Pu$ fission Xe was detected (Lunatic Asylum, 1970; Alexander, 1970): Finally, if 12013 was thrown to the Apollo 12 site from the Fra Mauro region, as has been suggested by Schnetzler and Nava (1971), Turner's direct comparison of light and dark phase in this rock indicates equal metamorphic ages. If it is assumed that the metamorphic ages of KREEP breccias and 12013 "granites" are equal at 3.95 AE and that no Rb/Sr fractionation occurred at this time, then the model $T_1$ becomes $0.47 \pm 0.04$ AE yielding a common formation age of 4.4? AE in agreement with the previously suggested KREEP formation age of Schonfeld (1971; 1972) and the model age $T_{BABI} = 4.43$ AE calculated for the average $^{87}Sr/^{86}Sr$ and $^{87}Rb/^{86}Sr$ for the data in Fig. 4.

It is also interesting to note the comparison of these ages with those of Tatsumoto *et al.* (1971) using U–Th–Pb systematics. The $^{207}Pb/^{204}Pb$–$^{206}Pb/^{204}Pb$ age of 12013 fragments is 4.37 AE; the concordia intercept age for 12013 materials is 4.4 AE and the model ages for concordant fragment 12013,10,45 are $4.36 \pm 0.04$ AE. U–Th–Pb data for soils 12033 and 12034, which contain more than 50% KREEP (Meyer *et al.*, 1971) also fall on the 4.4 AE line generated by the 12013 fragments (Fig. 5, p. 1539, Tatsumoto *et al.*, 1971). Although objections to the interpretation of a common formation age for KREEP and "granites" can be raised it nevertheless is an important consideration in view of the diminishing probability of finding a pristine crystalline lunar rock as old as the chondritic meteorites. If it is assumed that the moon originated at the time of formation of chondritic meteorites, i.e., about 4.6 AE ago, the ages derived here suggest that there may be a significant time interval between the accretion of the moon and the formation of KREEP basalts by partial melting.

*Acknowledgments*—We are especially grateful to G. J. Wasserburg and D. A. Papanastassiou for providing the C.I.T. seawater standard and for freely sharing their knowledge of analytical techniques. Discussions with I. L. Barnes and W. Compston were very helpful. M. Coscio suggested the use of the ZP-1 column and provided the procedure. C. E. Hedge provided the Eimer and Amend $SrCO_3$ standard. C. Meyer provided stimulating discussion, petrography, and microprobe analyses of some of the breccias. J. Warner classified the breccias according to metamorphic grade. E. Schonfeld provided programming assistance and much helpful discussion, particularly concerning KREEP and "granite" chronologies.

## REFERENCES

Alexander E. C. (1970) Rare gases from stepwise heating of lunar rock 12013. *Earth Planet. Sci. Lett.* **9,** 201–207.

Barnes I. L., Carpenter B. S., Garner E. L., Gramlich J. W., Kuehner E. C., Machlan L. A., Mainethal E. J., Moody J. R., Moore L. J., Murphy T. J., Paulsen P. J., Sappenfield K. M., and Shields W. R. (1972) The isotopic abundance ratio and assay analysis of selected elements in Apollo 14 samples (abstract). In *Lunar Science—III* (editor C. Watkins), pp. 41–43, Lunar Science Institute Contr. No. 88.

Bogard D. D. and Nyquist L. E. (1972) Noble gas studies on regolith materials from Apollo 14 and 15. *Proc. Third Lunar Sci. Conf., Geochim. Cosmochim. Acta* Suppl. 3, Vol. 2. MIT Press.

Bottino M. L., Fullagar P. D., Schnetzler C. C., and Philpotts J. A. (1971) Sr isotopic measurements in Apollo 12 samples. *Proc. Second Lunar Sci. Conf., Geochim. Cosmochim. Acta* Suppl. 2, Vol. 2, pp. 1487 1491. MIT Press.

Catanzaro E. J., Murphy T. J., Garner E. L., and Shields W. R. (1969) Absolute isotopic abundance ratio and atomic weight of terrestrial rubidium. *J. Res.* **73A,** 511–516.

Cliff R. A., Lee-Hu C., and Wetherill G. W. (1972) K, Rb, and Sr measurements in Apollo 14 and 15 material (abstract). In *Lunar Science—III* (editor C. Watkins), pp. 146–147, Lunar Science Institute Contr. No. 88.

Compston W., Berry H., Vernon M. J., Chappell B. W., and Kaye M. J. (1971a) Rubidium-Strontium chronology and chemistry of lunar material from the Ocean of Storms. *Proc. Second Lunar Sci. Conf., Geochim. Cosmochim. Acta* Suppl. 2, Vol. 2, pp. 1471–1485. MIT Press.

Compston W., Vernon M. J., Berry H., and Rudowski R. (1971b) The age of the Fra Mauro formation: A radiometric older limit. *Earth Planet. Sci. Lett.* **12,** 55–58.

Crozaz G., Drozel R., Graf H., Hohenberg C. M., Monnin M., Ragan D., Ralston C., Seitz M., Shirck J., Walker R. M., and Zimmerman J. (1972) Evidence for extinct $Pu^{244}$: Implications for the age of the pre-Imbrium crust (abstract). In *Lunar Science—III* (editor C. Watkins), pp. 164–166, Lunar Science Institute Contr. No. 88.

Drake M. J., McCallum I. S., McKay G. A., and Weill D. F. (1970) Mineralogy and petrology of Apollo 12 sample 12013: A progress report. *Earth Planet. Sci. Lett.* **9,** 103–123.

Funk H., Podosek F., and Rowe M. W. (1967) Fissiogenic xenon in the Renazzo and Murray meteorites. *Geochim. Cosmochim. Acta* **31,** 1721–1732.

Gancarz A. J., Albee A. L., and Chodos A. A. (1971) Petrologic and mineralogic investigation of some crystalline rocks returned by the Apollo 14 mission. *Earth Planet. Sci. Lett.* **12,** 1–18.

Gast P. W., Hubbard N. J., and Wiesmann H. (1970) Chemical composition and petrogenesis of basalts from Tranquility Base. *Proc. Apollo 11 Lunar Sci. Conf., Geochim. Cosmochim. Acta* Suppl. 1, Vol. 2, pp. 1143–1164. Pergamon.

Gibson E. K. and Hubbard N. J. (1972) Volatile element depletion investigation on Apollo 11 and 12 lunar basalts by means of thermal volatilization. *Proc. Third Lunar Sci. Conf., Geochim. Cosmochim. Acta* Suppl. 3, Vol. 2. MIT Press.

Hildreth R. A. and Henderson W. T. (1971) Comparison of $^{87}Sr/^{86}Sr$ for seawater strontium and the Eimer and Amend $SrCO_3$. *Geochim. Cosmochim. Acta* **35,** 235–238.

Hubbard N. J., Gast P. W., and Wiesmann H. (1970) Rare earth, alkaline and alkali metal and $^{87/86}Sr$ data for subsamples of lunar sample 12013. *Earth Planet. Sci. Lett.* **9,** 181–184.

Hubbard N. J., Meyer C. Jr., Gast P. W., and Wiesmann H. (1971) The composition and derivation of Apollo 12 soils. *Earth Planet. Sci. Lett.* **10,** 341–350.

Hubbard N. J., Rhodes J. M., Gast P. W., Bansal B., Wiesmann H., and Church S. E. (1972) Nonmare basalts: Part II. *Proc. Third Lunar Sci. Conf., Geochim. Cosmochim. Acta* Suppl. 3, Vol. 2. MIT Press.

Huneke J. C., Podosek F. A., Turner G., and Wasserburg G. J. (1972) $^{40}Ar$–$^{39}Ar$ systematics in lunar rocks and separated minerals of lunar rocks from Apollo 14 and Apollo 15 (abstract). In *Lunar Science—III* (editor C. Watkins) 413–414, Lunar Science Institute Contr. No. 88.

LSPET (1971) (Lunar Sample Preliminary Examination Team) Preliminary examination of lunar samples from Apollo 14. *Science* **173,** 681–693.

Lunatic Asylum (1970) Mineralogic and isotopic investigations on lunar rock 12013. *Earth Planet. Sci. Lett.* **9,** 137–163.

Meyer C. Jr., Brett R., Hubbard N. J., Morrison D. A., McKay D. S., Aitken F. K., Takeda H., and Schonfeld E. (1971) Mineralogy, chemistry, and origin of the KREEP component in soil samples from the Ocean of Storms. *Proc. Second Lunar Sci. Conf., Geochim. Cosmochim. Acta* Suppl. 2, Vol. 1, pp. 393–411. MIT Press.

Meyer C. Jr. (1972) Mineral assemblages and the origin of non-mare lunar rock types (abstract). In *Lunar Science—III* (editor C. Watkins), 542–544, Lunar Science Institute Contr. No. 88.

Murthy V. R., Evensen N. M., Jahn B., and Coscio M. R. Jr. (1971) Rb–Sr ages and elemental abundances of K, Rb, Sr, and Ba in samples from the Ocean of Storms. *Geochim. Cosmochim. Acta* **35,** 1139–1154.

NBS Tech. Note 546 (1970) (editor W. R. Shields), pp. 4–9. U.S. Govt. Printing Office.

Nyquist L. E., Hubbard N. J., Gast P. W., Wiesmann H., and Church S. E. (1972) Rb–Sr relationships for some chemically defined lunar materials. In *Lunar Science—III* (editor C. Watkins), pp. 584–586, Lunar Science Institute Contr. No. 88.

Papanastassiou D. A., Wasserburg G. J., and Burnett D. S. (1970a) Rb–Sr ages of lunar rocks from the Sea of Tranquility. *Earth Planet. Sci. Lett.* **8,** 1–19.

Papanastassiou D. A. and Wasserburg G. J. (1970b) Rb–Sr ages from the Ocean of Storms. *Earth Planet. Sci. Lett.* **8,** 269–278.

Papanastassiou D. A. and Wasserburg G. J. (1971a) Lunar chronology and evolution from Rb–Sr studies of Apollo 11 and 12 samples. *Earth Planet. Sci. Lett.* **11,** 37–62.

Papanastassiou D. A. and Wasserburg G. J. (1971b) Rb–Sr ages of igneous rocks from the Apollo 14 mission and the age of the Fra Mauro formation. *Earth Planet. Sci. Lett.* **12,** 36–48.

Papanastassiou D. A. and Wasserburg G. J. (1971c) Age of an Apollo 15 mare basalt: Lunar crust and mantle evolution. *Earth Planet. Sci. Lett.* **13,** 97–104.

Papanastassiou D. A. and Wasserburg G. J. (1972) Rb–Sr age of a Luna 16 basalt and the model age of lunar soils. *Earth Planet. Sci. Lett.* **8,** 368–374.

Reynolds J. H. (1963) Xenology. *J. Geophys. Res.* **68,** 2939–2956.

Roedder E. and Weiblen P. W. (1970) Lunar petrology of silicate melt inclusions, Apollo 11 rocks. *Proc. Apollo 11 Lunar Sci. Conf., Geochim. Cosmochim. Acta* Suppl. 1, Vol. 1, pp. 801–837. Pergamon.

Schnetzler C. C., Philpotts J. A., and Bottino M. L. (1970) Li, K, Rb, Sr, Ba, and rare-earth concentrations, and Rb–Sr age of lunar rock 12013. *Earth Planet. Sci. Lett.* **9,** 185–192.

Schnetzler C. C. and Nava D. F. (1971) Chemical composition of Apollo 14 soils 14163 and 14259. *Earth Planet. Sci. Lett.* **11,** 345–350.

Schonfeld E. (1971) The formation age of the lunar material KREEP. NASA Rep. MSC-4975. Rev. 1.

Schonfeld E. (1972) Component abundance and ages in soils and breccia (abstract). In *Lunar Science—III* (editor C. Watkins), pp. 683–685, Lunar Science Institute Contr. No. 88.

Tatsumoto M., Knight R. J., and Doe B. R. (1971) U–Th–Pb systematics of Apollo 12 lunar samples. *Proc. Second Lunar Sci. Conf., Geochim. Cosmochim. Acta* Suppl. 2, Vol. 2, pp. 1521–1546. MIT Press.

Tatsumoto M., Hedge C. E., Doe B. R., and Unruh D. (1972) U–Th–Pb and Rb–Sr measurements on some Apollo 14 lunar samples (abstract). In *Lunar Science—III* (editor C. Watkins), pp. 741–743, Lunar Science Institute Contr. No. 88.

Turner G. (1970) $^{40}Ar$–$^{39}Ar$ age determination of lunar rock 12013. *Earth Planet. Sci. Lett.* **9,** 177–180.

Turner G. (1971) $^{40}Ar$–$^{39}Ar$ ages from the lunar maria. *Earth Planet. Sci. Lett.* **11,** 169–191.

Turner G., Huneke J. C., Podosek F. A., and Wasserburg G. J. (1971) $^{40}Ar$–$^{39}Ar$ ages and cosmic ray exposure ages of Apollo 14 samples. *Earth Planet. Sci. Lett.* **12,** 19–35.

Warner J. L. (1972) Apollo 14 breccias: Metamorphic origin and classification (abstract). In *Lunar Science—III* (editor C. Watkins), pp. 782–784, Lunar Science Institute Contr. No. 88.

York D. (1966) Least-squares fitting of a straight line. *Can. J. Geophys. Res.* **44,** 1079–1086.

Proceedings of the Third Lunar Science Conference
(Supplement 3, *Geochimica et Cosmochimica Acta*)
Vol. 2, pp. 1531–1555
The M.I.T. Press, 1972

# U–Th–Pb and Rb–Sr measurements on some Apollo 14 lunar samples*

M. Tatsumoto, C. E. Hedge, B. R. Doe, and D. M. Unruh
U.S. Geological Survey, Denver, Colorado 80225

**Abstract**—The U–Th–Pb and Rb–Sr systems were studied on selected samples collected by the Apollo 14 mission. Concordia diagram treatment of U–Pb and U–Th–Pb of whole samples of breccias and rocks, and those of density fractions of a rock (basalt 14310) revealed that the 4650-m.y.-old source material was subjected to an event at about 3800 m.y. ago. This event resulted in formation of two igneous rocks (probably from melting of regolith) and numerous breccias. Support for the 4650-m.y. age of the source material was also obtained by use of the Rb–Sr method on whole-soil samples. An internal isochron was obtained for density fractions from the igneous rock 14310 on $^{206}Pb/^{207}Pb$ versus $^{238}U/^{207}Pb$ and $^{208}Pb/^{207}Pb$ versus $^{232}Th/^{207}Pb$ plots. The resultant isochron from the former method is 3800 $\pm$ 300 m.y. and agrees well with the Rb–Sr mineral isochron age of 3840 m.y. The most significant feature of Apollo 14 samples is that most whole-soil and whole-breccia samples (and basalt 14053) give evidence of significant increase of lead relative to uranium and thorium billions of years ago at the Fra Mauro site. This increase contrasts with evidence from Apollo 11 and Apollo 12 samples of lead loss or no change in lead content relative to those of uranium and thorium since the moon formed. Lead enrichment probably took place owing to two or more events. After lunar formation at about 4650 m.y., the second event is probably related to the time of basalt formation, interpreted to have occurred 3800 m.y. ago. In this event, the value of $^{238}U/^{204}Pb$ in basalt 14310 and breccia 14063 increased and decreased in basalt 14053, breccias 14307 and 14318, and soils 14003, 14163, and 14259 relative to their source materials. A "third event" may have occurred in the interval from 2000–3000 m.y. ago. There is no evidence that the basalts were affected by this event, nor is there clear evidence of any effect on 14163 soil, but soils 14259 and 14003 increased in $^{238}U/^{204}Pb$.

## Introduction

The U–Th–Pb and Rb–Sr systems have been studied on selected samples collected by the Apollo 14 mission to Fra Mauro: basalts 14053 and 14310; soils 14003 (contingency sample), 14163 (bulk fines), and 14259 (comprehensive fines); and breccias 14063, 14307, and 14318 (for sample descriptions, see Appendix). The data is presented on the $^{207}Pb/^{204}Pb$ versus $^{206}Pb/^{204}Pb$, $^{206}Pb/^{207}Pb$ versus $^{238}U/^{207}Pb$, U–Pb concordia, and U–Th–Pb concordia plots. The breccias were separated into matrices and clasts, and a basalt was also separated into density fractions. Previous studies of Apollo 11 (Tatsumoto, 1970a) and Apollo 12 (Tatsumoto, 1970b; Tatsumoto *et al.*, 1971; Cliff *et al.*, 1971) samples showed that the U–Pb concordia treatments of the whole-rock data indicate ages that are comparable with those obtained by Rb–Sr internal isochron and by $^{40}Ar/^{39}Ar$ whole-rock data. Furthermore, results from two attempts to obtain U–Pb internal isochron ages of Apollo 11 and 12 rocks (Tatsumoto,

* Publication authorized by the Director, U.S. Geological Survey.

1970; Tatsumoto *et al.*, 1971) were also comparable with those of other age determination methods, even though the isochron was poorly defined owing to low lead concentrations. Because of an extraordinarily great lead concentration in lunar basalt 14310, an internal isochron age of reasonably high quality was obtained on a $^{206}Pb/^{207}Pb$ versus $^{238}U/^{207}Pb$ diagram.

## DATA

### *Analytical procedures and accuracy*

Only aspects of the analytical procedures are covered here that were not previously presented in Tatsumoto *et al.* (1971) and papers cited therein.

*Sample preparation.* The techniques of sample preparation are basically as previously described. A matter of concern, however, has been the low ratios of $^{206}Pb/^{204}Pb$ that we observed in our density separations as compared with those of whole rocks. We have not been able to devise a satisfactory method of determining blanks on the reagents used other than for acetone. The lead found in the study of the acetone was equal to our chemical processing blank of 4 ng (nanograms). Therefore, the lead "blank" in the acetone is $<1$ ng per 50 ml. For bromoform used in obtaining fractions 1 and 2 and methylene iodide used in obtaining fractions 3 and 4, the heavy liquid was vacuum distilled and a 50-ml fraction was redistilled completely and the lead determined on the residue. These "blanks" were 17 ng/50 ml for bromoform and 135 ng for methylene iodide. The determinations are probably minimum values because if some lead distilled over the first time, some may have distilled over the second time. However, the blanks of these reagents cannot be directly applied to the mineral separates, because the mineral separates were washed thoroughly with acetone on a Teflon filter.

*Lead analysis.* We started our lead analyses with double electrodeposition, in which the anodic deposition step was followed by cathodic deposition instead of the barium coprecipitation step previously used. This procedure premitted us to do U, Th, Pb and K, Rb, Sr on one dissolution, without aliquoting for K, Rb, and Sr. Although the procedure appeared to work well on 14259a, difficulties, particularly with regard to yields, became progressively more apparent on other samples. We finally abandoned the method and returned to our previous procedure. The poor yields, about 23%, made the blank relatively more important. Reanalysis of 14259 as 14259b, 14003 as 14003b, and 14163 as 14163b by our former techniques demonstrated the magnitude of the problem.

Our Pb blanks fluctuated in the range of 2.8–17 ng and concentrated around 7–9 ng Pb. These blanks appear to be greater than our previous level of 4–7 ng. We found that this is due to a higher blank for $HClO_4$ kept in a quartz automatic pipette and blank contributions from $^{41}K$, $^{87}Rb$, and $^{84}Sr$ spikes. Although the "true" blank was probably as good or better than that previously attained, the "effective" blanks owing to the poor yields are properly counted as 9–17 ng Pb. In the Discussion section, we treat only data obtained by the previous method (barium coprecipitate and anodic deposition).

*Potassium, rubidium, and strontium analyses.* Potassium, rubidium, and strontium were determined by isotope dilution using spikes of $^{41}K$, $^{87}Rb$, and $^{84}Sr$. Separation of the elements was done by ion-exchange columns using Dowex 50-X12, 200–400 mesh (AG grade) resin. The solution used was either the residual from lead separation by the double electrodeposition or an aliquot taken immediately after dissolution by HF and $HClO_4$. The blank contributions were about 0.05% for the smallest samples (5–30 mg) and an order of magnitude less for the larger samples (200–300 mg). The mass spectrometry used triple-filament (rhenium) procedures. Precision of the potassium concentrations and the $^{87}Rb/^{86}Sr$ ratios is estimated to be $\pm 2\%$, while the precision of the $^{87}Sr/^{86}Sr$ ratios is $\pm 0.00005$ for the larger samples and $\pm 0.0002$ for the smaller samples. All samples were normalized to a value of 0.1194 for $^{88}Sr/^{86}Sr$.

### *Trace elements*

*Concentration.* The potassium, uranium, thorium, rubidium, and strontium contents of three soil samples and of basalt 14310 (Table 1) are similar to those contents in KREEP-rich soil and breccias collected by the Apollo 12 mission to Fra Mauro (12033 and 12034) (Table 2) and greater than the concentrations in soils and breccias collected by the Apollo 11 mission and the KREEP-poor soil 12070, as has been pointed out by LSPET (1971, p. 123). This similarity in trace element concentrations of the basaltic rock 14310 to those of soil and breccia is suggestive that 14310 is fused soil, as previously stated (Tatsumoto *et al.*, 1972). The lead contents of the Apollo 14 soils and 14310 basalt are greater than those of the KREEP-rich soils and breccia (12033 and 12034) from Apollo 12 probably because none of the Apollo 14 materials has been so extensively affected by the "third events" that resulted in lead depletion. Rock 12013, in terms of its great trace element contents, still remains unique among whole-rock lunar samples so far analyzed.

The other basalt analyzed, 14053, is perhaps closest in uranium, thorium, and lead contents to the Apollo 11 Group I (potassium-richer) basalts; however, 14053

Table 2. Lead, uranium, and thorium concentrations in Apollo 11, Apollo 12 and Apollo 14 samples.

| Samples | Pb (ppm) | U (ppm) | Th (ppm) | $Th^{232}/U^{238}$ |
|---|---|---|---|---|
| | | Apollo 11 | | |
| Basalts | | | | |
| Group II (3)* | 0.29 ~ 0.51 | 0.16 ~ 0.27 | 0.53 ~ 1.02 | 3.53 ~ 3.99 |
| Group I (3)* | 1.56 ~ 1.74 | 0.85 ~ 0.87 | 3.30 ~ 3.43 | 0.03 ~ 4.08 |
| Breccia (1)* | 1.7 | 0.67 | 2.6 | 3.94 |
| Fines (1)* | 1.4 | 0.54 | 2.1 | 3.97 |
| | | Apollo 12 | | |
| Basalts (8)* | 0.28 ~ 0.64 | 0.16 ~ 0.40 | 0.61 ~ 1.41 | 3.61 ~ 3.92 |
| Breccia (1)* | 4.2 | 3.6 | 13.0 | 3.76 |
| Fines (2)* | 3.2 ~ 4.4 | 1.6 ~ 3.3 | 6.0 ~ 12.1 | 3.75 ~ 3.79 |
| Igneous breccia [12013]† (4)‡ | 9.2 ~ 16.3 | 5.7 ~ 10.8 | 19.1 ~ 34.3 | 3.28 ~ 4.04 |
| | | Apollo 14 | | |
| Basalts (2)* | 1.7 ~ 6.2 | 0.59 ~ 3.10 | 2.1 ~ 10.4 | 3.67 ~ 3.70 |
| Breccia (3)* | 3.7 ~ 12.2 | 1.6 ~ 5.3 | 5.4 ~ 16.7 | 3.24 ~ 3.64 |
| Fines (3)* | 7.2 ~ 9.9 | 2.5 ~ 4.1 | 11.7 ~ 14.6 | 3.66 ~ 3.68 |

* Number of samples analyzed.
† Sample number.
‡ Number of chips analyzed.

Table 1. Concentrations of selected elements and elemental ratios in some Apollo 14 samples.

| Sample number | Fraction analyzed | Sample weight (mg) | Concentrations (ppm) K | U | Th | Pb | Rb | Sr | Ratios: Atomic $^{232}Th/^{238}U$ | Ratios: Weight K/U | K/Rb | Approx. initial Pb (ppb) |
|---|---|---|---|---|---|---|---|---|---|---|---|---|
| | | | | | Soil | | | | | | | |
| 14003,37a* | Whole | 242 | 2790 | 2.52 | — | 7.18 | 11.59 | 142.7 | — | 1100 | 242 | 755 |
| 14003,59b | Whole | 167 | — | 3.31 | 11.66 | 8.16 | — | — | 3.64 | — | — | 493 |
| 14003,37 | Acetone floats | 23 | — | 4.04 | 14.60 | 13.16 | — | — | 3.73 | — | — | 1604 |
| 14003,37 | $\rho$ = 2.9–3.3 | 222 | — | 4.05 | 14.64 | 11.16 | — | — | 3.74 | — | — | 569 |
| 14163,156 | Whole | 59 | 4250 | 3.51 | 12.46 | 9.97 | 14.61 | 182.1 | 3.66 | 1370 | 291 | 362 |
| 14259,16a | Whole | 150 | 3990 | 3.43 | 12.15 | 7.87 | 14.06 | 182.6 | 3.66 | 1160 | 284 | 265 |
| 14259,16b | | 215 | — | 3.47 | 12.35 | 7.52 | — | — | 3.68 | — | — | 185 |
| | | | | | Breccia | | | | | | | |
| 14063,37,M | Matrix | 303 | — | 3.40 | 12.64 | 7.95 | 20.65 | 180.9 | 3.84 | — | — | 117 |
| 14063,37,C | Clast | 29 | 3115 | 1.56 | 5.42 | 3.70 | 10.10 | 231.3 | 3.58 | 1997 | 308 | 194 |
| 14307,26,M | Matrix | 230 | 4930 | 3.36 | 11.85 | 11.45 | 16.44 | 171.1 | 3.64 | 1467 | 300 | 548 |
| 14307,26,C1 | Clast 1 | 178 | 3700 | 4.86 | 16.46 | 8.24 | 17.44 | 147.0 | 3.50 | 761 | 212 | 96 |
| 14307,26,C2 | Clast 2 | 36 | — | 4.99 | 17.29 | 9.60 | — | — | 3.58 | — | — | 119 |
| 14318,26,M | Matrix | 328 | — | 3.67 | 12.67 | 12.19 | 17.59 | 170.9 | 3.56 | — | — | 304 |
| 14318,26,C1 | Clast 1 | 76 | — | 5.32 | 16.69 | 10.70 | — | — | 3.24 | — | — | 236 |
| 14318,26,C2 | Clast 2 | 14 | 1490 | — | — | — | 5.35 | 141.7 | — | — | 279 | |
| 14318,40 | Sawdust | 419 | — | 3.29 | 11.35 | 13.9† | — | — | 3.56 | — | — | (271) |
| | | | | | Basalt | | | | | | | |
| 14053,27 | Whole | 103 | — | 0.592 | 2.101 | 1.71 | — | — | 3.67 | — | — | 92 |
| 14310,71a | Whole | 464 | 4010‡ | 3.10 | 10.42‡ | 6.18 | 9.79 | 154.8 | 3.43‡ | 1294 | 410 | 45 |
| 14310,71b | Whole | 35 | — | 3.22 | 11.36 | 6.10 | — | — | 3.64 | — | — | 45 |
| 14310,71,M1 | $\rho$ = 2.4–2.67 | 29 | — | 4.64 | 18.66 | 10.70 | — | — | 4.15 | — | — | — |
| 14310,71,M2 | $\rho$ = 2.67–2.89 | 154 | 3790 | 2.59 | 8.73 | 4.91 | 10.48 | 270.1 | 3.49 | 1463 | 362 | — |
| 14310,71,M3 | $\rho$ = 2.89–3.3 | 95 | — | 6.23 | 22.41 | 11.61 | — | — | 3.71 | — | — | — |
| 14310,71,M4 | $\rho$ > 3.3 | 192 | 1410 | 2.31 | 8.66 | 4.12 | 4.12 | 26.86 | 3.87 | 610 | 342 | — |

* a—lead was analyzed by double electroplating; b—analyzed by barium coprecipitation and anodic electroplating. M and C in sample numbers indicate matrix and clast, respectively. M1–M4 of 14310 indicate density fractions.

† The lead concentration corrected to provide reasonable U–Th–Pb systematics compared with other fractions of the sample would be 11.1 ppm.

‡ Uncertainty of the data is about 10 percent.

has a distinctly lesser value of $^{232}Th/^{238}U$ and K/U. Unlike uranium and thorium contents, potassium, rubidium, and strontium contents of 14053 are intermediate between those of the Group I and Group II basalts collected by the Apollo 11 mission but are similar to those for Apollo 11 soil and breccia.

*Th/U.* The value of Th/U in Apollo 14 whole samples, except a white clast in breccia 14318, is remarkably uniform, ranging from 3.5 to 3.7. The value of this ratio in lunar whole samples so far analyzed lies in the range from 3.5 to 4.1 (Table 2). Average ratios of Th/U in Apollo soils are 3.97 for the Apollo 11 mission, 3.77 for the Apollo 12 mission, and 3.67 for the Apollo 14 mission. The $^{232}Th/^{238}U$ values appear to decrease from Apollo 11 to Apollo 14 samples.

The trend in Th/U is in good agreement with that for $^{208}Pb/^{206}Pb$ (where $^{208}Pb$ is the stable daughter product of $^{232}Th$ and $^{206}Pb$ of $^{238}U$), which also decreases

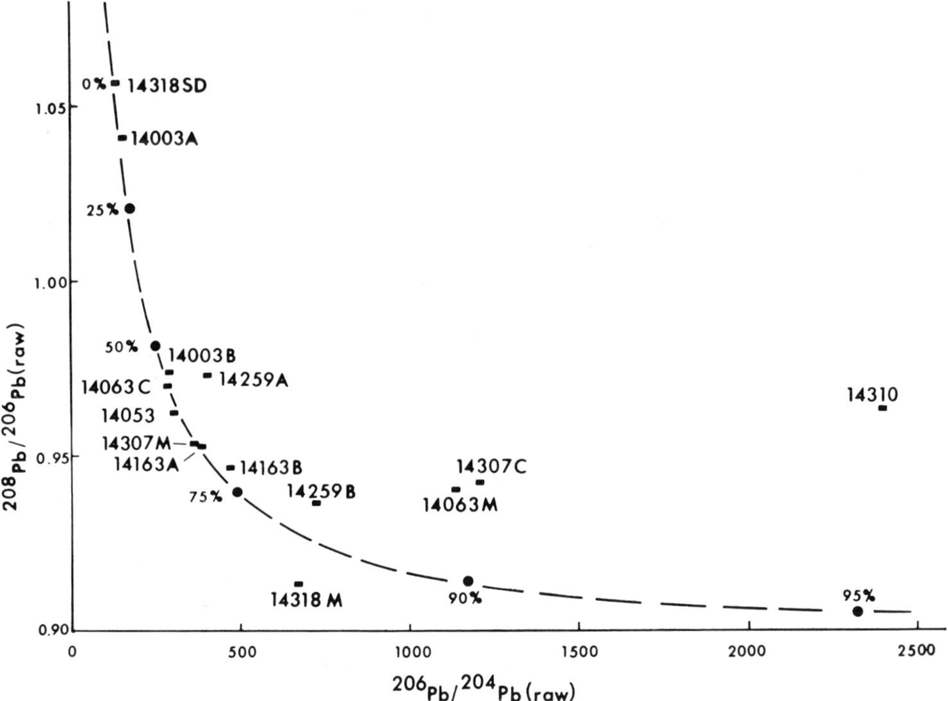


Fig. 1. Plot of the raw ratios of $^{208}Pb/^{206}Pb$ versus $^{206}Pb/^{204}Pb$ for samples analyzed from the Apollo 14 mission. Dashed line defines $^{208}Pb/^{206}Pb$ and $^{206}Pb/^{204}Pb$ for sample 14318SD corrected for variable amounts of its $^{204}Pb$ being ascribed to contamination. Several individual points are labeled in percent. The isotopic composition of the contaminating lead is assumed to be $^{206}Pb/^{204}Pb = 18.48$, $^{207}Pb/^{204}Pb = 15.73$, and $^{208}Pb/^{204}Pb = 38.38$. The limiting value of $^{208}Pb/^{206}Pb$ is 0.896 when all $^{204}Pb$ is assumed to be contaminant lead. Sample 14318M could be made to fit a 14318SD contamination line if the $^{208}Pb/^{206}Pb$ of the contaminant lead was about 5% greater than assumed. Size of the blocks approximates analytical uncertainties. Symbols A and B after a sample number refer to first and second analyses of a sample, C means clast and M matrix from a breccia, and SD means sawdust.

from unity in Apollo 11 samples to 0.95 for Apollo 14; however, precise determinations of $^{208}Pb/^{206}Pb$ values are difficult to make for most samples, a difficulty illustrated by Fig. 1. The figure shows how the values of $^{208}Pb/^{206}Pb$ and $^{206}Pb/^{204}Pb$ of 14318SD (sawdust) change as more and more of the $^{204}Pb$ is removed as a common lead contaminant. It is seen that corrected values of $^{208}Pb/^{206}Pb$ for the sawdust change rather rapidly up to values of $^{206}Pb/^{204}Pb$ of about 1000 (which includes most samples), whereas further removal of $^{204}Pb$ as a contaminant results in little change in $^{208}Pb/^{206}Pb$. Small differences in $^{208}Pb/^{206}Pb$ probably exist in the sequence basalt 14310 > breccia matrices 14307 and 14063 > breccia matrix 14318. Soil 14003, the contingency fines, has a greater value of $^{208}Pb/^{206}Pb$ and lesser value of $^{206}Pb/^{204}Pb$ than the other soil samples, and may have been contaminated on the lunar surface by rocket exhaust (Tatsumoto *et al.*, 1972). Breccia sawdust 14318 also has an unusually great value of $^{208}Pb/^{206}Pb$ and an unusually lesser value of $^{206}Pb/^{204}Pb$, probably accounted for by contamination from the vacuum cleaner bag used to gather up the sawdust at the Lunar Receiving Laboratory, such as was used to explain unusual isotopic characteristics in breccia 12013 sawdust (Tatsumoto *et al.*, 1971). The indigenous values of $^{206}Pb/^{204}Pb$ and $^{208}Pb/^{206}Pb$ of 14318SD and 14003 probably do not depart significantly from a grouping of samples including 14318 matrix, 14163 soil, 14063 clast, 14307 matrix, and basalt 14053. Soil 14259 does appear to belong to a different soil isotopic family than the other two in a $^{208}Pb/^{206}Pb$ versus $^{206}Pb/^{204}Pb$ grouping.

The narrow range in the values of $^{232}Th/^{238}U$ in lunar whole samples, 3.5–4.1, is rather remarkable compared with the internal heterogeneity for Th/U mineral separates. The Th/U varies from 2 to about 5 in density separates of basalt 10017 (Tatsumoto, 1970a), 2 to 4 in density separates of soil 12033 (Tatsumoto *et al.*, 1971), and 3.5 to 4.2 in density separates of basalt 14310 (Table 1). The ratio $^{232}Th/^{238}U$ is as little as 0.25 in zircon to as great as 20 in whitlockite (Burnett *et al.*, 1971; Andersen and Hinthorne, 1972). The uniformity in $^{232}Th/^{238}U$ in whole samples is one of the peculiarities of lunar samples as compared with terrestrial samples.

*K/U.* As has been pointed out by Tera *et al.* (1970) and Fanale and Nash (1971), the values of K/U for terrestrial rocks mostly lie within 0.5 orders of magnitude of $10^4$, whereas in lunar samples the ratio is closer to $10^3$. The K/U of Apollo 14 samples given in Table 1 is equal to or lower than that for samples from the Apollo 11 and 12 samples except for fragments from the breccia 12013. This relationship is in accord with that given by LSPET (1971, p. 123). According to the data of Wakita and Schmitt (1970), the K/U of 12013 covers the entire spectrum known for whole samples from the moon.

### *Isotope ratios*

*Lead.* The lead isotopic compositions of Apollo 14 samples (Table 3) are extremely radiogenic, as was found for other lunar samples from the Apollo 11 and 12 missions. The raw value of 2400 for $^{206}Pb/^{204}Pb$ of basalt 14310 is the greatest found so far for any whole-rock sample from the moon. Aside from basalt 14310, the only grouping readily apparent is the low value ($\sim$300) of $^{206}Pb/^{204}Pb$ for 14003, contingency fines previously suspected by Tatsumoto *et al.* (1972) to be contaminated.

Table 3. Isotopic composition of lead and strontium in some Apollo 14 samples.

| Sample number | Fraction analyzed | Sample weight for Pb (mg) | Raw data $^{206}Pb/^{204}Pb$ | Raw data $^{207}Pb/^{204}Pb$ | Raw data $^{208}Pb/^{204}Pb$ | Raw data $^{208}Pb/^{206}Pb$ | Corrected for blank $^{206}Pb/^{204}Pb$ | Corrected for blank $^{207}Pb/^{204}Pb$ | Corrected for blank $^{208}Pb/^{204}Pb$ | Corrected for blank $^{207}Pb/^{206}Pb$ | Corrected for blank $^{208}Pb/^{206}Pb$ | Raw $^{87}Rb/^{86}Sr$ | Normalized $^{87}Sr/^{86}Sr$ |
|---|---|---|---|---|---|---|---|---|---|---|---|---|---|
| Soil | | | | | | | | | | | | | |
| 14003,37a* | Whole | 490 | 155.3 | 112.1 | 161.7 | 1.0412 | 169.7 | 122.8 | 176.8 | 0.7236 | 1.0418 | 0.2345 | 0.71447 |
| 14003,69b | Whole | 222 | 293.3 | 208.3 | 285.3 | 0.9727 | 306.4 | 217.7 | 298.0 | 0.7107 | 0.9727 | — | — |
| 14003,37 | Acetone floats | 23 | — | — | — | — | 146.5 | 124.9 | — | 0.8528 | — | — | — |
| 14003,37 | $\rho$ = 2.9–3.3 | 318 | 108.6 | 70.4 | 122.6 | 1.1289 | 292.0 | 182.3 | 295.4 | 0.6242 | 1.0012 | — | — |
| 14163,a | Whole | 568 | 384.4 | 284.4 | 366.6 | 0.9537 | 406.7 | 301.2 | 387.8 | 0.7407 | 0.9536 | — | — |
| 14163,b | Whole | 301 | 469.8 | 347.8 | 444.6 | 0.9464 | 505.7 | 374.8 | 478.4 | 0.7411 | 0.9461 | 0.2324 | 0.71448 |
| 14259,a | Whole | 332 | 404.2 | 276.1 | 393.3 | 0.9730 | 557.8 | 379.2 | 537.6 | 0.6798 | 0.9638 | — | — |
| 14259,b | Whole | 850 | 745.4 | 501.6 | 697.2 | 0.9353 | 769.9 | 518.4 | 722.4 | 0.6734 | 0.9384 | 0.2232 | 0.71398 |
| Breccia | | | | | | | | | | | | | |
| 14063,37,M | Matrix | 555 | 1138 | 683.2 | 1070 | 0.9402 | 1323 | 794.4 | 1244 | 0.6007 | 0.9407 | 0.3308 | 0.71575 |
| 14063,37,C | Clast | 78 | 285.5 | 181.0 | 276.8 | 0.9695 | 364.7 | 230.5 | 348.9 | 0.6322 | 0.9568 | 0.1264 | 0.70698 |
| 14307,26,M | Matrix | 559 | 362.0 | 291.8 | 345.2 | 0.9536 | 374.9 | 302.6 | 357.9 | 0.8072 | 0.9546 | 0.2783 | 0.71588 |
| 14307,26,C1 | Clast 1 | 253 | 1207 | 636.9 | 1138 | 0.9428 | 1726 | 909.8 | 1623 | 0.5269 | 0.9402 | 0.3437 | 0.71906 |
| 14307,26,C2† | Clast 2 | 36 | — | — | — | — | 1628 | 858.3 | 1532 | 0.5271 | 0.9409 | — | — |
| 14318,26,M | Matrix | 559 | 678.1 | 568.6 | 619.2 | 0.9131 | 721.1 | 605.1 | 659.1 | 0.8397 | 0.9140 | 0.3041 | 0.71800 |
| 14318,26,C1† | Clast 1 | 76 | — | — | — | — | 911.6 | 509.1 | — | 0.5585 | — | — | — |
| 14318,26,C2 | Clast 2 | 14 | — | — | — | — | — | — | — | — | — | 0.1092 | 0.70570 |
| 14318,40 | Sawdust | 903 | 135.6 | 110.8 | 143.3 | 1.0568 | (738.7) | (602.4) | (686.8) | (0.8156) | (0.9299) | — | — |
| Basalt | | | | | | | | | | | | | |
| 14053,27 | Whole | 1272 | 305.0 | 236.6 | 293.6 | 0.9626 | 337.6 | 262.2 | 323.7 | 0.7765 | 0.9588 | — | — |
| 14310,71b | Whole | 742 | 2396 | 1177 | 2307 | 0.9629 | 2709 | 1332 | 2614 | 0.4917 | 0.9648 | 0.1832 | 0.71041 |

* a—lead analyzed by double electroplating; b—analyzed by barium coprecipitation and anodic electroplating. M and C in sample numbers indicate matrix and clast, respectively.

† $^{208}Pb$ spike run.

The single most unique feature of the Apollo 14 samples is their great values of $^{207}Pb/^{206}Pb$ which, except for basalt 14310, range from 0.67 to 0.92, exceeding values found on other lunar samples. Lead in the light-colored portions of the breccias is more radiogenic than that in the dark-colored portions.

## DISCUSSION

### *Rb–Sr ages*

Rb–Sr model ages of the three soil samples 14003, 14163, and 14259 are 4.60, 4.65, and 4.68 b.y., respectively—assuming a meteorite-initial $^{87}Sr/^{86}Sr$ (0.699). These ages are typical of those reported for lunar soils by both Rb–Sr and U–Pb methods (cf. Wasserburg *et al.*, 1972, and U–Pb results, this report) but these are higher than the 4.5 b.y. obtained by Rb–Sr total-rock "isochron" ages (Compston *et al.*, 1972, and Murthy *et al.*, 1972). A line corresponding to an age of 4.65 b.y. has been drawn through the soil samples in Fig. 2.

The basalt 14310 was analyzed for Rb–Sr as a whole-rock sample and two mineral phases, M2 (impure plagioclase) and M4 (olivine plus ilmenite). These three portions define an isochron (Fig. 2) whose age is $3.84 \pm 0.04$ b.y. with an initial $^{87}Sr/^{86}Sr$ of 0.7004. This age, presumably a crystallization age, is not analytically significantly different from that reported by Wasserburg *et al.* (1972, 3.89 b.y.) and Murthy *et al.* (1972, $3.93 \pm 0.06$ b.y.), but it is significantly lower than the 4-b.y. age reported by Compston *et al.* (1972).

The data for the breccia samples 14063, 14307, and 14318 are more difficult to interpret. The matrices of the breccias are isotopically similar in some respects to the soils, and it is conceivable that they are dominantly soil material to which Rb (and Pb) has been added. The location of matrices 14307 and 14318 in Fig. 2 is consistent with Rb having been added at about $3.8 \pm 0.1$ b.y. ago—in other words, at the time of basalt formation and presumably the time of major impact in this region. The matrix of breccia 14063 was affected at some later, indeterminate time, probably no earlier than about 3.5 b.y. ago. The alignment of the breccia clasts, in Fig. 2, along the dashed line, might be interpreted as indicating that the clasts are of one age, 4.01 b.y. However, there are no compelling reasons for assuming that the clasts are of a single age. If we assume that the time of breccia formation corresponds with the age of the basalts (as suggested by the position of breccia matrices 14307 and 14318 in Fig. 2), then we really only know the obvious about the age of the breccia clasts—namely, they are at least 3.8 b.y. old.

### *Isochron age methods of U–Th–Pb*

There are four isochron techniques of dating in the U–Th–Pb system, three of which are independent: two of three for the U–Pb system ($^{207}Pb/^{204}Pb$–$^{206}Pb/^{204}Pb$, $^{207}Pb/^{204}Pb$–$^{235}U/^{204}Pb$, $^{206}Pb/^{204}Pb$–$^{238}U/^{204}Pb$) and one for the Th–Pb system ($^{208}Pb/^{204}Pb$–$^{232}Th/^{204}Pb$). The Pb–Pb isochron, which was the first dating technique to give precise ages of meteorites, will be discussed first because of the simplicity of the method, its proven reliability on nonlunar samples, and the insensitivity of the method

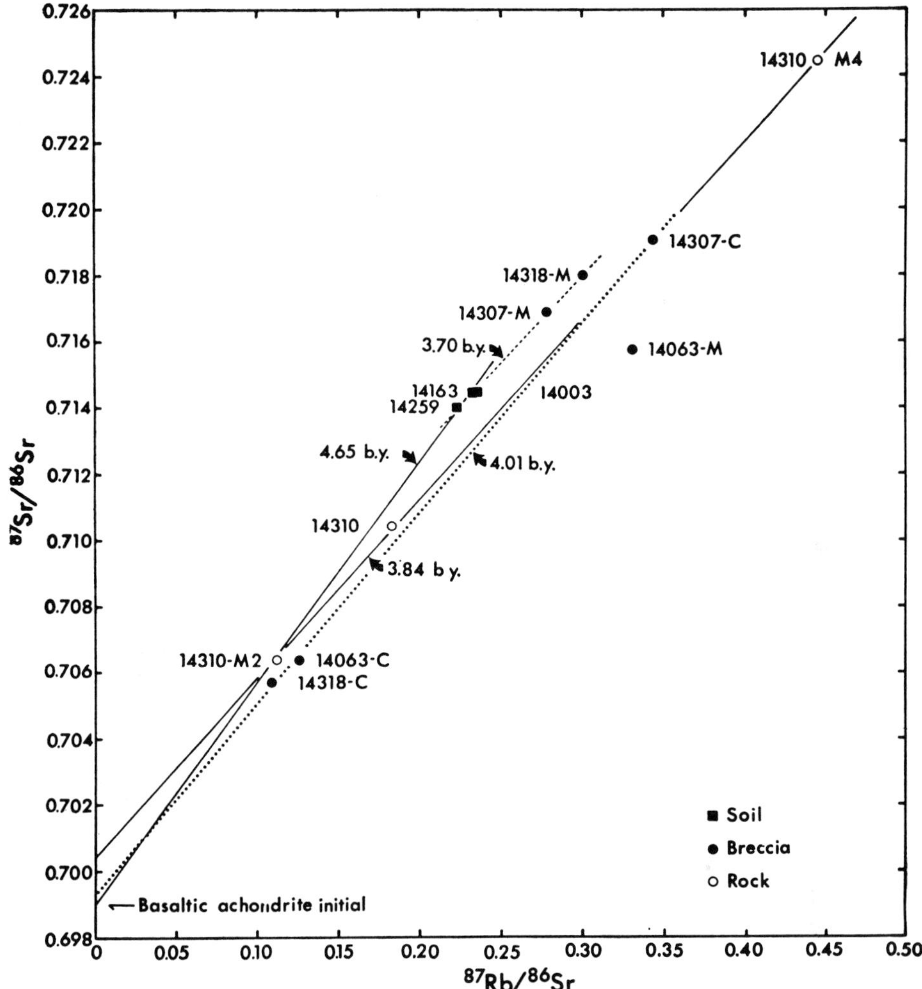


Fig. 2. Strontium evolution diagram for whole samples of Apollo 14 breccias and soils, and density fractions of rock 14310. Symbols after sample numbers are as follows: C for breccia clasts (white for 14318 and 14307, black for 14063); M for breccia matrix (white for 14063, medium gray for 14318, dark gray for 14307); M2 and M4 are density fractions from basalt 14310.

to various analytical uncertainties. Following will be discussions of the internal isochron expressed by $^{206}Pb/^{207}Pb$ versus $^{238}U/^{207}Pb$ and concordia diagrams.

*Primary Pb–Pb isochron*

The data for the analyzed Apollo 14 samples are shown in Fig. 3. Samples 14053, 14307M, 14318M, 14003, 14163, and 14259 have primary Pb–Pb isochron ages in excess of 4500 m.y. Rather than indicating such old ages for the samples, the data could be explained through a process whereby Pb was enriched relative to uranium

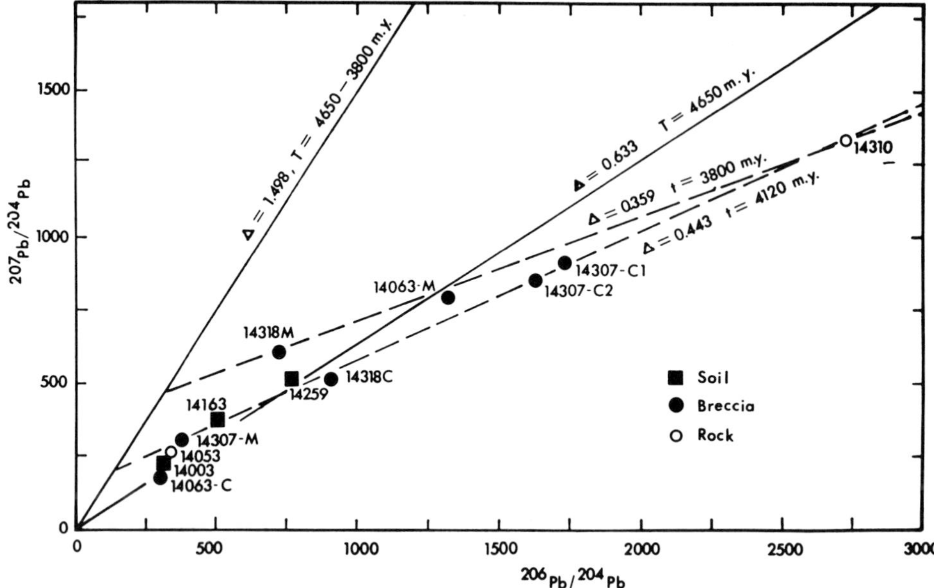


Fig. 3. $^{207}Pb/^{204}Pb$ versus $^{206}Pb/^{204}Pb$ for Apollo 14 samples analyzed in the study. Solid lines are primary isochrons given for reference. Δ is the slope of the primary isochron and T is the primary isochron age. Broken lines are secondary isochrons of slope Δ. Intersection of the 3800-m.y. secondary isochron with the 4650- to 3800-m.y. primary isochron represents the greatest "initial lead" found so far for lunar samples at their presumed time of formation in terms of $^{207}Pb/^{204}Pb$ and $^{206}Pb/^{204}Pb$, assuming a two-stage development. Significance of the 4120-m.y. secondary isochron is not known, but lead isochron "ages" of about 4100 m.y. repeatedly appear for many lunar samples (Tatsumoto, 1970; Doe and Tatsumoto, 1972). Symbols after sample numbers are as follows: C for breccia clast (white for 14318 and 14307, black for 14063); M for breccia matrix (white for 14063, medium gray for 14318, dark gray for 14307).

a long time ago. The validity of this explanation will become apparent in the discussion on concordia diagrams.

Fig. 4 summarizes the primary Pb–Pb isochron ages we have determined on basalts from the Apollo missions 11–14 plotted against the most reliable Rb–Sr mineral isochron ages determined on the same rocks. On the average, the primary Pb–Pb isochron age is about 630 m.y. more than the Rb-Sr mineral isochron age. This difference in age could well be accounted for by a problem of determining highly radiogenic initial leads in the magmas, which has been pointed out repeatedly by many authors (Tatsumoto, 1970; Compston *et al.*, 1970; Tatsumoto *et al.*, 1971; Wetherill, 1971). If the initial lead is radiogenic, then we are not dealing with primary systems but secondary or higher-order systems (with one or more disturbances of U/Pb).

*Secondary Pb–Pb isochron*

One explanation of the difference between the primary Pb–Pb isochron age and the Rb–Sr internal isochron and $^{40}Ar/^{39}Ar$ whole-rock ages involves a change in the

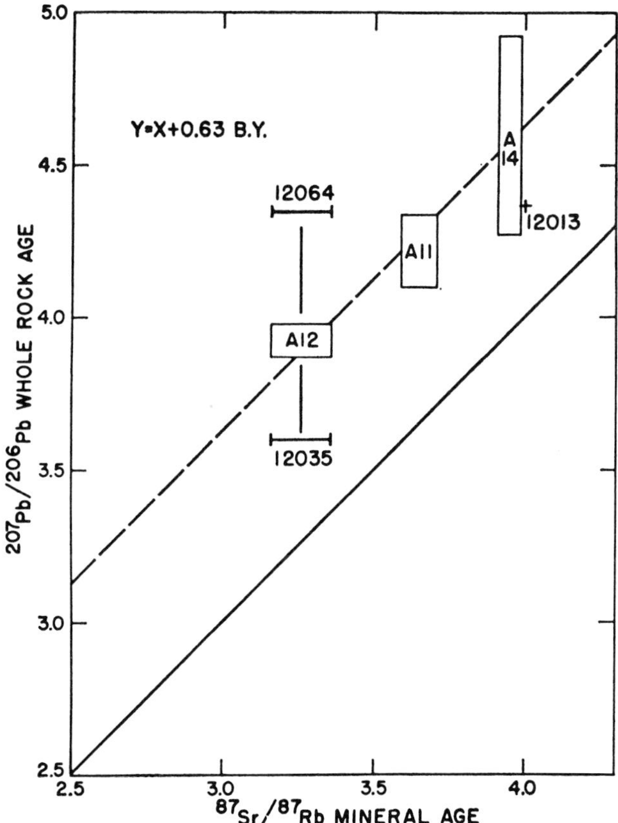


Fig. 4. $^{207}Pb/^{206}Pb$ or primary Pb–Pb isochron age versus Rb–Sr internal isochron age of the same lunar basalts. Lead isotope data are from Tatsumoto (1970a) for Apollo 11 (A11), Tatsumoto (1970b) for rock 12013, Tatsumoto *et al.* (1971) for Apollo 12 (A12, 12035 and 12064) basalts, and this paper for Apollo 14 basalts (A14). Rb–Sr internal isochron ages are from Papanastassiou *et al.* (1970) and Papanastassiou and Wasserburg (1970, 1971).

value of U/Pb. The mantle rock that melted to produce a lunar basalt might have one value of U/Pb, and through the partial melting of the mantle rock the melt itself would then have another value. Then the primary Pb–Pb isochron no longer indicates a true age. If the magma reservoir has a homogeneous value of lead isotopic composition and differentiates to form several different basalts each with its own value of U/Pb, the present-day data should lie along a line (a secondary isochron) of lesser slope than the primary isochrons, and we should be able to determine the age of the magmatic activity very precisely. Three conditions—difference in initial lead, third events, and contamination—might destroy a Pb–Pb whole rock secondary isochron or produce a spurious isochron. Secondary isochrons need not lie close to the value of primordial lead or contamination lead as primary isochrons do, and difficulties in correcting for contamination constitute a third manner in which secondary isochrons could be destroyed. One of the strengths of primary isochrons is that the value of $^{207}Pb/^{206}Pb$ of a lunar sample is little affected by contamination. The data points just move along the primary isochron when corrected for the contaminant. The old character of the $^{207}Pb/^{206}Pb$ in lunar samples is also little affected either by instrumental fractionation or by $^{204}Pb$ error (Tatsumoto *et al.*, 1971). None of this is true for secondary isochrons, and all corrections are likely to be at high angles to secondary isochrons (see Fig. 3), although correction of the sample for contamination is most likely the major problem. If the basalt was affected by some event (a "third event") subsequent to its crystallization that altered the value of U/Pb in the rock, the data might form a scatter array on the $^{207}Pb/^{204}Pb$ versus $^{206}Pb/^{204}Pb$ diagram. If the value of U/Pb was altered proportionally by the "third event" to the value originally in the basalt, or if the basalts were derived from many sources with different values of initial lead isotopic composition, a spurious isochron would be formed.

Basalt 14310 (Table 3 and Fig. 3) is truly remarkable in the radiogenic nature of its lead isotopic composition, even for lunar samples. If we accept the Rb-Sr internal age for this sample (this paper and Papanastassiou and Wasserburg, 1971b), assuming two-stage evolution, the initial lead (the intersection of the dashed line labeled t = 3800 m.y. with the solid line labeled T = 4650–3800 m.y.) for the rock is the greatest so far found for the moon (Fig. 3), even greater than for the breccia 12013. The trend in initial lead parallels that found for $^{87}Sr/^{86}Sr$ by Papanastassiou and Wasserburg (1971b) who reported 14310 to have the greatest initial $^{87}Sr/^{86}Sr$ for lunar basalts, although not so great as for the breccia 12013. Basalt 14053 appears to have a more normal initial lead isotopic composition for lunar basalts as compared with Group II basalts of Apollo 11 and the 12009 grouping from Apollo 12. The initial $^{87}Sr/^{86}Sr$ for this basalt is also more normal, according to Papanastassiou and Wasserburg (1971b).

Soils and breccias constitute a more complex problem with regard to secondary isochrons, not only because they are complex mixtures of materials but also because they commonly display evidence of "third events" (young impacts?). These complexities are such that further discussion of secondary isochrons on soils and breccias is not warranted here.

### *U–Pb and Th–Pb internal isochron*

Tatsumoto (1970a) obtained a $^{238}U/^{204}Pb$ versus $^{206}Pb/^{204}Pb$ isochron age for basalt 10017 of 3660 m.y., similar to that of the Rb–Sr internal isochron; however, the U–Pb internal isochron age was given little consideration because of problems involving contamination. Internal isochrons were also attempted for rock 12064 (Tatsumoto *et al.*, 1971, p. 1538) with a result of 3320 m.y. by the $^{238}U/^{204}Pb$ versus $^{206}Pb/^{204}Pb$ isochron, again in good agreement with the internal Rb–Sr ages, but again contamination problems persisted. The lead isotopic composition of the density fractions of 14310, corrected for analytical blank, is less radiogenic than the whole-rock value (Table 4). It is clear that there must be several contamination sources, especially heavy liquids used for the mineral separation. The lead contamination accounts for nearly all the $^{204}Pb$ content but much lesser percentages of the other isotopes; as a result ratios involving $^{204}Pb$ are greatly affected. To avoid this problem we subtracted a terrestrial common lead from the data corresponding to all the

Table 4. Lead and strontium isotopic data for density fractions of 14310.

| Sample | Raw $^{87}Rb/^{86}Sr$ | Normalized $^{87}Sr/^{86}Sr$ | $^{232}Th/^{238}U$ | Raw data $^{206}Pb/^{204}Pb$ | $^{207}Pb/^{204}Pb$ | $^{208}Pb/^{204}Pb$ | Corrected for analytical blank $^{206}Pb/^{204}Pb$ | $^{207}Pb/^{204}Pb$ | $^{208}Pb/^{204}Pb$ | Corrected for total $^{204}Pb$ $^{206}Pb/^{207}Pb$ | $^{208}Pb/^{207}Pb$ | $^{238}U/^{207}Pb$ | $^{232}Th/^{207}Pb$ |
|---|---|---|---|---|---|---|---|---|---|---|---|---|---|
| Whole rock | 0.1832 | 0.71041 | 3.64 | 2396 | 1177 | 2307 | 2701 | 1326 | 2597 | 2.0440 | 1.9564 | 2.2903 | 8.346 |
| M1 $\rho$ = 2.4–2.67* | — | — | 4.15 | — | — | — | 275.9 | 163.3 | — | 1.7372 | — | 1.8137 | 7.532 |
| M2 $\rho$ = 2.67–2.89 | 0.1132 | 0.70637 | 3.49 | 1080 | 552.3 | 1259 | 1259.3 | 643.8 | 1130.2 | 1.9752 | 1.7382 | 2.1959 | 7.656 |
| M3 $\rho$ = 2.89–3.3 | — | — | 3.71 | 669.6 | 318.1 | 673.9 | 983.0 | 464.5 | 982.8 | 2.1487 | 2.1040 | 2.4823 | 9.218 |
| M4 $\rho$ > 3.3 | 0.4447 | 0.72443 | 3.87 | 1436 | 624.0 | 1444 | 2069.4 | 897.4 | 2079.4 | 2.3257 | 2.3146 | 2.7608 | 10.691 |

* $^{208}Pb$, spike run.

$^{204}$Pb and have plotted the data for 14310 as $^{238}$U/$^{207}$Pb versus $^{206}$Pb/$^{207}$Pb (Fig. 5). This plotting is a variant of the representation first reported by Wasserburg and colleagues (Wasserburg *et al.*, 1972; Tera and Wasserburg, 1972b) in their construction of internal isochrons on 14053 and 14310 by a $^{238}$U/$^{206}$Pb versus $^{207}$Pb/$^{206}$Pb isochron method. In our representation, these particular data plot with a positive slope rather than the negative slope obtained by the method of Wasserburg and colleagues. The obtained "age" of 3815 ± 300 m.y. (Fig. 5) is close to the Rb–Sr internal isochron age of 3840 m.y. (this paper). The whole-rock value of 14053 has also been included for comparison in the plot and on an equivalent plot involving $^{232}$Th (Fig. 6). The proximity of the 14053 whole-rock sample to the 14310 isochron suggests that 14053 is of the same age as 14310, which is in agreement with the internal isochron given by Wasserburg *et al.* (1972) and Tera and Wasserburg (1972b) who also present a discussion of the systematics of these isochrons. Our data agree with the results on 14053 reported by these authors, who stated that the internal isochron results ob-

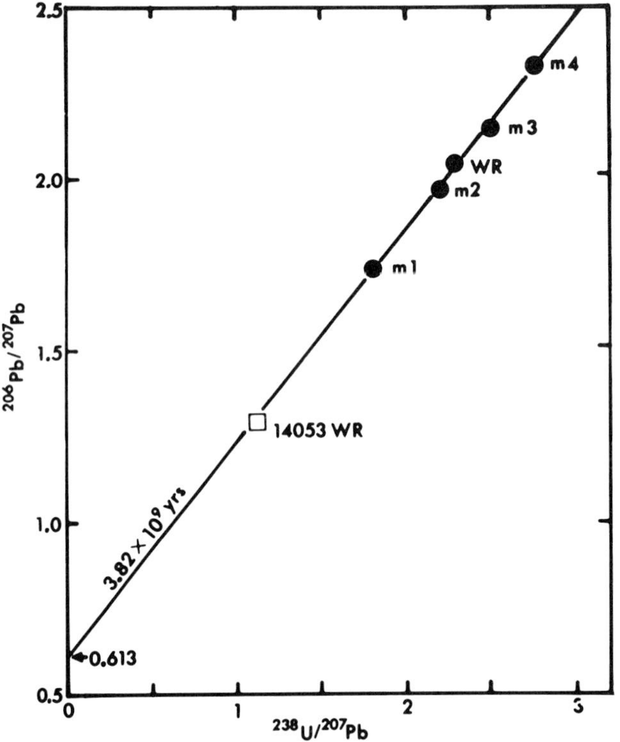


Fig. 5. $^{206}$Pb/$^{207}$Pb versus $^{238}$U/$^{207}$Pb for rock 14310.

$$\frac{d(^{206}\mathrm{Pb}/^{207}\mathrm{Pb})}{d(^{238}\mathrm{U}/^{207}\mathrm{Pb})} = (e^{\lambda 238} - 1) - (^{206}\mathrm{Pb}/^{207}\mathrm{Pb})_i \frac{e^{\lambda 235} - 1}{137.8}$$

where $(^{206}\mathrm{Pb}/^{207}\mathrm{Pb})_i$ is the initial ratio. Symbols are: whole rock, WR; density fractions, m1–m4.

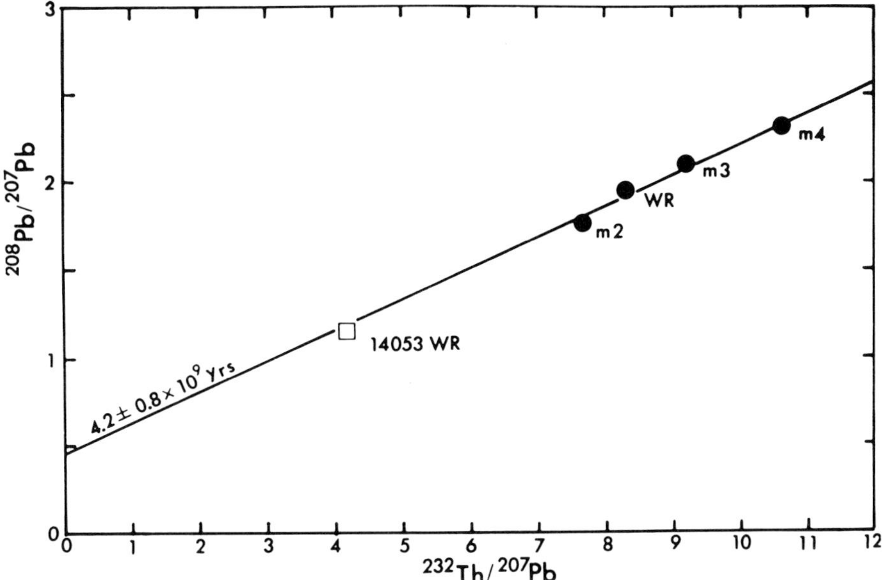


Fig. 6. $^{208}Pb/^{207}Pb$ versus $^{232}Th/^{207}Pb$ for rock 14310:

$$\frac{d(^{208}\mathrm{Pb}/^{207}\mathrm{Pb})}{d(^{232}\mathrm{Th}/^{207}\mathrm{Pb})} = (e^{\lambda_{232}} - 1) - (^{208}\mathrm{Pb}/^{207}\mathrm{Pb})_i \frac{(e^{\lambda_{235}} - 1)}{137.8 \times \kappa}$$

where $(^{208}Pb/^{206}Pb)_i$ is the initial ratio; $\kappa$ is $^{232}Th/^{238}U$. Symbols are: whole-rock WR; density fractions, m1–m4.

tained by the plotting of $^{238}U/^{206}Pb$ versus $^{207}Pb/^{206}Pb$ were completely compatible with the Rb–Sr internal isochron and the K–Ar ages. The initial value of $^{206}Pb/^{207}Pb$ of 0.613 as determined by us is different from that reported by Wasserburg *et al.* (1972) and Tera and Wasserburg (1972b) of 0.69. The reason for the discrepancy is yet to be evaluated. The Th–Pb data of density fractions of 14310 are plotted as $^{208}Pb/^{207}Pb$ versus $^{232}Th/^{207}Pb$ in Fig. 6. The obtained isochron age, 4.2 b.y., has a large uncertainty ($\pm 0.8$ b.y.).

*U–Pb concordia age relations*

Although "third events" were well demonstrated in lunar soil and breccia from the Apollo 12 mission (Tatsumoto *et al.*, 1971), they were not reported for basalts from either Apollo 11 or 12. This situation is best shown by a concordia diagram (Fig. 7). Plotted on the figure are the data on basalts from Apollo 12 (Tatsumoto *et al.*, 1971) for which precise Rb–Sr internal isochrons exist (Papanastassiou and Wasserburg, 1971a). The proximity of the U–Pb data to the appropriate discordia line is remarkable for all samples but 12035. This sample was coarse grained and friable. We therefore feel that the sample was subjected to a "third event"; however, it may be possible that the "third event" introduced into the sample was sample handling (loss of some lead-rich components) and that a proper sample would not show the lead loss. Correspondingly, the average Apollo 11 basalt (Tatsumoto, 1970a) is near a 4650- to

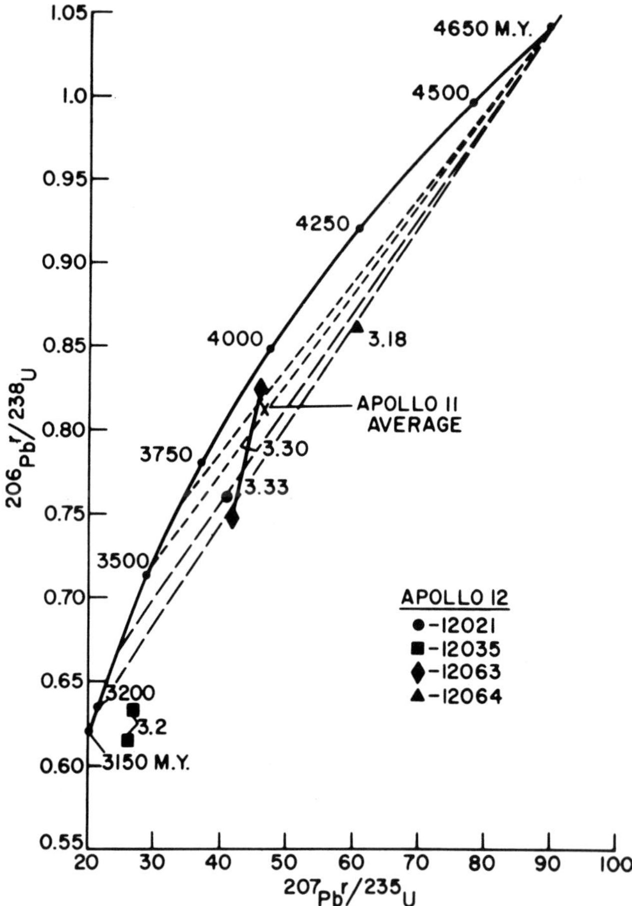


Fig. 7. Concordia diagram for basalts from the average Apollo 11 and Apollo 12 missions for which precise internal Rb–Sr isochrons (ages in billions of years shown next to data points) are available, showing good agreement between the discordia age and the Rb–Sr internal isochron age if source materials 4650 m.y. in age are used on the concordia diagram. Basalts 12063 and 12035 were analyzed twice, and both data points are shown for each sample. Internal Rb–Sr isochron ages are adopted from Papanastassiou and Wasserburg (1969 and 1971a, b).

3500-m.y.-old discordia line, again in good agreement with the Rb–Sr internal isochron as determined by Papanastassiou and Wasserburg (1970). We have no indication of any basalt being affected by a "third event" other than sample handling; so with appropriate sampling and precise data, rather good agreement should be possible between ages obtained by the concordia treatment of U–Pb in whole-rock basalts using 4650 m.y. as the age of the first stage and the Rb–Sr internal isochron age as the second.

The U–Pb concordia diagram for Apollo 14 samples is presented as Fig. 8. The

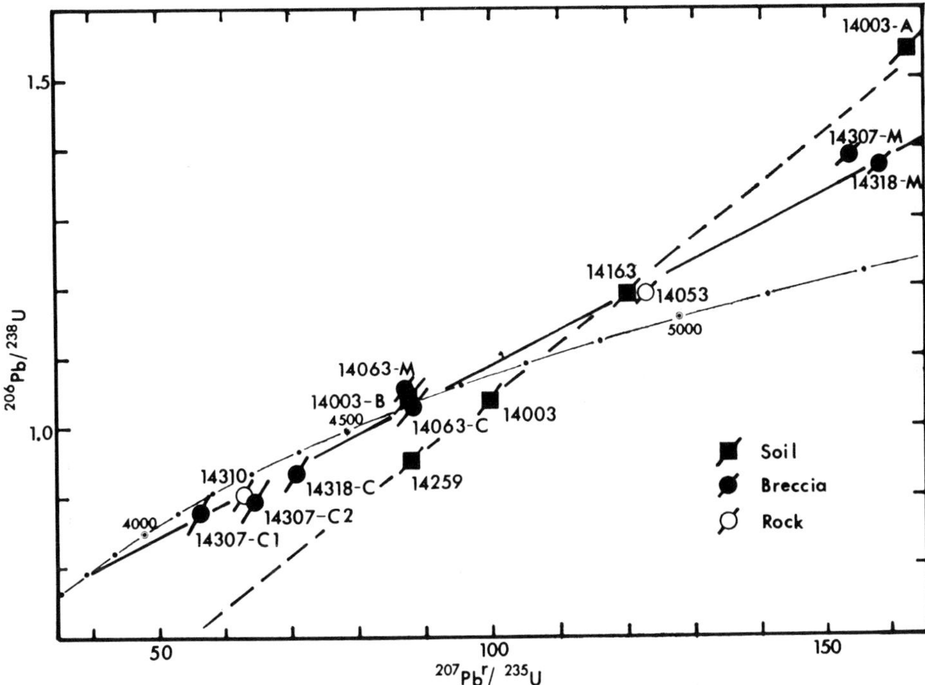


Fig. 8. U–Pb concordia diagram for samples collected by the Apollo 14 mission. The 14003A and B stand for the fractions of acetone floats and of density $\rho$ = 2.9–3.3, respectively. Symbols are as follows: C for breccia clasts (white from 14307 and 14318, black from 14063); M for matrix (white from 14063, medium gray from 14318, dark gray from 14037). Sample 14053 was misplotted in Tatsumoto *et al.* (1972).

whole-rock data from 14310 and 14053 are within a 3800-m.y. to 4650-m.y. discordia line within analytical uncertainties. The U–Pb data are in essential agreement with the Rb–Sr internal and whole-rock isochron (and $^{40}Ar/^{39}Ar$ whole-rock) data.

Tatsumoto *et al.* (1972) stated that the soils 14163, 14003, and 14259 lie along a discordia line that intersects concordia at about 2000 m.y. and about 4750 m.y. Data by Tera and Wasserburg (1972a) on 14163 and 14259 and by Barnes *et al.* (1972) on 14163 are in agreement with this conclusion. Silver (1972) obtained a discordia line for Apollo 14 soils between about 2900 and 4880 m.y. We interpret the low age intersection of 2000 m.y. to approximate the age of a "third event" represented in the lunar soil. We do not restrict the "third event" to just one episodic event; it might be multiple events. The upper intersection probably does not have an age significance. It may indicate that the soils (or source material for the soils) at about 3800 m.y. had a uniform value of $^{207}Pb/^{206}Pb$ in present-day terms of about 0.734 caused by a *decrease* in the value of $^{238}U/^{204}Pb$. The "third event" at roughly 2000 m.y. then resulted in an increase in the $^{238}U/^{204}Pb$ in 14259 and 14033 (second analysis) relative to its value at 3800 m.y.; however, 14163 appears to have been little affected by the "third event."

Inasmuch as samples from the Apollo 11 and 12 missions showed lead loss relative to uranium a long time ago or no change in $^{238}U/^{204}Pb$, localities with samples showing lead gain relative to uranium should be expected on the moon in order to maintain a material balance in uranium and lead. The Apollo 14 site at Fra Mauro appears to be the first such locality, and evidence of predominant lead gain relative to uranium at Fra Mauro is perhaps the single most significant discovery in the U–Th–Pb system in Apollo 14 samples.

The breccias studied—14063, 14307, and 14318—all possess U–Pb systems in the matrix that lie above the U–Pb concordia line and near a 3800- to 4650-m.y. discordia line. Clasts of these breccias lie below the concordia and around a 3800- to 4600-m.y. discordia line.

One way of grouping the Apollo 14 breccias is to place the breccias 14307 and 14318 with the three soils, inasmuch as all these samples come from an area of only 300 or 400 m in extent. Breccia 14063 comes from near Cone Crater some distance away.* Such a grouping suggests a "third-event" discordia line that intersects concordia at about 2900 m.y. The U–Pb plot of Silver (1972) and K–Ar data of Pepin *et al.* (1972) also suggest "third event(s)" in this time period. Neither continuous diffusion nor time-dependent diffusion can explain such a concordia time intersection as could be done for Apollo 12 soils and breccias, but diffusion with a "pulse" (Wetherill, 1963) could be invoked. The pulse could conceivably be at 3800 m.y. We prefer the "third-event" interpretation, which also gets support from the K–Ar data. The 1000-m.y. "third event" for Apollo 12 soils and breccias seems best explained by the impact that resulted in the Copernicus Crater (Tatsumoto *et al.*, 1971; Cliff *et al.*, 1971; Silver, 1971; Wetherill, 1971; Pepin *et al.*, 1972; Eberhardt *et al.*, 1972). The cause of the approximately 2500-m.y. "third event" is not known at this time, but it probably postdates the impact that formed the Imbrium Basin, which is judged on the basis of field geology to be older rather than younger than the mare sites visited by the Apollo 11 and 12 missions.

The data on density fractions of 14310 are shown on a U–Pb concordia diagram (Fig. 9). The data of 14310 lie nicely on the 3800- to 4650-m.y. discordia line with uncertainties of about 100 m.y. in the intercepts, in agreement with Rb–Sr internal isochron and whole-rock ages. We believe the primary isochron ages of 4900 m.y. and 4300 m.y. for 14053 and 14310 do not show the rock age but represent the age of 4650-m.y.-old source material (possibly soil) that was melted to produce these igneous rocks. Rock 14310 increased in $^{238}U/^{204}Pb$, but 14053 decreased in this ratio at the 3800-m.y. event.

These concordia plots of Figs. 8 and 9 clearly indicate that the original deduction given by Tatsumoto (1970a) for the age of outer portions of the moon as 4650 m.y. obtained in the study of Apollo 11 samples is still reasonable. The least such age from our work is the 4400-m.y. intersection for density fractions from 12064 (Tatsumoto *et al.*, 1971), and the greatest such age is 4750 m.y. for soils and breccias of Apollo 14 (this study).

* Breccias 14307 and 14318 are thought by Jackson and Wilshire (1972) to come from one stratigraphic unit (F2) and 14063 from a different one.

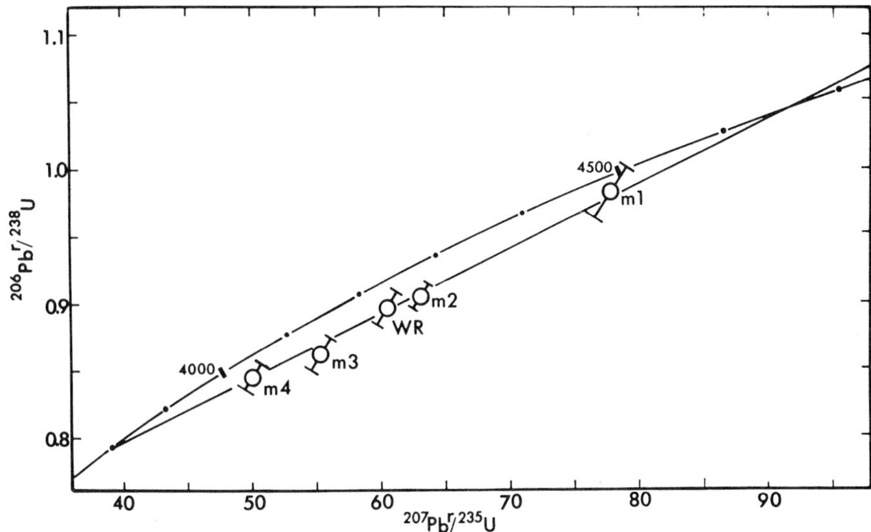


Fig. 9. U–Pb concordia diagram for density fractions of sample 14310. Whole rock is WR; m1–m4 are density fractions.

Table 5. Calculated parameters for concordant age; $t_1$, by two-stage model assuming $t_0 = 4.65 \times 10^9$ yr.

| Sample | Concordant $t_1$ ($10^9$ yr) | $(^{238}U/^{204}Pb)_{t_0-t_1}$ | $(^{206}Pb/^{204}Pb)_{t_1}$ | $(^{207}Pb/^{204}Pb)_{t_1}$ | Observed $(^{238}U/^{204}Pb)_{t=0}$ |
|---|---|---|---|---|---|
| | | Soil | | | |
| 14163 | 3.68 | 638.8 | 189.7 | 269.8 | 415.3 |
| | | Breccia | | | |
| 14063 Clast | 3.80 | 325.4 | 90.6 | 132.1 | 145.1 |
| 14307 Matrix | 3.38 | 514.3 | 196.1 | 253.7 | 262.8 |
| Clast 1 | 3.37 | 1084 | 405.2 | 525.2 | 1802 |
| Clast 2 | 3.99 | 369.6 | 82.3 | 127.0 | 1718 |
| 14318 Matrix | 3.76 | 1175 | 316.8 | 464.3 | 517.2 |
| Clast 1 | 3.37 | 674.0 | 254.6 | 329.6 | 966.7 |
| Sawdust | (3.45) | (1063) | (375.0) | (498.1) | (519.8) |
| | | Igneous rocks | | | |
| 14053 | 4.19 | 569.6 | 88.1 | 146.2 | 275.6 |
| 14310 | 3.71 | 1331 | 375.1 | 541.8 | 3036 |

If we assume two-stage development and concordance of all rocks, the initial lead and $^{238}U/^{204}Pb$ in the source material can be calculated (Table 5). Because the initial surface development of the moon occurred about 4650 m.y. ago, there is no solution for samples 14003, 14063 matrix, and 14259, and it consequently indicates "third event(s)."

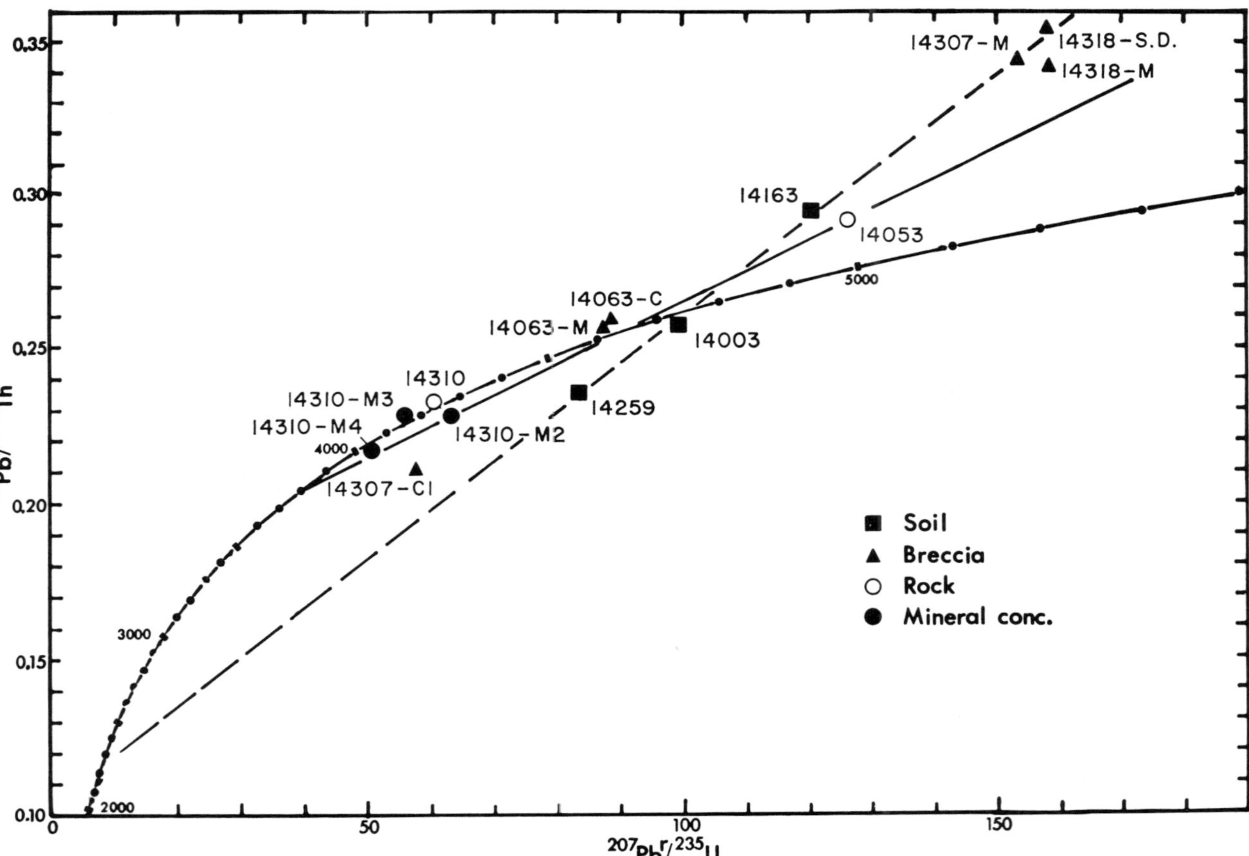


Fig. 10. U–Th–Pb concordia plot for Apollo 14 samples. Symbols after sample numbers are as follows: plain M for breccia matrix, C for breccia clast, SD for sawdust; m2–m4 are density fractions.

*U–Th–Pb concordia relations*

Figure 10 shows the data on a U–Th–Pb concordia plot developed by Allegre (1967) and Steiger and Wasserburg (1966). Because Th/U is not constant, as is $^{238}U/^{235}U$, the treatment of data on the U–Th–Pb concordia plot is more complex than is the U–Pb concordia diagram. Under lunar conditions, however, Th/U does not vary much and treatment of lunar samples on the U–Th–Pb concordia plot approaches the U–Pb concordia plot in simplicity.

As in the U–Pb concordia plot, basalt 14053 lies on a 3800-m.y. to 4650-m.y. discordia line, but even the second analysis of 14310 lies somewhat above the line. The Apollo 14 soils and breccias, with 14063 again being an exception, lie close to a discordia line suggesting an age for the third events of between 2000 and 3000 m.y. As in the U–Pb concordia diagram, samples 14259 and 14003 appear to have had lead depleted relative to thorium as well as uranium. The other samples have been enriched by the "third event."

In the U–Pb concordia plot, the present-day value of $^{207}Pb/^{206}Pb$ for the parent source of the soils and breccias disturbed by the "third event" is represented by the intersection of the 3800- to 4650-m.y. discordia line with the "third event" discordia line. The $^{207}Pb/^{206}Pb$ value is close to 0.72. From the U–Th–Pb isochron plot we can similarly estimate the present-day value of $^{208}Pb/^{207}Pb$ of the source of the soils and breccias disturbed by the "third event" to be about 1.26 (using a Th/U of about 3.65) or about 0.91 for $^{208}Pb/^{206}Pb$.

## CONCLUSIONS

The general chemistry as well as the U–Th–Pb and Rb–Sr systems of lunar soils is clearly complex. Evidence for the presence of lead-enriched components (Silver, 1970; Tatsumoto *et al.*, 1971) and for lead-depleted components (Tatsumoto *et al.*, 1971; this paper) clearly indicate that a lunar soil cannot be taken as a simple system. Despite Silver's arguments (1970, 1971, 1972), the U–Th–Pb evolution systems in the lunar soils, breccias, and rocks can however be clearly and easily explained by taking the age of the outer portions of the moon as 4650 m.y. The Rb–Sr evolution system in the soils, though it is not a closed system, seems to be more simple and also indicates a 4.6-b.y. "total" age. The complexities in the U–Th–Pb system of lunar soils as compared with the Rb–Sr system could be due to either a greater sensitivity of the U–Pb system for old events because of the short mean life of $^{235}U$, to a greater mobilizing character of lead than for rubidium, or to a combination of both causes. Considering all of these data of U–Th–Pb and Rb–Sr systems and those for meteorites (for example, Patterson, 1955; Burnett and Wasserburg, 1967; Gopalan and Wetherill, 1968), there is little doubt that the age of the moon and of the earth as well as that of the solar system can be taken to be about 4650 m.y. as the best working age. The matter of concern at present is what are the limits of the uncertainties in the age? As has previously been mentioned (Tatsumoto, 1970), the age of the moon, 4650 m.y., is meant to be the time of the initial development of the lunar surface. If the lunar gross structure formed in 200–200 m.y. by lunar global differentiation in the beginning stages of the moon (Urey *et al.*, 1971; Gast and McConnell, 1972), the original accretion age of the moon should be increased several hundred million years. On the other hand, if continuously volatilized lead was condensed back onto the lunar surface at the ending stage of lunar accretion from the solar nebula (Anders *et al.*, 1971), a lead greater in $^{207}Pb/^{206}Pb$ than that of primordial lead was added, and the age of the moon should be reduced by several hundred million years.

The great $^{207}Pb/^{206}Pb$ model ages of the soil and breccia must be related to the high volatility of lead and its radiogenic nature. The nature of the U–Th–Pb systematics of Apollo 11 soil and breccia is rather simple and could be explained by a two-stage process. Apollo 12 and 14 soils and breccias are more complex, and at least a three-stage evolution process is suggested. If concordant U–Pb ages (Tatsumoto, 1970; Tatsumoto *et al.*, 1971) and U–Pb internal isochrons are considered (Wasserburg *et al.*, 1972; Tera and Wasserburg, 1972b; and this paper), there is no fundamental difference between U–Th–Pb total-rock ages and Rb–Sr isochron ages for the lunar rocks. The difference is solely due to the highly radiogenic initial lead in the lunar rocks; however, the initial lead is shown to be no problem on lunar whole-rock basalts if the concordia diagrams are used, because there is no evidence of "third events" in lunar basalts and use of primordial lead as the initial lead still leaves the data on the proper discordia line between 4650 m.y. and the age of the basalt.

Perhaps the more significant features of U–Th–Pb system in Apollo 14 samples are: (1) Breccias and some soil samples as well as a basalt sample give evidence of significant enrichment of lead relative to uranium and thorium billions of years ago at the Fra Mauro site, contrasting to Apollo 11 and 12 samples, which give evidence of lead loss. (2) The Apollo 14 soils, 14003 and 14259, indicate clearly that a "third event(s)" occurred 2000–3000 m.y. ago. The $^{238}U/^{204}Pb$ ratios in most of the soils increased as a result of the event. (3) The highly radiogenic character of lead and the high concentration of U, Th, and Pb enabled us to construct a U–Pb internal isochron and U–Pb concordia diagram from density fractions of an igneous rock with reasonably small error. The obtained age from U–Pb internal isochron, 3820 m.y., is in good agreement with our Rb–Sr internal isochron age of 3840 m.y. and that of Wasserburg *et al.* (1972) of 3870 m.y. The U–Pb concordia diagram treatment for the density fractions of 14310 also agrees with a 3800-m.y. crystallization age for the basalt and its derivation from 4650-m.y.-old source material.

*Acknowledgments*—We benefited from discussions with Claude J. Allegre (University of Paris) and our colleagues in the U.S. Geological Survey, particularly Zell E. Peterman and John N. Rosholt. We thank Thomas Murphy (National Bureau of Standards) for advice on lead electroplating, Robert A. Hildreth and Philip A. Reed for laboratory assistance, John S. Stacey, Ernest E. Wilson, and Robert Terrazas for automation of mass spectrometer data collection and reduction, and F. Joseph Jurceka and Noel V. Carpenter for preparation of equipment used in this study. This study was supported in part by NASA Interagency Transfer Order T-2407A.

## References

Allegre C. J. (1967) Méthode de discussion géochronologique concordia généralisée. *Earth Planet. Sci. Lett.* **2,** 57–66.

Anders E., Ganapathy R., Keays R. R., Laul J. C., and Morgan J. W. (1971) Volatile and siderophile elements in lunar rocks: Comparison with terrestrial and meteoritic basalts. *Proc. Second Lunar Sci. Conf., Geochim. Cosmochim. Acta* Suppl. 2, Vol. 2, pp. 1021–1036. MIT Press.

## Appendix. Sample Description*

| Sample no. | Sub. no. | Description | Sample location |
|---|---|---|---|
| *Soils* | | | |
| 14003 | 37 and 59 | Fines, less than 1 mm | Contingency sample |
| 14163 | 156 | Fines, less than 1 mm | Bulk sample |
| 14259 | 16 | Fines, less than 1 mm | Comprehensive sample |
| *Breccias* | | | |
| 14063 | 37 | Moderately friable, light-colored feldspathic-rich, glass-poor, matrix and dark-colored shocked basaltic clasts (M. Tatsumoto consortium). | Station C1 |
| 14307 | 26 | Coherence is moderate; light-colored feldspathic clasts were separated from dark-colored glass-rich matrix. C1 and C2 were separated from different portions. Sample was obtained from the P. W. Gast consortium. | Station G |
| 14318† | 27 | Coherence: tough, fractured. Clast 1 (hornfelsed, microbrecciated basalt) and clast 2 (plagioclase-rich anorthositic fragment) were separated from dark-colored fine-grained matrix (M. Tatsumoto consortium). | Station H |
| *Basalts* | | | |
| 14053 | 27 | Subophitic, grain size 0.5–2 mm, 60% feldspar, 5% olivine, 30% pyroxene. | Station C2 |
| 140310 | 71 | Intergranular, grain size 0.1–0.5 mm, 45% feldspar, 45% pyroxene. | Station G |

* Detailed descriptions and sample locations may be obtained from NASA TM X-58062, *Apollo 14 Lunar Sample Information Catalog*, prepared by Curator's Office, Manned Spacecraft Center, and NASA SP-272, *Apollo 14 Preliminary Science Report*, prepared by NASA Manned Spacecraft Center.

† Detailed petrology of the breccia may be obtained from Chao *et al.* (1972).


Andersen C. A. and Hinthorne J. R. (1972) U, Th, Pb, and REE abundances and $Pb^{207}/Pb^{206}$ ages of individual minerals in returned lunar material by ion microprobe mass analysis. *Proc. Third Lunar Sci. Conf., Geochim. Cosmochim. Acta* Suppl. 3, Vol. 2. MIT Press.

Barnes I. L., Carpenter B. S., Garner E. L., Gramlich J. W., Kuehner E. C., Machlan L. A., Mainethal E. J., Moody J. R., Moore L. J., Murphy T. J., Paulsen P. J., Sappenfield K. M., and Shields W. R. (1972) The isotopic abundance ratio and assay analysis of selected elements in Apollo 14 samples (abstract). In *Lunar Science—III* (editor C. Watkins), pp. 41–43, Lunar Science Institute Contr. No. 88.

Burnett D. S. and Wasserburg G. J. (1967) $^{87}Rb$–$^{87}Sr$ ages of silicate inclusions in iron meteorites. *Earth Planet. Sci. Lett.* **2**, 397–408.

Burnett D., Monnin M., Seitz M., Walker R., and Yuhas D. (1971) Lunar astrology—U–Th distributions and fission-track dating of lunar samples. *Proc. Second Lunar Sci. Conf., Geochim. Cosmochim. Acta* Suppl. 2, Vol. 2, pp. 1503–1519. MIT Press.

Chao E. C. T., Minkin J. A., and Boreman J. A. (1972) The petrology of some Apollo 14 breccias. *Proc. Third Lunar Sci. Conf., Geochim. Cosmochim. Acta* Suppl. 3, Vol. 3. MIT Press.

Cliff R. A., Lee-Hu C., and Wetherill G. W. (1971) Rb–Sr and U, Th–Pb measurements on Apollo 12 material. *Proc. Second Lunar Sci. Conf., Geochim. Cosmochim. Acta* Suppl. 2, Vol. 2, pp. 1493–1502. MIT Press.

Compston W., Chappell B. W., Arriens P. A., and Vernon M. J. (1970) The chemistry and age of Apollo 11 lunar material. *Proc. Apollo 11 Lunar Sci. Conf., Geochim. Cosmochim. Acta* Suppl. 1, Vol. 2, pp. 1007–1027. Pergamon.

Compston W., Vernon M. J., Berry H., Rudowski R., Gray C. M., and Ware N. (1972) Age and petrogenesis of Apollo 14 basalts (abstract). In *Lunar Science—III* (editor C. Watkins), pp. 151–153, Lunar Science Institute Contr. No. 88.

Doe B. R. and Tatsumoto M. (1972) Volatilized lead from Apollo 12 and 14 soils. *Proc. Third Lunar Sci. Conf., Geochim. Cosmochim. Acta* Suppl. 3, Vol. 2. MIT Press.

Eberhardt P., Eugster O., Geiss J., Grögler N., Schwarzmüller J., Stettler A., and Weber L. (1972) When was the Apollo 12 KREEP ejected? (abstract). In *Lunar Science—III* (editor C. Watkins), pp. 206–208, Lunar Science Institute Contr. No. 88.

Fanale F. P., and Nash D. B. (1971) Potassium-uranium systematics of Apollo 11 and Apollo 12 samples: Implications for lunar material history. *Science* **171,** 282–284.

Gast P. W. and McConnell R. K. (1972) Evidence for initial chemical layering of the moon (abstract). In *Lunar Science—III* (editor C. Watkins), pp. 289–290, Lunar Science Institute Contr. No. 88.

Gopalan K. and Wetherill G. W. (1968) Rubidium-strontium age of hypersthene (L) chondrites. *J. Geophys. Res.* **73,** 7133–7136.

Jackson E. D. and Wilshire H. G. (1972) Classification of the samples returned from the Apollo 14 landing site (abstract). In *Lunar Science—III* (editor C. Watkins), pp. 418–420, Lunar Science Institute Contr. No. 88.

LSPET (Lunar Sample Preliminary Examination Team) (1971) Chap. 5. Preliminary examination of lunar samples. *Apollo 14 Preliminary Science Report*, NASA SP-272, pp. 109–131.

Murthy V. R., Evenson N. M., Jahn B., and Coscio M. R. Jr. (1972) Rb–Sr ages, trace elements, and speculations on lunar differentiation (abstract). In *Lunar Science—III* (editor C. Watkins), pp. 571–572, Lunar Science Institute Contr. No. 88.

Papanastassiou D. A. and Wasserburg G. J. (1970) Rb–Sr ages from the Ocean of Storms. *Earth Planet. Sci. Lett.* **8,** 269–278.

Papanastassiou D. A. and Wasserburg G. J. (1971a) Lunar chronology and evolution from Rb–Sr studies of Apollo 11 and 12 samples. *Earth Planet. Sci. Lett.* **11,** 37–62.

Papanastassiou D. A. and Wasserburg G. J. (1971b) Rb–Sr ages of igneous rocks from the Apollo 14 mission and the age of the Fra Mauro formation. *Earth Planet. Sci. Lett.* **12,** 36–48.

Patterson C. C. (1955) The $^{207}Pb/^{206}Pb$ ages of some stone meteorites. *Geochim. Cosmochim. Acta* **7,** 151–153.

Pepin R. O., Bradley J. G., Dragon J. C., and Nyquist L. E. (1972) K–Ar dating of lunar soils: Apollo 12, Apollo 14, and Luna 16 (abstract). In *Lunar Science—III* (editor C. Watkins), pp. 602–604, Lunar Science Institute Contr. No. 88.

Silver L. T. (1970) Uranium-thorium-lead isotopes in some Tranquility Base samples and their implications for lunar history. *Proc. Apollo 11 Lunar Sci. Conf., Geochim. Cosmochim. Acta* Suppl. 1, Vol. 2, pp. 1533–1574. Pergamon.

Silver L. T. (1971) U–Th–Pb isotope systems in Apollo 11 and 12 regolithic materials and a possible age for the Copernicus impact event (abstract). *Trans. Amer. Geophys. Union* **52,** 534.

Silver L. T. (1972) U–Th–Pb abundances and isotopic characteristics in some Apollo 14 rocks and soils and an Apollo 15 soil (abstract). In *Lunar Science—III* (editor C. Watkins), pp. 704–706, Lunar Science Institute Contr. No. 88.

Steiger R. H. and Wasserburg G. J. (1966) Systematics in the $Pb^{208}$–$Th^{232}$, $Pb^{207}$–$U^{235}$, and $Pb^{206}$–$U^{238}$ systems. *J. Geophys. Res.* **71,** 6065–6090.

Tatsumoto M. (1966) Genetic relations of oceanic basalts as indicated by lead isotopes. *Science* **153,** 1094–1101.

Tatsumoto M. (1970a) Age of the moon: An isotopic study of U–Th–Pb systematics of Apollo 11 lunar samples, II. *Proc. Apollo 11 Lunar Sci. Conf., Geochim. Cosmochim. Acta* Suppl. 1, Vol. 2, pp. 1595–1612. Pergamon.

Tatsumoto M. (1970b) U–Th–Pb age of Apollo 12 rock 12013. *Earth Planet. Sci. Lett.* **9,** 193–200.

Tatsumoto M., Knight R. J., and Doe B. R. (1971) U–Th–Pb systematics of Apollo 12 lunar samples. *Proc. Second Lunar Sci. Conf., Geochim. Cosmochim. Acta* Suppl. 2, Vol. 2, pp. 1521–1546. MIT Press.

Tatsumoto M., Hedge C. E., Doe B. R., and Unruh D. (1972) U–Th–Pb and Rb–Sr measurements on some Apollo 14 lunar samples (abstract). In *Lunar Science—III* (editor C. Watkins), pp. 741–743, Lunar Science Institute Contr. No. 88.

Tera F., Eugster O., Burnett D. S., and Wasserburg G. J. (1970) Comparative study of Li, Na, K, Rb, Cs, Ca, Sr, and Ba abundances in achondrites and in Apollo 11 lunar samples. *Proc. Apollo 11 Lunar Sci. Conf., Geochim. Cosmochim. Acta* Suppl. 1, Vol. 2, pp. 1637–1657. Pergamon.

Tera F. and Wasserburg G. J. (1972a) U–Th–Pb analyses from the Sea of Fertility. *Earth Planet. Sci. Lett.* **13,** 457–466.

Tera F. and Wasserburg G. T. (1972b) U–Th–Pb systematics in three Apollo 14 basalts and the problem of initial Pb in lunar rocks. *Earth Planet. Sci. Lett.* **14**, 281–304.

Urey H. C., Marti K., Hawkins J. W., and Liu M. K. (1971) Model history of the lunar surface. *Proc. Second Lunar Sci. Conf., Geochim. Cosmochim. Acta* Suppl. 2, Vol. 2, pp. 987–998. MIT Press.

Wakita H. and Schmitt R. A. (1970) Elemental abundances in seven fragments from lunar rock 12013. *Earth Planet. Sci. Lett.* **9,** 169–176.

Wasserburg G. J. (1963) Diffusion processes in lead-uranium systems. *J. Geophys. Res.* **68,** 4823–4846.

Wasserburg G. J., Turner G., Tera F., Podosek F. A., Papanastassiou D. A., and Huneke J. C. (1972) Comparison of Rb–Sr, K–Ar, and U–Th–Pb ages; lunar chronology and evolution (abstract). In *Lunar Science—III* (editor C. Watkins), pp. 788–790, Lunar Science Institute Contr. No. 88.

Wetherill G. W. (1963) Discordant uranium-lead ages—Pt. 2. Discordant ages resulting from diffusion of lead and uranium. *J. Geophys. Res.* **68,** 2957–2965.

Wetherill G. W. (1971) Of time and the moon. *Science* **173,** 383–392.

Proceedings of the Third Lunar Science Conference
(Supplement 3, *Geochimica et Cosmochimica Acta*)
Vol. 2, pp. 1557–1567
The M.I.T. Press, 1972

# The ages of lunar material from Fra Mauro, Hadley Rille, and Spur Crater

L. HUSAIN, O. A. SCHAEFFER, J. FUNKHOUSER, and J. SUTTER*

Department of Earth and Space Sciences, State University of New York, Stony Brook, New York 11790

**Abstract**—Twelve crystalline and fragmental rocks from Fra Mauro and six rocks from Hadley Rille and Spur Crater have been dated by the $^{40}Ar$–$^{39}Ar$ technique. Three characteristic types of thermal release patterns are observed that may influence the derived $^{40}Ar$–$^{39}Ar$ age by up to $0.1 \times 10^9$ yr. High-temperature ages from Fra Mauro suggest that the Imbrium Basin was excavated $(3.7 - 3.9) \times 10^9$ yr ago. An age of a Fra Mauro clastic fragment of $3.50 \times 10^9$ yr may represent the time of another major impact event. The crystallization age of the basalts from Hadley Rille and Spur Crater all group at $3.25 \pm 0.1 \times 10^9$ yr indicating that at least a portion of Mare Imbrium filled with basalt about $500 \times 10^6$ yr after its formation.

Cosmic-ray exposure ages were determined from $^{38}Ar_{sp}/^{37}Ar$ versus $^{37}Ar$ thermal release patterns. Such ages compare favorably with those calculated in the standard manner from unirradiated sample chips. A grouping at about $30 \times 10^6$ yr probably defines the Cone Crater event. The ages of the remaining samples from both landing sites vary widely, from 80 to $740 \times 10^6$ yr, with no apparent age groupings.

## INTRODUCTION

THE APPLICATION of the $^{40}Ar$-$^{39}Ar$ dating technique to a variety of lunar sample types has proved to be most useful in delineating periods of intense activity on the moon. The experimental uncertainties of the method are small and interlaboratory precision is remarkable. The major limitation in the accuracy of the $^{40}Ar$-$^{39}Ar$ dating is the interpretation of complex isotope release patterns. The most serious shortcoming to the *application* of $^{40}Ar$-$^{39}Ar$ ages, or any measured property of a sample for that matter, to site studies of the lunar surface involves the method of sampling. Limited by the lack of accessible outcrops, only loose, structurally undefined samples are collected, thus the knowledge gleaned from each sample can only be related to a specific region of the moon on a statistical basis. The probability of correctly assigning a set of properties to a particular event or location on the moon is increased, of course, with increasing numbers of samples analyzed.

The crystallization ages of a number of samples from Mare Tranquillitatis group into two units of 3.6 and $3.8 \times 10^9$ yr for the high-K and low-K rocks respectively (Papanastassiou and Wasserburg, 1971; Turner, 1970, 1971), while ages from Oceanus Procellarum lie about $3.3 \times 10^9$ yr (Compston *et al.*, 1971; Papanastassiou and Wasserburg, 1970, 1971; Turner, 1971). The $^{40}Ar$-$^{39}Ar$ ages of several highly brecciated fragmental and crystalline rocks from Fra Mauro suggest that the Imbrium Event occurred about $3.8 \times 10^9$ yr ago (Husain *et al.*, 1971; Sutter *et al.*, 1971). We

* Present address: Ohio State University, Columbus, Ohio.

report here additional $^{40}Ar$-$^{39}Ar$ formation ages of Apollo 14 samples, along with cosmic-ray exposures ages. Crystallization and exposure ages for six Apollo 15 samples are also presented, including 15415 and 15555 which have been reported elsewhere (Husain *et al.*, 1972).

## EXPERIMENTAL

A small portion of each lunar sample was removed for the preparation of thin sections. Another portion was used for a cursory mass spectrometric analysis to check for the presence of contaminating solar wind argon. Any sample with a measured $^{40}Ar/^{36}Ar$ ratio less than about 10 was not irradiated. Samples, weighing 10–180 mg each, along with hornblende standards (50–90 mg each) were then vacuum encapsulated in quartz tubing. The quartz tube along with a high-purity nickel wire 380 $\mu$ thick was packed in a water-tight aluminum can. The package was then irradiated in the core of the high flux beam reactor at Brookhaven National Laboratory. The fast neutron flux was monitored by the hornblende standard and by the nickel wire. The integrated fast neutron flux was 0.2 to 1 $\times$ $10^{19}$ n/cm$^2$. The neutron flux variation along the sample length of 3 cm was less than 1%. Irradiated samples were removed from the quartz tubes and loaded in a gas extraction system connected to a 15.2-cm, 60° sector, magnetic deflection mass spectrometer. Gases were extracted from a sample by radio-frequency induction heating in a series of successively higher temperatures ranging from 600° to 1500 or 1650°C. The argon was purified by means of hot titanium getters and a charcoal trap that was cooled with a mixture of dry ice and acetone. Samples and standards were held at each temperature for 1 hr, and each gas fraction was measured under static conditions. Procedural blanks were measured between each series of temperature runs and were generally less than 1% of the sample for every Ar isotope, except $^{39}Ar$, for which it was generally less than 2%. Instrument sensitivity was measured by means of an argon standard and was 1.8 $\times$ $10^{-11}$ cm$^3$ STP/mv. The correction for mass discrimination was 0.8% per mass for argon isotopes. The K–Ar age of the hornblende standard has been accurately measured as (2.61 $\pm$ 0.06) $\times$ $10^9$ yr (Hanson *et al.*, 1971; Hanson, 1971). The treatment of data has been described in detail by Husain *et al.* (1971; 1972) and will not be repeated here.

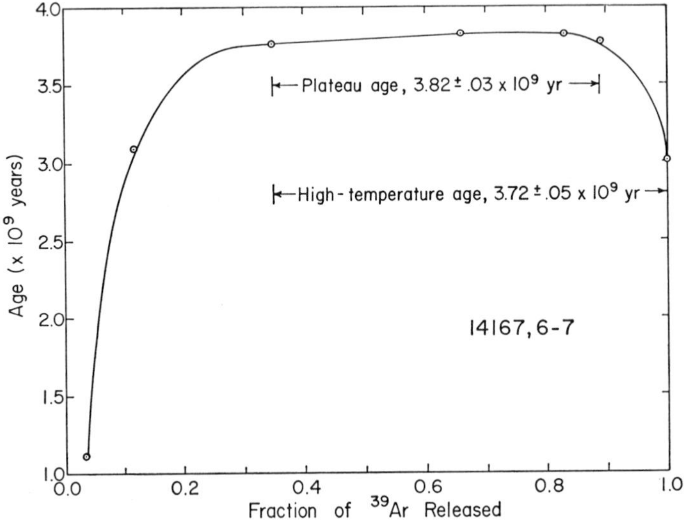


Fig. 1. The $^{40}Ar$-$^{39}Ar$ thermal release pattern for fragmental rock 14167,6,7, see text.

## $^{40}Ar$-$^{39}Ar$ AGES

We have now measured the $^{40}Ar$-$^{39}Ar$ ages for 12 samples from Fra Mauro and 6 from Hadley Rille and Spur Crater. Generally, the thermal release patterns of radiogenic argon and neutron-produced $^{39}Ar$ fall into three categories. Figures 1–3 represent $^{40}Ar$-$^{39}Ar$ ages of the individual temperature fractions plotted against the cumulative fraction of $^{39}Ar$ released for each of the three typical examples of release pattern.

Most samples from Fra Mauro exhibit a characteristic thermal release pattern in which the $^{40}Ar/^{39}Ar$ ratio increases to a stable plateau but then decreases in the highest temperature fraction, as shown in Fig. 1 for sample 14167,6,7. Most of the Apollo 15 basalts also show this behavior. No satisfactory explanation of this phenomenon has yet been advanced. The high-temperature dip raises the question just how to interpret the thermal release pattern in order to derive a valid formation age. The age calculated from the plateau is generally 0.05 to 0.10 $\times$ $10^9$ yr greater than the age determined

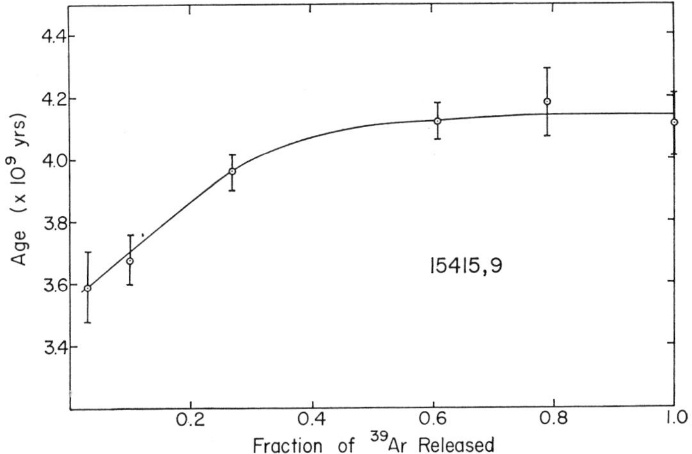


Fig. 2. The $^{40}Ar$-$^{39}Ar$ thermal release pattern for lunar anorthosite 15415,9.

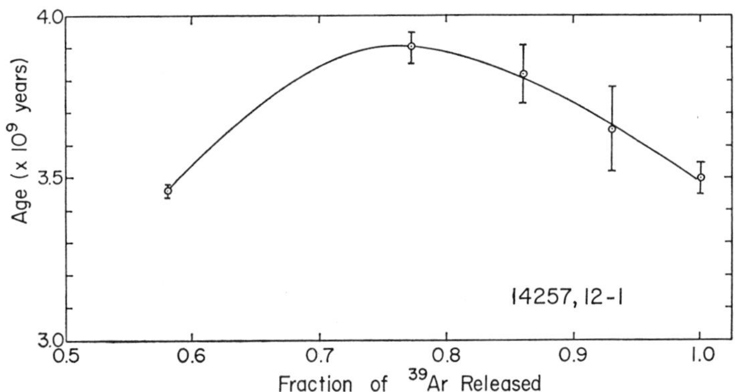


Fig. 3. The $^{40}Ar$-$^{39}Ar$ thermal release pattern for fragmental rock 14257,12,1.

Table 1. Age of Fra Mauro samples.

| Sample | K, %† | Formation Age, × $10^9$ yr | | | Exposure Age, × $10^6$ yr* |
|---|---|---|---|---|---|
| | | $^{40}Ar$–$^{39}Ar$ Plateau | $^{40}Ar$–$^{39}Ar$ High Temperature | K–Ar | |
| 14053,34 | 0.08 | 3.92 ± 0.08 | 3.83 ± 0.09 | 3.69 | 21 ± 5 (19, 14, 22, 19) |
| 14310,101 | 0.41 | 3.78 ± 0.03 | 3.74 ± 0.04 | 3.68 | 210 ± 50 (120, 140, 250, 260) |
| 14152,1,1 | 0.37 | 3.82 ± 0.10 | 3.77 ± 0.15 | 3.76 | 82 ± 12 |
| 14152,1,2 | 0.38 | 3.75 ± 0.05 | 3.75 ± 0.05 | 3.72 | 95 ± 10 |
| 14167,6,1 | 0.44 | 3.79 ± 0.03 | 3.76 ± 0.04 | 3.61 | 43 ± 9 |
| 14167,6,3 | 0.30 | 3.83 ± 0.05 | 3.80 ± 0.05 | 3.67 | ~50‡ |
| 14167,6,4 | 0.36 | 3.72 ± 0.03 | 3.70 ± 0.03 | 3.61 | 32 ± 3 |
| 14167,6,7 | 0.36 | 3.82 ± 0.03 | 3.72 ± 0.05 | 3.64 | 29 ± 5 (20, 17, 27, 36) |
| 14192,1,1 | 0.31 | No plateau | | 3.92 | 35 ± 5 |
| 14193,2,1 | 0.34 | 3.82 ± 0.10 | 3.75 ± 0.10 | 3.62 | 35 ± 5 |
| 14257,12,1 | 0.95 | 3.87 ± 0.06 | 3.78 ± 0.10 | 3.60 | ~40‡ |
| 14257,12,3 | 0.46 | 3.50 ± 0.12 | 3.50 ± 0.12 | 3.20 | 740 ± 130 |

* Exposure ages determined from $^{38}Ar_{sp}/^{37}Ar$ thermal release patterns. Uncertainties are statistical and do not reflect possible errors in K or the production rate of $^{38}Ar$ from Ca. The ages in parentheses were calculated from cosmogenic $^{3}He$, $^{21}Ne$, $^{38}Ar$, $^{126}Xe$, respectively, for unirradiated chips (see text and Table 4). The parenthetic group at sample 14167,6,7 was determined on a similar but different fragment, 14167,6,11 (Table 4).

† K determined by comparing the amount of $^{39}Ar$ in sample to that in hornblende standard of known K content. Accuracy ±15%.

‡ No plateau. Age calculated from total amount of $^{38}Ar_{sp}$ and $^{37}Ar$ released.

from the average $^{40}Ar/^{39}Ar$ ratio for all high-temperature fractions ("high-temperature age," see Fig. 1).

An ideal thermal release pattern is similar to that in Fig. 1, but with no high-temperature decrease. In this case, the plateau age and high-temperature age are identical. Fig. 2 closely approximates such a release pattern, yet even here some ambiguity exists as to which part of the plateau provides a valid age. The last three release points yield an age of 4.14 × $10^9$ yr, whereas the last four points give an age of 4.09 × $10^9$ yr. The ideal type of thermal release pattern is typical of those reported for most all Apollo 11 and 12 samples, excluding 12013 (Turner, 1971) but is not observed for any of the samples from Fra Mauro that we have analyzed.

A third type of thermal release pattern is depicted in Fig. 3 for sample 14257,12,1. Here there is no real plateau although we have calculated both plateau and high-temperature ages for this sample (Table 1). In extreme cases of this type of pattern, only a conventional K–Ar age based on the total $^{40}Ar$-$^{39}Ar$ ratio can be determined (14192,1,1; Table 1). Apparently there is a redistribution of Ar and/or K with an overall loss or gain of $^{40}Ar$; however, no petrologic or deformational characteristics have yet been correlated to either of the three types of thermal release patterns described here.

In Tables 1 and 2, we present both plateau and high temperature $^{40}Ar$-$^{39}Ar$ ages, as well as K–Ar ages (based on the total $^{40}Ar/^{39}Ar$ ratio) for all the Apollo 14 and 15 rocks we have analyzed. It is interesting to note that despite the apparent complex and violent history of the Fra Mauro rocks, they have lost far less radiogenic Ar than the crystalline rocks from Hadley Rille.

The plateau ages of the Fra Mauro rocks listed in Table 1 lie in the range (3.72–3.92) × $10^9$ yr with one sample at 3.50 × $10^9$ yr. Previously reported plateau

Table 2. Ages of Hadley Rille and Spur Crater samples.

| Sample | K, %† | Formation Age, × $10^9$ yr: $^{40}Ar$–$^{39}Ar$ | Formation Age, × $10^9$ yr: K–Ar | Exposure Age,* × $10^6$ yr |
|---|---|---|---|---|
| 15415,9 | 0.011 | 4.09 ± 0.19 | 3.95 | 90 ± 10 (3, 40, 95) |
| 15555,33 | 0.035 | 3.28 ± 0.06 | 2.87 | 80 ± 10 (77, 73, 81) |
| 15385,3 | 0.027 | 3.32 ± 0.06 | 3.28 | 270 ± 14 |
| 15668,5 | 0.040 | 3.15 ± 0.08 | 2.50 | 510 ± 32 |
| 15678,4 | 0.039 | 3.28 ± 0.06 | 2.51 | 150 ± 20 |
| 15683,3 | 0.037 | 3.27 ± 0.06 | 2.86 | 290 ± 19 |

* Age calculated from the total amount of $^{38}Ar_{sp}/^{37}Ar$ released. Uncertainties are statistical. Parenthetic ages were calculated from $^{3}He$, $^{21}Ne$, $^{38}Ar$, respectively, for unirradiated chips (see text and Table 4).

† K determined by comparing the amount of $^{39}Ar$ in sample to that in hornblende monitor of known K-content. Accuracy ±10%.

ages (Husain *et al.*, 1971; Schaeffer *et al.*, 1972; Sutter *et al.*, 1971; Turner *et al.*, 1971; York *et al.*, 1972) lie in the range (3.72 – 4.06) × $10^9$ yr. Since the Fra Mauro consists of an extensive ejecta blanket from the excavation of the Imbrium Basin, the ages of the rocks must be at least as old as the Imbrium event. On the other hand, as no Imbrium ejecta is observed on Mare Tranquillitatis, the Imbrium event must be older than the Mare Tranquillitatis basalts of 3.7 × $10^9$ yr age. Hence, we conclude that the Imbrium event occurred between 3.7 and 3.9 billion years ago.

The lower age of 3.50 × $10^9$ yr for the clastic rock chip, 14257,12,3, lies outside the experimental uncertainties of the Imbrium grouping and may be an ejecta from a neighboring basin. Two other rock chips, both classified as nonmare type, from sample 14161, give a similar age, (3.52 ± 0.03) and (3.59 ± 0.03) × $10^9$ yr (Kirsten *et al.*, 1972), thus giving credence to a possible large-scale impacting event at (3.5 ± 0.10) × $10^9$ yr ago.

The Hadley Rille basalt, 15555, provides a rare opportunity for comparing the age data from various laboratories using different methods. In Table 3, we list all published age measurements for this rock. With the exception of the 3.54 × $10^9$ yr age (Chappell *et al.*, 1972), all the results, including the analyses on mineral separates, are in excellent agreement. The $^{40}Ar$-$^{39}Ar$ age that we have reported earlier (Husain *et al.*, 1972) for the unique sample of lunar anorthosite, 15415, is in reasonable agreement with the ages obtained by other workers: (4.05 ± 0.15) × $10^9$ yr (Turner, 1972) and 4.00 × $10^9$ yr (Stettler *et al.*, 1972).

Table 3. Age measurements of 15555 by several investigators using different methods.

| Type of Sample | Age, × $10^9$ yr | Method | Reference |
|---|---|---|---|
| Whole Rock | 3.28 ± 0.06 | $^{40}Ar$–$^{39}Ar$ | Husain *et al*, 1972 |
| Whole Rock | 3.29 ± 0.05 | $^{40}Ar$–$^{39}Ar$ | Alexander *et al*, 1972 |
| Whole Rock | 3.31 ± 0.05 | $^{40}Ar$–$^{39}Ar$ | York *et al*, 1972 |
| Whole Rock | 3.219 ± 0.025 | $^{40}Ar$–$^{39}Ar$ | Podosek *et al*, 1972 |
| Plagioclase | 3.308 ± 0.025 | $^{40}Ar$–$^{39}Ar$ | Podosek *et al*, 1972 |
| Plagioclase | 3.32 ± 0.05 | K–Ar | Podosek *et al*, 1972 |
| Plagioclase | 3.31 ± 0.07 | K–Ar | Murthy *et al*, 1972 |
| Pyroxene | 3.28 ± 0.09 | $^{40}Ar$–$^{39}Ar$ | Podosek *et al*, 1972 |
| Isochron | 3.30 ± 0.08 | Rb–Sr | Murthy *et al*, 1972 |
| | 3.3 | Rb–Sr | Cliff *et al*, 1972 |
| | 3.54 ± 0.13 | Rb–Sr | Chappell *et al*, 1972 |

Excluding 15415, the high-temperature age of the five basalts from Hadley Rille and Spur Crater group at about (3.2–3.3) $\times$ $10^9$ yr, which defines a probable filling of at least a portion of Mare Imbrium, 500 $\times$ $10^6$ yr after its excavation.

## COSMIC-RAY EXPOSURE AGES

The calculation of exposure ages from the amount of spallation produced $^{38}$Ar ($^{38}Ar_{sp}$) for those samples on which $^{40}$Ar-$^{39}$Ar ages have been determined is complicated by the production of $^{38}$Ar and $^{36}$Ar from Ca, K, and Cl during neutron irradiation. In addition, the maximum temperature of 1500° employed for the release of $^{39}$Ar and $^{40}$Ar from most of the samples was not sufficient to extract all the $^{38}$Ar, $^{37}$Ar, $^{36}$Ar. In an effort to obviate this latter problem, and in order to learn more about the production and losses of spallogenic $^{38}$Ar, plots of $^{38}Ar_{sp}/^{37}Ar$ versus the fraction of $^{37}$Ar released were studied. The measured quantities of $^{38}$Ar and $^{36}$Ar at each release temperature were corrected for a neutron-induced production from Ca and K using the following measured relationships: $(^{38}Ar/^{37}Ar)_{Ca} = 5.9 \times 10^{-5}$, $(^{36}Ar/^{37}Ar)_{Ca} = 2.3 \times 10^{-4}$, $(^{38}Ar/^{39}Ar)_K = 1.0 \times 10^{-2}$. Spallation $^{38}$Ar was then calculated assuming $(^{36}Ar/^{38}Ar)_{solar\ wind} = 5.35$ and $(^{36}Ar/^{38}Ar)_{sp} = 0.54$, determined from 600 MeV proton interactions with calcium. Details of these experiments will be published elsewhere.

The bulk of $^{38}Ar_{sp}$ in lunar material is produced from Ca, thus it should be correlated to $^{37}$Ar. This is most apparent in Fig. 4 where $^{38}Ar_{sp}/^{37}Ar$ is plotted against the fraction of $^{37}$Ar released at each temperature for rock 14310. The exposure age at any point on the $^{38}Ar_{sp}/^{37}Ar$ thermal release diagram can be determined from the equation:

$$T = \kappa \frac{(^{39}Ar)\ (^{38}Ar)}{(K)\ \ (^{37}Ar)}$$

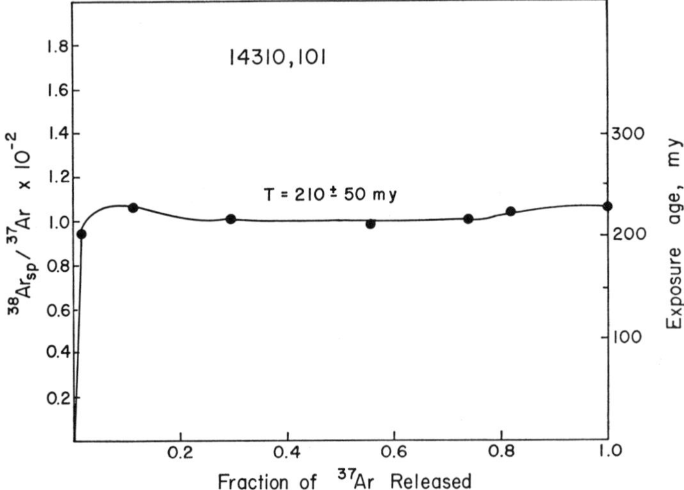


Fig. 4. $^{38}Ar_{sp}/^{37}Ar$ thermal release pattern for rock 14310,101. Stable plagioclase plateau yields an exposure age of 210 $\times$ $10^6$ yr.

where the ratio $^{39}Ar/K$ is determined in the total sample, and $\kappa = 3.0 \times 10^7$ g my/cm$^3$ STP using a production rate for $^{38}Ar$ of $1.7 \times 10^{-8}$ cm$^3$ STP/g Ca/my (Bogard *et al.*, 1971). The age so derived can only be related to the cumulative time the sample has spent in the top few meters of the lunar regolith so long as the points chosen correspond to $^{38}Ar_{sp}$ release from Ca sites.

When cosmogenic $^{38}Ar$ is significantly produced from other target elements, a thermal release pattern typified by that for 14053 results (Fig. 5). Here, the first hump may result from the $^{38}Ar$ produced from nuclear reactions on K. The coincidence of the first maximum release of $^{38}Ar_{sp}$ with that of $^{40}Ar_{rad}$ on a differential thermal release curve (Fig. 5) substantiates this. Additionally, a portion of this hump may

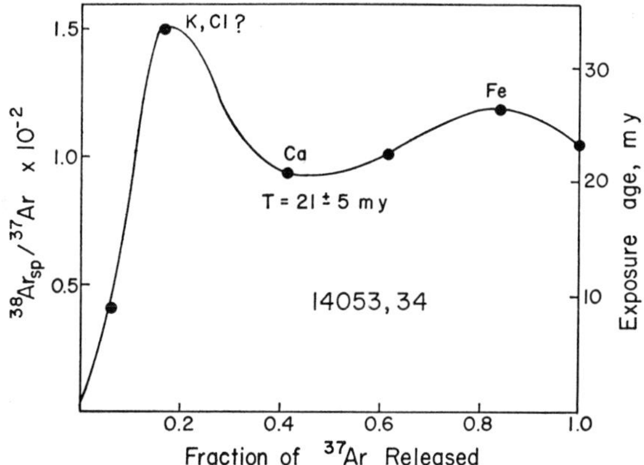


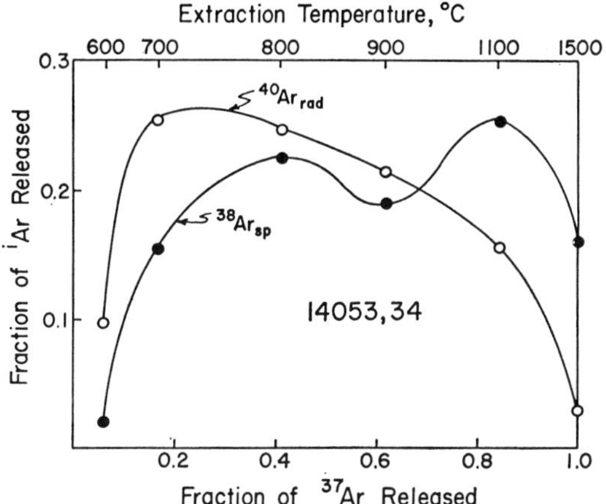


Fig. 5. $^{38}Ar_{sp}/^{37}Ar$ thermal release pattern for rock 14053,34 (top). The minimum is assumed to indicate release strictly from Ca sites and yields an exposure age consistent with those derived from an unirradiated sample. The first hump coincides with maximum release of radiogenic Ar (bottom figure) and represents production of cosmogenic $^{38}Ar$ mainly from K, whereas the other hump may correspond to release from Fe-containing minerals. The double-humped profile for differential $^{38}Ar_{sp}$ release in the bottom figure is typical of many meteorites and lunar samples.

represent $^{38}$Ar produced from Cl during neutron irradiation and released from relatively low-energy chlorine sites. The high-temperature hump could result from the release of cosmogenic $^{38}$Ar produced in low-Ca, high-Fe pyroxenes. Assuming that the valley represents release strictly from plagioclase, an exposure age of $(21 \pm 5) \times 10^6$ yr results, in agreement with exposure ages determined from an unirradiated portion of the same sample (see Table 1).

A more complicated $^{38}Ar_{sp}/^{37}Ar$ thermal release pattern is exemplified by that from 14257,12,1 (Fig. 6). No plagioclase plateau is apparent. An exposure age of

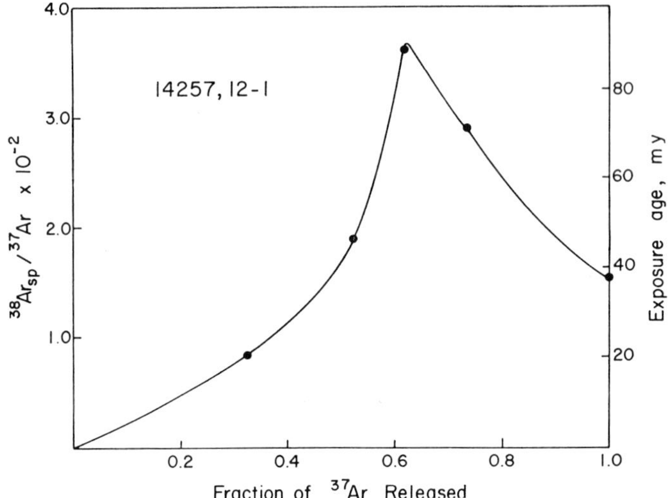


Fig. 6. $^{38}Ar_{sp}/^{37}Ar$ thermal release pattern for 14257,12,1. No plateau is apparent. An exposure age of about $40 \times 10^6$ yr is obtained from the total $^{38}Ar_{sp}/^{37}Ar$ released.

about $40 \times 10^6$ yr is estimated from the total $^{38}Ar_{sp}/^{37}Ar$ ratio, which, *if correct*, would indicate a relatively recent redistribution of cosmogenic $^{38}$Ar, perhaps coinciding with the possible redistribution of radiogenic argon (Fig. 3).

Exposure ages determined from $^{38}Ar_{sp}/^{37}Ar$ thermal release patterns for the samples for which we measured $^{40}$Ar-$^{39}$Ar ages are listed in Tables 1 and 2. Especially apparent for the Fra Mauro area is a grouping at about 30 my that has been suggested as the age of the Cone Crater event (LSPET, 1971; Turner *et al.*, 1971). The exposure age of the other samples are all older and range up to about $740 \times 10^6$ yr. No correlation to rock type or formation age is apparent.

The exposure ages for the Hadley-Rille–Spur-Crater area are variable and range from 80 to $560 \times 10^6$ yr. The exposure ages derived from $^3$He, $^{21}$Ne, and $^{38}$Ar for the anorthosite sample, 15415, are 3, 39, and $95 \times 10^6$ yr, respectively (see Table 2). Such a fractionation suggests diffusion losses, yet the $^{40}Ar/^{39}Ar$ thermal release pattern indicates $<4\%$ $^{40}Ar_{rad}$ loss. The differential temperature release curves for $^{38}Ar_{sp}$ and $^{40}Ar_{rad}$ are almost superimposable, thus little cosmogenic $^{38}$Ar has been lost and the age of 95 my, therefore, should be close to the true exposure age.

Table 4. Noble gas content of samples.

| | Noble Gas Abundances | | | | | | | | | |
|---|---|---|---|---|---|---|---|---|---|---|
| | Concentration, $10^{-8}$ cm$^3$ STP/g | | | | | | | | $10^{-12}$ cm$^3$ STP/g | |
| Sample | $^3$He | $^4$He | $^{20}$Ne | $^{21}$Ne | $^{22}$Ne | $^{36}$Ar | $^{38}$Ar | $^{40}$Ar | $^{84}$Kr | $^{132}$Xe |
| 14053,34 | 19.3 | 31,800 | 12.0 | 2.21 | 3.46 | 4.98 | 3.49 | 6420 | 93 | 50 |
| 14310,101 | 123 | 118,000 | 196 | 23.2 | 39.8 | 70.5 | 46.6 | 23,600 | 900 | 400 |
| 14167,6,11 | 23.1 | 221,000 | 303 | 3.38 | 26.8 | 93.1 | 20.6 | 26,300 | 740 | 150 |
| 15415,9 | 2.9 | 3980 | 118 | 4.85 | 15.0 | 38.2 | 26.6 | 681 | 270 | 37 |
| 15555,33 | 76.9 | 8970 | 79.7 | 11.5 | 19.0 | 18.5 | 11.3 | 913 | 140 | 25 |

*Note:* Elemental abundances are estimated to be accurate to $\pm 10\%$ except for Kr and Xe which are $\pm 25\%$.

## Noble Gas Content of Samples

The noble gas content of three unirradiated sample chips from Apollo 14 and two from Apollo 15 landing sites are presented in Table 4. The solar wind contribution to the gas content of clastic rock 14167,6,11 is low enough to allow calculation of spallation components. A good portion of the measured Kr–Xe spectra was also cosmogenic. Despite the low amounts of Kr and Xe, these spectra are typical of those from crystalline rocks except for a relatively low spallogenic $^{131}$Xe/$^{126}$Xe ratio of about 2. Since much of the $^{131}$Xe in lunar samples is produced from epithermal neutron capture by Ba (Eberhardt *et al.*, 1971; Kaiser and Berman, 1972), a low $^{131}$Xe/$^{126}$Xe ratio implies an unshielded location which is compatible with the fact that 14167,6,11 is probably part of the relatively recent surface layer of ejecta from Cone Crater.

Cosmic-ray exposure ages for the samples listed in Table 4 were determined from $^3$He, $^{21}$Ne, $^{38}$Ar and, where possible, $^{126}$Xe, in the usual manner using production rates from Bogard *et al.* (1971). Target element concentrations were taken from reported values (Chappell *et al.*, 1972; Compston *et al.*, 1972; Hubbard and Gast, 1972; LSPET, 1972; Schnetzler *et al.*, 1972; Stewart *et al.*, 1972) except for 14167,6,11. In this case, Mg + Si, Ca, and Ba were estimated to be similar to the values reported for bulk fines (LSPET, 1971).

The exposure ages calculated from each spallogenic isotope are given in parentheses in Tables 1 and 2. The agreement between these values and the ages determined from the plagioclase plateau of $^{38}$Ar$_{sp}$/$^{37}$Ar versus $^{37}$Ar thermal release patterns is quite good.

*Acknowledgments*—We wish to thank Professors J. J. Papike and A. E. Bence for many useful discussions, G. Barber, G. Dehner, and R. Warasila for their expert handling of the mass spectrometers, T. Ludkewycz, R. Schlott, and C. A. Landeen for assisting with the analyses, and the crew of the High Flux Beam Reactor for irradiations. We also would like to thank G. N. Hanson for providing the hornblende monitors. This work was supported by NASA Grant NGL-33-015-130.

## References

Alexander, E. C. Jr., Davis P. K., and Lewis R. S. (1972) Argon 40–Argon 39 dating of Apollo sample 15555. *Science* **175,** 417–419.

Bogard, D. D., Funkhouser J. G., Schaeffer O. A., and Zähringer J. (1971) Noble gas abundances in lunar material—Cosmic-ray spallation products and radiation ages from the Sea of Tranquility and the Ocean of Storms. *J. Geophys. Res.* **76,** 2757–2779.

Chappell, B. W., Compston W., Green D. H., and Ware N. G. (1972) Chemistry, geochronology, and petrogenesis of lunar sample 15555. *Science* **175,** 415–416.

Cliff R. A., Lee-Hu C., and Wetherill G. W. (1972) K, Rb, and Sr measurements in Apollo 14 and 15 material (abstract). In *Lunar Science—III* (editor C. Watkins), pp. 146–147, Lunar Science Institute Contr. No. 88.

Compston W., Berry H., Vernon M. J., Chappell B. W., and Kaye M. J. (1971) Rubidium-strontium chronology and chemistry of lunar material from the Ocean of Storms. *Proc. Second Lunar Sci. Conf., Geochim. Cosmochim. Acta* Suppl. 2, Vol. 2, pp. 1471–1485. M.I.T. Press.

Compston W., Vernon M. J., Berry H., Rudowski R., Gray C. M., and Ware N. (1972) Age and petrogenesis of Apollo 14 basalts (abstract). In *Lunar Science—III* (editor C. Watkins), pp. 151–153, Lunar Science Institute Contr. No. 88.

Eberhardt P., Geiss J., and Graf H. (1971) On the origin of excess $^{131}$Xe in lunar rocks. *Earth Planet. Sci. Lett.* **12,** 260–262.

Hanson G. N., Goldich S. S., Arth J. G., and Yardley D. H. (1971) Age of the early Precambrian rocks of the Saganaga Lake-Northern Light Lake Area, Ontario–Minnesota. *Can. J. Earth Sci.* **8,** 1110–1124.

Hanson, G. N. (1971) Private communication.

Hubbard N. J. and Gast P. W. (1972) Chemical composition of Apollo 14 materials and evidence for alkali volatilization (abstract). In *Lunar Science—III* (editor C. Watkins), pp. 407–409, Lunar Science Institute Contr. No. 88.

Husain L., Sutter J. F., and Schaeffer O. A. (1971) Ages of crystalline rocks from Fra Mauro. *Science* **173,** 1235–1236.

Husain L., Schaeffer O. A., and Sutter J. F. (1972) Age of a lunar anorthosite. *Science* **175,** 428–430.

Kaiser W. A. and Berman B. L. (1972) The average $^{130}$Ba $(n, \gamma)$ cross section and the origin of $^{131}$Xe on the moon (abstract). In *Lunar Science—III* (editor C. Watkins), p. 444, Lunar Science Institute Contr. No. 88.

Kirsten T., Deubner J., Ducati H., Gentner W., Horn P., Jessberger E., Kalbitzer S., Kancoka I., Kiko J, Kratschmer W., Muller H. W., Plieninger T., and Thio S. K. (1972) Rare gases and ion tracks in individual components in bulk samples of Apollo 14 and 15 fines and fragmental rocks (abstract). In *Lunar Science—III* (editor C. Watkins), pp. 452–454, Lunar Science Institute Contr. No. 88.

LSPET (Lunar Sample Preliminary Examination Team) (1971) Preliminary examination of lunar samples from Apollo 14. *Science* **173,** 681–693.

LSPET (Lunar Sample Preliminary Examination Team) (1972) The Apollo 15 samples: A preliminary description. *Science* **175,** 363–375.

Murthy V. R., Evensen N. M., Jahn B., Coscio M. R. Jr., Dragon J. C., and Pepin R. O. (1972) Rubidium–strontium and potassium–argon age of lunar sample 15555. *Science* **175,** 419–421.

Papanastassiou D. A. and Wasserburg G. J. (1970) Rb–Sr ages from the Ocean of Storms. *Earth Planet. Sci. Lett.* **8,** 269–278.

Papanastassiou D. A. and Wasserburg G. J. (1971) Lunar chronology and evolution from Rb–Sr studies of Apollo 11 and 12 samples. *Earth Planet. Sci. Lett.* **11,** 37–62.

Podosek F. A., Huneke J. C., and Wasserburg G. J. (1972) Gas-retention and cosmic-ray exposure ages of lunar rock 15555. *Science* **175,** 423–425.

Schaeffer O. A., Husain L., Sutter J., Funkhouser J., Kirsten T., and Kaneoka I. (1972) The ages of lunar material from Fra Mauro and the Hadley Rille–Apennine Front area (abstract). In *Lunar Science—III* (editor C. Watkins), pp. 675–677, Lunar Science Institute Contr. No. 88.

Schnetzler C. C., Philpotts J. A., Nava D. F., Schuhmann S., and Thomas H. H. (1972) Geochemistry of Apollo 15 basalt 15555 and soil 15531. *Science* **175,** 426–428.

Stettler A., Eberhardt P., Geiss J., and Grogler N. (1972) $Ar^{39}/Ar^{40}$ ages of Apollo 11, 12, 14, and 15 rocks (abstract). In *Lunar Science—III* (editor C. Watkins), pp. 724–725, Lunar Science Institute Contr. No. 88.

Stewart D. B., Ross M., Morgan B. A., Appleman D. E., Heubner J. S., and Commeau R. F. (1972) Mineralogy and petrology of lunar anorthosite 15415 (abstract). In *Lunar Science—III* (editor C. Watkins), pp. 726–728, Lunar Science Institute Contr. No. 88.

Sutter J. F., Husain L., and Schaeffer O. A. (1971) $^{40}Ar/^{39}Ar$ ages from Fra Mauro. *Earth Planet. Sci. Lett.* **11,** 249–253.

Turner G. (1970) Argon-40/Argon-39 dating of lunar rock samples. *Proc. Apollo 11 Lunar Sci. Conf., Geochim. Cosmochim. Acta* Suppl. 1, Vol. 2, pp. 1665–1684. Pergamon.

Turner G., Huneke J. C., Podosek F. A., and Wasserburg G. J. (1971) $^{40}Ar$–$^{39}Ar$ ages and cosmic-ray exposure ages of Apollo 14 samples. *Earth Planet. Sci. Lett.* **12,** 19–35.

Turner G. (1971) $^{40}Ar$–$^{39}Ar$ ages from the lunar maria. *Earth Planet. Sci. Lett.* **11,** 169–191.

Turner G. (1972) $^{40}Ar$–$^{39}Ar$ age and cosmic ray irradiation history of the Apollo 15 anorthosite, 15415. *Earth Planet. Sci. Lett.* **14,** 169–175.

York, D., Kenyon W. J., and Doyle R. J. (1972) $^{40}Ar$–$^{39}Ar$ ages of Apollo 14 and 15 samples (abstract). In *Lunar Science—III* (editor C. Watkins), pp. 822–824, Lunar Science Institute Contr. No. 88.

Proceedings of the Third Lunar Science Conference
(Supplement 3, *Geochimica et Cosmochimica Acta*)
Vol. 2, pp. 1569–1588
The M.I.T. Press, 1972

# K–Ar dating of lunar fines: Apollo 12, Apollo 14, and Luna 16

R. O. PEPIN, J. G. BRADLEY, J. C. DRAGON, and L. E. NYQUIST*

School of Physics and Astronomy, University of Minnesota,
Minneapolis, Minnesota 55455

**Abstract**—K–Ar ages of < 1 mm fines from the Apollo 12, Apollo 14, and Luna 16 sites range between 0.95 and 4.20 × $10^9$ yr (AE). The 0.95 AE age of the Apollo 12 KREEP soil component is slightly higher than but in reasonable agreement with the age of 0.85 AE deduced by other workers from the U/Pb and $^{40}Ar/^{39}Ar$ techniques. It appears that *two* chronologically distinct low-K components, of ages ~2.8 AE and ~4.0 AE, must be present in the Apollo 12 soils to accommodate the age data. One of these, at ~2.8 AE, could relate to local rock ages through diffusive Ar loss. Dating of grain-size separates of two Apollo 14 soils yields ages between 2.4 AE and 2.95 AE for one, and 4.20 AE for the other; in the latter sample, the trench-bottom fines, there is evidence for an unusual concentration of parentless $^{40}Ar$ in the < 37 $\mu$ size fractions. Ages of < 125 $\mu$ fine fines from both the ~6.5 cm layer and the ~30 cm layer of the Luna 16 core are less precise, but do appear to differ significantly; the values are ≈2.7 AE for the upper level and ~4.0 AE for the lower.

Ordinate intercept correlations for the Apollo 14 and Luna 16 samples give an average value for $^{38}Ar/^{36}Ar$ in the solar wind of 0.1849 ± 0.0008. There appears to be a difference of ~7‰ in this ratio between the upper and lower Luna 16 core samples. Cosmic ray exposure ages are close to 440 m.y. for both Apollo 14 samples and close to 840 m.y. for both Luna 16 samples.

When the K–Ar soil ages reported here are combined with approximate Apollo 11 soil ages deduced from literature data, and considered as a group, it is clear that they tend to fall in ~500 m.y. epochs centered around ~2.8 AE and ~4.0 AE, for all four lunar sites. Only the Apollo 12 KREEP component, probably dispersed by the Copernican impact, is clearly younger.

## INTRODUCTION

RATHER LITTLE IS KNOWN as yet about the detailed chronology of lunar soils, especially when compared with the comprehensive information now available on crystallization ages of lunar rocks. Rb–Sr *model* ages of lunar soils are consistently near 4.6 AE for all five sampled sites (Papanastassiou and Wasserburg, 1972; Murthy *et al.*, 1972a); the comparatively small departures from this number, down to about 4.3 AE (Murthy *et al.*, 1971), are undoubtedly significant but do not readily lend themselves to interpretation of soil history. The U–Th–Pb system, because of the comparative volatility of the daughter product, is an effective detector for soil formation and transport events for which the associated thermal energy input is sufficient to mobilize Pb but insufficient to fractionate Rb and Sr. Extensive studies of mobile lead in lunar soil systems have been carried out by Silver (see Silver, 1972a, for discussion and references). Silver derived a Pb remobilization age of 850 m.y. for the Apollo 12 KREEP

* Present address: Planetary and Earth Sciences Division, NASA Manned Spacecraft Center, Houston, Texas 77058.

soil component and thus, on the assumption of KREEP as Copernican ejecta, of the Copernicus impact (Silver, 1971). Tatsumoto *et al.* (1971) interpreted their U–Th–Pb data for Apollo 12 soils in terms of "second event" U/Pb alteration of 4.6 AE old material about 4 AE ago, with "third event" U/Pb differentiation, perhaps due to impact, about 800 m.y. ago. Both Silver (1972a, b) and Doe and Tatsumoto (1972), Tatsumoto *et al.* (1972) show that leads from Apollo 14 soils define linear arrays on U–Pb evolution diagrams, with intersection "ages" of 2.9 AE and about 2 AE, respectively. Silver considers that definition of his 2.9 AE number as an event in a precise geologic sense is premature; Tatsumoto *et al.* suggest that a "third event" U/Pb fractionation took place about 2 AE ago.

It is clear that K–Ar dating techniques, applied to the $<1$ mm fraction of lunar soils, should provide chronological information constituting a rather sensitive record of major impacts forming and shaping the regolith. The $^{40}Ar$–$^{39}Ar$ method, with its extraordinary capability for recording details of thermal history, is perhaps better suited to this application than any other dating technique. Unfortunately, use of either the $^{40}Ar$–$^{39}Ar$ or bulk K–Ar dating methods on lunar soils must contend with a formidable difficulty—the presence of surface-correlated, parentless $^{40}Ar$ in amounts that in the fine components of the soils can greatly exceed expected concentrations from *in situ* $^{40}K$ decay, even in the most ancient and K-rich samples. To our knowledge, no attempts have been made to obtain $^{40}Ar$-$^{39}Ar$ ages of $<1$ mm fines. It is not certain that the method is, in fact, inapplicable, despite the potential gross interference from parentless $^{40}Ar$; there is evidence from stepwise heating experiments on lunar fines that a large fraction of the $^{40}Ar$ from *in situ* $^{40}K$ decay is outgassed below the 700°C temperature where the correlated release of solar wind $^{36}Ar$ and parentless $^{40}Ar$ becomes significant (Pepin *et al.*, 1970).

The surficial siting of parentless $^{40}Ar$ in lunar materials, first noted by Heymann *et al.* (1970) in Apollo 11 fines and now thought to result from implantation by acceleration of atmospheric ions in the solar wind electric field (Manka and Michel, 1971), suggests the use of correlation techniques to separate the total $^{40}Ar$ in lunar fines into surface and volume components. This opens the possibility of bulk K–Ar dating using the volume-correlated $^{40}Ar$, which presumably arises from *in situ* $^{40}K$ decay. Heymann and Yaniv (1970) observed a correlation between $^{40}Ar$ and solar wind implanted $^{36}Ar$ in size separates of Apollo 11 fines 10084 and with this isolated a volume-correlated $^{40}Ar$ residual which yielded a bulk K–Ar age of $\sim$4.4 AE. Techniques utilized in the studies reported here are similar.

## EXPERIMENTAL TECHNIQUES

General sample preparation procedures and mass spectrometric techniques have been described elsewhere (Pepin *et al.*, 1970). The 6-inch double-focusing mass spectrometer used for these analyses (System I in Pepin *et al.*) has since been equipped for programmable electrostatic jump-scan operation employing digital voltmeter signal integration and magnetic tape data acquisition and was operated in this mode in the present studies. Grain-size separations of Apollo 14 and Luna 16 samples were carried out by M. R. Coscio, Jr.; aliquots of most of the Apollo 14 size fractions were taken for trace element determinations by Professor V. R. Murthy of the Department of Geology and Geophysics, and the particle separation techniques are described in the report by his group in these Proceedings (Murthy *et al.*, 1972a).

## Results

Ar data from Apollo 12, Apollo 14, and Luna 16 fines analyses are shown in Tables 1–3. He, Ne, Kr, and Xe data have also been obtained for all samples, and will be published separately. Errors in isotopic compositions are 1 $\sigma$ statistical errors, summed quadratically from all contributing sources, including blank corrections and drift in mass discrimination corrections between calibration runs. Errors in absolute concentrations of $^{36}$Ar include systematic components—uncertainties in the composition of the standard calibration gas (which has been cross-calibrated against research grade Ar) and in the volume of the gas pipette—as well as statistical errors. All samples (except for those in stepwise heating runs: for procedures see Pepin *et al.*, 1970) were volatilized by heating for 1 hour at 1650°C. $^{40}$Ar blanks for 1 hour, 1650°C extraction ranged from 1.2 to 3.6 $\times$ $10^{-8}$ ccSTP; isotopic composition of the blanks was approximately atmospheric. Blank corrections for $^{36}$Ar and $^{38}$Ar were always negligible; for $^{40}$Ar, corrections were $<1\%$ except for 12042,37 (2%), 14259,14; $\lesssim 1\ \mu$ (3%), and both series of Luna 16 samples ($<1\%$ to 6%). Mass discrimination corrections were based on Nier's (1950) isotopic composition of atmospheric Ar: ($^{38}$Ar/$^{36}$Ar) $= 0.1880 \pm 0.0005$, ($^{40}$Ar/$^{36}$Ar) $= 296 \pm 0.5$. There is a suggestion, in the departure of our instrumental Ar mass discrimination from linearity, that the atmospheric ($^{38}$Ar/$^{40}$Ar) calculated from Nier's numbers is $\sim 5‰$ low, but we continue to use Nier's analysis while obtaining additional discrimination data.

## Discussion

### *K–Ar dating*

As pointed out in the Introduction, the ideal dating technique, in theory, to apply in attempting to untangle the chronology of major impact events in the history of the regolith, as recorded in the lunar fines, is the $^{40}$Ar–$^{39}$Ar method. Experiments will eventually rule on whether parentless $^{40}$Ar contamination completely blocks its application. In the meantime, at the present stage of development of lunar science, particularly with respect to the origin and development of the regolith, it appears that the bulk K–Ar method is capable of yielding significant chronological information. Of particular interest is the comparison of K–Ar ages with current interpretations of Pb data. But in considering specific ages yielded by bulk K–Ar, the limitations and assumptions of the method must be kept in mind: (1) prior accumulation of radiogenic $^{40}$Ar is assumed to be completely lost in the dated event, and (2) subsequent retention of radiogenic $^{40}$Ar is assumed to be complete. Application to samples rich in parentless $^{40}$Ar imposes two additional limitations: (3) that the parentless $^{40}$Ar is present in the system entirely as a surface-correlated component, and (4) that the ratio of parentless $^{40}$Ar to solar wind $^{36}$Ar, represented by $(40/36)_{sc}$ in the following discussion, is constant over the grain-size distribution of a particular soil sample. There are no absolute cross checks on any of these assumptions, but reasonably convincing evaluations can be made of the extent to which the assumptions are violated in a particular sample, based on comparison with other data and on the internal coherence of the Ar data itself.

### *Apollo 12 fines*

Ar and literature K data for eight Apollo 12 fines samples analyzed in the Minnesota laboratory, including three from the 10 cm (12025,62), 21.5 cm (12028,101), and 37 cm (12028,140) levels of the Apollo 12 double core and for two samples analyzed by Hintenberger *et al.* (1971), are given in Table 1. Since for the Apollo 12 soils no K determinations were made on aliquots of the samples used for gas analyses, and since the spread in the literature K concentrations for many of the samples suggest some variation in the mixing ratio of the KREEP and basaltic soil components, it was thought best where possible to base all calculations on reasonable spreads defined by the literature values rather than on single analyses from one or two laboratories.

We assume the following component structure for Ar in lunar soils:

$$(^{40}\mathrm{Ar})_{\mathrm{measured}} = (^{40}\mathrm{Ar})_{\mathrm{radiogenic}} + (^{40}\mathrm{Ar})_{\mathrm{parentless}} + (^{40}\mathrm{Ar})_{\mathrm{spallation}} \quad (1)$$

$$(^{36}\mathrm{Ar})_{\mathrm{measured}} = (^{36}\mathrm{Ar})_{\mathrm{solar\,wind}} + (^{36}\mathrm{Ar})_{\mathrm{spallation}} \quad (2)$$

$(^{40}\mathrm{Ar})_{\mathrm{solar\,wind}}$ is assumed to be absent. With these equations, a little algebra yields:

$$(40)_m = [(40)_r + (36)_{sp}\{(40/36)_{sp} - (40/36)_{sc}\}] + (36)_m(40/36)_{sc} \quad (3)$$

where the ratio of parentless $^{40}$Ar to solar wind $^{36}$Ar is represented by $(40/36)_{sc}$. If a family of samples for which $(40/36)_{sc}$, $(40)_r$, $(36)_{sp}$, and the spallation isotopic composition are constant is plotted on a correlation diagram of $(40)_m$ versus $(36)_m$, a straight line results whose slope is $(40/36)_{sc}$ and whose intercept is related to $(40)_r$. Note that the intercept is not *equal* to $(40)_r$; however, in all of the samples considered in this work, the spallation term in the intercept expression in equation (3) ranges from less than to much less than the *error* involved in determining $(40)_r$; the calculation is discussed in a later section on spallation argon.

The family of samples which optimizes the chance of uniform $(40/36)_{sc}$ among its members and at the same time is characterized by a wide spread in the ratio of surface to volume correlated components, is the family consisting of grain-size separate of a single soil sample. While the Apollo 14 and Luna 16 data reported below were in fact obtained on size separates, the Apollo 12 fines were analyzed as bulk samples. Never-

Table 1. Argon and potassium contents and K–Ar ages of Apollo 12 fines.

| Sample | Weight (mg) | $^{36}$Ar ($\times 10^{-8}$ ccSTP/g) | $^{40}$Ar/$^{36}$Ar | $^{38}$Ar/$^{36}$Ar | $^{40}\mathrm{Ar}_{\mathrm{rad}}$ ($\times 10^{-8}$ ccSTP/g) | K§ (ppm) | K–Ar age‖ (AE) |
|---|---|---|---|---|---|---|---|
| 12032,32 | 39.9 | 4300 ± 150 | 1.023 ± 0.003 | 0.1969 ± 0.0004 | 2292 ± 82 | 2988–3487 | 1.25 ± 0.10 |
| 12033,42 | 45.0 | 2248 ± 47 | 1.429 ± 0.004 | 0.2054 ± 0.0006 | 2111 ± 44 | 2905–4050 | 1.12 ± 0.15 |
| 12034,21 | 97.8 | 624 ± 10 | 4.692 ± 0.016 | 0.2221 ± 0.0010 | 2622 ± 38 | 4320–4560 | 1.10 ± 0.04 |
| 12025,62(1)† | 14.5 | 22160 ± 460 | 0.582 ± 0.002 | 0.1922 ± 0.0006 | 2040 ± 125 | 2120–2750 | 1.41 ± 0.19 |
| *12025,62(2)† | 48.1 | 21960 ± 500 | 0.578 ± 0.002 | 0.1925 ± 0.0009 | 1930 ± 130 | 2120–2750 | 1.36 ± 0.19 |
| 12028,101(1)† | 20.1 | 11100 ± 200 | 1.067 ± 0.003 | 0.1929 ± 0.0007 | 2075 ± 50 | 2120–2750 | 1.43 ± 0.16 |
| *12028,101(2)† | 38.4 | 10070 ± 210 | 1.105 ± 0.004 | 0.1931 ± 0.0009 | 2266 ± 60 | 2120–2750 | 1.52 ± 0.16 |
| 12028,140† | 10.3 | 23980 ± 500 | 0.960 ± 0.003 | 0.1909 ± 0.0006 | 1920 ± 250 | 2120–2750 | 1.35 ± 0.35 |
| 12001,43‡ | — | 28500 ± 860 | 0.593 ± 0.012 | 0.199 ± 0.004 | 2940 ± 380 | 1980 | 2.06 ± 0.20 |
| 12042,37 | 42.3 | 27970 ± 780 | 0.596 ± 0.002 | 0.1924 ± 0.0006 | 2965 ± 175 | 1660–2160 | 2.11 ± 0.20 |
| 12044,11‡ | — | 24600 ± 740 | 0.626 ± 0.013 | 0.198 ± 0.004 | 3345 ± 345 | 2030 | 2.19 ± 0.15 |
| 12070,63 | 38.3 | 27860 ± 780 | 0.536 ± 0.006 | 0.1880 ± 0.0009 | 2815 ± 220 | 1760–2160 | 2.02 ± 0.22 |

* Integrated stepwise heating experiment
† Apollo 12 double core samples.
‡ Ar data from Hintenberger *et al.* (1971).
§ K data: 12032, 12033, 12034, 12042, 12044 from Warner (1971) compilation of literature values; 12001, 12025, 12028 from Schnetzler and Philpotts (1971); 12070 from Wänke *et al.* (1971) compilation.
‖ $^{40}$K decay constants: $\lambda_\beta = 4.72 \times 10^{-10}\ \mathrm{yr}^{-1}$, $\lambda_e = 0.584 \times 10^{-10}\ \mathrm{yr}^{-1}$; $^{40}\mathrm{K/K} = 1.19 \times 10^{-4}$.

theless, a $(40)_m$ versus $(36)_m$ plot of 12 analyses of 10 separate bulk Apollo 12 samples, shown in Fig. 1, reveals surprisingly coherent correlation patterns. To first order, all points fall on or near two distinct correlations, of slopes $(40/36)_{sc} = 0.49 \pm 0.02$ and $0.88 \pm 0.02$. From the point of view of the origin and history of parentless $^{40}$Ar implantation into regolith material, a subject that will not be considered here but which is under active investigation at Minneapolis and elsewhere (Heymann *et al.*, 1972a), it is interesting to note that the two samples collected near or below ~20 cm depth in the regolith are characterized by a decidedly higher value of $(40/36)_{sc}$ than the surface and near-surface samples.

The pair of correlations in Fig. 1 intersect $^{36}$Ar = 0 at a common $^{40}$Ar intercept of $\sim 2100 \times 10^{-8}$ ccSTP/g. Neglecting the spallation component in equation (3), the

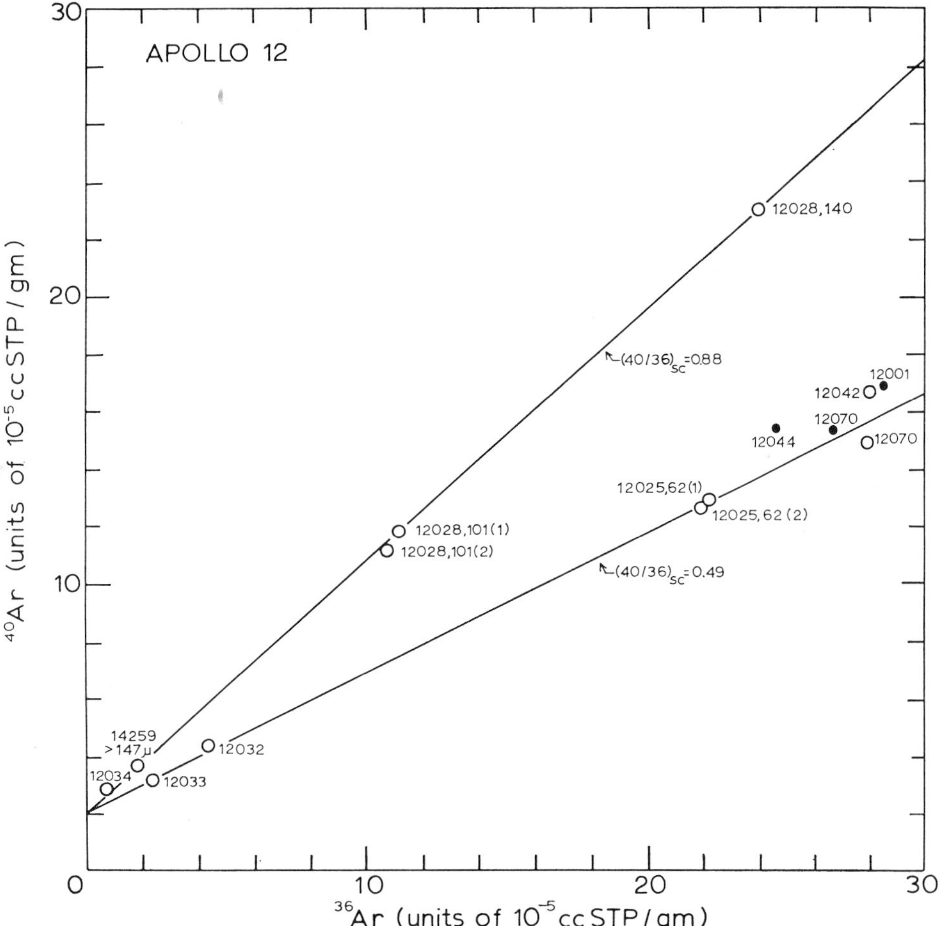


Fig. 1. $^{40}$Ar versus $^{36}$Ar correlation diagram for Apollo 12 < 1 mm fines. Data for 12001, 12044, and one sample of 12070 (filled symbols) from Hintenberger *et al.* (1971). Data for the 147-1000 μ fraction of Apollo 14 soil 14259 is also plotted.

average $(40)_r$ in all samples is then about $2100 \times 10^{-8}$ ccSTP/g; with an average K content of ~2600 ppm (Table 1), the mean K–Ar age of the Apollo 12 soils is about 1.4 AE. Study of equation (3) suggests an immediate way to improve on this crude estimate. If the slopes of the best fit lines in Fig. 1 give $(40/36)_{sc}$ for all samples on or near each line, then, with $(36)_{sp} = 0$, equation (3) may be rewritten as:

$$(40)_r = (36)_m[(40/36)_m - (40/36)_{sc}]. \quad (4)$$

Values of $(40)_r$ were calculated for each sample from $(36)_m$ and $(40/36)_m$, using equation (4) with $(40/36)_{sc} = 0.88$ for samples 12028,101 and 12028,140 and $(40/36)_{sc} = 0.49$ for all others except 12070,63 and are tabulated together with the corresponding K–Ar ages in Table 1. Errors in K–Ar ages are maximum errors, calculated using extreme values for $(40)_r$ and the end members of the K ranges. Hintenberger *et al.* (1971) have measured Ar in grain-size separates of 12070; their data yield $(40/36)_{sc} = 0.435$, and this value is used in equation (4) to determine $(40)_r$ in 12070.

If the Apollo 12 soils consisted of a single component with an assumed single, sharply defined "age," a standard interpretation of the age data in Table 1 would be to set a lower limit of ~2.2 AE on the age of the soils and attribute the suite of lower ages to diffusive loss of Ar. However, even if no other information on the soils or the site were available, the rough anticorrelation of K–Ar ages with K contents (Table 1) would suggest a system of at least two chemically distinct components with different ages. As is now well known, petrographic and chemical studies of the returned soils have firmly established the presence of two major soil components at Surveyor site; the photogeologic detection of a high albedo ray of Copernicus crossing the site and crew description and sampling of local areas of contrasting albedo in the regolith provide evidence for identifying the high-K KREEP soil component as Copernican ejecta and the low-K basaltic component as erosional detritus from local rocks. Of particular interest in the discussion of ages is the two-component mixing model of KREEP with average Apollo 12 basaltic material derived by Hubbard *et al.* (1971) from the trends of major and trace element chemical compositions over the suite of Apollo 12 soils and breccias. If each of the two components is characterized by a unique age as well as by a distinct chemical composition, the K–Ar ages given in Table 1 should fall along a smooth curve between the appropriate endpoint ages on a plot of age versus mixing ratio between the two components. Hubbard *et al.* give K(KREEP) $\simeq$ 6640 ppm and K(basalt) $\simeq$ 530 ppm as the approximate endpoint K concentrations in the <1 mm fines. We may then define the percentage of KREEP in the soils in Table 1 by $\%\text{KREEP} = 100(K - K_b)/(K_K - K_b)$, where $K = K(\text{soil})$, $K_K = 6640$ ppm and $K_b = 530$ ppm.

Figure 2 is a plot of K–Ar ages from Table 1 versus %KREEP; the error "rectangles" show the maximum ranges of K–Ar ages and K contents for each sample. The smooth curves connecting endpoint ages were derived by assuming the endpoint ages and calculating the apparent ages that appear along the mixing curve.

The salient features of Fig. 2 are readily apparent: (1) the K–Ar ages of the Apollo 12 soils do in fact trend smoothly between endpoint ages that are defined with reasonable precision by the data; (2) it appears that *two* low-K components, one

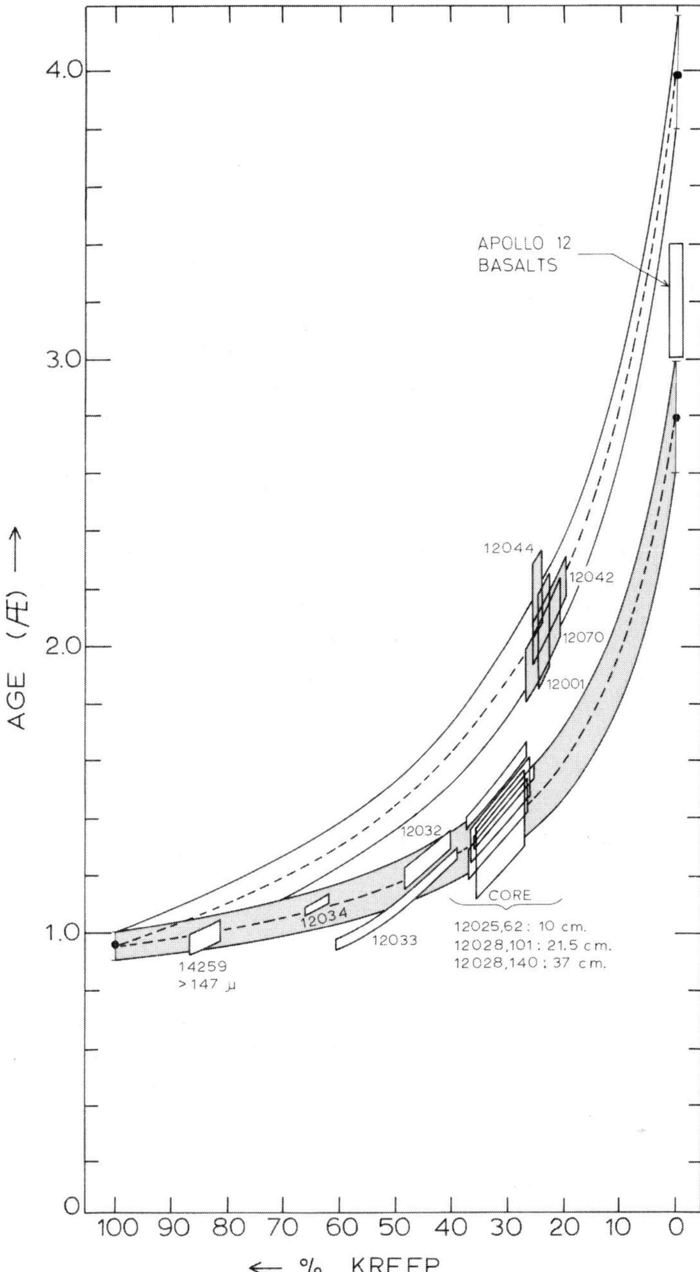


Fig. 2. Calculated K–Ar ages in AE (units of $10^9$ yr) versus % KREEP soil component for Apollo 12 soils. A possible age for the >147 $\mu$ fraction of 14259 is also plotted. Smooth curves are calculated mixing curves between the indicated endpoint component ages.

significantly older than the other, are required to accommodate the age data. In the absence of chemical evidence to the contrary, we have tentatively assumed that the same K content of $\sim 530$ ppm is characteristic of both low-K components.

The endpoint ages which best fit the data for 12032, 12033, 12034, and the three core samples are $t$(KREEP) $= 0.95 \pm 0.05$ AE and $t$(basalt) $= 2.8 \pm 0.2$ AE. The range of apparent ages which result from mixing of the two endpoint components is shown as the shaded lower belt in Fig. 2. The coarse grain size fraction of Apollo 14 soil 14259, which also falls on this mixing curve, is discussed below. Soils 12001, 12042, 12044, and 12070, all of which contain a low and virtually uniform proportion of the KREEP component, are tightly grouped significantly above this curve. The spread in mixing ratio among this group of soils is too narrow to permit an independent deduction of endpoint ages. Here, the age of the KREEP component is *assumed* to be that fixed by the lower correlation, $0.95 \pm 0.05$ AE; with this choice, and with K $= 530$ ppm for the low-K component, the age of this third component is fixed within reasonable limits at $4 \pm 0.2$ AE.

*There is therefore evidence that the Apollo 12 soils are mixtures of materials of at least three distinct K–Ar ages.* The K–Ar age of $0.95 \pm 0.05$ AE for the KREEP component in these soils, which may also be taken as the formation age of Copernicus on the reasonable (but still unproven) assumption that KREEP is Copernican ejecta, is marginally higher than Silver's (1971) value of 0.85 AE but the agreement is satisfactory. There are recent, additional data on the age of the KREEP material. While the $^{40}Ar$–$^{39}Ar$ method has not been attempted on $<1$ mm fines, it can be applied to directly date *coarse* fines fragments of diameters 1–2 mm and above, which are large enough to permit interior sample selection or removal of the parentless $^{40}Ar$-rich surface layers. Applying this technique, Eberhardt *et al.* (1972) have found low-temperature plateau $^{40}Ar$–$^{39}Ar$ ages of $0.85 \pm 0.05$ AE in two KREEP fragments separated from the 12033 coarse fines. In connection with the slightly higher age given by the bulk K–Ar method, it is interesting to note that the Eberhardt *et al.* data show a significant content of $^{40}Ar$ released *above* the low-temperature plateau. It seems clear that this is residual $^{40}Ar$, accumulated from $^{40}K$ decay during the pre-impact history of the material in sites of sufficiently high activation energy to prevent its complete loss during the impact thermal pulse. *Bulk* K–Ar dating of these fragments would thus yield ages *higher* than the event age. Funkhouser (1971) has obtained bulk K–Ar ages ranging up to 1.9 AE for individual KREEP fragments from 12033 and 12070 coarse fines; from these he deduces a KREEP deposition event $\geq 2$ AE ago, attributing all lower ages to subsequent, variable Ar loss. But on the evidence above, 2 AE is a lower limit *crystallization* age rather the lower limit age of a partial outgassing event. In fact, in complete reversal of the usual limit applied to K–Ar ages, the most accurate estimate of the age of a lunar impact event could well be to regard the *lowest* measured age in a distribution of fragments as an *upper* limit; the lowest K–Ar age measured by Funkhouser, except for one particle which was apparently outgassed in a very late event, is 0.85 AE.

In summary, if the K–Ar age of KREEP determined in the present study is slightly high, the reason is almost certainly residual $^{40}Ar$. As far as the future application of this technique is concerned, it is encouraging to note that the effect *in the $<1$ mm fines* appears to be very much less serious than in larger fragments. There is *no* evidence for significant diffusive loss of $^{40}$Ar from the KREEP soil component over a time period of roughly 1 AE.

In contrast, the most straightforward interpretation of the apparent 2.8 $\pm$ 0.2 AE age for the first basaltic component in Fig. 2 does imply Ar loss. Crystallization ages of the Apollo 12 basalts range from approximately 3 to 3.4 AE (Papanastassiou and Wasserburg, 1971; Murthy *et al.*, 1971). The basaltic soil component is apparently chemically indistinguishable from the large basalts and is presumably derived from them. Loss of about 30% of radiogenic $^{40}$Ar from the rock comminution products drops the apparent K–Ar age of the basaltic soil component from $\sim$3.2 to $\sim$2.8 AE; this loss is less than that shown by the rocks themselves, since their average bulk K–Ar age is about 2.3 AE (LSPET, 1970). We cannot rule out the possibility of an episodic outgassing event at the site $\sim$2.8 AE ago, and in fact the occurrence of approximately this age for Apollo 14 soil 14259 and perhaps for the A-level of the Luna 16 core, together with Silver's (1972a, b) U/Pb data on Apollo 14 soils, could suggest a large-scale event of some kind at $\sim$2.8 – 2.9 AE. But the most direct explanation as far as this Apollo 12 soil component is concerned is probably nonepisodic Ar leakage.

The primary assumption in the derivation of an age of $\sim$4 AE for the second low-K soil component in Fig. 2 is the use of $(40/36)_{sc} = 0.49$ in equation (4) for $(40)_r$ in 12001, 12042, and 12044; it is clear from equation (4) and from the $(40/36)_m$ values in Table 1 that $(40)_r$ in these three cases is extremely sensitive to the choice of $(40/36)_{sc}$. Grain-size separation experiments to directly measure $(40/36)_{sc}$ in each sample are required. However, the following point should be made: if the same value of $(40/36)_{sc} = 0.49$ is also assumed to apply to 12070, the sample falls far off the upper mixing curve in Fig. 2, in fact well below the lower curve. When the *directly determined* value 0.435 is used, 12070 falls as shown on the upper curve. So a direct case is made for the presence of $\sim$4 AE old material in one of these four similar soils, and it is probable that all four do group tightly in apparent age, as shown in Fig. 2. As far as the origin of this component is concerned, it is clear that since it appears to antedate the formation of local bedrock not only at this site but over large areas of Oceanus Procellarum (Murthy *et al.*, 1972b) by $\sim$700 m.y. (or more, if, as seems probable, diffusive Ar loss has occurred), it cannot be an erosional product of the Apollo 12 basalts but must instead belong to a more ancient generation of soils disseminated over the lunar surface by lateral transport.

### *Apollo 14 fines*

Ar data for five grain-size fractions of $<1$ mm fines 14259,14 are tabulated in Table 2 and plotted on a $(^{40}\text{Ar})_m$ versus $(^{36}\text{Ar})_m$ diagram in Fig. 3. The four finer fractions correlated linearly with a slope $(40/36)_{sc} = 0.93 \pm 0.01$, but the coarse 147–1000 $\mu$ fraction lies well off the correlation. Radiogenic $^{40}$Ar calculated from equation (4), and associated K–Ar ages for those fractions with known K (Murthy *et al.*, 1972a), are also given in Table 2; ages range from 2.39 to 2.95 AE for the $<147$ $\mu$ fractions. The coarse fraction, which is strikingly anomalous in both Ar and K contents, has a bulk K–Ar age, without *any* correction for parentless $^{40}$Ar, of

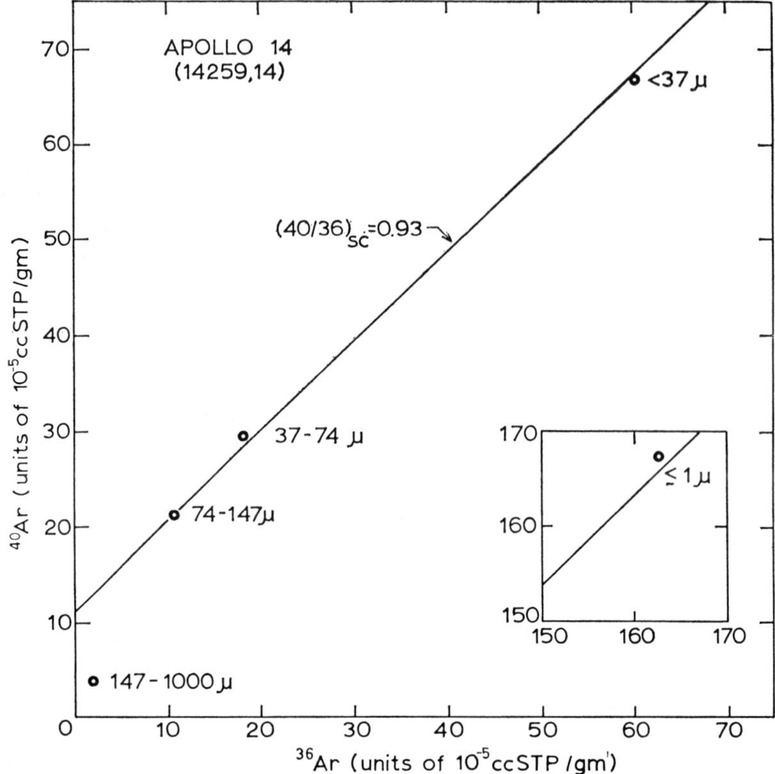


Fig. 3. $^{40}$Ar versus $^{36}$Ar correlation diagram for grain-size fractions of Apollo 14 soil 14259,14. The insert represents the line in the body of the diagram continued out to the coordinates of the $\lesssim 1$ μ fraction.

Table 2. Argon and potassium contents and K–Ar ages of grain-size separates of Apollo 14 fines 14259,14 and 14149,44.

| Sample | Weight (mg) | $^{36}$Ar ($\times 10^{-8}$ ccSTP/g) | $^{40}$Ar/$^{36}$Ar | $^{38}$Ar/$^{36}$Ar | $^{40}$Ar$_{rad}$ ($\times 10^{-8}$ ccSTP/g) | K* (ppm) | K–Ar age (AE) |
|---|---|---|---|---|---|---|---|
| *14259,14* | | | | | | | |
| 147–1000 μ | 84.8 | 1830 ± 95 | 2.070 ± 0.021 | 0.1886 ± 0.0009 | 2890 ± 160 (2090 ± 120) | 5660 ± 170 | 0.98 ± 0.05 (0.75 ± 0.04) |
| 74–147 μ | 69.0 | 10820 ± 315 | 1.957 ± 0.018 | 0.1905 ± 0.0009 | 11110 ± 390 | 4340 ± 130 | 2.80 ± 0.07 |
| 37–74 μ | 45.1 | 18710 ± 655 | 1.567 ± 0.014 | 0.1884 ± 0.0008 | 11920 ± 530 | 4190 ± 130 | 2.95 ± 0.08 |
| < 37 μ | 10.4 | 62000 ± 1800 | 1.075 ± 0.010 | 0.1865 ± 0.0007 | 8990 ± 920 | 4680 ± 140 | 2.39 ± 0.14 |
| ≲ 1 μ | 0.666 | 165900 ± 4800 | 1.008 ± 0.009 | 0.1858 ± 0.0009 | 12900 ± 2300 | — | — |
| *14149,44* | | | | | | | |
| 147–1000 μ | 17.75 | 3480 ± 150 | 8.88 ± 0.09 | 0.1935 ± 0.0004 | 27300 ± 1200 | 4480 ± 130 | 4.17 ± 0.09 |
| 74–147 μ | 14.25 | 9620 ± 210 | 3.933 ± 0.027 | 0.1909 ± 0.0004 | 28020 ± 820 | 4440 ± 130 | 4.23 ± 0.07 |
| 37–74 μ | 6.65 | 12080 ± 280 | 3.231 ± 0.020 | 0.1878 ± 0.0004 | 26710 ± 900 | 4250 ± 130 | 4.22 ± 0.07 |
| 16–37 μ | 7.69 | 22470 ± 490 | 2.796 ± 0.019 | 0.1876 ± 0.0004 | — | 4500 ± 130 | — |
| 4–16 μ | 7.56 | 50200 ± 1100 | 2.730 ± 0.017 | 0.1862 ± 0.0004 | — | 5430 ± 160 | — |
| < 4 μ | 2.57 | 102000 ± 2200 | 3.003 ± 0.022 | 0.1856 ± 0.0004 | — | — | — |
| ≲ 1 μ | 0.895 | 118800 ± 2600 | 3.120 ± 0.020 | 0.1856 ± 0.0004 | — | — | — |

* Murthy *et al.* (1972a).

1.20 AE. In this respect it resembles an Apollo 12 KREEP-rich soil, and in order to investigate this similarity its corrected K–Ar age was calculated in two ways, first assuming $(40/36)_{sc} = 0.49$, typical of Apollo 12 high-K soils, and second assuming $(40/36)_{sc} = 0.93$, characteristic of the finer components of 14259. The ages were 0.98 AE and 0.75 AE, respectively, and the first of these falls directly on the lower mixing curve for Apollo 12 soils in Fig. 2. Thus a possible interpretation of the data for the coarse fraction of 14259 is that the Apollo 12 KREEP component is present (but probably rare) at the Apollo 14 site. While this interpretation is reasonable in terms of the proximity of the site to Copernicus, Shoemaker (personal communication) has pointed out that there is no evidence for a Copernican ray across the site. There is, however, no question that this coarse fraction is an anomalous component in 14259, not only on chemical and chronological grounds but in exposure age as well. In a later section on spallation Ar, the exposure age of the $<147\ \mu$ fractions of 14259 is calculated to be 425 m.y., while that of the coarse fraction is 44 m.y. This exposure approaches that of 29 m.y. found by Turner *et al.* (1971) for coarse fines fragment 14167,8,1 collected from the same general area as 14259 and identified by them on the basis of its exposure age as ejecta from Cone Crater. These authors note evidence for both shock and substantial radiogenic $^{40}$Ar loss in 14167,8,1. If the coarse component of 14259 is also Cone ejecta, which seems probable, that event could have mobilized a major fraction of the radiogenic $^{40}$Ar; the bulk K–Ar age would then be low, and the position of the point in Fig. 2 merely fortuitous.

The ordinate intercept K–Ar age of the $<147\ \mu$ fractions of 14259, determined from the correlation line intercept of $11{,}000 \times 10^{-8}$ ccSTP/g and the total sample K of 4268 ppm (Murthy *et al.*, 1972a), is 2.8 AE. The more meaningful ages for the individual fractions, given in Table 2, fall roughly between the Tatsumoto *et al.* (1972) "third event" U/Pb fractionation age of $\sim 2$ AE for three Apollo 14 soils, among them 14259, and Silver's (1972a, b) intersection at $2.9 \pm 0.05$ AE. The three ages in Table 2 anticorrelate with K content, suggesting the possibility of a multi-component chronology in 14259. However, no clear indication of systematic multi-component mixing has emerged from chemical studies of the Apollo 14 soils, and the K–Ar data taken alone are too sparse and too tightly grouped to interpret in terms of possible endpoint component ages.

Sample 14149,44, from the bottom of the Apollo 14 trench, shows a significantly different chronology. Ar data for seven grain-size separates, with K contents for all but the two finest, are given in Table 2 and plotted in Fig. 4. The three fractions $>37\ \mu$, which include more than half the mass of the soil, show a linear correlation with $(40/36)_{sc} = 1.02 \pm 0.03$. Radiogenic Ar contents, calculated from equation (4), yield K–Ar ages closely grouped at $4.20 \pm 0.03$ AE (Table 2), the oldest and the most precisely determined soil ages in this study. However, the four finest grain sizes depart in an increasingly radical fashion from the correlation describing the three coarsest fractions, as may be clearly seen in Fig. 4. It is virtually certain that the effect cannot be due to increase in radiogenic $^{40}$Ar in these fractions, which would require either unreasonably high K or totally unrealistic ages. The most probable explanation is a smooth, systematic increase in $(40/36)_{sc}$ as a function of decreasing grain size. The effect is intriguing, and if we interpret its cause correctly, it should certainly provide

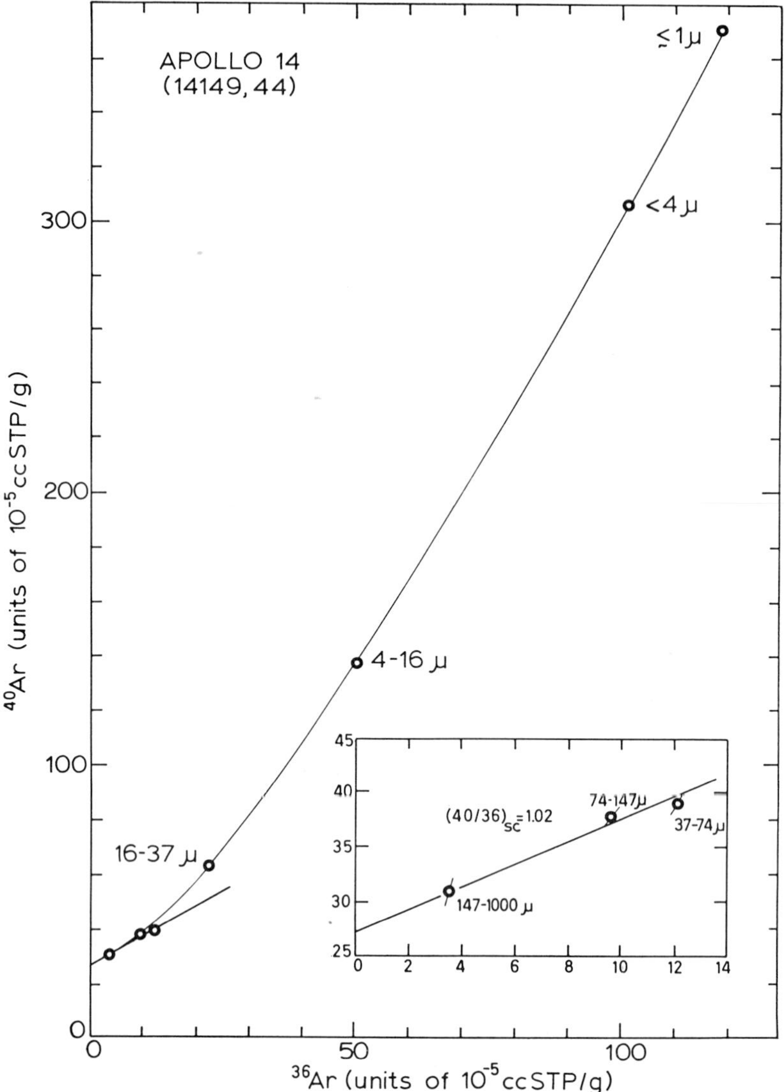


Fig. 4. $^{40}$Ar versus $^{36}$Ar correlation diagram for grain-size fractions of Apollo 14 soil 14149,44. The insert shows the correlation of the three coarse size fractions on an expanded scale.

clues to the mechanism of parentless $^{40}$Ar implantation and its variation with time. The pattern in Fig. 4 could relate to the depth of implantation of this component. It could also arise from $^{40}$Ar irradiation of fine material in the *absence* of solar wind irradiation, a process that does not conform well to current models of the implantation process. In any case, it is probably important that the effect appears in an extremely ancient soil (in the sense that its Ar content has not been seriously disturbed for more than 4 AE) but not in younger samples such as 14259.

*Luna 16 fines*

Ar data from grain-size separates of two 15 mg samples of <125 $\mu$ fine fines from the A and G levels of the Luna 16 core are tabulated in Table 3 and plotted in Fig. 5.

Table 3. Argon in grain-size separates of Luna 16 core fines: A-8 (~6.5 cm depth) and G-6 (~30 cm depth).

| Sample | Weight (mg) | $^{36}$Ar ($\times 10^{-8}$ ccSTP/g) | $^{40}$Ar/$^{36}$Ar | $^{38}$Ar/$^{36}$Ar |
|---|---|---|---|---|
| *A-8* | | | | |
| 74–125 $\mu$ | 1.135 | 28350 ± 600 | 1.057 ± 0.008 | 0.1887 ± 0.0003 |
| 37–74 $\mu$ | 4.508 | 31520 ± 720 | 1.018 ± 0.004 | 0.1874 ± 0.0003 |
| 16–37 $\mu$ | 3.971 | 39700 ± 1500 | 0.929 ± 0.007 | 0.1862 ± 0.0007 |
| 8–16 $\mu$ | 0.267 | 62600 ± 1900 | 1.141 ± 0.016 | 0.1858 ± 0.0006 |
| <8 $\mu$ | 2.103 | 102300 ± 3300 | 0.977 ± 0.004 | 0.1849 ± 0.0003 |
| Total* | | 44900 ± 1300 | 0.986 ± 0.006 | 0.1867 ± 0.0005 |
| *G-6* | | | | |
| 74–125 $\mu$ | 1.166 | 23600 ± 500 | 1.326 ± 0.010 | 0.1892 ± 0.0006 |
| 37–74 $\mu$ | 3.160 | 24590 ± 520 | 1.077 ± 0.005 | 0.1900 ± 0.0006 |
| 16–37 $\mu$ | 3.942 | 36960 ± 850 | 0.957 ± 0.004 | 0.1882 ± 0.0003 |
| 8–16 $\mu$ | 0.057 | 27900 ± 1200 | 14.79 ± 0.16 | 0.1890 ± 0.0004 |
| <8 $\mu$ | 1.338 | 130000 ± 3400 | 0.894 ± 0.012 | 0.1859 ± 0.0003 |

* Grain-size separation yield ≃100%. Total calculated by summing individual separates over the grain size versus mass distribution curve.

The data scatter rather more than has been the case for samples of larger mass, and in particular the 57 $\mu$g separate from the G-level material shows a wildly anomalous $^{40}$Ar content; no attempt is made to include this point in the correlations. Despite the scatter, correlations are evident in Fig. 5. Since no chemical determinations were made on the individual size separates, the ordinate intercept technique must be used to calculate the "average" radiogenic $^{40}$Ar in material from each level, Results are, for the A-level, $(40)_r = 2000 \pm 1000 \times 10^{-8}$ ccSTP/g, and for the G-level, $(40)_r = 5000 \pm 1000 \times 10^{-8}$ ccSTP/g. With K contents of 850 ppm and 880 ppm, respectively (Philpotts *et al.*, 1972), the corresponding K–Ar ages are $2.7^{+0.6}_{-0.9}$ AE for A-level fines and 4.05 ± 0.30 AE for G-level fines. The A-level age is too imprecise for anything more than to note that it could be roughly concordant with the ages between 2.5 AE and 3 AE that occur in regolith components at the Apollo 12 and 14 sites. The G-level soil is definitely old, and it seems clear that fine regolith materials with Ar-retention ages near or above 4 AE are widely distributed on the lunar surface.

*Spallation Argon*

Assuming that $(^{38}\mathrm{Ar})_m$ in lunar materials is a mixture of the same two components, solar wind and spallation, as $(^{36}\mathrm{Ar})_m$ (equation (2)), we may write

$$(38/36)_m = \frac{(36)_{sp}}{(36)_m}[(38/36)_{sp} - (38/36)_{sw}] + (38/36)_{sw} \tag{5}$$

where $(38/36)_{sw}$ and $(38/36)_{sp}$ are the isotopic ratios in the solar wind and spallation components respectively. Equation (5) is simply an inverted version of the ordinate

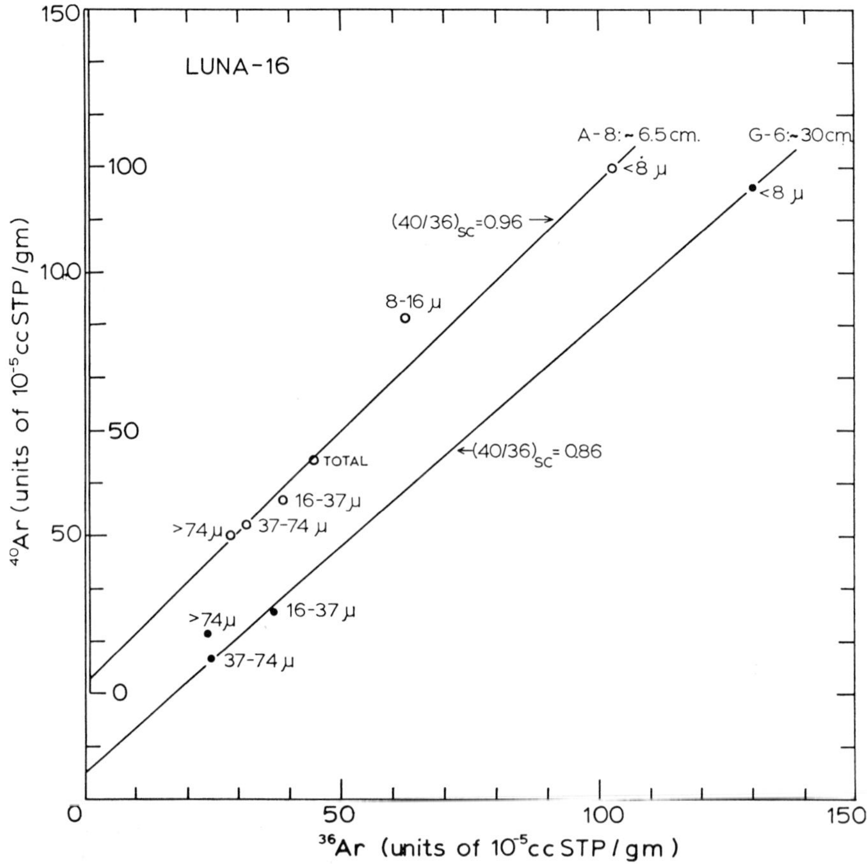


Fig. 5. $^{40}$Ar versus $^{36}$Ar correlation diagrams for grain-size fractions of the Luna 16 $<125\ \mu$ fine fines A-8, $\sim 6.5$ cm core depth, and G-6, $\sim 30$ cm core depth. The right side of the double ordinate scale applies to the A-8 data, the left side to the G-6 data. The 8–16 $\mu$ fraction of G-6 plots far above the diagram and is not shown.

intercept equation given by Eberhardt *et al.* (1970). For samples in which the solar wind and spallation isotopic ratios and the concentrations of spallogenic $^{36}$Ar are uniform, a plot of $(38/36)_m$ versus $1/(36)_m$ gives a straight line of slope $(36)_{sp}[(38/36)_{sp} - (38/36)_{sw}]$ and intercept $(38/36)_{sw}$. Variations from sample to sample in the concentrations of Ar spallation target elements (dominantly Ca, with smaller contributions from Ti, Fe, and Ni) or in exposure ages, or both, result in variations in $(36)_{sp}$ and thus lead to scatter away from the linear correlation given by equation (5). The isotopic ratio $(38/36)_{sp}$ is known to be remarkably constant at $\sim 1.5$ in meteorites and in lunar rocks (Huneke *et al.*, 1972). Whether the isotopic composition of solar wind rare gases in lunar materials is constant or variable is a moot question, and one which analyses such as these are designed to address.

It is clear, from assessments of probable variations in chemical compositions and exposure ages, that grain-size separates of single soil samples offer the best chance for successful application of equation (5) as a linear correlation technique. Both chemical compositions and exposure ages of Apollo 12 soils are known to be highly variable (Funkhouser, 1971), and at least for the present this technique is not applied to them. Data from the grain-size separates of the two Apollo 14 soils and the two Luna 16 core samples (Tables 2 and 3) are plotted on $(38/36)_m$ versus $1/(36)_m$ correlation diagrams in Fig. 6. Individual point corrections for target element variations (Eberhardt *et al.*, 1972b) cannot be carried out since the chemical data do not exist, but judging from the pronounced chemical similarity of a group of eight Apollo 14 soil samples (Compston *et al.*, 1972) and the nearly uniform chemical composition in $<125\ \mu$ and $125–425\ \mu$ size fractions of Luna 16 soils from both the A and G layers (Hubbard *et al.*, 1972), chances of large-scale chemical heterogeneity in the grain-size fractions

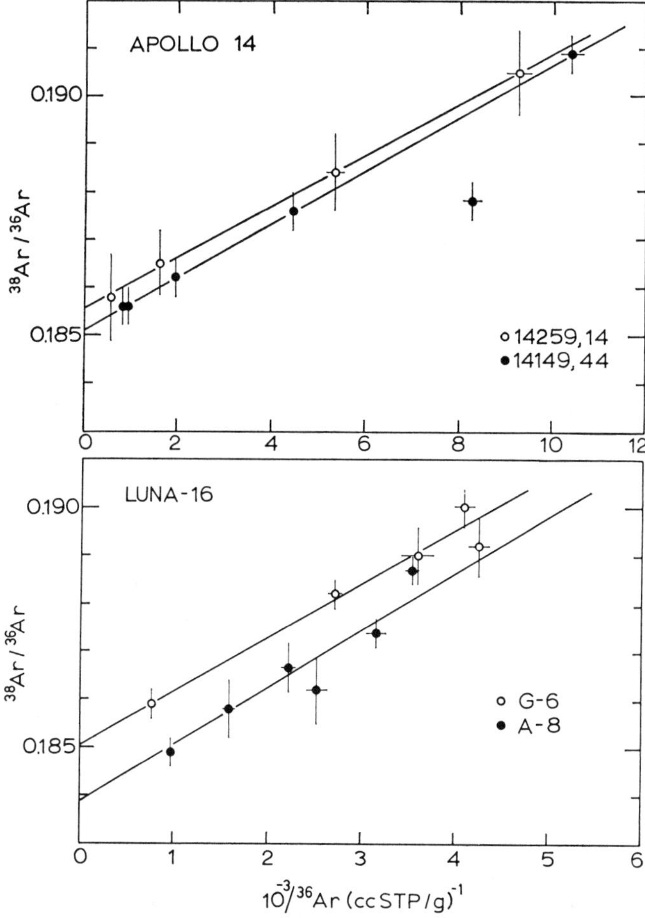


Fig. 6. Ordinate intercept plots of ($^{38}$Ar/$^{36}$Ar) versus 1/($^{36}$Ar) for two Apollo 14 and two Luna 16 soils.

appear to be small *if* the fractions are of reasonable mass. Clearly, the chemistry of a 57 $\mu$g sample could vary widely from a composition determined by Hubbard *et al.* on a sample 175 times larger.

Correlations for grain-size fractions $<147\ \mu$ are reasonably good for all four soils in Fig. 6, with the exception of the aberrant (but experimentally impeccable) 37–74 $\mu$ fraction of 14149. The coarse, 147–1000 $\mu$ fractions from both Apollo 14 soils, which are not shown in the figure, fall off the correlation lines in the direction of lower exposure age.

The correlations in Fig. 6 yield ordinate intercept values for $(38/36)_{sw}$ of 0.1856 (14259), 0.1851 (14149), 0.1838 (Luna A8), and 0.1850 (Luna G6); the average value is $0.1849 \pm 0.0008$. The difference between the Luna A8 and G6 intercepts appears to be real: $(38/36)_{sw}$ is $\sim 7‰$ lower in the near-surface layer than at the 30 cm depth.

With the $(38/36)_{sw}$ ratios above, and $(38/36)_{sp} = 1.53$ (Huneke *et al.*, 1972), the measured slopes of the correlations in Fig. 6 are used to calculate $(36)_{sp}$ in each soil, from the relation $\lambda_m = (36)_{sp}[(38/36)_{sp} - (38/36)_{sw}]$. Results, in units of $10^{-8}$ ccSTP/g, are: 39.4 (14259), 41.3 (14149), 89.1 (Luna A8), and 83.3 (Luna G6). Using the spallogenic $^{38}$Ar production rate equation given by Heymann *et al.* (1972b), together with chemical data from Compston *et al.* (1972) for the Apollo 14 soils and from Hubbard *et al.* (1972) for Luna 16, we calculate the cosmic ray exposure ages of these four soils to be: 430 m.y. (14259), 450 m.y. (14149), 865 m.y. (Luna A8), and 810 m.y. (Luna G6). If we consider probable errors in the Fig. 6 correlation line slopes, neither the two Apollo 14 nor the two Luna 16 soil exposure ages differ significantly from each other. That the Luna 16 soils have a much longer exposure age than the Apollo 14 soils, longer in fact that any other sampled lunar region, was reported recently by Kaiser (1972); his estimated $^{126}$Xe exposure age of $900 \pm 300$ m.y. is in agreement with the Ar ages reported here.

Known values of $(38/36)_{sw}$ (with $(38/36)_{sp} = 1.53$), together with $(38/36)_m$ and $(36)_m$ data from the tables, may be used in equation (5) to calculate $(36)_{sp}$ and thus exposure ages for individual samples and individual grain-size separates. Assuming that $(38/36)_{sw} = 0.1849$ applies to the Apollo 12 soils, we used the data in Table 1 to calculate the relative magnitudes of the spallation term and $(40)_r$ in equation (3) for each Apollo 12 sample. $(40/36)_{sp}$ was taken to be zero (a conservative assumption). In no case does the contribution to $(40)_r$ from this term exceed the *error* given for $(40)_r$ in Table 1. This correction is completely negligible for all Apollo 14 and Luna 16 samples.

Calculated exposure ages for the coarse, 147–1000 $\mu$ fractions of the two Apollo 14 soils are 44 m.y. for 14259 and 240 m.y. for 14149; both lie well below the $\sim$440 m.y. exposure characteristic of the finer components, particularly for 14259 where the integrated exposure is only 10% of that of the "parent" soil. The exposure age of the anomalous 37–74 $\mu$ fraction of 14149 (Fig. 6) is 265 m.y., close to that of the 147–1000 $\mu$ fraction.

It is interesting to observe that there is no characteristic of either the solar wind isotopic composition or the exposure age of 14149, both of which are essentially determined by correlations over the finest grain-size fractions, which reflects the fact that the soil is ancient and that $^{40}$Ar in these fractions seems to be greatly enriched (Fig. 4).

## Summary and Conclusions

A summary of the K–Ar ages of soils from the Apollo 12, Apollo 14, and Luna 16 sites is given in Fig. 7. Ages for Apollo 11 soils 10084 (3.8 ± 0.2 AE) and 10087 (2.7 ± 0.3) were calculated for Fig. 7 from equation (4), with data from the following sources: $(40/36)_{sc}$ = 0.95 (10084) and 1.51 (10087) from Eberhardt *et al.* (1970) and Hintenberger *et al.* (1971); $(36)_m$, $(40/36)_m$ from Hintenberger *et al.* (1971) and Pepin *et al.* (1970); K = 1120 ppm from Philpotts and Schnetzler (1970). The heavy vertical bars represent the spread of rock ages at the sites.

The horizontal shaded belts between 2.5 and 3 AE and between 3.75 and 4.25 AE in Fig. 7 seem to deliniate epochs in lunar chronology in which large-scale events deposited thermal energy in the fine regolith materials over a significant fraction of at least the nearside equatorial lunar surface. The upper belt may actually represent the end of an epoch of intense bombardment extending back to the formation of the original lunar surface, during which volatile species could have been continually

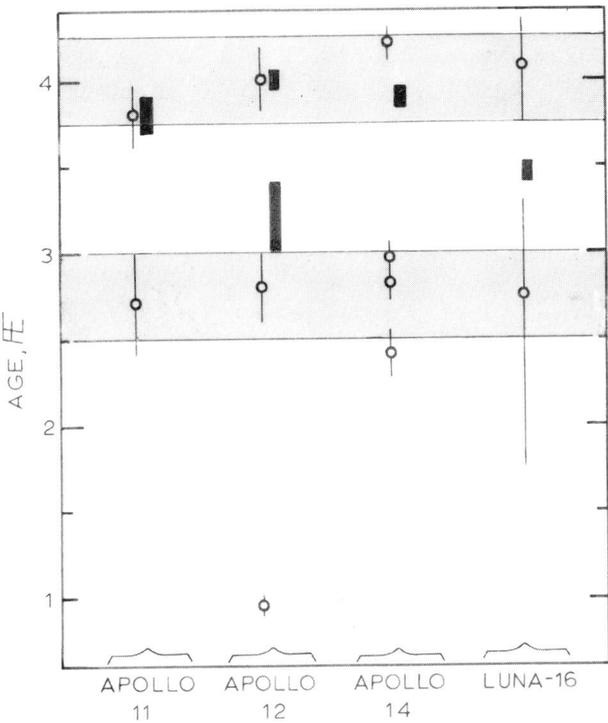


Fig. 7. K–Ar ages of lunar soils, in AE, plotted by site (open symbols). Sources for Apollo 11 data given in text; all others from this work. Heavy vertical bars represent spreads in rock ages at the sites. Rock age data from: Papanastassiou *et al.* (1970) and Turner (1970a) for Apollo 11; Papanastassiou and Wasserburg (1971), Murthy *et al.* (1971), Lunatic Asylum (1970) and Turner (1970b) for Apollo 12, including 12013 (upper bar at Apollo 12 site); Turner *et al.* (1971) and Murthy *et al.* (1972a) for Apollo 14; Papanastassiou and Wasserburg (1972) for Luna 16.

remobilized or lost. The site-by-site evidence for the younger epoch is not overwhelming: It rests in a single measurement on one sample for Apollo 11, a soil component which could have derived from local rocks by comparatively modest diffusive loss for Apollo 12, a spread of ages that could represent a two-component chronology with neither endmember age in this time-span for Apollo 14, and a single, highly imprecise age for Luna 16. But taken together, the four-site grouping is rather striking.

*Acknowledgments*—This research was supported by NASA Grant NGL 24-005-225 and in part by Contract N00014-67-A-0113-0018 with the U.S. Office of Naval Research. We are indebted to Mr. John Ho, Mr. T. R. Venkatesan, Mr. Jeffrey Basford, and Mr. Kin-Pong Leung for assistance with data reduction and analysis. The data analysis computer program is the creation of Mr. Neil Johnson. Grain-size separation techniques were developed and carried out by M. R. Coscio, Jr. We have enjoyed many rewarding conversations on this subject matter with Professor V. Rama Murthy and in addition wish to express our gratitude to him and his colleagues for their efforts and cooperation in this joint undertaking to measure both rare gas and trace element chemical concentrations in lunar fines. Dr. Bruce R. Doe has been the source of much vocal stimulation.

## References

Compston W., Vernon M. J., Berry H., Rudowski R., Gray C. M., Ware N., Chappell B. W., and Kaye M. (1972) Age and petrogenesis of Apollo 14 basalts (abstract). In *Lunar Science—III* (editor C. Watkins), pp. 151–153, Lunar Science Institute Contr. No. 88.

Doe B. R. and Tatsumoto M. (1972) Volatilized lead from Apollo 12 and 14 soils (abstract). In *Lunar Science—III* (editor C. Watkins), pp. 177–179, Lunar Science Institute Contr. No. 88.

Eberhardt P., Geiss J., Graf H., Grögler N., Krähenbühl U., Schwaller H., Schwarzmüller J., and Stettler A. (1970) Trapped solar wind noble gases, exposure age and K/Ar-age in Apollo 11 lunar fine material. *Proc. Apollo 11 Lunar Sci. Conf., Geochim. Cosmochim. Acta* Suppl. 1, Vol. 2, pp. 1037–1070. Pergamon.

Eberhardt P., Eugster O., Geiss J., Grögler N., Schwarzmüller J., Stettler A., and Weber L. (1972a) When was the Apollo 12 KREEP ejected? (abstract). In *Lunar Science—III* (editor C. Watkins), pp. 206–208, Lunar Science Institute Contr. No. 88.

Eberhardt P., Geiss J., Graf H., Grögler N., Mendia M. D., Mörgeli M., Schwaller H., Stettler A., Krähenbühl U., and von Gunten H. R. (1972b) Trapped solar wind gases in lunar fines and a breccia (abstract). In *Lunar Science—III* (editor C. Watkins), pp. 203–205, Lunar Science Institute Contr. No. 88.

Funkhouser J. (1971) Noble gas analysis of KREEP fragments in lunar soil 12033 and 12070. *Earth Planet. Sci. Lett.* **12,** 263–272.

Heymann D. and Yaniv A. (1970) $Ar^{40}$ anomaly in samples from Tranquility Base. *Proc. Apollo 11 Lunar Sci. Conf., Geochim. Cosmochim. Acta* Suppl. 1, Vol. 2, pp. 1261–1267. Pergamon.

Heymann D., Yaniv A., Adams J. A. S., and Fryer G. E. (1970) Inert gases in lunar samples. *Science* **167,** 555–558.

Heymann D., Yaniv A., and Walton J. (1972a) Inert gases in Apollo 14 fines and the case of parentless $^{40}Ar$ (abstract). In *Lunar Science—III* (editor C. Watkins), pp. 376–378, Lunar Science Institute Contr. No. 88.

Heymann D., Yaniv A., and Lakatos S. (1972b) Inert gases in twelve particles and one "dust" sample from Luna 16. *Earth Planet. Sci. Lett.* **13,** 400–406.

Hintenberger H., Weber H. W., and Takaoka N. (1971) Concentrations and isotopic abundances of the rare gases in lunar matter. *Proc. Second Lunar Sci. Conf., Geochim. Cosmochim. Acta* Suppl. 2, Vol. 2, pp. 1607–1625. MIT Press.

Hubbard N. J., Meyer C. Jr., Gast P. W., and Weismann H. (1971) The composition and derivation of Apollo 12 soils. *Earth Planet. Sci. Lett.* **10,** 341–350.

Hubbard N. J., Nyquist L. E., Rhodes J. M., Bansal B. M., Weismann H., and Church S. E. (1972) Chemical features of the Luna 16 regolith sample. *Earth Planet. Sci. Lett.* **13,** 423–428.

Huneke J. C., Podosek F. A., Burnett D. S., and Wasserburg G. J. (1972) Rare gas studies of the galactic cosmic ray irradiation history of lunar rock. *Geochim. Cosmochim. Acta* (in press).

Kaiser W. A. (1972) Rare gas studies in Luna-16-G-7 fines by stepwise heating technique. A low fission solar wind Xe. *Earth Planet. Sci. Lett.* **13,** 387–399.

LSPET (Lunar Sample Preliminary Examination Team) (1970) Preliminary examination of the lunar samples from Apollo 12. *Science* **167,** 1325–1339.

Lunatic Asylum (1970) Mineralogic and isotopic investigations on lunar rock 12013. *Earth Planet. Sci. Lett.* **9,** 137–163.

Manka R. H. and Michel F. C. (1971) Lunar atmosphere as a source of lunar surface elements. *Proc. Second Lunar Sci. Conf., Geochim. Cosmochim. Acta* Suppl. 2, Vol. 2, pp. 1717–1728. MIT Press.

Murthy V. Rama, Evensen N. M., Bor-Ming J., and Coscio M. R. Jr. (1971) Rb–Sr ages and elemental abundances of K, Rb, Sr, and Ba in samples from the Ocean of Storms. *Geochim. Cosmochim. Acta* **35,** 1139–1153.

Murthy V. Rama, Evensen N. M., Bor-Ming J., and Coscio M. R. Jr. (1972a) Apollo 14 and 15 samples: Rb–Sr ages, trace-elements and lunar evolution. *Proc. Third Lunar Sci. Conf., Geochim. Cosmochim. Acta* Suppl. 3, Vol. 2. MIT Press.

Murthy V. Rama, Evensen N. M., Bor-Ming J., Coscio M. R. Jr., Dragon J. C., and Pepin R. O. (1972b) Rubidium-strontium and potassium-argon age of lunar sample 15555. *Science* **175,** 419–421.

Nier A. O. (1950) A redetermination of the relative abundances of the isotopes of carbon, nitrogen, oxygen, argon, and potassium. *Phys. Rev.* **77,** 789–793.

Papanastassiou D. A., Wasserburg G. J., and Burnett D. S. (1970) Rb–Sr ages of lunar rocks from the Sea of Tranquility. *Earth Planet. Sci. Lett.* **8,** 1–19.

Papanastassiou D. A. and Wasserburg G. J. (1971) Lunar chronology and evolution from Rb–Sr studies of Apollo 11 and 12 samples. *Earth Planet. Sci. Lett.* **11,** 37–62.

Papanastassiou D. A. and Wasserburg G. J. (1972) Rb–Sr age of a Luna 16 basalt and the model age of lunar soils. *Earth Planet. Sci. Lett.* **13,** 368–374.

Pepin R. O., Nyquist L. E., Phinney D., and Black D. C. (1970) Rare gases in Apollo 11 lunar material. *Proc. Apollo 11 Lunar Sci. Conf., Geochim. Cosmochim. Acta* Suppl. 1, Vol. 2, pp. 1435–1454. Pergamon.

Philpotts J. A. and Schnetzler C. C. (1970) Apollo 11 lunar samples: K, Rb, Sr, Ba, and rare-earth concentrations in some rocks and separated phases. *Proc. Apollo 11 Lunar Sci. Conf., Geochim. Cosmochim. Acta* Suppl. 1, Vol. 2, pp. 1471–1486. Pergamon.

Philpotts J. A., Schnetzler C. C., Bottino M. L., Schuhmann S., and Thomas H. H. (1972) Luna 16: Some Li, K, Rb, Sr, Ba, rare-earth, Zr, and Hf concentrations. *Earth Planet. Sci. Lett.* **13,** 429–435.

Schnetzler C. C. and Philpotts J. A. (1971) Alkali, alkaline earth, and rare-earth element concentrations in some Apollo 12 soils, rocks, and separated phases. *Proc. Second Lunar Sci. Conf., Geochim. Cosmochim. Acta* Suppl. 2, Vol. 2, pp. 1101–1122. MIT Press.

Silver L. T. (1971) *Trans. Am. Geophys. Union* **52,** 534 (abstract).

Silver L. T. (1972a) Lead volatilization and volatile transfer processes on the moon (abstract). In *Lunar Science—III* (editor C. Watkins), pp. 701–703, Lunar Science Institute Contr. No. 88.

Silver L. T. (1972b) U–Th–Pb abundances and isotopic characteristics in some Apollo 14 rocks and soils and in an Apollo 15 soil (abstract). In *Lunar Science—III* (editor C. Watkins), pp. 704–706, Lunar Science Institute Contr. No. 88.

Tatsumoto M., Knight R. J., and Doe B. R. (1971) U–Th–Pb systematics of Apollo 12 lunar samples. *Proc. Second Lunar Sci. Conf., Geochim. Cosmochim. Acta* Suppl. 2, Vol. 2, pp. 1521–1546. MIT Press.

Tatsumoto M., Hedge C. E., Doe B. R., and Unruh D. (1972) U–Th–Pb and Rb–Sr measurements on some Apollo 14 lunar samples (abstract). In *Lunar Science—III* (editor C. Watkins), pp. 741–743, Lunar Science Institute Contr. No. 88.

Turner G. (1970a) Argon-40/argon-39 dating of lunar rock samples. *Proc. Apollo 11 Lunar Sci. Conf., Geochim. Cosmochim. Acta* Suppl. 1, Vol. 2, pp. 1665–1684. Pergamon.

Turner G. (1970b) $^{40}Ar$–$^{39}Ar$ age determination of lunar rock 12013. *Earth Planet. Sci. Lett.* **9,** 177–180.

Turner G., Huneke J. C., Podosek F. A., and Wasserburg G. J. (1971) $^{40}Ar$–$^{39}Ar$ ages and cosmic ray exposure ages of Apollo 14 samples. *Earth Planet. Sci. Lett.* **12,** 19–35.

Wänke H., Wlotzka F., Baddenhausen A., Balacescu A., Spettel B., Teschke F., Jagoutz E., Kruse H., Quijano-Rico M., and Rieder R. (1971) Apollo 12 samples: Chemical composition and its relation to sample location and exposure ages, the two component origin of the various soil samples and studies on lunar metallic particles. *Proc. Second Lunar Sci. Conf., Geochim. Cosmochim. Acta* Suppl. 2, Vol. 2, pp. 1187–1209. MIT Press.

Warner J. L. (1971) Apollo 12 data-base, a continuing compilation of Apollo 12 chemical data. Unpublished.

Proceedings of the Third Lunar Science Conference
(Supplement 3, *Geochimica et Cosmochimica Acta*)
Vol. 2, pp. 1589–1612
The M.I.T. Press, 1972

# $Ar^{40}$-$Ar^{39}$ systematics in rocks and separated minerals from Apollo 14

G. TURNER,* J. C. HUNEKE, F. A. PODOSEK, and G. J. WASSERBURG

Lunatic Asylum,† California Institute of Technology,
Pasadena, California 91109, U.S.A.

**Abstract**—The $Ar^{40}$-$Ar^{39}$ dating technique has been applied to separated minerals (plagioclase, pyroxene, quintessence and an "ilmenite" concentrate), and whole rock samples of Apollo 14 rocks 14310 and 14073. Plagioclase shows the best gas retention characteristics, with no evidence of anomalous behavior and only a small amount of gas loss in the initial release. Ages determined from the plagioclase of 14310 and 14073 are (3.87 ± 0.05) and (3.88 ± 0.05) AE respectively. Low apparent ages at low release temperatures, which are frequently observed in whole rock $Ar^{40}$-$Ar^{39}$ experiments on lunar basalts, are shown here to be principally due to gas loss in the high-K interstitial glass (quintessence) phase, confirming earlier suggestions. The decrease in apparent ages in the high temperature release previously observed in several total rock samples of Apollo 14 basalts has been identified with the pyroxene.

Plagioclase is also found to be the most suitable mineral for the determination of cosmic ray exposure ages, and exposure ages of 280 and 113 million years are found for 14310 and 14073, respectively, indicating that these rocks, which are very similar in many respects, have different exposure histories. The relative production rates of $Ar^{38}$ from Fe and Ca have been determined from a comparison of pyroxene and plagioclase measurements.

## INTRODUCTION

THE APPLICATION of the $Ar^{40}$–$Ar^{39}$ dating technique (Merrihue and Turner, 1966) to whole rock lunar samples has revealed significant variations in apparent K–Ar ages with release temperature in all of the samples so far analyzed (Turner, 1970; 1971; Turner *et al.*, 1971). On the basis of hypothetical explanations for the observed variations, attempts have been made to deduce crystallization ages. Agreement of the ages with Rb–Sr mineral isochron ages (Papanastassiou and Wasserburg, 1971a, b) indicates that for the most part the procedures adopted have been successful, although in some cases the detailed explanation of the observed $Ar^{40}$–$Ar^{39}$ release pattern is obscure (Turner *et al.*, 1971).

Features of previously observed $Ar^{40}$–$Ar^{39}$ release patterns are:

(1) An increase in $(Ar^{40*}/Ar^{39*})$ is frequently observed in the initial stages of argon release from lunar, as well as other, samples, usually interpreted in terms of prior diffusive loss of radiogenic $Ar^{40}$ from sites of low retentivity (cf. Turner, 1970).

(2) A decrease in the $(Ar^{40*}/Ar^{39*})$ ratio in argon released at high temperatures from several Apollo 14 basaltic rocks (Turner *et al.*, 1971) and samples from Apollo 12 rock 12013 (Turner, 1971).

* Present address: Department of Physics, University of Sheffield, Sheffield S3 7RH, England.
† Contribution No. 2136.

(3) Well-defined but different plateaux for both a total rock sample and a plagioclase separate from Apollo 15 rock 15555 (Podosek *et al.*, 1972).

(4) A peculiar pattern in which the apparent ages rise to very high values in the early stages of release and then decline steadily to very low values at higher temperatures (Turner *et al.*, 1971). The high ages do not appear to date any event. Ages calculated from the total Ar release agree with Rb–Sr ages and appear to be meaningful.

New data pertaining to all of these features have been discussed at the Third Lunar Science Conference (Huneke *et al.*, 1972a). Features (1), (2), and (3) are considered here in greater detail.

Comparison of $Ar^{40*}/Ar^{39*}$ age patterns obtained on plagioclase and total rock samples from rocks 14053 (Turner *et al.*, 1971) and 15555 (Podosek *et al.*, 1972) have already shown that plagioclase separates provide better defined gas retention ages. In both samples, the plagioclase separates showed considerably less loss of radiogenic $Ar^{40}$ than the total rock samples. In the case of 14053, the high temperature decrease in apparent age evident in the total rock sample was absent from the plagioclase. In the case of 15555, the plagioclase provided an $Ar^{40*}/Ar^{39*}$ plateau age significantly larger than the total rock sample, presumably due to large losses from the high-K quintessence phase in the total rock sample.

In an attempt to obtain a better understanding of the significance of $Ar^{40}$–$Ar^{39}$ release patterns for rocks which are complex assemblages of minerals, and to establish a sounder empirical basis for their interpretation, we have investigated separated minerals (plagioclase, pyroxene, K-rich glass (quintessence) and an ilmenite concentrate) as well as whole rock samples from Apollo 14 rocks 14310 and 14073.

## SAMPLE DESCRIPTION

Mineralogic, petrologic, and chemical descriptions of 14310 and 14073 are available in the literature (Gancarz *et al.*, 1971; Papanastassiou and Wasserburg, 1971b), as is information on the chronology of these rocks (Papanastassiou and Wasserburg, 1971b; Turner *et al.*, 1971; and Husain *et al.*, 1971). Samples were taken from the interior of 14310,100 and 14073,3. After gentle crushing fragments were picked for whole rock samples, and then mineral separations were carried out on finely crushed material using heavy liquids and a Frantz magnetic separator, with some hand picking (Papanastassiou and Wasserburg, 1971b).

The separates obtained from 14310,100 were:

(a) a high purity (>99%) plagioclase,

(b) a pyroxene (pyroxene 1) separate (>95%) which was mostly clear and light colored but with several percent darker pyroxenes and some plagioclase (<5%),

(c) a second pyroxene (pyroxene 2) separate (>97%) with highly variable pyroxenes (many of which were dark brown) and containing much less plagioclase than for (b),

(d) an ilmenite concentrate with ~30% ilmenite and a large proportion of light colored pyroxene and plagioclase and significant amounts of glassy material, and

(e) a quintessence concentrate comprised primarily of plagioclase, with ~20% quintessence grains.

Mineral separates obtained from 14073,3 were:

(a) a high purity (>99%) plagioclase,

(b) a pyroxene separate with some plagioclase (<5%), and

(c) a quintessence concentrate with about 50% quintessence, <40% plagioclase, <5% opaque minerals, <5% phosphate minerals.

These mineral separates and concentrates are not representative of the entire mineral assemblage in either of the rocks. The pyroxene, plagioclase and quintessence phases are heterogeneous and the mineral separation procedures by nature select for relatively uniform material which does not represent averages over these phases.

It should also be noted that the $Ar^{40*}/Ar^{39*}$ ratios in the later release stages from finely crushed separates will be lowered if the crushing procedure reduces the grain size sufficiently so that the diffusion path length during laboratory outgassing is comparable in size to the region in the crystal affected by loss on the moon. Furthermore, grains deriving from interior regions of crystals unaffected by lunar gas loss can contribute to the initial release and increase the ($Ar^{40*}/Ar^{39*}$) ratio for these release fractions which are normally low.

As a final caveat, we note that in the reaction $K^{39}(n,p)Ar^{39*}$, the recoil energy of the $Ar^{39*}$ is sufficient to displace the $Ar^{39*}$ from the K site by ~0.1 $\mu$ (Turner *et al.*, 1971). Thus if the K distribution in the sample is dominated by K-rich domains less than a micron in size included in K-poor material, the $Ar^{40*}$ and $Ar^{39*}$ will separately reflect the release characteristics of the different host sites, resulting in perturbations in the $Ar^{40*}/Ar^{39*}$ ratio and in the apparent age. This must be seriously considered both in the selection of suitable material for $Ar^{40}$–$Ar^{39}$ dating and in the understanding of the present results.

## Neutron Irradiation

The samples were irradiated for 4.4 days in the shuttle tube facility of the General Electric Test Reactor in Pleasanton, California. The nominal fluence was $7 \times 10^{18}$ $cm^{-2}$ ($E > 0.18$ MeV) and $4.8 \times 10^{19}$ $cm^{-2}$ ($E < 0.17$ eV). This irradiation is designated GTV-4.

The samples, wrapped in high purity Al foil, were irradiated on two levels within the rotating insert, with six samples on each level surrounding a central hornblende monitor at that level. Ni wire flux monitors were spaced around the central monitors and around the periphery of the sample clusters. It is clear from these Ni flux monitors that the container did not rotate continuously, resulting in a fluence variation of some 10% across the container. The fluence of individual samples relative to the centrally located monitors was estimated from a flux map established by the Ni flux wire measurements. The uncertainty in the relative fluence is estimated to be less than 1%.

All experimental procedure and data analyses employed in this study were the same as those detailed by Turner *et al.* (1971). All data discussed below have been corrected for interferences following Turner *et al.* (1971); use of the asterisk (*) in $Ar^{40*}$ and $Ar^{39*}$ denotes data corrected for everything but electron capture by $K^{40}$ or

Table 1. Argon isotope release pattern for Apollo sample 14310.

| Approximate Temperature (°C) | $Ar^{36}/Ar^{38}$ | $Ar^{38}/Ar^{37}$ | $Ar^{39*}/Ar^{37}$ | $Ar^{40*}/Ar^{39*}$ | $Ar^{39*}$[c] | Apparent Age (aeons) |
|---|---|---|---|---|---|---|
| 14310 Plagioclase (40 mg) | | | | | | |
| 610[a] | 0.827 ± 0.043 | 0.0376 ± 0.0015 | 0.0483 ± 0.0005 | 188.3 ± 1.4 | 1.57 | 3.67 ± 0.03 |
| 715 | 0.631 ± 0.013 | 0.0300 ± 0.0004 | 0.0277 ± 0.0003 | 206.7 ± 1.9 | 2.97 | 3.82 ± 0.02 |
| 810 | 0.633 ± 0.005 | 0.0305 ± 0.0002 | 0.0244 ± 0.0001 | 210.7 ± 0.8 | 7.28 | 3.85 ± 0.01 |
| 900 | 0.623 ± 0.003 | 0.0308 ± 0.0002 | 0.0231 ± 0.0001 | 211.9 ± 0.5 | 12.5 | 3.86 ± 0.01 |
| 980 | 0.606 ± 0.006 | 0.0314 ± 0.0002 | 0.0274 ± 0.0002 | 212.3 ± 1.6 | 5.35 | 3.87 ± 0.01 |
| 1385 | 0.620 ± 0.004 | 0.0314 ± 0.0001 | 0.0244 ± 0.0001 | 213.7 ± 0.7 | 7.28 | 3.88 ± 0.01 |
| Total | 0.628 | 0.0310 | 0.0251 | 210.7 | 36.9 | 3.86 |
| 14310 Quintessence concentrate (1.4 mg) | | | | | | |
| 460[b] | — | — | — | 103.8 ± 8.9 | 27.8 | 2.75 ± 0.12 |
| 540 | 1.00 ± 0.36 | 0.255 ± 0.086 | 3.08 ± 0.57 | 188.1 ± 1.9 | 83.4 | 3.67 ± 0.02 |
| 630 | 0.41 ± 0.10 | 0.234 ± 0.026 | 3.36 ± 0.14 | 208.5 ± 1.3 | 146.3 | 3.84 ± 0.01 |
| 740 | 0.49 ± 0.06 | 0.099 ± 0.006 | 0.753 ± 0.006 | 209.8 ± 1.3 | 178.2 | 3.85 ± 0.01 |
| 850 | 0.42 ± 0.08 | 0.061 ± 0.003 | 0.500 ± 0.005 | 210.5 ± 1.8 | 99.2 | 3.86 ± 0.01 |
| 930 | — | — | — | — | 100 | — |
| 1040 | 0.41 ± 0.08 | 0.058 ± 0.005 | 0.416 ± 0.005 | 208.9 ± 2.0 | 88.6 | 3.84 ± 0.02 |
| 1200 | 0.17 ± 0.15 | 0.073 ± 0.009 | 0.725 ± 0.011 | 212.5 ± 1.5 | 118.3 | 3.87 ± 0.01 |
| Total | 0.43 | 0.091 | 0.841 | 203.9 | 844 | 3.80 |
| 14310 Ilmenite concentrate (15 mg) | | | | | | |
| 500[a] | — | — | — | 107.2 ± 11.3 | 3.37 | 2.80 ± 0.15 |
| 610 | 0.45 ± 0.11 | 0.236 ± 0.039 | 2.81 ± 0.14 | 182.8 ± 1.3 | 12.7 | 3.626 ± 0.011 |
| 715 | 0.145 ± 0.053 | 0.198 ± 0.011 | 3.06 ± 0.06 | 212.9 ± 0.8 | 24.2 | 3.874 ± 0.006 |
| 810 | 0.260 ± 0.077 | 0.092 ± 0.004 | 1.35 ± 0.02 | 216.0 ± 0.6 | 30.0 | 3.898 ± 0.004 |
| 900 | 0.437 ± 0.029 | 0.0608 ± 0.0018 | 0.447 ± 0.003 | 212.1 ± 1.1 | 19.6 | 3.868 ± 0.008 |
| 980 | 0.480 ± 0.030 | 0.0642 ± 0.0022 | 0.148 ± 0.001 | 202.0 ± 1.8 | 9.21 | 3.788 ± 0.014 |
| 1385 | 0.555 ± 0.005 | 0.0532 ± 0.0003 | 0.0188 ± 0.0001 | 179.2 ± 1.9 | 9.74 | 3.594 ± 0.016 |
| Total | 0.504 | 0.0591 | 0.165 | 203.1 | 108.9 | 3.80 |
| 14310 Pyroxene 1 (22 mg) | | | | | | |
| 715[a] | 0.34 ± 0.07 | 0.090 ± 0.009 | 0.135 ± 0.005 | 179.2 ± 11.2 | 0.97 | 3.59 ± 0.10 |
| 900 | 0.49 ± 0.08 | 0.034 ± 0.002 | 0.0624 ± 0.0017 | 208.8 ± 8.9 | 1.36 | 3.84 ± 0.07 |
| 980 | 0.57 ± 0.13 | 0.039 ± 0.002 | 0.0509 ± 0.0024 | 173.5 ± 18.9 | 0.54 | 3.54 ± 0.17 |
| 1145 | 0.64 ± 0.04 | 0.0355 ± 0.0009 | 0.0137 ± 0.0004 | 171.0 ± 14.2 | 0.71 | 3.52 ± 0.13 |
| 1385 | 0.63 ± 0.01 | 0.0363 ± 0.0003 | 0.0042 ± 0.0002 | 201.8 ± 15.2 | 0.75 | 3.79 ± 0.12 |
| Total | 0.60 | 0.379 | 0.0161 | 196.5 | 4.47 | 3.74 |
| 14310 Pyroxene 2 (47 mg) | | | | | | |
| 715[a] | 0.37 ± 0.05 | 0.0604 ± 0.0032 | 0.412 ± 0.003 | 187.7 ± 1.3 | 3.95 | 3.67 ± 0.01 |
| 900 | 0.53 ± 0.04 | 0.0394 ± 0.0017 | 0.148 ± 0.001 | 219.2 ± 2.1 | 3.62 | 3.92 ± 0.02 |
| 980 | 0.44 ± 0.08 | 0.0379 ± 0.0028 | 0.081 ± 0.002 | 218.3 ± 7.0 | 0.90 | 3.92 ± 0.05 |
| 1385 | 0.628 ± 0.003 | 0.0357 ± 0.0002 | 0.00658 ± 0.00008 | 184.5 ± 3.4 | 2.38 | 3.64 ± 0.03 |
| Total | 0.607 | 0.0365 | 0.0266 | 200.1 | 10.8 | 3.78 |

[a] Based on earlier calibration with a thermocouple. The estimated uncertainty in absolute temperature is less than 50°C

[b] Temperatures above 800°C obtained by optical pyrometer (spectral emissivity of 0.7 assumed) and below 800°C calculated from heating currents. Different assumptions on the spectral emissivity change the absolute temperatures by up to 50°C.

[c] Amounts in units of $10^{-8}$ cc STP/g.

$K^{39}(n,p)$ $Ar^{39*}$. For the two hornblende monitors, we measured $(Ar^{40*}/Ar^{39*}) = 23.70 \pm 0.05$ and $(Ar^{40*}/Ar^{39*}) = 23.54 \pm 0.09$. This variation is consistent with the variability of $Ar^{40}/K$. With the previously measured ratio $(Ar^{40}/K) = (5.69 \pm 0.09) \times 10^{-3}$ ccSTP/gmK (Turner *et al.*, 1971), we have for the conversion ratio for the $K^{39}(n,p)$ $Ar^{39*}$ reaction

$$C_{39}(K) \equiv Ar^{39*}/K = 2.41 \pm 0.04 \times 10^{-4} \text{ ccSTP/gmK}. \tag{1}$$

From previous irradiations in the same facility (Turner *et al.*, 1971)

$$\frac{K}{Ca} = (0.54 \pm 0.02)\frac{Ar^{39*}}{Ar^{37}}, \tag{2}$$

so for the $Ca^{40}(n,\alpha)$ $Ar^{37}$ reaction on Ca, we have

$$C_{37}(Ca) \equiv \frac{Ar^{37}}{Ca} = 1.30 \pm 0.05 \times 10^{-4} \text{ ccSTP/gmCa}. \tag{3}$$

## Analytical Results

The results of the stepwise heating experiments are summarized in Tables 1 and 2. $Ar^{37}$ and $Ar^{39}$ have been corrected for decay during and after the irradiation. The ratios have been corrected for $Ar^{36}$, $Ar^{38}$, $Ar^{39}$ and $Ar^{40}$ produced by nuclear reactions on Ca and K during the irradiation using the production ratios $Ar^{36}/Ar^{37} = 3.05 \times 10^{-4}$, $Ar^{38}/Ar^{37} = 1 \times 10^{-4}$, $Ar^{39}/Ar^{37} = 7.32 \times 10^{-4}$, and $Ar^{40}/Ar^{39*} = 1 \times 10^{-2}$ (Turner *et al.*, 1971). Corrections for atmospheric $Ar^{40}$ of $4 \times 10^{-9}$ ccSTP $Ar^{40}$ and corresponding amounts of $Ar^{38}$ and $Ar^{36}$ have been applied to each release fraction on the basis of extraction system blanks determined at ~700°C, ~1000°C, and ~1400°C. An uncertainty of ±50% has been assumed for this correction and is included in the error figure for the $(Ar^{40*}/Ar^{39*})$ ratio. Measurements

Table 2. Argon isotope release pattern for Apollo sample 14073.

| Approximate Temperature (°C) | $Ar^{36}/Ar^{38}$ | $Ar^{38}/Ar^{37}$ | $Ar^{39*}/Ar^{37}$ | $Ar^{40*}/Ar^{39*}$ | $Ar^{39*c}$ | Apparent Age (aeons) |
|---|---|---|---|---|---|---|
| 14073 Whole rock (67 mg) | | | | | | |
| 540[b] | 0.375 ± 0.052 | 0.0476 ± 0.0017 | 0.908 ± 0.005 | 138.1 ± 0.4 | 11.1 | 3.185 ± 0.004 |
| 630 | 0.338 ± 0.025 | 0.0316 ± 0.0005 | 0.533 ± 0.002 | 208.5 ± 0.3 | 21.6 | 3.840 ± 0.002 |
| 740 | 0.429 ± 0.007 | 0.0207 ± 0.0001 | 0.2465 ± 0.0002 | 217.3 ± 0.2 | 29.8 | 3.908 ± 0.002 |
| 850 | 0.527 ± 0.007 | 0.0157 ± 0.0001 | 0.0897 ± 0.0003 | 217.9 ± 0.3 | 23.1 | 3.913 ± 0.002 |
| 930 | 0.477 ± 0.004 | 0.0175 ± 0.0001 | 0.0467 ± 0.0001 | 214.7 ± 0.4 | 12.5 | 3.889 ± 0.002 |
| 1040 | 0.237 ± 0.009 | 0.0376 ± 0.0003 | 0.0587 ± 0.0002 | 204.3 ± 0.9 | 4.31 | 3.807 ± 0.007 |
| 1200 | 0.431 ± 0.009 | 0.0224 ± 0.0002 | 0.0191 ± 0.0001 | 208.1 ± 1.5 | 2.72 | 3.837 ± 0.011 |
| 1450 | 0.453 ± 0.009 | 0.0204 ± 0.0001 | 0.0169 ± 0.0001 | 206.1 ± 1.5 | 2.66 | 3.821 ± 0.012 |
| Total | 0.430 | 0.0208 | 0.100 | 206.2 | 107.9 | 3.82 |
| 14073 Plagioclase (30 mg) | | | | | | |
| 715[a] | 0.637 ± 0.041 | 0.0132 ± 0.0006 | 0.0193 ± 0.0003 | 203.6 ± 4.0 | 2.48 | 3.80 ± 0.03 |
| 810 | 0.640 ± 0.011 | 0.0124 ± 0.0002 | 0.01578 ± 0.00008 | 211.6 ± 1.7 | 5.26 | 3.86 ± 0.01 |
| 900 | 0.631 ± 0.007 | 0.0124 ± 0.0001 | 0.01458 ± 0.00005 | 212.8 ± 1.2 | 6.99 | 3.87 ± 0.01 |
| 980 | 0.616 ± 0.016 | 0.0123 ± 0.0002 | 0.01608 ± 0.00024 | 216.4 ± 3.8 | 3.30 | 3.90 ± 0.03 |
| 1385 | 0.628 ± 0.012 | 0.0125 ± 0.0002 | 0.01496 ± 0.00007 | 215.4 ± 1.8 | 4.40 | 3.89 ± 0.01 |
| Total | 0.631 | 0.0125 | 0.01557 | 212.5 | 22.4 | 3.87 |
| 14073 Quintessence concentrate (1 mg) | | | | | | |
| 460[b] | 1.52 ± 0.46 | 0.242 ± 0.055 | 0.99 ± 0.11 | 138.1 ± 7.7 | 30.2 | 3.18 ± 0.09 |
| 540 | 1.13 ± 0.19 | 0.298 ± 0.039 | 2.99 ± 0.24 | 179.8 ± 2.4 | 93.5 | 3.60 ± 0.02 |
| 630 | 0.48 ± 0.22 | 1.00 ± 0.35 | 23.2 ± 5.6 | 206.5 ± 1.3 | 200 | 3.82 ± 0.01 |
| 740 | 1.52 ± 0.09 | 0.287 ± 0.013 | 3.39 ± 0.06 | 210.9 ± 0.8 | 326 | 3.86 ± 0.01 |
| 850 | 1.20 ± 0.19 | 0.107 ± 0.010 | 1.18 ± 0.05 | 206.0 ± 3.1 | 92.9 | 3.82 ± 0.03 |
| 930 | 0.58 ± 0.13 | 0.119 ± 0.014 | 1.09 ± 0.04 | 208.1 ± 2.6 | 97.9 | 3.84 ± 0.02 |
| 1040 | 0.69 ± 0.28 | 0.093 ± 0.026 | 1.12 ± 0.07 | 203.4 ± 4.6 | 58.8 | 3.80 ± 0.04 |
| 1200[b] | 0.30 ± 0.31 | 0.067 ± 0.023 | 1.76 ± 0.06 | 207.2 ± 2.1 | 109 | 3.83 ± 0.02 |
| Total | 1.01 | 0.181 | 2.25 | 203.4 | 1009 | 3.80 |
| 14073 Pyroxene (72 mg) | | | | | | |
| 715[a] | 0.64 ± 0.08 | 0.0383 ± 0.0034 | 0.350 ± 0.002 | 181.0 ± 1.3 | 2.95 | 3.61 ± 0.01 |
| 900 | 0.84 ± 0.05 | 0.0168 ± 0.0006 | 0.0893 ± 0.0005 | 216.1 ± 1.7 | 2.28 | 3.89 ± 0.01 |
| 980 | 0.44 ± 0.10 | 0.0190 ± 0.0025 | 0.0649 ± 0.0019 | 201.6 ± 7.4 | 0.59 | 3.78 ± 0.06 |
| 1385 | 0.64 ± 0.01 | 0.0151 ± 0.0001 | 0.00458 ± 0.00005 | 179.2 ± 2.7 | 1.58 | 3.59 ± 0.02 |
| Total | 0.65 | 0.0159 | 0.0191 | 193.1 | 7.40 | 3.72 |

[a] Based on earlier calibration with a thermocouple. The estimated uncertainty in absolute temperature is less than 50°C

[b] Temperatures above 800°C obtained by optical pyrometer (spectral emissivity of 0.7 assumed) and below 800°C calculated from heating currents. Different assumptions on the spectral emissivity change the absolute temperatures by up to 50°C.

[c] Amounts in units of $10^{-8}$ cc STP/g.

on both irradiated high purity Al foil and the Al foil into which the samples were repackaged after irradiation showed no significant $Ar^{40}$ or $Ar^{39}$ contributions from this source. A combined correction for cosmogenic and trapped $Ar^{40}$ has been applied on the basis of $(Ar^{40}/Ar^{36})_{c,t} = 1 \pm 0.5$.

The form chosen for presenting the results has a number of advantages. The $(Ar^{36}/Ar^{38})$ ratio is an indicator of the importance of the cosmogenic contribution to $Ar^{38}$. The $(Ar^{38}/Ar^{37})$ and $(Ar^{40*}/Ar^{39*})$ ratios are related to the exposure age and K–Ar age, respectively, while $(Ar^{39*}/Ar^{37})$ is proportional to the average (K/Ca) ratio of the sites releasing argon in each step (see equation 2). The $(Ar^{40*}/Ar^{39*})$ ratio (column 5 in Tables 1 and 2) for each stage of the gas release has been converted to an apparent K–Ar age (column 7) using the expressions

$$t = \frac{1}{\lambda} \ln \left[ \frac{\lambda}{\lambda_e} \left( \frac{Ar^{40*}}{K^{40}} \right) + 1 \right] \tag{4}$$

and

$$\left( \frac{Ar^{40*}}{K^{40}} \right) = C_{39}(K) \cdot \left( \frac{Ar^{40*}}{Ar^{39*}} \right) \cdot \left( \frac{K}{K^{40}} \right) \tag{5}$$

where $\lambda = 5.305 \times 10^{-10}/\mathrm{yr}$, $\lambda_e = 0.585 \times 10^{-10}/\mathrm{yr}$, $K^{40}/K = 0.0119\%$.

The apparent ages are plotted in the lower parts of Figs. 1–7 as a function of the cumulative fraction of $Ar^{39*}$ released. The errors in the apparent ages listed in Table 1 and illustrated in Figs. 1–7 are based only on the errors in $Ar^{40*}/Ar^{39*}$, as is appropriate for comparison of the various release fractions and thus for the evaluation of the release pattern of a single sample. Comparison of different samples must include an additional systematic uncertainty of $\pm 0.02$ AE (1 aeon = AE = $10^9$ years) due to fluence uncertainty. The absolute ages are uncertain by an additional $\pm 0.03$ AE, reflecting the uncertainty in the monitor $Ar^{40*}/K$ ratio. Uncertainties in the decay constants and K isotopic composition are not included in this error. The (K/Ca) ratios calculated from equation (3) are plotted in the upper part of Figs. 1–7 and may be directly correlated with the apparent age of each release fraction.

The absolute $Ar^{39*}$ and $Ar^{37}$ concentrations in all the samples analyzed have been converted to K and Ca concentrations which are summarized in Table 3.

Table 3. K and Ca abundances[a] of total rock samples and mineral separates of rocks 14310 and 14073

| | Sample | K (%) | Ca (%) |
|---|---|---|---|
| 14310 | Whole rock[b] | 0.39 | 8.3 |
| | Plagioclase | 0.158 | 11.6 |
| | Quintessence concentrate | 3.60 | 7.9 |
| | Pyroxene 1 | 0.019 | 2.19 |
| | Pyroxene 2 | 0.046 | 3.21 |
| | Ilmenite concentrate | 0.465 | 5.2 |
| 14073 | Whole rock | 0.46 | 8.5 |
| | Plagioclase | 0.096 | 11.4 |
| | Quintessence concentrate | 4.31 | 3.55 |
| | Pyroxene | 0.032 | 3.06 |

[a] Absolute abundances are accurate to better than 10%.
[b] See Turner *et al.* (1971).

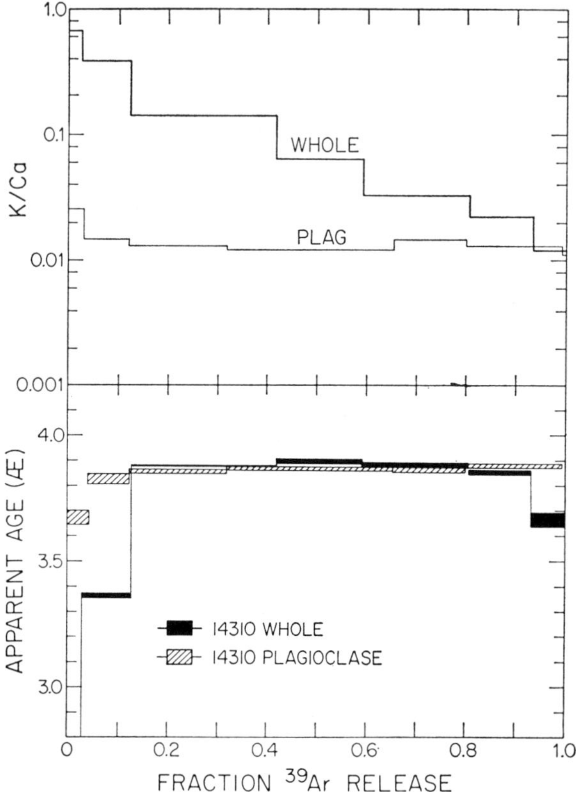


Fig. 1. Apparent age and K/Ca as a function of $Ar^{39*}$ release from a plagioclase separate and a whole rock sample from rock 14310. The whole rock data for 14310 are from Turner *et al.* (1971). The error in apparent age indicated by the height of the box is statistical only, and may only be used in comparison of different release fractions of the same sample. Since the whole rock was irradiated in a different irradiation than that for the mineral separates, the systematic uncertainty appropriate for comparison of whole rock and mineral separate absolute ages is $\pm 0.04$ AE.

## Sample 14310

Rock 14310, the largest crystalline rock returned by Apollo 14, is a basalt composed predominantly of plagioclase and pyroxene, with minor ilmenite, troilite, Fe-metal, and mesostasis (Gancarz *et al.*, 1971). Subsequent investigations (Bence and Papike, 1972; Hollister *et al.*, 1972; Ridley *et al.*, 1972; Gancarz, personal communication) indicate that a large proportion of the pyroxene is a relatively Ca-rich orthopyroxene.

A whole rock sample of 14310 was previously analyzed by Turner *et al.* (1971). The observed pattern showed both the effects of argon loss ($\sim 5\%$ of the total $Ar^{40*}$ content) in the initial part of the release pattern and a 13% decrease in the high temperature ($Ar^{40*}/Ar^{39*}$) ratio. On the basis of the high (K/Ca) ratio ($\sim 0.7$) associated

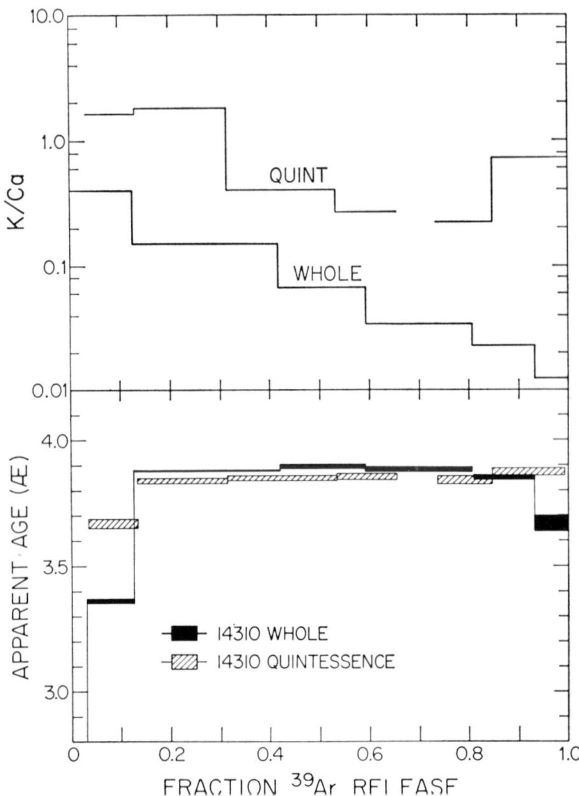


Fig. 2. Apparent age and K/Ca as a function of $Ar^{39*}$ release from a quintessence concentrate and a whole rock sample from rock 14310. Note the high K/Ca ratio for the quintessence. The decrease in $Ar^{40*}/Ar^{39*}$ for the whole rock occurs at a temperature at which the Ar extraction for the quintessence is essentially completed.

with the low temperature release, it was concluded that the $Ar^{40*}$ loss had occurred from a high-K phase in the rock, probably quintessence. In contrast, the unexplained high temperature decrease in $(Ar^{40*}/Ar^{39*})$ was associated with very low (K/Ca) ratios ($\sim 0.015$). The mineral separate data are compared in Figs. 1–4 with the whole rock data of Turner *et al.* (1971), since we wish to identify in the mineral separate data features contributing to the whole rock pattern. In this context it is very important to note that the whole rock and mineral separate samples were irradiated separately, and comparison must include uncertainties due to this fact.

It is also essential to note in the comparison that as various minerals will have different Ar release characteristics with respect to temperature, similar features may occur at different stages of $Ar^{39}$ release. A distinctive feature in the age pattern of a mineral which occurs at a given temperature is not necessarily observed in the whole rock age pattern at the same fractional $Ar^{39*}$ release point.

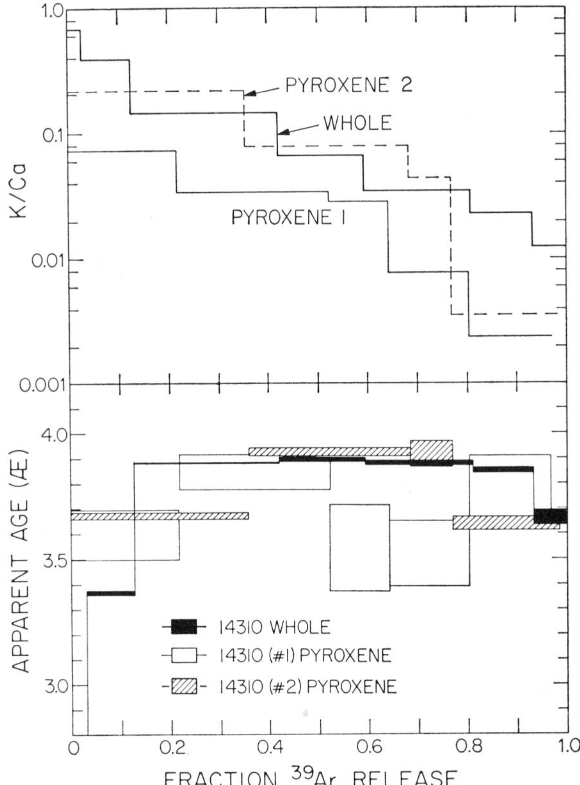


Fig. 3. Apparent age and K/Ca as a function of $Ar^{39*}$ release from pyroxene separates and a whole rock sample from rock 14310. A significant amount of the $Ar^{39*}$ release from the pyroxenes occurs at high temperatures, when the total rock $Ar^{39*}$ release is diminishing. Note that pyroxene 1 has a very low K/Ca ratio.

*Plagioclase*

The ($Ar^{40*}/Ar^{39*}$) ratio observed for plagioclase (Fig. 1) shows the least variation of the samples analyzed, and in this sense the plagioclase system appears to be the least disturbed by post-crystallization effects. The initial (610°C) ratio is only 10% low relative to the maximum observed in the 1385°C release. Above 810°C the ratio increases slightly throughout the release, though not outside error limits. The high temperature decrease apparent in the total rock age pattern (see Fig. 1) is absent from the plagioclase age pattern. The mean value of the 900°C to 1385°C releases corresponds to a high temperature plateau age of 3.87 AE. The difference between the total gas ($Ar^{40*}/Ar^{39*}$) ratio and the 1385°C value corresponds to an overall $Ar^{40*}$ loss from the plagioclase of only 1.4%.

The difference between the intermediate plateau of the whole rock sample and the high temperature plateau of the plagioclase separate is well within the uncertainties

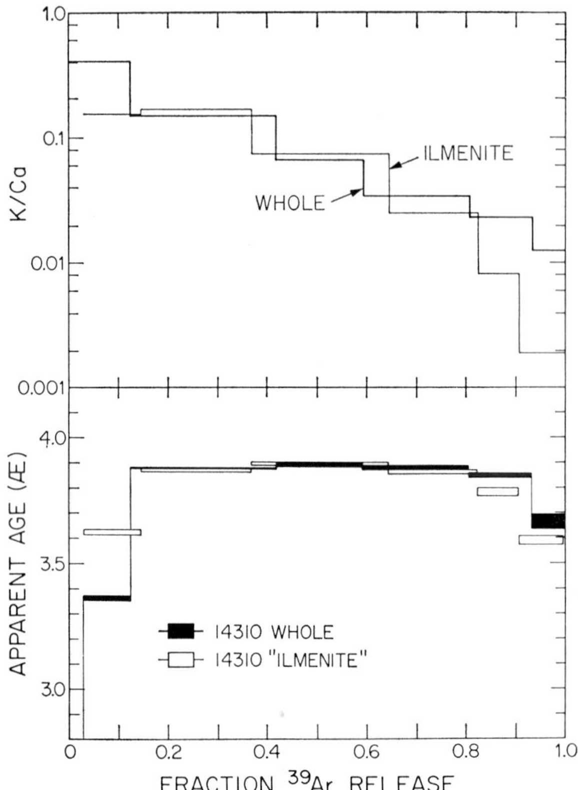


Fig. 4. Apparent age and K/Ca as a function of $Ar^{39*}$ release from an ilmenite concentrate and a whole rock sample from rock 14310. Comparison of any two samples must allow for the $\pm 0.02$ AE uncertainty for each sample (in addition to statistical uncertainty) corresponding to uncertainty in relative neutron fluence.

in the relative flux estimates and conversion coefficient $C_{39}(K)$ determinations of the two irradiations.

The variation in (K/Ca) is relatively small in the plagioclase. (K/Ca) is highest in the initial release, possibly as a result of zoning (K-rich rims) and the presence of minute amounts of quintessence. A small secondary maximum in K/Ca in the 980°C release is similar to that previously observed in a plagioclase separate from 14053 (Turner *et al.*, 1971). This plagioclase separate was of very high purity and contained only a small amount of quintessence, as shown by the K/Ca ratio variations.

*Quintessence concentrate*

The quintessence shows much lower ($Ar^{40*}/Ar^{39*}$) ratios than the plagioclase in the initial stages of the release pattern (Fig. 2). The 460°C and 540°C ratios are respectively 50% and 12% below the high temperature values. The high temperature release parallels that of the plagioclase although the experimental uncertainties are slightly

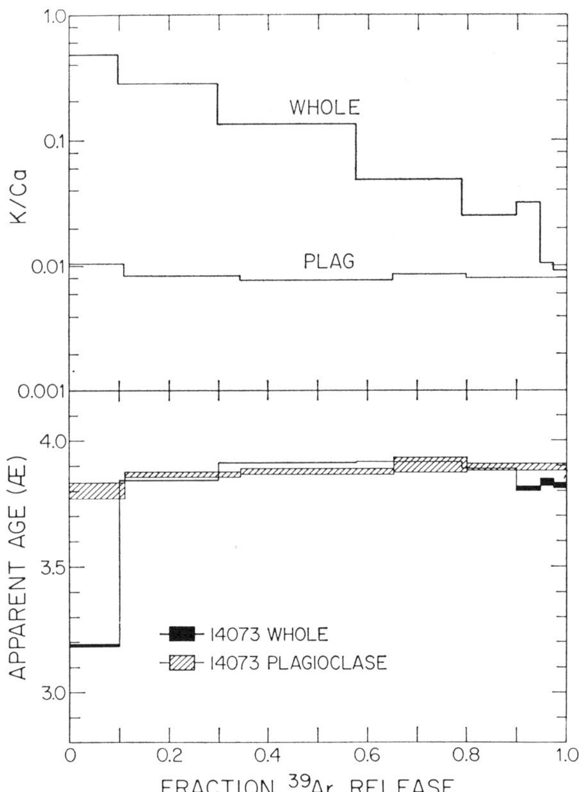


Fig. 5. Apparent age and K/Ca as a function of $Ar^{39*}$ release from a plagioclase separate and whole rock sample from rock 14073. Note the similarity in both release patterns with Figs. 1 and 8.

greater. As in the plagioclase the high temperature plateau is well-defined. The ages do not decrease in the last release stages. The vertical offset of the two curves is not significant in view of the uncertainty in the corrections applied for differences in neutron fluence for different samples. The mean high temperature plateau age deduced from the 740°C to 1200°C release is 3.86 AE.

The (K/Ca) ratio reflects the high K content of the quintessence. The (K/Ca) ratio varies by a factor of 6 and is highest in the material dominating the initial argon release. The dominant part of the Ca in this mineral separate is from the plagioclase. However, because of the very high concentration of K in the quintessence the $Ar^{40*}/Ar^{39*}$ results are almost exclusively a reflection of the behavior of the quintessence phases. These are chemically quite heterogeneous as has been shown by microprobe analyses (Gancarz *et al.*, 1971; Roedder and Weiblen, 1972).

Since the K and Ca are contained almost entirely in the quintessence and the plagioclase respectively, the relative abundances of the two minerals can be calculated from the Ar data. With ~0.5% Ca in the quintessence (Gancarz *et al.*, 1971) and

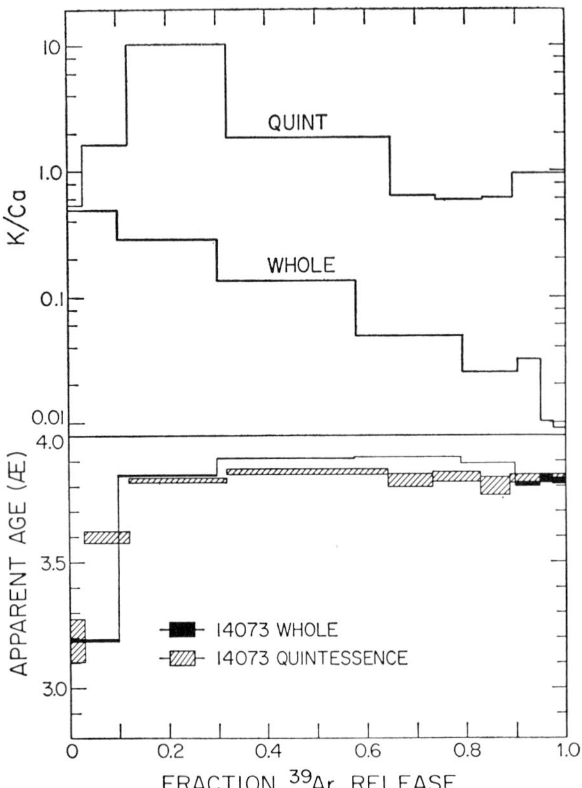


Fig. 6. Apparent age and K/Ca as a function of $Ar^{39*}$ release from a quintessence concentrate and whole rock sample from rock 14073. Note that the plateau for the quintessence is far below that of the total rock. As these were irradiated simultaneously there is an additional uncertainty of only 0.02 AE for each sample, and the difference is apparently real.

11.6% Ca in the plagioclase (Table 3), a plagioclase content of 67% and a quintessence content of 33% is indicated, in agreement with the visual estimates based on grain counts. Further, with 0.16% K in the plagioclase (Table 3), an 11% K concentration is calculated for the quintessence, on the K-rich end of the range observed by Gancarz *et al.* (1971).

### *Pyroxene*

Striking variations are observed in the $(Ar^{40*}/Ar^{39*})$ ratios measured for the pyroxenes, well in excess of the uncertainties. The pattern of variation (Fig. 3) is quite different from those of either plagioclase or quintessence. The characteristic feature of the pattern is a relatively low $(Ar^{40*}/Ar^{39*})$ ratio for *both* the *low and high* temperature release and high $(Ar^{40*}/Ar^{39*})$ ratio in the intermediate release. The high ratios at intermediate temperatures correspond to apparent ages of $3.84 \pm 0.07$ AE

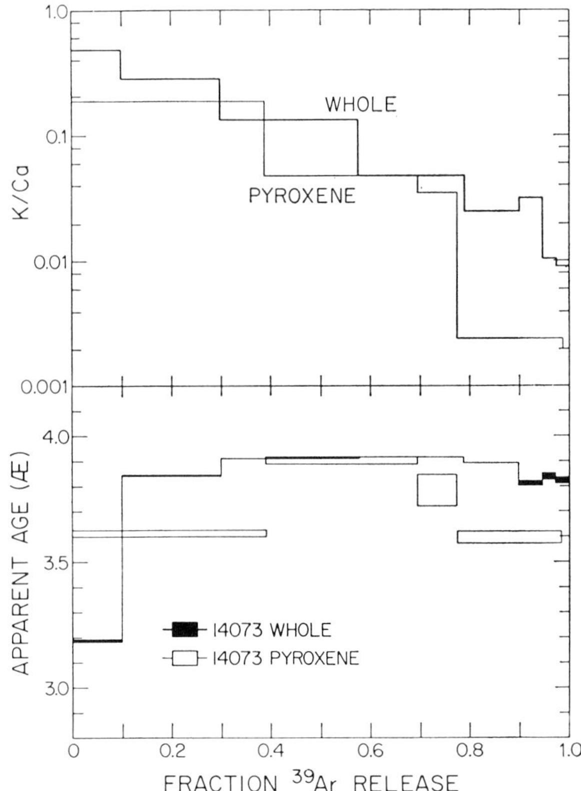


Fig. 7. Apparent age and K/Ca as a function of $Ar^{39*}$ release from a pyroxene separate and whole rock sample of rock 14073. The pyroxene again shows a low K/Ca ratio and a distinct decrease in the high temperature region where $Ar^{40*}/Ar^{39*}$ for the total rock decreases and $Ar^{39*}$ is relatively small in amount.

and 3.92 ± 0.02 AE for the pyroxenes 1 and 2, respectively, and are not significantly different from the high temperature ages of the plagioclase and quintessence. The high temperature points, however, define significantly lower apparent ages in the range 3.5–3.7 AE over a substantial fraction of the total release pattern.

The (K/Ca) ratios measured for the pyroxene separates show a steady decrease from 0.07 to 0.002 (pyroxene 1) and from 0.22 to 0.003 (pyroxene 2), almost two orders of magnitude. Lunar pyroxenes are characterized by extreme zoning and exsolution and this may be reflected in the observed variation in (K/Ca). Alternatively, in view of the very low K content of the pyroxene (0.02–0.05%) in comparison with the quintessence (8–13%), it may be that the initial argon release from the pyroxene (including the high ages of 3.9 AE) reflects a contribution from very small amounts of quintessence contamination.

The pyroxene separates analyzed in this work represent the low Ca population. Pyroxene 1 is the more homogeneous of the two separates analyzed and has an average

Ca content of 2.19%, as compared to the observed range in 14310 of from ~1 to ~12% Ca. We may thus not use the present analyses to represent the average pyroxene of this rock.

Comparison of the K/Ca ratio release pattern of the pyroxene with that of the K/Ca ratio of the total rock shows them to be roughly similar with monotonically decreasing values (see Fig. 3). It is quite evident that the pyroxene separates, particularly #1, have a very low-K content. The extent to which this K content reflects substitution in the pyroxene lattice as opposed to inclusions of quintessence is not evident. In contrast to the K/Ca pattern observed in the pyroxene, that in the plagioclase shows a rather constant K/Ca ratio over all temperature fractions. As a consequence, we can conclude that the plagioclase is essentially free of K-rich impurities and can interpret the release pattern as intrinsically due to the plagioclase phase. The pyroxenes show a wide variability in K/Ca, and it is not possible to conclude in a similar way that these separates are not a blend of phases including quintessence. However, the high temperature drop of $Ar^{40*}/Ar^{39*}$ is not present in the quintessence, so that part of the pyroxene release pattern must reflect an intrinsic character of the pyroxenes themselves.

We shall consider possible mechanisms for the observed ($Ar^{40*}/Ar^{39*}$) variations in pyroxene in a later section. For the present, however, we shall simply note the significant empirical observation that the high temperature decrease observed in ($Ar^{40*}/Ar^{39*}$) in the whole rock sample can be explained in terms of a pyroxene contribution.

*Ilmenite concentrate*

As described above, the ilmenite concentrate is of low quality and contains major contributions of pyroxene. The release pattern of the ilmenite concentrate is in many respects similar to that of the total rock. A comparison of the very close parallel in the initial ($Ar^{40*}/Ar^{39*}$) and (K/Ca) variations in quintessence and ilmenite suggests, in fact, that the "ilmenite" pattern is dominated by quintessence contamination, at least in the first part of the $Ar^{39*}$ release. The presence of K-rich quintessence inclusions in ilmenite has been documented by Rb–Sr studies (Papanastassiou and Wasserburg, 1971a). For the final stages of the $Ar^{39*}$ release from the "ilmenite," the ($Ar^{40*}/Ar^{39*}$) ratio decreases to a value typically observed for the pyroxene, with a parallel decrease in the (K/Ca) ratio. It is probable that little contribution arises from K within the ilmenite itself, and that the ilmenite acts as a dilutant only.

## SAMPLE 14073

Sample 14073 is an orthopyroxene basalt composed predominantly of plagioclase, clinopyroxene, and orthopyroxene, with minor ilmenite, troilite, Fe metal, mesostasis, and other phases (Gancarz *et al.*, 1971). In general terms, the features observed in the Ar release from the total sample and mineral separates of 14310, and described in detail in the preceding section, are also seen in the analysis of 14073, and much of the previous discussion applies.

*Total rock sample*

The whole rock release pattern for 14073 is quite similar to that of 14310. The ($Ar^{40*}/Ar^{39*}$) ratios in the 540°C and 630°C release are, respectively, 36% and 4% low relative to the maximum values observed in the 740°C and 850°C fractions, which account for 50% of the argon release. The ($Ar^{40*}/Ar^{39*}$) ratio decreases by a small but significant amount by 930°C and in the final three fractions (1040°C to 1450°C) is some 5% low. A comparison of the total ($Ar^{40*}/Ar^{39*}$) ratio with the maximum value indicates a 5% radiogenic $Ar^{40}$ loss overall.

*Plagioclase*

The ($Ar^{40*}/Ar^{39*}$) release pattern for 14073 plagioclase is very similar to that for 14310, with a small initial $Ar^{40*}$ loss in the 715°C fraction followed by a plateau in the higher temperature release. The mean ratio for the 900°C to 1385°C release corresponds to a high temperature plateau age of 3.88 AE and is identical to that obtained for 14310. The overall $Ar^{40*}$ loss from the plagioclase is around 2%.

The K/Ca ratio is 40% lower than that for 14310 plagioclase but shows a similar pattern of variation. The ratio is highest in the initial release and displays the secondary maximum at 980°C previously seen in plagioclase from 14053 and 14310.

*Quintessence*

The quintessence release pattern parallels the whole rock pattern in the low temperature release with 34% and 14% $Ar^{40*}$ loss in the first two extractions. Above 630°C the ($Ar^{40*}/Ar^{39*}$) ratio attains an approximate plateau corresponding to an age of 3.83 AE. This is, respectively, 0.05 AE and 0.07 AE below the plateaux of the plagioclase and total rock samples and is outside the limits estimated for uncertainty in relative fluence and represents a significant difference in $Ar^{40*}$ retention between the samples. Quintessence is very heterogeneous, particularly in rock 14073 (Gancarz *et al.*, 1971), and the sample is probably not completely representative.

The (K/Ca) variation is notable for the high ratio ($\sim$10) reached in the 630°C release. Again assuming essentially all the Ca is in the plagioclase and using the Ca concentrations (Table 3) to determine the relative mineral abundances, we calculate $\sim$25% plagioclase and $\sim$75% quintessence. The resulting K content of the quintessence is 6%.

*Pyroxene*

The variations of both ($Ar^{40*}/Ar^{39*}$) and (K/Ca) observed in the pyroxene of 14073 are essentially identical to those of pyroxene 2 from 14310. (The sole distinguishing feature of any significance between the two cases is in the ($Ar^{38}/Ar^{37}$) ratio, which reflects a factor of 2 difference in cosmic ray exposure age.) The high temperature decrease in the whole rock ($Ar^{40*}/Ar^{39*}$) ratio can again be understood in terms of a pyroxene contribution.

## ROCK 15555

For completeness of presentation of mineral separate results, we give here a brief discussion of the data for rock 15555 (Podosek *et al.*, 1972). The apparent ages and K/Ca ratios whole rock, plagioclase, and pyroxene separates are plotted in Fig. 8. A plateau is defined for both plagioclase and whole rock, but the corresponding ages are quite distinct and differ by an amount greater than can be attributed to uncertainty in neutron fluence (the 15555 samples were included in the LAV1 irradiation). As in 14310 and 14073, the degree of gas loss from the plagioclase is much smaller than from the whole rock. The inferred gas loss from the quintessence (presumed to represent about half of the total K) is quite extensive and is probably responsible for the difference between whole rock and plagioclase.

The 15555 pyroxene shows no evidence for a high temperature decrease such as that found in 14310 and 14053, although a moderate decrease would be hidden by

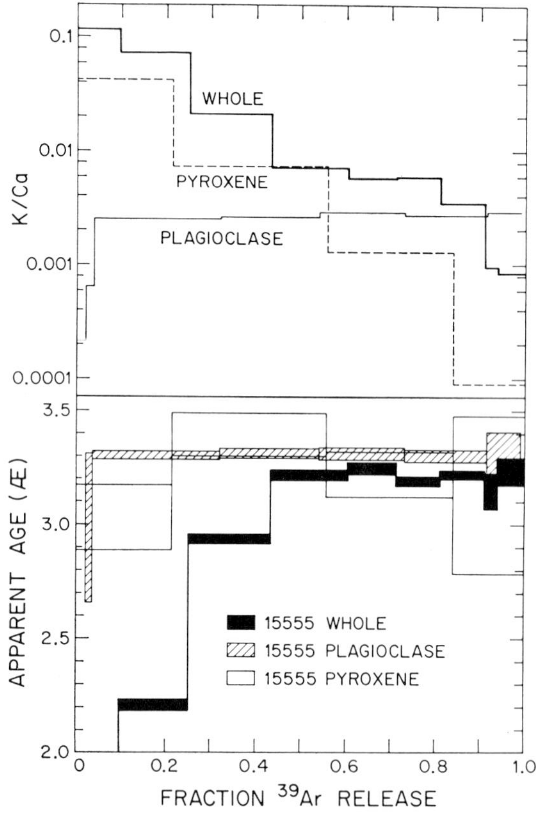


Fig. 8. Apparent ages and K/Ca as a function of $Ar^{39*}$ release from 15555 whole rock, plagioclase, and pyroxene. The lower part of the figure is reproduced from Podosek *et al.* (1972). The 15555 samples were part of a different irradiation from that of the 14310 and 14073 samples. The difference between the whole rock and plagioclase plateau ages is apparently real, since it is beyond the uncertainty attributed to variability of neutron fluence.

the large error limits. We note also that there is no high temperature decrease in the whole rock pattern.

As for 14310 and 14053, the K/Ca ratio in the plagioclase, after the first two stages of release, is quite uniform and also exhibits a small secondary maximum at $\sim 1100°C$. Both whole rock and pyroxene K/Ca ratios decrease through two orders of magnitude across the release pattern.

## Cosmic Ray Effects

The abundance of $Ar^{38}$ produced by the action of high energy cosmic rays is frequently used to calculate cosmic ray exposure ages, from which information can be obtained about the recent history of the rocks on the lunar surface. We have previously shown that the ($Ar^{38}/Ar^{37}$) ratio is particularly useful in this respect since it gives the ratio of cosmogenic $Ar^{38}$ to the principal target element Ca (Turner *et al.*, 1971). The major difficulty in using the ($Ar^{38}/Ar^{37}$) ratio to obtain precise information about the cosmic ray irradiation history is the production of additional $Ar^{38}$ from the target nuclei Ti and Fe. By measuring the ($Ar^{38}/Ar^{37}$) ratio in minerals of different chemical composition, but which have experienced the same exposure to cosmic rays, the present experiment permits the investigation of these contributions.

### *Calcium*

In plagioclase $Ar^{38}$ is produced almost entirely from Ca. Both the (K/Ca) ratios deduced from ($Ar^{39*}/Ar^{37}$) and the (Fe/Ca) ratios from microprobe analyses of 14310 and 14073 (Gancarz *et al.*, 1971) are of the order of 0.01. The production of $Ar^{38}$ in plagioclase solely from Ca is reflected in the constancy of the ($Ar^{38}/Ar^{37}$) ratio throughout the experiment (Tables 1 and 2). The situation is exactly analogous to the case of ($Ar^{40*}/Ar^{39*}$) that, apart from the initial release, reflects a constant ($Ar^{40*}$/K) ratio.

The $Ar^{38}$/Ca ratios calculated for 14310 and 14073 plagioclase are, respectively, $392 \times 10^{-8}$ and $158 \times 10^{-8}$ ccSTP/g. Assuming a nominal $Ar^{38}$ production rate of $1.4 \times 10^{-8}$ ccSTP/g/$10^6$ yr (Turner *et al.*, 1971), we calculate exposure ages of 280 and 113 m.y., respectively. The minimum ($Ar^{38}/Ar^{37}$) ratio measured previously on a total rock sample of 14310 (Turner *et al.*, 1971) corresponded to an exposure age of 290 m.y. The minimum value of the ($Ar^{38}/Ar^{37}$) ratio is selected to minimize contributions from $Cl^{37}(n,\gamma\beta)$ $Ar^{38}$ and spallation on Ti and Fe, but will overcompensate in the case of $Ar^{38}$ loss. From all considerations, the plagioclase provides a more reliable estimate and is to be preferred.

### *Potassium*

The correlation between ($Ar^{38}/Ar^{37}$) and ($Ar^{39*}/Ar^{37}$) can be used to determine the relative production rates of $Ar^{38}$ by cosmic rays on K and Ca (Turner *et al.*, 1971). Inspection of the tables will reveal that such a correlation does exist. High ($Ar^{38}/Ar^{37}$) ratios tend to be associated with high ($Ar^{39*}/Ar^{37}$) ratios. (The 630°C release from 14073 quintessence is the most extreme example.) The correlation is not perfect, however, and it is clear that some release fractions (e.g., 14073 460°C, 540°C, and 740°C) contain $Ar^{38}$ from some additional source. Apollo 14 samples are notably high in Cl contents (Reed *et al.*, 1972), and the high $Ar^{38}/Ar^{37}$ ratios may well reflect $Cl^{37}(n,\gamma\beta)$ $Ar^{38}$ contributions. In addition, the production of $Ar^{38}$ from neutrons on K may be greater than previously assumed. The slope of the correlations observed in 14310, 14073, 14001,7,1 (Turner *et al.*, 1971) and Luna 16 B-1 (Huneke *et al.*, 1972b) for data uncorrected for $Ar^{38}$ derived from K during the neutron irradiation are not significantly different despite a factor of 5 difference in exposure age. This is expected if the bulk of the K-correlated $Ar^{38}$ is produced during the neutron irradiation. With similar cosmic-ray production ratios $P_{38}(K)/P_{38}(Ca)$ the slope should be proportional to the exposure age if the data are properly corrected for contributions from K during the neutron irradiation and for variations in neutron flux between different irradiations. A fortuitous relationship between the production ratio and the exposure age of the sample would be required to explain the approximately constant slope observed. A simpler explanation is that the K-correlated $Ar^{38}$ observed has been largely produced by the reactor neutron irradiation. The ratio $(Ar^{38}/Ar^{39})_K$ for pile-produced Ar implied by the slope of the correlation is $\leq 0.04$. (The corresponding ratio for the core of the Herald reactor, A.W.R.E., Aldermaston, is 0.01 (Turner, 1970).) The credibility of relative cosmogenic production ratios calculated from the $Ar^{38}/Ar^{37}$ correlation with $Ar^{39*}/Ar^{37}$ is impaired until this question is resolved. In spite of this ambiguity, however, the observed correlation does permit a correction for *all* K-derived $Ar^{38}$ in the calculation of $Ar_c^{38}$ derived from Ca, providing the amounts of $Ar^{38}$ produced from $Cl^{37}$ are negligible.

### *Iron*

The target elements for the production of $Ar^{38}$ in the pyroxene are Ca, Fe, and Ti. The effect of Fe and Ti is to produce $(Ar^{38}/Ar^{37})$ ratios systematically higher than those observed in the plagioclase, and the production rate of $Ar^{38}$ from Fe relative to Ca can be estimated from the combined data on the plagioclase and pyroxene separates. $Ar^{36}/Ar^{38}$ ratios of 0.60–0.65 in these separates are typical for spallation Ar (cf. Huneke *et al.*, 1972c) and indicate only small contributions to $Ar^{38}$ from neutron reactions on $Cl^{37}$ or K and small contributions to $Ar^{36}$ and $Ar^{38}$ by trapped Ar ($Ar^{36}/Ar^{38} = 5.3$). With a small correction for K-derived $Ar^{38}$, assuming $(Ar^{38}/Ar^{39})_K = 0.02 \pm 0.02$, the ratio of $(Ar^{38}/Ar^{37})$ in the pyroxene relative to that in the plagioclase is calculated as $1.23 \pm 0.05$ in 14310 pyroxene 1, $1.19 \pm 0.03$ in 14310 pyroxene 2, and $1.27 \pm 0.02$ in 14073 pyroxene. The Ca contents are taken from Table 3. The Fe and Ti have not been independently measured.

Microprobe data on a large number of pyroxenes from 14310 (Gancarz *et al.*, unpublished data) show a high abundance of low Ca pyroxenes, and average contents of $11.5 \pm 4.0\%$ Fe and $0.3 \pm 0.15\%$ Ti are estimated from the microprobe data. While it is important to correct for the Ti contributions to $Ar_c^{38}$, these are not large enough to establish $Y_{38}(Ti)/Y_{38}(Fe)$ independently. The relative $Ar^{38}$ production rates per gram of target element from Ti and Fe are taken as $Y_{38}(Ti)/Y_{38}(Fe) = 2.9^{+2.9}_{-1.4}$ (cf. Huneke *et al.*, 1972c). With these data, we determine that $Y_{38}(Ca)/Y_{38}(Fe)$

is $26^{+21}_{-12}$ and $22^{+16}_{-10}$ for 14310 pyroxenes 1 and 2, respectively, where the errors are primarily due to the uncertainty in chemical composition.

The plagioclase and pyroxene data on rock 14073 can be combined in a similar calculation. The average composition of the pyroxenes from rock 14073 is determined from microprobe analysis of twenty grains to be 12.0 ± 3.0% Fe, 0.42 ± 0.12% Ti, 0.3% Cr, and 0.15% Mn. The average Ca content from the microprobe analyses is 2.9%, in very good agreement with the value of 3.05% derived from the $Ar^{37}$ concentration (Table 3). Assuming $Y_{38}(Ti)/Y_{38}(Fe)$ as before, a value of $Y_{38}(Ca)/Y_{36}(Fe) = 16.5^{+8}_{-5}$ is calculated for rock 14073.

Huneke *et al.* (1972c) have determined $Y_{36}(Ca)/Y_{36}$ (Fe) for Apollo 11 rocks 10017 and 10044 to be $15^{+8}_{-4}$ and $5^{+10}_{-3}$, respectively. The present measurements are in essential agreement, particularly with rock 10017 which appears to have acquired much of its exposure in a relatively shielded location. Whether the difference in relative production rates for 14310 and 14073 is significant or simply reflects an uncertainty in the Fe content is not possible to say. If the difference is real it would imply a harder effective flux spectrum for 14073 than 14310 since the production of $Ar^{38}$ from Fe ($\Delta A \approx 18$) requires considerably more energy than the production from Ca ($\Delta A \approx 2$). A soft flux spectrum implies some irradiation at depth, where the low energy secondaries become significant. If comparison with rock 10017 is meaningful the implication is that both 14073 and 14310 received some irradiation in a shielded location below the lunar surface.

*Titanium*

In principle, the plagioclase and pyroxene data can be combined with the data on the ilmenite concentrate to deduce the relative production ratio $Y_{38}(Ti)/Y_{38}(Ca)$. However, with uncertainties in the composition of both the pyroxene and the ilmenite concentrate and with large possible excesses in $Ar^{38}$ (evidenced by the very low $Ar^{36}/Ar^{38}$ ratios) the calculation is somewhat less definitive than desirable. Nevertheless, conservative limits on the production rate ratio can be set to determine consistency with the previous calculation. Assuming there is no trapped $Ar^{36}$ and that $(Ar^{36}/Ar^{38})_c = 0.63$, a value for $Ar^{38}_c$ can be estimated from the $Ar^{36}$. Relative to the plagioclase ratio, the corrected $Ar^{38}/Ar^{37}$ ratio for the ilmenite is 1.56.

The chemical composition is calculated from visual estimates of the ilmenite abundances (30 ± 10%) and electron microprobe studies of the non-opaque minerals. Analyses of 155 grains indicate 11% plagioclase and glass (±4% assumed), and the remaining 59 ± 14% pyroxene. The K content of the total concentrate (Table 3) implies a quintessence content of 4 ± 1%. Microprobe analyses of thirty pyroxene grains give average concentrations of 18 ± 2% Fe, 7 ± 2% Ca, 0.6% Ti, 0.3% Cr, and 0.3% Mn. The Cr and Mn are included with Fe for the purposes of the calculation. The ilmenite composition is taken from Gancarz *et al.* (1971). The Ca concentration calculated from the mineral composition and abundance data is 4.9%, in good agreement with the value of 5.2% derived from the $Ar^{37}$ (Table 3).

Combining the Ar data and estimated chemical compositions of all three samples, we calculate $Y_{38}(Ti)/Y_{38}(Ca) = 0.2^{+0.4}_{-0.1}$, where two-thirds of the positive error is due to the uncertainty in the amount of pile-produced $Ar^{38}$. This ratio is consistent with the previously assumed Y(Fe)/Y(Ca) ratio.

## SUMMARY AND DISCUSSION

Application of the $Ar^{40}$–$Ar^{39}$ technique to "total" rock samples in general yields a complex release pattern which is a reflection of the constituent mineral phases. Studies of some of the individual mineral phases show characteristic patterns which can explain some of the anomalies found in the whole rock release patterns of lunar basalts. These data serve to indicate when a plateau in the release pattern is to be expected and whether this plateau may be reasonably assigned a precise time meaning.

The investigations of separated minerals, as well as whole rock samples, from Apollo 14 rocks make it clear that the age resolution possible by the $Ar^{40}$–$Ar^{39}$ method is not limited by experimental errors, but by the interpretation of variations in the K–Ar age spectrum which far exceed experimental error. From the present work, earlier experiments on separated minerals from terrestrial rocks (Lanphere and Dalrymple, 1971), as well as the more numerous whole rock investigations, it is clear that ($Ar^{40*}/Ar^{39*}$) variations (and by implication ($Ar^{40}$/K) variations) are the rule rather than the exception. Insofar as these variations represent real variations in the distribution of $Ar^{40}$ relative to its parent K (as opposed to possible variations introduced by the irradiation itself) they indicate the lack of precision which is inherent in conventional K–Ar ages based only on total $Ar^{40}$/K ratios of separated minerals.

The general features which characterize a whole rock release pattern for unshocked rocks are:

(1) A steady increase in $Ar^{40*}/Ar^{39*}$ release in the initial release fractions. The low $Ar^{40*}/Ar^{39*}$ ratios obtained in the low temperature release is presumably due to argon loss in post crystallization processes such as thermal cycling from solar heating on the lunar surface (Turner, 1971), heating during impact processes transporting the rock to the lunar surface (Turner, 1970; Turner, 1972), or long term diffusive loss at relatively low temperatures.

(2) A broad flat region or peak over which a major fraction of the gas is released. This is usually taken to be the age "plateau" indicative of the formation time. However, in one out of four cases we have studied, this plateau is significantly below that found for plagioclase.

(3) A distinct decrease in ($Ar^{40*}/Ar^{39*}$) for the very high temperature release, which may comprise as much as 20% of the total gas release.

(4) A monotonically decreasing pattern for the K/Ca ratio throughout the whole degassing procedure.

In explaining the behavior of total rock ages as the sum of the individual contributions of the mineral phases, we are faced with the fact that the mineral separates are not in general representative of the "average" of that mineral species in the rock. The study of mineral separates is further complicated by the possibility that the behavior of even the purest separates obtainable is governed by microscopic and submicroscopic impurities (Roedder and Weiblen, 1972). Despite these limitations we can identify several critical empirical features which appear to explain the observations on total rocks in a phenomenological fashion and which also suggest plausible mechanisms responsible for the variations in the different minerals analyzed.

The mineral phases are characterized by the following release patterns.

*Plagioclase*

High purity separates of this mineral show a small but significant low temperature release with a low $Ar^{40*}/Ar^{39*}$ ratio. The $Ar^{40*}/Ar^{39*}$ rises rapidly to a very well-defined plateau, and no high temperature decrease is observed. The K/Ca ratio over most of the release is very uniform. The lowest temperature releases show somewhat higher values, and there is a small but distinct peak which appears at about 980°C. Except for the very low temperature behavior, the $Ar^{40*}/Ar^{39*}$ and K/Ca release patterns appear to reflect the specific mineral phase and are generally indicative of a well-behaved mineral system which should yield well-defined ages.

*Quintessence*

This heterogeneous assemblage of K-rich materials contains major portions of the total K and Ar in lunar basalts. Ar loss from this fine grained interstitial glass was recognized from the first as a primary cause of low total rock $Ar^{40*}/K$ ages (Lunatic Asylum, 1970). This study shows that the quintessence has an initial release with a very low $Ar^{40*}/Ar^{39*}$ ratio which then rises to a well-defined plateau with no high temperature decrease; the plateau may be distinctly below that of the plagioclase. Much of the argon release by the quintessence is at moderate temperatures (at 930°C or below). The $Ar^{39*}$ release from quintessence over a relatively wider temperature range than for the plagioclase may reflect in part a lower activation energy for diffusion.

There is a very good correlation between the variation of $Ar^{40*}/Ar^{39*}$ in the initial release from the whole rock and the quintessence, and these data show conclusively that the low temperature release pattern for total rocks is associated with the quintessence.

*Pyroxene*

This complex of mineral phases has a very low K content (see Table 3) but plays a significant role in the $Ar^{40*}/Ar^{39*}$ systematics. The $Ar^{40*}/Ar^{39*}$ ratio rises to a peak from the low initial values and then decreases markedly.

The K/Ca ratio shows a monotonic decrease throughout the total release curve and is generally parallel to that for the total rock. The most important characteristics are that the pyroxenes release a large fraction of their argon at higher temperatures (over 980°C) and that this gas has a low $Ar^{40*}/Ar^{39*}$ ratio. This release is in the high temperature regime in which there is little argon left in either the plagioclase or the quintessence. While these characteristics of the pyroxene are not understood, they are clearly responsible for the high temperature release pattern in the total rocks. Exsolution features are very evident in lunar pyroxenes and are possibly related to these characteristic $Ar^{40}/Ar^{39}$ patterns.

A quantitative reconstitution of the total rock release pattern in terms of constituent minerals is not possible with the limited information available. In particular, the pyroxene separates analyzed are surely not representative of average pyroxene compositions. Nevertheless, a semi-quantitative reconstruction indicates that all features observed in the whole rock patterns of 14310 and 14073 are directly attributable to features seen in the separates, so that no unknown phases need be invoked. On the basis of the modal abundances of Gancarz *et al.* (1971) and the elemental concentrations in Table 3, the quintessence is the dominant source of potassium in both 14310 ($\sim 75\%$) and 14073 ($\sim 85\%$); most of the remaining K is located in the plagioclase, with only minor amounts (3–5%) in the pyroxene. Over the majority of the release, these whole rock patterns should thus be dominated by the quintessence contributions. Inspection of Figs. 2 and 6 shows that the character of the release patterns is essentially the same, particularly the low apparent ages in the initial release. At intermediate temperatures, both plagioclase and quintessence define plateaux, as do the whole rock patterns. Both whole rock patterns show significant decreases in apparent age in the final 5–10% of their releases. Roughly half of the pyroxene release occurs in this temperature regime, however, so that the pyroxene contribution to these fractions is enhanced by factors of 5 to 10 above the average contribution. The observed high-temperature whole-rock pattern may thus be reasonably attributed to moderation of the observed pyroxene behavior with the residual release of quintessence and plagioclase.

These data clearly demonstrate major differences in the release patterns of mineral separates and offer an explanation on a phenomenological basis for the observed total rock patterns. However, the actual phases and the lattice sites which are contributing in each temperature regime are unknown.

The study of mineral separates has shown that the interpretation of total rock release patterns is complex and that "plateaux" may be found which are not simply related to the age. This may be seen most clearly for 15555. Thus care must be exercised in those cases where significant high quality data are required. Since a major part of lunar chronology is confined to the period 3.2 to 4.0 AE, it is clear that a possible level of uncertainty of 0.1 AE in the interpretation of whole rock $Ar^{40}$–$Ar^{39}$ release patterns is unsatisfactory. The requirement of greater precision will demand that in future investigations adequate attention be focused on the pertinent mineralogical characteristics, proper use of standards and, for self-consistency, a uniform use of reliable decay constants.

Of the minerals analyzed, the plagioclase has unquestionably the best gas retention characteristics with the ($Ar^{40*}/Ar^{39*}$) ratio having the least variation. For the four samples so far analyzed (14053, 14073, 14310, and 15555), the apparent $Ar^{40*}$ loss is least in the plagioclase and the *relative ages* indicated by the major part of the release are consistent to within 0.02 AE with the Rb–Sr measurements (Papanastassiou and Wasserburg, 1971). In addition, the correlation of $Ar^{40*}/Ar^{39*}$ with $Ar^{37}/Ar^{39*}$ for plagioclase shows a clear relationship, apparently well-defined by the mineralogy and the chemistry. Certainly any anomalies which would be present can be readily seen for this phase. At the present state of the art plagioclase must be regarded as a more reliable age indicator for lunar basalts than the other minerals analyzed, although one must clearly be wary of extrapolating this observation without further experiments on mineral separates.

Our measurements indicate that plagioclase, where the dominant target is Ca, is also an excellent mineral for determining exposure ages. This is particularly fortunate in view of the additional suitability of plagioclase for $Ar^{40}$–$Ar^{39}$ dating. Taken together these two considerations make a strong case for using plagioclase in preference to whole rock samples in future studies.

The present experiments have also provided useful data on the distribution of cosmic ray produced $Ar^{38}$ and its relationship to the principal target elements Ca, Fe, and Ti. Of these target elements Fe is potentially most useful in the sense of providing information on the energy spectrum of the bombarding particles which may be used to infer mean depths of burial in the regolith during the irradiation. The energies required to produce $Ar^{38}$ from Fe are typically hundreds of MeV in contrast to tens of MeV for production from Ca. Measurements of enhanced ($Ar^{38}$/Ca) ratios in pyroxene separates coupled with a knowledge of the Fe abundance thus provide energy spectrum information by way of the relative production rates, Y(Ca)/Y(Fe) (Huneke *et al.*, 1972). For the present experiments the production ratios determined in 14310 and 14073 were, respectively, 24 and 16.5. As we have previously pointed out (Turner *et al.*, 1971), high exposure ages at the Apollo 14 site are found in the Smooth Terrain away from Cone Crater and reflect a pre-Cone Crater exposure for this material. In comparison with the measurements of Huneke *et al.* (1972c) on 10017 and 10044 the Y(Fe)/Y(Ca) measurements indicate that a good part of this irradiation occurred in a shielded environment below the regolith surface.

*Acknowledgments*—One of us (G.T.) would like to thank the University of Sheffield for giving him the leave of absence necessary to commit himself to the Lunatic Asylum. We are grateful to A. Albee and A. Gancarz for many helpful discussions and assistance in mineralogical descriptions and to A. Chodos for valuable microprobe analyses. This work has been supported by NASA contract NGL-05-002-188.

## References

Bence A. E., Papike J. J. (1972) Crystallization histories of pyroxenes from lunar basalts (abstract). In *Lunar Science—III* (editor C. Watkins), pp. 59–61, Lunar Science Institute Contr. No. 88.

Gancarz A. J., Albee A. L., and Chodos A. A. (1971) Petrologic and mineralogic investigation of some crystalline rocks returned by the Apollo 14 mission. *Earth Planet. Sci. Lett.* **12,** 1–18.

Hollister L., Trzcienski W. Jr., Dymek R., Kulick C., Weigand P., and Hargraves R. (1972) Igneous fragment 14310,21 and the origin of the mare basalts (abstract). In *Lunar Science—III* (editor C. Watkins), pp. 386–388, Lunar Science Institute Contr. No. 88.

Huneke J. C., Podosek F. A., and Wasserburg G. J. (1972b) Gas retention and cosmic ray exposure ages of a basalt fragment from Mare Fecunditatis. *Earth Planet. Sci. Lett.* **13,** 375–383.

Huneke J. C., Podosek F. A., Burnett D. S., and Wasserburg G. J. (1972c) Rare gas studies of the galactic cosmic irradiation history of lunar rocks. *Geochim. Cosmochim. Acta,* to be published.

Huneke J. C., Podosek F. A., Turner G., and Wasserburg G. J. (1972a) $Ar^{40}$-$Ar^{39}$ systematics in lunar rocks and separated minerals of lunar rocks from Apollo 14 and 15 (abstract). In *Lunar Science—III* (editor C. Watkins), pp. 413–414, Lunar Science Institute Contr. No. 88.

Husain L., Sutter J. F., and Schaeffer O. A. (1971) Ages of crystalline rocks from Fra Mauro. *Science* **173,** 1235.

Lanphere M. A. and Dalrymple G. B. (1971) A test of the $^{40}$Ar-$^{39}$Ar age spectrum technique on some terrestrial minerals. *Earth Planet. Sci. Lett.* **12,** 359–373.

Lunatic Asylum (1970) Ages, irradiation history and chemical composition of lunar rocks from the Sea of Tranquility. *Science* **167,** 463.

Merrihue C. M. and Turner G. (1966) Potassium-argon dating by activation with fast neutrons. *J. Geophys. Res.* **71,** 2852.

Papanastassiou D. A. and Wasserburg G. J. (1971a) Lunar chronology and evolution from Rb–Sr studies of Apollo 11 and 12 samples. *Earth Planet. Sci. Lett.* **11,** 37–62.

Papanastassiou D. A. and Wasserburg G. J. (1971b) Rb–Sr ages of igneous rocks from the Apollo 14 mission and the age of the Fra Mauro formation. *Earth Planet. Sci. Lett.* **12,** 36–48.

Podosek F. A., Huneke J. C., and Wasserburg G. J. (1972). Gas-retention and cosmic-ray exposure ages of lunar rock 15555. *Science* **175,** 423–425.

Reed G. W. Jr., Jovanovic S., and Fuchs L. H. (1972) Concentrations and lability of the halogens, platinum metals and mercury in Apollo 14 and 15 samples (abstract). In *Lunar Science—III* (editor C. Watkins), pp. 637–639, Lunar Science Institute Contr. No. 88.

Ridley W. I., Williams R. J., Brett R., Takeda H., and Brown R. W. (1972) Petrology of lunar basalt 14310 (abstract). In *Lunar Science—III* (editor C. Watkins), pp. 648–650, Lunar Science Institute Contr. No. 88.

Roedder E. and Weiblen P. W. (1972) Petrographic and petrologic features of Apollo 14, 15, and Luna 16 samples (abstract). In *Lunar Science—III* (editor C. Watkins), pp. 657–659, Lunar Science Institute Contr. No. 88.

Turner G. (1970) Argon 40/argon 39 dating of lunar rock samples. *Geochim. Cosmochim. Acta* Suppl. 1, Vol. 2, 1665–1684.

Turner G. (1971) $^{40}$Ar–$^{39}$Ar ages from the lunar maria. *Earth Planet. Sci. Lett.* **11,** 169–191.

Turner G. (1972) $^{40}$Ar–$^{39}$Ar age and cosmic ray irradiation history of Apollo 15 anorthosite, 15415. *Earth Planet. Sci. Lett.* **14,** 169–175.

Turner G., Huneke J. C., Podosek F. A., and Wasserburg G. J. (1971) $^{40}$Ar–$^{39}$Ar ages and cosmic ray exposure ages of Apollo 14 samples. *Earth Planet. Sci. Lett.* **12,** 19–35.

Turner G. (1968) The distribution of potassium and argon in chondrites. In *Origin and Distribution of the Elements* (editor L. H. Ahrens), pp. 387–389. Pergamon.

Wasserburg G. J., Turner G., Tera F., Podosek F. A., Papanastassiou D. A., and Huneke J. C. (1972) Comparison of Rb–Sr, K–Ar, and U–Th–Pb ages; lunar chronology and evolution (abstract). In *Lunar Science—III* (editor C. Watkins), pp. 788–790, Lunar Science Institute Contr. No. 88.

Proceedings of the Third Lunar Science Conference
(Supplement 3, *Geochimica et Cosmochimica Acta*)
Vol. 2, pp. 1613–1622
The M.I.T. Press, 1972

# $^{40}$Ar–$^{39}$Ar ages of Apollo 14 and 15 samples

DEREK YORK, W. JOHN KENYON, and ROY J. DOYLE
Geophysics Division, Department of Physics,
University of Toronto, Toronto 5, Ontario, Canada

**Abstract**—The $^{40}$Ar–$^{39}$Ar method of whole-rock K–Ar dating has been applied to the following Apollo 14 material: two samples of fines (14167,9), two samples of basalt 14072,6, two fragments of basalt 14310,65, three basaltic clasts, one micro-breccia clast, and matrix material from breccia 14321,184. For most of the samples a characteristic evolution pattern is seen in which the $^{40}$Ar/$^{39}$Ar ratios rise quickly to a plateau around 3.9 b.y. but then display some tendency to drop at about the 50% evolution point. In contrast, basalt 14310 rises to virtually a steady plateau, while the micro-breccia clast and the matrix material from 14321 rise quickly to maxima of 4.2–4.3 b.y. and then drop off steadily in staircase fashion with increasing temperature. The results in general indicate a significant event occurred on the moon 3.9 $\pm$ 0.1 b.y. ago. Two fragments of Apollo 15 basalt 15555,26 gave an average plateau age of 3.31 $\pm$ 0.05 b.y. indicating the existence of a much younger event in the area sampled. Approximate $^{38}$Ar cosmic ray exposure ages of 27 m.y., 21 m.y., and 340 m.y. were found for the 14167,9 fines, 14072,6 basalt and 14310,65 basalt, respectively, and 76 m.y. for 15555,26.

## INTRODUCTION

THE $^{40}$Ar–$^{39}$Ar METHOD OF K–Ar DATING (Sigurgeirsson, 1962; Merrihue, 1965; Merrihue and Turner, 1966) has been applied with considerable success to lunar material (Turner, 1970a, b, 1971; Turner *et al.*, 1971). Most of the lunar samples have suffered some argon loss and the appeal of the $^{40}$Ar–$^{39}$Ar approach lies in its potential for allowing corrections for these losses. We present here such analyses of Apollo 14 and 15 material.

## EXPERIMENTAL PROCEDURE

Samples weighing 25–185 mg were wrapped in aluminium foil along with a nepheline standard, enclosed in an aluminium can and irradiated in the McMaster University reactor. The Gill Quarry nepheline standard used has been studied in this laboratory in detail with both the conventional and fast-neutron methods of potassium-argon analysis. The results of potassium and argon analyses of five different grain size fractions of this material have already been published (Macintyre *et al.*, 1969.) These give an age of 984 m.y. and interlaboratory comparisons indicate that this value should not be in error by more than 2% (Baksi *et al.*, 1967). The irradiated samples were loaded in a bakeable vacuum system and baked at 200°C for about forty hours to minimize terrestrial atmospheric argon contamination. Radio-frequency heating was used to raise the temperature of the samples in a degassed molybdenum crucible from about 500°C in a series of nine steps to about 1600°C. The temperature was held constant for one hour at each stage and was monitored with an optical pyrometer. The temperature readings quoted have an uncertainty of the order $\pm$30°C. Evolved argon fractions were purified with hot titanium sponge and analyzed on an MS10 mass spectrometer (Farrar *et al.*, 1964). Samples were allowed to come to equilibrium in the analyzer tube and the volumes were determined from the peak heights, the mass spectrometer being regularly calibrated during these analyses with $^{38}$Ar spikes. Atmospheric argon analyses were run regularly with the samples to monitor discrimination changes.

Several corrections were applied to the measured isotope ratios. System blanks indicated that each fraction was contaminated with about $7 \times 10^{-9}$ ccntp of terrestrial atmospheric argon and all analyses were adjusted to allow for this. Previous analyses in this laboratory on irradiated calcium and potassium salts showed that corrections for calcium-generated $^{39}Ar$ and $^{36}Ar$ and potassium-generated $^{40}Ar$ should be applied using the values

$$(^{39}Ar/^{37}Ar)_{calcium} = 8.5 \times 10^{-4}, \qquad (^{36}Ar/^{37}Ar)_{calcium} = 2.7 \times 10^{-4},$$

and

$$(^{40}Ar/^{39}Ar)_{potassium} = 2.7 \times 10^{-2}$$

(Berger and York, 1970). The effect of these corrections on the ages of the various fractions was usually less than 2%. Cosmogenic $^{40}Ar$ volumes were removed using the assumption that the $^{40}Ar/^{36}Ar$ ratio of this component is 1.00 (Turner, 1970b). This changed the ages of all fractions of basalts 14072 and 14310 and all the breccia samples by much less than 1%. In each run on the fines the effect was never more than 1% on the age. The effect again was less than 1% for basalt 15555,26 except in the last 15% of the gas evolved, where it reached a maximum correction of about 2% of the age calculated. No background peaks were visible in the mass spectrometer at masses 36, 37, 38, and 39 and that at mass 40 was negligible.

The reproducibility of our experimental procedure may be estimated from the results of those runs which are essentially duplicates. Thus, two complete runs were done on each of 14167,9; 14072,6; 14310,65, and 15555,26. In each case, what we later term the plateau ages of duplicates differ by about 1.5% or less.

## RESULTS

The data are displayed in Tables 1–7, and $^{40}Ar/^{39}Ar$ evolution spectra are shown in Figs. 1–3.

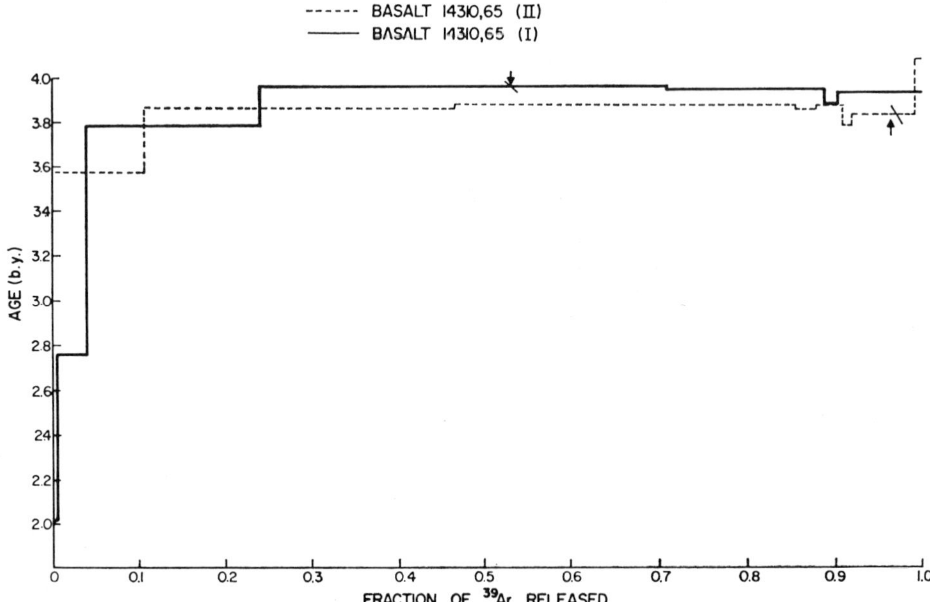


Fig. 1. Evolution spectra from duplicate analyses of basalt 14310,65. Little arrows indicate where two consecutive fractions were too close in age to be shown as distinct.

*14310,65*

Two fragments of basalt 14310,65 were analyzed. This was the largest igneous rock returned by the Apollo 14 mission. The evolution spectra may be seen in Fig. 1 where it is clear that good agreement was found between runs. This rock exhibits the least complicated evolution spectrum of all the samples shown in this paper. The usual low temperature ratios are quickly replaced by an almost perfect plateau in both analyses. A plateau age was computed for each run from the weighted mean of all the gas fractions released above 800°C, with the exception of the very small final fraction of run II. This gave 3.94 b.y. for run I and 3.88 b.y. for run II (Table 1). We conclude, therefore, that this basalt crystallized on the moon 3.91 $\pm$ 0.05 b.y. ago and subsequently lost only $\sim$3% of its radiogenic argon. Total gas ages of 3.88 b.y. and 3.85 b.y. were calculated for the two runs (Table 1)

*14072,6*

Analyses of two fragments of basalt samples 14072,6 gave the slightly more complex evolution spectra shown in Table 2. Argon loss is shown in only the first 5% of gas release, after which the pattern plateaus at just over 4 b.y. until about 50% of the gas has been evolved. After this there is a slight but significant sag in the $^{40}Ar/^{39}Ar$ ratios with some trend back towards the earlier plateau value in the last 20% of gas release. This spectrum is evidently an incipient version of those to be next described. The average ages for the two analyses calculated from the plateau area between about 5 and 50% argon release are 4.06 b.y. and 4.01 b.y. If these are interpretable as crystallization dates, then this basalt was formed 4.04 $\pm$ 0.05 b.y. ago. Total gas ages were 4.02 and 3.94 b.y.

*14167,9*

Two basaltic fragments of the 2–4 mm fines 14167,9 were examined. These show (Table 3) some radiogenic argon loss in the low temperature fractions, followed by a

Table 1. Analytical data for sample: 14310,65, basalt.

| Run I (Mass = 0.0838 g) | | | | Run II (Mass = 0.0987 g) | | | |
|---|---|---|---|---|---|---|---|
| T°C | $^{40}Ar^*$ ($10^{-8}$ cc/g) | $^{40}Ar^*/^{39}Ar^*$ | Age (b.y.) | T°C | $^{40}Ar^*$ ($10^{-8}$ cc/g) | $^{40}Ar^*/^{39}Ar^*$ | Age (b.y.) |
| 485 | 30 | 38.3 | 2.01 | 670 | 2369 | 114.7 | 3.59 |
| 600 | 437 | 66.7 | 2.76 | 795 | 9483 | 136.9 | 3.88* |
| 700 | 4860 | 129.3 | 3.79 | 900 | 10390 | 137.8 | 3.89* |
| 815 | 7671 | 143.6 | 3.96* | 965 | 665 | 134.4 | 3.85* |
| 870 | 4762 | 143.6 | 3.96* | 1060 | 747 | 135.4 | 3.86* |
| 990 | 4797 | 143.5 | 3.96* | 1150 | 109 | 131.7 | 3.81* |
| 1140 | 372 | 137.3 | 3.88* | 1260 | 1437 | 134.8 | 3.85* |
| 1420 | 2512 | 141.0 | 3.93* | 1480 | 752 | 135.2 | 3.86* |
| | | | | 1600 | 61 | 158.2 | 4.12 |
| Total | 25441 | 137.2 | 3.88 | | 26013 | 134.6 | 3.85 |

$(^{40}Ar^*/^{39}Ar^*)_{standard} = 13.77$.
$\lambda_\beta = 4.72 \times 10^{-10}\ yr^{-1}$; $\lambda_e = 0.584 \times 10^{-8}\ yr^{-1}$.
* The plateau ages were calculated from those fractions marked with * in Tables 1–6.

Table 2. Analytical data for sample 14072,6, basalt.

| Run I (Mass = 0.0419 g) | | | | Run II (Mass = 0.0442 g) | | | |
|---|---|---|---|---|---|---|---|
| T°C | $^{40}Ar^*$ ($10^{-8}$ cc/g) | $^{40}Ar^*/^{39}Ar^*$ | Age (b.y.) | T°C | $^{40}Ar^*$ ($10^{-8}$ cc/g) | $^{40}Ar^*/^{39}Ar^*$ | Age (b.y.) |
| 485 | 24 | 391.8 | 2.83 | 665 | 161 | 619.7 | 3.53 |
| 665 | 125 | 561.3 | 3.37 | 755 | 907 | 861.6 | 4.07* |
| 755 | 725 | 835.4 | 4.02* | 820 | 919 | 857.9 | 4.06* |
| 880 | 855 | 839.3 | 4.03* | 880 | 622 | 844.8 | 4.04* |
| 965 | 631 | 816.8 | 3.98* | 960 | 364 | 819.3 | 3.98 |
| 1080 | 385 | 795.1 | 3.94 | 1165 | 474 | 811.1 | 3.97 |
| 1165 | 326 | 756.5 | 3.85 | 1180 | 864 | 859.0 | 4.06 |
| 1280 | 578 | 806.2 | 3.96 | 1600 | 102 | 747.4 | 3.83 |
| 1600 | 57 | 847.8 | 4.04 | | | | |
| Total | 3706 | 798.5 | 3.94 | | 4413 | 834.3 | 4.02 |

$(^{40}Ar^*/^{39}Ar^*)_{standard} = 77.27$.

Table 3. Analytical data for sample 14167,9, 2–4 mm fines.

| Run I (Mass = 0.0255 g) | | | | Run II (Mass = 0.0332 g) | | | |
|---|---|---|---|---|---|---|---|
| T°C | $^{40}Ar^*$ ($10^{-8}$ cc/g) | $^{40}Ar^*/^{39}Ar^*$ | Age (b.y.) | T°C | $^{40}Ar^*$ ($10^{-8}$ cc/g) | $^{40}Ar^*/^{39}Ar^*$ | Age (b.y.) |
| 485 | 282 | 158.2 | 1.65 | 485 | 348 | 162.0 | 1.68 |
| 755 | 8843 | 776.7 | 3.90* | 665 | 2565 | 529.0 | 3.28 |
| 945 | 2352 | 803.1 | 3.95* | 755 | 8057 | 813.2 | 3.97* |
| 1060 | 931 | 634.3 | 3.57 | 780 | 5398 | 788.4 | 3.92* |
| 1185 | 409 | 361.8 | 2.71 | 850 | 2287 | 749.1 | 3.84 |
| 1240 | 618 | 521.8 | 3.26 | 970 | 675 | 529.8 | 3.28 |
| 1355 | 40 | 367.2 | 2.73 | 1065 | 329 | 308.5 | 2.49 |
| 1490 | 64 | 537.2 | 3.30 | 1100 | 357 | 321.3 | 2.54 |
| 1600 | 53 | 338.8 | 2.62 | 1600 | 229 | 517.5 | 3.25 |
| Total | 13592 | 670.5 | 3.66 | | 20245 | 659.1 | 3.63 |

$(^{40}Ar^*/^{39}Ar^*)_{standard} = 77.27$.
Fraction released at 835°C in run (I) lost.

plateau in the 20 to 80% gas release area. In the final 20% of the gas evolved, the $^{40}Ar/^{39}Ar$ ratios plunge dramatically to about 2.6 b.y. before recovering a little. The plateau ages are 3.91 and 3.95 b.y. for the two runs. One gas fraction was lost in run I. The remaining fractions yielded a whole gas age of 3.66 b.y. Run II gave a total gas age of 3.63 b.y.

*14321,184*

The authors were members of the "Goles Consortium" for breccia 14321,184. Dr. A Duncan provided three basaltic clasts, a microbreccia clast, and some matrix material. Evolution curves for the basalt clasts are shown in Fig. 2. The patterns are quite similar to those of 14072,6 discussed above. Plateau ages were computed for samples 1D, 12B, and 17B, yielding 3.91, 3.99, and 3.94 b.y., respectively. Correspondingly, whole gas ages were 3.84, 3.94, and 3.83 b.y. (Table 4).

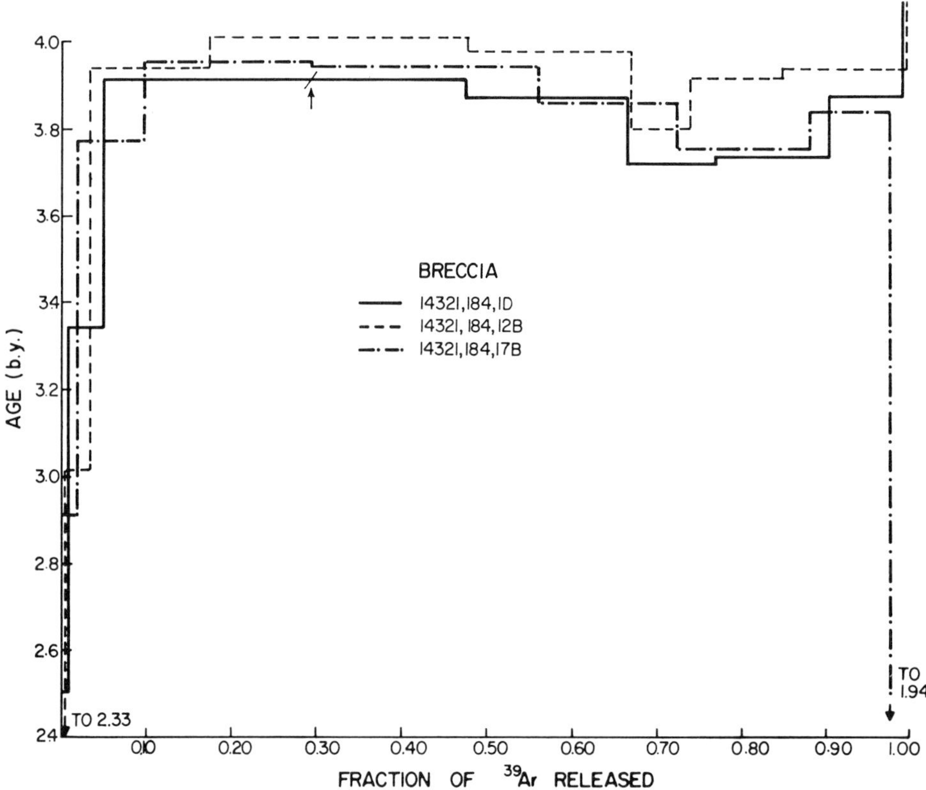


Fig. 2. Evolution spectra of basaltic clasts 1D, 12B, and 17B from breccia 14321,184. Little arrow on 1D pattern indicates where two consecutive fractions were too close in age to be shown as distinct.

Table 4. Analytical data for basalt clasts from sample 14321,184.

| Sample 14321,1D (Mass = 0.1170 g) | | | | Sample 14321,184,12B (Mass = 0.1830 g) | | | | Sample 14321,17B (Mass = 0.1170 g) | | | |
|---|---|---|---|---|---|---|---|---|---|---|---|
| T°C | $^{40}Ar^*$ ($10^{-8}$ cc/g) | $^{40}Ar^*/^{39}Ar^*$ | Age (b.y.) | T°C | $^{40}Ar^*$ ($10^{-8}$ cc/g) | $^{40}Ar^*/^{39}Ar^*$ | Age (b.y.) | T°C | $^{40}Ar^*$ ($10^{-8}$ cc/g) | $^{40}Ar^*/^{39}Ar^*$ | Age (b.y.) |
| 485 | 27 | 55.7 | 2.51 | 485 | 10 | 49.1 | 2.33 | 485 | 8 | 73.6 | 2.90 |
| 600 | 311 | 97.6 | 3.34 | 600 | 89 | 83.9 | 3.10 | 600 | 48 | 74.0 | 2.91 |
| 775 | 2489 | 139.8 | 3.91* | 700 | 731 | 142.5 | 3.94* | 700 | 410 | 128.1 | 3.77 |
| 825 | 1813 | 139.7 | 3.91* | 865 | 1662 | 148.1 | 4.01* | 790 | 1160 | 143.4 | 3.95* |
| 925 | 1880 | 136.0 | 3.87 | 985 | 1022 | 145.1 | 3.98* | 860 | 1533 | 141.1 | 3.93* |
| 1030 | 915 | 124.2 | 3.72 | 1085 | 336 | 130.3 | 3.80 | 950 | 904 | 135.7 | 3.86 |
| 1140 | 1172 | 125.2 | 3.73 | 1140 | 538 | 140.3 | 3.92 | 1085 | 774 | 126.7 | 3.75 |
| 1205 | 855 | 136.7 | 3.88 | 1240 | 738 | 142.4 | 3.94 | 1220 | 493 | 133.6 | 3.84 |
| 1600 | 109 | 227.7 | 4.74 | 1600 | 58 | 335.0 | 5.42 | 1600 | 38 | 36.0 | 1.94 |
| Total | 9571 | 133.4 | 3.84 | | 5184 | 142.4 | 3.94 | | 5368 | 132.8 | 3.83 |

$(^{40}Ar^*/^{39}Ar^*)_{standard} = 13.77$.

The microbreccia clast (26) and the matrix material (28) have strikingly different release patterns (Fig. 3). These give no semblance of a plateau. The $^{40}Ar/^{39}Ar$ ratios rise very quickly to brief peaks of 4.3 b.y. (microbreccia) and 4.2 b.y. (matrix) and thenceforward drop essentially monotonically in staircase fashion with rising temperature. Total gas ages are 3.9 b.y. for the microbreccia and 4.06 b.y. for the matrix (Table 5). Duncan *et al.* (1972) have described this matrix as a mixture of fragments

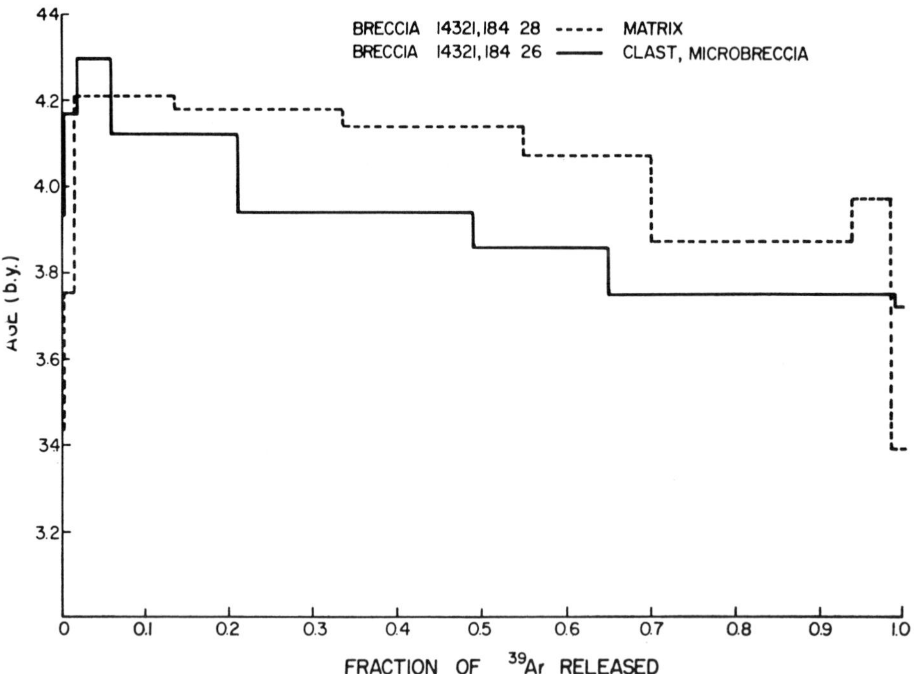


Fig. 3. Evolution spectra of microbreccia clast (26) and matrix material (28) from breccia 14321,184.

Table 5. Analytical data for sample 14321,184,26 microbreccia clast and 14321,184,28 matrix.

| | Sample 14321,184,26 (Mass = 0.1040 g) | | | | Sample 14321,184,28 (Mass = 0.0950 g) | | |
|---|---|---|---|---|---|---|---|
| T°C | $^{40}Ar^*$ ($10^{-8}$ cc/g) | $^{40}Ar^*/^{39}Ar^*$ | Age (b.y.) | T°C | $^{40}Ar^*$ ($10^{-8}$ cc/g) | $^{40}Ar^*/^{39}Ar^*$ | Age (b.y.) |
| 485 | 108 | 141.3 | 3.93 | 485 | 14 | 103.3 | 3.42 |
| 600 | 777 | 163.5 | 4.17 | 600 | 140 | 126.7 | 3.75 |
| 700 | 2529 | 176.7 | 4.30 | 700 | 1766 | 167.3 | 4.21 |
| 770 | 7911 | 156.6 | 4.10 | 780 | 2866 | 164.0 | 4.18 |
| 935 | 13020 | 140.9 | 3.93 | 855 | 3017 | 160.4 | 4.14 |
| 1025 | 7024 | 134.3 | 3.85 | 980 | 2013 | 153.8 | 4.07 |
| 1195 | 14190 | 125.1 | 3.73 | 1150 | 2778 | 136.2 | 3.87 |
| 1600 | 264 | 124.2 | 3.72 | 1300 | 555 | 144.9 | 3.97 |
| | | | | 1600 | 131 | 101.0 | 3.39 |
| Total | 45823 | 138.6 | 3.90 | | 13280 | 153.1 | 4.06 |

$(^{40}Ar^*/^{39}Ar^*)_{standard} = 13.77.$

from microbreccia and basalt clasts with the latter predominating. This is consistent with the three types of spectra involved. Thus the matrix spectrum is possibly what one might expect from a mixture of the microbreccia spectrum and the basalt spectrum. The fact that the matrix spectrum more nearly resembles that of the microbreccia is not inconsistent with Duncan *et al.*'s conclusion that there is more basalt than microbreccia in the matrix since the microbreccia contains from five to ten times as much argon as do the basalts.

### *15555,26*

Approximately 200 mg of basalt 15555,26 were received for analysis. The sample was divided into two for irradiation, and duplicates were run. The original basalt was collected near the rim of Hadley Rille. As may be seen from Table 6, this sample exhibited serious gas loss in the low temperature fractions. A plateau was reached after about 45% of the argon had been released. This was maintained until 90% evolution, after which a sharp decline in the $^{40}Ar/^{39}Ar$ ratios occurred. The plateaus obtained in the two runs corresponded to ages of 3.29 and 3.32 b.y. If these are crystallization ages, then this basalt formed about 3.31 ± 0.06 b.y. ago.

## Discussion

### *Apollo 14 material*

Three distinct types of $^{40}Ar/^{39}Ar$ evolution spectra have been observed. Firstly, basalt 14310 gave the simplest result, having a rapid rise in $^{40}Ar/^{39}Ar$ ratio to an almost perfect plateau. Basalt 14072, three basaltic clasts from breccia 14321, and two basaltic fragments of coarse soil 14167 show the second characteristic spectrum of a rapid rise in the low temperature $^{40}Ar/^{39}Ar$ ratios to a reasonable plateau followed by a mid- to high-temperature sag. Finally, the microbreccia clast and matrix material from breccia 14321 rose rapidly to a narrow peak in the age range 4.2–4.3 b.y. and then fell off steadily in age thereafter. These overall features are very similar

Table 6. Analytical data for sample 15555,26, basalt.

| Run I (Mass = 0.0981 g) | | | | Run II (Mass = 0.0972 g) | | | |
|---|---|---|---|---|---|---|---|
| T°C | $^{40}Ar^*$ ($10^{-8}$ cc/g) | $^{40}Ar^*/^{39}Ar^*$ | Age (b.y.) | T°C | $^{40}Ar^*$ ($10^{-8}$ cc/g) | $^{40}Ar^*/^{39}Ar^*$ | Age (b.y.) |
| 485 | 12 | 75.0 | 0.95 | 485 | 3 | 24.0 | 0.36 |
| 815 | 291 | 454.8 | 3.03 | 600 | 47 | 140.2 | 1.51 |
| 895 | 196 | 533.1 | 3.27* | 810 | 179 | 450.4 | 3.01 |
| 1015 | 162 | 598.0 | 3.45* | 860 | 125 | 557.4 | 3.34* |
| 1085 | 186 | 516.0 | 3.22* | 925 | 101 | 532.7 | 3.27* |
| 1130 | 117 | 515.0 | 3.22* | 1065 | 119 | 525.2 | 3.25* |
| 1220 | 20 | 423.5 | 2.92 | 1175 | 115 | 579.0 | 3.40* |
| 1600 | 14 | 426.0 | 2.93 | 1270 | 31 | 478.1 | 3.10 |
| | | | | 1615 | 12 | 301.2 | 2.44 |
| Total | 998 | 475.9 | 3.10 | | 732 | 420.6 | 2.91 |

$(^{40}Ar^*/^{39}Ar^*)_{standard}$ = 78.30.
Fraction released at 600°C in run I lost.

to those reported by Turner *et al.* (1971), who gave a detailed discussion of the various possible causes without finding a convincing explanation.

An interesting feature of the 14310 results is that we agree with Hussain *et al.* (1971) in finding an almost horizontal plateau whereas Turner *et al.* find a distinct high temperature sag in the $^{40}Ar/^{39}Ar$ ratios. It is not clear whether this is due to real differences among samples of 14310 or to differences in experimental technique. It appears, nonetheless, to be a very important feature, especially as our data in general are in close agreement with those of Turner *et al.*

Duncan *et al.* (1972), from a detailed textural study of breccia 14321, concluded there were several generations of breccias within breccias and developed a chronological flow chart describing the evolution of material within this sample. The three basaltic clasts analyzed here represented what they considered among the youngest material in the breccia. The matrix and microbreccia were expected to harbor older components. It is certainly true that the latter have quite different evolution spectra than the basaltic clasts. Furthermore, their low temperature maxima reach 4.2 and 4.3 b.y. This may be a reflection of the great age of these components partially reset in later brecciation events. Alternatively, it may be significant, as Turner *et al.* (1971) noted, that despite their curious argon evolution spectra, these staircase-type samples still give total gas ages which do not differ significantly from the general 3.9 b.y. ages found for most Apollo 14 material. The staircase may, therefore, result merely from a closed-system redistribution of $^{40}Ar$ in the samples.

The narrow range in plateau ages strongly suggests these ages have significant meaning in terms of the history of the rocks concerned. Their agreement with internal Rb–Sr isochron ages by Papanastassiou and Wasserburg (1971) and Compston *et al.* (1972) on the same or similar material indicates that they are in fact most probably formation ages. The common plateau age reported here of 3.95 $\pm$ 0.1 b.y. for samples of igneous rocks, soil fragments and breccia fragments suggests we are dealing with a significant event which occurred on the moon at this time. This may well be the time of excavation of the Imbrium Basin, although certainty in this regard is precluded by the obvious difficulty of ascertaining the geographical origin of the samples analyzed.

### *Apollo 15 material*

The plateau age of 3.3 b.y. for basalt 15555,26 indicates that this lava was almost certainly formed about 600 m.y. later than the event recorded above in the Apollo 14 materials. Interestingly, the evolution spectra again show some evidence of high-temperature sag in the $^{40}Ar/^{39}Ar$ ratio. Again it is difficult to relate the age measured to a specific lunar event. The date could refer to the time of filling of the Imbrium Basin, or it could be of relatively local significance. The age derived here, however, is in close agreement with those reported by several other groups at the Third Lunar Science Conference (e.g., Alexander *et al.*, 1972).

### *Cosmic ray ages*

Approximate $^{38}Ar$ exposure ages calculated according to Turner (1970b) are listed in Table 7 for those samples in which the neutron-generated $^{38}Ar$ from chlorine

Table 7. Analytical data for $^{38}Ar$ exposure ages.

| Sample | $^{36}Ar/^{38}Ar$ | Total $^{38}Ar$ ($10^{-8}$ cc/g) | $^{38}Ar$ exposure age (m.y.) |
|---|---|---|---|
| 14310,65 (I) | 0.63 | 43.3 | 333 |
| 14310,65 (II) | 0.66 | 45.4 | 347 |
| 14072,6 (I) | 1.13 | 2.7 | 19 |
| 14072,6 (II) | 1.22 | 3.2 | 22 |
| 14167,9 (I) | 4.67 | 26.2 | 26 |
| 14167,9 (II) | 4.82 | 37.3 | 27 |
| 15555,26 (I) | 1.03 | 11.3 | 79 |
| 15555,26 (II) | 0.92 | 9.9 | 72 |

Computations follow Turner (1970b).
$^{38}Ar$ production rate = $0.13 \times 10^{-8}$ cc/g/m.y.

was not important. The Apollo 14 exposure ages fall into two groups as noted by Turner *et al.* (1971), who found material collected near Cone Crater had exposure ages of about 26 m.y. Similar low exposure ages were given in the preliminary examination report LSPET (1971). Our average value of 21 m.y. for basalt 14072,6 is consistent with this, supporting an age of about 20 m.y. for the Cone Crater event. Our values of 26 and 27 m.y. for 14167 soil fragments agree with Turner *et al.*'s figure of 29 m.y. Basalt 14310 yielded ages of 333 and 347 m.y., again consistent with Turner *et al.* (1971), who quoted 300 m.y. Exposure ages of 79 and 72 m.y. were calculated for 15555 from the Hadley Rille locality; these ages may be compared with the 90 m.y. given by Burnett *et al.* (1972).

*Acknowledgments*—Useful discussions were enjoyed with Mr. G. W. Berger and Professor R. M. Farquhar. Drs. A. Duncan and G. Goles were extremely helpful in providing material from breccia 14321. This research was supported by a grant from the National Research Council of Canada.

## References

Alexander E. C. Jr., Davis P. K., Lewis R. S., and Reynolds J. H. (1972) Rare gas analyses on neutron irradiated lunar samples (abstract). In *Lunar Science—III* (editor C. Watkins), pp. 12–14, Lunar Science Institute Contr. No. 88.

Baksi A. K., York D., and Watkins N. D. (1967). The age of Steens Mountain geomagnetic polarity transition. *J. Geophys. Res.* **72,** 6299–6308.

Berger G. W. and York D. (1970) Precision of the $^{40}Ar/^{39}Ar$ dating technique. *Earth Planet. Sci. Lett.* **9,** 39–44.

Burnett D. S., Huneke J. C., Podosek F. A., Russ G. P. III, Turner G., and Wasserburg G. J. (1972) The irradiation history of lunar samples (abstract). In *Lunar Science—III* (editor C. Watkins), pp. 105–107, Lunar Science Institute Contr. No. 88.

Compston W., Vernon M. J., Berry H., and Rudowski R. (1971) The age of the Fra Mauro formation: A radiometric older limit. *Earth Planet. Sci. Lett.* **12,** 55–58.

Duncan A. R., Lindstrom M. M., Lindstrom D. J., McKay S. M., Stoeser J. W., Goles G. G., and Fruchter J. S. (1972) Comments on the genesis of breccia 14321 (abstract). In *Lunar Science—III* (editor C. Watkins), pp. 192–194, Lunar Science Institute Contr. No. 88.

Farrar E., Macintyre R. M., York D., and Kenyon W. J. (1964) A simple mass spectrometer for the analysis of argon at ultra-high vacuum. *Nature* **204,** 531–533.

Hussain L., Sutter J. F., and Schaeffer O. A. (1971) Ages of crystalline rocks from Fra Mauro. *Science* **173,** 1235–1236.

LSPET (Lunar Sample Preliminary Examination Team) (1970) Preliminary examination of lunar samples from Apollo 14. *Science* **173,** 681–693.

Macintyre R. M., York D., and Gittins J. (1969) K–Ar dating of nephelines. *Earth Planet. Sci. Lett.* **7,** 125–131.

Merrihue C. M. (1965) Trace-element determinations and potassium-argon dating by mass spectroscopy of neutron-irradiated samples (abstract). *Trans. Amer. Geophys. Union* **46,** 125.

Merrihue C. M. and Turner G. (1966) Potassium-argon dating by activation with fast neutrons. *J. Geophys. Res.* **71,** 2852–2857.

Papanastassiou D. A. and Wasserburg G. W. (1971) Rb–Sr ages of igneous rocks from the Apollo 14 mission and the age of the Fra Mauro formation. *Earth Planet. Sci. Lett.* **12,** 36–48.

Sigurgeirsson T. (1962) Dating recent basalt by the potassium-argon method. Dept. Physical Lab. of the Univ. Iceland, p. 9.

Turner G. (1970a) Argon-40/Argon-39 dating of lunar rock samples. *Science* **167,** 466–468.

Turner G. (1970b) Argon-40/Argon-39 dating of lunar rock samples. *Proc. Apollo 11 Lunar Science Conf., Geochim. Cosmochim. Acta* Suppl. 1, Vol. 2, pp. 1665–1684. Pergamon.

Turner G. (1971) $^{40}Ar$–$^{39}Ar$ ages from the lunar maria. *Earth Planet. Sci. Lett.* **11,** 169–191.

Turner G., Huneke J. C., Podosek F. A., and Wasserburg G. J. (1971) $^{40}Ar$–$^{39}Ar$ ages and cosmic ray exposure ages of Apollo 14 samples. *Earth Planet. Sci. Lett.* **12,** 19–35.

Proceedings of the Third Lunar Science Conference
(Supplement 3, *Geochimica et Cosmochimica Acta*)
Vol. 2, pp. 1623–1636
The M.I.T. Press, 1972

# Uranium and extinct $Pu^{244}$ effects in Apollo 14 materials

G. Crozaz,* R. Drozd, H. Graf, C. M. Hohenberg,
M. Monnin,† D. Ragan, C. Ralston, M. Seitz,‡
J. Shirck, R. M. Walker, and J. Zimmerman

Washington University, Laboratory for Space Physics,
St. Louis, Missouri 63130

**Abstract**—Xenon data from stepwise heating of rock 14301 show a large fission component that is attributed to the decay of $Pu^{244}$. A unique feature of 14301 is a clear separation in the release of the fission and spallation components. This allows a determination of the isotopic composition and the derived spectrum is essentially identical with that from spontaneous fission of $Pu^{244}$. The total quantity of fission xenon is also ~15 times that expected from $U^{238}$ spontaneous fission. If the Xe was produced by *in situ* decay, and if the same initial ratio of Pu/U is assumed as in Pasamonte, then portions of rock 14301 formed no later than 120 m.y. after Pasamonte. Whitlockite crystals in rocks 10057, 12040, 12063, and 12064 contain track densities much higher than expected from either slowing down cosmic rays or $U^{238}$ spontaneous fission. The anomalous track densities, which are based on total pit counts in the SEM, are shown to be due to spallation recoil tracks. Using an internal isochron method, a value of ~1 $t/cm^2/yr$ is found for the rate of production of spallation tracks during cosmic ray exposure. Using this spallation correction reasonable ages are obtained on the crystalline rock samples. This shows that whitlockite retains tracks under lunar environmental conditions. Spallation and fission tracks can also be separated on the basis of track length distributions measured on plastic replicas, but this method is difficult to apply at high track densities ($\gtrsim 10^8/cm^2$). Preliminary data on a whitlockite crystal from 14301 show a large track excess that could be due either to $Pu^{244}$ or an early irradiation. A phosphate crystal from breccia 14321 has an apparent excess of fission tracks that could also be attributed to $Pu^{244}$; however, the situation is confused in both samples by an inhomogeneous distribution of fossil tracks. Whitlockites in rock 14311 have fewer fission tracks than expected indicating a postformation annealing event.

## Introduction

In previous work (Burnett *et. al.*, 1970; Burnett *et al.*, 1971) we described the distribution of uranium and thorium in lunar samples and gave preliminary results of fission track dating of lunar phosphates. The purpose of the fission track work was to develop a method that would show fission track anomalies due to $Pu^{244}$ and hence (potentially) measure ages in the interesting region $>4 \times 10^9$ yr.

In this paper we discuss both rare gas and track data on Apollo 14 rocks. From the former we conclude that rock 14301 has a large fission Xe excess with the isotopic composition expected from $Pu^{244}$. The track method has now been developed to the point in lunar whitlockite where large anomalies due to Pu would be measurable. We

* Permanent Address: Université Libre de Bruxelles, Bruxelles, Belgium.
† Permanent Address: Université de Clermont, Clermont-Ferrand, France.
‡ Now at the Geophysical Laboratory, Carnegie Institution of Washington, Washington, D.C.

find no anomalies in Apollo 12 crystalline rocks or in rock 14311. One crystal from 14321 shows a potential anomaly, but the situation is confused by an inhomogeneous distribution of fossil tracks. Preliminary data on rock 14301 show a large track excess that could be due either to $Pu^{244}$, as suggested by the rare gas data, or an early irradiation by solar and galactic particles.

We first summarize the rare gas results and then discuss the track measurements. As we shall show, the essential problem in fission track dating is the correction for spallation recoil tracks. Two methods for making this correction are discussed.

## RARE GAS EVIDENCE FOR $Pu^{244}$ IN ROCK 14301

A possible excess of fission xenon in rock 14301 was first reported by the Preliminary Examination Team (LSPET, 1971). In confirmation of this report we find fission xenon in 14301 which is most likely due to the decay of extinct $Pu^{244}$. A detailed analysis of xenon from 14301 will be reported elsewhere (Drozd *et al.*, 1972) and is referred to for a more complete discussion.

Xenon data from stepwise heating of 14301 are shown in Figs. 1 and 2. Extraction temperatures in hundreds of degrees celsius are indicated by the numbers. The heavy xenon isotopes (Fig. 2) demonstrate a linear correlation indicative of two-component mixing between solar xenon (SUCOR) and a fission component lying on a northeast extension of the line. Figure 1 shows the light (fission shielded) xenon isotopes. The straight-line correlation suggests a two component mixing between solar xenon and cosmic-ray spallation xenon. It is therefore implied that each temperature fraction from 14301 can be constructed from various mixtures of the three basic components; solar, spallation, and fission.

A unique feature of 14301 is a clear separation in the release of the spallation and fission components, e.g., the most spallogenic xenon was extracted at 1200°C, a fraction low in fissiogenic xenon, while the most fissiogenic xenon was liberated at 500°C, the second lowest fraction in spallation. This property allows us to determine the isotopic composition of the pure spallation and pure fission components shown in Table 1. Those fractions richest in fission xenon provide the most precise spectrum. The computed fission spectrum is essentially identical with $Pu^{244}$ fission (Alexander *et al.*, 1971) and different from the spectrum of uranium spontaneous or particle-induced fission.

Additional evidence for $Pu^{244}$ comes from the quantity of fission xenon. If we attribute all the fission xenon to the decay of $U^{238}$ the required uranium content would be 74 ppm, in contrast with our measured uranium content for an aliquot of this sample of 5 $\pm$ 2 ppm (PET value for 14301 is 3.6 ppm). Other possible sources of fission-like xenon (high-energy particle induced fission of U and Th, resonance neutron fission and mass fractionation) can all be eliminated on the basis of both the quantity and isotope composition of the xenon produced (Drozd *et al.*, 1972).

Xenon in rock 14306 is dominated by spallation and fission components. The rare gases in the light and dark regions of this rock are virtually identical as are the uranium contents. Both regions have similar $Kr^{81}$–Kr exposure ages with an average value of 24.4 $\pm$ 1.6 m.y. The exposure age of 14301 is 102 $\pm$ 30 m.y. In contrast

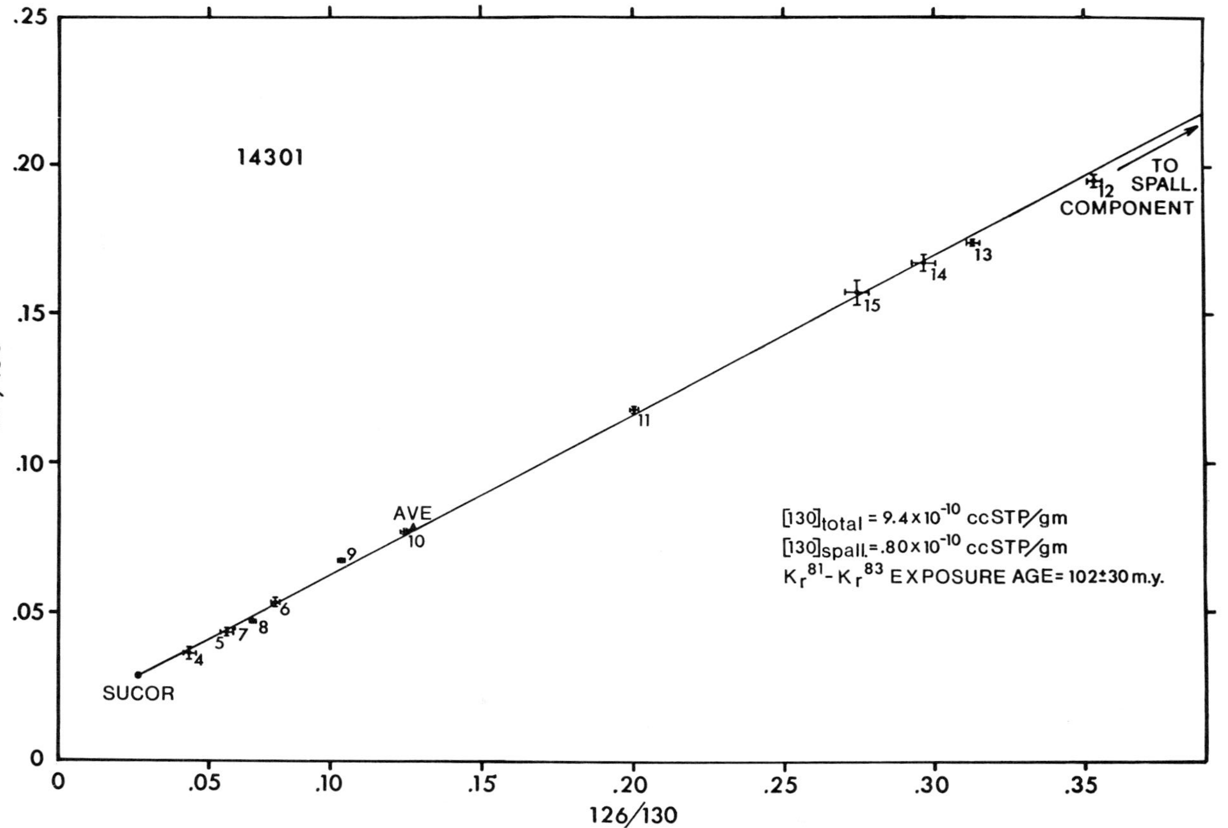


Fig. 1. Correlation diagram of $Xe^{124}/Xe^{130}$ versus $Xe^{126}/Xe^{130}$ released in step-wise heating of 14301. The straight line indicates two component mixing at solar xenon (SUCOR) (Podosek *et al.*, 1971) and spallation produced xenon. The $Kr^{81}$–$Kr^{83}$ exposure age is computed using the technique described by Marti and Lugmair (1971).

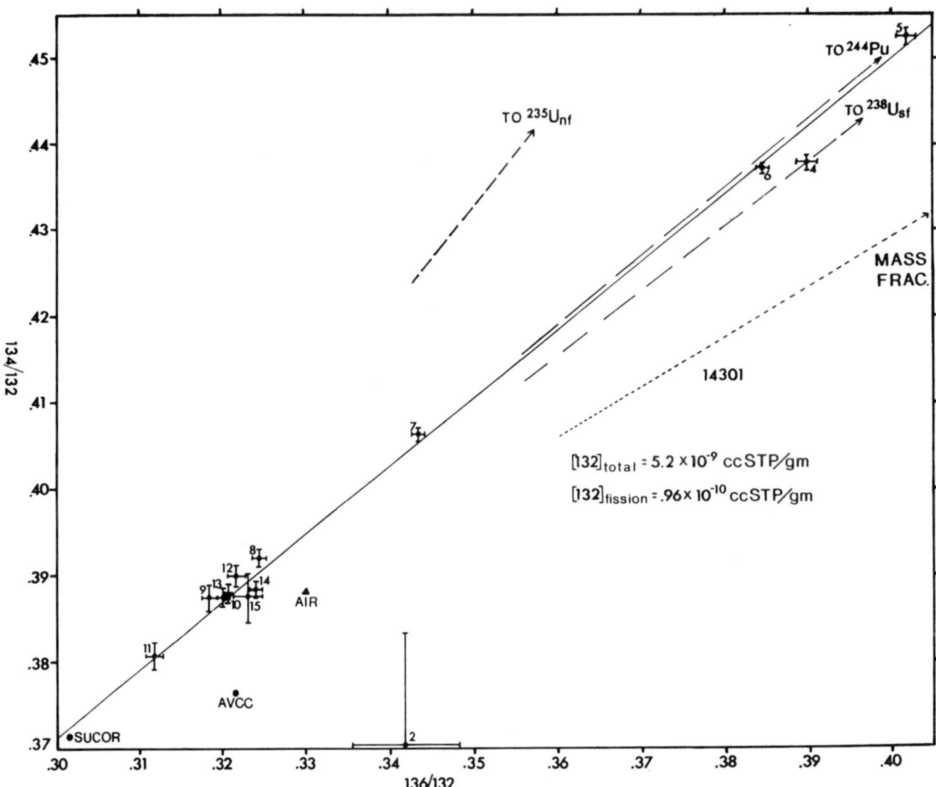


Fig. 2. Correlation diagram of $Xe^{134}/Xe^{132}$ versus $Xe^{136}/Xe^{132}$ representing mixing between solar xenon and fission-produced xenon. Correlation lines which would result from mixtures of solar and various fission sources are shown as dashed lines. The dotted line is mass fractionated solar xenon. The isotopic composition of air and carbonaceous chondrite xenon (AVCC) are shown for comparison.

Table 1. Calculated spectra for fission xenon and spallation xenon in 14301.

| *Fission spectrum* | | | | | |
|---|---|---|---|---|---|
| *14301* | *129* | *131* | *132* | *134* | *136* |
| 500–1200°C subtraction[(a)] | 25.6 ± 3.1 | 29.4 ± 3.3 | 88.1 ± 3.3 | 92.0 ± 2.0 | 100 |
| Rich component subtraction[(b)] | 27.1 ± 3.1 | 30.5 ± 1.3 | 87.6 ± 1.9 | 91.4 ± 0.9 | 100 |
| Average sample | 33.5 ± 10.6 | 28.6 ± 3.2 | 83.6 ± 6.5 | 91.4 ± 0.9 | 100 |
| $Pu^{244}$ [(c)] | — | 25.1 ± 2.2 | 88.0 ± 3.1 | 92.1 ± 2.7 | 100 |
| $U^{238}$ [(d)] | — | 7.6 | 59.5 | 83.2 | 100 |
| $U^{235}_{nf}$ [(d)] | — | 45.4 | 67.8 | 124.8 | 100 |

| *Spallation spectrum* | | | | | | | |
|---|---|---|---|---|---|---|---|
| *124* | *126* | *128* | *129* | *130* | *131* | *132* | *134* |
| 46.4 ± 0.2 | 87.8 ± 0.1 | 150 ± 2 | 230 ± 17 | 100 | 531 ± 12 | 87 ± 13 | 4 ± 4 |

[(a)] Fission component computed by subtraction of the 500°C fraction from a SUCOR–1200°C fraction mixture (Drozd *et al.*, 1972).
[(b)] Computed from the weighted average of all fission rich and spallation rich components.
[(c)] Alexander, *et al.* (1971).
[(d)] Hyde (1964).

with 14301 the total fission xenon content of 14306 is consistent with $U^{238}$ spontaneous fission.

The extensively studied meteorite Pasamonte has been found to contain $8^{+4}_{-2} \times 10^{-11}$ cc STP $Xe^{136}$ fission per microgram of uranium (Hohenberg *et al.*, 1967). The corresponding number of 14301 is $3 \times 10^{-11}$. If we assume that the initial ratio of $Pu^{244}$ to $U^{238}$ was uniform for the original material of these two objects, and if we assume that no plutonium fission xenon has been *added* to 14301 by processes other than *in situ* fission, we can compute the time of xenon retention in 14301 relative to that of Pasamonte. It would appear that portions of rock 14301 formed no later than 120 m.y. after Pasamonte. Therefore, if 14301 represents pre-Imbrian crustal material which was ejected over the Fra Mauro region by the Imbrian event, then the crust and the meteorites are nearly contemporaneous, having begun xenon retention early in the primitive solar system.

Even if extensive plutonium-uranium fractionation has occurred on the moon, it is unlikely that an *in situ* origin for the fission xenon in 14301 is compatible with an age of 3.9–4 b.y. typical of Apollo 14 breccias. Such a situation would require an average $Pu^{244}/U^{238}$ ratio of roughly 0.3 at 4.6 b.y. ago, a number too large to be taken seriously. However, if conventional dating techniques establish a relatively young age for 14301, other mechanisms for the incorporation of $Pu^{244}$ fission products will have to be carefully considered. Additional measurements on a variety of Apollo 14 breccias are clearly desirable. The study of xenon released in incremental heating of neutron irradiated sample should provide answers to the question of *in situ* origin for the fission component and perhaps rule on its usefulness as a method for dating the crustal formation.

## Track Studies

We previously reported (Burnett *et al.*, 1971) that the fossil track density in whitlockite crystals in rock 12040 was greatly in excess of the number expected from slowing down heavy cosmic rays (as measured in adjacent feldspar grains) or spontaneous fission from $U^{238}$. Similar track excesses are found in whitlockites from Apollo 12 and 14 rocks. We show below that these anomalous densities, which are based on total pit counts, result from a large production rate of spallation recoils in whitlockite. Two methods of estimating the spallation production rate are discussed.

The experimental techniques are similar to those previously reported. Polished samples of lunar rocks are covered with either mica or polished feldspar external detectors and then are neutron irradiated to produce a fission map. The external detectors are used both to locate interesting grains and to measure the induced track density. The neutron dose is measured with a U-1 glass standard placed next to mica. Calibration experiments (Seitz *et al.*, 1972) show that the neutron flux is equal to $3.35 \times 10^9\ \rho_{std}$ where $\rho_{std}$ is the density in the mica, as measured in an SEM.

An important point in this technique is to make certain that the polished section and detector are in intimate contact. When this is the case, the induced track density *in the exposed surface area of the whitlockite* can be obtained by tracing the outline of the grain on the external detector and counting tracks only in this region. An alternate approach is to count all the induced tracks in a fission star, including those that

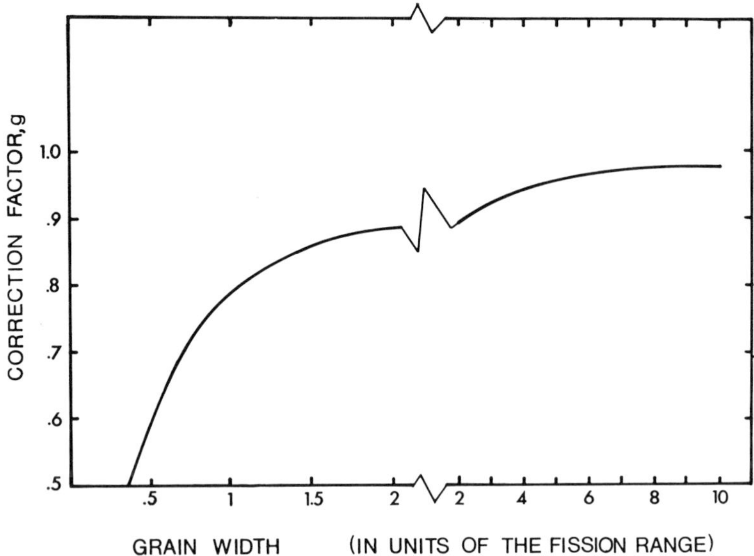


Fig. 3. Geometrical correction factor for long narrow grains as a function of grain width. The number of tracks passing through the exposed surface area of a grain of width $W$ is found by multiplying the total number of tracks in a fission star by $g$.

result from the spray of tracks from the sides of the buried whitlockite, and multiply by a calculated geometrical correction factor (see Fig. 3) to obtain the number that pass through the exposed surface. We employed both methods and most data re-reported here are for stars where reasonable agreement was found between the two approaches. In rock 12040 where a separation of the section and detector was found, the calculated induced densities were used.

Following irradiation the sections were etched in a 0.125% $HNO_3$ solution at 21.8°C between 20 sec and 6 min. The shorter time is appropriate for whitlockites and the latter for chloroapatites. Because the sample is etched after irradiation, the total track density $\rho_{tot}$ in the phosphates includes a neutron induced component $\rho_{iF}$ that must be subtracted to obtain the fossil density $\rho_f$, i.e., $\rho_f = \rho_{tot} - \rho_{iF}$. The induced track density in the whitlockite $\rho_{iF}$ is determined from the density in the external detector $[\rho_{iF}]_D$. The two densities are related by the expression $\rho_{iF} = \varepsilon[\rho_{iF}]_D$ where $\varepsilon$ is the ratio of the track registration efficiency of the whitlockite to that of the external detector. Separate calibration experiments on annealed, neutron irradiated samples of rock 12040 show that $\varepsilon = 1.0 \pm 10\%$ for mica detectors and $1.25 \pm 15\%$ for feldspar. The correction for the induced track density can be avoided if the samples are etched prior to irradiation, but this procedure runs the risk of altering the U concentration in the grains. Although auxiliary experiments indicate that the U loss is generally less than 15%, we prefer to use the postirradiation etching.

The fossil track density $\rho_f$ is the sum of two components: $\rho_{fF}$ due to fission, and $\rho_{fCR}$ due to cosmic rays. The cosmic ray background is itself composed of two components, one due to very heavy cosmic rays $\rho_{VH}$, the other, $\rho_{SR}$, due to spallation recoil. Track densities in lunar feldspar crystals are dominated by heavy cosmic rays and $\rho_{VH}$ can be determined by fossil track measurements in adjacent feldspar grains. Because whitlockite is more sensitive than feldspar, $\rho_{VH}$ will not be strictly the same in the two minerals. However, $\rho_{VH}$ is generally relatively small and this difference is not important.

Assuming that the fission component is due solely to the spontaneous fission of $U^{238}$ the fossil track density can be written as follows:

$$\rho_f = \rho_{\text{tot}} - \varepsilon[\rho_{iF}]_D = \rho_{fF} + \rho_{VH} + \rho_{SR}; \tag{1}$$

$$\rho_{fF} = \left(\frac{\lambda_F}{\lambda_D}\right)(e^{\lambda_D T} - 1)R_F N_0 C_U. \tag{2}$$

The symbols are defined below:

$\lambda_F$ = the spontaneous fission constant of $U^{238}$.
$\lambda_D$ = the total decay constant of $U^{238}$.
$N_0$ = the number of atoms/cm$^3$.
$R_F$ = the effective range of fission fragments.
$C_U$ = the atomic concentration of uranium.
$T$ = the total storage time of fission tracks.

The induced fission component is given by the following expression:

$$\rho_{iF} = \frac{R_F}{2} N_0 C_U \phi \sigma I \tag{3}$$

where $\phi$ is the neutron flux, $\sigma$ is the cross section for neutron induced fission on $U^{235}$, and $I$ is the isotopic abundance of $U^{235}$. Combining equations (2) and (3) and solving for the age, we obtain

$$T = \frac{1}{\lambda_D} \ln \left[\frac{(\rho_f - \rho_{VH} - \rho_{SR})\lambda_D \phi I \sigma}{2\lambda_F \rho_{iF}} + 1\right] \tag{4}$$

Since all the quantities in this expression are either constants or experimental quantities, it can be seen that determining the age reduces to determining $\rho_{fF}$ which in turn reduces to making the appropriate correction for $\rho_{CR} = \rho_{VH} + \rho_{SR}$. Since $\rho_{VH}$ can be measured in adjoining U-free grains, and in any event is small, the key problem is to determine $\rho_{SR}$.

The above equations are based on the assumption that the crystals have a thickness $\geq$ the range of fission fragments. They also assume that the polishing has removed at least a similar amount of material from the top of the grains. Neither assumption can be checked and occasional aberrant ages, both high and low, are expected. In expressing the variables in terms of track densities, it has further been assumed that the grains are large compared to the range of fission fragments. When the grains are small, the apparent track densities, obtained by measuring the number of tracks and dividing by the exposed area of the grains, must be divided by a geometric factor. A theoretical curve of this factor is shown in Fig. 3.

Different whitlockite grains in the same rock are found to contain different amounts of uranium. From equations (1) and (2) it can be seen that a plot of the fossil track density, $\rho_f$, versus $C_U$ should give a straight line assuming constant spallation recoil and cosmic ray track densities. The slope is a function of the age and the intercept represents $\rho_{SR} + \rho_{CR}$. In Fig. 4 we show such a plot for rock 12064. The data are uncorrected for a geometry factor because it has about the same value for all three grains. The slope gives an age of $4.2 \times 10^9$ yr and the intercept $\rho_{SR} + \rho_{CR} = 1.6 \times 10^8\ t/\text{cm}^2$. With a more appropriate age of $3.3 \times 10^9$ yr the intercept becomes $(2.3 \pm 0.2) \times 10^8\ t/\text{cm}^2$.

If we assume that the rate of spallation track production $P$ is constant in lunar whitlockites, the spallation density is given by the following expression:

$$\rho_{SR} = PT_{\text{exp}} \tag{5}$$

where $T_{\text{exp}}$ is the spallation exposure age of the rock. From the data in Fig. 4 and the published exposure age of $210 \times 10^6$ yr for 12064 (Hintenberger *et al.*, 1971) we find that $P = 1 \pm 0.1\ t/\text{cm}^2/\text{yr}$ (error includes only statistical uncertainty).

Application of equation (5) to other lunar rocks requires the assumptions that this production rate is constant and that changes in the spallation energy spectrum affect equally the measured exposure age and the rate of track production. In Table 2 we show that reasonable ages are obtained for rock 12063 and two grains of rock 12040 with this procedure. The corrections for grains 3 + 4 of rock 12040 and for rock 10057 are too large to make a meaningful test.

Correct and uncorrected ages are also shown in Fig. 5 for those cases where the spallation correction is smaller than 75%. It can be seen that the spallation correction adequately explains the previously noted excess track densities. The general agreement

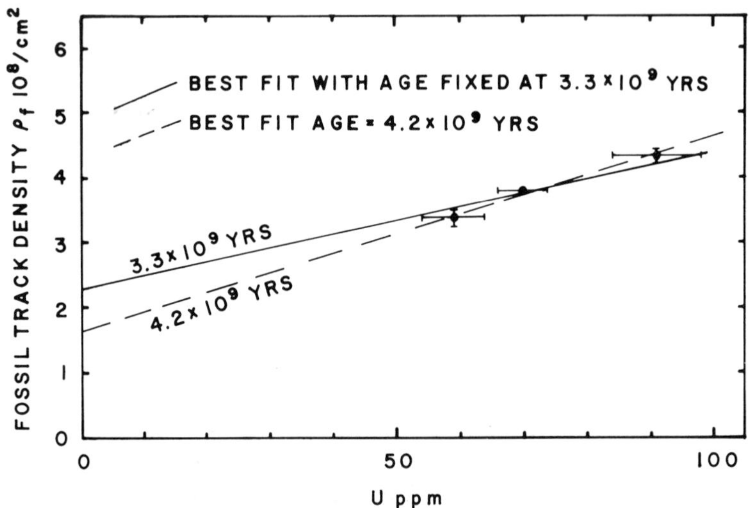


Fig. 4. Internal isochron for rock 12064. Since the dimensions of the grains are closely similar, no geometrical correction has been applied to the data.

Table 2. Results on phosphate grains using the pit counting method.
(Track densities are not corrected for a geometry factor. Statistical errors only are shown.)

| Sample | Grain W = whitlockite A = apatite | Detector | $(\rho_{iF})_D$ ($10^7$/cm$^2$) | $\phi$ ($10^{16}$ n/cm$^2$) | Geometry factor (g) | $\rho_{VH}$ ($10^7$/cm$^2$) | $\rho_{SR} = P \times T_{exp}$ ($10^7$/cm$^2$) | $\rho_f$ ($10^7$/cm$^2$) | $\phi\ (\rho_{fF}/\rho_{iF})$ ($10^{16}$/cm$^2$) | Age ($10^9$ yr) | $\frac{P \times T_{exp}}{\rho_f}$ (%) |
|---|---|---|---|---|---|---|---|---|---|---|---|
| 10057 | W | Mica | 0.765 ± 0.09 | 4.70 | 0.91 ± 0.05 | 1.6 | 5.2 | 7.4 ± 0.3 | 3.7 ± 1.9 | 1 | 70 |
| 12040 | 1W | Mica | 5.13 ± 0.35 | 5.43 | 0.90 ± 0.05 | 1.8 | 22 | 39.2 ± 1.0 | 16.3 ± 1.5 | 3.6 | 56 |
| | 2W | | 6.25 ± 0.23 | | 0.86 ± 0.05 | | | 36.5 ± 0.6 | 11.1 ± 0.7 | 2.7 | 60 |
| | 3W | | 3.82 ± 0.26 | | 0 90 ± 0 05 | | | 24.8 ± 0.7 | 1.4 ± 1.0 | <1 | 89 |
| | 4W | | 2.62 ± 0.16 | | 0.70 ± 0.15 | | | 25.4 ± 0.6 | 3.3 ± 1.3 | 1 | 88 |
| 12064 | 1W | Mica | 5.05 ± 0.27 | 4.90 | 0.87 ± 0.05 | 0.5 | | 38.0 ± 0.7 | | 4.2‡ | — |
| | 2W | | 6.51 ± 0.51 | | 0.85 ± 0.05 | | | 43.6 ± 0.8 | | 4.2‡ | — |
| | 3W | | 4.23 ± 0.34 | | 0.83 ± 0.10 | | | 33.8 ± 1.0 | | 4.2‡ | — |
| 12063 | 1W | Mica | 3.94 ± 0.28 | 6.26 | 0.75 ± 0.10 | 0.39 | 9.5 | 19.1 ± 0.6 | 14.6 ± 1.4 | 3.3 | 50 |
| | 2W | | 5.14 ± 0.51 | | 0.65 ± 0.20 | | | 24.7 ± 1.1 | 18.0 ± 2.2 | 3.9 | 38 |
| 14311 | 1W | Feldspar | 3.02 ± 0.06 | 0.365 | 1 | 0.07[a] | 66 | 90 ± 4 | 2.3 ± 0.4 | <1 | 73 |
| | 2W | | 2.19 ± 0.09 | | 1 | | | 19.5 ± 0.7 | — | — | — |
| | 3W | | 2.50 ± 0.18 | | 1 | | | 48 ± 4 | — | — | — |
| | 4A | | 6.95 ± 0.57 | 2.74 | 0.44 ± 0.30 | | | 11.3 ± 0.7 | — | — | — |
| | 5A | | 3.09 ± 0.15 | | 0.83 ± 0.10 | | | 16.5 ± 1.5 | — | — | — |
| | 6A | | 1.50 ± 0.15 | | 0.77 ± 0.10 | | | 34 ± 2 | — | — | — |
| 14321 | 1W* | Mica | 4.14 ± 0.15 | 8.05 | 1 | 0.15[b] | 2.7 | 15.9 ± 0.9 | 25.4 ± 1.8 | 5 | 17 |
| | 1W† | | 4.14 ± 0.15 | | 1 | | | 10.9 ± 0.6 | 15.6 ± 1.3 | 3.5 | 25 |

* "Hot spot" regions included. † "Hot spot" regions not included. ‡ Age derived from internal isochron.
[a] Hart *et al.*, 1972. [b] Hutcheon *et al.*, 1972.

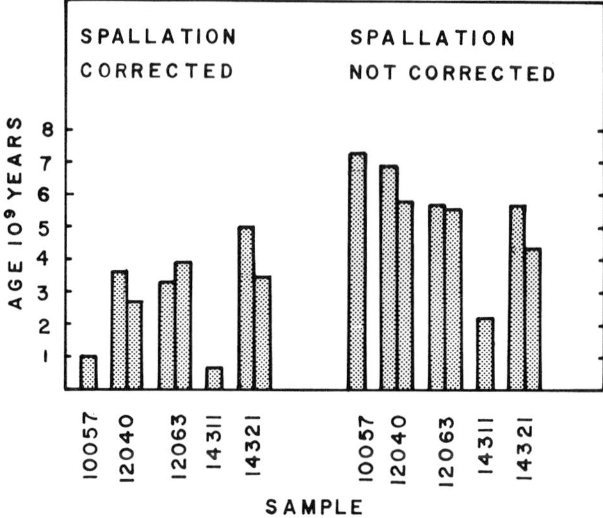


Fig. 5. Comparison between uncorrected and spallation corrected track ages. Only ages with spallation corrections smaller than 75% are shown.

with the previously published ages of crystalline lunar rocks leads us to conclude that fission tracks are stable in whitlockites under lunar environmental conditions and that fission track studies of this mineral would show any large fission anomalies if they exist.

In an independent measure of the spallation production rate, two samples of rock 12040 and one sample of the amphoterite St. Severin were annealed (12 hours at 700°C) and irradiated with the 28 GeV proton beam of the Brookhaven AGS accelerator. From the activity of $Na^{22}$ in the Al target holder the proton dose was determined to be $(5.5 \pm 1.1) \times 10^{15}$ $p/cm^2$, assuming a cross section for the reaction $Al^{27}$ $(p, 3p3n)$ $Na^{22}$ of 10 mb. The results are summarized in Table 3. An average production rate of $6.8 \times 10^{-8}$ $t/cm^2$ per $p/cm^2$ was obtained. Although this rate is about a factor of two higher than would be inferred from the data of 12064, we consider the agreement satisfactory in view of the differences in bombarding energy and difficulties in measuring the high energy proton flux.

The production rate corresponds to a high fraction of all nuclear interactions

Table 3. Results of annealed and proton irradiated samples.

| Sample | Method | Number of grains analyzed | U-conc. (ppm) | $\rho_{SR}$ ($10^8/cm^2$) | $\rho_{SR}$ Average ($10^8/cm^2$) | Proton flux ($10^{15}$ p/cm$^2$) |
|---|---|---|---|---|---|---|
| 12040 | Pit counts | 5 | 20–39 | 3.2–4.6 | 3.8 | $5.5 \pm 1.1$ |
| | | 1 | 65 | 3.5 | 3.5 | $5.5 \pm 1.1$ |
| | Replicas | 4 | 22–39 | 2.45–3.74 | 3.1 | $5.5 \pm 1.1$ |
| | | 2 | 50–60 | 3.1–3.42 | 3.3 | $5.5 \pm 1.1$ |
| St. Severin | Pit counts | 1 | 0.4 | 4.4 | 4.4 | $5.5 \pm 1.1$ |

($\gtrsim 2\%$) in the whitlockite and suggests that a major constituent, probably Ca, is responsible for the tracks.

Preliminary data on one large phosphate grain from an interior chip of 14301 show a large track excess. Unfortunately, both the uranium distribution and the fossil track densities are very inhomogeneous. Counts in corresponding areas give fission excesses ranging from 0.8 to 10. The *VH* contribution was taken from the work of Hart *et al.*, 1972 and the spallation correction from our measured spallation age of $102 \times 10^6$ yr. An inviting interpretation is to ascribe the excess tracks to $Pu^{244}$ decay. It must be remembered, however, that rock 14301 has a large solar gas content. Individual grains in the rock could retain a memory of an early solar flare or galactic irradiation leading to the apparent $Pu^{244}$ excess. More grains need to be studied and techniques need to be refined before a definitive assignment of these excess tracks to $Pu^{244}$ can be made.

In 14321 one grain $\sim 30 \times 40\mu$ was measured. This grain is unique in that it had several "hot spots" in the fossil track pattern. The uranium map on the other hand showed a relatively uniform distribution. If only tracks outside the anomalous regions were counted, we obtain an age of $3.5 \times 10^9$ yr. Inclusion of the hot spots however, gave 40% track excess over $U_{sf}^{238}$ tracks in $3.95 \times 10^9$ yr in qualitative agreement with Hutcheon *et al.*, 1972.

Although the identification of the track excess with $Pu^{244}$ contribution is the most plausible explanation, it must be kept in mind that other sources are also possible as in 14301.

Ages in 14311 are very low both before and after spallation correction showing that extensive track annealing has taken place. This may well have occurred in a single event when the rock was mechanically broken out of a larger object.

A different method for separating fission and spallation tracks is based on the difference in track lengths. In Fig. 6 we show replicas of two crystals of rock 12040, one that has been annealed and proton irradiated, the other containing fossil tracks. It can be seen that the spallation tracks are much shorter, a fact made more explicit in Fig. 7 that shows the measured track length distributions.

Although this method holds high promise, we have as yet been unable to get a satisfactory quantitative separation of spallation and fission at high track densities. Part of the problem appears to be a slight change in etching characteristics from one sample to the next, making it difficult to construct universal track length curves for the two components.

A by-product of the experimental procedure is a measurement of the U-distribution in different samples. Results for Apollo 12 rocks have been previously discussed (Burnett *et al.*, 1970; Burnett *et al.*, 1971). The rocks of Apollo 14 show a remarkable diversity. In the breccia 14306, which has a pronounced light-dark structure, the dark areas tend to give a fairly uniform pattern of distributed uranium with a concentration of $\sim 18 \pm 5$ ppm, however, some dark areas are noticeably lower in uranium. The white areas show regions that are largely devoid of distributed uranium but which contain intense fission stars. Qualitatively the pattern is similar to that in rock 12013. A section of rock 14311 on the other hand has a fairly uniform background of $\sim 5$ ppm with occasional zircon crystals up to $100\mu$ in size with uranium

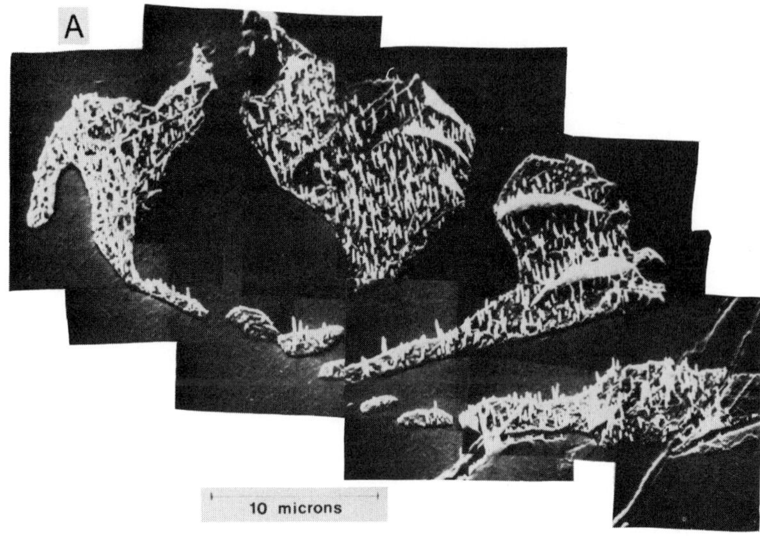


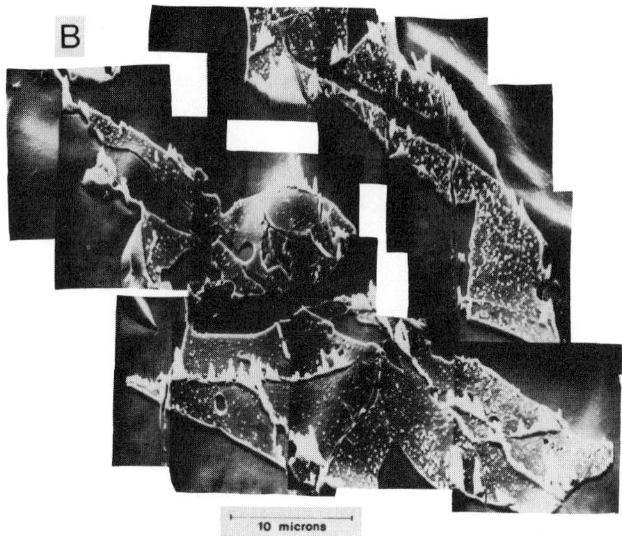


Fig. 6. Comparison between fossil tracks and spallation tracks in whitlockite grains of rock 12040: (a) SEM photograph of replica tracks in an as-received sample of 12040. (b) SEM photograph of spallation tracks in 12040. The sample was first annealed to remove fossil tracks and then irradiated with 28 GeV protons.

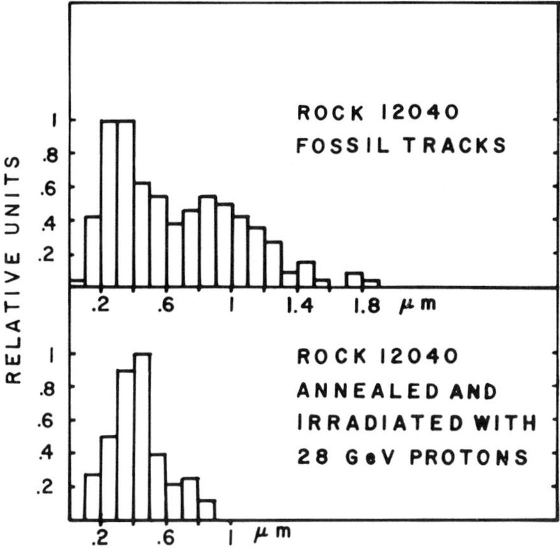


Fig. 7. Track length distributions of fossil tracks and spallation tracks in 12040. The measurements were made on plastic replicas similar to those in Fig. 6.

Table 4. Uranium-rich phases in lunar materials.

| Phase | 14306 U (ppm) | 14311 U (ppm) | 14310 U (ppm) | 14259† | 12028‡ U (ppm) | 10085,9§ |
|---|---|---|---|---|---|---|
| Phosphate | ≥ 200 | 40–80 | 20–200 | * | 50–200 | * |
| Zircon | — | 90–300 | — | * | — | — |
| Baddeylite | 80 | 270 | 400 | — | ≳ 500 | * |
| Zr–Ti | 250–> 500 | — | — | * | — | — |
| No. of stars | 12 | 10 | 14 | 16 | 4 | 8 |

* Mineral identified but concentration not measured.
† Comprehensive soil sample.
‡ Coarse-grained layer from Apollo 12 double core.
§ Two anorthosites removed from coarse fines of Apollo 11.

concentration of ~100 ppm. The large crystalline rock 14310 is studded with high uranium stars separated by regions with virtually no uranium. In all cases where the star-producing grains can be identified they are associated with P or Zr bearing minerals as in previous lunar samples. Table 4 gives a summary of the minerals that have been seen in various samples. The reader is cautioned that this does not represent a systematic study of the rocks, only a brief survey of some respresentative stars.

*Acknowledgments*—We acknowledge the assistance of P. Swan and S. Sutton in preparing and analyzing the samples and of F. Geisler for the uranium determinations. We also thank B. Drozd for the drafting and M. Daggett for manuscript preparation. This work was supported by NASA Contract NAS 9-8165.

## References

Alexander E. C. Jr., Lewis R. S., Reynolds J. H., and Michel M. C. (1971) Plutonium-244: Confirmation as an extinct radioactivity. *Science* **172,** 837–840.

Burnett D. S., Monnin M., Seitz M., Walker R., Woolum D., and Yuhas D. (1970) Charged particle track studies in lunar rock 12013. *Earth Planet. Sci. Lett.* **9,** 127–136.

Burnett D., Monnin M., Seitz M., Walker R., and Yuhas D. (1971) Lunar astrology: U–Th distributions and fission-track dating of lunar samples. *Proc. Second Lunar Sci. Conf., Geochim. Cosmochim. Acta* Suppl. 2, Vol. 2, pp. 1503–1519. MIT Press.

Drozd R., Hohenberg C. M., and Ragan D. (1972) Fission xenon from extinct $Pu^{244}$ in 14301: Age of the pre-Imbrium crust. *Earth Planet. Sci. Lett.* (submitted).

Hart H. R. Jr., Comstock G. M., and Fleischer R. L. (1972) The particle track record of Fra Mauro (abstract). In *Lunar Science—III* (editor C. Watkins), pp. 360–363, Lunar Science Institute Contr. No. 88.

Hintenberger H., Weber H., and Takaoka N. (1971) Concentrations and isotopic abundances of the rare gases in lunar matter. *Proc. Second Lunar Sci. Conf., Geochim. Cosmochim. Acta* Suppl. 2, Vol. 2, pp. 1607–1625. MIT Press.

Hohenberg C. M., Munk M. N., and Reynolds J. H. (1967) Spallation and fissiogenic xenon and krypton from stepwise heating of the pasamonte achondrite; the case for extinct plutonium 244 in meteorites; relative ages of chondrites and achondrites. *J. Geophys. Res.* **72,** 3139–3177.

Hutcheon I. D., Phakey P. P., Price P. B., and Rajan R. S. (1972) History of lunar breccias (abstract). In *Lunar Science—III* (editor C. Watkins), pp. 415–418, Lunar Science Institute Contr. No. 88.

Hyde E. K. (1964) The nuclear properties of heavy elements. In *Fission Phenomena*, Vol. 3. Prentice Hall.

LSPET (Lunar Sample Preliminary Examination Team) (1970) Preliminary examination of lunar samples from Apollo 14. *Science* **173,** 681–693.

Marti K. and Lugmair G. W. (1971) $Kr^{81}$–Kr and K–$Ar^{40}$ ages, cosmic-ray spallation products, and neutron effects in lunar samples from Oceanus Procellarum. *Proc. Second Lunar Sci. Conf., Geochim. Cosmochim. Acta* Suppl. 2, Vol. 2, pp. 1591–1605. MIT Press.

Podosek F. A., Huneke J. C., Burnett D. S., and Wasserburg G. J. (1971) Isotopic composition of xenon and krypton in the lunar soil and in the solar wind. *Earth Planet. Sci. Lett.* **10,** 199–216.

Seitz M. G., Walker R. M., and Carpenter S. (1972) Improved methods for measurement of thermal neutron dose by the fission track technique. *J. Appl. Phys.* (submitted).

Proceedings of the Third Lunar Science Conference
(Supplement 3, *Geochimica et Cosmochimica Acta*)
Vol. 2, pp. 1637–1644
The M.I.T. Press, 1972

# $^{237}$Np, $^{236}$U, and other actinides on the moon*

P. R. FIELDS, H. DIAMOND, D. N. METTA, D. J. ROKOP, and C. M. STEVENS
Chemistry Division, Argonne National Laboratory,
Argonne, Illinois 60439

**Abstract**—Widely varying amounts of $^{236}$U (2.4 × $10^7$ y half-life) have been observed in samples 12070,91, 12013,10,42, 12073,34, and 14305,80. In samples 10084,75 and 14163,135 the ratio $^{236}$U : $^{238}$U < 3 × $10^{-9}$. The $^{236}$U : $^{238}$U ratio in the whole 12070 sample is much higher than it is in the coarsest component of that sample. Arguments are presented to show that lunar $^{236}$U is produced mostly by solar flare protons irradiating the indigenous $^{238}$U. The $^{237}$Np (2.14 × $10^6$ y half-life) also predicted by this mechanism was sought and found in sample 12070. Preliminary calculations imply that the solar flare proton flux more than two million years ago was greater than it has been more recently.

Measurements of $^{235}$U : $^{238}$U mol ratios, total thorium and uranium contents, $^{234}$U : $^{238}$U activity ratios, and limits to $^{239}$Pu and $^{244}$Pu contents were made on samples 14305 and 14163.

## INTRODUCTION

THE FIRST OBSERVATION of naturally occuring $^{236}$U (2.4 × $10^7$ y half-life) was in sample 12070 by Fields *et al.* (1971). Six more measurements of $^{236}$U in lunar samples have been made, and in a separate paper by Rokop *et al.* (1972) $^{236}$U in terrestrial uranium has been reported. We suggest that most of the lunar $^{236}$U was made by the reaction of solar flare protons with the $^{238}$U in lunar soil. Since the same reaction should also produce $^{237}$Np, the neptunium fraction of the most $^{236}$U-rich sample was investigated, and $^{237}$Np was found.

Ultimately the $^{236}$U and $^{237}$Np contents of lunar samples may serve as long-lived solar proton flux monitors, but since the surface exposure and mixing history of the samples are not well defined and the pertinent cross sections are not known, only broad inferences are now possible. The very high $^{236}$U content of 12070 and some crude cross-section assumptions imply that the $^{236}$U was produced by a proton flux much more intense than that implied by the $^{26}$Al content of many lunar samples (Reedy and Arnold, 1972). This high proton flux exposure must have occurred at such a time that the $^{26}$Al would have had time to decay away, but not so long ago that $^{236}$U would be lost, that is, 2–100 million years ago.

Further actinide data has been collected for samples 14163,135 and 14305,80. Uranium and thorium contents and limits to $^{239}$Pu and $^{244}$Pu and other actinides are given. The mass spectrometric $^{235}$U : $^{238}$U ratios and the $^{234}$U : $^{238}$U activity ratios are reported.

* Work was performed under NASA Contract No. T76536 and U.S. Atomic Energy Commission.

## EXPERIMENTAL METHODS

The chemical procedures have been described earlier in Fields *et al.* (1970). Uranium, thorium, neptunium, and plutonium are all extracted into tricaporyl methyl ammonium chloride (Alequat 336) in xylene from $HNO_3$ solution and are resolved into their components with Dowex A–1 ion exchange columns with HCl and HI elutriants.

Isotopic dilution with $^{230}$Th ($8.0 \times 10^4$ y) and $^{233}$U [$(1.585 \pm 0.018) \times 10^5$ y] (Ellis, 1971) added to the dissolving sample was used for thorium and uranium analyses. $^{239}$Np, obtained from mass-separated $^{243}$Am, was added to the already isolated neptunium fraction of 12070,91. This was then reduced with HI to ensure exchange and then purified with another anion exchange column. The chemical yield of neptunium at the point of adding tracer was estimated from the yields of two dummy runs employing $^{237}$Np tracer. $^{236}$Pu tracer was also added to separated plutonium fractions whose chemical yields were estimated. The uncertainties of estimation were included in the stated errors and were incorporated into the upper limits of $^{239}$Pu and $^{244}$Pu that are reported.

The uranium and thorium determinations and the $^{235}$U : $^{238}$U ratios employed a 30 cm radius, 60° sector, permanent magnet mass spectrometer with a multiple filament surface ionization source and electron multiplier detection.

The very high sensitivity of the 100-inch mass spectrometer (Moreland *et al.*, 1967) was required for the $^{239}$Pu, $^{244}$Pu limits and for finding $^{237}$Np.

Because the $^{236}$U : $^{238}$U ratio can be very small, measurements were made with a 32.4 cm, 44° sector tandem magnetic mass spectrometer, specifically designed to measure widely disparate isotope ratios at adjacent mass peaks (Kaiser and Stevens, 1967; Moreland *et al.*, 1970). A high-pass energy filter exploited the 1.2 eV difference in the kinetic energy between $^{236}U^+$ and $^{235}UH^+$ to suppress the latter. Further discrimination was based upon the 5.7 MeV mass difference between $^{236}U^+$ and $^{235}UH^+$.

Interference at the $^{236}$U mass position by $^{235}UH^+$ can be estimated from the works of Moreland *et al.* (1970) to be less than $2 \times 10^{-13}$ of the $^{238}$U. Known mass 236 ions from surface ionization such as $^{39}K_5$ $^{41}K^+$, $^{204}Pb$ $^{16}O_2^+$, $^{187}Re$ $^{16}O_2$ $^{17}O^+$, and $^{12}C_{18}H_{20}^+$, have sufficiently different masses from $^{236}U^+$ to be displaced: the tandem mass spectrometer can resolve full peaks at mass 1500. Other isometic hydrocarbons such as $^{12}C_{19}H_8^+$ have also been ruled out by ancillary studies of their normal abundance ratios to more easily discerned peaks (Rokop *et al.*, 1972).

The data on $^{236}$U in terrestrial uranium reported elsewhere by Rokop *et al.* (1972) are included here to demonstrate the capability of the machine to achieve even greater sensitivity than was available with the much smaller amounts of uranium from lunar samples.

Each measurement reported in this paper was accompanied by one or more appropriate blanks. Except for one unreported result, the blanks were very small, and the corrections are incorporated into the results.

## RESULTS AND DISCUSSION

### $^{236}U$

$^{236}$U, with a half life of $2.4 \times 10^7$ y, is the longest-lived radioactive nuclide perceptible on the moon that could not have survived since the last known nucleogenetic event, $4.7 \times 10^9$ years ago. As such, its abundance reflects radionuclide generating conditions for a greater time span than do the abundances of other radioactive nuclei. In this section we will consider the possible sources of $^{236}$U in lunar material and will argue against the alternatives to production of $^{236}$U by solar flare proton reactions with the $^{238}$U, which is present at about one part per million in lunar surface material.

The ratios of $^{236}$U : $^{238}$U in six lunar samples are shown in Table 1. In a separate investigation (Fields *et al.*, 1971) the finest particles in 12070 (1/7 by volume) were

Table 1. $^{236}$U in lunar and terrestrial samples.

| Sample | $^{236}U/^{238}U$ ($\times 10^{-9}$) | $^{26}$Al (Literature) (c/m/Kg) |
|---|---|---|
| 10084,75 (soil) | <3 | 107–147 |
| 12070,91 (soil) | 233 ± 15 | 146–171 |
| 12070,91 (coarsest fraction of < 1 mm soil) | 31 ± 6 | |
| 12013–10,42* | 9.4 ± 2 | 115 |
| 12073,34 (breccia) | 47 ± 10 | 110 ± 10 |
| 14163,135 (soil) | <3 | 78 |
| 14305,80 (breccia) | 4 ± 2 | 85 ± 17 |
| Uraninite (Shinkolobwe Katanga, Congo)† | 0.23 ± 0.08 | |
| Shroeckingerite (Lost Creek, Wyoming)† | <0.2 | |
| Thucolite (Parry Sound, Ontario)† | 0.08 ± 0.15 | |
| Pitchblend (Great Bear Lake, Port Hope Refinery, Ontario)† | 0.62 ± 0.22 | |

* Supplied, already isolated and purified, by J. Rosholt.
† Rokop *et al.*, 1972.

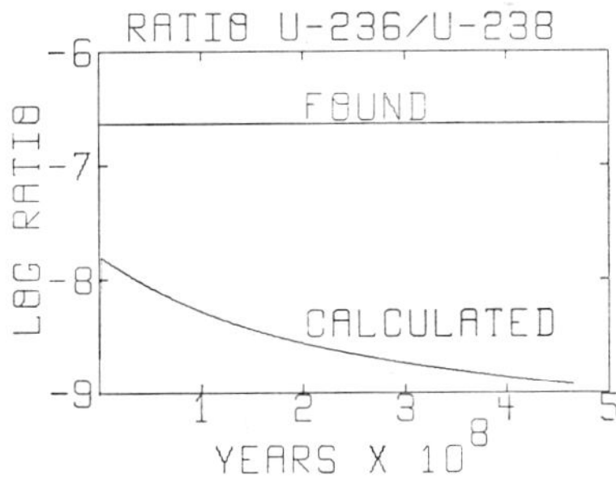


Fig. 1. $^{236}U : ^{238}U$ expectation from neutron capture by $^{235}$U if the nvt were $2.1 \times 10^{16}$ and the time of radiation were given by the abcissa. This is compared to the ratio found in 12070.

removed by collecting only those particles that settled faster than 0.17 cm/min. In this coarser portion of 12070 the $^{236}U : ^{238}U$ was nearly 8-fold lower than it was in the sample as a whole. This means that in the finest fraction of 12070 (less than 10 microns) the $^{236}U : ^{238}U$ ratio must have been greater than $2.33 \times 10^{-7}$. This places severe limits upon the possible modes of $^{236}$U formation.

The isotopic composition of the gadolinium in 12070 (Burnett *et al.*, 1971) indicates that the total number of neutrons (< 0.2 eV) to which 12070 was exposed is $2.1 \times 10^{16}$ n/cm$^2$. No information about the time distribution of this neutron exposure is available, but Fig. 1 shows the calculated $^{236}U : ^{238}U$ ratios one would expect if $^{236}$U were made from the capture of neutrons by $^{235}$U ($\sigma$ = 106 barns) and if the time of the irradiation were given by the abcissa and continued until now.

There are two problems with the above reasoning: (1) There might be a much larger flux of slightly higher energy neutrons that would not be reflected by the gadolinium isotopic distribution but could affect the $^{235}$U capture; (2) since 12070 is a mixture, the gadolinium might have acquired its irradiation history prior to its having been mixed with the uranium. The neutron flux estimated by Begemann *et al.* (1972) of 1 n/(cm$^2$ · sec) from $^{36}$Cl and by Fields *et al.* (1971) of less than 80 n/(cm$^2$ · sec) in 12070 from a $^{239}$Pu limit do not have the dependence upon low energy resonances of the gadolinium isotopes, and these also establish the inadequacy (at least recently) of the neutron flux for production of the required amount of $^{236}$U. Although separation of rare earths and uranium is possible, Hubbard and Gast (1971) and others have shown a fairly close correlation between the uranium and gadolinium contents in many lunar fractions.

The very wide variation of $^{236}$U content from sample to sample is more consistent with the source being solar flare protons, whose effective depth is only several centimeters, than it is with a neutron or galactic cosmic ray source, whose ranges would be meters. Just as is the case with neutrons, there are not enough galactic cosmic rays to produce the observed $^{236}$U (Reedy and Arnold, 1972; Armstrong and Alsmiller, 1971).

The steep decline of activity with depth within a lunar rock shows that all but about 60 d/min/Kg of $^{26}$Al activity is formed by solar flare protons (Finkel *et al.*, 1971). It can be seen in Table 1 that there is some correspondence between the high $^{236}$U contents and the high $^{26}$Al contents. Although the appropriate depth dependence measurements have not been made for $^{236}$U, the absence of reasonable alternative modes of formation, the high variability of $^{236}$U content, the crude correlation with $^{26}$Al, and the presence of $^{237}$Np in 12070 (discussed later) all point to the production of $^{236}$U by solar protons acting upon $^{238}$U. The pertinent reactions in order of importance would be $(p, t)$, $(p, 3n)$, $(p, p2n)$, and $(p, 2pn)$. Only 49% of the $^{236}$Np from $^{238}$U $(p, 3n)$ reactions decays into $^{236}$U.

There are very few measurements of proton spallation cross-sections for $^{238}$U. These are shown in Fig. 2 along with an assumed curve for all proton reactions transforming $^{238}$U to $^{236}$U. It is hoped that this curve might be within a factor of 2 or 3 of being correct—more sophisticated calculations are in progress. This curve was applied to a proton spectrum whose rigidity is 100 MV (Finkel *et al.*, 1971), and the flux required to produce $2.33 \times 10^{-7}$ $^{236}$U : $^{238}$U (in equilibrium) was calculated to be about 5000 protons/(sec · cm$^2$) above 10 MeV. This is about 2 orders of magnitude more protons than is invoked by Reedy and Arnold (1972) to explain the $^{26}$Al on the moon. These results suggest that there was a far greater proton flux 2–100 million years ago than at present, and hence the sun might have been much more active then than now.

The possibility that $^{236}$U was formed elsewhere and accreted upon the lunar surface cannot be excluded. However, such $^{236}$U could not have arisen from some supernova explosion, for then we would expect much more $^{244}$Pu, and we have reported a $^{244}$Pu content for 12070 of $< 2 \times 10^{-4}$ lower than $^{236}$U in that sample (Fields *et al.*, 1971).

### $^{237}$*Np*

The protons reacting with $^{238}$U can be expected to produce $2.2 \times 10^6$ y $^{237}$Np. The neptunium fraction in the chemical separations of the various Apollo samples

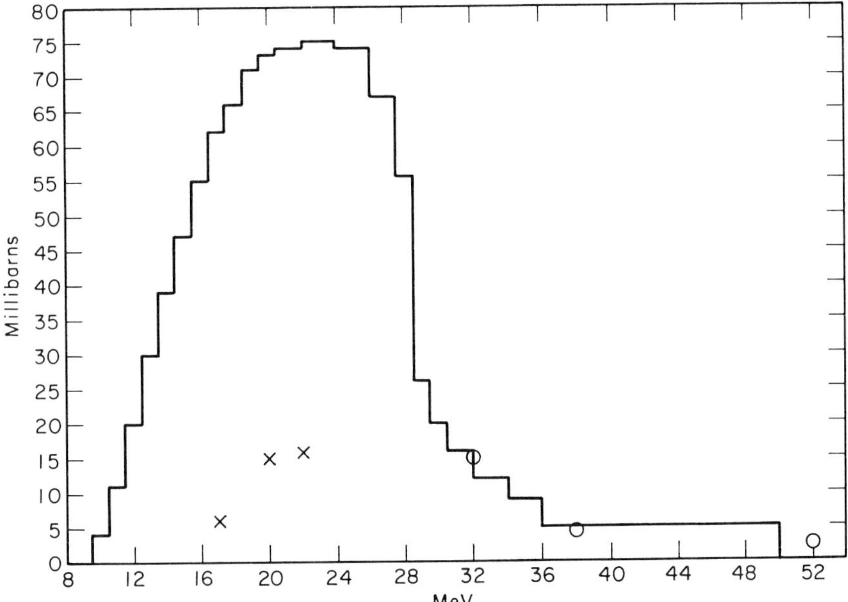


Fig. 2. Histogram of an assumed cross-section curve for all proton reactions transforming $^{238}U$ into $^{236}U$. This was employed in calculations leading to the conclusion that the proton flux creating $^{236}U$ was more intense than the flux used to explain the production of $^{26}Al$. The circles are from Lefort's (1961) $^{238}U$ ($p$, $3n$) cross sections. The crosses are from McCormick and Cohen's (1954) $^{238}U$ ($p$, $3n$). Forty-nine percent of these reactions result in $^{236}U$. See also Wade *et al.* (1957).

had previously been isolated, and tracer experiments had measured the chemical yield of neptunium in these fractions. $^{239}Np$ tracer was added to a fraction of 12070, and after further purification the $^{237}Np/^{239}Np$ ratio was measured with the 100-inch mass spectrometer. There are $1.1 \pm 0.3 \times 10^{-13}$ g $^{237}Np$ per g of 12070,91. This is 0.28 as much as the $^{236}U$ content; if one assumes a steady equilibrium bombardment for the last $10^8$ years, the cross section for formation of $^{237}Np$ would be 3.2 times the cross section for production of $^{236}U$.

Here, as in $^{236}U$, there appears to be no reasonable alternative to solar flare protons acting upon $^{238}U$ as a source for the observed $^{237}Np$.

## $^{235}U/^{238}U$

The isotopic ratios of $^{235}U : {}^{238}U$ in five samples are shown in Table 2. Although deviation from terrestrial standards as much as 2.6 $\sigma$ are reported, the lunar and terrestrial uraniums appear to have the same ratio.

Sample 14163,159 has been measured more precisely by Barnes *et al.* (1972) at 7257 and 7256 atoms of $^{235}U$ per $10^6$ atoms $^{238}U$.

Table 2. Isotopic ratios of $^{235}U/^{238}U(\times 10^6)$

| Sample | Ratio |
|---|---|
| 10084,75 | 7233 ± 15 |
| 12070,91 | 7153 ± 20 |
| 12073,34 | 7390 ± 200 |
| 14163,135 | 7230 ± 40 |
| 14305,80,41 | 7294 ± 15 |
| Terrestrial uranium (assumed) | 7257 |

### $^{244}$*Pu*, $^{239}$*Pu*, *Uranium, and Thorium*

The results of isotopic dilution analyses for two Apollo 14 samples are shown in Table 3. Two adjacent 1/3 g fragments of 14305,80,41 (taken from an outer portion of the breccia) exhibited varying amounts of thorium: the 17.2 ppm value in fragment b leads to the unusual thorium/uranium ratio of 4.16. The thorium/uranium ratio in 14163 is 3.88.

The uranium and thorium contents are in agreement with the less precise gamma-ray analyses from LSPET (1971).

The report by Hoffman *et al.* (1971) of $8.2 \times 10^7$ y $^{244}$Pu in terrestrial bastnesite is supported by some unpublished evidence at Argonne National Laboratory for $^{244}$Pu in terrestrial gadolinite (Metta *et al.*, 1971). In each case the observed $^{244}$Pu appears to be too abundant to have survived from the amount estimated to have been present during the condensation of the solar system. Despite the absence of $^{244}$Pu in five lunar samples, the lunar surface remains a promising collector of any $^{244}$Pu that might come from accretion of extrasolar sources.

The limits to $^{239}$Pu (24410 y) in Table 3 imply that recent thermal neutron fluxes were less than 33 n/(cm$^2$ · sec) and 370 n/(cm$^2$ · sec) for samples 14305,80 and 14163.

## COMMENT

The broad survey of lunar actinides has proven productive in finding new monitors for solar flare protons. Since $^{236}$U and $^{237}$Np both arise from $^{238}$U and protons, and since their half-lives differ by a factor of 11, one can reasonably expect to extract at least relative proton exposure histories by comparing $^{237}Np/^{236}U$ ratios in different samples.

Table 3. Thorium, uranium, $^{239}$Pu, and $^{244}$Pu content in two Apollo 14 samples.

| | 14163,135 | 14305,80,41 |
|---|---|---|
| Thorium (ppm) | 13.2 ± 0.4 | 15.6 ± 0.5[a]<br>17.2 ± 0.5[b] |
| Uranium (ppm) | 3.4 ± 0.1 | 4.13 ± 0.12[b] |
| $^{239}$Pu (ppm) | $<4.5 \times 10^{-9}$ | $<4 \times 10^{-10}$ |
| $^{244}$Pu (ppm) | $4^{+1}_{-4} \times 10^{-9}$ | $<2 \times 10^{-10}$ |
| $^{238}U/^{234}U$ activity ratio | 1.01 ± 0.01 | 1.01 ± 0.02 |

[a] Fragment a.
[b] Fragment b.

Although no $^{244}Pu$ or $^{239}Pu$ have been observed, one can now consider looking for them in larger samples whose radioactive contents indicate a long history of surface exposure.

Several schemes have been developed to isolate lunar superheavy elements ($Z = 110$–$114$), should they exist. However, no unambiguous scheme for identification of such elements now looks attractive enough to warrant the expenditure of large amounts of lunar samples. Here, too, samples whose radioactive contents indicate long exposure on the lunar surface would greatly enhance the probability of finding such elements.

## REFERENCES


Armstrong T. W. and Alsmiller R. G. Jr. (1971) Calculation of cosmogenic radionuclides in the moon and comparison with Apollo measurements. *Proc. Second Lunar Sci. Conf., Geochim. Cosmochim. Acta* Suppl. 2, Vol. 2, pp. 1729–1745. MIT Press.

Barnes I. L., Carpenter B. S., Garner E. L., Gramlich J. W., Keuhner E. C., Machlan L. A., Mainethal E. J., Moody J. R., Moore L. J., Murphy T. J., Paulsen P. J., Sappenfield K. M., and Shields W. R. (1972) Isotopic abundance ratio and assay analysis of selected elements in Apollo 14 samples (abstract). In *Lunar Science—III* (editor C. Watkins), pp. 41–43, Lunar Science Institute Contr. No. 88.

Begemann F., Born W., Palme H., Vilcsek E., and Wänke H. (1972) Cosmic ray produced radio isotopes in Apollo 12 and Apollo 14. *Third Lunar Science Conf.* (verbal presentation).

Burnett D. S., Huneke J. C., Podosek F. A., Russ G. P. III, and Wasserburg G. J. (1971) Irradiation history of lunar samples. *Proc. Second Lunar Sci. Conf., Geochim. Cosmochim. Acta* Suppl. 2, Vol. 2, pp. 1671–1679. MIT Press.

Ellis Y. A. (compiler) (1971) Nuclear Data Sheets **6,** 257.

Fields P. R., Diamond H., Metta D. N., Stevens C. M., Rokop D. J., and Moreland P. E. (1970) Isotopic abundances of actinide elements in lunar material. *Proc. Apollo 11 Lunar Sci. Conf., Geochim. Cosmochim. Acta* Suppl. 1, Vol. 2, pp. 1097–1102. Pergamon.

Fields P. R., Diamond H., Metta D. N., Stevens C. M., and Rokop D. J. (1971) Isotopic abundances of actinide elements in Apollo 12 samples. *Proc. Second Lunar Sci. Conf., Geochim. Cosmochim. Acta* Suppl. 2, Vol. 2, pp. 1571–1576. MIT Press.

Finkel R. C., Arnold J. R., Imamura M., Reedy R. C., Fruchter J. S., Loosli H. H., Evans J. C., Delany A. C., and Shedlovsky J. P. (1971) Depth variation of cosmogenic nuclides in a lunar surface rock and lunar soil. *Proc. Second Lunar Sci. Conf., Geochim. Cosmochim. Acta* Suppl. 2, Vol. 2, pp. 1773–1789. MIT Press.

Hoffman D. C., Lawrence F. O., Mewherter J. L., and Rourke F. M. (1971) Detection of Plutonium-244 in Nature, *Science* **234,** 132–134.

Hubbard N. J. and Gast P. W. (1971) Chemical composition and origin of nonmare basalts. *Proc. Second Lunar Sci. Conf., Geochim. Cosmochim. Acta* Suppl. 2, Vol. 2, pp. 999–1020. MIT Press.

Kaiser K. A. and Stevens C. M. (1967) Ion-retarding lens to improve the abundance sensitivity of tandem mass spectrometers. Argonne National Laboratory Report ANL-7393.

Leforte M. (1961) Fonction d'excitation de la réaction nucléaire $^{238}U$ $(p, 3n)$ $^{236}Np$ entre 30 et 150 MeV. *Compt. Rend.* **253,** 2221.

LSPET (Lunar Sample Preliminary Examination Team) (1971) Preliminary examination of lunar samples from Apollo 14. *Science* **173,** 681–693.

McCormick G. H. and Cohen B. L. (1954) Fission and total reaction cross sections for 22-MeV protons on $^{232}Th$, $^{235}U$ and $^{238}U$. *Phys. Rev.* **96,** 722.

Metta D. N., Rokop D. J., and Stevens C. M. (1971) (unpublished).

Moreland P. E. Jr., Stevens C. M., and Wahling D. B. (1967) Semiautomatic data-collection systems for the mass spectrometer. *Rev. Sci. Inst.* **38,** 760–764.

Moreland P. E. Jr., Rokop D. J., and Stevens C. M. (1970) Observations of uranium and plutonium hydrides formed by ion-molecule reactions. *Int. J. Mass Spectroscopy and Ion Physics* **5**, 127–136.

Reedy R. C. and Arnold J. R. (1972) Interaction of solar and galactic cosmic ray particles with the moon. *J. Geophys. Res.* (to be published).

Rokop D. J., Metta D. N., and Stevens C. M. (1972) $^{236}U/^{238}U$ measurements for three terrestrial minerals and one processed ore. *Int. J. of Mass. Spect. and Ion Physics* (to be published).

Wade W. H., Gazales-Vidal J., Glass R. A., and Seaborg G. T. (1957) Spallation-fission competition in the heaviest elements: triton production. *Phys. Rev.* **107**, 1311.

Proceedings of the Third Lunar Science Conference
(Supplement 3, *Geochimica et Cosmochimica Acta*)
Vol. 2, pp. 1645–1650
The M.I.T. Press, 1972

# $^{204}$Pb in Apollo 14 samples and inferences regarding primordial Pb lunar geochemistry*

R. O. ALLEN, JR.,† S. JOVANOVIC, and G. W. REED, JR.

Argonne National Laboratory, Argonne, Illinois 60439

**Abstract**—$^{204}$Pb has been measured in Apollo 14 samples and its aqueous (pH 5–6) leachability determined. Soils contain 5–9 ppb of $^{204}$Pb with 7–35% leachable; a fragmental rock contains 1.5 ppb $^{204}$Pb with ~30% leachable. $^{208}$Pb and Bi were also measured in some samples; both were more leachable on the average than $^{204}$Pb. Primordial lead is present in two forms, as a volatile, soluble salt on surfaces and as a constituent of a metal or metal related phase.

## INTRODUCTION

PRIMORDIAL Pb IN lunar samples can be studied only via its 204-isotope. The concentration and distribution of this element are of interest because of the volatility of its compounds, their presence as sublimates during terrestrial vulcanism, the tendency of Pb to be associated with terrestrial sulfides and with meteoritic sulfides and metal. Measurement of its distribution in lunar matter may help to determine, by analogy, its lunar geochemistry.

We have used fast neutron activation analysis to measure $^{204}$Pb as a method which avoids to a great extent the uncertainties arising from blank corrections of the same order as the ppb $^{204}$Pb concentrations found. The lability of $^{204}$Pb was also studied by its leachability in aqueous (pH 5–6) solution. $^{208}$Pb and Bi were measured in a few experiments along with $^{204}$Pb.

## EXPERIMENTAL

Neutron activation analysis using fast neutrons for $^{204}$Pb and slow neutrons for $^{208}$Pb and Bi was used in this work and has been adequately described by Turkevich *et al.* (1971). All previous irradiations were in reactors; in this work fast neutrons produced by $(d, n)$ reactions at the ANL 60-in. cyclotron were used as well. The advantage in the latter irradiation technique is that the samples are considerably less radioactive; the disadvantage is that Bi levels are too low to be detected. The $^{206}\text{Pb}\,(n, \alpha)\,^{203}\text{Hg}$ reaction makes the measurement of $^{206}$Pb feasible. Short reactor irradiations subsequent to the cyclotron run permit $^{208}$Pb and Bi measurements.

Samples prepared in a $N_2$ dry-box were sealed in supersil fused silica ampoules for irradiation. After irradiation all samples were leached for 10–15 min in a hot dilute $HNO_3$ solution at pH 5–6. The leached Pb as well as that remaining in the sample were determined. After the aqueous leach the samples were rinsed with alcohol and ether; these washes were discarded. They might be expected to remove any contaminant organic lead (lead tetraethyl) adsorbed on surfaces.

Bulk and sieved fractions of soils 14163,152 and 14259,119, the soil breccia 14049,35 and fragmental rock 14321,185 were determined. In the case of the latter sample a 620 mg piece, freed of cut surfaces, was crushed and 470 mg run as a "representative" sample.

*Work sponsored by USAEC and NASA.

†Permanent address: Chemistry Dept., University of Virginia, Charlottesville, Va.

## Results

The $^{204}Pb$ concentrations in the leach and the sample are given in Table 1. $^{204}Pb$

Table 1. Lead and bismuth in Apollo 14 samples.

| | $^{204}Pb$ (ppb) | | $^{208}Pb$ (ppm) | | Bi (ppb) | |
|---|---|---|---|---|---|---|
| Sample* | Residue | Leach | Residue | Leach | Residue | Leach |
| 14163,152 | 5.6 ± 1.7 | 3.0 ± 1.4 | 2.3 ± 0.3 | 1.0 ± 0.1 | — | — |
| 14163,152 > 150μ | 2.3 ± 0.14 | 0.05 ± 0.08 | — | — | — | — |
| < 150μ | 10 ± 1 | 2.8 ± 0.5 | — | — | — | — |
| Sum of fractions | 8.2 | 2.1 | — | — | — | — |
| 14259,119 | 4.4 ± 0.5 | 0.5 ± 0.2 | — | — | — | — |
| 14259,119 | 5.1 ± 0.6 | 0.4 ± 0.2 | 1.8 ± 0.5 | 0.5 ± 0.2 | 1.7 | <0.4 |
| 14259 (119,112) | | | | | | |
| >150μ | 7.0 ± 1.4 | 4.1 ± 1.0 | — | — | — | — |
| <150μ | 5.8 ± 0.7 | 1.1 ± 0.2 | — | — | — | — |
| Sum of fractions | 6.0 | 1.6 | — | — | — | — |
| 14049,35 | 4.7 ± 0.8 | 1.3 ± 0.2 | — | — | 1.1 | 4.8 |
| 14321,185 | 1.1 ± 0.2 | 0.45 ± 0.23 | — | — | 0.9 | 0.6 |

* Sample weights were 0.3–1.0 g.

contents range from about 1.5 ppb in fragmental rock 14321,185 to 5–9 ppb in the soil samples and the friable soil breccia 14049,35. Within the experimental uncertainties, which are essentially counting statistics, replicate determinations are in good agreement. From 7–35% of the total $^{204}Pb$ in soils and 29% in rock 14321,185 are leached under the mild conditions used.

The $^{204}Pb$ concentrations in the >150 μ and <150 μ sieved fractions of soil 14163 are 2.4 and 12.8 ppb, respectively, with essentially no leachable $^{204}Pb$ in the coarser grained fraction and about 22% leachable in the finer grained fraction. In soil 14259 the >150 μ sieve fraction contains 11.1 ppb $^{204}Pb$ and the <150 μ fraction contains 6.9 ppb; in these respective size fractions 37% and 16% are leachable. The summations of $^{204}Pb$ in the sieve fractions fall statistically within the amounts in the bulk samples but are higher by ~1 ppb in both residual samples and in the leach of one. This may be due to sampling since both residual and labile Pb would not be expected to be effected by contamination. Nevertheless, the possibility of contamination should be examined.

A number of observations which argue against contamination and in support of the $^{204}Pb$ in these samples being indigenous are: (1) The almost constant $^{204}Pb$ in the replicates of soil 14259,119, including the sieved sample, run at different times, and in the soil breccia; (2) the lack of a consistent grain size dependence of the labile Pb, even the apparent absence of any labile Pb in the 14163,152 >150 μ fraction; (3) the fact that the fragmental rock sample 14321,185, freshly broken out and crushed in our $N_2$-box, has a low total $^{204}Pb$ and an amount of leachable Pb similar to the 14259,119 soil sample.

$^{208}Pb$ concentrations measured in 14163,152 and 14259,119 are given in Table 1; the residual and leachable fractions are 2.3 ± 0.3 and 1.0 ± 0.1, and 1.8 ± 0.5 and 0.5 ± 0.2 ppm, respectively.

Table 2. Literature data comparison.

| Sample | Isotope or Element | This Work | Literature |
|---|---|---|---|
| 14163 | $^{204}Pb$ (ppb) | 8.6 | 9.40,[a] 10.3,[b] 7.05[c] |
| | $^{208}Pb$ (ppm) | 3.3 | 3.45,[a] 3.3,[b] 3.44[c] |
| 14259 | $^{204}Pb$ (ppb) | 5.2 | 4.08,[a] 4.3[d] |
| | $^{208}Pb$ (ppm) | 2.3 | 2.8,[a] 2.9,[a] 2.41[d] |
| | Bi (ppb) | 2.1 | 1.8[e] |
| 14321 | $^{204}Pb$ (ppb) | 1.6 | 1.91[c] |

[a] Tatsumoto *et al.*, 1972.
[b] Silver, 1972.
[c] Barnes *et al.*, 1972.
[d] Doe and Tatsumoto, 1972.
[e] Morgan *et al.*, 1972.

The Bi contents of three of the samples are given in Table 1 and range from 1.5 to 5.9 ppb. As in the case of Pb a significant fraction of the Bi is leachable, up to 81% in soil breccia 14049,35.

Results reported here and literature values are given in Table 2.

## Discussion

Measurements of the labile and nonlabile Pb permit us to attempt some deductions concerning how primordial lead is contained in lunar material. The amounts of $^{204}Pb$ released from the samples under the mild leaching conditions are fairly uniform at 0.4 to 1.3 ppb with the exception of 14163,152 in which leachable Pb is higher but statistically within this range. Soil 14259,119 and fragmental rock 14321,185 have the least amounts of leachable $^{204}Pb$. Other investigators have observed the lability of Pb in lunar samples under sometimes more drastic conditions than employed here, these include concentrated acid leaching (Silver, 1970) and volatilization (Silver, 1970, 1972; Huey *et al.*, 1971; Golpalan *et al.*, 1970; and Doe and Tatsumoto, 1972). In the strong acid treatment Silver (1970) reports 50% of the $^{204}Pb$ leached from soil 10084 and igneous rock 10017. Huey *et al.* (1971) find 50–80% $^{204}Pb$ volatilized from Apollo 11 and 12 soil at temperatures of 400–800°C, however, blank corrections were uncertain in these experiments. In other experiments Doe and Tatsumoto (1972) and Silver (1972) have found that on heating to the relatively low temperature of 600°C amounts of $^{204}Pb$ are volatile that are about the same as reported here in the aqueous leach. The results on the readily labile $^{204}Pb$ are summarized in Table 3. The two types

Table 3. Labile $^{204}Pb$.

| Sample | Conditions for Release | Fraction $^{204}Pb$ Released | Amount $^{204}Pb$ Released (ppb) | Investigator |
|---|---|---|---|---|
| 14163 | Leached, 15 min (pH 5–6) | 35% | 3.0 | This work |
| 14049 | Leached, 15 min (pH 5–6) | 19% | 1.3 | This work |
| 14321 | Leached, 15 min (pH 5–6) | 29% | 0.45 | This work |
| 14259 | Leached, 15 min (pH 5–6) | 7–21% | 0.4–0.5 | This work |
| 14259 | Heated, 600°C | 20% | 0.66 | Doe and Tatsumoto (1972) |
| 12033 | Heated, 600°C | 27% | 0.33 | Doe and Tatsumoto (1972) |
| 12070 | Heated, 600°C | 26% | 1.44 | Doe and Tatsumoto (1972) |
| 10084 | Heated, 615°C | 15% | 1.47 | Silver (1971) |

of observations suggest that the labile $^{204}Pb$ is both soluble and volatile, possibly $PbCl_2$, a common terrestrial volcanic sublimate. If the correspondence between water-leachable and readily volatilized Pb holds for all samples then all sites appear to contain this labile $^{204}Pb$ component in comparable amounts.

A clue to the site of the nonleachable Pb may be gleaned from the metal contents of samples. The metal contents determined by Gose *et al.* (1972) and nonleachable $^{204}Pb$ in soil, breccia, and fragmental rock from this work along with estimates of nonlabile $^{204}Pb$ based on measurements by other investigators are given in Table 4. A correlation appears to exist. The Pb in 12070 measured by Doe and Tatsumoto (1972) consists of 4.56 ppb $^{204}Pb$ that is not volatilized at 600°C. The 1.5 ppb $^{204}Pb$ extracted from 10084 by concentrated $HNO_3$ (Silver, 1970) could for the most part be in the metal and is consistent with the trend. This 1.5 ppb $^{204}Pb$ may represent the indigenous (noncontamination) amount; the lowest reported $^{204}Pb$ concentrations in 10084 are 2.16, 1.95, and 2.16 ppb (Gopalan *et al.*, 1970; Tatsumoto, 1970; and Turkevich *et al.*, 1971; respectively). Even the $^{204}Pb$ in Apollo 12 igneous rocks 12021 and 12063 reflect the trend.

We conclude then that (1) some primordial lead is present in lunar material as a readily volatile, soluble sublimate, possibly $PbCl_2$, and that (2) some of it is present in solid solution in metal phases or incorporated in phases closely correlated with metal.

We have less extensive data on radiogenic lead but shall examine our results on $^{208}Pb$ as well as relevant $^{204}Pb$ data for implications concerning the evolution of lead that are consistent with or complementary to mass spectrometric isotope results.

$^{208}Pb$, which is primarily radiogenic, is relatively more leachable than it is volatile at low temperatures. Silver (1972b) reports 9% ($\sim$0.18 ppm) of the total Pb in 14163,184 is released at low (600°C) temperatures; from his isotopic data we estimate this to correspond to 0.06 ppm $^{208}Pb$. Doe and Tatsumoto (1972) report 7% of the total Pb in 14259 volatilized at 600°C corresponding to 0.17 ppm $^{208}Pb$. We observe 20–30% ($1.0 \pm 0.2$ and $0.5 \pm 0.2$ ppm for 14163,152 and 14259,117, respectively) readily (pH 5–6) leachable $^{208}Pb$. Thus, in contrast to primordial Pb only a fraction of the leachable radiogenic Pb (as $^{208}Pb$) is volatile at 600°C. The enhanced solubility relative to volatility may be due to the fact that the radiogenic Pb is incorporated in phases which are soluble but relatively nonvolatile or it may be that the radiogenic Pb is not deposited as a relatively volatile volcanic sublimate.

A grain-size correlated, readily volatilized $^{207}Pb$ component has been reported

Table 4. Nonleachable $^{204}Pb$-metal content comparison.

| Sample | 14163 | 14049 | 14321 | 12070 | 10084 | 14310 | 12063 | 12021 |
|---|---|---|---|---|---|---|---|---|
| Nonleachable $^{204}Pb$ (ppb) | 5.6[a] | 4.7[a] | 1.1[a] | 4.56[b] | 1.5[c] | < 1.0[d] | < 0.26[e] | < 0.184[e] |
| Metal[f] (wt.%) | 0.58 | 0.59 | 0.19 | 0.31 | 0.4 | 0.1 | 0.06 | 0.04 |
| $^{204}Pb$ (ppb) / Metal (wt.%) | 10 | 8 | 5 | 15 | 4 | < 10 | < 4.3 | < 4.5 |

[a] This work; [b] Doe and Tatsumoto, 1972; [c] Silver, 1970; [d] Tatsumoto *et al.*, 1972; [e] Tatsumoto *et al.*, 1971; [f] Gose *et al.*, 1972.

by Tatsumoto (1970), Tatsumoto *et al.* (1971), and Silver (1972b). We find that $^{204}$Pb is not consistently grain-size correlated. As a possible explanation for this difference we can assume that the apparent relative amounts of $^{204}$Pb on surfaces may be related to the mechanics of regolith formation. Lateral transport as discussed by Huey *et al.* (1971), for instance, can introduce material from diverse source regions each with a unique amount of surface related Pb. Soil samples then are mixtures and may or may not exhibit a labile-Pb grain-size correlation. The $^{207}$Pb, on the other hand, may have been introduced during Tatsumoto *et al.* (1971) "third event" subsequent to regolith formation; in this case the deposited volatiles would be grain-size correlated.

*Acknowledgments*—We thank the crews of the ANL 60-in. cyclotron and the BNL High Flux Beam Reactor for irradiations. Milan C. Oselka of ANL modified the cyclotron target to make irradiations feasible. We are grateful to Dr. H. R. Heydegger, University of Chicago, who suggested several constructive changes in the manuscript. Association of G. W. Reed with and the use of facilities at the Enrico Fermi Institute, University of Chicago, are gratefully acknowledged.

## References

Barnes I. L., Carpenter B. S., Garner E. L., Gramlich J. W., Kuehner E. C., Machlan L. A., Mainethal E. J., Moody J. R., Moore L. J., Murphy T. J., Paulsen P. J., Sappenfield K. M., Shields W. R. (1972) The isotopic abundance ratio and assay analysis of selected elements in Apollo 14 samples (abstract). In *Lunar Science—III* (editor C. Watkins), pp. 41–43. Lunar Science Institute Contr. No. 88.

Cliff R. A., Lee-Hu C., Wetherill G. W. (1971) Rb–Sr and U, Th–Pb measurements on Apollo 12 material. *Proc. Second Lunar Sci. Conf., Geochim. Cosmochim. Acta* Suppl. 2, Vol. 2, pp. 1493–1502. M.I.T. Press.

Doe B. R. and Tatsumoto M. (1972) Volatilized lead from Apollo 12 and 14 soils (abstract). In *Lunar Science—III* (editor C. Watkins), pp. 177–179, Lunar Science Institute Contr. No. 88.

Gopalan K., Kaushal S., Lee-Hu C., and Wetherill G. W. (1970) Rb–Sr and U, Th–Pb ages of lunar materials. *Proc. Apollo 11 Lunar Sci. Conf., Geochim. Cosmochim. Acta* Suppl. 1, Vol. 2, pp. 1195–1205. Pergamon.

Gose W. A., Pearce G. W., Strangeway D. W., Larson E. E. (1972) On the magnetic properties of lunar breccias (abstract). In *Lunar Science—III* (editor C. Watkins), pp. 332–334, Lunar Science Institute Contr. No. 88.

Huey J. M., Inochi H., Black L. P., Ostic R. G., and Kohman T. P. (1971) Lead isotopes and volatile transfer in the lunar soil. *Proc. Second Lunar Sci. Conf., Geochim. Cosmochim. Acta* Suppl. 2, Vol. 2, pp. 1547–1564. M.I.T. Press.

Morgan J. W., Laul J. C., Krähenbühl U., Ganapathy R., and Anders E. (1972) Major impacts on the moon: Chemical characterization of projectiles (abstract). In *Lunar Science—III* (editor C. Watkins), pp. 552–554, Lunar Science Institute Contr. No. 88.

Silver L. T. (1970) Uranium–thorium–lead isotopes in some Tranquility base samples and their implications for lunar history. *Proc. Apollo 11 Lunar Sci. Conf., Geochim. Cosmochim. Acta* Suppl. 1, Vol. 2, pp. 1533–1574. Pergamon.

Silver L. T. (1972a) U–Th–Pb abundances and isotopic characteristics in some Apollo 14 rocks and soils and an Apollo 15 soil (abstract). In *Lunar Science—III* (editor C. Watkins), pp. 704–706, Lunar Science Institute Contr. No. 88.

Silver L. T. (1972b) Lead volatilization and volatile transfer processes on the moon (abstract). In *Lunar Science—III* (editor C. Watkins), pp. 701–703, Lunar Science Intsitute Contr. No. 88.

Tatsumoto M. (1970) Age of the moon: An isotopic study of U–Th–Pb systematics of Apollo 11 lunar samples—II. *Proc. Apollo 11 Lunar Sci. Conf., Geochim. Cosmochim. Acta* Suppl. 2, Vol. 2, pp. 1595–1612. M.I.T. Press.

Tatsumoto M., Knight R. J., and Doe B. R. (1971) U–Th–Pb systematics of Apollo 12 lunar samples. *Proc. Second Lunar Sci. Conf., Geochim. Cosmochim. Acta* Suppl. 2, Vol. 2, pp. 1521–1546. M.I.T. Press.

Tatsumoto M., Hedge C. E., Doe B. R., and Unruh D. (1972) U–Th–Pb and Rb–Sr measurements on some Apollo 14 lunar samples (abstract). In *Lunar Science—III* (editor C. Watkins), pp. 741–743, Lunar Science Institute Contr. No. 88.

Turkevich A., Reed G. W. Jr., Heydegger H. R., Collister J. (1971) Activation analysis determination of uranium and $^{204}$Pb in Apollo 11 lunar fines. *Proc. Second Lunar Sci. Conf., Geochim. Cosmochim. Acta* Suppl. 2, Vol. 2, pp. 1565–1570. M.I.T. Press.

Proceedings of the Third Lunar Science Conference
(Supplement 3, *Geochimica et Cosmochimica Acta*)
Vol. 2, pp. 1651–1658
The M.I.T. Press, 1972

# Abundances of primordial and cosmogenic radionuclides in Apollo 14 rocks and fines

JAMES S. ELDRIDGE, G. DAVIS O'KELLEY, and K. J. NORTHCUTT
Oak Ridge National Laboratory, Oak Ridge, Tennessee 37830

**Abstract**—Potassium, thorium, uranium, $^{26}Al$, and $^{22}Na$ concentrations were determined non-destructively by gamma-ray spectrometry in a group of breccias or clastic rocks and three samples of fines from Fra Mauro. Samples investigated were rocks 14169, 14170, 14265, 14271, 14272, 14273, 14321,38, and sawdust 14321,256. Fines samples 14148, 14156, and 14149,62 from the top, middle, and bottom, respectively, of the Soil Mechanics Experiment trench were measured in the same study. These samples were collected from the Comprehensive Sample and Station G from the smooth terrain, and from Station Cl in the blocky rim deposit of Cone Crater. There is a remarkable uniformity in the primordial radioelement content in all samples, and average concentrations exceed those of samples returned by Apollo 11 and 12 missions, with the exception of 12034 and 12013.

A simple two-component mixing model developed earlier for our K/U systematics yields KREEP contents of 60–85% for Apollo 14 soils and breccias.

Sawdust from extensive cutting of heterogeneous rocks is a valuable sample matrix for many analytical determinations, but dilution by saw-wire contamination must be accurately determined.

## INTRODUCTION

APOLLO 14 SAMPLES from the Fra Mauro formation were expected to be representative of the terra or highlands region as opposed to the mare regions of Apollo 11 and 12 and Luna 16 samples. Ejecta blanket material from the Imbrium Basin, ray material from Copernicus Crater, and Fra Mauro base material excavated by the 340 m diameter Cone Crater were expected to be present at the landing site. Stratigraphic data had indicated that the Apollo 14 landing site was an area considered to be older than the Apollo 11 and 12 mare sites.

LSPET (1971) reported gamma-ray analyses on seven clastic rocks, two crystalline rocks, and five samples of fines. A striking feature of the primordial radioelement content of those samples reported by LSPET (1971) is the high average values for K, Th, and U, compared to the content of these elements in samples from the Apollo 11 and 12 sites. In addition, the fragmental rock to basaltic rock ratio (by number of rocks) was found to be approximately 9 to 1, in contrast to Apollo 11 rocks, where the ratio is 1 : 1, and Apollo 12, where the ratio is 1 : 9.

Our previous gamma-ray spectrometry studies on Apollo 11 and 12 materials have shown distinct differences in the mass ratio K/U for the earth and moon (O'Kelley *et al.*, 1970a, 1970b, 1971a). With the availability of samples from the 43 kg of returned lunar material from the Apollo 14 landing site, we continued our characterization of primordial and cosmogenic radionuclide concentrations in these highland materials.

## EXPERIMENTAL METHODS

The gamma scintillation spectrometer used for these analyses contained two NaI(Tl) detectors, each 23 cm in diameter and 13 cm long, with 10 cm pure NaI light guides. The detectors were operated in a coincidence mode along with a large plastic scintillator in anticoincidence with the NaI(Tl) detectors. Further background reduction was achieved with a massive lead shield 20 cm thick. The data acquisition system included a coincidence-anticoincidence logic circuit interfaced with a 4096-channel analyzer containing dual 12-bit analog-to-digital converters. Coincident gamma-ray interactions were recorded in a 64 × 64 channel matrix configuration. Data analysis included the use of two IBM 360/91 programs; one to preprocess the matrix data and the second to perform the quantitative radionuclide determinations by the method of least squares (O'Kelley *et al.*, 1971a; Schonfeld, 1967).

Stainless steel cans of three different diameters and heights from 2.5 to 10 cm were used as sample containers for all rock and fine samples analyzed in this study. Sample weights ranged from 22 to 1100 grams. Calibrations of the system were performed for all the samples measured in this study by constructing exact replicas of them with a separate replica for each nuclide sought (typically eight), along with a blank replica. Electrolytically reduced iron powder was used as the dispersing medium for the nuclide standards in order to match the electronic density of the lunar materials. Bulk density adjustments were made by the addition of 3-mm polyurethane spheres.

Thus, all radionuclide determinations reported in this study were obtained from least-squares analysis of single—and coincident—gamma-ray spectra of all lunar samples using a library of calibration nuclides from exact replicas measured in the same stainless steel containers used for each lunar sample. This technique removes many of the uncertainties associated with the use of absorption corrections for geometrical inconsistencies associated with inexact replicas or with the use of corrections for the stainless-steel can attenuation of gamma rays. The least-squares method of quantitative radionuclide determination used in this study permits the choice of the matrix region used for the analysis. In general, the entire folded, processed matrix yields the highest statistical accuracy, but adjustment of the region of fit helps improve the quality of the data for some samples. Determinations of thorium, uranium, and potassium are based on the use of standards containing terrestrial isotopic abundances, and radioactive equilibrium is assumed for the thorium decay series through the $^{208}$Tl daughter and for the uranium decay chain through the $^{214}$Bi daughter. Errors reported in this study are conservative estimates of overall uncertainties, including counting statistics and calibration uncertainties.

## RESULTS AND DISCUSSION

### *Primordial radioelement content of rocks and fines*

Concentrations of K, Th, U, $^{26}$Al, and $^{22}$Na for seven clastic or brecciated rocks, three sieved samples of fines, and one composite sample of sawdust from extensive rock cutting of a fragmental rock are listed in Table 1. Two features of the tabular data are apparent from a cursory examination: (1) Levels of the primordial radionuclides are higher than those of the Apollo 11 and 12 samples by a factor of as much as 10; and (2) the spread in the concentrations of these radioelements is very narrow in all samples listed.

Figure 1 is a map of the Apollo 14 landing site and shows the location of samples measured in this study. Swann *et al.* (1971) described the local geologic setting and outlined three map units traversed during the EVA periods: a smooth terrain unit on which the LM landed, slopes of a cratered ridge of the Fra Mauro formation, and the blocky rim deposit of Cone Crater. The regolith in the Fra Mauro region is estimated to range from 10 to 29 m in thickness. All of the samples measured in this study, with the exception of 14321, came from the smooth terrain geologic unit,

Table 1. Gamma-ray analyses of rocks and fines from Apollo 14. (Concentration values have been corrected for decay to 1848 hours, GMT, 5 February 1971.)

| Sample No. | Weight (g) | K* (ppm) | Th* (ppm) | U* (ppm) | $^{26}$Al (dpm/kg) | $^{22}$Na (dpm/kg) |
|---|---|---|---|---|---|---|
| | | | Clastic Rocks | | | |
| 14169,0 | 78.66 | 5500 ± 300 | 14.2 ± 0.2 | 3.9 ± 0.1 | 82 ± 6 | 54 ± 7 |
| 14170,0 | 26.34 | 5850 ± 300 | 14.9 ± 0.5 | 4.1 ± 0.1 | 88 ± 6 | 39 ± 9 |
| 14265,0 | 65.79 | 4100 ± 200 | 10.9 ± 0.6 | 3.3 ± 0.2 | 102 ± 8 | 70 ± 7 |
| 14271,0 | 96.58 | 5250 ± 250 | 15.6 ± 0.2 | 4.5 ± 0.3 | 118 ± 6 | 61 ± 5 |
| 14272,0 | 46.20 | 4500 ± 200 | 11.3 ± 0.5 | 3.3 ± 0.2 | 94 ± 6 | 78 ± 9 |
| 14273,0 | 22.40 | 4560 ± 200 | 11.7 ± 0.5 | 3.1 ± 0.2 | 73 ± 7 | 66 ± 8 |
| 14321,38 | 1100.0 | 4050 ± 220 | 12.7 ± 0.5 | 3.9 ± 0.4 | 50 ± 20 | 35 ± 20 |
| 14321.256 | 200.2 | 3900 ± 200 | 10.8 ± 0.5 | 2.9 ± 0.4 | 70 ± 7 | 42 ± 5 |
| | | | Fines of less than 1 mm | | | |
| 14148,0 | 45.3 | 4150 ± 200 | 11.4 ± 0.5 | 3.3 ± 0.2 | 130 ± 10 | 74 ± 7 |
| 14149,62 | 50.0 | 4650 ± 200 | 11.4 ± 0.5 | 3.2 ± 0.2 | 105 ± 10 | 66 ± 6 |
| 14156,46 | 100.0 | 4410 ± 200 | 11.9 ± 0.5 | 3.3 ± 0.2 | 148 ± 12 | 68 ± 7 |

* Standardization for the assay of K, Th, and U made with reference standards of terrestrial isotopic abundances. Equilibrium of Th and U decay series is also assumed.

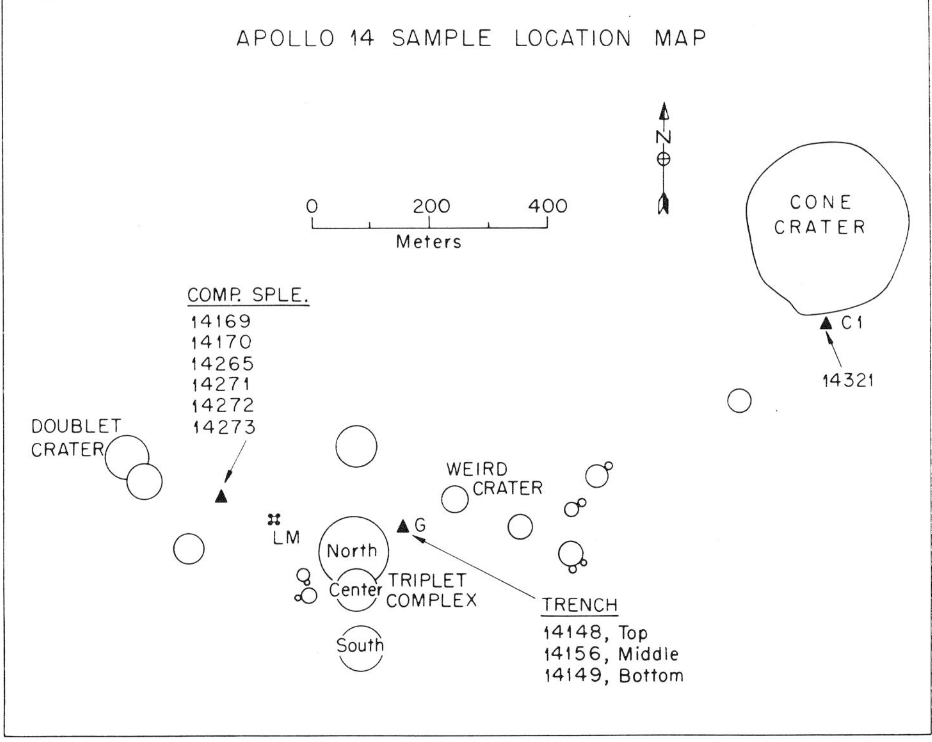


Fig. 1. Sample location map for rocks and fines described in this report. Open circles indicate craters, and sampling stations are shown as solid black triangles. The coordinates of the lunar module (LM) are 3.67°S 17.47°W.

which is densely populated with subdued crater forms several tens of meters to several hundred meters across and generally several meters to several tens of meters deep. Sample 14321 was collected at Station C1 on the second EVA and was collected from the hummocky ejecta blanket of Cone Crater. The rock was well rounded and partially buried (Swann *et al.* 1971).

Cosmic-ray exposure ages have been measured for samples 14053, 14063, 14066, 14167,8, 14305, and 14321 by LSPET (1971) and by Turner *et al.* (1971), yielding exposure ages in the range of 10 to 30 m.y. Turner *et al.* (1971) interpret the low exposure age and the sample location at the Cone Crater rim as evidence that samples 14053 and 14321 are Cone Crater ejecta; thus, the age of Cone Crater may be taken as $<30$ m.y. From the location on the rim, they conclude that sample 14321 was ejected from a depth $>10$ m.

LSPET (1971) calculated exposure ages of 10 to 20 m.y. for 14066 and 14305, whose locations were $\sim 300$ meters east and 100 meters west of the LM, respectively. Thus, if we make the assumption that rocks with exposure ages of 10–20 m.y. originated from the Cone Crater event, then it is obvious that the entire landing site is partially strewn with Cone Crater ejecta. From documentary photographs, Swann *et al.* (1971) predicted that 14305 landed in its position on the lunar surface as a result of a recent impact that produced a nearby crater and not as a result of the formation of any of the large older craters such as Cone. However, the exposure age is essentially the same as that of 14321, found partially buried on Cone Crater rim. Sample 14167,8,1, a 67 mg fragment from the comprehensive sample fines (smooth terrain unit) yielded an exposure age of 29 m.y. and was identified by Turner *et al.* (1971) as Cone Crater ejecta.

Exposure ages calculated for samples 14001, 14259, and 14310 fall in a range of 170 to 590 m.y. and are distinctively different from those of the Cone Crater age (LSPET 1971; Turner *et al.*, 1971). The primordial radionuclide content of at least two of these older exposure age samples (14310 and 14259) is high and is bracketed by the values shown in Table 1 (Keith *et al.*, 1972).

From the preceding discussion, it can be seen that the material returned from Fra Mauro Base sampled a wide area containing Cone Crater ejecta from depths $>10$ m, as well as surface materials with long exposure ages. From an analysis of our sample suite coupled with those of Keith *et al.* (1972) we can calculate an *average* thorium content of 12.9 ppm for 24 samples weighing 29 kg out of the total sample inventory of 43 kg. Thus, at least 68% by weight of the entire Fra Mauro sample collection contains $\sim 13$ ppm Th, and from the Th/U ratio of $\sim 3.6$ and the K/U ratio of $\sim 1400$, we calculate an *average* uranium content of $\sim 3.6$ ppm and an *average* potassium content of $\sim 5000$ ppm. To the extent that the Fra Mauro sample collection is considered a representative sample of the Fra Mauro formation and Imbrium Basin ejecta, we can speculate that these concentrations of K, Th, and U will be found in a wide area of the lunar surface in and around the Imbrium Basin.

Figure 2 shows a simple two-component mixing diagram fitting our K/U systematics, which were developed for Apollo 12 soils and breccias (O'Kelley *et al.*, 1971a). The solid lines are shown as previously described for the Apollo 12 samples with the additional data points added for Apollo 14 soils and breccias from this work and for

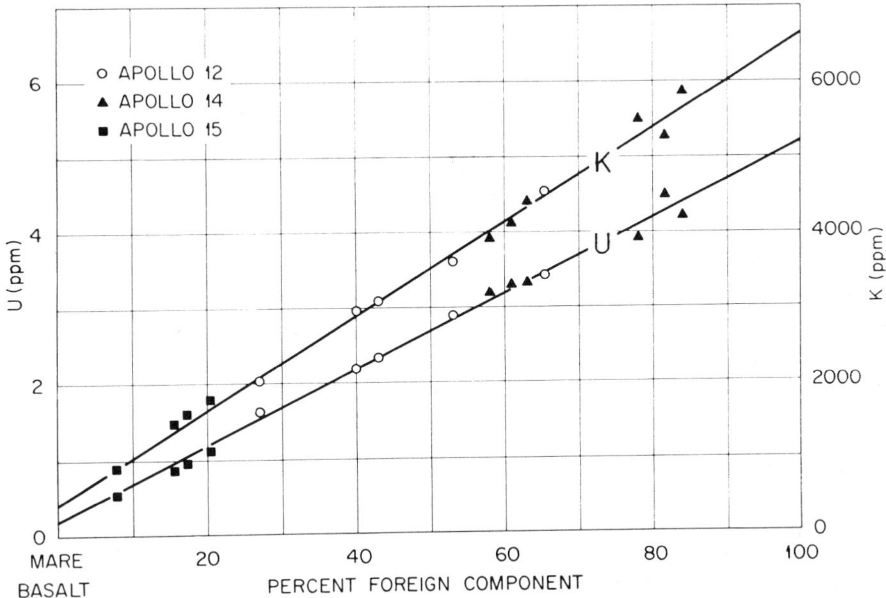


Fig. 2. Two-component mixing lines for K and U concentrations in lunar soils and breccias from Apollo 12, 14, and 15. KREEP has characteristics similar to the foreign component indicated here, with 6500 ppm K and 5.2 ppm U.

Apollo 15 samples as reported by O'Kelley *et al.* (1972a). This simple mixing diagram indicates ~60 to 85% content of foreign component (KREEP) in our collection of Apollo 14 soils and breccias. This finding is in excellent agreement with that of Schonfeld (1972), who used a 26-element linear mixing model and found a 60 to 95% KREEP content in Apollo 14 soils and breccias.

### *Rock 14321,38 and 14321,256 sawdust*

Sample 14321,38 is an 1100 g piece sawed from the north side of the 8996 g 14321 collected near Station C1 during the second EVA. The lunar orientation was well documented by lunar surface photographs (Swann *et al.*, 1971). Cutting diagrams prepared by the NASA-MSC Curator's staff indicate that section 14321,38 has a top surface area of ~60 $cm^2$ from a total top surface area of ~420 $cm^2$. Sample 14321,256 is a 200 g aliquot of ~1 kg of sawdust obtained from multiple slab- and wire-saw cuttings of rock 14321. Wrigley (1972) measured Th, U, $^{26}Al$, and $^{22}Na$ concentrations in 14321 sawdust and obtained excellent agreement with values shown in Table 1. Keith *et al.* (1972) measured radionuclide concentrations in 14321,38 and obtained good agreement with values shown in Table 1.

Showalter *et al.* (1972) measured 11 major and minor elements and 11 trace elements from two Apollo 12013 rock fragments and from a sample of 12013,17 sawdust. They found a dilution of the sawdust of 27% by contamination during the sawing process. They report that sawdust analyses should be quite valuable for overall whole rock compositions for complex materials such as the Apollo 14 clastic rocks if suitable corrections are made for dilutions by the saw-wire debris. From a comparison of our 14321 sawdust and 14321,38 rock values in Table 1, we can calculate an average dilution of the sawdust of $\sim 15\%$, based on the dilution of the thorium content from the rock.

Potassium and uranium dilutions yield different decrements, but the approximate 15% value is within the error range of the potassium and uranium dilutions. Rhodes (1972) found contamination of $\sim 15\%$ copper along with an unknown contamination of a fibrous material in the sawdust from the cutting of the breccia 14307. It is obvious that sawdust samples can be used for a variety of determinations to obtain whole rock or average concentrations for heterogeneous materials; however, the homogenized sawdust should be carefully analyzed in order to determine the type and quantity of diluents added by the cutting process.

*Cosmogenic radionuclide concentrations*

Galactic and solar cosmic rays produce many radionuclides in lunar samples by a variety of nuclear reactions. Many of these radionuclides may be determined by non-destructive gamma-ray spectrometry. In our suite of Apollo 11 and 12 samples, we determined nine radionuclides ranging in half-life from 5.7 d. ($^{52}$Mn) to 740,000 y. ($^{26}$Al) (O'Kelley *et al.*, 1970b, 1971b). The distribution schedule for Apollo 14 samples precluded any studies of short-lived radionuclides in our laboratory, since 115 days elapsed after the samples left the moon and 127 days after the intense solar flare of January 25, 1971, before we received our first sample (14321,38). In addition, the high concentrations of thorium and uranium masked the minor cosmogenic radionuclides. Concentrations of $^{26}$Al and $^{22}$Na were determined in all samples and are shown in Table 1. Cosmogenic radionuclide determinations reported here show little differences from those found in previous Apollo missions, with the exception of the three soil samples, which show unexpected results.

The Soil Mechanics Experiment trench was planned to be a 60 cm-deep trench about one crater diameter away from North Triplet Crater at Station G. The astronauts were instructed to dig the trench with one vertical sidewall to provide a means for sampling at depth. The trenching did not yield a vertical side wall; sloping occurred with walls of 60°–80°, and a maximum depth of 36 cm was achieved (Mitchell *et al.*, 1971). Samples 14148, 14149, and 14156, shown in Table 1, were taken from the top, bottom, and middle, respectively, of the trench and are all $<1$ mm sieved fractions. It was expected that there would be pronounced decreases in the concentrations of the cosmogenic species $^{26}$Al and $^{22}$Na with depth. Instead, all three samples show a surprising uniformity in concentrations of these nuclides. Calculations of $^{26}$Al and $^{22}$Na concentrations at depths of 36 cm show that sample 14149,62 should contain $\sim 40$ and $\sim 35$ dpm/kg for $^{26}$Al and $^{22}$Na, respectively (Armstrong and Alsmiller, 1971). Due to the uniform distribution of $^{26}$Al and $^{22}$Na and their high concentrations in the "bottom" sample, we must conclude that extensive mixing occurred and that sample 14149,62 is not representative of the soil at a 36 cm sampling depth. This also gives reason to question the uniformity of K, Th, and U concentrations in the different soil layers. In addition, the separation of the < 1 mm fraction from the trench bottom sample has further emphasized the sampling defect, because the bottom sample has a median grain size of 0.41 mm compared to 0.09 and 0.007 mm for the surface and middle trench samples (LSPET, 1971). Sample 14140, the 4–10 mm sieved fraction, is probably a more representative sample to characterize the trench bottom. The $^{26}$Al contents of 14150, 14151 (1–2 mm sieved fraction), and 14152 (2–4 mm sieved fraction) could be used to predict the most characteristic trench bottom sample. All three of these samples are ~11 g fractions from the trench bottom sample (Warner and Duke, 1971).

Our studies with similar trench samples from Hadley Base showed the expected decrease in $^{26}$Al and $^{22}$Na content with increasing depth in the trench (O'Kelley *et al.*, 1972a, 1972b).

## Summary

The Apollo 14 samples from Fra Mauro are unique in the uniformity of primordial radioelement content from all areas of the landing site and from samples ejected from depths ~10 m. These samples yielded average potassium, uranium, and thorium contents of 5000, 3.6, and 13 ppm, respectively. The K/U ratio of ~1400 compares favorably with that of Apollo 12 fines and breccias and to the value of 1250 predicted for KREEP (O'Kelley *et al.*, 1971a).

The K/U systematics developed for our Apollo 12 studies were used to estimate KREEP contents of Apollo 14 samples (O'Kelley *et al.*, 1971a). KREEP contents of 60–85% were determined for samples of this study.

Sawdust from cutting of lunar samples may be used for analytical purposes to obtain whole rock average concentrations of many elements if suitable corrections are made for dilution of the sawdust by saw-wire debris.

Care should be exercised in using sieved samples of fines or soils when there is a possibility that mixing of adjacent samples has occurred. This was found from trench samples collected at Station G. The bottom layer was known to be considerably more coarse-grained than the upper layers in the trench. Fine-grained material from the top layer contaminated the bottom of the trench; sieving of < 1 mm fractions from the coarse grains of the bottom thus enhanced the contamination effect.

*Acknowledgments*—The authors gratefully acknowledge contributions to the work reported here by R. S. Clark, M. B. Duke, R. E. Laughon, V. A. McKay, and E. Schonfeld. This research was carried out under Union Carbide's contract with the U.S. Atomic Energy Commission through interagency agreements with the National Aeronautics and Space Administration.

## References

Armstrong T. W. and Alsmiller R. G. Jr. (1971) Calculation of cosmogenic radionuclides in the moon and comparison with Apollo measurements. *Proc. Second Lunar Sci. Conf., Geochim. Cosmochim. Acta* Suppl. 2, Vol. 2, pp. 1729–1745. MIT Press.

Keith J. E., Clark R. S., and Richardson K. A. (1972) Gamma ray measurements of Apollo 12, 14, and 15 lunar samples (abstract). In *Lunar Science—III* (editor C. Watkins), pp. 445–448, Lunar Science Institute Contr. No. 88.

LSPET (Lunar Sample Preliminary Examination Team) (1971) Preliminary examination of lunar samples from Apollo 14. *Science* **173,** 681–693.

Mitchell J. K., Bromwell L. G., Carrier W. R. III, Costes N. C., and Scott R. F. (1971) Soil mechanics experiment. Sec. 4 of Apollo 14 Preliminary Science Report NASA SP-272, 1971.

O'Kelley G. D., Eldridge J. S., Schonfeld E., and Bell P. R. (1970a) Elemental compositions and ages of lunar samples by nondestructive gamma-ray spectrometry. *Science* **167,** 580–582.

O'Kelley G. D., Eldridge J. S., Schonfeld E., and Bell P. R. (1970b) Primordial radionuclide abundances, solar-proton and cosmic-ray effects and ages of Apollo 11 lunar samples by non-destructive gamma-ray spectrometry. *Proc. Apollo 11 Lunar Sci. Conf., Geochim. Cosmochim. Acta* Suppl. 1, Vol. 2, pp. 1407–1423. Pergamon.

O'Kelley G. D., Eldridge J. S., Schonfeld E., and Bell P. R. (1971a) Abundances of the primordial radionuclides K, Th, and U in Apollo 12 lunar samples by nondestructive gamma-ray spectrometry: Implications for origin of lunar soils. *Proc. Second Lunar Sci. Conf., Geochim. Cosmochim. Acta* Suppl. 2, Vol. 2, pp. 1159–1168. MIT Press.

O'Kelley G. D., Eldridge J. S., Schonfeld E., and Bell P. R. (1971b) Cosmogenic radionuclide concentrations and exposure ages of lunar samples from Apollo 12. *Proc. Second Lunar Sci. Conf., Geochim. Cosmochim. Acta* Suppl. 2, Vol. 2, pp. 1747–1755. MIT Press.

O'Kelley G. D., Eldridge J. S., Schonfeld E., and Northcutt K. J. (1972a) Concentrations of primordial radioelements and cosmogenic radionuclides in Apollo 15 samples by nondestructive gamma-ray spectrometry (abstract). In *Lunar Science—III* (editor C. Watkins), pp. 587–590, Lunar Science Institute Contr. No. 88.

O'Kelley G. D., Eldridge J. S., Schonfeld E., and Northcutt K. J. (1972b) Concentrations of primordial radioelements and cosmogenic radionuclides in Apollo 15 samples by nondestructive gamma-ray spectrometry. *Proc. Third Lunar Sci. Conf., Geochim. Cosmochim. Acta* Suppl. 3 (this volume).

Rhodes M. (1972) Private communication of unpublished results, NASA-MSC.

Schonfeld E. (1967) ALPHA M—an improved computer program for determining radioisotopes by least-squares resolution of the gamma-ray spectra. *Nucl. Instrum. Methods* **52,** 177–178.

Schonfeld E. (1972) Component abundance and ages in soils and breccia (abstract). In *Lunar Science—III* (editor C. Watkins), pp. 683–685, Lunar Science Institute Contr. No. 88.

Showalter D. L., Wakita H., Smith R. H., Schmitt R. A., Gillum D. E., and Ehmann W. D. (1972) Chemical composition of sawdust from lunar rock 12013 and comparison of a Java tektite with the rock. *Science* **175,** 170–172.

Swann G. A., Bailey N. G., Batson R. J., Eggleton R. E., Hait M. H., Holt, H. E., Larson K. B., McEwen M. C., Mitchell E. D., Schaber G. G., Schafer J. P., Shepard A. B., Sutton R. L., Trask N. J., Ulrich G. E., Wilshire H. G., and Wolfe E. W. (1971) Preliminary geologic investigations of the Apollo 14 landing site. Sec. 3 of Apollo 14 Preliminary Science Report, NASA SP-272, 1971.

Turner G., Huneke J. C., Podosek F. A., and Wasserburg G. J. (1971) $^{40}Ar$–$^{39}Ar$ ages and cosmic ray exposure ages of Apollo 14 samples. *Earth Planet. Sci. Lett.* **12,** 19–35.

Warner J. L. and Duke M. B. (1971) Apollo 14 lunar sample information catalog. NASA TM X-58062, 1971.

Wrigley R. C. (1972) Radionuclides at Fra Mauro (abstract). In *Lunar Science—III* (editor C. Watkins), pp. 814–815, Lunar Science Institute Contr. No. 88.

Proceedings of the Third Lunar Science Conference
(Supplement 3, *Geochimica et Cosmochimica Acta*)
Vol. 2, pp. 1659–1670
The M.I.T. Press, 1972

# Primordial radioelements and cosmogenic radionuclides in lunar samples from Apollo 15

G. DAVIS O'KELLEY, JAMES S. ELDRIDGE, and K. J. NORTHCUTT
Oak Ridge National Laboratory,
Oak Ridge, Tennessee 37830

and

E. SCHONFELD
Manned Spacecraft Center,
Houston, Texas 77058

**Abstract**—Gamma-ray spectrometers with low background were used to determine nondestructively the concentrations of K, Th, U, and cosmogenic radionuclides in Hadley Base basalts 15016, 15475, and 15495; in breccias 15285 and 15455; and in soils 15031, 15041, 15101, and 15601. Bulk densities were determined for six of the samples. The basalts of Apollo 15 are somewhat lower in K, Th, and U (respectively, about 400, 0.54, and 0.14 ppm) than basalts of Apollo 12 or the low-K basalts of Apollo 11 (about 520, 0.97, and 0.25 ppm). KREEP is ubiquitous in the Apollo 15 soils and breccias and ranges from 8 to 21% in the samples studied. Samples of soil or soil breccia from the mature, relatively undisturbed mare regolith and from the Apennine Front are highest in KREEP, while a sample of soil from the edge of Hadley Rille was significantly lower.

Two rocks have concentrations of $^{26}Al$ less than their saturation values; 15475 and 15495 may have been ejected onto the lunar surface as recently as 0.7 m.y. and 2.0 m.y. ago, respectively. Trench samples from the LM-ALSEP site show variations in radionuclide concentrations with depth similar to those of the Apollo 12 cores. The galactic cosmic-ray production rate of $^{48}V$ was determined as 57 ± 11 dpm/kg Fe. The concentration of $^{56}Co$ in rock 15016 leads to the conclusion that the solar flare of 24 January 1971 was 1.91 ± 0.45 times more intense than the flare of 3 November 1969, in agreement with other radiochemical data and with preliminary satellite measurements.

## INTRODUCTION

THE APOLLO 15 LANDING marked a notable scientific advance over previous Apollo missions. The increased astronaut mobility made possible the exploration of a large and diverse area of the Hadley-Apennine region. Well-documented rocks and rock fragments with a wide range of textures were returned, together with a variety of soil samples with significant geochemical differences from the igneous rocks and from each other. Thus, the samples from the selenological structures of the region offer unique opportunities to study the detailed geochemistry of the Apollo 15 landing site.

The techniques of nondestructive gamma-ray spectrometry have proved to be very useful for scanning a large number of samples to determine the concentrations of K, Th, and U. In addition to such chemical data, it is also possible to study in some detail the irradiation history of lunar samples by measuring the concentrations of radionuclides produced in the bombardment of lunar surface material by the solar and galactic cosmic rays. Our studies and related efforts by other investigators are included in papers on lunar samples from Apollo 11 (O'Kelley *et al.*, 1970a, 1970b),

Apollo 12 (O'Kelley *et al.*, 1971a, 1971b), and Apollo 14 (Eldridge *et al.*, 1972). A preliminary account of measurements on some of the Apollo 15 samples reported here was published by O'Kelley *et al.* (1972).

Samples from the Apollo 15 manned lunar landing have special significance to nuclear geochemistry because, unlike previous missions, sampling was not closely preceded by an intense solar flare. Thus, the Apollo 15 materials could be used to determine the galactic production rates of some short-lived radionuclides that previously were detected chiefly as products of solar-proton bombardment. Because no quarantine restrictions were imposed on samples from Apollo 15, it was possible to obtain samples for time-dependent studies in our laboratory rather soon after their arrival at the Lunar Receiving Laboratory (LRL).

## Description of Samples

The suite of nine samples to be discussed here includes specimens from the principal geological structures at the Hadley-Apennine landing site. Locations of the samples are shown in Fig. 1, which is a simplified map of Hadley Base drawn from the preliminary report of the Apollo Lunar Geology Investigation Team (1972). Unlike that of previous missions, documentation is rather complete on Apollo 15 samples. For the benefit of later discussion, brief sample descriptions follow. Further details may be found in the Sample Information Catalog (LRL, 1971).

Three of the samples are mare basalts. Sample 15016 is a porphyritic, highly vesicular basalt. Our measured bulk density of 2.4 (Table 1) suggests void space of

Table 1. Concentrations of primordial radionuclides in Apollo 15 samples.*

| Sample | Weight (g) | Density (g/cm$^3$) | K (ppm) | Th (ppm) | U (ppm) | K/U mass ratio | Th/U mass ratio |
|---|---|---|---|---|---|---|---|
| | | | | Crystalline rocks | | | |
| 15016,0 | 924 | 2.4 | 374 ± 20 | 0.52 ± 0.02 | 0.15 ± 0.01 | 2493 ± 213 | 3.47 ± 0.27 |
| 15475,0 | 288 | 2.9 | 354 ± 20 | 0.40 ± 0.02 | 0.12 ± 0.01 | 2950 ± 297 | 3.33 ± 0.32 |
| 15495,0 | 909 | 2.9 | 495 ± 25 | 0.60 ± 0.03 | 0.16 ± 0.01 | 3094 ± 248 | 3.75 ± 0.30 |
| Average, crystalline rocks | | | | | | 2846 | |
| | | | | Breccias | | | |
| 15285,0 | 251 | 2.4 | 1610 ± 80 | 3.4 ± 0.1 | 0.93 ± 0.05 | 1731 ± 127 | 3.66 ± 0.22 |
| 15455,0† | 833 | | 900 ± 150 | 2.0 ± 0.3 | 0.53 ± 0.08 | 1698 ± 382 | 3.77 ± 0.80 |
| | | | | Fines | | | |
| 15031,86‡ | 100 | | 1860 ± 95 | 4.3 ± 0.2 | 1.10 ± 0.05 | 1690 ± 116 | 3.91 ± 0.25 |
| 15041,100‡ | 100 | | 1740 ± 90 | 4.0 ± 0.2 | 1.10 ± 0.05 | 1582 ± 109 | 3.64 ± 0.24 |
| 15101,1 | 116 | ~1.2 | 1484 ± 74 | 3.1 ± 0.3 | 0.86 ± 0.08 | 1725 ± 182 | 3.60 ± 0.48 |
| 15601,2 | 204 | ~1.7 | 900 ± 45 | 1.8 ± 0.2 | 0.51 ± 0.05 | 1765 ± 194 | 3.53 ± 0.52 |
| Average, fines and breccias | | | | | | 1699 | |
| Average, all samples | | | | | | | 3.63 ± 0.15 |

* Calibration for assay of K, Th, and U assumed terrestrial isotopic abundances and equilibrium of Th and U decay series.

† "Black and white" breccia. Analysis given emphasizes the dark portion; determined with Ge(Li) detector.

‡ Trench samples near ALSEP (Station 8). Sample 15031 from bottom, 15041 from top.

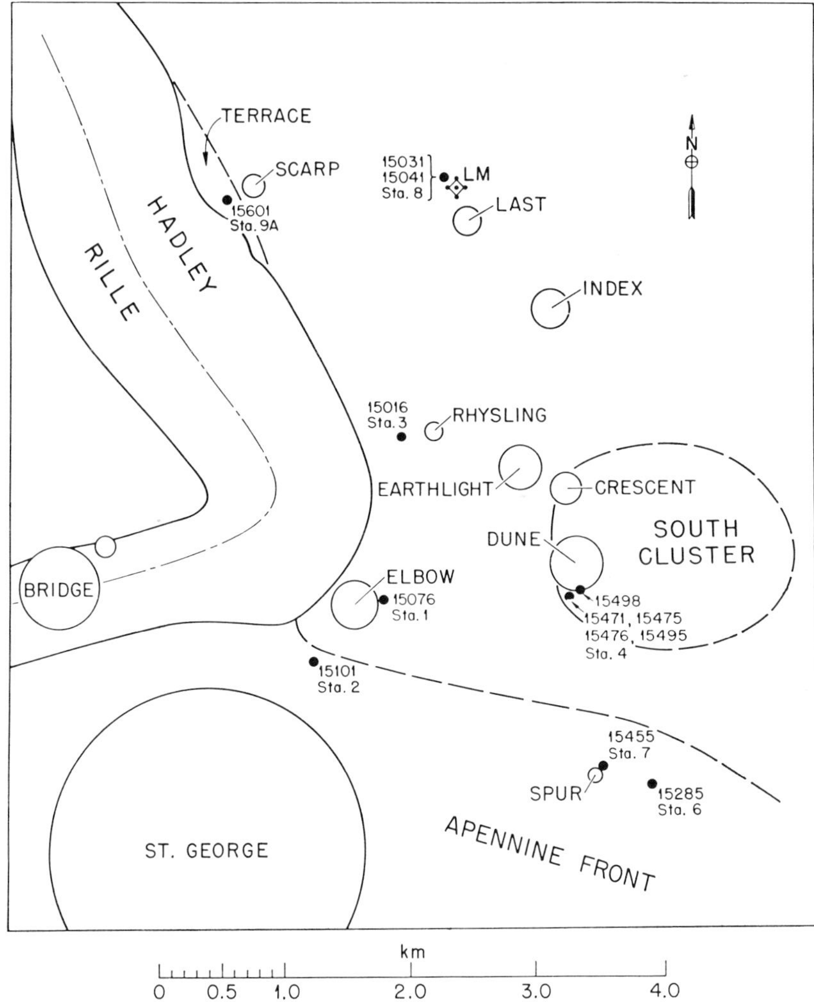


Fig. 1. Map showing original locations of samples measured in this study. Craters are shown as open circles, and sampling stations are shown as solid black circles. The coordinates of the lunar module (LM) are 26.43° N lat, 3.66° E long.

20%. Cavities are mostly vesicles, with about 2% void space as vugs. This sample can not be related clearly to the underlying bedrock. The ubiquitous porphyritic clinopyroxene basalts are typified by 15475 and 15495; sample 15475 represents the coarse basalts, while 15495, at the other extreme, may be classified as a gabbro by its grain size. Such variations suggest that these basalts may have been ejected from different depths of a thick, widespread unit that underlies the regolith at Hadley Base (LSPET, 1972).

Breccia sample 15285 from the Apennine Front is apparently very homogeneous in composition, with various amounts of glass coating on its surfaces. Its mineral content suggests a largely nonmare origin. The "black and white rock," sample 15455, is a dark breccia with white, norite clasts. It has been described by LSPET (1972).

Four soil samples were provided, all with particle sizes less than 1 mm. Photographs of the Hadley landing site show a south-trending ray that passes through the LM-ALSEP area (Station 8). Sample 15041 is from the top of the Soil Mechanics Trench in the LM-ALSEP area and is of chemical interest because of the possible presence of ray material. Sample 15031 is from the bottom of the trench, about 36 cm below the surface, and provides an opportunity to study cosmogenic radionuclide concentrations at depth. Soil sample 15101 consists of low-density fines ($\sim 1.2$ g/cm$^3$, Table 1) from an area off the ray, near St. George Crater. Sample 15601 from the "terrace" of Hadley Rille has the higher density ($\sim 1.7$ g/cm$^3$) associated with mare soils from the Apollo 12 site but shows interesting chemical differences discussed later.

## Experimental Methods

The gamma-ray spectrometer of low background used for most of the analyses reported in this study is located at Oak Ridge National Laboratory (ORNL). It consists of two large scintillation detectors at 180° with the sample between them, completely surrounded by an anticoincidence mantle of plastic scintillator. Background is further reduced by enclosing detectors and anticoincidence mantle inside a large lead shield. The detector and shield system at ORNL is identical to that described by O'Kelley *et al.* (1970b) for the Houston LRL.

Spectra are recorded by a suitable data acquisition system either as noncoincident, or "singles," data if an event occurs in one detector only, or as a gamma-gamma coincidence event if a signal is produced in both detectors simultaneously. Data reduction procedures for the ORNL scintillation spectrometer system were described previously by O'Kelley *et al.* (1971a).

The analysis of rock 15455 used a high-efficiency, low-background Ge(Li) semiconductor radiation detector at the NASA Manned Spacecraft Center, Houston, Texas.

All samples were enclosed in stainless steel containers for measurement. Libraries of standard spectra for quantitative analysis with the least-squares computer program were acquired with the aid of replicas that contained accurate additions of radionuclide standards. With the exception of 15455, all calibrations were carried out with exact replicas placed inside stainless-steel containers identical to those used for the lunar samples.

The internal consistency of the experimental procedures was checked with the help of standard radionuclide test mixtures incorporated in lunar sample replicas. Amounts taken and found agreed in each case within counting statistics, usually 2 to 5%. In his evaluation of 28 analyses of Apollo 12 soil 12070, Morrison (1971) showed that our analyses of that sample (O'Kelley *et al.*, 1971a) for K, Th, and U fall within an average deviation from the mean of $\pm 2.6\%$, which is smaller than our reported errors. Thus, the experimental method appears well established. As before, error statements given here are conservative estimates of the overall uncertainties, including counting statistics and calibration errors.

## Results and Discussion

Radionuclide concentrations of the samples described above are listed in Tables 1 and 2. Additional samples are under study in connection with specific geological problems and will be reported later. Preliminary results on these new samples are consistent with the discussion here. Except for a few exceptions noted later, agreement with other work (Keith *et al.*, 1972; LSPET, 1972; Rancitelli *et al.*, 1972; Wänke *et al.*, 1972) was obtained within the experimental errors in cases where measurements on the same samples could be compared.

Table 2. Concentrations (dpm/kg) of spallogenic radionuclides in Apollo 15 samples* (decays corrected to 1711 hours GMT, 2 August 1971).

| Sample | $^{22}Na$ | $^{26}Al$ | $^{46}Sc$ | $^{48}V$ | $^{54}Mn$ | $^{56}Co$ |
|---|---|---|---|---|---|---|
| | | | Crystalline rocks | | | |
| 15016,0 | 29 ± 2 | 82 ± 4 | 3 ± 1 | 10 ± 2 | 31 ± 4 | 16 ± 3 |
| 15475,0 | 32 ± 3 | 40 ± 3 | 3 ± 2 | — | 23 ± 3 | 11 ± 5 |
| 15495,0 | 29 ± 3 | 69 ± 3 | 3 ± 1 | trace | 25 ± 2 | 11 ± 2 |
| | | | Breccias and fines | | | |
| 15285,0 | 50 ± 4 | 85 ± 4 | 3 ± 2 | | 30 ± 5 | |
| 15031,86 | 33 ± 3 | 49 ± 3 | | | 40 ± 10 | |
| 15041,100 | 57 ± 4 | 99 ± 7 | 3 ± 2 | | 33 ± 10 | 17 ± 5 |
| 15101,1 | 44 ± 5 | 120 ± 12 | ≤4 | 9 ± 6 | 28 ± 8 | 11 ± 6 |
| 15601,2 | 55 ± 6 | 112 ± 11 | ≤4 | | 32 ± 8 | 28 ± 9 |

* Upper limits are $2\sigma$ evaluated from least-squares analysis.

### *Primordial radioelements K, Th, and U*

The primordial radioelement concentrations are listed in Table 1. A large number of samples from Apollo 11, 12, 14, and 15 have been determined by gamma-ray spectrometry, both by our group and by others. Thus, it is now possible to make some detailed comparisons between the chemistry of Hadley Base and that of previous landing sites.

Patterns of primordial radioelement distributions in the Apollo 15 materials show subtle differences from distributions in lunar material from other sites. The new data add significantly to the K, Th, and U systematics developed by O'Kelley *et al.* (1971a). Concentrations of K, Th, and U in Apollo 15 samples show trends similar to those observed for the rocks and soils of Apollo 12 but quite distinct from the distributions observed for Apollo 14 samples. The basalts of Apollo 15 are somewhat lower in K, Th, and U (respectively, about 400, 0.54, and 0.14 ppm) than the basalts of Apollo 12 or the low-K basalts of Apollo 11 (about 520, 0.97, and 0.25 ppm). Like the soils and breccias of the Apollo 12 site, the soils and breccias of Hadley Base have concentrations of K, Th, and U much higher and more variable than for the crystalline rocks; however, the amount of foreign component (KREEP) in the Apollo 15 soils and breccias is less than that for comparable Apollo 12 materials.

As we have shown earlier (O'Kelley *et al.*, 1970b, 1971a), mass ratios K/U for the moon (1000–3000) lie significantly below the terrestrial values (8000–20,000). These differences are believed to be characteristic of each planet. The relatively narrow range of values for the ratio K/U implies that this parameter was determined by early chemical fractionation and was little altered by later igneous processes. The low values of K/U are another result of the depletion of volatile elements, such as potassium (Ganapathy *et al.*, 1970), accompanied by an anomalously high content of refractory elements, such as uranium or thorium (Hubbard and Gast, 1971). This correlation between potassium and uranium concentrations also implies a similar relationship between potassium and thorium, because concentration ratios Th/U are nearly constant in lunar material (typically, 3.6–3.8). Schmitt (1972) has shown that potassium correlates similarly with hafnium, another refractory element. The most convincing

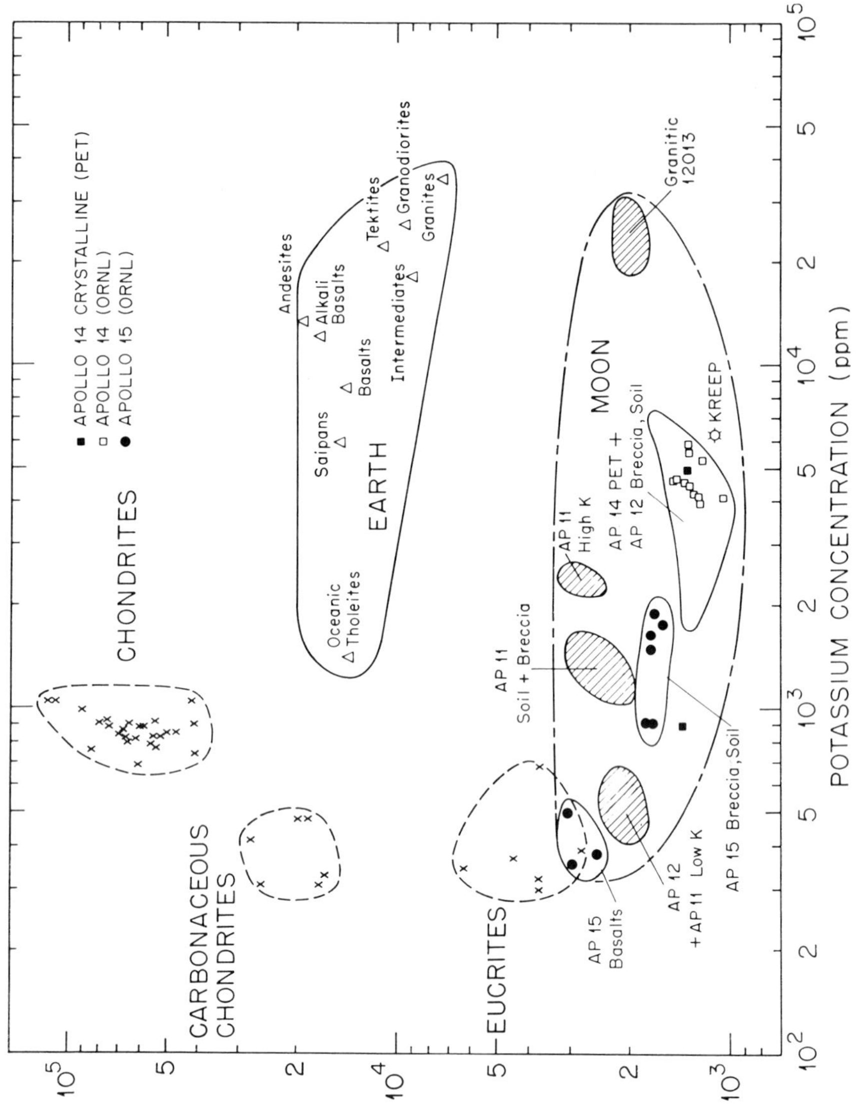


Fig. 2. Plot of concentration ratios K/U as a function of K concentration for terrestrial, meteoritic, and lunar materials. Data from this study are shown as solid circles.

interpretation of the situation is that the igneous liquids that formed the lunar samples were derived from an outer surface enriched in refractory elements and depleted in volatile elements (see Gast and McConnell, 1972).

Values of K/U from Hadley Base materials are displayed graphically in Fig. 2, which is redrawn from our systematics of Apollo 11 and Apollo 12 sample data (O'Kelley *et al.*, 1971a), with the addition of some recent results for Apollo 14 (LSPET, 1971; Eldridge *et al.*, 1972). Unlike our previous plots, the potassium concentrations of Fig. 2 were not normalized to 18% silicon.

A grouping according to sample type is apparent from Fig. 2. The points representing the data on Apollo 15 basaltic rocks overlap the eucrite zone and fall on a trend line for both meteorites and lunar materials. Data on 15085 and 15256 by Keith *et al.* (1972) and on 15556 by Rancitelli *et al.* (1972) are in good agreement with the boundaries of the Apollo 15 basalt zone. In their concentrations of potassium and uranium the Apollo 15 basalts show a remarkable similarity to the Nuevo Laredo eucrite. These gross similarities between the lunar mare basalts and the eucrites are further indications that both materials underwent similar genetic processes in the early history of the solar system (Silver and Duke, 1971).

The soils and breccias of Apollo 15 are much higher in primordial radioelements than the basalts from the same area. On a "trend line" connecting the Apollo 15 basalts and the lunar material KREEP (Meyer *et al.*, 1971) in Fig. 2, the Apollo 15 soils and breccias lie in a hitherto unfilled zone intermediate between the Apollo 12 basalts and the soils and breccias of Apollo 12 and 14. It is apparent that a simple, two-component mixing model that makes Apollo 15 soils and breccias from mare basalt as one end member requires as the other end member a material very similar to KREEP, because other materials shown in the figure would yield higher K/U ratios than those observed.

Two-component mixing diagrams that meet the requirements of our K/U systematics are shown in Fig. 3 for potassium and uranium in samples from Apollo 12 (O'Kelley *et al.*, 1971a), Apollo 14 (Eldridge *et al.*, 1972), and Apollo 15 (Table 1). Although such a two-component model is oversimplified in its neglect of anorthositic constituents, it yields the correct relative concentrations of KREEP, because anorthosites are very low in primordial radioelements and would simply act as diluents. It is seen that KREEP is ubiquitous at the Apollo 15 site but only varies from about 8 to 21%. The Apollo 12 samples show a range from 27 to 65%, but the Apollo 14 soils and breccias are very rich in KREEP, 58 to 84% for the samples shown.

Concentrations of KREEP in Hadley Base materials derived from the mixing lines of Fig. 3 compare favorably with the results of an extensive analysis by Schonfeld (1972a), who determined amounts of mare basalt, anorthosite, granitic material (light portion of 12013), ultramafic material, meteoritic components, and KREEP by use of a least-squares fit to a linear mixing model based on chemical concentrations of up to 26 elements. Schonfeld reports 18% KREEP for 15101 and 12% for 15601.

Although the breccias of the Apennine Front resemble the Fra Mauro breccias texturally, the Apennine Front breccias and soils we have measured are quite distinct chemically from their Fra Mauro counterparts. Table 1 and Fig. 3 show that the primordial radioelement and KREEP concentrations of soil 15101 and soil breccia

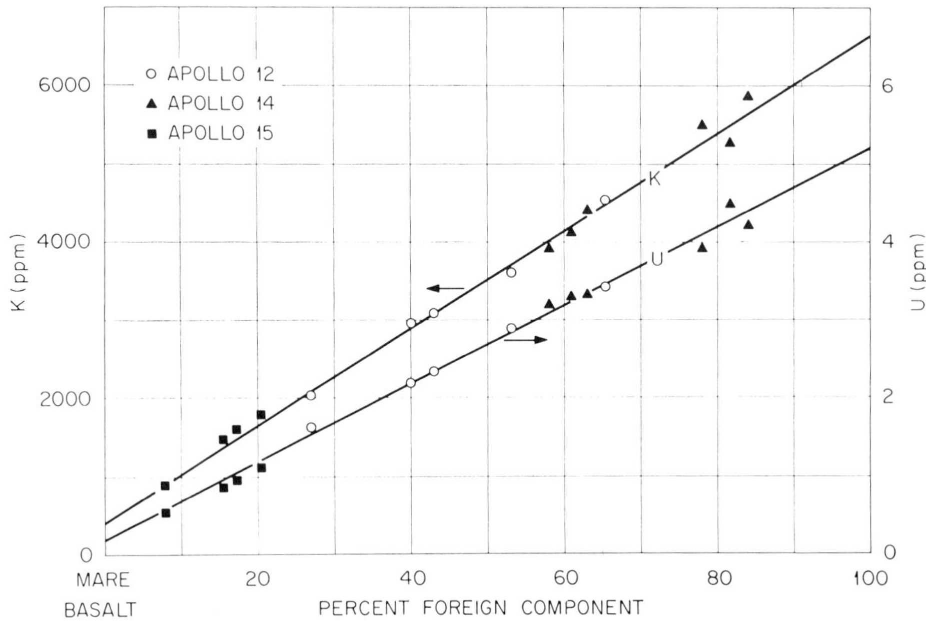


Fig. 3. Two-component mixing diagrams for K and U concentrations in lunar soils and breccias. Apollo 12 data from O'Kelley *et al.* (1971a), Apollo 14 data from Eldridge *et al.* (1972), Apollo 15 data from Table 1. Samples (and percent foreign component, or KREEP) are: 15601 and 15455 (8%), 15101 (16%), 15285 (17%), 15031 and 15041 (20%).

15285 lie well below the limits for Apollo 14 materials. This concentration of KREEP (~20%) is typical of soils and soil breccias along the Front. Further, Schonfeld (1972a) finds that these Front materials are lowest in basaltic component and highest in "anorthositic gabbro" of the materials from Hadley Base. Breccias 15455 (the "black and white" rock), 15426, and 15923 ("green clods") (Keith *et al.*, 1972; LSPET, 1972) all share a much lower concentration of primordial radioelements and, by inference, a lower concentration of KREEP. Two breccia samples (15205 and 15206) from the largest rock examined on the Front do approach the high primordial radioelement concentrations of the Apollo 14 samples (LSPET, 1972; Keith *et al.*, 1972; Rancitelli *et al.*, 1972). Such samples may prove to be Imbrium ejecta, like the Fra Mauro formation.

The mature, relatively undisturbed mare regolith appears to be high in KREEP. Soil sample 15041 from the surface at the LM-ALSEP site and 15031 from a depth of 36 cm have the highest concentrations of KREEP in our samples (~20%). The concentrations of potassium, thorium, and uranium do not show any significant variations with depth. It is not possible to associate the high concentration of KREEP at this location with ray material, although it may be significant that soil at the edge of Hadley Rille (15601, Table 1) is dramatically lower in radioelement concentrations and contains only about 8% KREEP.

The smaller amounts of potassium, thorium, and uranium in Rille soil 15601 may have originated *via* an erosion mechanism. Sample 15601 was collected only 20 m from the edge of the Hadley Rille. It is reasonable to assume that this area at the edge of the Rille was at one time covered by a layer of lunar material as rich in KREEP as the adjoining region (up to 20%) but that erosion by meteorite impacts at the margin of the Rille removed a large part of the layer. The effect of such impact erosion is much greater at an edge than on a horizontal plane, because the eroded material tends to fall away inside the Rille. It was reported that in the region under discussion, the regolith was almost absent 25 m from the edge of the Rille (Swann *et al.*, 1971). Thus, the soil of the terrace is believed to be thin, relatively young, and characteristic of the bedrock underlying this region.

### *Cosmogenic radionuclides*

The general concentration patterns for spallogenic radionuclides shown in Table 2 resemble those observed on previous Apollo missions. Because chemical analysis data are lacking for many of the samples reported here, detailed interpretations cannot be made in some cases. However, it is possible to discern important trends in the data in many cases and to calculate quantitative results in others.

The three crystalline rocks yielded especially precise results on cosmogenic species, because spectra of such weak components suffer less interference from the thorium and uranium decay series than spectra of lunar soils and breccias. The concentrations of $^{56}$Co, $^{54}$Mn, and $^{22}$Na for rocks 15016, 15475, and 15495 shown in Table 2 are close to the saturation values expected for basaltic rocks exposed to solar and galactic cosmic rays at the lunar surface. Photographic studies by Swann *et al.* (1971) noted the lack of extensive burial of these rocks. The $^{22}$Na concentration only measures exposure on about a 10-year time scale, while $^{26}$Al determines cosmic-ray exposure on a scale of a few million years. Rock 15016 appears to have a saturation value for $^{26}$Al. For 15475, the concentration of $^{26}$Al is only about one-half the saturation value; a detailed analysis indicates that this sample was ejected onto the lunar surface about $0.7 \times 10^6$ yr ago. The exposure of 15495 may also be low; if its chemical composition is the same as the other basalts in Table 2, its $^{26}$Al concentration may be consistent with an exposure as short as $2 \times 10^6$ yr.

Soil samples 15101 and 15601 appear to have been taken from relatively thin surface layers. The high concentrations of $^{22}$Na, $^{26}$Al, and $^{56}$Co are consistent with a mean sampling depth of about 3.5 cm for 15101 and a somewhat shallower depth of 2.3 cm for 15601, if it is assumed that their chemical compositions and radioactivity variations with depth are the same as found (Rancitelli *et al.*, 1971; Eldridge *et al.*, 1971) for Apollo 12 soils.

The trench samples from the LM-ALSEP area (Station 8) show the general trends expected for such samples. However, 15041 from the top of the trench seems to have been several cm thick; as in other Apollo surface samples, it seems difficult to obtain material representative of a thin surface layer. Our measurements of $^{22}$Na and $^{26}$Al in 15041 and those by Keith *et al.* (1972) and by Rancitelli *et al.* (1972) lead to mean depths of sampling of 2.9–2.1 cm. The spread in estimated sampling depth reflects a small, systematic difference in $^{22}$Na and $^{26}$Al concentrations of 15031 and 15041 between our measurements and those of Keith *et al.* (1972) and Rancitelli *et al.* (1972). The aliquots of samples measured by us were different from the aliquots measured by the other groups. Small differences in assays for thorium and uranium, together with the differences just noted, suggest that the aliquots may have been taken from samples that originally were slightly nonuniform in composition.

The radionuclide concentrations to be expected at a depth of 36 cm are not well established, but the amounts of $^{22}$Na and $^{26}$Al determined for 15031 are in approximate agreement with expectations. It is not possible to detect any difference in $^{54}$Mn content between 15041 and 15031. Calculations by Reedy and Arnold (1971) suggest a $^{54}$Mn concentration for 15041 of about 34 dpm/kg from solar and galactic cosmic-ray production, decreasing to about 30 dpm/kg for 15031, for which only galactic cosmic-ray production is significant.

During the preliminary examinations of the Apollo 11 and Apollo 12 samples in the LRL, we were able to detect two relatively short-lived radionuclides (O'Kelley *et al.*, 1970b, 1971b); 5.7-day $^{52}$Mn was determined in four rocks and 16-day $^{48}$V was determined in six rocks. Our studies showed that the concentrations of $^{48}$V were well correlated with the titanium concentrations of the rocks, as expected if most of the $^{48}$V was produced by solar-flare protons *via* the $^{48}$Ti $(p, n)$ reaction. From the $^{48}$V content of rock 12062, which appeared to have been buried, we inferred a yield from galactic proton bombardment of about 40 $\pm$ 20 dpm/kg Fe. Samples from Apollo 15 offered the best opportunity so far to determine the galactic production rate of $^{48}$V almost free from solar flare effects.

As shown in Table 2, we were able to determine $^{48}$V quantitatively in the first two samples received, 15016 and 15101. However, the results on 15016 were superior, because weak components of the gamma-ray spectra of mare basalts suffer less interference from the spectra of the thorium and uranium decay series than spectra of lunar soils and breccias. The concentration of $^{48}$V in 15016 leads to a galactic production rate for $^{48}$V of 57 $\pm$ 11 dpm/kg Fe, based on an FeO concentration of 22.6% (LSPET, 1972). This result agrees within the experimental errors with our earlier estimate, with the value of 90 $\pm$ 45 dpm/kg Fe determined by Honda and Arnold (1961) for the Aroos iron meteorite and with the value 75 $\pm$ 23 dpm/kg Fe obtained by Cressy (1970) for the St. Séverin amphoterite.

### *Solar flare of 25 January 1971*

Unlike previous Apollo manned lunar landings, no intense solar flare directly preceded the Apollo 15 mission; however, 77.3-day $^{56}$Co was detected in some of the samples. Because $^{56}$Co is almost totally produced by solar flare protons *via* the $^{56}$Fe $(p, n)$ reaction, the $^{56}$Co detected in our Apollo 15 samples was effectively produced by the intense solar flare of 24 January 1971, which preceded the collection of Apollo 14 samples. In our preliminary report on Apollo 15 samples (O'Kelley *et al.*, 1972) we compared the $^{56}$Co yields for our Apollo 15 samples with concentrations of $^{56}$Co in Apollo 12 samples and concluded that the flare of 24 January 1971 was at least 30% more intense than the well-characterized event of 3 November 1969. The

$^{56}Co$ concentration of 16 ± 3 dpm/kg for rock 15016 was corrected for iron content and for decay from 24 January 1971 and was then applied to a more careful estimate of the relative intensity of the solar flare that produced the $^{56}Co$ of the Apollo 14 and 15 samples.

The threshold of the reaction $^{56}Fe$ $(p, n)$ $^{56}Co$ is about 6 MeV, and the peak of the excitation function is about 13 MeV, so relative solar production rates for $^{56}Co$ relate rather well to solar proton fluxes of particles with energies above 10 MeV. Data on the concentration of $^{56}Co$ in rock 12002 (O'Kelley *et al.*, 1971b) were used to monitor the flare of 3 November 1969; a necessary correction for $^{56}Co$ already present from the flare of 12 April 1969 was estimated from our measurements on Apollo 11 rock 10017 (O'Kelley *et al.*, 1970a, 1970b). Differences in densities and geometries of the rocks were calculated. The new calculation concludes that the solar flare of 24 January 1971 was 1.91 ± 0.45 more intense than the flare of 3 November 1969. No absolute satellite measurements have been published; however, Bostrom (1972) obtained 2.0 for the above ratio from a preliminary analysis of satellite data. Also in excellent agreement are other radiochemical flux ratios of 1.6 ± 0.5 by Arnold *et al.* (1972), from a study of rock 14321; and 2.2 ± 0.3 by Schonfeld (1972b), from measurements on rock 14310.

*Acknowledgments*—The authors thank M. B. Duke and J. O. Annexstad for their rapid and efficient preparation of lunar material, the Lunar Sample Analysis Planning Team for their assistance and advice, and V. A. McKay and R. S. Clark for help with design and procurement of sample containers. Discussions with J. R. Arnold and P. W. Gast helped formulate our ideas. Research was carried out under Union Carbide's contract with the U.S. Atomic Energy Commission through interagency agreements with the National Aeronautics and Space Administration.

## References

Apollo Lunar Geology Investigation Team (1972) Geologic setting of the Apollo 15 samples. *Science* **175,** 407–415.

Arnold J. R., Finkel R. C., and Wahlen M. (1972) Personal communication.

Bostrom C. O. (1972) Unpublished data.

Cressy P. J. Jr. (1970) Multiparameter analysis of gamma radiation from the Barwell, St. Séverin, and Tatlith meteorites. *Geochim. Cosmochim. Acta* **34,** 771–779.

Eldridge J. S., Northcutt K. J., and O'Kelley G. D. (1971) Unpublished data.

Eldridge J. S., O'Kelley G. D., and Northcutt K. J. (1972) Abundances of primordial and cosmogenic radionuclides in Apollo 14 rocks and fines (abstract). In *Lunar Science—III* (editor C. Watkins), pp. 221–223, Lunar Science Institute Contr. No. 88.

Ganapathy R., Keays R. R., Laul J. C., and Anders E. (1970) Trace elements in Apollo 11 lunar rocks: Implications for meteorite influx and origin of moon. *Proc. Apollo 11 Lunar Sci. Conf., Geochim. Cosmochim. Acta* Suppl. 1, Vol. 2, pp. 1117–1143. Pergamon.

Gast P. W. and McConnell R. K. Jr. (1972) Evidence for initial chemical layering of the moon (abstract). In *Lunar Science—III* (editor C. Watkins), pp. 289–290, Lunar Science Institute Contr. No. 88.

Honda M. and Arnold J. R. (1961) Radioactive species produced by cosmic rays in the Aroos iron meteorite. *Geochim. Cosmochim. Acta* **23,** 219–232.

Hubbard N. J. and Gast P. W. (1971) Chemical composition and origin of nonmare lunar basalts. *Proc. Second Lunar Sci. Conf., Geochim. Cosmochim. Acta* Suppl. 2, Vol. 2, pp. 999–1020. MIT Press.

Keith J. E., Clark R. S., and Richardson K. A. (1972) Gamma ray measurements of Apollo 12,

14, and 15 lunar samples (abstract). In *Lunar Science—III* (editor C. Watkins), pp. 446–448, Lunar Science Institute Contr. No. 88.

LSPET (Lunar Sample Preliminary Examination Team) (1971) Preliminary examination of lunar samples from Apollo 14. *Science* **173,** 681–693.

LSPET (Lunar Sample Preliminary Examination Team) (1972) The Apollo 15 lunar samples: A preliminary description. *Science* **175,** 363–375.

Lunar Receiving Laboratory (1971) *Lunar Sample Information Catalog—Apollo 15.* NASA Manned Spacecraft Center Report MSC 03209.

Meyer C. Jr., Brett R., Hubbard N. J., Morrison D. A., McKay D. S., Aitken F. K., Takeda H., and Schonfeld E. (1971) Mineralogy, chemistry, and origin of the KREEP component in soil samples from the Ocean of Storms. *Proc. Second Lunar Sci. Conf., Geochim. Cosmochim. Acta* Suppl. 2, Vol. 1, pp. 393–411. MIT Press.

Morrison G. H. (1971) Evaluation of lunar elemental analyses. *Anal. Chem.* **43,** No. 7, 22A–31A.

O'Kelley G. D., Eldridge J. S., Schonfeld E., and Bell P. R. (1970a) Elemental compositions and ages of lunar samples by nondestructive gamma-ray spectrometry. *Science* **167,** 580–582.

O'Kelley G. D., Eldridge J. S., Schonfeld E., and Bell P. R. (1970b) Primordial radionuclide abundances, solar proton and cosmic-ray effects, and ages of Apollo 11 lunar samples by nondestructive gamma-ray spectrometry. *Proc. Apollo 11 Lunar Sci. Conf., Geochim. Cosmochim. Acta* Suppl. 1, Vol. 2, pp. 1407–1423. Pergamon.

O'Kelley G. D., Eldridge J. S., Schonfeld E., and Bell P. R. (1971a) Abundances of the primordial radionuclides K, Th, and U in Apollo 12 lunar samples by nondestructive gamma-ray spectrometry: Implications for origin of lunar soils. *Proc. Second Lunar Sci. Conf., Geochim. Cosmochim. Acta* Suppl. 2, Vol. 2, pp. 1159–1168. MIT Press.

O'Kelley G. D., Eldridge J. S., Schonfeld E., and Bell P. R. (1971b) Cosmogenic radionuclide concentrations and exposure ages of lunar samples from Apollo 12. *Proc. Second Lunar Sci. Conf., Geochim. Cosmochim. Acta* Suppl. 2, Vol. 2, pp. 1747–1755. MIT Press.

O'Kelley G. D., Eldridge J. S., Schonfeld E., and Northcutt K. J. (1972) Primordial radioelements and cosmogenic radionuclides in lunar samples from Apollo 15. *Science* **175,** 440–443.

Rancitelli L. A., Perkins R. W., Felix W. D., and Wogman N. A. (1972) Cosmic ray flux and lunar surface processes characterized from radionuclide measurements in Apollo 14 and 15 lunar samples (abstract). In *Lunar Science—III* (editor C. Watkins), pp. 630–632, Lunar Science Institute Contr. No. 88.

Reedy R. C. and Arnold J. R. (1971) Interaction of solar and galactic cosmic-ray particles with the moon. *J. Geophys. Res.* (in press).

Schmitt R. A. (1972) Personal communication.

Schonfeld E. (1972a) Component abundance and ages in soils and breccia (abstract). In *Lunar Science—III* (editor C. Watkins), pp. 683–685, Lunar Science Institute Contr. No. 88.

Schonfeld E. (1972b) Unpublished data.

Silver L. T. and Duke M. B. (1971) U–Th–Pb isotope relations in some basaltic achondrites (abstract). *EOS-Trans. Am. Geophys. Union* **52,** 269.

Swann G. A., Hait M. H., Schaber G. G., Freeman V. L., Ulrich G. E., Wolfe E. W., Reed V. S., and Sutton R. L. (1971) Preliminary description of Apollo 15 sample environments. U.S. Geological Survey Interagency Report 36.

Wänke H., Baddenhausen H., Balacescu A., Teschke F., Spettel B., Dreibus G., Quijano M., Kruse H., Wlotzka F., and Begemann F. (1972) Multielement analyses of lunar samples (abstract). In *Lunar Science—III* (editor C. Watkins), pp. 779–781, Lunar Science Institute Contr. No. 88.

Proceedings of the Third Lunar Science Conference
(Supplement 3, *Geochimica et Cosmochimica Acta*)
Vol. 2, pp. 1671–1680
The M.I.T. Press, 1972

# Gamma-ray measurements of Apollo 12, 14, and 15 lunar samples

J. E. Keith, R. S. Clark, and K. A. Richardson*

National Aeronautics and Space Administration, Manned Spacecraft Center,
Houston, Texas 77058

**Abstract**—Two Apollc 12 samples, eighteen Apollo 14 samples, and twenty-four Apollo 15 samples were analyzed by gamma ray spectroscopic methods. The measurements were made on the low level NaI(Tl) spectrometer at the Lunar Receiving Laboratory. Potassium, U, and Th levels in the fragmental rocks from Apollo 14 are higher (the K ranging from 0.2% to 0.7%) than those returned from the other lunar landing missions. Fragmental rocks from Apollo 15 show a wide range (in K, 86 ppm to 0.49%) in these elements. An Apollo 15 crystalline rock (15415) and an Apollo 15 breccia (15418) exhibit the lowest levels of the natural radioactivities seen in lunar materials.

Levels of cosmic ray induced radionuclides reflect their chemical composition and their exposure history. Three Apollo 15 samples (15205, 15206, and 15085) are shown to be unsaturated in $^{26}Al$, suggesting that they have only recently been exposed on the lunar surface. One Apollo 14 and two Apollo 15 (14045, 15426, and 15431) rocks show low $^{26}Al$ activity, probably as a result of a high erosion rate. Apollo 15 samples show a lower $^{56}Co$ activity than Apollo 12 or Apollo 14 samples, largely as a consequence of the absence of solar flares since January 1971.

## Introduction

Nondestructive gamma-ray spectroscopy has been shown to be an effective means of obtaining much information about the nature of lunar materials (O'Kelley *et al.*, 1970, 1971a, 1971b; Perkins *et al.*, 1970; Rancitelli *et al.*, 1971; Wrigley and Quaide, 1970; Wrigley, 1971). The method does not compromise the samples in any way and gives average concentrations of potassium, uranium, and thorium that may not be easily obtained in any other way in the case of inhomogeneous samples. The short range (about a centimeter) of solar particles and the meter to dekameter range of galactic cosmic rays combined with a suite of radionuclides whose half-lives range from a few days to a million years combine to preserve for us a record of the motions of the lunar regolith, the erosion rate of regolith materials, and the behavior of the sun during the last million years, both as a source of energetic particles and as a modulator of even more energetic particles in cislunar space.

In this paper we present the results of the measurement of the gamma-ray spectra of 44 lunar samples and some of the inferences that may be drawn from them.

## Procedure

### *Data acquisition*

The gamma-ray spectrometer data acquisition system in the Radiation Counting Laboratory has been previously described in detail by its developers (O'Kelley *et al.*, 1970, 1971a, 1971b), and the location and facility has been described by McLane *et al.* (1967).

* Now with the Geological Survey of Canada, Ottawa, Canada.

All of the samples were counted in stainless steel containers, the most common of which has been described by O'Kelley *et al.* (1971). One large Apollo 14 sample (14321) was counted in a stainless steel container of 10.2 cm height and 20.4 cm diameter. An additional type, the O'Kelley container, was employed for the measurement of some Apollo 15 samples. These stainless steel containers are of 10.2 cm diameter and of 2.5 cm or 5.1 cm height.

Soil samples were placed in cylindrical aluminum containers of 7.6 cm diameter and 2.0 cm height or 9.1 cm diameter and 5.0 cm height, which were then sealed in one of the above stainless steel containers.

Apollo 15 samples were sealed in two 2-mil teflon bags before being sealed in the stainless steel containers. The stainless steel container was then sealed in another 2-mil teflon bag.

The procedures used here are essentially the same as those described by O'Kelley *et al.* (1970, 1971a, 1971b); however, calculations have been added to the data acquisition program to facilitate gain and balance control. The gain is controlled manually to within $\pm 0.2\%$ by monitoring the centroid of the $^{40}$K peak. The balance refers to the relative number of events in the two crystals due to the sample, and in order for our data reduction schemes to work in the predetermined way, the number of events in the two crystals must be equal. By varying the spacing of the samples at a constant intercrystal distance, this equality is achieved to within $\pm 3\%$. The centroid of the $^{40}$K peak, its standard deviation, the ratio of the net number of counts in the upper detector to the lower detector over our standard energy regions (that is, the $^{40}$K peak and 0.10 to 2.0 MeV), and the ratio's standard deviation are all calculated by the data acquisition program.

### *Data reduction*

The sets of standard spectra (now 32 sets of 8 or more spectra) comprising the library with which we reduce our data were obtained before the measurement of the spectra of our samples. Most of the standards whose spectra comprise this library have been described by O'Kelley *et al.* (1970, 1971a, 1971b) or are very similar to them. To choose the most appropriate set, we first limit ourselves to those standards counted at the same gain and intercrystal spacing and choose that standard in which the ratio of the Compton backscatter area to the photopeak area for the 0.583 MeV gamma ray of $^{208}$Tl, measured in coincidence with its 2.62 MeV gamma ray, most closely approximates this ratio in the unknown spectrum. Since there is some uncertainty in this ratio, and since there may be some slight interference from other radionuclides in the unknown spectrum, several standards approximating this ratio are usually tried. The most appropriate is chosen by the examination of the residuals after data reduction.

The unknown spectrum is resolved into its components by sequentially comparing the unknown spectrum to a set of standard spectra. The bases of these comparisons are certain critical areas in the spectrum, one for each radionuclide, selected for minimum interference from other radionuclides. Those in the coincidence matrix are shown in Fig. 1. Those radionuclides emitting only one gamma ray are assigned the average of the areas of their photopeaks in the singles spectra of upper and lower detectors as their critical areas. The proportions of the standards resulting from these comparisons are subtracted from the unknown spectra and recorded, and the amount of the various radionuclides in the unknown are then calculated from the proportion subtracted, the counting times, the activity of the standard, and the sample weight. If the potassium content is greater than $0.2\%$, the order in which they are subtracted is Th, U, K, $^{26}$Al, $^{22}$Na, $^{56}$Co, $^{46}$Sc, and $^{54}$Mn. If the potassium content is less than $2\%$, it is subtracted after the $^{22}$Na.

This operation is currently being performed by a paper-tape-controlled floating point program STRIP, which can be called into the computer to operate in background while data acquisition continues in foreground. Since the program is controlled by paper-tape, it is possible to vary the order or the choice of critical regions easily (the standard areas are shown in Fig. 1) to meet changing conditions. STRIP stores the entire library, background, and unknown or its partial residual at any stage of the program on a 242 K word disc in floating point. A complete analysis currently takes about 22 minutes.

The appropriateness of the standard chosen is judged by the examination of the residuals. The quickest and most intuitively satisfying method is to look at the oscilloscope display of the residuals.

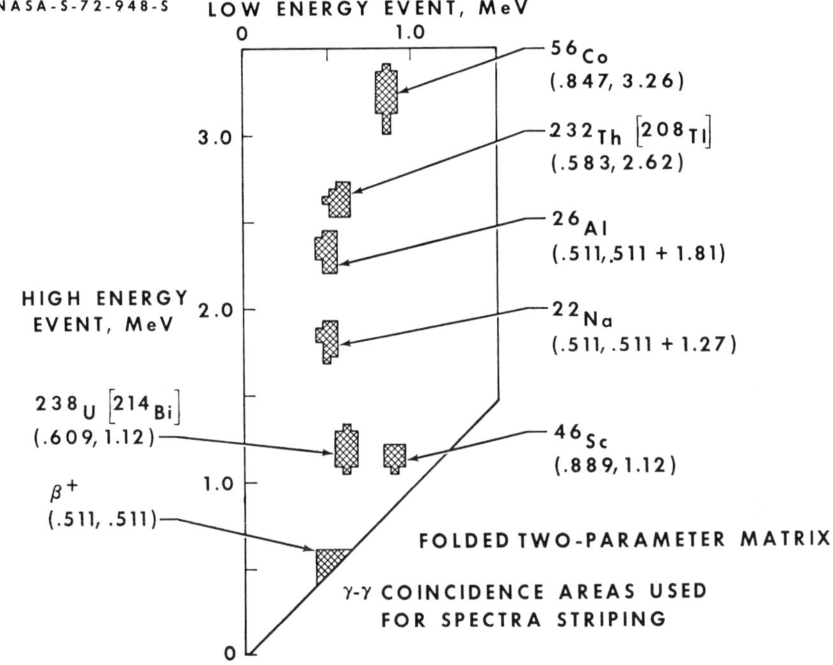


Fig. 1. Critical areas in the folded $\gamma$-$\gamma$ coincidence matrix used in STRIP. Data is taken as a 127 by 127 channel matrix folded along its minor diagonal so that in every event, the position to be incremented is formed by assigning the greater energy in the $Y$ direction, the lesser in the $X$ direction. Only part of the matrix is shown. The entire matrix is an isosceles right triangle stretching from 0 to 5.12 MeV (that is, 40 keV per channel).

A random, uncorrelated distribution symmetric about zero is persuasive, since all 8900 data points are manipulated by STRIP and appear in the residuals, and the sum of the data points in all the critical areas is a small fraction of these. The $0.511 \times 0.511$ MeV region is not used as a critical area in STRIP, but its area is compared with that due to the position emitters after stripping as a check. The algebraic sum of the residuals for each singles spectrum and the sum-coincidence spectrum is also calculated and is less than 2% of the corresponding sums for the original unknown spectrum if the standard is appropriate.

## Results

The results of these measurements and stripping of two Apollo 12 samples, eighteen Apollo 14 samples, and twenty-four Apollo 15 samples are presented in Tables 1 and 2. Those measurements made during preliminary examinations generally lasted about 2500 minutes; those made afterward generally lasted about 5000 minutes.

The sample spectra were resolved into their components by comparison with spectra obtained from various known distributed sources, as described previously. The errors listed in the tables are one standard deviation. They are derived from estimates of the errors due to counting statistics, errors in the standards, and the lack of fit due to inappropriateness of the standards. The errors due to counting

Table 1. Apollo 12 and 14 results.

| Sample | Weight (grams) | Th (ppm) | U (ppm) | K (%) | $^{26}$Al (dpm/kg) | $^{22}$Na (dpm/kg) | $^{54}$Mn (dpm/kg) | $^{56}$Co (dpm/kg) | $^{46}$Sc (dpm/kg) |
|---|---|---|---|---|---|---|---|---|---|
| | | | | Clastic Rocks | | | | | |
| 14066 | 497.5 | 15.3 ± 1.3 | 4.2 ± 0.2 | 0.72 ± 0.02 | 103 ± 6 | 43 ± 6 | 5 ± 12 | 31 ± 7 | 6 ± 3 |
| 14301 | 1370.0 | 13.2 ± 1.0 | 3.6 ± 0.5 | 0.604 ± 0.006 | 62 ± 18 | 27 ± 6 | < 33 | 8 ± 2 | 0.4 ± 0.5 |
| 14305,18 | 380.6 | 13.9 ± 1.7 | 3.8 ± 0.2 | 0.533 ± 0.010 | 74 ± 13 | 46 ± 11 | 4 ± 13 | 3 ± 6 | 0.6 ± 1.6 |
| 14318 | 600.2 | 12.0 ± 2.5 | 3.27 ± 0.14 | 0.49 ± 0.03 | 117 ± 7 | 41 ± 3 | 10 ± 11 | 28 ± 10 | 4 ± 3 |
| 14321,38 | 1100.0 | 12.7 ± 0.8 | 3.6 ± 0.2 | 0.402 ± 0.015 | 72 ± 11 | 38 ± 7 | 16 ± 6 | < 11 | < 4 |
| 14045 | 64.2 | 13.8 ± 1.3 | 3.6 ± 0.4 | 0.39 ± 0.03 | 139 ± 19 | 84 ± 9 | < 70 | 80 ± 20 | 5 ± 3 |
| 14315 | 115.0 | 8.8 ± 0.7 | 2.14 ± 0.08 | 0.328 ± 0.007 | 146 ± 16 | 58 ± 3 | < 28 | 52 ± 10 | 4 ± 3 |
| 14082 | 63.0 | 4.2 ± 0.3 | 1.24 ± 0.11 | 0.206 ± 0.009 | 120 ± 13 | 53 ± 4 | 6 ± 11 | 34 ± 10 | 1.6 ± 1.6 |
| | | | | Crystalline Rocks | | | | | |
| 14310,42 | 455.0 | 10.5 ± 0.8 | 3.0 ± 0.2 | 0.414 ± 0.013 | 97 ± 6 | 33 ± 9 | < 50 | 30 ± 30 | 1 ± 3 |
| 14053 | 251.3 | 2.29 ± 0.12 | 0.57 ± 0.05 | 0.0877 ± 0.0014 | 101 ± 4 | 57 ± 5 | 30 ± 2 | 44 ± 8 | 5 ± 1 |
| | | | | Fines | | | | | |
| 14259,8 | 496.4 | 14.4 ± 0.7 | 3.5 ± 0.3 | 0.416 ± 0.005 | 222 ± 9 | 91 ± 8 | 60 ± 20 | 60 ± 30 | 0.7 ± 1.5 |
| 14163 | 490.9 | 13.7 ± 0.7 | 3.9 ± 0.3 | 0.472 ± 0.011 | 79 ± 4 | 46 ± 5 | 4 ± 7 | 21 ± 6 | 0.7 ± 1.0 |
| 14160,11 | 100.0 | 14.2 ± 1.5 | 4.0 ± 0.5 | 0.52 ± 0.04 | 68 ± 10 | 44 ± 5 | < 60 | < 20 | 6 ± 3 |
| 14161,8 | 100.0 | 14.4 ± 1.2 | 3.9 ± 0.4 | 0.53 ± 0.08 | 73 ± 15 | 46 ± 5 | < 70 | 14 ± 19 | 4 ± 4 |
| 14162,10 | 100.0 | 14.3 ± 1.5 | 3.9 ± 0.5 | 0.52 ± 0.04 | 76 ± 9 | 49 ± 5 | 9 ± 9 | 60 ± 50 | 11 ± 5 |
| 14148 | 69.8 | 13.4 ± 1.0 | 3.7 ± 0.3 | 0.43 ± 0.02 | 170 ± 18 | 71 ± 5 | < 40 | 85 ± 16 | 1 ± 2 |
| 14156 | 136.0 | 13.9 ± 1.0 | 3.8 ± 0.3 | 0.40 ± 0.02 | 176 ± 17 | 66 ± 4 | 20 ± 20 | 64 ± 9 | 4.9 ± 1.7 |
| 14149 | 85.4 | 13.3 ± 1.0 | 3.5 ± 0.3 | 0.48 ± 0.02 | 132 ± 14 | 63 ± 4 | < 40 | 44 ± 9 | 0.1 ± 0.4 |
| | | | | Two Apollo 12 Rocks | | | | | |
| 12010 | 288.7 | 2.5 ± 0.6 | 0.60 ± 0.10 | 0.104 ± 0.013 | 83 ± 19 | 54 ± 14 | 42 ± 6 | < 70 | 5 ± 3 |
| 12031 | 185.0 | 0.94 ± 0.11 | 0.238 ± 0.013 | 0.0529 ± 0.0017 | 81 ± 5 | 54 ± 6 | 25 ± 8 | 20 ± 20 | 9 ± 3 |

*Note:* The errors listed include estimates of the errors due to counting statistics, the uncertainties of the standards, and the lack of fit in data reductions and are one standard deviation. Upper limits are three standard deviations above zero.

Table 2. Apollo 15 results.

| Sample | Weight (gms) | Th (ppm) | U (ppm) | K (%) | $^{26}$Al (dpm/kg) | $^{22}$Na (dpm/kg) | $^{54}$Mn (dpm/kg) | $^{56}$Co (dpm/kg) | $^{46}$Sc (dpm/kg) |
|---|---|---|---|---|---|---|---|---|---|
| | | | | Clastic Rocks | | | | | |
| 15206 | 85.5 | 12.0 ± 1.3 | 3.2 ± 0.4 | 0.487 ± 0.018 | 40 ± 5 | 49 ± 5 | 7 ± 8 | < 30 | < 8 |
| 15205 | 334.4 | 12.0 ± 0.4 | 2.9 ± 0.5 | 0.440 ± 0.008 | 48 ± 6 | 48 ± 4 | 50 ± 30 | < 13 | < 7 |
| 15465 | 364.9 | 5.9 ± 0.11 | 1.46 ± 0.13 | 0.234 ± 0.004 | 120 ± 30 | 56 ± 14 | 31 ± 18 | < 19 | < 5 |
| 15265 | 314.2 | 5.05 ± 0.12 | 1.27 ± 0.07 | 0.211 ± 0.08 | 72 ± 8 | 37 ± 3 | 12 ± 15 | 8 ± 6 | < 15 |
| 15558 | 1333.3 | 3.42 ± 0.18 | 1.01 ± 0.04 | 0.170 ± 0.006 | 84 ± 5 | 36 ± 5 | 23 ± 5 | 9 ± 3 | 3.0 ± 0.7 |
| 15255 | 240.4 | 3.5 ± 0.3 | 0.92 ± 0.07 | 0.156 ± 0.019 | 111 ± 7 | 43 ± 4 | 26 ± 3 | 11 ± 8 | 4 ± 4 |
| 15466 | 117.8 | 3.5 ± 0.2 | 0.86 ± 0.08 | 0.156 ± 0.004 | 79 ± 8 | 36 ± 4 | 4 ± 5 | 5 ± 4 | 0.5 ± 1.3 |
| 15086 | 172.1 | 3.2 ± 0.2 | 0.76 ± 0.03 | 0.143 ± 0.003 | 39 ± 6 | 50 ± 6 | 22 ± 6 | 11 ± 3 | 2.5 ± 0.7 |
| 15459 | 92.0 | 2.9 ± 0.4 | 0.70 ± 0.04 | 0.137 ± 0.004 | 120 ± 40 | 39 ± 3 | 16 ± 13 | 8 ± 9 | 3.9 ± 1.7 |
| 15455 | 881.1 | 2.0 ± 0.3 | 0.53 ± 0.16 | 0.106 ± 0.004 | 70 ± 30 | 42 ± 4 | 10 ± 6 | 6 ± 2 | 5 ± 11 |
| 15445 | 270.8 | 2.40 ± 0.18 | 0.63 ± 0.08 | 0.106 ± 0.014 | 81 ± 16 | 45 ± 5 | < 50 | < 18 | 0.8 ± 1.5 |
| 15426 | 125.7 | 1.89 ± 0.09 | 0.41 ± 0.02 | 0.090 ± 0.008 | 61 ± 5 | 39 ± 3 | 20 ± 20 | 7 ± 3 | 3.1 ± 1.2 |
| 15418 | 1127.5 | 0.102 ± 0.016 | 0.043 ± 0.002 | 0.0086 ± 0.0007 | 120 ± 5 | 26.6 ± 1.5 | 7.7 ± 0.9 | 1.9 ± 0.8 | 0.8 ± 0.2 |
| | | | | Crystalline Rocks | | | | | |
| 15085 | 471.0 | 0.57 ± 0.05 | 0.138 ± 0.010 | 0.0404 ± 0.0009 | 84 ± 10 | 37 ± 3 | 23 ± 3 | 12 ± 2 | 3.9 ± 1.1 |
| 15256 | 201.0 | 0.42 ± 0.04 | 0.139 ± 0.009 | 0.030 ± 0.005 | 97 ± 6 | 37 ± 3 | 25 ± 5 | 6 ± 7 | 3.6 ± 1.8 |
| 15415 | 269.4 | 0.028 ± 0.014 | 0.003 ± 0.005 | 0.0124 ± 0.0005 | 116 ± 9 | 36 ± 5 | 0.4 ± 0.9 | 3 ± 4 | < 8 |
| | | | | Fines | | | | | |
| 15021 | 132.3 | 5.0 ± 0.16 | 1.32 ± 0.04 | 0.161 ± 0.006 | 179 ± 9 | 51 ± 2 | 18 ± 6 | 15 ± 4 | 3.3 ± 1.5 |
| 15031 | 142.4 | 4.85 ± 0.12 | 1.25 ± 0.08 | 0.184 ± 0.006 | 55 ± 4 | 34 ± 3 | 10 ± 20 | < 7 | 6.8 ± 1.9 |
| 15041 | 145.7 | 4.64 ± 0.14 | 1.20 ± 0.05 | 0.174 ± 0.008 | 127 ± 8 | 61 ± 4 | 15 ± 30 | 35 ± 13 | 5.4 ± 1.9 |
| 15211 | 104.2 | 3.75 ± 0.17 | 0.98 ± 0.06 | 0.149 ± 0.002 | 130 ± 13 | 59 ± 6 | 19 ± 8 | 16 ± 4 | 3.4 ± 1.0 |
| 15271 | 527.9 | 4.1 ± 0.4 | 1.21 ± 0.04 | 0.162 ± 0.008 | 130 ± 30 | 37 ± 4 | 9 ± 6 | 5 ± 11 | 3 ± 2 |
| 15301 | 557.2 | 3.38 ± 0.19 | 0.80 ± 0.19 | 0.122 ± 0.002 | 104 ± 6 | 45 ± 6 | 22 ± 7 | < 12 | 3.6 ± 1.6 |
| 15401 | 86.3 | 3.4 ± 0.2 | 0.90 ± 0.11 | 0.143 ± 0.002 | 73 ± 13 | 58 ± 12 | 29 ± 17 | 12 ± 4 | 4.1 ± 1.3 |
| 15431 | 145.4 | 4.86 ± 0.15 | 1.12 ± 0.09 | 0.186 ± 0.005 | 66 ± 7 | 36 ± 4 | < 12 | 12 ± 12 | 3 ± 4 |

*Note:* The errors listed include estimates of the errors due to counting statistics, the uncertainties of the standards, and the lack of fit in data reductions and are one standard deviation. Upper limits are three standard deviations above zero.

statistics are calculated in the standard manner. The errors in the activities of the potassium, uranium, and thorium standards are estimated to be 1%, and all other standards, 5%. The error due to inappropriateness of the standards is in many cases the largest and is the product of the estimates of the degree of inappropriateness and the change of the value as a function of the degree of inappropriateness, as estimated by several strips with different sets of standards. In 12 cases, the sample was measured more than once. In these cases, the weighted average of the results of measurements appears in the table. In these measurements, no case of inappropriate dependence of activity on time was found. When a peak cannot be distinguished, an upper limit of three standard deviations is reported.

As an examination of the tables will reveal, high thorium contents impair the measurement of less abundant radionuclides, especially $^{54}$Mn. Since the details of the shape and thickness of the sample affect the composite thorium peak lying just above the $^{54}$Mn peak, this peak is zeroed separately in the residuals to estimate the $^{54}$Mn photopeak.

## Discussion

### *Apollo 12 samples*

Only three breccias that weighed over 50 g were returned in the Apollo 12 mission. Sample 12010 has a K content of 0.104% as compared to 0.46% and 0.3% for samples 12034 and 12073, respectively (O'Kelley *et al.*, 1971a). McKay *et al.* (1971) and Meyer *et al.* (1971) emphasize the heterogeneity of sample 12010 and the difficulty of getting a representative sample. Our value compares to 0.103% K for a piece of 12010 (Compston *et al.*, 1971), indicating that the value of McKay *et al.* (1971) of 9% KREEP (Hubbard *et al.*, 1971; Hubbard and Gast, 1971; Meyer *et al.*, 1971) based on the K mixing model is applicable to the entire rock. This is comparable to the $\leq$5% KREEP for total element mixing model used by Meyer *et al.* (1971), since the value for the Apollo 12 basalt component is 95 $\pm$ 5%.

### *Apollo 14 samples*

The K, U, and Th contents of the Apollo 14 samples are high, as compared to samples from previous missions (with the exception of sample 12013). Only two large crystalline rocks were collected during this mission, and they differ by a factor of 5 in these naturally occurring radioactive elements.

The clastic rocks show a variation of K from 0.2% to 0.7%. Most of the samples have approximately the same (0.40 to 0.53%) potassium content as the Apollo 14 soils. This is consistent with the formation of the soil mainly by breakdown of the breccias.

Various workers have calculated erosion rates of lunar rocks. Finkel *et al.* (1971) calculate the erosion rate for sample 12002 to be 0.5 mm/10$^6$ yr. Rancitelli *et al.* (1971) calculate an erosion rate of 1 mm/10$^6$ yr and suggest that the erosion rate for 12002 could be as high as 4 or 5 mm per m.y. Hörz *et al.* (1971) calculate a lower limit of 0.2 to 0.4 mm/10$^6$ yr based on pit concentrations in the surface of lunar rock. Different erosion rates can have a significant effect on the $^{26}$Al present near the surface of lunar rocks due to the steep gradient of the $^{26}Al$ (Finkel *et al.*, 1971) produced by solar protons.

With one exception, sample 14045, all of the Apollo 14 samples that we measured seem to be saturated with respect to $^{26}Al$. Sample 14045 is more friable than any Apollo 11 or Apollo 12 rock whose $^{26}Al$ concentration is known. This friability could cause the unsaturation by allowing the surface to be eroded fast enough to prevent saturation due to solar proton bombardment. Unfortunately, erosion rates of highly friable samples have not been investigated by physical methods, but considerable variation exists in the erosion rates of less friable samples (Crozaz *et al.*, 1971).

Sample 14310 has a high $^{26}Al/^{22}Na$ ratio. This can be explained by the high $Al_2O_3/MgO$ (LSPET, 1972) ratio in this particular sample as compared to the Apollo 14 basalt and the Apollo 14 clastic rocks.

It is obvious from the low levels of $^{26}Al$ and $^{22}Na$ in sample 14301 that this sample was buried to a considerable extent in the lunar regolith. Photographic location of the sample (Swann *et al.*, 1971a) shows that only a very small portion of this sample was exposed on the lunar surface and that portion had a considerable amount of dust cover.

Samples 14148, 14156, and 14149 were the $<1$ mm fraction of the top, middle, and bottom of the trench. The high values for the $^{26}Al$ and $^{22}Na$ in the middle and bottom trench samples indicate that there was considerable amount of contamination in the sampling of the trench. Cobalt-56, which can only be produced to any extent at the lunar surface by solar protons (Heydegger and Turkevich, 1971), shows this contamination and also indicates that approximately 75% of the middle trench sample and 50% of the bottom trench sample are equivalent to the top of the trench in surface exposure.

In the soil samples, the $^{26}Al$ concentrations for the $<1$ mm fines ranges from 79 to 222 dpm/kg, with the lower limit of the range representing the bulk fines and the upper limit representing the comprehensive fines that were collected from the top centimeter of the lunar surface. All of the sieve fractions of the bulk sample were analyzed, and no significant differences in the $^{26}Al$ and $^{22}Na$ were seen. This indicates that the different size fractions were evenly distributed within the depth sampled in the lunar surface.

### *Apollo 15 samples*

The Apollo 15 samples vary widely in their K, Th, and U concentrations. Two samples (15418 and 15415) contain the lowest concentrations of these elements yet found in a lunar sample. The clastic rocks seem to contain three distinct levels of naturally occurring radioactive elements, the majority of which resemble the fines, the low-K 15418, and two samples (15205 and 15206) of the same boulder, which have K, U, and Th contents more similar to Apollo 14 clastic rocks than the rest of the Apollo 15 samples.

In contrast to the trench sampling on Apollo 14, the Apollo 15 bottom trench sample (15031) was not appreciably contaminated with material from the surface. The Apollo 15 trench was about 35 to 40 cm deep (Swann *et al.*, 1971b), and no significant difference could be seen in $^{26}$Al and $^{22}$Na levels at the bottom of the trench or in levels of these nuclides seen in other lunar samples known to be buried at least a few centimeters. However, the concentrations of K found in the top and bottom of the trench, as estimated by the weighted averages of the values reported in this work and those of Rancitelli *et al.* (1972) and O'Kelley *et al.* (1972), are significantly different at the 90% confidence level and suggest vertical variations. The U and Th levels found by all three investigators also are consistent with this explanation. Layering has been shown to exist in core tubes and drill stems samples from the lunar regolith (LSPET 1970, 1971, and 1972).

The $^{22}$Na concentration in the top of the trench (15041) is significantly higher than that of the contingency sample (15021), while the $^{26}$Al concentration in 15041 is significantly lower than that of 15021. Rancitelli *et al.* (1971) showed that the $^{26}$Al gradient in the regolith is severely affected by gardening, and Swann *et al.* (1971b) has shown that an area within a meter of the contingency sampling area was disturbed by the LM landing. While no variation in depth of sampling can explain the discrepancy, it can be explained either by the removal of the top few millimeters of the regolith by the LM exhaust or by local differences in the gardening rate.

Sample 15401 is a sample of fines from the "saddle" of a large boulder sample at Station 6A. No chemical analysis for major elements is available to show if 15401 is derived by erosion from this boulder or not. Although the error limits are large, the sample appears to be grossly unsaturated in $^{26}$Al. The most likely possibility is that within the last one million years a nearby cratering event has thrown dust into the "saddle" of the boulder.

Forty percent of the samples that we analyzed were from Station 7, and a wide variety of samples were collected in this area. Rock 15415 has the lowest abundance of U and Th yet observed in any lunar sample and the lowest ratio of U/K and Th/K observed in lunar samples. The absence of Fe and Mg in this rock accounts for low activities of $^{54}$Mn, $^{56}$Co, and $^{22}$Na.

Sample 15418, a breccia with a very fine matrix, exhibits the lowest K content seen in any lunar sample. The ratios of U/K and Th/K are abnormally low compared to other lunar samples except 15415. The very low Fe and Mg contents of 15418 also account for the low levels of $^{54}$Mn, $^{56}$Co, and $^{22}$Na.

Other than the two unique samples just mentioned, the range of K for the breccia samples from Station 7 ran from 0.234% for 15465 to 0.09% for 15426.

Two samples from this area are very friable. Sample 15431 is the $<1$ mm fraction of the debris from the pedestal rock. Sample 15415 was sitting on top of this rock and is thought to be a clast from this very friable breccia (Swann *et al.*, 1971b). Samples 15431 to 15433 were portions of this breccia that were broken up during sampling and sample return. Sample 15426 represents a portion of the "green rocks." Both samples 15426 and 15431 are very friable and appear to be unsaturated in $^{26}$Al for the same reason as ascribed previously to sample 14045; that is, the erosion rate was high enough so that the $^{26}$Al could not build up to saturation on the existing surfaces.

Samples 15085 and 15086 were sampled from 60 m east of Elbow Crater. Since there are small fresh craters very close to each of these samples that are considered their own secondary craters (Swann *et al.*, 1971), it may be expected that these samples were thrown out by a common event. However, the induced radioactivity and other evidence do not support this hypothesis. The $^{26}$Al in 15086 is very low and is definitely unsaturated. The sample was probably moved from depth in the regolith in the last 200,000 to 500,000 years. Sample 15085 is apparently saturated in $^{26}$Al, so it would have to have been on the surface for at least the last few million years. Also, 15086 does not have a fillet, whereas 15085 does.

Rancitelli *et al.* (1972a, 1972b) have made an extensive study of the samples collected near the big boulder at Elbow Crater and have determined that the boulder was moved to this location about one million years ago. We also measured samples 15205 and 15206 and are in complete agreement with these data and conclusions. We also measured sample 15211 fines. This sample was thought to be derived from the boulder by erosion, but in the abundance of K, U, and Th it resembles the Apollo 15 fines and not the boulder.

## Summary

The lack of saturation of $^{26}$Al shown in three samples from Apollo 15 is probably due to the recent movement of the samples to the surface from some depth in the lunar regolith. Variation in cosmic ray induced radionuclide concentrations have been observed and shown to be due to burial, erosion, and chemical composition. Estimations of the concentration of the naturally occurring radioactive elements and cosmic ray induced radionuclides have made possible the recognition of genetic relationships among the samples. Most of these measurements were made with counting times of a few days, and survey by gamma-ray spectroscopy has been shown to be a valuable, rapid, noncompromising way to recognize interesting lunar samples.

*Acknowledgments*—Technical support in the operation and maintenance of the equipment in our laboratory was provided by Warren R. Portenier and Marshall K. Robbins of Brown and Root-Northrop. Miss Linda Bennett assisted us in the data reduction. Software for our computer systems was developed by C. Wendell Richardson and Ric C. Davies of Idaho Nuclear Corporation. Standards were prepared by Joe C. Northcutt and James S. Eldridge of Oak Ridge National Laboratory. Containers and tools used for our lunar samples were developed by Vern A. McKay of ORNL.

We wish to thank Louis A. Rancitelli, Richard W. Perkins of Battelle, Pacific Northwest Laboratories, G. Davis O'Kelley, and James S. Eldridge of Oak Ridge National Laboratory, and Donald A. Morrison of MSC for many helpful suggestions and discussions.

## References

Compston W., Berry H., Vernon M. J., Chappell B. W., and Kaye M. J. (1971) Rubidium–strontium chronology and chemistry of lunar material from the Ocean of Storms. *Proc. Second Lunar Sci. Conf., Geochim. Cosmochim. Acta* Suppl. 2, Vol. 2, pp. 1471–1485. M.I.T. Press.

Crozaz G., Walker R., and Woolum D. (1971) Nuclear track studies of dynamic surface processes on the moon and the constancy of solar activity. *Proc. Second Lunar Sci. Conf., Geochim. Cosmochim. Acta* Suppl. 2, Vol. 3, pp. 2543–2558. MIT Press.

Finkel R. C., Arnold J. R., Imamura M., Reedy R. C., Fruchter J. S., Loosli H. H., Evans J. C., and Delany A. C. (1971). Depth variation of cosmogenic nuclides in a lunar surface rock and lunar soil. *Proc. Second Lunar Sci. Conf., Geochim. Cosmochim. Acta* Suppl. 2, Vol. 2, pp. 1773–1789. MIT Press.

Heydegger H. R. and Turkevich A. (1970). Radioactivity induced in Apollo 11 lunar-surface material by solar flare protons. *Science* **168,** 575–576.

Hubbard N. J. and Gast P. W. (1971) Chemical composition and origin of nonmare lunar basalts. *Proc. Second Lunar Sci. Conf., Geochim. Cosmochim. Acta* Suppl. 2, Vol. 2, pp. 999–1020. MIT Press.

Hubbard N. J., Meyer C., Gast P. W., and Wiesmann H. (1971). The composition and derivation of Apollo 12 soils. *Earth Planet. Sci. Lett* **10,** 341–350.

LSPET (Lunar Sample Preliminary Examination Team) (1970) Preliminary examination of lunar samples from Apollo 12. *Science* **167,** 1325–1339.

LSPET (Lunar Sample Preliminary Examination Team) (1971) Preliminary examination of lunar samples from Apollo 14. *Science* **173,** 681–693.

LSPET (Lunar Sample Preliminary Examination Team) (1972) The Apollo 15 lunar samples: A preliminary description. *Science* **175,** 363–375.

McKay D. S., Morrison D. A., Clanton U. S., Ladle G. H., and Lindsay J. T. (1971) Apollo 11 soil and breccias. *Proc. Second Lunar Sci. Conf., Geochim. Cosmochim. Acta* Suppl. 2, Vol. 1, pp. 755–773. MIT Press.

McLane J. C., King E. A., Flory D. A., Richardson K. A., Dawson J. P., Kemmerer W. W., and Wooley B. C. (1967) Lunar Receiving Laboratory. *Science* **155,** 525–529.

Meyer C. Jr., Brett R., Hubbard N. J., Morrison D. A., McKay D. S., Aitken F. K., Takeda H., and Schonfeld E. (1971) Mineralogy, chemistry, and the origin of the KREEP component in soil samples from the Ocean of Storms. *Proc. Second Lunar Sci. Conf., Geochim. Cosmochim. Acta* Suppl. 2, Vol. 1, pp. 393–411. MIT Press.

O'Kelley G. D., Eldridge J. S., Schonfeld E., and Bell P. R. (1970) Primordial radionuclide abundance, solar proton and cosmic-ray effects and ages of Apollo 11 lunar samples by nondestructive gamma-ray spectrometry. *Proc. Apollo 11 Lunar Sci. Conf., Geochim. Cosmochim. Acta* Suppl. 1, Vol. 2, pp. 1407–1423. Pergamon.

O'Kelley G. D., Eldridge J. S., Schonfeld E., and Bell P. R. (1971a) Abundances of the primordial radionuclides K, Th, and U in Apollo 12 lunar samples by nondestructive gamma-ray spectrometry: Implications for origin of lunar soils. *Proc. Second Lunar Sci. Conf., Geochim. Cosmochim. Acta* Suppl. 2, Vol. 2, pp. 1159–1168. MIT Press.

O'Kelley G. D., Eldridge J. S., Schonfeld E., and Bell P. R. (1971b) Cosmogenic radionuclide concentrations and exposure ages of lunar samples from Apollo 12. *Proc. Second Lunar Sci. Conf., Geochim. Cosmochim. Acta* Suppl. 2, Vol. 2, pp. 1747–1755. MIT Press.

O'Kelley G. D., Eldridge J. S., Schonfeld E., and Northcutt K. J. (1972). Concentrations of primordial radioelements and cosmogenic radionuclides in Apollo 15 samples by nondestructive gamma-ray spectroscopy (abstract). In *Lunar Science—III* (editor C. Watkins), pp. 581–583, Lunar Science Institute Contr. No. 88.

Perkins R. W., Rancitelli L. A., Cooper J. A., Kaye J. H., and Wogman N. A. (1970) Cosmogenic and primordial radionuclide measurements in Apollo 11 lunar samples by nondestructive analysis. *Proc. Apollo 11 Lunar Sci. Conf., Geochim. Cosmochim. Acta* Suppl. 1, Vol. 2, pp. 1455–1471. Pergamon.

Rancitelli L. A., Perkins R. W., Felix W. D., and Wogman N. A. (1971) Erosion and mixing of the lunar surface from cosmogenic and primordial and radionuclide measurements in Apollo 12 lunar samples. *Proc. Second Lunar Sci. Conf., Geochim. Cosmochim. Acta* Suppl. 2, Vol. 2, pp. 1757–1772. MIT Press.

Rancitelli L. A., Perkins R. W., Felix W. D., and Wogman N. A. (1972a) Cosmic ray flux and lunar surface processes characterized from radionuclide measurements in Apollo 14 and 15 lunar samples (abstract). In *Lunar Science—III* (editor C. Watkins), pp. 631–633, Lunar Science Institute Contr. No. 88.

Rancitelli L. A., Perkins R. W., Felix W. D., and Wogman N. A. (1972b) (to be published).

Swann G. A., Bailey N. G., Batson R. M., Eggleton R. E., Hait M. H., Holt H. E., Larson K. B., McEwen M. C., Mitchell E. D., Schaber G. G., Schafer J. P., Shepard A. B., Sutton R. L., Trask N. J., Ulrich G. E., Wilshire H. G., and Wolfe E. W. (1971a) *Preliminary Geologic Investigations of the Apollo 14 Landing Site*, Interagency Report: 29, U.S. Geological Survey.

Swann G. A., Hait M. H., Schaber G. G., Freeman V. L., Ulrich G. E., Wolfe E. W., Reed V. S., and Sutton R. L. (1971b) *Preliminary Description of Apollo 15 sample environments*, Interagency Report: 36, U.S. Geological Survey.

Warner J. (1971) *A Summary of Apollo 11 Chemical, Age, and Modal Data.* Curator's Office, Manned Spacecraft Center.

Wrigley R. C. (1971) Some cosmogenic and primordial radionuclides in Apollo 12 lunar surface materials. *Proc. Second Lunar Sci. Conf., Geochim. Cosmochim. Acta* Suppl. 2, Vol. 2, pp. 1791–1796. MIT Press.

Wrigley R. C. and Quaide W. L. (1970) $Al^{26}$ and $Na^{22}$ in lunar surface materials: Implications for depth distribution studies. *Proc. Apollo 11 Lunar Sci. Conf., Geochim. Cosmochim. Acta* Suppl. 1, Vol. 2, pp. 1751–1757. Pergamon.

Proceedings of the Third Lunar Science Conference
(Supplement 3, *Geochimica et Cosmochimica Acta*)
Vol. 2, pp. 1681–1691
The M.I.T. Press, 1972

# Lunar surface processes and cosmic ray characterization from Apollo 12–15 lunar sample analyses*

L. A. RANCITELLI, R. W. PERKINS, W. D. FELIX, and N. A. WOGMAN

Battelle, Pacific Northwest Laboratories, Richland, Washington

**Abstract**—The Apollo 14 samples; 14310,187, 14321,0, and 14163,0 and the Apollo 15 samples 15205,0, 15206,0, 15211,2, 15221,2, 15231,1, 15091,15, 15261,15, 15271,19, 15031,14, 15041,14, 15501,2, 15505,0, 15535,0, 15557,1, and 15556,0 were analyzed for their cosmogenic and primordial radionuclide contents.

Samples from two stations at the Apollo 15 Hadley Apennine site were shown to have lunar surface exposure ages of less than one million years based on $^{26}$Al undersaturation. Two chips, 15205 and 15206, from a meter-sized boulder at Station 2 indicated the boulder had been ejected during a cratering incident about three-fourths of a million years ago. A soil clod from a 15 meter diameter crater and a breccia at Station 9 showed similar short exposure ages. Measurements of $^{56}$Co (77*d*) established that the proton flux above 10 MeV from the January 24, 1971 flare was about 1.4 $\times$ $10^9$ protons per cm$^2$ while $^{48}$V measurements indicated that the galactic flux is not significantly different between 1 and 2 AU. The solar flare proton energy distribution determined from the depth concentration gradients of both $^{26}$Al and $^{22}$Na is in accord, indicating constancy of the solar flare energy spectrum for the past few million years. The primordial radionuclide concentrations at the Apollo 15 site show a local range in concentrations comparable to the total concentration range from the previous Apollo landing sites.

## INTRODUCTION

IN ADDITION to the obvious benefits of direct geological lunar exploration, the lunar samples from the Apollo missions are permitting a great deal to be learned concerning the history of both the moon and the sun. The cosmogenic radionuclides reflect both the recent and long-term character of the solar and galactic cosmic ray flux, while primordial radionuclides help to describe the magmatic differentiation processes at the lunar surface. The Apollo 11 lunar sample studies provided our first look at the cosmogenic radionuclide production rates on the moon and confirmed the fact that the top centimeter of the lunar surface receives a relatively intense solar proton bombardment with a smaller galactic cosmic ray contribution (Perkins *et al.*, 1970; Shedlovsky *et al.*, 1970; O'Kelley *et al.*, 1970). Lunar samples returned by the Apollo 12 mission included core tubes which have allowed the concentration of cosmogenic radionuclides $^{26}$Al and $^{22}$Na to be characterized to depths of 40 cm, and thus provide a record on solar and galactic cosmic ray flux. Measurements on these samples have provided the basis for estimating the erosion rate of rocks, the mixing of soil, the long-term average galactic cosmic ray flux at 1 A.U. (Rancitelli *et al.*, 1971) and the recent and ancient solar cosmic ray intensity and energy spectrum (Rancitelli *et al.*, 1971;

* This paper is based on work supported by the National Aeronautics and Space Administration, Manned Spacecraft Center, Houston, Texas under Contract NAS 9-11712.

Finkel *et al.*, 1971). The Apollo 14 samples have not yet been analyzed by our laboratory to the extent of those from other missions, but have helped to confirm observations of samples from the Apollo 11, 12, and 15 missions. The Apollo 15 mission provided a remarkable collection that contains numerous unique samples (LSPET, 1972). This is due in large part to the very excellent lunar surface sample documentation (Swann *et al.*, 1971) provided by the astronauts and the fortuitous collection of material that has been shown by this present work to have a very young lunar surface exposure age. Of particular interest to the study of the character of the galactic cosmic ray flux was the fact that the last major solar flare occurred seven months before the Apollo 15 landing. This, coupled with the revocation of the quarantine requirements, permitted the galactic cosmic ray production rates of short-lived radionuclides such as $^{48}V$ to be measured directly.

Apollo 15 samples were received within 6 days of splashdown and subjected to our nondestructive gamma ray analysis (Perkins *et al.*, 1970) for both short-lived and long-lived cosmogenic radionuclides. The high degree of accuracy and precision attained in our earlier studies (Perkins *et al.*, 1970; Rancitelli *et al.*, 1971) was maintained in these analyses to permit the observation of subtle differences in radionuclide concentrations. This high precision and accuracy has been extremely helpful as it has permitted small, but very meaningful, variations of $^{26}Al$ in soil and rocks to be observed.

## Procedure

The multiple gamma coincidence counting techniques (Perkins *et al.*, 1970; Rancitelli *et al.*, 1971) described in our earlier work were used to analyze the Apollo 14 and 15 samples. Briefly, the technique involves counting the sample for 8000 to 15,000 minutes on an anticoincidence shielded multidimensional gamma-ray spectrometer. A comparison of the observed concentrations after background subtraction with the count rate of a sample mockup containing known radionuclide additions provides the basis for determining the radionuclide concentrations. The mockups are prepared from a mixture of casting resin, iron powder, and aluminum oxide that contains a precisely known radionuclide addition. These mockups reproduce the precise shape, size, physical and electron densities (of the samples). In most cases, the uncertainties associated with counting statistics were on the order of 1–2% for $^{22}Na$, $^{26}Al$, K, Th, and U while the absolute errors for these radionuclides based on all analytical uncertainties including the error in radioisotope standards ranged from 3 to 8%. The errors associated with the other radionuclides such as $^{46}Sc$, $^{48}V$, $^{56}Co$, and $^{60}Co$ are due predominantly to poor counting statistics.

## Results and Discussion

### *Radionuclide concentration and surface ages*

The summary of radionuclide concentrations in the Apollo 15 samples is contained in Tables 1 and 2. Table 1 is devoted to the unique set of samples collected at Station 2 during the first EVA, while Table 2 summarizes the samples collected at Stations 2, 6, 8, 9, and 9a. The samples collected at Station 2 include chips (15205 and

Table 1. Radionuclide content of Apollo 15 lunar samples from near St. George Crater, Station 2 (dpm/kg except as noted).

| | 15205,0 | 15206,0 | 15211,2 | 15221,1 | 15231,1 | 15091,15 |
|---|---|---|---|---|---|---|
| $^{22}$Na | 48 ± 2 | 50 ± 4 | 64 ± 3 | 72 ± 3 | 44 ± 3 | 50 ± 3 |
| $^{26}$Al | 52 ± 2 | 46 ± 4 | 152 ± 4 | 169 ± 5 | 104 ± 3 | 166 ± 5 |
| $^{46}$Sc | 3.3 ± 1.6 | <4 | <2.1 | 1.9 ± 1.9 | 2.8 ± 2.0 | 5 ± 4 |
| $^{48}$V | 6.7 ± 2.9 | — | — | 6 ± 6 | 9 ± 5 | — |
| $^{54}$Mn | 17 ± 20 | <30 | 12 ± 9 | <19 | 9 ± 7 | <40 |
| $^{56}$Co | <8 | <11 | 6 ± 6 | 5 ± 5 | <8 | <26 |
| $^{60}$Co | <6 | <3 | — | <2.7 | <3.1 | <3 |
| K (ppm) | 4680 ± 200 | 4980 ± 200 | 1440 ± 60 | 1360 ± 70 | 1410 ± 70 | 1440 ± 80 |
| Th (ppm) | 12.6 ± 0.3 | 12.4 ± 0.2 | 3.95 ± 0.08 | 3.57 ± 0.18 | 3.59 ± 0.18 | 3.97 ± 0.06 |
| U (ppm) | 3.28 ± 0.10 | 3.22 ± 0.10 | 1.02 ± 0.03 | 0.97 ± 0.03 | 0.94 ± 0.03 | 0.93 ± 0.04 |
| Sample weight (grams) | 334 | 85.5 | 104 | 96 | 96 | 96 |
| Sample type | Boulder-chips | | Boulder fillet | Soil | Soil | Soil |

15206) from the top of a meter size boulder, a soil sample from beneath the boulder (15231,1), a sample of soil adjacent to the boulder (15211,2) originally described by astronauts Scott and Irwin as a boulder fillet, and soil samples collected at about 0.5 m (15231,1) and 10 m (15091,15) from the boulder. The two boulder chips that were identified as breccias (lunar sample information catalog, Apollo 15) have high primordial radionuclide concentrations similar to the Apollo 14 breccias. The four soil samples collected at Station 2 showed little variation in their K, U, and Th concentrations, but contained only about one-third the primordial radionuclide contents of the boulder chips. The "soil fillet" (15211,2) adjacent to the boulder has primordial radionuclide concentrations which are essentially identical to the surrounding soil samples. This requires that the fillet sample originated almost entirely from soil rather than from erosion of the boulder.

The $^{22}$Na content of the two boulder chips (15205 and 15206) is at equilibrium as can be verified by comparison with the $^{22}$Na content of Apollo 11 (Perkins *et al.*, 1970), Apollo 12 (Rancitelli *et al.*, 1971) and Apollo 14 and 15 rocks. However, the $^{26}$Al content of these chips (46 to 50 dpm/kg) is considerably lower than any previously observed in the Apollo 11, 12, and 14 suites of samples. The fact that the $^{22}$Na is at saturation while the $^{26}$Al is at one-half to two-thirds of the saturation value indicates that this boulder has been in its present position for less than one million years. While firm conclusions as to the precise exposure age of the rocks must await chemical analysis data, it can be shown that the low $^{26}$Al content could not reasonably be the result of an artifact of the rock's composition. The range in $^{22}$Na and $^{26}$Al concentrations that could result from variations in chemical composition was determined by calculating production rates from solar protons using a model described in our earlier work (Rancitelli *et al.*, 1971). Basalts and breccias from the Apollo 11, 12, 14, and 15 collections with the highest and lowest reported Al, Na, Si, Ti, Fe, Mg, and Ca contents were chosen for the production rate calculations. The solar cosmic ray production rates were calculated for a sample integrated to a depth of 10 g/cm$^2$ (Rancitelli *et al.*, 1971), while galactic cosmic ray production rates through this depth were calculated by the method of Fuse and Anders (1969). These calculations show that

Table 2. Radionuclide content of Apollo 15 lunar samples from Stations 6, 8, 9, and 9a (dpm/kg except as noted).

| | 15261,15 (6) | 15271,19 (6) | 15031,14 (8)* | 15041,14 (8) | 15501,2 (9) | 15505,0 (9) | 15535,0 (9a) | 15557,1 (9a) | 15556,0 (9a) |
|---|---|---|---|---|---|---|---|---|---|
| $^{22}Na$ | 37 ± 3 | 50 ± 4 | 33 ± 2 | 65 ± 2 | 62 ± 3 | 44 ± 2 | 39 ± 2 | 39 ± 2 | 40 ± 2 |
| $^{26}Al$ | 50 ± 4 | 136 ± 3 | 60 ± 2 | 123 ± 4 | 74 ± 2 | 44 ± 2 | 61 ± 2 | 75 ± 2 | 103 ± 6 |
| $^{46}Sc$ | — | — | 6.3 ± 2.7 | 2.9 ± 1.6 | 7 ± 5 | < 3.0 | < 3.1 | 3.4 ± 1.7 | 6.5 ± 1.0 |
| $^{48}V$ | — | — | < 14 | < 11 | — | — | — | — | 12 ± 4 |
| $^{54}Mn$ | — | — | 24 ± 8 | 14 ± 8 | < 20 | 42 ± 30 | 21 ± 5 | 34 ± 9 | 41 ± 12 |
| $^{56}Co$ | — | — | 6 ± 6 | 27 ± 6 | < 31 | < 12 | < 16 | < 13 | 11 ± 4 |
| $^{60}Co$ | — | — | 4.3 ± 2.8 | < 1.5 | < 3.6 | < 1.2 | < 1.2 | < 0.8 | < 1.7 |
| K (ppm) | 1670 ± 70 | 1620 ± 70 | 1780 ± 60 | 1640 ± 60 | 1250 ± 100 | 1550 ± 70 | 490 ± 50 | 340 ± 20 | 440 ± 30 |
| Th (ppm) | 4.64 ± 0.09 | 4.87 ± 0.10 | 4.74 ± 0.12 | 4.56 ± 0.14 | 4.15 ± 0.12 | 3.64 ± 0.07 | 0.45 ± 0.03 | 0.44 ± 0.02 | 0.56 ± 0.00 |
| U (ppm) | 1.18 ± 0.04 | 1.22 ± 0.06 | 1.33 ± 0.04 | 1.28 ± 0.04 | 1.03 ± 0.03 | 0.94 ± 0.02 | 0.104 ± 0.010 | 0.131 ± 0.008 | 0.15 ± 0.01 |
| Sample weight (grams) | 104 | 104 | 142 | 146 | 46.6 | 862 | 230 | 408 | 1514 |
| Sample type | Trench soil | Soil | Trench soil | Trench soil | Soil clod | Breccia | Porphyritic basalt | Basalt | Basalt, vesicular |

* Sampling station.

even where the major target elements Al and Mg vary in concentration by a factor of 3, the production rate and thus the equilibrium ratios of $^{26}Al$ and $^{22}Na$ can only vary from about 1.4 to 2.5. Thus, the $^{26}Al$ to $^{22}Na$ ratio of 1.0 in the boulder chips 15205 and 15206 represent undersaturation of $^{26}Al$, even if the chemical composition proves to be as unusual as 12012 (low Si and high Mg), thus producing a low $^{26}Al$ to $^{22}Na$ ratio. The short exposure age of these boulder chips is further verified by comparing the $^{26}Al$ and $^{22}Na$ concentrations of the soil sample from beneath the boulder 15231,1 with those of soil samples adjacent to the boulders; 15211, 15221, and 15091. The $^{22}Na$ and $^{26}Al$ concentrations of sample 15231,1 are 44 dpm/kg and 104 dpm/kg, respectively, while those in the adjacent soil 15221 are on the order of 72 and 169 dpm/kg, respectively. Based upon the minimum shielding of 20 g/cm$^2$, subtended by the boulder over the sample 15231 (Swann *et al.*, 1971), we would expect both the $^{22}Na$ and the $^{26}Al$ concentrations to be about 45 dpm/kg. The excess $^{26}Al$ in this sample of about 60 dpm/kg would exist if the boulder had covered soil sample 15231 about three-fourths of a million years ago when the soil had a $^{26}Al$ content similar to that of the present-day nearby soil samples 15211 or 15221. During the last three-fourths million years, the $^{26}Al$ originally present would have decayed from 160 dpm/kg to 80 dpm/kg, while the galactic production of $^{26}Al$ would have reached one-half of the saturation value of 45 dpm/kg, producing a present-day total $^{26}Al$ content of about 105 dpm/kg. The short exposure age of this boulder is also supported by its physical appearance such as the sharp, well-defined features of its surface, including areas of bubbly glass. The origin of the boulder at Station 2 is evidently a fresh impact some distance from its final resting place. The region at Station 2 has been described in the Field Geology Report (Swann *et al.*, 1971) as being free of rock fragments larger than a few centimeters with the exception of two boulders. Large fragments without accompanying smaller debris are generally found beyond 30 crater diameters (Sutton, 1972). Thus, this young meter-sized boulder must be associated with a fresh crater at least a few hundred meters and more likely a kilometer or more from Station 2. A fresh crater that meets these requirements and may possibly be the source of the boulder is present on the northeast rim of St. George Crater, at a distance of about 1 kilometer from the boulder's location.

Samples collected from Station 9 (see Table 2) on the third EVA also proved to have a short exposure age as determined from $^{26}Al$ undersaturation. The soil clod 15501,2 was collected near the bench of a 15 m diameter crater which was described as having a very fresh appearance (Swann *et al.*, 1971). The rather high $^{22}Na$ content of 62 dpm/kg was consistent with that expected for a surface sample where the solar proton bombardment adds substantially to the total production of the cosmogenic radionuclides. On the basis of the $^{22}Na$ content of this soil clod, an $^{26}Al$ concentration of about 150 dpm/kg, which is similar to that in sample 15211 (see Table 1), would be predicted from our production rate model (Rancitelli *et al.*, 1971). The actual $^{26}Al$ content was only 74 dpm/kg, indicating that the sample has been exposed to solar protons for the past three-fourths million years. An alternate plausible explanation for the low $^{26}Al$ content is that this very soft soil clod was eroding at a relatively high rate. Our previous calculations (Rancitelli *et al.*, 1971) indicate that an erosion rate of greater than 3 cm per million years would be required to produce the observed

$^{22}Na$ and $^{26}Al$ concentrations. Such a high erosion rate is not supported by the very blocky appearance of the crater ejecta (Swann *et al.*, 1971). Thus, the young age, on the order of three-fourths million years, is supported. A second sample collected at Station 9, breccia 15505, also has a low $^{26}Al$ content of 44 dpm/kg and a normal $^{22}Na$ content of 49 dpm/kg, indicating extreme undersaturation of $^{26}Al$ and thus a recent lunar surface exposure age on the order of a few hundred thousand years. Although chemical analyses of these samples will permit a more accurate estimate of exposure ages, this will not change the general observation. The arguments employed in the case of boulder chips 15205 and 15206 can again be used to show that the observed $^{26}Al$ to $^{22}Na$ ratios of about 1 can only be the result of $^{26}Al$ undersaturation. Thus, the samples 15501 and 15505 together with the observations described in Swann *et al.* (1971) indicate a very recent cratering event near Station 9.

Three samples collected at Station 9a, approximately 300 meters from Station 9, were also analyzed. These samples consisted of a highly vesicular basalt 15556,0, a porphyritic basalt 15535, and a light gray basalt 15557. All of these samples contained normal amounts of $^{22}Na$ and $^{26}Al$, indicating surface exposure ages long compared with the $^{26}Al$ half-life. While the $^{22}Na$ content of all three basalt samples was essentially identical (39 dpm/kg), the highly vesicular basalt 15556,0 contained a 50% higher $^{26}Al$ concentration and an $^{26}Al$ to $^{22}Na$ ratio of 2.58. This ratio may be explained on the basis of the rock's chemical composition and the fact that it was only slightly recessed in the lunar soil with approximately 75% of its surface exposed to solar cosmic ray bombardment.

The radionuclide contents of two samples of soil collected from the top (15031,14) and bottom (15041,14) of a 35 cm deep trench at Station 8 are summarized in Table 2. The $^{26}Al$ and $^{22}Na$ contents of sample 15041 are compatible with an average sampling depth to 6 cm. Our previous observations of both rock and soil samples (Rancitelli *et al.*, 1971) have shown that where the chemical composition of the major target elements for $^{26}Al$ and $^{22}Na$ production are reasonably constant, the $^{26}Al$ to $^{22}Na$ ratio decreases by as much as twofold in the first 20 g/cm$^2$ of depth. The fact that the $^{26}Al$ to $^{22}Na$ ratio is the same at the surface as at the trench bottom requires either a very different chemical composition for the soil at the two levels, or that the soil at the top contains a very young component with an undersaturated $^{26}Al$ content. The latter hypothesis can be evaluated when the chemical analyses of these samples become available. The primordial radionuclide contents of the two soil samples are nearly identical; however, this does not insure the constancy of other elements. The approximate twofold increase in $^{54}Mn$ and $^{46}Sc$ concentrations with depth, and the fourfold decrease in the $^{56}Co$ with depth are in accord with the expected galactic cosmic ray buildup with depth and the solar cosmic ray attenuation with depth, respectively.

Two samples of soil were collected at Station 6: 15261,15 from a trench dug in a crater rim and 15271,19, a surface sample from the Lunar Rover track. Since the primordial radionuclide contents of these samples are nearly identical and they were collected in the same vicinity, it is not unreasonable to compare these samples in the same manner as the above trench samples from Station 8. The $^{26}Al$ to $^{22}Na$ ratio of the surface sample 15271 is 2.72, while the $^{26}Al$ to $^{22}Na$ ratio of the bottom trench sample 15261,15 is 1.73. This decrease in the ratio with depth is compatible with our

Table 3. Radionuclide content of Apollo 14 lunar samples (dpm/kg except as noted).

| | 14310,187 | 14321,40 | 14163,0 |
|---|---|---|---|
| $^{22}Na$ | 63 ± 5 | 32 ± 2 | 44 ± 4 |
| $^{26}Al$ | 165 ± 6 | 74 ± 4 | 89 ± 4 |
| $^{46}Sc$ | — | < 3.6 | < 5.3 |
| $^{48}V$ | — | — | — |
| $^{54}Mn$ | — | 37 ± 19 | < 38 |
| $^{56}Co$ | — | < 7 | < 13 |
| $^{60}Co$ | — | < 1.3 | — |
| K (ppm) | 3490 ± 160 | 4080 ± 120 | 4390 ± 130 |
| Th (ppm) | 11.3 ± 0.2 | 13.3 ± 0.3 | 14.6 ± 0.3 |
| U (ppm) | 2.85 ± 0.06 | 3.42 ± 0.07 | 3.65 ± 0.11 |
| Sample weight (grams) | 10.9 | 72.0 | 300 |
| Sample type | Rock slice | Rock | Soil |

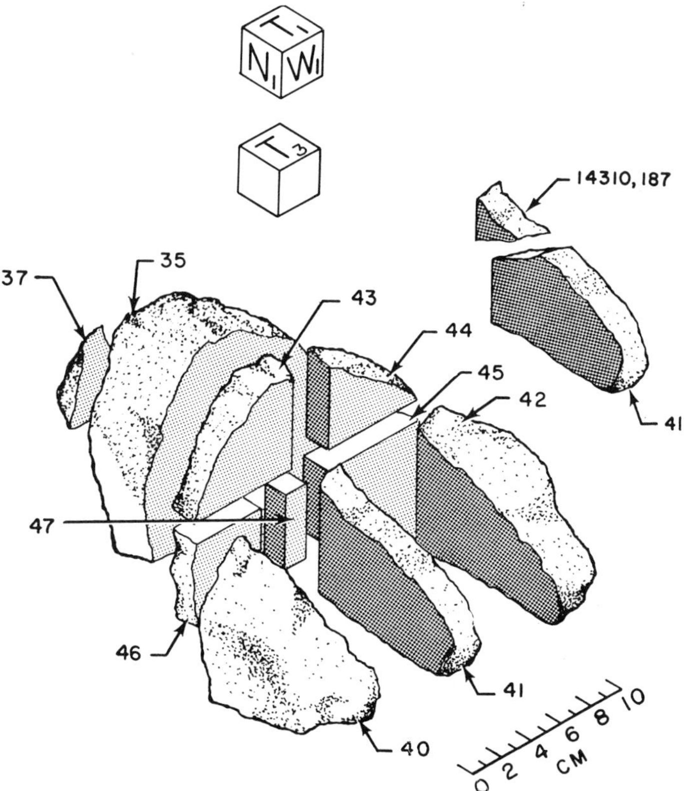


Fig. 1. Cutting diagram of rock 14310 showing the location of sample 14310,187.

observations in the Apollo 12 core tube and rock slices. The $^{26}Al$ content of 15261 is compatible with sampling depth to about 7 cm.

The radionuclide contents of three Apollo 14 samples are summarized in Table 3. Sample 14310,187 was a slice from the top surface of an igneous rock. The relationship of this sample to the original rock is presented in Fig. 1: Due to the small size of this sample the only radionuclides measured were the primordials K, U, and Th and the cosmogenics $^{22}Na$ and $^{26}Al$. The high $^{26}Al$ (165 dpm/kg) and $^{22}Na$ contents (63 dpm/kg) are compatible with the relatively unshielded position of this sample (5 g/cm$^2$). A predicted $^{26}Al$ content of this rock slice of 145 dpm/kg is obtained using the model described by Rancitelli *et al.* (1970) that employs the rock's chemical composition, the $^{26}Al$ excitation function from major target elements, and the average incident solar proton flux. The excellent agreement between the observed and predicted $^{26}Al$ contents of this rock slice indicates a saturation concentration and hence the rock was at the surface for at least the last two or three million years. This is of particular interest since preliminary track measurements (Walker, 1972) suggested the rock might have had a relatively short lunar surface age. Rock sample 14321,40 is a 700 gram surface specimen of the 9 kilogram football sized breccia. The $^{22}Na$ and $^{26}Al$ concentrations in this section indicate saturation; however, the $^{26}Al$ is somewhat low and may be the result of losses of surface material which are known to have occurred during the cutting of this specimen. Soil sample 14163,0 has substantially lower $^{22}Na$ and $^{26}Al$ concentrations than are present in the top two or three centimeters of lunar soil. From previously established concentration gradients for $^{26}Al$ it can be shown that the average sampling depth for these specimens exist from the surface to approximately 16 centimeters.

### *Solar and galactic fluxes*

The relatively short-lived cosmogenic radionuclides $^{48}V$ (16.1$d$) and $^{56}Co$ (77$d$) were measured in some of the Apollo 15 samples and provided a basis for determining the intensity of the January 24, 1971 solar flares and the relative galactic cosmic ray flux intensity at 1 AU compared with that at meteorite orbit distances. Cobalt-56 is produced in the first few centimeters of the lunar surface primarily from the interaction of solar protons on Fe. The galactic production rate of $^{56}Co$ has been estimated from meteorite data to be about 1 dpm/kg (Perkins *et al.*, 1970) while the solar proton produced $^{56}Co$ in lunar samples of a few grams per cm depth has been on the order of 20 to 50 dpm/kg. Since the last major solar flare prior to the Apollo 15 mission occurred January 24, 1971 seven months prior to the mission, the $^{56}Co$ (77$d$) produced by this flare had decayed through 2.5 half-lives by the time of collection, resulting in the low concentrations reported in Tables 1 and 2. In fact, the top section of trench soil sample 15041,14 was the only Apollo 15 sample which contained sufficient $^{56}Co$ (27 $\pm$ 6 dpm/kg) so it could be measured with a reasonable degree of accuracy. This, together with the fact that the average sampling depth of 6 cm was known from the $^{26}Al$ and $^{22}Na$ activities, permitted the recent solar flare intensity to be calculated. This $^{56}Co$ content was decay corrected to January 24, 1971, the time of the last major flare, and employed along with known excitation functions (Brodzinski *et al.*, 1971), the iron content of soil sample 15021 (LSPET, 1972), and a kinetic power law form of the solar proton energy distribution with a shape function equal to 3.1, see equation (1), to estimate the flare intensity. These computations indicate that the January 1971 flare emitted $14 \times 10^9$ protons above 10 MeV to produce the observed $^{56}Co$ content of 15041 in a $2\pi$ bombardment. Some uncertainty exists in these calculations since the precise chemical composition of 15041 and the shape function of the January flare were unavailable. However, they do indicate that the January 1971 flare and the April 1969 flare were approximately equal in intensity.

Vanadium-48 is produced in lunar material by high energy spallation of iron and low energy proton reactions on Ti. Due to its short half-life, the $^{48}V$ produced by the January 1971 solar flare decayed to insignificant levels during the seven-month interval prior to the Apollo 15 landing. Thus, the $^{48}V$ present in the Apollo 15 samples (see Tables 1 and 2) is the result of the steady state galactic cosmic ray bombardment of the lunar surface. The $^{48}V$ content of the Apollo 15 samples 15205 and 15556 was measured with sufficient accuracy to permit an estimation of the galactic cosmic ray production rate. Based upon an iron content of 18%, we calculate a production rate of $52 \pm 15$ dpm/kg Fe. This can be compared with a predicted production rate of $50 \pm 15$ dpm/kg Fe based on observed $^{48}V$ in the Allende meteorite (Rancitelli *et al.*, 1969), with appropriate corrections for the $2\pi$ bombardment on the moon. While the uncertainties are significant, this accord between the $^{48}V$ galactic production rate from iron in lunar samples and in meteorites supports our previous observations that the galactic cosmic ray flux is not significantly different between 1 AU and 2 AU (Rancitelli *et al.*, 1971). These earlier observations indicated that the galactic cosmic ray production rate of $^{26}Al$ was comparable in meteorites with that in the lunar surface if the $2\pi$ lunar irradiation was considered.

In our previous work (Rancitelli *et al.*, 1971), we described a model for calculating the depth dependence of $^{26}Al$ and $^{22}Na$ production from a representative incident solar proton spectrum, the appropriate excitation functions and the sample's target element composition. This model employs the kinetic power law form:

$$\frac{dJ}{dE} = kE^{-\alpha} \tag{1}$$

to describe the incident proton spectrum, where $J$ is the proton flux, $E$ is the proton energy, $\alpha$ is a shape function, and $k$ a constant related to the particle intensity. In this work on Apollo 12 samples, $^{26}Al$ and $^{22}Na$ concentrations as a function of depth were calculated for various lunar rock erosion rates. Concentrations versus depth were also calculated for lunar soil. The shape of the incident proton flux was chosen to conform with the average of the 1968–1969 solar flares measured by the IMP satellites (Simpson and Hsieh, 1970). These calculated gradients were then compared with measured gradients in lunar rocks and core tube materials to estimate the erosion rate of rocks and the mixing of soil.

In the present work, $^{26}Al$ and $^{22}Na$ concentration gradients were generated as a function of the incident proton spectrum shape while the erosion rates remained constant. The limits on the shape function ($\alpha$) for the incident proton spectrum were varied from 2.5 to 3.5. The best fit of the calculated to observed $^{26}Al$ gradients occurred for a shape function ($\alpha$) equal to 3.1; this would require an average proton flux above 10 MeV for the last few million years of 60 p/cm$^2$-sec. For comparison, as stated above, the January 1971 flare emitted $1.4 \times 10^9$ protons of greater than 10 MeV. This is equivalent to a two-day average flux of 7900 p/cm$^2$-sec and would contribute 43 p/cm$^2$-sec to the 1971 Average Annual Solar proton flux. The fact that the observed $^{26}$Al and $^{22}$Na concentration depth gradients in lunar material can be produced by the same proton energy and intensity spectra supports the idea that the solar flare activity has remained relatively unchanged for the past million years.

### *Primordial radionuclides*

The concentrations of K, U, and Th, the primordial radionuclides in lunar materials, are not only important to our understanding of the moon's geochemistry and its relationship to other objects in the solar system but can also serve as indicators of local mixing and transport of lunar soil. The primordial radionuclide contents of Apollo 15 soil samples fall in a range between those at the Apollo 11 and Apollo 12 sites (see Table 4). The K, U, and Th contents of the soil samples near Hadley Rille (Stations 2 and 9) are lower than those at the landing site (Station 9) and along the Apennine Front (Station 6) indicating that the wide differences in the primordial radionuclide contents of soil such as those between the Apollo 11 and 12 sites can also occur on a distance scale of a few kilometers. The K, U, and Th concentrations of the Apollo 14 soil sample (14163,0) are two- to fourfold higher than those of the Apollo 12 and 15 samples, while the K/U ratios (1200–1500) are comparable. The uniformity of K/U ratios for different lunar areas suggests that magnetic differentiation has not been sufficient to significantly alter their ratios over a large portion of the lunar surface.

The highest primordial radionuclide contents were found in the Apollo 14 rocks 14310,187 and 14321,40 and the Apollo 15 boulder chip 15205 and 15206. The K content (0.4 to 0.5%), the U content (2.8 to 3.4 ppm) and the Th content (11.3 to 13.3 ppm) were comparable to the Apollo 12 breccia 12034,9 (Rancitelli *et al.*, 1971). The three basalts from Apollo 15 sampling Station 9a had low K (0.034 to 0.049%), U (0.10 to 0.15 ppm), and Th (0.44 to 0.56 ppm) contents comparable to the Apollo 12 basalts. The low K basalts from Station 9a were also characterized by the highest K/U ratios (2600–4700) observed in the suite of Apollo 14 and 15 samples; generally

Table 4. Primordial radionuclide content of lunar soils.

| | K (%) | Th (ppm) | U (ppm) | K/U |
|---|---|---|---|---|
| Apollo 11 | 0.11 | 2.3 | 0.55 | 2000 |
| Apollo 12 | 0.20 | 6.7 | 1.7 | 1180 |
| Apollo 14 | 0.44 | 14.6 | 3.6 | 1220 |
| Apollo 15 | | | | |
| (2)* | 0.14 | 3.8 | 0.94 | 1500 |
| (6) | 0.16 | 4.7 | 1.2 | 1330 |
| (8) | 0.17 | 4.6 | 1.3 | 1310 |
| (9) | 0.12 | 4.2 | 1.0 | 1200 |

* Sampling station.

the K/U ratios of Apollo 15 rocks and soils ranged from 1200 to 1600, lower than the range of K/U values (1800–3100) we reported for the Apollo 12 samples.

The average relative atomic abundances of Th/U is 4.05 ± 0.04 in the two Apollo 14 rocks and 3.97 ± 0.28 in the fifteen Apollo 15 samples. These Th/U ratios are in excellent agreement with our previous measurements in Apollo 11 rocks (3.8 ± 0.2), Apollo 12 rocks (3.8 ± 0.1), and the calculated value of 3.8 ± 0.3 for the present-day solar system (Fowler and Hoyle, 1960).

*Acknowledgments*—We wish to thank R. M. Campbell, D. R. Edwards, J. G. Pratt, and J. H. Reeves of this Laboratory for their aid in standards preparation and in data acquisition. The unique and sensitive instrumentation that made this work possible was developed during the past decade under sponsorship of the United States Atomic Energy Commission, Division of Biology and Medicine.

## References

Brodzinski R. L., Rancitelli L. A., Cooper J. A., and Wogman N. A. (1971) High-energy proton spallation of iron. *Physical Review C.* **4,** 1257–1265.

Finkel R. C., Arnold J. R., Imamura M., Reedy R. C., Fruchter J. S., Loosli H. H., Evans J. C., and Delany A. C. (1971) Depth variation of cosmogenic nuclides in a lunar surface rock and lunar soil. *Proc. Second Lunar Sci. Conf., Geochim. Cosmochim. Acta* Suppl. 2, Vol. 2, pp. 1773–1789. MIT Press.

Fowler W. A. and Hoyle F. (1960) Nuclear cosmochronology. *Ann. Phys.* **10,** 280.

Fuse K. and Anders E. (1969) Aluminum-26 in meteorites—VI achondrites. *Geochim. Cosmochim. Acta* **33,** 653–670.

Lunar Sample Information Catalog, Apollo 15 (1971) National Aeronautics and Space Administration MSC 03209.

O'Kelley G. D., Eldridge J. S., Schonfeld E., and Bell P. R. (1970) Primordial radionuclide abundances, solar proton and cosmic-ray effects, and ages of Apollo lunar samples by nondestructive gamma ray spectrometry. *Proc. Apollo 11 Lunar Sci. Conf., Geochim. Cosmochim. Acta* Suppl. 1, Vol. 2, pp. 1407–1423. Pergamon.

Perkins R. W., Rancitelli L. A., Cooper J. A., Kaye J. H., and Wogman N. A. (1970) Cosmogenic and primordial radionuclide measurements in Apollo 11 lunar samples by nondestructive analysis. *Proc. Apollo 11 Lunar Sci. Conf., Geochim. Cosmochim. Acta* Suppl. 1, Vol. 2, pp. 1455–1469. Pergamon.

LSPET (Lunar Sample Preliminary Examination Team) (1972) Preliminary examination of lunar samples from Apollo 15. *Science* **175,** 363–375.

Rancitelli L. A., Perkins R. W., Cooper J. A., Kaye J. H., and Wogman N. A. (1969) Radionuclide composition of the Allende meteorite from nondestructive gamma-ray spectrometric analysis. *Science* **166,** 1269–1272.

Rancitelli L. A., Perkins R. W., Felix W. D., and Wogman N. A. (1971) Erosion and mixing of the lunar surface from cosmogenic and primordial radionuclide measurements in Apollo 12 lunar samples. *Proc. Second Lunar Sci. Conf., Geochim. Cosmochim. Acta* Suppl. 2, Vol. 2, pp. 1757–1772. MIT Press.

Shedlovsky J. P., Honda M., Reedy R. C., Evans J. C. Jr., Lal D., Lindstrom R. M., Delany A. C., Arnold J. R., Loosli H., Fruchter J. S., and Finkel R. C. (1970) *Proc. Apollo 11 Lunar Sci. Conf., Geochim. Cosmochim. Acta* Suppl. 1, Vol. 2, pp. 1503–1532. Pergamon.

Simpson J. and Hsieh J. (1970) Personal communication.

Sutton R. L. (1972) Personal communication.

Swann G. A., Hait M. H., Schaber G. G., Freeman V. L., Ulrich G. E., Wolfe E. W., Reed V. S., and Sutton R. L. (1971) Interagency Report: 36, Preliminary description of Apollo 15 sample environments. U.S. Department of the Interior Geological Survey, September.

Walker R. (1972) Personal communication.

Proceedings of the Third Lunar Science Conference
(Supplement 3, *Geochimica et Cosmochimica Acta*)
Vol. 2, pp. 1693–1702
The M.I.T. Press, 1972

# Cosmic-ray produced radioisotopes in Apollo 12 and Apollo 14 samples

F. BEGEMANN, W. BORN, H. PALME, E. VILCSEK, and H. WÄNKE

Max-Planck-Institut für Chemie (Otto-Hahn-Institut),
Mainz, Germany

**Abstract**—Results are reported for the $^{14}C$-, $^{22}Na$-, $^{26}Al$-, $^{36}Cl$-, and $^{39}Ar$-content of three lunar fines (12001, 14163, 14259) and three different depth regions in the igneous rock 12053.

$^{22}Na$ and $^{26}Al$ show the expected solar proton induced excess over the galactic cosmic ray produced levels. For the soil samples they allow qualitative statements about the mean sampling depth.

The $^{14}C$-data represent the first depth profile of this isotope in a lunar sample. The observed gradient is not reconcilable with a mean flux of 100 solar protons/cm$^2$ sec and $R_0 \approx 100$ MV. Instead, a 3–5 times higher flux is necessary or the presence of a surface correlated component. It is tentatively suggested that the required flux of $1 \times 10^{-3}$ $^{14}C$-atoms/cm$^2$ sec as well as the tritium flux of $5 \times 10^{-3}$ $^{3}H$/cm$^2$ sec discussed by D'Amico *et al.* is due to $(n, p)$-reactions of secondary neutrons on the surface of the sun.

Two grain size fractions of the 12001 soil were found to contain approximately equal amounts of $^{22}Na$, $^{26}Al$, and $^{36}Cl$ but grossly different $^{39}Ar$-activities. The data point to an almost complete loss of $^{39}Ar$ from the Ca-containing phases in the fine-grained fraction.

The $^{36}Cl$- and $^{39}Ar$-data obtained on rock 12053 as well as all others on rock samples available in the literature are discussed. The discrepancies between the predicted depth profiles and the measured ones are ascribed to the small size of the rocks analyzed. Model calculations for infinite flat surfaces neglect the enhanced flux of solar protons with ranges comparable to the rock radii.

## INTRODUCTION

THE CONTINUOUS BOMBARDMENT of the lunar surface by cosmic-ray particles produces a number of radioactive and stable isotopes. The measured amounts have been used with considerable success to unravel part of the history of the lunar surface and to derive the intensity as well as the energy and charge spectrum of the cosmic radiation in the near and distant past. In particular, these measurements have provided us with the first detailed information on the *mean solar* cosmic radiation, averaged over various time intervals. From these data it is apparent that in near-surface layers the production of a number of proton induced activities is dominated by low energy solar protons. Model calculations by Armstrong and Alsmiller, Rancitelli *et al.*, and Reedy and Arnold are indeed able to reproduce the steep gradients observed and, most important, such an analysis indicates the mean rigidity and intensity of the solar protons to have been the same during the past 10 yr as that during the last few million years.

As no data are available for intermediate time intervals of a few thousand years we continued our measurements of $^{14}C$ ($T_{1/2} = 5730$ yr). In addition, a number of samples has been analyzed for their content of $^{36}Cl$ and $^{39}Ar$, two isotopes that are mainly due to secondary neutrons produced by the galactic cosmic radiation. Although results are available for a number of samples, the situation is still rather confusing.

So far, neither the absolute amounts of these isotopes nor the depth gradients—or their absence—has been explained satisfactorily

## EXPERIMENTAL PROCEDURE AND RESULTS

Three samples of lunar fines (12001, 14163, 14259) and a 110 g slab of rock 12053 were available for the present investigation. Prior to any further treatment the $\gamma$-activity of sample 12001 was determined by means of a Ge (Li)-detector. (For specifications see Begemann *et al.*, 1970.) Subsequently, the sample was sieved into four grain size fractions ($<44\ \mu$, $44$–$75\ \mu$, $75$–$150\ \mu$, $>150\ \mu$). As the coarsest fraction consisted mostly of aggregates it was crushed gently in an agate mortar, and the resulting sieve fractions were combined with the ones obtained originally. Magnetic particles were separated by means of a hand magnet; for the $<44\ \mu$ fraction, this was done by agitating the powder with ultrasound in ethanol.

The surface of the rock slab was inspected for the presence of microcraters in order to determine the original orientation of rock 12053 on the lunar surface. Craters were found on about half the circumference, with a rather narrow transition region. A 4–6 mm layer was then removed from the top with a diamond wheel attached to a dental drill. The remainder was divided into two more depth regions of about equal weight (Fig. 1). These three portions were crushed to $<100\ \mu$ and the two deep samples were split into two portions each for the determination of $^{14}C$–$^{39}Ar$ and $^{22}Na$–$^{26}Al$–$^{36}Cl$–$^{39}Ar$, respectively.

For the extraction of $^{14}C$–$^{39}Ar$ the same procedure was followed as for the Apollo 11 samples; when samples were dissolved this was performed in one step as described for the *residues* of the Apollo 11 samples (Begemann *et al.*, 1970. See this reference for details of the counting procedure).

Our counting results are listed in Table 1. Whenever the same samples—or such from similar positions like for rock 12053—have been analyzed by other investigators their data are included for comparison. In all cases the agreement is seen to be good.

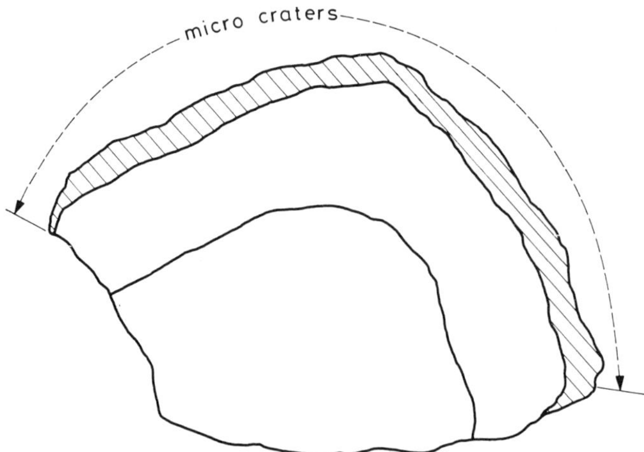


Fig. 1. Slab of rock 12053, showing the three depth regions analyzed.

Table 1. Specific activities determined in lunar soil and three depth regions in rock 12053. All activities are given in dpm/kg.

| Sample | Weight (g) | Depth (cm) | $^{14}C$ | $^{22}Na$ | $^{26}Al$ | $^{36}Cl$ | $^{39}Ar$ | $^{54}Mn$ | Author* |
|---|---|---|---|---|---|---|---|---|---|
| *Fines* | | | | | | | | | |
| 12001, bulk | | — | — | | 128 ± 24 | — | — | 83 ± 25 | — |
| < 44 μ | 20.85 | — | — | 84 ± 8 | 109 ± 11 | 16.2 ± 0.8 | 2.4 ± 0.6 | | — |
| > 75 μ | 22.98 | — | — | 78 ± 9 | 99 ± 12 | 13.4 ± 1.5 | 8.0 ± 0.6 | | — |
| 14163 | 15.36 | — | — | — | 86 ± 11 | 25.0 ± 2.0 | — | — | — |
| | — | — | — | 45 ± 9 | 78 ± 16 | — | — | | (1) |
| | — | — | — | — | 79 ± 6 | — | — | — | (2) |
| | — | — | — | 46 ± 5 | 79 ± 4 | — | — | 4 ± 7 | (3) |
| 14259 | 17.47 | 0–1 | — | — | 212 ± 25 | 16.5 ± 1.8 | — | — | — |
| | | | — | 84 ± 17 | 220 ± 40 | — | — | | (1) |
| | — | | — | 91 ± 8 | 222 ± 9 | — | — | 60 ± 20 | (3) |
| | — | | — | 95 ± 15 | 176 ± 22 | — | — | — | (4) |
| *Rock* | | | | | | | | | |
| 12053 | 8.3 | 0–0.5 | 72.4 ± 7 | — | — | — | 10.0 ± 1.0 | | — |
| | 15.0 | 0.5–2.0 | 32.8 ± 3 | — | — | — | 8.5 ± 0.7 | | — |
| | 29.88 | | — | 38 ± 6 | 69 ± 9 | 17.1 ± 0.8 | 7.7 ± 0.5 | | — |
| | | ≈ 1.5 | | 45 ± 13 | 75 ± 14 | — | — | — | (5) |
| | 15.0 | 2.0–6.5 | 29.7 ± 3 | — | — | — | 8.6 ± 0.5 | | — |
| | 27.72 | | — | 34 ± 5 | 54 ± 6 | 15.7 ± 0.7 | 7.7 ± 0.5 | | — |
| | — | ≈ 5.0 | — | 38 ± 8 | 58 ± 8 | — | — | — | (5) |

* For comparison the results obtained by other authors on the same samples are included. (1) LSPET (1971); (2) Wrigley (1972); (3) Keith *et al.* (1972); (4) Wahlen *et al.* (1972); (5) Rancitelli *et al.* (1971).

Table 2. Content of some relevant target elements in 12001 bulk soil and the two grain size fractions analyzed for activities. All entries are in weight percent.

| | Mg | Al | Si | K | Ca | Fe | Ti |
|---|---|---|---|---|---|---|---|
| 12001, bulk | 5.92 | 6.65 | 21.6 | 0.210 | 6.1 | 12.9 | 1.6 |
| 12001, < 45 μ | 5.8 | 7.4 | 21.8 | 0.221 | 7.4 | 11.7 | 1.7 |
| 75–150 μ | 6.7 | 6.0 | 21.8 | 0.171 | 6.7 | 12.7 | 1.3 |

Table 2 gives the chemical composition of the different 12001 fractions measured. Although there are distinct differences—especially in the Mg, Al, and K-content—these are only minor. They can barely be expected to result in significant changes of the activities of the radioisotopes investigated. The more surprising is the large difference in the $^{39}Ar$ content of the two grain size fractions (see below).

## Discussion

### *$^{22}Na$ and $^{26}Al$*

These isotopes show the, by now, well-established strong dependance on mean sampling depth (Shrelldallf, 1970; Perkins *et al.*, 1970; Rancitelli *et al.*, 1971; Finkel *et al.*, 1971). Taking into account the different chemistry of 12001 and the Apollo 14 fines, we conclude that 14163 represents the thickest surface layer and 14259 the shallowest one, with 12001 somewhere in between. For 14259 this is in agreement with the LSPET (1971) according to which this sample was "skimmed from the upper 1 cm of the surface." Our data are not numerous enough to either support or contradict the conclusion of Wahlen *et al.* (1972) that the layer must have been twice as thick or that the exposure history of the top 1 cm was rather complicated.

### *$^{14}C$*

According to the model calculations of Reedy and Arnold (1972) the concentration of $^{14}C$ should show a depth dependance similar to that of other proton induced activities, i.e., a decrease in activity with depth superimposed on a slowly increasing contribution produced by galactic cosmic rays. Hence, the initial decrease or the integrated "excess" activity in the surface layers should again allow to determine the average flux of solar protons. Due to the half-life of $^{14}C$ the average flux thus determined would be that over the past $10^4$ yr or so. This is an interesting time interval covered by no other isotopes.

The data obtained for rock 12053 are plotted in Fig. 2. The predicted decrease with depth of the activity is indeed observed. A quantitative analysis shows, however, that the results are not reconcilable with a $4\pi$ integral flux ($E > 10$ MeV) of 100 solar protons/cm$^2$ sec and a rigidity shape factor of $R_0 \approx 100$ MV as obatained from the analysis of various other isotopes (Finkel *et al.*, 1971). A rather good fit is obtained with $R_0 = 100$ MV and a three times higher flux, or $R_0 = 80$ MV and a five times higher flux. This, one could take to indicate that either the mean flux of solar protons over the past 10,000 yr has been 3–5 times higher than that averaged over the past 10 yr and the past million years, respectively, or that the cross sections used in the model calculations are completely wrong. Neither of these possibilities can be excluded with certainty. The fact, however, that the discrepancy between measured and calculated activities (with $R_0 = 100$ MV and a flux of 100 protons/cm$^2$ sec) is caused

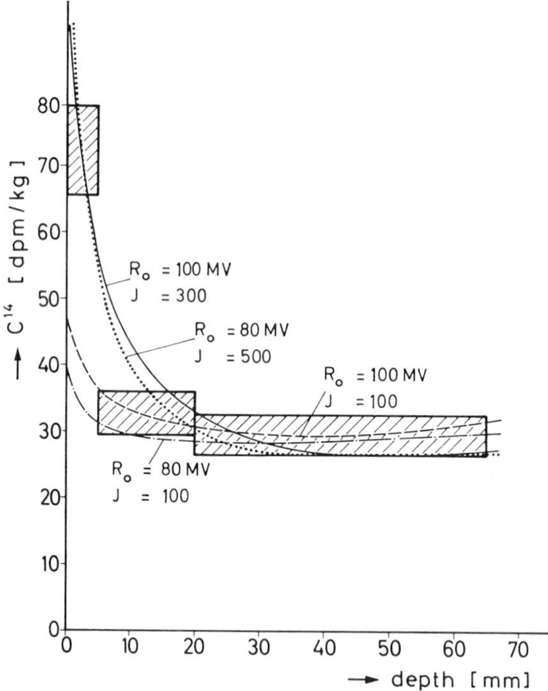


Fig. 2. $^{14}C$ activities measured in rock 12053. The curves are those predicted from the model calculations of Reedy and Arnold (1972) for different choices of $R_0$ and flux $J$ [solar protons with $E > 10$ MeV/cm$^2$ sec and $4\pi$ irradiation].

by the high activity of the surface sample suggests that the excess $^{14}C$ might be surface correlated. For the geometry of our sample a mean $^{14}C$ flux of about $1 \times 10^{-3}$ $^{14}C$ atoms/cm$^2$ sec would be required, assuming equilibrium conditions.

In this connection it is perhaps worth mentioning that D'Amico *et al.* (1971) suggested a $^3H$ flux of $5 \times 10^{-3}$ tritium atoms/cm$^2$ sec to account for a surface correlated $^3H$ component in rock 12002. While this tritium might presumably be produced by the *direct* interaction of solar flare protons with the photosphere of the sun and subsequently be removed as flare particles (Fireman, 1962) and implanted into the lunar surface, there is no chance to produce the required amounts of $^{14}C$ by a similar process. There is the possibility, however, that both isotopes are produced on the sun by secondary neutrons via the exothermic $(n, p)$-reactions on $^3He$ and $^{14}N$. If this were so the equilibrium ratio of the fluxes, barring any fractionation in the acceleration mechanism, would be

$$\frac{^3\mathrm{H}}{^{14}\mathrm{C}} = \left(\frac{^3\mathrm{He}}{^{14}\mathrm{N}}\right)_{\mathrm{sun}} \times \frac{T(^3\mathrm{H})}{T(^{14}\mathrm{C})} \times \frac{\sigma(3)}{\sigma(14)}.$$

Inserting the values for the solar abundance ratio (Cameron, 1968), the half-lives of tritium and $^{14}C$, and the ratio of the "thermal" cross sections yields a ratio of 1.6 as compared to the one of 5 deduced from the lunar samples.

We are aware, of course, that this close agreement is no proof for the processes suggested to take place on the required scale, especially since the assumption that equilibrium is established *at a rather high level of activity* needs special conditions. Vertical mixing rates on the solar surface are fast so that there is a huge dilution factor (mass in convective zone/mass in reaction zone). We understand, however, that solar physicists are discussing at present the possibility that following major solar flares the regions around the foot points are sucked up into the corona (Schmidt, private communication). Such a process would indeed obstruct the dilution and lead to conditions equivalent to equilibrium in the reaction zone.

### $^{36}$*Cl*

In lunar samples this isotope is produced by spallation reactions on Fe, Ti, and Ca and by secondary neutrons from Ca, with the latter being most important target element. A contribution from K or from the $^{35}Cl$ $(n, \gamma)$ reaction can only be expected for unusually high contents of K and Cl. This is shown in Fig. 3 for a sample with 25% (Fe + 4 Ti), 7% Ca, and 2000 ppm K. The differential energy spectra and fluxes of the nuclear active particles as well as the relevant cross sections used in this calculation are those of Reedy and Arnold (1972). From the computations of the same authors it is also apparent that *solar* protons contribute less then 1 dpm/kg even in the surface layer of $<1$ g/cm$^2$. Consequently, an increase in activity with depth must be expected.

Qualitatively, the soil samples listed in Table 1 conform to this expectation. The shallowest sample (14259) has an activity of $(16.5 \pm 1.8)$ dpm/kg, the deepest one (14163) $(25 \pm 2)$ dpm/kg. The intermediate sample 12001 (see above), corrected for its lower Ca content, falls into the same sequence. We do not believe, however, that

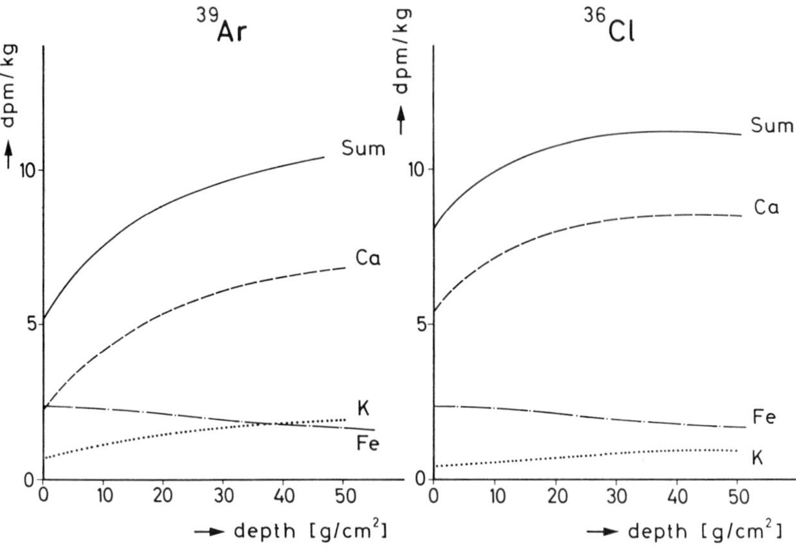


Fig. 3. Calculated $^{39}$Ar and $^{36}$Cl activities for a rock with 25% (Fe + 4 · Ti), 7% Ca, and 2000 ppm K.

the high activity of 14163 is entirely due to a greater mean depth. The only explanation we can put forward at the present time to account for the excess activity is that in this sample we have a major contribution of $^{35}$Cl $(n, \gamma)$-produced $^{36}$Cl. The Cl-content of an aliquot of this sample was found to be exceptionally high (280 ppm, Wänke *et al.*, 1972) and for this high Cl content a thermal neutron flux of about 1 neutron/cm$^2$ sec would be required to produce 10 dpm $^{36}$Cl/kg. It is somewhat disturbing, however, that Brunfelt *et al.* (1971) and Reed *et al.* (1972) report for the 14163 fines Cl-contents of only 47 and 51 ppm, respectively, although there is the definite possibility of a very inhomogeneous distribution of trace elements like Cl.

The two samples of rock 12053 (Table 1, column 7) show a decrease or perhaps constant activity with depth, contrary to expectations. Furthermore, the absolute activity is considerably higher than expected from the chemistry of this rock and the curves in Fig. 3. If one looks at the $^{36}$Cl data of all four rocks in which the activity has been measured for different layers (Shrelldallf, 1970; Finkel *et al.*, 1971; Wahlen *et al.*, 1972), it turns out that this is rather the rule (Fig. 4). Only for 14321 are the theoretical predictions in accord with the measurements although even here there might be an excess in the top layers. As the discrepancies appear to be the largest for the smallest rocks (10017, 0.98 kg; 12053, 0.88 kg) and smallest for 14321 (9.0 kg) we suggest that they might conceivably be due to applying a wrong model. Clearly, rocks with radii of only a few cm lying on top of the regolith do not have a $2\pi$ flat geometry for irradiation. Rather, the flux of solar protons with ranges comparable to the radii is considerably enhanced while the build-up of the secondary neutron flux is somewhat reduced.

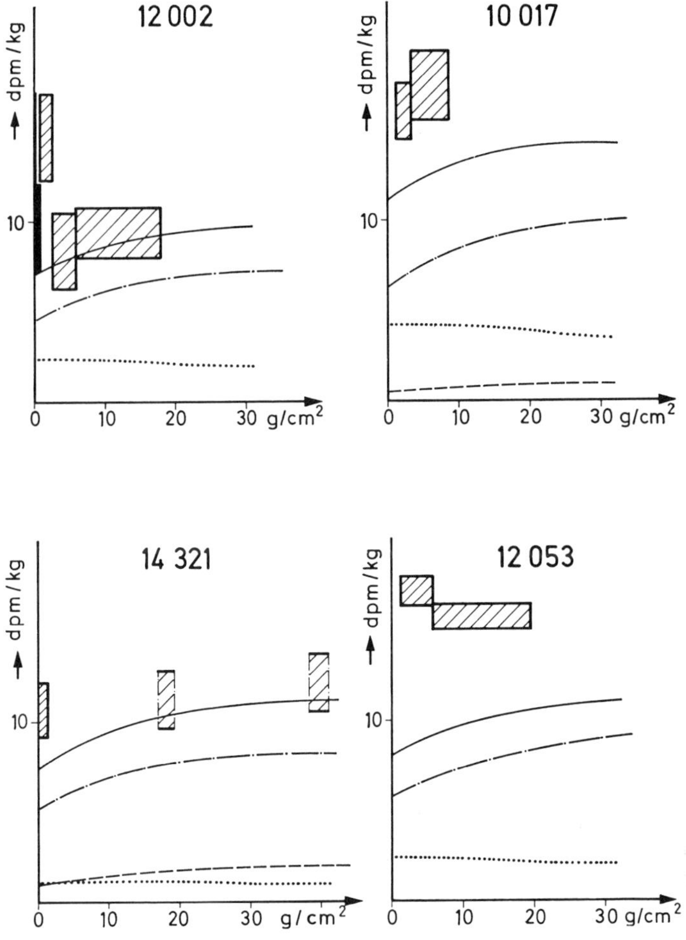


Fig. 4. Measured and predicted $^{36}$Cl activities in 4 lunar rocks. —·—·: Contribution from Ca; ····· contribution from Fe + 4·Ti; ———: contribution from K; —— sum.

### $^{39}$*Ar*

In our first paper on the $^{39}$Ar content of lunar samples (Begemann *et al.*, 1970) we were of the opinion that besides Fe and Ti, K is the most important target element. This conclusion was reached by estimating the contribution due to the $^{40}$Ca $(n, 2p)$-reaction from the measured $(n, 2p)$- cross sections on neighboring nuclides. In the meantime the cross section for the production of $^{39}$Ar from Ca has been measured with 14 MeV neutrons (Stoenner *et al.*, 1970) and Reedy and Arnold (1972) have constructed an excitation function for this reaction, with normalization at the one experimental point. Using these data it appears now that the contribution from Ca is dominant as the production per kg of K turns out to be only 10 times larger than that per kg Ca. This is somewhat surprising in view of the fact that for the enstatite chondrite Abee we found a ratio of 120 instead (Begemann *et al.*, 1967). It does remove some of the inconsistencies mentioned in our first paper, however.

The most striking result obtained on the two grain size fractions of the 12001 fines is the large difference in their $^{39}Ar$ content that is clearly *not* due to a different chemical composition (Table 2). If anything, from these differences one would rather expect the activity in the finest fraction to be higher. Instead, the activity found can almost be accounted for by the production from Fe and Ti. This points to a more or less complete loss of $^{39}Ar$ from the Ca-containing phases. Whether this is solely a grain size effect or due to a higher relative abundance of Ca-rich *glass* in the finest fraction is not clear at present. That such diffusion loss of $^{39}Ar$ from lunar fines does

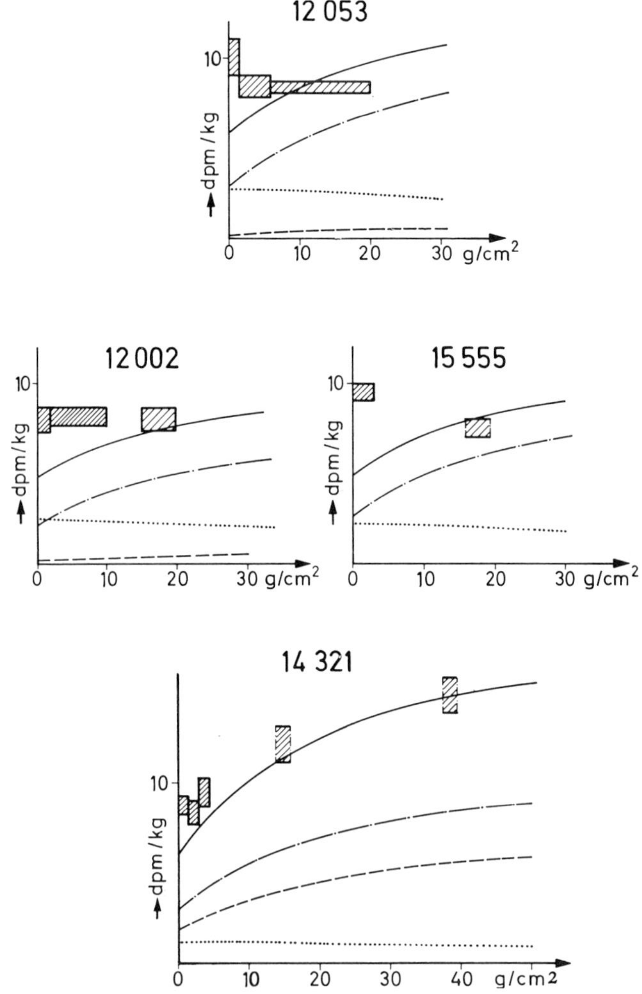


Fig. 5. Measured and predicted $^{39}Ar$ activities in 4 lunar rocks. For explanation of curves see legend to Fig. 4.

indeed occur has been shown by Stoenner *et al.* (1972) who report diffusion losses of $10^{-2}$–$10^{-3}$ %/day.

For the samples from the three layers of rock 12053 there is no increase with depth observed in the $^{39}$Ar activity, again in contradiction to the prediction of the model calculations. As in the case of $^{36}$Cl this is true for all lunar rocks analyzed so far, with the exception of 14321 (Fig. 5) (D'Amico *et al.*, 1971; Fireman *et al.*, 1972). Here, too, we suggest this to be due to the small size of the rocks analyzed. In the case of $^{39}$Ar, however, the target element for the production of the excess activity by solar protons is probably Ti (not Ca or K) as protons of any energy can produce $^{39}$Ar only from the rare isotopes of both elements.

*Acknowledgments*—We are grateful to NASA for making available the lunar material for this investigation. We wish to thank the staff of our Institute, in particular Miss G. Schmidt, Miss H. Prager and Mr. P. Keller. The financial support by the Bundesministerium für Bildung und Wissenschaft is gratefully acknowledged.

## References

Armstrong T. W. and Alsmiller R. G. Jr. (1971) Calculation of cosmogenic radionuclides in the moon and comparison with Apollo measurements. *Proc. Second Lunar Sci. Conf., Geochim. Cosmochim. Acta* Suppl. 2, Vol. 2, pp. 1729–1745. MIT Press.

Begemann F., Vilcsek E., and Wänke H. (1967) The origin of the "excess" argon-39 in stone meteorites. *Earth Planet. Sci. Lett.* **3,** 207–212.

Begemann F., Vilcsek E., Rieder R., Born W., and Wänke H. (1970) Cosmic-ray produced radioisotopes in lunar samples from the Sea of Tranquility (Apollo 11). *Proc. Apollo 11 Lunar Sci. Conf., Geochim. Cosmochim. Acta* Suppl. 1, Vol. 2, pp. 995–1005. Pergamon.

Brunfelt A. O., Heier K. S., Steinnes E., and Sundvoll B. (1971) Determination of 36 elements in Apollo 14 bulk fines 14163 by activation analysis. *Earth Planet. Sci. Lett.* **11,** 351–353.

Cameron A. G. W. (1968) A new table of abundances of the elements in the solar system. In *Origin and Distribution of the Elements* (editor L. H. Ahrens), pp. 125–143, Pergamon.

D'Amico J., DeFelice J., Fireman E. L., Jones C., and Spannagel G. (1971) Tritium and argon radioactivities and their depth variations in Apollo 12 samples. *Proc. Second Lunar Sci. Conf., Geochim. Cosmochim. Acta* Suppl. 2, Vol. 2, pp. 1825–1839. MIT Press.

Finkel R. C., Arnold J. R., Imamura M., Reedy R. C., Fruchter J. S., Lossli H. H., Evans J. C., and Delany A. C. (1971) Depth variation of cosmogenic nuclides in a lunar surface rock and lunar soil. *Proc. Second Lunar Sci. Conf., Geochim. Cosmochim. Acta* Suppl. 2, Vol. 2, pp. 1773–1789. MIT Press.

Fireman E. L. (1962) Tritium in meteorites and in recovered satellite material. In *Tritium in the Physical and Biological Sciences*, Vol. 1, pp. 69–74, IAEA, Vienna.

Fireman E. L., D'Amico J., DeFelice J., and Spannagel G. (1972) Radioactivities in Apollo 14 and 15 materials (abstract). In *Lunar Science—III* (editor C. Watkins), pp. 262–264, Lunar Science Institute Contr. No. 88.

Keith J. E., Clark R. S., and Richardson K. A. (1972) Gamma ray measurements of Apollo 12, 14, and 15 lunar samples (abstract). In *Lunar Science—III* (editor C. Watkins), pp. 446–448, Lunar Science Institute Contr. No. 88.

LSPET (Lunar Sample Preliminary Examination Team) (1971) Preliminary examination of lunar samples from Apollo 14. *Science* **173,** 681–693.

Perkins R. W., Rancitelli L. A., Cooper J. A., Kaye J. H., and Wogman N. A. (1970) Cosmogenic and primordial radionuclide measurements in Apollo 11 lunar samples by nondestructive analysis. *Proc. Apollo 11 Lunar Sci. Conf., Geochim. Cosmochim. Acta* Suppl. 1, Vol. 2, pp. 1455–1469. Pergamon.

Rancitelli L. A., Perkins R. W., Felix W. D., and Wogman N. A. (1971) Erosion and mixing of the lunar surface from cosmogenic and primordial radionuclide measurements in Apollo 12 lunar samples. *Proc. Second Lunar Science Conf., Geochim. Cosmochim. Acta* Suppl. 2, Vol. 2, pp. 1757–1772. MIT Press.

Reed G. W. Jr., Jovanovic S., and Fuchs L. H. (1972) Concentrations and lability of the halogens, platinum metals and mercury in Apollo 14 and 15 samples (abstract). In *Lunar Science—III* (editor C. Watkins), pp. 637–639, Lunar Science Institute Contr. No. 88.

Reedy R. C. and Arnold J. R. (1972) Interaction of solar and galactic cosmic-ray particles with the moon. *J. Geophys. Res.* **77,** 537–555.

Shrelldalff: Shedlovsky J. P., Honda M., Reedy R. C., Evans J. C. Jr., Lal D., Lindstrom R. M., Delany A. C., Arnold J. R., Loosli H. H., Fruchter J. S., and Finkel R. C. (1970) Pattern of bombardment-produced radionuclides in rock 10017 and in lunar soil. *Proc. Apollo 11 Lunar Science Conf., Geochim. Cosmochim. Acta* Suppl. 1, Vol. 2, pp. 1503–1532. Pergamon.

Stoenner R. W., Lyman W. J., and Davis R. Jr. (1970) Cosmic-ray production of rare-gas radioactivities and tritium in lunar material. *Proc. Apollo 11 Lunar Sci. Conf., Geochim. Cosmochim. Acta* Suppl. 1, Vol. 2, pp. 1583–1594. Pergamon.

Stoenner R. W., Lindstrom R. M., Lyman W., and Davis R. Jr. (1972) Argon, radon, and tritium radioactivities in the sample return container and the lunar surface (abstract). In *Lunar Science—III* (editor C. Watkins), pp. 729–730, Lunar Science Institute Contr. No. 88.

Wahlen M., Honda M., Imamura M., Fruchter J., Finkel R., Kohl C., Arnold J., and Reedy R. (1972) Cosmogenic nuclides in football-sized lunar rocks (abstract). In *Lunar Science—III* (editor C. Watkins), pp. 764–766, Lunar Science Institute Contr. No. 88.

Wänke H., Baddenhausen H., Balacescu A., Teschke F., Spettel B., Dreibus G., Quijano-Rico M., Kruse H., Wlotzka F., and Begemann F. (1972) Multielement analyses of lunar samples (abstract). In *Lunar Science—III* (editor C. Watkins), pp. 779–781, Lunar Science Institute Contr. No. 88.

Wrigley R. C. (1972) Radionuclides at Fra Mauro (abstract). In *Lunar Science—III* (editor C. Watkins), pp. 814–815, Lunar Science Institute Contr. No. 88.

Proceedings of the Third Lunar Science Conference
(Supplement 3, *Geochimica et Cosmochimica Acta*)
Vol. 2, pp. 1703–1717
The M.I.T. Press, 1972

# Argon, radon, and tritium radioactivities in the sample return container and the lunar surface

R. W. Stoenner, Richard M. Lindstrom,
Warren Lyman,* and Raymond Davis, Jr.

Chemistry Department, Brookhaven National Laboratory,
Upton, N.Y. 11973

**Abstract**—The radioactive rare gases $^{37}Ar$ (35 day), $^{39}Ar$ (269 yr), $^{222}Rn$ (3.8 day), and tritium (12.3 yr) were measured in the gas recovered from the Apollo 12, 14, and 15 sample return containers and in lunar rocks and soil. The activities observed in the sample return containers can be attributed to the diffusion of these activities from the fine material present in the box. An analysis of these results and the diffusion of argon activities and tritium from lunar fine material at various temperatures up to 900°C allows a determination of the heat of diffusion and the diffusion constant divided by the square of the mean grain size ($D/a^2$) for lunar material. An analysis is made of the loss of argon into the lunar atmosphere.

The cosmic ray production of $^{37}Ar$, $^{39}Ar$, and tritium in the lunar soil was measured in two deep samples, and in adjacent surface samples from the Apollo 15 mission. The $^{37}Ar$ activity was essentially the same at the surface and at depth (150 g/cm$^2$). This result is in accord with a 600-MeV proton bombardment of a thick target of lunar-like composition 100 g/cm$^2$ deep. Argon-37 activities from the Apollo 14 mission were high as a result of the January 1971 flare that occurred 12 days preceding the mission.

Measurements of the $^{222}Rn$ observed in the sample return container allow a determination of the $^{222}Rn$ emanation rate in lunar soil under conditions representative of the natural conditions on the moon. An analysis of the emanation rate permits an estimate of the amount of $^{222}Rn$ released to the lunar atmosphere.

## Introduction

The Sample Return Container (SRC) when closed on the surface of the moon includes an 18-liter sample of the lunar atmosphere. In addition, it contains all gaseous products, both radioactive and stable, that have been released from the contents of the SRC, in particular from the lunar fine material. The lunar fines have unique adsorptive and emanation properties that are important to an understanding of the lunar atmosphere, properties that are essentially preserved in the SRC up to the time of opening at the Lunar Receiving Laboratory. The concentration of gas in the lunar atmosphere or the amounts of gas emanating from a kilogram of lunar fines is extremely small, too small to be observed with a conventional high-sensitivity mass spectrometer. However, one has a very high sensitivity for observing radioactive rare gas atoms and tritium, and also the background amounts of these atoms present as contamination are extremely small. The detection limits of the radioactivities reported in this work are: 20 atoms of $^{222}Rn$ ($t_{1/2}$ 3.8 days), 50 atoms of $^{37}Ar$ ($t_{1/2}$

* Present address: Department of Earth and Space Sciences, State University of New York, Stony Brook, N.Y. 11790.

35 days), $1.5 \times 10^5$ atoms of $^{39}Ar$ ($t_{1/2}$ 269 years), and $10^5$ atoms of tritium ($t_{1/2}$ 12.3 years). In this paper we report measurements of the amounts of these radioactive gases present in the sample return container from the Apollo 12, 14, and 15 missions.

The results for argon and tritium will be interpreted in terms of the diffusion of these atoms from the fine material contained in the SRC. It will be shown that the loss of argon and tritium from the fines in the SRC at room temperature is quantitatively related to the loss of these gases from lunar fines heated to temperatures up to 1170°K, and also to the loss from bombarded simulated lunar fines over the temperature range 197 to 408°K. Combining these observations and the surface temperature of the moon, an estimate is made of the loss of argon and tritium to the lunar atmosphere.

The radon present in the lunar atmosphere and its daughter radioactivities have been the subject of direct observation from orbiting spacecraft and the Surveyor V landing spacecraft. It is of interest to measure the radon loss or emanating power of pristine lunar soil. The lunar fines present in the unopened SRC is the only available sample (the environmental sample is superior) of returned lunar material that approaches these requirements. The $^{222}Rn$ present in the SRC in the Apollo 12, 14, and 15 missions was measured, and the results are interpreted in terms of recoil processes and adsorption in the lunar soil. Using this measured emanating rate we estimate the amount of $^{222}Rn$ expected to appear in the lunar atmosphere.

After a brief description of our experimental procedures, we will discuss and interpret the results for individual radioactivities, argon, tritium, and radon in order.

## Experimental

### *Sample return container*

An effective seal was realized on one SRC from each of the Apollo 12, 14, and 15 missions, such that it was still under vacuum (30 to 70 $\mu$ Hg; see Table 1) upon arrival at the Lunar Receiving Laboratory. The second SRC returned from each of these missions could not be analyzed because the box was not properly closed on the lunar surface and therefore leaked to atmospheric pressure. The gas from the Apollo 15 SRC was removed in the same manner as described by Stoenner *et al.* (1971) except that to collect $H_2$ a trap with 10 g of activated powdered vanadium metal at room temperature was substituted for the liquid helium cooled charcoal trap used on Apollo missions 12 and 14. A liquid nitrogen cooled trap filled with 10 g of charcoal was used as before to collect all other gases. An absorption time of one hour was allowed in the box opening procedures, however this time was apparently sufficient to collect over 90% of the gases, as indicated by a thermocouple gauge on the pumping line between the SRC and the liquid nitrogen cooled charcoal trap. The argon, radon, and

Table 1. Argon and tritium radioactivities in the sample return container.*

| Mission | SRC Pressure (Torr) | Wt. fines in SRC (kg) | Observed activity (dpm $\times 10^3$) | | | Fraction of activity lost per day ($\times 10^5$) | | |
|---|---|---|---|---|---|---|---|---|
| | | | $^{37}Ar$ | $^{39}Ar$ | T | $^{37}Ar$ | $^{39}Ar$ | T |
| 12 | 0.030 | 2.7 | $40 \pm 2$ | $2.3 \pm 1.0$ | — | 6 | 2 | — |
| 14 | 0.070 | 0.60 | $40 \pm 2$ | $0.8 \pm 0.5$ | $280 \pm 20$ | 18 | 2 | 40 |
| 15 | 0.032 | 2.30 | $106 \pm 3$ | $3.0 \pm 1.2$ | $550 \pm 10$ | 18 | 2 | 10 |

* Errors quoted throughout this paper, unless otherwise noted, are one standard deviation based on counting statistics.

hydrogen collected were separated from each other, purified, and counted in small low-level proportional counters to obtain the amounts of $^{37}$Ar, $^{39}$Ar, $^{222}$Rn, and tritium activities. The separation procedures used are those described earlier (Stoenner *et al.*, 1970). In previous reports the counting efficiency for $^{222}$Rn and its daughters was only estimated. In the present work the radon counters were standardized and the results applied to the earlier measurements. Radon-222 was swept from NBS standard radium solutions with helium gas and the $^{222}$Rn trapped on charcoal. It was removed by heating to 250°C along with argon carrier gas and placed in the counter. Measured counting efficiencies were in the range of 1.9 counts per disintegration of $^{222}$Rn.

The $^{37}$Ar, $^{39}$Ar, tritium, and $^{222}$Rn observed in the Apollo 12, 14, and 15 missions is listed in Tables 1 and 2. The $^{37}$Ar activity is corrected to the time of liftoff from the lunar surface. The $^{222}$Rn activity given is the value at the time of box puncture corrected to an infinite accumulation time and therefore corresponds to the total equilibrium emanation rate for the mass of fine material in the container.

### *Argon and tritium activities in rocks and soils*

Argon radioactivities and tritium in lunar rocks and soils were measured using the procedures previously given (Stoenner *et al.*, 1970). Most of the samples were either surface chips of rocks or surface soil samples, with an estimated depth of a few grams per cm$^2$. Two Apollo 15 samples at depth were measured: (1) 15031,16, the bottom of a trench 35 cm deep dug at station 8 at a distance of 80 m from the Lunar Module, and (2) 15231,18, a sample of soil below a boulder approximately 60 cm in diameter and 1 m long at station 2.

It was of interest to observe the rate of loss of these radioactivities with temperature from the fine material to determine the variation of the diffusion constant with temperature. An experiment of this nature was performed during the Apollo 11 mission under the biological quarantine restrictions that necessitated preheating the sample a day at 120°C for sterilization purposes. Since the effect of the preheating was not known, we heated the two Apollo 15 trench soil samples (15031,16 and 15041,16) stepwise to check the earlier results. The temperatures used were 300°C, 600°C, and 900°C, for periods of one to four days. The experiments were concluded by melting the samples. Table 3 summarizes the results.

A long delay in receipt of the trench samples, and the division of activity into several heating fractions, resulted in rather large counting errors, especially in the $^{39}$Ar measurement. The fractions

Table 2. $^{222}$Radon in the sample return container.

| Mission | Total $^{238}$U Disintegration Rate in Fines (dpm) | $^{222}$Rn in SRC Gas (Saturation dpm) | Fraction Emanated ($\times 10^3$) |
|---|---|---|---|
| 12 | 3400 | 6.4 ± 0.4 | 1.9 |
| 14 | 1700 | 2.42 ± 0.03 | 1.4 |
| 15 | 1700 | 0.86 ± 0.05 | 0.50 |

Table 3. Thermal release of argon and tritium activities from lunar fines.

| Temp. (°C) | Period (days) | Trench Top, 15041,16 Percent Released | | | Period (days) | Trench Bottom, 15031,16 Percent Released | | |
|---|---|---|---|---|---|---|---|---|
| | | $^{37}$Ar | $^{39}$Ar | T | | $^{37}$Ar | $^{39}$Ar | T |
| 300 | 0.85 | 7 | 2 | 5 | 0.93 | 0 | 5 | 13 |
| 600 | 1.2 | 1 | 21 | 56 | 4.0 | 10 | 17 | 66 |
| 900 | 2.0 | 40 | 15 | 37 | 0.97 | 79 | 36 | 21 |
| Melt | — | 52 | 62 | 2 | — | 12 | 41 | 0 |
| Total dpm/kg | | 28.4 ± 3.6 | 12.9 ± 1.5 | 435 ± 8 | | 24.9 ± 4.2 | 7.6 ± 1.5 | 202 ± 6 |

were later combined to reassay the $^{39}$Ar activities; agreement was satisfactory. The temperature release of $^{37}$Ar and tritium in the Apollo 15 trench samples agrees well with the earlier Apollo 11 results, and indicates that the amount of tritium released in the sterilization treatment was small.

Table 2 summarizes the $^{37}$Ar, $^{39}$Ar, and tritium values on all Apollo missions measured at this laboratory.

## Discussion

### *The production of argon and tritium radioactivities by cosmic rays*

The argon radioactivities, $^{37}$Ar and $^{39}$Ar, are produced by high energy spallation processes on the abundant elements iron, titanium, calcium, and potassium. In addition, these activities are produced by the fast neutron reactions $^{40}$Ca$(n,\alpha)^{37}$Ar, $^{40}$Ca$(n,2p)^{39}$Ar, and $^{39}$K$(n,p)^{39}$Ar. In our previous reports the measured activities of lunar samples were compared to the 600-MeV proton thin target cross sections on iron, titanium, and calcium and fast neutron cross sections on potassium and calcium. Tritium is generally considered a high energy product because its production cross section increases with energy and target mass number. To obtain the relative production of these isotopes under more realistic conditions that would include the nucleon and meson cascade, a thick target (20 × 20 cm and 43 cm deep) of lunar-like material was irradiated with 600 MeV protons (Lyman, 1971). A detailed analysis of these results has not been made, but certain broad features of the results are clear. The total $^{37}$Ar, $^{39}$Ar, and tritium produced does not vary greatly with depth from 2 to 100 g/cm$^2$, and the activity ratio $^{37}$Ar/$^{39}$Ar is 2.7 and the ratio T/$^{39}$Ar is 20. The flat distribution with depth is a result of the nuclear cascade in the target; the target had lateral dimensions great enough to include about 80% of the cascade. The target contained 11.7% Fe, 3.9% Ti, 8.0% Ca, and 0.09% K, a composition similar to that of the lunar rocks and soils measured in this report. One might expect the $^{37}$Ar/$^{39}$Ar ratios to depend somewhat on the composition, and also upon the bombarding energy. The target measurements do not show the increase of a factor of two in $^{37}$Ar from the surface to a depth of 50 g/cm$^2$ expected in the calculations of Reedy and Arnold (1972). However, these calculations are based primarily on the $^{40}$Ca$(n,\alpha)^{37}$Ar reaction as the $^{37}$Ar production mechanism, and include a spectrum of cosmic ray particles interacting with the lunar surface with uniform angular distribution. These thick target results would indicate that lunar surface material under constant cosmic ray bombardment should exhibit little change in $^{37}$Ar, $^{39}$Ar, and tritium activity with depth, and the ratio $^{37}$Ar/$^{39}$Ar should be approximately 2.7, and the ratio T/$^{39}$Ar should be approximately 20.

It is clear that the surface of the moon is not subject to a constant bombardment but is exposed to periodic intense fluxes of solar flare particles. The $^{37}$Ar activity will suddenly increase during a flare event whereas the long-lived $^{39}$Ar will be little affected by any single event. These general features are seen in the argon activities observed in the four Apollo missions given in Table 2. Apollo 11 and 15 missions were relatively free of solar flare effects, and one observes that the $^{37}$Ar activities are similar and show no change with depth, as for example the surface (15031,16) versus the bottom of the trench (15041,16), and the surface (15221,17) and under the boulder (15231,18). It can also be noted that the $^{37}$Ar/$^{39}$Ar ratio in these samples varies considerably over the range 1.6 to 6.1, thus showing poor agreement with the ratio of 2.7 obtained in the thick target bombardment. The largest variation is in the $^{39}$Ar activities, and they appear to have lower values on the surface than at depth. This apparently anomalous result can be attributed to the loss of $^{39}$Ar from surface fine material by diffusion. An analysis of this phenomenon will be given below. The Apollo 12 and 14 missions were preceded by solar flare events 17 and 12 days prior to the landing. Three of the four samples analyzed showed high $^{37}$Ar levels. The value of 78 dpm/kg observed in the bulk fines sample 14259,84 is the highest $^{37}$Ar observed in a lunar sample. It is interesting to note that at the time of the flare this sample would have contained 90 dpm $^{37}$Ar/kg. The values reported by Fireman *et al.* (1972) for the Apollo 14 mission were also high (34 to 38 dpm $^{37}$Ar/kg) but not as high as those reported here. The variations we observe are somewhat inconsistent with the more or less uniform production of $^{37}$Ar expected from the target, consequently we can only attribute the variations observed to differences in chemical composition of the individual samples.

The tritium content in Apollo samples measured varies considerably. The Apollo 11 rock samples of all investigators appeared to be relatively uniform in tritium activity (Stoenner *et al.*, 1970; D'Amico *et al.*, 1970). If one takes an average of 226 dpm T/kg and an average of 10 dpm $^{39}$Ar/kg for rock samples, the T/$^{39}$Ar ratio of 23 agrees rather well with the value of 20 obtained from the 600-MeV proton irradiation. However, there are some large variations in the tritium content of fine material. The two studies of presumably adjacent samples, one from the surface and one at depth, gave results that are difficult to explain. The soil sample from under the boulder (15231,18) and a surface sample a few meters away (15221,17) had very low tritium contents, only 56 dpm T/kg, whereas one would expect a tritium content for both of these samples in the range of 200–300 dpm T/kg. We attribute these low values to the escape of tritium. In the beginning of the extraction procedure these samples were placed in the vacuum system and allowed to stand at room temperature 16 hours. During this period the samples lost $4.9 \pm 3.3$ (15231,18) and $10.7 \pm 3.2$ (15221,17) dpm T/kg. These tritium losses were regarded as small, close to line backgrounds. Following this experiment the samples were heated for several days at 150°C to measure the $^{222}$Rn emanation rate, but the tritium released during this period was not measured. It is entirely possible that these soil samples collected at Station 2 are not very retentive for hydrogen. It is interesting to note that Fireman *et al.* (1972) observed a 25% tritium loss in 3 months at room temperature in a sample of fines (15021,2). There apparently is a variation in the tritium retentivity of lunar fine samples, though the release patterns of the three fine samples that we studied in detail are very similar, as Table 4 shows.

The trench samples showed a higher tritium content at the bottom ($\sim 70$ g/cm$^2$) of the trench, 435 dpm T/kg, than at the surface, 202 dpm T/kg. It is to be noted that the trench bottom sample (15031,16) did not release a measurable amount of tritium when retained in a vacuum at room temperature for 16 hours, $0.5 \pm 2.5$ dpm T/kg. The higher tritium content of the trench bottom sample can be explained by noting that a sample at a depth in the lunar soil is shielded from the solar diurnal heating and hence retains most of the tritium produced in it whereas a surface sample is

Table 4. Argon and tritium radioactivities from lunar rocks and soils.

| Sample Information | | | Argon Radioactivities (dpm/Kg) | | | | Composition (%) | | | | |
|---|---|---|---|---|---|---|---|---|---|---|---|
| No. | Location | Depth ($g/cm^2$) | $^{37}Ar$ | $^{39}Ar$ | $^{37}Ar/^{39}Ar$ | Tritium (dpm/Kg) | Fe | Ti | Ca | K | Ref.* |
| 10002,6 | Surface fines | Surface | 18.7 ± 1.2 | 9.2 ± 0.4 | 2.0 ± 0.2 | 314 ± 13 | 12.4 | 4.2 | 8.6 | 0.10 | |
| 10057,3 | Exterior chip | Surface | 18.5 ± 0.9 | 11.2 ± 0.3 | 1.7 ± 0.1 | 224 ± 15 } | 15.5 | 7.5 | 7.1 | 0.15 | |
| 10057,27 | Interior chip | Surface | 23.7 ± 4.1 | 14.8 ± 0.4 | 1.6 ± 0.3 | 214 ± 20 } | | | | | |
| 12063,– | Exterior chip | Surface | 27.8 ± 0.6 | 7.8 ± 0.6 | 3.6 ± 0.2 | High on surface | 16.7 | 3.1 | 7.9 | 0.065 | |
| 12065,– | Exterior chip | Surface | 49.5 ± 0.6 | 7.8 ± 0.2 | 6.4 ± 0.2 | High on surface | 17.1 | 2.3 | 9.0 | 0.06 | (1) |
| 14259,84 | Bulk fines | Surface | 78.2 ± 3.4 | 9.1 ± 0.5 | 8.6 ± 0.6 | 203 ± 4 | 7.8 | 1.1 | 7.8 | 0.42 | (2) |
| 14163,116 | Bulk fines | Surface | 46.4 ± 3.4 | 10.2 ± 0.6 | 4.6 ± 0.4 | — | — | — | — | — | |
| 15221,17 | Surface fines, near boulder | Surface | 34 ± 4 | 5.6 ± 0.8 | 6.1 ± 1.1 | 56 ± 4 | 10.3 | 0.95 | 9.6 | 0.19 | (3) |
| 15231,18 | Fines, under boulder | ~150 | 35 ± 5 | 14 ± 1 | 2.6 ± 0.4 | 56 ± 4 | 10.3 | 0.95 | 9.6 | 0.19 | |
| 15031,16 | Trench, surface | Surface | 25 ± 4 | 5.9 ± 1.4 | 4.2 ± 1.2 | 202 ± 5 | 13.3 | 1.3 | 8.8 | 0.18 | (4) |
| 15041,16 | Trench, bottom | ~70 | 28 ± 4 | 14 ± 2 | 2.0 ± 0.4 | 435 ± 8 | 13.3 | 1.3 | 8.8 | 0.18 | |

* References for composition: (1) LSPET (1970); (2) LSPET (1971); (3) LSPET (1972), used 15101; (4) LSPET (1972), used 15021.

heated to 100 to 130°C during the lunar day. The diffusion loss will be discussed in a later section.

## Thermal Release

We now turn to a consideration of the diffusion of gaseous radioactivity from minerals in order to explain some of the anomalies observed above. The topic has been reviewed by Fechtig and Kalbitzer (1966), from whom the equations below are taken. The fraction $F$ of the total available gas diffusing from a mineral sample at absolute temperature $T$ for a time $t$ depends on the ratio $D/a^2$, where $D$ is the diffusion constant and $a$ is the mean grain size. For fractional losses $F < 0.9$,

$$\frac{D}{a^2} = \frac{2}{\pi t}\left(1 - \frac{\pi F}{6} - \sqrt{1 - \frac{\pi F}{3}}\right). \tag{1}$$

When ln $(D/a^2)$ is plotted against $1/T$, a straight line is obtained if a single, simply activated, process is operating; the slope of the line is $-Q/R$, where $Q$ is the heat of activation for diffusion and $R$ the gas constant.

We have three separate experiments to consider on the rate of diffusion of $^{37}$Ar, $^{39}$Ar, and T from lunar soil: the heating curves for the two Apollo 15 soils, the room-temperature diffusion from the Apollo 15 SRC, and our previous laboratory bombardment of simulated lunar fines (Stoenner *et al.*, 1971). Data for $T$, $F$, and $t$ from the trench fines are collected in Table 3. For the SRC we take $T = 25$°C, $t = 7.5d$ from box closure on the moon, and $^{37}$Ar, $^{39}$Ar, and T activities in the 2.3 kg of fines of 31.5, 9.9, and 192 dpm/kg, respectively. Data from the bombardment are given in Fig. 2 of Stoenner *et al.* (1971).

When we plot log $(D/a^2)$ for each nuclide on a common graph against $1/T$ we find a straight line of surprisingly good quality considering the quite different nature of the three experiments and the wide range of temperatures examined. The plot for $^{39}$Ar is given in Fig. 1; the other two are similar. Parameters obtained from least-squares fits for the three nuclides are collected in Table 5.

It is notable that the rate of diffusion loss of tritium into the Apollo 15 SRC is quite in line with the other points of the heating curve. If there were large amounts of loosely bound tritium in the Apollo 15 soil the SRC point would be high. We discuss the implications of this in the next section.

Table 5. Temperature release of Ar and T activities.

| | $^{37}$Ar | $^{39}$Ar | T |
|---|---|---|---|
| Temperature range (°C) | −76 to 900 | −76 to 900 | −76 to 600 |
| Number of points | 13 | 13 | 11 |
| Slope of log $D/a^2$ against $1/T$* | −3300 ± 230 | −2240 ± 210 | −2840 ± 280 |
| Intercept* | −4.00 ± 0.66 | −5.47 ± 0.60 | −3.71 ± 0.88 |
| Activation energy $Q$ (kcal/mole) | 15.1 ± 1.0 | 10.3 ± 1.0 | 13.0 ± 1.3 |

* Errors quoted for slope and intercept are standard deviations calculated from the fit, based on Williamson (1968).

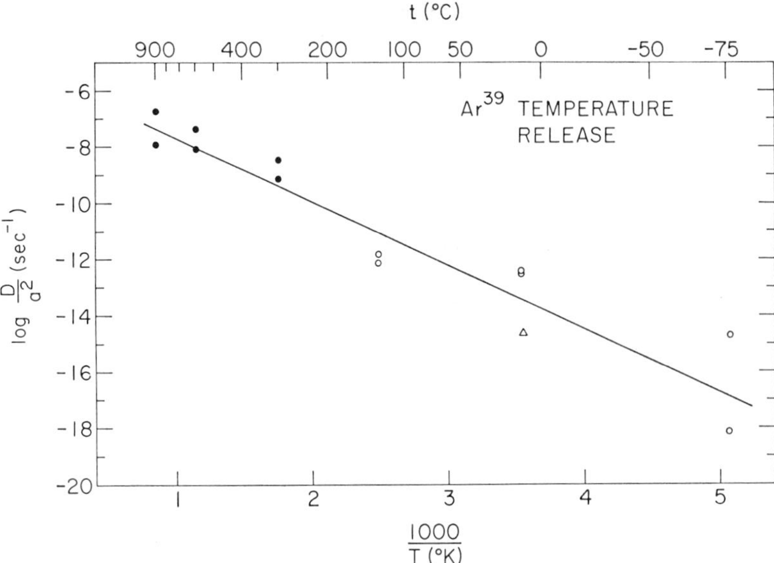


Fig. 1. Temperature release of $^{39}Ar$ from actual and simulated lunar fines. Solid points are the values obtained from stepwise heating of trench soils 15031,16 and 15041,16; open circles are the 600 MeV bombardment of minerals simulating lunar fines; and the open triangle is the Apollo 15 SRC. The vertical standard deviation of the points from the least-squares line is a factor 11.4.

The activation energies we give in Table 5 are twofold lower than the values found for the diffusion of Ar in meteorites (Fechtig and Kalbitzer, 1966) or hydrogen (Lord, 1967). There are several reasons for this.

At a given high temperature, particularly for the long heating times we employed, the finest particles may degas completely, with the result that as heating continues the gas is evolved from particles of larger $a^2$. There is independent evidence that the mean grain size in lunar soil is indeed very small. The surface area of 10087 fines has been measured to be 1 $m^2/g$ by adsorption of nitrogen by the BET method (Fuller *et al.*, 1971). A sample of uniform spheres of this specific area and density 3 $g/cm^3$ would have a diameter of 2 $\mu$, although the area obtained by summing over the sieve distribution determined by Heywood (1971) is 0.07 $m^2/g$. The difference is accounted for by adhesion of the finest particles to the surfaces of larger grains (McKay *et al.*, 1971; Arrhenius *et al.*, 1972).

A second reason for the low $Q$ we find is an increase of $D$ at low degassing rates by nonvolume diffusion along cracks and dislocations in the heavily shocked lunar soil. The curves in Fechtig and Kalbitzer (1966) clearly show this process overwhelming true volume diffusion at low temperatures. The coarse resolution of our experiments precludes separating the two components, with the result that we see a heat of activation for both volume and nonvolume diffusion.

Removal of rare gases and tritium by diffusion also takes place on the moon. An estimate of the loss of $^{39}Ar$, the longest lived nuclide we measure, may be approached by calculating an average lunar surface value of $D/a^2$. The exponential temperature dependence is given by the parameters in Table 5, and the variation of the lunar surface temperature with phase is approximated by a sine wave with minimum temperature 90°K and maximum 390°K. Then the average

$$\left(\frac{D}{a^2}\right)_{\text{av}} = \frac{1}{2\pi}\int_0^{2\pi} \exp\left[2.303\left(-5.47 - \frac{2240}{240 + 150\sin\phi}\right)\right] d\phi = 1.0 \times 10^{-12}\ \text{sec}^{-1}$$

for $^{39}$Ar. The fractional gas loss $F$ is then obtained from a rearrangement of equation (1). For $t = t_{1/2} = 269$ yr, $F_{39} = 0.29$. Similar calculations for $^{37}$Ar and T give fractional losses in one halflife of 0.0013 and 0.083, respectively. Clearly diffusive loss is important only for gases of long halflife.

Because of the strong damping of the diurnal heat wave with depth in the regolith, the loss rate just calculated is an upper limit. If we consider that the temperature range is decreased to $1/e$ of the surface value in 4 cm (Hoyt *et al.*, 1971; Robie and Hemingway, 1971), then the $^{39}$Ar deficiency we find is reasonably accounted for by a sampling depth comparable to the size of the scoop.

Reimplantation of escaping gas as a result of acceleration by the interplanetary electric field (Manka and Michel, 1970) and the temperate latitude of the Apollo 15 landing site both act further to reduce the importance of this degassing in our soil samples. The higher thermal conductivity and lower porosity of rocks reduce the effect still more, so that the observed $^{39}$Ar activity in these samples more nearly reflects the production rate.

## Tritium from the Sun

When surface lunar samples became available it was of great interest to search for tritium of solar origin. One recalls that Fireman and his associates (1961) observed excess amounts of tritium implanted in the Discoverer XVII satellite by the solar flare of November 12, 1960. Searches for implanted tritium in lunar materials have not revealed amounts in excess of that expected from cosmic ray interactions. One observation of extremely large amounts of lightly adsorbed tritium was reported by this laboratory on two Apollo 12 rocks, 12063 and 12065 (Stoenner *et al.*, 1971). This observation was made during the quarantine period, and unfortunately was not followed up immediately. The possibility of gross tritium contamination could not be eliminated at the time, and some additional searches for sources of tritium contamination will be described below.

If implanted, easily released, tritium is present in lunar samples, one might expect to find it present in the gas in the SRC. The Apollo 14 mission was the first opportunity to perform this measurement. In this mission a relatively small amount of charcoal was used as an adsorber. The stainless steel finger containing the charcoal and the valve assembly was sterilized at 160°C for 12 hours (a quarantine requirement) before removing the sample for counting. There was, of course, the possibility that tritium could be lost from the charcoal into the stainless steel during this heating. However, a test of this possibility was made at a later time with the trap assembly.

A known amount of hydrogen was adsorbed on the charcoal, the equivalent sterilization heating performed, and the hydrogen was recovered with greater than 90% yield. This experiment demonstrated that tritium losses during sterilization were small, and the measurements performed on the Apollo 14 SRC were valid. The measurements on Apollo 14 showed that very little tritium was present in the gas, even though the lunar surface was exposed to a solar flare 12 days prior to the mission. The small amount of tritium that was observed can easily be accounted for by diffusion from the fine material in the box. This was discussed quantitatively in an earlier section.

A second sample was obtained from the Apollo 15 SRC. In this experiment a larger amount of charcoal was used (10 g) and since there was no quarantine restriction, the assembly remained at room temperature (4 days) until the gas was removed. In addition, an activated vanadium metal hydrogen absorber (10 g) was used. This absorber will quantitatively absorb $H_2$ at room temperature. In this experiment $0.279 \pm 0.005$ dpm T was observed on the charcoal adsorber and $0.225 \pm 0.007$ dpm T was observed on the vanadium metal absorber. Again, it is clear that very little tritium is present in the gas phase of the sample return container. We can then safely conclude that there was little or no lightly adsorbed tritium present on the rock and soil samples returned by the Apollo 14 and 15 missions. Conditions for finding solar tritium were nearly ideal in the Apollo 14 mission. In the 12 days between a major flare and the landing, the sampling site was dark and cold, so tritium should have been retained well. We conclude that there was little or no tritium in the flare of January 24, 1971.

As mentioned earlier, large quantities of lightly adsorbed tritium was found on two rock samples measured during the Apollo 12 quarantine period. As much as 200 dpm of tritium was released at room temperature from a 10 g sample of rock, an enormous quantity. Because the amount of tritium was so large, we attributed it to gross surface contamination, though there was no obvious source of tritium in the spacecraft or at the Lunar Receiving Laboratory. Fireman and his associates (D'Amico *et al.*, 1971) examined one of these rocks (12065) for lightly adsorbed tritium 6 months later and found that no tritium was released from their sample at room temperature. This result could be explained easily if the tritium was only adsorbed on the surface. The question could be settled by finding a source of tritium contamination, but if none is found there is the real possibility that the large amount of tritium seen on these samples was of solar origin.

We would like now to describe some further efforts that were made to locate the sources of tritium contamination. Since these samples were returned from the lunar surface to the Lunar Receiving Laboratory not in a SRC but in the tote bag, the atmosphere inside the Command Module (CM) was considered as a possible source. A lithium hydroxide scrubber is used in the CM to remove carbon dioxide from the spacecraft atmosphere. Cosmic ray produced slow neutrons will yield tritium by the high cross section reaction $^6Li(n,\alpha)T$. One would expect that the tritium so formed would be retained as water in the lithium hydroxide, and consequently would be released to the atmosphere of the CM at a very low rate.

Samples of the air inside the CM from Apollo 14 were taken immediately after opening the hatch upon its delivery to the Lunar Receiving Laboratory. The CM had not been opened since the transfer of the astronauts to the Mobile Quarantine Facility aboard the recovery carrier, hence this should be a representative sample of the tritium level in the CM during the mission. Both tritium as hydrogen gas and tritium as water vapor were considered as possible chemical forms. Hydrogen gas was separated from the air by contacting with powdered vanadium metal at room temperature. The hydrogen was then evolved, diffused through hot palladium and counted. No tritium activity was observed in this fraction. The water vapor was separated from the air by freezing with liquid nitrogen and pumping away the air. The water was then treated with hot vanadium metal and the resulting hydrogen was purified and counted as above. The tritium activity in the CM was found to be $5.0 \pm 0.3$ dpm/liter of air, far too low to be considered a source of tritium contamination responsible for the large amount of tritium observed on rocks 12063 and 12065.

An additional sample was taken in the CM after the ventilation system had been started again, at a point near the lithium hydroxide cannister exhaust. Only the tritium in the water vapor was measured in this sample and it was found to be a little higher, $6.4 \pm 0.2$ dpm/liter, than the previous sample but still too low to contribute substantially to samples transported exposed in the CM. It should be remarked that the tritium levels observed in the water in these samples is very high compared to normal earth environmental samples. The T/H ratio corresponding to 5 dpm T per liter of air would be $50{,}000 \times 10^{-18}$, whereas terrestrial T/H ratios are from 20 to $50 \times 10^{-18}$. The CM levels are a factor of 1000 under maximum permissible concentrations for continuous exposure.

## $^{222}$Radon in the Sample Return Container and on the Moon

Material in the SRC should be, for the reasons mentioned above, the best sample on which to determine the emanation rate of $^{222}$Rn. Appreciable emanation should lead to an enrichment of daughter activities on the lunar surface. The topic has been widely studied (Yeh and Van Allen, 1969; Turkevich *et al.*, 1970; Economou and Turkevich, 1971; Stoenner *et al.*, 1971; Lindstrom *et al.*, 1971; Adams *et al.*, 1971; Yaniv and Heymann, 1972; Lambert *et al.*, 1972; Gorenstein and Bjorkholm, 1972) with considerable variation in the inferred flux of Rn into the lunar atmosphere. The present section is an attempt to rationalize this variation.

The $^{222}$Rn observed in the SRC is attributed to radon loss from the fine material contained in the box (Stoenner *et al.*, 1971). The $^{222}$Rn data from the Apollo 12, 14, and 15 missions are given in Table 2. To explain the quantities we observe, we discuss below the processes responsible for Rn emanation: recoil and stopping of daughter Rn from soil grains, and thermal diffusion through the pores of the soil.

Radon is first formed in the regolith by escape of recoil atoms resulting from $\alpha$ decay of $^{226}$Ra. The fraction of decays recoiling from a grain of diameter $a$ is

$$P = \frac{3}{2}\frac{L}{a} - \frac{1}{2}\left(\frac{L}{a}\right)^3$$

(Giletti and Kulp, 1955), where $L$ is the recoil range. Combination of this relation with the particle size distribution of Heywood (1971) shows that recoil from 5–15 $\mu$

grains accounts for nearly half the radon escape. For this size distribution the integrated escape is proportional to the range $L$, and $P = 1 \times 10^{-3}$ for $L = 200$ Å. This is indeed the fraction we find in the SRC (Table 2).

This calculation assumes that the size distribution of uranium-bearing phases is the same as that of the soil as a whole. In view of the observation of Burnett *et al.* (1971) and Haines *et al.* (1972) that U is concentrated in the finer grains of lunar rocks and soils, the calculation may be low and not all recoiling Rn atoms find their way into the atmosphere of the SRC.

The efficiency of release depends on thermalization and on diffusion of Rn atoms. The latter process is characterized by the diffusion constant $D$, which may be very different from values measured in terrestrial settings. In the absence of direct measurements we attempt to calculate $D$, following Friesen and Heymann (1972). From kinetic theory $D = x\bar{v}/3$, where $x$ is the mean free path and $\bar{v}$ the mean velocity of Rn atoms. In turn, $\bar{v} = x/(t_f + t_a)$, where $t_f = x\sqrt{\pi m/8kT}$ is the flight time across a distance $x$ for an atom of mass $m$ at temperature $T$. The second term $t_a$ is the time an atom spends adsorbed on a surface, and is given by $t_a = t_0 e^{Q/RT}$ (de Boer, 1953) where $Q$ is the heat of adsorption.

Values of these parameters may be estimated. The mean free path $x$ may be taken as $\bar{d}p$, where $\bar{d}$ is a typical grain size and $p$ the porosity. Taking $\bar{d} = 3\ \mu$ as the number median diameter (Görz *et al.*, 1971) and $p = 0.5$, a reasonable $x = 1.5\ \mu$. The characteristic sticking time $t_0$ is estimated from de Boer's (1953) equation $t_0 = 4.75 \times 10^{13}\sqrt{M_s V^{2/3}/T_s}$. Using the mean molecular weight $M$ and molar volume $V$ of lunar soil calculated from Morrison's (1971) compilation for 12070, and the melting temperature $T_s$ taken from Green *et al.* (1970), we calculate $t_0 = 1.1$–$1.2 \times 10^{-13}$ sec. For comparison, de Boer (1953) gives $t_0 = 0.75$, 0.95, and $0.67 \times 10^{-13}$ sec for $Al_2O_3$, $SiO_2$, and MgO, respectively. We take $t_0 = 0.9 \times 10^{-13}$ sec as a fair average. The heat of adsorption $Q$ of Rn on charcoal has been measured to be 7.51 kcal/mole from air (Gübeli and Störi, 1959) and estimated to be $13.5 \pm 1.4$ kcal/mole at low surface coverage in vacuum (Chackett and Tuck, 1957). In view of the low surface energy observed by Fuller *et al.* (1971) in adsorption of $N_2$ and CO on soil sample 10087, however, $Q$ in lunar soil may not be so high. The lower limit is the heat of vaporization of liquid Rn, which is 4.3 kcal/mole. The most likely range is $10 \pm 3$ kcal/mole. Measurement of this quantity is in progress.

The diffusion constant calculated from these numbers is $4 \times 10^{-1}$, $4 \times 10^{-3}$, and $3 \times 10^{-5}$ cm$^2$/sec for $T = 300$ K and $Q = 7$, 10, and 13 kcal/mole, respectively. These values are high enough that diffusion from the sample of soil in the rockbox is rapid during transport from the moon to the LRL. The exponential temperature dependence of the adsorption term, however, makes diffusion slow ($D = 10^{-5 \pm 3}$) at 225 K, the mean temperature of the regolith below a few decimeters in situ.

The flux $J$ of Rn emanating from the lunar surface is, from a simple diffusion model (Wilkening and Hand, 1960; Schroeder *et al.*, 1965),

$$J = C_\infty \sqrt{\lambda D}$$

where $C_\infty$ is the amount of Rn free to diffuse per unit volume of bulk soil, at a depth large compared to the diffusion length $\sqrt{2D/\lambda}$, and $D$ is the diffusion constant, and $\lambda$ is the decay constant of $^{222}$Rn.

Taking 1 ppm U, an emanating power (the product of recoil and stopping efficiencies) of $10^{-3}$ from our SRC measurements, and $D = 10^{-5}$ cm$^2$/sec, we calculate $J = 3 \times 10^{-7}$ atoms/cm$^2$ sec. This is to be compared with the lowest observed upper limit of $2 \times 10^{-5}$ alpha disintegrations/cm$^2$ sec (Lambert *et al.*, 1972). It should be clear that the present estimate is good to no better than an order of magnitude in $J$; the uncertainty in $Q$, and hence $D$, is the most important unknown at present.

*Acknowledgments*—We are indebted to the staff of the Lunar Receiving Laboratory, and especially the Gas Analysis Laboratory, for their assistance and the use of their facilities. The 600-MeV proton irradiation was performed by the staff of the NASA Space Radiation Effects Laboratory, Newport News, Va. We thank Don Bogard, Robert Reedy, and John Evans for useful discussions. This work was supported by NASA and the AEC.

## References

Adams J. A. S., Barretto P. M., Clark R. B., and Duval J. S. (1971) Radon-222 emanation and the high apparent lead isotope ages in lunar dust. *Nature* **231,** 174–175.

Arrhenius G., Asunmaa S. K., and Fitzgerald R. W. (1972) Electrostatic properties of lunar regolith (abstract). In *Lunar Science—III* (editor C. Watkins), pp. 30–32, Lunar Science Institute Contr. No. 88.

Burnett D., Monnin M., Seitz M., Walker R., and Yuhas D. (1971) Lunar astrology—U–Th distributions and fission-track dating of lunar samples. *Proc. Second Lunar Sci. Conf., Geochim. Cosmochim. Acta* Suppl. 2, Vol. 2, pp. 1503–1519. MIT Press.

Chackett K. F. and Tuck D. G. (1957) The heats of adsorption of the inert gases on charcoal at low pressure. *Trans. Farad. Soc.* **53,** 1652–1658.

D'Amico J., De Felice J., and Fireman E. L. (1970) The cosmic-ray and solar-flare bombardment of the moon. *Proc. Apollo 11 Lunar Sci. Conf., Geochim. Cosmochim. Acta* Suppl. 1, Vol. 2, pp. 1029–1036. Pergamon.

D'Amico J., De Felice J., Fireman E. L., Jones C., and Spannagel G. (1971) Tritium and argon radioactivities and their depth variations in Apollo 12 samples. *Proc. Second Lunar Sci. Conf., Geochim. Cosmochim. Acta* Suppl. 2, Vol. 2, pp. 1825–1839. MIT Press.

de Boer J. H. (1953) *The Dynamical Character of Adsorption.* Oxford Press.

Economou T. E. and Turkevich A. L. (1971) Examination of returned Surveyor III camera for alpha radioactivity. *Proc. Second Lunar Sci. Conf., Geochim. Cosmochim. Acta* Suppl. 2, Vol. 3, pp. 2699–2703. MIT Press.

Fechtig H. and Kalbitzer S. (1966) The diffusion of argon in potassium-bearing solids. In *Potassium Argon Dating* (editors: O. A. Schaeffer and J. Zahringer), pp. 68–107. Springer-Verlag.

Fireman E. L., De Felice J., and Tilles D. (1961) Solar flare tritium in a recovered satellite. *Phys. Rev.* **123,** 1935–1936.

Fireman E. L., D'Amico J., De Felice J., and Spannagel G. (1972) Radioactivities in Apollo 14 and 15 materials (abstract). In *Lunar Science—III* (editor C. Watkins), pp. 262–264, Lunar Science Institute Contr. No. 88.

Friesen L. J. and Heymann D. (1972) Model for radon diffusion through the lunar regolith. Proc. Conf. on Lunar Geophysics, *The Moon* (in press).

Fuller E. L., Holmes H. F., Gammage R. B., and Becker K. (1971) Interaction of gases with lunar materials: Preliminary results. *Proc. Second Lunar Sci. Conf., Geochim. Cosmochim. Acta* Suppl. 2, Vol. 3, pp. 2009–2019. MIT Press.

Giletti B. J. and Kulp J. L. (1955) Radon leakage from radioactive minerals. *Amer. Mineral.* **40,** 481–496.

Gorenstein P. and Bjorkholm P. (1972) Results of the Apollo 15 alpha particle spectrometer experiment (abstract). In *Lunar Science—III* (editor C. Watkins), pp. 326–328, Lunar Science Institute Contr. No. 88.

Görz H., White E. W., Roy R., and Johnson G. G. (1971) Particle size and shape distributions by CESEMI. *Proc. Second Lunar Sci. Conf., Geochim. Cosmochim. Acta* Suppl. 2, Vol. 3, pp. 2021–2025. MIT Press.

Green D. H., Ringwood A. E., Ware N. G., Hibberson W. O., Major A., and Kiss E. (1971) Experimental petrology and petrogenesis of Apollo 12 basalts. *Proc. Second Lunar Sci. Conf., Geochim. Cosmochim. Acta* Suppl. 2, Vol. 1, pp. 601–615. MIT Press.

Gübeli O. and Störi M. (1954) Zur Mischadsorption von Radon an Aktivkohle mit verschiedenen Trägergasen. *Helv. Chim. Acta* **37,** 2224–2230.

Haines E. L., Gancarz A. J., Albee A. L., and Wasserburg G. J. (1972) The uranium distribution in lunar soils and rocks 12013 and 14310 (abstract). In *Lunar Science—III* (editor C. Watkins), pp. 350–352, Lunar Science Institute Contr. No. 88.

Heywood H. (1971) Particle size and shape distribution for lunar fines sample 12057,72. *Proc. Second Lunar Sci. Conf., Geochim. Cosmochim. Acta* Suppl. 2, Vol. 3, pp. 1989–2001. MIT Press.

Hoyt H. P., Miyajima M., Walker R. M., Zimmerman D. W., Zimmerman J., Britton D., and Kardos J. L. (1971) Radiation dose rates and thermal gradients in the lunar regolith: Thermoluminescence and DTA of Apollo 12 samples. *Proc. Second Lunar Sci. Conf., Geochim. Cosmochim. Acta* Suppl. 2, Vol. 3, pp. 2245–2263.

Lambert G., Grejebine T., Le Roulley J. C., and Bristeau P. (1972) Alpha spectrometry of a surface exposed lunar rock (abstract). In *Lunar Science—III* (editor C. Watkins), pp. 472–474, Lunar Science Institute Contr. No. 88.

Lindstrom R. M., Evans J. C., Finkel R. C., and Arnold J. R. (1971) Radon emanation from the lunar surface. *Earth Planet. Sci. Lett.* **11,** 254–256.

Lord H. C. (1967) Ph.D. Thesis, University of California, San Diego.

LSPET (Lunar Sample Preliminary Examination Team) (1970) Preliminary examination of lunar samples from Apollo 12. *Science* **167,** 1325–1339.

LSPET (Lunar Sample Preliminary Examination Team) (1971) Preliminary examination of lunar samples from Apollo 14. *Science* **173,** 681–693.

LSPET (Lunar Sample Preliminary Examination Team) (1972) The Apollo 15 lunar samples: A preliminary description. *Science* **175,** 363–375.

Lyman W. J. (1971) Production of $^{37}Ar$, $^{39}Ar$, $^{42}Ar$, and T induced by 600-MeV proton bombardments on Ca, K, Ti and a simulated lunar target. BNL preliminary report, available on request from the author.

Manka R. H. and Michel F. C. (1970) Lunar atmosphere as a source of argon-40 and other lunar surface elements. *Science* **169,** 278–280.

McKay D. S., Morrison D. A., Clanton U. S., Ladle H. G., and Lindsay J. F. (1971) Apollo 12 soils and breccia. *Proc. Second Lunar Sci. Conf., Geochim. Cosmochim. Acta* Suppl. 2, Vol. 1, pp. 755–773. MIT Press.

Morrison G. H. (1971) Evaluation of lunar elemental analyses. *Anal. Chem.* **43** (7), 22A–31A.

Reedy R. C. and Arnold J. R. (1972) Interaction of solar and galactic cosmic-ray particles with the moon. *J. Geophys. Res.* **77,** 537–555.

Robie R. A. and Hemingway B. S. (1971) Specific heats of the lunar breccia (10021) and olivine dolerite (12018) between 90° and 350° Kelvin. *Proc. Second Lunar Sci. Conf., Geochim. Cosmochim. Acta* Suppl. 2, Vol. 3, pp. 2361–2365. MIT Press.

Schroeder G. L., Kraner H. W., and Evans R. D. (1965) Diffusion of radon in several naturally occurring soil types. *J. Geophys. Res.* **70,** 471–474.

Stoenner R. W., Lyman W. J., and Davis R. Jr. (1970) Cosmic-ray production of rare-gas radioactivities and tritium in lunar material. *Proc. Apollo 11 Lunar Sci. Conf., Geochim. Cosmochim. Acta* Suppl. 1, Vol. 2, pp. 1583–1594. Pergamon.

Stoenner R. W., Lyman W., and Davis R. Jr. (1971) Radioactive rare gases and tritium in lunar rocks and in the sample return container. *Proc. Second Lunar Sci. Conf., Geochim. Cosmochim. Acta* Suppl. 2, Vol. 2, pp. 1813–1823. MIT Press.

Turkevich A. L., Patterson J. H., Franzgrote E. J., Sowinski K. P., and Economou T. E. (1970) Alpha radioactivity of the lunar surface at the landing sites of Surveyors 5, 6, and 7. *Science* **167,** 1722–1724.

Wilkening M. H. and Hand J. E. (1960) Radon flux at the earth–air interface. *J. Geophys. Res.* **65,** 3367–3370.

Williamson J. H. (1968) Least-squares fitting of a straight line. *Can. J. Phys.* **46,** 1845–1847.

Yaniv A. and Heymann D. (1972) Radon emanation from Apollo 11, 12, and 14 fines (abstract). In *Lunar Science—III* (editor C. Watkins), pp. 816–818, Lunar Science Institute Contr. No. 88.

Yeh R. S. and Van Allen J. A. (1969) Alpha-particle emissivity of the moon: An observed upper limit. *Science* **166,** 370–372.

Proceedings of the Third Lunar Science Conference
(Supplement 3, *Geochimica et Cosmochimica Acta*)
Vol. 2, pp. 1719–1732
The M.I.T. Press, 1972

# Cosmogenic nuclides in football-sized rocks

M. WAHLEN, M. HONDA,* M. IMAMURA, J. S. FRUCHTER,† R. C. FINKEL, C. P. KOHL, J. R. ARNOLD, and R. C. REEDY

Department of Chemistry, University of California,
San Diego, La Jolla, California

**Abstract**—Short- and long-lived radionuclides were measured in rock 14321 at depths of $\sim 0.3$ g/cm$^2$, $\sim 1.1$ g/cm$^2$, 18 g/cm$^2$, and 40 g/cm$^2$. The effects of solar cosmic rays are seen in the upper samples of 14321. After correcting for an attrition to the rock (loss of surface material in handling) of $(0.4 \pm 0.2)$ mm, the $Co^{56}$ measurements imply, for the flare of January 24, 1971, a $J$ (>10 MeV) of 200 protons/cm$^2$ sec $(4\pi)$ averaged over the mean life of $Co^{56}$ and an $R_0$ (exponential rigidity parameter) of 100 MV, in good agreement with satellite data (Bostrom *et al.*). The $Fe^{55}$ and $Na^{22}$ values, which average over many flares, are consistent with our rock 12002 measurements (Finkel *et al.*). The near surface profiles of the longer-lived isotopes, $Mn^{53}$ and $Al^{26}$, suggest an erosion rate of about 2 mm/$10^6$ yr. $Mn^{54}$, $Cl^{36}$, and $Be^{10}$ were also measured in 14321.

$Fe^{55}$, $Be^{10}$, and $Mn^{53}$ were measured in a slice of rock 14310 from 24 g/cm$^2$ average depth. This sample was also separated into two density fractions enabling the production rates for $Na^{22}$, $Cl^{36}$, and $Al^{26}$ from different target elements to be determined.

The activities in our sample of 14310 and in our deepest sample of 14321 have no contribution from solar cosmic rays and give the beginning of the galactic cosmic-ray depth profile. A brief comparison between the data and the Reedy–Arnold model (Reedy and Arnold) is discussed.

We also measured $Co^{56}$, $Na^{22}$, $Al^{26}$, and $Mn^{53}$ in a sample of soil 14259. The observed solar cosmic ray produced activities are consistent with an average depth for the soil sample of 1 cm (1.8 g/cm$^2$).

## INTRODUCTION

THE STUDY OF lunar surface samples from the Apollo 11 and Apollo 12 missions has shown that the production of cosmogenic nuclides can be accounted for by bombardment by two classes of high-energy particles. The solar cosmic rays (SCR) produce significant effects on a depth scale of a few g/cm$^2$, while the galactic cosmic rays (GCR) penetrate to greater depths (Finkel *et al.*, 1971; Rancitelli *et al.*, 1971; O'Kelley *et al.*, 1971).

Rock 14321, a 9 kg fragmental rock whose orientation was photographically documented on the lunar surface, provided an opportunity to extend our study of the variation of production of radionuclides with depth down to about 40 g/cm$^2$, a region where according to present ideas only GCR effects should be important. We therefore measured the activity of long- and short-lived isotopes in a series of samples from a vertical column through the center of this rock, 14321. Also investigated was a sample from the lower portion of rock 14310 where, in order to study target effects, two different density fractions (mineral separates) were analyzed. In addition we

* Present address: Institute for Solid State Physics, University of Tokyo, Tokyo, Japan.

† Present address: Center for Volcanology, Department of Geology, University of Oregon, Eugene, Oregon.

measured a few nuclides in a sample from the comprehensive fines 14259, material reported to have been collected largely from the top centimeter of the lunar soil.

The study of the deep samples of 14321 and 14310 provided us with values for the activity of isotopes at points where only GCR effects are significant. At the same time, the results obtained from the surface samples of 14321 confirm the effects of solar cosmic rays observed earlier.

## EXPERIMENTAL TECHNIQUES

*Sample description*

Three categories of samples were used in this study. 14259,18 (FS) was a 5 g sample of lunar soil with a nominal depth of 0–1 cm (0–1.8 g/cm$^2$).

The second sample 14310,47 was a 5.5 cm long column taken from the bottom center of rock 14310. This rock is a fine grained igneous rock of density 2.9 g/cm$^3$ and weighs 3.4 kg. Its orientation on the lunar surface has been evaluated by counting in Houston (LSPET, 1971). The depth of our sample below the top of the rock was from 5.5–11 cm (Fig. 1).

Our most extensive set of samples was obtained from rock 14321, a complex fragmental rock of approximately 9 kg total mass and density 3.0 g/cm$^3$. The orientation of this "football-sized" rock on the lunar surface was documented photographically by the Apollo 14 astronauts. It was buried to ~4 cm in the lunar soil (cf. waterline Fig. 1). The three samples which we received (14321,91; 14321,92; 14321,123) were slices sawn from a large vertical column which was cut so that its base would be the most shielded position in the rock (cf. Fig. 1). The uppermost slice was further subdivided by grinding into samples of nominal depth 0–2 mm, FM-1, and 2–5 mm, FM-2, with an uncertainty of 0.5 mm. A large pit covered about 10% of the surface area. The middle sample, FM-3, has a mean

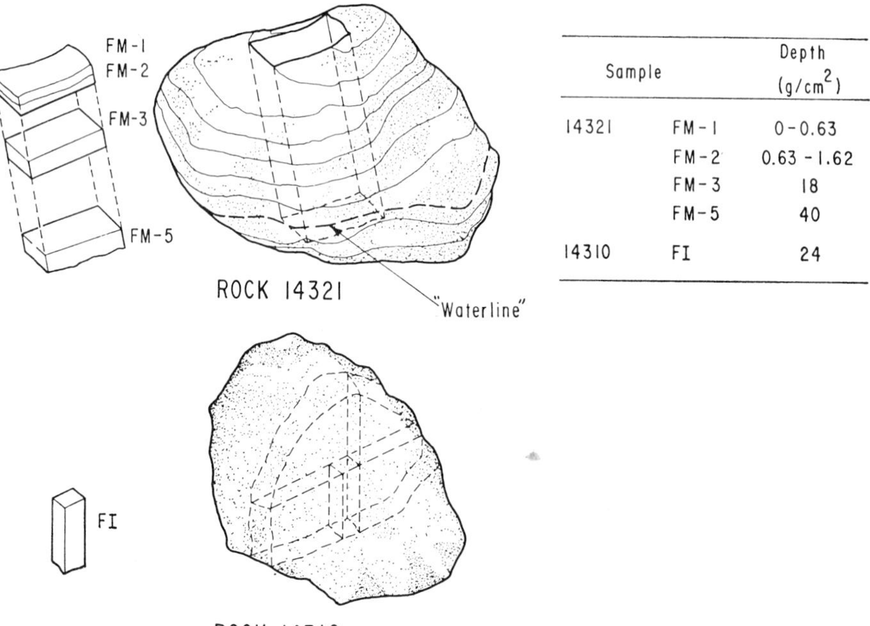


| Sample | | Depth (g/cm$^2$) |
|---|---|---|
| 14321 | FM-1 | 0-0.63 |
| | FM-2 | 0.63-1.62 |
| | FM-3 | 18 |
| | FM-5 | 40 |
| 14310 | FI | 24 |

Fig. 1. Location of the samples from rocks 14321 and 14310.

depth of about 18 g/cm$^2$ and a thickness of $\sim$3 g/cm$^2$; the bottom sample, FM-5, a mean depth of 40 g/cm$^2$ and a thickness of $\sim$3 g/cm$^2$.

*Chemical procedure*

Some mineral grains were picked from each sample for track analysis (Arrhenius, priv. comm.), and samples were also taken for rare gas studies and Gd-isotope analysis (Lugmair and Marti, 1971; Lugmair and Marti, 1972). The rock slices were then ground to below 120 $\mu$ size under isopropyl alcohol. Grinding was difficult because of the mixture of hard and soft regions, which are characteristic of such a fragmental rock.

In order to get information on target effects in the galactic cosmic ray production of certain radionuclides, 27 g of the sample of 14310 was subjected to a mineral separation procedure. The sample, ground to finer than 120 $\mu$ size, was separated into 3 density fractions using aqueous Tl-malonate-formate heavy liquid. The density fractions were selected with the goal of separating the two major mineral components, plagioclase ($\rho < 2.8$ g/cm$^3$) and clinopyroxene ($\rho > 3.2$ g/cm$^3$). The heavy density separate, FIH ($\rho > 3.22$), made up 36% of the total sample; the middle density separate, FIM ($3.0 \leq \rho \leq 3.22$), 14%; and the light density separate, FIL ($\rho < 3.0$), 50%. Only FIH and FIL were used, since the middle density fraction was regarded as a mixture of clinopyroxene and plagioclase. The separation of clinopyroxene and plagioclase obtained seemed to be satisfactory, as can be seen from the chemical compositions of FIH and FIL (Table 1). We were able to observe target effects in the induced activities of $Al^{26}$, $Na^{22}$, and $Cl^{36}$.

Table 1. Chemical composition of Apollo 14 samples determined by atomic absorption spectroscopy and colorimetry.

| | 14321 | | | | 14259 | 14310 | | |
|---|---|---|---|---|---|---|---|---|
| | FM-1 | FM-2 | FM-3 | FM-5 | FS | FIH | FIL | FI (calc.) |
| Na (ppm) | 4890 | 5260 | 5500 | 5330 | 4800 | 1540 | 7830 | 5560 |
| Mg (%) | 6.3 | 6.4 | 6.4 | | 5.6 | 11.5 | 1.0 | 4.8 |
| Al (%) | 7.8 | 8.0 | 8.3 | 8.2 | 9.2 | 3.0 | 16.3 | 11.5 |
| Si (%) | | | 21.9 | | 22.3 | 23.3 | 21.1 | 21.9 |
| K (ppm) | 3100 | 4000 | 4800 | 4000 | 4100 | 1580 | 5040 | 3800 |
| Ca (%) | 7.8 | 7.7 | 7.9 | 7.8 | 7.7 | 5.0 | 11.9 | 9.4 |
| Ti (%) | | | 1.4 | | 1.0 | 1.7 | 0.42 | 0.87 |
| Mn (ppm) | 1460 | 1440 | 1210 | 1280 | 1050 | 1830 | 278 | 840 |
| Fe (%) | 10.9 | 10.9 | 9.5 | 10.5 | 8.0 | 13.6 | 2.4 | 6.4 |

About 0.25 g of each sample was used for the determination of the concentration of the chief target elements. For the soil, 14259, where the size of our sample did not permit us to take such an aliquot, the data were adopted from analyses by other investigators. Measurements were done by atomic absorption spectrometry for all elements except Ti, which was determined by colorimetry. For Si, $LiBO_2$ fusion followed by atomic absorption spectrometry was applied using a separate 50 mg aliquot (Ingamells, 1966). For Cl the results of Reed *et al.* (1972) were used.

The chemical separation procedure for the radionuclides was essentially the same as in Apollo 11 and Apollo 12 (Shrelldalff, 1970; Finkel *et al.*, 1971) with some small modification in the Co and Mn chemistries. Be, Ar, Cl, Co, and I carriers were added at the stage of the still chemistry and Be, Na, Al, Cl, Mn, Fe, and Co were then separated and purified.

The counting sample for $Co^{56}$ was plated on a copper sample holder in an area of 1.8 cm $\times$ 3 cm from a plating solution of hydrazine sulfate (1 g/100 cc), ammonium tartrate (0.3 g/100 cc), and ammonium sulfate (5 g/100 cc). The $p$H was held between 5–7 with $NH_4OH$ and an applied voltage of 2 V was used.

$Mn^{53}$ was determined by neutron activation analysis following a procedure similar to that described in Finkel *et al.* (1971).

*Counting*

The counting was done with essentially the same detectors and equipment as previously reported (Shrelldalff, 1970; Finkel *et al.*, 1971), except for the modifications described below.

For isotopes such as $Fe^{55}$, which decay purely by electron capture, the method of pulse-rise-time discrimination has been applied (Davis *et al.*, priv. comm.) to the existing proportional counters in order to increase the precision of the measurements. This method makes use of the fact that in a given proportional counter the x-rays following an electron capture decay cause output signals different in rise-time from those triggered by "background events." X-rays of a few keV, because of their short ranged secondary ionization (interaction of the primary photo-electron with the counter gas) produce a response signal of short rise-time, typically $\sim$10 nsec. On the other hand, "background events" (charged cosmic ray particles, Compton electrons, interactions in the ineffective volume) lead to longer-range ionization, thus producing signals with considerably longer rise-times, typically $\sim$100 nsec. To take advantage of this difference the pulse-height analysis is performed in a two-parameter mode, using two separate amplifiers (with their differentiation-time-constants of about 10 nsec and 100 nsec, respectively) and two ADC's. In order to get sufficient resolution, a large memory array is necessary. This large memory has been obtained by interfacing the counter system on-line to an IBM-1800 computer (with large disc storage capacity) allowing us to obtain a maximum memory array of 256 $\times$ 256 channels. By this method the background was reduced in the case of $Fe^{55}$ by a factor 4 to 5, to 0.016 cpm without a substantial loss of efficiency.

A 40 cc Ge-Li well-detector (diameter of the well = 5 mm) was used for the measurement of $Mn^{54}$. For the 835 keV line of $Mn^{54}$ the resolution with this detector is 3.5 keV FWHM. The absolute total photopeak efficiency is 3.4%. The corresponding background, when operated in anticoincidence with a NaI guard annulus, is 0.07 cpm.

## RESULTS

*Counting*

The concentrations of radioisotopes found in the samples of rock 14321 and 14310 (including separate phases) and in the soil sample 14259 are given in Table 2. The activities are corrected to 6 February 1971. The errors quoted include (quadratically added) a one-standard deviation counting error, a 5% error due to calibration standards used for the determination of the detector efficiencies, a 5 to 10% error in the self-absorption correction where applicable, and a 5% error for the chemical yield. In the case of $Co^{56}$, where three commercially available standard sources differed widely among themselves, we allowed for a 10% error in the absolute efficiency.

Background counting was performed using blank samples made from Indian Ocean basalts (with appropriate amounts of K, Th, and U mixed in), which underwent the same chemical treatment as the actual lunar samples. In the case of rock 14310 the results for $Fe^{55}$, $Mn^{53}$, and $Be^{10}$ denoted as FI are from direct measurements of the combined elemental fractions of FIL and FIH, whereas those for $Al^{26}$, $Na^{22}$, and $Cl^{36}$ are calculated from the measurements on the separate phases.

Keith *et al.* (1972) measured several cosmogenic radionuclides in soil sample 14259,18. Their results and ours are, respectively [$Co^{56}$ (60 $\pm$ 30, 47 $\pm$ 24); $Al^{26}$ (222 $\pm$ 9, 170 $\pm$ 21); $Na^{22}$ (91 $\pm$ 8, 89 $\pm$ 14)]. The agreement is quite good except for $Al^{26}$ where there seems to be some discrepancy. Several groups (Keith *et al.*, 1972; Rancitelli *et al.*, 1972; Eldridge *et al.*, 1972) made measurements of cosmogenic radionuclides similar to ours in rocks 14321 and 14310. The data available at present do not allow a detailed comparison with our measurements. The results, however, do seem to be in general agreement.

Table 2. Counting results (dpm/kg rock), corrected to February 6, 1971.

| Sample | | | Isotope | | | | | | | |
|---|---|---|---|---|---|---|---|---|---|---|
| | Wt. (g) | Depth (g/cm$^2$) | $Co^{56}$ (77 days) | $Mn^{54}$ (303 days) | $Fe^{55}$ (2.6 yr) | $Na^{22}$ (2.6 yr) | $Al^{26}$ ($7.4 \times 10^5$ yr) | $Mn^{53}$ ($3.7 \times 10^6$ yr) | $Cl^{36}$ ($3.1 \times 10^5$ yr) | $Be^{10}$ ($2.5 \times 10^6$ yr) |
| 14321,91 | | | | | | | | | | |
| FM-1 | 6.9 | 0.0–0.63* | 220 ± 45 | 41 ± 11 | 310 ± 40 | 116 ± 13 | 148 ± 16 | 44 ± 3 | 10.7 ± 1.5 | 12.8 ± 2.3 |
| FM-2 | 19.0 | 0.63–1.62* | 78 ± 30 | | 135 ± 20 | 71 ± 7 | 109 ± 13 | 36 ± 2 | | |
| 14321,92 | | | | | | | | | | |
| FM-3 | 27.5 | 18 | | 21 ± 6 | 26 ± 7 | 39 ± 5 | 57 ± 7 | 23 ± 2 | 11.2 ± 1.6 | 13.8 ± 2.0 |
| 14321,123 | | | | | | | | | | |
| FM-5 | 42 | 40 | | 27 ± 5 | 34 ± 8 | 32 ± 4 | 54 ± 8 | 25 ± 2 | 12.1 ± 1.6 | 12.3 ± 2.0 |
| 14310,47 | | | | | | | | | | |
| FIL | 11.6 | 24 | | | | 28 ± 6 | 80 ± 13 | | 15.9 ± 2.2 | |
| FIH | 8.7 | 24 | | | | 38 ± 5 | 47 ± 7 | | 7.3 ± 1.5 | |
| Total | | | | | | | | | | |
| FI | | 24 | | | 22 ± 6 | 32 ± 4 calc. | 65 ± 9 calc. | 16 ± 1 | 12.3 ± 2.0 calc. | 13.1 ± 2.3 |
| 14259,18 | | | | | | | | | | |
| FS | 4.4 | 0.0–1.8† | 47 ± 24 | | | 89 ± 14 | 170 ± 21 | 44 ± 2 | | |

* This number does not include an attrition of 0.12 g/cm$^2$.

† Reported depth LSPET (1971); most probable depth 0–3.6 g/cm$^2$ derived from our measurements.

*Chemical composition*

The chemical composition of each sample is shown in Table 1. That of 14259 was taken from the data of several authors (Schnetzler *et al.*, 1971; Wänke *et al.*, 1972; Jackson *et al.*, 1972; Klein *et al.*, 1972). The variation of the chemical composition of 14321 from sample to sample was small but not negligible.

## DISCUSSION

*Solar cosmic ray effects*

The surface samples FM-1 and FM-2 (and FS) show the now familiar effects due to solar cosmic-ray bombardment.

$Co^{56}$ in these samples is essentially produced only by SCR's. Because of its short half-life, $Co^{56}$ production is dominated by recent solar flares. We know of only one flare of importance, that of 24 January 1971, that occurred close enough to the time of the Apollo 14 mission to have produced a significant amount of $Co^{56}$.

Comparison of the measured activities of $Co^{56}$, 220 $\pm$ 45 dpm/kg for FM-1 and 78 $\pm$ 30 dpm/kg for FM-2, with activities predicted by theoretical calculations based on satellite data (Bostrom *et al.*, 1971) for the January 1971 flare immediately implies that considerable attrition, i.e., loss of surface material due to handling and transit of the rock, has occurred. This fact is confirmed when one examines the activities observed for $Fe^{55}$ and $Na^{22}$. The best estimate for the amount of attrition as derived from these three isotopes is (0.4 $\pm$ 0.2) mm. This estimate is in excellent agreement with an estimate derived from comparison with the $Co^{56}$ measurements of Schonfeld (private communication) on rock 14310. After correction for the different Fe content, these measurements show a total activity (dpm/cm$^2$) about one-quarter greater in rock 14310 than in 14321. This excess is just what one would expect if rock 14321 had lost about 0.4 mm after its collection from the moon. Such a loss is very much in accord with the friability of 14321, a fragmental rock, as opposed to the strongly cohesive nature of 14310.

The best fit of our $Co^{56}$ data to the calculated values is obtained if we use an exponential rigidity shape with an $R_0$ of 100 MV and an equivalent steady state flux of $J$ ($E >$ 10 MeV) = 200 protons/cm$^2$ sec ($4\pi$) for the spectral shape and the intensity of the 24 January flare. The values reported from the data obtained by the solar proton monitor experiment (Bostrom *et al.*, 1971) correspond to an $R_0$ of about 70 MV and $J$ = 160 p/cm$^2$ sec ($4\pi$) [or $J$ ($E >$ 10 MeV) = $1.5 \times 10^9$ protons/cm$^2$ for the whole flare]. If we adopt the value of $R_0$ = 75 MV, to which our data also fit well within the limits of error, we find a $J$ of 175 p/cm$^2$ sec ($4\pi$) in good agreement with the satellite data. This flare seems to have been approximately 50% more intense and perhaps somewhat harder than the November 1969 flare responsible for the $Co^{56}$ production we observed in rock 12002.

In the light of this analysis the measurements of 35-day $Ar^{37}$ (Fireman *et al.*, 1972) on surface samples of 14321 similar to ours are quite puzzling. We can correct for GCR production by using measurements of $Ar^{37}$ in rock 15555 (Fireman *et al.*, 1972) which, because of the lack of flares immediately before the Apollo 15 mission, show essentially no SCR effect. The net solar cosmic-ray production profile in 14321 is then seen to be relatively flat, indicating for the 24 January 1971 flare a spectral shape $R_0 > 150$ MV and an equivalent steady state flux $J < 100$ p/cm$^2$ sec ($4\pi$). These results are in disagreement with the $Co^{56}$ data discussed above. At present the most likely reason for this discrepancy lies in the great uncertainty in the excitation functions for the production of $Ar^{37}$ from Ca, which are required for deducing flare characteristics from activity measurements. A more detailed discussion of this problem is given in Fireman *et al.* (1972a).

The depth gradients of $Fe^{55}$ and $Na^{22}$ ($t_{1/2} = 2.6$ yr) allow us to derive the mean spectral shape and intensity of the solar flare particle flux to which this rock was exposed, weighted according to the mean life of these isotopes (3.7 yr). In order to determine the characteristics of the bombarding SCR's we must first subtract that portion of the activity produced by GCR bombardment. The GCR contribution to each sample was determined by normalizing the theoretical depth profile (Reedy and Arnold, 1972) to the activity measured in the FM-5 sample. At the depth of this sample (40 g/cm$^2$) there is no appreciable production by SCR's. The shape of the GCR-produced profile will be discussed in more detail in the next section.

From both isotopes, $Fe^{55}$ and $Na^{22}$, we derive an $R_0 = 85$ MV for the spectral shape and a $J = 100$ p/cm$^2$ sec ($4\pi$) for the mean intensity of the solar flare particle flux responsible for the production of these nuclides. The loss of surface material by attrition was taken to be $(0.4 \pm 0.2)$ mm.

In order to extract the same information from the data on the long-lived nuclei, $Al^{26}$ and $Mn^{53}$, one has to consider two additional parameters: the surface irradiation age, which will affect the extent to which activity values have reached equilibrium with production rates, and the rate of erosion while on the surface of the moon. Undersaturation can be excluded by the Kr–Kr age of $27 \times 10^6$ yr (Lugmair and Marti, 1972), the track age of $(25 \pm 3) \times 10^6$ yr (Crozaz *et al.*, 1972), and the $Ar^{38}$ spallation age of $(24 \pm 2) \times 10^6$ yr (Burnett *et al.*, 1972). Using both $Al^{26}$ and $Mn^{53}$ one obtains, after correcting for attrition, the best fit to theoretical SCR produced activity gradients with a surface erosion rate of 1.5–3.0 mm/$10^6$ yr. The best fitting spectral shapes and intensities are then $R_0 = 85$–$100$ MV and $J = 70$ protons/cm$^2$ sec ($4\pi$) for $Al^{26}$, and $R_0 = 100$ MV and $J = 100$ protons/cm$^2$ sec ($4\pi$) for $Mn^{53}$. Figure 2 illustrates the range of values of $R_0$, $J$, and erosion rates which will fit the experimental profiles. The slight difference in the value of $J$ derived from the two "profiles" is not significant considering the precision of the data.

It is logical to assume that the rocks 12002 and 14321 both experienced the same cosmic ray particle flux, because both have long surface exposure ages. We can therefore compare the solar produced activity profiles obtained from 14321 to those of rock 12002 (Finkel *et al.*, 1971). Using our newly measured values for pure galactic cosmic ray production we can reexamine the solar profiles of rock 12002. Such a recalculation shows that the galactic production within this rock has been overestimated in the deeper samples OP-4, OP-5, and OP-6. This led to an underestimation of the solar production. The increased net solar production in the deeper samples thus would seem to imply that for rock 12002 the $R_0$'s should be higher than those given in Finkel *et al.* (1971). However, at least for $Mn^{53}$, this possibility is precluded

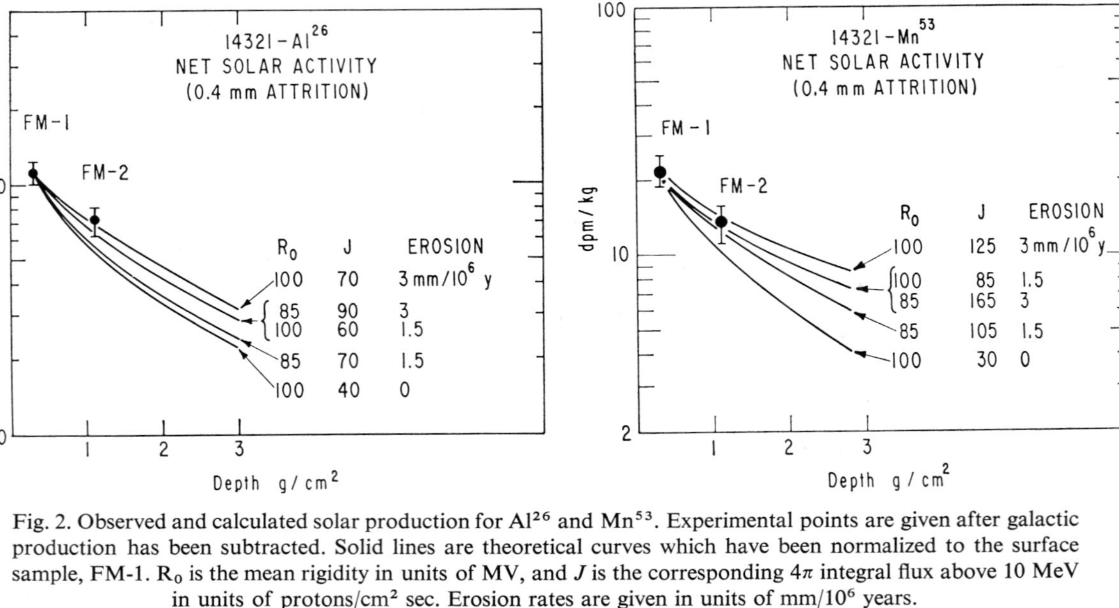


Fig. 2. Observed and calculated solar production for $Al^{26}$ and $Mn^{53}$. Experimental points are given after galactic production has been subtracted. Solid lines are theoretical curves which have been normalized to the surface sample, FM-1. $R_0$ is the mean rigidity in units of MV, and $J$ is the corresponding $4\pi$ integral flux above 10 MeV in units of protons/cm$^2$ sec. Erosion rates are given in units of mm/$10^6$ years.

by the $Be^{10}$ results. The complete lack of any observed solar cosmic ray produced $Be^{10}$ sets an upper limit for $R_0$ of about 150 MV for the average long-term parameter of the solar flare particle spectrum.

Therefore, a more likely explanation of the flat shape for both $Mn^{53}$ and $Al^{26}$, as well as for other isotopes, is that the infinite plane geometry previously used for assessing solar cosmic ray produced activities underestimates production below 3 g/cm$^2$. If we instead use a more realistic hemispherical geometry with a radius of 21 g/cm$^2$ for the calculation of solar cosmic ray production profiles, the best fit to the higher net solar values mentioned above is then found (assuming erosion rates of 0.5 mm/$10^6$ yr for 12002 and 2 mm/$10^6$ yr for 14321) with the following $R_0$'s and $J$'s:

$$\begin{array}{lll} Na^{22} & R_0 = 85 \text{ MV} & J(E > 10 \text{ MeV}) = 110 \text{ protons/cm}^2 \text{ sec } (4\pi) \\ Fe^{55} & R_0 = 100 \text{ MV} & J(E > 10 \text{ MeV}) = 100 \text{ protons/cm}^2 \text{ sec } (4\pi) \\ Al^{26} & R_0 = 100 \text{ MV} & J(E > 10 \text{ MeV}) = 80 \text{ protons/cm}^2 \text{ sec } (4\pi) \\ Mn^{53} & R_0 = 100 \text{ MV} & J(E > 10 \text{ MeV}) = 90 \text{ protons/cm}^2 \text{ sec } (4\pi). \end{array}$$

These values agree with those derived from 14321 above and are essentially the same as in Finkel *et al.* (1972).

Again one notices the striking fact that the fluxes and rigidities observed from $Al^{26}$ and $Mn^{53}$ are very similar to those obtained from the short-lived species. One exception to this conclusion is the $C^{14}$ results of Begemann *et al.* (1972). These results, which are in general agreement with unpublished results of Suess and Boeckl (private communication) on rock 12002, indicate an increase in $C^{14}$ production of about a factor of 4 over that predicted from a steady state solar activity as derived from other isotopic measurements. The meaning of this result is as yet uncertain. The result on another long-lived isotope, 24 my $U^{236}$, reported by Fields *et al.* (1972), may also be partially explained by cosmic ray bombardment.

The results on the soil sample 14259,18, which is reported to be material from the top centimeter of the surface LSPET (1971) look puzzling at first. The $Co^{56}$, $Al^{26}$, and $Na^{22}$ activities are all in good agreement with values obtained in the same sample by other workers. The low $Co^{56}$ value (although the error is quite large) together with the high $Al^{26}$ and $Mn^{53}$ activities could imply a rather complicated history for this sample. However, assuming an actual collection depth of 0 to about 2 cm (3.6 g/cm$^2$), the net solar activities for $Co^{56}$ (if this is taken at the upper edge of its error range), $Mn^{53}$, $Al^{26}$, and $Na^{22}$ are consistent with the results obtained from the surface samples of rock 14321 when corrected for the slightly different target chemistry.

*Galactic cosmic ray effects*

In contrast to the sharp decrease in activity between FM-1 ($\sim$0.3 g/cm$^2$) and FM-2 ($\sim$1.1 g/cm$^2$) which is evident in all isotopes except $Cl^{36}$ and $Be^{10}$, the gradient between FM-3 and FM-5 is quite small. The overlying mass of the rock has nearly extinguished any solar cosmic ray contribution at these depths. We are thus seeing only the effects of the GCR's. The Reedy–Arnold model predicts that galactic cosmic rays will produce an activity level that in general is low at the surface, rises gradually to a peak at about 40 to 70 g/cm$^2$ and then decreases exponentially with depth. However, in the case of FM-3, a small contribution of solar produced activities cannot be excluded. This would tend to obliterate any small positive gradient between FM-3 and FM-5 expected from GCR production. Table 3 shows the possible solar contribution to FM-3 calculated using two geometries: an infinite plane and a hemisphere with a radius $R$ of 21 g/cm$^2$. The second geometry seems to be more realistic for the top part of the rock. Both sets of calculations used the model of Reedy and Arnold (1972). The calculated production rates are higher for the hemisphere because the mean path for solar protons to any depth is smaller than for an infinite plane.

In particular the $Fe^{55}$ measurements (Fig. 3) illustrate a complete composite profile. Because of the short half-life, 2.6 yr, erosion and other peculiarities of a rock's

Table 3. Calculated solar cosmic ray production rates (dpm/kg rock) for sample FM-3 (18 g/cm$^2$ depth).

| | Model | | | |
|---|---|---|---|---|
| | Infinite plane | | Hemisphere with $R$ = 21 g/cm$^2$ | |
| | $J(>10$ MeV, $4\pi) =$ 100 p/cm$^2$ sec | | $J(>10$ MeV, $4\pi) =$ 100 p/cm$^2$ sec | |
| Isotope | $R_0$ = 75 MV | $R_0$ = 100 MV | $R_0$ = 75 MV | $R_0$ = 100 MV |
| $Fe^{55}$ | 0.6 | 2.6 | 1.9 | 7 |
| $Mn^{53}$ | 0.2 | 0.8 | 0.6 | 2.2 |
| $Na^{22}$ | 0.5 | 2.6 | 1.8 | 7.5 |
| $Al^{26}$ | 0.8 | 3.5 | 2.7 | 10.4 |

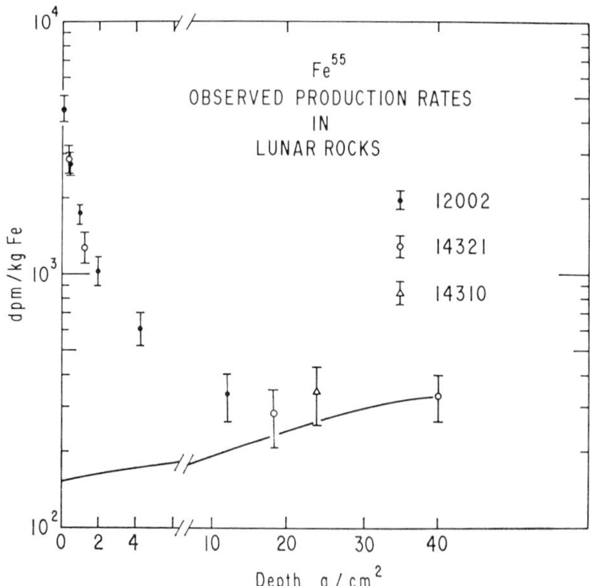


Fig. 3. Observed production rates (dpm/kg Fe) versus depth for $Fe^{55}$ in lunar rocks. The solid line is the calculated production from galactic cosmic rays alone, normalized at the deepest sample.

history while on the moon have no effect on this isotope. On the other hand, the half-life is long enough that any given solar flare event will have only a small effect on the total activity. These simplifications, coupled with the fact that iron is the only target of importance, allow us to compare directly our Apollo 14 measurements with our Apollo 12 measurements. As can be seen, the agreement between the samples is quite good. The line in Fig. 3 is the calculated galactic shape normalized at FM-5. There is no indication of a solar contribution to the FI sample, which integrates over a broad depth range from 16.5 g/cm$^2$ to 33 g/cm$^2$. The maximum possible SCR contribution to FM-3 is about 65 dpm/kg Fe or 7 dpm/kg rock 14321 (Table 3).

In the case of $Na^{22}$ the more complicated target chemistry makes direct inter-comparison of samples more difficult. The FM-3 and FM-5 values of 39 ± 5 dpm/kg and 32 ± 4 dpm/kg do again show the expected flat gradient.

The interpretation of the $Mn^{53}$ and $Al^{26}$ measurements must take into account the fact that rock history (erosion, exhumation, etc.) can have an important effect. This is especially illustrated for $Mn^{53}$ in Fig. 4, which presents measurements of the rocks 12002, 14321, and 14310, and Apollo 11 core samples 10004 and 10005. The lower values and flatter gradient of the 14321 surface samples compared with the 12002 measurements illustrate graphically the increased erosion which rock 14321 has experienced. The deeper samples, which are less affected by erosion, agree quite well in the different samples and again illustrate the flat nature of the galactic gradient in this region.

In the case of $Al^{26}$ we cannot as easily compare samples of different chemistry because several targets are important. However, measurements discussed below allow us to normalize the FI value to the chemistry of FM-3 and FM-5. The values then are

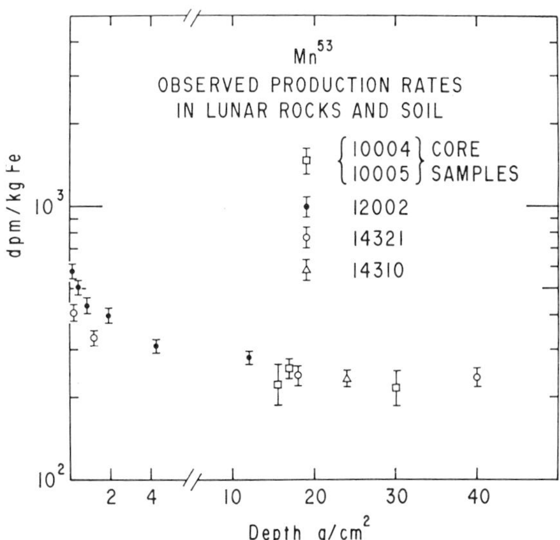


Fig. 4. Observed production rates (dpm/kg Fe) versus depth for $Mn^{53}$ in lunar rocks and soil.

57 ± 7 dpm/kg, 60 ± 8 dpm/kg, and 54 ± 8 dpm/kg for FM-3, FI, and FM-5, respectively. Again we see a flat spectrum as predicted.

Our sample of rock 14310, denoted FI, came from a mean depth of 24 g/cm$^2$. The separation of plagioclase and pyroxene which we were able to obtain on this sample allowed us to observe target effects in the induced activities of $Al^{26}$, $Na^{22}$, and $Cl^{36}$. The counting data, along with the chemical compositions, are given in Table 4.

$Al^{26}$ is the simplest case since production from only Al and Si is important. Production from Fe, Mg, and Ca is small enough to be ignored. A simple solution of the two equations involved give $P_{Al} = 270 \pm 60$ dpm/kg Al and $P_{Si} = 170 \pm 50$ dpm/kg Si. These values are in quite good agreement with the results calculated by Reedy and Arnold (1972) and with the values of Fuse and Anders (1969), obtained from studies of meteorites, after correcting for a $2\pi$ rather than a $4\pi$ bombardment. Also shown are values measured by Cressy (1971) in the meteorite Bruderheim and by Begemann *et al.* (1970) in lunar rocks.

In the case of $Na^{22}$ the situation is complicated by the fact that four targets make important contributions: Na, Mg, Al, and Si. This complication coupled with the fact that neither laboratory nor geochemical processes give useful separations of these targets makes it very difficult to calculate production rates from individual targets. If, however, we look at the Reedy–Arnold calculation for a sample of FIH and FIL chemistry we find good agreement between the predicted FIL/FIH ratio and that actually measured.

For $Cl^{36}$ we are in a somewhat better situation if we are willing to use measurements from meteorites and other lunar samples besides rock 14310. Using measurements from iron meteorites we can, after correcting for a $2\pi$ rather than a $4\pi$

Table 4. Galactic cosmic ray production rates for $Al^{26}$ and $Cl^{36}$ from various targets in rock 14310.

| | Al (%) | Si (%) | $Al^{26}$ dpm/kg rock |
|---|---|---|---|
| FIL | 16.3 | 21.1 | 80 ± 13 |
| FIH | 3.0 | 23.3 | 47 ± 7 |
| dpm/kg target | 270 ± 60 | 170 ± 50 | This work |
| | 540 | 170 | (Lunar rock)[1] |
| | 476 ± 54 | 310 ± 10 | (Meteorite)[2] |
| | 1130 ± 190 | 245 ± 31 | (Meteorite)[3] |
| | 300 | 160 | (Calculated)[4] |

| | Ca (%) | K (%) | Ti (%) | Fe (%) | $Cl^{36}$ dpm/kg rock |
|---|---|---|---|---|---|
| FIL | 11.9 | 0.50 | 0.42 | 2.4 | 15.9 ± 2.2 |
| FIH | 5.0 | 0.16 | 1.7 | 13.6 | 7.3 ± 1.5 |
| 10017 | 7.8 | 0.24 | 7.1 | 15.3 | 16.5 ± 1.8 |
| dpm/kg target | 115 ± 25 | — | 80 ± 35 | 8 (set) | This work |
| | 152 | | | | (Lunar rock)[1] |
| | 112 | 420 | 65 | 7 | (Calculated)[4] |

[1] Begemann *et al.*, 1970.
[2] Fuse and Anders, 1969.
[3] Cressy, 1971.
[4] Reedy and Arnold, 1972.

bombardment, set the production $P_{Fe} = 8$ dpm/kg Fe. Then using $Cl^{36}$ measurements from rock 10017 we can calculate $P_{Ti} = 80 \pm 35$ dpm/kg Ti and from the FIL and FIH determinations $P_{Ca} = 115 \pm 25$ dpm/kg Ca. Both numbers are in good agreement with Reedy–Arnold estimates.

$Fe^{55}$ and $Mn^{53}$ being produced from only one target can easily be compared with other samples as we have done above.

*Acknowledgments*—The authors have as always been helped by many persons. Norman Fong, Florence Kirchner and Lawrence Finnin supported our work in many ways. We have benefited from valuable discussions with Kurt Marti, Ernest Schonfeld, Edward Fireman, and other colleagues. We owe a special debt to the astronauts and others involved in collecting and documenting the large rocks which made this work possible. This research was supported by NASA Grant GNL 05-009-148.

## References

Begemann F., Born W., Palme H., Vilcsek E., and Wänke H. (1972) Cosmic ray produced radioisotopes in Apollo 12 and Apollo 14 samples (abstract). In *Lunar Science—III* (editor C. Watkins), p. 53, Lunar Science Institute Contr. No. 88.

Begemann F., Vilcsek E., Rieder R., Born W., and Wänke H. (1970) Cosmic ray produced radioisotopes in lunar samples from the Sea of Tranquility (Apollo 11). *Proc. Apollo 11 Lunar Sci. Conf., Geochim. Cosmochim. Acta* Suppl. 1, Vol. 2, pp. 995–1005. Pergamon.

Bostrom C. O., Williams D. J., and Arens J. R. (1971) Solar proton monitor experiment. *Solar Geophysical Data*, No. 328, Part II.

Burnett D. S., Huneke J. C., Podosek F. A., Russ G. P. III, Turner G., and Wasserburg G. J. (1972) The irradiation history of lunar samples (abstract). In *Lunar Science—III* (editor C. Watkins), p. 105, Lunar Science Institute Contr. No. 88.

Cressy P. J. Jr. (1971) The production rate of $Al^{26}$ from target elements in the Bruderheim chondrite. *Geochim. Cosmochim. Acta* **35,** 1283–1296.

Crozaz G., Drozd R., Hohenberg C. M., Hoyt H. P. Jr., Ragan D., Walker R. M., and Yuhas D. (1972) Solar flare and galactic cosmic ray studies of Apollo 14 samples (abstract). In *Lunar Science—III* (editor C. Watkins), p. 167, Lunar Science Institute Contr. No. 88.

Eldridge J. S., O'Kelley G. D., and Northcutt K. J. (1972) Abundances of primordial and cosmogenic radionuclides in Apollo 14 rocks and fines (abstract). In *Lunar Science—III* (editor C. Watkins), p. 221, Lunar Science Institute Contr. No. 88.

Fields P. R., Diamond H., Metta N. D., Rokop D. J., and Stevens C. M. (1972) $Np^{237}$, $U^{236}$, and other actinides on the moon (abstract). In *Lunar Science—III* (editor C. Watkins), p. 256, Lunar Science Institute Contr. No. 88.

Finkel R. C., Arnold J. R., Imamura M., Reedy R. C., Fruchter J. S., Loosli H. H., Evans J. C., Delany A. C., and Shedlovsky J. P. (1971) Depth variation of cosmogenic nuclides in a lunar surface rock and lunar soil. *Proc. Second Lunar Sci. Conf., Geochim. Cosmochim. Acta* Suppl. 2, Vol. 2, pp. 1773–1789. MIT Press.

Fireman E. L., D'Amico J., Defelice J., and Spannagel G. (1972) Radioactivities in Apollo 14 and 15 materials (abstract). In *Lunar Science—III* (editor C. Watkins), p. 262, Lunar Science Institute Contr. No. 88.

Fireman E. L., D'Amico J., Defelice J., and Spannagel G. (1972a) Radioactivities in returned lunar materials. *Proc. Third Lunar Sci. Conf., Geochim. Cosmochim. Acta* Suppl. 3, Vol. 2. MIT Press.

Fuse K. and Anders E. (1969) $Al^{26}$ in meteorites—VI. Achondrites. *Geochim. Cosmochim. Acta* **33,** 653–670.

Ingamells C. O. (1966) Absorptiometric methods in rapid silicate analysis. *Anal. Chem.* **38,** 1228.

Jackson P. F. S., Coetzee J. H. J., Strasheim A., Strelow F. W. E., Gricius A. J., Wybenga F., and Kokot M. L. (1972) The analysis of lunar material returned by Apollo 14 (abstract). In *Lunar Science—III* (editor C. Watkins), p. 424, Lunar Science Institute Contr. No. 88.

Keith J. E., Clark R. S., and Richardson K. A. (1972) Gamma ray measurements of Apollo 12, 14. and 15 lunar samples (abstract). In *Lunar Science—III* (editor C. Watkins), p. 446, Lunar Science Institute Contr. No. 88.

Klein C. Jr., Drake J. C., Frondel C., and Ito J. (1972) Mineralogy and petrology of several Apollo 14 rock types and chemistry of the soil (abstract). In *Lunar Science—III* (editor C. Watkins), p. 455, Lunar Science Institute Contr. No. 88.

LSPET (Lunar Sample Preliminary Examination Team) (1971) Preliminary examination of lunar samples from Apollo 14. *Science* **173,** 681–693.

Lugmair G. W. and Marti K. (1971) Neutron capture effects in lunar gadolinium and the irradiation histories of some lunar rocks. *Earth Planet. Sci. Lett.* **13,** 32.

Lugmair G. W. and Marti K. (1972) Neutron and spallation effects in Fra Mauro regolith (abstract). In *Lunar Science—III* (editor C. Watkins), p. 495, Lunar Science Institute Contr. No. 88.

O'Kelley G. D., Eldridge J. S., Schonfeld E., and Bell P. R. (1971) Cosmogenic radionuclide concentrations and exposure ages of lunar samples from Apollo 12. *Proc. Second Lunar Sci. Conf., Geochim. Cosmochim. Acta* Suppl. 2, Vol. 2, pp. 1747–1755. MIT Press.

Rancitelli L. A., Perkins R. W., Felix W. D., and Wogman N. A. (1971) Erosion and mixing of the lunar surface from cosmogenic and primordial radionuclide measurements in Apollo 12 lunar samples. *Proc. Second Lunar Sci. Conf., Geochim. Cosmochim. Acta* Suppl. 2, Vol. 2, pp. 1757–1772. MIT Press.

Rancitelli L. A., Perkins R. W., Felix W. D., and Wogman N. A. (1972) Cosmic ray flux and lunar surface processes characterized from radionuclide measurements in Apollo 14 and 15 lunar samples (abstract). In *Lunar Science—III* (editor C. Watkins), p. 630, Lunar Science Institute Contr. No. 88.

Reed G. W. Jr., Jovanovic S., and Fuchs L. H. (1972) Concentrations and lability of the halogens, platinum metals and mercury in Apollo 14 and Apollo 15 samples (abstract). In *Lunar Science—III* (editor C. Watkins), p. 637, Lunar Science Institute Contr. No. 88.

Reedy R. C. and Arnold J. R. (1972) Interaction of solar and galactic cosmic ray particles with the moon. *J. Geophys. Res.* **77,** No. 4, 537–555.

Schnetzler C. C. and Nava D. F. (1971) Chemical composition of Apollo 14 soils 14163 and 14259. *Earth Planet. Sci. Lett.* **11,** 345.

Shrelldalff: Shedlovsky J. P., Honda M., Reedy R. C., Evans J. C., Lal, D., Lindstrom R. M., Delany A. C., Arnold J. R., Loosli H. H., Fruchter J. S., and Finkel R. C. (1970) Pattern of bombardmentproduced radionuclides in rock 10017 and in lunar soil. *Proc. Apollo 11 Lunar Sci. Conf., Geochim. Cosmochim. Acta* Suppl. 1, Vol. 2, 1503–1532. Pergamon.

Wänke H., Baddenhausen H., Balacescu A., Teschke F., Spettel B., Dreibus G., Quijano-Rico M., Kruse H., Wlotzka F., and Begemann F. (1972) Multielement analyses of lunar samples (abstract). In *Lunar Science—III* (editor C. Watkins), p. 779, Lunar Science Institute Contr. No. 88.

Proceedings of the Third Lunar Science Conference
(Supplement 3, *Geochimica et Cosmochimica Acta*)
Vol. 2, pp. 1733–1746
The M.I.T. Press, 1972

# Cosmonuclides in lunar rocks

Y. YOKOYAMA, R. AUGER, R. BIBRON, R. CHESSELET, F. GUICHARD,
C. LEGER, H. MABUCHI,* J. L. REYSS, and J. SATO*

Centre des Faibles Radioactivités, C.N.R.S., 91-Gif-sur-Yvette, France

**Abstract**—The rates of production of cosmonuclides as a function of depth in the moon are calculated with a simplified method. The differential spectra of particles due to galactic cosmic ray and of solar protons are calculated as a function of depth. For solar protons, the range-energy relation of Wilson is used. For galactic cosmic rays, a three-steps cascade model is used. Integration as a function of incident angle is simplified by dividing zenith angles to have an equal solid angle. The calculated production rates are compared with experimental results and with the earlier calculations. Th, U, $^{26}$Al, and $^{55}$Fe are measured in lunar rock 14305.

## INTRODUCTION

LUNAR SAMPLES bear fossil records of cosmic radiation. From studies of Apollo 11 samples, clear evidence was obtained for the long term bombardment by solar particles in the region of tens of MeV per nucleon (Shrelldalff, 1970; O'Kelley *et al.*, 1970; Marti *et al.*, 1970; Crozaz *et al.*, 1970; Fleischer *et al.*, 1970; Lal *et al.*, 1970; Price and O'Sullivan, 1970). Studies of Apollo 12 samples (Finkel *et al.*, 1971) showed a similarity of the depth profile of $7.16 \times 10^5$ yr $^{26}$Al to that of 2.6 yr $^{22}$Na. Thus the flux of the solar particles averaged over $10^6$ yr has been similar to that observed recently. The profile for $3.7 \times 10^6$ yr $^{53}$Mn suggests a possibility of either an erosion rate of 0.2 mm per m.y. or the variation of the solar cosmic flux (Finkel *et al.*, 1971).

The aim of the work reported here comprises experimental measurements of the depth variation of radionuclides, theoretical calculations of the expected profiles, and comparison of these results. We report here mainly the results of the theoretical calculation which have been obtained by somewhat different approach using different input data from the previous work of Shrelldalff (1970), Armstrong and Alsmiller (1971), Reedy and Arnold (1971), and Tanaka *et al.* (1971).

## SOLAR PRODUCTION

We consider here only the interactions of the primary protons. The production by $\alpha$-particles will be reported later. The spectrum of incident solar protons is assumed to decrease exponentially with their rigidity:

$$N(R)\,dR = k \exp\left(-\frac{R}{R_0}\right) dR \qquad (1)$$

* Present address: Department of chemistry, Faculty of Science, University of Tokyo, Tokyo, Japan.

where $N(R)\, dR$ is the number of particles having rigidities between $R$ and $R + dR$, and $R_0$ is the characteristic slope (MV). By using the relation between the rigidity and the energy,

$$R^2 = E^2 + 1876E \tag{2}$$

the rigidity spectrum is converted in the energy spectrum:

$$N(E)\, dE = \frac{k(E + 938)}{R} \exp\left(-\frac{R}{R_0}\right) dE \tag{3}$$

where $N(E)\, dE$ is the number of particles having energies between $E$ and $E + dE$ (MeV), and 938 is the rest mass of the proton in MeV.

To calculate the depth variation of the proton flux, we have used a simplified relationship between range and energy, proposed by Wilson and Brobeck (Wilson, 1947):

$$P = \frac{2.21}{S} E^{1.8} \tag{4}$$

where $P$ is the range (mg/cm$^2$) of a proton of energy $E$ (MeV) and $S$ is the relative mass stopping power which depends on the chemical composition of the material and is 1 for air, 0.91 for the rock 14305, and 0.88 for the rock 12002. The relative mass stopping power was calculated by the equations (7-1, 7-2) of Friedlander and Kennedy (1955). The equation (4) is valid in the energy domain of about 10 to 200 MeV.

The incident energy $E$ of the particle coming with a zenith angle $\theta$ and having an energy $E_D$ at a given depth $D$ (mg/cm$^2$) is calculated by

$$E = \left(\frac{SD}{2.21 \cos\theta} + E_D{}^{1.8}\right)^{1/1.8} \tag{5}$$

The incident flux is assumed to be isotropic. Incident angles are divided in ten parts, of which the $\cos\theta$ are comprised between 0 to 0.1, 0.1, to 0.2 and so on, in order that each part has equal solid angle. Then each part is represented by its mean $\cos\theta$, namely $\cos\theta = 0.05$, 0.15, and so on.

The differential flux of particles having energy $E_D$ at a given depth $D$ is then obtained by integrating the flux of particles having initial energy $E$ ($E$ being a function of $\cos\theta$) over all angles:

$$\begin{aligned} N(E_D)\, dE_D &= N(E)\, dE = N(E) \frac{dE}{dE_D} dE_D = N(E) \left(\frac{E_D}{E}\right)^{0.8} dE_D \\ &= \frac{k}{20} \sum_{\cos\theta} \frac{E + 938}{R} \exp\left(-\frac{R}{R_0}\right) \left(\frac{E_D}{E}\right)^{0.8} dE_D \end{aligned} \tag{6}$$

where $E$ should be determined by the equation (5) as the function of $E_D$, $D$, $\cos\theta$; $R$ does by the equation (2) from $E$, and summation should be done for $\cos\theta = 0.05$, 0.15, . . . , to 0.95.

The depth variation of the proton flux thus calculated for several values of $R_0$ is shown in Fig. 1.

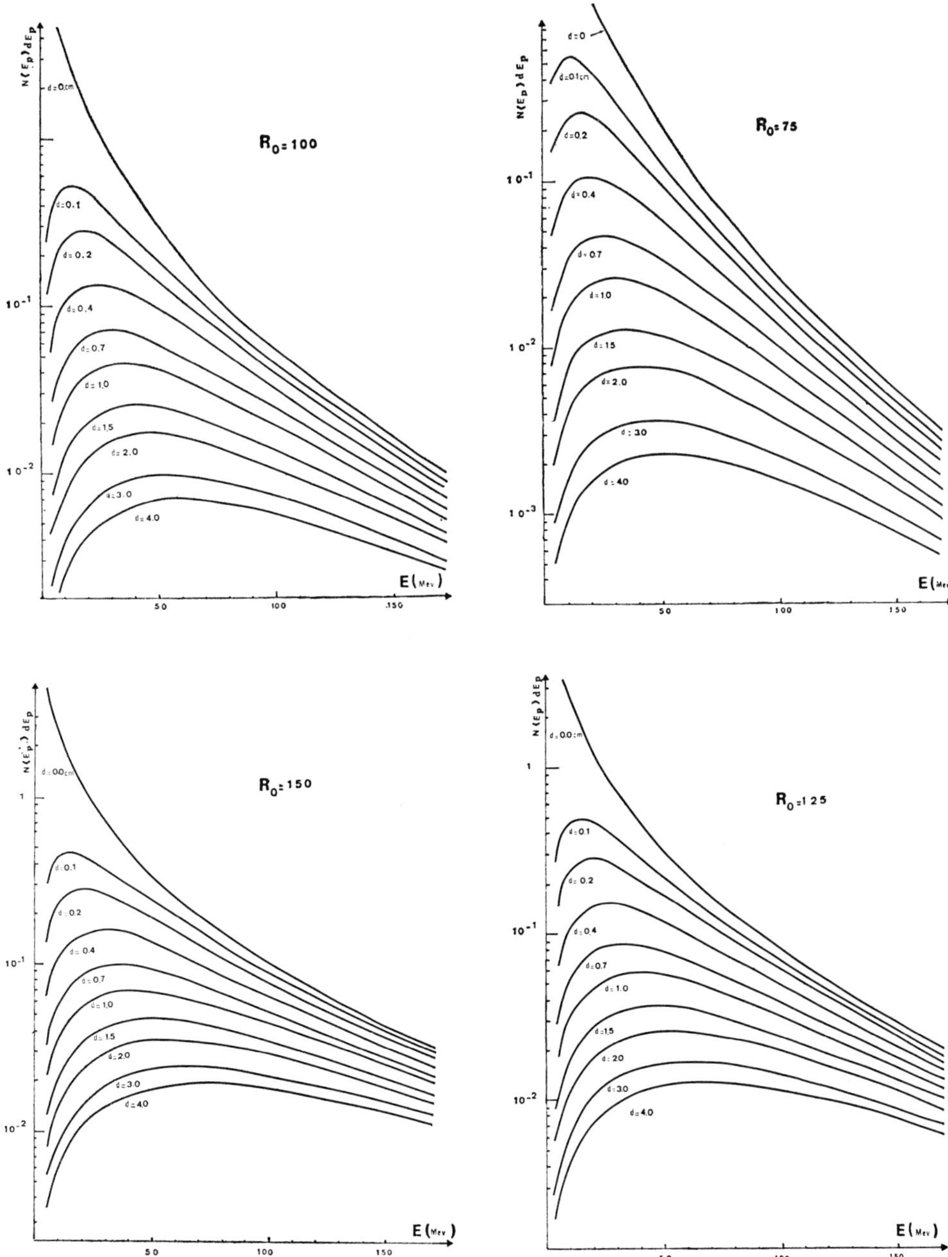

Fig. 1. Differential energy spectra of solar protons at various depth ($d$) in the moon (in units of proton/cm$^2$-sec-MeV). Here $J$ ($4\pi$ integral flux above 10 MeV in units of protons/cm$^2$-sec) is taken to be 100. $R_0$ is the mean rigidity in units of MV. The density of rock is taken to be 3.

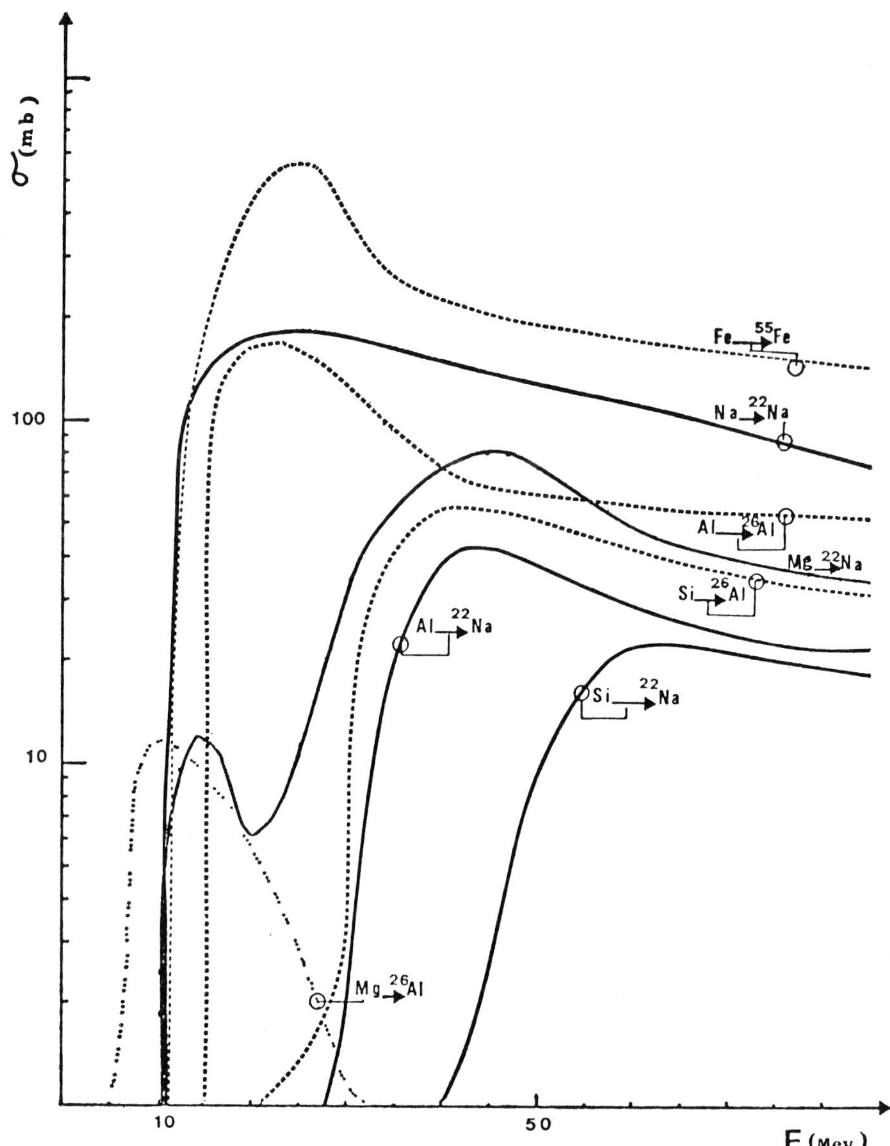


Fig. 2. Adopted excitation functions for the production of various radionuclides. The functions are for the proton reactions.

The production rates of nuclides are calculated by integrating $nN(E_D)\mathfrak{S}(E_D)\, dE_D$ for all energies, where $n$ is the number of target nuclei per kg, $\mathfrak{S}(E_D)$ is the excitation function of a given reaction. We have used the excitation functions selected from experiments, corrected for the adopted values of the monitor reactions, of decay schemes, and synthesized by Tobailem and de Lassus St-Genies (1971, 1972). The excitation functions of main reactions are shown in Fig. 2. For the productions of

$^{53}Mn$ and $^{54}Mn$, we have adopted the excitation functions of Reedy and Arnold (1971).

## Galactic Production

The depth variation of the particle flux due to galactic cosmic rays is estimated in the following way.

We divide particles in three components. The first component is the energetic incident particles. We assumed an exponential decrease of the flux of this component with the traversed length $L$:

$$N_1 = N_{1,0} \exp(-\mu_1 L) \tag{7}$$

where $N_{1,0}$ is the incident flux, $N_1$ is the flux at $L$ (cm), and $\mu_1$ is the attenuation coefficient (cm$^{-1}$).

The second component is the medium energy particles formed by the interaction of the first component with matter. Assuming $dN_2 = (m_1\mu_1N_1 - \mu_2N_2)\,dL$, we obtained

$$N_2 = C_1N_{1,0}(e^{-\mu_1L} - e^{-\mu_2L}) \tag{8}$$

where $N_2$ is the flux of the second component at $L$, $\mu_2$ is the attenuation length of this component, $m_1$ is the multiplication factor (a primary particle producing $m_1$ secondary particles), and $C_1$ is a constant equal to $m_1\mu_1/(\mu_2 - \mu_1)$.

The third component is the low energy particles (mainly neutrons) formed from the second component. The flux $N_3$ of the third component is obtained by assuming $dN_3 = (m_2\mu_2N_2 - \mu_3N_3)\,dL$,

$$N_3 = C_2N_{1,0}[e^{-\mu_1L} - e^{-\mu_2L} - (\mu_2 - \mu_1)Le^{-\mu_2L}] \tag{9}$$

where $m_2$ is the multiplication factor for the production of the third component and $C_2$ is a constant equal to $m_1m_2\mu_1\mu_2/(\mu_2 - \mu_1)^2$. We have assumed that the attenuation length of the third component $\mu_3$ is equal to $\mu_2$.

We have adopted five particles (nucleons)/cm$^2$/sec/4$\pi$ for $N_{1,0}$, and 0.02 and 0.08 cm$^{-1}$ for $\mu_1$ and $\mu_2$, respectively. The energy domains of the three components are assumed somewhat arbitrarily to be $E > 1$ BeV, 1 BeV $> E >$ 200 MeV, 200 MeV $> E >$ 2 MeV. The values of 3.3 and 13.3 were adopted for $C_1$ and $C_2$, respectively.

The incident flux is assumed to be isotropic. The effect of the incident angle was calculated in a similar way as that of solar production: by substituting $L$ in the equations (7, 8, 9) by $D/\cos\theta$, the fluxes $N_1$, $N_2$, $N_3$ are numerically integrated over $2\pi$. We assumed $L = 7$ cm if $L$ exceeds 7 cm, in order to fit approximately the shape of the rock 14305 (this was done also for the calculation of solar production, but the difference from the case of infinite plane was negligible).

The cross section in the energy domain over more than 200 MeV is approximately constant for a given reaction, if the difference of mass is small between target and product. We express this mean cross section by $\mathfrak{S}_H$.

On the other hand, if we assume the differential spectrum of the third component (2 to 200 MeV) to have a form of $E^{-1}$, we can calculate the mean cross section in this domain by an arithmetic mean of the cross sections corresponding to the energies taken as a geometric progression, such as 2, 2.8, 4, 5.7, 8, 11, 16, . . . , 128, 181 MeV. We express this mean by $\mathfrak{S}_L$, namely $\mathfrak{S}_L = (\mathfrak{S}_2 + \mathfrak{S}_{2.8} + \mathfrak{S}_4 + \cdots + \mathfrak{S}_{181})/14$ where $\mathfrak{S}_2$ is the cross section for 2 MeV and so on.

Then the production rate $A$ is calculated by

$$A = n[(N_{1,D} + N_{2,D})\mathfrak{S}_H + N_{3,D}\mathfrak{S}_L] \tag{10}$$

where $N_{1,D}$ is the flux of the first component at the depth $D$, $N_{2,D}$, that of the second and so on.

## RESULTS AND DISCUSSION

Figure 3 shows the calculated profiles for $^{26}$Al and the preliminary observed value obtained by us with nondestructive gamma-gamma coincidence spectrometry on the rock 14305 (Table 1). This rock was thought to have been turned recently,

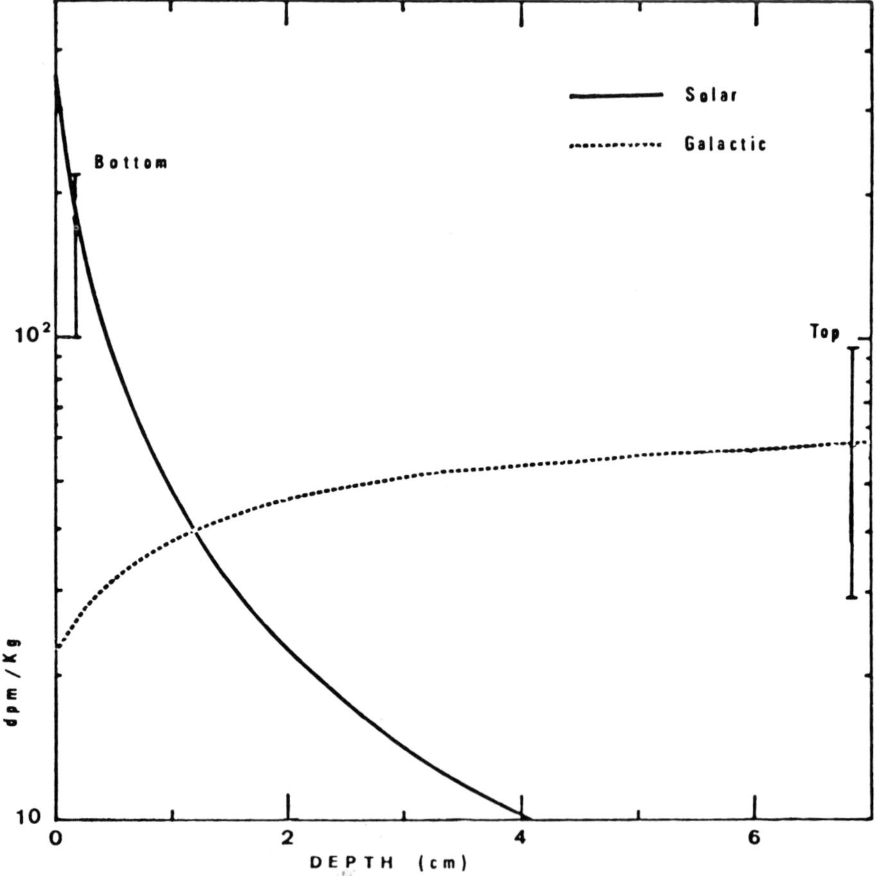


Fig. 3. Depth variation of $^{26}$Al in the lunar rock 14305. The calculated solar production is based on the values of $R_0 = 100$ and $J = 70$. Galactic contribution is not subtracted from experimental points.

Table 1. Nondestructive gamma-ray measurements of rocks 14305 and 14302.

| Samples | Weight (g) | Th (ppm) | U (ppm) | $^{26}$Al (dpm/kg) |
|---|---|---|---|---|
| This work 14305 | | | | |
| Top (0–3 mm) | 2.7 | 14.6 ± 0.9 | 3.8 ± 0.4 | 63 ± 34 |
| Bottom (0–3 mm) | 1.45 | 13.3 ± 0.9 | 3.8 ± 0.5 | 160 ± 60 |
| LSPET (1971) 14302* | 381 | 14.3 ± 1.4 | 3.8 ± 0.6 | 85 ± 17 |

* 14302 was a part of 14305 at the lunar surface.

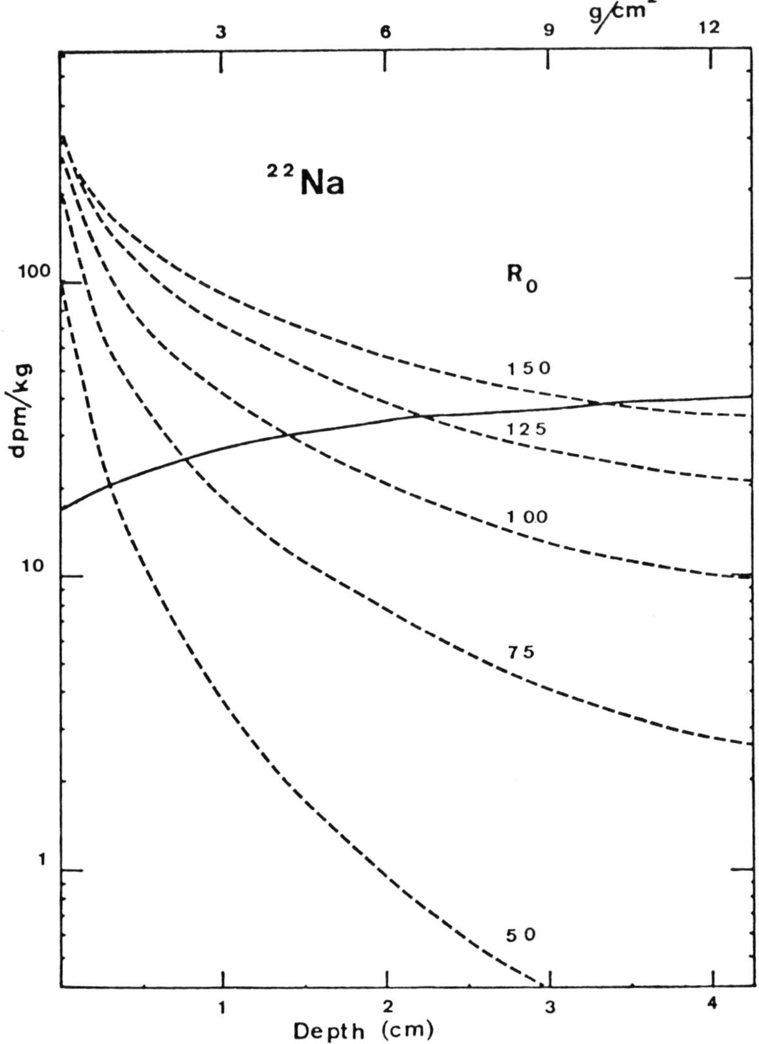


Figs. 4 to 8. Calculated depth variations of various radionuclides in the lunar rock 14305. The broken curves represent the solar production ($J$ = 100) and the full curves, the galactic production. Adopted contents of Si, Al, Mg, Fe, and Na are 23.0, 8.5, 7.8, 7.4, and 0.63%, respectively (LSPET, 1971).

from the photographic studies of its situation on the lunar surface and from the pits counting. This was confirmed by our results, showing more activities of $^{26}$Al at the bottom than at the top. The calculated profiles of $^{22}$Na, $^{26}$Al, $^{53}$Mn, $^{54}$Mn, and $^{55}$Fe are shown in Figs. 4 to 8. A preliminary result of the measurement of $^{55}$Fe of the rock 14305 is shown in Fig. 9.

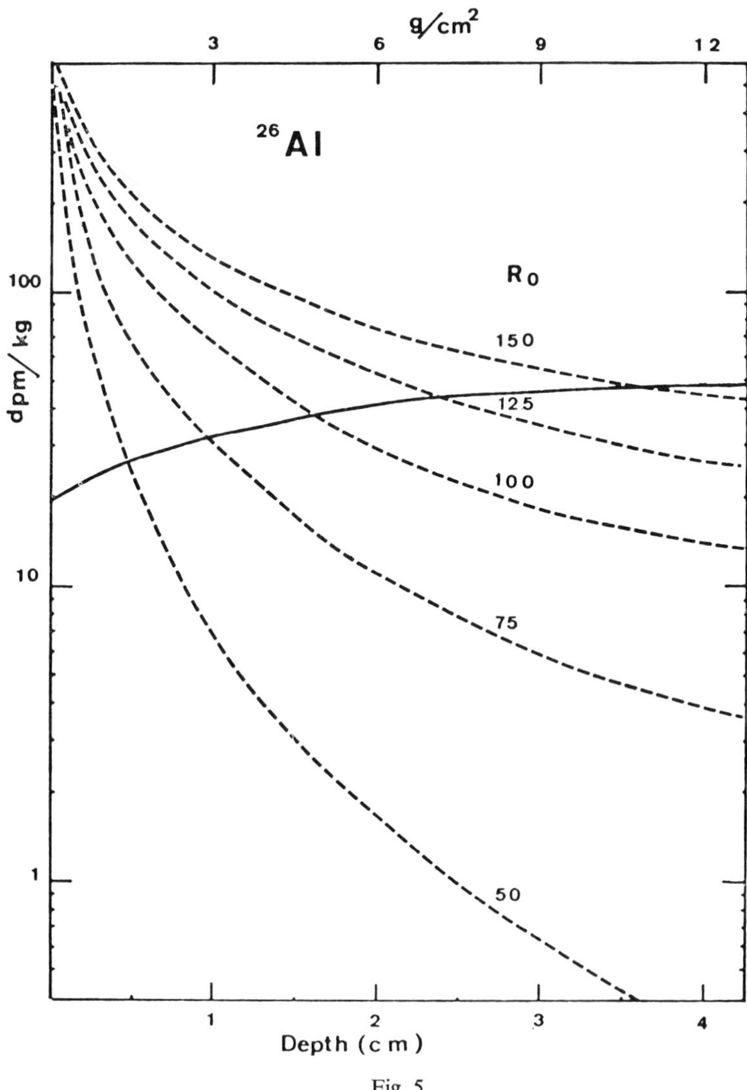


Fig. 5.

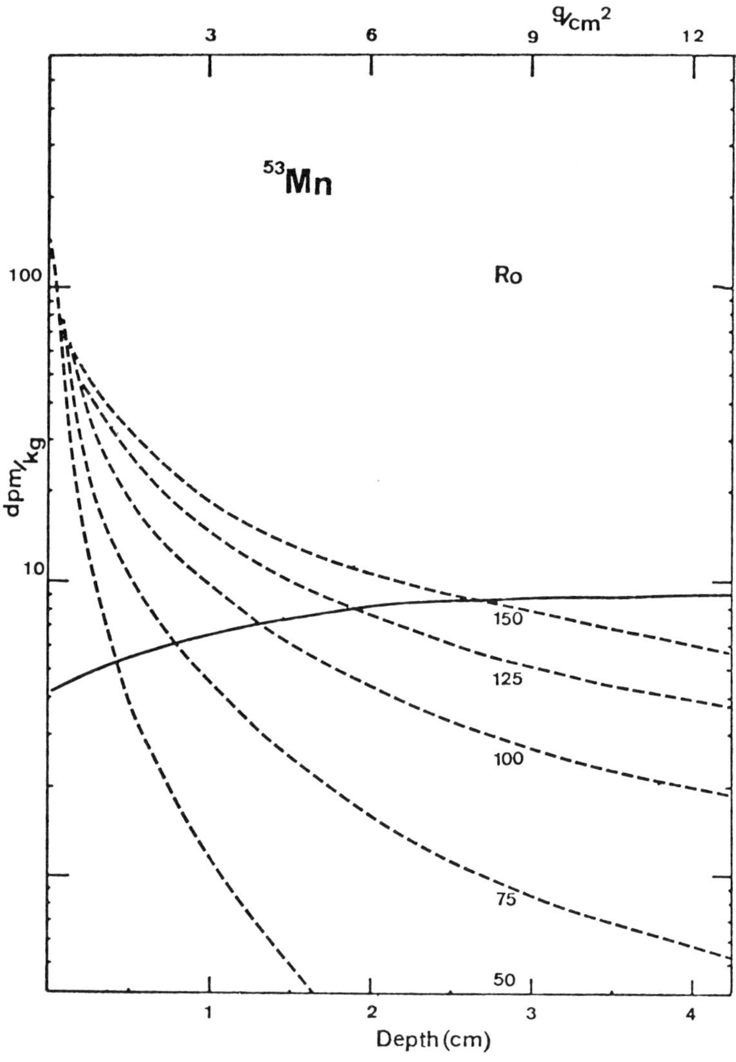


Fig. 6.

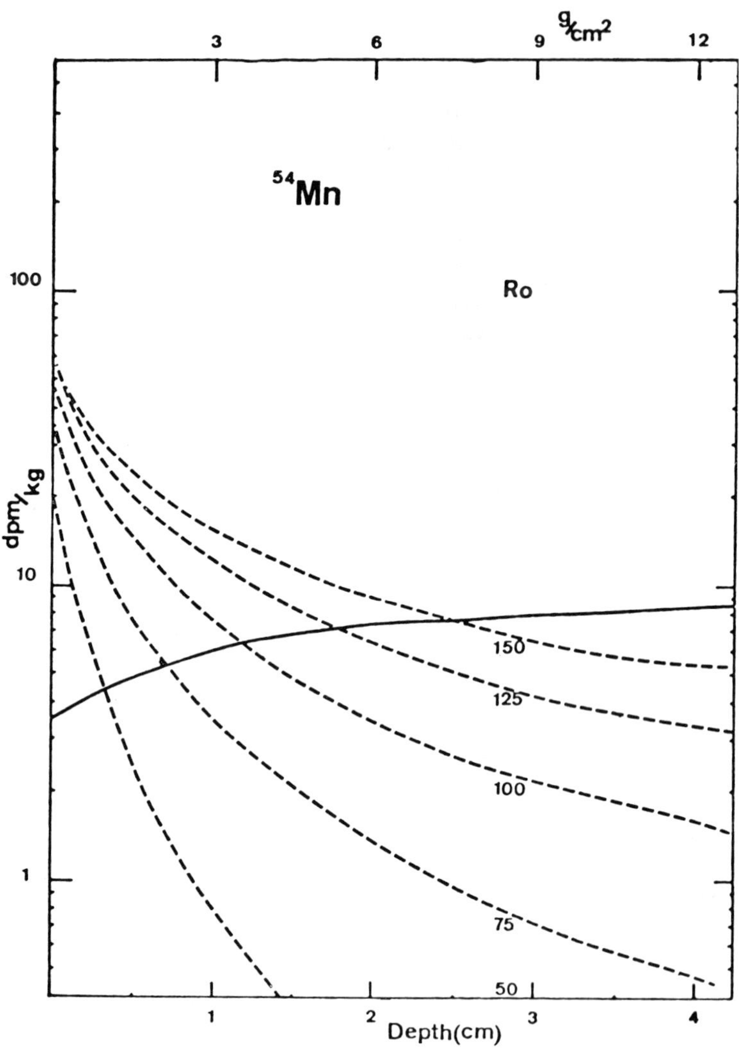
3
6
9
g/cm²
12
⁵⁴Mn
100
Ro
dpm/kg
10
150
125
100
1
75
50
1
2
Depth(cm)
3
4

Fig. 7.

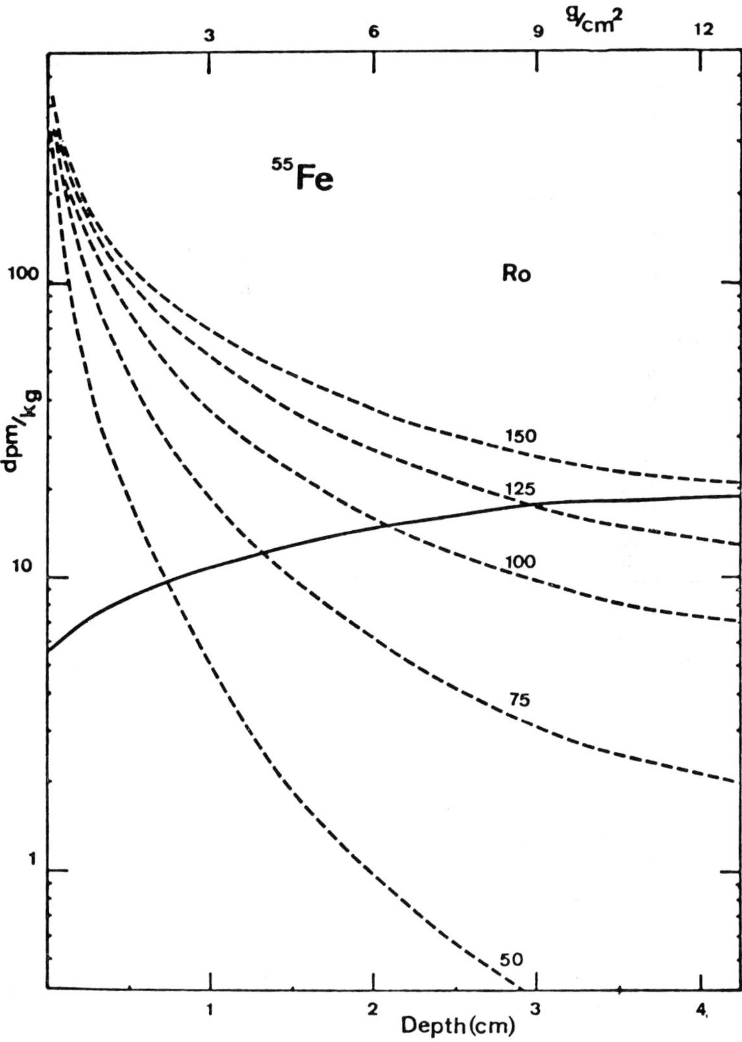

3
6
9
g/cm²
12
$^{55}$Fe
Ro
100
dpm/kg
150
125
100
10
75
1
50
1
2
3
4
Depth (cm)


Fig. 8.

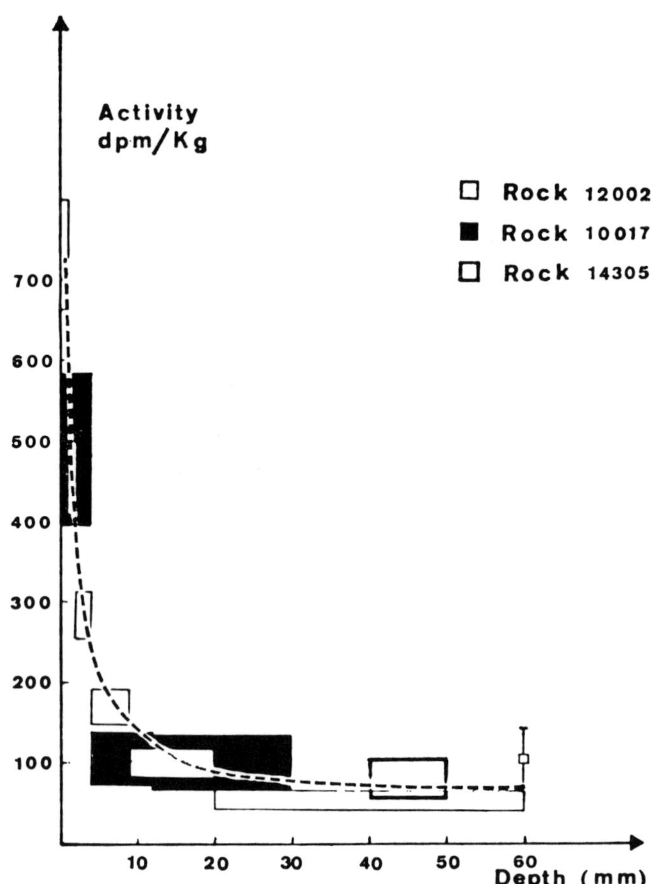


Fig. 9. Measurements of $^{55}Fe$ in the lunar rocks 14305 (this work), 12002 (Finkel *et al.*, 1971) and 10017 (Shrelldalff, 1970). Iron contents are normalized to 16.8%.

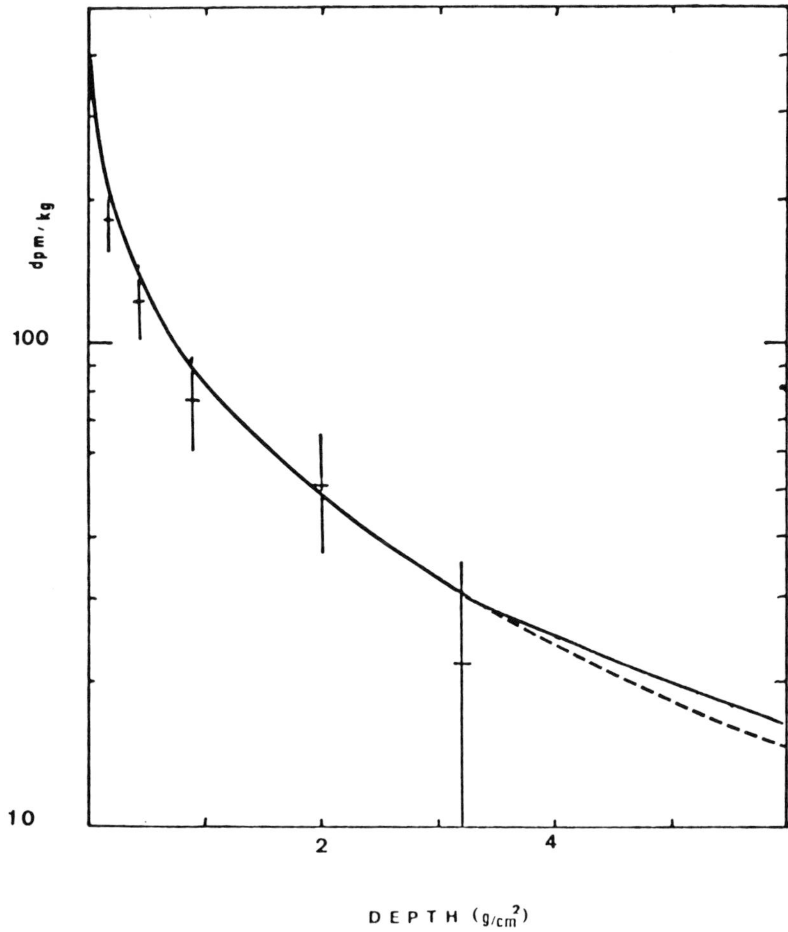


Fig. 10. Depth variation of $^{26}Al$ produced by solar protons in the lunar rock 12002. The full curve represents the calculated production rate by this work, and the broken curve, that of Reedy and Arnold (1971) for $R_0 = 100$ and $J = 70$. Experimental points are those of Finkel *et al.* (1971), and are given after subtracted for the galactic contributions calculated by this work.

The result of our calculation for the rock 12002 is in good agreement with that of Reedy and Arnold (1971) as shown in Fig. 10. It indicates the validity of our simplified method.

Natural radioactive elements, Th and U were also measured by nondestructive gamma-gamma coincidence spectrometry (Table 1). Our results are in good agreement with the results of LSPET (1971). It shows a possibility of measurement of these elements of small samples (of the order of grams) with a good precision by this method.

*Acknowledgments*—We are grateful to Dr. J. Labeyrie for his encouragement and helpful discussions. This work is done in the Consortium of rock 14305. We wish to thank Dr. J. Geiss, consortium leader and Dr. N. Grögler for their helpful discussions and their kindness of sawing and preparing of samples. We are also indebted to Drs. J. Tobailem and C. H. de Lassus St-Genies for their kindness of providing the cross-section data prior to publication.

## REFERENCES

Armstrong T. W. and Alsmiller R. G. (1971) Calculation of cosmogenic radionuclides in the moon and comparison with Apollo measurements. *Proc. Second Lunar Sci. Conf., Geochim. Cosmochim. Acta* Suppl. 2, Vol. 2. M.I.T. Press.

Crozaz G., Haack U., Hair M., Maurette M., Walker R., and Woolum D. (1970) Nuclear track studies of ancient solar radiations and dynamic lunar surface processes. *Proc. Apollo 11 Lunar Sci. Conf., Geochim. Cosmochim. Acta* Suppl. 1, Vol. 3, pp. 2051–2080. Pergamon.

Finkel R. C., Arnold J. R., Reedy R. C., Fruchter J. S., Loosli H. H., Evans J. C., Shedlovsky J. P., Imamura M., and Delany A. C. (1971) Depth variation of cosmogenic nuclides in a lunar surface rock. *Proc. Second Lunar Sci. Conf., Geochim. Cosmochim. Acta* Suppl. 2, Vol. 2, p. 1773. M.I.T. Press.

Fleischer R. L., Haines E. L., Hart H. R., Woods R. T., and Comstock G. M. (1970) The particle track record of the Sea of Tranquility. *Proc. Apollo 11 Lunar Sci. Conf., Geochim. Cosmochim. Acta* Suppl. 1, Vol. 3, pp. 2103–2120. Pergamon.

Friedlander G. and Kennedy J. W. (1955) *Nuclear and Radiochemistry*, p. 190. John Wiley.

Lal D., MacDougall D., Wilkening L., and Arrhenius G. (1970) Mixing of the lunar regolith and cosmic ray spectra: Evidence from particle track studies. *Proc. Apollo 11 Lunar Sci. Conf., Geochim. Cosmochim. Acta* Suppl. 1, Vol. 3, pp. 2295–2303. Pergamon.

LSPET (1971) (Lunar Sample Preliminary Examination Team) Preliminary examination of lunar samples from Apollo 14. *Science* **173,** 681–693.

Marti K., Lugmair G. W., and Urey H. C. (1970) Solar wind gases, cosmic ray spallation products, and the irradiation history of Apollo 11 samples. *Proc. Apollo 11 Lunar Sci. Conf., Geochim. Cosmochim. Acta* Suppl. 1, Vol. 2, pp. 1357–1367. Pergamon.

O'Kelley G. D., Eldridge J. S., Schofeld E., and Bell P. R. (1970) Primordial radionuclide abundances, solar proton and cosmic ray effects and ages Apollo 11 lunar samples by non-destructive gamma ray spectrometry. *Proc. Apollo 11 Lunar Sci. Conf., Geochim. Cosmochim. Acta* Suppl. 1, Vol. 2, pp. 1407–1422. Pergamon.

Price P. B. and O'Sullivan D. (1970) Lunar erosion rate and solar flare paleontology. *Proc. Apollo 11 Lunar Sci. Conf., Geochim. Cosmochim. Acta* Suppl. 1, Vol. 3, pp. 2351–2359. Pergamon.

Reedy R. C. and Arnold J. R. (1971) Interaction of solar and galactic cosmic ray particles with the moon. To be published in *J. Geophys. Res.*

Shrelldalff (Shedlovsky J. P., Honda M., Reedy R. C., Evans J. C., Lal D., Lindstrom R. M., Delany A. C., Arnold J. R., Loosli H. H., Fruchter J. S., and Finkel R. C.) (1970) Pattern of bombardment produced radionuclides in rock 10017 and in lunar soil. *Proc. Apollo 11 Lunar Sci. Conf., Geochim. Cosmochim. Acta* Suppl. 1, Vol. 2, pp. 1503–1532. Pergamon.

Tanaka S., Sakamoto K., and Komura K. (1971) $Al^{26}$ and $Mn^{53}$ production by solar flare particles in lunar rock and cosmic dust. To be published.

Tobailem J., de Lassus St-Genies C. H., and Leveque L. (1971) Sections efficaces des réactions nucléaires induites par protons, deutons, particules alpha, I. Réaction nucléaires moniteurs. Note CEA-N-1466(1) Commissariat à l'Energie Atomique, France.

Tobailem J. and de Lassus St-Genies C. H. (1972) Sections efficaces des réactions nucléaires induites par protons, deutons, particules alpha, II, III. In preparation.

Wilson R. R. (1947) Range, straggling, and multiple scattering of fast protons. *Phys. Rev.* **71,** 385–386.

Proceedings of the Third Lunar Science Conference
(Supplement 3, *Geochimica et Cosmochimica Acta*)
Vol. 2, pp. 1747–1762
The M.I.T. Press, 1972

# Radioactivities in returned lunar materials

E. L. FIREMAN, J. D'AMICO, J. DEFELICE, and G. SPANNAGEL

Smithsonian Institution, Astrophysical Observatory,
Cambridge, Massachusetts 02138

**Abstract**—The $H^3$, $Ar^{37}$, and $Ar^{39}$ radioactivities were measured at several depths in the large documented lunar rocks 14321 and 15555. The comparison of the $Ar^{37}$ activities from similar locations in rocks 12002, 14321, and 15555 gives direct measures of the amounts of $Ar^{37}$ produced by the 2 November 1969 and 24 January 1971 solar flares. From the $Ar^{37}$ measurements, the intensities of the 2 November 1969 and 24 January 1971 solar flares were estimated to be $(5.1 \pm 1.2) \times 10^6$ and $(5.9 \pm 1.0) \times 10^6$ protons ($> 50$ MeV)/cm$^2$ sr, respectively. Because of the large differences in the K and Fe contents of the documented rocks, the $Ar^{39}$ produced from the separate target elements, Fe + Ti, K, and Ca, at 1-, 5-, and 12-cm depths was obtained from the $Ar^{39}$ measurements. The high-energy proton flux ($> 200$ MeV) averaged over 1000 yr was obtained from the Fe + Ti $\rightarrow Ar^{39}$; the neutron flux ($\lesssim 1$ MeV) averaged over 1000 yr was obtained from the K $\rightarrow Ar^{39}$; and the neutron flux ($\lesssim 10$ MeV) averaged over 1000 yr was obtained from the Ca $\rightarrow Ar^{39}$. From the depth dependence of $Ar^{39}$, the intensity of solar-flare protons ($> 50$ MeV) averaged over the past 1000 yr was estimated to be $7 \times 10^8$/cm$^2$ yr. The tritium contents in the documented rocks decreased with increasing depth. The solar-flare intensity averaged over 30 yr obtained from the tritium depth dependence was approximately the same as the flare intensity averaged over 1000 yr obtained from the $Ar^{39}$ measurements. Radioactivities in two Apollo 15 soil samples, $H^3$ in several Surveyor 3 samples, and tritium and radon weepage were also measured.

## INTRODUCTION

THE RADIOACTIVITIES OF $Ar^{37}$, $Ar^{39}$, and $H^3$ were measured in a number of Apollo 11 and 12 samples (Fireman *et al.*, 1970; D'Amico *et al.*, 1970, 1971; Stoenner *et al.*, 1970a, b, 1971; Begemann *et al.*, 1970; Bochsler *et al.*, 1971). In the main, the measurements from different laboratories agree. D'Amico *et al.* (1971) measured the depth variation of $Ar^{37}$, $Ar^{39}$, and $H^3$ in rock 12002, a 6.4-cm-thick crystalline rock.

With samples from known locations in the large documented rocks, 14321 and 15555, we extended the $Ar^{37}$, $Ar^{39}$, and $H^3$ in a number of important ways. Apollo 14 material was subjected to a large solar flare on 24 January 1971, two weeks before sample recovery. No large solar flares occurred for several months before the Apollo 15 mission. The difference between the $Ar^{37}$ activities from similar locations in 12002 and 15555 gives direct measures of the $Ar^{37}$ activities produced by the 2 November 1969 flare, and the difference between the $Ar^{37}$ activities in 14321 and 15555 gives the $Ar^{37}$ activities produced by the 24 January 1971 flare. By use of measured $Ar^{37}$ cross sections in simulated lunar material, the intensities of these flares were determined.

The target elements important for $Ar^{39}$ production are Fe, Ti, K, and Ca. Rock 14321 is a breccia with low Fe and high K contents; rocks 12002 and 15555 have high Fe and low K contents; the Ca and Ti contents of the rocks differ only slightly. Since $Ar^{39}$ is produced by the action of high-energy protons ($> 200$ MeV) on Fe and Ti, by low-energy neutrons ($\lesssim 1$ MeV) on K, and by intermediate-energy neutrons ($\lesssim 10$

MeV) on Ca, the $Ar^{39}$ measurements from similar depths in the rocks determine the $Ar^{39}$ from the separate target elements and the fluxes of the corresponding particles that interact with the separate target elements. The $Ar^{39}$ has a 270-yr half-life so that the fluxes integrated over approximately 1000 yr are obtained from these determinations. The average high-energy proton flux for 1000 yr can be compared with measurements of the galactic rays; the neutron fluxes averaged over 1000 yr can be compared with the neutron fluxes that are obtained from $Ar^{37}$ measurements.

The tritium was released in a two-step heating process, 3-hour heating at 275°C and melting, in the Apollo 14 and 15 and Surveyor 3 materials. The tritium released at 275°C may be either tritium implanted by the solar wind as suggested by D'Amico *et al.* (1971) or terrestrial contamination as suggested by Bochsler *et al.* (1971). The tritium released at high temperatures is mainly caused by solar-flare and galactic cosmic-ray interactions. The depth dependence of the tritium in rocks 14321 and 15555, Apollo 15 soil samples, and Surveyor 3 samples and its temperature-release pattern give information about the sources of the tritium and information about the intensity of solar flares integrated over the past 30 yr.

## Sample Description

Both brecciated rock 14321 and crystalline rock 15555 are documented rocks, which means that their position and orientation on the lunar surface were determined from photographs taken by an astronaut on the lunar surface. We received samples 14321,81,* 14321,267,† and 14321,95 from rock 14321. A drawing of 14321, together with a slice taken from the center of the rock showing the locations of the samples, is given in Fig. 1. Sample 14321,81 extended from the top surface to 1.5-cm depth; the material from this sample was broken into three 0.5-cm-thick sections for separate analysis. The orientations of the sections were approximately preserved. Sample 14321,95 was from the bottom of the rock at approximately 12 cm or 40 g/cm² depth.

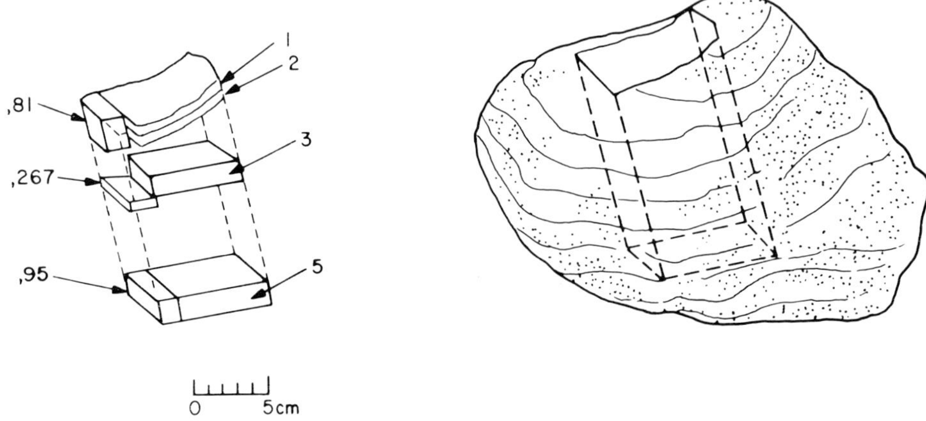


Fig. 1. Lunar rock 14321 and sample locations.

* Samples 14321,81 and 14321,83 were combined and called sample 14321,81.

† Samples 14321,261; 14321,262; and 14321,267 were combined and called 14321,267.

Sample 14321,267 was from 5-cm depth. Although samples 14321,81 and 14321,95 were received 78 days after the Apollo 14 mission, sample 14321,267 was received more than six months after the mission. Nearby samples of 14321 labeled 1, 2, 3, and 5 were analyzed by Wahlen *et al.* (1972) for a number of radioactive nuclides. Figure 1 was obtained from Wahlen *et al.* (1972). Rock 14321 was very fragile, so that samples could be broken with small hand tweezers without much difficulty.

Our samples of 15555 were taken from slice 15555,57 shown by solid lines in Fig. 2,

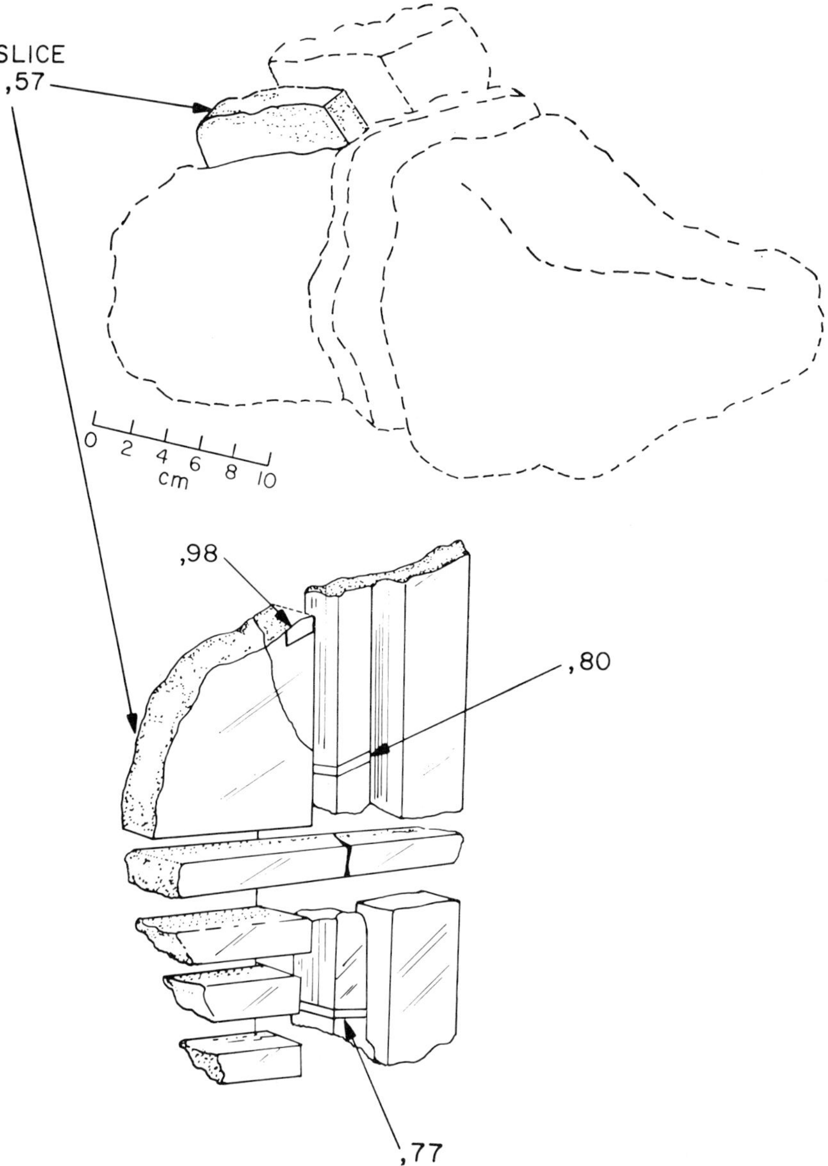


Fig. 2. Lunar rock 15555 and sample locations.

where the outline of rock 15555 is shown by dashed lines. This drawing was made by members of the Lunar Receiving Laboratory. Sample 15555,98 was a top-surface sample approximately 0.8-cm thick. Sample 15555,80 was from near the middle of an adjacent column cut from slice 15555,57 at approximately 6-cm depth, and sample 15555,77 was from near the bottom of the same column at approximately 14-cm depth. These samples were received 119 days after the Apollo 15 mission. Two soil samples, 15271,17 and 15261,14, of 1.97- and 2.04-g weight were received 51 days after the mission; their $Ar^{37}$ activities could be measured more accurately than those in the larger 15555 rock samples. Soil 15271,17 was taken from the compressed wheel track of the lunar rover, and soil 15261,14 was collected from a trench dug into the rim of a 12-m-diameter crater. The depth of the trench was estimated at approximately 30 cm.

## EXPERIMENTAL PROCEDURE

Since $Ar^{37}$ decays with a 35-day half-life, we measured the argon radioactivities as rapidly as possible and stored the tritium, except for that from the remelting of the samples, for later analysis. The tritium from the remelt was measured before proceeding to the next sample to ascertain that all tritium had been removed from the sample. The gamma-ray analysis of Apollo 14 samples done at LRL indicated that their uranium and thorium contents were ten-fold higher than those in most previous samples. We therefore put an additional charcoal trap in the purification system to ensure the complete removal of the larger amounts of radon expected from 14321 samples. Otherwise, the extraction and purification procedures for the gases from the 14321 samples were identical to those used for most previous lunar samples (D'Amico *et al.*, 1971). The additional charcoal trap was not used in the processing of the Apollo 15 samples.

The gas was removed in three heatings. After the sample had been placed in a molybdenum crucible that had been previously outgassed, the extraction system was pumped down to a pressure of several Torrs. Hydrogen carrier was added, and the sample was heated to 275°C with a resistance heater for 2 hr. The gases were removed with an automatic toepler pump and stored in a glass bulb for later purification and tritium analysis. The sample was then melted by induction heating in the presence of argon and hydrogen carrier. The gases from the molten Apollo 14 samples were collected on charcoal at liquid-nitrogen temperature and removed at dry-ice temperature, and then transferred to finely divided vanadium powder at 800°C. The gases from the molten Apollo 15 samples were transferred to the vanadium directly. The hot vanadium removed the chemically active constituents; the vanadium was then slowly cooled to room temperature to absorb hydrogen as vanadium hydride. The remaining gas was then removed from the cool vanadium, and its volume measured. If the volume of gas was larger than the amount of argon carrier used, the gas was repurified over vanadium. This procedure was repeated until the volume of gas did not change. At this stage, the volume was the same (within several percent) as that of the argon carrier used. This gas was repurified over hot titanium and condensed on charcoal at liquid-nitrogen temperature. The argon was removed from the charcoal at dry-ice temperature and placed in the same small low-level proportional counters used for Apollo 11 and 12 samples (Fireman *et al.*, 1970; D'Amico *et al.*, 1970, 1971). Methane (10%) was added to the argon. The counter was removed from the purification system and counted in the low-level system, where the counts between 0.4 and 8.0 keV were recorded on a 100-channel analyzer, and those above 7.4 keV, on scalers. Fig. 3 is a plot of the argon counting data for sample 14321,95 for an 8-day period from 27 April to 6 May 1971. Figure 4 is a plot of the argon counting data for 15555,80 for a 22-day period from 6 to 28 December 1971.

The hydrogen was recovered from the vanadium by reheating the vanadium to 800°C and pumping the released hydrogen into a storage bulb.

The sample was then remelted by induction heating, and wall deposits were heated with a torch in the presence of hydrogen carrier. The gas was transferred to finely divided vanadium power at

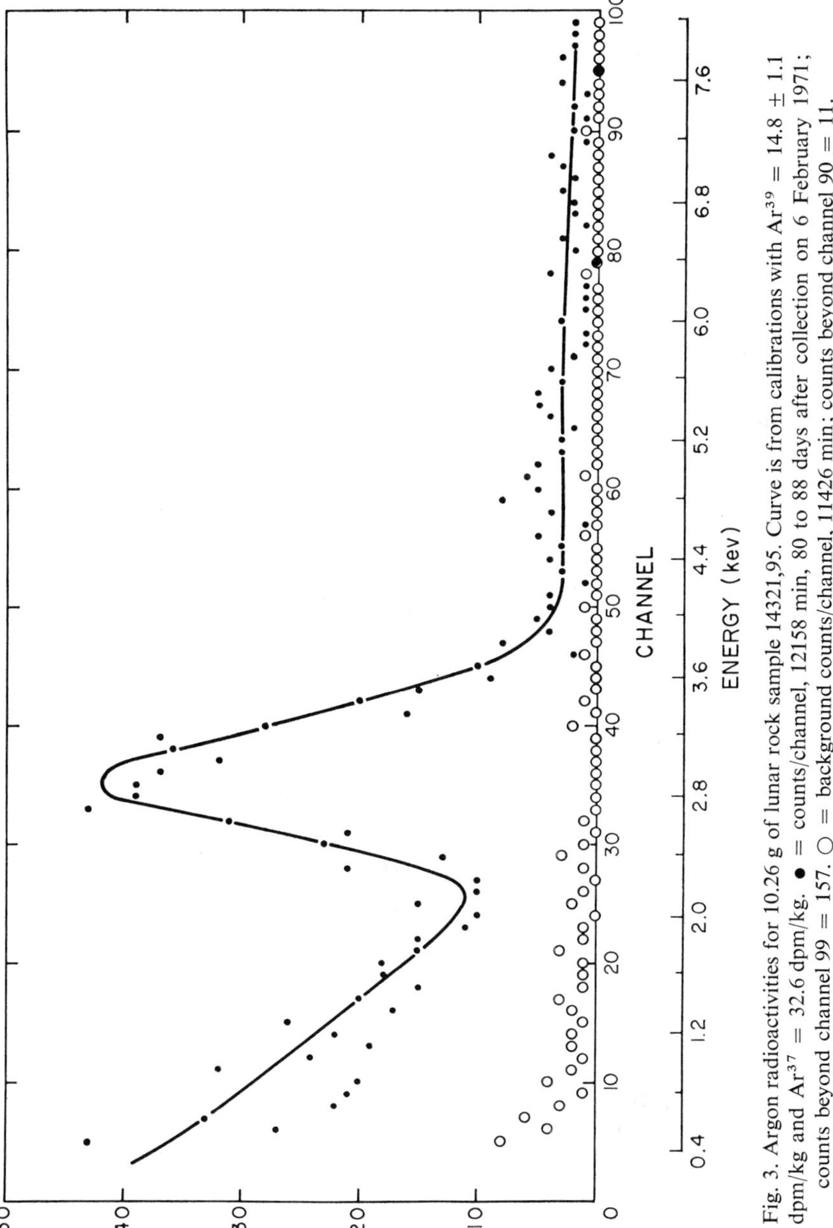


Fig. 3. Argon radioactivities for 10.26 g of lunar rock sample 14321,95. Curve is from calibrations with $Ar^{39}$ = 14.8 ± 1.1 dpm/kg and $Ar^{37}$ = 32.6 dpm/kg. ● = counts/channel, 12158 min, 80 to 88 days after collection on 6 February 1971; counts beyond channel 99 = 157. ○ = background counts/channel, 11426 min; counts beyond channel 90 = 11.

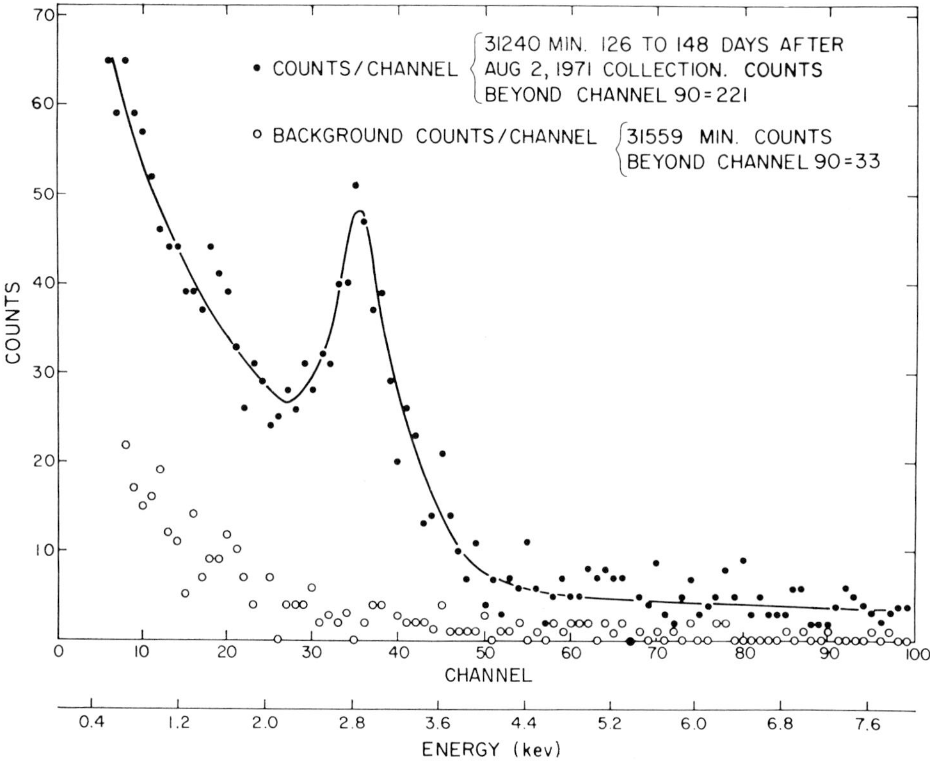


Fig. 4. Argon radioactivity for lunar rock sample 15555,80 (10.25 g).

800°C, which was slowly cooled to room temperature until no gas remained. The hydrogen was removed from the vanadium by reheating and passed through a charcoal trap at dry-ice temperature to remove any possible radon. After its volume had been measured, the hydrogen was added to a low-level proportional counter that contained 400-Torr pressure of P-10 gas. The resolution of the counter was checked with an $Fe^{55}$ source, and the counting was done in a low-level system.

## RESULTS

Table 1 gives the $Ar^{37}$ activities that we have measured and the dates of large solar flares that occurred within four months before the missions. Apollo 15 samples were not exposed to any such flares; Apollo 14 samples were exposed to the 24 January 1971 flare 13 days before recovery; and Apollo 12 samples were exposed to the 2 November 1969 flare 21 days before recovery. The $Ar^{37}$ activity in the top sample of 15555, 15.8 $\pm$ 3.0 dpm/kg is a factor of 2.5 lower than in the top sample of 14321 and a factor of 2 lower than in the top sample of 12002. The $Ar^{37}$ activities in 15555 are produced solely by galactic cosmic rays; those in 14321 and 12002 are produced by both galactic cosmic rays and a solar flare. $Ar^{37}$ activity is produced mainly by the action of protons and neutrons on Ca. The $Ar^{37}$ activity per kg of Ca in the top sample of 15555 is slightly lower than in soils 15271, 10017, and 10084, showing that

Table 1. $Ar^{37}$ activities and solar flares within 4 months of mission.

| Sample | Depth (cm) | Recovery date | Flare date | $Ar^{37}$* (dpm/kg) | Ca (%) | $Ar^{37}$ (dpm/kg Ca) | $\Delta Ar^{37}$† (dpm/kg Ca) | Solar-proton flux (1/cm$^2$ sr) | Solar-proton energies |
|---|---|---|---|---|---|---|---|---|---|
| 15555,98 | 0–0.8 | 2 August 1971 | none | 15.8 ± 2.8 | 6.7 | 236 ± 42 | — | — | — |
| 15555,80 | ~6 | 2 August 1971 | none | 31.2 ± 3.3 | 6.7 | 465 ± 49 | — | — | — |
| 15555,77 | ~14 | 2 August 1971 | none | 35.0 ± 6.5 | 6.7 | 520 ± 97 | — | — | — |
| 14321,81,83 | 0–0.5 | 6 February 1971 | 24 January 1971 | 38.5 ± 2.5 | 6.4 | 600 ± 39 | 364 ± 60 | (5.9 ± 1.0) × $10^6$ | (> 50 MeV) |
| 14321,81,83 | 0.5–1.0 | 6 February 1971 | 24 January 1971 | 37.6 ± 2.5 | 6.4 | 590 ± 39 | 354 ± 60 | — | — |
| 14321,81,83 | 1.0–1.5 | 6 February 1971 | 24 January 1971 | 35.0 ± 4.0 | 6.4 | 550 ± 63 | 324 ± 75 | — | — |
| 14321,95 | ~12 | 6 February 1971 | 24 January 1971 | 34.0 ± 2.7 | 6.4 | 530 ± 42 | ≤100 | (< 1.6) × $10^6$ | (> 200 MeV) |
| 12002,57 | 0–0.8 | 23 November 1969 | 2 November 1969 | 30.0 ± 4.0 | 5.4 | 556 ± 65 | 320 ± 70 | (5.1 ± 1.2) × $10^6$ | (> 50 MeV) |
| 12002,57 | 0.8–3.1 | 23 November 1969 | 2 November 1969 | 25.0 ± 1.5 | 5.4 | 463 ± 28 | 160 ± 50 | (2.6 ± 0.9) × $10^6$ | (> 70 MeV) |
| 12002,59 | 4.9–6.4 | 23 November 1969 | 2 November 1969 | 27.5 ± 2.5 | 5.4 | 510 ± 46 | 55 ± 65 | (0.9 ± 1.1) × $10^6$ | (> 100 MeV) |
| Soil 15271,17 | 0–5 | 2 August 1971 | none | 21.4 ± 2.0 | 8.1 | 264 ± 25 | — | — | — |
| Soil 15261,14 | ~30 | 2 August 1971 | none | 26.4 ± 3.0 | — | — | — | — | — |
| Soil 10017,14 | 0–5 | 21 July 1969 | none | 21.0 ± 2.0 | 7.4 | 284 ± 27 | 20 ± 37 | — | — |
| Soil 10084,24 | 0–5 | 21 July 1969 | none | 27.2 ± 2.2 | 8.8 | 309 ± 25 | 45 ± 35 | — | — |

* Activity at time of recovery.

† Solar-flare-produced $Ar^{37}$ from calcium, $\Delta Ar^{37} = (Ar^{37}/\%\ Ca) - [Ar^{37}(15)/\%\ Ca]$, where $Ar^{37}(15)$ is the $Ar^{37}$ in an Apollo 15 sample at a comparable depth.

these soil samples, although not affected by solar flares, were taken from slightly larger average depths than sample 15555,98 was. The solar-flare-produced $Ar^{37}$ in the top samples of 14321 and 12002 were 364 $\pm$ 60 dpm/kg Ca and 324 $\pm$ 70 dpm/kg Ca, respectively; however, at approximately 6-cm depth, the solar-flare-produced $Ar^{37}$ is only 55 $\pm$ 65 dpm/kg Ca.

Reedy and Arnold (1972) calculated the $Ar^{37}$ activities as a function of depth by galactic cosmic rays on lunar material. The $Ar^{37}$ activity measured at three depths in rock 15555 agrees quite well with Reedy and Arnold's calculations. It is not necessary to use Reedy and Arnold's calculations to obtain the intensities of the 24 January 1971 and the 2 November 1969 flares from the $Ar^{37}$ measurements. D'Amico *et al.* (1971) measured the $Ar^{37}$ production cross sections in a series of targets of simulated lunar material containing 6.7% Ca bombarded by 158-MeV protons. Interspersed between the lunar-type targets were iron absorbers so that the $Ar^{37}$ production cross sections for 50-, 85-, and 158-MeV protons could be determined. The protons of these energies in the flares should contribute more $Ar^{37}$ than those of higher energies if the differential energy spectrum varies as $E^{-3}$; the $Ar^{37}$ production cross section, $\sigma_{Ca}(Ar^{37})$, is 60 $\pm$ 3 mb for 158- and 85-MeV protons and 10.8 $\pm$ 1.0 mb for 50-MeV protons. The cross section falls steeply between 85 and 50 MeV; below 50 MeV, the cross section probably falls even more steeply. On the basis of these cross sections and the solar-flare-produced $Ar^{37}$ from calcium, $\Delta Ar^{37}$, which is defined in Table 1, we estimate the solar-flare-proton fluxes given in Table 1. The 24 January 1971 flare intensity is estimated to be $(5.9 \pm 1.0) \times 10^6$ protons/cm$^2$ sr greater than 50-MeV energy; the 2 November 1969 flare intensity is estimated to be $(5.1 \pm 1.2) \times 10^6$ protons/cm$^2$ sr (>50 MeV). These flare intensities obtained from lunar rocks can be compared with satellite measurements. From the Explorer 35 satellite, Van Allen (private communication, 1972) obtained $4.9 \times 10^6$ protons (>55 MeV/cm$^2$ sr) for the 24 January 1971 flare and $3.2 \times 10^6$ protons (>55 MeV/cm$^2$ sr) for the 2 November 1969 flare with 20% uncertainty. The intensities of the flares obtained from the $Ar^{37}$ activities in lunar rocks are slightly higher than those obtained by Van Allen from his Explorer 35 counters.

Table 2 gives the $Ar^{39}$ activities at several depths in three oriented lunar rocks and in four soil samples. An interesting feature of the $Ar^{39}$ activities is that the $Ar^{39}$ increases with depth in 14321, a high K rock, but is approximately constant with depth in low K rocks 15555 and 12002. $Ar^{39}$ is produced by high-energy protons on Fe (>300 MeV) and on Ti ($\lesssim$200 MeV); on the other hand, $Ar^{39}$ is produced from K and Ca by the action of neutrons through the $K^{39}$ $(n, p)$ and $Ca^{40}$ $(n, 2p)$ reactions. The $K^{39}$ reaction is exothermic; the $Ca^{40}$ reaction has an 8.1-MeV threshold energy. The $Ar^{39}$ production rates for four individual target elements can be calculated from $Ar^{39}$ measurements in four lunar samples at the same depth; however, such calculated rates would have considerable error and there would be no consistency checks. If the $Ar^{39}$ production rates from Fe and Ti are lumped together and the Fe + Ti $\rightarrow Ar^{39}$ production rate is estimated theoretically, then the K $\rightarrow Ar^{39}$ and Ca $\rightarrow Ar^{39}$ production rates are obtained from the data with relatively small errors and there are six samples that remain for consistency checks. We have therefore adopted this procedure for obtaining the $Ar^{39}$ production rates in K and Ca.

Table 2. $Ar^{39}$ activities, estimated productions from target elements with and without energetic solar flares, and neutron fluxes.

| | | | | | | | Situation 1 No solar-flare protons (> 200 MeV) past 1000 yr | | | Situation 2 Solar-flare protons (> 200 MeV) past 1000 yr equal galactic cosmic rays | | | Neutrons/cm$^2$ sec | |
|---|---|---|---|---|---|---|---|---|---|---|---|---|---|---|
| Sample | Depth (cm) | K (%)* | Fe (%)* | Ti (%)* | Ca (%)* | $Ar^{39}$ (dpm/kg) | Fe + Ti → $Ar^{39}$ (dpm/kg)† | K → $Ar^{39}$ (dpm/kg) | Ca → $Ar^{39}$ (dpm/kg) | Fe + Ti → $Ar^{39}$ (dpm/kg)‡ | K → $Ar^{39}$ (dpm/kg) | Ca → $Ar^{39}$ (dpm/kg) | > 1 MeV** | > 10 MeV†† |
| 15555,98 | 0–0.8 | 0.025 | 17.5 | 1.36 | 6.7 | 8.4 ± 0.5 | 1.8 | 0.13 | 6.5 | 3.6 | 0.17 | 4.7 | | |
| 15555,80 | ~ 6 | 0.025 | 17.5 | 1.36 | 6.7 | 7.5 ± 0.5 | 1.7 | 0.3 | 5.5 | 3.4 | 0.5 | 3.7 | | |
| 15555,77 | ~ 14 | 0.025 | 17.5 | 1.36 | 6.7 | 9.8 ± 0.8 | 1.4 | 0.4 | 8.0 | 2.8 | 0.6 | 6.3 | | |
| 14321,81 | 0–0.5 | 0.37 | 8.6 | 1.2 | 6.4 | 8.8 ± 0.5 | 1.0 | 2.0 ± 0.6 | 5.8 ± 0.7 | 2.0 | 2.6 ± 0.6 | 4.2 ± 0.6 | 3.7 | 2.0 |
| 14321,81 | 0.5–1.0 | 0.37 | 8.6 | 1.2 | 6.4 | 8.3 ± 0.7 | 1.0 | 2.0 ± 0.6 | 5.8 ± 0.7 | 2.0 | 2.6 ± 0.6 | 4.2 ± 0.6 | 3.7 | 2.0 |
| 14321,81 | 1.0–1.5 | 0.37 | 8.6 | 1.2 | 6.4 | 9.5 ± 0.8 | 1.0 | 2.0 ± 0.6 | 5.8 ± 0.7 | 2.0 | 2.6 ± 0.6 | 4.2 ± 0.6 | 3.7 | 2.0 |
| 14321,267 | ~ 5 | 0.37 | 8.6 | 1.2 | 6.4 | 12.1 ± 1.0 | 0.95 | 6.2 ± 1.1 | 5.0 ± 0.7 | 1.9 | 7.0 ± 1.2 | 3.2 ± 0.5 | 10.6 | 1.6 |
| 14321,95 | ~ 12 | 0.37 | 8.6 | 1.2 | 6.4 | 14.8 ± 1.0 | 0.84 | 6.8 ± 1.2 | 7.2 ± 0.9 | 1.7 | 7.5 ± 1.3 | 5.7 ± 0.8 | 12 | 2.7 |
| 12002,57 | 0–0.8 | 0.045 | 16.5 | 1.5 | 5.4 | 8.0 ± 0.7 | 1.8 | 0.3 | 5.2 | 3.6 | 0.4 | 3.7 | | |
| 12002,57 | 0.8–3.1 | 0.045 | 16.5 | 1.5 | 5.4 | 8.2 ± 0.5 | 1.8 | 0.4 | 5.2 | 3.5 | 0.5 | 3.7 | | |
| 12002,59 | 4.9–6.4 | 0.045 | 16.5 | 1.5 | 5.4 | 8.0 ± 0.6 | 1.7 | 0.7 | 4.4 | 3.4 | 0.9 | 3.0 | | |
| Soil: | | | | | | | | | | | | | | |
| 15271,17 | 0–5 | 0.17 | 9.5 | 0.88 | 8.1 | 10.0 ± 1.5 | 1.0 | 0.9 | 7.7 | 2.0 | 1.2 | 5.6 | | |
| 15261,14 | ~ 30 | — | — | — | — | 7.1 ± 1.0 | — | — | — | — | — | — | | |
| 10017,14 | 0–5 | 0.24 | 15.3 | 7.1 | 7.4 | 16.4 ± 0.9 | 3.0 | 1.3 | 7.2 | 6.0 | 1.8 | 5.2 | | |
| 10084,24 | 0–5 | 0.11 | 12.3 | 4.4 | 8.8 | 12.1 ± 0.7 | 2.0 | 0.6 | 8.4 | 4.0 | 0.8 | 6.2 | | |

* A weight percent obtained from various publications of chemical composition of lunar material.
† Reedy and Arnold's calculated galactic cosmic-ray production rate of $Ar^{39}$ from Fe + Ti (private communication, 1972).
‡ Twice calculated galactic cosmic-ray production rate of $Ar^{39}$ from Fe + Ti.
** Based on $Ar^{39}$ from $K^{39}$ with $\sigma(n,p) = 200$ mb.
‡‡ Based on $Ar^{39}$ from $Ca^{40}$ with $\sigma(n, 2p) = 40$ mb.

In Table 2, two situations are represented: (1) the $Ar^{39}$ from Fe + Ti is produced solely by galactic cosmic rays of the present average intensity and is given by Reedy and Arnold's calculation (private communication, 1972); and (2) extremely energetic solar flares have occurred during the past 1000 yr, which raised the cosmic-ray intensity to twice the present average, so that the $Ar^{39}$ from Fe + Ti is twice the Reedy and Arnold value.

The calculation proceeds by subtracting the $Ar^{39}$ produced from Fe + Ti from the measured $Ar^{39}$ to give the $Ar^{39}$ from Ca + K. The $Ar^{39}$ produced from Ca and the $Ar^{39}$ from K are regarded as unknowns, and a pair of simultaneous equations is set up with 14321 and 15555 data. These equations are solved for each depth. The results are given in Table 2. Because 15555 and 14321 samples differ by a factor of 15 in K contents, the procedure gives the $Ar^{39}$ produced from Ca and from K with errors of the same magnitude as the error in the $Ar^{39}$ measurement. $Ar^{39}$ from K for situation (1) is approximately 15% lower than from situation (2); the $Ar^{39}$ from Ca for situation (1) is 40% higher than from situation (2). With these production rates, the $Ar^{39}$ production rates from K and from Ca for rock 12002 samples and the soil samples are calculated. The sum of the productions from the individual target elements agrees fairly well with the measured $Ar^{39}$ activities for most cases. In soil 10017 the measured $Ar^{39}$ activity is too high. Situation (2) gives a better approximation to the measured activities for the six samples that serve as consistency checks than does situation (1). The K $\rightarrow$ $Ar^{39}$ production rate for the top of 14321 is lower than estimated by Begemann *et al.* (1970) for Apollo 11 fines.

The question arises as to how large can the Fe + Ti $\rightarrow$ $Ar^{39}$ rate be made before inconsistencies arise. The $Ar^{39}$ measured activity in sample 15555,80 limits the Fe + Ti $\rightarrow$ $Ar^{39}$ production rate to less than four times the Reedy and Arnold calculated rates. Therefore, the intensity of solar-flare protons of energy greater than ~200 MeV averaged over the past 1000 yr must be less than four times the intensity of the recent galactic cosmic rays if galactic cosmic rays have remained unchanged over 1000 yr.

The next question is whether or not there is any positive evidence from the $Ar^{39}$ measurements for the presence of lower energy protons (<200 MeV) from solar flares during the past 1000 yr. The depth variation of Ca $\rightarrow$ $Ar^{39}$ does provide positive evidence for low-energy solar flares. The Ca $\rightarrow$ $Ar^{39}$ production rates for all cases decrease slightly with increasing depth from 0 to 5 cm and then increase with increasing depth from 5 to 14 cm. Such a depth dependence is different than that expected for galactic cosmic rays. For example, Reedy and Arnold (1972) calculate that the Ca $\rightarrow$ $Ar^{39}$ from galactic cosmic rays should increase by a factor of 2 in the first 5-cm depth. A direct determination of the intensity of <200-MeV solar-flare protons averaged over 1000 yr from the Ca $\rightarrow$ $Ar^{39}$ production rates requires knowledge of the $Ar^{39}$ production cross section in thick Ca targets bombarded by protons of less than 200-MeV energy. These cross sections have not been measured. Since neutron-production cross sections for $Ar^{39}$ from $K^{39}$ and from $Ca^{40}$ are known, the neutron fluxes from the K $\rightarrow$ $Ar^{39}$ and Ca $\rightarrow$ $Ar^{39}$ determinations can be obtained and the neutron fluxes can be used to estimate the flux of solar-flare protons.

The $\sigma(n, p)$ cross section in $K^{39}$ decreases from 350 $\pm$ 50 mb for 14-MeV neutrons, to 270 $\pm$ 40 mb for 4-MeV neutrons, and to approximately 50 mb for 1.5-MeV neutrons (Stehn *et al.*, 1964). Although neutrons of lower energies produce $Ar^{39}$ from $K^{39}$, we call the neutron flux obtained from the $K^{39} \rightarrow Ar^{39}$ reaction the (>1 MeV) neutron flux in Table 2, because of energy dependence of the cross section. This flux is 3.7, 10.6, and 12/cm$^2$ sec at 1-, 5-, and 12-cm depths, respectively, with a $\sigma(n, p)$ cross section of 200 mb. The (>1-MeV) neutron flux increases with increasing depth. The $K \rightarrow Ar^{39}$ production rates in 14321 are almost twice the rates calculated by Reedy and Arnold for galactic cosmic-ray production (private communication, 1972).

The $\sigma(n, 2p)$ on $Ca^{40}$ has an 8.1-MeV threshold and is 35 mb for 14-MeV neutrons (Stoenner *et al.*, 1971). The contribution of $Ca^{42}$ to $Ar^{39}$ is small because of its low natural abundance, 0.6%, if the $(n, \alpha)$ $Ca^{42}$ cross section is similar to the $(n, \alpha)$ $Ca^{40}$ cross section. The $Ar^{39}$ from Ca is therefore mainly produced by greater than 10-MeV neutrons on $Ca^{40}$ with approximately a 40-mb cross section. The (>10 MeV) neutron fluxes obtained from the $Ca \rightarrow Ar^{39}$ column in Table 2 are 2.0, 1.6, and 2.7/cm$^2$ sec at approximately 1-, 5-, and 12-cm depths, respectively. The action of galactic cosmic rays on lunar material should produce a neutron flux (>10 MeV) that increases with increasing depth for the first 14 cm. Evidently, solar flares during the past 1000 yr have increased the >10-MeV neutron flux in the top centimeter of material by approximately one neutron per cm$^2$ sec relative to the flux at 5-cm depth. We shall assume that neutrons of greater than 10-MeV energy are produced by solar-flare protons of greater than 50-MeV energy with a cross section of 0.5 b; then 25 protons of (>50 MeV) energy are required to produce one neutron of (>10 MeV) energy in the top centimeter. With this estimate, a solar-flare-proton flux (>50 MeV) of 25/cm$^2$ sec or $7 \times 10^8$/cm$^2$ yr averaged over the past 1000 yr is obtained from $Ca \rightarrow Ar^{39}$ values in Table 2. On the bases of the $Ar^{37}$ depth dependence after the 24 January 1971 and 2 November 1969 flares, we estimated that approximately $6 \times 10^6$/cm$^2$ sr or $4 \times 10^7$/cm$^2$ of protons >50 MeV were in each of these flares. From the $Ar^{39}$ depth dependence, we estimate that during the past 1000 yr, the flare flux was equivalent to 15 such flares per year.

The measured tritium radioactivities from the samples of 14321 and 15555 and two Apollo 15 soil samples with their approximate depths are given in Table 3. Very little tritium was released at 275°C from the 14321 and 15555 samples. From soil 15271,17, 15% of the tritium was released at 275°C and from soil 15261,14, 8% of the tritium was released at 275°C. If the low-temperature tritium observed from the top of rock 12002 (D'Amico *et al.*, 1971) and from soil 15271 is caused by solar-wind implantation of tritium, then the top surface material from our 14321 and 15555 samples, which should contain solar wind, was lost during the sawings and handlings. This is quite likely for 14321, which was a fragile breccia. Rock 15555 is a crystalline rock; our top sample, 15555,98, was from a broken sloped section (see Fig. 2) and could also have had its surface material removed in handling and sawing; however, this is less likely. In both rocks 15555 and 14321, the tritium decreased with increasing depth; however, the decrease was quite small from 6- to 14-cm depth in rock 15555. The decrease of tritium contents with depth is evidence for tritium produced by solar-flare interactions. Figure 5 gives the $H^3$ depth dependence in the three rocks. The top samples of 15555 and of 14321 had 280 ± 14 dpm/kg compared to 392 ± 11 dpm/kg of tritium in the top of 12002. Differences in the topographies of the rocks probably

Table 3. Tritium activities in Apollo 15 and 14 samples.

| Sample | Wt (g) | Depth (cm) | Temperature (°C) | $H^3$ (dpm/kg) | Total $H^3$ (dpm/kg) | Extraction date |
|---|---|---|---|---|---|---|
| 15555,98 | 10.65 | 0–1 | 275 | $<3$ | | |
| | | surface | Melt | $260 \pm 9$ | $281 \pm 11$ | 12/10/71 |
| | | | Remelt | $21 \pm 5$ | | |
| 15555,80 | 10.25 | ~6 | 275 | $11 \pm 3$ | | |
| | | | Melt | $146 \pm 6$ | $171 \pm 8$ | 12/2/71 |
| | | | Remelt | $14 \pm 3$ | | |
| 15555,77 | 8.44 | ~14 | 275 | $<3$ | | |
| | | | Melt | $141 \pm 5$ | $166 \pm 8$ | 12/23/71 |
| | | | Remelt | $25 \pm 5$ | | |
| 14321,81 | 10.08 | 0–0.5 | 275 | $<3$ | | |
| | | surface | Melt | $270 \pm 12$ | $280 \pm 14$ | 5/10/71 |
| | | | Remelt | $10 \pm 2$ | | |
| 14321,81 | 8.49 | 0.5–1.0 | 275 | $5 \pm 3$ | | |
| | | | Melt | $197 \pm 6$ | $208 \pm 8$ | 5/3/71 |
| | | | Remelt | $6 \pm 3$ | | |
| 14321,81 | 9.46 | 1.0–1.5 | 275 | $6 \pm 3$ | | |
| | | | Melt | $164 \pm 8$ | $198 \pm 11$ | 4/28/71 |
| | | | Remelt | $28 \pm 5$ | | |
| 14321,267 | 4.7 | ~5 | 275 | $15 \pm 5$ | | |
| | | | Melt | $161 \pm 7$ | $176 \pm 11$ | 7/20/71 |
| | | | Remelt | $<7$ | | |
| 14321,95 | 10.08 | ~12 | 275 | $3 \pm 3$ | | |
| | | | Melt | $133 \pm 5$ | $145 \pm 8$ | 5/25/71 |
| | | | Remelt | $9 \pm 4$ | | |
| Soil 15271,17 | 1.97 | 0–5 | 275 | $45 \pm 7$ | | |
| | | | Melt | $255 \pm 10$ | $300 \pm 15$ | 10/7/71 |
| | | | Remelt | $<7$ | | |
| Soil 15261,14 | 2.03 | ~30 | 275 | $20 \pm 5$ | | |
| | | | Melt | $220 \pm 15$ | $240 \pm 15$ | 9/24/71 |
| | | | Remelt | $<4$ | | |

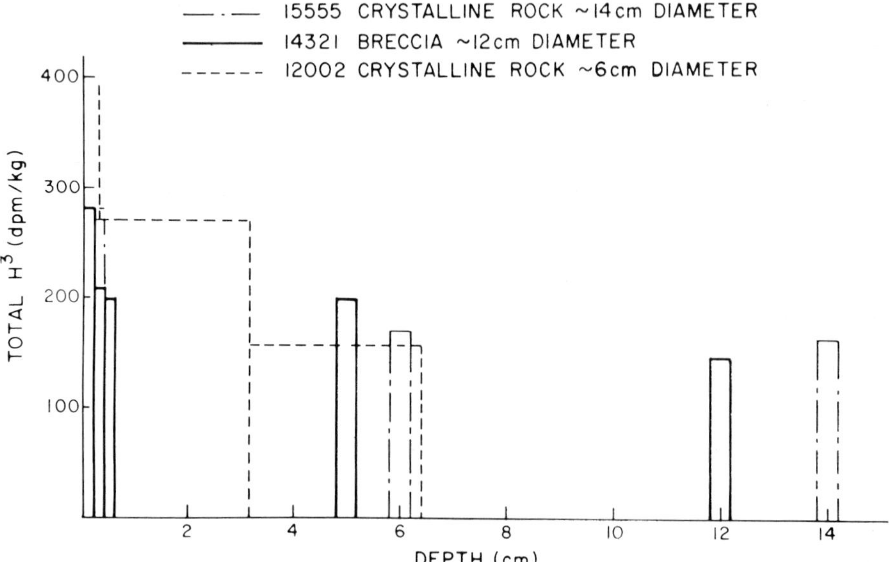


Fig. 5. Depth variation of tritium.

account for most of tritium differences between the samples at similar depths. We do not consider the differences in tritium contents between samples from 15555, 14321, and 12002 to indicate any serious discrepancies, except possibly for the high tritium in the top sample of 12002. Soil sample 15271,17 had 300 $\pm$ 15 dpm/kg, and soil 15261,14 had 240 $\pm$ 15 dpm/kg of tritium. An estimated 150 dpm/kg of tritium were produced by the action of solar flares in the top centimeter of 14321 and 15555, based on the tritium depth dependences. The solar-flare intensity calculated with this tritium excess is $1.7 \times 10^9$ protons ($>30$ MeV)/cm$^2$ yr or $6 \times 10^8$ ($>50$ MeV)/cm$^2$ yr with an $E^{-3}$ differential energy spectrum. The cross sections for this calculation are given by D'Amico *et al.* (1970). This intensity, which represents an average over the past 30 yr, is nearly the same as that obtained from the $Ar^{39}$.

The gases from the Apollo 15 soil samples were extracted approximately two months after recovery; the gases from rock 15555 samples were extracted 4 to $4\frac{1}{2}$ months after recovery. During most of these times, the samples were stored at room temperature with an atmosphere of either air or dry nitrogen. To examine whether or not this type of storage had any effect on the tritium contents, samples of an Apollo 15 soil, crystalline rock, and breccia were sealed in closed containers with air and hydrogen carrier on 9 August 1971, shortly after the returned sample box was opened. After approximately three months' storage at room temperature, the tritium and radon radioactivities in the gas were measured. The results are given in Table 5, where the tritium in the gas of the sealed container is called tritium weepage. No tritium weeped from the crystalline rock sample, which was a sample of 15555. We therefore conclude that the tritium activities measured in 15555 were actually the activities at the time of sample return. The amount tritium weeped from breccia 15565 was only 12.8 $\pm$ 1.0 dpm/kg, which is between 5 and 10% of the amounts observed in breccia 14321. If breccia 14321 is similar to breccia 15565 with regard to tritium weepage, then three months of storage only reduced its tritium contents by 5 to 10%. However, the tritium weepage from soil 15021, 51 $\pm$ 4 dpm/kg, was 20% of the amount observed in soil 15271. If the tritium content of 15271 soil is raised by 20%, its total tritium content is higher than in the top samples from rocks 14321 and 15555 and quite close to the tritium content observed in the top sample of 12002. This may add some support to the suggestion made by D'Amico *et al.* (1971) that the low-temperature tritium from the top of rock 12002 was solar-wind-implanted tritium.

Table 4 gives the hydrogen and tritium data from Surveyor 3 samples. Although most lunar samples were analyzed with the counter of 42-cm$^3$ volume and a background of 0.140 $\pm$ 0.005 count/min, most Surveyor 3 samples, because of their smaller hydrogen contents, were analyzed with the counter of 7-cm$^3$ volume and background of 0.0261 $\pm$ 0.0014 count/min. The Surveyor samples are very thin, $\sim$0.2 g/cm$^2$ thick. Two of the Surveyor samples were exposed to sunlight while on the moon; these released tritium at 270°C. One of the Surveyor samples was facing the soil and protected from sunlight; it released no tritium at 270°C. This sample has 198 $\pm$ 25 dpm/kg of tritium. Since 198 $\pm$ 25 dpm/kg of the tritium were produced during 2.4 yr, the tritium activity at saturation is 1580 $\pm$ 150 dpm/kg. The higher tritium contents in Surveyor 3 material exposed to sunlight than in lunar material indicates that those samples were either contaminated with terrestrial tritium or contain loosely bound tritium that was implanted at the surface by solar wind. Even the

Table 4. Tritium and hydrogen data from Surveyor 3 samples.

| Sample | Wt. (g) | Area ($cm^2$) | Extraction | Hydrogen (ccSTP) | Counter volume (efficiency) | Background (count/min) | Sample (count/min) | $H^3$ (dpm) | $H^3$ (dpm/kg) | $H^3$ (dpm/$cm^2$) | $H^3$ (sat.) (dpm/kg) |
|---|---|---|---|---|---|---|---|---|---|---|---|
| Painted aluminum blank | 0.320 | 1.3 | Melt | 0.45 | 42 $cm^3$ (62%) | 0.140 ± 0.005 | 0.139 ± 0.005 | < 0.012 | | | |
| Painted aluminum blank | 0.255 | 1.0 | Melt | 0.47 | 14 $cm^3$ (55%) | 0.048 ± 0.003 | 0.042 ± 0.003 | < 0.008 | | | |
| Painted aluminum blank | 0.335 | 1.3 | 20–270°C | 0.09 | 7 $cm^3$ (50%) | 0.0261 ± 0.0014 | 0.0269 ± 0.0017 | < 0.005 | | | |
| | | | 270°C-melt | 0.30 | | | | | | | |
| Surveyor 1011,2 | 0.335 | 1.3 | 20–270°C | 0.62 | 7 $cm^3$ (50%) | 0.0261 ± 0.0014 | 0.0353 ± 0.0016 | 0.018 ± 0.004 | 54 ± 12 | 0.014 ± 0.003 | 430 ± 95 |
| Some sunlight | | — | 270°C-melt | 0.82 | 7 $cm^3$ (50%) | 0.0261 ± 0.0014 | 0.0375 ± 0.0016 | 0.022 ± 0.004 | 66 ± 12 | 0.017 ± 0.003 | 530 ± 95 |
| | | — | Wall heating and remelt | 0.70 | 7 $cm^3$ (50%) | 0.0261 ± 0.0014 | 0.0383 ± 0.0016 | 0.023 ± 0.004 | 68 ± 12 | 0.018 ± 0.003 | 545 ± 96 |
| | | — | 2nd wall heating and remelt | 0.09 | 7 $cm^3$ (50%) | 0.0261 ± 0.0014 | 0.0252 ± 0.0016 | < 0.004 | < 12 | < 0.003 | < 95 |
| Total | | | | | | | | 0.063 ± 0.007 | 188 ± 21 | 0.049 ± 0.005 | 1505 ± 170 |
| Surveyor 1002 | 0.243 | 1.86 | 20–270°C | with $H_2$ carrier | 7 $cm^3$ (50%) | 0.0261 ± 0.0014 | 0.0270 ± 0.0016 | < 0.004 | < 16 | < 0.002 | |
| No sunlight | | | 270°C-melt | with $H_2$ carrier | 7 $cm^3$ (50%) | 0.0261 ± 0.0014 | 0.0432 ± 0.0016 | 0.034 ± 0.004 | 140 ± 16 | 0.018 ± 0.002 | 1120 ± 130 |
| | | | Wall heating and remelt | with $H_2$ carrier | 0.5 $cm^3$ (68%) | 0.0070 ± 0.0010 | 0.0166 ± 0.0014 | 0.014 ± 0.002 | 58 ± 8 | 0.008 ± 0.001 | 466 ± 50 |
| | | | 2nd wall heating and remelt | with $H_2$ carrier | 7 $cm^3$ (50%) | 0.0261 ± 0.0014 | 0.0284 ± 0.0016 | < 0.005 | < 20 | < 0.003 | < 150 |
| Total | | | | | | | | 0.048 ± 0.005 | 198 ± 25 | 0.026 ± 0.003 | 1580 ± 150 |
| Surveyor 1015 | 0.394 | 1.67 | 20–270°C | with $H_2$ carrier | 7 $cm^3$ (50%) | 0.0261 ± 0.0014 | 0.0651 ± 0.0044 | 0.078 ± 0.009 | 198 ± 23 | 0.047 ± 0.005 | 1580 ± 185 |
| Most sunlight | | | 270°C-melt | with $H_2$ carrier | 7 $cm^3$ (50%) | 0.0261 ± 0.0014 | 0.0515 ± 0.0028 | 0.051 ± 0.006 | 130 ± 15 | 0.030 ± 0.004 | 1040 ± 120 |
| | | | Remelt | with $H_2$ carrier | 42 $cm^3$ (62%) | 0.140 ± 0.004 | 0.147 ± 0.004 | 0.011 ± 0.009 | 28 ± 23 | 0.006 ± 0.005 | 210 ± 185 |
| Total | | | | | | | | 0.129 ± 0.014 | 328 ± 36 | 0.077 ± 0.009 | 2620 ± 240 |

Table 5. Tritium and radon weepage from Apollo 15 samples packaged with air on 9 August 1971.

| Sample | Type | Wt (g) | Extraction date | $H^3$ (dpm/kg) | $Rn^{222*}$ (dpm/kg) |
|---|---|---|---|---|---|
| 15555,1 | Crystalline rock | 18.7 | 23 November 1971 | < 3 | — |
| 15565 | Breccia | 9.7 | 28 October 1971 | 12.8 ± 1.0 | 25 ± 5 |
| 15021 | Soil | 10.0 | 2 November 1971 | 51 ± 4 | 5 ± 1 |
| | | | | | 2.5 ± 1† |

* Counter efficiency for $Rn^{222}$ is estimated to be 2.
† Under vacuum conditions.

Surveyor 3 sample that was not exposed to sunlight had a higher saturation $H^3$ content than the lunar samples did, which is somewhat of a mystery.

The radon emanation from the sealed containers was also measured. The soil under vacuum conditions had a radon emanation rate similar to that observed in the sample container (Stoenner *et al.*, 1971). The samples sealed with air had a higher rate of radon emanation.

*Acknowledgment*—This research was supported in part by Grant NGR 09-015-145 from the National Aeronautics and Space Administration.

## References

Begemann F., Vilcsek E., Rieder R., Born W., and Wänke H. (1970) Cosmic-ray produced radioisotopes in lunar samples from the Sea of Tranquility. *Proc. Apollo 11 Lunar Sci. Conf., Geochim. Cosmochim. Acta* Suppl. 1, Vol. 2, pp. 995–1007. Pergamon.

Bochsler P., Wahlen M., Eberhardt P., Geiss J., and Oeschger H. (1971) Tritium measurements of lunar material (fines 10084 and breccia 10046) from Apollo 11. *Proc. Second Lunar Sci. Conf., Geochim. Cosmochim. Acta* Suppl. 2, Vol. 2, pp. 1803–1812. MIT Press.

D'Amico J., DeFelice J., and Fireman E. L. (1970) The cosmic-ray and solar flare bombardment of the moon. *Proc. Apollo 11 Lunar Sci. Conf., Geochim. Cosmochim. Acta* Suppl. 1, Vol. 2, pp. 1029–1036. Pergamon.

D'Amico J., DeFelice J., Fireman E. L., Jones C., and Spannagel G. (1971) Tritium and argon radioactivities and their depth variations in Apollo 12 samples. *Proc. Second Lunar Sci. Conf., Geochim. Cosmochim. Acta* Suppl. 2, Vol. 2, pp. 1825–1839. MIT Press.

Fireman E. L., D'Amico J., and DeFelice J. (1970) Tritium and argon radioactivities in lunar material. *Science* **167,** 566–568.

Reedy R. C. and Arnold J. R. (1972) Interaction of solar and galactic cosmic-ray particles with the moon. *J. Geophys. Res.* **77,** 537–555.

Stehn J. R., Goldberg M. D., Magurno B. A., and Wiener-Chasman R. (1964) Neutron cross sections. Brookhaven Nat. Lab. Rep. 325, 2nd ed., Suppl. No. 2.

Stoenner R. W., Lyman W. J., and Davis R. Jr. (1970a) Cosmic-ray production of rare gas radioactivities and tritium in lunar material *Science* **167,** 553–555.

Stoenner R. W., Lyman W. J., and Davis R. Jr. (1970b) Cosmic-ray production of rare gas radioactivities and tritium in lunar material. *Proc. Apollo 11 Lunar Sci. Conf., Geochim. Cosmochim. Acta* Suppl. 1, Vol. 2, pp. 1029–1036. Pergamon.

Stoenner R. W., Lyman W. J., and Davis R. Jr. (1971) Radioactive rare gases in lunar rocks and in the lunar atmosphere. *Proc. Second Lunar Sci. Conf., Geochim. Cosmochim. Acta* Suppl. 2, Vol. 2, pp. 1813–1823. MIT Press.

Wahlen M., Honda M., Imamura M., Fruchter J., Finkel R., Kohl C., Arnold J., and Reedy R. (1972) Cosmogenic nuclides in football-sized lunar rocks (abstract). In *Lunar Science—III* (editor C. Watkins), pp. 764–766, Lunar Science Institute Contr. No. 88.

Proceedings of the Third Lunar Science Conference
(Supplement 3, *Geochimica et Cosmochimica Acta*)
Vol. 2, pp. 1763–1769
The M.I.T. Press, 1972

# Study on the cosmic ray produced long-lived Mn-53 in Apollo 14 samples

W. HERR and U. HERPERS
Institut für Kernchemie der Universitaet Köln

and

R. WOELFLE
Institut für Radiochemie der KFA Juelich, Germany

**Abstract**—Accurate determinations of long-lived spallation nuclides, are of current interest for the estimation of "radiation ages," depth profiles, and for the detection of possible variations in the cosmic flux. With respect to the previous unsatisfactorily known $^{53}$Mn half-life, we have performed a redetermination of the $^{53}$Mn-neutron cross section. Having irradiated meteoritic Mn for 392 days in a high neutron flux we find $\sigma_{\text{therm}} = 66 \pm 7$ barns. Based on the "Peace River" chondrite, the $^{53}$Mn content of which is well known, a product of $T \times \sigma = (260 \pm 32) \times 10^6$ barns $\times$ years is derived, thus leading to a $^{53}$Mn half-life of $(3.9 \pm 0.6) \times 10^6$ years. On the basis of this figure a number of Apollo 11 to 14 soil and rock samples are analyzed by neutron activation techniques for their $^{53}$Mn contents. The K-emitter $^{53}$Mn is converted by neutron capture to the $\gamma$-emitting $^{54}$Mn, which is sensitively detected by a 75 cc Ge(Li) well-type detector. The surface soil 14259 (comprehensive) with 644 dpm/kg Fe is found to be twice as high in $^{53}$Mn as the bulk soils. The $^{53}$Mn-depth effect on rock 10017 was reestablished. From the respective low value of the microbreccia 14305 (interior), pointing out that "saturation" activity is not reached, a remarkable short exposure age in the order of $\sim 6$ million years is concluded.

## INTRODUCTION

SINCE THE FIRST OBSERVATION of the existence of a long-lived manganese isotope in nature by Shedlovsky (1960), the interest in this spallation produced nuclide has steadily grown. It offers a possibility to get additional information about the complex radiation history of meteorites and of cosmic ray exposed matter. A supplement to $^{26}$Al, $^{36}$Cl, etc., $^{53}$Mn can be useful to reach back into the past, at least to $\sim$10–15 m.y. Obviously, the large production cross section (its main source is iron) and the high sensitivity in detection are of great importance. With respect to the latter the K-emitting nuclide can be transformed by neutron capture into the $\gamma$-emitting $^{54}$Mn isotope, raising the activity ratio by a factor of up to $> 10^4$ (Millard, 1965, Herpers *et al.*, 1967). We have developed this activation technique and shown on a larger number of iron meteorites that there are considerable advantages, as high accuracy and economy of sample need (Herpers *et al.*, 1969, Herr *et al.*, 1969). $^{53}$Mn depth profiles of lunar rocks were recently presented by Shedlovsky *et al.* (1970), Finkel *et al.* (1971), Wahlen *et al.* (1972) and also by our group (Herr *et al.*, 1971, 1972).

However, the decay constant of $^{53}$Mn was until recently not very well known, and more precise information was highly desirable, for being the basis of any dating work in cosmochemistry. Earlier attempts to establish the half-life were made on several approaches. On considerations of spin and nuclear reaction yields Sheline and Hooper (1957) came to $T = 2 \times 10^6$ yr. Kaye and Cressy (1965) estimated $T = (1.9 \pm 0.5) \times 10^6$ yr, based on counting meteoritic Mn. The latter value was considered the best for a long time. Matsuda *et al.* (1971) reported $T = (2.9 \pm 1.2) \times 10^6$ yr working with accelerator produced $^{53}Mn$ and $^{54}Mn$. Then mass-spectrometry on meteoritic Mn (Cañon Diablo) resulted in a figure of $T = (10.8 \pm 4.5) \times 10^6$ yr (Hohlfelder, 1969), pointing out the enormous technical difficulties. Obviously, more weight has the recent work of Honda and Imamura (1971) succeeding in a measurement of the specific activity and the abundance of naturally and artificially produced $^{53}Mn$. Their value is $T = (3.7 \pm 0.4) \times 10^6$ yr.

In view of the importance of this constant, we started independently three years ago an experiment, in order to establish the half-life via the $^{53}Mn(n, \gamma)^{54}Mn$ thermal neutron cross section ($\sigma_{53}$) (Herpers *et al.*, 1969). The principle is as follows: ~2 mg of Mn from the Duchesne meteorite (activity of $^{53}Mn = 412 \pm 34$ dpm/kg) were activated in a high flux reactor ($\Phi = 10^{14}$n cm$^{-2}$ s$^{-1}$), one aliquot was irradiated only for a "short" period (20*d*), the other for a "long" period (392*d*) together with suitable flux monitors (Fe, $^{55}Mn$, $^{54}Mn$, and Cu etc. standards). From the systematics of the activation process the thermal neutron capture cross section $\sigma_{53}$ was determined as 66 ± 7 *barns.*

The next step is the determination of the $^{53}Mn$ half-life. By courtesy of J. R. Arnold we received a sample of the meteorite "Peace River" having a well-known $^{53}Mn$ decay rate (measured by direct counting). With the "Peace River" manganese we performed an activation experiment and obtained a product of $T \times \sigma = (260 \pm 32) \times 10^6$ barns × yr.

On the basis of the above established $\sigma_{53} = 66 \pm 7$ b and the product $T \times \sigma$ we derived a $^{53}Mn$ half-life of $T = (3.9 \pm 0.6) \times 10^6$ yr. Obviously, this value is in excellent agreement with that published quite recently by Honda and Imamura (1971).

## EXPERIMENTAL (ANALYSIS OF LUNAR MATERIAL)

Amounts of 200–500 mg powdered Apollo 14 samples were fused in NaOH + $Na_2O_2$ and the Mn isolated and purified by ion exchange techniques as described earlier (Herr *et al.*, 1970, 1971). The $MnO_2$ probes were brought into an Al-foil sandwich and irradiated for 31 days (FRJ-2) at Juelich. $\Phi_{therm} = 5 \times 10^{12}$ n cm$^{-2}$ s$^{-1}$, $\Phi_{fast}/\Phi_{therm} \leq 0.001$. The latter ratio was measured by $^{31}P$ $(n, p)$ $^{31}Si$ and $^{59}Co$ $(n, \gamma)$ $^{60}Co$. Postirradiation treatment consisted of an $HMnO_4$ distillation followed by ion exchange separation. The radiometric essay was very much improved by a 75 cc Ge(Li) detector (well-type), kindly supplied by J. Eberth, Institut für Kernphysik, Köln. Resolution: 3.8 keV at 1.33 MeV and an efficiency of 1.4% at the $^{54}Mn$ 835 keV peak. By heavy shielding (Pb + Hg) the background was lowered to 0.3 cpm under the total line of $^{54}Mn$ at 835 ± 15.3 keV. No other interfering radioactivities were observed near, or at, the $^{54}Mn$ photo peak. Absolute neutron flux calibration was done by an internal Zn standard ($^{65}Zn$) against a standardized $^{22}Na$ sample on the basis of their annihilation radiation.

## RESULTS AND DISCUSSION

The importance of the knowledge of the $^{53}Mn$ half-life for precise estimations of exposure ages, sedimentation rates, and for the discovery of possible variations in the solar and cosmic fluxes is evident. Following these intentions, we have analyzed two Apollo 14 soils, 14163 (bulk) and 14259 (comprehensive soil) and also the interior part of rock 14305 (microbreccia) for their $^{53}$Mn contents. Based on our new half-life $T = 3.9 \times 10^6$ yr a number of Apollo 12 samples were also redetermined.

It has been already reported that the Fe and Mn contents of the minerals from the Apollo 14 landing site were lower by a factor of ~2 than those of the earlier missions (Waenke *et al.*, 1972) and others. Our Fe and Mn contents of the respective lunar samples are presented in Table 1. The Fe and Mn contents of the "Peace River" chondrite are given for comparison. They are by a factor of ~2.5 higher than the Apollo 14 samples. The results of our $^{53}$Mn determinations are listed in Table 2. The given errors include all systematic and counting ($1\sigma$) errors in the procedure. However, not included is the still remaining uncertainty in the product of the $^{53}$Mn half-life times cross section.

In the course of our work we were able to measure an aliquot of the $^{53}$Mn "Tokyo"-standard, kindly provided by Prof. M. Honda. A very good agreement was reached on this man-made sample.

It should be mentioned also that the Apollo 12 samples were recalculated. The results in the last column are given in dpm/kg iron. Most of the soil samples are in the range of 340–380 dpm/kg Fe, with the only exception of the comprehensive soil 14259 (~1 cm below surface), which is higher nearly by a factor of 2. Evidently, this extremely high value of 644 ± 21 dpm/kg Fe (the error presented corresponds to one standard deviation of counting) is due to a contribution of solar protons. This value is in concordance with that of J. R. Arnold and coworkers, presented at the Third

Table 1. Mn and Fe contents of Apollo 14 samples.

| Sample no. | Type | Mn (ppm) | Fe (%) |
|---|---|---|---|
| 14163 | Soil | 1117 ± 34 | 7.50 |
| 14259 | Soil | 1103 ± 33 | 7.74 |
| 14305,BD1 (interior) | Breccia | 1059 ± 31 | 7.91 |
| Peace River (chondrite) | | 2544 ± 52 | 21.35 |

Table 2. $^{53}$Mn content of lunar material.

| Sample no. | Type | Sample weight (mg) | $^{54}$Mn (total) (cpm) | Contribution of $(n,2n)$ plus $(n,p)$ (%) | $^{53}$Mn (dpm/kg) | Fe (%) | $^{53}$Mn (dpm/kg Fe) |
|---|---|---|---|---|---|---|---|
| 14163 | Soil | 248 | 0.377 ± 0.008 | 27.8 | 28 ± 3 | 7.5 | 372 ± 34 |
| 14259 | Soil | 231 | 0.563 ± 0.012 | 17.5 | 50 ± 5 | 7.7 | 644 ± 58 |
| 12070 | Soil | 501 | 14.1 ± 0.2 | 25.8 | 50 ± 4 | 13.2 | 380 ± 27 |
| 10084 | Soil | 990 | 17.7 ± 3.2 | 30.2 | 41 ± 7 | 12.0 | 341 ± 58 |
| 14305,BD1 (interior) | Breccia | 504 | 0.389 ± 0.007 | 49.4 | 13 ± 1 | 7.9 | 169 ± 16 |
| 12021 | Rock | 842 | 25.4 ± 0.3 | 38.8 | 33 ± 4 | 14.2 | 232 ± 28 |
| 12022 | Rock | 1038 | 35.7 ± 0.4 | 36.6 | 38 ± 4 | 16.4 | 232 ± 24 |
| 12053 | Rock | 1070 | 41.2 ± 0.4 | 34.1 | 41 ± 3 | 15.3 | 268 ± 20 |
| $^{53}$Mn* Tokyo—std. | 0.5 n HCl solution | 1000 | 16.89 ± 0.39 | 0.26 | (dpm/g sol) 0.392 ± 0.059 | | |

* $^{53}$Mn standard solution supplied by courtesy of Prof. M. Honda with a value of 0.395 dpm/g solution (private communication).

Lunar Science Conference (Wahlen *et al.*, 1972). Their figure of 44 ± 2 dpm/kg material corresponds with a value of 572 ± 27 dpm/kg Fe. It is notable, that in the comprehensive soil also the $^{22}$Na and $^{26}$Al contents are a factor of 1.9, respectively, 2.8 higher (LSPET, 1971, Rancitelli *et al.*, 1972) than those of the bulk soils (10084, 12070, and 14163). Since the latter are undefined in their depth position and may have suffered some "gardening," it is hard to estimate their former depth.

From earlier studies on Apollo 11 and 12 samples, clear evidence was obtained of a long-term bombardment of solar (wind) particles in the ~10 MeV per nucleon range (Finkel *et al.*, 1971). A certain similarity of the depth profile of $^{26}$Al ($T = 7.5 \times 10^5$ yr) to that of $^{22}$Na ($T = 2.6$ yr) was also observed, whereas the respective $^{53}$Mn profile seemed less steep. Evidently, from these measurements it followed that the solar particles, averaged over some $10^6$ yr, has been in the same range as those measured today. Furthermore, the precise estimation of activity profiles from short and longlived activities, especially $^{53}$Mn has led to the discussion of the still open question: Should there have happened (possible) variations in the solar and cosmic flux in the last $10^6$ yr, or is there either a rather large corrosion rate (of ~0.2 mm/$10^6$ yr) responsible for differences in the respective activity ratios? One has to work further on this problem. (We received the 14 material only very late in November 1971.)

In the diagram of Fig. 1, the depth dependence of $^{53}$Mn in rock 10017 together with data of Apollo 11 to 14 samples are shown. The rock 14305, a fist-sized micro-

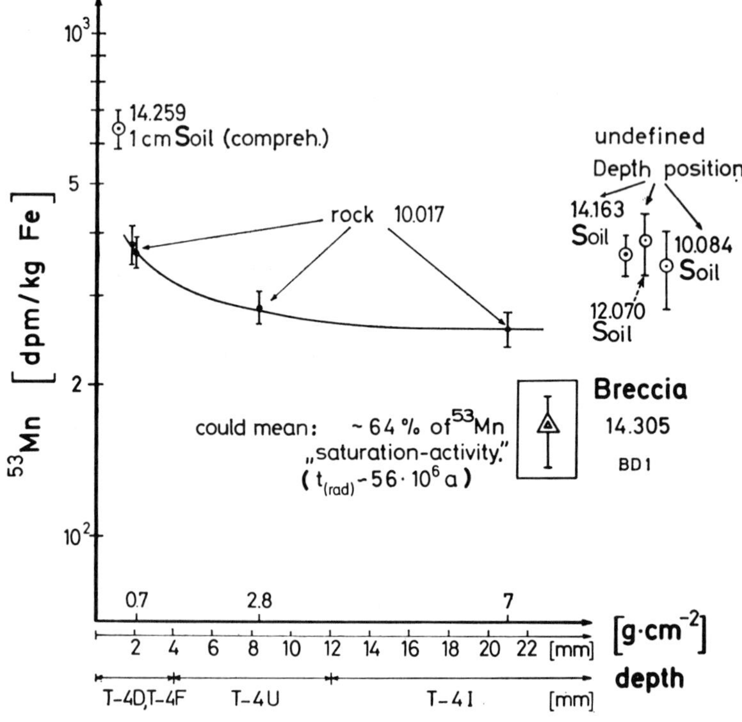


Fig. 1. $^{53}$Mn values in relation to the depth.

breccia, is thought to have had irradiations from the top and the bottom. There is evidence from photographic studies and also from the number of the microcraters on the surfaces that in this case we probably are dealing with a "moving" rock. The $^{26}$Al data from the top and the bottom recently published by Yokoyama *et al.* (1972) do support this view.

Some remarks to the $^{53}$Mn results on rock 10017, having a radiation age of $\sim 300 \times 10^6$ yr (D'Amico *et al.*, 1970, Marti *et al.*, 1970). Thus, it may be sure that its $^{53}$Mn concentration has reached "saturation," and only the depth effect reflects the differences in the specific activity. If we take this fact into consideration and transfer this model to the interior sample of breccia 14305 (taken from $\sim$2.5 cm below the top surface), we see that this sample has not yet reached the respective "saturation activity." However, we are simplifying and also neglecting the fact that this object may have "turned" around during its stay on the lunar surface. So, the assumed depth cannot be well defined.

Nevertheless, we conclude that the $^{53}$Mn radiation age of 14305 must be comparatively short and should be in the range of $T_{rad} \sim 6 \times 10^6$ yr only. This would mean an extraordinarily short irradiation period, but unless other detailed information on its $^{53}$Mn profile are available a more precise statement cannot be made. It appears that the breccia 14305 was buried on the average more deeply than the neighboring rocks, which give the 20–25 $\times 10^6$ yr radiation age.

There are some other independent investigations on the same microbreccia that strengthen our view that we are here really dealing with one of the youngest ejecta found in the Apollo 14 landing site. The rare gases were studied in these microbreccias and Bogard and Nyquist (1972) concluded that this material (it was recovered very near the Cone Crater), has a rare gas exposure age of max $T_{rad} \sim 20 \times 10^6$yr, which may reflect the age of the Cone Crater itself. This is also pointed out by Lugmair and Marti (1972). Independently, the group of Hart *et al.* (1972) came to comparable results by recording particle tracks. For a similar microbreccia from the rim of the Cone Crater, they evaluated a maximum surface age of $8 \times 10^6$ yr. It is, however, clear that these breccias have a very complicated history, at least compared with the basaltic rocks. The possibility that the major surface exposure occurred for a longer time (e.g., in the interior of a considerably larger rock or below the surface) cannot be excluded. It is noteworthy that the Apollo 14 breccias are by a factor of 10 to 20 shorter "exposed" compared with the basalts of Apollo 11 and 12. This may be also one of the results of the more friable nature of these 14—rocks that will allow even a more rapid and larger erosion and also a promoted catastrophic break up by nearby impacts.

*Acknowledgments*—We express our thanks to Prof. J. R. Arnold, La Jolla, for having forwarded a sample of the Peace River chondrite and to Prof. M. Honda, Tokyo, for sending his $^{53}$Mn standard. We are also indebted to Mr. J. Eberth, Köln, for placing the 75 cc Ge(Li) detector at our disposal. Our thanks are also due to NASA for supplying lunar material.

## References

Bogard D. D. and Nyquist L. E. (1972) Noble gas studies on regolith materials from Apollo 14 and 15 (abstract). In *Lunar Science—III* (editor C. Watkins), pp. 89–91, Lunar Science Institute Contr. No. 88.

D'Amico J., De Felice J., and Fireman E. L. (1970) The cosmic-ray and solar-flare bombardment of the moon. *Proc. Apollo 11 Lunar Sci. Conf., Geochim. Cosmochim. Acta* Suppl. 1, Vol. 2, pp. 1029–1036. Pergamon.

Finkel R. C., Arnold J. R., Imamura M., Reedy R. C., Fruchter J. S., Loosli H. H., Evans J. C., Delany A. C., and Shedlovsky J. P. (1971) Depth variation of cosmogenic nuclides in a lunar surface rock and lunar soil. *Proc. Second Lunar Sci. Conf., Geochim. Cosmochim. Acta* Suppl. 2. Vol. 2, pp. 1773–1789. MIT Press.

Hart H. R. Jr., Comstock G. M., and Fleischer R. L. (1972) The particle track record of Fra Mauro (abstract). In *Lunar Science—III* (editor C. Watkins), pp. 360–362, Lunar Science Institute Contr. No. 88.

Herpers U., Herr W., and Woelfle R. (1967) Determination of cosmic ray produced nuclides $^{53}$Mn, $^{45}$Sc, and $^{26}$Al in meteorites by neutron activation and gamma coincidence spectroscopy. In *Radioactive Dating and Methods of Low-Level Counting*, pp. 199–205, International Atomic Energy Agency, Vienna.

Herpers U., Herr W., and Woelfle R. (1969) Evaluation of $^{53}$Mn by $(n, \gamma)$ activation, $^{26}$Al, and special trace elements in meteorites by $\gamma$-coincidence techniques. In *Meteorite Research*, pp. 387–396, International Atomic Energy Agency, Vienna.

Herr W., Herpers U., and Woelfle R. (1969) Die Bestimmung von $^{53}$Mn, welches in meteoritischem Material durch kosmische Strahlung erzeugt wurde, mit Hilfe der Neutronenaktivierung. *J. Radioanal. Chem.* **2,** 197–203.

Herr W., Herpers U., Hess B., Skerra B., and Woelfle R. (1970) Determination of manganese-53 by neutron activation and other miscellaneous studies on lunar dust. *Proc. Apollo 11 Lunar Sci. Conf., Geochim. Cosmochim. Acta* Suppl. 1, Vol. 2, pp. 1233–1238. Pergamon.

Herr W., Herpers U., and Woelfle R. (1971) Spallogenic $^{53}$Mn ($T \sim 2 \times 10^6$ y) in lunar surface material by neutron activation. *Proc. Second Lunar Sci. Conf., Geochim. Cosmochim. Acta* Suppl. 2, Vol. 2, pp. 1797–1802. MIT Press.

Herr W., Herpers U., and Woelfle R. (1972) Study on the cosmic ray produced longlived Mn-53 in Apollo 14 samples (abstract). In *Lunar Science—III* (editor C. Watkins), pp. 373–375, Lunar Science Institute Contr. No. 88.

Hohlfelder J. J. (1969) Half-life of $^{53}$Mn. *Phys. Rev.* **186,** 1126–1131.

Honda M. and Imamura M. (1971) Half-life of $^{53}$Mn. *Phys. Rev.* C. **4,** 1182–1188.

Kaye J. H. and Cressy P. J. (1965) Half-life of manganese-53 from meteorite observations. *J. Inorg. Nucl. Chem.* **27,** 1889–1892.

LSPET (Lunar Sample Preliminary Examination Team) (1971) Apollo 14 Preliminary Science Report. NASA SP-272.

Lugmair G. W. and Marti K. (1972) Neutron and spallation effects in Fra Mauro regolith (abstract). In *Lunar Science—III* (editor C. Watkins), pp. 495–497, Lunar Science Institute Contr. No. 88.

Marti K., Lugmair G. W., and Urey H. C. (1970) Solar wind gases, cosmic-ray spallation products, and the irradiation history of Apollo 11 samples. *Proc. Apollo 11 Lunar Sci. Conf., Geochim. Cosmochim. Acta* Suppl. 1, Vol. 2, pp. 1357–1367. Pergamon.

Matsuda H., Umemoto S., and Honda M. (1971) Manganese-53 produced by 730 MeV proton bombardment of iron. *Radiochim. Acta* **15,** 51–53.

Millard H. T. Jr. (1965) Thermal neutron activation: Measurement of cross section for manganese-53. *Science* **147,** 503–504.

Rancitelli L. A., Perkins R. W., Felix W. D., and Wogman N. A. (1972) Cosmic ray flux and lunar surface processes characterized from radionuclide measurements in Apollo 14 and 15 lunar samples (abstract). In *Lunar Science—III* (editor C. Watkins), pp. 630–632, Lunar Science Institute Contr. No. 88.

Shedlovsky J. P. (1960) Cosmic-ray produced manganese-53 in iron meteorites. *Geochim. Cosmochim. Acta* **21,** 156–158.

Shedlovsky J. P., Honda M., Reedy R. C., Evans J. C. Jr., Lal D., Lindstrom R. M., Delany A. C., Arnold J. R., Loosli H. H., Fruchter J. S., and Finkel R. C. (1970) Pattern of bombardment-produced radionuclides in rock 10017 and in lunar soil. *Science* **167,** 574–576.

Shedlovsky J. P., Honda M., Reedy R. C., Evans J. C. Jr., Lal D., Lindstrom R. M., Delany A. C.,

Arnold J. R., Loosli H. H., Fruchter J. S., and Finkel R. C. (1970) Pattern of bombardment-produced radionuclides in rock 10017 and in lunar soil. *Proc. Apollo 11 Lunar Sci. Conf., Geochim. Cosmochim. Acta* Suppl. 1, Vol. 2, pp. 1503–1532. Pergamon.

Sheline R. K. and Hooper J. E. (1957) Probable existence of radioactive manganese-53 in iron meteorites. *Nature* **179,** 85–87.

Waenke H., Baddenhausen H., Balacescu A., Teschke F., Spettel B., Dreibus G., Quijano M., Kruse H., Wlotzka F., and Begemann F. (1972) Multielement analyses of lunar samples (abstract). In *Lunar Science—III* (editor C. Watkins), pp. 779–781, Lunar Science Institute Contr. No. 88.

Wahlen M., Honda M., Imamura M., Fruchter J., Finkel R., Kohl C., Arnold J., and Reedy R. (1972) Cosmogenic nuclides in football-sized lunar rocks (abstract). In *Lunar Science—III* (editor C. Watkins), pp. 764–766, Lunar Science Institute Contr. No. 88.

Yokoyama Y., Auger R., Bibron R., Chesselet R., Guichard F., Léger C., Mabuchi H., Reyss J. L., and Sato J. (1972) Cosmonuclides in lunar rocks (abstract). In *Lunar Science—III* (editor C. Watkins), pp. 819–821, Lunar Science Institute Contr. No. 88.

Proceedings of the Third Lunar Science Conference
(Supplement 3, *Geochimica et Cosmochimica Acta*)
Vol. 2, pp. 1771–1777
The M.I.T. Press, 1972

# Alpha spectrometry of a surface exposed lunar rock

GÉRARD LAMBERT, TOVY GRJEBINE, JEAN CLAUDE LE ROULLEY,
and PIERRE BRISTEAU

Centre des Faibles Radioactivités, CNRS,
Gif-sur-Yvette (91), France

**Abstract**—The $\alpha$ activity of the top surface of a moon rock has been measured in order to detect an eventual layer of lead 210 and polonium 210 that could be designated as descendant of the radon escaping from the lunar soil. The limit for this activity is proposed as $2.10^{-5}$ dis/sec/cm$^2$. The possibility for the existence of this process was meanwhile shown in a simulating experiment. The absence of the activity can be attributed to an apparently very low transmission coefficient of fines. This is also seen in another simulation experiment where a source emanating radon is covered with increasing layers of basalt powder.

## INTRODUCTION

ON EARTH, an average concentration of 4 ppm of uranium in the soil gives rise to an outgassing flux of radon 222 of 0.7 atoms per second and cm$^2$. If a layer of 7 cm of terrestrial soil were entirely degassed, it should produce such a flux. Therefore, if all processes involved were similar on the moon, an average concentration of 1 ppm of uranium would produce a flux in the range of 0.1 to 1 atoms of radon per second and cm$^2$. Kraner *et al.* (1966) have shown that, even with the day time heating of the moon's surface, the atoms do not have enough energy to leave the moon. It should be pointed out that, during the lunar night, the temperature reaches 90°K, i.e., a lower temperature than the melting point of radon (202°K).

After the $\alpha$ disintegration of radon 222 (energy 5.5 MeV) the residual nucleus of polonium 218 (RaA) has a recoil energy of about 100 keV. If the $\alpha$ particle were emitted toward the soil, this recoil energy would be sufficient to overcome the lunar gravity and the RaA would be able to leave the moon. If, on the contrary, the $\alpha$ particle were directed toward the outside, the RaA would be projected toward the lunar soil and become embedded in the external surface layer. Taking into account the mass and the recoil energy of the RaA, it may be assumed that its range is approximately $20.10^{-6}$ g/cm$^2$ (Domeij *et al.*, 1963). The effect of this depth is almost negligible for the $\alpha$ particles emitted by the further disintegrations.

Thus, we have the formation of a thin layer of lead 210 (half-life 20 yr) in equilibrium with the decay products bismuth 210 and polonium 210. This layer produces an $\alpha$ emission of definite energy, 5.30 MeV, but this is in contrast to the continuous spectrum of $\alpha$ radioactive decay products of uranium and thorium, which are directly yielded inside the rock.

## EXPERIMENTAL

In order to investigate the formation of such a layer of polonium 218 and its daughter products, several flat rocks were laid on a bed of pulverized sand mixed with 1 microcurie of radium 226.

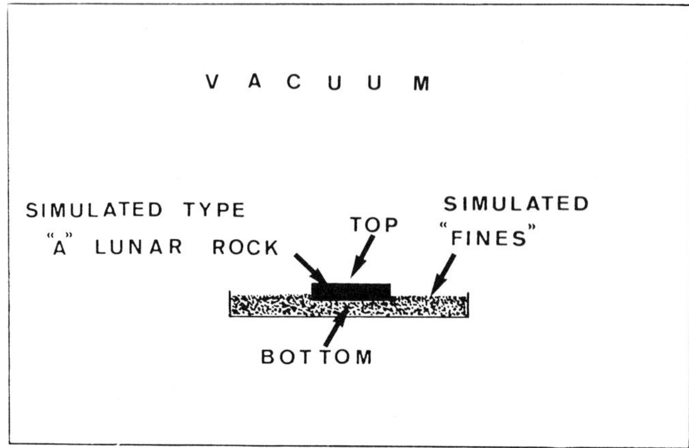


Fig. 1. Experimental device for simulating the embedding process of a recoiling atom, after an α disintegration.

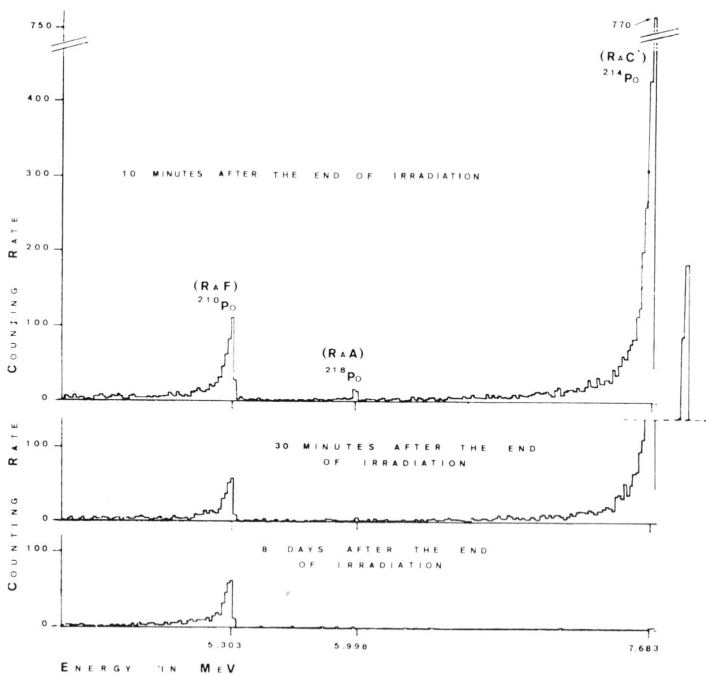


Fig. 2. α spectrum of the top face of a rock in a simulating experiment. Curve A: 10 minutes after removing from the vessel; Curve B: 30 minutes after removing from the vessel; Curve C: 8 days later after removing from the vessel.

This material was put in a vessel which was brought to a primary vacuum (Fig. 1). After 8 months, the rocks were removed and measured with an $\alpha$ spectrometer.

The $\alpha$ spectrum of the top face of a rock, 10 min after its removing from the vessel, is shown on Fig. 2, curve a. Characteristic peaks of polonium 218, 214, and 210 (RaA, C′, and F) are visible. Since there is no contamination by radium 226 or radon 222, this demonstrates that polonium 218 and its decay products were effectively embedded in the top face by the disintegration of the gaseous radon present in the vessel. Thirty minutes later, the polonium 218 peak disappeared (curve b). One day later, only the polonium 210 peak remained visible (curve c).

## Results and Discussion

Therefore, according to the theory proposed by Kraner *et al.* (1966) and to the results above, it seems possible to measure the flux of radon outgassing from the moon soil, by measuring the lead 210 deposited in the top layer of lunar rocks. Lunar sample 14321,212 provided by NASA was measured with an $\alpha$ semiconductor spectrometer. This sample was cut off the top of a rock. Two faces were measured: The first face was identified as an external face from the presence of a large meteoritic pit and available documentation. The second face was irregular, without any pits or erosion features.

The $\alpha$ spectrometer was comprised of a surface barrier detector (ORTEC 2 $cm^2$) connected to a linear amplifier and a 200 channel analyzer. It was located in an air conditioned room and fed with a battery stabilized power supply. The measurements were made on the external face for one and a half months (i.e., 1176 hours of effective counting) without changing the position of the sample. The apparatus was checked for stability every week by feeding calibrated pulses into the amplifier: no shift was noted.

The low activity of the sample suggested the regrouping of the channels four by four (100 keV intervals) in order to diminish the statistical fluctuations (Fig. 3). The negligible background, on the order of 2 counts per 100 keV in 1176 hours, was not subtracted. The smooth-curve can be interpreted as a spectrum of a powder. Uranium

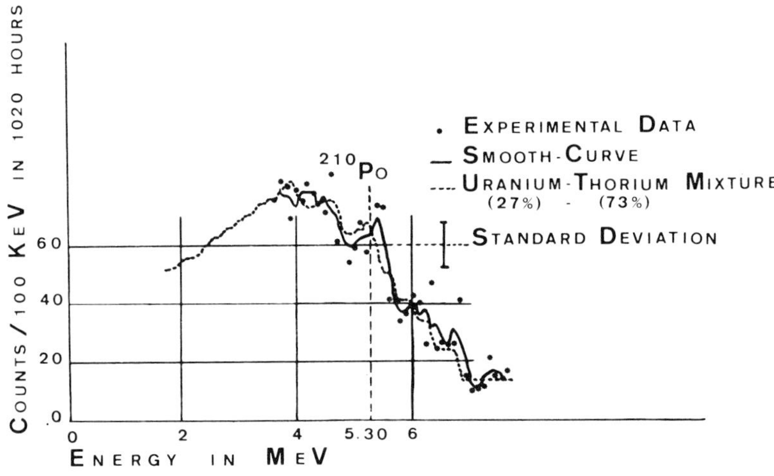


Fig. 3. The $\alpha$ spectrum of the top of the lunar rock 14321,212.

238 could account for 51% and thorium 232 for 49% of this activity, also keeping in secular equilibrium all their daughter products. Activities such as these are in agreement with the usually observed concentrations (LSPET, 1971). The observed data between 5.10 and 5.30 MeV show no significant activity above the continuous spectrum that could be attributed to excess polonium 210. The spectrum obtained with the internal face presents effectively the same features (Fig. 4). In the spectrum

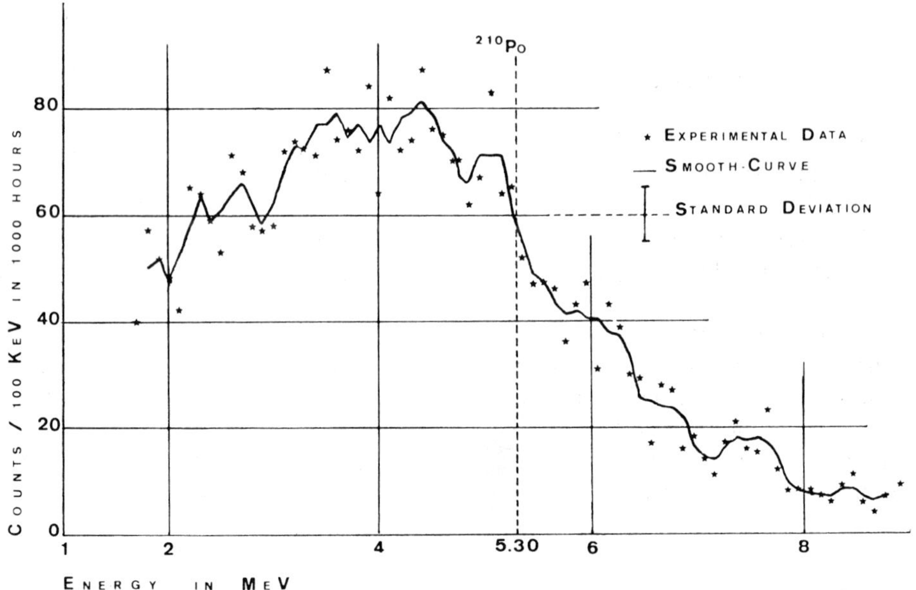


Fig. 4. The $\alpha$ spectrum of an internal face of lunar rock 14321,212.

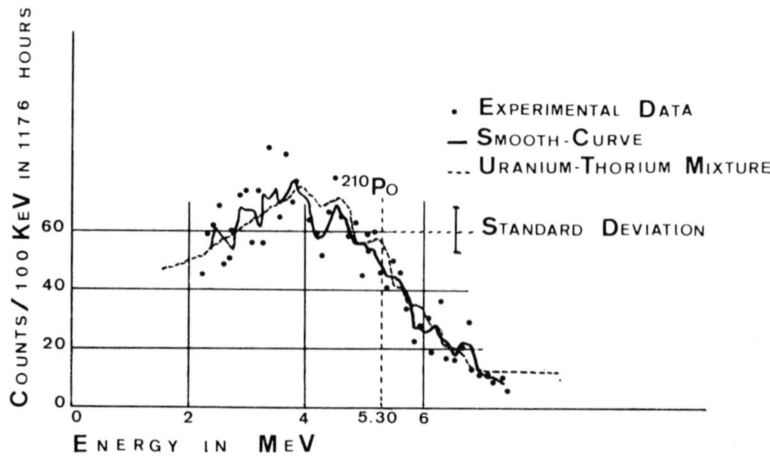


Fig. 5. The $\alpha$ spectrum of lunar fines 14163,92.

for the external face, the total number of counts between 5.10 and 5.30 MeV was $N = 124$ for 1776 hours. Setting as a detection limit 2 standard deviations of this rate ($2\sqrt{N} = 22.3$), and a counting efficiency of 13%, it is possible to assume that an excess of this rate would be detected. The upper limit is $4.10^{-5}$ dis/sec for the total sample. The measured surface of the sample was approximately 2 $cm^2$ therefore; the upper limit for a deposit of polonium 210 on the moon surface was $2.10^{-5}$ dis/sec/$cm^2$. A similar result obtained with fines sample 14163,92 (Fig. 5) confirms the extreme weakness of a possible excess of polonium 210 in the superficial lunar soil.

This result is in agreement with the measurements made by Economou and Turkevitch (1971) who found an upper limit of $5.10^{-3}$ dis/sec/$cm^2$ on the filter of the Surveyor 3 television camera, and of Lindstrom *et al.* (1971) who found 4.5 $10^{-4}$ dis/sec/$cm^2$ in Apollo 11 fines. All these limits are in agreement with the results obtained from Lunar orbiters by Yeh and Van Allen (1965) and Gorenstein and Bjokholm (1972). They did not detect any surplus in the lunar $\alpha$ activity. Only one experiment has shown different results: Turkevitch *et al.* (1970) using the Surveyor $\alpha$ spectrometer have reported at the Surveyor 5 landing site (Mare Tranquilitatis) an $\alpha$ activity that was described as polonium 210.

The lunar fines when brought to terrestrial conditions, on the other hand, apparently behave the same way as terrestrial soil. Adams *et al.* (1971) have found that 40% of the radon produced in 0.473 g of lunar fines was outgassed. This experiment was performed at room temperature and pressure. Yaniv and Heymann (1972), using a similar technique working in a moderate vacuum have lowered this emanating-to-production ratio between $2.10^{-3}$ and $5.10^{-3}$ for $Rn^{222}$ and $3.10^{-3}$ to $5.10^{-3}$ for $Rn^{220}$ in Apollo 14 fines. They did not observe any outgassing in Apollo 11 and Apollo 12 fines. Yaniv and Heymann (1972) have not therefore reached the same results as Adams *et al.* (1971). This suggests that the pressure may be of great importance. The absence of interstitial gas between the grains of the regolith may influence the outgassing in two different ways. First, Lindstrom *et al.* (1971) have suggested that, in the lunar soil, after the $\alpha$ emission of radium 226 the recoiling atom of radon 222 is not slowed down in the vacuum space between grains and therefore is reembedded in the next crystal.

In the previously described simulation experiment where the rocks were placed on the mixture of sand and radium 226, this effect may be seen on the bottom face. The $\alpha$ spectrum of this face shows (Fig. 6) although there is no contamination by radium 226 but the $\alpha$ peaks of radon 222 and daughter products are clearly apparent. Since the $\alpha$ spectrum remains almost unchanged after heating 15 min at 500°C, these atoms are strongly bound to the rock. These results can be interpreted as the reembedding process of radon atoms recoiling after the $\alpha$ disintegration of radium 226.

Grjebine *et al.* have calculated that, at a depth of a few centimeters, the temperature of the lunar soil is leveled to about 210°K. This calculation is in agreement with the measurements made during the Apollo 15 mission (Langseth *et al.*, 1972). These very low temperatures could mean that the adsorption possibilities of the lunar soil are very large for radon.

The combined effects of the reembedding and adsorption processes in fines

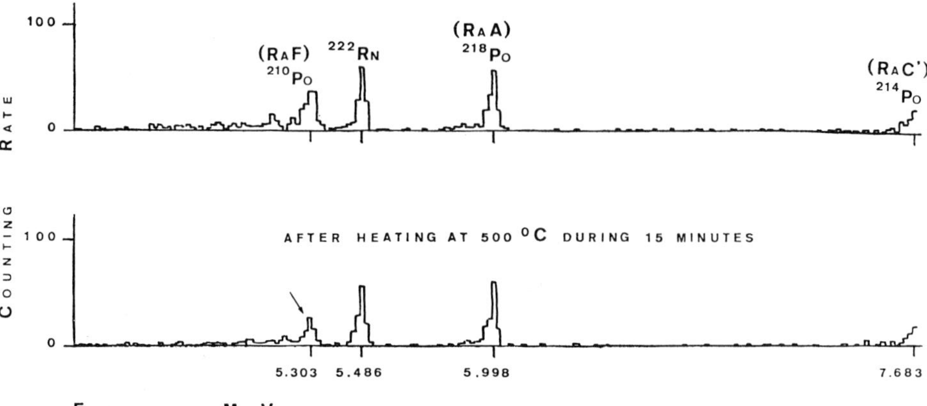


Fig. 6. The $\alpha$ spectrum of the bottom face of a rock in simulating experiment.

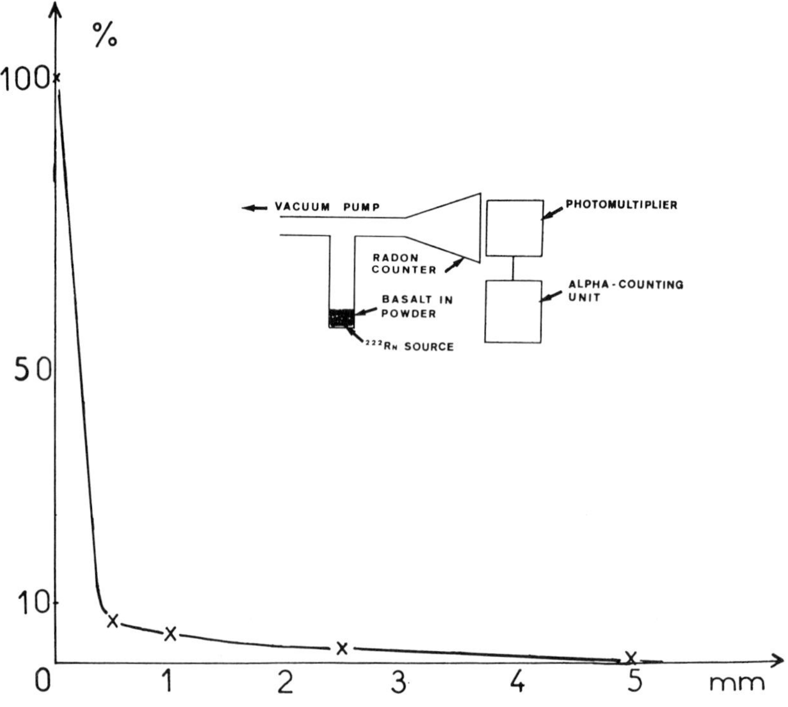


Fig. 7. Transmission of radon through basalt powder.

have been studied in the following experiment. A very thin source of radon was made by depositing radium on a platine plate. One sees in the $\alpha$ spectrum of this source that it loses half of its radon. This source was put in the bottom of a vessel connected to a scintillating bottle outside the view of the source (Fig. 7). The vessel was continuously pumped to maintain a primary vacuum. The radon emitted by the bare source was measured. Then the source was covered with successive layers of fine basalt powder (grain diameter $<50\mu$). It may be seen that the counting rate in the radon counter is lowered to 7% with half a mm of powder and to 1% with one cm. One sees the great importance of the first layers from the slope of the curve. This layer most probably combines the adsorption and reembedding processes.

The experimental result obtained in this work shows that there is $\leq 2.10^{-5}$ dis/sec/cm$^2$ lead 210 deposited layer on the moon surface. The measurement of the emanation-to-production ratio performed by Yaniv and Heymann (1972) do not support our simulation experiment of adsorption of radon. However, up to this point, there have been no experiments performed under exact lunar conditions. We must consider, in addition, other possible effects such as the partial erosion of volatile elements by solar winds.

## References

Adams J. A. S., Barretto M., Clark R. B., and Duvoljun J. S. (1971) Radon 222 emanation and the high apparent lead isotope ages in lunar dust. *Nature* **231,** 174–175.

Domeij I., Bergstrom I., Davies J. A., and Uhler J. (1963) A method of determining heavy ion range by analysis of α-line shapes. *Arkiv Fysik* **24,** 399–411.

Economou T. E. and Turkevitch A. L. (1971) Examination of returned Surveyor 3 camera visor for α radioactivity. Houston Conference.

Gorenstein P. and Bjokholm P. (1972) Results of the Apollo 15 α particle spectrometer experiment (abstract). In *Lunar Science—III* (editor C. Watkins), pp. 326–328, Lunar Science Institute Contr. No. 88.

Grjebine T., Lambert G., and Le Roulley J. C. (1972) α Spectrum of a surface exposed lunar rock. *Earth Planet. Sci. Lett.* **14,** 3, 322–324.

Kraner H. W., Schroeder G. L., Davidson G., and Carpentier J. W. (1966) Radioactivity of the lunar surface. *Science* **152,** 1235.

Langseth M. G. Jr., Clark S. P. Jr., Chute J. Jr., and Keihm S. (1972) The Apollo 15 lunar heat flow measurement (abstract). In *Lunar Science—III* (editor C. Watkins), pp. 475–477, Lunar Science Institute Contr. No. 88.

Lindstrom R. M., Evans J. C. Jr., Finkel R. C., and Arnold J. R. (1971) Radon emanations from the lunar surface. *Earth Planet. Sci. Lett.* **11,** 254.

LSPET (Lunar Sample Preliminary Examination Team) (1971) Apollo 14 Preliminary Science Report. NASA SP-272, pp. 123.

Turkevitch A. L., Patterson J. H., Franzgrote E. J., Sowinski K. P., and Economou T. E. (1970) Alpha radioactivity of the lunar surface at the landing sites of Surveyors 5, 6, and 7. *Science* **167,** 1722–1724.

Yaniv A. and Heymann D. (1972) Radon emanation from Apollo 11, 12, and 14 fines (abstract). In *Lunar Science—III* (editor C. Watkins), pp. 816–818, Lunar Science Institute Contr. No. 88.

Yeh R. S. and Van Allen J. A. (1969) Alpha particle emissivity of the moon: An observed upper limit. *Science* **166,** 370.

Proceedings of the Third Lunar Science Conference
(Supplement 3, *Geochimica et Cosmochimica Acta*)
Vol. 2, pp. 1779–1786
The M.I.T. Press, 1972

# Vanadium isotopic composition and the concentrations of it and ferromagnesian elements in lunar material

P. Rey

Department of Chemistry, Purdue University,
Lafayette, Ind. 47907, U.S.A.

H. Balsiger

Physikalisches Institut der Universität, Bern, Switzerland

and

M. E. Lipschutz

Departments of Chemistry and Geosciences, Purdue University and
Department of Physics and Astronomy, Tel-Aviv University,
Ramat Aviv, Tel-Aviv, Israel

**Abstract**—The isotopic composition of vanadium from three Apollo 14 samples, an Apollo 11 sample, the dark portion of a gas-rich chondrite and a standard rock are the same as that of previous lunar, meteoritic, and terrestrial samples within the limits of error. The maximum difference possible between the mean isotopic composition in lunar and meteoritic material is about 1%. The concentrations of iron, magnesium, titanium, chromium, and vanadium and the weight ratios of the first four of these elements to vanadium in Apollo 12 and 14 lunar materials suggest different primary source materials for different inclusions in rock 14321.

## Introduction

During the last several hundred million years lunar surface material and meteorites have been irradiated by energetic charged particles and the radiation history of these extraterrestrial samples has been investigated extensively by studies of a variety of monitors including a number of highly sensitive noble gas nuclides and radionuclides and tracks. Such monitors are not suitable for searches for the effects of a similar bombardment postulated to have taken place early in the history of the solar system (Fowler *et al.*, 1962; Bernas *et al.*, 1967) since this irradiation could have occurred under conditions such that these monitors would not be retained or would later be lost. The nongaseous elements provide several suitable radiation monitors, among the most sensitive of which is the vanadium isotopic composition (Shima and Honda, 1963; Burnett *et al.*, 1966; Balsiger *et al.*, 1969). An early charged particle irradiation should be revealed by comparison of the $^{50}V/^{51}V$ ratios in terrestrial, meteoritic and lunar samples unless either (a) there was thorough mixing of all matter now constituting these different objects, or (b) the integrated particle fluxes and the ratios of irradiated to shielded material were virtually identical in those parts of the solar system where these objects were formed (cf. Burnett *et al.*, 1965, 1966).

In previous studies (Balsiger *et al.*, 1969; Pelly *et al.*, 1970; Lipschutz *et al.*, 1971), we found that the vanadium isotopic composition in terrestrial, meteoritic and lunar samples was the same within the error limits (about 1%) and Albee *et al.* (1970) noted that the $^{50}V/^{51}V$ ratios in two Apollo 11 samples were the same as that of terrestrial vanadium to within 2 and 3%. However, our results showed that the $^{50}V/^{51}V$ ratios in lunar samples tended to be systematically higher than those in meteorites, the difference in the mean ratios lying just at the limits of confidence. We felt it worthwhile therefore to investigate additional samples to determine whether this difference persisted and, if so, the probable cause for this. Here we report results for four additional lunar samples, a chondrite and standard rock W-1. We also report determinations of some ferromagnesian elements (iron, magnesium, titanium, chromium) and vanadium in standard rocks and in these lunar samples and two others studied previously (Lipschutz *et al.*, 1971) which were undertaken to provide further information on the selenochemistry of vanadium.

## Experimental

The lunar materials, whose vanadium isotopic composition was studied here, included samples of 10084,247 (studied previously by Albee *et al.*, 1970) and <1 mm lunar fines 14163,129 and two portions of the football-sized rock 14321,184. This rock is quite complex petrographically as it consists of a variety of crystalline (lunar basaltic) and fragmental (microbreccia) fragments in a lighter-colored matrix (Grieve *et al.*, 1972). Inasmuch as we were particularly interested in determining the $^{50}V/^{51}V$ ratios in more primitive material we studied samples of the two inclusion types but not the matrix. These 1.5 g subsamples (1c and 20/22, respectively) were excavated from the interior of the rock with vanadium-free WC or TaC tools by A. R. Duncan (personal communication) and, like the fines samples and standard rock, were not etched prior to their solution. The chondrite (Leighton), whose vanadium isotopic composition is reported here, is one of six being studied concurrently, and is the only one for which we have data as yet. This H5 chondrite is a typical gas-rich meteorite with a light-dark structure and only the dark portion (which was etched briefly with 1:1 $HNO_3$ to remove surface contamination) was taken for analysis.

Each lunar sample was dissolved (Pelly *et al.*, 1970) and after centrifugation to remove insoluble material (i.e., sulfates, etc.) the supernate was divided into three aliquots. Vanadium was extracted from the largest (60–70%) of these aliquots by the chemical separation technique described by Pelly *et al.* (1970) and its isotopic composition was determined mass-spectrometrically by the surface-ionization technique described by Balsiger *et al.* (1969). We used the same chemical and mass-spectrometric techniques to determine the vanadium concentration by isotope dilution (Lipschutz *et al.*, 1971) in the second aliquot (20–33%) of this solution. The third aliquot (10–14%) was diluted to 25 ml with 2N $H_2SO_4$ containing 2% $NH_4Cl$ and portions of this solution were further diluted with appropriate reagents for determination of iron, magnesium, titanium, and chromium by atomic absorption spectroscopy (Slavin, 1968). Half-gram samples of standard rocks W-1 and BCR-1 and the 2 g sample of Leighton were treated similarly except that the sample solutions were each divided into only two aliquots, for determination of the vanadium isotopic composition and total vanadium concentration. Since we had not reserved appropriate aliquots of the Apollo 12 samples studied previously (Lipschutz *et al.*, 1971), 100–200 mg chips of two of the crystalline rocks (12021,49 and 12038,43) were processed chemically only for determination of iron, magnesium, titanium, and chromium. We also prepared separate stock solutions of standard rocks BCR-1 (for analysis of iron, magnesium, and titanium) and DTS-1 (for chromium), which for these elements have compositions similar to lunar samples, and we analyzed aliquots of these stock solutions whenever we determined these elements in lunar samples. The replicate analyses for each element in these aliquots differed by much less than 1%. We also attempted to determine the concentrations of these four elements in the precipitates removed prior to division of the lunar sample solutions, and we found only negligible quantities ($\ll$0.1%) of these elements in all cases save for magnesium in 12021,49 where the amount measured in the precipitate corresponded to ~2% of the total magnesium in the sample.

## Results and Discussion

The vanadium isotopic ratios for the samples included in this study are listed in Table 1 together with the uncertainty (the sum of the statistical error [three estimated standard errors] and those errors arising from correction of the mass-50 peak for contributions by $^{50}$Ti and $^{50}$Cr) associated with each measurement (Lipschutz *et al.*, 1971). Unfortunately the uncertainties associated with the measurements of lunar fines 10084,247 and 14163,129 were unusually large due to a substantial (~37%) correction for $^{50}$Cr in the former sample and low beam intensity in the latter. The remaining uncertainties are more in accord with our previous results. As before, the $^{50}$V/$^{51}$V ratios were not corrected for source or multiplier discrimination nor do the uncertainties listed include the error due to a possible variation in mass discrimination from one run to another.

The vanadium isotopic ratios and associated uncertainties listed in Table 1 are plotted in Fig. 1 together with our previous results (Pelly *et al.*, 1970; Lipschutz *et al.*, 1971). Within the uncertainty limits assigned to each of the $^{50}$V/$^{51}$V ratios none of the individual isotopic ratios differs from those of the other samples or from the mean isotopic ratio, $2.455 \times 10^{-3}$ (Table 2), for all samples. However the isotopic ratios for lunar samples still tend to be higher than those in meteoritic samples (Fig. 1) and the difference in the weighted means (obtained by inversely weighting each measurement by the square of its assigned uncertainty) for the two

Table 1. Vanadium isotopic ratios in samples investigated in this study.

| Material | Lunar Sample No. or Sample Name | Type | $^{50}$V/$^{51}$V* |
|---|---|---|---|
| Lunar | 10084,247 | fines | $(2.474 \pm 0.064) \times 10^{-3}$ |
| Lunar | 14163,129 | fines (< 1 mm) | $(2.477 \pm 0.065) \times 10^{-3}$ |
| Lunar | 14321,184(1c) | lunar basaltic inclusion | $(2.456 \pm 0.015) \times 10^{-3}$ |
| Lunar | 14321,184(20/22) | microbreccia inclusion | $(2.483 \pm 0.047) \times 10^{-3}$ |
| Terrestrial | W–1 | diabase | $(2.469 \pm 0.042) \times 10^{-3}$ |
| Meteoritic | Leighton | H5 (dark portion) | $(2.471 \pm 0.028) \times 10^{-3}$ |

* The uncertainty listed for each sample includes a statistical error (three estimated standard errors) as well as those errors arising from correction of the mass-50 peak (see text).

Table 2. Vanadium isotopic ratios in terrestrial, meteoritic, and lunar samples.

| Material | No. Investigated | Weighted $^{50}$V/$^{51}$V* |
|---|---|---|
| Lunar | 8 | $(2.461 \pm 0.002) \times 10^{-3}$ |
| Terrestrial | 4 | $(2.458 \pm 0.004) \times 10^{-3}$ |
| Meteoritic | 10 | $(2.449 \pm 0.004) \times 10^{-3}$ |
| All | 22 | $(2.455 \pm 0.002) \times 10^{-3}$ |

* The uncertainty listed for each group is the estimated weighted standard deviation of the mean calculated from the dispersions of the measurements of the individual samples.

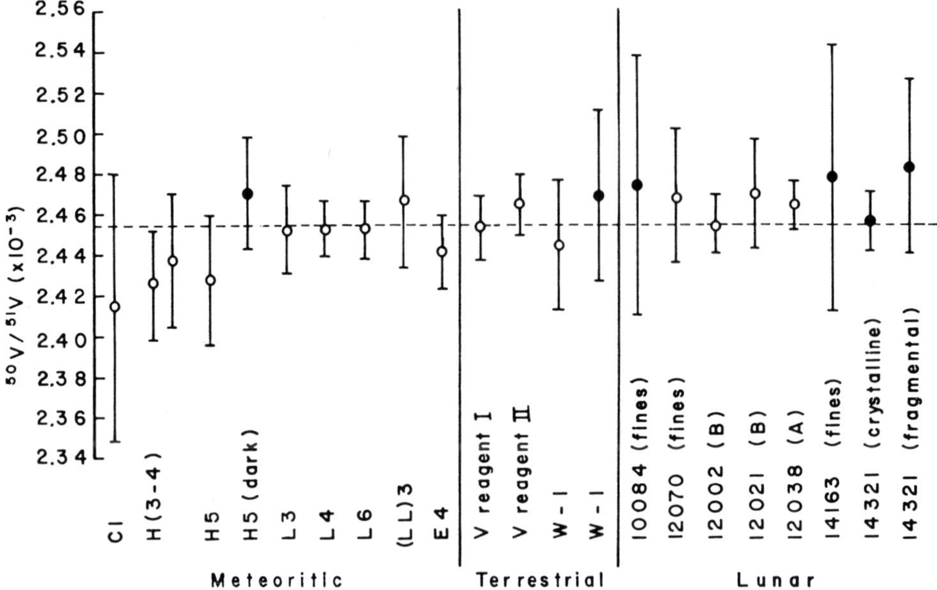


Fig. 1. $^{50}V/^{51}V$ ratios measured in this study (filled circles) compared with those measured in previous studies (open circles) of lunar samples, terrestrial diabase W-1 and reagent vanadium (I, II) and chondrites (uncertain classifications in parentheses). The weighted mean ratio of all samples is indicated by the dashed line.

groups persists (Table 2). The same weighting factors were used in calculating the estimated standard deviations of the group means (Table 2) from the dispersions of the measurements of the relevant individual samples, the standard deviation for the lunar-meteoritic difference, $0.013 \times 10^{-3}$, being $\pm 0.005 \times 10^{-3}$. It appears that the lunar-meteoritic difference may be marginally significant since it slightly exceeds two estimated standard deviations, whereas by the same criterion the terrestrial-meteoritic and lunar-terrestrial differences are not statistically significant.

If we treat the lunar-meteoritic difference as reflecting an irradiation effect, it cannot be ascribed to recent cosmic-ray or energetic solar particle bombardment since these should each alter the meteoritic and lunar $^{50}V/^{51}V$ ratios by less than 0.1% over times greater than the exposure ages of these materials (cf., Lipschutz *et al.*, 1971; Crozaz *et al.*, 1972). Since the bulk of the measurements of the lunar samples were made after the measurements of the meteorites and terrestrial samples it may well be that the lunar-meteoritic difference in the $^{50}V/^{51}V$ ratios merely reflects a long-term variation in mass-discrimination. Indeed the measurements of this ratio in Leighton and W-1 reported here tend to be higher than those of other meteoritic and terrestrial samples that we determined previously (Fig. 1). It will be necessary to carry out additional measurements, particularly of meteorites, to determine whether this tendency persists.

At this stage it seems safest to continue to assume that the samples constituting each of the groups in Table 2 have the same vanadium isotopic composition and that the lunar-meteoritic difference plus two estimated standard deviations, $(0.013 \pm 2 \times 0.005) \times 10^{-3}$, corresponds to the maximum difference possible between these two sample groups (Lipschutz *et al.*, 1971). This maximum difference of about 1% can be used to calculate upper limits for the possible integrated flux ($\Delta\phi$) differences experienced by these materials for two types of proton spectra (Balsiger *et al.*, 1969). The resulting $\Delta\phi$ limits are similar to those calculated by Lipschutz *et al.* (1971), i.e., $1.6 \times 10^{18}$ "hard" protons/cm$^2$ (cosmic-ray spectrum, $E > 30$ MeV) and $8 \times 10^{19}$ "soft" protons/cm$^2$ (solar flare spectrum, $E > 10$ MeV).

The concentrations of iron and vanadium in the samples studied must be utilized in calculating these $\Delta\phi$ limits, and we determined these concentrations as well as those of other elements expected to provide clues to the selenochemistry of vanadium. Our analyses for the ferromagnesian elements iron, magnesium, titanium, and chromium agree very well ($\leq 1\%$) with mean values for homogeneous standard rocks and the mutual agreement of the vanadium data is nearly as satisfactory (Table 3) especially when due regard is taken of the scatter of the data summarized by Fleischer (1969) and Flanagan (1969). Chemical inhomogeneities in 0.1–0.2 g lunar rock samples would be expected to result in less good agreement and indeed our analyses of these five elements in rocks 12021 and 12038 differ from the mean values reported by others by somewhat greater amounts, typically 5–10% (Table 3). Some results (notably iron, titanium and chromium in 10084 and chromium in 12021) exhibit even greater differences which may also be due to sample inhomogeneity. Our sample of fines 10084 was allocated after the Apollo 12 mission and quite possibly is not representative of samples prepared during the original allocation of these Apollo 11 fines. Indeed our results for fines 14163 are in much better agreement (generally ~5%) with the results of other investigations of this material (cf., Table 3 for references). We are at a loss to explain the unusually large discrepancy in the case of chromium in 12021 although it should be noted that this element appears to present serious analytical difficulties at lunar concentration levels, the wet chemical analyses (with which we generally agree) being markedly different in DTS-1 from those

Table 3. Concentrations of ferromagnesian elements in standard rocks and lunar samples.

| Sample | Fe (%) | | Mg (%) | | Ti (%) | | Cr (%) | | V (ppm) | |
|---|---|---|---|---|---|---|---|---|---|---|
| | This Study | Others* | This Study | Others* | This Study | Others* | This Study | Others* | This Study | Others* |
| BCR–1 | 9.48‡ | *9.45*(a) | 1.99‡ | *1.98*(a) | 1.35‡ | *1.34*(a) | — | *0.0016*(a) | 393 | *384*(a) 395(c) |
| DTS–1 | — | *6.19*(a) | — | *30.04*(a) | — | *0.014*(a) | 0.370 | *0.374*(a) | — | *19*(a) |
| W–1 | — | **7.76**(b) | — | **3.99**(b) | — | **0.641**(b) | — | **0.120**(b) | 259 | **240**(b) 256(c) |
| 10084,247 | 13.98 | *12.3*(d) | 4.96 | *4.73*(d) | 5.05 | *4.36*(d) | 0.255 | *0.20*(d) | † | — |
| 12021,49 | 15.71 | *15.0*(e) | 5.14 | *4.48*(e) | 2.08 | *2.01*(e) | 0.391 | *0.26*(e) | 160(e) | *175*(e) |
| 12038,43 | 13.09 | *13.8*(e) | 4.41 | 4.10(e) | 2.06 | *1.95*(e) | 0.219 | *0.20*(e) | 126(e) | *123*(e) |
| 14163,129 | 8.22 | *8.04*(f,g, h,j) | 5.77 | *5.58*(f,g, h,j) | 1.08 | *1.04*(f,g, h,j) | 0.153 | *0.13*(f,g, h,j) | 43 | *47*(h,l) |
| 14321,184(1c) | 12.07 | *12.8*(i,k) | 5.68 | 5.37(k) | 1.55 | 1.26(i) | 0.342 | *0.309*(i,k) | 133 | 96(i) |
| 14321,184 (20/22) | 7.75 | *8.8*(i,k) | 7.03 | 6.80(k) | 1.25 | 1.07(i) | 0.158 | *0.131*(i,k) | 44 | 48(i) |

* References: (a) Flanagan (1969); (b) Fleischer (1969); (c) Wyttenbach (1970); (d) Wänke *et al.* (1970) and references cited by them; (e) Lipschutz *et al.* (1971) and references cited by them; (f) Compston *et al.* (1972); (g) Hubbard *et al.* (1972); (h) Jackson *et al.* (1972); (i) Lindstrom and Duncan (personal communication); (j) Scoon (1972); (k) Wänke *et al.* (1972); (l) LSPET (1971). Entries in italics are mean values of different investigators, those in boldface are values recommended for standard rock W–1.

† Sample lost in laboratory mishap.

‡ Mean values of at least 6 replicate analyses.

obtained using other techniques (cf., Flanagan, 1969). Our results for the crystalline basaltic and fragmental microbreccia subsamples from rock 14321 generally agree quite well with the mean values listed in Table 3 (which were obtained from analyses only of similar type inclusions in this rock). In view of the heterogeneous nature of rock 14321 and, indeed, of the inclusions within it (Grieve *et al.*, 1972), those differences that do exist between our results and others may be easily ascribed to sample

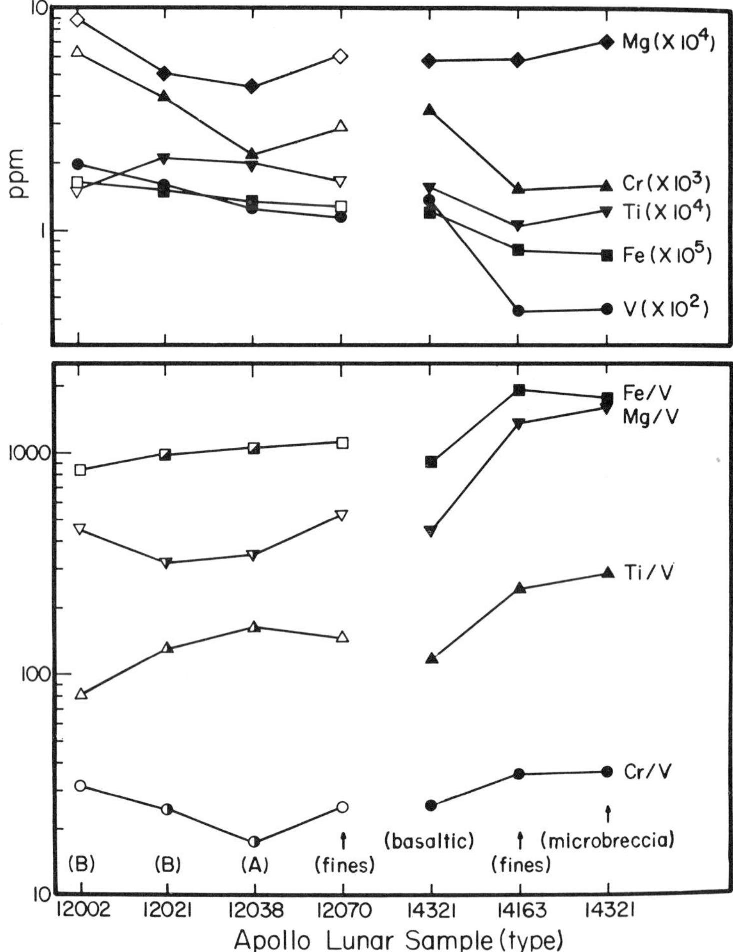


Fig. 2. Concentrations of four major and minor ferromagnesian elements and vanadium, and the weight ratios of these four elements relative to vanadium in Apollo 12 and 14 lunar samples. Filled and half-filled symbols represent our results (the former obtained from aliquots of the same sample solution, the latter obtained from different chips of the same subsample), and open symbols represent the average of other determinations of ferromagnesian elements (cf. Lipschutz *et al.* [1971] for references). For these elements the basaltic inclusion in rock 14321 resembles Apollo 12 crystalline rocks and fines and the microbreccia resembles fines 14163.

inhomogeneities. We should note that, overall, application of the Student's $t$ test shows that our data for each of the elements studied do not differ significantly from those reported by other investigators of the same samples.

Our results for the Apollo 12 and 14 lunar samples listed in Table 3, together with previous results for rock 12002 and fines 12070 (cf., Lipschutz *et al.*, 1971), are plotted in Fig. 2 in decreasing order of iron concentration. Even our very limited sampling yields results exhibiting the same gross trends noted by LSPET (1971), i.e., that iron, titanium, chromium and vanadium are less abundant in Apollo 14 than in Apollo 12 material whereas the amounts of magnesium are generally similar. Interestingly, the basaltic inclusion 184 (1c) in rock 14321 contains these five elements in amounts similar to those in 12070 fines. The contents of iron, chromium and vanadium in the microbreccia subsample 184 (20/22) are lower than those in the basaltic inclusion and are quite similar to those in the fines 14163 whereas magnesium exhibits the opposite trend, its concentrations in the basaltic inclusion and fines 14163 being quite similar. The titanium concentration in the basaltic inclusion is higher than that in the microbreccia, the latter having a content closer to that of fines 14163.

The trends exhibited by the concentrations of these five elements as well as the Fe/V, Mg/V, Ti/V and Cr/V concentration ratios illustrated in the bottom portion of Fig. 2 seemingly indicate that vanadium follows chromium in fines 14163 and rock 14321 rather than iron as in Apollo 12 fines and crystalline rocks (Lipschutz *et al.*, 1971). However, this conclusion contains the implicit assumption that the Apollo 14 material studied was largely, if not entirely, derived from a single source material. The mineralogy of rock 14321 (Grieve *et al.*, 1972) indicates a more likely alternative, i.e., that various subsamples of rock 14321 were derived from different source materials. In this event the concentrations and ratios illustrated in Fig. 2 suggest that basaltic inclusion 184 (1c) was derived primarily from mare basalt material similar in composition to fines 12070 and/or a mixture of Apollo 12 crystalline rocks whereas the primary source material for the microbreccia subsample 184 (20/22) may well have had a composition similar to that of fines 14163.

*Acknowledgments*—This research was supported by the U.S. National Aeronautics and Space Administration (Grant NGR-134) and the Swiss National Science Foundation (Grant 2-213-69). We thank Professor J. Geiss for his continuing aid and encouragement.

## References

Albee A. L., Burnett D. S., Chodos A. A., Eugster O. J., Huneke J. C., Papanastassiou D. A., Podosek F. A., Russ G. P. II, Sanz H. G., Tera F., and Wasserburg G. J. (1970) Ages, irradiation history, and chemical composition of lunar rocks from the Sea of Tranquillity. *Science* **167,** 463–466.

Balsiger H., Geiss J., and Lipschutz M. E. (1969) Vanadium isotopic composition in meteoritic and terrestrial matter. *Earth Planet. Sci. Lett.* **6,** 117–122.

Barnes I. L., Carpenter B. S., Garner E. L., Gramlich J. W., Kuehner E. C., Machlan L. A., Mainethal E. J., Moody J. R., Moore L. J., Murphy T. J., Paulsen P. J., Sappenfield K. M., and Shields W. R. (1972) The isotopic abundance ratio and assay analysis of selected elements in Apollo 14 samples (abstract). In *Lunar Science—III* (editor C. Watkins), pp. 41–43, Lunar Science Institute Contr. No. 88.

Bernas R., Gradsztajn E., Reeves H., and Schatzmann E. (1967) On the nucleosynthesis of lithium, beryllium and boron. *Ann. Phys.* **44,** 462–478.

Burnett D. S., Fowler W. A., and Hoyle F. (1965) Nucleosynthesis in the early history of the solar system. *Geochim. Cosmochim. Acta* **29,** 1209–1241.

Burnett D. S., Lippolt H. J., and Wasserburg G. J. (1966) The relative isotopic abundance of $K^{40}$ in terrestrial and meteoritic samples. *J. Geophys. Res.* **71,** 1249–1269, 3609.

Compston W., Vernon M. J., Berry H., Rudowski R., Gray C. M., Ware N., Chappell B. W., and Kaye M. (1972) Age and petrogenesis of Apollo 14 basalts (abstract). In *Lunar Science—III* (editor C. Watkins), pp. 151–153, Lunar Science Institute Contr. No. 88.

Crozaz G., Drozd R., Hohenberg C. M., Hoyt H. P. Jr., Ragan D., Walker R. M., and Yuhas D. (1972) Solar flare and galactic cosmic ray studies of Apollo 14 samples (abstract). In *Lunar Science—III* (editor C. Watkins), p. 165, Lunar Science Institute Contr. No. 88.

Flanagan F. J. (1969) U.S. Geological Survey Standards—II. First compilation of data for the new U.S.G.S. rocks. *Geochim. Cosmochim. Acta* **33,** 81–120.

Fleischer M. (1969) U.S. Geological Survey Standards—I. Additional data on rocks G-1 and W-1, 1965–1967. *Geochim. Cosmochim. Acta* **33,** 65–79.

Fowler W. A., Greenstein J. L., and Hoyle F. (1962) Nucleosynthesis during the early history of the solar system. *Geophys. J.* **6,** 148–220.

Grieve R., McKay G., Smith H., Weill D., and McCallum J. (1972) Mineralogy and petrology of polymict breccia 14321 (abstract). In *Lunar Science—III* (editor C. Watkins), pp. 338–340, Lunar Science Institute Contr. No. 88.

Hubbard N. J., Gast P. W., Rhodes M., and Wiesmann H. (1972) Chemical composition of Apollo 14 materials and evidence for alkali volatilization (abstract). In *Lunar Science—III* (editor C. Watkins), pp. 407–409, Lunar Science Institute Contr. No. 88.

Jackson P. F. S., Coetzee J. H. J., Strasheim A., Strelow F. W. E., Gricius A. J., Wybenga F., and Kokot M. L. (1972) The analysis of lunar material returned by Apollo 14 (abstract). In *Lunar Science—III* (editor C. Watkins), pp. 424–426, Lunar Science Institute Contr. No. 88.

Lipschutz M. E., Balsiger H., and Pelly I. Z. (1971) Vanadium isotopic composition and contents in lunar rocks and dust from the Ocean of Storms. *Proc. Second Lunar Sci. Conf., Geochim. Cosmochim. Acta* Suppl. 2, Vol. 2, pp. 1443–1450. M.I.T. Press.

LSPET (1971) (Lunar Sample Preliminary Examination Team) Preliminary examination of lunar samples from Apollo 14. *Science* **173,** 681–693.

Pelly I. Z., Lipschutz M. E., and Balsiger H. (1970) Vanadium isotopic composition and contents in chondrites. *Geochim. Cosmochim. Acta* **34,** 1033–1036.

Scoon J. H. (1972) Chemical analysis of lunar samples 14003, 14311, and 14321 (abstract). In *Lunar Science—III* (editor C. Watkins), pp. 690–691, Lunar Science Institute Contr. No. 88.

Shima M., and Honda M. (1963) Isotopic abundance of meteoritic lithium. *J. Geophys. Res.* **68,** 2844–2854.

Slavin W. (1968) *Atomic Absorption Spectroscopy*, Interscience.

Wänke H., Baddenhausen H., Balacescu A., Teschke F., Spettel B., Dreibus G., Quijano-Rico M., Kruse H., Wlotzka F., and Begemann F. (1972) Multielement analyses of lunar samples (abstract). In *Lunar Science—III* (editor C. Watkins), pp. 779–781, Lunar Science Institute Contr. No. 88.

Wänke H., Rieder R., Baddenhausen H., Spettel B., Teschke F., Quijano-Rico M., and Balacescu A. (1970) Major and trace elements in lunar material. *Proc. Apollo 11 Lunar Sci. Conf., Geochim. Cosmochim. Acta* Suppl. 1, Vol. 2, pp. 1719–1727.

Wyttenbach A. (1970) Die zerstörungsfreie Aktivierungsanalytische Bestimmung von Na, Mg, Al, Ca, V, und Mn in Gesteinen und Steinmeteoriten. Habilitationsschrift, EIR Bericht Nr. 177.

Proceedings of the Third Lunar Science Conference
(Supplement 3, *Geochimica et Cosmochimica Acta*)
Vol. 2, pp. 1787–1795
The M.I.T. Press, 1972

# Rare-gas analyses on neutron irradiated Apollo 12 samples

E. C. ALEXANDER, JR., P. K. DAVIS, and J. H. REYNOLDS
Department of Physics, University of California,
Berkeley, California 94720

**Abstract**—Argon, krypton, and xenon from stepwise heating of five Apollo 12 rocks that had been irradiated to a neutron fluence of $\sim 10^{19}$ neutrons/cm$^2$ were analyzed mass-spectrometrically. The $^{40}Ar$–$^{39}Ar$ ages were determined and range from $3.18 \pm 0.06 \times 10^9$ years to $3.32 \pm 0.06 \times 10^9$ years. Trace elements Ba, Br, I, and U were measured via the neutron induced reactions that produce isotopes of krypton and xenon. The following concentration ranges were observed: Ba, 46 to 70 ppm; Br, 48 to 146 ppb; I, 16 to 73 ppb; and U, 170 to 247 ppb.

## INTRODUCTION

THIS WORK REPRESENTS a continuation of our previous study (Davis *et al.*, 1971) of rare gases from neutron irradiated lunar samples. Five Apollo 12 crystalline rocks were analyzed: 12002,83; 12020,36; 12022,52; 12051,12; and 12065,33. The objectives are to obtain simultaneously K–Ar ages, via the $^{40}Ar$–$^{39}Ar$ technique (Merrihue and Turner, 1966) and concentrations of the trace elements Ba, Br, I, and U from each sample. Subtle changes in trace element concentrations with age (which normally would be obscured by sample inhomogeneity, interlaboratory biases, etc.) thus may be examined. Only those trace elements that produce rare-gas isotopes by neutron induced reaction are determinable, but both the volatile (Br and I) and refractory (U and Ba) elements are represented.

## EXPERIMENTAL TECHNIQUE

Experimental procedures for the Apollo 12 samples were modified from those used in our Apollo 11 work (Davis *et al.*, 1971). The samples, weighing about 90 mg, were sealed in evacuated quartz break-seal capsules for irradiation. This permitted the analysis of low temperature fractions, which previously were lost.

The neutron irradiation was carried out in the pool of the General Electric Test Reactor, Vallecitos Nuclear Center, Pleasanton, California, and is designated Vallecitos four. The Vallecitos center indicated the samples received the following fluence:

| | |
|---|---|
| Thermal ($E < 0.17$ eV) | $1.34 \times 10^{19}$ neutrons/cm$^2$ |
| Epithermal ($0.17$ eV $< E < 0.18$ MeV) | $4.21 \times 10^{18}$ neutrons/cm$^2$ |
| Fast ($E > 0.18$ MeV) | $1.16 \times 10^{18}$ neutrons/cm$^2$ |

Crystalline KI was irradiated with the samples to monitor the fluence. Isotope dilution of the $^{128*}Xe$ extracted from part of the KI gave $^{128*}Xe/^{127}I = (1.210 \pm 0.014) \times 10^{-4}$, which corresponds to a fluence of $1.92 \times 10^{19}$ neutrons/cm$^2$ using $\sigma_{127} = 6.3$ barns. This value, which is the value used in subsequent calculations, is larger than the thermal fluence indicated by the Vallecitos Center but agrees well with their total fluence. Fluence variations within the irradiation can were monitored by placing a Co-doped Al wire inside each quartz capsule. The maximum correction for fluence inhomogeneity was 3%.

After irradiation and cooling, the capsules were glass-blown to an extraction system. Any gases lost by the sample during or after the irradiation were measured as part of the 80°C fraction. This fraction and the 350°C fraction were obtained using an external resistance heater. After completion of the 350°C gas extraction, the samples were removed from the quartz capsules and placed in an induction heater extraction system for the high temperature analyses. All of the high temperature Kr and Xe fractions were accumulated and analyzed as a single total fraction after the Ar analyses were completed.

Details of data reduction are discussed in Alexander *et al.* (1972) and Podosek and Lewis (1972).

## $^{40}$Ar–$^{39}$Ar AGES

Table 1 lists the $^{37}Ar/^{39}Ar$, $^{40*}Ar/^{39}Ar$, and the concentration of $^{39*}Ar$ for the Apollo 12 rocks we analyzed. Complete sets of the isotopic data on which Table 1 is based have been prepared and are available on request from the authors. Data in the first two columns of Table 1 were used to construct $^{40*}Ar/^{39}Ar$ versus $^{37}Ar/^{39}Ar$ plots for each rock. Figure 1 shows a typical plot for sample 12051. Turner (1970b) has discussed the use and meaning of such plots, and Alexander *et al.* (1972) explain in detail our use of the plots. The "intercept" values listed in Table 1 are from cubic-least-squares fits (York, 1966) for all temperatures >800°C for each sample. The horizontal intercept is $(^{37}Ar/^{39}Ar)_{Ca}$ or calcium-derived argon, and the vertical intercept is $(^{40*}Ar/^{39}Ar)_K$ or potassium-derived argon. The latter values were used, after

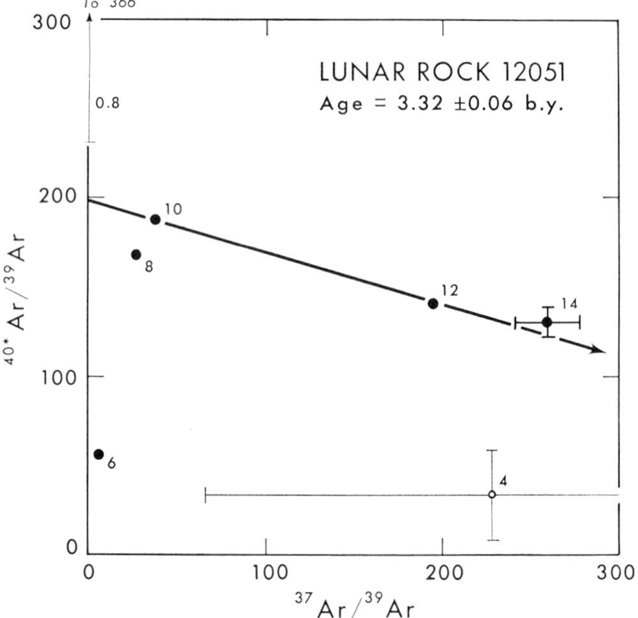


Fig. 1. Argon from neutron irradiated lunar rock 12051. The other rocks yield similar graphs. Errors of 1$\sigma$ are either indicated or are smaller than the points. The number by each point is the temperature of that fraction in hundreds of degrees Celsius. Low temperature points indicated in light symbols are included for completeness but represent negligible fractions of the gas. The figure is discussed more fully in the text.

Table 1. Argon data for five neutron irradiated Apollo 12 rocks.

| Temperature (°C) | $^{37}Ar/^{39}Ar$ | $^{40*}Ar/^{39}Ar$ | $^{39*}Ar$ ($\times 10^{-8}$ ccSTP/gm) |
|---|---|---|---|
| | Lunar rock 12002 (0.968 ± 0.0025)† | | |
| 400 | — | — | 0.55 |
| 600 | 9.1 ± 2.6 | 38.69 ± 0.85 | 4.48 |
| 800 | 36.0 ± 1.9 | 137.8 ± 1.1 | 6.30 |
| 1000 | 68.0 ± 1.8 | 177.3 ± 1.3 | 4.20 |
| 1200 | 243.2 ± 5.7 | 136.5 ± 2.7 | 2.21 |
| 1400 | 400 ± 112 | 57 ± 20 | 0.13 |
| Intercept | 811 ± 91 | 193.7 ± 3.7 | |
| | Lunar rock 12020 (0.991 ± 0.0026)† | | |
| 400 | — | 5.4 ± 3.4 | 0.52 |
| 600 | 9.0 ± 4.5 | 39.41 ± 0.40 | 5.13 |
| 800 | 28.1 ± 1.8 | 127.9 ± 1.9 | 6.57 |
| 1000 | 57.4 ± 2.0 | 166.5 ± 1.7 | 4.00 |
| 1200 | 191.6 ± 2.6 | 138.1 ± 1.5 | 3.43 |
| 1400 | 347 ± 17 | 106.1 ± 5.5 | 0.43 |
| Intercept | 848 ± 52 | 178.5 ± 2.4 | |
| | Lunar rock 12022 (0.989 ± 0.0026)† | | |
| 80 | — | 240 ± 73 | 0.11 |
| 350 | — | 200 ± 62 | 0.14 |
| 400 | — | 8.9 ± 4.3 | 0.78 |
| 600 | — | 46.45 ± 0.83 | 5.37 |
| 800 | 26.9 ± 2.4 | 151.2 ± 1.8 | 5.95 |
| 1000 | 42.8 ± 2.1 | 166.4 ± 1.1 | 6.11 |
| 1200 | 180.6 ± 2.3 | 134.0 ± 3.0 | 5.53 |
| 1400 | 325 ± 37 | 76 ± 13 | 0.15 |
| Intercept | 715 ± 99 | 177.1 ± 3.0 | |
| | Lunar rock 12051 (0.969 ± 0.0025)† | | |
| 80 | — | 366 ± 137 | 0.06 |
| 400 | 227 ± 160 | 33 ± 25 | 0.12 |
| 600 | 5.9 ± 3.3 | 59.4 ± 2.6 | 2.49 |
| 800 | 28.1 ± 1.3 | 167.7 ± 1.6 | 8.56 |
| 1000 | 38.8 ± 2.0 | 186.9 ± 2.6 | 9.60 |
| 1200 | 196.4 ± 2.8 | 140.6 ± 2.8 | 5.17 |
| 1400 | 261 ± 17 | 130.9 ± 7.9 | 0.45 |
| Intercept | 693 ± 64 | 197.7 ± 4.4 | |
| | Lunar rock 12065 (0.989 ± 0.0026)† | | |
| 350 | — | 271 ± 96 | 0.09 |
| 400 | — | 6.3 ± 4.0 | 0.69 |
| 600 | 14.1 ± 1.6 | 65.90 ± 0.71 | 4.50 |
| 800 | 29.0 ± 1.7 | 169.09 ± 0.87 | 7.18 |
| 1000 | 50.1 ± 2.7 | 171.5 ± 1.2 | 6.97 |
| 1200 | 224.7 ± 5.6 | 129.9 ± 3.1 | 3.40 |
| 1400 | 518 ± 69 | 84 ± 15 | 0.10 |
| Intercept | 791 ± 84 | 183.0 ± 2.7 | |

† Fluence corrections (normalized to one of the monitors). These factors are to be applied to the "intercept" values before the age calculation.

The tabulated amounts of $^{37}Ar$ and $^{39}Ar$ are corrected for radioactive decay that took place both during and after the irradiation. The additional error of 1.5% that this introduces into the $^{37}Ar/^{39}Ar$ ratio does not affect the age calculation and has not been included in the tabulated values.

Fractions in which there was no measurable $^{39}Ar$ have been omitted.

application of the fluence corrections listed in Table 1, to calculate the sample ages. The $(^{40*}Ar/^{39}Ar)_K$ intercepts (fluence corrected) for the monitors are 419.6 ± 2.8 and 410.1 ± 2.1 for an average of 414.8 ± 6.9.

Ages were calculated for the Apollo 12 samples using the equation:

$$T_s = (1/\lambda) \ln [1 + R(e^{\lambda T_m} - 1)], \quad (1)$$

where $T_s$ is the age of the sample, $\lambda$ is the total decay probability of $^{40}K$, $T_m$ is the age of the monitor, and $R = (^{40*}Ar/^{39}Ar)_K$ sample/$(^{40*}Ar/^{39}Ar)_K$ monitor. The age of the monitor is assumed to be 4.56 ± 0.05 b.y., based on Goplan and Wetherill's (1969) Rb–Sr age of the amphoterite chondrites. Any change in the age assigned to the monitor will propagate to the sample ages by a factor of $\delta T_s/\delta T_m = 0.914$.

Ages listed in the second column of Table 2 were calculated using $\lambda = 5.480 \times 10^{-10}$ yr$^{-1}$ from Beckinsale and Gale's (1969) summary of physical determinations of the $^{40}K$ decay parameters. Ages listed in the third column were calculated using $\lambda = 5.304 \times 10^{-10}$ yr$^{-1}$, which is the geochronological value (Smith, 1964) used by most laboratories. The former are our preferred values, and the values in the third column are included only to facilitate interlaboratory comparisons. Our results agree satisfactorily with those of other workers (Turner, 1971; Stettler *et al.*, 1972).

Figure 2 is a conventional "plateau plot" of $^{40*}Ar/^{39*}Ar$ and apparent age versus the fraction of $^{39}Ar$ released for sample 12051. This particular method of data display was first used by Podosek *et al.* (1972). The $^{39*}Ar$ is the $^{39}Ar$ derived from potassium and is given by:

$$^{39*}Ar = {}^{39}Ar - (^{39}Ar/^{37}Ar)_{Ca}\,{}^{37}Ar, \quad (2)$$

where $(^{39}Ar/^{37}Ar)_{Ca}$ is the inverse of the horizontal intercept in Fig. 1. Figure 2 is typical of the plots for the other samples.

## TRACE ELEMENTS

Krypton and xenon isotopic data for each of the temperature fractions are available on request from the authors. The complex Kr and Xe spectra were decomposed

Table 2. $^{40}Ar$–$^{39}Ar$ ages of Apollo 12 crystalline rocks.

| | Age (in units of $10^9$ yr) | | |
|---|---|---|---|
| Sample | Preferred values* (this work) | Comparison values† (this work) | Other workers‡ |
| 12002 | 3.29 ± 0.04 (±0.06) | 3.26 ± 0.06 | 3.24 ± 0.05[a] |
| 12020 | 3.20 ± 0.03 (±0.06) | 3.17 ± 0.06 | |
| 12022 | 3.18 ± 0.04 (±0.06) | 3.15 ± 0.06 | |
| 12051 | 3.32 ± 0.04 (±0.06) | 3.29 ± 0.06 | 3.27 ± 0.05[a] 3.15, 3.19[b] |
| 12065 | 3.23 ± 0.03 (±0.06) | 3.20 ± 0.06 | 3.24 ± 0.05[a] |

* Preferred values are calculated using $\lambda = 5.48 \times 10^{-10}$ yr$^{-1}$ (Beckinsale and Gale, 1969) and are relative to an assumed monitor age of 4.56 ± 0.05 × $10^9$ yr for St. Severin. The first error listed does not include the monitor error and is for internal comparison. The second error is the total error and is for interlaboratory comparisons.

† Comparison values are calculated using $\lambda = 5.304 \times 10^{-10}$ yr$^{-1}$ (Smith, 1964), which is the value used by the other workers.

‡ References: [a] Turner (1971), [b] Stettler *et al.* (1972).

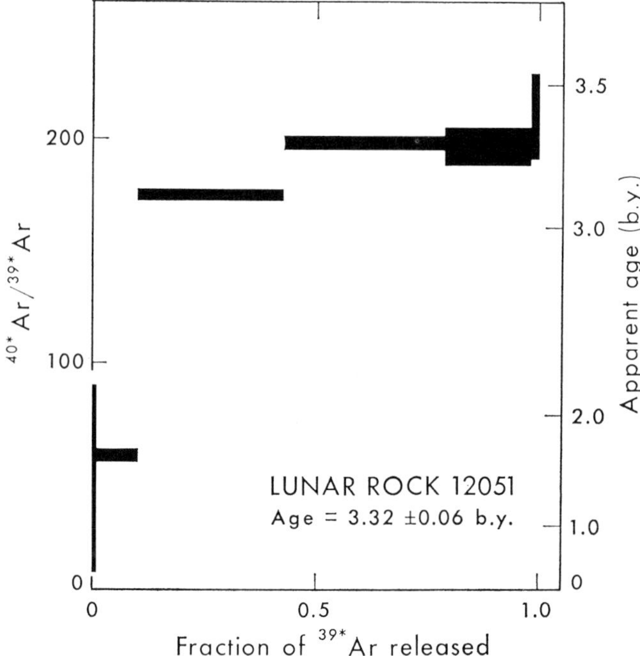


Fig. 2. The $^{40}$Ar–$^{39}$Ar release pattern for lunar rock 12051. The other rocks yield similar patterns. This particular form of data display was developed by Podosek *et al.* (1972). For each temperature fraction the calculated $^{40*}$Ar/$^{39*}$Ar ratio is at the center of a box the length of which is the fraction of $^{39*}$Ar released at that temperature. The height of the box is the statistical error associated with the $^{40*}$Ar/$^{39*}$Ar ratio.

into their various components using a multidimensioned matrix method outlined by Davis *et al.* (1971). The concentrations of I represented by the amount of $^{128*}$Xe in each temperature fraction were calculated using the $^{128*}$Xe/$^{127}$I ratio measured in a KI monitor. Br, Ba, and U concentrations were calculated using the KI monitor data, the ratios of $\sigma^{n,\gamma}_{80,82}$, $\sigma^{n,\gamma}_{130}$, and $\sigma^{n,f}_{235}$ to $\sigma^{n,\gamma}_{127}$, and the measured amounts of $^{80+82*}$Kr, $^{131*}$Xe, and $^{131-136f}$Xe. Table 3 lists the results of the calculations. The low temperature data in Table 3 were corrected for the contribution of the quartz capsule and aluminum foil to the observed neutron induced Kr and Xe isotopes. This correction was based on a "blank" quartz capsule, which contained only aluminum foil. An error equal to the size of this correction was applied to the data. Errors listed in Table 3 are the $1\sigma$ quadratic sum of all known sources of error, including sensitivity errors in the amount of rare gas.

Our Ba and U concentrations compare favorably with those measured for Apollo 12 rocks by many other workers. Our Br value for sample 12022, 48 ± 17 ppb, agrees well with that obtained by Reed and Jovanovic (1971), 43 ppb. Anders *et al.* (1971) reported Br values for samples 12002, 12020, and 12051 of 10, 16, and 16 ppb, respectively, while we observe values of 127 ± 23, 121 ± 20, and 146 ± 25 ppb, respectively. The reason for this (roughly order-of-magnitude) difference with Anders

Table 3. Trace element concentrations in Apollo 12 crystalline rocks.

| Sample | Temperature (°C) | Br (ppb) | Ba (ppm) | I (ppb) | U (ppb) |
|---|---|---|---|---|---|
| 12002 | 80 | 96 ± 23 | 0.08 ± 0.02 | 52.0 ± 7.8 | 0.8 ± 0.3 |
| | 350 | 20.2 ± 3.6 | < 0.03 | 1.9 ± 0.3 | < 1.0 |
| | 1730 | 11.7 ± 1.9 | 47.6 ± 5.7 | 0.6 ± 0.1 | 174 ± 21 |
| | Total | 127 ± 23 | 47.8 ± 5.7 | 54.5 ± 7.8 | 176 ± 21 |
| 12020 | 80 | 111 ± 20 | 0.18 ± 0.02 | 30.1 ± 4.3 | < 1.6 |
| | 350 | 5.1 ± 1.1 | < 0.04 | 0.8 ± 0.2 | < 0.6 |
| | 1730 | 5.7 ± 0.9 | 45.4 ± 5.4 | 0.4 ± 0.2 | 168 ± 21 |
| | Total | 121 ± 20 | 45.7 ± 5.4 | 31.3 ± 4.3 | 170 ± 21 |
| 12022 | 80 | 35 ± 17 | 0.17 ± 0.02 | 69.0 ± 9.4 | < 0.9 |
| | 350 | 6.1 ± 1.8 | < 0.2 | 3.2 ± 0.4 | < 4.0 |
| | 1730 | 7.6 ± 1.2 | 57.6 ± 6.9 | 0.4 ± 0.2 | 204 ± 25 |
| | Total | 48 ± 17 | 58.0 ± 6.9 | 72.6 ± 9.4 | 209 ± 26 |
| 12051 | 80 | 104 ± 24 | 0.08 ± 0.02 | 14.6 ± 5.1 | 0.3 ± 0.1 |
| | 350 | 27.6 ± 4.7 | 0.03 ± 0.02 | 1.0 ± 0.3 | < 1.5 |
| | 1730 | 14.0 ± 2.2 | 69.5 ± 8.3 | 0.7 ± 0.1 | 245 ± 30 |
| | Total | 146 ± 25 | 69.6 ± 8.3 | 16.3 ± 5.1 | 247 ± 30 |
| 12065 | 80 | † | † | † | † |
| | 350 | ~8.7 ± 1.8 | < 0.03 | ~0.4 ± 0.2 | < 1.2 |
| | 1730 | 8.3 ± 1.3 | 51.1 ± 6.1 | 0.5 ± 0.1 | 201 ± 24 |
| | Total | † | 51.2 ± 6.1 | † | 203 ± 24 |

† The 80° fraction of 12065 was lost because the "break seal" apparently was not gas tight. Some of the 350° fraction may have been lost, also.

*et al.*'s (1971) work is not clear, but we are encouraged by our agreement with Reed and Jovanovic's (1971) results. We conclude that the Br content of Apollo 12 crystalline rocks is higher than the average value of 13 ppb calculated by Anders *et al.* (1971).

The value of this unique (and laborious) method of trace element determination is particularly evident for I determinations. Reed and Jovanovic (1971) are the only other group to determine I in lunar samples, and their radiochemical method yields extremely variable results. Our I value for 12022, 72.6 ± 9.4 ppb, falls between the values reported by Reed and Jovanovic (1971) of 14 ppb (water extractable) and 508 ppb (total sample) for two samples of 12022. The rare gas I measurements show a smaller range of values than the radiochemical determinations and agree well with the range, 10 to 60 ppb, quoted by Reed and Jovanovic (1971) as their most reliable results.

Release patterns of the rare gases produced from each element are very similar for all five rocks but vary drastically from element to element. Figure 3 shows the release patterns for each of the elements in rock 12022. Gases derived from Ba and U are almost completely contained in the high temperature fractions. The $^{128*}$Xe from I is almost entirely contained in the low temperature fractions, and Br derived Kr isotopes, while mainly in the low temperature fractions, have a small but significant high temperature component. Similar release patterns for neutron irradiated meteorites have been inferred by Hohenberg (1968).

## DISCUSSION

Davis *et al.* (1971), in a comparison of the $^{40}$Ar release patterns from irradiated and unirradiated samples of 10044 and 10057, found that the release patterns were

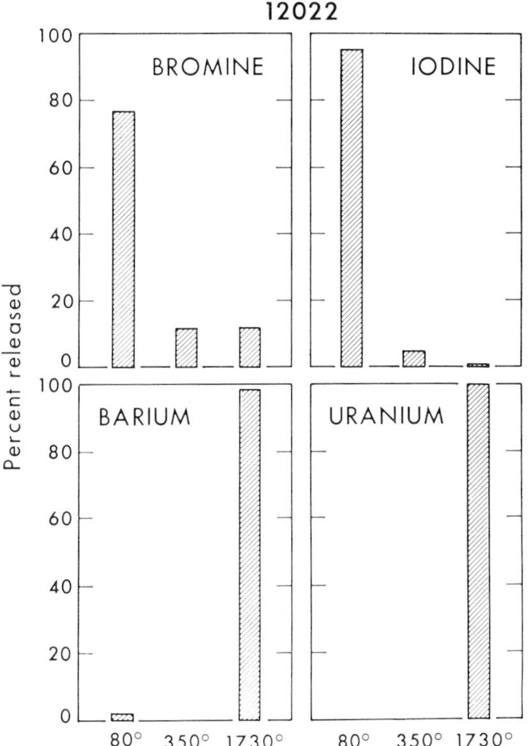


Fig. 3. Release patterns of neutron capture (or fission) rare gases from trace elements in lunar rock 12022. The other rocks have very similar release patterns. Bromine has a small but distinct high temperature component, while iodine does not.

clearly altered by the irradiation. One of the purposes of this work was to check for selective loss of $^{39}$Ar (due perhaps to recoil effects of the $^{39}$K $(n, p)$ $^{39}$Ar reaction) in lunar samples. The low temperature data indicate that loss of $^{39}$Ar was not a significant problem for the Apollo 12 samples. Potassium is apparently located in retentive sites in these samples. In this context it is interesting to note that all five Apollo 12 crystalline rocks yield high temperature plateaus and are datable using the $^{40}$Ar–$^{39}$Ar method. In contrast, three of the eight Apollo 11 crystalline rocks analyzed by the $^{40}$Ar–$^{39}$Ar method did not yield meaningful ages (Turner, 1970a, 1970b; Davis *et al.*, 1971). It will be interesting to see if $^{39}$Ar loss is significant in samples from other missions.

Figure 4 is a plot of trace element concentration versus age for the five Apollo 12 rocks. The spread in ages confirms the observation of Papanastassiou and Wasserburg (1971) that small but real differences exist in the ages of the Apollo 12 rocks. There appears to be significant variation of the I and Br concentrations with age. Although both I and Br are considered volatile elements, their trends are opposite. The Br increases with age, while the I decreases. The data for Ba and U do not define trends, but the oldest rock, 12051, is the highest in Ba and U.

The difference in the release patterns of Br and I described in the previous section

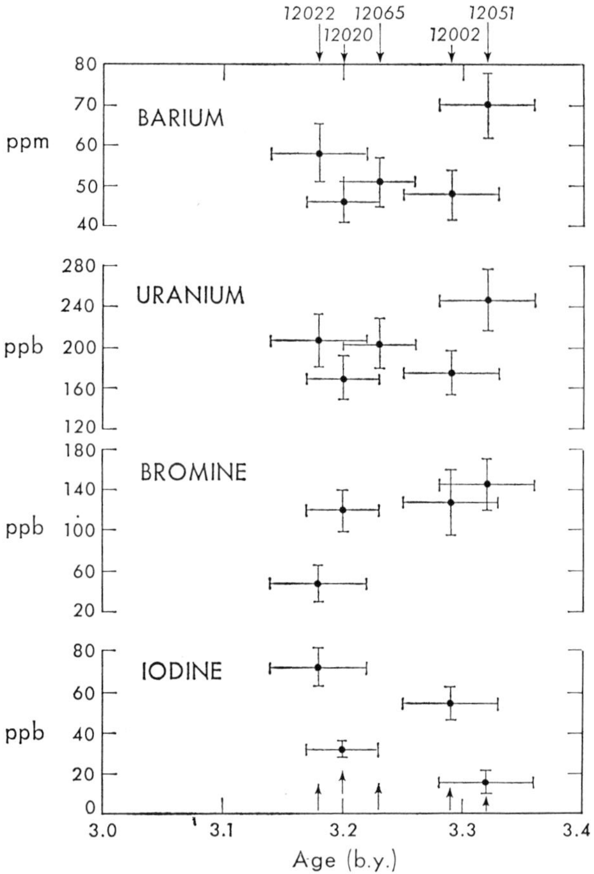


Fig. 4. Trace element content versus age for Apollo 12 samples. Bromine tends to increase with age, while iodine tends to decrease.

is relevant to the recent discovery of Br neutron capture anomalies in Apollo 14 rock 14310 (Lugmair and Marti, 1971). The high temperature fraction of the neutron capture Kr represents only about 10% of the total produced. As the low temperature fractions would be lost in the lunar surface environment, any attempt to calculate the lunar neutron fluence from the Br content and $^{80}$Kr and $^{82}$Kr excesses in unirradiated Apollo 14 samples will yield fluences that are an order of magnitude too low. The almost total absence of a high temperature $^{128*}$Xe component from neutron capture on $^{127}$I observed in the Apollo 12 samples probably explains why no neutron capture $^{128}$Xe was observed in those Apollo 14 samples that contained the Kr neutron capture anomalies. While most of the neutron capture Kr was lost on the lunar surface, all of the neutron capture $^{128}$Xe was lost.

*Acknowledgments*—We acknowledge many stimulating discussions of this data with Drs. R. S. Lewis, W. A. Kaiser, and B. Srinivasan. G. A. McCrory contributed invaluable assistance in maintaining the equipment. This work was supported by NASA under grant NGL 05-003-409.

## References


Alexander E. C. Jr., Davis P. K., and Lewis R. S. (1972) Argon 40–argon 39 dating of Apollo sample 15555. *Science* **175,** 417–419.

Anders E., Ganapathy R., Keays R. R., Laul J. C., and Morgan J. W. (1971) Volatile and siderophile elements in lunar rocks: Comparison with terrestrial and meteoric basalts. *Proc. Second Lunar Conf., Geochim. Cosmochim. Acta* Suppl. 2, Vol. 2, pp. 1021–1036. MIT Press.

Beckinsale R. D. and Gale N. H. (1969) A reappraisal of the decay constants and branching ratio of $^{40}$K. *Earth Planet. Sci. Lett.* **6,** 289.

Davis P. K., Lewis R. S., and Reynolds J. H. (1971) Stepwise heating analyses of rare gases from pile-irradiated rocks 10044 and 10057. *Proc. Second Lunar Sci. Conf., Geochim. Cosmochim. Acta* Suppl. 2, Vol. 2, pp. 1693–1703. MIT Press.

Goplan K. and Wetherill G. W. (1969) Rubidium–strontium age of amphoterite (LL) chondrites. *J. Geophys. Res.* **74,** 4349–4358.

Hohenberg C. A. (1968) Extinct radioactivities in meteorites. Unpublished paper written for the "Nininger Meteorite Award" competition.

Lugmair G. W. and Marti K. (1971) Neutron capture effects in lunar gadolinium and the irradiation histories of some lunar rocks. *Earth Planet. Sci. Lett.* **13,** 32–42.

Merrihue C. M. and Turner G. (1966) Potassium–argon dating by activation with fast neutrons. *J. Geophys. Res.* **71,** 2852–2857.

Papanastassiou D. A. and Wasserburg G. J. (1971) Lunar chronology and evolution from Rb–Sr studies of Apollo 11 and Apollo 12 samples. *Earth Planet. Sci. Lett.* **11,** 37–62.

Podosek F. A., Huneke J. C., and Wasserburg G. J. (1972) Gas-retention and cosmic ray exposure ages of lunar rock 15555. *Science* **175,** 423.

Podosek F. A. and Lewis R. S. (1972) $^{129}$I and $^{244}$Pu abundances in white inclusions of the Allende meteorite. *Earth Planet. Sci. Lett.*, in press.

Reed G. W. and Jovanovic S. (1971) The halogens and other trace elements in Apollo 12 samples and the implications of halides, platinum metals, and mercury on surfaces. *Proc. Second Lunar Sci. Conf., Geochim. Cosmochim. Acta* Suppl. 2, Vol. 2, pp. 1261–1276. MIT Press.

Smith A. G. (1964) Potassium–argon decay constants and age tables. *Quart. J. Geol. Soc. London* **120s,** 129.

Stettler A., Eberhardt P., Geiss J., and Grögler N. (1972) $Ar^{39}/Ar^{40}$ ages of Apollo 11, 12, 14, and 15 rocks (abstract). In *Lunar Science—III* (editor C. Watkins), p. 724, Lunar Science Institute Contr. No. 88.

Turner G. (1970a) Argon-40/argon-39 dating of lunar rock samples. *Science* **167,** 466–468.

Turner G. (1970b) Argon-40/argon-39 dating of lunar rock samples. *Proc. Apollo 11 Lunar Sci. Conf., Geochim. Cosmochim. Acta* Suppl. 1, Vol. 2, pp. 1665–1684. Pergamon.

Turner G. (1971) $^{40}$Ar–$^{39}$Ar ages from the lunar maria. *Earth Planet. Sci. Lett.* **11,** 169.

York D. (1966) Least-square fitting of a straight line. *Can. J. Phys.* **44,** 1079.

Proceedings of the Third Lunar Science Conference
(Supplement 3, *Geochimica et Cosmochimica Acta*)
Vol. 2, pp. 1797–1819
The M.I.T. Press, 1972

# Noble gas studies on regolith materials from Apollo 14 and 15

D. D. BOGARD and L. E. NYQUIST
NASA Manned Spacecraft Center,
Houston, Texas 77058

**Abstract**—Abundances and isotopic compositions of He, Ne, Ar, Kr, and Xe have been determined in a number of Apollo 14 and 15 samples, including bulk analyses on soils and breccias, a range of depths in a trench and a core, and grain size separates and stepwise temperature releases of Apollo 14 bulk fines. Several samples exhibit measured values of $^{84}Kr/^{36}Ar \simeq 3 \times 10^{-4}$ and $^{132}Xe/^{36}Ar \simeq 0.4 \times 10^{-4}$, and set new upper limits for these ratios in the solar wind. A few samples indicate a trapped Ne other than solar wind Ne with $^{20}Ne/^{22}Ne \leq 11.4$. Trapped solar wind, $^{20}Ne/^{22}Ne$ ($12.65 \pm 0.10$), and $^{36}Ar/^{38}Ar$ ($5.36 \pm 0.03$) obtained from grain size separates and stepwise temperature release data are identical to values found for Apollo 11 and 12; $^{4}He/^{3}He$ is variable (2000–2800). Both He and Ne exhibit a low-temperature enhancement of $^{3}He/^{4}He$ and $^{20}Ne/^{22}Ne$ which may be due to a lunar atmosphere implanted component. Trapped $^{40}Ar/^{36}Ar$ is variable among a number of samples. Low-temperature releases reveal a Kr component distinct from the cosmic ray spallation component defined by high temperature releases and grain size separates, and for which $^{82}Kr/^{86}Kr$, $^{83}Kr/^{86}Kr$, and $^{84}Kr/^{86}Kr$ are enriched over typical trapped values by up to 29%, 21%, and 3%, respectively. This low-temperature component is apparently not due to implantation of lunar atmosphere Kr, but may result from solar flare reactions on Rb. Xe isotope correlation diagrams for stepwise temperature releases and grain size separates exhibit linear correlations of slightly different slope, which apparently are not caused by mass fractionation during the stepwise heating. For both Kr and Xe, grain size separates and temperature releases above 900°C indicate different trapped components, which may be related as mixtures of trapped solar wind and other low temperature components. Spallation $^{131}Xe/^{126}Xe$ in many fines and gas-rich breccias varies by a factor of two, while $^{124}Xe/^{126}Xe$ is essentially constant at $0.53 \pm 0.02$. If much of the excess $^{131}Xe$ is due to neutron capture on Ba, the variations in spallation $^{131}Xe/^{126}Xe$ possibly indicate differences in average shielding. By this criterion, samples from core 14230 and the Apollo 14 trench samples show nearly the same average burial depth during cosmic ray exposure, the topmost samples having marginally greater $^{131}Xe/^{126}Xe$. A number of soils and breccia show differences in spallation xenon of $\geq$ a factor of three. indicating variable exposure times.

## INTRODUCTION AND TECHNIQUES

THE PURPOSE OF THE PRESENT investigation is to determine the nature of the various noble gas components in a number of lunar regolith samples in order to define solar wind implanted abundances of these elements, and to interpret lunar surface processes and history. Thus, we have determined abundances and isotopic compositions of the five stable noble gases, He, Ne, Ar, Kr, and Xe in soils and breccia from Apollo 14 and 15, including samples from a range of depths in a trench and a core tube, and grain size separates and stepwise temperature releases of the Apollo 14 bulk soil. Noble gas determinations were performed in a 6-inch all-metal mass spectrometer of high sensitivity (Bogard *et al.*, 1971a). Analog signal output was integrally digitized, and computations were performed on programmable calculators.

Table 1. Measured noble gas abundances.

| Sample | Weight (mg) | $\times 10^{-6}$ ccSTP/g | | | | $\times 10^{-9}$ ccSTP/g | | $^{20}Ne/^{22}Ne$ | $^{22}Ne/^{21}Ne$ | $^{36}Ar/^{38}Ar$ | $^{40}Ar/^{36}Ar$ |
|---|---|---|---|---|---|---|---|---|---|---|---|
| | | $^{3}He$ | $^{4}He$ | $^{22}Ne$ | $^{36}Ar$ | $^{84}Kr$ | $^{132}Xe$ | | | | |
| I. *Bulk samples* | | | | | | | | | | | |
| 14148,27 (trench top) | 5.31 | 22.7 | 55,040 | 74.6 | 267 | 170 | 20.1 | 12.52 ± 0.05 | 26.64 ± 0.21 | 5.32 ± 0.01 | 2.03 ± 0.02 |
| 14156,27 (trench middle) | 6.66 | 21.6 | 52,000 | 79.0 | 321 | 113 | 22.5 | 12.54 ± 0.05 | 27.27 ± 0.36 | 5.34 ± 0.01 | 1.77 ± 0.02 |
| 14149,43 (trench bottom) | 5.52 | 21.2 | 50,950 | 77.5 | 239 | 93 | 25.7 | 12.41 ± 0.04 | 26.42 ± 0.14 | 5.33 ± 0.02 | 3.12 ± 0.01 |
| 14003,6 (fines) | 9.32 | 25.5 | 62,560 | 84.0 | 253 | 159 | 13.8 | 12.60 ± 0.04 | 28.80 ± 0.16 | 5.38 ± 0.02 | 1.68 ± 0.01 |
| 14301,48 (breccia) | 28.78 | 2.98 | 9,260 | 11.3 | 3.51 | 0.17 | 0.098 | 12.39 ± 0.14 | 28.5 ± 0.7 | 5.54 ± 0.04 | 15.1 ± 0.1 |
| 14307,26,2 (breccia–dark) | 19.32 | 26.0 | 70,740 | 108 | 311 | 59 | 12.9 | 12.24 ± 0.04 | 29.4 ± 1.2 | 5.38 ± 0.02 | 5.67 ± 0.02 |
| 14307,26,2 (repeat–dark) | 10.76 | 16.3 | 42,020 | 101 | 83 | 36 | 8.7 | 12.38 ± 0.04 | 29.1 ± 0.9 | 5.38 ± 0.02 | 5.86 ± 0.02 |
| 14307,26 (light) | 35.62 | 1.0 | 1,240 | 0.26 | ≤0.12* | 0.33 | 0.33 | 2.73 ± 0.01 | 1.331 ± 0.002 | 1.45* | 109* |
| 14006,3 | 20.05 | 1.99 | 3,080 | 0.76 | 1.56 | 0.45 | 0.33 | 7.33 ± 0.08 | 2.74 ± 0.03 | 2.75 | 88 |
| 14068,3 (breccia) | 11.3 | 0.23 | 1,450 | 0.63 | 0.68 | 2 | 0.7 | 12.15 ± 0.03 | 13.70 ± 0.16 | 5.15 ± 0.05 | 41.7 ± 0.7 |
| 14230,113 (core–4 cm) | 16.65 | 15.3 | 35,110 | 47.4 | 136 | 66 | 6.85 | 12.57 ± 0.02 | 25.54 ± 0.27 | 5.37 ± 0.01 | 2.18 ± 0.01 |
| 14230,130 (core–8 cm) | 17.30 | 11.7 | 29,000 | 39.1 | 57 | 12 | 1.6 | 12.72 ± 0.04 | 25.3 ± 0.2 | 5.45 ± 0.02 | 2.48 ± 0.02 |
| 14230,121 (core–12 cm) | 9.85 | 16.4 | 40,730 | 61.1 | 170 | 72 | 10.2 | 12.69 ± 0.03 | 27.1 ± 0.2 | 5.37 ± 0.02 | 2.09 ± 0.02 |
| II. *Grain size separates* | | | | | | | | | | | |
| 14161,35 (2–4 mm) | 33.63 | 0.98 | 213 | 0.57 | 0.32 | 0.61 | 0.13 | 5.39 ± 0.03 | 2.00 ± 0.01 | 2.42 | 64 |
| 14162,22 (1–2 mm) (G) | 32.02 | 20.7 | 43,070 | 11.7 | 39.4 | 26 | 6.2 | 12.36 ± 0.04 | 19.17 ± 0.15 | 5.23 ± 0.03 | 4.50 ± 0.03 |
| 14163,97 | | | | | | | | | | | |
| 1000–250 μ (F) | 2.02 | 2.07 | 3,030 | 4.43 | 23.3 | 16* | 10* | 11.04 ± 0.05 | 8.70 ± 0.09 | 4.73 ± 0.02 | 32.6 ± 0.2 |
| 250–90 μ (E) | 1.33 | 7.02 | 14,200 | 22.7 | 94 | 45 | 15 | 12.11 ± 0.10 | 17.80 ± 0.11 | 5.15 ± 0.03 | 3.37 ± 0.2 |
| 90–60 μ (D) | 1.13 | 11.2 | 26,300 | 37.6 | 164 | 100 | 33 | 12.24 ± 0.04 | 22.82 ± 0.10 | 5.24 ± 0.03 | 2.94 ± 0.01 |
| 60–30 μ (C) | 1.97 | 19.5 | 49,200 | 63.3 | 285 | 130 | 25 | 12.8 ± 0.1 | 26.4 ± 0.1 | 5.32 ± 0.02 | 2.49 ± 0.03 |
| 30–20 μ (B) | 1.98 | 51.4 | 133,000 | 164 | 519 | 244 | 60 | 12.64 ± 0.10 | 29.17 ± 0.11 | 5.33 ± 0.01 | 2.39 ± 0.10 |
| 20 μ (A) | 2.54 | 62.6 | 166,500 | 187 | 533 | 170 | 41 | 12.64 ± 0.05 | 29.78 ± 0.11 | 5.33 ± 0.01 | 2.40 ± 0.05 |

Based on variations in machine sensitivity, abundances have an estimated uncertainty of ±5–10%, except for Kr which is ±15%. Blank corrections were usually small and have been applied. Values marked with an * possessed large blank corrections and are less accurate. Uncertainties for isotopic ratios are one sigma for multiple measurements. The $^{3}He/^{4}He$ was not measured directly, but can be obtained from the abundances with an uncertainty of about 2%. Letter codes listed for the grain size separates are for identification in Figs. 3 and 6 to 10.

Table 2. Measured noble gas isotopic composition for 14163,97 temperature release.

| Temperature (°C) | $^{4}He/^{3}He$ | $^{20}Ne/^{22}Ne$ | $^{21}Ne/^{22}Ne$ | $^{38}Ar/^{36}Ar$ | $^{40}Ar/^{36}Ar$ |
|---|---|---|---|---|---|
| 150 | 1720 | 14.46 ± 0.18 | — | — | — |
| 225 | 1770 | 14.30 ± 0.05 | 0.0365 ± 0.0006 | 0.180 ± 0.011 | 25.2 ± 1.5 |
| 275 | 1780 | 14.03 ± 0.03 | 0.0418 ± 0.0005 | 0.1818 ± 0.0006 | 28.9 ± 0.7 |
| 325 | 1900 | 13.30 ± 0.05 | 0.0462 ± 0.0005 | 0.1811 ± 0.0026 | 28.7 ± 0.8 |
| 400 | 2080 | 12.88 ± 0.05 | 0.0450 ± 0.0002 | 0.1836 ± 0.0013 | 30.2 ± 0.4 |
| 500 | 2165 | 12.71 ± 0.04 | 0.0354 ± 0.0004 | 0.1832 ± 0.0010 | 20.0 ± 0.1 |
| 600 | 2240 | 12.59 ± 0.02 | 0.0321 ± 0.0007 | 0.1812 ± 0.0007 | 10.40 ± 0.09 |
| 700 | 2365 | 12.79 ± 0.02 | 0.0341 ± 0.0002 | 0.1802 ± 0.0004 | 5.57 ± 0.03 |
| 800 | 2325 | 12.58 ± 0.03 | 0.0342 ± 0.0002 | 0.1858 ± 0.0004 | 2.32 ± 0.05 |
| 900 | 2675 | 12.61 ± 0.03 | 0.0348 ± 0.0001 | 0.1874 ± 0.0003 | 1.59 ± 0.01 |
| 1000 | 2570 | 12.58 ± 0.02 | 0.0388 ± 0.0004 | 0.1896 ± 0.0004 | 1.70 ± 0.01 |
| 1000 | 3105 | 12.31 ± 0.07 | 0.0418 ± 0.0002 | 0.1928 ± 0.0002 | 2.09 ± 0.01 |
| 1200 | 2555 | 11.91 ± 0.03 | 0.0452 ± 0.0006 | 0.1927 ± 0.0006 | 1.99 ± 0.01 |
| 1400 | 1305 | 10.81 ± 0.04 | 0.0797 ± 0.0008 | 0.1986 ± 0.0007 | 1.90 ± 0.07 |
| 1600 | 1175 | 12.0 ± 1.0 | 0.084 ± 0.02 | 0.208 ± 0.004 | 4.3 ± 0.1 |

Table 3. Measured krypton isotopic composition (relative to $^{86}Kr$).

| Sample | 78 | 80 | 82 | 83 | 84 |
|---|---|---|---|---|---|
| I. *Bulk samples* | | | | | |
| 14148,27 | 0.033 ± 0.003 | 0.164 ± 0.003 | 0.715 ± 0.006 | 0.721 ± 0.008 | 3.279 ± 0.021 |
| 14156,27 | 0.023 ± 0.003 | 0.152 ± 0.004 | 0.696 ± 0.012 | 0.703 ± 0.012 | 3.30 ± 0.05 |
| 14149,43 | 0.026 ± 0.003 | 0.151 ± 0.004 | 0.697 ± 0.008 | 0.701 ± 0.008 | 3.289 ± 0.032 |
| 14003,6 | 0.026 ± 0.003 | 0.148 ± 0.003 | 0.685 ± 0.006 | 0.692 ± 0.006 | 3.279 ± 0.021 |
| 14301,48 | 0.340 ± 0.032 | 0.182 ± 0.024 | 0.737 ± 0.073 | 0.741 ± 0.028 | 3.37 ± 0.09 |
| 14307,26,2 (dark) | 0.030 ± 0.003 | 0.163 ± 0.003 | 0.716 ± 0.004 | 0.716 ± 0.004 | 3.331 ± 0.009 |
| 14307,26 (light) | 1.18 | 3.22 | 4.99 | 6.34 | 4.44 |
| 14006,3 | 0.20 ± 0.02 | 0.572 ± 0.065 | 1.33 ± 0.12 | 1.53 ± 0.14 | 3.51 ± 0.09 |
| 14068,3 | 0.0328 ± 0.0033 | 0.162 ± 0.003 | 0.723 ± 0.005 | 0.720 ± 0.006 | 3.347 ± 0.020 |
| 14230,113,2 | 0.305 ± 0.0013 | 0.1553 ± 0.0010 | 0.703 ± 0.006 | 0.709 ± 0.006 | 3.311 ± 0.022 |
| 14230,130,2 | 0.044 | 0.179 ± 0.002 | 0.749 ± 0.009 | 0.757 ± 0.009 | 3.369 ± 0.034 |
| 14230,112,2 | 0.0303 ± 0.0006 | 0.1525 ± 0.0017 | 0.696 ± 0.003 | 0.701 ± 0.003 | 3.293 ± 0.009 |
| 15301,1 (fines) | 0.022 ± 0.002 | 0.135 ± 0.002 | 0.671 ± 0.009 | 0.670 ± 0.008 | 3.282 ± 0.024 |
| 15021,4 (fines) | 0.023 ± 0.007 | 0.138 ± 0.002 | 0.678 ± 0.008 | 0.678 ± 0.009 | 3.30 ± 0.03 |
| 15101,2 (fines) | 0.022 ± 0.001 | 0.139 ± 0.002 | 0.679 ± 0.007 | 0.679 ± 0.007 | 3.29 ± 0.03 |
| 15298,3 (breccia) | 0.028 ± 0.002 | 0.137 ± 0.002 | 0.675 ± 0.012 | 0.677 ± 0.013 | 3.28 ± 0.04 |
| 15558,3 (breccia) | 0.027 ± 0.002 | 0.146 ± 0.003 | 0.690 ± 0.006 | 0.693 ± 0.007 | 3.304 ± 0.025 |
| 15923,2 (matrix) | 0.017 ± 0.004 | 0.139 ± 0.007 | 0.692 ± 0.019 | 0.700 ± 0.020 | 3.280 ± 0.012 |
| 15923,2 (glass) | — | — | 0.712 ± 0.026 | 0.754 ± 0.028 | 3.35 ± 0.06 |
| 15498 (breccia) | 0.033 ± 0.003 | 0.155 ± 0.004 | 0.710 ± 0.008 | 0.711 ± 0.009 | 3.330 ± 0.017 |
| 15265 (breccia) | 0.031 ± 0.003 | 0.152 ± 0.003 | 0.696 ± 0.009 | 0.700 ± 0.008 | 3.306 ± 0.029 |
| 15601 (fines) | 0.026 ± 0.002 | 0.137 ± 0.002 | 0.674 ± 0.007 | 0.676 ± 0.008 | 3.284 ± 0.025 |
| II. *Grain size separates* | | | | | |
| 14161,35,7 (2–4 mm) | 0.45 ± 0.03 | 0.768 ± 0.028 | 1.53 ± 0.03 | 1.71 ± 0.03 | 3.62 ± 0.05 |
| 14162,22 (1–2 mm) | 0.042 ± 0.001 | 0.174 ± 0.002 | 0.725 ± 0.005 | 0.737 ± 0.004 | 3.291 ± 0.014 |
| 14163,97 | | | | | |
| 1000–250 μ | 0.050 ± 0.013 | 0.217 ± 0.004 | 0.793 ± 0.004 | 0.824 ± 0.004 | 3.338 ± 0.006 |
| 250–90 μ | 0.044 ± 0.003 | 0.183 ± 0.002 | 0.739 ± 0.007 | 0.760 ± 0.006 | 3.300 ± 0.019 |
| 90–60 μ | 0.036 ± 0.004 | 0.168 ± 0.001 | 0.714 ± 0.003 | 0.729 ± 0.003 | 3.287 ± 0.012 |
| 60–30 μ | 0.028 ± 0.002 | 0.149 ± 0.003 | 0.692 ± 0.012 | 0.697 ± 0.012 | 3.298 ± 0.045 |
| 30–20 μ | 0.043 ± 0.011 | 0.181 ± 0.004 | 0.797 ± 0.005 | 0.776 ± 0.003 | 3.537 ± 0.008 |
| 30–20 (repeat) | 0.041 ± 0.005 | 0.171 ± 0.004 | 0.729 ± 0.013 | 0.744 ± 0.014 | 3.31 ± 0.05 |
| III. *Stepwise temperature extractions* | | | | | |
| 14163,97 | | | | | |
| 325°C | — | — | 0.849 ± 0.035 | 0.800 ± 0.032 | 3.28 ± 0.09 |
| 500°C | — | — | 0.744 ± 0.027 | 0.712 ± 0.024 | 3.165 ± 0.040 |
| 700°C | 0.033 ± 0.013 | 0.187 ± 0.017 | 0.716 ± 0.008 | 0.699 ± 0.015 | 3.344 ± 0.034 |
| 800°C | 0.022 ± 0.003 | 0.134 ± 0.007 | 0.698 ± 0.019 | 0.689 ± 0.021 | 3.279 ± 0.064 |
| 900°C | 0.020 ± 0.007 | 0.139 ± 0.007 | 0.677 ± 0.004 | 0.673 ± 0.004 | 3.300 ± 0.011 |
| 1000°C | 0.023 ± 0.002 | 0.139 ± 0.002 | 0.670 ± 0.004 | 0.673 ± 0.004 | 3.268 ± 0.012 |
| 1100°C | 0.029 ± 0.002 | 0.152 ± 0.002 | 0.689 ± 0.003 | 0.701 ± 0.002 | 3.252 ± 0.004 |
| 1200°C | 0.032 ± 0.002 | 0.159 ± 0.002 | 0.704 ± 0.006 | 0.713 ± 0.006 | 3.257 ± 0.021 |
| 1400°C | 0.042 ± 0.002 | 0.183 ± 0.001 | 0.734 ± 0.004 | 0.761 ± 0.004 | 3.279 ± 0.011 |
| 1600°C | 0.08 ± 0.01 | 0.234 ± 0.004 | 0.828 ± 0.007 | 0.881 ± 0.004 | 3.300 ± 0.011 |

Uncertainties are one sigma for multiple measurements, except for $^{78}Kr$ and $^{80}Kr$, which include an additional uncertainty due to hydrocarbon corrections.

Table 4. Measured xenon isotopic compositions (relative to $^{136}Xe$).

| Sample | 124 | 126 | 128 | 129 | 130 | 131 | 132 | 134 |
|---|---|---|---|---|---|---|---|---|
| I. *Bulk samples* | | | | | | | | |
| 14148,27 | 0.075 ± 0.003 | 0.123 ± 0.003 | 0.458 ± 0.009 | 3.66 ± 0.08 | 0.658 ± 0.011 | 3.339 ± 0.065 | 3.390 ± 0.045 | 1.233 ± 0.019 |
| 14156,27 | 0.0532 ± 0.0014 | 0.084 ± 0.004 | 0.385 ± 0.003 | 3.56 ± 0.03 | 0.606 ± 0.010 | 3.067 ± 0.033 | 3.326 ± 0.022 | 1.224 ± 0.008 |
| 14149,43 | 0.0491 ± 0.0020 | 0.0776 ± 0.0011 | 0.371 ± 0.006 | 3.53 ± 0.03 | 0.595 ± 0.005 | 3.014 ± 0.025 | 3.317 ± 0.022 | 1.228 ± 0.009 |
| 14003,6 | 0.044 ± 0.002 | 0.065 ± 0.002 | 0.364 ± 0.004 | 3.59 ± 0.04 | 0.596 ± 0.010 | 3.002 ± 0.034 | 3.362 ± 0 033 | 1 234 ± 0.019 |
| 14301,48 | — | — | 0.380 ± 0.023 | 3.27 ± 0.13 | 0.567 ± 0.029 | 2.85 ± 0.12 | 3.10 ± 0.12 | 1.19 ± 0.07 |
| 14307,26,2 (dark) | 0.0327 ± 0.0010 | 0.0469 ± 0.0007 | 0.342 ± 0.005 | 3.75 ± 0.03 | 0.590 ± 0.004 | 2.960 ± 0.024 | 3.37 ± 0.02 | 1 236 ± 0.009 |
| 14307,26,2 (light) | 1.154 ± 0.014 | 2.163 ± 0.033 | 3.59 ± 0.05 | 5.06 ± 0.07 | 2.70 ± 0.03 | 17.05 ± 0.20 | 3.782 ± 0.024 | 1.43 ± 0.03 |
| 14006,3 | 0.375 ± 0.026 | 0.648 ± 0.039 | 1.307 ± 0.059 | 4.37 ± 0.17 | 1.20 ± 0.05 | 6.47 ± 0.26 | 3.75 ± 0.14 | 1.29 ± 0.06 |
| 14068,3 | 0.0314 ± 0.0011 | 0.0456 ± 0.0015 | 0.332 ± 0.005 | 3.65 ± 0.04 | 0.575 ± 0.007 | 2.90 ± 0.03 | 3.34 ± 0.04 | 1.234 ± 0.013 |
| 14230,113,2 | 0.0595 ± 0.0027 | 0.0930 ± 0.0024 | 0 408 ± 0.004 | 3.65 ± 0.03 | 0.628 ± 0.007 | 3.154 ± 0.023 | 3.38 ± 0.02 | 1.238 ± 0.120 |
| 14230,130,2 | 0.088 ± 0.015 | 0.135 ± 0.013 | 0.478 ± 0.016 | 3.73 ± 0.08 | 0.684 ± 0.017 | 3.34 ± 0.08 | 3.42 ± 0.07 | 1.24 ± 0.04 |
| 14230,121,2 | 0.0476 ± 0.0014 | 0.0773 ± 0.0024 | 0.370 ± 0.007 | 3.54 ± 0.04 | 0.600 ± 0.009 | 3.01 ± 0.04 | 3.33 ± 0.04 | 1.23 ± 0.02 |
| 15301,1 | 0.0211 ± 0.0010 | 0.0241 ± 0.0013 | 0.290 ± 0.003 | 3.47 ± 0.03 | 0.548 ± 0.011 | 2.745 ± 0.025 | 3.30 ± 0.02 | 1.23 ± 0.01 |
| 15021,4 | 0.0278 ± 0.0004 | 0.0356 ± 0.0008 | 0.316 ± 0.005 | 3.57 ± 0.04 | 0.575 ± 0.007 | 2.87 ± 0.04 | 3.39 ± 0.04 | 1.243 ± 0.015 |
| 15101,2 | 0.0270 ± 0.0010 | 0.0347 ± 0.0010 | 0.309 ± 0.009 | 3.51 ± 0.03 | 0.567 ± 0.006 | 2.83 ± 0.03 | 3.34 ± 0.03 | 1.233 ± 0.012 |
| 15298,3 | 0.0256 ± 0.0013 | 0.0338 ± 0.0010 | 0.300 ± 0.006 | 3.40 ± 0.03 | 0.551 ± 0.009 | 2.75 ± 0.03 | 3.28 ± 0.03 | 1.218 ± 0.012 |
| 15558,3 | 0.0287 ± 0.0010 | 0.0393 ± 0.0016 | 0.314 ± 0.003 | 3.50 ± 0.03 | 0.567 ± 0.008 | 2.831 ± 0.017 | 3.300 ± 0.014 | 1 220 ± 0.007 |
| 15923,2 (matrix) | — | — | 0.307 ± 0.015 | 3.51 ± 0.08 | 0.551 ± 0.016 | 2.80 ± 0.06 | 3.30 ± 0.07 | 1.19 ± 0.03 |
| 15923,2 (glass) | 0.050 ± 0.018 | 0.076 ± 0.009 | 0.395 ± 0.013 | 3.41 ± 0.13 | 0.610 ± 0.012 | 3.09 ± 0.06 | 3.32 ± 0.04 | 1.23 ± 0.04 |
| 15498,2 | 0.037 ± 0 003 | 0.052 ± 0.003 | 0.348 ± 0.007 | 3.63 ± 0.06 | 0.594 ± 0.012 | 2.92 ± 0.14 | 3.35 ± 0.04 | 1.24 ± 0.02 |
| 15265,3 | 0.045 ± 0.003 | 0.069 ± 0.006 | 0.361 ± 0.009 | 3.48 ± 0.07 | 0.595 ± 0.012 | 3.00 ± 0.05 | 3.32 ± 0.05 | 1.225 ± 0.023 |
| 15601,3 | 0.024 ± 0.003 | 0.0325 ± 0.0023 | 0.301 ± 0.005 | 3.48 ± 0.03 | 0.563 ± 0.014 | 2.78 ± 0.02 | 3.32 ± 0.02 | 1.224 ± 0.012 |
| II. *Grain size separates* | | | | | | | | |
| 14161,35,7 (2–4 mm) | 0.59 ± 0.05 | 1.06 ± 0.11 | 1.93 ± 0.14 | 4.82 ± 0.32 | 1.46 ± 0.09 | 7.19 ± 0.47 | 3.80 ± 0.23 | 1.28 ± 0.09 |
| 14162,22 (1–2 mm) | 0.0589 ± 0.0014 | 0.0927 ± 0.0022 | 0.372 ± 0.003 | 3.33 ± 0.02 | 0.575 ± 0.003 | 2.961 ± 0.021 | 3.200 ± 0.011 | 1.201 ± 0.007 |
| 14163,97 | | | | | | | | |
| 1000–250 μ | 0.1623 ± 0.0034 | 0.279 ± 0.004 | 0.710 ± 0.007 | 3.98 ± 0.03 | 0.826 ± 0.010 | 4.153 ± 0.047 | 3.545 ± 0.023 | 1.251 ± 0.010 |
| 250–90 μ | 0.0956 ± 0.0044 | 0.164 ± 0.003 | 0.517 ± 0.010 | 3.76 ± 0.06 | 0.699 ± 0.012 | 3.52 ± 0.06 | 3.44 ± 0.05 | 1.246 ± 0.022 |
| 90–60 μ | 0.0742 ± 0.0013 | 0.1226 ± 0.0010 | 0.451 ± 0.004 | 3.66 ± 0.03 | 0.651 ± 0.005 | 3.278 ± 0.026 | 3.38 ± 0.02 | 1.230 ± 0.010 |
| 60–30 μ | 0.0482 ± 0.0004 | 0.0747 ± 0.0007 | 0.375 ± 0.003 | 3.57 ± 0.02 | 0.601 ± 0.010 | 3.015 ± 0.016 | 3.321 ± 0.015 | 1.230 ± 0.008 |
| 30–20 μ | 0.058 ± 0.003 | 0.0862 ± 0.0024 | 0.423 ± 0.005 | 3.92 ± 0.02 | 0.651 ± 0.010 | 3.216 ± 0.018 | 3.519 ± 0.012 | 1.256 ± 0.019 |
| 30–20 (repeat) | 0.059 ± 0.003 | 0.0944 ± 0.0035 | 0.404 ± 0.007 | 3.62 ± 0.04 | 0.617 ± 0.010 | 3.114 ± 0.036 | 3.37 ± 0.03 | 1.228 ± 0.015 |

| Sample | 124 | 126 | 128 | 129 | 130 | 131 | 132 | 134 |
|---|---|---|---|---|---|---|---|---|
| III. *Stepwise temperature releases* | | | | | | | | |
| 14163,97 | | | | | | | | |
| 325°C | — | — | 0.271 ± 0.010 | 3.02 ± 0.08 | 0.473 ± 0.026 | 2.49 ± 0.05 | 3.05 ± 0.07 | 1.16 ± 0.04 |
| 525°C | — | — | 0.346 ± 0.004 | 3.30 ± 0.05 | 0.517 ± 0.013 | 2.64 ± 0.06 | 3.12 ± 0.03 | 1.171 ± 0.024 |
| 700°C | 0.025 | 0.044 | 0.324 ± 0.019 | 3.30 ± 0.19 | 0.53 ± 0.03 | 2.68 ± 0.15 | 3.14 ± 0.18 | 1.21 ± 0.07 |
| 800°C | 0.025 | 0.034 | 0.313 ± 0.007 | 3.48 ± 0.05 | 0.551 ± 0.008 | 2.76 ± 0.04 | 3.28 ± 0.04 | 1.23 ± 0.02 |
| 900°C | 0.0240 ± 0.0004 | 0.0297 ± 0.0007 | 0.307 ± 0.004 | 3.53 ± 0.04 | 0.563 ± 0.005 | 2.81 ± 0.03 | 3.33 ± 0.02 | 1.23 ± 0.01 |
| 1000°C | 0.0269 ± 0.0003 | 0.0359 ± 0.0004 | 0.314 ± 0.002 | 3.52 ± 0.03 | 0.568 ± 0.007 | 2.846 ± 0.016 | 3.324 ± 0.011 | 1.230 ± 0.007 |
| 1100°C | 0.0423 ± 0.0010 | 0.0650 ± 0.0010 | 0.356 ± 0.004 | 3.540 ± 0.023 | 0.593 ± 0.004 | 2.973 ± 0.019 | 3.333 ± 0.011 | 1.230 ± 0.005 |
| 1200°C | 0.0595 ± 0.0007 | 0.0953 ± 0.0014 | 0.406 ± 0.005 | 3.58 ± 0.04 | 0.620 ± 0.007 | 3.120 ± 0.035 | 3.344 ± 0.034 | 1.231 ± 0.016 |
| 1400°C | 0.0788 ± 0.0004 | 0.1290 ± 0.0014 | 0.454 ± 0.003 | 3.64 ± 0.02 | 0.653 ± 0.004 | 3.269 ± 0.015 | 3.367 ± 0.011 | 1.229 ± 0.005 |
| 1600°C | 0.086 ± 0.010 | 0.147 ± 0.011 | 0.493 ± 0.012 | 3.73 ± 0.09 | 0.685 ± 0.013 | 3.42 ± 0.08 | 3.42 ± 0.08 | 1.25 ± 0.03 |

## Results

Measured concentrations of all noble gases and isotopic abundances of He, Ne, and Ar in Apollo 14 bulk samples and grain size separates are given in Table 1. Isotopic compositions of He, Ne, and Ar from the stepwise temperature release of fines 14163,97 (30–60 microns) are given in Table 2. Tables 3 and 4 present the isotopic compositions of Kr and of Xe, respectively, for many of these samples and a number of Apollo 15 samples. Noble gas data on additional Apollo 14 samples were given in the PET report (LSPET, 1971); He, Ne, and Ar data and Kr and Xe concentrations for the Apollo 15 samples listed in Tables 3 and 4 were presented in the Apollo 15 PET report (LSPET, 1972). In the following figures and discussion use is made of these previously published data as well as those contained in Tables 1–4. Throughout much of the discussion we shall be especially interested in a comparison of the grain-size separates and stepwise heating data of 14163. In several important respects, the grain size separates and stepwise heating experimental methods give results that supplement each other, an informative approach which has been very little used.

## Solar Wind Derived Noble Gas Concentrations and Elemental Ratios

Most of the Apollo 14 and 15 soils analyzed exhibit elemental noble gas ratios which agree to within a factor of two, and show typical values of $^{4}He/^{36}Ar = 250$, $^{22}Ne/^{36}Ar = 0.30$, and $^{132}Xe/^{36}Ar = 0.5 \times 10^{-4}$. Previous work has shown the relative abundances of the lighter elements, He, Ne, and possibly Ar in bulk samples, to be mass fractionated compared to gases observed in ilmenite and the assumed solar wind abundances. However, relative abundances of the heavier noble gases are generally the same in silicate and ilmenite fractions, and upper limits to solar wind ratios have been given as $^{84}Kr/^{36}Ar \leq 4.3 \times 10^{-4}$ and $^{132}Xe/^{36}Ar \leq 0.62 \times 10^{-4}$ (Eberhardt *et al.*, 1970).

Figure 1 presents relative abundances for a number of the Apollo 14 and 15 sample analyses reported here and in LSPET (1972). Also given are the theoretical solar abundances of Aller (1961) and Cameron (1968), and the limits of the lowest measured $^{84}Kr/^{36}Ar$ and $^{132}Xe/^{36}Ar$ values in lunar soils by Eberhardt *et al.* (1970). Essentially all Ar, Kr, and Xe measurements by various laboratories on Apollo 11 and 12 materials fall above and to the right of this lower limit. Mass fractionation processes affecting the noble gases in these samples would tend to move the plotted points to the upper right. This is exemplified by the connected small points that represent the total gas content (T) and the gas remaining in fines 10084 during the early stages of a stepwise temperature release conducted on this sample (Pepin *et al.*, 1970). Several samples analyzed by us plot below the Bern limit, and establish new upper limits to the solar $^{84}Kr/^{36}Ar$ and $^{132}Xe/^{36}Ar$ ratios a factor of two below those quoted above. These limits are, however, still greater than the theoretical values by factors of two ($^{84}Kr/^{36}Ar$) and four ($^{132}Xe/^{36}Ar$). Likewise, measured $^{132}Xe/^{84}Kr$ ratios are variable in the range 0.1–0.5, and are all greater than the theoretical solar values by at least a factor of two.

Large relative fractionations in Ar, Kr, and Xe exhibited by these samples is

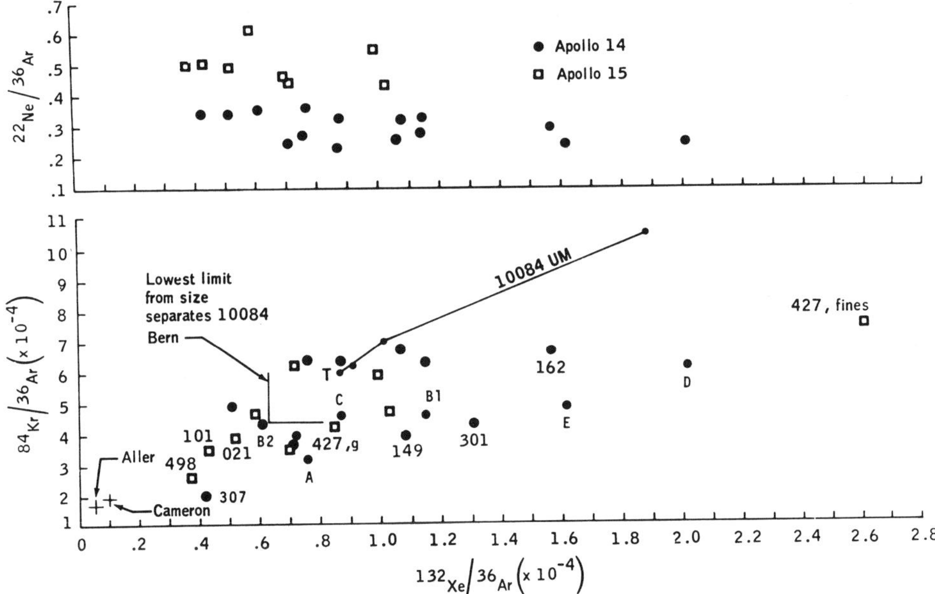


Fig. 1. Elemental abundance ratios for several Apollo 14 and 15 soils and gas rich breccia. Letters denote various grain size fractions of Table 1.

inconsistent with the much smaller variations in $^{22}Ne/^{36}Ar$ shown in the upper portion of Fig. 1. Note that most samples from a given landing site show $^{22}Ne/^{36}Ar$ ratios within a factor of two, with an obvious difference between Apollo 14 and 15, while $^{132}Xe/^{36}Ar$ varies by nearly a factor of ten. (One depth of core 14230 and fines 15427 gave much higher $^{22}Ne/^{36}Ar$ values of 0.68 and 2.45, respectively, and are not plotted.) Large fractionation for the heavier gases are also inconsistent with ilmenite results. Therefore, it appears that the relative noble gas abundances exhibited by many samples in Fig. 1 are inconsistent with simple, equilibrium fractionation processes. These samples showing the largest apparent fractionation in $^{132}Xe/^{36}Ar$ may represent mixtures of two trapped noble gas components, one representing severely fractionated solar gases for which Ne and Ar have been largely lost and the $^{132}Xe/^{84}Kr$ ratio increased to $\geq 0.5$. The second component represents much less fractionated trapped solar gas for which Ne and Ar are in equilibrium relative to one another and dominate the total concentrations in the sample. These two components may occur in separate soil particles, or alternatively may be produced in the same particles by severe outgassing followed by later gas loading.

Figure 2 presents the $^{4}He$, $^{22}Ne$, $^{36}Ar$, and $^{132}Xe$ abundances measured in size separates of soil 14163-14161, plotted against grain size (as separated by dry sieving). A straight line relationship of slope $-n$ results if $C \propto d^{-n}$ where $C \equiv$ gas concentration; $d \equiv$ grain diameter (Eberhardt *et al.*, 1970). Generally, Figure 2 shows well-defined linear relationships. The lower gas abundances at the smallest grain sizes have been noted on previous relationships of this type (Hintenberger *et al.*, 1971) and may

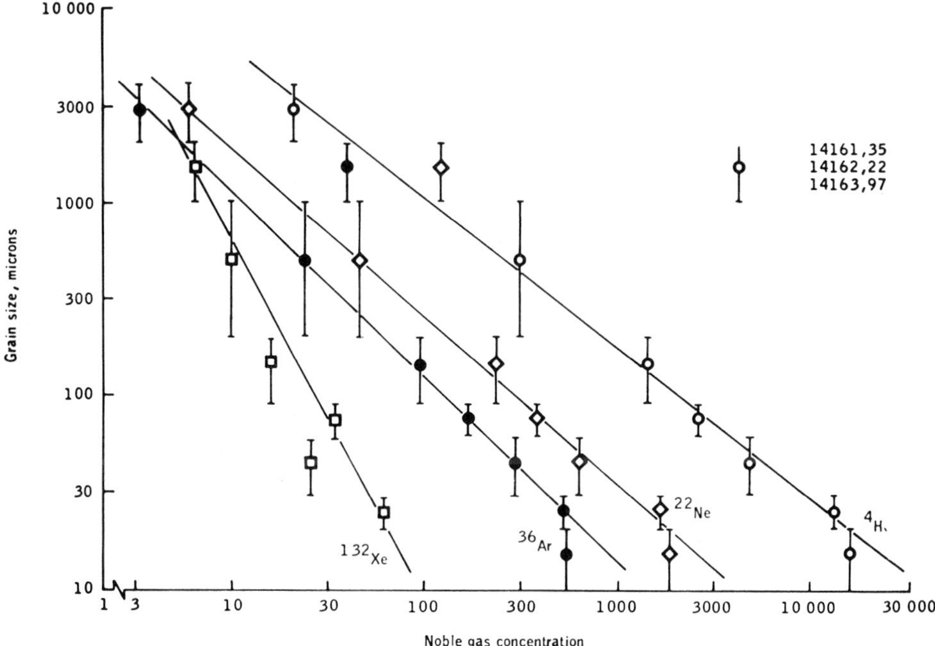


Fig. 2. Grain size effect on noble gas abundances for soil 14161 (2–4 mm), 14162 (1–2 mm), and 14163 (< 1 mm). Linear relationships indicate log gas concentration is proportional to log grain diameter. Grain sizes plotted are the median of sieve sizes.

be due to gas loss or different proportions of breccia particles. The 1–2 mm fraction possesses much higher He, Ne, and Ar contents than the simple grain size relationship allows, and is probably due to the presence of breccia particles with trapped interior gases. This size fraction also shows considerable relative enhancement of the lighter gases over the heavier (e.g., $^{4}$He/$^{22}$Ne, $^{22}$Ne/$^{38}$Ar) compared to the other size separates; however, these elemental ratios are still below those measured in ilmenite (Eberhardt *et al.*, 1970). While previous work has given nearly constant slopes for all the noble gases of about $-0.6$ for bulk samples and about $-1$ for ilmenite separates (Eberhardt *et al.*, 1970; Kirsten *et al.*, 1970; Hintenberger *et al.*, 1971; Kirsten *et al.*, 1971), the slopes in Figure 2 are variable and are: $^{4}$He $= -1.27$, $^{22}$Ne $= -1.13$, $^{36}$Ar $= -1.02$, and $^{132}$Xe $= -0.6$. A slope greater than $-1$ implies a higher gas concentration per unit surface area for the smaller grain sizes and may arise from longer irradiation times for smaller particles; a slope less than $-1$ probably indicates diffusion losses (Eberhardt *et al.*, 1965). If this explanation applies, the Xe grain size correlation is more suggestive of diffusion losses than the correlations for the lighter gases, implying a multistage irradiation of the type discussed in connection with Figure 1. We note that the grain size separates show a variation of a factor $2\frac{1}{2}$ in $^{132}$Xe/$^{36}$Ar and trend between the theoretical values and the 15427 point in Fig. 1. Alternatively, the pattern shown by Fig. 2 may arise from differences in the grain surface characteristics of the sized fractions resulting in a grain size dependency of the relative saturation concentrations for the various gases. In any case, we emphasize that elemental ratios can be grain size dependent.

## Isotopic Compositions of He, Ne, and Ar in the Bulk Samples, Grain Size Separates, and Stepwise Temperature Releases

### *Helium*

For most of these Apollo 14 and 15 fines and breccias which contain appreciable amounts of solar wind He, the measured $^4He/^3He$ ratio falls in the range 2000–2800 (Table 1; LSPET, 1971; LSPET, 1972). Many of the variations in $^4He/^3He$ among these samples can undoubtedly be explained by differences in radiogenic $^4He$ (about 10% of the $^4He$ in Apollo 14 fines and breccia is radiogenic, less in Apollo 15), and by differences in spallogenic $^3He$ (about 10% of the $^3He$ is probably spallogenic). However, as $^4He/^3He$ has been found to be quite variable in the solar wind (Geiss *et al.*, 1970; Bühler *et al.*, 1972) the possibility exists that this cause is also a major factor in the $^4He/^3He$ variations measured in these bulk samples. A plot (not shown) of $^3He/^4He$ vs. the inverse of the measured $^4He$ concentration for grain size separates of 14163 gives essentially a straight line for the five finer grain size fractions. From the ordinate intercept and the slope we obtain for 14163, $(^4He/^3He)$ trapped $= 2750 \pm 75$ and $(^3He)$ spallation $= 1.7 \times 10^{-6}$ cc/g. However, as spallation Xe is apparently grain size dependent in these separates (discussed later), spallation $^3He$ may be also, and thus the applicability of this technique is questionable.

### *Neon*

The isotopic composition of Ne for most bulk analyses is consistent with a two-component mixture of solar wind Ne and spallation Ne with a composition similar to that commonly found in meteorites. For the grain size separates of 14163, an ordinate-intercept plot of $^{22}Ne/^{21}Ne$ vs. the inverse of the $^{21}Ne$ concentration showed apparent scatter, due either to variations in abundances of the major target elements, Mg and Si, or to grain size differences in exposure age. However, the four finest grain size fractions define a straight line which gives $(^{22}Ne/^{21}Ne)$ trapped $= 31.8 \pm 0.5$ and $(^{21}Ne)$ spallation $= 5 \times 10^{-7}$ cc/g. This trapped $^{22}Ne/^{21}Ne$ is essentially identical with that obtained for other lunar fines (Eberhardt *et al.*, 1970; Kirsten *et al.*, 1970; Pepin *et al.*, 1970; Hintenberger *et al.*, 1971).

Several bulk analyses as well as the stepwise temperature release of 14163 reveal Ne components other than the simple trapped solar plus spallation mixture. Figure 3 is a neon isotope correlation diagram for these samples and omits other analyses which exhibit only two neon components and thus fall along the solar plus spallation mixing line. The 150–400°C temperature releases of 14163 show an enhancement in the $^{20}Ne/^{22}Ne$ and a structure in $^{21}Ne/^{22}Ne$ which has been seen in temperature releases on previous lunar samples (Pepin *et al.*, 1970; Hohenberg *et al.*, 1970; Kaiser, 1972). We interpret this as a lightly bound neon component arising from the lunar atmosphere and will discuss the phenomenon later in this paper. The 500–1000°C temperature releases and the finest grain size fractions cluster in a small region of Fig. 3,

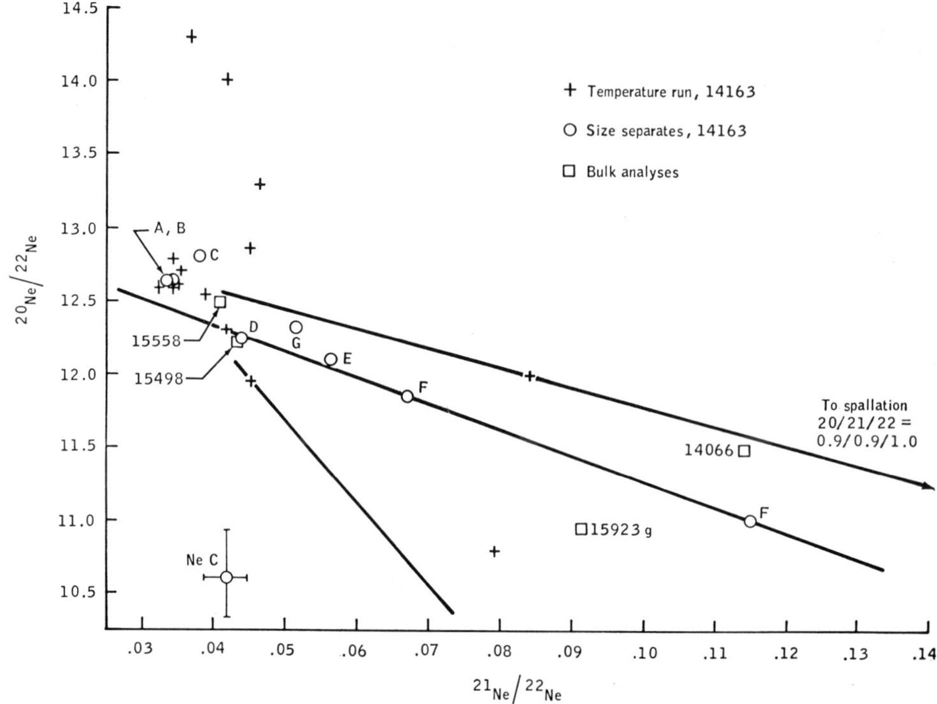


Fig. 3. Neon isotope correlation plot for grain size separates and stepwise temperature analyses of 14163,97 and bulk analyses of a few other fines and breccia. Letters denote various grain size fractions of Table 1.

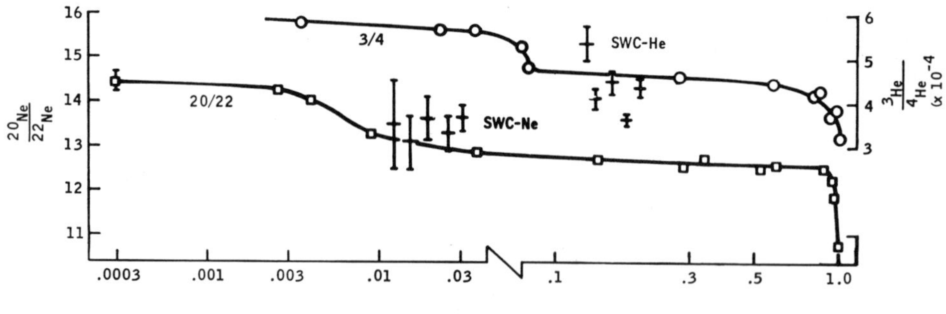


Fig. 4. Measured isotopic ratios of $^3He/^4He$ and $^{20}Ne/^{22}Ne$ as a function of fractional release of the heavier isotope for stepwise heating of fines 14163. Measured $^3He/^4He$ and $^{20}Ne/^{22}Ne$ ratios obtained from the Solar Wind Composition Experiment (SWC) from (plotted from left to right) Apollo 11, Apollo 12, Apollo 14, Surveyor, and Apollo 15 are shown (Eberhardt *et al.*, 1972; Bühler *et al.*, 1971).

and substantiate trapped solar $^{20}Ne/^{22}Ne = 12.65 \pm 0.10$ (also see Fig. 4). Bulk analyses of Apollo 14 and 15 samples not plotted on Fig. 3 show trapped Ne consistent with this $^{20}Ne/^{22}Ne$ ratio. Coarser grain size separates, the 1100–1400°C temperature releases, and bulk analyses on 15498 and the glass phase of 15923 all reveal the presence of either a trapped neon component with lower $^{20}Ne/^{22}Ne$ or a spallation neon with lower $^{21}Ne/^{22}Ne$. The line arbitrarily drawn through the grain size separates on Fig. 3 also passes very near the measured Ne isotopic compositions in two coarse-grained soils and a breccia from Apollo 12 (Nyquist and Pepin, unpublished data) and an amber glass soil fragment (Yaniv *et al.*, 1971). Neon from three high-temperature releases of Luna 16 soil (Kaiser, 1972a) would be essentially consistent with a line drawn through the trapped solar component and 15923. Thus, a variety of samples from at least four lunar sites display the existence of an additional neon component, which may also be present in minor amounts in other samples. The 1200°C and 1400°C neon data of 14163 establish an upper limit for any possible second spallation component as $^{21}Ne/^{22}Ne \leq 0.5$ (assuming spallation $^{20}Ne/^{22}Ne = 1$). However, there is no direct evidence for such a spallation Ne component in either meteorites or lunar samples. The second alternative, a trapped component with $^{20}Ne/^{22}Ne \leq 11.4$ (the limit set by our 1400°C point) is consistent with trapped Ne in several gas-rich meteorites. Black (1972) presents evidence for the existence in gas rich meteorites and lunar fines of Ne with $^{20}Ne/^{22}Ne = 10.6 \pm 0.3$ (Ne C in Figure 3), and infers this to be directly implanted solar flare Ne.

### *Argon*

An ordinate-intercept plot of $^{36}Ar/^{38}Ar$ vs. the inverse of the $^{38}Ar$ content for size separates of 14163 yields a trapped $^{36}Ar/^{38}Ar$ value of $5.36 \pm 0.03$, in agreement with previous determinations of this ratio (Eberhardt *et al.*, 1970; Kirsten *et al.*, 1970; Hintenberger *et al.*, 1971). Measured values of $^{40}Ar/^{36}Ar$ in various Apollo 14 and 15 fines and gas-rich breccia commonly range 0.7–3 (Table 1; LSPET, 1971; LSPET, 1972). Neither in situ decay of $^{40}K$ nor solar wind $^{40}Ar$ can account for $^{40}Ar/^{36}Ar$ ratios this large, and therefore much of the $^{40}Ar$ present in apparently orphan argon (Heymann and Yaniv, 1970; Manka and Michel, 1970; Manka and Michel, 1971; Baur *et al.*, 1972). Apollo 14 fines show larger $^{40}Ar/^{36}Ar$ ratios than Apollo 11 or 12 fines, but similar to those of Apollo 11 breccias. An $^{40}Ar/^{36}Ar$ ordinate-intercept plot (not shown) for the five finest grain size separates of 14163 form an essentially linear correlation whose ordinate intercept defines the trapped (surface correlated) $^{40}Ar/^{36}Ar = 2.16 \pm 0.08$ and whose slope defines the radiogenic (volume-correlated) $^{40}Ar$ (Eberhardt *et al.*, 1970; Heymann and Yaniv, 1970). Three different depths in core 14230 form a different linear relation with an intercept of $^{40}Ar/^{36}Ar = 1.9$ and a slope corresponding to the very low radiogenic $^{40}Ar$ value of $0.32 \times 10^{-4}$. The three trench samples 14149, 14148, and 14156 form a linear array which does not have a positive $^{40}Ar/^{36}Ar$ intercept, as is also the case for the fines samples 15301, 15101, and 15021. As measured K abundances are nearly identical for the trench samples (LSPET, 1971) and similar for the three Apollo 15 soils (LSPET, 1972), *differences in trapped* $^{40}Ar/^{36}Ar$ must produce this effect. Correcting for radiogenic $^{40}$Ar that would accumulate in these trench samples for $3.9 \times 10^9$ yr yields trapped $^{40}$Ar/$^{36}$Ar ratios of about 1.2 for 14148 and 14156 and 2.2 for 14149. Likewise, correcting 15021, 15101, and 15301 for radiogenic $^{40}$Ar that would accumulate in $3.5 \times 10^9$ yr yields trapped $^{40}$Ar/$^{36}$Ar ratios varying from 0.4 to 1.5. Thus, for both Apollo 14 and 15 samples there exist large variations in trapped $^{40}$Ar/$^{36}$Ar, even for samples collected in close proximity as were the trench fines. The explanation for these $^{40}$Ar/$^{36}$Ar variations apparently goes beyond lunar latitude effects on fluxes of lunar atmosphere $^{40}$Ar, and probably includes local shielding factors, chemical and mineralogical effects on trapping efficiencies, and possibly surface exposure histories.

The five grain size fractions of 14163 yield a radiogenic $^{40}$Ar content of $1.16 \times 10^{-4}$ cc/g, which with the measured K content of 0.48% (LSPET, 1971) gives the low K–Ar age of $2.7 \times 10^9$ yr. This age is essentially identical to the K–Ar age derived in the same manner for soil 14259 (Pepin *et al.*, 1972; Kirsten *et al.*, 1972) and is somewhat higher than the $2.0 \times 10^9$ yr event inferred from Pb–U data on these soils (Tatsumoto *et al.*, 1972).

*Stepwise temperature release of 14163.* Figure 4 shows $^3$He/$^4$He and $^{20}$Ne/$^{22}$Ne as a function of the fractional gas release during stepwise heating of the 30-60 micron fraction of fines 14163. Real variations also exist for $^{36}$Ar/$^{38}$Ar, $^{40}$Ar/$^{36}$Ar, and $^{83}$Kr/$^{86}$Kr (Tables 2 and 3). All of these ratios show sizeable changes above 80% gas release which is largely due to spallation-produced isotopes. The $^{40}$Ar/$^{36}$Ar ratio changes during gas release from a high of 30 to less than 2 above 70% argon release, presumably as a result of differential release of lunar atmosphere $^{40}$Ar and solar wind implanted $^{36}$Ar. Several ratios also show an enhancement of the lighter isotope over the heavier at very low fractional release. The maximum low-temperature enrichment of the lighter isotope compared to the ratios of 50% gas release are: $^3$He/$^4$He, 28%; $^{20}$Ne/$^{22}$Ne, 14%; $^{36}$Ar/$^{38}$Ar, 4%; $^{83}$Kr/$^{86}$Kr, 18% (also see Figs. 3 and 5). Similar low-temperature enrichments of $^3$He/$^4$He and $^{20}$Ne/$^{22}$Ne have been observed previously. In addition, stepwise heating and bulk analysis data from several laboratories show a correlation between high $^{20}$Ne/$^{22}$Ne and high $^{36}$Ar/$^{38}$Ar (this work; Pepin *et al.*, 1970; Hintenberger *et al.*, 1971; Yaniv *et al.*, 1971). These isotope enrichments have been variously attributed to mass fractionation during the heating experiment, a mass fractionation during solar wind implantation or subsequent gas loss, and a low-energy, directly implanted lunar atmosphere component (Hohenberg *et al.*, 1970; Pepin *et al.*, 1970; Black, 1972; Kaiser, 1972).

Simple models of laboratory-induced fractionation cannot explain the He and Ne release patterns of Fig. 4. For a gas initially uniformly distributed throughout a volume $V$ of surface area $S$, the fractional gas release $F$ as a function of time $t$ is given by

$$F = (2/\pi)(S/V)\sqrt{Dt}, \qquad (F \leq 0.25)$$

where $D$ is the diffusion coefficient (Lagerwall and Zimen, 1964). This equation approximates the solutions of Fick's law for small gas release. Although the exact solutions of Fick's law depend on the geometry of the reservoir, they are in general smooth functions. Thus it is difficult to see how the "break-over" at 7% of the $^4$He

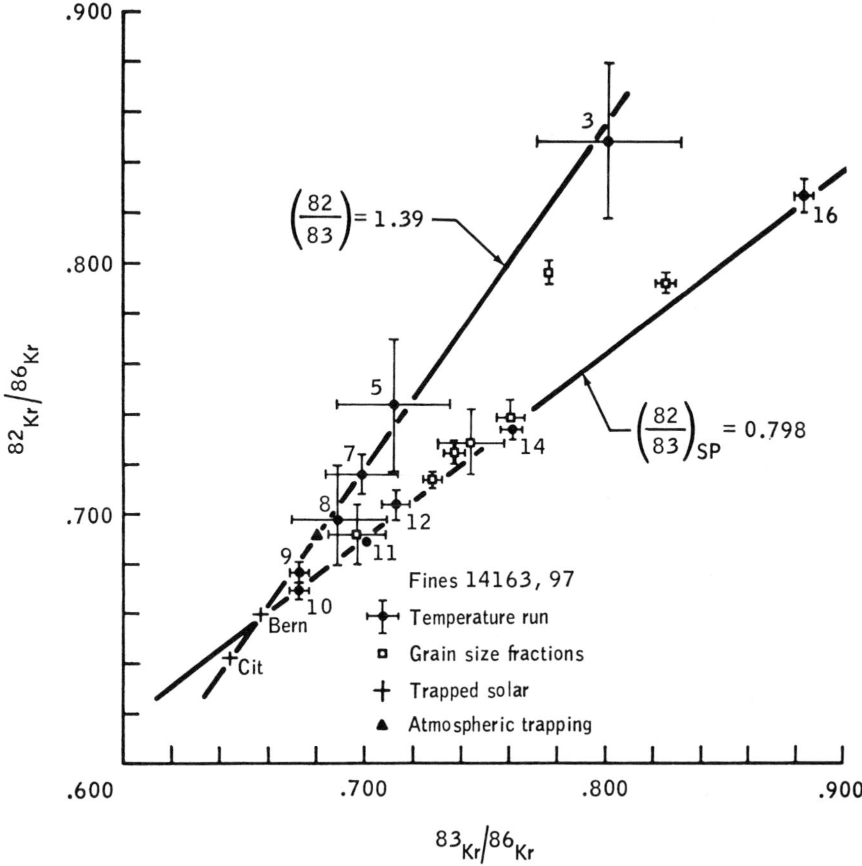


Fig. 5. $^{82}Kr/^{86}Kr$–$^{83}Kr/^{86}Kr$ isotope correlation plot for grain size separates and stepwise temperature release of 14163. Trapped solar compositions shown are from Eberhardt *et al.*, (1970) and Podosek *et al.* (1971). Letters denote various grain size fractions of Table 1. In addition to trapped solar Kr, a high temperature spallation component ($^{82}Kr/^{83}Kr = 0.80$) and a distinct low temperature component ($^{82}Kr/^{83}Kr = 1.39$) are apparent.

release and 0.6% of $^{22}Ne$ can occur. Assumption of gas release from multiple reservoirs would not materially affect this conclusion. One would expect the fractionation to persist at least until 25% of the "larger" reservoirs had been depleted. A possible mechanism might be fractionation during early gas release from radiation-damaged areas (P. Signer, pers. comm.); the break-over would then correspond to annealing at 300–400°C, a temperature at which gas release rates from various lunar mineral separates decrease (Baur *et al.*, 1972). It is nevertheless difficult to account for the magnitude of the sudden decrease in fractionation. The strongest reasonable mass dependence of the diffusion mechanism is $D \propto m^{-\frac{1}{2}}$, which produces fractionation proportional to $(m_2/m_1)^{1/4}$. Larger fractionations could be attributed to variations in the effective volume of diffusion for the two isotopes (Hohenberg *et al.*, 1970). While changes in the mass dependence of $D$ during annealing seem possible, changes in the relative concentration profiles of the isotopes do not. Thus we prefer the release of a low-energy trapped atmosphere component as the simplest explanation for the He and Ne release patterns of Fig. 4. As will be discussed later however, the low-temperature Kr component may have another origin.

*Krypton*

Isotope correlation plots of krypton measured in grain size separates and stepwise temperature extractions of 14163 are presented in Figs. 5 and 6. A two-component mixture of solar wind Kr and spallation-produced Kr will define a straight line, and deviations from this line will indicate other Kr components. Figure 5 clearly indicates that data from grain size separates and extraction temperatures above 900°C define a single spallation Kr with $^{82}Kr/^{83}Kr = 0.80$ (assuming $^{86}Kr$ spallation = 0), and the 300–900°C data define a distinct component with $^{82}Kr/^{83}Kr = 1.39$. Although less pronounced, there also appear to be a high-temperature $^{84}Kr/^{83}Kr$ component and a low-temperature $^{84}Kr/^{83}Kr$ component (Fig. 6).* A correlation plot of $^{80}Kr/^{86}Kr$ vs. $^{83}Kr/^{86}Kr$ (not shown) exhibits only one spallation $^{80}Kr/^{83}Kr$ component for both grain sizes and temperature releases and with a value of 0.49. Kr correlation plots for a number of bulk analyses of Apollo 14 and 15 fines and gas-rich breccia (Tables 3 and 4) tend to fall along two component straight lines with spallation $^{80}Kr/^{83}Kr = 0.51$ and $^{82}Kr/^{83}Kr = 0.79$. These ratios are identified as the galactic cosmic ray spallation Kr component in these samples and are similar to values found for Apollo 11 and 12 fines (Pepin *et al.*, 1970; Nyquist and Pepin, 1971; Podosek *et al.*, 1971). The similarity of the isotopic composition of this component in a number of samples indicates that no major shielding differences existed for many lunar soils and breccia.

Low-temperature Kr data of fines 10084 and 12028 (Pepin *et al.*, 1970; Nyquist and Pepin, 1971) also showed a third component for $^{82}Kr$, $^{83}Kr$, and $^{84}Kr$, but not $^{80}Kr$, with $^{82}Kr/^{83}Kr = 1.3$ and $^{84}Kr/^{83}Kr = 3.4$ (Fig. 6). The data presented here on 14163 have a factor of three greater measured enrichment of $^{82}Kr$ and $^{83}Kr$ than these previous measurements, and substantiate the occurrence of a lightly bound Kr component distinct from the typical cosmic ray spallation Kr. Podosek *et al.* (1971) have argued that the low-temperature Kr observed by the Minnesota laboratory is probably not due to neutron capture on bromine or to solar flare reactions, mainly because of the absence of any low-temperature $^{80}Kr$. A possible origin not considered by these workers is a lightly bound Kr component arising from low-energy implantation of lunar atmosphere Kr (Manka and Michel, 1971). The triangular point in Figs.

* The first analyses of the 20–30 $\mu$ fraction (B1) also shows the presence of this low-temperature component, as well as anomalous Xe isotopic ratios shown later. A second analysis of this grain size fraction (B2) performed much later possessed lower Kr and Xe contents by a factor of two and showed no evidence of anomalous Kr or Xe compared to other grain size fractions. As we have no reason to suspect the B1 analysis, we have included it in the figures.

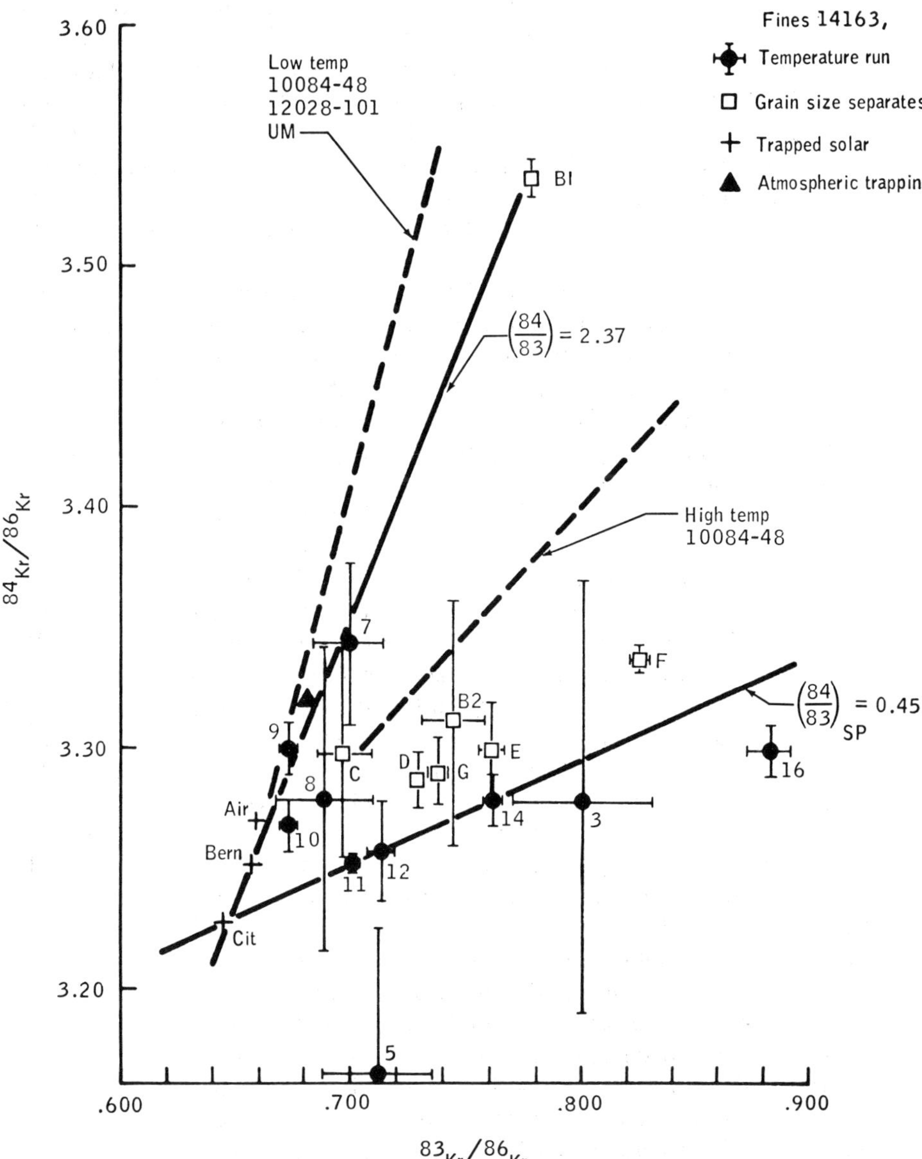


Fig. 6. $^{84}Kr/^{86}Kr$–$^{83}Kr/^{86}Kr$ isotope correlation plot for grain size separates and stepwise temperature release of 14163. Dotted lines are extrapolations of trends found by Pepin *et al.* (1970) and Nyquist and Pepin (1971).

5 and 6 represents the predicted simple mass-fractionated composition of this lunar atmosphere Kr, and in both figures falls on the lines defined by the low-temperature data. Lunar atmosphere $^{80}$Kr would not be apparent in the correlation plot mentioned earlier because mass fractionated lunar atmosphere $^{80}$Kr/$^{83}$Kr falls so close to the trapped solar plus spallation mixing line that it is not distinguishable from it. However, the Kr data for 14163 (Fig. 5) shows much greater $^{82}$Kr/$^{86}$Kr and $^{83}$Kr/$^{86}$Kr low temperature enrichments than predicted by the Manka and Michel model, while the $^{82}$Kr/$^{83}$Kr is the same. Thus the lunar atmosphere does not seem a probable source for this low-temperature Kr, as not even a multistage mass fractionation model would fractionate $^{82}$Kr/$^{86}$Kr and $^{83}$Kr/$^{86}$Kr to such a large extent compared to the $^{82}$Kr/$^{83}$Kr and $^{80}$Kr/$^{83}$Kr.

We believe the low-temperature Kr data could result from variations in the spallation spectra among different target elements. Spallation Kr ought to be produced in relatively large amounts from Rb by lower energy nuclear reactions. Only three common reactions are possible: $(p, \alpha)$ reactions on $^{85}$Rb and $^{87}$Rb to yield $^{82}$Kr and $^{84}$Kr and $(p, pn)$ reactions on $^{85}$Rb to yield $^{84}$Kr; reactions on Rb leading to $^{80}$Kr, $^{83}$Kr, and $^{86}$Kr are much less probable. After allowing for $^{84}$Kr and $^{82}$Kr produced by spallation reactions on Sr + Y + Zr, the remaining excess Kr in the low temperature fractions possesses a ratio $^{84}$Kr/$^{82}$Kr = 3. If $(p, \alpha)$ reactions leading to $^{82}$Kr and $^{84}$Kr possess roughly equal cross sections, then the $(p, pn)$ reaction on $^{85}$Rb would have to be roughly three times as probable as the $(p, \alpha)$ reaction on $^{87}$Rb. This cross-section ratio is quite reasonable in view of the uncertainty of the proton energies involved. To produce spallation $^{83}$Kr at low extraction temperatures, the $^{82}$Kr and $^{84}$Kr arising from Rb would have to be accompanied by spallation Kr from Sr. Thus, the low-temperature Kr may arise from the high Rb and K phases known to occur in Apollo 14 samples and where Rb/Sr can be increased by an order of magnitude (Papanastassiou and Wasserburg, 1971). Argon-39 dating has indicated that these phases lose argon at lower temperatures than other phases (Turner *et al.*, 1971). The excess $^{82}$Kr and $^{84}$Kr may also be produced from Rb on grain surfaces by solar flares of insufficient energy to produce normal amounts of spallation Kr from Sr. In this case the low-temperature Kr component would be characteristic of lunar soils but not of crystalline rocks.

Figures 5 and 6 also allow some distinction to be made as to the isotopic composition of trapped solar Kr in 14163 and the solar compositions reported by Eberhardt *et al.* (1970) and Podosek *et al.* (1971). The $^{82}$Kr/$^{83}$Kr mixing line for high temperature and grain size data and the $^{84}$Kr/$^{83}$Kr mixing line for grain size data tend to extrapolate through the Bern composition, while the $^{84}$Kr/$^{83}$Kr high temperature data extrapolate through the CIT composition. The low-temperature data extrapolates through both the Bern and CIT compositions. This suggests that Kr from bulk analyses may not extrapolate to the same trapped composition as temperature data, and that the Bern trapped component may be related to the CIT trapped Kr by the addition of a small amount of the low-temperature, lightly bound Kr. Isotope correlation plots for Xe discussed next also show a difference in trapped composition between grain size data and temperature data.

*Xenon*

Xenon isotope correlation plots for the 14163 grain size separates and stepwise temperature release are presented in Figs. 7 and 8. As no $^{124}Xe$ and $^{126}Xe$ data was obtained below 700°C, low-temperature ratios are not plotted for other isotopes. For temperatures above 700°C the data are consistent with a mixture of solar Xe plus spallation Xe. Correlation plots relative to $^{130}Xe/^{136}Xe$ (not shown) reveal a low-temperature third component, which is displaced in the direction of atmospheric Xe (also note the 700°C point of Figs. 7 and 8), and it may represent small amounts of adsorbed atmospheric Xe on our samples. As the presence of a fission Xe component would produce displacements in the same general direction on the Xe correlation diagrams, this may also be a source of the low-temperature Xe. Fission Xe is undoubtedly the explanation for similar displacements for Xe on bulk sample analyses (Table 4) which show 14301, 14307, and 14162 (G on Figs. 7, 8) to contain appreciable amounts of fission Xe (LSPET, 1971). As Crozaz *et al.* (1972) have recently attributed this fission Xe in 14301 partially to spontaneous fission of the extinct isotope $^{244}Pu$, the possibility exists of fission Xe of this type existing in these other samples and in the low-temperature release of 14163.

From Figs. 7 and 8 it can be seen that the 900–1400°C Xe data (containing 94% of the total Xe released) define a solar plus spallation mixing line which is slightly different in slope from the solar plus spallation mixing line defined by the grain size separates. (Analyses for B1 and G were not considered for reasons already given.) The fact that each set of data define a straight line indicates a single spallation component for the grain size separates lying on the line and for the temperature data. If we assume the CIT-trapped component, the derived spallation Xe for the 1400°C point and grain size fraction *F* are given in Table 5. For those isotopes showing large spallation enrichments the two spallation spectra are essentially the same. Differences in relative spallation yields for $^{129}Xe$, $^{131}Xe$, and $^{132}Xe$ are probably due to our choice of a single trapped solar Xe for these two samples. For the $^{131}Xe/^{136}Xe$, $^{130}Xe/^{136}Xe$, and $^{128}Xe/^{136}Xe$ correlation plots the 900–1400°C data extrapolate through the CIT-trapped solar composition, and the grain size separates extrapolate through the Bern trapped solar composition. This difference in slopes cannot be due to mass fractionation during the stepwise temperature release, for the high-temperature data (1400°C) show enrichment of the lighter Xe isotopes over the heavier, while the 900 and 1000°C data show depletion of the lighter Xe isotopes over the heavier. This is the opposite of the effect to be expected from mass fractionation during stepwise heating. Thus it appears that the grain size separates contain a sufficient

Table 5. Spallation Xe spectra for 14163.

| | 124 | 126 | 128 | 129 | 130 | 131 | 132 |
|---|---|---|---|---|---|---|---|
| 1400°C | 0.550 | 1.00 | 1.53 | 1.30 | 1.01 | 4.69 | 0.41 |
| ± | 0.008 | | 0.03 | 0.14 | 0.04 | 0.14 | 0.10 |
| 250–1000 $\mu$ | 0.553 | | 1.63 | 1.87 | 1.05 | 5.37 | 0.85 |
| ± | 0.015 | 1.00 | 0.03 | 0.12 | 0.04 | 0.20 | 0.10 |

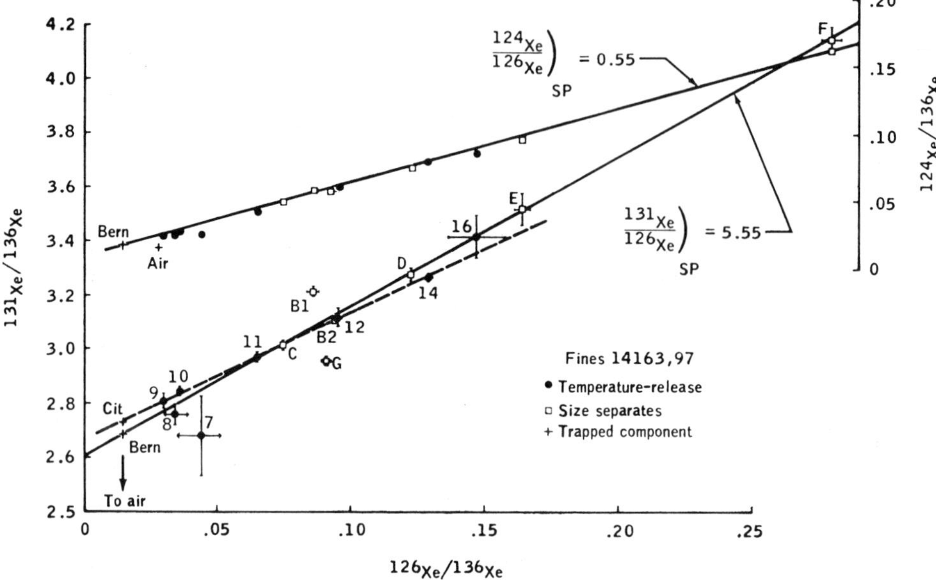


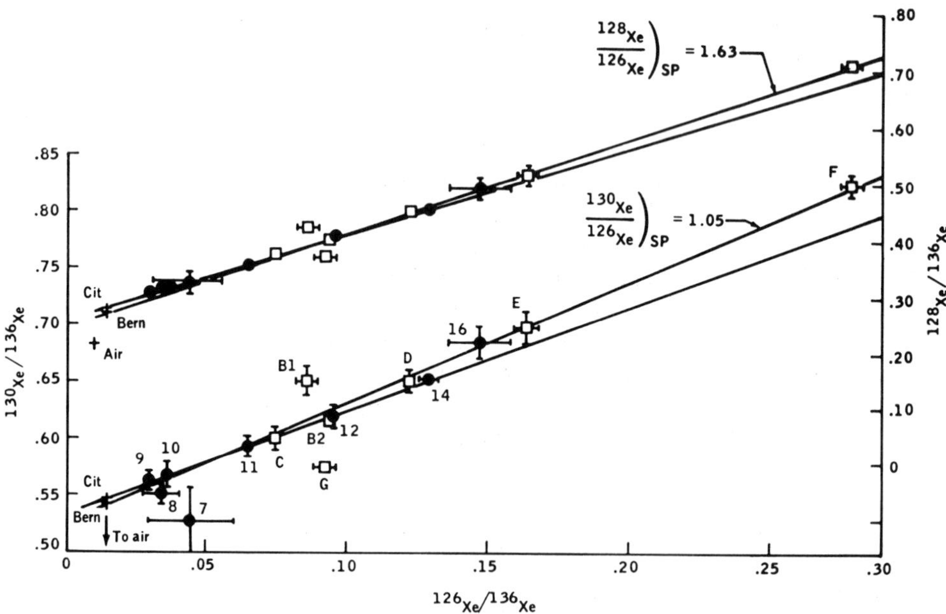


Fig. 7 and 8. Isotope correlation plots for $^{124}Xe/^{136}Xe$, $^{131}Xe/^{136}Xe$, $^{128}Xe/^{136}Xe$, and $^{130}Xe/^{136}Xe$ against $^{126}Xe/^{136}Xe$ for grain size separates and stepwise temperature release of fines 14163. Letters denote various grain size fractions of Table 1.

amount of Xe similar to the Xe released at low extraction temperatures to alter slightly the trapped Xe component. The trapped Xe composition given by CIT would then represent the more tightly bound solar wind component, while the Xe composition given by Bern would contain an additional small amount of the low temperature Xe component. This situation is analogous to the Kr correlation plots where the Bern trapped Kr composition appeared to be a mixture of the CIT high-temperature component and Kr released at low extraction temperatures.

Figures 5–8 demonstrate that grain size separates and stepwise heating of a lunar sample give nearly the same results for the isotopic composition of the spallation component. There is no evidence for a mass fractionation during the stepwise heating altering the spallation spectrum obtained. However, it is apparent that Kr and Xe in lunar fines are not simple mixtures of components, but that the trapped solar and spallation Kr and Xe components may vary in composition (also see Burnett *et al.*, 1971). The techniques used to obtain both the CIT and the Bern trapped Kr and Xe compositions (Podosek *et al.*, 1971; Eberhardt *et al.*, 1970) assume that Kr and Xe can be described as a two-component system for which the composition of the two components and the amounts of the spallation component are the same for all samples. The techniques for deriving the composition of the trapped component are thus sensitive to the types of isotope correlation variations seen for fines 14163.

Figure 9 presents the same Xe correlation diagram as Fig. 7 for a large number of

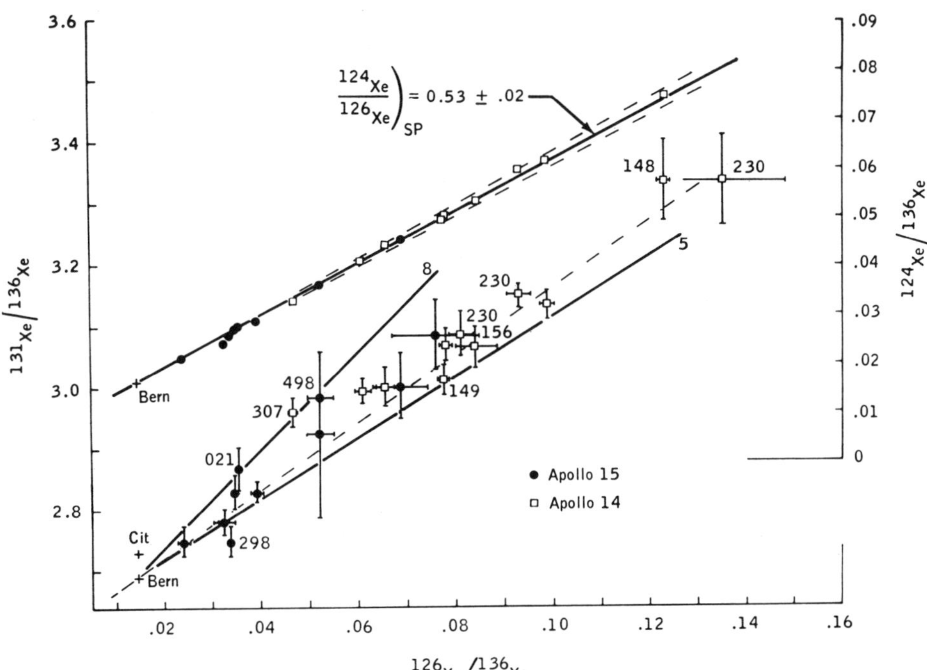


Fig. 9. Xenon isotope correlation plot for a number of Apollo 14 and 15 fines and gas rich breccia. The two solid lines represent spallation $^{131}Xe/^{126}Xe$ ratios of 5 and 8; the dashed line represents spallation $^{131}Xe/^{126}Xe$ determined for 14163 (Fig. 8).

Apollo 14 and 15 fines and gas-rich breccia. The spallation $^{124}Xe/^{126}Xe$ is essentially constant at 0.53 ± 0.02. This value contrasts with the range of $^{124}Xe/^{126}Xe$ spallation ratios found in meteorites of 0.52–0.65 (see Bogard *et al.*, 1971b, for references), and indicates that these particular lunar samples have not experienced major energy differences in the higher-energy cosmic-ray component that produces the light spallation isotopes of Xe. However, the spallation $^{131}Xe/^{126}Xe$ vary by a factor of two and possess values considerably higher than those commonly found for spallation Xe in meteorites. These high values for spallation $^{131}Xe/^{126}Xe$ have also been found to be typical of lunar crystalline rocks, and apparently are not due to chemical differences (Hohenberg *et al.*, 1970; Marti *et al.*, 1970; Pepin *et al.*, 1970; Bogard *et al.*, 1971a; Marti and Lugmair, 1971; Alexander, 1971; Huneke *et al.*, 1972). Recent target irradiation experiments (Eberhardt *et al.*, 1971; Kaiser, 1972b) have shown that this excess $^{131}Xe$ above what is expected from cosmic ray spallation is probably due to epithermal neutron capture on $^{130}Ba$. Thus, the spallation excess $^{126}Xe$ allows a quantitative determination of the integrated cosmic ray exposure, while the excess $^{131}Xe$ allows a qualitative determination of the integrated energy of that exposure (shielding). Samples which were physically associated on the lunar surface (e.g., core samples), but which do not exhibit the same spallation $^{131}Xe/^{126}Xe$ ratio on Figure 9 would have had different average burial depths in the regolith. Thus fines 15021 and breccia 15498 and 14307 possess excess $^{131}Xe/^{126}Xe = 8$, and should exhibit considerable evidence of low-energy neutron fluxes. By the same criterion, breccia 15298 must have existed on the surface for most of its exposure. The excess $^{131}Xe/^{126}Xe$ also allows examination of the various samples collected at depth. Three different depths of core 14230 and the three different trench samples show small apparent variations in this ratio. Trench sample 14148 and core 14230,113 both appear to have higher excess $^{131}Xe/^{126}Xe$, implying greater average shielding.

Calculated concentrations of spallation $^{126}Xe$ in many of these fines and gas-rich breccia are given in Table 6. Spallation gases in a number of gas-poor Apollo 14 breccia have been discussed elsewhere (Bogard and Nyquist, 1972; LSPET, 1971), and indicate that a number of these rocks were excavated by Cone Crater approximately $20 \times 10^6$ yr ago. The difference in $^{126}Xe$ concentrations between Apollo 14 and 15 fines and gas-rich breccia is largely due to differences in abundance of the main target element Ba. However, the differences in $^{126}Xe$ (>factor of three) which exist among the Apollo 14 samples apparently are not due to target element variations, but rather to real differences in cosmic ray exposure times. Ba concentrations in most Apollo 14 fines and gas-rich breccias are constant within ±20% (LSPET, 1971; Hubbard *et al.*, 1972, and personal comm.; Schnetzler and Nova, 1972; Wakita *et al.*, 1972). Utilizing $^{126}Xe$ production rates previously reported (Bogard *et al.*, 1971a; Huneke *et al.*, 1972) gives typical exposure times of Apollo 14 soils ranging from $200–500 \times 10^6$ yr. Very little Ba and rare earth data on Apollo 15 fines and breccia have been reported, and the factor of three variation in spallation $^{126}Xe$ for these samples may be due to either differences in target elements or integrated exposure times. Utilizing measured Ba contents of soils 15301, 15021, and 15101 (Hubbard, personal comm.) gives exposure ages of 190, 200, and $200 \times 10^6$ yr, respectively. The Apollo 14 bulk soil 14161-14163 shows an inverse correlation of spallation $^{126}Xe$

Table 6. Abundances of spallation $^{126}Xe$ in Apollo 14 and 15 fines and gas rich breccia ($10^{-10}$ cc/g).

| Sample | $^{126}Xe$ | Sample | $^{126}Xe$ |
|---|---|---|---|
| 14259,10 | 12.5 ± 1.7 | 14163 | |
| 14003,6 | 2.06 ± 0.29 | 1000–250 μ | 2.49 ± 0.29 |
| 14047,2 | 4.31 ± 0.52 | 250–90 μ | 4.07 ± 0.49 |
| 14049,3 | 5.31 ± 0.62 | 90–60 μ | 3.52 ± 0.38 |
| 14301,48* | 0.022 ± 0.007 | 60–30 μ | 5.56 ± 0.62 |
| 14307,26 (dark) | 1.25 ± 0.15 | 30–20 μ | 0.47 ± 0.06 |
| 14307,26 (light) | 1.83 ± 0.21 | 30–20 (repeat) | 0.45 ± 0.07 |
| 14230,113 | 1.57 ± 0.12 | 15301 | 0.47 ± 0.11 |
| 14230,130 | 0.56 ± 0.12 | 15021 | 0.73 ± 0.10 |
| 14230,121 | 1.93 ± 0.27 | 15101 | 0.53 ± 0.08 |
| 14148 (T) | 6.54 ± 0.84 | 15601 | 0.30 ± 0.07 |
| 14156 (M) | 4.71 ± 0.74 | 15298 | 0.85 ± 0.13 |
| 14149 (B) | 4.89 ± 0.58 | 15558 | 1.00 ± 0.17 |
| 14161 (2–4 mm) | 0.34 ± 0.07 | 15498 | 0.37 ± 0.07 |
| 14162 (1–2 mm) | 1.51 ± 0.19 | 15265 | 0.81 ± 0.17 |
| | | 15923 (matrix)* | 0.10 ± 0.06 |
| | | 15923 (glass) | 0.065 ± 0.016 |

Spallation abundances calculated from the data in Tables 1 and 4 and in LSPET (1971) and (1972) assuming the trapped solar composition given by Eberhardt *et al.* (1970). Uncertainties listed are 10% plus the fractional value given by the one sigma uncertainty in measured $^{126}Xe/^{136}Xe$ divided by the difference in this ratio between measured and trapped values, and thus seem conservatively estimated.
* Calculated on basis of (128/126) spallation = 1.55.

with grain size. Even after considering target element differences (Hubbard *et al.*, 1972, and personal comm.), at least a factor of five difference in exposure time remains between the largest grain size and the finer fractions. The reason for the abrupt drop in $^{126}Xe$ for the 30–20 μ fraction is not known. As the Ba content for this fraction was not measured, this decrease may reflect a gross chemical change for the smallest grain sizes. Alternatively, if the Ba content of this smallest grain size is the same as the bulk soil, the exposure time of this component would be $\sim 40 \times 10^6$ yr. The large differences in spallation gases shown by various soil and breccia samples and by grain sizes of the same soil are important considerations in the derivation of isotopic composition of the various trapped noble gas components.

*Acknowledgments*—We acknowledge the competent technical support of the following persons of the Brown and Root-Northrop support staff: W. Hart, W. Hirsch, C. Polo, and R. Wilkin.

## References

Alexander E. C. Jr. (1971) Spallogenic Ne, Kr, and Xe from a depth study of 12022. *Proc. Second Lunar Sci. Conf., Geochim. Cosmochim. Acta* Suppl. 2, Vol. 0, pp. 1643–1650. MIT Press.

Aller L. H. (1961) *The Abundance of the Elements.* Interscience.

Baur H., Frick U., Funk H., Schultz L., and Signer P. (1972) On the question of retrapped 40-Ar in lunar fines (abstract). In *Lunar Science—III* (editor C. Watkins), pp. 47–49, Lunar Science Institute Contr. No. 88.

Black D. C. (1972) On the origins of trapped He, Ne, and Ar isotopic variations in meteorites—I. Gas rich meteorites, lunar soil, and breccia. *Geochim. Cosmochim. Acta* **36,** 347–376.

Bogard D. D., Funkhouser J. G., Schaeffer O. A., and Zahringer J. (1971a) Noble gas abundances in lunar material—cosmic ray spallation products and radiation ages from the Sea of Tranquility and the Ocean of Storms. *J. Geophys. Res.* **76,** 2757–2779.

Bogard D. D., Huneke J. C., Burnett D. S., and Wasserburg G. J. (1971b) Xe and Kr analyses of silicate inclusions from iron meteorites. *Geochim. Cosmochim. Acta* **35,** 1231–1254.

Bogard D. D. and Nyquist L. E. (1972) Noble gas studies on regolith materials from Apollo 14 and 15 (abstract). In *Lunar Science—III* (editor C. Watkins), pp. 89–91, Lunar Science Institute Contr. No. 88.

Bühler F., Eberhardt P., Geiss J., and Schwarzmüller J. (1971) Trapped solar wind He and Ne in Surveyor 3 material. *Earth Planet. Sci. Lett.* **10,** 297–306.

Bühler F., Cerutti H., Eberhardt P., and Geiss J. (1972) Results of the Apollo 14 and 15 solar wind composition experiments (abstract). In *Lunar Science—III* (editor C. Watkins), pp. 102–104, Lunar Science Institute Contr. No. 88.

Burnett D. S., Huneke J. C., Podosek F. A., Russ G. P. III, and Wasserburg G. J. (1972) The irradiation history of lunar samples (abstract). In *Lunar Science—III* (editor C. Watkins), pp. 105–107, Lunar Science Institute Contr. No. 88.

Cameron A. G. W. (1968) A new table of abundances of the elements in the solar system. In *Origin and Distribution of the Elements*, pp. 125–143, Pergamon.

Crozaz G., Drozd H., Graf H., Hohenberg C. M., Mannin M., Rogan D., Ralston C., Seitz M., Shirck J., Walker R. M., and Zimmerman J. (1972) Evidence for extinct $Pu^{244}$: Implications for the age of the pre-Imbrium crust (abstract). In *Lunar Science—III* (editor C. Watkins), pp. 164–166, Lunar Science Institute Contr. No. 88.

Eberhardt P., Geiss J., and Grögler N. (1965) Über die Verteilung der Uredelgase in Meteoriten Khor Temiki. *Tschermaks Mineral. Petrogr. Mitt.* **10,** 535–551.

Eberhardt P., Geiss J., Graf H., Grögler N., Krähenbühl U., Schwaller H., Schwarzmüller J., and Stettler A. (1970) Trapped solar wind noble gases, exposure age and K/Ar age in Apollo 11 lunar fine material. *Proc. Apollo 11 Lunar Sci. Conf., Geochim. Cosmochim. Acta* Suppl. 1, Vol. 1, pp. 1037–1070. Pergamon.

Eberhardt P., Geiss J., and Graf H. (1971) On the origin of excess $^{131}Xe$ in lunar rocks. *Earth Planet. Sci. Lett.* **12,** 260–262.

Geiss J., Eberhardt P., Bühler F., Meister J., and Signer P. (1970) Apollo 11 and 12 solar wind composition experiments: Fluxes of He and Ne isotopes. *J. Geophys. Res.* **75,** 5972–5979.

Heymann D. and Yaniv A. (1970) $Ar^{40}$ anomaly in lunar samples from Apollo 11. *Proc. Apollo 11 Lunar Sci. Conf., Geochim. Cosmochim. Acta* Suppl. 1, Vol. 2, pp. 1261–1268. Pergamon.

Hintenberger H., Weber H. W., and Takaoka N. (1971) Concentrations and isotopic abundances of the rare gases in lunar matter. *Proc. Second Lunar Sci. Conf., Geochim. Cosmochim. Acta* Suppl. 2, Vol. 2, pp. 1607–1626. MIT Press.

Hohenberg C. M., Davis P. K., Kaiser W. A., Lewis R. S., and Reynolds J. H. (1970) Trapped and cosmogenic rare gases from stepwise heating of Apollo 11 samples. *Proc. Apollo 11 Lunar Sci. Conf., Geochim. Cosmochim. Acta* Suppl. 1, Vol. 2, pp. 1283–1310. Pergamon.

Hubbard N. J., Gast P. W., Rhodes M., and Wiesmann H. (1972) Chemical composition of Apollo 14 materials and evidence for alkali volatilization (abstract). In *Lunar Science—III* (editor C. Watkins), pp. 407–409, Lunar Science Institute Contr. No. 88.

Huneke J. C., Podosek F. A., Burnett D. S., and Wasserburg G. J. (1972) Rare gas studies of the galactic cosmic ray irradiation history of lunar rocks. *Geochim. Cosmochim. Acta* **36,** 269–302.

Kaiser W. A. (1972a) Rare gas studies in Luna 16 G-7 fines by stepwise heating technique. *Earth Planet. Sci. Lett.* **13,** 387–399.

Kaiser W. A. (1972b) The average $^{130}Ba$ $(n, \gamma)$ cross section and the origin of $^{131}Xe$ on the moon (abstract). In *Lunar Science—III* (editor C. Watkins), p. 44, Lunar Science Institute Contr. No. 88.

Kirsten T., Müller O., Steinbrum F., and Zähringer J. (1970) Study of distribution and variations of rare gases in lunar material by a microprobe technique. *Proc. Apollo 11 Lunar Sci. Conf., Geochim. Cosmochim. Acta* Suppl. 1, Vol. 2, pp. 1331–1344. Pergamon.

Kirsten T., Steinbrum F., and Zähringer J. (1971) Location and variation of trapped rare gases in Apollo 12 lunar samples. *Proc. Second Lunar Sci. Conf., Geochim. Cosmochim. Acta* Suppl. 2, Vol. 2, pp. 1651–1670. MIT Press.

Kirsten T., Deubner J., Ducati H., Gentner W., Horn P., Jessberger E., Kalbitzer S., Kaneoka I., Kiko J., Krätschmer W., Müller H., Plieninger T., and Thio S. (1972) Rare gases and ion tracks in individual components and bulk samples of Apollo 14 and 15 fines and fragmental rocks (abstract). In *Lunar Science—III* (editor C. Watkins), pp. 452–454, Lunar Science Institute Contr. No. 88.

Lagerwall T. and Zimen K. E. (1964) The kinetics of rare gas diffusion in solids. EURACE Report No. 772, Hahn-Meitner Institut für Kernforschung, Berlin.

LSPET (Lunar Sample Preliminary Examination Team) (1971) Preliminary examination of lunar samples from Apollo 14. *Science* **173,** 681–693.

LSPET (Lunar Sample Preliminary Examination Team) (1972) Preliminary examination of lunar samples from Apollo 15. *Science* **175,** 363–374.

Manka R. H. and Michel F. C. (1970) Lunar atmosphere as a source of argon-40 and other lunar surface elements. *Science* **169,** 278–280.

Manka R. H. and Michel F. C. (1971) Lunar atmosphere as a source of lunar surface elements. *Proc. Second Lunar Sci. Conf., Geochim. Cosmochim. Acta* Suppl. 2, Vol. 2, pp. 278–280. MIT Press.

Marti K., Lugmair G. W., and Urey H. C. (1970) Solar wind gases, cosmic-ray spallation products, and the irradiation history of Apollo 11 samples. *Proc. Apollo 11 Lunar Sci. Conf., Geochim. Cosmochim. Acta* Suppl. 1, Vol. 2, pp. 1357–1368. Pergamon.

Marti K. and Lugmair G. W. (1971) $Kr^{81}$–Kr and K–$Ar^{40}$ ages, cosmic ray spallation products, and neutron effects in lunar samples from Oceanus Procellarum. *Proc. Second Lunar Sci. Conf., Geochim. Cosmochim. Acta* Suppl. 2, Vol. 2, pp. 1591–1606. MIT Press.

Nyquist L. E. and Pepin R. O. (1971) Rare gases in Apollo 12 lunar materials. *Abstrs. Second Lunar Sci. Conf.* and unpublished data.

Papanastassiou D. A. and Wasserburg G. J. (1971) Rb–Sr ages of igneous rocks from the Apollo 14 mission and the age of the Fra Mauro formation. *Earth Planet. Sci. Lett.* **12,** 36–48.

Pepin R. O., Nyquist L. E., Phinney D., and Black D. C. (1970) Rare gases in Apollo 11 lunar material. *Proc. Apollo 11 Lunar Sci. Conf., Geochim. Cosmochim. Acta* Suppl. 1, Vol. 2, pp. 1435–1454. Pergamon.

Pepin R. O., Bradley J. G., Dragon J. C., and Nyquist L. E. (1972) K–Ar dating of lunar soils: Apollo 12, Apollo 14, and Luna 16 (abstract). In *Lunar Science—III* (editor C. Watkins), pp. 602–604, Lunar Science Institute Contr. No. 88.

Podosek F. A., Huneke J. C., Burnett D. S., and Wasserburg G. J. (1971) Isotopic composition of Xe and Kr in the lunar soil and in the solar wind. *Earth Planet. Sci. Lett.* **10,** 199–216.

Schnetzler C. C. and Nava D. F. (1971) Chemical composition of Apollo 14 soils. *Earth Planet. Sci. Lett.* **11,** 345–350.

Tatsumoto M., Hedge C. E., Doe B. R., and Unruh D. (1972) U–Th–Pb and Rb–Sr measurements on some Apollo 14 lunar samples (abstract). In *Lunar Science—III* (editor C. Watkins), pp. 741–743, Lunar Science Institute Contr. No. 88.

Turner G., Huneke J. C., Podosek F. A., and Wasserburg G. J. (1971) $^{40}$Ar–$^{39}$Ar ages and cosmic ray exposure ages of Apollo 14 samples. *Earth Planet. Sci. Lett.* **12,** 19–36.

Wakita H., Showalter D. L., and Schmitt R. A. (1972) Bulk, REE, and other abundances in Apollo 14 soils (abstract). In *Lunar Science—III* (editor C. Watkins), pp. 767–769, Lunar Science Institute Contr. No. 88.

Yaniv A., Taylor G. J., Allen S., and Heymann D. (1971) Stable rare gas isotopes produced by solar flares in single particles of Apollo 11 and Apollo 12 fines. *Proc. Second Lunar Sci. Conf., Geochim. Cosmochim. Acta* Suppl. 2, Vol. 2, pp. 1705–1716. MIT Press.

Proceedings of the Third Lunar Science Conference
(Supplement 3, *Geochimica et Cosmochimica Acta*)
Vol. 2, pp. 1821–1856
The M.I.T. Press, 1972

# Trapped solar wind noble gases in Apollo 12 lunar fines 12001 and Apollo 11 breccia 10046

P. EBERHARDT, J. GEISS, H. GRAF,* N. GRÖGLER, M. D. MENDIA,†
M. MÖRGELI, H. SCHWALLER, and A. STETTLER

Physikalisches Institut, University of Bern,
Sidlerstrasse 5, 3012 Bern, Switzerland

and

U. KRÄHENBÜHL‡ and H. R. VON GUNTEN

Institut für anorganische, analytische und physikalische Chemie,
University of Bern, Freiestrasse 3,
3012 Bern, Switzerland

**Abstract**—We have measured the concentrations and isotopic composition of the five noble gases in an unseparated bulk sample of the Apollo 12 fines 12001, in seven bulk and five ilmenite grain size fractions from 12001 and in four ilmenite grain size fractions separated from the Apollo 11 breccia 10046. In addition, we determined Sr, Zr, Ba, La, Ce, Sm, and Eu in several bulk grain size fractions from 12001.

Similar to previous investigations, a well defined anticorrelation between the concentration $C(Y, d)$ of trapped solar wind noble gases $Y$ and the grain size $d$ is found. The experimental data can be represented by a power law of the form $C(Y, d) = S_y\,(d/d_0)^{-n_y}$. For the ilmenite grain size fractions from soils 10084 (Eberhardt *et al.*, 1970) and 12001 the exponents $n_y$ are close to one, which is the value expected for solar wind implanted ions. Absolute gas concentrations ($S_y$) are also very similar in the two soil ilmenites. The 10084 and 12001 bulk grain size fractions have exponents $n_y \approx 0.6$. Several possible explanations for these low $n$ values are discussed, the most likely explanation being contamination of the coarser fractions with microbreccias or a similar volume correlated noble gas component. The He and Ne in the 10046 ilmenites show virtually the same grain size dependency as the two soil ilmenites. The grain size dependency of the heavier noble gases, especially of Ar, is considerably less steep, with exponents $n \approx 0.7$. The most probable explanation of this lower exponent is a volume correlated Ar, Kr, and Xe component, which probably originated during the compaction of the breccia from loose soil material.

$(Ar^{36}/Kr^{86})_{tr}$ and $(Kr^{86}/Xe^{132})_{tr}$ ratios are similar in the 10084 and 12001 bulk and ilmenite grain size fractions and also in the 10046 ilmenites. The average values are 5000 and 1.8, respectively. The $(He^4/Ne^{20})_{tr}$ is very similar in the 10084, 12001, and 10046 ilmenites with average values between 218 and 253. In the bulk samples from both soils this ratio is a factor 2 to 5 lower. Similarly, the $(Ne^{20}/Ar^{36})_{tr}$ ratio in the bulk samples is a factor of 5 lower than in the two soil ilmenites which gave an average value of 27. The $(Ne^{20}/Ar^{36})_{tr}$ ratio in the 10046 ilmenite is grain size dependent, which can be explained in terms of the volume component of trapped Ar, Kr, and Xe in the 10046 ilmenites.

The exposure ages for the 12001 soil are between $300 \times 10^6$ yr (ilmenite) and $440 \times 10^6$ yr (bulk grain size fractions). These ages are similar to the ones obtained for the 10084 soil. The ilmenite of breccia 10046 has a distinctly higher exposure age of $800 \times 10^6$ yr.

* Present address: Washington University, Laboratory for Space Physics, St. Louis, Mo. 63130.
† ESRO fellow, on leave from the University of Zaragoza, Spain.
‡ Present address: Enrico Fermi Institute, University of Chicago, Chicago, Ill. 60637.

The considerably lower $(He^4/Ne^{20})_{tr}$ and $(Ne^{20}/Ar^{36})_{tr}$ ratios of the bulks relative to the ilmenites show, that an evaluation of the isotopic composition of trapped He and Ne in the bulk is rather futile, as these ratios are certainly affected by the heavy diffusion loss. The 10084 and 12001 soil ilmenites have identical $(He^3/He^4)_{tr}$ ratios of $3.7 \times 10^{-4}$. This value, which corresponds to recent solar wind, is in basic agreement with the present solar wind $He^3/He^4$ ratio measured by the SWC experiment (Geiss *et al.*, 1971a). The breccia ilmenite, containing older solar wind, has a 10% lower $(He^3/He^4)$ ratio than the soil ilmenites. This result supports the proposed secular variation of the $He^3/He^4$ ratio in the solar wind and in the sun.

The isotopic composition of surface correlated trapped Ne is essentially identical in the 10084, 12001, and 10046 ilmenites. The average $(Ne^{20}/Ne^{22})_{sc} = 12.8$ is approximately 5% lower than the ratio measured in the present day solar wind. This difference might be the result of unequal trapping probabilities for the two Ne isotopes, of saturation effects and of diffusion losses.

The $(Ar^{36}/Ar^{38})_{sc}$ ratio in all investigated samples is identical and agrees excellently with the terrestrial value. The $(Ar^{40}/Ar^{36})_{sc}$ ratio is different in the five samples varying from 0.37 to 1.35. Differences between bulk samples could be explained in terms of differences in the mineralogy. The variability between ilmenites from different samples, however, must represent genuine variations of the $Ar^{40}/Ar^{36}$ ratio in the source of the surface correlated gas. It is well known that the trapped $Ar^{40}$ is not of direct solar wind origin but represents retrapped $Ar^{40}$ from a transient lunar atmosphere. We thus interpret the variability of the $(Ar^{40}/Ar^{36})_{sc}$ ratio in different ilmenites by time variations in the $Ar^{40}$ outgassing of the moon.

The isotopic composition of surface correlated Kr in the 12001 bulk grain size fractions (BEOC 12001) is intermediate between AVCC and atmospheric Kr and essentially related to these by a simple mass fractionation. However, to fully explain the difference, an excess $Kr^{86}$ component in atmospheric and AVCC Kr would be required. Such a 1.1% $Kr^{86}$ excess in the terrestrial atmosphere might be identified as a fission component, which however could not have originated from known terrestrial sources. Further experimental evidence seems necessary to establish the presence of excess $Kr^{86}$ in the terrestrial atmosphere.

The $Xe^{124}/Xe^{130}$, $Xe^{126}/Xe^{130}$, and $Xe^{128}/Xe^{130}$ ratios of the surface correlated Xe in the 12001 bulk grain size fractions (BEOC 12001) are identical within 1.5% with the isotopic composition of AVCC Xe. The well known excess in the heavy isotopes of AVCC Xe is also observed relative to BEOC 12001 Xe. The AVCC fission, derived from the comparison with BEOC 12001 Xe, is—within the rather large error limits—compatible with $Pu^{244}$ spontaneous fission Xe. BEOC 12001 and atmospheric Xe are not related by a mass fractionation process depending linearly on the mass difference. Our $Xe^{124,126,128}$ BEOC 12001 data are also not compatible with the 41‰ per mass unit fractionation proposed by Kaiser (1972).

## INTRODUCTION AND RESULTS

THE LUNAR SURFACE is directly exposed to the solar wind and a considerable fraction of the impinging solar wind particles is trapped in the surface of the exposed lunar material (typical ranges of solar wind ions in solids are a few hundred to two thousand Å). These trapped solar wind atoms can be detected in the lunar surface material for elements with low intrinsic abundance in lunar material, e.g., hydrogen, the noble gases and some other elements.

In this paper we report and discuss our investigations of trapped solar wind noble gases in the Apollo 12 fines 12001,38 and the Apollo 11 breccia 10046,45. The main aims of the present investigation were:

(1) To study the elemental and isotopic composition of the trapped solar wind particles in a soil sample from a locality of the moon different from Mare Tranquillitatis and to compare the results obtained at the two places.

(2) To establish the isotopic composition of trapped Kr and Xe with improved accuracy.

(3) To investigate possible time variations in the elemental and isotopic composition of the solar wind.

A single analysis of a bulk soil or breccia sample can give only limited and inconclusive evidence on the elemental and isotopic composition of the trapped solar wind particles. Interference from spallation produced isotopes and erratic diffusion losses of the lighter noble gases must be considered and corrected for, in order to obtain meaningful results. To date, the best experimental approach to solve these problems consists in measuring the noble gases in grain size fractions of the bulk material and of mineral separates, notably ilmenite. We have therefore prepared the following bulk and mineral grain size fractions and analyzed all noble gases therein:

An unseparated sample from 12001.
Seven bulk grain size fractions from 12001.
Five ilmenite grain size fractions from 12001.
Four ilmenite grain size fractions from 10046.

For a thorough evaluation of the data it is necessary to know the concentration of the relevant target elements for the production of spallation isotopes. Therefore, the elements Sr, Zr, Ba, La, Ce, Sm, and Eu were determined in aliquots of several bulk grain size fractions from 12001.

The experimental technique used for the mineral and grain size separation and the noble gas determinations has been previously described (cf. Eberhardt *et al.*, 1970). The mineralogical purity of the ilmenite fractions was estimated to be better than 90–95% (X-ray diffraction patterns and optical microscopy). Sr, Zr, and Ba were determined with the isotopic dilution technique, La, Ce, Sm, and Eu by neutron activation (cf. Krähenbühl *et al.*, 1972). All results are given in Tables 1 to 4.

## Grain Size Dependency of Trapped Gases

The $Ne^{20}$, $Ar^{36}$, $Kr^{86}$, and $Xe^{132}$ present in all investigated fractions is at least 99% trapped gas.* Corrections for spallation and other components can thus be neglected for the discussion of the elemental abundances of these isotopes. For $He^4$ a correction for radiogenic $He^4$ had to be applied in the coarser bulk grain size fractions ($He^4_r$ = 110,000 × $10^{-8}$ $cm^3$ STP/g, correction always smaller than 5%, see discussion of isotopic ratios). A distinct anticorrelation between trapped gas content and grain size down to 1$\mu$ is evident (Figs. 1 to 3). The trapped gas in the Apollo 12 fines 12001 and the ilmenite of breccia 11046 is thus surface correlated and represents most likely implanted low energy particles, such as solar wind ions. The 36$\mu$ ilmenite grain size fraction bb-51 does not fit onto the corresponding grain size-trapped gas correlation line. This is especially apparent for the heavier noble gases Ar, Kr, and

* As customary, we use the term trapped gas for the noble gas component that did not originate by *in situ* nuclear reactions (radioactive decay, spallation reactions, etc.). The following indices will be used to define different components: c: cosmogenic (includes spallation and low energy neutron reactions); f: fission; fs: spontaneous fission; fn: neutron induced fission; m: measured; n: produced by thermal or epithermal neutron capture; r: radiogenic, i.e., decay product of primordial nuclei; sc: surface correlated; sp: spallation; tr: trapped; vc: volume correlated.

Table 1. Results of He, Ne, and Ar measurements on an unseparated sample of Apollo 12 soil 12001, on bulk and ilmenite grain size fractions separated from 12001 and on ilmenite grain size fractions separated from Apollo 11 breccia 10046.

| Sample no. | Grain size ($\mu$) | Weight of analyzed sample (mg) | $He^4$ ($10^{-8}$ cm$^3$ STP/g) | $Ne^{20}$ ($10^{-8}$ cm$^3$ STP/g) | $Ar^{36}$ ($10^{-8}$ cm$^3$ STP/g) | $He^4/He^3$ | $Ne^{20}/Ne^{22}$ | $Ne^{22}/Ne^{21}$ | $Ar^{36}/Ar^{38}$ | $Ar^{40}/Ar^{36}$ |
|---|---|---|---|---|---|---|---|---|---|---|
| *Unseparated sample from 12001* | | | | | | | | | | |
| A-12 | — | 11.2 | 9,250,000 ± 320,000 | 129,000 ± 4,500 | 24,500 ± 2,000 | 2,315 ± 70 | 12.37 ± 0.20 | 29.0 ± 0.5 | 5.23 ± 0.06 | 0.57 ± 0.01 |
| *Bulk grain size fractions from 12001* | | | | | | | | | | |
| bb-12 | 151 | 10.4 | 2,270,000 ± 80,000 | 43,700 ± 1,800 | 9,300 ± 700 | 2,350 ± 70 | 12.37 ± 0.20 | 23.0 ± 0.4 | 5.18 ± 0.08 | 0.86 ± 0.01 |
| bb-13 | 82 | 9.5 | 2,930,000 ± 100,000 | 55,200 ± 2,000 | 13,300 ± 1,000 | 2,280 ± 70 | 12.17 ± 0.20 | 21.4 ± 0.4 | 5.20 ± 0.06 | 0.68 ± 0.01 |
| bb-14 | 44 | 6.3 | 4,270,000 ± 150,000 | 80,800 ± 3,000 | 16,800 ± 1,300 | 2,520 ± 70 | 12.42 ± 0.20 | 26.2 ± 0.3 | 5.14 ± 0.10 | 0.62 ± 0.01 |
| bb-15 | 20 | 5.3 | 9,970,000 ± 360,000 | 149,300 ± 5,500 | 23,500 ± 1,800 | 2,350 ± 80 | 12.33 ± 0.20 | 26.7 ± 0.8 | 5.20 ± 0.08 | 0.55 ± 0.01 |
| bb-16 | 10.4 | 4.4 | 16,900,000 ± 600,000 | 244,000 ± 9,000 | 48,000 ± 3,500 | 2,600 ± 70 | 12.52 ± 0.20 | 30.2 ± 0.6 | 5.29 ± 0.06 | 0.50 ± 0.01 |
| bb-17 | 2.0 | 3.0 | 44,100,000 ± 1,600,000 | 504,000 ± 30,000 | 89,000 ± 7,000 | 2,310 ± 70 | 12.52 ± 0.25 | 30.2 ± 0.6 | 5.28 ± 0.06 | 0.48 ± 0.01 |
| bb-18 | 1.3 | 2.0 | 58,000,000 ± 2,000,000 | 720,000 ± 25,000 | 142,000 ± 11,000 | 2,400 ± 70 | 12.66 ± 0.20 | 31.4 ± 1.0 | 5.31 ± 0.07 | 0.48 ± 0.01 |
| *Ilmenite grain size fractions from 12001* | | | | | | | | | | |
| bb-26 | 125 | 5.2 | 10,400,000 ± 400,000 | 40,800 ± 1,500 | 2,000 ± 170 | 2,600 ± 70 | 12.62 ± 0.30 | 29.3 ± 1.2 | 5.06 ± 0.07 | 0.88 ± 0.03 |
| bb-37 | 75 | 6.7 | 21,300,000 ± 2,300,000 | 89,000 ± 9,000 | 3,400 ± 250 | 2,600 ± 110 | 12.70 ± 0.45 | 30.0 ± 1.5 | 5.13 ± 0.06 | 0.57 ± 0.01 |
| bb-51 | 36 | 6.3 | 18,900,000 ± 700,000 | 105,700 ± 4,000 | 1,400 ± 130 | 2,615 ± 80 | 12.83 ± 0.20 | 30.8 ± 0.7 | 5.44 ± 0.06 | 0.83 ± 0.04 |
| bb-52 | 17.2 | 5.6 | 79,000,000 ± 3,000,000 | 309,000 ± 11,000 | 11,500 ± 900 | 2,590 ± 70 | 12.90 ± 0.20 | 31.6 ± 1.0 | 5.31 ± 0.06 | 0.45 ± 0.01 |
| bb-57 | 10.9 | 4.3 | 200,000,000 ± 8,000,000 | 760,000 ± 30,000 | 22,500 ± 1,800 | 2,700 ± 80 | 13.00 ± 0.20 | 32.2 ± 0.7 | 5.30 ± 0.06 | 0.44 ± 0.01 |
| *Ilmenite grain size fractions from 10046* | | | | | | | | | | |
| DA-14 | 114 | 1.6 | 13,900,000 ± 900,000 | 65,000 ± 3,500 | 6,700 ± 600 | 2,810 ± 80 | 12.61 ± 0.30 | 28.85 ± 0.8 | 5.07 ± 0.13 | 2.40 ± 0.09 |
| DA-15 | 78 | 4.8 | 31,000,000 ± 1,800,000 | 129,000 ± 6,000 | 8,800 ± 800 | 2,830 ± 70 | 12.54 ± 0.15 | 30.55 ± 0.6 | 5.18 ± 0.08 | 1.94 ± 0.05 |
| DA-16 | 28 | 5.0 | 66,000,000 ± 3,500,000 | 290,000 ± 16,000 | 15,100 ± 1,500 | 2,810 ± 100 | 12.49 ± 0.25 | 31.05 ± 0.8 | 5.21 ± 0.08 | 1.78 ± 0 .05 |
| DA-17 | 14 | 5.4 | 200,000,000 ± 10,000,000 | 825,000 ± 40,000 | 28,900 ± 3,000 | 3,010 ± 70 | 12.81 ± 0.15 | 30.85 ± 0.6 | 5.28 ± 0.08 | 1.61 ± 0.05 |

Grain size determined according to equation (4) of Eberhardt *et al.* (1965). The uncertainty in the average grain size is estimated to be always smaller than ±20%.

Table 2. Results of Kr measurements on an unseparated sample of Apollo 12 soil 12001, on bulk and ilmenite grain size fractions separated from 12001 and on ilmenite grain size fractions separated from Apollo 11 breccia 10046.

| Sample no. | $Kr^{86}$ ($10^{-8}$ $cm^3$ STP/g) | $Kr^{78}/Kr^{86}$ | $Kr^{80}/Kr^{86}$ | $Kr^{82}/Kr^{86}$ | $Kr^{83}/Kr^{86}$ | $Kr^{84}/Kr^{86}$ |
|---|---|---|---|---|---|---|
| | | | | × 100 | | |
| *Unseparated sample from 12001* | | | | | | |
| ba-2 | 6.0 ± 1.2 | 2.180 ± 0.07 | 13.38 ± 0.2 | 66.55 ± 0.7 | 66.95 ± 0.5 | 326.8 ± 1.7 |
| *Bulk grain size fractions from 12001* | | | | | | |
| bb-12 | 2.0 ± 0.4 | 2.510 ± 0.06 | 14.23 ± 0.18 | 67.20 ± 0.7 | 68.35 ± 0.6 | 325.6 ± 3 |
| bb-13 | 2.7 ± 0.9 | 2.440 ± 0.05 | 14.16 ± 0.18 | 67.45 ± 0.6 | 67.75 ± 0.6 | 324.5 ± 2.5 |
| bb-14 | 3.7 ± 0.7 | 2.295 ± 0.045 | 13.62 ± 0.18 | 66.45 ± 0.6 | 67.00 ± 0.45 | 324.9 ± 1.9 |
| bb-15 | 5.3 ± 1 | 2.205 ± 0.05 | 13.39 ± 0.2 | 66.25 ± 0.7 | 66.90 ± 0.45 | 325.4 ± 2 |
| bb-16 | 9.9 ± 2 | 2.095 ± 0.04 | 13.06 ± 0.17 | 65.80 ± 0.6 | 66.20 ± 0.4 | 325.3 ± 1.8 |
| bb-17 | 21 ± 4 | 2.040 ± 0.045 | 13.02 ± 0.2 | 66.20 ± 0.5 | 66.30 ± 0.4 | 328.1 ± 2 |
| bb-18 | 33 ± 6 | 1.970 ± 0.04 | 12.80 ± 0.14 | 65.65 ± 0.5 | 65.90 ± 0.45 | 327.3 ± 1.6 |
| *Ilmenite grain size fractions from 12001* | | | | | | |
| bb-26 | 0.39 ± 0.08 | 5.695 ± 0.09 | 21.44 ± 0.35 | 77.95 ± 0.9 | 84.30 ± 0.5 | 326.7 ± 3 |
| bb-37 | 0.59 ± 0.12 | 8.44 ± 0.3 | 27.06 ± 0.5 | 86.30 ± 0.9 | 93.45 ± 0.8 | 329.3 ± 3 |
| bb-51 | 0.099 ± 0.02 | 11.33 ± 0.25 | 34.20 ± 0.8 | 100.1 ± 1.7 | 108.9 ± 0.9 | 337.1 ± 4.5 |
| bb-52 | 2.3 ± 0.4 | 4.495 ± 0.1 | 18.54 ± 0.18 | 74.25 ± 0.7 | 76.80 ± 0.7 | 326.8 ± 1.7 |
| bb-57 | 4.8 ± 0.9 | 3.010 ± 0.035 | 15.26 ± 0.2 | 69.05 ± 0.6 | 70.80 ± 0.5 | 326.6 ± 1.8 |
| *Ilmenite grain size fractions from 10046* | | | | | | |
| DA-14 | 0.95 ± 0.20 | 3.90 ± 0.16 | 17.85 ± 0.40 | 74.8 ± 1.1 | 76.2 ± 0.9 | 329.0 ± 2.3 |
| DA-15 | 1.5 ± 0.3 | 3.69 ± 0.07 | 17.30 ± 0.25 | 73.9 ± 0.8 | 75.95 ± 0.8 | 330.0 ± 2.5 |
| DA-16 | 2.8 ± 0.5 | 2.93 ± 0.08 | 15.25 ± 0.25 | 70.35 ± 0.65 | 71.05 ± 0.65 | 328.6 ± 1.9 |
| DA-17 | 4.8 ± 0.9 | 2.57 ± 0.05 | 14.45 ± 0.20 | 69.1 ± 0.9 | 69.85 ± 0.75 | 329.9 ± 3 |

For grain sizes see Table 1.

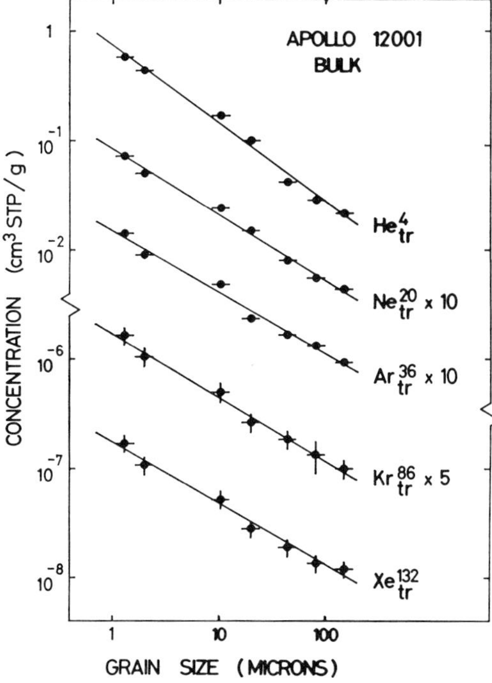


Fig. 1. Grain size dependency of the trapped $He^4$, $Ne^{20}$, $Ar^{36}$, $Kr^{86}$, and $Xe^{132}$ concentrations in unseparated Apollo 12 lunar fine material 12001.

Table 3. Results of Xe measurements on an unseparated sample of Apollo 12 soil 12001, on bulk and ilmenite grain size fractions separated from 12001 and on ilmenite grain size fractions separated from Apollo 11 soil 10046.

| Sample no. | $Xe^{132}$ ($10^{-8}$ $cm^3$ STP/g) | $Xe^{124}/Xe^{132}$ | $Xe^{126}/Xe^{132}$ | $Xe^{128}/Xe^{132}$ | $Xe^{129}/Xe^{132}$ | $Xe^{130}/Xe^{132}$ | $Xe^{131}/Xe^{132}$ | $Xe^{134}/Xe^{132}$ | $Xe^{136}/Xe^{132}$ |
|---|---|---|---|---|---|---|---|---|---|
| | | × 100 | | | | | | | |
| *Unseparated sample from 12001* | | | | | | | | | |
| ba-2 | 2.6 ± 0.5 | 0.862 ± 0.025 | 1.126 ± 0.025 | 9.50 ± 0.1 | 106.3 ± 1.1 | 16.96 ± 0.13 | 85.06 ± 0.7 | 36.91 ± 0.25 | 29.21 ± 0.3 |
| *Bulk grain size fractions from 12001* | | | | | | | | | |
| bb-12 | 1.21 ± 0.2 | 1.234 ± 0.05 | 1.810 ± 0.06 | 10.38 ± 0.16 | 106.3 ± 1.4 | 17.72 ± 0.3 | 88.35 ± 1.4 | 36.48 ± 0.4 | 29.80 ± 0.6 |
| bb-13 | 1.35 ± 0.25 | 1.073 ± 0.05 | 1.492 ± 0.05 | 9.85 ± 0.19 | 104.4 ± 1.8 | 17.23 ± 0.2 | 86.79 ± 0.35 | 37.00 ± 0.35 | 30.44 ± 0.6 |
| bb-14 | 1.91 ± 0.35 | 0.958 ± 0.025 | 1.355 ± 0.035 | 9.73 ± 0.2 | 105.1 ± 1.2 | 17.21 ± 0.15 | 86.23 ± 0.3 | 36.81 ± 0.35 | 30.01 ± 0.45 |
| bb-15 | 2.8 ± 0.5 | 0.832 ± 0.02 | 1.103 ± 0.045 | 9.33 ± 0.14 | 105.1 ± 1.2 | 17.09 ± 0.25 | 85.16 ± 0.9 | 36.90 ± 0.3 | 30.16 ± 0.45 |
| bb-16 | 5.2 ± 1 | 0.723 ± 0.02 | 0.892 ± 0.02 | 8.99 ± 0.16 | 104.5 ± 1.1 | 16.83 ± 0.25 | 83.92 ± 0.9 | 37.05 ± 0.45 | 30.12 ± 0.5 |
| bb-17 | 10.7 ± 2 | 0.625 ± 0.019 | 0.703 ± 0.02 | 8.78 ± 0.18 | 105.0 ± 1.2 | 16.70 ± 0.13 | 83.42 ± 0.5 | 36.88 ± 0.3 | 29.91 ± 0.45 |
| bb-18 | 17.1 ± 3 | 0.576 ± 0.025 | 0.625 ± 0.02 | 8.59 ± 0.12 | 105.0 ± 1.3 | 16.67 ± 0.17 | 83.30 ± 0.5 | 36.86 ± 0.3 | 30.04 ± 0.45 |
| *Ilmenite grain size fractions from 12001* | | | | | | | | | |
| bb-26 | 0.17 ± 0.03 | 1.564 ± 0.05 | 2.412 ± 0.09 | 11.00 ± 0.35 | 102.4 ± 2.2 | 17.46 ± 0.5 | 89.34 ± 1.4 | 36.94 ± 0.8 | 30.24 ± 0.7 |
| bb-37 | 0.22 ± 0.04 | 0.998 ± 0.045 | 1.409 ± 0.09 | 9.46 ± 0.35 | 103.5 ± 2.5 | 17.22 ± 0.45 | 85.17 ± 0.9 | 37.18 ± 0.45 | 30.18 ± 0.5 |
| bb-51 | 0.034 ± 0.007 | 0.796 ± 0.09 | 1.005 ± 0.1 | 8.76 ± 0.7 | 101.9 ± 4.5 | 17.34 ± 0.9 | 82.60 ± 0.6 | 37.69 ± 2 | 32.22 ± 2 |
| bb-52 | 0.96 ± 0.18 | 0.599 ± 0.018 | 0.626 ± 0.02 | 8.46 ± 0.45 | 104.2 ± 1.3 | 16.69 ± 0.45 | 82.49 ± 1 | 37.33 ± 0.35 | 30.54 ± 0.5 |
| bb-57 | 2.1 ± 0.4 | 0.524 ± 0.009 | 0.531 ± 0.02 | 8.00 ± 0.35 | 103.9 ± 0.9 | 16.35 ± 0.25 | 81.75 ± 0.9 | 37.10 ± 0.5 | 30.45 ± 0.3 |
| *Ilmenite grain size fractions from 10046* | | | | | | | | | |
| DA-14 | 0.60 ± 0.12 | 0.840 ± 0.050 | 1.082 ± 0.085 | 9.42 ± 0.30 | 104.8 ± 1.3 | 16.96 ± 0.1 | 84.85 ± 1.2 | 37.00 ± 0.45 | 30.98 ± 0.4 |
| DA-15 | 0.90 ± 0.18 | 0.695 ± 0.030 | 0.819 ± 0.050 | 8.90 ± 0.13 | 104.4 ± 2.2 | 16.88 ± 0.4 | 83.2 ± 1.2 | 37.30 ± 0.55 | 30.66 ± 0.5 |
| DA-16 | 2.1 ± 0.4 | 0.591 ± 0.025 | 0.633 ± 0.030 | 8.52 ± 0.13 | 103.8 ± 1.3 | 16.50 ± 0.2 | 82.55 ± 0.5 | 37.39 ± 0.3 | 30.87 ± 0.4 |
| DA-17 | 4.0 ± 0.8 | 0.5005 ± 0.015 | 0.494 ± 0.018 | 8.32 ± 0.11 | 104.1 ± 0.8 | 16.34 ± 0.2 | 81.85 ± 0.5 | 37.58 ± 0.25 | 30.78 ± 0.3 |

For grain sizes see Table 1.

Table 4. Results of Sr, Zr, Ba, La, Ce, Sm, and Eu determinations in bulk grain size fractions separated from Apollo 12 fines 12001.

| Sample no. | Sr | Zr | Ba | La | Ce | Sm | Eu |
|---|---|---|---|---|---|---|---|
| | | | | (ppm) | | | |
| bb-14 | 134 ± 7 | 405 ± 40 | 305 ± 20 | 22 ± 2 | 77 ± 5 | 10 ± 1 | 1.2 ± 0.1 |
| bb-15 | 141 ± 7 | 410 ± 40 | 315 ± 20 | — | — | — | — |
| bb-16 | 175 ± 9 | 460 ± 45 | 400 ± 25 | 38 ± 3 | 130 ± 8 | 18 ± 2 | 2.4 ± 0.2 |
| bb-17 | 182 ± 9 | 440 ± 45 | 490 ± 30 | — | — | — | — |
| bb-18 | 214 ± 11 | 400 ± 40 | 630 ± 40 | 53 ± 4 | 167 ± 10 | 28 ± 2 | 3.1 ± 0.3 |
| 12001 (a) | 145.5 | — | 370 | — | 87.2 | 16.1 | 1.78 |
| bulk (b) | 130 | — | 460 | 32.4 | 87 | 15.0 | 1.80 |

Measurements were done on aliquots of the samples used for the noble gas analyses. Sample sizes analyzed: 9 to 13 mg. For grain sizes see Table 1. Data for 12001 bulk sample from: (a) Schnetzler and Philpotts (1971); (b) Wänke *et al.* (1971).

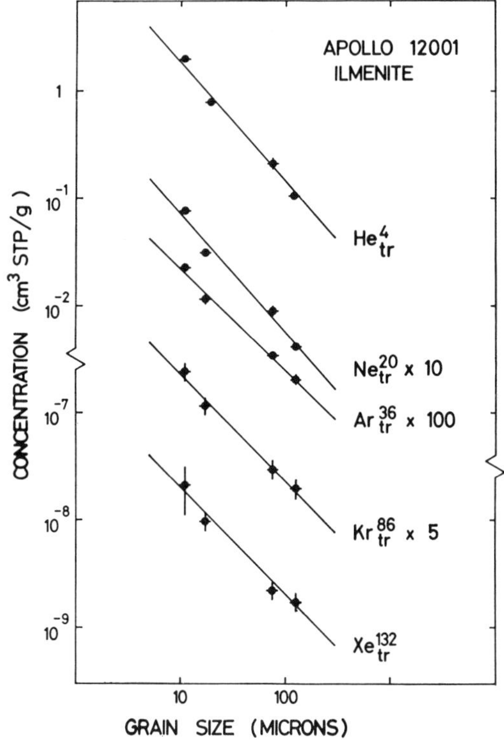


Fig. 2. Grain size dependency of the trapped $He^4$, $Ne^{20}$, $Ar^{36}$, $Kr^{86}$, and $Xe^{132}$ concentrations in ilmenite separated from the Apollo 12 lunar fine material 12001.

Xe. Not enough material was available to make a duplicate analysis and we can thus not decide whether this discrepancy is due to an analytical error or whether this peculiarity is inherent to this grain size fraction. Therefore, the results from this fraction will be disregarded in our further discussion.

Within the experimental uncertainties and limitations, the different grain size fractions fall on straight lines in the log grain size versus log concentration plots

(Figs. 1 to 3). The concentration $C(Y, d)$ of the trapped noble gas $Y$ in the grain size $d$ can be expressed as

$$C(Y, d) = S_y \left(\frac{d}{d_0}\right)^{-n_y} \tag{1}$$

where $d_0$ is an arbitrary reference grain size, $S_y$ the noble gas concentration in the grain size $d_0$, and $n_y$ the exponent defining the slope of the correlation line in the log-log plot. The $S_y$ and $n_y$ values obtained by a least square fit are given in Table 5. As reference grain size $d_0 = 10\ \mu$ was chosen.

For the ilmenite grain size fractions separated from the Apollo 11 and 12 fines, virtually identical values for the noble gas concentrations and the exponent $n$ are found. For $He^4_{tr}$ and $Ne^{20}_{tr}$ the ilmenite separated from breccia 10046 shows similar exponents and only slightly higher gas concentrations. The exponent $n$ for $Ar^{36}_{tr}$ and $Kr^{86}_{tr}$ is, however, distinctly smaller than for the other two ilmenites. This is also evident from Figs. 2 and 3. For strongly differing $n$ values the comparison of the reference trapped gas concentrations is not any more appropriate. The low $n_{36}$ value for the 10046 ilmenites will be discussed in more detail in a later section.

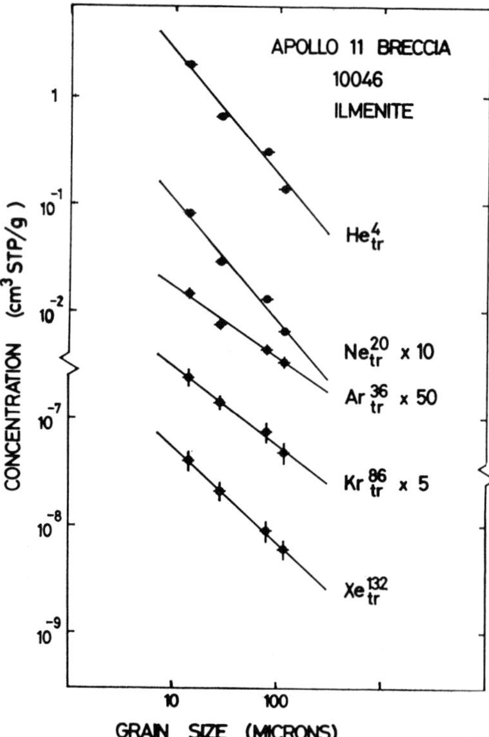


Fig. 3. Grain size dependency of the trapped $He^4$, $Ne^{20}$, $Ar^{36}$, $Kr^{86}$, and $Xe^{132}$ concentrations in ilmenite separated from the Apollo 11 breccia 10046.

Table 5. Best fit of the measured concentrations $C(Y,d)$ in the different grain sizes $d$ to the equation $C(Y, d) = S_y(d/d_0)^{-n_y}$ ($d_0 = 10\ \mu$).

| | $S_y$: trapped gas concentration in 10 $\mu$ fraction | | | | | $n_y$ | | | | |
|---|---|---|---|---|---|---|---|---|---|---|
| | $He^4$ | $Ne^{20}$ | $Ar^{36}$ | $Kr^{86}$ | $Xe^{132}$ | | | | | |
| Samples | ($10^{-3}$ cm$^3$ STP/g) | | | ($10^{-8}$ cm$^3$ STP/g) | | $He^4$ | $Ne^{20}$ | $Ar^{36}$ | $Kr^{86}$ | $Xe^{132}$ |
| *Bulk grain size fractions* | | | | | | | | | | |
| 10084 | 305 ± 15 | 3.15 ± 0.25 | 0.48 ± 0.05 | 9.5 | 5.7 | 0.64 ± 0.03 | 0.59 ± 0.04 | 0.60 ± 0.05 | 0.55 ± 0.06 | 0.56 ± 0.04 |
| 12001 | 145 ± 8 | 2.1 ± 0.1 | 0.41 ± 0.03 | 9.1 | 4.8 | 0.71 ± 0.03 | 0.60 ± 0.02 | 0.56 ± 0.03 | 0.58 ± 0.02 | 0.56 ± 0.03 |
| *Ilmenite grain size fractions* | | | | | | | | | | |
| 10084 | 1700 ± 170 | 7.3 ± 0.6 | 0.21 ± 0.06 | 4.8 | 2.1 | 1.15 ± 0.06 | 1.11 ± 0.04 | 0.96 ± 0.15 | 0.98 ± 0.11 | 0.82 ± 0.13 |
| 12001 | 1850 ± 350 | 7.1 ± 1.5 | 0.22 ± 0.02 | 4.6 | 1.95 | 1.12 ± 0.11 | 1.10 ± 0.12 | 0.95 ± 0.06 | 1.00 ± 0.07 | 1.02 ± 0.1 |
| 10046 | 2700 ± 700 | 11.1 ± 2.5 | 0.35 ± 0.03 | 6.1 | 5.3 | 1.16 ± 0.15 | 1.13 ± 0.12 | 0.67 ± 0.06 | 0.74 ± 0.06 | 0.89 ± 0.04 |

$S_y$ represents the trapped gas concentration in a (hypothetical) 10 $\mu$ grain size fraction; $-n_y$ the slope of the correlation lines in Figs. 1 to 3.

The bulk grain size fractions separated from 10084 and 12001 show remarkably similar grain size dependencies. However, the $n$ values are nearly a factor of 2 smaller than for the corresponding ilmenites. For the heavy noble gases the reference gas concentrations $S_y$ are similar in the two bulk samples, while for $Ne^{20}$ and especially for $He^4$ they are higher in 10084 than in 12001. For He and Ne the trapped gas concentrations in the bulk material are up to an order of magnitude lower than in the ilmenite for a grain size of 10 $\mu$. On the other hand, at $d = 10\ \mu$, the bulk material has twice as high Ar, Kr, and Xe concentrations as the ilmenite.

Presuming ideal conditions (e.g., no broken grains, no microbreccia, etc.) the exponent $n$ should be equal to unity for surface correlated gases. This holds quite well for all noble gases in the 10084 and 12001 ilmenite and for He and Ne in the 10046 ilmenite. The bulk samples have $n$ values considerably lower than one. Several explanations are possible:

(1) A systematically increasing dilution of the finer grain size fractions with broken larger grains or grains containing no trapped gases.
(2) A contamination of the coarser grain size fractions with microbreccias.
(3) The presence of a volume correlated component.
(4) A grain size dependent noble gas loss.

The ilmenite grain size fractions were prepared from the corresponding bulk grain size fractions. It is thus improbable that during our sample preparation a contamination of the finer bulk grain size fractions with fragmented grains occurred, without leading to a similar contamination in the ilmenite fractions. The same argument holds for the lunar surface. "Rapid" reworking of the regolith and breaking-up of the larger crystal should influence both the bulk grain size fractions and the ilmenite. A certain discrimination could occur if the ilmenite were more resistant than the bulk material against the specific break-up process. The exposure age of the ilmenite is then expected to be higher than that of the bulk material, which is not in agreement with our measurements (cf. Table 7). A continuous dilution of the lunar regolith by fine, fresh, ilmenite-poor dust can also be considered. No possible source and transport mechanism for such a virgin dust component is, however, apparent, especially considering the time scales involved. An exposure of a thousand years would already saturate the grain surface with trapped solar wind gases and rapid transport and burial (a few microns) of this virgin dust would be necessary.

Equivalent to a dilution of the finer grain sizes with a component free of trapped gases would be a contamination of the coarser grain size fractions with small grains, either adhering to larger grains or as microbreccias. The latter would essentially correspond to a volume correlated component. The coarser grain size fractions indeed contained some composite grains, whereas the finer fractions consisted essentially of single grains. We thus cannot exclude this contamination effect.

If a volume correlated component is present equation (1) must be replaced by:

$$C(Y, d) = S_y \left(\frac{d}{d_0}\right)^{-n_y} + V_y. \tag{2}$$

$V_y$ is the concentration of the grain size independent volume correlated component. In a log $C$–log $d$ plot equation (2) is not any more a straight line. The deviation from a straight line in a limited $d$ range is, however, relatively small. Figure 4 shows a best fit of the experimental data for $Ar^{36}_{tr}$ in the 10084 bulk grain size fractions both with equation (1) and (2). For equation (2) $n = 1$ was assumed. It is evident that equation (2), e.g., the assumption of a volume correlated component, gives at least as good a fit to the measured concentrations as the straight line of equation (1). $Ar^{36}_{tr}$ is a typical case for all the bulk grain size fractions since the value of $n_{36} = 0.6$ is very close to the $n$ values of all the other gases.

In analogy to the $Ar^{36}_{tr}$ in 10084 also the other noble gases in the bulk material of 10084 and 12001 could be explained as a mixture of a volume and a surface correlated ($n = 1$) component. A similarly satisfactory agreement of equation (2) with the observed grain size dependence as for $Ar^{36}$ in 10084 can be obtained. As the exponents $n_y$ of equation (1) are approximately identical for all the noble gases the volume correlated component would have to have similar relative abundance ratios as the surface correlated gas. A contamination by microbreccia, as discussed above, would satisfy this requirement.

A grain size dependent diffusion loss seems to us an unlikely explanation of the low $n$ values for the bulk grain size fractions. The similarity of the $n$ for all noble gases requires the same relative gas loss for all noble gases for any given grain size. All the available evidence on the diffusion behavior of the trapped noble gases in lunar material shows that the lighter noble gases have much higher diffusion constants than the heavier ones (Hohenberg *et al.*, 1970; Pepin *et al.*, 1970; Baur *et al.*, 1972). Diffusion loss by thermal heating will thus always give a strong mass fractionation. The loss pattern could, however, be changed if solar wind irradiation and

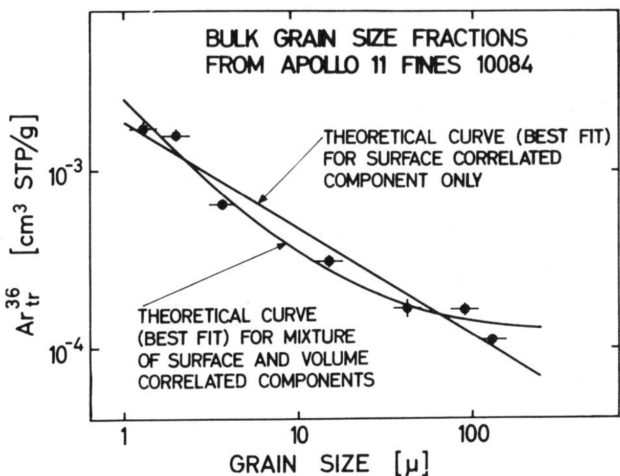


Fig. 4 Comparison of measured grain size dependency for $Ar^{36}_{tr}$ with the theoretical curves expected for purely surface correlated gas ($Ar^{36}_{tr} \propto (d/d_0)^{-0.6}$) and for a mixture of surface and volume correlated gas ($Ar^{36}_{tr} \propto 240\,(d/d_0)^{-1} + 12$). Experimental data from Eberhardt *et al.* (1970).

diffusion loss occurred simultaneously or alternately or if the loss is induced by saturation/sputtering effects.

Flaking-off of the amorphous surface layer (Borg *et al.*, 1971), saturated with trapped solar wind noble gases, could lead to a trapped gas loss independent of mass. If this flaking-off is grain size dependent, then such a process might be a possible explanation for the low $n$ values.

With the present available evidence we cannot decide which of the discussed processes is mainly responsible for making the trapped gas concentration less grain size dependent in the bulk fractions than in the ilmenites. It may well be that several of the above mentioned mechanisms, and perhaps some yet unrecognized ones, have contributed to the decrease in the slope of the grain size dependency curves.

## ELEMENTAL ABUNDANCE RATIOS OF TRAPPED GASES

In Table 6 the elemental abundance ratios—averaged over the grain size fractions—are given for the investigated samples (bulk and ilmenite). In two cases a systematic trend of the ratios with grain size is observed. Accordingly, no average value is then given, but the range of observed ratios is indicated in Table 6. In one case [$(Ne^{20}/Ar^{36})_{tr}$ for 12001 ilmenite] a weak grain size dependency may exist (cf. Fig. 5).

The abundance ratios between trapped Ar, Kr, and Xe are quite similar in all five investigated bulk and ilmenite samples. The $(Ar^{36}/Kr^{86})_{tr}$ agree within $\pm 15\%$, the $(Kr^{86}/Xe^{132})_{tr}$ within $\pm 30\%$. The trapping and diffusion behavior of the heavy noble gases in the different minerals must thus be fairly similar. Measurements on bulk grain size fractions ought to be no less reliable than measurements on separated minerals (e.g., ilmenite) for deriving the "true" isotopic and elemental abundances in the heavy trapped gases.

The $(He^4/Ne^{20})_{tr}$ and especially the $(Ne^{20}/Ar^{36})_{tr}$ ratios in the bulk material are 2 to 5 times lower than in the ilmenites. However, for the same type of material (bulk, ilmenite) the elemental abundances are remarkably similar (excluding the two cases of clear grain size dependency). Heavy diffusion loss of trapped He and Ne in the less retentive minerals of the bulk material must have occurred. Normalized to Ar, the bulk material has lost 90% of the trapped He and 80% of the trapped Ne if compared

Table 6. Average elemental abundance ratios of the trapped noble gases in different lunar grain size fractions.

| Sample | $He^4/Ne^{20}$ | $Ne^{20}/Ar^{36}$ | $Ar^{36}/Kr^{86}$ ($\times 10^3$) | $Kr^{86}/Xe^{132}$ |
|---|---|---|---|---|
| *Bulk grain size fractions* | | | | |
| 10084 | 96 ± 18 | 6.8 ± 2.0 | 5.1 ± 1.2 | 1.65 ± 0.55 |
| 12001 | 49 – 87 | 5.1 ± 0.7 | 4.5 ± 0.3 | 1.9 ± 0.1 |
| *Ilmenite grain size fractions* | | | | |
| 10084 | 218 ± 8 | 27.4 ± 4.3 | 4.7 ± 0.9 | 1.8 ± 0.25 |
| 12001 | 253 ± 10 | 26.8 ± 5.4 | 5.2 ± 0.5 | 2.4 ± 0.2 |
| 10046 | 231 ± 13 | 9.7 – 28.5 | 6.1 ± 0.7 | 1.45 ± 0.2 |

The errors are standard deviations and are thus a direct measure for the scattering of the measured ratios within the individual grain size fractions relative to the average value. A range of values indicates a systematic trend with grain size.

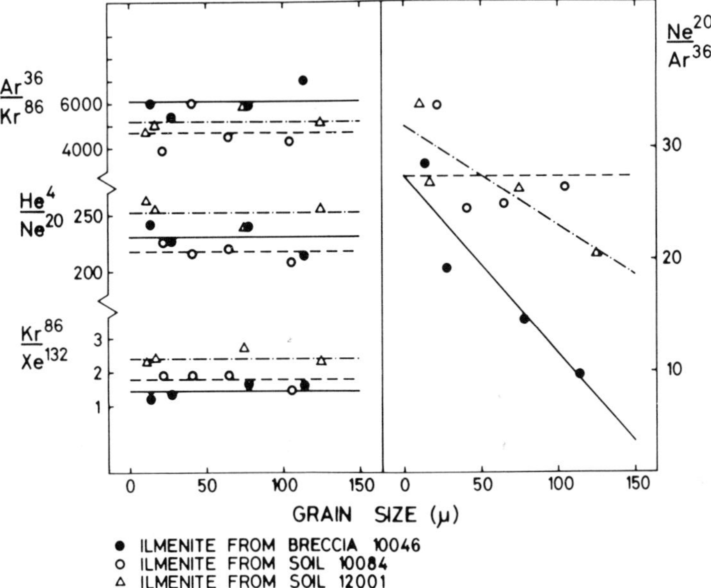


Fig. 5. Elemental abundance ratios of the trapped noble gases in the 10084, 12001, and 10046 ilmenite fractions as function of the grain size. For error see Table 1 to 3.

with the ilmenite. The ilmenite itself has probably also lost some He and perhaps Ne (see subsequent discussion). The heavy He and Ne diffusion loss in the bulk material is also evident when absolute gas concentrations in the same grain size of bulk and ilmenite are compared (cf. Table 5). It has to be expected that the large diffusion losses have altered the isotopic composition of trapped He and Ne in the bulk material. Thus data obtained on bulk lunar fine material are not representative for the elemental and isotopic composition of the light noble gases which were originally trapped in the lunar material.

### $(He^4/Ne^{20})_{tr}$ *variation in 12001 bulk grain size fractions*

A volume correlated component of trapped Ne, Ar, Kr, and Xe could explain the systematic decrease of $(He^4/Ne^{20})_{tr}$ with grain size.

More severe He diffusion loss in the larger grain sizes is the other possible explanation. The grain size dependent He loss does not require a grain size dependent thermal history. A more or less systematic change in the mineralogical composition or the glass content in the bulk grain size fractions bb-12 to bb-18 could lead to systematically varying loss factors.

### $(Ne^{20}/Ar^{36})_{tr}$ *variation in the 10046 ilmenite grain size fractions*

In Fig. 5 the grain size dependency of the trapped gas elemental abundance ratios in the investigated ilmenite fractions are shown. It is evident that these ratios, except for $(Ne^{20}/Ar^{36})_{tr}$, are essentially independent of grain size. The $(Ne^{20}/Ar^{36})_{tr}$ ratio in the 10046 ilmenite is strongly grain size dependent, increasing systematically by a factor of 3 from the coarsest to the finest fraction. A weaker grain size dependency is also present for the 12001 ilmenite. The 10084 ilmenite shows no systematic trend in the $(Ne^{20}/Ar^{36})_{tr}$ ratio, except that the finest fraction has a somewhat higher ratio. Two different models could explain the systematic trend in these elemental abundance ratios, and we would like to discuss them specifically for the 10046 ilmenite:

*Model A: A volume correlated* Ar, Kr, *and* Xe *component*

The trapped gas concentration in the different grain sizes is described by equation (2). The $(He^4/Ne^{20})_{tr}$ ratio is constant within $\pm 7\%$ and shows no grain size dependency, and thus no volume correlated $He_{tr}$ and $Ne_{tr}$ components are present ($V_4 = V_{20} = 0$). The constancy of the $(He^4/Ne^{20})_{tr}$ ratio also indicates that the grain size dependency of the surface correlated component is identical for $He^4_{tr}$ and $Ne^{20}_{tr}$ ($n_4 = n_{20}$, cf. Table 5). If we assume that the surface correlated component of $Ar^{36}_{tr}$, $Kr^{86}_{tr}$, and $Xe^{132}_{tr}$ have the same grain size dependency ($n_{36} = n_{86} = n_{132} = n_4 = n_{20}$), we obtain

$$\frac{C(Y, d)}{C(X, d)} = \frac{S_y}{S_x} + \frac{V_y}{C(X, d)}, \tag{3}$$

where $X$ is $He^4_{tr}$ or $Ne^{20}_{tr}$ and $Y$ is $Ar^{36}_{tr}$, $Kr^{86}_{tr}$ or $Xe^{132}_{tr}$. Equation (3) represents a straight line if $C(Y, d)/C(X, d)$ is plotted versus $1/C(X, d)$. The ratio $S_y/S_x$ is the ratio of the surface correlated component. For $X = Ne^{20}_{tr}$ and $Y = Ar^{36}_{tr}$ equation (3) reads:

$$\left(\frac{Ar^{36}}{Ne^{20}}\right)_{tr} = \left(\frac{Ar^{36}}{Ne^{20}}\right)_{sc} + \frac{V_{36}}{Ne^{20}_{tr}} \tag{4}$$

Figure 6 shows the corresponding correlation diagram. The experimental points define indeed a straight line within the experimental uncertainties. From Fig. 6 and the corresponding correlations for $Kr^{86}_{tr}$ and $Xe^{132}_{tr}$ the concentration of the volume correlated component and the elemental abundance ratios in the surface correlated component can be calculated.

*Model B: All trapped gases are surface correlated, but constant amounts of trapped* He *and* Ne *were lost*

In this model it is assumed that in all ilmenite fractions of rock 10046 a constant amount per unit weight of $He^4_{tr}$ and $Ne^{20}_{tr}$ was lost. Then the remaining concentration of a noble gas can be written as:

$$C(Y, d) = S_y\left(\frac{d}{d_0}\right)^{-n_y} - L_y. \tag{5}$$

The first term in equation (5) represents the surface correlated gas originally present, the second term, $L_y$, represents the lost gas. Since $(Ar^{36}/Kr^{86})_{tr}$ and $(Kr^{86}/Xe^{132})_{tr}$

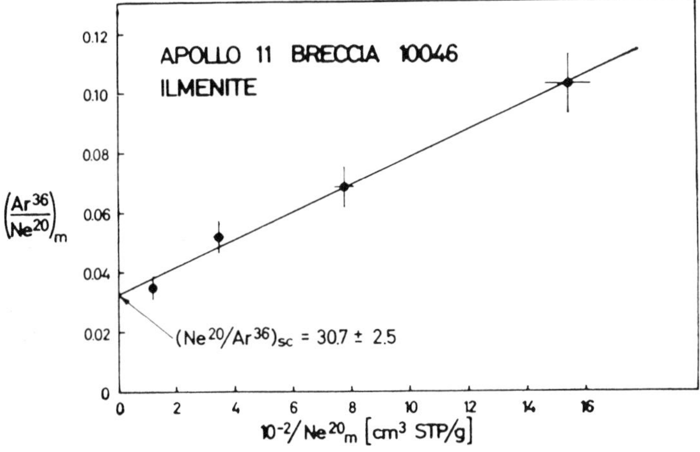


Fig. 6. Correlation between the $(Ar^{36}/Ne^{20})_m$ ratio and $1/Ne^{20}_m$ in the 10046 ilmenite grain size fractions. The excellent correlation proves that the grain size dependent $(Ne^{20}/Ar^{36})_{tr}$ ratio in these samples can be explained by the presence of a volume correlated $Ar^{36}_{tr}$ component.

are approximately constant, we take $L_{36} = L_{86} = L_{132} = 0$. We assume again that all $n_y$ are equal and obtain

$$\frac{C(X, d)}{C(Y, d)} = \frac{S_x}{S_y} - \frac{L_x}{C(Y, d)} . \tag{6}$$

Here $X$ is $He^4_{tr}$ or $Ne^{20}_{tr}$ and $Y$ is $Ar^{36}_{tr}$, $Kr^{86}_{tr}$ or $Xe^{132}_{tr}$. Equation (6) represents again a straight line if $C(X, d)/C(Y, d)$ is plotted versus $1/C(Y, d)$ and $S_x/S_y$ is the isotopic ratio of the surface correlated component. A detailed analyses of the data shows that also this model could give a satisfactory explanation of the experimental data.

## *Discussion*

The two models mentioned and the assumptions made have to be considered as limiting cases. Some assumptions, such as the absence of a He and Ne volume component, could be relaxed to some extent. Also, a mixture of the two models would equally well represent the experimental data.

Model B requires a constant He and Ne loss in all fractions. This gas loss could have occurred during the compaction process leading from the loose soil to the breccia. The surface correlated gases are concentrated in a layer approximately 1000 Å thick below the surface of the individual grains (Eberhardt *et al.*, 1970). The diffusion path is thus *not* grain size dependent, except for submicron sized grains. As Eberhardt *et al.* (1970) have shown, heavy saturation occurs in the regolith. The gas concentration in the surface layer is then not grain size dependent. A constant gas loss per gram of material would thus imply that the larger grains lost relatively more gas from the saturated surface layer than the smaller ones. Such a grain size dependent loss requires a grain size dependent temperature history, the larger grains having been subjected to a higher temperature or heated for a longer time.

Model B leads to an exponent $n \approx 0.65$ for the surface correlated gas, whereas model A gives $n \approx 1.15$ which is much closer to the expected value of 1, and very similar to the values observed in the ilmenite grain size fractions of the lunar fine material. Additional assumptions for explaining the low $n$ values of model B would be required. This is a serious drawback inherent to model B which does not arise with model A.

The Ar, Kr, and Xe volume component, required by model A, could be explained in several ways. The volume component could have originated during the crystallization and cooling or a later reheating of the ilmenite, provided a sufficient partial pressure of Ar, Kr, and Xe was present. Solar wind particles implanted at elevated temperatures, or alternating cycles of bombardment and heating might also lead to a volume component of the heavier noble gases. He and Ne diffuses much easier and could have been completely lost. The loss of the heavier noble gases will be less complete and part of the gases trapped at the surface will diffuse into the interior of the grains, slowly building up a volume component. Channeling effects (Piercy *et al.*, 1964), leading to much longer ranges for a very small fraction of the impinging solar wind ions, could also contribute to the build up of a volume component. Solar flare particles are another possible source for a volume component in millimeter and submillimeter sized lunar regolith particles. The present day average solar flare flux of $100\ cm^{-2}\ sec^{-1}$ (Finkel *et al.*, 1971) would lead, if quantitatively retained, to an $Ar^{36}_{tr}$ volume component of only $0.5 \times 10^{-8}\ cm^3$ STP/g ($8 \times 10^8$ yr exposure age, 20 cm average shielding, cf. discussion of $(Xe^{131}/Xe^{126})_{sp}$ ratios). This amount is 4 orders of magnitude smaller than the required volume component for 10046.

If we assume a genetic relationship between the regolith material and the breccia, then all the above discussed processes would also predict a volume correlated component in the regolith. No such component is evident in the 10084 ilmenite.

We must also consider the possibility that the volume correlated Ar, Kr, and Xe component is not contained in the ilmenite but associated with an impurity in our mineral separates. Our 10046 ilmenite fractions are at least 95% pure and this associated component would thus have to have at least concentrations of $90{,}000 \times 10^{-8}$ $cm^3$ STP $Ar^{36}$/g, $13 \times 10^{-8}\ cm^3$ STP $Kr^{86}$/g and $6 \times 10^{-8}\ cm^3$ STP $Xe^{132}$/g. Much higher trapped gas concentrations are found in the micron sized lunar fine material, but of course together with correspondingly higher trapped He and Ne concentrations. Unaltered, micron or submicron sized lunar fine material attached to the ilmenite grains could thus not explain the volume component. However, if some heating of this fine accessory material occurred most of the He and Ne might have been lost. A difference in the diffusion coefficients between the ilmenite and the accessory material would be required to explain the absence of diffusion loss in the ilmenite. This requirement would be fulfilled if this accessory phase consisted of glassy lunar material which shows much poorer gas retention properties.

## Exposure Ages and Spallation Produced Noble Gases

The calculation of the concentration and isotopic composition of the spallation produced noble gases and the exposure ages is hampered by the required large corrections for the trapped gas. Furthermore, for some isotopes the target element concentrations are not known. Our best estimates for the average exposure ages are given in Table 7. These exposure ages were calculated as follows:

### *12001, unseparated sample*

Assuming $(Xe^{126}/Xe^{136})_{tr} = 0.0141$ (this paper) and $Xe^{136}_{sp} \equiv 0$ we calculate $Xe^{126}_{sp} = 185 \times 10^{-12}$ cm$^3$ STP/g. From rocks 10017 and 10071 Eberhardt *et al.* (1972) obtained a production rate of $1.3 \times 10^{-21}$ cm$^3$ STP $Xe^{126}_{sp}$/yr "ppm Ba." This production rate includes the contribution from all target elements, i.e., Ba and the REE. For convenience, normalization to Ba alone was used. We denote this by writing ppm Ba in quotation marks. The REE contribute only approximately 25% of the total $Xe^{126}_{sp}$ (Eberhardt *et al.*, 1970a) and the production rate given above is valid within $\pm 10\%$ as long as the Ba/REE ratio is the same as in rocks 10017 and 10071 to within $\pm 40\%$. In rocks 10017 and 10071 the Ba/Ce ratios are 4.0 and 3.9, respectively (Gast *et al.*, 1970). In 12001 this ratio is 4.2 (Schnetzler and Philpotts, 1971). With a Ba content of 370 ppm (Schnetzler and Philpotts, 1971) we obtain for the 12001 bulk sample an exposure age of $390 \times 10^6$ yr.

### *12001, bulk grain size fractions*

From the $(Xe^{126}/Xe^{130})_m$ versus "Ba"/$Xe^{130}_m$ correlation (cf. Fig. 8) we obtain $Xe^{126}_{sp}$/"Ba" = $(0.57 \pm 0.015) \times 10^{-12}$ cm$^3$ STP/ppm. The average Ba/Ce ratio in the bulk grain size fractions is 3.6, again very similar to rocks 10017 and 10071. With the production rate given above we obtain $T_E = 440 \times 10^6$ yr.

### *12001, ilmenite grain size fractions*

From the slope of the correlation line in the $(Ar^{36}/Ar^{38})_m$ versus $1/Ar^{38}_m$ correlation diagram we obtain $Ar^{38}_{sp} = (24 \pm 4) \times 10^{-8}$ cm$^3$ STP/g. With an $Ar^{38}_{sp}$ production rate of $8.1 \times 10^{-16}$ cm$^3$ STP/yr g ilmenite (Eberhardt *et al.*, 1972) an exposure age of $300 \times 10^6$ yr is calculated.

### *10046, ilmenite grain size fractions*

From the slope of the correlation line in the $(Ar^{36}/Ar^{38})_m$ versus $1/Ar^{38}_m$ correlation diagram we obtain $Ar^{38}_{sp} = (70 \pm 15) \times 10^{-8}$ cm$^3$ STP/g and exposure age of $860 \times 10^6$ yr results.

An average $Kr^{83}_{sp}$ content of $1450 \times 10^{-12}$ cm$^3$ STP/g is obtained if the measured $(Kr^{83}/Kr^{86})_m$ ratios are individually corrected for trapped gas. The isotopic composition for trapped Kr as derived in this paper was used. With the average production rate of $1.95 \times 10^{-18}$ cm$^3$ $Kr^{83}_{sp}$/yr g ilmenite

Table 7. Average exposure ages (in million years) of the investigated fractions.

| | 10084 | | 12001 | | | 10046 |
|---|---|---|---|---|---|---|
| | Bulk grain size fractions | Ilmenite grain size fractions | Unseparated sample | Bulk grain size fractions | Ilmenite grain size fractions | Ilmenite grain size fractions |
| Average exposure age (m.y.) | 520 | 380 | 390 | 440 | 300 | 800 |

The exposure ages for the fractions separated from the Apollo 11 fines 10084 are from Eberhardt *et al.* (1970) (see also discussion in text). We estimate that the exposure ages are accurate within $\pm 20\%$ (neglecting the uncertainties in the production rates).

measured in ilmenite concentrates of rocks 10003 and 10017 (Eberhardt *et al.*, 1972 and yet unpublished results) we estimate from the average $Kr^{83}_{sp}$ content an exposure age of 740 × $10^6$ yr. The exposure ages derived from $Ar^{38}_{sp}$ and $Kr^{83}_{sp}$ agree fairly well. We adopt an exposure age of 800 × $10^6$ yr for breccia 10046.

*10084, ilmenite grain size fractions*

The $Ar^{38}_{sp}$ content given in Table 9 of Eberhardt *et al.* (1970) for the ilmenite separated from the Apollo 11 fines 10084 is incorrect. From the slope of the correlation line in Fig. 8 of Eberhardt *et al.* (1970) we calculate an $Ar^{38}_{sp}$ content of (31 ± 7) × $10^{-8}$ $cm^3$ STP/g ilmenite for the sample 10084. This corresponds to an exposure age of 380 × $10^6$ yr.

The exposure ages of the two soil samples are similar, with an indication that soil 12001 was, on the average, irradiated for a somewhat shorter time. Both ilmenite fractions have consistently lower exposure ages than the bulk fractions.

The ilmenite from breccia 10046 is older by a factor of more than two than the soil ilmenites. Assuming that the parent material of the breccia had a similar irradiation history as the 10084 and 12001 samples we can conclude that the compaction of breccia 10046 from loose soil material must have occurred at least 400 to 500 m.y. ago. Consequently, we estimate that the trapped solar wind particles are, on the average, at least 400 to 500 m.y. older than in the two soil samples.

Concentrations of spallation noble gases and the relative isotopic abundances of the spallation components are compiled in Table 8. Spallation concentrations were derived from the correlation between isotopic ratios and inverse concentrations. All isotopic compositions are derived from the slope of the correlation line in the usual two isotope ratio correlation diagrams [e.g., $(Xe^{124}/Xe^{126})_{sp}$ from the correlation between $(Xe^{124}/Xe^{132})_m$ and $(Xe^{126}/Xe^{132})_m$]. For error definition see next chapter.

The Kr spallation yields are similar to those observed in lunar rocks (Marti *et al.*, 1970; Pepin *et al.*, 1970; Hohenberg *et al.*, 1970; Eberhardt *et al.*, 1972). The observed variations are probably due to differences in the irradiation hardness and the Sr/Zr ratio (Eberhardt *et al.*, 1970a; Schwaller *et al.*, 1971). The $(Kr^{80}/Kr^{78})_c$ ratio could also be influenced by epithermal neutron capture on Br (Marti *et al.*, 1966; Eugster *et al.*, 1969; Lugmair and Marti, 1972).

Table 8. Concentrations and isotopic composition of the cosmogenic components in the investigated samples.

| | 12001 | | 10046 |
|---|---|---|---|
| | Bulk grain size fractions | Ilmenite grain size fractions | Ilmenite grain size fractions |
| $Ne^{21}_{sp}$ | 51 ± 10 | 11 ± 3 | 14 ± 4 |
| $Ar^{38}_{sp}$ | 58 ± 20 | 24 ± 4 | 70 ± 15 |
| $(Kr^{80}/Kr^{78})_c$ | 2.75 ± 0.15 | 2.18 ± 0.05 | 2.59 ± 0.06 |
| $(Kr^{82}/Kr^{78})_c$ | 3.1 ± 0.6 | 3.16 ± 0.08 | 4.4 ± 0.2 |
| $(Kr^{83}/Kr^{78})_{sp}$ | 4.3 ± 0.4 | 4.2 ± 0.4 | 5.2 ± 0.5 |
| $(Xe^{124}/Xe^{126})_{sp}$ | 0.556 ± 0.012 | 0.548 ± 0.01 | 0.575 ± 0.014 |
| $(Xe^{128}/Xe^{126})_c$ | 1.54 ± 0.04 | 1.58 ± 0.10 | 1.96 ± 0.08 |
| $(Xe^{130}/Xe^{126})_{sp}$ | 0.96 ± 0.09 | 0.66 ± 0.15 | 1.23 ± 0.27 |
| $(Xe^{131}/Xe^{126})_c$ | 5.0 ± 0.2 | 4.6 ± 0.2 | 5.6 ± 0.4 |

All concentrations in units of $10^{-8}$ $cm^3$ STP/g. For spallation Kr and Xe concentrations in some of the fractions see text.

The isotopic composition of the spallation Xe is in good agreement with the spectra observed in corresponding lunar rocks (Marti *et al.*, 1970; Pepin *et al.*, 1970; Hohenberg *et al.*, 1970; Eberhardt *et al.*, 1972). From the observed $(Xe^{131}/Xe^{126})_c$ ratio of approximately 5 it follows that these samples were not irradiated at the surface for most of their exposure time (cf. Eberhardt *et al.*, 1970a; Schwaller *et al.*, 1971; Eberhardt *et al.*, 1971). For surface irradiation a $(Xe^{131}/Xe^{126})_c$ ratio of 3 or below would be expected, for irradiation at considerable depth the ratio could be as high as 9. The intermediate ratio of 5 corresponds to an average shielding of approximately 20 g $cm^{-2}$ or an average burial depth in the regolith of approximately 10 cm.

## Isotopic Composition of Trapped Gases

Basically the same approach as used by Eberhardt *et al.* (1970) is utilized to make the necessary corrections in the measured isotope ratios in order to obtain the true trapped gas isotopic compositions. A more detailed discussion of the methods can be found in the above mentioned paper and only a short summary of the necessary corrections will be given here. If not otherwise stated, spallation isotope production rates used are from Eberhardt *et al.* (1972).

It is important to point out that the two basic approaches used by Eberhardt *et al.* (1970) are not completely equivalent. Correcting measured isotopic ratios individually for the cosmogenic, radiogenic, and fission components will give the isotopic composition of the total trapped component; i.e., the weighted average of the isotopic composition of the surface correlated and the volume correlated components. On the other hand the use of a correlation diagram, as introduced by Eberhardt *et al.* (1970), will give the isotopic composition of the surface correlated component. The two approaches are equivalent if no volume correlated trapped component is present ($n \approx 1$, i.e., ilmenite grain size fractions) or if the isotopic composition of the volume correlated and the surface correlated trapped gas are the same.

The errors given for a trapped gas isotopic composition obtained by individually correcting measured isotope ratios is the standard deviation of the individually corrected values, as determined from their distribution about the average. For the isotopic composition of the surface correlated gas derived by extrapolation, the standard deviation of the ordinate intercept is given as error. The same error definition was used in the case of spallation gas concentrations determined from the slope of the correlation diagrams. With this error definition the individual experimental errors are automatically taken into account, with the possible exception of systematic errors in the measured ratios or concentrations (possible systematic errors are, however, included in the errors given in Tables 1 to 3). As we mainly compare results obtained at Bern with identical experimental techniques, such systematic errors will essentially cancel out. Our error definition does not include possible uncertainties in the necessary assumptions, and this point will have to be discussed specifically for each result.

### *Helium*

*12001, unseparated sample.* With an exposure age of 390 × $10^6$ yr and a production rate of $10^{-14}$ $cm^3$ STP $He^3$/g yr we calculate $He^3_{sp}$ = 390 × $10^{-8}$ $cm^3$ STP/g. However, $He^3_{sp}$ diffusion loss is frequent in lunar material (Hintenberger *et al.*, 1971; Eberhardt *et al.*, 1972) and this concentration must be considered as an upper limit for the true $He^3_{sp}$ content. We estimate $He^4_r = 110{,}000 \times 10^{-8}$ cm³ STP/g, corresponding to an "age" of $2.5 \times 10^9$ yr (U and Th concentrations from Wänke *et al.*, 1971). With these assumptions we obtain $2290 \leq (He^3/He^4)_{tr} \leq 2630$.

*12001, bulk grain size fractions.* Similar to the unseparated sample we can only give an upper limit of $440 \times 10^{-8}$ cm³ STP/g for the spallation He³ content. This uncertainty makes an evaluation of the $(He^4/He^3)_{tr}$ ratio in the coarser grain size fractions impossible. For the two finest fractions bb-17 and bb-18 the correction for $He^3_{sp}$ is small ($He^3_{sp}/He^3_m < 2.5\%$) and we estimate $2210 \leq (He^4/He^3)_{tr} \leq 2540$.

*12001, ilmenite grain size fractions.* With an exposure age of $300 \times 10^6$ yr and a $He^3_{sp}$ production rate of $0.9 \times 10^{-14}$ cm³ STP/yr g ilmenite we obtain $He^3_{sp} = 270 \times 10^{-8}$ cm³ STP/g ilmenite. From the measurements on ilmenite separated from rocks 10071 and 10003 (Eberhardt *et al.*, 1972, and yet unpublished results) it is evident that $He^4_r$ is $<1\%$ even for the coarsest ilmenite fraction and can be neglected. After correcting for spallation He³ we obtain for the bb-26, bb-37, bb-52, and bb-57 grain size fractions $(He^4/He^3)_{tr}$ ratios of 2790; 2700; 2610; and 2710, respectively, and an average value of $2700 \pm 70$.

*10046, ilmenite grain size fractions.* The exposure age of $800 \times 10^6$ yr corresponds to $He^3_{sp} = 720 \times 10^{-8}$ cm³ STP/g ilmenite. Again $He^4_r$ can be neglected. After correcting for the spallation He³ we obtain for the four grain size fractions DA-14, DA-15, DA-16, and DA-17 ratios $(He^4/He^3)_{tr}$ of 3280; 3020; 2890; and 3040, respectively. The average value is $3060 \pm 160$.

*Neon*

*12001, unseparated sample.* The exposure age of the unseparated 12001 sample is 10% lower than the average age of the bulk grain size fractions (cf. Table 7). Thus we expect in the 12001 unseparated sample a 10% lower $Ne^{21}_{sp}$ content than in the grain size fractions. With $Ne^{21}_{sp} = 45 \times 10^{-8}$ cm³ STP/g we obtain $(Ne^{20}/Ne^{22})_{tr} = 12.43$ and $(Ne^{22}/Ne^{21})_{tr} = 33.0$.

*12001, bulk grain size fractions.* From the correlation between $(Ne^{22}/Ne^{21})_m$ and $1/Ne^{21}_m$, we obtain $(Ne^{22}/Ne^{21})_{sc} = 31.5 \pm 1.1$.

From the same correlation diagram an average $Ne^{21}_{sp}$ content of $(51 \pm 10) \times 10^{-8}$ cm³ STP/g is calculated. Correcting for the corresponding amount of $Ne^{22}_{sp}$ we calculate in the fractions bb-12 to bb-18 ratios $(Ne^{20}/Ne^{22})_{tr}$ of 12.56; 12.31; 12.52; 12.38; 12.55; 12.54; and 12.67, respectively. The correction is always smaller than 1.6% and less than 0.3% for the three finest fractions. From all fractions we obtain an average $(Ne^{20}/Ne^{22})_{tr} = 12.50 \pm 0.12$.

*12001, ilmenite grain size fractions.* From the correlation between $(Ne^{22}/Ne^{21})_m$ and $1/Ne^{21}_m$, we obtain $(Ne^{22}/Ne^{21})_{sc} = 32.0 \pm 0.4$. This value is in good agreement with the measured ratio in the two finest grain size fractions which contain negligible amounts of $Ne^{21}_{sp}$. From the slope of the correlation line $Ne^{21}_{sp} = 11 \times 10^{-8}$ cm³ STP/g is calculated. Correcting for the corresponding amounts of $Ne^{22}_{sp}$ we obtain for bb-26, bb-37, bb-52, and bb-57 ratios $(Ne^{20}/Ne^{22})_{tr}$ of 12.67; 12.72; 12.91; and 13.00, respectively. The average of $12.83 \pm 0.16$ is slightly lower than the value of $(Ne^{20}/Ne^{22})_{sc} = 12.96 \pm 0.06$ calculated from the correlation between $(Ne^{20}/Ne^{22})_m$ and $1/Ne^{22}_m$.

*10046, ilmenite grain size fractions.* From the correlation between $(Ne^{22}/Ne^{21})_m$ and $1/Ne^{21}_m$, we obtain $(Ne^{22}/Ne^{21})_{sc} = 31.4 \pm 0.4$. This is again in good agreement with the measured ratios of the finest two fractions which contain negligible amounts of $Ne^{21}_{sp}$. The slope of the correlation lines gives $Ne^{21}_{sp} = 14 \times 10^{-8}$ cm³ STP/g. Correcting for the corresponding amounts of $Ne^{22}_{sp}$ we obtain for DA-14 to DA-17 ratios $(Ne^{20}/Ne^{22})_{tr}$ of 12.64; 12.56; 12.50; and 12.81, respectively. The average is $12.63 \pm 0.13$.

*Argon*

*12001, bulk grain size fractions.* From the $(Ar^{36}/Ar^{38})_m$ versus $1/Ar^{38}_m$ correlation diagram, we obtain $(Ar^{36}/Ar^{38})_{sc} = 5.29 \pm 0.03$. From the correlation between $(Ar^{40}/Ar^{36})_m$ and $1/Ar^{36}_m$, a value of $(Ar^{40}/Ar^{36})_{sc} = 0.42 \pm 0.02$ is derived.

*12001, ilmenite grain size fractions.* From the $(Ar^{36}/Ar^{38})_m$ versus $1/Ar^{38}_m$ correlation diagram, we obtain $(Ar^{36}/Ar^{38})_{sc} = 5.33 \pm 0.03$. The correlation between $(Ar^{40}/Ar^{36})_m$ and $1/Ar^{36}_m$ gives $(Ar^{40}/Ar^{36})_{sc} = 0.37 \pm 0.05$.

*10046, ilmenite grain size fractions.* From the two corresponding correlations, we obtain $(Ar^{36}/Ar^{38})_{sc} = 5.33 \pm 0.03$ and $(Ar^{40}/Ar^{36})_{sc} = 1.35 \pm 0.14$.

*Krypton*

*12001, bulk grain size fractions.* The use of a correlation diagram, such as $(Kr^{78}/Kr^{86})_m$ versus $1/Kr^{86}_m$ for deducing the $(Kr^{78}/Kr^{86})_{sc}$ ratio, implies that the nontrapped $Kr^{78}$ and $Kr^{86}$ concentrations (e.g., $Kr^{78}_{sp}$) are the same in all grain size fractions. A necessary condition is thus the equality of the target element concentrations in all grain size fractions. This assumption is certainly valid for spallation argon in the ilmenites, as the important target elements Ti and Fe are main constituents. If trace elements are the major target elements, possible systematic or random variations with grain size must be considered.

In the five finest bulk grain size fractions we have measured the Sr and Zr concentrations. If [Sr] is the Sr abundance and [Zr] the Zr abundance, then we estimate from spallation theory (Geiss *et al.*, 1962; $n = 2$) that the concentration of a spallation Kr isotope is proportional to [Sr] + 0.8 [Zr] [assuming $Y/Zr = 0.3$; cf. Morrison *et al.* (1971); Wänke *et al.* (1971)]. A correlation diagram between a measured Kr isotope ratio $(Kr^M/Kr^{86})_m$ and $([Sr] + 0.8\,[Zr])/Kr^{86}_m$ will then give a straight line even if the target element concentration is different in different grain sizes. The ordinate intercept gives the isotopic composition of the surface correlated gas and the slope of the correlation line essentially the exposure age. This is true if two assumptions are fulfilled: (1) The gases in all fractions

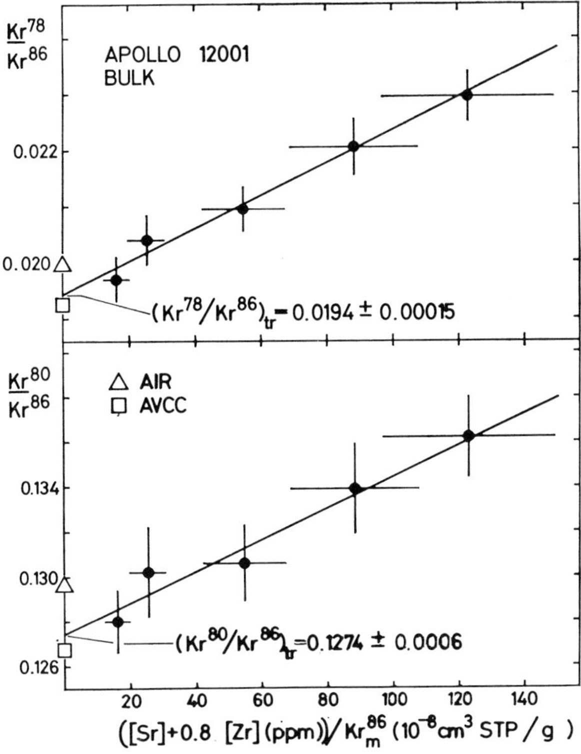


Fig. 7. Correlation between the measured $Kr^{78}/Kr^{86}$ and $Kr^{80}/Kr^{86}$ ratios for the 12001 bulk grain size samples bb-14 to bb-18 and $([Sr] + 0.8\,[Zr])/Kr^{86}_m$. The factor [Sr] + 0.8 [Zr] compensates the variable chemical composition in the different grain size fractions.

are a mixture with different mixing ratios of the same two components (trapped and "spallation"). (2) The exposure age of all fractions is the same.

Figure 7 shows the correlation diagram for the evaluation of the $(Kr^{78}/Kr^{86})_{sc}$ and $(Kr^{80}/Kr^{86})_{sc}$ ratios. The isotopic composition of the trapped Kr, as derived from such diagrams, is given in Table 11. It is important to point out that the factor 0.8 for the relative production cross section of Zr does not enter critically in the calculation for the trapped ratios. Assuming $Kr_{sp} \propto [Sr] + 1.6\,[Zr]$ would change the $(Kr^{78}/Kr^{86})_{sc}$ ratio by less than 0.2% and all other ratios by less than 0.1%.

### *Xenon*

*12001, bulk grain size fractions.* The isotopic composition of surface correlated Xe is obtained from correlation diagrams. Similar to the case of krypton, variations in the target element concentrations are taken into account by correlating the measured isotopic ratios with the $Ba/Xe^{130}_{m}$ ratio. Ba and the REE are the main target elements for the production of spallation Xe. The Ba/Ce ratio is the same within ±15% in the three fractions bb-14, bb-16, and bb-18. The Ba concentration is thus representative for the target element concentration (cf. discussion on exposure age of 12001 bulk sample).

The correction for the variability in the target element abundances is important for the 12001 bulk grain size fractions. This is evident from Fig. 8 where the correlations of the measured $(Xe^{126}/Xe^{130})_{m}$ ratio both with $1/Xe^{130}_{m}$ and with $Ba/Xe^{130}_{m}$ are shown. The two correlation lines are different and the use of the $(Xe^{126}/Xe^{130})_{m}$ versus $1/Xe^{130}_{m}$ correlation line would give a 30% higher $(Xe^{126}/Xe^{130})_{sc}$ ratio than the barium corrected diagram. The variable chemistry is of similar importance for the evaluation of $(Xe^{124}/Xe^{130})_{sc}$ and $(Xe^{128}/Xe^{130})_{sc}$. In Table 12 the isotopic composition of the trapped Xe in the 12001 bulk grain size fractions are compiled. All ratios were calculated from correlation with the $Ba/Xe^{130}_{m}$ ratio.

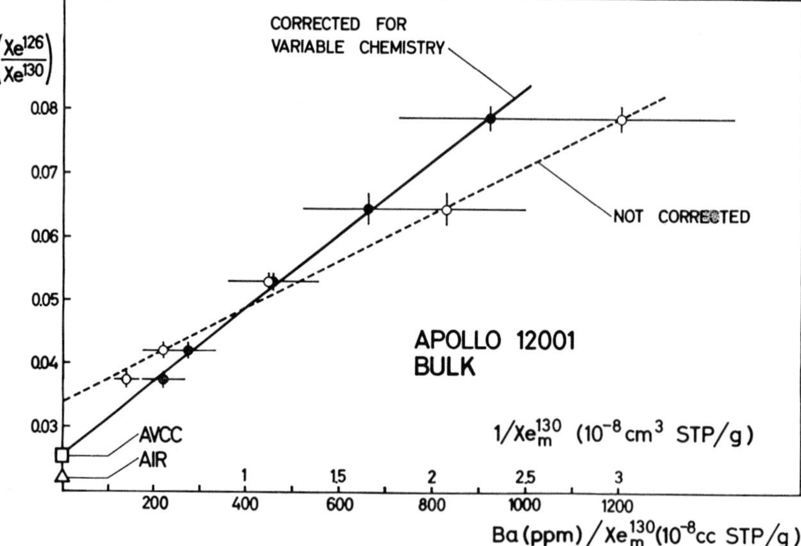


Fig. 8. Correlation between the measured $Xe^{126}/Xe^{130}$ ratios and $1/Xe^{130}_{m}$ (dashed line) and $[Ba]/Xe^{130}_{m}$ (solid line). The variable target element concentration in the different grain sizes is automatically compensated for by correlating the $(Xe^{126}/Xe^{130})_{m}$ ratio with the $[Ba]/Xe^{130}_{m}$ ratio. The importance of using this refined evaluation method for obtaining the isotopic composition of the surface correlated trapped gas from the measurements on the 12001 bulk grain size fractions is evident from this figure.

## Discussion of Isotopic Composition of Trapped Gases

### *Helium*

There is an excellent agreement between the $(He^3/He^4)_{tr}$ ratios in the 10084 and 12001 ilmenites (Table 9) (in this discussion we shall preferably use ratios with the heavier isotope in the denominator). The ilmenite of breccia 10046 shows a distinctly lower ratio. This systematic difference is also apparent if the measured ratios are directly compared. The $(He^4/Ne^{20})_{tr}$ ratio of 10046 is intermediate between the two soils. Therefore, helium diffusion losses, if any, must be similar in all three ilmenites. It is thus very unlikely that the low $(He^3/He^4)_{tr}$ ratio in 10046 is due to a different diffusion history of the trapped solar wind particles in this sample.

The $He^3/He^4$ ratio in the solar wind is variable (Bame *et al.*, 1968; Geiss *et al.*, 1970, 1971a). The most reliable estimate for the average ratio in the present day solar wind can be obtained from the observed correlation between the $He^3/He^4$ ratio and the magnetic index $K_p$ (Geiss *et al.*, 1971a). The soil ilmenite $(He^3/He^4)_{tr}$ is slightly lower than the solar wind value, but, within the experimental uncertainty, they would still agree. However, this comparison is rather futile as the influence of the unknown trapping probability of ilmenite for the solar wind ions, of the heavy saturation and of the possible He diffusion loss on the $(He^3/He^4)_{tr}$ ratio in the ilmenite has not yet been investigated. The low $(He^4/Ne^{20})_{tr}$ ratio in the ilmenite, as compared with the present day solar wind value, must be taken as an indication that the $(He^3/He^4)_{tr}$ ratio in the ilmenite might have been lowered by the same processes which changed the $(He^4/Ne^{20})_{tr}$ ratio. If such processes are taken into account, a still better agreement between the present day solar wind $He^3/He^4$ and the $(He^3/He^4)_{tr}$ in the ilmenite might result.

The trapped gases in aubritic meteorites are also surface correlated and located at the surface of the individual grains (Eberhardt *et al.*, 1965, 1965a). They most likely represent trapped old solar wind particles or similar low energy ions. The $(He^3/He^4)_{tr}$ ratio in the aubrites is distinctly higher than in the lunar ilmenites. The $(He^4/Ne^{20})_{tr}$ ratio is intermediate between the ilmenite and present day solar wind

Table 9. Comparison of elemental and isotopic abundance ratios in different reservoirs of trapped solar wind particles with average composition in present day solar wind (Geiss *et al.*, 1971a).

| Reservoir | $He^4/Ne^{20}$ | $Ne^{20}/Ar^{36}$ | $He^3/He^4$ ($\times 10^{-4}$) | $Ne^{20}/Ne^{22}$ | $Ne^{21}/Ne^{22}$ ($\times 10^{-2}$) |
|---|---|---|---|---|---|
| *Trapped solar wind in* | | | | | |
| 10084 ilmenite | 218 ± 8 | 27.4 ± 4 | 3.68 ± 0.12 | 12.85 ± 0.1 | 3.22 ± 0.08 |
| 12001 ilmenite | 253 ± 10 | 26.8 ± 5 | 3.70 ± 0.1 | 12.9 ± 0.1 | 3.13 ± 0.04 |
| 10046 ilmenite | 231 ± 13 | — | 3.27 ± 0.17 | 12.65 ± 0.15 | 3.18 ± 0.04 |
| Aubritic meteorites | 380 ± 70 | ~22 | 2.50 ± 0.30 | 12.5 ± 0.6 | — |
| *For comparison* | | | | | |
| Present day solar wind | 600 ± 150 | $37^{+10}_{-5}$* | 4.00 ± 0.45 | 13.6 ± 0.3 | 3.23 ± 0.4 |
| Terrestrial atmosphere | n.r. | 0.5 | n.r. | 9.80 ± 0.08 | 2.90 ± 0.08 |

Data for aubritic meteorites from Zähringer (1962) and Eberhardt *et al.* (1965); data for 10084 ilmenite from Eberhardt *et al.* (1970).

* Preliminary value from Apollo 14 SWC experiment only (Geiss *et al.*, 1971).

n.r. = not relevant.

values. In other meteorites, particularly in chondrites, trapped gases with low $(He^3/He^4)_{tr}$ ratios have been found (Hintenberger *et al.*, 1965). Strong evidence exists, that in these meteorites at least two components with different $(He^3/He^4)_{tr}$ ratio are present (Anders *et al.*, 1970; Black, 1970; Jeffery and Anders, 1970), a planetary component with a low $(He^3/He^4)_{tr}$ ratio and a solar component with a high ratio.

These at present available data on the $(He^3/He^4)_{tr}$ ratio in different reservoirs of trapped solar wind particles strongly suggests a secular variation of this ratio in the solar wind and probably also in the sun (Schatzman, 1970; Eberhardt *et al.*, 1970; Geiss *et al.*, 1970). Present day solar wind as directly measured or approximately preserved in the soil ilmenite shows the highest $He^3$ abundances. The solar wind particles in the breccia ilmenite are at least $500 \times 10^6$ yr older and have a $He^3$ abundance approximately 10% lower than in the soil ilmenite. A similar trend is also apparent in the measurements on bulk soil and breccia samples of Hintenberger *et al.* (1971). The trapped gas in aubrites would then represent perhaps the oldest solar wind reservoir, having a $He^3$ abundance which is nearly 40% lower. In our opinion the presently available data do not establish the secular increase of the $He^3/He^4$ ratio in the solar wind beyond doubt. However, our data from the 10084, 12001, and 10046 ilmenites clearly support a long-time variation. It will be most important to substantiate the $He^3/He^4$ secular trend as the solar $He^3/He^4$ ratio prevailing shortly after the formation of the solar system is an important parameter for deducing the D/H ratio in the solar nebula (Black, 1972; Geiss and Reeves, 1972).

### *Neon*

The isotopic composition of surface correlated Ne is very similar in the three ilmenites. The $(Ne^{20}/Ne^{22})_{sc}$ ratios are consistently lower than the ratio measured in the present day solar wind (Geiss *et al.*, 1970, 1971, 1971a) (cf. Table 9). The difference is small ($\approx 5\%$) and could well be due to fractionation in the trapping process of solar wind Ne ions or to the heavy saturation of the surface layer with solar wind particles. In the ilmenite there is no evidence for a large diffusion loss of trapped Ne. The $(Ne^{20}/Ar^{36})_{tr}$ ratio in the ilmenite is only 30% lower than the ratio measured in the solar wind during the Apollo 14 mission (Geiss *et al.*, 1971).

The systematic difference between the isotopic composition of terrestrial Ne and trapped lunar or solar wind Ne has already been discussed earlier (Eberhardt *et al.*, 1970).

### *Argon*

The surface correlated $Ar^{36}/Ar^{38}$ ratio is identical in the 12001 bulk and the three ilmenite samples (Table 10). The ratio agrees very well with the terrestrial $Ar^{36}/Ar^{38}$ abundance.

The $(Ar^{40}/Ar^{36})_{sc}$ ratio is different in all five samples. It has already been established that the $Ar^{40}_{sc}$ in lunar fine material is not of direct solar wind origin, but that it represents retrapped lunar $Ar^{40}$ from the decay of $K^{40}$ in lunar rocks (Heymann *et al.*, 1970; Heymann and Yaniv, 1970; Eberhardt *et al.*, 1970). Escaping $Ar^{40}$ atoms form a transient lunar atmosphere, which is ionized by solar uv or charge exchange with

Table 10. Comparison of argon isotopic composition in different reservoirs of trapped solar wind particles.

| Reservoir | $Ar^{36}/Ar^{38}$ | $Ar^{40}/Ar^{36}$ |
|---|---|---|
| *Trapped solar wind in* | | |
| 10084 bulk | — | $0.95 \pm 0.06$ |
| 12001 bulk | $5.29 \pm 0.03$ | $0.42 \pm 0.02$ |
| 10084 ilmenite | $5.32 \pm 0.08$ | $0.67 \pm 0.06$ |
| 12001 ilmenite | $5.33 \pm 0.03$ | $0.37 \pm 0.05$ |
| 10046 ilmenite | $5.33 \pm 0.03$ | $1.35 \pm 0.14$ |
| *For comparison* | | |
| Terrestrial atmosphere | $5.32 \pm 0.01$ | n.r. |

Error definition for 10084 is different from the one used in present paper (cf. Eberhardt *et al.*, 1970).
n.r. = not relevant.

the solar wind and subsequently accelerated. A fraction of these accelerated ions is trapped along with the solar wind particles in the surface of the regolith grains (cf. Manka and Michel, 1971). The $(Ar^{40}/Ar^{36})_{sc}$ ratio is thus a complicated function depending on the $Ar^{40}$ outgassing rate of the moon, the $Ar^{36}$ flux in the solar wind, the interplanetary magnetic field, the local lunar magnetic field, the lunar latitude and longitude of the sample and other parameters. Trapping probabilities, saturation and diffusion effects will also influence the $(Ar^{40}/Ar^{36})_{sc}$ ratio as the average energy and direction of the solar wind ions and the secondary ions from the lunar neutral atmosphere are significantly different. The observed variability in the $(Ar^{40}/Ar^{36})_{sc}$ ratio in the different grain size fractions is thus not unexpected and can have resulted from one or from a combination of any of the above mentioned factors. The variability of the $(Ar^{40}/Ar^{36})_{sc}$ ratio between different bulk samples has also been found and discussed by other investigator teams (LSPET, 1970; Heymann and Yaniv, 1971; Heymann *et al.*, 1972). For the three ilmenite fractions the influence of trapping probability, saturation, etc., should be relatively similar and we could interpret the lower $(Ar^{40}/Ar^{36})_{sc}$ ratio in the 12001 ilmenite relative to the 10084 ilmenite as a longitude effect. Similarly, the high $(Ar^{40}/Ar^{36})_{sc}$ ratio in the 10046 ilmenite may reflect a long time variation in the $Ar^{40}$ outgassing rate of the moon. About twice the present outgassing rate would be required $500 \times 10^6$ yr or longer ago.

Along with radiogenic $Ar^{40}$, the moon will loose other noble gases or volatiles by outgassing. These should be mainly radiogenic $He^4$, Rn, $Xe^{129}$, fission Kr and Xe, primordial trapped noble gases, volatile elements, and also trapped solar wind gases. For the lighter elements, the thermal escape from the lunar atmosphere will be fast compared with the lifetime for ionization. For heavier or less volatile elements the retrapping from the temporary lunar atmosphere may become of importance. As already discussed (Eberhardt *et al.*, 1970; Podosek *et al.*, 1971), this retrapping of outgassing products of the moon could have influenced the abundance and isotopic composition of the surface correlated heavier noble gases in lunar material.

### *Krypton and Xenon*

The isotopic composition of surface correlated trapped Kr and Xe derived from the 12001 bulk grain size fractions is compared in Tables 11 and 12 with other de-

Table 11. Isotopic composition of surface correlated trapped Kr in lunar fine material.

| | $Kr^{78}/Kr^{86}$ | $Kr^{80}/Kr^{86}$ | $Kr^{82}/Kr^{86}$ | $Kr^{83}/Kr^{86}$ | $Kr^{84}/Kr^{86}$ |
|---|---|---|---|---|---|
| | | | × 100 | | |
| BEOC 12001 (this paper) | 1.945 ± 0.015 | 12.74 ± 0.06 | 65.73 ± 0.2 | 65.86 ± 0.17 | 327.9 ± 0.7 |
| BEOC 10084 (Eberhardt *et al.*, 1970) | 2.00 ± 0.035 | 12.90 ± 0.2 | 66.0 ± 0.3 | 65.75 ± 0.3 | 325.2 ± 1.8 |
| SUCOR (Podosek *et al.*, 1971) | — | 12.42 ± 0.08 | 64.25 ± 0.26 | 64.55 ± 0.27 | 322.8 ± 0.8 |
| AVCC (Eugster *et al.*, 1967a) | 1.927 ± 0.014 | 12.65 ± 0.09 | 65.04 ± 0.2 | 65.11 ± 0.2 | 322.8 ± 0.8 |
| Atmosphere (Nief 1960; Eugster *et al.*, 1967) | 1.995 ± 0.008 | 12.96 ± 0.04 | 66.17 ± 0.16 | 66.00 ± 0.14 | 327.3 ± 0.7 |

The error definitions are not necessarily equivalent, see original papers.

terminations of the isotopic composition of surface correlated Kr and Xe in lunar material. Other authors have also published isotopic compositions of trapped lunar Kr and Xe (Hohenberg *et al.*, 1970; Marti *et al.*, 1970; Pepin *et al.*, 1970; Kaiser, 1972). However, all these authors had to assume at least one isotopic ratio in the trapped Kr in order to make the necessary corrections for spallation Kr and Xe.

In order to facilitate the following discussion we shall designate the determinations

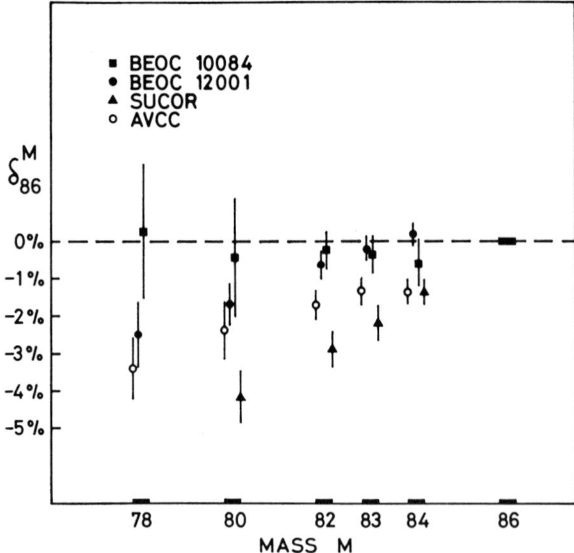


Fig. 9. Comparison between different determinations of the isotopic composition of surface correlated trapped Kr in lunar samples. Shown are the relative deviations from the composition of atmospheric Kr:

$$\delta_{86}^{M} = [(Kr^{M}/Kr^{86})_{sc}/(Kr^{M}/Kr^{86})_{atm} - 1].$$

Data from: BEOC 10084: Eberhardt *et al.* (1970); SUCOR: Podosek *et al.* (1971); BEOC 12001: this paper. The average isotopic composition of trapped Kr in carbonaceous chondrites (AVCC) is from Eugster *et al.* (1967a).

Table 12. Isotopic composition of surface correlated trapped Xe in lunar fine material.

| | $Xe^{124}/Xe^{130}$ | $Xe^{126}/Xe^{130}$ | $Xe^{128}/Xe^{130}$ | $Xe^{129}/Xe^{130}$ | $Xe^{131}/Xe^{130}$ | $Xe^{132}/Xe^{130}$ | $Xe^{134}/Xe^{130}$ | $Xe^{136}/Xe^{130}$ |
|---|---|---|---|---|---|---|---|---|
| | | | | × 100 | | | | |
| BEOC 12001 (this paper) | 2.90 ± 0.07 | 2.59 ± 0.09 | 50.38 ± 0.28 | 635.4 ± 1.7 | 498.8 ± 1.1 | 606.2 ± 1.6 | 223.9 ± 0.8 | 181.8 ± 0.6 |
| BEOC 10084 (Eberhardt *et al.*, 1970) | 2.99 ± 0.18 | 2.67 ± 0.2 | 50.45 ± 0.8 | 634.5 ± 3.5 | 495.5 ± 3 | 607 ± 5 | 225.8 ± 3.5 | 184.2 ± 3 |
| SUCOR (Podosek *et al.*, 1971) | 2.89 ± 0.04 | 2.63 ± 0.06 | 50.9 ± 0.3 | 637.1 ± 1.8 | 499.0 ± 2.2 | 606.6 ± 2.1 | 225.2 ± 0.9 | 182.7 ± 0.6 |
| Luna 16 (Kaiser, 1972) | — | — | — | 633.8 ± 6.5 | 494.7 ± 3 | 600.6 ± 3 | 220.1 ± 1.8 | 176.0 ± 2 |
| Pesyanoe 1000°C (Marti, 1969)* | — | — | — | 633.1 ± 9.5 | 495.0 ± 6.5 | 608 ± 7 | 222.2 ± 3.2 | 179.6 ± 2.8 |
| AVCC (Eugster *et al.*, 1967b) | 2.854 ± 0.065 | 2.550 ± 0.03 | 51.0 ± 0.45 | 634.4 ± 4† | 508.1 ± 3.9 | 621.9 ± 4.2 | 237.6 ± 2 | 199.6 ± 1.8 |
| Atmosphere (Nier, 1950; Podosek *et al.*, 1971) | 2.335 ± 0.014 | 2.176 ± 0.014 | 47.08 ± 0.17 | 650.5 ± 1.5 | 522.4 ± 1.2 | 661.4 ± 1.2 | 256.7 ± 0.7 | 218.2 ± 0.6 |

For Luna 16 and Pesyanoe no abundance data for the less abundant isotopes $Xe^{124-128}$ are given, as these depend critically on the spallation correction. The error definition for the different determinations are not equivalent, see original papers.

* The Xe in Pesyanoe is believed to be mainly trapped solar wind or low energy particles. For the discussion of the necessary spallation correction for Pesyanoe see Eberhardt *et al.* (1970).

† Abundance from Novo Urei (Marti, 1967).

made at Bern of the isotopic composition of the surface correlated trapped noble gases in the bulk material as BEOC 10084 and BEOC 12001 (BEOC = Bern Oberflächen-Correliert).

BEOC 10084, BEOC 12001, and SUCOR Kr do not agree within the error limits assigned by the respective authors. The deviations are however small and show a fairly systematic trend with the mass number (Fig. 9). For the BEOC 10084 and the SUCOR determinations the possibility of a variable target element composition in the different grain sizes was not investigated. Our experience with the BEOC 12001 evaluation has shown that a systematic variation of the spallation gas content with grain size can indeed lead to a nonnegligible shift of the ordinate intercept in the extrapolation method used in determining the BEOC 10084 and SUCOR isotopic compositions. Furthermore, mineral dependent mass fractionation could occur in the trapping of the solar wind particles in the lunar dust grains. The gas losses associated with the heavy saturation and the subsequent structural changes which occur at the surface of the grains (Borg *et al.*, 1971) and also possible thermal diffusion losses could give an additional mass fractionation. We may thus expect small differences in the isotopic composition of surface correlated trapped gases between lunar samples of different mineralogy, thermal history and solar wind irradiation history.

Figure 10 shows the isotopic composition of atmospheric Kr and AVCC Kr (AVCC = average carbonaceous chondrites) relative to BEOC 12001. All $\eta$ values, except for the reference isotope $Kr^{86}$, fall on straight lines. Our conclusions are:

(1) AVCC and BEOC 12001 Kr have identical isotopic composition except for a $Kr^{86}$ excess with $Kr^{86}_{excess}/Kr^{86} = 1.5\%$ in AVCC Kr and a very small mass fractionation of $\sim 1‰$ per mass unit.

(2) ATM Kr and BEOC 12001 Kr are related by a mass dependent fractionation process with a fractionation factor of 4.7‰ per mass unit (light isotopes enriched in atmospheric Kr). In addition, a $Kr^{86}$ excess with $Kr^{86}_{excess}/Kr^{86} = 1.1\%$ is present in atmospheric Kr.

The mass dependent fractionation between atmospheric and BEOC 12001 Kr could easily have occurred at the lunar surface (see above discussion). In our opinion no definite conclusion that solar wind Kr or solar Kr have an overall isotopic composition different from terrestrial Kr can be drawn with the presently available data.

Excess $Kr^{86}$ in the terrestrial atmosphere and in AVCC Kr could be interpreted as fission Kr. The presence of fission Kr in AVCC krypton has already been discussed (Eugster *et al.*, 1967). For the terrestrial atmosphere the required $Kr_f^{86}/Kr^{86} = 1.1\%$ corresponds to $Kr_f^{86} = 1.7 \times 10^{-3}$ cm$^3$ STP/cm$^2$ earth surface. The K abundance in the crust as given by Wasserburg *et al.* (1964) accounts for the production of atmospheric $Ar^{40}$. With the average U concentration of 9 g/cm$^2$ in the crust (Wasserburg *et al.*, 1964) a $Kr_f^{86}/U$ ratio of $2 \times 10^{-6}$ results. This ratio is much too high for spontaneous fission. For thermal neutron fission an integrated thermal flux of $3 \times 10^{17}$ to $2 \times 10^{19}$ neutrons cm$^{-2}$ would be necessary to give the required $Kr_f^{86}/U$ ratio. The lower flux corresponds to an irradiation early in the earth's history, the higher flux to a recent irradiation. Other neutron induced reactions, including some leading to noble gas isotopes, should also occur and at these high flux levels measurable effects should result. We estimate that with Cl/U and Br/U atomic ratios of 500

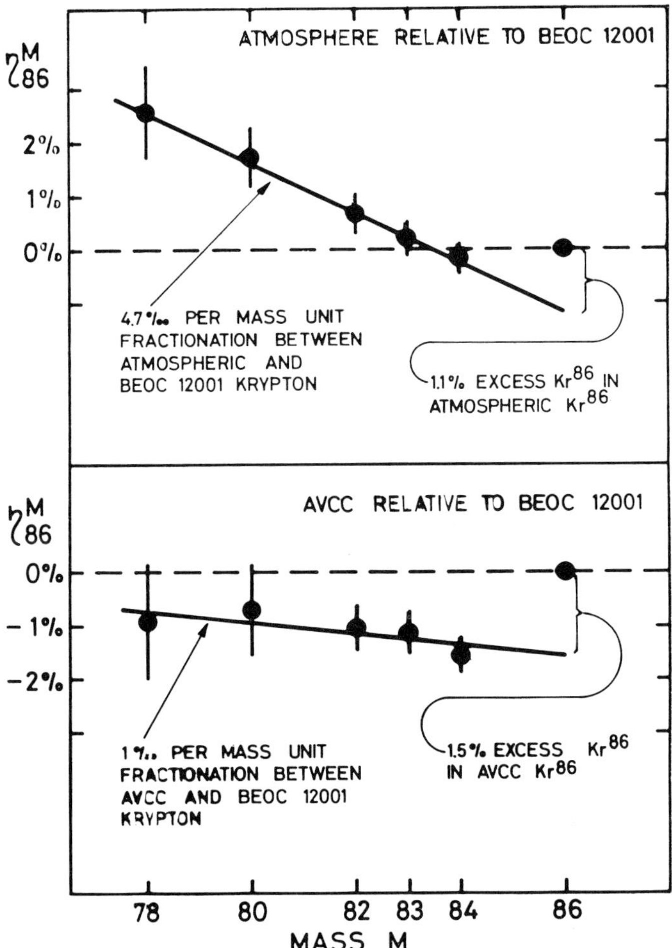


Fig. 10. Comparison of the isotopic composition of atmospheric Kr and average carbonaceous chondrite Kr (AVCC) with surface correlated (solar) trapped Kr in lunar material (BEOC 12001). Shown are the relative deviations

$$\eta_{86}^{M} = [(Kr^{M}/Kr^{86})_{sample}/(Kr^{M}/Kr^{86})_{BEOC\ 12001} - 1].$$

and 4 respectively, as are typical for the earth's crust (Mason, 1966), neutron capture on Cl and Br should produce between 4 and 300 $cm^3$ STP $Ar^{36}/cm^2$ and at least 0.05 $cm^3$ STP $Kr^{80}/cm^2$. These contributions would drastically alter the isotopic composition of atmospheric Ar and Kr. Fission Xe must be associated with fission $Kr^{86}$. Even the low $(Xe^{136}/Kr^{86})_f$ ratio of $U_{fn}^{235}$ would imply that 90% of the atmospheric $Xe^{136}$ is of fission origin, if $Kr_f^{86}$ is 1.1% of the $Kr^{86}$ in the atmosphere. For spontaneous fissions, with the correspondingly higher $(Xe^{136}/Kr^{86})_f$ ratios the resulting $Xe_f^{136}/Xe^{136}$ would be considerably greater than one. This problem could, however, be alleviated by either assuming incomplete Xe outgassing of the $Kr_f^{86}$ reservoir or by assuming completely different origins or histories for terrestrial Kr and Xe. The latter possibility is of course strongly supported by the large isotopic anomalies observed between terrestrial Xe and xenon present in other reservoirs of the solar system.

Our conclusion, that excess $Kr^{86}$ is present in the terrestrial atmosphere, hinges critically on the BEOC 12001 $(Kr^{84}/Kr^{86})_{sc}$ ratio. If this ratio would be in error by ~1%, i.e., if it would be ~324 instead of 327.9, then the difference between BEOC 12001 and atmospheric Kr could be explained by a simple mass fractionation. As the data in Table 2 show, the $Kr^{84}/Kr^{86}$ ratio increases with decreasing grain size. This is contrary to what would be expected from the spallation contribution and the negative slope in the $(Kr^{84}/Kr^{86})_m$ versus $([Sr] + 0.8[Zr])/Kr_m^{86}$ correlation diagram can only be explained by a volume component of $Kr^{86}$, i.e., a possible fission component in lunar krypton. A weak negative correlation is also indicated in the 10084 grain size data of Eberhardt *et al.* (1970). The amount of this volume component in the 12001 bulk grain size fractions is approximately $400 \times 10^{-12}$ $cm^3$ STP $Kr_{ff}^{86}$/g. An *in situ* origin of this fission Kr is not compatible with the U and Th abundances and estimated neutron fluxes. The small meteoritic component in the lunar fine material (Ganapathy *et al.*, 1970; Laul *et al.*, 1971) also cannot provide the necessary $Kr_f^{86}$ as in carbonaceous chondrites $Kr_f^{86}$ is only of the order of $20 \times 10^{-12}$ $cm^3$ STP/g (Eugster *et al.*, 1967). If this volume correlated $Kr^{86}$ component in the bulk 12001 grain size fractions is neglected, i.e., if the measured $Kr^{84}/Kr^{86}$ ratios are directly corrected for spallation $Kr^{84}$ only, an approximately 0.7% lower $(Kr^{84}/Kr^{86})_{tr}$ ratio would result.

We may summarize our conclusion regarding the presence of excess $Kr^{86}$ in atmospheric Kr as follows: With the presently available data we do not consider the presence of an excess $Kr^{86}$ component in the terrestrial atmosphere, presumably fission $Kr^{86}$, as firmly established. It cannot be fully excluded that the deduced excess is an artifact of spurious analytical uncertainties and/or of an oversimplified data treatment. It is mandatory to further study the isotopic composition of trapped Kr in lunar material in order to obtain firm conclusions.

BEOC 10084, BEOC 12001, and SUCOR xenon agree within the errors assigned by the respective authors. In our opinion, the BEOC 12001 data ought to be the most reliable of the three determinations, because the variable target element chemistry was taken into account in correcting for spallation. We shall therefore base our further discussion on the BEOC 12001 data.

In Fig. 11 the isotopic composition of BEOC 12001 is compared with AVCC-Xe (Eugster *et al.*, 1967). Excellent agreement in the abundance of the fission shielded isotopes is observed. For $Xe^{131-136}$ an excess is present in the AVCC-Xe. This excess relative to trapped solar wind has already been observed by many investigators (Hohenberg *et al.*, 1970; Marti *et al.*, 1970; Pepin *et al.*, 1970; Eberhardt *et al.*, 1970; Podosek *et al.*, 1971). Mass spectra of this presumed fission component in AVCC-Xe are compiled in Table 13. The presence of a fission component in AVCC xenon had already been recognized by several investigator teams before lunar samples were analyzed, either on the basis of temperature release experiments (Reynolds and Turner, 1964; Pepin, 1968) or from the direct comparison of fractionated terrestrial Xe with AVCC-Xe (Krummenacher *et al.*, 1962; Eugster *et al.*, 1967; Marti, 1967; Podosek *et al.*, 1971). These spectra are also compiled in Table 13 and compared with

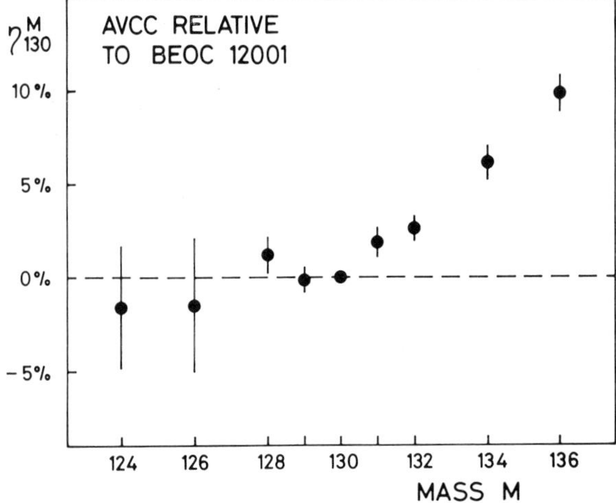


Fig. 11. Comparison of the isotopic composition of average carbonaceous chondrite Xe (AVCC) with surface correlated (solar) trapped Xe in lunar material (BEOC 12001). Shown are the relative deviations

$$\eta^{M}_{130} = [(Xe^{M}/Xe^{130})_{AVCC}/(Xe^{M}/Xe^{130})_{BEOC\ 12001} - 1].$$

Table 13. Isotopic composition of excess $Xe^{131-136}$ in AVCC Xenon. Results obtained by different experimental approaches are compared.

| Method | $Xe^{131}/Xe^{136}$ | $Xe^{132}/Xe^{136}$ | $Xe^{134}/Xe^{136}$ |
|---|---|---|---|
| *Temperature release experiment* | | | |
| Reynolds and Turner (1964) | 0.13 ± 0.1 | 0.16 ± 0.1 | 0.72 ± 0.04 |
| Pepin (1968) | 0.17 ± 0.08 | 0.33 ± 0.06 | 0.70 ± 0.03 |
| *AVCC-fractionated atmospheric Xe* | | | |
| Krummenacher *et al.* (1962) | 0.31 | 0.46 | 0.67 |
| Eugster *et al.* (1967) | 0.25 ± 0.15 | 0.38 ± 0.21 | 0.64 ± 0.1 |
| Marti (1967) | 0.12 ± 0.12 | 0.17 ± 0.17 | 0.57 ± 0.07 |
| Podosek *et al.* (1971) (Murray) | 0.19 ± 0.1 | 0.21 ± 0.18 | 0.62 ± 0.08 |
| *AVCC-surface correlated trapped Xe* | | | |
| Marti (1969) (Pesyanoe 1000° Xe)* | 0.65 ± 0.39 | 0.69 ± 0.42 | 0.77 ± 0.22 |
| Eberhardt *et al.* (1970) (BEOC 10084) | 0.80 ± 0.35 | 0.95 ± 0.45 | 0.75 ± 0.3 |
| Podosek *et al.* (1971) (SUCOR) | 0.38 ± 0.18 | 0.66 ± 0.21 | 0.66 ± 0.1 |
| Present results (BEOC 12001) | 0.52 ± 0.25 | 0.88 ± 0.33 | 0.77 ± 0.19 |
| Kaiser (1972)† | 0.57 ± 0.22 | 0.90 ± 0.24 | 0.74 ± 0.14 |
| *Known fission spectra* | | | |
| $U^{235}$ thermal neutron | 0.45 | 0.67 | 1.25 |
| $U^{238}$ spontaneous fission | 0.08 ± 0.01 | 0.60 ± 0.01 | 0.83 ± 0.01 |
| $Pu^{244}$ spontaneous fission | 0.25 ± 0.02 | 0.88 ± 0.03 | 0.92 ± 0.03 |

Known fission spectra from Hyde (1964) ($U^{235}_{fn}$), Wetherill (1953) ($U^{238}_{fs}$) and Alexander *et al.* (1971) ($Pu^{244}_{fs}$).

* Calculated using the data of Marti (1969), making the appropriate spallation correction (cf. Eberhardt *et al.*, 1970) and using the AVCC composition given by Eugster *et al.* (1967a).

† Calculated using AVCC composition given by Eugster *et al.* (1967a).

known fission spectra. It is apparent from this Table that the AVCC-fission spectra derived from temperature release experiments and from the direct comparison of fractionated atmospheric Xe with AVCC-Xe are quite similar. The comparison of AVCC-Xe with surface correlated Xe gives systematically higher $Xe^{131}$ and $X^{132}e$ yields. The dissimilarity of the fission spectra obtained from the comparison of AVCC-Xe with surface correlated Xe, and from the comparison of AVCC-Xe with fractionated terrestrial Xe is not necessarily surprising. Terrestrial Xe and BEOC 12001 xenon are most likely not related by a simple, linear mass fractionation (cf. Fig. 12). The two calculation methods are thus not equivalent. In our opinion, the comparison of the AVCC-Xe with the surface correlated Xe is preferable and more reasonable than the comparison with the hypothetical fractionated atmospheric Xe.

The discrepancy between the AVCC fission spectra derived from temperature release experiments and from the direct comparison of AVCC xenon with surface correlated Xe is difficult to understand. Either the surface correlated trapped Xe—even Kaiser's (1972) "low fission Xe"—is not representative for the nonfission Xe in carbonaceous chondrites or the difference results, at least partly, from the special conditions of the temperature release technique. The general agreement observed for the light Xe isotopes between AVCC and BEOC 12001 seems to rule out the first of the two possible explanations.

In Fig. 12 the isotopic composition of atmospheric Xe is compared with BEOC 12001 xenon. A very systematic trend of the $\eta$ values with the mass difference is

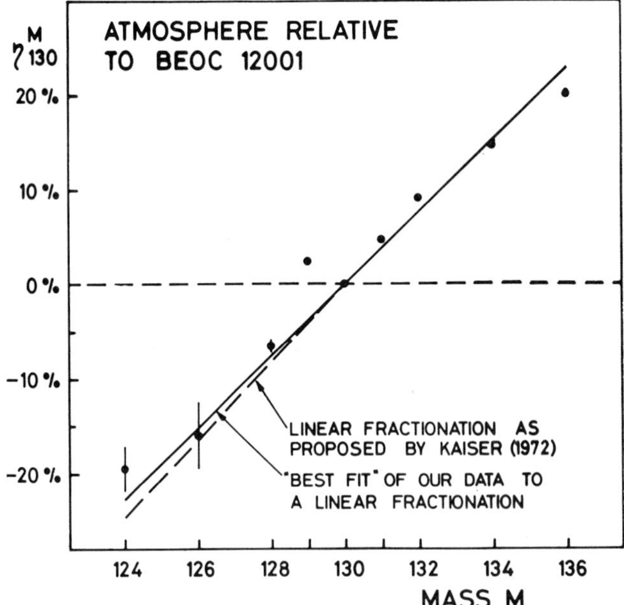


Fig. 12. Comparison of the isotopic composition of atmospheric Xe with surface correlated (solar) trapped Xe in lunar material (BEOC 12001). Shown are the relative deviations

$$\eta^{M}_{130} = [(Xe^{M}/Xe^{130})_{ATM}/(Xe^{M}/Xe^{130})_{BEOC\ 12001} - 1].$$

obvious. However, even pushing the error limits to the extreme, it is not possible to explain the difference between atmospheric and BEOC 12001 xenon by a simple, linear mass fractionation alone.

Recently, Kaiser (1972) observed Xe with somewhat lower $Xe^{131-136}$ abundances than BEOC 12001 in the stepwise degassing of a Luna 16 sample. He concludes that atmospheric Xe and his Luna 16 trapped Xe are related by a linear mass fractionation with a 4.1% per mass unit fractionation factor. Our BEOC 12001 data on the low mass Xe isotopes are not in agreement with Kaiser's suggestion (cf. Fig. 12).

A satisfactory explanation of the observed overall differences in the isotopic composition of xenon in the different reservoirs of the solar system (terrestrial atmosphere; surface correlated lunar Xe, e.g., solar Xe) is still lacking. A simple mass dependent fractionation as the sole source of the differences seems to be ruled out by our BEOC 12001 data. However, the question remains if BEOC 12001 corresponds indeed to the true solar Xe isotopic composition, or if secondary processes on the lunar surface could have significantly changed the isotopic composition of the trapped solar wind Xe.

*Acknowledgments*—We would like to thank the National Aeronautics and Space Administration for generously supplying the necessary lunar samples for this investigation. We would like to acknowledge stimulating discussions with Drs. E. Anders and O. Eugster. We are grateful to Carmen Geiss, Madeleine Thönen, W. Fahrer, V. Horvath, E. Lenggenhager, A. Schaller, and F. Schweizer for their help during the measurements and the preparation of the manuscript.

This research was supported by the Swiss National Science Foundation (grants NF 2.213.69, 2.190.69, 2.405.70, and 2.592.71).

## References

Alexander E. C. Jr. (1971) Spallogenic Ne, Kr, and Xe from a depth study of 12002. *Proc. Second Lunar Sci. Conf., Geochim. Cosmochim. Acta* Suppl. 2, Vol. 2, pp. 1643–1650. MIT Press.

Anders E., Heymann D., and Mazor E. (1970) Isotopic composition of primordial helium in carbonaceous chondrites. *Geochim. Cosmochim. Acta* **34,** 127–132.

Bame S. J., Hundhausen A. J., Asbridge J. R., and Strong I. B. (1968) Solar wind ion composition. *Phys. Rev. Lett.* **20,** 393–395.

Baur H., Frick U., Funk H., Schultz L., and Signer P. (1972) On the thermal release of helium, neon, and argon from lunar fines (abstract). In *Lunar Science—III* (editor C. Watkins), pp. 47–49, Lunar Science Institute Contr. No. 88.

Black D. C. (1970) Trapped helium-neon isotopic correlations in gas-rich meteorites and carbonaceous chondrites. *Geochim. Cosmochim. Acta* **34,** 132–140.

Black D. C. (1972) On the origins of trapped helium, neon, and argon isotopic variations in meteorites —I. Gas rich meteorites, lunar soil, and breccia. *Geochim. Cosmochim. Acta* **36,** 347–375.

Borg J., Maurette M., Durrieu L., and Jouret C. (1971) Ultramicroscopic features in micron-sized lunar dust grains and cosmophysics. *Proc. Second Lunar Sci. Conf., Geochim. Cosmochim. Acta* Suppl. 2, Vol. 3, pp. 2027–2040. MIT Press.

Eberhardt P., Geiss J., and Grögler N. (1965) Ueber die Verteilung der Uredelgase im Meteoriten Khor Temiki. *Tschermaks Mineral. Petrogr. Mitt.* **10,** 535–551.

Eberhardt P., Geiss J., and Grögler N. (1965a) Further evidence on the origin of trapped gases in the meteorite Khor Temiki. *J. Geophys. Res.* **70,** 4375–4378.

Eberhardt P., Geiss J., Graf H., Grögler N., Krähenbühl U., Schwaller H., Schwarzmüller J., and Stettler A. (1970) Trapped solar wind noble gases, exposure age and K/Ar-age in Apollo 11 lunar fine material. *Proc. Apollo 11 Lunar Sci. Conf., Geochim. Cosmochim. Acta* Suppl. 1, Vol. 2, pp. 1037–1070. Pergamon.

Eberhardt P., Geiss J., Graf H., Grögler N., Krähenbühl U., Schwaller H., Schwarzmüller J., and Stettler A. (1970a) Correlation between rock type and irradiation history of Apollo 11 igneous rocks. *Earth Planet. Sci. Lett.* **10,** 67–72.

Eberhardt P., Geiss J., Graf H., and Schwaller H. (1971) On the origin of excess $^{131}$Xe in lunar rocks. *Earth Planet. Sci. Lett.* **12,** 260–262.

Eberhardt P., Geiss J., Graf H., Grögler N., Krähenbühl U., Schwaller H., and Stettler A. (1972) Noble gas investigations of lunar rocks 10017 and 10071. *Geochim. Cosmochim. Acta* (to be submitted).

Eugster O., Eberhardt P., and Geiss J. (1967) The isotopic composition of krypton in unequilibrated and gas rich chondrites. *Earth Planet. Sci. Lett.* **2,** 385–393.

Eugster O., Eberhardt P., and Geiss J. (1967a) Krypton and xenon isotopic composition in three carbonaceous chondrites. *Earth Planet. Sci. Lett.* **3,** 249–257.

Eugster O., Eberhardt P., and Geiss J. (1969) Isotopic analysis of krypton and xenon in fourteen stone meteorites. *J. Geophys. Res.* **74,** 3874–3896.

Finkel R. C., Arnold J. R., Imamura M., Reedy R. C., Fruchter J. S., Loosli H. H., Evans J. C., Delany A. C., and Shedlovsky J. P. (1971) Depth variation of cosmogenic nuclides in a lunar surface rock and lunar soil. *Proc. Second Lunar Sci. Conf., Geochim. Cosmochim. Acta* Suppl. 2, Vol. 2, pp. 1773–1789. MIT Press.

Ganapathy R., Keays R. R., Laul J. C., and Anders E. (1970) Trace elements in Apollo 11 lunar rocks: Implications for meteorite influx and origin of moon. *Proc. Apollo 11 Lunar Sci. Conf., Geochim. Cosmochim. Acta* Suppl. 1, Vol. 2, pp. 1117–1142. Pergamon.

Gast P. W., Hubbard N. J., and Wiesman H. (1970) Chemical composition and petrogenesis of basalts from Tranquillity Base. *Proc. Apollo 11 Lunar Sci. Conf., Geochim. Cosmochim. Acta* Suppl. 1, Vol. 2, pp. 1143–1163. Pergamon.

Geiss J., Oeschger H., and Schwarz U. (1962) The history of cosmic radiation as revealed by isotopic changes in the meteorites and on the earth. *Space Sci. Rev.* **1,** 197–223.

Geiss J., Eberhardt P., Bühler F., Meister J., and Signer P. (1970) Apollo 11 and Apollo 12 solar wind composition experiments: Fluxes of He and Ne isotopes. *J. Geophys. Res.* **75,** 5972–5978.

Geiss J., Bühler F., Cerutti H., Eberhardt P., and Meister J. (1971) Solar wind composition experiment. *Apollo 14 Preliminary Science Report*, NASA SP-272, pp. 221–226.

Geiss J., Bühler F., Cerutti H., and Eberhardt P. (1971a) The solar wind composition experiment. *Apollo 15 Preliminary Science Report*, NASA SP, in press.

Geiss J. and Reeves H. (1972) Cosmic and solar system abundances of deuterium and helium-3. *Astron. Astrophys.*, in press.

Heymann D. and Yaniv A. (1970) $Ar^{40}$ anomaly in lunar samples from Apollo 11. *Proc. Apollo 11 Lunar Sci. Conf., Geochim. Cosmochim. Acta* Suppl. 1, Vol. 2, pp. 1261–1267. Pergamon.

Heymann D., Yaniv A., Adams J. A. S., and Fryer G. E. (1970) Inert gases in lunar samples. *Science* **167,** 555–558.

Heymann D. and Yaniv A. (1971) $Ar^{40}$ in meteorites, fines and breccias from the moon. *Chem. Erde.* **30,** 175–189.

Heymann D., Yaniv A., and Walton J. (1972) Inert gases in Apollo 14 fines and the case of parentless $Ar^{40}$ (abstract). In *Lunar Science—III* (editor C. Watkins), pp. 376–378, Lunar Science Institute Contr. No. 88.

Hintenberger H., Vilcsek E., and Wänke H. (1965) Ueber die Isotopenzusammensetzung und über den Sitz der leichten Uredelgase in Steinmeteoriten. *Z. Naturforsch.* **20a,** 939–945.

Hintenberger H., Weber H. W., and Takaoka N. (1971) Concentrations and isotopic abundances of the rare gases in lunar matter. *Proc. Second Lunar Sci. Conf., Geochim. Cosmochim. Acta* Suppl. 2, Vol. 2, pp. 1607–1625. MIT Press.

Hohenberg C. M., Davis P. K., Kaiser W. A., Lewis R. S., and Reynolds J. H. (1970) Trapped and cosmogenic rare gases from stepwise heating of Apollo 11 samples. *Proc. Apollo 11 Lunar Sci. Conf., Geochim. Cosmochim. Acta* Suppl. 1, Vol. 2, pp. 1283–1309. Pergamon.

Hyde E. K. (1964) The nuclear properties of heavy elements. In *Fission Phenomena*, Vol. 3, Prentice Hall.

Jeffery P. M. and Anders E. (1970) Primordial noble gases in separated meteoritic minerals—I. *Geochim. Cosmochim. Acta* **34,** 1175–1198.

Kaiser W. A. (1972) Rare gas studies in Luna 16 G-7 fines by stepwise heating technique. A low fission solar wind Xe. *Earth Planet. Sci. Lett.* **13,** 387–399.

Krähenbühl U., Rolli H. P., and von Gunten H. R. (1972) Aktivierungsanalytische Bestimmung von seltenen Erden in Gesteinsstandards und in Mondproben. *Helv. Chim. Acta* (submitted).

Krummenacher D., Merrihue C. M., Pepin R. O., and Reynolds J. H. (1962). Meteoritic krypton and barium versus the general isotopic anomalies in meteoritic xenon. *Geochim. Cosmochim. Acta* **26,** 231–249.

Laul J. C., Morgan J. W., Ganapathy R., and Anders E. (1971) Meteoritic material in lunar samples: Characterization from trace elements. *Proc. Second Lunar Sci. Conf., Geochim. Cosmochim. Acta* Suppl. 2, Vol. 2, pp. 1139–1158. MIT Press.

LSPET (Lunar Sample Preliminary Examination Team) (1970) Preliminary examination of the lunar samples from Apollo 12. *Science* **167,** 1325–1339.

Lugmair G. W. and Marti K. (1972) Neutron and spallation effects in Fra Mauro regolith (abstract). In *Lunar Science—III* (editor C. Watkins), pp. 495–497, Lunar Science Institute Contr. No. 88.

Manka R. H. and Michel F. C. (1971) Lunar atmosphere as a source of lunar surface elements. *Proc. Second Lunar Sci. Conf., Geochim. Cosmochim. Acta* Suppl. 2, Vol. 2, pp. 1717–1728. MIT Press.

Marti K. (1967) Isotopic composition of trapped krypton and xenon in chondrites. *Earth Planet. Sci. Lett.* **3,** 243–248.

Marti K. (1969) Solar-type xenon: A new isotopic composition of xenon in the Pesyanoe meteorite. *Science* **166,** 1263–1265.

Marti K., Eberhardt P., and Geiss J. (1966) Spallation, fission, and neutron capture anomalies in meteoritic krypton and xenon. *Z. Naturforsch.* **21a,** 398–413.

Marti K., Lugmair G. W., and Urey H. C. (1970) Solar wind gases, cosmic-ray spallation products, and the irradiation history of Apollo 11 samples. *Proc. Apollo 11 Lunar Sci. Conf., Geochim. Cosmochim. Acta* Suppl. 1, Vol. 2, pp. 1357–1367. Pergamon.

Mason B. (1966) *Principles of Geochemistry*, third edition. John Wiley.

Morrison G. H., Gerard J. T., Potter N. M., Gangadharam E. V., Rothenberg A. M., and Burdo R. A. (1971) Elemental abundances of lunar soil and rocks from Apollo 12. *Proc. Second Lunar Sci. Conf., Geochim. Cosmochim. Acta* Suppl. 2, Vol. 2, pp. 1169–1185. MIT Press.

Nief G. (1960) As reported in isotopic abundance ratios reported for reference samples stocked by the National Bureau of Standards, ed. Mohler F. (NBS Techn. Note 51).

Nier A. O. (1950) A redetermination of the relative abundances of the isotopes of neon, krypton, rubidium, xenon and mercury. Phys. Rev. **79,** 450–454.

Pepin R. O. (1968) Neon and xenon in carbonaceous chondrites. In *Origin and Distribution of the Elements* (editor L. H. Ahrens), pp. 379–386, Pergamon.

Pepin R. O., Nyquist L. E., Phinney D., and Black D. C. (1970) Rare gases in Apollo 11 lunar material. *Proc. Apollo 11 Lunar Sci. Conf., Geochim. Cosmochim. Acta* Suppl. 1, Vol. 2, pp. 1435–1454. Pergamon.

Piercy G. R., McCargo M., Brown F., and Davies D. A. (1964) Experimental evidence for the channeling of heavy ions in monocristalline aluminum. *Can. J. Phys.* **42,** 1116–1134.

Podosek F. A., Huneke J. C., Burnett D. S., and Wasserburg G. J. (1971) Isotopic composition of xenon and krypton in the lunar soil and in the solar wind. *Earth Planet. Sci. Lett.* **10,** 199–216.

Reynolds J. H. and Turner G. (1964) Rare gases in the chondrite Renazzo. *J. Geophys. Res.* **69,** 3263–3281.

Schatzmann E. (1970) CERN Lecture Notes.

Schnetzler C. C. and Philpotts J. A. (1971) Alkali, alkaline earth, and rare-earth element concentrations in some Apollo 12 soils, rocks, and separated phases. *Proc. Second Lunar Sci. Conf., Geochim. Cosmochim. Acta* Suppl. 2, Vol. 2, pp. 1101–1122. MIT Press.

Schwaller H., Eberhardt P., Geiss J., Graf H., and Grögler N. (1971) The $(^{78}Kr/^{83}Kr)_{sp}$–$(^{131}Xe/^{126}Xe)_{sp}$ correlation in Apollo 12 rocks. *Earth Planet. Sci. Lett.* **12,** 167–169.

Wänke H., Wlotzka F., Baddenhausen H., Balacescu A., Spettel B., Teschke F., Jagoutz E., Kruse H., Quijano-Rico M., and Rieder R. (1971) Apollo 12 samples: Chemical composition and its relation to sample locations and exposure ages, the two-component origin of the various soil samples and studies on lunar metallic particles. *Proc. Second Lunar Sci. Conf., Geochim. Cosmochim. Acta* Suppl. 2, Vol. 2, pp. 1187–1208. MIT Press.

Wasserburg G. J., MacDonald G. J. F., Hoyle F., and Fowler W. A. (1964) Relative contributions of uranium, thorium, and potassium to heat production in the earth. *Science* **143,** 465–467.

Wetherill G. W. (1953) Spontaneous fission yields from uranium and thorium. *Phys. Rev.* **92,** 907–912.

Zähringer J. (1962) Ueber die Uredelgase in den Achondriten Kapoeta und Staroe Pesyanoe. *Geochim. Cosmochim. Acta* **26,** 665–680.

Proceedings of the Third Lunar Science Conference
(Supplement 3, *Geochimica et Cosmochimica Acta*)
Vol. 2, pp. 1857–1863
The M.I.T. Press, 1972

# Inert gases from Apollo 12, 14, and 15 fines

D. HEYMANN, A. YANIV,* and S. LAKATOS

Departments of Geology and Space Science,
Rice University, Houston, Texas 77001

**Abstract**—He and Ne contents have been determined in five Apollo 14, five Apollo 12, and two Apollo 15 samples (Ar data are presented in the following paper). Kr and Xe contents were measured in all but the two Apollo 15 samples. The two "mini-rocks" from 14166 and 14167 contain radiogenic $He^3_R$. Their apparent U, Th–$He^4$ age, and in fact that of all the Apollo 14 fines is 3700 ± 800 m.y. If this age is real then it follows that either the (< 1 mm) ejecta from Cone Crater (about 1 km away) lost little if any $He^4_R$ during the formation of the crater, or that these ejecta constitute only a small proportion in the bulk, contingency, and comprehensive fines. The $He^3_C$ radiation ages of 14003, 14163, and 14259 are about 300 m.y.; the $Ne^{21}_C$ ages are about 400, 400, and 300 m.y., respectively. The $Ne^{21}_C$ age of 15601 is about 350 m.y.

## INTRODUCTION

IN THIS PAPER we report the results of our continuing studies of inert gases in lunar fines, i.e., from the Apollo 12, 14, and 15 missions. The inert gases are important geochemical tracers of the solar wind, the cosmic rays, and of the evolution of the regolith. The Apollo 15 fines are of particular interest in this respect because of the presence in them of significant amounts of green glass of high iron content. Because of our special interest in the $Ar^{40}$ problem, we will discuss it in a separate paper.

Our measurements were done by mass spectrometry; the details of which have been published elsewhere (Heymann and Yaniv, 1970a) hence need not be repeated here. Results are given in Table 1.

## RESULTS AND DISCUSSION

### (*1*) *Radiogenic $He^4$ and $Ar^{40}$*

Inert gases in lunar fines consist of the following principal constituents: trapped gases of solar wind origin, cosmogenic gases produced by galactic and solar cosmic rays, and radiogenic $He^4_R$ and $Ar^{40}_R$. $He^4_R$, if any is present, is difficult to detect in fines in general, because it is normally wholly masked by the solar wind component in these samples. However, samples 14166 and 14167 (1–2 mm and 2–4 mm fines, respectively) clearly contain $He^4_R$. These "mini-rocks" have $He^4/He^3$ ratios of 5500 and 8700, far greater than the largest directly measured solar-wind value of 2600 (Geiss *et al.*, 1970) or typical values seen in bulk fines, which are usually smaller than 3000. In addition, the $He^4/Ne^{20}$ ratios in 14166 and 14167, 160 and 390 respectively are far greater than the typical value of about 50 in the bulk Apollo 14 fines.

* On leave from the Department of Physics and Astronomy, Tel Aviv University, Ramat Aviv, Israel.

Table 1. Inert gas contents of Apollo 12, 14, and 15 samples.

| Sample | $He^3$ $10^{-5}$ | $He^4$ $10^{-3}$ | $Ne^{20}$ $10^{-3}$ | $Ne^{21}$ $10^{-6}$ | $Ne^{22}$ $10^{-4}$ | $Kr^{84}$ $10^{-8}$ | $Xe^{132}$ $10^{-8}$ |
|---|---|---|---|---|---|---|---|
| 14003 (bulk) | 2.95 | 84.3 | 1.31 | 3.64 | 1.01 | 19.0 | 4.6 |
| 14163 (bulk) | 3.20 | 80.6 | 1.40 | 4.06 | 1.08 | 20.0 | 3.8 |
| 14166 (1 fragment) ~1 mm | 0.0617 | 3.37 | 0.0207 | 0.201 | 0.0173 | — | — |
| 14167 (1 fragment) ~3 mm | 0.0372 | 3.25 | 0.0083 | 0.048 | 0.00656 | — | — |
| 14259 (bulk) | 3.48 | 85.7 | 1.52 | 4.16 | 1.15 | 18.9 | 3.6 |
| 14163 (<37 μ) | 4.41 | 108 | 2.02 | 5.58 | 1.56 | 20.1 | 5.2 |
| 14163 (37–63 μ) | 1.01 | 22.4 | 0.482 | 1.79 | 0.386 | 6.3 | 1.6 |
| 14163 (63–74 μ) | 0.693 | 15.4 | 0.312 | 1.31 | 0.251 | 4.9 | 1.4 |
| 14163 (74–88 μ) | 0.940 | 20.2 | 0.404 | 1.66 | 0.319 | 6.0 | 1.6 |
| 14163 (88–105 μ) | 0.798 | 17.3 | 0.332 | 1.44 | 0.264 | 5.6 | 1.5 |
| 14163 (105–250 μ) | 0.818 | 17.5 | 0.352 | 1.57 | 0.281 | 5.0 | 1.3 |
| 14163 (250–354 μ) | 0.599 | 12.3 | 0.293 | 1.20 | 0.178 | 4.2 | 0.93 |
| 14163 (354–500 μ) | 0.998 | 21.1 | 0.381 | 1.61 | 0.299 | 3.7 | 1.1 |
| 14163 (500–700 μ) | 0.369 | 5.49 | 0.0607 | 0.691 | 0.0553 | 0.86 | 0.36 |
| 14163 (700–1000 μ) | 0.281 | 3.96 | 0.0408 | 0.534 | 0.0361 | 0.82 | 0.35 |
| 14259 (<37 μ) | 4.83 | 123 | 2.13 | 5.61 | 1.65 | 26.1 | 5.8 |
| 14259 (37–53 μ) | 1.56 | 47.6 | 0.731 | 2.41 | 0.597 | 10.6 | 2.3 |
| 14259 (53–63 μ) | 1.94 | 44.9 | 0.852 | 2.40 | 0.651 | 12.2 | 2.6 |
| 14259 (63–74 μ) | 1.10 | 25.1 | 0.495 | 1.68 | 0.384 | 7.5 | 1.9 |
| 14259 (74–88 μ) | 1.05 | 27.0 | 0.504 | 1.70 | 0.390 | 9.6 | 3.1 |
| 14259 (88–105 μ) | 0.889 | 21.0 | 0.450 | 1.50 | 0.350 | 8.8 | 2.0 |
| 14259 (105–250 μ) | 0.757 | 16.6 | 0.322 | 1.25 | 0.254 | 6.5 | 2.2 |
| 14259 (250–354 μ) | 1.07 | 25.5 | 0.525 | 1.85 | 0.412 | 8.1 | 1.8 |
| 14259 (354–500 μ) | 0.616 | 12.1 | 0.205 | 0.988 | 0.162 | 3.3 | 0.91 |
| 14259 (500–700 μ) | 0.627 | 13.0 | 0.235 | 0.938 | 0.184 | 2.8 | 0.83 |
| 14259 (700–1000 μ) | 0.427 | 6.90 | 0.152 | 0.870 | 0.122 | 3.8 | 1.2 |
| 12042 (bulk) | 3.40 | 97.6 | 1.56 | 4.27 | 1.21 | 12.5 | 2.4 |
| 12073 (interior chip) | 1.10 | 27.7 | 0.494 | 1.67 | 0.383 | 4.2 | 1.0 |
| 12030 (bulk) | 16.0 | 382 | 6.14 | 16.8 | 4.76 | 4.2 | 1.2 |
| 12044 (bulk) | 3.27 | 84.2 | 1.49 | 4.06 | 1.14 | 11.5 | 1.9 |
| 12033 (>250 μ) | 0.0946 | 0.512 | 0.0197 | 0.315 | 0.0188 | 0.41 | 0.18 |
| 12033 (105–250 μ) strongly magnetic | 0.290 | 5.43 | 0.109 | 0.942 | 0.0882 | 1.2 | 0.33 |
| 12033 (105–250 μ) weakly magnetic | 0.117 | 0.786 | 0.0498 | 0.821 | 0.0463 | 0.31 | 0.089 |
| 12033 (<105 μ) | 0.303 | — | 0.164 | 0.786 | 0.134 | 1.6 | 0.55 |
| 12033 (<105 μ) | 0.197 | 3.59 | 0.0808 | 0.848 | 0.0658 | — | — |
| 15601 (bulk) | 2.59 | 61.9 | 1.33 | 3.88 | 1.04 | | |
| 15601 (500–700 μ) | 0.457 | 6.67 | 0.152 | 0.805 | 0.122 | | |
| 15601 (354–500 μ) | 0.594 | 7.75 | 0.162 | 1.06 | 0.132 | | |
| 15601 (250–354 μ) | 0.634 | 9.29 | 0.238 | 1.16 | 0.189 | | |
| 15601 (105–250 μ) | 0.859 | — | 0.324 | 1.29 | 0.257 | | |
| 15601 (88–105 μ) | 0.970 | 19.3 | 0.494 | 1.78 | 0.385 | | |
| 15601 (74–88 μ) | 1.06 | 22.1 | 0.588 | 2.03 | 0.444 | | |
| 15601 (63–74 μ) | 1.10 | 26.8 | 0.601 | 2.05 | 0.461 | | |
| 15601 (53–63 μ) | 0.702 | 15.6 | 0.552 | 1.84 | 0.431 | | |
| 15601 (44–53 μ) | 1.38 | 22.9 | 0.789 | 2.60 | 0.627 | | |
| 15601 (<44 μ) | 4.67 | 118 | 2.36 | 9.37 | 1.83 | | |
| 15091 (bulk) | 2.36 | 60.6 | 1.38 | 3.82 | 1.07 | | |

| Sample | 4/3 | 20/22 | 21/22 | 4/20 | 20/36 | 36/84 | 84/132 |
|---|---|---|---|---|---|---|---|
| 14003 (bulk) | 2900 | 13.0 | 0.036 | 64 | 3.7 | 1900 | 4.1 |
| 14163 (bulk) | 2500 | 12.9 | 0.038 | 58 | 4.0 | 1700 | 3.8 |
| 14166 (1 fragment) ~1 mm | 5500 | 11.9 | 0.116 | 160 | 3.0 | — | — |
| 14167 (1 fragment) ~3 mm | 8700 | 12.6 | 0.073 | 390 | 2.3 | — | — |
| 14259 (bulk) | 2500 | 13.2 | 0.036 | 56 | 3.5 | 2300 | 5.3 |
| 14163 (<37 μ) | 2400 | 13.0 | 0.036 | 54 | 4.0 | 2500 | 3.9 |
| 14163 (37–63 μ) | 2210 | 12.5 | 0.047 | 47 | 3.2 | 2400 | 3.9 |
| 14163 (63–74 μ) | 2200 | 12.4 | 0.052 | 49 | 3.0 | 2100 | 3.5 |
| 14163 (74–88 μ) | 2100 | 12.7 | 0.052 | 50 | 2.9 | 2400 | 3.8 |
| 14163 (88–105 μ) | 2200 | 12.6 | 0.055 | 52 | 2.4 | 2500 | 3.8 |
| 14163 (105–250 μ) | 2100 | 12.5 | 0.056 | 50 | 3.0 | 2400 | 3.9 |
| 14163 (250–354 μ) | 2100 | 13.5 | 0.068 | 52 | 2.4 | 2300 | 4.6 |
| 14163 (354–500 μ) | 2100 | 12.7 | 0.054 | 55 | 4.0 | 2600 | 3.2 |
| 14163 (500–700 μ) | 1500 | 11.0 | 0.125 | 90 | 3.7 | 1900 | 2.4 |
| 14163 (700–1000 μ) | 1400 | 11.3 | 0.148 | 97 | 3.9 | 1300 | 2.4 |
| 14259 (<37 μ) | 2500 | 13.0 | 0.034 | 58 | 3.6 | 2300 | 4.5 |
| 14259 (37–53 μ) | 3100 | 13.1 | 0.040 | 65 | 2.7 | 2600 | 4.5 |
| 14259 (53–63 μ) | 2300 | 13.1 | 0.037 | 53 | 2.9 | 2400 | 4.6 |
| 14259 (63–74 μ) | 2300 | 12.9 | 0.044 | 51 | 2.9 | 2300 | 3.9 |
| 14259 (74–88 μ) | 2600 | 12.9 | 0.044 | 54 | 2.5 | 2100 | 3.1 |
| 14259 (88–105 μ) | 2400 | 12.9 | 0.043 | 47 | 2.4 | 2100 | 4.4 |
| 14259 (105–250 μ) | 2200 | 12.7 | 0.049 | 52 | 2.6 | 1900 | 3.0 |
| 14259 (250–354 μ) | 2400 | 12.7 | 0.045 | 49 | 2.9 | 2200 | 4.6 |
| 14259 (354–500 μ) | 2000 | 12.7 | 0.061 | 59 | 3.2 | 2000 | 3.6 |
| 14259 (500–700 μ) | 2100 | 12.8 | 0.051 | 55 | 3.5 | 2400 | 3.3 |
| 14259 (700–1000 μ) | 1600 | 12.5 | 0.071 | 45 | 1.9 | 2100 | 3.2 |
| 12042 (bulk) | 2900 | 12.9 | 0.035 | 63 | 5.2 | 2400 | 5.3 |
| 12073 (interior chip) | 2500 | 12.9 | 0.044 | 56 | 4.9 | 2400 | 4.2 |
| 12030 (bulk) | 2400 | 12.9 | 0.039 | 63 | 4.9 | 3000 | 3.5 |
| 12044 (bulk) | 2600 | 13.1 | 0.036 | 57 | 4.9 | 2700 | 6.1 |
| 12033 (>250 μ) | 540 | 10.6 | 0.167 | 26 | 4.06 | 1200 | 2.3 |
| 12033 (105–250 μ) strongly magnetic | 1900 | 12.4 | 0.070 | 50 | 4.29 | 2000 | 3.7 |
| 12033 (105–250 μ) weakly magnetic | 670 | 10.8 | 0.177 | 16 | 5.64 | 1400 | 3.5 |
| 12033 (<105 μ) | — | 12.3 | 0.059 | — | 5.63 | 1800 | 2.9 |
| 12033 (<105 μ) | 1822 | 12.3 | 0.129 | 44 | 4.56 | — | — |
| 15601 (bulk) | 2390 | 12.85 | 0.035 | 47 | 6.33 | | |
| 15601 (500–700 μ) | 1370 | 12.44 | 0.066 | 44 | 5.26 | | |
| 15601 (354–500 μ) | 1220 | 12.26 | 0.080 | 48 | 5.70 | | |
| 15601 (250–354 μ) | 1470 | 12.60 | 0.061 | 39 | 6.15 | | |
| 15601 (105–250 μ) | — | 12.63 | 0.050 | — | 6.49 | | |
| 15601 (88–105 μ) | 1990 | 12.72 | 0.046 | 39 | 6.41 | | |
| 15601 (74–88 μ) | 1950 | 12.75 | 0.044 | 38 | 6.69 | | |
| 15601 (63–74 μ) | 2440 | 13.56 | 0.046 | 45 | 5.72 | | |
| 15601 (53–63 μ) | 2220 | 12.80 | 0.043 | 28 | 5.52 | | |
| 15601 (44–53 μ) | 1660 | 12.57 | 0.041 | 29 | 6.98 | | |
| 15601 (<44 μ) | 2530 | 12.88 | 0.039 | 50 | 7.11 | | |
| 15091 (bulk) | 2670 | 12.91 | 0.036 | 44 | 4.92 | | |

Notes: (1) $Ar^{36}$, $Ar^{38}$, and $Ar^{40}$ results are reported in the companion paper.
(2) Errors in absolute amounts are usually less than $\pm 5\%$.
(3) Typical system blanks were: $He^4 = 8 \times 10^{-8}$ cm³ STP; $Ne^{20} = 7 \times 10^{-10}$ cm³ STP; $Ar^{40} = 10^{-8}$ cm³ STP.
(4) Sample weights ranged from about 0.1–1 mg.

When we plot $He^4$ versus $Ne^{20}$ for all the Apollo 14 samples in Table 1, we find that most points fall close to a curve with $He^4/Ne^{20} \sim 50$ and a positive $He^4$ intercept of $(3 \pm 1) \times 10^{-3}$ cm$^3$ STP/g. We have assumed that *all* of the Apollo 14 samples in Table 1 contain this amount of $He^4$ and have assumed that it is, in fact, *in situ* produced radiogenic $He^4_R$ from U and Th decay. The average U and Th contents of Apollo 14 fines are 3.9 ppm and 13.7 ppm, respectively (LSPET, 1971). With these values we calculate an apparent U, Th-$He^4$ age of $3700 \pm 800$ m.y. This result is somewhat surprising in view of the fact that Cone Crater is located only a little more than 1 km to the east of the sites from which all five Apollo 14 fines were collected. The formation of Cone Crater has now been dated at 20–27 m.y. (Bogard and Nyquist, 1972; Lugmair and Marti, 1972). This means that either our assumptions are wrong, or that the Cone Crater materials lost little if any $He^4_R$ when they were ejected, or that these ejecta constitute only a relatively small proportion of the fines only 1 km away.

The $Ar^{40}$ intercepts from $Ar^{40}$–$Ar^{36}$ correlation plots have been used to obtain apparent K–$Ar^{40}$ ages of fines or certain constituents in fines (Heymann *et al.*, 1970; Pepin *et al.*, 1972). The $Ar^{40}$ intercepts of 14163 and 14259 (see companion paper) combined with K contents taken from LSPET (1971) give apparent K–$Ar^{40}$ ages of 3690 and 3630 m.y., respectively. These numbers seem to imply that the fines in question were not substantially heated since about 3600 m.y. ago, but this statement must be qualified in the light of our own Luna 16 results (Heymann *et al.*, 1972). We have found that breccia fragments in the Luna 16 soil contain "excess" $Ar^{40}$ of unknown origin. The breccia from Apollo 11 are exceedingly rich in $Ar^{40}$ (Funkhouser *et al.*, 1970). Certain Apollo 14 breccia rocks are also gas-rich (Bogard and Nyquist, 1972). Hence one must suspect that fines from nearly everywhere on the lunar surface contain variable proportions of gas-rich breccia fragments, whose $Ar^{40}$ consists in part of a component which cannot be accurately estimated from gas measurements on bulk fines.

### (2) *Cosmogenic gases*

For the Apollo 14 fines we have already made the assumption that these contain $(3 \pm 1) \times 10^{-3}$ cm$^3$ STP/g of $He^4_R$. Accordingly we have corrected all the $He^4$ measurements for this component. The corrected $He^4/He^3$ ratios range from 342 to 2860; the small values reflecting the presence of substantial quantities of cosmogenic $He^3_C$. For the Apollo 12 and 15 samples we could not make any correction for $He^4_R$; hence we have used the $He^4/He^3$ ratios as measured. We have further assumed that the trapped $(He^4/He^3)_T$ ratio in all of the samples is $2700 \pm 100$; i.e., we have calculated $He^3_C$ only when $(He^4/He^3)_T$ is less than 2600. The results are shown in Table 2. Perhaps the most striking result is that the amount of $He^3_C$ varies so little, and is typically around $0.3 \times 10^{-5}$ cm$^3$ STP/g, which corresponds to an apparent $He^3_C$ exposure age of about 300 m.y. For the few cases where $He^3_C$ exposure ages of fines in different locations on the moon can be directly compared, one finds that these ages are roughly the same: 300 m.y. for 10084 (Kirsten *et al.*, 1970); 150 m.y. for 12070 (Hintenberger *et al.*, 1971). But it is also known that ages determined via $Ne^{21}_C$, or $Kr^{81}_C - Kr^{83}_C$, or $Ba - Xe^{126}_C$ are nearly always considerably older than $He^3_C$ ages (Bogard *et al.*, 1971;

Table 2. Cosmogenic $He_C^3$ and $Ne_C^{21}$.

| Sample | $He_C^3$($10^{-5}$ cm$^3$ STP/g) | $Ne^{21}$($10^{-6}$ cm$^3$ STP/g) |
|---|---|---|
| 14003 (bulk) | — | 0.48 |
| 14163 (bulk) | 0.33 | 0.68 |
| 14166 (1 fragment) | — | 0.15 |
| 14167 (1 fragment) | — | 0.028 |
| 14259 (bulk) | 0.42 | 0.49 |
| 14259 (< 37 μ) | 0.39 | 0.45 |
| 14259 (37–53 μ) | — | 0.64 |
| 14259 (53–63 μ) | 0.39 | 0.34 |
| 14259 (63–74 μ) | 0.28 | 0.48 |
| 14259 (74–88 μ) | 0.16 | 0.48 |
| 14259 (88–105 μ) | 0.22 | 0.41 |
| 14259 (105–250 μ) | 0.25 | 0.47 |
| 14259 (250–354 μ) | 0.24 | 0.58 |
| 14259 (354–500 μ) | 0.28 | 0.49 |
| 14259 (500–700 μ) | 0.26 | 0.37 |
| 14259 (700–1000 μ) | 0.28 | 0.50 |
| 14163 (< 37 μ) | 0.52 | 0.72 |
| 14163 (37–63 μ) | 0.29 | 0.63 |
| 14163 (63–74 μ) | 0.23 | 0.56 |
| 14163 (74–88 μ) | 0.30 | 0.69 |
| 14163 (88–105 μ) | 0.27 | 0.64 |
| 14163 (105–250 μ) | 0.28 | 0.72 |
| 14163 (250–354 μ) | 0.25 | 0.50 |
| 14163 (354–500 μ) | 0.33 | 0.69 |
| 14163 (500–700 μ) | 0.28 | 0.55 |
| 14163 (700–1000 μ) | 0.25 | 0.44 |
| 12042 (bulk) | — | 0.47 |
| 12073 (interior chip) | 0.08 | 0.47 |
| 12030 (bulk) | 1.8 | 1.8 |
| 12044 (bulk) | 0.13 | 0.43 |
| 12033 (> 250 μ) | 0.08 | 0.03 |
| 12033 (105–250 μ) strongly magnetic | 0.08 | 0.68 |
| 12033 (105–250 μ) weakly magnetic | 0.09 | 0.70 |
| 12033 (< 105 μ) | — | 0.39 |
| 12033 (< 105 μ) | 0.06 | 0.65 |
| 15601 (bulk) | 0.30 | 0.44 |
| 15091 (bulk) | — | 0.45 |
| 15601 (500–700 μ) | 0.24 | 0.43 |
| 15601 (354–500 μ) | 0.35 | 0.67 |
| 15601 (250–354 μ) | 0.29 | 0.58 |
| 15601 (105–250 μ) | — | 0.50 |
| 15601 (88–105 μ) | 0.26 | 0.58 |
| 15601 (74–88 μ) | 0.32 | 0.60 |
| 15601 (63–74 μ) | 0.11 | 0.58 |
| 15601 (53–63 μ) | 0.12 | 0.49 |
| 15601 (44–53 μ) | 0.53 | 0.68 |
| 15601 (< 44 μ) | 0.30 | 3.61 |

Eberhardt *et al.*, 1970; Marti *et al.*, 1970). Apparently the $He_C^3$ ages are false, reflecting as they perhaps do a steady state between the production of $He_C^3$ by cosmic rays and diffusion losses from minerals of poor He retentivity, such as Ca-rich plagioclase.

The isotopic relationships of trapped neon ($Ne^{20}/Ne^{22}$; $Ne^{22}/Ne^{21}$)$_T$ were obtained for 14163 and 14259 by the method first used by Eberhardt *et al.* (1970). We obtained $(Ne^{22}/Ne^{21})_T = 31.1 \pm 1.0$ and $(Ne^{20}/Ne^{21})_T = 410 \pm 30$ for both fines.

This means that $(Ne^{20}/Ne^{22})_T = 13.2$. We have used the $(Ne^{20}/Ne^{21})_T$ ratio to calculate $Ne_C^{21}$, with the simple assumption that all $Ne^{20}$ is of the trapped variety. The results are given in Table 2. Again, the most salient result is that the amounts of $Ne_C^{21}$ are rather constant, e.g., 37 of 45 values (14166 and 14167 excluded) fall between 0.4 and $0.7 \times 10^{-6}$ $cm^3$ STP/g. The average $Ne_C^{21}$ contents of 14163 and 14259 (size fractions plus bulk) are $0.64 \times 10^{-6}$ and $0.47 \times 10^{-6}$ $cm^3$ STP/g, respectively. The $Ne_C^{21}$ production rate was calculated from the assumed $He_C^3$ rate of $1.0 \times 10^{-8}$ $cm^3$ STP/g per m.y. with the correlation of Bogard *et al.* (1971) for $He_C^3/Ne_C^{21}$ as a function of composition and the average composition of Apollo 14 fines as reported by LSPET (1971). We have thus calculated a production rate of $0.16 \times 10^{-8}$ $cm^3$ STP/g per m.y. with this value the $Ne_C^{21}$ exposure ages of 14163 and 14259 are 400 and 300 m.y., respectively. The average $Ne_C^{21}$ content of 15601 is $0.55 \times 10^{-6}$ $cm^3$ STP/g, which corresponds to an exposure age of 350 m.y.

The following results in Table 2 call for comments:

(1) The $Ne_C^{21}$ ages of the 14166 and 14167 fragments are well below the ages of the fines. But this probably reflects merely a statistical effect of ages determined on single particles as against ages determined on a large number of particles.

(2) $He_C^3$ contents of all 12033 samples are far below those of typical bulk fines; they correspond to an apparent exposure age of only about 80 m.y. Four of the five $Ne_C^{21}$ values on the other hand are in the same range as most of the values seen in the fines samples; i.e., the former correspond to ages of some 250–400 m.y. We think that the small $He_C^3$ in 12033 are indicative of substantial He diffusion losses from this sample.

(3) Both $He_C^3$ and $Ne_C^{21}$ in 12030 are unusually large, but this sample is also the most gas-rich in Table 1. The large numbers may well be an artifact because of our choice of $(He^4/He^3)_T = 2700$ and $(Ne^{20}/Ne^{21})_T = 410$. Sample 12030 is apparently not a genuine sample of fines but comes from glass covered fragments at the bottom of a 1-m diameter crater, which "don't seem to hold together very well" according to the astronaut's comments (Shoemaker *et al.*, 1970). From this we infer that 12030 is a loosely consolidated soil breccia. If the $He_C^3$ and $Ne_C^{21}$ shown in Table 2 are real, they may imply that this breccia was formed from pre-irradiated fines, which had been very near the top of the regolith for about 1800 m.y. ($He_C^3$ age), or that the crater itself was formed about 1800 m.y. ago. In the second case one must assume that the glass covered soil clod was produced by the impact, and was not covered since by a substantial amount of regolith to shield it from cosmic ray bombardment.

*Acknowledgment*—This work was supported by NASA Grant NGL 44-006-127.

## References

Bogard D. D., Funkhouser J. G., Schaeffer O. A., and Zähringer J. (1971) Noble gas abundances in lunar material-cosmic ray spallation products and radiation ages from the Sea of Tranquility and the Ocean of Storms. *J. Geophys. Res.* **76,** 2757–2779.

Bogard D. D. and Nyquist L. E. (1972) Noble gas studies on regolith materials from Apollo 14 and 15 (abstract). In *Lunar Science—III* (editor C. Watkins), pp. 89–91, Lunar Science Institute Contr. No. 88.

Eberhardt P., Geiss J., Graf H., Grögler N., Krähenbühl U., Schwaller H., Schwarzmüller J., and Stettler A. (1970) Trapped solar wind noble gases, exposure age and K/Ar-age in Apollo 11 lunar fine material. *Proc. Apollo 11 Lunar Sci. Conf., Geochim. Cosmochim. Acta* Suppl. 1, Vol. 2, pp. 1037–1070. Pergamon.

Funkhouser J. G., Schaeffer O. A., Bogard D. D., and Zähringer J. (1970) Gas analysis of the lunar surface. *Proc. Apollo 11 Lunar Sci. Conf., Geochim. Cosmochim. Acta* Suppl. 1, Vol. 2, pp. 1111–1116. Pergamon.

Geiss J., Eberhardt P., Signer P., Buehler F., and Meister J. (1970) The solar wind composition experiment. *Apollo 12 Preliminary Science Report*, NASA SP-235, pp. 99–102.

Heymann D., Yaniv A., Adams J. A. S., and Fryer G. E. (1970) Inert gases in lunar samples. *Science* **167,** 555–558.

Heymann D. and Yaniv A. (1970a) Inert gases in the fines from the Sea of Tranquility. *Proc. Apollo 11 Lunar Sci. Conf., Geochim. Cosmochim. Acta* Suppl. 1, Vol. 2, pp. 1247–1259. Pergamon.

Heymann D., Yaniv A., and Lakatos S. (1972) Inert gases in twelve particles and one "dust" sample from Luna 16. *Earth Planet. Sci. Lett.* **13,** 400–406.

Hintenberger H., Weber H. W., and Takaoka N. (1971) Concentrations and isotopic abundances of the rare gases in lunar matter. *Proc. Second Lunar Sci. Conf., Geochim. Cosmochim. Acta* Suppl. 2, Vol. 2, pp. 1607–1626. MIT Press.

LSPET (Lunar Sample Preliminary Examination Team) (1971) Preliminary examination of lunar samples. *Apollo 14 Preliminary Science Report*, NASA SP-272, pp. 109–132.

Lugmair G. W. and Marti K. (1972) Neutron and spallation effects in Fra Mauro regolith (abstract). In *Lunar Science—III* (editor C. Watkins), pp. 495–497, Lunar Science Institute Contr. No. 88.

Marti K., Lugmair G. W., and Urey H. C. (1970) Solar wind gases, cosmic ray spallation products and the irradiation history of Apollo 11 samples. *Apollo 11 Lunar Sci. Conf., Geochim. Cosmochim. Acta* Suppl. 1, Vol. 2, pp. 1357–1367. Pergamon.

Pepin R. O., Bradley J. G., Dragon J. C., and Nyquist R. E. (1972) K–Ar dating of lunar soils: Apollo 12, Apollo 14, and Luna 16 (abstract). In *Lunar Science—III* (editor C. Watkins), pp. 602–604, Lunar Science Institute Contr. No. 88.

Shoemaker E. M., Batson R. M., Bean A. L., Conrad C. Jr., Dahlem D. H., Goddard E. N., Hait M. H., Larson K. B., Schaber G. G., Schleicher D. L., Sutton R. L., Swann G. A., and Waters A. C. (1970) Preliminary geologic investigation of the Apollo 12 landing site. *Apollo 12 Preliminary Science Report*, NASA SP-235, pp. 113–156.

Proceedings of the Third Lunar Science Conference
(Supplement 3, *Geochimica et Cosmochimica Acta*)
Vol. 2, pp. 1865–1889
The M.I.T. Press, 1972

# The rare gas record of Apollo 14 and 15 samples

T. KIRSTEN, J. DEUBNER, P. HORN, I. KANEOKA, J. KIKO,
O. A. SCHAEFFER, and S. K. THIO

Max-Planck-Institut für Kernphysik,
Heidelberg, Germany

**Abstract**—Rare gases have been measured in numerous samples from Apollo 14 and 15 returns including bulk soils, individual soil grains, 10–50 mg rock fragments from the soil, clasts, pieces of breccia, and pieces of large crystalline rocks. Some of the samples were irradiated with fast neutrons before analysis, allowing a determination of $^{39}Ar$–$^{40}Ar$ crystallization ages and $^{37}Ar$–$^{38}Ar$ cosmic ray exposure ages. The ages show that the Fra Mauro formation at the Apollo 14 landing site contains mainly debris from the pre-Mare Imbrium terrain, with some fragments from neighboring maria and that there has been igneous activity at the pre-Mare Imbrium site for over 200 m.y. from 3.8 to $4.0 \times 10^9$ years ago. Exposure ages are found to lie between 15 m.y. and 770 m.y. for the rocks. Individual soil grains give ages as high as 1700 m.y. These ages reflect the gardening of the lunar soil. As the solar wind component in the soil is similar at all sites, it is inferred that the retained solar wind may be very ancient, possibly up to $4.0 \times 10^9$ years. Some single grains contain solar wind very little changed by diffusion. Several anomalies exist, the Ne is enriched in 15421 soil from near Spur Crater and the abundance ratio of "parentless" $^{40}Ar$ relative to implanted $^{36}Ar$ is variable and can be as high as 8.

## INTRODUCTION

THE RARE GAS CONTENTS of lunar samples provide a comprehensive record of their evolution. Radiogenic, cosmogenic, trapped, and dissolved rare gases have been used extensively to study formation ages, stratigraphic relationships, exposure ages, regolith formation, transport processes, and solar wind abundances.

Age determinations have shown that lunar rocks crystallized between $3.2 \times 10^9$ and $3.9 \times 10^9$ yr ago, except for rocks 12013 (Husain *et al.*, 1972) and 15415 (Turner, 1971) which show slightly older crystallization ages. Major events dated are one of the latest mare fillings at Oceanus Procellarum $3.3 \times 10^9$ yr ago, Mare Tranquilitatis $3.7 \times 10^9$ yr ago, and Imbrium Basin excavation $3.8 \times 10^9$ yr ago. Here we report on further age measurements using the $^{39}Ar$–$^{40}Ar$ method, shown by Turner (1970) to be particularly applicable to lunar rocks. Rock fragments from the soil and from clasts in 14303 have been studied. In addition, conventional K–Ar dating has been applied to determine model ages for lunar soils.

The time spent by rock fragments at the top layer of the regolith can be estimated from the amount of accumulated spallation products. In this paper we report particularly on spallation products in various lunar soils. Such data provide mean turnover rates for the sampled regolith sites. In addition, a search for extremely high exposure ages was performed by analyzing single soil particles. This could narrow the gap of recordable time between $\sim 1 \times 10^9$ and $3 \times 10^9$ yr.

The solar wind implanted gases in lunar soils and breccias are of interest with regard to elemental and isotopic solar abundances at present and in the past; they are, however, sensitive to the various temperatures within the lunar regolith. Apart from bulk soil analyses, we have therefore analyzed implanted gases in single Apollo 14 soil particles. This is a continuation of previous studies for Apollo 12 soil (Kirsten *et al.*, 1971). It is hoped then to find a particular specimen in which implanted gases display minimal alteration by solid state diffusion. Results on diffusion studies in single soil particles reported at the Third Lunar Science Conference (Kirsten *et al.*, 1972) will be published in Ducati *et al.*, 1972.

## SAMPLE DESCRIPTION AND PREPARATION

Bulk soils were measured in mg-sized samples from aliquots as obtained. Sieve fractions were prepared by wet sieving using anhydrous ethanol. Etched samples were exposed for 10 sec to a solution of 1 part $H_2SO_4$, 1 part HF, and 2 parts $H_2O$. Single soil particles were handpicked from the coarser sieve fractions under the microscope. They were identified by standard mineralogic techniques, occasionally controlled by XRD methods.

Single rock fragments from coarse fines were ultrasonically cleaned in ethanol. The fragmental rock 14303,13 was cut dry into slices of $\sim 20 \times 30 \times 2$ mm, using a wire saw. Two slices were gently crushed into fragments $\leq 4$ mm, the size of the largest clasts. Igneous and brecciated clasts were handpicked under the binocular, liberated from adhering ground mass particles, and ultrasonically cleaned. Different clast types were selected according to color, crystallinity, homogeneity, etc. Before analysis, small chips were broken off to prepare polished sections. A similar treatment was given to basalts handpicked from coarse fines 14161. The small chips used for polished sections may not always be representative of the whole sample. Occasionally very fine grained patches of mesostasis with a "sieve structure" enriched in K according to microprobe analysis (El Goresy, 1972) occur. In judging shock effects, the presence of milky white plagioclases, cataclastic deformations, and fragmentations in the minerals and planar elements in plagioclase were taken as indications. Local melting due to shock and subsequent recrystallization is rare.

*Sample description*

| | |
|---|---|
| 14001,5 | contingency soil, 2–4 mm, near LM |
| 14003,24 | contingency soil, < 1 mm, near LM |
| 14161,26 + 34 | coarse fines, composed of lithic clasts, pyroxene, plagioclase, glass, and opaques, 2–4 mm, near LM |
| 14257,9 | comprehensive soil, 2–4 mm, 100 m NW of LM |
| 14303,13 | fragmental rock, highly metamorphosed, group 6 of the classification by Warner (1972), 100 m NW of LM |
| 15021,94 | contingency soil < 1 mm, near LM |
| 15101,59 | soil < 1 mm from Apennine front (St. George Crater) |
| 15301,79 | soil < 1 mm, rich in green glass, rim of Spur Crater |
| 15421,21 | soil < 1 mm, very rich in green glass, rim of Spur Crater |
| 15556,25 | vesicular basalt, 60 m from the rim of Hadley Rille |
| 15601,63 | comprehensive soil < 1 mm, 20 m NE of rim of Hadley Rille, matured surface |

*Subsamples used for* $^{39}$Ar–$^{40}$Ar *dating*

| | |
|---|---|
| 14161,34,2 | ophitic basalt, shocked, interlocking plagioclase laths enclosed in pyroxenes; ilmenite, iron, chromite, ulvöspinel, troilite, very little tridymite; fine grained ($< 100\ \mu$); local patches of K-rich mesostasis |
| 14161,34,5 | intergranular basalt, rounded pyroxene grains in a matrix of plagioclase; ilmenite, iron, troilite, spinel, tridymite; relatively coarse grained ($\sim 100\ \mu$) |
| 14161,34,4 | vuggy intergranular basalt, shocked, plagioclase intergrown with pyroxenes; olivine; ilmenite, iron, chromite, troilite; little tridymite; troilite in microcracks of pyroxenes; coarse grained ($\sim 100\ \mu$); local patches of K-rich mesostasis |
| 14161,34,6 | basalt with cognate basalt inclusion, shocked, composed of two types of basaltic rocks, one being coarser grained than the other; both composed of pyroxene and plagioclase, laths of ilmenite, iron, large spinels, troilite, baddeleyite; resembling rock 14310 |
| 14303,13,R5,1 | intergranular basalt, shocked, plagioclase intergrown with pyroxenes; much tridymite and iron, little ilmenite, troilite and spinel; coarse grained ($\geq 100\ \mu$) |
| 14303,13,R5,21 | coarse grained annealed basaltic microbreccia, shocked, angular fragments of plagioclases ($\sim 100\ \mu$) in a fine grained matrix ($< 20\ \mu$) of plagioclase and pyroxene; ilmenite extremely fine grained and dispersed, iron, troilite, spinel; holocrystalline |
| 14303,13,R5,22 | fine grained annealed basaltic microbreccia, shocked as R5,21, finer grained; glass probably present |
| 14161,34,3 | feldspathic accumulate + pyroxene, shocked, very coarse grained plagioclases ($\sim 90\%$) and small aggregates of pyroxenes ($\sim 10\%$); no opaques |
| 14161,26,11 | intergranular basalt, pyroxenes intergrown with plagioclases; troilite, spinel, iron, ilmenite; ore content is small; coarse grained ($\geq 100\ \mu$) |
| 14161,26,12 | intergranular basalt, as 26,11, coarser |

*Subsamples used for rare gas analysis*

| | |
|---|---|
| 14001,5,1 + 2 | lithic clasts |
| 14003,24,1 + 2 | inhomogeneous sieve fraction $> 300\ \mu$ |
| 14161,26,1 | fine grained, coherent microbreccia |
| 14161,34,8 | "sintered soil" |
| 14161,34,9 | friable microbreccia, glass rich |
| 14161,34,10 | friable ground mass |
| 14257,9,1 | lithic clast |
| 14303,13,6 | annealed breccia, fine grained dark lithic clast |
| 14303,13,7 | coarse grained basaltic rock |
| 14303,13,8 | matrix material |
| 14303,13,9 | fine grained black basaltic rock |
| 14303,13,12 | anorthositic fragment |

## Experimental Procedures and Results

### $^{39}$*Ar*–$^{40}$*Ar dating*

Lunar samples (ranging from 10 to 60 mg) and terrestrial hornblende monitors were wrapped in aluminum foil, stacked alternatingly in quartz ampoules and vacuum-sealed together with a nickel wire to check the uniformity of the neutron flux from the induced $^{58}$Co activity. Samples marked "K" in Table 1 were irradiated for 14 days in the core of the Karlsruhe FR2 reactor behind Cd-shielding. The integrated fast neutron flux was $\sim 2.7 \times 10^{18}$ n/cm$^2$. The neutron flux gradient along the length of the ampoule (4.5 cm) was $< 2\%$. The "M" samples received a 10 days irradiation in the core of the BR-2 reactor at Mol, Belgium. Here, the integrated fast neutron flux

Table 1. Summary of $^{39}Ar$–$^{40}Ar$ ages and exposure ages.

| Sample | J value used | K (%) | Ca (%) | Total argon age ($\times 10^9$ yr) | High-temperature age ($\times 10^9$ yr) | Plateau age ($\times 10^9$ yr) | Exposure age $^{38}Ar$–Ca ($\times 10^6$ yr) |
|---|---|---|---|---|---|---|---|
| 14161,34,2 | 0.01610 K | 0.61 | 7.1 | $3.83 \pm 0.08$ | $3.85 \pm 0.08$ | $4.03 \pm 0.08$ | 380 |
| 14161,34,5 | 0.01610 K | 0.73 | 7.6 | $3.92 \pm 0.08$ | $3.93 \pm 0.08$ | $4.01 \pm 0.06$ | 430 |
| 14161,34,4 | 0.01610 K | 0.55 | 7.9 | $3.91 \pm 0.06$ | $3.96 \pm 0.06$ | $3.99 \pm 0.04$ | 300 |
| 14161,34,6 | 0.01740 K | 0.76 | 7.8 | $3.74 \pm 0.04$ | $3.86 \pm 0.04$ | $3.90 \pm 0.04$ | 390 |
| 14303,13,R5,1 | 0.01740 K | 1.10 | 7.1 | $3.75 \pm 0.02$ | $3.89 \pm 0.04$ | $3.91 \pm 0.04$ | 29 |
| 14303,13,R5,21 | 0.1502 M | 0.69 | 8.3 | $3.44 \pm 0.08$ | $3.45 \pm 0.08$ | ($3.78 \pm 0.14$) | — |
| 14303,13,R5,22 | 0.1502 M | 0.0057 | 7.9 | ($5.69 \pm 0.08$) | — | — | — |
| 14161,34,3 | 0.1502 M | 0.46 | 13.4 | $3.62 \pm 0.08$ | $3.63 \pm 0.10$ | $3.63 \pm 0.10$ | — |
| 14161,26,11 | 0.01740 K | 0.53 | 6.4 | $3.58 \pm 0.04$ | $3.59 \pm 0.04$ | $3.59 \pm 0.04$ | 360 |
| 14161,26,12 | 0.01740 K | 0.062 | 8.5 | $3.54 \pm 0.04$ | $3.52 \pm 0.04$ | $3.54 \pm 0.02$ | 320 |

Error figures indicate $2\sigma$ standard deviation. Errors for K and Ca $\leq 5\%$, for exposure ages $\leq 10\%$.
K = Karlsruhe irradiation.
M = Mol (Belgium) irradiation (see text).

was $2.1 \times 10^{19}$ n/cm$^2$. In this case, shielding from thermal neutrons turned out to be incomplete.

The irradiated samples were loaded into the high-vacuum, on-line extraction system of a high-sensitivity Nier-type mass spectrometer. Samples were inductively heated in a molybdenum crucible for periods of 1 hour, at successively higher temperatures ranging from 600°C to 1500°C, determined by a W–Re-alloy-thermocouple directly attached to the crucible. Temperature fluctuations were less than $\pm 5$°C.

After the purified gas was introduced into the spectrometer, the extraction furnace was isolated for the next extraction. Blanks determined after each sample ranged from $5 \times 10^{-10}$ cc STP $^{40}Ar$ at low temperatures to $10^{-8}$ cc STP $^{40}Ar$ at 1500°C.

$^{40}Ar/^{39}Ar$ ratios were determined from 15–20 recordings, other Ar-isotopes from three recordings, taking into account linear zero-time extrapolations. Errors given for the $^{40}Ar/^{39}Ar$ ratio are $2\sigma$ standard deviations, errors of other isotopes are estimated from the general spectrometer characteristics and reproducibility considerations. The following corrections were applied to the measured values:

(a) Blank corrections.
(b) $^{37}Ar$ decay between time of bombardment and time of measurement.
(c) $^{39}Ar$ produced from Ca, with $(^{39}Ar/^{37}Ar)_{Ca} = 0.001$.
(d) $^{40}Ar$ trapped, with $(^{40}Ar/^{36}Ar)_{tr} = 1$.
Using production ratios given by Brereton (1970), we corrected for:
(e) $^{40}Ar$ produced from K, with $(^{40}Ar/^{39}Ar)_K = (1.23 \pm 0.24) \times 10^{-2}$.
(f) $^{38}Ar$ produced from K, with $(^{38}Ar/^{39}Ar)_K = (1.14 \pm 0.03) \times 10^{-2}$.
(g) $^{38}Ar$ produced from Ca, with $(^{38}Ar/^{37}Ar)_{Ca} = (1.39 \pm 0.26) \times 10^{-4}$.
(h) $^{36}Ar$ produced from Ca, with $(^{36}Ar/^{37}Ar)_{Ca} = (2.47 \pm 0.09) \times 10^{-4}$.

K and Ca contents are determined from $^{39}Ar$ and $^{37}Ar$ in sample and monitor and the K and Ca content of the monitor (see Table 1).

A hornblende separate from the Northern Light Gneiss of Northern Minnesota was used as monitor. Its K–Ar age was found to be $2.60 \pm 0.04 \times 10^9$ yr (K content by neutron activation $3030 \pm 30$ ppm; $^{40}Ar_{rad}$ $(6.7 \pm 0.2) \times 10^{-5}$ cc STP/g; $\lambda_\beta = 4.72 \times 10^{-10}$ yr$^{-1}$; $\lambda_\kappa = 0.585 \times 10^{-10}$ yr$^{-1}$; $^{40}K/K = 1.19 \times 10^{-4}$). Atomic absorption analysis gave a Ca content of (8.00 ± 0.05)%. The $J$ values (Grasty and Mitchell, 1966) needed to convert the measured $^{40}Ar/^{39}Ar$ ratios into ages, are listed in Table 1.

The release patterns of all Ar isotopes for all samples and temperature steps will be presented in a more detailed discussion (Kaneoka, 1972). However, the corrected total Ar isotope concentrations are given in Table 2. Figures 1–5 show the radiogenic

Table 2. Total amount of argon isotopes.

| Sample | $^{40}Ar^*$ | $^{39}Ar_K$ | $10^{-8}$ccSTP/g $^{38}Ar_{corr}$ | $^{37}Ar_{corr}$ | $^{36}Ar_{corr}$ | $^{38}Ar_{sp}$ |
|---|---|---|---|---|---|---|
| 14161,34,2 | 30150 | 73.2 | 66.2 | 543 | 153 | 42.5 |
| 14161,34,5 | 38490 | 88.5 | 59.4 | 577 | 93.5 | 47.6 |
| 14161,34,4 | 29000 | 67.0 | 56.1 | 605 | 136 | 34.7 |
| 14161,34,6 | 35670 | 99.4 | 52.3 | 694 | 51.2 | 48.4 |
| 14303,13,R5,1 | 52400 | 144 | 4.20 | 629 | 6.69 | 3.34 |
| 14303,13,R5,21 | 27030 | 783 | 108† | 5090 | 3.91 | — |
| 14303,13,R5,22 | 539.9 | 4.18 | 40.3† | 3130 | 25.0 | — |
| 14161,34,3 | 20020 | 518 | 109† | 8170 | 6.10 | — |
| 14161,26,11 | 22770 | 69.6 | 49.8 | 572 | 72.8 | 41.0 |
| 14161,26,12 | 2591 | 8.10 | 51.6 | 755 | 40.7 | 49.9 |

† Partly due to improper shielding from thermal neutrons.
$^{36}Ar$ corrected for reactor produced $^{36}Ar$ from Ca.
$^{37}Ar$ corrected for decay after irradiation.
$^{38}Ar$ corrected for reactor produced $^{38}Ar$ from K and Ca.

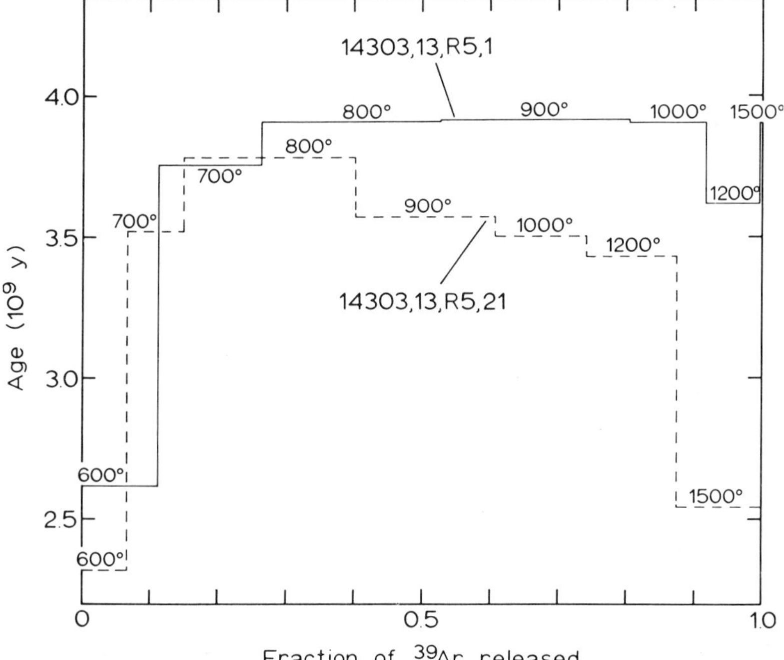


Fig. 1. $^{39}Ar$–$^{40}Ar$ age release patterns for basaltic fragment 14303,13,R5,1 showing a good plateau, and microbreccia 14303,13,R5,21 showing a steady decrease at higher temperatures.

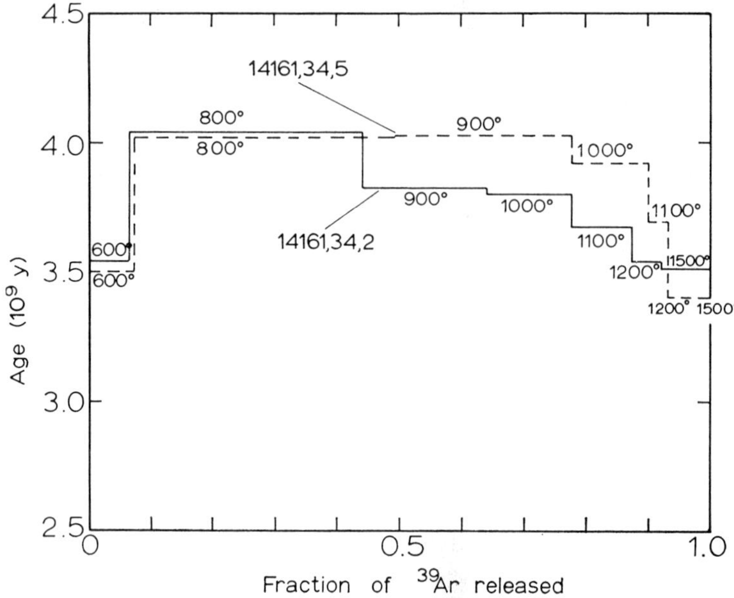


Fig. 2. $^{39}Ar$–$^{40}Ar$ age release patterns for basaltic fragments 14161,34,2 and 14161,34,5 with decrease in apparent ages at higher temperatures.

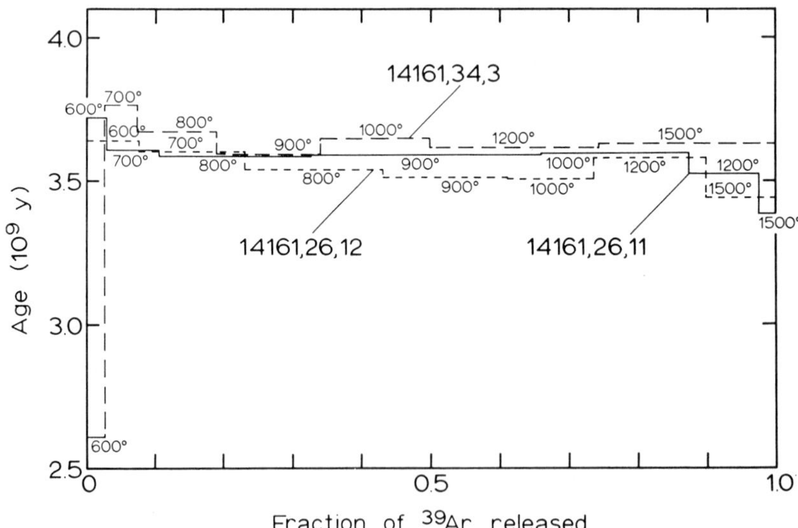


Fig. 3. $^{39}Ar$–$^{40}Ar$ age release patterns for basaltic fragments 14161,26,11 and 14161,26,12 and feldspathic accumulate, 14161,34,3 with very little loss of radiogenic argon and well defined plateaus.

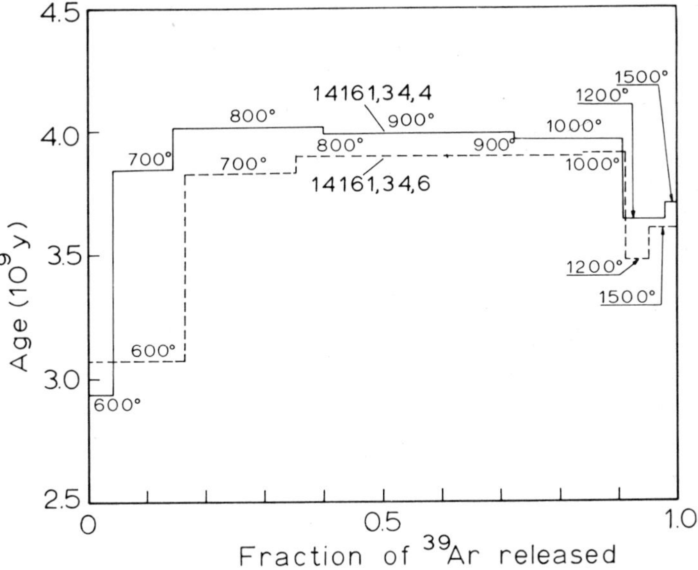


Fig. 4. $^{39}Ar$–$^{40}Ar$ age release patterns for basaltic fragments 14161,34,4 and 14161,34,6 with well-defined plateaus, but decrease at lower and higher temperatures.

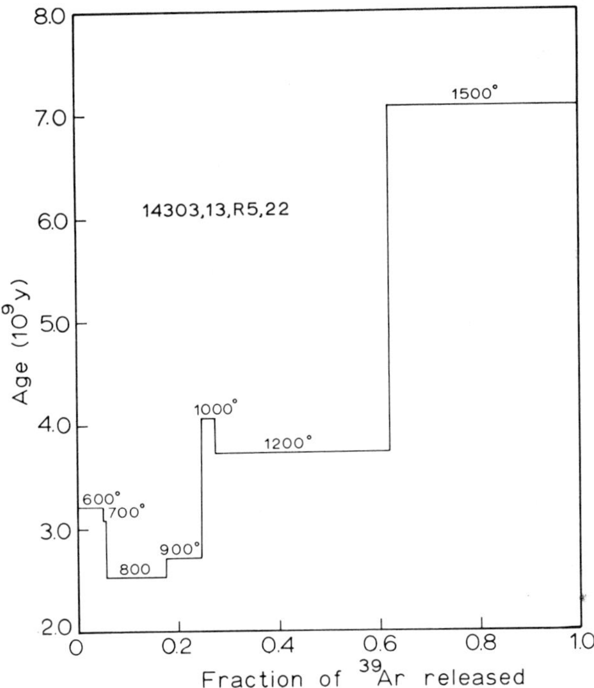


Fig. 5. $^{39}Ar$–$^{40}Ar$ age release pattern for microbreccia 14303,13,R5,22 with evidence for excess argon (even the total argon age is $5.7 \times 10^9$ yr).

gas release patterns converted into ages. As also observed by other authors (Turner *et al.*, 1971; Schaeffer *et al.*, 1972), the $^{40}Ar/^{39}Ar$ ratio in Apollo 14 rocks frequently decreases from the plateau value at the highest temperatures (e.g., Figs. 1, 2, and 4). Hence, we distinguish between total argon ages, plateau ages, and high-temperature ages. They are calculated from the Ar released between 600–1500°C, $T_{\text{Plateau, start}}$ to $T_{\text{Plateau, end}}$, and $T_{\text{Plateau, start}}$ to 1500°C, respectively. Table 1 is a summary of all ages so obtained. Error progression was calculated following Dalrymple and Lanphere (1971). Errors given are $2\sigma$ standard deviations. To calculate exposure ages from neutron-irradiated samples the corrected $^{38}Ar$ of each temperature fraction was split into a spallogenic and a trapped component assuming $(^{36}Ar/^{38}Ar)_{sp} = 0.63$; $(^{36}Ar/^{38}Ar)_{tr} = 5.35$. As shown in Figs. 6 and 7, the release pattern of $^{38}Ar$ is similar to that of Ca-derived $^{37}Ar$, especially at higher temperatures, indicating that spallogenic $^{38}Ar$ is mainly produced from Ca. Particularly, the lowest $^{38}Ar_{sp}/^{37}Ar$ ratio should represent spallogenic $^{38}Ar$ from Ca alone and is therefore used to determine the total Ca-derived $^{38}Ar_{sp}$. Since Ca contents are also known, $^{38}Ar$–Ca exposure ages can be calculated, adopting a production rate of $1.4 \times 10^{-8}$ cc STP/g Ca, m.y. (Turner *et al.*, 1971). Results are given in Table 6.

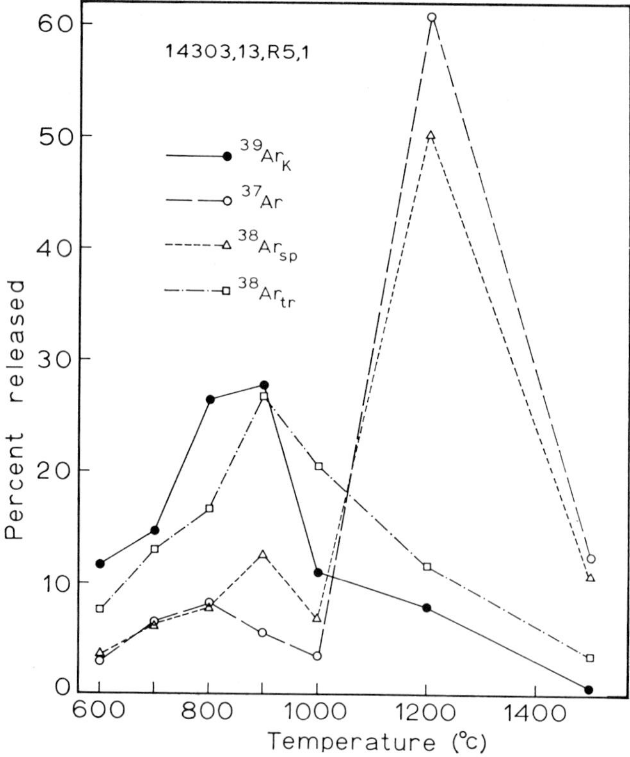


Fig. 6. Release pattern of different argon isotopes from basaltic fragment 14303,13,R5,1. Note correlation between $^{37}Ar$ and $^{38}Ar_{sp}$.

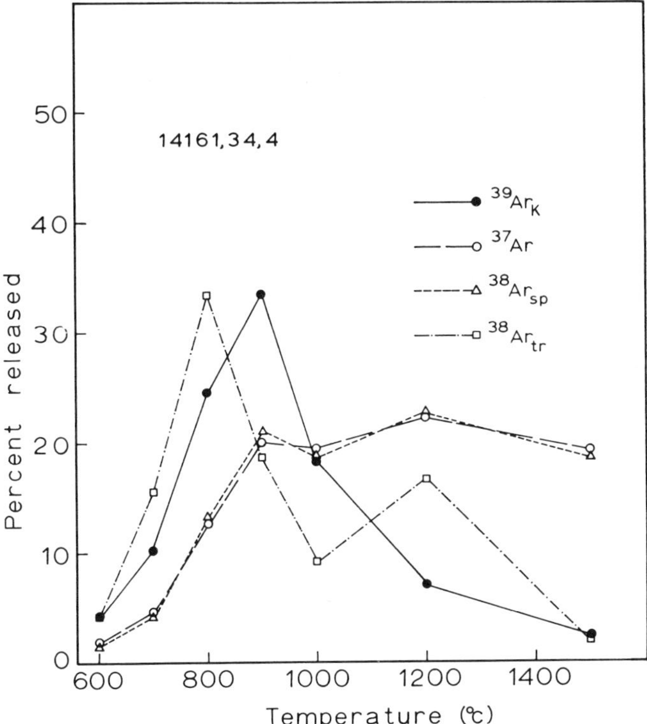


Fig. 7. Release pattern of each argon isotope for basaltic fragment 14161,34,4.

## *Rare gas analysis of bulk samples*

Standard procedures were applied in the mass spectrometric rare gas analysis of bulk samples. Blank values were (in $10^{-10}$ cc STP) $^{3}$He ~ 1, $^{4}$He ~ 40; $^{20}$Ne ~ 3, $^{21}$Ne ~ 0.5, $^{22}$Ne ~ 2, $^{36}$Ar ~ 0.1, $^{38}$Ar ~ 0.1, $^{40}$Ar ~ 150, $^{84}$Kr ~ 0.01, $^{132}$Xe ~ 0.01. Samples were predegassed for 6 hours at 180°C.

In performing conventional K–Ar dating of soils, aliquots of ~100 mg were split into two parts. One part was used to determine K and U by neutron activation, the other was slightly etched in a diluted $HF/H_2SO_4$ solution to reduce the concentration of surface correlated gases. Afterwards mg-sized samples were taken for the mass spectrometric analysis.

Table 3 contains the results for bulk soils, sieve-fractions of 14003, and etched sieve-fractions of 14003 and 15101. All these data are based on duplicate or triplicate analyses.

Results from the analysis of coarse fines and rocks are given in Table 4. Small sample weights (generally $\lesssim 1$ mg) were chosen for accurate He, Ne, and Ar determinations to avoid errors caused by volume dilutions. As a consequence, the less abundant Kr and Xe isotopes were determined only for the various fractions of soil 14003. These data are listed in Table 5.

Absolute errors for He, Ne, and Ar are estimated to be less than $\pm 6\%$ from the

Table 3. Rare gas concentrations in Apollo 14 and 15 soils (in $10^{-8}$ ccSTP/g).

| Sample | $^3He$ | $^4He$ | $^{20}Ne$ | $^{21}Ne$ | $^{22}Ne$ | $^{36}Ar$ | $^{38}Ar$ | $^{40}Ar$ | $^{84}Kr$ | $^{132}Xe$ | K (ppm) | U (ppm) |
|---|---|---|---|---|---|---|---|---|---|---|---|---|
| *Bulk fines* | | | | | | | | | | | | |
| 14003 | 3150 | 8880000 | 160000 | 448 | 12100 | 43300 | 7940 | 67800 | 24.3 | 3.9 | 4360 | 3.3 |
| 15021 | 2890 | 8942000 | 186000 | 497 | 13700 | 40500 | 7450 | 29800 | 23.7 | 3.15 | 1610* | 1.3* |
| 15101 | 3030 | 7900000 | 168000 | 456 | 12850 | 34600 | 6450 | 40800 | 12.0 | 1.8 | 1430 | 0.91 |
| 15601 | 2610 | 7633000 | 159000 | 452 | 12270 | 24400 | 4600 | 21100 | 8.3 | 1.4 | 820 | 0.51 |
| 15301 | 1820 | 5614000 | 183300 | 486 | 13620 | 31600 | 5850 | 52700 | 13 | 2.3 | 1220* | 0.8* |
| 15421 | 900 | 2540000 | 125000 | 368 | 9290 | 15130 | 2810 | 54800 | 5.5 | 1.6 | 940† | |
| *Sieve fractions* | | | | | | | | | | | | |
| 14003 < 25 μ | 5350 | 15340000 | 285000 | 753 | 21560 | 78200 | 14370 | 112900 | 53 | 7.3 | | |
| 25–60 μ | 1235 | 3700000 | 79500 | 240 | 5980 | 22700 | 4295 | 37740 | 16.2 | 2.0 | | |
| 60–109 μ | 990 | 2660000 | 55500 | 190 | 4250 | 20640 | 3805 | 39200 | 11.3 | 1.75 | | |
| 109–272 μ | 710 | 2092000 | 40900 | 147 | 3180 | 13540 | 2590 | 32800 | 9.7 | 1.65 | | |
| *Etched fines* | | | | | | | | | | | | |
| 14003 60–109 μ | 300 | 820000 | 10750 | 87 | 905 | 2790 | 569 | 14200 | 2.33 | 0.44 | 3780 | |
| 109–272 μ | 282 | 752000 | 15850 | 90 | 1290 | 7100 | 1375 | 23300 | 5.65 | 0.91 | 4240 | |
| 15101,59,1,3 60–109 μ | 334 | 572000 | 9900 | 87.2 | 830 | 2320 | 470 | 5800 | — | — | 1280 | 0.81 |
| 15101,59,1,2 109–272 μ | 298 | 548000 | 14000 | 90.3 | 1136 | 4200 | 800 | 7200 | — | — | 1190 | 0.74 |

* Keith *et al.* (1972). † LSPET (1972). Absolute errors < ±6%, for Kr and Xe < ±20%.

Table 4. Rare gas concentrations in coarse fragments, breccias and rocks (in $10^{-8}$ cc STP/g).

| Sample | $^{3}He$ | $^{4}He$ | $^{20}Ne$ | $^{21}Ne$ | $^{22}Ne$ | $^{36}Ar$ | $^{38}Ar$ | $^{40}Ar$ | $^{84}Kr$ | $^{132}Xe$ | K (ppm) | U (ppm) |
|---|---|---|---|---|---|---|---|---|---|---|---|---|
| *Coarse fines* | | | | | | | | | | | | |
| 14003,24,1 > 300 μ | 890 | 2500000 | 53000 | 177 | 4110 | 17200 | 3230 | 35000 | 19 | 4 | | |
| 14003,24,2 > 300 μ | 1200 | 4150000 | 60100 | 186 | 4640 | 10540 | 1990 | 86700 | 6.4 | 0.84 | | |
| 14001,5,1 | 510 | 504000 | 2045 | 109 | 278 | 482 | 161 | 28000 | 0.60 | 0.24 | | |
| 14001,5,2 | 490 | 420000 | 1200 | 144 | 199 | 240 | 114 | 28000 | 0.38 | 0.19 | | |
| 14257,9,1 | 66 | 262000 | 990 | 16.4 | 87 | 211 | 52.5 | 27600 | 0.18 | 0.1 | | |
| 14161,26,breccia | 1310 | 4720000 | 96000 | 283 | 7000 | 13540 | 2560 | 160000 | 6.4 | 1.3 | 4500* | |
| 14161,26,1 | 160 | 129000 | 309 | 40.8 | 66.7 | 78.9 | 51.1 | 31300 | | | for range of varia- | |
| 14161,34,8 | 228 | 351000 | 2520 | 42.7 | 247 | 640 | 157 | 22700 | | | tions see Table 1, | |
| 14161,34,9 | 3300 | 10900000 | 210000 | 632 | 15800 | 46000 | 8500 | 127000 | 27.7 | 5.7 | and Hubbard *et al.* | |
| 14161,34,10 | 2915 | 9700000 | 150000 | 409 | 11100 | 39000 | 7300 | 84700 | 24.8 | 4.4 | (1972). | |
| | | | | | | | | | | | 620–7600 | 3.3–5 |
| *Clastic rock fragments* | | | | | | | | | | | | |
| 14303, saw dust | 13.4 | 250000 | | 2.6 | | 9.4 | 3.66 | 24400 | 0.1 | 0.07 | 7450 | 4.0 |
| 14303,13,6 | 16.2 | 328000 | ≤4.7 | 2.65 | 3.18 | 13.6 | 4.4 | 40700 | | | | |
| 14303,13,7 | 16.8 | 78900 | | 3.0 | 3.8 | 9.9 | 5.5 | 64700 | | | | |
| 14303,13,8 | 10.2 | 246000 | | 2.3 | 4.2 | 11.4 | 4.3 | 58600 | | | | |
| 14303,13,9 | | 22700 | | | | 13.1 | 5.8 | 33500 | | | 10430 | |
| 14303,13,12 | 7.4 | 10680 | | 1.8 | | 6.9 | 3.9 | 2435 | | | | |
| *Crystalline rock* | | | | | | | | | | | | |
| 15556 | 490 | 24400 | 238 | 77.6 | 99 | 72.5 | 72 | 1030 | | | 255† | |

* Keith *et al.* (1972). † LSPET (1972). Absolute errors ≤ ±6%, for Kr and Xe ≤ ±20%.

Table 5. Isotopic composition of Kr and Xe in 14003 samples.

| Sample | $^{82}$Kr | $^{83}$Kr | $^{84}$Kr | $^{86}$Kr | $^{124}$Xe | $^{126}$Xe | $^{128}$Xe | $^{129}$Xe | $^{130}$Xe | $^{131}$Xe | $^{132}$Xe | $^{134}$Xe | $^{136}$Xe |
|---|---|---|---|---|---|---|---|---|---|---|---|---|---|
| | Normalized for $^{84}$Kr = 100 and $^{136}$Xe = 30 | | | | | | | | | | | | |
| 14003 bulk | 21 | 21 | =100 | 30.5 | 1.23 | 1.62 | 10.5 | 108 | 17.7 | 93 | 101 | 38 | =30 |
| <25 μ | 20 | 21 | =100 | 30.5 | 1.20 | 1.40 | 10.0 | 107 | 17.9 | 89 | 101 | 37 | =30 |
| 25–60 μ | 20 | 21 | =100 | 30 | | | 14.8 | 114 | 21.6 | 99 | 104 | 37.8 | =30 |
| 60–109 μ | 22 | 22 | =100 | 31 | | | 15.4 | 114 | 21.8 | 100 | 104 | 36.7 | =30 |
| 109–272 μ | 21 | 21 | =100 | 30 | | | 13.9 | 112 | 22.4 | 101 | 103 | 36 | =30 |
| 14003,24,2 > 300 μ | 21 | 21 | =100 | 30 | | | 18 | 117 | 21 | 92 | 105 | 38 | =30 |
| Etched 60–109 μ | 22 | 24 | =100 | 30 | | | 24 | 126 | 26.4 | 129 | 108.5 | 37 | =30 |
| Etched 109–272 μ | 22 | 24 | =100 | 30 | | | 18.1 | 122 | 23.7 | 116 | 105.5 | 36 | =30 |
| Solar* | 20 | 20 | =100 | 31 | 0.49 | 0.44 | 8.3 | 10.4 | 16.5 | 82 | 100 | 37 | =30 |
| Spallogenic† | 75 | =100 | 38 | — | 35 | 64 | =100 | 110 | 64 | 390 | 58 | — | — |

For absolute amounts, see Tables 3 and 4. Errors 0.1–2% depending on abundance.
Ba-content of 14003,24 aliquot was measured to be 825 ppm.
* Eberhardt *et al.* (1970). † Bogard *et al.* (1971).

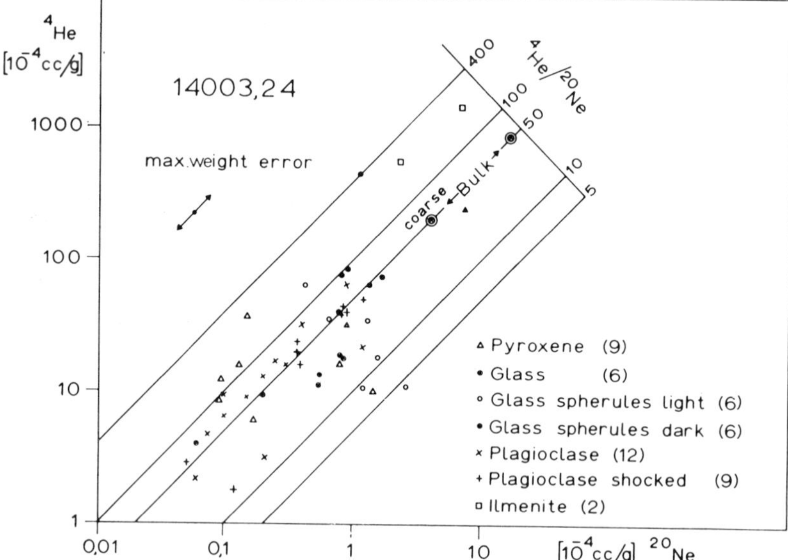


Fig. 8. $^4$He, $^{20}$Ne, and ($^4$He/$^{20}$Ne) in single grains of 14003,24 fines.

internal agreement of duplicate analyses and comparisons of the calibration standards with terrestrial and meteoritic interlaboratory standards. To calibrate Kr and Xe, Abee-meteorite standards were used instead of air spikes. In this case, absolute errors may be as large as 20%. Isotopic ratios have errors <2% for all gases.

K and U data required for the age determination are included in Tables 3 and 4. Their determination will be described in more detail by Müller (1972). Where comparisons are possible, our rare gas data compare favorably with those by Heyman *et al.* (1972) but tend to be generally higher than those reported by LSPET (1971, 1972). Sampling may account for some of the differences.

### *Single grain analysis*

Single minerals and glasses were handpicked from 14003 soil and identified by optical and x-ray methods. Fifty specimens were then weighed on a microbalance (range 10–300 μg) and measured in the Nier-type spectrometer, applying the single grain technique already described (Kirsten *et al.*, 1971). The results for $^4$He, $^{20}$Ne, and $^{36}$Ar are given in Figs. 8 and 9. The isotopic composition of these gases corresponds in general to the known ratios for solar wind implanted gases as measured in the bulk. Due to the high K content of Apollo 14 fines, individual $^{40}$Ar/$^{36}$Ar ratios are relatively high. They range from 0.4 to 20 and are centered around ~1.5. In 15 cases it was possible to determine the spallogenic contributions from the isotopic composition of Ne and Ar and to calculate the individual exposure ages given in Fig. 10.

For a survey of the variation of implanted $^4$He concentrations in a larger number of single grains we have applied the He-microprobe technique (for details, see Kirsten *et al.*, 1970) to 153 individual specimens with emphasis on differently colored glass spherules. The resulting $^4$He concentrations per cm$^3$ and per cm$^2$ surface area are given in Figs. 11 and 12. Grain surface areas were estimated by microscopic evaluation of shape parameters.

## Discussion

Radiogenic, spallogenic, and trapped components contribute to the rare gases in lunar samples. In general, they can be distinguished from each other on the basis of their isotopic composition, especially if data for grain size fractions and etched samples are also available. For bulk soils the $(^4He/^3He)_{tr}$ ratio is little affected by radiogenic $^4$He and spallogenic $^3$He. A plot of $^4$He versus $^3$He for all 14003 samples listed in Table 3 reveals that these contributions by chance compensate each other approximately. This prevents an exact determination of each component; however, an iteration which also involves the $^4$He/$^{20}$Ne ratio allows an estimate of the components.

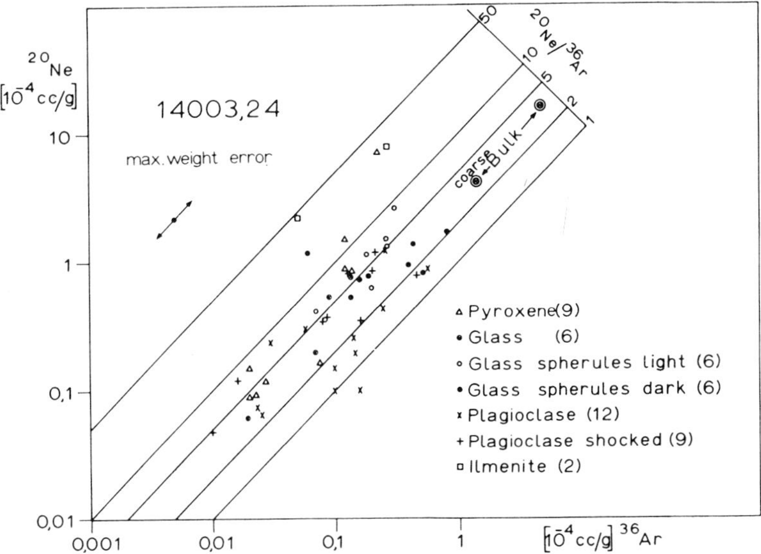


Fig. 9. $^{20}$Ne, $^{36}$Ar, and ($^{20}$Ne/$^{36}$Ar) in single grains of 14003,24 fines.

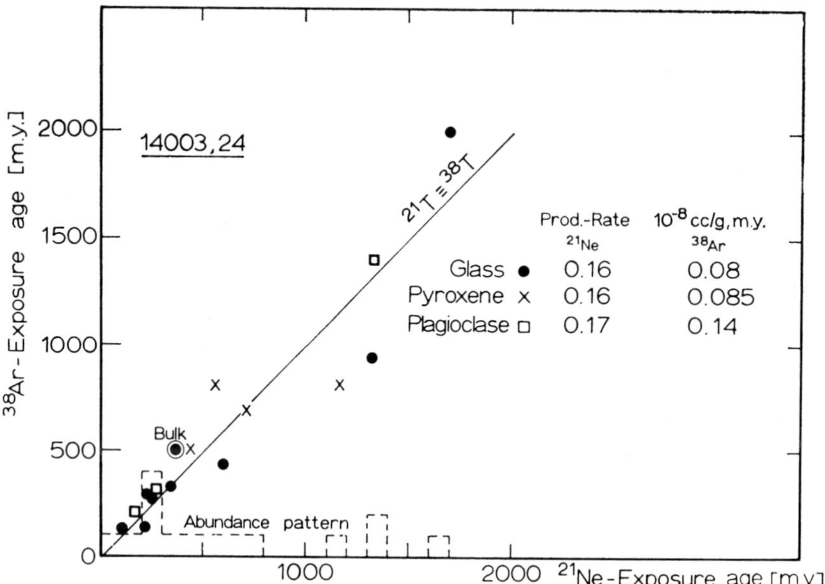


Fig. 10. $^{21}$Ne and $^{38}$Ar exposure ages of single particles from 14003,24 fines.

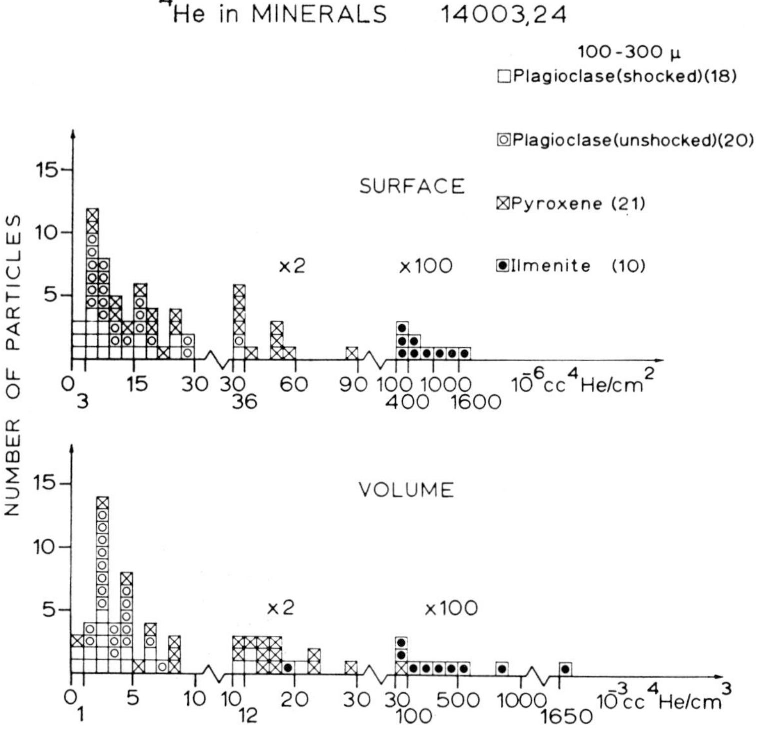


Fig. 11. $^4$He concentrations in single mineral grains of 14003,24 fines. Micro-He-probe analysis.

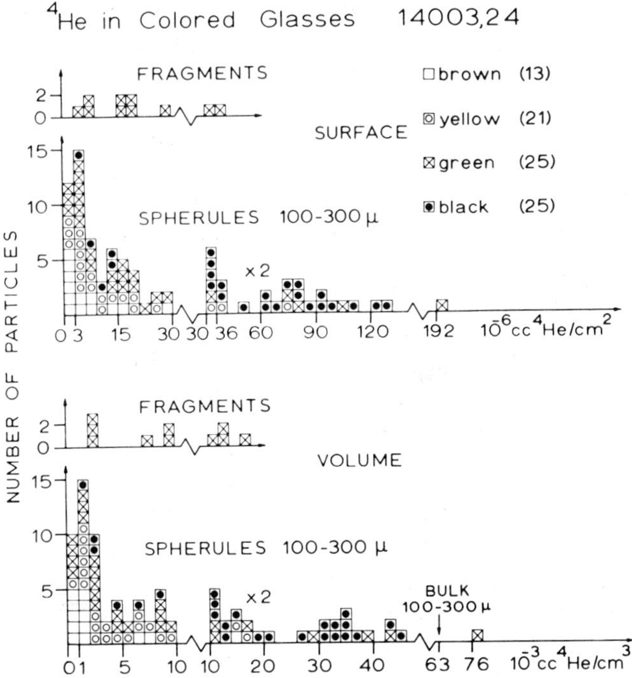


Fig. 12. $^4$He concentrations in single glass spherules and fragments of different color. Micro-He-probe analysis.

For 15101, the known U content limits the maximum contribution of radiogenic $^4$He. Similarly, spallogenic $^3$He could be estimated for some of the coarse fines listed in Table 4.

The Ne isotopes were divided into spallogenic and trapped components adopting a ratio of $(^{20}Ne/^{21}Ne/^{22}Ne)_{sp} = 0.95 : 1 : 1.05$ by an iteration which corresponds to the application of a two-isotope plot as introduced by Eberhardt *et al.* (1970). The spallogenic Ne, which is used to calculate the exposure ages (Table 6), is then subtracted to calculate the trapped component. A similar treatment was given to the Ar isotopes adopting $(^{36}Ar/^{38}Ar)_{sp} = 0.63$. A plot of $^{40}$Ar versus $^{36}$Ar allows a separation in favorable cases of the radiogenic $^{40}$Ar from the implanted $^{40}$Ar and a calculation of characteristic $(^{40}Ar/^{36}Ar)_{tr}$ ratios as well as total K–Ar model ages (Table 6). The Xe isotopic composition of 14003 samples (Table 5) is normalized to $^{136}Xe = 30$, since this corresponds approximately to the abundance of $^{136}$Xe relative to $^{132}Xe = 100$ in the trapped component and since there is no spallogenic $^{136}$Xe. Then, spallogenic Xe is derived by subtracting the next-to-last line of Table 5 from each sample.

### *Trapped gases*

The major source of the rare gases observed in lunar fines is solar wind implantation (e.g., Eberhardt *et al.*, 1970; Heymann *et al.*, 1970; Hintenberger *et al.*, 1970; Kirsten *et al.*, 1970). The absolute amounts of implanted solar wind gases turn out to be remarkably similar for the soils of all landing sites. Total He, Ne, and Ar concentrations of the analyzed Apollo 14 and 15 bulk soils differ only by a factor of 2, except for 15421 (Table 3), and are not too distinct from Apollo 11 and 12 soils.

Solar wind is surface correlated because of its low initial implantation depth ($\sim$400 Å). Our data on grain size fractions and etched samples support this result; however, it is also apparent that about 5 to 10% of these gases are related to the volume rather than to the surface. This is explained by solid state diffusion (Kirsten *et al.*, 1971). Other sources for this component may be stopped solar flare particles (Poupeau *et al.*, 1972) or inclusion of ancient solar wind gases during remelting (Megrue and Steinbrunn, 1972; Roedder and Weiblen, 1972).

The relatively uniform surface loading of different soils is mainly due to saturation effects. This is illustrated by the absence of single soil particles without implanted gases. According to micrometeorite influx data and present-day erosion rates, 3–4 $\times$ $10^9$ yr were required to form the lunar regolith. Sampling near medium-sized craters guarantees sampling from all regolith depths (e.g., Spur Crater, soil 15301). The rather uniform rare gas concentrations lead then to the conclusion that solar wind irradiation took place for at least 3 $\times$ $10^9$ yr with relatively little change.

On the other hand, large differences of solar wind concentrations are observed among individual soil particles (Figs. 8, 9, 11, and 12). They are best explained by subsequent diffusion loss at or near the lunar surface.

The relative diffusion losses are governed by the crystal lattice parameters of the various minerals. The lowest concentrations are observed in plagioclases (Fig. 11) in agreement with diffusion studies by Baur *et al.* (1972). Shock effects seem to have little influence on the diffusion parameters. The distributions for shocked and unshocked plagioclases are quite similar.

Ilmenites have the highest retentivity (Fig. 11), resulting in 10 to 100 times higher $^4$He concentrations. For a given mineral type, trapped gas concentrations vary by an order of magnitude since each grain has experienced its individual thermal history after it became saturated with solar wind.

Figure 12 exhibits a correlation between the solar wind content and the color of glass spherules. Mean concentrations increase in the order brown, yellow, green, black by a factor of about 10. Glass colors are due to different chemical composition (see, e.g., Bell and Mao, 1972) which may cause differences in retentivity. The color effect can be explained by glass production from different source materials (Giles and Nicholls, 1972). Darkening along with higher retentivity could also be due to opaque microcrystals within the glasses. Each black glass spherule we have looked at so far in polished sections contained fragments of ilmenite or newly crystallized phases (Ni–Fe and others).

*Elemental ratios.* We have measured $(^4\mathrm{He}/^{20}\mathrm{Ne})_{tr}$ ratios between 47 and 65 for Apollo 14 fines and between 20 and 48 for Apollo 15 fines. For soils from Apollo 11 and 12 this ratio varies between 30 and 120 (Bogard *et al.*, 1971). The large difference with the $^4$He/$^{20}$Ne ratios directly observed in the present-day solar wind (430–620, Bühler *et al.*, 1972) indicates loss of implanted $^4$He rather than a change of solar ratios. Similarly, $(^{20}\mathrm{Ne}/^{36}\mathrm{Ar})_{tr}$ ratios range from 3.7 to 8.3, much lower than the expected value of at least 43 which is observed in one single particle (see also Aller,

1961). The ($^{36}Ar : ^{84}Kr : ^{132}Xe$) ratios of the analyzed soils fall in the range previously observed for Apollo 11 and 12 soils but are also lower by factors of 3 to 5 than the expected solar abundances (Aller, 1961). Secondary fractionation of the lighter gases rather than solar abundance changes must be responsible.

Soil 15421, rich in green glass, is peculiar in its apparent Ne excess with ratios $(^{4}He/^{20}Ne)_{tr} = 20$ and $(^{20}Ne/^{36}Ar)_{tr} = 8.3$. Even higher $(^{20}Ne/^{36}Ar)_{tr}$ ratios are given by LSPET (1972) for fines and glass from soil 15923,2 which is derived from the same bulk sample as 15421. Apparently different portions of a particularly Ne-rich component (Bogard and Nyquist, 1972) are admixed to the samples. Single grain analysis will be performed to localize this component.

The basic uniformity of the composition of solar wind gases at various landing sites does not contradict the order of magnitude variations of elemental ratios observed in individual soil particles (Figs. 8 and 9). Similar results were obtained by Megrue and Steinbrunn (1972). In general, the depletion of light rare gases is enhanced for glasses and plagioclases compared to pyroxenes and ilmenites. The least fractionated gases are found in a dark glass spherule $(^{4}He/^{20}Ne)_{tr} = 400$ and in ilmenites $(^{4}He/^{20}Ne)_{tr} = 240$; $(^{20}Ne/^{36}Ar)_{tr} = 43$. Presumably, these particles were recently loaded with solar wind and suffered little subsequent diffusion. The $(^{4}He/^{20}Ne)_{tr}$ ratio of 400 is already close to the expected solar value. It is then indicated that with even better statistics, the maximum ratios obtained do closely represent the solar wind ratios.

The fragmental rock 14303 is almost free of solar wind. It has been almost completely outgassed during extensive metamorphism. It contains only traces of heavily fractionated gases [$8 \pm 3 \times 10^{-8}$ cc STP/g $^{36}Ar$; $(^{20}Ne/^{36}Ar) < 1$], which are probably remainders of ancient solar wind dissolved during the formation or subsequent annealing of this fragmental rock.

*Isotopic ratios.* $^{4}He/^{3}He$ ratios in the trapped component are $2850 \pm 100$ for 14003 and $2750 \pm 100$ for 15101. For the other fines radiogenic $^{4}He$ prevents an exact determination. U and age data (Tables 3 and 6) yield estimated values between 2600 and 3000. The present-day $^{4}He/^{3}He$ ratio in the solar wind falls distinctly below the observed ratios and ranges from 1860 to 2450 (Bühler *et al.*, 1972). It is probably not by chance that the only isotopic ratio for which a distinct deviation is observed is the one that is most sensitive to the evolution of the sun. Geiss *et al.* (1970) conclude a true change in the solar wind $^{4}He/^{3}He$ ratio. If this is true it would suggest that most of the solar wind collected in the lunar regolith is very ancient since it displays $^{4}He/^{3}He$ ratios higher than at present. The Ne-isotopic ratios measured in Apollo 14 and 15 soils range from 13 to 13.6 for $(^{20}Ne/^{22}Ne)_{tr}$ and from 30 to 32 for $(^{22}Ne/^{21}Ne)_{tr}$. This is similar to the ratios observed in the solar wind composition experiment (Bühler *et al.*, 1972). We observed that $(^{36}Ar/^{38}Ar)_{tr}$ ratios are $\sim 5.45$, except for 15601. This is slightly above the ratio observed by us in Apollo 11 and 12 samples and in 15601. Since LSPET (1972) reports normal ratios, it may be too early to assign significance to this deviation.

The so-called parentless $^{40}Ar$ (Heymann and Yaniv, 1970) is also observed in Apollo 14 and 15 fines. This component is surface related and correlated with the solar wind; however, variable $^{40}Ar/^{36}Ar$ ratios and nuclear abundance considerations exclude a solar wind origin. Heymann and Yaniv (1970) have proposed reimplantation of a latent lunar atmosphere by the solar wind. It is then interesting to compare the $(^{40}Ar/^{36}Ar)_{tr}$ ratios of different soils, since they may reflect the igneous outgassing processes in the early lunar history. In favorable cases (14003, 15101) a plot of $^{40}Ar$ versus $^{36}Ar$ allows one to distinguish in situ-radiogenic $^{40}Ar$ from the implanted $^{40}Ar$ (Table 6). For the other samples we have estimated the probable radiogenic contributions from the K concentrations and the K–Ar ages of Table 6 in order to calculate the $(^{40}Ar/^{36}Ar)_{tr}$ ratios given in Table 6.

Table 6. Ages and $(^{40}Ar/^{36}Ar)_{tr}$-ratios in Apollo 14 and 15 fines and rocks.

| | Exposure ages (m.y.) | | | | Gas retention ages ($10^9$ yr) | | |
|---|---|---|---|---|---|---|---|
| Sample | $^{3}He$ | $^{21}Ne$ | $^{38}Ar$ | $^{128}Xe$ | U/Th–He | Total K–Ar | $R = (^{40}Ar/^{36}Ar)_{tr}$ |
| *Soils < 1 mm* | | | | | | | |
| 14003 | 90 ± 30 | 350 ± 40 | 470 ± 80 | 580 ± 150 | 2.5 ± 0.2 | 3.1 ± 0.15 | 1.24 ± 0.04 |
| 15101 | 150 ± 50 | 390 ± 30 | 330 ± 50 | | | 2.6 ± 0.15 | 1.08 ± 0.04 |
| 15021 | | 370 ± 50 | | | | | 0.65 ± 0.08 |
| 15601 | | 380 ± 50 | | | | | 0.78 ± 0.08 |
| 15301 | | 310 ± 40 | | | | | 1.55 ± 0.1 |
| 15421 | | 460 ± 60 | | | | | 3.4 ± 0.2 |
| *Coarse fines* | | | | | | | |
| 14003 > 300 μ | | 250 ± 40 | | (470) | | | 14003,24,2: 4 < *R* < 8 |
| 14001 | 360 ± 40 | 700 ± 50 | 570 ± 30 | | | (~4) | |
| Various 14161 | 170 ± 30 | 300 ± 60 | 300 ± 40 | | | | Breccia: 9 < *R* < 12 |
| do, irrad. samples | | | 300–430 | | | 3.54–3.92 (Table 1) | |
| 14161,34,9 | | 770 ± 100 | 610 ± 150 | | | | 1.9 < *R* < 2.8 |
| 14257 | 50 ± 10 | 90 ± 10 | 110 ± 10 | | | (~4) | |
| *Rocks* | | | | | | | |
| 14303 | 15 ± 2 | 17 ± 2 | 22 ± 7 | | 2.25 ± 0.15 | 3.15 ± 0.1 | |
| do, irrad. samples | | | 29 ± 4 | | | 3.44–3.75 (Table 1) | |
| 15556 | 490 ± 50 | 525 ± 40 | 490 ± 50 | | | 3.4 ± 0.1 | |
| Production rate* in $10^{-8}$ cc/g (m.y.) | 1.0 | 0.154 | 0.14 | 0.000165 | | | |

* Based on chemical composition and specific production rates given by Bogard *et al.* (1971).

Variations between 0.65 and 3.4 exist for the various soils (<1 mm) even within one landing site. A soil breccia from coarse fines 14161 has a ratio $\gtrsim 8$, even after allowance is made for radiogenic $^{40}Ar$ produced from twice the maximum K amount observed in any of many investigated 14161 fragments (Tables 1 and 4, Hubbard *et al.* (1972)). A similar situation is encountered for 14003,24,2 with a minimum $(^{40}Ar/^{36}Ar)_{tr}$ ratio of 4.

The observed variations could be explained by a lunar atmosphere variable with time and incomplete mixing during transport of regolith material from various sites (Heymann *et al.*, 1972). However, it must be explained how *mean* differences persist within a relatively small, smooth area, while, on the other hand, a thorough mixing is required to explain the large variation of $^{40}Ar/^{36}Ar$ ratios in individual grains of one soil sample (Kirsten *et al.*, 1971) by a variable lunar atmosphere. The total $^{40}Ar/^{36}Ar$ ratios of the individual grains analyzed in this work vary from 0.4 to 20 and are centered around unity.

Poupeau *et al.* (1972) propose an early solar flare irradiation of Ca in highland materials as a source of the excess $^{40}Ar$. One could also consider direct adsorption of ascendent $^{40}Ar$ in the overlaying regolith. The loose correlation between $^{40}Ar$ and $^{36}Ar$ could be related to enhanced adsorption capability of material heavily damaged by solar wind irradiation. However, in diffusion experiments, Baur *et al.* (1972) have found little evidence for $^{40}Ar$ from a reimplanted lunar atmosphere. They consider adsorption of $^{40}Ar$ and subsequent migration caused by solar wind via a knock-on process or coatings of volatilized potassium as alternative explanations. Volatilization of K in crystallized material is likely to have occurred (Hubbard *et al.*, 1972; Biggar *et al.*, 1972; Nyquist *et al.*, 1972).

*Exposure ages*

The cosmic ray exposure ages given in Table 6 are based on the (relatively uniform) chemical composition of all analyzed samples and the production rates given by Bogard *et al.* (1971). It is known that $^{3}He$ exposure ages are often lowered by diffusion loss of spallogenic $^{3}He$. Most reliable are the $^{21}Ne$ ages, while $^{38}Ar$ ages are less accurate. Nevertheless, $^{21}Ne$ and $^{38}Ar$ ages mostly agree within the limits of error. The slightly higher $^{128}Xe$–Ba age of 14003 is not taken as indication for $^{21}Ne$ or $^{38}Ar$ losses.

Now that exposure ages for fines from four landing sites have been measured (for Apollo 11 and 12, see Kirsten *et al.*, 1970, 1971), it is remarkable that they are all around 400 m.y. This age probably reflects gardening in the regolith and indicates a relatively uniform mixing rate of $\sim$2 mm/m.y. for the whole moon. Contrary to this uniformity, particular fragments of coarse fines exhibit individual exposure histories and have exposure ages between 100 and 770 m.y. The $^{38}Ar$–Ca exposure ages derived from neutron irradiated 14161-fragments (300–430 m.y.) are similar to those of other 14161 samples. However, rock fragment 14161,9 has an exposure age of 770 m.y. The individual behavior of each fragment within the regolith is further illustrated by the exposure ages obtained for single soil fragments (Fig. 10). They range from $<$100 m.y. up to nearly $2 \times 10^9$ yr. This again is a proof for a regolith much older than $10^9$ yr. The exposure age of fragmental rock 14303 from the rim of Cone Crater is $\sim$25 m.y. This age is believed to be the age of Cone Crater (Bogard and Nyquist, 1972; Burnett *et al.*, 1972). Rock 15556 from the rim of Hadley Rille has an exposure age of $\sim$500 m.y.

*Total gas-retention ages*

Total K–Ar and U–He model ages for soil 14003 and the K–Ar age for soil 15101 have been calculated (Table 6).

According to soil mixing models (Hubbard *et al.*, 1972; Birck *et al.*, 1972; Wänke *et al.*, 1972; Schonfeld, 1972), soils are composed of at least three components with different ages and K contents. Apart from diffusion losses, the measured K–Ar ages of 3.1 and $2.6 \times 10^9$ yr. may reflect the mixing ratio of these components (Pepin *et al.*, 1972). A comparison of the K–Ar and the U–He ages of 14003 indicates only moderate He losses. The similarity of both K–Ar and U–He ages for 14003 and saw dust of fragmental rock 14303 seems to indicate that Apollo 14 soil is made of essentially the same material as the local clastic rocks with little gas loss during the grinding process. The crystalline fragment 14161,34,9 has a total K–Ar age near $4 \times 19^9$ yr and corresponds to samples 14161,34,4 and 14161,34,5 discussed in the next section.

The 15556 mare basalt gives an age of $3.4 \times 10^9$ yr and corresponds to the age obtained by other authors for mare flows at Imbrium basin near Hadley Rille (Schaeffer *et al.*, 1972; Cliff *et al.*, 1972).

*Crystallization ages*

The $^{39}Ar$–$^{40}Ar$ method of age determination has been very successful for lunar samples (Alexander *et al.*, 1972; Husain *et al.*, 1971; Turner, 1970, 1971; Turner *et al.*, 1971). In general, the gas release pattern obtained for measured $^{40}Ar/^{39}Ar$ ratios at different temperatures shows several features: (1) a plateau of high $^{40}Ar/^{39}Ar$ value which is ascribed to the gas release from the retentive minerals and from which an age is obtained; (2) lower $^{40}Ar/^{39}Ar$ values at the lower temperature which is ascribed to the gas released from less retentive minerals and from which the diffusion loss of Ar is obtained; and (3) lower $^{40}Ar/^{39}Ar$ values at the highest temperatures for which there is no clear explanation. For certain rocks (1) and/or (3) are of little importance (see, for example, Fig. 3), while in a number of cases, as seen in Figs. 1 and 2, the interpretation of the drop-off in the $^{40}Ar/^{39}Ar$ ratio is of importance in establishing a precise age. Several suggestions have been advanced (Huneke *et al.*, 1972; Turner *et al.*, 1971) to explain the low $^{40}Ar/^{39}Ar$ values at high temperature releases. One is that the high temperature minerals, such as the pyroxenes, do not retain their argon as well as feldspars. There is experimental evidence to show that in separated mineral phases plagioclase gave an age agreeing with the Rb–Sr age, while the pyroxene fraction gave a low age. The same authors suggest, in cases where the plateau age is apparently too high and the drop off in age is large at the high temperatures, that the rock represents a case where the K and/or the Ar have been redistributed. Assuming a closed system, the age then is given by the average over the high temperature release.

We should like to suggest as an alternative explanation the slow devitrification of a minor K-rich glass phase. The fall-off in age at high temperatures was not seen for Apollo 11 or 12 basalts but seems to be related to the explosive origin of the Fra Mauro formation. Although there is no observable glass present in the samples, there is petrological evidence that the Fra Mauro rocks contain a fine grained phase which crystallized from a glass. If this took place slowly, then that phase could exhibit a much younger age and at present be very retentive with high temperature melting minerals (see also Turner, 1971). This removes the burden of explaining gas loss from pyroxenes without corroborating evidence for diffusion loss, and with the uncertainty as to why Fra Mauro pyroxenes are different from those from Apollo 11 or 12 basalts.

At the present time, when none of the explanations is established, it may be better to calculate gas retention ages based both on the plateau and the whole high temperature portion. In cases where these ages agree, there is no uncertainty, while in cases where there is a discrepancy one should increase the error in the age.

In seven cases listed in Table 1 and shown in Figs. 1, 2, 3, and 4, the difference between the two ages is small, while in two cases (14161,34,2 and 14303,13,R5,21, Figs. 1 and 2), the difference between the two ages is larger.

For two basaltic samples (14161,26,11 and 14161,26,12) and a feldspathic accumulation (14161,34,3) the two ages are in perfect agreement, but much lower than those of the other fragments. From microscopic examination it was expected that this sample is anorthositic in composition as it is composed of ~90% plagioclase and ~10% pyroxene. The high Ca content (13.4%) is in accordance with a mainly feldspathic composition. But as the potassium content is rather high (0.46%), a K-feldspar phase may also be present. The age measured on this sample is very well defined at $3.63 \pm 0.05 \times 10^9$ yr total argon age, high temperature age, and the plateau age. The total argon loss is 2.4%. The slight shock effects observed in this sample evidently did not affect the ages.

The two basaltic rocks show equally well-defined ages of $3.59 \pm 0.02 \times 10^9$ yr (14161,26,11) and $3.54 \pm 0.01 \times 10^9$ yr (14161,26,12)—the argon losses being well below 1% (Table 1). These two basalts are different from the other basalts analyzed by their extremely low opaque mineral content and their relatively high pyroxene-to-feldspar ratio.

The ages of all three rocks are unusual for Apollo 14 samples. As mare material is rare (but present) in the Apollo 14 soils (Adams and McCord, 1972; Powell and Weiblein, 1972), it follows that we have measured three mare-derived rocks with a mean age of $3.58 \pm 0.05 \times 10^9$ yr. These ages are typically found for Mare Tranquilitatis basalts (Papanastassiou and Wasserburg, 1971; Turner, 1971). Therefore the question arises where these samples from the Apollo 14 soils originate.

As judged from crater counts, Sinus Medii is the nearest mare surface about equal in age to Mare Tranquilitatis (Gault, 1972). It is located at the geographical center of the moon's surface, about 500 km ENE from the Apollo 14 landing site. This distance is rather large. Therefore, our rocks more plausibly might be ejecta material from a crater in the nearby mare basalts of the Imbrium or Erathosthenian formation. There is no reason to doubt the existence of older lava flows below the surface flows which show ages from 3.36 to $3.16 \times 10^9$ yr (Papanastassiou and Wasserburg, 1971). We assume that such older material impacted at the Fra Mauro formation, contributing to the regolith but contributing very little to the total amount of material there. Probably the cosmic ray exposure ages of the two young basalt samples which are 320 m.y. and 360 m.y. approximately date such a secondary event. The ages listed in Table 1 fall into two classes, one group at 3.5 to $3.6 \times 10^9$ yr, presumably material not originating in the Imbrium event as discussed above, and another group at 3.7 to $4.0 \times 10^9$ yr, if we include all other ages (Cliff *et al.*, 1972; Compston *et al.*, 1972; Huneke *et al.*, 1972; Husain *et al.*, 1971, 1972; Papanastassiou and Wasserburg, 1971; Murthy *et al.*, 1972; Schaeffer *et al.*, 1972; Stettler *et al.*, 1972; Sutter *et al.*, 1971; Turner, 1970; Turner *et al.*, 1971; York *et al.*, 1972). Rocks of these ages are probably all derived from the Imbrium event.

Petrologic observations show that these basaltic rocks must have been crystallized within the lunar crust before they were excavated during the Imbrium impact event and embedded into the pyroclastic ejecta material. They were not even affected thermally or chemically to an extent which would have disturbed the Rb/Sr isotopic systems (Papanastassiou and Wasserburg, 1971). It seems likely that the Imbrium event did little in the way of releasing radiogenic $^{40}Ar$ so that the crystallization ages measured represent the times of crystallization of these rocks in the crust of the moon where Mare Imbrium is at present. In this case the event forming the Mare Imbrium basin is younger than the lowest age observed so far, $3.75 \pm 0.05 \times 10^9$ yr. On the other hand, as there is no evidence for pre-Mare Imbrium site ejecta on Mare Tranquilitatis, it is clear that Mare Tranquilitatis is younger than the Imbrium event. This, then, closely brackets the Imbrium event as being between $3.7 \times 10^9$ and $3.8 \times 10^9$ yr ago. As the rim of the Imbrium basin is penetrated by some craters which were flooded with basalts simultaneously with the mare one can assume that there is a time gap between basin formation and filling with basalts. The density of such craters is too small to allow an estimate for this time gap by crater counts (Gault, 1972).

In this respect the measurements on microbreccias 14303,R5,21 and 22 are probably meaningful (Table 1). The microbreccia-clast R5,21 shows a very badly defined "plateau age" of $3.78 \times 10^9$ yr (see Fig. 1). The $^{39}Ar$–$^{40}Ar$ release pattern gives an indication of the K/Ar system having been disturbed at $3.78 \times 10^9$ yr ago. The large argon loss of the sample (19.5%) indicates the formation of phases with poor Ar retentivities. The formation of this breccia is probably close to $3.78 \times 10^9$ yr ago. Sample R5,22 which also is a microbreccia but finer grained than R5,21 seems to be totally disturbed in respect of both K and Ar but not degassed. It therefore shows a large amount of excess argon in the high temperature range, possibly shock-implanted excess argon. It is interesting that this sample shows the smallest K content ever reported for a lunar rock. This low K content (57 ppm) most probably is due to potassium volatilization during the thermal spike caused by the large impact. A comparably low potassium content has recently been reported for the shock-molten gabbroic anorthosite 15418 (86 ppm, LSPET, Apollo 15, 1972).

In summary then, the upper crust of the moon at the pre-Mare Imbrium site was built up by an already complex sequence of rocks such as norite, granites, and anorthosites overlain by different basaltic lava flows. These were locally brecciated by impacts of meteorites—sometimes multiply. Multiple basalt flows or sills with distinct crystallization ages of 4.00 to $3.75 \times 10^9$ yr have been recognized. This complex was ejected at the time a large impact event formed the Mare Imbrium basin. The time of this event was between 3.70 and $3.80 \times 10^9$ yr ago. The ejecta material was deposited forming the Fra Mauro formation which was subsequently bombarded by meteorites leading to the formation of the regolith and impact craters.

Filling of the maria occurred some time after the Imbrium basin was excavated, starting with that of Mare Tranquilitatis at some time more than $3.70 \times 10^9$ yr ago and ending with the surface flows in the Oceanus Procellarum and the Mare Imbrium about $3.20 \times 10^9$ yr ago. Further bombardment of these surfaces led to cross contaminations of material from one place to the other over large areas and distances.

From this model it is expected that with increasing number of age determinations and refinements in their interpretation the clustering of ages for rocks from mare areas as it appears now to exist will disappear and continuous mare-filling processes will be recognized; these might vary in intensity within the time interval of 3.8 to $3.2 \times 10^9$ yr.

*Acknowledgments*—We are grateful to NASA for providing us with lunar samples. The skillful operation of the mass spectrometer by H. Richter is gratefully acknowledged. Thanks are given to Dr. O. Müller for making unpublished data available and for help in monitoring the neutron irradiations. The technical assistance of D. Dörflinger, W. Ehrhardt, R. Schwan, H. Urmitzer, and H. Weber is appreciated. Discussions with Drs. D. Gault, A. El Goresy, S. Kalbitzer, J. F. Lovering, and with Professor W. Gentner have substantially contributed to the interpretations. We appreciate valuable comments by Professor F. Begemann.

## References

Adams J. B. and McCord T. B. (1972) Optical evidence for regional cross-contamination of highland and mare-soils (abstract). In *Lunar Science—III* (editor C. Watkins), p. 1, Lunar Science Institute Contr. No. 88.

Alexander E. C. Jr., Davis P. K., and Lewis R. S. (1972) Argon-40–argon-39 dating of Apollo sample 15555. *Science* **175,** 417–419.

Aller L. H. (1961) *The Abundance of the Elements.* Interscience, New York.

Baur H., Frick U., Funk H., Schultz L., and Signer P. (1972) On the question of retrapped 40-Ar in lunar fines (abstract). In *Lunar Science—III* (editor C. Watkins), p. 47, Lunar Science Institute Contr. No. 88.

Bell P. M. and Mao H. K. (1972) Initial findings of a study of chemical composition and crystal field spectra of selected grains from Apollo 14 and 15 rocks, glasses and fine fractions (less than 1 mm) (abstract). In *Lunar Science—III* (editor C. Watkins), p. 55, Lunar Science Institute Contr. No. 88.

Biggar G. M., Ford C. E., Humphries D. J., Wilson G., and O'Hara M. J. (1972) Melting relations of more primitive mare-type basalt 14053 and of breccia 14321 and soil 14162 (average lunar crust?) (abstract). In *Lunar Science—III* (editor C. Watkins), p. 74, Lunar Science Institute Contr. No. 88.

Birck J. L., Loubet M., Manhes G., Provost A., Tatsumoto M., and Allegre C. J. (1972) Age and origin of lunar soils (abstract). In *Lunar Science—III* (editor C. Watkins), pp. 80–81, Lunar Science Institute Contr. No. 88.

Bogard D. D., Funkhouser J. G., Schaeffer O. A., and Zähringer J. (1971) Noble gas abundances in lunar material-cosmic-ray spallation products and radiation ages from the Sea of Tranquility and the Ocean of Storms. *J. Geophys. Res.* **76,** 2757–2779.

Bogard D. D. and Nyquist L. E. (1972) Noble gas studies on regolith materials from Apollo 14 and 15 (abstract). In *Lunar Science—III* (editor C. Watkins), pp. 89–91, Lunar Science Institute Contr. No. 88.

Brereton N. R. (1970) Corrections for interfering isotopes in the $^{40}$Ar/$^{39}$Ar dating method. *Earth Planet. Sci. Lett.* **8,** 427–433.

Bühler F., Cerutti H., Eberhardt P., and Geiss J. (1972) Results of the Apollo 14 and 15 solar wind composition experiments (abstract). In *Lunar Science—III* (editor C. Watkins), pp. 102–104, Lunar Science Institute Contr. No. 88.

Burnett D. S., Huneke J. C., Podosek F. A., Russ G. P., Turner G., and Wasserburg G. J. (1972) The irradiation history of lunar samples (abstract). In *Lunar Science—III* (editor C. Watkins), pp. 105–107, Lunar Science Institute Contr. No. 88.

Cliff R. A., Lee-Hu C., and Wetherill G. W. (1972) K, Rb, and Sr measurements in Apollo 14 and 15 material (abstract). In *Lunar Science—III* (editor C. Watkins), pp. 146–147, Lunar Science Institute Contr. No. 88.

Compston W., Vernon M. J., Berry H., Rudowski R., Gray C. M., Ware N., Chappell B. W., and Kaye M. (1972) Age and petrogenesis of Apollo 14 basalts (abstract). In *Lunar Science—III* (editor C. Watkins), pp. 151–153, Lunar Science Institute Contr. No. 88.

Dalrymple G. B. and Lanphere M. A. (1971) $^{40}$Ar/$^{39}$Ar technique of K–Ar dating: a comparison with the conventional technique. *Earth Planet. Sci. Lett.* **12,** 300–308.

Ducati H., Kalbitzer S., Kiko J., and Kirsten T. (1972) Rare gas diffusion studies in individual lunar soil particles. To be published.

Eberhardt P., Geiss J., Graf H., Grögler N., Krähenbühl U., Schwaller H., Schwarzmüller J., and Stettler A. (1970) Trapped solar wind noble gases, exposure age, and K/Ar-age in Apollo 11 lunar fine material. *Proc. Apollo 11 Lunar Sci. Conf., Geochim. Cosmochim. Acta* Suppl. 1, Vol. 2, pp. 1037–1070. Pergamon.

El Goresy A. (1972) Private communication.

Gault D. E. (1972) Private communication.

Geiss J., Eberhardt P., Bühler F., Meister J., and Signer P. (1970) Apollo 11 and Apollo 12 solar wind composition experiments: fluxes of He and Ne isotopes. *J. Geophys. Res.* **75,** 5972–5979.

Giles H. N. and Nicholls G. D. (1972) Preliminary results of mass spectrometric analysis of individual grains from lunar samples (abstract). In *Lunar Science—III* (editor C. Watkins), pp. 306–308, Lunar Science Institute Contr. No. 88.

Grasty R. L. and Mitchell J. G. (1966) Single sample potassium-argon ages using the omegatron. *Earth Planet. Sci. Lett.* **1,** 121–122.

Heymann D. and Yaniv A. (1970) $Ar^{40}$ anomaly in lunar samples from Apollo 11. *Proc. Apollo 11 Lunar Sci. Conf., Geochim. Cosmochim. Acta* Suppl. 1, Vol. 2, pp. 1261–1267. Pergamon.

Heymann D., Yaniv A., Adams J. A., and Fryer G. (1970) Inert gases in lunar samples. *Science* **167,** 555–558.

Heymann D., Yaniv A., and Walton J. (1972) Inert gases in Apollo 14 fines and the case of parentless $Ar^{40}$ (abstract). In *Lunar Science—III* (editor C. Watkins), pp. 376–378, Lunar Science Institute Contr. No. 88.

Hintenberger H., Weber H. W., Voshage H., Wänke H., Begemann F., Vilcsek E., and Wlotzka F. (1970) Rare gases, hydrogen, and nitrogen: Concentrations and isotopic composition in lunar material. *Science* **167,** 543–545.

Hubbard N. J., Gast P. W., Rhodes M., and Wiesmann H. (1972) Chemical composition of Apollo 14 materials and evidence for alkali volatilization (abstract). In *Lunar Science—III* (editor C. Watkins), pp. 407–409, Lunar Science Institute Contr. No. 88.

Huneke J. C., Podosek F. A., Turner G., and Wasserburg G. J. (1972) $^{40}Ar$–$^{39}Ar$ systematics in lunar rocks and separated minerals of lunar rocks from Apollo 14 and 15 (abstract). In *Lunar Science—III* (editor C. Watkins), pp. 413–414, Lunar Science Institute Contr. No. 88.

Husain L., Sutter J. F., and Schaeffer O. A. (1971) Ages of crystalline rocks from Fra Mauro. *Science* **173,** 1235–1236.

Husain L., Schaeffer O. A., and Sutter J. F. (1972) Age of a lunar anorthosite. *Science* **175,** 428–430.

Kaneoka I. (1972) $^{40}Ar/^{39}Ar$ age studies of Apollo 14 samples. To be published.

Keith J. E., Clark R. S., and Richardson K. A. (1972) Gamma ray measurements of Apollo 12, 14, and 15 lunar samples (abstract). In *Lunar Science—III* (editor C. Watkins), pp. 446–448, Lunar Science Institute Contr. No. 88.

Kirsten T., Müller O., Steinbrunn F., and Zähringer J. (1970) Study of distribution and variations of rare gases in lunar material by a microprobe technique. *Proc. Apollo 11 Lunar Sci. Conf., Geochim. Cosmochim. Acta* Suppl. 1, Vol. 2, pp. 1331–1343. Pergamon.

Kirsten T., Steinbrunn F., and Zähringer J. (1971) Location and variation of trapped rare gases in Apollo 12 lunar samples. *Proc. Second Lunar Sci. Conf., Geochim. Cosmochim. Acta* Suppl. 2, Vol. 2, pp. 1651–1669. MIT Press.

Kirsten T., Deubner J., Ducati H., Gentner W., Horn P., Jessberger E., Kalbitzer S., Kaneoka I., Kiko J., Krätschmer W., Müller H. W., Plieninger T., and Thio S. K. (1972) Rare gases and ion tracks in individual components and bulk samples of Apollo 14 and 15 fines and fragmental rocks (abstract). In *Lunar Science—III* (editor C. Watkins), pp. 452–454, Lunar Science Institute Contr. No. 88.

LSPET (Lunar Sample Preliminary Examination Team) (1971) Preliminary examination of lunar samples from Apollo 14. *Science* **173,** 681–693.

LSPET (Lunar Sample Preliminary Examination Team) (1972) The Apollo 15 lunar samples: A preliminary description. *Science* **175,** 363–375.

Megrue G. H. and Steinbrunn F. (1972) Classification and source of lunar soils; clastic rocks; and individual mineral, rock, and glass fragments from Apollo 12 and 14 samples as determined by the concentration gradients of the helium, neon, and argon isotopes (abstract). In *Lunar Science—III* (editor C. Watkins), pp. 532–534, Lunar Science Institute Contr. No. 88.

Müller O. (1972) Alkali and alkaline earth elements, La and U in Apollo 14 and Apollo 15 samples. To be published.

Murthy Rama V., Evensen N. M., Bor-ming Jahn, and Coscio M. R. (1972) Rb–Sr ages, trace elements, and speculations on lunar differentiation (abstract). In *Lunar Science—III* (editor C. Watkins), pp. 571–572, Lunar Science Institute Contr. No. 88.

Nyquist L. E., Hubbard N. J., Gast P. W., Wiesmann H., and Church S. E. (1972) Rb–Sr relationships for some chemically defined lunar materials (abstract). In *Lunar Science—III* (editor C. Watkins), pp. 584–586, Lunar Science Institute Contr. No. 88.

Papanastassiou D. A. and Wasserburg G. J. (1971) Lunar chronology and evolution from Rb–Sr studies of Apollo 11 and 12 samples. *Earth Planet Sci. Lett.* **11,** 37.

Papanastassiou D. A. and Wasserburg G. J. (1971) Rb–Sr ages of igneous rocks from the Apollo 14 mission and the age of the Fra Mauro formation. *Earth Planet. Sci. Lett.* **12,** 36–48.

Pepin R. O., Bradley J. G., Dragon J. C., and Nyquist L. E. (1972) K–Ar dating of lunar soils: Apollo 12, Apollo 14, and Luna 16 (abstract). In *Lunar Science—III* (editor C. Watkins), pp. 602–604, Lunar Science Institute Contr. No. 88.

Poupeau G., Berdot J. L., Chetrit G. C., and Pellas P. (1972) Predominant trapping of solar-flare gases in lunar soils (abstract). In *Lunar Science—III* (editor C. Watkins), pp. 613–615, Lunar Science Institute Contr. No. 88.

Powell B. N. and Weiblen P. W. (1972) Petrology and origin of rocks in the Fra Mauro formation (abstract). In *Lunar Science—III* (editor C. Watkins), pp. 616–618, Lunar Science Institute Contr. No. 88.

Roedder E. and Weiblen P. W. (1972) Petrographic and petrologic features of Apollo 14, 15, and Luna 16 samples (abstract). In *Lunar Science—III* (editor C. Watkins), pp. 657–659, Lunar Science Institute Contr. No. 88.

Schaeffer O. A., Husain L., Sutter J., Funkhouser J. G., Kirsten T., and Kaneoka I. (1972) The ages of lunar material from Fra Mauro and the Hadley Rille-Apennine front area (abstract). In *Lunar Science—III* (editor C. Watkins), pp. 675–677, Lunar Science Institute Contr. No. 88.

Schönfeld E. (1972) Component abundance and ages in soils and breccia (abstract). In *Lunar Science—III* (editor C. Watkins), pp. 683–685, Lunar Science Institute Contr. No. 88.

Stettler A., Eberhardt P., Geiss J., and Grögler N. (1972) $Ar^{39}/Ar^{40}$ ages of Apollo 11, 12, 14, and 15 rocks (abstract). In *Lunar Science—III* (editor C. Watkins), pp. 724–725, Lunar Science Institute Contr. No. 88.

Sutter J. F., Husain L., and Schaeffer O. A. (1971) $^{40}Ar/^{39}Ar$ ages from Fra Mauro. *Earth Planet. Sci. Lett.* **11,** 249–253.

Turner G. (1970) Argon-40/argon-39 dating of lunar rock samples. *Proc. Apollo 11 Lunar Sci. Conf., Geochim. Cosmochim. Acta* Suppl. 1, Vol. 2, pp. 1665–1684. Pergamon.

Turner G. (1971) $^{40}Ar/^{39}Ar$ ages from the lunar maria. *Earth Planet. Sci. Lett.* **11,** 169–191.

Turner G., Huneke J. C., Podosek F. A., and Wasserburg G. J. (1971) $^{40}Ar$–$^{39}Ar$ ages and cosmic ray exposure ages of Apollo 14 samples. *Earth Planet. Sci. Lett.* **12,** 19–35.

Wänke H., Baddenhausen H., Balacescu A., Teschke F., Spettel B., Dreibus G., Quijano M., Kruse H., Wlotzka F., and Begemann F. (1972) Multielement analyses of lunar samples (abstract). In *Lunar Science—III* (editor C. Watkins), pp. 779–781, Lunar Science Institute Contr. No. 88.

Warner J. L. (1972) Apollo 14 breccias: Metamorphic origin and classification (abstract). In *Lunar Science—III* (editor C. Watkins), pp. 782–784, Lunar Science Institute Contr. No. 88.

York D., Kenyon W. J., and Doyle R. J. (1972) $^{40}Ar/^{39}Ar$ ages of Apollo 14 and 15 samples (abstract). In *Lunar Science—III* (editor C. Watkins), pp. 822–824, Lunar Science Institute Contr. No. 88.

Proceedings of the Third Lunar Science Conference
(Supplement 3, *Geochimica et Cosmochimica Acta*)
Vol. 2, pp. 1891–1897
The M.I.T. Press, 1972

# Exposure ages and neutron capture record in lunar samples from Fra Mauro

G. W. LUGMAIR and KURT MARTI

University of California, San Diego Chemistry Department,
La Jolla, California 92037

**Abstract**—Cosmic-ray exposure ages of Apollo 14 rocks and rock fragments obtained by the $^{81}Kr$–$^{83}Kr$ method range from 27 to 700 million years. Rock 14321, collected near the Cone crater rim, is one of the many ~27 m.y. old ejecta which were reported at the Third Lunar Science Conference. All the other rocks have considerably higher exposure ages. Isotopic anomalies from neutron capture in gadolinium, in bromine and in barium are used to obtain information on the lunar neutron spectrum at various depths below the lunar surface. The flux ratio of resonance and slow (< 0.3 eV) neutrons is found to be nearly constant in the topmost ~100 g/cm$^2$.

## INTRODUCTION

COSMIC RAYS AND SOLAR WIND are continuously bombarding the lunar surface. Nuclear transformation induced by primary or secondary particles yield information on surface processes and the time of exposure to cosmic rays. Information on the average irradiation depth of a sample can be obtained from measurements of spallogenic noble gases and neutron produced isotopic anomalies in gadolinium (Eugster *et al.*, 1970; Burnett *et al.*, 1971; Marti and Lugmair, 1971; Lugmair and Marti, 1971; Burnett *et al.*, 1972; Lugmair and Marti, 1972). Neutron capture anomalies were recently found in lunar Sm (Russ *et al.*, 1971), as well as in Kr from neutron capture in Br (Lugmair and Marti, 1971). This paper reports on the extension of this work to some Apollo 14 samples. The investigated rock fragments (4–10 mm) from sample 14160 were allocated to a consortium, and our results can be correlated with track data (Macdougall *et al.*, 1972). Sample 14257,10E consists of three basaltic fragments (2–4 mm) from the comprehensive fines.

## EXPOSURE AGES

The time intervals of exposure to cosmic rays on or near the lunar surface have been determined by the $^{81}Kr$–$^{83}Kr$ method (Marti, 1967). This method has been used to date Apollo 11 and 12 materials and has given reliable ages whenever the irradiation history is not too complex. Since $^{81}Kr$–$^{83}Kr$ ages obtained from surface material may be affected by recent solar flares, we have only sampled the center portion of the soil fragments. The age obtained from a surface sample of rock 14321 (FM1 + 2) is about 10% lower than that obtained from a bottom sample (FM5). The $^{81}Kr$–$^{83}Kr$ ages are listed in Table 1 together with the relevant ratios. In rock 14321, no evidence is found for neutron capture anomalies from Br and, therefore, $P_{81}/P_{83}$ is calculated from $P_{81}/P_{83} = 0.95(^{80}Kr + {}^{82}Kr)/2\ {}^{83}Kr$ (Marti, 1967). However, generally, in Apollo

Table 1. $^{81}$Kr–$^{83}$Kr ages of Apollo 14 samples.

| Sample | $^{83}$Kr/$^{81}$Kr | $^{83}$Kr$_{sp}$/$^{81}$Kr | $P_{81}/P_{83}$ | $^{81}$Kr–$^{83}$Kr age (m.y.) |
|---|---|---|---|---|
| 14160,4 | 3765 ± 130 | 3754 | 0.615 | 700 ± 24 |
| 14160,6 | 1879 ± 84 | 1872 | 0.618 | 351 ± 16 |
| 14160,8 | 1377 ± 59 | 1376 | 0.634 | 264 ± 12 |
| 14160,10 | 2311 ± 145 | 2297 | 0.605 | 421 ± 27 |
| 14257,10E | 858 ± 36 | 661 | 0.618 | 124 ± 5 |
| 14310,47 | 1445 ± 40 | 1427 | 0.600 | 259 ± 7 |
| 14321,FM1 + 2 | 264 ± 6 | 122 | 0.645 | 23.8 ± 0.6 |
| 14321,FM5 | 263 ± 5 | 142 | 0.633 | 27.2 ± 0.5 |

$\lambda_{81} = 3.3 \times 10^{-6}$ yr$^{-1}$ is used; for the ratio $P_{81}/P_{83}$ see text.
Errors in the ages do not include uncertainties in $\lambda_{81}$ and in $P_{81}/P_{83}$.

14 samples, this production ratio cannot be obtained by the interpolation method, because both $^{80}$Kr and $^{82}$Kr show anomalies from neutron capture in Br. $P_{81}/P_{83}$ ratios were calculated from

$$\frac{P_{81}}{P_{83}} = 0.850\left(\frac{^{78}\text{Kr}}{^{83}\text{Kr}}\right)_{\text{spall}} + 0.442 \tag{1}$$

a relation which is analogous to equation (5) of Lugmair and Marti (1971), applied to rock 14310. The data of rocks 14321, 10017, 10057, and 10071 are used to obtain relation (1), since the abundance ratios of the major target elements Sr/Zr are similar in these rocks.

## NEUTRON CAPTURE EFFECTS

Enrichments in the $^{158}$GdO/$^{157}$GdO due to neutron capture range from <0.02% in rock 14321 up to 0.7% in fragment 14160,4. The analytical techniques and data reduction have been discussed in detail (Lugmair and Marti, 1971). Isotopic ratios of Gd in some additional fragments from sample 14160 are given in Table 2. The $^{158}$GdO/$^{157}$GdO ratios and the calculated fluences ($\Phi_s$) and fluxes ($\phi_s$) of slow (<0.3 eV) neutrons are compiled in Table 3. Russ *et al.* (1971), from a comparison of Gd and Sm isotopic anomalies, have shown that the lunar neutron energy spectrum is hardened and that, therefore, the effective cross sections may be lower by a factor of about 2.5 (Lingenfelter *et al.*, 1971). The fluxes listed in Table 3 are based on 2200 m/sec cross sections ($\sigma_{157} = 2.54 \times 10^5$ barn) and have not been adjusted.

Rock 14310 was found to be the first lunar sample to show clear evidence for neutron capture anomalies in $^{80}$Kr and $^{82}$Kr, due to resonance (30–300 eV) neutron capture in Br (Lugmair and Marti, 1971). With the exception of rock 14321, Kr of all our Apollo 14 samples is found to have neutron produced anomalies. This evidence reflects larger concentrations of Br in Apollo 14 material (Morgan *et al.*, 1972; Reed

Table 2. Isotopic ratios of GdO from Apollo 14 samples (corrected for mass fractionation[a])

| Sample | $^{152}$GdO/$^{160}$GdO* | $^{154}$GdO/$^{160}$GdO* | $^{154}$GdO/$^{160}$GdO$_{corr}$ | $^{155}$GdO/$^{160}$GdO | $^{156}$GdO/$^{160}$GdO | $^{157}$GdO/$^{160}$GdO | $^{158}$GdO/$^{160}$GdO |
|---|---|---|---|---|---|---|---|
| Rock fragments: | | | | | | | |
| 14160,4 | 0.009 28 | 0.099 63 | 0.099 63 | 0.674 63 | 0.935 14 | 0.713 37 | 1.138 51 |
| | ± 1 | ± 3 | ± 3 | ± 4 | ± 7 | ± 5 | ± 7 |
| 14160,6 | 0.009 30 | 0.099 64 | 0.099 61 | 0.675 02 | 0.934 65 | 0.714 98 | 1.136 70 |
| | ± 3 | ± 3 | ± 3 | ± 2 | ± 7 | ± 8 | ± 4 |
| 14160,8 | 0.009 24 | 0.099 62 | 0.099 62 | 0.675 21 | 0.934 47 | 0.715 88 | 1.135 73 |
| | ± 2 | ± 3 | ± 3 | ± 1 | ± 1 | ± 7 | ± 9 |
| 14160,10 | 0.009 29 | 0.099 61 | 0.099 61 | 0.675 06 | 0.934 69 | 0.714 98 | 1.136 64 |
| | ± 6 | ± 3 | ± 3 | ± 1 | ± 6 | ± 4 | ± 7 |
| Rocks: | | | | | | | |
| 14310,47 | 0.009 28 | 0.099 59 | 0.099 59 | 0.674 86 | 0.934 87 | 0.714 28 | 1.137 57 |
| | ± 2 | ± 3 | ± 3 | ± 4 | ± 3 | ± 6 | ± 8 |
| 14321,FM3 | 0.009 27 | 0.099 59 | 0.099 59 | 0.675 35 | 0.934 52 | 0.716 40 | 1.135 60 |
| | ± 2 | ± 3 | ± 3 | ± 2 | ± 4 | ± 8 | ± 12 |
| Terr. GdO | 0.009 27 | | 0.099 60 | 0.675 33 | 0.934 41 | 0.716 37 | 1.135 30 |
| | ± 3 | | ± 1 | ± 1 | ± 3 | ± 2 | ± 8 |

* Not corrected for SmO and $^{151}$Eu$(n,\beta^-)^{152}$Gd contributions.
[a] For method see (Lugmair and Marti, 1971). The errors given are $2\sigma_{mean}$ (statistical errors only).

Table 3. Neutron capture anomalies in Apollo 14 samples and neutron fluences $\Phi_s$ of slow (<0.3 eV) neutrons.

| Sample | $^{158}$GdO/$^{157}$GdO | $\Phi_s$ ($10^{15}$ n/cm$^2$) | $\phi_s = \frac{\Phi_s}{T_r}$ (n/cm$^2$ sec) | $\frac{^{131}Xe_n}{^{126}Xe_{sp}}$ | $\frac{^{80}Kr_n{}^*}{^{83}Kr_{sp}}$ | $\frac{^{82}Kr_n{}^*}{^{83}Kr_{sp}}$ | $\left(\frac{^{78}Kr}{^{83}Kr}\right)_{sp}$ |
|---|---|---|---|---|---|---|---|
| 14160,4 | 1.595 97 | 17.00 | 0.77 | 2.59 | 0.1707 | 0.0635 | 0.204 |
| | ± 15 | ±0.29 | | | ± 19 | ± 17 | ± 1 |
| 14160,6 | 1.589 82 | 7.66 | 0.70 | 2.64 | 0.0649 | 0.0240 | 0.207 |
| | ± 18 | ±0.33 | | | ± 13 | ± 16 | ± 1 |
| 14160,8 | 1.586 48 | 2.58 | 0.31 | 0.92 | 0.0496 | 0.0206 | 0.226 |
| | ± 20 | ±0.35 | | | ± 14 | ± 17 | ± 1 |
| 14160,10 | 1.589 75 | 7.55 | 0.57 | — | 0.0360 | 0.0100 | 0.192 |
| | ± 12 | ±0.25 | | | ± 16 | ± 20 | ± 1 |
| 14310,47 | 1.592 62 | 11.92 | 1.45 | 3.60 | 0.1668 | 0.0563 | 0.186 |
| | ± 18 | ±0.33 | | | ± 18 | ± 21 | ± 1 |
| Terr. GdO | 1.584 79 | ≡0 | | | | | |
| | ± 12 | | | | | | |

* Errors include experimental uncertainties only but do not include uncertainties in the spallation correction.

*et al.*, 1972). Relative $^{80}Kr_n$ and $^{82}Kr_n$ neutron capture anomalies are compiled in Table 3. Neutron capture excesses are calculated from

$$\frac{^{m}Kr_n}{^{83}Kr_{sp}} = \left(\frac{^{m}Kr}{^{83}Kr}\right)_{c.r.} - \left(\frac{^{m}Kr}{^{83}Kr}\right)_{sp} \tag{2}$$

and the spallation ratios $(^{m}Kr/^{83}Kr)_{sp}$ ($m$ = 80, 82) are obtained by interpolation between those found in rock 14321,FM5 and in Apollo 11A rocks 10017, 10057, and 10071. Figure 1 shows the correlation between the relative $^{80}Kr_n$ and $^{82}Kr_n$ excesses. The slope is determined by the ratio $0.92\sigma_{79}/\sigma_{81}$ which is smaller than the thermal

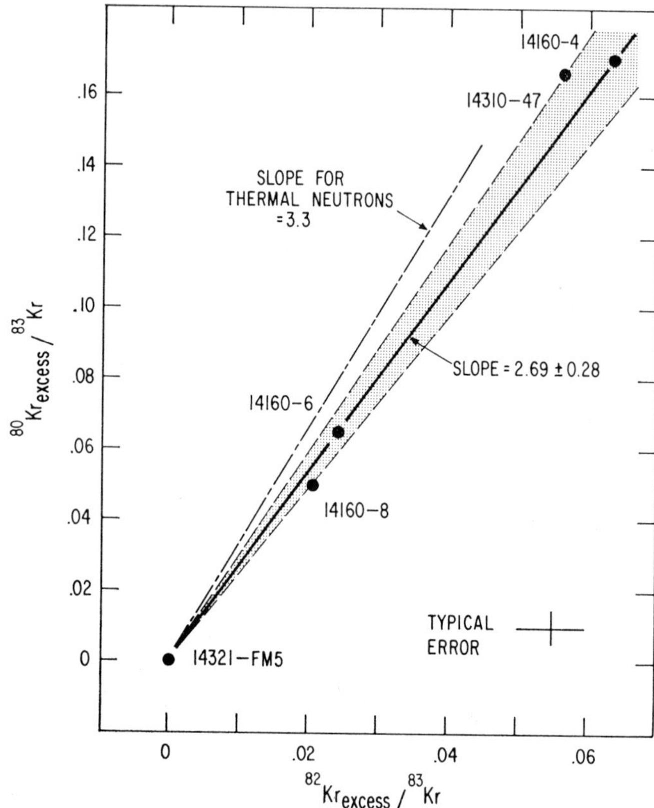


Fig. 1. $^{80}Kr_{excess}/^{83}Kr_{sp}$ versus $^{82}Kr_{excess}/^{83}Kr_{sp}$. The slope of the correlation line is 2.69 ± 0.28 and corresponds to 0.92 $\sigma_r$ ($^{79}$Br)/$\sigma_r$ ($^{81}$Br). Sub $r$ indicates the resonance energy range 30–300 eV.

cross-section ratio of Br but in agreement with ratios calculated from neutron resonance parameters of Br (Goldberg *et al.*, 1966).

Recent measurements show that the $^{130}$Ba cross section for resonance neutrons is 200–300 barn (Kaiser and Berman, 1972; Eberhardt *et al.*, 1971), and that, therefore, the neutron capture contribution to the $^{131}$Xe yield is quite large in many lunar samples. We have attempted to calculate the neutron excesses from

$$\frac{^{131}Xe_n}{^{126}Xe_{sp}} = \left(\frac{^{131}Xe}{^{126}Xe}\right)_{c.r.} - \left(\frac{^{131}Xe}{^{126}Xe}\right)_{sp} \quad (3)$$

analogous to the Kr data. The spallation ratios $(^{131}Xe/^{126}Xe)_{sp}$ cannot be obtained by an interpolation method because no data for spallation *only* are currently known. We have calculated the $^{131}Xe_{sp}$ yield by extrapolation from the best (power) fits to the $^{126,128,129}Xe_{sp}$ yields. The $^{131}Xe_n/^{126}Xe_{sp}$ ratios are also listed in Table 3. A detailed discussion of the rare gas data will be given in a separate paper (Marti *et al.*, 1972).

## Discussion

The exposure age of rock 14321 (27 m.y.) is very much shorter than those of the other samples. Rock 14321 was collected on the Cone Crater flank. Similar exposure ages were found in many other rocks from this location (Turner *et al.*, 1971; G. Crozaz *et al.*, 1972) and, therefore, 14321 appears to be one of the Cone Crater ejecta.

Data for three different neutron capture anomalies are given in Table 3. Rock fragment 14160,8 experienced a very small neutron flux of $\phi_s = 0.31$ n/cm$^2$ sec. This, together with the very high $^{78}$Kr/$^{83}$Kr spallation ratio and the high $P_{81}/P_{83}$ production ratio indicates a close to surface irradiation during most of the time of exposure to cosmic rays. The steep track gradient found in this sample by MacDougall *et al.* (1972) supports this contention.

It is possible to study the correlation between slow and resonance neutron fluxes. Resonance fluxes $\phi_r$ can be calculated from either $^{80,82}$Kr$_n$ or $^{131}$Xe$_n$ data. Flux calculations from the Kr$_n$ data require knowledge of Br abundances. Unfortunately, only Br data for rock 14310 are available. Furthermore, the data obtained by Morgan *et al.* (1972) and Reed *et al.* (1972) differ by a factor of 3.6. If we combine the Kr$_n$ data of 14310 with the Br concentration of Morgan *et al.* (0.235 ppm) a flux $\phi_r(1) = 3.68$ n/cm$^2$ sec is obtained, and the Reed *et al.* value (0.85 ppm) yields a flux $\phi_r(2) = 1.02$ n/cm$^2$ sec. This is to be compared to the slow neutron flux $\phi_s = 1.45$ n/cm$^2$ sec or 3.6 n/cm$^2$ sec, if the thermal cross section is adjusted according to Lingenfelter *et al.* (1971). The possibility that the resonance flux at or below the lunar surface is smaller than the thermal flux seems very unlikely. Therefore, the Br value by Reed *et al.* (1972) is not applicable to our sample.

If we use $\sigma = 220$ barn for $^{130}$Ba (Kaiser and Berman, 1971) and a Ba concentration of 630 ppm, a resonance flux for rock 14310 of $\sim 3.5$ n/cm$^2$ sec is calculated. This value is consistent with $\phi_r(1)$ obtained above from neutron produced Kr.

Ba does not only account for the neutron capture excess $^{131}$Xe$_n$, but it is also the major target element for spallation. We have, therefore, a favorable situation inasmuch as knowledge of the Ba concentration is not required:

$$^{131}\mathrm{Xe}_n = [\mathrm{Ba}]\frac{^{130}\mathrm{Ba}}{\mathrm{Ba}}\,\phi_r T_r \sigma(^{130}\mathrm{Ba}), \qquad \text{with } \phi_r \equiv \frac{\Phi_r}{T_r} \tag{4}$$

$$^{126}\mathrm{Xe}_{sp} = \mathrm{p}_{126}(E)T_r([\mathrm{Ba}] + f(E)[\mathrm{La} + \mathrm{Ce} + \mathrm{Nd}])$$

$$^{126}\mathrm{Xe}_{sp} = \mathrm{p}_{126}(E)T_r[\mathrm{Ba}](1 + f(E)R), \qquad \text{with } R = \frac{[\mathrm{La} + \mathrm{Ce} + \mathrm{Nd}]}{[\mathrm{Ba}]} \tag{5}$$

$$\frac{^{131}\mathrm{Xe}_n}{^{126}\mathrm{Xe}_{sp}} = \frac{0.001\sigma(^{130}\mathrm{Ba})}{(1 + f(E)R)\mathrm{p}_{126}(E)}\,\phi_r\,. \tag{6}$$

In a plot of $^{131}\mathrm{Xe}_n/^{126}\mathrm{Xe}_{sp}$ versus $\phi_s = \Phi_s/T_r$ as calculated from the $^{158}$GdO/$^{157}$GdO data, therefore, essentially $\phi_r$ and $\phi_s$ are correlated, and the energy dependence of $\mathrm{p}_{126}$ is considered to be small over the depth of $\lesssim 100$ g/cm$^2$. Figure 2 shows that the two fluxes correlate nearly linearly, which in turn indicates that the ratio of resonance and slow neutron fluxes is not strongly depth dependent. This is in accord with

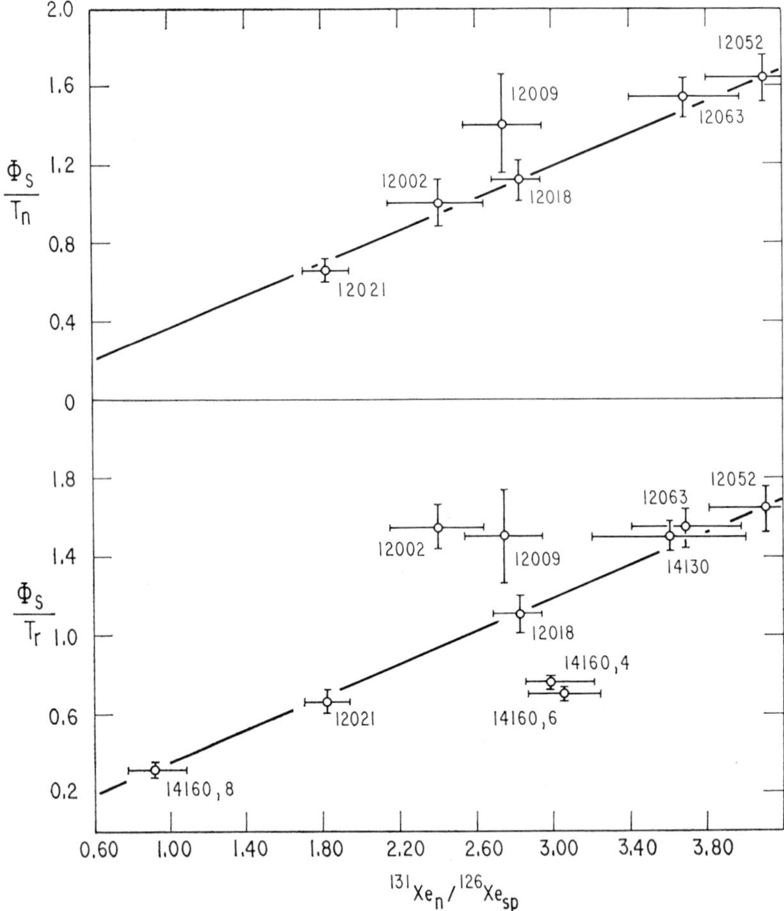


Fig. 2. Correlation of the slow neutron flux $\phi_s = \Phi_s/T$ (in n/cm² sec) and the resonance neutron parameter $^{131}Xe_n/^{126}Xe_{sp}$. The uncertainties of the $^{131}Xe_n/^{126}Xe_{sp}$ ratios shown were compounded including uncertainties in amounts of trapped Xe, fission produced Xe, and of possible neutron produced $^{128}Xe_n$. Assumed excesses $^{128}Xe_n$ are consistent with $^{80}Kr_n$ excesses. Identical isobaric fraction yields were assumed for spallation produced $^{126}Xe$, $^{128}Xe$, and $^{129}Xe$. The use of calculated effective neutron ages $T_n$ (Marti and Lugmair, 1971) instead of $^{81}Kr$–$^{83}Kr$ ages $T_r$ for Apollo 12 rocks improves the correlation of the flux parameters (top).

theoretical calculations (Armstrong and Alsmiller, 1971). From spallation systematics (Marti *et al.*, 1966; Burnett *et al.*, 1971), we estimate $^{131}Xe_n/^{126}Xe_{sp}$ ratios for Apollo 14 material to be some 8% larger than for Apollo 12 because of ~20% smaller [La + Ce + Nd]/[Ba] ratio.

*Acknowledgments*—We gratefully acknowledge the help of K. R. Goldman, B. D. Lightner, T. W. Osborn, and A. Schimmel in the analytical work and in the data reduction. This research was supported by NASA Grant NGR 05-009-150.

## REFERENCES

Armstrong T. W. and Alsmiller R. G. Jr. (1971) Calculation of cosmogenic radionuclides in the moon and comparison with Apollo measurements. *Proc. Second Lunar Sci. Conf., Geochim. Cosmochim. Acta* Suppl. 2, Vol. 2, pp. 1729–1745. MIT Press.

Burnett D. S., Huneke J. C., Podosek F. A., Russ G. P. III, and Wasserburg G. J. (1971) The irradiation history of lunar samples. *Proc. Second Lunar Sci. Conf., Geochim. Cosmochim. Acta* Suppl. 2, Vol. 2, pp. 1671–1679. MIT Press.

Burnett D. S., Huneke J. C., Podosek F. A., Russ G. P. III, Turner G., and Wasserburg G. J. (1972) The irradiation history of lunar samples (abstract). In *Lunar Science—III* (editor C. Watkins), pp. 105–107, Lunar Science Institute Contr. No. 88.

Crozaz G., Drozd R., Graf H., Hohenberg C. M., Monnin M., Rajan D., Ralston C., Seitz M., Shirck J., Walker R. M., and Zimmerman J. (1972) Evidence for extinct $^{244}$Pu: Implications for the age of the pre-Imbrium crust (abstract). In *Lunar Science—III* (editor C. Watkins), pp. 164–166, Lunar Science Institute Contr. No. 88.

Eberhardt P., Geiss J., Graf H., and Schwaller H. (1971) On the origin of excess $^{131}$Xe in lunar rocks. *Earth Planet. Sci. Lett.* **12,** 260–262.

Eugster O., Tera F., Burnett D. S., and Wasserburg G. J. (1970) The isotopic composition of Gd and neutron capture effects in Apollo 11 samples. *Earth Planet. Sci. Lett.* **8,** 20–30.

Goldberg M. D., Mughabghab S. F., Magurno B. A., and May V. M. (1966) Neutron cross-sections, Vol. 11A, BNL 325, Second Edition, Suppl. No. 2.

Kaiser W. A. and Berman B. L. (1972) The average $^{130}$Ba $(n, \gamma)$ cross-section and the origin of $^{131}$Xe on the moon (abstract). In *Lunar Science—III* (editor C. Watkins), p. 444, Lunar Science Institute Contr. No. 88.

Lingenfelter R. E., Canfield E. H., and Hampel V. H. (1971) The lunar neutron flux revisited. *Earth Planet. Sci. Lett.* (to be submitted).

Lugmair G. W. and Marti K. (1971) Neutron capture effects in lunar Gd and the irradiation histories of some lunar rocks. *Earth Planet. Sci. Lett.* **13,** 32–42.

Lugmair G. W. and Marti K. (1972) Neutron and spallation effects in Fra Mauro regolith (abstract). In *Lunar Science—III* (editor C. Watkins), pp. 495–497, Lunar Science Institute Contr. No. 88.

Macdougall D., Martinek B., and Arrhenius G. (1972) Regolith dynamics (abstract). In *Lunar Science—III* (editor C. Watkins), pp. 498–500, Lunar Science Institute Contr. No. 88.

Marti K. (1967) Mass-spectrometric detection of cosmic-ray-produced $^{81}$Kr in meteorites and the possibility of Kr–Kr dating. *Phys. Rev. Lett.* **18,** 264–266.

Marti K., Eberhardt P., and Geiss J. (1966) Spallation, fission, and neutron capture anomalies in meteoritic krypton and xenon. *Z. Naturforsch.* **21,** Ser. a, 398–413.

Marti K. and Lugmair G. W. (1971) $Kr^{81}$–Kr and K–$Ar^{40}$ ages, cosmic-ray spallation products, and neutron effects in lunar samples from Oceanus Procellarum. *Proc. Second Lunar Sci. Conf., Geochim. Cosmochim. Acta* Suppl. 2, Vol. 2, pp. 1591–1605. MIT Press.

Marti K., Lightner B. D., and Osborn T. W. (1972) In preparation.

Morgan J. W., Laul J. C., Krähenbühl U., Ganapathy R., and Anders E. (1972) Major impacts on the moon: Chemical characterization of projectiles (abstract). In *Lunar Science—III* (editor C. Watkins), pp. 552–554, Lunar Science Institute Contr. No. 88.

Reed G. W. Jr., Jovanovic S., and Fuchs L. H. (1972) Concentrations and lability of the halogens, platinum metals and mercury in Apollo 14 and 15 samples (abstract). In *Lunar Science—III* (editor C. Watkins), pp. 637–639, Lunar Science Institute Contr. No. 88.

Russ G. P. III, Burnett D. S., Lingenfelter R. E., and Wasserburg G. J. (1971) Neutron capture on $^{149}$Sm in lunar samples. *Earth Planet. Sci. Lett.* **13,** 53–60.

Turner G., Huneke J. C., Podosek F. A., and Wasserburg G. J. (1971) $^{40}$Ar–$^{39}$Ar ages and cosmic-ray exposure ages of Apollo 14 samples. *Earth Planet. Sci. Lett.* **12,** 19–35.

Proceedings of the Third Lunar Science Conference
(Supplement 3, *Geochimica et Cosmochimica Acta*)
Vol. 2, pp. 1899–1916
The M.I.T. Press, 1972

# Classification and source of lunar soils; clastic rocks; and individual mineral, rock, and glass fragments from Apollo 12 and 14 samples as determined by the concentration gradients of the helium, neon, and argon isotopes

G. H. MEGRUE and F. STEINBRUNN

Smithsonian Institution, Astrophysical Observatory,
Cambridge, Massachusetts 02138

**Abstract**—Concentration gradients of the helium, neon, and argon isotopes from various lunar materials were determined *in situ* by laser probe-mass spectrometry. The results demonstrate that many lunar glasses, soil agglomerates, and breccias contain solar gases throughout the grains at a depth greater than their expected range of a few microns. The element ratios of the solar gases are used to classify uniquely lunar soil, agglomerates, and breccias. No evidence for thermal diffusion of radiogenic $^{4}He$ and $^{40}Ar$ has been found in the profile analysis of one norite (KREEP) fragment. The potassium concentrations of 13 glasses correlate with $^{40}Ar/^{36}Ar$, $^{4}He/^{20}Ne$, and $^{4}He/^{36}Ar$ with coefficients of 0.98, 0.81, and 0.78, respectively, which clearly demonstrate the close affinity of radiogenic $^{4}He$ and $^{40}Ar$. Furthermore, $^{4}He/^{3}He$, $^{4}He/^{20}Ne$ variations are shown to result from differing radiogenic $^{4}He$ abundances. K–Ar ages of the glasses in the Apollo 14 regolith indicate that melting events occurred at 17 m.y., $<600$ m.y., and 2–2.6, and 3.2–3.7 $\times$ $10^{9}$ yr. Cosmic-ray exposure ages $<40$ m.y. and the high frequency of soil agglomerates in the lunar regolith similar to breccia 14047 indicate that breccias 14047 and 14321 were put on the lunar surface quite recently and most likely are derived from the formation of Cone Crater. Breccia 14301, which contains an abnormally high radiogenic $^{40}Ar$ within the matrix, is similar to the chemistry and inferred structure of the Fra Mauro formation. A fragment within this breccia has an exposure age of 200 m.y. Our analyses show no evidence for reimplanted $^{40}Ar$ onto the surface of lunar grains by solar-wind irradiation.

## INTRODUCTION

THE SPATIAL DISTRIBUTION of the helium, neon, and argon isotopes, which have been incorporated into the lunar surface by solar-wind irradiation, cosmic-ray spallation reactions, and *in situ* decay of the radioactive isotopes of uranium, thorium, and potassium, has been determined by laser probe-mass spectrometry for the purpose of (1) classifying lunar soils, fragments, and clastic rocks, (2) determining ultimately the source and temporal frequency of impact events on the lunar surface, and (3) studying the temporal variation of solar-wind irradiation.

The two soil samples that were analyzed in this study were collected at the Apollo 14 LM site (Swann *et al.*, 1971), whereas the three rock breccias were collected from the following: (1) approximately 320 m east of the LM site at Station B (sample 14047), (2) the ridge of Cone Crater at Station C1 (sample 14321), and (3) on the return traverse from Cone Crater to the LM at Station G1 approximately 150 m east of the LM site (sample 14301).

## EXPERIMENTAL RESULTS

The lunar samples were prepared for laser probe-mass spectrometric analysis by either sieving, handpicking, or cutting with a wire saw. The particular method used for sample preparation will be described more fully with the experimental results for each individual sample. After sample preparation, the individual samples were photographed, mounted in an ultra-high-vacuum system, and analyzed by laser probe-mass spectrometry as previously described (Megrue, 1971). Each laser pulse removed approximately 1 $\mu$g of material from the sample, as calibrated by destruction of the Cumberland Falls meteorite. Generally, between 5 and 25 pulses were used in the gas extraction of an individual sample. The mass spectrometer was calibrated before and after each analysis by addition of known aliquots of $^3$He, $^4$He, and atmospheric neon and argon. The sensitivity of the analytical system for a signal/noise = 2 is in units of $10^8$ atoms: $^3$He = 3.7, $^4$He = 340, $^{20}$Ne = 21, $^{21}$Ne = 1.3, $^{22}$Ne = 1.3, $^{36}$Ar = 1.2, $^{38}$Ar = 0.7, and $^{40}$Ar = 20.

*Soil 14163,93*

This sample of soil was sieved into a fine fraction (<320 mesh) and a coarse fraction (>60 mesh). The fine fraction was analyzed without further sample handling. The coarse fraction of individual glass, feldspar, norite (KREEP), and chondrule fragments was washed in acetone, and each individual particle was mounted in Indium metal (Fig. 1). Laser probe-mass spectrometric analysis of these grains was performed in successive steps by first removing and analyzing the surface material with a defocused laser beam, and then penetrating and analyzing to successively deeper and deeper levels in the grain. Alternatively, we analyzed an internal horizontal section

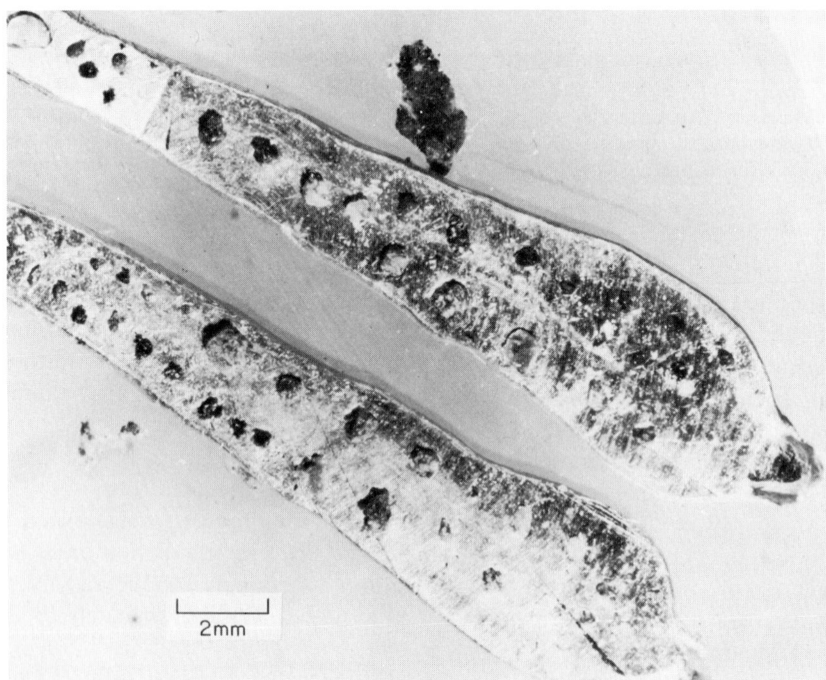


Fig. 1. Individual glass, mineral, and rock fragments from 14163,93 mounted in indium metal.

from the periphery to the center of the grain after the defocused beam had removed the upper portion of the grain.

*Fine fraction.* The results of laser probe-mass spectrometric analysis from five samples of lunar fines (14163,93) are given in Table 1. Ten laser irradiations were used to destroy approximately 10 $\mu$g for each analysis. The assigned errors to the abundance ratios include a 5% uncertainty that is inherent in calibration of the spike system, although the relative errors from each sample measurement were generally <5%.

*Individual glass, feldspar, norite (KREEP), and chondrule fragments.* Relative isotopic abundances of helium, neon, and argon from individual glass, feldspar, norite, and chondrule fragments that were separated from the coarse fraction (>60 mesh) of lunar sample 14163,93 are highly variable between the surface and interior of each individual particle (Table 2). More importantly, solar gases within the glass samples occur at depths greater than their expected range of a few microns.

Table 1. $^4$He abundances and ratios of helium, neon, and argon isotopes released from 10 laser irradiations into lunar fines 14163,93.

| Sample | $^4$He ($10^{10}$ atm) | $^4$He/$^3$He | $^4$He/$^{20}$Ne | $^4$He/$^{36}$Ar | $^{20}$Ne/$^{22}$Ne | $^{21}$Ne/$^{22}$Ne | $^{36}$Ar/$^{38}$Ar | $^{40}$Ar/$^{36}$Ar |
|---|---|---|---|---|---|---|---|---|
| 1 | 7770 | 2100 ± 210 | 85 ± 11 | 700 ± 80 | 12.4 ± 0.6 | 0.035 ± 0.006 | 5.26 ± 0.19 | 2.5 ± 1 |
| 2 | 5630 | 2160 ± 200 | 82 ± 10 | 660 ± 75 | 13.6 ± 0.7 | 0.031 ± 0.006 | 5.22 ± 0.19 | 2.7 ± 0.2 |
| 3 | 7770 | 2300 ± 230 | 101 ± 13 | 840 ± 95 | 13.3 ± 0.7 | 0.038 ± 0.006 | 5.30 ± 0.2 | 2.8 ± 0.2 |
| 4 | 4020 | 2250 ± 225 | 82 ± 10 | 570 ± 63 | 12.5 ± 0.6 | 0.033 ± 0.006 | 5.36 ± 0.2 | 2.2 ± 0.2 |
| 5 | 3480 | 2150 ± 215 | 81 ± 10 | 630 ± 70 | 12.0 ± 0.6 | 0.039 ± 0.006 | 5.6 ± 0.2 | 2.5 ± 0.2 |

Table 2. Relative abundances of helium, neon, and argon isotopes released by laser irradiation from individual glass, feldspar, norite, and chondrule fragments in lunar sample 14163,93.*

| Sample | $^4$He ($10^{10}$ atm/10 $\mu$g) | $^4$He/$^3$He | $^4$He/$^{20}$Ne | $^4$He/$^{36}$Ar | $^{20}$Ne/$^{22}$Ne | $^{36}$Ar/$^{38}$Ar | $^{40}$Ar/$^{36}$Ar |
|---|---|---|---|---|---|---|---|
| *Ropy glass* | | | | | | | |
| Surface | 777 | 1900 | 98 | 640 | 11.3 | 4.4 | — |
| Interior | 80 | 1040 | 340 | 1296 | 3.7 | 1.5 | 2.8 |
| *Light-green glass* | | | | | | | |
| Surface | 536 | 2460 | 69 | 266 | 11.4 | 5.3 | 1.1 |
| Interior | 322 | 2330 | 73 | 414 | 12.5 | 5.3 | 2.0 |
| *Yellow glass* | | | | | | | |
| Surface | 268 | 3280 | 86 | 756 | 11.0 | 4.1 | 21 |
| Interior | 24 | 560 | — | — | — | — | — |
| *Dark-brown glass* | | | | | | | |
| Surface | 1072 | 2430 | 63 | 567 | 11.7 | 5.3 | 14 |
| Interior | 16 | 630 | 19 | 88 | 9.9 | 5.0 | 2.9 |
| *Feldspar* | | | | | | | |
| Surface | 295 | 2380 | 76 | 403 | 11.9 | 5.0 | 3.3 |
| Interior | 19 | 380 | — | 195 | — | 2.4 | 3.9 |
| *Norite* | | | | | | | |
| Surface | 536 | 1300 | 190 | 592 | 9.1 | 4.7 | 49 |
| Interior | 188 | 1130 | — | 7680 | <1 | 0.67 | 960 |
| *Chondrule* | | | | | | | |
| Surface | 107 | 2250 | 77 | 298 | 11.3 | 5.3 | 0.95 |

* Analytical errors for this and the following tables are estimated to be: $^4$He/$^3$He = ±10%, $^4$He/$^{20}$Ne = ±15%, $^4$He/$^{36}$Ar = ±10%, $^{20}$Ne/$^{22}$Ne = ±5%, $^{21}$Ne/$^{22}$Ne = 20%, $^{36}$Ar/$^{38}$Ar = ±5%, and $^{40}$Ar/$^{36}$Ar = ±10%.

From the interior of a norite grain, we find a cosmogenic $^{36}Ar/^{38}Ar = 0.67$ with a corresponding $^{4}He/^{3}He = 1130$ and $^{4}He/^{36}Ar = 7680$. These ratios must reflect the high radiogenic $^{4}He$ content of the sample and most emphatically demonstrate that large isotopic variations exist from the surface to the interior of the grain. Moreover, these isotopic variations are averaged when measuring single grains by induction melting (Table 3) (Kirsten *et al.*, 1972).

Table 3. Relative abundances of helium, neon, and argon isotopes released by induction melting of individual glass, feldspar, and norite fragments from lunar sample 14163,93.

| Sample | Weight ($\mu$g) | $^{4}He$ ($10^{10}$ atm/10 $\mu$g) | $^{4}He/^{3}He$ | $^{4}He/^{20}Ne$ | $^{4}He/^{36}Ar$ | $^{20}Ne/^{22}Ne$ | $^{21}Ne/^{22}Ne$ | $^{36}Ar/^{38}Ar$ | $^{40}Ar/^{36}Ar$ |
|---|---|---|---|---|---|---|---|---|---|
| Brown glass | 340 | 88 | 2207 | 48 | 300 | 11.7 | 0.15 | 4.9 | 19 |
| Feldspar | 520 | 55 | 1695 | 31 | 792 | 11.7 | 0.14 | 3.7 | 17 |
| Norite | 3730 | 126 | 1345 | 123 | 790 | 10.8 | 0.19 | 4.3 | 26 |

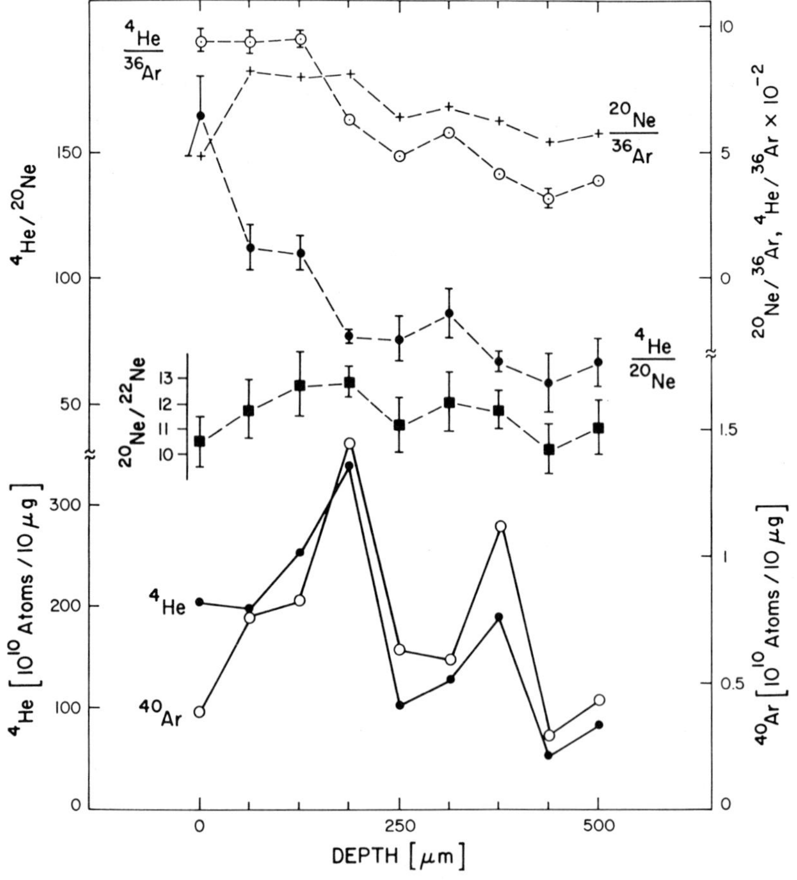


Fig. 2. Relative isotopic abundances of helium, neon, and argon as a function of depth in a cylindrical spheroid.

*Profile analyses.* In order to investigate more thoroughly the distribution of the gaseous constituents within individual rock and glass grains, we measured *in situ* with the laser probe-mass spectrometer the isotopic abundances of helium, neon, and argon in finite steps from the surface to the interior of a glass cylindrical spheroid and a norite (KREEP) grain. The results for the glass fragment (Fig. 2) demonstrate that the $^4$He and $^{40}$Ar are distributed radically symmetric about the center as measured from the near surface at 0 depth to the far surface at a depth of 500 $\mu$. The $^4$He varies between 100 and 350 $\times$ $10^{10}$ atoms/10 $\mu$g. From the $^{20}$Ne/$^{22}$Ne $= 11.6 \pm 0.9$ and the average $^{36}$Ar/$^{38}$Ar $= 5.1 \pm 0.2$, we identify a solar-type gas to be present throughout the grain, which must reflect incomplete outgassing of the glass when it is formed by melting of lunar soil or of previously solar-irradiated material. Variations in the $^4$He/$^{20}$Ne and $^{20}$Ne/$^{36}$Ar between the surface and the interior are currently not completely understood.

The concentration gradients of the radiogenic gases of $^4$He and $^{40}$Ar were measured in a KREEP grain (Fig. 3). The results show a rather constant $(^4\text{He})_r/(^{40}\text{Ar})_r$ as a function of depth within the grain and indicate that thermal diffusion of radioactive gases from this grain has been negligible. A radiogenic $^4$He/$^3$He cosmogenic $=$ 1300 is observed to be constant throughout the interior of the grain as the $^{36}$Ar/$^{38}$Ar varies from a solar value of 4.8 at the surface to a cosmogenic value of 0.65 in the interior.

*Chemical composition versus radiogenic gas concentrations.* In an attempt to understand whether radiogenic $^4$He and $^{40}$Ar can be correlated with the color or, more specifically, the chemical composition of glasses, we had 13 glasses analyzed by electron-probe fluorescence spectrometry before analyzing the samples by laser probe-mass spectrometry. The results of this investigation (Table 4) show that the potassium contents of the glasses vary by about two orders of magnitude. Five of the analyzed glasses gave sufficiently reliable chemical data that a mineral norm was calculated. As one might expect, the increasing K content of the glasses yields an

Table 4. $^4$He abundances and isotopic ratios of helium, neon, and argon from glasses of 14163,93 with calculated mineral norms and arranged in order of increasing potassium abundances.

| | | | | | | | Mineral norm | | |
|---|---|---|---|---|---|---|---|---|---|
| Glass type | $^4$He ($10^{10}$ atm/10 $\mu$g) | $^4$He/$^{20}$Ne | $^4$He/$^{36}$Ar | $^{36}$Ar/$^{38}$Ar | $^{40}$Ar/$^{36}$Ar | $K_2O$ (wt.%) | Pyroxene (wt.%) | Feldspar (wt.%) | Oxide (wt.%) |
| Light green | 118 | 29 | 344 | 5.2 | 0.62 | 0.02 | | | |
| Dark green | 38 | 7 | 28 | 5.2 | 3.5 | 0.17 | | | |
| Dark amber | 166 | 49 | 208 | 5.3 | 15 | 0.18 | | | |
| Olive green | 27 | 5 | 15 | 5.3 | 3.9 | 0.23 | | | |
| Brown | 160 | 13 | 80 | 5.2 | 0.42 | 0.25 | En 22.5<br>Fs 15.5 | Or 1.5<br>Al 2.1<br>An 50.8 | Il 3.8<br>Qz 2.7 |
| Light amber | 70 | 35 | 139 | 4.4 | 35 | 0.28 | | | |
| Brown | 13 | 52 | 82 | — | ≤1.2 | 0.29 | En 22.3<br>Fs 23.2<br>Wo 5.6 | Or 1.8<br>Al 1.6<br>An 35.8 | Il 6.0<br>Cr 0.4<br>Qz 2.7 |
| Dark brown | 200 | 58 | 194 | 5.8 | 4.8 | 0.62 | | | |
| Dark brown | 140 | 57 | 125 | 5.2 | 0.43 | 0.70 | | | |
| Amber | 37 | 29 | 109 | 4.7 | ≤0.4 | 0.90 | En 22.2<br>Fs 13.0<br>Wo 2.1 | Or 5.5<br>Al 4.8<br>An 41.0 | Il 3.8<br>Cr 0.3 |
| Dark brown | 305 | 72 | 382 | 5.7 | 15 | 1.1 | | | |
| Yellow | 32 | 45 | 308 | 4.1 | 75 | 1.6 | En 20.7<br>Fs 14.1 | Or 10.2<br>Al 9.5<br>An 21.9 | Il 4.1<br>Qz 5.4<br>Cr 5.7 |
| White | 175 | 119 | 596 | 5.9 | 193 | 10.7 | | Or 63.6 | Qz 31.4 |

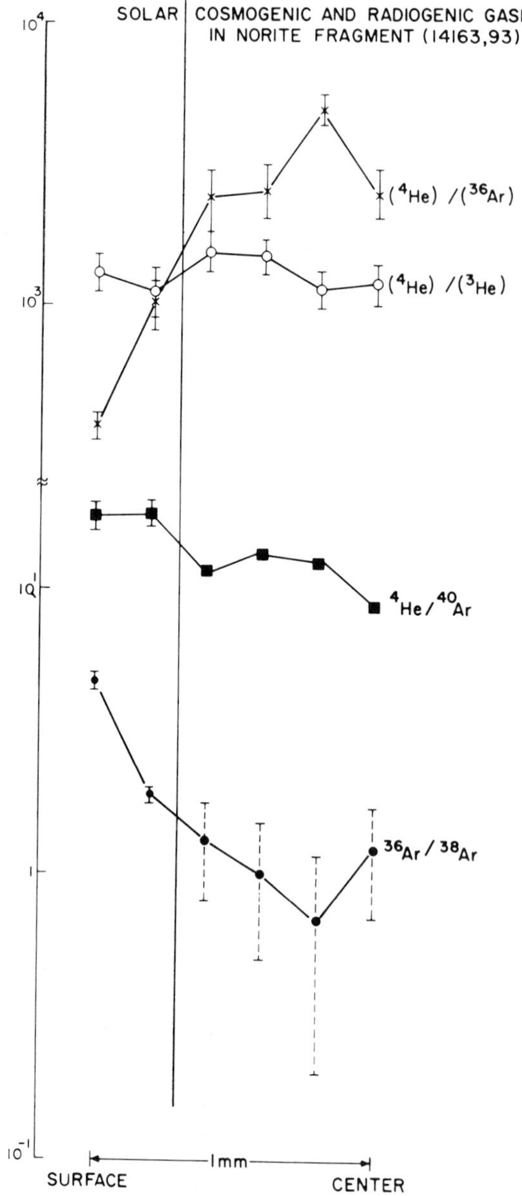


Fig. 3. Relative concentration gradients of solar, radiogenic, and cosmogenic gases across a norite fragment.

increasingly higher calculated mineral norm of orthoclase, as well as albite, with a correspondingly lower abundance of anorthite. The calculated enstatite and ferrosilite mineral abundances of the glasses do not appear to be correlated with the varying K abundances.

Correlation coefficients were calculated for the $K_2O$ abundances and the other measured isotopic abundance ratios of helium, neon, and argon for the 13 glass samples. The correlation coefficient between $K_2O$ and the $^{40}Ar/^{36}Ar$ is 0.98, whereas the correlation coefficients between $K_2O$ and $^{4}He/^{20}Ne$ and $^{4}He/^{36}Ar$ are 0.81 and 0.78, respectively. Furthermore, the correlation coefficient between $^{4}He/^{36}Ar$ and $^{40}Ar/^{36}Ar$ is 0.77. Consequently, the above observations indicate that radiogenic $^{40}Ar$ and $^{4}He$ are associated together within the glasses, and their abundances vary proportionally with the potassium content of the sample (Burnett *et al.*, 1971). More importantly, these observations demonstrate that variable $^{4}He/^{20}Ne$ abundances, which are sometimes attributed to solar gas, can, in fact, result from the contamination of radiogenic $^{4}He$.

*Breccia 14047,37*

This sample is a friable fine-grained clastic rock with a small percentage of subangular leucocratic clasts in a medium-gray matrix (Swann *et al.*, 1971). Well-developed slickensides are present on one surface of the sample. The purpose of our experiments was to ascertain whether solar-wind gases were distributed homogeneously throughout the surface and interior of the sample, and whether large amounts of radiogenic and cosmogenic gases were contained within the white crystalline inclusions.

We split the sample with a chisel; one-half of the sample was left untreated before laser probe-mass spectrometric analysis (Fig. 4), whereas the other half was cut into slices with a wire saw for the purpose of exposing for analysis the interior matrix and

Fig. 4. Slickensided and irregular surface of 14047,37 with glass-lined craters produced by laser irradiation.

fragments. One interior portion of the sample (Fig. 5) was analyzed from the surface (a) inward 600 $\mu$ toward the center of the section in order to determine a concentration gradient of the solar-wind-implanted gases.

The relative abundances of the helium, neon, and argon isotopes (Table 5) do not differ significantly between the fine-grained surface matrix and the corresponding interior matrix. The solar $^{4}He/^{20}Ne$ and $^{4}He/^{36}Ar$ from these matrices differ approx-

Table 5. Distribution of helium, neon, and argon isotopes released by laser irradiation from the surface and interior section of lunar breccia 14047,37.

| Sample | $^{4}He$ ($10^{10}$ atm/10 $\mu$g) | $^{4}He/^{3}He$ | $^{4}He/^{20}Ne$ | $^{4}He/^{36}Ar$ | $^{20}Ne/^{22}Ne$ | $^{21}Ne/^{22}Ne$ | $^{36}Ar/^{38}Ar$ | $^{40}Ar/^{36}Ar$ |
|---|---|---|---|---|---|---|---|---|
| *Surface—matrix* | | | | | | | | |
| Slickenside | 3750 | 2364 | 71 | 280 | 11.2 | 0.036 | 4.9 | 1.3 |
| Slickenside | 2470 | 2322 | 73 | 337 | 11.0 | 0.032 | 5.0 | 1.3 |
| Irregular | 1180 | 2580 | 69 | 310 | 11.3 | 0.031 | 5.1 | 1.1 |
| Irregular | 2950 | 2262 | 77 | 355 | 11.5 | 0.035 | 5.2 | 1.5 |
| *Surface—white inclusion* | | | | | | | | |
| | 54 | 2258 | 137 | 699 | — | — | 3.4 | <2 |
| *Interior—matrix* | | | | | | | | |
| Margin (a) | 2950 | 2400 | 82 | 355 | 11.1 | — | 5.2 | 2.5 |
| 200 $\mu$ from (a) | 1640 | 2375 | 67 | 310 | 10.8 | — | 5.0 | 3.4 |
| 400 $\mu$ from (a) | 4020 | 2266 | 72 | 252 | 11.3 | — | 5.1 | 2.5 |
| 600 $\mu$ from (a) | 5630 | 2598 | 79 | 304 | 11.5 | — | 5.1 | 1.9 |

Fig. 5. Interior section of 14047,37 with aligned, laser-produced, glass craters that were used to measure the concentration gradients of the helium, neon, and argon isotopes.

imately 10–15% and 50%, respectively, from the corresponding ratios of the fine-grained lunar soil 14163,93 (Table 1).

The white inclusions from the surface and interior portion of 14047,37 contain different $^{4}He/^{20}Ne$, $^{4}He/^{3}He$, and $^{36}Ar/^{38}Ar$ than the corresponding ratios within the fine-grained matrix. These differences are attributed to the presence of a higher proportion of cosmogenic and radiogenic gases within the inclusions than within the fine-grained matrix.

*Breccia 14301,61*

This rock is a coherent, medium-gray clastic rock with subangular leucocratic clasts and less abundant melanocratic clasts in a fine-grained matrix (Swann *et al.*, 1971). We prepared this rock for laser probe-mass spectrometric analysis in a similar manner as the previously described breccia (i.e., the surface of one-half of the sample (Fig. 6) was analyzed without any alteration of the sample as received from the Lunar Receiving Laboratory, whereas the other half of the sample was sectioned into six slices with a wire saw before selecting one sample for detailed analysis (Fig. 7)).

From the surface sample and from different regions of the interior of breccia 14301,61 (Fig. 7), we find a $^{4}He/^{3}He$ that is 25% higher than what we measured in the lunar fines fraction 14163,93 and a $^{4}He/^{20}Ne$ that is 25–30% less than corresponding results from the lunar fines (see Table 6). The above observations along with a $^{40}Ar/^{36}Ar > 10$ strongly suggest that this breccia contains an abnormally high radiogenic $^{4}He$ and $^{40}Ar$ component in conjunction with the solar-wind gases that are

Fig. 6. Exterior surface of 14301,61 showing analyzed regions of gray and white areas.

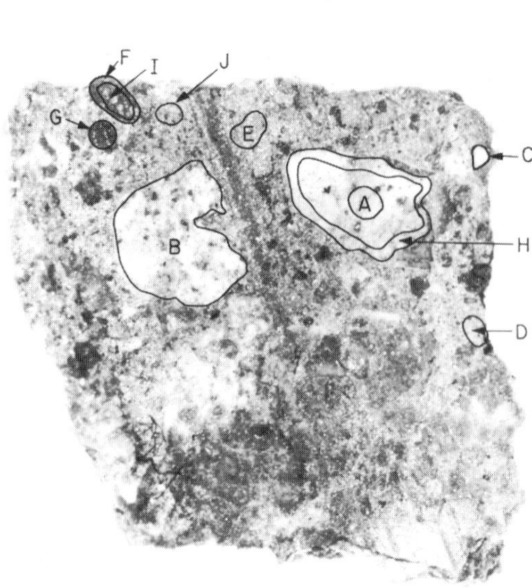


Fig. 7. Interior section of 14301,61 showing specific regions analyzed *in situ* for relative isotopic abundances of helium, neon, and argon.

Table 6. Distribution of helium, neon, and argon isotopes from the surface and interior of lunar breccia 14301,61.

| Sample | $^{4}He$ ($10^{10}$ atm/10 μg) | $^{4}He/^{3}He$ | $^{4}He/^{20}Ne$ | $^{4}He/^{36}Ar$ | $^{20}Ne/^{22}Ne$ | $^{21}Ne/^{22}Ne$ | $^{36}Ar/^{38}Ar$ | $^{40}Ar/^{36}Ar$ |
|---|---|---|---|---|---|---|---|---|
| *Surface* | | | | | | | | |
| Gray | | | | | | | | |
| Matrix | 7880 | 3170 | 67 | 916 | 12.6 | 0.030 | 5.5 | 16 |
| Matrix | 4930 | 3060 | 55 | 670 | 13.1 | 0.038 | 5.3 | 16 |
| Matrix | 4100 | 2920 | 65 | 846 | 12.5 | 0.036 | 5.2 | 18 |
| White | | | | | | | | |
| Matrix | 780 | 3950 | 79 | 1120 | 12.7 | 0.044 | 5.0 | 34 |
| Crystalline | 1580 | > 3100 | > 1000 | > 7700 | — | — | — | > 230 |
| Crystalline | 160 | > 4600 | > 2000 | > 15700 | — | — | — | > 490 |
| *Interior* | | | | | | | | |
| D matrix | 3860 | 2850 | 53 | 522 | 12.1 | 0.034 | 5.1 | 17 |
| E matrix | 2840 | 3140 | 60 | 505 | 11.0 | 0.041 | 5.0 | 16 |
| J matrix | 3080 | 2800 | 50 | 587 | 11.7 | 0.035 | 5.1 | 19 |
| K matrix | 1980 | 2730 | 62 | 576 | 12.6 | 0.040 | 5.4 | 16 |
| *Fragments* | | | | | | | | |
| B xenobreccia | 730 | 2880 | 82 | 692 | 11.9 | 0.041 | 5.0 | 23 |
| C white | 1470 | 2690 | 53 | 514 | 11.0 | 0.037 | 4.9 | 17 |
| *Glass teardrop* | | | | | | | | |
| F rim | 1020 | 2850 | 57 | 549 | 10.8 | 0.033 | 5.2 | 15 |
| I core | 560 | 3050 | 54 | 452 | 10.8 | 0.037 | 5.1 | 14 |
| G spherule | 46 | — | 30 | 128 | 10.6 | — | 5.1 | 17 |
| *Rock* | | | | | | | | |
| H rim | 100 | 2026 | 62 | 450 | 10.2 | 0.051 | 5.0 | 20 |
| A core | 26 | — | 83 | 303 | 7.6 | 0.133 | 5.5 | 26 |

found distributed throughout the matrix both on the surface and in the interior of the breccia.

No solar gas was found within the crystalline sample of the white surface. A glass teardrop from the interior of the breccia gives results that are consistent with the glasses that were analyzed in sample 14163,93, namely, that solar-type gases exist at a depth greater than their expected ranges. A xenobreccia (sample B) (Fig. 7) from the interior of the prepared breccia sections contains solar $^{4}He/^{20}Ne$ and $^{4}He/^{36}Ar$ that are consistent with the results of the lunar fines fraction 14163,93. However, the $^{4}He/^{3}He$ from this xenobreccia is 30% higher than the ratio measured for the lunar fines; this higher value reflects the increased radiogenic $^{4}He$ concentration, as evidenced by the higher $^{40}Ar/^{36}Ar = 23$, which is higher than that in the matrix of the breccia and lunar fines (14163,93).

### *Breccia 14321,244*

This rock is a coherent breccia with about 40% melanocratic clasts and a very light gray matrix (Swann *et al.*, 1971). Since this sample was received in many pieces from the Lunar Receiving Laboratory, we selected a few samples of different petrographic type for cutting with the wire saw. The prepared slices and surfaces of some untreated samples were analyzed by laser probe-mass spectrometry. In contrast to the other breccia samples, we could not detect any solar-type gases from the interior matrix or fragments of this sample. However, from one external surface, after subtracting a 30% correction for radiogenic $^{4}He$ and a 50% correction for radiogenic $^{40}Ar$, which were equivalent to what we measured below the surface of the sample at the identical spot, we measured the following ratios: $^{4}He/^{3}He = 2310$, $^{4}He/^{20}Ne = 102$, $^{4}He/^{36}Ar = 786$, $^{20}Ne/^{22}Ne = 13.9$, $^{36}Ar/^{38}Ar = 5.1$, and $^{40}Ar/^{36}Ar = 53$. All these except the latter are consistent with the corresponding ratios for the solar gases from the lunar soil 14163,93. From the interior of the sample, we measured a radiogenic $^{4}He/^{40}Ar = 16 \pm 4$.

### *Coarse fines 14161,31*

Coarse fragments from this sample were individually mounted in aluminum foil before incorporation into the vacuum system. The fragments appear petrographically to consist of: (1) fine-grained breccias, resembling soil clods, (2) slag or vuggy fine-grained particles, and (3) coarsely crystalline rock or breccia fragments. Two of the fine-grained breccias had glass coatings, both of which were individually analyzed; furthermore, a slickensided surface of a fine-grained breccia fragment was analyzed so as to compare the gas content of the slickensided surface with the gas content of the underlying material.

The results of the distribution of the helium, neon, and argon isotopes from eight particles (Table 7) show that solar-wind gases and radiogenic gases are the predominant constituents of these samples. Measurements from a slickensided surface (A-1) and matrix (A-2) of a fine-grained breccia, which resembles in petrographic appearance lunar breccia 14047,37, gave similar helium, neon, and argon isotopic ratios (Table 5 and Table 7), which substantiates the fact that these samples were part

Table 7. $^4$He abundances and relative isotopic abundances of helium, neon, and argon isotopes from coarse fragments of lunar sample 14161,31.

| Sample | $^4$He ($10^{10}$ atm/10 $\mu$g) | $^4$He/$^3$He | $^4$He/$^{20}$Ne | $^4$He/$^{36}$Ar | $^{20}$Ne/$^{22}$Ne | $^{21}$Ne/$^{22}$Ne | $^{36}$Ar/$^{38}$Ar | $^{40}$Ar/$^{36}$Ar |
|---|---|---|---|---|---|---|---|---|
| Breccia (fine-grained) | | | | | | | | |
| A-1 | 8470 | 2403 | 69 | 345 | 11.3 | 0.033 | 5.1 | 3.4 |
| A-2 | 6750 | 2388 | 66 | 345 | 11.7 | 0.034 | 5.1 | 3.3 |
| B | 2572 | 2456 | 65 | 320 | 11.8 | 0.046 | 5.1 | 1.3 |
| B′ | 54 | 1831 | 27 | 164 | 10.9 | — | 5.4 | 1.2 |
| C | 22800 | 2450 | 79 | 455 | 11.2 | 0.034 | 5.0 | 2.0 |
| D | 388 | 2680 | 133 | 1288 | 11.3 | 0.066 | 4.6 | 14 |
| D′ | 47 | 1386 | 148 | 1225 | 11.3 | 0.21 | 4.5 | 308 |
| D″ | 128 | 3483 | 337 | 1942 | 9.8 | 0.38 | 3.5 | 53 |
| Breccia (slag) | | | | | | | | |
| E | 3100 | 2761 | 51 | 138 | 11.6 | 0.043 | 5.1 | 1.5 |
| F | 536 | 3229 | 48 | 153 | 12.1 | 0.17 | 5.1 | 3.2 |
| Crystalline | | | | | | | | |
| G | 344 | 1850 | 102 | 626 | 10.0 | 0.078 | 4.9 | 15.0 |
| H | 140 | 1520 | 203 | 1040 | 9.8 | — | 3.9 | — |

of the same soil sample or region of the moon at some previous time. Likewise, sample B, a soil breccia similar to sample A and lunar breccia 14047,37, contains similar isotopic abundances of helium, neon, and argon; however, a glass coating (B′) on the surface of sample B contains completely different isotopic abundances of helium, neon, and argon than do the soil breccia samples. Sample C of lunar sample 14161,31 contains a much higher absolute abundance of $^4$He than the breccia fragments A and B of similar petrographic appearance. Whether this reflects a higher radiogenic $^4$He component or a sample that was thermally less metamorphosed when it was agglomerated is yet unknown.

Sample D is a fine-grained soil clod that is physically associated with two different glasses (samples D′ and D″). From the results of the relative isotopic abundances of helium, neon, and argon from these samples, we would say that this soil clod is unrelated to the soil agglomerates (samples A, B, and C) because of its higher $^4$He/$^{20}$Ne and $^4$He/$^{36}$Ar and higher concentration of cosmogenic gases (indicated by the higher $^{21}$Ne/$^{22}$Ne and lower $^{36}$Ar/$^{38}$Ar) than the corresponding ratios for the other soil agglomerates. Moreover, glass samples D′ and D″ contain unusually large abundances of radiogenic $^{40}$Ar and $^4$He in comparison to the soil clod D. This indicates that the sources of the glasses and soil clod were unrelated (i.e., the glasses are not merely melted underlying lunar soil).

A comparison of the relative isotopic abundances of helium, neon, and argon from two samples (E and F, Table 7) of breccia slag with the corresponding ratios from lunar soil 14163,93 indicates that the solar-type gases within these breccia slag samples have been selectively lost by heating and outgassing of previously solar-irradiated lunar soil.

Crystalline fragments (G and H) of 14161,31 (Table 7) contain cosmogenic gases, solar gases, and radiogenic gases that yield relative isotopic abundances of helium, neon, and argon that are more similar to norite (KREEP) fragments (Table 2) than to any other type of particle that we have analyzed.

## Summary and Discussion

### *Solar gases*

A summary of the relative $^4He/^{36}Ar$ and $^{20}Ne/^{36}Ar$ measured *in situ* from the fine-grained matrices of different breccias and soil samples from Apollo 12 and 14 samples (Fig. 8) displays a variation that can be used ultimately to provide a genetic classification for lunar soil and breccia samples. In particular, the relative $^4He/^{36}AR$ and $^{20}Ne/^{36}AR$ from the surface of breccia 14321,244 and from the soils of 12070,67 and 14163,93 are similar within $\pm 1\ \sigma$. The $^4He/^{36}Ar$ and $^{20}Ne/^{36}Ar$ from the exterior and interior surface of the fine-grained matrix of breccia 14047,37 are similar to ratios measured in the fine-grained matrix of two 2-mm-sized breccia particles of 14161,31. This observation leads us to infer that fine-grained breccia particles in the regolith of Apollo 14 that have analogous counterparts in the rock samples of the Apollo 14 regolith have a source that is quite local since the sorting of impact-produced material has been shown to be a strong function of distance from the crater (Shoemaker *et al.*, 1970).

Solar $^4He/^{36}Ar$ and $^{20}Ne/^{36}Ar$ from the surface and interior of two breccias, 14301,61 and 12010,13, show a similar and interesting variation, namely, that the relative $^4He/^{36}Ar$ and $^{20}Ne/^{36}Ar$ from the surfaces of these breccias are consistently higher than the corresponding ratios measured within the interior of the sample. This result suggests that the surface has been reirradiated by the solar wind in contrast to the interior matrix, which carries the history of previous solar irradiation in addition to possible thermal heating and gas fractionation.

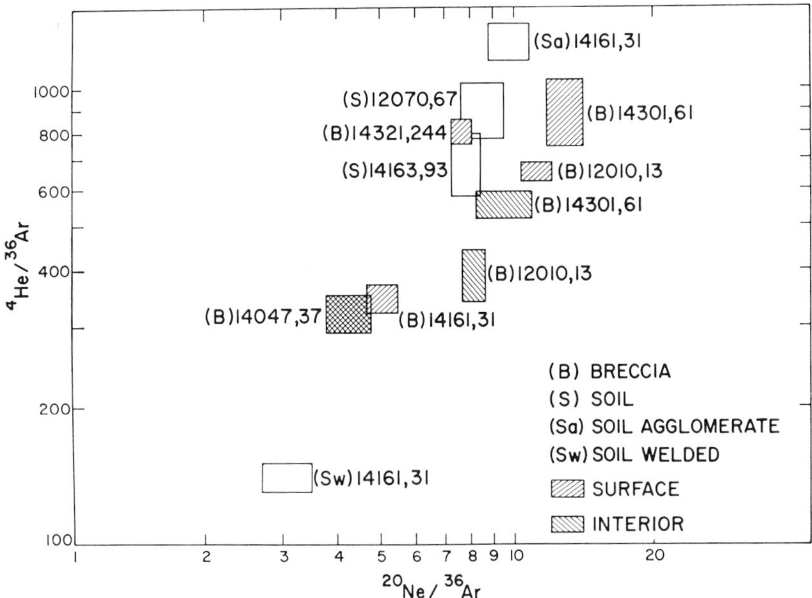


Fig. 8. Summary of the solar $^4He/^{36}Ar$ and $^{20}Ne/^{36}Ar$ from fine-grained soil and matrix of breccias from Apollo 12 and Apollo 14 samples.

Welded or slag breccias from 14161,31 yield the lowest solar $^4He/^{36}Ar$ and $^{20}Ne/^{36}Ar$ and indicate that the origin of this material was previously solar-irradiated fine-grained material that was heated and partially outgassed. A soil agglomerate from 14161,31 has $^4He/^{36}Ar$ and $^{20}Ne/^{36}Ar$ that are distinctly higher than those of other fine-grained breccias or soils from Apollo 12 and 14. This observation implies that lunar soil contains soil agglomerates equivalent to large breccias from nearby localities and that it is possible to differentiate by *in situ* rare-gas analysis lunar soil contained within other lunar soils. This differentiation should provide an understanding of the sources that contribute to the composition of the lunar regolith at other sites on the moon.

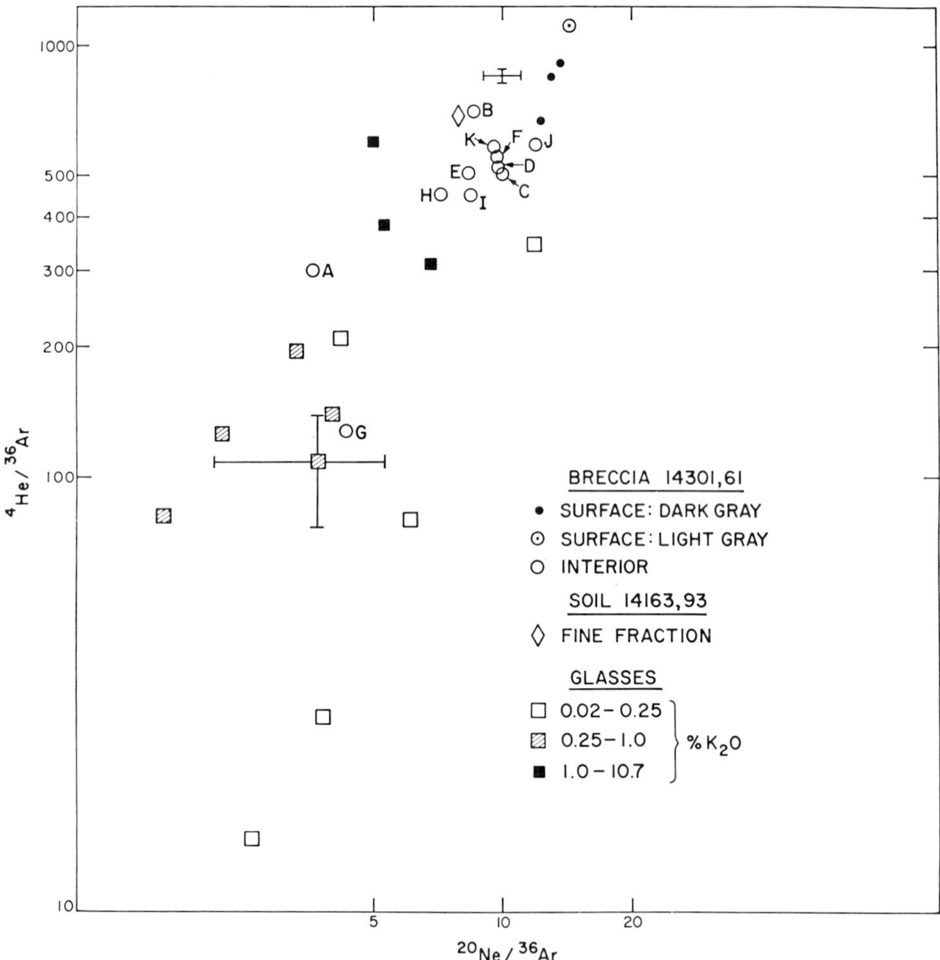


Fig. 9. Summary of $^4He/^{36}Ar$ and $^{20}Ne/^{36}Ar$ from glasses of 14163,93 and from interior and exterior of 14301,61.

A summary of the $^4He/^{36}Ar$ and $^{20}Ne/^{36}Ar$ from glasses of varying $K_2O$ (Fig. 9) shows that glasses of low $K_2O$ (i.e., <0.25%) have distinctly lower gas ratios than do those observed in the fine fraction of soil. This result implies that the glasses are melted lunar soil or regolith that had been previously irradiated by the solar wind and that upon melting produced a fractionation of the previously incorporated solar gas. As the $K_2O$ content of the glasses increases (Fig. 9), the $^4He/^{36}Ar$ increases, whereas the $^{20}Ne/^{36}Ar$ remains relatively constant. Since we have shown that in glasses the radiogenic $^4He$ is correlated with $K_2O$ as well as with radiogenic $^{40}Ar$, we can understand the abrupt increase of $^4He/^{36}Ar$ with increasing $K_2O$ content as being the result of increasing contents of radiogenic $^4He$. This result suggests that high $^4He/^{36}Ar$ or $^4He/^{20}Ne$ from various mineral separates (i.e., "ilmenite"), which has been attributed to undersaturated or nonthermally metamorphosed solar-gas-bearing minerals, may, in fact, contain abnormally high radiogenic $^4He$ abundances that result from the decay of abnormally high uranium-thorium concentrations (Eberhardt *et al.*, 1970; Kirsten *et al.*, 1971).

The surface $^4He/^{36}Ar$ and $^{20}Ne/^{36}Ar$ from the light-gray and dark-gray surface of breccia 14301,61 are systematically higher than the ratios observed from various interior portions of the sample (Fig. 9). Whether this observation reflects contamination from solar-irradiated fine-grained particles similar to the relative rare-gas abundances found in soil agglomerate Sa, 14161,31 (Fig. 8) or can be attributed to increasing radiogenic $^4He$ content as is observed for the potassium-rich glasses is a subject for future investigation.

### *Radiogenic and cosmogenic gases*

The distribution of the radiogenic gases ($^4He$ and $^{40}Ar$) can give us insight into the location and distribution of the potassium- and uranium-bearing minerals on the lunar surface. As we have previously shown, the distribution of radiogenic $^4He$ can confuse the measurements of the solar $^4He/^{20}Ne$ and $^4He/^{36}Ar$ from different minerals.

More importantly, if $^{40}Ar$ has been reimplanted into solar-irradiated minerals or rock surfaces, either recently or in the past, we should be able to detect this reimplantation experimentally by *in situ* analysis with a laser probe-mass spectrometer. So far, our profile measurements of one norite grain do not show an excessive abundance of radiogenic $^4He$ or $^{40}Ar$ on the surface of the grain. Furthermore, we do not find a loss of $^4He$ or $^{40}Ar$ from the interior of the grain, which must be assumed to have occurred if radiogenic gases are to be reimplanted into mineral surfaces.

Calculation of the potassium-argon ages of glasses separated from 14163,93 displays a variation in age from 17 m.y. upward to $3.7 \times 10^9$ yr (Table 8). Although a large uncertainty may exist in the absolute ages because of the possibilities of inherent radiogenic argon and of an extreme heterogeneity in the potassium abundances, the fact remains that discrete melting events may have produced these glasses at various intervals over the lifetime of the moon. If we assume that complete degassing of radiogenic $^{40}Ar$ has occurred from these glasses at the time of formation, then melting events occurred at the following intervals: 17 m.y., <600 m.y., and 2–2.6 and

Table 8. K–Ar ages and exposure ages of Apollo 14 samples.

| Sample | $^{40}Ar/^{36}Ar$ | $^{40}Ar \times 10^{-6}$ (cc STP/g) | K (ppm) | K–Ar age ($10^9$ yr) | Exposure age $^3He$ or $^{21}Ne$ ($10^6$ yr) |
|---|---|---|---|---|---|
| *Glasses* | | | | | |
| 14163,93 | | | | | |
| Light green | 0.62 | 7.9 | 166 | 3.7 | |
| Dark green | 3.5 | 177 | 1410 | — | |
| Dark amber | 15 | 447 | 1500 | — | |
| Olive green | 3.9 | <260 | 1910 | — | |
| Brown | 0.42 | 31 | 2080 | 2.0 | |
| Light amber | 35 | 658 | 2320 | — | |
| Brown | <1.2 | <7 | 2400 | <0.6 | |
| Dark brown | 4.8 | 185 | 5150 | 3.3 | |
| Dark brown | 0.43 | 18 | 5800 | — | |
| Amber | <0.4 | 5 | 7460 | 0.017 | |
| Dark brown | 15 | 447 | 9150 | 3.7 | |
| Yellow | 75 | 291 | 13300 | 2.5 | |
| White | 193 | 2100 | 89000 | 2.6 | |
| *Breccias* | | | | | |
| 14301,61 | 16 | 3850 | — | — | <200 |
| 14321,244 | 53 | 160 | 3500 | 3.7 | <30 |
| 14047,37 | 1.3 | 410 | — | — | <45 |
| *Fragments* | | | | | |
| 14161,31 | | | | | |
| Glass | 53 | 29 | — | — | 60 |
| Breccia | 15 | 14 | — | — | 18 |
| Soil | 11 | 287 | — | — | 34 |
| Soil glass | 254 | 800 | — | — | 36 |
| 14163,93 | | | | | |
| KREEP | 26 | 160 | — | — | 135 |
| | 620 | 850 | — | — | 320 |
| Plagioclase | 15 | 44 | — | — | 40 |

$3.3–3.7 \times 10^9$ yr (which is in agreement with Pepin *et al.*, 1972). Three glasses of low-K abundance (<2500 ppm) give abnormally high $^{40}Ar/^{40}K$ ($\gg 1$), which must be attributed either to an inhomogeneous distribution of potassium and argon within the glass or to the incorporation of "parentless" radiogenic $^{40}Ar$ (Manka and Michel, 1970; Heymann and Yaniv, 1970).

From the light matrix of breccia 14301,61, we measure a radiogenic $^{40}Ar$ of $3.8 \times 10^{-3}$ cc STP/g, which would require a potassium content >5% for the K–Ar age $<4 \times 10^9$ yr. A potassium abundance <5% would indicate that this material is the source of "parentless" $^{40}Ar$ (Heymann and Yaniv, 1971; Bauer *et al.*, 1972). Whether this "parentless" $^{40}Ar$ is correlated with the excess fission xenon from decay of $^{244}Pu$ (Crozaz *et al.*, 1972) awaits further research. A K–Ar age of $3.7 \times 10^9$ yr is calculated for breccia 14321, using the average K abundance of 3500 ppm (Preliminary Examination Team, 1971) and an average radiogenic $^{40}Ar = 1.6 \times 10^{-4}$ cc STP/g from the surface and interior of the sample. Atmospheric argon contamination was detected in 14047,37 as seen by the factor of 2 higher $^{40}Ar/^{36}Ar$ from the interior sections as contrasted to the surface measurements (Table 5). This atmospheric argon contamination is attributed to the fact that we prepared this rock section by cutting in a water medium and observed that the $^{40}Ar/^{36}Ar$ decreased as a function of time when the sample was placed in a vacuum system.

Cosmogenic exposure ages have been calculated for various lithic fragments within breccias and soil samples by using the production rates for $^{3}He = 1 \times 10^{-8}$ cc STP/g/m.y. and $^{21}Ne = 0.2 \times 10^{-8}$ cc STP/g/m.y. The results (Table 8) demonstrate that soil and glass fragments from the coarse fraction of 14161,31 have relatively recent exposure ages $<40$ m.y. Fragments from breccia samples 14047,37 and 14301,61 have exposure ages of 45 and 200 m.y., respectively. Since we can assume that the emplacement of the breccia on to the lunar surface is younger than the youngest exposure age of its constituents, we place only upper limits on the surface age of the breccia. Plagioclase and norite (KREEP) fragments have exposure ages of 40 and $>100$ m.y., respectively.

## Classification and Source of Analyzed Lunar Samples

The relative concentration gradients of the helium, neon, and argon isotopes from returned lunar samples provide a means of uniquely classifying these samples. This is because once the indigenous radiogenic gases and the gases from the external sources of the moon (solar wind and cosmic-ray bombardment) have been incorporated into the lunar material, then subsequent relocation or heating of the lunar samples can alter the initial gas distribution by mere gas fractionation. Consequently, thermal metamorphism of lunar soils, breccias, or individual particles can be recognized by ascertaining the pattern of gas fractionation for these different grain-sized materials. Furthermore, the source of these materials can be inferred from the radiogenic gas contents and cosmic-ray exposure ages of the materials. The temporal variations of the external solar and cosmic-ray sources can be studied, provided that the radiogenic gases (in particular, $^{4}He$) can be uniquely identified.

If we assume that solar-wind irradiation is constant as a function of time on the lunar surface, then the $^{4}He/^{36}Ar$ and $^{20}Ne/^{36}Ar$ variations, which we see in fine-grained lunar soil and breccia matrices (Fig. 8), must result from thermal diffusion of the gases from the sample. Consequently, we would define a relative thermal metamorphic scale from the most thermally metamorphosed to the least thermally metamorphosed as: glass or welded soil $>$ soil breccia 14047 $>$ breccia 14301 $>$ soil agglomerate 14161. This scale defines thermal metamorphism in terms of gas loss rather than the petrographic description of crystallinity or observed shock features (Warner, 1972).

The common occurrence of soil agglomerates in the lunar regolith that are similar in relative noble gas contents to those contents found in breccia 14047, as well as the relatively recent exposure ages of 14047 and 14321 $<40$ m.y., indicates that the source of these rocks is probably quite local and of very recent origin. Consequently, we would identify the genesis of these rocks on the lunar surface as resulting from the formation of Cone Crater. Moreover, the abundant radiogenic $^{4}He$ and $^{40}Ar$ in the fine-grained matrix of 14301 indicate that this rock has an unusually high content of KREEPy material, which is consistent with the chemistry of the Fra Mauro formation. If this rock were ejected from Cone Crater along with the other breccias, then some very special conditions must be invoked to understand why it contains fragments with exposure ages of $\sim 200$ m.y.

*Acknowledgments*—We wish to thank J. Bower for assistance in the chemical analysis of the lunar glasses and G. J. Taylor for assistance in mounting the samples and collection of the mineralogical data. This research was supported in part by Grant NGR 09-015-144 from the National Aeronautics and Space Administration.

## REFERENCES

Bauer H., Frick N., Funk H., Schultz L., and Signer P. (1972) On the thermal release of helium, neon, and argon from lunar fines (abstract). In *Lunar Science—III* (editor C. Watkins), p. 47, Lunar Science Institute Contr. No. 88.

Burnett D., Monin M., Seitz M., Walker R., and Yuhas D. (1971) Lunar astrology—U–Th distributions and fission track dating of lunar samples. *Proc. Second Lunar Sci. Conf., Geochim. Cosmochim. Acta* Suppl. 2, Vol. 2, pp. 1503–1519. MIT Press.

Crozaz G., Drozd R., Graf H., Hohenberg C. M., Monnin M., Ragan D., Ralston C., Seitz M., Shirck J., Walker R. M., and Zimmerman J. (1972) Evidence for extinct $^{244}$Pu: Implications for the age of the pre-Imbrium crust (abstract). In *Lunar Science—III* (editor C. Watkins), p. 164, Lunar Science Institute Contr. No. 88.

Eberhardt P., Geiss J., Graf H., Grogler N., Krähenbühl U., Schwaller H., Schwarzmuller J., and Stetter A. (1970) Trapped solar wind noble gases, $^{81}$Kr/$^{80}$Kr exposure ages and K/Ar ages in Apollo 11 lunar material. *Science* **167,** 558–560.

Heymann D. and Yaniv A. (1970) $^{40}$Ar anomaly in lunar samples from Apollo 11. *Proc. Apollo 11 Lunar Sci. Conf., Geochim. Cosmochim. Acta* Suppl. 1, Vol. 2, pp. 1261–1267. Pergamon.

Heymann D. and Yaniv A. (1971) Ar-40 in meteorites, fines and breccias from the moon. *Chem. Erde Journ.* **175.**

Kirsten T., Deubner H., Ducati H., Gentner W., Horn P., Jessberger E., Kalbitzer S., Kaneoka I., Kiko J., Kratschmer W., Muller H. W., Plieninger T., and Thio S. K. (1972) Rare gases and ion tracks in individual components and bulk samples of Apollo 14 and 15 fines and fragmented rocks (abstract). In *Lunar Science—III* (editor C. Watkins), p. 452, Lunar Science Institute Contr. No. 88.

Kirsten T., Steinbrunn F., and Zähringer J. (1971) Location and variation of trapped rare gases in Apollo 12 lunar samples. *Proc. Second Lunar Sci. Conf., Geochim. Cosmochim. Acta* Suppl. 2, Vol. 2, pp. 1651–1669. MIT Press.

LSPET (Lunar Science Preliminary Examination Team) (1971) *Apollo 14 Preliminary Science Report*, NASA SP-272.

Manka R. H. and Michel F. C. (1970) Lunar atmosphere as a source of argon-40 and other lunar surface elements. *Science* **169,** 278–280.

Megrue G. H. (1971) Distribution and origin of helium, neon, and argon isotopes in Apollo 12 samples by *in situ* analysis with a laser-probe mass spectrometer. *J. Geophys. Res.* **76,** 4956–4968.

Pepin R. O., Bradley J. G., Dragon J. C., and Nyquist L. E. (1972) K–Ar dating of lunar soils: Apollo 12, Apollo 14, and Luna 16 (abstract). In *Lunar Science—III* (editor C. Watkins), p. 602, Lunar Science Institute Contr. No. 88.

Shoemaker E. M., Hait M. H., Swann G. A., Schleicher D. L., Schaber G. G., Sutton R. L., Dahlen D. H., Goddard E. N., and Waters A. D. (1970) Origin of the lunar regolith at Tranquility Base. *Proc. Apollo 11 Lunar Sci. Conf., Geochim. Cosmochim. Acta* Suppl. 1, Vol. 3, pp. 2399–2412. Pergamon.

Swann G. A., Bailey N. G., Batson R. M., Eggleton R. E., Hait M. H., Holt H. E., Larson K. B., McEwen M. C., Mitchell E. D., Schaber G. G., Schafer J. B., Shepard A. B., Sutton R. L., Trask N. J., Ulrich G. E., Wilshire H. G., and Wolfe E. W. (1971) Preliminary geologic investigations of the Apollo 14 landing site. *Apollo 14 Preliminary Science Report*, NASA SP-272, pp. 39–85.

Warner J. L. (1972) Apollo 14 breccias: Metamorphic origin and classification (abstract). In *Lunar Science—III* (editor C. Watkins), p. 782, Lunar Science Institute Contr. No. 88.

Proceedings of the Third Lunar Science Conference
(Supplement 3, *Geochimica et Cosmochimica Acta*)
Vol. 2, pp. 1917–1925
The M.I.T. Press, 1972

# Isotopic anomalies in lunar rhenium

R. MICHEL, U. HERPERS, H. KULUS, and W. HERR

Institut fuer Kernchemie der Universitaet Koeln, Germany

**Abstract**—Apollo 14 soils 14163 (bulk) and 14259 (comprehensive) and breccias from 14321 and 14305 were analyzed for their Re- and W-contents. Simultaneously, the Re isotopic composition was determined by means of neutron activation. The latter was achieved by comparing the 137 keV line of $^{186}$Re with the 155 keV line of $^{188}$Re. The Apollo 14 soils were considerably enriched in Re (0.95–1.4 ppb) relative to basaltic rocks (0.01–0.09 ppb for Apollo 12). The higher abundance is explained by a stronger contribution of a meteoritic component, about twice as intensive as that at the Apollo 12 landing place. Tungsten behaves in a similar way (Apollo 14 soils: 1.3–1.9 ppm, in contrast to basaltic rock 12021 with only 0.28 ppm W). We have also established that the $^{187}$Re is enriched in lunar material relative to the terrestrial value by 1.4 to 1.8% (in 14 regoliths) and by as much as 29% in one of the breccias. The W/Re ratios varied between 1,100 (soils) and 41,000 (breccia 14321). Thus suggesting that the isotopic anomaly is mainly caused by neutron capture in $^{186}$W resulting in a surplus of $^{187}$Re. However, it should be pointed out, that neutron capture in $^{186}$W has taken place not only on the moon, but also in the course of the analytical procedure. Therefore the excess of $^{187}$Re has to be corrected for the effect of enhancement from the "terrestrial" pile neutrons. This was done by calculation and comparison with suitable Re and W standards. The lunar neutron contribution is between 20 and 60% of the total excess and is considered to be primarily due to epithermal neutrons up to $\sim$100 eV.

## INTRODUCTION

THE NATURAL $\beta^-$- RADIOACTIVITY of $^{187}$Re is of geochemical and cosmochemical interest, since it has been shown to be useful in dating minerals, ore deposits, and iron meteorites (Herr *et al.*, 1961; Herr *et al.*, 1967). Although some open problems still exist concerning the decay mode (bound state $\beta$-decay) (Gilbert, 1958; Brodzinski and Conway, 1965), it was pointed out, that $^{187}$Re could still be a valuable tool in cosmochronology (Clayton, 1964). As was also discussed earlier, the weak $\beta$-emitter $^{187}$Re ($E_{max}$ $\sim$2.6 keV) may possibly be depleted in "solar wind" Re by a temperature dependant "faster" decay in the sun (Clayton, 1969). In 1971 we reported on attempts to measure the isotopic abundance of lunar rhenium by neutron activation techniques (Herr *et al.*, 1971). Contrary to our expectation we observed a slight enrichment in $^{187}$Re for Apollo 12 samples. However, the error of the former analytical procedure was too large to prove clearly the significance of the effect. In order to improve the precision of our procedure we found it necessary to redetermine the half lifes of $^{188}$Re and of $^{186}$Re since these showed considerable variation in the literature. Our results are $T_{1/2}$ = (90.64 ± 0.09) h for $^{186}$Re and $T_{1/2}$ = (16.98 ± 0.02) h for $^{188}$Re (Michel and Herpers, 1971). The errors are by a factor of 2 and 10, respectively, lower than for earlier determinations.

With a 25 cc Ge(Li)-detector and appropriate shielding the efficiency and resolution of the system was highly improved, and amounts of Re, as small as $10^{-12}$ g, were isotopically analysed. The very low concentrations of Re in lunar material (0.01 ppb Re for Apollo 12 igneous rocks, Herr *et al.*, 1971) makes it difficult to analyze by mass spectrometry. However, the high sensitivity of activation analysis combined with the radiochemical techniques today is a more realistic approach. The isotopic ratios can be measured by comparison of the respective radioactivities of $^{186}$Re and $^{188}$Re. After an extensive chemical treatment counting was done by $\gamma$-spectroscopy. Although the 155 keV-$\gamma$-radiation of $^{188}$Re has an intensity of only 10% and also the 137 keV-$\gamma$-line of $^{186}$Re has 9%, $\gamma$-spectroscopy has considerable advantages over the more sensitive $\beta^-$-counting-technique suggested by Morgan (1970). The comparison of the two $\gamma$-lines allows one to record both Re isotopes simultaneously with high precision ($\pm 0.3\%$).

The question of the cause of a possible enrichment of $^{187}$Re led us to search for a responsible nuclear reaction. The high abundance of the neighboring element tungsten (Wänke *et al.*, 1970, 1971; Brunfelt *et al.*, 1971) brought to our mind the consideration of the neutron capture reaction of $^{186}$W (Herpers *et al.*, 1971). This $(n, \gamma)$-process has a cross section of $\sigma_{\text{therm}} = 38$ barns, producing $^{187}$Re after the $^{187}$W ($T_{1/2} = 24$ h) has decayed. This formation of $^{187}$Re is favored by high W abundances. All other elements near to Re in the periodic table are found only in the ppb of ppm region, so spallation and other nuclear reactions are less probable. Therefore our interest was directed to the W/Re ratios, which were found to be between 1,100 (soil) and 41,000 (breccia 14321).

With respect to the computation of the experimental Re-isotopic abundances, the following point must be considered. One has to prove to what extent the observed isotopic effect is due to the lunar neutron irradiation or due to the neutron bombardment in the reactor. The latter cannot be avoided, because it is the basis of our analytical procedure. However, it can be corrected for by calculation and by comparison with suitable Re and W standards, which were pile activated under the same conditions.

## EXPERIMENTAL

The following samples were analyzed: 14163,118 (<1 mm, bulk soil), 14259,86 (<1 mm, comprehensive soil), 14321,174 (breccia) and chips of breccia 14305 (,36 and ,45) from different positions. The samples were irradiated in the FRJ-2-reactor $\Phi_{\text{therm}} = 8 \times 10^{13}\ n\ \text{cm}^{-2}\ \text{sec}^{-1}$ for 12 to 70 hr. Our chemical procedures were very similar to those described elsewhere (Herr *et al.*, 1971). It should be mentioned that now the last step includes an ion exchange separation from interfering $^{99m}$Tc. Tungsten was separated several days later by precipitation with conc. $HNO_3$ as tungstic acid and repeatedly purified. It was measured in the form of $WO_3$. As standards $KReO_4$ and $WO_3$ on high purity Al-foils were used, each of which was specpure (Johnson and Mattey). They were chemically treated in the same way as the mineral samples.

Our first task was to establish the contribution of the reaction $^{186}\text{W}(n, \gamma)\ ^{187}\text{W} \xrightarrow{\beta^-} {}^{187}\text{Re}(n, \gamma)$ $^{188}$Re taking place during the reactor irradiation. Therefore, aliquots of our $WO_3$ standards were first analyzed for their own Re contents and, thereafter, the isotopic composition of Re was also checked. $^{186}$W has been previously used as a monitor for intermediate neutrons (Zijp, 1970). But since the resonance integrals of $^{186}$W were not precisely known for the respective irradiation positions, it seemed not to be permitted for us to make the necessary correction for $^{188}$Re breeding on the basis of calculation only.

A 10 cm lead-shielded 25 cc Ge(Li) detector combined with a 4096 channel analyzer was used. The $^{186}$Re and $^{188}$Re decay was followed for at least 6 half lifes for each radioisotope. The $\gamma$-spectra were stored on papertape and processed by a PDP 9 computer. The Re elemental abundances were evaluated via the 3.8 $d$ activity of $^{186}$Re, assuming the terrestrial abundance of $^{185}$Re = 37.07% Re (White and Cameron, 1948). The errors presented correspond to one standard deviation. The W content was determined on the basis of three different $\gamma$ energies of $^{187}$W (134,479,686 keV).

In Fig. 1 a typical Re $\gamma$ spectrum of the breccia 14321 is compared with that of a terrestrial Re standard. From the deviations in the actual counting and decay times the difference in the isotopic composition of lunar Re is evident. The $^{188}$Re peak (155 keV) is several times higher than the "terrestrial" $^{188}$Re peak. In addition spectra of a pure W standard and of a mixed W plus Re standard are presented in Fig. 2. Obviously, it is possible to simulate the "lunar Re" $\gamma$-spectrum of 14321 by the preparation of a suitable mixed W plus Re standard. On the other hand, the Fig. 2 shows that one is able to determine precisely the yield of the reaction $^{186}W\,(n, \gamma)\,^{187}W \xrightarrow{\beta-} {}^{187}Re\,(n, \gamma)\,^{188}Re$ ($T_{1/2}$ = 17 h) during the pile activation. The production of $^{186}$Re from $^{184}$W is in this case negligible, since the cross section of $^{184}$W is small ($\sigma_{therm}$ = 1.9 barn) and the 70 days half life of $^{185}$W causes a rather slow growth of $^{186}$Re.

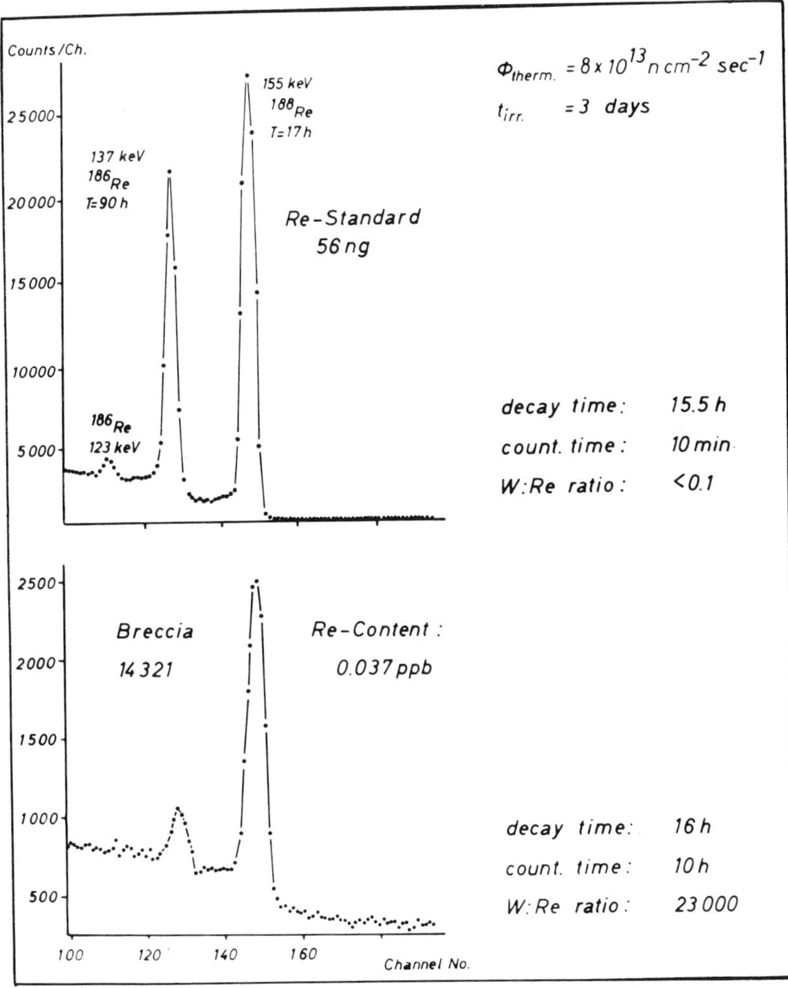


Fig. 1. Comparison of neutron activated "terrestrial" and "lunar" rhenium $\gamma$-ray-spectra.

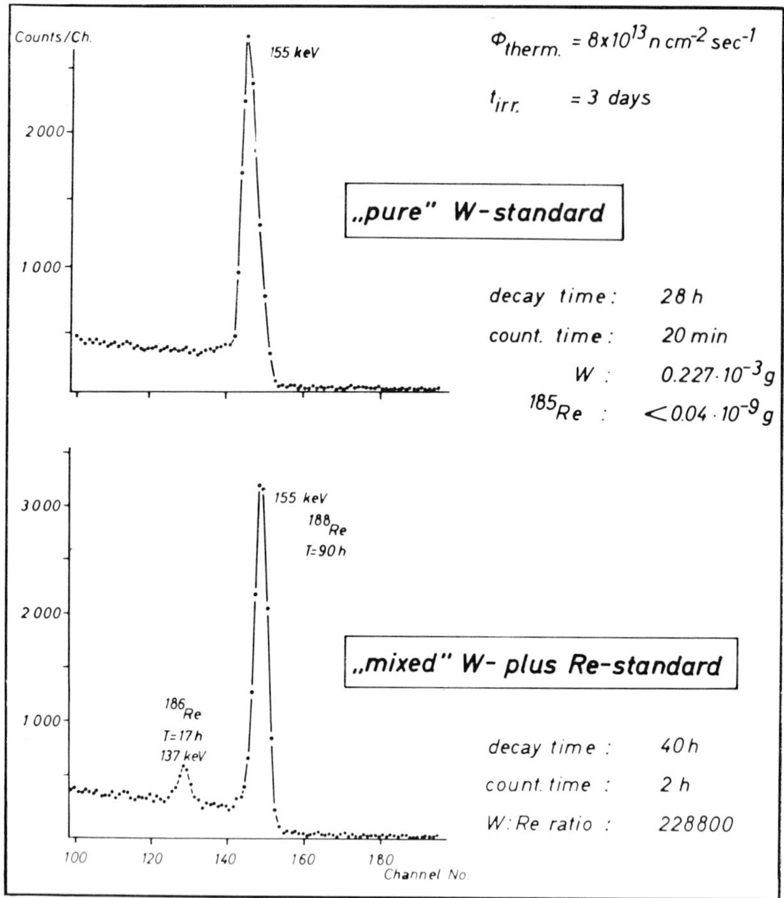


Fig. 2. Gamma-ray spectra of Re from irradiated W and W plus Re standards demonstrating the production of $^{188}$Re by the reaction $^{186}W\,(n, \gamma)\,^{187}W \xrightarrow{\beta-} {}^{187}Re\,(n, \gamma)\,^{188}Re$.

## RESULTS AND DISCUSSION

In Table 1 Re and W contents of our Apollo 14 samples are compared with those from Apollo 12 igneous rocks. As already mentioned in earlier investigations, lunar basaltic rocks are strikingly depleted in Re and W and other siderophile elements like Au, Os, Ir, etc. The enrichment in the soil samples in Re, etc., is evidently due to a meteoritic component (Herr *et al.*, 1971; Keays *et al.*, 1970; Anders *et al.*, 1971; Lovering and Butterfield, 1970; Lovering and Hughes, 1971). The enhancement of Re is considerably higher for Apollo 14 soils (by a factor of 2) than for the former Apollo missions. For the soils 14163 and 14259 we find our Re figures to be in excellent agreement with those, which were recently presented by Ganapathy *et al.* (1972). It should be mentioned, however, that there are certain variations (up to $\sim$30%) for Re in individual soil probes. This indicates strong inhomogenities,

Table 1. Rhenium and Tungsten contents of lunar material.

| Sample | | Type | W (ppm) | Re* (ppb) | Ratio W:Re (× $10^3$) |
|---|---|---|---|---|---|
| 14163 | a | soil | $1.32 \pm 0.02$ | $1.316 \pm 0.004$ | 1.1 |
| | b | (< 1 mm) | $1.94 \pm 0.03$ | $0.957 \pm 0.007$ | 2.0 |
| 14259 | a | soil | $1.56 \pm 0.02$ | $1.41 \pm 0.03$ | 1.1 |
| | b | (< 1 mm) | $1.85 \pm 0.03$ | $1.16 \pm 0.04$ | 1.6 |
| 14321 | a | breccia | $0.82 \pm 0.01$ | | |
| | b | | $0.86 \pm 0.01$ | $0.021 \pm 0.002$ | 40.9 |
| | c | | $0.87 \pm 0.01$ | $0.037 \pm 0.004$ | 23.3 |
| 14305 BD1 int. | | breccia | $1.74 \pm 0.03$ | $0.620 \pm 0.003$ | 2.8 |
| 14305 AE1 surf. | | | $2.39 \pm 0.03$ | $0.834 \pm 0.003$ | 2.9 |
| 12053 | | rock | 0.12† | $0.033 \pm 0.002$<br>$0.030 \pm 0.003$<br>$0.026 \pm 0.002$ | 4.0 |
| 12021 | | rock | $0.28 \pm 0.01$ | $0.077 \pm 0.007$<br>$0.078 \pm 0.007$ | 3.6 |

* Re contents based on $^{185}$Re $(n, \gamma)$ $^{186}$Re ($T = 90.6$ h).
† Given by Wänke *et al.* (1971).

probably due to statistically distributed iron grains. The behavior of the breccias falls between soil samples and igneous rocks.

The basaltic rocks are found to be relatively poor in tungsten, on the other hand the breccias and soils contain considerably more W, by as much as a factor of 10 more. We find remarkable agreement among our three independant analyses for W in the breccia 14321. This is not the case with the breccia 14305, where a difference of ~30% is found for samples from the surface and from the interior. The absolute W content of this rock is a factor 2–3 higher than 14321. Wlotzka *et al.* (1972) have pointed out the interesting fact, that W in soil is strongly attached to the metallic phase (Fe grains). This would support the presumed inhomogenity in the W distribution.

Of particular interest for the Re isotopic analysis is the W/Re ratio, which is in all cases > 1,100. This ratio is even higher in breccias and igneous rocks and reaches in 14321 the extremely high value of 41,000 (cosmic ratio W/Re = 3.63, Suess and Urey, 1956). In our isotopic analysis of Re we found a striking enrichment of $^{187}$Re. The excess in the soils amounts to 1.4 to 1.8%, going up to 29% for the breccia 14321. Because the W/Re ratio differs rather strongly, we have plotted in Fig. 3 the $^{187}$Re/$^{185}$Re ratios against the $^{186}$W/$^{185}$Re ratios in three separate diagrams. Presented are the measured ratios as well as the final values corrected for the terrestrial neutron activation. The errors given are those of the complete analysis. As can be seen from these diagrams the $^{187}$Re enhancement, which we attribute to the effect of lunar neutrons, is significant.

Assuming this to be true, i.e., that lunar neutrons have produced the excess of $^{187}$Re, we are faced with the following problems:

(1) How large is the response integral of the $^{186}$W $(n,\gamma)$ process for the respective lunar neutron energy spectrum?

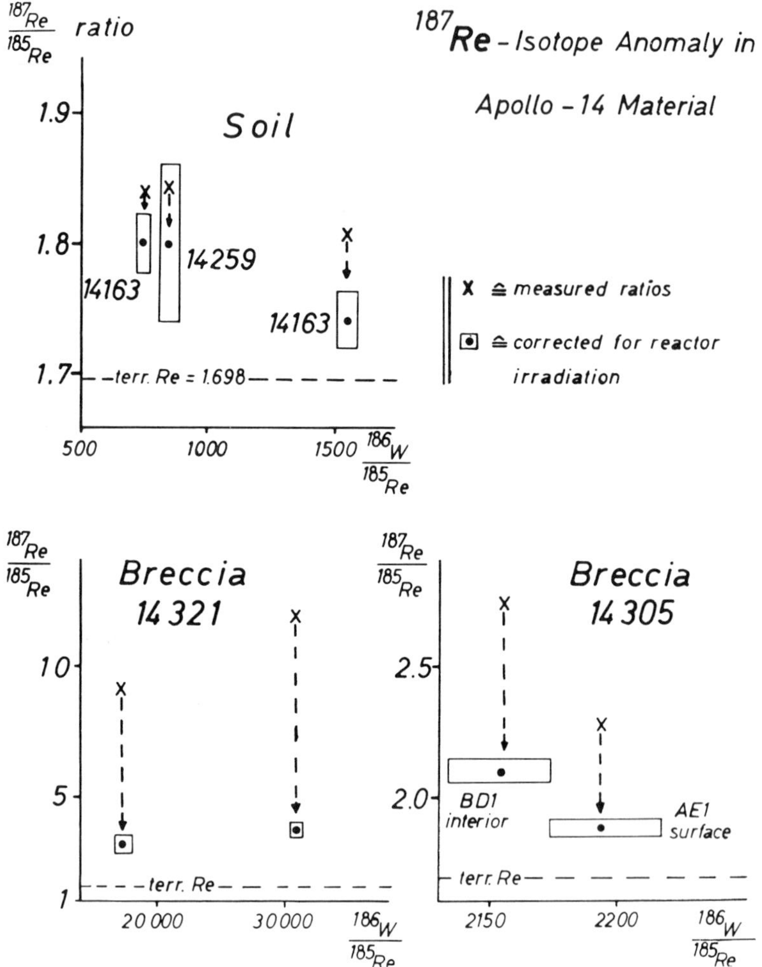


Fig. 3. Results of the Re isotopic analysis of Apollo 14 material.

(2) What are the time integrated fluxes the lunar samples have received and

(3) How are our results correlated with the already better known (lunar) neutron monitors, as for instance, Gd, Sm, Ba, and Br (Albee *et al.*, 1970; Burnett *et al.*, 1971; Russ *et al.*, 1972; Lugmair and Marti, 1972; Kaiser and Berman, 1971; Eberhardt *et al.*, 1970)?

From the experiment we can calculate the number of neutrons ($\varepsilon_w$) captured for atom $^{186}$W according to equation (1)

$$\varepsilon_w = \frac{(^{187}\text{Re}/^{186}\text{W})_M - (^{185}\text{Re}/^{186}\text{W})_M(62.93/37.07)}{1 + (^{185}\text{Re}/^{186}\text{W})_M(62.93/37.07)} . \quad (1)$$

If $\Delta^{187}$Re(lun.) is the number of $^{187}$Re atoms produced on the lunar surface the equation can be simplified to a good approximation by

$$\varepsilon_w = \frac{\Delta^{187}\text{Re(lun.)}}{^{186}\text{W}} \,. \tag{2}$$

In order to enable us to perform a more quantitative treatment, we have put together all data, which are needed to produce the ratio $\Delta^{187}$Re(lun.)/$^{186}$W in Table 2.

Table 2. Contribution of the reactor irradiation to the $^{187}$Re anomaly.

| Apollo 14 Samples | $\Delta^{187}$Re Brutto ($10^{-12}$ g) | $\Delta^{187}$Re Reactor ($10^{-12}$ g) | $\Delta^{187}$Re Lunar ($10^{-12}$ g) | $^{186}$W ($10^{-9}$ g) | $\Delta^{187}$Re (lun.)/$^{186}$W ($\times 10^{-3}$) |
|---|---|---|---|---|---|
| 14163 | 3.5 | 0.94 | 2.6 ± 0.3 | 18.8 | 0.14 ± 0.01 |
| | 1.9 | 1.21 | 0.7 ± 0.3 | 27.6 | 0.027 ± 0.01 |
| 14259 | 3.8 | 1.13 | 2.6 ± 1.4 | 22.2 | 0.12 ± 0.06 |
| 14321 | 3.9 | 3.18 | 0.7 ± 0.09 | 10.7 | 0.071 ± 0.007 |
| | 5.1 | 4.13 | 1.0 ± 0.09 | 12.4 | 0.081 ± 0.006 |
| 14305 BD1 int. | 11.9 | 7.39 | 4.5 ± 0.3 | 24.8 | 0.18 ± 0.01 |
| 14305 AE1 surf. | 8.9 | 6.21 | 2.7 ± 0.2 | 33.9 | 0.081 ± 0.005 |

All data normalized to 50 mg weight (individ. weights 49–104 mg).

The absolute amounts of $^{186}$W and excess $-^{187}$Re in column 2–4 are normalized to a 50 mg sample weight, to allow the comparison of our samples analyzed with each other.

From column 2 it follows that the $^{187}$Re measured excess ranges from ($\sim$2 to $\sim$12) $\times$ $10^{-12}$ g. The lunar fraction (column 5) is between 20–60% of this amount. In the last column the ratio $\Delta^{187}$Re(lun.)/$^{186}$W shows a relatively wide spread from (0.027 to 0.18) $\times$ $10^{-3}$.

Attention should be paid to the rather low value 0.027 $\times$ $10^{-3}$ for soil 14163, indicating a small neutron dosage for this particular sample. A possible explanation is, that tungsten is concentrated in metallic grains of the soil. W contents up to 223 ppm have been detected here by Wlotzka *et al.* (1972). One can imagine that such a metallic fragment, with low neutron exposure might change the apparent neutron dose of the 50 mg sample.

The number of neutrons captured per $^{186}$W atom is equal to the product of the cross section weighted over the energy spectrum and the fluence. However, the mean $(n, \gamma)$ cross section $\langle\sigma\rangle$ of $^{186}$W for the respective lunar neutron spectrum is not yet known. Recently, calculations of the lunar neutron spectrum have been carried out by Lingenfelter *et al.* (1961; 1972) considering now the chemical composition. As soon as these results are available $\langle\sigma\rangle$ will be evaluated. The neutron capture in $^{186}$W is mainly sensitive to epithermal neutrons as can be derived from Fig. 4. The differential cross section of $^{157}$Gd shows that preferentially it reacts with low energy neutrons ($E_n$ <0.1 eV), while $^{186}$W has a pronounced maximum at $\sim$20 eV and is still high up to $\sim$100 eV.

Evidently, $^{187}$Re is breeded via neutron capture, but simultaneously $^{185}$Re is

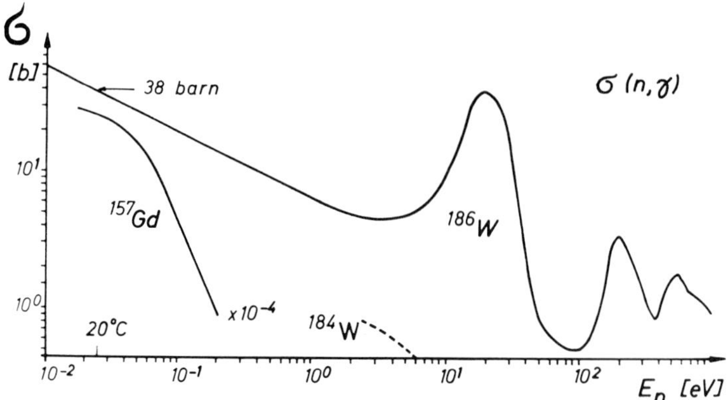


Fig. 4. Comparison of the $^{186}W$ and $^{157}Gd$ $(n, \gamma)$ cross sections (compilation of Goldberg *et al.*, 1966).

produced from $^{184}W$. However, here the thermal cross section is lower by a factor of 20 and the intermediate neutron capture is comparingly small (Goldberg *et al.*, 1966). As conclusion we think that the Re isotopic anomaly may be a useful tool for obtaining more detailed information on radiation profiles, neutron energy spectra, and the radiation history of the lunar surface.

## REFERENCES

Albee A. L., Burnett D. S., Chodos A. A., Eugster O. J., Hunecke J. C., Papanastassiou D. A., Podosek F. A., Russ G. P. III, Sanz H. G., Tera F., and Wasserburg G. J. (1970) Ages, irradiation history and chemical composition of lunar rocks from the Sea of Tranquility. *Science* **167,** 463–466.

Anders E., Ganapathy R., Keays R., Laul J. C., and Morgan J. W. (1971) Volatile and siderophile elements in lunar rocks: Comparison with terrestrial and meteoritic basalts. *Proc. Second Lunar Sci. Conf., Geochim. Cosmochim. Acta* Suppl. 2, Vol. 2, pp. 1021–1036. M.I.T. Press.

Brodzinski R. L. and Conway D. C. (1965) Decay of $^{187}$rhenium. *Phys. Rev.* **138B,** 1368–1371.

Brunfelt A. O., Heier K. S., and Steinnes E. (1971) Determination of 40 elements in Apollo 12 materials by neutron activation analysis. *Proc. Second Lunar Sci. Conf., Geochim. Cosmochim. Acta* Suppl. 2, Vol. 2, pp. 1281–1290. M.I.T. Press.

Burnett D. S., Huneke J. C., Podosek F. A., Russ G. P. III, and Wasserburg G. J. (1971) The irradiation history of lunar samples. *Proc. Second Lunar Sci. Conf., Geochim. Cosmochim. Acta* Suppl. 2, Vol. 2, pp. 1671–1679. M.I.T. Press.

Clayton D. D. (1964) Cosmoradiogenic chronologies of nucleosynthesis. *Astrophys. J.* **139,** 637–663.

Clayton D. D. (1969) Isotopic composition of cosmic importance. *Nature* **224,** 56–57.

Eberhardt P., Geiss J., Graf H., Groegler N., Kraehenbuehl U., Schwaller H., Schwarzmueller J., and Stettler A. (1970) Correlation between rock type and irradiation history of Apollo 11 igneous rocks. *Earth Planet. Sci. Lett.* **10,** 67–72.

Ganapathy R., Laul J. C., Morgan J. W., and Anders E. (1972) Moon: Possible nature of the body that produced the Imbrian basin, from the composition of Apollo 14 samples. *Science* **175,** 55–59.

Gilbert N. (1958) Etude théorique de la desintegration du rhenium 187. *Compt. Rend.* **247,** 868–871.

Goldberg M. D., Nughabghal S. F., Purohit S. N., Maurno B. A., and May V. M. (1966) Neutron cross sections Vol. II C, Z = 61 – 87. BNL – 325.

Herpers U., Michel R., and Herr W. (1971) Search for "solar wind" rhenium in moon samples. *Angewandte Chemie* (International Edition in English) **10,** 818–867.

Herr W., Hoffmeister W., Hirt B., Geiss J., and Houtermans F. G. (1961) Versuch zur Datierung von Eisenmeteoriten nach der Re/Os-Methode. *Z. Naturforsch.* **16a,** 1053–1058.

Herr W., Woelfle R., Eberhardt P., and Kopp E. (1967) Development and recent application of the Re/Os dating method. In *Radioactive Dating and Methods of Low-Level Counting*, pp. 499–508, International Atomic Energy Agency, Vienna.

Herr W., Herpers U., Michel R., Rassoul Abdel A., A. and Woelfle R. (1971) Search for rhenium isotopic anomalies in lunar surface material by neutron bombardment. *Proc. Second Lunar Sci. Conf., Geochim. Cosmochim. Acta* Suppl. 2, Vol. 2, pp. 1337–1341. M.I.T. Press.

Kaiser W. A. and Berman B. L. (1971) The integral $^{130}$Ba $(n, \gamma)$ cross section and the origin of $^{131}$Xe on the moon. To be submitted to the *Earth Planet. Sci. Lett.*

Keays R. R., Ganapathy R., Laul J. C., Anders E., Herzog G. F., and Jeffery P. M. (1970) Trace elements and radioactivity in lunar rocks: Implications for meteorite infall, solar wind flux and formation conditions of moon. *Science* **167,** 490–493.

Lingenfelter R. E., Canfield E. H., and Hess W. N. (1961) The lunar neutron flux. *J. Geophys. Res.* **66,** 2665–2671.

Lingenfelter R. E., Canfield E. H., and Hampel V. H. (1972) The lunar neutron flux revisited. To be submitted to the *Earth Planet. Sci. Lett.*

Lovering J. F. and Butterfield D. (1970) Neutron activation analysis of rhenium and osmium in Apollo 11 lunar material. *Proc. Apollo 11 Lunar Sci. Conf., Geochim. Cosmochim. Acta* Suppl. 1, Vol. 2, pp. 1351–1355. Pergamon.

Lovering J. F. and Hughes T. C. (1971) Rhenium and osmium abundance determinations and meteoritic contamination levels in Apollo 11 and Apollo 12 lunar samples. *Proc. Second Lunar Sci. Conf., Geochim. Cosmochim. Acta* Suppl. 2, Vol. 2, pp. 1331–1335. M.I.T. Press.

Lugmair G. W. and Marti K. (1972) Neutron capture effects in Gadolinium and the irradiation histories of some lunar rocks. *Earth Planet. Sci. Lett.* **13,** 32–42.

Michel R. and Herpers U. (1971) Praezisionsbestimmungen der Halbwertszeiten von $^{186}$Re und $^{188}$Re. *Radiochim. Acta* **16,** 115.

Morgan J. W. (1970) Anomalous rhenium isotopic ratio in the solar wind: Detection at the nanogram level. *Nature* **225,** 1037–1038.

Russ G. P. III, Burnett D. S., Lingenfelter R. E., and Wasserburg G. J. (1972) Neutron capture on $^{149}$Sm in lunar samples. *Earth Planet. Sci. Lett.* **13,** 53–60.

Suess H. and Urey H. C. (1956) Abundance of the elements. *Rev. Mod. Phys.* **28,** 53–74.

Wänke H., Rieder R., Baddenhausen H., Spettel B., Teschke F., Quijano-Rico M., and Balacescu A. (1970) Major and trace elements in lunar material. *Proc. Apollo 11 Lunar Sci. Conf., Geochim. Cosmochim. Acta* Suppl. 1, Vol. 2, pp. 1719–1727. Pergamon.

Wänke H., Wlotzka F., Baddenhausen H., Balacescu A., Spettel B., Teschke F., Jagoutz E., Kruse H., Quijano-Rico M., and Rieder R. (1971) Apollo 12 samples: Chemical composition and its relation to sample locations and exposure ages, the two component origin of the various soil samples and studies on lunar metallic particles. *Proc. Second Lunar Sci. Conf., Geochim. Cosmochim. Acta* Suppl. 2, Vol. 2, pp. 1187–1208. M.I.T. Press.

White J. R. and Cameron A. E. (1948) The natural abundance of isotopes of the stable elements. *Phys. Rev.* **74,** 991–1000.

Wlotzka F., Jagoutz E., Spettel B., Baddenhausen H., Balacescu A., and Wänke H. (1972) On lunar metallic particles and their contribution to the trace element content of the Apollo 14 and 15 soils (abstracts). In *Lunar Science—III* (editor C. Watkins), pp. 806–808, Lunar Science Institute Contr. No. 88.

Zijp W. L. (1970) Intermediate neutrons. In *Neutron Fluence Measurements*, pp. 77–140, International Atomic Energy Agency, Vienna.

Proceedings of the Third Lunar Science Conference
(Supplement 3, *Geochimica et Cosmochimica Acta*)
Vol. 2, pp. 1927–1945
The M.I.T. Press, 1972

# A comparison of noble gases released from lunar fines (#15601.64) with noble gases in meteorites and in the earth

B. SRINIVASAN*, E. W. HENNECKE, D. E. SINCLAIR, and O. K. MANUEL

Division of Earth and Planetary Science,
Department of Chemistry, University of Missouri,
Rolla, Missouri 65401

**Abstract**—The abundance and isotopic composition of helium, neon, argon, krypton, and xenon which had been released by stepwise heating of lunar soil (15601,64) were measured mass spectrometrically. The extraction temperatures ranged from 100°C to 1500°C in 100° intervals. The bulk of the noble gases were released between 700°C and 1500°C, and the release pattern over this temperature range showed a preferential release of the lighter weight noble gases at low extraction temperatures. This release pattern was also observed across the isotopes of the individual noble gases. Very little gas was released below 700°C, but the gas release pattern from the very low extraction temperatures was unlike the general release trend, showing an enrichment of the heavier noble gases. This anomalous release pattern at low extraction temperatures ($T \leq 700°C$) was also observed across the isotopes of individual noble gases and is attributed to one, or a combination, of two effects: (a) adsorption of atmospheric gases on the lunar soil during its residence on earth, or (b) selective release of the lighter weight noble gases from the low-temperature sites of lunar soil during its residence on the lunar surface. Isotopic release patterns of noble gases from lunar soil are compared with atmospheric and meteoritic noble gases. Our results indicate that major differences between the isotopic composition of He, Ne, Ar, and Kr in terrestrial, meteoritic and lunar samples can be accounted for by a combination of mass fractionation and *in situ* nuclear processes in solid planetary material. However, it is evident that additional processes, perhaps nuclear reactions in the sun, have altered the isotopic composition of xenon. Relative to fractionated atmospheric xenon. the spallation-free xenon in lunar fines appears to be enriched in the heavy isotopes, $^{134}Xe$ and $^{136}Xe$, and depleted in $^{124}Xe$ and $^{129}Xe$.

## INTRODUCTION

THE NOBLE GASES TRAPPED IN solid planetary material contain a record of many geologic events. Due to their chemical inertness and volatile nature, these elements were almost completely lost from other chemical elements of the planetary system when solid bodies formed. Studies of the abundance of noble gases in meteorites and in the earth show that the heavier noble gases have been preferentially retained, and the fractionation pattern across the noble gases provides a record of the conditions which prevailed when these gases separated from more condensable matter in the early history of the solar system (Suess, 1949; Brown, 1949; Canalas *et al.*, 1968).

Over the history of the solar system, induced nuclear reactions and natural radioactive decay of the more abundant elements have produced many changes in the isotopic composition of the residual noble gases in planetary material. The present isotopic composition of noble gases shows the cumulative effect of many nuclear

* Present address: Department of Physics, University of California, Berkeley, California 94720.

processes, and in some meteorites almost the entire inventory of noble gases is the result of nuclear reactions or radioactive decay. Trapped noble gas isotopes which could not be accounted for by nuclear processes in a meteorite were discovered by Gerling and Levskii (1956). Subsequent studies on noble gases trapped in meteorites showed that the relative abundance of the noble gases followed the same general fractionation pattern observed for atmospheric noble gases, and the possibility of isotopic anomalies resulting from mass fractionation were noted (Reynolds, 1960a; Zähringer and Gentner, 1960; Krummenacher *et al.*, 1962; Signer and Suess, 1963). Further studies showed that many brecciated-type meteorites contained large amounts of the light weight noble gases (Koenig *et al.*, 1961; Hintenberger *et al.*, 1962) which had been implanted in the dark-colored matrix by the solar wind (Suess *et al.*, 1964; Eberhardt *et al.*, 1965a, b; Hintenberger *et al.*, 1965).

Analyses on the first lunar samples returned to earth (LSPET, 1969) demonstrated that lunar fines and breccias contain large amounts of noble gases from the solar wind. This study of noble gases released from lunar fines (15601,64) was undertaken in order to obtain quantitative information on noble gas isotopes which have been implanted by the solar wind. A comparison of the isotopic composition of noble gases in planetary material with the isotopic composition of noble gases presently being implanted by the solar wind is useful in identifying the effects of (a) radioactive decay and induced nuclear reactions which have occurred in planetary bodies, (b) mass fractionation, and (c) solar nuclear reactions which have occurred since the separation of planetary material from the elements in the sun.

## EXPERIMENTAL PROCEDURE

An aliquot of soil sample 15601,64 weighing 0.9861 g was used in this study. This soil consists of fine $<1$ mm. The sample was wrapped in aluminum foil and mounted in a quartz side-arm chamber of an extraction bottle. After the pressure had been reduced to $10^{-8}$ Torr, a blank Al foil in another quartz side-arm chamber was heated stepwise from 100°C through 600°C by resistance heating. Following analyses of these blanks, the sample was heated stepwise through the same temperatures. Higher extraction temperatures were achieved by dropping the sample into a previously degassed molybdenum crucible heated by rf induction. Each extraction temperature was maintained for thirty minutes.

The gases released from the sample were cleaned on Cu–CuO at 550° and on Ti at 850°C, then transferred into a secondary clean-up system by adsorption on charcoal cooled with liquid nitrogen. After 15 minutes the secondary clean-up system was isolated and the gases cleaned on a fresh Ti surface at 850°C. The gases were admitted into a Reynolds' (1956) high-sensitivity mass spectrometer as four fractions: helium and neon, argon, krypton, and finally xenon.

The procedures for gas analyses and the data reduction have been described by Hennecke and Manuel (1971), except for the following modification. In order to measure the isotope ratios for extraction temperatures that released large amounts of gas, small aliquots ($\approx \frac{1}{100}$ and $\approx \frac{1}{30}$) of the gases in the secondary clean-up system were isolated and analyzed after the regular gas analysis procedure was complete.

The errors reported in the isotopic ratios for noble gases represent one standard deviation ($\sigma$) from the least squares line through the observed ratios plotted as a function of time. The ratios reported are calculated from the least squares lines at the point of entry of the gases into the mass spectrometer. Mass discrimination across the heavier noble gases was determined in the air spikes. The discrimination across the neon isotopes was less than 0.5% and the discrimination across the two helium isotopes was not measured due to the low atmospheric abundance of $^{3}$He. Except for the three lightest noble gases, all of the isotopic ratios have been corrected for mass discrimination. The concentration of each noble gas was determined from its peak height relative to the average peak height observed in two air spikes analyzed before and after the sample analysis.

## Results and Discussion

In comparing the isotopic composition of noble gases extracted from our lunar soil sample with the isotopic composition of noble gases observed in meteorites or in the earth's atmosphere, the following equation for diffusive fractionation is used (Aston, 1933; Kuroda *et al.*, 1971):

$$\log r = [(m_2 - m_1)/(m_2 + m_1)][\log (V_i/V_f)] \tag{1}$$

This equation refers to a particular noble gas, and $r$ is the enrichment of the heavy isotope of mass $m_2$ relative to a lighter isotope of mass $m_1$ that results when the volume of a gas is reduced from some initial volume, $V_i$, to some final volume, $V_f$, by diffusive gas loss. Our use of equation (1) should not be considered as an indication that we believe all isotopic fractionation effects in noble gases are due to diffusive fractionation. We suspect that many different mass-dependent processes may account for part of the complex relationship between planetary and solar-implanted gases in lunar soil. Equation (1) is used in order to make our comparisons quantitative (and at the same time to indicate our bias for diffusive gas loss as a major effect in the relationship between planetary and solar noble gases).

Changes in the relative abundance of three or more isotopes of a noble gas due to mass fractionation can be calculated from equation (1). However, one cannot use equation (1) to distinguish between mass fractionation and nuclear reactions for gases, such as helium and argon, that have less than three nonradiogenic isotopes.

### *Helium, neon, and argon*

The results of our analyses for helium, neon, and argon are shown in Table 1. The isotopic ratios for argon released at 100°C and 200°C and the isotopic ratios for neon released at 200°C have been corrected for signals observed in blanks, but the

Table 1. Helium, neon, and argon from stepwise heating of lunar fines 15601,64.

| Temp. (°C) | Isotope ratios | | | | | Gas content ($\times 10^{-8}$ ccSTP/g) | | |
|---|---|---|---|---|---|---|---|---|
| | $^4$He/$^3$He | $^{20}$Ne/$^{22}$Ne | $^{21}$Ne/$^{22}$Ne | $^{38}$Ar/$^{36}$Ar | $^{40}$Ar/$^{36}$Ar | $^4$He | $^{22}$Ne | $^{36}$Ar |
| 100° | — | — | — | — | 215.8 ± 0.8 | 1.8 | 0.20 | 0.015 |
| 200° | 1922 ± 7 | 12.66 ± 0.03 | 0.0307 ± 0.0004 | — | 29.8 ± 0.1 | 38 | 0.69 | 0.60 |
| 300° | 1838 ± 14 | 13.49 ± 0.02 | 0.0329 ± 0.0003 | 0.185 ± 0.002 | 28.2 ± 0.2 | 300 | 3.2 | 1.5 |
| 400° | 1864 ± 23 | 13.48 ± 0.05 | 0.0325 ± 0.0012 | 0.186 ± 0.001 | 33.1 ± 0.2 | 400 | 4.0 | 3.9 |
| 500° | 1911 ± 9 | 13.70 ± 0.02 | 0.0336 ± 0.0001 | 0.183 ± 0.001 | 31.8 ± 0.2 | 870 | 7.0 | 6.4 |
| 600° | 2273 ± 3 | 13.47 ± 0.02 | 0.0337 ± 0.0001 | 0.181 ± 0.001 | 26.7 ± 0.1 | 6,400 | 60 | 15 |
| 700° | 2280 ± 10 | 13.01 ± 0.02 | 0.0327 ± 0.0001 | 0.178 ± 0.003 | 4.14 ± 0.05 | 450,000 | 15,000 | 5,200 |
| 800° | 2093 ± 16 | 12.76 ± 0.03 | 0.0320 ± 0.0002 | 0.177 ± 0.001 | 2.50 ± 0.01 | 28,000 | 640 | 1,400 |
| 900° | 2215 ± 12 | 12.80 ± 0.05 | 0.0337 ± 0.0002 | 0.185 ± 0.001 | 0.764 ± 0.001 | 54,000 | 2,500 | 25,000 |
| 1000° | 2268 ± 5 | 12.82 ± 0.02 | 0.0372 ± 0.0001 | 0.186 ± 0.001 | 0.558 ± 0.001 | 34,000 | 2,700 | 28,000 |
| 1100° | 1355 ± 3 | 11.91 ± 0.01 | 0.0656 ± 0.0002 | 0.194 ± 0.001 | 0.835 ± 0.005 | 260 | 260 | 2,300 |
| 1200° | 2108 ± 23 | 12.26 ± 0.01 | 0.0557 ± 0.0001 | 0.204 ± 0.002 | 0.811 ± 0.007 | 950 | 270 | 1,800 |
| 1300° | 1928 ± 27 | 12.13 ± 0.01 | 0.0673 ± 0.0003 | 0.254 ± 0.002 | 0.855 ± 0.001 | 430 | 84 | 580 |
| 1400° | 2208 ± 18 | 12.41 ± 0.01 | 0.0461 ± 0.0001 | 0.196 ± 0.001 | 0.711 ± 0.001 | 1,300 | 290 | 3,100 |
| 1500° | 2169 ± 22 | 12.66 ± 0.08 | 0.0434 ± 0.0004 | 0.194 ± 0.001 | 0.700 ± 0.001 | 130 | 65 | 700 |
| Air | 7,700,000 | 9.81 | 0.0290 | 0.187 | 296 | — | — | — |

signals observed from blank analyses were negligible relative to the sample signals for all other data shown in Table 1.

The $^4He/^3He$ ratios increase with increasing extraction temperature from 300°C to 700°C, where the bulk of the helium is released. Hohenberg *et al.* (1970) previously observed a preferential release of $^3He$ from lunar soil 10084 at low extraction temperatures, and suggested that the separation was due to mass fractionation. In Table 1 the highest value for $^4He/^3He$ occurs at 700°C, and the $^4He/^3He$ ratio in the 700° fraction is 24% higher than the $^4He/^3He$ ratio released at 300°C. In the stepwise heating of Apollo 11 lunar soils Hohenberg *et al.* (1970) and Pepin *et al.* (1970b) observed that the $^4He/^3He$ ratio for the total helium was higher than the $^4He/^3He$ ratios in the helium released at low extraction temperatures by 36% and 30%, respectively. Fig. 1 compares the $^4He/^3He$ ratios in sample 15601,64 with the amount of $^4He$ released at each extraction temperature. The ratios $^4He/^3He$ correlate with the amount of $^4He$ released over the entire range of extraction temperatures. The correlation below 700°C may be due to isotopic fractionation, and the correlation above 700°C may be due to the release of cosmogenic $^3He$ at high extraction temperatures. However, it is not possible to unambiguously distinguish between mass fractionation and nuclear processes from the helium results alone.

The neon isotopes released from sample 15601,64 are shown in Fig. 2, where the effects of isotopic mass fractionation and spallation reactions can be separated in the manner employed by Manuel (1967). The approximately vertical lines extrapolate to cosmogenic neon, and the line through atmospheric neon shows the relationship between the neon isotopes which results from fractionation. This approximately

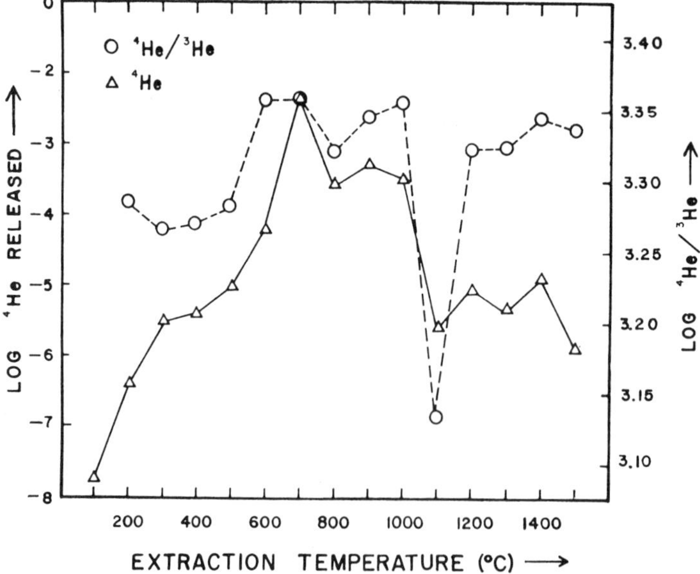


Fig. 1. The release pattern of helium from sample 15601,64. The $^4He$ content on the left ordinate is in units of cc STP/g.

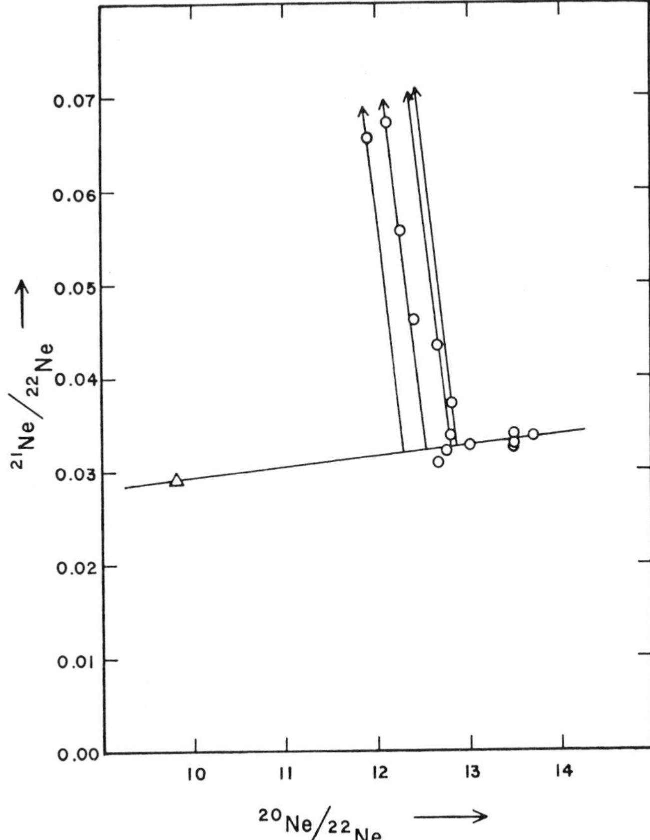


Fig. 2. Correlation of the neon isotopes released from sample 15601,64. The triangle represents atmospheric neon and the line through atmospheric neon results from isotopic fractionation. The arrows point toward pure cosmogenic neon, and the intersections of the lines correspond to the isotopic composition of spallation-free neon (Manuel, 1967).

horizontal line is defined by equation (1), and all of the neon isotopes released at extraction temperatures $<1000°C$ lie along this line.

From Fig. 2 we note that isotopic mass fractionation is responsible for the high $^{20}Ne/^{22}Ne$ released at 500°C and that fractionation and spallation effects are both present in the neon released at high extraction temperatures. The neon released at extraction temperatures $\leq 400°C$ is not as enriched in the light neon isotopes as that released at 500°C, suggesting that fractionation observed in the neon isotopes released from lunar soil may be a mixture of several mass-dependent processes. The pattern of neon isotopes released from 15601,64 is very similar to the release pattern of neon isotopes reported from the gas-rich regions of Fayetteville (Manuel, 1967), where the highest $^{20}Ne/^{22}Ne$ ratio was observed at 600°C and the neon isotopes released at lower extraction temperatures fit the fractionation line shown in Fig. 2. The neon isotopes released from lunar soil in this study can be explained in terms of isotopic fractionation and spallation products. A combination of these two processes has previously been proposed to account for the neon isotopes released by stepwise heating of the gas-rich regions of the Fayetteville meteorite (Manuel, 1967) and Apollo 11 fines (Hohenberg *et al.*, 1970). Since lunar soil and the gas-rich meteorites have both acquired neon from solar-wind implantation, one should not overlook the possible role of other mechanisms which have been proposed to account for variations of the neon isotopes in gas-rich meteorites (Pepin, 1967, 1968; Black and Pepin, 1969; Black, 1969, 1970).

Table 1 shows the maximum release of argon from lunar fines 15601,64 in the 900°–1000° fractions, and the argon released in the three earlier temperature fractions (600°–800°C) shows lower $^{38}Ar/^{36}Ar$ ratios than the argon released at 900°–1000°C. This enrichment of the lighter isotopes in temperature fractions preceeding the release of the bulk of the argon parallels the release pattern observed for neon and helium. Thus, it appears that the argon isotopes released by stepwise heating of lunar fines also show evidence of mass dependent fractionation, as has been suggested earlier by Hohenberg *et al.* (1970). The $^{38}Ar/^{36}Ar$ ratios in argon released at 300°–500°C show less enrichment of the lighter isotope, $^{36}Ar$, than does argon in the 600°–800° fractions. This release pattern of argon isotopes from the 300°–500° fractions is similar to the release pattern of neon isotopes from the 200°–400° fractions and the release pattern of helium isotopes from the 200° fraction. Variations of the $^{38}Ar/^{36}Ar$ ratios in the high temperature fractions are probably due to both spallation and fractionation, as was observed in neon. However, a three-isotope correlation plot cannot be used to distinguish spallation and fractionation effects in the isotopes of argon because of the presence of $^{40}Ar$ from $^{40}K$ decay.

In our analyses, the $^{40}Ar/^{36}Ar$ ratios in the different temperature fractions of lunar soil show the greatest variation of any ratio of two isotopes of a single noble gas. This is undoubtedly due to the presence of radiogenic $^{40}Ar$ in the soil sample. Heymann *et al*: (1970) and Heymann and Yaniv (1970) have shown that some of the radiogenic $^{40}Ar$ in lunar fines is surface-correlated and suggested that this excess radiogenic $^{40}Ar$ may be due to secondary implantation of argon from the lunar atmosphere. The lowest $^{40}Ar/^{36}Ar$ ratio from our sample is observed in the argon released at 1000°C, the temperature fraction that released the greatest amount of $^{36}Ar$. This sets an upper limit of $^{40}Ar/^{36}Ar \lesssim 0.56$ in the solar wind.

## *Krypton*

The content and isotopic composition of krypton released from 15601,64 is shown in Table 2. Marti (1967) and Eugster *et al.* (1967) suggested that krypton in carbonaceous chondrites (AVCC Kr) may be related to atmospheric krypton (Nief, 1960) by isotopic fractionation, and Manuel (1970) noted that some of the variations in the isotopic composition of krypton released from lunar soil fits the pattern expected from mass fractionation. Isotope data on krypton released in several low temperature fractions are not shown in Table 2. The amount of krypton released at extraction temperatures 100°–400°C was too low for reliable isotopic analysis. The initial mass spectrometer signal for krypton in the 600° fraction was higher than that observed from any of the lower temperature fractions but, for reasons which are not understood,

Table 2. Krypton released from stepwise heating of lunar fines 15601,64.

| Temperature (°C) | $^{86}$Kr released ($\times 10^{-10}$ ccSTP/g) | $^{78}$Kr | $^{80}$Kr | $^{82}$Kr | $^{83}$Kr | $^{84}$Kr | $^{86}$Kr |
|---|---|---|---|---|---|---|---|
| 500° | 0.33 | — | — | 73 ± 1 | 70 ± 1 | 342 ± 4 | ≡100 |
| 700° | 2.5 | 2.35 ± 0.02 | 13.8 ± 0.1 | 69.4 ± 0.7 | 68.7 ± 0.5 | 336 ± 2 | ≡100 |
| 800° | 0.35 | — | 12.8 ± 0.2 | 67.9 ± 0.5 | 67.2 ± 0.5 | 331 ± 1 | ≡100 |
| 900° | 8.0 | 2.10 ± 0.01 | 13.3 ± 0.1 | 67.8 ± 0.2 | 67.8 ± 0.3 | 332 ± 1 | ≡100 |
| 1000° | 25 | 2.10 ± 0.01 | 13.3 ± 0.1 | 66.9 ± 0.3 | 66.6 ± 0.3 | 328 ± 1 | ≡100 |
| 1100° | 11 | 2.76 ± 0.03 | 15.1 ± 0.1 | 69.9 ± 0.2 | 70.2 ± 0.2 | 331 ± 1 | ≡100 |
| 1200° | 8.2 | 2.34 ± 0.02 | 13.7 ± 0.1 | 66.3 ± 0.3 | 67.6 ± 0.4 | 327 ± 1 | ≡100 |
| 1300° | 3.7 | 2.71 ± 0.04 | 14.5 ± 0.3 | 69.1 ± 0.5 | 70.8 ± 0.5 | 332 ± 2 | ≡100 |
| 1400° | 14 | 2.38 ± 0.03 | 13.7 ± 0.1 | 66.8 ± 0.4 | 67.4 ± 0.3 | 329 ± 1 | ≡100 |
| 1500° | 3.2 | 2.25 ± 0.02 | 13.4 ± 0.1 | 66.0 ± 0.3 | 66.3 ± 0.3 | 324 ± 1 | ≡100 |
| Atmosphere | — | 2.02 | 13.0 | 66.2 | 66.0 | 327 | ≡100 |
| AVCC | — | 1.92 | 12.6 | 64.9 | 65.1 | 323 | ≡100 |

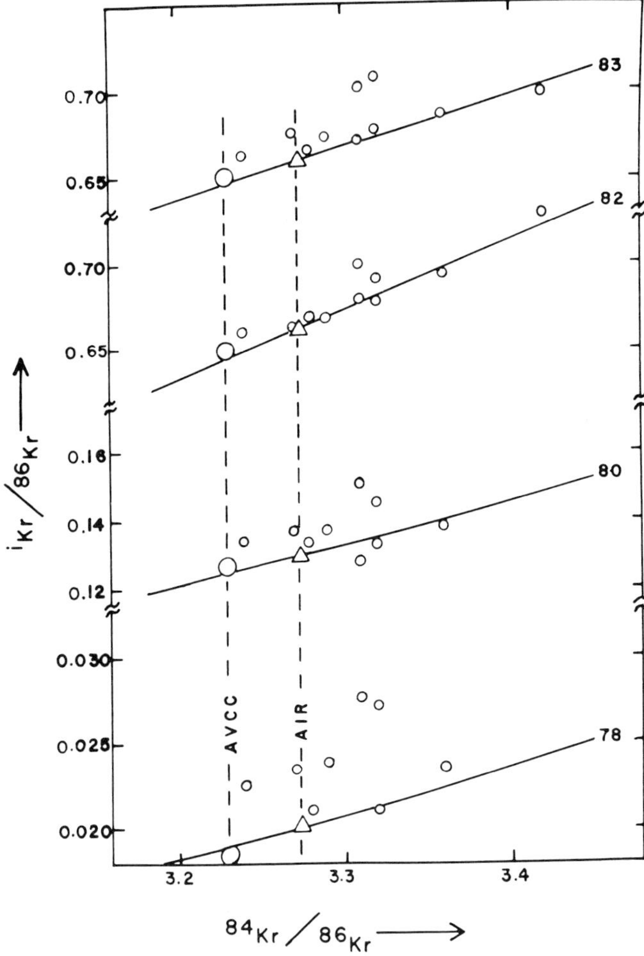


Fig. 3. Correlation of the krypton isotopes released from sample 15601,64. The solid lines passing through atmospheric krypton are defined by equation (1). The scatter of isotope ratios from the correlation lines is due to cosmogenic krypton.

the "peak heights" decreased very rapidly with time rendering isotopic analysis impossible.

Figure 3 shows a plot of the data from Table 2, with mass fractionation lines passing through atmospheric krypton in the manner described by equation (1). The isotopic ratios of krypton released from lunar fines at extraction temperatures $<1100°C$ and the isotopic ratios of krypton from average carbonaceous chondrites (AVCC Kr) cluster along these fractionation lines. The krypton isotopes released from lunar fines at extraction temperatures $\geq 1100°C$ do not follow the fractionation lines, suggesting the presence of a component of krypton isotopes from nuclear processes. Using the spallation yields reported by Marti and Lugmair (1971) for krypton isotopes in rock 12021, we can subtract the contribution of spallation-

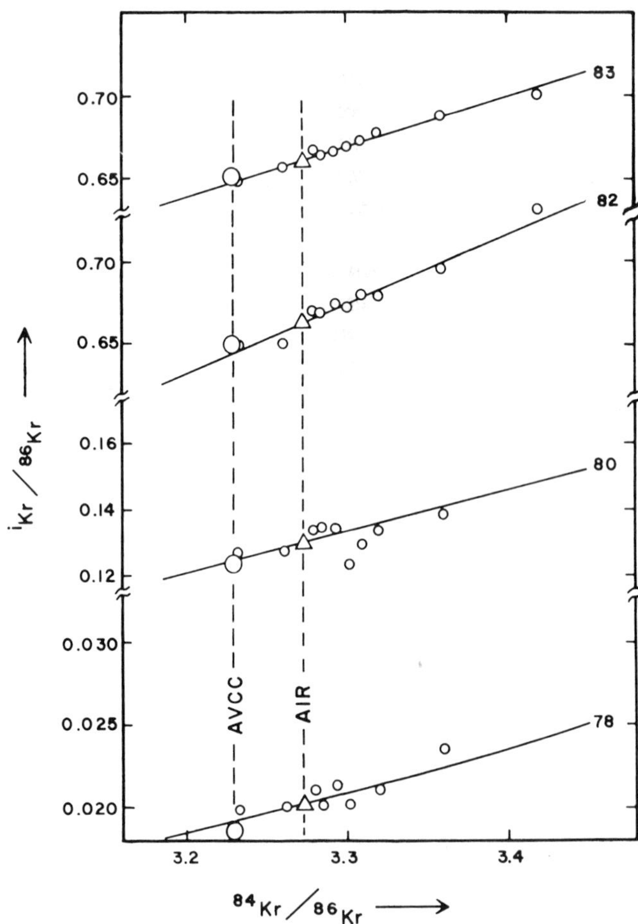


Fig. 4. Correlation of the spallation-free krypton isotopes released from 15601,64. The solar-wind implanted krypton isotopes correlate in the manner expected from isotopic fraction, lying on the same correlation line which relates atmospheric and AVCC krypton (Marti, 1967; Eugster *et al.*, 1967).

produced krypton isotopes in a manner analogous to that shown for subtracting the spallation-produced neon in Fig. 2.

The krypton isotope ratios, corrected for spallation, are shown in Fig. 4. From Table 2 it can be seen that the maximum release of krypton from our sample occurred at 1000°C. The krypton isotopes released in the 500°, 700°–900° fractions are enriched in the lighter isotopes in the manner expected from isotopic mass fractionation, and the krypton released at 1500°C resembles AVCC type krypton. Since all of the lighter noble gases show evidence for an enrichment of the light-mass isotopes in the temperature fractions preceeding the maximum release of that gas, we suggest that this enrichment of lighter krypton isotopes in the 500°, 700°–900° fractions results from isotopic mass fractionation. A similar low temperature release pattern of krypton isotopes from lunar soil 10084,48 has been reported earlier by Pepin *et al.* (1970), who suggested that a new type of spallation krypton might be released in the low temperature fractions.

*Xenon*

The abundance and isotopic composition of xenon released from lunar fines 15601,64 are shown in Table 3, together with the isotopic composition of xenon in the

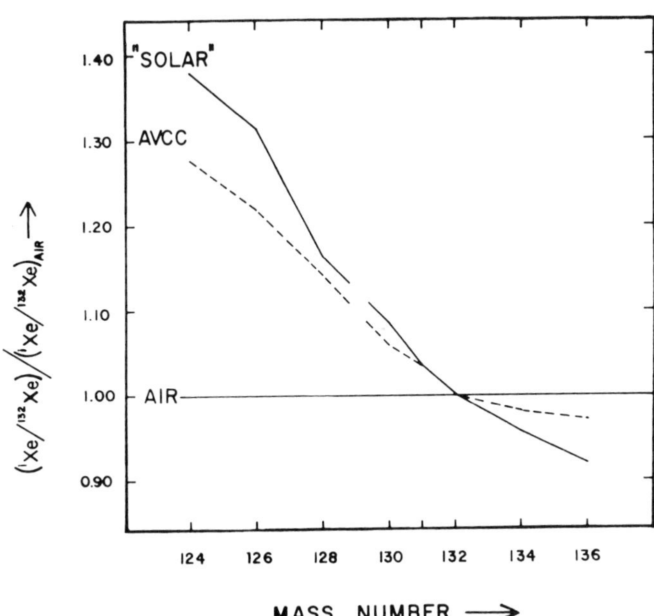


Fig. 5. A comparison of the abundances of xenon isotopes in average carbonaceous chondrites (Eugster *et al.*, 1967; Marti, 1967) and in solar-type xenon (Eberhardt *et al.*, 1971) with the abundance of xenon isotopes in the earth's atmosphere (Nier, 1950). Fractionation across the heavy isotopes $^{132-136}$Xe appears to be less steep than fractionation across the light isotopes $^{124-132}$Xe, and the $^{132}$Xe isotope is common to both fractionation patterns.

Table 3. Xenon released from stepwise heating of lunar fines 15601,64.

| Temperature (°C) | $^{132}Xe$ released $\times 10^{-10}$ ccSTP/g | $^{124}Xe$ | $^{126}Xe$ | $^{128}Xe$ | $^{129}Xe$ | $^{130}Xe$ | $^{131}Xe$ | $^{132}Xe$ | $^{134}Xe$ | $^{136}Xe$ |
|---|---|---|---|---|---|---|---|---|---|---|
| 400° | 0.04 | — | — | — | 99.8 ± 0.08 | 15.3 ± 0.2 | 79.4 ± 0.9 | ≡100 | 39.6 ± 0.6 | 33.6 ± 0.7 |
| 500° | 0.03 | — | — | — | 99.1 ± 1.6 | 15.6 ± 0.2 | 79.8 ± 1.0 | ≡100 | 39.4 ± 0.4 | 32.0 ± 0.5 |
| 700° | 0.27 | — | — | 9.35 ± 0.12 | 101.1 ± 1.3 | 15.9 ± 0.1 | 80.9 ± 1.0 | ≡100 | 38.5 ± 0.5 | 32.6 ± 0.4 |
| 800° | 0.18 | — | — | 7.24 ± 0.04 | 98.2 ± 0.5 | 15.0 ± 0.1 | 78.4 ± 0.4 | ≡100 | 38.4 ± 0.2 | 32.5 ± 0.2 |
| 900° | 1.2 | 0.487 ± 0.008 | 0.467 ± 0.005 | 8.48 ± 0.07 | 104.9 ± 0.5 | 16.5 ± 0.1 | 82.2 ± 0.5 | ≡100 | 37.3 ± 0.2 | 30.5 ± 0.2 |
| 1000° | 6.8 | 0.518 ± 0.003 | 0.526 ± 0.003 | 8.35 ± 0.04 | 103.4 ± 0.8 | 16.2 ± 0.1 | 82.4 ± 0.5 | ≡100 | 37.2 ± 0.2 | 30.3 ± 0.2 |
| 1100° | 6.2 | 1.00 ± 0.01 | 1.38 ± 0.01 | 9.71 ± 0.03 | 103.8 ± 0.4 | 17.2 ± 0.1 | 86.0 ± 0.3 | ≡100 | 36.9 ± 0.1 | 30.3 ± 0.1 |
| 1200° | 4.9 | 0.794 ± 0.009 | 1.06 ± 0.01 | 9.14 ± 0.10 | 103.2 ± 1.1 | 16.9 ± 0.1 | 84.3 ± 0.9 | ≡100 | 37.6 ± 0.4 | 31.0 ± 0.3 |
| 1300° | 1.6 | 0.899 ± 0.007 | 1.22 ± 0.01 | 9.42 ± 0.05 | 103.7 ± 0.5 | 17.0 ± 0.1 | 85.4 ± 0.6 | ≡100 | 37.2 ± 0.2 | 30.8 ± 0.1 |
| 1400° | 6.1 | 0.758 ± 0.006 | 0.959 ± 0.007 | 9.05 ± 0.06 | 104.2 ± 0.5 | 16.9 ± 0.1 | 84.3 ± 0.5 | ≡100 | 37.2 ± 0.2 | 30.4 ± 0.1 |
| 1500° | 1.7 | 0.692 ± 0.004 | 0.857 ± 0.008 | 8.92 ± 0.06 | 103.3 ± 0.7 | 16.6 ± 0.1 | 83.7 ± 0.5 | ≡100 | 37.2 ± 0.3 | 30.3 ± 0.2 |
| Atmosphere | — | 0.357 | 0.335 | 7.14 | 98.3 | 15.2 | 78.8 | ≡100 | 38.8 | 33.0 |
| AVCC | — | 0.452 | 0.406 | 8.09 | — | 16.1 | 81.5 | ≡100 | 38.1 | 32.0 |

atmosphere (Nier, 1950) and in average carbonaceous chondrites (Eugster *et al.*, 1967; Marti, 1967). We have used $^{132}$Xe as the reference isotope because this isotope is common to two different fractionation lines which seem to relate atmospheric and AVCC xenon (Reynolds, 1960a; Krummenacher *et al.*, 1962; Eugster *et al.*, 1967; Marti, 1967), as shown in Fig. 5. Also shown is the relationship between "solar" xenon, as estimated by Eberhardt *et al.* (1970) from analysis of xenon in lunar fines 10084,47, and atmospheric xenon. The abundance of the $^{129}$Xe isotope is not shown in Fig. 5 due to the existence of radiogenic $^{129}$Xe in meteorites (Reynolds, 1960b) and in the earth (Boulos and Manuel, 1971).

The greatest amount of $^{132}$Xe is released from lunar fines 15601,64 at 1000°C. The xenon released in the preceeding temperature fraction, 900°C, shows a greater enrichment of the lighter isotopes, $^{128-130}$Xe, and a greater depletion of the heavier isotopes, $^{134,136}$Xe, than does the xenon released at 1000°C. The $^{124,126}$Xe isotopes are not considered now because of the presence of a sizable contribution of spallation products in the 1000° fraction, as will be discussed later. Since the release pattern of xenon isotopes in the temperature fraction preceeding the maximum release of xenon parallels the release pattern observed for all other noble gases in the temperature fractions prior to the release of the maximum amount of each gas, it appears that isotopic fractionation may be responsible for some differences in the xenon released in our stepwise heating experiment.

Prior to analysis of gases released at 800°C, the spectrometer tube was accidently exposed to the atmosphere. Subsequent analyses of the lighter noble gases seemed to show no evidence for contamination with atmpspheric-type noble gases, and it was thought that analyses of these lighter noble gases from the 800° fraction would act to "scrub" the spectrometer of any heavier atmospheric gases. However, before letting the 800° fraction of xenon into the spectrometer we observed instrument background xenon equivalent to about $\frac{1}{5}$ of the xenon in the 800° fraction. Due to the possibility of atmospheric contamination in the 800° fraction of xenon, we will not include this temperature fraction in our discussion of the isotope release pattern of xenon.

The isotopic composition of xenon released at 400°, 500°, and 700°C appears to be enriched in the heavy isotopes. Isotopic ratios for xenon released at 600°C are not shown because of a rapid decrease of the spectrometer signal during analysis of these isotopes. No enrichment of heavy krypton isotopes was observed in low-temperature fractions but the 200° fraction of helium, the 300°–400° fractions of neon, and the 300°–500° fractions of argon are more enriched in the heavy isotopes than are the gases released in the next few higher temperature fractions. The reasons for this enrichment of heavy isotopes in low-temperature fractions is not clear, but may result from adsorption of atmospheric-type gases on the sample prior to analysis or to the selective loss of the lighter noble gas isotopes (from sites which are degassed at low laboratory extraction temperatures) during the time that the soil resided on the lunar surface.

In Table 4 we have shown the cumulative percent release of each noble gas as the extraction temperatures are increased. For extraction temperatures $\leq$500°C, the cumulative percent release of Xe and Kr exceeds that of the three lighter weight noble gases. This release pattern may be related to the enrichment of heavy isotopes

Table 4. Cumulative percent of each noble gas released from stepwise heating of lunar fines 15601,64.

| Temperature (°C) | $^{4}$He | $^{22}$Ne | $^{36}$Ar | $^{84}$Kr | $^{132}$Xe |
|---|---|---|---|---|---|
| 100° | 0.00031 | 0.00090 | 0.000021 | 0.0057 | 0.0094 |
| 200° | 0.00070 | 0.0040 | 0.00090 | 0.026 | 0.085 |
| 300° | 0.059 | 0.018 | 0.0030 | 0.111 | 0.157 |
| 400° | 0.128 | 0.036 | 0.0088 | 0.288 | 0.294 |
| 500° | 0.278 | 0.069 | 0.019 | 0.735 | 0.393 |
| 600° | 1.388 | 0.338 | 0.040 | 1.783 | 0.831 |
| 700° | 79.388 | 69.521 | 7.621 | 5.114 | 1.763 |
| 800° | 84.208 | 72.412 | 9.684 | 5.566 | 2.395 |
| 900° | 93.618 | 83.770 | 46.876 | 15.840 | 6.470 |
| 1000° | 99.488 | 95.835 | 87.585 | 48.363 | 29.992 |
| 1100° | 99.534 | 96.983 | 90.934 | 63.179 | 51.240 |
| 1200° | 99.698 | 98.138 | 93.533 | 73.670 | 67.958 |
| 1300° | 99.772 | 98.516 | 94.482 | 78.474 | 73.487 |
| 1400° | 99.993 | 99.809 | 98.981 | 95.893 | 94.318 |
| 1500° | 100 | 100 | 100 | 100 | 100 |

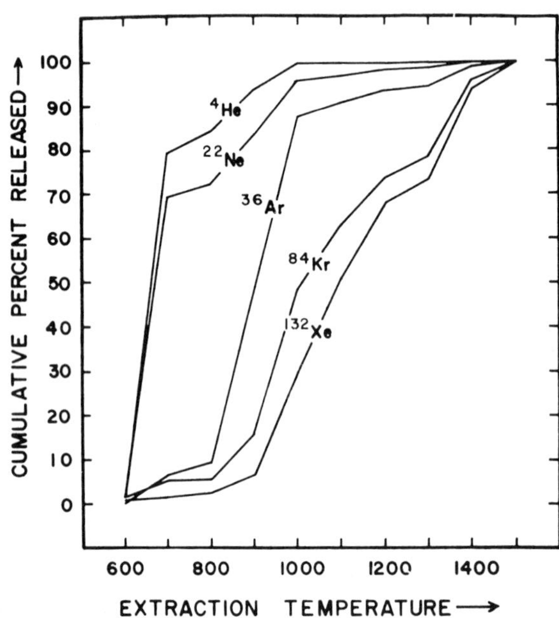


Fig. 6. The release pattern of noble gases from 15601,64 at extraction temperatures ≥600°C. The preferential release of light weight noble gases at low extraction temperatures is paralleled by isotopic fractionation across the isotopes of each noble gas.

observed in the low extraction temperatures, but the mechanism involved is not understood. For extraction temperatures >600°C, the gas release pattern clearly demonstrates the preferential release of the lighter noble gases. This is shown in Fig. 6, where the cumulative percent yield is plotted against extraction temperature.

The isotopic composition of xenon from Table 3 is shown in Fig. 7. Solid lines passing through atmospheric xenon fit the fractionation pattern defined by equation

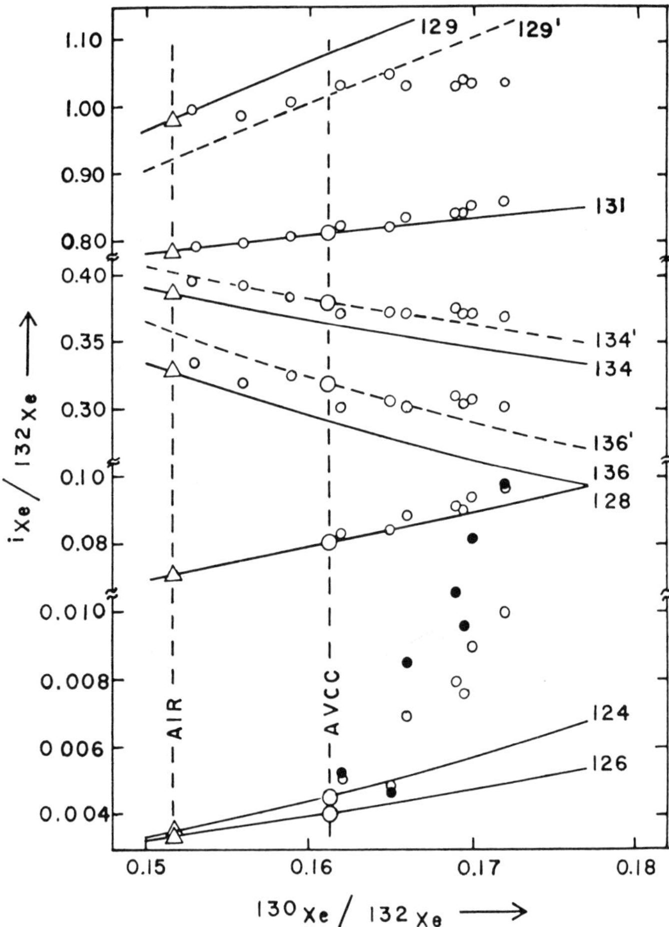


Fig. 7. Correlation of the xenon isotopes released from sample 15601,64. The solid lines passing through atmospheric xenon are defined by equation (1). The broken lines through AVCC $^{134}Xe/^{132}Xe$ and $^{136}Xe/^{132}Xe$ ratios also fit isotopic fractionation, and the broken line labeled 129′ would pass through atmospheric $^{129}Xe/^{132}Xe$ which had been corrected for radiogenic $^{129}Xe$ in air (Boulos and Manuel, 1971). The $^{126}Xe$ isotope is shown as darkened points. This isotope shows the largest scatter from the fractionation line, due to the large effect of spallation products on the $^{126}Xe/^{132}Xe$ ratio.

(1), and dashed fractionation lines are shown through $^{134}Xe$ and $^{136}Xe$ in AVCC xenon. The $^{124}Xe$ and $^{126}Xe$ isotopes show a large scatter due to the presence of spallation components, and the $^{126}Xe$ data are shown as dark points in order to distinguish them from $^{124}Xe$ and $^{128}Xe$. The dashed fractionation line labeled 129′ assumes that about 5.7% of atmospheric $^{129}Xe$ is due to radiogenic $^{129}Xe$ from the decay of extinct $^{129}I$ (Boulos and Manuel, 1971).

Using the spallation yields reported by Marti and Lugmair (1971), the contribution of spallation products to each xenon isotope has been subtracted in the manner employed for neon and krypton. For this correction we first used deviations of the points ($^{126}$Xe/$^{132}$Xe, $^{130}$Xe/$^{132}$Xe) from the fractionation line to compute the amount of spallation-produced $^{126}$Xe. The contribution of spallation to each xenon isotope was then calculated from its spallation relative to that of $^{126}$Xe. The xenon spectrum, corrected for spallation products, is shown in Fig. 8.

From the isotope correlations shown in Fig. 8 we note that there are only small fractionation effects in xenon released from our sample at extraction temperatures $\geq$900°C. The behavior of $^{129}$Xe, $^{136}$Xe, and $^{134}$Xe isotopes in the 400° and 500°

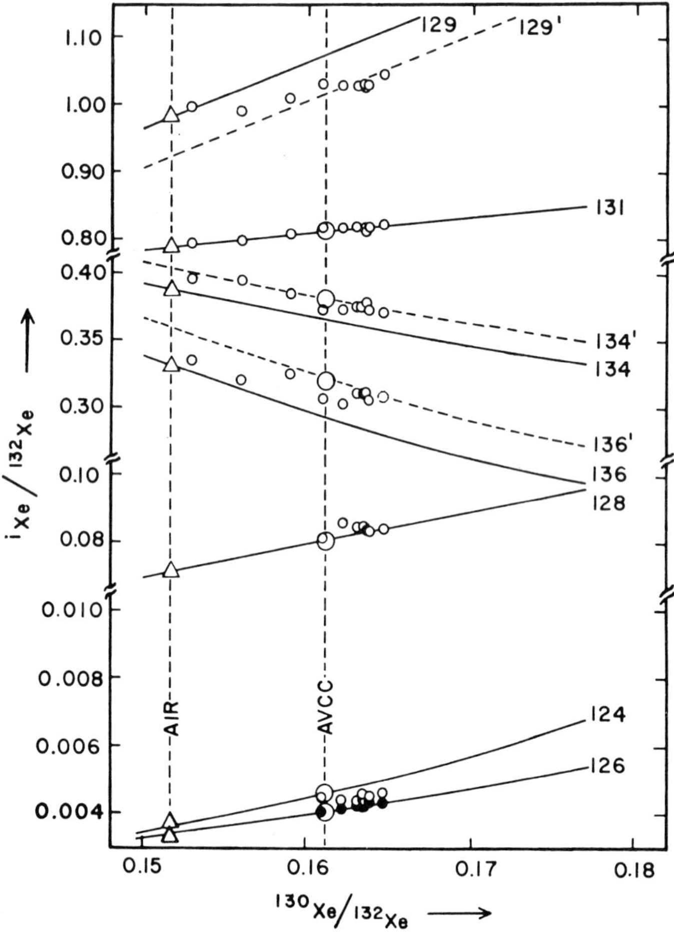


Fig. 8. Correlation of the spallation-free xenon isotopes from sample 15601,64. The heavy xenon isotopes are enriched above the atmospheric fractionation line, as are the heavy xenon isotopes in AVCC xenon. The $^{124}$Xe isotope appears to be depleted relative to fractionated atmospheric xenon.

fractions seems to better fit a mixture of atmospheric and AVCC-type xenon than the dashed fractionation lines. The $^{124}$Xe seems depleted relative to fractionated AVCC or atmospheric xenon in all temperature fractions. The apparent depletion of $^{124}$Xe may be due to (a) an excess of $^{126}$Xe in the solar wind, perhaps due to the $^{127}$I($\gamma$, $n\beta$)$^{126}$Xe reaction in the sun; (b) a depletion of $^{124}$Xe in the solar wind, perhaps due to the $^{124}$Xe($n$, $\gamma$)$^{125}$Xe reaction in the sun; or (c) a smaller $^{124}$Xe:$^{126}$Xe spallation yield in our sample than in lunar rock 12021.

The heavy xenon isotopes, $^{134-136}$Xe, released from our sample seem to better fit mass-fractionated AVCC xenon than mass-fractionated atmospheric xenon. Our results do not support the view of Marti (1969) that solar-wind xenon can be related to atmospheric xenon by mass fractionation, nor the view of Eberhardt *et al.* (1970) that solar-wind xenon confirms the presence of a fission component in AVCC xenon. To account for the enrichment of $^{136}$Xe and $^{134}$Xe in our sample above the solid fractionation line through atmospheric xenon by fission of a volatile, superheavy element (Anders and Heymann, 1969) would require unreasonably high abundances of superheavy elements in the early solar system. Kuroda and Manuel (1970) and Sabu and Manuel (1971) have previously discussed the difficulties encountered in explaining AVCC xenon by the fission hypothesis.

An alternate way of viewing the relationship between solar, AVCC, and atmospheric xenon has been presented by Kuroda and his associates (Kuroda and Manuel, 1970; Manuel *et al.*, 1970; Kuroda *et al.*, 1971; Kuroda, 1971). These authors have used the heaviest isotope, $^{136}$Xe, as a reference and compared the other isotopic ratios of xenon with the ratios calculated by assuming that variations in the $^{134}$Xe/$^{136}$Xe ratio result from fractionation. When the xenon data are examined in this manner one finds that xenon in both average carbonaceous chondrites and lunar soil appears enriched in the light isotopes, $^{124-132}$Xe, relative to fractionated atmospheric xenon. After subtracting the contribution of cosmogenic xenon from the excess $^{124-132}$Xe, Kuroda (1971) has suggested that the remaining excess $^{130}$Xe and $^{132}$Xe may de due to neutron capture processes, $^{129}$Xe($n$, $\gamma$)$^{130}$Xe and $^{131}$Xe($n$, $\gamma$)$^{132}$Xe, in the sun. We have normalized the isotopic ratios of xenon from lunar soil, 15601,64, to $^{136}$Xe and find that there remains an excess of $^{128}$Xe, $^{130}$Xe, and $^{131}$Xe after correcting for spallation and fractionation effects. Any excess $^{132}$Xe or $^{129}$Xe is very small and within the error limits of the measurement. The amount of excess $^{131}$Xe contains a large uncertainty due to the highly variable spallation yield of this isotope in lunar samples. However, the excess $^{128}$Xe and $^{130}$Xe, which occurs in a ratio of approximately 1:1 in all temperature fractions, is outside the error limits of the measurement or the spallation yields assumed. These anomalies may contain quantitative information on nuclear processes which have occurred in the evolution of the sun.

Thus, a comparison of our results on xenon from lunar soil with the results of earlier analyses of xenon in meteorites and in the atmosphere shows that either (a) the xenon in chondrites and the xenon implanted in lunar soil contains an enrichment of $^{134}$Xe and $^{136}$Xe from some unknown source, or (b) the xenon in chondrites and in lunar fines contains the products of nuclear reactions in the sun, or (c) the xenon in the earth's atmosphere came from a different nucleosynthesis source than the xenon in meteorites and in the solar wind. Since the other elements of lunar, meteoritic, terrestrial, and solar material appear to have been derived from a common batch of chemicals, it appears that the earth should have had an original inventory of xenon isotopes similar to that in meteoritic and lunar material. If the observed xenon in meteorites and in lunar fines has not been appreciably altered by nuclear reactions in the sun, then the possibility that atmospheric xenon is from some extra-solar source [item (c) above] could be experimentally verified if AVCC or solar-type xenon were found within the earth. Although recent analyses in our laboratory (Boulos and Manuel, 1971, 1972) of terrestrial xenon from deep gas wells have shown the decay products of two now extinct nuclides, $^{129}I$ and $^{244}Pu$, the nonradiogenic xenon isotopes in these wells more closely resembled atmospheric xenon than AVCC or solar-type xenon.

## CONCLUSIONS

The results of this investigation on the abundance and isotopic composition of noble gases released by stepwise heating of lunar soil, sample 15601,64, lead us to the following conclusions:

(1) The fractional amount of each gas released above 600°C shows preferential release of the light weight noble gases.

(2) Due to fractionation effects the isotopic composition of solar wind implanted gases cannot be unambiguously identified from analysis of gases in lunar soils. However, we estimate the isotopic composition of the noncosmogenic noble gases trapped in our sample to be $^{4}He:^{3}He = 2280$; $^{20}Ne:^{21}Ne:^{22}Ne = 13.0:0.033:1.00$; $^{36}Ar:^{38}Ar:^{40}Ar = 1.00:0.185:\leq 0.558$; $^{78}Kr:^{80}Kr:^{82}Kr:^{83}Kr:^{84}Kr:^{86}Kr = 0.021:0.132:0.665:0.665:3.29:1.00$; $^{124}Xe:^{126}Xe:^{128}Xe:^{129}Xe:^{130}Xe:^{131}Xe:^{132}Xe:^{134}Xe:^{136}Xe = 0.00445:0.00425:0.0824:1.03:0.163:0.817:1.00:0.374:0.305$.

(3) At very low extraction temperatures there appears to be a gas enriched in the heavy isotopes and in the heavy noble gases. This is due to adsorption of atmospheric gases or to selective release of the light isotopes from these low-temperature sites during the period the sample was on the moon.

(4) The isotopic composition of the noble gases released from lunar fines at different extraction temperatures shows variations due to mass fractionation. Differences in the isotopic composition of He, Ne, Ar, and Kr in meteorites, in the earth's atmosphere and in the solar wind can also be accounted for by mass fractionation and reasonable nuclear reactions. The results of this analysis of noble gases from 15601,64 and recent results from our studies of noble gases in terrestrial samples (Boulos and Manuel, 1970, 1972) and meteorites (Srinivasan and Manuel, 1971; Hennecke and Manuel, 1971) substantiate the view that mass fractionation is responsible for many of the isotopic anomalies of noble gases in nature.

(5) The xenon isotopes released from our sample cannot be accounted for by spallation plus fractionation of atmospheric xenon. Unreasonably high abundances of superheavy elements would be required to account for the "excess" $^{134}Xe$ and $^{136}Xe$ in solar-type xenon relative to fractionated atmospheric xenon.

(6) The relationship between atmospheric, solar, and AVCC xenon is complex.

Additional studies on the isotopic composition of xenon in a wide variety of lunar, meteoritic, and terrestrial samples are needed to decipher the xenon record of events in the early history of the solar system.

*Acknowledgments*—The authors are grateful to Mr. R. D. Beaty, Mr. P. H. Ruehle, Mr. R. L. Stranghoener and Ms. Phyllis Johnson for assistance with calculations and preparation of the manuscript and figures. This research was supported by funds from the University of Missouri-Rolla and the National Aeronautics and Space Administration, NASA NGR 26-003-057.

## References

Anders E. and Heymann D. (1969) Elements 112 to 119: were they present in meteorites? *Science* **164,** 821–823.

Aston F. W. (1933) *Mass-Spectra and Isotopes.* Edward Arnold and Co., London.

Black D. C. (1969) Isotopic variations in trapped meteoritic argon. *Meteoritics* **4,** 260.

Black D. C. (1970) Trapped helium-neon isotopic correlations in gas rich meteorites and carbonaceous chondrites. *Geochim. Cosmochim. Acta* **35,** 230–235.

Black D. C. and Pepin R. O. (1969) Trapped neon in meteorites, II. *Earth Planet. Sci. Lett.* **6,** 395–405.

Boulos M. S. and Manuel O. K. (1970) Isotopic anomalies of xenon and krypton in deep gas wells. Abst. 696 of Combined S.E.-S.W. Meeting Am. Chem. Soc., New Orleans, Lousiana, Dec. 2–4,

Boulos M. S. and Manuel O. K. (1971) The xenon record of extinct radioactivities in the earth. *Science* **174,** 1334–1336.

Boulos M. S. and Manuel O. K. (1972) Extinct radioactive nuclides and production of xenon isotopes in natural gas. *Nature Phys. Sci.* **235,** 150–152.

Brown H. (1949) *The Atmosphere of the Earth and the Planets.* University of Chicago Press, Chicago.

Canalas R. A., Alexander E. C. Jr., and Manuel O. K. (1968) Terrestrial abundance of noble gases. *J. Geophys. Res.* **73,** 3331–3334.

Eberhardt P., Eugster O., and Marti K. (1965) A determination of the isotopic composition of atmospheric neon. *Z. Naturforsch.* **20,** Ser. a, 623.

Eberhardt P., Geiss J., and Grögler N. (1965a) Ueber die verteilung der Uredelgase im Meteoriten Khor Temiki. *Tschermak's mineral petrogr. Mitt.* **10,** 1–4.

Eberhardt P., Geiss J., and Grögler N. (1965b) Further evidence on the origin of trapped gases in the meteorite Khor Temiki. *J. Geophys. Res.* **70,** 4375–4378.

Eberhardt P., Geiss J., Graf H., Grögler N., Krähenbühl U., Schawaller H., Schwarzmüller J., and Stettler A. (1970) Trapped solar wind noble gases, exposure age and K/Ar-age in Apollo 11 lunar fine material. *Proc. Apollo 11 Lunar Sci. Conf., Geochim. Cosmochim. Acta* Suppl. 1, Vol. 2, pp. 1037–1070. Pergamon.

Eugster O., Eberhardt P., and Geiss J. (1967) Krypton and xenon isotopic composition in three carbonaceous chondrites. *Earth Planet. Sci. Lett.* **3,** 249–257.

Gerling E. K. and Levskii L. K. (1956) On the origin of the rare gases in stony meteorites. *Dokl. Akad. Nauk. SSSR* **110,** 750–754.

Hennecke E. W. and Manuel O. K. (1971) Mass fractionation and the isotopic anomalies of xenon and krypton in ordinary chondrites. *Z. Naturforsch.* **16,** Ser. a, 1980–1986.

Heymann D. and Yaniv A. (1970) $Ar^{40}$ anomaly in lunar samples from Apollo 11. *Proc. Apollo 11 Lunar Sci. Conf., Geochim. Cosmochim. Acta* Suppl. 1, Vol. 2, pp. 1261–1267. Pergamon.

Heymann D., Yaniv A., Adams J. A. S., and Fryer G. E. (1970) Inert gases in lunar samples. *Science* **167,** 555–558.

Hintenberger H., Koenig H., and Waenke H. (1962) Uredelgase im Meteoriten Breitscheid. *Z. Naturforsch.* **17,** Ser. a, 306–309.

Hintenberger H., Vilcsek E., and Waenke H. (1965) Ueber die Isotopenzusammensetzung und ueber den Sitz der leichten Uredelgase in Steinmeteoriten. *Z. Naturforsch.* **20,** Ser. a, 939–945.

Hohenberg C. M., Davis P. K., Kaiser W. A., Lewis R. S., and Reynolds J. H. (1970) Trapped and cosmogenic rare gases from stepwise heating of Apollo 11 samples. *Proc. Apollo 11 Lunar Sci. Conf., Geochim. Cosmochim. Acta* Suppl. 1, Vol. 2, pp. 1283–1309. Pergamon.

Koenig H., Keil K., Hintenberger H., Wlotzka F., and Begemann F. (1961) Untersuchungen an Steinmeteoriten mit extrem hohen Edelgasgehalt, I. Der Chondrite Pantar. *Z. Naturforsch.* **16,** Ser. a, 1124–1130.

Krummenacher D., Merrihue C. M., Pepin R. O., and Reynolds J. H. (1962) Meteoritic krypton and barium versus the general isotopic anomalies in meteoritic xenon. *Geochim. Cosmochim. Acta* **26,** 231–249.

Kuroda P. K. (1971) Temperature of the sun in the early history of the solar system. *Nature* **230,** 40–42.

Kuroda P. K. and Manuel O. K. (1970) Mass fractionation and isotope anomalies in neon and xenon. *Nature* **227,** 1113–1116.

Kuroda P. K., Reynolds M. A., Sakamoto K., and Miller D. K. (1971) Isotope anomalies in rare gases. *Nature* **230,** 42.

LSPET (Lunar Sample Preliminary Examination Team) (1969) Preliminary examination of lunar samples from Apollo 11. *Science* **165,** 1211–1227.

LSPET (Lunar Sample Preliminary Examination Team) (1972) The Apollo 15 lunar samples: A preliminary description. *Science* **175,** 363–375.

Manuel O. K. (1967) Noble gases in the Fayetteville meteorite. *Geochim. Cosmochim. Acta* **31,** 2413–2431.

Manuel O. K. (1970) Isotopic anomalies of noble gases from mass fractionation. *Meteoritics* **5,** 207–208.

Manuel O. K., Wright R. J., Miller D. K., and Kuroda P. K. (1970) Heavy noble gases in Leoville: The case for mass fractionated xenon in carbonaceous chondrites. *J. Geophys. Res.* **75,** 5693–5701.

Marti K. (1967) Isotopic composition of trapped krypton and xenon in chondrites. *Earth Planet. Sci. Lett.* **3,** 243–248.

Marti K. (1969) Solar-type xenon; a new isotopic composition of xenon in the Pesyanoe meteorite. *Science* **166,** 1263–1265.

Marti K. and Lugmair G. W. (1971) $Kr^{81}$–Kr and Kr–$Ar^{40}$ ages, cosmic-ray spallation products and neutron effects in lunar samples from Oceanus Procellarum. *Proc. Second Lunar Sci. Conf., Geochim. Cosmochim. Acta* Suppl. 2, Vol. 2, pp. 1591–1605. MIT Press.

Nief G. (1960) (as reported in isotopic abundance ratios given for reference samples stocked by the National Bureau of Standards) *NBS* Tech. Note 51, F. Mohler, ed.

Nier A. O. (1950) A redetermination of the relative abundances of the isotopes of krypton, rubidium, xenon, and mercury. *Phys. Rev.* **97,** 450–454.

Pepin R. O. (1967) Trapped neon in meteorites. *Earth Planet. Sci. Lett.* **2,** 13–18.

Pepin R. O. (1968) Neon and xenon in carbonaceous chondrites. In *Origin and Distribution of the Elements* (editor L. H. Ahrens), Pergamon.

Pepin R. O., Nyquist L. E., Phinney D., and Black D. C. (1970a) Isotopic composition of rare gases in lunar samples. *Science* **167,** 550–553.

Pepin R. O., Nyquist L. E., Phinney D., and Black D. C. (1970b) Rare gases in Apollo 11 lunar material. *Proc. Apollo 11 Lunar Sci. Conf., Geochim. Cosmochim. Acta* Suppl. 1, Vol. 2, pp. 1435–1454. Pergamon.

Reynolds J. H. (1956) High sensitivity mass spectrometer for noble gas analysis. *Rev. Sci. Instrum.* **27,** 928–934.

Reynolds J. H. (1960a) Isotopic composition of primordial xenon. *Phys. Rev. Lett.* **4,** 351–354.

Reynolds J. H. (1960b) Determination of the age of the elements. *Phys. Rev. Lett.* **4,** 8–10.

Sabu D. D. and Manuel O. K. (1971) Superheavy elements: were they present in meteorites? *Trans. Mo. Acad. Sci.* **5,** pp. 16–21.

Signer P. and Suess H. E. (1963) Rare gases in the sun, in the atmosphere, and in meteorites. In *Earth Science and Meteoritics* (editors J. Geiss and E. D. Goldberg), North Holland.

Srinivasan B. and Manuel O. K. (1971) On the isotopic composition of trapped helium and neon in carbonaceous chondrites. *Earth Planet. Sci. Lett.* **12,** 282–286.

Suess H. E. (1949) Die Haufigkeit der Edelgase auf der Erde und im Kosmos. *J. Geolog. Res.* **57,** 600–607.

Suess H. E., Waenke H., and Wlotzka F. (1964) On the origin of gasrich meteorites. *Geochim. Cosmochim. Acta* **28,** 595–607.

Zähringer J. (1962) Isotopie-Effekt und Häufigkeiten der Edlegas in Steinmeteoriten und auf der Erde. *Z. Naturforsch.* **17,** Ser. a, 460–471.

Zähringer J. and Genter W. (1960) Uredelgase in einigen Steinmeteoriten. *Z. Naturforsch.* **15,** Ser. a, 600–602.

Proceedings of the Third Lunar Science Conference
(Supplement 3, *Geochimica et Cosmochimica Acta*)
Vol. 2, pp. 1947–1966
The M.I.T. Press, 1972

# Thermal release of helium, neon, and argon from lunar fines and minerals

H. BAUR, U. FRICK, H. FUNK, L. SCHULTZ, and P. SIGNER

Swiss Federal Institute of Technology,
Sonneggstr. 5, 8006 Zürich, Switzerland

**Abstract**—He, Ne, and Ar from (1) grain-size fractions of bulk lunar soils (10084,48; 14163,166; 15101,88; 15321,54) and (2) sieved mineral separates of lunar soil 14163,166 were studied in stepwise and linear heating experiments. The gases were monitored continuously during release in a specially constructed system. The results are compared to simulation experiments and model predictions. The release patterns of He, Ne, and Ar from different grain size fractions of bulk soils appear to be quite similar. The He, Ne, and Ar retentivities of the separated minerals are different: (in increasing order) plagioclase, magnetic glass, pyroxene. Diffusive alteration of the implantation profiles is detected and correlates with the degree of retentivity. Comparison of simple model calculations with release patterns of gases implanted as ions with energies in the keV region has met with only limited success in that certain portions, but not the totality, of the observed release profiles can be described. Particularly intractable in this regard are (1) the low temperature enhancement observed in most release patterns and (2) the shape of the high temperature release, which may (in some cases) be an earmark of release of gas from greater than normal penetration depths. Only minor amounts of $^{40}Ar$ were detected in the low temperature release range, a feature that will require reexamination of theories which ascribe "orphan" $^{40}Ar$ to a low energy retrapping of $^{40}Ar$ from the lunar atmosphere. Detailed consideration of the results have led to a search for additional sources of "orphan" $^{40}Ar$.

## INTRODUCTION

SEVERAL INVESTIGATORS OF lunar material carried out stepwise heating experiments to resolve different gas components via isotope correlation diagrams (Hohenberg *et al.*, 1970; Marti *et al.*, 1970; Pepin *et al.*, 1970; Davis *et al.*, 1971). In connection with differential thermal analysis experiments gas releases for linear temperature increase have been reported by Hanneman (1970), Wachi *et al.* (1971), and Gibson and Johnson (1971). Concentration profiles within the grains under investigation have been studied in etching experiments (Eberhardt *et al.*, 1970; Hintenberger *et al.*, 1970; Kirsten *et al.*, 1970, 1971), in experiments with different grain sizes (Eberhardt *et al.*, 1970; Hintenberger *et al.*, 1970, 1971; Kirsten *et al.*, 1970, 1971; Heymann and Yaniv, 1970), and in experiments on single crystals using techniques with high resolution (Kirsten *et al.*, 1970, 1971; Megrue and Steinbrunn, 1972).

Our experiments aim at the determination of concentration profiles of the noble gases in lunar fines. Conclusions can then be drawn about origins of noble gas components and the history of the samples.

Two analytical techniques have been used in the present investigation: (1) Determination of effective activation energies in stepwise heating experiments with rapid temperature changes between steps (Huneke *et al.*, 1969; Nyquist *et al.*, 1972). (2) Linear heating experiments, to obtain refined release patterns as compared to stepwise heating experiments.

The correlation of the release patterns with the concentration profiles depends on the complex diffusion properties in an unevenly radiation-damaged and possibly shock-affected target. Furthermore, the present-day concentration profiles, particularly of the light noble gases, may differ from the initial, instantaneous implantation profiles. Possible erosion, sputtering during exposure, and diffusion may alter the initial concentration profiles. Hence, a straightforward interpretation of our data can not yet be given. Further experimental investigations and also extensive simulation experiments are needed.

A particular aspect of this investigation is the resolution of a possible retrapped argon component (Heymann and Yaniv, 1970, 1971; Eberhardt *et al.*, 1970) from the radiogenic argon of *in situ* decay of potassium. We have, therefore, separated minerals from Apollo 14 soil to obtain samples of homogeneous diffusion characteristics and varying potassium content. The study of the mineral separates gives information about the noble gas retentivity in different minerals.

## EXPERIMENTAL PROCEDURE

### *Sample preparation*

The following lunar samples ($< 1000\ \mu$) were available to us:

(1) Apollo 11 soil 10084,48 (0.89 g)
(2) Apollo 14 soil 14163,166 (13.98 g)
(3) Apollo 14 soil 14240,15 (0.054 g)
(4) Apollo 15 soil 15101,88 (1.0 g)
(5) Apollo 15 soil 15231,54 (1.0 g)

In addition, grain size fractions from the dark material of the chondrite Weston were investigated.

To reduce the influence of different grain sizes on the concentration of noble gases, we investigated certain grain size fractions. For what we call "bulk sample analyses" we chose the $< 25\ \mu$ fraction. Samples from this fraction have high gas concentrations, requiring the use of small sample weights. The coarser grains were preserved for mineral separation.

Grain size fractions from the lunar soils were obtained by sieving in acetone. From 14163 a mineral separation was carried out on a grain size fraction of 30–48 $\mu$. This size fraction contains few lithic grains and can still be handled relatively easily during the separation process. Repeated passes through a covered Frantz magnetic separator gave a clean feldspar fraction. For further separation the plagioclase component was treated with heavy liquids. The fraction with $\rho < 2.65$ g/cm$^3$ (90% of the plagioclases) is denoted as *plagioclase A*. Single crystals were examined by x-ray investigations and identified as bytownite.

The magnetic fraction richest in pyroxenes (*pyroxene A*) was treated with heavy liquids ($\rho > 3.2$ g/cm$^3$) and further refined by handpicking. Optical examinations showed that this fraction contains at most 10% impurities (olivine, ilmenite, coated feldspars).

The most magnetic fraction with density smaller than 3.2 g/cm$^3$ is denoted *magnetic glass*, because this sample consisted predominantly of black glass fragments. However, x-ray examinations showed that the sample also contains various mineral fragments. This sample is therefore considered to be rather ill-defined. To enhance the ratio of *in situ*-produced radiogenic $^{40}$Ar to the surface-correlated $^{40}$Ar, a feldspar fraction of grain size 200–350 $\mu$ was crushed and sieved. The 30–48 $\mu$ fraction is called *plagioclase B*.

### *Analytical procedure*

The analytical system consists of a diffusion oven connected to two 6″, 60° single-focusing mass spectrometers (Fig. 1). The whole system is operated under static vacuum conditions. Up to 6 samples,

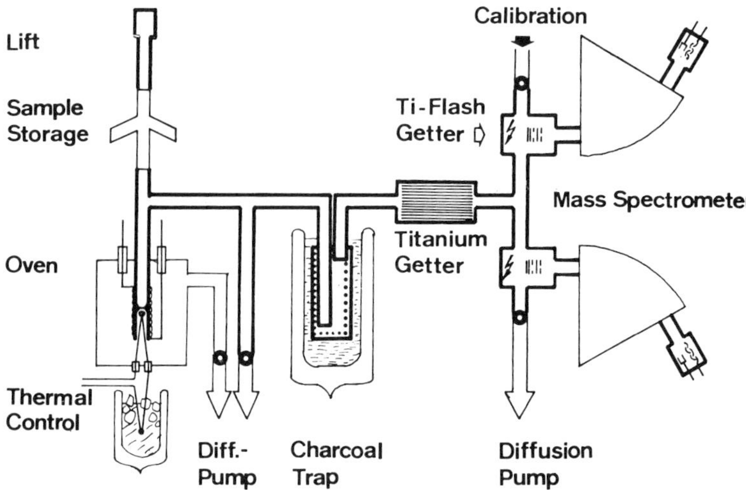


Fig. 1. Outline of the analytical system used for stepwise and linear heating experiments. The purified nobel gases enter both spectrometers. Each spectrometer is used to record two isotopes continuously under static conditions. The accumulated gases are pumped periodically.

wrapped in Ni-foil for magnetic maneuverability, are stored in glass fingers from where they can be dropped individually into the crucible. After a degassing analysis, the sample can be lifted back into the glass finger by means of a magnetic elevator system. The furnace is resistance heated and has a fast thermal reaction. Only the inside of the crucible belongs to the sample vacuum system. The temperature is determined by a thermocouple welded into the bottom of the crucible close to the location of the sample. The emf of the thermocouple is used to control the power applied to the heater by means of a proportional-differential control unit. For linear heating analyses a constant temperature increase (100°/h) is approximated by comparing the emf of the thermocouple to a linearly increasing reference voltage. The linearity of the emf increase is maintained to better than 1%. For stepwise heating the temperature of the oven is raised over a temperature interval of 100°C in less than one minute without overshooting the preset value. The constancy of the temperature is controlled to better than 0.5°C. The absolute temperature of the oven is believed to be known within 1%. The released gases pass through a charcoal trap and a cold titanium getter to the two statically operated mass spectrometers. Ti flash getters are not used during the runs. Both mass spectrometers are equipped with independent multiplier and Faraday detection systems. The sensitivities of the system were determined before and after an analysis. They remained constant within 3% over a period of four months. The detection limits, defined by the level of noise of the amplification system were:

$$^{4}\text{He:}\ 0.03 \times 10^{-10}\ \text{ccSTP}$$
$$^{20}\text{Ne:}\ 0.01 \times 10^{-10}\ \text{ccSTP}$$
$$^{40}\text{Ar:}\ 0.01 \times 10^{-10}\ \text{ccSTP}.$$

Blanks for $^{3}$He, $^{20}$Ne, and $^{22}$Ne were fairly constant. Blank rates for $^{4}$He, $^{36}$Ar, and $^{40}$Ar were of the order of 0.1, 0.01, and $2 \times 10^{-10}$ ccSTP/min, respectively.

In each run only 4 isotopes were registered continuously, either $^{3}$He, $^{4}$He, $^{20}$Ne, and $^{22}$Ne or $^{4}$He, $^{22}$Ne, $^{36}$Ar, and $^{40}$Ar. The two particular combinations were chosen since $^{40}Ar^{++}$ interferes with $^{20}Ne^{+}$. In the case of the analyses of He and Ne the charcoal trap is cooled with liquid nitrogen. If argon is analyzed, the trap is cooled with a $CO_2$-acetone mixture. To measure the 6 isotopes two independent runs were necessary. In this way $^4He$ and $^{22}Ne$ were determined twice and the reproducibility of the data could be checked. The agreement of two corresponding release patterns is excellent. The deviations are less than 1% and can not be resolved in our graphical representations.

In a programmed sequence with an automatic scanning device two mass peaks per minute were registered in each mass spectrometer. In linear heating experiments properly averaged readings were taken every 5 minutes. Thus, a release pattern between 100° and 900°C is documented by 97 data points for each isotope.

## Results and Discussion

### *Simulation experiments*

Noble gases in lunar soils consist predominantly of trapped solar particles. If the trapped gases have been injected with different energies, then one would expect that the release patterns reflect the energy of implantation. In a simulation experiment we bombarded an aluminum foil with $^4He$ ions of different energies and then observed the release patterns in a stepwise heating experiment (Figs. 2a, b). The temperature of maximum release correlates indeed with the energy of implantation.

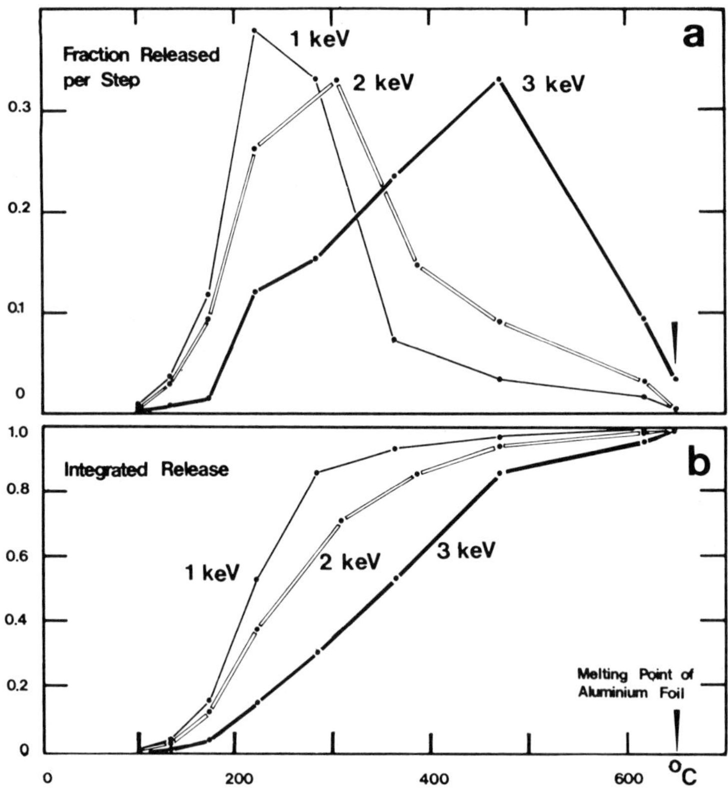


Fig. 2. Release of $^4He$ implanted in aluminium foil with different implantation energies. (a) Shows the release rates in stepwise heating; (b) the corresponding integral gas loss.

In another experiment, quartz crystals were doped with helium in two ways: (1) by 2.5 keV ion bombardment and (2) by pressure impregnation at 800°C and 3.5 atm for up to one month. For the latter case we assume a saturation concentration with a rectangular concentration profile (Funk *et al.*, 1971). The results (Fig. 3) are quite unanticipated and are presented here as an illustration of the complexity of the diffusion processes.

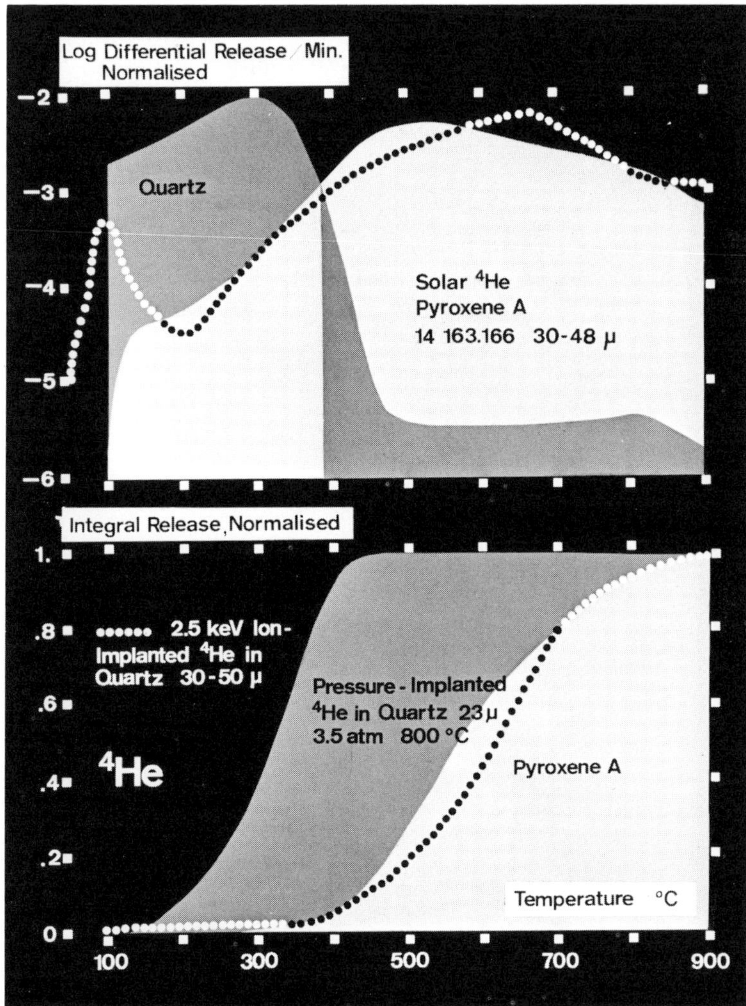


Fig. 3. Comparison of $^4$He release from doped quartz (ion bombarded and pressure introduced, respectively). Note the drastic difference of the release pattern for ion-implanted and pressure-introduced He from quartz, and the similarity for ion-implanted He from quartz and He from lunar pyroxene above 200°C.

*Activation energies*

The release rates and effective activation energies as a function of temperature for He and Ne of 10084 (grain size: <25 μ, 25–42 μ, and 64–100 μ) and 14163 (<25 μ) are shown in Fig. 4. Within the limits of errors a dependence on grain size is not detectable.

Therefore, the values are shown in a shaded area. For comparison, the release of the trapped gases from material of Weston obtained in a similar experiment (grain size: 1–5 μ and 48–100 μ) is shown in Fig. 4 (Schultz *et al.*, 1971). The activation energy increases with temperature from 15 kcal/mol to 60 kcal/mol for $^4$He. The effective activation energies of $^{20}$Ne show a similar picture; they increase from 20 kcal/mol to 80 kcal/mol. For a grain size fraction of 14163 a linear heating with two different heating rates were carried out to obtain activation energies by a different method (Fechtig and Kalbither, 1966). The calculated values of 40 kcal/mol for $^4$He and 55

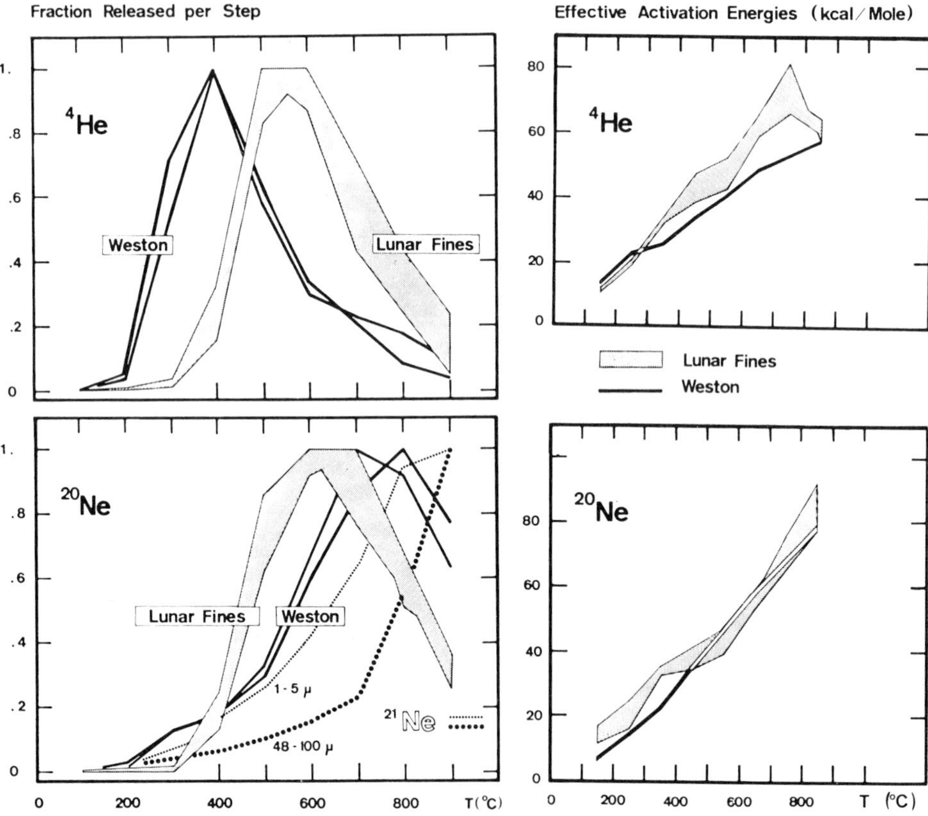


Fig. 4. Comparison of release patterns and effective activation energies obtained from stepwise heating of different grain sizes samples of lunar soils (10084: <25 μ, 25–42 μ, 64–100 μ; 14163: <25 μ, 64–100 μ) and the noble gas rich meteorite Weston (1–5 μ, 48–100 μ). Note the similarity of the release patterns and activation energies for all lunar soil grain size fractions.

kcal/mol for $^{20}Ne$ agree very well with those found at the temperature of maximal gas release by stepwise heating. Furthermore, these values are in agreement with recently published values by Kirsten *et al.* (1972).

The absence of a grain-size effect in the release pattern of noble gases from lunar fines is an additional proof for the surface correlation of the trapped gases. In Weston, however, a grain-size effect is observed for the spallogenic $^{21}Ne$ because spallogenic $^{21}Ne$ is homogeneously distributed throughout the grains. The shift of the maximum in the release pattern to higher temperatures is due to diffusion from larger grains. The maximum release rate for trapped $^{20}Ne$ from Weston appears about 100° later than that for lunar material, but for helium it is the other way around. The $^{4}He/^{20}Ne$ ratio of trapped solar gases in Weston is 455 (Schultz *et al.*, 1971) and is comparable to the ratios found in the solar wind from the SWC experiments (i.e., Bühler *et al.*, 1972). Assuming similar diffusion characteristics for He and Ne in Weston and lunar fines, the release pattern of He in lunar fines can be explained by loss of helium during the lunar day at temperatures around 120°C. We note that Weston shows a loss of about 1.5% of $^{4}He$ in 45 min at 200°C.

*Retention properties*

Our investigation offers three sets of data pertinent to the evaluation of gas retention properties: (1) total gas concentrations, (2) gas retained by the minerals throughout the degassing sequence, and (3) release patterns.

The gas amounts released between 100° and 900°C and those from the subsequent total extractions are given in Table 1. All concentrations are corrected for blanks. The errors of the concentration determinations from the diffusion runs are believed to be smaller than 20%; those for the total extractions are smaller than 5% with exception of $^{40}Ar$ from samples No. 1, 2, and 10. A comparison of these data with the results from other laboratories is only possible in a few cases. This is because (1) of the dependence of the concentrations of noble gases on grain size, and (2) small samples of unsieved bulk material may not be representative for the total sample. Nevertheless, our results agree well with those given by Eberhardt *et al.* (1970) and Heymann *et al.* (1972).

The $^{20}Ne/^{36}Ar$ ratios of the total gas concentrations in the grain size separates of bulk soils agree to better than 20%. However, the $^{4}He$ in the 10084 is two to three times more abundant than in the Apollo 14 and 15 soils. This may reflect different mineralogic compositions or different conditions during and/or after irradiation. Since gas retention depends on mineral types the total gas concentration of mineral separates of 14163 (plagioclase A, magnetic glass, and pyroxene A) should not contain equal amounts of solar gases. The $^{4}He$ and $^{20}Ne$ contents (Table 1) increase in the sequence plagioclase A, magnetic glass, pyroxene A. The $^{4}He$ concentrations in the plagioclase are lower by a factor of 30 than those of the pyroxene. The normalized ratios of the trapped gases in the minerals with respect to solar abundance (Aller, 1961) are shown in Fig. 5. The abundances of the trapped gases in ilmenite (Eberhardt *et al.*, 1970) are also shown. Pyroxene is clearly the most retentive mineral investigated here, but even

Table 1. Rare gas concentrations in $10^{-6}$ cm$^3$ STP/g observed in linear heating experiments (LH) and total extractions (TE).

| Sample | | Size ($\mu$) | Weight (mg) | $^3$He | $^4$He | $^{20}$Ne | $^{22}$Ne | $^{36}$Ar | $^{40}$Ar |
|---|---|---|---|---|---|---|---|---|---|
| | | | | (in $10^{-6}$ ccSTP/g) | | | | | |
| 1. 10084,48 | LH | 25–42 | 2.1 | — | 190 000 | — | 96 | 184 | 205 |
| | TE | | | — | 670 | 95 | 8 | 24 | 83 |
| Total | | | | — | 190 670 | — | 104 | 208 | 288 |
| 2. 10084,48 | LH | < 25 | 3.16 | — | 370 000 | 3170 | 250 | — | — |
| | TE | | | — | 1 700 | 440 | 37 | 92 | 96 |
| Total | | | | — | 371 700 | 3610 | 287 | — | — |
| 3. 14163,166 | ME | < 1000 | 1.00 | 38.4 | 72 000 | 12400 | 99.8 | 299 | 788 |
| 4. 14163,166 | LH | < 25 | 9.9 | 100 | 230 000 | 3100 | 246 | — | — |
| | TE | | | 1 | 1 600 | 380 | 31 | 123 | 210 |
| Total | | | | 101 | 231 600 | 3480 | 277 | 565* | 1290* |
| 5. 14163,166 | LH | < 25 | 16.15 | | 234 000 | — | 240 | 442 | 1080 |
| 6. 14163,166 Plag. B | LH | 30–48 | 5.75 | — | 1 200 | — | 2 | 5.6 | 201 |
| | TE | | | — | 11 | 1 | 0.1 | 0.3 | 24 |
| Total | | | | — | 1 211 | — | 2.1 | 5.9 | 225 |
| 7. 14163,166 Plag. A | LH | 30–48 | 2.77 | 1 | 2 330 | 54 | 4.7 | — | — |
| 8. 14163,166 Plag. A | LH | 30–48 | 10.22 | — | 2 210 | — | 4.6 | 29.7 | 220 |
| | TE | | | — | 47 | 5.4 | 0.4 | 5 | 25 |
| Total | | | | — | 2 257 | 59 | 5 | 35 | 245 |
| 9. 14163,166 Pyrox. A | LH | 30–48 | 3.45 | 32 | 75 000 | 1030 | 83 | — | — |
| | TE | | — | — | 2 050 | 280 | 25 | 34.5 | 83 |
| Total | | | — | — | 77 050 | 1310 | 108 | — | — |
| 10. 14163,166 Pyrox. A | LH | 30–48 | 11.63 | — | — | — | — | 109 | 194 |
| | TE | | — | — | 1 620 | 226 | 22 | 35 | 60 |
| Total | | | — | — | — | — | — | 144 | 159 |
| 11. 14163,166 Magn. glass | LH | 30–48 | 3.0 | 14.3 | 37 000 | 646 | 52.4 | — | — |
| | TE | | — | — | 680 | 136 | 11.3 | — | — |
| Total | | | — | — | 37 680 | 782 | 63.7 | — | — |
| 12. 14163,166 Magn. glass | LH | 30–48 | 20.0 | — | 36 000 | — | 51.8 | 214 | 335 |
| | TE | | — | 0.22 | 620 | 153 | 12.8 | 123 | 210 |
| Total | | | — | — | 36 620 | — | 64.6 | 337 | 545 |
| 13. 14240,15 | ME | < 1000 | 0.88 | 26.2 | 60 500 | 1090 | 87.5 | 286 | 677 |
| 14. 15101,88 | LH | < 25 | 5.6 | 56 | 160 000 | 2500 | 226 | — | — |
| | TE | | | 0.6 | 1 900 | 420 | 34 | 81 | 29 |
| Total | | | — | 57 | 162 900 | 2920 | 260 | 391* | 409* |
| 15. 15101,88 | LH | < 25 | 10.15 | — | 156 000 | — | 230 | 310 | 380 |
| 16. 15231,54 | LH | < 25 | 1.63 | 64 | 182 000 | 2730 | 218 | — | — |
| 17. 15231,54 | LH | < 25 | 5.75 | — | 176 000 | — | — | 368 | 345 |
| | TE | | | 0.6 | 1 920 | 470 | 38 | 87 | 70 |
| Total | | | — | 65* | 178 000 | 3200* | 256* | 455 | 415 |

LH = linear heating (100–900°C). TE = total extraction (> 900°C). ME = melt extraction (1700°C without previous heat treatment).
* Calculated with values from the parallel measurement.

this mineral is still less retentive than ilmenite. The $^4$He from pyroxene is about a factor of 10 lower than an ilmenite sample of similar grain size. We also note a $^{36}$Ar excess in the magnetic glass samples in comparison to the pyroxene samples (Table 1 and Fig. 5). This excess may be due to an incomplete degassing of $^{36}$Ar at the time the glass was formed by impact. The amounts of gas in the final extractions (1700°C) are also listed in Table 1. Only 1–2% of the $^4$He, some 10–15% of $^{20}$Ne and about 10–40% of the argon are retained.

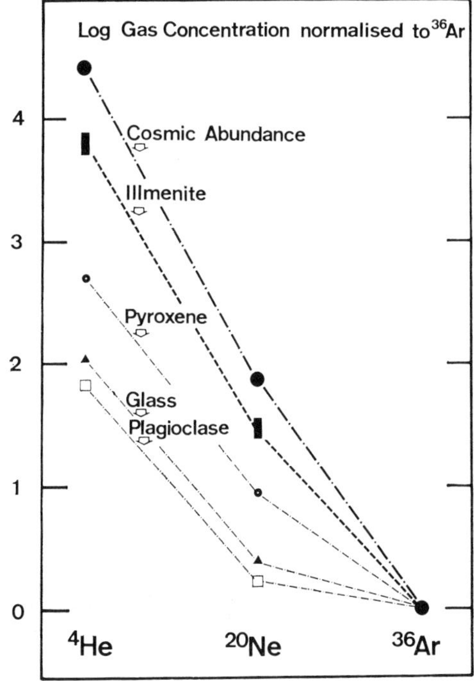


Fig. 5. Abundance of light noble gases in plagioclase, magnetic glass, and pyroxene separates (14163,166, grain sizes 30–48 $\mu$) compared to ilmenite from 10084,47 (Eberhardt *et al.*, 1970) and cosmic abundances (Aller, 1961).

The release patterns for the 25 $\mu$ grain size fraction of unseparated material from the four soils 10084, 14163, 15101, and 15231 are presented in Fig. 6. The amounts of gas released per gram and minute are plotted on a logarithmic scale versus the temperature. The upper boundary of each shaded field represents the actual release curve. Crossovers of curves are shown by intersecting fields.

In the following discussion, the release patterns of $^{40}$Ar are excluded; they will be discussed in a separate section. First, we note the overall similarity of the release patterns of the four nuclides from the different soils. They agree with those observed in stepwise heating experiments by Pepin *et al.* (1970). Characteristic differences exist between the nuclides in each soil (Fig. 7), but only minor variations are observed between the different soils. The temperature of maximum gas release occurs generally in the sequence $^{4}$He, $^{20}$Ne, $^{36}$Ar. Second, we observe a more or less pronounced peculiarity of all release patterns below about 300°C (Fig. 6). This feature is also observed for ion implanted $^{4}$He in quartz (Fig. 3). Therefore, we interpret this peculiarity as an enhanced low temperature release, characteristic for the release of noble gases embedded in solids by ion implantation. The amount of $^{4}$He released from quartz at low temperatures is about two orders of magnitude larger than that released from lunar fines. This can be explained in terms of the thermal history of the lunar fines, because the weakly bound rare gas constituent diffuses out during the lunar day.

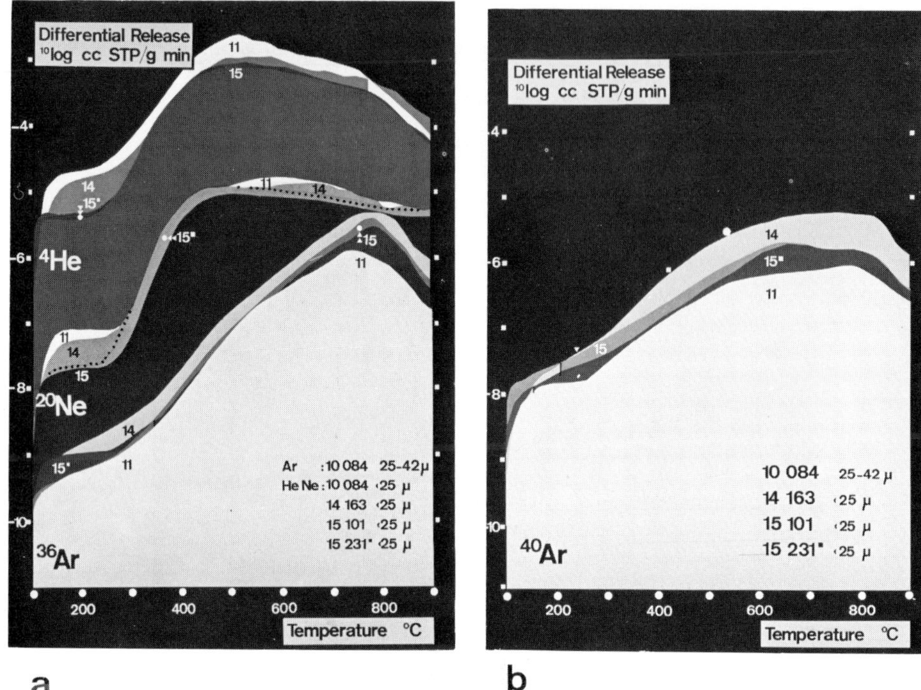


Fig. 6. Release rates of $^{4}$He, $^{20}$Ne, $^{36}$Ar, and $^{40}$Ar versus temperature from $<25\ \mu$ grain size bulk samples of lunar soil on a logarithmic scale. Symbols on the curves indicate integrated gas loss (▲ = 0.1%; ■ = 1%; ● = 10%).

The release patterns of the four soils investigated are complex superpositions of the release patterns of the mineral constituents. A quantitative discussion of these release patterns is not possible. However, for the mineral separates a more quantitative discussion and examination with respect to their different retentivities is possible.

Figure 8 shows the differential release as a function of temperature for $^{4}$He, $^{20}$Ne, $^{36}$Ar, and $^{40}$Ar from pyroxene, magnetic glass, and plagioclase A. The release patterns are, on the whole, similar to those observed from the bulk grain size fractions. Also striking is the great similarity of the He releases in spite of the large differences in gas retention. Another interesting feature is the occurrence of the enhancement of the release at low temperatures observed already on the bulk fractions. This clearly shows that this release is not due to degassing of low retentive minerals. The fraction of gas released at low temperatures is relatively constant for all minerals. This implies that the low temperature release is independent of chemistry and structure.

Additional information concerning the different retentivities of noble gases can be obtained from the examination of Figs. 8 and 9. Figure 9 shows the integrated losses from the different minerals as a function of temperature. Interesting are the similarity of the integrated He releases and the clear differences of the $^{20}$Ne and $^{36}$Ar release. The $^{20}$Ne is released much more readily from plagioclase than from the other

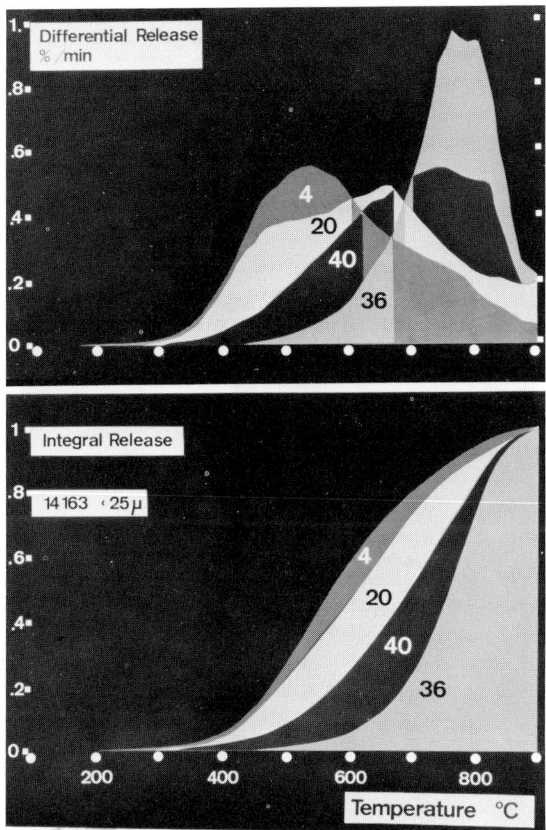


Fig. 7. Integral and differential release from $<25\ \mu$ grain size bulk samples of lunar soil 14163,166 as a function of temperature. Amounts are normalized to total gas released in diffusion run up to 900°C.

minerals. In fact, the plagioclase release maximum of the $^{20}$Ne even occurs before that of $^{4}$He. Excluding the $^{4}$He releases, the release patterns correlate to the retentivity sequence noted already from the consideration of gas concentrations.

In summary, the release patterns of the mineral separate are understood to show: (1) A substantial amount of $^{4}$He has been lost from sites of low retention from the plagioclase. This leads to the inversion of the release sequence of He as compared to the other releases. (2) Plagioclase has a high mean diffusion constant as compared to the other minerals, which causes large fractions of gas to be lost and considerable alterations of the concentration profiles. (3) Helium and neon show considerable high temperature tailing, which indicates that neon has diffused into the grains. In the glass, the Ne tailing is most pronounced, which may indicate that a substantial fraction of Ne is distributed throughout the volume of the grains perhaps because of incomplete degassing of the glass parent material. The same conclusion has been reached above on the basis of the high $^{36}$Ar concentration and is supported by the Ar release patterns.

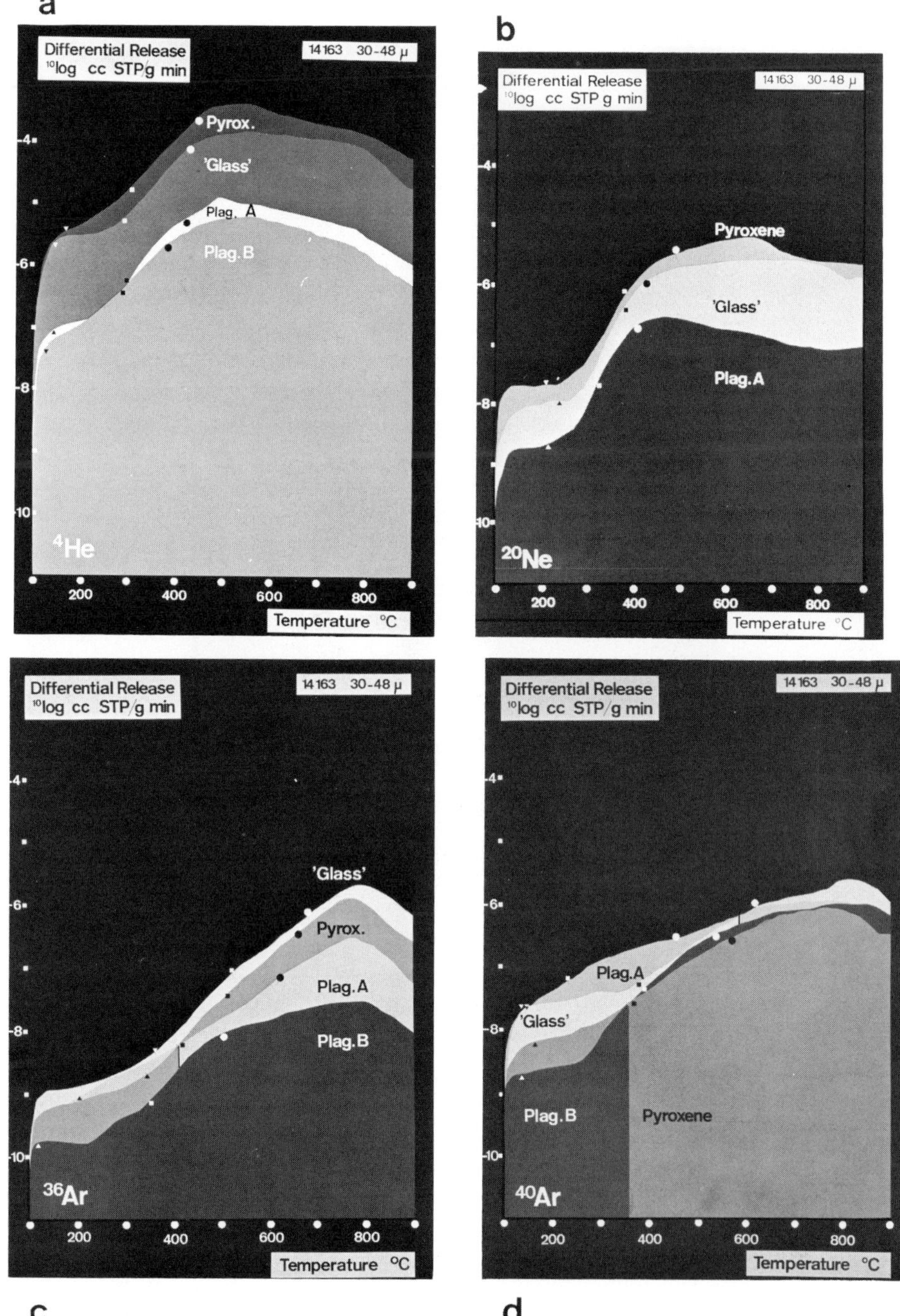


Fig. 8. Release rates from 30–48 $\mu$ grain size mineral separates plagioclase A, magnetic glass, and pyroxene. Presentation as in Fig. 6. Symbols on the curves indicate integrated gas loss (▲ = 0.1%, ■ = 1%, ● = 10%).

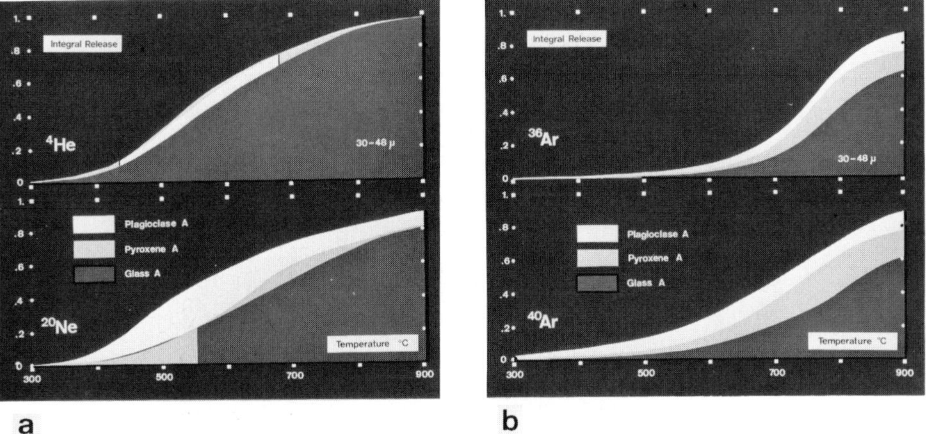


Fig. 9. Integral loss for mineral separates.

These conclusions are to be regarded as tentative until simulation experiments to examine the release patterns from gases of known distributions are carried out.

The low plagioclase retentivity observed here stands in contradiction to the observations made by Huneke *et al.* (1972). In a study of $^{40}Ar/^{39}Ar$ ages of minerals from lunar rocks they found that the plagioclase retains $^{40}Ar$ better than pyroxene. We tend to explain this divergence as due to different degrees of lattice damage in the respective plagioclases investigated.

*Isotopic ratios.* As examples for the isotope ratios as a function of temperature $^4He/^3He$, $^{20}Ne/^{22}Ne$, and $^{40}Ar/^{36}Ar$ ratios of sample 14163 observed during linear heating are shown as function of temperature in Fig. 10. Isotopic ratios of the other bulk samples behave in a virtually identical fashion. The $^4He/^3He$ ratios observed in the low temperatures range ($<300°C$) have relatively large errors due to the small amounts of $^3He$ released. Therefore, the low temperature maximum needs to be confirmed. The $^4He/^3He$ pattern is determined at least by three facts: (1) The initial concentration profiles of $^3He$ and $^4He$ are different because the isotopes are implanted with different energies. (2) The initial concentration profile is changed by thermal effects on the lunar surface. (3) Mass fractionation occurs because of the mass dependence of diffusion.

At low temperatures ($<300°C$) the $^{20}Ne/^{22}Ne$ values lie for all samples between 14 and 15 and above about 400°C between 12 and 13. We note that the occurrence of the high ratio is correlated with the low temperature enhancement. To clarify this point, linear heating experiments on larger lunar samples and minerals with artificially implanted Ne are underway.

*Model release patterns*

Model calculations were carried out based on (1) Fick's second law

$$\frac{\partial c}{\partial t} = D\,\Delta c$$

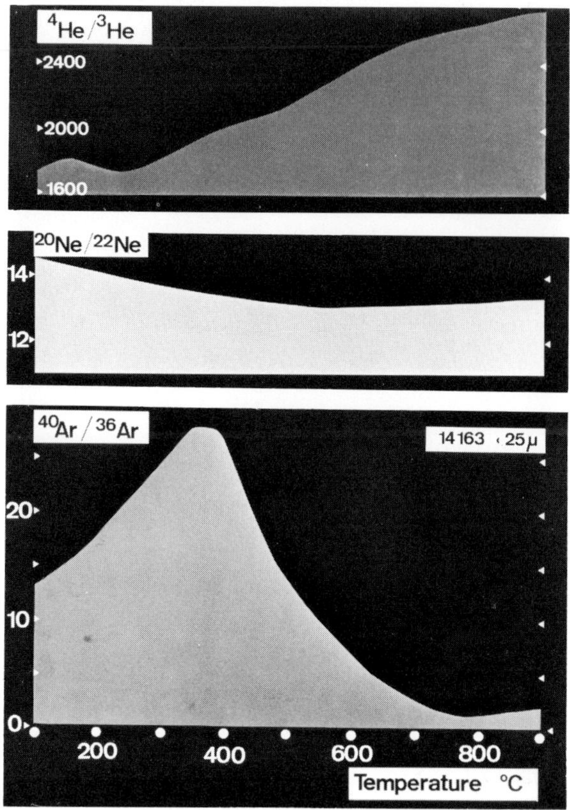


Fig. 10. Isotope ratios as a function of temperature during linear heating of $<25\ \mu$ grain size fraction from 14163,166.

where $c$ is the concentration of the diffusing gas, $t$ the time, and $D$ the diffusion coefficient; and (2) the Arrhenius equation, modified for "linear heating,"

$$D = D_0 \exp\left(-\frac{Q}{R\alpha t}\right)$$

where $D_0$ is the frequency factor, $Q$ the activation energy, $R$ the gas constant, and $\alpha$ the rate of temperature increase.

The fractional loss of gas was determined for three sets of initial and boundary conditions. (1) "Sphere," radius $a$, with initially homogeneous gas distribution. This model was chosen for its simplicity as a first approach. (2) "Plane sheet," thickness $a$, with initially homogeneous gas distribution. In this case the fractional loss as a function of time is identical with that of a plane sheet bounded by an impermeable surface on one side. This represents an approximation for a grain, where the gas is located only in an outer layer and the diffusion coefficient in this layer is much greater than in the rest of the grain. (3) "Semi-infinite medium" with an initial gas concentration

$C = C_0 \cdot \exp(-x/a)$ where $x$ is the distance from the surface, $C_0$ the initial concentration at the surface, and $a$ a characteristic length.

The resulting formula for the fractional loss contains 2 parameter, $Q$ and $D_0/a^2$ that are used for fitting the model values to the experimental data. The influence of these parameters is demonstrated in Fig. 11. One notes that both parameters can independently define the temperature of maximum release. With the temperature of maximum release taken as constant, higher values for $Q$ or $D_0/a^2$ give higher maximum release rates.

The noble gases observed in this study are predominantly trapped gases of solar origin. A high concentration near the surface and monotonic decrease with depth are, therefore, appropriate assumptions, so that models (2) and (3) can be regarded as reasonable approximations. Model (3) is justified if one keeps in mind that solar particles can have a range of energies up to those of solar flares. Higher energies occur with smaller probabilities.

In Fig. 12 differential loss of $^{36}Ar$ in pyroxene A is plotted together with curves based on the above models. This particular isotope of this particular sample was chosen because it should be the least affected by diffusion on the moon. It is apparent that the low temperature release of the naturally implanted gases is not fitted by any of the models. In the high temperature release, differences between the natural release patterns and the model releases are observed, as seen, for example, from a comparison of the Ne release from glass in Fig. 8 and the model release in Figs. 11 and 12.

All three models fit the experimental data equally well, at least in the temperature region of the major loss. Model (3) gives the best approximation for the high temperature range because in this model diffusion into the grain has been considered and the high temperature fraction should consist of surface distant gas.

It should be noted that the model calculations can be extended for $Q$ as a linear function of temperature and the resulting fitted curves are the same as for $Q =$ constant, not changing (improving) the poor approximation for the high temperature release.

None of the models fits the low temperature release which we believe is due to a weakly bound gas component.

### *The $^{40}Ar$ problem*

An unknown fraction of the $^{40}Ar$ has been produced *in situ* from the decay of radiogenic $^{40}K$. Only a very small fraction of the remaining $^{40}Ar$ is of direct solar origin, because the solar $^{40}Ar/^{36}Ar$ ratio is expected to be $10^{-4}$ (Cameron, 1968). Thus a substantial part of the $^{40}Ar$ has a nontrivial origin (Eberhardt *et al.*, 1970; Heymann and Yaniv, 1970, 1971; Heymann *et al.*, 1972). These authors, as well as Manka and Michel (1970, 1971), propose a retrapping of $^{40}Ar$ from the lunar atmosphere to account for this excess in the lunar fines.

The energy of retrapped $^{40}Ar$ is low in comparison with that of the solar wind $^{36}Ar$. Manka and Michel (1970) predict energies up to 2 keV for reimplanted $^{40}Ar$ ions (solar wind velocity of 400 km/sec). Consequently, the initial concentration profiles of solar $^{36}Ar$ and retrapped $^{40}Ar$ should be different: $^{36}Ar$ is embedded in a

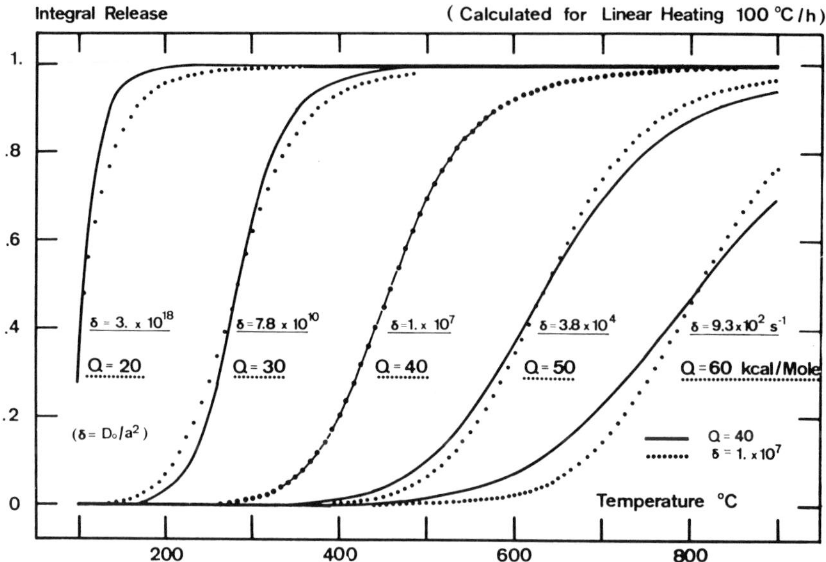


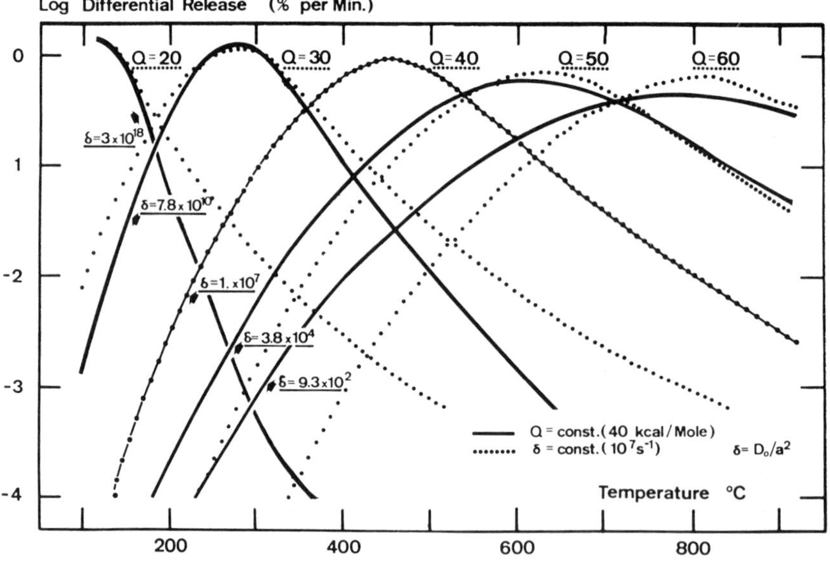


Fig. 11. Release rates from model calculation with an exponential concentration profile. The upper part shows the integrated gas loss and the lower part the release rates on a logarithmic scale versus the temperature. The solid curves are calculated with constant activation energy Q (40 kcal/mol) and with varying $\delta = D_0/a^2$, whereas for the dotted curves the $D_0/a^2$ is kept constant ($1 \times 10$ $\sec^{-1}$) and the activation energy varies.

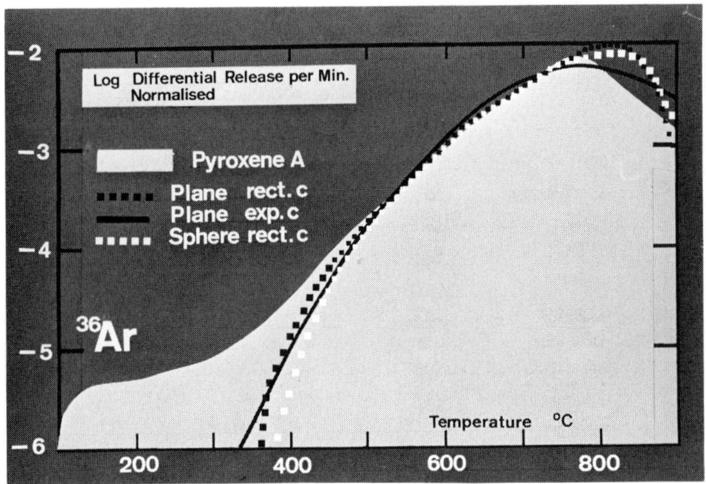


Fig. 12. Comparison of $^{36}$Ar release rate as function of temperature, with those computed from different models.

layer of some 500 Å, whereas $^{40}$Ar should be distributed in a much thinner surface layer.

From our simulation experiment with $^{4}$He implanted into Al (Fig. 2), we expected that $^{40}$Ar retrapped from the lunar atmosphere should be released at lower temperatures than the $^{36}$Ar from the solar wind. However, in agreement with observations by Pepin *et al.* (1970), no substantial fraction of $^{40}$Ar is released at low temperatures.

The $^{36}$Ar, $^{40}$Ar releases, and the $^{40}$Ar/$^{36}$Ar ratios of the mineral separates are shown in Figs. 7c, d, and 13. The release of the $^{40}$Ar differs significantly from that of $^{36}$Ar; especially at lower temperatures. The points on the release curves indicating 0.1, 1, and 10% loss show, nevertheless, that the enhancement of $^{40}$Ar over $^{36}$Ar comprises a small fraction of the total $^{40}$Ar only. Unfortunately, it was not possible to evaluate the release of *in situ*-produced radiogenic $^{40}$Ar, possibly because the $^{40}$Ar distribution is not homogeneous throughout the grains. It is known that the potassium in plagioclase is not homogeneous (Brown and Gay, 1971).

In summary, we conclude that the concentration profiles of solar wind $^{36}$Ar and "orphan" $^{40}$Ar in the grains are—at least at present—rather similar. If the orphan $^{40}$Ar is indeed due to retrapping from the lunar atmosphere, its initial distribution must have been severely altered during the grain's history. Such an alteration of the concentration profile can be caused by diffusion, by ablation, and/or sputtering, and by "knock-on processes." Knock-on processes certainly occur, but it is difficult to account for an equalization of the $^{40}$Ar and $^{36}$Ar profiles by such processes. Ablation and/or sputtering processes probably do not uniformize the concentration profiles of the two nuclides. Thus, we are left with diffusion. However, in this case the fraction of $^{40}$Ar lost from the grain is considerably higher than that of the $^{36}$Ar. Therefore, a rather substantial production of argon ions is required to explain the "orphan" Ar as retrapped from the lunar atmosphere. If our interpretation of the release patterns is

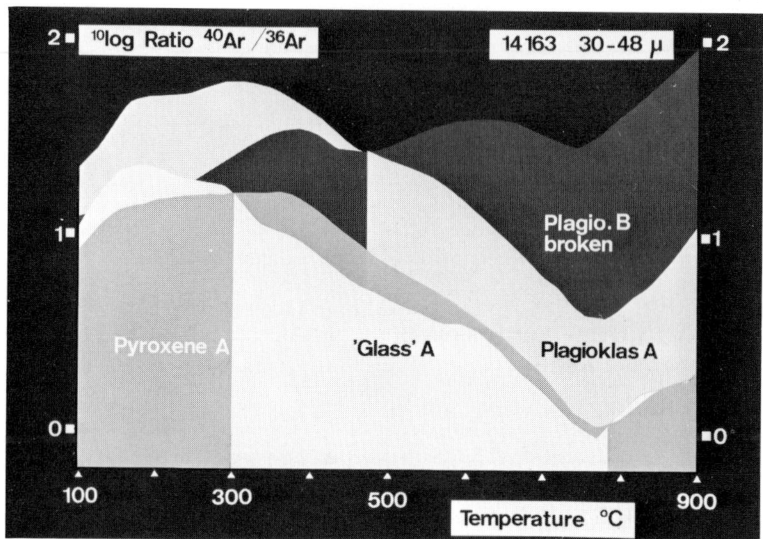


Fig. 13. $^{40}Ar/^{36}Ar$ as a function of temperature during linear heating for different minerals.

correct, the retention of retrapped $^{40}Ar$ under lunar conditions may be even lower than Manka and Michel (1970) assumed.

One is thus led to search for other possible sources of the "orphan" argon, if one does not postulate a much higher $^{40}Ar/^{36}Ar$ ratio in the solar wind.

*Potassium-coated regolith.* Silver (1972) reported that part of the radiogenic lead in the lunar fines is readily volatilized. Other authors (De Maria *et al.*, 1971; Naughton *et al.*, 1971; Gibson, 1972) conclude that lunar material can lose alkalis. In analogy, large quantities of potassium could have been volatilized during extrusion of mare material and during impacts. K and/or $K_2O$ liberated would be distributed over large areas of the moon, coating existing regolith material. If the radiation damaged surface of this material would allow diffusion of potassium into the grains, surface correlated $^{40}Ar$ would be produced. If this is the origin of the "orphan" argon, the site-correlation of the bulk $^{40}Ar/^{36}Ar$ ratio noted by Heymann and Yaniv (1972) and the $^{40}Ar$ increase with depth in a core tube noted by Marti *et al.* (1971) can be understood. Indications for the potassium coating could still exist in the lunar regolith and especially lunar breccias. A positive grain size correlation of the potassium content of mineral separates or a depth correlation in single grains would support the hypothesis. A negative result of such an investigation would not necessarily disprove it because loss of potassium could have occurred. After receiving the proceedings of the second lunar science conference we discovered that Naughton *et al.* (1971) independently had considered similar processes to explain the orphan argon in lunar soils.

In conclusion we suggest that the retrapping of $^{40}Ar$ from the lunar atmosphere as proposed by Heymann *et al.* (1970) is not the only source of orphan Ar.

*Acknowledgments*—We are grateful to NASA for supplying the lunar samples. For helpful discussions and advice we thank Drs. G. Bayer, M. Grünenfelder, and R. Schmid. We are indebted to Dr. D. Phinney who contributed in experimental work and discussions, and Dr. A. B. Harnik who carried out the x-ray investigations. This work was supported by the Swiss National Science Foundation Grant NF 2.386.70.

## REFERENCES

Aller L. H. (1961) *The Abundance of the Elements*. Interscience.

Brown M. G. and Gay P. (1971) Lunar Antiperthites. *Earth Planet. Sci. Lett.* **11,** 23–27.

Buehler F., Cerutti H., Eberhardt P., and Geiss J. (1972) Results of the Apollo 14 and 15 solar wind composition experiment (abstract). In *Lunar Science—III* (editor C. Watkins), pp. 102–104, Lunar Science Institute Contr. No. 88.

Cameron A. G. W. (1968) A new table of abundances of the elements in the solar system. In *Origin and Distribution of the Elements* (editor L. H. Ahrens), pp. 125–143, Pergamon.

Davis P. K., Lewis R. S., and Reynolds J. H. (1971) Stepwise heating analyses of rare gases from pile-irradiated rocks 10044 and 10057. *Proc. Second Lunar Sci. Conf., Geochim. Cosmochim. Acta* Suppl. 2, Vol. 2, pp. 1693–1703. MIT Press.

De Maria G., Balducci G., Guido M., and Piacente V. (1971) Mass spectrometric investigation of the vaporization process of Apollo 12 lunar samples. *Proc. Second Lunar Sci. Conf., Geochim. Cosmochim. Acta* Suppl. 2, Vol. 2, pp. 1367–1380. MIT Press.

Eberhardt P., Geiss J., Graf H., Groegler N., Kraehenbuehl U., Schwaller H., Schwarzmueller J., and Stettler A. (1970) Trapped solar wind noble gases, exposure age and K/Ar-age in Apollo 11 lunar fine material. *Proc. Apollo 11 Lunar Sci. Conf., Geochim. Cosmochim. Acta* Suppl. 1, Vol. 2, pp. 1037–1070. Pergamon.

Fechtig H. and Kalbitzer S. (1966) The diffusion of argon in potassium-bearing solids. In *Potassium Argon Dating*, pp. 68–107, Springer.

Funk H., Baur H., Frick U., Schultz L., and Signer P. (1971) On the diffusion of helium in quartz. *Z. Kristallogr.* **133,** 225–233.

Gibson E. K. (1972) Volatile element depletion investigations on Apollo 11 and 12 lunar basalts via thermal volatilizations (abstract). In *Lunar Science—III* (editor C. Watkins), pp. 303–305, Lunar Science Institute Contr. No. 88.

Gibson E. K. Jr. and Johnson S. M. (1971) Thermal analysis-inorganic gas release studies of lunar samples. *Proc. Second Lunar Sci. Conf., Geochim. Cosmochim. Acta* Suppl. 2, Vol. 2, pp. 1351–1366. MIT Press.

Hanneman R. E. (1970) Thermal and gas evolution behavior of Apollo 11 samples. *Proc. Apollo 11 Lunar Sci. Conf., Geochim. Cosmochim. Acta* Suppl. 1, Vol. 2, pp. 1207–1211. Pergamon.

Heymann D. and Yaniv A. (1970) $Ar^{40}$ anomaly in lunar samples from Apollo 11. *Proc. Apollo 11 Lunar Sci. Conf., Geochim. Cosmochim. Acta* Suppl. 1, Vol. 2, pp. 1261–1267. Pergamon.

Heymann D. and Yaniv A. (1971) $^{40}Ar$ in meteorites, fines, and breccias from the moon. *Chem. Erde.* **30,** 175–190.

Heymann D., Yaniv A., and Walton J. (1972) Inert gases in Apollo 14 fines and the case of parentless $Ar^{40}$ (abstract). In *Lunar Science—III* (editor C. Watkins), pp. 376–378, Lunar Science Institute Contr. No. 88.

Hintenberger H., Weber H. W., Voshage H., Waenke H., Begemann F., and Wlotzka F. (1970) Concentrations and isotopic abundance of the rare gases, hydrogen and nitrogen in Apollo 11 lunar matter. *Proc. Apollo 11 Lunar Sci. Conf., Geochim. Cosmochim. Acta* Suppl. 1, Vol. 2, pp. 1268–1282. Pergamon.

Hintenberger H., Weber H. W., and Takaoka N. (1971) Concentrations and isotopic abundances of the rare gases in lunar matter. *Proc. Second Lunar Sci. Conf., Geochim. Cosmochim. Acta* Suppl. 2, Vol. 2, pp. 1607–1625. MIT Press.

Hohenberg C. M., Davis P. K., Kaiser W. A., Lewis R. S., and Reynolds J. H. (1970) Trapped and cosmogenic rare gases from stepwise heating of Apollo 11 samples. *Proc. Apollo 11 Lunar Sci. Conf., Geochim. Cosmochim. Acta* Suppl. 1, Vol. 2, pp. 1283–1309. Pergamon.

Huneke J. C., Nyquist L. E., Funk H., Koeppel V., and Signer P. (1969) The thermal release of rare gases from separated minerals of the Mocs meteorite. In *Meteorite Research* (editor P. Millman), pp. 901–921. Reidel.

Huneke J. C., Podosek F. A., Turner G., and Wasserburg G. J. (1972) $^{40}$Ar–$^{39}$Ar systematics in lunar rocks and separated minerals of lunar rocks from Apollo 14 and 15 (abstract). In *Lunar Science—III* (editor C. Watkins), pp. 413–414, Lunar Science Institute Contr. No. 88.

Kirsten T., Mueller O., Steinbrunn F., and Zaehringer J. (1970) Study of distribution and variations of rare gases in lunar material by a microprobe technique. *Proc. Apollo 11 Lunar Sci. Conf., Geochim. Cosmochim. Acta* Suppl. 1, Vol. 2, pp. 1331–1343. Pergamon.

Kirsten T., Steinbrunn F., and Zaehringer J. (1971) Location and variation of trapped rare gases in Apollo 12 lunar samples. *Proc. Second Lunar Sci. Conf., Geochim. Cosmochim. Acta* Suppl. 2, Vol. 2, pp. 1651–1669. MIT Press.

Kirsten T., Deubner J., Ducati H., Gentner W., Horn P., Jessberger E., Kalbitzer S., Kaneoka I., Kiko J., Kraetschmer W., Mueller H. W., Plieninger T., and Thio S. K. (1972) Rare gases in individual components and bulk samples of Apollo 14 and 15 fines and fragmental rocks (abstract). In *Lunar Science—III* (editor C. Watkins), pp. 452–454, Lunar Science Institute Contr. No. 88.

Manka R. H. and Michel F. C. (1970) Lunar atmosphere as a source of argon-40 and other lunar surface elements. *Science* **169,** 278–280.

Manka R. H. and Michel F. C. (1971) Lunar atmosphere as a source of lunar surface elements. *Proc. Second Lunar Sci. Conf., Geochim. Cosmochim. Acta* Suppl. 2, Vol. 2, pp. 1717–1728. MIT Press.

Marti K., Lugmair G. W., and Urey H. C. (1970) Solar wind gases, cosmic-ray spallation products and the irradiation history of Apollo 11 samples. *Proc. Apollo 11 Lunar Sci. Conf., Geochim. Cosmochim. Acta* Suppl. 1, Vol. 2, pp. 1357–1367. Pergamon.

Marti K. and Lugmair G. W. (1972) Kr$^{81}$–Kr and K–Ar$^{40}$ ages, cosmic-ray spallation products and neutron effects in lunar samples from Oceanus Procellarum. *Proc. Second Lunar Sci. Conf.. Geochim. Cosmochim. Acta* Suppl. 2, Vol. 2, pp. 1591–1605. MIT Press.

Megrue G. H. and Steinbrunn F. (1972) Classification and source of lunar soils; clastic rocks; and individual mineral, rock, and glass fragments from Apollo 12 and 14 samples as determined by the concentration gradients of the helium, neon, and argon isotopes (abstract). In *Lunar Science—III* (editor C. Watkins), pp. 532–534, Lunar Science Institute Contr. No. 88.

Naughton J. J., Derby J. V., and Lewis V. A. (1971) Vaporization from heated lunar samples and the investigation of lunar erosion by volatilized alkalis. *Proc. Second Lunar Sci. Conf., Geochim. Cosmochim. Acta* Suppl. 2, Vol. 1, pp. 449–457. MIT Press.

Nyquist L. E., Huneke J. C., Funk H., and Signer P. (1972) Thermal release characteristics of spallogenic He, Ne, and Ar from the Carbo iron meteorite. *Earth Planet. Sci. Lett.* **14,** 207–214.

Pepin R. O., Nyquist L. E., Phinney D., and Black D. C. (1970) Rare gases in Apollo 11 lunar material. *Proc. Apollo 11 Lunar Sci. Conf., Geochim. Cosmochim. Acta* Suppl. 1, Vol. 2, pp. 1435–1454. Pergamon.

Schultz L., Frick U., and Signer P. (1971) Nachweis des Diffusionsverlustes von Sonnenwind $^4$He im Mondstaub. *Helv. Phys. Acta* **44,** 614–616.

Schultz L., Signer P., Lorin J. G., and Pellas P. (1972) Complex irradiation history of the Weston chondrite. *Earth Planet. Sci. Lett.* (submitted).

Silver L. T. (1970) Lead volatilization and volatile transfer processes on the moon (abstract). In *Lunar Science—III* (editor C. Watkins), pp. 701–703, Lunar Science Institute Contr. No. 88.

Walchi F. M., Gilmartin D. E., Oro J., and Updegrove W. S. (1971) Differential thermal analysis and gas release studies of Apollo 11 samples. *Icarus* **15,** 304–313.

Proceedings of the Third Lunar Science Conference
(Supplement 3, *Geochimica et Cosmochimica Acta*)
Vol. 2, pp. 1967–1980
The M.I.T. Press, 1972

# Atmospheric $Ar^{40}$ in lunar fines

A. YANIV* and D. HEYMANN

Departments of Geology and Space Science, Rice University,
Houston, Texas 77001

**Abstract**—The criticism by Baur *et al.* of the solar wind implantation mechanism of atmospheric (lunar) $Ar^{40}$, which is mainly based on thermal release studies of argon isotopes from sample 14163, is rebutted. The solar wind mechanism could produce atmospheric $Ar^{40}$ ions with kinetic energies comparable to those of typical solar-wind $Ar^{36}$ ions if the magnetic field near the moon was stronger in the *past* than it is today or if the average solar wind velocity was greater in the *past* than it is today. The $Ar^{40}/Ar^{36}$ ratio tends to become smaller for "younger" regolith. One interpretation is that the concentration of neutral $Ar^{40}$ in the moon's atmosphere decreased between 4000 and 3000 m.y. ago. The trend suggests that most of the atmospheric $Ar^{40}$ *now* seen in fines was trapped before 3000 m.y. ago and that the concentration of neutral $Ar^{40}$ in the lunar atmosphere had become very small by the time of the Copernican event, 850 m.y. ago. Hence, calculations of the kinetic energies of atmospheric $Ar^{40}$ ions with today's values of solar wind velocities and near-lunar magnetic field values could be irrelevant. The preservation of small-scale $Ar^{40}$ provinces within the Apollo landing sites probably reflects the ejecta blankets of sporadic, relatively recent impacts in the areas. These blankets apparently were not greatly disturbed after they formed. The regolith seems to be quite inhomogeneous both vertically as well as horizontally. The Apollo 15 fines apparently do not agree with the $Ar^{40}/Ar^{36}$ versus age trend. The reason for this is not known at this time.

## INTRODUCTION

ATMOSPHERIC $Ar^{40}$, that is, $Ar^{40}$ that has emanated from the solid moon into its atmosphere, has become a geochemical tracer of considerable interest and a subject of controversy. Several investigators (Heymann and Yaniv, 1970a; Manka and Michel, 1970; Eberhardt *et al.*, 1970) have suggested that most of this gas has been removed from the moon and that about 10% of it has been retrapped in the regolith by the interaction of Ar-ions with the solar wind. At the 3rd Lunar Science Conference, Baur *et al.* (1972) criticized this theory, contending that their thermal release studies of argon isotopes from sample 14163 gave no evidence for the presence of substantial amounts of atmospheric argon in these fines.

In the present paper we report Ar measurements in five Apollo 12, five Apollo 14, and two Apollo 15 samples. With the data from five locations on the moon before us we will review the case of $Ar^{40}$ in lunar fines, in particular the known variation of $Ar^{40}/Ar^{36}$ ratios.

Our measurements were done mass-spectrometrically, as before (Heymann and Yaniv, 1970b). The results are given in Table 1.

---

* On leave of absence from the Department of Physics and Astronomy, Tel Aviv University, Ramat Aviv, Israel.

Table 1. Argon contents of 14003, 14163, 14166, 14167, 14259, 12042, 12073, 12030, 12044, 12033, 14091, and 15601 (units: $cm^3$ STP/g).*

| Sample | $Ar^{36}$ $10^{-4}$ | $Ar^{38}$ $10^{-5}$ | $Ar^{40}$ $10^{-4}$ | $Ar^{36}/Ar^{38}$ | $Ar^{40}/Ar^{36}$ |
|---|---|---|---|---|---|
| 14003 (bulk) | 3.54 | 6.68 | 6.16 | 5.30 | 1.74 |
| 14163 (bulk) | 3.47 | 6.58 | 8.49 | 5.29 | 2.45 |
| 14166 (1 fragment) | 0.0684 | 0.190 | 3.29 | 3.90 | 48.1 |
| 14167 (1 fragment) | 0.0357 | 0.0686 | 1.08 | 5.20 | 30.2 |
| 14259 (bulk) | 4.39 | 8.19 | 5.58 | 5.35 | 1.27 |
| 14163 (< 37 μ) | 5.10 | 9.44 | 12.4 | 5.40 | 2.43 |
| (37–63 μ) | 1.53 | 2.89 | 4.25 | 5.30 | 2.77 |
| (63–74 μ) | 1.03 | 1.96 | 3.61 | 5.26 | 3.49 |
| (74–88 μ) | 1.42 | 2.74 | 4.70 | 5.18 | 3.30 |
| (88–105 μ) | 1.39 | 2.53 | 4.67 | 5.50 | 3.35 |
| (105–250 μ) | 1.19 | 2.25 | 4.00 | 5.28 | 3.35 |
| (250–354 μ) | 0.980 | 1.88 | 3.98 | 5.21 | 4.06 |
| (354–500 μ) | 0.963 | 1.85 | 5.76 | 5.20 | 5.98 |
| (500–700 μ) | 0.163 | 0.355 | 3.38 | 4.59 | 20.8 |
| (700–1000 μ) | 0.106 | 0.235 | 3.27 | 4.51 | 30.9 |
| 14259 (< 37 μ) | 5.98 | 11.3 | 6.92 | 5.30 | 1.16 |
| (37–53 μ) | 2.75 | 5.14 | 3.98 | 5.35 | 1.45 |
| (53–63 μ) | 2.90 | 5.46 | 4.05 | 5.32 | 1.39 |
| (63–74 μ) | 1.69 | 3.22 | 2.80 | 5.25 | 1.66 |
| (74–88 μ) | 2.02 | 3.84 | 3.44 | 5.26 | 1.70 |
| (88–105 μ) | 1.87 | 3.54 | 3.15 | 5.28 | 1.69 |
| (105–250 μ) | 1.26 | 2.39 | 2.88 | 5.26 | 2.28 |
| (250–354 μ) | 1.80 | 3.40 | 6.02 | 5.30 | 3.34 |
| (354–500 μ) | 0.649 | 1.25 | 2.67 | 5.20 | 3.16 |
| (500–700 μ) | 0.676 | 1.29 | 2.98 | 5.23 | 4.41 |
| (700–1000 μ) | 0.789 | 1.53 | 2.41 | 5.20 | 3.02 |
| 12042 (bulk) | 3.01 | 5.69 | 2.06 | 5.29 | 0.69 |
| 12073 (interior chip) | 1.01 | 1.92 | 0.797 | 5.28 | 0.79 |
| 12030 (bulk) | 1.25 | 2.38 | 3.11 | 5.26 | 2.49 |
| 12044 (bulk) | 3.06 | 5.78 | 1.97 | 5.30 | 0.64 |
| 12033 (> 250 μ) | 0.0485 | 0.120 | 0.243 | 4.06 | 5.01 |
| (105–250 μ) strongly magnetic | 0.254 | 0.470 | 0.419 | 5.08 | 1.76 |
| (105–250 μ) weakly magnetic | 0.0434 | 0.105 | 0.236 | 4.21 | 5.32 |
| (< 105 μ) | 0.292 | 0.582 | 0.348 | 5.01 | 1.19 |
| (< 105 μ) | 0.177 | 0.359 | 0.473 | 3.59 | 2.67 |
| 15091 (bulk) | 2.81 | 5.18 | 3.36 | 5.42 | 1.20 |
| 15610 (bulk) | 2.10 | 3.85 | 1.91 | 5.45 | 0.910 |
| 15601 (< 44 μ) | 3.32 | 6.97 | 2.97 | 4.76 | 0.897 |
| (44–53 μ) | 1.13 | 2.17 | 1.06 | 5.21 | 0.941 |
| (53–63 μ) | 1.00 | 1.90 | 0.952 | 5.29 | 0.950 |
| (63–74 μ) | 1.05 | 1.81 | 0.910 | 5.27 | 0.953 |
| (74–88 μ) | 0.879 | 1.70 | 0.834 | 5.17 | 0.949 |
| (88–105 μ) | 0.771 | 1.46 | 0.745 | 5.28 | 0.970 |
| (105–250 μ) | 0.499 | 0.968 | 0.537 | 5.16 | 1.08 |
| (250–354 μ) | 0.387 | 0.755 | 0.513 | 5.13 | 1.33 |
| (354–500 μ) | 0.284 | 0.586 | 0.387 | 4.84 | 1.33 |
| (500–700 μ) | 0.289 | 0.569 | 0.403 | 5.07 | 1.40 |

* Errors in gas contents are always less than 5%.

## RESULTS

### *Apollo 14 samples*

The size fractions of 14163 and 14259 were obtained by washing the fines with acetone through standard mesh sieves. In the case of the three coarsest fractions the samples for analysis consisted of handpicked particles. For each of the remaining fractions the samples contained at least 50 particles.

All of the samples contain small amounts of cosmogenic argon. However, our calculations show that the amounts of cosmogenic $Ar^{36}$ and $Ar^{40}$ are always less than 1% of the total amount present; hence we will use the uncorrected data from Table 1 in the following discussion.

Figure 1 shows $Ar^{40}$ versus $Ar^{36}$. The 14163 results fall on, or near, the curve:

$$Ar^{40} = 1.87\ Ar^{36} + (2.21 \pm 0.10) \times 10^{-4}\ cm^3\ STP/g;$$

those of 14259 yield:

$$Ar^{40} = 0.81\ Ar^{36} + (1.87 \pm 0.10) \times 10^{-4}\ cm^3\ STP/g.$$

*Apollo 12 samples: 12033 size fractions*

Three size fractions and magnetic separates of the 105–250 $\mu$ fraction were prepared for analysis. The most striking characteristic of 12033 is its small trapped gas content relative to the concentrations seen in other Apollo 12 fines, such as 12070.

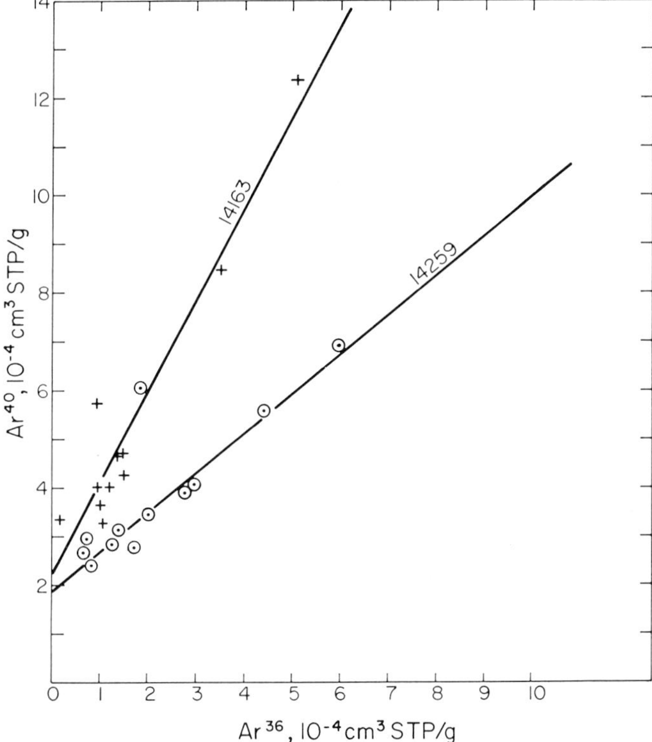


Fig. 1. $Ar^{40}$–$Ar^{36}$ correlations for Apollo 14 fines 14163 and 14259. Slopes $\varphi$ are 1.87 and 0.81; $Ar^{40}$-intercept values I are 2.21 and 1.87 $\times$ $10^{-4}$ $cm^3$ STP/g, respectively.

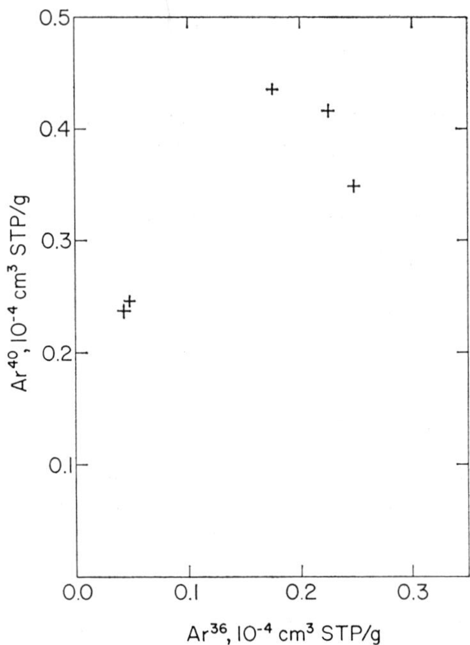


Fig. 2. $Ar^{40}$–$Ar^{36}$ correlation for Apollo 12 sample 12033. There is an increase of $Ar^{40}$ with $Ar^{36}$, but we suggest that this is due to contamination of an argon-poor component with argon-rich fines such as 12042, 12044, or 12070.

Funkhouser *et al.* (1971) have reported similarly small trapped gas contents in a bulk sample of 12033. The $Ar^{40}$–$Ar^{36}$ correlation is shown in Fig. 2. Although data points do not fall closely near a straight line, there is a general trend of increasing $Ar^{40}$ with increasing $Ar^{36}$. We believe that this trend is real but is simulated by the mixing (on the lunar surface) of a gas-poor component, perhaps most closely represented by the $>250\ \mu$ and 150–250 $\mu$ (weakly magnetic) fractions with small amounts of gas-rich fines in the Apollo 12 landing area.

*Apollo 15 samples*

The size fractions of 15601 were obtained in the same way as the Apollo 14 size fractions. Green glasses were *not* removed; however, the three coarsest fractions contained no such particles. Figure 3 shows the $Ar^{40}$–$Ar^{36}$ correlation:

$$Ar^{40} = 0.86\ Ar^{36} + (0.12 \pm 0.02) \times 10^{-4}\ \text{cm}^3\ \text{STP/g}.$$

## DISCUSSION

Linear correlations between $Ar^{40}$ and $Ar^{36}$ are now known to occur in fines from all four Apollo sites and in the Luna 16 soil (Table 2). Apparently such correlations are a moon-wide characteristic of regolith fines. Hence, the trapping of $Ar^{40}$ in the regolith must have occurred for certain between about 3300 and 3900 m.y. ago (the ages of Apollo 12 and Apollo 14 basaltic rocks) and possibly during most of the

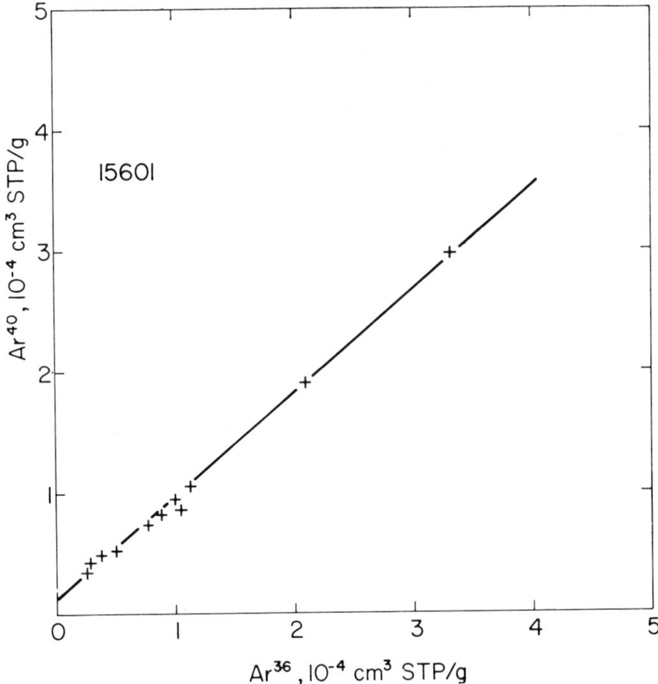


Fig. 3. $Ar^{40}$–$Ar^{36}$ correlation from Apollo 15 sample 15601. Slope is 0.86; $Ar^{40}$ intercept value is $0.12 \times 10^{-4}$ cm³ STP/g.

Table 2. Slope and intercept values of $Ar^{40}$–$Ar^{36}$ correlations.*

| Sample | $\phi$ | $\phi_{corr}$ | I | Reference |
|---|---|---|---|---|
| 10084 (< 1000 μ) | 1.06 | 1.11 | 0.71 | Heymann and Yaniv (1970a) |
| 10084 (< 300 μ) | 0.947 | | 0.747 | Hintenberger *et al.* (1970) |
| 10084 (< 130 μ) | 0.95 | | 0.40 | Eberhardt *et al.* (1970) |
| 10084 (ilmenite 22–105 μ) | 0.66 | | 0.094 | Eberhardt *et al.* (1970) |
| 10087 (< 200 μ) | 1.53 | 1.45 | 0.33 | Hintenberger *et al.* (1971) |
| 12070 (< 500 μ) | 0.41 | 0.28 | 0.37 | Hintenberger *et al.* (1971) |
| Luna 16 (12 particles) | 0.88 | 0.92 | 0.49 | Heymann *et al.* (1972) |
| Luna 16 (breccia only) | 0.65 | | 1.59 | Heymann *et al.* (1972) |
| Luna 16 (basalts only) | 0.65 | | 0.47 | Heymann *et al.* (1972) |
| 14163 (< 1000 μ) | 1.87 | 1.89 | 2.21 | This work |
| 14259 (< 1000 μ) | 0.81 | 0.89 | 1.87 | This work |
| 15601 (< 700 μ) | 0.86 | 0.75 | 0.12 | This work |

* $Ar^{40} = \phi Ar^{36} + I(10^{-4}$ cm³ STP/g); $\phi_{corr} = [Ar^{40}_{measured} - 4 \times 10^{-8}$ (K, ppm)]/$Ar^{36}_{measured}$.

moon's entire history. Which are the trapping mechanisms? What is their relative importance? And what, if any, are their time variations?

The first theory on the subject was given by Heymann *et al.* (1970) at the Apollo 11 Lunar Science Conference. They confirmed the conclusion reached by all other investigators that the apparent $K^{40}$–$Ar^{40}$ age of the 10084 fines was much greater than 4600 m.y., the most widely accepted age of the moon. This implies that the fines contain "excess" $Ar^{40}$ of some kind and that the $K^{39}$, complementary with $K^{40}$ from which the "excess" was produced, is not *now* present in 10084. Heymann *et al.* proposed that this potassium was, and still is, elsewhere in the moon. They suggested further that a small fraction of the $Ar^{40}$, produced by $K^{40}$ decay in the moon, had diffused into the lunar atmosphere, whence about 10% of it had become implanted into the regolith by the solar-wind mechanism (see Heymann and Yaniv, 1970a; Manka and Michel, 1970, 1971).

At the same conference, Dr. A. G. W. Cameron brought to our attention a paper by Murray *et al.* (1969), in which the transport of dust along the lunar surface by "nuée ardente"-like events had been discussed. A similar mechanism was recently advocated by Pai *et al.* (1972). One may argue that if the nuees contained $Ar^{40}$ in their gas phases, dust fragments carried along with them could have acquired some of this gas. This is a viable hypothesis; perhaps a fraction of the $Ar^{40}$ in fines is of this origin. But we doubt that this fraction could be very significant. Why? Because the $Ar^{36}$-correlated $Ar^{40}$ nearly always comprises by far the largest fraction of $Ar^{40}$. This gas is clearly surface correlated, which implies that near-equilibrium between dust grains and gasphase was never reached in any of the nuees, and this we find difficult to understand, especially for the very fine dust (diameter $\lesssim 40\ \mu$).

Baur *et al.* (1972) have raised a basic question about the solar-wind mechanism. They have extracted argon from the 14163 fines by stepwise heating and have shown that the bulk of solar wind $Ar^{36}$ and of atmospheric $Ar^{40}$ is given off together at about 750°C. Similar release patterns had been reported earlier for 10084 by Hohenberg *et al.* (1970) and Pepin *et al.* (1970). This, Baur *et al.* argue, is unexpected because "the retrapped $Ar^{40}$ implanted with energies in the 1 keV range should be released more easily (that is, at lower temperatures) than the solar wind $Ar^{36}$" (which impacts with energies of about 30 keV). From this Baur *et al.* conclude "that our experiments do not give evidence for *substantial* amounts of $Ar^{40}$ implanted at low energies." They propose an alternative theory: "Such an Ar distribution would result if the solar $Ar^{40}/Ar^{36}$ ratio is higher than expected and/or if an in situ decay of $K^{40}$ contributed to the surface correlated $Ar^{40}$, whereby at least some of the K was later redistributed."

The following comments are in order:

(1) That the surface-correlated $Ar^{40}$ has come to the moon with the solar wind has been questioned by several investigators because it seemed improbable that the $Ar^{40}/Ar^{36}$ ratio in the sun could be as large as 1.0 (see Heymann and Yaniv, 1970a; Eberhardt *et al.*, 1970). The data in Table 2 would seem to imply furthermore that the solar-wind $Ar^{40}/Ar^{36}$ ratio had varied *randomly* by nearly an order of magnitude between 3900 and 3300 m.y. ago. This is so unlikely that the theory of the solar origin of the bulk of the $Ar^{40}$ in lunar fines should be laid to rest.

(2) Nothing is known about the thermal release of argon, implanted at any energy, from silicate surfaces that are so heavily radiation-damaged that they are virtually amorphous (Bibring *et al.*, 1972).

(3) The solar wind mechanism *can* accelerate ions up to 30 keV or greater, provided that the near-lunar electric field gradient *in the past* has been roughly an order of magnitude greater than it is *today*, either because the strength of the magnetic field near the moon was greater, or because the average solar wind velocity was greater, or both. (However, the kinetic energy of $Ar^{36}$ increases as $v^2$, whereas that of at-

mospheric $Ar^{40}$ increases as $v$. The main point is that $Ar^{40}$ energies of tens of keV are possible.)

(4) The hypothesis of Baur *et al.* (1972) requires that most of the "missing" potassium either must have been wholly lost from the moon or must have diffused out of the regolith to places unknown.

Thus we see no reason to abandon the solar wind mechanism now in favor of another one that calls for a greater number of unproven assumptions.

Let us consider argon measurements in *bulk* fines as reported in the literature, because these can be added to the list of $Ar^{40}/Ar^{36}$ ratios in Table 2. The measured $Ar^{40}/Ar^{36}$ ratio of a *bulk* sample is always greater than the slope obtained from size fractions, because of the positive $Ar^{40}$ intercept (Figs. 1 and 3). However, Fig. 4

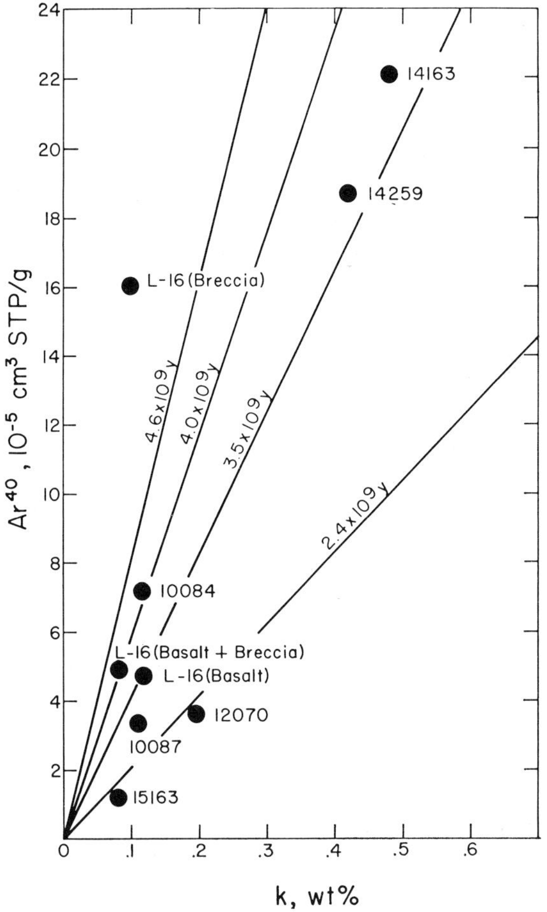


Fig. 4. $Ar^{40}$ intercept values plotted against K-contents. The curves represent loci of all points having the same apparent K–Ar age as indicated. Note the rough correlation between $Ar^{40}$ and K, which we have used to correct *bulk* argon measurements in the literature (correction used = $-4.0 \times 10^{-8}$ cm$^3$ $Ar^{40}$ g$^{-1}$ ppm$^{-1}$). Note also that the Luna 16 breccia fragments have apparent K–Ar ages greater than 4600 m.y.

shows that there exists a *gross* correlation between K content and $Ar^{40}$ intercept value. The most notable exceptions are the breccia fragments from Luna 16. The other points lie close to a line with slope of $4.0 \times 10^{-8}$ $cm^3$ STP of $Ar^{40}$ per gram and ppm of K. We have used this value to correct the bulk $Ar^{40}$ data. The corrected $Ar^{40}/Ar^{36}$ slopes are given in Table 3.

The Luna 16 breccias are exceptional because their $Ar^{40}$ intercept corresponds to an *apparent* $K^{40}$–$Ar^{40}$ age greater than 4600 m.y. We suggest that these fragments contain $Ar^{40}$ that was trapped by the parent breccia rocks when and if these formed in a base-surge type event (see Heymann and Yaniv, 1971).

The distinct difference in $Ar^{40}/Ar^{36}$ between the Apollo 11 fines (0.76–1.45) and the Apollo 12 fines (0.25–0.45) suggests the existence of large-scale "provinces" of atmospheric $Ar^{40}$ on the moon, "provinces" that are perhaps mare-wide. The Apollo 14 fines show that there appear to be also "provinces" of perhaps $10^4$–$10^6$ $m^2$ concomitant with the length of the traverse. The work of Marti and Lugmair (1971) and Nyquist and Pepin (1971) shows that there are $Ar^{40}/Ar^{36}$ variations with depth in the regolith, too.

It is possible that all of these $Ar^{40}/Ar^{36}$ variations are merely caused by differences

Table 3. Slope values $\phi_{corr}$ from bulk measurements of lunar fines.

| Sample | $Ar^{40}/Ar^{36}$ | K(ppm)* | $\phi_{corr}$ | Argon references† |
|---|---|---|---|---|
| 10004 | 1.17 | 1200 | 0.76 | (1) |
| 10005 | 1.4 | *1200* | 1.20 | (1) |
| 10010 | 1.1 | *1200* | 0.96 | (2) |
| 10010 | 1.2 | *1200* | 1.06 | (2) |
| 10084 | 1.23 | 1160 | 1.11 | (3) |
| 10084 | 1.13 | 1160 | 1.01 | (4) |
| 10084 | 1.125 | 1160 | 1.03 | (5) |
| 10084 | 1.09 | 1160 | 0.95 | (6) |
| 10087 | 1.53 | *1200* | 1.45 | (7) |
| 12001 | 0.615 | 2100 | 0.31 | (7) |
| 12032 | 1.71 | 3230 | negative | (2) |
| 12033 | 1.65 | 3400 | negative | (2) |
| 12042 | 0.73 | 2160 | 0.45 | (1) |
| 12042 | 0.685 | 2160 | 0.40 | (8) |
| 12044 | 0.543 | 2110 | 0.25 | (7) |
| 12060 | 0.81 | 1660 | 0.26 | (1) |
| 12070 | 0.57 | 1950 | 0.28 | (7) |
| 12070 | 0.65 | 1950 | 0.29 | (1) |
| Luna 16 | 0.98 | 830 | 0.92 | (9) |
| 14003 | 1.74 | 4200 | 1.27 | (8) |
| 14163 | 2.45 | 4800 | 1.89 | (8) |
| 14259 | 1.27 | 4200 | 0.89 | (8) |
| 15091 | 1.20 | *1330* | 1.01 | (8) |
| 15601 | 0.910 | 830 | 0.86 | (8) |

* Potassium data were taken from the literature, principally from the Apollo 11 and 12 proceedings, or from Apollo 11, 12, 14, 15 LSPET reports. When several investigators had reported on the same fines, an average was taken. When no potassium data were available, an average of all fines from one site was used; these numbers are printed in italics.

† Argon references: (1) Funkhouser *et al.* (1970); (2) Funkhouser *et al.* (1971); (3) Heymann and Yaniv (1970b); (4) Hintenberger *et al.* (1970); (5) Hohenberg *et al.* (1970); (6) Marti *et al.* (1970); (7) Hintenberger *et al.* (1971); (8) this work; (9) Vinogradov (1971).

in trapping efficiencies and $Ar^{40}$ retentivity of the different fines. Too little is known about the pertinent properties to allow any firm conclusions.

It is also possible that one "province" became "contaminated" with soils either poor or rich in atmospheric $Ar^{40}$, while the others did not. Thus the Apollo 12 $Ar^{40}/Ar^{36}$ values could be low because at that site the regolith is replete with materials such as 12032 and 12033, which contain little, if any, atmospheric $Ar^{40}$. But why are these fines themselves comparatively poor in atmospheric $Ar^{40}$?

We propose that variations in the concentration of neutral $Ar^{40}$ in the lunar atmosphere with time constitute one of the most important factors responsible for $Ar^{40}/Ar^{36}$ variations now seen in fines. The lifetime of neutral $Ar^{40}$ against ionization was probably never much greater than two weeks (Manka and Michel, 1970), so that fluctuations or long-term variations in the concentration of neutral $Ar^{40}$ was followed by the $Ar^{40}$ ion flux on a time scale that was very short compared to the rate of regolith evolution and turnover. Let us now test this working hypothesis.

In Fig. 5 we have plotted the $Ar^{40}/Ar^{36}$ ratios from Table 3 against the ages of basaltic rocks. We assume that these ages represent the date of *onset* of regolith formation in each landing area. This is to say that the exterior surfaces of Apollo 11 regolith fragments, produced from local rocks, are *on the average* older than surfaces of Apollo 12 regolith fragments and that Apollo 11 surfaces began to trap atmospheric $Ar^{40}$ some 350 m.y. earlier than those of the Apollo 12 fines. Our assumptions are probably safe for the Apollo 11, 12, and Luna 16 sites, which are geologically quite similar; but they are questionable for areas of much greater geologic diversity, such as the Apollo 14 and 15 sites. In the case of Apollo 14, Turner *et al.* (1971) have cautioned

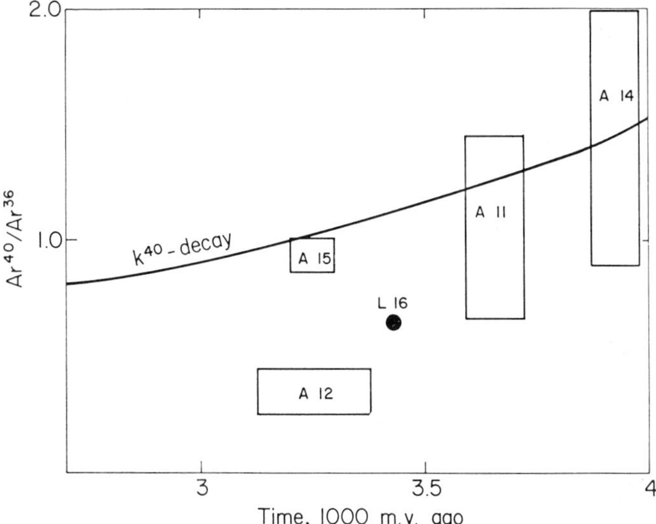


Fig. 5. $Ar^{40}/Ar^{36}$ versus "age" of regolith. The rock ages have been adopted to represent the onset of regolith formation at each location. Age-ranges from Wasserburg *et al.* (1972) and Schaeffer *et al.* (1972). $Ar^{40}/Ar^{36}$-ranges from Table 3. Note decrease of ratio for "younger" regolith. Apollo 15 fines do not agree with the trend.

that the basaltic rocks could be Imbrian ejecta or rocks produced by local intrusions. Despite the ambiguity of the time scale and the paucity of data points, we would like to speculate that Fig. 5 does show a trend in which $Ar^{40}/Ar^{36}$ ratios tend to become smaller for a "younger" regolith. Note that the decrease is considerably faster than what one would expect from $K^{40}$ decay in the moon alone. The light-grey, KREEP-rich fines 12032 and 12033 are consistent with the trend, because these soils have been associated by several authors with the Copernican ray, which traverses the Apollo 12 landing site (LSPET, 1970; Adams and McCord, 1971; Dollfus *et al.*, 1971; Morgan *et al.*, 1971). The Copernican event is now placed at about 850 m.y. ago (Eberhardt *et al.*, 1972; Pepin *et al.*, 1972).

Another trend is seen among the four samples of 10084 in Table 2. The $Ar^{40}/Ar^{36}$ ratio decreases with grain size and is particularly small in the ilmenite separates. But this is precisely what one would expect if the *average* neutral $Ar^{40}$ concentration in the atmosphere decreased during the evolution of the Apollo 11 regolith, because the exterior surfaces of finer particles should be, on the average, younger than those of coarser ones; and the exterior surfaces of nearly-pure mineral fragments should probably be younger still.

A number of inferences can be drawn from our hypothesis. First, the near-surface regions of the moon were apparently cooling between 4000 and 3000 m.y. ago; or alternatively, the extent and frequency of extrusions (for example, mare fills) decreased during this time. We have seen that the average neutral $Ar^{40}$ content of the lunar atmosphere seems to have decreased on a time scale faster than $K^{40}$ decay. Cooling of the moon would almost certainly retard the emanation of $Ar^{40}$ into the lunar atmosphere. However, this should not imply that the moon began to cool suddenly about 4000 m.y. ago and stopped cooling 3000 m.y. ago; it could well have already been cooling before 4000 m.y. ago and have continued to cool after 3000 m.y. ago. Also, since we do not know where in the moon the atmospheric $Ar^{40}$ was produced, we cannot say whether the whole moon or only parts of it were cooling. Second, judging from Fig. 5, regolith evolution took place on a time scale concomitant with the age differences of the various landing sites, that is, a few times $10^8$ yr. Third, mixing of regolith along the lunar surface over distances of hundreds of kilometers has always been relatively limited; otherwise, the differences in average $Ar^{40}/Ar^{36}$ ratios would have been ironed out.

The Apollo 14 samples 14163 and 14259 illustrate well the variations within a given landing area and the problems connected with their interpretation. The first sample has an $Ar^{40}/Ar^{36}$ slope of 1.87, the second, 0.81. Sample 14163 ("bulk sample") was scooped up a few tens of meters west of the LM, 14259 ("comprehensive sample"), about 120 meters west of the LM (Swann *et al.*, 1971). Both samples come from the smooth terrain material of the Fra Mauro formation. The only known significant difference between them is that the comprehensive sample was taken by the skimming off of roughly the top one centimeter of the regolith; the bulk sample contains fines from a greater, but unknown, depth. The area of Cone Crater materials begins about 700 meters to the east. Compositionally the two fines are very similar, in both major and trace element concentrations (Wänke *et al.*, 1972). The main difference seems to be that 14259 is significantly richer in glass than 14163 (Carr and Meyer, 1972).

Apparently the $Ar^{40}/Ar^{36}$ ratio is much smaller in the upper centimeter of the regolith than the average of roughly the first 10 centimeters, but there is no clear-cut evidence that the upper centimeter is derived from any particular crater nearby, such as Cone Crater. It is unlikely that the difference could be due to the removal of the top layer of fines, low in $Ar^{40}/Ar^{36}$, from the vicinity of the LM by the gases of the LM descent engine, because the "contingency fines" 14003 have an $Ar^{40}/Ar^{36}$ ratio intermediate between 14163 and 14259 (1.27); and they were collected from a spot even closer to the LM than 14163.

The most detailed depth information is available from the Apollo 12 double core 12025–12028 (Marti and Lugmair, 1971; Nyquist and Pepin, 1971). In this core, $Ar^{36}$ increases with depth, reaches a maximum at about 12 cm, then decreases again all the way to the bottom at about 40 cm. The $Ar^{40}$ content, on the other hand, increases continuously. From 0–12 cm the (bulk) $Ar^{40}/Ar^{36}$ ratio decreases, then it increases. Marti and Lugmair (1971) have reported apparent K–Ar ages of 3400 m.y. and 4800 m.y. for the topmost and deepest sample, respectively.

The problem with the interpretation of this core is that the apparent age of the topmost sample (depth, 3 cm) is in such good agreement with the ages of basaltic rocks from the Apollo 12 site that one cannot accurately estimate the amount of atmospheric $Ar^{40}$. The upper portion of the double core (to an unknown depth) may contain very little atmospheric $Ar^{40}$, indeed. Because of this we cannot tell whether the ratio atmospheric $Ar^{40}$/solar wind $Ar^{36}$ increases or decreases in the upper portion. In the deeper portions of this core the $Ar^{40}/Ar^{36}$ ratio convincingly increases with depth. However, it is much too early to say whether increasing $Ar^{40}/Ar^{36}$ ratios are a general characteristic of the regolith everywhere on the moon.

The working hypothesis developed earlier in this section predicts an increase in $Ar^{40}/Ar^{36}$ ratio with depth but only for a special model of regolith evolution, one in which the regolith was formed layer upon layer, with vertical mixing restricted at any given time to the upper few centimeters. For such a regolith, the "oldest" particle surfaces are at the bottom, the "youngest" at the top. It is tempting to explain the Apollo 12 double core along this line of reasoning, but Burnett *et al.* (1971) have shown that this core could not have been stratified in its present configuration for more than 50 m.y. Nyquist and Pepin (1971) have reported that the $Ar^{40}/Ar^{36}$ ratio at the bottom of the core is larger than the typical values in Apollo 12 surface fines of 0.3–0.4. This would seem to imply that the lowest portions of this core were already in place much longer than 50 m.y. ago, at a time when the $Ar^{40}/Ar^{36}$ implantation ratio was fairly large. Thus, we do not understand at this time how the regular $Ar^{40}$–$Ar^{36}$ stratification could have become established in this core during approximately the last 50 m.y.

Samples 12042 (northwest of Halo Crater), 12044 (southern rim of Surveyor Crater), 12060 (fines from Surveyor 3 scoop inside Surveyor Crater), and 12070 (contingency fines, near LM) show that the $Ar^{40}/Ar^{36}$ ratio in surface fines is fairly uniform (0.25–0.45) to the south and east of the LM. At least in this area (roughly 500 $m^2$) the ejecta blankets of nearby craters have apparently not produced any substantial $Ar^{40}/Ar^{36}$ variation at the surface, such as seen in the Apollo 14 landing site and, to a lesser degree, in the Apollo 11 landing site (Table 3). However, if the regolith is stratified with respect to $Ar^{40}/Ar^{36}$, one may expect a surface expression of such stratification, if not everywhere in a given landing site, then perhaps occasionally. The $Ar^{40}/Ar^{36}$ ratio in fines collected from the very top of the regolith could well be dominated by the most recent impacts on the moon. In certain cases local impacts could have been the major contributors (Apollo 11, Apollo 14?); in other cases distant impacts could have contributed much to the uppermost portion of the regolith (Apollo 12?). In the first case one could understand the $Ar^{40}/Ar^{36}$ variations of surface fines from a given landing site in terms of depth of excavation by the impact: the deeper the crater, the greater the proportion of regolith fines with "older" surfaces and the larger the $Ar^{40}/Ar^{36}$ ratio in the ejecta blanket. Whether this is basically correct or wrong cannot, unfortunately, be firmly settled with the information now at hand.

The Apollo 15 fines do not fit the trend of Fig. 5; their $Ar^{40}/Ar^{36}$ ratios are significantly greater than what one would expect on the basis of the ages of basaltic rocks, which are similar to those of the Apollo 12 basalts. Perhaps the regolith in a geologically complex area, such as Hadley Rille, is much older than the ages shown by the rocks dated thus far.

*Acknowledgment*—This work was supported by NASA Grant NGL 44-006-127.

## REFERENCES


Adams J. B. and McCord T. B. (1971) Alteration of lunar optical properties: Age and composition effects. *Science* **71,** 567–571.

Baur H., Frick U., Funk H., Schultz L., and Signer P. (1972) On the question of retrapped $Ar^{40}$ in lunar fines (abstract). In *Lunar Science—III* (editor C. Watkins), pp. 47–49, Lunar Science Institute Contr. No. 88.

Bibring J. P., Maurette M., Meunier R., Durieu L., Jouret C., and Eugster O. (1972) Solar wind implantation effects in the lunar regolith (abstract). In *Lunar Science—III* (editor C. Watkins), pp. 71–73, Lunar Science Institute Contr. No. 88.

Burnett D. S., Huneke J. C., Podosek F. A., Price Russ G. III, and Wasserburg G. J. (1971) The irradiation history of lunar samples. *Proc. Apollo 12 Lunar Sci. Conf., Geochim. Cosmochim. Acta* Suppl. 2, Vol. 2, pp. 1671–1680. MIT Press.

Carr M. H. and Meyer C. E. (1972) Petrologic and chemical characterization of soils from the Apollo 14 landing area (abstract). In *Lunar Science—III* (editor C. Watkins), pp. 116–118, Lunar Science Institute Contr. No. 88.

Dollfus A., Geake J. E., and Titula C. (1971) Polarimetric properties of the lunar surface and its interpretation. Part 3: Apollo 11 and Apollo 12 lunar samples. *Proc. Apollo 12 Lunar Sci. Conf., Geochim. Cosmochim. Acta* Suppl. 2, Vol. 3, pp. 2285–2300. MIT Press.

Eberhardt P., Geiss J., Graf H., Grögler N., Krühenbühl U., Schwaller H., Schwarzmüller J., and Stettler A. (1970) Trapped solar wind noble gases, exposure age and K/Ar-age in Apollo 11 lunar fine material. *Proc. Apollo 11 Lunar Sci. Conf., Geochim. Cosmochim. Acta* Suppl. 1, Vol. 2, pp. 1037–1070. Pergamon.

Eberhardt P., Eugster O., Geiss J., Grögler N., Schwarzmüller J., Stettler A., and Weber L. (1972) When was the Apollo 12 KREEP ejected? (abstract). In *Lunar Science—III* (editor C. Watkins), pp. 206–208, Lunar Science Institute Contr. No. 88.

Funkhouser J. G., Schaeffer O. A., Bogard D. D., and Zähringer J. (1970) Gas analysis of the lunar surface. *Proc. Apollo 11 Lunar Sci. Conf., Geochim. Cosmochim. Acta* Suppl. 1, Vol. 2, pp. 1111–1116. Pergamon.

Funkhouser J. G., Schaeffer O. A., Bogard D. D., and Zähringer J. (1971) Noble gas abundances in lunar material, I. Solar wind implanted gases in Mare Tranquilitatis and Oceanus Procellarum (submitted to *J. Geophys. Res.*).

Heymann D. and Yaniv A. (1970a) $Ar^{40}$ anomaly in lunar samples from Apollo 11. *Proc. Apollo 11 Lunar Sci. Conf., Geochim. Cosmochim. Acta* Suppl. 1, Vol. 2, pp. 1261–1267. Pergamon.

Heymann D. and Yaniv A. (1970b) Inert gases in the fines from the Sea of Tranquility. *Proc. Apollo 11 Lunar Sci. Conf., Geochim. Cosmochim. Acta* Suppl. 1, Vol. 2, pp. 1247–1259. Pergamon.

Heymann D., Yaniv A., Adams J. A. S., and Fryer G. E. (1970) Inert gases in lunar samples. *Science* **167,** 555–558.

Heymann D. and Yaniv A. (1971) Breccia 10065: Release of inert gases by vacuum crushing at room temperature. *Proc. Apollo 12 Lunar Sci. Conf., Geochim. Cosmochim. Acta* Suppl. 2, Vol. 2, pp. 1681–1692. MIT Press.

Heymann D., Yaniv A., and Lakatos S. (1972) Inert gases in twelve particles and one "dust" sample from Luna 16. *Earth Planet. Sci. Lett.* **13,** 400–406.

Hintenberger H., Weber H. W., Voshage H., Wänke H., Begemann F., and Wlotzka F. (1970) Concentrations and isotopic abundances of the rare gases, hydrogen and nitrogen in lunar matter. *Proc. Apollo 11 Lunar Sci. Conf., Geochim. Cosmochim. Acta* Suppl. 1, Vol. 2, pp. 1269–1282. Pergamon.

Hintenberger H., Weber H. W., and Takaoa N (1971) Concentrations and isotopic abundances of the rare gases in lunar matter *Proc. Apollo 12 Lunar Sci. Conf., Geochim. Cosmochim. Acta* Suppl. 2, Vol. 2, pp. 1607–1626. MIT Press.

Hohenberg C. M., Davis P. K., Kaiser W. A., Lewis R. S., and Reynolds J. H. (1970) Trapped and cosmogenic rare gases from stepwise heating of Apollo 11 samples. *Proc. Apollo 11 Lunar Sci. Conf., Geochim. Cosmochim. Acta* Suppl. 1, Vol. 2, pp. 1283–1310. Pergamon.

LSPET (Lunar Sample Preliminary Examination Team) (1970) Preliminary examination of lunar samples from Apollo 12. *Science* **167,** 1325–1339.

Manka R. H. and Michel F. C. (1970) Lunar atmosphere as a source of argon-40 and other lunar surface elements. *Science* **169,** 278–280.

Manka R. H. and Michel F. C. (1971) Lunar atmosphere as a source of lunar surface elements. *Proc. Apollo 12 Lunar Sci. Conf., Geochim. Cosmochim. Acta* Suppl. 2, Vol. 2, pp. 1717–1728. MIT Press.

Marti K., Lugmair G. W., and Urey H. C. (1970) Solar wind gases, cosmic ray spallation products, and the irradiation history of Apollo 11 samples. *Apollo 11 Lunar Sci. Conf., Geochim. Cosmochim. Acta* Suppl. 1, Vol. 2, pp. 1357–1367. Pergamon.

Marti K. and Lugmair G. W. (1971) $Kr^{81}$–Kr and K–$Ar^{40}$ ages, cosmic-ray spallation products, and neutron effects in lunar samples from Oceanic Procellarum. *Proc. Apollo 12 Lunar Sci. Conf., Geochim. Cosmochim. Acta* Suppl. 2, Vol. 2, pp. 1591–1606. MIT Press.

Morgan J. W., Laul J. C., Ganapathy R., and Anders E. (1971) Glazed lunar rocks: Origin by impact. *Science* **172,** 556–558.

Murray J. D., Spiegel E. A., and Theys J. (1969) Fluidization on the Moon (?). *Comments on Astrophysics and Space Physics* **1,** 165–171.

Nyquist L. E. and Pepin R. O. (1971) *Second Lunar Sci. Conf.* (unpublished proceedings).

Pai S. I., Hsieh T., and O'Keefe J. A. (1972) Lunar ash flows: How they work (abstract). In *Lunar Science—III* (editor C. Watkins), pp. 593–594, Lunar Science Institute Contr. No. 88.

Pepin R. O., Nyquist L. E., Phinney D., and Black D. C. (1970) Rare gases in Apollo 11 lunar material. *Apollo 11 Lunar Sci. Conf., Geochim. Cosmochim. Acta* Suppl. 1, Vol. 2, pp. 1435–1454. Pergamon.

Pepin R. O., Bradley J. G., Dragon J. C., and Nyquist L. E. (1972) K–Ar dating of lunar soils: Apollo 12, Apollo 14, and Lunar 16 (abstract). In *Lunar Science—III* (editor C. Watkins), pp. 602–603, Lunar Science Institute Contr. No. 88.

Schaeffer O. A., Husain L., Sutter J., Funkhouser J., Kirsten T., and Kaneoka I. (1972) The ages of lunar material from Fra Mauro and the Hadley Rille–Apennine front area (abstract). In *Lunar Science—III* (editor C. Watkins), pp. 675–677, Lunar Science Institute Contr. No. 88.

Swann G. A., Bailey N. G., Batson R. M., Eggleton R. E., Hait M. H., Holt H. E., Larson K. B., McEwen M. C., Mitchell E. D., Schaber G. G., Schafer J. B., Shepard A. B., Sutton R. L., Trask N. J., Ulrich G. E., Wilshire H. G., and Wolfe E. W. (1971) Preliminary geologic investigation of the Apollo 14 landing site. NASA SP-272, pp. 39–86.

Turner G., Huneke J. C., Podosek F. A., and Wasserburg G. J. (1971) $Ar^{40}$–$Ar^{39}$ ages and cosmic ray exposure ages of Apollo 14 samples. *Earth Planet. Sci. Lett.* **12,** 19–35.

Vinogradov A. P. (1971) Preliminary data on lunar ground brought to Earth by automatic probe "Luna-16." *Proc. Apollo 12 Lunar Sci. Conf., Geochim. Cosmochim. Acta* Suppl. 2, Vol. 2, pp. 1–16. MIT Press.

Wänke H., Baddenhausen H., Balacescu A., Teschke F., Spettel B., Dreibus G., Quijano M., Kruse H., Wlotzka F., and Begemann F. (1972) Multielement analysis of lunar samples (abstract). In *Lunar Science—III* (editor C. Watkins), pp. 779–781, Lunar Science Institute Contr. No. 88.

Wasserburg G. J., Turner G., Tera F., Podosek F. A., Papanastassiou D. A., and Huneke J. C. (1972) Comparison of Rb–Sr, K–Ar, and U–Th–Pb ages; lunar chronology and evolution (abstract). In *Lunar Science—III* (editor C. Watkins), pp. 788–790, Lunar Science Institute Contr. No. 88.

Proceedings of the Third Lunar Science Conference
(Supplement 3, *Geochimica et Cosmochimica Acta*)
Vol. 2, pp. 1981–1988
The M.I.T. Press, 1972

# Volatilized lead from Apollo 12 and 14 soils*

BRUCE R. DOE and MITSUNOBU TATSUMOTO

U.S. Geological Survey, Denver, Colorado 80225

**Abstract**—Lead has been volatilized from lunar soils 12033, 12070, and 14259 under vacuum at the following temperature intervals: 0°–600°, 600°–800°, 800°–1000°, and 1000°–1350°C. These studies confirm observations on previous volatilizations from soils 10084 and 12070 by Silver (1970, 1971) and Huey *et al.* (1971) that $^{206}Pb/^{204}Pb$ progressively increases and $^{207}Pb/^{206}Pb$ progressively decreases in the temperature range 600°–1000°C. In addition, we have found that the lead released from the three soils has values of $^{207}Pb/^{206}Pb$ in the range 0.57–0.59 in the 1350°C fraction, independent of whether the values for this ratio are greater or less than this value in the lower temperature fractions. Values of $^{208}Pb/^{206}Pb$ generally decrease from the 600°C to the 1350°C fractions from as great as 2 to as little as 0.7. These 1350°C fraction data suggest the presence of a highly refractory phase in the soils that may be about 4500 m.y. in age and that appears to have values of Th/U ($\sim$2.8–3.6), as calculated from $^{208}Pb/^{206}Pb$, that are less than those for the whole-soil (Th/U $\sim$3.7–3.9). The data that are available on Th/U in individual minerals from the studies by Burnett *et al.* (1971) and Haines *et al.* (1971, 1972) suggest that the highly refractory phase may be Zr–Ti minerals such as tranquilityite.

Volatilization data for soils 12033 and 14259 in the temperature range 600°–1000°C suggest a Pb–Pb secondary isochron "age" of about 4200 m.y. and indicate that the value of $^{238}U/^{204}Pb$ in these temperature fractions in the interval between about 4500 and 4200 m.y. was about 1000 for soil 14259 but more on the order of 100 for soil 12033.

## INTRODUCTION

SILVER (1970, 1972) AND TATSUMOTO (1970a) have made a case for mobile lead in the lunar soils. Studies of the isotopic composition of lead volatilized from lunar soils 14259, 12070, and 12033 as a function of temperature (in stepwise increments: 0–600°C, 600–800°C, 800–1000°C, and 1000–1350°C) were undertaken in an attempt to better understand this mobilized lead. (Preliminary data were given in Doe and Tatsumoto, 1972.) All investigations of 12070 showed the whole-soil to lie below concordia (Tatsumoto *et al.*, 1971; Cliff *et al.*, 1971; Wetherill, 1971; Silver, 1971), which, coupled with data on whole-soil 12033 and breccia 12034, confirmed Tatsumoto's (1970b) findings on breccia 12013 that materials collected by Apollo 12 had been subjected to "third events" in the age range of 1000 ± 1000 m.y. Despite the known complexity of the U–Th–Pb system in the soil 12070, both for whole-soil and for density fractions, the lead-isotope volatilization data already given by Huey *et al.* (1971) and by Silver (1971) look no more complex than those observed on soil 10084, whose whole-soil U–Th–Pb system gave concordant ages.

We performed additional volatilization experiments on 12033, which has a whole-soil analysis that lies even more below concordia than does that of 12070, and in which all the density fractions are strongly affected by the "third events" (Tatsumoto

*Publication authorized by the Director, U.S. Geological Survey.

*et al.*, 1971). The prime differences between 12070 and 12033 appear to be contents of more basalt and of very much more cindery glass in 12070 and more homogeneous and very much more ropy, brown glass in 12033 (Meyer *et al.*, 1971; Marvin *et al.*, 1971). The brown, ropy glass resembles the dark phase of breccia 12013 (Meyer *et al.*, 1971; Marvin *et al.*, 1971), and therefore perhaps the position of the whole-soil sample on the same discordia line for "third events" as for 12013 (between about 800 and about 4400 m.y.; Tatsumoto *et al.*, 1971) and the fact that 12070 whole-soil, poor in ropy brown glass, does not lie on the same "third-events" discordia line are of little surprise. In conducting volatilization experiments on 12033, a kind of soil distinctly different from soil 12070 is being investigated.

We performed volatilization studies on soil 14259 because the whole-soil has a $^{207}Pb/^{206}Pb$ of $>0.57$ (Tatsumoto *et al.*, 1972). Like the other two samples, however, it also is "normally" discordant; that is, it lies below concordia.

## Analytical Procedure

The given amount of sample was placed in a doubly preoutgassed graphite crucible, which, in turn, was placed in a vacuum chamber and evacuated to about $10^{-5}$ mm Hg. No powdered graphite was mixed with the sample. The sample was then heated by induction up to 600°C over a period of 4 hr and was held at that temperature overnight. Material volatilized was condensed on a fused silica "cold finger." The cold finger was replaced, and the sample was heated in a similar manner to each of the other temperature intervals. In the 1350°C fraction, the sample was heated to 1350° over a period of 4 hr and then held at 1350°C for 1 hr. The condensed material was then dissolved in doubly vacuum distilled concentrated $HNO_3$ and split into two fractions. The lead was coprecipitated with purified $Ba(NO_3)$, electroplated, and analyzed in the silica gel-phosphoric acid mixture as given in Tatsumoto (1970a, 1970b).

The crucible blank at 1500°C, including the blank in the chemical processing of about 1 ng (nanogram), was 28 ng for the first outgassing and 13 ng after the second outgassing. The blank should be somewhat lower after the second outgassing, and we have assumed that there was only chemical processing blank in the 600°, 800°, and 1000°C volatilizations (1 ng). The lead collected on the cold finger of the volatilization unit is aliquoted into two portions—one portion for lead-isotopic composition measurement and one portion for isotope-dilution determination of lead content. The crucible blank should therefore be apportioned between the two aliquots (4 ng for each) of the 1350°C volatilization runs. The resultant blank corrections are initially somewhat unsatisfying, because the data that looks the best before blank correction, the data with the least $^{204}Pb$, receive the largest correction. This relationship arises because the samples with the lesser values of $^{206}Pb/^{204}Pb$ tend to have the greater values of $^{207}Pb/^{206}Pb$. If the relationship were reversed (that is, if the samples with the greater $^{206}Pb/^{204}Pb$ had the greater values of $^{207}Pb/^{206}Pb$), then the samples with the greater values of $^{206}Pb/^{204}Pb$ would receive smaller blank corrections. Blank corrections have little effect on the values of $^{207}Pb/^{206}Pb$, as has been discussed previously by Tatsumoto *et al.* (1971). Some blank correction is certainly required, even though the raw data appear to be quite acceptable for most purposes, but it turns out that the correction is not critical to any of our conclusions.

## Data

Data from volatilization experiments are given for samples 14259, 12070, and 12033 in Table 1 and are plotted on Figs. 1, 2, 3, and 4. Certain features are conspicuous:

(1) In all three samples, the values of $^{206}Pb/^{204}Pb$ at 600°C are lower than the values at greater temperatures and are in accord with the data from similar studies on 10084 and 12070 by Huey *et al.* (1971) and Silver (1970, 1971).

Table 1. Data for volatilization extractions of lead from lunar soils 12033, 12070, and 14259. (600, 800, and 1000°C extractions are overnight; 1350°C extraction is for 1 hour).

| | Temp. (°C) | Type of Data* | Total Pb Recovered (ppm) | Atomic Ratios | | | | |
|---|---|---|---|---|---|---|---|---|
| | | | | $^{206}Pb/^{204}Pb$ | $^{207}Pb/^{204}Pb$ | $^{208}Pb/^{204}Pb$ | $^{207}Pb/^{206}Pb$ | $^{208}Pb/^{206}Pb$ |
| | | | Sample 12033 (0.889 gm) | | | | | |
| | 600 | raw | | 278.1 | 151.4 | 387.8 | 0.5445 | 1.3943 |
| | | *corr.* | *0.304* | *294.6* | *160.3* | *411.3* | *0.5441* | *1.3961* |
| | 800 | raw | | 786.1 | 388.3 | 916.9 | 0.4940 | 1.1664 |
| | | *corr.* | *0.470* | *885.3* | *437.1* | *1033.5* | *0.4937* | *1.1674* |
| | 1000 | raw | | 1825.7 | 870.7 | 1488.1 | 0.4769 | 0.8151 |
| | | *corr.* | *1.620* | *1931.6* | *922.2* | *1577.9* | *0.4774* | *0.8169* |
| | 1350 | raw | | 1042.9 | 600.6 | 874.0 | 0.5759 | 0.8380 |
| | | *corr.* | *0.920* | *1627.8* | *936.3* | *1355.6* | *0.5752* | *0.8328* |
| Total | | *corr.* | *3.314* | — | — | — | — | — |
| Whole soil | | *corr.* | *†4.22* | *1134* | *577.6* | *1028* | *0.5093* | *0.9065* |
| | | | Sample 12070 (0.844 gm) | | | | | |
| | 600 | raw | | 47.9 | 37.7 | 97.2 | 0.7876 | 2.0298 |
| | | *corr.* | *0.273* | *48.6* | *39.3* | *98.9* | *0.7881* | *2.0350* |
| | 800 | raw | | 100.9 | 68.8 | 145.4 | 0.6819 | 1.4415 |
| | | *corr.* | *0.614* | *102.2* | *69.8* | *147.6* | *0.6830* | *1.4442* |
| | 1000 | raw | | 431.6 | 237.7 | 431.5 | 0.5507 | 0.9998 |
| | | *corr.* | *1.686* | *440.0* | *242.6* | *440.9* | *0.5514* | *1.0020* |
| | ‡1350 | raw | | 459.8 | 264.9 | 404.7 | 0.5761 | 0.8801 |
| | | *corr.* | *0.763* | *555.6* | *320.1* | *486.6* | *0.5761* | *0.8758* |
| Total | | *corr.* | *3.336* | — | — | — | — | — |
| Whole soil | | *corr.* | *†3.51* | *433.4* | *251.7* | *415.6* | *0.5808* | *0.9589* |
| | | | Sample 14259 (0.735 gm) | | | | | |
| | ‡600 | raw | | 242.1 | 223.4 | 262.5 | 0.9218 | 1.0836 |
| | | *corr.* | *0.502* | *251.7* | *232.7* | *273.0* | *0.9245* | *1.0846* |
| | ‡800 | raw | | 495.9 | 366.7 | 584.2 | 0.7382 | 1.1820 |
| | | *corr.* | *3.173* | *503.2* | *372.4* | *594.4* | *0.7401* | *1.1812* |
| | ‡1000 | raw | | 1007.2 | 578.3 | 706.5 | 0.5745 | 0.7013 |
| | | *corr.* | *1.980* | *1045.0* | *600.7* | *734.3* | *0.5748* | *0.7027* |
| | 1350 | raw | | 903.1 | 538.0 | 615.7 | 0.5958 | 0.6817 |
| | | *corr.* | *1.202* | *1179.2* | *702.1* | *798.4* | *0.5954* | *0.6771* |
| Total | | *corr.* | *6.857* | — | — | — | — | — |
| Whole soil | | *corr.* | *†7.78* | *757.7* | *518.8* | *721.6* | *0.6847* | *0.9524* |

* Raw ratios are not corrected for laboratory blank or instrumental fractionation; corrected ratios (*corr.*) have been corrected for laboratory blank, 1 ng for each fraction; graphite crucible blank of 8 ng was applied to the 1350°C fraction.
† Concentration of lead in whole soil is an average of replicated determinations.
‡ These samples were analyzed twice on the mass spectrometer from one sample loading.

(2) The $^{206}Pb/^{204}Pb$ value at 1000°C is greater than that at 800°C.

(3) The value of $^{207}Pb/^{206}Pb$ at 600°C is greater than that ratio at 800°C, which, in turn, is greater than that ratio at 1000°C, in accord with the trend given for 10084 and 12070 by Huey *et al.* (1971) and Silver (1970, p. 1571, 1971).

(4) The value of $^{207}Pb/^{206}Pb$ at 1350°C, however, is always greater than that ratio at 1000°C. (Data on a 1350°C volatilization, where the sample is molten, have not previously been published.)

(5) The volatilization data for all samples overlap at a $(\Delta^{207}Pb)/(\Delta^{206}Pb)$ model age of about 4500 m.y. (see Figs. 1 and 2), and the $(\Delta^{207}Pb)/(\Delta^{206}Pb)$ model age for the leads released in the 1350°C volatilization is always close to 4500 m.y.

(6) The older the $(\Delta^{207}Pb)/(\Delta^{206}Pb)$ model age of the whole-soil, the greater the span of the values of $^{207}Pb/^{206}Pb$ for the volatilization extractions of each sample.

(7) Although the data from the volatilization experiments, in general, are similar

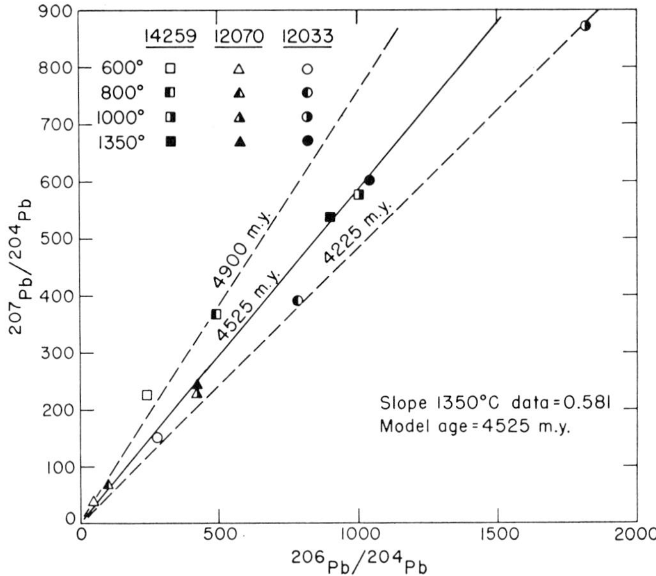


Fig. 1. $^{207}Pb/^{204}Pb$ versus $^{206}Pb/^{204}Pb$ for leads volatilized from lunar soils at various temperatures as shown. The lines are *primary* $\Delta(^{207}Pb/^{206}Pb)/\Delta(^{206}Pb/^{204}Pb)$ isochrons radiating from primordial lead.

to data obtained in the density fractions by Tatsumoto *et al.* (1971) and acid leaches of 10084 by Silver (1970), some important departures occur. For example, the spread in $^{207}Pb/^{206}Pb$ of the volatilization results on soil 12033 (Table 1) is greater than was found in the density fractions, primarily owing to the 1350°C data; therefore, the sample is internally more heterogeneous than might have been guessed from the density separation experiments. However, the spread in $^{207}Pb/^{206}Pb$ for volatilized leads from soil 12070 is less than was found in the density fractions.

(8) The values of $^{208}Pb/^{206}Pb$ are generally greater in the 600°C fractions and lesser in the greater temperature fractions, with a range from 2 for the 12070 fraction at 600°C to 0.7 for the 14259 fraction at 1350°C. This range is in marked contrast to the whole-soil samples, all of which are in the range 1.0 to 0.9.

## Discussion

### *Possible age and Th/U value of refractory phase*

As illustrated in Figs. 1 and 2, the values of $^{207}Pb/^{206}Pb$ for the 1350°C fractions are all close to an "age" of 4500 m.y., regardless of the variations in $^{207}Pb/^{206}Pb$ for the other temperature fractions when primordial lead from meteorites is used as the initial lead. This observation is of possible great importance and may reflect the presence of a refractory component in the three soils that is about 4500 m.y. old. Although he had not performed a volatilization experiment involving melting, L. T. Silver (oral communication, 1971) obtained analyses of the residues left from

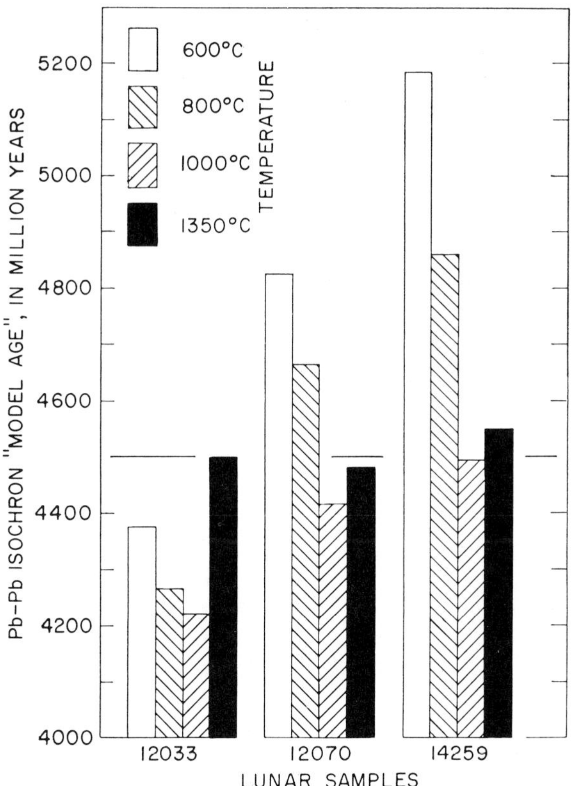


Fig. 2. The *primary* $\Delta(^{207}Pb/^{204}Pb)/\Delta(^{206}Pb/^{204}Pb)$ isochron "model age" for leads volatilized from lunar soils is shown. The initial lead isotopic compositions used in the model are those of primordial lead.

his 1000°C volatilizations that generally agree with our 1350°C volatilization data concerning a refractory component about 4500 m.y. in age.

If we assume that the refractory phase(s) were not altered in Th/U during the 4500 m.y. period, which seems reasonable if U/Pb was not altered, then the Th/U in the refractory phase seems to be less than that of the whole-soil in the three samples:

| Sample No. | Th/U Whole-Soil Observed | Th/U Refractory Phase(s) Calculated |
|---|---|---|
| 12033 | ~3.7 | ~3.4 |
| 12070 | ~3.9 | ~3.6 |
| 14259 | ~3.8–3.9 | ~2.8 |

Burnett *et al.* (1971), in their study of Th/U in the individual minerals of breccia 12013 and basalt 12040, reported values of >5 for whitlockite and <5 for apatite, both phosphates, mostly <1 for zircon (in general agreement with values of Th/U

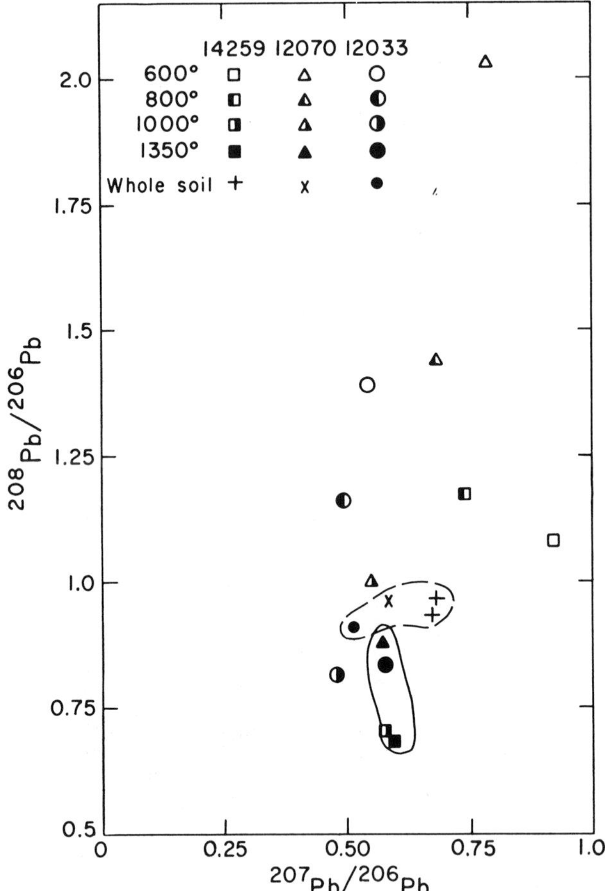


Fig. 3. $^{208}Pb/^{206}Pb$ versus $^{207}Pb/^{206}Pb$ for leads extracted from lunar soils at various temperatures and for the whole-soil samples. The field enclosed by a solid line encompasses the data on lead released between 1000° and 1350°C. The field enclosed by a dashed line includes the whole-soil data.

in terrestrial zircons), and 2.9 for a Zr–Ti phase. Haines *et al.* (1971) gave values of Th/U between 0.75 and 3.32 for phase $\beta$, a Zr–Ti phase in breccia 12013. Haines *et al.* (1972) also identify Zr–Ti phases as being major uranium-bearing phases in lunar soil. Although more work of this nature is clearly called for, the data available indicate that Zr–Ti phases such as *tranquilityite* appear to be the best candidates for the refractory phase.

*Possible homogenizing event at 4200 m.y. for soils 12033 and 14259*

There is some suggestion for 12033 and 14259 that—if the 1350°C refractory fraction is dropped from consideration—the U–Pb system was re-equilibrated by a

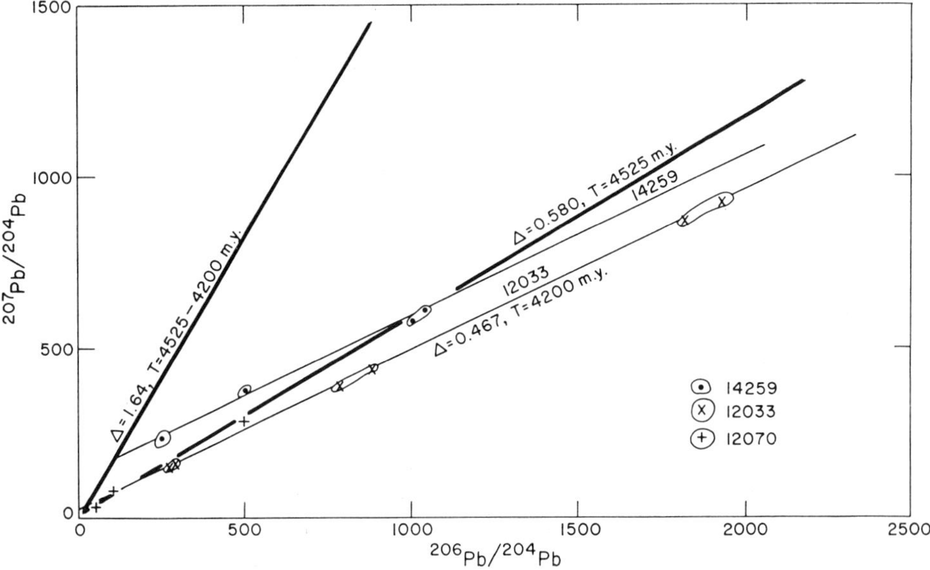


Fig. 4. $^{207}Pb/^{204}Pb$ versus $^{206}Pb/^{204}Pb$ for leads released from lunar soils in the increments 0°–600°C, 600°–800°C, and 800°–1000°C. Heavy solid lines are the *primary isochron* shown for the 1000°–1350°C fraction and soil 12070 in Fig. 1 of 4525 m.y. and the hypothetical *primary isochron* for lead generated in the interval from 4525 to 4200 m.y. ago. Light lines are *secondary isochrons* (that is, isochrons that do not pass through primordial lead), with slopes equivalent to 4200 m.y. in age, which would be an age or re-equilibration for the observed leads. The intersection of the light lines with the 4525-m.y.-old *primary isochron* is the isotopic composition for all the fractions if the sample was undisturbed for 4525 m.y. The intersections of *secondary isochrons* with the hypothetical *primary isochron* for leads generated in the interval between 4525 m.y. and4 200 m.y. is the isotopic composition of the 4525-m.y.-old sample at the time of its re-equilibration 4200 m.y. ago. Left symbol in each enclosed field is raw data, right symbol is corrected for blank.

"second event" at about 4200 m.y. (Fig. 4), because on each sample the data lie near a straight line of lesser slope than the *primary* isochrons (Fig. 1). Soil 12070 apparently was not subjected to this re-equilibration, because the slope of a line through the lower temperature data is not of lesser slope. For 14259, the value of $^{238}U/^{204}Pb$ in the soil material would appear to have been about 1100 during the period from 4525 m.y. to 4200 m.y., at which time $^{238}U/^{204}Pb$ was dropped to roughly 750. This value of 750 compares reasonably well with the value of about 850 actually observed in the soil today. The comparison is probably within the uncertainties of the blank correction and the complications of the "third event" observed in the whole-soil data (Tatsumoto *et al.*, 1971). For 12033, the value of $^{238}U/^{204}Pb$ would have been roughly 130 between 4525 and 4200 m.y. and subsequently 1210, as compared with a value of about 1580 actually observed in the soil; again, we find agreement possibly within the scope of the blank correction and the complication of the third event seen in the whole-soil data (Tatsumoto *et al.*, 1972).

*Acknowledgments*—The study benefited from discussions with L. T. Silver (California Institute of Technology) and Z. E. Peterman (U.S. Geological Survey). Ernest Wilson, Robert Terrazas, and John Stacey (U.S. Geological Survey) automated the data collection and reduction of the mass spectrometer. We thank C. J. Leistner (Ultra Carbon Company) for supplying us with the preoutgassed graphite crucibles and T. Murphy (National Bureau of Standards) for his suggestions on lead electroplating. We appreciate the laboratory assistance of D. M. Unruh (Colorado School of Mines) and P. Reed (Denver University). This research was supported in part by NASA Interagency Transfer Order No. T-2407A.

## References

Burnett D., Monnin M., Seitz M., Walker R., and Yuhas D. (1971) Lunar astrology—U–Th distributions and fission-track dating of lunar samples. *Proc. Second Lunar Sci. Conf., Geochim. Cosmochim. Acta* Suppl. 2, Vol. 2, pp. 1503–1519. MIT Press.

Cliff R. A., Lee-Hu C., and Wetherill G. W. (1971) Rb–Sr and U, Th–Pb measurements on Apollo 12 material. *Proc. Second Lunar Sci. Conf., Geochim. Cosmochim. Acta* Suppl. 2, Vol. 2, pp. 1493–1502. MIT Press.

Doe B. R. and Tatsumoto M. (1972) Volatilized lead from Apollo 12 and 14 soils (abstract). In *Lunar Science—III* (editor C. Watkins), pp. 177–179, Lunar Science Institute Contr. No. 88.

Haines E. L., Albee A. L., Chodos A. A., and Wasserburg G. J. (1971) Uranium-bearing minerals of lunar rock 12013. *Earth Planet. Sci. Lett.* **12,** 145–154.

Haines E. L., Gancarz A. J., Albee A. L., and Wasserburg G. J. (1972) The uranium distribution in lunar soils and rocks 12013 and 14310. In *Lunar Science—III* (editor C. Watkins), pp. 350–352, Lunar Science Institute Contr. No. 88.

Huey J. M., Ihochi H., Black L. P., Ostic R. G., and Kohman T. P. (1971) Lead isotopes and volatile transfer in the lunar soil. *Proc. Second Lunar Sci. Conf., Geochim. Cosmochim. Acta* Suppl. 2, Vol. 2, pp. 1547–1564. MIT Press.

Marvin U. B., Wood J. A., Taylor G. J., Reid J. B. Jr., Powell B. N., Dickey J. S. Jr., and Bower J. F. (1971) Relative proportions and probable sources of rock fragments in the Apollo 12 soil samples. *Proc. Second Lunar Sci. Conf., Geochim. Cosmochim. Acta* Suppl. 2, Vol. 1, pp. 679–699. MIT Press.

Meyer C. Jr., Aitken F. K., Brett R., McKay D., and Morrison D. (1971) Rock fragments and glasses rich in K, REE, P in Apollo 12 soils: Their mineralogy and origin. Second Lunar Science Conference (unpublished proceedings).

Silver L. T. (1970) Uranium–thorium–lead isotopes in some Tranquility Base samples and their implications for lunar history. *Proc. Apollo 11 Lunar Sci. Conf., Geochim. Cosmochim. Acta* Suppl. 1, Vol. 2, pp. 1533–1574. Pergamon.

Silver L. T. (1971) U–Th–Pb relations in Apollo 11 and Apollo 12 lunar samples. Second Lunar Science Conference (unpublished proceedings).

Silver L. T. (1972) Lead volatilization and volatile transfer (abstract). In *Lunar Science—III* (editor C. Watkins), pp. 701–703, Lunar Science Institute Contr. No. 88.

Tatsumoto M. (1970a) Age of the moon: An isotopic study of U–Th–Pb systematics of Apollo 11 lunar samples, II. *Proc. Apollo 11 Lunar Sci. Conf., Geochim. Cosmochim. Acta* Suppl. 1, Vol. 2, pp. 1595–1612. Pergamon.

Tatsumoto M. (1970b) U–Th–Pb age of Apollo 12 rock 12013. *Earth Planet. Sci. Lett.* **9,** 193–200.

Tatsumoto M., Knight R. J., and Doe B. R. (1971) U–Th–Pb systematics of Apollo 12 lunar samples. *Proc. Second Lunar Sci. Conf., Geochim. Cosmochim. Acta* Suppl. 2, Vol. 2, pp. 1521–1546. MIT Press.

Tatsumoto M., Hedge C. E., Doe B. R., and Unruh D. (1972) U–Th–Pb and Rb–Sr measurements on some Apollo 14 lunar samples (abstract). In *Lunar Science—III* (editor C. Watkins), pp. 741–743, Lunar Science Institute Contr. No. 88.

Wetherill G. W. (1971) Of time and the moon. *Science* **173,** no. 3995, 383–392.

Proceedings of the Third Lunar Science Conference
(Supplement 3, *Geochimica et Cosmochimica Acta*)
Vol. 2, pp. 1989–2001
The M.I.T. Press, 1972

# Trace element relations between Apollo 14 and 15 and other lunar samples, and the implications of a moon-wide Cl-KREEP coherence and Pt-metal noncoherence*

G. W. REED, JR., S. JOVANOVIC, and L. FUCHS

Chemistry Division, Argonne National Laboratory, Argonne, Illinois 60439

**Abstract**—The halogen, Hg, Ru, Os, U, Li, and Te contents in Apollo 14 soils 14259 and 14163, fragmental rocks 14321 and 14305, crystalline rock 14310, and some Apollo 15 soils are given. Br concentrations as high as 1 ppm are observed for the first time. The Cl remaining after aqueous leaching of samples is correlated with the total $P_2O_5$ in soils from all sites. Ru and Os are not coherent with the other Pt-metals, nor with each other; a lunar fractionation process appears to be necessary.

## INTRODUCTION

THE CONCENTRATION and, in some cases, aqueous leachability and volatility of a number of trace elements have been determined in Apollo 14 soils—including trench samples, fragmental rocks, and crystalline rock 14310. A few Apollo 15 soil samples have also been measured. Concentrations of the halogens, Cl, Br, and I, and a few F results are reported. The significance of the leachable and nonleachable fractions are discussed, and the correlation with total phosphate content is explored. The Pt-metals, Ru and Os, were determined, and on the basis of the concentration variations observed, inferences concerning lunar versus extralunar origins are considered. Hg was volatilized from samples by stepwise heating; this has proved to be an effective way of studying how this volatile element is incorporated in rocks and soil. U, Li, and, in a few cases, Te were also determined.

Of the Apollo 14 samples, the fragmental rocks present the greatest problem in sampling but also opportunities to sample an assortment of rocks of different geneses and thermal histories. In our work so far we have treated these rocks as consisting of two easily separated fractions. A dark part, which consisted of $\sim 90\%$ dark clasts, and a light part, which constituted the matrix and was composed of a mixture of $\sim 50\%$ dark and $50\%$ gray and white fragments. The mineralogy of the dark fractions of 14305,64 and 14321,185 is generally similar to that of the light fractions, but there are marked contrasts between the texture of the dark and that of the light. The dark clasts of both samples consist of relatively large subangular fragments of plagioclase, pyroxene, and olivine, up to 0.1 mm in diameter, embedded in a fine-grained (0.01 mm) matrix of silicates and abundant needles or plates of small (1–3 $\mu$) opaque grains. The hard and vuggy dark clasts appear to be partially recrystallized breccias. The light fraction, interstitial to the well-defined dark clasts, contains an assortment of less consolidated angular mineral and glass grains and both recrystallized and unaltered rock fragments.

---

* Work performed under the auspices of the USAEC and NASA.

Table 1. Halogens, ruthenium, osmium, mercury, and other trace elements in Apollo 14 and 15 samples. (All data on a line are for a single aliquant.)

| Sample | Cl $r$† (ppm) | Cl $l$† (ppm) | Br $r$ (ppm) | Br $l$ (ppm) | I†† (ppb) | Ru (ppb) | Os (ppb) | Hg ≤130°C (ppb) | Hg Total (ppb) | U $r$ (ppm) | U $l$ (ppm) | Li $r$ (ppm) | Li $l$ (ppm) | Te (ppb) |
|---|---|---|---|---|---|---|---|---|---|---|---|---|---|---|
| 14163,108 | 42 | 9.4 | 1.1 | 0.23 | 9.2 | | | | | 1.8 | 0.02 | 19 | < 0.07 | 0.34 |
| 14163,108 | | | 0.29 | | | 17 | 45 | 0.12 | 2.7 | 4.1 | | | | |
| 14259,96 | 37 | 18 | 0.17 | 0.03 | 4.0 | | | | | 4.0 | < 0.02 | 19 | < 0.01 | nd |
| 14259,96 > 150μm | 32 | 3.7 | 0.70 | 0.21 | 23 | | | | | 4.6 | 0.03 | 20 | < 3.5 | nd |
| 14259,96 < 74μm | 36 | 7.9 | 0.26 | 0.30 | 172 | | | | | 6.0 | 0.19 | 12 | ≤ 0.11 | nd |
| 14259,96 | | | 0.19 | | | 19 | 50 | 0.14 | 1.6 | 3.3 | | | | |
| 14259,96 > 150μm | | | 0.21 | | | 26 | 70 | | | 4.0 | | | | |
| 14259,96 74–150μm | | | 0.19 | | | 14 | 46 | | | 2.7 | | | | |
| 14259,96 < 74μm | | | 0.43 | | | 37 | 14 | | | 8.9 | | | | |
| Sum of fractions | | | 0.26 | | | 22 | 77 | | | 4.7 | | | | |
| 14148,13 trench top | | | | | | (17) | 107 | 1.2 | 7.3 | | | | | |
| 14156,3 middle | | | | | | (8.8) | 40 | 1.4 | 5.9 | | | | | |
| 14149,21 bottom | | | | | | (38) | 82 | 1.2 | 7.3 | | | | | |
| 14321,185* (light) | 46 | 3.8 | 0.33 | 0.10 | 82 | | | | | 2.0 | 0.01 | 32 | < 1.3 | ≲54 |
| 14321,185 (dark) | 61 | 3.9 | 0.91 | 0.11 | 51 | | | | | 6.3 | 0.01 | 40 | < 0.4 | ≲19 |
| 14305,84–1** | | | | | | ≤ 23 | 31 | 0.83 | 2.4 | | | | | |
| 14305,84–2 | | | | | | < 8.3 | 20 | 0.28 | 1.5 | | | | | |
| 14305,64 (light) | | | | | | ≤ 33 | 22 | 0.80 | 11 | | | | | |
| 14305,64 (dark) | | | | | | ≤ 14 | 22 | 0.11 | 0.68 | | | | | |
| 14305,64 (light) | 9.3 | 2.0 | 0.11 | 0.08 | 30 | | | | | 5.8 | 0.03 | 50 | < 1 | ≲62 |
| 14305,64 (dark) | 20 | 1.3 | 0.38 | 0.59 | 41 | | | | | 3.7 | 0.03 | 56 | < 1 | ≲65 |
| 14310,124–1** | | | | | | ~ 15 | 10 | 0.35 | 0.76 | | | | | |
| 14310,124–2 | | | | | | < 13 | 11 | 0.22 | 0.94 | | | | | |
| 14310,124 | 4.2 | 1.7 | 0.75 | 0.10 | 4.7 | | | | | 3.5 | 0.01 | 27 | 0.14 | ~ 51 |
| 14310,122–C | | | | | | 11 | 12 | 18 | 42 | | | | | |
| 15091,32 sta 2 | 11 | 7.3 | 0.42 | 0.29 | 2.1 | | | | | 0.96 | 0.01 | 16 | < 0.2 | ~ 9 |
| 15091,32 sta 2 | | | | | | 25 | 24 | 0.5 | 10 | | | | | |
| 15251,28 sta 6 | 21 | 3.4 | 0.52 | 0.25 | 8.3 | | | | | 2.0 | 0.02 | 16 | — | nd |
| 15041,32 sta 8 | 20 | — | 0.29 | 0.28 | 2.8 | | | | | 1.7 | 0.02 | 15 | ~ 0.2 | nd |
| 15041,32 sta 8 | | | | | | 15 | 28 | 1.3 | 3.0 | | | | | |
| 15231,29 sta 2 | | | | | | 16 | 14 | 7.8 | 11 | | | | | |

* 14321,185 and 14305,64 are interior samples.
** 14310,124 and 14305,84 are exterior samples: 1 = outer 2 mm and 2 = 1 cm below surface and just below surface, respectively.
† $r$ = residue after leaching; $l$ = leach solution.
†† I detected in leach only.
( ) = lower limits due to possible volatilization losses.
nd = not detected.

The experimental procedures are essentially those used previously (Reed and Jovanovic, 1969). Samples were handled only in a $N_2$ dry-box to avoid possible contamination, a major hazard to measurements of some of the elements. The restrictions on separation of phases were largely imposed by the necessity of operating inside this box.

## RESULTS

### *Halogens*

The halogen contents of the Apollo 14 and 15 samples measured are given in Table 1. The Cl contents of soils 14163,108 and 14259,96 are about 50 ppm, and its distribution does not appear to be grain size dependent. Wänke *et al.* (1972) report similar 14259 Cl contents. With the exception of one 14163,108 measurement of 1.3 ppm, the Br contents of these two soils average about 0.24 ppm. The light and dark fractions of fragmental rock 14321,185 have Cl contents of 50 and 65 ppm, respectively; these are similar to the soil values (~56 ppm), but 14305,64 has much less Cl, 11 and 21 ppm in the light and dark fractions, respectively. The Br in the 14321,185 and 14305,64 light fractions is 0.43 and 0.19 ppm, respectively; these are similar to the soil values. The dark fractions in both rocks contain ~1 ppm Br. The dark fractions from both breccias are compositionally quite uniform compared to the diverse mixture in our lighter fractions. Nonleachable Cl and Br are correlated in the dark samples; Cl decreases from 61 to 20 ppm and Br from 0.91 to 0.38 ppm in 14321,185 and 14305,64, respectively. The Cl/Br ratios are not constant. As will be discussed later, it appears that nonleachable Cl is present in a phosphate mineral; nonleachable Br may be also but in variable amounts.

The crystalline rock 14310,124 also contains a high Br concentration (0.85 ppm), while its Cl content is quite low, ~6 ppm. The high Br concentrations in Apollo 14 samples are in the direction needed, though inadequate, to explain the amounts of neutron capture-produced $^{80}Kr$ and $^{82}Kr$ reported by Lugmair and Marti (1971).

The leachability of Cl and Br are different. In general Br is more leachable, and this is probably consistent with its characterization as a dispersed element.

The total Cl and Br contents in Apollo 15 soils from Stations 2, 6, and 8, the front and maria, are almost constant at ~20 ppm Cl and ~0.7 ppm Br. Morgan *et al.* (1972a, 1972b) report Br contents of Apollo 14 soils and fragmental rocks in the ranges we observed, but all of their Apollo 15 soil values are lower (0.09–0.16 ppm) than those reported here.

Some "moon-wide" observations on Cl and Br are suggested:

(1) The Apollo 14 and 15 Cl and Br contents are compared with those reported for Apollo 11, 12, and Luna 16 samples in Fig. 1. We have used our data (Reed and Jovanovic, 1970, 1971), with the exception of the Luna 16 results of Vinogradov (1971), in order to minimize systematic interlaboratory differences. For soils, Apollo 11 and 15 have about the same average Cl contents, while Apollo 12, 14, and Luna 16 are higher. Cl contents are lower for rocks than for soils, with the exception of Apollo 11. Br contents at the various sites overlap, with the exception of Apollo 15 values, which tend to be higher. In contrast to Cl, with the exception of Apollo 12, the Br in rocks is comparable to that in soils.

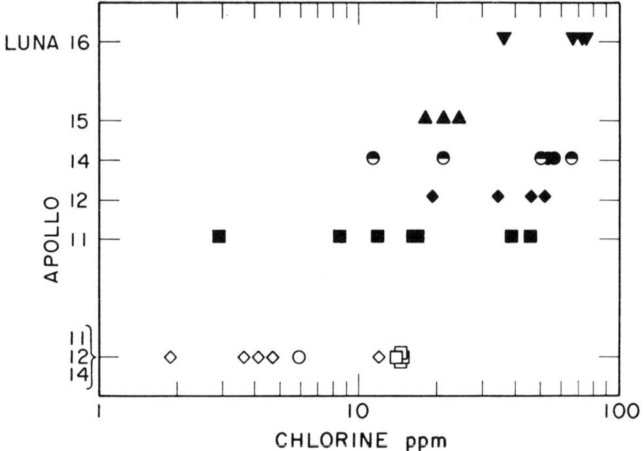


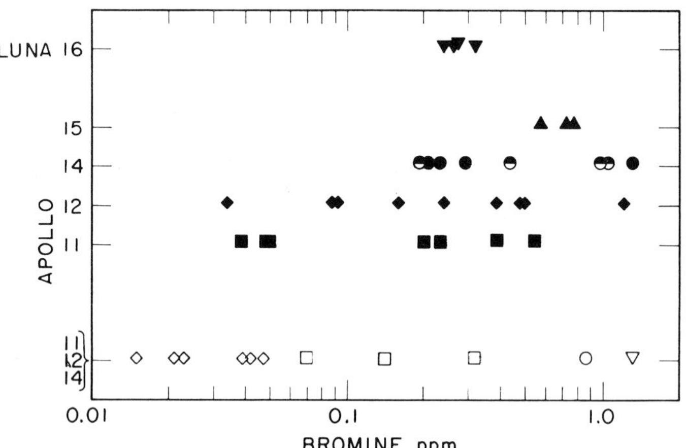


Fig. 1. Chlorine and bromine contents in soil and rock samples from different sites. All Apollo data are from our laboratory. Luna 16 data are from Vinogradov (1971). Symbols: closed = soil and breccia; half-open = fragmental rocks; open = igneous rocks.

(2) The larger variability of the Cl contents in comparison with Br contents may be caused by its association with phosphate minerals. Nonleachable Cl contents in soils and breccias from four sites are compared with total $F_2O_5$ in Fig. 2. A correlation exists with a mean $Cl/P_2O_5$ ratio of 0.008 and a spread of about 12%. The ratio from Luna 16-B (Vinogradov, 1971) of 0.013 is an upper limit,since it is for total Cl. Compared to the soils, the igneous rocks have lower $Cl/P_2O_5$ ratios: 0.001 (14310,124), 0.004 (10044, 12021), 0.006 (10017,72 and 10072,24). The higher ratio for the latter two Apollo 11 high-Rb rocks may be due to the presence of chlorapatite. An apatite with a ratio of 0.008 has been reported in rock 10017 (French *et al.*, 1970). Thus the soils and breccias for all sites should contain a phosphate, or a phase with associated

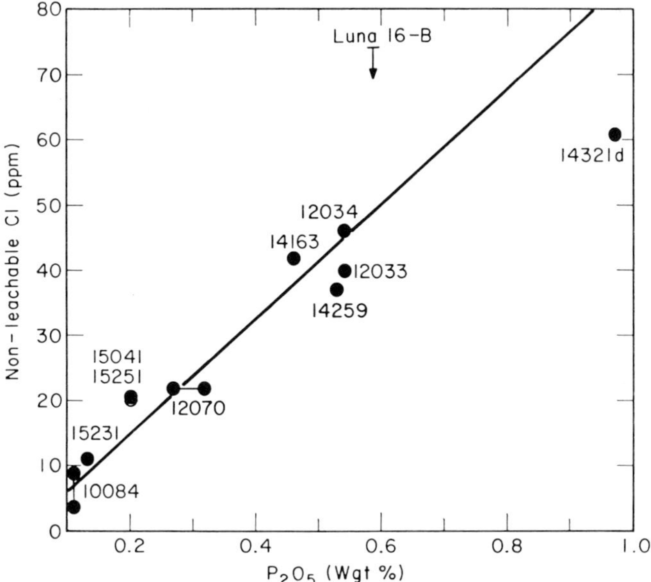


Fig. 2. Correlation of nonleachable chlorine to $P_2O_5$ in soils from different sites indicates a moon-wide coherence of Cl/P ratios. $P_2O_5$ data from: 10084—Maxwell *et al.* (1970); 15041, 15251, and 15231—LSPET (1972); 12070—J. Warner, MSC, Apollo 12 data compilation; 14163—LSPET (1972); 14259—Rose *et al.* (1972); 12033—Cuttitta *et al.* (1971); Luna 16—Vinogradov (1971).

Cl and $P_2O_5$, which has a higher Cl/P ratio than found in most crystalline rocks; moreover, it must be present in amounts sufficient to maintain the constant Cl/P ratio. Of the two known lunar phosphates, whitlockite and apatite, the latter appears as the more likely candidate. Lunar apatites with appropriate $Cl/P_2O_5$ ratios in the range greater than 0.008 have been reported in Luny Rock #1 (Albee and Chodos, 1970) and in 12013 (Lunatic Asylum, 1970). Lunar whitlockites have much too low ratios; to quote a few examples, values as high as 0.001 but usually $<0.0005$ are reported (Keil *et al.*, 1971; Albee and Chodos, 1970; Lunatic Asylum, 1970; Fuchs, 1971). It is rather surprising that the incidence of reported chlorapatite in the soils and breccias is not larger.

We have found that literature values for the $K_2O$ and REE contents for those samples plotted in Fig. 2 are also correlated with Cl and P. It appears that P, Cl, K, and REE must have coexisted in a sympathetic relationship at the time of soil and breccia formation. The source magma supplied specific amounts of these components, which have not been seriously altered by subtraction or addition of material.

It is interesting to note that although the Apollo 11 high-Rb rocks such as 10017 and 10072 have a $Cl/P_2O_5$ ratio near that of soils and relatively high REE and $K_2O$ contents compared to low-Rb rocks, their total KREEP-Cl contents are too low to permit this type of rock to account for the amounts of these elements found in 12033 and 12034 and Apollo 14 soils and breccia samples.

(3) The nonleachable/leachable Cl and Br ratios were shown to be independent of sample type and site for Apollo 11 and 12 (Reed and Jovanovic, 1971). Histograms of the ratios matched those of terrestrial basalts but not those of meteorites. Apollo 14 and 15 sample data conform to this same pattern, suggesting a parallel in the evolution of basalts on a moon-wide basis and on the earth.

F has been measured in Apollo 14 samples. Some uncertainly in the reliability of the monitoring technique used in earlier experiments has been under investigation, and only recent results will be given. The F contents of Apollo 14 and 15 soils fall in the range of 100–200 ppm, in good agreement with those reported by Wänke (1972). The most interesting results, however, are for the Apollo 15 anorthosite 15415,42 and gabbroic anorthosite 15418,30-08. They contain a few ppm or less of F; these results are lower than values reported earlier for a plagioclase concentrate from 10084,75 and an anorthosite inclusion in breccia 12073,34 (Reed *et al.*, 1971) and illustrate the uniqueness of the high F observed at Tycho by Surveyor VII (Patterson *et al.*, 1970).

The I in Apollo 14 and 15 samples is observed only in the aqueous leach as discussed previously for other samples (Reed and Jovanovic, 1971). The fact that I appears to be almost exclusively in the aqueous leach may be consistent with a trend suggested by the increased leachability of Br relative to Cl. Concentrations in Apollo 14 and 15 soils fall between 2 and 10 ppb (Table 1), although two grain size fractions from 14259,96 gave considerably higher values. Both fragmental rocks, 14321,185 and 14305,64, contain appreciably more I (30–80 ppb) than the soils; similar concentrations were found in Apollo 11 and 12 soils and some rocks by Reed and Jovanovic (1971) and by Alexander *et al.* (1972).

### *Mercury*

Hg concentrations in Apollo 14 samples usually are less than 2–3 ppb, with the notable exception of the trench samples, which contain $\sim 7$ ppb Hg. The trench samples also contain the higher concentrations of labile Hg (volatilized at $\leq 130°C$). These trench samples do not show an expected variation with depth probably because of mixing during sample collection, as also suggested by cosmogenic radionuclide data (Eldridge *et al.*, 1972; Keith *et al.*, 1972) and rare gas data (Bogard *et al.*, 1972). The 15231,29 soil sample obtained from under a boulder should have been cold, and it indeed has a large concentration of labile Hg compared to that in 15091,32, a nearby soil sample exposed to lunar daytime temperatures. Their total Hg concentrations of $\sim 10$ ppb are greater than in most other soil samples.

The most striking aspect of the Apollo 14 Hg measurements is the large release above 450°C during stepwise heating. In samples from other sites only a small fraction of the Hg remained above this temperature; in Apollo 14 most samples still retained 30% or more Hg (Fig. 3). We interpret the presence of Hg in highly retentive sites as being due to high temperature equilibration (Jovanovic and Reed, 1968). Since these samples are known to be associated with the impact that excavated the Imbrium Basin, both high temperature quenching and shock implantation, as well as

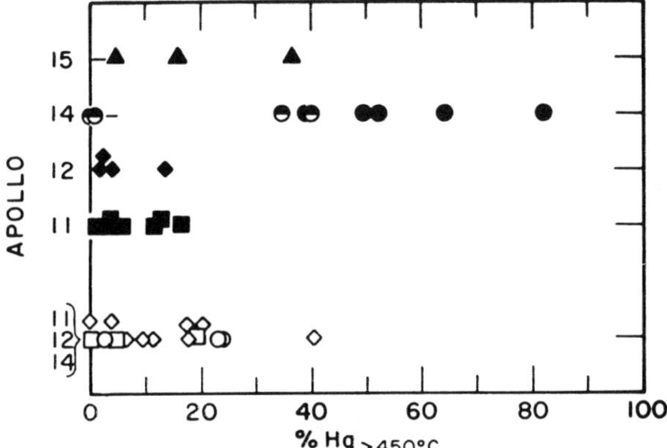


Fig. 3. Comparison of Hg fractions released at >450°C in stepwise heating experiments from Apollo 11–15 samples. The large fraction of Hg in the retentive sites in Apollo 14 samples is attributed to the thermal and shock effects associated with the Imbrium event. Symbols: closed = soil and breccia; half-open = fragmental rocks; open = igneous rocks.

pre-existing high temperatures at the impact site may account for the Hg in retentive sites.

*Ruthenium and osmium*

Our data for these elements are given in Table 1 and Fig. 4. Ru concentrations are constant at about 16 ppb in the Apollo 14 and 15 samples measured. Similar concentrations are found in Apollo 12 rocks (Reed and Jovanovic, 1971) and Luna 16 samples (Vinogradov, 1971). It is of interest to note the very high Ru concentrations of 2.13 ppm reported in 12053,31 basalt (Reed and Jovanovic, 1971) and 6 ppm in Luna 16 basalt (Vinogradov, 1971).

The Os concentrations in Apollo 14 samples vary considerably, decreasing from soil to fragmental to igneous rock in the order $\sim 67 \rightarrow 24 \rightarrow 11$ ppb. The Os content in Apollo 15 soil 15231,29 from Station 2 appears to be lower than that in soils from Stations 6 and 8, but the determination of whether this is significant must await further Apollo 15 sample measurements.

In contrast to our Os results on the Apollo 12 rocks in which there were significant differences between interior and exterior samples, no difference is observed in Os concentrations for interior and exterior samples of either fragmental rock 14305 or crystalline rock 14310. Lovering and Hughes (1971) found an exterior-surface enrichment in rock 12020 in contrast to our observation of a depletion in rocks 12022, 12052, and 12053 (Reed and Jovanovic, 1971).

In Table 2 we have summarized the Pt-metal results, from our own work and that of a number of other investigators, in the form of metal/metal ratios. The abundances reported for the other Pt-metals, Ir, Re, and Pd, appear to be correlated to one another

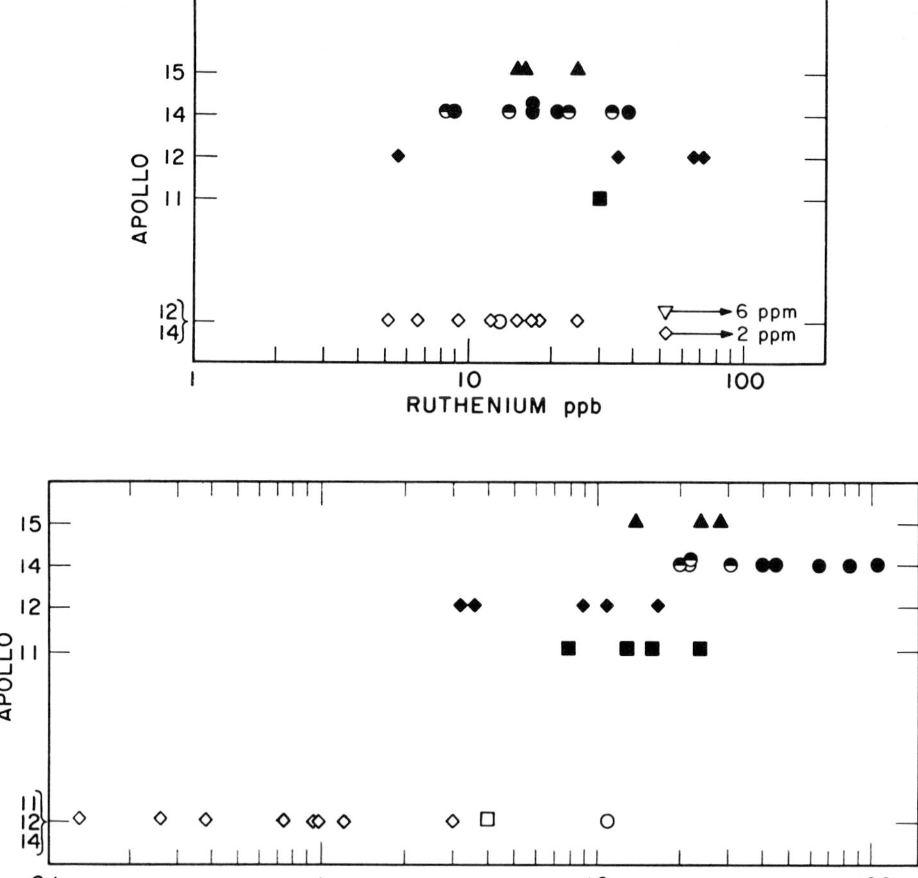


Fig. 4. Os and Ru contents in soil and rock samples from different sites. Luna 16 data are from Vinogradov (1971). Symbols: closed = soil and breccia: half-open = fragmental rocks; open = igneous rocks.

and to meteoritic (solar system) abundances; as a consequence, their abundances in the lunar surface have been associated with extralunar (meteoritic) sources (Morgan *et al.*, 1972b; Baedecker *et al.*, 1972). This association cannot be applied to Ru and Os without qualification. The relations between Os and components identified as largely extralunar, namely, Ni content and metal content, are plotted in Fig. 5. There appear to be at least two, possibly exogenous, populations of Os bearing phases and an indigenous contribution represented by the Os contents of igneous rocks and anorthosite fractions of soils and breccias. The other platinum metals, for example Ir, do not fit the pattern seen for Os, although they are correlated with the Ni and metal contents.

Table 2. Intercomparisons of Pt-Metals.*

| Sample | Ir/Re | Pd/Ir | Ru/Ir | Os/Ir | Ru/Os |
|---|---|---|---|---|---|
| 14163 | 11, 18 | 2.5, 1.5 | 1.5, 0.9 | 4.0, 2.4 | 0.4 |
| 14259 | 8 | 1.3 | 1.4 | 4.3 | 0.3 |
| 14305 | 6 | 1.5 | $\lesssim 1$ | 2.4 | $\leq 1$ |
| 14310 | — | — | 1.4 | 1.3 | 1.1 |
| Apollo 12 soils | 10 | — | 6.5, 7.9, 0.65 | 0.9, 1.0, 3.8 | 7.8, 20, 0.5 |
| Apollo 12 rocks | 7 | — | 88 | 4.3, 1.6, 5.3, 13 | 3.1, 26, 20 |
| Apollo 11 soils | — | 1.1 | 1.4, 3.9 | 1.8, 1.4, 0.9, 2.7 | 1.6 |
| Apollo 11 rocks | — | — | — | 8.7 | — |
| Apollo 11 anorthosite concentrates | — | — | — | 1.3, 2.1 | — |
| Meteorite | 9 | 1.9 | 2.8 | 1.7 | 1.7 |

* Ir: Ganapathy *et al.* (1970), Laul *et al.* (1971), Morgan *et al.* (1972), Wänke *et al.* (1971, 1972), Wasson and Baedecker (1971), Baedecker *et al.* (1972).
Re: Herr *et al.* (1971), Herpers *et al.* (1972).
Pd: Wänke *et al.* (1971, 1972), Ganapathy *et al.* (1970).
Os and Ru: Reed and Jovanovic (1971), Reed, Jovanovic, and Fuchs (1971, 1972).
Direct comparisons of data in the same experiment were usually not possible; hence some discrepancies will appear when ratios are used to calculate other ratios, for example, Ru/Os from Ru/Ir/Os/Ir, since some selection and averaging of data were necessary.

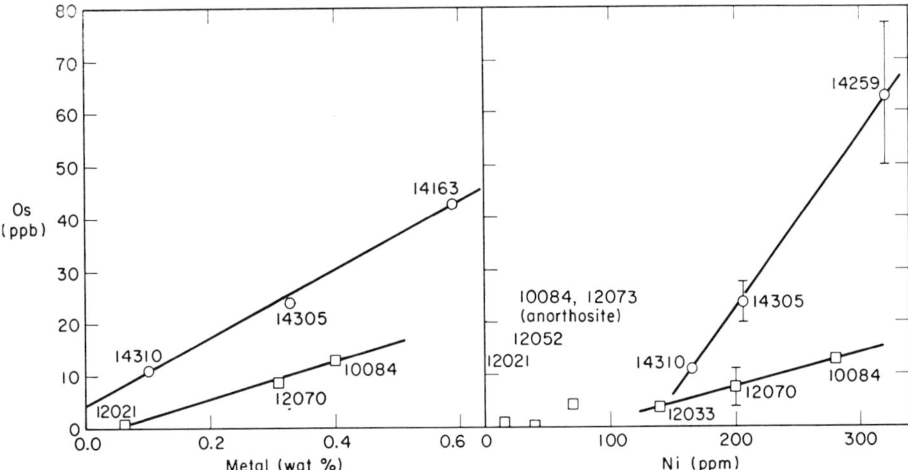


Fig. 5. Correlation of Os and metal content, and Os and Ni content in soils and rocks from different sites. Ni contents are from LSPET (1970, 1971), Wänke *et al.* (1971), Cuttitta *et al.* (1971), and Laul (1971). The metal contents are from Gose *et al.* (1972). The metal content for 14305 is not reported. We have used an approximate average for 8 fragmental rocks; a single high value was not included.

Ru and Os are the only Pt-metals that might be readily fractionated, provided they can be oxidized to their higher oxidation states. The large enhancement of Ru and Os relative to Ir in Apollo 11 and 12 rocks in comparison with soil may be an indication of the incorporation of oxidized species. Data on the relative state of oxidation of lunar samples can be examined to see whether any correlations may exist. There is no direct or inverse relationship between Ru and Os contents and the oxygen deficiency of Ehmann and Morgan (1971), the excess reducing capacity of Cuttitta *et al.* (1971), or the Eu depletion reported by a number of investigators.

Prolonged concentrated nitric acid leaching of 10084 soil by Laul *et al.* (1971) and metal separates from 14163 soil by Wlotzka *et al.* (1972) extracted in each case ~50% of the Ir; 50 and 73% of the Ni was also extracted. We have observed even higher extractions of Os from the Apollo 12 breccia 12073,34 and the anorthosite inclusion in the breccia under conditions of alkaline and brief (2-minute), dilute (5%) acid leaching (Reed *et al.*, 1971). If Os is contained in lunar matter in the same way as Ir, the mild conditions employed would not have extracted the Os. There is some evidence for Ru extraction, but the data are too poor to warrant consideration.

The Os content in the 12073,34 anorthosite inclusion is 4.4 ppb, about half that in Apollo 12 soils and considerably higher (3–35 times) than that in the Apollo 12 igneous rocks. Comparison of Ir from 10084 anorthosite separates (Laul *et al.*, 1971) to Os from 10084,75 anorthositic-rich phases and 12073,34 anorthosite (Reed *et al.*, 1971) yields Os/Ir ratios 1.3 and 2.1, respectively, which are meteoritic (1.7) within the uncertainties of the measurements. The suggestion, then, is that Os is apparently meteoritic, as are other Pt-metal elements studied by Laul *et al.* (1971). However, not only are the leach results contrary to such a conclusion, but the extreme purity of the 12073,34 inclusion, 99% anorthite and 1% clinopyroxene (Reed *et al.*, 1971) does not permit accommodation of an 0.5–1% meteoritic component (Laul *et al.*, 1971). It appears that most of the Pt-metals in these plagioclase samples are indigenous.

In spite of the obvious contribution of extralunar matter, which, in fact, we have observed in fragmental rocks (Reed *et al.*, 1972), it is not inconsistent with the data that some of the Os, and probably Ru, are indigenous components fractionated from one another and from other Pt-metals and between soil and rocks by lunar processes.

The apparent variation in Os concentrations in exterior versus interior samples of Apollo 12 igneous rocks is further evidence for the occurrence of a mobilizing process, and in spite of the lack of a relationship between Os contents and the reducing capacity of lunar samples discussed earlier, processes that involve oxidation and transport of Os and Ru seem to be required. A possible oxidizing agent could have been $H_2O$. It may have been effective very early in the history of the moon and might have caused Os fractionation.

Three observations, though possibly coincidental, may support the suggestion that a significant fraction of the Os, at least (~50% of Apollo 11 and 12 soils and ≲10% of the Apollo 14 samples) has been processed on the moon and is probably indigenous: (1) The two curves for Os versus Ni (Fig. 5) intersect at [Os] ~4.5 ppb, close to the Os content of 12033,18, a soil similar to Apollo 14 soils and breccia; (2) the Os content of two anorthosite samples in which 12033 soil and Apollo 14 samples are relatively enriched is ~4.4 ppb; (3) the Os content of metal-free Apollo 14 samples extrapolates to ~4 ppb (Fig. 5).

### *Li, U, and Te*

The Apollo 14 and 15 results on these elements are listed in Table 1. The uniformly high (several ppm) U values in all Apollo 14 samples are consistent with results reported by LSPET (1971) and a number of other investigators. The Apollo 15 results span the values given by LSPET (1972) and are in the 1–2 ppm range reported by most investigators for 12070 soil.

The Li contents of Apollo 14 soils have been reported by several laboratories and appear to be variable within a single sample. For instance, 14259 Li values of 19 ppm (this work), 35.6 ppm (Eugster, 1972), and 14 ppm (Wänke *et al.*, 1972) were reported. The average Li content (including Taylor *et al.*, 1972) for 14163 and 14259 soils is $\sim$24 ppm. More Li is found in Apollo 14 fragmental rocks and in 14141 Cone Crater soil (Eugster, 1972) than in any other samples, except 12033 soil. To a crude approximation the Li contents appear to parallel the relative amounts of exotic components in soils; that is, Li concentrations decrease in the order: 12033 > Ap 14 > Ap 15 > Ap 12 $\cong$ Ap 11, based on data from our laboratory. Eugster (1972) has already noted this trend and has suggested a two-component mixing model with the feed material (14141 soil) from Fra Mauro.

The Te results given in Table 1, as well as those previously reported for Apollo 11 and 12, merely indicate the magnitude of the concentration of this element.

## Conclusions

(1) $Cl/P_2O_5$ ratios are essentially constant at 0.008 in soils from all sites sampled. Igneous rocks, with the exception of Apollo 11 high-Rb rocks, have variable ratios but lower than the soils.

(2) Nonleachable/leachable Cl and Br ratios are similar at all sites; the distribution of these ratios in lunar samples is similar to those in terrestrial, but not meteoritic, samples.

(3) Hg volatilized at subsolar temperatures is found in samples from all sites. The amount of this Hg adsorbed on surfaces is sensitive to the temperatures the particular samples were exposed to.

(4) The Apollo 14 samples are unique among those measured in that they contain large amounts of Hg in retentive sites. This is attributed to thermal and shock effects associated with the Imbrium impact or to pre-impact thermal conditions at the site.

(5) Ru and Os are not coherent relative to the Pt-metals Ir, Re, and Pd. Whereas these latter are present in meteoritic ratios, Os and Ru deviate from these ratios relative to the other Pt-metals and to each other. Os appears to have been fractionated by processes, probably magmatic, occurring on the moon.

*Acknowledgments*—We thank the chemistry department and the operating staff of the High Flux Reactor, Brookhaven National Laboratory, for making their facilities available. We are especially grateful to J. Hudis and Mrs. E. Rowland for their assistance and to the Argonne National Laboratory Linac staff for their cooperation. The association of G. W. Reed with the Enrico Fermi Institute, University of Chicago, and the use of their facilities are gratefully acknowledged.

## References

Albee A. L. and Chodos A. A. (1970) Microprobe investigations on Apollo 11 samples. *Proc. Apollo 11 Lunar Sci. Conf., Geochim. Cosmochim. Acta* Suppl. 1, Vol. 1, pp. 135–157. Pergamon.

Alexander E. C. Jr., Davis P. K., Lewis R. S., and Reynolds J. H. (1972) Rare gas analyses on neutron irradiated lunar samples (abstract). In *Lunar Science—III* (editor C. Watkins), pp. 12–14, Lunar Science Institute Contr. No. 88.

Baedecker P. A., Chou C. L., Kimberlin J., and Wasson J. T. (1972) Trace element studies of lunar rocks and soils (abstract). In *Lunar Science—III* (editor C. Watkins), pp. 35–37, Lunar Science Institute Contr. No. 88.

Bogard D. D. and Nyquist L. E. (1972) Noble gas studies on regolith materials from Apollo 14 and 15 (abstract). In *Lunar Science—III* (editor C. Watkins), pp. 89–91, Lunar Science Institute Contr. No. 88.

Cuttitta F., Rose H. J. Jr., Annell C. S., Carron M. K., Christian R. P., Dwornik E. J., Greenland L. P., Helz A. W., and Ligon D. T. Jr. (1971) Elemental composition of some Apollo 12 lunar rocks and soils. *Proc. Second Lunar Sci. Conf., Geochim. Cosmochim. Acta* Suppl. 2, Vol. 2, pp. 1217–1229. MIT Press.

Ehmann W. D. and Morgan J. W. (1971) Major element abundances in Apollo 12 rocks and fines by 14 MeV neutron activation. *Proc. Second Lunar Sci. Conf., Geochim. Cosmochim. Acta* Suppl. 2, Vol. 2, pp. 1237–1245. MIT Press.

Eldridge J. S., O'Kelley G. D., and Northcutt K. J. (1972) Abundances of primordial and cosmogenic radionuclides in Apollo 14 rocks and fines (abstract). In *Lunar Science—III* (editor C. Watkins), pp. 221–223, Lunar Science Institute Contr. No. 88.

Eugster O. (1972) Li, Be, and B abundances in fines from the Apollo 11, Apollo 12, Apollo 14, and Luna 16 missions (abstract). In *Lunar Science—III* (editor C. Watkins), pp. 247–249, Lunar Science Institute Contr. No. 88.

French B. M., Walter L. S., and Heinrich K. J. F. (1970) Quantitative mineralogy of an Apollo 11 lunar sample. *Proc. Apollo 11 Lunar Sci. Conf., Geochim. Cosmochim. Acta* Suppl. 1, Vol. 1, pp. 433–444. Pergamon.

Fuchs L. H. (1971) Orthopyroxene and orthopyroxene-bearing rock fragments rich in K, REE, and P in Apollo 14 soil sample 14163. *Earth Planet. Sci. Lett.* **12,** 170–174.

Ganapathy R., Keays R. R., Laul J. C., and Anders E. (1970) Trace elements in Apollo 11 lunar rocks: Implications for meteoritic influx and origin of moon. *Proc. Apollo 11 Lunar Sci. Conf., Geochim. Cosmochim. Acta* Vol. 2, pp. 1117–1142. Pergamon.

Gose W. A., Pearce G. W., Strangway D. W., and Larson E. E. (1972) On the magnetic properties of lunar breccias (abstract). In *Lunar Science—III* (editor C. Watkins), pp. 332–334, Lunar Science Institute Contr. No. 88.

Grieve G., McKay G., Smith H., and Weill D. (1972) Mineralogy and petrology of polymict breccia 14321 (abstract). In *Lunar Science—III* (editor C. Watkins), pp. 338–340, Lunar Science Institute Contr. No. 88.

Herpers U., Herr W., Kulus H., and Michel R. (1972) Isotopic-anomalies in lunar rhenium (abstract). In *Lunar Science—III* (editor C. Watkins), pp. 370–372, Lunar Science Institute Contr. No. 88.

Herr W., Herpers U., Michel R., Abdel Rassoul A. A., and Woelfle R. (1971) Search for rhenium isotopic anomolies in lunar surface material by neutron bombardment. *Proc. Second Lunar Sci. Conf., Geochim. Cosmochim. Acta* Suppl. 2, Vol. 2, pp. 1337–1341. MIT Press.

Jovanovic S. and Reed G. W. Jr. (1968) Hg in metamorphic rocks. *Geochim. Cosmochim. Acta* **32,** 341–346.

Keil K., Prinz M., and Bunch T. E. (1971) Mineralogy, petrology, and chemistry of some Apollo 12 samples. *Proc. Second Lunar Sci. Conf., Geochim. Cosmochim. Acta* Suppl. 2, Vol. 1, pp. 319–341. MIT Press.

Keith J. E., Clark R. S., and Richardson K. A. (1972) Gamma ray measurements of Apollo 12, 14, and 15 lunar samples (abstract). In *Lunar Science—III* (editor C. Watkins), pp. 446–448, Lunar Science Institute Contr. No. 88.

Laul J. C., Morgan J. W., Ganapathy R., and Anders E. (1971) Meteoritic material in lunar samples: Characterization from trace elements. *Proc. Second Lunar Sci. Conf., Geochim. Cosmochim. Acta* Suppl. 2, Vol. 2, pp. 1139–1158. MIT Press.

Lovering J. F. and Hughes T. C. (1971) Rhenium and osmium abundance determinations and meteoritic contamination levels in Apollo 11 and Apollo 12 lunar samples. *Proc. Second Lunar Sci. Conf., Geochim. Cosmochim. Acta* Suppl. 2, Vol. 2, pp. 1331–1335. MIT Press.

LSPET (Lunar Sample Preliminary Examination Team) (1970) Preliminary examination of lunar samples from Apollo 12. *Science* **167,** 1325–1339.

LSPET (Lunar Sample Preliminary Examination Team) (1971) Preliminary examination of lunar samples. NASA-MSC, Apollo 14 Preliminary Science Report SP-272, 117-1180.

LSPET (Lunar Sample Preliminary Examination Team) (1972) The Apollo 15 lunar samples: A preliminary description. *Science* **175,** 363–375.

Lugamair G. W. and Marti K. (1972) Neutron and spallation effects in Fra Mauro regolith (abstract). In *Lunar Science—III* (editor C. Watkins), pp. 495–497, Lunar Science Institute Contr. No. 88.

Lunatic Asylum (1970) Mineralogy and isotopic investigations on lunar rock 12013. *Earth Planet. Sci. Lett.* **9,** 137–163.

Maxwell J. A., Peck L. C., and Wiik H. B. (1970) Chemical composition of Apollo 11 lunar samples 10017, 10020, 10072, and 10084. *Proc. Apollo 11 Lunar Sci. Conf., Geochim. Cosmochim. Acta* Suppl. 1, Vol. 2, pp. 1369–1374. Pergamon.

Morgan J. W., Laul J. C., Krähenbühl U., Ganapathy R., and Anders E. (1972a) Major impacts on the moon: Chemical characterization of projectiles (abstract). In *Lunar Science—III* (editor C. Watkins), pp. 552–554, Lunar Science Institute Contr. No. 88.

Morgan J. W., Laul J. C., Krähenbühl U., Ganapathy R., and Anders E. (1972b) Volatile and siderophile elements in Apollo 14 and 15 rocks (abstract). In *Lunar Science—III* (editor C. Watkins), pp. 555–557, Lunar Science Institute Contr. No. 88.

Patterson J. A., Turkevich A. L., Franzgrote E. J., Economou T. E., and Sowinski K. P. (1970) Chemical composition of the lunar surface in the terra region near the crater Tycho. *Science* **168,** 825.

Reed G. W. Jr. and Jovanovic S. (1969) Some halogen measurements on achondrites. *Earth Planet. Sci. Lett.* **6,** 316–320.

Reed G. W. Jr. and Jovanovic S. (1970) Halogens, mercury, lithium, and osmium in Apollo 11 samples. *Proc. Apollo 11 Lunar Sci. Conf., Geochim. Cosmochim. Acta* Suppl. 1, Vol. 2, pp. 1487–1492. Pergamon.

Reed G. W. Jr. and Jovanovic S. (1971) The halogens and other trace elements in Apollo 12 samples and the implications of halides, platinum metals, and mercury on surfaces. *Proc. Second Lunar Sci. Conf., Geochim. Cosmochim. Acta* Suppl. 2, Vol. 2, pp. 1261–1276. MIT Press.

Reed G. W. Jr., Jovanovic S., and Fuchs L. H. (1971) Fluorine and other trace elements in lunar plagioclase concentrates. *Earth Planet. Sci. Lett.* **11,** 354–358.

Reed G. W. Jr., Jovanovic S., and Fuchs L. H. (1972) Concentrations and lability of the halogens, platinum metals and mercury in Apollo 14 and 15 samples (abstract). In *Lunar Science—III* (editor C. Watkins), pp. 637–639, Lunar Science Institute Contr. No. 88.

Rose J. R., Cuttitta F., Annell C. S., Carron M. K., Christian R. P., Dwornik E. J., and Ligon D. T. Jr. (1972) Compositional data for fifteen Fra Mauro lunar samples (abstract). In *Lunar Science—III* (editor C. Watkins), pp. 660–662, Lunar Science Institute Contr. No. 88.

Taylor S. R., Muir P., Nance W., Rudowski R., and Kaye M. (1972) Composition of the lunar uplands, I. Chemistry of Apollo 14 samples from Fra Mauro (abstract). In *Lunar Science—III* (editor C. Watkins), pp. 744–746, Lunar Science Institute Contr. No. 88.

Vinogradov A. P. (1971) Preliminary data on lunar ground brought to earth by automatic probe "Luna-16." *Proc. Second Lunar Sci. Conf., Geochim. Cosmochim. Acta* Suppl. 2, Vol. 1, pp. 1–16. MIT Press.

Wänke H., Wlotzka I., Baddenhausen H., Balacescu A., Spettel B., Teschke I., Jagoutz E., Kruse H., Quijano-Rico M., and Rieder R. (1971) Apollo 12 samples: Chemical composition and its relation to sample locations and exposure ages, the two component origin of the various soil samples and studies of lunar metallic particles. *Proc. Second Lunar Sci. Conf., Geochim. Cosmochim. Acta* Suppl. 2, Vol. 2, pp. 1187–1208. MIT Press.

Wänke H., Baddenhausen H., Balacescu A., Teschke F., Spettel B., Dreibus G., Quijano M., Kruse H., Wlotzka F., and Begemann F. (1972) Multielement analyses of lunar samples (abstract). In *Lunar Science—III* (editor C. Watkins), pp. 779–781, Lunar Science Institute Contr. No. 88.

Wasson J. T. and Baedecker P. A. (1970) Ga, Ge, In, Ir, and Au in lunar, terrestrial, and meteoritic basalts. *Proc. Apollo 11 Lunar Sci. Conf., Geochim. Cosmochim. Acta* Vol. 2, pp. 1741–1750. Pergamon.

Wlotzka F., Jagoutz E., Spettel B., Baddenhausen H., Balacescu A., and Wänke H. (1972) On lunar metallic particles and their contribution to the trace element content of the Apollo 14 and 15 soils (abstract) .In *Lunar Science—III* (editor C. Watkins), pp. 806–808, Lunar Science Institute Contr. No. 88.

Proceedings of the Third Lunar Science Conference
(Supplement 3, *Geochimica et Cosmochimica Acta*)
Vol. 2, pp. 2003–2014
The M.I.T. Press, 1972

# Thermal volatilization studies on lunar samples

EVERETT K. GIBSON, Jr., and NORMAN J. HUBBARD
Geochemistry Branch, NASA Manned Spacecraft Center,
Houston, Texas 77058

**Abstract**—Thermal volatilization studies on lunar basalts, breccias, and soil have been carried out to measure the depletion of selected elemental abundances during heating under vacuum. The study revealed that the easily volatilized alkali element Rb is lost from lunar samples at temperatures of 950°C or less. The alkali elements K and Na are lost at higher temperatures while the refractory elements (Li, Ba, Sr, REE) are not volatilized below 1400°C. The vaporization temperature of Na in basaltic lunar samples is near 1050°C and Na is slowly lost from Apollo 14 soil at 1050°C. This study indicates that thermal volatilization of basaltic lunar materials is restricted to selected alkali elements and other elements of similar or greater volatility.

The naturally occurring changes in K/Rb and K/Ba ratios of KREEP breccia and glass samples have been duplicated during laboratory heating experiments at 1050°C and indicate that some KREEP breccias have been subjected to a metamorphic event with temperatures of 1050°C or below. The kinetics of K, Rb, and Pb loss are sufficiently fast that these elements can be vaporized and lost during production of impact glasses, and in fact vapor deposited K, Rb, and Pb have been found in lunar soils by other investigators.

It is improbable that lunar basalt flows have lost a significant percentage of K, Rb, and Na by volatilization unless the flow was extensively stirred. The uniformly low Na concentrations of mare basalts are not caused by post eruption vaporization and loss of Na.

## INTRODUCTION

ONE OF THE DISTINCTIVE geochemical features of lunar basalts is the low concentrations of volatile elements with respect to terrestrial basalts. Sodium is especially depleted, while elements such as K and Rb are depleted to a lesser degree (Hubbard and Gast, 1971). DeMaria *et al.* (1971) have investigated the volatilization of alkalis from lunar samples using a Knudsen cell–mass spectrometer and reported volatilization of sodium and potassium from lunar basalt 12022 at temperatures as low as 1025°C. No other elements studied by them, except perhaps Fe, volatilized within the temperature range where lunar basalts crystallize. Naughton *et al.* (1971) reported vaporization of alkalis from the surfaces of lunar soils and exterior pieces of lunar basalts at temperatures as low as 527°C. They suggested that the K lost at low temperatures had been vapor deposited on the surfaces of regolith particles. Silver (1970, 1972) found that Pb is lost from soil samples at temperatures as low as 500°C and possibly other volatile elements such as Rb could also be volatilized at low temperatures from soil samples. Given the evidence for vapor deposited K and Pb in soil samples it is desirable to understand how the K and Pb were vaporized. The abundance of impact produced glass in lunar soils is direct evidence that a fraction of lunar soils have been heated to temperatures above the vaporization temperatures of K, Rb, Na oxides (Brewer, 1953) and could have lost K, Rb, Na, and other volatile elements if the kinetics of vaporization are favorable.

The vaporization temperatures for K, Rb, and Na oxides can be roughly extrapolated from the $10^{-3}$ atmosphere data of Brewer (1953) to lunar conditions (Table 1). Use of Brewer's data for simple oxides is a reasonable approximation, because lunar samples consist mainly of silicates that have the elements of interest bonded to oxygen atoms. The data of Naughton *et al.* (1971) show that lunar samples lose alkalis as a variety of species, especially oxides (K, $K_2$, KO, $K_2O$, Na, $Na_2$, NaO, $Na_2O$), thus differences in volatilization temperatures for Na, K, and Rb oxides measured by Brewer (1953) are expected to be applicable to lunar samples.

In order to better understand the relative depletion patterns of the alkali elements (Na, K, Rb, Li), Ba, Sr, and selected rare earth elements, thermal volatilization studies were carried out under a vacuum of $2 \times 10^{-6}$ Torr and $P_{O_2} < 10^{-10}$ atm on lunar samples at different temperatures and for varying periods of time. Sample residues were analyzed for the elements of interest. Changes in elemental concentrations are used to evaluate previous hypotheses about volatile element depletion in lunar samples, to identify samples that have undergone natural thermal volatilization events, and to approximate boundary conditions within which selective volatilization may have produced observed elemental abundance patterns in lunar samples.

## EXPERIMENTAL PROCEDURES

Thermal volatilization was carried out using techniques similar to those described by Gibson and Johnson (1971). Powdered samples were placed in a platinum crucible in a Mettler recording vacuum thermal analyzer as described by Gibson and Johnson (1972). Samples were evacuated to $2 \times 10^{-6}$ Torr while the sample weight, temperature, and in some experiments the composition of evolved gases were continuously recorded. The partial pressures of oxygen were calculated from the mass spectra obtained during each volatilization experiment. Samples were heated at 4° and 6°C/min until the desired temperature (950°C, 1050°C, 1200°C, and 1400°C) was reached. For most runs the furnace was held at temperature for 2 hours and then allowed to cool along its normal cooling curve (from 1400°C to room temperature required 4 hours). One set of runs was held at the same temperature (950°C and 1050°C) for varying lengths of time between 1 and 72 hours. The sample residues then were analyzed for Na, K, Rb, Li, Ba, Sr, and REE using stable isotope dilution mass spectrometry and atomic absorption (Na only) as previously described by Gast *et al.* (1970).

## RESULTS

The measured depletions of Na, K, and Rb concentrations in lunar samples during thermal volatilization are listed in Table 2. The relative depletions are consistent with

Table 1. Vaporization temperatures of elemental oxides.*

| Oxide | Vaporization temperatures (°C) | |
|---|---|---|
| | $10^{-3}$ atm | 1 atm |
| $Na_2O$ | 1057 | 1767 |
| $K_2O$ | 877 | 1477 |
| $Rb_2O$ | 827 | 1327 |
| SrO | 2127 | 3227 |
| BaO | 1847 | 2723 |
| $Li_2O$ | 1557 | 2327 |

* After Brewer (1953).

Table 2. Elemental abundances measured in this work.

| Sample | 10017 Basalt | | 10073 Breccia | | 12022 Basalt | | | | | 14163 Soil | | | | |
|---|---|---|---|---|---|---|---|---|---|---|---|---|---|---|
| Element (ppm) | Initial | 1400°C 2 hr | Initial | 1400°C 2 hr | Initial | 950°C 2 hr | 1050°C 2 hr | 1200°C 2 hr | 1400°C 2 hr | Initial† | 1050°C instan-taneous | 1050°C 1 hr | 1050°C 6 hr | 1050°C 72 hr |
| K | 2610 | 641 | 1200 | 385 | 536 | 449 | 360 | 300 | 179 | 4840 | 4220 | 3027 | 1437 | 88 |
| Rb | 5.63 | 1.05 | 2.84 | 0.69 | 0.738 | 0.482 | 0.286 | 0.288 | 0.173 | 15.3 | 12.75 | 7.54 | 2.25 | 0.10 |
| Na | 3800 | 500 | 3500 | — | 1824 | 1820 | 1800 | 1200 | 673 | 5100 | 4660 | 4110 | 3140 | 1730 |
| Ba | — | — | — | — | 59.5 | — | — | — | 59.3 | 873 | — | 822 | 826 | 829 |
| Li | 18.1* | 19.3 | — | — | 9.5 | — | 9.0 | 10.1 | 11.8 | — | — | — | — | — |
| Ce | 77.3 | — | 46.5 | — | 17.4 | — | — | — | 17.7 | — | — | — | — | — |
| Nd | 59.5 | — | 35.4 | — | 14.8 | — | — | — | 15.0 | — | — | — | — | — |
| Sm | 20.9 | — | 12.4 | — | 5.58 | — | — | — | 5.59 | — | — | — | — | — |
| Eu | 2.14 | — | 1.70 | — | 1.28 | — | — | — | 1.28 | — | — | — | — | — |
| Gd | 27.4 | — | 15.9 | — | 7.90 | — | — | — | 7.85 | — | — | — | — | — |
| K/Rb | 464 | 610 | 422 | 558 | 723 | 932 | 1259 | 1041 | 1035 | 316 | 332 | 401 | 639 | 880 |
| K/Ba | — | — | — | — | 9.0 | 7.5 | 6.1 | 5.0 | 3.0 | 5.5 | ~5.1 | 3.7 | 1.7 | 0.1 |

* Tera *et al.* (1970). †Composition of separate aliquot of 14163 soil (Hubbard *et al.*, 1972).

those expected from the Brewer (1953) data for simple oxide systems, indicating that silicates at $10^{-6}$ Torr and simple oxides at $10^{-3}$ atm are comparable systems when discussing alkali volatilization. Figure 1 shows the loss of alkali elements as a function of temperature for lunar basalt 12022. The refractory elements Li, Sr, Ba, and REE have not been depleted. Loss of Rb after 2 hours at 1050°C is surprisingly large, in fact only 30% of the original Rb concentration remains. Depletions of K are considerably less than for Rb at 1050° and 950°C. Depletion of Na does not begin until above 1050°C (Figs. 1 and 2) in agreement with the Knudsen cell data of DeMaria *et al.* (1971) and the $Na_2O$ data ($10^{-3}$ atm) of Brewer. The large depletions of Rb and lesser depletions of K in 12022 during heating at 950° and 1050°C is consistent with the lower vaporization temperatures of $Rb_2O$ and $K_2O$ relative to $Na_2O$ (Brewer, 1953) and show that the vaporization temperatures of K and Rb are below 950°C for lunar basalt 12022. Also, the vaporization temperature of Rb is lower than for K and results in larger losses of Rb, relative to K, at low temperatures (900–1100°C) and therefore produces changes in K/Rb ratio at these temperatures. For example, the K/Rb ratio of 12022 changed from 723 in the unheated sample to 932 (950°C), 1259 (1050°C), 1041 (1200°C), and 1035 (1400°C) demonstrating this effect (Fig. 2). The lower vaporization temperature of Rb is also seen in the meteorite volatilization data of Smales *et al.* (1964) where the maximum Rb loss occurred at slightly lower temperature than the maximum K loss.

A set of constant temperature-variable time volatilization runs were made on lunar soil sample 14163, homogenized by grinding to 74 $\mu$, to determine the time dependence of Na, K, and Rb losses at the temperature of maximum K/Rb (1050°C)

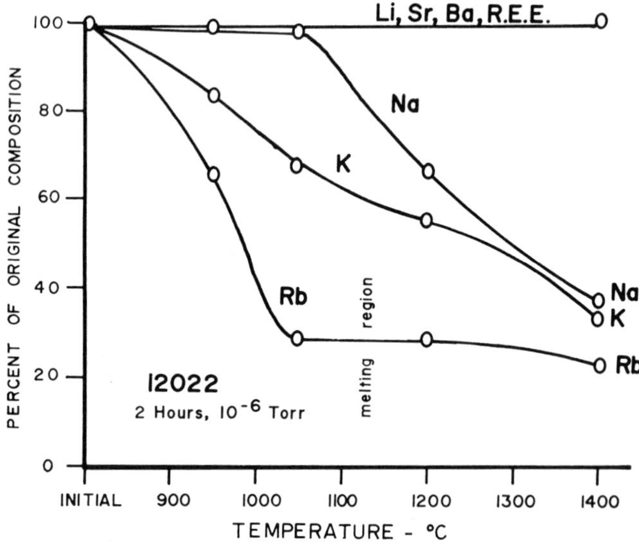


Fig. 1. Thermal volatilization of lunar basalt 12022 showing the depletion of Na, K, and Rb after heating under a vacuum of $2 \times 10^{-6}$ Torr for 2 hr. Note that the abundances of the refractory elements are not altered.

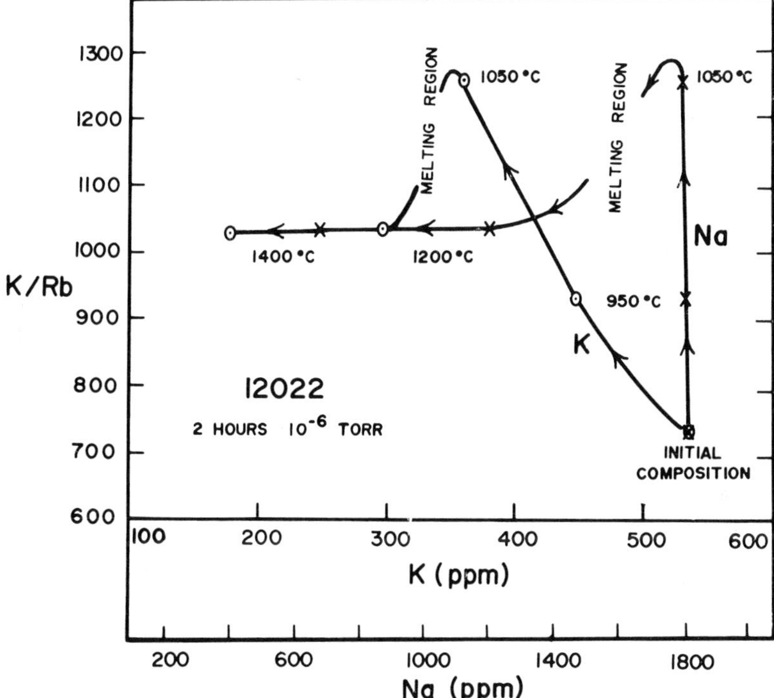


Fig. 2. Changes in the K/Rb ratio versus K and Na concentrations after thermal volatilization of lunar basalt 12022. Samples were heated at the different temperatures for two hours under a vacuum of $2 \times 10^{-6}$ Torr. Note that the K/Rb ratio increases to a maximum and then with further heating falls to a value that remains essentially constant.

ratio for mare basalt sample 12022. Results are shown in Fig. 3 and Table 2. The results demonstrate that almost total K and Rb loss can occur at 1050°C and $10^{-6}$ Torr if the sample is held for 72 hours under vacuum, thus placing the vaporization temperature of K and Rb well below 1050°C. Sodium loss occurs slowly at 1050°C (Table 2), showing that 1050°C is above the vaporization temperature of Na, as indicated by the DeMaria *et al.* (1971) data that gives 1025°C. Shorter periods of time are sufficient for sizeable alkali losses, even a few minutes at 1050°C is sufficient to change the K/Rb ratios. Note that the 1050°C temperature selected for the volatilization studies is almost 100°C below the initial melting point of the major soil components (Gibson and Moore, 1972).

Three lunar samples held for 2 hours at 1400°C show a nearly constant degree of K, Rb, and Na loss and change in K/Rb ratios, suggesting that identical thermal treatments cause nearly identical results (Table 2). Some explanation of this portion of the experiment is necessary because in our experimental apparatus 2 hours at 1400°C is not equivalent to instantaneous heating to 1400°C, holding for 2 hours, and instant quenching. Instead, the region of K and Rb loss above 900°C is crossed in about 150 min at a heating rate of 6°C/min during heating and crossed again in about

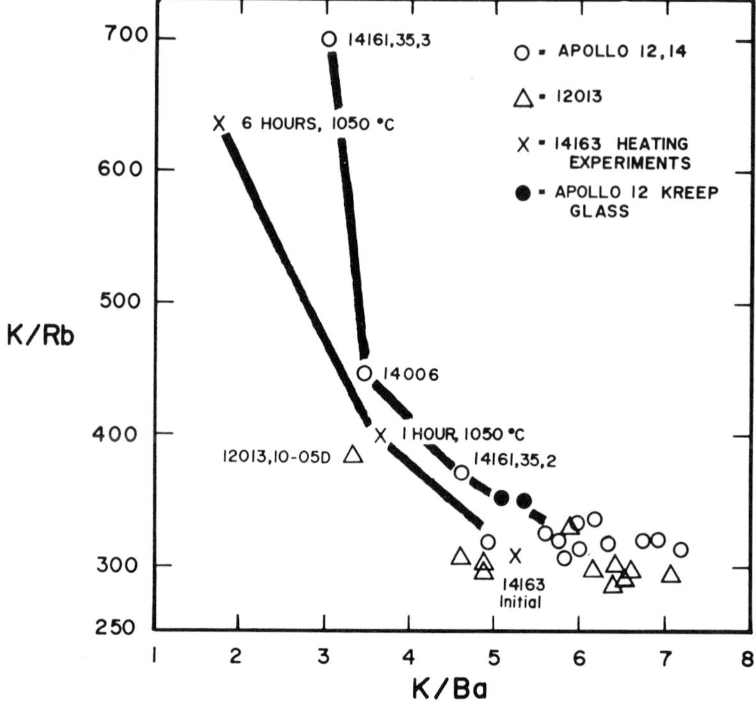


Fig. 3. Natural changes and experimentally produced changes in the K/Rb versus K/Ba ratios. Data are from Schnetzler *et al.* (1970), Hubbard and Gast (1971), and this report. Note the heating experiments on sample 14163 causes the K/Ba ratio to decrease while the K/Rb ratio increases. Both of these changes are similar to the ratios observed in certain lunar samples which have undergone thermal alteration. Differences in the analytical data of Schnetzler *et al.* (1970) (the lower triangles in the figure) and those of Hubbard and Gast (1971) (upper triangles) are due to 10% interlab difference in K.

60 min during cooling. Thus enough time has been spent above the K and Rb vaporization temperatures so that the 1400°C runs, and all other runs above the vaporization temperatures, produce results integrated over a range of temperatures. This integration effect is reduced at the lower temperatures where the greatest changes in K/Rb ratio can be generated and where kinetics of K and Rb loss are slow relative to heating rates of the furnace.

## Discussion

Some thermally metamorphosed breccia samples from Apollo 14 have features indicating that they have lost K and Rb during thermal metamorphism. Samples 14161,35,3 (a coarse fines fragment) and 14006 (a walnut-sized breccia fragment) have the typical major and trace element chemistry of Apollo 14 basaltic breccias of KREEP composition (Hubbard *et al.*, 1972) except that both have low K and Rb concentrations accompanied by increases in K/Rb ratios. These samples are of rather high metamorphic grade (Warner, 1972, personal communication) and have been subjected to the temperatures required for vaporization of K and Rb. Scanning electron microscope studies (McKay *et al.*, 1972) reveal that 14006 has cavities lined with vapor deposited crystals. The vapor deposited crystals, the low K and Rb concentrations, the high K/Rb ratio, and the high metamorphic grade combine to identify this sample as having lost K and Rb by thermal volatilization. Neither of these samples have lost an appreciable amount of Na, placing a limit of about 1050°C on the temperature to which these samples were heated.

Given two samples (14161,35,3 and 14006,3) exhibiting K and Rb loss by volatilization from KREEP basaltic breccia, it is necessary to establish the extent to which K and Rb volatilization has produced the measured K and Rb values in these materials and the extent to which volatilization experiments using Apollo 14 soil 14163 duplicate the results. Before the effects of K and Rb volatilization can be precisely studied, other causes of K and Rb concentration changes must be normalized. A common and major cause of K and Rb variations in KREEP materials is the presence of K-feldspar + quartz "granophyre" patches, veins, and cavity fillings. This "granophyre" material is also enriched in Ba. The K/Rb and K/Ba ratios of this "granophyre" material in 12013 are identical to or overlap the K/Rb and K/Ba ratios for KREEP breccias with "normal" amounts of this high K, Rb and Ba material (Hubbard *et al.*, 1972; Nyquist *et al.*, 1972). Thus K and Rb variations due to varying amounts of this material can be normalized by plotting K/Rb versus K/Ba (Fig. 3). The data plotted in Fig. 3 show that most KREEP breccias have K/Ba ratios between 4.5 and 7.2 and K/Rb ratios between 285–340. Three samples have "normal" K/Ba ratios but slightly high K/Rb ratios. Additional samples have both high K/Rb ratios and low K/Ba ratios. Two of the samples with "normal" K/Ba ratios, but slightly high K/Rb ratios are samples of KREEP glass produced by impact melting (Hubbard *et al.*, 1971; Hubbard and Gast, 1971; Meyer *et al.*, 1971). If one makes the reasonable assumption that these samples had premelting K/Rb ratios of 325 and K/Ba ratios of 6.0 then the K and Rb losses are near 10%. The third sample has a K/Ba ratio of 4.6 (i.e., only barely within the range of "normal"), is of the same metamorphic grade as 14161,35,3 and 14006 (Warner, personal communication) and is therefore expected to have lost K and Rb.

We have subjected aliquots of Apollo 14 soil samples 14163 to thermal volatilization at 1050°C for varying lengths of time and plotted the experimental results along with the Apollo 12 and Apollo 14 KREEP breccia data (Fig. 3). The experimental results parallel the observed data for the lower K/Rb ratios but do not exactly match the K/Rb and K/Ba values of 14161,35,3. By matching the naturally occurring K/Rb and K/Ba ratios with experimental results, the past thermal histories of these samples can be defined in terms of boundary temperatures and times. For samples 14161,35,3 and 14006,3 that were open to alkali loss, one can say the following: (1) temperatures were below 1050°C because little or no Na has been lost, and (2) observed K/Rb ratios for these two samples (set by the metamorphic event) have been reproduced experimentally by heating experiments at temperatures of 1050°C for no longer than 6 hours at $P = 10^{-6}$ Torr and $P_{O_2} < 10^{-10}$ atm. The actual temperature was lower than 1050°C and thus the time was longer because of the small Na loss. Preliminary results at 950°C indicate that temperatures above 950°C or times much longer than 72 hours are required for K loss.

Not all highly metamorphosed KREEP breccias have been treated to an identical thermal history. Sample 14161,35,3 has a higher K/Rb and K/Ba than produced by our 1050°C experimental runs for times less than 6 hours. This difference is explained if this sample was heated to a lower temperature, where the loss of Rb relative to K was more rapid than at 1050°C. Further experiments are being conducted to determine the relative thermal history of this sample.

Some samples (in particular 14068) have been metamorphosed to about the same degree as 14161,35,3 and 14006 but do not show elevated K/Rb ratios and decreased K and Rb concentrations. Others (14161,35,2) have less elevated K/Rb ratios (Fig. 3) but similar thermal histories. As noted above, our experimental data show that such variations in K/Rb, K, and Rb can be a function of time spent above the vaporization temperatures of K and Rb in a system open to vapor loss. A system closed to vapor loss will by definition lose no K or Rb and experience no change in K/Rb. In effect, the observed and natural changes in K/Rb are a function of both temperature and the time that the system is open to loss of vaporized K and Rb. The physical state of Fra Mauro KREEP breccias prior to the thermal metamorphic event can be considered to be that of a typical lunar regolith. Lunar regoliths are porous and permeable systems until highly compacted, and thus could provide subsystems with all degrees of openness to vapor loss. Thus variations in natural K/Rb ratios exhibited by KREEP breccias are more consistantly explained by noting that K/Rb ratios, K, and Rb concentrations are a composite result of temperature, time, and the degree of openness of the sample to loss of vaporized K and Rb. This restricts volatile element loss from thermal volatilization to shallow depths where porosity and permeability are maintained.

The KREEP glass samples for Apollo 12 allow an assessment of the extent to which volatile elements other than K and Rb have been lost during an impact melting event. The amount of Pb loss for KREEP glass can be calculated graphically using the Pb–U data of Tatsumoto *et al.* (1971), a U–Pb evolution diagram, the $0.85 \times 10^9$ yr $Ar^{39/40}$ age for KREEP glass (Eberhardt *et al.*, 1972), and the $4.4 \times 10^9$ yr Rb–Sr model age for KREEP materials (Nyquist *et al.*, 1972). The result is that the KREEP glass rich soil, 12033, has been moved along the 4.4 to $0.85 \times 10^9$ yr discordia line (Silver, 1971) a distance proportional to ~30% Pb loss. Note that ~10% Rb loss requires only a few minutes (<20) at temperatures near 1050°C according to our experimental data and that the approximate time for 30% Pb loss can be calculated using the Silver (1972) data for Pb loss rates at 970°C (~1.8 μg/g/hour) for Apollo 14 soil. The calculated result is ~45 min and represent a rough maximum because 9% of the total Pb was lost at lower temperatures and KREEP glass was heated to >970°C. A factor of 3 increase in the effective Pb loss rate for KREEP glass is probably permissible and neatly reconciles the Pb and Rb losses.

KREEP glass demonstrates that sizeable volumes of volatilized Pb can be produced when lunar rocks are converted to glass by impact and that the amount of volatilized Rb produced is about 1/3 the amount of Pb. For Apollo 14 soils the volatilized Pb is ~9% of the total (Silver, 1972) and we calculate that ~3% of the Rb is vapor deposited. Nyquist and Bansal (personal communication) report that 1.5% Rb is easily leachable from soil 14163, essentially the amount that we calculate to be vapor deposited within the soils. A similar amount of K should also be present as vapor deposited material on soil fragments, thus accounting for the low temperature K release found by Naughton *et al.* (1971).

Lunar lava flows seem to provide ideal conditions for alkali loss by volatilization because they attain the required temperatures and low pressures. In order to consider the volatilization process within a lunar basalt flow, the effects of lunar gravity, lithostatic pressure, and vapor pressures of the species of interest must all be taken into consideration. Assuming the lunar atmospheric pressure to be $10^{-9}$ atm and the mean density of Apollo 11 and 12 lunar basalt to be 3.3 g/cc, we can calculate the lithostatic pressure within a basalt flow and apply this pressure information to vaporization processes within the lunar basalt during formation. If we take the vapor pressure data of DeMaria *et al.* (1971) for the alkali elements, alkaline earth elements, iron, chromium, titanium, and aluminum for lunar basalt 12022, we find that any volatilization process (neglecting stirring) would cause only a "skin-effect" depletion of a volatile element. The lithostatic pressure below $10^{-3}$ cm in the lunar basalt exceeds the vapor pressures at reasonable temperatures of all volatile species found in the lunar basalts. This constraint implies that any volatilization process will deplete volatile elements only on the outermost surfaces of lunar basalt flows. For any large-scale loss by volatilization from lunar lava flows to occur on the surface, some means of mixing or exposing almost the entire volume of the lava flow to the lunar vacuum must be found. Such mechanism could involve gas bubbling within the lava, convection, splattering, or fountaining. In any case, it must be very efficient to be very effective.

O'Hara *et al.* (1970) have claimed that the depletion of Na and other volatile elements occurred during cooling of the lunar lava flows after extrusion, and hence the relative depletions observed in the Apollo 11 basalts are not characteristic of the source regions. In opposition, Ringwood and Essene (1970) and others have argued that Na depletions in the lunar materials are characteristic of the source regions of the lunar basalts and accordingly provide clues to the chemical process by which the moon was formed. Ringwood and Essene (1970) have argued that alkali loss by volatilization was ineffective during crystallization because alkali granophyre patches are preserved in the basalts. Further, if thermal volatilization of K and Na had been predominant at subliquids temperatures, the Apollo 11 and 12 plagioclases should have been zoned with K and Na contents decreasing outward from the core (Essene *et al.*, 1970). Plagioclases in lunar basalts show exactly the opposite zoning, and also the lunar basalts are remarkably constant in their sodium contents despite their wide range of cooling histories (Ringwood and Essene, 1970).

If the low Na concentrations in lunar basalts are due to Na loss early in the cooling of the lunar lava flow and the loss mechanism was strongly temperature-dependent, the Na loss could have preceeded plagioclase crystallization. In view of the greater volatility of K and Rb, the nearly tenfold lower Na concentration in lunar basalts, relative to terrestrial basalts, should be accompanied by almost total K and Rb loss, which is not observed. It must be concluded that the low Na concentrations of lunar basalts are not due to the simple thermal volatilization of Na. Further, basaltic lava flows on the moon must have temperatures above 1050°C to be fluid and thus would lose Na and other volatile elements from their surfaces and also from other exposed parts of the flow if the lava is exposed to lunar vacuum. It is highly improbable that all of the lava in the flow, and all lava flows, will be exposed to lunar vacuum for exactly the time required to produce the observed low and nearly constant Na concentrations in lunar mare basalts. If one assumes that the highly improbable did happen and that the initial lava had tenfold more Na, then the lost Na must be accounted for. Thus far, no high Na materials have been found on the moon although high K-materials (K-granophyres) are commonly associated with KREEP breccias. We conclude that the relatively low Na concentrations are a primary feature of lunar basalts and hence their source.

## CONCLUSIONS

The naturally occurring changes in K/Rb and K/Ba ratios of KREEP breccias and glass have been duplicated during laboratory heating experiments at 1050°C. The metamorphic event that caused some K and Rb loss from KREEP breccias caused no discernible Na loss, indicating that temperatures were below 1050°C. Almost total loss of K and Rb results when these materials are held at 1050°C for 72 hours under a vacuum of $10^{-6}$ Torr. Thus, we conclude that the KREEP breccias were subjected to a lower temperature metamorphic event or that the analyzed samples were open to alkali loss for shorter periods of time.

The Apollo 12 KREEP glasses demonstrate that small percentages of K and Rb ($\sim$10%), much larger percentages of Pb ($\sim$30%), and nearly all $Ar^{40}$ can be lost when small glass samples are produced by impact melting, even though temperatures do not exceed the liquidus temperatures and times are short ($<$20 min).

Loss of alkali elements from lunar basalts by thermal volatilization appears to be generally absent or insignificant and improbable on a large scale. The low Na abundances in lunar basalts are a primary feature and reflect low Na abundances in the moon.

*Acknowledgments*—The assistance of Henry Wiesmann and Brij Bansal in carrying out the chemical separations and alkali element determinations is acknowledged. Suzanne M. Johnson and Gary W. Moore assisted in operation of the thermal analysis equipment. Helpful discussions with Paul W. Gast, Richard J. Williams, David S. McKay, Jeff Warner, and Larry Nyquist are also acknowledged. The bedroom critique by G. J. Wasserburg of an earlier version of this manuscript is acknowledged.

## REFERENCES

Brewer L. (1953) The thermodynamic properties of the oxides and their vaporization processes. *Chem. Rev.* **52,** 1–75.

DeMaria G., Balducci B., Guido M., and Piacente V. (1971) Mass spectrometric investigation of the vaporization process of Apollo 12 lunar samples. *Proc. Second Lunar Sci. Conf., Geochim. Cosmochim. Acta* Suppl. 2, Vol. 2, pp. 1367–1380. MIT Press.

Eberhardt P., Eugster O., Geiss J., Grogler N., Schwarzmuller J., Stettler A., and Weber L. (1972) When was the Apollo 12 KREEP ejected? (abstract). In *Lunar Science—III* (editor C. Watkins), pp. 206–208, Lunar Science Institute Contr. No. 88.

Essene E., Ringwood A. E., and Ware N. (1970) Petrology of lunar rocks from the Apollo 11 landing site. *Proc. Apollo 11 Lunar Sci. Conf., Geochim. Cosmochim. Acta* Suppl. 1, Vol. 1, pp. 385–397. Pergamon.

Gast P. W., Hubbard N. J., and Wiesmann H. (1970) Chemical composition and petrogenesis of basalts from Tranquility Base. *Proc. Apollo 11 Lunar Sci. Conf., Geochim. Cosmochim. Acta* Suppl. 1, Vol. 2, pp. 1143–1163. Pergamon.

Gibson E. K. Jr. and Johnson S. M. (1971) Thermal analysis-inorganic gas release studies of lunar samples. *Proc. Second Lunar Sci. Conf., Geochim. Cosmochim. Acta* Suppl. 2, Vol. 2, pp. 1351–1366. MIT Press.

Gibson E. K. Jr. and Johnson S. M. (1972) Thermogravimetric-quadrupole mass spectrometric analysis of geochemical samples. *Thermochimica Acta* **4,** 49–56.

Gibson E. K. Jr. and Moore G. W. (1972) Inorganic gas release and thermal analysis study of Apollo 14 and 15 soils. *Proc. Third Lunar Sci. Conf. Geochim. Cosmochim. Acta* Suppl. 3, Vol. 2. MIT Press.

Hubbard N. J. and Gast P. W. (1971) Chemical composition and origin of nonmare lunar basalts. *Proc. Second Lunar Sci. Conf., Geochim. Cosmochim. Acta* Suppl. 2, Vol. 2, pp. 999–1020. MIT Press.

Hubbard N. J., Meyer C. Jr., Gast P. W., and Wiesmann H. (1971) The composition and derivation of Apollo 12 soils. *Earth Planet. Sci. Lett.* **10,** 341–350.

Hubbard N. J., Rhodes J. M., Gast P. W., and Church S. E. (1972) Three lunar rock associations, their chemical compositions, chronology and possible modes of origin. *Proc. Third Lunar Sci. Conf., Geochim. Cosmochim. Acta* Suppl. 3, Vol. 2. MIT Press.

McKay D. S., Clanton U. S., Heiken G. H., Morrison D. A., Taylor R. M., and Ladle G. (1972) Vapor phase crystallization in Apollo 14 breccias and size analysis of Apollo 14 soils (abstract). In *Lunar Science—III* (editor C. Watkins), pp. 529–531, Lunar Science Institute Contr. No. 88.

Meyer C. Jr., Brett P. R., Hubbard N. J., Morrison D. A., McKay D. S., Aitken F. K., Takeda H., and Schonfeld E. (1971) Mineralogy, chemistry, and origin of the KREEP component in soil samples from the Ocean of Storms. *Proc. Second Lunar Sci. Conf., Geochim. Cosmochim. Acta* Suppl. 2, Vol. 1, pp. 393–411. MIT Press.

Naughton J. J., Derby J. V., and Lewis V. A. (1971) Vaporization from heated lunar samples and the investigation of lunar erosion by volatilized alkalis. *Proc. Second Lunar Sci. Conf., Geochim. Cosmochim. Acta* Suppl. 2, Vol. 1, pp. 449–457. MIT Press.

Nyquist L. E., Hubbard N. J., and Gast P. W. (1972) Rb–Sr relationships for some chemically defined lunar materials (abstract). In *Lunar Science—III* (editor C. Watkins), pp. 584–586, Lunar Science Institute Contr. No. 88.

O'Hara M. J., Biggar G. M., Richardson S. W., Ford C. E., and Jamieson B. G. (1970) The nature of seas, mascons, and the lunar interior in the light of experimental studies. *Proc. Apollo 11 Lunar Sci. Conf., Geochim. Cosmochim. Acta* Suppl. 1, Vol. 1, pp. 695–710. Pergamon.

Ringwood A. E. and Essene E. (1970) Petrogenesis of Apollo 11 basalts, internal constitution and origin of the moon. *Proc. Apollo 11 Lunar Sci. Conf., Geochim. Cosmochim. Acta* Suppl. 1, Vol. 1, pp. 769–799. Pergamon.

Schnetzler C. C., Philpotts J. A., and Bottino M. L. (1970) Li, K, Rb, Sr, Ba, and rare-earth concentrations, and Rb–Sr age of lunar rock 12013. *Earth Planet. Sci. Lett.* **9,** 185–192.

Silver L. T. (1970) Uranium-thorium-lead isotopes in some Tranquility Base samples and their implications for lunar history. *Proc. Apollo 11 Lunar Sci. Conf., Geochim. Cosmochim. Acta* Suppl. 1, Vol. 2, pp. 1533–1574. Pergamon.

Silver L. T. (1971) U–Th–Pb isotope systems in Apollo 11 and 12 regolithic materials and a possible age for the Copernicus impact event. *Trans. A.G.U.* **52,** 534.

Silver L. T. (1972) Lead volatilization and volatile transfer processes on the moon (abstract). In *Lunar Science—III* (editor C. Watkins), pp. 701–703, Lunar Science Institute Contr. No. 88.

Smales A. A., Hughes T. C., Mapper D., McInnes C. A. S., and Webster R. K. (1964) The determination of rubidium and caesium in stony meteorites by neutron activation analysis and by mass spectrometry. *Geochim. Cosmochim. Acta* **28,** 209–233.

Tatsumoto M., Knight R. J., and Doe B. R. (1971) U–Th–Pb systematics of Apollo 12 lunar samples.

*Proc. Second Lunar Sci. Conf., Geochim. Cosmochim. Acta* Suppl. 1, Vol. 2, pp. 1521–1546. MIT Press.

Tera R., Eugster O., Burnett D. S., and Wasserburg G. J. (1970) Comparative study of Li, Na, K, Rb, Cs, Ca, Sr, and Ba abundances in achondrites and in Apollo 11 lunar samples. *Proc. Apollo 11 Lunar Sci. Conf., Geochim. Cosmochim. Acta* Suppl. 1, Vol. 2, pp. 1637–1657. Pergamon.

Warner J. L. (1972) Apollo 14 breccias: Metamorphic origin and classification (abstract). In *Lunar Science—III* (editor C. Watkins), pp. 782–784, Lunar Science Institute Contr. No. 88.

Proceedings of the Third Lunar Science Conference
(Supplement 3, *Geochimica et Cosmochimica Acta*)
Vol. 2, pp. 2015–2024
The M.I.T. Press, 1972

# The nature and effect of the volatile cloud produced by volcanic and impact events on the moon as derived from a terrestrial volcanic model*

J. J. NAUGHTON, D. A. HAMMOND, S. V. MARGOLIS, and D. W. MUENOW

Chemistry Department and Hawaii Institute of Geophysics,
University of Hawaii, Honolulu, Hi. 96822

**Abstract**—The ejection of volatiles from Hawaiian volcanoes has been used as a model for the emission of the volatile cloud that might be produced during volcanic or impact events on the moon. From the known composition of terrestrial and lunar lava, and that of terrestrial volatiles, an estimate can be made of the elemental composition of a lunar volatile cloud. If it is assumed that the volatiles approach equilibrium, as occurs in terrestrial volcanism, then the equilibrium molecular composition of the lunar cloud can be calculated for a variety of temperatures. CO is the main gaseous species, while $S_2$ and elemental substances would be important components as sublimates in the highly reducing environment.

A further study was made of the erosive action of alkali metal vapors, which laboratory experiments and equilibrium calculation show to be important components of the vapor emitted from heated lunar basalts. Electron microscope scans of lunar rocks exposed to K vapors demonstrate attack along grain boundaries of the crystalline portion, while glass-lined impact pits remain as unattacked protrusions. This is characteristic of alkali-eroded rock surfaces and has been noted as a feature of lunar rock surfaces. The production of a plasma of K vapor produced no enhancement of the erosive attack. The mechanism of emission of vapor from high-temperature silicate glasses and melts into a vacuum was shown to consist of episodic bursts released by breaking bubbles superimposed on a much slower diffusive release.

## INTRODUCTION

RESEARCHES ON THE LUNAR SAMPLES returned by the Apollo missions have led to an increasing realization that surface volatilization could play an important role as a later-stage process of material transfer on the moon. Thus, it would be expected that in the lunar vacuum the alkali metals, and lead, mercury, and other elements or compounds known to exert high vapor pressures, would be vaporized from high-temperature regions created by meteoritic impact or volcanism. Some fraction of the light elements and compounds would be lost to space, but much volatilized material would be deposited in the cooler zones. Localized heating would occur as a continuing process from meteorite impact effects, as is evidenced by the many glass-lined craters of all ages noted on the lunar surface. However, the very extensive flows of the mare regions that occurred at one time must have led to the evaporative movement of large amounts of material. Evidence for this type of transfer has been noted in reports of the surface association of some volatile elements (Kurat and Keil, 1970; Naughton *et al.*, 1971; Reed *et al.*, 1972) and, recently, in evidence for the

* Hawaii Institute of Geophysics Contribution No. 472.

volatile-supplied growth of crystals on protected surfaces (McKay *et al.*, 1972; Skinner and Winchell, 1972).

The effects of such vaporization as a differentiation process should be significant for lava being spewed forth in fountains, for persistently hot lava or ash flows, and for lava lakes. Other effects have been suggested such as the vaporization loss of rubidium and its consequences in Rb–Sr age dating (Gibson and Hubbard, 1972), the isotopic fractionation that occurs during the volatilization of lead and its effects on lead system chronologies (Silver, 1972), and the action of vaporized alkalis as a lunar rock eroding agent (Naughton *et al.*, 1971). Thus, an investigation of the elements and compounds that possibly would be present in a vapor cloud produced under lunar condition should be of interest.

## MODEL FOR LUNAR VAPORIZATION

Terrestrial volcanism has provided a useful comparative model for the study of features produced by volcanic eruptions and lava flows on the lunar surface. This has ranged all the way from the use of eruptive areas as training grounds for astronauts and their equipment to attempts to identify certain lunar surface features as analogs of structures associated with volcanism on the earth. Such comparisons must take into account the differences between lunar and terrestrial lava composition, and the unique effects of the lunar high vacuum on volcanic processes.

The presence of dissolved gases or of atoms that can volatilize, or associate to form high vapor pressure compounds, has a major effect on the properties of fluid lavas. Viscosity is modified, vesicularity is controlled, and the expansion of released volatiles contributes the main force in lava fountaining. On the earth these vaporizing elements supply components to the crust not provided by igneous rock weathering—Rubey's (1951) "excess volatiles" that make up an important part of the sediments, the atmosphere, the biosphere, and the oceans. Oceanic volcanics of the Hawaiian type probably provide the best model available of the degassing of a primitive earth crust and would be the most suitable for extrapolation to similar processes on the moon.

From laboratory studies on lunar rocks heated under vacuum, a loss of certain components has been noted by several workers (O'Hara *et al.*, 1970; Gibson and Hubbard, 1972), and studies of the vaporized species have been made by De Maria *et al.* (1971) and by Naughton *et al.* (1971). In the latter case, a Knudsen cell-mass spectrometric study of alkali volatilization was made so that species containing these elements were reported, but other substances also were noted, though many escaped detection because of the sporadic type of vapor release that takes place and the time needed to make mass scans. Also, reaction between silicate samples and containers at high temperatures is a particular problem in this type of work because it releases spurious vapor species. Because a general knowledge of the mechanism of vaporization from silicate species is lacking, it was believed that a fruitful approach to the evaluation of this phenomenon under lunar conditions could be made through a study of the analogous vaporization found in terrestrial situations—mainly in volcanoes.

Collection and research on volatiles from terrestrial volcanoes has been going on for over 150 yr, hampered by the obvious difficulties of sampling. At one extreme there is the explosive degassing that is accompanied by a rapidly moving high-temperature cloud of gas, debris, and ash (*nuée ardente*), of the kind involved in many instances of destructive volcanism. The types of welded breccias and ignimbrites that result are well-known products of volcanism and may have direct analogs among lunar breccias. On the other hand we have Hawaiian volcanoes, which, because of their relatively quiescent nature, are particularly favored for the difficult and dangerous task of collecting and experimenting with the volatiles that accompany eruptions. Only recently has some order and understanding been achieved in these studies by dealing with the volatile system as one in thermodynamic equilibrium, an approach first suggested by Ellis (1957).

## Evaluation of the Vaporization Model

From observations of the degassing from nonexplosive lava fountains or from lava lakes, we have noted that volatiles are emitted in the initial stage as gases, which in part condense to particles ("blue haze") and ultimately to acidic steam clouds. Analyses of the main gaseous components in volcanic fountains have been accomplished using infrared techniques (Naughton *et al.*, 1969). Also recently, it has been possible to collect volatiles from directly above lava fountains. Analyses of these samples give us the composition of the volatile phase emanating from oceanic basaltic lava at the erupting temperature (circa 1200°C). Details of these latter collections and studies will be published elsewhere.

Lunar basalts have been analyzed extensively so that the content of elements that would contribute to a volatile phase is known. For this work we have used the compilation of analyses given by Mason and Melson (1970) based mainly on Apollo 11 materials. From the elemental compositions of terrestrial lavas and volatiles, and of lunar lavas, then, using a simple proportionality, one may arrive at an approximation for the elemental composition of a lunar volatile phase. This assumes that similar temperatures would obtain during lunar and terrestrial volatile emission, that the vaporizing molecules in the two systems would be the same, and that the partial vapor pressures would be proportional to the atomic fractional composition in the respective lava systems (Raoult's law). The gram-atom elemental composition of a lunar volatile phase calculated by this method is given in Table 1.

As mentioned earlier, volcanic volatiles have been studied successfully by treating

Table 1. Calculated elemental composition of a lunar vapor phase. Composition in log gram atoms.

| Element | Log gram atoms | Element | Log gram atoms |
|---|---|---|---|
| H | −1.70 | Ca | −0.82 |
| O | 0.18 | Mg | −1.00 |
| C | 0.11 | Al | −0.96 |
| S | 0.40 | Fe | −1.07 |
| N | −0.68 | Si | −2.24 |
| Cl | −1.62 | Ti | −1.16 |
| F | −1.07 | P | −2.36 |
| Na | −1.13 | Pb | −4.70 |
| K | −1.85 | | |

them as an equilibrium system (Heald *et al.*, 1963). If we assume that this same condition may obtain or be approached in the lunar situation, then from the elemental composition we can calculate the relative equilibrium concentrations of all likely volatile molecular compounds of the elements for which thermodynamic data are available. One hundred and twenty-two compounds were used in this case (JANAF, 1971). This is done by finding the composition corresponding to the minimum free energy for this system, using well-known computer-assisted procedures (Heald and Naughton, 1962). The resulting calculated composition of a lunar vapor phase emitted at high temperatures is shown in Table 2. The respective content of each substance

Table 2. Mole fractions ($f_M$) of major equilibrium components of lunar vapor phase. Same compounds in terrestrial volcanic vapor listed for comparison.*

| | Lunar | | | Terrestrial |
|---|---|---|---|---|
| Compound | $\log f_M$ (1200°K) | $\log f_M$ (1400°K) | $\log f_M$ (1600°K) | $\log f_M$ (1400°K) |
| $H_2O$ | −3.9 | −3.6 | −3.4 | −0.028 |
| $H_2$ | −3.7 | −3.2 | −3.0 | −2.5 |
| $O_2$ | −16.4 | −13.7 | −12.0 | −7.8 |
| $CO_2$ | −1.1 | −1.2 | −1.3 | −1.4 |
| CO | −0.63 | −0.50 | −0.46 | −3.5 |
| S | −6.6 | −5.2 | −4.2 | −8.9 |
| $S_2$ | −0.47 | −0.43 | −0.42 | −7.8 |
| $H_2S$ | −2.5 | −2.7 | −2.8 | −5.6 |
| HS | −4.6 | −4.0 | −3.5 | −7.3 |
| $SO_2$ | −4.7 | −4.0 | −3.3 | −2.0 |
| COS | −1.0 | −1.4 | −1.8 | −8.1 |
| $CS_2$ | −1.6 | −2.2 | −2.9 | −15.5 |
| CS | −4.6 | −4.2 | −4.0 | −13.8 |
| C | −2.7 | −3.4 | −4.2 | −9.4 |
| CHO | −8.9 | −8.0 | −7.4 | −10.7 |
| $N_2$ | −1.5 | −1.5 | −1.5 | −3.3 |
| $NH_3$ | −9.9 | −9.7 | −9.6 | −11.3 |
| HCl | −9.7 | −8.1 | −7.0 | −2.1 |
| HF | −7.6 | −6.4 | −5.6 | −2.9 |
| Na | −1.6 | −1.7 | −1.7 | −7.1 |
| NaCl | −3.3 | −3.2 | −3.1 | −3.0 |
| K | −2.4 | −2.4 | −2.4 | −8.7 |
| KCl | −3.3 | −3.2 | −3.2 | −3.9 |
| $Al_2O$ | −1.8 | −1.8 | −1.9 | −19.4 |
| AlS | −3.8 | −3.3 | −3.0 | −17.3 |
| AlOF | −2.7 | −2.5 | −2.3 | −6.6 |
| Ca | −1.8 | −1.8 | −1.7 | −14.0 |
| CaCl | −3.1 | −2.9 | −2.7 | −9.2 |
| CaF | −1.6 | −1.6 | −1.7 | −10.5 |
| Mg | −5.2 | −5.8 | −6.3 | −12.9 |
| MgS | −1.5 | −1.5 | −1.5 | −12.2 |
| Fe | −1.5 | −1.6 | −1.6 | −5.8 |
| FeO | −4.9 | −4.5 | −4.1 | −5.9 |
| FeF | −5.8 | −5.5 | −5.2 | −6.7 |
| TiO | −1.6 | −1.7 | −1.7 | −8.6 |
| $TiO_2$ | −3.2 | −3.4 | −3.5 | −7.5 |
| TiOF | −4.4 | −4.5 | −4.5 | −8.3 |
| SiO | −6.1 | −5.3 | −4.7 | −4.0 |
| SiS | −2.7 | −2.7 | −2.8 | −9.0 |
| $P_4O_6$ | −3.4 | −3.5 | −3.5 | −4.9 |
| Pb | −5.2 | −5.2 | −5.2 | −7.8 |
| PbO | −9.0 | −8.5 | −8.0 | −8.0 |

* Mole fractions differ from values previously listed in revised abstracts (Naughton, *et al.*, 1972) due to calculation error in the latter.

is listed as log mole or volume fraction (log $f_M$). Because of space considerations, we have limited the listing to the major gaseous components and sublimates, with two or three of the main compounds that would be found in a vapor phase given for each element. Similar calculations for terrestrial systems (Table 2) show that the same compounds also do occur in the primary volcanic gaseous phase, though, of course, in very different proportions. The components shown for comparison are those of importance in the lunar system, and not necessarily the major volatile compounds found in terrestrial volcanics. All values listed are for gaseous mole fractions with a total pressure of one atmosphere.

The very highly reduced nature of the system as calculated is evident in the very low fugacity of $O_2$, about 5 orders of magnitude lower than that observed in terrestrial basaltic lava systems at corresponding temperatures. This is in agreement with estimates made from other studies (Cameron, 1971; Sato and Helz, 1971). The pressure under which a lunar vapor would be confined would change as it is volatilized, varying from a high value within a vesicle or bubble to a very low value after release to the lunar vacuum. Wellman (1970) has estimated the pressure within lunar vesicles and vugs to be between 0.01 and 100 atm. Calculations of the change of equilibrium concentrations with increasing pressure for terrestrial systems made by Heald *et al.* (1963) show that for the majority of components the mole fraction composition remains relatively constant, although the partial pressure of each compound increases. We have chosen 1 atm as a convenient value for demonstrating the general composition of the vapor, but calculation also could be made for any other pressure.

The main components in a vaporizing system would be CO and $S_2$, comprising about 75% of the total pressure. It will be noted that a surprisingly large contribution to the gaseous phase would be made by condensibles, the "sublimates" of volcanology. These would produce about 50% of the total pressure. Elemental substances would be important components under the reducing conditions that prevail. The high concentration of the alkali elements is noteworthy in reference to their effects as eroding agents. The absolute value used for the lead content is arbitrary, because we have found lead in terrestrial sublimates in highly variable amounts, ranging from zero to a few tenths percent. It was introduced here to demonstrate the probable form in which it would be transported as a vapor, and because of its importance in geochronology. The possible transport of some of the refractory elements normally considered as nonvolatile, such as Al, Mg, and Ti, as gases in the form of lower valence oxides (TiO), or oxyhalides (AlOF), or sulfides (MgS), is also noteworthy. Differentiation by this process may to some degree account for the variability in the average content of some elements in rock samples obtained from different areas of the moon.

The vapor compositions at three temperatures, 1200°K, 1400°K, and 1600°K, are listed in Table 2. These encompass the eruption temperature of terrestrial basaltic lava (about 1400°K–1500°K), and would include the minimum temperature of extrusion of lunar basalts (1350°K–1400°K) (Ringwood and Essene, 1970; Roedder and Weiblen, 1970; Muan *et al.*, 1971). In this range the relative concentrations of the prevalent vapor molecules containing an element will be noted to change with temperature, but only to a minor degree.

A more important process that would occur during the cooling of the gas cloud would be the appearance of solid species as the vaporized sublimates condense. Also, at this stage, reactions between vapor components would produce certain silicates, sulfides, oxides, and other forms which have greater stability at lower temperatures. Unfortunately, thermodynamic data are lacking for many of these substances, and it is therefore not possible at present to calculate the complete equilibrium compositional history of all phases—gases, solids, and liquids—that would be present as the high temperature gas cloud cools to the ambient lunar conditions.

## Mechanism of Vapor Release

The release of vapors from high-temperature solid surfaces or nonagitated viscous liquids is a slow process, limited by replenishment of evaporated material by diffusion. In natural terrestrial systems, the main release takes place through bubbles generated within the lava, which agitate and stir the lava and vent to the atmosphere quietly, or in a more spectacular form as lava fountains. Such bubbles can be observed frozen into crystalline or glassy lavas as vesicles and have been noted as near-surface features within lunar glasses (Bayer and Wiedemann, 1972).

We have studied the mechanism of vapor release by progressively heating glassy droplets (Pele's tears) within a vacuum system and noting the gas release pattern by means of a mass spectrometer. Samples were removed after each heating step in the softening range of the glass (about 700°C), where major gas release was noted, and were examined microscopically and photographed. Examples of such release behavior are shown in Fig. 1, where the stages of bubble formation and breakage are illustrated in (a) and (b). Similar droplet forms, with vesicle breakage holes, are found in lunar fines (Bayer and Wiedemann, 1972). Some of the bubblet craters left after breakage bear a superficial resemblance to lunar impact craters but can be easily differentiated by the absence of a halo. A degree of gas and vapor release was observed throughout the heating cycle, but very large spikes of gas were noted at and above the softening point. Almost certainly these were due to the venting of vapor from bubbles as they broke at the surface and were easily missed unless a mass peak was being held or scanned when the event occurred. A continuously recording, rapid-scan mass spectrometer is needed to most effectively study the vapors being emitted, and future work will be done with an instrument of this type.

The release of gas from lunar samples was studied using glass spherules handpicked from sample 14163. This was done by photographing the spheres as they were heated within a vacuum system. Sequential gas release events are shown in Fig. 1 (c and d), and the similarity to the bubble-type vapor release process that takes place in terrestrial samples is evident.

The release and expansion of gases from dust and glassy particles carried in the vapor cloud from a lunar cratering event would supply a high-temperature gaseous envelope that would maintain the mobility of such a cloud over large distances, similar to the mechanism that is believed to operate in volcanic *nuées ardente*. The bubble-type release from lava lakes or eruption sites on the moon would provide stirring,

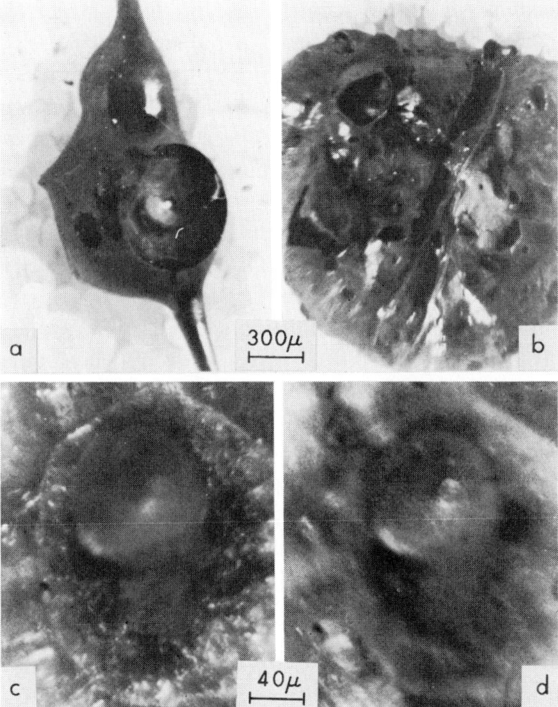


Fig. 1. Photomicrographs illustrating the chief mode of gas release from natural silicate glasses through bubble formation and breakage. (a) and (b), Terrestrial volcanic glass droplets (Pele's tears). (c) and (d), Sequential formation and breakage of gas bubble in spherule from lunar fines (14163). Latter photographed within vacuum system during heating and release. All heated at about 800°C in vacuum.

and prevent formation of a cooled, nonemitting crust, again analogous to the behavior observed in terrestrial situations.

## Erosion by Alkali Vapor

The surface association of some volatile elements in lunar rocks has been reported (Kurat and Keil, 1970), and the special case of the alkalis has been investigated by us (Naughton *et al.*, 1971). Also we have studied the action of volatilized alkali elements as eroding agents for crystalline rocks (Naughton *et al.*, 1965). In previous work we had noted the particular effectiveness of potassium in this regard and had found that the eroding effect seemed to constitute a detachment of individual crystals by loosening at grain boundaries. It was deduced that an exchange between potassium and sodium takes place in these regions, with a resultant expansion of the lattice which caused the individual grains to pop out of the structure.

Further studies have been conducted in an effort to better understand the process that causes this breakdown. Glasses and glassy rocks were not attacked. The ionization form of potassium vapor in a plasma seemed not to be effective at all. Lunar crystalline rocks seemed to be particularly susceptible to attack and erosion by the alkali metals. A carefully controlled attack of the surface of rock 14310 was undertaken in equipment where the surface could be observed microscopically during the process. After a limited amount of erosion was achieved, the sample was removed and examined in the scanning electron microscope (SEM). The microscopic and SEM observations showed that the glass-lined micrometeorite pit craters present on the surface were not attacked but were left as protruding islands or pedestal-like structures with the removal of the surrounding crystalline material by the erosion attack. Additional indication of grain boundary attack also was observed. These effects can be seen in Fig. 2, (a) and (b), respectively. The survival of protruding glass-splashed surfaces or glass-lined craters after alkali metal attack is the one phenomenon observed thus far that is diagnostic of this erosive action. Areas of splashed glass around craters also survive. Such pedestal craters or stylus pits have been noted on the surface of lunar rocks and are believed by some to be evidence of erosion by means other than impact (LSPET, 1970). However, Hörz *et al.* (1971) believe them to be primary effects of the impact that produced the craters, but laboratory experiments simulating micrometeorite pit-forming events have not as yet succeeded in producing this type of crater.

## Conclusion

From researches that have been conducted on the degassing of terrestrial volcanoes, it is possible to construct a model of degassing under lunar conditions that gives us a reasonable picture of the nature of the components of a lunar vapor cloud that would be produced in high-temperature areas. Such areas would be the result of meteorite impacts, or of isolated or regional volcanism. Reduced compounds of the elements would be major components of such a gas cloud under the low-oxygen fugacity of the lunar surface, with the elemental forms playing an important role. The presence of specific vapor molecules of the elements normally thought of as "volatile," is demon-

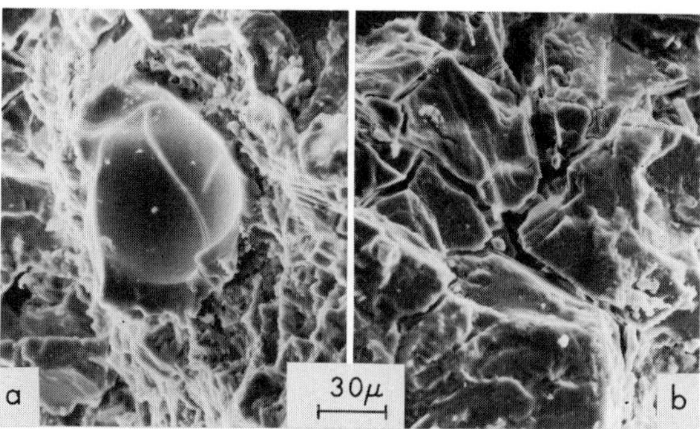


Fig. 2. Scanning electron micrographs showing the effects of a limited erosive attack by potassium vapor on a lunar rock (14310). (a) Unattacked, glass-lined micrometeorite crater left as stylus pit after removal of surrounding rock by alkali erosion. (b) Indication of erosive action as attack and separation along grain boundaries on crystalline portion of rock surface.

strated for H, O, C, S, Cl, F, N, Na, K, P, and Pb. Also, vapor forms can exist for "refractory" elements such as Al, Si, Ti, Ca, and Mg, and these may play a role in the transport and differentiation of such elements in lunar systems.

The mechanism of degassing has been investigated, and it has been shown that the major contribution to the gaseous phase is made by bubbles or vesicles produced internally, which release bursts of vapor on breaking at the surface of the melt, or that of viscous high-temperature glass droplets. Thus, degassing may be continuous in a hot particulate cloud, and would contribute to the low viscosity and ready transport over great distances, as in terrestrial glowing cloud (*nuées ardente*) eruptions.

One of the effects of a volatile cloud that is of interest is the erosive action that would be exerted by the alkali metals condensing from the cloud. It is believed that alkali attack takes place by a loosening of crystals from the rock matrix at grain boundaries. Evidence of this is seen in SEM studies of rock surfaces that have been subjected to this attack. Also noted are the resistance of glasses to alkali attack, and the importance of this peculiarity in preserving glass-lined pit craters as protrusions or pedestals ("stylus pits") on surfaces which have been subjected to this type of attack. Such pedestal pits are observed on lunar surfaces, and are taken as evidence that alkali erosion may have played a significant role in the erosion of these areas *in situ*.

*Acknowledgment*—The work reported here was supported in part by Grant No. NGR 12-001-081 from the National Aeronautics and Space Administration.

## References

Bayer G. and Wiedemann H. G. (1972) Microstructure, melting, and crystallization characteristics of lunar, vitreous fines (abstract). In *Lunar Science—III* (editor C. Watkins), pp. 50–52, Lunar Science Institute Contr. No. 88.

Cameron E. N. (1971) Opaque minerals in certain lunar rocks from Apollo 12. *Proc. Second Lunar Sci. Conf., Geochim. Cosmochim. Acta* Suppl. 2, Vol. 1, pp. 193–206. MIT Press.

De Maria G., Balducci G., Guido M., and Piacente V. (1971) Mass spectrometric investigation of the vaporization process of Apollo 12 lunar samples. *Proc. Second Lunar Sci. Conf., Geochim. Cosmochim. Acta* Suppl. 2, Vol. 2, pp. 1367–1380. MIT Press.

Ellis A. J. (1957) Chemical equilibrium in magmatic gases. *Amer. J. Sci.* **255,** 416–431.

Gibson E. K. and Hubbard N. J. (1972) Volatile element depletion investigations on Apollo 11 and 12 lunar basalts via thermal volatilization (abstract). In *Lunar Science—III* (editor C. Watkins), pp. 303–305, Lunar Science Institute Contr. No. 88.

Heald E. F. and Naughton J. J. (1962) Calculation of chemical equilibria in volcanic systems by means of computers. *Nature* **193,** 642–644.

Heald E. F., Naughton J. J., and Barnes I. L. Jr. (1963) The chemistry of volcanic gases. *J. Geophys. Res.* **68,** 545–557.

Hörz F., Hartung J. B., and Gault D. E. (1971) Micrometeorite craters on lunar rock surfaces. *J. Geophys. Res.* **76,** 5770–5798.

JANAF *Thermochemical Tables*, 2nd edition (1971) NSRDS-NBS 37, Nat. Bur. Stds.

Kurat G and Keil K. (1970) Effects of vaporization and condensation on Apollo 11 glass spherules: Implications for cooling rates. Personal communication preprint.

LSPET (Lunar Sample Preliminary Examination Team) (1970) Preliminary examination of lunar samples from Apollo 12. *Science* **167,** 1325–1339.

Mason B. and Melson W. G. (1970) *The Lunar Rocks*. Wiley-Interscience.

McKay D. S., Clanton U. S., Heiken G. H., Morrison D. A., and Taylor R. M. (1972) Vapor phase crystallization in Apollo 14 breccias and size analysis of Apollo 14 soils (abstract). In *Lunar Science—III* (editor C. Watkins), pp. 529–531, Lunar Science Institute Contr. No. 88.

Muan A., Hauck J., Osborn E. F., and Schairer J. F. (1971) Equilibrium relations among phases occurring in lunar rocks. *Proc. Second Lunar Sci. Conf., Geochim. Cosmochim. Acta* Suppl. 2, Vol. 1, pp. 497–505. MIT Press.

Naughton J. J., Barnes I. L., and Hammond D. A. (1965) Rock degradation by alkali metals: A possible lunar erosion mechanism. *Science* **149,** 630–632.

Naughton J. J., Derby J. V., and Glover R. B. (1969) Infrared measurements on volcanic gas and fume: Kilauea eruption, 1968. *J. Geophys. Res.* **74,** 3273–3277.

Naughton J. J., Derby J. V., and Lewis V. A. (1971) Vaporization from heated lunar samples and the investigation of lunar erosion by volatilized alkalis. *Proc. Second Lunar Sci. Conf., Geochim. Cosmochim. Acta* Suppl. 2, Vol. 1, pp. 449–457. MIT Press.

Naughton J. J., Hammond D. A., Margolis S. V., and Muenow D. W. (1972) A study of the nature of the gas cloud produced by volcanic and impact events on the moon and its relation to alkali erosion (abstract). In *Lunar Science—III* (editor C. Watkins), pp. 578–581, Lunar Science Institute Contr. No. 88.

O'Hara M. J., Biggar G. M., and Richardson S. W. (1970) Experimental petrology of lunar materials: The nature of mascons, seas, and the lunar interior. *Science* **167,** 605–607.

Reed G. W. Jr., Jovanovic S., and Fuchs L. H. (1972) Concentrations and lability of the halogens, platinum metals, and mercury in Apollo 14 and 15 samples (abstract). In *Lunar Science—III* (editor C. Watkins), pp. 637–639, Lunar Science Institute Contr. No. 88.

Ringwood A. E. and Essene E. (1970) Petrogenesis of Apollo 11 basalts, internal constitution and origin of the moon. *Proc. Apollo 11 Lunar Sci. Conf., Geochim. Cosmochim. Acta* Suppl. 1, Vol. 1, pp. 769–799. Pergamon.

Roedder E. and Weiblen P. W. (1970) Lunar petrology of silicate melt inclusions, Apollo 11 rocks. *Proc. Apollo 11 Lunar Sci. Conf., Geochim. Cosmochim. Acta* Suppl. 1, Vol. 1, pp. 801–837. Pergamon.

Rubey W. W. (1951) Geologic history of sea water. *Bull. Geol. Soc. Amer.* **62,** 1111–1147.

Sato M. and Helz R. T. (1971) Oxygen fugacity studies of Apollo 12 basalts by the solid-electrolyte methods (abstract). Abstracts of the Second Lunar Sci. Conf. (unpublished proceedings).

Silver L. T. (1972) Lead volatilization and volatile transfer processes on the moon (abstract). In *Lunar Science—III* (editor C. Watkins), pp. 701–703, Lunar Science Institute Contr. No. 88.

Skinner B. J. and Winchell H. (1972) Vapor phase growth of feldspar crystals and fractionation of alkalis in feldspar crystals from 12038.22 (abstract). In *Lunar Science—III* (editor C. Watkins), pp. 710–712, Lunar Science Institute Contr. No. 88.

Wellman T. R. (1970) Gaseous species in equilibrium with the Apollo 11 holocrystalline rocks during their crystallization. *Nature* **225,** 716–717.

Proceedings of the Third Lunar Science Conference
(Supplement 3, *Geochimica et Cosmochimica Acta*)
Vol. 2, pp. 2025–2027
The M.I.T. Press, 1972

# Analysis of single particles of lunar dust for dissolved gases

F. M. Ernsberger
PPG Industries, Inc.

The inert-gas analyses of Heymann, *et al.* (1970) are among the few that have been made on the gas content of the glass component of lunar fines. Apparently there are no cases in which glass alone has been analyzed for other than inert gases. Such data are needed, because the amount and distribution of gases within the glassy particles can tell us something of the conditions under which the particles were melted.

Special techniques are required to secure an uncontaminated release and then a quantitative analysis of the extremely small absolute quantity (of the order of $10^{-6}$ cm$^3$ STP) of gas from a particle 50–100 $\mu$ in diameter. A combination of the neodymium laser and the omegatron type of bakeable mass spectrometer has been found to be suitable for this task.

Figure 1 shows a block diagram of the apparatus, omitting power supplies and the strip-chart recorder for the output spectrum. The laser and microscope were collimated to deliver a focused pulse (about 5 joules) at a point defined by the cross-hairs in the reticle of the microscope, and an $X$, $Y$, $Z$ positioner permitted the successive disintegration of individual preselected particles. The mass spectrometer was operated in a static mode; that is, with the pump valve closed. The accumulation of background gases during the few minutes necessary to record the spectrum was made negligible by a previous bake-out of the all-glass system. The small volume of the system (about 0.3 liter) and the high ion-collection efficiency of the omegatron made the detection limits about $10^{-9}$ cm$^3$ for He, $2 \times 10^{-10}$ cm$^3$ for $N_2$, and $10^{-10}$ cm$^3$ for Ar. Calibration for these three gases was accomplished by breaking within the system a capillary containing a known mixture of helium and air.

Entrance and exit windows for the laser pulse, and in particular the support for the particles, must be made of a material that does not appreciably absorb the 1.06 $\mu$ laser radiation. Ordinary glasses are suitable, provided they are quite thin ($<0.1$ mm). Microscope cover-slips, further thinned by etching, were used as the sample support, and thin blisters blown in the Pyrex walls of the system were used as windows.

The platinum black exposed within the system had the important function of selectively adsorbing $O_2$ and CO. Oxygen is the gas released in largest amount by the laser pulse, but it has no meaning in the present context because it arises from thermal decomposition of the glass. Likewise CO is largely an artifact; and unless it is removed, it seriously complicates the determination of $N_2$.

Samples were selected from fines returned from both Apollo 11 (sample 10089, 156) and Apollo 14 (14163). The finest particles were removed by elutriation with a

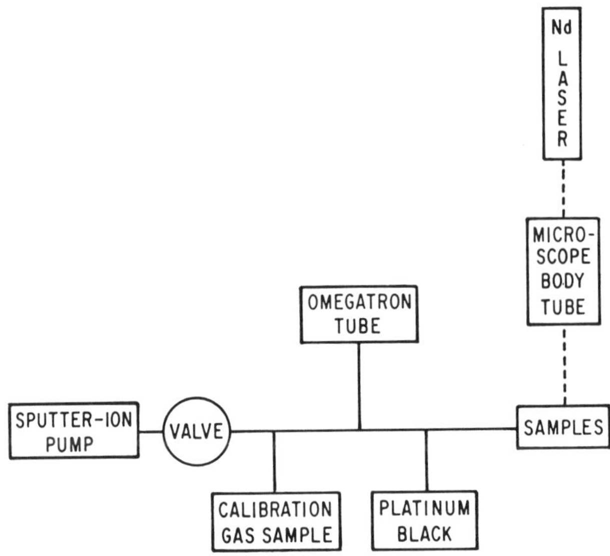


Fig. 1. System for analysis of gases occluded in single particles.

very dilute solution of sodium silicate. The silicate functioned as a dispersing agent, and was used instead of an organic surfactant, to avoid carbonaceous contamination. The remaining coarse particles were spread on a microscope slide, and single particles were selected under a microscope.

Table 1 shows data relating to a particular group of six particles, all of which were done with one loading of the sample carrier. All the particles in this group were essentially spherical and were selected to be similar in visual characteristics such as color. Half of the group was etched 30 seconds with 2.5% aqueous HF.

This limited amount of data is already sufficient to answer certain qualitative questions. It is apparent that the lunar glasses do contain dissolved helium and nitrogen. The relation between the amounts of helium and nitrogen is the reverse of that reported by Hintenberger *et al.* (1970) for Apollo 11 breccias, but a part of the helium content of our samples was probably lost during the bakeout procedure, even though special coolers were applied to keep the samples below 200°C.

Table 1. Gases in glassy lunar microspheres. All particles were selected from Apollo 11 fines. Quantities are given in units of $10^{-9}$ cm$^3$ STP.

| Diameter, $\mu$ | Helium | Nitrogen |
|---|---|---|
| 82 | 3 | 25 |
| 75 | 6 | 28 |
| 50 | 7 | 52 |
| 82 (etched) | 2 | 140 |
| 73 (etched) | 1 | 61 |
| 91 (etched) | 3 | 43 |

Etching seems to remove a portion of the helium content, confirming what others have noted, that helium in the lunar fines is largely of solar-wind origin though its distribution has been modified by subsequent diffusion. Nitrogen, however, seems to be as plentiful in the interior as at the surface. Apparently nitrogen is indigenous in the regolith, because Hintenberger, *et al.* (1970) and Moore, *et al.* (1970) found it within Type A rock, where other light gaseous elements were absent.

Methane is not reported in the table, but it was observed in amounts comparable to that of nitrogen in every sample analyzed. No attempt was made to record quantitative data on methane, because it is rapidly cracked by the filament of the omegatron and in any event does not represent the total carbon content of the particle. An unknown proportion was converted to carbon monoxide and adsorbed by the platinum black.

Hydrogen was never observed. It was probably present but would be partly converted to water and adsorbed on the walls of the vacuum system, while the remainder would be adsorbed by the platinum black. Neon and argon were detectable only in a few of the larger particles.

Exploratory measurements on crystalline particles and on Apollo 14 fines established that all dust particles contain the same dissolved gases in roughly the same proportions.

The salient conclusions from this work are first, that the laser microprobe, mass-spectrograph combination is a useful technique for investigating the nitrogen content of single particles of lunar dust; and second, that lunar glasses are apparently formed under conditions such that indigenous nitrogen is largely retained. This is consistent with an impact theory, if we postulate a large enough impact to provide a transient gaseous atmosphere in which the droplets of molten rock may cool without loss of their dissolved nitrogen.

## References

Heymann D., Yaniv A., Adams J. A. S., and Fryer G. E. (1970) Inert gases in lunar samples. *Science* **167,** 555–558.

Hintenberger H., Weber H. W., Voshage H., Wänke H., Begemann F., and Wlotzka F. (1970) Concentrations and isotopic abundances of the rare gases, hydrogen and nitrogen in Apollo 11 lunar matter. *Proc. Apollo 11 Lunar Sci. Conf., Geochim. Cosmochim. Acta* Suppl. 1, Vol. 2, pp. 1269–1282. Pergamon.

Moore C. B., Lewis C. F., Gibson E. K., and Nichiporuk W. (1970) Total carbon and nitrogen abundances in lunar samples. *Science* **167,** 495–497.

Proceedings of the Third Lunar Science Conference
(Supplement 3, *Geochimica et Cosmochimica Acta*)
Vol. 2, pp. 2029–2040
The M.I.T. Press, 1972

# Inorganic gas release and thermal analysis study of Apollo 14 and 15 soils

EVERETT K. GIBSON, Jr.

Geochemistry Branch, NASA Manned Spacecraft Center,
Houston, Texas 77058

and

GARY W. MOORE

Lockheed Electronics Corp.,
Houston, Texas 77058

**Abstract**—Inorganic gas release studies have been made on Apollo 14 and Apollo 15 soils. The samples were heated at 6°C/min to 1400°C under vacuum, and the abundances, temperature ranges, and sequences of released gases were determined. The gases measured included $H_2$, He, $H_2O$, CO, $N_2$, $O_2$, $H_2S$, $CO_2$, and $SO_2$. The evolved gases are from several sources: (1) atmospheric contaminants, (2) solar wind derived species, (3) chemical reaction products, and (4) gases from vesicles and inclusions and/or gases exsolving from the melt. Soil samples lost less than 1.7 wt.% when heated to 1400°C under vacuum, while a sample heated in air to 1400°C gained 4.8 wt.%. Initial melting of the Apollo 14 bulk soil and two Apollo 15 soils occurs at 1140 ± 20°C.

Proton irradiation experiments with 100 keV protons have been carried out on samples of terrestrial olivine in an attempt to produce low molecular weight species which might be generated from solar wind irradiation of lunar samples. Irradiated samples of olivine contain 300 ppm $H_2O$ above the control samples. The quantity of water produced and its temperature release profile is similar to that found in some lunar samples.

## INTRODUCTION

THE STUDY OF VOLATILES in lunar samples is important to the understanding of past thermal conditions on the lunar surface. The redox state of lunar soils can also be inferred from the volatile gaseous species released by heating the soils under vacuum. The purpose of our investigation was to determine the gaseous species released from the Apollo 14 and 15 lunar soils during vacuum pyrolysis of the samples during programmed linear heating. This type of experiment is important in order to determine which volatiles are present and their abundance, as well as the temperature release range and sequence of release of these volatiles. A review of previous investigations of the volatile gaseous species from the Apollo 11 and 12 lunar samples has recently been made (Gibson and Moore, 1972).

Zeller *et al.* (1970) proposed that $H_2O$, $CH_4$, and possibly other hydrocarbons could be formed from irradiation of the lunar surface by solar wind protons. To test this model, lunar analogs of minerals, rocks, and meteorite samples have been irradiated with 100 keV protons for 8 hours and some of the initial analytical results for the production of water are presented in this paper.

## Experimental Procedures

The analytical techniques employed were similar to those previously described by Gibson and Johnson (1971, 1972) with one major modification. The quadrupole mass spectrometer has recently been interfaced with a small laboratory computer that controls the operation of the mass spectrometer and collects data as a function of signal strength.

The lunar samples were analyzed with a Mettler recording vacuum thermal analyzer interfaced with a Finnigan 1015S/L quadrupole mass spectrometer. The source of the mass spectrometer was placed directly in the reaction chamber. With this arrangement the evolved gases are analyzed without requiring any gas transfer procedures. Soil samples used in this study were placed into a previously outgassed 16 mm platinum or alumina crucible and evacuated to $2 \times 10^{-6}$ torr. The sample weight, temperature, and chamber pressure were continuously recorded. The sensitivity of the thermal balance used for the weight-loss studies is 0.05 mg. The samples were heated from ambient temperature to 1400°C at a heating rate of 6°C/min. Sample temperatures were measured with calibrated Pt/Pt-10% Rh thermocouples located at the base of the sample crucible. Spectra were obtained every 5°C during the heating cycle by the automatic mass spectrometer-computer control. The analytical data were stored on magnetic tape until processing after the programmed heating cycle was completed. Reproducible background spectra were obtained during the bakeout procedure with an empty crucible before sample analysis and were later subtracted from the spectra obtained for the samples.

The lunar soil samples were neither crushed nor homogenized before analyses, under the assumption that previous sample splitting during sample description and processing within the Lunar Receiving Laboratory had homogenized the samples. The sample sizes used for the analyses were between 200 and 250 mg. After the thermal analysis-gas release study was completed, the sample residues were chemically analyzed by atomic absorption and isotopic dilution analysis for their K, Rb, Na, Li, Sr, Ba, and REE concentrations. The analytical results from the volatilization study will be reported in a separate paper (Gibson and Hubbard, 1972).

Samples selected for the proton irradiation experiments included the following materials: San Carlos Arizona olivine, meteoritic cohenite, iron carbide ($Fe_3C$), graphite, quartz, and slices of the Allende meteorite. The six samples selected for the experiment were cleaned and divided into three equal pieces. One split was retained as a control, the second was subjected to the thermal vacuum of the irradiation chamber, and the third split was subjected to the proton exposure in addition to the environment of the irradiation facility. Samples were irradiated with protons in the Deep Space Materials Engineering Laboratory, NASA, Houston, Texas. The exposure conditions were as follows: chamber pressure less than $1 \times 10^{-8}$ Torr, chamber wall temperature of room temperature (ambient) to $-179$°C, proton energy of $100 \pm 5$ keV, proton particle density of $2 \times 10^{8}$ particles/$cm^2$/sec, and an exposure time of $8.0 \pm 0.1$ hours per sample. Extreme caution was used in handling the cleaned samples in order to prevent contamination. After the sample irradiation, the samples were placed in a desiccator until time for analysis. Samples of the irradiated and control olivine were analyzed within one week of the irradiation for their $H_2O$ contents using the procedures of Gibson and Johnson (1971). Analyses of the samples for $CH_4$ and hydrocarbon contents will be carried out in the near future using the procedures of Chang *et al.* (1971).

## Experimental Results

The soil samples analyzed included: (1) Apollo 14 bulk soil 14163,178 (in a platinum crucible), (2) Apollo 14 bulk soil 14163,178 (in an alumina crucible), (3) Apollo 15 LM soil 15021,21, and (4) Apollo 15 comprehensive soil 15601,31 from the edge of the Hadley Rille.

### *Weight loss investigations*

The results of the weight loss measurements for the Apollo 14 and 15 soils are given in Table 1. The experimental results obtained are similar to those reported previously

Table 1. Apollo 14 and Apollo 15 summary of weight-loss measurements.

| | 14163,178 (Pt cruc.) 202.5 mg | | 14163,178 ($Al_2O_3$ cruc.) 216.78 mg | | 15021,21 202.56 mg | | 15601,31 205.7 mg | |
|---|---|---|---|---|---|---|---|---|
| | (mg loss) | (% loss) | (mg loss) | (% loss) | (mg loss) | (% loss) | (mg loss) | (% loss) |
| 800° | 0.13 | 0.064 | 0.13 | 0.06 | 0.08 | 0.039 | 0.06 | 0.029 |
| 1000° | 0.25 | 0.12 | 0.26 | 0.12 | 0.15 | 0.074 | 0.11 | 0.054 |
| 1100° | 0.68 | 0.34 | 0.75 | 0.35 | 0.35 | 0.17 | 0.21 | 0.10 |
| 1200° | 1.06 | 0.52 | 1.30 | 0.60 | 0.58 | 0.28 | 0.38 | 0.18 |
| 1400° | 3.43 | 1.70 | 3.59 | 1.65 | 2.32 | 1.14 | 1.63 | 0.79 |
| Initial melting temp. | | 1150°C | | 1150°C | | 1130°C | | 1130°C |

for the Apollo 11 and 12 soils (Gibson and Johnson, 1971). The results of two separate analyses of the Apollo 14 soil 14163 are identical despite the fact that the analyses were carried out in platinum and alumina crucibles. The lunar soils analyzed lose less than 0.35 ± 0.10 wt.% when heated to 1100°C under vacuum, and even after 1400°C the soils lost less than 1.70 ± 0.5 wt.%. A sample of Apollo 14 soil 14163,178 heated to 1400°C in air gained 4.8 wt.%. The weight gain resulted from the oxidation of iron and other reduced phases in the soil sample. The weight gain from heating soil samples in air could perhaps be used as an indicator of relative degrees of reduction for the various soil samples. Hanneman (1970) previously reported a net weight gain when the samples were heated under oxidizing conditions.

Initial melting of the Apollo 14 soil occurred at 1150 ± 10°C while the two Apollo 15 soils from the LM and Hadley Rille began to melt at a slightly lower temperature of 1130 ± 10°C. The melting regions are similar to those reported previously (Gibson and Johnson, 1971). Sintering of the soil samples occurs near 1050°C.

### *Gas release patterns for Apollo 14 and 15 soils*

The gas release patterns for the four soils are shown in Figs. 1–4. The Apollo 14 and 15 soils which were analyzed in platinum crucibles show only minor differences in the volatiles released and the gas release patterns were similar to previously obtained patterns for the Apollo 11 and 12 soils (Gibson and Johnson, 1971). The patterns (Figs. 1 and 2) for the two Apollo 14 analyses carried out in platinum and alumina crucibles are identical with the exception of the $O_2$ release. The gases released from the soils result from terrestrial atmospheric contamination, solar wind components, chemical reaction products, and gaseous components from vesicles and inclusions and/or gases exsolving from the melt. The gas release patterns at elevated temperatures (>700°C are similar to those previously reported for synthetic lunar analogs (Gibson and Johnson, 1971).

### *Atmospheric contamination of soil samples*

Evidence for terrestrial contamination of the lunar soils with gases from the atmosphere is seen in the release patterns for the two separate samples of Apollo 14 soil 14163,178. For the initial analysis the sample was not exposed to the atmosphere

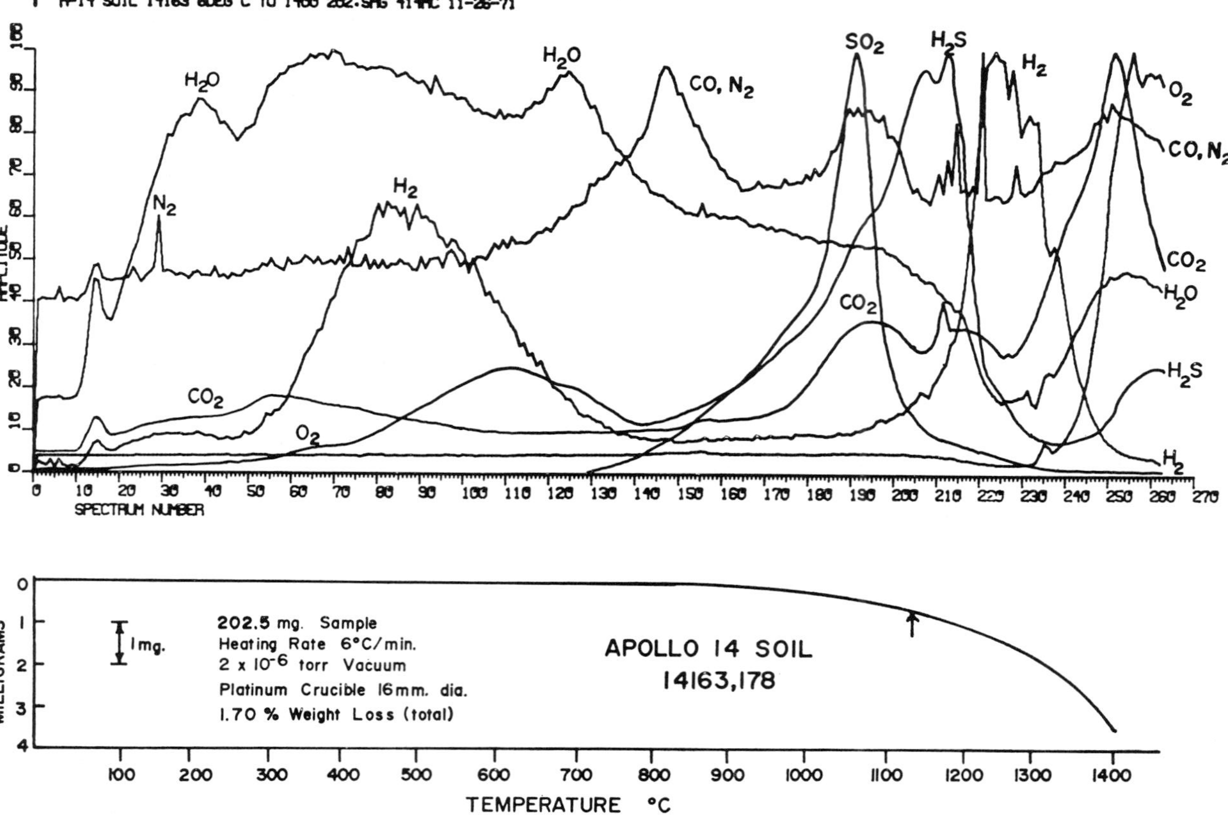


Fig. 1. Gas release pattern and weight-loss curve for Apollo 14 soil 14163,178 heated in a platinum crucible. Heating rate 6°C/min. Each of the released gases have been plotted so that their temperature of greatest abundance has been normalized to 100% amplitude. The arrow shown on the weight loss curve indicates the initial melting of the soil sample.

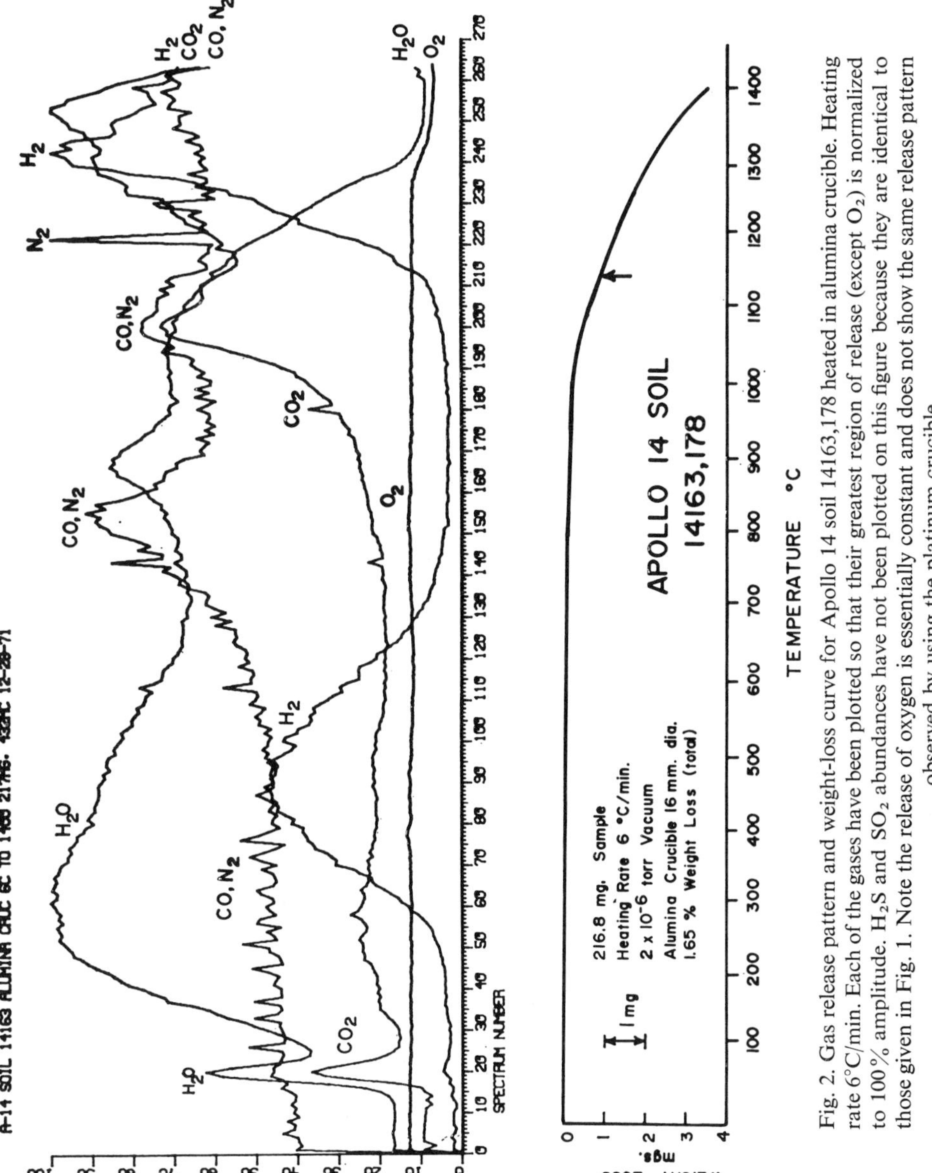


Fig. 2. Gas release pattern and weight-loss curve for Apollo 14 soil 14163,178 heated in alumina crucible. Heating rate 6°C/min. Each of the gases have been plotted so that their greatest region of release (except $O_2$) is normalized to 100% amplitude. $H_2S$ and $SO_2$ abundances have not been plotted on this figure because they are identical to those given in Fig. 1. Note the release of oxygen is essentially constant and does not show the same release pattern observed by using the platinum crucible.

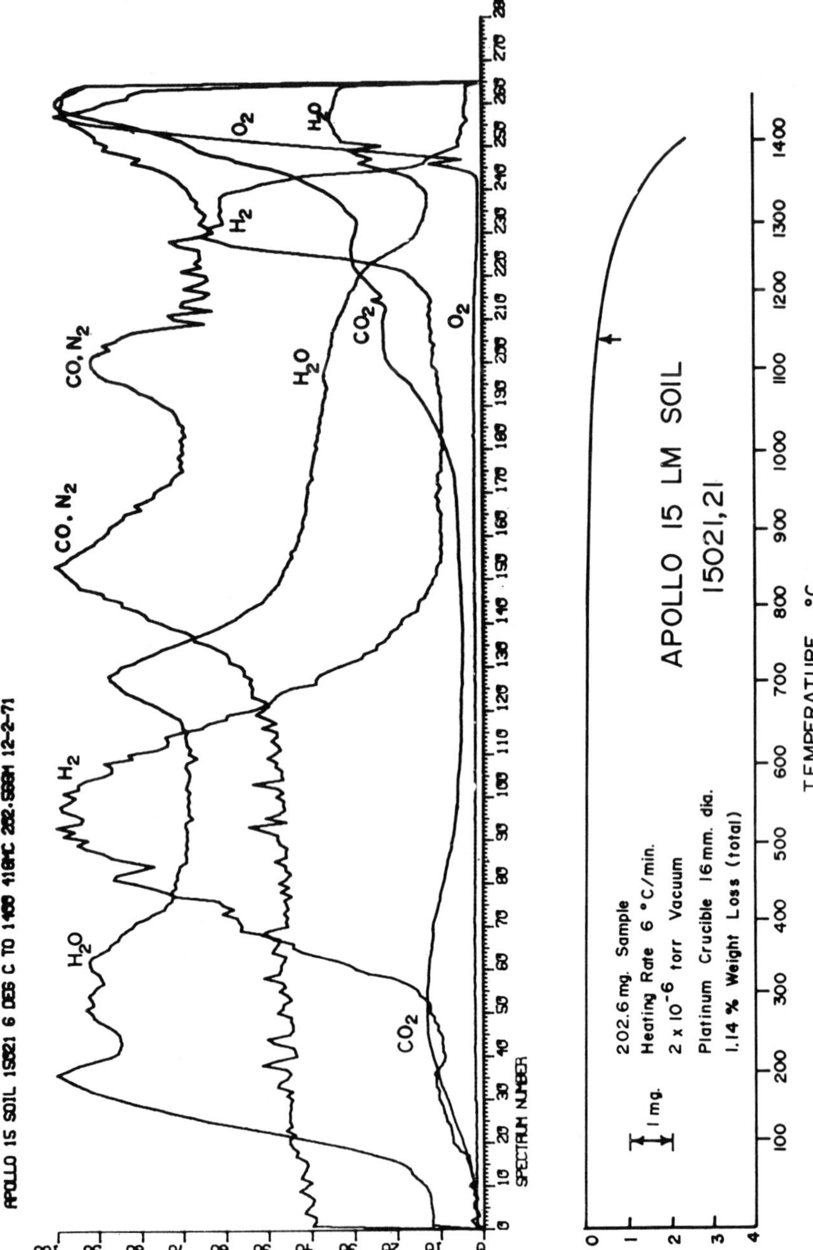


Fig. 3. Gas release pattern and weight-loss curve for Apollo 15 soil 15021,21. Sample heated in platinum crucible at 6°C/min. Each gas shown has been normalized to its greatest release temperature (100% amplitude). $H_2S$ and $SO_2$ release patterns have not been plotted because they are identical to the release patterns shown in Fig. 1.

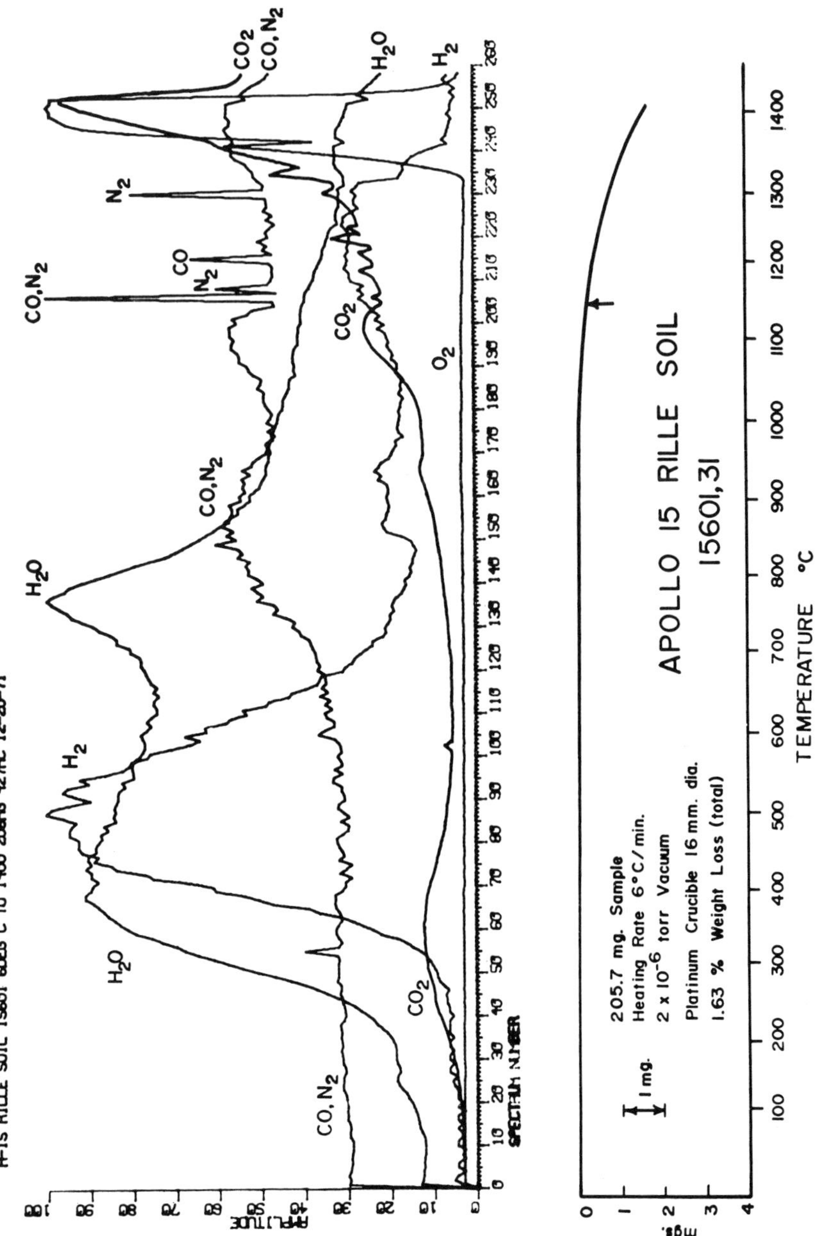


Fig. 4. Gas release pattern and weight-loss curve for Apollo 15 rille soil 15601,31. Sample heated in platinum crucible at 6°C/min. Each gas shown has been normalized to its greatest release temperature (100% amplitude). $H_2S$ and $SO_2$ release patterns have not been plotted because they are identical to the release patterns shown in Fig. 1.

until analysis was ready to begin. The first analysis of the sample 14163,178 (in a platinum crucible) indicated that only trace amounts of $H_2O$ and $CO_2$ were released during the heating of the sample below 150°C (Fig. 1). No major contamination by atmospheric $N_2$ is evidenced from our analysis of this soil and additional Apollo 14 and 15 soils. The second sample of Apollo 14 soil 14163,178 (in an alumina crucible) showed considerably more $H_2O$ and $CO_2$ contamination than did the earlier analysis of sample 14163 (Fig. 2). The second sample had been exposed to the room atmosphere for a period of one month. Oró *et al.* (1971) previously noted that $CO_2$ was adsorbed by the lunar soil samples and our experimental results support their observation.

*Low temperature release of solar wind products and other gaseous species*

Carbon dioxide is the major carbon containing gaseous phase released below 500°C from the Apollo 14 and 15 soils. Similar results were obtained for the Apollo 11 and 12 soil samples (Gibson and Johnson, 1971; Chang *et al.*, 1971; and others). The origin of the low temperature $CO_2$ is still unknown. Cadogen *et al.* (1972) have suggested that the low temperature $CO_2$ is not derived entirely from the solar wind. Hayes (1972) has suggested that the low temperature $CO_2$ results from cometary and/or meteorite gases driven back into the lunar soil by the solar wind after the impacting event. Such a process could be similar to the process noted by Heymann and Yaniv (1970) for the enrichment of $Ar^{40}$ in the lunar fines.

Hydrogen and helium derived from the solar wind are released between 300 and 700°C. The release temperatures are similar to those reported by other investigators for the noble gases He and Ne released from step-wise heating of the lunar fines (D. Bogard, private communication, 1972). The release of helium is identical to that of hydrogen in the 300 to 700°C temperature range and has not been plotted in Figs. 1–4. Similar release patterns for H and He were noted for the Apollo 11 and 12 soils (Gibson and Johnson, 1971).

The release of water vapor from the lunar soils was noted previously for the Apollo 11 and 12 soils (Gibson and Johnson, 1971; Oró *et al.*, 1971). The wide temperature range over which the $H_2O$ is released is similar to the water released from the Apollo 14 soils reported by Holland *et al.* (1972). We reported previously (Gibson and Johnson, 1971) that the majority of the water found in the lunar samples has been produced from the solar wind proton irradiation of the lunar fines and the water resulted from the mechanisms proposed by Zeller *et al.* (1970).

*Proton irradiation of lunar analogs*

Proton irradiation experiments with 100 keV protons have been carried out on samples of terrestrial olivine in an attempt to produce low molecular weight species, especially water, that might be generated from solar wind irradiation of lunar samples. After the proton bombardment with 100 keV protons with a flux of $2 \times 10^8$ particles/$cm^2$/sec for 8 hours, the irradiated olivine sample and its control samples were analyzed for their water contents (Gibson and Johnson, 1971). The irradiated olivine sample contained 300 ppm $H_2O$ above the two control samples which had not been subjected to the irradiation. However, it has been pointed out (C. T. Pillinger, private communication, 1972) that the proton flux and exposure time used were not sufficient (by an order of magnitude) to generate the 300 ppm $H_2O$ measured on the olivine samples. The measured water can possibly be accounted for from the irradiation process in addition to a large quantity of terrestrial water vapor adsorbed on the fresh surfaces of the olivine. The proton irradiation can damage the crystal surfaces and increase the water vapor adsorption capacity of the olivine. However, the $H_2O$ release pattern is not identical with those produced by adsorbed water vapor. The $H_2O$ release pattern and concentration from the proton irradiated sample were almost identical to the $H_2O$ release profile obtained from the thermal analysis-mass spectrometric analysis of lunar soils from the four Apollo missions.

The model proposed by Zeller *et al.* (1970) for the production of hydroxyl and water on the lunar surface has been partially confirmed by the experimental results obtained from our proton irradiation experiments on olivine. More detailed experimental work is required before the process is completely understood with regard to the flux rates and energies involved which could produce additional low molecular weight species in the lunar samples by proton bombardment. Our experimental results only approximate the conditions of the solar wind solar flares. However, the preliminary experimental evidence from Cadogan *et al.* (1972) for the production of hydrocarbons and our work indicates that indeed $H_2O$, in either the hydroxyl form or as $H_2O$, and hydrocarbons can be synthesized from the interactions of the solar wind protons with components found in the soils.

### *High temperature release of gaseous species*

The most abundant gaseous phase released above 700°C is mass 28, CO and/or $N_2$. The mass 28 is released in three distinct regions: (1) between 700 and 900°C, (2) between 950°C and the melting region of the soil (around 1130°C), and (3) immediately above the melting region of the soil. The CO and $N_2$ are produced in almost equal amounts in the first two regions. Examination of the fragment ions of CO and $N_2$ (e.g. masses 12 ($C^+$) and 14 ($N^+$)) provided the information concerning the relative amounts of the two mass 28 species. Holland *et al.* (1972) obtained similar results for the relative amounts of CO and $N_2$.

The source of the CO is from the reaction products at elevated temperatures of phases containing carbon such as cohenite and/or a "carbide" with the silicate minerals and glass phases found in the soils. The CO is identical to the reaction products produced from synthetic lunar analogs (Gibson and Johnson, 1971).

The mass 28 peaks or spikes produced at the melting region of the soil sample provide important information concerning gaseous species which may be trapped within gas-rich inclusions and/or vesicles or simply gases exsolving from the melt. The gases suddenly released from the Apollo 14 and 15 soils are both CO and $N_2$. Several of the spikes have been examined in detail and found to contain in some cases only $N_2$ or CO but in a few cases both CO and $N_2$ are present (Fig. 4). Funkhouser *et al.* (1971) also reported the presence of $N_2$ when lunar soil samples were crushed. The gases analyzed in our study are similar to those predicted by Wellman (1970).

Carbon dioxide was released at elevated temperatures from the lunar soils. We believe that the $CO_2$ produced, like a portion of the CO, results from chemical reactions of carbon bearing phases with the mineral and glass components found in the lunar soils. Similar concentrations of $CO_2$ have been produced from synthetic lunar analogs at elevated temperatures (Gibson and Johnson, 1971). The amount of $CO_2$ produced from the lunar soils at temperatures above 700°C is generally less than one-half the amount of CO produced. Holland *et al.* (1972), Chang *et al.* (1971), Oró *et al.* (1971), and others have found similar results for the analysis of lunar soils.

The $H_2S$ and $SO_2$ released between 900° and 1300°C are chemical reaction products produced from sulfides, mainly FeS, found in the lunar soils and the silicate and glass matrix of the soil (Fig. 1). The releases of $H_2S$ and $SO_2$ from the Apollo 14 soil (Fig. 2) and the two Apollo 15 soils (Figs. 3 and 4) are essentially identical with respect to both the temperature release profiles and relative abundances as found for the Apollo 14 soil shown in Fig. 1. The $H_2S$ and $SO_2$ species were not plotted in Fig. 2 to 4 in order to prevent further cluttering of the figures. Approximately three times more $H_2S$ than $SO_2$ is evolved from both the Apollo 14 and 15 soils. Identical gas release patterns have been generated from synthetic lunar analogs and meteoritic troilite mixtures (Gibson and Johnson, 1971).

Small amounts of hydrogen are released between 1150 and 1350°C (Figs. 1 and 3). This high temperature release of $H_2$ might result from chemical reactions similar to the water-gas (or town-gas) reaction. When $H_2$ concentrations increase, $H_2O$ abundances simultaneously decrease, providing further evidence of a reaction of the water-gas type. It should be noted that the concentrations of $H_2$ in the Apollo 14 samples is lower than the amount of $H_2$ in the two Apollo 15 soil samples.

Previously we noted that $O_2$ was released from the lunar samples when they were heated in a platinum crucible at temperatures greater than 1300°C (Gibson and Johnson, 1971). We proposed that the $O_2$ release resulted from the iron in the sample dissolving into the platinum crucible and liberating $O_2$. The experiments carried out in an alumina crucible confirm our earlier proposal. No significant amount of $O_2$ above the system background was released from the lunar soils when they were heated in an alumina crucible to temperatures of 1400°C. No other differences were noticed between the two gas release patterns for soil 14163 carried out in platinum and alumina crucibles (Figs. 1 and 2).

## CONCLUSIONS

Thermal analysis-inorganic gas release studies of lunar soils has shown that there is very little volatile rich material or hydrous mineral phases found in the lunar soils. The gases released during the heating of the soils results from four separate and distinct sources. (1) $H_2O$ vapor and $CO_2$ have been adsorbed by the lunar soils since return to earth. These two gaseous species are loosely bound to the lunar soils and are lost from the samples when they are heated under vacuum at temperatures below 150°C. (2) The solar wind components such as $H_2$ and He are trapped within the outer surfaces of individual mineral and glass grains and these two gases are released between 300 and 700°C. Additional low molecular weight species such as $H_2O$ (possibly in the hydroxyl form) can be formed on the outer surfaces of rocks and mineral grains as the result of proton irradiation from the solar wind. Initial laboratory experimental results from irradiation of lunar analogs with 100 keV protons indicates that the quantities of $H_2O$ and the temperature release profiles produced are similar to those of the lunar samples. The Zeller *et al.* (1970) mechanism for $H_2O$ and low molecular weight hydrocarbon production on the lunar surface is believed to operate, however, at the present time we do not know how efficient such a process operates. (3) The release of the gaseous species CO, $CO_2$, $H_2S$, and $SO_2$ result from the reaction products of carbon and sulfur containing phases found in the soils with mineral and glass components found in the soils. Identical gas release profiles have been produced using lunar analogs (Gibson and Johnson, 1971). (4) We have found the presence of a gas phase composed of $N_2$ and/or CO which may be trapped within vesicles and gas-rich inclusions and/or simply exsolving from the melt after the initial melting of the samples. These gas phases are released as sudden bursts of gas at temperatures immediately above the melting point ($1130 \pm 20°C$) of the samples. The identity of the gases found are similar to those that should be expected from theoretical considerations of the lunar gaseous environment during the time of the crystallization of the lunar samples (Wellman, 1970).

*Acknowledgments*—Assistance from the personnel of the Deep Space Materials Engineering Laboratory, NASA-MSC is acknowledged. Suzanne M. Johnson contributed significantly to the analysis of the irradiated lunar analogs and her assistance is acknowledged. Vic Borgnis and Pete Olsen assisted with the mass spectrometer-computer interfacing and required computer programming which allowed significant time saving procedures to be employed during this investigation. Discussions with C. T. Pillinger on the proton irradiation experiments are acknowledged.

## References

Cadogan P. H., Eglinton G., Firth J. N. M., Maxwell J. R., Mays B. J., and Pillinger C. T. (1972) Survey of lunar carbon compounds, II: The carbon chemistry of Apollo 11, 12, 14, and 15 samples (abstract). In *Lunar Science—III* (editor C. Watkins), pp. 113–115, Lunar Science Institute Contr. No. 88.

Chang S., Kvenvolden K. A., Lawless J., Ponnamperuma C., and Kaplan I. R. (1971) Carbon in an Apollo 12 sample: Concentration, isotopic composition, pyrolysis products, and evidence for indigenous carbides and methane. *Science* **171,** 474–477.

Funkhouser J., Jessberger E., Müller O., and Zahringer J. (1971) Active and inert gases in Apollo 12 and Apollo 11 samples released by crushing at room temperature and by heating at low temperatures. *Proc. Second Lunar Sci. Conf., Geochim. Cosmochim. Acta* Suppl. 2, Vol. 2, pp. 1381–1396. MIT Press.

Gibson E. K. Jr. and Hubbard N. J. (1972) Volatilization studies on lunar samples. *Proc. Third Lunar Sci. Conf., Geochim. Cosmochim. Acta* Suppl. 3, Vol. 2. MIT Press.

Gibson E. K. Jr. and Johnson S. M. (1971) Thermal analysis-inorganic gas release studies of lunar samples. *Proc. Second Lunar Sci. Conf., Geochim. Cosmochim. Acta* Suppl. 2, Vol. 2, pp. 1351–1366. MIT Press.

Gibson E. K. Jr. and Johnson S. M. (1972) Thermogravimetric-quadrupole mass spectrometric analysis of geochemical samples. *Thermochimica Acta* **4,** 49–56.

Gibson E. K. Jr. and Moore C. B. (1972) Compounds of the organogenic elements in Apollo 11 and 12 lunar samples—A review. *Space Life Sci.* (in press).

Hanneman R. E. (1970) Thermal and gas evolution behavior of Apollo 11 samples. *Proc. Apollo 11 Lunar Sci. Conf., Geochim. Cosmochim. Acta* Suppl. 1, Vol. 2, pp. 1207–1211. Pergamon.

Hayes J. M. (1972) Extralunar sources for carbon on the moon. *Space Life Sci.* (in press).

Heymann D. and Yaniv A. (1970) $Ar^{40}$ anomaly in lunar samples from Apollo 11. *Proc. Apollo 11 Lunar Sci. Conf., Geochim. Cosmochim. Acta* Suppl. 1, Vol. 2, pp. 1261–1267. Pergamon.

Holland P. T., Simoneit B. R., Wszolek P. C., McFadden W. H., and Burlingame A. L. (1972) Carbon compounds in Apollo 12, 14, and 15 samples (abstract). In *Lunar Science—III* (editor C. Watkins), pp. 383–385, Lunar Science Institute Contr. No. 88.

Oró J., Flory D. A., Gibert J. M., McReynolds J., Lichtenstein H. A., and Wikstrom S. (1971) Abundances and distribution of organogenic elements and compounds in Apollo 12 lunar samples. *Proc. Second Lunar Sci. Conf., Geochim. Cosmochim. Acta* Suppl. 2, Vol. 2, pp. 1913–1925. MIT Press.

Wellman T. (1970) Gaseous species in equilibrium with the Apollo 11 holocrystalline rocks during their crystallization. *Nature* **225**, 716–717.

Zeller E. J., Dreschhoff G., and Kevan L. (1970) Chemical alterations resulting from proton irradiation of the lunar surface. *Modern Geology* **1**, 141–148.

Proceedings of the Third Lunar Science Conference
(Supplement 3, *Geochimica et Cosmochimica Acta*)
Vol. 2, pp. 2041–2050
The M.I.T. Press, 1972

# Total nitrogen contents of some Apollo 14 lunar samples by neutron activation analysis

P. S. GOEL and B. K. KOTHARI

Department of Chemistry, Indian Institute of Technology,
Kanpur, India

**Abstract**—Total nitrogen was measured by neutron activation in one basaltic rock, one fines sample, and several breccias from Apollo 14. The basalt and all but one breccias contained 20–25 ppm nitrogen. One breccia (14049) and the fines had 70–80 ppm nitrogen. Nitrogen distribution in the latter two samples among different grain sizes shows that most of their nitrogen is surface correlated and is therefore likely to have come from solar wind implantation. Some leaching experiments have been performed, demonstrating surface residence of excess nitrogen in fines. The solar wind C : N ratio is found to be 10 : 6 and is in good agreement with estimates based on spectroscopic data. Our data generally give lower nitrogen contents than reported earlier. Classical methods apparently introduce a contamination of 20 to 50 ppm nitrogen.

## INTRODUCTION

IN LUNAR SAMPLES returned by missions Apollo 11 and Apollo 12, extensive determinations of nitrogen contents have been made by Moore *et al.* (1970, 1971). Measurements on some samples have also been reported by Hintenberger *et al.* (1970), and Morrison *et al.* (1970, 1971). Several laboratories presented data on nitrogen contents of Apollo 14 and Apollo 15 samples at the Third Lunar Science Conference (Goel and Kothari, 1972; Moore *et al.*, 1972; Müller, 1972; Sakai *et al.*, 1972). The results, even though disagreeing in details among various groups, clearly show that the lunar fines contain more nitrogen in comparison with the igneous rocks. Presumably, nitrogen is introduced from solar wind impact (Moore *et al.*, 1970, 1971). Some of the published data, however, do seem to indicate serious terrestrial contamination.

We ran nitrogen analysis on several samples from Apollo 14 mission using activation analysis technique developed for meteorite studies (Goel, 1970; Kothari and Goel, 1972). In our technique the laboratory contamination is minimized since most of the handling is done after the samples have been irradiated. Our objectives were to establish that the excess nitrogen in fines is of solar wind origin and to learn more about the chemical history of lunar materials by studying this volatile compound-forming element.

## EXPERIMENTAL

A lunar sample, 0.2 to 1 g, along with several pieces of stone and iron meteorites, were packed in a quartz irradiation vial. If necessary, a sample was fragmented in an agate mortar. The vial also contained some standard NBS steels (33*d* and 12*h*) whose nitrogen contents were known and which served as monitors for the neutron flux. During packing the handling was kept to a minimum in a clean room free from nitrogen compounds in the air to avoid contamination. The quartz vial was evacuated below 25 $\mu$ of pressure for 10 hr and sealed under vacuum. The irradiation lasted for

2 to 4 weeks and was carried out at Bhabha Atomic Research Center, Bombay, in the CIRUS reactor. The neutron flux was (2 to 10) × $10^{12}$ n $cm^{-2}$ $sec^{-1}$.

An irradiated sample was powdered (if necessary) and mixed with five times its mass of an oxidizing flux of $PbCrO_4$ and fused $K_2CrO_4$. Some $CaCO_3$ carrier was added to provide enough $BaCO_3$ for convenient handling. The sample was fused in a flowing oxygen atmosphere at about 1100°C for two hours in an arrangement as shown in Fig. 1. The evolved gases were swept through a set of oxidants and purifiers, as shown, for complete oxidation of carbon to carbon dioxide that was absorbed in 2N NaOH solution and was subsequently recovered as $BaCO_3$ by precipitation with $BaCl_2$. Fusion was carried out in a hot lab. The $BaCO_3$ was decomposed in a cold lab with dilute HCl in a closed system for purification. The evolved $CO_2$ was bubbled through a set of purifiers and was absorbed in 2N NaOH. It was recovered as $BaCO_3$ which was counted against an end window low-background counter (Lowbeta system manufactured by Sharp Labs). The counter background was about 0.3 cpm. The sample counting rates ranged from four to several hundred cpm. In each case we collected a total of about 2000 counts to keep the statistical error below 3%. Self-absorption correction was experimentally determined. In most cases the amount of $BaCO_3$ from NBS steels and from samples were of comparable mass requiring almost the same correction factors for self-absorption. The radiocarbon from NBS steel standard was isolated, purified, and counted in the same way.

Before being applied to lunar materials, the technique was established for its reliability and reproducibility by a number of control experiments (Goel, 1970). Replicate measurements on irradiated and homogenized stone meteorites gave reproducible results. A second fusion of once-fused meteorite and also of lunar sample gave less than 3% radiocarbon in the second fusion. In some meteorites, we measured less than 5 ppm nitrogen, indicating a low laboratory contamination blank. We have assumed a conservative 10% error in our data because of the combined effect of spread in chemical recovery, nonuniform sample mounting, statistical error of counting, etc. More details about the technique will be given in a separate publication (Kothari and Goel, 1972).

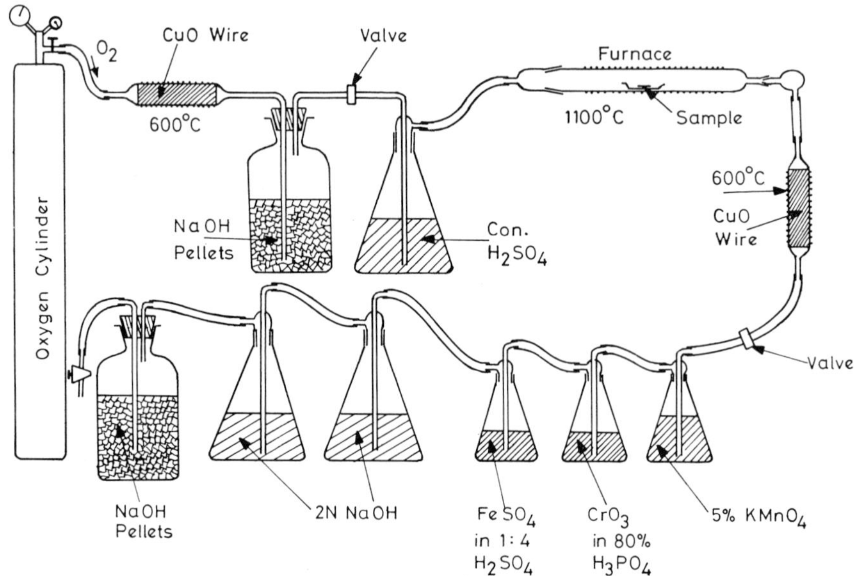


Fig. 1. Experimental arrangement for fusion.

## Results and Discussion

Our data, obtained from three irradiations, are given in Table 1. Good agreement between replicate measurements on 14163, a fairly homogenized sample, demonstrates the reliability and reproducibility of our technique. Since this sample was sealed at different times and in different ways for the two irradiations, the data also indicate an absence of any serious contamination by laboratory handling of the samples. It is interesting to note that in the breccias the nitrogen distribution is uniform, despite their gross structural inhomogeneity.

In Table 2 we compare the published nitrogen data from various laboratories. The results of our samples 14049 and 14321 are significantly lower than those of Moore *et al.* (1972). Since both laboratories use the same reference standard (NBS steel, 33*d*), the discrepancy can not arise from a calibration or standardization error. It appears that the nitrogen data of Moore *et al.* (1972) may have a terrestrial contamination of about 20 to 50 $\mu$g N/g. The results of Müller (1972) also appear to be a bit high, particularly in view of the fact that he measures *only* the chemically bound nitrogen. However, a direct comparison with his data can not be made because of the absence of a common sample between the two laboratories. The fragmentary data by others are generally of exploratory nature and need not be discussed seriously. High nitrogen values reported by Sakai *et al.* (1972) for 14321 breccia may be due to contamination.

The data of Table 2, however, clearly bring out the point first suggested by Moore *et al.* (1970) that the lunar fines in general contain more nitrogen than the rocks. This arises from solar wind implantation on the surface of dust grains. In order to understand more about the nature of this excess nitrogen in lunar fines, we carried out two sets of experiments. We did an analysis of nitrogen distribution versus grain size (Müller, 1972) demonstrating that nitrogen in fines is surface correlated. We also carried out a series of leaching experiments that showed that the solar wind nitrogen can be "peeled off" the grain surfaces by mild chemical attack.

Table 1. Nitrogen contents of Apollo 14 samples.

| Sample | Description | Mass (mg) | Nitrogen* (ppm) | Weighted mean* (ppm) |
|---|---|---|---|---|
| 14163,117 | Fines | 74 | 78 | |
| | | 61 | 80 | |
| | | 19.2 | 82 | |
| | | 16.3 | 90 | |
| | | 15.5 | 72 | |
| | | 32 | 81 | 80 |
| 14049,34 | Fragmental | 109 | 71 | |
| | (soil clod) | 106 | 70 | 71 |
| 14305,78 | Fragmental | 103.5 (one piece) | 20 | |
| | | 73 (white clast) | 26 | |
| | | 41 (five pieces) | 20 | 22 |
| 14321,224 | Fragmental | 110 (dark clasts) | 23 | |
| | | 70 (one piece) | 25 | 24 |
| 14310,121 | Basaltic | 105 (one piece) | 22 | |
| | | 126 (one piece) | 21 | 21 |

* Uncertainty $\pm 10\%$.

Table 2. Summary of nitrogen data on lunar samples (ppm).

| Mission | Fines | Breccia | Rocks |
|---|---|---|---|
| Apollo 11 | 10084 110 Morrison *et al.* (1970) | 10021 131 Hintenberger *et al.* (1970) | 10049 116 Moore *et al.* (1970) |
| | 10086 102 Moore *et al.* (1970) | 10061 125 Hintenberger *et al.* (1970) | 10050 30 Moore *et al.* (1970) |
| | 10086 153 Moore *et al.* (1970) | 10002 125 Moore *et al.* (1970) | 10020 40 Morrison *et al.* (1970) |
| | 10084 93 Müller (1972) | 10044 98 Moore *et al.* (1970) | 10057 70 Morrison *et al.* (1970) |
| | | 10056 70 Morrison *et al.* (1970) | 10072 110 Morrison *et al.* (1970) |
| | | 10060 20 Morrison *et al.* (1970) | 10058 40 Morrison *et al.* (1970) |
| | | 10046 131 Müller (1972) | 10046 260 Morrison *et al.* (1970) |
| | | | 10057 64 Müller (1972) |
| Apollo 12 | 12001 110 Moore *et al.* (1971) | 12073 130 Morrison *et al.* (1971) | 12002 43 Moore *et al.* (1971) |
| | 12003 85 Moore *et al.* (1971) | | 12022 44 Moore *et al.* (1971) |
| | 12023 120 Moore *et al.* (1971) | | 12053 36 Morrison *et al.* (1971) |
| | 12032 48 Moore *et al.* (1971) | | 12051 54 Morrison *et al.* (1971) |
| | 12033 46 Moore *et al.* (1971) | | 12063 < 10 Müller (1972) |
| | 12037 96 Moore *et al.* (1971) | | 12075 < 10 Müller (1972) |
| | 12042 130 Moore *et al.* (1971) | | |
| | 12070 40 Morrison *et al.* (1971) | | |
| | 12070 48 Müller (1972) | | |
| Apollo 14 | 14163 80 Goel and Kothari (1972) | 14049 71 Goel and Kothari (1972) | 14310 21 Goel and Kothari (1972) |
| | 14003 92 Müller (1972) | 14321 24 Goel and Kothari (1972) | |
| | 14298 164 Sakai *et al.* (1972) | 14305 22 Goel and Kothari (1972) | |
| | | 14049 130 Moore *et al.* (1972) | |
| | | 14321 57 Moore *et al.* (1972) | |
| | | 14303 31 Müller (1972) | |
| | | 14321 181 Sakai *et al.* (1972) | |
| Apollo 15 | 15101 109 Müller (1972) | | 15556 < 10 Müller (1972) |
| | 15601 80 Müller (1972) | | |
| | 15271 111 Sakai *et al.* (1972) | | |

Table 3. Grain size dependence of nitrogen in fines, 14163 and soil clod, 14049.

| Grain size ($\mu$) | Mass recovered (mg) | | Mass analyzed (mg) | | Nitrogen (ppm) | | Mean (ppm) | | Excess (ppm) | |
|---|---|---|---|---|---|---|---|---|---|---|
| | 14163 | 14049 | 14163 | 14049 | 14163 | 14049 | 14163 | 14049 | 14163 | 14049 |
| < 38 | 122.2 | 117.7 | 63.7<br>58.5 | 35<br>41.5 | 101<br>88 | 145<br>127 | 95 | 136 | $73 \pm 10$ | $114 \pm 14$ |
| 38–75 | 211.2 | 227.6 | 96.5<br>114.7 | 90.3<br>46.5 | 92<br>72 | 82<br>80 | 82 | 81 | $60 \pm 9$ | $59 \pm 9$ |
| 75–151 | 96.3 | 86.8 | 28.5 | 86.8 | 47<br>50*, 47* | 55 | 48 | 55 | $26 \pm 5$ | $33 \pm 6$ |
| 151–270 | 93.5 | 92.8 | 93.5 | 62.8 | 42<br>44*, 41* | 41 | 42 | 41 | $20 \pm 5$ | $19 \pm 5$ |
| > 270 | 62.1 | 41.0 | 62.1 | 41 | 20<br>27*, 27* | 30 | 25 | 30 | $5 \pm 3$ | $8 \pm 4$ |

* Separate irradiation.

### *Grain size analysis*

Irradiated samples of 14163 (fines) and 14049 (fragmental, soil clod) were subjected to grain size separation by sieving through standard sieves. Nitrogen contents of different sieve fractions were measured. Often replicate measurements were done. The results are presented in Table 3 and are shown graphically in Fig. 2. The data for both 14163 and 14049 obey the relation

$$c \propto d^{-n}$$

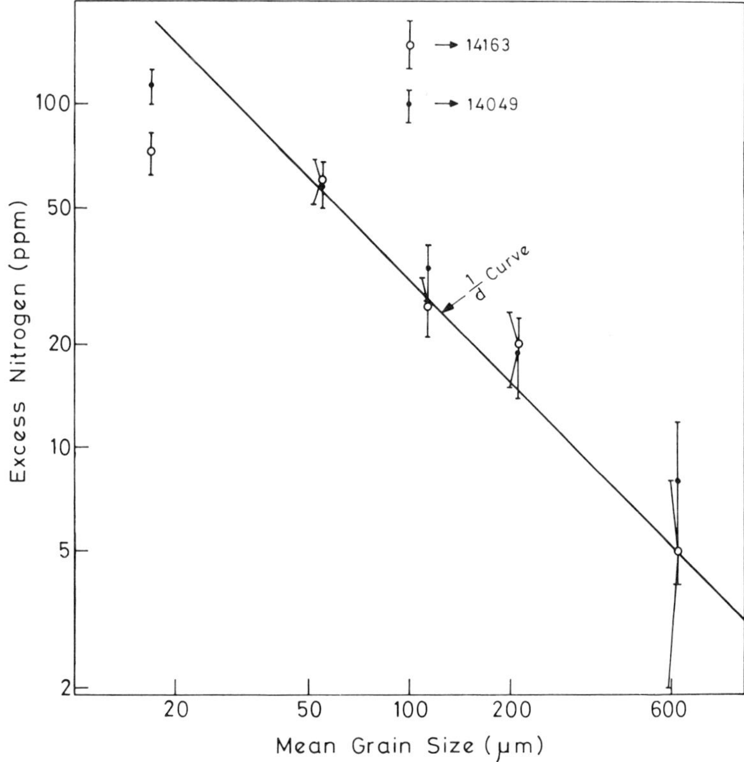


Fig. 2. Grain size dependence of nitrogen in fines, 14163 and soil clod, 14049.

where $c$ is the concentration of excess nitrogen and $d$ is the mean grain diameter. The value of exponent $n$ is seen to be unity indicating that the nitrogen concentration in all grain sizes is proportional to the surface area of the grains. The points for the smallest grain size fraction are lower than the expected values. This deficiency could result if the finer particles settle on the lunar surface below the coarser ones due to gravity sorting and are thus shielded from the solar wind part of the time (Goel and Kothari, 1972). Another process that could explain these observations might be the thermal effects on the surface of grains by solar wind impact (Bibring *et al.*, 1972). Rare gases implanted by solar wind in lunar fines show similar grain size dependences (Eberhardt *et al.*, 1970; Hintenberger *et al.*, 1971) and are known to be concentrated in the outer parts of the grains (Megrue, 1971).

*Leaching experiments*

We devised a differential chemical dissolution scheme as shown in Fig. 3 for the removal of surface implanted nitrogen. The irradiated powdered sample was attacked with 1% NaOH and the radiocarbon was isolated from the gases. The sample was decanted, the NaOH solution was dried and fused with the oxidizing flux, and the

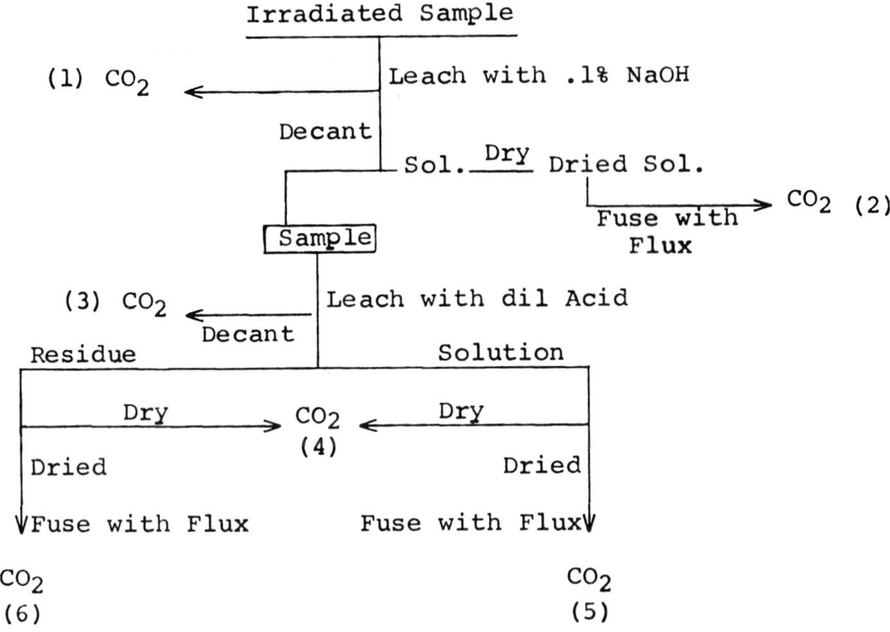


Fig. 3. Scheme for stepwise removal of nitrogen.

Table 4. Stepwise removal of nitrogen.

| Sample | Mass (mg) | Nitrogen (ppm) in various steps (1) | (2) | (3) | (4) | (5) | (6) | sum | Total nitrogen (ppm) |
|---|---|---|---|---|---|---|---|---|---|
| 14163, Fines | 77 | 1.3 | 4 | 22 | — | 6 | 15 | 48.3 | 80 |
| | 120 | — | — | 8* | — | 19.4 | 18.6 | 46 | 80 |
| | 19.2 | — | — | 25.4 | 11.8 | 2.3 | 13.9 | 53.4 | 80 |
| Shalka hypersthene achondrite | 107 | $<0.5$ | $<1$ | $<0.5$ | — | $<1$ | 14.3 | 14.3 | 19 |

* Sample was not heated.

residue was leached with 3M $HNO_3$. The nitric acid was decanted from the residue and both were evaporated in a closed system with a sweeping by purified air to recover any evolving $CO_2$. The dried solution and the residue were separately fused with flux. Since the NaOH attack did not show any radiocarbon evolution, this step was omitted in the later runs. The $CO_2$ evolved during the drying of acid solution and residue was not recovered in the earlier runs since its significance was realized only after some runs had been made.

The results of the leaching experiments are presented in Table 4. The data show that from lunar fines leached with dilute $HNO_3$ about 20 to 30% nitrogen (as radiocarbon) is released in the gases; the acid solution traps about 20% nitrogen and the residue has about 20 ppm nitrogen—the same as in crystalline rocks. Since the acid can dissolve part of most minerals from the surface layers (Begemann *et al.*, 1970) the nitrogen (as radiocarbon) removed as gases and found in solution is most likely from the surface of the grains. The results are quite striking in comparison with the data on Shalka achondrite where one does not find significant "peeling off" effects for radiocarbon.

*Solar wind composition*

Our experiments clearly show that (1) the lunar fines contain about 60 ppm nitrogen in excess over crystalline rocks, (2) this excess nitrogen in various grains is proportional to their surface area, and (3) it can be leached away from the grains. These results support the view, originally put forward by Moore *et al.* (1970) that the lunar fines contain nitrogen from solar wind implantation. Two other possibilities for the origin of this excess nitrogen may be considered.

First we discuss the possibility of atmospheric nitrogen adsorbed on the surface of grains. Our data on fragmental rocks show that nitrogen from the atmosphere is not adsorbed on surfaces, because we found the same nitrogen contents in a single piece as in five smaller pieces. A powdered sample of Shalka achondrite was bombarded and was found to contain 15 ppm nitrogen, the same as the average value for whole rock pieces—an indication of absence of surface adsorption for the meteorite case. The surfaces of lunar fines are not very active for adsorption of gases. Laboratory studies have shown that the gases can be desorbed rapidly by pumping (Grossman *et al.*, 1972; Fuller *et al.*, 1971). Our technique is such that the adsorbed gases, if any, are pumped off before bombardment.

The other possible source of excess nitrogen may be the 1.5 to 2% influx of carbonaceous chondritic C1 type material that has been suggested to be present in lunar fines to explain the abundance pattern of volatile elements (Ganapathy *et al.*, 1970). This would add about 700 $\mu$g carbon per gram of fines. The observed carbon contents are only 100 to 150 $\mu$g/g (Moore *et al.*, 1971) indicating that the incoming micrometeoritic particles loose their carbon (and nitrogen) by volatilization. Melting of these on impact heating would result in such a behavior (Laul *et al.*, 1971). The C/N ratio in C1 type meteorites is 10 : 1 (Gibson *et al.*, 1971) against the observed value of 10 : 6 in the lunar fines. Moreover the carbon contents in lunar materials parallel their neon contents (LSPET, 1971) suggesting that both arise from solar wind. All these lines of evidences show that the excess nitrogen in lunar fines can not be accounted for in terms of a micrometeoritic influx.

In Table 5 we compare the relative abundances of solar wind C, N, and Ne in

Table 5. Solar system abundances of C, N, and Ne (C = 100).

| Element | Planetary nebulae[1] | Sun[2] | Solar cosmic rays[3] | Trapped on moon dust |
|---|---|---|---|---|
| C | 100 | 100 | 100 | 100 |
| N | 60 | 40 | 32 | 60 |
| Ne | 80 | — | 22 | 0.36 |

[1] Unsöld (1969).
[2] Goldberg *et al.* (1960); Müller (1968).
[3] Biswas *et al.* (1966).

lunar fines with the solar abundances as inferred from spectroscopic and other direct measurements. The C/N ratio in fines is in good agreement with the astrophysical data. Neon is low, probably indicating its poor retention. From the distribution of grain size, we calculated the surface concentration of solar wind nitrogen as $1 \times 10^{16}$ atoms $cm^{-2}$. Based on data from foil experiments on solar wind composition (Buehler *et al.*, 1972), Banks (1971) has calculated a flux of solar wind nitrogen to be $2 \times 10^{4}$ atoms $cm^{-2}$ $sec^{-1}$ at the lunar surface. The observed concentration would accumulate in about $2 \times 10^{4}$ yr. The surface exposure ages of lunar fines are of the order of few millions of years (Bhandari *et al.*, 1972). The low nitrogen concentration, by two orders of magnitude, may be due to the effects of surface erosion. Shielding by a layer of one micron would prevent the solar wind ions from reaching the surface. The energetic track forming particles from the sun would still penetrate giving a high exposure age. It is also possible that the surface does not retain 100% of the nitrogen.

### *Breccia formation*

Jackson *et al.* (1972) have given a classification of Apollo 14 breccias. The fragmental rocks 14321 and 14305 belong to the class F4, "thermally metamorphosed microbreccias," while 14049 belongs to the class F1, "porous unshocked microbreccias" (Chao *et al.*, 1971). The absence of any excess (solar wind) nitrogen in F4 breccias means that either the solar wind nitrogen is lost during brecciation or that the matter forming these breccias was not exposed to the influence of solar wind. Particle track studies on breccias in general give the same picture (Bhandari *et al.*, 1972). The fragmental rock 14049 which is a soil clod and is an agglomerate of dust grains has a solar wind nitrogen at about the same level as the fines. This rock also contains higher concentration of solar wind carbon (Moore *et al.*, 1972; Holland *et al.*, 1972). The solar wind nitrogen has remained intact in this rock. Studies on a wider variety of breccias may help to give a better understanding of brecciation processes.

## Conclusions

The present investigations provide a highly sensitive technique for total nitrogen determination in lunar and other materials. One can make measurements on 10 mg samples without any serious laboratory or atmospheric contamination problem. The results obtained on Apollo 14 lunar materials show that (1) the indigeneous nitrogen in lunar rocks is about 20 $\mu$g/g; (2) the lunar fines contain an excess nitrogen of about 60 $\mu$g/g which is shown to be of solar wind origin; (3) the elemental abundance ratio of C : N in the solar wind is 10 : 6; and (4) the effective exposure time of lunar grains to the solar wind is about $2 \times 10^{4}$ yr. If some erosion takes place, the exposure time would be still longer.

We plan to study nitrogen contents in a wider variety of rocks and also make mineral separations on irradiated rocks. Simultaneous measurements of $^{6}$Li via the reaction $^{6}$Li $(n, \alpha)$ $^{3}$H are planned to get more information in the same runs.

*Acknowledgments*—We are grateful to NASA for making it possible for us to participate in the lunar investigation program by providing us the precious lunar materials. We appreciate the technical assistance of Mr. R. K. Sharma and Mr. H. O. Shankar during this work. B. K. Kothari is grateful to the Department of Atomic Energy, Bombay, for a fellowship support.

## REFERENCES

Banks P. M. (1971) Interplanetary hydrogen and helium from cosmic dust and the solar wind. *J. Geophys. Res.* **76,** 4341–4348.

Begemann F., Vilcsek E., Rieder R., Born W., and Wänke H. (1970) Cosmic-ray produced radioisotopes in lunar samples from the Sea of Tranquility (Apollo 11). *Proc. Apollo 11 Lunar Sci. Conf., Geochim. Cosmochim. Acta* Suppl. **1**, Vol. 2, pp. 995–1005. Pergamon.

Bhandari N., Bhat S. G., Goswami J. N., Gupta S. K., Krishnaswami S., Lal D., Tamhane A. S., and Venkatavaradan V. S. (1972) Collision controlled radiation history of lunar regolith (abstract). In *Lunar Science—III* (editor C. Watkins), pp. 68–70, Lunar Science Institute Contr. No. 88.

Bibring J. P., Maurette M., Meunier R., Durieu I., Jouret C., and Eugster O. (1972) Solar wind implantation effects in the lunar regolith (abstract). In *Lunar Science—III* (editor C. Watkins), pp. 71–73, Lunar Science Institute Contr. No. 88.

Biswas S., Fichtel C. E., and Guss D. E. (1966) Solar cosmic-ray multiply charged nuclei and the July 18, 1961 solar event. *J. Geophys. Res.* **71,** 4071–4077.

Buehler F., Cerutti H., Eberhardt P., and Geiss J. (1972) Results of the Apollo 14 and 15 solar wind composition experiments (abstract). In *Lunar Science—III* (editor C. Watkins), Lunar Science Institute Contr. No. 88.

Chao E. C. T., Boreman J. A., and Desborough G. A. (1971) The petrology of unshocked and shocked Apollo 11 and Apollo 12 microbreccias. *Proc. Second Lunar Sci. Conf., Geochim. Cosmochim. Acta* Suppl. 2, Vol. 1, pp. 797–816. MIT Press.

Fuller E. L., Holmes H. F., Gammage R. B., and Becker K. (1971) Interaction of gases with lunar materials: Preliminary results. *Proc. Second Lunar Sci. Conf., Geochim. Cosmochim. Acta* Suppl. 2, Vol. 3, pp. 2009–2019. MIT Press.

Ganapathy R., Keays R. R., Laul J. C., and Anders E. (1970) Trace elements in Apollo 11 lunar rocks: Implications for meteorite influx and origin of moon. *Proc. Apollo 11 Lunar Sci. Conf., Geochim. Cosmochim. Acta* Suppl. 1, Vol. 2, pp. 1117–1142. Pergamon.

Gibson E. K., Moore C. B., and Lewis C. F. (1971) Total nitrogen and carbon abundances in carbonaceous chondrites. *Geochim. Cosmochim. Acta* **35,** 599–604.

Goel P. S. (1970) Determination of nitrogen in iron meteorites. *Geochim. Cosmochim. Acta* **34,** 932–935.

Goel P. S. and Kothari B. K. (1972) Nitrogen abundances in lunar samples by neutron activation analysis (abstract). In *Lunar Science—III* (editor C. Watkins), pp. 315–317, Lunar Science Institute Contr. No. 88.

Goldberg L., Müller E. A., and Aller L. H. (1960) Abundances of the elements in the solar atmosphere. *Astrophys. J. Suppl. Ser.* **5,** 1–137.

Grossman J. J., Mukherjee N. R., and Ryan J. A. (1972) Microphysical, microchemical, and adhesive properties of lunar material, III: Gas interaction with lunar material (abstract). In *Lunar Science—III* (editor C. Watkins), pp. 344–346, Lunar Science Institute Contr. No. 88.

Hintenberger H., Weber H. W., Voshage H., Wänke H., Begemann F., and Wlotzka F. (1970) Concentrations and isotopic abundances of the rare gases, hydrogen, and nitrogen in Apollo 11 lunar matter. *Proc. Apollo 11 Lunar Sci. Conf., Geochim. Cosmochim. Acta* Suppl. 1, Vol. 2, pp. 1269–1282. Pergamon.

Holland P. T., Simoneit B. R., Wszolek P. C., McFadden W. H., and Burlingame A. (1972) Carbon compounds in Apollo 12, 14, and 15 samples (abstract). In *Lunar Science—III* (editor C. Watkins), pp. 383–385, Lunar Science Institute Contr. No. 88.

Jackson E. D. and Wilshire H. G. (1972) Classification of the samples returned from the Apollo 14 landing site (abstract). In *Lunar Science—III* (editor C. Watkins), pp. 418–420, Lunar Science Institute Contr. No. 88.

Kothari B. K. and Goel P. S. Nitrogen and lithium measurements in rocks and metals by neutron activation analysis. To appear.

Laul J. C., Morgan J. W., Ganapathy R., and Anders E. (1971) Meteoritic material in lunar samples: Characterization from trace elements. *Proc. Second Lunar Sci. Conf., Geochim. Cosmochim. Acta* Suppl. 2, Vol. 2, pp. 1139–1158. MIT Press.

LSPET (Lunar Sample Preliminary Examination Team) (1971) Preliminary examination of lunar samples from Apollo 14. *Science* **173,** 681–693.

Megrue G. H. (1971) Distribution and origin of helium, neon, and argon isotopes in Apollo 12 samples measured by in situ analysis with a laser-probe mass spectrometer. *J. Geophys. Res.* **76,** 4956–4968.

Moore C. B., Gibson E. K., Larimer J. W., Lewis C. F., and Nichiporuk W. (1970) Total carbon and nitrogen abundances in Apollo samples and selected achondrites and basalts. *Proc. Apollo 11 Lunar Sci. Conf., Geochim. Cosmochim. Acta* Suppl. 1, Vol. 2, pp. 1375–1382. Pergamon.

Moore C. B., Lewis C. F., Larimer J. W., Delles F. M., Gooley R. C., Nichiporuk W., and Gibson E. K. Jr. (1971) Total carbon and nitrogen abundances in Apollo 12 lunar samples. *Proc. Second Lunar Sci. Conf., Geochim. Cosmochim. Acta* Suppl. 2, Vol. 2, pp. 1343–1350. MIT Press.

Moore C. B., Lewis C. F., Cripe J., Kelly W. R., and Delles F. (1972) Total carbon, nitrogen, and sulfur abundances in Apollo 14 lunar samples (abstract). In *Lunar Science—III* (editor C. Watkins), pp. 550–552, Lunar Science Institute Contr. No. 88.

Morrison G. H., Gerard J. T., Kashuba A. T., Gangadharam E. V., Rothenberg A. M., Potter N. M., and Miller G. B. (1970) Elemental abundances of lunar soil and rocks. *Proc. Apollo 11 Lunar Sci. Conf., Geochim. Cosmochim. Acta* Suppl. 1, Vol. 2, pp. 1383–1392. Pergamon.

Morrison G. H., Gerard J. T., Potter N. M., Gangadharam E. V., Rothenberg A. M., and Burdo R. A. (1971) Elemental abundances of lunar soil and rocks from Apollo 12. *Proc. Second Lunar Sci. Conf., Geochim. Cosmochim. Acta* Suppl. 2, Vol. 2, pp. 1169–1185. MIT Press.

Müller E. A. (1968) The solar abundances. In *Origin and Distribution of the Elements* (editor L. H. Ahrens), pp. 155–176. Pergamon.

Müller O. (1972) Chemically bound nitrogen abundances in lunar samples, and active gases released by heating at lower temperatures (250 to 500°C) (abstract). In *Lunar Science—III* (editor C. Watkins), pp. 568–570, Lunar Science Institute Contr. No. 88.

Sakai H., Petrowski C., Goldhaber M. B., and Kaplan I. R. (1972) Distribution of carbon, sulfur, and nitrogen in Apollo 14 and 15 material (abstract). In *Lunar Science—III* (editor C. Watkins), pp. 672–674, Lunar Science Institute Contr. No. 88.

Unsöld A. O. J. (1969) Steller abundances and the origin of the elements. *Science* **163,** 1015–1025.

Proceedings of the Third Lunar Science Conference
(Supplement 3, *Geochimica et Cosmochimica Acta*)
Vol. 2, pp. 2051–2058
The M.I.T. Press, 1972

# Total carbon, nitrogen, and sulfur in Apollo 14 lunar samples

C. B. MOORE, C. F. LEWIS, J. CRIPE,
F. M. DELLES, and W. R. KELLY

Arizona State University, Tempe, Arizona 85281

and

E. K. GIBSON, JR.

NASA Manned Spacecraft Center, Houston, Texas 77058

**Abstract**—The total carbon abundances in ten samples of Apollo 14 lunar fines ranged from 90 to 190 $\mu$g/g. A light-gray-colored fines sample 14141 from Cone Crater had from 42 to 80 $\mu$g/g total carbon. These samples are similar to the normal dark- and light-colored lunar fines from Apollo 11 and 12. Apollo 14 basalt 14310 had a total carbon abundance of 35 $\mu$g/g. Apollo 14 fragmental rock and breccia samples ranged from 21 to 225 $\mu$g/g total carbon. The total carbon contents showed a correlation with each rock's petrologic classification. Breccias in the first group (F-1) with friable matrices and light-colored clasts had total carbon contents of about 150 $\mu$g/g. The F-2 breccias with light-colored clasts and moderate coherency had variable total carbon contents (60 to 170 $\mu$g/g). The third group (F-3) of fragmental rocks with dark-colored clasts and coherent matrices had low total carbon contents ranging from 21 to 80 $\mu$g/g. Four individual samples from the large F-3 breccia 14321 had consistent total carbon contents of 21, 25, 32, and 43 $\mu$g/g. It is suggested that the fragmental rocks with higher carbon contents have normal dark fines material in their matrices, while the lower carbon fragmental rocks have no lunar surface exposed material.

Total sulfur contents in the Apollo 14 breccias ranged from 700 to 1000 $\mu$g/g. Significant fractionation among petrologic groups was not evident. The total nitrogen content of an F-1 fragmental rock was 130 $\mu$g/g, and of an F-3 fragmental rock, 60 $\mu$g/g. These values correlate directly with the total carbon content for these samples.

## INTRODUCTION

TOTAL CARBON, nitrogen, and sulfur contents of Apollo 14 lunar samples were determined on nine lunar fines and seven lunar breccias. The analytical techniques utilized for carbon and nitrogen were the same as used for Apollo 11 and 12 samples as reviewed by Moore *et al.* (1971). For the Apollo 14 samples, total sulfur was also determined. The analytical method used for sulfur was based on the combustion of the sample splits in an induction furnace in an oxygen atmosphere followed by an iodometric titration of the $SO_2$ produced.

Investigations of total carbon in the Apollo 11 and 12 samples indicated that it has a higher concentration in lunar fines and breccias than in the lunar basalts (Moore *et al.* 1970a, 1970b, 1971). Studies of the chemical species of carbon released from the lunar materials by pyrolysis, leaching, and crushing indicated that the carbon was mainly present as inorganic carbon. A detailed review of these studies by all investigators has been compiled by Gibson and Moore (1972). Detailed studies of the total nitrogen contents of Apollo 11 and 12 samples were fewer in number. Nitrogen-like carbon had a higher reported concentration in lunar fines than in rocks. A major difference in the data appeared when Hintenberger *et al.* (1971) reported that the basalts from Apollo 11, analyzed by a vacuum fusion-mass spectrometric method, contained less than 3 $\mu$g/g of total nitrogen. The results reported by Moore *et al.* (1970b, 1971) in Apollo 11 and 12 basalts were in the range of 30 to 120 $\mu$g/g of total nitrogen. The difference was tentatively attributed to atmospheric contamination of the crushed samples. The analyses of lunar fines gave vacuum fusion concentrations in the range of 100 $\mu$g/g, which are similar to the values of 100 to 150 $\mu$g/g reported by Moore *et al.* (1970b, 1971).

Because of these reported differences and studies made on related systems, total nitrogen was determined in only two Apollo 14 breccias. We have concluded that the adsorption of atmospheric nitrogen on the surface of freshly crushed lunar rocks is, in fact, responsible for the nitrogen detected in some lunar samples.

The total sulfur content in lunar samples from Apollo 11 and 12 is significantly higher than either carbon or nitrogen. Total sulfur analyses in the Apollo 11 samples have been reported by Compston *et al.* (1970), Kaplan and Smith (1970), Maxwell *et al.* (1970), and Agrell *et al.* (1970). The basalts are reported to contain from 0.15 to 0.24 weight percent total sulfur, the breccias 0.11 to 0.15 weight percent, and the fines 0.07 to 0.14 weight percent. In a model for the distribution of volatile elements in the lunar fines Moore *et al.* (1970b) proposed that the difference in concentrations between the basalts and the fines was due to volatile element depletion during meteoroid impact. On the basis of data from the Apollo 12 and 14 lunar samples we now prefer to attribute the major depletion to the mixing in of low sulfur lunar material, rather than to sulfur loss. Total sulfur values in the Apollo 12 samples as measured by Kaplan and Petrowski (1971), Compston *et al.* (1971), and Maxwell and Wiik (1971) are lower than those from Apollo 11. The Apollo 12 basalts ranged from 0.04 to 0.09 weight percent total sulfur, while the breccias and fines had values of 0.07 to 0.12 weight percent.

## Experimental Methods

The techniques for the total nitrogen and carbon contents are the same as those used for Apollo 11 and 12. Included in each set of analyses are two samples of standard rock sample BCR–1. The total carbon content of BCR–1 is 68 $\pm$ 10 $\mu$g/g. The use of this standard is particularly important in assuring analytical consistency between the carbon analyzer utilized at the Lunar Receiving Laboratory during the preliminary examination and subsequent analyses done at Arizona State University. The presently accepted value differs from that reported by Moore *et al.* (1970b) due to a computational error in the earlier publication.

After the initial report of low nitrogen contents in lunar basalts by Hintenberger *et al.* (1971), experiments were made to show the effect of atmospheric contamination on crushed material. A sample of syenite was analyzed as a single chip of approximately 50 mg, a group of coarse grains, and a sample of finely powdered material. The total nitrogen contents were 67 $\mu$g/g, 95 $\mu$g/g, and 950 $\mu$g/g, respectively, illustrating the effect of atmospheric adsorption. Heating and vacuum cleaning experiments did not give evidence for the complete removal of the adsorbed nitrogen. It does appear likely that part of the reported total nitrogen in lunar samples is of terrestrial origin. Müller (1972) has initiated a program of determining chemically bound nitrogen in lunar samples. His Kjeldahl analysis discriminates between molecular and chemically labile forms of nitrogen. His data support the evidence that some basalts are very low in combined nitrogen but also support our earlier evidence that some fine grained vesicular basalts and the lunar fines and breccias do have moderate total nitrogen contents.

The total sulfur contents of the lunar samples reported in this paper were determined on samples ranging from 50 mg to 100 mg in weight. The samples were analyzed using a LECO 532-000 sulfur analyzer. The samples are mixed with iron chips and a copper-tin accelerator and fused at a minimum of 1400°C in a flowing oxygen atmosphere to release $SO_2$. The combustion products are carried by the oxygen stream to an automatic titration vessel containing diluted HCl, KI, and arrowroot starch indicator. A powdered antimony trap is used to remove any halides that would interfere with the titration. The solution is titrated with a $KIO_3$ solution with an automatic buret to a preselected blue endpoint. Each analytical run is standardized with a silicate matrix sulfur standard. Studies have shown that the initial form of the sulfur does not affect the analytical result. The precision of the method as determined by replicate standard analyses ranges from 4% to 10% of the total sulfur volume reported depending upon the initial sample weight taken.

## RESULTS

The results of the Apollo 14 analyses are given in Table 1. Total carbon analyses indicated with an asterisk (*) were run at the Lunar Receiving Laboratory as a part

Table 1. Carbon, nitrogen, and sulfur in Apollo 14 samples (μg/g).

| Sample No. | Sample Description | C | N | S |
|---|---|---|---|---|
| 14003 | Fines | 140 | — | — |
| 14042* | F-1 | 225 | — | — |
| 14047 | F-1 | 140 | — | 990 |
| 14047* | F-1 | 210 | — | — |
| 14049 | F-1 | 135 | 130 | 970 |
| 14049* | F-1 | 190 | — | — |
| 14063* | F-3 | 80 | — | — |
| 14066* | F-4 | 90 | — | — |
| 14141 | Fines, Cone Crater | 42 | — | 1000 |
| 14141* | Fines, Cone Crater | 80 | — | — |
| 14148* | Fines, trench surface | 160 | — | — |
| 14149* | Fines, trench bottom | 135 | — | — |
| 14156 | Fines, trench middle | 110 | — | — |
| 14156* | Fines, trench middle | 186 | — | — |
| 14163 | Fines | 110 | — | 930 |
| 14163* | Fines | 120 | — | — |
| 14230 | Fines | 150 | — | 960 |
| 14240 | Fines | 90 | — | 900 |
| 14259* | Fines | 160 | — | — |
| 14298 | Fines | 150 | — | — |
| 14301 | F-2 | 70 | — | 910 |
| 14301* | F-2 | 50 | — | — |
| 14305* | F-4 | 32 | — | — |
| 14310* | Basalt | 35 | — | — |
| 14311 | F-4 | 40 | — | 950 |
| 14313 | F-2 | 170 | — | 1020 |
| 14313* | F-2 | 130 | — | — |
| 14318 | F-2 | 86 | — | — |
| 14321 | F-4 | 22 | 57 | 580 |
| 14321* | F-4 | 28 | — | — |
| 14321A | F-4 | 32 | — | 700 |
| 14321B | F-4 | 21 | — | 650 |
| 14321C | F-4 | 43 | — | — |
| 14421 | Fines | 160 | — | 910 |
| 14422 | Fines | 120 | — | — |

* Samples were analyzed at the Lunar Receiving Laboratory during preliminary examination. Fragmental rocks are classified after Jackson and Wilshire (1972).

of the preliminary examination (LSPET, 1971). Each of the LRL results is from a single analysis. The other values are weighted means from two or three individual analyses. No attempt was made to homogenize individual lunar samples. This minimized the possibility of contamination. Rock samples were crushed with a single stroke in a clean diamond mortar. To guard against contamination they were not sieved or run through a mechanical splitter. The analytical precision for the carbon, nitrogen, and sulfur analyses taken from the 90% confidence levels on the standard analytical curves has a maximum of 10% of the value reported.

It may be noted that the total carbon contents on replicate samples run both at the LRL and Arizona State University often show significant differences. We attribute this to variations in the size distribution of the grab samples taken for analysis. It has been shown by Moore *et al.* (1970a, 1970b) and Kaplan and Smith (1970) that the finest regolith material has a higher carbon content than the coarser fraction, so that size variations could produce a noticeable difference in their carbon contents.

In addition to the samples listed in Table 1 the carbon content of sample 14311 was reported to be 200 $\mu$g/g in the Apollo 14 Preliminary Science Report (LSPET, 1971). This rock classified as an F-4 fragmental rock by Jackson and Wilshire (1972) appears to have an anomalously high carbon content. We attribute the high carbon content to fine lunar dust adhering to the sample provided during the preliminary examination. Rock samples taken during the preliminary examination consisted of chips broken from the surface of the main mass in the sample container. These chips, together with fines dislodged from the rock's surface, provided an unreliable sample.

The carbon contents of fines or "D" samples from Apollo 14 show similarities to those from Apollo 11 and 12. They approach a maximum of 200 $\mu$g/g total carbon. In this sense they are normal dark lunar fines. Sample 14141 from near Cone Crater is similar in carbon content to samples 12032 and 12033 from Apollo 12. Like the Apollo 12 samples, it is from the ejecta regolith on a crater rim that has not been exposed at the immediate lunar surface or mixed appreciably with surface fines.

From Table 1 it may be seen that the F-1 breccias have high total carbon contents. The three samples analyzed contained over 135 $\mu$g/g C. The F-4 breccias are uniformly low with less than 40 $\mu$g/g total carbon, with the exception of rock 14066 run during the preliminary examination. Five samples of the F-4 fragmental rock 14321 had relatively constant carbon contents in the range 22 to 43 $\mu$g/g. The individual samples from this large fragmental rock were selected to show different lithologies as indicated by their color and coherency. There is apparently little, if any, fractionation of carbon in this material. A portion of the 90 $\mu$g/g total carbon reported for 14066 may be due to surface fines. The intermediate value for the F-3 fragmental rock 14063 may also be due to the same cause. Care obviously should be exercised in the use of the preliminary examination results. The F-2 fragmental rocks appear to fill an intermediate position, with values ranging between 50 and 170 $\mu$g/g total carbon.

The large selection of breccias or fragmental rocks sampled during Apollo 14 are different than those from Apollo 11. While the breccias from Apollo 11 appear to be indurated normal lunar regolith, many of the Apollo 14 breccias appear to have another mode of origin. The total carbon contents of the Apollo 14 fragmental rocks range from 22 to 225 $\mu$g/g C. In Table 1 each of these rocks has been classified according to the hand specimen classification of Jackson and Wilshire (1972). This fourfold classification divides the fragmental rocks according to their clast color and degree of coherency. They note that the F-1 rocks correspond to the "porous, unshocked microbreccias," the F-2 rocks to the "shock compressed microbreccias," and the F-3 and F-4 rocks to the "thermally metamorphosed microbreccias" of Chao *et al.* (1971).

The two Apollo 14 fragmental rocks analyzed for total nitrogen were from the high carbon F-1 group and the low carbon F-4 group. The nitrogen contents found followed the carbon trend but also indirectly supported our conclusion that the samples have adsorbed atmospheric nitrogen.

Total sulfur values for the Apollo 14 samples appear to be intermediate between the higher sulfur samples from Apollo 11 and the lower values from Apollo 14. The normal fines have a range of 760 to 930 $\mu$g/g S. The light-colored low carbon fines from Cone Crater have 1000 $\mu$g/g S. The fragmental rocks have similar, but slightly higher, values in the range of 910 to 1020 $\mu$g/g S, except for the F-4 fragmental rock 14321 from which three different samples had lower total sulfur contents of 580 to 700 $\mu$g/g.

## Discussion

The total carbon, nitrogen, and sulfur contents of the Apollo 14 rocks are generally similar to those for Apollo 11 and 12. The normal dark lunar fines have higher total carbon contents than do the basalts. In our earlier papers (Moore *et al.*, 1970a, 1970b, 1971) we attributed this higher carbon content to the addition of carbon by the solar wind. In addition we developed a model in which the volatile elements, carbon, nitrogen, and sulfur, were depleted on the moon surface by impact events. This was primarily done to account for the lower sulfur contents in the Apollo 11 fines than in the Apollo 11 basalts. Further work on the petrologic makeup of the lunar fines by Reid *et al.* (1972) indicates that to the first approximation, the total sulfur content of the fines may be determined by rock mixing models rather than by impact loss. Table 2 illustrates the total sulfur balance for Apollo 14 fines. The total sulfur values for the mare-derived component were taken from analyses by

Table 2. Mixing model for Apollo 14 lunar fines.*

| Rock Type | Sulfur Observed ($\mu$g/g) | Reid Model | |
|---|---|---|---|
| | | Relative Abundance (%) | Sulfur Contribution ($\mu$g/g) |
| Mare-derived | 1200 | 11 | 132 |
| Fra Mauro basalt | 1000 | 59 | 590 |
| Highland basalt | 300 | 27 | 81 |
| Anorthositic | 200 | 1 | 2 |
| Potash granite | — | 1 | — |

* Total calculated fines = 800; observed fines = 850.

Compston *et al.* (1972) and Hubbard *et al.* (1972). Fra Mauro sulfur values were determined by the authors utilizing unpublished Apollo 12 results in a mixing diagram following the method of Wänke *et al.* (1971). The highland basalt and anorthositic material values were selected from the Apollo 15 LSPET (1972). This preliminary calculation does not take into account sulfur from the solar wind (Moore *et al.* 1971) or recycled meteoritic sulfur, but it does indicate that large amounts of sulfur cannot have been added to the lunar surface from meteorities unless some major loss mechanism has taken place.

A critical discussion of the origin of carbon in the lunar fines has recently been prepared by Hayes (1972). In this review he considers that, in addition to the solar wind, meteorites and comets may also contribute to the lunar carbon via a cycle of carbon through the lunar atmosphere. Such a contribution is supported by the work of Heymann and Yaniv (1970), who showed an implantation on the lunar surface of Ar-40 from a tenuous lunar atmosphere. At the present time there is not enough evidence to consider quantitatively the origin and distribution of the lunar carbon. A portion of it does come from the lunar rocks, with probable additions from extralunar sources. The carbon in the lunar fines appears to be surface correlated. Calculations of the mean surface size of sieved fractions of lunar fines indicate that they have a direct correlation with total carbon content after corrections for indigenous carbon are made. In addition to the accumulation mechanism, loss mechanisms, such as proton stripping and impact volatilization, must also be operative to some degree.

A major point of variation between Apollo 14 carbon contents and those from Apollo 11 and 12 is found in the rock samples commonly referred to as breccias. The Apollo 14 rocks are largely in this category and have commonly been attributed to an impact origin during the Imbrian event (LSPET, 1971). We interpret the low carbon abundances in the F-4 breccias to indicate that they do not contain solar wind-(extralunar) contributed carbon and, hence, differ from normal regolith breccias. The high carbon contents in the F-1 breccias indicate the presence of lunar fines and, hence, support the idea that they are compacted breccias that originated from a normal lunar surface environment. This conclusion is supported by noble gas distribution patterns (LSPET, 1971). Neon-20 contents attributed to a solar wind source show a direct and significant correlation with the total carbon contents for lunar fines and fragmental rocks.

From the data in Table 1, the increase in total carbon in the fines and F-1 rocks is readily apparent. The evolution of the composition of the individual samples appears to include an increase in total carbon content away from a starting material approximated by the F-4 fragmental rocks and the Cone Crater fines sample 14141. Also apparent is the low carbon and sulfur content of the F-4 fragmental rock 14321. This may be indigenous to the pre-Imbrian starting material or indicative of depletion of sulfur during heating and compaction during the Imbrian event. This may be supported by the sulfur analysis by Compston *et al.* (1972) for the rock 14310, which is 600 $\mu$g/g. Ehrlich *et al.* (1972) support the possibility that basalt 14310 may in fact be a melted soil and, hence, closely related to the family of fragmental rocks. If these conclusions are correct, some F-2 fragmental rocks appear to contain lunar surface material, while others do not. Jackson and Wilshire (1972) emphasize that their classification is descriptive and that a similar genesis of rocks in a given group is not required. This seems to be true for the F-2 fragmental rocks 14301, 14318, and 14313. The high carbon content of 14313 indicates that it is an indurated fines sample. Quaide (1972) also concludes, on the basis of abundance glassy aggregates and spherules, that rock 14313 must have originated by lithification of local regolith deposits.

The data from Apollo 14 analyses indicate that care must be taken in the use of the inclusive terms "breccia" and "fragmental rocks." Some distinction should be made to identify local regolith breccias or normal fines breccias and those that Quaide (1972) terms "annealed breccias."

*Acknowledgments*—This work was supported principally by NASA grants NGL 03-001-001 and NGR 03-001-057. The assistance of Dr. Michael Reynolds in facilitating our work in the Lunar Receiving Laboratory is sincerely appreciated.

## References

Agrell S. O., Scoon J. H., Muir I. D., Jong J. V. P., McConnell J. D. C., and Peckett A. (1970) Mineralogy and petrology of some lunar samples. *Science* **167**, 583–586.

Chao E. C. T., Boreman J. A., and Desborough G. A. (1971) The petrology of unshocked and shocked Apollo 11 and Apollo 12 microbreccias. *Proc. Second Lunar Sci. Conf., Geochim. Cosmochim. Acta* Suppl. 2, Vol. 1, pp. 797–816. MIT Press.

Compston W., Chappell B. W., Arriens P. A., and Vernon M. J. (1970) The chemistry and age of Apollo 11 lunar material. *Proc. Apollo 11 Lunar Sci. Conf., Geochim. Cosmochim. Acta* Suppl. 1, Vol. 2, pp. 1007–1027. Pergamon.

Compston W., Berry H., Vernon M. J., Chappell B. W., and Kaye M. J. (1971) Rubidium–strontium chronology and chemistry of lunar material from the Ocean of Storms. *Proc. Second Lunar Sci. Conf., Geochim. Cosmochim. Acta* Suppl. 2, Vol. 2, pp. 1471–1485. MIT Press.

Compston W., Vernon M. J., Berry H., Rudowski R., Gray C. M., and Ware N. (1972) Age and Petrogenesis of Apollo 14 Basalts (abstract). In *Lunar Science—III* (editor C. Watkins), pp. 151–153, Lunar Science Institute Contr. No. 88.

Ehrlich R., Vogel T. A., and Weinberg B. (1972) Inferences from Fourier Shape Analysis of Plagioclase from Apollo 14 Rocks and Soil (abstract). In *Lunar Science—III* (editor C. Watkins), pp. 212–213, Lunar Science Institute Contr. No. 88.

Gibson E. K. and Moore C. B. (1972) Compounds of the organogenic elements in Apollo 11 and 12 lunar samples—A review. *Space Life Sciences* (in press).

Hayes J. M. (1972) Extralunar sources for carbon on the moon. *Space Life Sciences* (in press).

Heymann D. and Yaniv A. (1970) $Ar^{40}$ anomaly in lunar samples from Apollo 11. *Proc. Apollo 11 Lunar Sci. Conf., Geochim. Cosmochim. Acta* Suppl. 1, Vol. 2, pp. 1261–1267. Pergamon.

Hintenberger H., Voshage H., and Specht S. (1971) Investigations of non-rare gases in lunar fines, breccias and rocks (unpublished proceedings presented at the Second Lunar Sci. Conf.).

Hubbard N. J. and Gast P. W. (1972) Chemical Composition of Apollo 14 Materials and Evidence for Alkali Volatilization (abstract). In *Lunar Science—III* (editor C. Watkins), pp. 407–409, Lunar Science Institute Contr. No. 88.

Jackson E. D. and Wilshire H. G. (1972) Classification of the Samples Returned from Apollo 14 Landing Site (abstract). In *Lunar Science—III* (editor C. Watkins), pp. 418–420, Lunar Science Institute Contr. No. 88.

Kaplan I. R. and Smith J. W. (1970) Concentration and isotopic composition of carbon and sulfur in Apollo 11 lunar samples. *Science* **167**, 541–543.

Kaplan I. R. and Petrowski C. (1971) Carbon and sulfur isotopic studies on Apollo 12 lunar samples.

*Proc. Second Lunar Sci. Conf., Geochim. Cosmochim. Acta* Suppl. 2, Vol. 2, pp. 1397–1406. MIT Press.

LSPET (Lunar Sample Preliminary Examination Team) (1971) Preliminary examination of lunar samples from Apollo 14. *Science* **173,** 681–693.

LSPET (Lunar Sample Preliminary Examination Team) (1972) The Apollo 15 lunar samples: A preliminary description. *Science* **175,** 363–375.

Maxwell J. A., Peck L. C., and Wiik H. B. (1970) Chemical composition of Apollo 11 lunar samples 10017, 10020, 10072, and 10084. *Proc. Apollo 11 Lunar Sci. Conf., Geochim. Cosmochim. Acta* Suppl. 1, Vol. 2, pp. 1369–1374. Pergamon.

Maxwell J. A. and Wiik H. B. (1971) Chemical composition of Apollo 12 lunar samples 12004, 12033, 12051, 12052, and 12065. *Earth Planet. Sci. Lett.* **10,** 285–288.

Moore C. B., Lewis C. F., Gibson E. K., and Nichiporuk W. (1970a) Total carbon and nitrogen abundances in lunar samples. *Science* **167,** 496–497.

Moore C. B., Gibson E. K., Larimer J. W., Lewis C. F., and Nichiporuk W. (1970b) Total carbon and nitrogen abundances in Apollo 11 lunar samples and selected achondrites and basalts. *Proc. Apollo 11 Lunar Sci. Conf., Geochim. Cosmochim. Acta* Suppl. 1, Vol. 2, pp. 1375–1382. Pergamon.

Moore C. B., Lewis C. F., Larimer J. W., Delles F. M., Gooley R., and Nichiporuk W. (1971) Total carbon and nitrogen abundances in Apollo 12 lunar samples. *Proc. Second Lunar Sci. Conf., Geochim. Cosmochim. Acta* Suppl. 2, Vol. 2, pp. 1343–1350. MIT Press.

Müller O. (1972) Chemically Bound Nitrogen Abundances in Lunar Samples, and Active Gases Released by Heating at Lower Temperatures (250 to 500°C) (abstract). In *Lunar Science—III* (editor C. Watkins), pp. 568–570, Lunar Science Institute Contr. No. 88.

Quaide W. (1972) Mineralogy and Origin of Fra Mauro Fines and Breccias (abstract). In *Lunar Science—III* (editor C. Watkins), pp. 627–629, Lunar Science Institute Contr. No. 88.

Reid A. M., Ridley W. I., Warner J., Harmon R. S., Brett R., Jakes P., and Brown R. (1972) Chemistry of Highland and Mare Basalts as Inferred from Glasses in the Lunar Soils (abstract). In *Lunar Science—III* (editor C. Watkins), pp. 640–642, Lunar Science Institute Contr. No. 88.

Wänke H., Baddenhausen H., Balacescu A., Teschke F., Spettel B., Dreibus G., Quijano M., Kruse H., Wlotzka F., and Begemann F. (1972) Multielement Analyses of Lunar Samples (abstract). In *Lunar Science—III* (editor C. Watkins), pp. 779–781, Lunar Science Institute Contr. No. 88.

Proceedings of the Third Lunar Science Conference
(Supplement 3, *Geochimica et Cosmochimica Acta*)
Vol. 2, pp. 2059–2068
The M.I.T. Press, 1972

# Chemically bound nitrogen abundances in lunar samples, and active gases released by heating at lower temperatures (250 to 500°C)

O. Müller
Max-Planck-Institut für Kernphysik, 69 Heidelberg, Germany

**Abstract**—Chemically bound nitrogen concentrations have been determined in lunar samples from different Apollo landing sites using the Kjeldahl method. This technique discriminates against molecular nitrogen, $N_2$, which may be a possible cause for contamination of lunar material by terrestrial atmospheric nitrogen. The results show that the main amount of nitrogen is present in a bound state. Nitrogen analyses on grain size fractions of Apollo 14 and Apollo 15 fines have revealed that nitrogen is clearly correlated with the grain surfaces and derives most likely from the solar wind. Nitrogen can be used as a reference element in terms of solar elemental abundances because it is chemically active and, therefore, is better retained on grain surfaces than the lighter solar wind noble gases.

The predominant gases released from fines 14003 and 15101 and fragmental rock 14303 by heating between 250 and 500°C are $H_2$, CO, $CO_2$, and $CH_4$. Minor amounts of $C_2H_6$, $C_3H_8$, $C_2H_4(C_2H_2)$, and $C_3H_6$ were detected only at the higher temperatures. No molecular nitrogen was detected during the heating experiments; this is probably due to the high portion of chemically bound nitrogen.

Crushing experiments have been performed on some vesicular lunar rock and glass samples to study trapped gases. Only traces of methane have been detected, and no other active gases have been found.

## Introduction

In most nitrogen work on lunar material of previous Apollo missions the total nitrogen content has been determined. Moore *et al.* (1970, 1971, 1972) have used a high temperature extraction method by which all nitrogen compounds are transformed to $N_2$. Goel and Kothari (1972) reported on total nitrogen in Apollo 14 material using the neutron activation reaction $^{14}N$ $(n, p)$ $^{14}C$. These analytical techniques do not distinguish between different chemical forms in which nitrogen may be present. It is known that nitrogen can occur in crystalline material and glasses either as gaseous, molecular $N_2$ or as chemically bound nitrogen; the latter may be present as ammonium- and nitride-nitrogen.

In an earlier investigation on trapped gases in Apollo 11 and 12 samples (Funkhouser *et al.*, 1971) we have detected molecular $N_2$, as well as other gases, by crushing breccias and igneous rocks in a steel apparatus. The amounts of nitrogen released were only in the $10^{-9}$ g $N_2$/g range, and, by far, too low compared with total nitrogen contents of the samples. On heating various lunar samples sealed under vacuum in small glass vials, gases like $H_2$, CO, $CO_2$, and hydrocarbons were detected, but no $N_2$. These results also indicated that the major portion of nitrogen present is not in a gaseous form, but rather in a chemically bound state.

There exist only few measurements of chemically bound nitrogen in lunar samples by Hintenberger *et al.* (1970), Chang *et al.* (1971), and Sakai *et al.* (1972). The analyses of lunar fines and breccias by these authors showed that the major portion of total nitrogen can be hydrolyzed to ammonia by acid treatment.

The aims of the present investigation were: (1) to determine the abundance of chemically bound nitrogen in a larger number of lunar samples and to compare it with total nitrogen content, (2) to study the surface dependence of bound nitrogen in grain size fractions of Apollo 14 and Apollo 15 fines to learn more about nitrogen derived from solar wind, (3) to study the release of indigenous and generated gases from lunar material by means of heating experiments, and (4) to study trapped gases in vesicular lunar samples by crushing at room temperature.

## Analytical Procedure

Chemically bound nitrogen concentrations have been determined in lunar samples and some standard materials by the Kjeldahl method, which discriminates against molecular nitrogen. Bound nitrogen was converted to $NH_4^+$ by acid hydrolysis with HF and $H_2SO_4$ in a platinum crucible. No residue was observed after sample decomposition. The sample break-up was performed in a lucite box that was continuously ventilated with ammonia-free air. Parallel to the samples, blank analyses were run. After a Kjeldahl distillation (Fig. 1) the ammonia was determined spectrophotometrically with Nessler's reagent ($K_2HgI_4$).

The ammonia distillations were performed as follows: First, any ammonia present in the NaOH solution was expelled by water steam distillation. Then, the blank was rinsed into the 250 ml flask and distilled, and, finally, the ammonia of the lunar sample was distilled. To neutralize ammonia, 1 ml of N/50 $H_2SO_4$ was present in the 50 ml recipient at the end of the condenser (see Fig. 1).

The extinction of the yellow ammonia complex formed with Nessler's reagent was measured against $H_2O$ in 2 cm quartz cuvettes at 425 $\mu$ in a ZEISS spectrophotometer, model PMQ II. The dependence of ammonia concentration and extinction was determined by processing ammonium sulfate solutions in the same manner as the samples. A linear calibration curve was obtained for ammonia concentrations between 4 $\mu$g and 80 $\mu$g $NH_3$/50 ml distillate.

Sample weights of about a 100 mg were used for a single analysis of lunar fines, their grain size fractions, and breccias. For lunar crystalline rock analyses, about 200 mg material was used. For some lunar samples, duplicate analyses were performed, which agreed to about $\pm 5$ ppm N. The results are compiled in Table 1.

Table 1. Chemically bound nitrogen (chem.b.N) concentrations in lunar bulk fines and grain size fractions, breccia, fragmental and igneous rocks determined as ammoniacal nitrogen by the Kjeldahl method.*

| Sample Type | and Number | Chem.b.N (ppm) | Sample Type | and Number | Chem.b.N (ppm) |
|---|---|---|---|---|---|
| Fines | 10084,31-bulk | 93 | Fines | 15601,63-bulk | 80 |
| | < 48 $\mu$ | 148 | Grain size | < 24 $\mu$ | 157 |
| Fines | 12070,76-bulk | 80 | fractions | 24–48 $\mu$ | 83 |
| Fines | 14003,24-bulk | 92 | | | |
| | < 48 $\mu$ | 194 | Breccia | 10046,1 | 131 |
| Grain | < 24 $\mu$ | 226 | Fragmental | | |
| size | 24–48 $\mu$ | 149 | rock | 14303,13 | 31 |
| fractions | 48–60 $\mu$ | 73 | | | |
| | 60–109 $\mu$ | 60 | Vesicular | | |
| Fines | 15101,59-bulk | 109 | basalt | 10057,80 | 64 |
| Grain | < 24 $\mu$ | 272 | Microgabbro | 12063,112 | < 10 |
| size | 24–48 $\mu$ | 113 | Basalt | 12075,13 | < 10 |
| fractions | 48–60 $\mu$ | 86 | Vesicular | | |
| | 60–109 $\mu$ | 65 | basalt | 15556,25 | < 10 |

* Error $\pm$ 5 ppm.

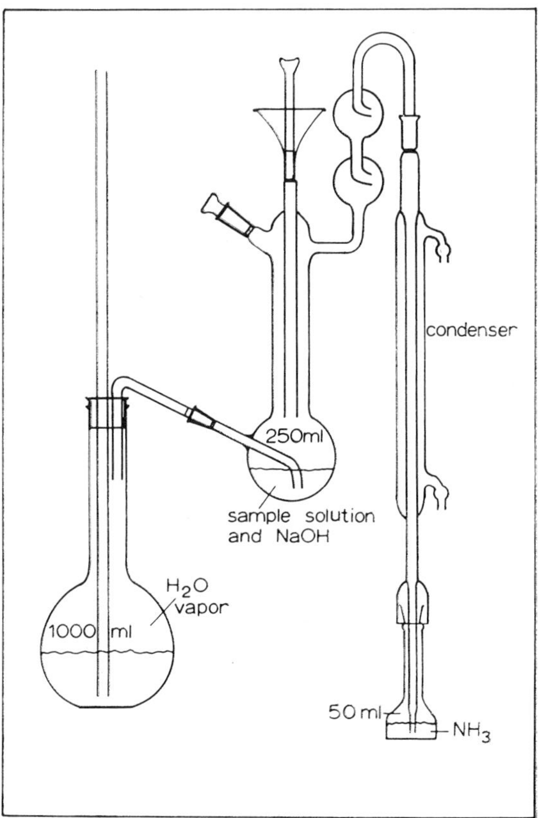


Fig. 1. Kjeldahl distillation apparatus for the determination of chemically bound nitrogen (ammonium- and nitride-nitrogen). The part with the 250 ml flask consists of quartz-glass.

To test the precision and accuracy of the nitrogen procedure we have analyzed various materials. In U.S.G.S. standard diabase W–1 we found a mean value of 14 ppm N, which is in excellent agreement with the 14 ppm found by Wlotzka (1961). As further reference we analyzed a soda-lime-silica glass that contained 2.06 weight percent chemically bound nitrogen. This large amount of nitrogen could be dissolved in the glass by bubbling ammonia gas into the molten glass. This glass was synthesized a few years ago at the Max-Planck-Institut für Silikatforschung in Würzburg, Germany, and recommended as a standard for nitrogen determination. The above value was obtained by Kjeldahl distillation and subsequent titration of the ammonia. With the Kjeldahl-Nessler method we could confirm this result within $\pm 5$ percent. Titanium nitride, TiN, was also analyzed and yielded the expected nitrogen content of $(22.6 \pm 1.0)$ weight percent.

Heating experiments of lunar fines 14003 and 15101 and fragmental rock 14303 have been performed to examine the gas release at different temperatures. Samples of about 5 mg weight were preheated at 160°C in small glass vials under vacuum, sealed and heated at 250, 300, 350, 400, 450, and 500°C for 15 hours, respectively. The volume of the vials was about 35 mm$^3$. The gases released from the samples were measured by breaking the vials in a stainless steel crushing apparatus connected to a sensitive gas chromatograph (Varian MAT, model 1532, 2B). This model is equipped with two columns, a zeolite molecular sieve and an organic polymer, Porapak Q, column, and two helium

Table 2. Gas chromatographic results of heating experiments.*

| Sample and Temperature | $H_2$ | CO | $CO_2$ | $N_2$ | $CH_4$ | $C_2H_6$ | $C_3H_8$ | $C_2H_4(C_2H_2)$ | $C_3H_6$ |
|---|---|---|---|---|---|---|---|---|---|
| Fines 14003,24 | | | | | | | | | |
| 250°C | 1150 | 350 | 1410 | — | 23 | — | — | — | — |
| 300 | 5790 | 1810 | 2070 | — | 45 | — | — | — | 20 |
| 350 | 5720 | 1330 | 2160 | — | 47 | 12 | — | — | 40 |
| 400 | 49400 | 1870 | 2100 | — | 425 | 260 | 153 | 128 | 300 |
| 450 | 52070 | 1870 | 1530 | — | 492 | 160 | 41 | 50 | 170 |
| 500 | 50000 | 5000 | 380 | — | 1800 | 530 | 240 | 120 | 260 |
| Fragmental rock 14303,13 | | | | | | | | | |
| 250°C | 34 | 250 | 960 | — | — | — | — | — | — |
| 300 | 174 | 1010 | 2190 | — | 20 | — | — | — | 21 |
| 350 | 305 | 1760 | 4150 | — | 80 | — | — | — | 51 |
| 400 | 9340 | 4970 | 15860 | — | 730 | 247 | 167 | 10 | 420 |
| 450 | 8550 | 9520 | 18260 | — | 1270 | 425 | 280 | 24 | 800 |
| Fines 15101,59 | | | | | | | | | |
| 250°C | 257 | 130 | 50 | — | — | — | — | — | — |
| 300 | 6500 | 480 | 600 | — | 20 | — | — | — | 10 |
| 350 | 8550 | 630 | 1000 | — | 60 | 7 | — | — | 26 |
| 400 | 62900 | 780 | 600 | — | 170 | 50 | 23 | 10 | 54 |
| 450 | 110400 | 1720 | 1000 | — | 780 | 140 | 45 | 35 | 120 |
| 500 | 105000 | 4400 | 230 | — | 2700 | 600 | 160 | 80 | 100 |

* Gas amounts are given in $10^{-8}$ cc STP/5 mg sample.
*Note:* Gases released at 250, 300, 350, 400, 450, and 500°C from Apollo 14 and 15 samples sealed in small vacuum glass vials. Sample weights are about 5 mg.—below detection limits.

ionization detectors. The results of the heating experiments are compiled in Table 2. Blanks were determined from processing in the same manner empty vials, and vials containing powdered quartz, which was predegassed at 850°C. The blanks yielded only negligible amounts of atmospheric gases and traces of methane.

Several vesicular Apollo 14 and 15 samples were crushed at room temperature, and the gas release of trapped gases was studied by gas chromatography as described by Funkhouser *et al.* (1971).

## Results and Discussion

### *Chemically bound nitrogen abundances*

The results of the chemically bound nitrogen determinations for various lunar samples of different Apollo landing sites are given in Table 1. Fig. 2 illustrates the nitrogen content found in bulk fines, breccias, fragmental and igneous rocks. The length of the bars represents the approximate error of $\pm 5$ ppm N.

The analyzed bulk fines 10084, 12070, and 14003 (Apollo 11, 12, and 14) have a chemically bound nitrogen concentration of 93, 80, and 92 ppm N, respectively. Apollo 15 fines 15101 and 15601 contain 109 and 80 ppm N, respectively. The concentrations of *chemically bound nitrogen* found in this work are somewhat, but not much, lower compared with the *total nitrogen* values for Apollo 11 fines (range 102 to 153 ppm N) found by Moore *et al.* (1970, 1971) and some Apollo 12 fines (range 85 to 140 ppm N). It is an important result that the major portion of nitrogen in fines is present in a chemically bound state. Dr. C. B. Moore recently pointed out

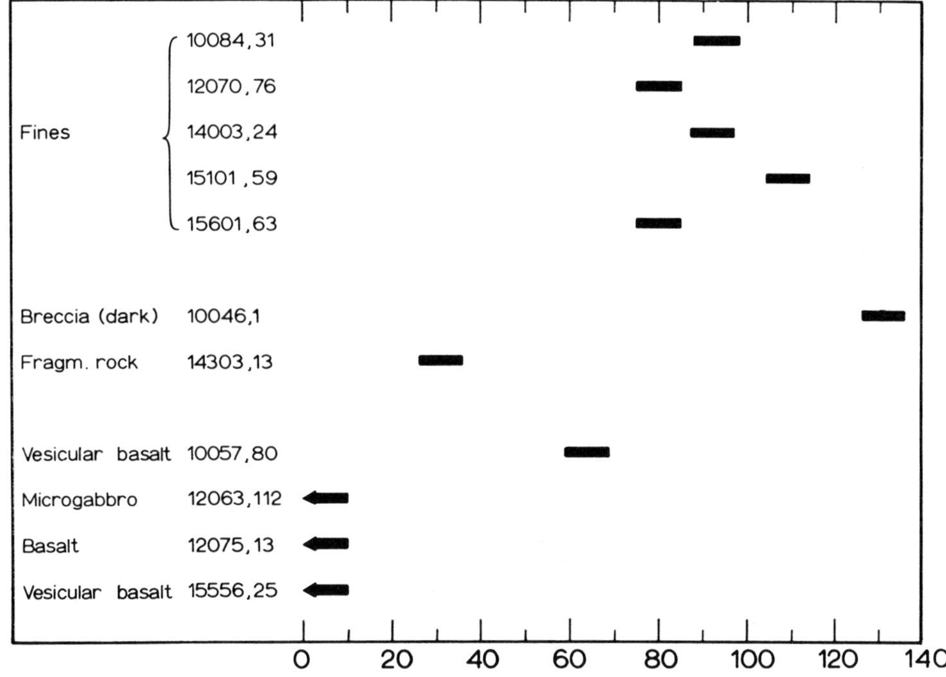


Fig. 2. Chemically bound nitrogen content (in ppm N) in lunar bulk fines, breccia, fragmental and igneous rocks. The length of the bars corresponds to the error of ±5 ppm N.

(written communication) that the fines are probably contaminated by a strongly adsorbed atmospheric nitrogen component (about 30 ppm N). The Kjeldahl method applied by us discriminates against this component. Taking this amount into account, the nitrogen data of both laboratories agree quite well. Goel and Kothari (1972) reported, besides other data, 80 ppm total nitrogen (mean of 5 replicate analyses) in fines 14163 using $^{14}N$ $(n, p)$ $^{14}C$ activation analysis. We found 92 ppm N in fines 14003. Sakai *et al.* (1972) found 111 ppm hydrolyzable nitrogen in fines 15271.

To learn more about nitrogen derived from the solar wind, we have separated fines 14003, 15101, and 15601 into grain size fractions and determined the chemically bound nitrogen content. The data given in Table 1 show a distinct increase of nitrogen content with decreasing grain size. Because of lack of material of 15601, only two grain size fractions could be analyzed. Goel and Kothari (1972) found a similar dependence in grain size fractions of fines 14163. Hintenberger *et al.* (1970) found an enrichment of bound nitrogen on grain surfaces of fines 10084 by using a stepwise dissolution technique.

In Fig. 3 the nitrogen concentrations (in ppm) of grain size fractions of fines 14003, 15101, and 15601 are plotted versus the mean grain diameter (in $\mu$). As approximate mean grain diameter of the fractions 60–109 $\mu$, 48–60 $\mu$, 24–48 $\mu$, and $<24$ $\mu$, we have used 79, 53, 32, and 8 $\mu$, respectively, in this plot. The results for 14003 and

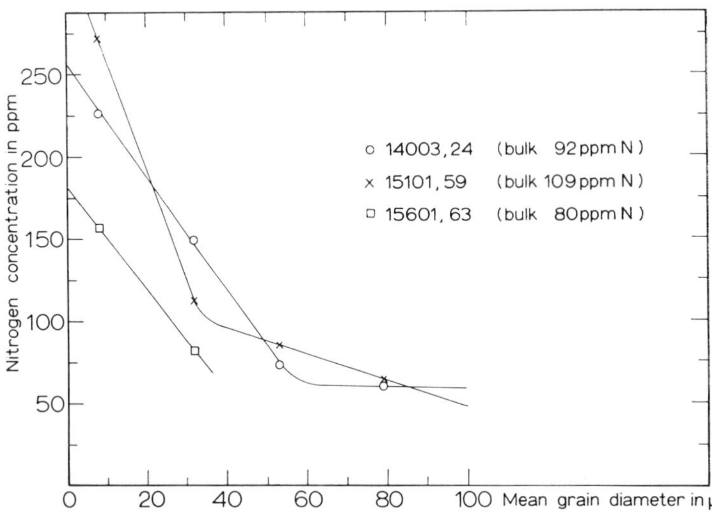


Fig. 3. Grain size dependence of chemically bound nitrogen (ammonium- and nitride-nitrogen) in sieve fractions of Apollo fines 14003, 15101, and 15601. Because of lack of material, only the two finest sieve fractions of fines 15601 could be analyzed.

15101 show a distinct increase of nitrogen content by a factor of about 4 with decreasing grain size, that is, nitrogen is clearly correlated with the grain surfaces and derives most likely from the solar wind. With increasing grain size of fines 14003 the slope of the nitrogen curve turns sharply and becomes almost horizontal. This indicates a volume component of nitrogen that can be interpreted as indigenous lunar nitrogen. For fines 15101 the course of the nitrogen curve is somewhat different. The slope at smaller grain sizes is steeper, at larger grain sizes not as horizontal compared to that of fines 14003.

Besides the solar wind, as source of the surface correlated nitrogen, one has also to consider a meteoritic contribution to lunar surface material. Ganapathy *et al.* (1970) found an admixture of about 2% of carbonaceous-chondrite-like material in Apollo 11 fines and breccias. Assuming a nitrogen content of 2600 ppm N (mean value for C–1 chondrites; Moore, 1971) in this material, the addition is not sufficient to account for the nitrogen content in the finest grain size fractions of lunar fines. Moreover, the C/N ratio of about 1.5 in lunar fines and breccias is distinctly different from that of C–1 chondrites (C/N ~12).

The amount of chemically bound nitrogen of 131 ppm in a dark-colored chip of breccia 10046 is similar to total nitrogen in breccias 10002 and 12057 found by Moore *et al.* (1970, 1971).

Fragmental rock 14303, however, has a relatively low nitrogen content of 31 ppm. Typical solar wind constituents, such as light rare gases (Kirsten *et al.*, 1972) and hydrogen (see heating experiments, following section), are much less abundant in this clastic rock than in lunar fines and breccias. Therefore, most of the nitrogen is probably not derived from the solar wind, but is, rather, indigenous. Goel and Kothari

(1972) found for the fragmental rocks 14049 (soil clod), 14305, and 14321 mean values of 71, 22, and 24 ppm total N, respectively. Moore *et al.* (1972) reported 130 ppm and 57 ppm total N for fragmental rocks 14049 and 14321, respectively. The total carbon data of a large number of Apollo 14 fragmental rocks show a wide range between 210 and 21 ppm C (Moore *et al.*, 1972). The same is probably true for the nitrogen contents of these rocks. The vesicular basalt sample 10057 and the two dense igneous rocks 12063 and 12075 differ significantly in chemically bound nitrogen content. The sample 10057 has a relatively high N value of 64 ppm, whereas in the two Apollo 12 rocks practically no nitrogen was detected. We suggest that part of the gases present during the formation of the vesicles was ammonia and/or nitrogen, which interacted with the magma to form chemically bound nitrogen. However, as a chip of Apollo 15 vesicular basalt 15556 has a N content below 10 ppm, this suggestion is questionable. More data on vesicular and dense igneous rocks are needed to reach a conclusive solution to this problem.

The finding of surface correlated nitrogen in lunar fines and its presence in a bound state suggest a comparison of nitrogen with solar wind-derived noble gases in terms of solar elemental abundances. It is known that the lighter solar wind noble gases are partly lost from grain surfaces by sputtering and diffusion. As distinct from noble gases, solar wind nitrogen impinges on the lunar surface in a chemically active form (plasma state) and probably reacts with hydrogen and other elements forming ammonium- and nitride-nitrogen. Therefore, nitrogen should be less prone to be lost from grain surfaces and may thus be useful as a reference element for solar abundances. In Table 3 the abundance of chemically bound nitrogen in the $<24\ \mu$ fraction of fines 14003* is compared with the abundances of trapped solar wind noble gases. Noble gas data are from Kirsten *et al.* (1972). Our measured $^4He/N$ ratio of 1.2 is lower by a factor of about 500 than the solar ratios of Cameron (1968) and of Suess and Urey (1956). Our $^{20}Ne/N$ ratio of 0.02 is lower by a factor of about 50 than the solar ratio. We infer from these data that nitrogen is, as expected, more tightly retained than the noble gases and may thus be used as a reference. The loss of

Table 3. The significance of solar wind-derived nitrogen as reference element in terms of solar abundances deduced from its relation with solar wind noble gases.

| | $^4He/^{20}Ne$ | $^{20}Ne/^{36}Ar$ | $^4He/N$ | $^{20}Ne/N$ |
|---|---|---|---|---|
| Fines 14003,24 Grain size fraction $<24\ \mu$ | 54 | 3.6 | 1.2 | 0.02 |
| Solar wind composition experiment, Apollo 14 (Geiss *et al.*, 1971) | 550 | 37 | — | — |
| Solar abundances (Cameron, 1968) | 980 | 11 | 860 | 1.1 |
| Solar abundances (Suess and Urey, 1956) | 394 | 67 | 466 | 0.8 |

* From the value given in Table 1 the volume component of nitrogen (60 ppm N) was subtracted (see Fig. 3).

noble gases from grain surfaces is evident when comparing $^{4}He/^{20}Ne$ and $^{20}Ne/^{36}Ar$ ratios of fines 14003 with the ratios found in the Solar Wind Composition experiment by Geiss *et al.* (1971).

### *Heating experiments*

The results of the gas release by heating fines 14003, fragmental rock 14303, and fines 15101 at temperatures between 250 and 500°C using gas chromatography are given in Table 2. The predominant gases are $H_2$, CO, $CO_2$, and $CH_4$. Minor amounts of $C_2H_6$, $C_3H_8$, $C_2H_4(C_2H_2)$, and $C_3H_6$ were detected only at higher temperatures. From fragmental rock 14303 distinctly less hydrogen is evolved compared to the two fines samples, indicating a lower content of solar wind. In all three samples no molecular nitrogen was detected during the heating experiments. This is probably due to the high portion of chemically bound nitrogen, the diffusion rate of which is negligible compared to that of molecular nitrogen.

It was shown in numerous papers dealing with the organic chemistry of lunar samples that carbon is present in various chemical forms. The studies of Chang *et al.* (1971) and Abell *et al.* (1971) revealed that a significant portion of the carbon is in a nonvolatile form. This is supported by a previous study (Funkhouser *et al.*, 1971), where we did not detect any indigenous CO or $CO_2$ in crushed unheated Apollo 11 and 12 samples. It is, therefore, plausible to assume that the CO and $CO_2$ detected in Apollo 14 and 15 material are generated from some sort of carbonaceous material by heating (Table 2). Part of the $CO_2$ may be due to terrestrial contamination (Kaplan and Petrowski, 1971).

How far the *hydrocarbons* are indigenous to fines 14003 and 15101 and rock 14303 is difficult to evaluate. At least part of the hydrocarbons is produced from CO and $H_2$ during heating, possibly by Fischer-Tropsch synthesis. Distinction between indigenous and generated hydrocarbons can be made only by the isotopic labeling method, which was first applied to Apollo 11 fines by Abell *et al.* (1970). The heating experiments show that at relatively low temperatures a variety of volatile chemical compounds are produced in lunar surface material. Similar chemical reactions should take place on the moon's surface during the lunar day. Volatile compounds as carbon oxides and hydrocarbons will gradually diffuse away from the lunar surface. Such a mechanism may be partly responsible for the observed isotopic fractionation of carbon (enrichment of $^{13}C$ relative to $^{12}C$).

### *Crushing experiments*

To study trapped gases in vesicular lunar material we have performed some crushing experiments on a vesicular basalt specimen of 15556, on a small piece of a glass-coated breccia, and on a bubble-containing glass cylinder (5 × 2 mm); these last two were both picked from coarse fines 14257. The eight chips of 15556 we obtained from NASA show open bubbles of about millimeter size on the surface. Unfortunately, no interior closed vesicles are present as x-ray photographs have shown. On crushing a 400 mg chip only a trace of methane ($<10^{-7}$ cc) was detected by gas chromatography. The two glass samples possibly also contain a trace of methane ($<10^{-7}$ cc) and no other active gases.

*Acknowledgments*—We thank the National Aeronautics and Space Administration for providing the lunar samples. Valuable discussions with Prof. W. Gentner, Dr. T. Kirsten, and Prof. O. A. Schaeffer are highly appreciated. The collaboration of Mrs. S. Hasse and D. Kaether is gratefully acknowledged. I thank Dr. P. Horn and R. Schwan for making the grain size fractions of lunar fines and Dr. H. E. Schmid, Research Laboratory of Brown-Boveri Company, Heidelberg, for performing the x-ray photography.

## References

Abell P. I., Eglinton G., Maxwell J. R., Pillinger C. T., and Hayes J. M. (1970) Indigenous lunar methane and ethane. *Nature* **226,** 251–252.

Abell P. I., Cadogan P. H., Eglinton G., Maxwell J. R., and Pillinger C. T. (1971) Survey of lunar carbon compounds, I. The presence of indigenous gases and hydrolysable carbon compounds in Apollo 11 and 12 samples. *Proc. Second Lunar Sci. Conf., Geochim. Cosmochim. Acta* Suppl. 2, Vol. 2, pp. 1843–1863. MIT Press.

Cameron A. G. W. (1968) A new table of abundances of the elements in the solar system. In *Origin and Distribution of the Elements* (editor L. H. Ahrens), pp. 125–143. Pergamon.

Chang S., Kvenvolden K., Lawless J., Ponnamperuma C., and Kaplan I. R. (1971) Carbon, carbides, and methane in an Apollo 12 sample. *Science* **171,** 474–477.

Funkhouser J., Jessberger E., Müller O., and Zähringer J. (1971) Active and inert gases in Apollo 12 and Apollo 11 samples released by crushing at room temperature and by heating at low temperatures. *Proc. Second Lunar Sci. Conf., Geochim. Cosmochim. Acta* Suppl. 2, Vol. 2, pp. 1381–1396. MIT Press.

Ganapathy R., Keays R. R., Laul J. C., and Anders E. (1970) Trace elements in Apollo 11 lunar rocks: Implications for meteorite influx and origin of moon. *Proc. Apollo 11 Lunar Sci. Conf., Geochim. Cosmochim. Acta* Suppl. 1, Vol. 2, pp. 1117–1142. Pergamon.

Geiss J., Buehler F., Cerutti H., Eberhardt P., and Meister J. (1971) The solar-wind composition experiment. Apollo 14 Preliminary Science Report, NASA SP-272, pp. 221–226.

Goel P. S. and Kothari B. K. (1972) Nitrogen abundances in lunar samples by neutron activation analysis (abstract). In *Lunar Science—III* (editor C. Watkins), pp. 315–317, Lunar Science Institute Contr. No. 88.

Hintenberger H., Weber H. W., Voshage H., Wänke H., Begemann F., and Wlotzka F. (1970) Concentrations and isotopic abundances of the rare gases, hydrogen and nitrogen in Apollo 11 lunar matter. *Proc. Apollo 11 Lunar Sci. Conf., Geochim. Cosmochim. Acta* Suppl. 1, Vol. 2, pp. 1269–1282. Pergamon.

Kaplan I. R. and Petrowski C. (1971) Carbon and sulfur isotope studies on Apollo 12 lunar samples. *Proc. Second Lunar Sci. Conf., Geochim. Cosmochim. Acta* Suppl. 2, Vol. 2, pp. 1397–1406. MIT Press.

Kirsten T., Deubner J., Ducati H., Gentner W., Horn P., Jessberger E., Kalbitzer S., Kaneoka I., Kiko J., Krätschmer W., Müller H. W., Plieninger T., and Thio S. K. (1972) Rare gases and ion tracks in individual components and bulk samples of Apollo 14 and 15 fines and fragmental rocks (abstract). In *Lunar Science—III* (editor C. Watkins), pp. 452–454, Lunar Science Institute Contr. No. 88.

Moore C. B. (1971) Nitrogen. In *Handbook of Elemental Abundances in Meteorites* (editor B. Mason), pp. 93–98. Gordon and Breach Science Publishers.

Moore C. B., Gibson E. K., Larimer J. W., Lewis C. F., and Nichiporuk W. (1970) Total carbon and nitrogen abundances in Apollo 11 lunar samples and selected achondrites and basalts. *Proc. Apollo 11 Lunar Sci. Conf., Geochim. Cosmochim. Acta* Suppl. 1, Vol. 2, pp. 1375–1382. Pergamon.

Moore C. B., Lewis C. F., Larimer J. W., Delles F. M., Gooley R. C., Nichiporuk W., and Gibson E. K. Jr. (1971) Total carbon and nitrogen abundances in Apollo 12 lunar samples. *Proc. Second Lunar Sci. Conf., Geochim. Cosmochim. Acta* Suppl. 2, Vol. 2, pp. 1343–1350. MIT Press.

Moore C. B., Lewis C. F., Cripe J., Kelly W. R., and Delles F. (1972) Total carbon, nitrogen and sulfur abundances in Apollo 14 lunar samples (abstract). In *Lunar Science—III* (editor C. Watkins), pp. 550–551, Lunar Science Institute Contr. No. 88.

Sakai H., Petrowski C., Goldhaber M. B., and Kaplan I. R. (1972) Distribution of carbon, sulfur and nitrogen in Apollo 14 and 15 material (abstract). In *Lunar Science—III* (editor C. Watkins), pp. 672–674, Lunar Science Institute Contr. No. 88.

Suess H. E. and Urey H. C. (1956) Abundances of the elements. *Rev. Modern Phys.* **28,** 53–74.

Wlotzka F. (1961) Untersuchungen zur Geochemie des Stickstoffs. *Geochim. Cosmochim. Acta* **24,** 106–154.

Proceedings of the Third Lunar Science Conference
(Supplement 3, *Geochimica et Cosmochimica Acta*)
Vol. 2, pp. 2069–2090
The M.I.T. Press, 1972

# Survey of lunar carbon compounds: II. The carbon chemistry of Apollo 11, 12, 14, and 15 samples

P. H. CADOGAN, G. EGLINTON, J. N. M. FIRTH, J. R. MAXWELL,
B. J. MAYS, and C. T. PILLINGER

School of Chemistry, University of Bristol,
Bristol BS8 1TS, U.K.

**Abstract**—The methane and carbide concentrations of a number of Apollo 11, 12, 14, and 15 samples of fines and breccias have been examined by the deuterated acid dissolution method. Location studies indicate that these carbon compounds are concentrated in the outer surfaces of the fines particles of 48–152 $\mu$ diameter; for larger particles a volume-related component may contribute. In individual samples the methane and carbide concentrations correlate with parameters indicative of lunar surface exposure. The data provide further evidence that solar wind implantation is the major source of the methane in the fines and that the carbide originates from both solar wind implantation and meteorite impacts. The carbon chemistry can be used as an exposure and reworking parameter and provides evidence for mixing in the Apollo 14 fines. The $CH_4$ and carbide concentrations in the Apollo 12 double core show the complex nature of the regolith at the Apollo 12 site. The metamorphic history of the Apollo 14 breccias is reflected in their carbon chemistry.

## INTRODUCTION

GIBSON AND MOORE (1972) HAVE recently reviewed the carbon chemistry of Apollo 11 and 12 samples in terms of the variety of low molecular weight carbon compounds that can be released from lunar samples by vacuum pyrolysis, acid etching and dissolution, and vacuum crushing. Application of these techniques, notably acid dissolution with deuterium-labeled reagents, suggests that the carbon chemistry of the regolith is highly dependent on phenomena occurring at the lunar surface, including solar wind implantation and meteorite impacts (Cadogan *et al.*, 1971; Abell *et al.*, 1971). Previously we have reported preliminary data suggesting that both the methane and carbide in the lunar fines might be located at the surfaces of the grains (Abell *et al.*, 1970a) and that there is a correlation between the concentrations of these species and other parameters indicative of lunar surface exposure (Cadogan *et al.*, 1971; Abell *et al.*, 1971). The work reported herein describes a detailed study of the concentration dependence of these carbon compounds on grain size, and of the relationships between their concentrations in individual samples and surface exposure parameters. Demonstration of such relationships within the fines and breccias should provide an additional parameter in understanding the history of the lunar regolith.

The analytical method used has the capability of measuring $C_1$ to $C_3$ species; we have concentrated, however, on $CH_4$ and $CD_4$ because (1) these are the most abundant undeuterated and deuterated species released by deuterated acid dissolution; (2) they can be considered as representative of the endogenous hydrocarbons and the carbides, respectively, and (3) analyses can be performed routinely on 5-mg aliquots of typical samples of fines and breccias.

## EXPERIMENTAL

The sample handling, storage, acid dissolution, and gas chromatographic procedures, and the gas fractionation and concentration system have been previously described in detail (Abell *et al.*, 1970a, 1971). Only minor modifications are specified herein.

### *General*

Since adsorbed terrestrial methane is desorbed by the standard degassing procedure (see below), samples are stored at atmospheric pressure in stoppered glass tubes equipped with ground-glass joints and Teflon sleeves. The tubes are stored inside larger glass vessels with ground-glass flanges. Storage in this way decreases sample processing and laboratory exposure times. The vacuum system for degassing samples and acid is now equipped with an additional liquid nitrogen trap.

### *Acid dissolution*

Previously, samples were exchanged with $D_2O$ prior to degassing (to remove adsorbed $H_2O$ after isotopic exchange); however, we have found that degassing at $5 \times 10^{-6}$ Torr makes $D_2O$ exchange unnecessary. Samples of different acids afforded varying yields of $CD_4$ from dissolution of aliquots of the same sample of 10086 fines (see Table 1 and Results section). The standard procedure adopted, namely dissolution (100°C, 2 hr) with DCl in $D_2O$ (38% Ciba, isotopic purity 99.5 atom percent D), afforded reproducible and optimum recoveries of hydrocarbons and deuterocarbons.

### *Gas fractionation and concentration system*

The revised system (Fig. 1) incorporates a gas sampling valve (Type GIL, Jones Chromatography Co.) in place of the gas-tight syringe used previously, to prevent sample losses and improve reproducibility. Two fractionation traps replace the Töepler pump used for gas concentration, to eliminate reactions between hydrocarbons and mercury (Johnson *et al.*, 1970); $CH_4$, $CD_4$, and more volatile species are condensed (77°K) on 5 A molecular sieve in one trap (B); $CO_2$, $C_2$, and $C_3$ species can be condensed (77°K) in the other trap, which also serves as sample loop (1.1 ml) (C) of the gas sampling valve (A). Residual gases present in the system immediately prior to each analysis are concentrated and analyzed by gas chromatography as a background control.

### *Gas chromatography*

Analyses were carried out either with a Varian 1200 series gas chromatograph as previously described, or with a Perkin Elmer F11 Mark 2 instrument equipped with a flame ionization detector alone. Gases released from lunar samples were analyzed on 10 m × 1 mm i.d. stainless steel columns packed with 40–60 mesh Graphon (efficiences about 15,000 theoretical plates for $CH_4$ or $CD_4$).

Table 1. Methane evolved from Apollo 11 fines by different acid reagents.

| Acid | Strength (%) | $CD_4$ (μg/g) | $CH_4$ (μg/g) | Total methane (μg/g) |
|---|---|---|---|---|
| HF | 40 | — | 20–32 | 20–32 |
| HF/HCl | 20/18 | — | 24 | 24 |
| DF (batch 1) | 20* | 20 | 5.3 | 25 |
| DF (batch 2) | 20† | 2.0 | 2.9 | 5 |
| DF | 30‡ | 8.2 | 4.0 | 12 |
| DF/DCl | 15/19 | 10 | 3.5 | 14 |
| DCl | 38 | 17–22 | 3.6–5.1 | 21–27 |

* As specified by manufacturer.
† As above but estimated also by titration.
‡ Courtesy of B. R. Simoneit.

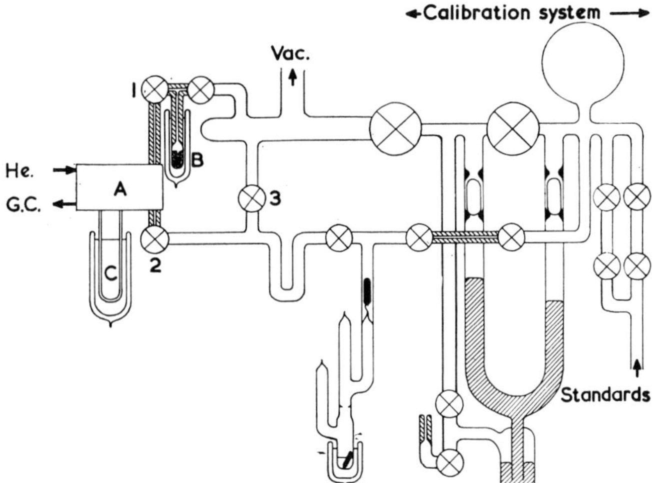


Fig. 1. Gas fractionation and concentration system. Methane and permanent gases are condensed in trap B. The less volatile gases are condensed in the stainless steel sample loop (C). Valve 2 is closed and the less volatile gases may be analyzed by gas chromatography. The gas sampling valve (A) is returned to the "fill" position and residual gases pumped away through valves 2 and 3. Valve 2 is closed and valve 1 opened to allow the volatile fraction to be analyzed (see Abell *et al.*, 1971).

At regular intervals a standard mixture of $CD_4$ and $CH_4$ was analyzed to ensure that the calibration factors remained constant. The overall error in the analytical sequence was $\pm 20\%$.

*Dry sieving of fines*

The sieving assemblage comprised a series of open stainless steel barrels (each 2.0 cm deep × 1.25 cm i.d.) with threaded outer surfaces. Stainless steel mesh of various aperture sizes (152, 108, 77, and 48 $\mu$, in sequence) and aluminum foil (penultimate barrel) was clamped between the barrels by means of threaded brass collars, the entire assemblage containing seven barrels. An aliquot of Apollo 11 fines (10086 bulk fines D, 0.5294 g) was sieved into the five fractions by vibrating (30 min) the assemblage with a vibrator (engraving tool, Burgess Products Ltd.). The sieves were dismantled and the separated fractions weighed (Table 2). The $> 152\ \mu$ fraction was further separated in a similar manner (0.5 mm mesh) into two fractions (152–500 $\mu$ and 500–about 2000 $\mu$).

## Results

*Comparison of methane concentrations released from Apollo 11 fines by different acids*

Table 1 lists the concentrations of $CH_4$ and $CD_4$ released by dissolution of aliquots of 10086 bulk fines in different mineral acids. HF (40%), a mixture (1:1) of HF (40%), and HCl (36%), and one batch (Batch 1) of DF in $D_2O$ (20%, Merck, Sharp, and Dohme; isotopic purity $>99.5\%$) afforded total methane concentrations in the range 20–32 $\mu$g/g (Abell *et al.*, 1971). A second batch of 20% DF (Batch 2, specification as above) whose strength was confirmed by titration, afforded a drastically reduced estimate of the carbide concentration as $CD_4$; in contrast, the recovery of $CH_4$ and $C_2$ hydrocarbons was only slightly lower. A stronger solution of DF in

Table 2. Concentrations of $CH_4$ and $CD_4$ released by acid dissolution (DCl in $D_2O$, 38%, 100°C, 2 hr) of lunar fines and breccias.

| Sample | Total carbon (ppm)* | Weight (mg) | $CD_4$ ($\mu$g/g) | $CH_4$ ($\mu$g/g) | $CD_4$:$CH_4$ |
|---|---|---|---|---|---|
| Fines | | | | | |
| 10086 | 140, 225 | 16–34 | 17.5–18.7 | 4.2–4.9 | 3.7–4.5 |
| 12032 | 25–86 | 57.5 | 2.5 | 0.34 | 7.4 |
| 12033 | 23–60 | 56.4 | 1.4 | 0.11 | 12.7 |
| 12070 | n.m. | 9.7 | 6.7 | 1.1 | 5.9 |
| 12025/12028 | | | | | |
| 59 9.0–9.4 cm (VIII) | 150 | 13.8 | 10 | 2.1 | 4.8 |
| 173 20.8–21.8 cm (IV) | 150 | 10.5 | 5.0 | 1.3 | 3.9 |
| 176 25.4–26.1 cm (III) | 160 | 17.6 | 5.7 | 1.7 | 3.4 |
| 180 28.8–30.0 cm (III) | 100 | 10.6 | 7.6 | 1.7 | 4.5 |
| 183 30.0–30.6 cm (III) | 160, 140 | 11.9 | 7.7 | 2.0 | 3.9 |
| 185 30.6–31.2 cm (III) | n.m. | 9.8 | 10 | 2.6 | 3.9 |
| 187 36.7–37.2 cm (II) | 170 | 24.1 | 8.0 | 2.0 | 4.0 |
| 188 39.2–39.8 cm (II) | n.m. | 17.5 | 3.6 | 0.96 | 3.8 |
| 10086† | | | | | |
| 500–about 2000 $\mu$ | n.m. | 23.2 | 24 | 6.9 | 3.6 |
| 152–500 $\mu$ | n.m. | 19.2 | 13 | 2.3 | 5.6 |
| 108–152 $\mu$ | n.m. | 4.9 | 11 | 2.3 | 4.9 |
| 77–108 $\mu$ | n.m. | 11.8 | 17 | 3.4 | 5.0 |
| 48–77 $\mu$ | n.m. | 19.0 | 23 | 5.2 | 4.5 |
| $<48\ \mu$ | n.m. | 4.9, 5.9 | 25.5, 23 | 6.7, 6.1 | 3.8 |
| 14240 | 89 | 9.4 | 2.6 | 1.0 | 2.6 |
| 14148 | 160 | 9.8 | 11.6 | 3.0 | 3.8 |
| 14149 | 135 | 18.1 | 8.7 | 3.0 | 2.9 |
| 14156 | 105, 186 | 8.7 | 13.0 | 4.0 | 3.2 |
| 14163 | 112, 120 | 18.2 | 7.7 | 2.6 | 3.0 |
| 14298 | 138, 145 | 11.5 | 13.0 | 3.6 | 3.8 |
| 14141 | 42, 80 | 53.1 | 1.1 | 0.50 | 2.2 |
| 15231 | n.m. | 6.0 | 14.8 | 5.1 | 2.9 |
| 15261 | n.m. | 5.8 | 14.6 | 4.9 | 3.0 |
| 15041 | 160 | 5.5 | 23.5 | 4.6 | 5.1 |
| 15031 | 130 | 12.6 | 16.4 | 4.0 | 4.1 |
| 15431 | n.m. | 23.1 | 9.5 | 3.2 | 3.0 |
| 15471 | 90 | 8.4 | 11.5 | 3.2 | 3.6 |
| 15501 | 130 | 12.5 | 14.7 | 4.1 | 3.6 |
| Breccias | | | | | |
| 10059 | n.m. | 19.3, 30.1 | 20, 17 | 5.0, 5.2 | 4.0, 3.3 |
| 14063 | 80 | 41.5 | 0.07 | $<0.04$ | n.m. |
| 14267 | n.m. | 80.5, 64.6 | 0.11, 0.15 | n.d. | n.m. |
| 14311 | 44 | 14.2, 19.4 | 0.80, 0.37 | n.d. | n.m. |
| 14313 | 130, 170 | 12.6 | 9.9 | 3.3 | 3.0 |
| 14321 | 21–43 | 167.0 | 0.29 | n.d. | n.m. |

* Moore *et al.* (1970, 1971); Kaplan and Petrowski (1971); Epstein and Taylor (1971); Sakai *et al.* (1972); LSPET (1971, 1972).

†Weighted average of sieved fractions: $CH_4$, 4.9 $\mu$g/g; $CD_4$, 21 $\mu$g/g.

n.d. = not detected. n.m. = not measured.

$D_2O$ (30%, courtesy of B. R. Simoneit, University of California, Berkeley) and a mixture (1:1) of DF (30%) and DCl (38%) in $D_2O$ released intermediate concentrations of $CD_4$.

## *Contaminants and artifacts*

A series of experiments was carried out to investigate possible sources of contamination via hydrolysis of lunar carbides by adsorbed terrestrial $H_2O$, or adsorption of $CH_4$ from the terrestrial atmosphere. First, exposure (24 hours) of 10086 fines to an atmosphere of $CD_4$ (about 20 Torr) followed by dissolution in HF showed that only a trace of the $CD_4$ ($<0.2$ $\mu$g/g) was not desorbed by the standard degassing procedures. Second, analysis by the standard procedure of the gases released from freshly exposed interior chips of a breccia (10059) afforded concentrations of $CH_4$ and $CD_4$ similar to those observed for 10086 fines (Table 2). The breccia, therefore, contains $CH_4$ and carbide in the interior at concentrations similar to those in the fines. Third, immersion of 10086 fines in $D_2O$ for 41 days did not generate $CD_4$; dissolution in HF of another aliquot (after removal of $D_2O$) of fines exposed to $D_2O$ for 90 days also failed to reveal any $CD_4$.

### *Size fractionation of Apollo 11 fines*

Dry sieving of the aliquot of 10086 fines was carried out to minimize contamination of the sample. The cumulative particle-size distribution curve obtained was similar to those obtained by LSPET (1969) and King *et al.* (1971) over the particle-size range investigated. The mean grain diameter was 74 $\mu$. The concentrations of $CD_4$ and $CH_4$ released by dissolution of the fractions are shown in Table 2. When these concentrations are plotted against the reciprocal of the mean grain radius (half of the arithmetic mean of the limiting mesh opening) of the fractions there exists a linear relationship for the three fractions in the range 48–152 $\mu$ diameter (Fig. 2). This surface area relationship, however, is no longer valid for particles $>152$ $\mu$ diameter (Fig. 2). It is possible that this is a result of the presence of particles having composite grain surfaces (see below).

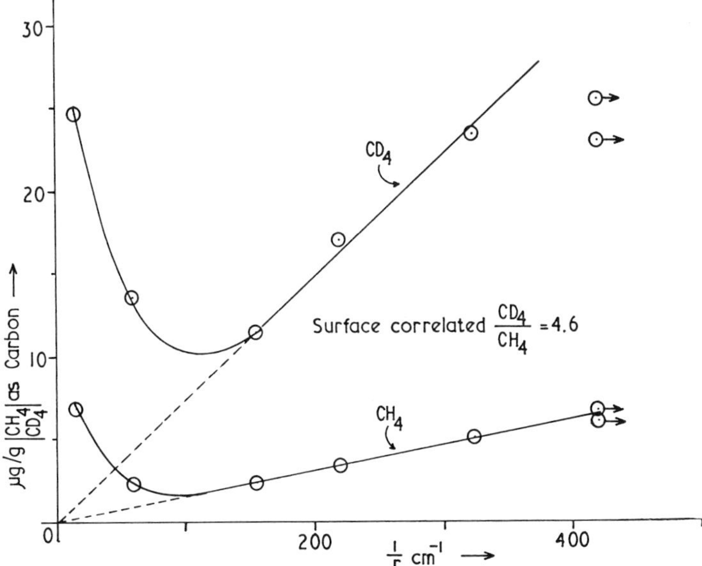


Fig. 2. Plot of the concentrations of $CD_4$ and $CH_4$ released by DCl dissolution versus the reciprocal of the mean grain radius of sieved fractions (10086 fines).

*Survey of $CH_4$ and carbide in fines and breccias*

Table 2 lists the concentrations of $CD_4$ and $CH_4$ released by DCl dissolution of aliquots of Apollo 11, 12, 14, and 15 fines, and of Apollo 11 and 14 breccias. The significance of the observed variations between samples is discussed below.

## Discussion

*Variations in $CH_4$ and $CD_4$ released by dissolution in different acids*

The data in Table 1 indicate that it is likely that the ionic strength of the mineral acid (and possibly also the concentrations of various ionic impurities present) affects the efficiency of carbide hydrolysis (Putnam and Kobe, 1937). The yields of $CH_4$ are less affected, as expected for a trapped species. Holland *et al.* (1972a), using 30% DF, have reported concentrations of $CD_4$, released from fines samples 12001, 12042, 12023, and 12032, which are lower by a factor of 6 than our previous results (DF, 20%; manufacturer's specification, Batch 1) obtained with the same samples (Abell *et al.*, 1971). They have suggested that this is a consequence of the increased reaction time used by us. We suggest that a more likely explanation is that the manufacturer's specification of our DF (Batch 1) was inaccurate and that the ionic strength was greater than 20% (cf. HF, 40%; result, Table 1). This explanation is in agreement with the yields of $CD_4$ obtained in the present study from dissolution of Apollo 11 fines with an aliquot of 30% DF as used by Holland *et al.*, (1972b; Table 1). The concentration of $CD_4$ released was higher than that from dissolution in DF (20%, Batch 2, checked by titration) but lower than that from dissolution in HF (40%). These data indicate that care should be taken in comparing the results obtained by different investigators. In the present study, repeated analyses of Apollo 11 fines and duplicate analyses of an Apollo 11 breccia (10059) with different batches of DCl (38%) afforded efficient and more reproducible recoveries of hydrocarbons and deuterocarbons (Tables 1 and 2). Thus, dissolution with this reagent at 100°C was adopted as the standard method for estimating $CH_4$ and carbide concentrations in all the samples examined (Table 2). The development of new reagents, which will afford a more quantitative conversion of lunar carbide into volatile compounds, is desirable. Even with the standard method of dissolution, meteoritic cohenite afforded only a 6% yield of the carbon as $CD_4$. Small variations in the $CD_4$ to $CH_4$ ratios measured for the same samples of fines and breccias are probably real and reflect sample inhomogeneity, rather than any effect of the acid.

*Contaminants and artifacts*

Oró *et al.* (1971) have suggested that the $CH_4$ released from lunar fines by deuterated acid dissolution could arise from hydrolysis of lunar carbide by adsorbed terrestrial water. The two experiments involving exposure of Apollo 11 fines to $D_2O$ indicate, however, that adsorbed $H_2O$ does not react with the samples at ambient to form $CH_4$ which is released from the particles immediately or by subsequent acid dissolution. Also, the chips of an Apollo 11 regolith breccia (10059) used were freshly exposed and were not crushed prior to dissolution. Thus, although the surface area to mass ratio was low, $CH_4$ and $CD_4$ were released in concentrations similar to those from the Apollo 11 fines (Table 2). Two other groups of investigators have reached similar conclusions using different methods. Flory *et al.* (1972) exposed a sample of 14240 fines (SESC) to $D_2O$ vapor and examined the products of volatilization by combined gas chromatography–mass spectrometry. No $CD_4$ was detected in the products. Holland *et al.* (1972b) found that DF treatment of an aliquot of the same sample, which had received minimum exposure to terrestrial $H_2O$, afforded $CH_4$ in concentrations similar to those observed from pyrolysis in the temperature range 500°–800°C. Also, electron microscopic examination of Apollo 14 fines does not reveal any evidence of reaction after exposure to water vapor (Cadenhead *et al.*, 1972) although the presence of certain major elements (at ppm concentrations) in deionized water after exposure of 12070 fines for 81 days shows that some dissolution must occur (Keller and Huang, 1971).

Another possible source of contamination is adsorption of $CH_4$ from the terrestrial atmosphere. This is unlikely because $CD_4$ can be efficiently removed from Apollo 11 fines by the standard degassing procedure, after exposure to concentrations vastly in excess of those of $CH_4$ in the terrestial atmosphere. Flory *et al.* (1972) believe, however, that adsorbed $CH_4$ is a source of contamination since samples of fines with long terrestrial exposure times showed a decrease in $CD_4/CH_4$ ratios. Our data do not support this hypothesis; aliquots of fines (12032, 12033) with the *lowest* $CH_4$ contents showed no significant increase in $CH_4$ when analyzed at an interval of one year (Abell *et al.*, 1971, and Table 2). Also there is no reason to suppose that $CH_4$ is irreversibly adsorbed by the fines since a number of other gases (Ar, $O_2$, $N_2$, and CO) show completely reversible adsorption isotherms (Fuller *et al.*, 1971).

Vigorous heating of DCl (200°C, 2 hours) with fragments of laboratory glass released CO, indicating that the single high concentration of CO previously reported for an aliquot of 10086 fines was an artifact, as suspected (Abell *et al.*, 1971). No CO was generated in the standard reaction procedure blank (100°C, 2 hours). The quantities of CO released by the standard DCl procedure from a variety of Apollo 11, 12, and 14 samples certainly do not exceed 3 $\mu$g/g, and are usually of the order 1 $\mu$g/g.

### *Location studies*

The linear relationships between the $CH_4$ and $CD_4$ concentrations and the reciprocal of the radius for Apollo 11 particles in the size range 48–152 $\mu$ mean grain diameter show that there is a surface relationship for both carbon species ($CH_4$ and carbide); the surface concentration is constant for both species (Fig. 2). This is in agreement with our earlier sieving experiment (Abell *et al.* 1970a). Holland *et al.* (1972a) have reached a similar conclusion from an examination of the gases released by DF dissolution of a sample of Apollo 14 fines (14240, SESC); $CH_4$ was linearly related to the reciprocal of the radius for particles from 37 to 420 $\mu$ diameter. The $C_2H_6$ released was dependent on the reciprocal of a higher power of the radius, suggesting a bimolecular synthetic process. Moore *et al.* (1970) and Kaplan *et al.* (1970) have shown that total carbon in the Apollo 11 samples is also surface correlated. A number of other investigators have observed a similar surface area relationship for rare gases attributed to solar wind implantation (Eberhardt *et al.*, 1970; Heymann and Yaniv, 1970a; Hintenberger *et al.*, 1970, 1971; Kirsten *et al.*, 1970, 1971). The surface concentrated $CH_4$ in Apollo 11 and 14 fines is therefore consistent with a solar wind origin, the carbide being consistent with both a solar wind and a meteoritic origin (see below). A similar method has been used by Muller (1972) to suggest that chemically bound nitrogen is solar wind in origin.

It appears that for the finest particles, the concentrations of $CH_4$ and carbide may be less than those expected from extrapolation of the linear relationships, the grain size plotted (48 $\mu$) for the finest fraction being the maximum diameter and not the mean grain diameter of the particles (Fig. 2). A similar effect has been observed by Eberhardt *et al.* (1970), Heymann and Yaniv (1970a), Hintenberger *et al.* (1970, 1971), and Kirsten *et al.* (1971) with rare gases. Further investigation of the effect requires wet sieving methods, which should also minimize contributions from fine particles adhering to larger grains.

The coarser particles ($>152\ \mu$) show methane and carbide concentrations in excess of the anticipated surface-related component, indicating that there is a volume-related component which increases with increasing grain size (Fig. 2). Again, this effect has been observed in Apollo 12 fines (12070) for solar wind rare gases (Hintenberger *et al.*, 1971). There is no evidence to suggest that there should be higher concentrations of $CH_4$ and carbide in the coarser igneous fragments. Indeed the quantity of $CH_4$ and of carbide measured in an igneous rock (12022) was very small (Abell *et al.*, 1971). The volume related $CH_4$ and $CD_4$ components probably arise from the presence of particles with composite grain surfaces (microbreccias and glassy aggregates). One explanation would be a higher proportion of these particles in the coarser fractions; this, however, does not appear to be the case for the particle-size range investigated (Quaide and Bunch, 1970; Duke *et al.*, 1970; McKay *et al.*, 1970). It appears likely, therefore, that these larger composite particles (152–2000 $\mu$ diameter) contain concentrations of $CH_4$ and carbide which are higher than those in the finer fractions. In Apollo 11 samples, LSPET (1969) and Funkhouser *et al.* (1970), and Moore *et al.* (1970) and Kaplan *et al.* (1970) have observed higher concentrations of rare gases and total carbon, respectively, for some of the breccias compared to the fines. Also, a 4 mm microbreccia selected from Apollo 11 fines afforded very high concentrations of $CH_4$ and $CD_4$ (5.6 and 24 $\mu$g/g). Several mechanisms can be invoked to explain these observations; e.g. (1) these breccias were formed from fines containing higher concentrations of $CH_4$ and carbide. The soil could derive from a site not yet investigated or represent "fossil" material from the Apollo 11 site. The high concentrations may have arisen either from increased exposure of a thinner regolith or from increased activity in the past of the major processes giving rise to both carbon species. (2) Enrichment could have occurred during breccia formation in the presence of a gas phase (Heymann and Yaniv, 1970a) or losses of $CH_4$ and carbide may have occurred during comminution processes. (3) The coarse particles were aggregated from very fine particles which had experienced a long surface exposure.

At present it is difficult to distinguish among these possibilities and detailed analyses of individual microbreccias and breccias, including determination of "formation ages" (Fleischer *et al.*, 1972), are necessary.

*Correlation studies*

Analysis of the $CH_4$ and carbide from a wide variety of fines and breccias shows large variations in concentrations between samples. This allows extension of our previous preliminary correlations between the concentrations measured and parameters indicative of lunar surface exposure of the samples (Abell *et al.*, 1971; Cadogan *et al.*, 1971). All available measurements of $^{36}Ar$ content, fractions of grains with track densities $>10^8$ tracks $cm^{-2}$, and fractions of grains in the finest fines with amorphous coatings, have been utilized for the samples listed in Table 2 and those previously analyzed. A discussion of the applicability of these correlations has been given previously (Abell *et al.*, 1971; Cadogan *et al.*, 1971). Two groups of samples, namely the Apollo 12 double core (12025/12028) and the Apollo 14 breccias, are discussed separately.

Figure 3 shows the relationships between $CH_4$ and $CD_4$ and $^{36}Ar$ concentrations for Apollo 11, 12, and 14 samples. There is an approximately linear relationship

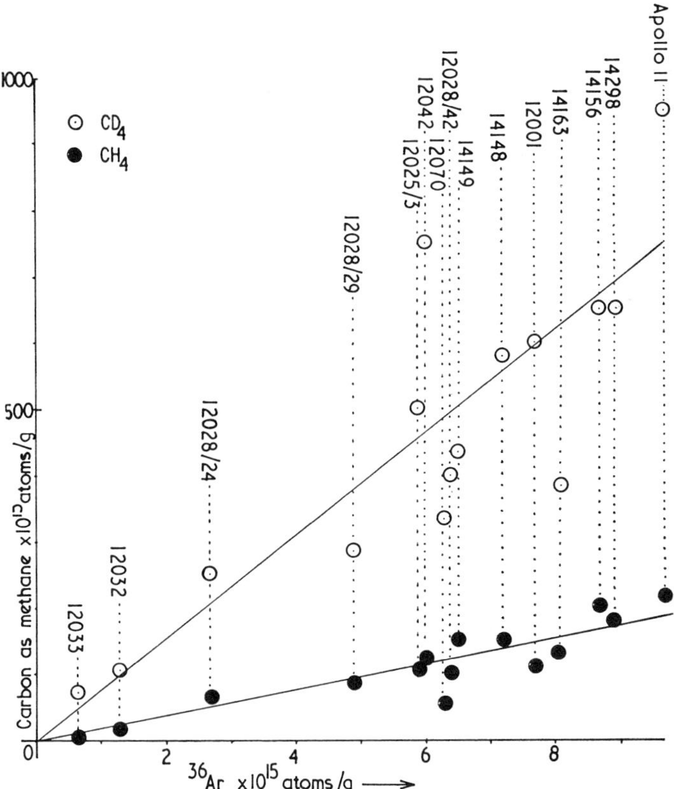


Fig. 3. Concentration of $CD_4$ and $CH_4$ as carbon ($10^{15}$ atoms/g) versus concentration of $^{36}Ar$ ($10^{15}$ atoms/g) in Apollo 11, 12, and 14 samples ($^{36}Ar$ data from Hintenberger *et al.*, 1971; Marti, 1971; LSPET, 1971, 1972; Pepin, 1971; Bogard, 1972; Funkhouser, 1971).

between $CH_4$ and solar wind $^{36}Ar$, consistent with a solar wind origin for the $CH_4$. The relationship for $CD_4$ also appears consistent with the proposed extralunar origins for the carbide (see below), although there is a greater degree of scatter. A tentative correlation between total carbon and solar wind $^{20}Ne$ in Apollo 14 fines and breccias has been reported by LSPET (1971).

Particles with very high solar flare cosmic ray track densities have a high probability of having experienced unshielded surface exposure during their history; that is, a high probability of exposure to the solar wind (Arrhenius *et al.*, 1971). Figure 4 shows a plot of $CH_4$ concentrations against the fraction of grains having these high track densities in a particular sample of fines. Samples showing a high proportion of grains with high track densities also have high $CH_4$ concentrations. The proportions of grains showing high track densities measured by Bhandari (1972) for the Apollo 14 samples (14141, 14148, 14163, 14298) are consistently lower than those measured by other investigators for the same samples. The relationship in Fig. 4 shows a tendency toward track density saturation, as would be expected, i.e., the $CH_4$ concentration continues to increase, whereas the proportion of grains with track densities greater than $10^8$ tracks $cm^{-2}$ cannot exceed 100%. Optical counting methods do not allow resolution of track densities > about $10^8$ tracks $cm^{-2}$; clearly, a more accurate estimate of unshielded surface exposure is the proportion of particles with extremely high track densities (>10-in. tracks $cm^{-2}$, as measured by electron microscopy) which themselves correlate with the proportions of particles showing solar wind radiation damage in the form of amorphous coatings (see below; Borg *et al.*, 1972). At present insufficient data are available for correlation of these track densities with $CH_4$ concentration.

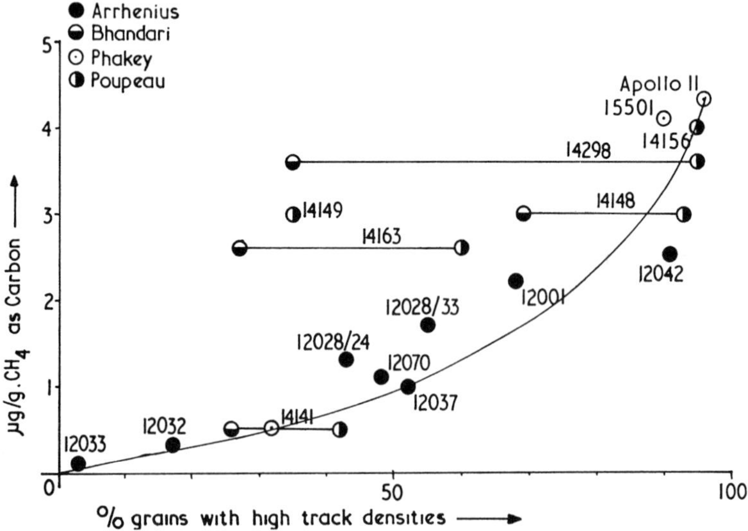


Fig. 4. $CH_4$ as carbon ($\mu g/g$) in total sample versus percentage mineral grains having track densities $>10^8$ $cm^{-2}$. (Track density data from Arrhenius *et al.*, 1971; Bhandari, 1972; Phakey *et al.*, 1972; Poupeau *et al.*, 1972.)

The $CD_4$ concentrations again show a greater degree of scatter although a similar trend to that shown in Fig. 4 is evident.

Sample 14149, from the bottom of the soil mechanics trench, is a notable exception (Fig. 4); it has a relatively high $CH_4$ and $^{36}Ar$ concentrations (Fig. 3) but contains only a small proportion of particles having high track densities (Crozaz *et al.*, 1972; Poupeau *et al.*, 1972; Phakey *et al.*, 1972). It also has a high median grain size and a low proportion of glass, indicating that the degree of reworking is low (LSPET, 1971). It has been suggested (Crozaz *et al.*, 1972) that this sample constitutes Cone Crater ejecta and was covered by older, more irradiated samples (14148, 14156). However, sample 14149 contains high concentrations of short-lived spallogenic nuclides ($^{26}Al$ and $^{22}Na$) which have been attributed to sample mixing during collection (Eldridge *et al.*, 1972). Measurements made on bulk samples (e.g., $CH_4$, $^{36}Ar$, $^{26}Al$, $^{22}Na$) are more likely to be affected by sample mixing contamination than are measurements carried out on selected particles (e.g., track densities). It appears that sample 14149 probably constitutes low exposure Fra Mauro fines (typified by Cone Crater ejecta) cross contaminated with more exposed fine material from the surface of the trench, as a result of crumbling of the sampling face. Sample 14240 (SESC), also from the trench bottom, and having a high median grain size (Holland *et al.*, 1972a, b) is probably more representative of Cone Crater ejecta (Table 2), but at present measurements are not available for track density or for $^{36}Ar$, $^{26}Al$, and $^{22}Na$; these measurements should correlate with $CH_4$ and $CD_4$ concentrations and indicate low surface exposure. The $CD_4$ to $CH_4$ ratios provide further evidence for sample mixing (Table 2). Sample 14141 (Cone Crater ejecta) has the lowest ratio (2.2) measured for any sample at the Apollo 14 site, whereas the upper layers of the trench (14148 and 14156) have ratios of 3.8 and 3.2, respectively. Sample 14149 has an intermediate ratio (2.9), whereas 14240 is still intermediate (2.6) but tends towards the ratio for 14141.

Sample 14163 (the bulk fines surface sample scooped down to a depth of several centimeters) shows some deviation from the proposed correlation in Fig. 4, which can also be explained by sample mixing. The $CD_4$ to $CH_4$ ratio (3.0) is lower than that (3.8) of the comprehensive surface sample (14298) collected nearby from the top 1 cm of the regolith. This is explicable in terms of mixing with a sample such as 14141, having a lower ratio (cf. 14149 above). This suggests that the unexposed Fra Mauro fine material may be close to the surface at this sampling site. It appears, therefore, that a two-component mixing model (based on 14298 and 14141) can explain the carbon chemistry of the Apollo 14 soil samples.

The available measurements (Maurette, 1972) of the fractions of grains in the 1–2 $\mu$ size range having amorphous coatings are compared in Fig. 5 with the $CH_4$ concentrations in the unsieved samples for Apollo 11, 12, and 14 fines; a similar, but less well defined, trend is evident for $CD_4$. The data support our previous preliminary hypothesis (Abell *et al.*, 1971; Cadogan *et al.*, 1971) that samples having high proportions of particles with amorphous coatings, attributed (Borg *et al.*, 1971) to solar wind radiation damage, have high $CH_4$ contents. The new data provide further evidence for the participation of the solar wind in $CH_4$ and carbide synthesis. Samples 14163 and 14149 show a deviation for the correlation in Fig. 5. This can be rationalized by the mixing proposed above.

Factors related to the degree of reworking of the regolith (median grain size,

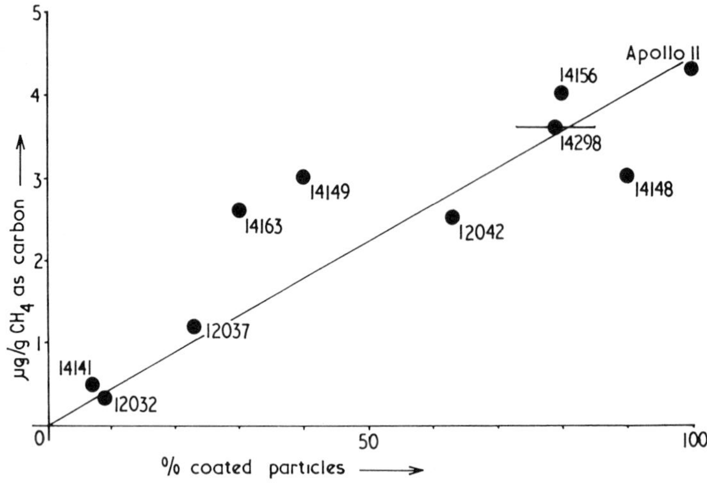


Fig. 5. Bulk concentration of $CH_4$ as carbon ($\mu$g/g) versus percentage of grains of 1–2 $\mu$ diameter in the 400 mesh residues showing amorphous coatings (Maurette, 1972) for Apollo 11, 12, and 14 fines.

sorting index, proportion of glass, microbreccias, and glassy aggregates, etc.) provide a general indication of the maturity of the soil, and have not been plotted against $CH_4$ or $CD_4$. The information available, however, about the degree of reworking of the Apollo 14 fines samples confirms the trends observed previously for various Apollo 11 and 12 samples (Abell *et al.*, 1971; Cadogan *et al.*, 1971). Samples releasing relatively high $CH_4$ and $CD_4$ concentrations (e.g., 14148, 14163, and 14298; Table 2) are the most reworked (McKay *et al.*, 1972). The sample of Cone Crater fines (14141) is the least reworked and releases low concentrations of $CH_4$ and $CD_4$ (Table 2). Another trend reported previously, relating $CH_4$ concentration to the fractionation of carbon isotopes, cannot be extended at present because insufficient data are available (Cadogan *et al.*, 1971). Few data for $^{36}$Ar, track densities, amorphous coatings, and degree of reworking are available for the Apollo 15 samples listed in Table 2. The $CH_4$ (and possibly $CD_4$) concentrations appear to indicate, in general, that the regolith at the Apollo 15 site shows a high degree of maturity in terms of turnover and concomittant exposure. Samples collected at the edge of Hadley Rille (Stations 9A and 10) or from the Apennine Front (e.g., 15231, 15261, 15431) may be reworked by downslope gravity transport (Phakey *et al.*, 1972), in addition to the processes of meteorite bombardment (McKay *et al.*, 1971; Quaide *et al.*, 1971) and electrostatic redistribution effects (Gold, 1971). Thus, some samples (e.g., 15231, 15261) may be found to have high values for other exposure parameters similar to those observed for the Apollo 11 fines.

### *Origins of $CH_4$ and carbide in the fines*

The detailed correlations between $CH_4$ and carbide and exposure parameters, and the location studies with sieved fractions confirm our earlier data which indicated an extralunar origin for both species (Abell *et al.*, 1971; Cadogan *et al.*, 1971).

The major contribution to the $CH_4$, and presumably to the smaller concentrations of other hydrocarbons present, arises from reactions involving species contributed by the solar wind. Possible mechanisms have been discussed previously (Abell *et al.*, 1970a, b). Simulation studies of solar wind implantation suggest that at least two of these mechanisms operate, namely hydrogen from the solar wind reacts with carbon, either from the solar wind or already present at the surface of the grains, to produce $CH_4$ (Pillinger *et al.*, 1972). A contribution to the $CH_4$ from $CH_4$ in meteorites is unlikely (Abell *et al.*, 1971; Cadogan *et al.*, 1971). We have also suggested previously that there might be a small contribution from a primordial (igneous rock) source, but we are unable to estimate gas losses that might occur during the formation of fines. Since no measurable quantities of $CH_4$ were detected in the metamorphic Fra Mauro breccias, any $CH_4$, if present, must have been lost during impact events. A similar severe impact giving rise to fine material, would also result in extensive loss of primordial $CH_4$. Also, an investigation of the depletion of alkali elements during the formation of lunar glasses suggests that similarly volatile elements, including carbon, might be expected to be lost (Gibson and Hubbard, 1972). Other processes giving rise to fines are less severe but, overall, primordial $CH_4$ would be greatly depleted.

The situation with respect to the carbide is more complex. Two extralunar sources are likely, namely solar wind implantation and meteorite impact. Laboratory-based simulation studies show that both are feasible. Material reacting as carbide can be synthesized by implantation of carbon at solar wind energies into metal targets. Vacuum vaporization of the iron carbide in an iron filament gives rise to material reacting as carbide in the deposited iron film, indicating that impacting meteorites containing carbide could contribute "carbides" to the lunar surface (Pillinger *et al.*, 1972). Alternatively, the iron phase could arise by impact heating under reducing conditions (Carter, 1972; Housley *et al.*, 1972); "carbide" could thus be formed by dissolution of siderophilic carbon in the iron.

Micrometeorite impact is a major reworking process leading to exposure of lunar material to the solar wind. Contributions from the meteorites and from the solar wind should increase in parallel. The correlation and location studies, therefore do not distinguish between the two origins for carbide. The carbide concentration tends to increase with the concentrations of a number of volatile and siderophile elements (Ir, Au, Se, Ag, Bi) indicative of the micrometeorite contributions (Ganapathy *et al.*, 1970). Figure 6 shows a plot of $CD_4$ concentration against Bi concentration for a number of Apollo 11, 12, 14, and 15 samples. A trend is evident but again it is impossible to distinguish origins. A high Bi content indicates an increased meteorite contribution and greater probability of exposure to the solar wind. The situation may be further complicated either by direct enrichment from components rich in siderophile and volatile elements (Morgan *et al.*, 1972) or by vapor transfer of volatile elements (Silver, 1972; Allen *et al.*, 1972; Epstein and Taylor, 1972).

Previously, two pieces of evidence indicated that the meteorite contribution to carbide might be small (Cadogan *et al.*, 1971). First, DF dissolution of meteoritic cohenite afforded only 1% to 2% of the carbon as $CD_4$. A similar yield from lunar carbide would require total carbon contents an order of magnitude in excess of those observed. Dissolution of cohenite by the standard DCl method, however, affords a 6% yield of the carbon as $CD_4$. The carbon as lunar carbide, calculated on this

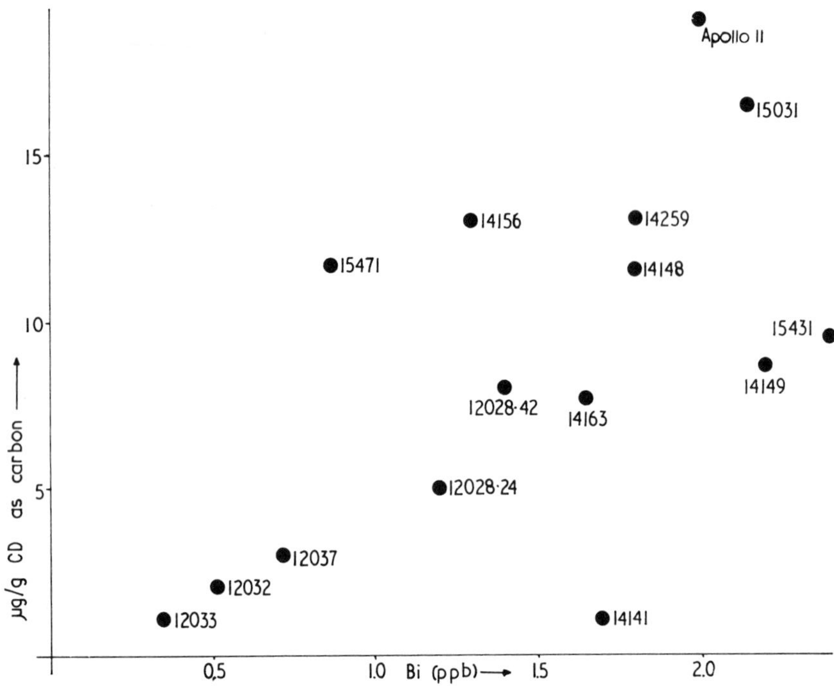


Fig. 6. Plot of $CD_4$ as carbon ($\mu$g/g) versus bismuth concentrations (ppb) in Apollo 11, 12, 14, and 15 fines (bismuth data from Laul *et al.*, 1971, and Morgan *et al.*, 1972).

increased yield, does not greatly exceed the observed total carbon measurements, and in some cases (e.g., 14163, 14148) is equivalent to total carbon. Second, isotope ratio measurements suggested that, overall, lunar carbide is distinguishable from meteoritic carbide (Chang *et al.*, 1971). Two $\delta^{13}C$ values measured for the total hydrocarbons released by HCl dissolution of 12023 fines (+5, +14‰) are greater than those reported for meteoritic carbide (−4 to −8‰). Meteoritic carbide, however, could be fractionated on the lunar surface and two processes have been suggested: (1) A solar wind hydrogen stripping process, involving synthesis and removal of volatile hydrogenated species, could lead to an enrichment in $^{13}C$ (Chang *et al.*, 1971); (2) Enrichment in the heavier isotopic species, could occur by preferential condensation during vaporization and deposition of meteoritic carbide (Pillinger *et al.*, 1972). Simulation studies show, in favor of these processes, that carbon in lunar fines could be converted to methane by solar wind hydrogen and that carbide can be vaporized and deposited by impact conditions on the lunar surface (Pillinger *et al.*, 1972), although the isotopic fractionation processes taking place during these simulations require investigation. On the other hand, progressive etching of 14298 fines with $H_2SO_4$ shows a parallel enrichment in $^{13}C$ in the gases released (Sakai *et al.*, 1972).

The contributions to lunar carbide from the solar wind and meteorites are unknown. Although the relative abundances of carbon and rare gases in the solar wind have not been determined accurately and nothing is known about relative diffusion losses of carbon-containing species and rare gases, calculations have been made of the solar wind carbon contribution to lunar fines. Perhaps the best estimate, 50 ppm, is that of Hayes (1972). Thus, solar wind implantation alone would not be expected to account for all the lunar carbide (based on a 6% yield of $CD_4$ on acid hydrolysis).

Perhaps the best evidence of a meteoritic contribution to lunar carbide is the mineralogical observation in Apollo 11 fines of occasional grains of carbide with a meteoritic composition. Carbide with the same elemental composition as the cohenite in the Odessa meteorite has recently been observed in breccia 14321 (Reed *et al.*, 1972). Observations of meteoritic cohenite are restricted to iron meteorites (Brett, 1967). The maximum content of metallic iron in the fines is about 0.5% (Housley *et al.*, 1972); assuming that all the iron is meteoritic (unlikely) and that the maximum content of carbon of iron meteorites is 0.2% (Moore *et al.*, 1969), then the *maximum* contribution of carbide from this source is about 10 ppm. Acid hydrolysis (assuming a 6% yield) would afford only 0.6 $\mu$g/g of carbon as $CD_4$. It is likely, therefore, that if meteorites contribute significantly to lunar carbide the process must be an indirect one, such as the carbon of carbonaceous chondrites reacting with metallic iron.

Reimplantation of ions from a lunar atmosphere has been proposed as a possible explanation of the excess of $^{40}Ar$ in lunar fines and regolith breccias (Heymann and Yaniv, 1970b). If such a process occurs, it could have applicability to lunar carbon chemistry in that carbon compounds could be contributed to the atmosphere by sputtering processes, diffusion, meteorite impact, and volcanism. Ionization processes (exchange with solar wind) should result in fragmentation and formation of carbon ions which could be implanted into the fines. Carbon species synthesized in this way would be located at the surfaces of the fines and thus be indistinguishable (except by accurate depth measurement) from carbon species implanted by the solar wind. Processes contributing to the lunar atmosphere would fractionate in favor of $^{12}C$; the observation (Sakai *et al.*, 1972) of lighter carbon at the surfaces of grains could constitute evidence for a lunar atmosphere contribution. Hayes (1972) has calculated, by a circuitous route, the amount of carbon contributed from the lunar atmosphere to the fines. The calculation, which affords a value of 10 ppm, is misleading since it assumes a reimplantation "sticking factor" based on Heymann and Yaniv's (1970b) factor (1/50) for $^{40}Ar$. It is impossible to estimate the "sticking factor" for carbon by this method and there are other deficiencies in the approach.

### *Apollo 12 double core* (*12025/12028*)

The core has been described by LSPET (1971) and comprises 10 major stratigraphic units. The concentrations of $CD_4$ and $CH_4$ released from eight samples from various depths are listed in Table 2. In all cases, the ratio of $CD_4$ to $CH_4$ was about 4:1 although there was a threefold variation in the quantities released. There is no direct relationship between the concentration of either species with depth, in agreement with the concentrations of total methane released by undeuterated mineral acid (Henderson *et al.*, 1971). The highest values for $CH_4$ and $CD_4$ were measured in units VIII and III at depths of 9 to 9.4 cm and 30.6 to 31.2 cm, respectively. These are of the same order as the highest values measured for surface samples (12001, 12042) collected at the same site (Abell *et al.*, 1971; Cadogan *et al.*, 1971) and are indicative of higher surface exposure and more extensive reworking than for the other depths of the core (Table 2). The data of other investigators show that the core contains two units of minimal surface exposure (units VI and I) in terms of $^{36}Ar$ content (Funkhouser, 1971), proportions of grains with high track densities (Arrhenius *et al.*, 1971), and proportions of glassy aggregates (Sellers *et al.*, 1971; McKay *et al.*, 1971). Our allocation did not include samples from these two units but we expect that the concentrations of $CH_4$ and carbide would also be very low. These carbon species have, however, been measured for a number of samples between units VI and I; the concentrations increase from unit IV to a maximum at the 30.6 to 31.2 cm depth (about the middle of unit III) and fall to a minimum at the bottom of unit II (Table 2). This is explicable in terms of varying surface residence times for each layer after deposition and before burial, and mixing across the stratigraphic boundaries. Radionuclide measurements have shown that soil mixing can occur to a depth of about 3 cm (Rancitelli *et al.*, 1972). Thus, relatively unexposed material from unit I could be mixed with exposed material from units II and III; methane and carbide values are lower at depth 39.2 to 39.8 cm than depth 36.7 to 37.2 cm in unit II (Table 2). If mixing did occur the recognizable stratigraphy was not destroyed. This suggests that analysis of core samples, selected solely on a stratigraphic basis, is insufficient to obtain a detailed history of the deposition of the regolith. Layer by layer deposition, defining the stratigraphy, is the dominant process but small-scale vertical mixing processes may be equally important.

Unit I, possibly representative of the Copernican ejecta blanket of KREEP material (typified by sample 12033), is probably widespread and intermixed throughout the area sampled (Sellers *et al.*, 1971). This is reflected in the overall concentrations of $CH_4$ and $CD_4$ observed in the surface samples at the Apollo 12 site in comparison with those at the Apollo 11 mare site. The concentrations observed are probably low, due to the presence of a component of unexposed KREEP material. Thus there is an anticorrelation between $CH_4$ and carbide and the KREEP content estimated by Meyer *et al.* (1971). A similar effect might be expected from material typified by unit VI but there is no mineralogical evidence to suggest that this material, thought to be the ejecta from Bench Crater, is widely distributed at the site (Sellers *et al.*, 1971).

### *Apollo 14 breccias*

The Apollo 14 breccias have been classified into four basic types (F-1 to F-4) from visual examination of the exterior, and in some cases interior, surfaces (LSPET, 1971; Jackson and Wilshire, 1972). The classification, based on the character and distribution of the clasts present, describes the F-1 breccias as being porous and unshocked, F-2 as shock compressed, and both F-3 and F-4 as thermally metamorphosed but having different glass contents. Sample 14313 (F-2) contains relatively high concentrations of $CH_4$ and carbide, of the same order as 10059, a typical example of consolidated mare regolith (Table 2). This reflects their mild thermal and shock history. Samples 14063 (F-3), 14311, and 14321 (F-4), however, contain low concentrations of carbide and no $CH_4$; this indicates that the material constituting these samples either had minimal exposure to the solar wind or was drastically heated during formation of the breccias. It is generally accepted that the metamorphic breccias were transported to the Fra Mauro site in a hot turbulent jet (base surge) at the time of the impact that produced the Imbrium Basin, and there is evidence that the temperature reached in this dust and vapor cloud was about 1000°C (e.g., Anderson *et al.*, 1972). Hydrocarbons would certainly be lost at this temperature (Abell *et al.*, 1970b) and a high proportion of the other carbon species, including carbide, might be expected to be lost as gaseous species (Gibson and Johnson, 1971). Breccia 14267 (European Consortium rock) however, may have been incorrectly classified as F-2 by Jackson and Wilshire (1972). The $CH_4$ and carbide contents (Table 2) suggest that it should be classified as F-3 or F-4. The Apollo 14 breccias can also be classified according to solar rare gas content (Megrue and Steinbrunn, 1972) and track densities (Dran *et al.*, 1972); these classifications are in agreement with the carbon chemistry measurements. Perhaps the most comprehensive classification is that of Warner (1972) in which petrological description of 27 samples distinguished eight groups within three metamorphic facies. In contrast, classification based on the $CH_4$ and carbide concentrations in Table 2 can only distinguish the degree of metamorphism.

## Conclusions

The carbon chemistry of the lunar regolith has been established as a significant indicator of exposure and reworking. In conjunction with other parameters indicative of exposure of the fines on the lunar surface, it should contribute to an understanding of the complex history of the regolith. A variety of correlations, including those described herein, have shown that these parameters are inter-relatable (Poupeau *et al.*, 1972; Bibring *et al.*, 1972; Holland *et al.*, 1972a). A second approach involves laboratory-based simulation of the effects of lunar surface phenomena. Thus, irradiation of the fines with protons decreases the albedo (Dollfus *et al.*, 1972; Gold *et al.*, 1972), implantation with rare gases results in the formation of amorphous outer layers with concomittant rounding of the particles (Bibring *et al.*, 1972), and bombardment with electrons causes electrostatic transportation (Gold, 1972). Further, implantation of lunar fines and metal films with carbon and hydrogen species at solar wind energies results in the formation of methane and carbide (Pillinger *et al.*, 1972) and heating of simulated fines with sulfur, carbon, and hydrogen produces iron and iron sulfide mounds similar to those observed in the fines themselves (Carter, 1972).

The quantities of indigenous total carbon present in the igneous rocks are small. The contribution from this source to the carbon species in the fines is uncertain, although it is likely to be even smaller as a result of losses incurred during the formation of the fines. The mechanisms of generation of the carbon species observed require elucidation, as do the relative contributions made to lunar carbide in the fines and breccias by the solar wind and meteorites.

The carbon content of the species released by deuterated acid dissolution of lunar fines and breccias varies with sample from 2% to 20% of the total carbon as measured by pyrolysis and combustion. The major uncertainty in expressing the experimental data as carbon content of the species present in the samples is associated with defining the yield of $CD_4$ from lunar carbide. Our data for meteoritic cohenite provide a multiplication factor of 16 to convert $CD_4$ carbon to carbide total carbon. When applied to the available data for lunar samples this factor leads to estimates of indigenous hydrocarbons and carbides which account for 40% to 230% of the total carbon. Such a calculation is misleading because the precise nature of lunar carbides and their chemical environment remains unknown, and the $CD_4$ yield on deuterated acid dissolution may be very different from that observed with meteoritic carbide. However, at present, the factor calculated from cohenite represents the best available approximation. For a number of samples, the total carbon measured by combustion and pyrolysis approximately equals the total carbon calculated from the yield of acid dissolution products (hydrocarbons and carbides). Consequently, carbide in the fines and breccia may account for a substantial portion of the lunar carbon.

*Acknowledgments*—We thank the Science Research Council for financial assistance and the award of a Fellowship (C.T.P.), the Petroleum Research Fund of the American Chemical Society for a Research Studentship (P.H.C.), and the National Aeronautics and Space Administration for a Research Studentship (B.J.M.) from a subcontract for organic geochemical studies (NGL-05-003-003) made through the University of California, Berkeley. We are grateful to the following for communicating unpublished data: Professor E. Anders, University of Chicago; Dr. N. Bhandari, Tata Institute of Fundamental Research, Bombay; Dr. D. D. Bogard, Manned Spacecraft Center, Houston; Professor K. Marti, University of California, San Diego; Dr. M. Maurette, Centre de Spectrométrie de Masse du C.N.R.S., Orsay; Professor R. O. Pepin, University of Minnesota, Minneapolis; Professor P. Signer, Swiss Federal Institute of Technology, Zurich. We also thank M. S. Reid and I. A. Manning for technical assistance.

## REFERENCES

Abell P. I., Eglinton G., Maxwell J. R., Pillinger C. T., and Hayes J. M. (1970a) Indigenous lunar methane and ethane. *Nature* **226,** 251–252.

Abell P. I., Draffan G. H., Eglinton G., Hayes J. M., Maxwell J. R., and Pillinger C. T. (1970b) Organic analysis of the returned Apollo 11 lunar sample. *Proc. Apollo 11 Lunar Sci. Conf., Geochim. Cosmochim. Acta* Suppl. 1, Vol. 2, pp. 1757–1773. Pergamon.

Abell P. I., Cadogan P. H., Eglinton G., Maxwell J. R., and Pillinger C. T. (1971) Survey of lunar carbon compounds, I. The presence of indigenous gases and hydrolysable carbon compounds in Apollo 11 and Apollo 12 samples. *Proc. Second Lunar Sci. Conf., Geochim. Cosmochim. Acta* Suppl. 2, Vol. 2, pp. 1843–1863. MIT Press.

Allen R. O., Jovanovic S., and Reed G. W. (1972) $^{204}Pb$ in Apollo 14 samples (abstract). In *Lunar Science—III* (editor C. Watkins), pp. 15–17, Lunar Science Institute Contr. No. 88.

Anderson A. T., Braziunas T. F., Jacoby J., and Smith J. V. (1972) Breccia populations and thermal history: Nature of the pre-Imbrium crust and impacting body (abstract). In *Lunar Science—III* (editor C. Watkins), pp. 24–26, Lunar Science Institute Contr. No. 88.

Arrhenius G., Liang S., Macdougall D., Wilkening L., Bhandari N., Bhat S., Lal D., Rajagopalan G., Tamhane A. S., and Venkatavaradan V. S. (1971) The exposure history of the Apollo 12 regolith. *Proc. Second Lunar Sci. Conf., Geochim. Cosmochim. Acta* Suppl. 2, Vol. 3, pp. 2583–2598. MIT Press.

Bhandari N. (1972) Personal communication.

Bibring J. P., Maurette M., Meunier R., Durrieu L., Jouret C., and Eugster O. (1972) Solar wind implantation effects in the lunar regolith (abstract). In *Lunar Science—III* (editor C. Watkins), pp. 71–73, Lunar Science Institute Contr. No. 88.

Bogard D. D. (1972) Personal communication.

Borg J., Maurette M., Durrieu L., and Jouret C. (1971) Ultramicroscopic features in micron-sized lunar dust grains and cosmophysics. *Proc. Second Lunar Sci. Conf., Geochim. Cosmochim. Acta* Suppl. 2, Vol. 3, pp. 2027–2040. MIT Press.

Borg J., Maurette M., Durrieu L., Jouret C., Lacaze J., and Peter P. (1972) Search for low energy ($10 \lesssim E \lesssim 300$ keV/amu) nuclei in space: Evidence from track and electron diffraction studies in lunar dust grains and in Surveyor 3 material (abstract). In *Lunar Science—III* (editor C. Watkins), pp. 92–94, Lunar Science Institute Contr. No. 88.

Brett R. (1967) Cohenite: Its occurrence and proposed origin. *Geochim. Cosmochim. Acta* **31,** 143–159.

Cadenhead D. A., Wagner N. J., Jones B. R., and Stetter J. R. (1972) Some surface characteristics and gas interactions of Apollo 14 fines and rock fragments (abstract). In *Lunar Science—III* (editor C. Watkins), pp. 110–111, Lunar Science Institute Contr. No. 88.

Cadogan P. H., Eglinton G., Maxwell J. R., and Pillinger C. T. (1971) Carbon chemistry of the lunar surface. *Nature* **231,** 29–31.

Carter J. L. (1972) Morphology and chemical composition of metallic mounds produced by $H_2$ and C reduction of simulated lunar composition (abstract). In *Lunar Science—III* (editor C. Watkins), pp. 125–127, Lunar Science Institute Contr. No. 88.

Chang S., Kvenvolden K. A., Lawless J., Ponnamperuma C.; and Kaplan I. R. (1971) Carbon in an Apollo 12 sample; concentration, isotopic composition, pyrolysis products, and evidence for indigenous carbides and methane. *Science* **173,** 474–477.

Crozaz G., Drozd R., Hohenberg C. M., Hoyt H. P., Ragan D., Walker R. M., and Yuhas D. (1972) Solar flare and galactic cosmic ray studies of Apollo 14 samples (abstract). In *Lunar Science—III* (editor C. Watkins), pp. 167–169, Lunar Science Institute Contr. No. 88.

Dollfus A., Bowell E., Geake J. E., and Maurette M. (1972) Optical polarimetric and photometric studies (abstract). In *Lunar Science—III* (editor C. Watkins), pp. 181–182, Lunar Science Institute Contr. No. 88.

Dran J. C., Durand J. P., Maurette M., Durrieu L., Jouret C., and Legressus C. (1972) The high resolution track and texture record of lunar breccias and gas rich meteorites (abstract). In *Lunar Science—III* (editor C. Watkins), pp. 183–185, Lunar Science Institute Contr. No. 88.

Duke M. B., Woo C. C., Sellers G. A., Bird M. L., and Finkelman R. B. (1970) Genesis of lunar soil at Tranquility Base. *Proc. Apollo 11 Lunar Sci. Conf., Geochim. Cosmochim. Acta* Suppl. 1, Vol. 1, pp. 347–361. Pergamon.

Eberhardt P., Geiss J., Graf H., Grögler N., Krähenbühl U., Schwaller H., Schwarzmüller J., and Stettler A. (1970) Trapped solar wind noble gases, exposure age and K/Ar-age in Apollo 11 lunar fine material. *Proc. Apollo 11 Lunar Sci. Conf., Geochim. Cosmochim. Acta* Suppl. 1, Vol. 2, pp. 1037–1070. Pergamon.

Eldridge J. S., O'Kelley G. D., and Northcutt K. J. (1972) Abundances of primordial and cosmogenic radionuclides in Apollo 14 rocks and fines (abstract). In *Lunar Science—III* (editor C. Watkins), pp. 221–223, Lunar Science Institute Contr. No. 88.

Epstein S. and Taylor H. P. (1971) $O^{18}/O^{16}$, $Si^{30}/Si^{28}$, D/H, and $C^{13}/C^{12}$ ratios in lunar samples. *Proc. Second Lunar Sci. Conf., Geochim. Cosmochim. Acta* Suppl. 2, Vol. 2, pp. 1421–1441. MIT Press.

Epstein S. and Taylor H. P. (1972) $O^{18}/O^{16}$, $Si^{30}/Si^{28}$, $C^{13}/C^{12}$, and D/H studies of Apollo 14 and 15 samples (abstract). In *Lunar Science—III* (editor C. Watkins), pp. 236–238, Lunar Science Institute Contr. No. 88.

Fleischer R. L., Hart H. R., and Comstock G. M. (1972) Particle track dating of mechanical events (abstract). In *Lunar Science—III* (editor C. Watkins), pp. 265–267, Lunar Science Institute Contr. No. 88.

Flory D. A., Wikstrom S., Gupta S., Gibert J. M., and Oró J. (1972) Analysis of organogenic elements in Apollo 11, 12, and 14 lunar samples (abstract). In *Lunar Science—III* (editor C. Watkins), pp. 271–273, Lunar Science Institute Contr. No. 88.

Fuller E. L., Holmes H. F., Gammage R. B., and Becker K. (1971) Interaction of gases with lunar materials; preliminary results. *Proc. Second Lunar Sci. Conf., Geochim. Cosmochim. Acta* Suppl. 2, Vol. 3, pp. 2009–2019. MIT Press.

Funkhouser J. G. (1971) Personal communication.

Funkhouser J. G., Schaeffer O. A., Bogard D. D., and Zähringer J. (1970) Gas analysis of the lunar surface. *Proc. Apollo 11 Lunar Sci. Conf., Geochim. Cosmochim. Acta* Suppl. 1, Vol. 2, pp. 1111–1116. Pergamon.

Ganapathy R., Keays R. R., Laul J. C., and Anders E. (1970) Trace elements in Apollo 11 lunar rocks: Implications for meteorite influx and origin of moon. *Proc. Apollo 11 Lunar Sci. Conf., Geochim. Cosmochim. Acta* Suppl. 1, Vol. 2, pp. 1117–1142. Pergamon.

Gibson E. K. and Johnson S. M. (1971) Thermal analysis—inorganic gas release studies of lunar samples. *Proc. Second Lunar Sci. Conf., Geochim. Cosmochim. Acta* Suppl. 1, Vol. 2, pp. 1351–1366. MIT Press.

Gibson E. K. and Hubbard N. J. (1972) Volatile element depletion investigations on Apollo 11 and 12 lunar basalts via thermal volatilization (abstract). In *Lunar Science—III* (editor C. Watkins), pp. 303–305, Lunar Science Institute Contr. No. 88.

Gibson E. K. and Moore C. B. (1972) Compounds of the organogenic elements in Apollo 11 and 12 lunar samples. A review. *Space Life Sci.*, in press.

Gold T. (1971) Evolution of mare surface. *Proc. Second Lunar Sci. Conf., Geochim. Cosmochim. Acta* Suppl. 2, Vol. 3, pp. 2675–2680. MIT Press.

Gold T. (1972) The depth of the lunar dust layer (abstract). In *Lunar Science—III* (editor C. Watkins), pp. 321–322, Lunar Science Institute Contr. No. 88.

Gold T., Bilson E., and Yerbury M. (1972) Grain size analysis, optical reflectivity measurements, and determination of high frequency electrical properties for Apollo 14 lunar samples (abstract). In *Lunar Science—III* (editor C. Watkins), pp. 318–320, Lunar Science Institute Contr. No. 88.

Hayes J. M. (1972) Extralunar sources for carbon on the moon. *Space Life Sci.*, in press.

Henderson W., Kray W. C., Newman W. A., Reed W. E., Simoneit B. R., and Calvin M. (1971) Study of carbon compounds in Apollo 11 and Apollo 12 returned lunar samples. *Proc. Second Lunar Sci. Conf., Geochim. Cosmochim. Acta* Suppl. 2, Vol. 2, pp. 1901–1912. MIT Press.

Heymann D. and Yaniv A. (1970a) Inert gases in the fines from the Sea of Tranquility. *Proc. Apollo 11 Lunar Sci. Conf., Geochim. Cosmochim. Acta* Suppl. 1, Vol. 2, pp. 1247–1259. Pergamon.

Heymann D. and Yaniv A. (1970b) $Ar^{40}$ anomaly in lunar samples from Apollo 11. *Proc. Apollo 11 Lunar Sci. Conf., Geochim. Cosmochim. Acta* Suppl. 1, Vol. 2, pp. 1261–1267. Pergamon.

Hintenberger H., Weber H. W., Voshage H., Wänke H., Begemann F., and Wlotzka F. (1970) Concentrations and isotopic abundances of the rare gases, hydrogen and nitrogen in Apollo 11 lunar matter. *Proc. Apollo 11 Lunar Sci. Conf., Geochim. Cosmochim. Acta* Suppl. 1, Vol. 2, pp. 1269–1282. Pergamon.

Hintenberger H., Weber H. W., and Takaoka N. (1971) Concentrations and isotopic abundances of the rare gases in lunar matter. *Proc. Second Lunar Sci. Conf., Geochim. Cosmochim. Acta* Suppl. 2, Vol. 2, pp. 1607–1625. MIT Press.

Holland P. T., Simoneit B. R., Wszolek P. C., McFadden W. H., and Burlingame A. L. (1972a) Carbon chemistry of Apollo 14 size-fractionated fines. *Nature* **235,** 106–108.

Holland P. T., Simoneit B. R., Wszolek P. C., and Burlingame A. L. (1972b) Study of carbon compounds in Apollo 12 and 14 lunar samples. *Space Life Sci.*, in press.

Housley R. M., Grant R. W., and Abdel-Gawad M. (1972) Study of excess iron metal in the lunar fines by magnetic separation (abstract). In *Lunar Science—III* (editor C. Watkins), pp. 392–394, Lunar Science Institute Contr. No. 88.

Jackson E. D. and Wilshire H. G. (1972) Classification of the samples returned from the Apollo 14 landing site (abstract). In *Lunar Science—III* (editor C. Watkins), pp. 418–420, Lunar Science Institute Contr. No. 88.

Johnson J. H., Knipe R. H., and Gordan A. S. (1970) Chemical reactions in a Toepler pump. *Can. J. Chem.* **48,** 3604–3605.

Kaplan I. R., Smith J. W., and Ruth E. (1970) Carbon and sulfur concentration and isotopic composition in Apollo 11 lunar samples. *Proc. Apollo 11 Lunar Sci. Conf., Geochim. Cosmochim. Acta* Suppl. 1, Vol. 2, pp. 1317–1329.

Kaplan I. R. and Petrowski C. (1971) Carbon and sulfur isotope studies on Apollo 12 lunar samples. *Proc. Second Lunar Sci. Conf., Geochim. Cosmochim. Acta* Suppl. 2, Vol. 2, pp. 1397–1406. MIT Press.

Keller W. D. and Huang W. H. (1971) Response of Apollo 12 lunar dust to reagents simulative of those in the weathering environment of Earth. *Proc. Second Apollo Lunar Sci. Conf., Geochim. Cosmochim. Acta* Suppl. 2, Vol. 1, pp. 973–981. MIT Press.

King E. A., Butler J. C., and Carman M. F. (1971) The lunar regolith as sampled by Apollo 11 and Apollo 12: Grain size analyses, modal analyses, and origins of particles. *Proc. Second Lunar Sci. Conf., Geochim. Cosmochim. Acta* Suppl. 2, Vol. 1, pp. 737–746. MIT Press.

Kirsten T., Müller O., Steinbrunn F., and Zähringer J. (1970) Study of distribution and variations of rare gases in lunar material by a microprobe technique. *Proc. Apollo 11 Lunar Sci. Conf., Geochim. Cosmochim. Acta* Suppl. 1, Vol. 2, pp. 1331–1343. Pergamon.

Kirsten T., Steinbrunn F., and Zähringer J. (1971) Location and variation of trapped rare gases in Apollo 12 lunar samples. *Proc. Second Lunar Sci. Conf., Geochim. Cosmochim. Acta* Suppl. 2, Vol. 2, pp. 1651–1669. MIT Press.

Laul J. C., Morgan J. W., Ganapathy R., and Anders E. (1971) Meteorite material in lunar samples: Characterization from trace elements. *Proc. Second Lunar Sci. Conf., Geochim. Cosmochim. Acta* Suppl. 2, Vol. 2, pp. 1139–1158. MIT Press.

LSPET (Lunar Sample Preliminary Examination Team) (1969) Preliminary examination of lunar samples of Apollo 11. *Science* **165,** 1211–1227.

LSPET (Lunar Sample Preliminary Examination Team) (1971) Preliminary examination of lunar samples from Apollo 14. *Science* **173,** 681–693.

LSPET (Lunar Sample Preliminary Examination Team) (1972) Preliminary examination of lunar samples of Apollo 15. *Science* **175,** 407–443.

Marti K. (1971) Personal communication.

Maurette M. (1972) Personal communication.

McKay D. S., Greenwood W. R., and Morrison D. A. (1970) Origin of small lunar particles and breccia from the Apollo 11 site. *Proc. Apollo 11 Lunar Sci. Conf., Geochim. Cosmochim. Acta* Suppl. 1, Vol. 1, pp. 673–694. Pergamon.

McKay D. S., Morrison D. A., Clanton U. S., Ladle G. H., and Lindsay J. F. (1971) Apollo 12 soil and breccia. *Proc. Second Lunar Sci. Conf., Geochim. Cosmochim. Acta* Suppl. 2, Vol. 1, pp. 755–773. MIT Press.

McKay D. S., Clanton U. S., Heiken G. H., Morrison D. A., and Taylor R. M. (1972) Vapor phase crystallization in Apollo 14 breccias and size analysis of Apollo 14 soils (abstract). In *Lunar Science—III* (editor C. Watkins), pp. 529–531, Lunar Science Institute Contr. No. 88.

Megrue G. H. and Steinbrunn F. (1972) Classification and source of lunar soils; clastic rocks; and individual mineral, rock, and glass fragments from Apollo 12 and 14 samples as determined by the concentration gradients of the helium, neon, and argon isotopes (abstract). In *Lunar Science—III* (editor C. Watkins), pp. 532–534, Lunar Science Institute Contr. No. 88.

Meyer C., Brett R., Hubbard N. J., Morrison D. A., McKay D. S., Aitken F. K., Takeda H., and Schonfeld E. (1971) Mineralogy, chemistry, and origin of the KREEP component in soil samples from the Ocean of Storms. *Proc. Second Lunar Sci. Conf., Geochim. Cosmochim. Acta* Suppl. 2, Vol. 2, pp. 393–411. MIT Press.

Moore C. B., Lewis C. F., and Nava D. (1969) Superior analyses of iron meteorites. In *Meteorite Research* (editor P. M. Millman), pp. 738–748, D. Reidel Publishing Co., Holland.

Moore C. B., Lewis C. F., Gibson E. K., and Nichiporuk W. (1970) Total carbon and nitrogen abundances in Apollo 11 lunar samples and selected achondrites and basalts. *Proc. Apollo 11 Lunar Sci. Conf., Geochim. Cosmochim. Acta* Suppl. 1, Vol. 2, pp. 1375–1382. Pergamon.

Moore C. B., Lewis C. F., Larimer J. W., Delles F. M., Gooley R. C., and Nichiporuk W. (1971) Total carbon and nitrogen abundances in Apollo 12 lunar samples. *Proc. Second Lunar Sci. Conf., Geochim. Cosmochim. Acta* Suppl. 2, Vol. 2, pp. 1343–1350. MIT Press.

Morgan J. W., Laul J. C., Krähenbühl U., Ganapathy R., and Anders E. (1972a) Major impacts on the moon: Chemical characterization of projectiles (abstract). In *Lunar Science—III* (editor C. Watkins), pp. 552–554, Lunar Science Institute Contr. No. 88.

Morgan J. W., Krähenbühl U., Ganapathy R., and Anders E. (1972b) Volatile and siderophile elements in Apollo 14 and 15 rocks (abstract). In *Lunar Science—III* (editor C. Watkins), pp. 555–557, Lunar Science Institute Contr. No. 88.

Müller O. (1972) Chemically bound nitrogen abundances (abstract). In *Lunar Science—III* (editor C. Watkins), pp. 568–570, Lunar Science Institute Contr. No. 88.

Oró J., Flory D. A., Gibert J. M., McReynolds J., Lichtenstein H. A., and Wikstrom S. (1971) Abundances and distribution of organogenic elements and compounds in Apollo 12 lunar samples. *Proc. Second Lunar Sci. Conf., Geochim. Cosmochim. Acta* Suppl. 2, Vol. 2, pp. 1913–1925. MIT Press.

Pepin R. O. (1971) Personal communication.

Phakey P. P., Hutcheon D., Rajan R. S., and Price P. B. (1972) Radiation damage in soils from five lunar missions (abstract). In *Lunar Science—III* (editor C. Watkins), pp. 608–610, Lunar Science Institute Contr. No. 88.

Pillinger C. T., Cadogan P. H., Eglinton G., Maxwell J. R., Mays B. J., Grant W. A., and Nobes M. J. (1972) Simulation study of lunar carbon chemistry. *Nature* **235,** 108–109.

Poupeau G., Berdot J. L., Chetrit G. C., and Pellas P. (1972) Predominant trapping of solar-flare gases in lunar soils (abstract). In *Lunar Science—III* (editor C. Watkins), pp. 613–615, Lunar Science Institute Contr. No. 88.

Putnam G. L. and Kobe K. A. (1937) Hydrocarbons from carbides. *Chem. Rev.* **20,** 131–143.

Quaide W. and Bunch T. (1970) Impact metamorphism of lunar surface materials. *Proc. Apollo 11 Lunar Sci. Conf., Geochim. Cosmochim. Acta* Suppl. 1, Vol. 1, pp. 711–729. Pergamon.

Quaide W., Oberbeck V., and Bunch T. (1971) Investigations of the natural history of the regolith at the Apollo 12 site. *Proc. Second Lunar Sci. Conf., Geochim. Cosmochim. Acta* Suppl. 2, Vol. 1, pp. 701–718. MIT Press.

Rancitelli L. A., Perkins R. W., Felix W. D., and Wogman N. A. (1972) Cosmic ray flux and lunar surface processes characterized from radionuclide measurements in Apollo 14 and 15 lunar samples (abstract). In *Lunar Science—III* (editor C. Watkins), pp. 630–632, Lunar Science Institute Contr. No. 88.

Reed G. W., Jovanovic S., and Fuchs L. H. (1972) Concentrations and lability of the halogens, platinum metals, and mercury in Apollo 14 and 15 samples (abstract). In *Lunar Science—III* (editor C. Watkins), pp. 637–639, Lunar Science Institute Contr. No. 88.

Sakai H., Petrowski C., Goldhaber M. B., and Kaplan I. R. (1972) Distribution of carbon, sulfur, and nitrogen in Apollo 14 and 15 material (abstract). In *Lunar Science—III* (editor C. Watkins), pp. 672–674, Lunar Science Institute Contr. No. 88.

Sellers G. A., Woo C. C., and Bird M. L. (1971) Composition and grain-size characteristics of fines from the Apollo 12 double-core tube. *Proc. Second Lunar Sci. Conf., Geochim. Cosmochim. Acta* Suppl. 2, Vol. 1, pp. 665–678. MIT Press.

Silver L. T. (1972) U–Th–Pb abundances and isotopic characteristics in some Apollo 14 rocks and soils and in an Apollo 15 soil (abstract). In *Lunar Science—III* (editor C. Watkins), pp. 704–706, Lunar Science Institute Contr. No. 88.

Warner J. L. (1972) Apollo 14 breccias: Metamorphic origin and classification (abstract). In *Lunar Science—III* (editor C. Watkins), pp. 782–784, Lunar Science Institute Contr. No. 88.

Proceedings of the Third Lunar Science Conference
(Supplement 3, *Geochimica et Cosmochimica Acta*)
Vol. 2, pp. 2091–2108
The M.I.T. Press, 1972

# Analysis of organogenic compounds in Apollo 11, 12, and 14 lunar samples

D. A. Flory, S. Wikstrom, S. Gupta, J. M. Gibert, and J. Oró

Department of Biophysical Sciences, University of Houston,
Houston, Texas 77004

**Abstract**—Quadrupole mass spectrometry (QMS) and gas chromatography mass spectrometry (GC–MS) have been applied to the study of volatiles released by acid and thermal treatments of lunar samples. DCl was used in order to distinguish hydrocarbons generated by the acid treatment from those present as such. Multiple ion plotting permitted separation of deuterated from nondeuterated hydrocarbons in the DCl products. Stepwise thermal treatment was used to aid in distinguishing the source of the volatile compounds.

The acid treatment released $H_2$, $N_2$, CO, $CO_2$, $CH_4$, $C_2H_6$, $C_3H_6$, $C_3H_8$, and $H_2S$ from all samples. $CO_2$ ranged from about 2 ppm to 50 ppm and was the major carbon compound released. $CH_4$ ranged from less than 0.1 ppm to several ppm and total hydrocarbon contents were slightly higher. The deuterated to nondeuterated hydrocarbon ratios indicate more than two-thirds of the hydrocarbons are produced by the reaction of the DCl with reactive carbon species. Evidence was obtained for partially deuterated hydrocarbons and could mean that partially hydrogenated carbon species are present.

The thermal treatments also released the above volatiles in similar amounts with the exception of $H_2S$. Significant amounts of $CO_2$ were released at higher temperatures ($>500°C$) indicating that indigenous $CO_2$ may be present as trapped gas or mineral carbonates. Hydrocarbon release peaked in the 500–900°C temperature range and showed some correlation with $CO_2$, indicating release of the compounds as such and/or synthesis may be significant contributors. The total water observed in unexposed samples ranged from about 50 to 150 ppm with 10–20% being released at temperatures above 700°C. CO was evolved from some samples at 200–600°C but the greater portion came off above 800°C. Evidence was obtained for the presence of LM rocket exhaust products in two samples. Additional evidence was obtained to indicate a substantial portion of the $CO_2$ evolved below 400°C, and water evolved below 500°C is terrestrial contamination but does not exclude the presence of indigenous carbonates in the samples. Temperature release data for $N_2$ indicate that high temperature stable nitrides are present in some samples.

## Introduction

Our previous examinations of Apollo 11 and 12 samples, as well as the work of other investigators, have indicated that organogenic and organic compounds are released upon hydrolysis with selected acids (Cadogan *et al.*, 1971; Chang *et al.*, 1971; Oro *et al.*, 1971), by thermal extraction methods (Burlingame *et al.*, 1971; Chang *et al.*, 1971; Henderson *et al.*, 1971; Nagy *et al.*, 1971; Oró *et al.*, 1971; Preti *et al.*, 1971), and by crushing (Abell *et al.*, 1971; Chang *et al.*, 1971; Funkhouser *et al.*, 1971). Compounds released have been summarized in a review by Gibson and Moore (1971) and include $H_2$, CO, $CO_2$, $N_2$, $H_2O$, $H_2S$, methane, ethane, ethylene, acetylene, benzene, toluene, and higher molecular weight organic compounds.

In general the origin of these compounds appears to be multiple while the chemical and/or physical state of the organic elements and their simple compounds indigenous to the lunar samples remains to be determined. Sources of these elements and compounds that we have considered include (1) primordial lunar material (e.g. in vesicles or grain boundaries); (2) extralunar, such as the solar wind, comets, and meteorites; (3) artifacts of the analytical procedure employed (e.g. synthesis of hydrocarbons from CO and $H_2$, reaction of indigenous species with terrestrial water); and (4) terrestrial contamination (Flory and Simoneit, 1971).

In our work we have placed particular emphasis on obtaining information to allow us to identify which source(s) is the principal contributor for the compounds evolved by thermal treatment and hydrolysis. Quadrupole mass spectrometry (QMS) and gas chromatography-mass spectrometry (GC-MS) were used to analyze gases released by thermal treatment. Gases released by acidolysis were analyzed by GC-MS. One sample was also extracted with benzene-methanol and the extract analyzed by GC-MS with no extractable compounds being found at a sensitivity of $10^{-9}$ g/g. A list of the samples analyzed and brief description of their history and properties is given in Table 1.

## Acidolysis

The gases generated upon treatment of the samples with acid in a 10 cc reaction vessel were swept into and separated in a Porapak Q (2.5 mm i.d. × 2 m) packed column following our previously described method (Oró *et al.*, 1970). Both HCl and DCl were used in order to distinguish hydrocarbons generated by reaction between the acid and the sample from hydrocarbons present as such in the sample. At the time the generated gases were swept into the column the temperature was maintained isothermally near $-50°C$ for 6 min and then programmed at a rate of 8°C/min to 150°C. These conditions proved to give a better separation of the different com-

Table 1. Brief description of samples analyzed.

| Sample no. | Type | Carbon content (ppm) | Remarks |
|---|---|---|---|
| 14003 | Fines | 140[a] | Collected near LM, should be one of most contaminated samples (< 1 mm). |
| 14156 | Fines | 186[a] | Trench soil—middle (< 1 mm) ≈ 50 m from LM. |
| 14240 | Fines | 89[a] | Collected in a trench 1200 m from LM and returned in SESC. Cleanest sample available. Unsieved. |
| 14311 | Clastic rock | 44[a] | Fine grained crystalline groundmass, 600 m from LM. |
| 14421 | Fines | 160[a] | Reserve from unsieved comp. sample, collected 120 m from LM. |
| 14422 | Fines | 120[a] | < 1 mm, reserve from 14163 bulk sample collected ≈ 40 m from LM. |
| 12023 | Fines | 150[b] | SESC sample, collected ≈ 440 m from LM. |
| 10059 | Breccia rock | ND | Processed for distribution by Calvin's group at UCB. |
| 10086 | Bulk fines | 109–353[c] | Organic reserve sample. |

[a] C. B. Moore *et al.*, 1972.
[b] C. B. Moore *et al.*, 1971.
[c] C. B. Moore *et al.*, 1970.
ND = not determined.

ponents evolved from the sample than did our previous work, although separation of all components was not achieved. A chromatogram from an Apollo 14 sample treated with HCl is shown in Fig. 1.

In order to distinguish between deuterated and nondeuterated hydrocarbons, to resolve $C_2H_2$ from $C_2H_6$, and to resolve $C_3H_6$ from $C_3H_8$ semicontinuous mass scans were recorded to provide mass spectrometric evidence for the presence of the different species. Multiple ion plots were then constructed to improve the gas chromatographic separation as reported earlier (Oró *et al.*, 1971).

Figure 2 shows the multiple ion plot of the $C_2H_4$, $C_2H_2$, $C_2H_6$, $C_3H_6$, and $C_3H_8$ region of sample 14422. It is possible to see how the ions m/e 25 and 26 of acetylene arrive at their peak values 10 sec earlier than the m/e 30, 28, and 27 peaks of ethane. Similarly, the separation of propane and propylene can also be seen. It should be noted that the several ions are plotted against different scales and, therefore, do not represent relative quantities of the ions. In this particular case acetylene is only about 1/100 the intensity of propane while propylene is approximately twice the intensity of propane.

The results of the acidolysis runs with DCl are presented in Table 2. The values given for the various volatiles were obtained by integrating the areas under the gas chromatogram peaks obtained with the mass spectrometer total ion current monitor, normalizing them to sample weight, and multiplying by a calibration factor. The calibration factor for methane was determined by injecting different quantities of

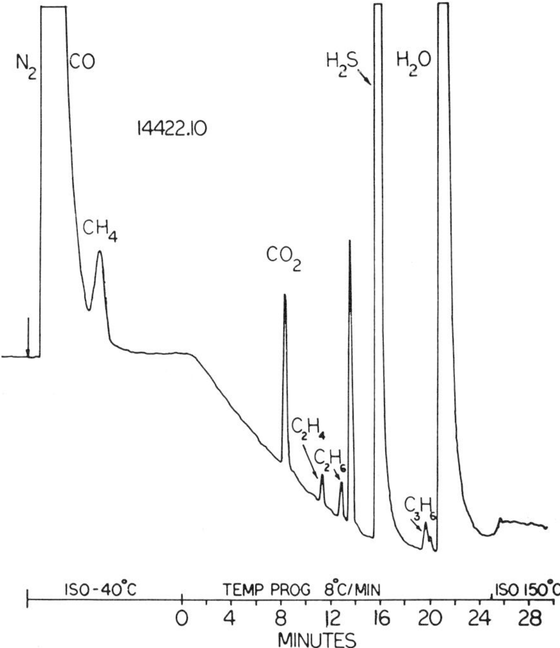


Fig. 1. Typical gas chromatogram of products released by acid treatment (HCl) Apollo 14 sample 14,422,10. Porapak Q. (2.5 mm i.d. × 2 m) packed column.

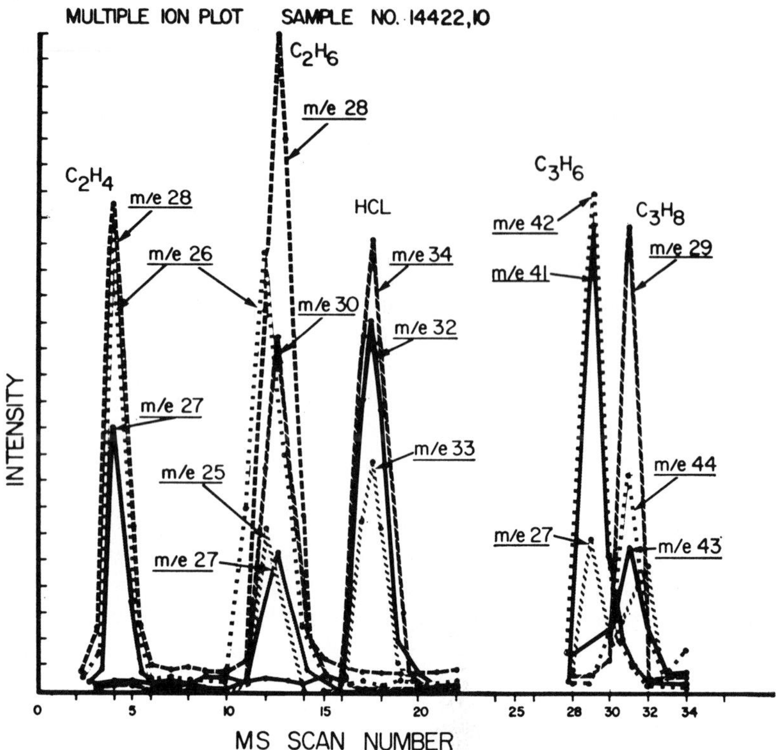


Fig. 2. Multiple ion plot of $C_2H_4$, $C_2H_2$, $C_2H_6$, and $C_3H_6$–$C_3H_8$ regions of the gas chromatogram. Ion intensities taken from mass scans taken at 10 sec intervals. Sample treated with HCl.

Table 2. Acidolysis results (DCI treatment).

| Sample | Compound | $CO_2$ | $CH_4$ | $C_3H_6$ + $C_3H_8$ | $C_2H_4$ | $C_2H_6$ | $H_2S$ | $CD_4$/$CH_4$ |
|---|---|---|---|---|---|---|---|---|
| 10059,2,3 | Crushed breccia | 17 | 20 | T | 2.9 | 5.9 | 98 | 2.1 |
| 10059,2,4 | Solid breccia | 0.9 | 0.5 | T | 0.1 | 0.1 | 8 | 3.3 |
| 12023,9,2 | Fines | 2.4 | 3.7 | 1.0 | 0.7 | 1.0 | T | 3.3 |
| 12023,9,2 | Second sweep | 1.6 | 2.2 | 0.9 | 0.5 | 0.5 | T | ND |
| 14311 | Clastic rock | 0.55 | 0.1 | T | 0.1 | 0.1 | T | 4.9 |
| 14311 | Crushed | 1.6 | 0.05 | 0.1 | 0.15 | 0.15 | T | 5.4 |
| 14422 | Fines | 0.96 | 1.3 | 0.4 | 0.2 | 0.3 | 1.2 | 4.0 |

T = Trace, present in mass spectra taken at appropriate retention time but not in sufficient quantities to produce an integratable GC peak.

Note: Hydrocarbon peaks actually include deuterated and nondeuterated species combined since they are not separated chromatographically.

$H_2$, $N_2$ and CO evolved by all samples, but not resolved in chromatograms. Values in ppm.

methane with a gas syringe and that for $CO_2$ by volatilizing a known amount of $Na_2CO_3$. The MS ion source was operated at 20 eV so ionization efficiencies will vary for the various gases as a function of their ionization potential. The calibration factor for $CH_4$ was used for all species with equal or lower ionization potentials, including $H_2O$ and the other hydrocarbons, even though it is not strictly valid due to the differences in ionization efficiency.

The factor for $CO_2$ was used for the $H_2S$ results. Samples from Apollo 11 and 12 were again analyzed for comparison. Samples of crushed and solid rock were analyzed. In the case of sample 12023,9,2 a second sweep or injection is shown to give an indication of the efficiency of this sampling method. It can be seen that a single sweep (of 60 sec duration) as was used to obtain these data may be only about 50% efficient. In the case of the carbonate standard the single sweep efficiency was better than 90%. These values, therefore, can be considered only semi-quantitative and perhaps should be doubled when compared to other investigators' results. Hydrogen, nitrogen, and CO were released by all samples but were not completely resolved in the chromatograms. The final column in Table 2 gives the ratios of deuterated and nondeuterated methane determined from the multiple ion plots. Similar values were observed for these ratios in the other hydrocarbons evolved. Figure 3 shows a multiple ion plot for

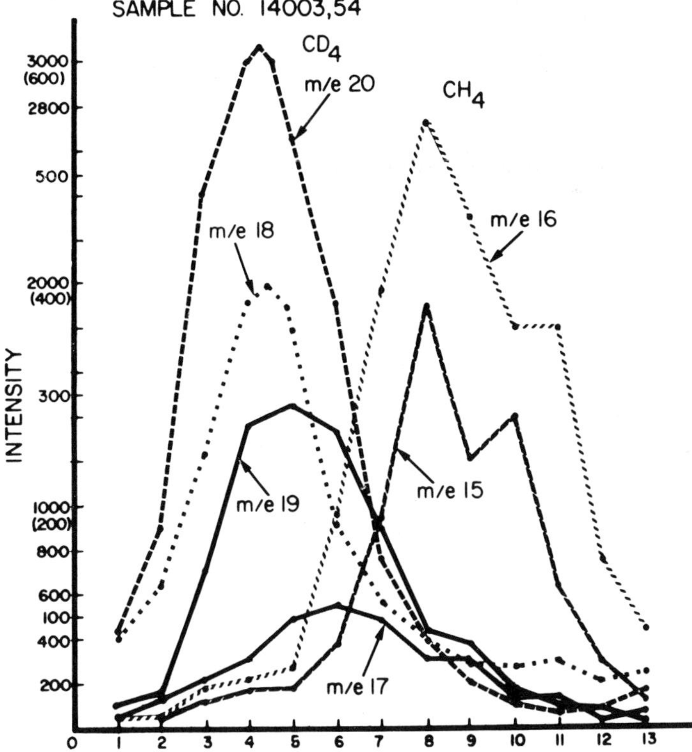


Fig. 3. Multiple ion plot of $CH_4$–$CD_4$ region of gas chromatogram from DCl treatment of sample 14003. Ion intensities taken from mass scans at 10 sec intervals.

sample 14422 over the time period that methane and its deuterated homologs would emerge from the column and indicates the presence of all possible deuterated species: m/e 20-$CD_4$, m/e 19-$CD_3H$, m/e 18-$CD_2H_2$, m/e 17-$CDH_3$, and m/e 16-$CH_4$. Similar evidence was obtained for sample 10059.

As would be expected, the crushed rocks evolved larger quantities of gas than did the solid rocks, but the increase is much more pronounced in the case of the Apollo 11 breccia than for the Apollo 14 clastic rock. The Apollo 11 breccia also evolved much larger quantities of gases when crushed than any other sample analyzed. It should also be noted that this breccia sample evolved a much larger proportion of nondeuterated hydrocarbons than any other sample. The Apollo 12 fines sample displayed the second largest proportion of nondeuterated hydrocarbons. It is evident that the major portion (70% or more) of the hydrocarbons observed result from the acid treatment in agreement with the range of $CD_4/CH_4$ values reported by others (Cadogan *et al.*, 1971; Cadogan *et al.*, 1972; Chang *et al.*, 1971). The deuterated hydrocarbons produced by acid treatment can be considered reaction products with reactive or "carbide-like" carbon atoms in the lunar material, while the nondeuterated species evolved should be present as such in the sample if there is no significant D-H exchange during the times involved in the procedure. An experiment was performed to determine the exchange rate by loading a previously DCl treated lunar sample, adding fresh DCl and methane, and measuring the $CD_4 : CH_4$ ratio with time. Three percent exchange was noted over an 18-hour period, which should not be significant over the 15–20-min experiment time. The nondeuterated species can then be truly indigenous to the sample and derived from either extralunar sources or trapped primordial lunar material; alternately, they can be the result of contamination from terrestrial exposure. The deuterated : nondeuterated ratio does not allow one to distinguish between these two sources, but the higher amounts of nondeuterated hydrocarbons found in the older, more exposed samples (average 4.8 in contrast to 2.9 for Apollo 14 samples) suggest either significant heterogeneity or that contamination may be a factor. These ratios, along with the amounts detected, indicate that there are substantially less hydrocarbons present as such in Apollo 14 samples than in earlier samples.

The observation of the several deuterated species of methane (Abell *et al.*, 1971, reported $CD_3H$ as an acidolysis product from DCl treatment of Apollo 11 fines) may be interpreted as evidence for the presence of some partially hydrogenated carbon atoms which react with the DCl to produce partially deuterated methane. Alternately, the implanted hydrogen evolving from the sample near to the reactive carbon species could be competing favorably with the $D^+$ ions from the DCl for the active carbon sites. Note that the relative concentrations of methane species having an even number of hydrocarbons ($CH_4$, $CD_2H_2$) are substantially higher than those having an odd number of hydrogens (CH, $CH_3$). This could be due to the longer lifetime of diradicals.

The small amounts of $CO_2$ produced by acid hydrolysis can in part be due to terrestrial contamination and in part due to reaction or indigenous material. We have previously obtained evidence (Oró *et al.*, 1971) for adsorption of a few ppm $CO_2$ in prepyrolyzed samples from terrestrial atmospheric exposure and additional evidence is presented in the QMS section of this paper. This adsorbed $CO_2$ should be released by acid treatment. Crushing has also been shown to release $CO_2$ from breccia sample 10009 (Chang *et al.*, 1971). Funkhouser *et al.* (1971) did not observe any $CO_2$ during crushing of several Apollo 11 and 12 samples but these same samples evolved a few ppm of $CO_2$ at a temperature of 300°C. These data indicate that adsorbed $CO_2$, presumably from the terrestrial atmosphere, can account for a substantial fraction of the $CO_2$ released by acidolysis.

## Volatilization-Gas Chromatograph-Mass Spectrometry

The experimental procedure for GC-MS analysis of volatiles released by stepwise heating of the sample to 950°C was similar to that described in our previous analyses of Apollo 11 and 12 samples (Oró *et al.*, 1970). The procedure involves the use of a quartz pyrolysis tube (4 mm i.d. × 25 cm) that can be heated in several steps to 980°C which is connected in line to a modified LKB-9000. The physical construction of the system and the gas chromatographic conditions have been improved to give a better separation of the different components evolved. Several samples of both fines and rocks have been analyzed by this volatilization GC-MS system following two slightly different sets of analytical conditions. Sample 14422,10,6 was evacuated to $10^{-2}$ Torr before the initial heating and between each step in the heating. All other samples were only evacuated prior to the initial heating and remained at 40 psi He (the carrier gas used to sweep the volatiles out of the quartz tube) throughout successive heating steps. Gas chromatographic conditions were identical to those for acidolysis.

Sample 14240 (SECS, fines), which should be the cleanest and driest sample provided to us, was given special handling. This sample was loaded into the volatilization apparatus in an ultrapure dry nitrogen filled cabinet at the NASA Manned Spacecraft Center, exposed to $D_2O$ vapors for 72 hours, and then the volatiles released by heating were analyzed by GC-MS in a manner which excluded any exposure to the terrestrial atmosphere. The special handling of this sample was intended to aid in determining amounts of light hydrocarbons which may be produced by hydrolysis of indigenous material due to terrestrial atmospheric exposure.

Table 3 gives the amounts of volatiles (in ppm) released as a function of temperature for the various samples analyzed. Number values were obtained by applying the same calibration factors to the GC peak areas as in acidolysis. The symbols $T$ and + are used to indicate the components present in the mass spectra taken at the appropriate retention time, even though there was no integratable GC peak. The − symbol denotes no gas sample taken over that particular temperature range. The symbol NF indicates that spectra taken at the appropriate retention time gave no indication of that compound being present. Limitations in data handling capacity, however, preclude taking mass spectra at every component retention time during analysis of the gas released at each temperature step. The term NS denotes the cases where no spectra were available. $H_2$, $N_2$, and CO were detected in all samples but are not shown in Table 3, because they were not resolved completely in the chromatograms. In general the CO maximum evolution occurred at the highest temperatures. Acetylene was also

Table 3. Amounts of gases evolved at various temperatures in VOL-GC-MS analyses.

| | | | Temperature range (°C) | | | | | | | |
|---|---|---|---|---|---|---|---|---|---|---|
| Gas | Sample | Type | Ambient | Amb-210 | 210–350 | 350–430 | 430–570 | 570–740 | 740–980 | Amb-980 |
| $H_2O$ | 14003,54,6 | Fines | – | 4.2 | 4.4 | – | 3.2 | – | 29 | 41 |
| | 14240,19,2 | Fines | 9.2 | 11 | – | – | – | 8 | 5.2 | 33 |
| | 14240,19,2[1] | Fines | – | 52 | – | – | 11 | 13 | 8.3 | 84 |
| | 14240,19,3 | Fines | 57 | – | 49 | – | – | 31 | 48 | 181 |
| | 14311,57,5 | Crushed rock | – | – | 7.1 | – | 9.2 | – | 14 | 30 |
| | 14422,10,6[2] | Fines | – | 6.7 | 8.1 | 11 | 10 | – | 13 | 49 |
| | 14422,10,7 | Fines | – | 11 | 16 | 13 | 18 | 10 | 22 | 89 |
| $CO_2$ | 14003,54,6 | Fines | – | 0.045 | 0.10 | – | 0.58 | – | 15 | 16 |
| | 14240,19,2 | Fines | 0.066 | 0.26 | – | – | – | 0.05 | 0.05 | 0 .41 |
| | 14240,19,2[1] | Fines | – | 1.7 | – | – | 0.25 | 0.08 | 0.21 | 2.2 |
| | 14240,19,3 | Fines | NS | – | 0.038 | – | – | 0.92 | 2.0 | 3.1 |
| | 14311,57,5 | Crushed rock | – | – | 5.3 | – | 2.2 | – | 0.13 | 7.6 |
| | 14422,10,6[2] | Fines | – | 0.12 | 0.36 | 0.76 | 0.89 | – | 0.59 | 2.7 |
| | 14422,10,7 | Fines | – | 0.27 | 1.9 | 2.7 | 4.0 | 6.0 | 1.7 | 17 |
| $CH_4$ | 14003,54,6 | Fines | – | – | + | – | + + | – | 4.1 | 4.1 |
| | 14240,19,2 | Fines | + | + | – | – | – | + + | + | 0.02 |
| | 14240,19,2[1] | Fines | – | NS | – | – | NS | – | 0.10 | 0.10 |
| | 14240,19,3 | Fines | NS | – | 0.016 | – | – | 0.34 | 0.056 | 0.41 |
| | 14311,57,5 | Crushed rock | – | – | + + + | – | T | – | + | 0.03 |
| | 14422,10,6[2] | Fines | – | + | + | + + | + | – | + + | 0.04 |
| | 14422,10,7 | Fines | – | + | + + | 0.53 | 2.4 | 8.0 | 0.53 | 11 |
| $C_2H_4$ | 14003,54,6 | Fines | – | – | – | – | 0.14 | – | 2.6 | 2.7 |
| | 14240,19,2 | Fines | NS | NS | – | – | – | 0.035 | T | 0.035 |
| | 14240,19,2[1] | Fines | – | T | – | – | NS | 0.10 | 0.023 | 0.12 |
| | 14240,19,3 | Fines | NS | – | NS | – | – | NS | + | 0.005 |
| | 14311,57,5 | Crushed rock | – | – | 0.07 | – | + + + | – | NS | 0.07 |
| | 14422,10,6[2] | Fines | – | NS | T | + | + + | – | T | 0.010 |
| | 14422,10,7 | Fines | – | NS | 0.062 | 0.72 | 2.9 | 1.8 | 1.2 | 6.7 |
| $C_2H_6$ | 14003,54,6 | Fines | – | NS | NS | – | + | – | + + | 0.015 |
| | 14240,19,2 | Fines | NS | NS | – | – | – | T | T | T |
| | 14240,19,2[1] | Fines | – | NS | – | – | – | NS | NS | |
| | 14240,19,3 | Fines | NS | – | NS | – | – | NS | NS | |
| | 14311,57,5 | Crushed rock | – | – | NS | – | + | – | NS | 0.005 |
| | 14422,10,6[2] | Fines | – | NS | NS | + | + | – | T | 0.010 |
| | 14422,10,7 | Fines | – | NS | T | 0.30 | 0.58 | 0.075 | T | 0.96 |
| $C_3H_6$ | 14003,54,6 | Fines | – | NS | NS | – | T | – | + + + | 0.015 |
| | 14240,19,2 | Fines | NS | NS | – | – | – | NF | NS | |
| | 14240,19,2[1] | Fines | – | NS | – | – | 0.055 | 0.04 | NS | 0.095 |
| | 14240,19,3 | Fines | NS | – | NS | – | – | NS | + | 0.005 |
| | 14311,57,5 | Crushed rock | – | – | 0.1 | – | 0.12 | – | NF | 0.22 |
| | 14422,10,6[2] | Fines | – | T | NS | + | + + | – | NS | 0.015 |
| | 14422,10,7 | Fines | – | NS | + | 1.9 | 3.3 | 0.71 | 0.14 | 6.1 |
| $C_3H_8$ | 14003,54,6 | Fines | – | NS | NS | – | T | – | + + + | 0.015 |
| | 14240,19,2 | Fines | NS | NS | – | – | – | + | NS | 0.005 |
| | 14240,19,2[1] | Fines | – | NS | – | – | T | NF | NS | |
| | 14240,19,3 | Fines | NS | – | NS | – | – | NS | NS | |
| | 14311,57,5 | Crushed rock | – | – | NS | – | + | – | NS | 0.005 |
| | 14422,10,6[2] | Fines | – | NS | NS | + | + + | – | NS | 0.015 |
| | 14422,10,7 | Fines | – | NS | T | 0.42 | 0.80 | 0.06 | + | 1.3 |

[1] After three weeks exposure to laboratory air.
[2] Sample chamber evacuated between each temperature step.
Notes: Sample 14240 exposed to water ($D_2O$) prior to analyses.
T: Trace detected in mass spectra only, equivalent to $< 1$ ppb.
+ : Detected in mass spectra only, equivalent to $\approx 5$ ppb.
– : Denotes no gas sample taken at this temperature.
NF: Denotes component not detected in spectra taken at appropriate retention time.
NS: Denotes no spectra available for appropriate retention time.
All numerical values in ppm.

detected in trace quantities in the samples where multiple ion plots were generated. In the case of sample 14240 the values listed for water are actually $H_2O + D_2O$ since they are not resolved chromatographically.

The data for sample 1422,7 indicate that the pressurized condition produces considerably higher amounts of all compounds. Contaminants in the carrier gas would seem to be ruled out by the fact that other runs with the same tank of helium do not produce nearly such quantities. Leakage into the system can be ruled out, because it would most certainly be greater for the evacuated condition. This same behavior was observed in our previously reported (Oró *et al.*, 1971) investigation of Apollo 12 samples and the increase in hydrocarbons may be due to enhanced conditions for synthesis. We cannot offer any explanation for the increase in $CO_2$ and CO other than some type of exchange phenomena. Fines sample 14003 and 14422 analyzed under pressurized conditions evolved hydrocarbons totalling several micrograms comparable to our previous observations on Apollo 12 (Oró *et al.*, 1971). The crushed clastic rock 14311 gave only trace amounts of methane and the $C_2$ hydrocarbons under the pressurized conditions. In general the samples display the maximum hydrocarbon release above 700°C which could be attributed to enhanced conditions for synthesis.

The temperature release data for water in samples 14003, 14311, and 14422 given in Table 3 indicate a substantial portion of the total water is given off in the final temperature step. One would initially interpret this strange behavior as an artifact of the system, but blank runs on the system and reheating of samples do not give measurable GC peaks for water. It is difficult to imagine adsorption processes (physical or chemical) that would retain terrestrial water to these temperatures. Even though $N_2$ was not usually resolved in the GC the $N_2$ intensity in the mass spectra also appears to increase at the highest temperature step in samples 14003 and 14311. Other sources could be bursting of vesicles or vugs containing trapped water, and/or decomposition of hydrated minerals. The most probable explanation, however, is that water is formed on the outer surfaces of rocks and mineral grains as a result of proton irradiation from the solar wind as suggested by Gibson and Moore (1972). The mineral goethite, FeO(OH) reported in breccia sample 14301 (Argell *et al.*, 1972), which should break down at 200–400°C to give water and hematite (Pollack *et al.*, 1970), could produce water in quantities of 10–100 ppm if present in concentrations of 0.01 to 0.1%.

Sample 14240 (SECS sample), which we attempted to volatilize without exposure to terrestrial atmospheric water, gave unique results compared to all other samples of fines that were investigated. This sample was analyzed at temperature steps of 20°C, 200°C, 740°C, and 950°C. Nitrogen was, of course, very large in the first analysis (20°C) but only trace amounts of $CH_4$ could be detected. There was no detectable $CD_4$ (from exposure to $D_2O$) at 20°C, even though consecutive scans over the mass range m/3 12–60 with the mass spectrometer were made during the possible elution time of $CD_4/CH_4$ from the gas chromatographic column. $CH_4$ increased to a maximum at 740°C as did $C_2H_4$ and $C_2H_6$. $N_2$ and CO had a second peak at 950°C while $CH_4$, $CO_2$, $C_2H_4$, $C_2H_6$, and $H_2$ decreased. $CD_4$ could be detected only as a trace with a $CH_4 : CD_4$ ratio greater than 15 throughout the temperature range. The total amounts of all types of gases released were considerably less than for the other sample of fines. These surprising results led us to repeat the entire analysis procedure including the loading in nitrogen and exposure to $D_2O$. Results were similar to those for the first run although larger amounts of some gases were evolved. After exposure to the laboratory environment for three weeks the sample was rerun.

During the second pyrolysis of the exposed sample, adsorbed $CO_2$ and water were detected at 200°C in several times larger amounts than compared to the unexposed sample, but no $CH_4$ was detected at this temperature. Comparatively small amounts of $N_2$ and CO were evolved at 200°C. $N_2$ increased slightly at 540°C while $CO_2$ and $H_2O$ decreased three to four times. Methane, ethylene, and propane were first detected at 540°C and $CH_4$ had its maximum evolution at this temperature. Only very few volatiles were evolved in sufficient quantities after the 200°C step to produce GC peaks. A fluorinated hydrocarbon in equal amount to $CO_2$ was detected at 540°C, presumably caused by pyrolysis of teflon from the system seals. At 750°C, the $N_2$, $CH_4$, $CO_2$, and $H_2O$ decreased, $C_2H_4$ and $C_3H_6$ remained constant, and CO increased. At 920°C, the CO, $N_2$, $CH_4$, $CO_2$, and $H_2O$ increased compared to 750°C, but the $C_2H_4$ and $C_3H_6$ decreased. The mass spectra intensities indicated the hydrocarbons released during the second heating at all temperatures were significantly lower than for the unexposed sample.

The results obtained for sample 14240 indicate that hydrolysis of reactive "carbide like" substances by terrestrial water adsorbed prior to the experiment does not contribute significantly to the amounts of methane found as such in the DCl treatment. The very low amounts of volatiles, especially methane and the other hydrocarbons, can be attributed to either true differences in the samples or the effects of exposure to the terrestrial environment. Table 3 shows that many samples are evolving $CO_2$ at temperatures above that where one would expect terrestrial contaminant $CO_2$ to be evolved. These data indicate that there is some indigenous source of $CO_2$ that may be in the form of carbonates or gas trapped in vesicles and interstitial grains. Gay *et al.* (1970) have in fact reported the presence of the carbonate mineral aragonite in an Apollo 11 sample. Table 4 gives the total amounts of $CO_2$, $CH_4$, and total hydrocarbon released by heating in order of decreasing $CO_2$ content. There appears to be a correlation between the hydrocarbons and the $CO_2$. This correlation could be due to simultaneous contamination of $CO_2$ and the hydrocarbons or synthesis involving $CO_2$. It should be noted that this sample of fines was not sieved whereas all others analyzed were less than 1 mm sieve fractions. However, the $<1$ mm sieve fraction would contain better than 90% of the unsieved fraction based on particle size distributions reported for lunar samples (LSPET, 1971), therefore sieving should not greatly affect the results. The data from the second run of the already pyrolyzed sample 14240, combined with the general observation of increased hydrocarbons at higher temperatures for all samples, indicate that synthesis (such as Fischer-Tropsch type) is an important factor in pyrolysis.

## Quadrupole Mass Spectrometry

The basic technique used in the quadrupole-mass spectrometric analyses was essentially the same as reported (Oró *et al.*, 1970) for our previous work on Apollo 11

Table 4. Correlation of total amounts of evolved gases in VOL-GC-MS analyses.

| | $CO_2$ | $CH_4$ | Total hydrocarbons |
|---|---|---|---|
| 14422,10,7 | 17 | 11 | 24.8 |
| 14003,54,6 | 16 | 4.1 | 6.8 |
| 14311,57,5 | 7.6 | 0.03 | 0.32 |
| 14240,19,3 | 3.1 | 0.41 | 0.41 |
| 14240,19,2[1] | 2.2 | 0.10 | 0.27 |
| 14240,19,2 | 0.41 | 0.02 | 0.05 |

Values in ppm.
[1] Rerun after three days exposure to laboratory atmosphere.

and 12 samples. However, three important modifications in the experimental approach have significantly improved the accuracy and performance of the equipment. (1) The stainless steel sample tube was replaced by quartz in order to minimize the adsorption of $CO_2$ and moisture, etc., from the atmosphere. (2) The ion source filament mounting was modified to increase instrument sensitivity. (3) The sample heating technique was improved by constructing a well insulated cylindrical quartz oven that could be heated to 950°C at a constant rate by means of a motor-driven powerstat. This system was adjusted to provide a regulated heating of the sample at a rate of approximately 5°/min. Apart from its convenience, this method provides more reproducible temperature read-out values and greatly improved thermal release data. The relative contributions of $N_2$ and CO in the spectra were obtained by determining the fragmentation patterns for the pure gases and solving simultaneous equations using the m/e 12, 14, 16, and 28 intensities. The following Apollo 14 samples were run: (1) 14003, (2) 14156, (3) 14311 (rock), (4) 14421, (5) 14422. For comparison purposes we also repeated previous analyses of one sample each from Apollo 11 (10086) and Apollo 12 (12023,9).

Figure 4 is a plot of the $H_2O$, CO, $N_2$, and $CO_2$ temperature release profiles for sample 14003,59. The behavior of these gases in this sample is very similar to that observed for sample 14156 and 14422. although peak locations and maxima do shift around somewhat. In addition to the above sample 14003 also evolved $CH_4$ in trace

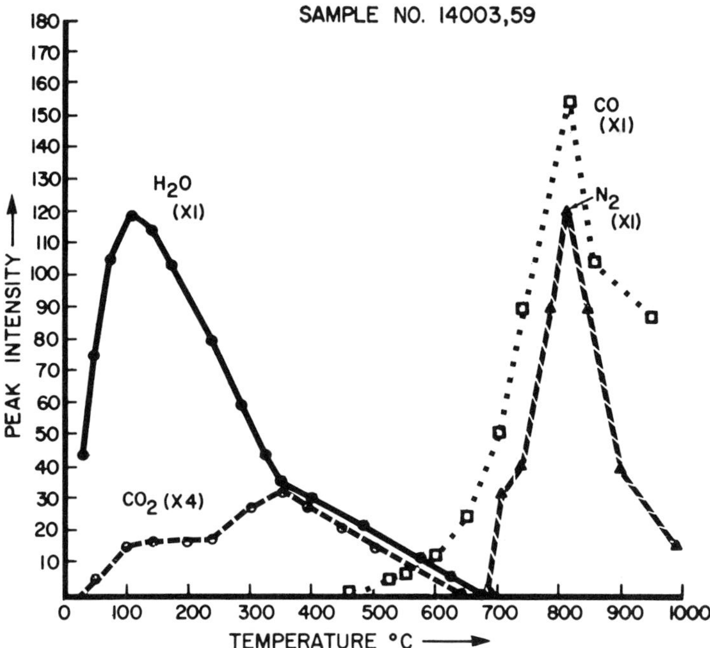


Fig. 4. Temperature release data for sample 14003. Temperature range: ambient–1000°C. Heating rate 5°C/min. Sample weight 500 mg. Quartz sample holder. Compounds identified by quadrupole mass spectrometry.

quantities with the maximum release at 160°C (no other hydrocarbons were detected); argon with a maximum at 820°C, and several compounds identified previously (Simoneit *et al.*, 1969) as lunar exhaust products including $CH_2 = C = O$ (m/e 41), $C_3H_3$ (m/e 39), HCN (m/e 27), $HN = C = O$ (m/e 45), and $NH_3$.

Figure 5 is a plot of the $H_2O$ and $CO_2$ and Fig. 6 of the CO and $N_2$ for sample 14311, a crushed breccia rock. Note that both the CO and $N_2$ exhibit more than one peak in contrast to sample 14003 and that there is an extremely large $CO_2$ peak above 400°C. The total $CO_2$ evolved was greater in this sample than any other examined, which is not in agreement with VOL-GC-MS results given in Table 4. This could be caused by inhomogeneities in the samples. Sakai *et al.* (1972) reported severe inhomogeneity of sulfur distribution in a sample of Apollo 14 fines. This sample also gave mass spectra indicative of $C_3H_3$ and $CH_2 = C = O$ even though collected at a distance of 600 m from the LM.

Sample 14156 was obtained from a trench about 50 m from the LM and gave no evidence of rocket exhaust products. The evolution of $H_2O$, CO, $N_2$, $CO_2$, and argon was similar to that of 14003.

Figures 7 and 8 present results obtained for $H_2O$ and $CO_2$ evolved from sample 14421 in quartz and stainless steel sample tubes, respectively. This sample was heated in both the quartz and stainless steel sample tubes to determine if the sample tube

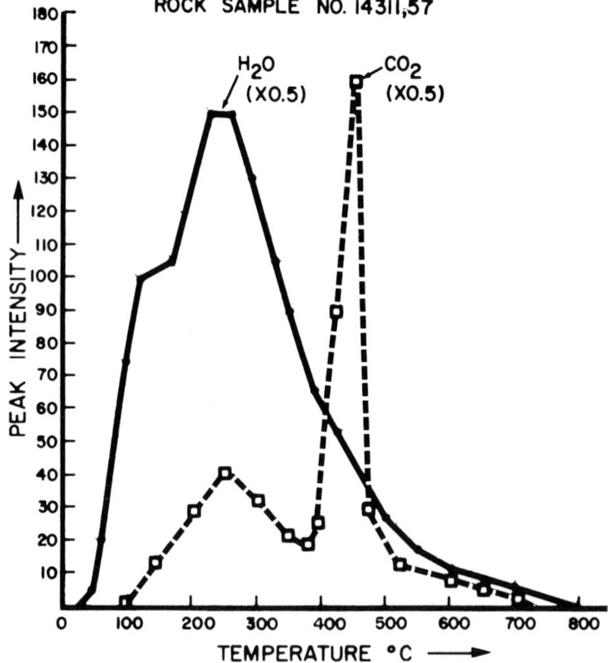


Fig. 5. Temperature release data for sample 14311. The relative peak intensities for $CO_2$ and $H_2O$ are 0.5 × the actual experimental values. Conditions identical to those for Fig. 4.

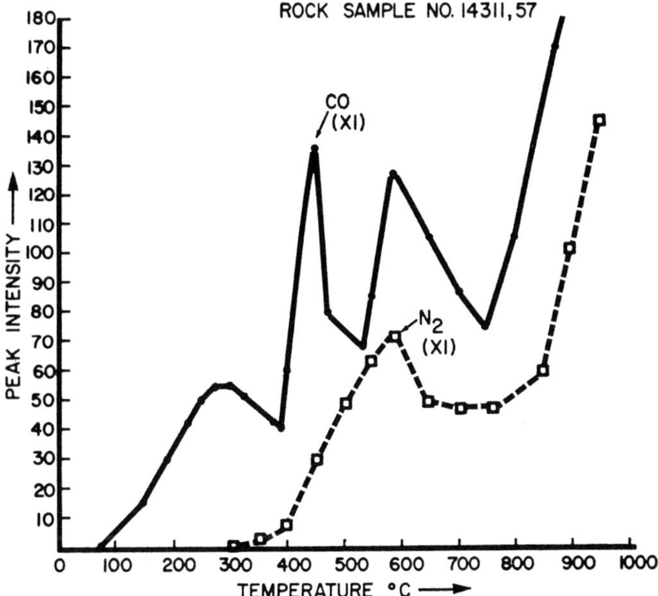


Fig. 6. Temperature release data for $N_2$ and CO in sample 14311. Conditions identical to those of Fig. 4.

itself contributed significantly to the evolved gases. The plots labelled first heating, or I, are the initial heating of the sample, and those labelled second heating, or II, are from a second heating made after the indicated exposure to laboratory atmosphere. No gases other than $H_2O$ and $CO_2$ were detectable upon heating an exposed prepyrolyzed sample. (Loosely bound adsorbed gases such as $N_2$ would be removed upon initial evacuation prior to beginning the analysis.) Immediate reheating while still maintaining vacuum was carried out with every sample run and never produced peaks greater than 0.1% of the initial heating. These results show that the water evolved from the initial heating of the sample behaves very similar to adsorbed atmospheric water and that $CO_2$ evolution occurs as if carbonates were being formed in the sample upon exposure to atmospheric $CO_2$. Similar results were observed for other samples exposed to atmosphere. $CaCO_3$ and $Na_2CO_3$ were volatilized under identical conditions and were found to release over 90% of their $CO_2$ content in narrow peaks (about 50°C width at half-height) at 550 and 525°C, respectively. The $CO_2$ peak at 750°C, however, is not generated upon atmospheric exposure and could result from indigenous carbonate in the lunar sample. These results also show a slight increase in adsorption of the stainless steel versus the quartz sample holder but good reproducibility between the two aliquots of the same sample.

Chang *et al.* (1971) reported that pyrolysis of iron carbide produces both CO and $CO_2$. We mixed calcium carbide with a prepyrolyzed sample and upon heating again did not observe any CO in the temperature region where the lunar samples release their CO. This observation does not support this calcium carbide as a source of CO in

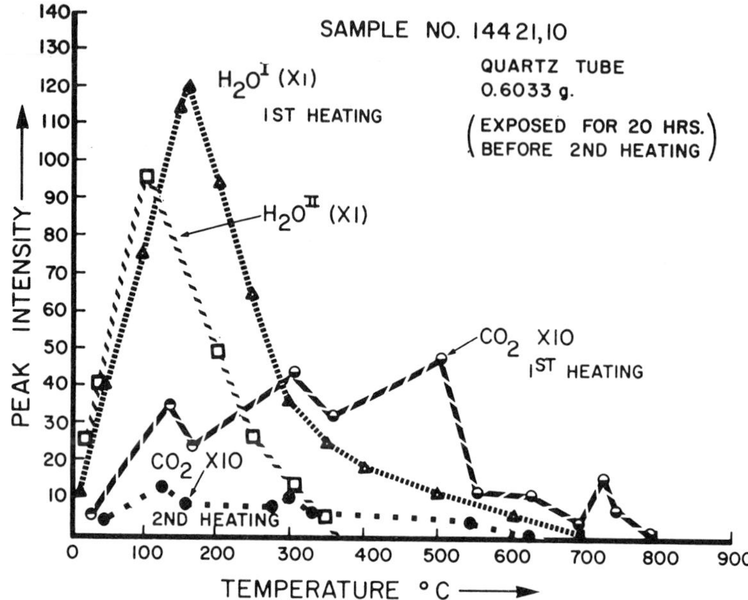


Fig. 7. Temperature release data for $H_2O$ and $CO_2$ in sample 14421. Same conditions as Fig. 4. $CO_2$ plots are × 10 actual values.

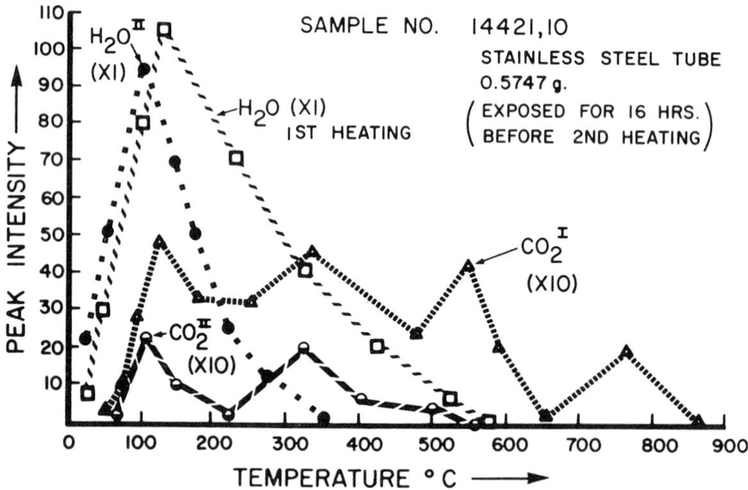


Fig. 8. Temperature release data for $H_2O$ and $CO_2$ in sample 14421. Same conditions as Fig. 4 with exception of stainless steel sample holder. $CO_2$ plots are × 10 actual values.

lunar samples but does not rule out the possibility that iron, nickel, and cobalt carbides may produce CO. Further work will be needed to clarify this point.

Samples 10086 and 12023 gave results similar to those for sample 14311 in that all three samples gave more than one peak for some of the detectable species. The major peaks for sample 10086 were one $H_2O$ peak at 100°C; four CO peaks at 235°C, 300°C, 660°C, and 875°C; two $N_2$ peaks at 625°C and 900°C; two $CO_2$ peaks at 220°C and 440°C. The major peaks for sample 12023 were two $H_2O$ peaks at 95°C and 260°C; through CO peaks at 280°C, 380°C, and 760°C; one $CO_2$ peak at 365°C; and one $N_2$ peak at 735°C. Both samples gave indications for the presence of small amounts of lunar exhaust products (m/e 39 and 41).

The results of the QMS analyses of the various samples can be summarized as follows. The presence of $H_2O$, CO, $CO_2$, $N_2$, and Ar was confirmed in all samples. The number of peaks for the individual gases released as a function of temperature as well as the peak maximum temperatures can vary considerably from sample to sample. The water released below 500°C and the $CO_2$ released below 400°C contain a substantial terrestrial atmospheric component, but the behavior of certain samples indicates the possible presence of indigenous hydrates (sample 12023) and carbonates (samples 14221 and 10086). The release behavior for nitrogen could be caused by bursting of vesicles or other traps at high temperatures or, alternately, by the presence of high temperature nitrides. The CO released at lower temperatures may be due to mineral reactions in the sample, but further work is needed to establish this fact.

The QMS results indicate the absence of any hydrocarbons with the exception of $CH_4$ in sample 14003 in contrast to the VOL-GC-MS results. This can probably be attributed to the fact that conditions are much less favorable for synthesis of hydrocarbons in these experiments than in other experimental techniques. Sensitivity factors cannot be ruled out, however, as the QMS method does not have the concentration effect of the time-temperature integration and gas chromatographic separation present in the VOL-GC-MS technique.

The detection of LM rocket exhaust products is also significant. The presence of these compounds in samples 10086 and 14003 was expected since they were collected near the LM. It is rather surprising to learn that exhaust products reached sample 14311 which is thought to have been collected about 600 m from the LM. Cross contamination during return to earth may be a factor here.

## Summary

Quadrupole-mass spectrometry and gas chromatography-mass spectrometry have been used to analyze gases released by thermal and acid treatment of lunar samples. The acid and thermal treatments released $H_2$, $N_2$, CO, $CO_2$, $CH_4$, $C_2H_6$, $C_3H_6$, $C_3H_8$, and Ar from the samples. $H_2S$ was also released by the acid treatment. These results and their significance are summarized below.

### *Carbon dioxide*

$CO_2$ contents ranged from 2–50 ppm. The small amounts produced by acid hydrolysis can in part be due to terrestrial contamination and in part to reaction of indigenous material. Temperature release data indicate that the $CO_2$ released below 400°C contains a substantial terrestrial atmospheric component, but the evolution of $CO_2$ at higher temperatures from certain samples indicates the possible presence of indigenous carbonates.

*Water*

Water contents in samples not exposed to water ranged from 30–90 ppm. Samples intentionally exposed to water ($D_2O$) evolved somewhat higher amounts. The temperature release patterns indicated that a substantial fraction of the water evolved below 500°C is terrestrial contamination, but one sample gave evidence for the possible presence of a low temperature hydrate. A significant fraction (10–20%) was released about 700°C in several samples. The high temperature water is most likely formed by solar wind proton irradiation.

*Nitrogen*

Temperature release data for nitrogen show rather sharp peaks with the maxima in the 500–900°C range. This behavior could be due to the rupture of vesicles and other traps and/or the presence of nitrides up to these stable high temperatures.

*Hydrocarbons*

Methane and other hydrocarbons were released by both acid and thermal treatments in amounts of 0.1 to a few ppm. The ratio of deuterated to nondeuterated hydrocarbons upon DCl treatment indicate more than 70% of the hydrocarbons are produced by reaction of the acid with indigenous reactive carbon species. Partially deuterated species are found indicating partially hydrogenated indigenous carbon species. The remaining 30% can be present as such or result from contamination. Temperature release data indicate maximum hydrocarbon evolution above 500°C. Reheating of an intentionally exposed prepyrolyzed sample produced small quantities of hydrocarbons. These results indicate contamination and/or synthesis from other gases may be important as sources of the observed hydrocarbons.

*LM exhaust compounds*

Compounds previously identified as lunar module engine exhaust products were identified in Apollo 12 and 14 samples. One sample was collected at a distance of 600 m from the LM.

*Carbon balance*

We are not able to make a carbon balance with the various volatiles released and account for the total carbon as reported by Moore *et al.* (1971, 1972) because of the difficulties in calibration mentioned. It is clear, however, that the major portion of the lunar carbon is released as CO above 700°C and is not affected by acid treatment. This indicates the bulk of the lunar carbon is either in a very stable form or is not accessible to acid attack due to matrix effects. The indigenous $CO_2$ evolved indicates that a few percent of the total carbon may be present as carbonates. Indigenous hydrocarbons cannot represent more than a fraction of 1% of the total carbon in the samples we have analyzed.

*Acknowledgments*—We wish to thank the lunar sample curator's staff, especially Mr. Richard Furman, for their help in loading sample 14240. We also thank Dr. Daryl Nooner for extracting one sample with benzene-methanol. This work was supported by NASA Grant NGR-44-005-125.

## REFERENCES

Abell P. I., Cadogan P. H., Eglinton G., Maxwell J. R., and Pillinger C. T. (1971) Survey of lunar carbon compounds, I. The presence of indigenous gases and hydrolysable carbon compounds in Apollo 11 and Apollo 12 samples. *Proc. Second Lunar Sci. Conf., Geochim. Cosmochim. Acta* Suppl. 2, Vol. 2, pp. 1843–1863. MIT Press.

Agrell S. O., Scoon J. H., Long J. V. P., and Coles J. N. (1972) The occurrence of goethite in a microbreccia from the Fra Mauro formation (abstract). In *Lunar Science—III* (editor C. Watkins), pp. 7–9, Lunar Science Institute Contr. No. 88.

Burlingame A. L., Hauser J. S., Simoneit B. R., Smith D. H., Biemann K., Mancuso N., Murphy R., Flory D. A., and Reynolds M. A. (1971) Preliminary organic analysis of the Apollo 12 cores. *Proc. Second Lunar Sci. Conf., Geochim. Cosmochim. Acta* Suppl. 2, Vol. 2, pp. 1891–1899. MIT Press.

Cadogan P. H., Eglinton G., Maxwell J. R., and Pillinger C. T. (1971) Carbon chemistry of the lunar surface. *Nature* **281,** 29–31.

Cadogan P. H., Eglinton G., Firth J. N. M., Maxwell J. R., Mays B. J., and Pillinger C. T. (1972) Survey of lunar carbon compounds, II. The carbon chemistry of Apollo 11, 12, 14, and 15 samples (abstract). In *Lunar Science—III* (editor C. Watkins), pp. 113–115, Lunar Science Institute Contr. No. 88.

Chang S., Kvenvolden K., Lawless J., Ponnamperuma C. and Kaplan I. R. (1971) Carbon carbides and methane in an Apollo 12 sample. *Science* **171,** 474–477.

Flory D. A. and Simoneit B. (1971) Terrestrial contamination in Apollo samples. *Space Life Sci.*, in press.

Funkhouser J., Jessberger E., Muller O., and Zahringer J. (1971) Active and inert gases in Apollo 12 and 11 samples released by crushing at room temperature and by heating at low temperatures. *Proc. Second Lunar Sci. Conf., Geochim. Cosmochim. Acta* Suppl. 2, Vol. 2, pp. 1381–1396. MIT Press.

Gay P., Bancroft G. M., and Brown M. G. (1970) Diffraction and Mössbauer studies of minerals from lunar soils and rocks. *Proc. Apollo 11 Lunar Sci. Conf., Geochim. Cosmochim. Acta* Suppl. 1, Vol. 1, pp. 481–497. Pergamon.

Gibson E. K. Jr. and Moore C. B. (1971) Compounds of the organogenic elements in Apollo 11 and 12 lunar samples—a review. *Space and Life Sci.*, in press.

Gibson E. K. and Moore G. W. (1972) Inorganic gas release and thermal analysis study of Apollo 14 and 15 studies. *Proc. Third Lunar Sci. Conf., Geochim. Cosmochim. Acta* Suppl. 3, Vol. 2, pp. 2003–2014. MIT Press.

Henderson W., Kray W. C., Newman W. A., Reed W. E., Simoneit B. R., and Calvin M. (1971) Study of carbon compounds in Apollo 11 and 12 returned lunar samples. *Proc. Second Lunar Sci. Conf., Geochim. Cosmochim. Acta* Suppl. 2, Vol. 2, pp. 1901–1913. MIT Press.

LSPET (Lunar Sample Preliminary Examination Team) (1971) Preliminary examination of lunar samples from Apollo 14. *Science* **173,** 681–693.

Moore C. B., Gibson E. K., Larimer J. W., Lewis C. F., and Nichiporuk W. (1970) Total carbon and nitrogen abundances in Apollo 11 lunar samples and selected acondrites and basalts. *Proc. Apollo 11 Lunar Sci. Conf., Geochim. Cosmochim. Acta* Suppl. 1, Vol. 2, pp. 1375–1382. Pergamon.

Moore C. B., Lewis C. F., Larimer J. W., Delles F. M., Gooley R. C., and Nichiporuk W. (1971) Total carbon and nitrogen abundances in Apollo 12 lunar samples. *Proc. Second Lunar Sci. Conf., Geochim. Cosmochim. Acta* Suppl. 2, Vol. 2, pp. 1343–1350. MIT Press.

Moore C. B., Lewis C. F., Cripe J., Kelly W. R., and Delles F. (1972) Total carbon, nitrogen, and sulfur abundances in Apollo 14 lunar samples (abstract). In *Lunar Science—III* (editor C. Watkins), pp. 550–551, Lunar Science Institute Contr. No. 88.

Nagy B., Modzeleski J. E., Modzeleski V. E., Jabbar Mohammed M. A., Nagy L. H., Scott W. M.,

Drew C. M., Thomas J. E., Ward R., Hamilton P. B., and Urey H. C. (1971) Carbon compounds in Apollo 12 samples. *Nature* **232,** 94–98.

Oró J., Updegrove W. S., Gibert J., McReynolds J., Gil-Av E., Ibanez J., Zlatkis A., Flory D. A., Levy R. L., and Wolf C. (1970) Organogenic elements and compounds type C and D lunar samples from Apollo 11. *Proc. Apollo 11 Lunar Sci. Conf., Geochim. Cosmochim. Acta* Suppl. 1, Vol. 2, pp. 1901–1920. Pergamon.

Oró J., Flory D. A., Gibert J. M., McReynolds J., Lichtenstein H. A., and Wikstrom S. (1971) Abundances and distribution of organogenic elements and compounds in Apollo 12 lunar samples. *Proc. Second Lunar Sci. Conf., Geochim. Cosmochim. Acta* Suppl. 2, Vol. 2, pp. 1913–1925. MIT Press.

Pollack J. B., Pitman D., Bishun N. K., and Sagan C. (1970) Goethite on Mars: A laboratory study of physically and chemically bound water in ferric oxides. *J. Geophys. Res.* **75,** 7480–7490.

Preti G., Murphy R. C., and Biemann K. (1971) The search for organic compounds in various Apollo 12 samples by mass spectrometry. *Proc. Second Lunar Sci. Conf., Geochim. Cosmochim. Acta* Suppl. 2, Vol. 2, pp. 1879–1889. MIT Press.

Saki H., Petrowski C., Goldhaber M. B., and Kaplan I. R. (1972) Distribution of carbon, sulfur, and nitrogen in Apollo 14 and 15 material (abstract). In *Lunar Science—III* (editor C. Watkins), pp. 672–673, Lunar Science Institute Contr. No. 88.

Simoneit B. R., Burlingame A. L., Flory D. A., and Smith I. D. (1969) Apollo lunar module engine exhaust products. *Science* **166,** 733–738.

Proceedings of the Third Lunar Science Conference
(Supplement 3, *Geochimica et Cosmochimica Acta*)
Vol. 2, pp. 2109–2118
The M.I.T. Press, 1972

# Amino acid precursors in lunar fines from Apollo 14 and earlier missions

SIDNEY W. FOX, KAORU HARADA, and P. E. HARE

Institute for Molecular and Cellular Evolution, University of Miami
Coral Gables, Florida 33134

and

Geophysical Laboratory, Carnegie Institution of Washington,
Washington, D.C. 20008

**Abstract**—Analyses by two investigative teams within one laboratory of samples obtained by hydrolysis of aqueous extracts of specially collected lunar fines (SESC) from Apollo 14 (14240,2) have verified a common pattern of five to six amino acids obtained by Fox *et al.* by IEC (ion exchange chromatography) on this and other samples. The publicized negative results by Ponnamperuma *et al.* by GLC (gas liquid chromatography) of volatile derivatives of amino acids on earlier samples from the moon have been explained as largely due to inappropriate methods of preparation and processing of the materials, e.g., direct hydrolysis of the samples. The method of hydrolysis of aqueous extracts has, however, been applied by the GLC team to fruitful examination of meteoritic samples. The results have to a late date been spoken of as amino acids whereas such values represent, in fact, chemical *precursors* hydrolyzable to amino acids in either lunar or meteoritic material.

The positive findings of amino acid precursors have been systematically examined for inhomogeneities in samples, for terrestrial and human contamination, and experiments have been performed to assess the possibility of contamination by the products of oxidation of rocket fuel. The partial evidence is against the last explanation; the two other interpretations mentioned for the positive IEC results on samples from Apollo 11 and Apollo 12 have been rigorously excluded.

Accordingly, hydrolyzable precursors of amino acids have been demonstrated in fines from seven collections from Apollo 11, 12, and 14 by the chemically discriminatory and sensitive assay technique of ion exchange chromatography. The range of amounts found corresponded to 20–70 ng/g of lunar soil.

The finding of amino acid precursors in lunar fines is of significance to a theory of molecular evolution.

## INTRODUCTION

EVIDENCE FOR THE PRESENCE or absence of organic compounds on the moon may indirectly provide insight into the state of molecular and cellular evolution in the solar system. Aside from laboratory experiments performed under geologically relevant conditions in pursuit of this objective (Fox, 1971), information is being sought from extraterrestrial sources. Evidence from life-bearing bodies (so far only the earth) and from nonlife-bearing bodies may be meaningfully compared. Samples of the surface of the moon which have now been returned from three missions (Apollo 11, 12, and 14) have been examined in the context of the prebiotic emergence of biomolecules.

The first reports of free amino acids from Apollo 11 fines (Fox *et al.*, 1970a; Nagy *et al.*, 1970) and of amino acid precursors obtained by hydrolysis of aqueous extracts

(Fox *et al.*, 1970a) were disputed (Ponnamperuma *et al.*, 1970). Examination of the published details in the methods of those failing to find amino acids suggested the possibility that some of the failure was due to the use of cold water instead of hot water for extraction. A controlled experiment on basaltic material is reported in this paper relevant to this question.

Published details (Ponnamperuma *et al.*, 1970) suggested to us that a failure to find amino acid precursors in extracts of lunar fines was due to direct hydrolysis of the samples, in contrast to hydrolysis of an aqueous extract, which would contain less mineral to accelerate decomposition during hydrolysis, and which would also minimize losses and contamination in any necessary desalting. A controlled experiment comparing direct hydrolysis with hydrolysis of aqueous extracts is reported in detail in this paper.

After amino acid precursors were again found in lunar dust from two collections from the Apollo 12 mission, arrangements were made to test Apollo 14 SESC samples side-by-side at the Ames Research Laboratory by representative analysts of two of the teams (Ponnamperuma *et al.*, 1972,* and Fox *et al.*, 1972). This choice also placed two of the assay techniques, i.e., IEC (ion exchange chromatography) and GLC (gas liquid chromatography) under close comparative observation. In attempting to assure that the time available would be conserved to this last mentioned comparison, the plan called for one investigator (K. Harada) to prepare all samples as necessary, and for an examination of details of each method of sample preparation as remaining time would allow.

Principal purposes of this paper are to report on the results of the intralaboratory comparison of assays by two groups of investigators, and then to compare these with accumulated results of samples from Apollo 11, 12, and 14 missions by the authors of this paper. Confidence in any one set of such analyses is, of course, partly a function of its repeatability. Seven collections of lunar fines have now been examined in our studies.

## Experimental Techniques

*Cleanroom and hood*

All operations in IEC analyses other than in the intralaboratory comparison were conducted in an Edcraft laminar flow hood under positive pressure in a carefully maintained cleanroom in the Institute of Molecular Evolution at the University of Miami. All of the operations for IEC analysis, and all of the preparations and partition of samples by K. Harada in the comparison at the Ames Research Center were carried out within a laminar flow hood at the latter site. The hood was first cleaned for the purpose. This hood, operated under positive pressure, was employed to minimize the possibility of atmospheric contamination within the area set aside as a cleanroom.

*Preparation of water*

The water used was freshly distilled, after which it was filtered through millipore filters (GSWPO 4700, pore size 0.22 $\mu$) washed with purified water, again in the laminar flow hood. The final distillation was carried out also in the hood, in a 2 ft wrapped Vigreux column.

---

* The abstract of the paper by Ponnamperuma *et al.* (1972) described analyses first carried out by Harada and Hare without acknowledgment to the latter.

For the intralaboratory comparison, the water prepared in Miami was selected for use in the analyses at the Ames Research Center; this water revealed no significant contamination.

*Purification of HCl*

The hydrochloric acid used for catalyzed hydrolysis was Baker's Analytical Grade mixed with an equal volume of purified water, and distilled slowly twice through a short column in the laminar flow hood. The purified HCl was examined for content of amino acid precursors and found to contain significantly lower amounts than those found in the samples. Also, such HCl was used in blanks for the entire analytical procedure; the results confirmed the absence of significant contamination, as also established in the direct examination of the 6 N HCl. In the intralaboratory comparison, the HCl was prepared by E. Peterson of the Ames Research Center.

*Preparation of glassware*

Glassware was initially cleaned by immersion in dichromate-sulfuric acid, copious washing with tap water, then with purified water, and baking overnight at 500°C. During the analyses of Apollo 12 samples, the cleaning of glassware was altered to employ washing with Haemo-Sol in almost boiling water instead of cleaning in the dichromate reagent. This glassware was subsequently rinsed copiously with hot water, then cold water, and finally with distilled water. The glassware was stored in a laminar flow hood, and was baked at 525°C for at least 3 hours before use. The glassware in the hood was used as soon as it was cool.

*Extraction of lunar and other samples*

The difference between cold water and hot water in their efficiency for extraction of amino acid precursors is revealed in Table 1. This experiment was performed on samples of cooled lava collected from a flowing molten stream at Mauna Ulu on the Island of Hawaii in September 1970, which hot

Table 1. Comparison of efficiency of cold and hot water extraction of Mauna Ulu lava* (as determined by amino acid composition in hydrolyzates).

| Amino acid | Cold (nmole) | Hot (nmole) | Hot − cold (nmole) |
|---|---|---|---|
| Aspartic acid | 0.0 | 37.5 | 38 |
| Threonine | 0.0 | 7.0 | 7 |
| Serine | 0.0 | 5.8 | 6 |
| Glutamic acid | 0.0 | 80.7 | 81 |
| Proline | 0.0 | 22.3 | 22 |
| Glycine | 0.0 | 50.4 | 50 |
| Alanine | 0.0 | 29.5 | 30 |
| Valine | 0.0 | 10.5 | 11 |
| Methionine | 0.0 | 3.7 | 4 |
| Isoleucine | 3.3 | 11.0 | 8 |
| Leucine | 5.4 | 16.6 | 11 |
| Tyrosine | 0.7 | 2.3 | 2 |
| Phenylalanine | 1.9 | 6.5 | 5 |
| Histidine | 0.4 | 6.0 | 6 |
| Lysine | 3.1 | 5.6 | 3 |
| Arginine | 0.8 | 3.6 | 3 |
| Total | 15.6 | 299.0 | 283 |
| Ammonia | 130 | 174 | 44 |
| Unknowns (leucine equivalent) | 93 | 0.0 | |

* Experiment performed by Mr. C. R. Windsor. Each analysis corresponds to 10.0 g of lava.

samples were placed directly into horizontally opened autoclaved Mason jars. By hot water, 299 nmole of amino acid precursors were extracted, whereas only 16 nmole were extracted by cold water.

Thirty grams of lava was mechanically shaken in an Erlenmeyer with 200 ml of cold distilled water for 24 hr. Another 30 gm (the hot extraction) was refluxed with 200 ml of water for 24 hr. Each extract was dried at 45°C in a rotary evaporator, and the residue hydrolyzed with 6 N HCl under reflux for 24 hr. The HCl solution was again evaporated, and the residue dissolved in 0.30 ml of water. Of this 0.10 ml was then analyzed by the method of Hare (1969).

*Direct hydrolysis of lunar fines compared to hydrolysis of aqueous extracts of lunar fines*

Aqueous extraction of lunar fines and effective temperatures for that extraction, as just described, are significant for two reasons. One is that free glycine and alanine were found in hot aqueous extracts of Apollo 11 fines by two groups (Nagy *et al.*, 1970; Fox *et al.*, 1970a). (An alternative explanation for the finding of free amino acids is that amino acid precursors were partly hydrolyzed by the hot aqueous extraction in the presence of mineral salts.)

The special significance of aqueous extraction to this analytical program is that hydrolysis of aqueous extracts has been found to yield amino acids, whereas direct hydrolysis of the solid samples of fines (Oró *et al.*, 1970; Ponnamperuma *et al.*, 1970) has not done so. The difference in results from these two methods (Table 2) was earlier observed on geochemical samples (Fox *et al.*, 1970b; Hare *et al.*, 1970).

*Extraction and hydrolysis in intralaboratory comparison*

Typically, 6 g of lunar fines was extracted by refluxing 20 ml of water for 12 to 24 hr in a 50 ml ground joint flask under a 30 cm condenser. The supernatant was removed by centrifugation, or filtration through a millipore filter, and the extracts were evaporated over flake sodium hydroxide in a vacuum desiccator. For the experiment of Table 2, water-extracted lunar fines were fortified with amino acid precursors by heating with formaldehyde and ammonia (Fox and Windsor, 1970).

The residues from the extracts were each washed three times with 1.0 ml each of purified 6 N HCl, the washings being transferred into tubes, which were evacuated with an aspirator and sealed in a flame. The contents were heated for 24 hr in an oilbath at 110–112°C.

Direct hydrolysis was performed in the usual way (Oró *et al.*, 1970; Ponnamperuma *et al.*, 1970).

The hydrolyzates were dried in the vacuum desiccator over flake sodium hydroxide. The residue was appropriately diluted with 0.01 N HCl. At this stage, samples were divided for analysis by IEC and by GLC.

All operations were carried out in the laminar flow hood. All analyses were controlled by carrying purified water through the entire sequence of operations, including hydrolysis.

Table 2. Recovery of amino acids* from fortified water-extracted lunar fines (in $\mu$mole).

| | Indirect hydrolysis | Direct hydrolysis |
|---|---|---|
| Aspartic acid | 0.04 | 0.00 |
| Glutamic acid | 0.35 | 0.00 |
| Glycine + Alanine | 1.12 | 0.00 |
| Valine | 0.35 | 0.00 |
| Isoleucine | 0.43 | 0.00 |
| Leucine | 0.27 | 0.00 |
| Unknowns | 0.78 | 0.28 |

* As judged by RT (retention time).

*Estimation of amino acids by IEC and GLC*

The nanogram analyzer of Hare (1969) and the GLC method of Gehrke *et al.* (1971) were employed, the latter by C. W. Gehrke, R. W. Zumwalt, and K. Kuo. This close comparison emphasized that, while the ninhydrin reagent employed for automatic IEC is relatively highly specific for amino acids, the GLC method reveals a multiplicity of peaks, without discriminating between them chemically. Only some of the peaks represent amino acids. The GLC assay on an aliquot from Apollo 14 material included, however, peaks corresponding to the amino acids identified by IEC, both in species and amount and was thus confirmatory. Three such confirmatory analyses were recorded; one of these three is presented in Table 3.

Table 3. Analyses of SESC samples from Apollo 14.

| | Analysis by Miami group (IEC)[a,b] ($10^{-2}$ nmole/g[d]) | Analysis by Ames group (GLC)[a,c] ($10^{-2}$ nmole/g) |
|---|---|---|
| | Sample 14240 received at Miami | |
| *Amino acid* | | |
| Aspartic acid | 0.7 | 0.2 |
| Serine | 0.3 | 0.3 |
| Glutamic acid | 1 | 0.2 |
| Glycine | 4 | 3–4 |
| Alanine | 1 | 0.1 |
| | Sample 14240 received at Ames | |
| Aspartic acid | 1 | 0.4 |
| Serine | 0.3 | 0.9 |
| Glutamic acid | 0.7 | 0.3 |
| Glycine | 6 | 4–5 |
| Alanine | 1.3 | 0.4 |

[a] For these results, both Ames and Miami groups used aliquots of material prepared by K. Harada. Samples were, however, dried by methods used only at Ames.

[b] IEC (ion exchange chromatography) is part of Miami method.

[c] GLC (gas liquid chromatography) is part of Ames method.

[d] The water blank was found to contain $0.17 \times 10^{-2}$ nmole of glycine by RT in the volume of water used for extraction. Other peaks were smaller. Since this value was negligible compared to the amount in the sample, no correction was employed for the water used.

Close examination of patterns of amino acids resulting from deliberate contamination (Harada *et al.*, 1971), hydrolyzed or unhydrolyzed, indicates that the procedures of collection and handling are practically free of human contaminants. Although the Apollo 14 SESC sample was specially collected to minimize the possibility of terrestrial contaminants, the pattern of results is similar to those obtained on samples collected without special effort.

For the Apollo 14 sample examined most recently, No. 14298, drying was carried out after a few drops of purified 6 N HCl had been added to the aqueous extract before evaporation. The water in the control blank for this analysis was of high purity, exhibiting only $\frac{1}{200}$ as much amino acid precursor as the extract of the sample itself. The methods of analysis are being examined for possible further improvement.

Some clues as to the chemical nature of the precursors have been obtained. The extraction of lunar fines with hot water, for example, releases some basic volatile material which can be hydrolyzed to glycine and alanine. Different details in the drying of volatile precursors in the two laboratories may explain why figures in Table 3 are smaller than those in other analyses. Because of these differences, the data of Table 3 are presented separately from the combined data of Table 4. The figures of Table 3 are not included in the range of 20–70 ng/g in Table 4, which presents the more quantitatively reliable results.

Table 4. Amino acid contents of hydrolyzates of extracts of lunar samples (% molar composition)[a]

| Amino acids in hydrolyzate | Apollo 11 analysis no. 1 No. 10086 | Apollo 11 analysis no. 2 No. 10086 | Apollo 12 Trench No. 12033 | Apollo 12 Surface No. 12001 | Apollo 14 No. 14003 | Apollo 14 No. 14163 | Apollo 14 No. 14298 |
|---|---|---|---|---|---|---|---|
| Aspartic acid | 5 | 5 | < 1 | 2 | 2 | 2 | 7 |
| Threonine | 2 | 3 | < 1 | | 1 | 1 | 2 |
| Serine | 9 | 10 | < 1 | 3 | 4 | 6 | 10 |
| Glutamic acid | 9 | 11 | 20 | 6 | 12 | 20 | 13 |
| Glycine | 50 | 52 | 37 | 70 | 62 | 47 | 57 |
| Alanine | 25 | 19 | 12 | 3 | 20 | 26 | 7 |
| Valine | | < 1 | < 1 | < 1 | | | |
| Isoleucine | | < 1 | < 1 | < 1 | | | 1 |
| Leucine | | < 1 | 3 | < 1 | | | 3 |
| Tyrosine | | | | 2 | | | |
| Phenylalanine | | | | 2 | | | |
| BAA | | | 25 | 10 | | | |
| Total ppb $\frac{\text{ng. amino acids}^{b}}{\text{g. lunar soil}}$ | 53 | 37 | 19 | 69 | 19 | 30 | 37 |

[a] Calculated without ammonia.
[b] A typographical error in the text, but not the tables, of the revised abstract misstates ng/g as mg/g.

## Results

Table 1 shows a marked difference in efficiency between extracting basalt with cold water as compared to extracting it with hot water. The conditions of extraction by cold water and the sample of lava are not the same conditions and material as used in extractions by others (Kvenvolden *et al.*, 1970) on lunar fines, but experiments of the type recorded in Table 1 illustrate why our group (Fox *et al.*, 1970a) and Nagy's (Nagy *et al.*, 1970) used hot water on lunar samples. These results strongly suggest futility in extraction of lunar fines by cold water.

The recovery of amino acids in the experiment of Table 2 reveals such recovery only in the case of the extract, contrasted with total loss during direct hydrolysis of the fortified lunar material. All positive results on Apollo 11, 12, and 14 samples have been obtained by this method of hydrolysis of the contents of the hot aqueous extract.

The results of one comparison of IEC with GLC are presented in Table 3.

### *Development of interpretations of the data*

The degree of confidence in the analyses is related to the number of amino acid profiles that has been accumulated from varied lunar sites. These have displayed a generally characteristic pattern. With analyses of a single sample from Apollo 11, the results were, however, equivocally interpreted (Fox *et al.*, 1970a) as due to "terrestrial contaminants, fuel exhaust products, or indigenous lunar material."

By the time three collections had been analyzed (Harada *et al.*, 1971), the main positive result was corroborated, and the evidence permitted stating that the possibility that the amino acids observed are due to oxidized rocket fuel has been largely ruled out. It was also possible to state at that time that the amino acid profiles obtained were not typical of terrestrial contamination.

Meanwhile, two groups (Ponnamperuma *et al.*, 1970; Oró *et al.*, 1970) had recorded no amino acids from lunar dust after hydrolysis. As suggested in our first papers and supported further in this one, the failure of others to find amino acid precursors is attributable especially to their use on the lunar samples of direct hydrolytic conditions. Indeed, in their findings of amino acids from the Murchison meteorite reported subsequently to our analyses of samples from Apollo 11 and 12, the Ponnamperuma group reported that they also used extraction by hot water, followed by hydrolysis of the material in the extract (Kvenvolden *et al.*, 1971).

A view that Ponnamperuma *et al.* had repeated our method without obtaining our results on lunar fines was however held recently (Anonymous, 1970). As this and earlier papers establish, the Ponnamperuma group had in fact not used our method on Apollo 11 or 12 samples, although they subsequently used close to the same method, based on IEC, on a meteorite. No *documented* evidence is at hand, even yet, that they have used our modes of sample preparation and analysis *in toto* on any lunar sample.

The intralaboratory comparison on the SESC sample from Apollo 14 (14240), as explained in detail in this paper, revealed that two analytical methods could be applied to aqueous extracts to verify the findings in this Apollo 14 sample, and by inference, our findings in Apollo 11 and 12 samples. This study showed, however, that the GLC method at these levels is most useful for confirmation, since it lacks the necessary specificity, as provided by ninhydrin for the IEC method. After programmed heating, peaks corresponding to the N-trifluoroacetylamino acid n-butyl esters of alanine, glycine, serine, aspartic acid, and glutamic acid in that order can be observed in the GLC analysis, but often they are found in the midst of a number of other mainly unidentified peaks. The order of elution on IEC is aspartic acid, serine, glutamic acid, glycine, and alanine. The finding of these five amino acids by two methods thus provides strong assurance of the identities and amounts of these five amino acids.

The SESC sample (14240) plus three other Apollo 14 samples have contributed four to now a total of seven collections which have yielded a characteristic amino acid profile. These seven collections have been subjected to analysis more than once each (note two analyses of sample 10086 in Table 4) but no negative results have been observed within the entire series.

## Discussion

Table 4 summarizes the results of analyses of six collections, exclusive of the SESC samples. The BAA of the two Apollo 12 samples occurs at an RT in the region of the basic amino acids (BAA), but it does not coincide with the RT of any proteinous basic amino acid. The presence of tyrosine and phenylalanine in one sample (12001), only, correlates with the fact that a small amount of unfiltered air entered the flask in which this sample was prepared. The values recorded for these two amino acids begin to provide some sense of the quantitative effect of contamination, when it is known to have occurred.

Questions of inhomogeneity and human contamination of lunar samples have been discussed elsewhere (Harada *et al.*, 1971). The possibility of contamination by oxidized rocket fuel has been examined indirectly and evaluated as unlikely. Examination of Apollo 15 samples from near to and far from the landing vehicle will permit a more direct test of this question. The identity of the five amino acids, in species and amount, is virtually assured by their being observed by both IEC of free amino acids and GLC of derivatives. In this regard, also, finding of sets of amino acids by ninhydrin is much more unambiguous evidence than is obtainable, for example, from a comparison of a single RT with that of a standard amino acid.

The fact that the amino acids are obtained by acid-catalyzed hydrolysis renders them of special interest in the context of evolution (Fox, 1972). Free amino acids are not as likely chemically to be found in situ on a dry moon as are polyamino acids or chemical precursors of amino acids or polymers of the precursors. Free amino acids in aqueous solution, however, are in a relatively stable state. Indeed, the hydrolytic conversion to amino acids in these analyses indicates such stability. The chemical nature of the precursors or polymers can thus be of value in clarifying a pathway of molecular evolution that proceeds through intermediates closely related to amino acids. The fact that the unknown compounds are precursors in an operational laboratory sense is analytically significant. The fact that such precursors are removed from free amino acids by the single, geologically relevant, step of hydrolysis is significant to understanding evolution. The fact that they are precursors rather than amino acids, which are unstable under some conditions, may contribute to the survival of the line of molecular evolution.

The repeated finding of amino acid precursors in the 20–70 ppb range, when the carbon and nitrogen contents are in the $10^2$ ppm range (Moore *et al.*, 1972), indicates that amino acid precursors tend to form extraterrestrially when organic matter of suitably related type is present. The source and range of suitable carbon-nitrogen compounds are questions for future investigation.

We can explain the fact that molecular evolution on the moon has not proceeded to amino acids as owing to the absence of solvent water. A general theoretical flowsheet for molecular evolution is

$$\text{Cosmic reactants} \xrightarrow[\text{(a)}]{} \text{amino acid precursors} \xrightarrow[\text{(b)}]{+\mathrm{H_2O}}$$

$$\text{amino acids} \xrightarrow[\text{(c)}]{-\mathrm{H_2O}} \text{polyamino acids} \xrightarrow[\text{(d)}]{\mathrm{H_2O}} \text{protocells.}$$

On the moon, the sequence appears to have terminated before (b) because of lack of water. Should further investigation reveal that the compounds found in lunar fines are chemical precursors rather than polyamino acids, the results will favor the interpretation of the absence of water from the moon during the entire period in which the amino acid-related compounds have been present there.

Since various mineral assemblages on the surface of the moon are suggestive of earlier melting (Gibson *et al.*, 1972; Stewart *et al.*, 1972), the amino acid precursors found are difficult to explain as products of indigenous organic synthesis, and are likely to have been formed by onfall of appropriate organic compounds subsequent to some cooling of the surface. Candidate sources are meteorites (Kvenvolden *et al.*, 1971), comets (Oró, 1961), interstellar clouds of organic matter (Snyder and Buhl,

1970), meteorites, and solar wind (Chang *et al.*, 1971; Cadogan *et al.*, 1972). Further chemical examination of the amino acid precursors is desirable in order to attempt to identify the source and chemical nature of the organic reactants and to obtain answers to other related questions such as the cosmochemical pathways.

*Acknowledgments*—This work was aided by Grant No. NGR 10-007-088 from the National Aeronautics and Space Administration. Contribution No. 214 of the Institute for Molecular and Cellular Evolution. We thank Mr. Charles R. Windsor for assistance.

## References

Anonymous (1970) Scientists disagree on lunar amino acids. *Chem. Eng. News* **48** (40), 37–38.

Cadogan P. H., Eglinton G., Firth J. N. M., Maxwell J. R., Mays B. J., and Pillinger C. T. (1972) Survey of lunar carbon compounds II: The carbon chemistry of Apollo 11, 12, 14, and 15 samples (abstract). In *Lunar Science—III* (editor C. Watkins), pp. 113–115, Lunar Science Institute Contr. No. 88.

Chang S., Kvenvolden K., Lawless J., Ponnamperuma C., and Kaplan I. R. (1971) Carbon, carbides, and methane in an Apollo 12 sample. *Science* **171,** 474–477.

Fox S. W. (1971) Chemical origins of cells—2. *Chem. Eng. News* **49** (50), 46–53.

Fox S. W. (1972) Evolution from amino acids: Lunar occurrence of their precursors. *Ann. N.Y. Acad. Sci.*, in press.

Fox S. W., Harada K., and Hare P. E. (1972) Amino acid precursors in lunar fines from Apollo 11, Apollo 12, and Apollo 14 (abstract). In *Lunar Science—III* (editor C. Watkins), pp. 277–279, Lunar Science Institute Contr. No. 88.

Fox S. W., Harada K., Hare P. E., Hinsch G., and Mueller G. (1970a) Bio-organic compounds and glassy microparticles in lunar fines and other materials. *Science* **167,** 767–770.

Fox S. W., Hare P. E., Harada K., and Windsor C. R. (1970b) *Sixth Ann. Rept. Inst. Molec. Evol'n.* pp. 2–3, Univ. of Miami.

Fox S. W. and Windsor C. R. (1970) Synthesis of amino acids by the heating of formaldehyde and ammonia. *Science* **170,** 984–985.

Gehrke C. W., Zumwalt R. W., Stalling D. L., Roach D., Aue W. A., Ponnamperuma C., and Kvenvolden K. A. (1971) A search for amino acids in Apollo 11 and Apollo 12 lunar fines. *J. Chromatog.* **59,** 305–319.

Gibson E. K. Jr. and Hubbard N. J. (1972) Volatile element depletion investigations on Apollo 11 and 12 lunar basalts via thermal volatilization (abstract). In *Lunar Science—III* (editor C. Watkins), pp. 303–305, Lunar Science Institute Contr. No. 88.

Harada K., Hare P. E., Windsor C. R., and Fox S. W. (1971) Evidence for compounds hydrolyzable to amino acids in aqueous extracts of Apollo 11 and Apollo 12 lunar fines. *Science* **173,** 433–435.

Hare P. E. (1969) (Geochemistry of proteins, peptides, and amino acids) Amino acid analysis. In *Organic Geochemistry* (editors G. Eglinton and M. T. J. Murphy), p. 452, Springer-Verlag.

Hare P. E., Harada K., and Fox S. W. (1970) Analyses for amino acids in lunar fines. *Proc. Apollo 11 Lunar Sci. Conf., Geochim. Cosmochim. Acta* Suppl. 1, Vol. 2, pp. 1799–1803. Pergamon.

Kvenvolden K. A., Chang S., Smith J. W., Flores J., Pering K., Saxinger C., Woeller F., Keil K., Breger I., and Ponnamperuma C. (1970) Carbon compounds in lunar fines from Mare Tranquillitatis —I. Search for molecules of biological significance. *Proc. Apollo 11 Lunar Sci. Conf., Geochim. Cosmochim. Acta* Suppl. 1, Vol. 2, pp. 1813–1828. Pergamon.

Kvenvolden K. A., Lawless J. G., and Ponnamperuma C. (1971) Nonprotein amino acids in the Murchison Meteorite. *Proc. Nat. Acad. Sci. U.S.* **68,** 486–490.

Moore C. B., Lewis C. F., Cripe J., Kelly W. R., and Delles F. (1972) Total carbon, nitrogen and sulfur abundances in Apollo 14 lunar samples (abstract). In *Lunar Science—III* (editor C. Watkins). pp. 550–552, Lunar Science Institute Contr. No. 88.

Nagy B., Drew C. M., Hamilton P. B., Modzeleski V. E., Murphy S. M. E., Scott W. M., Urey H. C., and Young M. (1970) Organic compounds in lunar samples: Pyrolysis products, hydrocarbons, amino acids. *Science* **167,** 770–773.

Oró J. (1961) Comets and the formation of biochemical compounds on the primitive earth. *Nature* **190,** 389–390.

Oró J., Updegrove W. S., Gibert J., McReynolds J., Gil-Av E., Ibanez J., Zlatkis A., Flory D. A., Levy R. L., and Wolf C. (1970) Organogenic elements and compounds in surface samples from the Sea of Tranquillity. *Science* **167,** 765–767.

Ponnamperuma C., Gehrke C., and Kvenvolden K. (1972) The search for amino acids in the Apollo 12 and Apollo 14 samples (abstract). In *Lunar Science—III* (editor C. Watkins), p. 611, Lunar Science Institute Contr. No. 88.

Ponnamperuma C., Kvenvolden K., Chang S., Johnson R., Pollock G., Philpott D., Kaplan I., Smith J., Schopf J. W., Gehrke C., Hodgson G., Breger I. A., Halpern B., Duffield A., Krauskopf K., Barghoorn E., Holland H., and Keil K. (1970) Search for organic compounds in the lunar dust from the Sea of Tranquillity. *Science* **167,** 760–762.

Snyder L. E. and Buhl D. (1970) Molecules in the interstellar medium. *Sky and Telescope* **40,** 267–271, 345–348.

Stewart D. B., Ross M., Morgan B. A., Appleman D. E., Huebner J. S., and Commeau R. F. (1972) Mineralogy and petrology of lunar anorthosite 15415 (abstract). In *Lunar Science—III* (editor C. Watkins), pp. 726–728, Lunar Science Institute Contr. No. 88.

Proceedings of the Third Lunar Science Conference
(Supplement 3, *Geochimica et Cosmochimica Acta*)
Vol. 2, pp. 2119–2129
The M.I.T. Press, 1972

# Amino acid analyses of Apollo 14 samples

CHARLES W. GEHRKE, ROBERT W. ZUMWALT, KENNETH KUO, and WALTER A. AUE

Experiment Station Chemical Laboratories, University of Missouri, Columbia, Missouri 65201

DAVID L. STALLING

Fish Pesticide Research Laboratory, U.S. Bureau of Sport Fisheries and Wildlife, Columbia, Missouri 65201

KEITH A. KVENVOLDEN

Exobiology Division, NASA, Ames Research Center, Moffett Field, California 94035

and

CYRIL PONNAMPERUMA

Laboratory of Chemical Evolution, Department of Chemistry, University of Maryland, College Park, Maryland 20742

**Abstract**—Water extracts of lunar fines from the Apollo 14 mission were analyzed for amino acids by a gas-liquid chromatograph technique wherein amino acids are converted to their N-trifluoroacetyl-*n*-butyl esters prior to analysis. Detection limits were established at 300 pg (picograms) to 1 ng (nanogram) for different amino acids. Initial analyses of water extracts and acid-hydrolyzed (6N HCl) water extracts of sample 14240 showed no amino acids above background. Also the acid-hydrolyzed water extracts of 14298 did not contain amino acids at levels greater than 1 ng/g of lunar fines. In collaborative experiments at Ames Research Center, two groups of investigators found extremely low concentrations of at least two amino acids in other samples of 14210. The group from Ames Research Center used gas-liquid chromatography while the group from Miami University used ion-exchange chromatography. Glycine was reported at concentration of 3 to 4 ng/g and alanine was less than 1 ng/g. Because of the minute quantities, these identifications could not be confirmed by gas chromatography-mass spectrometry, and therefore should still be considered as tentative even though the identifications have been collaborated by two independent chromatographic techniques. The evidence is yet insufficient to suggest that these amino acids are indigenous to the lunar surface.

Other studies included the analyses of performance standards at the 2 to 6 ng level for each of 17 amino acids, and the analyses of 5 ml of $H_2O$ containing 2 ppb of each amino acid. Recovery of amino acids added to lunar fines was conducted at the 10, 50, and 70 ng level of each amino acid with 50 to 70 mg of lunar material. The recoveries varied from as high as 80% for some of the neutral and acidic amino acids to complete loss of ornithine and lysine.

## INTRODUCTION

THE EXAMINATION OF the lunar material acquired by the Apollo missions for indigenous, biologically significant substances has been of primary concern to several research groups. These studies are of obvious importance, as they provide a unique possibility for gaining insight into the nature of chemical evolution. The important roles played by amino acids in terrestrial life and the possibility to determine them at a trace level by recently developed methodologies have made these compounds prime targets in our investigation of extraterrestrial materials.

This study was primarily directed toward the examination of Apollo 14 lunar fines for indigenous amino acids or materials which could be converted to amino acids on hydrolysis with 6N hydrochloric acid.

Previous gas chromatographic studies conducted by Gehrke *et al.* (1971a) on unhydrolyzed and hydrolyzed water extracts of Apollo 11 and 12 samples have resulted in none of the protein amino acids being detected. Hare *et al.* (1970) and Nagy *et al.* (1970) reported the presence of glycine and alanine in unhydrolyzed water extracts of Apollo 11 fines, though analyses of a subsurface Apollo 12 sample (12023) resulted in no free amino acids being detected by either Harada *et al.* (1971) or Nagy *et al.* (1971), both of whom used ion-exchange chromatographic techniques. However, glycine, alanine, glutamic acid, and leucine were reported by Harada *et al.* (1971) to be present after hydrolysis of the aqueous extract of sample 12023 at a total concentration of 19 ppb.

In this investigation, a special sample of fines brought back by the Apollo 14 astronauts was made available to the three groups for analysis for amino acids. This sample (14240 SESC) was opened and distributed in a clean facility at the Space Sciences Laboratory in Berkeley, California. Earlier experiments had shown this facility to be free of any detectable amino acid contamination.

## THE EXPERIMENT

### *Possible sources of contamination*

Gas-liquid chromatographic methods have been developed by Zumwalt *et al* (1971) that permit $10^{-11}$ mole quantities of amino acids to be detected. At this level of analysis, the problems of background become even more serious. Contamination by organic compounds may occur at any time from the collection of the lunar sample on the surface of the moon to injection of processed lunar extracts on the chromatographic column. An organic-clean room is essential with rigorous control of cleanliness. Special hoods with filtered laminar-flow air should be used in the working area to exclude particulate matter.

In our work on amino acids, we surveyed sources of contamination such as hair, dust, cigarette smoke, dandruff, flakes of skin, saliva, $N_2$ as carrier gas, ion-exchange resin solubles, fingerprints, and the ghosting effect of certain chromatographic columns. All of the above may be very important and can confuse a chromatogram. Indeed, it is a difficult experimental exercise to analyze for "nothing," as always something will be found at the nanogram level. Other serious sources of contamination are the protective coverings commonly used in the laboratory. Latex gloves can be a major source of contamination, from the drying powder to the plasticizers in the gloves. This source can be eliminated by using white cotton gloves or polyethylene gloves.

Cleaning glassware for organic analysis has been one of the most difficult problems in the removal of contaminants from the analytical system. Hot sulfuric chromic acid has been commonly used as the cleaning agent, followed by copious rinsing with distilled water and organic solvents. Perhaps the best procedure that has been evolved is the method of Hare *et al.* (1970) in which the glassware is thoroughly washed with an alkaline reagent, copiously rinsed with distilled water, wrapped in aluminum foil, placed in stainless steel containers and heated at 500 to 600°C for 3 to 20 hours. This removes all organics that interfere in GC and GC–MS experiments.

In our work, an instrument modification permitted venting in the injection port of the excess solvents and reagents resulting from the derivatization reactions (Gehrke *et al.*, 1971; Zumwalt *et al.*, 1971). This allows one to inject much larger amounts of the final derivatized sample into the GC (up to 100 $\mu$l; see chromatograms). With venting, the solvents and reagents do not traverse the chromatographic column, interact with it, and change its performance. This saves the columns and substrates and allows hundreds of injections to be made. In fact, we used our EGA and OV–17 columns for months with a high integrity of reproducibility and reliability. Only the derivatized molecule or molecules of similar volatility are permitted to traverse the chromatographic column. In this way, an improvement of at least one and perhaps two orders of magnitude in sensitivity is achieved. Furthermore, the total sample can be placed on the column for analysis. This is an important advance in instrumentation and should be useful in further GLC and GC–MS investigations at the nanogram level of analysis.

*Apparatus and reagents*

A Varian 2100 Series gas chromatograph with a Varian Model 20 dual pen strip chart recorder was used. The electrometer output was attached to a voltage divider which allowed the signal to be displayed on both pens of the recorder with a fourfold difference of amplification. The ethylene glycol adipate (EGA) column packing and solvent vent system described earlier by Zumwalt *et al.* (1971) were used.

The derivatization reagents, *n*-butanol · 3N HCl, dichloromethane, and trifluoroacetic anhydride (TFAA) were of the quality described by Gehrke *et al.* (1971). The derivatization of the amino acids to the N-trifluoroacetyl *n*-butyl (N-TFA) esters was conducted in pyrex microreaction vials with all-teflon screw caps. The derivatization reagents were added to the samples in the vials via 25, 50, and 100 $\mu$l micropipets attached to glass syringes with teflon tubing.

The following precautions were observed to reduce difficulties during the derivatization and chromatography of the samples.

(1) Aqueous samples were placed under an infrared lamp and allowed to dry without excessive heating. Dichloromethane was added, then evaporated, to azeotropically remove the last traces of $H_2O$.

(2) After formation of the amino acid *n*-butyl esters with *n*-butanol · 3N HCl, the butanol · HCl was also evaporated under the IR lamp. Care was taken to prevent excess heating of the sample, as the more volatile aliphatic amino acid esters are subject to losses at this point. However, all the butanol · HCl must be removed, as injection of butanol · HCl may present problems in the chromatographic analysis.

(3) Careful preparation of the chromatographic column proved essential, with special attention devoted to obtaining a uniform coating on the support material and a minimum of fractured particles after coating. Conditioning of the column to reduce liquid phase "bleed" was conducted at 220°C with a $N_2$ carrier flow of about 50 ml/min until the "bleed" rate had been reduced to a satisfactory level. When not in use, the columns were kept at 200°C in the chromatograph with a carrier flow of 20 to 50 ml/min. If the columns had to be removed, the ends were tightly capped to exclude moisture. Generally, EGA columns should not be subjected at any time to temperatures in excess of 225°C for longer than one to two hours.

### *Preparation of extracts and hydrolysates*

The samples of lunar fines (0.5 to 2 g) were placed in pyrex glass conical centrifuge tubes (13 ml), water added (5 ml/g of fines), and the tubes were closed with all-teflon screw caps. The tubes were shaken manually, then placed in a 100°C sand bath for 24 hours with periodic shaking. At the end of this period, the tubes were allowed to cool, then centrifuged at 3300 rpm for 10 min. The supernatant was then decanted into small (10 to 20 ml) beakers for evaporation, followed by hydrolysis and derivatization.

After evaporation of the extracts to about 1 ml under the IR lamp, they were transferred to pyrex micro reaction vials and taken to dryness. Then, 100 $\mu$l of 6N HCl were added, and the vials capped with teflon "Mininert" caps with stopcock-like apertures for attachment to a laboratory vacuum line.

After partial evacuation, the vials were closed and heated for 22 hours at 110°C in a sand bath. The samples were then allowed to cool, and the HCl evaporated. The samples were then derivatized as described by Gehrke *et al.* (1971a).

## RESULTS AND DISCUSSION

Initial experiments were conducted to confirm the integrity of the derivatization reactions and reagents and to optimize the GLC instrumental and chromatographic system for the separation and flame ionization detection of the amino acid derivatives (Gehrke *et al.*, 1972).

Interest was then centered on the analysis of the water extract of the special Apollo 14 sample (ARC 14240 SESC). This extract was divided into two equal portions; one portion was derivatized and analyzed for free amino acids, the other was hydrolyzed with 6N HCl, then derivatized and analyzed. Figure 1 shows the resultant chromatogram from the unhydrolyzed extract and the corresponding procedural blank. None of the protein amino acids were observed above background. Compare the corresponding positions on the two chromatograms. The arrows mark the elution positions for the amino acids, and the retention temperatures were reproducible to better than 0.5°C. The EGA chromatographic column gave a good background over most of the temperature scale, i.e., less than 500 pg to 180°C, and 3 ng at a temperature of 195°C. However, this background did not interfere with the detection of glutamic acid as it eluted at 197°C and was completely resolved from the interference. Although unidentified chromatographic peaks could have partially obscured valine, isoleucine, threonine, serine, methionine, and hydroxyproline, the positions of the

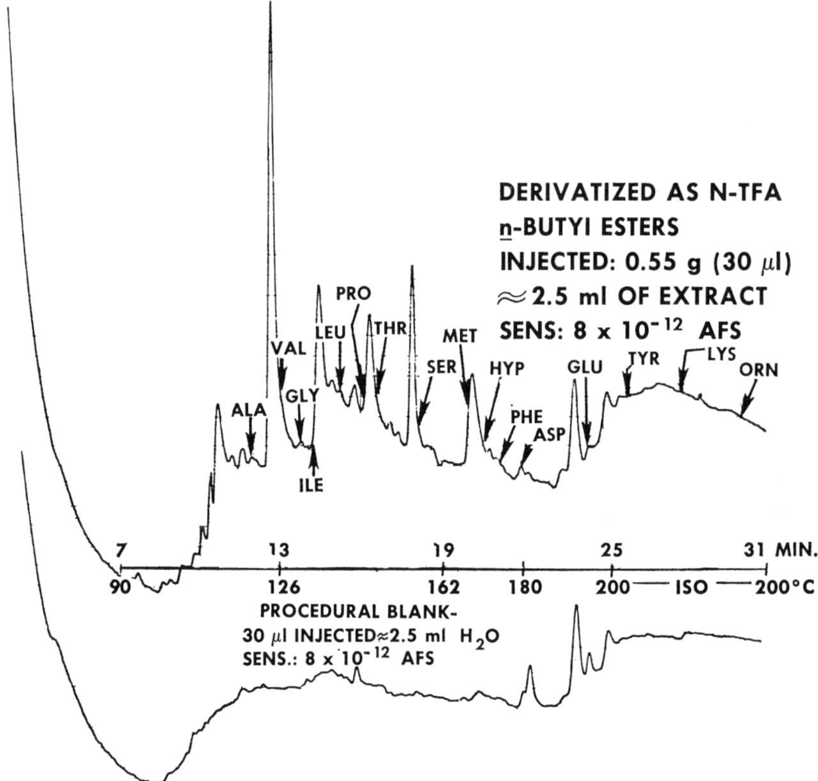


Fig. 1. GLC of water extract of Apollo 14 (ARC 14240, SESC). Derivatized as N–TFA *n*-butyl esters—injected: 0.55 g (30 $\mu$l) $\simeq$ 2.5 ml of extract; sens: 8 $\times$ $10^{-12}$ AFS. Procedural blank—injected: 30 $\mu$l $\simeq$ 2.5 ml $H_2O$; sens: 8 $\times$ $10^{-1}{}_2$ AFS.

other amino acids were free from interferences with a background of <500 pg. This is especially noteworthy in the case of alanine and glycine, amino acids of primary interest because they had been reported to occur in relatively large amounts in lunar fines by Harada *et al.* (1971) and Nagy *et al.* (1970). The chromatographic peaks for the hydrolyzed portion of the extract were of the following magnitudes: alanine, about 1 ng; glycine, about 3 ng; serine, about 1 ng; aspartic acid, about 2 ng; and glutamic acid, about 2 ng; the same amino acids were present at comparable levels in the hydrolyzed water blank.

A further series of analyses for amino acids were conducted on a separate 0.5 g Apollo 14 sample (ARC 14298). Alanine, glycine, serine, aspartic acid, and glutamic acid were observed at concentrations ranging from 6 to 12 ng/g and were identified by cochromatography of a standard amino acid mixture with the remainder of the sample. Since the 0.5 g suggested the presence of traces of amino acids, a second 1.0 g sample was analyzed as a "scale-up" study. No increase in the size of the amino acid peaks was noted on comparison of the chromatograms for the 0.5 and 1.0 g samples.

After subtraction of amino acids present in the hydrolyzed procedural blank, it was concluded that these amino acids were not present at levels >1 ng/g of lunar fines.

*Collaborative experiments at Ames*

In an effort to resolve the divergent results which have originated from the analyses of the lunar fines with regard to amino acids, a special collaborative investigation was conducted on the Apollo 14 fines (14240 SESC) at the Ames Research Center, Moffett Field, California, during October, 1971. The major objective was to resolve the question of whether amino acids were or were not present in hydrolyzed water extracts of the lunar fines and whether the different techniques of ion-exchange chromatography and gas-liquid chromatography would yield the same results.

The two participating teams were composed of the following: Ponnamperuma and his associates Gehrke, Zumwalt, Kuo, and Kvenvolden on the one hand, and Fox and coworkers Hare and Harada on the other. Gas-liquid chromatographic methods were used in the analyses by Ponnamperuma and coworkers, and Fox and associates used ion-exchange chromatography in their studies. The samples analyzed were: (a) a 6-gram sample of Apollo 14 SESC assigned to Ponnamperuma, (b) a 6-gram sample of Apollo 14 SESC assigned to Fox, and (c) another portion of the Apollo SESC sample extracted and hydrolyzed in Miami and brought to the Ames Research Center. In all of the experiments the two teams worked together in the extraction, preparation, and hydrolysis of the extracts. The final analysis for amino acids was made by GLC and IEC on the divided hydrolyzed lunar extracts by the two participating teams.

All the glassware used for the extraction, evaporation, and hydrolysis was cleaned by washing with Alconox in hot water, and thorough rinsing with hot water and distilled $H_2O$. The glassware was then placed in a furnace at 520 to 550°C for 3 to 15 hours to remove organic contaminants.

Prior to the extraction of the fines, blanks of the extraction system, distilled water, and reagents were prepared and analyzed both by GLC and IEC.

The water extracts were prepared by refluxing 6 g of Apollo 14 fines (14240) with about 22 ml of water for 12 hours. At the end of this period, the samples were allowed to cool and the water decanted into pyrex glass centrifuge tubes and centrifuged at 3500 rpm for 15 min. The supernatants were then decanted into small beakers, which were placed in a desiccator over NaOH, and a partial vacuum was applied (<25 mm). In most instances, the aqueous extracts became frozen during this step and sublimation occurred.

After vacuum drying of the aqueous extracts, the samples were dissolved in 3 ml of 6N HCl and transferred to a pyrex glass hydrolysis tube. A vacuum was applied to the samples for a few minutes to remove air and the tubes were sealed with a propane-oxygen torch. The hydrolysis tubes were wrapped in aluminum foil to ensure even heat distribution throughout the tube, then placed in 110°C oven for 20 hours. After removal of the tubes from the oven, they were placed in a desiccator over NaOH pellets with partial vacuum for removal of the 6N HCl. After drying, the residue was taken up in 200 $\mu$l of 0.01 N HCl; then one-half (100 $\mu$l) was taken for analysis by GLC and the other half for analysis by IEC.

GLC analyses of three consecutive water extracts of the samples (a) (Ames Research Center) are presented in Fig. 2. In extract 1, 3 ng/g of glycine were observed, less than 0.5 ng/g of alanine, about 1 ng/g of serine, and about 0.5 ng/g of both aspartic and glutamic acids. Extract 2 contained a similar amount of alanine, about 1 ng/g of glycine, and lesser amounts of serine, aspartic acid, and glutamic acid. In extract 3, the amount of glycine was similar to extract 2 (1 ng/g), alanine was well below the 0.5 ng/g level, serine at about 0.5 ng/g or less, and smaller quantities of aspartic and glutamic acids.

Thus, the amino acid found by GLC in the largest concentration was glycine, at 3 ng/g in the Apollo 14 SESC lunar fines (Ames). The GLC data obtained for this sample (extract 1) were in excellent agreement with IEC analysis on the other half of the sample.

Also of interest in Fig. 2 are the chromatographic peaks which do not correspond to the common protein amino acids. Analysis of extract 1 resulted in a large peak (30 to 40 ng/g) between the alanine and glycine elution positions, a second large peak near

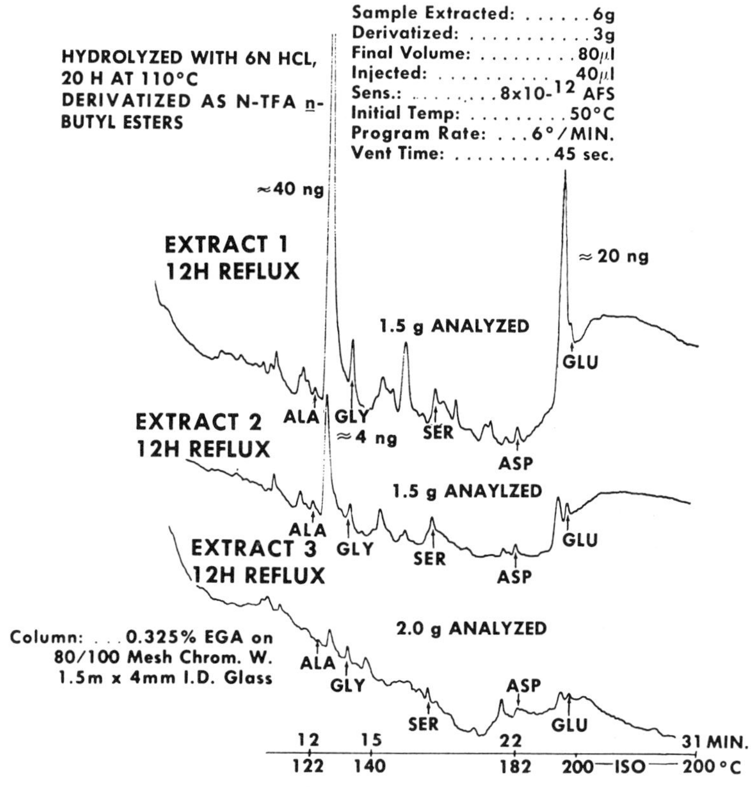


Fig. 2. GLC of water extract of Apollo 14 (14240 SESC), Ames sample joint study. Hydrolyzed with 6N HCl, derivatized as N-TFA *n*-butyl esters—injected: 40 μl; sens: $8 \times 10^{-12}$ AFS.

glutamic acid, and numerous smaller peaks throughout the chromatogram. Essentially all of these peaks decreased in intensity as the sample was re-extracted.

Figure 3 presents the chromatogram obtained on GLC analysis of the hydrolyzed water extract of sample (b) (Miami). The top GLC chromatogram shows about 2 to 3 ng/g of glycine present, and less than 0.5 ng/g of alanine, serine, aspartic acid, and glutamic acid. Again, these results were in excellent agreement with IEC data on the remaining half of the extract. Also, these results are in agreement with those obtained from the Ames sample, with the most abundant amino acid, glycine, being present at the 3 to 4 ng/g level.

Figure 4 shows the chromatogram which results from the analysis of a hydrolyzed water extract of Apollo 12 sample (ARC 12023,22). Again, although small amounts of amino acid peaks were observed (3 to 4 ng of glycine, serine, aspartic acid), similar quantities were observed in the corresponding procedural blank. The large chromatographic peak in the valine region of the chromatogram was identified as derivatized oxalic acid by GC–MS.

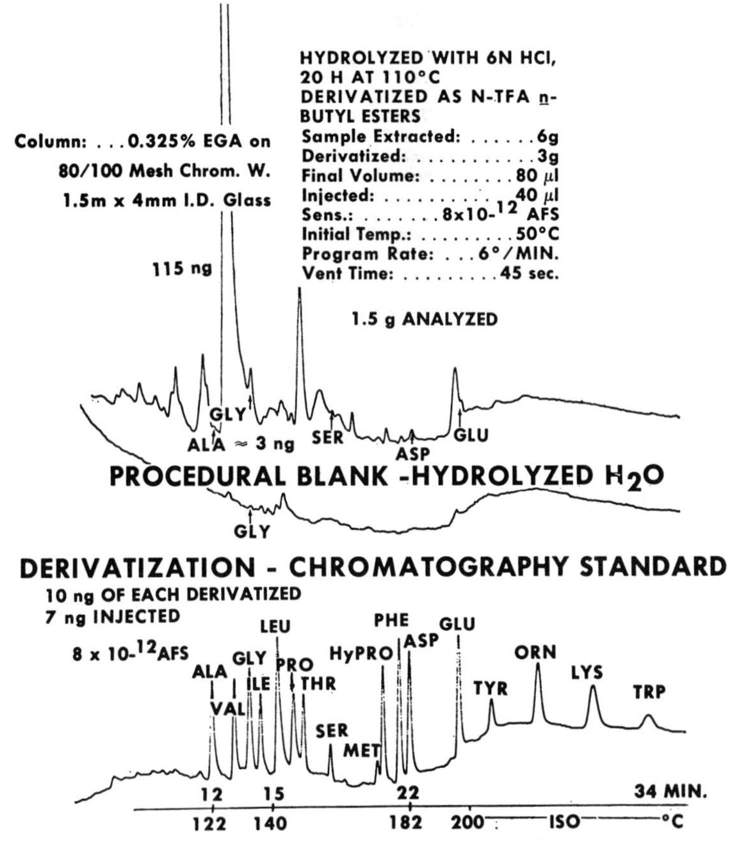


Fig. 3. GLC of water extract of Apollo 14 (14240 SESC), Miami sample joint study. Hydrolyzed with 6N HCl, derivatized as N–TFA *n*-butyl—injected: 80 μl; sens: $8 \times 10^{-12}$ AFS.

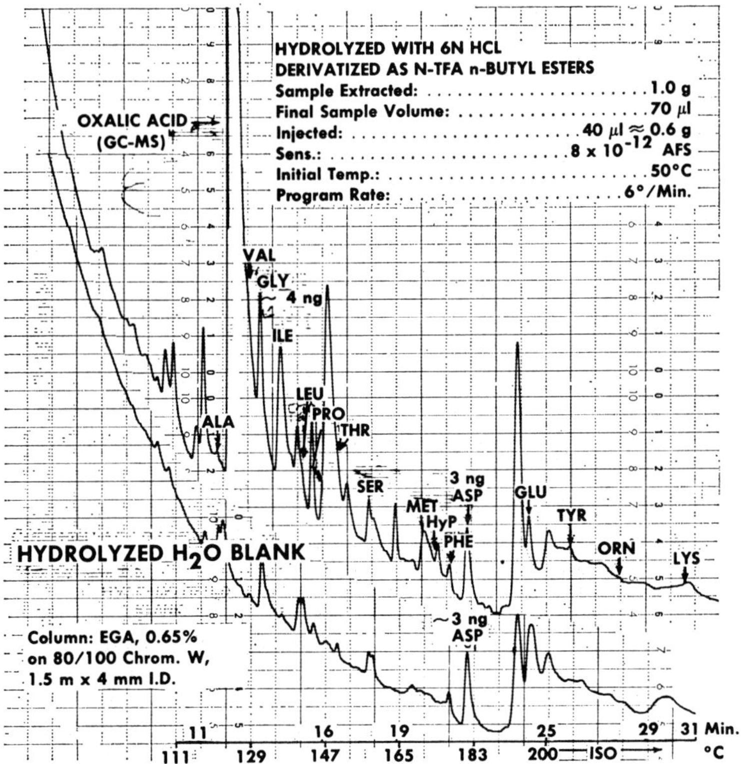


Fig. 4. GLC of water extract of Apollo 12 (ARC 12023,22). Hydrolyzed with 6N HCl, derivatized as N-TFA *n*-butyl esters—injected: 40 μl (0.6 g); sens: $8 \times 10^{-12}$ AFS.

A series of experiments was then conducted to gain information on the recovery of amino acids added to lunar fines. Ten (10) ng of each amino acid were added to 50 mg of lunar fines. The amino acids were added to the lunar fines in the following manner. Five microliters (5 μl) of an aqueous standard solution (2 ng/μl of each amino acid) were added to the sample and dried for a few minutes under the IR lamp, and 1 ml of water was added. After water extraction and derivatization, the recoveries were low (10 to 20%). To this extracted sample an additional 50 ng of each amino acid were added, followed by re-extraction and derivatization. This experiment yielded much higher recoveries, of 60 to 80%. Another experiment was then conducted in which 70 ng of each amino acid were added to 70 mg of Apollo 14 fines.

The resultant chromatogram is seen in Fig. 5, with recovery of the amino acids varying greatly. The aliphatic amino acids gave the highest recoveries (40 to 50%), while phenylalanine yielded only 19% and ornithine and lysine were completely lost. These recovery experiments are of considerable significance in the interpretation of the extractability of free amino acids at pH 4.6 from lunar fines. Further studies on the extraction of amino acids and peptides must be conducted.

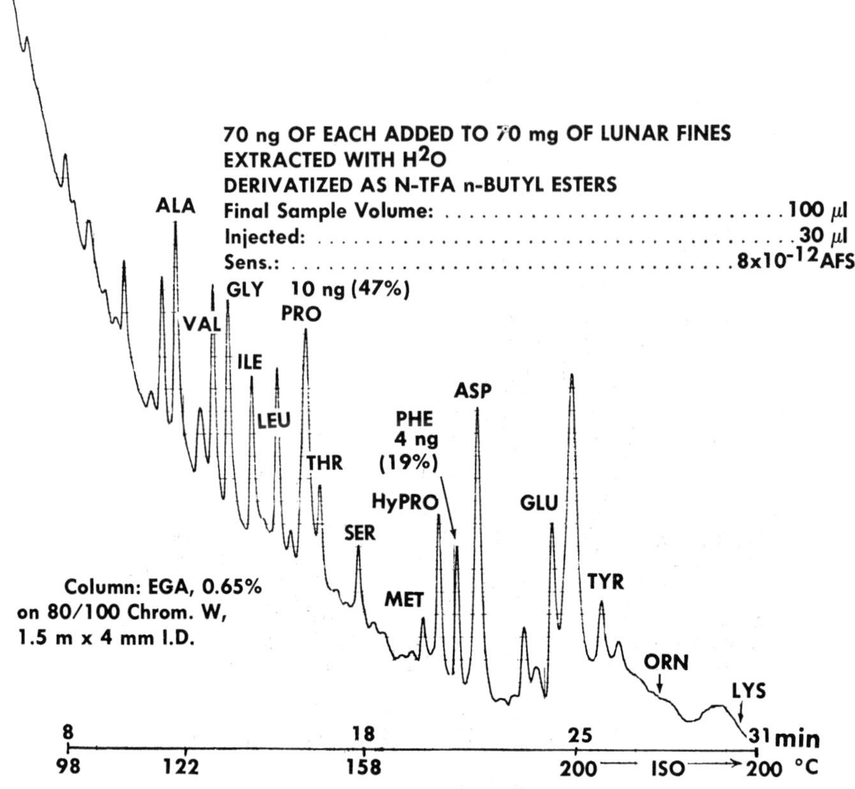


Fig. 5. Recovery of amino acids added to lunar fines (ARC 14240, 1.01 SFSC). 70 ng of each added to 70 mg of lunar fines, derivatized as N-TFA *n*-butyl esters—injected: 30 $\mu$l; sens: $8 \times 10^{-12}$ AFS.

## SUMMARY AND CONCLUSIONS

Analyses of both unhydrolyzed and hydrolyzed water extracts of Apollo 12 and 14 lunar fines were conducted by gas-liquid chromatography. The Apollo 14 sample 14240 SESC had been specially designated for amino acid studies. This SESC sample had been exposed to a minimum of manipulative steps prior to analysis and was considered to be of primary importance. On examination of reflux aqueous extracts of sample 14240 SESC, the common protein amino acids were not extracted or observed in concentrations above background in either the unhydrolyzed or hydrolyzed extracts. Another Apollo 14 sample (14298) was also studied, and again extractable indigenous amino acids were not found above the procedural blank values.

A hydrolyzed water extract of an Apollo 12 sample (12023) presented no indication of indigenous protein amino acids. Oxalic acid dibutyl ester was identified by GC–MS as the compound responsible for a large peak observed on GLC analysis of this sample.

To resolve the divergent data which have been reported for amino acids in lunar fines, a special collaborative study was conducted on Apollo 14 fines (14240 SESC) at the Ames Research Center, Moffett Field, California, during October, 1971.

By both GLC and IEC, glycine was observed at 2 to 4 ng/g; serine at 1 ng/g; and alanine, and aspartic, and glutamic acids at less than 0.5 ng/g. Of interest are the GLC chromatographic peaks which do not correspond to the common protein amino acids. A large unknown peak (30 to 40 ng/g) eluted between the alanine and glycine peaks. As the quantities of the amino acids are so minute, GLC–MS data were not obtained. The source of the amino acids may be synthesis during extraction and hydrolysis, rocket exhaust, low-level contamination, or indigenous.

In studies on the recovery of amino acids added to lunar fines, 10 to 20% when 10 ng of each amino acid were added to 50 mg of virgin fines, but the subsequent addition of 50 ng of each to the previously extracted sample resulted in much higher recoveries (60 to 80%). These results show that there is a need for an in-depth study of the extractability of amino acids and peptides from lunar material.

The quantities of amino acids in these studies are so minute (about 1 to 3 ng) that mass spectral data were not obtained. Until such information is forthcoming, the identifications must be considered tentative. Besides, the source of the amino acids is not clear. Identification does not establish the origin of the amino acids. Synthesis during the extraction and hydrolysis procedure from rather simple compounds such as cyanides (Biemann, 1972) which may be indigenous to the sample cannot be overlooked. Low levels of contamination are distinct possibilities. Information on the isomeric composition of the amino acids would be useful in clarifying the situation. There is also reason to believe that free amino acids are unlikely to survive extended periods of time in the harsh conditions of the lunar surface (Sagan, 1972). Although very small concentrations of amino acids were tentatively identified in Apollo 14 samples, the evidence is yet insufficient to suggest that these amino acids are indigenous to the lunar surface.

## References

Biemann, K. (1972) In situ synthesis during organic analysis of lunar samples. *Space Life Sci.,* in press.

Gehrke C. W., Kuo K. C., Zumwalt R. W., and Stalling D. L. (1970) Solvent-vent-chromatographic system, patent pending.

Gehrke C. W., Zumwalt R. W., Aue W. A., Stalling D. L., and Rash J. J. (1971) Search for organics in hydrolysates of lunar fines. *J. Chromatogr.* **54,** 169.

Gehrke C. W., Zumwalt R. W., Stalling D. L., Roach D., Aue W. A., Ponnamperuma C., and Kvenvolden K. A. (1971a) Search for amino acids in Apollo 11 and 12 lunar fines. *J. Chromatogr.* **59,** 305.

Gehrke C. W., Ponnamperuma C., Zumwalt R., Kuo K., Kvenvolden R. A. (1972) Amino acids in lunar samples? (abstract). In *Lunar Science—III* (editor C. Watkins), pp. 10–12, Lunar Science Institute Contr. No. 88.

Harada K., Hare P. E., Windsor C. R., and Fox S. W. (1971) Evidence for compounds hydrolyzable to amino acids in aqueous extracts of Apollo 11 and Apollo 12 lunar fines. *Science* **173,** 433.

Hare P. E., Harada K., and Fox S. W. (1970) Analyses for amino acids in lunar fines. *Proc. Apollo 11 Lunar Sci. Conf., Geochim. Cosmochim. Acta* Suppl. 2, Vol. 2, p. 1799. MIT Press.

Nagy B., Drew C. M., Hamilton P. B., Modzeleski V. E., Murphy M. E., Scott W. M., Urey H. C., and Young M. (1970) Organic compounds in lunar samples; pyrolysis products, hydrocarbons, amino acids. *Science* **167,** 770.

Sagan, C. (1972) The search for indigenous lunar organic matter. *Space Life Sci.,* in press.

Zumwalt R. W., Kuo K. C., and Gehrke C. W. (1971) A nanogram and picogram method for amino acid analysis. *J. Chromatogr.* **57,** 193.

Proceedings of the Third Lunar Science Conference
(Supplement 3, *Geochimica et Cosmochimica Acta*)
Vol. 2, pp. 2131–2147
The M.I.T. Press, 1972

# Compounds of carbon and other volatile elements in Apollo 14 and 15 samples

P. T. HOLLAND,* B. R. SIMONEIT, P. C. WSZOLEK,
and A. L. BURLINGAME

Space Sciences Laboratory, University of California,
Berkeley, California 94720

**Abstract**—Several aspects of the carbon chemistry of a variety of Apollo 14 and 15 samples have been investigated by DF dissolution and pyrolysis. The yields of indigenous methane and ethane for the suite of samples correlate with the amounts of solar wind gases ($^{20}Ne$ and $^{36}Ar$) and the deuterocarbon reaction products. The lunar fines and F1 breccias examined yield relatively higher amounts of $CH_4$ and $CD_4$ than more complex breccias and fragmental rocks. The yields of higher-weight deuterocarbons were significant for some samples.

Pyrolysis of the samples examined evolved mainly CO and $N_2$ with some $CO_2$. The CO and $N_2$ exhibit a similar bimodal evolution pattern. Terrestrial contaminants ($H_2O$, hydrocarbons and some $CO_2$) are volatilized at lower temperatures. In the intermediate temperature ranges some apparent pyrolytic reaction products are released. The bulk of the nonlabile carbon is released at higher temperatures to sample fusion.

Other compounds observed and quantitated by these techniques are DCN, HCN, $CS_2$, $D_2S$, and $PD_3$. The source of a major portion of the C and N appears to be solar wind and meteoritic input.

## INTRODUCTION

TOTAL CARBON MEASUREMENTS on crystalline lunar rocks returned by the Apollo missions have revealed only low concentrations of carbon (10–50 ppm) (Moore *et al.*, 1970, 1971, 1972; LSPET, 1971, 1972). The breccias and fines have, in general, higher carbon concentrations, ranging up to 200 ppm. This carbon is more concentrated in the finer fractions (Moore *et al.*, 1970) and correlated with the solar-wind noble gas content (Moore *et al.*, 1971). It has been postulated that most of the carbon associated with the fines and breccias is of solar-wind origin (Moore *et al.*, 1970; Kaplan and Smith, 1970; Kaplan *et al.*, 1970; Hayes, 1972).

Dissolution of lunar material by mineral acids has been the principal chemical technique, apart from direct pyrolysis, for getting data on lunar carbon (Abell *et al.*, 1970; Cadogan *et al.*, 1971; Chang *et al.*, 1971a, b; Henderson *et al.*, 1971). The varieties of hydrocarbons and other organic materials found have important implications for lunar carbon chemistry. In particular, experiments on Apollo 12 samples using

* Present address: Ruakura Soil Research Station, Department of Agriculture, Hamilton, New Zealand.

deuterated reagents have shown the presence of indigenous $CH_4$ and $C_2H_6$ (Abell *et al.*, 1970; Cadogan *et al.*, 1971; Chang *et al.*, 1971a, b; Henderson *et al.*, 1971). The $CD_4$, $C_2D_4$, and $C_2D_6$ also released are evidence for the presence of hydrolyzable carbide-like material in the samples. The yields of these gases from several samples were shown to correlate with various parameters related to solar-wind exposure (Cadogan *et al.*, 1971; Abell *et al.*, 1971). Shallow etching of fines with NaOD released more than 60% of the $CH_4$ found by complete dissolution (Abell *et al.*, 1971), indicating that most of the $CH_4$ is in the surface of the grains and may therefore be connected with solar-wind implantation (Abell *et al.*, 1970). Work from this laboratory (Holland *et al.*, 1972a) has shown that the yields of $CH_4$ and $CD_4$ released by 40% DF from a size fractionated sample 14240,5 (SESC trench sample) correlate with the inverse mean particle radius and also with the yields of solar-wind $^{20}Ne$ and $^{36}Ar$ released. These results firmly demonstrate the surface correlation of the indigenous $CH_4$ and provide direct evidence for a solar-wind origin of the indigenous $CH_4$ and $C_2H_6$. Recent solar-wind simulation experiments (Pillinger *et al.*, 1972) show that carbon and hydrogen bombardment of substrates can reproduce some of the features of lunar carbon chemistry attributed to solar-wind interactions. We have continued these dissolution experiments on a variety of Apollo 14 and 15 samples in order to extend our knowledge of lunar carbon and in particular the interesting chemistry induced in the regolith by the solar wind.

Pyrolysis of lunar samples directly into the ion source of a mass spectrometer can provide extensive information on the quantity and state of the indigenous compounds and other volatile material, as well as on contaminants. However, past approaches to this technique have either used low-resolution mass measurements (LSPET, 1969, 1970; Burlingame *et al.*, 1971b; Gibson and Johnson, 1971; Oró *et al.*, 1971; Nagy *et al.*, 1971) or have concentrated on high-resolution measurements over limited temperature, mass, or intensity ranges (Oró *et al.*, 1971; Preti, *ct al.*, 1971). Preliminary results on a new pyrolysis/mass spectrometer/computer system developed at this laboratory have been presented (Holland *et al.*, 1972b) and show the advantages of taking high-resolution data continuously over a wide dynamic range while pyrolyzing the sample from ambient to above melting (1200–1300°C) to obtain the maximum information of pyrolysis behavior. Pyrolysis of samples exposed only to low levels of contamination exhibit only a minor release of organic material, with $CH_4$ and benzene the principal components at the ppm level (Preti *et al.*, 1971; Oró *et al.*, 1971; Burlingame *et al.*, 1971b; Holland *et al.*, 1972b) and a variety of higher medium weight organic compounds at the ppb level. The main portion of the 100–200 ppm carbon in the fines is evolved as CO and $CO_2$ at high temperatures (Oró *et al.*, 1971; Nagy *et al.*, 1971; Holland *et al.*, 1972b). We have continued these pyrolysis experiments in some detail on several Apollo 14 and 15 samples.

## Experimental Procedures

The experimental procedures for the dissolution and pyrolysis experiments have been described elsewhere in some detail (Holland *et al.*, 1972b). The dissolutions were carried out using 40% DF for 1 hour in an all-glass evacuated system. The gases were collected and quantitated by high-resolution mass spectrometry and gas chromatography. Pyrolyses were carried out from 0 to 1300°C over a 40-minute period with continuous analysis of the evolved gases by high-resolution mass spectrometry (resolution 8000) using LOGOS computer data acquisition (Smith *et al.*, 1971). The only change in method is the use of quartz pyrolysis tubes which are used only once (after blanking) and give lower and more reproducible backgrounds than the alumina tubes used previously.

All handling of samples was carried out in the dry nitrogen atmosphere of a glove box in the UCB-SSL clean room (Burlingame *et al.*, 1971a). A magnetic separation was carried out on sample 14421,3 using a hand magnet. Four grams of fines yielded 160 mg of material that was strongly attracted to the magnet (fraction 14421,3,3). The average grain size of this fraction is 300 $\mu$, with approximately 10–20% by weight of particulates of diameters $< 10\ \mu$.

## Results and Discussion

### *Dissolutions*

The yields of various gases released by DF dissolution of some Apollo 14 and 15 samples are listed in Tables 1 and 2 respectively. In general the results are similar to those reported previously for Apollo 11 and 12 material (Abell *et al.*, 1971; Cadogan *et al.*, 1971; Chang *et al.*, 1971a, b; Holland *et al.*, 1972b). In those samples where substantial yields of solar-wind $^{20}Ne$ and $^{36}Ar$ were detected, the yields of indigenous $CH_4$ and $C_2H_6$ and of deuterocarbons are correspondingly high. Figs. 1 and 2 plot the yields of $^{20}Ne$ and $^{36}Ar$ respectively against the yields of methane and ethane expressed as $\mu$g/g carbon for all the samples we have examined so far. It is seen that the

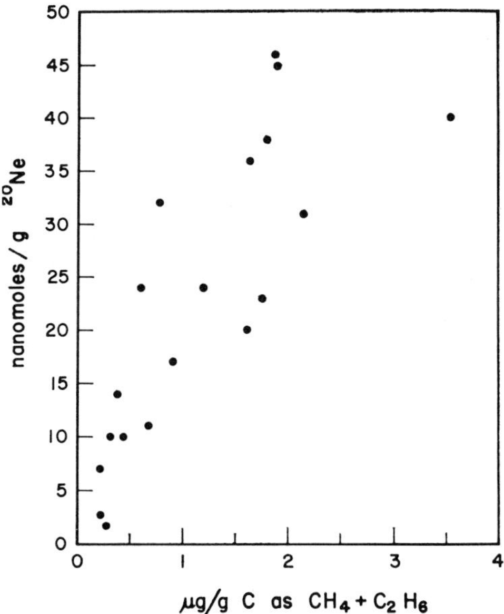


Fig. 1. Correlation plot of methane and ethane versus neon-20 yields from DF dissolution of Apollo 14 and 15 fines and breccias.

Table 1. Gases released on DF hydrolysis of Apollo 14 samples (nanomoles/g).

| | $CD_4$ + $CD_3H$ | $CH_4$ | $C_2H_6$ | Deuterocarbons $C_2$ | $C_3$ | $C_4$ | DCN + HCN | $D_2S$ + HDS | $CS_2$ | $^{20}Ne$ | $^{36}Ar$ | $PD_3$ | Total carbon (ppm)† |
|---|---|---|---|---|---|---|---|---|---|---|---|---|---|
| *Fines* | | | | | | | | | | | | | |
| 14148,3 (trench surface) | 908 | 290 | 3.5 | 158 | 61 | 3.8 | 50 | 760 | 110 | 40 | 25 | n.d. | 160 |
| 14156,6 (trench middle) | 663 | 160 | 10 | 62 | 9 | 2.1 | 20 | 4700 | 8 | 31 | 15 | n.d. | 180 |
| 14149,25 (trench bottom) | 410 | 150 | 1.2 | 20 | 2 | 2.0 | 13 | 3700 | 4 | 38 | 12 | n.d. | 135 |
| 14141,18 (Cone Crater) | 40 | 30 | 0.3 | 6 | 0.1 | * | 6 | 15300 | 4 | 14 | 2.5 | 6 | 42–80 |
| 14003,45 (LM contingency) | 620 | 156 | n.d. | n.d. | n.d. | n.d. | n.d. | n.d. | n.d. | 46 | 17 | n.d. | 140 |
| 14421,3 (bulk sample) | 410 | 123 | 4 | 53 | 23 | 5 | 2 | n.d. | 12 | 55 | 17 | 185 | 160 |
| 14421,3,3 (magnetic separate) | 1700 | 550 | n.d. | n.d. | n.d. | n.d. | n.d. | n.d. | n.d. | n.d. | n.d. | n.d. | n.d. |
| *Breccias* | | | | | | | | | | | | | |
| 14047,16 (interior) | 448 | 98 | 1.1 | 38 | 8 | 1.1 | 24 | 12300 | 11 | 24 | 10 | 150 | 140–210 |
| 14049,13 (interior) | 356 | 92 | 1.1 | 20 | 5.6 | 1.2 | 25 | 13700 | 11 | 32 | 9 | 48 | 135–190 |
| 14318,17 (interior) | 4 | 19 | * | * | * | * | 3 | 29000 | 9 | 3 | 1 | 3 | 86 |
| 14301,67 (interior) | 18 | 22 | 0.4 | 4 | 0.3 | * | 7 | 4100 | 3 | 1.6 | 0.2 | 1 | 70 |
| 14066,21 (clasts) | < 1 | * | n.d. | n.d. | n.d. | n.d. | n.d. | n.d. | n.d. | * | * | n.d. | 90 (matrix and |
| 14066,21 (matrix) | < 1 | * | n.d. | n.d. | n.d. | n.d. | n.d. | n.d. | n.d. | * | * | n.d. | clasts) |
| 14083,14 (white rock) | * | * | * | * | * | * | * | n.d. | n.d. | * | * | * | n.d. |
| 14321,48 (interior) | 3 | * | * | * | * | * | 30 | 1300 | 14 | * | 3 | * | 22–28 |

* = not detected. † Moore *et al.* (1972). n.d. = not determined.
All quantitation by mass spectrometry. Gas chromatographic determinations of $CH_4$ and $CD_4$ are generally within 30% of the listed values.

Table 2. Gases released on DF hydrolysis of Apollo 15 samples (nanomoles/g).

| | $CD_4$ + $CD_3H$ | $CH_4$ | $C_2H_6$ | Deuterocarbons | | | DCN + HCN | $D_2S$ + HDS | $CS_2$ | $^{20}Ne$ | $^{36}Ar$ | $PD_3$ | Total carbon (ppm)† |
|---|---|---|---|---|---|---|---|---|---|---|---|---|---|
| | | | | $C_2$ | $C_3$ | $C_4$ | | | | | | | |
| 15021,37 (fines) | 746 | 155 | 2.2 | 82 | 15 | 3 | 26 | 14400 | 6.9 | 45 | 9.6 | 107 | 140 |
| 15261,32 (fines) | 960 | 305 | 7.0 | 67 | 14.5 | n.d. | 57 | 12300 | 10 | 42 | 16 | 55 | n.d. |
| 15301,29 (fines) | 361 | 129 | 3.7 | 87 | 18 | 3 | 34 | 3700 | 5.5 | 37 | 5.2 | 112 | 160 |
| 15401,12 (fines) | 126 | 35 | 1.0 | 22 | 4 | 1 | 11 | 5950 | 9.2 | 10 | 2.2 | 17 | n.d. |

n.d. = not determined. † LSPET (1972).

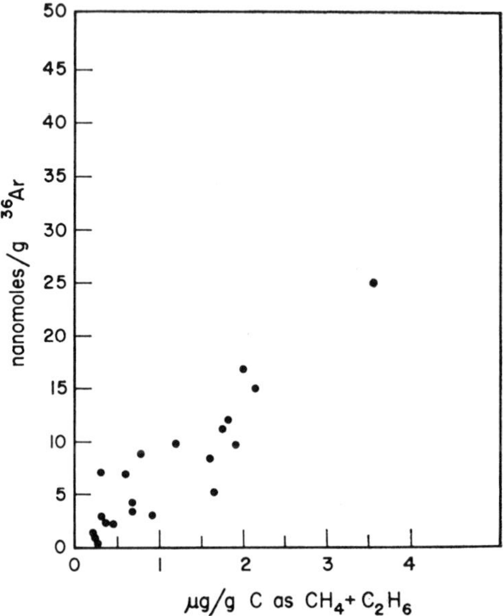


Fig. 2. Correlation plot of methane and ethane versus argon-36 yields from DF dissolution of Apollo 14 and 15 fines and breccias.

quantities are reasonably well correlated and provide direct support for the correlations made previously using the noble gas measurements of other workers (Abell *et al.*, 1971; Cadogan *et al.*, 1971). The plots are strong circumstantial evidence for a solar-wind source of the indigenous methane and ethane, as substantiated by the NaOD etching (Abell *et al.*, 1971) and size fractionation experiments (Holland *et al.*, 1972a) which show the surface-related nature of these gases. The exposure of the surface of the regolith grains to solar wind has been observed to give rise to an amorphous coating of highly disordered material (Borg *et al.*, 1971; Bibring *et al.*, 1972). This layer is presumably the site of solar-wind carbon and production of methane by proton bombardment.

Abell *et al.* (1971) have observed that the $CH_4$ and $CD_4$ yields correlate with the percentage of grains with very high track densities. The fines samples examined here seem to exhibit a similar correlation. Typical "mature" fines such as 14003, 14148, and 15021 give large $CD_4$ and $CH_4$ yields upon dissolution and appear to have surface exposure ages, estimated from track density, of greater than $2 \times 10^7$ yr (Bhandari *et al.*, 1972; Dran *et al.*, 1972; Hart *et al.*, 1972). The Cone Crater fines sample (14141) has been calculated to have an exposure age of 7–8 $\times 10^6$ yr (Bhandari *et al.*, 1972; Hart *et al.*, 1972). This sample contains less $CH_4$ (Table 1) than the "mature" soils, supporting an exposure-controlled mechanism for $CH_4$ formation and indicating that equilibrium for methane production is not achieved in a time period of $10^7$ yr. It has been previously argued (Holland *et al.*, 1972a) that solar-wind destruction of methane would be sufficient to cause the methane concentration to reach an equilibrium value in about $10^3$ yr if the gas were retained in the layer accessible to solar protons. Since equilibrium is not achieved in considerably longer time periods, the $CH_4$ contents measured are probably only a fraction of the total $CH_4$ production. A large fraction may be lost by a mechanism which increases the $\delta\ ^{13}C$ values of the regolith (Epstein and Taylor, 1970; Kaplan and Petrowski, 1971). The $CH_4$ which is retained probably results from a buildup of gas that has been able to diffuse deeper into the grains, where it is protected from solar radiation.

The rock fillet sample (15401) is interesting because it too exhibits low methane yields which correlate with its low exposure age (Hart *et al.*, 1972). This sample must originate primarily from microerosion of the parent rock without much horizontal transport of regolith. The methane yields for the Station G trench samples (14148, 14156, 14149) decrease with depth and a corresponding decrease in track density has been noted (Hart *et al.*, 1972; Bhandari *et al.*, 1972) confirming that vertical turnover of the regolith at ~10 cm depth may be operating over longer time scales than $10^7$ yr.

The carbon chemistry of the breccias divides them into two classes. The friable F1 breccias (14047 and 14049) exhibit similar total carbon contents and dissolution behavior (Table 1) to the mature fines samples. Bibring *et al.* (1972) have observed that the grains of this type of breccia have retained their amorphous coatings and high track densities. It appears that this type of sample represents soil compacted under relatively mild conditions which do not cause substantial loss of carbon or the track record. Other more complex Apollo 14 breccias and fragmental rocks such as 14066, 14318, 14301, 14083, and 14321 give very little $CH_4$ or $CD_4$ upon dissolution, and thus, if any of their components contained substantial acid labile carbon in the past,

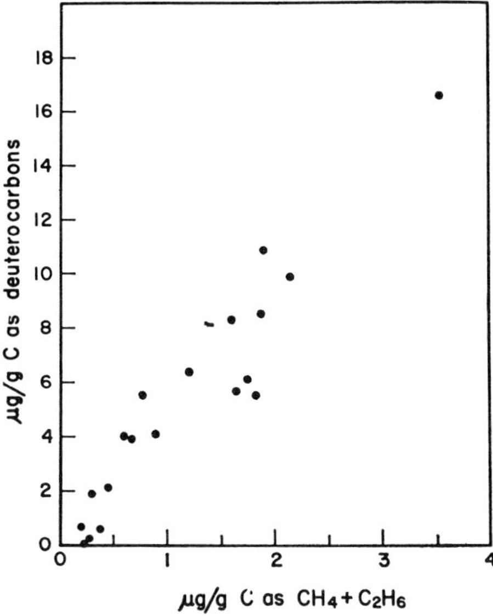


Fig. 3. Correlation plot of methane and ethane versus deuterocarbon yields from DF dissolution of Apollo 14 and 15 fines and breccias.

the processes operating to bring about brecciation or recrystallization have been severe enough to remove it or change its form. The nature and origin of the acid labile carbon in the fines and F1 breccias will be considered next.

The yields of deuterocarbons released by DF from our suite of samples are also well correlated with the yields of indigenous methane and ethane (Fig. 3). The slope of the correlation gives a ratio of 4.5 for carbons released as deuterocarbons to those released as hydrocarbons. This figure is not available from other workers because they do not analyze above $C_2$ (Abell *et al.*, 1971), so only the ratios of $CD_4/CH_4$ can be compared. It should be noted that the higher deuterocarbons make an important contribution to the fraction of total carbon released on acid dissolution. As we have commented previously (Holland *et al.*, 1972b) our $CD_4$ yields are consistently lower, giving rise to a $CD_4/CH_4$ ratio of about 3.5 versus the values in the region of 5.5 obtained by Abell *et al.* (1971). While most of this discrepancy may be due to different reagents and reaction conditions, we have found that our dissolution procedure was losing some $CD_4$.

Table 3 presents some results of experiments on sample 15261,32 carried out under different conditions. The first set of results shows that some $CD_4$ is being released from the sample by freeze/pump/thaw cycling of the DF in the same flask as the sample (our previous method of degassing the DF). The second set of results were obtained by degassing the DF in a Teflon valved tube prior to sample introduction and show the yields of $CD_4$, but not $CH_4$, to be enhanced when the sample is not exposed to DF vapor prior to the reaction period. Exposure of the sample to DF vapor only via the Teflon valve also gives rise to $CD_4$ evolution. These results firmly indicate that a solid/gas phase reaction is operative on the surface of the sample giving rise to $CD_4$, but no $CH_4$ or noble gases are lost within the accuracy of the measurements. Our previous procedure was therefore losing some $CD_4$, and the deuterocarbon results for lunar fines reported previously (Holland *et al.*, 1971a, b) and in Tables 1 and 2 of this paper are 5–10% low but remain qualitatively valid. The fact that DF vapor can yield $CD_4$ from lunar fines is strong evidence for the presence of carbides or carbide-like material in the surface layer of the grains.

A model for lunar carbon chemistry must accommodate the following experimental evidence on the methanes released by DF dissolution of the samples:

(1) Crystalline rocks yield very little $CD_4$ or $CH_4$.

(2) $CD_4$ and $CH_4$ yields are proportional to the surface area of the grains.

(3) DF vapor can give $CD_4$, probably by reacting with surface-related carbide material.

(4) $CD_4$ and $CH_4$ yields are correlated with solar wind exposure (Figs. 1 to 3).

Solar-wind implantation of carbon into the surface of the regolith grains has been proposed as the major source of lunar carbon (Moore *et al.*, 1970; Epstein and Taylor, 1970). This simple mechanism can explain the above facts by assuming implantation results in some carbide formation and that the solar proton flux can convert carbon to $CH_4$. Meteoritic carbon has previously been assumed to be lost on impact with the lunar surface (Moore *et al.*, 1970; Abell *et al.*, 1971) but many aspects of lunar carbon

Table 3. Effect of the DF degassing procedure on yields from sample 15261,32 (nanomoles/g).*

| | $CD_4$ | $CH_4$ | $^{20}Ne$ | $^{36}Ar$ |
|---|---|---|---|---|
| | *Run I. DF degas in presence of sample* | | | |
| 2nd degas | 4 | < 0.2 | n.d. | n.d. |
| 3rd degas | 10 | < 0.2 | n.d. | n.d. |
| 4th degas | 9 | 0.5 | n.d. | n.d. |
| Dissolution | 882 | 305 | 61 | 18 |
| | *Run II. DF degas separate from sample* | | | |
| Degas | 0 | < 0.2 | n.d. | n.d. |
| Dissolution | 960 | 305 | 42 | 16 |

n.d. = not determined.
* Quantitation by gas chromatography for degas and by mass spectrometry for dissolution.

chemistry are not inconsistent with a meteoritic contribution to the lunar regolith supplementing the solar-wind source of carbon.

Meteorites have brought at least 400 μg/g C to Mare Tranquillitatis (Epstein and Taylor, 1970) and even if all this carbon were lost by impact volatilization, about 5–10 μg/g may be returned by reimplantation from a lunar atmosphere (Hayes, 1972). If all the meteoritic carbon were not lost on impact, the following arguments could explain the experimental evidence. A direct contribution of meteoritic carbide could give the correlation of hydrocarbons to deuterocarbons because the level of $CH_4$ and $C_2H_6$ is governed by the sample's solar-wind exposure history, and this exposure also means the exposure to micrometeorites, the main source of meteoritic material in the regolith (Hartung *et al.*, 1972). The grain-size dependence of the $CD_4$ yields could be explained either by a concentration of meteoritic material in the finer size fractions or by redistribution of meteoritic carbon and carbide by volatilization and then condensation on the surfaces of nearby fines material. The observation of carbide transfer during vapor deposition of iron (Pillinger *et al.*, 1972) shows the latter process to be possible if the meteorite is vaporized upon impact (Hörz *et al.*, 1971). Our direct observation that lunar carbon is released by acid vapor can be interpreted as deriving from redeposited meteoritic carbide. Laul *et al.* (1971) have observed that trace elements attributable to meteorite influx to the regolith are selectively leached from Apollo 12 fines by dilute acid. This also implies surface correlated meteoritic material. In the solar-wind simulation experiments of Pillinger *et al.* (1972) methane formation seemed a more favored process than that for carbide. If the implantation conditions used by the Bristol group can be taken as valid, then meteoritic carbide may be necessary to explain the additional acid labile carbon in lunar samples.

In order to further investigate the possible importance of meteoritic material to lunar carbon chemistry, the magnetic fraction of sample 14421,3 was examined. The magnetic fraction of lunar fines contains a high proportion of particles with meteoritic metal inclusions (Mason *et al.*, 1970; Goldstein and Yakowitz, 1971; Goldstein *et al.*, 1972) and metallic particles from Apollo 12, 14, and 15 fines are also principally of meteoritic origin (Wänke *et al.*, 1971; Wlotzka *et al.*, 1972). The dissolution results from sample 14421,3 and its magnetic separate 14421,3,3 are given in Table 1. The substantially higher $CD_4$ and $CH_4$ yields from the magnetic fraction support a direct meteoritic contribution, since the increases are too large to be attributed to a size distribution effect. Carbide enrichment in the magnetic material may also be due to a more efficient formation process in the metallic phase by solar-wind implantation, resulting in higher $CD_4$ yields. A higher carbon content in the magnetic fraction that can be converted to methane by solar protons can explain the increased $CH_4$ yields. Pyrolysis experiments in progress should clarify the latter point and show if a meteoritic contribution to lunar carbon is probable.

Other volatile elements that we have observed in various combinations in the DF dissolution products are N, P, and S (Tables 1 and 2). From 0.1 to 0.5 $\mu$g C/g is evolved as DCN from the fines and the F1 breccias (14047 and 14049), but is a factor of 10 lower in the fragmental rocks. It presumably arises from hydrolysis of cyanide in the sample. The yields from size-fractionated fines increased with decreasing grain size. A solar-wind origin for this component seems likely and illustrates the complex chemistry that the solar wind can induce on the lunar surface. The small amounts of $CS_2$ observed may also arise from an implantation of solar-wind carbon into troilite (FeS). Based on the sulfur contents of Apollo 11, 12, and 14 samples (Kaplan and Smith, 1970; Kaplan *et al.*, 1970; Kaplan and Petrowski, 1971) of about 700 $\mu$g/g and the typical carbon content of fines samples of about 160 $\mu$g/g, the amount of carbon in association with sulfur may be about 0.1 $\mu$g/g, which is comparable to the $CS_2$ yields measured. The $D_2S$ evolved arises from hydrolysis of troilite in the samples but the yields are not very reproducible, as has been noted by other workers (Sakai *et al.*, 1972). Whether this represents sample inhomogeneity or is an experimental artifact has not been determined.

We report here the release of $PD_3$ from the samples by DF which indicates the presence of phosphides in the samples. The Apollo 14 and 15 fines exhibit yields corresponding to about 2 to 5 $\mu$g P/g while the Apollo 14 fragmental rocks give amounts 100 times lower. These yields are undoubtedly very qualitative as phosphine also forms the phosphonium ion with halogen acids and is rather unstable to oxidation. Soil samples contain more phosphorus than the crystalline rocks because of an input of KREEP material (Meyer *et al.*, 1971; Meyer, 1972) but the increased $PD_3$ yields from fines would appear to be due to a surface exposure effect because the smaller size fractions of sample 14240 gave higher yields. The solar-wind input of phosphorus should be on the order of $10^3$ times lower than that of carbon, making this an insufficient source (by about a factor of 10) for phosphide formation. It seems probable that the phosphides are due in part to the input of meteoritic material. Non-chondritic meteorites contain significant quantities of the phosphide mineral schreibersite $(Fe, Ni,Co)_3P$ (Mason, 1962). The chondrites contain about 0.05% P in the form of various phosphates and these may also be reduced to phosphides during impact melting (Taylor and Heymann, 1972). Goldstein and Yakowitz (1971) have observed a melted meteoritic metal particle containing schreibersite.

### *Pyrolyses*

The yields of the major carbon species and nitrogen from the pyrolysis of various samples are listed in Table 4. Also given are the total carbon contents of the samples,

Table 4. Pyrolysis of Apollo 14 and 15 samples (yields in $\mu$g/g as C or N).

| Sample | Description | CO | $CO_2$ | Benzene | Total C | Total N | Total C* | Total N* |
|---|---|---|---|---|---|---|---|---|
| 14240,5 | < 37 $\mu$ SESC fines | 244 | 109 | 1.8 | 370 | 255 | bulk | bulk |
| 14240,5 | 55–106 $\mu$ SESC fines | 70 | 33 | 0.8 | 110 | 60 | 89 | n.d. |
| 14240,5 | 106–420 $\mu$ SESC fines | 58 | 29 | 0.9 | 90 | 48 | | |
| 14003,45 | Contingency fines (LM) | 148 | 19 | 0.3 | 170 | 152 | 140 | n.d. |
| 14083,14 | White rock, Station C1 | 21 | 17 | 0.1 | 38 | 0.5 | n.d. | n.d. |
| 14049,13 | F1 breccia, Station Bg (interior chip) | 154 | 28 | 0.5 | 185 | 82 | 135,190 | 130 |
| 15261,32 | Trench fines, Station 6 | 337 | 42 | 1.1 | 385 | 400 | n.d. | n.d. |
| 15021,37 | Contingency fines (LM) | 118 | 27 | 0.5 | 135 | 122 | 160 | n.d. |

* Moore *et al.* (1972); LSPET (1972). n.d. = not determined.

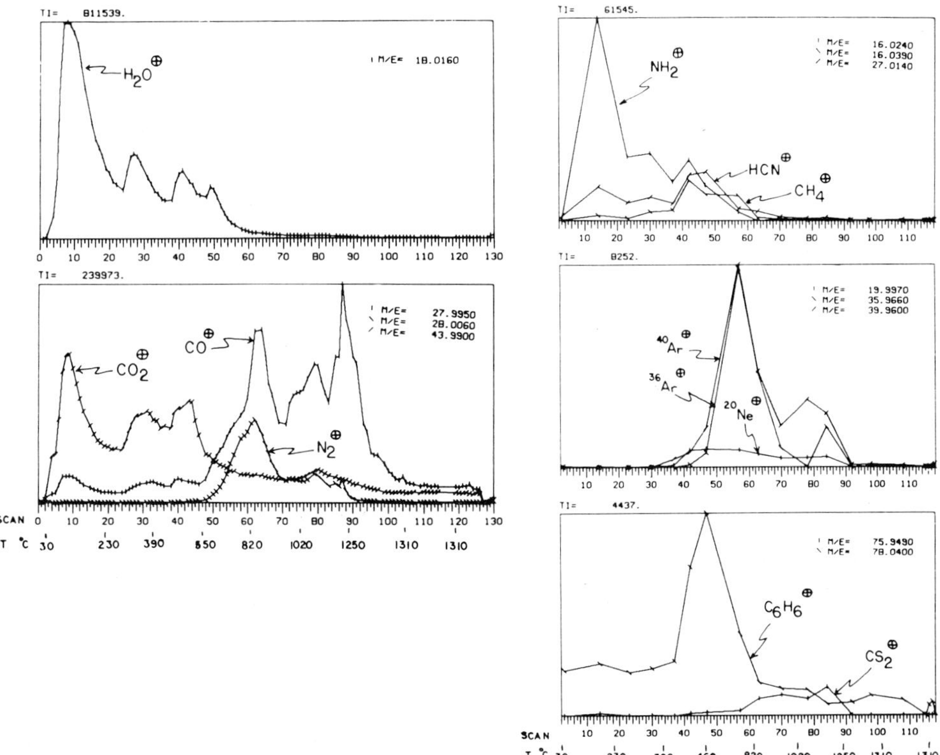


Fig. 4. Selected pyrolysis/high-resolution mass-spectrometric ion sum plots for sample 14049,13.

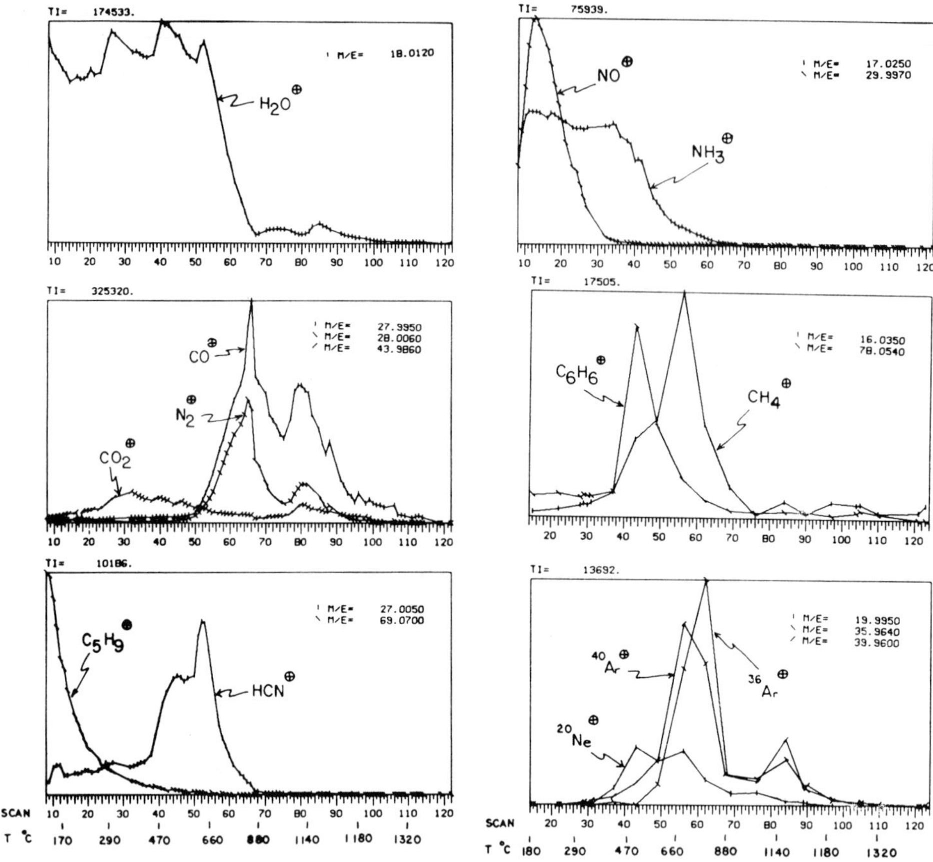


Fig. 5. Selected pyrolysis/high-resolution mass-spectrometric ion sum plots for sample 15261,32.

which include the minor contributions from carbonaceous components such as HCN, $CS_2$, toluene, and other hydrocarbons. The values for the total carbon and nitrogen content are corroborated by the data reported elsewhere (Moore *et al.*, 1972; LSPET, 1972). The crystalline white rock (14083) is low in volatilizable matter, whereas the fines and F1 breccias exhibit high values. The major constituents evolved at high temperatures are $N_2$ and CO with some $CO_2$ for all samples analyzed from the Apollo 14 and 15 missions. In the case of the SESC sample (14240,5) both the carbon and nitrogen correlate with the surface area parameter, $r^{-1}$, the values increasing as the grain size decreases. The pyrolytic gas-release patterns for all the size fractions were identical in shape, indicating that the volatile components are present in similar environments in the grains of various sizes.

In order to illustrate the pyrolysis/high-resolution mass-spectrometric data two examples were chosen, sample 14049,13 in Fig. 4 and sample 15261,32 in Fig. 5. The

$N_2$ and CO correlate with each other at high temperatures (above 550°C). The bimodal evolution of these gases indicates a surface release followed by a melting release. Gibson and Johnson (1971) observed a similar bimodal evolution, but were not able to resolve the CO and $N_2$ contributions. Water is evolved mainly at the beginning to a temperature of approximately 800°C and appears to be mainly adsorbed contamination. The $CO_2$ is also evolved mainly at lower temperatures, again indicating terrestrial contamination.

Benzene and methane are generated at levels of 1–2 $\mu$g/g from the fines samples, with the releases peaking at 500° and 650°C respectively. A variety of other low molecular weight C/H/N/S compounds (e.g., HCN, $CS_2$, toluene, etc.) are released in this temperature range, similar to those observed by Preti *et al.* (1971). These compounds may be synthesized by heat-promoted reaction of simple solar-wind products in the surface layer and are evolved after surface contamination has volatilized, but before sample fusion. This may be seen in the data of sample 15261,32 (Fig. 5). The hydrocarbon contamination, as indicated by the representative mass-spectrometric fragment ion of composition $C_5H_9$, is volatilized below 470°C. The benzene is not evolved below 450°C and this volatilization maximum at higher pyrolysis temperatures is strong evidence that the benzene is not contamination. Similar results were reported by Burlingame *et al.* (1971b) for some Apollo 12 double-core samples. Methane is evolved from 400–800°C, as are $^{20}$Ne, $^{36}$Ar, and $^{40}$Ar.

Ammonia is released in the low temperature range (below 600°C). It is probably a product of the reaction of $H_2O$ with nitrides (Müller, 1972). Nitrides would derive from solar-wind implantation of N and possibly from meteoritic contribution of such minerals as, for example, carlsbergite, CrN (Buchwald and Scott, 1971), and osbornite, TiN (Mason, 1962). The high release of NO, HCN, and $NH_3$ from sample 15261,32 (Fig. 5) is of interest, since this is a trench sample and it should not be contaminated by LM exhaust. Carbon disulfide was also released from several samples and may be a pyrolysis product of carbon implanted in sulfide minerals.

## CONCLUSIONS

From the methanes released by DF dissolution of the samples it can be concluded that: the crystalline rocks release small amounts of $CD_4$ and $CH_4$; the yields of these gases are proportional to the grain surface area and correlate with solar-wind exposure; and DF vapor yields $CD_4$, probably by reacting with surface-related carbide material. This carbide-like material in the amorphous surface layer of the regolith grains may be derived from both solar-wind implantation of C plus direct and/or redeposited meteoritic carbide. The indigenous methane and ethane are reasonably well correlated with the yields of solar-wind-derived $^{20}$Ne and $^{36}$Ar. These gases are probably trapped in the amorphous outer layer of the grains. The yields of higher-weight deuterocarbons were significant for some samples and indicate the presence of alkaline earth carbides.

Compounds of the volatile elements N, P, and S were also observed. DCN was evolved from the fines and F1 breccias, probably from acid hydrolysis of cyanide in the sample. Small amounts of $CS_2$ were observed, which may arise from solar-wind carbon implantation into the sulfide minerals. The $D_2S$ is released from acid hydrolysis of troilite. The fines samples yielded from 2–5 $\mu$g P/g as phosphine, indicating the presence of phosphides. Solar-wind input of P into the regolith is insufficient to generate the phosphide content observed. Meteoritic input of such minerals as for example schreibersite probably also occurs.

Pyrolysis of various samples indicated the major constituents evolved at high temperatures to be CO and $N_2$ with some $CO_2$. The $N_2$ and CO correlate with each other and their bimodial evolution pattern indicates a surface release, followed by a melting release. Water, some $CO_2$ and hydrocarbon contamination are volatilized mainly at lower temperatures, confirming their terrestrial origin. In the intermediate temperature region compounds such as methane, benzene, toluene, HCN, $CS_2$, are evolved. They may be synthesized by heat-promoted reaction of solar-wind products in the surface layer. The ammonia which is released in the low temperature range may be a hydrolysis product of nitrides.

The total carbon and nitrogen contents are in agreement with the values reported by Moore *et al.* (1972) and LSPET (1972). These pyrolysis results show that the bulk of the carbon is in a nonlabile form and is not released completely until sample fusion. The source of most of this carbon appears to be solar-wind and meteoritic input.

*Acknowledgments*—We thank Dr. J. Chang, Miss R. Jackson, Dr. W. H. McFadden, Mr. H. Melling, Dr. J. O'Connor, Mr. J. Wilder, Mr. C. Wong, and Mrs. E. Yang for their valuable contributions. Financial support from the National Aeronautics and Space Administration (Grants NGR 05-003-418 and NGR 05-003-435) is gratefully acknowledged.

This paper represents Part XXXVIII in the series High Resolution Mass Spectrometry in Molecular Structure Studies. For Part XXXVII, see B. R. Simoneit and A. L. Burlingame, in *Advances in Organic Geochemistry 1971* (editors von Gaertner H. R. and Wehner H.), Pergamon-Vieweg, in press.

A.L.B. is a John Simon Guggenheim Memorial Fellow, 1970–1972.

## REFERENCES

Abell P. I., Eglinton G., Maxwell J. R., Pillinger C. T., and Hayes J. M. (1970) Indigenous lunar methane and ethane. *Nature* **226,** 251–252.

Abell P. I., Cadogan P. H., Eglinton G., Maxwell J. R., and Pillinger C. T. (1971) Survey of lunar carbon compounds, I. The presence of indigenous gases and hydrolysable carbon compounds in Apollo 11 and Apollo 12 samples. *Proc. Second Lunar Sci. Conf., Geochim. Cosmochim. Acta* Suppl. 2, Vol. 2, pp. 1843–1863. MIT Press.

Bhandari N., Bhat S. G., Goswami J. N., Gupta S. K., Krishnaswami S., Lal D., Tamhane A. S., and Venkatavaradan V. S. (1972) Collision controlled radiation history of lunar regolith (abstract). In *Lunar Science—III* (editor C. Watkins), pp. 68–70, Lunar Science Institute Contr. No. 88.

Bibring J. P., Maurette M., Meunier R., Durrieu L., Jouret C., and Eugster O. (1972) Solar wind implantation effects in the lunar regolith (abstract). In *Lunar Science—III* (editor C. Watkins), pp. 71–73, Lunar Science Institute Contr. No. 88.

Borg J., Maurette M., Durrieu L., and Jouret C. (1971) Ultramicroscopic features in micron-sized lunar dust grains and cosmophysics. *Proc. Second Lunar Sci. Conf., Geochim. Cosmochim. Acta* Suppl. 2, Vol. 3, pp. 2027–2040. MIT Press.

Buchwald V. F. and Scott E. R. D. (1971) First nitride (CrN) in iron meteorites. *Nature* (Phys. Sci.) **233,** 113–114.

Burlingame A. L., Holland P., McFadden W. H., Simoneit B. R., Wilder J. T., and Wszolek P. C. (1971a) UCB Space Sciences Laboratory organic clean room and lunar material transfer facilities. Space Sciences Laboratory Report, University of California, Berkeley, California, May 18.

Burlingame A. L., Hauser J. S., Simoneit B. R., Smith D. H., Biemann K., Mancuso N., Murphy R., Flory D. A., and Reynolds M. A. (1971b) Preliminary organic analysis of the Apollo 12 cores. *Proc. Second Lunar Sci. Conf., Geochim. Cosmochim. Acta* Suppl. 2, Vol. 2, pp. 1891–1899. MIT Press.

Cadogan P. H., Eglinton G., Maxwell J. R., and Pillinger C. T. (1971) Carbon chemistry of the lunar surface. *Nature* **231,** 29–31.

Chang S., Kvenvolden K. A., Lawless J., Ponnamperuma C. A., and Kaplan I. R. (1971a) Carbon, carbides, and methane in an Apollo 12 sample. *Science* **171,** 474–477.

Chang S., Smith J. W., Kaplan I., Lawless J., Kvenvolden K. A., and Ponnamperuma C. (1971b) Carbon compounds in lunar fines from Mare Tranquillitatis—IV. Evidence for oxides and carbides. *Proc. Apollo 11 Lunar Sci. Conf., Geochim. Cosmochim. Acta* Suppl. 1, Vol. 2, pp. 1857–1869. Pergamon.

Dran J. C., Duraud J. P., Maurette M., Durrieu L., Jouret C., and Legressus C. (1972) The high resolution track and texture record of lunar breccias and gas-rich meteorites (abstract). In *Lunar Science—III* (editor C. Watkins), pp. 183–185, Lunar Science Institute Contr. No. 88.

Epstein S. and Taylor H. P. Jr. (1970) The concentration and isotopic composition of hydrogen, carbon, and silicon in Apollo 11 lunar rocks and minerals. *Proc. Apollo 11 Lunar Sci. Conf., Geochim. Cosmochim. Acta* Suppl. 1, Vol. 2, pp. 1085–1096. Pergamon.

Gibson E. K. Jr. and Johnson S. M. (1971) Thermal analysis—inorganic gas release studies of lunar samples. *Proc. Second Lunar Sci. Conf., Geochim. Cosmochim. Acta* Suppl. 2, Vol. 2, pp. 1351–1366. MIT Press.

Goldstein J. I. and Yakowitz H. (1971) Metallic inclusions and metal particles in the Apollo 12 lunar soil. *Proc. Second Lunar Sci. Conf., Geochim. Cosmochim. Acta* Suppl. 2, Vol. 2, pp. 177–191. MIT Press.

Goldstein J. I., Yen F., and Axon H. J. (1972) Metallic particles in the Apollo 14 lunar soil (abstract). In *Lunar Science—III* (editor C. Watkins), pp. 323–325, Lunar Science Institute Contr. No. 88.

Hart H. R. Jr., Comstock G. M., and Fleischer R. L. (1972) The particle track record of Fra Mauro (abstract). In *Lunar Science—III* (editor C. Watkins), pp. 360–362, Lunar Science Institute Contr. No. 88.

Hartung J. B., Hörz F., and Gault D. E. (1972) The origin and significance of lunar microcraters (abstract). In *Lunar Science—III* (editor C. Watkins), pp. 364–366, Lunar Science Institute Contr. No. 88.

Hayes J. M. (1972) Extralunar sources for carbon on the moon. *Space Life Sci.* (in press).

Henderson W., Kray W. C., Newman W. A., Reed W. E., Simoneit B. R., and Calvin M. (1971) Study of carbon compounds in Apollo 11 and 12 returned lunar samples. *Proc. Second Lunar Sci. Conf., Geochim. Cosmochim. Acta* Suppl. 2, Vol. 2, pp. 1901–1912. MIT Press.

Holland P. T., Simoneit B. R., Wszolek P. C., McFadden W. H., and Burlingame A. L. (1972a) Carbon chemistry of Apollo 14 size-fractionated fines. *Nature* (Phys. Sci.) **235,** 106–108.

Holland P. T., Simoneit B. R., Wszolek P. C., and Burlingame A. L. (1972b) Study of carbon compounds in Apollo 12 and 14 lunar samples. *Space Life Sci.* (in press).

Hörz F., Hartung J. B., and Gault D. E. (1971) Micrometeorite craters on lunar rock surfaces. *J. Geophys. Res.* **76,** 5770–5798.

Kaplan I. R. and Smith J. W. (1970) Concentration and isotopic composition of carbon and sulfur in Apollo 11 lunar samples. *Science* **167,** 541–543.

Kaplan I. R., Smith J. W., and Ruth E. (1970) Carbon and sulfur concentration and isotopic composition in Apollo 11 lunar samples. *Proc. Apollo 11 Lunar Sci. Conf., Geochim. Cosmochim. Acta* Suppl. 1, Vol. 2, pp. 1317–1329. Pergamon.

Kaplan I. R. and Petrowski C. (1971) Carbon and sulfur isotope studies in Apollo 12 lunar samples. *Proc. Second Lunar Sci. Conf., Geochim. Cosmochim. Acta* Suppl. 2, Vol. 2, pp. 1397–1406. MIT Press.

Laul J. C., Morgan J. W., Ganapathy R., and Anders E. (1971) Meteoritic material in lunar samples: Characterization from trace elements. *Proc. Second Lunar Sci. Conf., Geochim. Cosmochim. Acta* Suppl. 2, Vol. 2, pp. 1139–1158. MIT Press.

LSPET (Lunar Sample Preliminary Examination Team) (1969) Lunar sample information catalog —Apollo 11. NASA-MSC, August 31.

LSPET (Lunar Sample Preliminary Examination Team) (1970) Lunar sample information catalog —Apollo 12. NASA Technical Report R-353.

LSPET (Lunar Sample Preliminary Examination Team) (1971) Preliminary examination of lunar samples from Apollo 14. *Science* **173,** 681–693.

LSPET (Lunar Sample Preliminary Examination Team) (1972) The Apollo 15 lunar samples: A preliminary description. *Science* **175,** 363–375.

Mason B. (1962) *Meteorites.* John Wiley & Sons.

Mason B., Fredriksson K., Henderson E. P., Jarosewich E., Melson W. G., Towe K. M., and White J. S. Jr. (1970) Mineralogy and petrography of lunar samples. *Proc. Apollo 11 Lunar Sci. Conf., Geochim. Cosmochim. Acta* Suppl. 1, Vol. 1, pp. 655–660. Pergamon.

Meyer C. Jr., Brett R., Hubbard N. J., Morrison D. A., McKay D. S., Aitken F. K., Takeda H., and Schonfeld E. (1971) Mineralogy, chemistry, and origin of the KREEP component in soil samples from the Ocean of Storms. *Proc. Second Lunar Sci. Conf., Geochim. Cosmochim. Acta* Suppl. 2, Vol. 1, pp. 393–411. MIT Press.

Meyer C. Jr. (1972) Mineral assemblages and the origin of the non-mare lunar rock types (abstract). In *Lunar Science—III* (editor C. Watkins), pp. 542–544, Lunar Science Institute Contr. No. 88.

Moore C. B., Gibson E. K., Jr., Larimer J. W., Lewis C. F., and Nichiporuk W. (1970) Total carbon and nitrogen abundances in Apollo 11 lunar samples and selected achondrites and basalts. *Proc. Apollo 11 Lunar Sci. Conf., Geochim. Cosmochim. Acta* Suppl. 1, Vol. 2, pp. 1375–1382. Pergamon.

Moore C. B., Lewis C. F., Larimer J. W., Delles F. M., Gooley R. C., Nichiporuk W., and Gibson E. K. Jr. (1971) Total carbon and nitrogen abundances in Apollo 12 lunar samples. *Proc. Second Lunar Sci. Conf., Geochim. Cosmochim. Acta* Suppl. 2, Vol. 2, pp. 1343–1350. MIT Press.

Moore C. B., Lewis C. F., Cripe J., Kelly W. R., and Delles F. (1972) Total carbon, nitrogen, and sulfur abundances in Apollo 14 lunar samples (abstract). In *Lunar Science—III* (editor C. Watkins), pp. 550–551, Lunar Science Institute Contr. No. 88.

Müller O. (1972) Chemically bound nitrogen abundances in lunar samples, and active gases released by heating at lower temperatures (250 and 500°C) (abstract). In *Lunar Science—III* (editor C. Watkins), pp. 568–570, Lunar Science Institute Contr. No. 88.

Nagy B., Modzeleski J. E., Modzeleski V. E., Mohammad M. A., Nagy L. A., Scott W. M., Drew C. M., Thomas J. E., Ward R., Hamilton P. B., and Urey H. C. (1971) Carbon compounds in Apollo 12 lunar samples. *Nature* **232,** 94–98.

Oró J., Flory D. A., Gibert J. M., McReynolds J., Lichtenstein H. A., and Wikstrom S. (1971) Abundances and distribution of organogenic elements and compounds in Apollo 12 lunar samples. *Proc. Second Lunar Sci. Conf., Geochim. Cosmochim. Acta* Suppl. 2, Vol. 2, pp. 1913–1925. MIT Press.

Pillinger C. T., Cadogan P. H., Eglinton G., Maxwell J. R., Mays B. J., Grant W. A., and Nobes M. J. (1972) Simulation study of lunar carbon chemistry. *Nature* (Phys. Sci.) **235,** 108–109.

Preti G., Murphy R. C., and Biemann K. (1971) The search for organic compounds in various Apollo 12 samples by mass spectrometry. *Proc. Second Lunar Sci. Conf., Geochim. Cosmochim. Acta* Suppl. 2, Vol. 2, pp. 1879–1889. MIT Press.

Sakai H., Petrowski C., Goldhaber M. B., and Kaplan I. R. (1972) Distribution of carbon, sulfur, and nitrogen in Apollo 14 and 15 material (abstract). In *Lunar Science—III* (editor C. Watkins), pp. 672–674, Lunar Science Institute Contr. No. 88.

Smith D. H., Olsen R. W., Walls F. C., and Burlingame A. L. (1971) Real-time organic mass spectrometry: LOGOS—A generalized laboratory system for high and low resolution, GC/MS and closed loop applications. *Anal. Chem.* **43,** 1796–1806.

Taylor G. J. and Heymann D. (1972) Post shock thermal history of reheated chondrites. *J. Geophys. Res.* (in press).

Wänke H., Wlotzka F., Baddenhausen H., Balacescu A., Spettel B., Teschke F., Jagoutz E., Kruse H., Quijano-Rico M., and Rieder R. (1971) Apollo 12 samples: Chemical composition and its relation to sample locations and exposure ages, the two component origins of various soil samples, and studies on lunar metallic particles. *Proc. Second Lunar Sci. Conf., Geochim. Cosmochim. Acta* Suppl. 2, Vol. 2, pp. 1187–1208. MIT Press.

Wlotzka F., Jagoutz E., Spettel B., Baddenhausen H., Balacescu A., and Wänke H. (1972) On lunar metallic particles and their contribution to the trace element content of the Apollo 14 and 15 soils (abstract). In *Lunar Science—III* (editor C. Watkins), pp. 806–808, Lunar Science Institute Contr. No. 88.

Proceedings of the Third Lunar Science Conference
(Supplement 3, *Geochimica et Cosmochimica Acta*)
Vol. 2, pp. 2149–2155
The M.I.T. Press, 1972

# Spectrofluorometric search for porphyrins in Apollo 14 surface fines

JOON H. RHO, EDWARD A. COHEN, and A. J. BAUMAN

Jet Propulsion Laboratory, California Institute of Technology,
Pasadena, California 91103

**Abstract**—Benzene-methanol extracts of surface fines 14163,179,181 from Fra Mauro have been analyzed for porphyrins by means of fluorescence spectroscopy augmented by computer data treatment. Signal averaging and background removal revealed no porphyrin fluorescence features under conditions expected to produce a signal to noise ratio of 2 : 1 for $1.6 \times 10^{-14}$ moles of extractable metalloporphyrin per gram of fines. These results are based upon the assumption that if metalloporphyrins were present they would be demetallated with the same efficiency as is Ni-mesoporphyrin IX.

## INTRODUCTION

TEN GRAMS OF Fra Mauro fines, sample 14163,179,181 (designated sample A), were Soxhlet-extracted in a microporous Teflon thimble with fluorometric grade benzene-methanol (3/2, by volume) for 24 hours and prepared for fluorometric analysis as described previously (Rho *et al.*, 1970, 1971). Samples of 200 mesh optical quartz (designated sample B) identical in weight to that of the lunar sample were similarly extracted and examined. The samples were placed in the thimble as 5 mm thick layers separated by equally thick layers of borosilicate glass wool in order to promote free solvent drainage. Each thimble was then placed in a double-jacketed Soxhlet apparatus so designed that extraction was effected efficiently by solvent at its boiling point. The extract was first taken to a small volume in a flash evaporator then to near dryness under a stream of nitrogen. It was then dissolved in 0.4 ml of benzene and treated with methanesulfonic acid to remove any metals present (Rho *et al.*, 1970), as metallated porphyrins are often nonfluorescent.

Nickel mesoporphyrin IX (0.2 ng) dissolved in the same solvent system as that used with samples A and B was demetallated similarly and used as a porphyrin standard (sample C). Sixty percent of sample A or the equivalent of 6 grams of fines was carried through fluorometric analysis. The remainder was not processed due to accidental possible contamination. In addition, we were given the opportunity to record the spectra of a blank (D) and an extract of Apollo 14 sample 14163,169,176 (E) which had been prepared independently by Hodgson and co-workers. They had determined that these latter solutions contained no porphyrins by fluorometry following incremental demetallation (Hodgson *et al.*, 1972). Spectra of D and E provided us with additional characterization of our instrumental background.

An examination of the excitation spectra of samples A and B revealed no porphyrin features. Emission scans with excitation at 405 nm were then recorded for all samples. These emission spectra and their analysis are described below.

*Fluorescence spectra and data processing*

Extracts of samples A and B as well as sample C were each examined in a single quartz microcuvette with lightpath of 0.5 cm. Averages of 128 traces (10 sec/trace, time constant ~0.5 sec) for each solution were recorded in the region of 500–750 nm with excitation at 405 nm. Extracts of samples D and E were examined in separate 1 cm lightpath quartz cuvettes and averages of 64 traces were recorded under the conditions described above.

One hundred and eighty points from 555 to 734 nm for each sample were fitted to an eight term polynomial by the least squares procedure described by Cohen and Rho (1971). Neither lunar sample spectrum showed any appreciable differences from that of its corresponding blank. Table 1 shows the standard deviations of the computed traces from the experimental data for samples A, B, and C. The standard deviations are computed by taking the point-by-point deviations between the observed and calculated traces $d_i$ and using the usual formula

$$\delta = \left\{\frac{\sum d_i^2}{N}\right\}^{1/2}$$

Table 1.

| | Standard deviations |
|---|---|
| Porphyrin standard (C) | 0.120 |
| Blank (B) | 0.039 |
| Apollo 14 (A) | 0.036 |

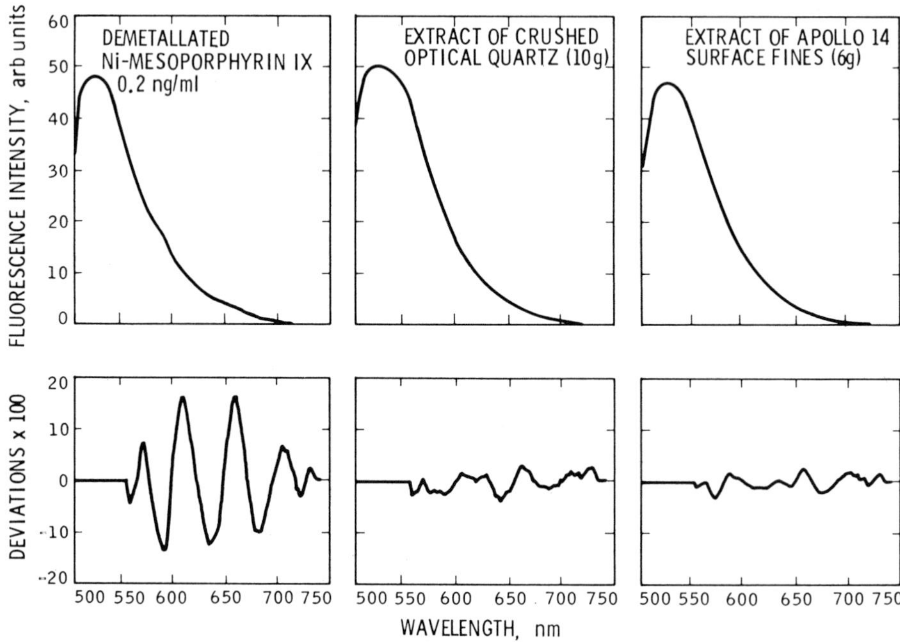


Fig. 1. Top: Averaged traces of fluorescence spectra of samples C, B, and A. Bottom: deviations of top curves from best fitting eight term polynomials.

where $N$ is the number of points and the $d_i$ are in arbitrary fluorescence intensity units. Figure 1 shows the original traces and plots of the deviations from the computed background for these three samples.

We have shown previously (Cohen and Rho, 1971) that a truncated polynomial can be chosen which will fit the smooth background to within the noise level. This same polynomial will not fit a sharp feature superimposed upon the background. The standard deviation is thus a measure of the intensity of sharp features in the region of interest. The difference between the standard deviations of the blank, B, and sample A shown in Table 1 is insignificant and similar to differences we have observed among blanks. The traces shown in Fig. 1 also provide no basis for distinguishing the sample spectrum from the blank.

If the sharp features on the background are random noise, then the standard deviations provide a measure of the signal-to-noise ratio. However, after all the data were processed, it was noted that the residual features in the lunar sample spectra and blanks were similar. In order to demonstrate that the source of these features was not a computational artifact and to determine the level of the random noise, the data were reduced as follows:

The value of the intensity $I_0$ at each wavelength $\lambda_0$ was replaced by the difference between itself and the average of the $I^+$, intensity at $\lambda_0 + 20$ nm, and $I^-$, the intensity

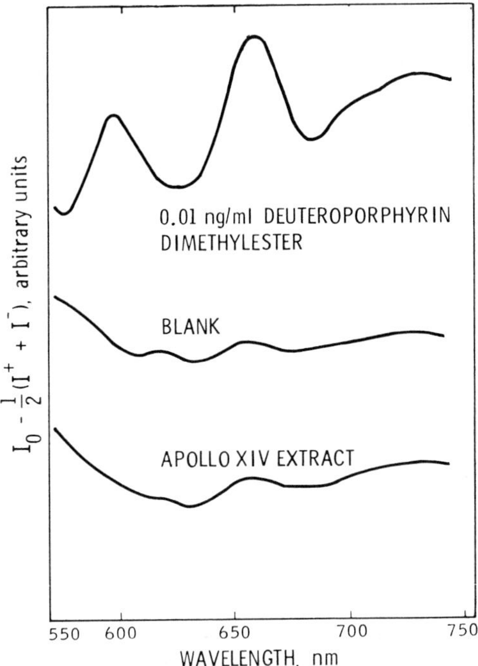


Fig. 2. Processed traces of a porphyrin reference spectrum and those of samples D and E. The reference, which required no demetallation and could be contained in a low scattering medium, had its spectrum recorded at ten times the fluorometer gain used for the other spectra.

at $\lambda_0 - 20$ nm. The interval of 20 nm is approximately the half-width of a typical porphyrin fluorescence feature and gave optimum enhancement of the spectrum relative to the background.

The logic behind the method is straightforward. The quantity

$$I' = I_0 - \tfrac{1}{2}(I^+ + I^-)$$

is zero for a linear function of wavelength. For a monotonic, smooth function such as the background light scattering from our instrument, $I'$ is always much smaller than $I_0$ for intervals of the order of a typical line width. However, if one considered only the intensity contribution of a feature whose width is equal to the interval between $I^+$ and $I^-$ then the maximum value of $I'$ is half the maximum value of $I_0$, and the peak-to-peak deflection of $I'$ will be approximately 0.75 the maximum value of $I_0$. The trace of $I'$ versus $\lambda$ is the same as would be observed if one conducted an experiment similar to that described by Verdieck and Cornwell (1961).

Since the background for the present case is not linear, it is not entirely removed, but the extent of its removal is sufficient to allow scaling up of the data to reveal weak features. Figure 2 shows the processed data of samples D and E as well as that of an 0.01 ng/ml deuteroporphyrin dimethylester solution. Notice that there is a coincidence between the features at 665 nm in the spectra of the blank of the lunar sample and of the porphyrin. Figure 3 shows the processed data for samples A, B, and C after reduction in the same manner. Again, anomalous features at 662 nm are present in the blank and lunar samples.

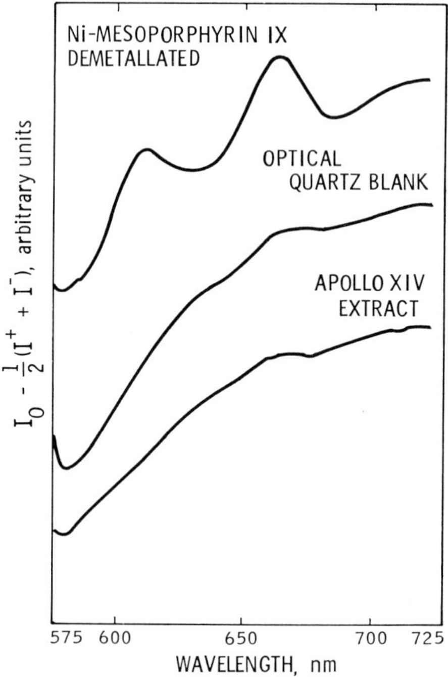


Fig. 3. Processed fluorescence spectra of samples A, B, and C (from bottom to top).

Because the lunar sample extracts and blanks showed the 662–665 nm features, it was assumed that the results were due to an instrumental anomaly and were possibly proportional to total light intensity. To test this assumption, the processed spectra of samples D and E were scaled in proportion to scattered light intensity. An overlay of these spectra and their difference is shown in Fig. 4. Sample A and B spectra were similarly scaled. Figure 5 shows the difference between the blank and the standard porphyrin solution with the difference between the lunar sample and the blank. The fact that the differences between appropriately scaled spectra of samples D and E, as well as of A and B, show no distinguishable features would seem to confirm our assumption. The residual low frequency noise in Fig. 5 is no more than 14% of the intensity of the standard porphyrin spectrum. This leads to a signal-to-noise ratio of about 7 : 1 for the standard.

An additional experiment suggested that the anomalous peaks were artifacts. A polished aluminum mirror was placed in the sample compartment, excitation was set at 405 nm and a Wratten #4 filter was used to block the light below 500 nm. The resulting trace in Fig. 6 shows that a considerable amount of undispersed light from the xenon excitation lamp is passed by the excitation nonochromator. In particular the features at 620 and 658 nm are possible sources of the unwanted peaks in the fluorescence data. Distortion of the peaks by the background and recalibration of the instrument can account for the slight differences in apparent wavelength from the

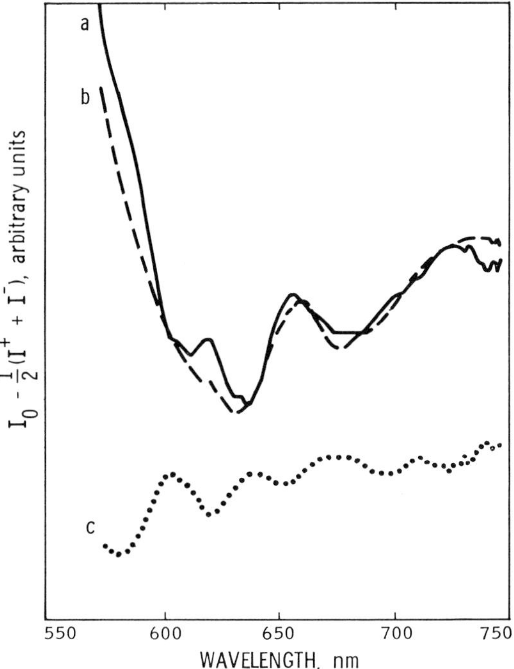


Fig. 4. Overlay of processed spectra of sample D — and E - - - when scaled in proportion to total light intensity. Trace c is the difference b − a.

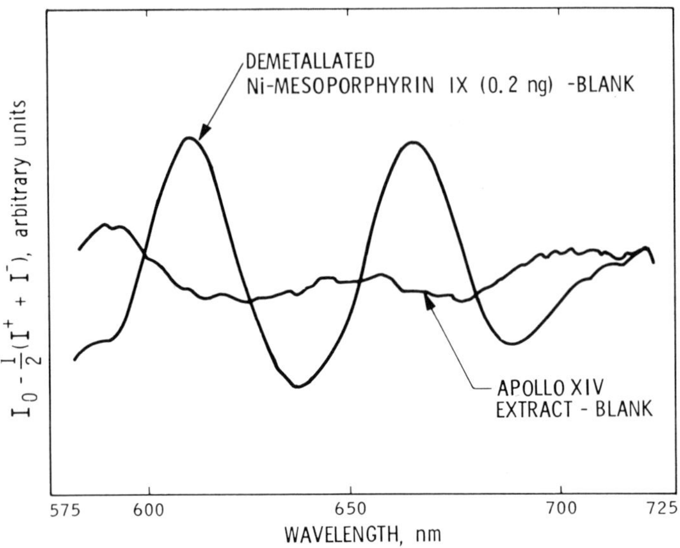


Fig. 5. A comparison of the processed spectra of the sample blank and porphyrin standard blank.

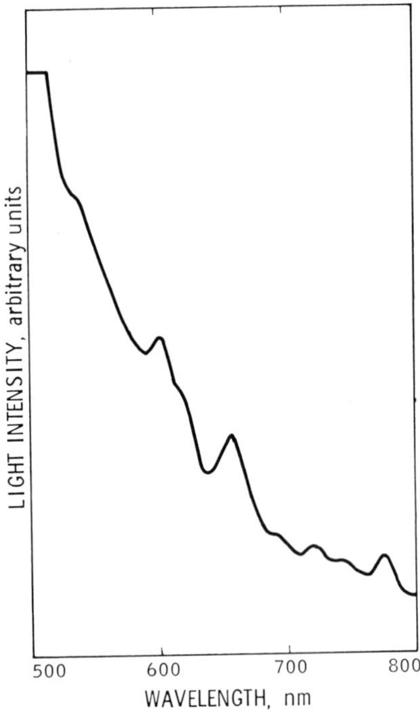


Fig. 6. Spectrum of light passed by the excitation monochrometer when set at 405 nm.

processed spectra. This finding illustrates one of the pitfalls of trace analysis as this effect did not become apparent until the instrumentation was pushed well beyond its specifications.

These results demonstrate that:

(a) The standard deviation of the lunar sample fluorescence spectrum does not differ significantly from that of the blank and is well below that of the porphyrin standard.

(b) The features of two independently processed lunar samples and two independently constructed blanks are essentially identical, proportional to scattered light intensity, and apparently caused by undispersed xenon light.

(c) Two different numerical procedures cannot distinguish the blank from the lunar sample.

## Conclusion

The signal-to-noise ratio for the standard solution was determined to be approximately 7 : 1. This solution contained the demetallation products of 0.2 ng or $3.26 \times 10^{-13}$ mole of Ni-mesoporphyrin IX. If a metalloporphyrin of similar structure were present in the sample and if it were to be demetallated with the same efficiency as that of Ni-mesoporphyrin IX to yield an equivalent fluorescence per mole, then $9.3 \times 10^{-14}$ mole would have given a spectrum with a signal-to-noise ratio of 2 : 1. This spectral feature would have been easily discernible; however, no spectral features were observed from the equivalent extracts of 6 gm of lunar fines.

On the basis of the amount of lunar sample extracted and the spectral intensity resulting from demetallation of known amounts of Ni-mesoporphyrin IX, we conclude that less than $1.6 \times 10^{-14}$ moles of extractable metalloporphyrin is present per gram of these lunar fines.

*Acknowledgment*—The authors thank G. W. Hodgson and E. Peterson for providing us with the results of their analysis. Thanks are also due J. Bonner and T. F. Yen for very useful discussions and suggestions. The technical assistance of J. R. Thompson is acknowledged. This work represents one phase of research carried out at the Jet Propulsion Laboratory, California Institute of Technology, under Contract No. NAS7-100, sponsored by the National Aeronautics and Space Administration.

## References

Cohen E. A. and Rho J. H. (1971) Computer enhancement of weak porphyrin fluorescence spectra. Amer. Chem. Soc. Meeting, Abstract No. 25, Sept. 1971. Washington, D.C.

Hodgson G. W., Kvenvolden K., Peterson E., and Ponnamperuma C. (1972) Search for porphyrins in lunar soil: Samples from Apollo 11, 12, and 14. *Space Life Sci.* In press.

Rho J. H., Bauman A. J., Yen T. F., and Bonner J. (1970) Fluorometric examination of a lunar sample. *Science* **167,** 754–755.

Rho J. H., Bauman A. J., Yen T. F., and Bonner J. (1971) Absence of porphyrins in an Apollo 12 lunar surface sample. *Proc. Second Lunar Sci. Conf., Geochim. Cosmochim. Acta* Suppl. 2, Vol. 2, pp. 1875–1877. MIT Press.

Verdieck J. F. and Cornwell C. D. (1961) Radio-frequency spectrometer with bidirectional square wave frequency modulation. *Rev. Sci. Inst.* **32,** 1382–1386.

# Lunar Sample Cross Reference
# (1970 and 1971 Proceedings)

Two sample cross references are presented in this Proceedings set. This sample cross reference refers to the *Proceedings of the Apollo 11 Lunar Science Conference* (set 1), and the *Proceedings of the Second Lunar Science Conference* (set 2). The sample cross reference to the *Proceedings of the Third Lunar Science Conference* (set 3) is presented at the end of Volume 3. The format of the sample cross reference is as follows:

Sample Number
(set number, volume number) page numbers . . . .

(2, 2) 1021, 1261, 1351, 1503, 1521, 1591, 1607, 1627, 1671, 1729, 1747, 1773, 1797, 1813
(2, 3) 2103, 2125, 2165, 2337, 2345, 2367, 2477, 2515, 2543, 2559, 2611, 2705

10018
(1, 1) 1, 267, 287, 315, 673, 749, 763, 801, 849
(1, 2) 991, 1007, 1029, 1071, 1111, 1117, 1143, 1165, 1177, 1239, 1407, 1471, 1685, 1719
(2, 1) 797, 949

10019
(1, 1) 1, 561, 627, 673, 763, 801
(1, 2) 1007, 1071, 1117, 1165, 1177, 1195, 1239, 1331, 1351, 1407, 1487, 1493, 1685, 1793, 1933
(1, 3) 2413
(2, 2) 1331
(2, 3) 2041, 2639

10020
(1, 1) 1, 81, 135, 267, 315, 419, 513, 599, 655, 661, 711, 801, 849, 927
(1, 2) 1007, 1071, 1111, 1117, 1143, 1213, 1311, 1369, 1383, 1429, 1435, 1499, 1533, 1595, 1637, 1659, 1685, 1729
(1, 3) 1937, 2013, 2127, 2221, 2243, 2289, 2341, 2413
(2, 1) 117, 247, 285, 413, 469, 507, 775
(2, 2) 1063, 1301, 1417
(2, 3) 2235, 2317, 2337, 2345, 2381

10021
(1, 1) 1, 419, 513, 801, 927
(1, 2) 991, 1071, 1103, 1111, 1117, 1143, 1165, 1177, 1239, 1269, 1407, 1471, 1659, 1685, 1741, 1775, 1805
(1, 3) 2325
(2, 2) 1037, 1301, 1607, 1901
(2, 3) 2361, 2461

10022
(1, 1) 1, 55, 195, 221, 247, 419, 607, 661, 801, 897, 937
(1, 2) 1007, 1071, 1111, 1165, 1177, 1195, 1213, 1311, 1393, 1429, 1493, 1637, 1665, 1685
(1, 3) 1953, 2093, 2155, 2199, 2213, 2221
(2, 1) 17, 143, 413, 469, 529, 559, 923
(2, 2) 1021, 1063, 1237, 1381
(2, 3) 2173, 2223, 2477, 2485

10023
(1, 1) 1, 267, 287, 315, 673, 711, 849
(1, 2) 1111
(1, 3) 1953, 2051, 2093, 2127, 2435
(2, 1) 17, 797

10024
(1, 1) 1, 539, 607, 749, 801
(1, 2) 1007, 1071, 1111, 1117, 1143, 1165, 1195, 1493, 1637, 1665, 1685, 1805
(1, 3) 1953, 2093, 2325
(2, 1) 413, 559
(2, 2) 1063, 1237
(2, 3) 2461, 2477, 2515

10025
(1, 1) 65
(2, 1) 797

10026
(1, 1) 673
(1, 2) 1239

10027
(1, 1) 363
(1, 2) 1111

(2, 1) 17, 469, 507
(2, 2) 1187
12013
(2, 1) 17, 319, 377, 393, 413, 431, 439, 459, 469, 507, 583, 617, 645, 727, 755, 817, 833, 949
(2, 2) 987, 999, 1021, 1083, 1101, 1123, 1139, 1159, 1169, 1187, 1231, 1237, 1319, 1421, 1487, 1503, 1521, 1577, 1627, 1747
(2, 3) 2103, 2153, 2183, 2223, 2235, 2543
12014
(2, 1) 17, 469, 507, 679
(2, 2) 1037, 1187
12015
(2, 1) 17, 469, 617
(2, 2) 1187
12016
(2, 1) 17
(2, 2) 1757
12017
(2, 1) 17, 469, 817
(2, 2) 1021, 1063, 1139
(2, 3) 2559, 2629
12018
(2, 1) 17, 39, 91, 219, 247, 343, 413, 469, 481, 497, 507, 583, 601, 617, 701, 855, 909
(2, 2) 1063, 1101, 1187, 1217, 1247, 1281, 1307, 1417, 1471, 1487, 1591, 1607
(2, 3) 2125, 2137, 2345, 2361, 2461, 2599, 2611
12019
(2, 1) 17, 469, 583
12020
(2, 1) 17, 151, 207, 265, 413, 469, 481, 497, 601, 617
(2, 2) 999, 1021, 1063, 1101, 1187, 1217, 1231, 1301, 1307, 1319, 1331, 1471, 1607
(2, 3) 2103, 2125, 2137, 2223, 2285, 2461, 2599, 2611
12021
(2, 1) 17, 39, 59, 91, 109, 117, 141, 247, 265, 285, 301, 343, 413, 439, 469, 481, 497, 529, 559, 601, 617, 645
(2, 2) 987, 1063, 1083, 1101, 1159, 1169, 1187, 1217, 1237, 1261, 1281, 1301, 1307, 1337, 1381, 1407, 1417, 1421, 1443, 1487, 1493, 1521, 1577, 1591, 1797, 1843
(2, 3) 2083, 2137, 2451, 2501, 2515, 2529, 2559, 2621, 2629, 2639
12022
(2, 1) 17, 151, 193, 219, 301, 413, 439, 449, 459, 469, 481, 497, 601, 617
(2, 2) 987, 999, 1037, 1063, 1083, 1101, 1187, 1237, 1253, 1261, 1307, 1331, 1337, 1343, 1351, 1367, 1397, 1407, 1461, 1521, 1577, 1797, 1843
(2, 3) 2165, 2317, 2327, 2381, 2485, 2543, 2705
12023
(2, 2) 1343, 1351, 1397, 1407, 1843, 1865, 1875, 1879, 1901, 1913, 1929
12024
(2, 2) 1343, 1757
(2, 3) 2501
12025
(2, 1) 665, 701, 755
(2, 2) 1063, 1101, 1139, 1261, 1343, 1407, 1487, 1591, 1671, 1681, 1757, 1891
(2, 3) 1959, 2057, 2125, 2183, 2235, 2245, 2543, 2559, 2569, 2593, 2599
12026
(2, 1) 701
(2, 2) 1929, 1939
(2, 3) 1959

# Author Index

# Subject Index

Pages 1–1132: Volume I, *Mineralogy and Petrology*
Pages 1133–2156: Volume 2, *Chemical and Isotope Analyses/Organic Chemistry*
Pages 2157–3263: Volume 3, *Physical Properties*

(Where an index entry refers to the opening page of an article, the entry includes the entire article.)